D0510046

Chicago &
Cook County

StreetFinder®

page	Contents
A	Using the StreetFinder
B	PageFinder Map
C	Vicinity Map
1	Contents
2	Downtown Chicago Map
4	Quick Reference Guide
9	Chicago Block Number System
10	StreetFinder Maps
97	StreetFinder Index

Photo credit: Chicago Skyline by Peter J. Schulz / Chicago Office of Tourism

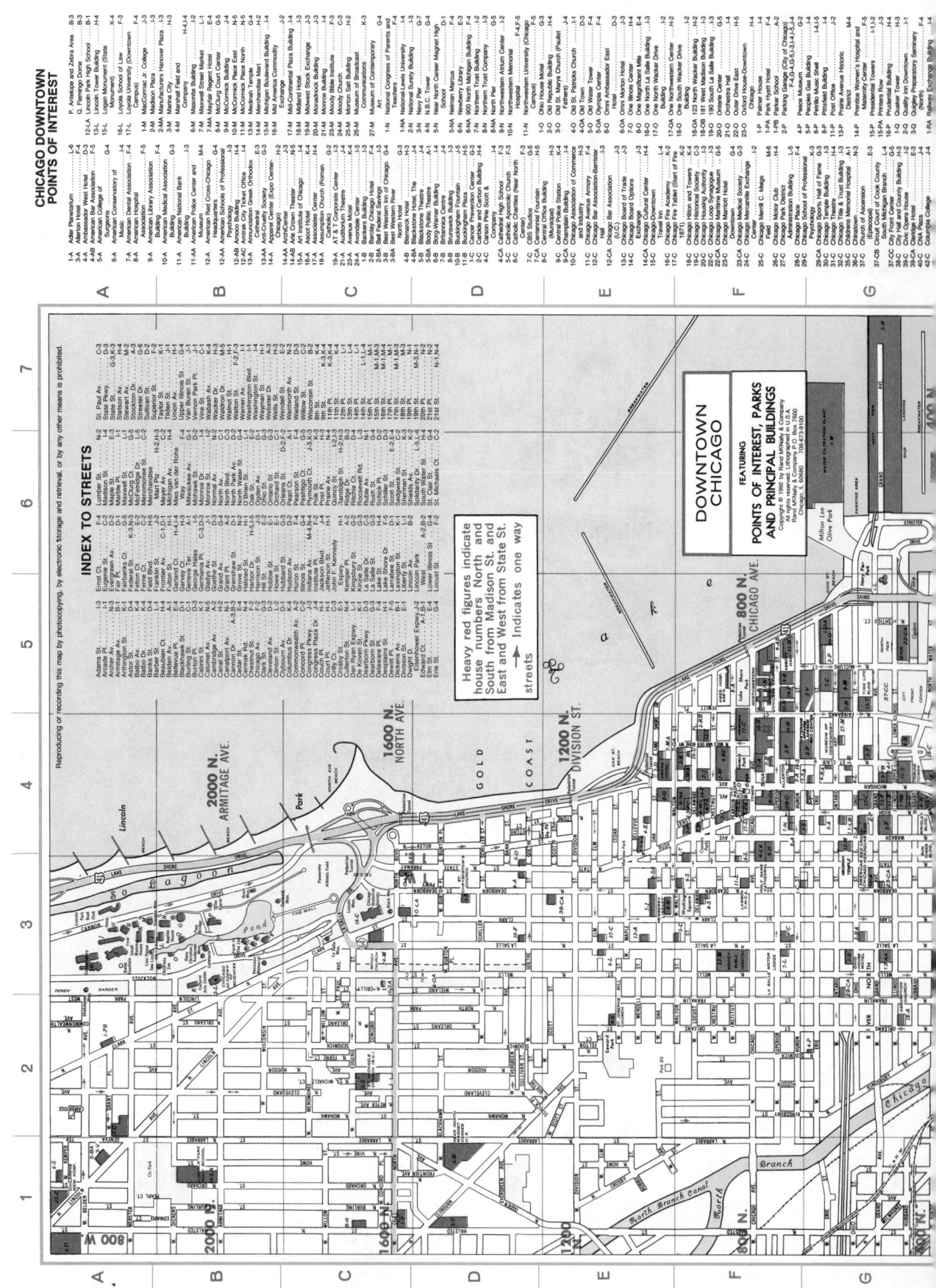

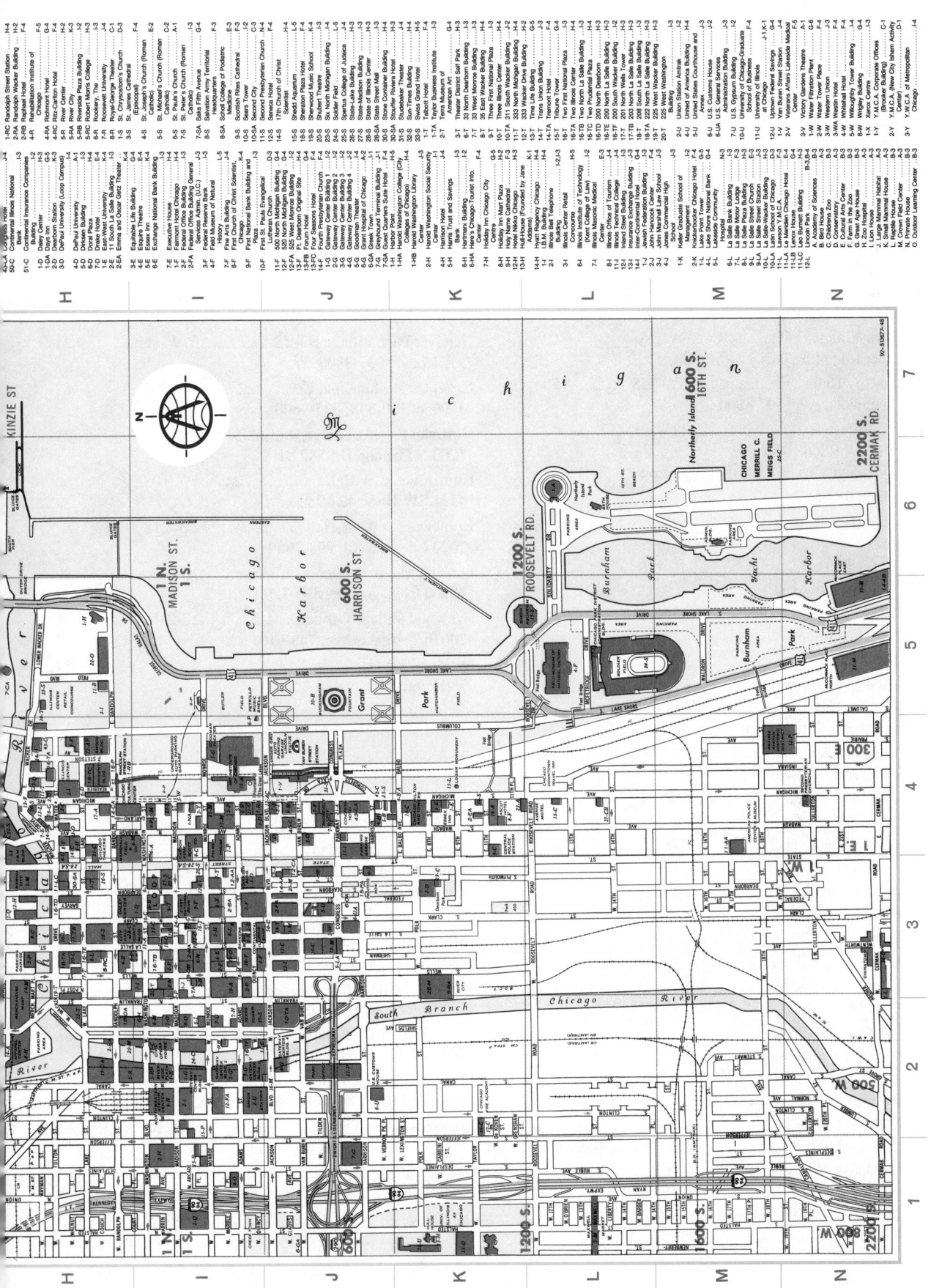

Chicago downtown map with alphanumeric grid (H–N columns, 1–7 rows), showing Lake Michigan, Chicago River, Grant Park, Burnham Park, Northerly Island, Merrill C. Meigs Field, and downtown streets including Kinzie St, Madison St, Harrison St, Roosevelt Rd, Cermak Rd, Lake Shore Drive, and the numbered block markers (1 N., 1 S., 600 S., 1200 S., 1600 S., 2200 S., 300 E, 500 W, 1200 W, 1600 W, 2200 W).

92-51307J-46

Index of points of interest with grid references:

- 1-RC Randolph Street Station — H-2
- 2-RA Randolph Wacker Building — H-2
- 2-RB Raphael Hotel — F-4
- 4-R Rehabilitation Institute of Chicago — F-5
- 4-RA Richmont Hotel — G-4
- 4-RC Ritz-Carlton Hotel — F-4
- 5-R River Center — H-2
- 5-RA River City — J-1
- 5-RB Riverside Plaza Building — H-3
- 5-RC Robert Morris College — H-3
- 7-R Rookery, The — J-3
- 7-RA Roosevelt University — J-4
- 8-R Royal-George Theater — F-3
- 3-S St. James Cathedral — F-4
- 4-S St. Joseph's Church (Roman Catholic) — E-2
- 5-S St. Michael's Church (Roman Catholic) — A-1
- 6-S St. Paul's Church — I-3
- 7-S St. Peter's Church (Roman Catholic) — G-4
- 7-SA Saks Fifth Avenue — F-3
- 8-S Salvation Army Territorial Headquarters — E-3
- 8-SA Scholl College of Podiatric Medicine — F-3
- 9-S Scottish Rite Cathedral — E-4
- 10-S Sears Tower — I-3
- 10-SA Second City — C-3
- 12-S Seneca Hotel — F-4
- 15-S Shedd Aquarium — L-5
- 15-SA Sheraton Plaza Hotel — K-4
- 18-S Sherwood Music School — L-5
- 19-S Soldier Field — L-5
- 20-S Spertus College of Judaica — J-4
- 24-S Spertus Museum — J-4
- 25-S State-Lake Building — H-3
- 26-S State-Madison Building — H-3
- 28-S State Street Mall — H-3
- 28-SA Stone Container Suite Hotel — H-4
- 30-S Stouffer Riviera Hotel — H-4
- 30-SA Studebaker Theater — H-5
- 33-S Sun-Times Building — H-3
- 33-SA Swiss Grand Hotel — H-4
- 1-T Talbot Hotel — G-4
- 1-TA Taylor Business Institute — G-4
- 2-T Teets Museum of American Art — J-3
- 3-T Theater District Self Park — G-4
- 7-T 33 West Monroe Building — H-3
- 7-TA 35 East Wacker Building — H-3
- 10-T Three Illinois Center — H-4
- 10-TA Three First National Plaza — H-3
- 11-T 201 North Wells Tower — H-3
- 17-T Three Illinois Center — J-2
- 17-TA 311 South Wacker Building — J-2
- 17-TB 333 North Michigan Building — H-3
- 18-T 203 North La Salle Building — H-3
- 18-TA 208 South La Salle Building — H-3
- 19-T 222 North La Salle Building — H-3
- 20-T 225 West Washington Building — H-3
- 2-U Union Station Amtrak — J-2
- 5-U United States Courthouse and Annex — J-2
- 6-U U.S. Customs House — G-5
- 6-UA U.S. General Services Administration Building — N-3
- 7-U U.S. Gypsum Building — H-3
- 10-U University of Chicago—Graduate School of Business at Chicago — J-4
- 12-U Uptown Federal Savings — H-3
- 2-V Van Buren Street Station — E-4
- 3-V Veterans Affairs, Lakeside Medical Center — M-4
- 3-VA Victory Gardens Theatres — B-3, B-4
- 3-W Water Filtration Plant — A-1
- 3-WA Water Tower Place — F-4
- 4-W Western Union — F-4
- 4-WA Westin Hotel — F-4
- 1-X Xerox Centre — H-3
- 1-Y Y.M.C.A. Corporate Offices — F-5
- 2-Y Y.M.C.A. (New City Isham Activity Center) — A-3
- 3-Y Y.W.C.A. of Metropolitan Chicago — L-4

- 45-LA Congress Hotel — J-4
- 5C-C Continental Illinois National — J-4
- 51-C Continental Insurance Companies — G-5
- 1-D Days Inn — F-4
- 1-DA DePaul Station — G-5
- 2-D Dearborn Station — J-4
- 3-D DePaul University (Loop Campus) — H-4
- 4-D DePaul University — A-1
- 6-D Dirksen Building — H-3
- 6-DA Drake Hotel — F-4
- 7-D Drake Hotel — E-4
- 1-E East-West University — J-4
- 4-E Emma and Oscar Getz Theater — K-4
- 2-EA Equitable Life Building — K-4
- 3-E Esquire Theatre — F-3
- 4-E Essex Inn — K-4
- 6-E Exchange National Bank Building — K-3
- 7-E Executive House — H-4
- 1-F Fairmont Hotel Chicago — H-4
- 2-F Field Museum of Natural History — L-5
- 2-FA Federal Office Building General — I-3
- 3-F Federal Reserve Bank — J-3
- 3-FA 525 West Monroe Building — I-2
- 3-FC Ft. Dearborn Original Site — F-4
- 4-F Four Seasons Hotel — F-4
- 7-F Fourth Presbyterian Church — F-4
- 8-F Fine Arts Building — J-4
- 9-F First National Bank and First Church of Christ Scientist, Chicago — J-3
- 10-F First St. Paul's Evangelical Lutheran Church — D-3
- 11-F 500 North Michigan Building — G-4
- 12-F 520 North Michigan Building — F-4
- 2-G Gateway Center Building 1 — I-2
- 3-G Gateway Center Building 2 — I-2
- 3-GA Gateway Center Building 3 — I-2
- 4-G Gateway Center Building 4 — J-2
- 5-G Goodman Theater — H-3
- 6-G Grant Hospital of Chicago — A-2
- 6-GA Greek Town — J-1
- 7-GA Greyhound Terminal Building — H-5
- 7-GA Guest Quarters Suite Hotel — G-4
- 1-H Harbor Point — H-4
- 1-HA Harold Washington College (City Colleges of Chicago) — H-4
- 1-HB Harold Washington Library Center — J-3
- 2-H Harold Washington Social Security Center — J-1
- 4-H Harris Trust and Savings Bank — J-3
- 5-H Harris Bank — I-3
- 6-H Harmers Building — G-4
- 6-HA Heller College—Tourist Info. Center — F-4
- 7-H Holiday Inn Chicago City Centre — G-4
- 8-H Holiday Inn Mart Plaza — H-2
- 9-H Holy Name Cathedral — F-4
- 12-H Hotel Nikko Chicago — H-3
- 14-H Hyatt Regency Chicago — H-4
- 3-I Illinois Bell Telephone Building — J-3
- 3-J Illinois Institute of Technology (Kent College of Law) — J-2
- 6-I Illinois Masonic Medical Center — A-3
- 7-I Illinois Office of Tourism — E-3
- 8-I Illinois Technical College — G-4
- 12-I Insurance Exchange Hotel — J-3
- 13-I Inter-Continental Hotel — G-4
- 14-I John Hancock Center — E-4
- 2-J John Marshall Law School (J. Addams) — J-3
- 1-K Keller Graduate School of Management — J-3
- 6-L La Salle Hotel — I-2
- 8-L La Salle Plaza — I-2
- 9-L La Salle Plaza Building — H-3
- 9-LA La Salle Street Church — I-3
- 10-L La Salle-Wacker Building — H-3
- 10-LA Latin School of Chicago — E-4
- 11-LA Le Meridien Chicago Hotel — H-3
- 11-LB Le Meridien Chicago Hotel — G-4
- 11-LC Lenox House — F-4
- 12-L Lincoln Park — B-3, B-4

Cook County Municipal Offices

	Location	Page
Alsip Village	5W-14S	74
4500 123rd St; 385-6902		
Arlington Heights		
Village Hall	17W-14N	24, 25
33 S Arlington Heights Rd; 253-2340		
Barrington Village Hall	25W-18N	12
206 S. Hough; 381-2141		
Barrington Hills Village Hall	28W-16N	11
112 Algonquin Rd; 428-1200		
Bartlett Village Hall	29W-8N	30, 31
228 Main St; 837-0800		
Bedford Park Village Hall	9W-7S	60
6701 Archer Rd.; 458-2067		
Blue Island City Hall	3W-15S	75
13051 Greenwood Ave; 597-8600		
Broadview Village Hall	11W-0S	47
1600 Roosevelt Rd; 681-3600		
Burbank City Hall	8W-9S	65
6530 W 79th; 559-5500		
Burr Ridge Village Hall	15W-8S	58
220 75th; 325-0420		
Calumet City City Hall	4E-17S	83
204 Pulaski Rd; 891-8100		
Calumet Park Village Offices	1W-14S	76
12409 Throop; 388-0850		
Chicago City Hall	0W-0N	51
121 N. LaSalle Ave; 744-5000		
Chicago Heights City Hall	1W-25S	93
1601 Chicago Rd; 756-5317		
Chicago Ridge Village Hall	7W-12S	66
10655 S Oak Ave; 425-7700		
Cicero Town Offices	6W-2S	55
4936 25th Pl; 656-3600		
Country Club Hill City Hall	4W-21S	87
3700 W 175th Pl; 798-2616		
Countryside City Hall	12W-5S	58
Des Plaines City Hall	13W-11N	26
1420 Miner; 391-5300		
East Hazel Crest Village Hall	1W-20S	88
17223 Throop; 798-0213		
Elk Grove Village Hall	19W-8N	33
901 Wellington Ave; 439-3900		
Elmwood Park Village Hall	9W-3N	41
11 Conti Pkwy; 452-7302		
Evanston Civic Center	2W-12N	29
2100 Ridge Ave; 328-2100		
Evergreen Park Village Hall	4W-10S	66, 67
9418 Kedzie Ave; 422-1551		
Flossmoor Village Hall	3W-23S	87
2800 Flossmoor Rd; 798-2300		

	Location	Page
Forest Park Village Hall	9W-0S	48
517 Des Plaines Ave; 366-2323		
Franklin Park Village Hall	11W-3N	41
9554 Belmont; 671-4800		
Glencoe Village Ctr	6W-17N	18, 19
325 Hazel; 835-4111		
Glenview Village Hall	8W-13N	27
1225 Waukegan Rd; 724-1700		
Glenwood Village Hall	0E-22S	88
13 S Rebecca; 758-5150		
Golf Village Hall	8W-12N	27
1 Briar Rd; 998-8852		
Hanover Park Village Hall	27W-8N	31
2121 Lake; 837-3800		
Hazel Crest Municipal Ctr	3W-20S	87
3000 W 170th Pl; 335-9600		
Hoffman Estates Village Hall	24W-12N	22
1200 N Gannon Dr; 882-9100		
Hometown City Hall	5W-10S	66
4331 SW Hwy; 424-7500		
Homewood Village Hall	2W-21S	87
2020 Chestnut Rd; 798-3000		
Lansing Mayor & Clerks Office	3E-21S	89
18200 Chicago Ave; 895-7200		
Lynwood Village Office	2E-24S	95
20636 Torrence Ave; 758-6101		
Lyons Village Hall	9W-4S	54
7801 Ogden Ave; 447-8886		
Markham City Hall	3W-19S	81
16313 Kedzie Ave; 331-4905		
Maywood Village Hall	10W-0N	47
115 S 5th Ave; 344-1200		
Merrionette Park Village Hall	3W-13S	75
3165 115th St.; 396-3180		
Midlothian Village Hall	4W-17S	81
14801 Pulaski Rd; 389-0200		
Morton Grove Village Hall	7W-10N	37
6101 Capulina Ave; 965-4100		
Niles Village Admin Bldg	9W-9N	37
7601 Milwaukee Ave; 967-6100		
Norridge Village Offices	9W-5N	41
4020 Olcott Ave; 453-0800		
Northbrook Village Hall	10W-17N	17
1225 Cedar Lane; 272-5050		
Northfield Village Hall	6W-15N	18, 19
361 Happ Rd; 446-9200		
North Riverside Village Hall	10W-2S	53
2400 Desplaines Ave; 447-4211		
Oak Forest City Hall	7W-18S	80
15440 Central Ave; 687-4050		
Oak Lawn Village Hall	6W-10S	66
5252 Dumke Dr; 636-4400		

	Location	Page
Oak Park Village Hall	8W-0N	48
1 Village Hall Plaza; 383-6400		
Orland Park Village Hall	12W-17S	78
14415 Beacon Ave; 349-5400		
Park Forest Village Hall	3W-26S	92
200 Forest Blvd; 748-1112		
Park Ridge City Hall	10W-8N	36, 37
505 Park Pl; 399-5200		
Phoenix Village Hall	0W-18S	82
15240 Vincennes Rd; 331-2636		
Posen Village Hall	3W-17S	81
2440 W 147th; 385-0139		
Richton Park Village Hall	5W-27S	91
4455 Sauk Tr; 841-2200		
Riverdale Village Hall	0W-16S	82
325 W 142d; 841-2200		
River Forest Village Hall	9W-0N	48
400 Park Ave; 366-5481		
River Grove Village Hall	10W-3N	42
2601 Thatcher Ave; 453-8000		
Robbins Village Hall	4W-16S	75
3329 137th; 385-8940		
Rolling Meadows City Hall	20W-13N	23
3600 Kirchoff Rd; 394-8500		
Schaumburg Municipal Bldg	22W-10N	32, 33
101 S Schaumburg Ct; 894-4500		
Schiller Park Village Hall	12W-5N	40, 41
9526 Irving Park Rd; 678-2550		
Skokie Village Hall	6W-9N	38
5127 Oakton; 673-0500		
South Chicago Heights Village Hall	1W-26S	93
2729 Jackson Ave; 755-1880		
South Holland Village Hall	0E-19S	82
16226 Wausau Ave; 333-0572		
Steger Village Hall	1W-27S	93
35 W 34th; 754-3395		
Streamwood Village Hall	28W-9N	31
401 Irving Park Rd; 837-0200		
Tinley Park Village Hall	8W-20S	85,86
17355 68th Ct; 532-7700		
Westchester Village Hall	12W-0S	46,47
10240 Roosevelt Rd; 345-0020		
Wheeling Village Hall	14W-18N	16
255 W Dundee Rd; 459-2600		
Wilmette Village Hall	3W-13N	29
1200 Wilmette Ave; 251-2700		

Colleges & Universities

	Location	Page
American Islamic Col	0W-5N	45
Brisk Rabbinical Col	3W-7N	39
Catholic Theological Union	2E-5S	63
Chicago Babtist Institute	0E-5S	62
Chicago Col of Osteopathic Medicine	1E-5S	62, 63
Chicago Natural Col of Naprapathy	5W-4N	44
Chicago State Univ	0E-11S	68
Chicago Theological Sem	1E-6S	62, 63
Christ Sem- Seminex	1E-5S	62, 63
City Col of Chicago-Kennedy-King	0W-7S	62
City Col of Chicago-Richard J Daley Col	5W-8S	60
City Col of Chicago-Truman Col	1W-5N	45
City Col of Chicago-Malcolm X Col	2W-0S	50
City Col of Chicago-Olive-Harvey	1E-11S	68
City Col of Chicago-Wright Col	7W-4N	42, 43
Columbia Col	0E-0S	51
Concordia Col	9W-1N	48
Cortez W Peters Bus Col	2W-11S	67
DePaul Univ	1W-2N	51
DePaul Univ-Lincoln Park	1W-2N	51
DeVry Inst of Tech	3W-4N	44
Garrett Evangelical Theological Sem	2W-12N	29
Harper College NE Ctr	13W-15N	16
Hebrew Theological Col	6W-8N	38
Ill Col of Optometry	0E-3S	57
Ill Inst of Technology	0E-3S	57
Kendall Col	2W-12N	29
Loyola Univ	1W-8N	39
Loyola Univ-Downtown	0E-1N	51
Lutheran School of Theology at Chicago	1E-5S	62, 63
McCormick Theological Sem	1E-6S	62, 63
Meadville-Lombard Theologiçal School	1E-6S	62, 63
Moody Bible Inst	0W-1N	51
Moraine Valley Col	11W-12S	65
Morton Jr Col	6W-3S	55
Mundelein Col	1W-8N	39
National Col	2W-13N	29
North Park Col & Theological Seminary	4W-6N	38, 44
Northeastern Ill Univ	4W-6N	38, 44
Northwestern Bus Col	6W-6N	43
Northwestern Univ	2W-12N	29
Northwestern Univ-Chicago	0E-0N	51
Oakton Comm Col	12W-12N	26
Prairie State Col	1W-23S	88
Roosevelt Univ	0E-0S	51
Rosary Col	9W-1N	48
St Augustine Col	1W-6N	45
St Xavier Col	4W-11S	66, 67
Seabury-Western Theological Sem	2W-12N	29
Thornton Comm Col	0W-18S	82
Triton Col	10W-2N	47, 48
Univ of Chicago	1E-6S	62, 63
Univ of Illinois at Chicago	1W-0S	51
Vander Cook Col of Music	0E-3S	57
Willam Rainey Harper Col	22W-13N	23

Forest Preserves

Forest preserves may cover an extensive area.
Only one set of map coordinates is given for each page
of the StreetFinder on which a particular woods or
preserve may be found.

	Location	Page
Algonquin Woods	12W-9N	35, 36
Allison Woods	12W-15N	16, 26
Bachelor Grove Woods	8W-16S	73, 79
Beaubien Woods	1E-15S	76, 77
Black Partridge FP	20W-13S	70
Bunker Hill Woods	8W-8N	37
Burnham Woods	4E-16S	83
Calumet Grove	2W-16S	75
Che-che-pin-qua Woods	10W-4N	41, 42
Chippewa Woods	11W-8N	36
Clayton F Smith Woods	7W-8N	37, 38
Crabtree Nature Center	27W-16N	11
Dam No 1 Woods	13W-17N	16
Dam No 4 Woods	11W-8N	36
Dan Ryan Woods	2W-10S	67
Deer Grove FP	21W-18N	12, 13
Edgebrook Woods	6W-7N	37
Eggers Woods	4E-13S	77
Forest Preserves	7W-17N	18
	7W-21S	86
	10W-1S	53
	13W-12N	26
	12W-11N	26
	12W-2S	52
	9W-0N	48
Fullerton Woods East	10W-2N	48
Fullerton Woods West	10W-2N	48
GAR Woods	10W-0N	47, 48
Harms Woods	7W-12N	28
Indian Road Woods	6W-7N	38
Iroquois Woods	12W-9N	36
Kickapoo Meadows	1W-16S	82
La Bagh Woods	5W-6N	43
Lansing Woods	2, 3E-22S	89
Linne Woods	7W-11N	27, 28
Linne Woods West	8W-11N	27, 28
Lion Woods	13W-11N	26
Miami Woods	8W-10N	37
Midlothian Meadow	5W-18S	81, 82
Miller Meadows	10W-1S	54
Ned Brown FP	19W-11N, 18W-10N	33, 34
Paul Douglas FP	24W-13N	22
Palos Hills Forest Preserve	14W-13S	71, 72
Poplar Creek FP	27W-12N, 30W-11N	21, 30, 31
Potawatomi Woods	13W-18N	16

	Location	Page
Rubio Woods	7W-16S	74
St Paul Woods	8W-10N	37
Sauk Trail Woods	2W-27S	93
Schuth's Grove	10W-1S	53
Seymour Simon Preserve	10, 11W-4N	41
Spring Creek Valley FP	29W-15N	20, 21
Spring Valley Nature Sanct	22W-10N	33
Tampier Slough FP	14, 15W-15S	71
Thatcher Woods	10W-1N	47, 48
Thatcher Woods Glen	10W-0N	47, 48
Thomas Jefferson Woods	10W-0N	47, 48
Tinley Creek Woods	8W-16S	73
Whistler Woods	0W-15S	76
Yankee Woods	7W-19S	80

Country Clubs & Golf Courses

	Location	Page
Anetsberger GC	9W-16N	18
Arlington Lakes GC	18, 19W-12N	24
Bartlett Hills G&CC	29W-8N	30, 31
Beverly CC	3W-9S	67
Billy Caldwell GC	6W-7N	37, 38
Bryn Mawr CC	5W8N	38
Burnham Woods GC	4E-16S	83
Calumet CC	2W-20S	87
Cherry Hill CC	4W-22S	87
Chicago Heights CC	2W-24, 25S	93, 94
Chick Evans GC	7W-11N	27, 28
Cog Hill CC	16, 17W-13S	71
Columbus Park GC	7W-0S	48, 49
Edgebrook GC	7W-7N	37, 38
Edgewood Valley CC	13W-8S	58
Evanston GC	5W-10N	28, 38
Evergreen Park GC	3W-10S	67
Flossmoor CC	2W-23S	87
Fresh Meadow G&CC	14W-1S	52
Glencoe GC	7W-18N	18
Gleneagles CC	16W-14S	71
Glenview GC	7W-12N	27, 28
Glenview Naval Air Station GC	9, 10W-13N	27
Glenwoodie CC	0E-23S	88
Golden Acres CC	23W-10, 11N	32
Golden Acres CC	8W-18N	18
Hickory Hills GC	10W-11S	65
Highland Woods GC	22, 23W-13N	22, 23
Hilldale CC	25W-12N	22
Idlewild CC	2W-22, 23S	87
Indian Boundary GC	10W-4N	41
Indian Hill GC	4, 5W-14N	28, 29
Inverness GC	23W-16N	12, 13
Jackson Park GC	2E-7S	63
Joe Louis GC	1W-15S	75, 76
La Grange CC	13W-5S	58

	Location	Page
Lake Shore CC	6W-18N	18, 19
Lakewood CC	3E-26S	95
Marquette Park GC	4W-7S	61
Midlothian CC	6W-17S	80
Mission Hills GC	12W-16N	16
Mt Prospect CC	6W-12N	25
North Shore CC	7, 8W-13N	27, 28
Northwestern GC	7W-11N	27, 28
Oak Park CC	10W-2N	47
Old Orchard CC	15W-14N	25
Olympia Fields CC	3W-24S	92, 93
Palatine Hills GC	21W-17N	13
Palos CC	11W-15S	72, 73
Park Ridge CC	10W-9N	36
Poplar Creek GC	26W-12N	31, 32
Poplar Lake GC	26W-12N	31, 32
Ravisloe CC	2, 3W-21S	87
Ridge CC	3W-12S	67
Ridgemoor CC	8W-5N	42
River Forest GC	15W-4N	40
River Oaks GC	2E-19S	83
Riverside GC	10W-2S	53
Robert A Black GC	2W-8N	39
Rob Roy GC	14W-14N	25
Rolling Green CC	16W-14N	25
Rolling Knolls GC	31W-11N	30
Sheridan Shore Yacht Club	2W-13N	29
Silver Lake CC	10W-17S	79
Skokie CC	6W-17N	18
Skokie Playfield GC	6W-15, 16N	18, 19
South Shore GC	3E-7S	63
Sportsmans CC	11, 12W-17N	16
Sunset Ridge CC	8W-16N	18
Thunderbird CC	24W-18N	12
Villa Olivia G & CC	30, 31W-9N	30
Walnut Hills CC	10W-11S	65
Waveland Ave. GC	0W-5N	45
Westgate Valley CC	8W-15S	73, 74
Westmore-land CC	5W-12, 13N	28, 29
Wilmette GC	6, 7W-14N	28

Hospitals

	Location	Page
Alexian Brothers Medical Ctr	19W-8N	33
Augustana Hosp	0W-2N	45, 51
Belmont Comm Hosp	5W-4N	44
Bethany Methodist Hosp	2W-6N	45
Bethany Hosp	4W-0S	50
Bethesda Hosp	3W-9N	39
Central Comm Hosp	2W-6S	61
Charter Barclay Hosp	1W-5N	45
Chicago Lakeshore Hosp	1W-6N	45
Chicago-Read Mental Health Center	8W-5N	42
Childrens Mem Hosp	0W-2N	45, 51
Christ Hosp & Medical Ctr	5W-10S	66
Columbus Hosp	0W-3N	45
Cook County Hosp	2W-0S	50
Edgewater Hosp	2W-7N	39
Elizabeth Hosp	2W-1N	50
Evanston Hosp	2W-13N	29
Forest Hosp	12W-11N	26
Frank Cuneo Mem Hosp	1W-5N	45
Franklin Blvd Comm Hosp	4W-0N	50
Glenbrook Hosp	11W-14N	26
Gottlieb Mem Hosp	10W-2N	47
Grant Hosp	0W-2N	51
Hartgrove Hosp	4W-0N	50
Holy Cross Hosp	3W-7S	61
Holy Family Hosp	13W-12N	26
Hospital of Englewood	1W-6S	61, 62
Humana Hosp Hoffman Ests	26W-12N	22
Hyde Park Comm Hosp	1E-6S	63
Ill Masonic Med Ctr	1W-3N	45
Ill State Psychiatric Inst	2W-0S	50
Ingalls Memorial Hosp	1W-18S	81, 82
Jackson Park Hosp	2E-8S	63
La Rabida Children's Hosp & Res Ctr	0W-1S	63
Lakeside Comm Hosp	0W-1S	57
Lincoln West Medical Ctr	3W-5N	44
Little Company of Mary Hosp	3W-10S	67
Loretto Hosp	7W-0S	48, 49
Loyola Univ Med Ctr	10W-0S	47
Lutheran General Hosp	11W-10N	36
MacNeal Hosp	8W-3S	54
Martha Washington Hosp	2W-5N	44, 45
Mary Thompson Hosp	2W-0N	50
Mercy Hosp & Med Ctr	0E-2S	57
Michael Reese Hosp & Med Ctr	0E-2S	57
Northwest Community Hosp	18W-13N	24
Northwest Hosp	7W-4N	42, 43
Northwest Mem Hosp	0E-0N	51
Norwegian American Hosp	3W-1N	50
Oak Forest Hosp	5W-19S	80, 81
Oak Park Hosp	8W-0S	47
Old Orchard Hosp	5W-12N	28
Olympia Fields Osteopathic Medical Ctr	4W-24S	92
Palos Comm Hosp	8W-14S	73
Provident Med Ctr	0E-5S	62
Ravenswood Hosp Med Ctr	2W-5N	45
Rehab Institute of Chicago	0E-0N	51
Resurrection Hosp	9W-7N	37
Riveredge Hosp	10W-0S	47
Roseland Comm Hosp	0W-13S	76
Rush Presbyterian-St Lukes Med Ctr	2W-0S	50

	Location	Page
St Annes Hosp	6W-1N	49
St Annes Hosp West	13W-1N	46
St Anthony Hosp	3W-1S	56
St Bernard Hosp	0W-7S	62
St Frances X Cabrini Hosp	1W-0S	51
St Francis Hosp	2W-9N	39
St Francis Hosp	2W-15S	75
St James Hosp	1W-25S	93
St Joseph Hosp	0W-3N	45
St Mary of Nazareth Hosp Ctr	2W-1N	50
Schwab Rehab Ctr	3W-1S	56
Sheridan Road Hosp	1W-7N	39
Shriners Hosp for Crippled Children	8W-2N	48
Skokie Valley Hosp	5W-12N	28, 29
South Chicago Comm Hosp	2E-10S	69
South Shore Hosp	2E-9S	69
South Suburban Hosp	4W-21S	87
Swedish Covenant Hosp	3W-6N	44
Thorek Hosp & Med Ctr	0W-5N	45
Tinley Park Mental Health Ctr	9W-21S	85
Univ of Chicago Hosps & Clins	1E-6S	62, 63
Univ of Ill Hosp	2W-0S	50
VA Hosp (Hines)	10W-1S	53
VA Lakeside Med Ctr	0E-0N	51
VA West Side Med Ctr	2W-0S	50
Walther Memorial Hosp	4W-1N	50
Weiss Mem Hosp	0W-5N	45
Westlake Comm Hosp	11W-0N	47
West Suburban Hosp Med Ctr	7W-0N	51, 52
Woodlawn Hosp	1E-6S	62, 63

Shopping Centers

	Location	Page
Brickyard SC	8W-3N	42
Canterbury SC	3W-19S	81
Deerbrook SC	10W-18N	17
Dunbrook SC	10W-18N	17
Edens Plaza SC	6W-14N	28
Evergreen Plaza SC	3W-11S	67
Flossmoor Commons East SC	4W-23S	87
Flossmoor Commons West SC	4W-23S	87
Ford City SC	5W-8S	60
Forest Park Mall	9W-0S	48
Four Flags SC	10W-11N	27
French Vill SC	4W-21S	87
Golf Mill SC	10W-11N	36, 47
Golf Rose SC	23W-11N	32
Golfview SC	8W-12N	27, 28
Greenlake SC	11W-12N	26
Grenoble Sq SC	4W21S	87
Grove Mall SC	19W-8N	33
Highland Plaza SC	4W-20S	87
Hillside Mall SC	13W-0S	46

	Location	Page
Hillside Mall SC	28W-10N	31
Hoffman Plaza SC	23W-11N	32
Lake Cook Plaza SC	10W-18N	17
Lincoln Mall SC	5W-25S	91
Market Place SC	15W-11N	25
Matteson Town Ctr SC	6W-25S	91
McDade SC	10W-11N	27
Meadow Plaza SC	9W-17N	17
Millbrook SC	10W-11N	27
Mt Prospect SC	14W-13N	25
Northbrook Court SC	9W-18N	18
Northbrook Shop Plaza SC	9W-17N	18
Northpoint SC	17W-16N	14,
North Riverside Park Mall SC	9W-2S	53
Norwood SC	3W-26S	93
Oak Park Mall	8W-0N	48
Old Orchard SC	6W-12N	28
Orland Court SC	11W-18S	78, 79
Orland Sq SC	11W-17S	78, 79
Palwaukee SC	13W-15N	16
Park & Shop SC	18W-10N	34
Park Plaza SC	25W-8N	32
Plaza SC	3W-27S	92, 93
Plaza-del-Prado SC	10W-15N	17
Prospect Heights SC	15W-14, 15N	15, 25
Randhurst SC	15W-14N	25
River Oaks SC	3E-19S	83
River Oaks West SC	2E-19S	83
Sanders Court SC	12W-17N	16
Sauk Trail SC	4W-27S	92
Shoppers Walk	10W-11N	27
Stonebridge SC	3W-21S	87
Streamwood SC	28W-10N	31
Town-N-Country Mall SC	17W-15N	14
Town Sq. SC	23W-10N	32
Tradewinds SC	26W-9N	31
Venture SC	10W-12N	27
Weathersfield Commons SC	25W-10N	32
White Plains SC	11W-17N	16, 17
Wiseway Plaza SC	25W-9N	32
Woodfield Commons SC	21W-11N	33
Woodland Hts SC	28W-9N	31

Getting Around Chicago

For the most part, streets in Chicago lie in a grid pattern, with the city center at the intersection of State and Madison Streets in downtown Chicago. Madison Street, running east and west, and State Street, running north and south, are the base lines from which all block and house numbers are determined.

Generally, block numbers in Chicago ascend in all directions by multiples of 100. Chicago street names and block numbers may not extend into neighboring suburbs. Here is a list of the major streets in Chicago along with their distance from the baseline in both miles and block numbers:

North of Madison Street		Block No.
Chicago Ave	1 mile	800
North Ave	2 miles	1600
Fullerton Ave	3 miles	2400
Belmont Ave	4 miles	3200
Irving Park Rd	5 miles	4000
Lawrence Ave	6 miles	4800
Bryn Mawr Ave	7 miles	5600
Devon Ave	8 miles	6400
Touhy Ave	9 miles	7200
Howard St	9 1/2 miles	7600

South of Madison Street		
Roosevelt Rd	1 mile	1200
Cermak Rd	2 miles	2200
31st St	3 miles	3100
Pershing Rd	4 miles	3900
47th St	5 miles	4700
55th St	6 miles	5500
63rd St	7 miles	6300
71st St	8 miles	7100
79th St	9 miles	7900
87th St	10 miles	8700
95th St	11 miles	9500

West of State Street		
Halsted St	1 mile	800
Ashland Ave	2 miles	1600
Western Ave	3 miles	2400
Kedzie Ave	4 miles	3200
Pulaski Rd	5 miles	4000
Cicero Ave	6 miles	4800
Central Ave	7 miles	5600
Narragansett Ave	8 miles	6400
Harlem Ave	9 miles	7200

East of State Street		
South Park Ave	1/2 mile	400
Cottage Grove Ave	1 miles	800
Woodlawn Ave	1 1/2 miles	1200
Stony Island Ave	2 miles	1600
Jeffrey Ave	2 1/2 miles	2000
Yates Ave	3 miles	2400
Brandon Ave	4 miles	3200

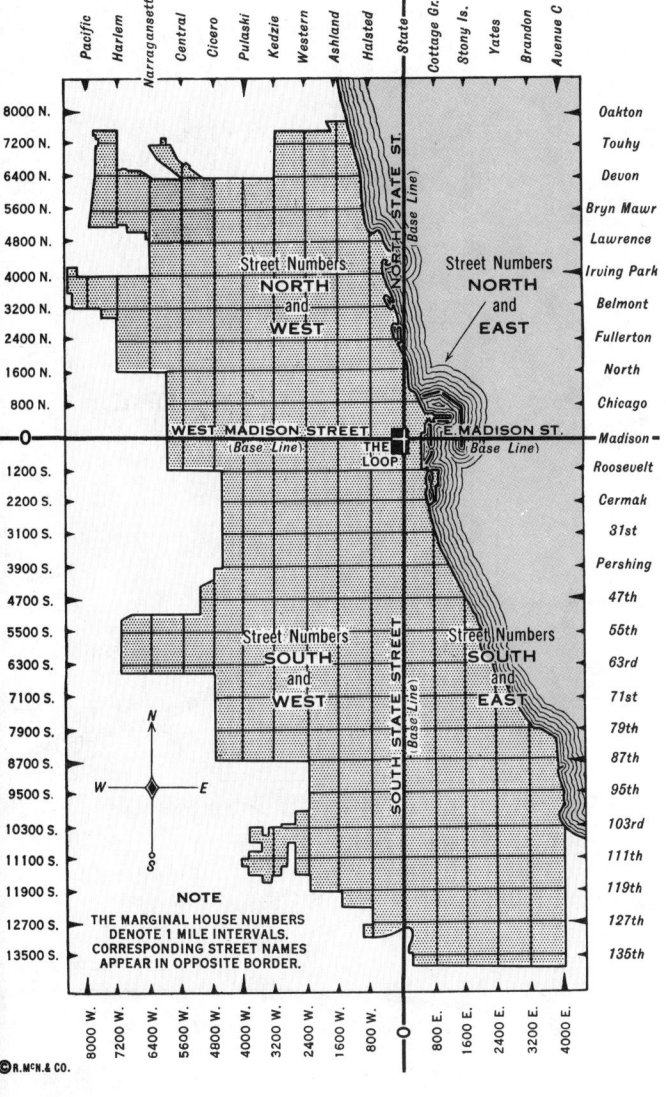

NOTE
THE MARGINAL HOUSE NUMBERS
DENOTE 1 MILE INTERVALS.
CORRESPONDING STREET NAMES
APPEAR IN OPPOSITE BORDER.

© R. McN. & CO.

SEE McHENRY COUNTY PAGES 48-49

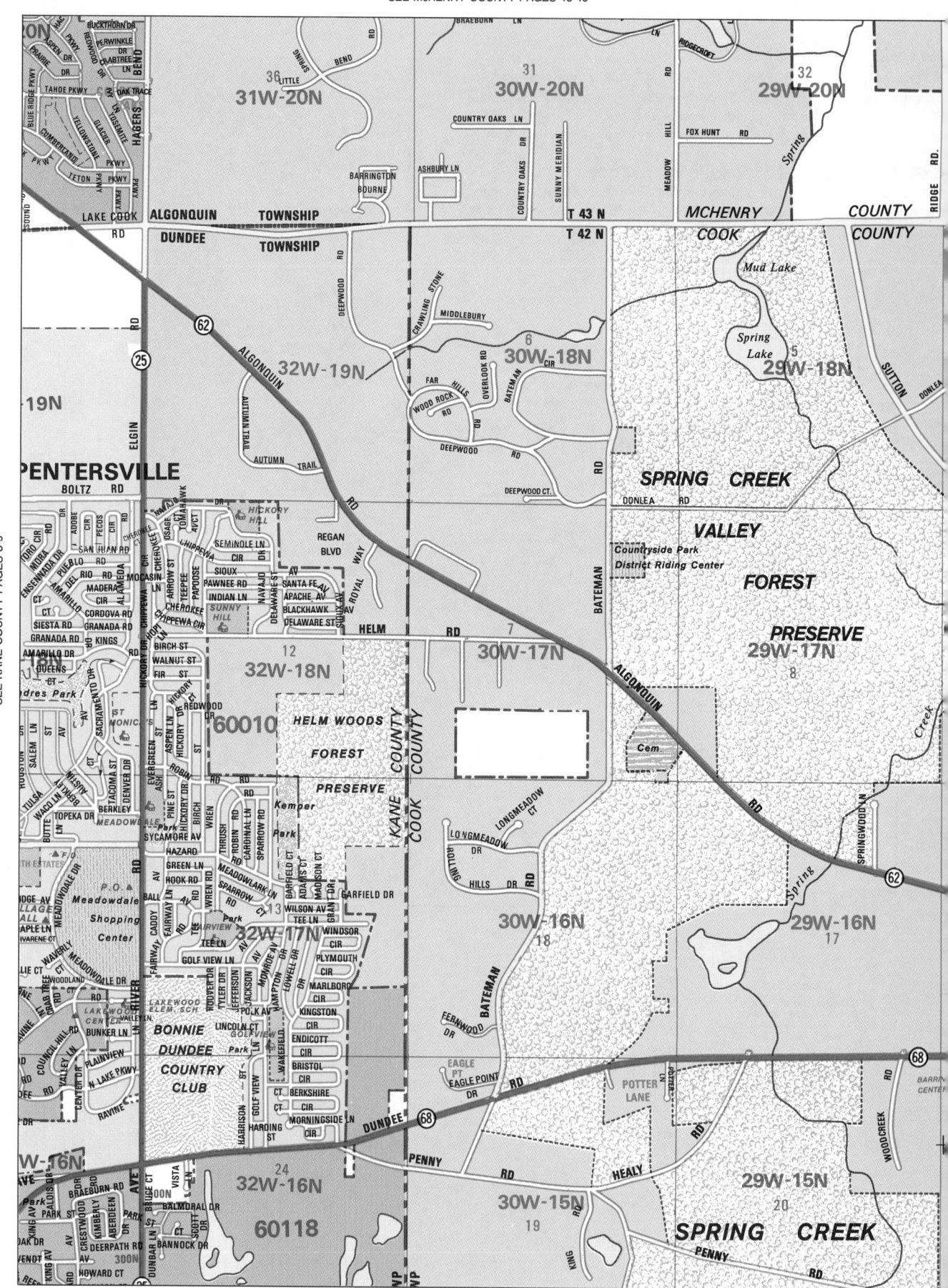

SEE LAKE COUNTY PAGES 44-45

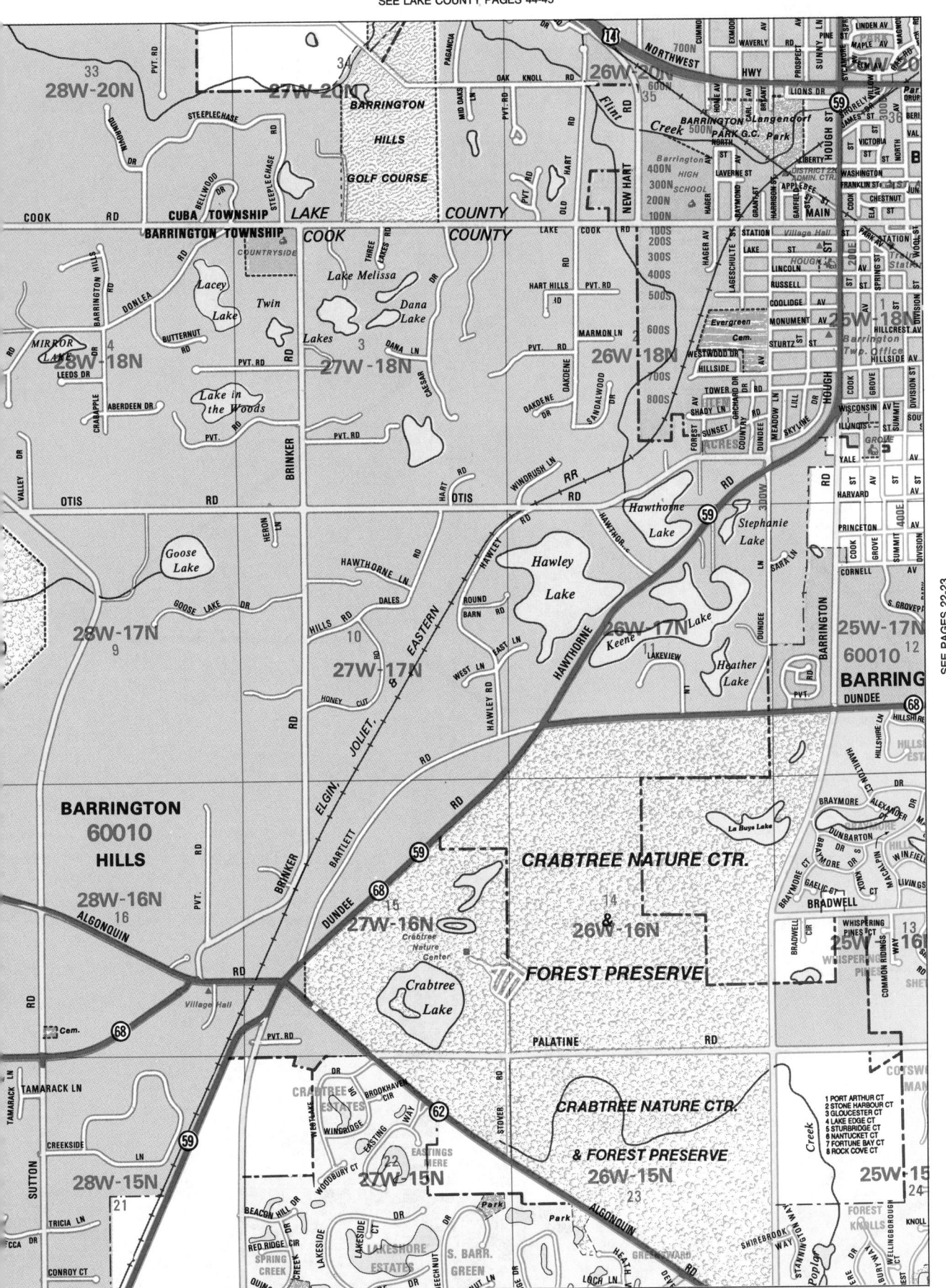

SEE PAGES 22-23

SEE LAKE COUNTY PAGES 44-45

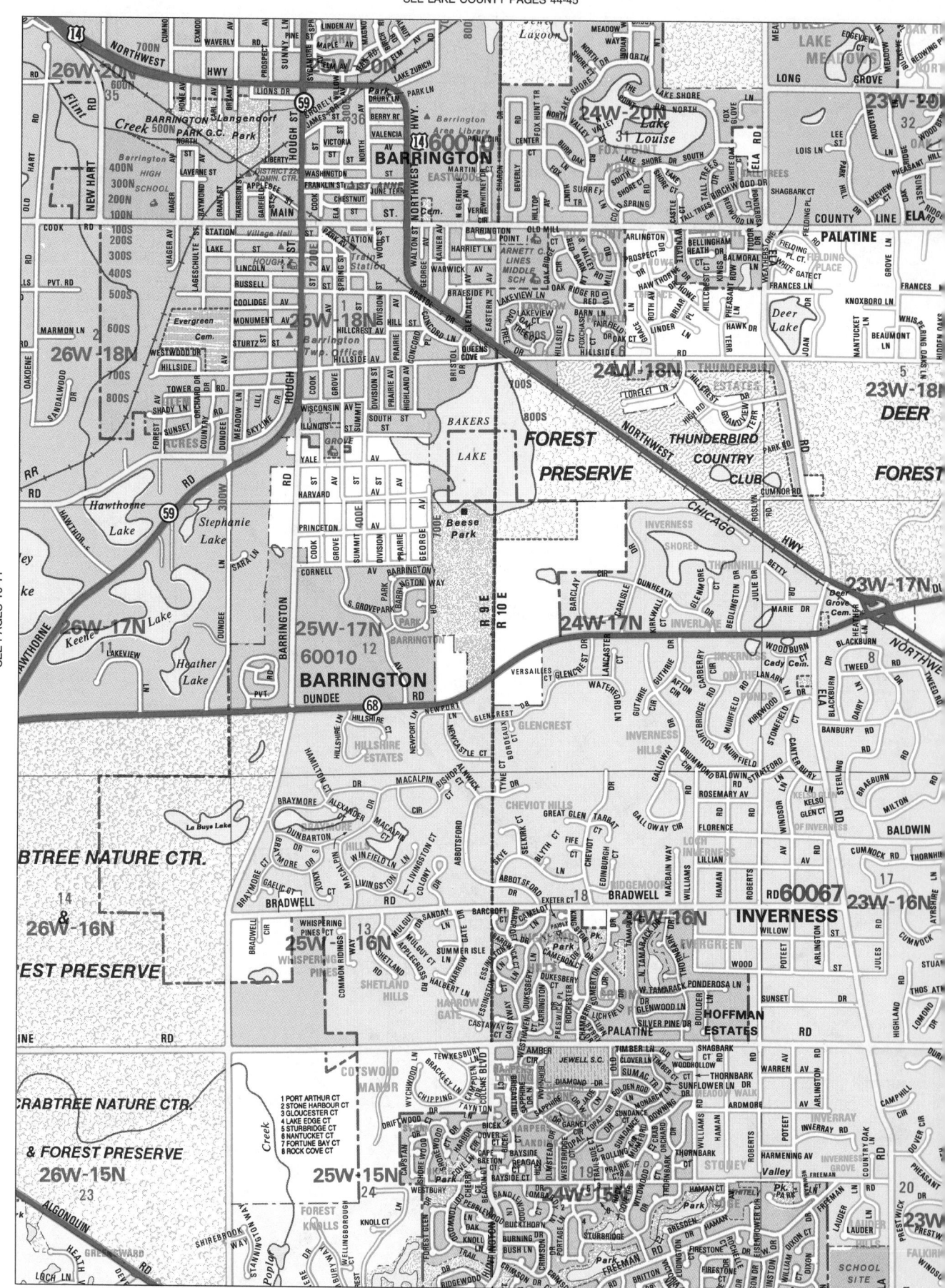

SEE PAGES 22-23

SEE LAKE COUNTY PAGES 46-47

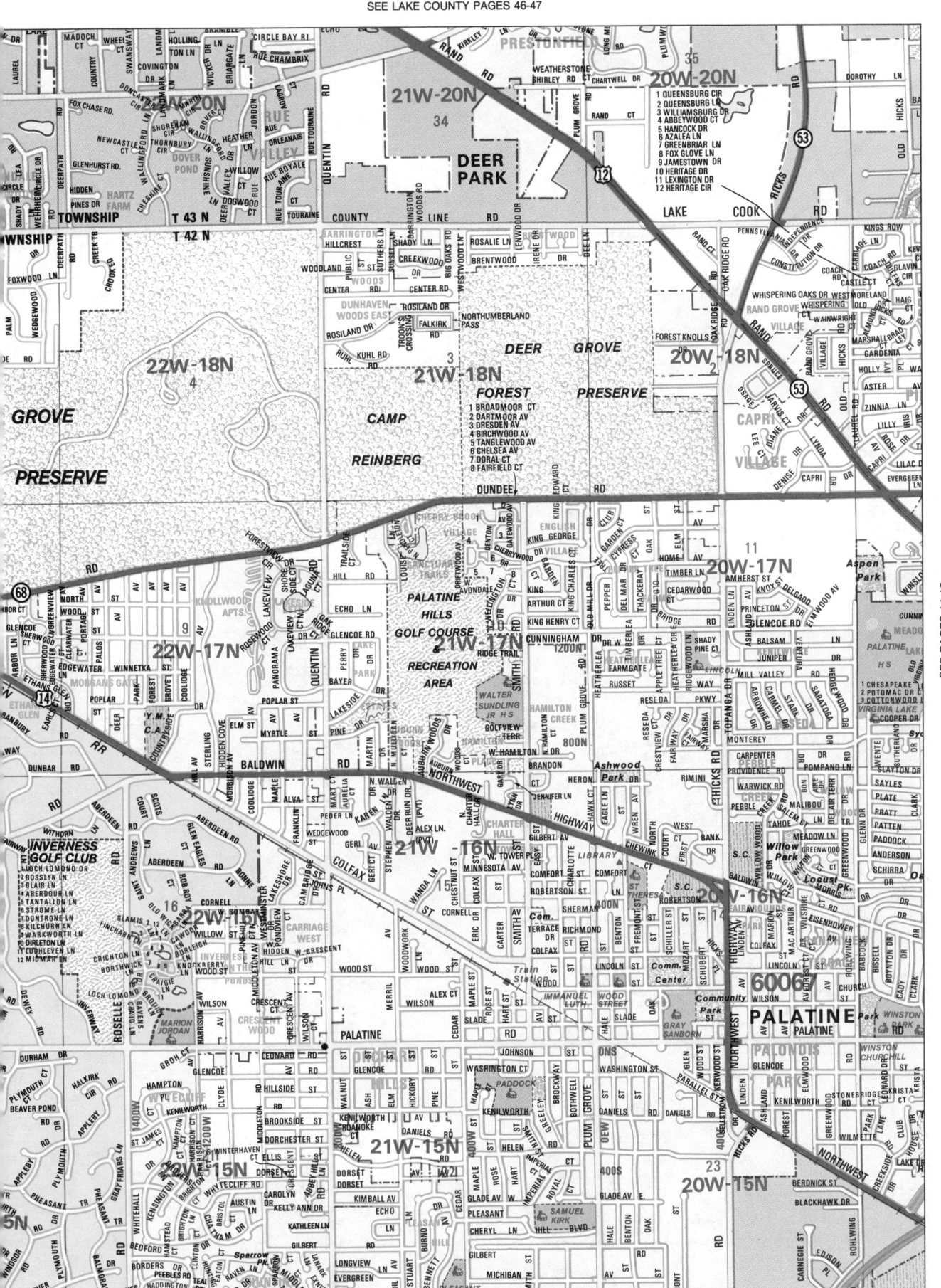

SEE PAGES 14-15

SEE PAGES 22-23

SEE LAKE COUNTY PAGES 46-47

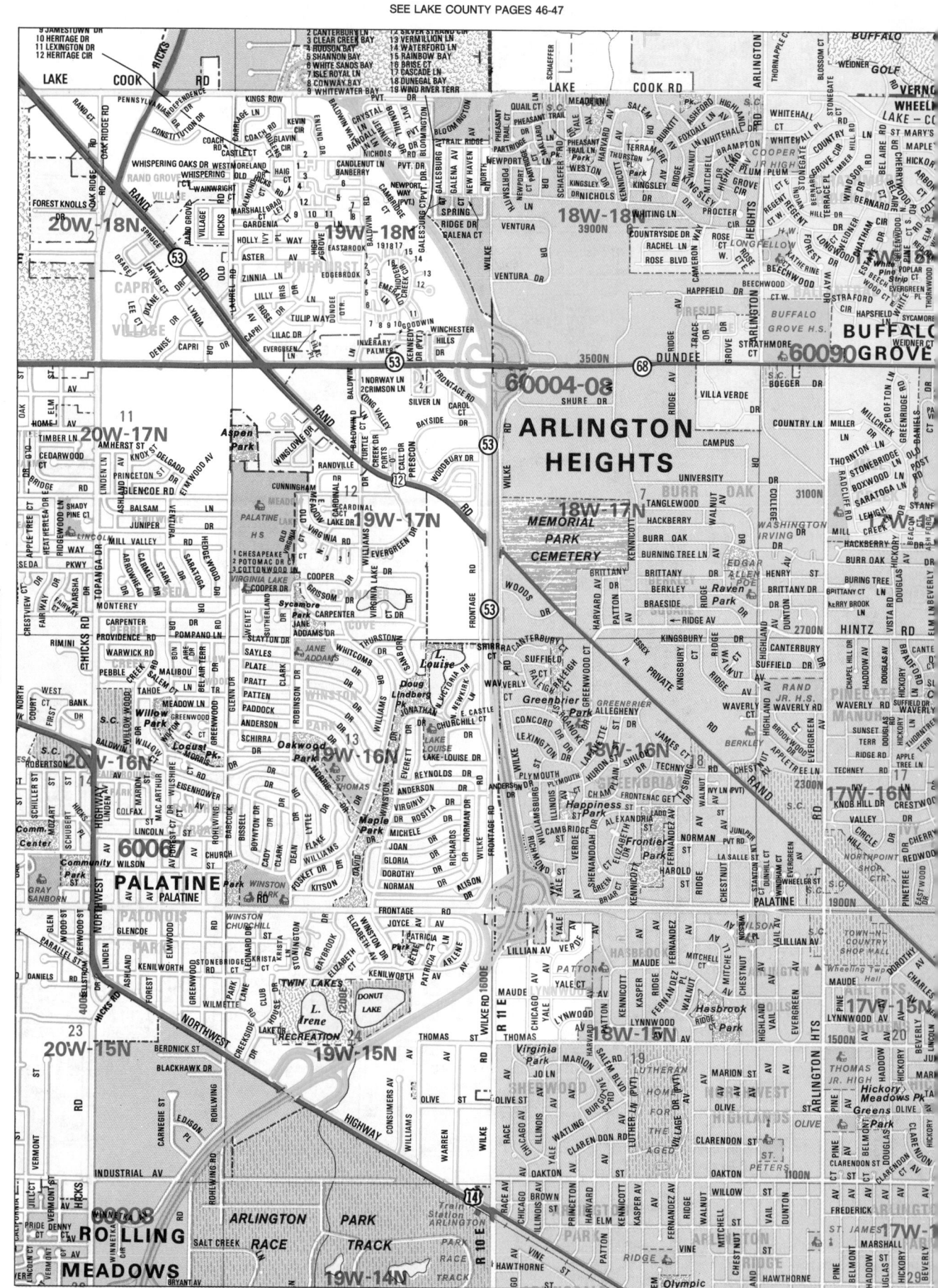

SEE PAGES 12-13

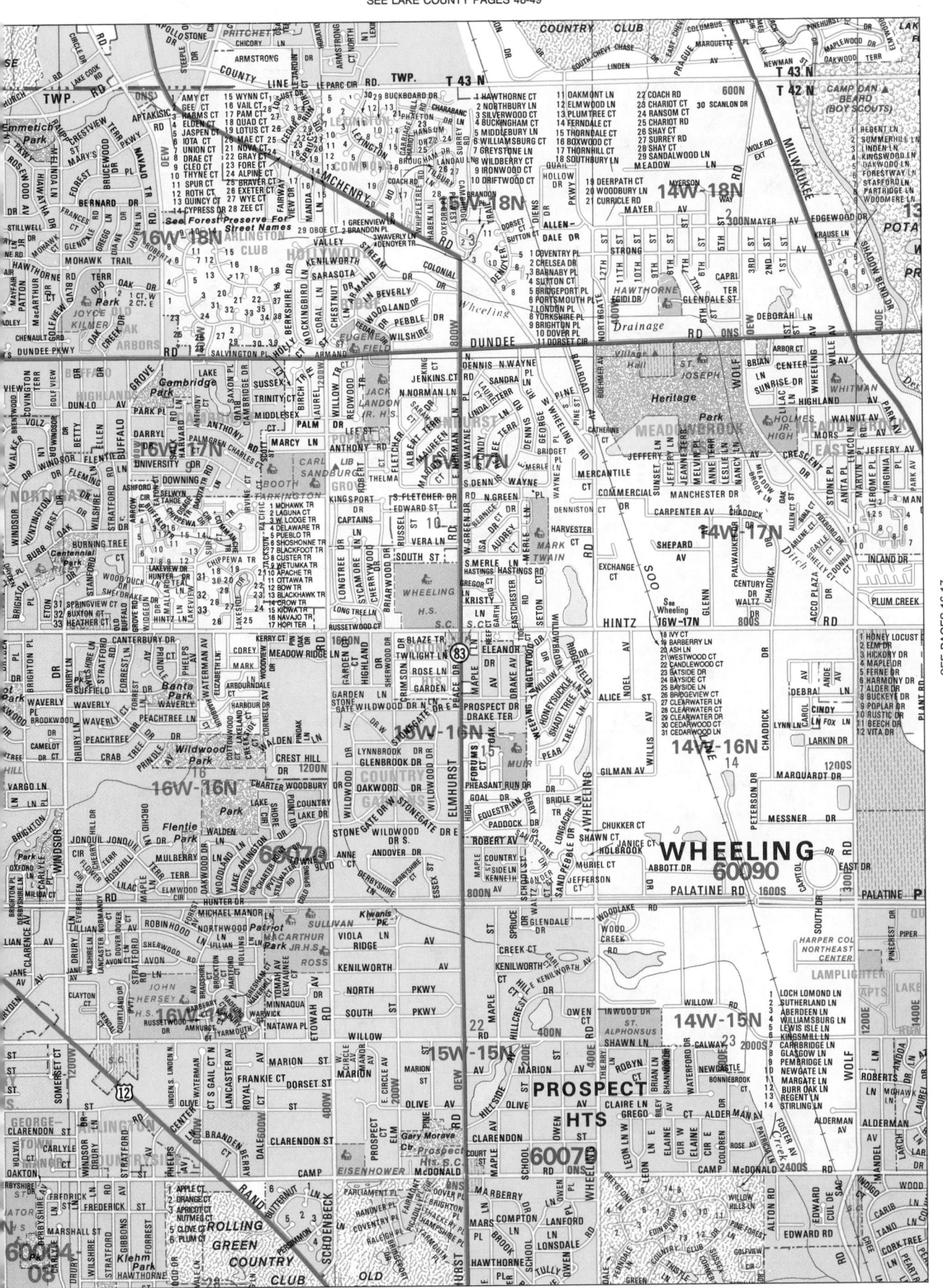

SEE LAKE COUNTY PAGES 48-49

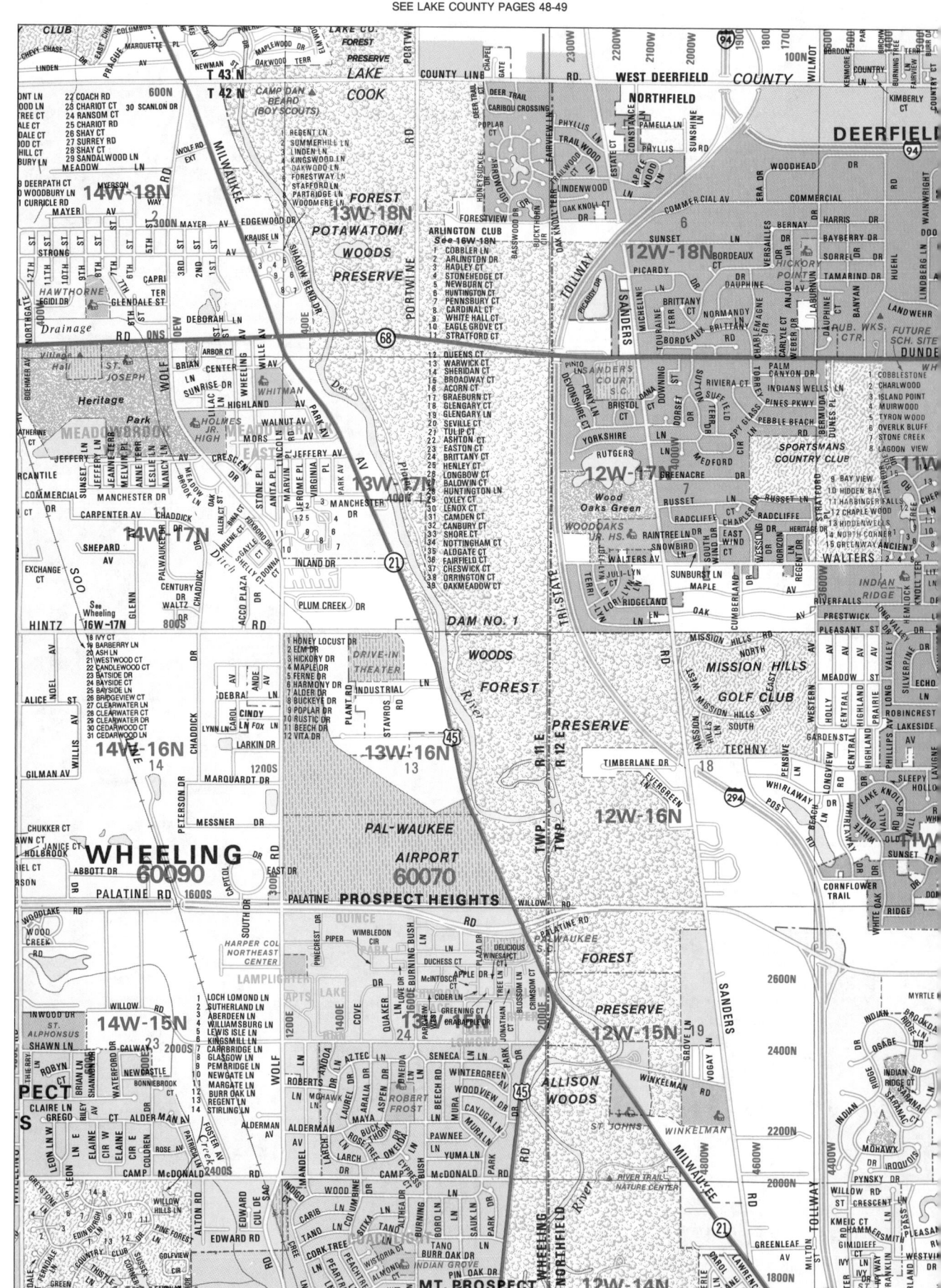

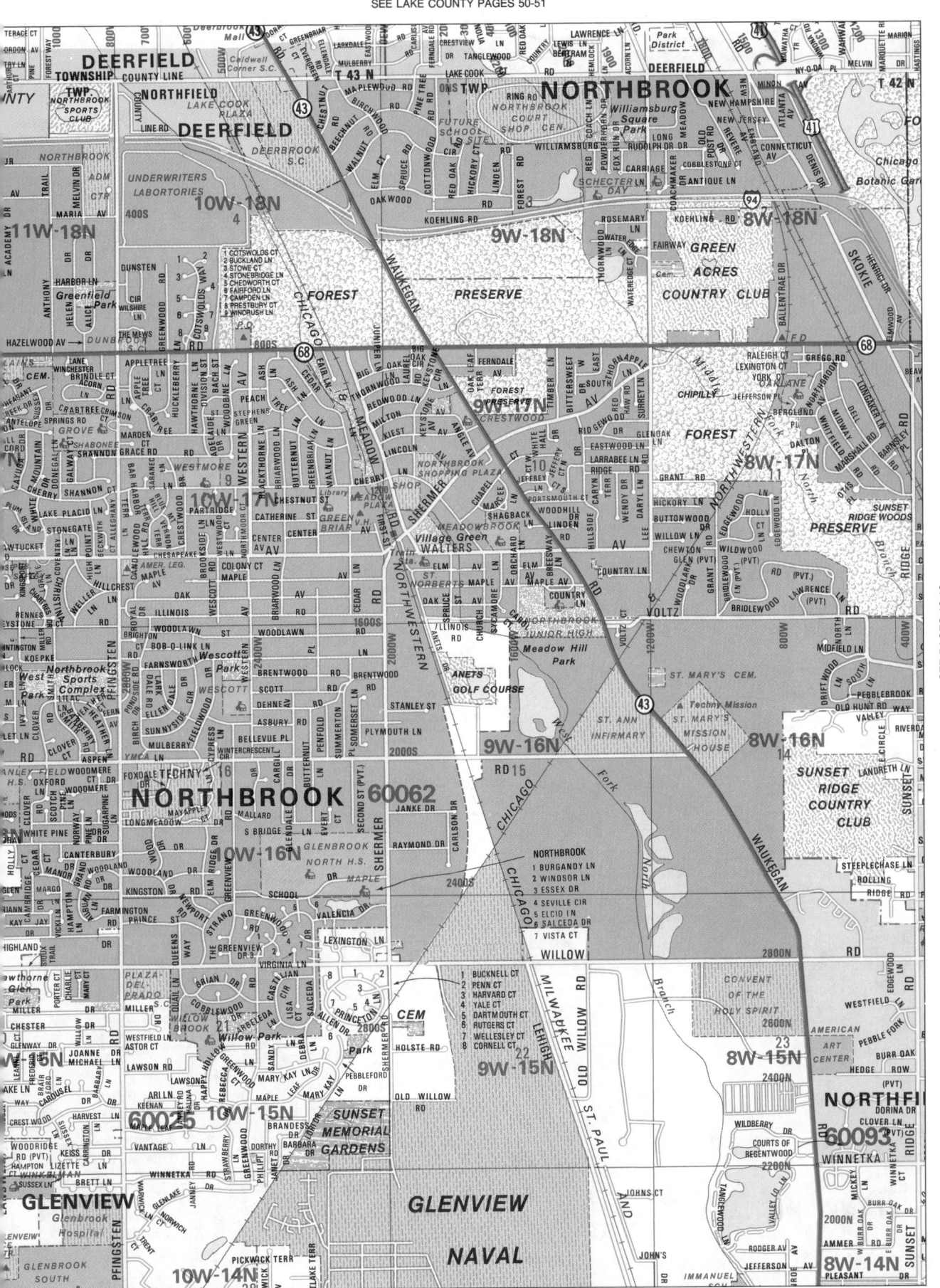

SEE PAGES 18-19

SEE LAKE COUNTY PAGES 50-51

SEE PAGES 16-17

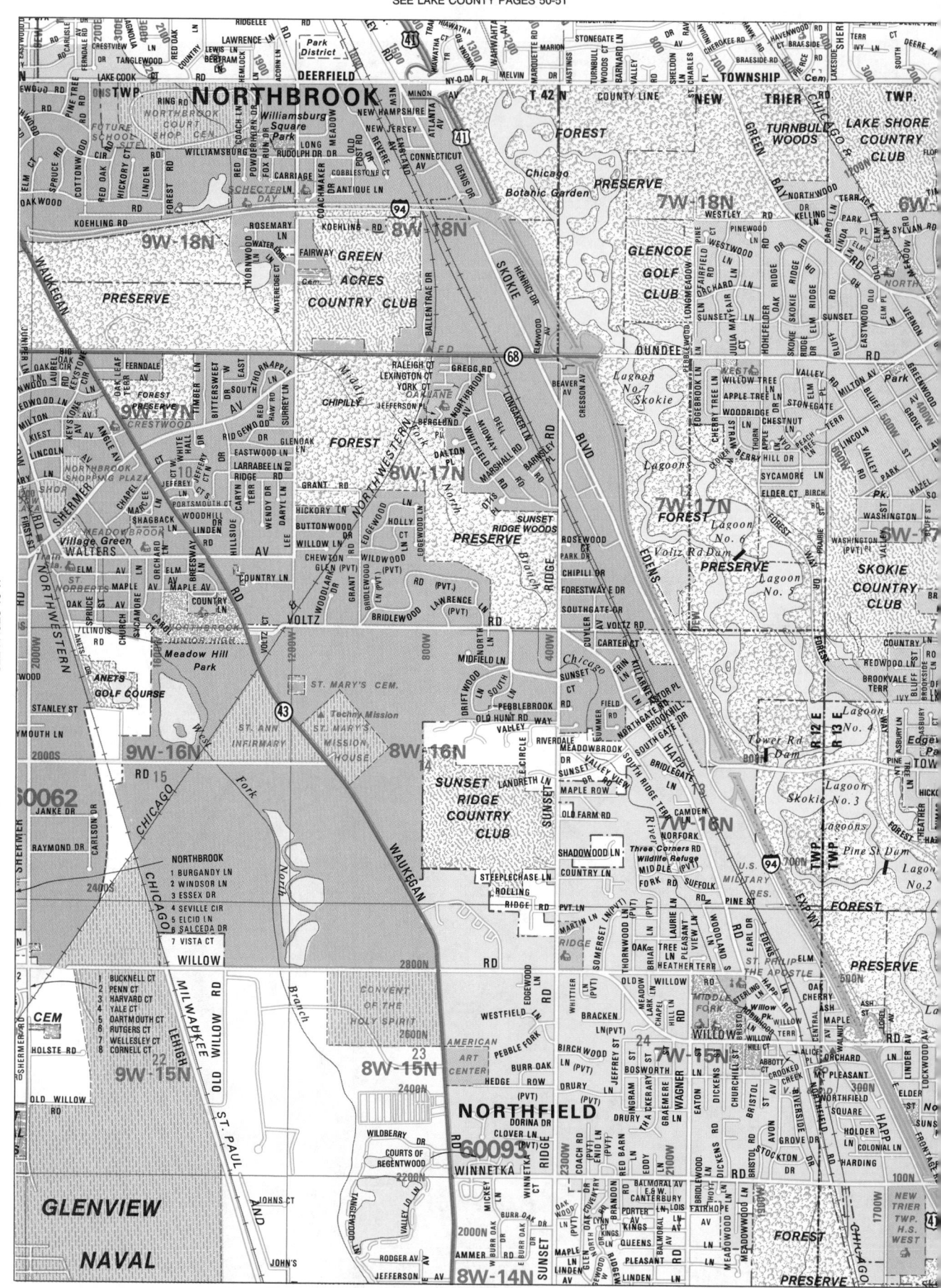

SEE PAGES 28-29

GLENCOE
60022

WINNETKA
5W-15N   60093

KENILWORTH
60043

LAKE MICHIG

SEE PAGES 10-11

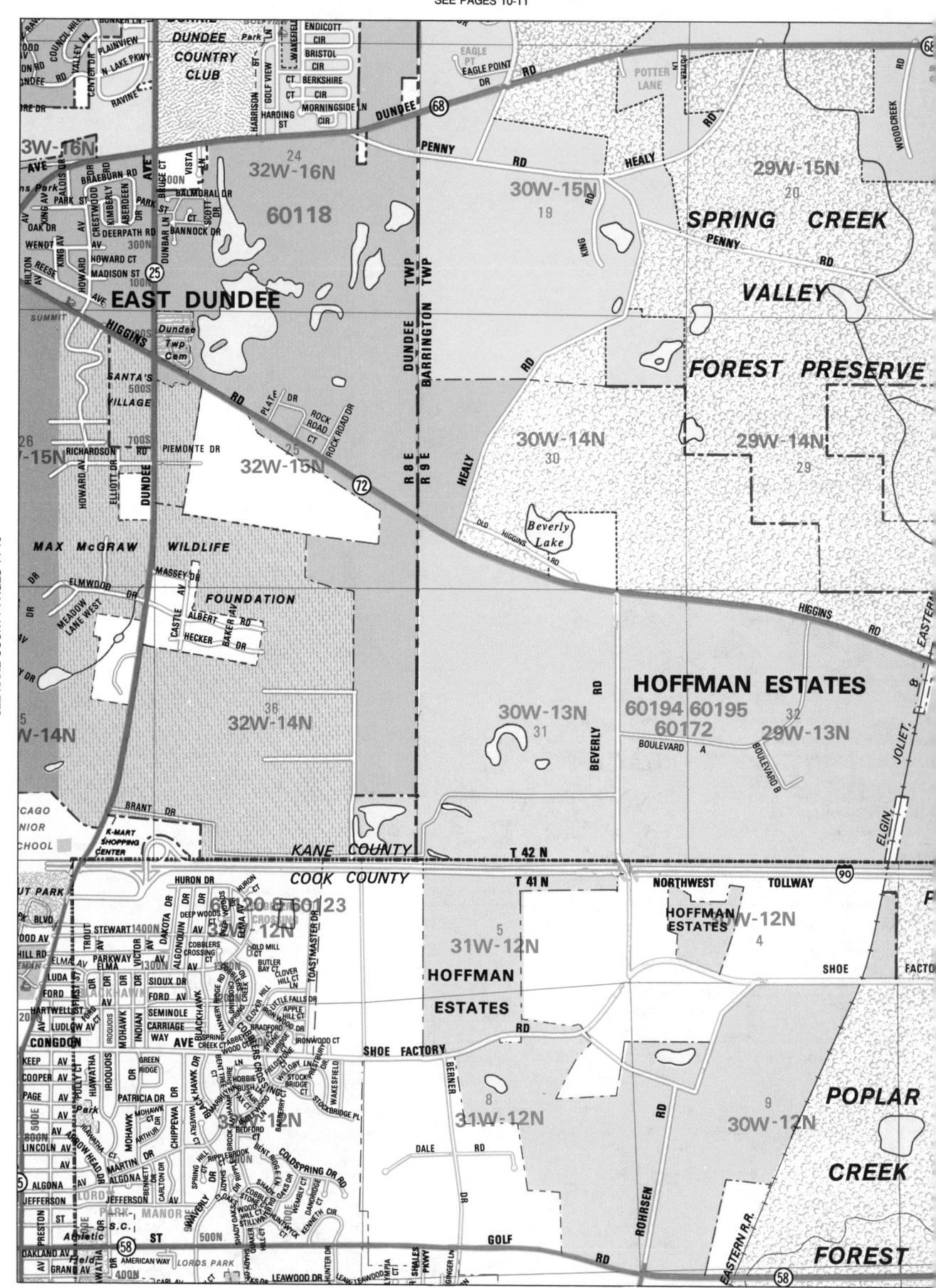

SEE PAGES 30-31

SEE PAGES 10-11

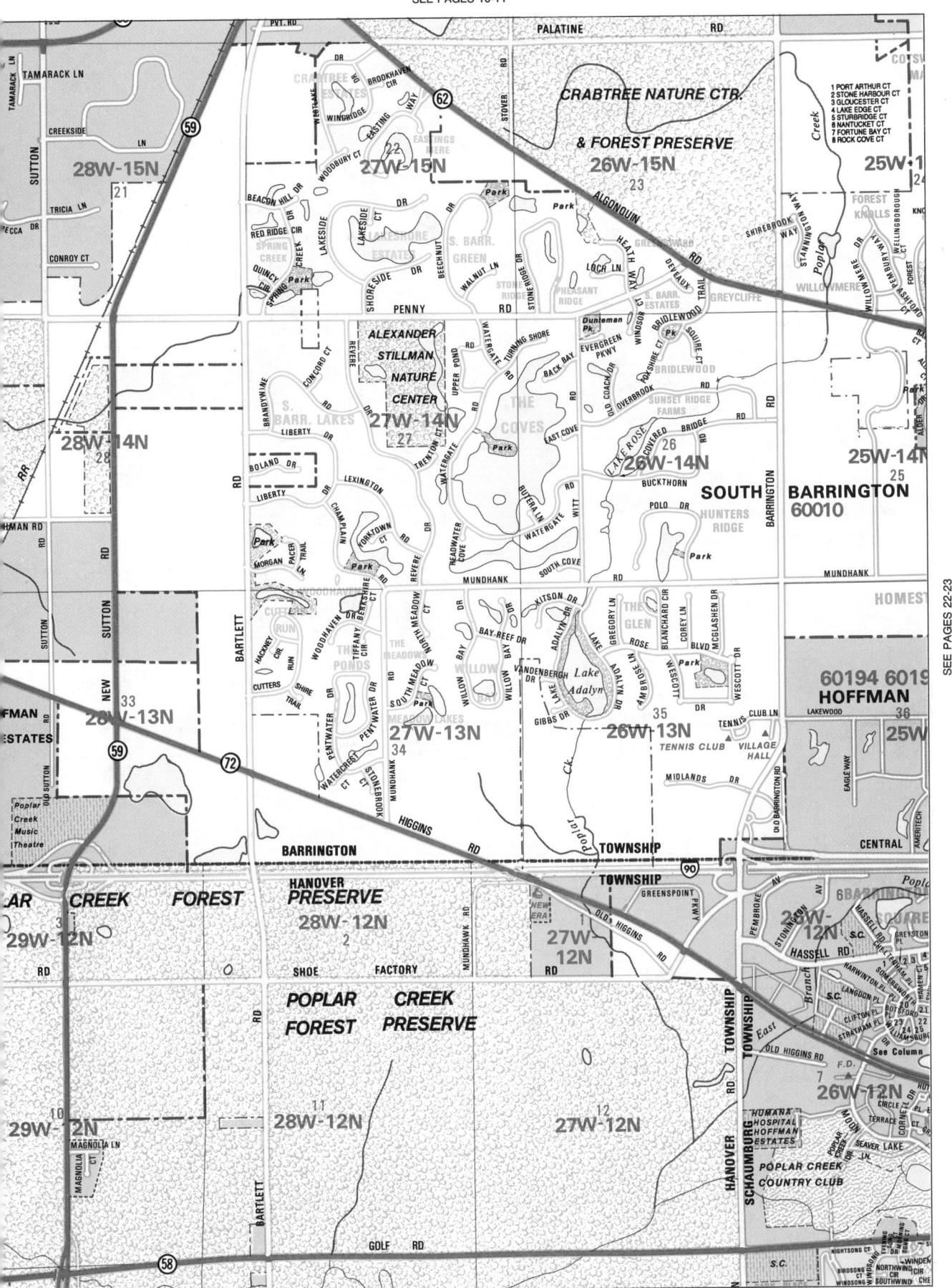

SEE PAGES 22-23

SEE PAGES 12-13

SEE PAGES 20-21

HOFFMAN ESTATES
PALATINE
SOUTH BARRINGTON
SOUTH BARRINGTON 60010
HOFFMAN ESTATES
60194 60195 60172
PAUL DOUGLAS FOREST PRESERVE
PALATINE TOWNSHIP
SCHAUMBURG TOWNSHIP
HANOVER TOWNSHIP
HIGHLAND WOODS GOLF COURSE
POPLAR CREEK COUNTRY CLUB
HILLDALE COUNTRY CLUB
HUMANA HOSPITAL HOFFMAN ESTATES
60172 60193-95
25W-15N
24W-15N
23W-15N
25W-14N
24W-14N
23W-14N
25W-13N
25W-12N
24W-12N
26W-12N
1 PORT ARTHUR CT
2 STONE HARBOUR CT
3 GLOUCESTER CT
4 LAKE EDGE CT
5 STURBRIDGE CT
6 NANTUCKET CT
7 FORTUNE BAY CT
8 ROCK COVE CT
COLUMN "A"
1 HOLBROOK LN.
2 CARDIGAN PL.
3 BRIGHTON LN.
4 OXFORD LN.
5 KETTERING RD.
6 GEORGETOWN LN.
7 ERIE LN.
8 DUNMORE PL.
9 FRANKLIN PL.
10 RALEIGH PL.
11 STOCKTON DR.
12 HASTINGS DR.
13 HANCOCK DR.
14 MARQUETTE LN.
15 LIBERTY PL.
16 RALEIGH LN.
17 BAYBERRY LN.
18 GARDEN TERR.
19 SUDBURY DR.
20 WHITINGHAM LN.
21 DANBURY PL.
22 SUTHERLAND PL.
23 SMETHWICK LN.
24 GRANTHAM PL.
25 WELLINGTON PL.
COLUMN "B"
1 MIDDLEBURY CT.
2 SANDHURST CT.
3 OAK KNOLL CT.
4 BOXWOOD CT.
5 DRIFTWOOD CT.
6 IRONWOOD CT.
7 DOGWOOD CT.
8 BIRCHWOOD CT.
9 ARROWOOD CT.
10 SANDALWOOD CT.
11 WHITEHALL CT.
12 CLARIDGE CT.
13 OAK MEADOW CT.
14 OAKMONT CT.
15 PLUM TREE CT.
16 SOUTHBURY CT.
17 WOODBURY CT.
18 GREYSTONE CT.
19 WARWICK CT.
20 FERNWOOD CT.
21 SIEVERWOOD CT.
22 LEXINGTON CT.
COLUMN "C"
1 ANDOVER CT.
2 DENTON CT.
3 HUNTLY CT.
4 SHAW CT.
5 NEWTON CT.
6 RAMSEY CIR.
7 HYDE CT.
8 KENDALL CT.
9 BURGESS CT.
10 CLAYTON CIR.
11 ACADEMY CT.
12 MANOR CIR.
13 BRYN MAWR CT.
14 BRITTANY CT.
15 CARDINAL CT.
16 DORCHESTER CT.
17 DEERFIELD CT.
18 EAGLE CT.
19 FLOWER CT.
20 GLENVIEW CT.
21 OXHILL CT.
22 PALACE CT.
23 OLD KINGS CT.
24 ONYX CT.
25 LIBERTY CT.
26 LEAR CT.
27 KAVALIER CT.
28 KNOLLWOOD CT.
29 HAMILTON CIR
30 HAMILTON PL
COLUMN "D"
1 FISKEVILLE LN.
2 DANVERS CT.
3 STONINGTON CT.
4 SWANSEA CT.
5 WOONSOCKET CT.
6 TAUNTON CT.
7 BELMONT CT.
8 MATFIELD CT.
9 GROTON CT.
10 RUSKIN CT.
11 HOLLISTON CT.
12 WALPOLE LN.
13 SUDBURY CT.
14 GLASTONBURY LN.
15 SAYLESVILLE LN.
16 SHANNOCK LN.
17 MOHEGAN LN.
COLUMN "E"
1 HASTINGS CT.
2 FINCHLEY CT.
3 RICHMOND CT.
4 CARNABY CT.
5 DUNBAR CT.
6 BISHOP CT.
7 ANGUS CT.
8 CARLISLE CT.
9 MANSFIELD CT.
10 FENWICK CT.
11 HANLEY CT.
12 LARGO CT.
13 BROMLEY CT.
14 CARDIFF CT.
15 DUNSFORD CT.
16 LANCASHIRE CT.
17 STOCKBRIDGE CT.
18 CLARENDON SPRINGS CT.
1 WILLOW TREE CT
2 CEDAR TREE CT
3 JUNIPER TREE CT
ALGONQUIN RD
NORTHWEST TOLLWAY
HIGGINS RD
GOLF RD
CENTRAL RD
MUNDHANK RD
FREEMAN RD
BARRINGTON RD
ELA RD
HASSELL RD
HOMESTEAD
CENTER DR
LAKEWOOD BLVD

SEE PAGES 32-33

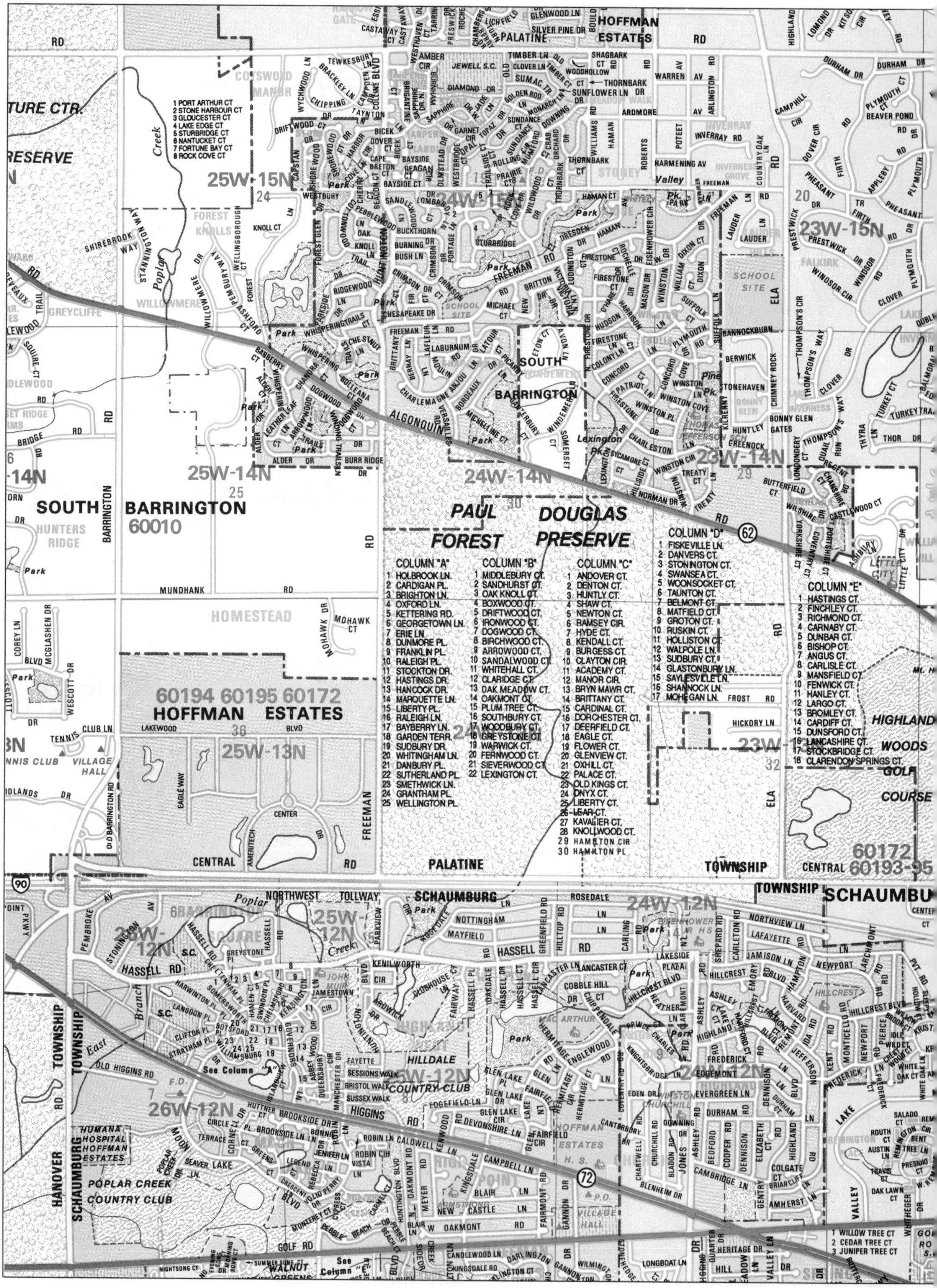

SEE PAGES 12-13

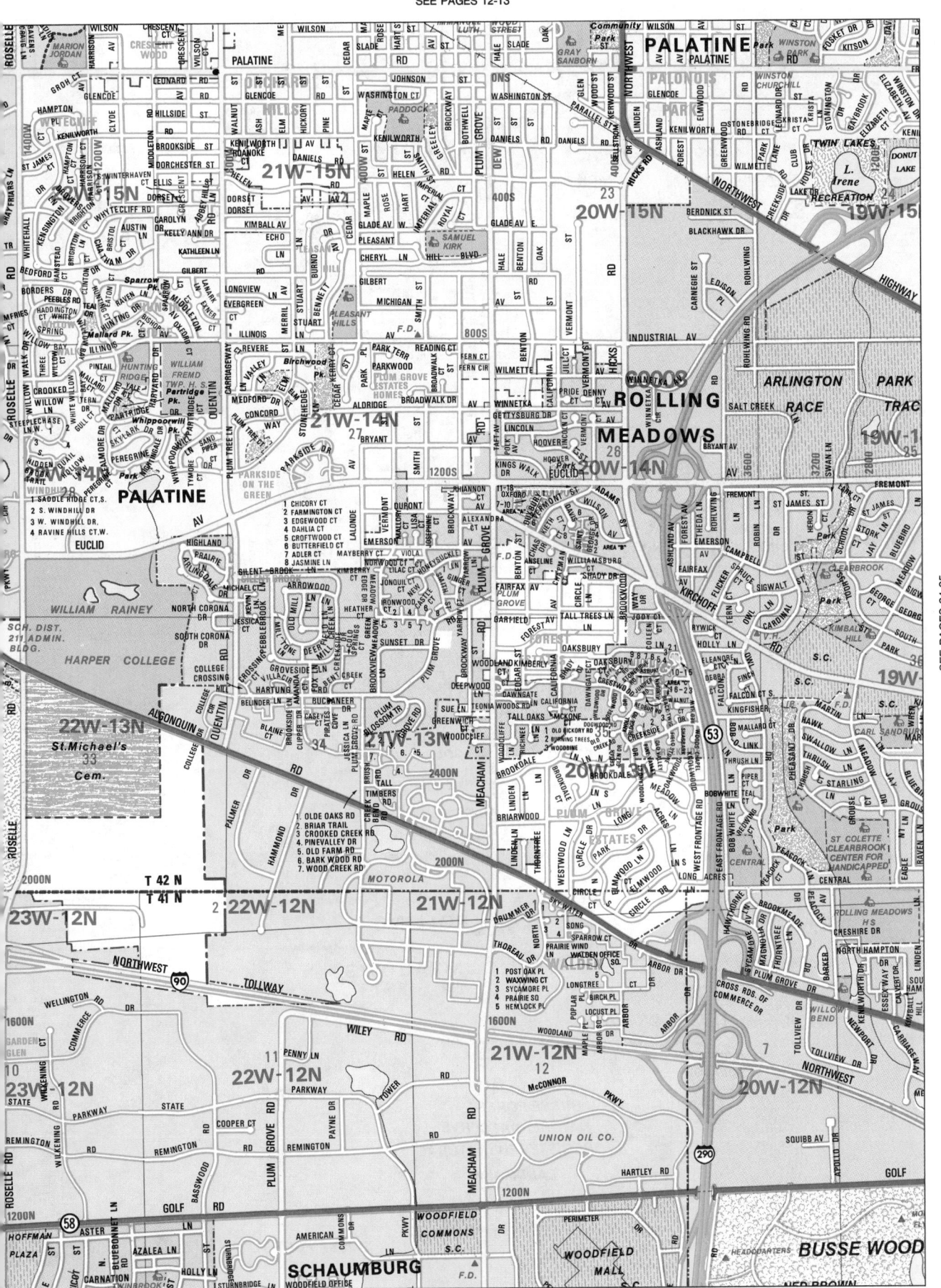

SEE PAGES 24-25

SEE PAGES 14-15

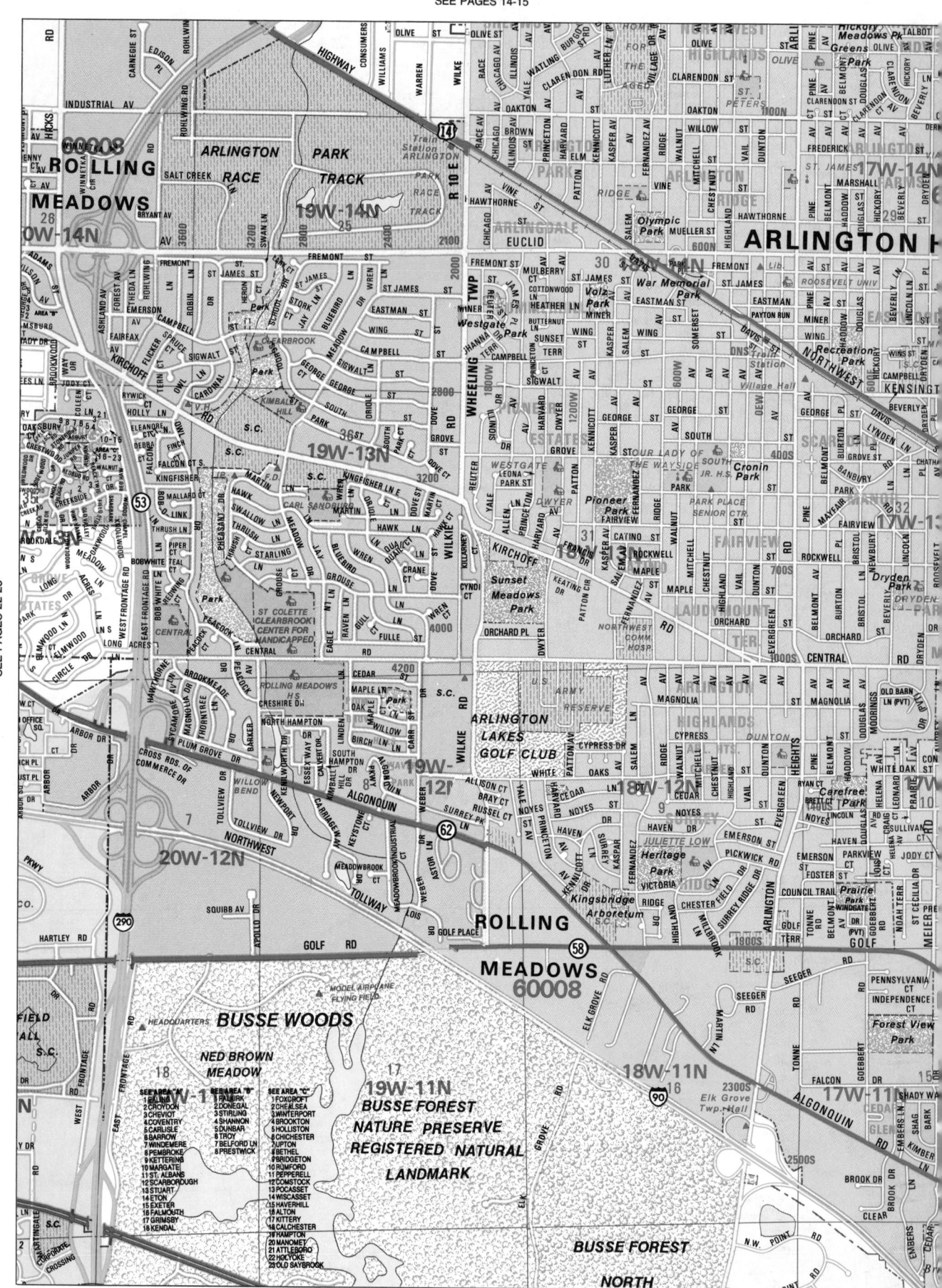

SEE PAGES 22-23

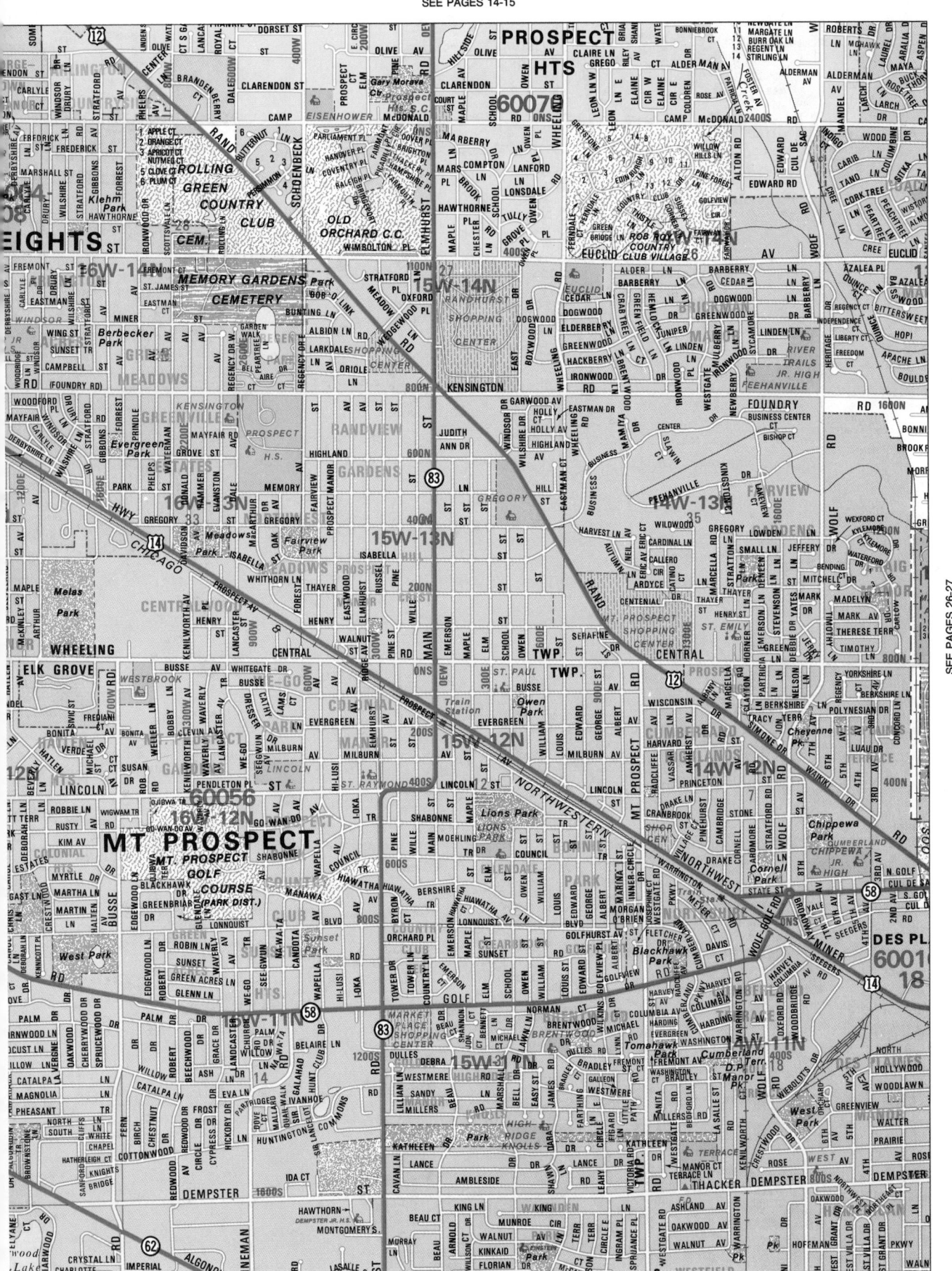

SEE PAGES 26-27

SEE PAGES 16-17

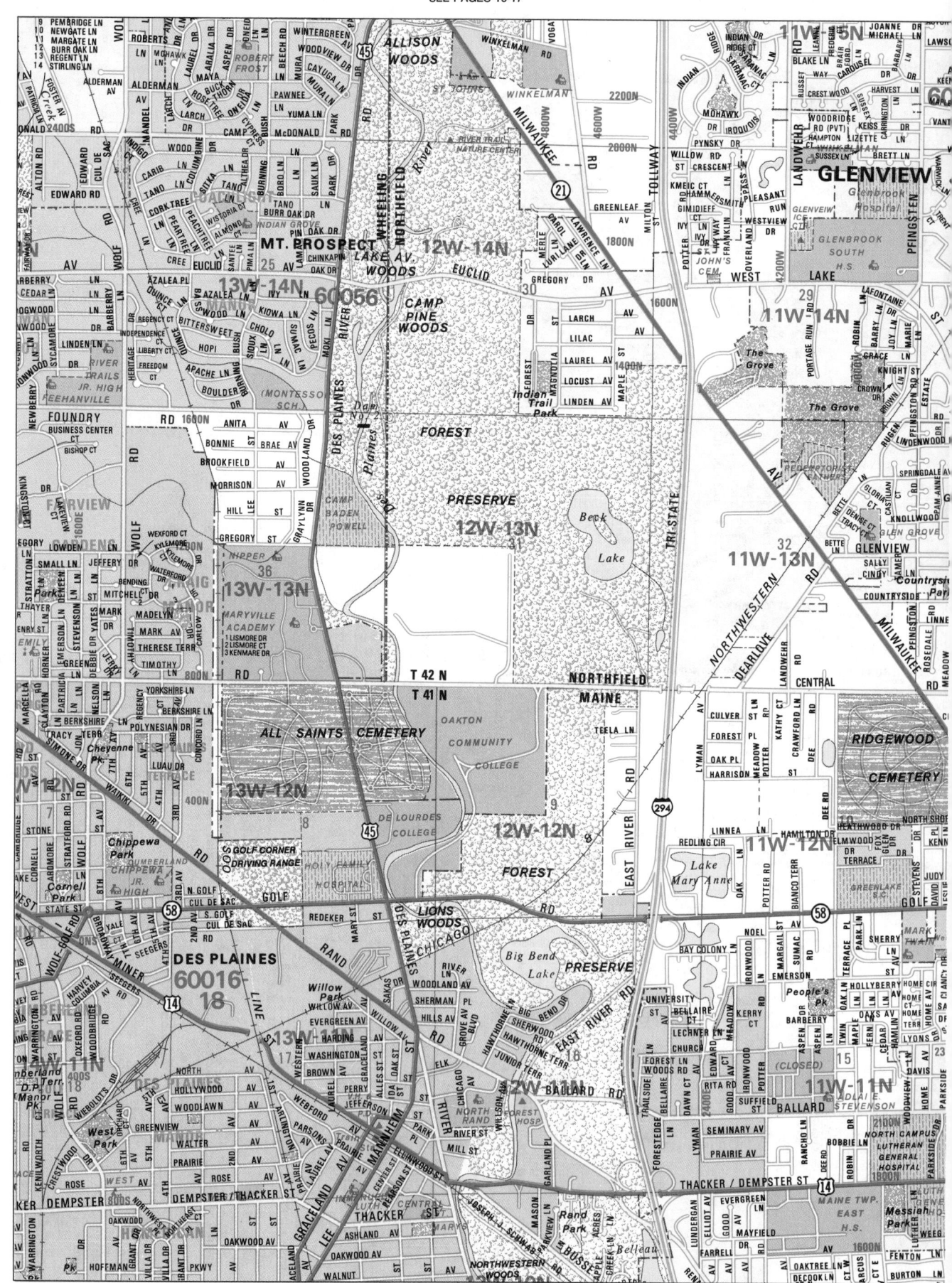

SEE PAGES 24-25

SEE PAGES 16-17

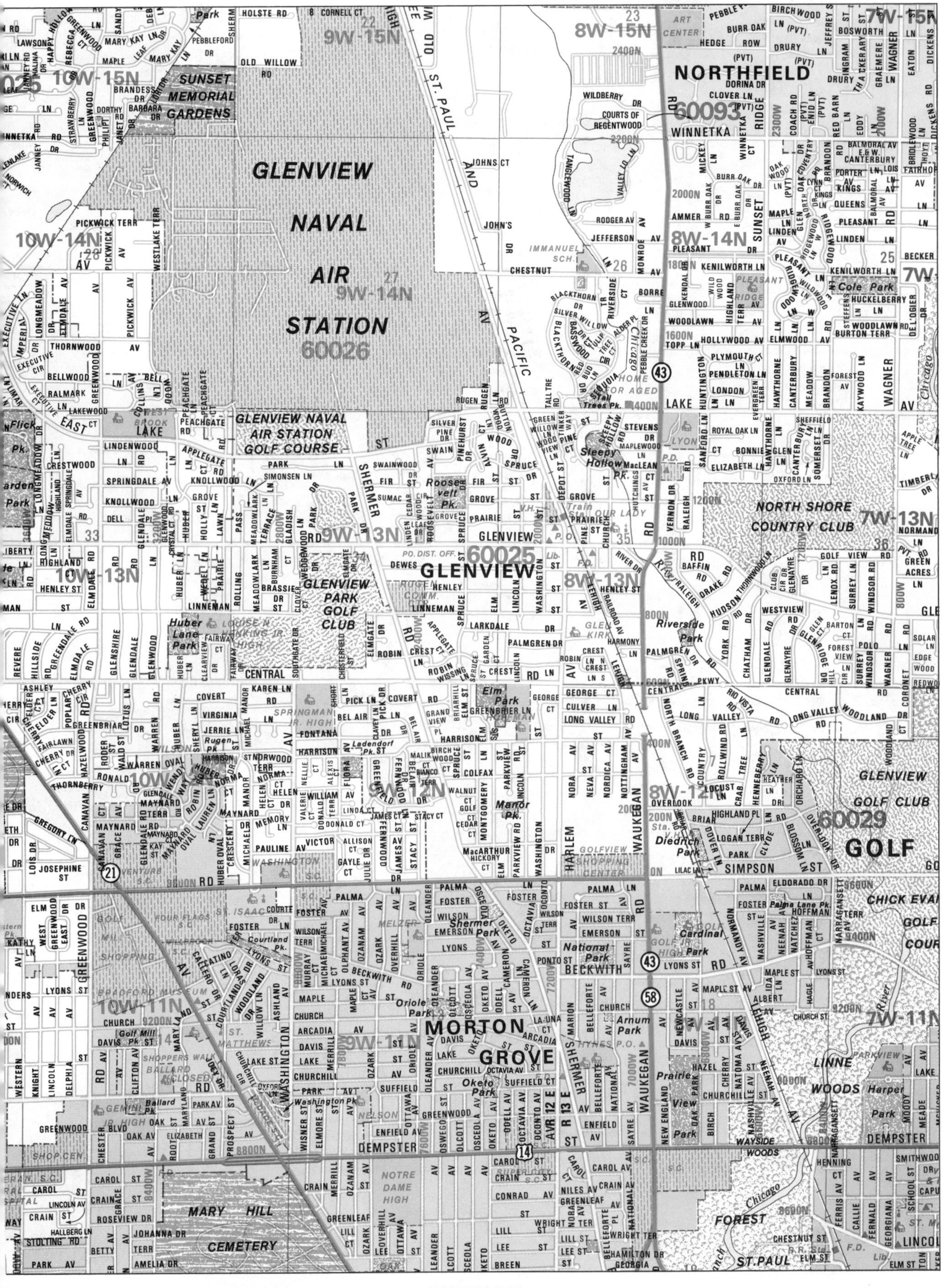

GLENVIEW
NAVAL
AIR
STATION
60026

NORTHFIELD
60093

GLENVIEW NAVAL
AIR STATION
GOLF COURSE

GLENVIEW
PARK GOLF
CLUB

GLENVIEW
60025

NORTH SHORE
COUNTRY CLUB

SUNSET
MEMORIAL
GARDENS

GLENVIEW
GOLF CLUB
60029
GOLF

CHICK EVANS
GOLF
COURSE

MORTON
GROVE

LINNE
WOODS

MARY HILL
CEMETERY

NOTRE
DAME
HIGH

FOREST

ST. PAUL

9W-15N
8W-15N
7W-15N
10W-15N
10W-14N
9W-14N
8W-14N
7W
9W-13N
10W-13N
8W-13N
7W-13N
9W-11N
8W-12N
10W
9W-11N
7W-11N

SEE PAGES 28-29

SEE PAGES 18-19

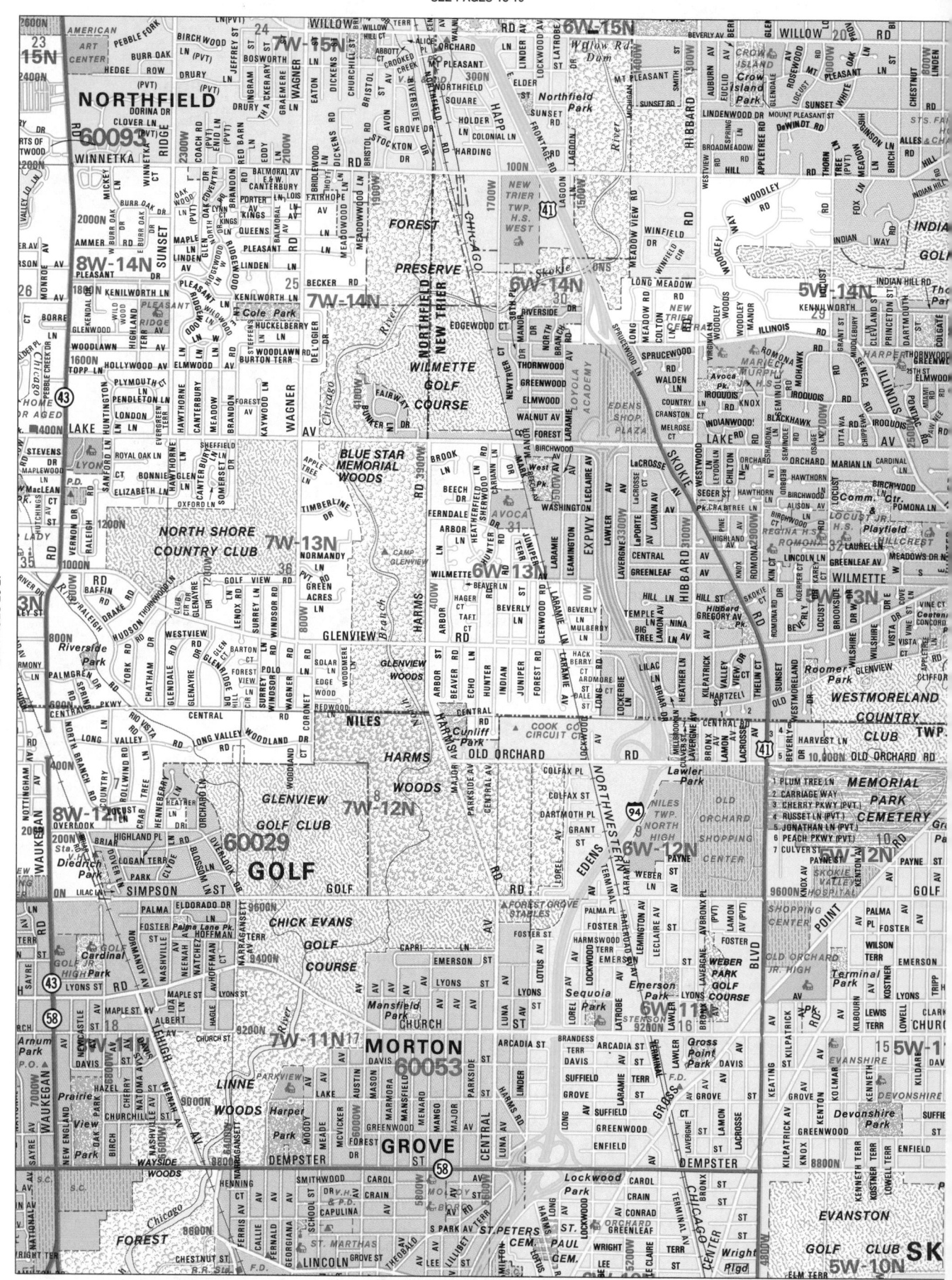

SEE PAGES 18-19

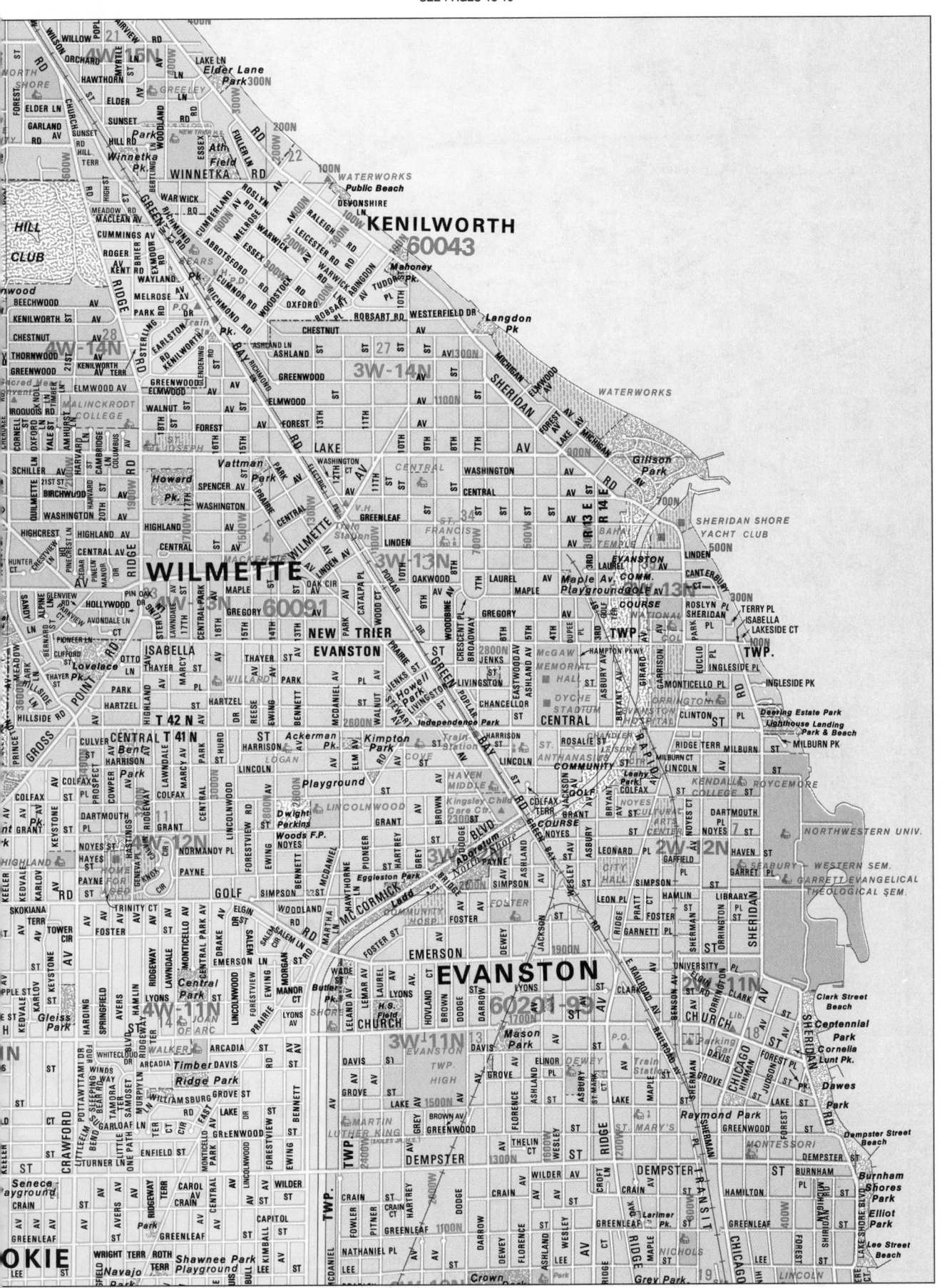

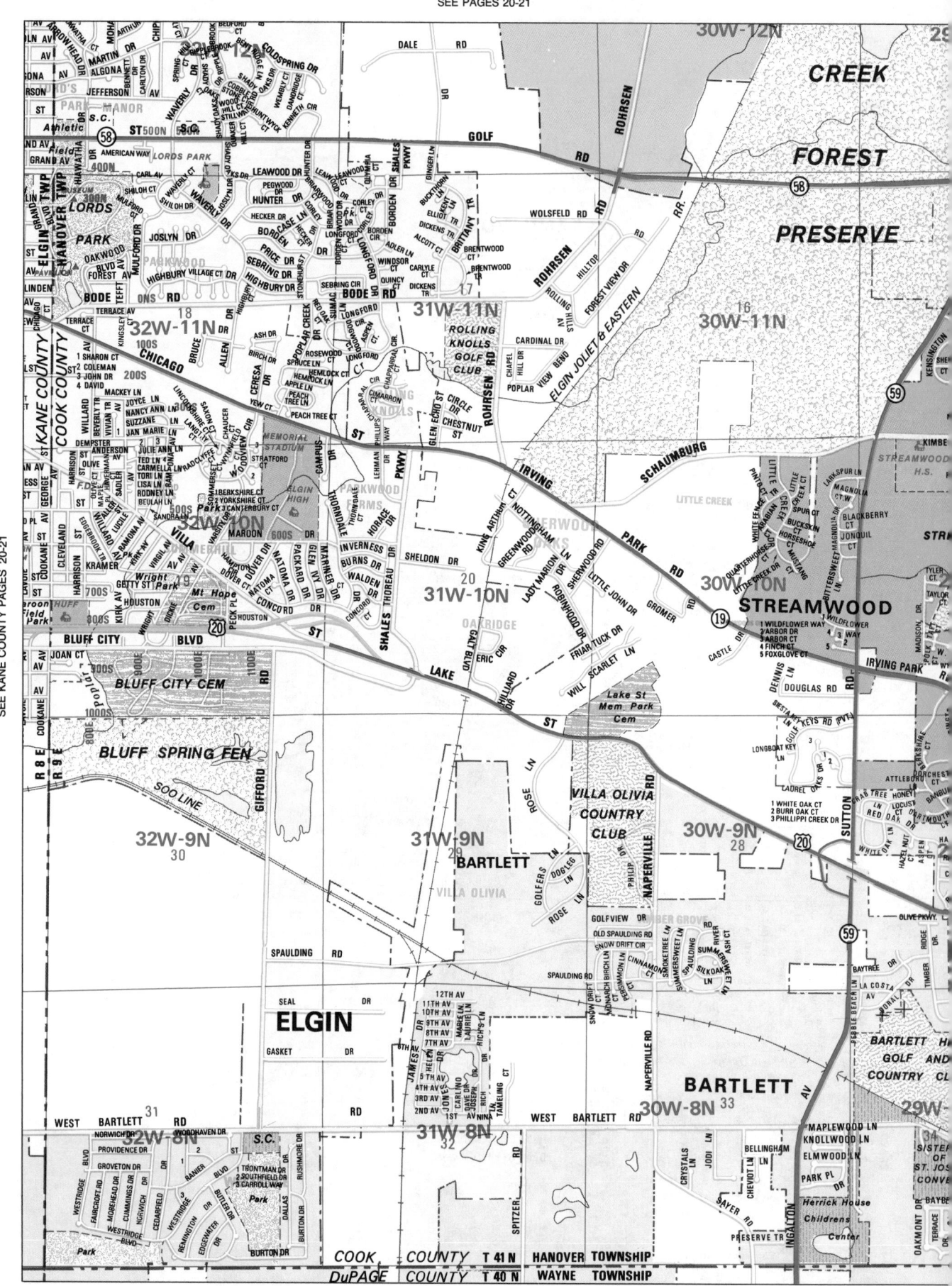

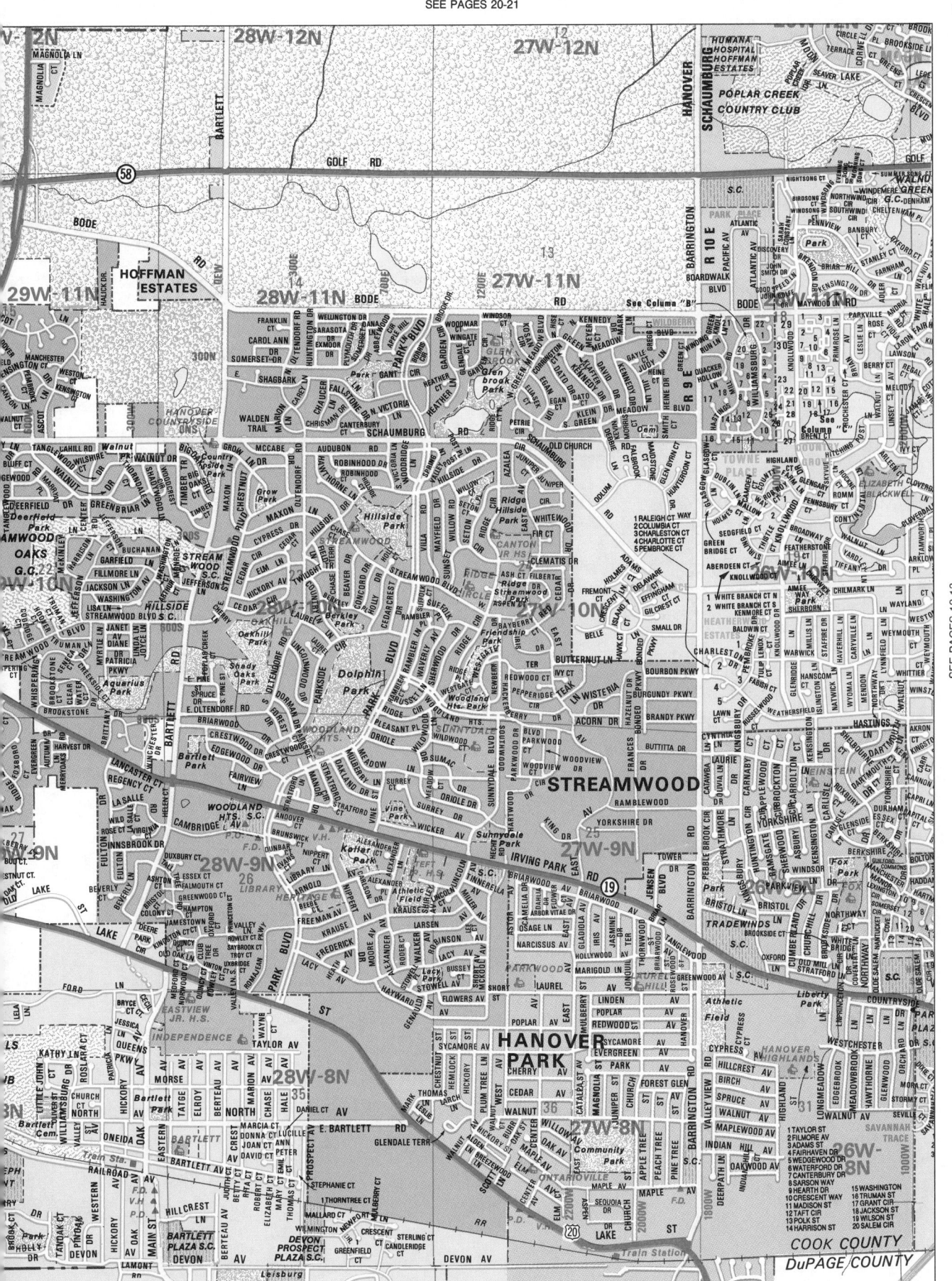

SEE PAGES 22-23

SEE PAGES 30-31

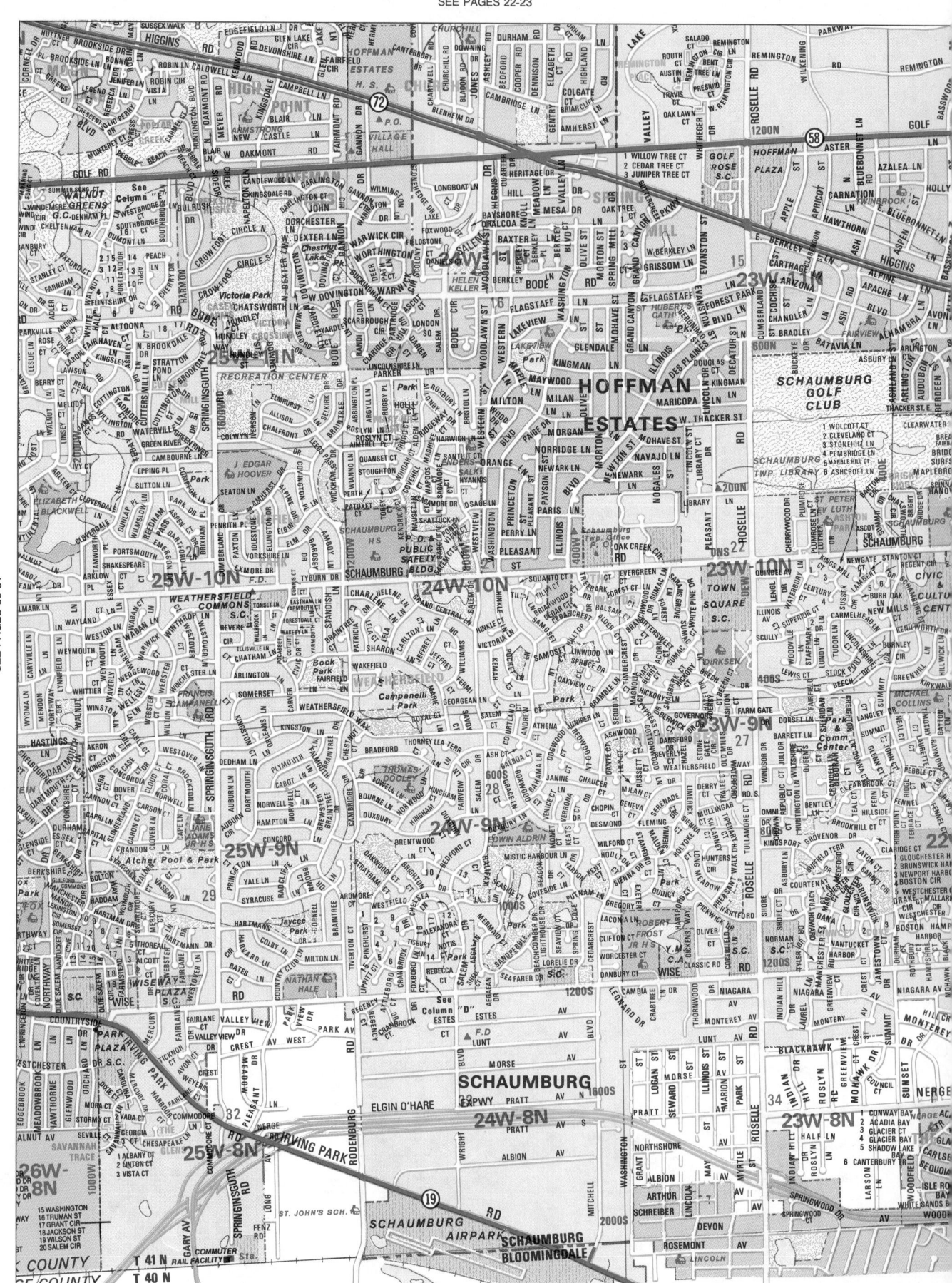

SEE PAGES 22-23

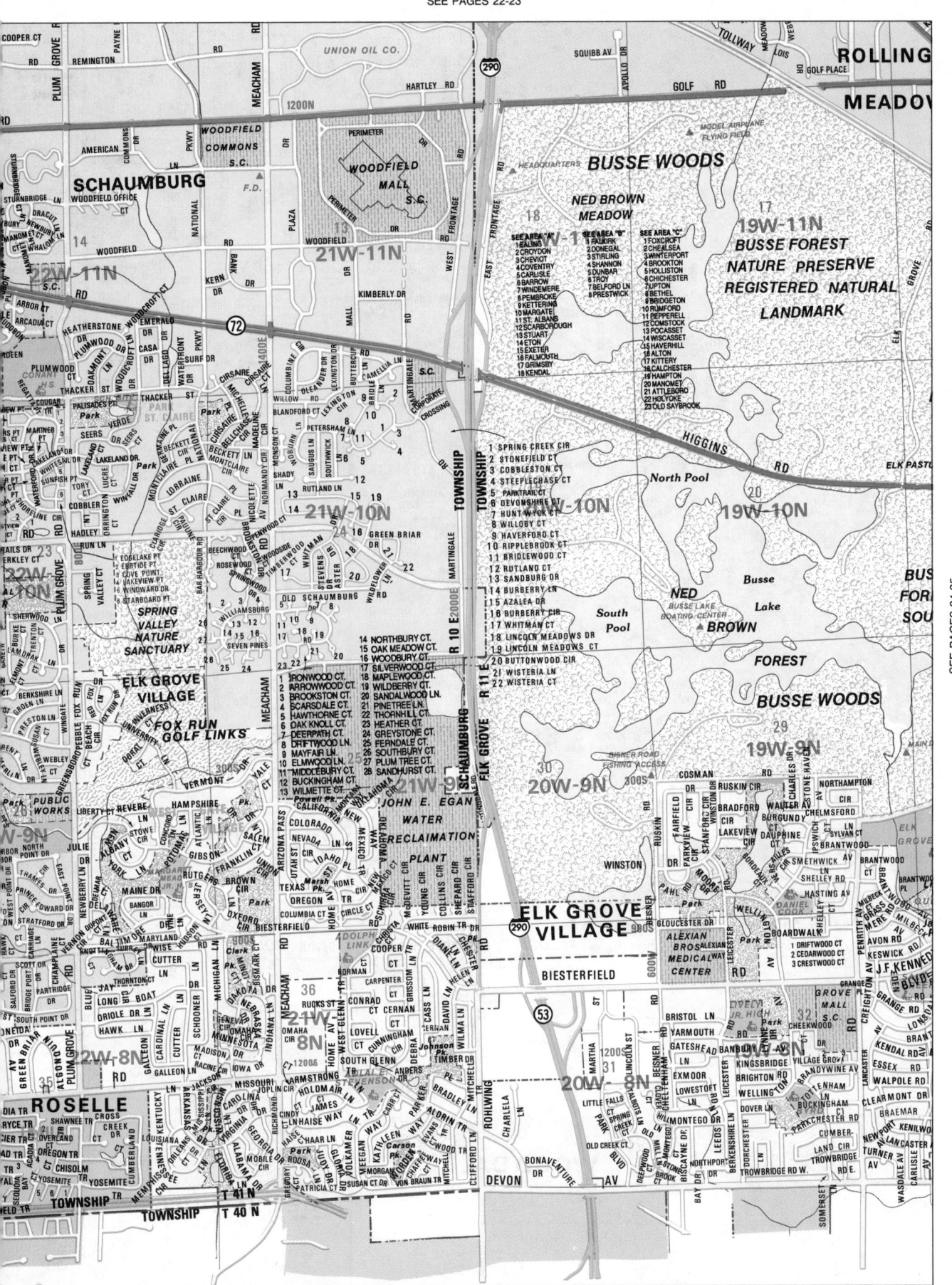

SEE PAGES 34-35

SEE PAGES 24-25

SEE PAGES 32-33

BUSSE FOREST NORTH

BUSSE FOREST SOUTH

ELK PASTURE

18W-10N

17W-10N

16W-10N

DEVONSHIRE WEST

18W-9N

17W-9N

16W-9N

HIGGINS IND. PARK

HIGGINS

COMMERCE

BRUMMEL

18W-8N

17W-8N

16W-8N

CENTEX IND. PARK

ELK GROVE VILLAGE

60007

18W-7N

17W-7N

16W-7N

COOK COUNTY

DuPAGE COUNTY

ELK GROVE TWP

ADDISON TWP

WOOD DALE

BENSENVILLE

ARLINGTON

Lions Park

Audubon Pk.

Morrison

Morton Pleasanton

Burbank Pk.

Clearmont

Muir Park

Appleseed Park

Osborn Park

Briarwood Lake

DEMPSTER

ALGONQUIN

NORTHWEST

HIGGINS

OAKTON

LANDMEIER

TOLLWAY

DEVON

THORNDALE

BRYN MAWR

INDUSTRIAL DR

SEE DU PAGE COUNTY PAGES 8-9

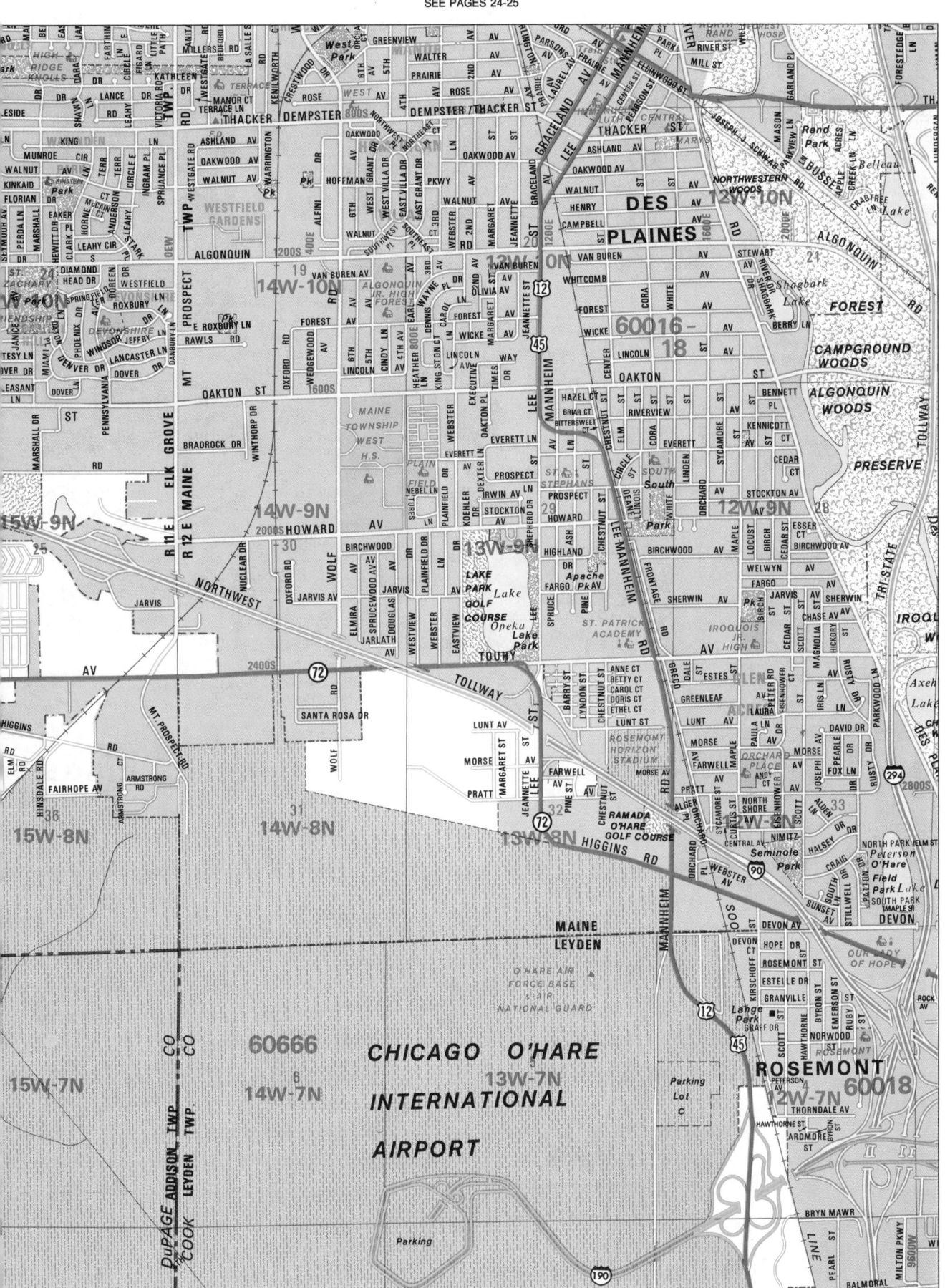

SEE PAGES 36-37

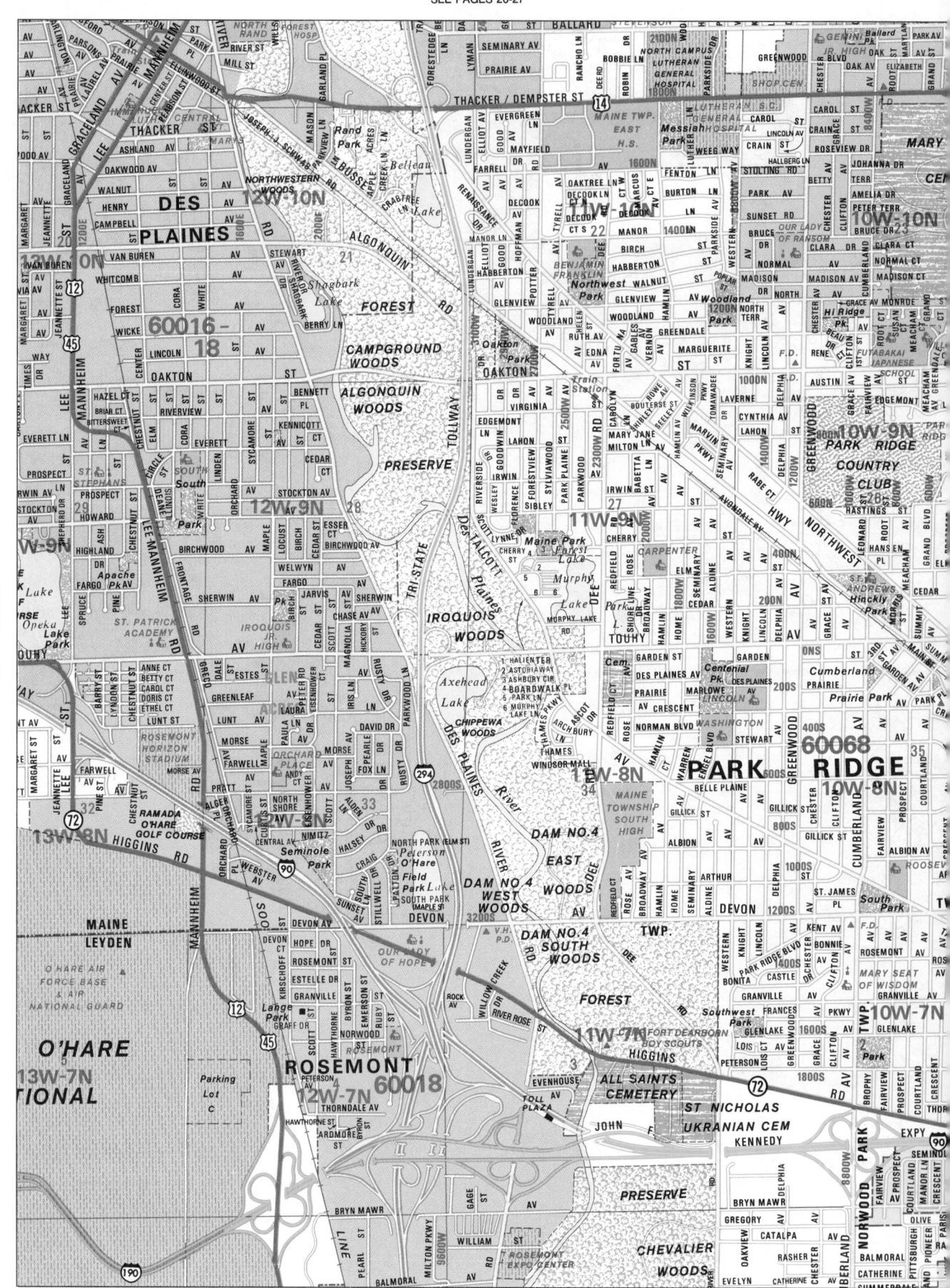

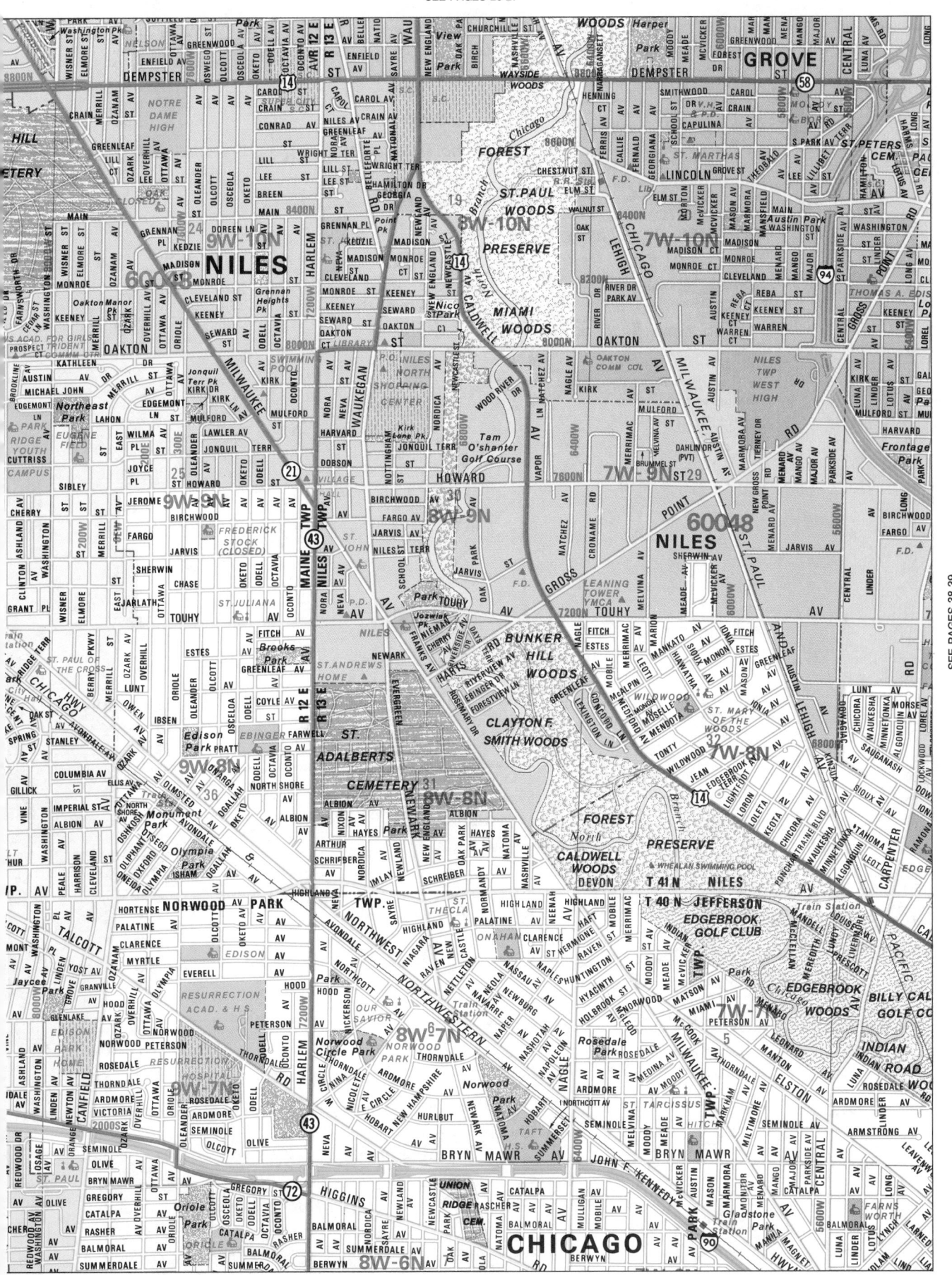

SEE PAGES 28-29

SEE PAGES 36-37

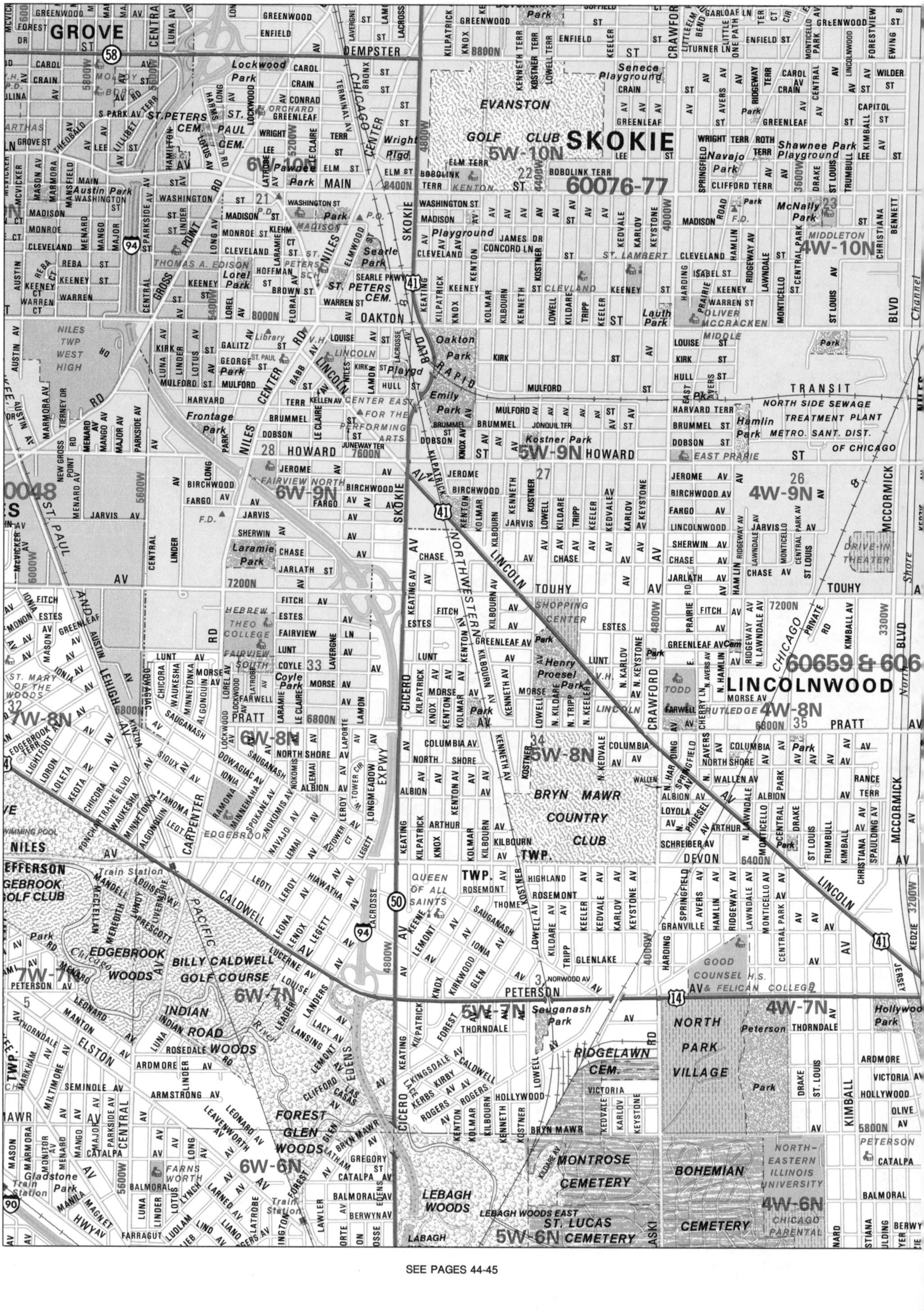

GROVE ST

EVANSTON GOLF CLUB

SKOKIE

60076-77

5W-10N

4W-10N

6W-10N

NILES TWP WEST HIGH

SKOKIE

5W-9N

6W-9N

4W-9N

NORTH SIDE SEWAGE TREATMENT PLANT METRO. SANT. DIST. OF CHICAGO

TRANSIT

DRIVE-IN THEATER

7W-8N

6W-8N

5W-8N

4W-8N

LINCOLNWOOD

60659 & 606

HEBREW THEO COLLEGE

BRYN MAWR COUNTRY CLUB

ST. MARY OF THE WOODS

JEFFERSON

EDGEBROOK GOLF CLUB

BILLY CALDWELL GOLF COURSE

7W-7N

6W-7N

5W-7N

4W-7N

QUEEN OF ALL SAINTS

TWP.

NORTH PARK VILLAGE

GOOD COUNSEL H.S. & FELICAN COLLEGE

RIDGELAWN CEM.

INDIAN ROAD WOODS

FOREST GLEN WOODS

6W-6N

5W-6N

4W-6N

MONTROSE CEMETERY

BOHEMIAN CEMETERY

LEBAGH WOODS

ST. LUCAS CEMETERY

NORTH-EASTERN ILLINOIS UNIVERSITY

CHICAGO PARENTAL

SEE PAGES 44-45

SEE PAGES 28-29

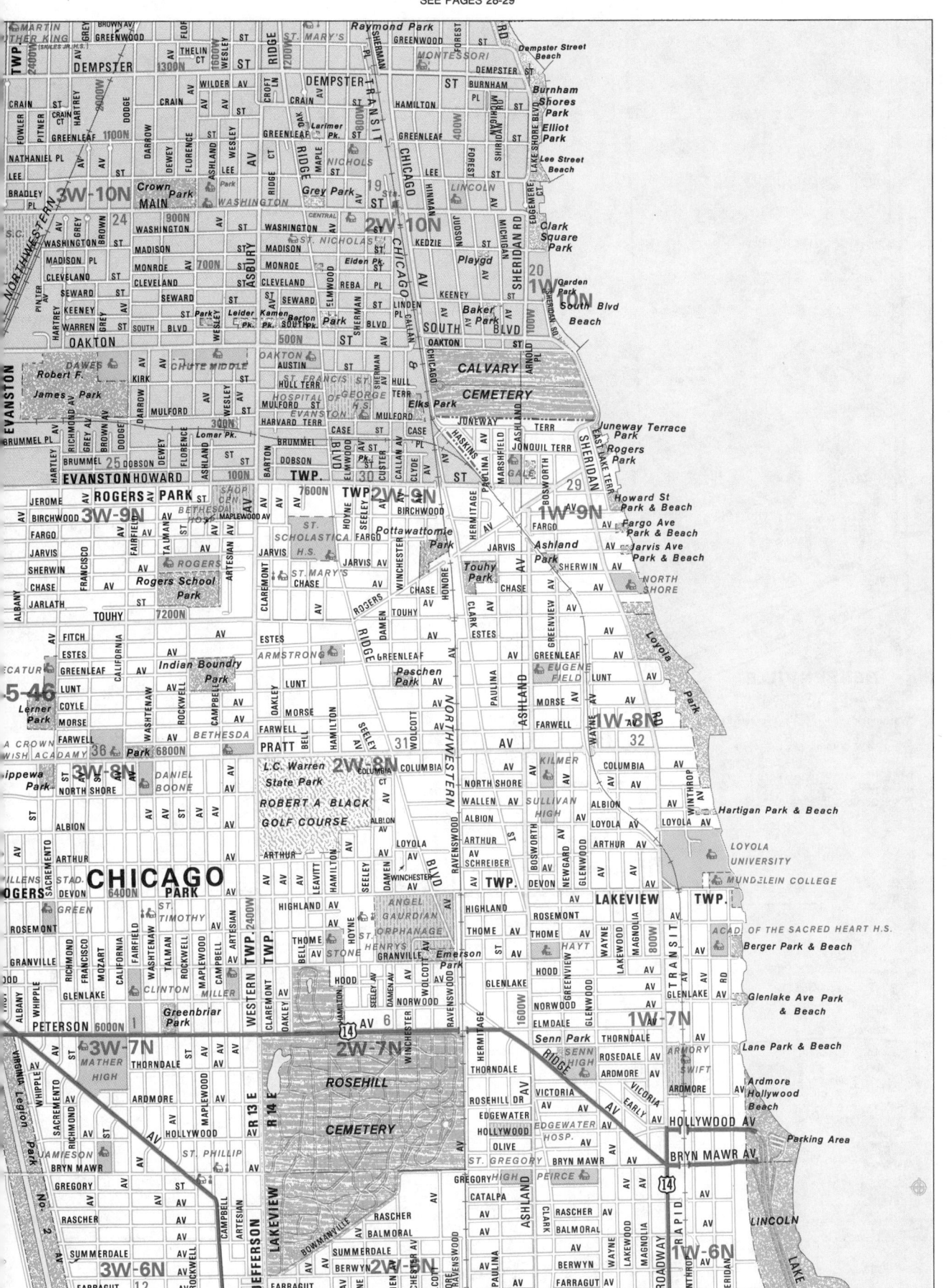

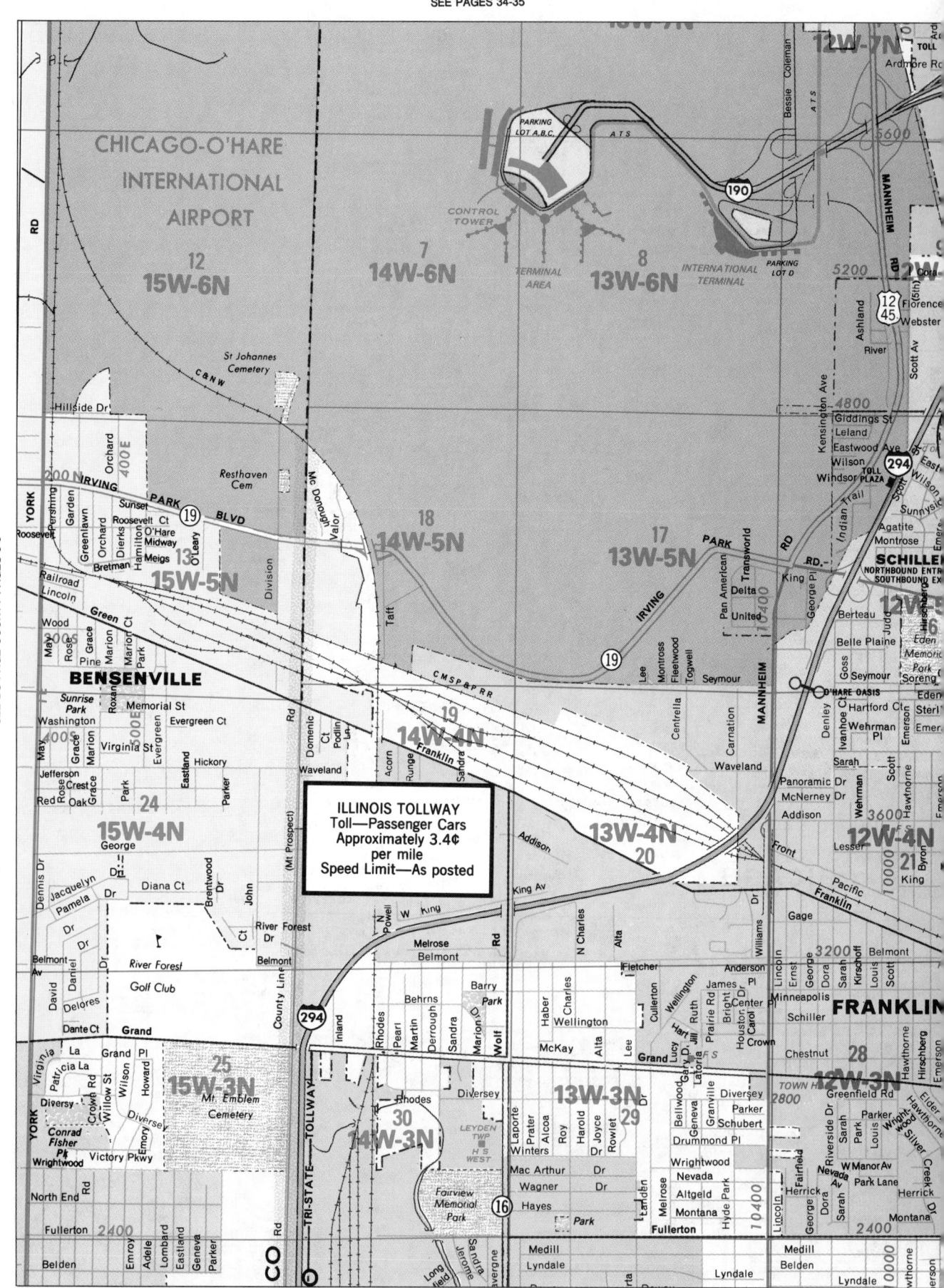

CHICAGO-O'HARE
INTERNATIONAL
AIRPORT

12
15W-6N

7
14W-6N

8
13W-6N

PARKING LOT A.B.C.

ATS

CONTROL TOWER

TERMINAL AREA

INTERNATIONAL TERMINAL

PARKING LOT D

190

MANNHEIM

5600

5200

12W

12
45
Florence
Webster

294
TOLL PLAZA

4800

Giddings St
Leland
Eastwood Ave
Wilson
Windsor

Indian Trail

Agatite
Montrose

SCHILLE
NORTHBOUND ENTR
SOUTHBOUND EX

St Johannes Cemetery

C&NW

Resthaven Cem

Mc Donough

18
14W-5N

17
13W-5N

PARK

IRVING

Kensington Ave

Pan American
Delta
United

Transworld

King Pl
George Pl

RD.

Berteau
Belle Plaine
Eden

12W
16

200 N IRVING PARK BLVD
400 E
Orchard
Sunset
Roosevelt Ct
O'Hare Midway
Meigs

19

13
O'Leary

15W-5N

Division

Taft

19

Lee
Montross
Fleetwood
Togwell
Seymour

MANNHEIM

Goss
Seymour
Soreng

O'Hare Oasis
Hartford Ct
Ivanhoe Ct
Wehrman Pl

Emer
Sterl

Emer

YORK
Roosevelt
Pershing
Garden
Greenlawn
Orchard
Dierks
Hamilton
Bretman
Railroad
Lincoln
Wood
May
Rose
Grace
Marion
Marion Ct
Pine
Park

Green

Grace

CMSP&PRR

Centrella
Carnation

Waveland

Denley

3600

BENSENVILLE

Sunrise Park
Washington
May
Marion
Grace
Rose
Crest
Oak
Grace
Red

Memorial St
Evergreen Ct
Virginia St
Hickory
Eastland
Parker

Roxann
Evergreen

Jefferson
Park

600 E
400 S

19
14W-4N

Domenic
Podlin
Ln
Acorn
Runge
Franklin
Sandra

Waveland

Waveland

Panoramic Dr
McNerney Dr
Addison

Wehrman
Sarah
Scott

12W-4N

Lesser

21
King

24
15W-4N
George

ILLINOIS TOLLWAY
Toll—Passenger Cars
Approximately 3.4¢
per mile
Speed Limit—As posted

13W-4N
20

Addison

10000

Front
Pacific
Franklin
Gage

Williams Dr

3200

Lincoln
Ernst
George
Dora
Sarah
Kirschoff
Louis
Scott

Belmont

Dennis Dr
Jacquelyn
Pamela
Dr
Dr
David
Daniel
Delores
Dante Ct

Diana Ct
Brentwood Dr
Dr
John
Ct

River Forest Dr
Belmont

King Av

N Powell
W King

Melrose
Belmont
Belmont

N Charles
Alta

Fletcher

Anderson
Minneapolis
Schiller

FRANKLIN

River Forest Golf Club

Belmont Av

Grand

County Line Rd

294

Inland

Rhodes
Pearl
Martin
Derrough
Sandra
Marion Dr
Wolf

Behrns

Barry Park

Haber
Charles
McKay

Wellington
Alta
Lee

Culleton
Lucy
Hart
Jill
Ruth
Prairie Rd
Bright

James Pl

Houston Dr
Center

Carol
Crown

Chestnut

28
12W-3N

Hawthorne
Hirschberg
Emerson

Virginia
La
Patricia La
Diversy
Conrad Fisher Pk
Wrightwood

Grand
La
Wilson Pl
Willow St
Howard
Crown Rd
Victory Pkwy

25
15W-3N
Mt. Emblem Cemetery

Diversey

Rhodes

30
14W-3N

Diversey

LEYDEN TWP
H S
WEST

Laporte
Prater
Winters

Diversey
Alcoa
Roy
Harold
Joyce
Rowlet

Granville
Bellwood
Geneva
Greenfield Rd

Schubert
Parker
Schubert
Drummond Pl

TOWN HALL
2800

Greenfield Rd

29

Riverside Dr
Nevada
Dora
Sarah
Fairfield

Park
Louis

Wright
Hawthorne
Elder
wood

Park Lane

Hirschberg
Silver
Creek Rd

North End
Rd

Fullerton
2400

16

Fairview Memorial Park

Sandra
Jerome
Evergreen

Mac Arthur
Dr
Wagner
Dr
Hayes

Park

Wrightwood
Nevada
Altgeld
Montana

Fullerton

Medill
Lyndale

Melrose
Lincoln
Hyde Park
George
Sarah

W Manor Av
Herrick

Herrick
Montana

2400

Belden

Emroy
Adele
Lombard
Geneva
Eastland
Parker

CO
RD

0

TRI-STATE TOLLWAY

Long field

Medill
Lyndale

Medill
Belden
Lyndale

0000

Hawthorne

YORK

SEE PAGES 36-37

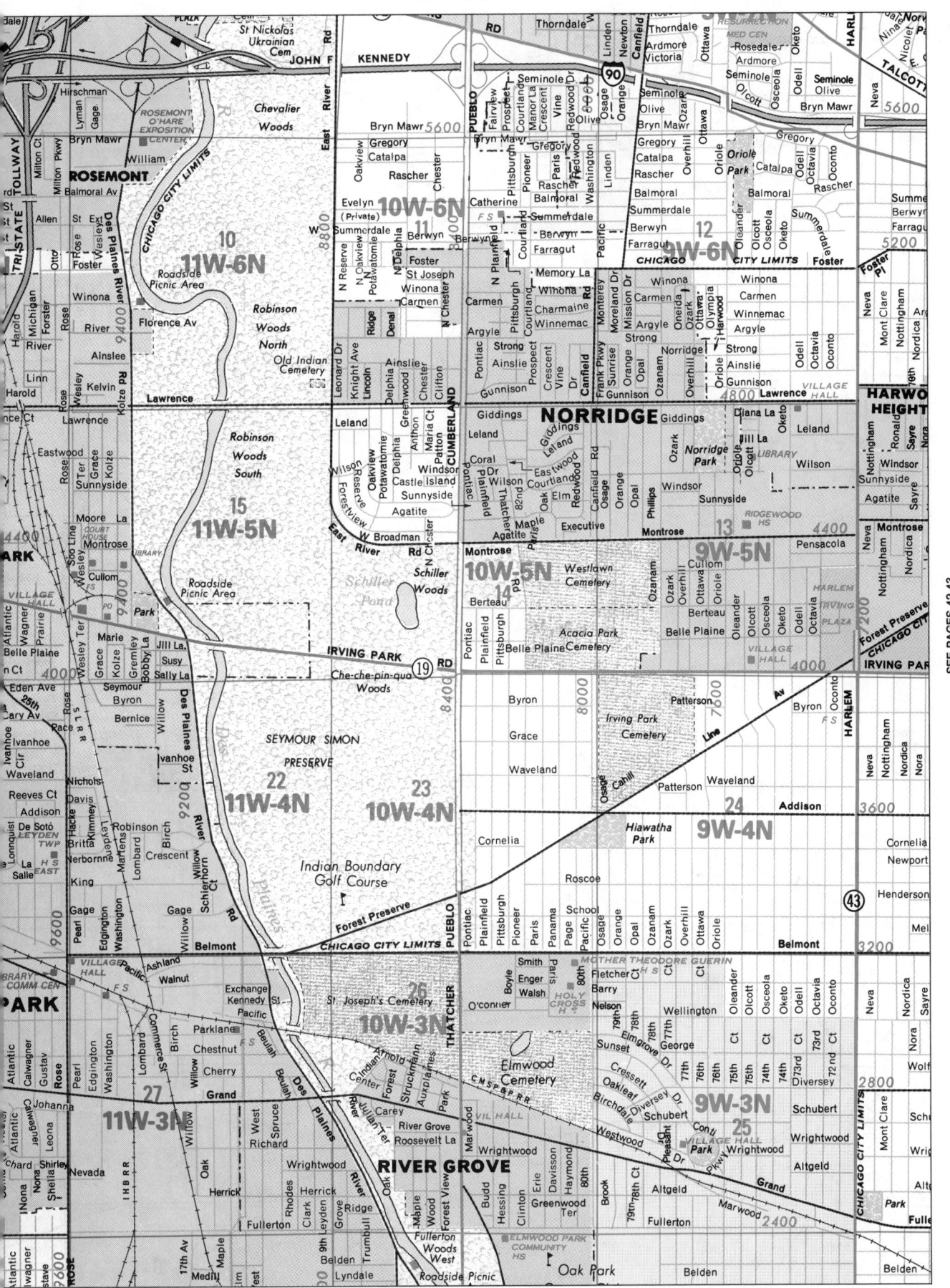

SEE PAGES 42-43

SEE PAGES 46-47

SEE PAGES 36-37

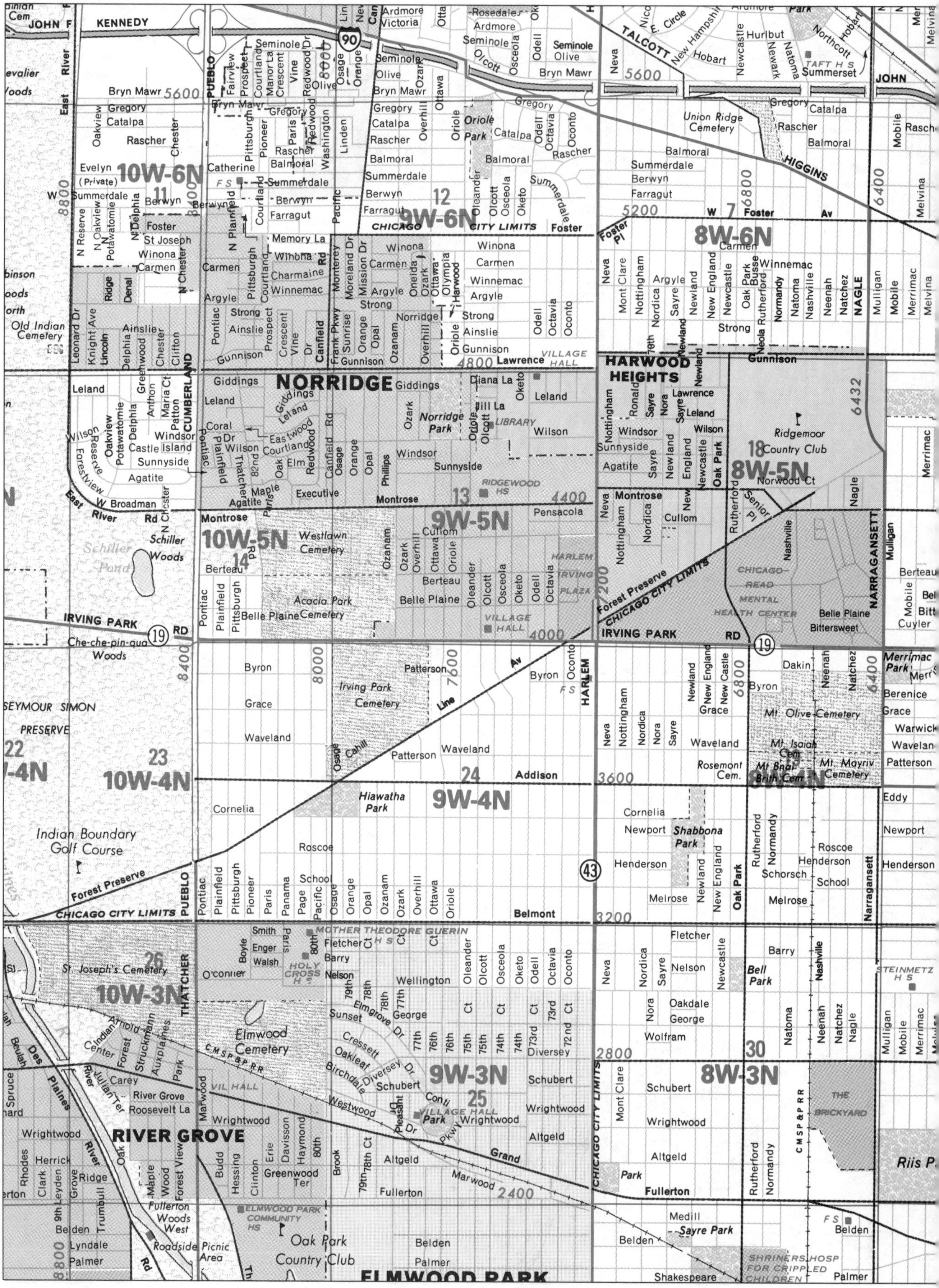

SEE PAGES 40-41

JOHN F. KENNEDY

10W-6N 11

9W-6N

8W-6N

CHICAGO CITY LIMITS

NORRIDGE

HARWOOD HEIGHTS

9W-5N

10W-5N 14

8W-5N 18

CHICAGO CITY LIMITS

IRVING PARK RD

SEYMOUR SIMON PRESERVE

10W-4N 23

9W-4N 24

Indian Boundary Golf Course

CHICAGO CITY LIMITS

10W-3N 26

9W-3N 25

8W-3N 30

THE BRICKYARD

RIVER GROVE

ELMWOOD PARK

SEE PAGES 38-39

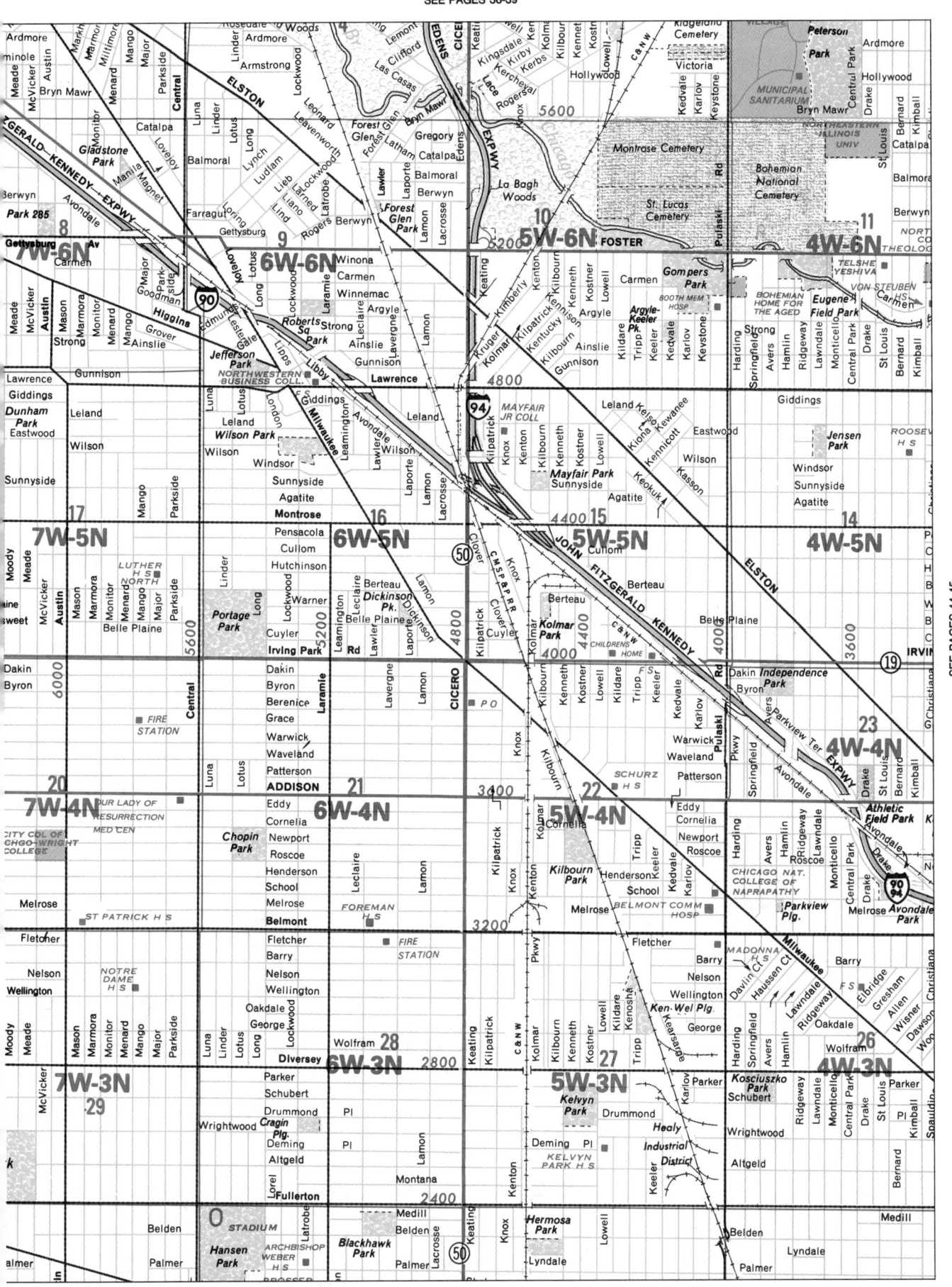

SEE PAGES 44-45

SEE PAGES 38-39

SEE PAGES 42-43

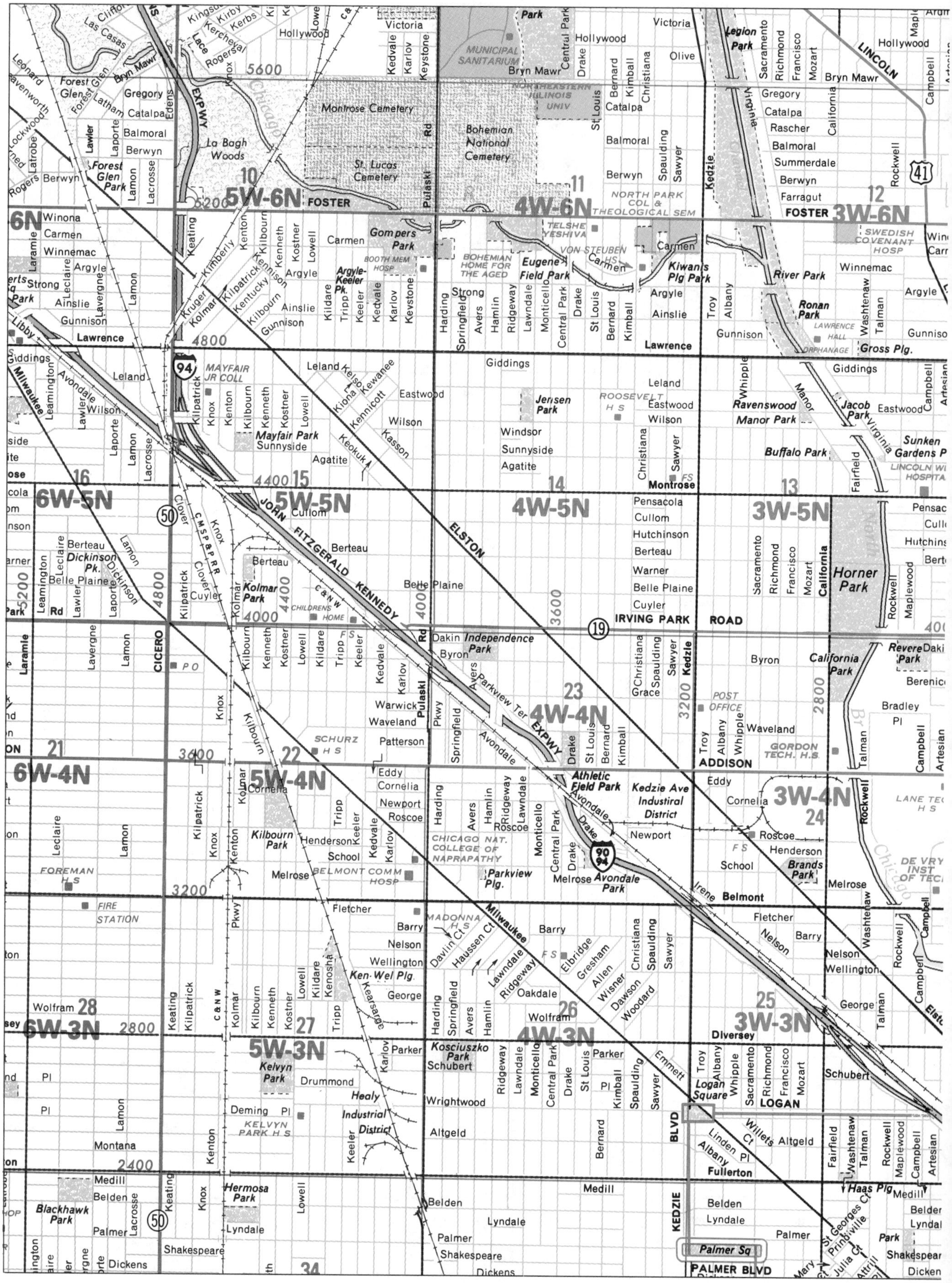

5W-6N · 10 · FOSTER · 4W-6N · 11 · FOSTER · 3W-6N · 12

6N

Montrose Cemetery
St. Lucas Cemetery
Bohemian National Cemetery
MUNICIPAL SANITARIUM
NORTHEASTERN ILLINOIS UNIV
NORTH PARK COL & THEOLOGICAL SEM
SWEDISH COVENANT HOSP

Forest Glen Park
La Bagh Woods
Gompers Park
Argyle-Keeler Pk.
BOOTH MEM HOSP
BOHEMIAN HOME FOR THE AGED
Eugene Field Park
TELSHE YESHIVA
VON STEUBEN H.S.
Kiwanis Plg Park
River Park
Ronan Park
LAWRENCE HALL ORPHANAGE
Gross Plg.

Winona, Carmen, Winnemac, Argyle, Ainslie, Gunnison, Lawrence

6W-5N · 16 · 5W-5N · 15 · 4W-5N · 14 · 3W-5N · 13

MAYFAIR JR COLL
Mayfair Park
Sunnyside
JOHN FITZGERALD KENNEDY EXPWY
Jensen Park
ROOSEVELT H.S.
Ravenswood Manor Park
Jacob Park
Buffalo Park
Sunken Gardens P.
LINCOLN VILLAGE HOSPITAL
Horner Park

Leland, Wilson, Windsor, Sunnyside, Agatite, Montrose
Eastwood, Giddings, Leland, Eastwood, Wilson, Giddings, Eastwood

Pensacola, Cullom, Hutchinson, Berteau, Warner, Belle Plaine, Cuyler

6W-4N · 21 · 5W-4N · 22 · 4W-4N · 23 · 3W-4N · 24

Dickinson Pk.
Belle Plaine
Kolmar Park
CHILDRENS HOME
C & N W
Independence Park
Warwick, Waveland
SCHURZ H.S.
Patterson
GORDON TECH. H.S.
POST OFFICE
Horner Park
California Park
Revere Park
Berenice
Bradley Pl.
LANE TECH H.S.

IRVING PARK ROAD · (19)
ADDISON

Kilbourn Park
School
Melrose
BELMONT COMM HOSP
CHICAGO NAT. COLLEGE OF NAPRAPATHY
Parkview Plg.
Athletic Field Park
Kedzie Ave Industrial District
Avondale Park
Brands Park
DE VRY INST OF TECH

FIRE STATION
FOREMAN H.S.

Eddy, Cornelia, Newport, Roscoe
Henderson, Melrose, Fletcher, Barry

6W-3N · 28 · 5W-3N · 27 · 4W-3N · 26 · 3W-3N · 25

Belmont
Fletcher, Barry, Nelson, Wellington, George, Diversey

MADONNA H.S.
Ken-Wel Plg.
Kosciuszko Park
Schubert
Kelvyn Park
KELVYN PARK H.S.
Healy
Industrial District
Hermosa Park
Blackhawk Park

LOGAN SQUARE
LOGAN BLVD
Willets
Haas Plg

Wolfram, Drummond, Deming Pl., Altgeld, Montana, Medill, Belden, Palmer, Dickens
Wrightwood, Fullerton

KEDZIE BLVD

Palmer Sq
PALMER BLVD

SEE PAGES 50-51

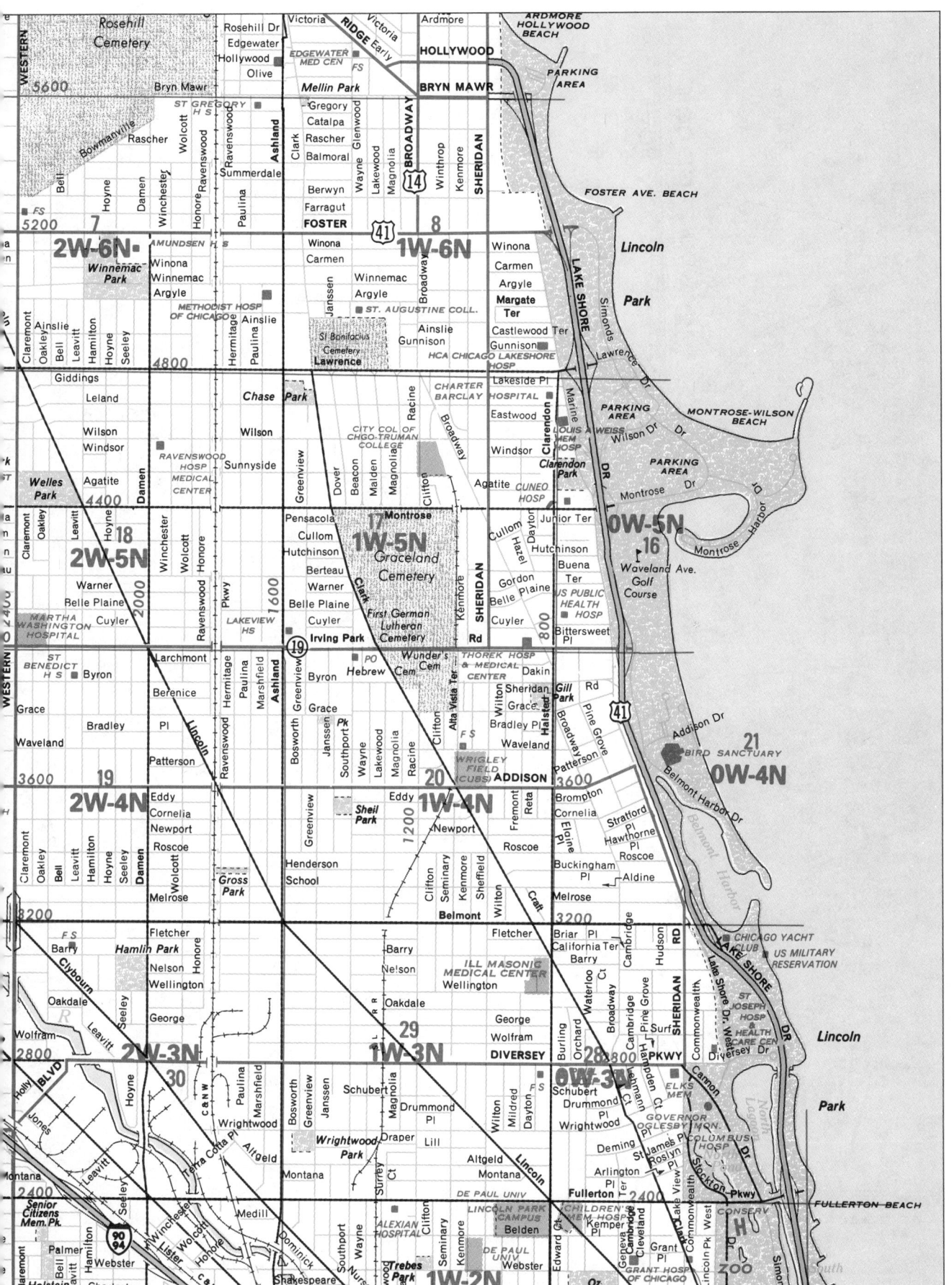

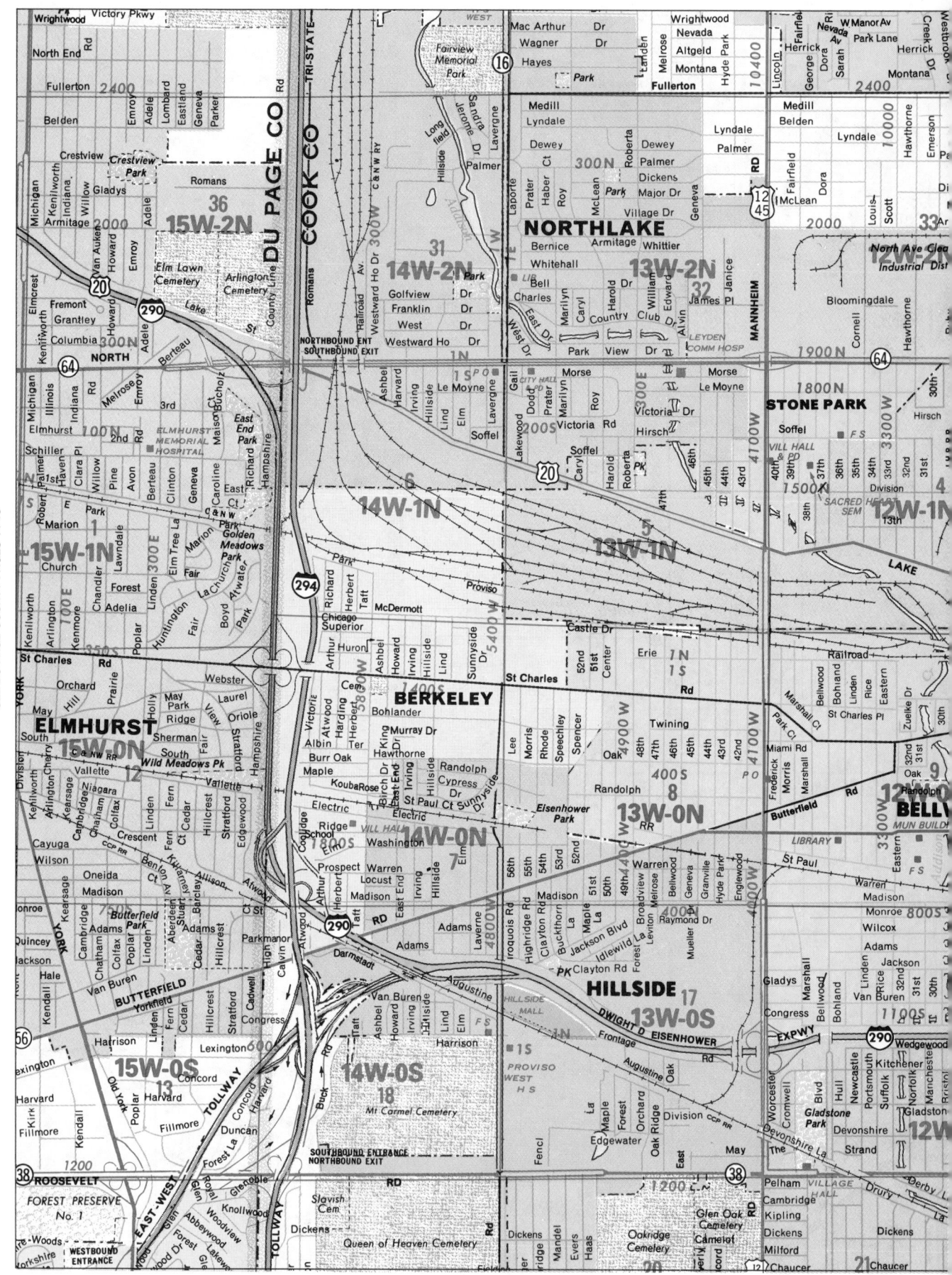

SEE PAGES 40-41

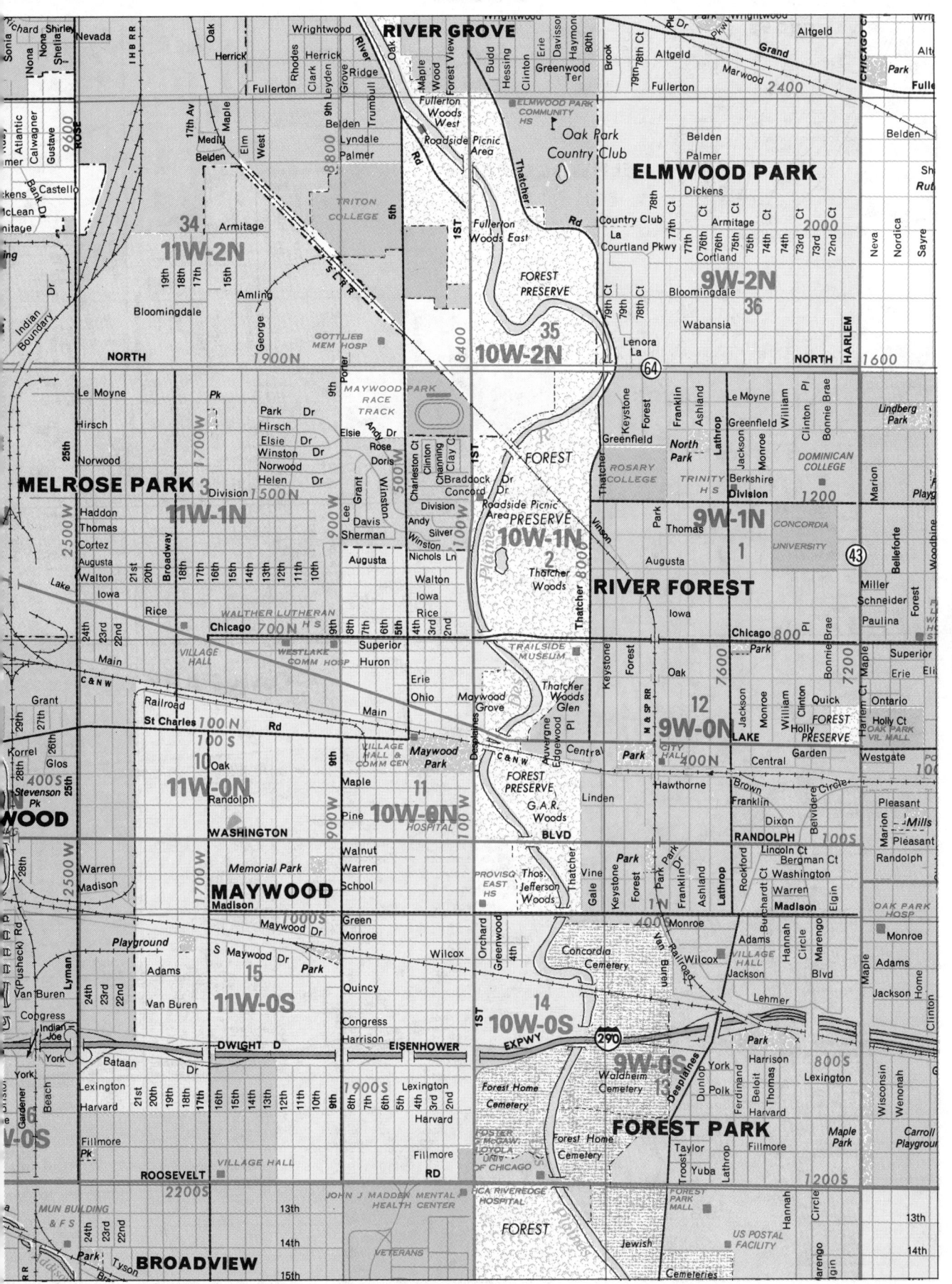

SEE PAGES 48-49

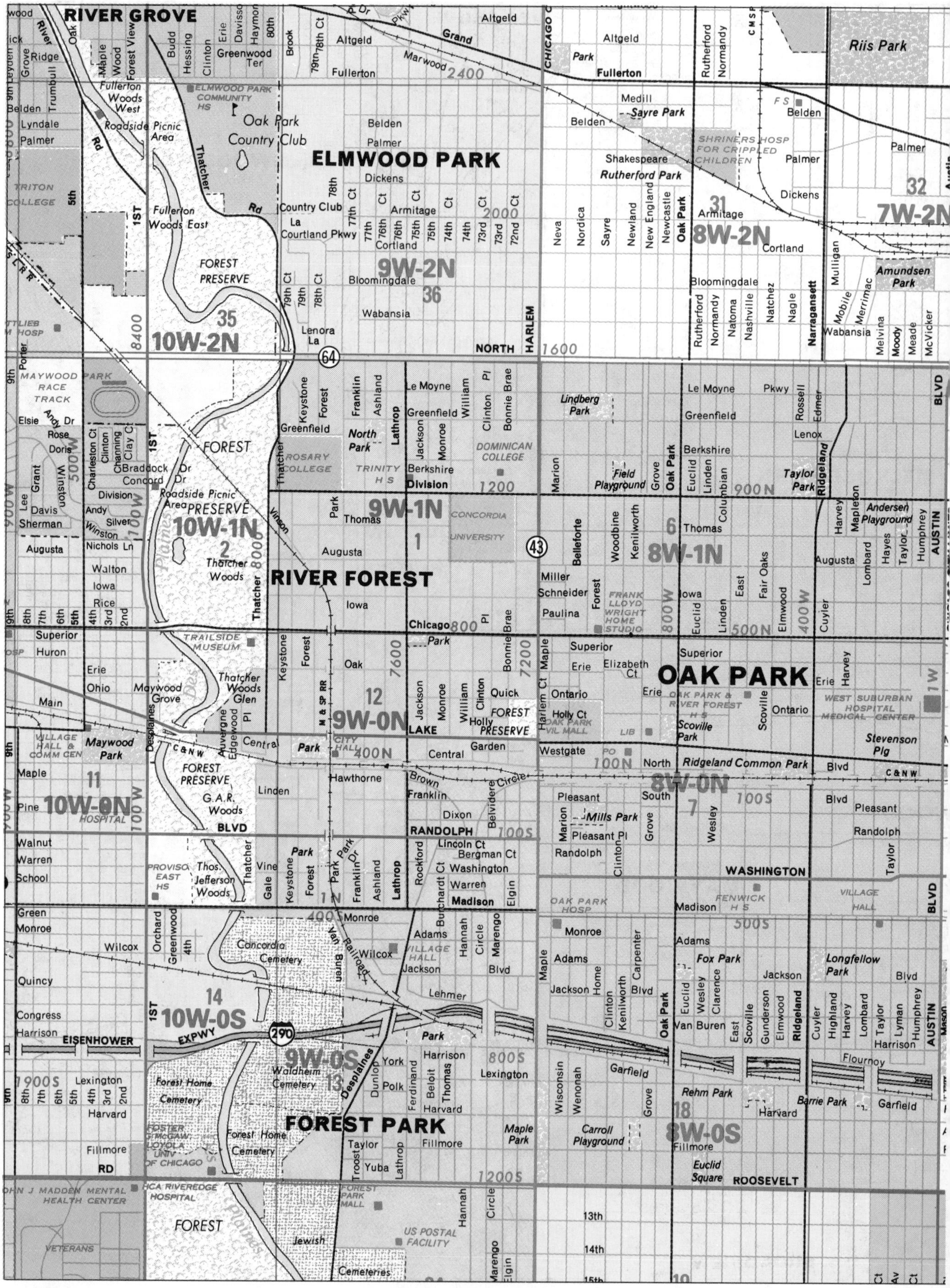

RIVER GROVE

ELMWOOD PARK

FOREST PRESERVE

10W-2N 35

9W-2N 36

8W-2N 31

7W-2N 32

Riis Park

Sayre Park

Rutherford Park

Shriners Hosp for Crippled Children

Amundsen Park

NORTH 1600 HARLEM

64

North Park

Rosary College

Trinity HS

Dominican College

Lindberg Park

Field Playground

Taylor Park

Andersen Playground

9W-1N 1

10W-1N 2 Thatcher Woods

RIVER FOREST

Concordia University

8W-1N 6

Frank Lloyd Wright Home Studio

Trailside Museum

Thatcher Woods Glen

9W-0N 12

OAK PARK

West Suburban Hospital Medical Center

10W-0N 11

Maywood Park

Forest Preserve

G.A.R. Woods

Oak Park & River Forest HS

Mills Park

Ridgeland Common Park

8W-0N 7

Proviso East HS

Thos. Jefferson Woods

RANDOLPH

WASHINGTON

Fenwick HS

Village Hall

Oak Park Hosp

Fox Park

Longfellow Park

Concordia Cemetery

Village Hall

10W-0S 14

9W-0S 13

290 EISENHOWER EXPWY

Waldheim Cemetery

Forest Home Cemetery

FOREST PARK

Rehm Park

Barrie Park

18 8W-0S

Carroll Playground

Euclid Square

ROOSEVELT

Forest Park Mall

John J Madden Mental Health Center

Riveredge Hospital

FOREST

US Postal Facility

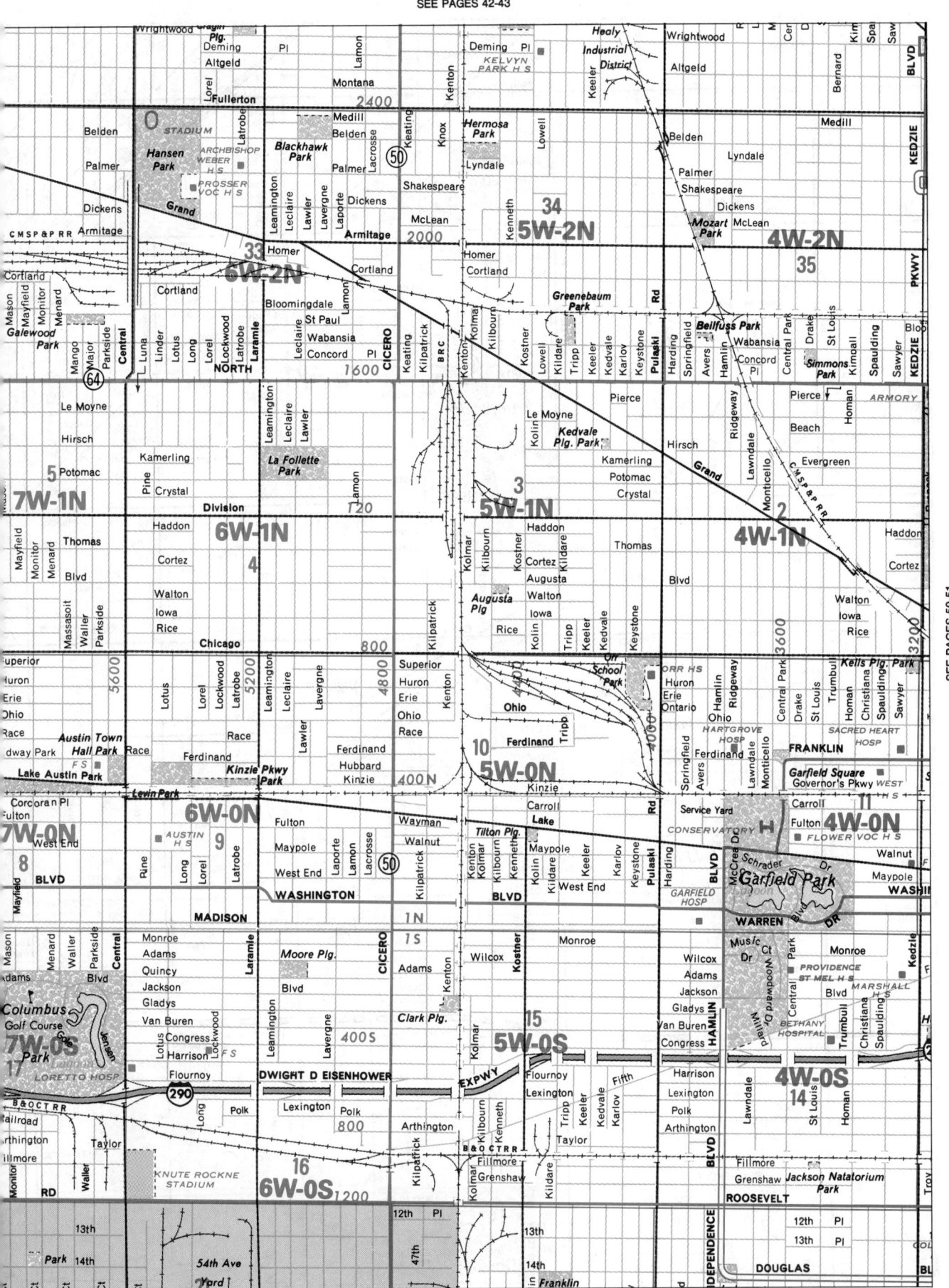

SEE PAGES 50-51

SEE PAGES 48-49

Industrial District

Altgeld · Bernard · Willets · Altgeld · Fairfield · Washtenaw · Talman · Rockwell · Maplewood · Campbell · Artesian · Montana · Wrightwood · Terra Cotta Pl · Altgeld

Keeler · Medill · Linden Pl · BLVD · KEDZIE · Albany · Fullerton · Haas Plg · Medill · Senior Citizens Mem. Pk · Winchester · Wolcott · Honore · C.&N.W. · Medill

Belden · Lyndale · Belden · Lyndale · Belden · Lyndale · Claremont · Oakley · Bell · Leavitt · Hamilton · Palmer · Charleston · Lister · Hobson · 90 94

Palmer · Shakespeare · St Georges Ct · Prindiville · Palmer · Park · Shakespeare · Holstein Park · Dickens · Avondale · Domini

Dickens · Mozart Park · McLean · PALMER BLVD · Dickens · St Mary · Henry Ct · Chanay Julia Ct · St Helen · Attrill · Francis Pl · Bingham · McLean · 2000 · McLean · Mozart Park

4W-2N · 35 · Armitage · BLVD · 3W-2N · 36 · Homer · 2W-2N · 31

Homer · Cortland · Moffat · Homer · Cortland · Moffat · Churchill · Cortland

Greenebaum Park · Kildare · Tripp · Keeler · Kedvale · Karlov · Keystone · Pulaski · Harding · Springfield · Hamlin · Central Park · Drake · St Louis · Spaulding · Sawyer · Bellfuss Park · Wabansia · Concord Pl · Simmons Park · KEDZIE · Troy · Albany · Whipple · HUMBOLDT · Richmond · Francisco · Mozart · California · Fairfield · Washtenaw · Talman · Rockwell · Campbell · Artesian · St Paul · Wabansia · Bell · WESTERN · Canton · Concord Pl · Damen · Winchester · Wolcott · Honore · Hermitage · Paulina · Marshfield · Ashland · Bloomingdale · Walsh Plg · Willow · St Paul · S.L.R.R · Bloomingdale · 1600

Pierce · Pierce · Homan · ARMORY · NORTH · 64 · Grower · Maplewood · Plg. Park · Le Moyne · Bell · JOSEPHINUM HS · Pierce · Le Moyne · Julian · Bosworth

Kedvale Plg. Park · Hirsch · Ridgeway · Lawndale · Monticello · Beach · HUMBOLDT DR · Le Moyne · St Elizabeth's Hosp · Hirsch · Claremont · Schiller · Wicker Park · Elk Grove · Beach · Wicker Park · Dean Plg

Kamerling · Potomac · Crystal · Grand · Evergreen · Humboldt Park · Evergreen · Potomac · Clemente Park · 1200 · Oakley · Leavitt · Schiller · Evergreen · Crystal · Marion Ct · Ellen · Moorman · Dean · Bauwans · Blackha · Mautene Ct

2 · 4W-1N · Haddon · Division · Crystal · 2800 · Crystal · 3W-1N · CLEMENTE HS · 2W-1N · Haddon · Po · Thor

Thomas · Haddon · Cortez · NORWEGIAN-AMERICAN HOSP · Haddon · Thomas · Cortez · ST MARY OF NAZARETH HOSP CENTER · Augusta · 6 · Winchester · Honore · WELLS HS · Cortez

Blvd · Walton · Iowa · Rice · Richmond · 2400 · Walton · Iowa · Rice · Blvd · Pearson · Commercial Club Plg · Fry · Greenview

Tripp · Keeler · Kedvale · Keystone · Chicago · 3200 · 3600 · Chicago · 800 · Rice · 2000 · 1600 · Su

Off School Park · ORR HS · Huron · Erie · Ontario · Superior · Huron · Smith Park · Erie · Ohio · Lee Pl · Hartland Ct · Hermitage · Ontario · Armour

4000 · HARTGROVE HOSP · Hamlin · Ridgeway · Central Park · Drake · St Louis · Trumbull · Christiana · Spaulding · Sawyer · Troy · Albany · Huron · Race · Jessie Ct · Hoyne · Hart

Tripp · Springfield · Avers · Ferdinand · SACRED HEART HOSP · FRANKLIN · BLVD · Ohio · Grand · Ferdinand · Hubbard · Anson Pl · Seeley

Garfield Square · Governor's Pkwy · WEST · Sacramento Square · Kinzie · Grand

12 · 3W-0N · C&NW

Service Yard · CONSERVATORY · Carroll · Fulton · 4W-0N · FLOWER VOC HS · Walnut · FS · Lake · Francisco · Mozart · California · Fairfield · Washtenaw · Talman · Maplewood · Campbell · Artesian · WESTERN · Claremont · Oakley · Bell · Leavitt · Carroll · Fulton · Walnut · Wolcott · Wood · Paulina · Marshfield · Oswego · Arbor · Union Park · Justine

H · GARFIELD HOSP · McCrea Dr · Schrader Dr · Garfield Park · Maypole · WASHINGTON · Lake · 7 · Maypole · SPALDING HS

West End · Keeler · Karlov · Keystone · Pulaski · Harding · GARFIELD HOSP · Garfield Park · WARREN · WASHINGTON · WARREN · BLVD · 2W-0N

Monroe · Music Dr Ct · Woodard Dr · Central Park · Monroe · Madison · Monroe · ARMORY · Sain Plg. Park · 1N · 1S · MARION ADULT & CAREER TRAINING · Young Park

Wilcox · Adams · Jackson · Gladys · Van Buren · Congress · PROVIDENCE · ST MEL HS · Blvd · MARSHALL HS · Fifth · Wilcox · Adams · Jackson · Gladys · Van Buren · Rockwell · Artesian · Blvd · 400S · CRANE HS · Touhy-Herbert Park · Hamilton · Damen · CREGIER VOC HS · Winchester · CITY COL OF CHGO MALCOLM X COL · CHICAGO STADIUM · Arcade · DWIGHT D

HAMLIN · BETHANY HOSPITAL · Trumbull · Christiana · Spaulding · Horan Plg · 290 · Congress · Altgeld Park · Harrison · Flournoy · RUSH UNIV PRESBYTERIAN ST LUKES MED CENTER

4W-0S · 14 · St Louis · Homan · 13 · 3W-0S · Albany · California · Lexington · Polk · Arthington · Maplewood · Campbell · Denvir · 800 · 2W-0S · 18 · Claremont · Oakley · Bell · Leavitt · Bowler · Hoyne · VA WEST SIDE MED CENTER · COOK CO HOSP · UNIV OF ILLINOIS HOSPITAL · Wolcott · Hermitage · Paulina · Marshfield · Garibaldi Plg · ILL ST PSYCH INST

Harrison · Lexington · Polk · Arthington · Taylor · MANLEY HS · Taylor · Fillmore · 1200 · Fillmore · Taylor

Fillmore · Grenshaw · Jackson Natatorium Park · ROOSEVELT · Troy · RD · Richmond · Francisco · Fairfield · Washtenaw · Grenshaw · 1200 · Washburne · Ada

Independence · DOUGLAS · BLVD · 12th Pl · 13th Pl · Thompson · Burkhardt Dr · 12th Pl · 13th · Talman · COOK CO JUVENILE COURT · CHICAGO LIGHTHOUSE FOR THE BLIND · 13th · Hastings

Franklin Park · Independence Square · Douglas · SACRAMENTO · Burkhardt · SCHWAB REHABILITATION CENTER · 14th · 14th Pl · Heath · BLVD · 14th · Pl · 15th · 15th

SEE PAGES 44-45

SEE PAGES 56-57

SEE PAGES 46-47

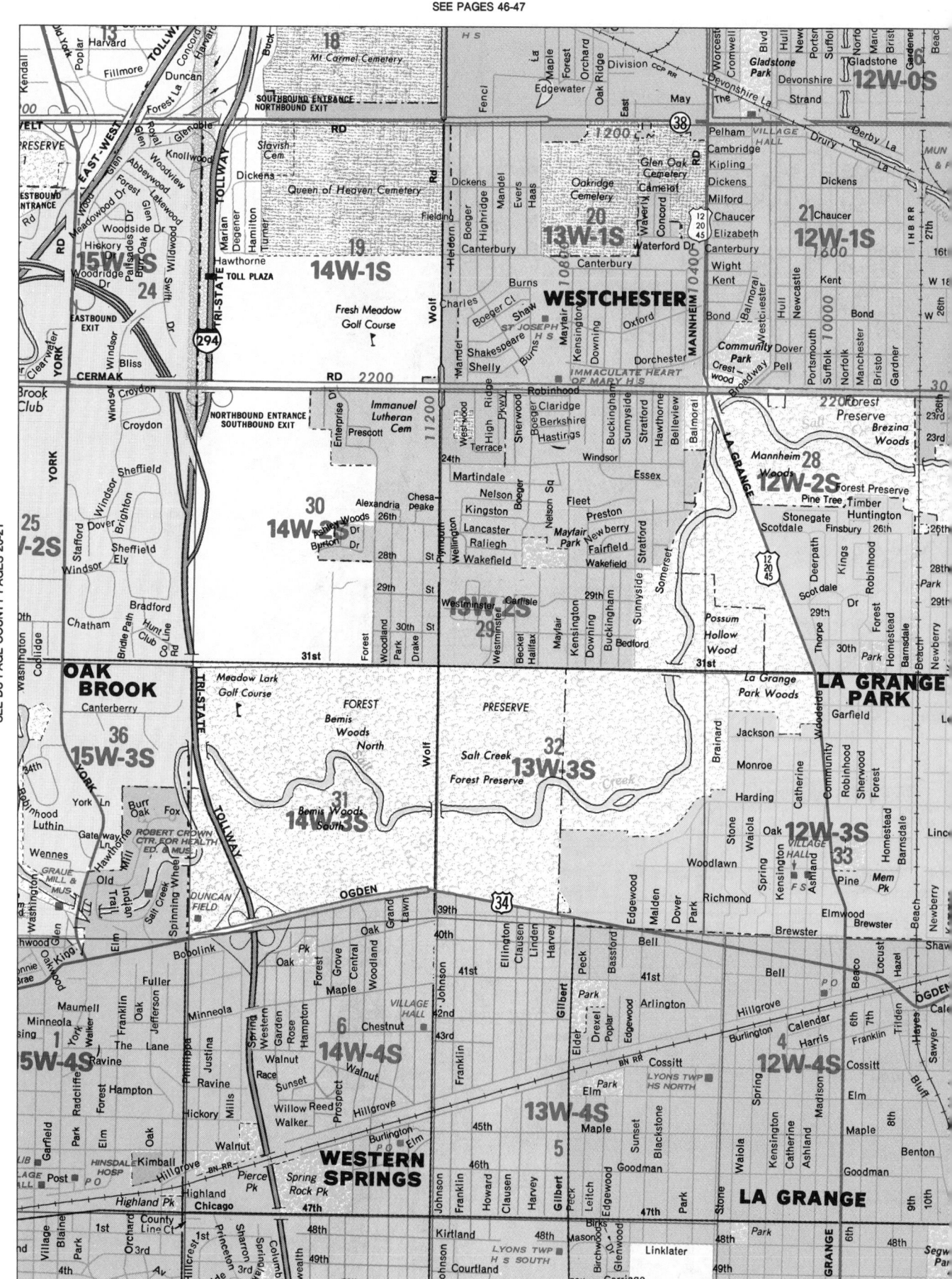

SEE PAGES 58-59

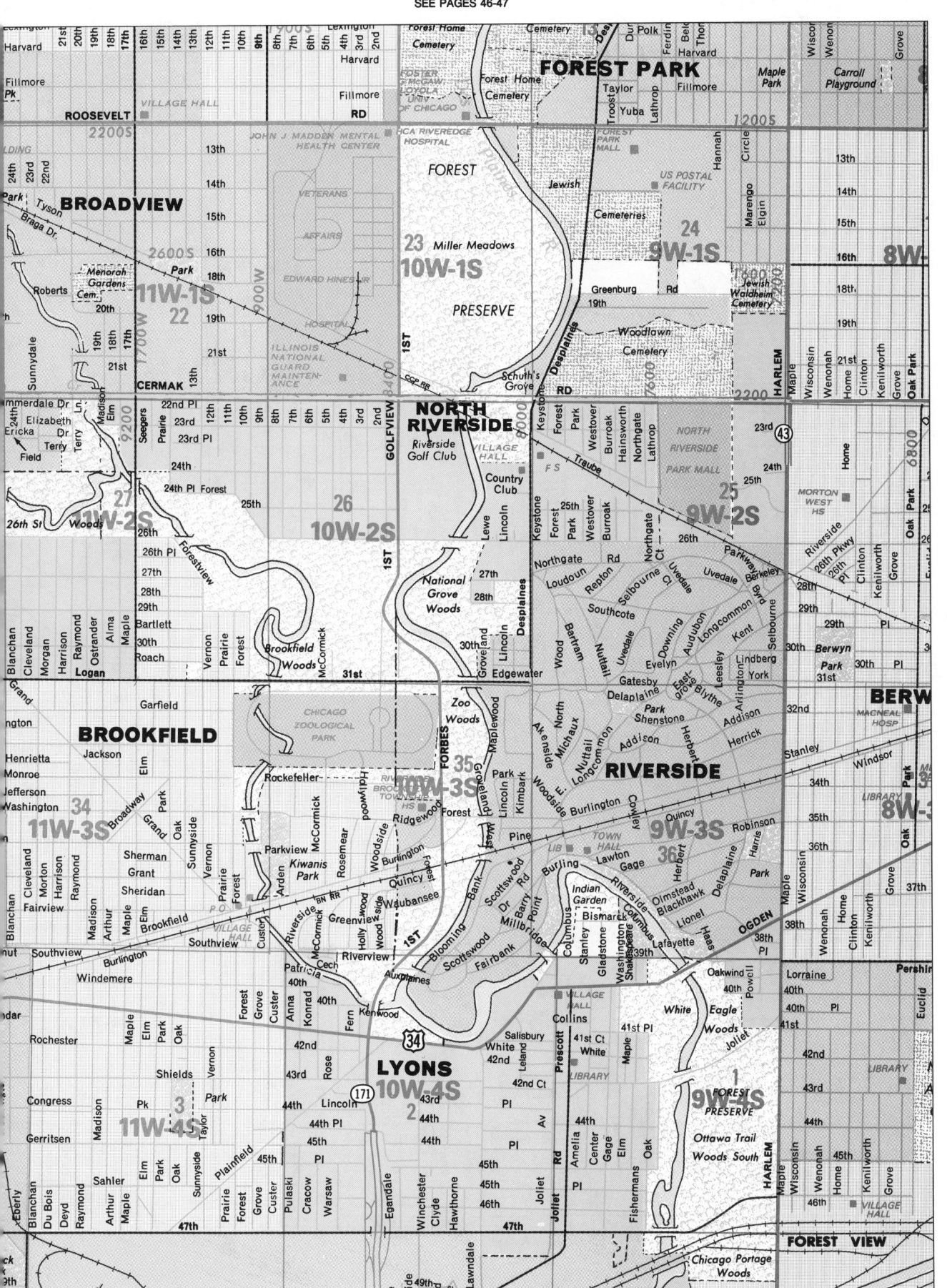

SEE PAGES 48-49

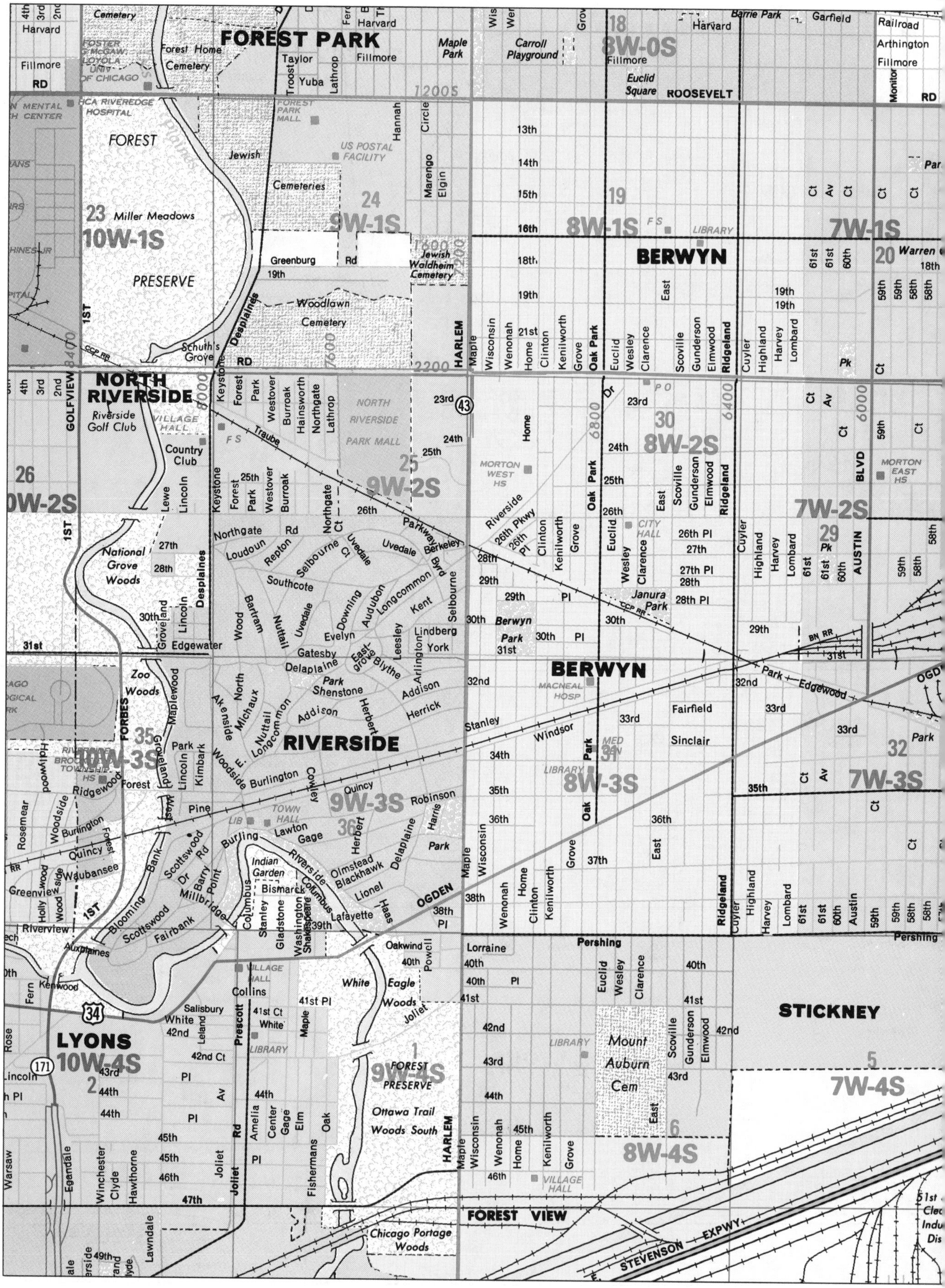

SEE PAGES 52-53

FOREST PARK

BERWYN

NORTH RIVERSIDE

RIVERSIDE

BERWYN

LYONS

STICKNEY

FOREST VIEW

FOREST PRESERVE

Miller Meadows

10W-1S 9W-1S 8W-0S 8W-1S 7W-1S

10W-2S 9W-2S 8W-2S 7W-2S

10W-3S 9W-3S 8W-3S 7W-3S

10W-4S 9W-4S 8W-4S 7W-4S

Roosevelt Rd
Ogden
Harlem
Desplaines
1st
Golfview
Austin
Ridgeland
Oak Park Blvd
Pershing

Ottawa Trail Woods South
Chicago Portage Woods
Mount Auburn Cem
Woodlawn Cemetery
Jewish Waldheim Cemetery
Forest Home Cemetery
Forest Park Mall
North Riverside Park Mall
National Grove Woods
Zoo Woods
Eagle Woods
White Eagle Woods

MacNeal Hosp
Riveredge Hospital
Mental Health Center
Riverside Golf Club / Country Club
Riverside Brookfield HS
Morton West HS
Morton East HS

STEVENSON EXPWY

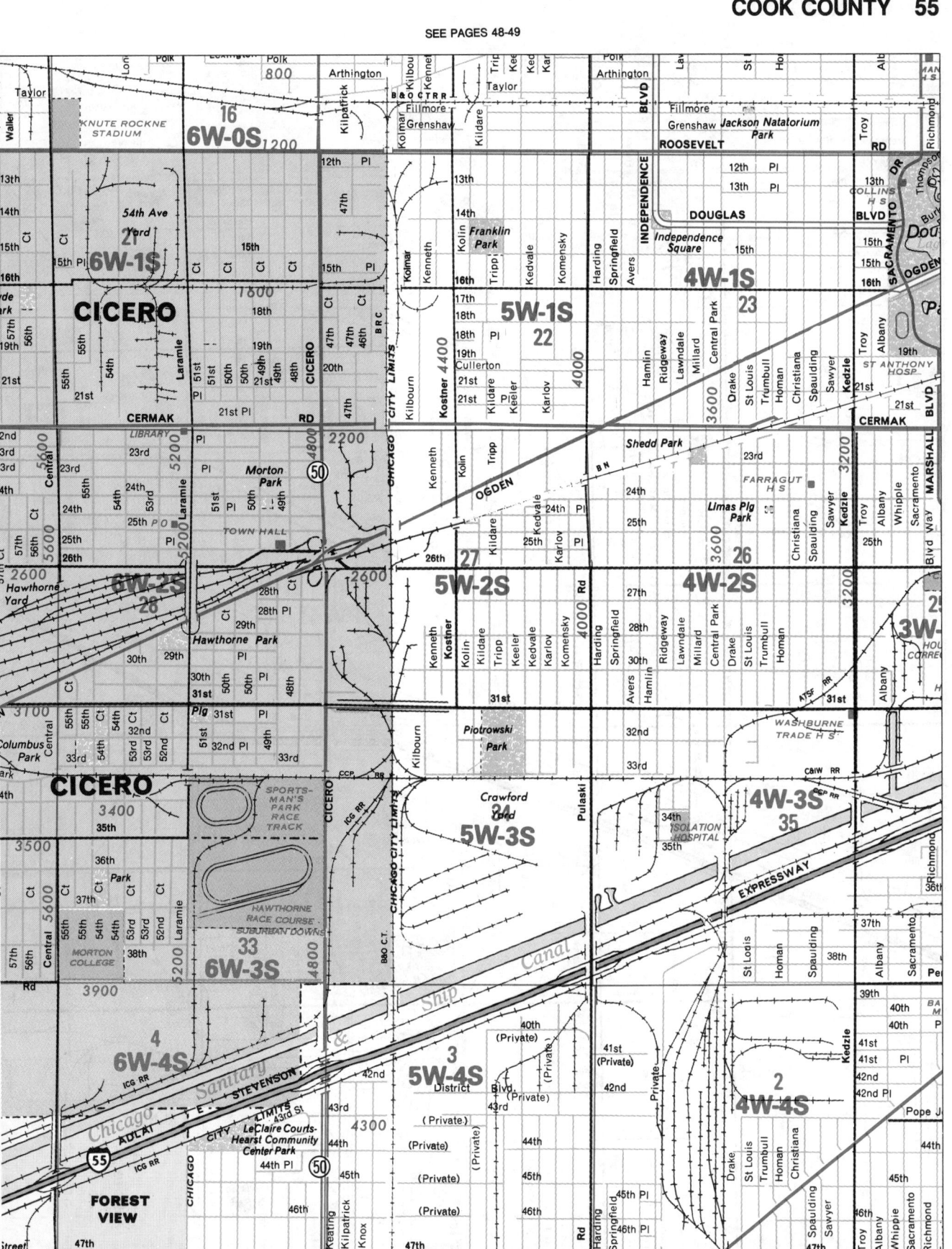

SEE PAGES 56-57

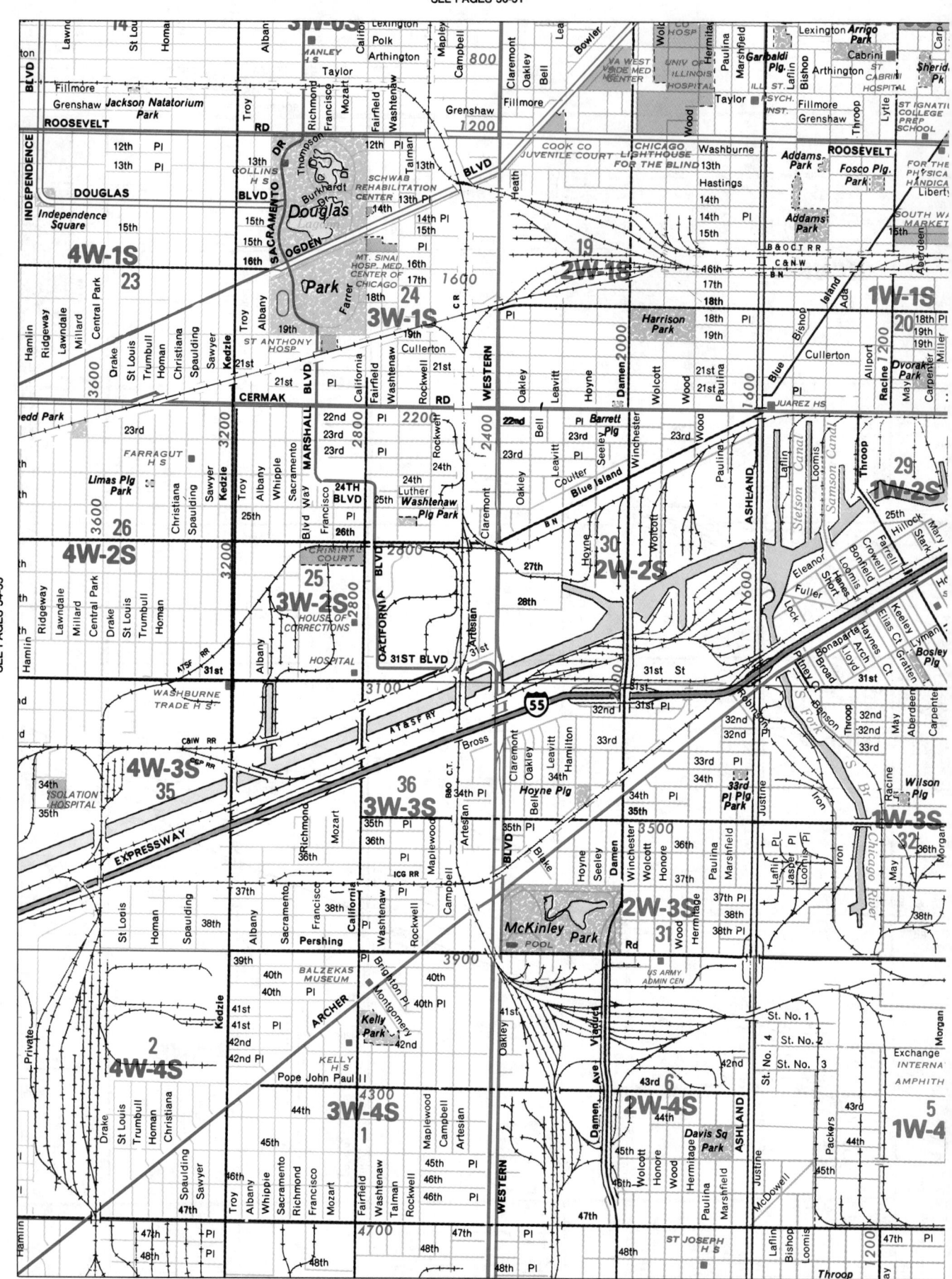

SEE PAGES 50-51

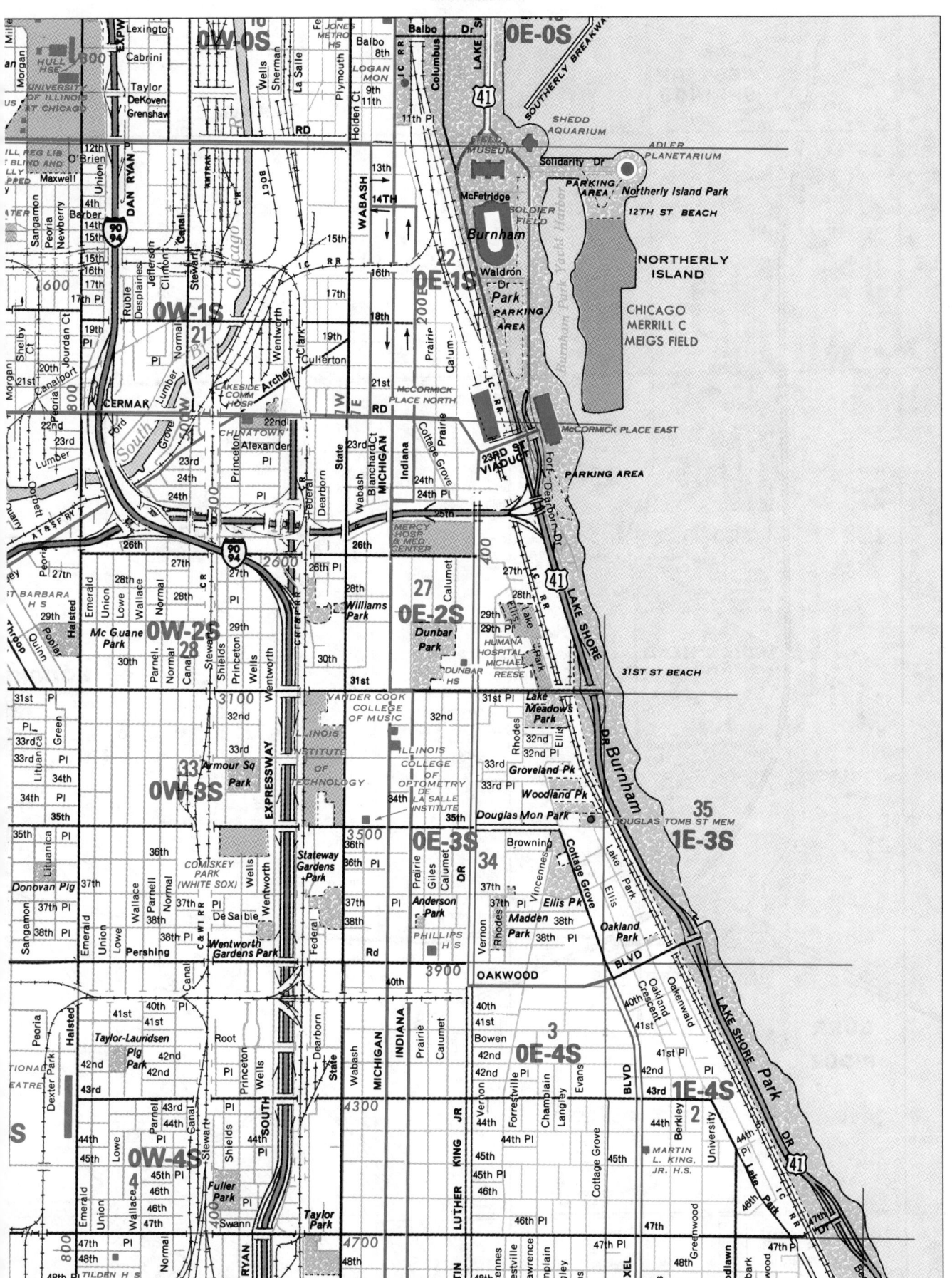

SEE PAGES 62-63

SEE PAGES 52-53

SEE DU PAGE COUNTY PAGES 26-27

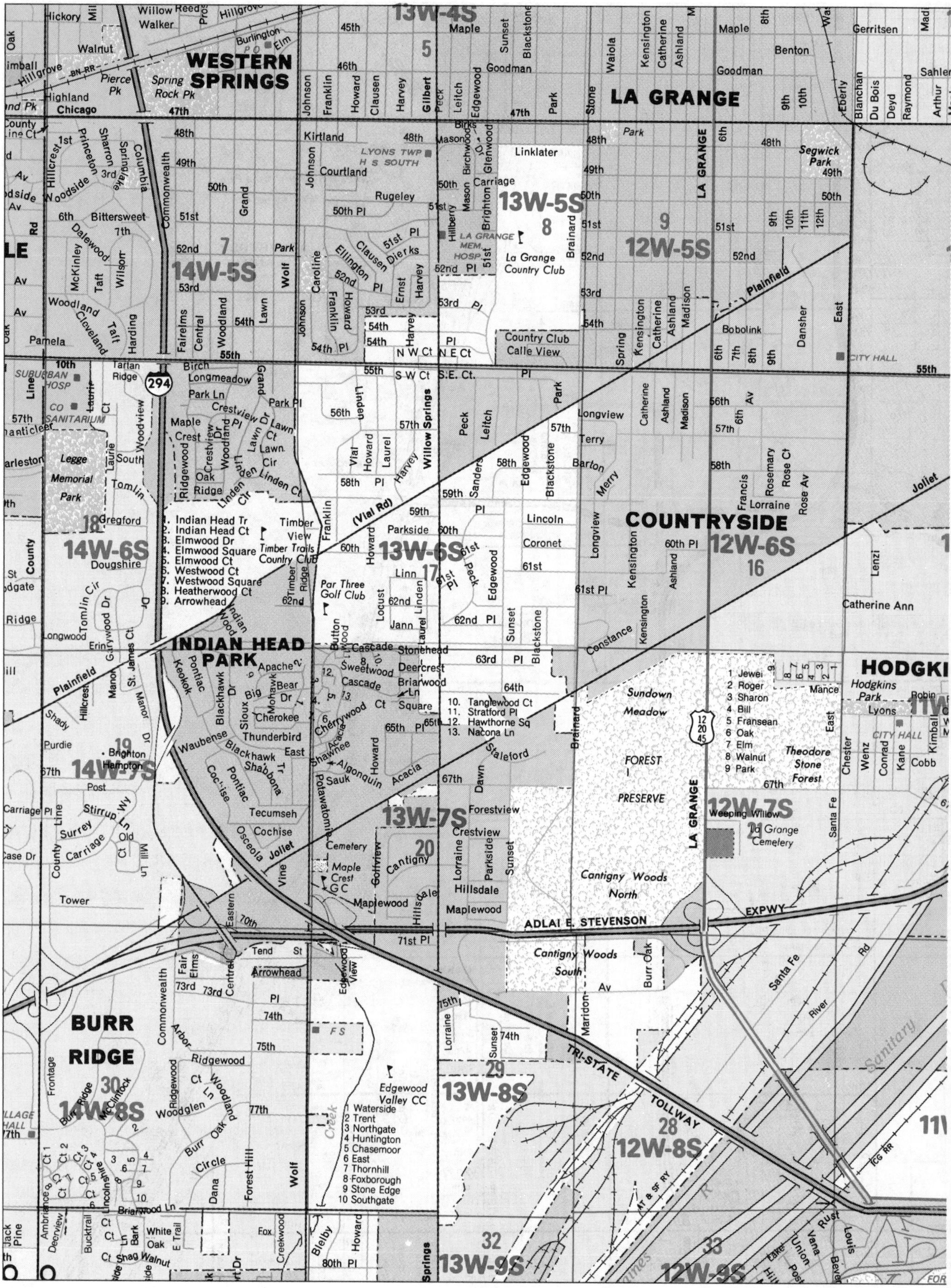

WESTERN SPRINGS

LA GRANGE

13W-4S

13W-5S

12W-5S

14W-5S

COUNTRYSIDE

14W-6S

13W-6S

12W-6S

INDIAN HEAD PARK

14W-7S

13W-7S

12W-7S

HODGKI

BURR RIDGE

14W-8S

13W-8S

12W-8S

13W-9S

12W-9S

LYONS TWP H S SOUTH

Linklater

La Grange Country Club

LA GRANGE MEM HOSP

Segwick Park

CITY HALL

SUBURBAN HOSP

CO SANITARIUM

Legge Memorial Park

Timber Trails Country Club

Par Three Golf Club

Sundown Meadow

FOREST PRESERVE

Theodore Stone Forest

Hodgkins Park

CITY HALL

Cantigny Woods North

Cantigny Woods South

Weeping Willow

La Grange Cemetery

Maple Crest G C

Edgewood Valley CC

ADLAI E. STEVENSON EXPWY

TRI-STATE TOLLWAY

1. Indian Head Tr
2. Indian Head Ct
3. Elmwood Dr
4. Elmwood Square
5. Elmwood Ct
6. Westwood Ct
7. Westwood Square
8. Heatherwood Ct
9. Arrowhead

10. Tanglewood Ct
11. Stratford Pl
12. Hawthorne Sq
13. Nacona Ln

1. Jewel
2. Roger
3. Sharon
4. Bill
5. Fransean
6. Oak
7. Elm
8. Walnut
9. Park

1. Waterside
2. Trent
3. Northgate
4. Huntington
5. Chasemoor
6. East
7. Thornhill
8. Foxborough
9. Stone Edge
10. Southgate

SEE PAGES 64-65

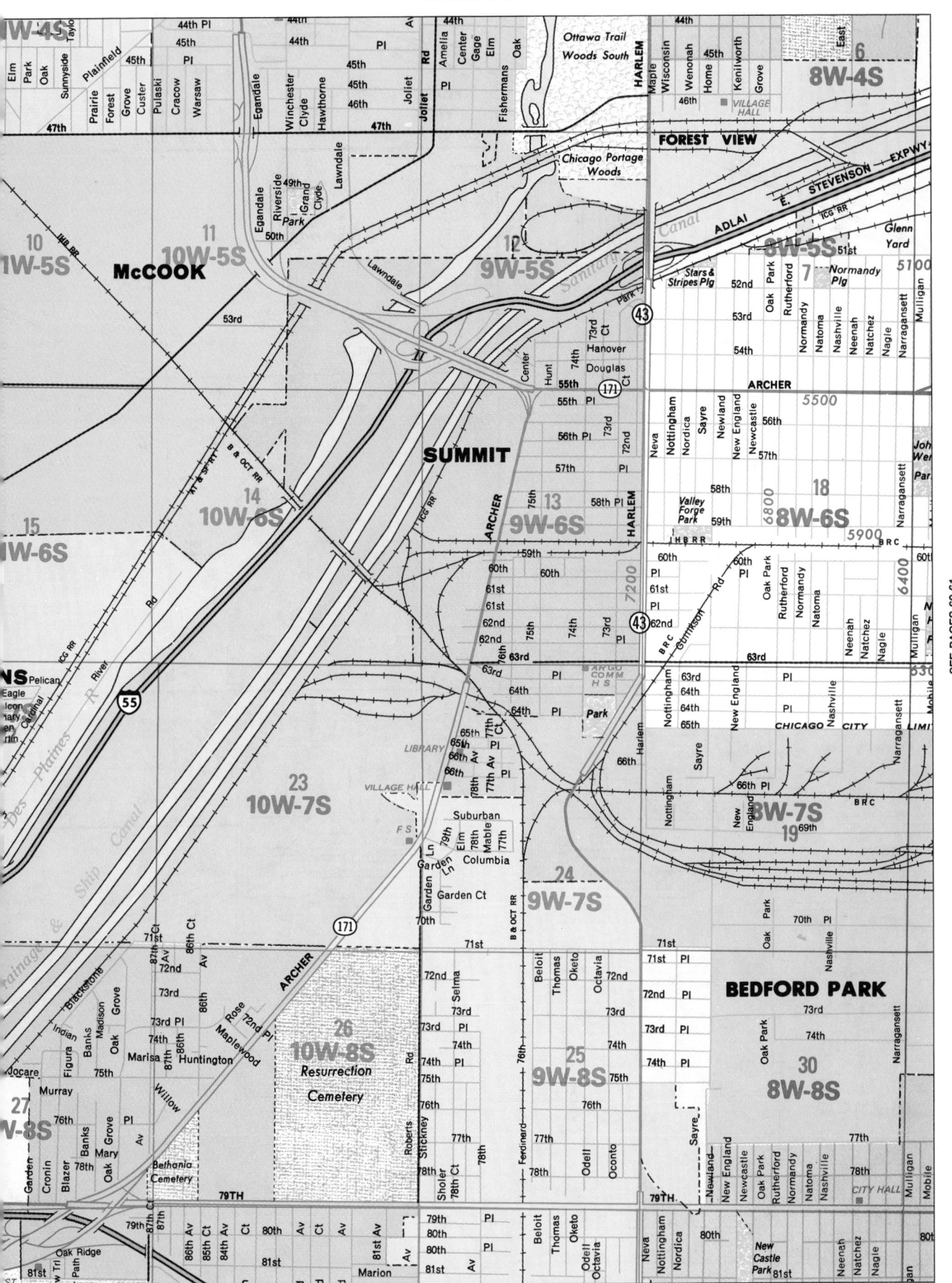

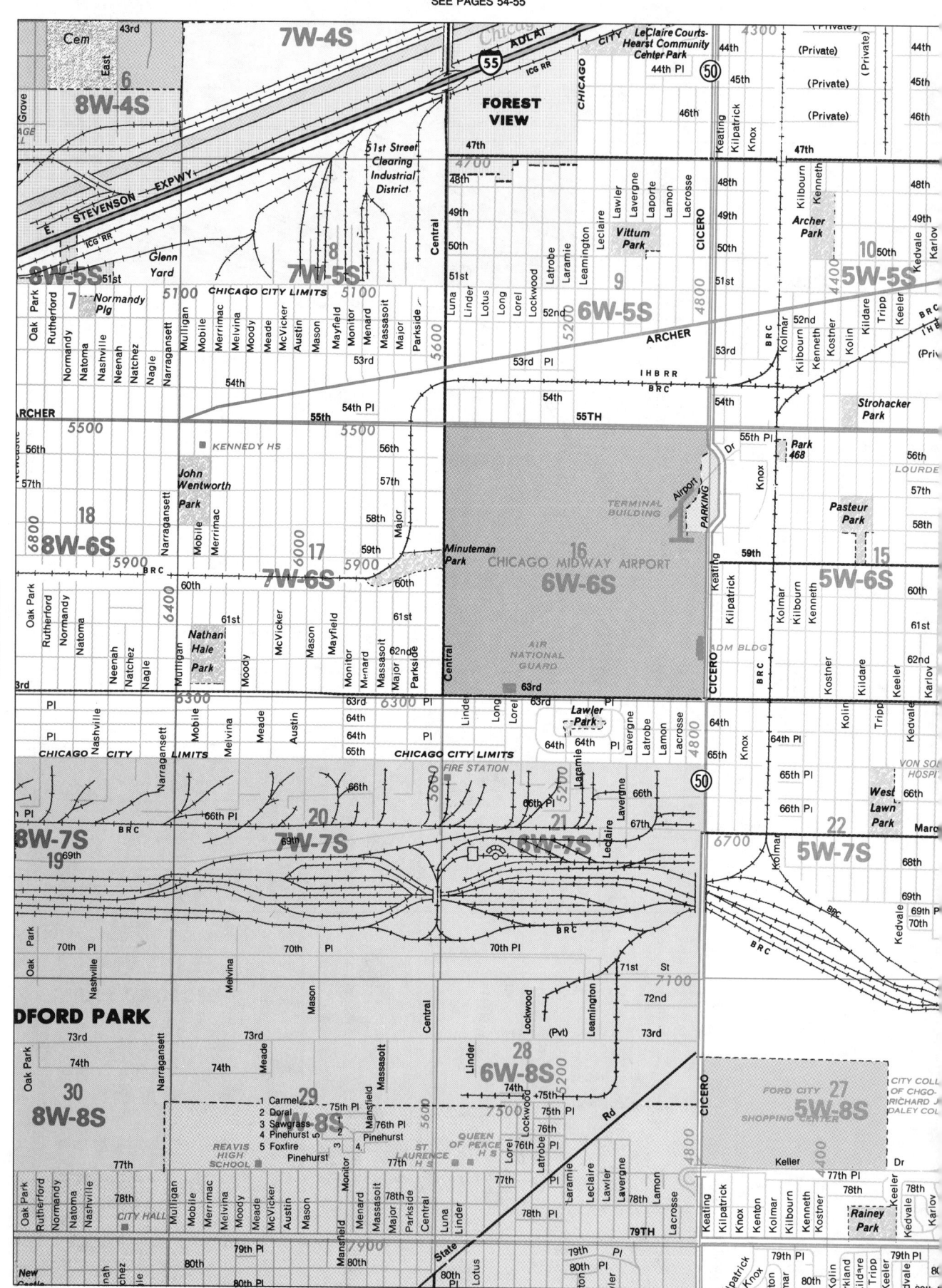

SEE PAGES 58-59

SEE PAGES 56-57

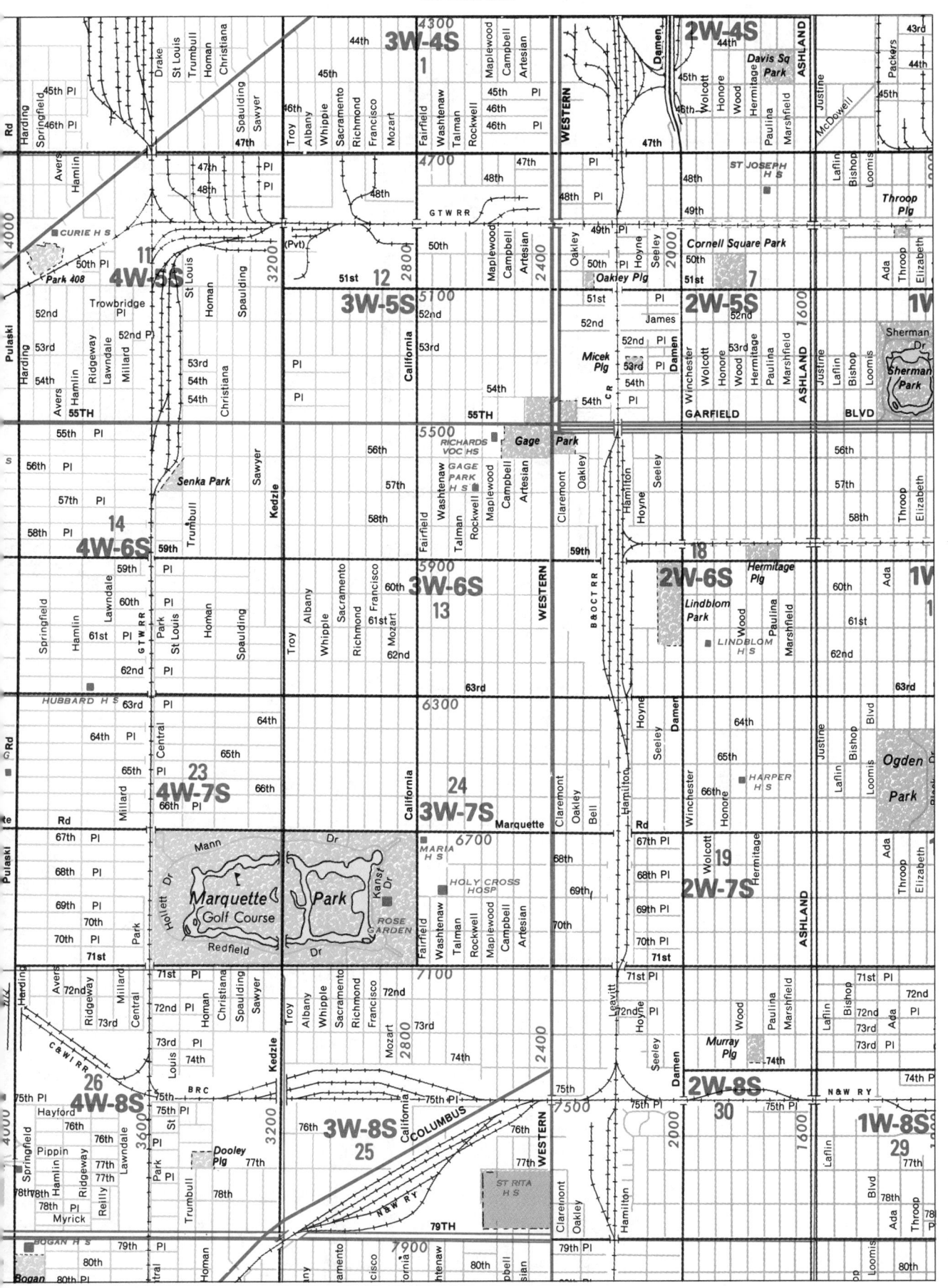

SEE PAGES 62-63

SEE PAGES 56-57

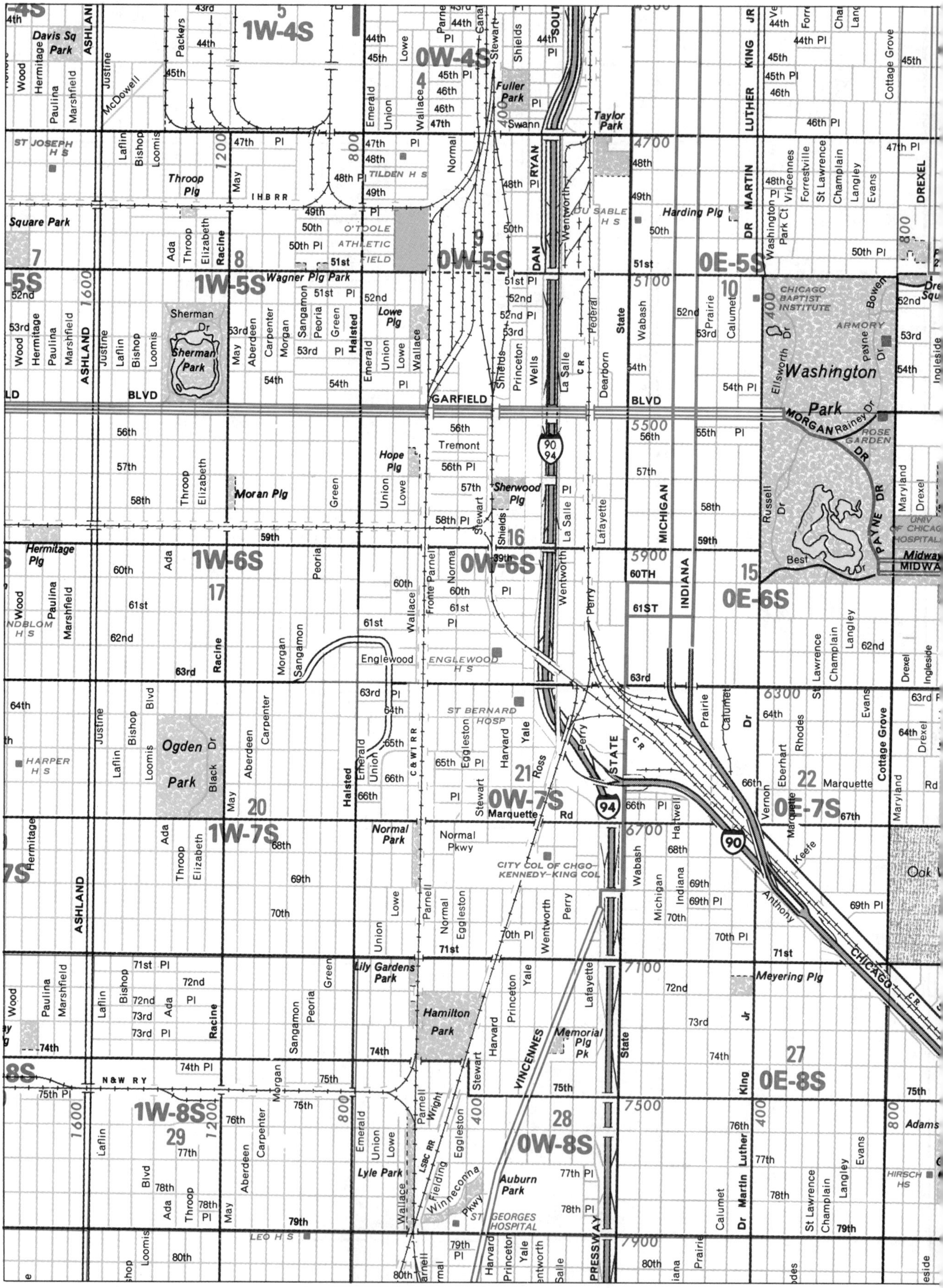

SEE PAGES 60-61

Davis Sq Park

1W-4S

0W-4S

Fuller Park

Taylor Park

ST JOSEPH H S

Throop Plg

TILDEN H S

DU SABLE H S

Harding Plg

Square Park

O'TOOLE ATHLETIC FIELD

0W-5S

0E-5S

CHICAGO BAPTIST INSTITUTE

ARMORY

Wagner Plg Park

1W-5S

Sherman Dr

Sherman Park

Lowe Plg

Washington Park

GARFIELD BLVD

MORGAN Rainey Dr

ROSE GARDEN

Hope Plg

Tremont

Sherwood Plg

Russell Dr

Best

Hermitage Plg

Moran Plg

1W-6S

0W-6S

0E-6S

Midway

UNIV OF CHICAGO HOSPITAL

Ogden Park

Englewood

ENGLEWOOD H S

HARPER H S

ST BERNARD HOSP

0W-7S

0E-7S

LINDBLOM H S

1W-7S

Normal Park

Normal Pkwy

CITY COL OF CHGO KENNEDY-KING COL

Lily Gardens Park

Meyering Plg

Hamilton Park

Memorial Plg Pk

N & W RY

0E-8S

1W-8S

Lyle Park

0W-8S

Auburn Park

ST GEORGES HOSPITAL

HIRSCH H S

LEO H S

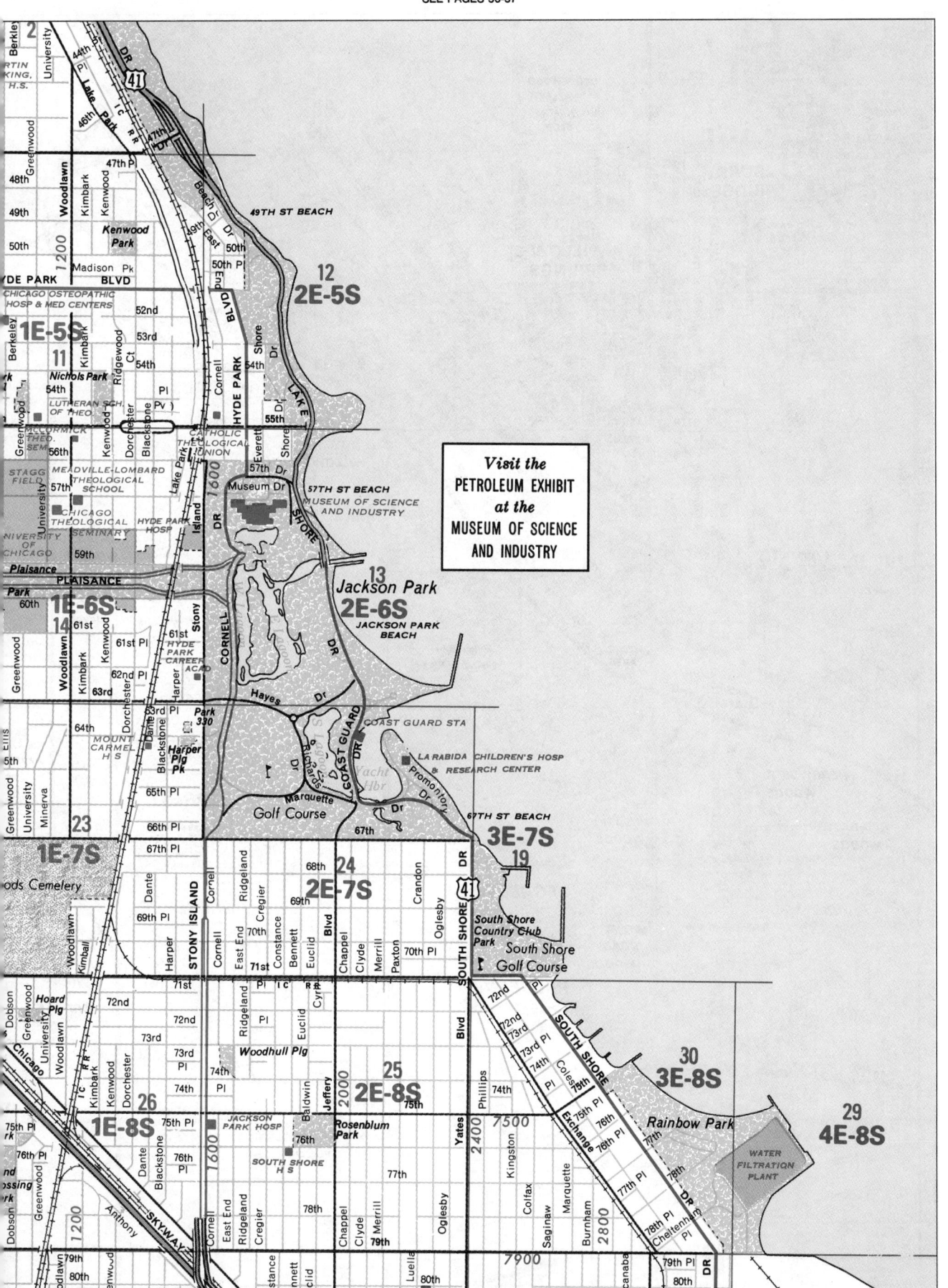

Visit the
PETROLEUM EXHIBIT
at the
MUSEUM OF SCIENCE
AND INDUSTRY

SEE PAGES 58-59

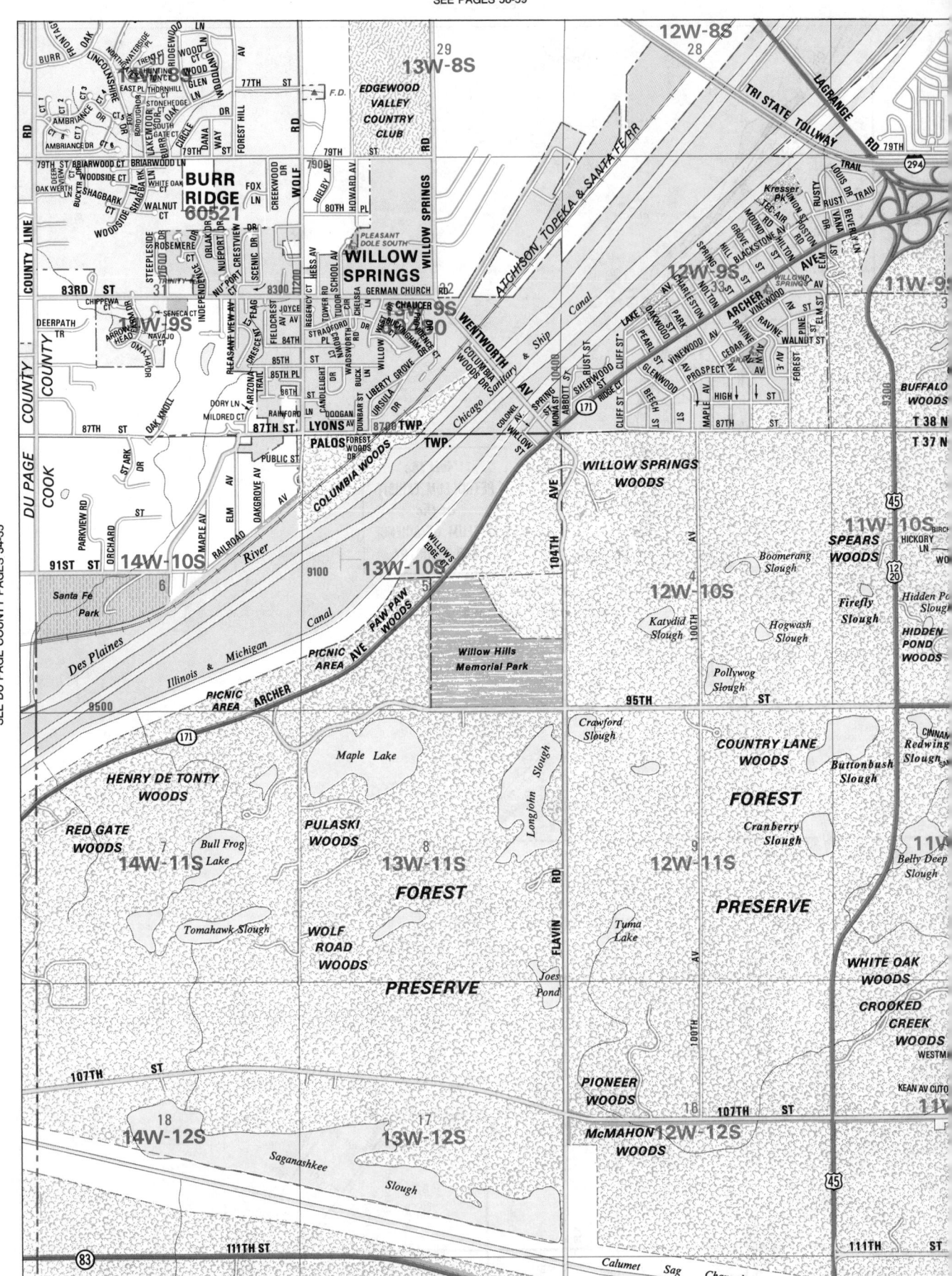

SEE PAGES 72-73

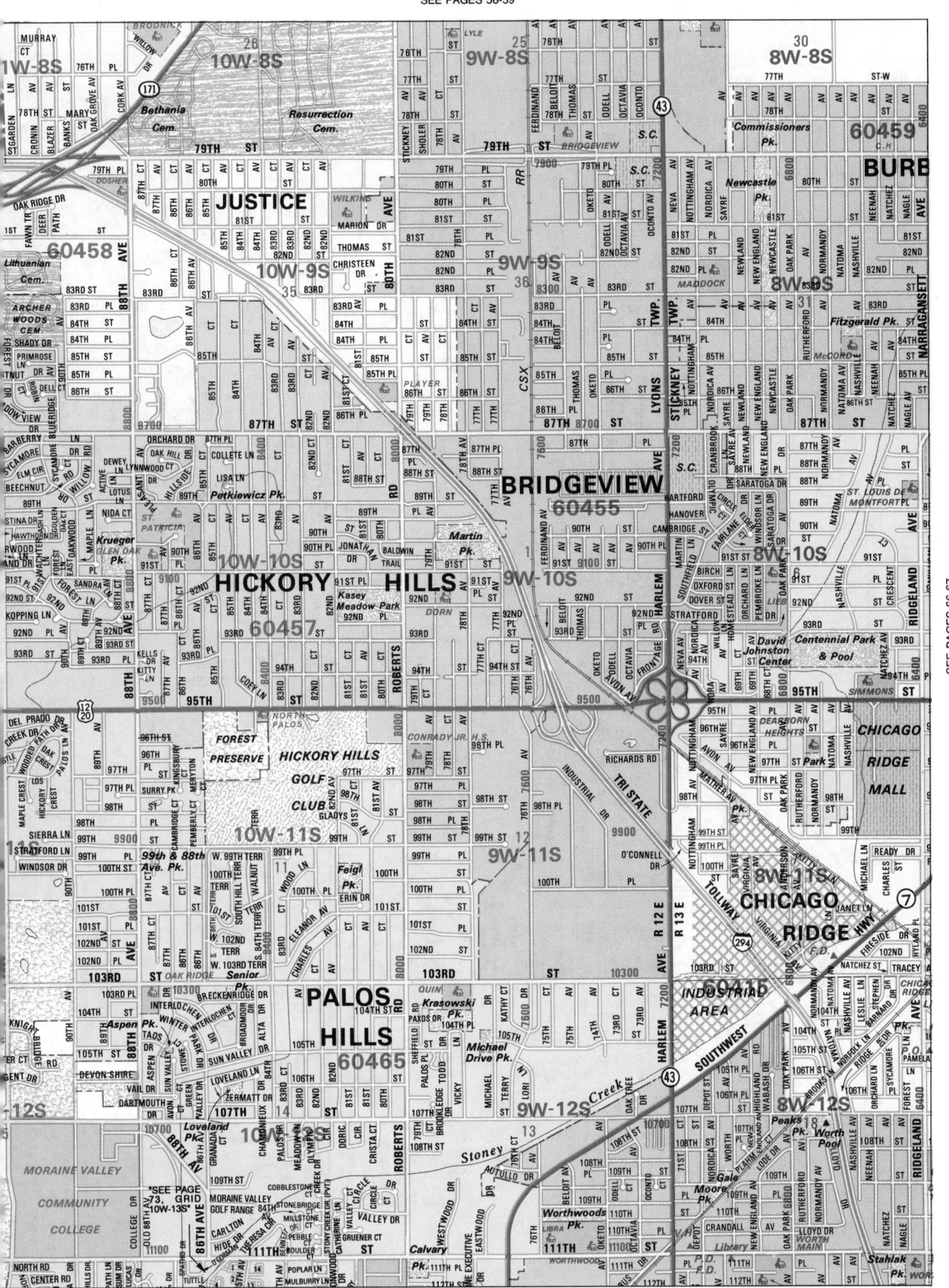

SEE PAGES 60-61

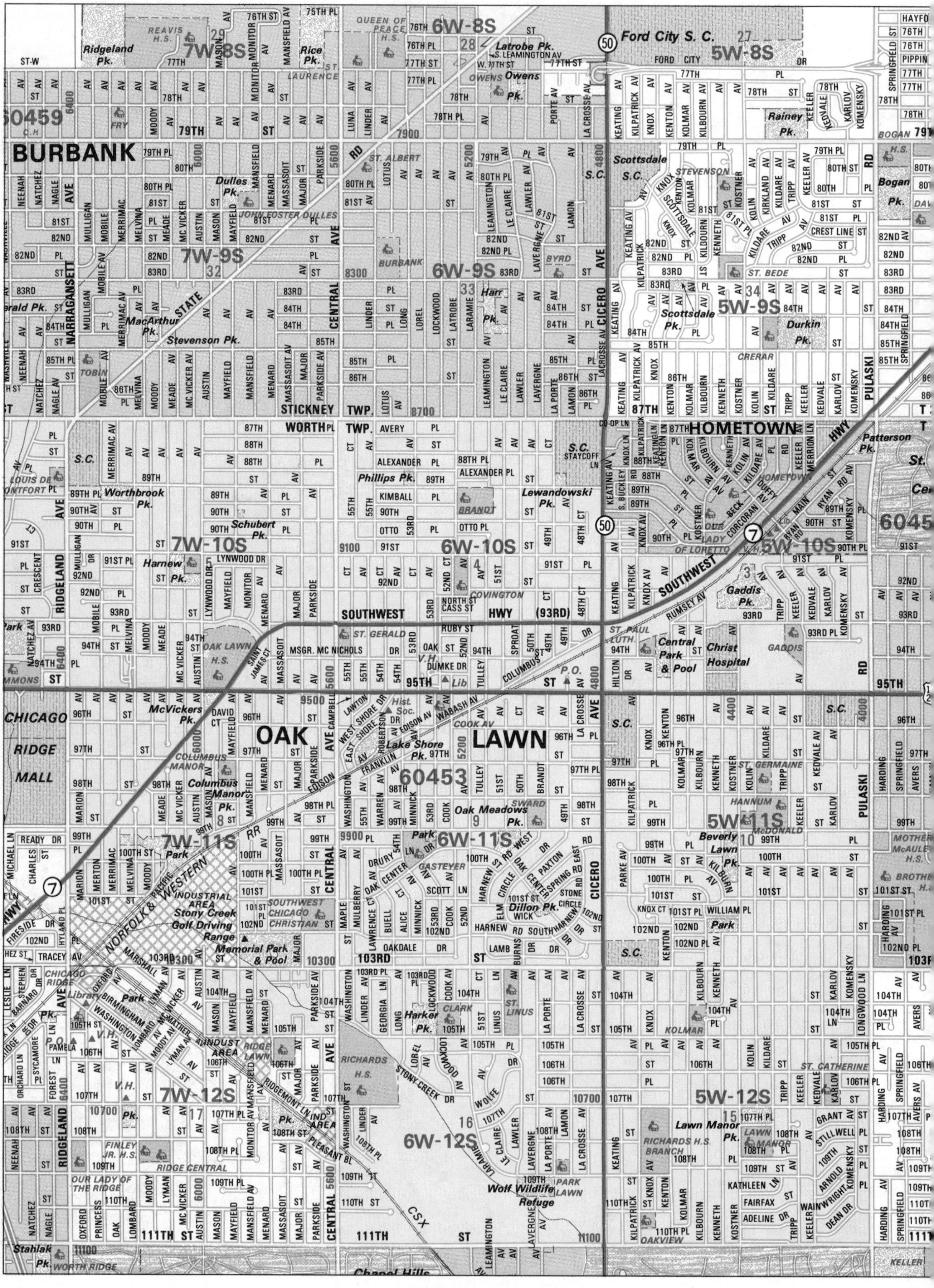

BURBANK

OAK LAWN

CHICAGO RIDGE MALL

HOMETOWN

Ford City S. C.

60459

60453

60451

Grid references: 7W-8S 6W-8S 5W-8S 7W-9S 6W-9S 5W-9S 7W-10S 6W-10S 5W-10S 7W-11S 6W-11S 5W-11S 7W-12S 6W-12S 5W-12S

STICKNEY TWP.

WORTH TWP.

SEE PAGES 64-65

SEE PAGES 60-61

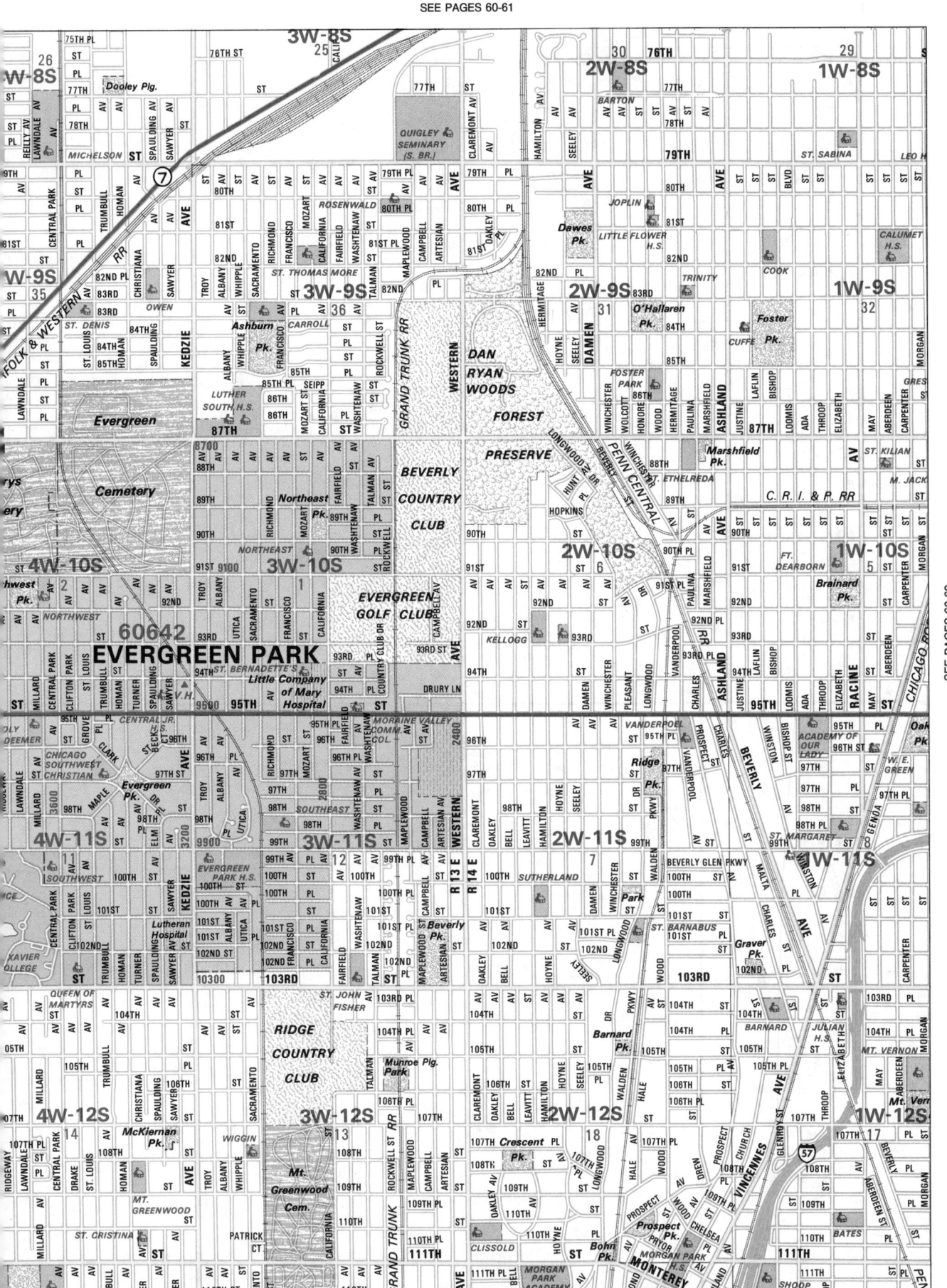

SEE PAGES 68-69

SEE PAGES 66-67

Grid references: 1W-8S, 0W-8S, 0E-8S, 1W-9S, 0W-9S, 0E-9S, 1E-9S, 1W-10S, 0W-10S, 0E-10S, 1E-10, 1W-11S, 0W-11S, 0E-11S, 1E-11S, 1W-12S, 0W-12S, 0E-12S, 1E-12S

Major streets and landmarks:

- OGLESBY, ST. SABINA, LEO H.S., ST. LEO
- WINNECONNA PKWY, EGGLESTON AV, WESCOTT, VINCENNES, HOOKWAY
- Grand Crossing Pk., ST. DOROTHY, RUGGLES, HIRSCH H.S.
- 76TH, 77TH, 77TH PL, 78TH, 78TH PL, 79TH, 80TH, 81ST, 82ND, 83RD, 84TH, 85TH, 86TH, 87TH, 88TH, 89TH, 90TH, 91ST, 92ND, 93RD, 94TH, 95TH
- CALUMET H.S., SIMEON VOC. H.S., West Chatham Pk., MORGAN ELEM.
- GRESHAM, ST. KILIAN, M. JACKSON
- MERCY H.S., DIXON, NEIL, McDADE
- AVALON PK., SBARBARO, PIRIE, UNIVERSITY
- DAN RYAN EXPRESSWAY, HOLLAND RD
- MARTIN LUTHER KING JR., COTTAGE GROVE, BURNSIDE
- FT. DEARBORN, Brainard Pk., Wallace Pk., Harvard Pk.
- TURNER DREW LANGUAGE ACADEMY, KIPLING, GILLESPIE
- Tuley Pk., Abbott Pk., ST. JOACHIMS
- C.R.I. & P. RR (CHICAGO ROCK ISLAND & PACIFIC RR)
- ACADEMY OF OUR LADY, Oakdale Pk., W.E. GREEN, WACKER
- Euclid Pk., Evers Pk., HARLAN H.S., Abbott Pk.
- ST. MARGARET, CHICAGO STATE UNIVERSITY, SHEDD, CALUMET EXPRESSWAY
- FERNWOOD, BENNETT, Gately Stadium Pk., W. SMITH, SCHMID, CORLISS H.S.
- M. GARVEY, JULIAN H.S., MT. VERNON, Mt. Vernon Pk., L. HUGHES, KOHN
- POE PL, CULLEN, All Saints
- DUNNE, BOOKER T. WASHINGTON (BR. VAN VLISSINGEN), VAN VLISSINGEN, BATES, MENDEL H.S.
- SHOOP, ROSELAND HOSP, Palmer
- 96TH, 97TH, 98TH, 99TH, 100TH, 101ST, 102ND, 103RD, 104TH, 105TH, 106TH, 107TH, 108TH, 109TH, 110TH, 111TH, 112TH

Avenue names (north-south): BISHOP, LOOMIS, ADA, THROOP, ELIZABETH, MAY, ABERDEEN, CARPENTER, SANGAMON, PEORIA, GREEN, HALSTED, RACINE, EMERALD, UNION, LOWE, WALLACE, PARNELL, NORMAL, EGGLESTON, HARVARD, PRINCETON, YALE, WENTWORTH, LA SALLE, PERRY, LAFAYETTE, STEWART, STATE, WABASH, MICHIGAN, INDIANA, PRAIRIE, FOREST, CALUMET, VERNON, EBERHART, RHODES, ST. LAWRENCE, CHAMPLAIN, LANGLEY, EVANS, COTTAGE GROVE, MARYLAND, DREXEL, INGLESIDE, ELLIS, DOBSON, GREENWOOD, UNIVERSITY, WOODLAWN, AVALON, KIMBARK, WINSTON, GENOA, PENN

Route markers: 1, 57, 94, 95

SEE PAGES 62-63

LAKE MICHIGAN

SEE PAGES 76-77

SEE DU PAGE COUNTY PAGES 34-35

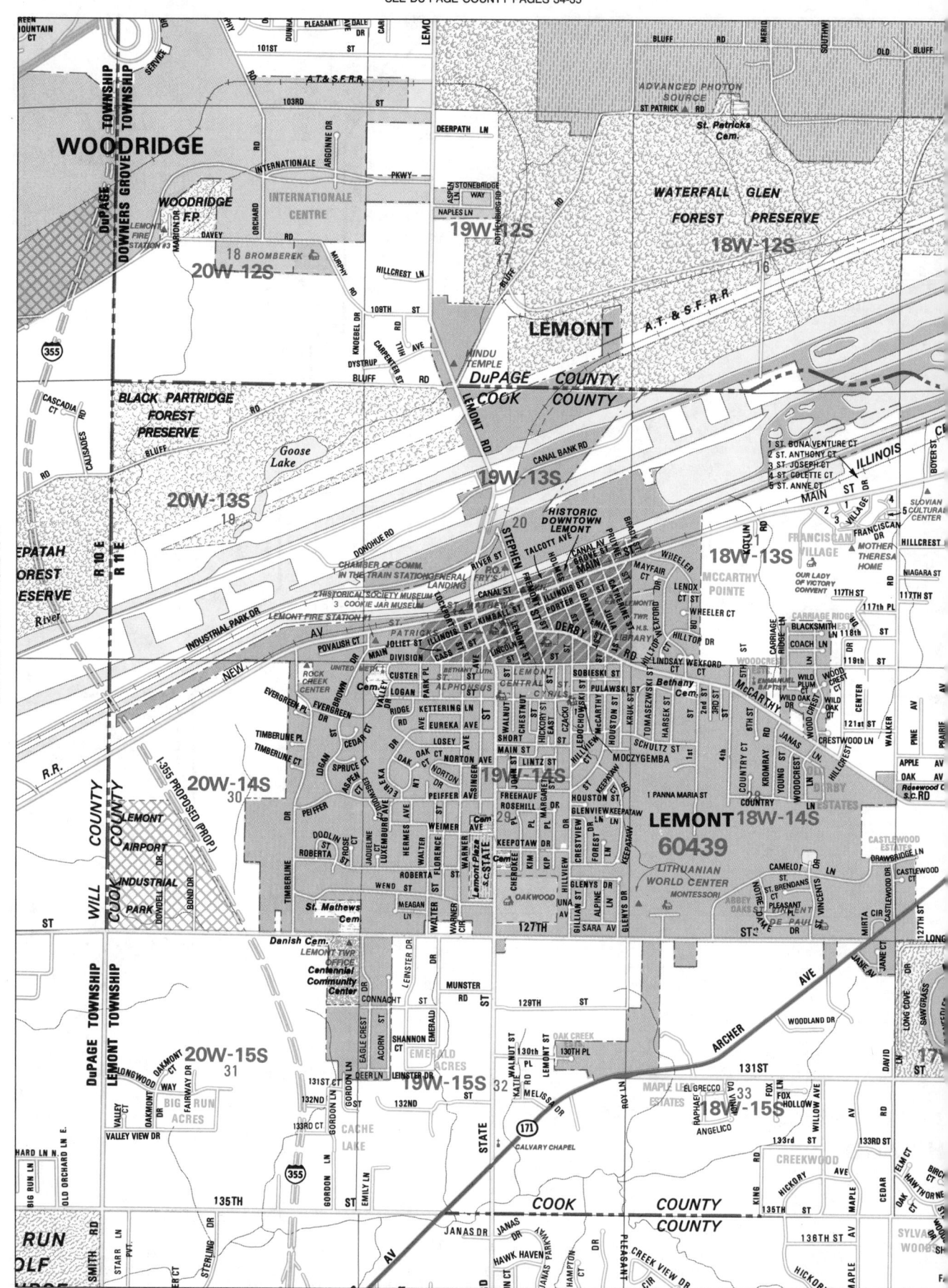

SEE WILL COUNTY PAGES 10-11

SEE DU PAGE COUNTY PAGES 34-35

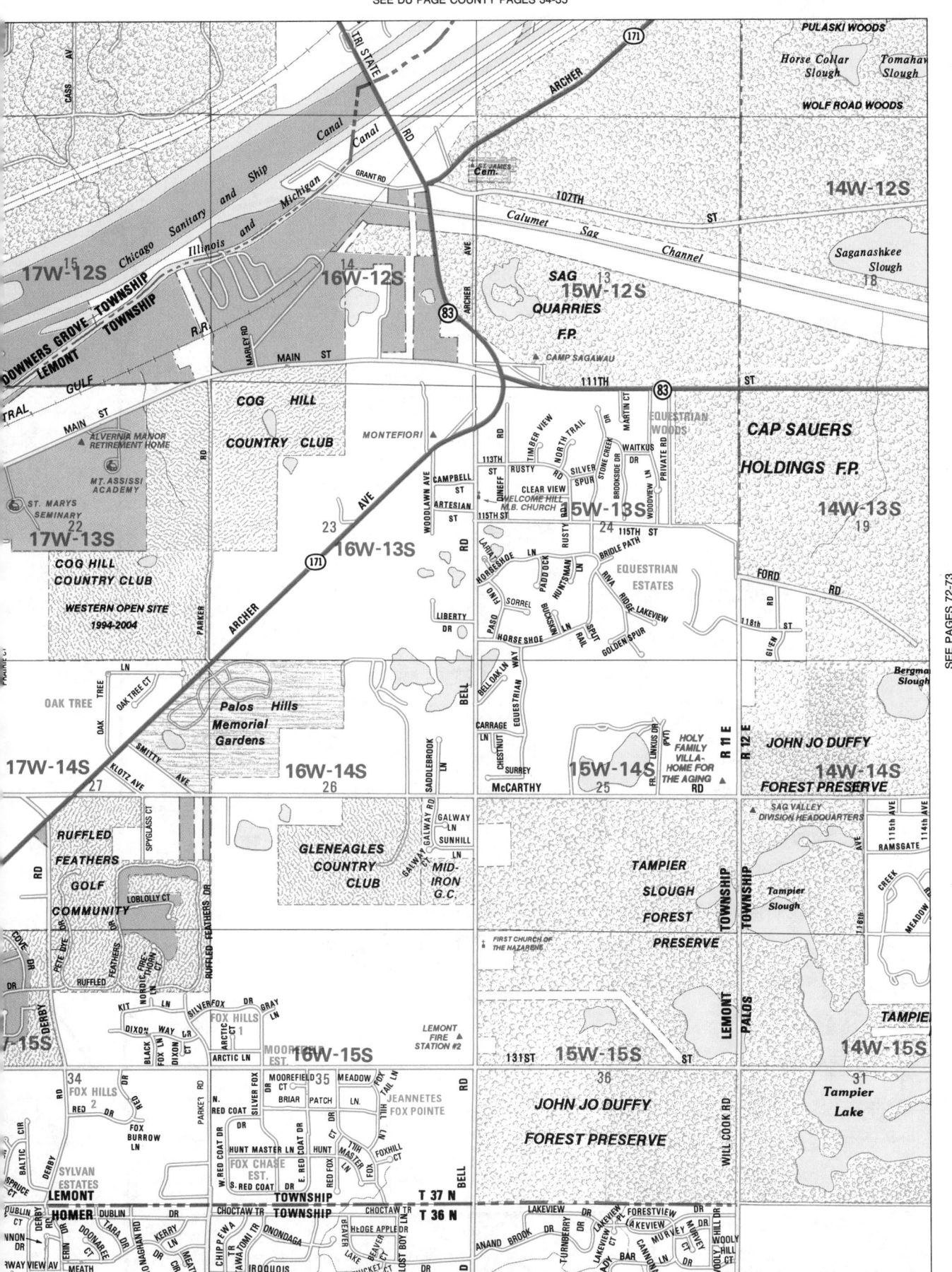

PULASKI WOODS

Horse Collar Slough

Tomahawk Slough

WOLF ROAD WOODS

171

ARCHER

ST JAMES Cem.

107TH

Calumet

Sag

ST

Channel

GRANT RD

Canal

Canal

Ship

and

Michigan

and

Illinois

Chicago Sanitary and Ship

14W-12S

Saganashkee Slough

18

17W-12S

16W-12S

SAG

15W-12S

QUARRIES

F.P.

13

DOWNERS GROVE TOWNSHIP

LEMONT TOWNSHIP

R.R.

MAIN ST

GULF

83

CAMP SAGAWAU

ARCHER AVE

111TH

83

ST

CENTRAL

MAIN ST

ALVERNIA MANOR RETIREMENT HOME

MT. ASSISSI ACADEMY

COG HILL COUNTRY CLUB

MONTEFIORI

EQUESTRIAN WOODS

CAP SAUERS

HOLDINGS F.P.

14W-13S

19

ST. MARYS SEMINARY

22

17W-13S

COG HILL COUNTRY CLUB

WESTERN OPEN SITE 1994-2004

23

16W-13S

171

ARCHER

WOODLAWN AVE

CAMPBELL ST

ARTESIAN ST

CLEAR VIEW WELCOME HILL M.B. CHURCH

113TH ST

DINEFF

115TH ST

TIMBER VIEW

NORTH TRAIL

RUSTY RD

SILVER SPUR

STONE CREEK DR

BROOKSIDE DR

WAITKUS LN

WOODVEW LN

MARTIN CT

PRIVATE RD

15W-13S

24 115TH ST

BRIDLE PATH

HUNTSMAN LN

RUSTY

RNA RIDGE LAKEVIEW

EQUESTRIAN ESTATES

LIBERTY DR

LAFIA

HORSESHOE LN

PINO

PASO

SORREL

PADDOCK

BUCKSKIN LN

SPLIT RAIL

GOLDEN SPUR

FORD RD

118th ST

GLENN

PARKER RD

PRAIRIE

LN

OAK TREE CT

OAK TREE

OAK TREE

Palos Hills Memorial Gardens

HORSE SHOE

BELL

BELL OAK LN

EQUESTRIAN WAY

CARRAGE LN

CHESTNUT LN

SURREY LN

McCARTHY

HOLY FAMILY VILLA-HOME FOR THE AGING RD

FR. LINUS DR

R 11 E

R 12 E

JOHN JO DUFFY

14W-14S

FOREST PRESERVE

Bergman Slough

17W-14S

27

SMITTY AVE

KLOTZ AVE

16W-14S

26

SADDLEBROOK LN

15W-14S

25

SAG VALLEY DIVISION HEADQUARTERS

115TH AVE

114TH AVE

RAMSGATE

RUFFLED FEATHERS GOLF COMMUNITY

SPYGLASS CT

LOBLOLLY CT

PETE DYE DR

RUFFLED FEATHERS DR

GALWAY RD

GALWAY LN

SUNHILL LN

GALWAY

MID-IRON G.C.

GLENEAGLES COUNTRY CLUB

FIRST CHURCH OF THE NAZARENE

TAMPIER

SLOUGH

FOREST

PRESERVE

Tampier Slough

CREEK

MEADOW

RD

COVE RD

NORDIC FIRE LN

THORN CT

RUFFLED

KIT LN

DIXON WAY

SILVERFOX DR

GRAY LN

ARCTIC CT 1

DIXON LR

FOX HILLS CT

LEMONT FIRE STATION #2

131ST ST

15W-15S

36

LEMONT TOWNSHIP

PALOS TOWNSHIP

WILL COOK RD

14W-15S

TAMPIER

Tampier Lake

DERBY

15S

BLACK LN

DIXON LN

ARCTIC LN

MOOREFIELD EST.

16W-15S

SILVER FOX CT

MOOREFIELD CT

BRIAR

MEADOW LN

PATCH LN

JEANNETES FOX POINTE

TAIL LN

TAIL LN

34

FOX HILLS 2

RED LN

PARKER RD

N. RED COAT DR

W. RED COAT DR

RED COAT DR

35

HUNT MASTER LN

E. RED COAT DR

HUNT

RED FOX

FOXHILL CT

131ST

JOHN JO DUFFY

FOREST PRESERVE

31

FOX BURROW LN

BALTIC CIR

DERBY

RED

DERBY

SPRUCE CT

SYLVAN ESTATES

FOX CHASE EST.

S. RED COAT DR

HUNT MASTER LN

BELL RD

LEMONT TOWNSHIP

T 37 N

T 36 N

DUBLIN CT

LEMONT HOMER

ERIN LN

MONAGHAN DR

DUBLIN DR

TARA DR

KERRY LN

DOONAREE CT

MEATH

CHIPPEWA TR

TAWATOMI TR

IROQUOIS

CHOCTAW TR

BEAVER LN

H.LOGE APPLE DR

BEAVER LAKE THICKET CT

LOST BOY LN

DR

CHOCTAW TR TOWNSHIP

ANAND BROOK DR

LAKEVIEW DR

TURNBERRY

LAKEVIEW CT

LADY BAR

MURVEY

FORESTVIEW DR

LAKEVIEW DR

MURVEY CT

CANNON

WOOLY HILL DR

WOOLY HILL

NNON DR

MEATH

RWAY VIEW AV

SEE PAGES 72-73

SEE WILL COUNTY PAGES 16-17

SEE PAGES 64-65

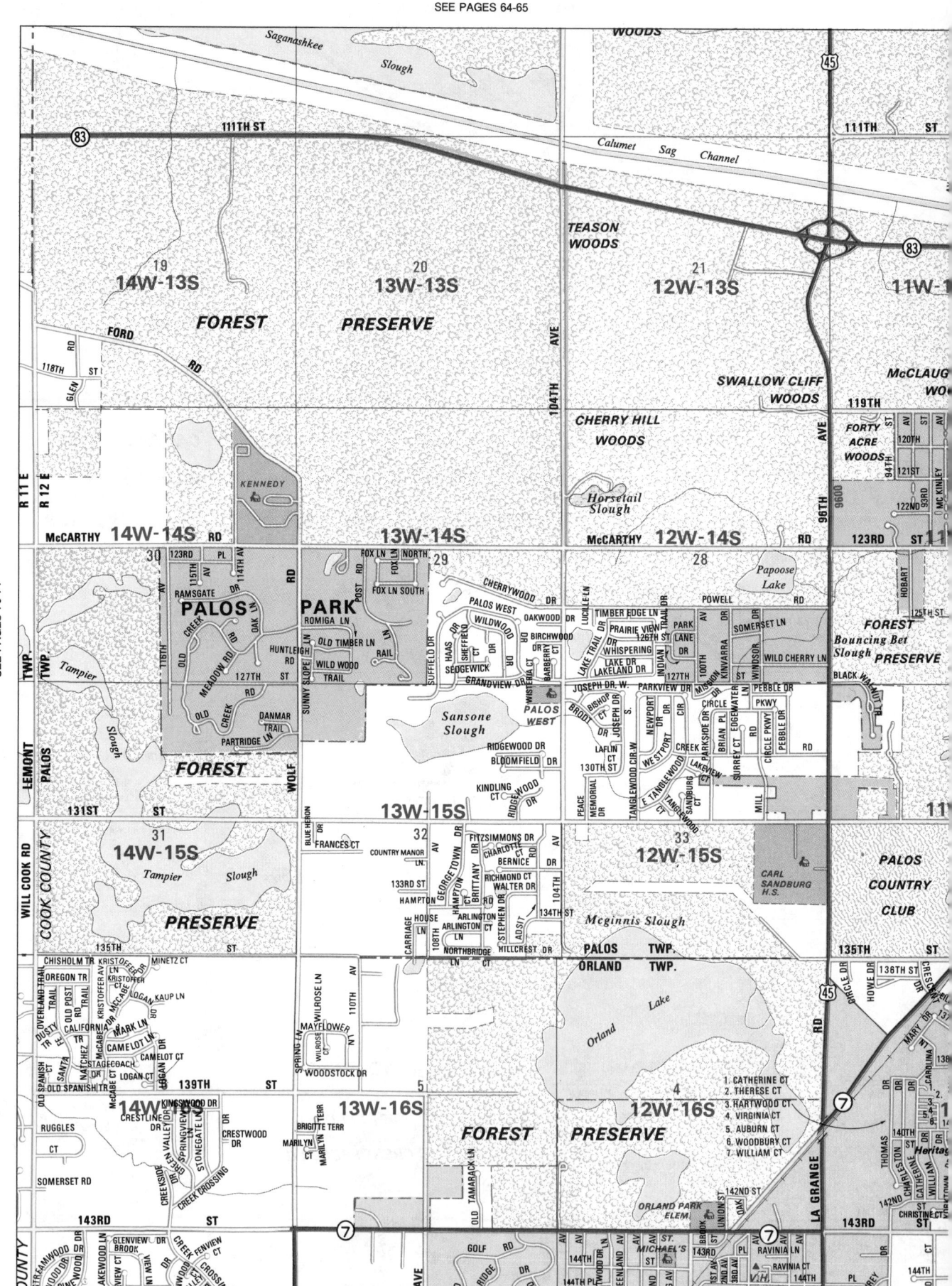

SEE PAGES 70-71

SEE PAGES 64-65

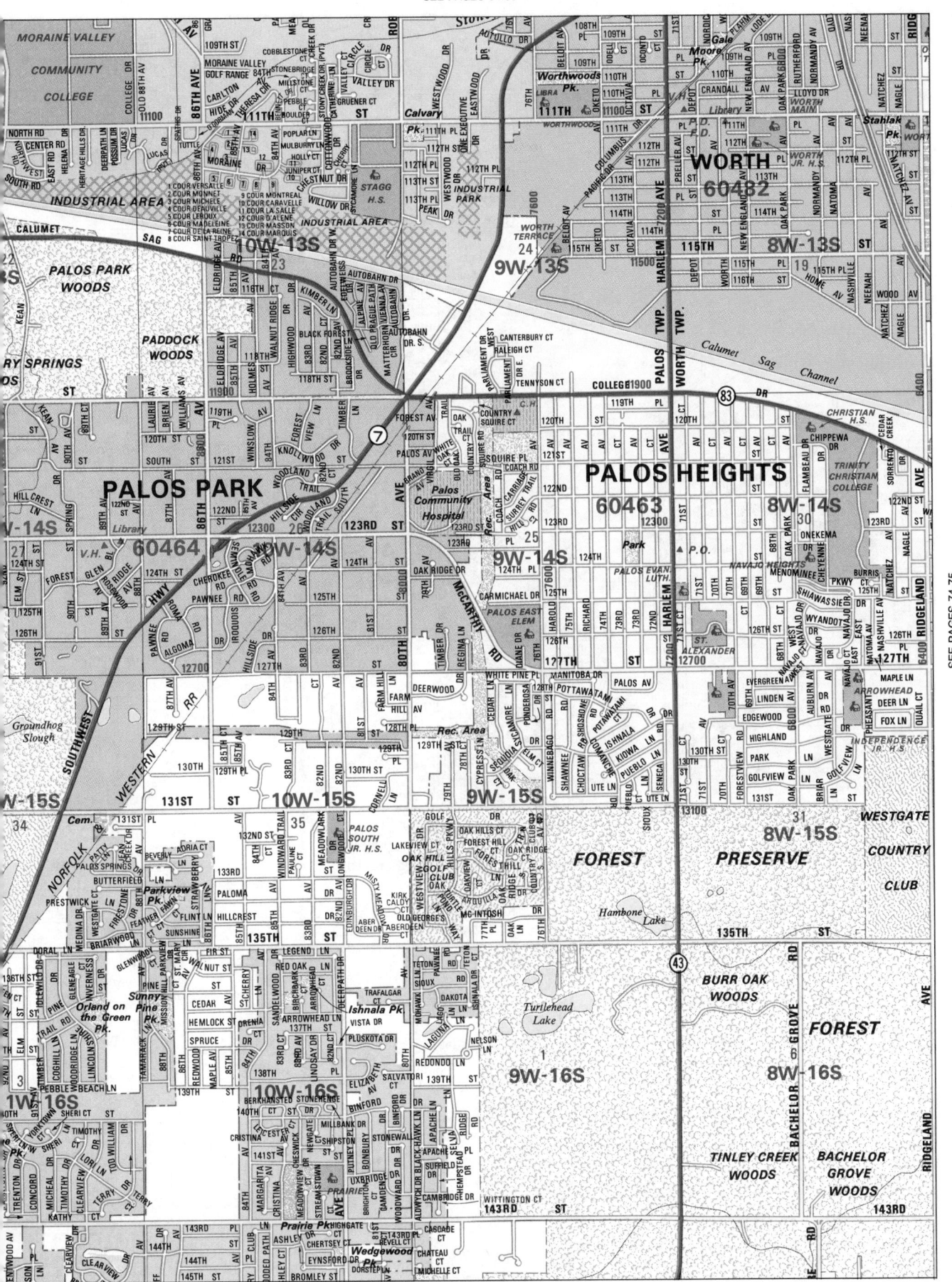

SEE PAGES 74-75

SEE PAGES 78-79

SEE PAGES 66-67

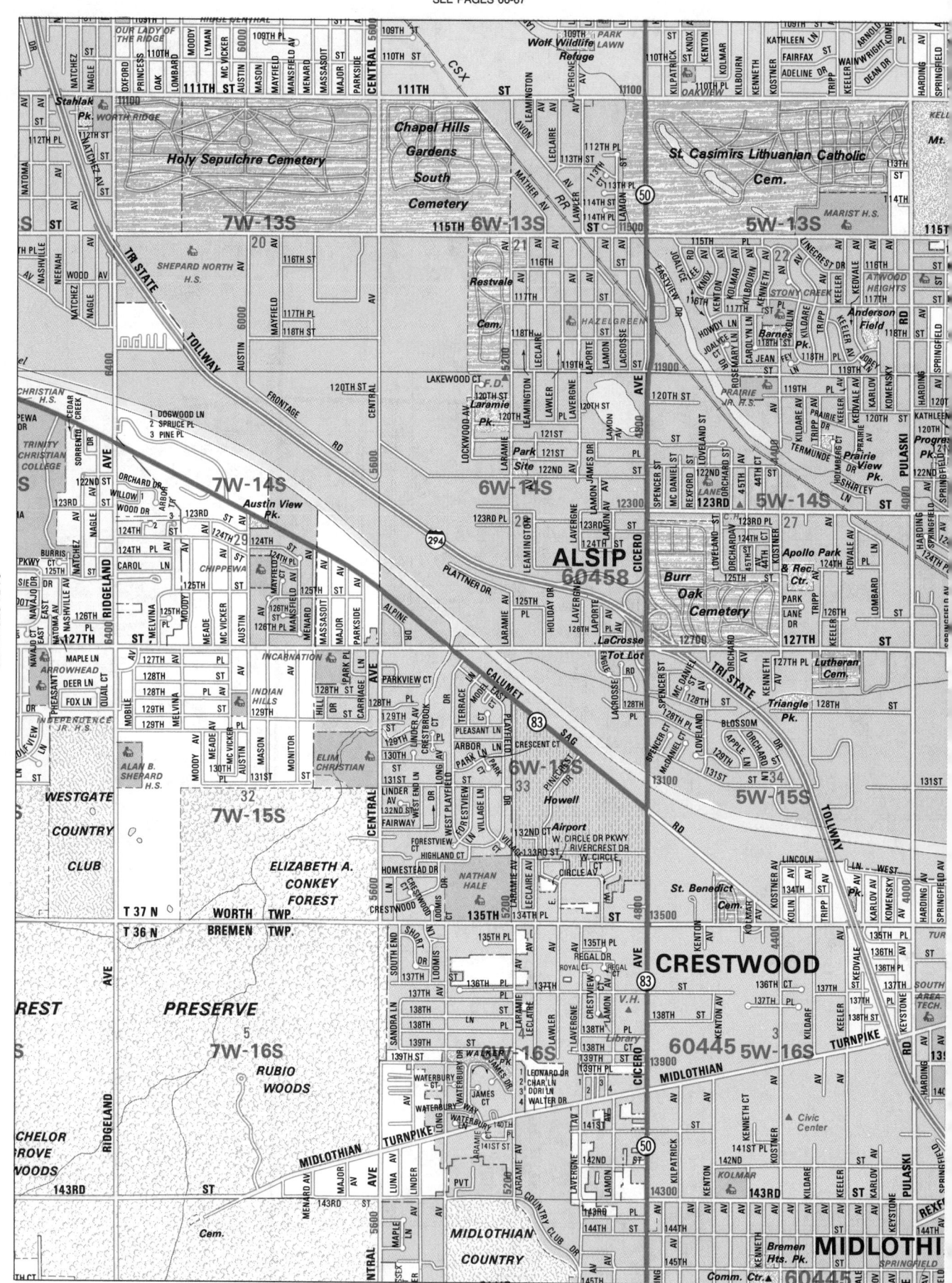

SEE PAGES 80-81

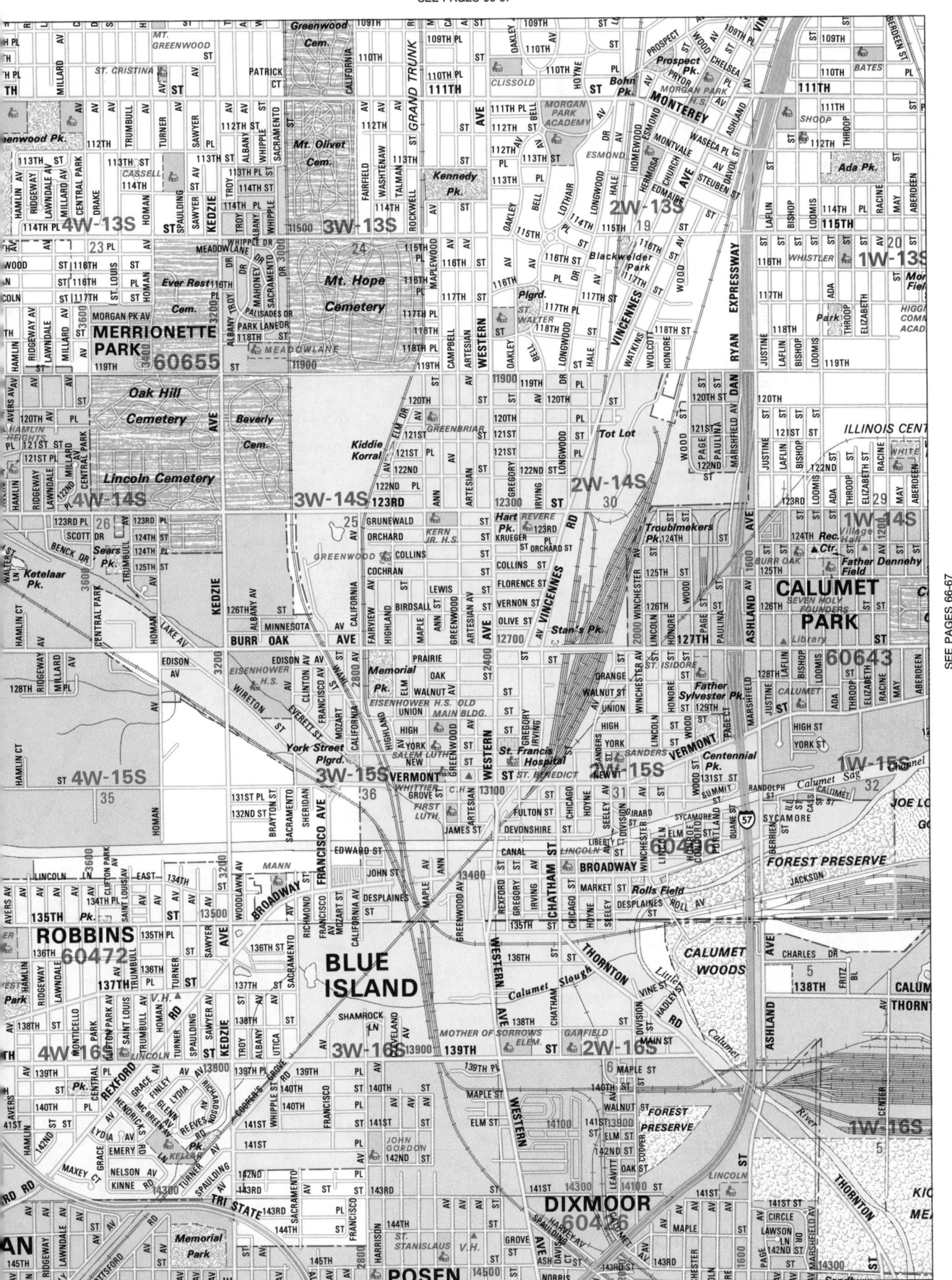

SEE PAGES 68-69

SEE PAGES 74-75

1W-12S
0W-12S
1E-12S
0E-12S
1E-13S
0W-13S
0E-13S
1W-13S
0W-14S
0E-14S
1W-14S
1W-15S
0W-15S
0E-15S

MARTIN LUTHER KING JR.
HALSTED ST
CENTRAL
PENN
STATE ST
INDIANA AVE
LINCOLN AVE
CHICAGO & EASTERN ILLINOIS RR
ILLINOIS CENTRAL RR
CALUMET EXPRESSWAY
COTTAGE GROVE RR
CSX

Fernwood Pk.
L. Hughes
Kohn
Mt. Vernon Pk.
Cullen
Poe Pl
Corliss
Maryland
Woodlawn

Julian H.S.
Mt. Vernon
Booker T. Washington (Br. Van Vlissingen)
All Saints
Dunne
Van Vlissingen
Mendel H.S.
Shoop
Roseland Hosp
Palmer Pk.
Holy Rosary
Pullman
Ellis
Ada Pk.
Fenger H.S.
St. Willbrord H.S.
St. Nicholas
Brenan
Yale
Curtis
St. Anthonys
Kensington
Whistler
Morgan Field
Higgins Comm. Acad.
St. Catherine
Scanlan
Kensington Pk.
St. Salameas
Lake
Port of Chicago
Lake Calumet
Habor Facilities
West Pullman
Assumption
Gompers
Jesse Owens Magnet
Metcalfe
West Pullman Pk.
Lion Field
Kiawanis Field
Brayton St
Illinois Central RR
White Park
Village Hall
Father Dennehy Field
Calumet Park Pool & Tot Lot
CALUMET PARK
Library
60643
Cedar Park Cemetery
Seven Holy Founders
Private Rds
Aldridge
Geo. Washington Carver Primary
Dubois
Carver Pk.
Drexel
Calumet Sag
Joe Louis The Champ Golf Course
Whistler Preserve
Forest View
Carver Middle
Rest Preserve
Jackson
Community Ctr. & Pool
Blue Island-Riverdale
Mohawk Pk.
Memorial Fieldhouse
Riverdale Pk.
St. Mary's
St. Paul
Sunshine Pk.
Calumet Twp.
Thornton Twp.
RIVERDALE
Washington
Little Calumet River
13100
13400
13800

Elizabeth, May, Aberdeen, Sangamon, Peoria, Green, Morgan, Throop, Bishop, Loomis, Racine, Carpenter, Normal, Wallace, Lowe, Union, Emerald, Eggleston, Stewart, Harvard, Princeton, Wentworth, Perry, State, Michigan, Indiana, LaSalle, Lafayette, Wabash, Prairie, Forest, Calumet, Vernon, Eberhart, Front, St. Lawrence, Forestville, Champlain, Langley, Cottage Grove, Rhodes, Ellis, Corliss, Maryland, Evans, Ingleside, Greenwood, Drexel

103RD, 104TH, 105TH, 106TH, 107TH, 108TH, 109TH, 110TH, 111TH, 112TH, 113TH, 114TH, 115TH, 116TH, 117TH, 118TH, 119TH, 120TH, 121ST, 122ND, 123RD, 124TH, 125TH, 126TH, 127TH, 128TH, 129TH, 130TH, 131ST, 132ND, 133RD, 134TH, 135TH, 136TH, 137TH, 138TH, 139TH, 140TH

T 37 N
T 36 N

SEE PAGES 82-83

SEE PAGES 68-69

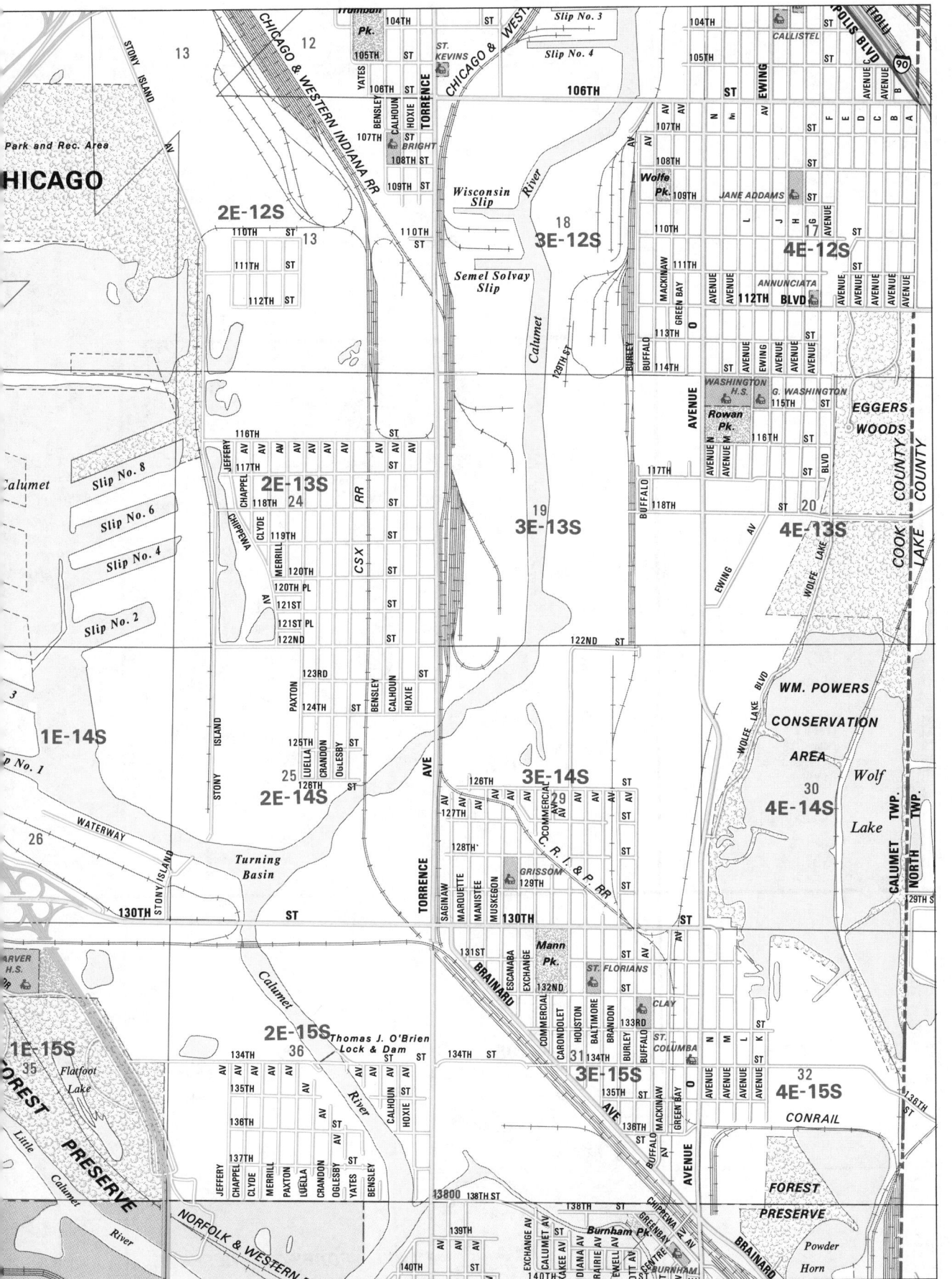

SEE PAGES 72-73

PRESERVE

COOK

PALOS   TWP.
ORLAND   TWP.

FOREST   PRESERVE

Orland Lake

1. CATHERINE CT
2. THERESE CT
3. HARTWOOD CT
4. VIRGINIA CT
5. AUBURN CT
6. WOODBURY CT
7. WILLIAM CT

Meginnis Slough

CLUB

14W   13W-16S   12W-16S   4

CHISHOLM TR KRISTOFFER MINETZ CT
OREGON TR KRISTOFFER AV
OVERLAND TRAIL DUSTY TR OLD POST RD TRAIL LOGAN DR KAUP LN
SANTA CALIFORNIA DR MCCABE DR CAMELOT LN
NATCHEZ TR STAGECOACH CAMELOT CT
OLD SPANISH TR LOGAN CT
OLD SPANISH TRABE MCCABE DR

MAYWILROSE LN
WILROSE CT
MAYFLOWER
WOODSTOCK DR

139TH   ST

CRESTLINE KINGSWOOD DR
CRESTWOOD DR GREEN VALLEY DR
SPRINGVIEW DR STONEGATE LN
CREEKSIDE DR CREEK CROSSING
BRIGITTE TERR
MARILYN CT MARILYN TERR

RUGGLES CT

SOMERSET RD

143RD   ST

STREAMWOOD DR GLENVIEW DR CREEK FENVIEW CT
CREEKWOOD DR LAKEWOOD LN VIEW LN CREEK CROSSING
PINEWOOD DR PINEVIEW DR FENWOOD DR FENVIEW
PINECREEK CT WOODLAKE LN VIEW DR MESQUITE DR
PINE CREEK DR N CREEKVIEW DR VALLEY VIEW LN CREEK CROSSING
SPRUCE LN MAPLE DR CREEK CROSSING
PINECREEK DR S OAK DR

GOLF RD
LAKE RIDGE DR
RIDGE DR MORNINGSIDE RD GREENLAND AV WESTWOOD DR HIGHLAND AV IRVING AV
HOLLOW TREE DR OAKLEY DR RANEY'S WOODLAND
CRYSTAL TREE WESTWOOD DR
GOLF RD

13W-17S   12W-17S

14W-17S
HARVEST CROSSING SINGLETREE RD
WIND MILL WINDMILL RD WINDMILL TURN
SILO RIDGE RD WEST SILO RIDGE RD EAST
BRAMLETT CT

R 11 E   R 12 E

151ST   ST
SILO RIDGE RD

ST. MICHAEL'S V.H.
143RD PL RAVINIA LN
1ST ST 3RD AV RAVINIA CT
145TH ST BEACON P.D.
145TH PL RAVINIA AV
146TH ST
147TH BROWN PK.
147TH DOOGAN HUMPHREY COMPLEX
LIBRARY HIGH POINT
ORLAND AVENIDA DEL NORTE
JR. H.S. LA REINA DR AVENIDA DEL ESTE
HAWTHORNE LA REINA
HYACINTH DR HOPKINS
HIBISCUS DR HALE LA REINA REAL
151ST HOLIDAY WEST
151ST ST

ORLAND
PARK
60462

ORLAND SQUARE SHOPPING CENTER

ORLAND SQUARE DR

ORLAND CENTER

WOLF   RD
108TH   AVE

SHAKER CT POPLAR CREEK LN
PINE GROVE DR VAN ARBOR DR SILO RIDGE DR GRANDVIEW DR STINGER CREEK LN LAKE LAWN SPRING
ALPINE DR SHADE COVE CT COTTONWOOD DR WILLOW CREEK LN
TIMBER RIDGE CT TIMBER EDGE LN ROYAL CREEK LN

CREEK LN POPLAR CREEK CT

HIAWATHA FRANCHESCA EL CAMENO LN
HILLTOP HIGHLAND CORDOBA EL CAMENO CT
HILLTOP CT CAMINO REAL PK. EL CAMENO
ORLAND COURT S.C.
BILL YOUNG PK. WEST AV SHADY LN
HUNTINGTON WILSHIRE DR
TREETOP RANQUIST PK.
CENTENNIAL CT
WEST AV CONSTITUTION DR
RAVINIA AV

14W-18S   13W-18S   12W-18S

LYNCH
118TH
155TH 156TH 157TH 158TH
117TH 116TH 115TH 114TH 113TH 112TH

PEMBROOK LN MAYFAIR LN OXFORD STRADFORD
BEDFORD CT
WHERRY LN BRADFORD LN WATERFORD DR SUNRISE AV

159TH   ST

DOCTOR US 6

JUANITA DR

14W-19S   13W-19S   12W-19S

HOMER TWP. ORLAND TWP.

KINGSPORT RD DEER ISLAND 107TH AV 163RD PL 105TH CT 164TH CT
MEADE ST W. HAZEL 165TH ST 106TH PL
HANCOCK SHERMAN TOUCHSTONE RD
ORANGE WILLOW PAW PAW PALM CEDAR BEECH GRANTS TR

167TH

SIR RALEIGH DR HUNTERS CT
CHAUSER DR VICTORIA PL 162ND
TOWER DR 163RD
LANDINGS DR ROBIN LN SPICEBUSH RD CARDINAL RD LINDSAY ST
LEE ST 167TH PL 168TH ST 169TH BLUE HERON DR WOOD DUCK LN YEARLING CROSSING

ORLAND HILLS 604

11W   ORLAND HILLS

FOREST   PRESERVE

WILL COUNTY COOK COUNTY

SEE WILL COUNTY PAGES 16-17

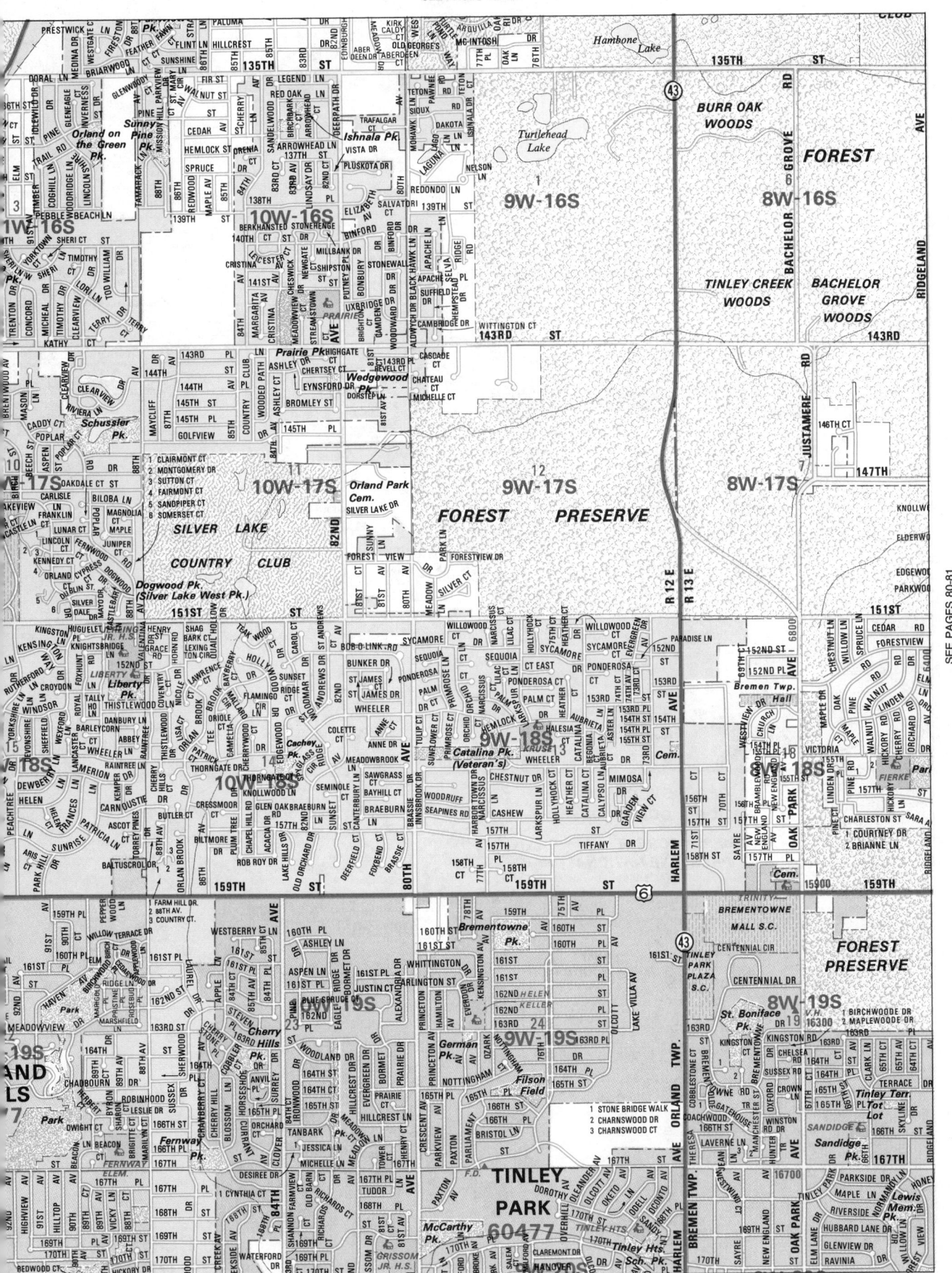

FOREST PRESERVE
7W-16S
RUBIO WOODS

BACHELOR GROVE WOODS

W-16S

143RD ST

MIDLOTHIAN TURNPIKE AVE

Cem.

7W-17S ST

JUSTAMERE RD
146TH CT

147TH ST

MIDLOTHIAN COUNTRY CLUB
6W-17S

OUR LADY OF SORROWS CONVENT

KNOLLWOOD DR DANIEL LN
ELDERWOOD CT HARBOR
CARRIAGE WAY
EDGEWOOD CT
PARKWOOD CT
151ST ST

S.W. CASE

OAK FOREST H.S.

CEDAR RD EL MORRO LN
FORESTVIEW
CHESTNUT LN WILLOW LN SPRUCE LN
MAPLE DR PINE RD OAK RD WALNUT RD LINDEN RD
ELM LAS ROBLES RIO VERDE CATALINA
AVALON COLINA AV CHAUCER DR
VENTURA ST FLORES HILLSIDE AV EL MORRO
ARROYO ST OAKLAND FOSTER ROY DR GEOFFREY DR
BOCA RIO SEQUOIA ST BRIAR CONDADO DR
VICTORIA HICKORY CHERRY RD ORCHARD RD
LA GRANGE ROB VICTORIA CHARLES LANCASTER MERLIN DR
Park OROGRANDE ALAMEDA ORANGE JAMES DR BETTY ANN LN
DOLORES DR EDWARD DOVER RD V.H.
VISTA ARROYO ALBERT ADELINE
CHARLESTON ST SARA ANN LN JILL ANN LIBERTY NATALIE
157TH ST JOANN LN ROY DR INDEPENDENCE AV
JON RD ROBHERON DR COREY LN
1 COURTNEY DR LA PAZ BROOKWOOD DR REYNOLDS
2 BRIANNE LN BRET DR TERRACE DR PEGGY
15900 159TH ST

7W-18S
FIERKE Park
La Paz
SCARLET OAK
INDEPENDENCE AV
SCENE
Library
CAROL BELLE TR

6W-18S
OAK FOREST
60452

OAK PARK AVE

SEE PAGES 78-79

GEORGE W. DUNNE NATIONAL GOLF COURSE
(FORMERLY FOREST PRESERVE NATIONAL)
7W-19S

SEE PAGES 79 & 85, GRID 9W-19S
1 BIRCHWOODE DR
2 MAPLEWOODE DR

FOREST PRESERVE
W-19S

JAMES ST

YANKEE WOODS

FOREST PRESERVE
6W-19S

JODY LN
MARK AV LESLIE ANN LN
KIMBERLY
GAYNELLE
KIMBERLY HEIGHTS
166TH ST PATRICIA FULTON TERR AUSTIN
167TH ST

Tinley Terr. Tot Lot
SANDIDGE Sandidge Pk.
165TH PL
SKYLINE

PARKSIDE DR
MAPLE LN HONEY LANE LN LAURA LN JACQUELIN PL
NORMANDY LN ARCADIA LN JENNIFER LN HILLSIDE DR
RIVERSIDE LN CARLSBAD ANDREE AV BRITTNEY LN
HUBBARD LANE LN BARBARA SCOTT CT DEBRA CT
GLENVIEW DR GAYNELLE BARBARA LN CHRISTOPHER LN
RAVINIA WILLOW LN FOREST VIEW LAKEOUT DR
PINE POINT DR AUTUMN DR BELLE LAKE BLUFF DR
-20S 171ST ST Vogt Woods Pk. 7W-20S
FULTON ST FOREST
VOGT PRESERVE
173RD PL

SANDRA LN 138TH LN LAMON V.H.
138TH 139TH LN PL LAWLER Library
WATERBURY 139TH PL LEONARD DR CREST 7W-16S 60445 5W-16S
WATERBURY WAY CHAR LN DORI LN
JAMES CT WALTER DR CICERO MIDLOTHIAN
140TH 141ST 13900
141ST ST 50 TURNPIKE
LUNA AV LINDER 142ND Civic Center
143RD PVT 14300 KOLMAR
MAPLE LN 144TH **MIDLO**
SUSSEX CT 144TH Bremen Hts. Pk.
SCARBOROUGH 145TH KENNETH
CROMWELL LN Comm. Ctr. 60445
14700 147TH ST 14700 St. Christo

GRANGE AV SUNSET 147TH CT 5W-17S
FERN AV 148TH 148TH
PARKSIDE AV 149TH FOREST EDGE LN 149TH
MASSASOIT TERR TEMPLE ST JEAN CT BECKY CT Pool
CALETTA DR MISSION Pk. 150TH SCHOOL SITE
150TH ST LA PALM LEClaire LAVERGNE CONCORDIA CT
150TH PL VINE ST LORAN LN 151ST
VINE ST 16100 151ST Kostner Pk.
CAROLYN CT 152ND
152ND GREEN FAWN CT CAROLYN 153RD
CRESCENT DIANE LAPORTE LACROSSE AV
NATALIE DR 153RD KEATING KILPATRICK KNOX AV KOLMAR
ST. DAMIEN 154TH WOODLAND
155TH 16 155TH ST WAVERLY
156TH LOREL AV LATROBE AV LAMON
ANN MARIE DR LONG AV Park
157TH LOCKWOOD DR 157TH
158TH 158TH ST
15900 159TH ST 6 159TH ST

FOREST PRESERVE
5W-18S
MIDLOTHIAN MEADOW

LESLIE LN 160TH ARBOR PARK 160TH ST County
VERA LN JAMIE LN 160TH Oak Forest Hospital
MARGIE LN AIMEE 161ST 5W-19S
BABETTE CT DEBBIE CT JANET 162ND ST
MARY ANN CT PAMELA CT LATROBE WAGMAN
ELLEN CT DEBRA LONG AV LAVERGNE HENRY MANN ST
163RD ST LOREL LOCKWOOD GROVE TERRY SCOTT BLAIR
16300 6W-19S OAK St. Gabriel Cem. HAROLD RICHARD BARRY
163RD CICERO GEORGE ADELE LN LAURA LN
16700 BRENDEN LN BROCKTON LN BARTON LN LISA LN
FIELDCREST DR WILLOWICK DR Industrial Park
166TH ST

KILPATRICK
ALDERSYDE RD ORCHARD HILL
BELLAIRE FOREST KIMBERLEY CT 16700
COULTER TIMBER CT
NEWPORT MEADOWDALE RD Park GINGERWOOD
169TH ST FAWNLEY LECLAIRE AV KARA
170TH ST DEERPATH ELMWOOD LANGLEY 5W-20S
FOREST TR GREENTREE CYPRESS CT 1 TIMERL
170TH HAWTHORNE RD 2 HOLLY C
BONNIE MARTHA JUDY CT SYCAMORE LN 3 PEARTR
JESSICA LN BELLE TR 28 171ST ELIZABETH LN
173RD ST 6W-20S 57 173RD **COUNTRY CLUB HILLS**
173RD PL HWY LECLAIRE

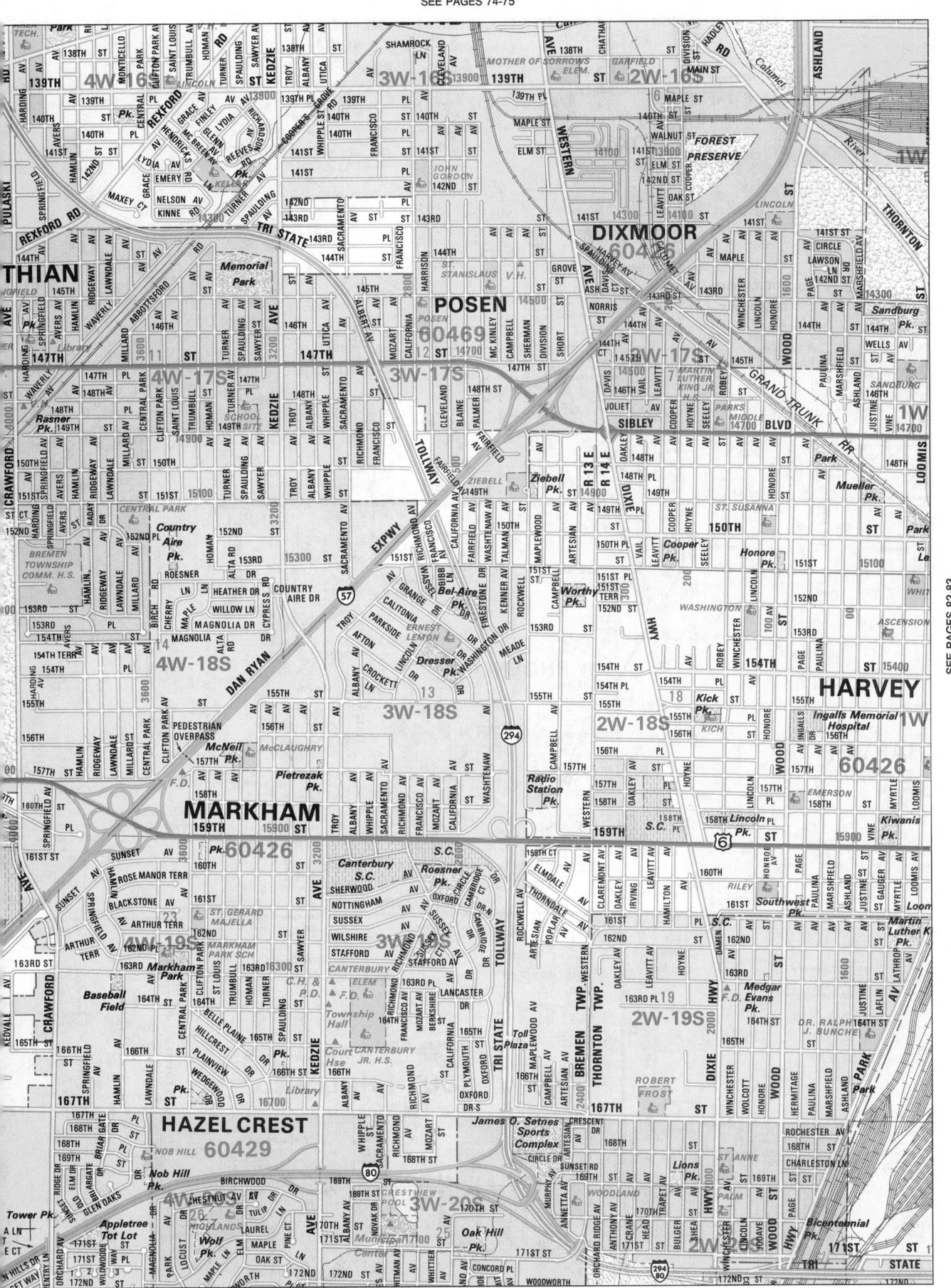

SEE PAGES 76-77

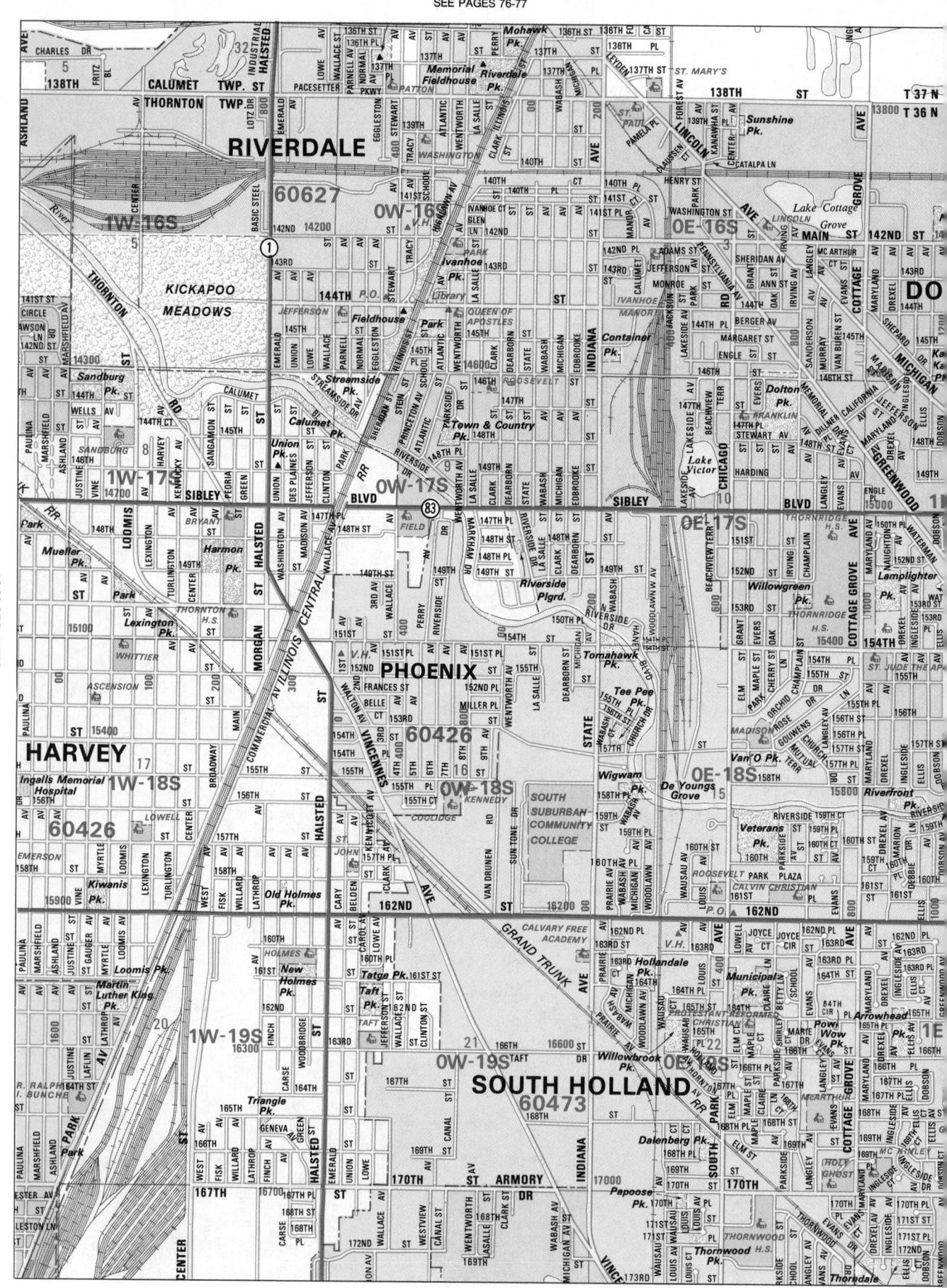

SEE PAGES 88-89

SEE PAGES 80-81

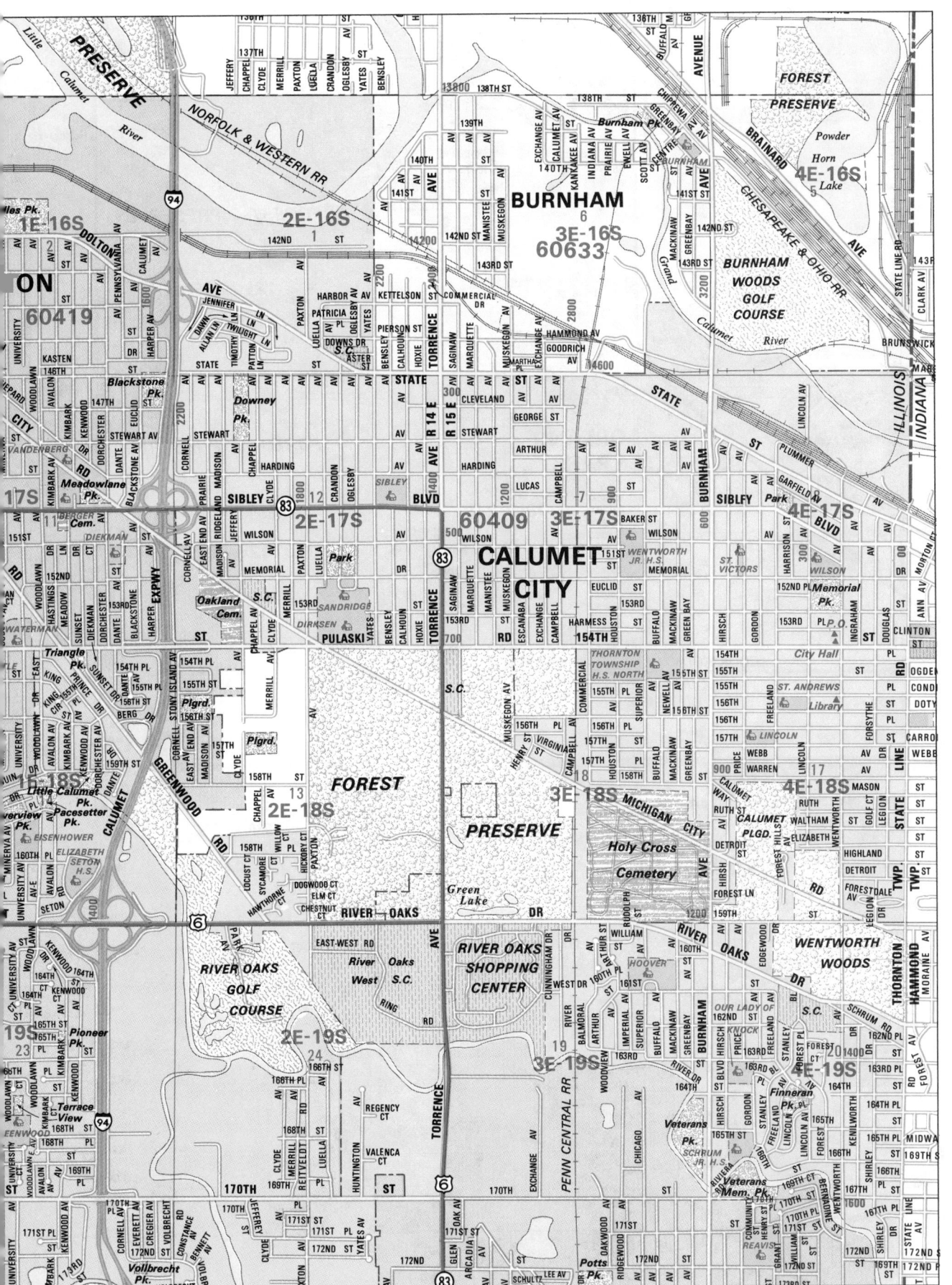

SEE PAGES 78-79

SEE WILL COUNTY PAGES 22-23

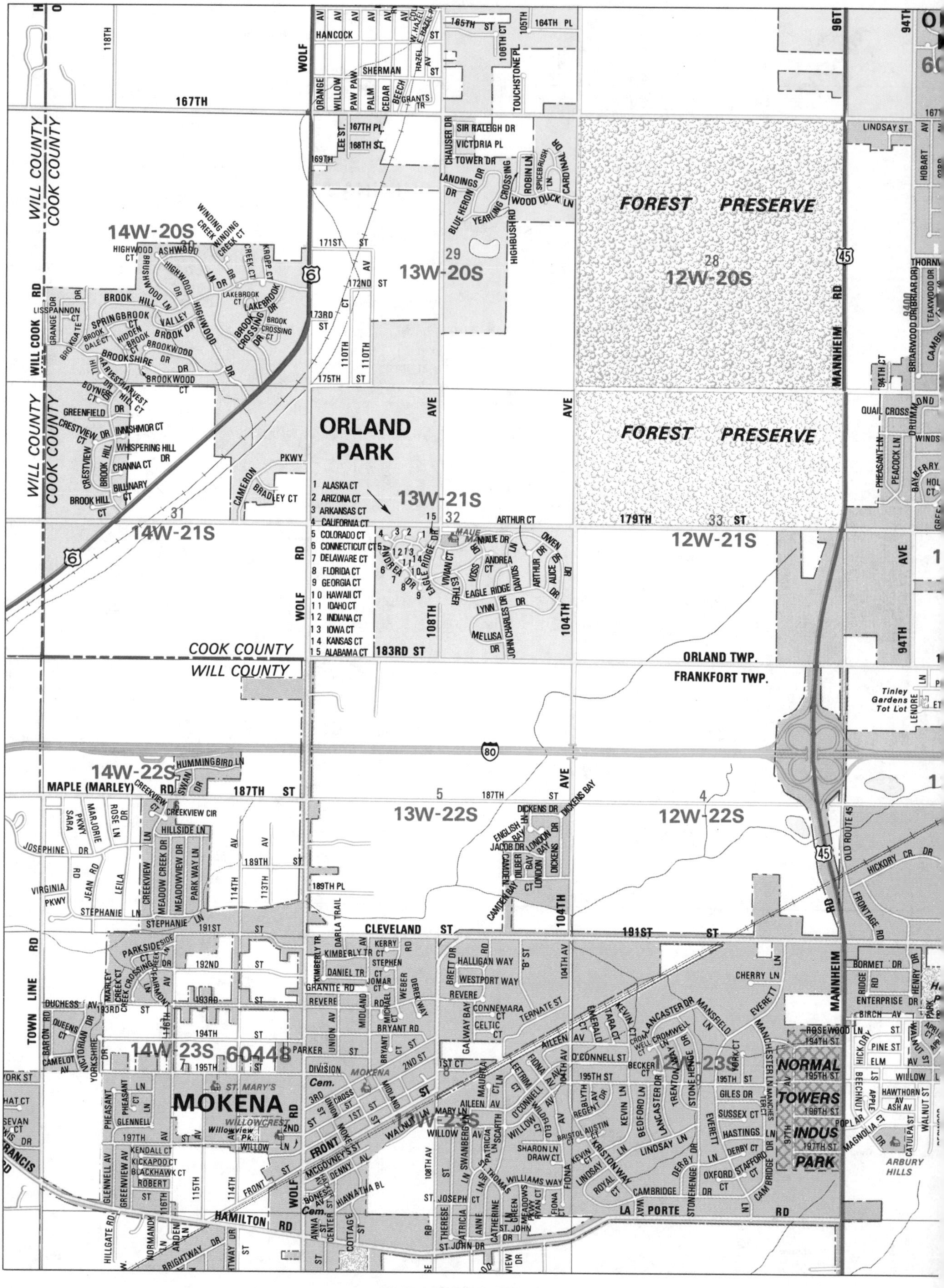

14W-20S

13W-20S

12W-20S

FOREST PRESERVE

28

ORLAND PARK

13W-21S

14W-21S

12W-21S

FOREST PRESERVE

179TH ST

33

1 ALASKA CT
2 ARIZONA CT
3 ARKANSAS CT
4 CALIFORNIA CT
5 COLORADO CT
6 CONNECTICUT CT
7 DELAWARE CT
8 FLORIDA CT
9 GEORGIA CT
10 HAWAII CT
11 IDAHO CT
12 INDIANA CT
13 IOWA CT
14 KANSAS CT
15 ALABAMA CT

183RD ST

COOK COUNTY
WILL COUNTY

ORLAND TWP.
FRANKFORT TWP.

Tinley Gardens Tot Lot

14W-22S

13W-22S

12W-22S

MAPLE (MARLEY) RD

187TH ST

CLEVELAND ST

191ST ST

14W-23S 60448

MOKENA

13W-23S

12W-23S

NORMAL TOWERS INDUS PARK

ARBURY HILLS

HAMILTON RD

LA PORTE RD

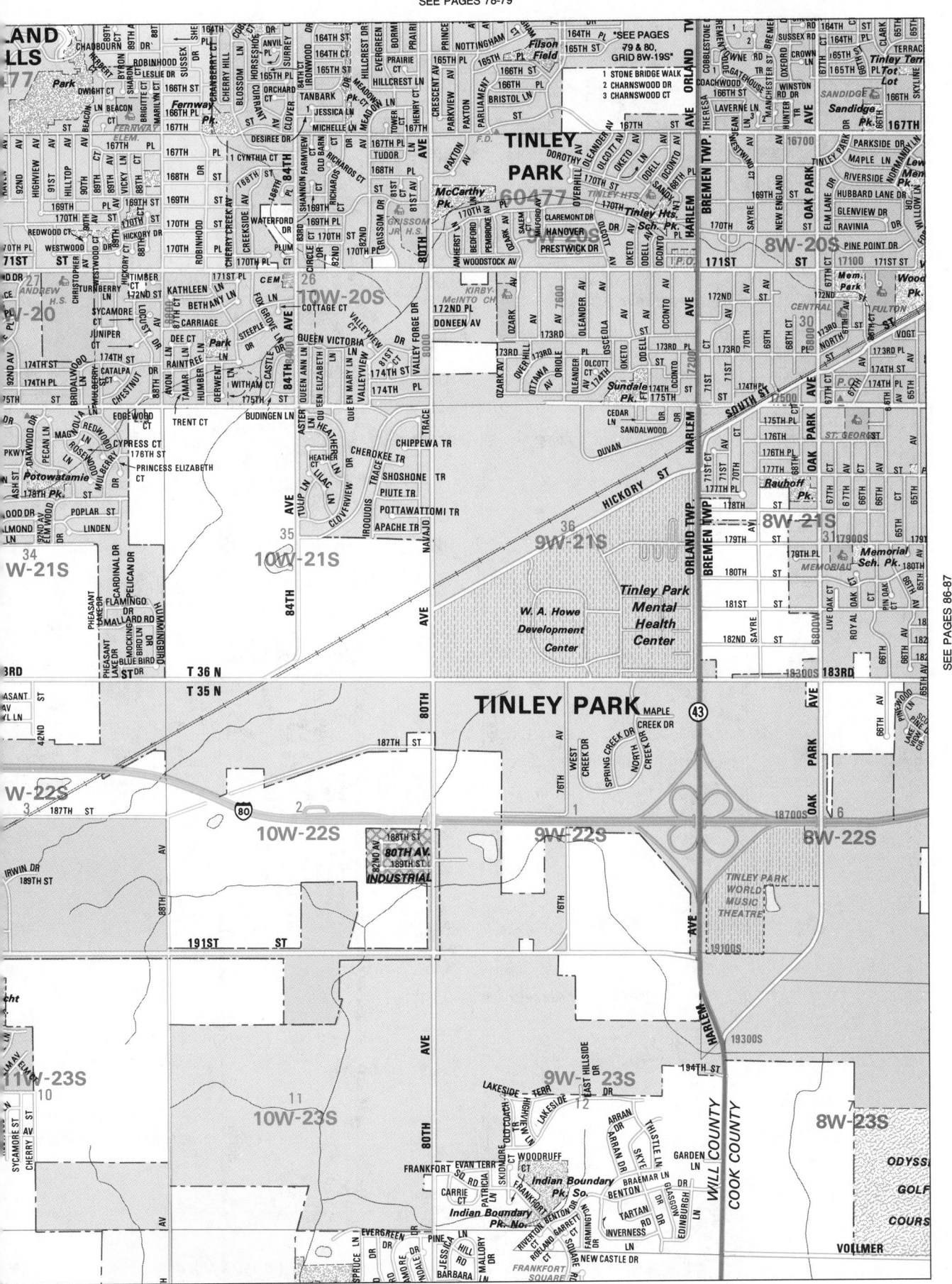

SANDIDGE
Sandidge Pk.
Lot 166TH SKYL
GAYNELLE
166TH ST
KIMBERLY HEIGHTS
BEVERLY ST
RIDGELAND
PATRICIA
FULTON
AUSTIN
166TH ST
167TH ST

PARKSIDE DR
TINLEY PARK DR
NORMANDLA
MAPLE LN
HONEY LANE DR
ARCADIA DR
RIVERSIDE
JENNIFER DR
CARLSBAD
SCOTT CT
BARBARA
LAURA LN
ANNE MARIE CT
ANDRES
BRITTNEY LN
HILLSIDE
GLENVIEW DR
HUBBARD LANE DR
RAVINIA DR
FOREST VIEW
GAYNELLE
DEBRA AV
CHRISTOPHER CT
ANNE MARIE CT
LAKEWOOD
GENTRY

ALDERSYDE
BELLAIRE
OAKWOOD
NEWPORT
MEADOWDALE
FOREST
TIMBER CT
KIMBERLEY CT
169TH ST
COULTER
170TH ST
DEERPATH
FOREST TR
170TH
ELMWOOD
GREENTREE
FARNSLEY AV
LANGLEY
CYPRESS CT
GINGERWOOD Park
KILPATRICK
ORCHARD HILL

FIELDCREST DR
WILLOWICK DR
16700
Industrial Park
166TH ST
KILBOURN
16700

PINE POINT DR
OAK PARK AVE
ELM LANE DR
WILLOW LN
171ST ST
Lewis Mem. Pk.
Vogt Woods Pk.
20S
17100
MARTHA
THACKERY AV
LAKEWOOD
HAWTHORNE RD
SYCAMORE LN
FOREST
ELIZABETH
5W-20S
27
1. TIMERLE
2. HOLLY C
3. PEARTRE

CENTRAL AVE
FULTON
172ND
VOGT ST
173RD AV
174TH ST
174TH PL
175TH ST
6800
6700
66TH
65TH
64TH
NORTH
Mem. Park
P.O.
BONNIE AV
BELLE TR
JESSICA DR
JUDY CT
LARAMIE
LOCKWOOD
171ST PL
171ST
173RD ST
174TH ST
28
6W-20S

7W-20S
29
FOREST PRESERVE
6000
PARKSIDE
173RD ST

BRENNAN HWY
LECLAIRE
174TH ST
FRONTAGE RD
LAVERGNE
173RD
175TH
MOCOMBO WAY
Wilshire
CICERO
174TH ST
175TH
COUNTRY CLUB HILLS
60477
5W-20S

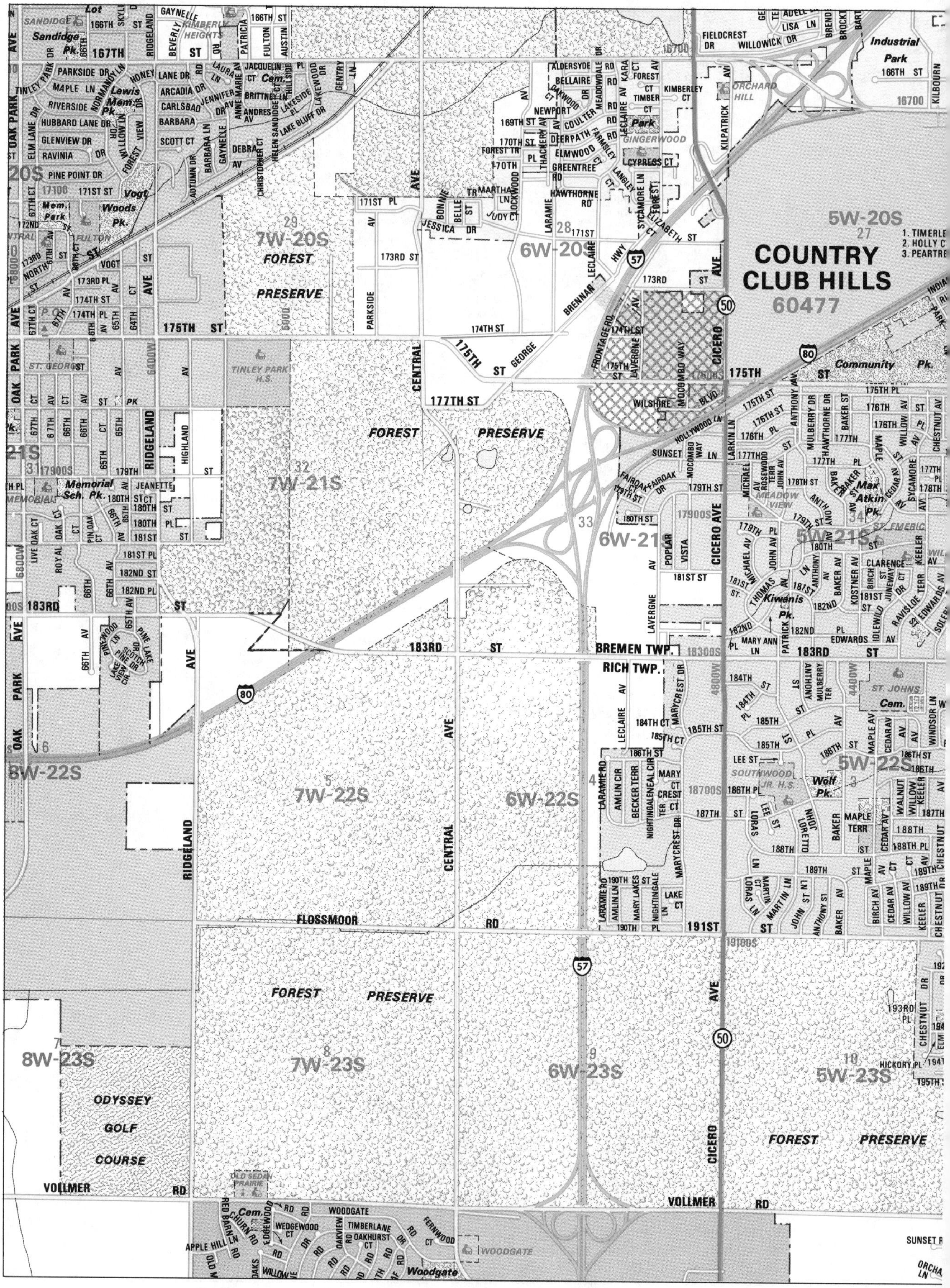

OAK PARK AVE
RIDGELAND
HIGHLAND
21S
17900S
67TH CT
66TH
65TH
179TH ST
Memorial Sch. Pk.
JEANETTE
180TH STCT
180TH ST
181ST
St. GEORGE
6400
Tinley Park H.S.
PK
FOREST PRESERVE
7W-21S
32
177TH ST

175TH ST
Community Pk.
175TH PL
176TH ST
176TH ST
177TH
HOLLYWOOD LN
SUNSET
MOCOMBO WAY
FAIROAK
FAIROAK DR
179TH ST
180TH ST
17900S
MICHAEL AV
LARKIN AV
ANTHONY AV
HAWTHORNE DR
BAKER ST
MAPLE
WILLOW AV
CHESTNUT
Max Atkin Pk.
MEADOW VIEW
33
6W-21S
5W-21S
St. EMERIC
34

6800
6700
LIVE OAK CT
ROYAL
PIN OAK
65TH
66TH
183RD ST
181ST PL
182ND ST
182ND PL
SEE PAGES 84-85
PINWOOD LN
PINE LAKE
LAKE VIEW CIR
PINE DR
POPLAR
VISTA
CICERO AVE
181ST ST
179TH PL
THOMAS AV
MICHAEL AV
BAKER AV
ANTHONY
181ST ST
182ND
Kiwanis Pk.
KOSTNER AV
CLARENCE AV
JUNEWAY
RAVISLOE TERR
EDWARDS AV
SOLE
IDLEWILD
183RD ST
4400
5W-21S

8W-22S
OAK
80
6
RIDGELAND AVE
CENTRAL AVE
7W-22S
5
6W-22S
4
LECLAIRE AV
AMLIN CIR
BECKER TERR
NIGHTINGALE CIR
MARYCREST DR
NEAL CIR
MARY CT
MARYCREST DR
184TH ST
184TH PL
185TH CT
LEE ST
186TH PL
BREMEN TWP.
RICH TWP.
18300S
184TH
184TH PL
185TH
185TH
186TH ST
MULBERRY AV
ANTHONY
JOHN AV
St. JOHNS Cem.
CEDAR AV
MAPLE AV
186TH
Southwood JR. H.S.
Wolf Pk.
WALNUT
WILLOW
KEELER
WINDSOR LN
3
5W-22S

FLOSSMOOR RD
LARAMIE RD
AMLIN LN
MARY LAKES
NIGHTINGALE
LAKE CT
187TH
JOHN
LORAS
MARTIN LN
BAKER
MAPLE TERR
188TH
188TH PL
BIRCH AV
CEDAR AV
WILLOW AV
KEELER AV
189TH
189TH
191ST ST
18700S
18300S
JOHN ST N
ANTHONY LN
CHESTNUT DR
190TH ST
191S
190TH

8W-23S
7
7W-23S
8
6W-23S
9
5W-23S
10
57
FOREST PRESERVE
50
CICERO AVE
193RD
CHESTNUT DR
HICKORY PL
194T
195TH
194

ODYSSEY GOLF COURSE
VOLLMER RD
OLD SEDAN PRAIRIE
Cem.
RED BARN RD
RED CHURN RD
EDGEWOOD
APPLE HILL LN
OLD M
OAKS
WILLOW
WOODGATE CT
WEDGEWOOD CT
TIMBERLANE DR
OAKVIEW RD
OAKHURST DR
FERNWOOD
WOODGATE CT
Woodgate
VOLLMER RD
FOREST PRESERVE
SUNSET R
ORCHA LN

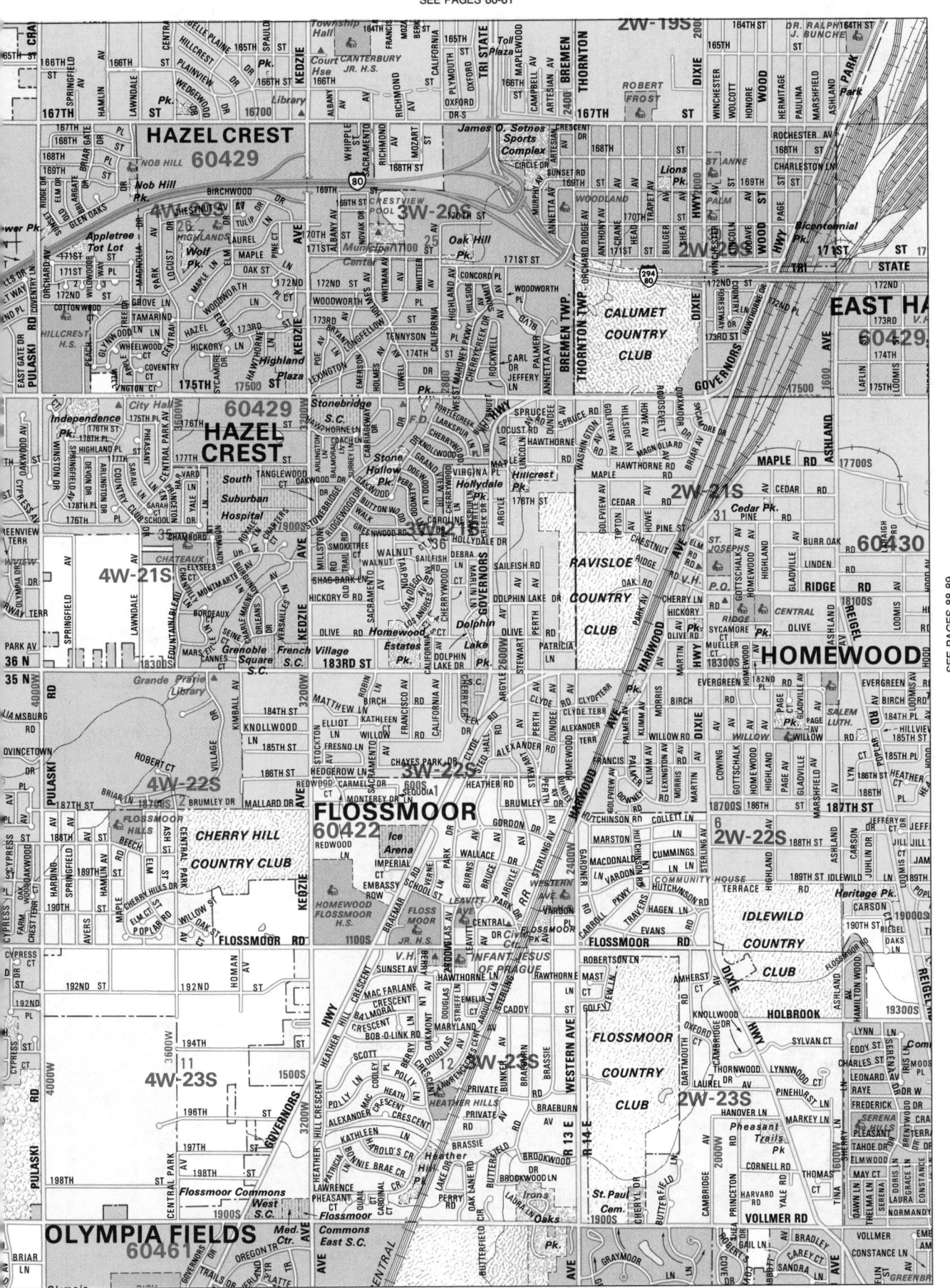

SEE PAGES 88-89

SEE PAGES 82-83

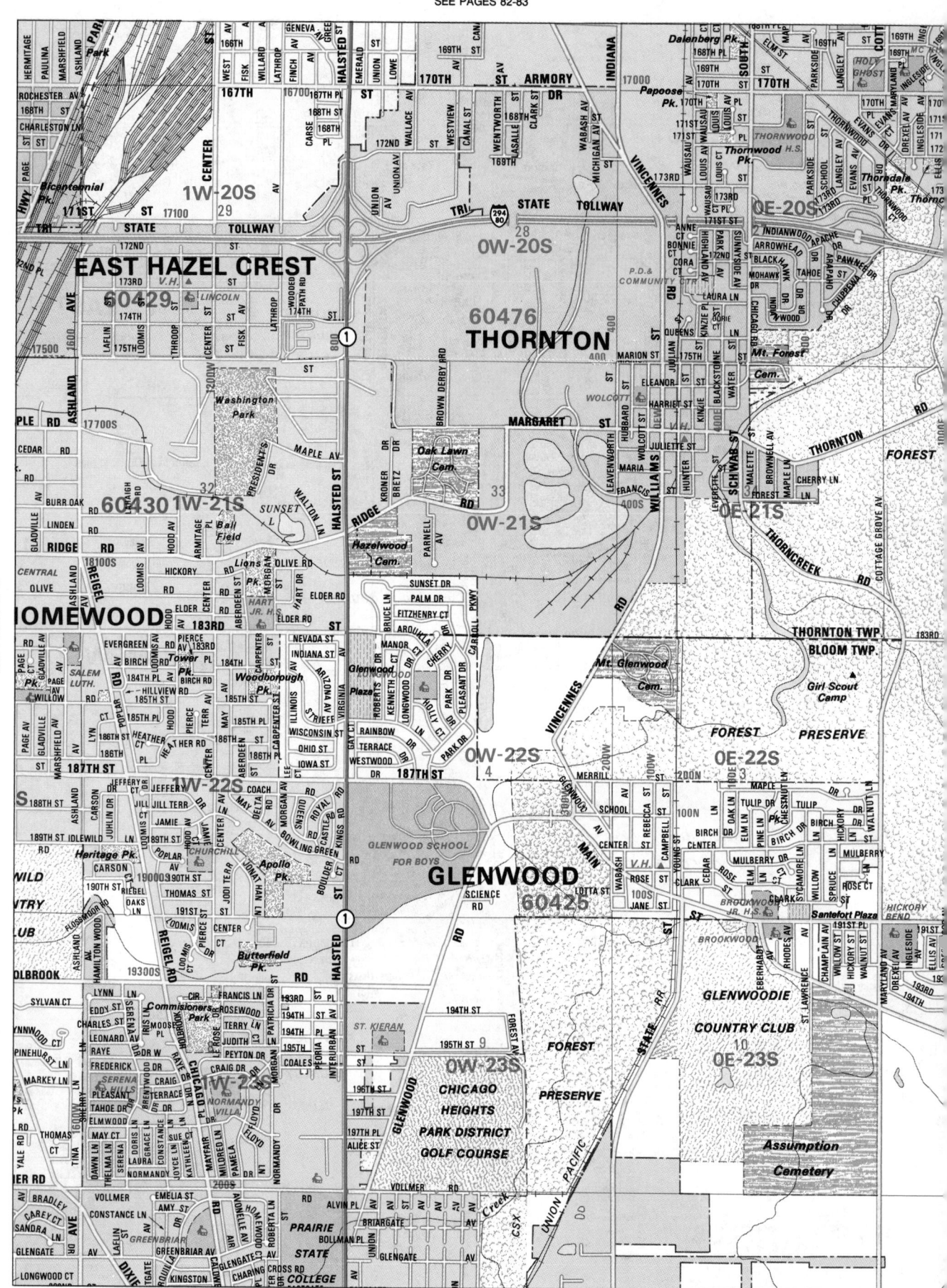

SEE PAGES 86-87

SEE PAGES 82-83

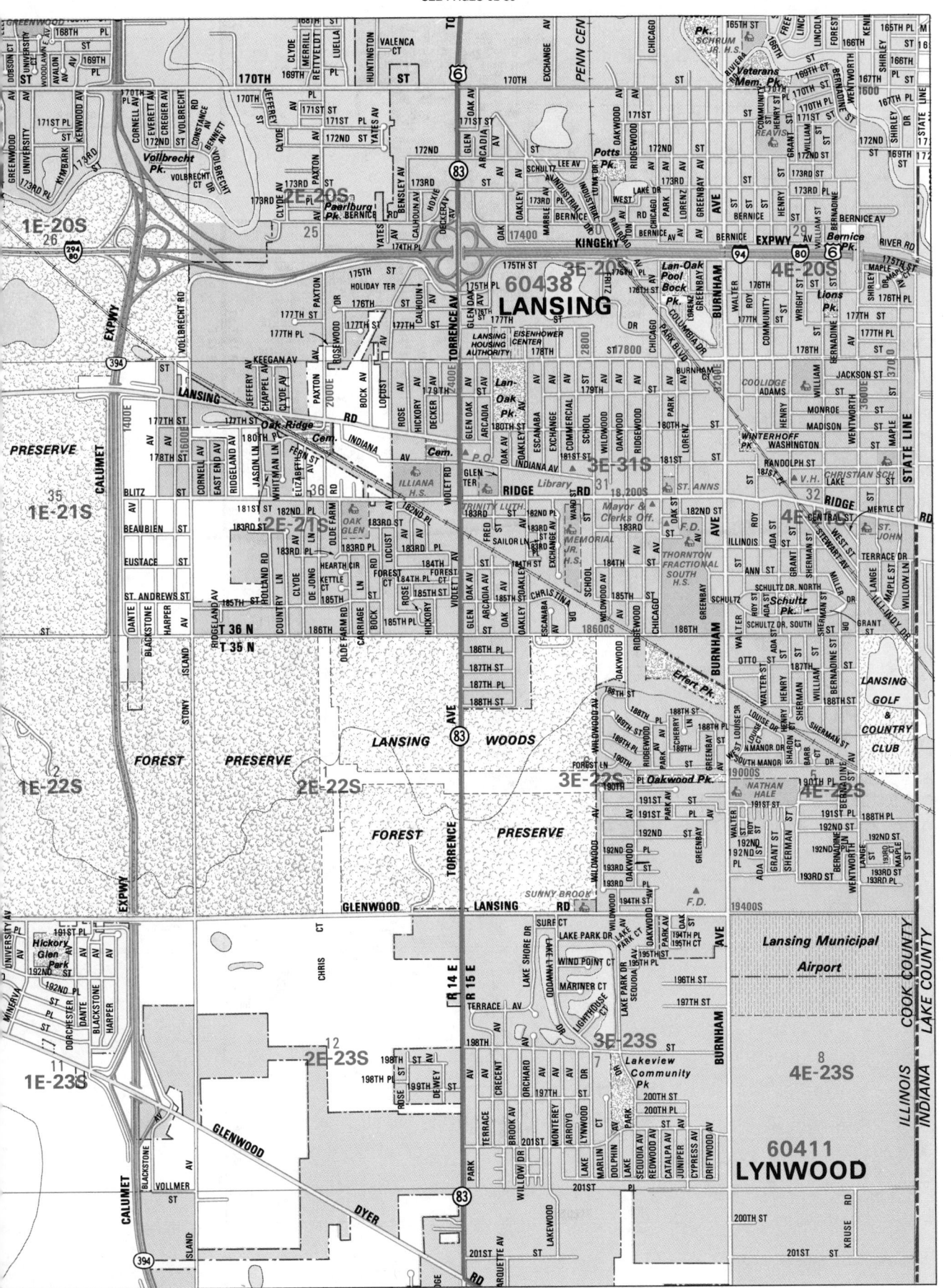

SEE PAGES 84-85

SEE PAGES 84-85

ODYSSEY GOLF COURSE

VOLLMER

ST. FRANCIS RD

9W-24S

8W-24S

10W-24S

W-24S

15

14

18

1W-25S

LINCOLN HWY

10W-25S

9W-25S

8W-25S

22

23

24

19

SEE WILL COUNTY PAGES 32-33

LIBRARY

Old Plank Road Trail
(Under development)

1. WILDROCK TER.
2. TULLAMORE TER.
3. BRIARBRANCH TER.
4. PLEASANT TER.
5. N. WINDMERE CIR.
6. S. WINDMERE CIR.
7. KNOLLWOOD CIR.
8. HEDGEWICK CT.
9. BURLWOOD CT.
10. CANDLEGATE CIR.
11. BRUSHWOOD DR.
12. THISTLE CT.
13. HEARTSIDE RD.
14. HEATHERMEAD RD.
15. PRAIRIE RD.
16. GREEN SWARD WAY
17. HICKORY GLEN
18. CHAPPARAL TER.
19. WOODBINE TER.
20. IVYLOG TER.
21. THORNTREE TER.

TIMBER

HUNTSBRIDG

PRESTWICK COUNTRY CLUB

FRANKFORT TWP.
RICH TWP.

8W-26S

27

26

25

9W-26S

11W-26S

10W-26S

SAUK TRAIL

30

FRANKFORT JR.H.S.

HICKORY CREEK

LARAWAY FOREST RD SAUK TRAIL

PRESERVE

34

35

36

31

10W-27S

9W-27S

8W-27S

27S

T 35 N STEGER RD

T 34 N

SEE PAGES 86-87

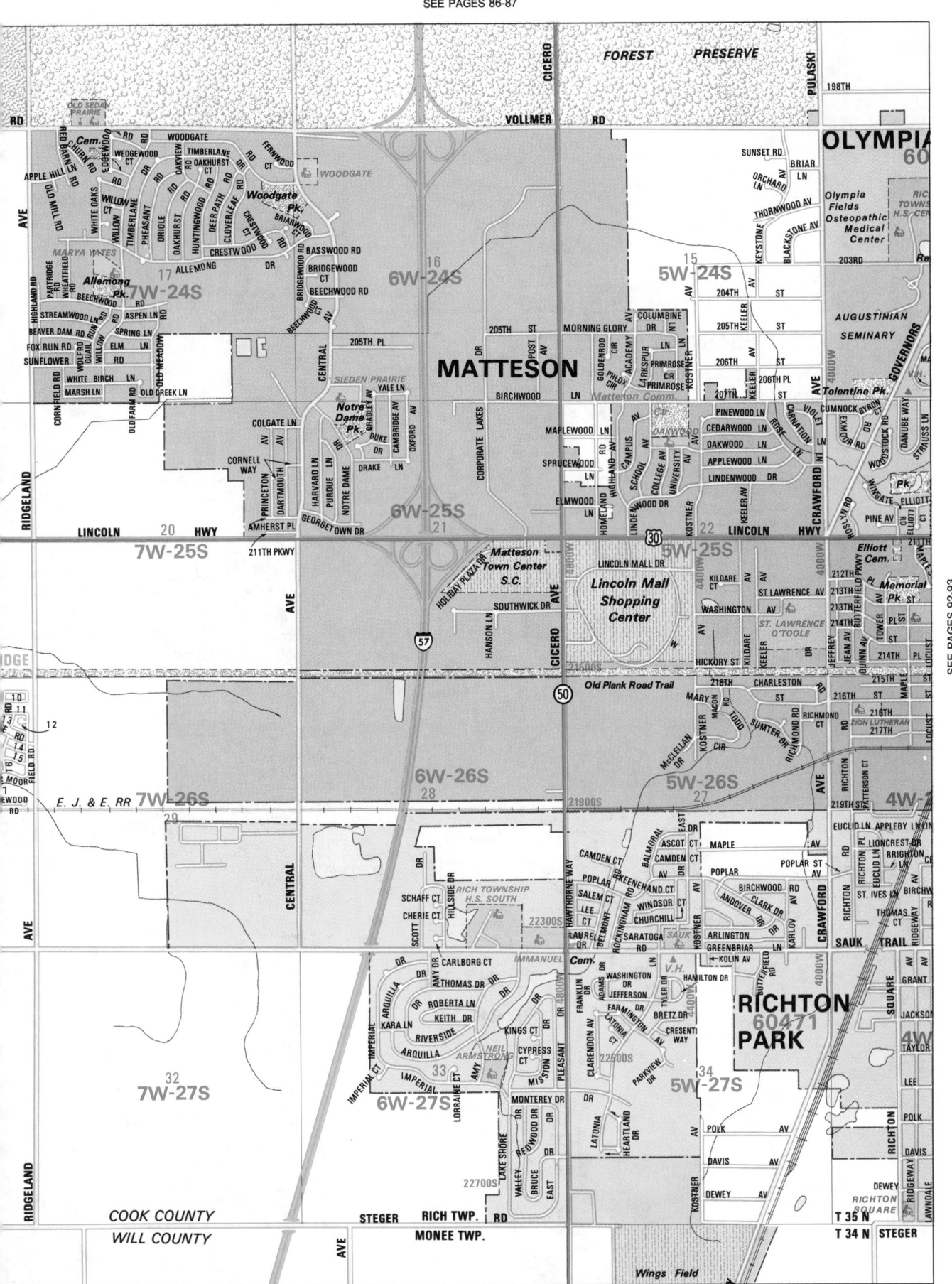

SEE PAGES 92-93

SEE WILL COUNTY PAGES 42-43

SEE PAGES 86-87

SEE PAGES 90-91

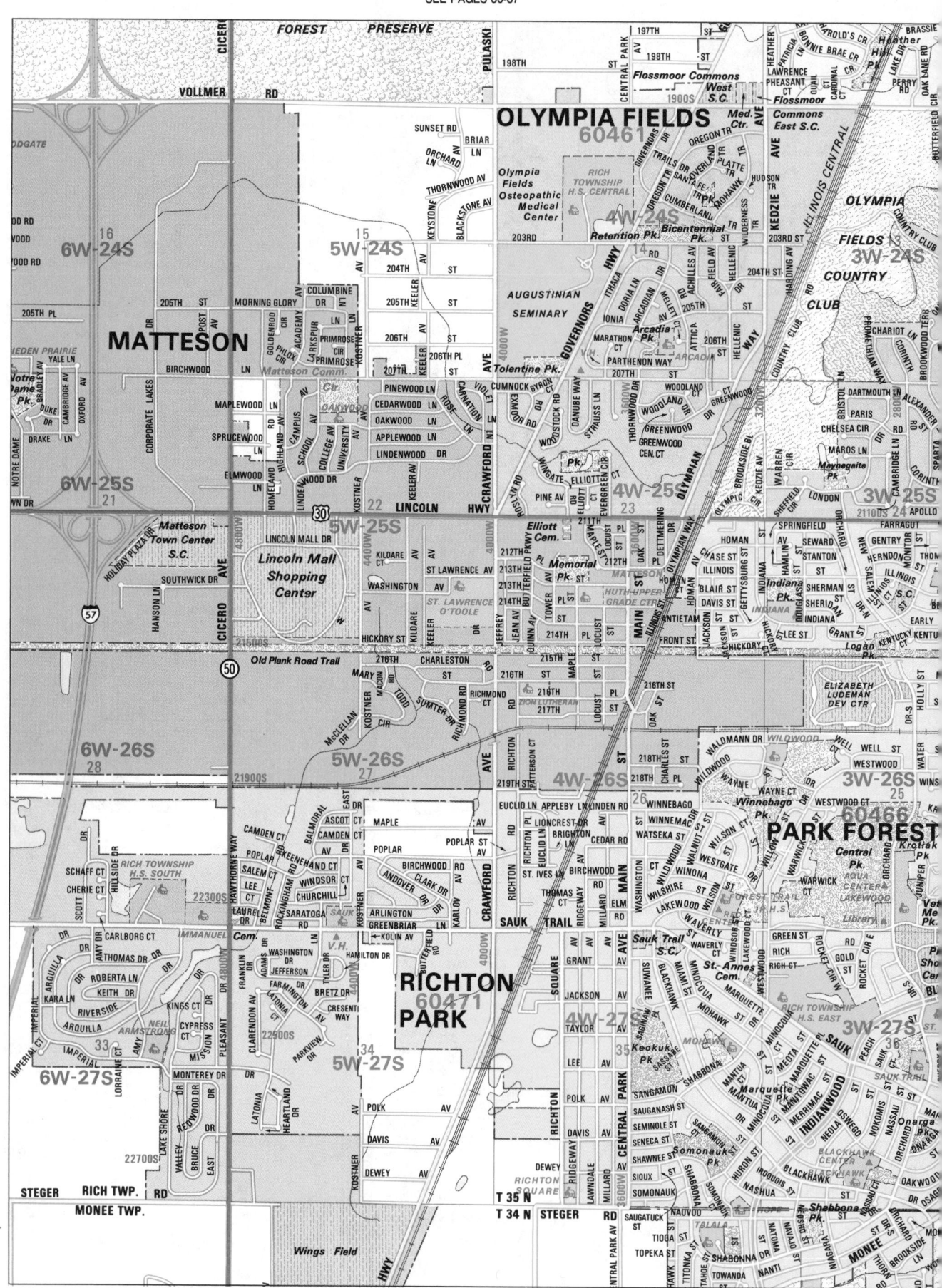

SEE WILL COUNTY PAGES 42-43

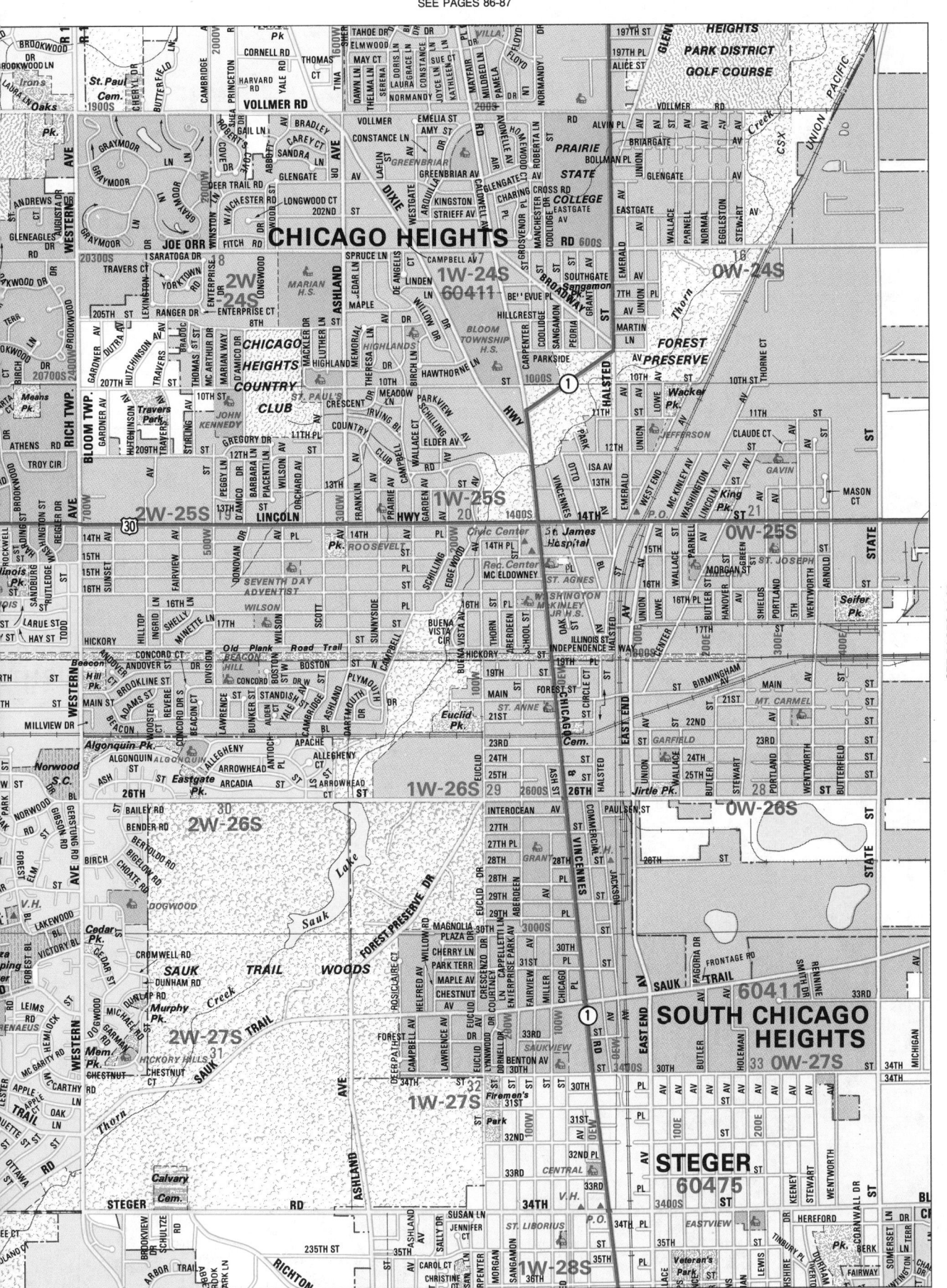

SEE PAGES 88-89

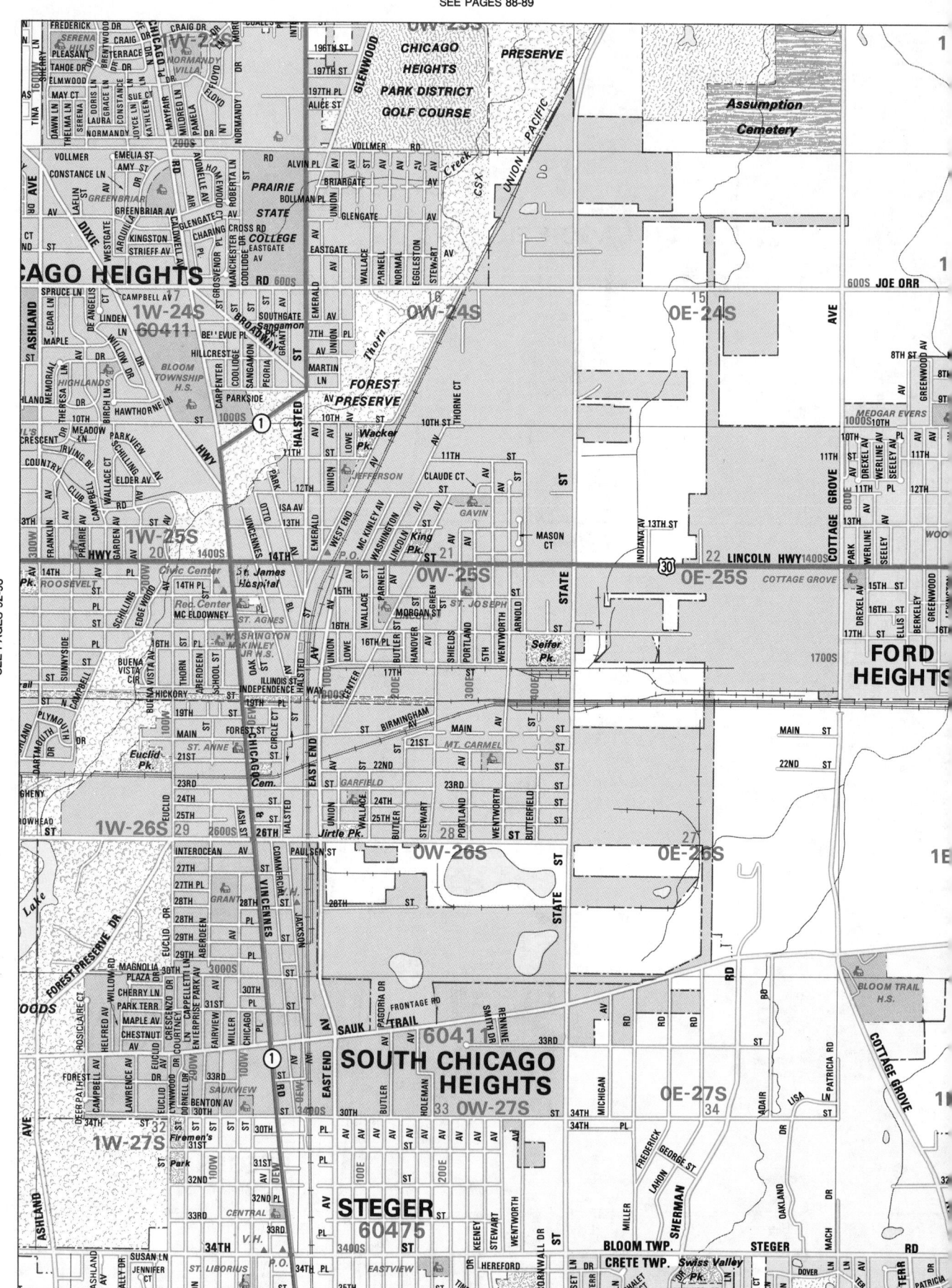

SEE WILL COUNTY PAGES 44-45

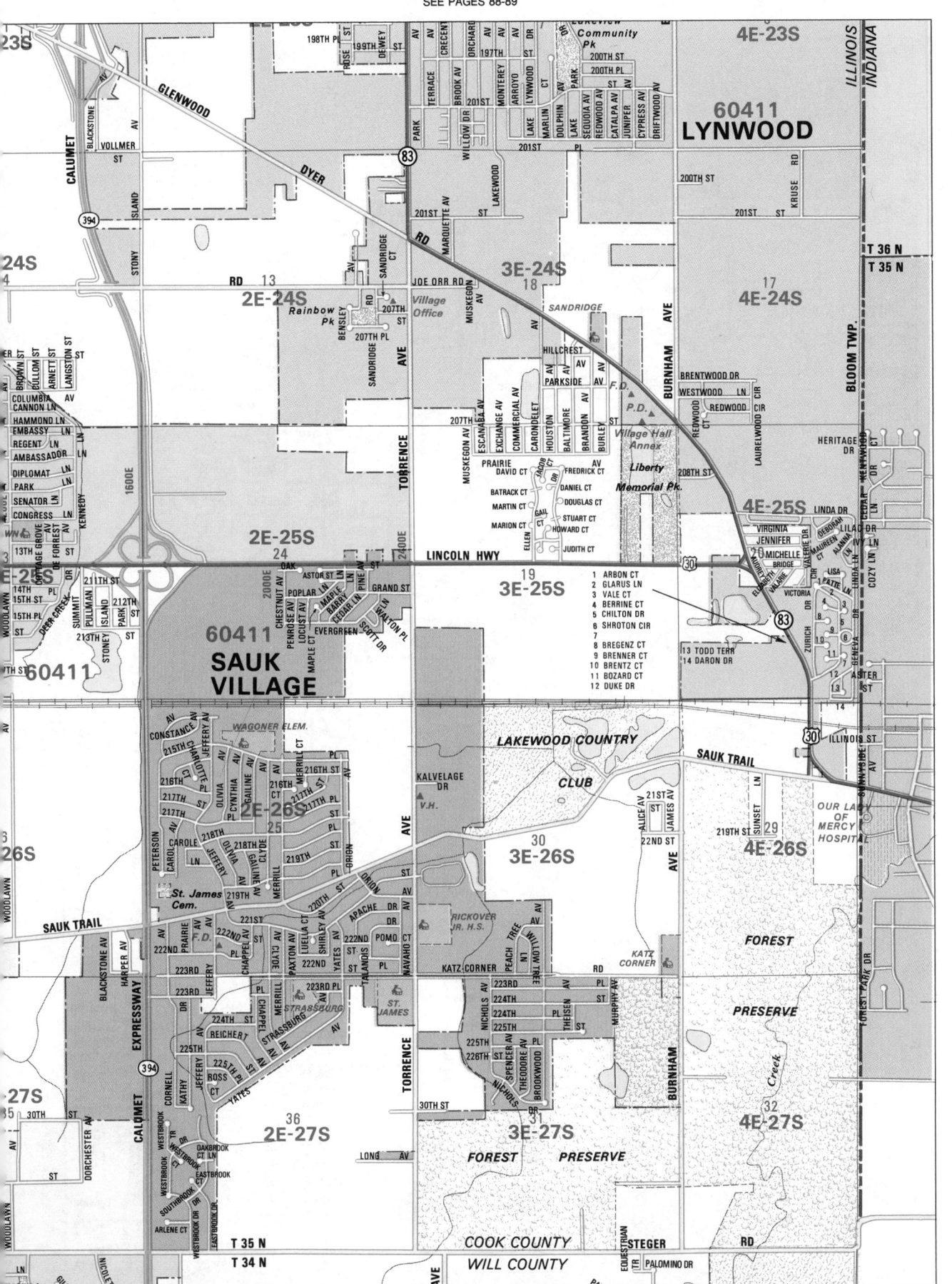

Chicago & Cook County

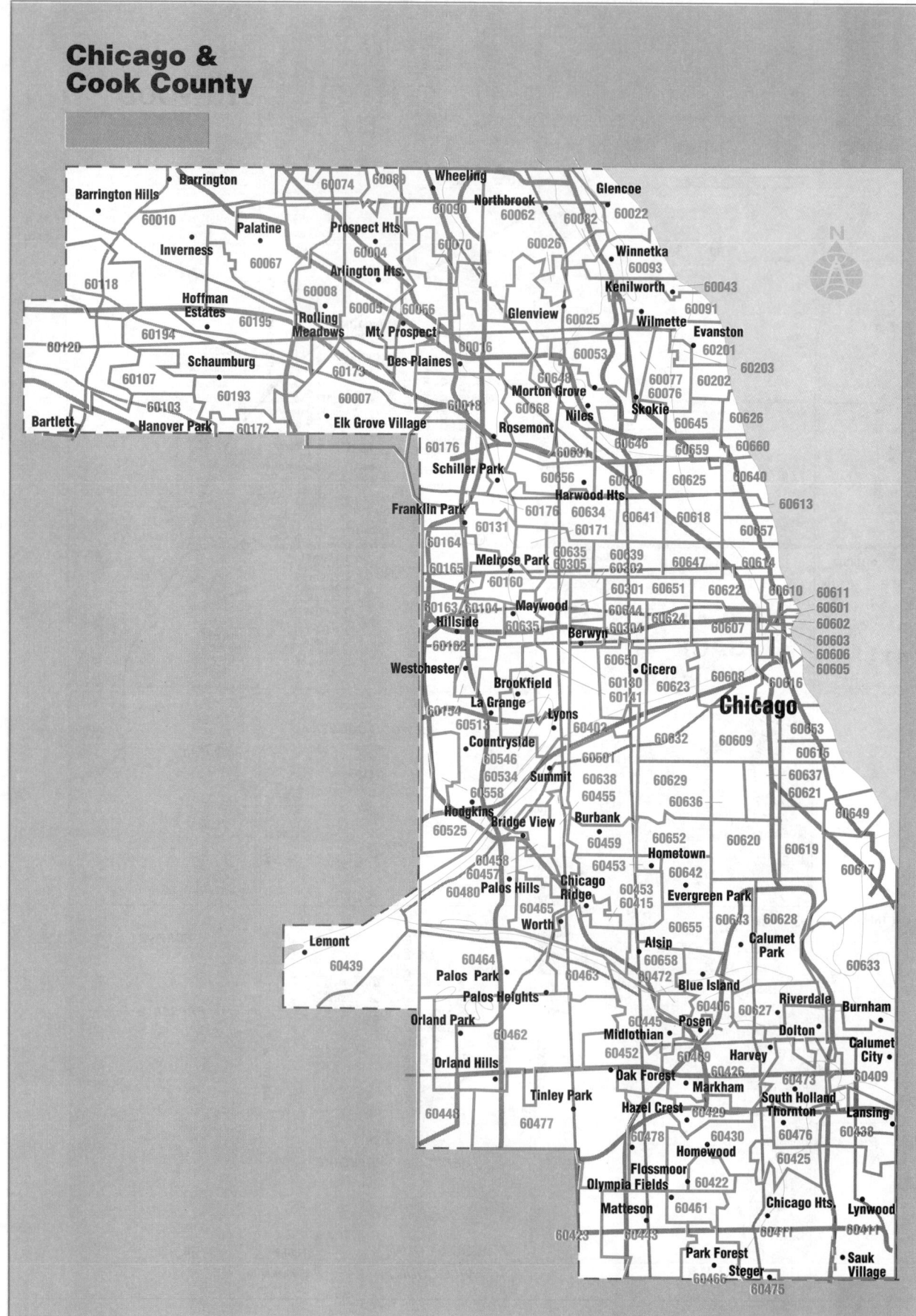

Street Index

	Grid	Page
Chicago		
Abbott St. N&S	0W-10S	68
Aberdeen St. N&S		
	1W-14S,1W-0N	76
Ada St. N&S	1W-14S,1W-2N	76
Adams St. E&W		
	6W-0S,0W-0S	49
Addison Dr. N&S	0W-4N	45
Addison St. E&W		
	9W-4N,1W-4N	42
Agatite Av. E&W		
	10W-5N,1W-5N	41
Ainslie St. E&W		
	7W-6N,1W-6N	42
Airport Dr. NE&SW	5W-6S	60
Albany Av. N&S		
	3W-13S,3W-9N	75
Albion Av. E&W		
	9W-8N,1W-8N	37
Aldine Av. E&W	0W-4N	45
Alexander St. E&W	0W-2S	57
Algonquin Av. N&S	6W-8N	38
Allen Av. NE&SW	4W-3N	44
Allport St. N&S	1W-1S	56
Alta Vista Ter. N&S	1W-4N	45
Altgeld St. E&W		
	8W-3N,1W-3N	42
Ancona St. E&W	1W-0N	51
Anson Pl. E&W	2W-0N	50
Anthon Av. E&W	10W-5N	41
Anthony Av. NW&SE		
	0E-7S,1E-8S	62
Arbor Pl. E&W	1W-0N	51
Arcade Pl. E&W		
	2W-0S,0W-0S	50
Arch St. NW&SE	1W-2S	56
Archer Av. NE&SW		
	8W-5S,0W-1S	59
Ardmore Av. E&W		
	9W-7N,1W-7N	37
Argyle St. E&W		
	8W-6N,1W-6N	42
Arlington Pl. E&W	0W-3N	45
Armitage Av. E&W		
	8W-2N,0W-2N	48
Armour St. N&S	1W-0N	51
Armstrong Av. E&W	6W-7N	38
Artesian Av. N&S		
	3W-13S,3W-9N	75
Arthington St. E&W		
	7W-0S,1W-0S	48
Arthur Av. E&W		
	8W-8N,1W-8N	37
Ashland Av. N&S		
	2W-13S,2W-9N	75
Astor St. N&S	0E-1N	51
Attrill St. NE&SW	3W-2N	50
Augusta Blvd. E&W		
	5W-1N,2W-1N	49
Austin Av. N&S		
	7W-7S,7W-6N	60
Avalon Av. N&S		
	1E-11S,1E-9S	68
Avenue A N&S	4E-12S	69
Avenue B N&S		
	4E-12S,4E-11S	69
Avenue C N&S		
	4E-12S,4E-11S	69
Avenue D N&S	4E-12S	69
Avenue E N&S	4E-12S	69
Avenue F N&S	4E-12S	69
Avenue G N&S		
	4E-12S,4E-11S	69
Avenue H N&S		
	4E-12S,4E-10S	69
Avenue J N&S		
	4E-12S,4E-10S	69
Avenue K N&S	4E-15S	77
Avenue L N&S		
	4E-15S,4E-10S	77
Avenue M N&S		
	4E-15S,4E-10S	77
Avenue N N&S		
	4E-15S,4E-10S	77
Avenue O N&S		
	3E-15S,4E-10S	77
Avers St. N&S		
	4W-12S,4W-7N	67
Avondale Av. NW&SE		
	9W-8N,2W-2N	37
Balbo Dr. E&W	0E-0S	51
Baldwin Av. N&S	2E-8S	63
Balmoral Av. E&W		
	10W-6N,1W-6N	41
Baltimore Av. N&S		
	3E-15S,3E-10S	77
Banks St. E&W	0E-1N	51
Barber St. E&W	0W-1S	57
Barry Av. E&W		
	8W-3N,0W-3N	42
Bauwans St. N&S	2W-1N	50
Beach Av. E&W		
	4W-1N,2W-1N	50
Beach Dr. NW&SE	2E-8S	63
Beacon St. N&S	1W-5N	45
Beaubien Ct. N&S	0E-0N	51
Belden Av. E&W		
	8W-2N,1W-2N	48

	Grid	Page
Bell Av. N&S	2W-13S,2W-8N	75
Belle Plaine Av. E&W		
	10W-5N,1W-5N	41
Bellevue Pl. E&W	0E-1N	51
Belmont Av. E&W		
	9W-4N,1W-4N	42
Belmont Harbor Dr. NW&SE		
	0W-4N	45
Bennett Av. N&S		
	2E-11S,2E-7S	69
Bensley Av. N&S		
	2E-15S,3E-11S	77
Berenice Av. E&W		
	7W-4N,2W-4N	42
Berkeley Av. N&S		
	1E-5S,1E-4S	62
Bernard St. N&S		
	4W-3N,4W-7N	44
Berteau Av. E&W		
	10W-5N,1W-5N	41
Berwyn Av. E&W		
	10W-6N,1W-6N	41
Besly Ct. N&S	1W-2N	51
Bessie Coleman Dr. N&S		
	12W-7N	40
Best Dr. E&W	0E-6S	62
Beverly Av. NW&SE	1W-11S	68
Beverly Glen Pkwy. E&W		
	2W-11S	67
Beverly St. NW&SE		
	1W-12S,2W-10S	68
Bingham St. NW&SE	3W-2N	50
Birchwood Av. E&W		
	9W-9N,2W-9N	37
Birkhoff Av. NW&SE	0W-9S	68
Bishop St. N&S		
	1W-14S,1W-1N	76
Bissell St. N&S	1W-2N	51
Bittersweet Pl. E&W		
	8W-5N,0W-5N	42
Black Dr. N&S	1W-7S	62
Blackhawk St. E&W		
	1W-1N,0W-1N	51
Blackstone Av. N&S		
	1E-10S,1E-5S	68
Blake St. NW&SE	2W-3S	56
Bliss St. NE&SW	1W-1N	51
Bloomingdale Av. E&W		
	8W-2N,2W-2N	48
Blue Island Av. NE&SW		
	2W-2S,1W-1S	56
Bonaparte St. NE&SW		
	1W-2S	56
Bonfield St. NW&SE	1W-2S	56
Bosworth Av. N&S		
	1W-1N,1W-9N	51
Boulevard Way N&S	3W-2S	56
Bowen Av. E&W	0E-4S	57
Bowler St. NE&SW	2W-0S	50
Bowmanville NE&SW	2W-6N	45
Bradley Pl. E&W		
	8W-4N,1W-4N	44
Brainard Av. NW&SE	3E-15S	77
Brandon Av. N&S		
	3E-15S,3E-9S	77
Brayton Av. E&W	0W-14S	76
Brayton St. NW&SE	0W-14S	76
Brennan Av. N&S	2E-11S	69
Brenock Dr. N&S	3W-1N	50
Briar Pl. E&W	0W-3N	45
Brighton Pl. NW&SE	3W-4S	56
Broad St. NW&SE	1W-2S	56
Broadway St. N&S		
	0W-3N,1W-6N	45
Brompton Av. E&W	0W-4N	45
Bross Av. E&W	3W-3S	56
Browning Av. E&W	0E-3S	57
Bryn Mawr Av. E&W		
	10W-7N,1W-7N	36
Buckingham Pl. E&W	0W-4N	45
Buena Ter. E&W	0W-5N	45
Buffalo Av. N&S		
	3E-15S,4E-9S	77
Burkhardt Dr. E&W	3W-1S	56
Burley Av. N&S		
	3E-15S,4E-9S	77
Burling St. N&S		
	0W-1N,0W-3N	51
Burnham Av. N&S		
	3E-15S,3E-8S	69
Burnside Av. NW&SE		
	0E-10S	68
Burton Pl. E&W	0E-1N	51
Busse Av. N&S	8W-6N	42
Byron St. E&W		
	10W-4N,1W-4N	41
Cabrini St. E&W		
	1W-0S,0W-0S	51
Cahill Ter. NE&SW	3W-4N	42
Caldwell Av. NW&SE		
	7W-8N,5W-7N	37
Calhoun Av. N&S		
	3E-15S,3E-11S	77
California Av. N&S		
	3W-12S,3W-9N	67
California Blvd. N&S	3W-2S	56
California Ter. E&W	0W-3N	45
Calumet Av. N&S		
	0E-15S,0E-1S	76

	Grid	Page
Calumet Expwy. N&S		
	0E-11S,1E-14S	68
Cambridge Av. N&S		
	0W-1N,0W-3N	51
Campbell Av. N&S		
	3W-13S,3W-9N	75
Canal St. N&S		
	0W-4S,0W-0S	57
Canalport Av. SW&NE		
	1W-1S	56
Canfield Av. N&S		
	9W-6N,9W-7N	42
Cannon Dr. NW&SE	0W-3N	45
Canton St. E&W	2W-2N	50
Carmen Av. E&W		
	10W-6N,1W-6N	41
Carondolet Av. N&S		
	3E-15S,3E-14S	77
Carpenter Rd. N&S	6W-8N	38
Carpenter St. N&S		
	1W-14S,1W-0N	76
Carroll Av. E&W		
	2W-0N,0W-0N	50
Carver Dr. NW&SE	1E-15S	76
Castle Island Av. E&W		
	10W-5N	41
Castlewood Ter. E&W		
	1W-6N	45
Catalpa Av. E&W		
	10W-6N,1W-6N	41
Catherine Av. E&W	10W-6N	41
Cedar St. E&W	0E-1N	51
Central Av. N&S		
	6W-6S,7W-7N	60
Central Park Av. N&S		
	4W-13S,4W-7N	75
Central Park Blvd. N&S		
	4W-0S,4W-0N	50
Cermak Rd. E&W		
	3W-1S,0W-1S	56
Champlain Av. N&S		
	0E-13S,0E-4S	76
Chanay St. NE&SW	3W-2N	50
Chappel Av. N&S		
	2E-15S,2E-7S	77
Charles St. N&S		
	1W-11S,2W-10S	68
Charleston St. E&W	2W-2N	50
Chase Av. E&W		
	9W-9N,1W-9N	37
Chelsea Pl. NW&SE	2W-12S	67
Cheltenham Pl. NE&SW		
	3E-8S	63
Cherry St. NW&SE	1W-1N	51
Chester Av. N&S	10W-6N	41
Chestnut St. E&W		
	1W-1N,0W-1N	51
Chicago Av. E&W		
	6W-1N,0W-0N	49
Chicago Skyway N&S		
	0E-7S,4E-11S	62
Chicora Av. NW&SE		
	7W-8N,6W-8N	37
Chippewa St. N&S	2E-13S	77
Christiana Av. N&S		
	4W-12S,4W-7N	67
Church St. NE&SW		
	2W-13S,1W-12S	75
Churchill St. E&W	2W-2N	50
Cicero Av. N&S		
	5W-8S,5W-7N	60
Claremont Av. N&S		
	2W-12S,2W-9N	67
Clarence Av. E&W		
	9W-7N,8W-7N	37
Clarendon Av. N&S	1W-5N	45
Clark St. N&S	0W-15S,2W-8N	57
Cleaver St. N&S	1W-1N	51
Cleveland Av. N&S		
	0W-1N,0W-2N	51
Clifford Av. NE&SW	6W-7N	38
Clifton Av. N&S		
	1W-2N,1W-5N	51
Clinton St. N&S		
	0W-1S,0W-0N	57
Clover Av. NE&SW	5W-5N	43
Clybourn Av. NW&SE		
	2W-3N,1W-2N	51
Clyde Av. N&S		
	2E-15S,2E-7S	77
Coast Guard Dr. N&S	2E-7S	63
Coles Av. N&S	3E-9S,3E-8S	69
Colfax Av. N&S		
	3E-10S,3E-8S	69
Columbia Av. E&W		
	2W-8N,1W-8N	39
Columbus Av. N&S		
	4W-9S,3W-8S	67
Columbus Dr. N&S	0E-0S	51
Commercial Av. N&S		
	3E-15S,3E-9S	77
Commonwealth Av. N&S		
	0W-2N,0W-3N	51
Concord Pl. E&W		
	6W-2N,0W-2N	49
Congress Pkwy. E&W		
	6W-0S,1W-0S	49
Constance Av. N&S		
	2E-10S,2E-7S	69

	Grid	Page
Corcoran Pl. E&W	7W-0N	48
Corliss Av. N&S		
	0E-15S,1E-12S	76
Cornelia Av. E&W		
	10W-4N,1W-4N	41
Cornell Av. N&S		
	2E-10S,2E-5S	69
Cortez Dr. E&W		
	0E-14S,0E-2S	76
Cortez St. E&W		
	6W-1N,1W-1N	49
Cortland Pkwy. E&W	1W-2N	51
Cottage Grove Av. N&S		
	0E-14S,0E-2S	76
Couch Pl. E&W	0W-0N	51
Coulter St. NE&SW	2W-2S	56
Court Pl. E&W	0W-0N	51
Courtland Av. N&S	10W-6N	41
Coyle Av. E&W		
	9W-8N,3W-8N	37
Craft St. NW&SE	1W-4N	45
Crandon Av. N&S		
	2E-15S,2E-7S	77
Cregier Av. N&S		
	2E-10S,2E-7S	69
Crestline St. E&W	5W-9S	66
Crilly Ct. N&S	0W-2N	51
Crosby St. NW&SE	0W-1N	51
Crowell St. NW&SE	1W-2S	56
Crystal St. E&W		
	6W-1N,1W-1N	49
Cullerton St. E&W		
	5W-1S,0W-1S	55
Cullom Av. E&W		
	6W-5N,1W-5N	43
Cumberland Av. N&S		
	10W-5N,10W-6N	41
Cuyler Av. E&W		
	7W-5N,1W-5N	42
Cyril Ct. N&S	2E-8S	63
Dakin St. E&W		
	8W-4N,1W-4N	42
Damen Av. N&S		
	2W-11S,2W-8N	67
Dante Av. N&S		
	1E-10S,1E-7S	68
Dauphin Av. NE&SW	1E-10S	68
Davlin Ct. NE&SW	4W-3N	44
Davol St. NE&SW	2W-13S	75
Dawson Av. N&S	4W-3N	44
Dayton St. N&S		
	1W-2N,1W-5N	51
DeKoven St. E&W	0W-0S	51
Dean St. NW&SE	2W-1N	50
Dearborn St. N&S		
	0W-5S,0W-1N	62
Delaware Pl. E&W	0W-1N	51
Delphia Av. N&S		
	10W-5N,10W-6N	41
Deming Pl. E&W		
	6W-3N,0W-3N	43
Denvir Av. NE&SW	3W-0S	50
DesPlaines St. N&S		
	0W-1S,0W-0N	57
Devon Av. E&W		
	9W-8N,1W-8N	37
Dewitt Pl. N&S	0E-1N	51
Dexter Park Av. N&S	1W-4S	56
Dickens Av. E&W		
	8W-2N,1W-2N	48
Dickinson St. NW&SE		
	6W-5N	43
District Blvd. E&W	5W-4S	55
Diversey Av. E&W		
	6W-3N,1W-3N	43
Diversey Dr. E&W	0W-3N	45
Diversey Pkwy. E&W	1W-3N	45
Division St. E&W		
	6W-1N,0W-1N	49
Dix St. NW&SE	1W-1N	51
Dobson St. N&S		
	1E-15S,1E-8S	76
Dominick St. NW&SE	1W-2N	51
Dorchester Av. N&S		
	1E-10S,1E-5S	68
Doty Av. NW&SE	1E-15S	76
Douglas Blvd. E&W	4W-1S	55
Dover St. N&S	1W-5N	45
Dowagiac Av. NW&SE		
	6W-8N	38
Drake Av. N&S		
	4W-13S,4W-7N	75
Draper St. E&W	1W-3N	45
Drew St. N&S	2W-12S	67
Drexel Av. N&S		
	1E-11S,1E-6S	68
Drexel Blvd. N&S	1E-5S	62
Drummond Pl. E&W		
	6W-3N,0W-3N	43
Early Av. NW&SE	1W-7N	39
Cir.	8W-7N	37
East End Av. N&S		
	1E-11S,1E-6S	68
East River Rd. NS&EW		
	10W-5N	41
Eastlake Ter. N&S	1W-9N	39
Eastman St. E&W	1W-1N	51
Eastwood Av. E&W		
	7W-5N,1W-5N	42
Eberhart Av. N&S		
	0E-15S,0E-7S	76

	Grid	Page
Edbrooke Av. N&S		
	0W-14S,0E-12S	76
Eddy St. E&W		
	7W-4N,1W-4N	42
Edens Expwy. N&S	6W-6N	43
Edens Expwy. N&S	6W-7N	38
Edgebrook Ter. NW&SE		
	7W-8N	37
Edgewater Av. E&W	2W-7N	39
Edmaire St. NW&SE	2W-13S	75
Edmunds St. NE&SW		
	6W-6N	43
Edward Ct. N&S	0W-2N	51
Eggleston Av. N&S		
	0W-15S,0W-7S	76
Eisenhower, Dwight Expy.		
E&W	7W-0S,1W-0S	48
Elaine Pl. N&S	0W-4N	45
Elbridge Av. NE&SW	4W-3N	44
Eleanor St. E&W	1W-2S	56
Elias St. NW&SE	1W-2S	56
Elizabeth St. N&S		
	1W-14S,1W-0N	76
Elk Grove Av. NW&SE		
	2W-1N	50
Ellen St. E&W	2W-1N	50
Elliott Av. NE&SW	2E-9S	69
Ellis Av. N&S	1E-15S,0E-2S	76
Ellis Av. E&W	9W-8N	37
Ellsworth Dr. N&S	0E-5S	62
Elm St. E&W	0W-1N	51
Elmdale Av. E&W	1W-7N	39
Elston Av. NE&SW		
	7W-7N,1W-1N	37
Emerald Av. N&S		
	0W-14S,0W-2S	76
Emmett St. NW&SE	4W-3N	44
Englewood Av. E&W	0W-6S	62
Erie St. E&W	7W-0N,0W-0N	48
Ernst Ct. N&S	0E-1N	51
Escanaba Av. N&S		
	3E-15S,3E-9S	77
Esmond St. NE&SW		
	2W-13S,2W-12S	75
Essex Av. N&S		
	3E-10S,3E-9S	69
Estes Av. E&W		
	9W-8N,1W-8N	37
Euclid Av. N&S		
	2E-10S,2E-7S	69
Eugenie St. E&W	0W-2N	51
Evans Av. N&S		
	0E-15S,0E-4S	76
Evelyn Ln. E&W	10W-6N	41
Everell Av. E&W	9W-7N	37
Everett Av. N&S	2E-6S	63
Evergreen Av. E&W		
	4W-1N,0W-1N	50
Ewing Av. N&S		
	4E-13S,4E-10S	77
Fairbanks Ct. N&S	0E-0N	51
Fairfield Av. N&S		
	3W-13S,3W-9N	75
Fargo Av. E&W		
	9W-9N,1W-9N	37
Farragut Av. E&W		
	10W-6N,1W-6N	41
Farrell St. NW&SE	1W-2S	56
Farrer Dr. NE&SW	3W-1S	56
Farwell Av. E&W		
	9W-8N,1W-8N	37
Federal St. N&S		
	0W-5S,0W-0S	62
Felton Ct. N&S	0W-2N	51
Ferdinand St. E&W		
	6W-0N,2W-0N	49
Fern Ct. N&S	0W-2N	51
Field Blvd. N&S	0E-0N	51
Fielding Av. NE&SW	0W-8S	62
Fifth Av. NE&SW		
	5W-0S,3W-0S	49
Fillmore St. E&W		
	7W-0S,1W-0S	48
Fitch Av. E&W		
	9W-8N,3W-8N	37
Fleetwood St. N&S	13W-5N	40
Fletcher St. E&W		
	8W-3N,1W-3N	42
Flournoy St. E&W		
	6W-0S,1W-0S	49
Ford Av. N&S	0W-2S	57
Ford City Dr. E&W	5W-8S	60
Foreman Dr. E&W	4E-10S	69
Forest Av. N&S		
	0E-15S,0E-10S	76
Forest Glen Av. NE&SW		
	6W-6N,5W-7N	43
Forest Preserve Dr. N&S		
	10W-4N	41
Forestview Av. N&S	10W-5N	41
Forestville Av. N&S		
	0E-15S,0E-4S	76
Fort Dearborn Dr. N&S		
	0E-2S	57
Foster Av. E&W		
	9W-6N,1W-6N	42
Foster Pl. E&W	8W-6N	42
Francis Pl. NE&SW	3W-2N	50

	Grid	Page
Francisco Av. N&S	3W-9S,3W-9N	67
Franklin Blvd. E&W	4W-0N,3W-0N	50
Franklin St. N&S	0W-0S,0W-1N	51
Fremont St. N&S	1W-2N,1W-4N	51
Front Av. N&S	0E-14S	76
Front St. N&S	0E-13S,0E-12S	76
Fronte Av. N&S	0W-6S	62
Frontier Av. N&S	0W-1N	51
Fry St. E&W	1W-1N	51
Fuller St. E&W	1W-2S	56
Fullerton Av. E&W	8W-3N,1W-3N	42
Fullerton Pkwy. E&W	0W-3N	45
Fulton St. E&W	7W-0N,1W-0N	48
Gale St. NE&SW	6W-6N	43
Garfield Blvd. E&W	2W-5S,0W-5S	61
Garland Ct. N&S	0E-0N	51
Geneva Ter. N&S	0W-2N	51
Genoa Av. NE&SW	1W-11S,1W-10S	68
George St. E&W	8W-3N,1W-3N	42
Germania Pl. E&W	0E-1N	51
Gettysburg Av. E&W	6W-6N	43
Giddings St. E&W	7W-5N,2W-5N	42
Gilbert Ct. NW&SE	0W-9S	68
Giles Av. N&S	0E-3S	57
Givins Ct. NW&SE	0W-9S	68
Gladys Av. E&W	6W-0S,0W-0S	49
Glenlake Av. E&W	5W-7N,1W-7N	38
Glenroy St. N&S	1W-12S	68
Glenwood Av. N&S	1W-6N,1W-8N	45
Goethe St. E&W	0W-1N	51
Gole Dr. N&S	7W-0S	49
Goodman St. E&W	7W-6N	42
Gordon Ter. E&W	1W-5N	45
Governor's Pkwy. E&W	4W-0N	49
Grace St. E&W	10W-4N,1W-4N	41
Graham Ct. N&S	0E-1N	51
Grand Av. NW&SE	6W-2N,0E-0N	49
Grant Pl. E&W	0W-2N	51
Granville Av. E&W	4W-7N,1W-7N	38
Gratten Av. NW&SE	1W-2S	56
Green Bay Av. N&S	3E-15S,4E-9S	77
Green St. N&S	1W-15S,1W-0N	76
Greenleaf Av. E&W	9W-8N,1W-8N	37
Greenview Av. N&S	1W-1N,1W-9N	51
Greenwood Av. N&S	1E-15S,1E-4S	76
Gregory St. E&W	10W-6N,1W-6N	41
Gresham Av. NE&SW	4W-3N	44
Grilly Dr. NW&SE	4E-10S	69
Grove St. NE&SW	0W-2S	57
Grover St. NW&SE	7W-6N	42
Grower Dr. N&S	3W-1N	50
Gullikson Rd. NE&SW	8W-6S	59
Gunnison St. E&W	7W-6N,1W-6N	42
Haddon Av. E&W	6W-1N,2W-1N	49
Haft St. NE&SW	7W-7N	37
Haines St. NE&SW	1W-1N	51
Halas, George Dr. N&S	0E-0N	51
Hale Av. N&S	2W-13S,2W-12S	75
Halsted St. N&S	1W-14S,1W-4N	76
Hamilton Av. N&S	2W-12S,2W-8N	67
Hamlin Av. N&S	4W-13S,4W-7N	75
Hampden Ct. N&S	0W-3N	45
Hanes St. NW&SE	1W-2S	56
Harding Av. N&S	4W-12S,4W-7N	67
Harlem Av. N&S	9W-4N,9W-7N	42
Harper Av. N&S	1E-10S,1E-6S	68
Harrison St. E&W	6W-0S,0W-0S	49
Hart St. N&S	2W-0N	50
Hartland Ct. N&S	2W-0N	50
Hartwell Av. N&S	0E-7S	62
Harvard Av. N&S	0W-14S,0W-7S	76
Haskins Av. NW&SE	2W-9N	39
Hastings St. E&W	2W-1S	56
Haussen Ct. NE&SW	4W-3N	44
Hawthorne Pl. E&W	0W-4N	45
Hayes Av. E&W	8W-8N	37
Hayes Dr. E&W	2E-6S	63
Hayford St. E&W	4W-8S	61
Haynes Ct. NW&SE	1W-2S	56
Hazel St. N&S	1W-5N	45
Heath Av. N&S	2W-1S	56
Henderson St. E&W	8W-4N,1W-4N	42
Henry Ct. NE&SW	3W-2N	50
Hermione St. NE&SW	7W-7N	37
Hermitage Av. N&S	2W-9S,2W-9N	67
Hermosa Av. NE&SW	2W-13S,2W-12S	75
Hiawatha Av. NW&SE	7W-8N,5W-7N	37
Hickory St. NW&SE	1W-1N	51
Higgins Av. NW&SE	8W-6N	42
Higgins Rd. NW&SE	7W-6N	43
Highland Av. E&W	8W-7N,2W-7N	37
Hill St. E&W	0W-1N	51
Hillock Av. NE&SW	1W-2S	56
Hirsch Dr. E&W	7W-1N,3W-1N	48
Hobart Av. NE&SW	8W-7N	37
Hobbie St. E&W	0W-1N	51
Hobson Av. NE&SW	2W-2N	50
Hoey St. NE&SW	1W-2S	56
Holbrook St. NE&SW	7W-7N	37
Holden Ct. N&S	0E-0S	51
Holland Rd. NW&SE	0W-10S	68
Hollett Dr. N&S	4W-7S	61
Holly Av. NW&SE	2W-2N	50
Hollywood Av. E&W	5W-7N,1W-7N	38
Homan Av. N&S	4W-13S,4W-1N	75
Homer St. E&W	6W-2N,2W-2N	49
Homewood Av. N&S	2W-13S	75
Honore St. N&S	2W-9S,2W-9N	67
Hood Av. E&W	9W-7N,1W-7N	37
Hooker St. NW&SE	1W-1N	51
Hopkins Pl. E&W	2W-10S	67
Hortense Av. E&W	9W-7N	37
Houston Av. N&S	3E-15S,3E-9S	77
Howard St. E&W	3W-9N	39
Howe St. N&S	0W-1N,0W-2N	51
Hoxie Av. N&S	2E-15S,2E-11S	77
Hoyne Av. N&S	2W-12S,2W-9N	67
Hubbard St. E&W	6W-0N,0W-0N	49
Hudson Av. N&S	0W-1N,0W-3N	51
Huguelet Pl. N&S	0E-1N	51
Humboldt Blvd. N&S	3W-2N	50
Humboldt Dr. N&S	3W-1N	50
Hunt Pl. NE&SW	2W-10S	67
Huntington St. NE&SW	7W-7N	37
Hurlbut Av. E&W	8W-7N	37
Huron St. E&W	7W-0N,0W-1N	48
Hutchinson St. E&W	6W-5N,0W-5N	43
Hyacinth St. NE&SW	7W-7N	37
Hyde Park Blvd. E&W	1E-5S,2E-5S	62
Ibsen St. E&W	9W-8N	37
Illinois St. E&W	0W-0N	51
Imlay Av. E&W	8W-8N	37
Independence Blvd. E&W	4W-1S,4W-0S	55
Indian Rd. NW&SE	7W-7N,6W-7N	37
Indiana Av. N&S	0W-14S,0E-2S	76
Indianapolis Blvd. NW&SE	4E-11S	69
Ingleside Av. N&S	0E-15S,1E-5S	76
Institute Pl. E&W	0W-1N	51
Interstate 290 E&W	7W-0S,1W-0S	48
Interstate 55 NE&SW	2W-3S	56
Interstate 57 N&S	0W-11S,1W-13S	68
Interstate 90 NW&SE	4E-11S,10W-7N	69
Interstate 90-94 N&S	0W-6S,4W-4N	62
Interstate 94 N&S	1E-12S,6W-7N	68
Ionia Av. NE&SW	5W-7N,7W-8N	38
Iowa St. E&W	6W-1N,2W-1N	49
Irene Av. NW&SE	3W-4N	44
Iron St. N&S	1W-3S	56
Irving Park Rd. E&W	10W-5N,1W-5N	41
Isham Av. E&W	9W-8N	37
Jackson Blvd. E&W	0W-0S,7W-0S	49
Jackson Dr. E&W	0E-0S	51
James St. E&W	2W-5S	61
Janssen Av. N&S	1W-3N,1W-6N	45
Jarlath Av. E&W	9W-9N	37
Jarlath St. E&W	3W-9N	39
Jarvis Av. E&W	9W-9N,2W-9N	37
Jasper Pl. N&S	1W-3S	56
Jean Av. NE&SW	7W-8N	37
Jefferson St. N&S	0W-1S,0W-0N	57
Jeffery Av. N&S	2E-15S,2E-10S	77
Jeffery Blvd. N&S	2E-9S,2E-7S	69
Jerome St. E&W	9W-9N,3W-9N	37
Jersey Av. N&S	4W-7N	38
Jessie Ct. N&S	2W-0N	50
Jones St. NW&SE	2W-3N	45
Jonquil Ter. E&W	1W-9N	39
Jourdan Ct. N&S	1W-1S	56
Julia Ct. SE&NW	3W-2N	50
Julian St. E&W	2W-1N	50
Juneway Ter. E&W	2W-9N	39
Junior Ter. E&W	0W-5N	45
Justine St. N&S	1W-14S,1W-0N	76
Kamerling Av. E&W	6W-1N,5W-1N	49
Kanst Dr. N&S	3W-7S	61
Karlov Av. N&S	5W-9S,5W-7N	66
Kasson Av. NW&SE	5W-5N	43
Kearsarge Av. NW&SE	5W-3N	43
Keating Av. N&S	5W-9S,5W-7N	66
Kecte Av. NE&SE	0E-7S	62
Kedvale Av. N&S	5W-9S,5W-7N	66
Kedzie Av. N&S	4W-13S,4W-7N	75
Kedzie Blvd. N&S	4W-2N,4W-3N	50
Kedzie Pkwy. N&S	4W-2N	50
Keeler Av. N&S	5W-9S,5W-7N	66
Keeley St. NW&SE	1W-2S	56
Keene Av. NE&SW	5W-7N	38
Keller Dr. E&W	5W-8S	60
Kelso Av. NW&SE	5W-5N	43
Kemper Pl. E&W	0W-2N	51
Kenmore Av. N&S	1W-2N,1W-6N	51
Kennedy, John F. Expy. NW&SE	10W-7N,1W-1N	36
Kenneth Av. N&S	5W-9S,5W-7N	66
Kennicott Av. NE&SW	5W-5N	43
Kennison Av. NW&SE	5W-6N	43
Kenosha Av. N&S	5W-3N	43
Kensington Av. E&W	0E-13S	76
Kenton Av. N&S	5W-9S,5W-7N	66
Kentucky Av. NE&SW	5W-6N	43
Kenwood Av. N&S	1E-10S,1E-5S	68
Keokuk Av. NW&SE	5W-5N	43
Keota Av. NE&SW	7W-8N	37
Kerbs Av. NE&SW	5W-7N	38
Kercheval Av. NW&SE	5W-7N	44
Kerfoot Av. NW&SE	0W-9S	68
Kewanee Av. NE&SW	5W-5N	43
Keystone Av. N&S	5W-0N,5W-7N	49
Kilbourn Av. N&S	5W-9S,5W-7N	66
Kildare Av. N&S	5W-9S,5W-7N	66
Kilpatrick Av. N&S	5W-9S,5W-7N	66
Kimball Av. N&S	4W-2N,4W-7N	50
Kimbark Av. N&S	1E-10S,1E-5S	68
Kimberly Av. NE&SW	5W-6N	43
Kingsbury St. NW&SE	1W-2N,0W-1N	51
Kingsdale Av. NE&SW	5W-7N	38
Kingston Av. N&S	3E-10S,3E-8S	69
Kinzie St. E&W	6W-0N,0E-0N	49
Kinzua Av. NW&SE	7W-8N	37
Kiona Av. NE&SW	5W-5N	43
Kirby Av. NE&SW	5W-7N	38
Kirkland Av. N&S	5W-9S	66
Kirkwood Av. NE&SW	5W-7N	38
Knox Av. N&S	5W-9S,5W-7N	66
Kolin Av. N&S	5W-9S,5W-1N	66
Kolmar Av. N&S	5W-9S,5W-7N	66
Komensky Av. N&S	5W-9S,5W-1S	66
Kostner Av. N&S	5W-9S,5W-7N	66
Kruger Av. NE&SW	5W-6N	43
LaCrosse Av. N&S	6W-7S,6W-7N	60
LaSalle St. N&S	0W-14S,0W-1N	76
Lacey Av. NW&SE	6W-7N	38
Lafayette Av. N&S	0W-14S,0W-6S	76
Laflin Pl. N&S	1W-3S	56
Laflin St. N&S	1W-14S,1W-0N	76
Lake Park Av. NW&SE	1E-6S,0E-2S	62
Lake Shore Dr. N&S	2E-6S,0W-6N	63
Lake Shore Dr. West E&W	0W-3N	45
Lake St. E&W	7W-0N,0W-0N	48
Lakeside Pl. E&W	1W-5N	45
Lakeview Av. N&S	0W-3N	45
Lakewood Av. N&S	1W-2N,1W-7N	51
Lamon Av. N&S	6W-7S,6W-6N	60
Landers Av. NE&SW	6W-7N	38
Langley Av. N&S	0E-15S,0E-4S	76
Langley Pl. N&S	0E-9S	68
Lansing Av. NW&SE	6W-7N	38
Laporte Av. N&S	6W-0N,6W-6N	49
Laramie Av. N&S	6W-5S,6W-6N	60
Larchmont Av. E&W	2W-4N	45
Larned Av. NW&SE	6W-6N	43
Larrabee St. N&S	0W-1N,0W-2N	51
Las Casas Av. NW&SE	6W-7N	38
Latham Av. NW&SE	6W-6N	43
Latrobe Av. N&S	6W-7S,6W-6N	60
Lavergne Av. N&S	6W-7S,6W-6N	60
Lawler Av. N&S	6W-5S,6W-6N	60
Lawndale Av. N&S	4W-13S,4W-6N	75
Lawrence Av. E&W	6W-6N,1W-6N	43
Lawrence Dr. E&W	0W-6N	45
Le Mai Av. NE&SW	6W-8N	38
Le Moyne Dr. E&W	7W-1N,1W-1N	48
Leader Av. NE&SW	6W-7N	38
Leamington Av. N&S	6W-5S,6W-5N	60
Leavenworth Av. NW&SE	6W-6N	43
Leavitt St. N&S	2W-12S,2W-8N	67
Leclaire Av. N&S	6W-5S,6W-6N	60
Lee St. N&S	13W-5N	40
Legett Av. NE&SW	6W-7N	38
Lehigh Av. NW&SE	7W-8N	37
Lehmann Ct. NW&SE	0W-3N	45
Leland Av. E&W	10W-5N,2W-5N	41
Lemai NE&SW	6W-8N	38
Lemont Av. NE&SW	6W-7N,5W-7N	38
Lenox Av. NE&SW	6W-7N	38
Leona Av. NE&SW	6W-7N	38
Leonard Av. NW&SE	6W-6N,7W-7N	43
Leoti Av. NW&SE	6W-7N,7W-8N	38
Leroy Av. NE&SW	6W-7N	38
Lessing St. N&S	1W-1N	51
Lester Av. NW&SE	6W-6N	43
Lexington St. E&W	6W-0S,0W-0S	49
Leyden Av. NW&SE	0E-15S	76
Liano Av. NW&SE	6W-6N	43
Libby Av. NW&SE	6W-6N	43
Liberty E&W	1W-1S	56
Lieb Av. NE&SW	6W-6N	43
Lightfoot Av. NE&SW	7W-8N	37
Lill Av. E&W	1W-3N	45
Lincoln Av. NW&SE	1W-3N,4W-7N	45
Lincoln Park West St. N&S	0W-2N	51
Lind Av. NW&SE	6W-6N	43
Linden Pl. NW&SE	3W-3N	44
Linder Av. N&S	6W-7S,6W-7N	60
Line Av. NE&SW	9W-4N	42
Lipps Av. NW&SE	6W-6N	43
Lister Av. NW&SE	2W-2N	50
Lituanica Av. N&S	1W-3S	56
Livermore Av. N&S	6W-7N	38
Lloyd Av. NW&SE	1W-2S	56
Lock St. NW&SE	1W-2S	56
Lockwood Av. N&S	6W-5S,6W-7N	60
Locust St. E&W	0W-1N	51
Logan Blvd. E&W	3W-3N,2W-3N	44
Loleta Av. NE&SW	7W-8N	37
London Av. N&S	6W-5N	43
Long Av. N&S	6W-7S,6W-6N	60
Longwood Dr. N&S	2W-13S,2W-10S	75
Loomis Blvd. N&S	1W-9S,1W-7S	68
Loomis Pl. N&S	1W-3S	56
Loomis St. N&S	1W-14S,1W-0N	76
Lorel Av. N&S	6W-7S,6W-3N	60
Loring Av. NW&SE	6W-6N	43
Loron Av. NE&SW	7W-8N	37
Lothair Av. N&S	2W-13S	75
Lotus Av. N&S	6W-5S,6W-6N	60
Louise Av. NW&SE	6W-7N	38
Lovejoy Av. NW&SE	7W-6N,6W-6N	42
Lowe Av. N&S	0W-14S,0W-2S	76
Lowell Av. N&S	5W-2N,5W-7N	49
Loyola Av. E&W	2W-8N,1W-8N	39
Lucerne Av. NW&SE	6W-7N	38
Ludlam Av. NE&SW	6W-6N	43
Luella Av. N&S	2E-15S,2E-7S	77
Lumber St. E&W	1W-2S,0W-1S	56
Luna Av. N&S	6W-5S,6W-7N	60
Lundy Av. NE&SW	6W-7N	38
Lunt Av. E&W	9W-8N,1W-8N	37
Luther St. E&W	3W-2S	56
Lyman St. NE&SW	1W-2S	56
Lynch Av. NE&SW	6W-6N	43
Lyndale Av. E&W	5W-2N,3W-2N	49
Lytle St. N&S	1W-0S	51
Mackinaw Av. N&S	4E-15S,4E-9S	77
Madison Park E&W	1E-5S	62
Madison St. E&W	6W-0N,1W-0N	49
Magnet Av. NW&SE	7W-6N	42
Magnolia Av. N&S	1W-1N,1W-7N	51
Major Av. N&S	7W-6S,7W-7N	60
Malden St. N&S	1W-5N	45
Malta St. NW&SE	1W-11S	68
Mandell Av. NW&SE	7W-7N	37
Mango Av. N&S	7W-2N,7W-6N	48
Manila Av. NE&SW	7W-6N	42
Manistee Av. N&S	3E-14S,3E-9S	77
Mankato Av. NE&SW	7W-8N	37
Mann Dr. E&W	4W-7S	61
Manor Av. NW&SE	3W-5N	44
Manton Av. NW&SE	7W-7N	37
Maple St. E&W	0W-1N	51
Maplewood Av. N&S	3W-13S,3W-9N	39
Marble Pl. E&W	0W-0S	51
Marcey St. NW&SE	1W-2N	51
Margate Ter. E&W	1W-6N	45
Maria Ct. N&S	10W-5N	41
Marine Dr. NW&SE	0W-5N	45
Marion Ct. N&S	2W-1N	50
Market St. E&W	0E-15S	76
Markham Av. N&S	7W-7N	37
Marmora Av. N&S	7W-3N,7W-7N	42
Marquette Av. N&S	3E-14S,3E-8S	77
Marquette Dr. E&W	2E-7S	63
Marquette Rd. E&W	5W-7S,0E-7S	60
Marshall Blvd. N&S	3W-2S,3W-1S	56
Marshfield Av. N&S	2W-10S,2W-9N	67
Martin Luther King Jr. Dr N&S	0E-12S,0E-4S	68
Mary St. NW&SE	1W-2S	56
Maryland Av. N&S	0E-15S,1E-6S	76
Mason Av. N&S	7W-6S,7W-6N	60
Massasoit Av. N&S	7W-6S,7W-1N	60
Matson Av. NE&SW	7W-7N	37
Maud Av. NW&SE	1W-2N	51
Mautene Ct. NE&SW	2W-1N	50
Maxwell St. E&W	1W-1S	56
May St. N&S	1W-13S,1W-1N	76
Mayfield Av. N&S	7W-6S,7W-2N	60
Maypole Av. E&W	6W-0N,2W-0N	49
McAlpin Av. NE&SW	7W-8N	37
McClellan Av. NW&SE	7W-7N	37
McClurg Ct. N&S	0E-0N	51
McCook Av. N&S	7W-7N	37
McCormick Blvd. NE&SW	4W-8N	38
McCrea Dr. N&S	4W-0N	50
McDowell Av. E&W	1W-4S	56
McFetridge Dr. E&W	0E-1S	57
McLean Av E&W	5W-2N,1W-2N	49
McLeod Av. NW&SE	7W-7N	37
McVicker Av. N&S	7W-6S,7W-9N	60
Meade Av. N&S	7W-7S,7W-9N	60
Medford Av. NW&SE	7W-8N	37
Medill Av. N&S	8W-2N,2W-2N	48
Medina Av. NE&SW	7W-7N	37
Melrose St. E&W	8W-4N,0W-4N	42
Melvina Av. N&S	7W-7S,7W-7N	60
Memory Ln. E&W	10W-6N	41

	Grid	Page
Menard Av. N&S	7W-6S,7W-6N	60
Mendell St. N&S	1W-2N	51
Mendota Av. NE&SW	7W-8N	37
Menomonee St. E&W	0W-2N	51
Meredith Av. N&S	7W-7N	37
Merrill Av. N&S	2E-15S,2E-7S	77
Merrimac Av. N&S	7W-6S,7W-8N	60
Merrion Av. N&S	2E-11S	69
Meyer Ct. N&S	0W-2N	51
Miami Av. E&W	7W-7N	37
Michaels Ct. N&S	0W-2N	51
Michigan Av. N&S	0W-14S,0E-0N	76
Midway Park E&W	7W-0N	48
Midway Plaisance E&W	1E-6S	62
Mies van der Rohe Way N&S	0E-1N	51
Mildred Av. N&S	1W-3N	45
Millard Av. N&S	4W-13S,4W-0S	75
Miller St. NE&SW	1W-1S,1W-0S	56
Miltimore Av. N&S	7W-7N	37
Milwaukee Av. NW&SE	7W-7N,0W-0N	37
Minerva Av. N&S	1E-7S	62
Minnehaha Av. NE&SW	6W-8N	38
Minnetonka Av. NE&SW	7W-8N,6W-8N	37
Mobile Av. N&S	7W-7S,7W-8N	60
Moffat St. E&W	3W-2N,2W-2N	50
Mohawk St. N&S	0W-2N	51
Monitor Av. N&S	7W-6S,7W-7N	60
Monon Av. NE&SW	7W-8N	37
Monroe Dr. E&W	0E-0S	51
Monroe St. E&W	6W-0S,1W-0S	49
Mont Clare Av. N&S	8W-3N,8W-6N	42
Montana St. E&W	6W-3N,1W-3N	43
Monterey Av. NW&SE	2W-13S	75
Montgomery Av. NW&SE	3W-4S	56
Monticello Av. N&S	4W-0N,4W-6N	50
Montrose Av. E&W	10W-5N,1W-5N	41
Montrose Dr. E&W	0W-5N	45
Montrose Harbor Dr. E&W	0W-5N	45
Montrose St. N&S	13W-5N	40
Montvale Av. NW&SE	2W-13S	75
Moody Av. N&S	7W-6S,7W-7N	60
Moorman St. NW&SE	2W-1N	50
Morgan Dr. NW&SE	0E-6S	62
Morgan St. N&S	1W-15S,1W-0N	76
Morse Av. E&W	6W-8N,1W-8N	38
Moselle Av. NE&SW	7W-8N	37
Mozart St. N&S	3W-9S,3W-7N	67
Mulligan Av. N&S	7W-6S,7W-7N	60
Museum Dr. E&W	2E-6S	63
Music Ct. E&W	4W-0S	50
Cir.	4W-0S	50
Muskegon Av. N&S	3E-14S,3E-9S	77
Myrick St. E&W	4W-8S	61
Myrtle Av. E&W	9W-7N	37
Nagle Av. N&S	8W-6S,8W-8N	59
Naper Av. NE&SW	8W-7N	37
Naples Av. NW&SE	8W-7N	37
Napoleon Av. NE&SW	8W-7N	37
Narragansett Av. N&S	8W-7S,8W-5N	59
Nashotah Av. NE&SW	8W-7N	37
Nashville Av. N&S	8W-7S,8W-3N	59
Nashville Av. N&S	8W-8N	37
Nashville Av. N&S	8W-7S,8W-2N	59
Nashville Av. N&S	8W-3N,8W-6N	42
Nassau Av. NW&SE	8W-7N	37
Natchez Av. N&S	8W-6S,8W-6N	59
Natoma Av. N&S	8W-6S,8W-8N	59
Navajo Av. NE&SW	6W-8N	38
Navarre Av. NW&SE	8W-7N	37
Neenah Av. N&S	8W-6S,8W-7N	59
Nelson St. E&W	8W-3N,1W-3N	42
Neola Av. N&S	8W-6N,8W-7N	42
Nettleton Av. NE&SW	8W-7N	37
Neva Av. N&S	8W-6S,8W-8N	59
New England Av. N&S	8W-6S,8W-8N	59
New Hampshire Av. NE&SW	8W-7N	37
Newark Av. N&S	8W-7N,8W-8N	37
Newberry Av. N&S	1W-1S	56
Newburg Av. NW&SE	8W-7N	37
Newcastle Av. N&S	8W-6S,8W-7N	59
Newgard Av. N&S	1W-8N	39
Newland Av. N&S	8W-6S,8W-8N	59
Newport Av. E&W	8W-4N,1W-4N	42
Niagara Av. NE&SW	8W-7N	37
Nickerson Av. N&S	8W-7N	37
Nicolet Av. N&S	8W-7N	37
Nina Av. NE&SW	8W-7N	37
Nixon Av. N&S	8W-8N	37
Noble St. E&W	1W-0N,1W-1N	51
Nokomis Av. NE&SW	6W-8N	38
Nora Av. N&S	8W-3N	42
Nordica Av. N&S	8W-6S,8W-8N	59
Normal Av. N&S	0W-15S,0W-1S	76
Normal Pkwy. E&W	0W-7S	62
Normandy Av. N&S	8W-6S,8W-8N	59
North Av. E&W	6W-2N,0W-2N	49
North Blvd. E&W	0W-2N	51
North Branch St. N&S	1W-1N	51
North Park Av. N&S	0W-1N	51
North Shore Av. E&W	9W-8N,2W-8N	37
North Water St. E&W	0E-0N	51
Northcott Av. NW&SE	8W-7N	37
Northwest Hwy. NW&SE	8W-7N	37
Norwood Av. E&W	9W-7N,1W-7N	37
Nottingham Av. N&S	8W-7S,8W-6N	59
Nursery St. NW&SE	1W-2N	51
O'Brien St. E&W	0W-1S	57
Oak Park Av. N&S	8W-6S,8W-8N	59
Oak St. E&W	0W-1N	51
Oakdale Av. E&W	8W-3N,1W-3N	42
Oakenwald Av. NW&SE	1E-4S	57
Oakley Av. N&S	2W-13S,2W-8N	75
Oakley Blvd. N&S	2W-0S,2W-1N	50
Oakview St. N&S	10W-5N,10W-6N	41
Oakwood Blvd. E&W	0E-4S	57
Oconto Av. N&S	9W-3N,9W-9N	42
Octavia Av. N&S	9W-3N,9W-9N	42
Odell Av. N&S	9W-3N,9W-9N	42
Ogallah Av. NE&SW	9W-8N	37
Ogden Av. NE&SW	5W-2S,1W-0N	55
Oglesby Av. N&S	2E-15S,2E-7S	77
Ohio St. E&W	7W-0N,0W-0N	48
Oketo Av. N&S	9W-3N,9W-9N	42
Olcott Av. N&S	9W-3N,9W-8N	42
Old Mannheim Rd. N&S	13W-7N	40
Oleander Av. N&S	9W-3N,9W-8N	42
Oliphant Av. NE&SW	9W-8N	37
Olive Av. E&W	9W-7N,2W-7N	37
Olmsted Av. NW&SE	9W-8N	37
Olympia Av. N&S	9W-6N,9W-8N	42
Onarga Av. NW&SE	9W-8N	37
Oneida Av. NW&SE	9W-8N	37
Ontario St. E&W	4W-0N,0W-0N	50
Opal Av. N&S	9W-4N	42
Orange Av. N&S	9W-4N,9W-7N	42
Orchard St. N&S	0W-2N,0W-3N	51
Oriole Av. N&S	9W-4N,9W-8N	42
Orleans St. N&S	0W-1N	51
Osage Av. N&S	9W-4N,9W-7N	42
Osceola Av. N&S	9W-3N,9W-8N	42
Oshkosh Av. NE&SW	9W-8N	37
Oswego St. N&S	2W-0N	50
Otsego Av. NW&SE	9W-8N	37
Ottawa Av. N&S	9W-4N,9W-9N	42
Otto Av. N&S	12W-6N	40
Overhill Av. N&S	9W-4N,9W-8N	42
Owen Av. NW&SE	9W-8N	37
Oxford Av. NE&SW	9W-8N	37
Ozanam Av. N&S	9W-4N,9W-7N	42
Ozark Av. N&S	9W-4N,9W-9N	42
Pacific Av. N&S	10W-4N,10W-6N	41
Packers Av. N&S	1W-4S	56
Page Av. N&S	10W-4N	42
Palatine Av. E&W	9W-7N,8W-7N	37
Palmer Blvd. E&W	3W-2N	50
Palmer St. E&W	8W-2N,2W-2N	48
Panama Av. N&S	10W-4N	41
Paris Av. N&S	10W-4N,10W-6N	41
Parker Av. E&W	6W-3N,4W-3N	43
Parkside Av. N&S	7W-6S,7W-7N	60
Parkview Ter. NW&SE	4W-4N	44
Parnell Av. N&S	0W-14S,0W-2S	76
Patrick Ct. E&W	3W-12S	67
Patterson Av. E&W	9W-4N,0W-4N	42
Patton Dr. N&S	10W-5N	41
Paulina St. N&S	2W-10S,2W-9N	67
Paxton Av. N&S	2E-15S,2E-7S	77
Payne Dr. N&S	0E-6S	62
Pearson St. E&W	2W-1N,0E-1N	50
Pensacola Av. E&W	6W-5N,1W-5N	43
Peoria St. N&S	1W-15S,1W-0N	76
Perry Av. N&S	0W-14S,0W-6S	76
Pershing Rd. E&W	3W-3S,0W-3S	56
Peterson Av. E&W	9W-7N,3W-7N	37
Phillips Av. N&S	3E-10S,3E-8S	69
Pierce Av. E&W	5W-1N,2W-1N	49
Pine Av. N&S	6W-0N,6W-1N	49
Pine Grove Av. N&S	0W-3N,0W-4N	45
Pioneer Av. N&S	10W-4N,10W-6N	41
Pippin St. E&W	4W-8S	61
Pitney Ct. NW&SE	1W-2S	56
Pittsburgh Av. N&S	10W-4N,10W-6N	41
Plainfield Av. N&S	10W-4N,10W-5N	41
Pleasant Av. N&S	2W-10S	67
Plymouth Ct. N&S	0W-0S	51
Poe St. NW&SE	1W-2N	51
Point St. NW&SE	3W-2N	50
Polk St. E&W	6W-0S,3W-0S	49
Ponchartrain Blvd. NE&SW	7W-8N	37
Pontiac Av. N&S	10W-4N,10W-5N	41
Pope John Paul II Dr. E&W	3W-4S	56
Poplar Av. NW&SE	1W-2S	56
Potawatomie St. N&S	10W-5N,10W-6N	41
Potomac Av. E&W	7W-1N,1W-1N	48
Prairie Av. N&S	0E-15S,0E-1S	76
Pratt Av. E&W	9W-8N,2W-8N	37
Prescott Av. NW&SE	6W-7N	38
Princeton Av. N&S	0W-14S,0W-2S	76
Prindiville St. NE&SW	3W-2N	50
Promontory Dr. N&S	2E-7S	63
Prospect Av. N&S	2W-12S,2W-11S	67
Pryor Av. NW&SE	2W-12S,2W-13S	67
Pueblo Av. N&S	10W-4N,10W-7N	41
Pulaski Rd.(Crawford) N&S	5W-9S,5W-7N	66
Quarry St. NW&SE	1W-2S	56
Quincy St. E&W	6W-0S,0W-0S	49
Quinn St. NW&SE	1W-2S	56
Race Av. E&W	7W-0N,1W-0N	48
Racine Av. N&S	1W-13S,1W-5N	76
Railroad Av. E&W	7W-0S	48
Rainey Dr. E&W	0E-6S	62
Randolph St. E&W	0W-0N	51
Rascher Av. E&W	10W-6N,1W-6N	42
Rascher Av. E&W	10W-6N,9W-6N	42
Raven St. E&W	8W-7N,7W-7N	37
Ravenswood Av. N&S	2W-4N,2W-8N	45
Redfield Dr. E&W	4W-7S	61
Reilly Ter. N&S	4W-8S	61
Reserve Av. N&S	10W-5N,10W-6N	41
Reta Av. N&S	1W-4N	45
Rhodes Av. N&S	0E-15S,0E-3S	76
Rice St. E&W	6W-1N,2W-1N	49
Richards Dr. N&S	2E-7S	63
Richmond St. N&S	3W-9S,3W-7N	67
Ridge Av. NW&SE	1W-7N	39
Ridge Blvd. NW&SE	2W-8N	39
Ridge Dr. N&S	0W-2N	51
Ridgeland Av. N&S	2E-10S,2E-7S	69
Ridgeway Av. N&S	4W-13S,4W-7N	75
Ridgewood Ct. N&S	1E-5S	62
Ritchie Ct. N&S	0E-1N	51
Riverdale NW&SE	0E-15S	76
Robinson St. NW&SE	2W-3S	56
Rockwell St. N&S	3W-13S,3W-8N	75
Rogers Av. NE&SW	6W-6N,2W-9N	43
Roosevelt Rd. E&W	4W-0S,1W-1S	50
Root St. E&W	0W-4S	57
Roscoe St. E&W	10W-4N,0W-4N	41
Rosedale Av. E&W	9W-7N,6W-7N	37
Rosehill Dr. E&W	2W-7N	39
Rosemont Av. E&W	5W-7N,1W-7N	38
Roslyn Pl. E&W	0W-3N	45
Ross Av. NE&SW	0W-7S	62
Rt. 19 E&W	10W-5N,1W-5N	41
Rt. 43 N&S	8W-6S,9W-9N	59
Rt. 50 N&S	5W-7S,5W-7N	60
Rt. 64 E&W	7W-2N,0W-2N	48
Rt. 72 NW&SE	10W-7N	36
Ruble St. N&S	0W-1S	57
Rundell Pl. E&W	4W-0S	50
Rush St. N&S	0E-0N,0E-1N	51
Russell Dr. N&S	0E-6S	62
Rutherford Av. N&S	8W-6S,8W-6N	59
Ryan, Dan Expwy. N&S	2W-13S,0W-1S	75
Sacramento Av. N&S	3W-12S,3W-8N	67
Sacramento Blvd. N&S	3W-0S,3W-0N	50
Sacramento Dr. N&S	3W-1S	56
Saginaw Av. N&S	3E-14S,3E-8S	77
Sangamon St. N&S	1W-15S,1W-1N	76
Sauganash Av. NW&SE	6W-8N,5W-7N	38
Sawyer Av. N&S	4W-13S,4W-6N	75
Sayre Av. N&S	8W-6S,8W-7N	59
Schiller St. E&W	2W-1N,0W-1N	50
School St. E&W	10W-4N,1W-4N	41
Schorsch St. E&W	8W-4N	42
Schrader Dr. E&W	4W-0N	50
Schreiber Av. E&W	8W-8N,2W-8N	37
Schubert Av. E&W	8W-3N,0W-3N	42
Scott St. E&W	0W-1N	51
Scottsdale Av. NW&SE	5W-9S	66
Sedgwick St. N&S	0W-1N	51
Seeley Av. N&S	2W-12S,2W-9N	67
Seipp St. E&W	3W-9S	67
Seminary Av. N&S	1W-2N,1W-4N	51
Seminole Av. E&W	9W-7N,7W-7N	37
Senour Av. NW&SE	1W-2S	56
Service Yard Dr. E&W	4W-0N	50
Shakespeare Av. E&W	8W-2N	48
Sheffield Av. N&S	1W-2N,1W-4N	51
Shelby Ct. N&S	1W-1S	56
Sheridan Rd. N&S	0W-3N,1W-9N	45
Sherman Av. N&S	0W-0S	51
Sherman Dr. E&W	1W-5S	62
Sherman St. N&S	0W-0S	51
Sherwin Av. E&W	9W-9N,1W-9N	37
Shields Av. N&S	0W-5S,0W-2S	62
Shore Dr. N&S	2E-6S,2E-5S	63
Short St. NW&SE	1W-2S	56
Simonds Dr. N&S	0W-2N,0W-6N	51
Sioux Av. NW&SE	7W-8N,6W-8N	37
Solidarity Dr. E&W	0E-1S	57
South Chicago Av. NW&SE	1E-8S,2E-9S	62
South Exchange Av. N&S	3E-15S,3E-8S	77
South Shore Dr. N&S	3E-9S,2E-7S	69
Southport Av. N&S	1W-2N,1W-4N	51
Spaulding Av. N&S	4W-13S,4W-6N	75
Spokane Av. NE&SW	6W-8N	38
Springfield Av. N&S	4W-12S,4W-7N	67
St. Clair St. N&S	0E-0N	51
St. Georges Ct. NE&SW	3W-2N	50
St. Helen Ct. NE&SW	3W-2N	50
St. James Pl. E&W	0W-3N	45
St. Joseph Av. E&W	10W-6N	41
St. Lawrence Av. N&S	0E-15S,0E-5S	76
St. Louis Av. N&S	4W-13S,4W-7N	75
St. Mary St. NE&SW	3W-2N	50
St. Paul Av. E&W	6W-2N,0W-2N	49
Stark St. NW&SE	1W-2S	56
State Rt. 19 E&W	10W-5N,1W-5N	41
State Rt. 43 N&S	8W-6S,8W-4N	59
State Rt. 50 N&S	5W-7S,6W-7N	60
State Rt. 64 E&W	7W-1N,0W-1N	48
State St. N&S	0W-14S,0W-1N	76
Stave St. NW&SE	3W-2N	50
Stetson Av. N&S	0E-0N	51
Steuben St. NW&SE	2W-13S	75
Stevenson Expy. NE&SW	6W-4S	55
Stewart Av. N&S	0W-14S,0W-1S	76
Stockton Dr. N&S	0W-3N	45
Stone Ct. N&S	0E-1N	51
Stony Island Av. N&S	2E-14S,1E-6S	77
Stratford Pl. NE&SW	0W-4N	45
Streeter Dr. NS&EW	0E-0N	51
Strong St. E&W	8W-6N,4W-6N	42
Sullivan St. E&W	0W-1N	51
Summerdale Av. E&W	10W-6N,2W-6N	41
Summerset Av. NE&SW	8W-7N	37
Sunnyside Av. E&W	10W-5N,2W-5N	41
Superior St. E&W	7W-0N,0W-0N	48
Surf St. E&W	0W-3N	45
Surrey Ct. N&S	1W-3N	45
Swann St. E&W	0W-4S	57
Tahoma Av. NW&SE	6W-8N	38
Talcott Av. NW&SE	9W-7N	37
Talman Av. N&S	3W-13S,3W-9N	75
Taylor St. E&W	7W-0S,0W-0S	48
Terra Cotta Pl. NE&SW	2W-3N	45
Thomas St. E&W	7W-1N,1W-1N	48
Thome Av. E&W	5W-7N,1W-7N	38
Thompson Dr. N&S	3W-1S	56
Thorndale Av. E&W	9W-7N,1W-7N	37
Throop St. N&S	1W-14S,1W-2N	76
Togwell St. N&S	13W-5N	40
Tonty Av. NE&SW	7W-8N	37
Torrence Av. N&S	2E-14S,2E-11S	77
Touhy Av. E&W	9W-9N,2W-9N	37
Tremont St. E&W	0W-6S	62
Tripp Av. N&S	5W-9S,5W-7N	66
Troy Av. N&S	3W-13S,3W-12S	75
Troy St. N&S	3W-9S,3W-7N	67
Trumbull Av. N&S	4W-13S,4W-0N	75
Turner Av. N&S	4W-13S	75
U.S. Rt. 12 E&W	2E-10S,4E-11S	69
U.S. Rt. 14 N&S	1W-6N,7W-8N	45
U.S. Rt. 20 E&W	2E-10S,4E-11S	69
U.S. Rt. 41 NW&SE	4E-11S,4W-7N	69
Union Av. N&S	0W-14S,0W-0N	76
University Av. N&S	1E-11S,1E-4S	68
Van Buren St. E&W	6W-0S,1W-0S	49
Van Vlissingen Rd. NW&SE	2E-11S	69
Vanderpool Av. N&S	2W-11S,2W-10S	67
Vermont Av. NE&SW	1W-15S,0W-15S	76
Vernon Av. N&S	0E-15S,0E-3S	76
Victoria Av. E&W	9W-7N,1W-7N	37
Vincennes Av. NW&SE	2W-13S,0E-3S	75
Vine Av. N&S	0W-2N	51
Virginia Av. NW&SE	3W-5N,3W-6N	44

	Grid	Page
Wabansia Av. E&W	7W-2N,2W-2N	48
Wabash Av. N&S	0W-14S,0E-0N	76
Wacker Dr. N&S	0W-0S,0W-0N	51
Walden Pkwy. N&S	2W-12S,2W-11S	67
Waldron Dr. E&W	0E-1S	57
Wallace St. N&S	0W-14S,0W-2S	76
Wallen Av. E&W	2W-8N	39
Waller Av. N&S	7W-0S,7W-1N	48
Walnut St. E&W	5W-0N,0W-0N	49
Walton Dr. N&S	4E-10S	69
Walton St. E&W	6W-1N,1W-1N	49
Warner Av. E&W	6W-5N,1W-5N	43
Warren Av. E&W	0W-0N	51
Warren Blvd. E&W	3W-0N	50
Warren Dr. E&W	4W-0N	50
Warwick Av. E&W	7W-4N,5W-4N	42
Waseca Pl. NW&SE	2W-13S	75
Washburne Av. E&W	2W-1S	56
Washington Blvd. E&W	6W-0N,0W-0N	49
Washington Park Ct. N&S	0E-5S	62
Washtenaw Av. N&S	3W-13S,3W-8N	75
Water St. E&W	0E-0N	51
Waterloo Ct. N&S	0W-3N	45
Waterway NW&SE	1E-14S	76
Watkins Av. NE&SW	2W-13S	75
Waukesha Av. N&S	6W-8N	38
Waveland Av. E&W	10W-4N,1W-4N	41
Wayman St. E&W	5W-0N,1W-0N	49
Wayne Av. N&S	1W-2N,1W-8N	51
Webster Av. E&W	2W-2N,1W-2N	50
Weed St. E&W	1W-1N,0W-1N	51
Wellington Av. E&W	9W-3N,1W-3N	42
Wells St. N&S	0W-5S,0W-1N	62
Wendell St. E&W	0W-1N	51
Wentworth Av. N&S	0W-14S,0W-1S	76
West Broadman Av. E&W	10W-5N	41
Cir.	8W-7N	37
West End Av. E&W	7W-0N	48
West Exchange Av. E&W	1W-4S	56
Western Av. N&S	3W-13S,3W-9N	75
Western Blvd. N&S	2W-4S	56
Whipple St. N&S	3W-12S,3W-7N	67
Wicker Park Av. NW&SE	2W-1N	50
Wieland St. N&S	0W-1N	51
Wilcox St. E&W	5W-0S,3W-0S	49
Wildwood Av. NE&SW	7W-8N	37
Willard Ct. N&S	1W-0N	51
Willets Ct. NW&SE	3W-3N	44
Williams Av. N&S	0E-10S	68
Willow St. E&W	2W-2N,0W-2N	50
Wilmot Av. NW&SE	2W-2N	50
Wilson Av. E&W	10W-5N,2W-5N	41
Wilson Dr. E&W	0W-5N	45
Wilton Av. N&S	1W-3N,1W-4N	45
Winchester Av. N&S	2W-11S,2W-9N	67
Windsor Av. E&W	10W-5N,1W-5N	41
Winnebago Av. NW&SE	2W-2N	50
Winneconna Pkwy. NE&SW	0W-8S	62
Winnemac Av. E&W	8W-6N,1W-6N	42
Winona St. E&W	10W-6N,1W-6N	41
Winston Av. N&S	1W-11S	68
Winthrop Av. N&S	1W-6N,1W-8N	45
Wisconsin St. E&W	1W-2N,0W-2N	51
Wisner Av. NE&SW	4W-3N	44
Wolcott Av. N&S	2W-13S,2W-8N	75
Wolfe Lake Blvd. N&S	4E-14S,4E-13S	77
Wolfram St. E&W	8W-3N,1W-3N	42
Wood St. N&S	2W-13S,2W-2N	75
Woodard St. NE&SW	4W-3N	44
Woodlawn Av. N&S	1E-12S,1E-5S	68
Woodward Dr. N&S	4W-0S	50
Wright St. NE&SW	0W-8S	62
Wrightwood Av. E&W	8W-3N,0W-3N	42
Yale Av. N&S	0W-14S,0W-7S	76
Yates Av. N&S	2E-15S,2E-11S	77
Yates Blvd. N&S	2E-10S,2E-8S	69
9th St. E&W	0E-0S	51
11th Pl. E&W	0E-0S	51
11th St. E&W	0E-0S	51
12th Pl. E&W	4W-1S,0W-1S	55
13th Pl. E&W	4W-1S,3W-1S	55
13th St. E&W	5W-1S,0E-1S	55
14th Pl. E&W	3W-1S,2W-1S	56
14th St. E&W	5W-1S,0E-1S	55
15th Pl. E&W	3W-1S,0W-1S	56
15th St. E&W	4W-1S,0W-1S	55
16th St. E&W	5W-1S,0E-1S	55
17th Pl. E&W	0W-1S	57
17th St. E&W	5W-1S,0W-1S	55
18th Pl. E&W	5W-1S,1W-1S	55
18th St. E&W	5W-1S,0E-1S	55
19th Pl. E&W	1W-1S	56
19th St. E&W	5W-1S,0W-1S	55
20th Pl. E&W	1W-1S	56
21st Pl. E&W	5W-1S,2W-1S	55
21st St. E&W	5W-1S,0E-1S	55
22nd Pl. E&W	3W-2S,0W-2S	56
22nd St. E&W	1W-2S,0W-2S	56
23rd Pl. E&W	3W-2S,0W-2S	56
23rd St. E&W	4W-2S,0E-2S	55
23rd St. Viaduct E&W	0E-2S	57
24th Blvd. E&W	3W-2S	56
24th Pl. E&W	5W-2S,0E-2S	55
24th St. E&W	4W-2S,0E-2S	55
25th Pl. E&W	5W-2S,3W-2S	55
25th St. E&W	4W-2S,0E-2S	55
26th Pl. E&W	0W-2S	57
26th St. E&W	5W-2S,0E-2S	55
27th St. E&W	4W-2S,0E-2S	55
28th Pl. E&W	0W-2S	57
28th St. E&W	4W-2S,0E-2S	55
29th Pl. E&W	0E-2S	57
29th St. E&W	1W-2S,0E-2S	56
30th St. E&W	4W-2S,0W-2S	55
31st Blvd. E&W	3W-2S	56
31st Pl. E&W	2W-3S,0E-3S	56
31st St. E&W	5W-2S,0E-2S	55
32nd Pl. E&W	1W-3S,0E-3S	56
32nd St. E&W	4W-3S,0E-3S	55
33rd Pl. E&W	2W-3S,0E-3S	56
33rd St. E&W	4W-3S,0E-3S	55
34th Pl. E&W	3W-3S,1W-3S	56
34th St. E&W	4W-3S,0E-3S	55
35th Pl. E&W	3W-3S,1W-3S	56
35th St. E&W	4W-3S,0E-3S	55
36th Pl. E&W	3W-3S,0E-3S	56
36th St. E&W	3W-3S,0E-3S	56
37th Pl. E&W	3W-3S,0E-3S	56
37th St. E&W	2W-3S,0E-3S	56
38th Pl. E&W	4W-3S,0E-3S	55
38th St. E&W	3W-3S,0E-3S	56
39th Pl. E&W	3W-4S,0W-6S	56
40th Pl. E&W	3W-4S,0W-4S	56
40th St. E&W	5W-4S,1E-4S	55
41st Pl. E&W	3W-4S,1E-4S	56
41st St. E&W	4W-4S,1E-4S	55
42nd Pl. E&W	3W-4S,1E-4S	56
42nd St. E&W	5W-4S,0E-4S	55
43rd Pl. E&W	0W-4S	57
43rd St. E&W	5W-4S,1E-4S	55
44th Pl. E&W	6W-4S,1E-4S	55
44th St. E&W	5W-4S,1E-4S	55
45th Pl. E&W	4W-4S,0E-4S	55
45th St. E&W	5W-4S,1E-4S	55
46th Pl. E&W	4W-4S,0E-4S	55
46th St. E&W	6W-4S,1E-4S	55
47th Dr. E&W	1E-4S	57
47th Pl. E&W	4W-5S,1E-5S	61
47th St. E&W	5W-4S,1E-4S	55
48th Pl. E&W	4W-5S,0E-5S	61
48th St. E&W	6W-5S,1E-5S	60
49th Dr. NE&SW	2E-5S	63
49th Pl. E&W	2W-5S,1W-5S	61
49th St. E&W	5W-5S,1E-5S	60
50th Pl. E&W	4W-5S,2E-5S	61
50th St. E&W	5W-5S,2E-5S	60
51st Pl. E&W	2W-5S,0W-5S	61
51st St. E&W	5W-5S,0E-5S	60
52nd Pl. E&W	4W-5S,0W-5S	61
52nd St. E&W	8W-5S,1E-5S	59
53rd Pl. E&W	6W-5S,1W-5S	60
53rd St. E&W	8W-5S,1E-5S	59
54th Pl. E&W	7W-5S,1E-5S	60
54th St. E&W	8W-5S,2E-5S	59
55th Pl. E&W	5W-6S,0E-6S	60
55th St. E&W	7W-5S,2E-5S	60
56th Pl. E&W	4W-6S,0W-6S	61
56th St. E&W	8W-6S,1E-6S	59
57th Dr. E&W	2E-6S	63
57th Pl. E&W	4W-6S,0W-6S	61
57th St. E&W	8W-6S,1E-6S	59
58th Pl. E&W	4W-6S,0W-6S	61
58th St. E&W	8W-6S,1E-6S	59
59th Pl. E&W	4W-6S	61
59th St. E&W	8W-6S,1E-6S	59
60th Pl. E&W	8W-6S,0W-6S	59
60th St. E&W	8W-6S,1E-6S	59
61st Pl. E&W	4W-6S,1E-6S	61
61st St. E&W	8W-6S,1E-6S	59
62nd Pl. E&W	4W-6S,1E-6S	61
62nd St. E&W	8W-6S,0E-6S	59
63rd Pl. E&W	8W-7S,1E-7S	59
63rd St. E&W	8W-6S,1E-6S	59
64th Pl. E&W	8W-7S,1E-7S	59
64th St. E&W	8W-7S,1E-7S	59
65th Pl. E&W	5W-7S,1E-7S	60
65th St. E&W	8W-7S,1E-7S	59
66th Pl. E&W	5W-7S,1E-7S	60
66th St. E&W	5W-7S,0E-7S	60
67th Pl. E&W	4W-7S,1E-7S	61
67th St. E&W	0E-7S,2E-7S	62
68th Pl. E&W	4W-7S,2W-7S	61
68th St. E&W	5W-7S,2E-7S	60
69th Pl. E&W	4W-7S,1E-7S	61
69th St. E&W	5W-7S,2E-7S	60
70th Pl. E&W	4W-7S,2E-7S	61
70th St. E&W	5W-7S,2E-7S	60
71st Pl. E&W	4W-8S,1E-8S	61
71st St. E&W	4W-7S,2E-7S	61
72nd Pl. E&W	4W-8S,3E-8S	61
72nd St. E&W	4W-8S,3E-8S	61
73rd Pl. E&W	4W-8S,3E-8S	61
73rd St. E&W	4W-8S,3E-8S	61
74th Pl. E&W	1W-8S,3E-8S	62
74th St. E&W	4W-8S,3E-8S	61
75th Pl. E&W	4W-8S,3E-8S	61
75th St. E&W	4W-8S,3E-8S	61
76th Pl. E&W	4W-8S,3E-8S	61
76th St. E&W	4W-8S,3E-8S	61
77th Pl. E&W	5W-8S,3E-8S	60
77th St. E&W	4W-8S,3E-8S	61
78th Pl. E&W	4W-8S,3E-8S	61
78th St. E&W	5W-8S,3E-8S	60
79th Pl. E&W	5W-9S,4E-9S	66
79th St. E&W	5W-8S,2E-8S	60
80th Pl. E&W	5W-9S,3E-9S	66
80th St. E&W	5W-9S,3E-9S	66
81st Pl. E&W	5W-9S,3E-9S	66
81st St. E&W	5W-9S,3E-9S	66
82nd Pl. E&W	5W-9S,1E-9S	66
82nd St. E&W	5W-9S,3E-9S	66
83rd Pl. E&W	5W-9S,2E-9S	66
83rd St. E&W	5W-9S,4E-9S	66
84th Pl. E&W	5W-9S,2E-9S	66
84th St. E&W	5W-9S,4E-9S	66
85th Pl. E&W	4W-9S,2E-9S	67
85th St. E&W	5W-9S,4E-9S	66
86th Pl. E&W	4W-9S,2E-9S	67
86th St. E&W	5W-9S,4E-9S	66
87th Pl. E&W	0E-10S,2E-10S	68
87th St. E&W	5W-9S,3E-9S	66
88th Pl. E&W	0E-10S,1E-10S	68
88th St. E&W	2W-10S,4E-10S	67
89th Pl. E&W	0E-10S,1E-10S	68
89th St. E&W	2W-10S,4E-10S	67
8th St. E&W	0E-0S	51
90th Pl. E&W	2W-10S,1E-10S	67
90th St. E&W	2W-10S,3E-10S	67
91st Pl. E&W	2W-10S,2E-10S	67
91st St. E&W	2W-10S,3E-10S	67
92nd Pl. E&W	2W-10S,2E-10S	67
92nd St. E&W	2W-10S,3E-10S	67
93rd Pl. E&W	2W-10S	67
93rd St. E&W	2W-10S,3E-10S	67
94th St. E&W	2W-10S,3E-10S	67
95th Pl. E&W	2W-11S,2E-11S	67
95th St. E&W	1W-10S,3E-10S	68
96th Pl. E&W	0W-11S,1E-11S	68
96th St. E&W	2W-11S,4E-10S	67
97th Pl. E&W	1W-11S,3E-11S	68
97th St. E&W	2W-11S,4E-10S	67
98th Pl. E&W	1W-11S,2E-11S	68
98th St. E&W	2W-11S,4E-10S	67
99th Pl. E&W	3W-11S,0E-11S	67
99th St. E&W	2W-11S,4E-11S	67
100th Blvd. E&W	3E-11S,4E-11S	69
100th Pl. E&W	3W-11S,0E-11S	67
100th St. E&W	3W-11S,3E-11S	67
101st Pl. E&W	3W-11S,0E-11S	67
101st St. E&W	3W-11S,4E-11S	67
102nd Pl. E&W	3W-11S,0E-11S	67
102nd St. E&W	3W-11S,4E-11S	67
103rd Pl. E&W	3W-12S,0E-12S	67
103rd St. E&W	3W-11S,4E-11S	67
104th Pl. E&W	4W-12S,0E-12S	67
104th St. E&W	4W-12S,4E-11S	67
105th Pl. E&W	4W-12S,0E-12S	67
105th St. E&W	4W-12S,4E-11S	67
106th Pl. E&W	3W-12S,0W-12S	67
106th St. E&W	4W-12S,4E-11S	67
107th Pl. E&W	4W-12S,0W-12S	67
107th St. E&W	4W-12S,4E-12S	67
108th Pl. E&W	4W-12S,0W-12S	67
108th St. E&W	4W-12S,4E-12S	67
109th Pl. E&W	4W-12S,0W-12S	67
109th St. E&W	4W-12S,4E-12S	67
110th Pl. E&W	4W-12S,0W-12S	67
110th St. E&W	4W-12S,4E-12S	67
111th Pl. E&W	2W-13S,0E-13S	75
111th St. E&W	4W-12S,4E-12S	67
112th Blvd. E&W	4E-12S	69
112th Pl. E&W	4W-13S,0W-13S	75
112th St. E&W	3W-13S,2E-12S	75
113th Pl. E&W	2W-13S,0W-13S	75
113th St. E&W	4W-13S,4E-12S	75
114th Pl. E&W	4W-13S,0E-13S	75
114th St. E&W	5W-13S,4E-12S	74
115th Pl. E&W	4W-13S,0W-13S	75
115th St. E&W	4W-13S,4E-13S	75
116th Pl. E&W	4W-13S,0W-13S	75
116th St. E&W	4W-13S,4E-13S	75
117th Pl. E&W	3W-13S,0E-13S	75
117th St. E&W	4W-13S,4E-13S	75
118th Pl. E&W	3W-13S,0E-13S	75
118th St. E&W	4W-13S,4E-13S	75
119th Pl. E&W	0E-14S	76
119th St. E&W	3W-13S,2E-13S	75
120th Pl. E&W	0E-14S,2E-13S	76
120th St. E&W	1W-14S,2E-13S	76
121st Pl. E&W	0E-14S,2E-13S	76
121st St. E&W	1W-14S,2E-13S	76
122nd Pl. E&W	0E-14S	76
122nd St. E&W	1W-14S,3E-13S	76
123rd St. E&W	1W-14S,2E-14S	76
124th Pl. E&W	0W-14S	76
124th St. E&W	0W-14S,2E-14S	76
125th Pl. E&W	0W-14S	76
125th St. E&W	0W-14S,2E-14S	76
126th Pl. E&W	0W-14S	76
126th St. E&W	0W-14S,3E-14S	76
127th Pl. E&W	0W-15S	76
127th St. E&W	0W-14S,3E-14S	76
128th Pl. E&W	1W-15S,0W-15S	76
128th St. E&W	0W-15S,3E-14S	76
129th Pl. E&W	1W-15S,0W-15S	76
129th St. E&W	3E-14S	77
130th Pl. (Pvt.) E&W	0E-15S,1E-15S	76
130th St. E&W	0E-14S,3E-14S	76
131st St. (Pvt.) E&W	0E-15S,3E-15S	76
132nd Pl. (Pvt.) E&W	1E-15S	76
132nd St. (Pvt.) E&W	0E-15S,3E-15S	76
133rd Pl. E&W	0E-15S	76
133rd St. (Pvt.) E&W	0E-15S,3E-15S	76
134th Pl. E&W	0E-15S	76
134th St. E&W	0E-15S,3E-15S	76
135th Pl. E&W	0E-15S	76
135th St. E&W	0E-15S,3E-15S	76
136th Pl. E&W	0E-15S	76
136th St. E&W	0E-15S,3E-15S	76
137th St. E&W	0E-15S,2E-15S	76
138th St. E&W	0E-15S	76
138th St. E&W	3E-15S	77

Alsip

	Grid	Page
Albany Av. N&S	3W-14S	75
Alpine Dr. NW&SE	6W-14S	74
Apple Ln. N&S	5W-15S	74
Austin Av. N&S	7W-14S,7W-13S	74
Avers Av. N&S	4W-14S	75
Avon Av. NW&SE	6W-13S	74
Benck Dr. NW&SE	4W-14S	75
Blossom Dr. E&W	5W-15S	74
Carolyn Ln. N&S	5W-13S	74
Central Av. N&S	7W-14S,7W-13S	74
Central Park Av. N&S	4W-14S	75
Cicero Av. N&S	6W-14S,6W-13S	74
Eastview Dr. NW&SE	5W-13S	74
Engle Rd. NS&EW	6W-15S	74
Fey Ln. E&W	5W-13S	74
Frontage Rd. NW&SE	7W-14S	74
Hamlin Av. N&S	4W-14S,4W-13S	75
Hamlin Ct. N&S	4W-15S,4W-14S	75
Harding Av. N&S	4W-14S,4W-13S	75
Holiday Dr. N&S	6W-14S	74
Holmberg Ct. N&S	5W-14S	74
Homan Av. N&S	4W-15S,4W-14S	75
Howdy Ln. E&W	5W-13S	74
Interstate 294 NW&SE	6W-14S	74
James Dr. N&S	6W-14S	74
Jean St. E&W	5W-13S	74
Joalyce Ct. NE&SW	5W-13S	74
Joalyce Dr. NW&SE	5W-13S	74
Joan St. E&W	4W-13S	75
Jobey Ln. NW&SE	5W-13S	74
Karlov Av. N&S	5W-14S,5W-13S	74
Kathleen Ct. N&S	4W-14S	75
Kedvale Av. N&S	5W-14S,5W-13S	74
Kedzie Av. N&S	4W-15S,4W-14S	75
Keeler Av. N&S	5W-14S,5W-13S	74
Kenneth Av. N&S	5W-15S,5W-13S	74
Kenton Av. N&S	5W-13S	74
Kilbourn Av. N&S	5W-13S	74
Kildare Av. N&S	5W-14S,5W-13S	74
Knox Av. N&S	5W-13S	74
Kolin Av. N&S	5W-13S	74
Kolmar Av. N&S	5W-13S	74
Komensky Ave. N&S	5W-14S	74
Kostner Av. N&S	5W-14S	74
Lacrosse Av. N&S	6W-15S,6W-13S	74
Lake Av. NW&SE	4W-14S	75
Lamon Av. N&S	6W-14S,6W-13S	74
Laporte Av. N&S	6W-14S,6W-13S	74
Laramie Av. N&S	6W-14S,6W-14S	74
Lavergne Av. N&S	6W-14S	74
Lawler Av. N&S	6W-14S,6W-13S	74
Lawndale Av. N&S	4W-14S	75
Leamington Av. N&S	6W-14S,6W-13S	74
Leclaire Av. N&S	6W-13S	74
Lee Rd. N&S	5W-13S	74
Linecrest Dr. E&W	5W-13S	74
Lockwood Av. N&S	6W-14S	74
Lockwood Ct. E&W	6W-14S	74
Lombard Ln. N&S	5W-14S	74
Loveland Av. N&S	5W-14S	74
Loveland St. N&S	5W-14S	74
Mansfield Av. N&S	7W-14S	74
Mather Av. NW&SE	6W-14S	74
Mayfield Av. N&S	7W-13S	74
Mayfield Ct. N&S	7W-14S	74
McDaniels Ct. NE&SW	5W-15S	74
McDaniels St. N&S	5W-15S,5W-14S	74
Menard Av. N&S	7W-14S	74
Millard Av. N&S	4W-15S,4W-14S	75
Minnesota Av. E&W	3W-14S	75
Orchard Av. N&S	5W-15S,5W-14S	74
Orchard Ln. N&S	5W-15S	74
Orchard St. N&S	5W-14S	74
Park Lane Dr. E&W	5W-14S	74
Prairie Av. N&S	5W-14S	74
Prairie Dr. E&W	5W-14S	74
Pulask Rd. N&S	5W-15S,5W-13S	74
Rexford St. E&W	5W-14S	74
Ridgeway Av. N&S	4W-15S,4W-14S	75
Rosemary Ln. N&S	5W-13S	74
Rt. 50 N&S	5W-13S	74
Scott Dr. E&W	4W-14S	75
Shirley Ln. NW&SE	5W-14S	74
Spencer Ct. NE&SW	5W-15S	74

	Grid	Page
Spencer St. N&S		
	5W-15S,5W-14S	74
Springfield Av. N&S		
	4W-14S,4W-13S	75
Termunde Dr. NW&SE		
	5W-14S	74
Tri State Tollway NW&SE		
	7W-13S,5W-15S	74
Tripp Av. N&S		
	5W-14S,5W-13S	74
Trumbull Av. N&S	4W-14S	75
Walter Ln. E&W	4W-14S	75
44th Ct. N&S	4W-14S	74
45th Av. N&S	5W-14S	74
111th St. E&W	6W-12S	66
112th Pl. E&W	6W-13S	74
113th Ct. NE&SW	6W-13S	74
113th Pl. E&W	6W-13S	74
113th St. E&W	6W-13S	74
114th Pl. E&W	6W-13S	74
114th St. E&W	6W-13S	74
115th Pl. E&W	6W-13S	74
115th St. E&W		
	7W-13S,5W-13S	74
116th Pl. E&W	5W-13S	74
116th St. E&W		
	7W-13S,5W-13S	74
117th Pl. E&W	7W-13S	74
117th St. E&W		
	6W-13S,5W-13S	74
118th Pl. E&W	5W-13S	74
118th St. E&W		
	7W-13S,5W-13S	74
119th Pl. E&W	5W-14S	74
119th St. E&W		
	6W-13S,4W-13S	74
120th Pl. E&W		
	6W-14S,4W-14S	74
120th St. E&W		
	7W-14S,5W-14S	74
121st Pl. E&W		
	6W-14S,4W-14S	74
121st St. E&W		
	6W-14S,4W-14S	74
122nd St. E&W		
	6W-14S,4W-14S	74
123rd Pl. E&W		
	6W-14S,4W-14S	74
123rd St. E&W		
	6W-14S,5W-14S	74
124th Pl. E&W		
	7W-14S,4W-14S	74
124th St. E&W		
	6W-14S,5W-14S	74
125th Pl. E&W	6W-14S	74
125th St. E&W		
	7W-14S,4W-14S	74
126th Pl. E&W		
	7W-14S,6W-14S	74
126th St. E&W		
	7W-14S,3W-14S	74
127th Pl. E&W	5W-15S	74
127th St. E&W		
	7W-14S,5W-14S	74
128th Pl. E&W		
	6W-15S,4W-15S	74
128th St. E&W	5W-15S	74
129th St. E&W	5W-15S	74
131st St. E&W		
	5W-15S,4W-15S	74

Arlington Heights

	Grid	Page
Alexandria St. E&W		
	18W-16N	14
Algonquin Rd. NW&SE		
	18W-11N,17W-11N	24
Allegheny Dr. E&W	18W-16N	14
Allen Ln. N&S	18W-13N	24
Allison Ct. E&W	19W-12N	24
Amhurst Ct. NW&SE		
	16W-15N	15
Anderson Dr. E&W	18W-16N	14
Appletree Ln. N&S	17W-16N	14
Arlington Heights Rd. N&S		
Walter	18W-11N,18W-18N	24
Arthur Av. N&S	17W-13N	24
Ashbury Ct. NW&SE		
	16W-15N	15
Ashford Ln. E&W	18W-18N	14
Avon Ct. E&W	16W-15N	15
Avon Ln. E&W	16W-15N	15
Banbury Rd. E&W	17W-13N	24
Barberry Ct. E&W	16W-15N	15
Bel Aire Dr. E&W	16W-14N	25
Belmont Av. N&S		
	17W-12N,17W-15N	24
Berkley Dr. E&W	18W-17N	14
Best Dr. NW&SE		
	17W-17N,16W-17N	14
Beverly Av. N&S		
	17W-14N,17W-17N	24
Beverly Ln. E&W		
	17W-13N,17W-15N	24
Bloomington Av. NE&SW		
	19W-18N	14
Boeger Dr. N&S	17W-17N	14
Bradford Dr. E&W	17W-16N	14
Bradford Dr. N&S	18W-17N	14
Bradshire Ct. N&S	16W-15N	15
Braeside Dr. E&W	18W-17N	14
Brampton Ln. E&W		
	18W-18N	14
Brandenberry Ct. NW&SE		
	16W-15N	15
Bray Ct. E&W	19W-12N	24
Brett Ct. E&W	17W-12N	24

	Grid	Page
Brighton Pl. N&S		
	17W-13N,17W-17N	24
Bristol Ln. N&S	17W-13N	24
Brittany Ct. E&W	17W-17N	14
Brittany Dr. E&W		
	18W-17N,17W-17N	14
Brockton Ct. N&S	16W-15N	15
Brook Dr. E&W	17W-11N	24
Brookwood Ct. N&S		
	17W-16N	14
Brookwood Dr. E&W		
	17W-16N	14
Brown St. E&W	18W-14N	24
Buffalo Grove Rd. N&S		
	16W-17N	15
Burgoyne Rd. N&S	18W-15N	14
Burke N&S	17W-16N	14
Burkitt NE&SW	18W-18N	14
Burning Tree Ct. E&W		
	16W-17N	15
Burning Tree Ln. E&W		
	18W-17N,17W-17N	14
Burr Oak Dr. E&W		
	18W-17N,17W-17N	14
Burton Pl. N&S	17W-13N	24
Butternut Ln. E&W	18W-14N	24
Cambridge St. E&W		
	17W-16N	14
Camelot Ct. E&W	17W-16N	14
Campbell St. E&W		
	18W-14N,16W-14N	24
Campus Dr. N&S		
	18W-17N	14
Canterbury Ct. E&W		
	18W-16N	14
Canterbury Dr. E&W		
	17W-16N,16W-16N	14
Carlyle NW&SE	17W-13N	24
Carlyle Ct. NS&EW		
	17W-15N	14
Carlyle Ln. N&S	17W-14N	24
Carlyle Pl. N&S	17W-16N	14
Carriage Way Dr. N&S		
	17W-17N	14
Catino St. E&W	18W-13N	24
Cedar Ct. E&W	18W-12N	24
Cedar Glen Dr. N&S		
	18W-11N	24
Cedar Ln. E&W	18W-12N	24
Cedar St. E&W	18W-12N	24
Central Rd. E&W	17W-13N	24
Champlain St. NS&EW		
	18W-16N	14
Chapel Hill Dr. N&S		
	17W-16N	14
Charles NW&SE	17W-15N	14
Charter Point Dr. NW&SE		
	16W-16N	15
Charter Point Dr. (Pvt) NE&SW		
	16W-16N	15
Chatham Pl. NE&SW		
	17W-13N	24
Cherry Hill Dr. N&S		
	16W-16N	15
Cherry Ln. E&W	17W-16N	14
Chesterfield Dr. E&W		
	18W-12N	24
Chestnut Av. N&S		
	18W-12N,18W-16N	24
Chicago Av. N&S		
	18W-14N,18W-16N	24
Circle Hill Dr. NS&EW		
	17W-16N	14
Clarence Av. N&S	17W-15N	14
Clarendon Av. NW&SE		
	17W-15N	14
Clarendon Ct. NE&SW		
	17W-15N	14
Clarendon Rd. E&W		
	18W-15N	14
Clarendon St. E&W		
	18W-15N,17W-15N	14
Clayton Ct. N&S	18W-15N	15
Clearbrook Dr. NS&EW		
	17W-11N	24
Cleveland Av. N&S	17W-13N	24
Coldspring Rd. (Pvt.) NE&SW		
	16W-16N	15
College Dr. NS&EW		
	17W-17N	14
Concord Dr. NS&EW		
	18W-16N	14
Cottonwood Ln. E&W		
	18W-14N	24
Council Trail E&W	17W-12N	24
Country Lake Rd. E&W		
	16W-16N	15
Country Ln. E&W	17W-17N	14
Courtland Dr.(Pvt.) N&S		
	16W-15N	15
Crabtree Dr. NE&SW		
	17W-16N,16W-16N	14
Crestwood Dr. E&W		
	17W-16N	14
Cyndi Ct. E&W		
	17W-12N,18W-12N	24
Cypress Dr. E&W	18W-12N	24
Cypress St. E&W	18W-12N	24
Dale Av. N&S		
	16W-13N,16W-16N	25
Daniels St. N&S	17W-7N	14
Davis St. E&W		
	16W-13N,18W-14N	25
Derbyshire Av. N&S		
	17W-14N	24
Derbyshire Dr. N&S		
	17W-16N	14

	Grid	Page
Derbyshire Ln. NW&SE		
	16W-13N,17W-16N	25
Donald Av. N&S		
	16W-13N,16W-14N	25
Dorothy Av. NE&SW		
	17W-15N	14
Douglas N&S	17W-17N	14
Douglas Av. N&S	17W-16N	14
Douglas Av. N&S		
	5W-14N,17W-15N	24
Douglas Rd. N&S		
	17W-16N,17W-17N	14
Dover Ct. N&S	16W-15N	15
Dover Ln. N&S	16W-15N	15
Drury Ln. NW&SE		
	16W-13N,16W-16N	25
Dryden Av. N&S		
	17W-14N,17W-15N	24
Dryden Pl. N&S		
	17W-13N,17W-17N	24
Dundee Rd. E&W	18W-18N	14
Dunhill Ct. N&S	17W-16N	14
Dunton Av. N&S		
	17W-12N,18W-17N	24
Dwyer Av. N&S		
	18W-13N,18W-14N	24
Eastman Ct. E&W	16W-14N	25
Eastman St. E&W		
	18W-14N,16W-14N	25
Eastwood Dr. N&S	17W-16N	14
Elizabeth Dr. NE&SW		
	18W-16N	14
Elk Grove Rd. N&S		
	18W-11N	24
Elm Ln. N&S	17W-17N	14
Elm St. E&W	18W-14N	24
Elmwood Cir. E&W		
	16W-16N	15
Embers Ln. N&S	17W-11N	24
Emerson St. E&W		
	18W-12N,17W-12N	24
Essex Pl. NE&SW	18W-16N	14
Eton Dr. E&W	17W-17N	14
Euclid Av. E&W		
	16W-14N,18W-12N	25
Evanston Av. N&S		
	16W-13N,16W-14N	25
Evergreen Av. N&S		
	18W-12N,18W-16N	24
Evergreen Ter. NS&EW		
	16W-16N	15
Fairview St. N&S		
	18W-14N,16W-14N	25
Falcon Dr. E&W	17W-11N	24
Fernandez Av. N&S		
	18W-12N,18W-16N	24
Fernandez Ct. E&W		
	18W-12N	24
Fernandez Pl. 18W-15N	14	
Fleming Dr. NS&EW		
	16W-17N	15
Flentie Ln. E&W	16W-17N	15
Flower Cir. NW&SE		
	16W-16N	15
Forest Av. N&S		
	16W-13N,16W-15N	25
Forrest Ln. N&S	16W-16N	15
Foster St. E&W	17W-12N	24
Foxdale Ln. NE&SW		
	18W-18N	14
Francis Dr. E&W	18W-13N	24
Frederick St. E&W		
	17W-14N,16W-14N	24
Fremont Ct. E&W	16W-14N	25
Fremont St. E&W		
	16W-14N,18W-14N	25
Frontenac Dr. E&W		
	18W-16N	14
Galena Dr. E&W	18W-18N	14
Galesburg Ct. N&S	19W-18N	14
Garden Walk E&W	16W-14N	25
George St. E&W		
	17W-13N,18W-13N	24
Gettysburg Dr. NS&EW		
	18W-16N	14
Gibbons Av. N&S		
	16W-13N,16W-14N	25
Goebbert Rd. N&S		
	17W-11N,17W-12N	24
Golf Rd. E&W	18W-12N	24
Golf Ter. E&W	17W-12N	24
Green Briar Ct. NS&EW		
	18W-16N	14
Greenwood Ct. N&S		
	18W-16N	14
Gregory St. N&S	16W-13N	25
Gresham Ct. NE&SW		
	16W-15N	15
Grove St. E&W		
	16W-13N,18W-13N	25
Hackberry Ln. E&W		
	18W-17N,17W-17N	14
Haddow Av. N&S		
	17W-12N,17W-16N	24
Happfield Dr. E&W	18W-11N	24
Harold St. E&W	18W-16N	14
Hartford Ct. N&S	19W-15N	14
Harvard Av. N&S		
	18W-12N,18W-18N	24
Haven Dr. E&W		
	17W-12N,18W-12N	24
Haverhill Ct. NE&SW		
	16W-15N	15
Hawthorne St. E&W		
	18W-14N,16W-14N	24
Heather Ln. E&W	18W-14N	24
Henry St. E&W	17W-17N	14

	Grid	Page
Hickory Av. N&S		
	17W-14N,17W-17N	24
Hickory Ln. N&S	17W-16N	14
Highland Av. N&S	18W-12N	24
Hintz Rd. E&W	17W-16N	14
Hunter Dr. E&W	16W-16N	15
Hunter Dr. NE&SW		
	16W-16N	15
Huntington Dr. E&W		
	17W-16N	14
Huron St. NE&SW	18W-16N	14
Illinois Av. N&S		
	18W-14N,18W-16N	24
Illinois Dr. NS&EW	18W-13N	24
Independence Ct. E&W		
	17W-11N	24
Interstate 90 NW&SE		
	18W-11N	24
Ironwood Dr. N&S	16W-14N	25
Ivy Ln. E&W	17W-16N	14
Ivy Ln. (Pvt.) E&W	18W-15N	14
James Ct. NW&SE	18W-16N	14
Jane Av. E&W	17W-15N	14
Jennifer Ct. E&W		
	17W-15N	14
Jo Ln. N&S	18W-15N	14
Johanna Ter. NE&SW		
	18W-14N	24
Jonquil Cir. E&W	16W-16N	15
Jonquil Ter. E&W	16W-16N	15
Jules E&W	17W-15N	14
Juniper Ln. NS&EW		
	18W-16N	14
Kaspar Av. N&S		
	18W-12N,18W-15N	24
Keating Ct. E&W	18W-13N	24
Kendal Ct. NW&SE		
	16W-15N	15
Kennicott Av. N&S		
	18W-13N,18W-18N	24
Kennicott Dr. N&S	18W-16N	14
Kensington (Foundry) E&W		
	17W-16N,16W-14N	24
Kerrybrook Ln. E&W		
	17W-17N	14
Kimber Ln. E&W	17W-11N	24
Kingsbury Ct. N&S	18W-16N	14
Kingsbury Dr. N&S		
	18W-16N	14
Kingsley Dr. E&W	18W-16N	14
Kinley Av. N&S	17W-13N	24
Kirchoff Rd. NW&SE		
	17W-13N,18W-13N	24
Knob Hill Dr. E&W	17W-16N	14
LaSalle St. E&W	18W-16N	14
Ladd St. E&W	18W-16N	14
Lafayette St. N&S	18W-16N	14
Lake Arlington Dr. E&W		
	16W-16N	15
Lake Cook Rd. E&W		
	17W-18N	14
Lake Shore Cir. NS&EW		
	16W-16N	15
Lancaster Ct. N&S	16W-15N	15
Leona Ter. E&W	18W-13N	24
Lexington Dr. NW&SE		
	16W-13N,16W-15N	25
Lilac Ter. E&W	16W-16N	15
Lillian Av. N&S		
	18W-15N,16W-15N	14
Lillian Ln. N&S	16W-15N	15
Lincoln Ct. N&S	17W-15N	14
Lincoln Ln. N&S		
	17W-13N,17W-15N	24
Lincoln Rd. E&W	17W-12N	24
Luther Ln. (Pvt.) N&S		
	18W-15N	14
Lynden Ln. NW&SE		
	17W-13N	24
Lynnwood Av. E&W		
	18W-15N,17W-15N	14
Magnolia St. E&W		
	18W-12N,17W-12N	24
Mallard Ln. N&S	18W-18N	14
Maple Ln. N&S	16W-16N	15
Maple St. E&W		
	18W-13N,17W-13N	24
Margaret Av. NE&SW		
	17W-15N	14
Marion Rd. E&W	18W-15N	14
Marion St. E&W		
	18W-15N,17W-15N	14
Marshall St. E&W	17W-14N	24
Martin Ln. N&S	18W-11N	24
Maude Av. E&W		
	18W-15N,17W-15N	14
Mayfair Rd. N&S		
	17W-13N,16W-13N	24
McKinley Av. N&S	17W-13N	24
Meade St. E&W	18W-18N	14
Michael Manor Ln. E&W		
	16W-15N	15
Milida E&W	17W-16N	14
Millbrook Ln. N&S	18W-12N	24
Miner St. E&W		
	18W-14N,16W-14N	24
Mitchell Av. N&S		
	18W-12N,18W-16N	24
Mitchell Ct. N&S	18W-15N	14
Mitchell St. N&S	18W-16N	14
Moorings Dr. (Pvt.) NS&EW		
	17W-12N	24
Mueller St. E&W	18W-14N	24
Mulberry Ct. N&S	18W-14N	24
Mulberry St. N&S	18W-14N	24
Newbury Pl. N&S	17W-13N	24
Newport Ct. N&S	18W-18N	14
Newport Ln. E&W	18W-18N	14

	Grid	Page
Newport Way (Pvt.) NW&SE		
	19W-18N	14
Nichols E&W	18W-18N	14
Normandy Ct. N&S		
	16W-15N	15
Northwest Hwy NW&SE		
	18W-15N,16W-13N	14
Northwest Tollway NW&SE		
	18W-11N,17W-10N	24
Northwood Ln. E&W		
	16W-15N	15
Noyes St. E&W		
	19W-12N,17W-12N	24
Oakton St. E&W	18W-15N,16W-15N	14
Oakwood Dr. N&S	16W-16N	15
Old Barn Ln. (Pvt.) E&W		
	17W-12N	24
Olive St. E&W		
	18W-15N,16W-15N	14
Orchard Pl. E&W	18W-13N	24
Orchard St. E&W		
	18W-13N,17W-13N	24
Orchid Ln. N&S	16W-16N	15
Oxford Ct. E&W	17W-16N	14
Page N&S	17W-17N	14
Palatine Rd. E&W		
	18W-16N,16W-16N	14
Par Ct. N&S	18W-18N	14
Park Pl. NW&SE	18W-14N	24
Park St. E&W		
	18W-13N,18W-16N	24
Parkview Ct. E&W	17W-12N	24
Partridge Ct. N&S	18W-18N	14
Partridge Ln. NE&SW		
	18W-18N	14
Patton Av. N&S		
	18W-13N,18W-17N	24
Patton Cir. N&S	18W-13N	24
Payton Run NW&SE		
	18W-14N	24
Peachtree Dr. E&W		
	16W-16N	15
Peachtree Ln. E&W		
	16W-16N	15
Peartree Ln. N&S	16W-14N	25
Pennsylvania Ct. E&W		
	17W-11N	24
Pheasant Tr. Ct. N&S		
	18W-18N	14
Pheasant Trail Dr. NE&SW		
	18W-18N	14
Pheasant Trail Ln. N&S		
	18W-18N	14
Phelps Av. N&S		
	16W-13N,16W-16N	25
Phelps Ct. E&W	16W-16N	15
Pickwick Rd. E&W	18W-12N	24
Pine Av. N&S		
	17W-12N,17W-16N	14
Pinetree Ln. N&S	17W-16N	14
Plum Grove Cir. E&W		
	18W-16N	14
Plymouth Ct. E&W	18W-16N	14
Plymouth Dr. E&W	18W-16N	14
Portsmith Ln. NS&EW		
	18W-16N	14
Princeton Av. N&S		
	18W-12N,18W-15N	24
Princeton Ct. N&S	18W-14N	24
Prindle Av. N&S		
	16W-13N,16W-16N	25
Private Rd. NW&SE		
	18W-16N	14
Proctor Cir. NW&SE		
	18W-18N	14
Pvt. Rd. E&W	18W-16N	14
Quail Ct. E&W	18W-18N	14
Race Av. N&S		
	18W-14N,18W-15N	24
Radcliff Rd. N&S	17W-17N	14
Radford Ct. E&W	16W-15N	15
Raleigh Ct. NE&SW		
	16W-14N	25
Raleigh St. NW&SE		
	18W-16N	14
Rammer Av. N&S		
	16W-13N,16W-14N	25
Rand Rd. NE&SW		
	18W-16N,17W-15N	14
Redwood Ln. E&W		
	17W-16N	14
Regency Ct., East N&S		
	16W-14N	25
Regency Ct., West N&S		
	16W-14N	25
Regency Dr., East N&S		
	16W-14N	25
Regency Dr., West N&S		
	16W-14N	25
Reuter Dr. N&S		
	18W-13N,18W-14N	24
Richmond St. E&W		
	18W-16N	14
Ridge Av. N&S		
	18W-12N,18W-18N	24
Ridge Ct. N&S	18W-15N	14
Ridge Dr. NS&EW	18W-17N	14
Ridge Rd. E&W	17W-16N	14
Roanoke Dr. NW&SE		
	18W-16N	14
Robinhood Ln. NW&SE		
	16W-15N	15
Rockwell St. E&W		
	18W-13N,16W-13N	24
Rolling Ln. N&S	16W-15N	15
Roosevelt Av. N&S		
	17W-13N	24

	Grid	Page
Rosehill Dr. E&W	16W-16N	15
Rt. 53 N&S	18W-15N,18W-17N	14
Rt. 58 E&W	18W-12N,17W-12N	24
Rt. 62 NW&SE	18W-11N,17W-11N	24
Rt. 68 E&W	18W-17N	14
Russell Ct. E&W	19W-12N	24
Russetwood Dr. NW&SE	16W-15N	15
Ryan Ct. E&W	17W-12N	24
Salem Av. N&S	18W-13N,18W-14N	24
Salem Blvd. NW&SE	18W-15N	14
Salem Dr. N&S	18W-18N	14
Salem Ln. N&S	18W-12N	24
Schaefer N&S	18W-18N	14
Schoenbeck Rd. N&S	16W-16N	15
Scottsvale Ln. N&S	16W-14N	25
Seeger Rd. E&W	18W-11N,17W-11N	24
Shady Way E&W	17W-11N	24
Shag Bark N&S	17W-11N	24
Shenandoah Dr. N&S	18W-16N	14
Sherwood Rd. NW&SE	16W-15N	15
Shiloh Dr. E&W	18W-16N	14
Shirra Ct. E&W	18W-16N	14
Shure Dr. E&W	18W-17N	14
Sigwalt St. E&W	18W-14N,17W-14N	24
Somerset Ct. N&S	17W-18N	14
Somerset Ln. N&S	18W-14N	24
South St. E&W	18W-13N	24
Spring Ridge Dr. NS&EW	19W-18N	14
Spruce Tr. N&S	17W-16N	14
St. James Pl. N&S	18W-14N	24
St. James St. E&W	18W-14N,16W-14N	24
Stanford Ln. N&S	17W-17N	14
Stanton Ct. N&S	17W-16N	14
Stillwater Rd. NE&SW	16W-16N	15
Stratford Rd. N&S	16W-13N,16W-17N	25
Stuart Dr. N&S	17W-16N,16W-16N	14
Suffield Ct. E&W	18W-16N	14
Suffield Dr. E&W	17W-16N,16W-16N	14
Sunset Dr. E&W	17W-16N	14
Sunset Ter. E&W	18W-14N	24
Sunset Tr. E&W	16W-14N	25
Surrey Park Ln. NW&SE	19W-12N	24
Surrey Ridge Dr. NE&SW	18W-12N	24
Talbot St. E&W	17W-15N	14
Tanglewood Dr. NS&EW	17W-17N,17W-18N	14
Techny Rd. E&W	17W-16N	14
Techny Rd. E&W	18W-16N,17W-16N	14
Terramere Av. NS&EW	18W-18N	14
Thomas Av. E&W	18W-15N,17W-15N	14
Thorntree Ter. NE&SW	17W-16N	14
Three States Blvd. N&S	17W-17N	14
Thurston Pl. NW&SE	18W-18N	14
Tonne Rd. N&S	17W-11N,17W-12N	24
Towne Blvd. NW&SE	16W-16N	15
U.S. Route 12 NW&SE	18W-17N,16W-15N	14
U.S. Route 14 NW&SE	18W-14N,16W-13N	24
University Dr. E&W	18W-17N,17W-17N	14
Vail Av. N&S	18W-12N,18W-15N	24
Valley Ln. E&W	17W-16N	14
Vargo Ln. E&W	17W-16N	14
Ventura Dr. NS&EW	18W-18N	14
Verde Av. N&S	18W-16N	14
Verde Dr. N&S	18W-16N	14
Viator Ct. NS&EW	17W-15N	14
Victoria Ln. E&W	18W-12N	24
Villa Verde Dr. E&W	18W-17N	14
Village Dr. (Pvt.) N&S	18W-15N	14
Vine St. E&W	18W-14N	24
Vista Rd. N&S	17W-17N	14
Voltz Dr. NS&EW	17W-17N	14
W. Whiting Ln. E&W	18W-18N	14
Walden Ln. E&W	16W-16N	15
Walker Ln. NS&EW	17W-17N	14
Walnut NS&EW	18W-12N,18W-18N	24
Walnut Ct. NE&SW	18W-16N	14
Warwick Ct. E&W	16W-15N	15
Waterman Av. N&S	16W-13N,16W-16N	25
Watling Rd. NE&SW	18W-15N	14
Waverly Ct. NS&EW	16W-16N,18W-16N	15
Waverly Dr. NW&SE	17W-16N,16W-16N	14
Waverly Pl. E&W	17W-16N	14
Waverly Rd. E&W	17W-16N	14
Weston Dr. E&W	18W-18N	14
Wheeler St. E&W	17W-16N	14
White Oak St. E&W	18W-12N,17W-12N	24
Whitehall Dr. E&W	18W-18N	14
Wilke Rd. N&S	19W-18N,18W-18N	14
Wilke Rd. N&S	19W-12N,19W-18N	24
Williamsburg St. N&S	18W-16N	14
Willow St. E&W	18W-14N,17W-14N	24
Wilshire Ln. NS&EW	16W-13N,16W-17N	25
Windham Ct. N&S	17W-16N	14
Windsor Dr. NS&EW	16W-13N,17W-17N	25
Wing St. E&W	18W-14N,16W-14N	24
Woodbury Dr. E&W	16W-16N	15
Woodford Pl. E&W	17W-13N	24
Woodland Ln. N&S	16W-16N	15
Woodridge Ln. N&S	17W-14N	24
Woods Dr. NS&EW	18W-17N	14
Yale Av. N&S	18W-13N,18W-18N	24
Yale Av. N&S	19W-12N	24
Yale Ct. E&W	18W-15N	14
Yarmouth Ct. NE&SW	16W-15N	15

Barrington*
(Also see Lake County)

	Grid	Page
Balmoral Ln. E&W	24W-18N	12
Barrington Point E&W	25W-18N	12
Barrington Rd. N&S	25W-17N,25W-18N	12
Bellingham Dr. E&W	24W-18N	12
Braeside Pl. E&W	25W-18N	12
Bristol Dr. NW&SE	25W-18N	12
Concord Ln. NW&SE	25W-18N	12
Concord Pl. N&S	25W-18N	12
Cook St. N&S	25W-18N,25W-20N	12
Coolidge Av. E&W	25W-18N	12
Country Dr. N&S	26W-18N	11
Division St. N&S	25W-18N	12
Dundee Av. N&S	26W-18N	11
Dundee Rd. E&W	25W-17N	12
Eastern Av. N&S	25W-18N	12
Ela Rd. N&S	24W-18N,24W-20N	12
Fairfield Dr. NS&EW	24W-18N	12
Forest Av. N&S	26W-18N	11
George Av. N&S	25W-18N	12
Glendale Av. N&S	25W-18N	12
Grove Av. N&S	25W-18N	12
Hager Av. N&S	26W-18N,26W-20N	11
Harriet Ln. E&W	25W-18N	12
Heath Ct. E&W	24W-18N	12
Highland Av. N&S	25W-18N	12
Hill St. E&W	25W-18N	12
Hillcrest Av. E&W	25W-18N	12
Hillside Av. E&W	26W-18N,25W-18N	11
Hillside Ct. N&S	24W-18N	12
Hough St. N&S	25W-18N	12
Illinois St. E&W	25W-18N	12
Kainer Av. N&S	25W-18N	12
Kings Row N&S	24W-18N	12
Lageschulte St. N&S	26W-18N	11
Lake St. E&W	26W-18N,25W-18N	11
Lakewood Ct. E&W	24W-18N	12
Lakewood Dr. NS&EW	24W-18N	12
Lill St. N&S	25W-18N	12
Lincoln Av. E&W	25W-18N	12
Meadow Ln. N&S	25W-18N	12
Monument Av. E&W	25W-18N	12
Newport Ln. E&W	25W-17N	12
Northwest Hwy. NW&SE	24W-18N	12
Oak Ridge Cir. NS&EW	24W-18N	12
Oak Ridge Rd. E&W	24W-18N	12
Oakland Ct. NW&SE	24W-18N	12
Oakland Dr. NS&EW	24W-18N	12
Old Mill Ct. E&W	24W-18N	12
Old Mill Rd. N&S	24W-18N	12
Orchard Dr. N&S	26W-18N	11
Park Av. NW&SE	25W-18N	12
Park Barrington Dr. NS&EW	25W-17N	12
Park Barrington Way NS&EW	25W-17N	12
Park Rd. E&W	23W-18N	12
Prairie Av. N&S	25W-18N	12
Private Rd. (Pvt.) NS&EW	25W-17N	12
Queens Cove E&W	25W-18N	12
Red Barn Ln. NS&EW	24W-18N	12
Rt. 59 N&S	25W-18N,25W-20N	12
Rt. 68 E&W	25W-17N	12
Russell St. E&W	25W-18N	12
S. Grove Av. NW&SE	25W-17N	12
Shady Ln. E&W	26W-18N	11
Skyline Dr. NE&SW	25W-18N	12
South St. E&W	25W-18N	12
South Valley Rd. NW&SE	24W-18N	12
Spring St. N&S	25W-18N	12
Station St. E&W	26W-18N,25W-18N	11
Sturtz St. E&W	25W-18N	12
Summit St. N&S	25W-18N	12
Sunset Rd. NE&SW	26W-18N	11
Tower Rd. E&W	26W-18N	11
Tudor Dr. N&S	24W-18N	12
Walton St. N&S	25W-18N	12
Warwick Av. E&W	25W-18N	12
Westwood Dr. E&W	26W-18N	11
Wisconsin Av. E&W	25W-18N	12
Wool St. N&S	25W-18N	12
Wyngate Dr. NS&EW	24W-18N	12

Barrington Hills*
(Also see Kane, Lake and McHenry Counties)

	Grid	Page
Aberdeen Dr. E&W	28W-18N	11
Algonquin Rd. NW&SE	28W-16N	11
Barrington Hills Rd. N&S	28W-18N	11
Bartlett Rd. NE&SW	27W-16N,26W-17N	11
Bateman Cir. NS&EW	30W-18N	10
Bateman Circle Rd. NS&EW	30W-18N	10
Bateman Rd. N&S	30W-16N,30W-18N	10
Brinker Rd. N&S	28W-16N,27W-18N	11
Butternut Rd. NS&EW	28W-18N	11
Caesar Dr. N&S	27W-18N	11
Conroy Ct. E&W	28W-15N	21
Crabapple Dr. N&S	28W-18N	11
Crawling Stone NE&SW	30W-18N	10
Creekside Ln. NS&EW	28W-15N	21
Dales Rd. NS&EW	27W-17N	11
Dana Ln. E&W	27W-18N	11
Deepwood Ct. E&W	30W-18N	10
Deepwood Rd. NW&SE	30W-18N	10
Donlea Rd. NE&SW	28W-18N	11
Dundee Ln. N&S	26W-17N,27W-16N	10
Dundee Rd. NE&SW	30W-15N,26W-17N	20
Eagle Point Dr. NS&EW	30W-15N	20
East Ln. NE&SW	27W-17N	11
Far Hills Rd. NS&EW	30W-18N	10
Fernwood Dr. E&W	30W-16N	10
Fernwood Ln. E&W	30W-16N	10
Goose Lake Dr. E&W	28W-17N,27W-17N	11
Hart Hills Rd. E&W	26W-18N	11
Harts Rd. N&S	27W-18N	11
Hawley Rd. NE&SW	27W-17N,26W-17N	11
Hawthorne NW&SE	26W-17N	11
Hawthorne Ln. NW&SE	27W-17N	11
Hawthorne Rd. NE&SW	26W-17N	11
Healy Rd. NE&SW	30W-15N,29W-15N	20
Helm Rd. E&W	30W-17N	10
Heron Ln. N&S	27W-17N	11
Hills Rd. NE&SW	27W-17N	11
Honey Cut Rd. E&W	27W-17N	11
King Rd. N&S	30W-15N	20
Lakeview Ln. NS&EW	26W-17N	11
Leeds Dr. E&W	28W-18N	11
Long Meadow Ct. NE&SW	30W-16N	10
Long Meadow Dr. E&W	30W-16N	10
Marmon Ln. E&W	26W-18N	11
Middlebury E&W	30W-18N	10
Oakdene Rd. NS&EW	26W-18N	11
Otis Rd. E&W	28W-18N,26W-18N	11
Overlook Rd. NS&EW	30W-18N	10
Palatine Rd. E&W	26W-16N	11
Penny Rd. E&W	30W-15N	20
Potter Ln. NW&SE	30W-15N	20
Private Rd. E&W	28W-18N,27W-18N	11
Private Rd. N&S	28W-16N	11
Rebecca Dr. E&W	28W-15N	21
Rock River Rd. N&S	29W-21N	56
Rolling Hills Dr. NS&EW	30W-16N	10
Round Barn Rd. E&W	27W-17N	11
Rt. 59 NE&SW	28W-15N,26W-17N	21
Rt. 62 NW&SE	30W-18N,27W-16N	10
Rt. 68 E&W	30W-15N,28W-16N	11
Sandlewood Dr. NS&EW	26W-18N	11
Sarah Ln. NE&SW	25W-17N	12
Springwood Ln. N&S	29W-16N	10
Sutton Rd. NW&SE	29W-18N	10
Sutton Rd. NW&SE	28W-13N,29W-18N	21
Tamarack Ln. NS&EW	28W-15N	21
Three Lakes Rd. N&S	27W-18N	11
Tricia Ln. E&W	28W-15N	21
Valley Dr. N&S	28W-18N	11
West Ln. E&W	27W-17N	11
Wichman Rd. E&W	28W-14N	21
Wichman Rd. NS&EW	29W-14N	20
Windrush Ln. NE&SW	26W-18N	11
Wood Creek Rd. N&S	29W-15N	20
Wood Rock Rd. NE&SW	30W-18N	10
Woodcreek Rd. N&S	29W-15N	20

Barrington Township

	Grid	Page
Algonquin Rd. NW&SE	26W-15N	21
Bartlett Rd. N&S	28W-13N	21
Boland Dr. E&W	27W-14N	21
Bradwell Cir. N&S	25W-16N	12
Cook St. N&S	25W-17N,25W-18N	12
Division St. N&S	25W-17N,25W-18N	12
George Av. N&S	25W-17N,25W-18N	12
Grove Av. N&S	27W-17N,25W-18N	11
Harvard Av. E&W	25W-18N	12
Old Higgins Rd. NW&SE	30W-14N	20
Palatine Rd. E&W	26W-16N	11
Prairie Av. N&S	25W-17N,25W-18N	12
Princeton Av. E&W	25W-17N	12
Stover Rd. N&S	27W-15N	21
Summit St. N&S	25W-17N,25W-18N	12
Yale Av. E&W	25W-18N	12

Bartlett*
(Also see Du Page County)

	Grid	Page
Acorn Ct. E&W	28W-8N	31
Ann Ct. E&W	28W-8N	31
Aspen Ct. N&S	29W-9N	30
Bartlett Av. E&W	29W-8N,28W-8N	30
Bay Tree Dr. E&W	29W-9N	30
Bayberry Dr. E&W	29W-8N	30
Baytree Dr. NE&SW	29W-9N	30
Berteau Av. N&S	28W-8N	31
Betty Ct. N&S	28W-8N	31
Bryce Ct. E&W	29W-8N	30
Burton Dr. N&S	32W-8N	30
Butler Dr. NW&SE	32W-8N	30
Candleridge Ct. E&W	28W-8N	31
Capalino N&S	31W-8N	30
Carroll Way* E&W	31W-8N	30
Cecil Ct. NW&SE	29W-8N	30
Cedarfield Dr. N&S	32W-8N	30
Chase Av. N&S	28W-8N	31
Chestnut Ct. E&W	29W-9N	30
Church Ct. E&W	29W-8N	30
Cobblewood Ln. E&W	28W-8N	31
Crab Tree Ln. NS&EW	29W-9N	30
Crescent Ct. E&W	28W-8N	31
Crescent Ct. NW&SE	28W-8N	31
Crest Av. N&S	28W-8N	31
Crystals Ln. N&S	30W-8N	30
Daniel Ct. N&S	28W-8N	31
Dave Dr. N&S	31W-8N	30
David Ct. E&W	28W-8N	31
Deere Park Cir. NS&EW	29W-9N	30
Devon Ave. E&W	29W-8N,28W-8N	30
Dogleg Ln. NE&SW	31W-9N	30
Donna Ct. E&W	28W-8N	31
Doral Dr. NE&SW	29W-8N,29W-9N	30
Eastern Av. N&S	28W-8N	31
Elizabeth Ct. N&S	28W-8N	31
Elmwood Ln. E&W	30W-8N	30
Elroy Av. N&S	28W-8N	31
Emil Ct. N&S	28W-8N	31
Ford Ln. E&W	29W-8N	30
Golfer's Ln. NE&SW	31W-9N	30
Golfview Dr. E&W	31W-9N,30W-9N	30
Greenfield Ct. E&W	28W-8N	31
Hackberry Ct. E&W	29W-9N	30
Hale Av. N&S	28W-8N	31
Hazelnut Ct. N&S	29W-9N	30
Helen Dr. N&S	31W-8N	30
Hickory Av. N&S	29W-8N	30
Hillcrest Ln. E&W	28W-8N	31
Holly Dr. E&W	29W-8N	30
Honey Locust Ct. E&W	29W-9N	30
Ingalton Av. N&S	30W-8N	30
James Dr. N&S	31W-8N	30
Jessica Ln. E&W	29W-8N	30
Joan Ct. E&W	28W-8N	31
Jodi Ln. N&S	30W-8N	30
Jones Dr. N&S	31W-8N	30
Joseph Dr. N&S	31W-8N	30
Judith Ct. N&S	28W-8N	31
Kathy Ln. E&W	29W-8N	30
Knollwood Ln. E&W	30W-8N	30
La Costa Av. E&W	29W-8N	30
Lake St. E&W	29W-9N,28W-8N	30
Laurie Ln. N&S	31W-8N	30
Lela Ln. N&S	31W-8N	30
Little John Ct. N&S	29W-8N	30
Lucille Ct. E&W	28W-8N	31
Mable Ln. N&S	31W-8N	30
Main St. N&S	29W-8N	30
Mallard Ct. NE&SW	28W-8N	31
Maplewood Ln. E&W	30W-8N	30
Marcia Ct. E&W	28W-8N	31
Marion Av. N&S	28W-8N	31
Mary Ct. N&S	28W-8N	31
Monarch Birch Ct. E&W	30W-9N	30
Monarch Birch Ln. E&W	30W-9N	30
Morse Av. E&W	29W-8N,28W-8N	30
Mulberry Ct. E&W	28W-8N	31
Mulberry Ct. NE&SW	28W-8N	31
Newport Ln. N&S	28W-8N	31
Nina Ln. NS&EW	31W-8N	30
North Av. E&W	29W-8N,28W-8N	30
Oak Av. N&S	29W-8N	30
Oakbrook Ct. N&S	29W-8N	30
Oakmont Dr. N&S	29W-8N	30
Oakwood Dr. N&S	29W-8N	30
Olive Pkwy. E&W	29W-9N	30
Oliver Valley N&S	29W-8N	30
Oneida Av. E&W	29W-8N,28W-8N	30
Park Pl. E&W	30W-8N	30
Patricia Ln. N&S	29W-8N	30
Pebble Beach Ln. N&S	29W-8N,29W-9N	30
Persimmon Ct. E&W	30W-9N	30
Persimmon Ln. E&W	30W-9N	30
Peter Ct. E&W	28W-8N	31
Philip Dr. NE&SW	30W-9N	30
Pinoak Dr. N&S	29W-8N	30
Preserve Tr. E&W	30W-8N	30
Prospect Av. N&S	28W-7N,28W-8N	31
Queens Pkwy. E&W	29W-8N	30
Railroad Av. E&W	29W-8N	30
Ranier St. NE&SW	32W-8N	30
Red Bud Ct. E&W	29W-9N	30
Red Oak Ct. E&W	29W-9N	30
Red Oak Dr. NW&SE	29W-9N	30
Rich Dr. N&S	31W-8N	30
Rich's Ln. N&S	31W-8N	30
Rita Ct. N&S	28W-8N	31
Robert Ct. N&S	28W-8N	31
Rose Ln. N&S	31W-9N	30
Rose Ln. NW&SE	31W-9N	30
Roslara Ct. N&S	29W-8N	30
Rt. 59 N&S	30W-8N	30
Snow Drift Cir. E&W	30W-9N	30
Snow Drift Ct.* E&W	30W-9N	30
Southfield Dr.* E&W	31W-8N	30
Spaulding Av. E&W	31W-9N,30W-9N	30
Spaulding Rd. E&W	30W-9N	30

	Grid	Page
Spitzer Rd. N&S	31W-8N	30
Stephani N&S	28W-8N	31
Sterling Ct. E&W	28W-8N	31
Taige Av. N&S	28W-8N	31
Tanoak Ct. N&S	29W-8N	30
Taylor Av. E&W	28W-8N	31
Thomas Ct. N&S	28W-8N	31
Thorn Tree Ct. NE&SW	28W-8N	31
Timber Ridge Dr. N&S	29W-9N	30
Trontman Dr.* E&W	31W-8N	30
Valley St. N&S	29W-8N	30
Wayne Ct. N&S	28W-8N	31
West Bartlett Rd. E&W	30W-8N,29W-8N	30
Western Av. N&S	29W-8N	30
White Oak Ln. NS&EW	29W-9N	30
Williamsburg Dr. N&S	29W-8N	30
Wilmington Dr. NW&SE	28W-8N	31
1st Av. E&W	31W-8N	30
2nd Av. E&W	31W-8N	30
3rd Av. E&W	31W-8N	30
4th Av. E&W	31W-8N	30
5th Av. E&W	31W-8N	30
6th Av. E&W	31W-8N	30
7th Av. E&W	31W-8N	30
8th Av. E&W	31W-8N	30
9th Av. E&W	31W-8N	30
10th Av. E&W	31W-8N	30
11th Av. E&W	31W-8N	30
12th Av. E&W	31W-8N	30

Bedford Park

	Grid	Page
Belt Cir. N&S	8W-7S	59
Central N&S	7W-8S	60
Harlem St. N&S	8W-7S	59
Interstate 55 SW&NE	11W-7S	58
Laramie N&S	6W-7S	60
Lavergne N&S	6W-7S	60
Leamington N&S	6W-8S	60
Leclaire N&S	6W-7S	60
Linder N&S	6W-8S	60
Lockwood N&S	6W-8S	60
Mason N&S	7W-8S	60
Massasoit N&S	7W-8S	60
Meade N&S	7W-8S	60
Narragansett N&S	8W-8S	59
Nashville Av. N&S	8W-7S	59
New England St. N&S	8N-7S	59
Oak Park N&S	8W-8S	59
Rt. 171 SW&NE	10W-7S	59
Rt. 50 N&S	6W-7S	60
Sayre N&S	8W-6S,8W-7S	59
65th E&W	9W-7S	59
65th Pl. E&W	9W-7S	59
66th E&W	9W-7S,6W-7S	59
66th Pl. E&W	9W-7S,6S-7S	59
67th E&W	6W-7S	60
68th E&W	8W-7S	59
70th Pl. E&W	6W-7S	60
71st E&W	11W-7S,8W-7S	58
71st Pl. E&W	8W-8S	59
72nd E&W	6W-8S	60
72nd Pl. E&W	8W-8S	59
73rd E&W	8W-8S,6W-8S	59
73rd Pl. E&W	8W-8S	59
74th E&W	8W-8S,6W-8S	59
74th Pl. E&W	8W-8S	59
75th E&W	6W-8S	60
77th E&W	8W-8S	59
77th Av. N&S	9W-7S	59
77th Ct. N&S	9W-7S	59
78th Av. N&S	9W-7S	59

Bellwood

	Grid	Page
Adams E&W	12W-0S	46
Bellwood N&S	12W-0S,12W-0N	46
Bohland N&S	12W-0S,12W-0N	46
Butterfield Rd. E&W	12W-0N	46
Castle Dr. E&W	13W-0N	46
Center N&S	13W-0N	46
Congress E&W	12W-0S	46
Eastern N&S	12W-0N	46
Erie E&W	13W-0N	46
Frederick N&S	12W-0N	46
Gladys E&W	12W-0S	46
Glos E&W	12W-0N	46
Grant E&W	12W-0N	46
Jackson E&W	12W-0S	46
Korel E&W	12W-0N	46
Linden N&S	12W-0S,12W-0N	46
Lyman N&S	12W-0S	46
Madison E&W	12W-0N	46
Marshall N&S	12W-0N,12W-0S	46
Marshall Ct. E&W	12W-0N	46
Miami Rd. E&W	12W-0N	46
Monroe E&W	12W-0S	46
Morris N&S	12W-0N	46
Oak E&W	13W-0N	46
Park Ct. E&W	12W-0N	46
Pusheck Rd. N&S	12W-0S	46
RR E&W	13W-0N,12W-0N	46
Railroad E&W	12W-0N	46
Randolph E&W	13W-0N	46
Rice N&S	12W-0S,12W-0N	46
St. Charles Pl. E&W	12W-0N	46
St. Paul E&W	12W-0N	46
Twining E&W	13W-0N	46
Van Buren E&W	12W-0S	46
Warren E&W	12W-0N	46
Wilcox E&W	12W-0S	46
Zuelke Dr. N&S	12W-0N	46
22nd N&S	11W-0S	47
23rd N&S	11W-0S	47
24th N&S	11W-0S	47
25th N&S	12W-0N	46
26th N&S	12W-0N	46
27th N&S	12W-0N	46
28th N&S	12W-0N	46
29th N&S	12W-0N	46
30th N&S	12W-0S,12W-0N	46
31st N&S	12W-0S,12W-0N	46
32nd N&S	12W-0S,12W-0N	46
42nd N&S	13W-0N	46
43rd N&S	13W-0N	46
44th N&S	13W-0N	46
45th N&S	13W-0N	46
46th N&S	13W-0N	46
47th N&S	13W-0N	46
48th N&S	13W-0N	46
51st N&S	13W-0N	46
52nd N&S	13W-0N	46

Berkeley

	Grid	Page
Albin Ter. E&W	14W-0N	46
Arthur N&S	14W-0S,14W-1N	46
Ashbel N&S	14W-0N	46
Atwood N&S	14W-0S,14W-0N	46
Bohlander E&W	14W-0N	46
Burr Oak E&W	14W-0N	46
Calvin N&S	14W-0S	46
Chicago E&W	14W-0N	46
Coolidge N&S	14W-0N	46
Electric E&W	14W-0N	46
Elm E&W	14W-0N	46
Harding N&S	14W-0N	46
Hawthorne E&W	14W-0N	46
Herber N&S	14W-0S,14W-1N	46
High N&S	14W-0S	46
Hillside N&S	14W-0N	46
Howard N&S	14W-0N	46
Huron E&W	14W-0N	46
Interstate 294 N&S	14W-1N	46
Irving N&S	14W-0N	46
KubaRose E&W	14W-0N	46
Lee N&S	13W-0N	46
Lind N&S	14W-0N	46
Maple E&W	14W-0N	46
McDermott E&W	14W-1N	46
Morris N&S	13W-0N	46
Murray Dr. E&W	14W-0N	46
Park E&W	14W-1N	46
Prospect E&W	14W-0N	46
Proviso E&W	14W-1N	46
Rhode N&S	13W-0N	46
Ridge E&W	14W-0N	46
School E&W	14W-0N	46
Speechley N&S	13W-0N	46
Spencer N&S	13W-0N	46
St. Charles E&W	13W-1N	46
Sunnyside Dr. N&S	14W-0N	46
Superior E&W	14W-0N	46
Taft N&S	14W-0S,14W-1N	46
Victoria N&S	14W-0N	46

Berwyn

	Grid	Page
Clarence N&S	8W-2S,8W-1S	54
Clinton N&S	8W-3S,8W-1S	54
Cuyler N&S	7W-3S,7W-1S	54
East N&S	8W-3S,8W-1S	54
Elmwood N&S	8W-2S,8W-1S	54
Euclid N&S	8W-2S,8W-1S	54
Fairfield E&W	8W-3S	54
Grove N&S	8W-3S,8W-3S	54
Gunderson N&S	8W-2S,8W-1S	54
Harvey N&S	7W-3S,7W-1S	54
Highland N&S	7W-3S,7W-1S	54
Home N&S	8W-3S,8W-1S	54
Kenilworth N&S	8W-3S,8W-1S	54
Lombard N&S	7W-3S,7W-1S	54
Maple N&S	8W-3S,8W-1S	54
Oak Park N&S	8W-3S,8W-1S	54
Park E&W	7W-3S	54
Ridgeland N&S	8W-3S,8W-1S	54
Riverside Dr. N&S	8W-2S	54
Rt. 43 N&S	8W-2S	54
Scoville N&S	8W-2S,8W-1S	54
Sinclair E&W	8W-3S	54
Stanley E&W	8W-3S	54
Wenonah N&S	8W-3S,8W-1S	54
Wesley N&S	8W-2S,8W-1S	54
Windsor E&W	8W-3S	54
Wisconsin N&S	8W-3S,8W-1S	54
13th E&W	8W-1S	54
14th E&W	8W-1S	54
15th E&W	8W-1S	54
16th E&W	8W-1S	54
18th E&W	8W-1S	54
19th E&W	8W-1S,7W-1S	54
21st E&W	8W-1S	54
23rd E&W	8W-2S	54
24th E&W	8W-2S	54
25th E&W	8W-2S	54
26th E&W	8W-2S	54
26th Pkwy. E&W	8W-2S	54
26th Pl. E&W	8W-2S	54
27th E&W	8W-2S	54
27th Pl. E&W	8W-2S	54
28th E&W	8W-2S	54
28th Pl. E&W	8W-2S	54
29th E&W	8W-2S,7W-2S	54
29th Pl. E&W	8W-2S	54
30th E&W	8W-2S	54
30th Pl. E&W	8W-2S	54
32nd E&W	8W-3S,7W-3S	54
33rd E&W	8W-3S,7W-3S	54
34th E&W	8W-3S	54
35th E&W	8W-3S,7W-3S	54
36th E&W	8W-3S	54
37th E&W	8W-3S	54
38th E&W	8W-3S	54

Bloom Township

	Grid	Page
Alice Av. N&S	3E-26S	95
Arnett St. N&S	1E-24S	94
Ashland Av. N&S	1W-27S,1W-26S	93
Audrey NW&SE	4W-25S	92
Baltimore Av. N&S	3E-25S,3E-24S	95
Batrack Ct. E&W	3E-25S	95
Blackstone Av. N&S	1E-26S,1E-23S	94
Brandon Av. N&S	3E-25S,3E-24S	95
Bridge E&W	4E-25S	95
Brown St. N&S	1E-24S	94
Burley Av. N&S	3E-25S,3E-24S	95
Burnham Av. N&S	3E-27S,3E-24S	95
Butler St. E&W	1E-24S	94
Calumet Expwy. N&S	1E-27S,1E-22S	94
Carondelet Av. N&S	3E-25S,3E-24S	95
Chicago Av. N&S	3E-22S	89
Chris Ct. N&S	2E-23S	89
Columbia Av. E&W	1E-24S	94
Commercial Av. N&S	3E-25S	95
Cottage Grove Av. N&S	0E-24S,1E-27S	94
Cullom St. N&S	1E-24S	94
Daniel Ct. E&W	3E-25S	95
David Ct. E&W	3E-25S	95
Douglas Ct. E&W	3E-25S	95
Dutra Av. NE&SW	2W-24S	93
Elizabeth NE&SW	4E-25S	95
Ellen Dr. N&S	3E-25S	95
Escanaba Av. N&S	3E-25S	95
Exchange Av. N&S	3E-25S	95
Frederick Ct. E&W	3E-25S	95
Gail Ct. E&W	3E-25S	95
Glenwood Dyer Rd. NW&SE	2E-23S,2E-24S	89
Glenwood Lansing Rd. E&W	2E-22S	89
Glenwood-Dyer Rd. NW&SE	1E-23S,3E-24S	89
Harper Av. N&S	1E-26S	94
Hillcrest Av. N&S	3E-24S	95
Houston Av. N&S	3E-25S,3E-24S	95
Howard Ct. N&S	3E-25S	95
Indiana Av. N&S	0E-25S	94
Jacob Ct. N&S	3E-25S	95
James Av. N&S	3E-26S	95
Jennifer E&W	4E-25S	95
Joe Orr Rd. E&W	1E-24S,2E-24S	94
Judith Ct. E&W	3E-25S	95
Langston St. N&S	1E-24S	94
Lincoln Hwy. E&W	3E-25S	95
Long Av. E&W	2E-27S	95
Main St. E&W	0E-26S	94
Marion Ct. E&W	3E-25S	95
Martin Ct. E&W	3E-25S	95
Michelle E&W	4E-25S	95
Muskegon Av. N&S	3E-25S	95
Oakwood Av. N&S	3E-22S	89
Park N&S	1E-25S	94
Parkside Av. E&W	3E-24S	95
Prairie Av. E&W	3E-25S	95
Pullman N&S	1E-25S	94
Sauk Trail E&W	2E-27S,4E-26S	95
State St. N&S	0E-23S	88
Steger Rd. E&W	2W-27S,4E-27S	93
Stirling Av. N&S	2W-25S	93
Stoney Island N&S	1E-25S	94
Stony Island Av. N&S	2E-24S,2E-22S	95
Stuart Ct. E&W	3E-25S	95
Summit N&S	1E-25S	94
Sunset Ln. N&S	4E-26S	95
Torrence Av. N&S	2E-27S,2E-22S	95
Travers Av. N&S	2W-25S,2W-24S	93
Valarie NE&SW	4E-25S	95
Valerie Dr. N&S	4E-25S	95
Victoria Cir. NS&EW	4E-25S	95
Virginia E&W	4E-25S	95
Vollmer St. E&W	1E-23S	89
13th St. E&W	0E-25S	94
21st St. E&W	3E-26S	95
22nd St. E&W	0E-26S	94
22nd St. E&W	3E-26S	95
30th St. E&W	1E-27S	94
190th St. E&W	1W-22S	88
205th St. E&W	2W-24S	93
207th St. E&W	2W-24S,3E-24S	93
211th St. E&W	1E-25S	94
212th St. E&W	1E-25S	94
213th St. E&W	1E-25S	94
219th St. E&W	4E-26S	95

Blue Island

	Grid	Page
Albany Av. N&S	3W-14S	75
Ann St. N&S	3W-15S,3W-14S	75
Artesian Av. N&S	3W-15S,3W-14S	75
Ashland Av. N&S	2W-15S	75
Berrien St. N&S	1W-15S	76
Birdsall St. E&W	3W-14S	75
Broadway E&W	3W-15S,2W-15S	75
Burr Oak Av. E&W	3W-14S	75
California Av. N&S	3W-16S,3W-14S	75
Calumet St. N&S	1W-15S	76
Canal St. E&W	2W-15S	75
Chatham St. N&S	2W-16S,2W-15S	75
Chicago St. N&S	2W-15S	75
Cleveland Av. N&S	3W-16S	75
Clinton Av. N&S	3W-15S	75
Cochran St. E&W	3W-14S	75
Collins St. E&W	3W-14S,2W-14S	75
Concord St. N&S	2W-15S	75
Coopers Grove Rd. NE&SW	3W-16S	75
Dan Ryan Expressway N&S	2W-15S	75
Des Plaines St. E&W	3W-15S,2W-15S	75
Devonshire St. E&W	2W-15S	75
Division St. N&S	2W-16S,2W-15S	75
Duane St. NS&EW	2W-15S	75
Edison Av. E&W	4W-15S,3W-15S	75
Edward St. E&W	3W-15S	75
Elim E&W	2W-15S	75
Elm Dr. NE&SW	3W-14S	75
Elm St. N&S	3W-15S,3W-14S	75
Everett St. NW&SE	3W-15S	75
Fairview Av. N&S	3W-14S	75
Florence St. E&W	2W-14S	75
Francisco Av. N&S	3W-16S,3W-15S	75
Fulton St. E&W	2W-15S	75
Girard St. E&W	2W-15S	75
Greenwood Av. N&S	3W-15S,3W-14S	75
Gregory St. N&S	2W-15S,2W-14S	75
Grove St. E&W	3W-15S,2W-15S	75
Grunewald St. E&W	3W-14S	75
Hadley St. NE&SW	2W-16S	75
High St. E&W	3W-15S,2W-15S	75
Highland Av. N&S	3W-15S,3W-14S	75
Homan Av. N&S	4W-14S	75
Honore St. N&S	2W-15S	75
Hoyne Av. N&S	2W-15S	75
Interstate 57 N&S	2W-15S	75
Irving Av. N&S	2W-15S,2W-14S	75
James St. E&W	3W-15S	75
John St. E&W	3W-15S	75
Kedzie Av. N&S	3W-15S	75
Krueger St. E&W	2W-14S	75
Lewis St. E&W	3W-14S	75
Liberty St. NW&SE	2W-15S	75
Lincoln St. N&S	2W-15S	75
Longwood Dr. N&S	2W-14S	75
Main St. E&W	2W-16S	75
Maple Av. N&S	3W-15S,3W-14S	75
Market St. E&W	2W-15S	75
Minnesota Av. E&W	3W-14S	75
Mozart St. N&S	3W-15S	75
New St. E&W	3W-15S	75
Oak St. E&W	3W-15S	75
Olive St. E&W	2W-14S	75
Orange St. E&W	2W-15S	75
Orchard St. E&W	3W-14S,2W-14S	75
Portland St. N&S	2W-15S	75
Prairie St. E&W	3W-15S,2W-15S	75
Randolph St. E&W	1W-15S	75
Rexford St. N&S	2W-15S	75
Roll Av. E&W	2W-15S	75
Sacramento Av. N&S	3W-15S	75
Sanders St. N&S	2W-15S	75
Seeley Av. N&S	2W-15S	75
Shamrock Ln. N&S	3W-16S	75
Summit St. E&W	2W-14S	75
Sycamore St. E&W	1W-15S	76
Thornton Rd. NW&SE	2W-16S	75
Union St. E&W	3W-15S,2W-15S	75
Vermont St. E&W	3W-15S,2W-15S	75
Vernon St. E&W	2W-14S	75
Vincennes Rd. NE&SW	2W-14S	75
Vine St. E&W	2W-16S	75
Wahl St. NW&SE	3W-15S	75
Walnut St. E&W	3W-15S,2W-15S	75
Western Av. N&S	3W-16S,3W-14S	75
Whipple St. N&S	3W-16S	75
Winchester Av. N&S	2W-15S	75
Wireton St. NW&SE	3W-15S	75
Wood St. N&S	2W-15S,2W-14S	75
York St. E&W	3W-15S,2W-15S	75
119th Pl. E&W	2W-14S	75
120th Pl. E&W	2W-14S	75
120th St. E&W	3W-14S,2W-14S	75
121st Pl. E&W	3W-14S,2W-14S	75
121st St. E&W	3W-14S,2W-14S	75
122nd Pl. E&W	3W-14S	75
122nd St. E&W	3W-14S,2W-14S	75
123rd Pl. E&W	2W-14S	75
123rd St. E&W	3W-14S,2W-14S	75
127th St. E&W	3W-15S,2W-15S	75
131st St. E&W	2S-14S	75
135th St. E&W	2W-16S	75
136th St. E&W	2W-16S	75
138th St. E&W	2W-16S	75
139th Pl. E&W	3W-16S	75
139th St. E&W	3W-16S,2W-16S	75
140th Pl. E&W	3W-16S	75
140th St. E&W	3W-16S	75
141st Pl. E&W	3W-16S	75
141st St. E&W	3W-16S	75

Bremen Township

	Grid	Page
Artesian Av. N&S	3W-19S	81
Austin Blvd. N&S	7W-19S	80
Bachelor Grove Rd. N&S	8W-16S	79
Beverly Av. N&S	7W-19S	80
Bramblewood Rd. N&S	8W-18S	79
California Av. N&S	3W-18S,3W-17S	81
Central Av. N&S	7W-21S,7W-20S	86
Elmdale Av. NW&SE	3W-18S,2W-18S	81
Fairfield Av. NW&SE	3W-17S	81
Forest Av. N&S	6W-20S	86
Francisco Av. N&S	3W-18S,3W-17S	81
Fulton Terr. N&S	7W-19S	80
Gaynelle Rd. NS&EW	7W-19S	80
George Brennan Hwy. NE&SW	6W-20S	86
Highland Av. N&S	7W-21S	86
Interstate 57 NE&SW	6W-21S,6W-20S	86
James St. E&W	7W-19S	80
Jody Ln. E&W	7W-19S	80
Justamere Rd. N&S	8W-17S	79
Kilpatrick Av. N&S	6W-20S	86
Kimberly Ln. E&W	7W-19S	80
Lamon Av. N&S	6W-17S	80
Laramie Av. N&S	6W-20S	86
Lavergne Av. N&S	6W-21S	86
Lawndale Av. N&S	4W-21S	87
Leclaire Av. N&S	6W-20S	86
Leslie Ann Dr. N&S	7W-19S	80
Linder Av. N&S	6W-16S	74
Lockwood Av. N&S	6W-20S	86
Luna Av. N&S	6W-16S	74
Major Av. N&S	7W-16S	74
Mark Ln. N&S	7W-19S	80
Menard Av. N&S	7W-16S	74
Midlothian Turnpike NE&SW	7W-16S,6W-16S	74
New England Av. N&S	8W-18S	79
Oak Park Av. N&S	8W-18S	79
Parkside Av. N&S	7W-20S	86
Patricia Av. N&S	7W-19S	80
Poplar Av. N&S	3W-19S	81
Richmond Av. N&S	3W-18S	81
Rt. 43 N&S	8W-17S,8W-16S	79
Sacramento Av. N&S	3W-17S,3W-16S	81
Sayre Av. N&S	8W-18S	79
Sayre Av. N&S	8W-21S	85
Springfield Av. N&S	4W-21S	87
Thorndale Av. NW&SE	3W-18S,2W-18S	81
139th St. E&W	6W-16S	74
142nd Pl. E&W	3W-16S	75
143rd Pl. E&W	3W-17S	81
143rd St. E&W	8W-16S,3W-16S	79
144th St. E&W	3W-17S	81
147th Ct. E&W	6W-17S	80
147th St. E&W	8W-17S,7W-17S	79
150th St. E&W	6W-17S	80
151st St. E&W	3W-17S	81
155th Pl. E&W	8W-18S	79
155th St. E&W	8W-18S	79
156th Pl. E&W	8W-18S	79
157th Pl. E&W	8W-18S	79

Name	Grid	Page
157th St. E&W	8W-18S	79
159th Ct. E&W	2W-18S	81
163rd St. E&W	5W-19S	80
166th St. E&W	7W-19S	80
171st Pl. E&W	7W-20S	86
171st St. E&W	6W-20S	86
173rd St. E&W	7W-20S,6W-20S	86
174th St. E&W	6W-21S,6W-20S	86
175th St. E&W	7W-20S,6W-21S	86
177th St. E&W	6W-21S	86
178th St. E&W	8W-21S	86
179th St. E&W	7W-21S	86
180th St. E&W	8W-21S	85
181st St. E&W	8W-21S	85
182nd St. E&W	8W-21S	85
183rd St. E&W	7W-21S,6W-21S	86

Bridgeview

Name	Grid	Page
Avon Av. NW&SE	9W-10S	65
Beloit Av. N&S	9W-10S,9W-8S	65
Birch E&W	8W-10S	65
Cambridge St. E&W	9W-10S,8W-10S	65
Circle Ct. NW&SE	8W-10S	65
Columbia E&W	9W-7S	59
Cranbrook Ln. N&S	8W-10S	65
Dover E&W	8W-10S	65
Elder Ct. NW&SE	8W-10S	65
Elm N&S	9W-7S	59
Fairlane Dr. NE&SW	8W-10S	65
Ferdinand Av. N&S	9W-10S,9W-8S	65
Frontage Rd. N&S	9W-10S	65
Garden Ct. E&W	9W-7S	59
Garden Ln. E&W	9W-7S	59
Garden Ln. N&S	9W-7S	59
Hanover Dr. E&W	8W-10S	65
Harlem Av. N&S	9W-10S,9W-8S	65
Hartford E&W	9W-10S,8W-10S	65
Homestead Ln. N&S	8W-10S	65
Industrial Dr. NW&SE	9W-11S	65
Mable N&S	9W-7S	59
Martin Ln. N&S	8W-10S	65
Oak Park Av. N&S	8W-10S	65
Oconto N&S	9W-8S	59
Oconto Av. N&S	9W-9S,9W-8S	65
Octavia Av. N&S	9W-10S,9W-8S	65
Odell Av. N&S	9W-10S,9W-8S	65
Oketo Av. N&S	9W-10S,9W-8S	65
Olympic Dr. NS&EW	8W-10S	65
Orchard Ln. N&S	8W-10S	65
Oxford St. E&W	8W-10S	65
Pembroke Ln. N&S	8W-10S	65
Richards Rd. E&W	9W-11S	65
Roberts Rd. N&S	9W-8S	59
Saratoga Dr. NS&EW	8W-10S	65
Selma N&S	9W-8S	59
Sholer Av. N&S	9W-8S	59
Southfield Dr. NE&SW	8W-10S	65
Stickney N&S	9W-8S	59
Stickney Av. N&S	9W-8S	59
Stratford Dr. E&W	8W-10S	65
Suburban E&W	9W-7S	59
Thomas Av. N&S	9W-10S,9W-8S	65
Tri-State Tollway NW&SE	9W-11S,9W-10S	65
Willow Ln. N&S	8W-10S	65
Windsor Ln. N&S	8W-10S	65
70th E&W	9W-7S	59
71st E&W	9W-7S	59
72nd E&W	9W-8S	59
73rd E&W	9W-8S	59
73rd Pl. E&W	9W-8S	59
74th E&W	9W-8S	59
74th Pl. E&W	9W-8S	59
75th E&W	9W-8S	59
75th St. E&W	9W-8S	59
76th E&W	9W-8S	59
76th St. E&W	9W-8S	59
77th E&W	9W-8S	59
77th Av. N&S	9W-10S,9W-9S	65
77th Ct. N&S	9W-9S	65
77th St. N&S	9W-8S	59
78th E&W	9W-8S	59
78th Av. N&S	9W-9S,9W-8S	65
78th Ct. N&S	9W-8S	59
78th St. E&W	9W-8S	59
79th N&S	9W-7S	59
79th Pl. E&W	9W-8S	59
79th St. E&W	9W-8S	59
80th Pl. E&W	9W-9S	65
80th St. E&W	9W-8S	59
81st Pl. E&W	9W-9S	65
81st St. E&W	9W-9S	65
82nd Pl. E&W	9W-9S	65
82nd St. E&W	9W-9S	65
83rd Pl. E&W	9W-9S	65
83rd St. E&W	9W-9S	65
84th Pl. E&W	9W-9S	65
84th St. E&W	9W-9S	65
85th Pl. E&W	9W-9S	65
85th St. E&W	9W-9S	65
86th Pl. E&W	9W-9S	65
86th St. E&W	9W-9S	65
87th Pl. E&W	9W-10S	65
87th St. E&W	9W-9S	65
88th St. E&W	9W-10S	65
90th Pl. E&W	9W-10S	65
90th St. E&W	9W-10S	65
91st St. E&W	9W-10S	65
92nd Pl. E&W	9W-10S	65
92nd St. E&W	9W-10S	65
93rd St. E&W	9W-10S	65
98th Pl. E&W	9W-11S	65
100th Pl. E&W	9W-11S	65

Broadview

Name	Grid	Page
Beach N&S	12W-0S	46
Braga Dr. NW&SE	11W-1S	53
Cermak E&W	11W-1S	53
Elizabeth Dr. E&W	11W-2S	53
Ericka E&W	11W-2S	53
Field Av. N&S	11W-2S	53
Fillmore E&W	11W-0S	47
Gardner N&S	12W-0S	46
Roberts E&W	11W-1S	53
Roosevelt Rd. E&W	11W-0S	47
Summerdale Dr. E&W	11W-2S	53
Sunnydale N&S	11W-2S	53
Terry Ln. E&W	11W-1S	53
Tyson E&W	11W-1S	53
York E&W	12W-0S	46
13th E&W	11W-0S	47
13th N&S	11W-1S	53
14th E&W	11W-0S	47
14th N&S	11W-0S	47
15th E&W	11W-1S	53
15th N&S	11W-0S	47
16th E&W	11W-1S	53
16th N&S	11W-0S	47
17th N&S	11W-1S	53
18th E&W	11W-1S	53
18th N&S	11W-1S	53
19th E&W	11W-1S	53
19th N&S	11W-1S	53
20th E&W	11W-1S	53
21st E&W	11W-1S	53
22nd N&S	11W-1S	53
23rd N&S	11W-1S	53
23rd Pl. E&W	12W-2S	52
23rd St. E&W	12W-2S	52
24th N&S	11W-2S	53
24th St. E&W	11W-1S	53
26th N&S	12W-2S	52
27th N&S	12W-1S	52

Brookfield

Name	Grid	Page
Arden N&S	10W-3S	53
Arthur N&S	11W-4S,11W-3S	53
Bartlett E&W	11W-2S	53
Blanchan N&S	11W-4S,11W-3S	53
Broadway SW&NE	11W-3S	53
Brookfield E&W	11W-3S	53
Burlington E&W	11W-4S	53
Cleveland N&S	11W-3S	53
Congress Park E&W	11W-4S	53
Custer E&W	11W-4S,11W-3S	53
Deyd N&S	11W-4S	53
DuBois N&S	11W-4S	53
Elm N&S	11W-4S,11S-3S	53
Fairview E&W	11W-3S	53
Forest	11W-4S,11W-3S	53
Forestview E&W	11W-2S	53
Garfield E&W	11W-3S	53
Gerritsen E&W	11W-4S	53
Grand NW&SE	11W-3S	53
Grant E&W	11W-3S	53
Greenview N&S	11W-4S	53
Grove N&S	11W-4S	53
Harrison N&S	11W-3S	53
Henrietta E&W	11W-3S	53
Hollywood N&S	10W-3S	53
Jackson N&S	11W-3S	53
Jefferson E&W	11W-3S	53
Lexington E&W	11W-3S	53
Madison N&S	11W-4S,11W-3S	53
Maple N&S	11W-4S,11W-3S	53
McCormick N&S	10W-3S	53
Monroe E&W	11W-3S	53
Morton N&S	11W-3S	53
Oak N&S	11W-4S,11W-3S	53
Park N&S	11W-4S,11S-3S	53
Parkview E&W	11W-4S	53
Plainfield E&W	11W-4S	53
Prairie N&S	11W-4S,11W-2S	53
Raymond N&S	11W-4S,11W-3S	53
Riverside N&S	10W-3S	53
Riverview N&S	10W-3S	53
Roach E&W	11W-2S	53
Rochester E&W	11W-3S	53
Rockfeller E&W	10W-3S	53
Rosemear N&S	11W-3S	53
Sahler E&W	11W-4S	53
Sheridan N&S	11W-3S	53
Sherman N&S	11W-3S	53
Shields E&W	11W-4S	53
Southview N&S	11W-3S	53
Sunnyside N&S	11W-4S,11W-3S	53
Taylor N&S	11W-4S	53
Vernon N&S	11W-4S,11W-2S	53
Washington E&W	11W-3S	53
Windemere E&W	11W-4S	53
Woodside N&S	10W-3S	53
26th E&W	11W-2S	53
26th Pl. E&W	11W-2S	53
27th E&W	11W-2S	53
28th E&W	11W-2S	53
29th E&W	11W-2S	53
30th E&W	11W-2S	53
31st E&W	10W-2S	53
45th E&W	11W-4S	53
47th E&W	11W-4S	53

Buffalo Grove*
(Also see Lake County)

Name	Grid	Page
Anthony Ct. N&S	16W-17N	15
Anthony Rd. E&W	16W-17N	15
Arbor Ct. NW&SE	17W-18N	14
Arlington Heights Rd. N&S	18W-18N	14
Ashford Ct. N&S	17W-17N	14
Beacon Ct. N&S	17W-17N	14
Beechwood Ct., E. NW&SE	17W-18N	14
Beechwood Rd. E&W	17W-18N	14
Bel Aire Ct. N&S	17W-18N	14
Bel Aire Rd. N&S	17W-18N	14
Bernard Ct. N&S	17W-18N	14
Bernard Dr. E&W	17W-18N,16W-18N	14
Betty Dr. N&S	16W-17N	15
Boxwood Ln. NE&SW	17W-17N	14
Bradley Rd. E&W	17W-18N	14
Brentwood N&S	17W-17N	14
Brucewood Dr. N&S	16W-18N	15
Buffalo Grove Rd. N&S	16W-17N,16W-20N	15
Cambridge Dr. N&S	16W-17N	15
Cameron Way N&S	18W-18N	14
Carriage Way N&S	17W-17N	14
Charles Ct. NW&SE	16W-17N	15
Chatham Cir. N&S	17W-18N	14
Chenault Rd.	17W-18N	14
Cherrywood Rd. N&S	17W-18N	14
Church Rd. NE&SE	17W-18N	14
Cottonwood Rd. NE&SE	17W-18N	14
Country Ln. NE&SW	17W-18N	14
Countryside Dr. E&W	18W-18N	14
Covington Ter. N&S	17W-17N	14
Crestview Ter. NW&SE	16W-18N	15
Crofton Ln. N&S	17W-17N	14
Ct. E. E&W	16W-18N	15
Ct. W. E&W	16W-18N	15
Darryl Dr. E&W	16W-17N	15
Diane Dr. N&S	16W-18N	15
Downing Rd. E&W	16W-17N	15
Dun-Lo Av. E&W	16W-17N	15
Dundee Pkwy. E&W	18W-18N,17W-18N	14
Dundee Rd. E&W	18W-18N,17W-18N	14
Eaton Ct. N&S	16W-18N	15
Ellen Dr. N&S	16W-17N	15
Elmwood Dr. NS&EW	17W-18N	14
Estate Dr. N&S	17W-18N	14
Evergreen Pl. E&W	16W-18N	15
Forest Pl. NS&EW	16W-18N	15
Forest Way Dr. N&S	17W-18N	14
Frances Ct. NW&SE	16W-18N	15
Glendale Rd. NS&EW	16W-18N	15
Golf View Terr. N&S	17W-18N	14
Golfview Ter. NS&EW	16W-18N	15
Greenridge Rd. N&S	17W-17N	14
Gregg Ln. N&S	16W-18N	14
Grove Cir. N&S	17W-18N	14
Grove Dr. N&S	18W-18N	14
Hapsfield Ln. E&W	17W-18N	14
Harvard Ln. N&S	16W-17N	15
Hawthorne Rd. E&W	17W-18N	14
Hiawatha Dr. N&S	17W-18N	14
Hickory Av. N&S	17W-17N	14
Hickory Dr. NW&SE	17W-18N	14
Indian Hill Dr E&W	17W-18N	14
Juniper Ct. NE&SE	17W-18N	14
Katherine Ct. N&S	17W-18N	14
Lake Blvd. NS&EW	16W-17N	15
Lake Cook Rd. E&W	17W-18N	14
Lauren Ln. N&S	16W-18N	15
Lehigh Ln. NE&SW	17W-17N	14
Lincoln Ter. N&S	17W-18N	14
Longwood Ct. NE&SW	17W-18N	14
Longwood Dr. NS&EW	17W-18N	14
MacArthur Ct. E&W	17W-18N	14
MacArthur Dr. N&S	17W-18N	14
Maple Dr. E&W	17W-18N	14
Marylu Ln. N&S	17W-18N	14
Mayfair Ct., North E&W	17W-18N	14
Mayfair Ct., South NW&SE	17W-18N	14
Mayfair Ln. NS&EW	17W-18N	14
Melinda Ln. N&S	17W-18N	14
Middlesex Ct. E&W	16W-17N	15
Mill Creek Dr. NS&EW	17W-17N	14
Miller Ln. E&W	17W-17N	14
Mohawk Ct. E&W	16W-18N	15
Mohawk Tr. E&W	16W-18N	15
Mulberry Ct. E&W	17W-18N	14
N. Greenwood Ct. N&S	17W-18N	14
Navajo Tr. N&S	16W-18N	15
Oak Creek Dr. NE&SW	16W-18N	15
Old Oak Dr. E&W	16W-18N	15
Old Post Rd. NE&SW	17W-17N	14
Palmgren Ct. E&W	16W-17N	15
Park Pl. E&W	16W-17N	15
Park View Ter. N&S	17W-17N	14
Parkview Ter. E&W	17W-17N	14
Patton Dr. N&S	17W-18N	14
Plum Grove Cir. E&W	17W-18N	14
Poplar St. E&W	17W-18N	14
Rachel Ln. NS&EW	18W-18N	14
Raupp Blvd. N&S	17W-18N,16W-18N	14
Red Oak Ct. NW&SE	16W-18N	15
Regent Ct., East NE&SW	16W-18N	15
Regent Ct., West NE&SW	17W-18N	14
Regent Dr. NS&EW	17W-18N	14
Roberta Ct. NW&SE	16W-18N	15
Rose Blvd. NS&EW	18W-18N	14
Rose Ct., East E&W	18W-18N	14
Rose Ct., West E&W	18W-18N	14
Rosewood Av. N&S	17W-18N	14
S. Greenwood Ct. N&S	17W-18N	14
Saratoga Ln. NE&SW	17W-17N	14
Saxon Pl. N&S	16W-17N	15
Selwyn Ln. N&S	16W-17N	15
St. Mary's Pkwy. E&W	17W-18N,16W-18N	14
Stanford Ln. E&W	17W-17N	14
Stillwell Dr. E&W	17W-18N	14
Stonebridge Ln. E&W	17W-17N	14
Stonegate Ct. N&S	17W-18N	14
Stonegate Rd. NS&EW	17W-18N	14
Straford Cir. E&W	17W-18N	14
Strathmore Ct. E&W	18W-18N	14
Sussex Ct. NS&EW	16W-17N	15
Sycamore Rd. E&W	17W-18N	14
Terrace Pl. N&S	17W-18N	14
Thornton Ln. E&W	17W-17N	14
Thornwood Rd. N&S	17W-18N	14
Timber Hill Rd. N&S	17W-18N	14
Trace Dr. E&W	18W-18N	14
Trinity Ct. E&W	16W-17N	15
University Dr. E&W	16W-17N	15
Vernon Ct., North E&W	17W-18N	14
Vernon Ct., South E&W	17W-18N	14
Vernon Ln. N&S	17W-18N	14
W. Beechwood Ct. E&W	17W-18N	14
Weidner Ct. E&W	17W-18N	14
Weidner Rd. N&S	17W-18N	14
White Pine Rd. N&S	17W-18N	14
Whitehall Ct. E&W	17W-18N	14
Whitehall Pl. E&W	17W-18N	14
Windsor Dr. N&S	17W-18N	14

Burbank

Name	Grid	Page
Austin Av. N&S	7W-9S,7W-8S	66
Carmel St. E&W	7W-8S	60
Central Av. N&S	6W-9S,6W-8S	66
Cicero Av. N&S	6W-9S,6W-8S	66
Doral St. E&W	7W-8S	60
Foxfire St. N&S	7W-8S	60
Harlem Av. N&S	8W-9S	65
La Crosse Av. N&S	6W-9S,6W-8S	66
La Porte Av. N&S	6W-9S,6W-8S	66
LaCrosse N&S	6W-8S	60
Lamon Av. N&S	6W-9S	66
Laramie Av. N&S	6W-9S,6W-8S	66
Latrobe Av. N&S	6W-9S,6W-8S	66
Lavergne Av. N&S	6W-9S,6W-8S	66
Lawler Av. N&S	6W-9S,6W-8S	66
Le Claire Av. N&S	6W-9S,6W-8S	66
Leamington Av. N&S	6W-9S,6W-8S	66
Linder Av. N&S	6W-9S,6W-8S	66
Lockwood Av. N&S	6W-9S,6W-8S	66
Long Av. N&S	6W-9S,6W-8S	66
Lorel Av. N&S	6W-9S,6W-8S	66
Lotus Av. N&S	6W-9S,6W-8S	66
Luna Av. N&S	6W-9S	60
Major Av. N&S	7W-9S,7W-8S	66
Mansfield Av. N&S	7W-9S,7W-8S	66
Mason Av. N&S	7W-9S,7W-8S	66
Massasoit Av. N&S	7W-9S,7W-8S	66
Mayfield Av. N&S	7W-9S,7W-8S	66
McVicker Av. N&S	7W-9S,7W-8S	66
Meade Av. N&S	7W-9S	66
Melvina Av. N&S	7W-9S,7W-8S	66
Menard Av. N&S	7W-9S,7W-8S	66
Merrimac Av. N&S	7W-9S,7W-8S	66
Mobile Av. N&S	7W-9S,7W-8S	66
Monitor N&S	7W-8S	60
Monitor Av. N&S	7W-8S	60
Moody Av. N&S	7W-9S,7W-8S	66
Mulligan Av. N&S	7W-9S,7W-8S	66
Nagle Av. N&S	8W-9S,8W-8S	65
Narragansett Av. N&S	8W-9S,8W-8S	65
Nashville Av. N&S	8W-9S,8W-8S	65
Natchez Av. N&S	8W-9S,8W-8S	65
Natoma Av. N&S	8W-9S,8W-8S	65
Neenah Av. N&S	8W-9S,8W-8S	65
Neva Av. N&S	8W-9S	65
New England Av. N&S	8W-9S	65
Newcastle Av. N&S	8W-9S,8W-8S	65
Newland Av. N&S	8W-9S,8W-8S	65
Nordica Av. N&S	8W-9S	65
Normandy Av. N&S	8W-9S,8W-8S	65
Nottingham Av. N&S	8W-9S	65
Oak Park Av. N&S	8W-9S,8W-8S	65
Parkside Av. N&S	7W-8S	60
Parkside Av. N&S	7W-9S,7W-8S	66
Pinehurst St. E&W	7W-8S	60
Rt. 43 N&S	8W-9S	65
Rt. 50 N&S	6W-9S,6W-8S	66
Rutherford Av. N&S	8W-9S,8W-8S	65
Sawgrass St. E&W	7W-8S	60
Sayre Av. N&S	8W-9S	65
State Rd. NE&SW	7W-9S,6W-8S	66
75th Pl. E&W	7W-8S,6W-8S	60
76th E&E	7W-8S,6W-8S	60
76th Pl. E&W	6W-8S	60
76th St. E&W	7W-8S,6W-8S	60
77th E&W	7W-8S	60
77th Pl. E&W	7W-8S	60
77th St. E&W	8W-8S,6W-8S	59
78th E&W	8W-8S,6W-8S	59
78th Pl. E&W	6W-8S	60
78th St. E&W	8W-8S,6W-8S	59
79th E&W	8W-8S,6W-8S	59
79th Pl. E&W	8W-9S,6W-8S	59
79th St. E&W	8W-8S,6W-8S	59
80th Pl. E&W	8W-9S,7W-9S	65
80th St. E&W	8W-9S,6W-9S	65
81st Pl. E&W	8W-9S,7W-9S	65
81st St. E&W	8W-9S,6W-9S	65
82nd Pl. E&W	8W-9S,6W-9S	65
82nd St. E&W	8W-9S,6W-9S	65

CHICAGO AND COOK COUNTY 105

Grid Page
83rd Pl. E&W .. 8W-9S,6W-9S 65
83rd St. E&W .. 8W-9S,6W-9S 65
84th Pl. E&W .. 8W-9S,6W-9S 65
84th St. E&W .. 8W-9S,6W-9S 65
85th Pl. E&W .. 8W-9S,6W-9S 65
85th St. E&W .. 8W-9S,6W-9S 65
86th Pl. E&W .. 7W-9S,6W-9S 66
86th St. E&W .. 8W-9S,6W-9S 65
87th St. E&W .. 8W-9S,6W-9S 65

Burnham

Bensley Av. N&S 2E-16S 83
Brainard Av. NW&SE 4E-16S 83
Burnham Av. N&S 3E-16S 83
Calhoun Av. N&S 2E-16S 83
Calumet Av. N&S 3E-16S 83
Centre Av. NE&SW .. 3E-16S 83
Chippewa Av. NW&SE 3E-16S 83
Commercial Dr. NW&SE 2E-16S 83
Ewell Av. N&S 3E-16S 83
Exchange Av. N&S .. 3E-16S 83
Goodrich Av. E&W 3E-16S 83
Greenbay Av. N&S .. 3E-16S 83
Hammond Av. E&W .. 3E-16S 83
Hoxie Av. N&S 2E-16S 83
Indiana Av. N&S 3E-16S 83
Kankakee Av. N&S .. 3E-16S 83
Kettelson St. E&W 2E-16S 83
Mackinaw St. N&S .. 3E-16S 83
Manistee Av. N&S 3E-16S 83
Marquette Av. N&S .. 3E-16S 83
Muskegon Av. N&S .. 3E-16S 83
Pierson St. E&W 2E-16S,3E-16S 83
Prairie Av. N&S 3E-16S 83
Saginaw Av. N&S 3E-16S 83
Scott Av. N&S 3E-16S 83
State Line Rd. N&S .. 4E-16S 83
State St. E&W 2E-16S,3E-16S 83
Torrence Av. N&S 2E-16S 83
Yates Av. N&S 2E-16S 83
138th St. E&W 3E-16S 83
139th St. E&W 3E-16S 83
140th St. E&W 2E-16S,3E-16S 83
141st St. E&W 2E-16S,3E-16S 83
142nd St. E&W 3E-16S 83
143rd St. E&W 3E-16S 83

Burr Ridge*
(Also see Du Page County)

Ambriance St. N&S .. 14W-9S 58
Arbor NW&SE 14W-8S 58
Arrowhead Farm Dr. NS&EW 14W-9S 64
Bastern Rd. N&S 14W-7S 58
Briarwood Ct. E&W .. 14W-9S 64
Briarwood Ln. E&W .. 14W-9S 64
Bridle Path E&W 14W-7S 58
Brighton E&W 14W-7S 58
Brighton Pl. NW&SE 14W-7S 58
Buck Trail Dr. N&S .. 14W-9S 64
Burr Oak NE&SW 14W-8S 58
Burr Ridge N&S 14W-8S 58
Carriage Pl. E&W 14W-7S 58
Carriage Wy. NE&SW 14W-7S 58
Central N&S 14W-8S 58
Chasemoor Dr. NS&EW 14W-8S 58
Chippewa Ct. E&W .. 14W-9S 64
Cir. 14W-8S 58
Circle Dr. NE&SW 14W-8S 58
Commonwealth N&S 14W-8S 58
County Line N&S 14W-7S 58
County Line Ln. N&S 14W-7S 58
County Line Rd. N&S 14W-9S,14W-6S 64
Creekwood Dr. (pvt) N&S 14W-9S 64
Ct. 1 N&S 14W-8S 58
Ct. 2 N&S 14W-8S 58
Ct. 3 N&S 14W-8S 58
Ct. 4 NE&SW 14W-8S 58
Ct. 5 E&W 14W-8S 58
Ct. 6 E&W 14W-8S 58
Ct. 7 N&S 14W-8S 58
Ct. 8 E&W 14W-8S 58
Dana N&S 14W-8S 58
Dana Wy. N&S 14W-8S 58
Deerview Rd. N&S .. 14W-9S 58
Dougshire E&W 14W-6S 58
Dougshire Ct. E&W .. 14W-6S 58
East St. E&W 14W-8S 58
East Trail Ln. N&S .. 14W-9S 64
Emco Dr. N&S 14W-8S 58
Emro N&S 14W-8S 58
Erin E&W 14W-7S 58
Fair Elms Av. N&S .. 14W-7S 58
Forest Hill N&S 14W-8S 58
Forest Hill Av. N&S .. 14W-8S 58
Fox Borough Dr. N&S 14W-8S 58
Fox Ln. E&W 14W-9S 64
Frontage Rd. NE&SW 14W-8S 58
Garywood Dr. N&S 14W-7S,14W-6S 58
Gregford E&W 14W-6S 58
Hampton E&W 14W-7S 58

Grid Page
Hillcrest N&S 14W-7S 58
Huntington Ct. E&W 14W-8S 58
Laurie Ct. N&S 14W-6S 58
Laurie Ln. N&S 14W-6S 58
Lincolnshire N&S 14W-8S 58
Longwood Dr. E&W .. 14W-7S 58
Manor N&S 14W-7S 58
Manor Dr. N&S 14W-7S 58
McClintock N&S 14W-8S 58
Navajo Ct. N&S 14W-9S 64
Northgate Pl. NS&EW 14W-8S 58
Omaha Dr. NW&SE .. 14W-9S 64
Plainfield SW&NE 14W-7S 58
Purdie St. E&W 14W-7S 58
Ridgewood E&W 14W-8S 58
Ridgewood Ln. NS&EW 14W-8S 58
Seneca Ct. NE&SW .. 14W-9S 64
Shady St. NW&SE .. 14W-7S 58
Shagbark Ct. E&W .. 14W-9S 64
Shagbark Ln. NE&SW 14W-9S 64
South E&W 14W-6S 58
South Dr. E&W 14W-6S 58
Southgate Ct. E&W .. 14W-8S 58
St. James Ct. N&S .. 14W-7S 58
Steeplechase Ln. N&S 14W-9S 64
Steepleside Dr. N&S 14W-9S 64
Stirrup Ln. E&W 14W-7S 58
Stonehedge Ct. E&W 14W-8S 58
Surrey SW&NE 14W-7S 58
Surrey Ln. E&W 14W-7S 58
Tartan Ridge Rd. NW&SE 14W-6S 58
Tend St. E&W 14W-8S 58
Thornhill Ct. E&W 14W-8S 58
Tomlin N&S 14W-6S 58
Tomlin Cir.,Dr. NS&EW 14W-6S 58
Tower E&W 14W-7S 58
Trent Ct. NE&SW 14W-9S 64
Walnut Ct. E&W 14W-9S 64
Waterside Pl. NE&SW 14W-9S 64
White Oak Ct. E&W .. 14W-9S 64
Wolf N&S 14W-8S 58
Wolf Rd. N&S 14W-9S,14W-8S 64
Woodglen E&W 14W-8S 58
Woodland Ct. E&W .. 14W-8S 58
Woodland Ln. NW&SE 14W-8S 58
Woodside Ct. E&W .. 14W-9S 64
Woodside Ln. N&S .. 14W-9S 64
Woodview N&S 14W-6S 58
63rd E&W 14W-6S 58
67th E&W 14W-7S 58
72nd St. E&W 14W-8S 62
73rd Pl. E&W 14W-8S 58
73rd St. E&W 14W-8S 62
74th St. E&W 14W-8S 58
75th St. E&W 14W-8S 62
77th St. E&W 14W-8S 62
79th St. E&W 14W-8S 62
83rd St. E&W 14W-9S 64

Calumet City

Allan Ln. NE&SW 2E-16S 83
Arthur Av. E&W 3E-17S 83
Arthur St. N&S 3E-19S 83
Aster St. E&W 2E-16S 83
Baker St. E&W 3E-17S 83
Balmoral Av. N&S 3E-19S 83
Bensley Av. N&S 2E-17S 83
Blackstone Av. N&S .. 1E-17S 82
Buffalo Av. N&S 3E-19S,3E-17S 83
Burnham Av. N&S 3E-19S,3E-17S 83
Calhoun Av. N&S 2E-17S 83
Calumet Way NW&SE 4E-18S 83
Campbell Av. N&S 3E-18S,3E-17S 83
Chappel Av. N&S 2E-17S 83
Chestnut Ct. E&W 2E-18S 83
Cleveland Av. E&W .. 3E-17S 83
Clyde Av. N&S 2E-18S,2E-17S 83
Commercial Av. N&S 3E-18S,3E-17S 83
Cornell Av. N&S 2E-18S,2E-17S 83
Crandon Av. N&S 2E-17S 83
Cunningham Dr. N&S 3E-19S 83
Dante Av. N&S 1E-17S 82
Dawn Ln. E&W 2E-16S 83
Detriot St. E&W 4E-18S 83
Dogwood Ct. E&W .. 2E-18S 83
Dolton Av. NW&SE .. 2E-16S 83
Dorchester Av. N&S .. 1E-17S 82
Douglas Av. N&S 4E-17S 83
Downs Dr. E&W 2E-16S 83
East End Av. N&S 2E-18S,2E-17S 83
East-West Rd. E&W .. 2E-19S 83
Edgewood Dr. N&S .. 4E-19S 83
Elm Ct. E&W 2E-18S 83
Escanaba Av. N&S .. 3E-17S 83
Euclid St. E&W 3E-17S 83
Exchange Av. N&S .. 3E-17S 83
Forest Av. N&S 4E-17S 83
Forest Ct. E&W 4E-19S 83

Grid Page
Forest Hills Av. N&S .. 4E-18S 83
Forest Ln. E&W 4E-18S 83
Forest Pl. N&S 4E-19S 83
Forestdale Av. E&W .. 4E-18S 83
Forsythe Av. N&S 4E-18S,4E-17S 83
Freeland Av. N&S 4E-19S,4E-18S 83
Garfield Av. NW&SE .. 4E-17S 83
George St. E&W 3E-17S 83
Golf Ct. N&S 4E-18S 83
Gordon Av. N&S 4E-19S,4E-17S 83
Green Bay Av. N&S 3E-19S,3E-17S 83
Greenwood Rd. NW&SE 2E-18S 83
Harbor Av. E&W 2E-16S 83
Harding Av. E&W 2E-17S,4E-17S 83
Harmess St. E&W 3E-17S 83
Harrison St. N&S 4E-17S 83
Hawthorne Ct. NE&SW 2E-18S 83
Henry St. NE&SW 3E-18S 83
Hickory Ct. N&S 2E-18S 83
Highland St. E&W 4E-18S 83
Hirsch Av. N&S 4E-19S,4E-17S 83
Hirsch Blvd. N&S 4E-19S 83
Houston Av. N&S 3E-18S,3E-17S 83
Hoxie Av. N&S 2E-17S 83
Huntington Av. N&S .. 2E-19S 83
Imperial Av. N&S 3E-19S 83
Ingram Av. N&S 4E-17S 83
Jeffery Av. N&S 2E-17S 83
Jennifer Ln. EW&NS .. 2E-16S 83
Kenilworth Dr. N&S .. 4E-19S 83
Legion Dr. N&S 4E-18S 83
Lincoln Av. N&S 4E-19S,4E-17S 83
Lincoln Pl. N&S 4E-19S 83
Locust Ct. N&S 2E-18S 83
Lucas St. E&W 3E-17S 83
Luella Av. N&S 2E-17S,2E-16S 83
Mackinaw Av. N&S 3E-19S,3E-17S 83
Madison Av. N&S 2E-18S,2E-17S 83
Manistee Av. N&S 3E-17S 83
Marquette Av. N&S .. 3E-17S 83
Mason St. E&W 4E-18S 83
Memorial Dr. E&W 2E-17S,4E-17S 83
Merril Av. N&S 2E-17S 83
Michigan City Rd. NW&SE 2E-17S,4E-18S 83
Muskegon Av. N&S 3E-18S,3E-17S 83
Newell Av. N&S 3E-18S 83
Oglesby Av. N&S 2E-17S,2E-16S 83
Park Av. NW&SE 2E-19S 83
Patricia Pl. E&W 2E-16S 83
Patton Ln. N&S 2E-16S 83
Paxton Av. N&S 2E-18S,2E-16S 83
Plummer Av. NW&SE 4E-17S 83
Prairie Av. N&S 2E-17S 83
Price Av. N&S 4E-19S,4E-17S 83
Pulaski Rd. E&W 2E-17S 83
Regency Ct. E&W 2E-19S 83
Ridgeland Av. N&S .. 2E-17S 83
Ring Rd. NS&EW 2E-19S 83
River Dr. N&S 3E-19S 83
River Oaks Dr. NW&SE 2E-19S,4E-19S 83
Rt. 83 NS&EW 2E-17S 83
Rudolph St. N&S 3E-18S 83
Ruth St. E&W 4E-18S 83
Saginaw Av. N&S 3E-17S 83
Schrum Rd. NW&SE .. 4E-19S 83
Shirley Dr. N&S 4E-20S,4E-19S 89
Sibley Blvd. E&W 2E-17S,4E-17S 83
Stanley Blvd. N&S .. 4E-19S 83
State Line Rd. N&S .. 4E-18S 83
State St. E&W 2E-17S,4E-17S 83
Stewart Av. E&W 2E-17S,3E-17S 83
Superior Av. N&S 3E-19S,3E-18S 83
Sycamore Ct. N&S .. 2E-18S 83
Timothy Ln. N&S 2E-16S 83
Torrence Av. N&S 2E-19S,2E-17S 83
Twilight Ln. E&W 2E-16S 83
Valenca Ct. E&W 2E-19S 83
Virginia St. E&W 3E-18S 83
Waltham St. E&W 4E-18S 83
Warren Av. E&W 4E-18S 83
Webb Av. E&W 4E-18S 83
Wentworth Av. N&S 4E-20S,4E-17S 89
West Dr. E&W 3E-19S 83
William St. E&W 3E-19S 83
Willow Ct. N&S 2E-18S 83
Wilson Av. E&W 2E-17S,4E-17S 83
Woodview Av. N&S .. 3E-19S 83
Yates Av. N&S 2E-17S,2E-16S 83

Grid Page
142nd St. E&W 2E-16S 83
151st St. E&W 3E-17S 83
152nd Pl. E&W 4E-17S 83
153rd Pl. E&W 4E-17S 83
153rd St. E&W 2E-17S,4E-17S 83
154th Pl. E&W 4E-18S 83
154th St. E&W 3E-17S,4E-17S 83
155th Pl. E&W 3E-18S,4E-18S 83
155th St. E&W 3E-18S,4E-18S 83
156th Pl. E&W 3E-18S,4E-18S 83
156th St. E&W 3E-18S,4E-18S 83
157th Pl. E&W 3E-18S 83
157th St. E&W 3E-18S,4E-18S 83
158th Pl. E&W 2E-18S 83
158th St. E&W 3E-18S 83
159th St. E&W 3E-18S,4E-18S 83
160th Pl. E&W 3E-19S 83
160th St. E&W 3E-19S 83
161st St. E&W 3E-19S,4E-19S 83
162nd Pl. E&W 4E-19S 83
162nd St. E&W 3E-19S,4E-19S 83
163rd Pl. E&W 4E-19S 83
163rd St. E&W 3E-19S,4E-19S 83
164th Pl. E&W 4E-19S 83
164th St. E&W 3E-19S,4E-19S 83
165th Pl. E&W 4E-19S 83
165th St. E&W 4E-19S 83
166th Pl. E&W 4E-19S 83
166th St. E&W 4E-19S 83
167th Pl. E&W 4E-20S 89
167th St. E&W 4E-19S 83
169th St. E&W 4E-20S 89
170th St E&W 2E-19S,4E-19S 83

Calumet Park

Aberdeen St. N&S 1W-15S,1W-14S 76
Ada St. N&S 1W-15S,1W-14S 76
Ashland Av. N&S 2W-15S,2W-14S 75
Bishop St. N&S 1W-15S,1W-14S 76
Calumet St. E&W 1W-15S 75
Carpenter St. N&S 1W-15S,1W-14S 76
Cass St. N&S 1W-15S 75
Elizabeth St. N&S 1W-15S,1W-14S 76
Green St. N&S 1W-14S 76
Halsted St. N&S 1W-14S 76
High St. E&W 1W-15S 76
Honore St. N&S 2W-14S 75
Illinois St. N&S 1W-15S 75
Justine St. N&S 1W-15S,1W-14S 76
Laflin St. N&S 1W-15S,1W-14S 76
Lincoln St. N&S 2W-14S 75
Loomis St. N&S 1W-15S,1W-14S 76
Marshfield Av. N&S 2W-15S,2W-14S 75
May St. N&S 1W-15S,1W-14S 76
Morgan St. N&S 1W-14S 76
Page Ct. N&S 2W-15S 75
Page St. N&S 2W-14S 75
Paulina St. N&S 2W-14S 75
Peoria St. N&S 1W-14S 76
Racine Av. N&S 1W-15S,1W-14S 76
Sangamon St. N&S .. 1W-14S 76
Throop St. N&S 1W-15S,1W-14S 76
Vermont St. E&W 1W-15S 76
Winchester Av. N&S 2W-14S 75
Wood St. N&S 2W-14S 75
York St. E&W 1W-15S 76
119th St. E&W 2W-14S 75
120th St. E&W 2W-14S 75
121st St. E&W 2W-14S 75
122nd St. E&W 2W-14S 75
123rd St. E&W 1W-14S 76
124th St. E&W 2W-14S,1W-14S 75
125th St. E&W 2W-14S,1W-14S 75
126th St. E&W 2W-14S,1W-14S 75
127th St. E&W 2W-14S 75
128th St. E&W 1W-15S 76
129th Pl. E&W 1W-15S 76
129th St. E&W 2W-15S 75

Chicago Heights

Abbott Av. N&S 2W-24S 93
Aberdeen St. N&S .. 1W-25S 93
Adams St. NE&SW .. 2W-26S 93
Alden Ct. N&S 2W-26S 93
Alice St. E&W 0W-23S 88
Alvin Pl. E&W 0W-24S 94
Amy St. E&W 1W-24S 93

Grid Page
Andover Ct. NW&SE 2W-26S 93
Andover Dr. E&W 2W-26S 93
Arnold St. N&S 0W-25S 94
Arquilla Dr. N&S 1W-24S 93
Ash St. NW&SE 1W-26S 93
Ashland Av. N&S 2W-25S,2W-23S 93
Avonelle Av. N&S 1W-24S 93
Barbara Ln. N&S 2W-25S 93
Beacon Blvd. E&W .. 2W-26S 93
Beacon Ct. N&S 2W-26S 93
Bellevue Pl. E&W 1W-24S 93
Birch Ln. N&S 1W-24S 93
Birmingham Av. E&W 0W-26S 94
Bollman Pl. E&W 0W-24S 94
Boston St. N. E&W .. 2W-26S 93
Boston St. W. N&S .. 2W-26S 93
Bradley Dr. E&W 2W-24S 93
Bradoc St. N&S 2W-24S 93
Brentwood Dr. N&S .. 1W-23S 88
Briargate Av E&W .. 0W-24S 94
Broadway NW&SE .. 1W-24S 93
Brookline St. E&W .. 2W-26S 93
Buena Vista Av. N&S 1W-25S 93
Buena Vista Cir. N&S 1W-25S 93
Bunker St. N&S 2W-26S 93
Butler St. N&S 0W-26S,0W-25S 94
Butterfield St. N&S .. 0W-26S 94
Caldwell Av. N&S 1W-24S 93
Cambridge St. N&S .. 2W-26S 93
Campbell Av. N&S 1W-25S,1W-24S 93
Carey Ct. E&W 2W-24S 93
Carpenter St. N&S .. 1W-24S 93
Cedar Ln. N&S 1W-24S 93
Center Av. NE&SW .. 0W-25S 94
Charing Cross Rd. E&W 1W-24S 93
Charles St. E&W 1W-23S 88
Chicago Rd. N&S 1W-24S,1W-23S 93
Circle Ct. N&S 1W-26S 93
Claude Ct. N&S 0W-25S 94
Coales St. E&W 1W-23S 88
Concord Ct. NW&SE 2W-26S 93
Concord Dr. S. N&S 2W-26S 93
Concord Dr. W. E&W 2W-26S 93
Constance Ln. N&S 1W-24S,1W-23S 93
Coolidge St. N&S 1W-24S 93
Cottage Grove Av. N&S 0E-25S 94
Country Club Rd. E&W 1W-25S 93
Craig Dr. E&W 1W-23S 88
Crescent Dr. E&W .. 1W-25S 93
D'amico Dr. N&S 2W-25S,2W-24S 93
Dartmouth Dr. N&S .. 1W-26S 93
Dawn Ln. N&S 1W-23S 88
De Angelis Ct. N&S .. 1W-24S 93
Deer Trail Rd. E&W .. 2W-24S 93
Division St. N&S 2W-26S,2W-25S 93
Dixie Hwy. NW&SE 1W-25S,1W-24S 93
Donovan Dr. N&S 2W-25S 93
Doris Ln. N&S 1W-23S 88
East End Av. N&S .. 0W-26S 94
Eastgate Av. E&W 1W-24S,0W-24S 93
Eddy St. E&W 1W-23S 88
Edgewood Av. N&S .. 1W-25S 93
Elder Av. E&W 1W-25S 93
Elmwood Dr. E&W .. 1W-23S 88
Emelia St. E&W 1W-24S 93
Emerald Av. N&S 0W-25S,0W-24S 94
Engleston Av. N&S .. 0W-24S 94
Enterprise Ct. NW&SE 2W-24S 93
Enterprise Dr. N&S .. 2W-24S 93
Euclid Av. N&S 1W-26S 93
Fairview Av. N&S 2W-25S 93
Fitch Rd. E&W 2W-24S 93
Flossmoor Rd. NE&SW 1W-23S 88
Floyd Ln. N&S 1W-23S 88
Forest St. E&W 1W-26S 93
Francis Ln. E&W 1W-23S 88
Franklin Av. N&S 1W-25S 93
Frederick Dr. E&W .. 1W-23S 88
Frontage Rd. E&W .. 0W-27S 94
Gail Ln. E&W 2W-24S 93
Garden Av. N&S 1W-25S 93
Glengate Av. E&W 2W-24S,0W-24S 93
Glenwood Rd. NE&SW 0W-23S 88
Grace Ln. N&S 1W-23S 88
Grant Av. N&S 1W-24S 93
Green St. N&S 0W-25S 94
Greenbriar Av. E&W .. 1W-24S 93
Gregory Dr. E&W 2W-25S 93
Grosvenor Pl. N&S .. 1W-24S 93
Halsted St. N&S 1W-26S,1W-24S 93
Hamilton Wood N&S 1W-23S 88
Hanover St. N&S 0W-25S 94

Street	Grid	Page
Hawthorne Ln. E&W	1W-24S	93
Hickory St. E&W	2W-25S,1W-25S	93
Highland Dr. E&W	2W-24S	93
Hillcrest Av. E&W	1W-24S	93
Hilltop Av. N&S	2W-25S	93
Holbrook Cir. NS&EW	1W-23S	88
Holbrook Rd. E&W	1W-23S	88
Homewood Ct. N&S	1W-24S	93
Illinois St. E&W	1W-23S	93
Independence Way E&W	1W-25S	93
Ingrid Ln. N&S	2W-25S	93
Iris Ln. N&S	1W-23S	88
Irving Blvd. E&W	1W-25S	93
Isa Av. E&W	1W-25S	93
Joe Orr Rd. E&W	2W-24S,0E-24S	93
Joyce Ln. N&S	1W-23S	88
Judith Ln. E&W	1W-23S	88
Kathleen Ln. N&S	1W-23S	88
Kingston Pl. E&W	1W-24S	93
Laflin St. N&S	1W-24S	93
Laura St. E&W	1W-23S	88
Lawrence Dr. N&S	2W-26S	93
Le Rose Dr. N&S	1W-23S	88
Leonard Av. E&W	1W-23S	88
Lexington Dr. N&S	2W-24S	93
Lincoln Av. N&S	0W-25S	94
Lincoln Hwy. E&W	2W-25S,0E-25S	93
Linden Ln. E&W	2W-24S	93
Longwood Ct. E&W	2W-24S	93
Longwood Dr. N&S	2W-24S	93
Lowe Av. N&S	0W-25S,0W-24S	94
Luther Ln. N&S	2W-24S	93
Lynn Ln. E&W	1W-23S	88
Mackler Dr. N&S	2W-24S	93
Main St. E&W	2W-26S,0W-26S	93
Manchester Dr. N&S	1W-24S	93
Maple Dr. E&W	1W-24S	93
Marian Way N&S	2W-24S	93
Martin Ln. E&W	0W-24S	94
Mason Ct. N&S	0W-25S	94
May Ct. E&W	1W-23S	88
Mayfair Pl. N&S	1W-24S,1W-23S	93
McArthur Dr. N&S	2W-24S	93
McEldowney Pl. E&W	1W-25S	93
McKinliey Av. NE&SW	0W-25S	94
Meadow Ln. E&W	1W-25S	93
Memorial Dr. NS&EW	1W-24S	93
Mildred Ln. N&S	1W-23S	88
Millview Dr. N&S	3W-26S	92
Minette Ln. NE&SW	2W-25S	93
Morgan St. E&W	0W-25S	94
Morgan St. N&S	1W-23S	88
Normal Av. N&S	0W-24S	94
Normandy Dr. NS&EW	1W-23S	88
Oak St. NW&SE	1W-25S	93
Orchard Av. N&S	2W-25S	93
Otto Blvd. NW&SE	1W-25S	93
Pagoria Dr. N&S	0W-27S	94
Pamela Ln. N&S	1W-23S	88
Park Av. NW&SE	1W-25S	93
Parkside Av. E&W	1W-24S	93
Parkview Av. NW&SE	1W-25S	93
Parnell Av. N&S	0W-25S,0W-24S	94
Patricia Dr. N&S	1W-23S	88
Peggy Ln. N&S	2W-25S	93
Peoria St. N&S	1W-24S	93
Peyton Dr. N&S	1W-23S	88
Piacenti Ln. N&S	1W-23S	93
Pleasant Dr. E&W	1W-23S	88
Plymouth Dr. NW&SE	1W-26S	93
Portland Av. N&S	0W-25S	94
Prairie Av. N&S	1W-25S	93
Ranger Dr. E&W	2W-24S	93
Raye Dr. N. N&S	1W-23S	88
Raye Dr. W. E&W	1W-23S	88
Revere St. N&S	2W-26S	93
Roberta Ln. N&S	1W-24S	93
Rosewood Ln. E&W	1W-23S	88
Rt. 1 N&S	1W-25S	93
Sandra Ln. E&W	2W-24S	93
Sangamon St. N&S	1W-24S	93
Saratoga Dr. N&S	2W-24S	93
Schilling Av. N&S	1W-25S	93
School St. N&S	1W-25S	93
Scott Av. N&S	2W-25S	93
Serena Dr. N&S	1W-23S	88
Shea Dr. N&S	2W-24S	93
Shelly Ln. N&S	2W-25S	93
Sherry Ln. N&S	1W-23S	88
Shields St. N&S	0W-25S	94
Southgate Av. E&W	1W-24S	93
Spruce Ln. E&W	1W-24S	93
Standish St. N&S	1W-24S	93
State St. N&S	0W-26S,0W-24S	94
Stewart Av. N&S	0W-26S,0W-24S	94
Stirling Av. N&S	2W-25S	93
Strieff Av. N&S	1W-23S	88
Sue Ct. E&W	1W-23S	88
Sunnyside Av. N&S	1W-25S	93
Sunset Av. N&S	2W-25S	93
Tahoe Dr. E&W	1W-23S	88
Terrace Dr. E&W	1W-23S	88
Terry Ct. E&W	1W-23S	88
Thelma Ln. N&S	1W-23S	88
Theresa Ln. N&S	1W-24S	93
Thomas St. N&S	1W-23S	93
Thorn St. N&S	1W-23S	93
Thorne Ct. NE&SW	1W-23S	94
Travers Ct. NE&SW	2W-24S	93
Union Av. N&S	0W-26S,0W-24S	94
Vincennes Av. NW&SE	1W-25S	93
Vollmer Rd. E&W	2W-23S,0W-23S	87
Wallace Ct. N&S	1W-25S	93
Wallace St. N&S	0W-26S,0W-24S	94
Washington Av. NE&SW	0W-25S	94
Wentworth Av. N&S	0W-26S,0W-25S	94
West End Av. NE&SW	0W-25S	94
Western Av. N&S	2W-25S	93
Westgate Av. N&S	1W-24S	93
Willow Dr. NW&SE	1W-24S	93
Wilson Av. N&S	2W-25S	93
Winchester Rd. E&W	2W-24S	93
Winston Ln. N&S	2W-24S	93
Wood St. N&S	2W-24S	93
Wooster Ct. N&S	2W-26S	93
Yale Av. N&S	2W-24S	93
Yorktown Rd. E&W	2W-24S	93
5th Av. N&S	0W-25S	94
7th Pl. E&W	2W-24S	94
8th St. E&W	2W-24S	93
10th St. E&W	2W-25S,0W-24S	93
11th Pl. E&W	2W-25S	93
11th St. E&W	1W-25S,0W-25S	93
12th St. E&W	2W-25S,0W-25S	93
13th St. E&W	1W-25S	93
14th Pl. E&W	2W-25S,1W-25S	93
14th St. E&W	1W-25S,0W-25S	93
15th Pl. E&W	2W-25S,1W-25S	93
15th St. E&W	2W-25S,0W-25S	93
16th Pl. E&W	2W-25S,0W-25S	93
16th St. E&W	2W-25S,0W-25S	93
17th St. E&W	2W-25S,0W-25S	93
19th Pl. E&W	1W-26S	93
19th St. E&W	1W-26S	93
21st St. E&W	1W-26S,0W-26S	94
22nd St. E&W	0W-26S	94
23rd St. E&W	1W-26S,0W-26S	93
24th St. E&W	1W-26S,0W-26S	93
25th St. E&W	1W-26S,0W-26S	93
26th St. E&W	1W-26S,0W-26S	93
28th St. E&W	0W-26S	94
197th Pl. E&W	0W-23S	88
197th St. E&W	0W-23S	88
202nd St. E&W	2W-24S	93
205th St. E&W	2W-24S	93
209th St. E&W	2W-24S	93

Chicago Ridge

Street	Grid	Page
Anderson Av. N&S	8W-11S	65
Austin Av. N&S	7W-12S	66
Avon Av. NW&SE	8W-11S	65
Barnard Dr. NE&SW	8W-12S	65
Birmingham Av. NW&SE	7W-12S	66
Brooks Ln. NE&SW	8W-12S	65
Central Av. N&S	7W-12S	66
Charles St. N&S	8W-11S	65
Fireside Dr. E&W	8W-11S	65
Forest Ln. N&S	8W-12S	65
Harlem Av. N&S	8W-11S	65
Hyland Pl. N&S	8W-11S	65
Interstate 294 NW&SE	8W-11S	65
Janet Ln. E&W	8W-11S	65
Kitty Ln. NW&SE	8W-11S	65
Leslie Ln. N&S	8W-11S	65
Lombard Av. N&S	7W-12S	66
Lyman Av. E&W	7W-12S	66
Lyman Av. NE&SW	7W-12S	66
Major Av. N&S	7W-12S	66
Mansfield Av. N&S	7W-12S	66
Marshall Av. NW&SE	7W-12S	66
Mason Av. N&S	7W-12S	66
Massasoit Av. N&S	7W-12S	66
Mather Av. NW&SE	8W-11S,7W-11S	65
Mayfield Av. N&S	7W-12S	66
McVicker Av. N&S	7W-12S	66
Menard Av. N&S	7W-12S	66
Michael Ln. N&S	8W-11S	65
Monitor Av. N&S	7W-12S	66
Moody Av. N&S	7W-12S	66
Nashville Av. N&S	8W-12S	65
Natchz St. N&S	8W-12S	65
Natoma Av. N&S	8W-12S	65
New England Av. N&S	8W-11S	65
Norfolk Ln. NE&SW	8W-12S	65
Normandy Av. N&S	8W-12S	65
Nottingham Av. N&S	8W-11S	65
O'Connell Dr. N&S	8W-11S	65
Oak Av. N&S	7W-12S	66
Oak Park Av. N&S	8W-12S,8W-11S	65
Orchard Av. N&S	8W-11S	65
Oxford Av. N&S	7W-12S	66
Pamela Ln. E&W	8W-12S	65
Parkside Av. N&S	7W-12S	66
Pleasant Blvd. NW&SE	7W-12S	66
Princess Av. N&S	7W-12S	66
Ready Dr. E&W	8W-11S	65
Ridge Dr. NE&SW	8W-11S	65
Ridgeland Av. N&S	7W-12S,7W-11S	66
Ridgemont Ln. NW&SE	7W-12S	66
Rt. 43 N&S	8W-11S	65
Rt. 7 NE&SW	8W-11S	65
Rutherford Av. N&S	8W-11S	65
Sayre Av. N&S	8W-11S	65
Southwest Hwy. NE&SW	8W-12S,8W-11S	65
Stephen Dr. N&S	8W-11S	65
Sycamore Dr. N&S	8W-11S	65
Tracey Av. E&W	8W-11S	65
Virginia Av. NW&SE	8W-11S	65
Washington St. NW&SE	7W-12S	66
97th Pl. E&W	8W-11S	65
98th St. E&W	8W-11S	65
99th Pl. E&W	8W-11S	65
99th St. E&W	8W-11S	65
100th St. E&W	8W-11S	65
101st St. E&W	7W-11S	66
102nd Pl. E&W	8W-11S	65
103rd St. E&W	8W-11S,7W-11S	65
104th Pl. E&W	8W-12S	65
104th St. E&W	8W-12S	65
105th St. E&W	8W-12S,7W-12S	65
106th Pl. E&W	8W-12S	65
106th St. E&W	8W-12S,7W-12S	65
107th Pl. E&W	7W-12S	66
107th St. E&W	8W-12S,7W-12S	65
108th Pl. E&W	7W-12S	66
108th St. E&W	7W-12S	66
109th Pl. E&W	7W-12S	66
109th St. E&W	7W-12S	66
110th St. E&W	7W-12S	66
111th St. E&W	7W-12S	66

Cicero

Street	Grid	Page
Austin N&S	7W-3S	54
Austin Blvd. N&S	7W-3S	54
Central N&S	7W-3S,7W-2S	54
Cermak Rd. E&W	6W-1S	55
Cicero Av. N&S	5W-3S,6W-1S	55
Edgewood E&W	7W-3S	54
Laramie N&S	6W-3S,6W-1S	55
Ogden E&W	7W-3S	54
Park E&W	7W-3S	54
Rt. 50 N&S	6W-2S	55
12th Pl. E&W	6W-1S	55
13th E&W	7W-1S	54
14th E&W	7W-1S	54
15th E&W	7W-1S,6W-1S	55
15th Pl. E&W	7W-1S,6W-1S	55
16th E&W	7W-1S	54
18th E&W	7W-1S,6W-1S	54
19th E&W	7W-1S,6W-1S	54
20th E&W	5W-1S	55
21st E&W	7W-1S,6W-1S	54
21st Pl. E&W	6W-1S	55
22nd E&W	7W-2S	54
23rd E&W	7W-2S,6W-2S	54
24th E&W	6W-2S	55
25th E&W	6W-2S	55
25th Pl. E&W	6W-2S	55
26th E&W	6W-2S	55
28th E&W	6W-2S	55
28th Pl. E&W	6W-2S	55
29th E&W	6W-2S	55
29th Pl. E&W	6W-2S	55
30th E&W	6W-2S	55
30th Pl. E&W	6W-2S	55
31st E&W	7W-2S,6W-2S	55
31st Pl. E&W	6W-2S	55
32nd E&W	6W-3S	55
32nd Pl. E&W	6W-3S	55
33rd E&W	7W-3S,6W-3S	55
34th E&W	7W-3S	54
35th E&W	6W-3S	55
36th E&W	6W-3S	55
37th E&W	6W-3S	55
38th E&W	6W-3S	55
46th Ct. N&S	5W-1S	55
47th N&S	5W-1S	55
47th Ct. N&S	5W-1S	55
48th Ct. N&S	6W-2S,6W-1S	55
49th N&S	6W-3S,6W-1S	55
49th Ct. N&S	6W-1S	55
50th N&S	6W-2S,6W-1S	55
50th Ct. N&S	6W-3S,6W-1S	55
51st N&S	6W-3S,6W-1S	55
51st Ct. N&S	6W-1S	55
52nd Ct. N&S	6W-3S	55
53rd N&S	6W-3S,6W-2S	55
53rd Ct. N&S	6W-3S	55
54th N&S	6W-3S,6W-1S	55
54th Ct. N&S	6W-3S	55
55th N&S	6W-3S,6W-1S	55
55th Ct. N&S	6W-3S	55
56th Ct. N&S	7W-3S,7W-1S	54
57th N&S	7W-3S,7W-1S	54
57th Ct. N&S	7W-3S,7W-1S	54
58th N&S	7W-3S,7W-1S	54
58th Ct. N&S	7W-3S,7W-1S	54
59th N&S	7W-3S,7W-1S	54
59th Ct. N&S	7W-3S,7W-1S	54
60th N&S	7W-3S	54
60th Ct. N&S	7W-3S,7W-1S	54
61st N&S	7W-3S,7W-1S	54
61st Ct. N&S	7W-3S,7W-1S	54

Country Club Hills

Street	Grid	Page
Amlin Cir. N&S	6W-22S	86
Amlin N&S	6W-22S	86
Anthony Av. N&S	5W-21S	86
Anthony St. N&S	5W-22S	86
Apple Tree Dr. N&S	4W-20S	87
Arlington Dr. N&S	4W-21S	87
Baker Av. N&S	5W-22S,5W-21S	86
Baker St. NE&SW	5W-21S	86
Becker Tr. N&S	6W-22S	86
Birch Av. N&S	5W-22S	86
Birch St. N&S	5W-21S	86
Brairgate Dr. N&S	4W-20S	87
Cedar Av. N&S	5W-22S,5W-21S	86
Cedar St. N&S	5W-21S	86
Central Pk. Av. N&S	4W-21S	87
Chestnut Av. N&S	5W-21S,5W-22S	86
Chestnut Dr. N&S	5W-23S,5W-22S	86
Cicero Av. N&S	6W-21S,6W-20S	86
Clarence Av. E&W	5W-21S	86
Country Club Dr. N&S	4W-21S	87
Coventry Ln. N&S	5W-20S	86
Crest Ct. E&S	6W-22S	86
Cypress Av. N&S	5W-22S,5W-21S	86
Cypress Ct. E&W	5W-23S	86
Cypress Dr. N&S	5W-23S,5W-22S	86
Devon Dr. N&S	4W-21S	87
East Gate Dr. N&S	5W-20S	86
Edwards Av. E&W	5W-21S	86
Elm Dr. N&S	5W-23S	86
Fairoak Ln. N&S	6W-21S	86
Fairoak Dr. E&W	6W-21S	86
Fairway Terr. NW&SE	5W-21S	86
Farm Crest Ter. N&S	5W-22S	87
Frontage Rd. N&S	6W-20S	86
Glen Oaks Ct. NW&SE	4W-20S	87
Glen Oaks Dr. NE&SE	4W-20S	87
Greenview Terr. N&S	5W-21S	86
Harvard Ln. N&S	4W-21S	87
Hawthorne Dr. N&S	5W-21S	86
Hickory Pl. N&S	5W-23S	86
Highland Pl. E&W	4W-21S	87
Holly Ct.* N&S	4W-20S	87
Hollywood Ln. NE&SE	6W-21S	86
Idlewild Dr. N&S	5W-21S	86
Indian Hills Dr. NE&SW	5W-20S	86
Interstate 57 NE&SW	6W-20S	86
Interstate 80 NE&SW	5W-21S	86
John Av. N&S	5W-21S	86
John St. N&S	5W-22S	86
Juneway Ct. N&S	5W-21S	86
Keeler Av. N&S	5W-22S,5W-21S	86
Kostner Av. N&S	5W-21S	86
Lake Ct. E&W	5W-21S	86
Laramie Rd. N&S	6W-22S	86
Larkin Ln. N&S	5W-22S	86
Lavergne Av. N&S	6W-20S,6W-21S	86
Laverne Av. N&S	5W-22S	86
Lee St. N&S	5W-22S	86
Loras Ln. N&S	5W-22S	86
Loretto Ln. E&W	5W-22S	86
Maple Av. N&S	5W-22S,5W-20S	86
Maple Ter. E&W	5W-22S	86
Martin Ct. E&W	5W-22S	86
Martin Ln. N&S	5W-22S	86
Mary Ann Ln. NE&SW	5W-22S	86
Mary Ct. E&W	6W-22S	86
Marycrest Dr. E&W	6W-22S	86
Marylakes N&S	6W-22S	86
Michael Av. N&S	5W-22S	86
Mocombo Way N&S	6W-21S	86
Mulberry Dr. N&S	5W-21S	86
Mulberry Terr. N&S	5W-22S	86
Neal Cir. N&S	6W-22S	86
Nightingale Ln. N&S	6W-22S	86
Nightingdale Ter. N&S	6W-22S	86
Oakwood Av. N&S	5W-22S,5S-21S	86
Oakwood Dr. N&S	5W-20S	86
Old Elm Dr. N&S	4W-20S	87
Olympia Dr. N&S	5W-21S	86
Orchard Av. N&S	4W-20S	87
Park Ln. N&S	5W-22S	86
Patrick Av. N&S	5W-21S	86
Peartree Ct.* N&S	4W-21S	87
Pheasant Dr. N&S	4W-21S	87
Poplar N&S	5W-21S	86
Princeton Av. N&S	4W-21S	87
Provincetown Dr. E&W	5W-22S	86
Pulaski Rd. N&S	5W-20S	86
Pulaski Rd. N&S	5W-20S,5W-23S	86
Ravisloe Terr. N&S	5W-21S	86
Rosewood Terr. N&S	5W-21S	86
Rt. 50 N&S	5W-22S,5W-20S	86
Russet Way NE&SW	5W-20S	86
Sarah Ct. NW&SE	4W-21S	87
Sarah Ln. N&S	4W-21S	87
School Dr. E&W	4W-21S	87
Soleri Dr. NE&SW	5W-21S	86
Springfield Av. N&S	4W-21S	87
Sunset Ln. E&W	6W-21S	86
Sunset Ridge Dr. N&S	4W-20S	87
Sycamore Av. N&S	5W-21S	86
Thomas Ln. NE&SW	5W-21S	86
Timerlea Ct. N&S	4W-20S	87
Timerlea Ln.* N&S	4W-20S	87
Vista N&S	6W-21S	86
Walnut Av. N&S	5W-22S	86
Wildwood Way N&S	4W-20S	87
Williamsburg Rd. E&W	5W-22S	86
Willow Av. N&S	5W-22S,5W-21S	86
Willow Ct. N&S	5W-21S	86
Wilshire Blvd. E&W	6W-21S	86
Windsor Ln. N&S	5W-22S	86
Winston Ct. NW&SE	4W-21S	87
Winston Dr. N&S	4W-21S	87
Yale Ln. N&S	4W-21S	87
19th Pl. E&W	6W-22S	86
167th Pl. E&W	4W-20S	87
167th St. E&W	5W-19S,4W-19S	81
168th Pl. E&W	4W-20S	87
168th St. E&W	4W-20S	87
169th St. E&W	4W-20S	87
171st Pl. E&W	4W-20S	87
171st St. E&W	4W-20S	87
172nd Pl. NE&SW	5W-20S	86
172nd St. E&W	4W-20S	87
173rd St. E&W	6W-20S	86
175th Pl. E&W	5W-21S,4W-21S	86
175th St. E&W	5W-20S,4W-21S	86
176th St. E&W	5W-21S,4W-21S	86
177th Pl. E&W	5W-21S	86
177th St. E&W	5W-21S,4W-21S	86
178th Pl. E&W	5W-21S	87
178th St. E&W	5W-21S	86
179th Pl. E&W	5W-21S	86
179th St. E&W	6W-21S,5W-21S	86
180th St. E&W	5W-21S	86
181st St. E&W	5W-21S	86
182nd Pl. E&W	5W-21S	86
182nd St. E&W	5W-21S	86
183rd St. E&W	5W-21S	86
184th Ct. E&W	5W-22S	86
184th Pl. NE&SW	5W-22S	86
184th St. E&W	5W-22S	86
185th Ct. E&W	5W-22S	86
185th Pl. E&W	5W-22S	86
185th St. E&W	6W-22S,6W-21S	86
186th Pl. E&W	5W-22S	86
186th St. E&W	5W-22S	86
187th Pl. E&W	5W-22S	86
187th St. E&W	6W-22S,5W-22S	86
188th Pl. E&W	5W-22S	86
188th St. E&W	5W-22S	86
189th Pl. E&W	5W-22S	86
189th St. E&W	5W-23S	86
190th St. E&W	5W-23S	86
191st St. E&W	6W-22S,6W-21S	86
192nd Pl. E&W	5W-23S	86
192nd St. E&W	5W-23S	86
193rd Pl. E&W	5W-23S	86
194th Ct. E&W	5W-23S	86
194th St. E&W	5W-23S	86
195th St. E&W	5W-23S	86

Countryside

Street	Grid	Page
Adlai E. Stevenson Expwy. E&W	13W-7S,12W-7S	58
Ashland N&S	12W-6S,12W-5S	58
Barton E&W	12W-6S	58
Bobolink E&W	12W-5S	58

	Grid	Page
Brainard N&S	12W-7S	58
Cantigny E&W	13W-7S	58
Catherine N&S	12W-6S,12W-5S	58
Constance SW&NE	12W-7S,12W-6S	58
Crestview E&W	13W-7S	58
Dansher N&S	12W-5S	58
Dawn N&S	13W-7S	58
East N&S	12W-5S	58
Forestview E&W	13W-7S	58
Francis Av. N&S	12W-6S	58
Golfview N&S	13W-7S	58
Hillsdale E&W	13W-7S	58
Kensington N&S	12W-6S	58
Leitch N&S	13W-6S	58
Longview E&W	12W-6S	58
Longview N&S	12W-6S	58
Lorraine E&W	12W-6S	58
Lorraine N&S	13W-7S	58
Madison N&S	12W-6S,12W-5S	58
Maplewood E&W	13W-7S	58
Merry N&S	12W-6S	58
Park N&S	13W-6S	58
Parkside N&S	13W-7S	58
Peck N&S	13W-6S	58
Rose Av. N&S	12W-6S	58
Rose Ct. N&S	12W-6S	58
Rosemary N&S	12W-6S	58
S.E. Ct. Pl. E&W	13W-6S	58
Staleford E&W	13W-7S	58
Sunset N&S	13W-7S	58
Terry E&W	12W-6S	58
6th N&S	12W-6S,12W-5S	58
8th N&S	12W-5S	58
9th N&S	12W-5S	58
56th E&W	12W-6S	58
57th St. E&W	12W-6S	58
58th E&W	12W-6S	58
60th Pl. E&W	12W-6S	58
61st Pl. E&W	12W-6S	58
67th E&W	13W-7S	58
71st Pl. E&W	13W-8S	58
75th E&W	13W-8S	58
7th N&S	12W-5S	58

Crestwood

	Grid	Page
Arbor Ln. E&W	6W-15S	74
Avers Av. N&S	4W-16S	75
Calumet Sag Rd. NW&SE	6W-15S,5W-15S	74
Carriage Ln. N&S	7W-15S	74
Central Av. N&S	7W-15S	74
Char Ln. N&S	6W-16S	74
Cicero Av. N&S	6W-16S,6W-15S	74
Circle Ct. E&W	6W-15S	74
Crescent Ct. NW&SE	6W-15S	74
Crestbrook Ct. N&S	6W-15S	74
Crestview Ct. N&S	6W-16S	74
Crestwood Ct. NW&SE	6W-15S	74
Crestwood Dr. E&W	6W-15S	74
Dori Ln. N&S	6W-16S	74
E. Circle Av. N&S	6W-15S	74
East Playfield Dr. N&S	6W-15S	74
Fairway Dr. NS&EW	6W-15S	74
Forestview Ct. E&W	6W-15S	74
Forestview Ln. N&S	6W-15S	74
Hamlin Av. N&S	4W-16S	75
Harding Av. N&S	4W-16S	75
Highland Ct. E&W	6W-15S	74
Hill Dr. N&S	7W-15S	74
Homestead Dr. E&W	6W-15S	74
Interstate 294 NW&SE	5W-16S,4W-16S	74
James Ct. E&W	6W-16S	74
James Dr. NW&SE	6W-16S	74
Karlov Av. N&S	5W-16S	74
Keeler Av. N&S	5W-16S	74
Kenneth Ct. N&S	6W-15S	74
Kenton Av. N&S	6W-15S,5W-16S	74
Kildare Av. N&S	5W-16S	74
Kilpatrick Av. N&S	5W-16S	74
Kostner Av. N&S	5W-16S	74
Lamon Av. N&S	6W-16S	74
Laramie Av. N&S	6W-16S,6W-15S	74
Laramie Ct. N&S	6W-16S	74
Lavergne Av. N&S	6W-16S	74
Lawler Av. N&S	6W-16S	74
LeClaire Av. N&S	6W-15S	74
Leclaire Av. N&S	6W-16S	74
Leonard Dr. N&S	6W-16S	74
Linder Av. E&W	6W-15S	74
Long Av. N&S	6W-16S,6W-15S	74
Loomis Ct. N&S	6W-15S	74
Loomis Ln. N&S	6W-16S	74
Midlothian Turnpike NE&SW	6W-16S,5W-16S	74
Model Ct. NE&SW	6W-15S	74
Park Ct. NW&SE	6W-15S	74
Park Ln. E&W	6W-15S	74
Park Pl. N&S	7W-15S	74
Parkview Ct. E&W	6W-15S	74
Pinecrest Dr. N&S	6W-15S	74
Pleasant Ln. E&W	6W-15S	74
Pulaski Rd. N&S	5W-16S	74
Regal Ct. N&S	6W-16S	74
Regal Dr. E&W	6W-16S	74
Rexford Rd. NE&SW	4W-16S	75
Rivercrest Dr. E&W	6W-15S	74
Royal E&W	6W-16S	74
Rt. 50 N&S	5W-16S	74
Sandra Ln. N&S	6W-16S,6W-15S	74
Short Dr. N&S	6W-16S	74
South End Ln. N&S	6W-16S	74
Springfield Av. N&S	4W-16S	75
Terrace Ln. N&S	6W-15S	74
Tri State Tollway NW&SE	5W-16S,4W-16S	74
Village Ct. NW&SE	6W-15S	74
Village Ln. N&S	6W-15S	74
W. Circle Av. N&S	6W-15S	74
W. Circle Dr. Pkwy. E&W	6W-15S	74
Walter Dr. N&S	6W-16S	74
Waterbury Ct. E&W	6W-16S	74
Waterbury Dr. N&S	6W-16S	74
Waterbury Ln. E&W	6W-16S	74
Waterbury Way E&W	6W-16S	74
West End Ln. N&S	6W-15S	74
West Playfield Dr. N&S	6W-15S	74
128th Pl. E&W	6W-15S	74
128th St. E&W	7W-15S	74
129th Pl. E&W	6W-15S	74
129th St. E&W	7W-15S	74
130th St. E&W	6W-15S	74
131st St. E&W	6W-15S	74
132nd Ct. E&W	6W-15S	74
132nd St. E&W	6W-15S	74
133rd St. E&W	6W-15S	74
134th Pl. E&W	6W-15S	74
135th Pl. E&W	6W-16S	74
135th St. E&W	6W-15S,5W-15S	74
136th Ct. E&W	5W-16S	74
136th Pl. E&W	6W-16S	74
137th Pl. E&W	6W-16S,5W-16S	74
137th St. E&W	6W-16S,5W-16S	74
138th Ct. E&W	6W-16S	74
138th Pl. E&W	6W-16S	74
138th St. E&W	6W-16S,5W-16S	74
139th Pl. E&W	6W-16S	74
139th St. E&W	6W-16S	74
140th Pl. E&W	6W-16S	74
141st St. E&W	6W-16S,4W-16S	74
142nd St. E&W	6W-16S,5W-16S	74
143rd Pl. E&W	6W-17S	80
143rd St. E&W	5W-16S	74

Deerfield*

(Also see Lake County)

	Grid	Page
County Line Rd. NS&EW	10W-18N	17
Interstate 94 E&W	11W-18N	16
Route 43 NW&SE	10W-18N	17

Des Plaines

	Grid	Page
Acres Ln. N&S	12W-10N	36
Albany Ln. NE&SW	14W-12N	25
Alden Ln. NW&SE	12W-8N	36
Alfini Dr. NS&EW	14W-10N	35
Alger E&W	12W-8N	36
Algonquin Rd. E&W	14W-10N	35
Alles St. N&S	13W-11N	26
Ambleside Rd. E&W	15W-11N	25
Amherst Av. N&S	14W-12N	25
Anderson Ter. N&S	15W-10N	34
Andrea Ln. N&S	15W-10N	34
Andy Ct. E&W	12W-8N	36
Anita N&S	14W-11N	25
Applecreek Ln. N&S	12W-10N	36
Ardmore Rd. N&S	14W-12N	25
Arlington Av. N&S	13W-11N	26
Armstrong Rd. NS&EW	15W-8N	34
Arnold Ct. N&S	15W-10N	34
Ash Ln. N&S	13W-9N	35
Ashland Av. E&W	14W-10N,13W-10N	35
Ballard Rd. E&W	12W-11N	26
Beau Ct. E&W	15W-10N,15W-11N	34
Beau Dr. N&S	15W-10N,15W-11N	34
Bedford Ln. N&S	14W-11N	25
Bell Dr. N&S	15W-11N	25
Bellaire Av. N&S	12W-11N	26
Bellaire Ct. E&W	11W-11N	26
Bennett Ln. N&S	15W-11N	25
Bennett Pl. E&W	12W-9N	36
Berkshire Ct. N&S	14W-12N	25
Berkshire Ln. E&W	14W-12N,13W-12N	25
Berry Ln. E&W	12W-10N	36
Big Bend Dr. NW&SE	12W-11N	26
Birch St. N&S	12W-9N	36
Birchwood Av. E&W	14W-9N,12W-9N	35
Bittersweet Ct. NW&SE	13W-9N	35
Bradley Ct. NE&SW	15W-11N	25
Bradley St. E&W	15W-11N,14W-11N	25
Bradrock Dr. E&W	14W-9N	35
Brentwood Dr. E&W	15W-11N	25
Briar Ct. E&W	13W-9N	35
Broadway N&S	14W-12N	25
Brown St. E&W	13W-11N	26
Busse Hwy. NW&SE	12W-10N	36
Cambridge Rd. N&S	14W-12N	25
Campbell Av. E&W	13W-10N,12W-10N	35
Carol Ln. NS&EW	13W-10N	35
Cavan Ln. N&S	15W-11N	25
Cedar Ct. E&W	12W-9N	36
Cedar St. N&S	12W-9N	36
Center St. NE&SW	13W-10N,13W-11N	35
Central Av. E&W	12W-8N	36
Central Rd. E&W	14W-12N	25
Chase Av. E&W	12W-9N	36
Chestnut St. N&S	13W-9N	35
Chicago Av. NS&EW	12W-11N	26
Church St. E&W	11W-11N	26
Cindy Ln. N&S	14W-10N	35
Circle St. NE&SW	13W-9N	35
Clark Ln. N&S	15W-10N	34
Clayton Ln. N&S	14W-12N	25
Columbia Av. E&W	14W-11N	25
Concord Ln. NS&EW	13W-12N	26
Cora St. N&S	12W-9N,12W-10N	36
Cordial Ln. E&W	15W-10N	34
Cornell Av. N&S	14W-12N	25
Courtesy Ln. E&W	15W-10N	34
Crabtree Ln. N&S	12W-10N	36
Craig Dr. NS&EW	12W-8N	36
Cranbrook St. E&W	14W-12N	25
Crestwood Dr. NS&EW	14W-11N	25
Cumberland Pkwy. N&S	14W-11N,14W-12N	25
Curtis St. N&S	12W-8N	36
Dale St. N&S	12W-8N	36
Danbury Ln. N&S	15W-10N	34
Dara James Rd. N&S	15W-11N	25
Daton Pl. N&S	15W-10N	34
David Dr. E&W	12W-8N	36
Davis Ct. E&W	14W-11N	25
Dawn Ct. N&S	12W-11N	26
Deane St. N&S	13W-9N	35
Debra Dr. E&W	15W-11N	25
Dempster St. E&W	14W-10N,12W-10N	35
Dennis Pl. NS&EW	13W-10N	35
Denver Dr. E&W	15W-10N	34
Des Plaines River Rd. N&S	12W-8N,13W-11N	36
Devon Av. E&W	12W-8N	36
Devonshire Dr. E&W	15W-10N	34
Dexter Ln. N&S	13W-9N	35
Diamond Head Dr. E&W	15W-10N	34
Doreen Dr. N&S	15W-10N	34
Dorthy Dr. E&W	15W-10N	34
Douglas Av. N&S	14W-9N	35
Dover Dr. E&W	15W-10N	34
Dover Ln. E&W	15W-10N	34
Drake Ln. E&W	14W-12N	25
Dulles Rd. E&W	15W-11N	25
Eaker Pl. E&W	15W-10N	34
Earl Av. N&S	13W-10N	35
East Grant Dr. N&S	13W-10N	35
East River Rd. NE&SW	12W-11N	26
East Villa Dr. N&S	13W-10N	35
Eastview Dr. N&S	13W-9N	35
Easy St. N&S	15W-11N	25
Edward Ct. N&S	11W-11N	26
Eisenhower Ct. N&S	12W-8N	36
Eisenhower Dr. N&S	12W-8N	36
Elizabeth Ln. E&W	15W-10N	34
Elk Blvd. NS&EW	12W-11N	26
Ellinwood St. NW&SE	13W-11N	26
Elm St. N&S	13W-9N	35
Elmhurst Rd. N&S	15W-10N	34
Elmira Av. N&S	14W-9N	35
Esser Ct. E&W	12W-9N	36
Estes Av. E&W	12W-8N	36
Everett Av. E&W	13W-9N,12W-9N	35
Everett Ln. E&W	13W-9N	35
Evergreen Av. E&W	14W-11N,13W-11N	25
Executive Way NS&EW	13W-10N	35
Fairhope Av. E&W	15W-8N	34
Fargo Av. E&W	13W-9N,12W-9N	35
Farthing Ln. N&S	15W-11N	25
Farwell Av. E&W	12W-8N	36
Figard Ln. NS&EW	15W-11N	25
Fletcher Dr. E&W	14W-11N	25
Florian Dr. E&W	15W-10N	34
Forest Av. E&W	14W-10N,12W-10N	35
Forest Ln. E&W	12W-11N	26
Forestedge Ln. N&S	12W-11N	26
Fox Ln. E&W	12W-8N	36
Fremont Av. E&W	14W-11N	25
Fremont Ct. E&W	15W-11N	25
Frontage Rd. N&S	13W-9N	35
Galleon N&S	15W-11N	25
Garland Pl. N&S	12W-11N	26
Golf Cul De Sac, North E&W	13W-12N	26
Golf Cul De Sac, South E&W	13W-12N	26
Golf Rd. E&W	13W-12N	26
Good Av. N&S	11W-11N	26
Graceland Av. NE&SW	13W-10N,13W-11N	35
Greco Av. N&S	12W-8N	36
Greenleaf Av. E&W	12W-8N	36
Greenview Av. E&W	14W-11N,13W-11N	25
Gregory St. E&W	13W-13N	26
Grove Av. N&S	12W-11N	26
Halsey Dr. NE&SW	12W-8N	36
Harding Av. E&W	14W-11N,13W-11N	25
Harvard St. E&W	14W-12N	25
Harvey Av. NS&EW	14W-11N	25
Hawthorne Ln. N&S	12W-11N	26
Hawthorne Ter. NW&SE	12W-11N	26
Hazel Ct. E&W	13W-9N	35
Heather Ln. N&S	13W-10N	35
Henry Av. E&W	13W-10N,12W-10N	35
Hewitt Dr. N&S	15W-10N	34
Hickory St. N&S	12W-9N	36
Higgins Rd. E&W	15W-8N	34
Highland Dr. E&W	13W-9N	35
Hills Av. E&W	13W-11N,12W-11N	26
Hinsdale Rd. N&S	15W-8N	34
Hoffman Pkwy. E&W	14W-10N,13W-10N	35
Holiday Ln. E&W	15W-10N	34
Hollywood Av. E&W	13W-11N	26
Horne Ter. N&S	15W-10N	34
Howard Av. E&W	14W-9N,12W-9N	35
Ida St. N&S	13W-11N	26
Illinois St. N&S	13W-9N	35
Ingram Pl. N&S	15W-10N	34
Inner Circle Dr. N&S	14W-12N	25
Interstate 294 N&S	12W-8N	36
Interstate 90 NW&SE	16W-9N,12W-8N	34
Iris Ln. N&S	12W-8N	36
Ironwood N&S	11W-11N	26
Irwin Av. E&W	13W-9N	35
James Rd. N&S	15W-11N	25
Janice Av. N&S	15W-10N	34
Jarlath Av. E&W	14W-9N	35
Jarvis Av. E&W	15W-9N,12W-9N	34
Jeanette St. N&S	13W-10N	35
Jefferson St. E&W	13W-11N	26
Jeffry Ln. E&W	15W-10N	34
Jill Ct. E&W	15W-10N	34
Jon Ct. N&S	15W-11N	25
Jon Ln. N&S	14W-12N	25
Joseph Av. N&S	12W-8N	36
Joseph J. Schwab Rd. NW&SE	12W-10N	36
Josephine Ct.* N&S	14W-12N	25
Joyce N&S	15W-10N	34
Junior Ter. NW&SE	12W-11N	26
Kathleen Dr. E&W	15W-11N,14W-11N	25
Kenilworth Ct. N&S	14W-11N	25
Kennicott Ct. E&W	12W-9N	36
Kerry Ct. E&W	11W-11N	26
King Ln. E&W	15W-10N	34
Kingston Ct. N&S	13W-10N	35
Kinkaid Ct. E&W	15W-10N	34
Koehler Dr. N&S	13W-9N	35
Kylemore Ct. NW&SE	13W-13N	26
Kylemore Dr. E&W	13W-13N	26
LaSalle St. N&S	14W-11N	25
Lancaster Ln. E&W	15W-10N	34
Lance Dr. E&W	15W-11N	25
Laura Ln. E&W	12W-8N	36
Laurel Av. N&S	13W-11N	26
Lawn Ln. N&S	15W-11N	25
Leahy Cir. E. N&S	15W-10N,15W-11N	34
Leahy Cir., S. E&W	15W-10N	34
Lechner Ln. E&W	11W-11N	26
Lee St. N&S	13W-8N,13W-10N	35
Leslie Ln. N&S	15W-10N	34
Lillian Ln. N&S	15W-11N	25
Lincoln Av. E&W	15W-10N,12W-10N	34
Linden St. N&S	12W-9N	36
Lismore Ct. E&W	13W-13N	26
Lismore Dr. E&W	13W-13N	26
Little Path Rd. N&S	15W-11N	25
Locust St. N&S	12W-9N	36
Luau Dr. E&W	14W-12N	25
Lunt Av. E&W	12W-8N	36
Lyman Av. N&S	12W-11N	26
Lynn Ct. N&S	15W-11N	25
Madelyn Dr. NS&EW	13W-13N	26
Magnolia St. N&S	12W-9N	36
Mannheim Rd. N&S	13W-9N,13W-11N	35
Manor Ct. N&S	14W-11N	25
Maple St. N&S	12W-8N,12W-9N	36
Marcella Rd. N&S	14W-12N	25
Margaret St. N&S	13W-10N	35
Marina St. N&S	15W-12N	25
Mark Av. E&W	13W-13N	26
Marshall Dr. N&S	15W-9N,15W-11N	34
Mary St. N&S	13W-12N	26
Mason Ln. N&S	12W-10N	36
McCain Ct. E&W	15W-10N	34
Meadow Ln. N&S	11W-11N	26
Meyer Ct. NW&SE	14W-11N,14W-12N	25
Miami Ln. NW&EW	15W-10N	34
Michael Ct. E&W	15W-11N	25
Michael Rd. E&W	15W-11N	25
Mill St. E&W	12W-11N	26
Millers Rd. E&W	15W-11N,14W-11N	25
Miner (Northwest Hwy) NW&SE	14W-11N	25
Morgan O'brien St. E&W	14W-12N	25
Morray Ct. E&W	15W-10N	34
Morse Av. E&W	12W-8N	36
Mt. Prospect Rd. N&S	15W-8N,14W-12N	34
Munroe Cir. E&W	15W-10N	34
Murray Ln. E&W	15W-10N	34
Nebel Ln. E&W	13W-9N	35
Nelson Ln. N&S	14W-12N	25
Nimitz Dr. E&W	12W-8N	36
Norman Ct. E&W	15W-11N	25
North Av. E&W	13W-11N	26
North Shore Av. E&W	12W-8N	36
Northeast Pl. NE&SW	13W-10N	35
Northwest Hwy. NW&SE	14W-12N	25
Northwest Pl. NW&SE	14W-10N	35
Northwest Tollway NW&SE	16W-9N,12W-8N	34
Nuclear Dr. N&S	14W-9N	35
Oak St. N&S	13W-11N	26
Oakton Pl. N&S	13W-9N	35
Oakton St. E&W	15W-10N,12W-10N	34
Oakwood Av. E&W	14W-10N,13W-10N	35
Oakwood Ct. E&W	14W-10N,13W-10N	35
Olivia Av N&S	13W-10N	35
Orchard Ct. N&S	14W-11N	25
Orchard Pl. N&S	12W-8N	36
Orchard St. N&S	12W-9N,12W-10N	36
Oxford Rd. N&S	14W-9N,14W-11N	35
Park Pl. NW&SE	13W-11N	26
Parkview Ln. N&S	12W-10N	36
Parkwood Ln. N&S	12W-8N	36
Parsons Av. NW&SE	13W-11N	26
Partricia Ln. N&S	14W-12N	25
Patton Dr. N&S	12W-8N	36
Paula Ln. NS&EW	12W-8N	36
Pearle Dr. N&S	12W-8N	36
Pearson St. NE&SW	13W-10N,13W-11N	35
Pennsylvania Av. N&S	15W-10N	34
Perda Ln. N&S	15W-10N	34
Perry St. E&W	13W-11N	26
Peter Rd. N&S	12W-8N	36
Phoenix Dr. N&S	15W-10N	34
Pine St. N&S	13W-9N	35
Pinehurst Dr. N&S	14W-12N	25
Plainfield Dr. N&S	13W-9N	35
Pleasant Ln. E&W	15W-10N	34
Polynesian Dr. E&W	14W-12N	25
Potter Rd. N&S	11W-11N	26
Prairie Av. NS&EW	13W-11N,11W-11N	26
Pratt Av. E&W	12W-8N	36
Princeton St. E&W	14W-12N	25
Prospect Av. E&W	13W-9N,12W-9N	35
Prospect Ln. E&W	13W-9N	35
Radcliff Av. N&S	14W-11N,14W-12N	25
Rand Rd. NW&SE	13W-11N	26
Rawls Rd. E&W	14W-10N	35
Redeker Rd. E&W	13W-12N	26
Regency N&S	14W-12N	25
Ridge Ln. N&S	15W-10N	34
Rita Rd. E&W	11W-11N	26
River Dr. N&S	12W-10N	36
River Ln. E&W	12W-11N	26
River St. E&W	12W-11N	26

	Grid	Page
Riverview Av. E&W ..	13W-9N	35
Rose Av. E&W		
	14W-11N,13W-11N	25
Roxbury Ln. E&W ..	15W-10N	34
Roxbury Ln., East E&W		
	14W-10N	35
Rt. 58 E&W		
	14W-11N,13W-12N	25
Rt. 72 E&W		
	14W-9N,12W-8N	35
Rt. 83 N&S		
	15W-10N,15W-11N	34
Rusty Dr. N&S	12W-8N	36
Sakas Dr. N&S	13W-11N	26
Sandy Ln. E&W ...	15W-11N	25
Santa Rosa Dr. E&W		
	14W-8N	35
Scott St. N&S		
	12W-8N,12W-9N	36
Seegers Rd. E&W ..	14W-11N	25
Seminary Av. E&W	11W-11N	26
Seymour St. N&S ..	15W-10N	34
Shagbark Dr. N&S ..	12W-10N	36
Shannon Ct. NS&EW		
	15W-11N	25
Shawn Ln. N&S	15W-11N	25
Shepherd Dr. N&S ..	13W-9N	35
Sherman Pl. E&W		
	13W-11N,12W-11N	26
Sherwin Av. E&W	12W-9N	36
Sherwood Rd. NW&SE		
	12W-11N	26
Short Ter. N&S ..	15W-10N	34
Simone Dr. NW&SE		
	14W-12N	25
South Ln. N&S ..	12W-8N	36
Southeast Pl. NE&SW		
	13W-10N	35
Southwest Pl. NE&SW		
	14W-10N	35
Springfield Ter. E&W		
	15W-10N	34
Spruance Pl. N&S ..	15W-10N	34
Spruce Av. N&S	13W-9N	35
Sprucewood Av. N&S		
	14W-9N	35
Stark Pl. N&S	15W-10N	34
State St. N&S	14W-12N	25
Stewart Av. E&W ..	12W-10N	36
Stillwell Dr. N&S ...	12W-8N	36
Stockton Av. E&W		
	12W-9N,13W-9N	36
Stone St. E&W ..	14W-12N	25
Stratford Rd. N&S ..	14W-12N	25
Suffield St. E&W ..	11W-11N	26
Sunset Av. E&W ..	12W-8N	36
Susan Dr. N&S	15W-10N	34
Sycamore St. N&S		
	12W-8N,13W-9N	36
Terrace Ln. E&W ..	14W-11N	25
Thacker St. E&W		
	14W-11N,13W-10N	25
Therese Ter. E&W ..	13W-13N	26
Times Dr. N&S	13W-10N	35
Timothy Ln. NS&EW		
	14W-13N,13W-13N	25
Touhy Av. E&W	13W-9N	35
Tracy Ter. E&W ..	14W-12N	25
Trailside Dr. N&S ..	12W-11N	26
Tri-State Tollway N&S		
	12W-8N,12W-11N	36
Tures Ln. N&S ..	13W-9N	35
U.S. Route 12 SE&NW		
	13W-10N,14W-12N	35
U.S. Route 14 SE&NW		
	14W-11N	25
U.S. Route 45 N&S		
	13W-10N	35
University Dr. E&W	11W-11N	26
Van Buren Av. E&W		
	14W-10N,13W-10N	35
Vassar Ln. N&S ..	14W-12N	25
Victoria Rd. N&S ..	15W-11N	25
Village Ct. NE&SW	14W-12N	25
Waikiki Dr. NW&SE		
	14W-12N	25
Walnut Av. E&W		
	15W-10N,12W-10N	34
Walnut Ct. E&W		
	14W-10N,13W-10N	35
Walter Av. E&W	13W-11N	26
Warrington Rd. NW&SE		
	14W-10N,14W-12N	35
Washington Ct. E&W		
	14W-11N	25
Washington St. E&W		
	14W-11N,13W-11N	25
Wayne Dr. NS&EW		
	13W-10N	35
Webford Av. NW&SE		
	13W-11N	26
Webster Av. E&W ..	12W-8N	36
Webster Ln. N&S		
	13W-9N,13W-10N	35
Wedgewood Av. N&S		
	14W-10N	35
Welwyn Av. E&W ..	12W-9N	36
West Grant Dr. N&S		
	15W-10N	35
West Villa Dr. N&S	14W-10N	35
Western Av. N&S ..	13W-11N	26
Westfield Ln. E&W ..	15W-10N	34
Westgate N&S		
	14W-10N,14W-12N	35
Westmere Rd. E&W		
	15W-11N	25
Westview Dr. N&S ..	13W-9N	35
Wexford Ct. E&W ..	13W-13N	26

	Grid	Page
Whitcomb Av. E&W		
	13W-10N,12W-10N	35
White St. N&S		
	12W-9N,12W-10N	36
Wicke Av. E&W		
	13W-10N,12W-10N	35
Wieboldts Dr. NW&SE		
	14W-11N	25
Wilkins Dr. N&S ..	15W-11N	25
Wille Rd. E&W	15W-9N	34
Willow Av. E&W ..	13W-11N	26
Wilson Av. N&S ..	12W-11N	26
Wilson Dr. N&S ..	15W-10N	34
Windsor Dr. NE&SW		
	15W-10N	34
Winthrop Dr. N&S....	14W-9N	35
Wisconsin Dr. E&W		
	14W-12N	25
Wolf Rd. N&S		
	14W-9N,13W-13N	35
Woodbridge Rd. N&S		
	14W-11N	25
Woodland Av. E&W		
	13W-11N,12W-11N	26
Woodlawn Av. E&W		
	13W-11N	26
Woods Rd. E&W ..	12W-11N	26
Yale Ct. N&S	14W-12N	25
Yorkshire Ln. E&W		
	14W-12N,13W-12N	25
1st Av. N&S.........	13W-11N	26
2nd Av. N&S		
	13W-10N,13W-12N	35
3rd Av. N&S		
	13W-10N,13W-12N	35
4th Av. N&S		
	14W-10N,14W-12N	35
5th Av. N&S		
	14W-10N,14W-12N	35
5th Ct. NE&SW	14W-11N	25
6th Av. N&S		
	14W-10N,14W-12N	35
7th Av. N&S		
	14W-11N,14W-12N	25
8th Av. N&S	14W-12N	25

Dixmoor

	Grid	Page
Ash St. E&W ..	2W-16S	75
Ashland Av. N&S ..	1W-17S	75
Calumet Av. N&S ..	2W-16S	81
Circle Dr. NS&EW....	2W-16S	75
Cooper Av. N&S ..	2W-17S	81
Cooper St. N&S ..	2W-16S	75
Dan Ryan Expressway		
NE&SW ..	2W-17S,2W-16S	81
Davis Av. N&S	2W-17S	81
Davis Ct. N&S	2W-16S	75
Elm St. E&W	2W-16S	75
Harvey Av. NW&SE ..	2W-16S	81
Honore Av. N&S		
	2W-17S,2W-16S	81
Hoyne Av. N&S	2W-17S	81
Interstate 57 NE&SW		
	2W-17S,2W-16S	81
Joliet Av. E&W ..	2W-17S	81
Lawson Ln. N&S ..	2W-16S	75
Leavit Av. N&S		
	2W-17S,2W-16S	81
Lincoln Av. N&S		
	2W-17S,2W-16S	81
Maple St. E&W ..	2W-16S	75
Marshfield Av. N&S ..	1W-17S	82
Norris St. E&W ..	2W-17S	81
Oak St. E&W	2W-16S	75
Oakley Av. N&S ..	2W-17S	81
Page Av. N&S	2W-16S	75
Paulina Av. N&S ..	1W-17S	82
Robey Av. N&S		
	2W-17S,2W-16S	81
Seeley Av. N&S....	2W-17S	81
Sibley Blvd. E&W ..	2W-17S	81
Spaulding Av. NW&SE		
	2W-16S	75
Thornton Rd. NW&SE		
	2W-17S	81
Vail Av. N&S	2W-17S	81
Walnut St. E&W ..	2W-16S	75
Western Av. N&S		
	2W-17S,2W-16S	81
Winchester Av. N&S		
	2W-17S,2W-16S	81
Wood St. N&S		
	2W-17S,2W-16S	81
140th St. E&W ..	2W-16S	75
141st St. E&W ..	2W-16S	75
142nd St. E&W ..	2W-16S	75
143rd St. E&W ..	2W-17S	81
144th Ct. E&W ..	2W-17S	81
144th St. E&W		
	2W-17S,1W-17S	81
145th St. E&W ..	2W-17S	81
146th St. E&W ..	2W-17S	81

Dolton

	Grid	Page
Adams St. E&W	0E-16S	82
Ann St. E&W	0E-16S	82
Atlantic St. NE&SW ..	0W-17S	82
Avalon Av. N&S ..	1E-16S	82
Beachview Terr. N&S		
	0E-17S	82
Berg Dr. E&W	1E-18S	82
Berger Av. E&W ..	1E-18S	82
Blackstone Av. N&S ..	1E-16S	82
Blouin St. E&W ..	1E-18S	82
California Av. NE&SW		
	0E-17S,1E-17S	82

	Grid	Page
Calumet Expressway N&S		
	1E-17S	82
Calumet Av. N&S		
	0E-16S,1E-16S	82
Catalpa Ln. E&W ..	0E-16S	82
Center Av. N&S	0E-16S	82
Champlain Av. N&S ..	1E-17S	82
Chicago Rd. N&S		
	0E-17S,0E-16S	82
Clark St. N&S	0W-17S	82
Claussen Ct. NE&SW		
	0E-16S	82
Cornell Av. N&S		
	2E-18S,2E-17S	83
Cottage Grove Av. N&S		
	0E-18S,0E-16S	82
Dante Av. N&S		
	1E-17S,1E-16S	82
Dearborn St. N&S..	0W-17S	82
Diekman St. N&S ..	1E-17S	82
Dillner Pl. NE&SW ..	0E-17S	82
Dobson Av. N&S		
	1E-18S,1E-16S	82
Dolton Av. NW&SE ..	1E-16S	82
Dorchester Av. N&S		
	1E-17S,1E-16S	82
Drexel Av. N&S		
	1E-18S,1E-16S	82
East End Av. N&S....	2E-18S	83
Edbrooke Av. N&S ..	0W-17S	82
Ellis Av. N&S	1E-18S,1E-16S	82
Engle Pl. E&W	1E-17S	82
Engle St. N&S	1E-17S	82
Euclid Av. N&S ..	1E-17S	82
Evans Ct. N&S	0E-17S	82
Evans St. N&S	0E-16S	82
Evers St. N&S........	0E-17S	82
Forest Av. N&S ..	0E-16S	82
Grant St. N&S		
	0E-17S,0E-16S	82
Greenwood Rd. NW&SE		
	0E-16S,1E-18S	82
Harding Av. E&W ..	0E-17S	82
Harper Av. N&S		
	1E-17S,1E-16S	82
Hastings Dr. N&S ..	1E-17S	82
Henry St. E&W	0E-16S	82
Indiana Av. N&S		
	0W-17S,0W-16S	82
Ingleside Av. N&S		
	1E-18S,1E-16S	82
Interstate 94 N&S ..	1E-16S	82
Irving Av. N&S		
	0E-17S,0E-16S	82
Jackson St. N&S ..	0E-16S	82
Jefferson St. NW&SE		
	1E-16S,1E-17S	82
Kanawha St. N&S ..	0E-16S	82
Kasten Dr. E&W	1E-17S	82
Kenwood Av. N&S		
	1E-17S,1E-16S	82
Kimbark Av. N&S		
	1E-17S,1E-16S	82
La Salle St. N&S	0W-17S	82
Lakeside Av. N&S		
	0E-17S,0E-16S	82
Langley Av. N&S		
	0E-17S,0E-16S	82
Lincoln Av. NW&SE ..	0E-16S	82
Madison Av. N&S ..	2E-18S	83
Madison St. NW&SE ..	1E-17S	82
Main St. E&W	0E-16S	82
Manor Ct. N&S ..	0E-16S	82
Margaret St. N&S ..	0E-16S	82
Maryland Av. N&S		
	1E-17S,1E-16S	82
Mc Arthur Ln. N&S ..	0E-16S	82
Meadow Ln. N&S	1E-17S	82
Memorial Dr. NW&SE		
	0E-17S	82
Michigan Av. N&S....	0W-17S	82
Michigan City Rd. NW&SE		
	0E-16S,2E-17S	82
Minera Av. N&S		
	1E-18S,1E-17S	82
Monroe St. E&W ..	0E-16S	82
Murray Av. N&S......	0E-16S	82
Oak St. N&S..	0E-17S,0E-16S	82
Pamela Pl. NE&SW ..	0E-16S	82
Park Av. N&S	0E-16S	82
Parkside Dr. N&S....	0W-17S	82
Pennsylvania Av. N&S		
	1E-16S	82
Pennsylvania Av. NW&SE		
	0E-16S	82
Princeton Av. NE&SW		
	0W-17S	82
Riverside Dr. NW&SE		
	0W-17S	82
Sanderson Av. N&S ..	0E-16S	82
Shepard Dr. N&S		
	0E-16S,1E-17S	82
Sheridan Av. NW&SE....	0E-16S	82
Sheridan St. NE&SW		
	0W-17S	82
Sibley Blvd. E&W		
	0W-17S,1E-17S	82
State St. N&W	0W-17S	82
Stein St. NE&SW	0W-17S	82
Stewart Av. E&W		
	1E-17S,0E-17S	82
Stoney Island Av. N&S		
	2E-18S	83
Sunset Dr. N&S		
	1E-17S,1E-16S	82
University Av. N&S		
	1E-18S,1E-16S	82
Van Buren St. N&S ..	0E-16S	82

	Grid	Page
Wabash Av. N&S	0W-17S	82
Washington St. E&W	0E-16S	82
Wentworth St. N&S..	0W-17S	82
Woodlawn Av. N&S		
	1E-17S,1E-16S	82
138th St. E&W		
	0E-16S,1E-16S	82
139th Pl. E&W	0E-16S	82
140th Pl. NS&EW ..	0E-16S	82
141st Pl. E&W	0E-16S	82
141st St. E&W	0E-16S	82
142nd Pl. E&W	0E-16S	82
142nd St. E&W		
	0E-16S,1E-16S	82
143rd St. E&W		
	0E-16S,1E-16S	82
144th Ct. E&W	0E-16S	82
144th St. E&W		
	0E-16S,1E-16S	82
145th St. E&W		
	0E-16S,1E-16S	82
146th St. E&W		
	0E-16S,1E-16S	82
147th Pl. E&W	0E-17S	82
147th St. E&W		
	0W-17S,1E-17S	82
148th Pl. E&W	0W-17S	82
148th St. E&W		
	0W-17S,1E-17S	82
149th St. E&W		
	0W-17S,1E-17S	82
151st St. E&W		
	0W-17S,1E-17S	82
152nd St. E&W		
	0W-17S,1E-17S	82
153rd St. E&W		
	0W-17S,1E-17S	82
154th Pl. E&W		
	0W-17S,1E-17S	82
154th St. E&W		
	0W-17S,1E-17S	82
155th Pl. E&W	1E-18S	82
155th St. E&W		
	1E-18S,2E-18S	82
156th St. E&W		
	1E-18S,2E-18S	82
157th St. E&W ..	1E-18S	82

Downers Grove Township

	Grid	Page
Deerpath Tr. E&W ..	14W-9S	64
Maple Av. N&S	14W-10S	64
Orchard St. NS&EW		
	14W-10S	64
Public St. E&W ..	14W-10S	64
Railroad Av. (91st St.)		
NE&SW	14W-10S	64
114th Av. N&S ..	14W-10S	64

East Dundee*
(Also see Kane County)

	Grid	Page
Healy Rd. NE&SW ..	30W-14N	20
Higgins Rd. NW&SE		
	30W-14N	20

East Hazel Crest

	Grid	Page
Ashland Av. N&S	2W-20S	87
Center St. N&S......	1W-20S	88
Country Ln. NS&EW		
	2W-20S	87
Dixie Hwy. N&S ..	2W-20S	87
Fisk Av. N&S......	1W-20S	88
Forestway Dr. NS&EW		
	2W-20S	87
Governors Hwy. N&S		
	2W-20S	87
Halsted St. N&S ..	1W-20S	88
Hawthorne Dr. NE&SW		
	2W-20S	87
Laflin St. N&S ..	1W-20S	88
Lathrop Av. N&S ..	1W-20S	88
Loomis St. N&S ..	1W-20S	88
Rt. 1 N&S	1W-20S	88
Throop St. N&S ..	1W-20S	88
Tri State Tollway N&S		
	2W-20S,1W-20S	87
Union Av. N&S ..	0W-20S	88
Westview Av. N&S ..	0W-20S	88
Wooded Path Rd. N&S		
	1W-20S	88
171st St. E&W		
	2W-20S,1W-20S	87
172nd Pl. NW&SE ..	2W-20S	87
172nd St. E&W		
	2W-20S,1W-20S	87
173rd St. E&W		
	2W-20S,1W-20S	87
174th St. E&W ..	1W-20S	88
175th St. E&W	1W-20S	88

Elgin*
(Also see Kane County)

	Grid	Page
Abbeywood Ct. E&W		
	32W-12N	20
Adler Ln. NW&SE ..	31W-11N	30
Alcott Ln. NW&SE ..	31W-11N	30
Alder Ln. NW&SE ..	31W-11N	30
Algona Av. E&W ..	32W-12N	20
Algonquin Dr. N&S..	32W-12N	20
Allen Dr. N&S......	32W-10N	30
Allen Dr. NE&SW ..	32W-10N	30
Aller St. N&S......	32W-10N	30
American Way E&W		
	32W-12N	20

	Grid	Page
Anderson E&W	32W-10N	30
Anderson St. N&S ..	32W-10N	30
Apple Hill Ct. E&W ..	32W-12N	20
Apple Ln. NW&SE ..	32W-11N	30
Arrowhead Dr. N&S		
	32W-12N	20
Arthur Dr. NS&EW ..	32W-12N	20
Ash Dr. E&W	32W-11N	30
Aspen Ct. N&S ..	31W-11N	30
Barberry Ct. N&S ..	32W-12S	20
Beaulah E&W	32W-10N	30
Bedford Ct. E&W ..	32W-12S	20
Bend Ridge Ln. E&W		
	32W-12N	20
Bennett Dr. N&S ..	32W-12N	20
Bent Tree Ct. N&S ..	32W-12N	20
Berkshire Ct. NW&SE		
	32W-12N	30
Beulah E&W	32W-10N	30
Beverly Tr. N&S....	32W-11N	30
Birch Dr. E&W	32W-11N	30
Blackhawk Dr. N&S		
	32W-12N	20
Bluff City Blvd. E&W		
	32W-10N	30
Bode Rd. E&W		
	32W-11N,31W-11N	30
Borden Ct. E&W ..	31W-11N	30
Borden Ct. N&S......	32W-11N	30
Borden Dr. NS&EW		
	32W-11N,31W-11N	30
Bradford Ct. NE&SW		
	32W-12N	20
Brentwood Tr. NS&EW		
	31W-11N	30
Briarwood Ct. NW&SE		
	32W-11N	30
Briarwood Dr. N&S		
	31W-11N	30
Brittany Tr. NE&SW		
	31W-11N	30
Bruce Dr. E&W ..	32W-11N	30
Brunswick Ct. E&W		
	32W-12N	20
Burns Dr. E&W ..	31W-10N	30
Butler Bay Ct. E&W		
	32W-13N	20
Campus Dr. N&S ..	32W-10N	30
Canterbury Ct. E&W		
	32W-10N	30
Cardinal Dr. E&W ..	31W-11N	30
Carl Av. E&W	32W-12N	20
Carl Av. N&S	32W-11N	30
Carlton Dr. N&S ..	32W-12N	20
Carlyle Ct. E&W ..	31W-11N	30
Carmella Ln. E&W ..	32W-10N	30
Carriage Way E&W		
	32W-12N	20
Case Ln. NE&SW ..	32W-11N	30
Ceresa Dr. N&S ..	32W-12N	20
Chapel Hill Dr. N&S		
	31W-11N	30
Chaucer Ct. N&S ..	32W-11N	30
Chicago St. NW&SE		
	32W-11N	30
Chippewa Dr. N&S ..	32W-12N	20
Christopher Ct. NE&SW		
	32W-12N	30
Cleveland Av. N&S ..	32W-10N	30
Clover Hill Ct. N&S	32W-13N	20
Cloverhill Ln. NS&EW		
	32W-12N	20
Cobblers Crossing NS&EW		
	32W-12N	20
Cobblers Crossing Ct. N&S		
	32W-13N	20
Cobblestone Ct. NW&SE		
	32W-12N	20
Coldspring Dr. NS&EW		
	32W-12N	20
Coleman* N&S ..	32W-11N	30
Concord Ct. NE&SW		
	31W-10N	30
Concord Dr. E&W		
	32W-10N,31W-10N	30
Congdon Av. E&W..	32W-12N	20
Congress St. E&W ..	32W-10N	30
Cookane Av. N&S ..	32W-12N	20
Corley Ct. E&W ..	31W-11N	30
Corley Ct. NW&SE ..	31W-11N	30
Corley Dr. NW&SE		
	32W-11N,31W-11N	30
Dakota Dr. N&S	32W-12N	30
Danridge Ct. N&S ..	32W-12N	20
David* N&S	32W-11N	30
Deepwoods Ct. E&W		
	32W-13N	20
Deepwoods Dr. N&S		
	32W-13N	30
Dempster St. E&W	32W-11N	30
Dickens Tr. NW&SE		
	31W-11N	30
Dickie Av. NE&SW..	32W-10N	30
Dogwood Ct. N&S ..	31W-11N	30
Dover Ct. NE&SW ..	32W-10N	30
Dover Dr. NE&SW ..	32W-10N	30
Edgebrook Tr. NW&SE		
	32W-10N	30
Eliot Tr. NW&SE	31W-11N	30
Elma Av. E&W	32W-12N	20
Elma Av. NS&EW ..	32W-13N	20
Eric Cir. NS&EW ..	31W-10N	30
Fairfax Ct. N&S ..	32W-12S	20
Fairfax Ct. NW&SE		
	32W-12S	20
Fieldstone Ct. NE&SW		
	32W-12N	20
First St. N&S	32W-13N	20

	Grid	Page
Ford Av. E&W	32W-13N	20
Ford Ct. NE&SW	32W-12N	20
Forest Av. E&W	32W-12N,32W-11N	20
Galt Blvd. N&S	31W-10N	30
Gaskett Dr. E&W	32W-8N,31W-8N	30
Getty St. E&W	32W-10N	30
Gifford Rd. N&S	32W-8N,32W-10N	30
Glen Ivy Dr. N&S	32W-10N	30
Green Ridge NS&EW	32W-12N	20
Hampshire Ln. E&W	32W-12N	20
Harrison St. N&S	32W-10N	30
Hastings St. E&W	32W-10N	30
Hathaway Ct. N&S	32W-12N	20
Haverman Ct. N&S	32W-10N	30
Hecker Ct. E&W	32W-11N	30
Hecker Dr. E&W	32W-11N	30
Hemlock Ct. E&W	32W-11N	30
Hemlock Ln. NW&SE	32W-11N	30
Hiawatha Ct. NW&SE	32W-12N	20
Hiawatha Dr. N&S	32W-12N	20
Hiawatha Ln. N&S	32W-12N	20
Highbury Ct. NE&SW	32W-11N	30
Highbury Dr. E&W	32W-11N	30
Hilliard Dr. NE&SW	31W-10N	30
Hobble Bush NS&EW	32W-12N	20
Horace Dr. NE&SW	31W-10N	30
Houston E&W	32W-10N	30
Hunt Wyck Ct. NW&SE	32W-12N	20
Hunter Dr. NS&EW	32W-11N,32W-12N	30
Huron Ct. NE&SW	32W-13N	20
Huron Dr. E&W	32W-12N	20
Indian Dr. N&S	32W-12N	20
Inglewood Ln. NE&SW	32W-12N	20
Inverness E&W	31W-10N	30
Ironwood Ct. N&S	32W-12N	20
Ironwood Dr. E&W	32W-12N	20
Iroquois Dr. N&S	32W-12N	20
Jan Marie Ln. E&W	32W-11N	30
Jay St. E&W	32W-10N	30
Jefferson Av. E&W	32W-12N	20
Joan Ct. E&W	32W-10N	30
John Dr.* N&S	32W-11N	30
Joslyn Dr. NE&SW	32W-11N	30
Joyce Ln. E&W	32W-11N	30
Julie Ann Ln. E&W	32W-10N	30
Cir.	32W-12N	20
Kingsley Ct. N&S	32W-11N	30
Kirk Av. NE&SW	32W-10N	30
Kramer Av. E&W	32W-10N	30
Lake St. NW&SE	32W-10N,31W-10N	30
Langtry Ct. NW&SE	32W-11N	30
Leawood Ct. NE&SW	31W-11N	30
Leawood Dr. E&W	32W-12N	20
Lehman Dr. NS&EW	31W-10N	30
Lincolnshire Ct. N&S	32W-11N	30
Lisa Ln. E&W	32W-10N	30
Little Falls Dr. NS&EW	32W-12N	20
Long Ford Cir. NS&EW	31W-11N	30
Longford Ct. E&W	31W-11N	30
Longford Dr. N&S	31W-11N	30
Lovell Rd. N&S	31W-10N	30
Lucile Av. N&S	32W-10N,32W-11N	30
Lucile Av. NE&SW	32W-10N	30
Luda St. E&W	32W-12N	20
Ludlow Av. E&W	32W-12N	20
Mackey Ln. E&W	32W-11N	30
Maple Pl. N&S	32W-10N	30
Maribil NE&SW	32W-12N	20
Mariner Ct. E&W	31W-10N	30
Mariner Dr. N&S	32W-10N	30
Maroon Dr. E&W	32W-10N	30
Martin Dr. NE&SW	32W-12N	20
Mohawk Ct. E&W	32W-12N	20
Mohawk Dr. N&S	32W-12N	20
Mulford Ct. NW&SE	32W-11N	30
Mulford Dr. NS&EW	32W-11N	30
Nancy Ann Ln. E&W	32W-11N	30
Natoma Ct. E&W	32W-10N	30
Natoma Dr. NW&SE	32W-10N	30
Nicola Dr. N&S	32W-12N	20
Oakwood Blvd. NE&SW	32W-11N	30
Old Mill Ct. N&S	32W-13N	20
Olive Ct. N&S	32W-10N	30
Olive St. E&W	32W-10N	30
Olympia Ct. N&S	31W-11N	30
Packard Dr. N&S	32W-10N	30
Parkway Av. E&W	32W-13N	20
Patricia Dr. E&W	32W-12N	20
Peach Tree Ct. N&S	32W-11N	30
Peach Tree Ln. E&W	32W-11N	30
Peck Pl. N&S	32W-10N	30
Pegwood Dr. E&W	32W-11N	30
Polly Ct. N&S	32W-12N	20
Poplar Creek Dr. NE&SW	32W-11N	30
Poplar View Bend NE&SW	31W-11N	30
Prestbury Dr. NE&SW	32W-12N	20
Price Dr. E&W	32W-11N	30
Providence Cir. E&W	32W-12N	20
Quaker Hill Ct. N&S	32W-12N	20
Quincy Ct. E&W	31W-11N	30
Radclyffe Ct. NE&SW	32W-10N	30
Ramona Av. NE&SW	32W-10N	30
Red Oak Ct. NE&SW	32W-11N	30
Ripplebrook Ct. NE&SW	32W-12N	20
Ripplebrook Ln. N&S	32W-12N	20
Ripplebrook Ln. NW&SE	32W-12S	20
Rodney Ln. E&W	32W-10N	30
Rose Ln. N&S	31W-9N	30
Rosewood Ct. E&W	32W-11N	30
Rt. 19 E&W	32W-11N	30
Rt. 58 E&W	32W-12N	20
Sadler Av. N&S	32W-10N,32W-11N	30
Sandra Ln. E&W	32W-10N	30
Saxon Ct. N&S	32W-11N	30
Seal Dr. E&W	32W-8N,31W-8N	30
Sebring Dr. E&W	32W-11N	30
Seminole Dr. E&W	32W-12N	20
Shady Oaks Ct. NE&SW	32W-12N	20
Shady Oaks Dr. NS&EW	32W-11N,32W-12N	30
Sharon Ct.* N&S	32W-11N	30
Shawford Way N&S	32W-12N	20
Sheldon Dr. E&W	31W-10N	30
Shiloh Ct. E&W	32W-11N	30
Shiloh Ln. NS&EW	32W-11N	30
Shoe Factory Rd. E&W	32W-12N	20
Sioux Dr. E&W	32W-13N	20
Sommerset Ct. N&S	32W-10N	30
Spaulding Rd. E&W	32W-9N,31W-9N	30
Spring Creek Ct. E&W	32W-12N	20
Spring Creek Rd. N&S	32W-12N	20
Spring Hill Ct. N&S	32W-12N	20
Spruce Ln. E&W	32W-11N	30
Stewart Av. E&W	32W-12N	20
Stillwater Rd. NE&SW	32W-12N	20
Stockbridge Ct. N&S	32W-12N	20
Stockbridge Pl. NW&SE	32W-12N	20
Stonebridge Ct. NE&SW	32W-12N	20
Stonehurst Dr. N&S	32W-11N	30
Stratford Ct. E&W	32W-10N	30
Sumac Ln. NS&EW	31W-11N	30
Summit Av. E&W	32W-12N	20
Suzzane Ln. E&W	32W-11N	30
Tannery Ridge Rd. N&S	32W-12N	20
Ted Ln. E&W	32W-10N	30
Tefft Av. N&S	32W-11N	30
Terrace Av. E&W	32W-11N	30
Terrace Ct. E&W	32W-11N	30
Thoreau Dr. N&S	31W-10N	30
Thorndale Ct. NE&SW	31W-10N	30
Thorndale Dr. NW&SE	32W-10N,31W-10N	30
Toastmaster Dr. N&S	32W-13N	20
Tori Ln. E&W	32W-10N	30
Trout Av. N&S	32W-12N	20
U.S. Route 20 NW&SE	32W-10N,31W-10N	30
Varsity Dr. NE&SW	32W-10N	30
Victor Av. N&S	32W-13N	20
Victoria Av. N&S	32W-12N	20
Villa Av. NW&SE	32W-10N	30
Village Ct. E&W	32W-11N	30
Virgil Av. NE&SW	32W-10N	30
Vivian Tr. N&S	32W-11N	30
Wakesfield N&S	32W-12N	20
Walden Dr. E&W	31W-10N	30
Waverly Ct. N&S	32W-12N	20
Waverly Ct. NE&SW	32W-11N	30
Waverly Ct. N. NE&SW	32W-11N	30
Waverly Ct. S. NE&SW	32W-11N	30
Waverly Dr. NW&SE	32W-11N,32W-12N	30
Wembly Ct. N&S	32W-12N	20
Willard Av. N&S	32W-10N,32W-11N	30
Willoby Ln. NS&EW	32W-12N	20
Windsor Ct. E&W	31W-11N	30
Woodhill Ct. E&W	32W-12N	20
Woodview Cir. N&S	32W-11N	30
Wright Av. NE&SW	32W-10N	30
Wynnfield Ct. NE&SW	32W-10N	30
Yew Ct. E&W	32W-11N	30
Yorkshire Ct. NW&SE	32W-10N	30

Elk Grove Township

	Grid	Page
Crest Av. N&S	18W-9N	34
Dierking Ter. E&W	16W-9N	34
Elizabeth Dr. N&S	16W-9N	34
Forestview Av. E&W	18W-9N	34
Hamilton Rd. E&W	16W-9N	34
Higgins Rd. NW&SE	20W-10N,16W-9N	33
Howard St. N&S	18W-9N	34
Kenneth Dr. E&W	16W-10N	34
Laurel St. N&S	18W-9N	34
Lee Ln. N&S	16W-9N	34
Lela St. E&W	16W-9N	34
Lindale St. E&W	17W-10N	34
Linneman Rd. N&S	16W-10N	34
Lund Rd. N&S	16W-9N	34
Malmo Rd. N&S	16W-10N	34
Northwest Tollway NW&SE	16W-9N	34
Roppolo Dr. E&W	16W-9N	34
Rt. 72 NW&SE	20W-10N,15W-9N	33
Thorndale Av. E&W	18W-9N	34
Tonne N&S	17W-12N	24
Touhy Av. E&W	15W-8N	34
Vera Ln. N&S	16W-9N	34
Weiler Rd. N&S	16W-9N	34

Elk Grove Village*
(Also see Du Page County)

	Grid	Page
Alabama Dr. NW&SE	21W-8N	33
Albany Ct. E&W	22W-9N	33
Aldrin Tr. NW&SE	21W-8N	33
Alexian Way E&W	19W-8N	33
Anders Dr. N&S	21W-8N	33
Arizona Pass N&S	21W-9N	33
Arkansas Dr. N&S	22W-8N	33
Armstrong Ct. NW&SE	21W-8N	33
Armstrong Ln. E&W	21W-8N	33
Arthur Av. E&W	17W-8N,16W-8N	34
Ash St. E&W	18W-8N	34
Aspen Ln. N&S	18W-8N	34
Atlantic Ln. N&S	22W-9N	33
Avon Rd. E&W	19W-8N	33
Baltimore Dr. E&W	22W-9N	33
Banbury Av. E&W	19W-8N	33
Bangor Ln. NS&EW	22W-9N	33
Banyan Dr. N&S	18W-8N	34
Basswood Ct. E&W	17W-10N	34
Basswood Dr. N&S	17W-9N	34
Bay Dr. N&S	19W-8N	33
Beisner Rd. N&S	20W-8N,20W-9N	33
Bennett Rd. E&W	18W-10N,17W-10N	34
Berkenshire Ln. N&S	19W-8N	33
Bianco Dr. E&W	18W-9N	34
Biesterfield Pkwy. E&W	18W-8N	34
Biesterfield Rd. E&W	21W-9N,19W-8N	33
Birchwood Av. E&W	18W-8N	34
Biscayne Dr. N&S	19W-8N	33
Bismark Ct. N&S	21W-8N	33
Blue Jay Cir. NS&EW	22W-8N	33
Boardwalk E&W	19W-8N	33
Bonaventure Dr. E&W	20W-8N	33
Bond St. N&S	17W-10N	34
Bonita Av. N&S	18W-9N	34
Bonnie Ln. N&S	17W-9N	34
Bordeaux Ct. NW&SE	19W-9N	33
Borman Ct. NS&EW	21W-8N	33
Bosworth Ln. N&S	18W-8N	34
Bradford Cir. E&W	19W-9N	33
Bradley Ln. NW&SE	21W-8N	33
Braemer Dr. E&W	19W-8N,18W-8N	33
Brandywine Av. E&W	19W-8N	33
Brantwood Av. NS&EW	19W-9N,18W-8N	33
Brantwood Ct. E&W	19W-9N	33
Brantwood Pl. NW&SE	19W-9N	34
Briarwood Ln. NW&SE	18W-10N	34
Brighton Rd. E&W	19W-8N	33
Bristol Ln. NS&EW	19W-8N	33
Brookhaven Dr. E&W	18W-8N	34
Brown Cir. NE&SW	21W-9N	33
Brummel St. E&W	17W-9N,16W-9N	34
Brynhaven Ct. N&S	18W-10N	34
Brynhaven St. E&W	18W-10N	34
Buckingham Ct. E&W	19W-8N	33
Burgundy Ct. E&W	19W-9N	33
Busse Rd. N&S	16W-9N	34
California St. NS&EW	21W-9N	33
Cambridge Dr. NW&SE	16W-9N	34
Canyon Dr. N&S	18W-8N	34
Cardinal Ln. N&S	22W-8N	33
Carlisle Av. N&S	19W-8N,18W-8N	34
Carmen Dr. N&S	15W-8N	34
Carolina Dr. E&W	21W-8N	33
Carpenter Ct. NS&EW	21W-8N	33
Carr Ct. NE&SW	21W-8N	33
Carroll Sq. N&S	17W-10N	34
Carswell Av. N&S	18W-8N	34
Carswell Ct. E&W	18W-8N	34
Carswell Dr. N&S	18W-8N	34
Cass Ln. N&S	21W-8N	33
Cedar Ln. NS&EW	18W-8N	34
Cedarwood Ct.* NS&EW	20W-8N	33
Center St. E&W	17W-8N	34
Cernan Ct. E&W	21W-8N	33
Chaffee Ct. N&S	21W-8N	33
Charing Cross Rd. N&S	18W-9N	34
Charlela Ln. N&S	20W-8N	33
Charles Rd. N&S	19W-9N	33
Chase Av. E&W	17W-9N	34
Cheekwood Ct. N&S	19W-8N	33
Cheekwood Dr. E&W	19W-8N	33
Chelmsford Ln. NS&EW	19W-9N	33
Cheltenham Rd. N&S	19W-8N	33
Chester Ln. NW&SE	21W-8N	33
Cindy Ln. E&W	21W-8N	33
Circle Ct. E&W	21W-9N	33
Clearmont Dr. E&W	19W-8N,18W-8N	33
Clifford Ln. N&S	21W-8N	33
Clover Hill Ct. NS&EW	20W-8N	33
Clover Hill Ln. NS&EW	20W-8N,19W-8N	33
Collins Cir. N&S	21W-9N	33
Colorado Ln. E&W	21W-9N	33
Columbia Ct. E&W	21W-9N	33
Commerce Dr. E&W	16W-9N	34
Concord Ln. N&S	22W-9N	33
Conrad Ct. E&W	21W-8N	33
Cooper Ct. NS&EW	21W-8N	33
Corrinthia Ct. N&S	18W-9N	34
Corrinthia Dr. E&W	18W-9N	34
Cosman Rd. E&W	19W-9N	33
Cottonwood Dr. N&S	17W-9N	34
Coyle Ln. E&W	15W-8N	34
Creighton Av. N&S	19W-8N	33
Crest Av. NS&EW	18W-9N	34
Crestwood Ct.* NS&EW	20W-8N	33
Criss Cir. E&W	17W-10N	34
Crossen Av. N&S	17W-9N,17W-10N	34
Cumberland Cir. NS&EW	19W-8N	33
Cunningham Cir. NE&SW	21W-8N	33
Cutter Ln. NS&EW	22W-8N	33
Cypress Ln. N&S	18W-8N	34
Dakota Dr. E&W	21W-8N	33
Dauphine Ct. E&W	19W-9N	33
David Ln. N&S	21W-8N	33
Debra Ln. N&S	21W-8N	33
Deepwood Ct. N&S	20W-8N	33
Delaware Ln. NS&EW	22W-9N	33
Delmar Cir. N&S	22W-9N	33
Delphia Av. E&W	18W-9N	34
Delphia Ct. N&S	18W-9N	34
Devon Av. E&W	20W-8N,19W-8N	33
Devon Av. E&W	17W-8N,16W-8N	34
Diane Ln. NW&SE	21W-8N	33
Dierking Terr. N&S	16W-9N	34
Dogwood Tr. E&W	18W-8N	34
Doral Ct. NE&SW	22W-9N	33
Dorchester Ln. N&S	19W-8N	33
Dover Ln. E&W	19W-8N	33
Driftwood Ct.* NS&EW	20W-8N	33
Dupoint Cir. NE&SW	22W-9N	33
Eagle Dr. N&S	18W-9N	34
Eden Rd. NS&EW	19W-8N,18W-8N	33
Edgeware Rd. NW&SE	18W-9N	34
Edgewood Ln. NW&SE	18W-9N	34
Elk Grove Blvd. N&S	18W-8N,18W-9N	34
Elmhurst Rd. N&S	16W-8N,16W-9N	34
Elmwood Ln. N&S	18W-8N	34
Essex Rd. E&W	19W-8N	33
Estes Av. E&W	17W-8N,16W-8N	34
Evans Ct. NE&SW	21W-8N	33
Evergreen Cir E&W	18W-10N	34
Evergreen St. N&S	18W-10N	34
Exmoor Rd. NS&EW	19W-8N	33
Fairfield Cir. N&S	19W-9N	33
Fargo Av. E&W	17W-9N	34
Fern Dr. E&W	18W-8N	34
Fleetwood Ln. N&S	17W-9N	34
Florida Ln. NS&EW	21W-8N	33
Forest Ln. N&S	18W-10N	34
Forestview Av. N&S	18W-9N	34
Fox Run Dr. NS&EW	22W-9N	33
Franklin Ln. NS&EW	21W-9N	33
Galleon Ln. NS&EW	22W-9N	33
Garlisch Dr. N&S	17W-10N	34
Gateshead Ln. E&W	19W-8N	33
Gaylord St. N&S	18W-10N	34
Geneva Cir. E&W	21W-8N	33
Georgia Dr. NW&SE	21W-8N	33
Germaine Ln. E&W	18W-9N	34
Germaine Pl. N&S	18W-9N	34
Gibson Dr. NE&SW	22W-9N,21W-9N	33
Gloria Dr. N&S	21W-8N	33
Gloucester Dr. E&W	19W-8N	33
Gordon St. N&S	18W-10N	34
Grange Pl. NW&SE	19W-8N	33
Grange Rd. NW&SE	19W-8N	33
Grassmere Rd. NE&SW	19W-8N	33
Green Briar St. E&W	18W-9N	34
Greenleaf Av. E&W	17W-8N,15W-8N	34
Greensboro Ct. N&S	22W-8N	33
Gregory Ct. N&S	21W-8N	33
Grissom Tr. N&S	21W-8N	33
Grosvener Ct. NE&SW	18W-9N	34
Grosvener Ln. N&S	18W-9N	34
Grove Dr. N&S	17W-9N	34
Gullo Av. N&S	17W-10N	34
Haar Ln. E&W	21W-8N	33
Haise Ct. NE&SW	21W-8N	33
Haise Ln. E&W	21W-8N	33
Hampshire Dr. NS&EW	22W-9N,21W-9N	33
Harmony Ln. E&W	18W-8N	34
Hartford Ln. N&S	18W-8N	34
Hastings Av. N&S	19W-9N	33
Hawk Ln. E&W	22W-8N	33
Helen Ln. NW&SE	21W-9N	33
Hemlock N&S	18W-8N	34
Hickory Ln. N&S	18W-8N	34
Higgins Rd. E&W	18W-10N,17W-10N	34
Holdmair Ln. E&W	21W-8N	33
Holly Ln. N&S	18W-9N	34
Home Av. N&S	21W-8N,21W-9N	33

Elk Grove Village*
(Also see Du Page County)

	Grid	Page
Home Cir. E&W	21W-9N	33
Howard St. E&W	18W-9N,17W-10N	34
Hudson Ct. NW&SE	22W-9N	33
Idaho Pl. E&W	21W-9N	33
Indiana Ln. N&S	21W-8N	33
Interstate 90 SE&NW	17W-10N	34
Inverness Ct. SW&NE	22W-9N	33
Iowa Dr. E&W	21W-8N	33
Ipswich Ct. N&S	19W-9N	33
Ironwood Dr. E&W	17W-10N	34
Ironwood Ln. E&W	17W-10N	34
J.F. Kennedy Blvd. NE&SW	19W-8N,18W-8N	33
Jackson Cir E&W	22W-8N	33
James Way E&W	21W-8N	33
Jarvis St. E&W	17W-9N	34
Jefferson Sq. E&W	18W-8N	34
Jersey Ln. N&S	22W-9N,21W-9N	33
Joey Ln. N&S	17W-10N	34
Joplin Cir. NE&SW	21W-8N	33
Judy Dr. N&S	21W-8N	33
Julie Dr. E&W	22W-9N	33
Kathleen Way NS&EW	21W-8N	33
Kelly St. N&S	18W-10N	34
Kendal Rd. E&W	19W-8N	33
Kenilworth Av. NS&EW	19W-8N,18W-8N	33
Kent Av. N&S	17W-9N	34
Kentucky Ln. NS&EW	22W-8N	33
Keswick Rd. E&W	19W-8N	33
King St. N&S	17W-10N	34
Kingsbridge Rd. E&W	19W-8N	33

	Grid	Page
Knottingham Dr. NS&EW	22W-8N	33
Lakeview Cir. E&W	19W-9N	33
Lancaster Av. NS&EW	19W-8N	33
Landmeir Rd. E&W	17W-9N,16W-9N	34
Larchmont Dr. N&S	18W-8N	34
Laurel St. E&W	18W-9N	34
Lee Ln. E&W	16W-9N	34
Lee St. E&W	17W-10N	34
Leeds Ln. N&S	19W-8N	33
Leicester Rd. N&S	19W-8N,19W-9N	33
Lela St. NE&SW	16W-9N	34
Liberty Ct. E&W	22W-9N	33
Lilac Ln. E&W	18W-9N	34
Lincoln N&S	18W-8N	34
Lindale St. E&W	17W-10N	34
Little Falls Ct. E&W	20W-8N	33
Long Boat Dr. E&W	22W-8N	33
Lonsdale Rd. NS&EW	19W-8N,18W-8N	33
Louis Av. E&W	17W-9N	34
Louisiana Dr. NW&SE	22W-8N	33
Love St. N&S	18W-8N,18W-9N	34
Lovell Ct. NS&EW	21W-8N	33
Lowestoft Ln. E&W	19W-8N	33
Lund Rd. N&S	16W-9N	34
Lunt Av. E&W	17W-8N,16W-8N	34
Lynne Dr. N&S	19W-8N	33
Madison Ct. NW&SE	21W-8N	33
Magnolia Ln. E&W	18W-8N	34
Maine Dr. E&W	22W-9N	33
Maple Ct. E&W	18W-8N	34
Maple Ln. NS&EW	18W-8N	34
Martin Ln. N&S	17W-10N	34
Maryland Dr. E&W	22W-9N	33
McCabe Av. N&S	16W-9N	34
Mcdevitt Cir. N&S	21W-9N	33
Meacham Rd. N&S	21W-8N,21W-9N	33
Meegan Way NS&EW	21W-8N	33
Memphis Cir. NE&SW	22W-8N	33
Michigan Ln. N&S	21W-8N	33
Middlebury Ln. N&S	18W-9N	34
Midway Ct. NW&SE	17W-10N	34
Milbeck Av. NW&SE	19W-8N	33
Milbeck Ct. NS&EW	19W-8N	33
Mimosa Ln. E&W	18W-8N	34
Minnesota Dr. NW&SE	21W-8N	33
Minot Ct. N&S	21W-8N	33
Mississippi Ln. N&S	22W-8N	33
Missouri Dr. E&W	21W-8N	33
Mitchell Tr. N&S	21W-8N	33
Mobile Cir. E&W	21W-8N	33
Montana Way NE&SW	21W-9N	33
Montego Ct. N&S	19W-8N	33
Montego Dr. E&W	19W-8N	33
Moore Dr. NW&SE	19W-9N	33
Morgan Dr. NS&EW	21W-8N	33
Morse Av. E&W	17W-8N	34
Mulberry Ln. E&W	18W-8N	34
N.W. Point Rd. N&S	18W-10N,17W-10N	34
Nebraska Dr. N&S	21W-8N	33
Nerge Rd. E&W	22W-8N,20W-8N	33
Nevada Ln. E&W	21W-9N	33
New Mexico Ct. NE&SW	21W-9N	33
New Mexico Tr. N&S	21W-9N	33
New York Ln. NS&EW	22W-9N	33
Newberry Ln. N&S	22W-9N	33
Newport Av. E&W	19W-8N	33
Nicholas Blvd. N&S	16W-8N,16W-9N	34
North Hampton Av. E&W	19W-9N	33
North Pkwy. E&W	17W-8N	34
Northampton Cir. E&W	19W-9N	33
Northport Dr. E&W	19W-8N	33
Northwest Tollway NW&SE	17W-10N	34
Oak St. N&S	18W-9N	34
Oakton St. E&W	17W-10N	34
Oakwood Dr. N&S	18W-10N	34
Oklahoma Cir. NE&SW	21W-9N	33
Oklahoma Way N&S	21W-9N	33
Old Creek Ct. NW&SE	20W-8N	33
Old Mill Ln. NS&EW	20W-8N	33
Oregon Tr. E&W	21W-9N	33
Oriole Dr. E&W	22W-8N	33
Orleans Ct. NE&SW	22W-8N	33
Oxford Cir. NW&SE	21W-9N	33
Pagni Dr. E&W	17W-10N	34
Pahl Rd. NS&EW	19W-9N	33
Park Blvd. NS&EW	20W-8N	33
Parkchester Rd. E&W	19W-8N	33
Parker Pl. NS&EW	21W-8N	33
Parkview Cir. N&S	19W-9N	33

	Grid	Page
Patricia Ct. E&W	21W-8N	33
Pauly Dr. E&W	17W-10N	34
Peach Tree Ln. E&W	18W-8N	34
Pebble Beach Cir. NS&EW	22W-9N	33
Penrith Av. N&S	19W-8N	33
Perrie Dr. N&S	17W-9N	34
Pinewood Dr. E&W	18W-8N	34
Placid Way E&W	18W-8N	34
Pleasant Dr. E&W	18W-8N	34
Potomac Ln. N&S	22W-9N	33
Pratt Blvd. E&W	17W-8N,16W-8N	34
Racine Cir. E&W	21W-8N	33
Randell St. N&S	18W-10N	34
Red Fox Ln. N&S	22W-9N	33
Redwood Av. N&S	17W-9N	34
Rev. Morrison Blvd NE&SW	18W-8N	34
Revere Ln. E&W	22W-9N	33
Richmond Ct. E&W	21W-8N	33
Ridge Av. N&S	18W-8N,18W-9N	34
Ridge Ct. E&W	18W-8N	34
Ridge Sq. E&W	18W-8N	34
Ridgewood Rd. N&S	18W-10N	34
Rockwood Dr. N&S	17W-9N	34
Rohlwing Rd. N&S	20W-18N,20W-19N	13
Roosa Ln. NW&SE	21W-8N	33
Roppolo Dr. N&S	16W-9N	34
Rt. 290 N&S	20W-8N,20W-9N	33
Rt. 72 E&W	17W-10N	34
Rt. 83 N&S	17W-9N	34
Ruskin Cir. E&W	19W-9N	33
Ruskin Dr. NS&EW	19W-9N	33
Rutgers Ln. E&W	22W-9N,21W-9N	33
Salem Ct. E&W	21W-9N	33
Schirra Cir. N&S	21W-9N	33
Schooner Ln. N&S	22W-8N	33
Scott St. N&S	17W-10N	34
Seegars Av. N&S	17W-10N	34
Shadywood Ln. E&W	18W-9N	34
Shelley Ct. N&S	19W-8N	33
Shelley Rd. E&W	19W-9N	33
Shepard Cir. N&S	21W-9N	33
Smethwick Ln. E&W	19W-9N	33
Somerset Ln. N&S	19W-8N	33
South Glenn Tr. E&W	21W-8N	33
Spring Creek Ct. NS&EW	20W-8N	33
Springdale Ln. NS&EW	18W-8N	34
Spruce Ln. E&W	18W-8N	34
Stafford Cir. N&S	21W-9N	33
Stanford Cir. N&S	19W-9N	33
Stanley St. N&S	17W-10N	34
Stone Brook Ct. E&W	20W-8N	33
Stonehaven Av. N&S	19W-9N	33
Stowe Cir. E&W	22W-9N	33
Susan Ct. E&W	21W-8N	33
Sussex Ct. N&S	18W-9N	34
Sycamore Dr. E&W	18W-8N	34
Sylvan Ct. E&W	19W-9N	33
Talbots Ln. N&S	20W-8N	33
Tanglewood Ct. N&S	18W-9N	34
Tanglewood Dr. N&S	18W-9N	34
Tennessee Ln. NS&EW	22W-8N	33
Texas St. E&W	21W-9N	33
Thorndale Av. N&S	18W-9N	34
Thornton Ct. E&W	22W-8N	33
Timber Ln. E&W	21W-8N	33
Tonne Rd. N&S	17W-8N,17W-9N	34
Tottenham Ln. E&W	19W-8N	33
Touhy Av. E&W	17W-8N,16W-8N	34
Tower Ln. E&W	18W-9N	34
Trowbridge Rd., East NS&EW	19W-8N	33
Trowbridge Rd., West E&W	19W-8N	33
Turner Av. E&W	19W-8N	33
Union Cir. NW&SE	21W-9N	33
University Ln. NW&SE	22W-9N	33
Utah Cir. E&W	21W-9N	33
Utah St. N&S	21W-9N	33
Vera Ln. E&W	16W-9N	34
Verde Ln. E&W	18W-9N	34
Vermont Dr. E&W	22W-9N,21W-9N	33
Vernon Cir. NE&SW	22W-9N	33
Versailles Cir. N&S	19W-9N	33
Victoria Ln. E&W	18W-8N,18W-9N	34
Village Grove Dr. E&W	19W-8N	33
Vine Ln. N&S	22W-9N	33
Virginia Dr. NE&SW	21W-8N	33
Volkamer Tr. NS&EW	21W-8N	33
Von Braun Ct. NE&SW	21W-8N	33
Von Braun Tr. E&W	21W-8N	33

	Grid	Page
Walnut Ln. E&W	18W-8N	34
Walpole Rd. E&W	19W-8N	33
Walter Av. E&W	19W-8N	33
Warwick Ln. N&S	18W-8N	34
Wasdale Av. N&S	19W-8N	33
Washington Sq. N&S	18W-8N	34
Wellington Av. NS&EW	20W-8N,19W-9N	33
West Gate Av. N&S	18W-9N	34
West Glenn Tr. N&S	21W-8N	33
Westview Dr. N&S	18W-8N	34
White Robin Dr. E&W	21W-8N	33
White Robin Tr. E&W	21W-8N	33
White Tr. E&W	21W-9N	33
Wildwood Dr. N&S	18W-9N	34
Wildwood Pl. E&W	18W-9N	34
Wildwood Rd. N&S	18W-9N	34
Willow Ln. E&W	18W-9N	34
Wilma Ln. N&S	21W-8N	33
Wilshire Av. N&S	18W-8N	34
Winston Dr. NS&EW	19W-9N	33
Wisconsin Ln. N&S	21W-8N	33
Wise Rd. E&W	22W-9N	33
Wood Tr. NW&SE	21W-8N	33
Woodcrest Ln. N&S	18W-10N	34
Woodview Av. N&S	18W-9N	34
Worden Way NS&EW	21W-8N	33
Yale Ct. NE&SW	21W-9N	33
Yarmouth Rd. E&W	19W-8N	33
Young Cir. N&S	21W-9N	33

Elmwood Park

	Grid	Page
Altgeld E&W	9W-3N	42
Armitage E&W	9W-2N	48
Barry E&W	9W-3N	42
Belden E&W	9W-2N	48
Cir.	9W-3N	42
Bloomingdale E&W	9W-2N	48
Brook N&S	9W-3N	42
Cortland E&W	9W-2N	48
Cortland Pkwy. E&W	9W-2N	48
Country Club Ln. E&W	9W-2N	48
Cir.	9W-3N	42
Dickens E&W	9W-2N	48
Diversey E&W	9W-3N	42
Cir.	9W-3N	42
Fletcher E&W	9W-3N	42
Fullerton E&W	9W-3N	42
George E&W	9W-3N	42
Grand E&W	9W-3N	42
Harlem Av. N&S	9W-2N	48
Lenora Ln. E&W	9W-2N	48
Marwood E&W	9W-3N	42
Nelson E&W	9S-3N	42
North E&W	9W-2N	48
Cir.	9W-3N	42
Palmer E&W	9W-2N	48
Schubert E&W	9W-3N	42
Cir.	9W-3N	42
Wabansia E&W	9W-2N	48
Wellington E&W	9W-3N	42
Westwood Dr. E&W	9W-3N	42
Wrightwood E&W	9W-3N	42
72nd Ct. N&S	9W-2N,9W-3N	48
73rd N&S	9W-2N,9W-3N	48
73rd Ct. N&S	9W-2N,9W-3N	48
74th N&S	9W-2N,9W-3N	48
74th Ct. N&S	9W-2N,9W-3N	48
75th N&S	9W-2N,9W-3N	48
75th Ct. N&S	9W-2N,9W-3N	48
76th N&S	9W-2N,9W-3N	48
76th Ct. N&S	9W-2N,9W-3N	48
77th N&S	9W-2N,9W-3N	48
77th Ct. N&S	9W-2N,9W-3N	48
78th N&S	9W-2N,9W-3N	48
78th Ct. N&S	9W-2N,9W-3N	48
79th N&S	9W-2N,9W-3N	48
79th Ct. N&S	9W-2N	48

Evanston

	Grid	Page
Arnold Pl. N&S	2W-10N	39
Ashbury Av. N&S	3W-10N,2W-13N	39
Ashland Av. N&S	3W-9N,3W-13N	39
Austin St. E&W	2W-9N	39
Barton Av. N&S	2W-9N	39
Bennett Av. N&S	4W-12N,4W-13N	29
Benson Av. N&S	2W-11N	29
Bernard St. N&S	4W-13N	29
Bradley Pl. E&W	3W-10N	39
Bridge St. NW&SE	3W-12N	29
Broadway N&S	3W-13N	29
Brown Av. N&S	3W-9N,3W-12N	39
Brummel Pl. E&W	3W-9N	39
Brummel St. E&W	3W-9N,2W-9N	39
Bryant Av. N&S	2W-12N,2W-13N	29
Burnham Pl. E&W	2W-10N	39
Callan Av. N&S	2W-9N,2W-10N	39
Calvin Cir. N&S	4W-12N	29
Case Pl. E&W	2W-9N	39
Case St. E&W	2W-9N	39
Central Park Av. N&S	4W-12N	29
Central St. E&W	4W-12N,3W-12N	29
Chancellor St. E&W	3W-13N	29
Chicago Av. N&S	2W-10N,2W-11N	39
Church St. E&W	3W-11N,2W-11N	29
Clark St. E&W	2W-11N	29
Cleveland St. E&W	3W-10N,2W-10N	39
Clifford St. E&W	4W-13N	29
Clinton Pl. E&W	2W-13N	29
Clyde Av. N&S	2W-9N	39
Colfax Pl. E&W	4W-12N	29
Colfax St. E&W	4W-12N,2W-12N	29
Colfax Ter. N&S	3W-12N	29
Cowper Av. N&S	4W-12N	29
Craford Av. NW&SE	5W-13N	28
Crain St. E&W	3W-10N,2W-10N	39
Croft Ln. N&S	2W-10N	39
Culver St. E&W	4W-12N	29
Custer Av. N&S	2W-9N	39
Darrow Av. N&S	3W-9N,3W-11N	39
Dartmouth Pl. E&W	4W-12N,2W-12N	29
Davis St. E&W	3W-11N,2W-11N	29
Dempster St. E&W	3W-11N,2W-10N	29
Dewey Av. N&S	3W-9N,3W-11N	39
Dobson St. E&W	3W-9N,2W-9N	39
Dodge Av. N&S	3W-9N,3W-12N	39
East Railroad Ave. NW&SE	2W-11N	29
Eastwood Av. N&S	3W-13N	29
Edgemere Ct. N&S	1W-10N	39
Elgin Rd. NW&SE	2W-11N	29
Elinor Pl. E&W	3W-11N	29
Elm Av. N&S	3W-12N	29
Elmwood Av. N&S	2W-9N,2W-10N	39
Emerson St. E&W	3W-11N	29
Euclid Park Pl. N&S	2W-13N	29
Ewing Av. N&S	4W-12N,4W-13N	29
Florence Av. N&S	3W-9N,3W-11N	39
Forest Av. N&S	2W-10N,2W-11N	39
Forest Pl. N&S	2W-11N	29
Forestview Rd. N&S	4W-12N	29
Foster St E&W	3W-11N,2W-11N	29
Fowler Av. N&S	3W-10N	39
Gaffield Pl. E&W	2W-12N	29
Garnett Pl. E&W	2W-11N	29
Garret Pl. E&W	2W-12N	29
Garrison Av. N&S	2W-13N	29
Geneva Pl. N&S	4W-12N	29
Girard Av. N&S	2W-13N	29
Golf Rd. E&W	4W-12N	29
Grant St. E&W	4W-12N,2W-12N	29
Greeley Av. N&S	5W-12N	28
Green Bay Rd. NW&SE	3W-12N	29
Greenleaf St. E&W	3W-10N,2W-10N	39
Greenwood St. E&W	3W-11N,2W-11N	29
Grey Av. N&S	3W-9N,3W-12N	39
Gross Point Rd. NE&SW	4W-13N	29
Grove St. E&W	3W-11N,2W-11N	29
Hamilton St. E&W	2W-10N	39
Hamlin St. E&W	2W-12N	29
Hampton Pkwy. N&S	2W-13N	29
Harrison St. E&W	4W-12N,3W-12N	29
Hartrey Av. N&S	3W-9N,3W-12N	39
Hartzel St. E&W	4W-13N	29
Harvard Ter. E&W	2W-9N	39
Hastings Av. N&S	4W-12N	29
Haven St. E&W	2W-12N	29
Hawthorne Ln. N&S	3W-12N	29
Hayes St. E&W	4W-12N	29
Highland Av. N&S	4W-13N	29
Hillside Ln. NW&SE	4W-13N	29
Hillside Rd. E&W	4W-13N	29
Hinman Av. N&S	2W-10N,2W-11N	39
Hovland Ct. N&S	3W-11N	29
Howard St. E&W	3W-9N	39
Hull Ter. E&W	2W-9N	39
Hurd Av. N&S	4W-12N	29
Ingleside Pk. E&W	2W-13N	29
Ingleside Pl. E&W	2W-13N	29
Isabella E&W	4W-13N,2W-13N	29
Jackson Av. N&S	3W-12N	29
Jenks St. NE&SW	3W-13N	29
Judson Av. N&S	2W-10N,2W-11N	39
Kedzie St. E&W	2W-10N	39
Keeney St. E&W	3W-10N,2W-10N	39
Kirk St. E&W	3W-9N	39
Knox Cir. N&S	4W-12N	29
Lake Shore Blvd. N&S	1W-10N	39
Lake St. E&W	3W-11N,2W-11N	29
Lakeside Ct. NW&SE	2W-13N	29
Laurel Av. N&S	3W-11N	29
Lawndale Av. N&S	4W-12N	29
Lee St. E&W	3W-10N,2W-10N	39
Leland Av. N&S	3W-11N	29
Lemar Av. N&S	3W-11N	29
Leon Pl. E&W	2W-12N	29
Leonard Pl. E&W	2W-12N	29
Library Pl. E&W	2W-12N	29
Lincoln St. E&W	4N-12N,2W-12N	29
Lincolnwood Dr. N&S	4W-12N	29
Livingston St. E&W	3W-13N	29
Lyons St. E&W	3W-11N	29
Madison Pl. E&W	3W-10N	39
Madison St. E&W	3W-10N,2W-10N	39
Main St. E&W	3W-10N	39
Maple Av N&S	2W-10N,2W-11N	39
Marcy Av. N&S	4W-12N,4W-13N	29
Martha Ln. N&S	4W-11N	29
McCormick Blvd. NE&SW	3W-11N,3W-12N	29
McDaniel Av. N&S	3W-10N,4W-13N	39
Meadowlark Ln. N&S	4W-13N	29
Michigan Av. N&S	2W-10N	39
Milburn Pk. E&W	2W-12N	29
Milburn St. E&W	2W-12N	29
Monroe St. E&W	3W-10N,2W-10N	39
Monticello St. E&W	2W-13N	29
Mulford St. E&W	3W-9N,2W-9N	39
Nathaniel Pl. E&W	3W-10N	39
Normandy Pl. E&W	4W-12N	29
Noyes Ct. N&S	2W-12N	29
Noyes St. E&W	4W-12N,2W-12N	29
Oak Av. N&S	2W-10N	39
Oakton St. E&W	3W-10N,1W-10N	39
Orrington Av. N&S	2W-11N	29
Otto Ln. E&W	4W-13N	29
Park Pl. E&W	4W-13N,3W-13N	29
Payne St. E&W	4W-12N,3W-12N	29
Pioneer Rd. N&S	3W-12N	29
Pitner Av. N&S	3W-10N	39
Poplar St. NW&SE	3W-13N	29
Prairie Av. NW&SE	3W-13N	29
Pratt Ct. N&S	2W-11N	29
Princeton Av. N&S	4W-12N	29
Prospect Av. N&S	4W-12N	29
Railroad Ave NW&SE	2W-11N	29
Reba Pl. E&W	2W-10N	39
Reese Av. N&S	4W-13N	29
Richmond Av. N&S	3W-9N	39
Ridge Av. N&S	2W-11N,2W-12N	29
Ridge Blvd. N&S	2W-10N	39
Ridge Ct. N&S	2W-10N	39
Ridge Ter. E&W	2W-12N	29
Ridgeway Av. N&S	4W-12N	29
Rosalie St. E&W	3W-12N	29
Roslyn Pl. E&W	2W-13N	29
Seward St. E&W	3W-10N,2W-10N	39
Sheridan Pl. E&W	2W-13N	29
Sheridan Rd. N&S	2W-10N,2W-11N	39
Sheridan Sq. N&S	1W-10N	39
Sherman Av. N&S	2W-10N,2W-12N	39
Sherman Pl. N&S	2W-11N	29
Simpson St. E&W	4W-12N,2W-12N	29
South Blvd. E&W	3W-10N,2W-10N	39
St. Mark Ct. N&S	2W-11N	29
Stewart Av. NW&SE	3W-13N	29
Terry Pl. E&W	2W-13N	29
Thayer St. E&W	4W-13N	29
Thelin Ct. E&W	3W-11N	29
University Pl. E&W	2W-11N	29
Wade St. E&W	3W-11N	29
Walnut Av. N&S	3W-13N	29
Warren St. E&W	3W-10N	39
Washington St. E&W	3W-10N,2W-10N	39
Wesley Av. N&S	3W-9N,3W-12N	39
Wilder Av. E&W	3W-10N	39
Woodbine Av. N&S	3W-13N	29

Evergreen Park

	Grid	Page
Albany Av. N&S	3W-11S,3W-10S	67
Artesian Av. N&S	3W-11S	67
Avers Av. N&S	4W-11S,4W-10S	67
Beck Ct. N&S	4W-11S	67

	Grid	Page
California Av. N&S	3W-11S,3W-10S	67
Campbell Av. N&S	3W-11S	67
Central Park Av. N&S	4W-11S,4W-10S	67
Clark Dr. NW&SE	4W-11S	67
Clifton Park Av. N&S	4W-11S,4W-10S	67
Country Club Dr. N&S	3W-10S	67
Drury Ln. E&W	3W-10S	67
Elm Pl. N&S	4W-11S	67
Fairfield Av. N&S	3W-11S,3W-10S	67
Francisco Av. N&S	3W-11S,3W-10S	67
Grove Pl. N&S	4W-11S	67
Hamlin Av. N&S	4W-11S,4W-10S	67
Harding Av. N&S	4W-11S,4W-10S	67
Homan Av. N&S	4W-11S,4W-10S	67
Kedzie Av. N&S	4W-11S,4W-10S	67
Lawndale Av. N&S	4W-11S,4W-10S	67
Maple St. NE&SW	4W-11S	67
Maplewood Av. N&S	3W-11S	67
Millard Av. N&S	4W-11S,4W-10S	67
Mozart St. N&S	3W-11S,3W-10S	67
Pulaski Rd. N&S	5W-11S,5W-10S	66
Richmond Av. N&S	3W-11S,3W-10S	67
Ridgeway Av. N&S	4W-11S,4W-10S	67
Rockwell St. N&S	3W-10S	67
Sacramento Av. N&S	3W-11S,3W-10S	67
Sawyer Av. N&S	4W-11S,4W-10S	67
Spaulding Av. N&S	4W-11S,4W-10S	67
Springfield Av. N&S	4W-11S,4W-10S	67
St. Louis Av. N&S	4W-11S,4W-10S	67
Talman Av. N&S	3W-10S	67
Troy Av. N&S	3W-11S,3W-10S	67
Trumbull Av. N&S	4W-11S,4W-10S	67
Turner Av. N&S	4W-11S,4W-10S	67
U.S. Route 12-20 E&W	4W-11S,3W-11S	67
Utica Av. N&S	3W-11S,3W-10S	67
Washtehaw Av. N&S	3W-11S,3W-10S	67
Western Av. N&S	3W-11S,3W-10S	67
87th St. E&W	3W-10S	67
88th St. E&W	3W-10S	67
89th Pl. E&W	3W-10S	67
89th St. E&W	3W-10S	67
90th Pl. E&W	3W-10S	67
90th St. E&W	3W-10S	67
91st St. E&W	4W-10S,3W-10S	67
92nd St. E&W	4W-10S,3W-10S	67
93rd Pl. E&W	3W-10S	67
93rd St. E&W	4W-10S,3W-10S	67
94th Pl. E&W	3W-10S	67
94th St. E&W	4W-10S,3W-10S	67
95th Pl. E&W	4W-11S,3W-11S	67
95th St. E&W	4W-10S,3W-10S	67
96th Pl. E&W	3W-11S	67
96th St. E&W	4W-11S,3W-11S	67
97th Pl. E&W	3W-11S	67
97th St. E&W	4W-11S,3W-11S	67
98th Pl. E&W	4W-11S,3W-11S	67
98th St. E&W	4W-11S,3W-11S	67
99th Pl. E&W	3W-11S	67
99th St. E&W	4W-11S,3W-11S	67
100th Pl. E&W	3W-11S	67
100th St. E&W	4W-11S,3W-11S	67
101st Pl. E&W	3W-11S	67
101st St. E&W	4W-11S,3W-11S	67
102nd Pl. E&W	4W-11S,3W-11S	67
102nd St. E&W	4W-11S,3W-11S	67
103rd St. E&W	4W-11S	67

Flossmoor

	Grid	Page
Alexander Crescent E&W	3W-23S	87
Amherst Ct. E&W	2W-23S	87
Argyle Av. N&S	3W-22S	87
Arquilla Av. NE&SW	3W-23S	87
Ash St. N&S	4W-22S	87
Avers Av. N&S	4W-22S	87
Balmoral Crescent E&W	3W-22S	87
Beech St. E&W	4W-22S	87
Berry Ln. N&S	3W-23S	87
Bob-o-link Rd. E&W	3W-23S	87
Bonnie Brae Crescent E&W	3W-23S	87
Braeburn Av. NS&EW	3W-23S	87
Braemar Rd. NE&SW	3W-22S	87
Brassie Av. N&S	3W-23S	87
Brookwood Dr. E&W	3W-23S	87
Bruce Av. N&S	3W-22S	87
Brumley Dr. E&W	3W-22S	87
Bunker Av. N&S	3W-23S	87
Burns Av. N&S	3W-22S	87
Butterfield Cir. N&S	3W-24S,3W-23S	92
Butterfield Ln. N&S	2W-23S	87
Butterfield Rd. N&S	3W-23S	87
Caddy St. E&W	3W-23S	87
Cambridge Av. N&S	2W-23S	87
Cardinal Ct. N&S	3W-23S	87
Carmel Dr. E&W	3W-22S	87
Carrol Pkwy. NE&SW	2W-22S	87
Central Dr. E&W	3W-22S	87
Central Park Av. N&S	4W-23S,4W-22S	87
Cherry Hills Dr. E&W	4W-22S	87
Cheryl Dr. N&S	2W-23S	87
Collett Ln. E&W	2W-22S	87
Cooley Pl. N&S	3W-23S	87
Cornell Rd. E&W	2W-23S	87
Cummings Ln. E&W	2W-22S	87
Dartmouth Rd. N&S	2W-23S	87
Dixie Hwy. NW&SE	2W-23S	87
Douglas Av. N&S	3W-23S,3W-22S	87
Dundee Av. N&S	3W-22S	87
Elm Ct. SW&NE	4W-22S	87
Elm St. N&S	4W-22S	87
Embassy Row E&W	3W-22S	87
Emelia Ct. E&W	3W-23S	87
Evans Rd. E&W	2W-22S	87
Flossmoor Rd. E&W	4W-22S,2W-22S	87
Gardner Rd. N&S	2W-22S	87
Golfview Ln. N&S	2W-23S	87
Gordon Dr. E&W	3W-22S	87
Governors Hwy NE&SW	4W-23S,3W-22S	87
Hagen Ln. E&W	2W-22S	87
Hamlin Av. N&S	4W-22S	87
Hanover Ln. E&W	2W-23S	87
Harding Av. N&S	4W-22S	87
Harold's Crescent E&W	3W-23S	87
Harvard Rd. E&W	2W-23S	87
Hawthorne Ln. E&W	3W-23S	87
Heather Hill Crescent N&S	3W-23S	87
Heather Rd. E&W	3W-22S	87
Holbrook Rd. E&W	2W-23S	87
Hutchinson Rd. NS&EW	2W-22S	87
Imperial Ct. E&W	3W-22S	87
Kathleen Ln. E&W	3W-23S	87
Kedzie Av. N&S	4W-23S,4W-22S	87
Knollwood Dr. E&W	2W-23S	87
Lake Dr. NE&SW	3W-23S	87
Cir.	3W-23S	87
Laurel St. E&W	2W-23S	87
Lawrence Crescent NE&SW	3W-23S	87
Leavitt Av. N&S	3W-22S	87
Lynnwood Ct. E&W	2W-23S	87
MacDonald Ln. E&W	2W-22S	87
MacFarlane Crescent E&W	3W-23S	87
MacHeath Crescent E&W	3W-23S	87
Maple Rd. N&S	4W-22S	87
Markey Ln. E&W	2W-23S	87
Marston Ln. E&W	2W-22S	87
Maryland Av. E&W	3W-23S	87
Mast Ct. E&W	2W-23S	87
Monterey Dr. E&W	3W-22S	87
Oak Ct. NE&SW	4W-22S	87
Oak Lane Rd. N&S	3W-23S	87
Oak St. E&W	4W-22S	87
Oakmont Av. N&S	3W-23S	87
Oxford Ct. E&W	2W-23S	87
Park Dr. NS&EW	3W-22S	87
Patricia Ln. NE&SW	3W-23S	87
Perry Rd. E&W	3W-23S	87
Perth Av. N&S	3W-22S	87
Pheasant Ct. N&S	3W-23S	87
Pinehurst Ln. E&W	2W-23S	87
Polly Ln. E&W	3W-23S	87
Poplar Rd. E&W	4W-22S	87
Princeton Rd. N&S	2W-23S	87
Private Rd. E&W	3W-23S	87
Pulaski Rd. N&S	4W-23S,4W-22S	87
Quail Ct. N&S	3W-23S	87
Redwood Ct. E&W	3W-22S	87
Redwood Ln. N&S	3W-22S	87
Robertson Ln. E&W	2W-23S	87
School St. E&W	3W-22S	87
Scott Crescent E&W	3W-23S	87
Springfield Av. N&S	4W-22S	87
Sterling Av. N&S	2W-22S	87
Sterling Av. NE&SW	3W-22S	87
Strieff Ln. N&S	3W-23S	87
Sunset Av. E&W	3W-23S	87
Sylvan Ct. E&W	2W-23S	87
Thomas Ct. E&W	2W-23S	87
Thornwood Dr. E&W	2W-23S	87
Tina Ln. N&S	2W-23S	87
Travers Ln. N&S	2W-22S	87
Vardon Ln. E&W	2W-22S	87
Vardon Pl. E&W	3W-22S	87
Verne Ln. N&S	3W-22S	87
Vollmer Rd. E&W	4W-23S,2W-23S	87
Wallace Dr. E&W	3W-22S	87
Western Av. N&S	3W-23S	87
Willow St. E&W	4W-22S	87
Yale Rd. N&S	2W-23S	87
187th St. E&W	4W-22S	87
188th St. E&W	4W-22S	87
189th St. E&W	4W-22S	87
190th St. E&W	4W-22S	87
198th St. E&W	4W-23S	87

Ford Heights

	Grid	Page
Ambassador Ln. E&W	1E-25S	94
Berkeley Av. N&S	1E-25S	94
Calumet Expressway N&S	1E-25S	94
Cannon Ln. E&W	1E-24S	94
Columbia Av. E&W	1E-24S	94
Congress Ln. E&W	1E-25S	94
Cottage Grove Av. N&S	0E-25S	94
Cottage Grove Av. N&S	1E-25S	94
DeForest Av. N&S	1E-25S	94
Deer Creek Dr. NE&SW	1E-25S	94
Diplomat Ln. E&W	1E-25S	94
Drexel Av. N&S	1E-25S	94
Ellis Av. N&S	1E-25S	94
Embassy Ln. E&W	1E-25S	94
Greenwood Av. N&S	1E-25S,1E-24S	94
Hammond Ln. E&W	1E-24S	94
Kennedy Ln. N&S	1E-25S	94
Lexington Av. N&S	1E-25S	94
Lincoln Hwy. E&W	1E-25S	94
Park Av. N&S	1E-25S	94
Park Ln. E&W	1E-25S	94
Regent Ln. E&W	1E-25S	94
Rt. 394 N&S	1E-25S	94
Seeley Av. N&S	1E-25S	94
Senator Ln. E&W	1E-25S	94
U.S. Rt. 30 E&W	1E-25S	94
Werline Av. N&S	1E-25S	94
Woodlawn Av. N&S	1E-25S	94
8th Pl. E&W	1E-24S	94
8th St. E&W	1E-24S	94
9th St. E&W	1E-24S	94
10th Pl. E&W	1E-25S	94
10th St. E&W	1E-24S	94
11th Pl. E&W	1E-25S	94
11th St. E&W	1E-25S	94
12th St. E&W	1E-25S	94
13th St. E&W	1E-25S	94
14th Pl. E&W	1E-25S	94
15th Pl. E&W	1E-25S	94
15th St. E&W	1E-25S	94
16th St. E&W	1E-25S	94
17th St. E&W	1E-25S	94

Forest Park

	Grid	Page
Adams E&W	9W-0S	48
Beloit N&S	9W-0S	48
Belvidere N&S	9W-0N	48
Bergman Ct. E&W	9W-0N	48
Brown E&W	9W-0N	48
Burchardt N&S	9W-0N	48
Circle E&W	9W-1S,9W-0N	54
Des Plaines N&S	9W-0S	48
Dixon E&W	9W-0N	48
Dunlop N&S	9W-0S	48
Elgin N&S	9W-1S,9W-0N	54
Ferdinand N&S	9W-0S	48
Fillmore E&W	9W-0S	48
Franklin E&W	9W-0N	48
Hannah N&S	9W-1S,9W-0S	54
Harlem N&S	9W-1S	54
Harrison E&W	9W-0S	48
Harvard E&W	9W-0S	48
Jackson Blvd. E&W	9W-0S	48
Lathrop N&S	9W-0S	48
Lehmer E&W	9W-0S	48
Lexington E&W	9W-0S	48
Lincoln Ct. E&W	9W-0N	48
Madison E&W	9W-0N	48
Marengo N&S	9W-1S,9W-0S	54
Monroe E&W	9W-0S	48
Polk E&W	9W-0S	48
Railroad N&S	9W-0S	48
Randolph E&W	9W-0N	48
Rockford N&S	9W-0N	48
Taylor E&W	9W-0S	48
Thomas N&S	9W-0S	48
Troost N&S	9W-0S	48
Van Buren N&S	9W-0S	48
Warren E&W	9W-0N	48
Washington E&W	9W-0N	48
Wilcox E&W	9W-0S	48
York E&W	9W-0S	48
Yuba E&W	9W-0S	48

Forest View

	Grid	Page
Adlai E. Stevenson Expy. E&W	8W-5S	59
Grove N&S	8W-4S	54
Home N&S	8W-4S	54
Interstate 55 E&W	6W-4S	55
Kenilworth N&S	8W-4S	54
Maple N&S	8W-4S	54
Wenonah N&S	8W-4S	54
Wisconsin N&S	8W-4S	54
46th St. E&W	8W-4S	53
46th St. E&W	8W-4S	59
47th E&W	6W-4S	55

Franklin Park

	Grid	Page
Acorn N&S	14W-4N	40
Addison E&W	12W-4N	40
Addison E&W	13W-4N	40
Alta N&S	13W-4N	40
Anderson Pl. E&W	13W-3N	40
Ashland E&W	11W-3N	41
Atlantic N&S	12W-3N	40
Belden E&W	11W-2N	47
Belmont E&W	14W-4N,11W-4N	40
Birch N&S	11W-3N,11W-4N	41
Bright N&S	13W-3N	40
Britta E&W	11W-4N	41
Calwagner N&S	12W-3N	40
Carnation N&S	13W-4N	40
Carol N&S	13W-3N	40
Center Pl. E&W	13W-3N	40
Centrella N&S	13W-4N	40
Cherry E&W	11W-3N	41
Chestnut E&W	12W-3N,11W-3N	40
Commerce St. N&S	11W-3N	41
Crescent E&W	11W-4N	41
Crown E&W	13W-3N	40
Cullerton St. N&S	13W-3N	40
Davis E&W	11W-4N	41
Des Plaines River Rd. N&S	11W-4N	41
Desoto St. E&W	12W-4N	41
Dodge E&W	12W-4N	40
Dora N&S	12W-3N,12W-4N	40
Edgington N&S	11W-3N,11W-4N	41
Elder Ln. N&S	12W-3N	40
Emerson N&S	12W-3N,12W-4N	40
Ernst N&S	12W-2N,12W-4N	46
Exchange Pl. E&W	11W-3N	41
Fairfield N&S	12W-3N	40
Fletcher E&W	13W-3N	40
Franklin E&W	12W-4N	40
Front E&W	12W-4N	40
Gage E&W	12W-4N,11W-4N	40
Gary N&S	13W-3N	40
George N&S	12W-3N,12W-4N	40
Grand E&W	13W-3N,11W-3N	40
Greenfield Rd. E&W	12W-3N	40
Gustav N&S	12W-3N	40
Hacke N&S	11W-4N	41
Hart E&W	13W-3N	40
Hawthorne N&S	12W-3N,12W-4N	40
Herrick E&W	12W-3N,11W-3N	40
Hirschberg N&S	12W-3N	40
Houston Dr. N&S	13W-3N	40
Iona St. E&W	12W-4N	41
James E&W	13W-3N	40
Jill N&S	13W-3N	40
Johanna E&W	12W-3N	40
Kennedy St. E&W	11W-3N	41
Kimmey N&S	11W-4N	41
King E&W	12W-4N,11W-4N	40
King Av. E&W	13W-4N	40
Kirshoff N&S	12W-3N,12W-4N	40
LaSalle St. E&W	12W-4N	41
Laidria N&S	13W-3N	40
Leona N&S	12W-3N	40
Lesser E&W	12W-4N	40
Leyden N&S	11W-4N	41
Lincoln N&S	12W-3N,12W-4N	40
Lombard N&S	11W-3N,11W-4N	41
Lonnquist Dr. N&S	12W-4N	40
Louis N&S	12W-3N	40
Lucy N&S	13W-3N	40
Mannheim N&S	13W-4N	40
Maple N&S	11W-2N	47
Martens N&S	11W-4N	41
McNerney Dr. E&W	12W-4N	40
Medill E&W	11W-2N	47
Melrose E&W	14W-4N	40
Minneapolis E&W	12W-3N	40
Montana E&W	12W-3N	40
Nerbonne E&W	11W-4N	41
Nevada E&W	12W-3N,11W-3N	40
Nichols E&W	11W-4N	41
Nona N&S	12W-3N	40
North Charles N&S	13W-4N	40
North Powell N&S	14W-4N	40
Oak N&S	11W-3N	41
Pacific E&W	12W-4N,11W-3N	40
Park Lane E&W	12W-3N	40
Parklane E&W	11W-3N	41
Pearl N&S	11W-3N,11W-4N	41
Prairie Rd. N&S	13W-3N	40
Reeves Ct. E&W	12W-4N	40
Reuter N&S	12W-3N	40
Richard E&W	12W-3N	40
Riverside Dr. N&S	12W-3N	40
Robinson E&W	11W-4N	41
Rose N&S	12W-3N	40
Rose N&S	11W-2N	47
Ruby N&S	12W-3N,12W-4N	40
Runge N&S	14W-4N	40
Ruth N&S	13W-3N	40
Sandra N&S	14W-4N	40
Sarah N&S	12W-3N,12W-4N	40
Schierhorn Ct. N&S	11W-4N	41
Schiller E&W	12W-3N	40
Scott N&S	12W-3N	40
Seymour E&W	13W-5N	40
Shelia N&S	12W-3N	40
Shirley E&W	12W-3N	40
Silver Creek Dr. N&S	12W-3N	40
Sonia N&S	12W-3N	40
Wagner N&S	12W-3N	40
Walnut E&W	11W-3N	41
Washington N&S	11W-3N,11W-4N	41
Waveland E&W	13W-4N	40
Wehrman N&S	12W-4N	40
Wellington St. NE&SW	13W-3N	40
West King E&W	14W-4N	40
West Manor Av. E&W	12W-3N	40
Westbrook Dr. N&S	12W-3N	40
Williams Dr. N&S	13W-4N	40
Willow N&S	11W-3N,11W-4N	41
Wrightwood N&S	12W-3N	40
17th Av. N&S	11W-2N	47

Glencoe

	Grid	Page
Adams Av. E&W	6W-17N	18
Apple Tree Ln. E&W	7W-17N	18
Aspen Ln. E&W	6W-18N	18
Beach Rd. NW&SE	6W-17N,6W-18N	18
Birch Rd. E&W	7W-17N	18
Bluff Rd. N&S	6W-18N	18
Bluff St. N&S	6W-17N,6W-16N	18
Brentwood Dr. E&W	6W-18N	18
Briar Ln. E&W	6W-17N	18
Brookside Ln. N&S	6W-16N	18
Brookvale Ter. E&W	6W-16N	18
Carol Ln. N&S	6W-18N	18
Cedar Ln. NE&SW	6W-17N	18
Cherry Tree Ln. N&S	7W-17N	18
Chestnut Ln. E&W	7W-17N	18
Clover Ln. NE&SW	7W-17N	18
Country Ln. E&W	6W-16N	18
Crescent Dr. E&W	6W-18N	18
Crescent Ln. E&W	5W-17N	19
Dell Pl. NE&SW	5W-17N	19
Dennis Ln. NW&SE	6W-18N	18
Drexel Av. N&S	6W-16N	18
Drexel Ln. N&S	6W-16N	18
Dundee Rd. E&W	7W-18N	18
Eastwood Rd. N&S	6W-18N	18
Edens Expwy. NW&SE	7W-16N	18
Edgebrook Ln. N&S	7W-17N	18
Elder Ct. E&W	7W-17N	18
Elm Ct. NE&SW	6W-18N	18
Elm Pl. N&S	7W-17N	18
Elm Ridge Dr. N&S	7W-18N	18
Estate Dr. E&W	6W-18N	18
Euclid Av. N&S	5W-16N	19
Fairfield Rd. N&S	7W-18N	18
Fairview Rd. N&S	5W-17N	19
Forest Av. NE&SW	6W-18N	18
Franklin Rd. N&S	6W-18N	18
Glade Rd. (Pvt.) N&S	6W-18N	18
Glencoe Dr. N&S	6W-17N,6W-18N	18
Glencoe Rd. NW&SE	6W-17N	18
Glenwood Av. NE&SW	5W-16N	19
Green Bay Rd. NW&SE	6W-18N,5W-17N	18
Green Leaf Av. N&S	6W-17N,6W-18N	18
Greenwood Av. N&S	6W-17N	18
Grove St. N&S	6W-17N	18
Harbor St. NE&SW	5W-17N	19
Hawthorn Av. NE&SW	6W-17N	18
Hazel Av. NE&SW	6W-17N	18
Hillcrest Dr. N&S	6W-18N	18
Hogarth Ln. (Pvt.) NW&SE	6W-18N	18
Hohlfelder Rd. N&S	7W-18N	18
Ida Pl. N&S	6W-17N	18
Ivy Ln. E&W	6W-16N	18
Jackson Av. E&W	6W-17N	18
Jefferson Av. E&W	6W-17N	18
Julia Ct. N&S	7W-18N	18
Kelling Ln. E&W	7W-18N	18

Grid   Page

Knollwood Ln. E&W.. 6W-18N 18
Lake Cook Rd. E&W
.... 7W-18N 18
Lake St. N&S 6W-16N 18
Lakeside Ter. N&S .. 5W-17N 19
Lakewood Dr. (Pvt.) NE&SW
.... 6W-18N 18
Lapier St. NE&SW .. 5W-17N 19
Lincoln Av. NE&SW .. 6W-17N 18
Lincoln Dr. E&W .. 6W-17N 18
Linda Ln. N&S 6W-17N 18
Linden Av. NW&SE .. 5W-17N 19
Longmeadow Ln. N&S
.... 7W-18N 18
Longwood Av. N&S. 6W-18N 18
Madison Av. E&W .. 6W-17N 18
Maple Hill Rd. E&W. 6W-18N 18
Mary St. N&S 5W-17N 19
Mayfair Ln. N&S .. 7W-18N 18
Meadow Rd. N&S .. 6W-18N 18
Milton Av. NE&SW .. 6W-18N 18
Monroe NE&SW 5W-17N 19
Monroe Av. E&W .. 6W-17N 18
Montgomery St. E&W
.... 5W-17N 19
Mortimer Ln. .. 5W-17N 19
North Way NE&SW .. 6W-17N 18
Northwood Dr. E&W
.... 7W-18N 18
Oak Dr. N&S 7W-17N 18
Oak Ridge Dr. N&S .. 7W-18N 18
Oak Ter. N&S 7W-18N 18
Oakdale Av. E&W.. 6W-16N 18
Old Elm Ln. N&S .. 6W-18N 18
Old Elm Pl. N&S .. 6W-18N 18
Orchard Ln. E&W .. 7W-18N 18
Palos Rd. NW&SE .. 5W-17N 19
Park Av. NW&SE .. 6W-17N 18
Park Pl. E&W 6W-18N 18
Peach Tree Ln. NE&SW
.... 7W-17N 18
Pebblewood Ln. N&S
.... 7W-18N 18
Pine Ct. N&S 7W-18N 18
Pinewood Ln. N&S .. 7W-18N 18
Prairie St. N&S 7W-17N 18
Randolph St. N&S
.... 6W-16N,6W-17N 18
Ravine Glade St. E&W
.... 5W-17N 19
Redwood Ln. E&W .. 6W-16N 18
Robin Ln. E&W 6W-16N 18
Rockgate Ln. (Pvt.) NW&SE
.... 6W-18N 18
Scott Av. NE&SW .. 5W-16N 19
Scott Ln. NW&SE... 5W-16N 19
Sheridan Rd. NW&SE
.... 6W-18N 18
Skokie Ln., North E&W
.... 6W-16N 18
Skokie Ln., South E&W
.... 6W-16N 18
Skokie Ridge Dr. N&S
.... 7W-18N 18
Skokie Ridge Rd. N&S
.... 7W-18N 18
South Av. E&W .. 6W-17N 18
South Way N&S 6W-17N 18
Stonegate Ter. NE&SW
.... 7W-17N 18
Strawberry Hill Dr. N&S
.... 7W-17N 18
Sunrise Cir. NW&SE
.... 5W-17N 19
Sunset Ln. E&W
.... 7W-18N,6W-18N 18
Surfside Pl. NW&SE 5W-17N 19
Sycamore Ln. E&W . 7W-17N 18
Sylvan Rd. E&W 6W-18N 18
Temple Ct. NE&SW .. 6W-17N 18
Terrace Ct. N&S 6W-18N 18
Thorn Apple Ln. N&S
.... 7W-17N 18
Timber Ln. E&W 6W-18N 18
Tudor Ct. NE&SW .. 6W-18N 18
Valley Rd. N&S
.... 7W-17N,6W-17N 18
Valley St. N&S 6W-17N 18
Vernon Av. N&S
.... 6W-16N,6W-17N 18
Village Ct. NW&SE .. 6W-17N 18
Walden Dr. E&W 6W-18N 18
Washington Av. E&W
.... 6W-17N 18
Washington Pl. (Pvt.) E&W
.... 6W-17N 18
Wentworth St. NE&SW
.... 5W-17N 19
Westley Rd. E&W... 7W-18N 18
Westwood Ln. E&W . 7W-18N 18
Whitebridge Hill St. NE&SW
.... 5W-17N 19
Wilgate Ter. NW&SE
.... 6W-17N 18
Willow Tree Ln. E&W
.... 7W-18N 18
Woodlawn Av. NE&SW
.... 6W-16N 18
Woodlawn St. NE&SW
.... 5W-17N 19
Woodridge Ln. E&W
.... 7W-17N 18

Glenview

Alder Pl. E&W 8W-14N 27
Alvin Ct. N&S 9W-13N 27
Ammer Rd. E&W .. 8W-14N 27

Grid   Page

Apple Tree Ln. E&W
.... 7W-13N 28
Applegate Ct. NW&SE
.... 9W-13N 27
Arbor Ln. N&S 6W-13N 28
Arbor St. N&S 6W-13N 28
Ardmore St. E&W... 6W-13N 28
Ari Ln. NS&EW 10W-15N 17
Astor Ct. N&S 10W-15N 17
Baffin Rd. E&W.... 8W-13N 27
Balmoral Av. N&S... 7W-14N 28
Barbara Dr. E&W .. 10W-15N 17
Barbary Ln. N&S .. 11W-15N 16
Barry Ln. N&S...... 11W-14N 26
Barton Ct. N&S 7W-13N 28
Basswood Cir N&S.. 9W-13N 27
Basswood Ct. NE&SW
.... 8W-14N 27
Beaver Ln. N&S .. 6W-13N 28
Beaver Rd. N&S ... 6W-13N 28
Becker Rd. E&W .. 7W-14N 28
Beech Dr. E&W.... 6W-13N 28
Bel Air Dr. N&S... 9W-12N 27
Bel Air Ln. E&W .. 9W-12N 27
Bellwood Ln. N&S.. 10W-14N 27
Bette Ln. N&S.... 11W-13N 26
Beverly Ln. E&W .. 6W-13N 28
Bianco Ter. E&W .. 9W-12N 27
Birchwood Ct. E&W.. 9W-12N 27
Blackthorn Dr. N&S.. 8W-14N 27
Blake Ln. E&W .. 11W-15N 16
Bonnie Glen Ln. E&W
.... 8W-13N,7W-13N 27
Borre Ln. N&S 8W-14N 27
Brairford Ln. N&S .. 11W-15N 16
Brandess Dr. E&W. 10W-15N 17
Brandon Rd. N&S.... 7W-13N 28
Brassie Dr. E&W .. 9W-13N 27
Brett Ln. E&W.... 11W-14N 26
Briarhill N&S 9W-12N 27
Brook Ln. E&W.... 6W-13N 28
Brookdale Ln. NW&SE
.... 11W-15N 16
Bucknell Ct. N&S.... 9W-13N 27
Burnham Ct. N&S... 9W-13N 27
Burr Oak Dr. E&W .. 8W-14N 27
Burton Ter. E&W .. 7W-14N 28
Button Wood Ln. NW&SE
.... 9W-13N,9W-14N 27
Canterbury Ln. N&S
.... 7W-13N,7W-14N 28
Cariann Ln. N&S .. 6W-13N 28
Carol Ln. NW&SE .. 12W-14N 26
Carousel Dr. E&W .. 11W-15N 16
Carrington Ln. N&S
.... 11W-15N 16
Castle Dr. N&S .. 11W-13N 26
Cedar Ct. E&W...... 9W-12N 27
Cedar Wood N&S.... 9W-13N 27
Central Pkwy. E&W. 8W-13N 27
Central Rd. E&W
.... 10W-13N,6W-13N 27
Charlie Ct. N&S .. 11W-15N 16
Chatham Dr. N&S... 8W-13N 27
Chester Dr. E&W .. 11W-15N 16
Chesterfield St. N&S
.... 9W-13N 27
Chestnut Av. E&W .. 8W-14N 27
Church St. N&S 8W-13N 27
Cindy Ln. N&S .. 11W-13N 26
Clavey Ln. N&S.... 9W-12N 27
Clearview Dr. N&S .. 10W-13N 27
Clover Ct. N&S 9W-13N 27
Cir. 7W-13N 28
Colfax St. N&S...... 9W-12N 27
Collins Av. N&S 10W-14N 27
Cornell Ct.* N&S 9W-15N 17
Coronet Rd. N&S
.... 7W-12N,7W-13N 28
Country Ln. N&S .. 8W-12N 27
Covert Rd. E&W
.... 10W-12N,9W-12N 27
Crab Tree Ln. N&S .. 8W-12N 27
Crescent Ln. E&W .. 11W-14N 26
Crestwood Dr. E&W
.... 11W-15N 16
Crestwood Ln. E&W
.... 10W-13N 27
Crown Dr. N&S .. 11W-14N 26
Crown Ln. N&S .. 11W-14N 26
Crystal Ct. N&S .. 10W-13N 27
Culver Ln. E&W 8W-12N 27
Dale St. E&W 6W-13N 28
Dartmouth Ct.* N&S 9W-15N 17
De L'ogier Dr. N&S .. 7W-14N 28
Dearlove Rd. N&S .. 11W-13N 26
Debra Ln. N&S .. 10W-15N 17
Dell Pl. E&W 10W-13N 27
Denise Ct. NW&SE
.... 11W-13N 26
Depot St. N&S 8W-13N 27
Dewes St. E&W .. 9W-13N 27
Dorthy Rd. E&W... 10W-15N 17
Drake Rd. E&W .. 8W-13N 27
East Burr Oak Dr. N&S
.... 8W-14N 27
East Lake St. E&W
.... 10W-13N 27
Echo Ln. N&S 6W-13N 28
Edgewood E&W .. 7W-13N 28
Elizabeth Ln. N&S .. 8W-13N 27
Elm St. N&S
.... 9W-12N,9W-13N 27
Elmdale Ln. N&S .. 10W-14N 27
Elmdale Rd. NS&EW
.... 10W-13N 27
Elmgate Dr. N&S .. 9W-13N 27
Elmwood Av. E&W .. 7W-14N 28

Grid   Page

Estate Ln. N&S 10W-14N 27
Evergreen Ter. N&S.. 8W-14N 27
Executive Cir. N&S
.... 10W-14N 27
Executive Ct. N&S .. 10W-14N 27
Executive Ct. N&S.. 10W-14N 27
Fairhope Av. E&W .. 7W-14N 28
Fairway Ct. E&W .. 10W-13N 27
Fairway Dr. N&S .. 10W-13N 27
Ferndale Dr. E&W .. 6W-13N 28
Fernwood Ln. N&S .. 9W-12N 27
Fir St. N&S 9W-13N 27
Flora Av. N&S 9W-12N 27
Fontana Dr. E&W .. 9W-12N 27
Forest Av. E&W .. 7W-14N 28
Forest Rd. N&S 6W-13N 28
Forestview Ln. E&W
.... 8W-14N 28
Franklin Ln. N&S .. 11W-14N 26
Frederic Ct. N&S .. 11W-15N 16
Garden Ct. N&S .. 9W-13N 27
George Ct. E&W 8W-12N 27
Gimidieff Ct. N&S .. 11W-14N 26
Gladish Ln. N&S .. 9W-13N 27
Glen Ct. N&S 8W-13N 28
Glen Oak Dr. N&S .. 7W-14N 28
Glenayre Dr. N&S .. 7W-14N 28
Glendale Rd. N&S .. 7W-13N 28
Glenlake Dr. NE&SW
.... 10W-14N 27
Glenridge Dr. N&S .. 8W-13N 28
Glenview Rd. E&W
.... 11W-13N,7W-13N 26
Glenway Dr. E&W .. 11W-15N 16
Glenwood Av. E&W. 8W-14N 27
Glenwood Rd. N&S
.... 10W-13N 27
Glenwood Rd. N&S.. 6W-13N 28
Gloria Ct. N&S .. 11W-13N 26
Golf Ct. E&W 9W-12N 27
Golf Rd. E&W
.... 9W-12N,8W-12N 27
Golf View Ln. N&S .. 7W-13N 28
Golf View Rd. E&W .. 7W-13N 28
Grace Ln. E&W .. 11W-14N 26
Grandview Pl. E&W.. 9W-12N 27
Green Acres Ln. E&W
.... 7W-13N 28
Green Willow Ln. E&W
.... 8W-13N 27
Greenbrier Dr. E&W
.... 10W-12N 27
Greenbrier Ln. E&W .. 9W-12N 27
Greenfield Dr. N&S .. 9W-12N 27
Greenwood Av. N&S
.... 10W-14N 27
Greenwood Ct. E&W
.... 10W-15N 17
Greenwood Rd. N&S
.... 10W-15N 17
Gregory Dr. E&W .. 12W-14N 26
Grove St. E&W
.... 9W-13N,8W-13N 27
Hackberry Ln. E&W .. 6W-13N 28
Hager Ct. E&W.... 6W-13N 28
Hammersmith E&W
.... 11W-14N 26
Happy Hollow Rd. N&S
.... 10W-15N 17
Harlem Av. N&S 8W-12N 27
Harmony Ln. E&W .. 8W-13N 27
Harms Rd. N&S 7W-13N 28
Harrison St. E&W .. 8W-13N 27
Harvard Ct.* E&W .. 9W-15N 17
Harvest Ln. E&W .. 11W-15N 16
Hawthorne Ln. N&S
.... 7W-13N,7W-14N 28
Heather Ln. E&W .. 8W-12N 27
Heatherfield Ln. N&S
.... 6W-13N 28
Henley St. E&W
.... 9W-13N,8W-13N 27
Henneberry Ln. N&S
.... 8W-12N 27
Hickory Ct. E&W .. 9W-12N 27
Highland Ln. N&S .. 10W-13N 27
Highland Ter. N&S .. 8W-14N 27
Hill Cir. N&S 7W-13N 28
Hollywood Av. E&W. 8W-14N 27
Huber Ln. N&S 10W-13N 27
Huber Ln. N&S 10W-12N 27
Huckelberry Ln. E&W
.... 7W-14N 28
Hudson Rd. E&W... 8W-13N 27
Hunter Rd. N&S 6W-13N 28
Huntington Ln. N&S.. 8W-14N 27
Hutchings Av. N&S .. 8W-13N 27
Imperial Dr. N&S.... 10W-14N 27
Indian Rd. N&S 8W-14N 27
Indian Ridge Ct. N&S
.... 11W-15N 16
Indian Ridge Dr. NS&EW
.... 11W-15N 16
Iroquois Dr. E&W .. 11W-15N 16
Ivy Dr. N&S...... 11W-14N 26
Ivy Ln. E&W 11W-14N 26
Ivy Way N&S 11W-14N 26
Janet Dr. N&S 10W-15N 17
Janney Rd. N&S.... 10W-15N 17
Jefferson Av. E&W .. 8W-14N 27
Jerrie Ln. E&W 8W-13N 27
Joanne Dr. E&W .. 11W-15N 16
John's Dr. N&S.... 9W-14N 27
Johns Ct. N&S 9W-14N 27
Joy Ln. N&S 11W-14N 26
Juniper St. N&S .. 6W-13N 28
Juniper Ter. N&S .. 6W-13N 28
Karen Ln. E&W .. 10W-12N 27

Grid   Page

Kaywood Ln. N&S .. 7W-14N 28
Keenan Ln. E&W .. 10W-15N 17
Keiss Dr. E&W 11W-15N 16
Kendal Dr. N&S.... 8W-14N 27
Kenilworth Ln. E&W
.... 8W-14N,7W-14N 27
Kings Av. N&S 7W-14N 28
Kings Ln. E&W 7W-14N 28
Kmeic Ct. N&S 11W-14N 26
Knight St. N&S 11W-14N 26
Knollwood Ln. E&W
.... 10W-13N 27
LaFontaine Dr. NW&SE
.... 11W-14N 26
Lake Av. E&W 8W-14N 27
Lakewood Ct. E&W
.... 10W-13N 27
Landwehr Rd. N&S
.... 7W-14N 28
.... 11W-14N,11W-15N 26
Laramie Av. N&S .. 6W-13N 28
Laramie Ln. E&W .. 6W-13N 28
Larkdale Dr. N&S .. 9W-13N 27
Lawrence Ln. NW&SE
.... 12W-14N 26
Lawson Rd. E&W .. 10W-15N 17
Leaf Dr. E&W 9W-13N 27
Leanne Ct. N&S .. 11W-15N 16
Lehigh Av. NW&SE .. 8W-13N 27
Lenox Rd. N&S.... 7W-13N 28
Lexington Ln. E&W
.... 10W-13N 27
Liberty Ln. N&S .. 10W-13N 27
Limar Ln. N&S 10W-14N 27
Lincoln St. N&S
.... 9W-12N,9W-13N 27
Linden N&S 7W-14N 28
Linden Av. N&S 7W-14N 28
Linden Leaf Dr. N&S
.... 9W-13N 27
Linden Ln. E&W .. 7W-14N 28
Lindenwood Ln. N&S
.... 11W-13N,10W-13N 26
Linneman St. N&S .. 9W-13N 27
Lizette Ln. N&S .. 11W-15N 16
Locust Ln. N&S .. 8W-12N 27
Lois Ln. N&S 7W-14N 28
London Ln. E&W .. 8W-14N 27
Long Rd. N&S 6W-13N 28
Long Valley Rd. E&W
.... 8W-12N 27
Longmeadow N&S
.... 10W-13N,10W-14N 27
Lori Dr. N&S 10W-15N 17
Lotus Ln. N&S 10W-15N 17
Luri Lane Dr. NE&SW
.... 12W-14N 26
Lynn Ct. E&W 7W-14N 28
MacArthur Dr. E&W. 9W-12N 27
MacLean Ct. E&W .. 8W-13N 27
Malik Ct. E&W 9W-12N 27
Maple Leaf Dr. E&W
.... 9W-13N 27
.... 7W-14N 28
Maple Ln. E&W .. 7W-14N 28
Maplewood Ln. E&W
.... 8W-13N 27
Marie Ln. N&S 11W-14N 26
Mark Dr. N&S 6W-13N 28
Mary Ct. N&S .. 11W-15N 16
Mary Kay Ln. E&W
.... 10W-15N 17
Meadow Ln. N&S... 7W-14N 28
Meadowlark Ln. N&S
.... 9W-13N 27
Merle Ln. N&S .. 12W-14N 26
Michael Ln. E&W .. 11W-15N 16
Michael Manor N&S
.... 9W-12N 27
Mickey Ln. N&S .. 8W-13N 27
Miller Dr. E&W
.... 11W-15N,10W-15N 16
Mohawk Dr. E&W .. 11W-15N 16
Monroe Av. E&W .. 8W-14N 27
Montgomery Ln. N&S
.... 9W-12N 27
Mulberry Ln. E&W .. 6W-13N 28
Myrtle Ln. E&W .. 11W-15N 16
Neva Av. N&S 8W-12N 27
Nora Av. N&S 8W-12N 27
Nordica Av. N&S .. 8W-12N 27
Normandy Ln. E&W.. 7W-13N 28
North Branch Rd. N&S
.... 8W-12N 27
North Dr. N&S 7W-14N 28
Norwich Ct. NW&SE
.... 10W-13N 27
Norwood Ter. E&W
.... 10W-12N 27
Nottingham Av. N&S
.... 8W-12N 27
Oakwood Ln. NE&SW
.... 7W-14N 28
Old Willow Rd. E&W
.... 9W-13N 27
Old Willow Rd. N&S 9W-15N 17
Osage Dr. E&W .. 11W-15N 16
Overland Pass N&S
.... 11W-14N 26
Overlook Dr. E&W .. 8W-12N 27
Oxford Ln. E&W .. 7W-13N 28
Palmgren Dr. E&W
.... 9W-13N,8W-13N 27
Pam-Anne Dr. N&S
.... 10W-13N 27
Park Dr. N&S 9W-13N 27
Park Ln. N&S 9W-13N 27
Parkview N&S 9W-12N 27
Pebble Creek Dr. N&S
.... 8W-14N 27

Grid   Page

Pebbleford Dr. N&S
.... 10W-15N 17
Pendleton Ln. E&W.. 8W-14N 27
Penn Ct.* NE&SW .. 9W-15N 17
Pfingsten Rd. N&S
.... 11W-13N,11W-14N 26
Philipi Ln. N&S .. 10W-15N 17
Pick Dr. N&S 9W-12N 27
Pick Ln. E&W 9W-12N 27
Pickwick Av. N&S .. 10W-14N 27
Pickwick Ter. E&W
.... 10W-13N 27
Pine Ct. E&W 8W-13N 27
Pine St. N&S 8W-13N 27
Pinehurst Dr. N&S .. 9W-13N 27
Pleasant Dr. E&W.. 8W-14N 27
Pleasant Ln. NW&SE
.... 7W-14N 28
Pleasant Run E&W .. 11W-14N 26
Plymouth Ct. E&W .. 8W-13N 27
Plymouth Ln. E&W .. 8W-13N 27
Polo Rd. E&W 7W-13N 28
Portage Run Rd. N&S
.... 11W-14N 26
Porter Av. E&W 11W-15N 16
Porter Ct. N&S 11W-15N 16
Potter Rd. N&S 11W-14N 26
Prairie Lawn Rd. N&S
.... 9W-13N 27
Prairie St. E&W
.... 9W-13N,8W-13N 27
Princeton NS&EW
.... 10W-15N 17
Pvt. Rd. E&W 7W-13N 28
Pynsky Dr. E&W... 11W-15N 16
Queens Ln. E&W .. 7W-14N 28
Railroad Av. NW&SE
.... 8W-13N 27
Raleigh Rd. NW&SE
.... 8W-13N 27
Ralmark Ln. E&W .. 10W-14N 27
Rebecca Ln. N&S .. 10W-15N 17
Red Bud Ln. NE&SW
.... 8W-14N 27
Ridgewood Ln., East N&S
.... 7W-14N 28
Ridgewood Ln., West N&S
.... 7W-14N 28
Rio Vista Rd. NW&SE
.... 8W-12N 27
River Dr. NW&SE... 8W-13N 27
Riverside Ct. N&S .. 8W-14N 27
Robin Crest Ln. E&W
.... 9W-13N 27
Robin Crest Ln., North E&W
.... 8W-13N 27
Robin Crest Ln., South E&W
.... 8W-13N 27
Robin Ln. N&S 11W-14N 26
Roder St. N&S 10W-12N 27
Rogers Av. E&W .. 8W-14N 27
Rolling Pass N&S .. 9W-13N 27
Rollwind Rd. N&S... 8W-12N 27
Roosevelt Av. N&S .. 9W-13N 27
Royal Oak Ln. E&W. 8W-13N 27
Rugen Ln. N&S 9W-14N 27
Rugen Rd. E&W
.... 11W-13N,9W-14N 26
Russet Way N&S .. 11W-15N 16
Rutgers Ct.* NE&SW
.... 9W-15N 17
Sally E&W 10W-13N 27
Sandy Ln. E&W .. 10W-15N 17
Sanford Ct. E&W .. 8W-13N 27
Sanford Ln. N&S .. 8W-13N 27
Saranac Ct. E&W .. 11W-15N 16
Saranac Ln. N&S .. 11W-15N 16
Sequoia Tr. NE&SW 8W-14N 27
Sheffield Ln. E&W .. 7W-13N 28
Shermer Rd. NW&SE
.... 9W-13N 27
Sherwood Rd. N&S. 6W-13N 28
Sheryl Ln. N&S 10W-12N 27
Short Ln. N&S 9W-13N 27
Silver Pine Dr. N&S. 9W-13N 27
Silver Willow Dr. NW&SE
.... 8W-14N 27
Simonsen Ln. E&W .. 9W-13N 27
Sioux Trail N&S .. 11W-15N 16
Sleepy Hollow Rd. NE&SW
.... 8W-13N 27
Solar Ln. E&W 7W-13N 28
Somerset Dr. N&S .. 7W-13N 28
Southgate Dr. N&S .. 9W-13N 27
Spring Rd. N&S 8W-13N 27
Springdale Av. NS&EW
.... 11W-13N,10W-13N 26
Spruce Ct. N&S .. 9W-13N 27
Spruce St. N&S
.... 9W-12N,9W-13N 27
Steffens Ln. N&S.... 7W-14N 28
Stevens Dr. E&W .. 8W-13N 27
Strawberry Ln. N&S
.... 10W-15N 17
Sumac Cir. N&S .. 9W-13N 27
Sunset Ridge Rd. N&S
.... 8W-14N 27
Surrey Ln. N&S.... 7W-13N 28
Sussex Ln. E&W .. 11W-14N 17
Sussex Ln. N&S .. 11W-15N 16
Swainwood Dr. N&S
.... 8W-14N 27
Taft Ct. E&W...... 6W-13N 28
Tall Tree Ln. N&S .. 7W-13N 28
Tamer N&S 11W-13N 26
Tanglewood Ln. NW&SE
.... 8W-14N 27
Terrace Ln. N&S .. 9W-13N 27

Street	Grid	Page
Thornwood Av. N&S	10W-14N	27
Thornwood Ln. NE&SW	8W-13N	27
Timberline Dr. E&W	7W-13N	28
Tinkerway N&S	8W-13N	27
Topp Ln. E&W	8W-14N	27
Tracy Ct. NW&SE	11W-13N	26
Trent Ct. NW&SE	10W-14N	26
Tri-state Tollway N&S	11W-13N	26
Tulip Tree Ct. NE&SW	8W-14N	27
Valley Lo Ln. N&S	8W-14N	27
Vantage Ln. E&W	10W-15N	17
Vernon Dr. N&S	8W-13N	27
Virginia Ln. E&W	10W-12N	27
Virginia Ln. N&S	10W-15N	17
Wagner Rd. N&S	7W-13N,7W-14N	28
Wald St. N&S	10W-12N	27
Walnut Ct. E&W	9W-12N	27
Warren Rd. N&S	10W-12N	27
Washington St. N&S	9W-12N,9W-13N	27
Waukegan Rd. N&S	8W-12N	27
Wedgewood Dr. N&S	9W-13N	27
Wellesley Ct.* NE&SW	9W-15N	17
West Burr Oak Dr. N&S	8W-14N	27
West Lake Av. E&W	11W-14N	26
Westfield Ln. E&W	10W-15N	17
Westlake Ter. N&S	10W-14N	27
Westview Dr. E&W	11W-14N	26
Westview Rd. E&W	7W-13N	28
Wild Wood N&S	8W-14N	27
Wildberry Dr. E&W	8W-15N	18
Wildwood Ln. NW&SE	7W-14N	28
Willow Ln. N&S	11W-15N	16
Willow Rd. E&W	11W-15N,9W-15N	16
Willow Rd. E&W	11W-14N	26
Wilmette Av. E&W	6W-13N	28
Windsor Rd. N&S	7W-13N	28
Winnetka Rd. E&W	10W-15N	17
Wissing Ln. NE&SW	9W-13N	27
Wood Dr. E&W	9W-13N	27
Woodland Ct. N&S	7W-12N	28
Woodland Dr. N&S	7W-12N	28
Woodland Rd. E&W	7W-14N	28
Woodlawn Av. E&W	8W-14N	27
Woodmere Ln. N&S	7W-13N	28
Woodridge Rd. (Pvt.) E&W	11W-15N	16
Woodview Ln. E&W	8W-13N	27
Yale Ct.* E&W	9W-15N	17
York Rd. N&S	8W-13N	27

Glenwood

Street	Grid	Page
Arizona Av. N&S	1W-22S	88
Arquilla Dr. N&S	0W-22S,0W-21S	88
Birch Dr. E&W	0E-22S	88
Blackstone Av. N&S	1E-23S	89
Bruce Ln. N&S	0W-21S	88
Calumet Expwy. N&S	1E-23S	89
Campbell St. N&S	0E-22S	88
Carroll Pkwy. N&S	0W-22S,0W-21S	88
Cedar Ln. N&S	0E-22S	88
Center St. E&W	0W-22S,0E-22S	88
Champlain Av. N&S	0E-23S	88
Cherry Dr. N&S	0W-22S	88
Chestnut Ln. N&S	0E-22S	88
Clark St. E&W	0E-22S	88
Coales St. E&W	1W-23S	88
Cottage Grove Ave. N&S	0E-23S,0E-22S	88
Dante Av. N&S	1E-23S	89
Dorchester Av. N&S	1E-23S	89
Drexel Av. N&S	1E-23S	89
Eberhardt Av. N&S	0E-23S	88
Ellis Av. N&S	1E-23S	89
Elm Ln. N&S	0E-22S	88
Fitzhenry Ct. E&W	0W-21S	88
Forest Av. N&S	0W-23S	88
Gay Ct. N&S	0W-22S	88
Glenwood Av. NW&SE	0W-22S	88
Glenwood Dyer Rd. NW&SE	0E-23S,1E-23S	88
Glenwood Rd. NE&SW	0W-23S	88
Glenwood-Lansing Rd. E&W	1E-23S	89
Greenwood Av. N&S	1E-23S	89
Halsted St. N&S	0W-22S	88
Harper Av. N&S	1E-23S	89
Hickory Ln. N&S	0E-22S	88
Hickory St. N&S	0E-23S	88
Holly Ct. N&S	0W-22S	88
Illinois Av. N&S	1W-22S	88
Indiana St. E&W	1W-22S	88
Ingleside Av. N&S	1E-23S	89
Interurban Av. N&S	1W-23S	88
Iowa St. E&W	1W-22S	88
Jane St. E&W	0E-22S	88
Kenneth Ct. N&S	0W-22S	88
Lee Ct. N&S	1W-22S	88
Longwood Dr. N&S	0W-22S	88
Lotta St. E&W	0W-22S	88
Main St. NW&SE	0W-22S,0E-22S	88
Manor Ct. E&W	0W-22S	88
Maple Dr. E&W	0E-22S	88
Maryland Av. N&S	1E-23S	89
Merril St. E&W	0W-22S,0E-22S	88
Minerva Av. N&S	1E-23S	89
Mulberry Ct. N&S	0E-22S	88
Mulberry Dr. E&W	0E-22S	88
Nevada St. E&W	1W-22S	88
Oak Ln. N&S	0E-22S	88
Ohio St. E&W	1W-22S	88
Palm Dr. E&W	0W-21S	88
Park Dr. N&S	0W-22S	88
Peoria St. N&S	1W-23S	88
Pine Ln. N&S	0E-22S	88
Pleasant Dr. N&S	0W-22S	88
Rainbow Dr. E&W	0W-22S	88
Rebecca St. N&S	0E-22S	88
Rhodes Av. N&S	0E-23S	88
Roberts Dr. N&S	0W-22S	88
Rose Ct. E&W	0E-22S	88
Rose St. E&W	0E-22S	88
Rt. 1 N&S	0W-22S	88
School St. E&W	0W-22S,0E-22S	88
Science Rd. E&W	0W-22S	88
Spruce Ln. N&S	0E-22S	88
St. Lawrence Av. N&S	0E-23S	88
State St. N&S	0E-23S,0E-22S	88
Strieff Ln. E&W	1W-22S,0W-22S	88
Sunset Dr. E&W	0W-21S	88
Sycamore Ln. N&S	0E-22S	88
Terrace Dr. E&WN&	0W-22S	88
Tulip Dr. E&W	0E-22S	88
University Av. N&S	1E-23S	89
Virginia Av. N&S	1W-22S	88
Wabash Av. N&S	0E-22S	88
Walnut Ln. N&S	0E-22S	88
Walnut St. N&S	0E-23S	88
Westwood Dr. E&W	0W-22S	88
Willow Ln. N&S	0E-22S	88
Willow St. N&S	0E-23S	88
Wisconsin St. E&W	1W-22S	88
Young St. N&S	0E-22S	88
183rd St. E&W	1W-22S	88
187th St. E&W	0W-22S	88
191st Pl. E&W	0E-23S,1E-23S	88
192nd Pl. E&W	1E-23S	89
192nd St. E&W	1E-23S	89
193rd Pl. E&W	1W-23S,1E-23S	88
193rd St. NW&SE	1E-23S	89
194th Pl. E&W	1W-23S	88
194th St. E&W	1W-23S,1E-23S	88
195th St. E&W	1W-23S,0W-23S	88
196th St. E&W	0W-23S	88

Golf

Street	Grid	Page
Blossom Ln. NW&SE	8W-12N	27
Briar Rd. NW&SE	8W-12N	27
Clyde Ln. N&S	8W-12N	27
Dover Ln. NW&SE	8W-12N	27
Highland Pl. N&S	8W-12N	27
Lilac Ln. NW&SE	8W-12N	27
Logan Ter. E&W	8W-12N	27
Orchard Ln. N&S	8W-12N	27
Overlook Dr. NW&SE	7W-12N	28
Park Ln. E&W	8W-12N	27
Simpson St. E&W	8W-12N	27

Hanover Park*

(Also see Du Page County)

Street	Grid	Page
Adams St. E&W	25W-9N	32
Alden Ln. NW&SE	27W-8N	31
Apple Tree St. N&S	27W-8N	31
Applewood Ct. N&S	26W-9N	31
Arbor Vitae Dr. E&W	27W-9N	31
Asbury Cir. N&S	26W-9N	31
Astor Av. N&S	27W-9N	31
Bamberg Ct. NW&SE	26W-9N	31
Barrington Rd. N&S	27W-8N,27W-9N	31
Berkshire Ct. NW&SE	26W-9N	31
Berkshire Dr. E&W	26W-9N	31
Birch Av. E&W	26W-8N	31
Bolton Way E&W	25W-9N	32
Breezewood Ln. SE&NW	27W-8N	31
Briar Ln. NE&SW	27W-9N	31
Briarwood Av. NW&SE	27W-9N	31
Briarwood St. N&S	27W-9N	31
Bristol Ln. E&W	26W-9N	31
Brockton Ct. N&S	26W-9N	31
Brookside Ct. E&W	26W-9N	31
Brookside Dr. N&S	26W-9N	31
Burr Oak St. NE&SW	27W-8N	31
Camelia Dr. N&S	27W-9N	31
Canterbury Dr. E&W	25W-9N	32
Carlisle Ct. E&W	26W-9N	31
Carlisle Dr. N&S	26W-9N	31
Carnaby Ct. N&S	26W-9N	31
Carrolton Ct. N&S	26W-9N	31
Catalpa St. N&S	27W-8N	31
Catawba Ln. N&S	26W-9N	31
Cedar Av. E&W	27W-8N	31
Center Av. N&S	27W-8N	31
Cherry Av. E&W	27W-8N	31
Chestnut St. N&S	27W-8N	31
Church St. N&S	27W-8N	31
Churchill Dr. N&S	26W-9N	31
Countryside Dr. E&W	26W-8N	31
Coventry Ln. N&S	26W-9N	31
Crescent Way NE&SW	26W-9N	31
Cumberland Dr. N&S	26W-9N	31
Cynthia Ln. E&W	26W-9N	31
Cypress Av. E&W	26W-8N	31
Cypress Ct. N&S	26W-8N	31
Dahlia Dr. N&S	27W-9N	31
Dartmouth Ct. NE&SW	26W-9N	31
Dartmouth Ln. NS&EW	26W-9N	31
Deerpath Ln. N&S	26W-8N	31
Durham Ct. E&W	26W-9N	31
East Av. N&S	27W-8N,27W-9N	31
Edgebrook Ln. N&S	26W-8N	31
Elm Av. NW&SE	27W-8N	31
Elm Ct. N&S	27W-8N	31
Essex Ct. E&W	26W-9N	31
Evergreen Av. E&W	27W-8N	31
Fairhaven Dr. E&W	25W-9N	32
Farmstead Ln. N&S	25W-9N	32
Filmore Av.* E&W	25W-9N	32
Flower Ct. N&S	27W-9N	31
Forest Glen E&W	27W-8N	31
Gladiola Av. N&S	27W-9N	31
Glendale Ter. NE&SW	27W-8N	31
Glenside Ct. E&W	26W-9N	31
Glenwood Ln. N&S	26W-8N	31
Grant Cir. E&W	25W-9N	32
Greenwood Av. E&W	27W-9N	31
Guilford Commons E&W	26W-9N	31
Hadom Way E&W	25W-9N	32
Hanover St. N&S	27W-8N,27W-9N	31
Harrison St.* E&W	26W-9N	31
Hartman Dr. E&W	25W-9N	32
Hartmann NW&SE	25W-9N	32
Hastings Ln. E&W	26W-9N	31
Hawthorne Ln. N&S	26W-8N	31
Hearth Dr. N&S	25W-9N	32
Hearth Dr. NE&SW	25W-9N	32
Hemlock St. N&S	27W-9N	31
Hickory St. N&S	27W-8N	31
Highland St. N&S	26W-8N	31
Hillcrest Av. E&W	26W-8N	31
Hollywood Av. E&W	27W-9N	31
Huntington Cir. N&S	26W-9N	31
Indian Hill Av. E&W	26W-8N	31
Indian Hill Ct. N&S	26W-8N	31
Iris Av. N&S	27W-9N	31
Irving Park Rd. NW&SE	27W-8N	31
Jackson St.* E&W	25W-9N	32
Jasmine Dr. N&S	27W-9N	31
Jensen Blvd. N&S	27W-9N	31
Jonquil Ter. N&S	27W-9N	31
Juniper St. N&S	27W-8N	31
Kensington Ln. N&S	26W-9N	31
Kent Ct. NW&SE	26W-9N	31
Kingsbury Dr. N&S	26W-9N	31
Lake St. NW&SE	28W-8N,27W-8N	31
Larch Ln. NE&SW	27W-8N	31
Laurel Av. E&W	27W-9N	31
Laurie Ln. E&W	26W-9N	31
Leslie Ln. NW&SE	28W-8N,27W-8N	31
Lexington Cir. E&W	26W-9N	31
Lincoln Cir.* NW&SE	26W-8N	31
Linden Av. E&W	27W-8N	31
Longmeadow Ln. N&W	26W-8N	31
Madison St.* N&S	26W-9N	31
Madrid Ct. E&W	26W-9N	31
Magnolia St. N&S	27W-8N	31
Manchester Manor NW&SE	26W-9N	31
Maple Av. NW&SE	27W-8N	31
Maplewood Av. E&W	26W-8N	31
Marigold Ln. E&W	27W-9N	31
Mark Thomas Ln. E&W	28W-8N,27W-8N	31
Meadowbrook Ln. N&S	26W-8N	31
Mulberry St. N&S	27W-8N,27W-9N	31
Nantucket Cove N&S	26W-9N	31
Narcissus Av. E&W	27W-9N	31
Northway Ct. E&W	26W-9N	31
Northway Dr. N&S	26W-9N	31
Oak Av. NW&SE	27W-8N	31
Oakwood Av. E&W	26W-8N	31
Old Mill Ln. E&W	26W-9N	31
Olde Salem Cir. N&S	25W-9N	32
Olde Salem Dr. NE&SW	26W-9N	31
Olde Salem Rd. N&S	26W-9N	31
Olivia Ln. N&S	26W-9N	31
Orchard Ln. N&S	26W-8N	31
Osage Ln. E&W	27W-9N	31
Oxford Ln. E&W	26W-9N	31
Park Av. E&W	27W-8N	31
Parkview Dr. E&W	26W-9N	31
Peach Tree St. N&S	27W-8N	31
Pebblebrook Cir. N&S	26W-9N	31
Pine Tree St. N&S	27W-8N	31
Plum Tree Ln. N&S	27W-8N	31
Polk St.* N&S	26W-8N	31
Poplar Av. E&W	27W-8N	31
Princeton Circle Dr. N&S	26W-9N	31
Ramblewood Dr. E&W	26W-9N	31
Ramsgate Cir. N&S	26W-9N	31
Redwood Av. E&W	27W-8N	31
Roosevelt Rd.* E&W	26W-8N	31
Rosewood St. N&S	27W-9N	31
Roxbury Ct. NW&SE	26W-9N	31
Rt. 19 E&W	27W-9N,26W-9N	31
Salem Cir. N&S	26W-9N	31
Sarson Way NE&SW	25W-9N	32
Scott Ln. NE&SW	27W-8N	31
Sequoia Dr. E&W	27W-8N	31
Shelbourne Ct. NW&SE	26W-9N	31
Sherwood Cir. N&S	26W-9N	31
Sommerset Cir. E&W	26W-9N	31
Spruce Av. E&W	26W-8N	31
Stratford Ln. E&W	26W-9N	31
Strathmore Ln. N&S	26W-9N	31
Sycamore Av. E&W	27W-8N	31
Taft Cir.* E&W	26W-8N	31
Tanglewood Av. NW&SE	27W-9N	31
Taylor St.* E&W	25W-9N	32
Tea Party Ln. N&S	26W-9N	31
Thornwood St. N&S	27W-9N	31
Tower Dr. E&W	27W-9N	31
Truman St.* NW&SE	25W-9N	32
U.S. Route 20 NW&SE	27W-8N	31
Valley View Rd. N&S	26W-8N	31
Walnut Av. E&W	27W-8N,26W-8N	31
Walnut Ct. N&S	27W-8N	31
Washington St.* E&W	26W-9N	31
Waterford Dr. E&W	25W-9N	32
Wedgewood Dr. E&W	25W-9N	32
West Av. N&S	27W-8N	31
Westchester Dr. E&W	26W-8N	31
Weymouth Cir. NE&SW	26W-9N,25W-9N	31
Weymouth Cir. NS&EW	25W-9N	32
White Bridge Ct. N&S	26W-9N	31
White Bridge Ln. E&W	26W-9N	31
Willow Av. NW&SE	27W-8N	31
Wilson St.* E&W	25W-9N	32
Windsor Ln. E&W	26W-9N	31
Wise Rd. E&W	25W-9N	32
Yorkshire Ct. N&S	26W-9N	31
Yorkshire Dr. E&W	26W-9N	31

Hanover Township

Street	Grid	Page
Adler Ln. NW&SE	31W-11N	30
Bartlett Rd. N&S	28W-11N,28W-12N	31
Bellingham Ln. E&W	30W-8N	30
Berner Dr. N&S	31W-12N	20
Beverly Rd. N&S	31W-12N	20
Bode Rd. NW&SE	29W-11N,27W-11N	30
Borden Dr. N&S	31W-11N	30
Cardinal Dr. E&W	31W-11N	30
Chapel Hill Dr. N&S	31W-11N	30
Chestnut St. E&W	31W-11N	30
Cheviot Ln. N&S	30W-8N	30
Circle Dr. NW&SE	31W-11N	30
Dale Dr. E&W	31W-12N	20
Dale Rd. E&W	31W-12N	20
Dennis Ln. N&S	30W-10N	30
Douglas Rd. E&W	30W-10N	30
Forest View Dr. NE&SW	31W-11N,30W-11N	30
Friar Tuck Dr. NE&SW	31W-10N	30
Gifford Rd. N&S	32W-9N	30
Glen Echo St. N&S	31W-11N	30
Golf Keys Rd. (Pvt) NS&EW	30W-9N	30
Golf Rd. E&W	31W-12N,27W-12N	20
Golfview Dr. E&W	30W-9N	30
Greenfeather Ln. NE&SW	30W-10N	30
Greenwood Rd. NE&SW	31W-10N	30
Gromer Rd. NE&SW	30W-10N	30
Hillard Dr. NE&SW	31W-10N	30
Hilltop Rd. NE&SW	31W-11N,30W-11N	30
Irving Park Rd. NW&SE	31W-10N,30W-10N	30
King Arthur Ct. NE&SW	31W-10N	30
Lady Marion Dr. NE&SW	31W-10N	30
Lake St. NW&SE	31W-10N,29W-9N	30
Little John Dr. NW&SE	31W-10N	30
Longboat Key Ln. E&W	30W-9N	30
Magnolia Ct. N&S	29W-12N	20
Magnolia Ln. E&W	29W-12N	20
Mundhawk Rd. N&S	28W-12N	21
Naperville Rd. N&S	30W-9N	30
Northwest Tollway E&W	31W-12N,27W-12N	20
Nottingham Ln. NW&SE	31W-10N	30
Old Higgins Rd. NW&SE	30W-14N,27W-12N	20
Old Lake St. NE&SW	29W-9N	30
Philip Dr. NE&SW	30W-9N	30
Phillippi Creek Dr. NS&EW	30W-9N	30
Poplar Creek Dr. N&S	32W-11N	30
Poplar View Bend NE&SW	31W-11N	30
Quincey Ct. E&W	31W-11N	30
Robinhood Dr. NW&SE	31W-10N	30
Rohrson Rd. N&S	31W-11N	30
Rolling Hills Av. NS&EW	31W-11N	30
Rosewood Ct. NE&SW	32W-11N	30
Sayer Rd. NW&SE	30W-8N	30
Schaumburg Rd. E&W	30W-10N	30
Sherwood Rd. NE&SW	31W-10N	30
Shoe Factory Rd. E&W	31W-12N,27W-12N	20
Siesta Key Ln. NW&SE	30W-9N	30
Spaulding Rd. E&W	30W-9N	30
Stonehurst Dr. N&S	32W-11N	30
Sutton Rd. N&S	30W-9N	30
Tameling Ct. N&S	31W-8N	30
West Bartlett Rd. E&W	32W-8N,30W-8N	30
Will Scarlet Ln. NE&SW	31W-10N,30W-10N	30
Windsor Ct. E&W	31W-11N	30
Wolsfeld Rd. E&W	31W-11N	30

Harvey

Street	Grid	Page
Artesian Av. N&S	3W-17S	81
Ashland Av. N&S	2W-19S,2W-17S	82
Belden Av. N&S	0W-18S	82
Broadway Av. N&S	1W-18S,1W-17S	82
California Av. N&S	3W-17S	81
Calumet Blvd. E&W	1W-17S	82
Campbell Av. N&S	2W-18S	81
Carol Av. N&S	0W-19S	82
Carse Av. N&S	1W-20S,1W-19S	88
Cary Av. N&S	0W-19S,0W-18S	82
Center Av. N&S	1W-18S,1W-17S	82
Center St. N&S	1W-20S,1W-19S	88
Clark Av. N&S	0W-18S	82
Clinton St. N&S	0W-19S,0W-17S	82
Commercial Av. NE&SW	1W-18S	82
Cooper Av. N&S	2W-17S	81
Des Plaines St. N&S	0W-17S	82
Dixie Hwy. NW&SE	2W-18S,2W-17S	81
Emerald Av. N&S	0W-19S	82
Fairfield Av. N&S	3W-17S	81
Finch Av. N&S	1W-19S,1W-18S	82
Fisk Av. N&S	1W-19S,1W-18S	82
Gauger Av. N&S	1W-19S	82
Geneva Av. E&W	1W-19S	82
Green St. N&S	1W-19S,1W-17S	82
Halsted St. N&S	1W-19S,1W-17S	82
Harvey Av. N&S	1W-17S	82
Honore Av. N&S	2W-19S,2W-17S	81
Hoyne Av. N&S	2W-18S,2W-17S	81

	Grid	Page
Ingalls Dr. N&S	1W-18S	82
Jefferson St. N&S	0W-19S,0W-17S	82
Justine St. N&S	1W-19S,1W-17S	82
Kennicott Av. N&S	0W-18S	82
Kentucky Av. N&S	1W-17S	82
Lathrop Av. N&S	1W-19S,1W-18S	82
Leavitt Av. N&S	2W-17S	81
Lexington Av. N&S	1W-18S,1W-17S	82
Lincoln Av. N&S	2W-19S,2W-17S	81
Loomis St. N&S	1W-19S,1W-17S	82
Lowe Av. N&S	0W-19S	82
Madison N&S	0W-17S	82
Main St. N&S	2W-18S,1W-17S	82
Maplewood Av. N&S	3W-17S	81
Markham Dr. N&S	0W-17S	82
Marshfied Av. N&S	2W-19S,2W-17S	81
Morgan St. N&S	1W-17S	82
Myrtle Av. N&S	1W-19S,1W-17S	82
Normal Av. N&S	0W-17S	82
Oakley Av. N&S	2W-17S	81
Page Av. N&S	1W-19S,1W-17S	82
Park Av. NE&SW	0W-17S	82
Parnell Av. N&S	0W-17S	82
Paulina Av. N&S	1W-19S,1W-17S	82
Peoria St. N&S	1W-17S	82
Robey St. N&S	2W-17S	81
Rockwell St. N&S	2W-18S	81
Rt. 83 E&W	0W-17S	82
Sangamon St. N&S	1W-17S	82
Seeley Av. N&S	2W-17S	81
Sibley Blvd. E&W	2W-17S,0W-17S	81
Streamside Dr. NW&SE	1W-17S,0W-17S	82
Talman Av. N&S	3W-17S	81
Thornton Rd. NW&SE	1W-17S	82
Turlington Av. N&S	2W-18S,1W-17S	82
U.S. Route 6 E&W	2W-18S	81
Union St. N&S	0W-19S,0W-17S	82
Vail Av. N&S	2W-17S	81
Vine Av. N&S	1W-18S,1W-17S	82
Wallace Rd. N&S	0W-17S	82
Wallace St. N&S	0W-19S,0W-17S	82
Washtehaw Av. N&S	3W-17S	81
Wells Av. E&W	1W-17S	82
West Av. N&S	1W-19S,1W-18S	82
Western Av. N&S	2W-18S	81
Willard Av. N&S	1W-19S,1W-18S	82
Winchester Av. N&S	2W-18S,2W-17S	81
Wood St. N&S	2W-19S,2W-17S	81
Woodbridge Dr. N&S	1W-19S,1W-18S	82
144th Ct. E&W	1W-17S	82
144th St. E&W	1W-17S	82
145th St. E&W	2W-17S,1W-17S	81
146th St. E&W	2W-17S,1W-17S	81
147th Pl. E&W	0W-17S	82
148th Pl. E&W	2W-17S	81
148th St. E&W	2W-17S,0W-17S	81
149th Pl. E&W	2W-17S	81
149th St. E&W	3W-17S,1W-17S	81
150th Pl. E&W	2W-17S	81
150th St. E&W	3W-17S,1W-17S	81
151st Pl. E&W	2W-18S	81
151st St. E&W	3W-17S,1W-17S	81
151st Terr. E&W	2W-18S	81
152nd St. E&W	2W-18S,1W-18S	81
153rd St. E&W	3W-18S,1W-18S	81
154th Pl. E&W	2W-18S	81
154th St. E&W	2W-18S,1W-18S	81
155th Pl. E&W	2W-18S	81
155th St. E&W	2W-18S,1W-18S	81
156th Pl. E&W	2W-18S	81
156th St. E&W	2W-18S,1W-18S	81
157th Pl. E&W	2W-18S,0W-18S	81
157th St. E&W	2W-18S,0W-18S	81
158th Pl. E&W	2W-18S	81
158th St. E&W	2W-18S,0W-18S	81
159th St. E&W	2W-18S,1W-18S	81
160th Pl. E&W	0W-9S	68

	Grid	Page
160th St. E&W	2W-19S,0W-19S	81
161st St. E&W	2W-19S,0W-19S	81
162nd St. E&W	1W-19S,0W-19S	82
163rd St. E&W	1W-19S,0W-19S	82
164th St. E&W	1W-19S,0W-19S	82
165th St. E&W	1W-19S,0W-19S	82
166th St. E&W	1W-19S,0W-9S	82
167th Pl. E&W	1W-20S	88
167th St. E&W	1W-20S,0W-20S	88
168th Pl. E&W	1W-20S	88
168th St. E&W	1W-20S	88
171st E&W	1W-20S	88

Harwood Heights

	Grid	Page
Ainslie E&W	9W-6N	42
Argyle E&W	9W-6N	42
Carmen E&W	9W-6N	42
Cullom E&W	8W-5N	42
Diana Ln. E&W	9W-5N	42
Gunnison E&W	9W-6N,8W-5N	42
Jill Ln. E&W	9W-5N	42
Lawrence E&W	9W-6N,8W-5N	42
Leland E&W	9W-5N,8W-5N	42
Nagle N&S	8W-5N	42
New Castle N&S	8W-5N	42
New England N&S	8W-5N	42
Newland N&S	8W-5N	42
Nora N&S	8W-5N	42
Norwood Ct. E&W	8W-5N	42
Oak Park N&S	8W-5N	42
Oconto N&S	9W-6N	42
Octavia N&S	9W-6N	42
Odell N&S	9W-6N	42
Oketo N&S	9W-5N	42
Olcott N&S	9W-5N	42
Oriole N&S	9W-6N	42
Ronald N&S	8W-5N	42
Rutherford N&S	8W-5N	42
Sayre N&S	8W-5N	42
Senior Pl. NW&SE	8W-5N	42
Strong E&W	9W-6N	42
Wilson N&S	9W-5N,8W-5N	42
Winnemac E&W	9W-6N	42
Winona E&W	9W-6N	42

Hazel Crest

	Grid	Page
Albany Av. N&S	3W-20S	87
Annetta Av. N&S	3W-20S	87
Anthony Av. N&S	2W-20S	87
Apple Tree Dr. N&S	4W-20S	87
Argyle Av. N&S	3W-21S	87
Arlington Ln. N&S	3W-20S	87
Artesian Av. N&S	3W-20S	87
Balmoral Ln. N&S	3W-21S	87
Birchwood Dr. E&W	4W-20S	87
Bordeaux Ct. E&W	4W-21S	87
Briar Ln. E&W	4W-22S	87
Brumley Dr. E&W	4W-22S	87
Bryant Ln. NW&SE	3W-20S	87
Bulger Av. N&S	2W-20S	87
Burgandy Ln. NW&SE	4W-21S	87
Burr Oak Ln.	4W-20S	87
Butternut Rd. N&S	3W-21S	87
Buttonwood Walk E&W	3W-20S	87
California Av. N&S	3W-20S	87
Cannes Ct. E&W	4W-21S	87
Carl Dr. E&W	3W-20S	87
Carraigeway Dr. N&S	3W-21S	87
Central Park Av. N&S	4W-20S	87
Chambord Ln. N&S	4W-21S	87
Chantilly Ln. N&S	4W-21S	87
Charlemagne Av. N&S	4W-21S	87
Charleston Ln. E&W	2W-20S	87
Charters Ct. NE&SW	4W-21S	87
Cherry Creek Dr. NE&SW	3W-20S	87
Cherrywood Ln. NW&SE	3W-21S	87
Chestnut Av. E&W	3W-20S	87
Circle Dr. E&W	3W-20S	87
Coach Ln. E&W	3W-21S	87
Concord Pl. E&W	3W-20S	87
Cottonwood Ct. NW&SE	4W-20S	87
Coventry Ct. E&W	4W-20S	87
Crane Av. N&S	2W-20S	87
Crescent Dr. E&W	3W-20S	87
Dixie Hwy E&W	2W-20S,2W-19S	87
Dogwood Ln. N&S	3W-21S	87
Elm Dr. N&S	4W-20S	87
Elysees Ln. E&W	4W-21S	87
Emerson Av. N&S	3W-20S	87
Fountainbleau Dr. NE&SW	4W-21S	87
Glynwood Ln. E&W	4W-20S	87
Governors Hwy. NE&SW	2W-20S	87
Grandview Dr. N&S	3W-21S	87

	Grid	Page
Greenwood Rd E&W	3W-21S	87
Grove Ln. E&W	4W-20S	87
Hawthorne Ln. E&W	3W-21S	87
Hazel Ln. E&W	4W-20S	87
Head Av. N&S	2W-20S	87
Hickory Ln. E&S	4W-20S	87
Highland Av. N&S	3W-20S	87
Hillside Av. N&S	3W-20S	87
Holmes Av. N&S	3W-20S	87
Interstate 80 E&W	4W-20S,3W-20S	87
Jeffery Ln. E&W	4W-20S	87
Jodave Av. N&S	2W-20S	87
Kedzie Av. N&S	4W-21S,4W-20S	87
Kimball Av. N&S	4W-20S	87
Knollwood Pl. E&W	3W-21S	87
Larkspur Ln. NW&SE	3W-21S	87
Laurel Ln. NE&SW	4W-20S	87
Lexington Dr. E&W	3W-20S	87
Lincoln St. N&S	2W-20S	87
Locust Dr. N&S	4W-20S	87
Longfellow Av. NE&SW	3W-20S	87
Lowell Av. N&S	3W-20S	87
Magnolia Dr. N&S	4W-20S	87
Maple Ln. E&W	4W-20S	87
Marseille Ln. NS&EW	4W-21S	87
Millstone Rd. N&S	3W-21S	87
Montmarte Av. NE&SW	4W-21S	87
Murphy Av. N&S	3W-20S	87
Normandy Ln. N&S	4W-21S	87
Novak Dr. N&S	3W-20S	87
Oak St. N&S	4W-20S	87
Oakwood Dr. E&W	3W-20S	87
Old Trail Rd. N&S	3W-21S	87
Orchard Ridge Av. N&S	2W-20S	87
Orleans Dr. N&S	4W-21S	87
Page St. N&S	2W-20S	87
Palmer Blvd. N&S	3W-20S	87
Peach Grove Ln. EW&NS	4W-20S	87
Pebblewood Ln. N&S	3W-21S	87
Pine Ct. N&S	4W-20S	87
Poe Av. N&S	3W-20S	87
Pulaski Rd. N&S	4W-22S	87
Ridgewood Dr. NE&SW	3W-21S	87
Rochester Av. E&W	2W-20S	87
Rockwell Av. N&S	3W-20S	87
Royale Ln. NE&SW	4W-21S	87
Sacramento Av. N&S	3W-20S	87
Seine St. NW&SE	4W-20S	87
Shagbark Ln.	3W-21S	87
Shea Av. N&S	2W-20S	87
Smoketree Ct. N&S	3W-21S	87
Stonebridge Dr. NE&SW	3W-21S	87
Summit Av. N&S	3W-20S	87
Sunset Rd. E&W	3W-20S	87
Surrey Ln. N&S	3W-21S	87
Sycamore Dr. N&S	4W-20S	87
Tamarino Ln. E&W	4W-20S	87
Tanglewood Ct.* N&S	4W-21S	87
Tennyson Pl. E&W	3W-20S	87
Trapet Av. N&S	2W-20S	87
Tri-State Tollway E&W	3W-20S,2W-20S	87
Tulip Dr. NE&SW	4W-20S	87
Turtlecreek Dr. NW&SE	3W-21S	87
Versailles Ln. N&S	4W-21S	87
Village Dr. N&S	4W-22S	87
Wellington Ct. E&W	4W-20S	87
West Mahoney Pkwy. NE&SW	3W-20S	87
Wheelwood Ct. E&W	4W-20S	87
Whipple St. N&S	3W-20S	87
Whitman Av. N&S	3W-20S	87
Whittier Av. N&S	3W-20S	87
Winchester Av. N&S	2W-20S	87
Wood St. N&S	2W-20S	87
Woodworth Pl. E&W	4W-20S,3W-20S	87
167th St. E&W	4W-20S,1W-20S	87
168th St. E&W	3W-20S,2W-20S	87
169th St. E&W	3W-20S,2W-20S	87
170th St. E&W	3W-20S,2W-20S	87
171st St. E&W	3W-20S,2W-20S	87
172nd Ct. E&W	4W-20S	87
172nd St. E&W	3W-20S,2W-20S	87
173rd St. E&W	4W-20S,3W-20S	87
174th St. E&W	3W-20S,2W-20S	87
175th St. E&W	4W-20S,2W-20S	87
176th St. E&W	4W-21S	87
177th St. E&W	4W-21S	87
183rd St. E&W	4W-22S	87

Hickory Hills

	Grid	Page
Active Ln. N&S	11W-10S	64
Ash Ln. N&S	11W-10S	65
Baldwin Trail NS&EW	10W-10S	65
Barberry Ln. E&W	11W-10S	64
Bath Av. N&S	10W-12S	65
Beechnut Rd. E&W	11W-10S	64
Birch Ln. N&S	11W-10S	65
Blueridge Av. N&S	11W-9S	64
Briarwood Ln. E&W	11W-10S	64
Chestnut Dr. E&W	11W-9S	64
Christina Dr. E&W	11W-10S	65
Coey Ln. NW&SE	10W-10S	65
Collette Ln. E&W	10W-10S	65
Dell Ct. E&W	11W-9S	64
Dewey Ln. E&W	11W-10S	64
Elm Cir. E&W	11W-10S	64
Forest Dr. NS&EW	11W-9S	64
Forest Ln. NW&SE	11W-10S	64
Forest Ln., East N&S	11W-10S	64
Gladys Ln. NS&SE	10W-11S	65
Goldenoak Ct. N&S	11W-10S	65
Hawthorn Dr. E&W	11W-10S	65
Hickory Ln. N&S	11W-10S	64
Hillside Dr. N&S	10W-10S	65
Interstate 294 SE&NW	9W-10S	65
Jonathan Dr. E&W	10W-10S	65
Kean Av. N&S	11W-10S,11W-9S	64
Kells Dr. E&W	10W-10S	65
Kitty Ln. E&W	10W-10S	65
Kopping Ln. E&W	11W-10S	64
Lisa Ln. E&W	11W-10S	64
Lotus Ln. E&W	11W-10S	64
Lynnwood Dr. E&W	10W-10S	65
Maple Ln. N&S	11W-10S	64
Meadowview Dr. E&W	11W-9S	64
Nida Ct. E&W	11W-10S	64
Northwest Rd. NW&SE	11W-13S	72
Oak Hill Ct. E&W	10W-10S	65
Oakwood Dr. N&S	11W-10S	64
Orchard Dr. E&W	10W-10S	65
Pleasant Av. N&S	11W-10S	64
Primrose Ln. E&W	11W-9S	64
Roberts Rd. N&S	11W-9S	64
Robin Ct. E&W	11W-9S	64
Sandra Ln. E&W	11W-10S	64
Shady Dr. E&W	11W-9S	64
Sycamore Ct. N&S	11W-10S	65
Sycamore Dr. E&W	11W-10S	64
U.S. Route 12-20 E&W	11W-10S	64
Willow Rd. N&S	11W-10S	64
Woodland Dr. E&W	11W-10S	64
76th Av. N&S	9W-11S,0W-10S	65
76th Ct. N&S	9W-10S	65
77th Av. N&S	9W-10S	65
77th Ct. N&S	9W-10S	65
78th Av. N&S	9W-11S,9W-10S	65
78th Ct. N&S	9W-11S,9W-10S	65
79th Av. N&S	9W-11S,9W-10S	65
79th Ct. N&S	9W-10S	65
80th Ct. N&S	10W-10S	65
81st Av. N&S	10W-10S	65
81st Ct. N&S	10W-10S,10W-9S	65
82nd Av. N&S	10W-10S,10W-9S	65
82nd Ct. N&S	10W-10S,10W-9S	65
83rd Av. N&S	10W-10S,10W-9S	65
83rd Ct. N&S	10W-10S,10W-9S	65
84th Av. N&S	10W-10S,10W-9S	65
84th Ct. N&S	10W-10S,10W-9S	65
85th Av. N&S	10W-10S	65
85th Ct. N&S	10W-10S,10W-9S	65
85th Av. N&S	10W-10S	65
86th Av. N&S	10W-10S,10W-9S	65
86th Ct. E&W	10W-10S	65
86th Pl. E&W	10W-10S	65
86th St. E&W	10W-10S	64
87th Av. N&S	10W-10S	65
87th Ct. N&S	10W-10S	65
87th Pl. E&W	9W-10S	65
87th St. E&W	11W-9S,10W-9S	64
88th Av. N&S	11W-10S	64
88th Ct. N&S	11W-10S	64
88th St. E&W	10W-10S,9W-10S	65
89th Av. N&S	11W-10S	64
89th Ct. N&S	11W-10S	65
89th Pl. E&W	10W-10S,9W-10S	65
89th St. E&W	11W-10S,9W-10S	64

	Grid	Page
90th Av. N&S	11W-10S,11W-9S	64
90th Pl. E&W	10W-10S	65
90th St. E&W	10W-10S	65
91st Pl. E&W	11W-10S,9W-10S	64
91st St. E&W	10W-10S,9W-10S	65
92nd Pl. E&W	11W-10S,10W-10S	64
92nd St. E&W	11W-10S,9W-10S	64
93rd Pl. E&W	11W-10S,9W-10S	65
93rd St. E&W	11W-10S,9W-10S	65
94th St. E&W	10W-10S,9W-10S	65
95th St. E&W	11W-10S,9W-10S	64
96th Pl. E&W	9W-11S	65
97th Pl. E&W	9W-11S	65
97th St. E&W	9W-11S	65
98th Pl. E&W	9W-11S	65
98th St. E&W	9W-11S	65
99th St. E&W	9W-11S	65
104th Pl. E&W	9W-12S	65

Hillside

	Grid	Page
Adams E&W	14W-0S	46
Ashbel N&S	14W-0S	46
Augustine NW&SE	14W-0S,13W-0S	46
Bellwood N&S	14W-0S	46
Broadview N&S	13W-0N	46
Buckthorn N&S	13W-0N	46
Clayton Rd. E&W	13W-0S	46
Clayton Rd. N&S	13W-0S	46
Cypress Dr. E&W	14W-0N	46
Darmstadt E&W	14W-0S	46
Division E&W	13W-0S	46
East N&S	13W-0S	46
East End N&S	14W-0N	46
Edgewater E&W	13W-0S	46
Electric E&W	14W-0N	46
Elm N&S	14W-0S,14W-0N	46
Englewood N&S	13W-0N	46
Fencl N&S	13W-0S	46
Forest N&S	13W-0S	46
Frontage Rd. NW&SE	13W-0S	46
Geneva N&S	13W-0N	46
Granville N&S	13W-0N	46
Highridge Rd. N&S	13W-0S	46
Hillside N&S	14W-0S,14W-0N	46
Howard N&S	14W-0S	46
Hyde Park N&S	13W-0N	46
Idlewild Ln. SW&NE	13W-0S	46
Iroquis Rd. N&S	13W-0S	46
Irving S	14W-0S,14W-0N	46
Jackson Blvd. SW&NE	13W-0S	46
Laverne N&S	14W-0S	46
Leviton N&S	13W-0S	46
Lind N&S	14W-0S	46
Locust N&S	14W-0N	46
Madison E&W	14W-0N,13W-0N	46
Maple Ln. N&W	13W-0S	46
May E&W	13W-0S	46
Melrose N&S	13W-0S	46
Mueller N&S	13W-0S	46
Oak N&S	13W-0S	46
Oak Ridge N&S	13W-0S	46
Orchard N&S	13W-0S	46
Randolph E&W	14W-0N	46
Raymond Dr. E&W	13W-0S	46
St. Paul Ct. E&W	14W-0N	46
Sunnyside Dr. E&W	14W-0N	46
Taft St. N&S	14W-0S	46
Van Buren E&W	14W-0S	46
Warren E&W	14W-0N,13W-0N	46
Washington E&W	14W-0N	46
49th N&S	13W-0N	46
50th N&S	13W-0N	46
51st N&S	13W-0N	46
52nd N&S	13W-0N	46
53rd N&S	13W-0N	46
54th N&S	13W-0N	46
55th N&S	13W-0N	46

Hinsdale*
(Also see Du Page County)

	Grid	Page
Bittersweet E&W	14W-5S	58
Bittersweet Ln. E&W	14W-5S	58
Bobolink E&W	14W-4S	52
Chicago E&W	14W-4S	52
Cleveland Dr. NW&SE	14W-5S	58
Cleveland Rd. NW&SE	14W-5S	58
Columbia N&S	14W-5S	58
County Line Ct. E&W	14W-5S	58
Dalewood NW&SE	14W-5S	58
Fuller Rd. E&W	14W-4S	45
Harding E&W	14W-4S	52
Hickory E&W	14W-4S	52
Highland E&W	14W-4S	52
Hill Grove E&W	14W-4S	52
Hillcrest N&S	14W-5S	58
Hillcrest Cir. N&S	14W-5S	58
Interstate 294 N&S	14W-6S	58

	Grid	Page
Justina N&S	14W-4S	52
McKinley N&S	14W-5S	58
Mills N&S	14W-5S	58
Minneola E&W	14W-4S	52
Pamela E&W	14W-5S	58
Pamela Cir. E&W	14W-5S	45
Phillippa St. N&S	14W-4S	52
Phillipps N&S	14W-4S	52
Princeton N&SE	14W-4S	58
Ravine E&W	14W-4S	52
Sharron N&S	14W-5S	58
Springlake N&S	14W-5S	58
Taft N&S	14W-5S	58
Walnut E&W	14W-4S	52
Wilson N&S	14W-5S	58
Woodland Av. E&W	14W-5S	58
Cir.	14W-5S	58
Woodside E&W	14W-4S	58
Woodside Av.	14W-4S	45
1st E&W	14W-5S	58
3rd E&W	14W-5S	58
6th E&W	14W-5S	58
7th E&W	14W-5S	58
10th E&W	14W-6S	58

Hodgkins

	Grid	Page
Bill Rd. N&S	12W-7S	58
Canary Rd. E&W	11W-7S	59
Cardinal Rd. NE&SW		
	11W-7S	59
Catherine Ann E&W	11W-6S	58
Chester N&S	11W-7S	58
Cobb E&W	11W-7S	58
Conrad N&S	11W-7S	58
Eagle Rd. E&W	11W-7S	59
East N&S	12W-7S	58
Elm N&S	12W-7S	58
Falcon Rd. E&W	11W-7S	59
Franbean Rd. N&S	11W-7S	58
Jewel Rd. N&S	12W-7S	58
Kane N&S	11W-7S	58
Kimball N&S	11W-7S	58
La Grange N&S	12W-7S	58
Lenzi N&S	11W-6S	58
Lyons E&W	11W-7S	58
Mance Rd. E&W	12W-7S	58
Martin Rd. E&W	11W-7S	59
Oak Rd. N&S	12W-7S	58
Park St. N&S	12W-7S	58
Pelican Rd. E&W	11W-7S	59
River Rd. NE&SW	12W-8S	58
Roger Rd. N&S	12W-7S	58
Santa Fe SW&NE	12W-8S	58
Sharon Rd. N&S	12W-7S	58
U.S. Rt. 12 N&S	12W-7S	58
U.S. Rt. 20 N&S	12W-7S	58
U.S. Rt. 45 N&S	12W-7S	58
Walnut St. N&S	12W-7S	58
Weeping Willow Pl. NS&EW		
	12W-7S	58
Wenz N&S	11W-7S	58
Wren Rd. E&W	11W-7S	59
67th E&W	12W-7S	58

Hoffman Estates

	Grid	Page
Abbeywood Dr. N&S		
	25W-12N	22
Aberdeen Ct. E&W	22W-11N	33
Aberdeen St. N&S	22W-11N	33
Adler Ct. N&SE	25W-14N	22
Alcoa Ln. E&W	24W-11N	32
Alder Dr. NS&EW	25W-14N	22
Algonquin Rd. NW&SE		
	25W-14N,23W-14N	22
Alhambra Ln. NE&SW		
	22W-11N	33
Almond Ln. N&S	22W-11N	33
Alpine Ln. NW&SE		
	23W-11N,22W-11N	32
Amber Cir. NS&EW		
	24W-15N	22
Ameritech Center Dr. NS&EW		
	25W-13N	22
Amherst Ln. E&W	24W-12N	22
Anjou St. N&S	24W-14N	22
Apache Ln. NW&SE		
	23W-11N,22W-11N	32
Apple St. N&S	23W-11N	32
Apricot St. N&S	23W-11N	32
Arbor Ct. N&S	22W-11N	33
Arcadia Ct. E&W	22W-11N	33
Ardwick Dr. NW&SE		
	25W-12N	22
Arizona Blvd. NW&SE		
	23W-11N	32
Arlington Ct. NW&SE		
	22W-11N	33
Arlington St. N&S	22W-11N	33
Arrowwood Ln. N&S		
	25W-14N	22
Asbury Ln. E&W	23W-11N	32
Ascot Ct. NE&SW	24W-11N	32
Ash Rd. N&S	23W-11N	32
Ashland St. N&S	22W-11N	33
Ashley Ct. NW&SE	24W-12N	22
Ashley Rd. N&S	24W-12N	22
Aspen St. N&S	22W-11N	33
Aster Ln. E&W		
	23W-11N,22W-11N	32
Atlantic Av. N&S		
	26W-11N	31
Audobon Ct. NW&SE		
	22W-11N	33
Audobon Pl. N&S	22W-11N	33
Audobon St. N&S	22W-11N	33
Avondale Ln. E&W	22W-11N	33

	Grid	Page
Azalea Ln. E&W		
	23W-11N,22W-11N	32
Barberry Ct. NW&SE		
	25W-14N	22
Barcroft Ct. NW&SE		
	24W-16N	12
Barcroft Dr. NS&EW		
	24W-16N	12
Bardwick Ct. N&S	24W-16N	12
Barrington Rd. N&S		
	27W-11N,27W-12N	31
Bartlett Rd. N&S	28W-12N	21
Basswood St. N&S		
	22W-11N	33
Batavia Ln. E&W	23W-11N	32
Baxter Ln. E&W	24W-11N	32
Bayberry Ln.* E&W		
	24W-13N	22
Bayside Cir. NS&EW		
	24W-15N	22
Bayside St. E&W	24W-15N	22
Beacon Ct. NW&SE		
	25W-15N	22
Bedford Ct. NW&SE		
	24W-12N	22
Bedford Rd. N&S	24W-12N	22
Berkley Ct. N&S	24W-11N	32
Berkley Ln. E&W	24W-11N	32
Berkley Pl. N&S	24W-11N	32
Bernay Ln. N&S	24W-14N	22
Beverly Rd. N&S	30W-13N	20
Bicek Ct. N&S	24W-15N	22
Bicek Dr. NE&SW	24W-15N	22
Blair Ln. E&W	25W-12N	22
Boardwalk Blvd. E&W		
	26W-11N	31
Bode Ln. N&S	24W-11N	32
Bode Rd. NS&EW		
	26W-11N,23W-11N	31
Bolleana Ct. NW&SE		
	25W-14N	22
Bonnie Ln. E&W	25W-12N	22
Bordeaux Dr. NE&SW		
	24W-12N	22
Botsford Pl. E&W	26W-12N	21
Boulder Ln. N&S	24W-16N	12
Bradley Ln. E&W	23W-11N	32
Briarcliff Ln. E&W	24W-12N	22
Brigantine Ct. N&S	24W-15N	22
Brigantine Ln. N&S		
	24W-15N	22
Brigton Ln.* N&S	24W-14N	22
Bristol Walk E&W	25W-12N	22
Brittany Ln. N&S	24W-14N	22
Brookside Dr. N&S		
	25W-12N	22
Brookside Ln. E&W		
	25W-12N	22
Buckeye Dr. E&W	23W-11N	32
Buckingham Ct. NW&SE		
	24W-11N	32
Buckthorn Dr. E&W		
	24W-15N	22
Burnham Dr. E&W		
	24W-15N	22
Burning Bush Ln. E&W		
	24W-15N	22
Burr Ridge Dr. NS&EW		
	25W-14N	22
Buttercreek Ct. NW&SE		
	23W-11N	32
Butterfield Ct. E&W		
	23W-14N	22
Caldwell Ln. N&S	25W-12N	22
Cambridge Ln. NW&SE		
	24W-12N	22
Camelot Ln. NE&SW		
	24W-16N	12
Cameron Dr. N&S		
	24W-16N	12
Campbell Ln. E&W	25W-12N	22
Candlewood Ln. E&W		
	25W-11N	32
Cape Breton Ct. E&W		
	24W-15N	22
Capstan Dr. NS&EW		
	25W-15N	22
Cardigan Pl.* N&S	24W-14N	22
Carleton Rd. N&S	24W-12N	22
Carling Rd. N&S	24N-12N	22
Carmel Ct. NE&SW		
	24W-12N	22
Carnation Ln. E&W		
	23W-11N	32
Carthage Ln. E&W	23W-11N	32
Castaway Ct. E&W		
	24W-16N	12
Castaway Ln. E&W	24W-16N	12
Castlewood Ct. E&W		
	23W-14N	22
Cedar Tree Ct. E&W		
	23W-11N	32
Central Rd. E&W	25W-13N	22
Chambers Dr. N&S		
	24W-16N	12
Chandler Ln. E&W	23W-11N	32
Chanora Ct. NE&SW		
	25W-14N	22
Charlemagne Dr. NE&SW		
	24W-14N	22
Charleston Ln. E&W		
	24W-14N,23W-14N	22
Chatsworth Ln. E&W		
	25W-12N	22
Chelmsford Pl. NE&SW		
	25W-12N	22
Cheltenham Pl. NW&SE		
	26W-12N	21

	Grid	Page
Cherry Ct. N&S	25W-15N	22
Chesapeake Ct. N&S		
	24W-15N	22
Chesapeake Dr. E&W		
	24W-15N	22
Chestnut Ln. E&W	25W-14N	22
Chippendale Rd. E&W		
	24W-12N	22
Circle Pl. E&W	26W-12N	21
Claremont Rd. N&S		
	24W-12N	22
Clarendon St. NE&SW		
	23W-11N	32
Claridge Cir. NE&SW		
	24W-11N	32
Clifton Pl. E&W	26W-12N	21
Clover Ln. E&W	24W-15N	22
Clubhouse Ln. E&W		
	25W-12N	22
Cobble Hill Ct. E&W		
	24W-12N	22
Cobble Hill Dr. E&W		
	24W-12N	22
Cochise St. N&S	23W-11N	32
Colgate Ct. E&W	24W-12N	22
Colony Ct. NW&SE		
	24W-14N	22
Colony Ln. NE&SW		
	24W-14N	22
Concord Cove NE&SW		
	23W-14N	22
Concord Ln. E&W		
	24W-14N,23W-14N	22
Cooper Rd. N&S	24W-12N	22
Cornell Cir. E&W	26W-12N	21
Cornell Ct. N&S	26W-12N	21
Cornell Dr. N&S	26W-12N	21
Cornell Pl. E&W	26W-12N	21
Cornell Terr. N&S	26W-12N	21
Cottonwood Tr. NS&EW		
	24W-15N	22
Cove Ln. N&S	25W-15N	22
Coventry Ct. N&S	23W-14N	22
Crab Orchard Dr. N&S		
	24W-15N	22
Cranshire Ct. NW&SE		
	23W-14N	22
Crescent Ct. NE&SW		
	25W-12N	22
Crescent Ln. NW&SE		
	25W-12N	22
Crimson Ct. NW&SE		
	24W-15N	22
Crimson Dr. NS&EW		
	24W-15N	22
Cumberland St. N&S		
	23W-11N	32
Cypress Ct. N&S	25W-12N	22
Danbury Pl.* N&S	24W-13N	22
Darien St. E&W	24W-11N	32
Darlington Cir. NS&EW		
	23W-11N	32
Darlington Ct. E&W		
	25W-11N	32
Decatur St. N&S	23W-11N	32
Deer Path Ct. N&S	23W-11N	32
Dennison Rd. N&S	24W-12N	22
Des Plaines Ln. E&W		
	23W-11N	32
Devonshire Ln. NW&SE		
	24W-12N	22
Diamond Rd. N&S	24W-15N	22
Dixon Ct. NE&SW	23W-15N	22
Dixon Dr. N&S	23W-15N	22
Dogwood Ct. NE&SW		
	25W-14N	22
Dogwood Dr. NW&SE		
	25W-14N	22
Dorchester Ln. E&W		
	23W-11N	32
Douglas Ct. N&S	23W-11N	32
Dover Ct. E&W	24W-15N	22
Dovington Ct. N&S	25W-11N	32
Downing Dr. NE&SW		
	24W-15N	22
Dresden Ct. NW&SE		
	24W-15N	22
Dresden Dr. E&W	24W-15N	22
Driftwood Ct. E&W		
	25W-15N	22
Dukesbery Ct. E&W		
	24W-16N	12
Dukesbery Ln. N&S		
	24W-16N	12
Dunmore Pl.* E&W	24W-13N	22
Durham Ct. N&S		
	24W-12N	22
Durham Ln. E&W	24W-12N	22
E. Berkley Ln. N&S		
	23W-11N	32
E. Bluebonnet Ln. NW&SE		
	22W-11N	33
Eagle Way N&S	25W-13N	22
Edgefield Ln. E&W	25W-12N	22
Edgemont Ln. E&W		
	24W-12N	22
Eisenhower Cir. N&S		
	24W-14N	22
Elizabeth Ct. N&S	24W-12N	22
Emory Rd. N&S	24W-12N	22
Englewood Rd. NE&SW		
	24W-12N	22
Eric Dr. NE&SW	26W-12N	21
Erie Ln.* N&S	24W-13N	22
Essington Ct. N&S	24W-16N	12
Essington Ln. NE&SW		
	24W-16N	12

	Grid	Page
Evanston St. N&S	23W-11N	32
Evergreen Ln. E&W		
	24W-12N	22
Exeter Ct. E&W	24W-16N	12
Fairfield Cir. E&W	25W-12N	22
Fairfield Ln. NS&EW		
	25W-12N	22
Fairmont Rd. N&S	25W-12N	22
Fairway Ct. N&S	25W-12N	22
Fayette E&W	25W-12N	22
Fir Ct. N&S	24W-15N	22
Firestone Ct. N&S	24W-15N	22
Firestone Dr. NW&SE		
	24W-15N,24W-14N	22
Firestone Ln. N&S	24W-14N	22
Flagstaff Ln. N&S		
	24W-11N,23W-11N	32
Forest Glen Dr. N&S		
	25W-15N	22
Forest Park Ln. E&W		
	23W-11N	32
Fortune Bay Ct. NS&EW		
	25W-15N	22
Franklin Pl.* E&W	24W-13N	22
Frederick Ct. NW&SE		
	23W-12N	22
Frederick Ln. E&W		
	24W-12N,23W-12N	22
Freeman Rd. NW&SE		
	25W-13N,23W-15N	22
Fremont Ct. NW&SE		
	24W-12N	22
Fremont Rd. N&S		
	24W-12N	22
Gannon Ct. NW&SE		
	24W-11N	32
Gannon Dr. N&S		
	24W-11N,24W-12N	32
Garden Ter.* E&W	24W-11N	32
Garnet Cir. E&W	24W-15N	22
Gentry Rd. N&S	24W-12N	22
Georgetown Ln.* NW&SE		
	24W-13N	22
Geronimo St. NW&SE		
	23W-11N	32
Glen Lake Cir. E&W		
	25W-12N	22
Glen Lake Dr. E&W		
	25W-12N	22
Glen Lake Pl. E&W		
	25W-12N	22
Glen Lake Rd. N&S		
	25W-12N	22
Glen Ln. NS&EW		
	25W-12N,24W-12N	22
Glendale Ln. E&W	24W-11N	32
Glenwood Ln. E&W		
	24W-16N	12
Gloucester Ct. E&W		
	25W-15N	22
Golden Rod Ln. E&W		
	24W-15N	22
Golf Rd. E&W		
	26W-12N,24W-12N	21
Governors Ln. N&S		
	25W-12N	22
Grand Canyon Pkwy. NE&SW		
	23W-11N	32
Grand Canyon St. N&S		
	23W-11N	32
Grantham Pl.* NE&SW		
	24W-13N	22
Greenfield Rd. N&S		
	23W-11N	32
Greens Ct. NW&SE		
	26W-12N	21
Greenspoint Pkwy. NS&EW		
	27W-12N	21
Greystone Pl. E&W		
	26W-12N	21
Grissom Ct. N&S	23W-11N	32
Grissom Ct. E&W	23W-11N	32
Haman Av. NE&SW		
	24W-15N	22
Haman Ct. N&S	24W-15N	22
Hampton Rd. N&S	24W-12N	22
Hancock Dr.* NE&SW		
	24W-13N	22
Harbor Ct. NS&EW		
	25W-15N	22
Harrison Ln. NW&SE		
	24W-15N,23W-15N	22
Harvard Ct. NW&SE		
	24W-12N	22
Harwinton Pl. NW&SE		
	26W-12N	21
Hassell Cir. N&S	25W-12N	22
Hassell Ct. N&S	25W-12N	22
Hassell Dr. N&S	25W-12N	22
Hassell Pl. N&S	25W-12N	22
Hassell Rd. E&W		
	26W-12N,24W-12N	21
Hassell Rd. N&S	26W-12N	21
Hastings Dr.* NW&SE		
	24W-13N	22
Hawthorn Ln. E&W		
	23W-11N,22W-11N	32
Heather Ln. E&W		
	24W-11N	32
Heritage Dr. E&W	24W-11N	32
Heritage Cir. N&S		
	24W-12N	22
Hermitage Ct. N&S		
	24W-12N	22
Hermitage Ln. E&W		
	24W-12N	22
Higgins Quarter Dr. N&S		
	24W-11N	32

	Grid	Page
Higgins Rd. NW&SE		
	30W-14N,22W-11N	20
Highland Blvd. NS&EW		
	24W-12N	22
Hill Dr. E&W	24W-11N	32
Hillcrest Blvd. NE&SW		
	24W-12N,23W-12N	22
Hillcrest Ct. N&S	24W-12N	22
Hillside Ct. NE&SW		
	24W-14N	22
Hilltop Rd. N&S	24W-12N	22
Holbrook Ln.* NE&SW		
	24W-14N	22
Holly Ln. E&W	22W-11N	33
Hudson Dr. NE&SW		
	24W-15N	22
Huntington Blvd. N&S		
	25W-12N,25W-15N	22
Huttner Ct. E&W		
	26W-12N	21
Ida Rd. NE&SW	24W-12N	22
Illinois Blvd. N&S		
	24W-10N,23W-11N	32
Interstate 90 E&W		
	31W-12N,24W-12N	20
Islandview Ct. N&S		
	25W-12N	22
Jade Ln. NE&SW	24W-15N	22
Jamestown Cir. NS&EW		
	25W-12N	22
Jamison Ln. E&W	24W-12N	22
Jefferson Rd. NW&SE		
	25W-12N	22
Jenifer Ln. E&W	25W-12N	22
Jody Ln. N&S	24W-11N	32
John Dr. E&W		
	25W-11N,24W-11N	32
Jones Rd. N&S		
	24W-11N,24W-12N	32
Juniper Tree Ct. E&W		
	23W-11N	32
Kenilworth Cir. NS&EW		
	25W-12N	22
Kensington Ln. NE&SW		
	25W-12N	22
Kent Rd. E&W	25W-12N	22
Kenwood Rd. N&S	25W-12N	22
Kettering Rd.* E&W		
	24W-13N	22
Kingman Ln. E&W		
	24W-11N,23W-11N	32
Kingsdale Rd. NS&EW		
	25W-11N,25W-12N	32
Kingston Dr. N&S	24W-16N	12
Knoll Ln. N&S	24W-11N	32
Laburnum Rd. E&W		
	24W-14N	22
Lafayette Ln. E&W		
	24W-12N,23W-12N	22
Lafleur Ln. E&W	24W-14N	22
Lake Ct. NW&SE	24W-11N	32
Lake Edge Ct. E&W		
	25W-15N	22
Lakeside Plaza NS&EW		
	24W-12N	22
Lakeview Ln. E&W	24W-11N	32
Lakewood Blvd. E&W		
	25W-13N	22
Lancaster Ct. E&W		
	24W-12N	22
Lancaster Ln. E&W		
	25W-12N,24W-12N	22
Langdon Pl. E&W	26W-12N	21
Larchmont Rd. NE&SW		
	24W-13N	22
Latour Ct. NE&SW	24W-14N	22
Leather Leaf Ct. E&W		
	25W-14N	22
Leather Leaf Ln. E&W		
	25W-14N	22
Legend Ln. E&W	25W-12N	22
Lexington Dr. NE&SW		
	24W-14N,23W-14N	22
Liberty Pl.* E&W	24W-13N	22
Lichfield Dr. NS&EW		
	24W-16N	12
Lincoln Dr. N&S	23W-11N	32
Lincoln St. N&S	23W-10N	32
Lincolnshire E&W		
	24W-11N	32
Lombardy Ct. N&S	24W-15N	22
Lombardy Ln. NS&EW		
	24W-15N	22
London Sq. E&W	24W-11N	32
Londondery Ct. N&S		
	23W-14N	22
Ludington Ct. N&S	23W-15N	22
Manchester Dr. N&S		
	25W-12N	22
Maple Ln. NW&SE	24W-11N	32
Maricopa Ln. E&W	23W-11N	32
Marquette Ln.* N&S		
	24W-13N	22
Mason Dr. N&S	23W-15N	22
Mayfield Ln. E&W		
	26W-12N,25W-12N	21
Maywood Dr. E&W		
	24W-11N	32
Meadow Ln. N&S	24W-12N	32
Mesa Dr. E&W	24W-12N	32
Meyer Rd. N&S	25W-12N	22
Michael Ct. NW&SE		
	24W-15N	22
Michelline Ct. NW&SE		
	24W-14N	22
Milan Ln. E&W	24W-11N	32
Milton Ln. E&W	24W-11N	32

	Grid	Page
Mohave St. NS&EW	23W-10N,23W-11N	32
Monarch St. NE&SW	24W-15N	22
Monterey Ct. NE&SW	25W-12N	22
Monticello Rd. N&S	23W-12N	22
Moon Lake Rd. NW&SE	26W-12N,25W-12N	21
Morgan Ln. E&W	24W-10N	32
Morton St. N&S	24W-10N	32
Moulin Ln. NW&SE	24W-14N	22
Mumford Ct. NW&SE	24W-15N	22
Mumford Dr. N&S	24W-15N	22
N. Bluebonnet Rd. N&S	23W-11N	32
N. Dexter Ln. N&S	25W-11N	32
N. Dovington Dr. N&S	23W-11N	32
N. Firestone Dr. NW&SE	23W-14N,24W-15N	22
N. Oakmont Rd. N&S	25W-12N	22
N. Sapphire Dr. N&S	24W-15N	22
N. Tamarack Dr. N&S	24W-16N	12
N. Victoria Ln. N&S	24W-15N	22
Nantucket Ct. E&W	25W-15N	22
Nantucket Ct. NW&SE	25W-15N	22
Navajo Ln. E&W	23W-10N	32
New Britton Dr. NS&EW	24W-15N	22
New Castle Ln. E&W	25W-12N	22
New Sutton Rd. N&S	28W-13N	21
Newark Ln. N&S	24W-10N,23W-10N	32
Newman Pl. N&S	23W-15N	22
Newport Rd. NS&EW	24W-12N,23W-12N	22
Newton St. NE&SW	23W-10N	32
Nogales St. N&S	23W-10N	32
Norman Dr. E&W	24W-14N,23W-14N	22
Norridge Ln. E&W	24W-10N	32
Northview Ln. NW&SE	24W-12N	22
Northwest Tollway E&W	31W-12N,24W-12N	20
Nottingham Ln. E&W	26W-12N,25W-12N	21
O'Hare Dr. N&S	24W-15N	22
Oak Knoll Ln. NW&SE	25W-15N	22
Oak Tree Ct. E&W	23W-14N	32
Oakdale Rd. N&S	25W-12N	22
Old Barrington Rd. N&S	25W-13N	22
Old Higgins Rd. E&W	26W-12N	21
Old Sutton Rd. N&S	28W-13N	21
Old Timber Ct. NW&SE	24W-15N	22
Old Timber Ln. NS&EW	24W-15N	22
Olive St. N&S	24W-11N	32
Olmstead Dr. E&W	24W-15N	22
Olmstead Dr. NE&SW	24W-15N	22
Opal Ln. E&W	24W-10N	32
Orange Ln. E&W	24W-10N	32
Osage Ln. E&W	24W-10N	32
Oxford Ln.* N&S	24W-13N	22
Pacific Av. N&S	26W-11N	31
Paige Dr. NE&SW	24W-10N	32
Paisly Ct. E&W	24W-16N	12
Palatine Rd. E&W	24W-16N	12
Paris Ln. E&W	24W-10N	32
Park Ln. E&W	23W-15N	22
Parkside Dr. NS&EW	25W-15N	22
Parkview Cir. N&S	25W-13N	22
Partridge Hill Dr. NS&EW	24W-11N	32
Patriot Ct. NE&SW	24W-14N	22
Patriot Ln. E&W	24W-14N	22
Payson Dr. N&S	24W-10N	32
Pebble Beach Ct. NW&SE	25W-12N	22
Pebble Beach Dr. NS&EW	25W-12N	22
Pebblewood Ln. NW&SE	25W-15N	22
Pembroke Av. NS&EW	26W-12N	21
Perry Ln. E&W	24W-10N	32
Picardy Ln. NW&SE	24W-12N	22
Pierce Rd. N&S	23W-14N	22
Pleasant St. N&S	24W-10N	32
Plymouth Rd. NE&SW	23W-14N	22
Ponderosa Ln. N&S	24W-16N	12
Poplar Creek Dr. NE&SW	26W-12N	21
Port Arthur Ct. E&W	24W-15N	22
Portage Ln. N&S	24W-15N	22
Portshire St. N&S	23W-14N	22
Preswick Pl. NS&EW	24W-16N	12
Princeton St. N&S	24W-10N	32
Queensbury Cir. N&S	25W-12N	22
Raleigh Ln.* E&W	24W-13N	22
Raleigh Pl.* E&W	24W-13N	22
Randi Ln. N&S	24W-11N	32
Rebecca Dr. NE&SW	25W-12N	22
Regent Dr. NS&EW	23W-14N	22
Ridgewood Ln. E&W	24W-15N	22
Robin Cir. N&S	24W-12N	22
Robin Ln. E&W	24W-12N	22
Rochelle Ct. N&S	24W-15N	22
Rochester Dr. N&S	24W-16N	12
Rock Cove Ct. N&S	24W-15N	22
Rock Cove Ln. N&S	24W-15N	22
Rolling Prairie Ct. E&W	24W-15N	22
Rosedale Ln. E&W	25W-12N,24W-12N	22
Roselle Rd. N&S	23W-11N	32
Rt. 58 E&W	26W-12N	21
Rt. 59 N&S	28W-13N	21
Rt. 62 E&W	23W-14N	22
Rt. 72 NW&SE	30W-14N,22W-11N	20
Rubin Ln. N&S	25W-12N	22
Salem Dr. N&S	24W-11N,24W-12N	32
Salem Dr. N&S	25W-11N	32
Sandlewood Ln. NS&EW	24W-15N	22
Scarbrough Cir. E&W	24W-11N	32
Seaver Ln. E&W	26W-12N	21
Sessions Walk E&W	24W-15N	22
Shagbark Ct. E&W	24W-15N	22
Shephard Rd. N&S	24W-15N	22
Shoe Factory Rd. E&W	31W-12N,27W-12N	20
Shorewood Ct. N&S	25W-15N	22
Shorewood Dr. N&S	25W-15N	22
Silver Pine Dr. NS&EW	24W-16N	12
Smethwick Ln.* E&W	24W-13N	22
Somersworth Pl. E&W	26W-12N	21
Somerton Dr. N&S	24W-16N	12
Spring Mill Dr. N&S	24W-11N	32
Stockton Dr.* NW&SE	24W-13N	22
Stone Harbor Ct. N&S	25W-15N	22
Stonington Av. NE&SW	26W-12N	21
Stratham Pl. NE&SW	26W-12N	21
Sturbridge Ct. E&W	25W-15N	22
Sturbridge Dr. NE&SW	25W-15N	22
Sudbury Dr.* N&S	24W-13N	22
Suffolk Ct. NW&SE	23W-15N	22
Suffolk Ln. N&S	23W-14N,23W-15N	22
Sumac Tr. NW&SE	24W-15N	22
Sundance Cir. NE&SW	24W-15N	22
Sundance Ct. NW&SE	24W-15N	22
Sunflower Ln. NS&EW	24W-15N	22
Sussex Walk E&W	25W-12N	22
Sutherland Pl.* N&S	24W-13N	22
Swindon Pl. N&S	26W-12N	21
Sycamore Ct. NW&SE	24W-14N	22
Tamarack Ct. N&S	24W-16N	12
Tarrington Dr. N&S	24W-16N	12
Terrace Ct. E&W	26W-12N	21
Thacker St. E&W	22W-11N	33
Thornbark Ct. E&W	24W-15N	22
Thornbark Dr. N&S	24W-15N,24W-16N	22
Tolouse Ct. NW&SE	24W-14N	22
Topaz Dr. NE&SW	24W-15N	22
Trailside Ct. N&S	24W-15N	22
Treaty Ct. N&S	23W-14N	22
Treaty Ln. N&S	24W-14N	22
Turnberry Dr. NW&SE	24W-16N	12
Valley Ln. N&S	24W-14N	22
Versailles Rd. N&S	24W-14N	22
Victoria Dr. NS&EW	24W-15N	22
Vista Ln. E&W	25W-12N	22
Volid Dr. NE&SW	25W-12N	22
W. Berkley Ln. E&W	23W-11N	32
W. Dexter Ln. E&W	25W-11N	32
W. Dovington Dr. E&W	25W-11N,24W-11N	32
W. Firestone Dr. E&W	24W-15N,23W-15N	22
Cir.	24W-15N	22
W. Oakmont Rd. E&W	25W-12N	22
W. Tamarack Dr. E&W	24W-16N	12
W. Thacker St. E&W	23W-11N	32
Wainsford Dr. NS&EW	25W-11N	32
Warington Ln. NS&EW	24W-11N	32
Warwick Cir., North E&W	24W-11N	32
Warwick Cir., South E&W	24W-11N	32
Washington Blvd. N&S	24W-10N,24W-11N	32
Wellington Pl.* N&S	24W-13N	22
Westbridge Ct. N&S	24W-15N	22
Westbury Dr. E&W	25W-15N	22
Western St. N&S	24W-11N	32
Westhaven Ct. N&S	24W-16N	12
Westview St. N&S	24W-10N,24W-11N	32
Whispering Trails Ct. E&W	25W-15N	22
Whispering Trails Dr. NS&EW	25W-14N,25W-15N	22
Whitepine St. N&S	24W-16N	12
Whitingham Ln.* E&W	24W-13N	22
Wildwood Ct. NE&SW	24W-15N	22
William Ct. N&S	23W-15N	22
Williamsburg Dr. E&W	26W-12N	21
Willow Tree Ct. E&W	23W-11N	32
Wilmington Ln. NS&EW	24W-11N	32
Wilshire Dr. NW&SE	23W-14N	22
Winding Trail Ln. N&S	25W-14N	22
Winston Cir. NE&SW	23W-14N	22
Winston Cove E&W	23W-14N	22
Winston Dr. N&S	23W-14N,23W-15N	22
Winston Ln. E&W	23W-14N	22
Winston Pl. E&W	23W-14N	22
Wood Dr. NS&EW	24W-11N	32
Woodhollow Ct. E&W	24W-15N	22
Woodlawn St. N&S	24W-11N	32
Worthington Dr. E&W	24W-11N	32
Yardley Ct. E&W	25W-11N	32
Yardley Ln. N&S	25W-11N	32
Yorkshire Ct. N&S	23W-14N	22

Hometown

	Grid	Page
Beck Pl. NE&SW	5W-10S	66
Cicero Av. N&S	5W-10S	66
Co-op Ln. E&W	5W-10S	66
Corcoran Rd. NE&SW	5W-10S	66
Duffy Av. NW&SE	5W-10S	66
Keating Av. N&S	5W-10S	66
Keating Ln. N&S	5W-10S	66
Keeler Av. N&S	5W-10S	66
Kenneth Av. N&S	5W-10S	66
Kenton Ln. N&S	5W-10S	66
Kilbourn Av. N&S	5W-10S	66
Kildare Av. N&S	5W-10S	66
Kilpatrick Ln. N&S	5W-10S	66
Knox Av. N&S	5W-10S	66
Knox Ln. N&S	5W-10S	66
Kolin Av. N&S	5W-10S	66
Kolmar Av. N&S	5W-10S	66
Komensky Av. N&S	5W-10S	66
Kostner Av. N&S	5W-10S	66
Main St. NE&SW	5W-10S	66
Merrion Ln. N&S	5W-10S	66
Pulaski Rd. N&S	5W-10S	66
Rt. 50 N&S	5W-10S	66
Rt. 7 NE&SW	5W-10S	66
Ryan Rd. NE&SW	5W-10S	66
South Buckley Rd. N&S	5W-10S	66
Southwest Hwy. NE&SW	5W-10S	66
Staycoff Ln.* NE&SW	5W-10S	66
87th Pl. E&W	5W-10S	66
87th St. E&W	5W-9S	66
88th Pl. E&W	5W-10S	66
88th St. E&W	5W-10S	66
89th Pl. E&W	5W-10S	66
89th St. E&W	5W-10S	66
90th Pl. E&W	5W-10S	66
90th St. E&W	5W-10S	66
91st St. E&W	5W-10S	66

Homewood

	Grid	Page
Aberdeen St. N&S	1W-22S,1W-21S	88
Alexander Rd. E&W	3W-22S	87
Alexander Terr. E&W	2W-22S	88
Argyle Av. N&S	3W-22S,3W-21S	87
Armitage Pl. N&S	1W-21S	88
Ashland Av. N&S	2W-21S	87
Birch Rd. E&W	3W-22S,3W-21S	87
Boulder Ct. N&S	1W-22S	88
Bowling Green Rd. E&W	1W-22S	88
Bretz Dr. N&S	0W-21S	88
Briar Av. NE&SW	2W-21S	87
Burr Oak Rd. E&W	2W-21S,1W-21S	87
California Av. N&S	3W-22S,3W-21S	87
Capenter St. N&S	1W-21S	88
Caroline Dr. E&W	3W-21S	87
Carson Ct. E&W	1W-22S	88
Carson Rd. N&S	2W-21S	87
Castle Rd. N&S	1W-22S	88
Cedar Rd. E&W	2W-21S	87
Center Av. N&S	1W-22S,1W-21S	88
Center Ct. E&W	1W-23S	88
Chayes Park Dr. E&W	3W-22S	87
Cherry Creek Dr. N&S	3W-22S	87
Cherry Ln. E&W	2W-21S	87
Cherrywood Ct. E&W	3W-21S	87
Cherrywood Ln. N&S	3W-21S	87
Chestnut Rd. NW&SE	2W-21S	87
Clyde Rd. NE&SW	3W-22S	87
Clyde Terr. E&W	2W-22S	87
Coach Rd. E&W	1W-22S	88
Cowing Av. N&S	2W-21S	87
Debra Ln. E&W	3W-22S	87
Delta Rd. N&S	1W-22S	88
Dixie Hwy. N&S	2W-22S,2W-21S	87
Dixmoor Dr. N&S	2W-21S	87
Dolphin Lake Dr. NS&EW	3W-21S	87
Downey Rd. E&W	2W-22S	87
Dundee Av. N&S	3W-22S,3W-21S	87
Elder Rd. E&W	1W-21S	88
Elliot Ln. E&W	2W-22S	87
Elm Rd. NE&SW	2W-21S	87
Evergreen Rd. E&W	2W-22S,1W-22S	87
Francis Rd. E&W	3W-22S	87
Francisco Av. N&S	3W-22S	87
Fresno Rd. NS&EW	3W-22S	87
Gladville Av. N&S	2W-22S,2W-21S	87
Golfview Av. N&S	2W-22S,2W-21S	87
Gottschalk Av. N&S	2W-21S	87
Governors Hwy. N&S	3W-21S	87
Halsted St. N&S	1W-23S,1W-21S	88
Hart Dr. N&S	1W-21S	88
Harwood Av. NE&SW	2W-22S,2W-21S	87
Hawthorne Rd. E&W	3W-21S,3W-10S	87
Heather Ct. E&W	1W-22S	88
Heather Rd. E&W	1W-22S	88
Hedgerow Ln. E&W	2W-22S	87
Hickory Rd. E&W	3W-21S,3W-10S	87
Highland Av. N&S	2W-22S,2W-21S	87
Hillside Av. N&S	2W-21S	87
Hillview Rd. E&W	1W-22S	88
Holbrook Rd. E&W	1W-23S	88
Hollydale Dr. E&W	1W-22S	87
Homewood Av. N&S	2W-22S,2W-21S	87
Homewood Av. N&S	3W-22S	87
Hood Av. N&S	1W-22S,1W-21S	88
Howe Av. N&S	2W-21S	87
Idlewild Ln. E&W	1W-22S	88
Jamie Ct. E&W	1W-22S	88
Jamie Ln. E&W	1W-22S	88
Jefferey Ct. E&W	1W-22S	88
Jefferey Rd. E&W	1W-22S	88
Jill Ct. E&W	1W-22S	88
Jill Terr. E&W	1W-22S	88
Jodi Terr. N&S	1W-22S	88
Jonathan Ln. N&S	1W-22S	88
Juhlin Dr. E&W	1W-22S	88
Kathleen Ct. E&W	3W-22S	87
Kings Rd. N&S	1W-22S	88
Klimm Av. N&S	2W-22S	87
Knollwood Ln. N&S	2W-22S	87
Kroner Rd. N&S	0W-21S	88
L.A. Ct., N. NE&SW	3W-22S	87
L.A. Ct., S. SE&NW	3W-22S	87
Lahaigh Rd. N&S	1W-22S	88
Larkspur Ln. N&S	1W-22S	88
Lexington Av. N&S	2W-22S	87
Lincoln Av. N&S	3W-21S	87
Linden Rd. E&W	2W-21S,1W-21S	87
Locust Rd. E&W	3W-21S	87
Loomis Av. N&S	1W-22S,1W-21S	88
Loomis Ct. N&S	1W-23S	88
Loomis Dr. N&S	1W-23S,1W-22S	88
Los Angeles Av. N&S	3W-21S	87
Lyn Ct. N&S	1W-22S	88
Magnolia Rd. E&W	2W-22S	87
Maple Av. N&S	1W-21S	88
Maple Rd. E&W	3W-21S,2W-21S	87
Marlin Ln. E&W	3W-22S	87
Marshfield Av. N&S	2W-22S	87
Martin Av. N&S	2W-22S,2W-21S	87
Matthew Ln. E&W	3W-22S	87
May Av. N&S	1W-22S	88
Morgan Av. N&S	1W-22S,1W-21S	88
Morris Av. N&S	2W-22S	88
Mueller Ct. E&W	2W-21S	87
Oak Rd. N&S	2W-22S	87
Olive Rd. E&W	3W-21S,1W-21S	87
Page Av. N&S	1W-22S	88
Page Ct. N&S	2W-22S	87
Palmer Av. N&S	2W-22S	87
Palmer Rd. NW&SE	2W-22S	87
Park Av. NE&SW	2W-21S	87
Patricia Ln. E&W	3W-21S	87
Perth Av. N&S	3W-22S,3W-21S	87
Pierce Av. N&S	2W-22S	88
Pierce St. N&S	1W-23S	88
Pierce Terr. N&S	1W-23S	88
Pine Rd. E&W	2W-21S	87
Pine St. N&S	1W-23S	88
Poplar Av. NW&SE	1W-22S	88
Presidents Dr. NE&SW	1W-21S	88
Queens Rd. N&S	1W-21S	88
Reigel Rd. NW&SE	1W-23S,1W-21S	88
Ridge Rd. E&W	2W-21S,1W-21S	87
Riegel Oaks Ln. E&W	1W-22S	88
Robin Ln. NE&SW	2W-22S	87
Roosevelt Av. N&S	2W-22S	87
Royal Rd. N&S	1W-22S	88
Sacramento Av. N&S	3W-22S,3W-21S	87
Sailfish Rd. E&W	1W-21S	88
San Diego Av. N&S	3W-21S	87
Spruce Rd. E&W	3W-21S,1W-21S	87
Stedhall Rd. N&S	3W-21S	87
Stewart Av. N&S	3W-22S,3W-21S	87
Stockton Av. N&S	3W-22S	87
Sycamore Ct. E&W	2W-21S	87
Sycamore Dr. N&S	2W-21S	87
Tarpon Ct. N&S	1W-22S	88
Terrace Rd. E&W	2W-22S	87
Thomas St. E&W	1W-22S	88
Tipton Av. N&S	2W-22S	87
Turtle Creek Dr. N&S	3W-21S	87
Virginia Pl. E&W	3W-21S	87
Walnut Rd. E&W	3W-21S	87
Walton Ln. NW&SE	1W-22S	88
Washington Av. N&S	2W-21S	87
Willow Rd. E&W	3W-22S,1W-22S	88
175th St. E&W	1W-22S	88
178th St. E&W	3W-21S	87
182nd Pl.* E&W	2W-22S	87
183rd Pl. E&W	1W-22S	88
183rd St. E&W	3W-21S,1W-21S	88
184th Pl. E&W	1W-22S	88
184th St. E&W	4W-22S,1W-22S	87
185th Pl. E&W	1W-22S	88
185th St. E&W	4W-22S,1W-22S	87
186th Pl. E&W	1W-22S	88
186th St. E&W	4W-22S,1W-22S	87
187th St. E&W	2W-22S,1W-22S	87
188th St. E&W	2W-22S	87
190st St. E&W	1W-22S	88
191st St. E&W	1W-22S	88

Indian Head Park

	Grid	Page
Acacia NS&EW	13W-7S	58
Algonquiin N&S	13W-7S	58
Apache E&W	14W-7S	58
Arrowhead* N&S	14W-6S	58
Big Bear Dr. N&S	13W-7S	58
Blackhawk Dr. N&S	14W-7S	58
Briarwood Ln. E&W	13W-7S	58
Button Wood N&S	13W-7S	58
Cascade Ct. E&W	13W-7S	58
Cherokee E&W	14W-7S	58
Cherrywood NE&SW	13W-7S	58
Cochise E&W	14W-7S	58
Deercrest E&W	13W-7S	58
East E&W	14W-7S	58
Elmwood Ct.* E&W	14W-6S	58

	Grid	Page
Elmwood Square* N&S	14W-6S	58
Hawthorne Sq.* N&S	13W-7S	58
Heatherwood Ct.* N&S	14W-6S	58
Howard N&S	13W-7S	58
I-294 N&S	14W-7S	58
Indian Head Ct.* N&S	14W-6S	58
Indian Head Tr.* N&S	14W-6S	58
Joliet SW&NE	14W-7S	58
Keokok NW&SE	14W-7S	58
Mohawk Ct. NW&SE	14W-7S	58
Nacona Ln.* E&W	13W-7S	58
Osceola NE&SW	14W-7S	58
Pontiac NW&SE	14W-7S	58
Potawatomie N&S	13W-7S	58
Sauk E&W	13W-7S	58
Cir.	14W-7S	58
Shawnee E&W	13W-7S	58
Sioux N&S	14W-7S	58
Square E&W	13W-7S	58
Stonehead E&W	13W-7S	58
Stratford Pl.* N&S	13W-7S	58
Sweetwood E&W	13W-7S	58
Tanglewood Ct.* N&S	13W-7S	58
Tecumseh N&S	14W-7S	58
Thunderbird E&W	14W-7S	58
Timber Ridge Rd. N&S	14W-6S	58
Vine N&S	14W-7S	58
Waubanse NE&SW	14W-7S	58
Westwood Ct.* E&W	14W-6S	58
Westwood Square* N&S	14W-6S	58
Wood N&S	14W-6S	58
62nd E&W	14W-6S	58
65th E&W	13W-7S	58
65th Pl. E&W	13W-7S	58

Inverness

	Grid	Page
Abbotsford Dr. NS&EW	25W-16N,24W-16N	12
Aberdeen Rd. E&W	22W-16N	13
Aberdour Ln.* N&S	23W-16N	12
Afton Cir. NW&SE	24W-17N	12
Alexander Ct.	25W-16N	12
Alnwick Ct. NW&SE	25W-16N	12
Appleby Rd. N&S	23W-15N,22W-15N	22
Applecross Rd. NW&SE	25W-16N	12
Ardmore Av. N&S	24W-15N,23W-15N	22
Arlington Rd. N&S	23W-15N,23W-16N	22
Ayshire Ln. N&S	23W-16N	12
Baldwin Rd. E&W	24W-16N	12
Balmoral Cir. NE&SW	22W-14N	23
Balmoral Dr. E&W	22W-14N	23
Balmoral Ln. N&S	22W-14N,22W-15N	23
Banbury Rd. N&S	23W-17N,22W-17N	12
Bannockburn Rd.	23W-14S	22
Barclay Cir. NS&EW	24W-17N	12
Barra Ln. NE&SW	22W-16N	13
Beaver Pond Rd. E&W	23W-15N	22
Bedlington Dr. N&S	24W-17N	12
Berwick E&W	23W-14N	22
Betty Dr. NW&SE	23W-16N	12
Bishop Ct. NE&SW	25W-16N	12
Blackburn Rd. NS&EW	23W-17N	12
Blair Ln.* E&W	23W-16N	12
Blythe Ct. NE&SW	24W-16N	12
Bonne Ln. NW&SE	22W-16N	13
Bonny Glen Gates E&W	23W-14N	22
Bordeaux Ct. N&S	24W-17N	12
Borthwick Ln. E&W	22W-16N	13
Bradwell Rd. E&W	25W-16N,23W-16N	12
Braeburn Rd. N&S	23W-16N,23W-17N	12
Braymore Ct. NE&SW	25W-16N	12
Braymore Dr. E&W	25W-16N	12
Braymore Dr., South NS&EW	25W-16N	12
Brodick Ln. NW&SE	22W-16N	13
Burleigh Ln. E&W	22W-16N	13
Camphill Cir. N&S	23W-15N	22
Canterbury Ln. NW&SE	23W-16N,23W-17N	12
Carberry Cir. NS&EW	24W-17N	12
Carlisle Dr. NS&EW	24W-17N	12

	Grid	Page
Cawdor Ln. NE&SW	22W-16N	13
Cheviot Dr. N&S	24W-16N	12
Chimney Rock N&S	23W-14N	22
Clover Dr. NE&SW	23W-14N,22W-15N	22
Colonial Pkwy. NS&EW	23W-14N	22
Colony Dr. N&S	25W-16N	12
Common Ridings Way N&S	25W-16N	12
Cornell Av. E&W	22W-16N	13
Country Oak Ln. N&S	23W-15N	22
Courtbridge Rd. N&S	24W-17N	12
Craigie Ln. E&W	22W-16N	13
Crichton Ln. E&W	22W-16N	13
Cumnock Rd. NS&EW	23W-16N	12
Dairy Ln. N&S	23W-17N	12
Dewey Rd. E&W	23W-16N	12
Dirleton Ln.* E&W	23W-16N	12
Dover Cir. NE&SW	23W-15N	22
Drummond Cir. NW&SE	24W-17N	12
Dublin Ct. N&S	22W-15N	23
Dumfries Ct. NE&SW	22W-15N	23
Dunbar Rd. E&W	23W-17N,22W-17N	12
Dunbarton Dr. N&S	25W-16N	12
Dunheath Dr. N&S	24W-17N	12
Duntrone Ln.* E&W	23W-16N	12
Durham Dr. E&W	23W-16N	12
Edinburgh Ct. NS&EW	24W-17N	12
Ela Rd. N&S	23W-14N,23W-17N	22
Fairway Ln. N&S	23W-16N	12
Fife Ct. E&W	24W-16N	12
Fincharn Ln. N&S	24W-16N	12
Firth Rd. NS&EW	23W-15N	22
Florence Av. E&W	24W-16N,23W-16N	12
Fox Trail NW&SE	22W-14N	23
Freeman Ct. E&W	23W-15N	22
Freeman Rd. E&W	23W-15N	22
Gaelic Ct. E&W	25W-16N	12
Cir.	24W-16N	12
Glamis Ln. NW&SE	22W-16N	13
Glen Eagles Ct. NW&SE	22W-16N	13
Glencrest Dr. NS&EW	24W-17N	12
Glenmore Ct. N&S	24W-17N	12
Grayfriars Ln. N&S	22W-15N	23
Great Glen E&W	24W-16N	12
Greenock NS&EW	22W-16N	13
Guthrie Cir. N&S	24W-17N	12
Guthrie Dr. NE&SW	24W-17N	12
Halbert Ln. NW&SE	25W-16N	12
Halkirk Cir. E&W	22W-15N	23
Haman Rd. N&S	24W-15N,24W-16N	22
Hamilton Ct. N&S	25W-16N	12
Harmening Av. E&W	23W-16N	22
Harrow Gate Dr. N&S	25W-16N	12
Heather Ln. N&S	23W-17N	12
Highland Rd. E&W	23W-16N	12
Hillshire Ct. E&W	25W-17N	12
Hillshire Ln. NS&EW	25W-17N	12
Huntley E&W	23W-14N	22
Inverray Rd. E&W	23W-15N	22
Jules Rd. N&S	23W-16N	12
Julie Dr. N&S	23W-16N	12
Keise Glen Ct. (Pvt.) E&W	23W-16N	12
Kilchurn Ct. N&S	23W-16N	12
Kilrenny N&S	23W-14N	22
Kirkwall Ct. N&S	24W-17N	12
Kirkwood Dr. NE&SW	23W-17N	12
Kitson Cir. N&S	24W-16N	12
Knockderry Ln. E&W	22W-16N	13
Knox Ct. NS&EW	25W-16N	12
Lanark Ln. NW&SE	23W-17N	12
Lancaster Ct. N&S	24W-17N	12
Lauder Ln. NS&EW	23W-16N	22
Lillian Av. E&W	24W-16N,23W-16N	12
Livingston Ln. E&W	25W-16N	12
Livingston Ln. E&W	25W-16N	12
Loch Lomond Dr. NS&EW	22W-16N	13
Lochleven Ln.* NW&SE	23W-16N	12
Lomond Ln. N&S	23W-16N	12
Cir.	25W-16N	12

	Grid	Page
MacAlpin Ct. NW&SE	25W-16N	12
MacAlpin Dr. NS&EW	25W-16N	12
Macbain Way N&S	24W-16N	12
Marie Dr. E&W	23W-17N	12
Midmar Ln.* E&W	23W-16N	12
Milton Rd. E&W	23W-16N,23W-17N	12
Muirfield Rd. NW&SE	24W-17N	12
Mulguy Ct. NW&SE	25W-16N	12
Mulguy Dr. NE&SW	25W-16N	12
Newcastle Ct. NW&SE	25W-17N	12
Newport Ct. E&W	25W-17N	12
Newport Ln. NS&EW	25W-17N	12
Northwest Highway NW&SE	24W-17N,22W-16N	12
Old Wick Ln. NW&SE	22W-16N	13
Palatine Rd. E&W	25W-16N,22W-16N	12
Peregrine Dr. E&W	22W-14N	23
Pheasant Tr. NW&SE	23W-15N,22W-15N	22
Plymouth Ct. NE&SW	23W-15N	22
Plymouth Dr. N&S	23W-15N	22
Poteet Av. N&S	23W-15N,23W-16N	22
Prestwick Dr. NS&EW	23W-15N	22
Quail Run N&S	23W-14N	22
Ravenscraig Ln. N&S	22W-16N	13
Rob Roy Ct. NW&SE	22W-16N	13
Roberts Rd. N&S	24W-15N,24W-17N	22
Roselle Rd. N&S	22W-14N,22W-16N	23
Rosemary Av. E&W	24W-16N	12
Rosslyn Ln.* NE&SW	23W-16N	12
Rt. 68 E&W	25W-17N	12
Sanday Ln. E&W	25W-16N	12
Scots Ct. N&S	22W-16N	13
Selkirk Ct. NS&EW	24W-16N	12
Shetland Rd. NW&SE	23W-16N	12
Shire Ct. N&S	22W-14N	23
Skye E&W	24W-16N	12
St. Andrews Ln. N&S	23W-16N	13
Sterling Rd. N&S	23W-16N,23W-17N	12
Stonefield Ct. NE&SW	23W-17N	12
Stonehaven E&W	23W-14N	22
Stratford Ln. NE&SW	23W-16N,23W-17N	12
Strome Ln.* E&W	23W-16N	12
Stuart Ln. E&W	23W-16N	12
Summer Isle Ln. E&W	25W-16N	12
Sunset Dr. E&W	23W-16N	12
Tantallon Ln.* E&W	23W-16N	12
Tarbat Ct. NS&EW	24W-16N	12
Thompson's Way N&S	23W-14N	22
Thompsons Cir. E&W	23W-14N	22
Thor Dr. E&W	23W-14N,22W-14N	22
Thornhill Rd. E&W	23W-16N	12
Thos Atkinson Rd. E&W	23W-16N	12
Thyra N&S	23W-16N	12
Turkey Ct. NE&SW	23W-14N	22
Turkey Trail E&W	23W-14N,22W-14N	22
Tweed Rd. NS&EW	23W-16N	12
Tyne Ct. N&S	24W-16N	12
Valley Lake Dr. NE&SW	23W-14N	22
Versailles Ct. N&S	24W-17N	12
Warkworth Ln.* N&S	23W-16N	12
Warren Av. E&W	23W-15N	22
Waterford Ln. E&W	23W-17N	12
Whispering Pines Ct. E&W	25W-16N	12
Williams Rd. N&S	24W-15N,24W-16N	22
Willow Ln. E&W	24W-16N,23W-16N	12
Windsor Cir. NW&SE	23W-15N	22
Windsor Rd. E&W	23W-15N	22
Winfield Ln. N&S	25W-16N	12
Withorn Ln. E&W	22W-16N	13
Wood St. E&W	24W-16N,23W-16N	12
Woodburn Ct. N&S	23W-17N	12

Justice

	Grid	Page
Archer SW&NE	11W-8S	58
Archer Av. N&S	11W-9S,10W-8S	64
Banks N&S	11W-8S	58
Blackstone SW&NE	11W-8S	58
Blazer N&S	11W-8S	58
Christeen Dr. NS&EW	10W-9S	65
Cork Av. N&S	11W-8S	58
Cronin N&S	11W-8S	58
Deer Path N&S	11W-9S	64
Fawn Tr. N&S	11W-9S	64
Figura N&S	11W-8S	58
Garden Ln. N&S	11W-8S	58
Huntington N&S	10W-9S	59
Indian NW&SE	11W-8S	58
Interstate 294 SE&NW	10W-9S	65
Jocare E&W	11W-8S	58
Madison N&S	11W-8S	58
Maple NW&SE	10W-9S	59
Marion Dr. E&W	10W-9S	65
Marisa N&S	10W-9S	59
Mary E&W	11W-8S	58
Murray E&W	11W-8S	58
Oak Grove N&S	11W-8S	58
Oak Ridge Dr. E&W	11W-9S	64
Pawtrail Rd. N&S	11W-9S	59
Rose SW&NE	10W-8S	59
Rt. 171 NE&SW	11W-9S,11W-8S	64
Thomas St. E&W	10W-9S	65
Tri State Tollway SE&NW	10W-9S	65
Willow N&S	10W-9S	59
72nd E&W	10W-9S	59
72nd Pl. NW&SE	10W-9S	59
73rd E&W	10W-9S	59
73rd Pl. NW&SE	10W-9S	59
74th E&W	10W-9S	59
75th E&W	11W-8S	58
78th E&W	11W-8S	58
78th Ct. N&S	9W-8S	59
78th St. E&W	11W-8S	58
79th E&W	10W-9S	59
79th Av. N&S	9W-9S	65
79th Ct. N&S	9W-9S	65
79th Pl. E&W	11W-9S	64
79th St. E&W	11W-8S,10W-8S	58
80th Av. N&S	10W-9S	65
80th St. E&W	10W-9S	65
81st Av. N&S	10W-9S	65
81st St. E&W	11W-9S,10W-9S	64
82nd Av. N&S	10W-9S	65
82nd Ct. N&S	10W-9S	65
82nd St. E&W	10W-9S	65
83rd Av. N&S	10W-9S	65
83rd Pl.	11W-9S,10W-9S	64
83rd St. E&W	11W-9S,10W-9S	64
84th Av. N&S	11W-9S	64
84th Ct. N&S	10W-9S	65
84th Pl. E&W	11W-9S,10W-9S	64
84th St. E&W	11W-9S,10W-9S	64
85th Av. N&S	11W-9S	64
85th Ct. N&S	10W-9S	65
85th Pl. E&W	11W-9S,10W-9S	64
85th St. E&W	11W-9S,9W-9S	64
86th Av. N&S	10W-9S,10W-9S	65
86th Ct. N&S	11W-9S	64
86th St. E&W	11W-9S,9W-9S	64
87th Ct. N&S	10W-9S	65
87th St. E&W	10W-9S,9W-9S	65
88th Av. N&S	11W-9S	58
90th Av. N&S	11W-9S	64

Kenilworth

	Grid	Page
Abbotsford Rd. NW&SE	4W-14N	29
Abingdon Av. NE&SW	3W-14N	29
Brier N&S	4W-14N	29
Cumberland Av. NE&SW	4W-14N	29
Cummings Av. E&W	4W-14N	29
Cumnor Rd. NW&SE	4W-14N	29
Devonshire Ln. N&S	3W-14N	29
Earlston Rd. NE&SW	4W-14N	29
Essex Rd. N&S	4W-14N	29
Exmoor Rd. N&S	4W-14N	29
Glendening Rd. N&S	4W-14N	29
Green Bay Rd. NW&SE	4W-14N	29
Greenwood Av. E&W	4W-14N	29
Ivy Ct. E&W	4W-14N	29
Kenilworth Av. N&S	4W-14N	29
Kenilworth Terr. E&W	4W-14N	29
Kent Rd. E&W	4W-14N	29
Leicester Rd. NW&SE	3W-14N	29
McLean Av. E&W	4W-14N	29
Melrose Av. NE&SW	4W-14N	29
Oxford Rd. NE&SW	3W-14N	29
Park Dr. E&W	4W-14N	29
Raleigh Rd. NW&SE	3W-14N	29
Richmond Rd. N&S	4W-14N	29
Ridge Rd. NW&SE	4W-14N	29
Robsart Pl. E&W	3W-14N	29
Robsart Rd. NE&SW	4W-14N	29
Rogers Av. E&W	3W-14N	29
Roslyn Rd. NW&SE	4W-14N	29
Sterling Rd. N&S	4W-14N	29
Tudor Pl. E&W	3W-14N	29
Warwick Ln. E&W	3W-14N	29
Warwick Rd. NW&SE	4W-14N,3W-14N	29
Wayland Av. E&W	4W-14N	29
Woodstock N&S	4W-14N,3W-14N	29
10th St. N&S	3W-14N	29

La Grange

	Grid	Page
Arlington E&W	12W-4S	52
Ashland N&S	12W-4S	52
Bassford N&S	13W-4S	52
Beaco N&S	12W-4S	52
Bell E&W	13W-4S,12W-4S	52
Benton E&W	12W-4S	52
Birchwood N&S	13W-5S	58
Birks E&W	13W-5S	58
Blackstone N&S	12W-4S	52
Bluff N&S	12W-4S	52
Brighton N&S	13W-5S	58
Burlington SW&NE	12W-4S	52
Calendar E&W	12W-4S	52
Calendar SW&NE	12W-4S	52
Calle View E&W	12W-5S	58
Carriage E&W	13W-5S	58
Catherine N&S	12W-4S	52
Cossitt E&W	13W-4S,12W-4S	52
Country Club E&W	12W-5S	58
Drexel N&S	13W-4S	52
Eberly N&S	12W-4S	52
Edgewood N&S	13W-4S	52
Elder N&S	13W-4S	52
Elm E&W	13W-4S,12S-4S	52
Franklin St. N&S	12W-4S	52
Glenwood N&S	13W-5S	58
Goodman E&W	13W-4S,12W-4S	52
Haris SW&NE	12W-4S	52
Hayes N&S	12W-4S	52
Hazel N&S	12W-4S	52
Hill Grove SW&NE	12W-4S	52
Hillbery N&S	13W-5S	58
Kensington N&S	12W-5S,12W-4S	52
LaGrange N&S	12W-5S	58
Leitch E&W	13W-4S	52
Linklater E&W	12W-5S	58
Locust N&S	12W-4S	52
Madison N&S	12W-4S	52
Maple E&W	13W-4S,12W-4S	52
Mason N&S	13W-5S	58
Mason Dr. E&W	13W-5S	52
Ogden E&W	13W-4S	52
Park N&S	13W-4S	52
Peck N&S	13W-4S	52
Plainfield NE&SW	13W-6S	58
Poplar N&S	13W-4S	52
Sawyer N&S	12W-4S	52
Shawmut E&W	12W-4S	52
Spring N&S	12W-5S,12W-4S	58
Stone N&S	12W-4S	52
Sunset N&S	12W-4S	52
Tilden N&S	12W-4S	52
Waiola N&S	12W-4S	52
Washington N&S	12W-4S	52
6th N&S	12W-5S,12W-4S	58
7th N&S	12W-4S	52
8th N&S	12W-4S	52
9th N&S	12W-5S,12W-4S	58
10th N&S	12W-5S,12W-4S	58
11th N&S	12W-5S	58
12th N&S	12W-5S	58
41st E&W	13W-4S	52
47th E&W	13W-4S	52
48th E&W	12W-4S	52
49th E&W	12W-5S	58
50th E&W	13W-5S,12W-4S	58
51st E&W	12W-4S	52
51st N&S	13W-4S	58
52nd E&W	12W-5S	58
52nd Pl. E&W	13W-5S	58
53rd E&W	12W-5S	58
54th E&W	12W-5S	58

La Grange Park

	Grid	Page
Alma N&S	11W-2S	53
Ashland N&S	12W-3S	52
Barnsdale St. N&S	12W-3S	52
Beach N&S	12W-3S,12W-2S	53
Blanchan N&S	11W-2S	53
Brainard N&S	12W-3S	52
Brewster E&W	12W-3S	52
Catherine N&S	12W-3S	52
Cleveland N&S	11W-2S	53
Community N&S	12W-3S	52
Deerpath N&S	12W-2S	52
Dover N&S	13W-3S	52
Edgewood N&S	13W-3S	52

	Grid	Page
Elmwood E&W	12W-3S	52
Finsbury E&W	12W-2S	52
Forest N&S	12W-3S,12W-2S	52
Garfield E&W	12W-3S	52
Harding E&W	12W-3S	52
Harrison E&W	11W-2S	53
Homestead N&S	12W-3S,12W-2S	52
Huntington E&W	12W-2S	52
Jackson E&W	12W-3S	52
Kemman N&S	12W-3S,12W-2S	52
Kensington N&S	12W-3S	52
Kings N&S	12W-2S	52
Lincoln E&W	12W-3S	52
Logan E&W	11W-2S	53
Malden N&S	13W-3S	52
Maple N&S	11W-2S	53
Monroe E&W	12W-3S	52
Morgan N&S	11W-2S	53
Newberry N&S	12W-3S,12W-2S	52
Oak E&W	12W-3S	52
Ostrander N&S	11W-2S	53
Park N&S	13W-3S	52
Pine E&W	12W-3S	52
Pine Tree E&W	12W-2S	52
Raymond N&S	11W-2S	53
Richmond E&W	12W-3S	52
Robinhood N&S	12W-3S,12W-2S	52
Scotdale Dr. E&W	12W-2S	52
Spring N&S	12W-2S	52
Stone N&S	12W-3S	52
Stonegate E&W	12W-2S	52
Thorpe N&S	12W-2S	52
Timber N&S	12W-2S	52
U.S. Rt. 12 N&S	12W-2W	52
U.S. Rt. 40 N&S	12W-2S	52
U.S. Rt. 45 N&S	12W-2S	52
Waiola N&S	12W-3S	52
Woodlawn E&W	13W-3S,12W-2S	52
Woodside N&S	12W-2S	52
26th E&W	12W-2S	52
28th E&W	12W-2S	52
29th E&W	12W-2S	52
30th E&W	12W-2S	52
31st E&W	13W-2S,12W-2S	52

Lansing

Ada St. N&S	4E-22S,4E-21S	89
Adams St. E&W	4E-21S	89
Ann St. E&W	4E-21S	89
Arcadia Av. N&S	3E-21S,3E-20S	89
Barb Ct. N&S	4E-22S	89
Bensley Av. N&S	2E-20S	89
Bernadine Ln. N&S	4E-22S	89
Bernadine St. N&S	4E-22S,4E-20S	89
Bernice Av. E&W	3E-20S,4E-20S	89
Bernice Rd. E&W	2E-20S,4E-20S	89
Bock Rd. N&S	2E-21S	89
Burnham Av. N&S	3E-20S,3E-20S	89
Burnham Ct. E&W	3E-20S	89
Burnham St. N&S	3E-21S	89
Calhoun Av. N&S	2E-20S	89
Calhoun St. N&S	3E-20S	89
Carriage Ln. N&S	2E-21S	89
Central E&W	4E-21S	89
Cherry Ln. N&S	3E-22S	89
Chicago Av. N&S	3E-22S,3E-19S	89
Christina Dr. E&W	3E-21S	89
Clyde Av. N&S	2E-21S	89
Columbian Dr. NW&SE	3E-20S	89
Commercial Av. N&S	3E-21S	89
Community St. N&S	4E-20S	89
Dash St. N&S	3E-20S	89
De Jong Ln. N&S	2E-21S	89
Decker Av. N&S	3E-21S	89
Escanaba Av. N&S	3E-21S	89
Exchange Av. N&S	3E-21S,3E-19S	89
Forest E&W	2E-21S	89
Forest Ln. E&W	3E-22S	89
Fred St. N&S	3E-21S	89
Fritz Dr. N&S	3E-20S	89
Glen Oak Av. N&S	3E-21S,3E-20S	89
Glen Ter. N&S	3E-21S	89
Glenwood Lansing Rd. E&W	3E-22S	89
Grant St. N&S	4E-22S,4E-20S	89
Greenbay Av. N&S	3E-22S,3E-20S	89
Hearth Cir. E&W	2E-21S	89
Henry Ct. N&S	4E-21S	89
Henry St. N&S	4E-22S,4E-20S	89
Hickory Av. N&S	2E-21S	89
Holiday Ter. E&W	2E-20S	89
Hoxie Av. N&S	2E-20S	89
Illi-Indy Dr. NE&SW	4E-21S	89
Illinois Av. N&S	4E-21S	89
Indiana Av. E&W	2E-21S,3E-21S	89
Industrial Dr. NW&SE	3E-20S	89
Jackson St. E&W	4E-21S	89
Keegan Av. N&S	2E-20S	89
Kettle Ct. E&W	2E-21S	89
Kingery Expwy. E&W	2E-20S,4E-20S	89
Lake St. E&W	4E-21S	89
Lange St. N&S	4E-21S,4E-22S	89
Lee Av. E&W	3E-20S	89
Locust Av. N&S	2E-21S,2E-20S	89
Lorenz Av N&S	3E-21S,3E-20S	89
Louise Ct. NE&SW	4E-22S	89
Louise Dr. NW&SE	4E-22S	89
Luna Dr. N&S	3E-20S	89
Madison St. E&W	4E-21S	89
Maple Av. NW&SE	4E-22S	89
Maple St. NE&SW	4E-20S	89
Maple St. N&S	4E-20S	89
Marble Av. N&S	3E-20S	89
Mertle St. NW&SE	4E-21S	89
Miller Dr. N&S	4E-21S	89
Monroe St. E&W	4E-21S	89
N. Manor Dr. E&W	4E-22S	89
Oak Av. N&S	3E-21S,3E-20S	89
Oak St. N&S	3E-21S,3E-23S	89
Oakley Av. N&S	3E-21S,3E-20S	89
Oakwood Av. N&S	3E-22S,3E-21S	89
Old Farm Rd. N&S	2E-21S	89
Otto St. E&W	4E-22S	89
Park Av. N&S	3E-20S	89
Park Av. N&S	3E-22S,3E-20S	89
Park Blvd. NW&SE	3E-20S	89
Paxton Av. N&S	3E-20S	89
Railroad Av. NW&SE	3E-20S	89
Randolph St. E&W	4E-21S	89
Ridge Rd. E&W	2E-21S,3E-21S	89
Ridgewood Av. N&S	3E-22S,3E-21S	89
River Rd. E&W	4E-20S	89
Riviera Rd. NE&SW	4E-19S	83
Rose Av. N&S	3E-20S	89
Rosewood Dr. N&S	2E-20S	89
Roy St. N&S	4E-22S,4E-20S	89
Roy St. N&S	4E-20S	89
Rt. 94-80 N&S	2E-20S,4E-20S	89
Sailor Ln. N&S	4E-20S	89
School St. N&S	3E-21S	89
Schultz Av. E&W	3E-20S	89
Schultz Dr. E&W	4E-21S	89
Schultz Dr., North E&W	4E-21S	89
Schultz Dr., South E&W	4E-21S	89
Sharon Ct. N&S	4E-22S	89
Sherman St. N&S	4E-22S,4E-21S	89
Shirley St. E&W	4E-20S	89
South Manor Dr. E&W	4E-22S	89
State Line Rd. N&S	4E-21S,4E-20S	89
Stewart Av. NW&SE	4E-21S	89
Terrace Dr. E&W	4E-20S	89
Ton Av. N&S	3E-20S	89
Torrence Av. N&S	2E-22S,2E-20S	89
Violet Av. N&S	2E-21S	89
Violet Rd. N&S	2E-21S	89
Vollbrecht Rd. N&S	1E-20S	89
Walter St. N&S	4E-22S,4E-20S	89
Ward St. N&S	3E-21S	89
Washington St. E&W	4E-21S	89
Wentworth Av. N&S	4E-22S,4E-20S	89
West Lake Dr. E&W	3E-20S	89
West Louise Dr. N&S	4E-22S	89
West St. NW&SE	4E-22S	89
Wildwood Av. N&S	3E-22S,3E-21S	89
William St. N&S	4E-22S,4E-20S	89
Williams St. N&S	4E-22S,40E-20S	89
Willow Ln. N&S	4E-21S	89
Wright St. N&S	4E-20S	89
Yates Av. N&S	2E-20S	89
169th St. E&W	4E-19S	83
170th Pl. E&W	4E-20S	89
170th St. E&W	3E-19S,4E-20S	83
171st St. E&W	3E-20S,4E-20S	89
172nd St. E&W	2E-20S,4E-20S	89
173rd Pl. E&W	3E-20S,4E-20S	89
173rd St. E&W	2E-20S,4E-20S	89
174th Pl. E&W	2E-20S	89
175th Pl. E&W	3E-20S	89
175th St. E&W	2E-20S,4E-20S	89
176th Av. E&W	3E-20S	89
176th Pl. E&W	4E-20S	89
176th St. E&W	2E-20S,4E-20S	89
177th Pl. E&W	2E-20S,4E-20S	89
177th St. E&W	2E-20S,4E-20S	89
178th St. E&W	3E-20S,4E-20S	89
179th St. E&W	2E-21S,3E-21S	89
180th St. E&W	4E-21S	89
181st Pl. E&W	4E-21S	89
181st St. E&W	4E-21S	89
182nd Pl. E&W	2E-21S,3E-21S	89
182nd St. E&W	3E-21S	89
183rd Pl. E&W	2E-21S,3E-21S	89
183rd St. E&W	2E-21S,3E-21S	89
184th Pl. E&W	2E-21S	89
184th St. E&W	2E-21S,3E-21S	89
185th Pl. E&W	3E-21S	89
185th St. E&W	2E-21S,3E-21S	89
186th Pl. E&W	3E-22S	89
186th St. E&W	2E-21S,4E-21S	89
187th Pl. E&W	3E-22S	89
187th St. E&W	3E-22S,4E-22S	89
188th Pl. E&W	3E-22S	89
188th St. E&W	3E-20S,4E-22S	89
189th Pl. E&W	3E-22S	89
189th St. E&W	3E-22S	89
190th Pl. E&W	3E-22S,4E-22S	89
190th St. E&W	3E-22S	89
191st Pl. E&W	3E-22S,4E-22S	89
191st St. E&W	3E-22S,4E-22S	89
192nd Pl. E&W	3E-22S,4E-22S	89
192nd St. E&W	3E-22S,4E-22S	89
193rd Pl. E&W	3E-22S	89
193rd St. E&W	3E-22S,4E-22S	89
194th Pl. E&W	3E-23S	89
194th St. E&W	3E-22S	89
195th Ct. E&W	3E-23S	89

Lemont

Acorn St. N&S	19W-15S	70
Alpine Av. N&S	19W-14S	70
Aspen Ct. NE&SW	20W-14S	70
Brady St. NW&SE	19W-14S	70
Brown Dr. NE&SW	20W-14S	70
Canal Av. N&S	19W-14S	70
Cass St. NE&SW	19W-13S	70
Catherine St. NW&SE	19W-13S	70
Cedar St. N&S	20W-14S	70
Cherokee Pl. N&S	19W-14S	70
Chestnut Dr. N&S	19W-14S	70
Connacht St. N&S	19W-15S	70
Country Ct. E&W	18W-14S	70
Country Ln. E&W	18W-14S	70
Crestview Dr. N&S	19W-14S	70
Crestview Rd. NW&SE	19W-14S	70
Custer St. E&W	19W-14S	70
Czacki St. N&S	19W-14S,19W-13S	70
Deer Ln. E&W	19W-15S	70
Derby Rd. N&S	19W-13S,18W-13S	70
Division St. E&W	19W-14S	70
Eaglecrest N&S	19W-15S	70
East St. N&S	19W-14S	70
Edgewood Ct. N&S	19W-14S	70
Emil St. E&W	19W-13S	70
Eureka Av. E&W	19W-14S	70
Eureka Dr. N&S	19W-14S	70
Evergreen Dr. NW&SE	20W-14S	70
Evergreen Pl. NW&SE	20W-14S	70
Florence St. N&S	19W-14S	70
Forest Ln. N&S	19W-14S	70
Freehauf St. E&W	19W-14S	70
Fremont St. NW&SE	19W-14S	70
Gillian St. N&S	19W-14S	70
Glenview Ln. E&W	19W-14S	70
Glenys Dr. NS&EW	19W-14S	70
Grant St. N&S	19W-13S	70
Grove St. N&S	19W-13S	70
Harasek St. N&S	18W-14S	70
Hermes Av. N&S	19W-14S	70
Hickory St. N&S	19W-14S	70
Hilltop Ct. NE&SW	18W-13S	70
Hilltop Dr. N&S	18W-13S	70
Hillview Ct. NE&SW	19W-14S	70
Hillview Dr. N&S	19W-14S	70
Holmes St. NW&SE	19W-14S	70
Houston St. N&S	19W-14S	70
Illinois St. NE&SW	19W-13S	70
Jacqueline Ct. N&S	19W-14S	70
Jands Ln. N&S	18W-14S	70
Jane Av. N&S	18W-15S	70
John St. N&S	19W-14S	70
Joliet St. NW&SE	19W-13S	70
Julia St. N&S	19W-13S	70
Keepotaw Ct. E&W	19W-14S	70
Keepotaw Dr. E&W	19W-14S	70
Keepotaw Ln. E&W	19W-14S	70
Kettering Ln. E&W	19W-14S	70
Kim Pl. N&S	19W-14S	70
Kimball St. NE&SW	19W-13S	70
Kip Pl. N&S	19W-13S	70
Kromray Rd. N&S	18W-14S	70
Kruz St. NS&EW	18W-14S	70
Ledochowski St. N&S	19W-14S	70
Lemont St. NW&SE	19W-13S	70
Lenox Ct. E&W	18W-13S	70
Lenox St. E&W	18W-13S	70
Lincoln NE&SW	19W-13S	70
Lindsay Ct. E&W	18W-13S	70
Lintz St. E&W	19W-14S	70
Lockport St. E&W	19W-13S	70
Logan St. N&S	20W-14S,19W-14S	70
Losey St. E&W	19W-14S	70
Luna Av. E&W	19W-14S	70
Luxemburg Av. N&S	19W-14S	70
Main St. E&W	19W-14S	70
Main St. N&S	19W-14S	70
Margret St. N&S	19W-14S	70
Mayfair St. NW&SE	18W-13S	70
McCarthy St. N&S	19W-14S,19W-13S	70
Meagan Ln. NS&EW	19W-14S	70
Cir.	18W-14S	70
Moczygemba St. N&S	19W-14S	70
Norton Av. N&S	19W-14S	70
Norton Dr. NW&SE	19W-13S	70
Oak Ct. NE&SW	19W-14S	70
Oak Ln. N&S	19W-14S	70
Para Maria St. N&S	18W-14S	70
Park Pl. N&S	19W-14S	70
Peiffer Av. E&W	20W-14S,19W-14S	70
Porter St. N&S	19W-13S	70
Povalish Ct. E&W	20W-13S	70
Pruyne St. NW&SE	19W-13S	70
Przewolki St. N&S	18W-14S	70
Pulawski St. N&S	19W-14S	70
Ridge Rd. NS&EW	19W-13S	70
River St. NE&SW	19W-13S	70
Roberta St. E&W	19W-15S,19W-14S	70
Rosehill Dr. E&W	19W-14S	70
S. Canal Av. NE&SW	19W-13S	70
Sara Av. E&W	19W-14S	70
Schultz St. N&S	18W-14S	70
Short St. E&W	19W-14S	70
Singer Av. N&S	19W-14S	70
Sobieski St. N&S	19W-14S	70
Spruce Ct. E&W	20W-14S	70
State St. N&S	19W-14S	70
Stephen St. NW&SE	19W-13S	70
Talcott St. NE&SW	19W-13S	70
Timberline Ct. NW&SE	20W-14S	70
Timberline Dr. N&S	20W-14S	70
Timberline Pl. E&W	20W-14S	70
Tomasezwski St. N&S	19W-14S	70
Una Dr. E&W	19W-4S	70
Walnut St. N&S	19W-14S	70
Walter St. N&S	19W-14S	70
Warner Av. N&S	19W-14S	70
Weimer Av. E&W	19W-14S	70
Wend St. E&W	19W-15S,19W-14S	70
Wexford Ct. E&W	18W-13S	70
Wexford St. E&W	19W-14S	70
Wheeler Ct. E&W	18W-13S	70
Wheeler Dr. NW&SE	18W-13S	70
Woodcrest Ln. N&S	18W-14S	70
Young St. N&S	18W-14S	70
2nd St. N&S	18W-14S	70
4th St. N&S	18W-14S	70
6th St. N&S	18W-14S	70
127th St. E&W	19W-14S	70

Lemont Township

Angelico E&W	18W-15S	70
Apple Av. N&S	17W-14S	71
Archer Av. NE&SW	18W-15S,19W-11S	70
Arctic Ct. N&S	16W-15S	71
Arctic Ln. N&S	16W-15S	71
Artesian St. E&W	16W-13S	71
Baltic Cir. N&S	17W-15S	71
Bell Oak Ln. N&S	15W-14S	71
Bell Rd. N&S	16W-15S,16W-14S	71
Birch Ct. NW&SE	17W-15S	71
Black Fox Ln. N&S	17W-15S	71
Bluff Rd. E&W	20W-13S	70
Boyer St. N&S	17W-15S	71
Briar Patch Ln. E&W	16W-15S	71
Bridle Path NE&SW	15W-13S	71
Brookside Dr. N&S	15W-13S	71
Buckskin NW&SE	15W-13S	71
Campbell St. E&W	16W-13S	71
Carriage Ln. E&W	15W-14S	71
Cedar Rd. N&S	18W-14S	70
Center Dr. N&S	18W-14S	70
Chestnut N&S	15W-14S	71
Clear View E&W	15W-13S	71
Connacht St. E&W	15W-14S	70
Crestwood Ln. E&W	18W-15S	70
DaVinci N&S	18W-15S	70
David Ln. N&S	18W-15S	70
Derby Rd. N&S	17W-15S,17W-14S	71
Dineff Rd. E&W	18W-15S	71
Dixon Way Dr. E&W	17W-15S	71
El Grecco E&W	18W-15S	70
Elm Ct. NE&SW	17W-15S	71
Emerald Dr. N&S	19W-15S	70
Emily Ln. NS&EW	19W-15S	70
Equestrian Way N&S	15W-14S	71
Fairway Dr. N&S	20W-15S	70
Fox Burrow Ln. E&W	17W-15S	71
Fox Hollow E&W	18W-15S	70
Fox Ln. N&S	18W-15S	70
Fox Tail Ln. N&S	16W-15S	71
Foxhill Ct. E&W	16W-15S	71
Foxhill Dr. N&S	16W-15S	71
Fr. Linkus Dr. (Pvt) N&S	15W-14S	71
Franciscan Dr. E&W	18W-13S	70
Galway Ct. NE&SW	16W-14S	71
Galway Ln. E&W	16W-14S	71
Galway Rd. N&S	16W-14S	71
Golden Spur NE&SW	15W-13S	71
Gordon Ln. N&S	20W-15S	70
Gray Ln. N&S	18W-15S	71
Hawthorne Dr. NS&EW	17W-15S	71
Hickory Av. E&W	17W-14S	70
Hillcrest Dr. N&S	18W-14S,18W-13S	71
Hillcrest Rd. E&W	17W-13S	71
Horseshoe Ln. NE&SW	15W-13S	71
Hunt Master Ln. E&W	16W-15S	71
Huntsman N&S	15W-13S	71
Katie Rd. N&S	19W-15S	70
King Rd. N&S	18W-15S	70
Kit Ln. E&W	17W-15S	71
Klotz Av. N&S	17W-14S	71
Lakeview NW&SE	15W-13S	71
Lariat Dr. N&S	15W-13S	71
Leinster Dr. NS&EW	19W-13S	70
Lemont Rd. NW&SE	19W-13S	70
Lemont St. N&S	19W-15S	70
Liberty Dr. E&W	16W-13S	71
Longwood Way N&S	20W-15S	70
Main St. E&W	16W-12S	71
Maley Rd. N&S	16W-12S	71
Maple St. N&S	18W-15S	70
Martin Ct. N&S	15W-13S	71
McCarthy Rd. N&S	17W-14S,15W-14S	71
Meadow Hill Ln. NS&EW	16W-15S	71
Melissa Dr. E&W	19W-15S	70
Moorefield Ct. E&W	16W-15S	71
Munster Rd. N&S	19W-15S	70
N. Red Coat Dr. E&W	16W-15S	71
Nordic Ln. N&S	17W-15S	71
North Trail N&S	15W-13S	71
Oak Av. E&W	17W-14S	71
Oak Ct. NE&SW	17W-15S	71
Oak Tree Ct. NE&SW	17W-14S	71
Oak Tree Ln. NS&EW	17W-14S	71
Oakmont Ct. NE&SW	20W-15S	70
Oakmont Dr. N&S	20W-15S	70
Paddock N&S	15W-13S	71
Parker Rd. N&S	17W-15S,17W-13S	71
Paso Fino N&S	15W-13S	71
Pine Av. N&S	17W-14S	71
Prairie Av. N&S	17W-14S	71
Raphael N&S	18W-15S	70
Red Dr. NS&EW	17W-15S	71
Red Fox N&S	16W-15S	71
Riva Ridge NW&SE	15W-13S	71
Roy Ln. N&S	18W-15S	70
Rusty Rd. NS&EW	17W-15S	71
S. Red Coat Dr. E&W	16W-15S	71

Name	Grid	Page
Saddlebrook Ln. N&S	16W-14S	71
Shannon Ct. E&W	19W-15S	70
Silver Spur E&W	15W-13S	71
Silverfox Dr. NS&EW	17W-15S,16W-15S	71
Smitty Av. NW&SE	17W-14S	71
Sorrel Ct. E&W	15W-13S	71
Split Rail NW&SE	15W-13S	71
Spruce Ct. NW&SE	17W-15S	71
St. Anne Ct.* N&S	17W-15S	70
St. Anthony Ct.* E&W	17W-13S	70
St. Bonaventure Ct.* E&W	17W-13S	70
St. Colette Ct.* N&S	17W-13S	70
St. Joseph Ct.* E&W	17W-13S	70
State St. N&S	19W-15S	70
Stone Creek Dr. N&S	15W-13S	71
Sunhill Ln. E&W	16W-14S	71
Surrey E&W	15W-13S	71
Sylvan Woods Dr. NW&SE	17W-15S	71
Timber View N&S	15W-13S	71
Valley View Ct. N&S	20W-15S	70
Valley View Dr. E&W	20W-15S	70
Village Dr. N&S	18W-13S	70
W. 113th St. E&W	15W-13S	71
W. 115th St. E&W	15W-13S	71
W. Red Coat Dr. N&S	16W-15S	71
Waitkus Dr. E&W	15W-13S	71
Walker Rd. N&S	18W-14S,18W-13S	70
Walnut St. N&S	19W-15S	70
Wild Oak Ct. E&W	18W-14S	70
Wildplum Ct. E&W	18W-14S	70
Wildwood Dr. NW&SE	18W-14S	70
Will Cook Rd. N&S	15W-15S,15W-13S	71
Willow Av. N&S	18W-15S	70
Woodcrest Ct. E&W	18W-14S	70
Woodcrest Ln. N&S	18W-14S	70
Woodcrest Ln. E&W	18W-14S,18W-13S	70
Woodland Dr. E&W	18W-15S	70
Woodlawn Av. N&S	16W-13S	71
Woodview Ln. N&S	15W-13S	71
107th St. E&W	15W-12S	71
111th St. E&W	15W-13S	71
117th Pl. E&W	18W-13S	70
117th St. E&W	18W-13S,17W-13S	70
117th St. E&W	17W-13S	71
118th St. E&W	17W-13S	70
119th St. E&W	18W-13S,15W-13S	70
121st St. E&W	18W-14S	70
127th St. E&W	20W-14S,18W-14S	70
127th St. E&W	17W-14S	71
129th St. E&W	19W-15S	70
130th St. E&W	19W-15S	70
131st Ct. E&W	19W-15S	70
131st St. E&W	18W-15S,15W-15S	70
132nd St. E&W	19W-15S	70
133rd Ct. E&W	19W-15S	70
133rd St. E&W	18W-15S	71

Leyden Township

Name	Grid	Page
Alcoa N&S	13W-3N	40
Alta N&S	13W-3N	40
Altgeld E&W	13W-3N	40
Armitage E&W	12W-2N	46
Atlantic N&S	12W-2N	46
Bank Dr. N&S	12W-2N	46
Barry E&W	14W-3N	40
Behrns Dr. E&W	14W-3N	40
Belden E&W	12W-2N	46
Bellwood N&S	13W-3N	40
Calwagner N&S	12W-2N	46
Castello E&W	12W-2N	46
Charles N&S	13W-3N	40
Derrough N&S	14W-3N	40
Dickens E&W	12W-2N	46
Diversey E&W	13W-3N	40
Dora N&S	12W-2N	46
Drummond Pl. E&W	13W-3N	40
Emerson N&S	12W-2N	46
Fairfield N&S	12W-2N	46
Fullerton E&W	13W-3N	40
Geneva N&S	13W-3N	40
Granville N&S	13W-3N	40
Gustave N&S	12W-2N	46
Haber N&S	13W-3N	40
Harold N&S	13W-3N	40
Hawthorne N&S	12W-2N	46
Hyde Park N&S	13W-3N	40
Inland N&S	14W-3N	40
Joyce N&S	13W-3N	40
Landen N&S	13W-3N	40
Laporte N&S	13W-3N	40
Lee N&S	13W-3N	40
Louis N&S	12W-2N,12W-3N	46

Name	Grid	Page
Lyndale E&W	13W-2N,12W-2N	46
Mac Arthur Dr. E&W	13W-3N	40
Marion N&S	14W-3N	40
Martin N&S	14W-3N	40
McDonough Rd. N&S	14W-5N	40
McKay E&W	14W-3N	40
McLean N&S	12W-2N	46
Medill E&W	12W-2N	46
Melrose N&S	13W-3N	40
Montana E&W	13W-3N	40
Nevada E&W	13W-3N	40
Palmer E&W	13W-2N,12W-2N	46
Parker E&W	13W-3N,13W-2N	40
Pearl N&S	14W-3N	40
Prater N&S	14W-3N	40
Rhodes N&S	14W-3N	40
Rowlet N&S	13W-3N	40
Roy N&S	13W-3N	40
Ruby N&S	12W-2N	46
Sandra N&S	13W-3N	40
Schubert E&W	13W-3N	40
Scott N&S	12W-2N,12W-3N	46
Wellington E&W	13W-3N	40
Winters Dr. E&W	13W-3N	40
Wolf N&S	14W-3N	40
Wrightwood E&W	13W-3N	40

Lincolnwood

Name	Grid	Page
Albion Av. E&W	6W-8N,4W-8N	38
Arthur Av. E&W	5W-8N,4W-8N	38
Avers Av., North N&S	4W-8N	38
Carpenter Rd. N&S	6W-8N	38
Central Park Av. N&S	4W-8N	38
Chase Av. E&W	5W-8N,4W-9N	38
Cherry Av. N&S	4W-8N	38
Christiana Av. N&S	4W-8N	38
Cicero Av. N&S	5W-8N	38
Columbia Av. E&W	5W-8N,4W-8N	38
Crawford Av. N&S	5W-8N	38
Devon Av. E&W	4W-8N	38
Dowagiac Av. NW&SE	6W-8N	38
Drake Av. N&S	4W-8N	38
Estes Av. E&W	5W-8N	38
Farwell Av. E&W	4W-8N	38
Fitch Av. E&W	5W-8N,4W-8N	38
Greenleaf Av. E&W	5W-8N,4W-8N	38
Hamlin Av., North N&S	4W-8N	38
Harding Av., North N&S	4W-8N	38
Interstate 94 N&S	6W-8N	38
Ionia Av. NW&SE	6W-8N	38
Jarlath Av. E&W	4W-9N	38
Karlov Av., North N&S	5W-8N	38
Keating Av. N&S	6W-8N,5W-8N	38
Kedvale Av., North N&S	5W-8N	38
Keeler Av., North N&S	5W-8N	38
Kenneth Av. N&S	5W-8N	38
Kenton Av. N&S	5W-8N	38
Keystone Av., North N&S	5W-8N	38
Kilbourn Av., North N&S	5W-8N	38
Kildare, Av., North N&S	5W-8N	38
Kilpatrick Av. N&S	5W-8N	38
Kimball Av. E&W	4W-8N	38
Knox Av. N&S	5W-8N	38
Kolmar Av. N&S	5W-8N	38
Kostner Av., North N&S	5W-8N	38
LaPorte Av. N&S	6W-8N	38
Lawndale Av. North N&S	4W-8N	38
Legett NE&SW	6W-8N	38
Lemai Av. N&S	6W-8N	38
Leroy Av. N&S	6W-8N	38
Lincoln Av. NW&SE	5W-8N	38
Longmeadow Av. N&S	6W-8N	38
Lowell Av. N&S	5W-8N	38
Loyola Av. E&W	4W-8N	38
Lunt Av. E&W	5W-8N	38
McCormick Blvd. N&S	4W-8N	38
Minnehaha Av. NE&SW	6W-8N	38
Monticello Av. N&S	4W-8N	38
Morse Av. E&W	5W-8N,4W-8N	38
Navajo Av. NE&SW	6W-8N	38
Nokomis Av. NE&SW	6W-8N	38
North Shore Av. E&W	6W-8N,4W-8N	38
Prairie Rd., East N&S	4W-8N	38
Pratt Av. E&W	4W-8N	38
Private Rd. NE&SW	4W-8N	38
Proesel Av., North NE&SW		
Ramona Av. NE&SW	6W-8N	38
Rance Ter. E&W	4W-8N	38

Name	Grid	Page
Ridgeway Av. N&S	4W-8N	38
Rt. 50 N&S	5W-8N	38
Sauganash Av. NW&SE	6W-8N	38
Schreiber Av. E&W	4W-8N	38
Sherwin Av. E&W	4W-9N	38
Spaulding Av. N&S	4W-8N	38
Spokane Av. NE&SW	6W-8N	38
Springfield Av. N&S	4W-8N	38
St. Louis Av. N&S	4W-8N	38
Touhy Av. E&W	5W-9N	38
Tower Cir., West N&S	6W-8N	38
Tower Ct. NE&SW	6W-8N	38
Tripp Av., North N&S	5W-8N	38
Trumbull Av. N&S	4W-8N	38
U.S. Route 41 SE&NW	5W-8N,4W-8N	38
Wallen Av. E&W	4W-8N	38

Lynwood

Name	Grid	Page
Alanna Ln. N&S	4E-25S	95
Arbon Ct.* NW&SE	3E-23S	95
Arroyo Av. N&S	3E-23S	89
Bensley Av. N&S	2E-24S	95
Berrine Ct.* N&S	3E-25S	95
Bozard Ct.* E&W	3E-25S	95
Bregenz Ct.* E&W	3E-25S	95
Brenner Ct.* E&W	3E-25S	95
Brentwood Dr. E&W	4E-24S	95
Brentz Ct.* E&W	3E-25S	95
Brook Av. N&S	3E-23S	89
Burnham Av. N&S	3E-24S,3E-23S	95
Catalpa Av. N&S	3E-23S	89
Chilton Dr.* NS&EW	3E-25S	95
Crescent Av. N&S	3E-23S	89
Cypress Av. N&S	3E-23S	89
Daron Dr. NS&EW	4E-25S	95
Deborah Ct. NE&SW	4E-25S	95
Dewey Av. N&S	2E-23S	89
Dolphin Av. N&S	3E-23S	89
Driftwood Av. N&S	3E-25S	95
Duke Dr.* N&S	3E-25S	95
Geneva Dr. N&S	3E-23S	89
Glarus Ln.* NW&SE	3E-25S	95
Glenwood Dyer Rd. NW&SE	2E-23S,3E-24S	89
Joe Orr Rd. E&W	3E-24S	95
Juniper Av. N&S	3E-23S	89
Kruse Rd. N&S	4E-24S	95
Lake Lynwood Dr. N&S	3E-23S	89
Lake Park Ct. NE&SW	3E-23S	89
Lake Park Dr. NS&EW	3E-23S	89
Lake Shore Dr. N&S	3E-23S	89
Lakewood Av. N&S	3E-24S	95
Laurelwood Cir. N&S	4E-24S	95
Lighthouse Ct. NE&SW	3E-23S	89
Linda Dr. E&W	4E-25S	95
Linda Ln. N&S	4E-25S	95
Lisa Ln. NS&EW	4E-25S	95
Mariner Ct. E&W	3E-23S	89
Marlin Ct. N&S	3E-23S	89
Marquette Av. N&S	3E-24S	95
Maureen Ct. NE&SW	4E-25S	95
Monterey Av. N&S	3E-23S	89
Muskegon Av. N&S	3E-24S	95
Oakwood Av. N&S	3E-23S	89
Orchard Av. N&S	3E-23S	89
Park Av. N&S	3E-23S	89
Pattie Ln. NW&SE	4E-25S	95
Redwood Av. N&S	3E-23S	89
Redwood Cir. E&W	3E-23S	89
Redwood Ct. N&S	4E-25S	95
Rose St. N&S	2E-23S	89
Rt. 83 SE&NW	2E-23S,4E-25S	89
Sandridge Ct. N&S	2E-24S	95
Sandridge Rd. N&S	2E-24S	95
Sequoia Av. N&S	3E-25S	95
Cir.	3E-25S	95
Stony Island Av. N&S	1E-24S	95
Surf Ct. E&W	3E-23S	89
Terrace Av. NS&EW	3E-23S	89
Todd Terr. N&S	4E-25S	95
Torrence Av. N&S	3E-23S	89
Vale Ct.* N&S	3E-25S	95
Vollmer St. E&W	1E-24S	95
Westwood Ln. E&W	4E-24S	95
Willow Dr. N&S	3E-23S	89
Wind Point Ct. E&W	3E-23S	89
Zurich Dr. N&S	3E-23S	89
195th Pl. E&W	3E-23S	89
195th St. E&W	3E-23S	89
196th St. E&W	3E-23S	89
197th St. E&W	3E-23S	89
198th Pl. E&W	2E-23S	89
198th St. E&W	2E-23S,3E-23S	89
199th Pl. E&W	2E-23S	89
200th Pl. E&W	2E-23S	89
200th St. E&W	3E-23S,4E-23S	89
201st Pl. E&W	3E-23S	89
201st St. E&W	3E-23S,4E-24S	89
207th Pl. E&W	2E-24S	95
207th St. E&W	2E-24S	95
208th St. E&W	4E-25S	95

Lyons

Name	Grid	Page
Amelia N&S	9W-4S	54
Anna N&S	10W-4S	53
Cech N&S	10W-4S	53
Center N&S	9W-4S	54
Clyde N&S	10W-4S	53
Collins E&W	9W-4S	54
Columbus N&S	10W-3S	53
Cracow N&S	10W-4S	53
Egandale N&S	9W-4S	53
Elm N&S	9W-4S	54
Fern N&S	10W-4S	53
Fishermans N&S	10W-4S	53
Gage N&S	9W-4S	54
Harlem N&S	10W-4S	54
Hawthorne N&S	10W-4S	53
Joliet E&W	9W-4S	54
Joliet Av. N&S	10W-4S	53
Joliet Rd. N&S	9W-4S	54
Kenwood E&W	10W-4S	53
Konrad N&S	10W-4S	53
Lafayette E&W	9W-3S	54
Leland E&W	10W-4S	53
Lincoln E&W	10W-4S	53
Maple N&S	9W-4S	54
Oak N&S	10W-4S	54
Oakwind E&W	9W-4S	54
Patricia E&W	10W-4S	53
Powell N&S	9W-4S	54
Prescott N&S	9W-4S	54
Pulaski N&S	10W-4S	53
Rose N&S	10W-4S	53
Salisbury St. E&W	10W-4S	53
U.S. Rt. 34 E&W	10W-4S	53
Warsaw N&S	10W-4S	53
White N&S	9W-4S	54
Winchester N&S	10W-4S	53
38th Pl. E&W	9W-4S	54
39th E&W	9W-3S	54
40th E&W	10W-4S,9W-4S	53
41st Ct. E&W	9W-4S	54
41st Pl. E&W	9W-4S	54
42nd E&W	10W-4S	53
42nd Ct. E&W	10W-4S	53
43rd E&W	10W-4S	53
43rd Pl. E&W	10W-4S	53
44th E&W	10W-4S,9W-4S	53
44th Pl. E&W	10W-4S	53
45th E&W	10W-4S	53
45th Pl. E&W	10W-4S	53
46th E&W	10W-4S	53
47th E&W	10W-4S	53

Lyons Township

Name	Grid	Page
Bielby Av. N&S	13W-9S	64
Blackstone N&S	13W-6S	58
Burr Oak N&S	12W-8S	58
Coronet E&W	13W-6S	58
Deerpath Tr. NS&EW	14W-9S	64
Degener E&W	14W-1S	52
Dickens E&W	14W-1S	52
Dory N&S	14W-9S	64
Edgewood N&S	13W-6S	58
Edgewood View N&S	13W-8S	58
Franklin E&W	13W-6S	58
Hamilton N&S	14W-1S	52
Howard N&S	13W-6S	58
Howard Av. N&S	13W-9S	64
Interstate 294 N&S	14W-1S	52
Jann E&W	13W-6S	58
Laurel N&S	13W-6S	58
Lincoln E&W	13W-6S	58
Linden N&S	13W-6S	58
Linn E&W	13W-6S	58
Locust E&W	13W-6S	58
Lorraine N&S	13W-8S	58
Marian N&S	14W-1S	52
Maridon N&S	12W-8S	58
Mt. Lored Ct. E&W	14W-9S	64
Northeast Ct. E&W	13W-5S	58
Northwest Ct. E&W	13W-5S	58
Oak Knoll NE&SW	14W-9S	64
Parkside E&W	13W-6S	58
Peck N&S	13W-6S	58
Sanders N&S	13W-6S	58
Southwest Ct. E&W	13W-5S	58
Sunset N&S	13W-6S	58
Tri-State Tollway N&S	14W-1S	52
Tri-State Tollway NW&SE	12W-8S	58
Turner N&S	14W-1S	52
Vial N&S	13W-6S	58
Willow Springs N&S	13W-6S	58
53rd Pl. E&W	13W-5S	58
54th E&W	13W-5S	58
54th Pl. E&W	13W-5S	58
55th E&W	13W-6S	58
56th E&W	13W-6S	58
57th E&W	13W-6S	58
58th E&W	13W-6S	58
58th Pl. E&W	13W-6S	58
59th E&W	13W-6S	58
59th Pl. E&W	13W-6S	58
60th E&W	13W-6S	58
61st E&W	13W-6S	58
61st Pl. E&W	13W-6S	58
62nd E&W	13W-6S	58
62nd Pl. E&W	13W-6S	58
63rd Pl. E&W	13W-7S	58
64th E&W	13W-7S	58
74th E&W	13W-8S	58
79th E&W	13W-8S	58

Name	Grid	Page
79th St. E&W	13W-8S,12W-8S	58
80th Pl. E&W	13W-9S	64
87th St. E&W	14W-9S	64

Maine Township

Name	Grid	Page
Acres Ln. N&S	12W-10N	36
Alexis Ct. N&S	9W-12N	27
Allison Ct. E&W	9W-12N	27
Apple Creek Ln. N&S	12W-10N	36
Ashley Ct. E&W	10W-12N	27
Aspen Dr. N&S	11W-11N	26
Aspen Ln. N&S	11W-11N	26
Barberry Ln. E&W	11W-11N	26
Bay Colony E&W	11W-11N	26
Bianco Ter. N&S	11W-11N	26
Bobbie E&W	11W-11N	26
Burton Ln. E&W	12W-10N	36
Busse Hwy. NW&SE	12W-10N	36
Canavan Rd. N&S	10W-12N	27
Cedar Ln. N&S	11W-11N	26
Cherry Cir. E&W	11W-12N,10W-12N	26
Cherry Ct. NE&SW	10W-12N	27
Cherry Ln. NW&SE	10W-12N	27
Clancy Dr. N&S	11W-11N	26
Crawford Ln. N&S	11W-12N	26
Crescent Dr. N&S	10W-12N	27
Culver St. E&W	11W-11N	26
David Pl. N&S	11W-12N	26
Decook Ct., North E&W	11W-10N	36
Decook Ct., South E&W	11W-10N	36
Decook Ln. E&W	11W-10N	36
Dee Rd. N&S	11W-10N,11W-12N	26
Delphia Av. N&S	10W-11N	27
Diane Rd. N&S	16W-10N	34
Donald Ct. E&W	9W-12N	27
Donald Ter. N&S	9W-12N	27
East River Rd. N&S	12W-10N	26
Elder Ct. N&S	10W-12N	27
Elder Ln. N&S	10W-12N	27
Elm Dr. E&W	10W-11N	27
Elm Rd. N&S	15W-10N	34
Emerson St. E&W	11W-11N	26
Fairlawn Dr. E&W	10W-12N	27
Farwell Av. E&W	13W-8N	35
Fenton Ln. E&W	11W-10N	36
Fern Ln. N&S	11W-11N	26
Fernwood Dr. NS&EW	9W-12N	27
Forest Pl. E&W	11W-11N	26
Gayle Ct. E&W	10W-12N	27
Greenwood Dr., East N&S	11W-11N	27
Greenwood Dr., West N&S	10W-11N	27
Greenwood Rd. N&S	10W-11N	27
Gregory Ln. NW&SE	10W-12N	27
Hamilton Dr. E&W	11W-12N	26
Hamlin Ln. N&S	11W-11N	26
Harrison St. E&W	11W-12N	26
Hazelwood Dr. E&W	10W-12N	27
Helen Ct. N&S	10W-12N	27
Helen Dr. E&W	10W-12N	27
Hollyberry Av. E&W	11W-11N	26
Home Av. N&S	11W-11N	26
Home Cir. E&W	11W-11N	26
Home Ct. E&W	11W-11N	26
Home Ter. N&S	11W-11N	26
Ironwood Ln. N&S	11W-11N	26
James Ct. N&S	9W-12N	27
James St. N&S	9W-12N	27
Josephine St. E&W	10W-12N	27
Judy Dr. N&S	11W-12N	26
Julie E&W	9W-12N	27
Kathy Ct. N&S	11W-11N	26
Kathy Ln. E&W	11W-11N	26
Kenneth Dr. E&W	11W-11N	26
Knight Av. N&S	10W-11N	27
Lee Ln. E&W	16W-9N	34
Leslie Ln. N&S	11W-12N	26
Lincoln Av. N&S	10W-11N	27
Linda Ct. E&W	9W-12N	27
Linnea N&S	11W-12N	26
Lois Dr. N&S	10W-12N	27
Lunt Av. E&W	13W-8N	35
Lyman Av. N&S	11W-12N	26
Lyons St. E&W	11W-11N,10W-10N	26
Manor Ln. N&S	11W-11N	26
Maple Ln. N&S	11W-11N	26
Maple Rd. N&S	15W-8N	34
Marcus Ct., East N&S	11W-10N	36
Marcus Ct., West N&S	11W-10N	36
Margail St. N&S	11W-11N	26
Margaret St. N&S	13W-8N	35
Maynard Ter. E&W	10W-12N	27
Meadow Ln. N&S	11W-11N,11W-12N	26
Memory Ln. E&W	10W-12N	27
Michael Manor N&S	10W-12N	27

	Grid	Page
Morse Av. E&W	13W-8N	35
Nellie Ct. N&S	9W-12N	27
Noel Av. E&W	11W-11N	26
Norma Ct. E&W	10W-12N	27
North Shore Dr. E&W	11W-12N	26
Oak Ln. N&S	11W-12N	26
Oak Pl. E&W	11W-12N	26
Oaks Av. E&W	11W-11N	26
Park Ln. N&S	11W-11N	26
Parkside Dr. N&S	11W-11N	26
Pauline Av. E&W	10W-12N	27
Poplar N&S	10W-12N	27
Potter Rd. N&S	11W-12N	26
Pratt Av. E&W	13W-8N	35
Rancho Ln. N&S	11W-11N	26
Redling Cir. NS&EW	11W-12N	26
Robin Dr. N&S	11W-11N	26
Ronald Rd. NS&EW	10W-12N	27
Roppollo Dr. N&S	16W-9N	34
Sanders Dr. E&W	11W-11N	26
Sell Rd. E&W	15W-8N	34
Sherry Ln. E&W	11W-11N	26
Stacy Ct. N&S	9W-12N	27
Stacy St. N&S	9W-12N	27
Stevens Dr. N&S	11W-12N	26
Sumac Rd. N&S	11W-11N	26
Teela Ln. E&W	12W-12N	26
Terrace Dr. E&W	11W-12N	26
Terrace Pl. N&S	11W-11N	26
Thornberry E&W	10W-12N	27
Tri-state Tollway N&S	12W-9N	36
Twin Oak Ln. N&S	11W-11N	26
Tyrell Av. N&S	11W-10N	36
Valerie Ct. N&S	9W-12N	27
Victor Av. E&W	9W-12N	27
Western Av. N&S	10W-11N	27
William Av. E&W	9W-12N	27

Markham

	Grid	Page
Afton Dr. NW&SE	3W-18S	81
Albany Av. N&S	3W-19S,3W-17S	81
Alta Rd. N&S	4W-18S	81
Artesian Av. N&S	3W-19S	81
Arthur Terr. E&W	4W-19S	81
Ashland Av. N&S	2W-19S	81
Belleplaine Dr. NW&SE	4W-19S	81
Berkshire Dr. N&S	3W-19S	81
Bibb Ln. N&S	3W-18S	81
Birch Rd. N&S	4W-18S	81
Blackstone Av. E&W	4W-19S	81
California Av. N&S	3W-19S,3W-18S	81
Calitonia Dr. NW&SE	3W-18S	81
Cambridge Ct. N&S	3W-19S	81
Cambridge Dr. N&S	3W-19S	81
Campbell Av. N&S	3W-19S	81
Central Park Av. N&S	4W-19S,4W-18S	81
Cherry Ln. N&S	4W-18S	81
Cir.	3W-19S	81
Clifton Park Av. N&S	4W-19S,4W-18S	81
Country Aire Dr. E&W	4W-18S	81
Crawford Av. N&S	4W-19S	81
Crockett Ln. NW&SE	3W-18S	81
Cypress Rd. N&S	4W-18S	81
Dan Ryan Expwy. NE&SW	4W-18S,3W-18S	81
Dixie Hwy. N&S	2W-19S	81
Firestone Dr. N&S	3W-18S	81
Francisco Av. N&S	3W-19S,3W-18S	81
Grange Dr. NW&SE	3W-18S	81
Hamlin Av. N&S	4W-19S,4W-18S	81
Harding Av. N&S	4W-18S	81
Heather Dr. E&W	4W-18S	81
Hermitage Av. N&S	2W-19S	81
Hillcrest Dr. NW&SE	4W-19S	81
Homan Av. N&S	4W-19S,4W-17S	81
Honore St. N&S	2W-19S	81
Hoyne Av. N&S	2W-19S	81
Interstate 294 N&S	3W-18S	81
Interstate 57 NE&SW	4W-18S	81
Justine Av. N&S	1W-19S	82
Kedvale Av. N&S	5W-19S	80
Kedzie Av. N&S	4W-19S,4W-17S	81
Kenner Av. N&S	3W-18S	81
Laflin St. N&S	1W-19S	82
Lancaster Dr. E&W	3W-19S	81
Lathop Av. N&S	1W-19S	82
Lawndale Av. N&S	4W-19S,4W-18S	81
Leavitt Av. N&S	2W-19S	81
Lincoln Dr. NE&SW	3W-18S	81
Magnolia Dr. E&W	4W-18S	81
Maple Ln. NS&EW	4W-18S	81
Maplewood Av. N&S	3W-19S	81
Marshfield Av. N&S	2W-19S	81
Meade Ln. NE&SW	3W-18S	81
Millard Av. N&S	4W-18S	81
Mozart Av. N&S	3W-20S,3W-18S	87
Nottingham Av. E&W	3W-19S	81
Oakley Av. N&S	2W-19S	81
Oxford Dr., North NS&EW	3W-19S	81
Oxford Dr., South NS&EW	3W-19S	81
Park Av. NE&SW	1W-19S	82
Parkside Dr. NW&SE	3W-18S	81
Paulina Av. N&S	2W-19S	81
Plainview Dr. NW&SE	4W-19S	81
Plymouth Dr. N&S	3W-19S	81
Richmond Av. N&S	3W-20S,3W-18S	87
Ridgeway Av. N&S	4W-18S	81
Rockwell Av. N&S	3W-19S	81
Rockwell St. N&S	3W-18S	81
Roesner Dr. E&W	4W-18S	81
Rose Manor Terr. E&W	4W-19S	81
Sacramento Av. N&S	3W-18S,3W-17S	81
Sawyer Av. N&S	4W-19S,4W-17S	81
Sherwood Av. E&W	3W-19S	81
Spaulding Av. N&S	4W-19S,4W-17S	81
Springfield Av. N&S	4W-19S	81
St. Louis Av. N&S	4W-19S,4W-17S	81
Stafford Av. E&W	3W-19S	81
Sunset Av. E&W	4W-19S	81
Sussex Av. NS&EW	3W-19S	81
Sussex Ct. NE&SW	3W-19S	81
Tri-State Tollway N&S	3W-19S,3W-18S	81
Troy Av. N&S	3W-18S,3W-17S	81
Trumbull Av. N&S	4W-19S,4W-17S	81
Turner Av. N&S	4W-19S,4W-17S	81
Washington Dr. NE&SW	3W-18S	81
Washtenaw Av. N&S	3W-18S	81
Wassel Dr. NW&SE	3W-18S	81
Wedgewood Dr. NW&SE	4W-19S	81
Western Av. N&S	3W-19S	81
Whipple Av. N&S	3W-18S,3W-17S	81
Willow Ln. E&W	4W-18S	81
Wilshire Av. E&W	3W-19S	81
Winchester Av. N&S	2W-19S	81
Wolcott Av. N&S	2W-19S	81
Wood St. N&S	2W-19S	81
145th Ct. E&W	6W-17S	80
149th St. E&W	4W-17S	81
150th St. E&W	4W-17S	81
151st St. E&W	3W-17S	81
152nd St. E&W	4W-18S	81
153rd St. E&W	4W-18S	81
154th Pl. E&W	4W-18S	81
154th St. E&W	4W-18S	81
154th Ter. E&W	4W-18S	81
155th St. E&W	4W-18S,3W-18S	81
156th St. E&W	4W-18S,3W-18S	81
157th Pl. E&W	2W-18S	81
157th St. E&W	4W-18S,3W-18S	81
158th St. E&W	4W-18S,2W-18S	81
159th Pl. E&W	4W-19S	81
159th St. E&W	4W-18S,3W-18S	81
160th St. E&W	4W-19S	81
161st Pl. E&W	2W-19S	81
161st St. E&W	5W-19S,2W-19S	80
162nd Pl. E&W	4W-19S	81
162nd St. E&W	4W-19S,2W-19S	81
163rd Pl. E&W	3W-19S,2W-19S	81
163rd St. E&W	4W-19S,2W-19S	81
164th St. E&W	4W-19S,1W-19S	81
165th St. E&W	4W-19S,2W-19S	81
166th St. E&W	4W-19S,3W-19S	81
167th St. E&W	4W-19S,2W-19S	81
168th St. E&W	3W-20S	87

Matteson

	Grid	Page
Academy Av. N&S	5W-24S	92
Allemong Dr. E&W	7W-24S	91
Amherst Pl. E&W	7W-25S	91
Apple Hill Ln. E&W	7W-24S	91
Applewood Ln. E&W	5W-25S	92
Aspen Ln. E&W	7W-24S	91
Basswood Rd. E&W	6W-24S	91
Beaver Dam Rd. E&W	7W-24S	91
Beechwood Ct. NE&SW	6W-24S	91
Beechwood Rd. E&W	7W-24S,6W-24S	91
Birchwood Ln. E&W	6W-24S,5W-24S	91
Bradley Av. N&S	6W-25S	91
Briarwood Ct. NW&SE	7W-24S,6W-24S	91
Bridgewood Ct. E&W	6W-24S	91
Bridgewood Rd. N&S	6W-24S	91
Butterfield Pkwy. N&S	4W-25S	92
Cambridge Av. N&S	6W-25S	91
Campus Av. N&S	5W-25S	92
Carnation Ln. N&S	5W-25S	92
Cedarwood Ln. E&W	5W-25S	92
Central Av. N&S	7W-26S,6W-24S	91
Charles St. N&S	4W-26S	92
Charleston Rd. E&W	5W-26S	92
Churn Rd. NW&SE	7W-24S	91
Cicero Av. N&S	6W-26S,6W-24S	91
Cloverleaf Rd. N&S	7W-24S	91
Colgate Ln. E&W	7W-25S	91
College Av. N&S	5W-25S	92
Columbine Ln. NS&EW	5W-24S	92
Cornell Way E&W	7W-25S	91
Cornfield Rd. N&S	7W-24S	91
Corporate Lakes Dr. N&S	6W-25S,6W-24S	91
Crawford Av. N&S	5W-26S,5W-25S	92
Crestwood Ct. NW&SE	7W-24S	91
Crestwood Rd. NE&SW	7W-24S	91
Dartmouth Av. N&S	7W-25S	91
Deerpath Rd. N&S	7W-24S	91
Dettmering Dr. N&S	4W-25S	92
Drake Ln. E&W	6W-25S	91
Duke Dr. E&W	6W-25S	91
Edgewood Rd. N&S	7W-24S	91
Elm Ln. E&W	7W-24S	91
Elmwood Ln. E&W	5W-25S	92
Fernwood Ct. E&W	7W-24S	91
Fox Run Rd. E&W	7W-24S	91
Front St. E&W	4W-25S	92
Georgetown Dr. E&W	6W-25S	91
Goldenrod Cir. N&S	5W-24S	92
Hanson Ln. N&S	6W-25S	91
Harvard Ln. N&S	6W-25S	91
Highland Av. N&S	5W-25S	92
Highland Rd. N&S	7W-24S	91
Holiday Plaza Dr. NE&SW	6W-26S	91
Homeland Rd. N&S	5W-25S	92
Huntingwood Rd. N&S	7W-24S	91
Interstate 57 N&S	6W-26S,6W-24S	91
Jean Av. N&S	4W-25S	92
Jeffrey Dr. N&S	4W-25S	92
Keeler Av. N&S	5W-25S,5W-24S	92
Kildare Av. N&S	5W-25S	92
Kildare Ct. E&W	5W-25S	92
Kostner Av. N&S	5W-26S,5W-25S	92
Larkspur Ln. NS&EW	5W-24S	92
Lincoln Hwy. E&W	6W-25S,4W-25S	91
Lincoln Mall Dr. NS&EW	5W-25S	92
Lindenwood Dr. E&W	5W-25S	92
Locust St. N&S	4W-26S,4W-25S	92
Macon Rd. N&S	5W-26S	92
Main St. N&S	4W-26S,4W-25S	92
Maple St. N&S	4W-26S,4W-25S	92
Marsh Ln. E&W	7W-24S	91
Mary Todd Cir. E&W	5W-26S	92
McClellan Dr. NE&SW	5W-26S	92
Morning Glory Dr. E&W	5W-24S	92
Notre Dame Dr. N&S	6W-25S	91
Oak St. N&S	4W-26S,4W-25S	92
Oakhurst Ct. N&S	7W-24S	91
Oakhurst Rd. N&S	7W-24S	91
Oakview Rd. N&S	7W-24S	91
Oakwood Ln. E&W	5W-25S	92
Old Creek Ln. E&W	7W-24S	91
Old Farm Rd. N&S	7W-24S	91
Old Meadow Rd. N&S	7W-24S	91
Old Mill Rd. NW&SE	7W-24S	91
Olympian Way NE&SW	4W-25S	92
Oriole Rd. N&S	7W-24S	91
Oxford Av. N&S	6W-25S	91
Partridge Rd. N&S	7W-24S	91
Pheasant Rd. N&S	7W-24S	91
Phlox Cir. NW&SE	5W-24S	92
Pinewood Ln. E&W	5W-25S	92
Post Av. N&S	6W-24S	91
Primrose Cir. E&W	5W-24S	92
Primrose Ln. NS&EW	5W-24S	92
Princeton Av. N&S	7W-25S	91
Purdue Ln. N&S	6W-25S	91
Quail Run Rd. N&S	7W-24S	91
Quinn Av. N&S	4W-25S	92
Red Barn Rd. N&S	7W-24S	91
Richmond Ct. E&W	5W-26S	92
Richmond Rd. N&S	5W-26S	92
Richton Rd. N&S	4W-26S	92
Ridgeland Av. N&S	7W-24S	91
Rose Ln. NW&SE	5W-25S	92
Rt. 50 N&S	6W-26S,6W-24S	91
School Av. N&S	5W-25S	92
Southwick Dr. E&W	6W-26S	91
Spring Ln. E&W	7W-24S	91
Sprucewood Ln. E&W	5W-25S	92
St. Lawrence Av. E&W	5W-25S	92
Streamwood Ln. E&W	7W-24S	91
Sumter Dr. E&W	5W-26S	92
Sunflower Rd. E&W	7W-24S	91
Timberlane Dr. NS&EW	7W-24S	91
Tower Av. N&S	4W-25S	92
U.S. Route 30 E&W	6W-25S,4W-25S	91
University Av. N&S	5W-25S	92
Violet Ln. N&S	5W-25S	92
Vollmer Rd. E&W	7W-23S,6W-23S	86
Washington Av. E&W	5W-25S	92
Wedgewood Ct. E&W	7W-24S	91
Wheatfield Rd. N&S	7W-24S	91
White Birch Ln. E&W	7W-24S	91
White Oaks Rd. N&S	7W-24S	91
Willow Ct. E&W	7W-24S	91
Willow Rd. N&S	7W-24S	91
Wolf Rd. N&S	7W-24S	91
Woodgate Rd. E&W	7W-24S	91
Yale Ln. E&W	6W-24S	91
205th Pl. E&W	6W-24S	91
207th St. E&W	5W-24S	92
211th Pkwy. E&W	7W-25S	91
211th Pl. E&W	4W-25S	92
212th Pl. E&W	4W-25S	92
213th Pl. E&W	4W-25S	92
213th St. E&W	4W-25S	92
214th Pl. E&W	4W-25S	92
214th St. E&W	4W-25S	92
215th St. E&W	4W-26S	92
216th Pl. E&W	4W-26S	92
216th St. E&W	5W-26S,4W-26S	92
217th St. E&W	4W-26S	92
218th Pl. E&W	4W-26S	92
218th St. E&W	4W-26S	92
219th St. E&W	4W-26S	92

Maywood

	Grid	Page
Adams E&W	11W-0S	47
Augusta Ln. E&W	10W-1N	47
Bataan E&W	11W-0S	47
Congress E&W	10W-0S	47
DesPlaines N&S	10W-0N	47
Dwight D. Eisenhower Expy E&W	11W-0S,10W-0S	47
Erie E&W	10W-0N	47
Fillmore E&W	11W-0S	47
Green E&W	10W-0S	47
Greenwood N&S	10W-0S	47
Harrison E&W	10W-0S	47
Harvard E&W	11W-0S,10W-0S	47
Huron E&W	10W-0N	47
Iowa E&W	10W-1N	47
Lexington E&W	11W-0S,11W-0S	47
Madison E&W	11W-0N	47
Main E&W	10W-0N	47
Maple E&W	10W-0N	47
Maywood Dr. E&W	11W-0S	47
Monroe E&W	10W-0S	47
Oak E&W	11W-0N	47
Ohio E&W	10W-0N	47
Orchard N&S	10W-0S	47
Pine E&W	10W-0N	47
Quincy E&W	10W-0S	47
Railroad E&W	11W-0N	47
Randolph E&W	11W-0N	47
Rice E&W	10W-1N	47
School E&W	10W-0N	47
South Maywood Dr. E&W	11W-0S	47
St. Charles Rd. E&W	11W-0N	47
Superior E&W	10W-0N	47
Van Buren E&W	11W-0S	47
Walnut E&W	10W-0N	47
Walton E&W	10W-1N	47
Warren E&W	10W-0N	47
Washington Blvd. E&W	11W-0N	47
Wilcox E&W	10W-0S	47
1st N&S	10W-0S	47
2nd N&S	10W-0S,10W-1N	47
3rd N&S	10W-0S,10W-1N	47
4th N&S	10W-0S	47
4th N&S	10W-0S,10W-1N	47
5th N&S	10W-0S,10W-1N	47
6th N&S	10W-0S,10W-1N	47
7th N&S	10W-0S,10W-1N	47
8th N&S	10W-0S,10W-1N	47
9th N&S	11W-0S,11W-0N	47
10th N&S	11W-0S	47
11th N&S	11W-0S	47
12th N&S	11W-0S	47
17th N&S	11W-0S	47
18th N&S	11W-0S	47
19th N&S	11W-0S	47
20th N&S	11W-0S	47
21st N&S	11W-0S	47

McCook

	Grid	Page
Clyde N&S	10W-5S	59
Egandale N&S	10W-5S	59
Grand N&S	10W-5S	59
Joliet SW&NE	11W-6S	58
Lawndale N&S	10W-5S	59
Riverside N&S	10W-5S	59
49th E&W	10W-5S	59
50th E&W	10W-5S	59
53rd E&W	10W-5S	59
55th E&W	11W-5S	58

Melrose Park

	Grid	Page
Amling E&W	11W-2N	47
Andy Dr. E&W	10W-1N	47
Armitage E&W	11W-2N	47
Augusta E&W	11W-1N	47
Bloomingdale E&W	12W-1N,11W-2N	46
Braddock E&W	10W-1N	47
Broadway N&S	11W-1N	47
Cary N&S	13W-1N	46
Channing Ct. N&S	10W-1N	47
Charleston Ct. N&S	10W-1N	47
Chicago E&W	11W-1N	47
Clay Ct. N&S	10W-1N	47
Clinton Ct. N&S	10W-1N	47
Concord E&W	10W-1N	47
Cornell N&S	12W-2N	46
Cortez E&W	11W-1N	47
Davis E&W	10W-1N	47
Division E&W	12W-1N,11W-1N	46
Doris St. E&W	10W-1N	47
Elsie Dr. E&W	11W-1N	47
George N&S	11W-2N	47
Grant N&S	10W-1N	47
Haddon E&W	11W-1N	47
Harold N&S	13W-1N	46
Hawthorne N&S	12W-2N	46
Helen Dr. E&W	11W-1N	47
Hirsch E&W	12W-1N,11W-1N	46
Indian Boundary Dr. N&S	12W-2N	46
Iowa E&W	11W-1N	47
James Pl. E&W	13W-2N	46
Janice N&S	13W-2N	46
Lake E&W	12W-1N	46
Le Moyne E&W	11W-1N	47
Lee N&S	10W-1N	47
Main E&W	11W-0N	47
Mannheim N&S	13W-2N	46
Nichols Ln. E&W	10W-1N	47
North Av. E&W	11W-2N	47
Norwood E&W	11W-1N	47
Park Dr. E&W	11W-1N	47
Porter St. N&S	10W-2N	47
Rice E&W	11W-1N	47
Roberta N&S	13W-1N	46
Rose St. E&W	10W-1N	47
Rt. 64 E&W	12W-2N	46
Ruby N&S	12W-2N	46
Sherman E&W	10W-1N	47
Silver St. E&W	10W-1N	47
Thomas E&W	11W-1N	47
Walton E&W	11W-1N	47
Winston Dr. NS&EW	11W-1N,10W-1N	47
9th N&S	11W-1N	47
10th N&S	11W-1N	47
11th N&S	11W-1N	47
12th N&S	11W-1N	47
13th E&W	12W-1N	46
13th N&S	11W-1N	47
14th N&S	11W-1N	47
15th N&S	11W-1N,11W-2N	47
16th N&S	11W-1N	47
17th N&S	11W-1N,11W-2N	47
18th N&S	11W-1N,11W-2N	47
19th N&S	11W-2N	47
20th N&S	11W-1N	47
21st N&S	11W-1N	47
22nd N&S	11W-0N,11W-1N	47
23rd N&S	11W-0N,11W-1N	47
24th N&S	11W-0N,11W-1N	47
25th N&S	12W-1N	46
30th N&S	12W-1N	46
31st N&S	12W-1N	46
32nd N&S	12W-1N	46
33rd N&S	12W-1N	46
34th N&S	12W-1N	46
35th N&S	12W-1N	46
36th N&S	12W-1N	46
46th N&S	13W-1N	46
47th N&S	13W-1N	46

Merrionette Park

	Grid	Page
Albany Dr. N&S	3W-13S	75
Central Park Av. N&S	4W-14S	75
Homan Av. N&S	4W-13S	75
Kedzie Av. N&S	4W-13S	75
Mahoney Dr. N&S	3W-13S	75

Street	Grid	Page
Meadowlane Dr. E&W	3W-13S	75
Morgan Park Av. E&W	4W-13S	75
Palisades Dr. E&W	3W-13S	75
Park Lane Dr. E&W	3W-13S	75
Sacramento Dr. N&S	3W-13S	75
Troy Dr. N&S	3W-13S	75
Whipple Dr. E&W	3W-13S	75
Whipple St. N&S	3W-13S	75
113th Pl. E&W	3W-13S	75
113th St. E&W	3W-13S	75
114th Pl. E&W	3W-13S	75
114th St. E&W	3W-13S	75
115th St. E&W	3W-13S	75
116th Pl. E&W	3W-13S	75
118th St. E&W	3W-13S	75
119th St. E&W	4W-13S,3W-13S	75

Midlothian

Street	Grid	Page
Abbottsford Rd. NE&SW	4W-17S	81
Avers Av. N&S	4W-18S,4W-17S	81
Central Av. N&S	7W-17S	80
Central Park Av. N&S	4W-17S	81
Cicero Av. N&S	5W-17S	80
Clifton Park Av. N&S	4W-17S	81
Concordia Ct. E&W	5W-17S	80
Country Club Dr. NW&SE	6W-17S	80
Crawford Av. N&S	5W-17S	80
Hamlin Av. N&S	4W-18S,4W-17S	81
Harding Av. N&S	4W-18S	81
Homan Av. N&S	4W-17S	81
Interstate 294 E&W	4W-17S	81
Karlov Av. N&S	5W-17S	80
Keating Av. N&S	5W-17S	80
Kedvale Av. N&S	5W-17S	80
Kedzie Av. N&S	4W-17S	81
Keeler Av. N&S	5W-17S	80
Kenneth Av. N&S	5W-17S	80
Kenton Av. N&S	5W-17S	80
Keystone Av. N&S	5W-17S	80
Kilbourn Av. N&S	5W-18S,5W-17S	80
Kildare Av. N&S	5W-17S	80
Kilpatrick Av. N&S	5W-17S	80
Knox Av. N&S	5W-17S	80
Kolin Av. N&S	5W-17S	80
Kolmar Av. N&S	5W-17S	80
Kostner Av. N&S	5W-17S	80
Lacrosse Av. N&S	6W-17S	80
Lamon Av. N&S	6W-17S	80
Laporte Av. N&S	6W-17S	80
Lawndale Av. N&S	4W-18S,4W-17S	81
Linder Av. N&S	6W-17S	80
Long Av. N&S	6W-17S	80
Maple Ln. NS&EW	6W-17S	80
Midlothian Turnpike NE&SW	6W-16S	74
Millard Av. N&S	4W-18S,4W-17S	81
Raday Dr. N&S	4W-18S	81
Rexford Rd. NE&SW	4W-17S	81
Ridgeway Av. N&S	4W-18S,4W-17S	81
Rt. 50 N&S	5W-17S	80
Rt. 83 N&S	5W-17S	80
Saint Louis Av. N&S	4W-17S	81
Sawyer Av. N&S	4W-17S	81
Spaulding Av. N&S	4W-17S	81
Springfield Av. N&S	4W-18S,4W-17S	81
Terrace Ln. NE&SW	5W-17S	80
Tri State Tollway E&W	4W-17S	81
Tripp Av. N&S	5W-17S	80
Trumbull Av. N&S	4W-17S	81
Turner Av. N&S	4W-17S	81
Waverly Av. NE&SW	5W-18S,4W-18S	80
143rd Pl. E&W	6W-17S	80
143rd St. E&W	5W-16S	74
144th St. E&W	6W-17S,4W-17S	80
145th St. E&W	6W-17S,4W-17S	80
146th St. E&W	5W-17S,4W-17S	80
147th Ct. E&W	6W-17S	80
147th Pl. E&W	4W-17S	81
147th St. E&W	6W-17S,4W-17S	80
148th Ct. E&W	6W-17S	80
148th Pl. E&W	4W-17S	81
148th St. E&W	5W-17S,4W-17S	80
149th St. E&W	5W-17S,4W-17S	80
150th St. E&W	5W-17S,4W-17S	80
151st Ct. E&W	5W-18S	80
151st St. E&W	5W-17S,4W-17S	80
152nd Pl. E&W	4W-18S	81
152nd St. E&W	5W-18S,4W-18S	80
153rd Pl. E&W	4W-18S	81
153rd St. E&W	4W-18S	81
154th St. E&W	4W-18S	81

Morton Grove

Street	Grid	Page
Albert Av. E&W	8W-11N	27
Arcadia Av. E&W	9W-11N	27
Austin Av. N&S	7W-10N,7W-11N	37
Beckwith Rd. E&W	9W-11N,8W-11N	27
Belleforte Av. N&S	8W-11N	27
Birch Av. N&S	8W-11N	27
Callie Av. N&S	7W-10N	37
Cameron Ln. NW&SE	9W-11N	27
Capri Ln. E&W	7W-11N	28
Capulina Av. E&W	7W-10N	37
Carol Av. E&W	7W-10N	37
Central Av. N&S	7W-10N,7W-11N	37
Cherry Av. N&S	8W-11N	27
Chestnut St. E&W	8W-10N	37
Church St. E&W	9W-11N,7W-11N	27
Church Terr. E&W	9W-11N	27
Churchill Av. E&W	9W-11N	27
Churchill St. E&W	8W-11N	27
Cleveland St. E&W	7W-10N	37
Crain St. E&W	7W-10N	37
Davis St. E&W	9W-11N,7W-11N	27
Dempster St. E&W	9W-11N,7W-11N	27
Edens Expy. N&S	7W-10N	37
Eldorado Dr. E&W	8W-11N	27
Elm St. E&W	8W-10N,7W-10N	37
Emerson St. E&W	9W-11N,7W-11N	27
Enfield Av. E&W	9W-11N,8W-11N	27
Fernald Av. N&S	7W-10N	37
Ferris Av. N&S	7W-10N	37
Forest Dr. E&W	7W-11N	28
Foster Av. E&W	9W-11N	27
Georgiana Av. N&S	7W-10N	37
Golf Rd. E&W	9W-11N	27
Greenwood Av. E&W	9W-11N,7W-11N	27
Gross Point Rd. NE&SW	6W-10N	38
Grove St. E&W	7W-10N	37
Hamilton N&S	6W-10N	38
Harms Rd. NW&SE	6W-11N	28
Hazel St. E&W	8W-11N	27
Henning Ct. E&W	7W-10N	37
Hoffman E&W	8W-11N	27
Hoffman Ct. N&S	8W-11N	27
Ida Ln. N&S	8W-11N	27
Interstate 94 N&S	7W-10N	37
Keeney Ct. E&W	7W-10N	37
Keeney St. E&W	7W-10N,6W-10N	37
Kirk St. E&W	7W-9N	37
Lake St. E&W	9W-11N,7W-11N	27
Launa Ct. E&W	9W-11N	27
Lee St. E&W	7W-10N	37
Lehigh Av. NW&SE	8W-11N,7W-10N	27
Lillibet Ter. NE&SW	7W-10N	37
Lincoln Av. E&W	7W-10N	37
Linder Av. N&S	6W-9N,6W-11N	38
Long Av. N&S	6W-10N	38
Lotus Av. N&S	6W-9N	38
Luna Av. N&S	6W-9N,6W-11N	38
Lyons Av. E&W	6W-11N	28
Lyons St. E&W	9W-11N,6W-11N	27
Madison Ct. E&W	7W-10N	37
Madison St. E&W	7W-10N	37
Main St. E&W	7W-10N	37
Major Av. N&S	7W-10N,7W-11N	37
Mango Av. N&S	7W-10N,7W-11N	37
Mansfield Av. N&S	7W-10N,7W-11N	37
Maple Ct. N&S	9W-11N	27
Maple St. E&W	9W-11N,8W-11N	27
Marion Av. N&S	8W-11N	27
Marmora Av. N&S	7W-10N,7W-11N	37
Mason Av. N&S	7W-10N,7W-11N	37
McVicker Av. N&S	7W-10N,7W-11N	37
Meade Av. N&S	7W-11N	28
Menard Av. N&S	7W-10N,7W-11N	37
Merrill N&S	9W-11N	27
Michael Av. N&S	9W-11N	27
Michael Ct. N&S	9W-11N	27
Monroe Ct. E&W	7W-10N	37
Monroe St. E&W	7W-10N	37
Moody Av. N&S	7W-11N	28
Morton N&S	7W-10N	37
Mulford St. E&W	6W-9N	38
Murray Ct. N&S	9W-11N	27
Nagle Av. N&S	8W-9N	37
Narragansett Av. N&S	7W-10N,7W-11N	37
Nashville Av. N&S	8W-11N	27
Natchez N&S	8W-11N	27
Natchez Av. N&S	8W-9N	37
National Av. N&S	8W-11N	27
Natoma Av. N&S	8W-11N	27
Neenah Av. N&S	8W-11N	27
New England Av. N&S	8W-11N	27
Newcastle Av. N&S	8W-11N	27
Normandy Av. NW&SE	8W-11N	27
Oak Park Av. N&S	8W-11N	27
Oak St. E&W	8W-10N	37
Oakton St. E&W	7W-10N	37
Oconto Av. N&S	9W-11N	27
Octavia Av. N&S	9W-11N	27
Odell Av. N&S	9W-11N	27
Oketo Av. N&S	9W-11N	27
Olcott N&S	9W-11N	27
Oleander N&S	9W-11N	27
Olphant Av. N&S	9W-11N	27
Oriole Av. N&S	9W-11N	27
Osceola Av. N&S	9W-11N	27
Oswego St. N&S	9W-11N	27
Ottawa Av. N&S	9W-11N	27
Overhill Av. N&S	9W-11N	27
Ozanam Av. N&S	9W-11N	27
Ozark Av. N&S	9W-11N	27
Palma Ln. E&W	9W-11N,8W-11N	27
Park Av. E&W	7W-10N	37
Park Av., South E&W	7W-10N	37
Parkside Av. N&S	7W-10N,7W-11N	37
Ponto St. E&W	9W-11N	27
Reba Ct. N&S	7W-10N	37
Reba St. E&W	7W-10N	37
River Dr. NS&EW	7W-10N	37
Rt. 43 N&S	8W-11N	27
Rt. 58 E&W	8W-11N,7W-11N	27
Sayre Av. N&S	8W-11N	27
School St. N&S	7W-10N	37
Shermer Rd. NW&SE	8W-11N	27
Smithwood Dr. E&W	7W-10N	37
Suffield Ct. E&W	9W-11N	27
Suffield St. E&W	9W-11N	27
Theobald Rd. NE&SW	7W-10N	37
U.S. Route 14 E&W	9W-11N	27
Warren Ct. E&W	7W-10N	37
Warren St. E&W	7W-10N	37
Washington St. E&W	7W-10N,6W-10N	37
Waukegan Rd. N&S	8W-11N	27
Wilson Ter. E&W	9W-11N,8W-11N	27

Mt. Prospect

Street	Grid	Page
Addison Ct. E&W	16W-10N	34
Albert St. N&S	15W-11N,15W-12N	25
Albion Ln. E&W	15W-14N	25
Alder Ln. E&W	14W-14N	25
Algonquin Rd. NW&SE	17W-11N,16W-10N	24
Algonquin Tr. N&S	17W-11N	24
Almond Ct. E&W	13W-14N	26
Althea Dr. N&S	13W-14N	26
Andoa Ln. N&S	13W-15N	16
Ann Marie N&S	17W-10N	34
Apache Ln. E&W	13W-14N	26
Apple Ct. N&S	16W-14N	25
Apricot Ct. NW&SE	16W-14N	25
Aralia Dr. N&S	13W-15N	16
Arbor Ct. N&S	17W-11N	24
Ardyce Ln. E&W	14W-13N	25
Ash Dr. E&W	16W-11N	25
Aspen Dr. N&S	13W-15N	16
Audrey Ln. N&S	17W-12N	24
Autumn Ln. NW&SE	14W-13N	25
Azalea Ln. E&W	13W-14N	26
Azalea Pl. E&W	13W-14N	26
Aztec Ln. E&W	13W-15N	16
Badger Rd. N&S	16W-10N	34
Barberry Ln. N&S	14W-14N	25
Basswood Ln. N&S	13W-14N	26
Beech Rd. N&S	13W-15N	16
Beechwood Dr. N&S	16W-11N	25
Belaire Ln. E&W	16W-11N	25
Bershire Ln. E&W	15W-12N	25
Beverly Ln. N&S	17W-12N	24
Birch Dr. N&S	16W-11N	25
Bittersweet E&W	13W-14N	26
Blackhawk Dr. E&W	16W-12N	25
Bob-o-link Rd. E&W	15W-14N	25
Bobby Ln. N&S	16W-12N	25
Bonita Av. E&W	17W-12N,16W-12N	24
Boro Ln. N&S	13W-14N	26
Boulder Dr. E&W	13W-14N	26
Boxwood Dr. N&S	15W-14N	25
Brentwood Ln. N&S	14W-14N	25
Briarwood Dr., East N&S	17W-10N	34
Bridgeport Dr. N&S	15W-14N	25
Brighton Pl. E&W	15W-14N	25
Brownstone Ln. N&S	17W-11N	24
Buckthorn Dr. N&S	13W-15N	16
Bunting Ln. E&W	15W-14N	25
Burning Bush Ln. N&S	13W-14N,13W-15N	26
Burr Oak Dr. E&W	13W-14N	26
Business Center Ct. E&W	14W-13N	25
Business Center Dr. N&S	14W-13N	25
Busse Av. E&W	16W-12N,15W-12N	25
Busse Rd. N&S	17W-10N,17W-12N	34
Butternut Ln. NS&EW	16W-14N	25
Byron Ct. N&S	15W-12N	25
Callero Cir. E&W	14W-13N	25
Camp McDonald Rd. E&W	13W-15N	16
Candota Av. N&S	16W-11N	25
Cannon Dr. N&S	16W-10N	34
Carboy Rd. E&W	16W-10N	34
Cardinal Ln. E&W	14W-13N	25
Carib Ln. E&W	13W-14N	26
Carol Ln. N&S	17W-11N,17W-12N	24
Catalpa Ln. E&W	17W-11N,16W-11N	24
Cathy Ln. N&S	16W-12N	25
Catino Ct. N&S	14W-13N	25
Cayuga Ln. NW&SE	13W-15N	16
Cedar Glen N&S	17W-11N	24
Cedar Ln. E&W	14W-14N	25
Centenial Dr. E&W	14W-13N	25
Central Rd. E&W	16W-13N,14W-13N	25
Chapel E&W	17W-11N	24
Chariot Ct. E&W	17W-10N	34
Charlotte Rd. E&W	17W-10N	34
Cherrywood Dr. N&S	17W-11N	24
Chestnut Dr. N&S	16W-11N	25
Chinkapin Oak Dr. E&W	13W-14N	26
Chold Ln. N&S	13W-14N	26
Chris Ln. N&S	17W-12N	24
Church Rd. N&S	16W-11N	25
Circle Dr. N&S	16W-11N	25
Clevin Av. E&W	16W-12N	25
Clove Ct. NE&SW	16W-14N	25
Colony Dr. N&S	16W-10N	34
Columbine Dr. N&S	13W-14N	26
Connie Ln. E&W	17W-12N	24
Corktree Ln. E&W	13W-14N	26
Cottonwood Ln. E&W	16W-11N	25
Council Tr. E&W	16W-12N,15W-12N	25
Country Ln. N&S	15W-11N	25
Coventry Pl. E&W	15W-14N	25
Crabtree Ln. N&S	14W-14N	25
Craig Ct. N&S	17W-12N	24
Cree Ln. NS&EW	13W-14N	26
Crestwood Ln. N&S	17W-12N	24
Crystal Ln. N&S	17W-10N	34
Cypress Ct. NW&SE	13W-15N	16
Cypress Dr. N&S	16W-11N	25
Dale Av. N&S	16W-13N	25
Debbie Dr. N&S	14W-13N	25
Deborah Ln. N&S	17W-11N,17W-12N	24
Delaware Dr. N&S	16W-10N	34
Dempster St. E&W	15W-11N	25
Deneen Ln. N&S	14W-13N	25
Dennis Dr. E&W	17W-10N	34
Des Plaines River Rd. N&S	13W-14N,13W-15N	26
Diane Rd. N&S	16W-10N	34
Dogwood Ln. E&W	15W-14N,14W-14N	25
Dove Ct. N&S	16W-11N	25
Dover Pl. E&W	15W-14N	25
Dresser Dr. NS&EW	16W-12N	25
East Dr. N&S	15W-14N	25
Eastman Ct. N&S	15W-13N	25
Eastman Dr. N&S	15W-13N	25
Eastwood Av. N&S	15W-13N	25
Edgewood Ln. N&S	16W-11N,16W-12N	25
Edward St. N&S	15W-11N,15W-12N	25
Elderberry Ln. E&W	14W-14N	25
Elm St. N&S	15W-11N,15W-13N	25
Elmhurst Av. N&S	16W-12N,16W-13N	25
Elmhurst Rd. N&S	16W-10N	34
Elyane Ct. N&S	17W-10N	34
Emerson Ct. NW&SE	15W-11N	25
Emerson Ln. N&S	14W-13N	25
Emerson St. N&S	15W-11N,15W-13N	25
Enterprise Dr. N&S	16W-10N	34
Eric Av. N&S	14W-13N	25
Eric Ct. E&W	14W-13N	25
Estates Dr. E&W	17W-12N	24
Euclid Av. E&W	13W-14N	26
Eva Ln. E&W	16W-11N	25
Evergreen Av. E&W	16W-12N,15W-12N	25
Fairmont Pl. N&S	15W-14N	25
Fairview St. N&S	15W-13N	25
Falcon Dr. E&W	17W-11N	24
Feehanville Dr. NS&EW	14W-13N	25
Fern Dr. N&S	16W-11N	25
Forest Av. N&S	16W-13N	25
Forest Cove Dr. E&W	17W-10N	34
Foster St. E&W	17W-12N	24
Foundry Rd. E&W	14W-13N	25
Franklin Dr. N&S	16W-10N	34
Frediani Ct. E&W	17W-12N	24
Freedom Ct. E&W	13W-14N	26
Frost Dr. E&W	16W-11N	25
Garwood Av. E&W	15W-13N	25
George St. N&S	15W-11N,15W-12N	25
Glendale Ln. N&S	16W-12N	25
Glenn Ln. E&W	16W-11N	25
Go-Wan-Do Av. E&W	16W-12N	25
Golf Rd. E&W	17W-11N,15W-11N	24
Golfhurst Av. E&W	15W-11N	25
Golfview Dr. E&W	15W-11N	25
Golfview Pl. N&S	15W-11N	25
Grace Dr. N&S	16W-11N	25
Green Acres Ln. E&W	16W-11N	25
Green Ln. E&W	14W-13N	25
Greenbriar Dr. E&W	16W-12N	25
Greenfield Ln. N&S	14W-14N	25
Greenwood Dr. E&W	14W-14N	25
Gregory Ln. E&W	14W-13N	25
Gregory St. E&W	16W-13N,15W-13N	25
Grindel Dr. E&W	17W-12N	24
Grove Dr. N&S	17W-11N	24
Hackberry Ln. E&W	14W-14N	25
Hampshire Pl. E&W	15W-14N	25
Hanover Pl. E&W	15W-14N	25
Harvest Ln. E&W	14W-13N	25
Hatherleigh Ct. E&W	17W-11N	24
Hatlen Av. N&S	17W-12N	24
Haven St. E&W	17W-12N	24
Hawthorn NS&EW	16W-10N	34
Hazelhill N&S	17W-10N	34
Helena Av. N&S	17W-12N	24
Hemlock Ln. N&S	14W-14N	25
Henry St. E&W	16W-13N,14W-13N	25
Heritage Dr. N&S	13W-14N	26
Hi-lusi Av. N&S	16W-11N,16W-12N	25
Hiawatha Av. E&W	16W-12N,15W-12N	25
Hiawatha Ct. N&S	15W-12N	25
Hiawatha Tr. E&W	15W-12N	25
Hickory Dr. N&S	16W-11N	25
Highland Av. E&W	15W-13N	25
Hill St. E&W	15W-13N	25
Holly Av. E&W	15W-13N	25
Holly Ct. E&W	15W-13N	25
Hopi Ln. E&W	13W-14N	26
Horner Ln. N&S	14W-13N	25
Hunt Club Dr. N&S	16W-11N	25
Huntington Commons Rd. E&W	16W-11N	25
I-oka Av. N&S	16W-11N,16W-12N	25
Ida Ct. N&S	16W-11N	25
Imperial Ct. E&W	16W-10N	34
Independence Ct. E&W	13W-14N	26
Indigo Ct. N&S	13W-14N	26
Inner Circle Dr. N&S	15W-12N	25
Interstate 90 NW&SE	17W-10N	34
Ironwood Dr. E&W	14W-14N	25
Ironwood Pl. N&S	14W-14N	25
Isabella St. E&W	16W-13N,15W-13N	25
Ivanhoe E&W	16W-11N	25
Ivy Ln. E&W	13W-14N	26
Jeffrey Dr. E&W	14W-13N	25
Jerry Dr. N&S	14W-13N	25
Jody Ct. E&W	17W-12N	24
Judith Ann Dr. E&W	15W-13N	25
Juniper Ln. NS&EW	14W-14N	25
Kenilworth Av. N&S	16W-12N,16W-13N	25
Kennicott Pl. N&S	17W-11N	24
Kensington Rd. E&W	15W-14N	25
Kim Av. E&W	17W-12N	24
Kingston Ct. N&S	14W-13N	25
Kiowa Ln E&W	13W-14N	26
Knights Bridge E&W	17W-11N	24
La Vergne Dr. N&S	17W-11N	24
LaSalle St. E&W	16W-10N	34
Lakeview Ct. N&S	14W-13N	25
Lama Ln. N&S	13W-14N	26
Lams Ct. N&S	16W-12N	25

	Grid	Page
Lancaster Av. N&S	16W-11N,16W-13N	25
Larch Dr. NS&EW	13W-15N	16
Larkdale Ln. E&W	15W-14N	25
Laurel Dr. N&S	13W-15N	16
Leonard Ln. N&S	17W-12N	24
Lexington Dr. N&S	16W-10N	34
Liberty Ct. E&W	13W-14N	26
Lincoln St. E&W	17W-12N,15W-12N	24
Linden Ln. E&W	14W-14N	25
Linneman Rd. N&S	16W-10N,16W-11N	34
Locust Ln. E&W	17W-11N	24
Lois Ct. N&S	17W-12N	24
Lonnquist Blvd. E&W	16W-12N	25
Louis St. N&S	15W-11N,15W-12N	25
Lowden Ln. E&W	14W-13N	25
Lynn Ct. N&S	17W-10N	34
MacArthur Dr. N&S	16W-13N	25
Magnolia Ln. E&W	17W-11N	24
Main St. N&S	15W-13N	25
Mallard N&S	16W-11N	25
Malmo Rd. N&S	16W-10N	34
Mamiya Dr. N&S	14W-13N	25
Manawa Tr. E&W	16W-12N	25
Mansard Ln. NS&EW	17W-10N	34
Maple St. N&S	15W-11N,15W-13N	25
Marcella Rd. N&S	14W-13N	25
Mark Dr. E&W	14W-13N	25
Mark Ter. E&W	17W-12N	24
Martha Ln. E&W	17W-12N	24
Martin Ln. E&W	17W-12N	24
Maya Ln. E&W	13W-15N	16
Meadow Ln. NW&SE	15W-14N	25
Meier Rd. N&S	17W-12N	24
Memory Ln. E&W	16W-13N,15W-13N	25
Michael St. N&S	17W-12N	24
Midway Dr. E&W	16W-10N	34
Milburn Av. E&W	16W-12N,15W-12N	25
Mitchell Dr. E&W	14W-13N	25
Moehling Dr. E&W	15W-12N	25
Mohawk Ln. E&W	13W-15N	16
Moki Ln. N&S	13W-14N	26
Montgomery St. E&W	16W-10N	34
Morrishill* N&S	16W-10N	34
Mt. Prospect Rd. N&S	15W-8N,14W-12N	34
Mulberry Ln. N&S	14W-14N	25
Mura Ln. N&S	13W-15N	16
Myrtle Dr. E&W	17W-12N	24
Na-wa-ta Av. N&S	16W-11N,16W-12N	25
Neil Av. N&S	14W-13N	25
Newberry Ln. N&S	14W-14N	25
Noah Ter. N&S	17W-12N	24
Nordic Rd. N&S	16W-10N	34
North Ln. E&W	17W-11N	24
Northwest Hwy. NW&SE	16W-13N,15W-12N	25
Northwest Tollway NW&SE	17W-10N	34
Nutmeg Ct. NW&SE	16W-14N	25
Oak Av. N&S	16W-13N	25
Oak Dr. E&W	13W-14N	26
Oakwood Dr. N&S	17W-11N	24
Ojibwa Tr. E&W	16W-12N	25
Ojibwa Tr. N&S	16W-12N	25
Ojibway Ter. N&S	16W-12N	25
Oneida Ln. N&S	13W-15N	16
Orchard Pl. E&W	15W-11N	25
Oriole Ln. E&W	15W-14N	25
Owen St. N&S	15W-11N,15W-13N	25
Oxford Pl. E&W	15W-14N	25
Palm Dr. E&W	17W-11N,16W-11N	24
Park Dr. N&S	13W-14N,13W-15N	26
Parliament Pl. E&W	15W-14N	25
Partridge E&W	16W-11N	25
Pawnee Ln. E&W	13W-15N	16
Peachtree Ln. N&S	13W-14N	26
Peartree Ln. N&S	13W-14N	26
Pecos Ln. N&S	13W-14N	26
Pendleton Pl. E&W	16W-12N	25
Persimmon Ln. E&W	16W-14N	25
Pheasant Tr. E&W	17W-11N	24
Phillip* N&S	16W-10N	34
Picadilly Cir. N&S	15W-14N	25
Pima Ln. N&S	13W-14N	26
Pin Oak Dr. E&W	13W-14N	26
Pine St. N&S	15W-12N,15W-13N	25
Plum Ct. NW&SE	16W-14N	25
Prairie Av. N&S	17W-12N	24
Prendegast Ln. E&W	17W-12N	24
Prospect Av. NW&SE	16W-13N,15W-12N	25
Prospect Manor Av. N&S	15W-13N	25
Quail Walk N&S	16W-11N	25
Quince Ct. E&W	13W-14N	26
Quince Ln. N&S	13W-14N	26
Raleigh Pl. E&W	15W-14N	25
Rand Rd. NW&SE	16W-14N,14W-13N	25
Redwood Av. N&S	16W-11N	25
Redwood Dr. N&S	16W-11N	25
Regency Ct. E&W	13W-14N	26
Ridge Av. N&S	15W-13N	25
Rob Rd. N&S	16W-12N	25
Robbie Ln. E&W	17W-12N	24
Robert Dr. N&S	16W-11N	25
Robin Ln. E&W	16W-11N	25
Rosetree Ln. NW&SE	13W-15N	16
Rt. 58 E&W	16W-11N	25
Rt. 62 SE&NW	16W-10N	34
Rt. 83 N&S	15W-13N	25
Russel St. N&S	15W-13N	25
Rusty Rd. E&W	17W-12N	24
Sable Ln. E&W	17W-10N	34
Sanford Cliffs N&S	17W-11N	24
Santee Ln. N&S	13W-14N	26
Sauk Ln. N&S	13W-14N	26
School St. N&S	15W-11N,15W-13N	25
Scott Ter. E&W	17W-12N	24
See-gwun Av. N&S	16W-11N,16W-12N	25
Seneca Ln. E&W	13W-15N	16
Shabonne Tr. E&W	16W-12N,15W-12N	25
Sioux Ln. N&S	13W-14N	26
Sir Galahad N&S	16W-11N	25
Sir Lancelot N&S	16W-11N	25
Sitka Ln. N&S	13W-14N	26
Sivic St. N&S	17W-12N	24
Slawin Ct. NW&SE	14W-13N	25
Small Ln. E&W	14W-13N	25
South Ln. E&W	17W-11N	24
Sprucewood Dr. N&S	17W-11N	24
St. Cecilia Dr. N&S	17W-12N	24
Stevenson Ln. N&S	14W-13N	25
Stratford Pl. E&W	15W-14N	25
Stratton N&S	14W-13N	25
Sullivan Ct. E&W	17W-12N	24
Sumac Ln. N&S	13W-14N	26
Sunset Rd. E&W	16W-11N,15W-11N	25
Susan Dr. E&W	16W-12N	25
Sycamore Ln. N&S	14W-14N	25
Tamarack Dr. N&S	17W-11N	24
Tano Ln. E&W	13W-14N	26
Terminal Dr. E&W	16W-10N	34
Thackery Pl. E&W	15W-14N	25
Thayer St. E&W	15W-13N,14W-13N	25
Thornwood Ln. E&W	17W-11N	24
Tower Dr. N&S	15W-11N	25
Tower Ln. N&S	15W-11N	25
Verde Dr. E&W	17W-12N	24
Victoria Dr. E&W	17W-10N	34
W. Briarwood Dr. NW&SE	17W-10N	34
Wall St. N&S	16W-10N	34
Walnut St. E&W	15W-13N	25
Wapella Av. N&S	16W-11N,16W-12N	25
Waverly Av. N&S	16W-11N	25
Waverly Pl. N&S	16W-12N,16W-13N	25
We-go Tr. N&S	16W-11N,16W-12N	25
Wedgewood NE&SW	15W-14N	25
Weller Ln. N&S	16W-12N	25
Westgate Rd. N&S	14W-14N	25
Wheeling Rd. N&S	14W-13N,15W-14N	25
White E&W	17W-11N	24
White Oak St. E&W	17W-12N	24
Whitegate Dr. E&W	16W-12N	25
Whithorn Ln. E&W	16W-13N	25
Wigwam Tr. E&W	16W-12N	25
Wildwood E&W	14W-13N	25
Wille St. N&S	15W-12N,15W-13N	25
William St. N&S	15W-11N,15W-13N	25
Willow Ln. E&W	17W-11N,16W-11N	24
Willow Rd. E&W	13W-15N	16
Wilshire Dr. N&S	15W-13N	25
Wimbolton Pl. E&W	15W-14N	25
Windsor Dr. N&S	15W-13N	25
Wintergreen Av. E&W	13W-15N	16
Wistoria Ct. E&W	13W-14N	26
Wolf Rd. N&S	14W-14N	25
Wood Ln. E&W	13W-14N	26
Woodview Dr. E&W	13W-15N	16
Yarmouth Pl. N&S	15W-14N	25
Yates Ln. N&S	14W-13N	25
Yuma Ln. E&W	13W-15N	16

New Trier Township

	Grid	Page
Colton Ln. N&S	6W-14N	28
Forest Way Dr. N&S	6W-16N,6W-15N	18
Golf Rd. N&S	5W-14N	28
Holly Ln. E&W	6W-16N	18
Indian Hill Rd. E&W	5W-14N	28
Locust Rd. N&S	5W-14N	28
Long Meadow Rd. N&S	8W-14N	27
Woodley Manor N&S	5W-14N	28
Woodley Rd. NE&SW	5W-14N	28
Woodley Way N&S	5W-14N	28
Woodley Woods N&S	5W-14N	28

Niles

	Grid	Page
Albion Av. E&W	8W-8N	37
Amelia Dr. E&W	10W-10N	36
Ashland Av. N&S	10W-11N	27
Austin Av. N&S	7W-8N	37
Belleforte Pl. N&S	8W-10N	37
Betty Ter. E&W	10W-10N	36
Birchwood Av. E&W	8W-9N	37
Breen St. E&W	9W-10N	37
Bruce Dr. E&W	10W-10N	36
Callero Dr. NW&SE	10W-11N	27
Canavan Ct. N&S	10W-12N	27
Canavan Rd. N&S	10W-12N	27
Carol Av. E&W	8W-10N	37
Carol Ct. NW&SE	8W-10N	37
Carol St. E&W	10W-9N,10W-10N	36
Catino NE&SW	10W-11N	27
Cedar Ln. NE&SW	10W-10N	36
Cherry Av. NE&SW	8W-8N	37
Chester Av. N&S	10W-10N,10W-11N	36
Church St. E&W	10W-11N	27
Churchill Cir. NW&SE	10W-11N	27
Clara Ct. E&W	10W-10N	36
Clara Dr. E&W	10W-10N	36
Cleveland St. E&W	9W-10N,8W-10N	37
Clifton Av. N&S	10W-10N,10W-11N	36
Concord Ln. NW&SE	7W-8N	37
Conrad Av. E&W	9W-10N	37
Courte Dr. E&W	10W-11N	27
Courtland Dr. NE&SW	10W-11N	27
Crain Av. E&W	8W-10N	37
Crain St. E&W	10W-9N,10W-10N	36
Croname Rd. N&S	7W-9N	37
Cumberland Av. N&S	10W-10N	36
Davis St. E&W	10W-11N	27
Days Ter. NW&SE	8W-8N	37
Dempster St. E&W	10W-11N	27
Dobson St. E&W	8W-9N	37
Doreen Ln. E&W	9W-10N	37
Ebinger Dr. NE&SW	8W-8N	37
Elizabeth Av. E&W	10W-11N	27
Elmore St. N&S	9W-10N,9W-11N	37
Elmwood Dr. E&W	11W-12N	26
Fargo Av. E&W	8W-9N	37
Farnsworth Dr. N&S	10W-10N	36
Field Dr. NE&SW	10W-10N	36
Forestview Ln. NE&SW	8W-8N	37
Foster Ln. E&W	10W-11N	27
Fox Glen Dr. N&S	11W-12N	26
Franks Av. NW&SE	8W-8N	37
Georgia Dr. E&W	8W-10N	37
Glendale Rd. N&S	10W-12N	27
Grace Av. N&S	10W-12N	27
Grand Ct. N&S	10W-10N	36
Grand St. N&S	10W-11N	27
Greendale St. N&S	10W-10N	36
Greenleaf Av. E&W	8W-10N	37
Greenleaf Dr. E&W	8W-8N,7W-8N	37
Greenleaf St. E&W	9W-10N	37
Greenwood Blvd. E&W	10W-11N	27
Greenwood Rd. N&S	10W-10N,10W-11N	36
Grennan Pl. E&W	9W-10N,8W-10N	37
Gross Point Rd. SW&NE	8W-9N,7W-9N	37
Hamilton Dr. E&W	8W-10N	37
Harlem Av. N&S	9W-10N	37
Harts Rd. NE&SW	8W-8N	37
Harvard St. E&W	8W-9N	37
Heathwood Dr. E&W	11W-12N	26
Howard St. E&W	9W-9N,7W-9N	37
Huber Oval NW&SE	10W-12N	27
Jarvis Av. E&W	8W-9N,7W-9N	37
Jarvis St. E&W	8W-9N	37
Joey Dr. NW&SE	10W-11N	27
Johanna Dr. E&W	10W-10N	36
Jonquil Ter. E&W	9W-9N,8W-9N	37
Kay St. E&W	10W-12N	27
Kedzie St. E&W	9W-10N	37
Keeney St. E&W	9W-10N,8W-10N	37
Kirk Dr. E&W	9W-9N	37
Kirk Ln. E&W	9W-9N	37
Kirk St. E&W	9W-8N,9W-9N	37
Lake St. E&W	10W-11N	27
Lauren Ln. NE&SW	10W-12N	27
Lawler Av. E&W	9W-9N	37
Lee St. E&W	9W-10N,8W-10N	37
Lehigh Av. NW&SE	7W-8N	37
Lexington Ln. NW&SE	7W-8N	37
Lill Ct. E&W	9W-10N	37
Lill St. E&W	9W-10N,8W-10N	37
Loras Ln. NW&SE	10W-11N	27
Lyons St. NW&SE	10W-11N	27
Madison Av. E&W	10W-10N	36
Madison Ct. E&W	10W-10N	36
Madison Dr. E&W	10W-10N	36
Madison St. E&W	9W-10N,8W-10N	37
Main St. E&W	9W-10N,8W-10N	37
Maryland St. N&S	10W-11N	27
Maynard Dr. E&W	10W-12N	27
Maynard Oval NE&SW	10W-13N	27
Maynard Rd. E&W	10W-12N	27
Maynard Ter. N&S	10W-12N	27
McVicker Av. N&S	7W-9N	37
Meacham Ct. N&S	10W-10N	36
Meade Av. N&S	7W-9N	37
Melvina Av. N&S	7W-9N	37
Merard Av. N&S	7W-9N	37
Merrill St. N&S	9W-10N,9W-11N	37
Merrimac Av. N&S	7W-9N	37
Milwaukee Av. NW&SE	10W-11N,8W-8N	27
Monroe Ct. E&W	8W-10N	37
Monroe St. E&W	10W-10N,8W-10N	36
Mulford Av. E&W	9W-9N	37
Mulford St. E&W	9W-9N,7W-9N	37
Natchez Av. N&S	8W-9N	37
National Av. N&S	8W-10N	37
Neva Av. N&S	8W-9N,8W-10N	37
New England Av. N&S	8W-10N	37
Newark Av. N&S	8W-8N	37
Newcastle N&S	8W-10N	37
Newland Av. N&S	8W-10N	37
Nieman NE&SW	8W-8N	37
Niles Av. E&W	8W-10N	37
Niles Ter. E&W	8W-9N	37
Nora Av. N&S	8W-9N,8W-10N	37
Nordica Av. N&S	8W-9N	37
Norma Ct. NE&SW	10W-12N	27
Normal Av. E&W	10W-10N	36
Normal Ct. E&W	10W-10N	36
North Av. E&W	10W-10N	36
Nottingham St. N&S	8W-9N	37
Oak Av. E&W	10W-11N	27
Oak Ln. E&W	10W-10N	36
Oak Park Av. N&S	8W-9N	37
Oakton Ct. E&W	8W-10N	37
Oakton St. E&W	9W-10N,8W-10N	37
Oconto Av. N&S	9W-9N	37
Octavia Av. N&S	9W-10N	37
Odell Av. N&S	9W-9N,9W-10N	37
Oketo Av. N&S	9W-9N,9W-10N	37
Olcott Av. N&S	9W-10N	37
Oleander Av. N&S	9W-9N,9W-10N	37
Oriole N&S	9W-10N	37
Osceola Av. N&S	9W-10N	37
Ottawa Av. N&S	9W-10N	37
Overhill Av. N&S	9W-10N	37
Ozanam Av. N&S	9W-10N	37
Ozark Av. N&S	9W-10N	37
Park Av. E&W	10W-11N,9W-11N	27
Park Av. E&W	10W-10N	36
Peter Ter. E&W	10W-10N	36
Prospect Ct. E&W	10W-10N	36
Prospect St. N&S	10W-11N	27
Riverside Dr. NE&SW	8W-8N	37
Riverview Av. NE&SW	8W-8N	37
Robin Rd. NE&SW	10W-12N	27
Root Ct. N&S	10W-10N	36
Root St. N&S	10W-11N	27
Rosemary Dr. NW&SE	8W-8N	37
Roseview Dr. E&W	10W-10N	36
Rt. 21 SE&NW	9W-9N	37
Rt. 43 N&S	9W-9N	37
School St. N&S	8W-9N	37
Seward St. E&W	9W-10N,8W-10N	37
Sherwin Av. E&W	7W-9N	37
Stolting Rd. E&W	10W-10N	36
Sunset Rd. E&W	10W-10N	36
Susan Ct. N&S	10W-10N	36
Touhy Av. E&W	8W-9N,7W-9N	37
Vapor Ln. N&S	8W-9N	37
Warren Oval E&W	10W-12N	27
Washington St. N&S	10W-10N	36
Waukegan Rd. N&S	8W-9N	37
Wendy Way NE&SW	10W-12N	27
Willow Ln. N&S	10W-11N	27
Wisner St. N&S	9W-10N,9W-11N	37
Wood River Dr. NE&SW	8W-9N	37
Woodland Dr. NE&SW	10W-11N	27
Wright Ter. E&W	8W-10N	37

Norridge

	Grid	Page
Agatite E&W	10W-5N,8W-5N	41
Ainslie E&W	10W-6N	41
Argyle E&W	10W-6N,9W-6N	41
Belle Plaine E&W	9W-5N	42
Berteau E&W	9W-5N	42
Berwyn E&W	10W-6N	41
Canfield Rd. N&S	10W-5N,10W-6N	41
Carmen E&W	10W-6N,9W-6N	41
Charmaine E&W	10W-6N	41
Chester N&S	10W-6N	41
Clifton N&S	10W-6N	41
Coral E&W	10W-5N	41
Courtland N&S	10W-5N,10W-6N	41
Crescent N&S	10W-6N	41
Cullom E&W	9W-5N	42
Delphia N&S	10W-6N	41
Denal N&S	10W-6N	41
Eastwood E&W	10W-5N	41
Elm E&W	10W-5N	41
Executive E&W	10W-5N	41
Foster E&W	10W-6N	41
Frank Pkwy. N&S	9W-6N	42
Giddings E&W	10W-5N,9W-5N	41
Gunnison E&W	10W-6N,9W-6N	41
Knight Av. N&S	10W-6N	41
Leland E&W	10W-5N	41
Leonard Dr. N&S	10W-6N	41
Lincoln N&S	10W-6N	41
Maple E&W	10W-5N	41
Mission Dr. N&S	9W-6N	42
Monterey N&S	9W-6N	42
Montrose E&W	9W-5N,8W-5N	42
Moreland Dr. N&S	9W-6N	42
Neva N&S	8W-5N	42
Nordica N&S	8W-5N	42
Norridge E&W	9W-6N	42
North Chester N&S	10W-6N	41
North Plainfield N&S	10W-6N	41
Nottingham N&S	8W-5N	42
Oak E&W	10W-5N	41
Octavia N&S	9W-5N	42
Odell N&S	9W-5N	42
Oketo N&S	9W-5N	42
Olcott N&S	9W-5N	42
Oleander N&S	9W-6N	42
Oneida N&S	9W-6N	42
Opal N&S	9W-5N,9W-6N	42
Orange N&S	9W-5N,9W-6N	42
Oriole N&S	9W-5N	42
Osage N&S	9W-5N	42
Osceola N&S	9W-5N	42
Ottawa N&S	9W-5N	42
Overhill N&S	9W-5N,9W-6N	42
Ozanam N&S	9W-5N,9W-6N	42
Ozark N&S	9W-5N,9W-6N	42
Paris N&S	10W-5N	41
Pensacola E&W	9W-5N	42
Philips N&S	9W-5N	42
Pittsburgh N&S	10W-6N	41
Plainfield N&S	10W-5N	41
Pontiac N&S	10W-5N,10W-6N	41
Prospect N&S	10W-6N	41
Redwood Dr. N&S	10W-5N	41
Ridge N&S	10W-6N	41
Sayre N&S	8W-5N	42
Strong E&W	10W-6N,9W-6N	41
Sunnyside E&W	9W-5N,8W-5N	42
Sunrise N&S	9W-6N	42
Thatcher N&S	10W-5N	41
Vine N&S	10W-6N	41
Wilson E&W	10W-5N	41
Windsor E&W	9W-5N,8W-5N	42
Winnemac E&W	10W-6N	41
Winona E&W	10W-6N,9W-6N	41
82nd N&S	10W-5N	41

North Riverside

	Grid	Page
Burroak N&S	9W-2S	54
Country Club E&W	10W-2S	53
Des Plaines N&S	10W-2S	53
Edgewater E&W	10W-2S	53
Elm N&S	11W-2S	53
Forest E&W	11W-2S	53
Forest N&S	9W-2S	54
Golfview N&S	10W-2S	53
Groveland N&S	10W-2S	53
Hainsworth N&S	9W-2S	54
Keystone N&S	9W-2S	54
Lathrop N&S	9W-2S	54
Lewe N&S	10W-2S	53
Lincoln N&S	10W-2S	53
Madison N&S	11W-2S	53
Northgate N&S	9W-2S	54
Park N&S	9W-2S	54
Prairie N&S	11W-2S	53

	Grid	Page
Seegers N&S	11W-2S	53
Traube N&S	9W-2S	54
Westover N&S	9W-2S	54
1st N&S	10W-2S	53
2nd N&S	10W-2S	53
3rd N&S	10W-2S	53
4th N&S	10W-2S	53
5th N&S	10W-2S	53
6th N&S	10W-2S	53
7th N&S	10W-2S	53
8th N&S	10W-2S	53
9th N&S	11W-2S	53
10th N&S	11W-2S	53
11th N&S	11W-2S	53
12th N&S	11W-2S	53
22nd Pl. E&W	11W-2S	53
23rd E&W	11W-2S	53
23rd Pl. E&W	11W-2S	53
24th N&S	11W-2S,9W-2S	53
24th Pl. E&W	11W-2S	53
25th E&W	11W-2S,9W-2S	53
26th E&W	9W-2S	54
27th E&W	10W-2S	53
28th E&W	10W-2S	53
30th E&W	10W-2S	53

Northbrook

	Grid	Page
Academy Dr. N&S	11W-18N	16
Acorn Ln. E&W	11W-17N	16
Adelaide N&S	10W-17N	17
Adirondack Dr. N&S	11W-17N	16
Alice Rd. N&S	11W-18N	16
Alleghany Ln. N&S	11W-17N	16
Allen Dr. NW&SE	10W-15N	17
Ancient Tree Rd. NS&EW	11W-17N	16
Angle Av. NW&SE	9W-17N	16
Anjou Dr. N&S	12W-18N	16
Antelope Springs Rd. E&W	11W-17N	16
Anthony Trail N&S	11W-18N	16
Antique Ln. E&W	8W-18N	18
Appletree Ln. N&S	10W-17N	17
Appletree Ln. E&W	10W-17N	17
Applewood Ln. NS&EW	12W-18N	16
Arbeleda Ln. NE&SW	10W-15N	17
Arnold Ln. E&W	11W-18N	16
Arrowood Dr. N&S	13W-18N	16
Asbury Rd. E&W	10W-16N	17
Ash Ln. NW&SE	10W-17N	17
Aspen Dr. E&W	11W-16N	16
Astor Pl. NE&SW	7W-16N	18
Atlanta Av. N&S	8W-18N	18
Auburn Ln. N&S	11W-15N,11W-16N	16
Bach St. N&S	10W-17N	17
Ballantrae Dr. N&S	8W-18N	18
Banyan Dr. N&S	11W-18N	16
Bar Harbor Ter. N&S	10W-17N	17
Barberry Ln. N&S	11W-16N	16
Barnsley Pl. NE&SW	8W-17N	18
Basswood Dr. N&S	13W-18N	16
Bay View* E&W	11W-17N	16
Bayberry Dr. E&W	11W-18N	16
Beaver Av. E&W	7W-17N	18
Beckwith Ct. N&S	11W-17N	16
Beechnut Rd. NE&SW	10W-18N	17
Bellevue Pl. E&W	10W-16N	17
Berglund Pl. E&W	8W-17N	18
Bermuda Dunes Pl. N&S	12W-17N	16
Bernay Dr. E&W	12W-18N	16
Big Oak Cir. E&W	9W-17N	17
Big Oak Ln. E&W	10W-17N,9W-17N	17
Bindle Ct. E&W	11W-17N	16
Birch Rd. N&S	10W-16N	17
Birchwood Cir. N&S	10W-18N	17
Bittersweet Dr., East N&S	9W-17N	17
Bittersweet Dr., South E&W	9W-17N	17
Bittersweet Dr., West N&S	9W-17N	17
Blackthorn N&S	10W-17N	17
Blue Hill Ter. NW&SE	10W-17N	17
Bob-o-link Ln. E&W	10W-16N	17
Bordeaux Ct. NW&SE	12W-18N	16
Bordeaux Dr. NE&SW	12W-18N	16
Brairwood Ln. N&S	12W-18N	16
Brees Way Ln. N&S	9W-17N	17
Brentwood Rd. E&W	10W-16N	17
Brian Dr. E&W	10W-15N	17
Briarwood Ln. E&W	10W-17N	17
Bridlewood Ln. (Pvt.) N&S	8W-17N	18
Bridlewood Rd. (Pvt.) E&W	8W-17N	18
Brighton Ct. E&W	10W-16N	17
Brindle Ct. N&S	11W-17N	16

	Grid	Page
Bristol Ct. E&W	12W-17N	16
Brittany Ct. NW&SE	12W-18N	16
Brittany Rd. E&W	12W-18N	16
Brookside Ln. N&S	10W-17N	17
Buckland Ln. NE&SW	10W-18N	17
Buckthorn Cir. NS&EW	13W-18N	16
Burgundy Ln.* NE&SW	9W-16N	17
Butternut Ln. N&S	10W-16N,10W-17N	17
Buttonwood Dr. E&W	8W-17N	18
Cambridge Dr. N&S	11W-16N	16
Campden Ln. E&W	10W-18N	17
Candlewood Hill Rd. N&S	10W-17N	17
Canterbury Dr. E&W	11W-16N	16
Cargill Dr. N&S	10W-16N	17
Caribou Crossing NS&EW	13W-18N	16
Carlson Dr. N&S	9W-16N	17
Carlyle Ct. N&S	11W-17N	16
Carol Ct. NW&SE	9W-17N	17
Carriage Av. E&W	9W-17N	17
Carter Ct. E&W	7W-16N	18
Caryn Ter. N&S	9W-17N	17
Cir.	10W-15N	17
Catherine St. E&W	10W-17N	17
Cayuga Ln. N&S	11W-17N	16
Cedar Ct. N&S	11W-17N	16
Cedar Ln. NW&SE	10W-17N	17
Center Av. E&W	10W-17N	17
Chapel Ct. NE&SW	9W-17N	17
Chapel Wood* E&W	11W-17N	16
Charlemagne Dr. N&S	12W-18N	16
Charles Dr. NE&SW	12W-18N	16
Charlwood* N&S	11W-17N	16
Chartres Dr. NW&SE	11W-17N	16
Chedworth Ct. E&W	10W-18N	17
Cherry Ln. E&W	11W-17N,9W-17N	16
Chesapeake Ln. E&W	10W-17N	17
Chestnut Rd. N&S	10W-17N	17
Chestnut St. E&W	10W-17N	17
Chewton Glen (Pvt.) E&W	8W-17N	18
Chipili Dr. E&W	7W-17N	18
Christina Ln. N&S	11W-17N	16
Church N&S	9W-17N	17
Clover Ct. NE&SW	11W-16N	16
Clover Rd. N&S	11W-16N	16
Coachmaker Dr. N&S	8W-18N	18
Cobblestone Ct. E&W	8W-18N	18
Cobblestone* N&S	11W-17N	16
Cobblewood Dr.	10W-15N	17
Colonade Dr. N&S	8W-18N	18
Colony Ct. E&W	10W-17N	17
Commercial Dr. E&W	12W-18N,11W-18N	36
Concord Ct. E&W	11W-17N	16
Connecticut Av. E&W	8W-18N	18
Constance Ln. N&S	12W-18N	16
Cornflower Trail E&W	11W-16N	16
Cotswolds Ct. NW&SE	10W-18N	17
Cotswolds Way N&S	10W-18N	17
Cottonwood Rd. N&S	9W-18N	17
Country Ln. E&W	9W-17N	17
County Line Rd. N&S	10W-18N	17
Coventry Ln. N&S	11W-17N	16
Crabtree Ln. N&S	11W-17N,10W-17N	16
Cresson Av. N&S	7W-17N	18
Crestwood Dr. N&S	10W-17N	17
Crimson Ct. E&W	11W-17N	16
Cuyler Av. N&S	7W-17N	18
Cypress Ln. N&S	10W-17N	17
Dalton Pl. E&W	8W-17N	18
Dana Ct. NE&SW	12W-17N	16
Daryl Ln. N&S	9W-17N	17
Dauphine Av. NS&EW	12W-18N	16
Dauphine Ln. N&S	11W-18N	16
Deer Trail E&W	13W-18N	16
Deer Trail Ct. N&S	13W-18N	16
Dehne Rd. E&W	10W-16N	17
Dell Rd. NW&SE	8W-18N	18
Dennis Dr. E&W	8W-18N	18
Devonshire Ct. N&S	11W-16N	16
Division St. N&S	10W-17N	17
Donegal Ln. N&S	10W-17N	17
Donovan Glen Ct. E&W	11W-16N	16
Doolittle Dr. E&W	11W-18N	16

	Grid	Page
Doriann Dr. E&W	11W-15N	16
Dorset Ln. N&S	12W-17N	16
Downing St. N&S	12W-17N	16
Driftwood Ln. NS&EW	8W-16N	18
Dundee Rd. E&W	11W-18N	16
Cir.	11W-18N	17
East Wind Ct. E&W	12W-17N	16
Eastwood Ln. E&W	9W-17N	17
Echo Ln. E&W	11W-16N	16
Edens Expwy N&S	7W-17N	18
Edgewood Ln. NS&EW	8W-17N	18
Elcid Ln.* N&S	9W-15N	17
Elder Ct. E&W	11W-16N	16
Ellendale Dr. NS&EW	10W-16N	17
Elm Av. E&W	9W-17N	17
Elm Ct. N&S	10W-18N	17
Elm Ridge Dr. N&S	10W-17N	17
Era Dr. N&S	12W-18N	16
Erin Ln. E&W	7W-16N	18
Essex Dr.* E&W	9W-16N	17
Estate Ct. N&S	12W-18N	16
Evert Ct. E&W	9W-17N	17
Fair Ln. NW&SE	10W-17N	17
Fairford Ln. E&W	10W-18N	17
Fairview Ln. N&S	12W-18N	16
Fairway E&W	8W-18N	18
Farmington Rd. E&W	11W-15N,11W-16N	16
Farnsworth Ln. E&W	10W-16N	17
Fern Av. E&W	9W-17N	17
Ferndale Av. E&W	9W-17N	17
Fieldwood Dr. N&S	10W-16N	17
First St. NW&SE	10W-17N	17
Floral Dr. E&W	11W-16N	16
Forest Rd. N&S	9W-18N	17
Forestview Dr. E&W	13W-18N	16
Forestway Dr., East E&W	7W-17N	18
Foxdale Ct. E&W	10W-16N	17
Foxrun Dr. N&S	9W-18N	17
Galway Ct. N&S	11W-16N	16
Glen Oak Ln. E&W	9W-17N	17
Glendale Av. N&S	10W-16N	17
Grace Rd. N&S	10W-17N	17
Grand Rd. N&S	11W-16N	16
Grant Rd. N&S	8W-17N	18
Greenacre Dr. E&W	12W-17N	16
Greenbriar Ln. N&S	10W-15N	17
Greenview Dr. E&W	10W-15N	17
Greenview Rd. N&S	11W-17N	16
Greenway* NS&EW	11W-17N	16
Greenwood Rd. N&S	10W-15N	17
Greenwood Rd. N&S	10W-18N	17
Gregg Rd. E&W	8W-18N	18
Hampton Ln. N&S	11W-17N,11W-16N	16
Harbinger Falls* E&W	11W-18N	16
Harbor Ln. E&W	11W-18N	16
Harborside Rd. N&S	11W-18N	16
Harris Dr. E&W	11W-18N	16
Hawthorne Ln. N&S	11W-18N	16
Hazelwood Av. N&S	11W-17N	16
Heather Ct. NE&SW	11W-16N	16
Heather Ln. NW&SE	11W-16N	16
Helen Rd. N&S	11W-18N	16
Hemlock Knoll Ter. N&S	11W-17N	16
Hemlock Ln. E&W	11W-16N	16
Henrici Dr. E&W	8W-18N	18
Heritage Dr. E&W	12W-17N	16
Hickory Ct. N&S	9W-18N	17
Hickory Ln. E&W	8W-17N	18
Hidden Bay* N&S	11W-18N	16
Hiddenwells* NE&SW	11W-17N	16
High Point Dr. N&S	11W-17N	16
Highland Dr. E&W	10W-15N	17
Hillcrest Ln. NW&SE	11W-17N	16
Hillside Rd. N&S	9W-17N	17
Holly Ct. N&S	8W-17N	18
Holly Ln. N&S	11W-16N	16
Holste Rd. E&W	10W-15N	17
Honeysuckle Dr. N&S	13W-18N	16
Horizon Ln. N&S	12W-17N	16
Huckleberry Ln. N&S	10W-17N	17
Huehl Rd. N&S	11W-16N	16
Huntington Ln. N&S	11W-16N	16
Illinois Rd. N&S	10W-17N,9W-17N	17
Indian Wells Ln. E&W	12W-17N	16
Interstate 294 N&S	12W-16N	16

	Grid	Page
Interstate 94 E&W	11W-18N,8W-18N	16
Iris Ct. E&W	11W-16N	16
Island Point* N&S	11W-17N	16
Ivy Ln. N&S	11W-16N	16
Janke Dr. E&W	9W-16N	17
Jefferson Pl. NW&SE	8W-17N	18
Jeffrey Ct., North N&S	9W-17N	17
Jeffrey Ct., South N&S	9W-17N	17
Jeffrey Ct., West N&S	9W-17N	17
Jeffrey Ln. E&W	9W-17N	17
Juli-lyn Ct. N&S	12W-17N	16
Juli-lyn Ln. N&S	12W-17N	16
Juniper Ln. N&S	10W-17N	17
Kay Jay Dr. E&W	11W-15N	16
Keystone Av. N&S	9W-17N	17
Keystone Cir. N&S	9W-17N	17
Keystone Rd. E&W	11W-17N	16
Kiest Av. N&S	10W-17N,9W-17N	17
Killarney Ln. NW&SE	7W-16N	18
Kingsport Ct. N&S	11W-17N	16
Kingston Ct. E&W	10W-16N	17
Koehling Rd. E&W	9W-18N,8W-18N	17
Koepke Rd. E&W	11W-16N	16
Laburnun Dr. N&S	12W-18N	16
Lagoon View* E&W	11W-17N	16
Cir.	11W-16N	16
Lake Placid Ln. E&W	11W-17N	16
Lake-Cook Rd. E&W	9W-18N,8W-18N	17
Lakeside Av. E&W	11W-16N	16
Landwehr Rd. N&S	11W-17N,11W-18N	16
Lane Winchester E&W	11W-17N	16
Larkdale Rd. N&S	10W-16N	17
Larrabee Ln. E&W	9W-17N	17
Laurel Rd. N&S	9W-17N	17
Lavigne Ln. N&S	11W-16N	16
Lawrence Ln. (Pvt.) E&W	8W-17N	18
Lee Rd. N&S	9W-17N	17
Lexington Ct. E&W	8W-17N	18
Lilac Ln. E&W	11W-16N	16
Lincoln Av. E&W	10W-17N	17
Lindberg Ln. N&S	11W-18N	16
Linden Rd. E&W	9W-17N	17
Linden Rd. NS&EW	9W-18N	17
Lindenwood Ln. E&W	12W-18N	16
Lisa Ct. N&S	10W-15N	17
Little Ln. E&W	11W-17N	16
Long Meadow N&S	8W-18N	18
Long Valley Dr. NS&EW	11W-16N	16
Longaker Ln. NW&SE	8W-17N	18
Longmeadow Rd. NS&EW	10W-16N	17
Lorete Ln. N&S	11W-17N	16
Lori-lyn Ln. NE&SW	12W-17N	16
Mallard NS&EW	10W-16N	17
Manor Dr. NW&SE	11W-16N	16
Maple Av. E&W	10W-17N,9W-17N	17
Maplewood Dr. E&W	10W-18N,9W-18N	17
Marcee Ln. NS&EW	9W-17N	17
Marden Ct. E&W	11W-16N	16
Margo Ln. E&W	11W-18N	16
Maria Av. E&W	11W-18N	16
Marshall Rd. E&W	8W-17N	18
Mayapple Ct. NS&EW	10W-16N	17
Mayapple Ln. NE&SW	10W-16N	17
McArthur Blvd. E&W	11W-18N	16
Meadow Rd. NW&SE	10W-17N	17
Cir.	12W-17N	16
Melvin Dr. N&S	11W-18N	16
Micheline Ln. NS&EW	12W-18N	16
Midfield Ln. E&W	8W-16N	18
Midway Rd. NW&SE	8W-17N	18
Miller Rd. N&S	11W-16N	16
Milton Av. E&W	10W-17N,9W-17N	17
Moon Hill Rd. E&W	10W-17N	17
Mt. Vernon Ter. NW&SE	11W-17N	16
Muirwood* N&S	11W-17N	16
Mulberry Ln. N&S	10W-17N	17
New England Av. N&S	8W-18N	18
New Hampshire E&W	8W-18N	18
New Jersey Av. N&S	8W-18N	18
Newport Rd. NW&SE	10W-15N	17
Norfolk Rd. NS&EW	7W-16N	18
Normandy Ln. E&W	12W-18N	16

	Grid	Page
North Corner* NE&SW	11W-17N	16
North Ln. N&S	8W-16N	18
Northbrook Av. NE&SW	8W-17N	18
Northmoor Ct. N&S	10W-17N	17
Norway Pine Ln. N&S	11W-16N	16
Oak Av. E&W	10W-17N,9W-17N	17
Oak Knoll Ct. E&W	12W-18N	16
Oak Knoll Ter. N&S	12W-18N	16
Oak Leaf Ter. N&S	9W-18N	17
Oakwood Rd. E&W	10W-18N,9W-18N	17
Old Hunt Rd. E&W	8W-16N	18
Old Mill Rd. NS&EW	11W-16N	16
Old Post Rd. N&S	8W-18N	18
Orchard Ln. N&S	11W-16N	16
Otis Pl. NE&SW	8W-17N	18
Overlook Bluff* E&W	11W-17N	16
Oxford Ln. E&W	11W-16N	16
Palm Canyon Dr. NS&EW	12W-17N	16
Palm Ln. E&W	11W-17N	16
Pamela Ln. E&W	12W-18N	16
Park Dr. E&W	7W-17N	18
Partridge Ln. E&W	11W-17N	16
Pawtucket Rd. E&W	11W-17N	16
Peachtree Ln. N&S	10W-17N	17
Pebble Beach Rd. E&W	12W-17N	16
Pebblebrook Rd. E&W	8W-16N	18
Penfold Pl N&S	10W-16N	17
Pfingsten Rd. N&S	11W-16N	16
Pheasant Creek Dr. NE&SW	11W-17N	16
Phillips Av. N&S	11W-16N	16
Phyllis Ln. NW&SE	12W-18N	16
Picardy Ct. E&W	12W-18N	16
Picardy Dr. NS&EW	12W-18N	16
Pine Tree Rd. N&S	9W-18N	17
Pinto Ln. E&W	8W-16N	18
Pleasant St. E&W	11W-17N	16
Plum Island Dr. E&W	11W-17N	16
Plymouth Ln. N&S	10W-16N	17
Pondside Dr. N&S	10W-16N	17
Poplar Ct. NW&SE	13W-18N	16
Portsmouth Ct. E&W	9W-17N	17
Powderhorn Dr. N&S	9W-18N	17
Prestbury Ct. NS&EW	10W-18N	17
Prestwick Ln. E&W	11W-17N	16
Prince St. N&S	10W-15N	17
Quail N&S	10W-15N	17
Queens Way N&S	10W-15N	17
Radcliffe Dr. E&W	12W-17N	16
Raintree Ln. E&W	12W-17N	16
Raliegh Ct. E&W	8W-17N	18
Raymond Dr. E&W	9W-16N	17
Red Coach Ln. N&S	9W-18N	17
Red Oak Rd. N&S	9W-18N	17
Redhaw Rd. N&S	9W-17N	17
Redwood Ln. E&W	10W-17N,9W-17N	17
Regent Dr. N&S	12W-17N	16
Rennes Ct. E&W	11W-17N	16
Revere Dr. NW&SE	11W-18N	16
Ridge Rd. E&W	11W-18N	16
Ridge Rd. N&S	8W-17N	18
Ridgeland Ln. E&W	12W-17N	16
Ridgewood Dr. E&W	9W-17N	17
Ring Rd. N&S	9W-18N	17
Riverdale E&W	7W-16N	18
Riverfalls Dr. E&W	11W-17N	16
Riviera Ct. E&W	12W-17N	16
Robincrest Dr. E&W	11W-17N	16
Rosemary Ln. E&W	9W-18N	17
Rosewood Ct. E&W	7W-17N	18
Royal Dr. N&S	12W-17N	16
Rt. 43 NW&SE	9W-16N	17
Rt. 68 E&W	10W-18N,8W-18N	17
Rudolph Dr. E&W	9W-18N	17
Russet Ct. NE&SW	12W-17N	16
Russet Ln. E&W	12W-17N	16
Rutgers Ln. E&W	12W-17N	16
Salceda Dr. N&S	10W-15N,9W-15N	17
Sanders Rd. N&S	12W-17N,12W-18N	16
Saranec Ln. N&S	10W-17N	17
School Dr. E&W	11W-16N	16
Scotch Pine Ln. N&S	11W-16N	16
Scott Rd. E&W	9W-17N	17
Second St.(pvt.) N&S	10W-16N	17
Seville Cir.* E&W	9W-15N	17
Shagback Ln. E&W	9W-17N	17
Shannon Ct. E&W	11W-17N	16

	Grid	Page
Shannon Rd. E&W	11W-17N,10W-17N	16
Shawnee Tr. N&S	11W-16N	16
Shermer Av. N&S	10W-16N,9W-17N	17
Silverpine Dr. NS&EW	11W-16N	16
Skokie Blvd. NW&SE	8W-18N	18
Sleepy Hollow N&S	11W-16N	16
Smith Rd. N&S	11W-16N	16
Snowbird Ln. E&W	12W-17N	16
Somerset Ln. N&S	10W-16N	17
Sorrel Dr. E&W	11W-16N	16
South Bridge Ln. E&W	10W-16N	17
South Ln. NE&SW	8W-16N	18
South Wind Dr. N&S	12W-17N	16
Southgate Dr. N&S	7W-17N	18
Spring Hill Dr. NS&EW	11W-17N	16
Spruce Rd. N&S	9W-18N	17
Spruce St. N&S	9W-17N	17
Spyglass Ln. NE&SW	12W-17N	16
Stanley St. N&S	9W-16N	17
Stephens Green E&W	12W-17N	17
Stone Creek* NW&SE	10W-17N	17
Stonebridge Ln. E&W	10W-18N	17
Stonegate E&W	11W-17N	16
Stowe Ct. NE&SW	10W-18N	17
Stratford Pl. N&S	12W-17N	16
Suffield Ter. NS&EW	12W-17N	16
Sugarpine Ln. N&S	11W-16N	16
Summerfield Rd. NS&EW	7W-16N	18
Summerton Pl. N&S	10W-16N	17
Sunburst Ln. N&S	12W-17N	16
Sunnyside Cir. N&S	10W-16N	17
Sunset Ct. E&W	7W-16N	18
Sunset Ln. E&W	12W-18N	16
Sunset Ridge Rd. N&S	8W-16N,8W-17N	18
Surrey Ln. N&S	9W-17N	17
Sussex Dr. N&S	11W-17N	16
Sutton Dr. N&S	12W-17N	16
Sycamore Ln. N&S	9W-17N	17
Tamarind Dr. E&W	11W-18N	16
Techny Rd. E&W	11W-16N,8W-16N	16
Terri-lyn N&S	12W-17N	16
The Mews Ln. E&W	10W-18N	17
The Strand NE&SW	10W-15N	17
Thornapple Ln. NS&EW	9W-17N	17
Thornwood Ln. E&W	10W-17N,9W-17N	17
Thornwood Ln.	9W-18N	17
Timber Ln. N&S	9W-17N	17
Toulon Dr. NS&EW	11W-17N	16
Touraine Ter. N&S	12W-18N	16
Trailwood Ct. NW&SE	12W-18N	16
Trailwood Ln. NW&SE	12W-18N	16
Tri-state Tollway N&S	12W-17N	16
Tyron Wood* N&S	11W-17N	16
Valencia Dr. N&S	11W-16N	17
Valley Rd. N&S	11W-16N	16
Vermont Av. E&W	8W-18N	18
Versailles Dr. N&S	12W-18N	16
Vicki N&S	11W-15N	16
Violet Ln. E&W	11W-16N	16
Vista Ct.* N&S	9W-15N	17
Voltz Ct. N&S	9W-16N	17
Voltz Rd.	8W-17N,7W-17N	18
Wainwright Dr. N&S	11W-16N	16
Walnut Cir. NE&SW	10W-18N	17
Walnut Ln. N&S	10W-17N	17
Walters Av. E&W	12W-17N,9W-17N	16
Water Edge Ct. N&S	9W-18N	17
Water Edge Ln. NW&SE	9W-18N	17
Waukegan Rd. NW&SE	8W-16N,10W-18N	18
Weber Dr. N&S	12W-18N	16
Weller Ln. NE&SW	11W-17N	16
Wendy Dr. N&S	9W-17N	17
Wescott Rd. N&S	9W-17N	17
Wessling Dr. N&S	11W-16N	16
Western Av. N&S	10W-16N,10W-17N	16
Westwood Ct. N&S	10W-17N	17
Whisperwoods Ct. E&W	11W-16N	16
White Mountain Dr. N&S	11W-17N	16

	Grid	Page
White Oak Dr. N&S	11W-16N	16
White Pine Dr. E&W	11W-16N	16
Whitehall Dr. NS&EW	9W-17N	17
Whitfield Rd. NW&SE	8W-17N	18
Wildwood Ln. (Pvt.) NS&EW	8W-17N	18
Williamsburg Dr. E&W	9W-18N	17
Willow Ln. E&W	8W-17N	18
Willow Rd.	10W-15N,8W-15N	17
Wilshire Ln. E&W	10W-18N	17
Windrush Ln. NS&EW	10W-18N	17
Windsor Ln.* NW&SE	9W-16N	17
Wintercrescent Cir. E&W	10W-16N	17
Wood Dr. N&S	10W-16N	17
Woodbine Ln. N&S	10W-17N	17
Woodhead Dr. N&S	12W-18N	16
Woodhill Dr. E&W	9W-17N	17
Woodland Dr. E&W	11W-16N,10W-16N	17
Woodlark Dr. N&S	8W-17N	18
Woodlawn Ct. N&S	10W-16N	17
Woodlawn St. E&W	10W-16N	17
Woodmere Ct. N&S	11W-16N	16
Woodmere Dr. E&W	11W-16N	16
York Ct. E&W	8W-17N	18
Yorkshire Ln. E&W	12W-17N	16

Northfield

	Grid	Page
Abbot Ct. E&W	7W-15N	18
Alice Pl. E&W	7W-15N	18
Ash E&W	6W-15N	18
Avon Av. N&S	7W-15N	18
Balmoral Av., East N&S	7W-14N	28
Balmoral Av., West N&S	7W-14N	28
Birchwood Ln. (Pvt.) E&W	7W-15N	18
Bosworth Ln. E&W	7W-15N	18
Bracken Ln. (Pvt.) NS&EW	7W-15N	18
Brandon Rd. N&S	7W-14N	28
Briar Ln. (Pvt.) N&S	7W-15N	18
Bridlegate Ln. NW&SE	7W-15N	18
Bridlewood Ln. NS&EW	7W-14N	28
Bristol Rd. N&S	7W-15N	18
Bristol St. N&S	7W-15N	18
Brookhill Dr. NE&SW	7W-14N	28
Burr Oak E&W	8W-15N	18
Camden Ln. E&W	7W-15N	18
Canterbury Ln. E&W	7W-14N	28
Central Av. N&S	7W-15N	18
Chapel Hill Ln. N&S	7W-15N	18
Cherry E&W	6W-15N	18
Churchill St. N&S	7W-15N	18
Clover Ln. (Pvt.) E&W	8W-15N	18
Coach Rd. (Pvt.) N&S	7W-15N	18
Colonial Ln. E&W	6W-15N	18
Country Ln. E&W	7W-16N	18
Courts of Regentwood NS&EW	8W-15N	18
Coventry Rd. N&S	7W-14N	28
Crooked Creek E&W	7W-15N	18
Dickens Rd. N&S	7W-15N	18
Dickens St. N&S	7W-15N	18
Dorina Dr. E&W	8W-15N	18
Drury Ln. (Pvt.) E&W	7W-15N	18
Earl Dr. N&S	7W-15N	18
Eaton St. N&S	7W-15N	18
Eddy Ln. E&W	7W-15N	18
Edens Expwy. N&S	7W-16N,6W-15N	18
Edens Ln. N&S	7W-15N	18
Edgewood Ln. N&S	8W-15N	18
Elder St. E&W	6W-15N	18
Elm E&W	7W-15N	18
Enid Ln. (Pvt.) N&S	7W-15N	18
Frontage Rd. N&S	6W-15N	18
Graemere St. N&S	7W-15N	18
Grove Dr. E&W	7W-15N	18
Happ Rd. N&S	7W-16N,6W-15N	18
Harding Rd. N&S	6W-15N	18
Heather Ter. E&W	7W-15N	18
Hedge Row (Pvt.) E&W	8W-15N	18
Holder Ln. E&W	6W-15N	18
Hoyt Ln. N&S	7W-14N	28
Ingram St. N&S	7W-15N	18
Jeffrey St. N&S	7W-15N	18
Lagoon Dr. N&S	6W-15N	18
Lagoon Ln. N&S	6W-14N	28
Latrobe Av. N&S	7W-15N	18
Laurie St. N&S	7W-15N	18

	Grid	Page
Linder Av. N&S	6W-15N	18
Lockwood Av. N&S	6W-15N	18
Maple E&W	6W-15N	18
Maple Row E&W	7W-15N	18
Martin Ln. E&W	7W-15N	18
Meadow Lark Ln. N&S	7W-15N	18
Meadowbrook Dr. E&W	7W-16N	18
Meadowview Rd. N&S	6W-14N	28
Meadowwood Ln. N&S	7W-14N	28
Middle Fork Rd. (Pvt.) E&W	7W-16N	18
Mt. Pleasant St. E&W	6W-15N	18
Norfork Rd. E&W	7W-16N	18
Northfield Rd. N&S	6W-15N	18
Northfield Sq. E&W	6W-15N	18
Northgate Rd. NE&SW	7W-16N	18
Oak E&W	7W-15N	18
Oak Tree Ln. E&W	7W-15N	18
Old Farm Rd. E&W	7W-16N	18
Old Willow Rd. E&W	7W-15N	18
Orchard Ln. E&W	6W-15N	18
Pebble Fork E&W	8W-15N	18
Pine St. E&W	7W-16N	18
Pleasant View Ln. N&S	7W-15N	18
Pvt. Ln. E&W	7W-15N	18
Red Barn Ln. N&S	7W-15N	18
Ridge Rd. N&S	8W-15N	18
Riverside Dr. N&S	7W-15N	18
Robinhood Ln. NW&SE	7W-15N	18
Rolling Ridge Rd. E&W	8W-16N	18
Rt. 43 N&S	8W-15N	18
Shadowood Ln. E&W	7W-16N	18
Somerset Ln. (Pvt.) N&S	7W-15N	18
South Ridge Ter. NW&SE	7W-16N	18
Southgate Rd. NE&SW	7W-16N	18
Sterling Ln. NE&SW	7W-15N	18
Stockton Dr. N&S	7W-15N	18
Suffolk Rd. NS&EW	7W-15N	18
Sunset Dr. E&W	7W-16N	18
Sunset Rd. E&W	6W-15N	18
Sunset Ridge Rd. N&S	8W-16N	18
Thackerary St. N&S	7W-15N	18
Thornwood Ln.(Pvt.) N&S	7W-15N	18
Valley View Rd. NW&SE	7W-16N	18
Wagner Rd. N&S	7W-15N	18
Walnut N&S	6W-15N	18
Westfield Ln. E&W	8W-15N	18
Whittier Ln. (Pvt.) N&S	7W-15N	18
Willow Hill Ct. N&S	7W-15N	18
Willow Rd. E&W	7W-15N	18
Willow Ter. E&W	7W-15N	18
Winfield Cir. N&S	6W-14N	28
Winfield Dr. E&W	6W-14N	28
Winnetka Ct. N&S	8W-15N	18
Winnetka Rd. E&W	8W-15N	18
Woodland Ln., South N&S	7W-15N	18

Northfield Township

	Grid	Page
Anets Dr. NW&SE	9W-16N	17
Applegate Ct. N&S	10W-13N	27
Applegate Ln. N&S	10W-13N	27
Applegate Rd. E&W	10W-13N	27
Beach Ln. N&S	12W-16N	16
Central Av. N&S	11W-16N	16
Central Rd. E&W	11W-13N	26
Constance Ln. N&S	12W-18N	16
Countryside Ln. E&W	11W-13N,10W-13N	26
Cumberland Dr. N&S	12W-17N	16
East Circle N&S	8W-16N	18
Elmdale Rd. N&S	10W-13N	27
Euclid Av. E&W	12W-14N	26
Evergreen Ln. NW&SE	12W-16N	16
Forest Dr. N&S	12W-14N	26
Forest Way Dr. N&S	7W-17N	18
Garden St. E&W	11W-16N	16
Glendale Rd. N&S	10W-13N	27
Glenshire Rd. N&S	10W-13N	27
Glenwood N&S	10W-13N	27
Greendale Rd. NE&SW	10W-13N	27
Greenleaf Av. E&W	12W-14N	26
Greenwood Rd. N&S	10W-13N	27
Grove Ln. N&S	10W-13N	27
Grove St. N&S	10W-13N	27
Henley St. E&W	10W-13N	27
Highland Av. N&S	11W-16N	16
Highland Ct. N&S	10W-13N	27
Hillside Rd. N&S	10W-13N	27

	Grid	Page
Holly Av. N&S	11W-16N	16
Holly Ln. N&S	10W-13N	27
Huber Ln. N&S	10W-13N	27
Knollwood Ln. E&W	10W-13N	27
Landreth Ln. NW&SE	8W-16N	18
Landwehr Rd. N&S	11W-13N	26
Larch Av. E&W	12W-14N	26
Laurel Av. E&W	12W-14N	26
Liberty Ln. E&W	10W-13N	27
Lilac Av. E&W	12W-14N	26
Linden Av. E&W	10W-13N	27
Linneman St. E&W	10W-13N	27
Locust Av. E&W	12W-14N	26
Long Meadow Ln. N&S	10W-13N	27
Longview Rd. N&S	11W-16N	16
Magnolia St. N&S	12W-14N	26
Maple Av. E&W	12W-17N	16
Maple St. N&S	12W-14N	26
Meadow Ln. NS&EW	10W-13N	27
Meadow St. N&S	11W-16N	16
Milton St. N&S	12W-14N	26
Milwaukee Av. NW&SE	12W-15N,11W-13N	16
Oak Av. E&W	12W-17N	16
Pamela Ln. E&W	12W-18N	16
Peachgate Ct. N&S	10W-13N	27
Peachgate Ln. N&S	10W-13N	27
Peachgate Rd. E&W	10W-13N	27
Pensive Ln. N&S	12W-16N	16
Pfingston Rd. N&S	11W-13N	26
Pfingston Rd. N&S	11W-13N	26
Phyllis Rd. E&W	12W-18N	16
Post Rd. N&S	11W-16N	16
Prairie Av. N&S	11W-16N	16
Revere Rd. N&S	10W-13N	27
Ridge Dr. N&S	11W-16N	16
Riverdale E&W	7W-16N	18
Rolling Ridge Rd. E&W	8W-16N	18
Rosedale Rd. N&S	10W-13N	27
Steeplechase Ln. E&W	8W-16N	18
Sunset Trail E&W	11W-16N	16
Sunshine Ln. N&S	12W-18N	16
Techny Rd. E&W	12W-16N	16
Timberlane Dr. E&W	12W-16N	16
Tri-state Tollway N&S	11W-13N,12W-17N	26
Valley Way E&W	8W-16N	18
Vogay Ln. N&S	12W-16N	16
Wedel Ln. N&S	10W-13N	27
Western Av. N&S	12W-16N	16
Whirlaway Dr. NS&EW	12W-16N,11W-16N	16
Winkleman Rd. NW&SE	12W-15N	16

Northlake

	Grid	Page
Alvin N&S	13W-2N	46
Armitage E&W	13W-2N	46
Ashbel N&S	14W-1N	46
Bell Dr. E&W	13W-2N	46
Bernice E&W	13W-2N	46
Caryl E&W	13W-2N	46
Charles E&W	13W-2N	46
Country Club Dr. E&W	13W-2N	46
Dewey E&W	13W-2N	46
Dickens E&W	13W-2N	46
Diversey E&W	14W-3N	40
Dodd N&S	13W-1N	46
East Dr. N&S	13W-2N	46
Edward N&S	13W-2N	46
Elm N&S	14W-1N	46
Franklin Dr. E&W	14W-2N	46
Gail N&S	13W-1N	46
Geneva N&S	13W-2N	46
Golfview Dr. E&W	14W-2N	46
Haber Ct. N&S	13W-2N	46
Harold N&S	13W-2N	46
Harvard N&S	14W-1N	46
Hayes E&W	13W-3N	40
Hillside N&S	14W-1N	46
Hirsch E&W	13W-1N	46
Irving N&S	14W-1N	46
Jerome Dr. N&S	14W-2N	46
Lakewood N&S	13W-1N	46
Laporte N&S	14W-1N	46
Lavergne N&S	14W-1N,14W-2N	46
Le Moyne E&W	14W-1N,13W-1N	46
Lind N&S	14W-1N	46
Longfield E&W	14W-2N	46
Lyndale E&W	13W-2N	46
Major E&W	13W-2N	46
Marilyn N&S	13W-1N,13W-2N	46
McLean N&S	13W-2N	46
Medill E&W	13W-2N	46
Morse E&W	13W-1N	46
Palmer E&W	13W-2N	46
Park View Dr. N&S	13W-1N	46
Prater N&S	13W-1N,13W-2N	46
Railroad Av. N&S	14W-1N	46
Rhodes E&W	14W-3N	40
Roberta N&S	13W-2N	46
Romans N&S	14W-2N	46

	Grid	Page
Roy N&S	13W-1N,13W-2N	46
Rt. 16 N&S	14W-3N	40
Sandra N&S	14W-2N	46
Soffel E&W	14W-1N,13W-1N	46
Tri-State Tollway N&S	14W-3N	40
U.S. Rt. 12 N&S	13W-2N	46
U.S. Rt. 20 E&W	13W-1N	46
U.S. Rt. 45 N&S	13W-2N	46
Victoria Dr. E&W	13W-1N	46
Victoria Rd. N&S	13W-1N	46
Village Dr. E&W	13W-1N	46
Wagner Dr. N&S	13W-3N	40
West Dr. E&W	14W-2N	46
West Dr. N&S	14W-2N	46
Westward Ho Dr. E&W	14W-2N	46
Whitehall E&W	13W-2N	46
Whittier E&W	13W-2N	46
William N&S	13W-2N	46
44th N&S	13W-1N	46
45th N&S	13W-1N	46

Norwood Park Township

	Grid	Page
Balmoral E&W	10W-6N	41
Bryn Mawr E&W	10W-7N	41
Courtland N&S	10W-7N	41
Fairview N&S	10W-7N	41
Gregory E&W	10W-6N	41
Linden N&S	9W-6N	41
Manor Ln. N&S	10W-7N	41
Olive E&W	10W-7N	41
Prospect N&S	10W-7N	41
Redwood N&S	10W-7N	41
Seminole E&W	10W-7N	41
Summerdale E&W	10W-6N	41
Vine N&S	10W-7N	41
Washington N&S	10W-6N	41

Oak Forest

	Grid	Page
Adele Ln. E&W	5W-19S	80
Adeline Pl. E&W	6W-18S	80
Aimee Ct. N&S	6W-19S	80
Alameda Av. NS&EW	7W-18S	80
Albert Dr. NW&SE	7W-18S	80
Aldersyde Rd. E&W	6W-20S	86
Ann Marie Dr. N&S	6W-18S	80
Ann Marie Ln. N&S	6W-18S	80
Anne Ct. N&S	6W-18S	80
Arroyo Dr. N&S	7W-18S	80
Avalon E&W	7W-18S	80
Babette Ct. E&W	6W-19S	80
Barry Ln. E&W	5W-19S	80
Barton Ln. N&S	6W-18S	80
Becky Ct. N&S	6W-17S	80
Bellaire Rd. E&W	6W-20S	86
Belle St. N&S	6W-20S	86
Beth Ct. N&S	6W-19S	80
Betty Ann Ln. N&S	6W-18S	80
Blair Ln. N&S	5W-19S	80
Boca Rio Dr. N&S	7W-18S	80
Bonnie Tr. NS&EW	6W-20S	86
Brenden Ln. N&S	5W-19S	80
Bret Dr. E&W	7W-18S	80
Brianne Ln. N&S	8W-18S	79
Briar Ln. N&S	5W-19S	80
Brockton Ln. N&S	5W-19S	80
Brookwood Dr. E&W	7W-18S	80
Caletta Terr. E&W	7W-17S	80
Carol Belle Tr. E&W	6W-18S	80
Carolyn Ct. E&W	6W-18S	80
Carriage Way E&W	7W-17S	80
Catalina Av. E&W	7W-18S	80
Cedar Rd. E&W	8W-18S	79
Central Av. N&S	7W-18S,7W-17S	80
Charles E&W	7W-18S	80
Charleston St. E&W	8W-18S	79
Chaucer Dr. NS&EW	7W-18S	80
Cherry Rd. N&S	8W-18S	79
Chestnut Ln. N&S	8W-18S	79
Church Rd. NE&SW	8W-18S	79
Cicero Av. N&S	6W-20S,6W-17S	86
Colina Av. N&S	7W-18S	80
Concha Ct. NW&SE	7W-18S	80
Condado Dr. N&S	7W-18S	80
Corey Ln. E&W	7W-18S	80
Coulter Rd. E&W	6W-20S	86
Courtney E&W	8W-18S	79
Craig Dr. N&S	5W-19S	80
Creekside Dr. NE&SW	6W-18S	80
Crescent Green Ln. NS&EW	6W-18S	80
Crescent Ln. E&W	6W-18S	80
Cromwell Av. E&W	6W-17S	80
Cypress Ct. E&W	6W-20S	86
Dan Ryan Expwy. NE&SW	5W-19S	80
Daniel Ln. E&W	7W-17S	80
David Ln. N&S	5W-19S	80
Debbie Ct. NE&SW	6W-19S	80
Debra Dr. N&S	6W-19S	80
Deerpath Rd. E&W	6W-20S	86
Diane Ct. E&W	6W-18S	80
Dolores Dr. E&W	7W-18S	80
Dover Rd. NW&SE	7W-18S	80
Duncan Rd. N&S	7W-18S	80
Edgewood Ct. E&W	7W-17S	80
Edward Dr. E&W	7W-18S	80
El Morro Ct. N&S	7W-18S	80

	Grid	Page
El Morro Ln. E&W	7W-18S	80
El Vista Av. N&S	6W-17S	80
Elderwood Ct. E&W	7W-17S	80
Elizabeth Ct. NW&SE	6W-20S	86
Ellen Ct. E&W	6W-19S	80
Elm Ln. E&W	8W-18S	79
Elmwood Rd. E&W	6W-20S	86
Essex Rd. E&W	7W-18S	80
Fairfax Rd. E&W	7W-18S	80
Farmsley Ct. NW&SE	6W-20S	86
Fawn Ct. E&W	6W-18S	80
Fern Av. E&W	7W-17S,6W-17S	80
Fieldcrest Dr. E&W	5W-19S	80
Forest Ave. N&S	6W-20S,6W-19S	86
Forest Ct. E&W	6W-20S	86
Forest Edge Ln. N&S	6W-17S	80
Forest Tr. E&W	6W-20S	86
Forestview Dr. E&W	8W-18S,7W-18S	79
Geoffrey Rd. N&S	7W-18S	80
George Dr. N&S	5W-19S	80
Grange Av. E&W	7W-17S,6W-17S	80
Greentree Rd. E&W	6W-20S	86
Grove Av. N&S	6W-19S	80
Harbor Dr. N&S	7W-17S	80
Harold St. N&S	5W-19S	80
Hawthorne Rd. E&W	6W-20S	86
Henry St. E&W	5W-19S	80
Heron Dr. E&W	7W-18S	80
Hickory Ln. N&S	8W-18S	79
Hickory Rd. N&S	8W-18S	79
Hillside Av. N&S	7W-18S	80
Independence Av. E&W	7W-18S	80
Independence Ct. N&S	7W-18S	80
Interstate 57 NE&SW	5W-19S	80
James Dr. NW&SE	7W-18S	80
Jamie Ct. N&S	6W-19S	80
Jamie Ln. E&W	6W-19S	80
Janet Ct. NW&SE	6W-19S	80
Jean Ct. N&S	6W-18S	80
Jessica Dr. E&W	6W-20S	86
Jill Ann Dr. E&W	7W-18S	80
Joann Ln. N&S	7W-18S	80
Jon Rd. N&S	7W-18S	80
Jones Ct. N&S	6W-17S	80
Judy Ct. E&W	6W-20S	86
Kara Ct. N&S	6W-20S	86
Keating N&S	5W-18S	80
Kedvale Av. N&S	5W-19S	80
Kenton Av. N&S	5W-18S	80
Kilbourn Av. N&S	5W-20S,5W-18S	86
Kilpatrick Av. N&S	5W-18S	80
Kimberly Ct. E&W	6W-20S	86
Knollwood Dr. E&W	7W-17S	80
Knox Av. N&S	5W-18S	80
Kolmar N&S	5W-18S	80
La Grange Ct. E&W	7W-18S	80
La Palm Ct. E&W	6W-17S	80
La Palm Dr. NS&EW	7W-17S	80
La Paz Ct. N&S	7W-18S	80
La Paz Dr. E&W	7W-18S	80
Lacrosse Av. N&S	6W-18S	80
Lamon Av. N&S	6W-18S,6W-17S	80
Lancaster Ln. NE&SW	7W-18S	80
Landings Ln. N&S	7W-17S	80
Langley Ct. NW&SE	6W-20S	86
Laporte Av. N&S	6W-18S	80
Laramie Av. N&S	6W-19S,6W-17S	80
Laramie Ct. E&W	6W-18S	80
Las Flores Ln. N&S	7W-18S	80
Las Robles Ct. NW&SE	7W-18S	80
Las Robles St. N&S	7W-18S	80
Latrobe Av. N&S	6W-19S,6W-18S	80
Laura Ln. N&S	5W-19S	80
Lavergne Av. N&S	6W-19S,6W-17S	80
Lawler Av. N&S	6W-17S	80
Leclaire Av. N&S	6W-20S,6W-17S	86
Leslie Ln. E&W	6W-19S	80
Liberty Square E&W	7W-18S	80
Linden Dr. NE&SW	8W-18S	79
Linder Av. N&S	6W-17S	80
Lisa Ln. E&W	5W-19S	80
Lockwood Dr. N&S	6W-19S,16W-18S	80
Long Av. N&S	6W-19S,6W-17S	80
Loran Ln. E&W	6W-17S	80
Lorel Av. N&S	6W-19S,6W-18S	80
Lynne Ct. N&S	6W-17S	80
Mann St. E&W	5W-19S	80
Maple Ct. NW&SE	8W-18S	79
Maple Dr. N&S	8W-18S	79
Margie Ln. E&W	6W-19S	80
Martha Ln. E&W	6W-20S	86
Mary Ann Ct. E&W	6W-19S	80
Massasoit Av. N&S	7W-17S	80
Meadowdale Dr. N&S	6W-20S	86
Menard Av. N&S	7W-17S	80
Merlin Ct. E&W	7W-18S	80
Michaele Dr. N&S	6W-18S	80
Mission Av. N&S	6W-17S	80
Moorings Ln. N&S	7W-17S	80
Natalie Dr. NS&EW	6W-18S	80
Newport Dr. NE&SW	6W-20S	86
Oak Park Av. N&S	8W-18S	79
Oak Rd. NE&SW	8W-18S	79
Oak St. N&S	6W-19S	80
Oakland E&W	7W-18S	80
Oakwood Ct. NW&SE	6W-20S	86
Orange Ln. E&W	7W-18S	80
Orchard Rd. N&S	8W-18S	79
Orogrande Ct. NW&SE	7W-18S	80
Orogrande St. N&S	7W-18S	80
Oxford Dr. N&S	7W-18S	80
Pamela Ct. NW&SE	6W-19S	80
Park Av. N&S	6W-17S	80
Parkside Av. N&S	7W-17S	80
Parkwood Ct. E&W	7W-17S	80
Peggy Ln. N&S	7W-18S	80
Pine Ct. N&S	8W-18S	79
Pine Rd. N&S	8W-18S	79
Reynolds Ln. N&S	7W-18S	80
Richard Av. E&W	5W-19S	80
Ridgeland Av. N&S	8W-18S	79
Ridgewood Dr. N&S	7W-17S	80
Rio Verde Dr. E&W	7W-18S	80
Rob Roy Cir. NW&SE	7W-18S	80
Rob Roy Dr. NS&EW	7W-18S	80
Roy St. N&S	5W-19S	80
Sara Ann Ln. NW&SE	7W-18S	80
Scarborough Ln. NW&SE	6W-17S	80
School E&W	7W-18S	80
Scott St. E&W	5W-19S	80
Sequoia St. N&S	7W-18S	80
Sierra Dr. N&S	7W-18S	80
Spruce Ln. N&S	8W-18S	79
Stuart Ln. E&W	7W-18S	80
Sunset Av. N&S	6W-17S	80
Sunset Ct. N&S	6W-17S	80
Sussex Ct. N&S	6W-17S	80
Sycamore Ln. N&S	6W-20S	86
Temple St. N&S	6W-17S	80
Terrace Dr. N&S	7W-18S	80
Terry Ln. N&S	5W-19S	80
Thackery Av. N&S	6W-20S	86
Timber Ct. E&W	6W-20S	86
Tudor Rd. N&S	7W-18S	80
Ventura St. N&S	7W-18S	80
Vera Ct. N&S	6W-19S	80
Victoria Ct. N&S	7W-18S	80
Victoria Dr. E&W	8W-18S,7W-17S	79
Vine St. E&W	7W-17S,6W-17S	80
Vista Dr. N&S	7W-18S	80
Wagman St. E&W	5W-19S	80
Walnut Rd. NE&SW	8W-18S	79
Waverly Av. NE&SW	5W-18S	80
Westview Dr. NS&EW	8W-18S	79
Willow Ln. N&S	8W-18S	79
Willowick Dr. E&W	5W-19S	80
Woodland Dr. E&W	6W-18S	80
69th Ct. N&S	8W-18S	79
147th St. E&W	7W-17S,6W-17S	80
149th St. E&W	6W-17S	80
150th Pl. E&W	7W-17S	80
150th St. E&W	7W-17S,6W-17S	80
151st St. E&W	8W-17S,6W-17S	79
152nd Pl. E&W	8W-18S	79
152nd St. E&W	8W-18S,5W-18S	79
153rd St. E&W	6W-18S,5W-18S	80
154th Pl. E&W	8W-18S	79
154th St. E&W	6W-18S	80
155th Pl. E&W	8W-18S	79
155th St. E&W	6W-18S	80
156th St. E&W	6W-18S	80
157th Pl. E&W	7W-18S	80
157th St. E&W	8W-18S,6W-18S	79
158th St. E&W	6W-18S	80
159th St. E&W	8W-18S,6W-18S	79
160th St. E&W	6W-19S	80
161st St. E&W	6W-19S	80
162nd St. E&W	6W-19S	80
163rd St. E&W	6W-19S	80
165th St. E&W	5W-19S	80
166th St. E&W	5W-19S	80
167th St. E&W	7W-19S,6W-19S	80
169th St. E&W	6W-20S	86
170th Pl. E&W	6W-20S	86
170th St. E&W	6W-20S	86

Oak Lawn

	Grid	Page
Adeline Dr. E&W	5W-12S	66
Alexander Pl. E&W	6W-10S	66
Alice Ct. N&S	6W-11S	66
Arnold Pl. NE&SW	5W-12S	66
Austin Av. N&S	7W-11S,7W-10S	66
Avery Pl. E&W	6W-10S	66
Avon Av. NW&SE	8W-11S	65
Brandt Av. N&S	6W-11S	66
Buell Av. N&S	6W-11S	66
Burns Dr. N&S	6W-11S	66
Campbell N&S	6W-11S	66
Cass St. E&W	6W-10S	66
Central Av. N&S	6W-12S,6W-10S	66
Cicero Av. N&S	6W-12S,6W-10S	66
Columbus Dr. NE&SW	6W-10S	66
Cook Av. N&S	6W-12S,6W-11S	66
Crescent Ct. N&S	8W-10S	65
David Ct. E&W	7W-11S	66
Dean Dr. NE&SW	5W-12S	66
Drury Ln. E&W	6W-11S	66
Dumke Dr. E&W	6W-10S	66
East Shore Dr. NE&SW	6W-11S	66
Edison Av. NE&SW	7W-11S,6W-11S	66
Elm Circle Dr. NE&SW	6W-11S	66
Fairfax St. E&W	5W-12S	66
Franklin Av. NE&SW	6W-11S	66
Georgia Ln. N&S	6W-12S	66
Grant St. E&W	5W-12S	66
Harnew Rd., East N&S	6W-11S	66
Harnew Rd., South E&W	6W-11S	66
Harnew Rd., West NE&SW	6W-11S	66
Hilton Dr. N&S	5W-10S	66
Karlov Av. N&S	5W-12S,5W-10S	66
Kathleen Ln. E&W	5W-12S	66
Keating Av. N&S	5W-12S,5W-10S	66
Kedvale Av. N&S	5W-12S,5W-10S	66
Keeler Av. N&S	5W-12S,5W-10S	66
Kenneth Av. N&S	5W-12S,5W-11S	66
Kenton Av. N&S	5W-12S,5W-10S	66
Kilbourn Av. N&S	5W-12S,5W-10S	66
Kildare Av. N&S	5W-12S,5W-11S	66
Kilpatrick Av. N&S	5W-12S,5W-10S	66
Kimball Pl. E&W	6W-10S	66
Knox Av. N&S	5W-12S,5W-10S	66
Knox Ct. E&W	5W-11S	66
Kolin Av. N&S	5W-12S,5W-11S	66
Kolmar Av. N&S	5W-12S,5W-10S	66
Komensky Av. N&S	5W-12S,5W-10S	66
Kostner Av. N&S	5W-12S,5W-10S	66
La Crosse Av. N&S	6W-12S,6W-11S	66
La Porte Av. N&S	6W-12S	66
Lamb Dr. E&W	6W-11S	66
Lamon Av. N&S	6W-12S	66
Laramie Av. N&S	6W-12S,6W-11S	66
Lavergne Av. N&S	6W-12S	66
Lawler Av. N&S	6W-12S	66
Lawrence Ct. N&S	6W-11S	66
Lawton Dr. NE&SW	6W-11S	66
Le Claire Av. N&S	6W-12S	66
Linder Av. N&S	6W-12S	66
Linus Ln. N&S	6W-12S	66
Lockwood Av. N&S	6W-12S	66
Long Av. N&S	6W-12S	66
Longwood Ln. N&S	5W-12S	66
Lorel Av. N&S	6W-12S	66
Lynwood Dr. NS&EW	7W-10S	66
Major Av. N&S	7W-11S,7W-10S	66
Mansfield Av. N&S	7W-12S,7W-11S	66
Maple Av. N&S	6W-11S	66
Marion Av. N&S	7W-11S	66
Mason Av. N&S	7W-12S,7W-11S	66
Massasoit Av. N&S	7W-11S,7W-10S	66
Mayfield Av. N&S	7W-12S,7W-10S	66
Meade Av. N&S	7W-11S,7W-10S	66
Melvina Av. N&S	7W-11S,7W-10S	66
Menard Av. N&S	7W-12S,7W-10S	66
Merrimac Av. N&S	7W-11S,7W-10S	66
Merton Av. N&S	7W-11S	66
Minnick Av. N&S	6W-11S	66
Mobile Av. N&S	7W-10S	66
Monitor Av. N&S	7W-10S	66
Moody Av. N&S	7W-11S,7W-10S	66
Msgr. Mc Nichols Dr. E&W	7W-10S,7W-9S	66
Mulberry Av. N&S	6W-11S	66
Mulligan Dr. N&S	7W-10S	66
Nashville Av. N&S	8W-11S,8W-10S	65
Natchez Av. N&S	8W-10S	65
Natoma Av. N&S	8W-11S,8W-10S	65
Neva Av. N&S	8W-10S	65
New England Av. N&S	8W-11S,8W-10S	65
Newland Av. N&S	8W-10S	65
Nora Av. N&S	8W-10S	65
Nordica Av. N&S	8W-10S	65
Normandy Av. N&S	8W-11S,8W-10S	65
North St. E&W	6W-10S	66
Nottingham Av. N&S	8W-11S	65
Oak Center Dr. E&W	6W-11S	66
Oak Park Av. N&S	8W-11S,8W-10S	65
Oak St. E&W	6W-10S	66
Oakdale Dr. E&W	6W-11S	66
Otto Pl. E&W	6W-10S	66
Pacific NE&SW	7W-11S	66
Pacific SW&NE	7W-11S	66
Parke Av. N&S	5W-11S	66
Parkside Av. N&S	7W-12S,7W-10S	66
Paxton Rd. NE&SW	6W-11S	66
Pulaski Rd. N&S	5W-12S,5W-10S	66
Ridgeland Av. N&S	8W-11S,8W-10S	65
Robertson Av. N&S	6W-11S	66
Rt. 7 E&W	8W-11S,5W-10S	65
Ruby St. E&W	6W-10S	66
Rumsey Av. NE&SW	5W-10S	66
Rutherford Av. N&S	8W-11S	65
Sayre Av. N&S	8W-11S,8W-10S	65
Scott Ln. E&W	6W-11S	66
Southwest Hwy. E&W	7W-11S,5W-10S	66
Spring Rd. NE&SW	6W-11S	66
Sproat Av. N&S	6W-10S	66
St. James Ct. NE&SW	7W-10S	66
Stillwell Pl. E&W	5W-12S	66
Stone Circle E&W	6W-11S	66
Stony Creek Dr. NW&SE	6W-12S	66
Tripp Av. N&S	5W-12S,5W-10S	66
Tulley Av. N&S	6W-11S,6W-10S	66
Wabash Av. NE&SW	6W-11S	66
Wainwright Pl. NE&SW	5W-12S	66
Warren Av. N&S	6W-11S	66
Washington St. N&S	6W-12S,6W-11S	66
West Shore Dr. NE&SW	6W-11S	66
Wick Dr. NW&SE	6W-11S	66
William Pl. E&W	5W-11S	66
Wolfe Dr. NE&SW	6W-12S	66
48th Ct. N&S	6W-10S	66
49th Av. N&S	6W-11S,6W-10S	66
49th Ct. N&S	6W-10S	66
50th Av. N&S	6W-10S	66
50th Ct. N&S	6W-11S	66
51st Av. N&S	6W-11S,6W-10S	66
51st Ct. N&W	6W-12S	66
52nd Av. N&S	6W-11S,6W-10S	66
52nd Ct. N&S	6W-10S	66
53rd Av. N&S	6W-11S,6W-10S	66
53rd Ct. N&S	6W-10S	66
54th Av. N&S	6W-11S,6W-10S	66
54th Ct. N&S	6W-10S	66
55th Av. N&S	6W-11S,6W-10S	66
55th Ct. N&S	6W-10S	66
68th Ct. N&S	8W-10S	65
69th Av. N&S	8W-10S	65
69th Ct. N&S	8W-10S	65
87th Pl. E&W	8W-10S,7W-10S	65
87th St. E&W	8W-10S,6W-10S	65
88th Pl. E&W	8W-10S,6W-10S	65
88th St. E&W	8W-10S,6W-10S	65
89th Pl. E&W	8W-10S,7W-10S	65
89th St. E&W	7W-10S,6W-10S	66
90th Pl. E&W	7W-10S	66
90th St. E&W	8W-10S,6W-10S	65
91st Pl. E&W	8W-10S,5W-10S	65
91st St. E&W	8W-10S,6W-10S	65
92nd Pl. E&W	7W-10S	66
92nd St. E&W	8W-10S,5W-10S	65
93rd Pl. E&W	8W-10S,5W-10S	65
93rd St. E&W	8W-10S,5W-10S	65
94th Pl. E&W	8W-10S	65
94th St. E&W	8W-10S,5W-10S	65
95th Pl. E&W	8W-11S	65
95th St. E&W	8W-10S,5W-10S	65
96th Pl. E&W	8W-11S,5W-11S	65
96th St. E&W	8W-11S,5W-11S	65
97th Pl. E&W	6W-11S,5W-11S	66
97th St. E&W	8W-11S,5W-11S	65
98th Pl. E&W	7W-11S,5W-11S	66
98th St. E&W	8W-11S,5W-11S	65
99th Pl. E&W	7W-11S,5W-11S	66
99th St. E&W	8W-11S,5W-11S	65
100th Pl. E&W	7W-11S,5W-11S	66
100th St. E&W	7W-11S,5W-11S	66
101st Pl. E&W	7W-11S,5W-11S	66
101st St. E&W	7W-11S,5W-11S	66
102nd Pl. E&W	5W-11S	66
102nd St. E&W	7W-11S,5W-11S	66
103rd Pl. E&W	6W-12S	66
103rd St. E&W	7W-11S,5W-11S	66
104th Ln. E&W	5W-12S	66
104th Pl. E&W	5W-12S	66
104th St. E&W	7W-12S,5W-12S	66
105th Pl. E&W	6W-12S,5W-12S	66
105th St. E&W	7W-12S,5W-12S	66
106th Pl. E&W	6W-12S,15W-12S	66
106th St. E&W	6W-12S,5W-12S	66
107th Pl. E&W	5W-12S	66
107th St. E&W	6W-12S,5W-12S	66
108th Pl. E&W	6W-12S,5W-12S	66
108th St. E&W	6W-12S,5W-12S	66
109th St. E&W	6W-12S,5W-12S	66
110th Pl. E&W	5W-12S	66
110th St. E&W	6W-12S,5W-12S	66
111th St. E&W	6W-12S,5W-12S	66

Oak Park

	Grid	Page
Adams E&W	8W-0S	48
Augusta E&W	7W-1N	48
Austin Blvd. N&S	7W-0S,7W-1N	48
Belleforte N&S	8W-1N	48
Berkshire E&W	8W-1N	48
Carpenter N&S	8W-0S	48
Clarence N&S	8W-0S	48
Clinton N&S	8W-0S,8W-0N	48
Columbian N&S	8W-1N	48
Cuyler N&S	7W-0S,7W-1N	48
East N&S	8W-0S,8W-1N	48
Edmer N&S	7W-1N	48
Elizabeth Ct. E&W	8W-0N	48
Elmwood N&S	8W-0S,8W-1N	48
Erie E&W	8W-0N,7W-0N	48
Euclid N&S	8W-0S,8W-1N	48
Fair Oaks N&S	8W-1N	48
Fillmore E&W	8W-0S	48
Flournoy E&W	7W-0S	48
Forest N&S	8W-1N	48
Garfield E&W	8W-0S,7W-0S	48
Greenfield E&W	8W-1N	48
Grove N&S	8W-0S,8W-1N	48
Gunderson N&S	8W-0S	48
Harlem Ct. N&S	8W-0N	48
Harrison E&W	7W-0S	48
Harvard E&W	8S-0S	48
Harvey N&S	7W-0S,7W-1N	48
Hayes N&S	7W-1N	48
Highland N&S	7W-0S	48
Holly Ct. E&W	8W-0N	48
Home N&S	8W-0S	48
Humphrey N&S	7W-0S,7W-1N	48
Iowa E&W	8W-1N	48
Jackson Blvd. E&W	8W-0S	48
Kenilworth N&S	8W-0S,8W-1N	48
Le Moyne Pkwy. E&W	8W-1N	48
Lenox E&W	8W-1N	48
Linden N&S	8W-1N	48
Lombard N&S	7W-0S,7W-1N	48
Lyman N&S	7W-0S	48
Madison E&W	8W-0N	48
Maple N&S	8W-0S,8W-0N	48
Mapleton N&S	7W-1N	48

	Grid	Page
Marion N&S	8W-0N,8W-1N	48
Miller E&W	8W-1N	48
Monroe E&W	8W-0S	48
North Blvd. E&W	8W-0N	48
Oak Park N&S	8W-0S,8W-1N	48
Ontario E&W	8W-0N	48
Paulina E&W	8W-1N	48
Pleasant E&W	8W-0N,7W-0N	48
Pleasant Pl. E&W	8W-0N	48
Randolph E&W	8W-0N,7W-0N	48
Ridgeland N&S	8W-0S,8W-1N	48
Roosevelt E&W	8W-0S	48
Rossell N&S	8W-1N	48
Schneider E&W	8W-1N	48
Scoville N&S	8W-0S,8W-0N	48
South Blvd. E&W	8W-0N	48
Superior E&W	8W-0N	48
Taylor N&S	7W-0S,7W-1N	48
Thomas E&W	8W-1N	48
Van Buren E&W	8W-0S	48
Washington E&W	8W-0N	48
Wenonah N&S	8W-0S	48
Wesley N&S	8W-0S,8W-0N	48
Westgate E&W	8W-0N	48
Wisconsin N&S	8W-0S	48
Woodbine N&S	8W-1N	48

Olympia Fields

	Grid	Page
Achilles Av. N&S	4W-24S	92
Alexander St. NW&SE	3W-25S	92
Apollo Cir. E&W	3W-25S	92
Arcadian Dr. NE&SW	4W-24S	92
Athens Rd. E&W	3W-25S	92
Attica Rd. N&S	4W-24S	92
Birch Ln. N&S	3W-24S	92
Bristol Ln. N&S	3W-25S	92
Brookside Blvd. N&S	4W-25S	92
Brookwood Dr. NS&EW	3W-25S,3W-24S	92
Brookwood Terr. N&S	3W-24S	92
Byron Ct. NE&SW	4W-25S	92
Cambridge Ln. N&S	3W-25S	92
Chariot Ln. E&W	3W-24S	92
Chelsea Cir. E&W	3W-25S	92
Corinth Rd. N&S	3W-25S,3W-24S	92
Country Club Rd. NE&SW	3W-24S	92
Crawford Av. N&S	4W-25S,4W-24S	92
Cumberland Tr. NW&SE	4W-24S	92
Cumnock Rd. E&W	4W-25S	92
Danube Way N&S	4W-25S	92
Dartmouth Ln. E&W	3W-25S	92
Doria Ln. NE&SW	4W-24S	92
Dutra Av. N&S	2W-24S	93
Elliott Ct. NS&EW	4W-25S	92
Evergreen Cir. N&S	4W-25S	92
Exmoor Rd. N&S	4W-25S	92
Fairfield Av. N&S	4W-24S	92
Gardner Av. N&S	2W-25S,2W-24S	93
Gleneagles Dr. E&W	3W-24S	92
Governors Dr. NE&SW	4W-24S	92
Governors Hwy. NE&SW	4W-25S,4W-24S	92
Graymoor Ln. NS&EW	2W-24S	93
Greenwood Cen. Ct. N&S	4W-25S	92
Greenwood Ct. E&W	4W-25S	92
Greenwood Dr. E&W	4W-25S	92
Harding Av. N&S	3W-24S	92
Harding St. N&S	3W-25S	92
Hellenic Dr. N&S	4W-24S	92
Hutchinson Av. N&S	2W-25S,2W-24S	93
Ionia Av. E&W	4W-24S	92
Ithaca Rd. NE&SW	4W-24S	92
Joe Orr Rd. E&W	2W-24S	93
Kedzie Av. N&S	4W-25S,4W-24S	92
Lincoln Hwy. E&W	4W-25S,3W-25S	92
London Dr. E&W	3W-25S	92
Marathon Ct. N&S	4W-24S	92
Maros Ln. E&W	3W-25S	92
Mellett Ct. E&W	4W-24S	92
Mohawk Tr. NE&SW	4W-24S	92
Oakwood Dr. E&W	3W-24S	92
Oakwood Terr. NE&SW	3W-24S	92
Olympian Way NE&SW	4W-25S,4W-24S	92
Olympic Cir. E&W	4W-25S	92
Orchard Dr. N. N&S	3W-25S	92
Oregon Tr. NE&SW	4W-24S	92
Overland Trail NE&SW	4W-24S	92
Paris Rd. E&W	3W-25S	92
Parthenon Way E&W	4W-24S	92
Pine Av. E&W	4W-25S	92
Platte Tr. E&W	4W-24S	92
Promethian Way N&S	3W-24S	92
Reigler Dr. N&S	3W-25S	92
Roslyn Rd. N&S	4W-25S	92
Santa Fe Trail NW&SE	4W-24S	92
Sheffield Cir. NE&SW	3W-25S	92
Sparta Ct. NE&SW	3W-25S	92
Sparta Ln. N&S	3W-25S	92
St. Andrews Ct. N&S	3W-24S	92
St. Andrews Dr. E&W	3W-24S	92
Strauss Ln. N&S	4W-25S	92
Tenata Ct. N&S	3W-24S	92
Thornwood Dr. N&S	4W-25S	92
Trails Dr. NW&SE	4W-24S	92
Travers Av. N&S	2W-25S,2W-24S	93
Troy Cir. E&W	3W-25S	92
U.S. Rt. 30 E&W	4W-25S,3W-25S	92
Vollmer Rd. E&W	4W-24S,2W-24S	92
Warren Cir. N&S	3W-25S	92
Washington St. N&S	3W-25S	92
Western Av. N&S	3W-25S,3W-24S	92
Wilderness Tr. N&S	4W-24S	92
Wingate Rd. N&S	4W-25S	92
Woodland Ct. E&W	4W-25S	92
Woodland Dr. E&W	4W-25S	92
Woodstock Rd. N&S	4W-25S	92
203rd St. E&W	4W-24S,3W-24S	92
204th St. E&W	4W-24S,3W-24S	92
205th St. E&W	4W-24S	92
206th St. E&W	4W-24S	92
207th St. E&W	4W-24S	92

Orland Hills

	Grid	Page
Applewood Ln. N&S	11W-19S	78
Beacon Ct. E&W	11W-19S	78
Beacon Ln. N&S	11W-19S	78
Birch Ct. N&S	11W-19S	78
Birchwood Dr. E&W	11W-19S	78
Brigitte Ct. N&S	11W-19S	78
Cedarwood Dr. NW&SE	11W-19S	78
Christopher Av. N&S	11W-20S	85
Dwight Ct. E&W	11W-19S	78
Elm Pl. E&W	11W-19S	78
Fox Ct. E&W	11W-19S	78
Haven Av. N&S	11W-20S	85
Haven Av. NE&SW	11W-19S	78
Herbert Ct. NW&SE	11W-19S	78
Hickory Ct. N&S	11W-20S	85
Hickory Dr. E&W	11W-20S	85
Highview Av. N&S	11W-20S	85
Hilltop Av. N&S	11W-20S	85
Hobart Av. N&S	11W-20S	85
Hunter's Ct. E&W	11W-19S	78
Leslie Dr. E&W	11W-19S	78
Marigold Pl. N&S	11W-19S	79
Marilyn Ct. N&S	11W-19S	78
Marshfield Ln. E&W	11W-19S	79
Meadowview Dr. E&W	11W-19S	79
Pepperwood Ln. N&S	11W-19S	78
Pristine Pl. N&S	11W-19S	79
Quail Ct. E&W	11W-19S	78
Redwood Ct. N&S	11W-20S	85
Rioge Ln. E&W	11W-19S	79
Rosebud Pl. N&S	11W-19S	79
Sharon Ct. N&S	11W-19S	78
Vicky Ln. N&S	11W-20S	85
Westwood Ct. N&S	11W-20S	85
Westwood Dr. E&W	11W-20S	85
Willow Terrace Dr. E&W	11W-19S	78
88th Ct. N&S	11W-20S	85
88th St. N&S	11W-20S,11W-19S	85
89th Av. N&S	11W-20S	85
89th Ct. N&S	11W-20S	85
90th Av. N&S	11W-20S	85
90th Ct. N&S	11W-19S	78
91st Av. N&S	11W-20S,11W-19S	85
92nd Av. N&S	11W-20S,11W-19S	85
93rd Av. N&S	11W-20S	85
94th Av. N&S	11W-20S,11W-19S	85
159th Pl. E&W	11W-19S	78
160th Pl. E&W	11W-19S	78
161st Pl. E&W	11W-19S	78
161st St. E&W	11W-19S	78
162nd St. E&W	11W-19S	78
163rd St. E&W	11W-19S	78
167th Pl. E&W	11W-20S	85
167th St. E&W	11W-19S	78
169th Pl. E&W	11W-20S	85
169th St. E&W	11W-20S	85
170th Pl. E&W	11W-20S	85
170th St. E&W	11W-20S	85
171st St. E&W	11W-20S	85

Orland Park

	Grid	Page
Abbey Ln. E&W	11W-18S	78
Acacia Dr. N&S	10W-18S	79
Adria Ct. E&W	10W-15S	73
Alabama Ct.* NE&SW	13W-22S	84
Alaska Ct.* NE&SW	13W-22S	84
Aldwych Dr. N&S	9W-16S	79
Alice Dr. N&S	13W-21S	84
Alpine Dr. N&S	14W-18S	78
Alveston St. E&W	12W-17S	78
Andrea Ct. E&W	13W-21S	84
Andrea Dr. NS&EW	13W-21S	84
Anne Ct. NE&SW	10W-18S	79
Anne Dr. E&W	10W-18S	79
Apache Ln. N&S	9W-16S	79
Apache Pl. E&W	9W-16S	79
Arbor Dr. E&W	14W-18S	78
Aris Ct. NW&SE	11W-18S	78
Arizona Ct.* N&S	13W-22S	84
Arkansas Ct.* NE&SW	13W-22S	84
Arrowhead Ct. N&S	10W-16S	79
Arrowhead Ln. E&W	10W-16S	79
Arthur Ct. E&W	13W-21S	84
Arthur Dr. N&S	13W-21S	84
Ascot Ct. E&W	11W-18S	78
Ash St. N&S	11W-17S	78
Ashley Ct. N&S	10W-17S	79
Ashley Dr. E&W	10W-17S	79
Ashwood Ln. NW&SE	14W-20S	84
Aspen St. N&S	11W-17S	78
Aster Ln. N&S	9W-18S	79
Aubrieta Ct. N&S	9W-18S	79
Aubrieta Ln. E&W	9W-18S	79
Auburn Ct.* E&W	12W-16S	78
Avenida Del Este N&S	12W-17S	78
Avenida Del Norte E&W	12W-17S	78
Baltuscrol Dr. E&W	11W-18S,10W-18S	78
Barleycorn Ct. E&W	11W-18S	78
Bayberry Ct. N&S	10W-18S	79
Bayhill Ct. E&W	10W-18S	79
Beacon Av. N&S	12W-17S	78
Bedford Ln. E&W	11W-18S	78
Beech St. N&S	11W-17S	78
Begonia Ct. N&S	9W-18S	79
Berkhansted Ct. E&W	10W-16S	79
Beverly Ln. E&W	10W-15S	73
Billinary Ct. E&W	14W-21S	84
Biloba Ln. E&W	11W-17S	78
Biltmore Dr. E&W	10W-18S	79
Binford Dr. E&W	10W-16S	79
Birch St. NS&EW	11W-17S	78
Birchbark Ct. N&S	10W-16S	79
Blackhawk Ln. N&S	9W-16S	79
Blue Heron Dr. NE&SW	13W-20S	84
Bob-o-link Rd. NS&EW	10W-18S	79
Bonbury Ln. N&S	10W-16S	79
Boyne Ct. NW&SE	14W-21S	84
Bradford Ln. E&W	11W-18S	78
Bradley Ct. E&W	14W-21S	84
Braeburn Ln. E&W	10W-18S	79
Bramlett Ct. E&W	14W-17S	78
Brassie Ct. N&S	10W-18S	79
Brassie Dr. N&S	10W-18S	79
Brentwood Av. N&S	11W-17S	78
Briarwood Ln. E&W	11W-15S	72
Brighton Ct. NS&EW	10W-16S	79
Bromley St. E&W	10W-17S	79
Brook Crossing Ct. NE&SW	14W-20S	84
Brook Crossing Dr. NE&SW	14W-20S	84
Brook Hill Ct. E&W	14W-21S	84
Brook Hill Dr. NS&EW	14W-21S,14W-20S	84
Brook St. N&S	12W-17S	78
Brookdale Ct. E&W	14W-20S	84
Brookgate Dr. N&S	14W-20S	84
Brookshire Dr. E&W	14W-20S	84
Brookwood Ct. N&S	14W-20S	84
Brookwood Dr. NS&EW	14W-20S	84
Brushwood Ln. NW&SE	14W-20S	84
Bunker Dr. NS&EW	10W-18S	79
Butler Ct. E&W	10W-18S	79
Butterfield Ln. E&W	11W-15S,10W-15S	72
Byron Dr. N&S	11W-19S	78
Caddy Ct. E&W	11W-17S	78
California Ct.* NW&SE	13W-22S	84
Calypso Ln. N&S	9W-18S	79
Cambridge Dr. E&W	10W-16S,9W-16S	79
Camden Dr. N&S	10W-16S	79
Camelia Ln. NS&EW	10W-18S	79
Cameron Pkwy. NS&EW	14W-21S	84
Canterbury Ln. N&S	10W-18S	79
Cardinal Dr. N&S	13W-20S	84
Carlisle Ln. E&W	11W-17S	78
Carnoustie Dr. E&W	11W-18S,10W-18S	78
Carol Ct. N&S	10W-18S	79
Carolina Ln. N&S	11W-16S	78
Cascade Ct. E&W	9W-17S	79
Cashew Dr. E&W	9W-18S	79
Castlebar Ln. N&S	11W-17S	78
Catalina Ct. N&S	9W-18S	79
Catalina Dr. N&S	9W-18S	79
Cathrine Ct.* E&W	12W-16S	78
Cathrine Dr. N&S	11W-16S	78
Centennial Ct. N&S	12W-18S	78
Chadburn Dr. E&W	11W-19S	78
Chapel Hill Rd. N&S	10W-18S	79
Charleston Dr. N&S	11W-16S	78
Chateau Ct. E&W	9W-17S	79
Chauser Dr. N&S	13W-20S	84
Cherry Hills Ct. N&S	10W-18S	79
Cherrywood Ct. N&S	10W-18S	79
Chertsey Ct. E&W	10W-17S	79
Chesterfield Ln. N&S	11W-18S	78
Chestnut Dr. E&W	9W-18S	79
Cheswick Dr. N&S	10W-16S	79
Christine Ct. E&W	11W-16S	78
Clairmont Ct.* E&W	10W-17S	79
Clearview Dr. NS&EW	11W-17S,11W-16S	78
Coghill Ln. N&S	11W-16S	78
Coleman Dr. N&S	13W-19S	78
Colette Ct. E&W	10W-18S	79
Colorado Ct.* E&W	13W-22S	84
Concord Dr. N&S	11W-16S	78
Connecticut Ct.* NE&SW	13W-22S	84
Constitution Ct. E&W	12W-18S	78
Constitution Dr. E&W	12W-18S	78
Cordoba Ct. E&W	12W-18S	78
Cottonwood Ct. N&S	14W-18S	78
Country Club Ln. N&S	10W-17S	79
Country Ct. NW&SE	10W-18S	79
Coventry Ct. N&S	10W-18S	79
Cranna Ct. E&W	14W-21S	84
Cressmoor Ct. E&W	10W-18S	79
Crestview Ct. NW&SE	14W-21S	84
Crestview Dr. NS&EW	14W-21S	84
Cristina Av. NS&EW	10W-16S	79
Croydon Ln. E&W	11W-18S	78
Crystal Tree Dr. N&S	13W-17S	78
Cypress Ct. NE&SW	11W-17S	78
Dakota Ln. E&W	9W-16S	79
Danbury Ln. E&W	11W-18S	78
Davids Ln. N&S	13W-21S	84
Dear Island Av. E&W	13W-19S	78
Deerfield Ct. N&S	10W-18S	79
Deerpath Dr. N&S	10W-16S	79
Delaware Ct.* NE&SW	13W-22S	84
Devonshire Ln. N&S	11W-18S	78
Dewberry Ln. E&W	11W-18S	78
Dexter Ct. E&W	11W-17S	78
Dogwood Dr. NW&SE	11W-17S	78
Doral Ln. E&W	11W-16S	78
Dorstep Ln. E&W	10W-17S	79
Dublin St. E&W	11W-17S	78
Eagle Ridge Dr. NS&EW	13W-21S	84
Edgewood Dr. N&S	10W-18S	79
Eileen Ct. NW&SE	11W-16S	78
El Cameno Ct. E&W	12W-18S	78
El Cameno Ln. E&W	12W-18S	78
El Cameno Re'al N&S	12W-18S,12W-17S	78
El Cameno Terr. N&S	12W-18S	78
Elizabeth Av. NS&EW	10W-16S	79
Esther N&S	13W-21S	84
Evergreen Av. N&S	9W-18S	79
Evergreen Dr. N&S	9W-18S	79
Eynsford Dr. E&W	10W-17S	79
Fairmont Ct.* NE&SW	10W-17S	79
Fairway Dr. E&W	11W-17S	78
Farm Hill Dr. NE&SW	10W-18S	79
Fawn Ct. NE&SW	10W-15S	73
Feather Ct. NE&SW	10W-15S	73
Fernwood Ct. NW&SE	11W-17S	78
Firestone Dr. N&S	11W-15S	72
Flamingo Cir. E&W	10W-18S	79
Flint Ln. E&W	10W-15S	73
Florida Ct.* NE&SW	13W-22S	84
Foxbend Ct. N&S	10W-18S	79
Frances Ln. N&S	11W-18S	78
Franchesca Ct. E&W	12W-18S	78
Franklin Ct. E&W	11W-17S	78
Garden View Ct. N&S	9W-18S	79
Georgia Ct.* N&S	13W-22S	84
Ginger Creek Ln. N&S	14W-18S	78
Glen Oak Rd. E&W	10W-18S	79
Gleneagle Ct. N&S	11W-16S	78
Glenwoody Ct. E&W	10W-16S	79
Golf Rd. NS&EW	13W-17S	78
Golfview Dr. E&W	10W-17S	79
Grace Rd. E&W	10W-18S	79
Grandview Dr. N&S	14W-18S	78
Grange Dr. N&S	14W-20S	84
Green View Rd. N&S	13W-17S	78
Greencastle Ln. E&W	11W-17S	78
Greenfield Dr. E&W	14W-21S	84
Greenland Av. N&S	12W-17S	78
Hale Dr. N&S	12W-17S	78
Halesia Ct. E&W	9W-18S	79
Harbor Town Dr. N&S	9W-18S	79
Harlem Av. N&S	9W-18S	79
Hartwood Ct.* E&W	12W-16S	78
Harvest Crossing E&W	14W-17S	78
Harvest Hill Ct. NW&SE	14W-21S	84
Harvest Hill Dr. NW&SE	14W-20S	84
Hawaii Ct.* N&S	13W-22S	84
Hawthorne Dr. E&W	12W-17S	78
Hazel Ct. NW&SE	12W-17S	78
Heather Ct. N&S	9W-18S	79
Helen Ct. N&S	11W-18S	78
Helen Ln. E&W	11W-18S	78
Hemlock Dr. E&W	9W-18S	79
Hempstead Dr. N&S	9W-16S	79
Henry St. E&W	10W-18S	79
Hiawatha Trail NS&EW	12W-18S	78
Hibiscus Dr. E&W	12W-17S	78
Hidden Brook Ct. NE&SW	14W-20S	84
Highbrush Rd. N&S	13W-20S	84
Highgate Ct. E&W	10W-17S	79
Highland Av. N&S	12W-18S,12W-17S	78
Highwood Ct. NW&SE	14W-20S	84
Highwood Dr. NW&SE	14W-20S	84
Hill Creek Ct. N&S	14W-20S	84
Hilltop Ct. E&W	12W-18S	78
Hilltop Dr. N&S	12W-18S	78
Holiday Ct. N&S	12W-17S	78
Hollow Tree Ct. NW&SE	13W-17S	78
Hollow Tree Rd. NS&EW	13W-17S	78
Holly Ct. N&S	12W-17S	78
Hollyhock Ct. N&S	9W-18S	79
Hollywood Dr. N&S	10W-18S	79
Hopkins Ct. N&S	12W-17S	78
Horn Rd. N&S	10W-18W	79
Huguelet Pl. E&W	11W-18S	78
Huntington Ct. N&S	12W-18S,12W-17S	78
Huntington Pl. NW&SE	12W-17S	78
Hyacinth Dr. E&W	12W-17S	78
Idaho Ct.* NE&SW	13W-22S	84
Idlewild Dr. N&S	11W-16S	78
Indiana Ct.* NE&SW	13W-22S	84
Innsbrook Dr. N&S	9W-18S	79
Innshmor Ct. E&W	14W-21S	84
Inverness Dr. N&S	11W-16S	78
Iowa Ct.* NE&SW	13W-22S	84
Irving Av. N&S	12W-17S	78
Ishnala Dr. N&S	9W-16S	79
Jean Creek Dr. N&S	11W-15S	73
Jefferson Av. N&S	12W-17S	78
John Charles Dr. NW&SE	13W-21S	84
John Humphrey Dr. N&S	11W-17S	78

	Grid	Page
Juniper Ct. E&W....11W-17S		78
Kansas Ct.* NE&SW		
..............13W-22S		84
Kathy Ct. E&W....11W-16S		78
Kemper Dr. N&S....11W-18S		78
Kennedy Ct. E&W..11W-17S		78
Kensington Way E&W		
..............11W-18S		78
Kingsport Rd. N&S 13W-19S		78
Kingston Ln. E&W..11W-18S		78
Knightsbridge Ln. E&W		
..............11W-18S		78
Knollwood Ln. E&W		
..............10W-18S		79
Kropp Ct. N&S....14W-20S		84
La Grange Rd. N&S		
......12W-18S,12W-16S		78
La Reina Re'al N&S..12W-17S		78
La Reina Re'al N&S		
......12W-18S,12W-17S		78
Lake Brook Dr. NE&SW		
..............14W-20S		84
Lake Hills Dr. N&S..10W-18S		79
Lake Lawn Ln. NE&SW		
..............14W-18S		78
Lake Ridge Rd. NE&SW		
..............13W-17S		78
Lakebrook Ct. N&S		
..............14W-20S		84
Lakebrook Dr. NE&SW		
..............14W-20S		84
Lakeview Dr. NS&EW		
..............11W-17S		78
Lancaster Ct. N&S..11W-18S		78
Landings Dr. E&W..13W-20S		84
Larkspur Ln. N&S....9W-18S		79
Laurel Dr. N&S....10W-19S		79
Lawrence Ct. NE&SW		
..............10W-18S		79
Lee St. N&S........13W-20S		84
Legend Ln. E&W..10W-16S		78
Leicester Ct. E&W..10W-16S		79
Lexington Cir. E&W		
..............10W-18S		79
Lilac Ct. N&S........9W-18S		79
Lincoln Ct. E&W..11W-17S		79
Lincolnshire Dr. N&S		
..............11W-16S		78
Lindsay Dr. N&S..10W-16S		79
Lisa Ct. N&S......10W-18S		79
Lissp Annon Ct. N&S		
..............14W-20S		84
Lori Ln. NW&SE....11W-16S		78
Lunar Ct. E&W....11W-17S		78
Lynn Dr. E&W......13W-21S		84
Magarita Av. N&S..10W-16S		79
Magnolia Ct. E&W..11W-17S		78
Mallard Cir. NW&SE		
..............10W-18S		79
Maple E&W......11W-17S		78
Mary Dr. NE&SW..11W-16S		78
Mason Ln. N&S....11W-17S		78
Maude Dr. E&W....13W-21S		84
Maycliff Dr. N&S....10W-17S		79
Mayfair Ln. E&W..11W-18S		78
Mayo Dr. N&S......11W-17S		78
Meadowbrook Ln. E&W		
..............10W-18S		79
Meadowview Ct. N&S		
..............10W-16S		79
Medina Dr. N&S..11W-15S		72
Melissa Dr. N&S..13W-21S		84
Merion Dr. E&W....11W-18S		78
Micheal Dr. N&S....11W-16S		78
Michelle Ct. E&W....10W-17S		79
Cir. 10W-16S		79
Mimosa Dr. E&W...9W-18S		79
Mission Hill Ct. N&S		
..............10W-16S		79
Misty Hill Rd. NE&SW		
..............13W-17S		78
Mohawk Ln. N&S...9W-16S		79
Montgomery Dr.* E&W		
..............10W-17S		79
Morningside Rd. N&S		
..............13W-17S		78
Narcissus Ct. N&S..9W-18S		79
Narcissus Ln. N&S..9W-18S		79
Newgate Ct. N&S..10W-16S		79
Nicole Ct. N&S....10W-16S		79
Oak N&S..........12W-16S		78
Oak St. E&W....11W-17S		78
Oakdale Ct. E&W..11W-17S		78
Oakley Av. N&S....12W-17S		78
Old Orchard Dr. N&S		
..............10W-18S		79
Orchid Ct. N&S......9W-18S		79
Orchid Ln. N&S......9W-18S		79
Orenia Ct. E&W....10W-16S		79
Oriole Ct. E&W....10W-18S		79
Orlan Brook Dr. N&S		
..............10W-18S		79
Orland Ct. E&W....11W-18S		78
Cir. 11W-17S		78
Oxford Ln. N&S....11W-18S		78
Palm Ct. N&S......9W-18S		79
Palm Dr. E&W......9W-18S		79
Palos Springs Dr. N&S		
..............11W-15S		73
Paradise Ln. E&W..9W-18S		79
Park Hill Dr. N&S..11W-18S		78
Park Ln. N&S......11W-18S		78
Parkview Dr. N&S..10W-16S		79
Patricia Ln. NW&SE		
..............11W-18S		78
Patrick Ct. NW&SE		
..............10W-18S		79
Patty Ln. NE&SW..11W-15S		73

	Grid	Page
Pawnee Rd. N&S....9W-16S		79
Peachtree Dr. N&S..11W-18S		78
Pebble Beach Ln. E&W		
..............11W-16S		79
Pembrooke Ln. E&W		
..............11W-18S		78
Pine Grove Dr. E&W		
..............14W-18S		78
Pine St. E&W		
......11W-16S,10W-16S		78
Pine Tree Rd. N&S 13W-17S		78
Plum Tree Dr. N&S		
..............10W-18S		79
Pluskota Dr. E&W..10W-18S		79
Ponderosa Ct. E&W..9W-18S		79
Poplar Creek Ct. N&S		
..............14W-18S		78
Poplar Creek Ln. NS&EW		
..............14W-18S		78
Poplar Ct. NE&SW..11W-17S		78
Poplar Rd. NS&EW		
..............11W-17S		78
Prestwick Ln. E&W		
..............11W-15S		72
Primrose Ct. NW&SE..9W-18S		79
Primrose Ln. N&S....9W-18S		79
Putney Pl. N&S....10W-16S		79
Quail Hollow Dr. N&S		
..............10W-18S		79
Raintree Dr. N&S....10W-18S		79
Raintree Ln. E&W..11W-18S		78
Raney's Ln. N&S....12W-17S		78
Ravinia Av. N&S		
......12W-18S,12W-17S		78
Ravinia Dr. E&W..11W-18S		78
Ravinia Ln. E&W..12W-17S		78
Red Oak Ln. E&W..10W-16S		79
Regent Dr. N&S..11W-18S		78
Revell Ct. E&W....10W-17S		79
Ridge Av. N&S....12W-17S		78
Riviera Ln. E&W..11W-18S		78
Rob Roy Dr. E&W..10W-18S		79
Robin Ln. N&S....13W-20S		84
Robinhood Dr. NS&EW		
......10W-20S,10W-19S		85
Royal Creek Ln. N&S		
..............14W-18S		78
Royal Foxhunt Rd. N&S		
..............11W-18S		78
Rt. 43 N&S........9W-18S		79
Rt. 7 E&W		
......12W-17S,11W-16S		78
Rutherford Ln. NE&SW		
..............11W-18S		78
Ryehill Cir. N&S..13W-19S		78
Salvatori St. E&W..9W-16S		79
Sandlewood Dr. N&S		
..............10W-16S		79
Sandpiper Ct.* E&W		
..............10W-17S		79
Sawgrass Ct. E&W		
..............10W-18S		79
Sayre N&S........8W-18S		79
Seapines Rd. E&W..9W-18S		79
Seminole Ct. E&W..9W-18S		79
Sequoia Ct. N&S....9W-18S		79
Shade Cove Ct. E&W		
..............14W-18S		78
Shady Ln. E&W..12W-18S		78
Shagbark Ct. E&W..10W-18S		79
Shaker Dr. N&S....14W-18S		78
Sheffield Ln. N&S..11W-18S		78
Sheri Ct. NW&SE..11W-16S		78
Sheri Lane West E&W		
..............11W-18S		78
Sheri Ln. E&W....11W-18S		78
Sherwood Dr. N&S 10W-19S		79
Shipston St. E&W..10W-16S		79
Silo Ridge Dr. N&S 14W-17S		78
Silo Ridge Rd. East N&S		
..............14W-17S		78
Silo Ridge Rd. North E&W		
..............14W-17S		78
Silo Ridge Rd. South E&W		
..............14W-17S		78
Silo Ridge Rd. West N&S		
..............14W-17S		78
Silverdale Dr. E&W 11W-17S		78
Singletree Dr. NE&SW		
..............14W-17S		78
Sioux Rd. E&W......9W-16S		79
Sir Raleigh Dr. E&W		
..............13W-20S		84
Somerset Ct.* NS&EW		
..............10W-17S		79
Southwest Hwy. SW&NE		
..............11W-16S		78
Spicebrush Ln. N&S		
..............13W-20S		84
Spring Creek Ln. E&W		
..............14W-18S		78
Springbrook Ct. E&W		
..............14W-20S		84
Spyglass Cir. E&W..10W-18S		79
St. Andrews Ct. N&S		
..............10W-18S		79
St. Andrews Dr. N&S		
..............10W-18S		79
St. James Ct. E&W		
..............10W-18S		79
St. James Dr. N&S		
..............10W-18S		79
St. Mary Cir. N&S..10W-18S		79
Stonehenge Dr. E&W		
..............10W-18S		79
Stonewall Ln. E&W..10W-18S		79
Stradford Ln. N&S..11W-18S		78

	Grid	Page
Strawberry Ln. N&S		
..............10W-15S		73
Streamstown Ct. N&S		
..............10W-16S		79
Suffield Ct. E&W....9W-16S		79
Sunflower Ct. N&S..9W-18S		79
Sunny Ln. N&S....10W-17S		79
Sunrise Ln. NS&EW		
..............11W-18S		78
Sunset Ridge Ct. N&S		
..............10W-16S		79
Sunset Ridge Dr. N&S		
..............10W-18S		79
Sunshine Ln. N&S..10W-15S		73
Sussex Dr. N&S....10W-18S		79
Sutton Ct.* E&W..10W-17S		79
Sycamore Ct. E&W..9W-18S		79
Sycamore Dr. N&S..9W-18S		79
Tally Ho Ln. E&W..11W-18S		78
Tamarack Ln. N&S 10W-16S		79
Teakwood Ct. E&W		
..............10W-16S		79
Tee Brook Dr. N&S		
..............10W-18S		79
Terry Ct. NW&SE..11W-16S		78
Terry Dr. N&S....11W-16S		78
Teton Ct. N&S......9W-16S		79
Teton Rd. E&W......9W-16S		79
Therese Ct.* E&W..12W-16S		78
Thistlewood Dr. N&S		
..............10W-18S		79
Thistlewood Ln. E&W		
..............11W-18S		78
Thomas Dr. N&S..11W-16S		78
Thorngate Ct. E&W		
..............10W-18S		79
Thorngate Dr. E&W		
..............10W-18S		79
Tiffany Dr. E&W....9W-18S		79
Timber Edge Ln. E&W		
..............14W-18S		78
Timber Trail Rd. N&S		
..............11W-16S		78
Timberidge Ct. N&S		
..............14W-18S		78
Timberline Ct. NW&SE		
..............13W-17S		78
Timothy Ct. E&W..11W-16S		78
Timothy Dr. N&S..11W-16S		78
Tod William Dr. N&S		
..............11W-16S		78
Torrey Pines Dr. N&S		
..............11W-18S		78
Tower Dr. E&W....13W-20S		84
Trafalgar Ct. E&W..10W-16S		79
Treetop Dr. NS&EW		
..............12W-18S		78
Trenton Dr. N&S....11W-16S		78
Tulip Ct. N&S......9W-18S		79
Union St. N&S....12W-16S		78
Uxbridge Dr. E&W..10W-18S		79
Vail Ct. NW&SE....14W-18S		78
Valentina Dr. N&S..10W-18S		79
Valley Brook Dr. E&W		
..............14W-20S		84
Valley Ct. NE&SW..13W-17S		78
Virginia Ct.* E&W..12W-16S		78
Vista Dr. E&W....10W-18S		79
Vivian Ct. N&S....13W-21S		84
Voss Dr. N&S......13W-21S		84
Waterford Ln. E&W		
..............11W-18S		78
West Av. N&S		
......12W-18S,12W-17S		78
Westgate Ct. N&S..11W-15S		72
Westminster Dr. N&S		
..............11W-18S		73
Westwood Dr. N&S		
..............12W-17S		78
Wexford Ln. N&S..11W-17S		78
Wheeler Dr. E&W		
......11W-18S,9W-18S		78
Wherry Ln. N&S....11W-18S		78
Whispering Hill Dr. N&S		
..............14W-18S		78
Whitehall Ln. N&S..11W-18S		78
Wildflower Rd. E&W		
..............13W-17S		78
Will Cook Rd. N&S 14W-20S		84
William Ct.* E&W..12W-16S		78
William Dr. N&S....11W-16S		78
Willow Creek Ln. E&W		
..............14W-18S		78
Willow St. N&S....11W-17S		78
Willowood Ct. E&W..9W-18S		79
Wind Mill Rd. E&W 14W-17S		78
Winding Creek Ct. N&S		
..............14W-20S		84
Winding Creek Dr. NW&SE		
..............14W-20S		84
Windmill Turn N&S..14W-17S		78
Windsor Dr. EW&NS		
......11W-18S,9W-18S		79
Wittington Ct. N&S..9W-18S		79
Woburn Ct. E&W..12W-17S		78
Wood Duck Ln. N&S		
..............13W-20S		84
Woodbury Ct.* N&S		
..............12W-16S		78
Wooded Path N&S..10W-17S		79
Woodland Av. N&S		
..............12W-17S		78
Woodmar Dr. N&S..10W-18S		79
Woodridge Ln. N&S		
..............11W-16S		78
Woodruff Dr. E&W..9W-18S		79
Woodward Dr. N&S		
..............10W-16S		79

	Grid	Page
Yearling Crossing NE&SW		
..............13W-20S		84
Yorkshire Ln. N&S..11W-18S		78
Yorktown Ct. NE&SW		
..............11W-16S		78
Yorktown Dr. N&S..11W-16S		78
1st Av. N&S....12W-17S		78
2nd Av. N&S....12W-17S		78
3rd Av. N&S....12W-17S		78
70th Ct. N&S....8W-18S		79
71st Ct. E&W....8W-18S		79
73rd Av. N&S....9W-18S		79
73rd Ct. E&W....9W-18S		79
74th Av. N&S....9W-18S		79
74th Ct. E&W....9W-18S		79
75th Ct. E&W....9W-18S		79
77th Av. N&S....9W-18S		79
80th Av. N&S		
......10W-18S,10W-16S		79
81st Av. N&S....10W-17S		79
81st Ct. N&S....10W-17S		79
82nd Av. N&S		
......10W-18S,10W-16S		79
82nd Ct. N&S....10W-16S		79
83rd Av. N&S....10W-16S		79
83rd Ct. E&W....10W-16S		79
84th Av. N&S		
......10W-17S,10W-16S		79
85th Av. N&S....10W-17S		79
86th Av. N&S		
......10W-18S,10W-15S		79
87th Av. N&S....10W-17S		79
88th Av. N&S		
......10W-18S,11W-19S		79
88th St. E&W....11W-19S		78
89th Av. N&S....11W-18S		78
89th Ct. E&W....11W-18S		78
91st Av. N&S....11W-16S		78
92nd Av. N&S....11W-18S		78
92nd Ct. N&S....11W-17S		78
94th Ave. N&S		
......11W-19S,11W-17S		78
96th Av. N&S....12W-19S		78
104th Av. N&S		
......13W-20S,13W-19S		84
105th Ct. N&S....13W-19S		78
106th Ct. N&S....13W-19S		78
107th Av. N&S....13W-19S		78
108th Av. N&S		
......13W-20S,13W-21S		84
108th Av. N&S		
......13W-21S,13W-16S		84
131st St. E&W....11W-15S		72
135th St. E&W		
......11W-15S,10W-15S		72
137th St. E&W		
......11W-16S,10W-16S		78
138th Pl. E&W....10W-16S		79
138th St. E&W		
......11W-16S,10W-16S		78
139th St. E&W		
......11W-16S,10W-16S		78
140th St. E&W		
......11W-16S,10W-16S		78
141st St. E&W....10W-16S		78
142nd St. E&W..11W-16S		78
143rd Pl. E&W		
......12W-17S,10W-17S		78
143rd St. E&W		
......13W-16S,10W-16S		78
144th Pl. E&W		
......12W-17S,10W-17S		78
144th St. E&W		
......12W-17S,10W-17S		78
145th Pl. E&W		
......12W-17S,10W-17S		78
145th St. E&W		
......12W-17S,10W-17S		78
146th St. E&W....12W-17S		78
147th St. E&W		
......12W-17S,11W-17S		78
149th St. E&W....12W-17S		78
151st St. E&W		
..............14W-17S,9W-17S		78
152nd St. E&W		
......10W-18S,8W-18S		79
153rd Pl. E&W		
......12W-18S,9W-18S		78
153rd St. E&W		
......13W-18S,9W-18S		78
154th Pl. E&W....9W-18S		79
154th St. E&W....9W-18S		79
155th St. E&W....9W-18S		79
156th St. E&W....8W-18S		79
157th Pl. E&W....9W-18S		79
157th St. E&W		
......10W-18S,8W-18S		79
158th Pl. E&W....9W-18S		79
158th St. E&W....8W-18S		79
159th St. E&W		
......12W-18S,9W-18S		78
160th St. E&W....13W-19S		78
161st Pl. E&W....10W-19S		78
161st St. E&W....13W-19S		78
162nd St. E&W....10W-19S		78
163rd Pl. E&W....13W-19S		78
163rd St. E&W		
......11W-19S,10W-19S		78
164th Pl. E&W		
......13W-19S,11W-19S		78
164th St. E&W		
......11W-19S,10W-19S		78
165th St. E&W		
......13W-19S,10W-19S		78
166th Pl. E&W....10W-19S		78
166th St. E&W....10W-19S		78
167th Pl. E&W....13W-20S		84

	Grid	Page
167th Pl. E&W....10W-20S		85
167th St. E&W		
......13W-19S,10W-19S		78
168th St. E&W....13W-20S		84
168th St. E&W....10W-20S		85
169th St. E&W....10W-20S		85
170th Pl. E&W....10W-20S		85
170th St. E&W....10W-20S		85
171st St. E&W....11W-20S		85
179th St. E&W....13W-21S		84
183rd St. E&W....13W-21S		84

Orland Township

	Grid	Page
Beech Av. N&S....13W-19S		78
Brigitte Terr. E&W..13W-16S		78
Brook View Ln. NS&EW		
..............14W-17S		78
California Tr. E&W..14W-16S		78
Camelot Ct. N&S..14W-16S		78
Camelot Ln. E&W..14W-16S		78
Cedar Av. N&S....13W-19S		78
Cedar St. E&W....10W-16S		79
Cherry Ln. N&S....10W-16S		79
Chisholm Tr. E&W..14W-16S		78
Circle Dr. N&S......11W-16S		78
Creek Crossing Ct. N&S		
..............14W-17S		78
Creek Crossing Dr. N&S		
..............14W-16S		78
Creekside Dr. N&S..14W-16S		78
Creekview Dr. N&S		
..............14W-17S		78
Creekwood Dr. NE&SW		
..............14W-17S		78
Crescent Dr. N&S..11W-16S		78
Crestline Dr. E&W..14W-16S		78
Crestwood Dr. E&W		
..............14W-16S		78
Dusty Tr. N&S....14W-16S		78
E. Hazel Pl. N&S..13W-19S		78
Elm St. N&S......11W-16S		78
Fenview Ct. NE&SW		
..............14W-17S		78
Fenwood Ct. NE&SW		
..............14W-17S		78
Fir St. E&W........14W-16S		79
Forest View Dr. E&W		
......10W-17S,9W-17S		79
Glenview Dr. E&W..14W-17S		78
Grants Tr. N&S....13W-19S		78
Green Valley Dr. N&S		
..............14W-16S		78
Hancock St. E&W..13W-19S		78
Hazel Av. N&S....13W-19S		78
Hemlock St. E&W..10W-16S		79
Howe Dr. N&S....11W-16S		78
Juanita Dr. E&W....14W-19S		78
Kaup Ln. E&W....14W-16S		78
Kingswood Dr. E&W		
..............14W-16S		78
Kristoffer Av. N&S..14W-16S		78
Kristoffer Ct. E&W..14W-16S		78
Kristoffer Ln. E&W..14W-16S		78
Lago Ln. N&S......9W-16S		79
Laguna Ln. NE&SW..9W-16S		79
Lakewood Ln. N&S		
..............14W-17S		78
Logan Ct. E&W....14W-16S		78
Logan Dr. E&W....14W-16S		78
Logan Dr. N&S....14W-16S		78
Lynch E&W........14W-18S		78
Mannheim Rd. N&S		
..............12W-20S		84
Maple Av. N&S....10W-16S		79
Maple Dr. N&S....14W-17S		78
Marilyn Ct. E&W..14W-16S		78
Marilyn Terr. N&S..13W-16S		78
Mark Ln. E&W....14W-16S		78
Mayflower Ln. N&S		
..............13W-16S		78
McCabe Ct. N&S..14W-16S		78
McCabe Dr. E&W..14W-16S		78
McCabe St. E&W..14W-16S		78
Meade St. E&W....13W-19S		78
Meadow Ln. N&S..9W-17S		79
Mesquite Dr. N&S..14W-16S		78
Minetz Ct. E&W..14W-16S		78
Natchez Trail N&S..14W-16S		78
Nelson Ln. E&W..14W-16S		78
Oak Ct. NW&SE....14W-17S		78
Old Post Rd. N&S..14W-16S		78
Old Spanish Ct. N&S		
..............14W-16S		78
Old Spanish Tr. E&W		
..............14W-16S		78
Old Tamarack Ln. N&S		
..............13W-16S		78
Orange Av. N&S....13W-19S		78
Oregon Tr. E&W..14W-16S		78
Overland Trial N&S 14W-16S		78
Palm Av. N&S....13W-19S		78
Park Ln. N&S......9W-17S		79
Paw Paw Av. N&S..14W-16S		78
Pine Dr. NE&SW...14W-17S		78
Pine St. E&W......14W-16S		78
Pinecreek Dr., South E&W		
..............14W-17S		78
Pinecreek Dr.,North E&W		
..............14W-17S		78
Pineview Ct. N&S..14W-17S		78
Pineview Dr. E&W..14W-17S		78
Pinewood Ct. NE&SW		
..............14W-17S		78
Pinewood Dr. NS&EW		
..............14W-17S		78
Redondo Av. N&S..9W-16S		79
Redwood Av. N&S..10W-16S		79

	Grid	Page
Ridge Rd. N&S	9W-16S	79
Rt. 43 N&S	9W-17S,9W-16S	79
Rt. 45 N&S	12W-21S,12W-20S	84
Rt. 6 NE&SW	14W-21S,13W-19S	84
Ruggles Ct. E&W	14W-16S	78
Sante Fe Trail N&S	14W-16S	78
Selva Ln. N&S	9W-16S	79
Sherman St. E&W	13W-19S	78
Silver Ct. E&W	9W-17S	79
Silver Lake Dr. E&W	10W-17S	79
Somerset Rd. E&W	14W-16S	78
Spring Ln. N&S	13W-16S	78
Springview Ln. N&S	14W-16S	78
Spruce Dr. E&W	10W-16S	79
Spruce Ln. NS&EW	14W-17S	78
Stagecoach Dr. E&W	14W-16S	78
Stonegate Dr. N&S	14W-16S	78
Streamwood Dr. NE&SW	14W-17S	78
Touchstone Pl. N&S	13W-19S	78
Valley View Ct. NW&SE	14W-17S	78
Valley View Dr. NE&SW	14W-17S	78
W. Hazel Pl. E&W	13W-19S	78
Walnut St. E&W	10W-16S	79
Will Cook Rd. N&S	14W-18S,14W-16S	78
Willow Av. N&S	13W-19S	78
Wilrose Ct. N&S	13W-16S	78
Wilrose Ln. N&S	13W-16S	78
Wolf Rd. N&S	14W-21S,14W-16S	84
Woodlake Ln. N&S	14W-17S	78
Woodstock Dr. E&W	13W-16S	78
80th Av. N&S	10W-21S,10W-17S	85
81st Av. N&S	10W-17S	79
81st Ct. N&S	10W-17S	79
84th Av. N&S	10W-21S	85
85th Av. N&S	10W-16S	79
86th Av. N&S	10W-16S	79
94th Av. N&S	11W-21S,11W-20S	85
94th Ct. N&S	11W-20S	85
104th Av. N&S	13W-20S,13W-19S	84
108th Av. N&S	13W-21S,13W-16S	84
109th Av. N&S	13W-18S	78
110th Av. N&S	13W-20S,13W-16S	84
110th Ct. N&S	13W-20S	84
112th Ct. N&S	14W-18S	78
113th Av. N&S	14W-18S	78
113th Ct. N&S	14W-18S	78
114th Av. N&S	14W-18S	78
114th Ct. N&S	14W-18S	78
115th Av. N&S	14W-18S	78
115th Ct. N&S	14W-18S	78
116th Av. N&S	14W-18S	78
116th Ct. N&S	14W-18S	78
117th Av. N&S	14W-18S	78
117th Ct. N&S	14W-18S	78
118th Av. N&S	14W-19S,14W-18S	78
135th St. E&W	14W-15S,10W-15S	72
136th St. E&W	11W-16S	78
139th St. E&W	14W-16S,9W-16S	78
143rd St. E&W	14W-16S,9W-16S	78
144th St. E&W	11W-17S	78
145th Pl. E&W	10W-17S	79
151st St. E&W	14W-17S	78
152nd St. E&W	13W-18S	78
155th St. E&W	14W-18S	78
156th St. E&W	14W-18S	78
157th St. E&W	14W-18S	78
158th St. E&W	14W-18S	78
159th St. E&W	14W-18S,12W-18S	78
167th St. E&W	14W-19S,9W-19S	78
169th St. E&W	13W-20S	84
171st St. E&W	13W-20S,11W-20S	84
172nd St. E&W	13W-20S	84
173rd St. E&W	13W-20S	84
175th St. E&W	13W-20S,11W-20S	84
179th St. E&W	12W-21S	84
183rd St. E&W	13W-21S,10W-21S	84
Palatine		
Abbey Hill Ln. N&S	22W-15N	23
Aldridge Av. E&W	21W-14N	23
Alex Ct. E&W	21W-16N	13
Alex Ln. (Pvt) E&W	21W-16N	13
Alison Dr. E&W	19W-16N	14
Amherst St. E&W	20W-17N	13
Anderson Dr. NW&SE	19W-16N,18W-16N	14
Arbor Ct. E&W	23W-17N	12
Arbor Ln. N&S	23W-17N	12
Arlene Ln. NS&EW	19W-15N	14
Arrowhead Dr. N&S	20W-17N	13
Ash St. N&S	21W-15N	23
Ashland Av. N&S	20W-15N,20W-17N	23
Auburn Woods Ct. NS&EW	21W-17N	13
Auburn Woods Dr. NS&EW	21W-16N,21W-17N	13
Aurelia Ct. N&S	21W-16N	13
Austin Ln. NS&EW	22W-15N	23
Babcock Dr. NS&EW	19W-16N	14
Baldwin Rd. E&W	22W-17N	13
Baldwin Rd. NW&SE	20W-16N,19W-16N	13
Balsam Ln. E&W	20W-17N	13
Baybrook Dr. N&S	19W-15N	14
Beacon Dr. NS&EW	19W-16N	14
Bedford Dr. E&W	22W-15N	23
Bel Air Terr. N&S	20W-16N	13
Belle Av. N&S	19W-15N	14
Bennett Av. N&S	21W-15N	23
Benton St. N&S	20W-15N,20W-16N	23
Bishop Ct. NW&SE	22W-15N	23
Bissell Dr. NS&EW	19W-16N	14
Bon Aire Dr. NS&EW	20W-16N	13
Borders Dr. E&W	22W-15N	23
Bothwell St. N&S	21W-15N,21W-16N	23
Boynton Dr. N&S	19W-16N	14
Brandon Ct. NS&EW	21W-16N	13
Brighton Ct. NE&SW	22W-15N	23
Brighton Ln. N&S	22W-15N	23
Bristol Ct. N&S	22W-15N	23
Broadwalk Ct. N&S	21W-14N	23
Broadwalk Dr. E&W	21W-14N	23
Brockway St. N&S	21W-14N,21W-16N	23
Bryant Av. E&W	21W-14N	23
Burno Dr. NS&EW	21W-15N	23
Cady Dr. N&S	19W-16N	14
Cambridge Dr. N&S	22W-16N	13
Cardinal NS&EW	19W-17N	14
Cardinal Dr. N&S	19W-17N	14
Carmel Dr. N&S	20W-17N	13
Carolyn Dr. E&W	22W-15N	23
Carpenter Dr. E&W	20W-17N,19W-17N	13
Carriageway Ct. NW&SE	21W-14N	23
Carriageway Ln. N&S	21W-14N	23
Carter St. N&S	21W-16N	13
Cedar St. N&S	21W-14N,21W-16N	23
Charlotte N&S	21W-16N	13
Chatham Dr. E&W	22W-15N	23
Cherrywood Dr. NW&SE	21W-17N	13
Cheryl Ln. E&W	21W-15N	23
Chesapeake Ct. N&S	19W-17N	14
Chesapeake Ln. N&S	19W-17N	14
Chestnut St. N&S	21W-16N	13
Chewink Ct. NW&SE	20W-16N	13
Church St. E&W	19W-16N	14
Churchill Dr. E&W	19W-16N	14
Clark Dr. N&S	19W-16N,19W-17N	14
Clearwater Ct. N&S	22W-17N	13
Clinton Ct. N&S	22W-15N	23
Club House Dr. N&S	19W-15N	14
Colfax St. NS&EW	21W-16N,20W-16N	13
Comfort Ln. NS&EW	20W-16N	13
Comfort St. E&W	21W-16N	13
Concord Way E&W	21W-14N	23
Consumers Av. N&S	19W-15N	14
Coolidge Av. N&S	22W-17N	13
Cooper Dr. E&W	19W-17N	14
Cornell Av. E&W	21W-16N	13
Cottonwood Ln. N&S	19W-17N	14
Countryside Dr. NS&EW	22W-17N	13
Creekside Dr. NE&SW	19W-15N	14
Creekwood Dr. E&W	21W-18N	13
Crestview Dr. N&S	20W-17N	13
Crooked Willow Ln. E&W	22W-14N	23
Cunningham Dr. E&W	21W-17N	13
Daniels Rd. E&W	20W-15N	23
David Dr. N&S	19W-16N	14
Dean Dr. N&S	19W-16N	14
Deer Run Dr. (Pvt) N&S	21W-16N	13
Delgado Dr. NW&SE	20W-17N	13
Dorest Av. E&W	21W-15N	23
Dorothy Dr. E&W	19W-16N	14
Drovers Ln. NW&SE	22W-15N	23
DuPont Av. E&W	21W-14N	23
Dundee Rd. E&W	21W-18N	13
E. Castle Ct. E&W	19W-16N	14
Eagle Ln. N&S	20W-16N	13
East Meadow Lake Dr. NS&EW	19W-17N	14
Easy St. N&S	21W-16N	13
Eaton Ct. N&S	22W-15N	23
Echo Ln. NS&EW	21W-15N	23
Edgewater Ct. N&S	23W-17N	12
Edgewater Ln. E&W	23W-17N,22W-17N	12
Eisenhower Av. E&W	20W-16N	13
Eisenhower Ct. N&S	20W-16N	13
Elizabeth Av. NW&SE	19W-15N	14
Elizabeth Ct. NE&SW	19W-15N	14
Elm Ct. NW&SE	21W-14N	23
Elm St. E&W	22W-17N	13
Elm St. N&S	21W-14N,21W-15N	23
Elmwood Av. N&S	20W-15N,20W-16N	23
Eric Dr. N&S	21W-16N	13
Euclid Av. E&W	22W-14N,21W-14N	23
Everett Dr. N&S	19W-16N	14
Evergreen Ct. E&W	21W-15N	23
Evergreen Dr. NS&EW	19W-17N	14
Exner Ct. NW&SE	22W-15N	23
Fairway Ct. NW&SE	20W-17N	13
Fairway Dr. NE&SW	20W-17N	13
Falmore Dr. N&S	22W-14N	23
Fern Cir. E&W	21W-14N	23
Fern Ct. E&W	21W-14N	23
First Bank Dr. NS&EW	20W-16N	13
Flake Dr. NS&EW	19W-16N	14
Forest Av. N&S	20W-15N,20W-16N	23
Forest Ct. N&S	20W-16N	13
Forest Knolls Dr. E&W	20W-18N	13
Forestview Cir. NW&SE	22W-17N	13
Fosket Dr. NE&SW	19W-16N	14
Fremont St. N&S	20W-16N	13
Frontage Rd. NS&EW	19W-15N,19W-16N	14
Gary Dr. N&S	21W-16N	13
Geri Av. E&W	21W-16N	13
Geri Ct. N&S	21W-16N	13
Gilbert Av. E&W	21W-16N	13
Gilbert Av. E&W	21W-15N	23
Gilbert Rd. E&W	21W-15N,20W-15N	23
Gilbert St. E&W	22W-15N	23
Glade Av., East E&W	20W-15N	23
Glade Av., West E&W	21W-15N	23
Glencoe Rd. E&W	20W-17N	13
Glencoe Rd. E&W	21W-15N	23
Glencoe St. E&W	23W-17N	13
Glencoe St. E&W	20W-15N,19W-15N	23
Glenn Dr. N&S	19W-16N,19W-17N	14
Glennwood St. N&S	20W-15N,20W-16N	23
Gloria Dr. E&W	19W-16N	14
Golfview Ter. E&W	21W-17N	13
Goodwin E&W	19W-18N	14
Greeley Av. N&S	21W-15N,21W-16N	23
Greenwood Ct. E&W	20W-16N	13
Greenwood Dr. N&S	20W-16N	13
Grissom Dr. E&W	19W-17N	14
Groh Ct. E&W	22W-15N	23
Gull Ct. NE&SW	22W-14N	23
Haddington Ct. E&W	22W-15N	23
Hale St. N&S	20W-15N,20W-16N	23
Hampton Ct. N&S	22W-15N	23
Hampton Pl. E&W	22W-15N	23
Hamstead Ct. N&S	22W-15N	23
Harrison Av. N&S	22W-15N	23
Harrison Ct. N&S	22W-15N	23
Hart St. N&S	21W-15N,21W-16N	23
Harvard Dr. N&S	22W-14N	23
Hawk Ct. N&S	20W-16N	13
Heatherlea Dr., East N&S	20W-17N	13
Hedgewood Dr. N&S	20W-17N	13
Helen Rd. E&W	21W-15N,20W-15N	23
Heron Dr. E&W	20W-16N	13
Hickory St. N&S	21W-15N	23
Hicks Pl. NW&SE	20W-16N	13
Hicks Rd. N&S	20W-15N,20W-17N	23
Hill Av. N&S	22W-16N	13
Hunting Ct. NW&SE	22W-15N	23
Hunting Dr. NE&SW	22W-15N	23
Illinois Av. E&W	22W-15N,21W-15N	23
Imperial Ct. NE&SW	21W-15N	23
Inverary E&W	19W-18N	14
Jane Addams Dr. E&W	19W-17N	14
Jennifer Ln. NS&EW	21W-16N	13
Joan Dr. E&W	19W-16N	14
Johnson St. E&W	21W-15N	23
Jonathan Dr. E&W	19W-16N	14
Joyce Av. E&W	19W-15N	14
Juniper Dr. E&W	20W-17N	13
Karen Ln. NE&SW	21W-16N	13
Kathleen Dr. E&W	22W-15N	23
Kelly Ann Dr. E&W	22W-15N	23
Kenilworth Av. E&W	22W-15N,19W-15N	23
Kenilworth Rd. E&W	20W-15N	23
Kensington Ct. NE&SW	22W-15N	23
Kerry Ct. NS&EW	21W-14N	23
Kerwood St. N&S	20W-16N,20W-15N	13
Kimball Av. E&W	21W-15N	23
King Edward Ct. N&S	21W-17N,21W-18N	13
Kitson Dr. NS&EW	19W-16N	14
Knox St. E&W	20W-17N	13
Krista Ct. E&W	19W-15N	14
Krista Ln. N&S	19W-15N	14
Lake Dr. E&W	19W-15N	14
Lake Louise Dr. E&W	19W-16N	14
Lakeshore Dr. N&S	22W-16N	13
Lalonde Av. N&S	21W-14N	23
Lanark Ln. N&S	22W-15N	23
Lanark St. E&W	22W-15N,21W-15N	23
Leonard Dr. N&S	19W-15N	14
Lincoln St. E&W	20W-16N	13
Linden Av. N&S	20W-15N,20W-17N	23
Linden St. N&S	20W-15N,20W-16N	23
Longview Ln. E&W	21W-15N	23
Louise Ln. N&S	21W-17N	13
Lynn Dr. NS&EW	21W-16N	13
Lytle Dr. N&S	19W-16N	14
Mac Arthur Dr. N&S	20W-16N	13
Malibou Ln. E&W	20W-16N	13
Mallard Ct. NW&SE	22W-14N	23
Mallard Dr. NE&SW	22W-14N	23
Maple Ct. N&S	21W-15N	23
Maple St. N&S	21W-15N,21W-16N	23
Marion St. N&S	20W-16N	13
Marsha Dr. N&S	20W-17N	13
Mart Ct. N&S	21W-16N	13
Meadow Ln. E&W	20W-16N	13
Medford Dr. E&W	21W-14N	23
Merril Av. N&S	21W-15N,21W-16N	23
Michele Dr. E&W	19W-16N	14
Michigan Av. E&W	21W-15N,20W-15N	23
Middleton Av. NW&SE	22W-15N	23
Middleton Ct. NE&SW	22W-15N	23
Mill Valley Rd. E&W	20W-17N	13
Minnesota Av. E&W	21W-16N	13
Monterey Rd. E&W	20W-17N	13
Morris Dr. E&W	20W-16N,19W-16N	13
Morrison Av. N&S	22W-16N,22W-17N	13
Mozart St. N&S	20W-16N	13
Myrtle St. E&W	22W-17N	13
N. Hamilton Ct. N&S	21W-17N	13
N. Mulligan Ct. N&S	21W-17N	13
N. Newkirk Ln. N&S	19W-16N	14
N. Victoria Dr. NS&EW	19W-16N	14
N. Westminster Dr. N&S	22W-16N	13
Nightingale Dr. N&S	22W-14N	23
Norman Dr. E&W	19W-16N	14
North Court N&S	20W-16N	13
Northwest Hwy. NW&SE	21W-16N,19W-15N	13
Oak Ridge Rd. N&S	20W-18N	13
Oak St. N&S	20W-15N,20W-16N	23
Old Northwest Highway NW&SE	22W-16N,21W-16N	13
Old Virginia Ct. N&S	19W-17N	14
Old Virginia Rd. NS&EW	19W-17N	14
Olive St. E&W	19W-15N	14
Oxford Ct. NW&SE	22W-15N	23
Paddock Dr. E&W	19W-16N	14
Palatine Rd. E&W	22W-15N,19W-15N	23
Palmer E&W	19W-18N	14
Panorama Dr. NS&EW	22W-17N	13
Parallel St. NW&SE	20W-15N	23
Park Lane N&S	19W-15N	14
Park Pl. N&S	21W-14N	23
Park Terr. E&W	21W-14N	23
Parkside Dr. NS&EW	21W-14N	23
Parkwood E&W	21W-14N	23
Partridge Ct. N&S	22W-14N	23
Partridge Dr. E&W	22W-14N	23
Patricia Ct. E&W	19W-15N	14
Patricia Ln. NE&SW	19W-15N	14
Patten Dr. E&W	19W-16N	14
Pebble Creek Rd. NS&EW	20W-16N	13
Pebbles Dr. N&S	22W-15N	23
Peder Ln. E&W	21W-16N	13
Peregrine Ct. NE&SW	22W-14N	23
Pine St. N&S	21W-15N	23
Pintail Ct. E&W	22W-14N	23
Plate Dr. E&W	19W-16N	14
Pleasant Hill Blvd. E&W	21W-15N	23
Plum Grove Rd. N&S	21W-15N,21W-17N	23
Plum Tree Ct. NW&SE	21W-14N	23
Plum Tree Ln. N&S	21W-14N	23
Pompano Ln. E&W	20W-16N	13
Poplar St. E&W	22W-17N	13
Potomac Ct. N&S	19W-17N	14
Potomac Dr. N&S	19W-17N	14
Pratt Dr. E&W	19W-16N	14
Princeton St. E&W	20W-17N	13
Providence Rd. E&W	20W-16N	13
Quail Hollow Ln., West NE&SW	22W-14N	23
Quentin Rd. N&S	22W-14N,22W-17N	23
Rand Rd. NW&SE	19W-17N	14
Raven Ln. E&W	22W-15N	23
Ravine Hills Ct., West NW&SE	22W-14N	23
Reading Ct. E&W	21W-14N	23
Reseda Ct. N&S	20W-17N	13
Reseda Pkwy. E&W	20W-17N	13
Revere Ln. E&W	21W-14N	23
Reynolds Dr. E&W	19W-16N	14
Richards Dr. N&S	19W-16N	14
Richmond St. E&W	21W-16N	13
Ridge Trail E&W	21W-17N	13
Rimini Ct. NS&EW	20W-16N	13
Roanoke Ct. E&W	21W-15N	23
Robertson St. E&W	21W-16N	13
Robinson Dr. N&S	19W-16N	14
Rohlwing Rd. N&S	20W-15N,20W-17N	23
Rose St. N&S	21W-15N,21W-16N	23
Rosita Dr. E&W	19W-16N	14
Route 68 E&W	23W-17N	12
Royal Ct. NE&SW	21W-15N	23
Russet Way E&W	20W-17N	13
S. Hidden Brook Trail NS&EW	22W-14N	23
S. Windhill Dr. N&S	22W-14N	23
Saddle Ridge Ct., South N&S	22W-14N	23
Salem Ct. NW&SE	20W-16N	13
Sanborn Dr. NS&EW	19W-16N,19W-17N	14
Sandpiper Ct. E&W	22W-14N	23
Saratoga Dr. N&S	20W-17N	13
Sayles Dr. E&W	19W-16N	14
Schiller St. N&S	20W-16N	13
Schirra Dr. E&W	19W-16N	14
Schubert St. N&S	20W-16N	13
Sellstrom Dr. NW&SE	20W-15N	23
Shady Ln. E&W	21W-18N	13
Sherman St. E&W	21W-16N,20W-16N	13
Sherwood Ct. E&W	23W-17N	12
Sherwood Ln. NS&EW	23W-17N	12
Skylark Ct. N&S	22W-14N	23
Skylark Dr. E&W	22W-14N	23
Slade St. E&W	21W-16N,20W-16N	13
Slayton Dr. E&W	19W-16N	14
Smith Rd. N&S	21W-17N	13

	Grid	Page
Smith St. N&S	21W-15N,21W-16N	23
Sparrow Ct. N&S	22W-15N	23
Spring Willow Bay E&W	22W-15N	23
St. James Ct. E&W	22W-15N	23
Stark Dr. N&S	20W-16N,20W-17N	13
Steeplechase Ln., West E&W	22W-14N	23
Stephen Dr. N&S	21W-16N	13
Sterling Av. N&S	22W-17N	13
Stonebridge Ct. E&W	19W-15N	14
Stonehedge Ln. N&S	21W-14N	23
Stonington Dr. N&S	19W-15N	14
Stuart Ln. NS&EW	21W-15N	23
Sunset Ln. E&W	22W-17N	13
Sutherland Dr. N&S	19W-17N	14
Tahoe Tr. E&W	20W-16N	13
Teal E&W	22W-15N	23
Tern Dr. E&W	22W-14N	23
Terrace Dr. NS&EW	21W-16N	13
Thomas St. E&W	19W-15N	14
Three Willow Ct. N&S	22W-14N	23
Thurston Dr. E&W	19W-16N	14
Topanga Dr. N&S	20W-17N	13
Tymore Ct. N&S	22W-14N	23
U.S. Route 14 N&S	23W-17N,19W-14N	12
Valley Ln. NE&SW	21W-14N	23
Ventura Dr. N&S	20W-17N	13
Vermont St. N&S	20W-15N	23
Virginia Dr. E&W	19W-16N	14
Virginia Lake Ct. N&S	19W-17N	14
W. Bombay Way E&W	22W-16N	13
W. Crescent Dr. E&W	22W-16N	13
W. Hamilton Dr. E&W	21W-17N	13
W. Hidden Hill Ln. E&W	22W-16N	13
W. Windhill Dr. E&W	22W-14N	23
Walnut St. N&S	21W-15N	23
Wanda Ln. NE&SW	21W-16N	13
Warren Av. N&S	19W-15N	14
Warwick Rd. E&W	20W-16N	13
Washington Ct. E&W	21W-15N	23
Washington St. E&W	20W-15N	23
Wente N&S	19W-17N	14
West Ct. E&W	20W-16N	13
Westminster Dr. N&S	22W-16N	13
Whippoorwill Ln. NE&SW	22W-14N	23
Whitcomb Dr. NW&SE	19W-16N	14
White Willow Bay N&S	22W-14N,22W-15N	23
Whitehall Dr. N&S	22W-15N	23
Whytecliff Rd. E&W	22W-15N	23
Wilke Rd. N&S	19W-15N,19W-16N	14
Williams Av. N&S	19W-15N	14
Williams Dr. N&S	19W-16N,19W-17N	14
Willmette Av. E&W	20W-14N	23
Willow Ct. NW&SE	20W-16N	13
Willow Walk Dr. N&S	22W-14N	23
Willow Wood Dr. N&S	20W-16N	13
Wilmette Rd. E&W	20W-15N,19W-15N	23
Wilshire Ct. N&S	20W-16N	13
Wilson St. E&W	21W-16N,20W-16N	13
Wilton Ct. NE&SW	20W-16N	13
Winnetka St. E&W	22W-17N	13
Winslowe Dr. NE&SW	19W-17N	14
Winston Dr. N&S	19W-15N,19W-16N	14
Wood St. E&W	21W-16N,20W-16N	13
Woodwork Ln. N&S	21W-16N	13
Wren Av. N&S	20W-16N	13
Yale Ct. NW&SE	22W-14N	23

Palatine Township

	Grid	Page
Abbeywood Ct.* NS&EW	20W-20N	47
Aldridge Av. E&W	21W-14N	23
Algonquin Rd. NW&SE	23W-14N,20W-13N	22
Almond Ct. N&S	19W-18N	14
Alva St. E&W	22W-16N	13
Anna Ct. E&W	23W-18N	12
Apple Tree Ct. N&S	20W-17N	13
Arlington Dr. NS&EW	24W-18N	12
Ashbury Ln. NS&EW	23W-14N	22
Aster Av. E&W	19W-18N	14
Azalea Ln.* E&W	20W-20N	47
Baldwin Ct. N&S	19W-17N	14
Baldwin Ln. N&S	19W-17N,19W-18N	14
Baldwin Rd. N&S	19W-18N	14
Baldwin Way NW&SE	19W-18N	14
Banberry E&W	19W-18N	14
Barrington Woods Rd. N&S	21W-18N	13
Bayer Dr. E&W	21W-17N	13
Bayside Dr. NS&EW	19W-17N	14
Beaumont Ln. E&W	23W-18N	12
Benton St. N&S	20W-14N,20W-15N	23
Big Oaks Rd. N&S	21W-18N	13
Birchwood Av.* E&W	21W-18N	13
Bloomington Av. N&S	19W-18N	14
Blossom Tr. NE&SW	21W-13N	23
Bonhill Dr. NW&SE	19W-18N	14
Bradley Ct. NS&EW	19W-18N	14
Brentwood Dr. E&W	21W-18N	13
Briar Pl. NE&SW	24W-18N	12
Briarwood Ln. E&W	20W-13N	23
Brise Ct.* N&S	19W-18N	14
Broadmore Ct.* NW&SE	21W-18N	13
Brookdale Ct. N&S	20W-13N	23
Brookdale Ln. E&W	20W-13N	23
Brookdale Ln., North E&W	20W-13N	23
Brookdale Ln., South E&W	20W-13N	23
Brookside St. E&W	22W-15N	23
Brookview Ln. N&S	21W-13N	23
Bryant Av. E&W	21W-14N	23
California Av. N&S	20W-13N,20W-14N	23
Cambridge Ct. NW&SE	19W-18N	14
Candlenut Ln. E&W	19W-18N	14
Canterbury Ln.* N&S	19W-20N	47
Canterbury Ter.* N&S	19W-20N	47
Capri Dr. NE&SW	20W-18N,19W-18N	13
Carol Ct. NS&EW	19W-17N	14
Carolyn Dr. E&W	22W-15N	23
Carriage Ln. NS&EW	19W-18N	14
Cascade Ln.* N&S	19W-18N	14
Castle Ct. E&W	19W-18N	14
Cedarwood Ct. E&W	20W-17N	13
Center Rd. E&W	21W-18N	13
Central Rd. E&W	23W-13N,22W-13N	22
Chelsea Av.* N&S	21W-18N	13
Cherry Wood Dr. NW&SE	21W-17N	13
Circle Ct. E&W	21W-12N	23
Circle Ln. N&S	20W-14N	23
Clear Creek Bay* E&W	19W-18N	14
Club Ct. N&S	20W-17N	13
Clyde Av. N&S	22W-15N	23
Coach Rd. NE&SW	20W-18N	13
Constitution Dr. E&W	20W-18N	13
Conway Bay* N&S	19W-18N	14
Coolidge Av. N&S	22W-16N,22W-17N	13
County Line Rd. E&W	23W-18N,21W-18N	12
Creekwood Dr. E&W	21W-18N	13
Crescent Av. N&S	22W-15N,22W-16N	23
Crimson Ln. NE&SW	19W-17N	14
Crooked Creek Tr. NS&EW	22W-18N	13
Crystal Ln. NE&SW	19W-18N	14
Cumnor Rd. E&W	23W-17N	12
Cunningham Rd. E&W	21W-17N	13
Cypress Ct. NE&SW	20W-17N	13
Daniels Rd. E&W	21W-15N	23
Dartmoor Av.* E&W	21W-18N	13
Dee Ln. N&S	21W-18N	13
Deer Av. N&S	22W-17N	13
Deerpath Rd. N&S	22W-18N	13
Del Mar Dr. N&S	20W-17N	13
Denise Dr. NE&SW	20W-18N	13
Denton Av. NS&EW	21W-17N	13
Diane Dr. NE&SW	20W-18N	13
Diane Ln. NS&EW	20W-18N	13
Doe Rd. N&S	23W-17N	12
Doral Ct.* E&W	21W-18N	13
Dorchester St. E&W	22W-15N	23
Dorset Av. E&W	22W-15N	23
Dresden Av.* E&W	21W-18N	13
Driftwood Av. N&S	21W-17N	13
Dundee Qtr. NS&EW	19W-18N	14
Dundee Rd. E&W	24W-17N,19W-18N	12
Dunegal Bay* N&S	19W-18N	14
Dupont Av. E&W	21W-14N	23
Eastbrook E&W	19W-18N	14
Echo Ln. E&W	21W-17N	13
Edgebrook E&W	19W-18N	14
Ela Rd. N&S	23W-13N,23W-18N	22
Ellis St. E&W	22W-15N	23
Elm St. NS&EW	22W-17N,20W-17N	13
Elmwood Av. NE&SW	20W-17N	13
Elmwood Ct. NW&SE	20W-13N	23
Elmwood Ln., North NE&SW	20W-13N	23
Elmwood Ln., South NE&SW	20W-13N	23
Emerald Ln. NW&SE	19W-18N	14
Emerson Av. E&W	21W-14N	23
Enlund Dr. N&S	19W-18N	14
Evergreen Ln. N&S	19W-18N	14
Fairfax Av. E&W	20W-14N	23
Fairfield Ct.* E&W	21W-18N	13
Farmgate E&W	20W-17N	13
Fern Cir. E&W	21W-14N	23
Fern Ct. E&W	21W-14N	23
Fielding Pl. NW&SE	23W-18N	12
Fielding Place Ct. NW&SE	23W-18N	12
Forest Av. N&S	22W-17N	13
Forest Ln. E&W	20W-13N	23
Fox Chase Ct. N&S	24W-18N	12
Fox Glove Ln.* E&W	20W-20N	47
Foxwood Ln. E&W	23W-18N	12
Frances Ln. E&W	23W-18N	12
Franklin Av. N&S	22W-16N,22W-17N	13
Frontage Rd. N&S	19W-17N	14
Frost Rd. E&W	23W-13N	22
Galena Av. N&S	19W-18N	14
Galesburg Av. N&S	19W-18N	14
Garden Av. E&W	21W-17N,20W-17N	13
Garden Ct. N&S	21W-17N	13
Gardenia Ln. E&W	19W-18N	14
Garfield Av. E&W	20W-14N	23
Gary Dr. N&S	21W-16N,21W-17N	13
Gatewood Av. N&S	21W-17N	13
Glavin Cir. E&W	19W-18N	14
Glencoe Rd. E&W	23W-17N,20W-17N	12
Glencoe Rd. E&W	22W-15N	23
Glencoe St. E&W	22W-17N	13
Goodwin E&W	19W-18N	14
Grace Ln. N&S	24W-18N	12
Grandview Ter. NS&EW	24W-18N	12
Greenbrier Ln.* E&W	20W-20N	47
Greenview Av. N&S	23W-17N	12
Grove Av. N&S	22W-17N	13
Grove Ln. N&S	23W-18N	12
Haig Ct. E&W	19W-18N	14
Hancock Dr. NS&EW	20W-20N	47
Harrison Av. N&S	22W-15N,22W-16N	23
Hawk Dr. E&W	24W-18N	12
Hawthorne Dr. NE&SW	24W-18N	12
Heatherlea Dr., West NS&EW	20W-17N	13
Helen Rd. E&W	22W-15N	23
Heritage Cir.* NS&EW	20W-18N	14
Heritage Dr.* E&W	20W-18N	14
Hickory Ln. E&W	23W-13N	22
Hicks Rd. N&S	20W-17N,20W-18N	13
Cir.	19W-18N	14
Hidden Oaks Dr. N&S	23W-18N	12
High Grove Ln. N&S	19W-18N	14
High Rd. SW&NE	24W-18N	12
Hill Av. N&S	22W-16N	13
Hill Rd. E&W	21W-17N	13
Hillcrest E&W	21W-18N	13
Hillcrest Ct. N&S	24W-18N	12
Hillcrest Dr. NW&SE	24W-18N,23W-18N	12
Hillside Rd. E&W	24W-18N,24W-18N	12
Hillside St. E&W	22W-15N	23
Holly Way E&W	19W-18N	14
Home Av. E&W	20W-17N	13
Howe Ter. NW&SE	24W-18N	12
Hudson Bay* E&W	19W-18N	14
Independence Dr. NE&SW	20W-18N	13
Irene Dr. N&S	21W-18N	13
Iris Dr. N&S	19W-18N	14
Isle Royal Ln.* N&S	19W-18N	14
Ivy Pl. N&S	19W-18N	14
Jamestown Dr.* NS&EW	20W-20N	47
Jarvis Ct. NW&SE	20W-18N	13
Jennifer Ln. NW&SE	19W-18N	14
Joan Dr. N&S	23W-18N	12
Kenilworth Av. E&W	22W-15N,21W-15N	23
Kennedy Dr. (Pvt) N&S	19W-18N	14
Kevin Cir. E&W	19W-18N	14
King Arthur Ct. E&W	21W-17N	13
King Charles Ct. N&S	21W-17N	13
King George Dr. NS&EW	21W-17N	13
King Henry Ct. E&W	21W-17N	13
Kings Row NS&EW	19W-18N	14
Knollwood N&S	20W-13N	23
Knoxboro Ln. E&W	23W-18N	12
Kuhl Rd. E&W	21W-18N	13
Lakeside Dr. NE&SW	21W-17N	13
Laurel Rd. N&S	19W-18N	14
Lee Ct. N&S	20W-18N	13
Lenwood Dr. N&S	21W-18N	13
Leonard Rd. E&W	22W-15N	23
Lexington Dr.* N&S	20W-18N	14
Lilac Dr. E&W	19W-18N	14
Lilac Pl. N&S	19W-18N	14
Lilly Ln. E&W	19W-18N	14
Linden Ln. N&S	20W-13N	23
Linder Ln. E&W	24W-18N	12
Little City Dr. N&S	23W-14N	22
Long Acres Ln. NS&EW	20W-13N	23
Long Ln. E&W	20W-13N	23
Long Valley Ln. NW&SE	19W-17N	14
Lorelei Ln. EW&NS	24W-18N	12
Louise Ln. N&S	21W-17N	13
Lynda Dr. NW&SE	20W-18N	13
Maple Av. N&S	22W-16N,22W-17N	13
Marshall Ct. E&W	19W-18N	14
Martin Dr. N&S	21W-17N	13
Meacham Rd. N&S	21W-13N,21W-14N	23
Meadow Ln. NW&SE	20W-13N	23
Middleton Av. N&S	22W-15N,22W-16N	23
Morrison Av. N&S	22W-16N	13
Myrtle St. E&W	22W-17N	13
N. Circle Dr. N&S	20W-13N	23
Nantucket Ln. N&S	23W-18N	12
New Haven Av. N&S	19W-18N	14
Nichols Rd. E&W	19W-18N	14
North Rd. N&S	19W-18N	14
North St. E&W	22W-17N	13
Norway Ln. NS&EW	19W-17N	14
Oak Ct. E&W	24W-18N	12
Oak Ridge Rd. N&S	20W-18N	13
Oak St. N&S	20W-17N	13
Oaksbury Ln. E&W	20W-13N	23
Oakwood N&S	20W-13N	23
Old Bridge Rd. NS&EW	20W-17N	13
Old Hicks Rd. NS&EW	19W-18N	14
Old Mill Dr. N&S	20W-17N	13
Osage NW&SE	20W-18N	13
Palm Dr. NS&EW	23W-18N	12
Palos Av. N&S	22W-17N	13
Panorama Dr. NS&EW	22W-17N	13
Park Av. N&S	22W-17N	13
Park Dr. NE&SW	20W-13N	23
Pennsylvannia Dr. NW&SE	20W-18N	13
Pepper Tree Dr. N&S	20W-17N	13
Perry Dr. N&S	21W-17N	13
Pheasant Ln. N&S	24W-18N	12
Pheasant Ter. N&S	24W-18N	12
Pine Dr. E&W	21W-17N	13
Plum Grove Rd. N&S	21W-13W,21W-17N	23
Poplar St. E&W	22W-17N	13
Portage Av. N&S	22W-17N	13
Ports 'O' Call Dr. N&S	19W-17N	14
Prescon N&S	19W-17N	14
Private Dr. NS&EW	19W-18N	14
Prospect Dr. E&W	24W-18N	12
Public St. N&S	21W-18N	13
Queens Ct. NW&SE	19W-18N	14
Queensburg Cir.* NS&EW	20W-20N	47
Queensburg Ln.* E&W	20W-20N	47
Quentin Rd. N&S	22W-15N,22W-17N	23
Rainbow Bay* N&S	19W-18N	14
Rand Ct. NW&SE	20W-18N	13
Rand Grove Village N&S	20W-18N	13
Rand Rd. NW&SE	20W-18N,19W-17N	13
Randall Ln. NE&SW	19W-18N	14
Randville Dr. E&W	19W-17N	14
Ridgewood Ln. N&S	20W-17N	13
Rosalie Ln. E&W	21W-18N	13
Rose Av. NW&SE	19W-18N	14
Rosiland Rd. E&W	21W-18N	13
Roslyn Rd. N&S	24W-17N	12
Roth Av. N&S	24W-18N	12
Rt. 53 NW&SE	20W-18N,19W-17N	13
Rt. 68 E&W	23W-17N,20W-17N	12
S. Circle Dr. E&W	21W-12N	23
Shadow Lake Ter.* N&S	19W-20N	47
Shady Dr. E&W	20W-14N	23
Shady Lane E&W	21W-18N	13
Shady Pine Ct. E&W	20W-17N	13
Shagbark Ct. NW&SE	23W-18N	12
Shannon Bay* E&W	19W-18N	14
Silver Ln. NS&EW	19W-17N	14
Silver Strand Cir.* E&W	19W-20N	47
Smith Rd. N&S	21W-17N	13
Smith St. N&S	21W-14N,21W-15N	23
Spruce NW&SE	20W-18N	13
Sue Ln. E&W	21W-13N	23
Sunset Dr. E&W	21W-13N	23
Sunset Rd. N&S	21W-18N	13
Suthers Ln. N&S	21W-18N	13
Tall Trees Ln. E&W	20W-14N	23
Tanglewood Av.* NS&EW	21W-18N	13
Thackeray Dr. N&S	20W-17N	13
Thorntree Ln. N&S	20W-13N	23
Timber Ln. E&W	20W-17N	13
Timberlea Dr. N&S	20W-17N	13
Trail Ridge E&W	19W-18N	14
Tulip Way E&W	19W-18N	14
Turtle Creek Dr. N&S	19W-17N	14
Vermillion Ln.* E&W	19W-18N	47
Vermont St. N&S	21W-14N	23
Wainwright Ct. N&S	19W-18N	14
Waterford Ln. NE&SW	19W-18N	14
Wedgewood Dr. N&S	23W-18N	12
West Frontage Rd. N&S	20W-13N	23
Westmoreland Dr. N&S	19W-18N	14
Westwood Ln. N&S	20W-13N,21W-18N	23
Wheatherstone Rd. NS&EW	23W-18N	12
Whispering Oaks Ln. NS&EW	23W-18N	12
Whispering Springs* NW&SE	19W-20N	47
White Gate Ct. NW&SE	23W-18N	12
White Sands Bay* N&S	19W-18N	14
Whitewater Bay* N&S	19W-18N	14
Willamsburg Dr.* NS&EW	20W-20N	47
Willow St. E&W	22W-16N	13
Wilmette Av. E&W	20W-14N	23
Wilson St. E&W	22W-16N	13
Winchester Hills Dr. NS&EW	19W-18N	14
Wind River Ter.* N&S	19W-18N	14
Winnetka Av. E&W	20W-14N	23
Winnetka St. E&W	22W-17N	13
Wood St. E&W	22W-17N	13
Wood St. E&W	22W-16N	13
Woodbury Dr. NE&SW	19W-17N	14
Woodland Ct. N&S	20W-13N	23
Woodland St. E&W	21W-18N	13
Zinnia Ln. E&W	19W-18N	14

Palos Heights

	Grid	Page
Arbor Tr. N&S	7W-14S	74
Arquilla Dr. E&W	9W-15S	73
Auburn Av. N&S	8W-15S	73
Briar Ln. N&S	8W-15S	73

Street	Grid	Page
Burris Ct. E&W	8W-14S	73
Canterberry Ct. E&W	9W-13S	73
Carmicheal Dr. E&W	9W-14S	73
Carol Ln. E&W	7W-14S	74
Carriage Trail NE&SW	9W-19S	79
Cedar Creek N&S	8W-14S	73
Cedar Ln. N&S	9W-15S	73
Cheyenne Dr. N&S	8W-14S	73
Chippewa Dr. E&W	8W-14S	73
Choctaw Rd. N&S	9W-15S	73
Coach Rd. NS&EW	9W-14S	73
College Dr. E&W	9W-14S,9W-13S	73
Comanche Dr. N&S	9W-13S	73
Country Club Ct. N&S	9W-15S	73
Country Squire Ct. NE&SW	9W-14S	73
Country Squire Rd. N&S	9W-14S	73
Cypress Ln. N&S	9W-15S	73
Deer Ln. E&W	8W-14S	73
Diane Dr. N&S	8W-14S	73
Dogwood Ln.* N&S	7W-14S	74
Edgewood Rd. E&W	8W-14S	73
Elm Ct. N&S	9W-15S	73
Evergreen Av. E&W	8W-14S	73
Flambeau Dr. N&S	8W-14S	73
Forest Hill Ct. NS&EW	9W-15S	73
Forest Hill Ln. E&W	9W-15S	73
Forestview Rd. N&S	8W-15S	73
Fox Ln. E&W	8W-15S	73
Golf Dr. E&W	9W-15S	73
Golfview Ln. E&W	8W-15S	73
Harlem Av. N&S	9W-15S,9W-14S	73
Harold Av. N&S	9W-14S	73
Highland Dr. E&W	8W-15S	73
Ishnala Dr. NE&SW	9W-15S	73
Kiowa Ln. NE&SW	9W-15S	73
Lakeview Ct. E&W	9W-15S	73
Linden Dr. E&W	8W-15S	73
Manitoba Dr. E&W	9W-15S	73
Maple Ln. E&W	8W-14S	73
McCarthy Rd. NW&SE	9W-14S	73
McIntosh Dr. E&W	9W-14S	73
Melvina Av. N&S	7W-14S	74
Menominee Pkwy. E&W	8W-14S	73
Nagle Av. N&S	8W-14S	73
Nashville Av. N&S	8W-14S	73
Natchez Av. N&S	8W-14S	73
Natoma Av. N&S	8W-14S	73
Navaho Dr. N&S	8W-14S	73
Navajo Ct., East N&S	8W-14S	73
Navajo Ct., West NE&SW	8W-14S	73
Navajo Dr., East N&S	8W-14S	73
Navajo Dr., West N&S	8W-14S	73
Oak Ct. NW&SE	9W-15S	73
Oak Hills Ct. E&W	9W-15S	73
Oak Hills Pkwy. N&S	9W-15S	73
Oak Ln. N&S	9W-15S	73
Oak Park Av. N&S	8W-15S,8W-14S	73
Oak Ridge Ct. E&W	9W-15S	73
Oak Ridge Trail N&S	9W-15S	73
Oak Trail Ct. N&S	9W-14S	73
Oakview Ct. N&S	9W-15S	73
Old George's Way NS&EW	9W-15S	73
Old Oak Trail N&S	9W-14S	73
Onekema Dr. E&W	8W-14S	73
Orchard Dr. E&W	7W-14S	74
Palos Av. E&W	9W-15S	73
Park Ln. E&W	8W-15S	73
Parliament Dr., East N&S	9W-13S	73
Parliament Dr., West N&S	9W-13S	73
Pheasant N&S	9W-14S	73
Pine Pl.* NS&EW	7W-14S	74
Ponderosa Dr. NS&EW	9W-15S	73
Pottawatami Ct. NE&SW	9W-15S	73
Pottawatami Dr. E&W	9W-15S	73
Pueblo Ct. N&S	9W-15S	73
Pueblo Ln. N&S	9W-15S	73
Quail Ct. N&S	8W-15S	73
Raleigh Ct. N&S	9W-15S	73
Richard Av. N&S	9W-14S	73
Ridgeland Av. N&S	8W-15S,8W-14S	73
Seneca Rd. N&S	9W-15S	73
Sequoia Dr. NE&SW	9W-15S	73
Shawnee Rd. N&S	9W-15S	73
Shiawassie Dr. E&W	8W-14S	73
Shoshone Rd. N&S	9W-15S	73
Sioux Ln. N&S	9W-15S	73
Sorrento Dr. N&S	8W-14S	73
Spruce Pl.* NS&EW	7W-14S	74
Squire Pl. E&W	9W-14S	73
Surrey Hill Ct. NW&SE	9W-14S	73
Surrey Hill Rd. NE&SW	9W-14S	73
Sycamore Ln. N&S	9W-15S	73
Tennyson Ct. E&W	9W-13S	74
Turtle Pond Ln. N&S	9W-15S	73
Ute Ln. E&W	9W-15S	73
Westgate Av. N&S	8W-15S	73
Westview Dr. N&S	9W-15S	73
White Oak Ct. NE&SW	9W-14S	73
White Pine Pl. E&W	9W-15S	73
Willow Wood Dr. E&W	7W-14S	74
Winnebago Rd. N&S	9W-15S	73
Wolf Rd. N&S	14W-15S,14W-14S	72
Wyandot Dr. E&W	8W-14S	73
68th Ct. N&S	8W-14S	73
69th Av. N&S	8W-14S	73
69th Ct. N&S	8W-15S,8W-14S	73
70th Av. N&S	8W-15S,8W-14S	73
70th Ct. N&S	8W-15S,8W-14S	73
71st Av. N&S	8W-15S,8W-14S	73
71st Ct. N&S	8W-15S,8W-14S	73
72nd Ct. N&S	8W-14S	73
73rd Av. N&S	9W-14S	73
73rd Ct. N&S	9W-14S	73
74th Av. N&S	9W-14S	73
75th Av. N&S	9W-14S	73
76th Av. N&S	9W-15S,9W-14S	73
78th Ct. N&S	9W-15S	73
79th Ct. N&S	9W-15S	73
80th Av. N&S	9W-15S	73
96th Av. N&S	11W-14S	72
119th Pl. E&W	9W-14S	73
120th St. E&W	9W-14S,8W-14S	73
121st St. E&W	9W-14S,8W-14S	73
122nd St. E&W	9W-14S,8W-14S	73
123rd Pl. E&W	9W-14S	73
123rd St. E&W	9W-14S,8W-14S	73
124th Pl. E&W	9W-14S,7W-14S	73
124th St. E&W	9W-14S,7W-14S	73
125th Pl. E&W	7W-14S	74
125th St. E&W	9W-14S,8W-14S	73
126th Pl. E&W	8W-14S	73
126th St. E&W	9W-14S,8W-14S	73
127th St. E&W	9W-14S,8W-14S	73
128th St. E&W	9W-15S	73
129th St. E&W	9W-15S	73
130th St. E&W	8W-14S	73
131st St. E&W	9W-15S,8W-15S	73

Palos Hills

Street	Grid	Page
Alta Dr. N&S	10W-12S	65
Amee Ct. (Pvt.) E&W	10W-12S	65
Aspen Dr. N&S	10W-12S	65
Aum Ridge Dr. (Pvt) E&W	10W-12S	65
Avon Ct. N&S	10W-12S	65
Bennett Dr. (Pvt.) E&W	10W-12S	65
Boulder Ct.* E&W	10W-13S	73
Breckenridge Dr. E&W	10W-12S	65
Broadmoor Dr. N&S	10W-12S	65
Brookledge Dr. N&S	9W-12S	65
Cambridge Ct. N&S	10W-11S	65
Carlton Hide Dr. E&W	10W-12S	65
Catherine Ln.* N&S	10W-13S	73
Center Rd. E&W	11W-13S	72
Chamonieux Dr. N&S	10W-12S	65
Charles Av. NE&SW	10W-11S	65
Cherry Ln. N&S	10W-13S	73
Chestnut Dr. N&S	10W-13S	73
Cinnamon Creek Dr. E&W	11W-11S	64
Circle Ct. N&S	10W-12S	65
Circle Dr. E&W	10W-12S	65
Cobblestone Ct.* E&W	10W-13S	73
College Dr. N&S	10W-12S	65
Cottonwood Dr. N&S	10W-13S	73
Cour Caravelle E&W	10W-13S	73
Cour D'Alene N&S	10W-13S	73
Cour De La Reide N&S	10W-13S	73
Cour Deauville N&S	10W-13S	73
Cour La Salle E&W	10W-13S	73
Cour Leroux N&S	10W-13S	73
Cour Madeleine N&S	10W-13S	73
Cour Marquis E&W	10W-13S	73
Cour Masson E&W	10W-13S	73
Cour Michele E&W	10W-13S	73
Cour Monnet N&S	10W-13S	73
Cour Montreal N&S	10W-13S	73
Cour Saint Tropez N&S	10W-13S	73
Cour Versaille E&W	10W-13S	73
Crista Ct. N&S	10W-12S	65
Dartmouth Dr. E&W	10W-12S	65
Deerpath Ln. N&S	11W-13S	72
Del Prado Dr. E&W	11W-11S	64
Devon Shire Ln. E&W	11W-12S	64
Doric Cir. N&S	10W-13S	65
East Rd. N&S	11W-13S	72
Eastwood Dr. N&S	9W-12S	65
Eleanor Av. NE&SW	10W-11S	65
Erin Dr. E&W	10W-11S	65
Granada Ct. N&S	10W-11S	65
Green Valley Dr. NS&EW	10W-12S	65
Gruener Ct. (Pvt.) E&W	10W-11S	65
Gruener Ct.* E&W	10W-13S	73
Harlem Av. N&S	10W-12S	65
Helena Dr. N&S	11W-12S	72
Heritage Hills Dr. N&S	11W-13S	72
Hickory Crest N&S	11W-11S	64
Holly Ct. E&W	10W-13S	73
Interlochen Ct. NE&SW	10W-12S	65
Interlochen Dr. NW&SE	10W-11S	65
Juniper Ct. E&W	10W-13S	73
Kathy Ct. N&S	9W-12S	65
Kean Av. N&S	11W-13S,11W-11S	72
Kingsbury Ct. N&S	10W-11S	65
Lori Ln. N&S	9W-12S	65
Los Palos Ln. NE&SW	11W-11S	64
Loveland Ln. E&W	10W-12S	65
Lucas Dr. (Pvt.) NS&EW	11W-13S,10W-13S	72
Maple Crest N&S	11W-11S	64
Meadow Ln. N&S	10W-12S	65
Mehta Ct. (Pvt.) E&W	10W-11S	65
Meryton Ct. N&S	10W-11S	65
Michael Dr. N&S	9W-12S	65
Millstone Ct.* E&W	10W-13S	73
Moraine Dr. E&W	10W-13S	73
Moraine Valley Golf Range E&W	10W-12S	65
Mullberry Ct. E&W	10W-13S	73
North Rd. E&W	11W-13S	72
Northwest Rd. NW&SE	11W-13S	73
O'Gorman N&S	10W-13S	73
O'Gorman N&S	10W-13S	73
Oak Crest NW&SE	11W-11S	64
Olympic Cir. N&S	10W-12S	65
One Executive Dr. N&S	10W-13S	73
Palos Dr. N&S	10W-12S	65
Paxos Dr. N&S	9W-12S	65
Peak Dr. NW&SE	9W-13S	73
Pebble Ct.* E&W	10W-13S	73
Pemberly Ct. N&S	10W-11S	65
Poplar Ct. E&W	10W-13S	73
Possum Dr. N&S	11W-13S	72
Roberts Rd. N&S	10W-12S,10W-11S	65
Sand Castle Dr. N&S	11W-11S	64
Sheffield Rd. N&S	9W-12S	65
Sierra Ln. E&W	11W-11S	64
South 84th Terr. N&S	10W-11S	65
South 86th Terr. N&S	10W-11S	65
South Rd. E&W	11W-13S	72
Spathis Dr. N&S	10W-13S	73
Stonebridge Ct.* E&W	10W-13S	73
Stoney Creek Dr. (Pvt.) N&S	10W-12S	65
Stony Creek Dr. N&S	10W-12S	65
Stowe Ct. N&S	10W-12S	65
Stratford Dr. E&W	11W-11S	64
Sun Valley Ct. N&S	10W-12S	65
Sun Valley Dr. E&W	10W-12S	65
Surray Pk. E&W	10W-11S	65
Sycamore Ct. N&S	10W-13S	73
Sycamore Ln. N&S	10W-13S	73
Taos Dr. N&S	10W-12S	65
Terry Dr. N&S	9W-12S	65
Theresa Cir. NE&SW	10W-12S	65
Todd Dr. N&S	9W-12S	65
Tuttle Dr. N&S	10W-13S	73
Urmi Ct. (Pvt.) E&W	10W-12S	65
Vail Dr. E&W	10W-12S	65
Valley Ct. N&S	10W-12S	65
Valley Dr. E&W	10W-12S	65
Vickey Ln. N&S	9W-12S	65
Walnut Terr. N&S	10W-11S	65
West 102nd Ter. E&W	10W-11S	65
West 103rd Ter. E&W	10W-11S	65
West 99th Ter. E&W	10W-11S	65
Westwood Dr. N&S	9W-13S,9W-12S	65
Willow Dr. N&S	10W-13S	73
Windsor Dr. E&W	11W-11S	64
Winter Park Dr. NW&SE	10W-12S	65
Wood Ln. N&S	10W-11S	65
Wooded Path Dr. NE&SW	11W-11S	64
Zermatt Dr. E&W	10W-12S	65
73rd Av. N&S	9W-12S	65
73rd Ct. N&S	9W-12S	65
74th Av. N&S	9W-12S	65
75th Av. N&S	9W-12S	65
75th Ct. N&S	9W-12S	65
76th Av. N&S	9W-12S	65
78th Av. N&S	9W-12S	65
79th Ct. N&S	9W-12S	65
80th Ct. N&S	10W-12S,10W-11S	65
81st Av. N&S	10W-12S,10W-11S	65
81st Ct. N&S	10W-12S,10W-11S	65
82nd Av. N&S	10W-12S,10W-11S	65
82nd Ct. N&S	10W-12S,10W-12S	65
83rd Av. N&S	10W-12S	65
83rd Ct. N&S	10W-12S,10W-12S	65
84th Av. N&S	10W-13S,10W-12S	73
85th Av. N&S	10W-13S,10W-12S	73
86th Av. N&S	10W-13S	65
86th Ct. N&S	10W-11S	65
87th Av. N&S	10W-11S	65
87th Ct. N&S	10W-11S	65
88th Av. N&S	10W-12S,11W-11S	65
89th Av. N&S	11W-12S,11W-11S	64
90th Av. N&S	11W-12S,11W-11S	64
96th Pl. E&W	10W-11S	65
96th St. E&W	10W-11S	65
97th Pl. E&W	11W-11S	64
97th St. E&W	11W-11S,10W-11S	64
98th Pl. E&W	11W-11S,10W-11S	64
98th St. E&W	11W-11S,10W-11S	64
99th Pl. E&W	11W-11S,9W-11S	64
99th St. E&W	11W-11S,9W-11S	64
100th Pl. E&W	11W-11S,9W-11S	64
100th St. E&W	11W-11S,9W-11S	64
101st Pl. E&W	11W-11S,9W-11S	64
101st St. E&W	10W-11S,9W-11S	65
101st Terr. E&W	10W-11S	65
102nd Pl. E&W	11W-11S	64
102nd St. E&W	11W-11S,9W-11S	64
103rd Pl. E&W	11W-12S	64
103rd St. E&W	11W-11S,9W-11S	64
104th St. E&W	11W-12S,10W-12S	64
105th St. E&W	11W-12S,9W-12S	64
106th St. E&W	10W-12S	65
107th St. E&W	10W-12S,9W-12S	65
108th St. E&W	9W-12S	65
109th St. E&W	10W-12S	65
111th Pl. E&W	10W-13S	73
111th St. E&W	11W-12S,9W-12S	64
112th Pl. E&W	9W-13S	73
112th St. E&W	10W-13S	73
113th Pl. E&W	9W-13S	73
113th St. E&W	9W-13S	73

Palos Park

Street	Grid	Page
Alpine Av. (Pvt.) N&S	10W-13S	73
Autobahn Dr. (Pvt.) NW&SE	10W-13S	73
Autobahn Dr. S. E&W	10W-13S	73
Autobahn Dr. W. NW&SE	10W-13S	73
Autobahn Dr.,East (Pvt.) N&S	10W-13S	73
Black Forest Ln. (Pvt.) NW&SE	10W-13S	73
Brand Av. NE&SW	9W-14S	73
Brien Av. N&S	10W-14S	73
Brookside Dr. N&S	10W-13S	73
Calumet Sag. Rd. E&W	11W-13S,10W-13S	72
Cherokee Rd. E&W	10W-14S	73
Danmar Trail E&W	14W-15S	72
Edelweiss Dr. (Pvt.) N&S	10W-13S	73
Eldridge Av. N&S	10W-13S	73
Elm St. N&S	11W-14S	72
Forest Av. N&S	9W-14S	73
Forest Glen Blvd. E&W	11W-14S	72
Forest View Ln. N&S	10W-14S	73
Fox Ln. N&S	13W-14S	72
Fox Ln., South E&W	13W-14S	72
Fox Ln.,North E&W	10W-14S	73
Highwood Dr. N&S	10W-13S	73
Hillcrest Ln. E&W	11W-14S	72
Hillside Cir. NE&SW	10W-14S	73
Hillside Dr. NS&EW	10W-14S	73
Hobart Av. N&S	10W-13S	73
Holmes Av. N&S	10W-13S	73
Huntleigh Rd. N&S	13W-14S	72
Iroqois Rd. N&S	10W-14S	73
Kean Av. N&S	11W-14S	72
Kimber Ln. NW&SE	10W-13S	73
Kinvarra Dr. N&S	12W-14S	72
Knollwood Dr. E&W	10W-14S	73
Laurie Av. N&S	10W-14S	73
Matterhorn Cir. (Pvt.) N&S	10W-13S	73
Mc Kinley Av. N&S	11W-14S	72
McCarthy Rd. E&W	14W-14S,9W-14S	73
Meadow Rd. NE&SW	14W-14S	72
Mohawk Rd. NE&SW	10W-14S	73
Oak Ln. N&S	10W-14S	72
Oak Ridge Dr. E&W	9W-14S	73
Old Creek Rd. NS&EW	14W-15S,14W-14S	72
Old Prague Path (Pvt.) N&S	10W-13S	73
Old Timber Ln. E&W	13W-14S	72
Palos Av. E&W	9W-14S	73
Park Lane Dr. E&W	12W-14S	72
Partridge Ln. E&W	14W-15S	72
Pawnee Rd. NS&EW	10W-14S	73
Post Rd. N&S	13W-14S	72
Rail Ln. NS&EW	13W-14S	72
Ramsgate Dr. E&W	14W-14S	72
Regina Ln. E&W	14W-14S	73
Ridge Av. NE&SW	11W-14S	73
Roma Rd. NW&SE	10W-14S	73
Romiga Ln. NS&EW	13W-14S	72
Rosewood Av. NW&SE	11W-14S	72
Rt. 7 SW&NE	11W-15S,10W-14S	72
Rt. 83 SE&NW	11W-13S,10W-13S	72
Seminole Rd. N&S	10W-14S	73
Somerset E&W	12W-14S	72
South St. E&W	11W-14S,10W-14S	72
Southwest Hwy. NE&SW	11W-15S,10W-14S	72
Spring Dr. N&S	11W-14S	72
Sunny Slope Ln. N&S	13W-14S	72
Timber Dr. N&S	9W-14S	73
Timber Ln. N&S	10W-14S	73
Vienna Av. (Pvt.) N&S	10W-13S	73
Virgil Av. N&S	9W-14S	73
Walnut Ridge Dr. N&S	10W-13S	73
Wild Cherry Ln. E&W	12W-14S	72
Wild Wood Tr. E&W	13W-14S	72
Williams Av. N&S	10W-14S	73
Windsor Dr. N&S	12W-14S	72
Winslow Rd. N&S	11W-14S	72
Woodland Trail E&W	10W-14S	73
Woodland Trial, South N&S	10W-14S	73
79th Av. N&S	9W-14S	73
80th Av. N&S	10W-14S	73
81st Ct. N&S	10W-14S	73
82nd Av. N&S	10W-14S,10W-13S	73
82nd Ct. N&S	10W-14S,10W-13S	73
83rd Av. N&S	10W-14S,10W-13S	73
84th Av. N&S	10W-14S	73
85th Av. N&S	10W-14S,10W-13S	73
86th St. N&S	10W-14S	73
87th Av. N&S	10W-14S	73
88th Av. N&S	11W-14S	72
89th Av. N&S	11W-14S	72

Street	Grid	Page
89th Ct. N&S	11W-14S	72
90th Av. N&S	11W-14S	72
90th St. N&S	11W-14S	72
91st St. N&S	11W-14S	72
92nd St. N&S	11W-14S	72
93rd St. N&S	11W-14S	72
94th St. N&S	11W-14S	72
114th Av. N&S	14W-14S	72
115th Av. N&S	14W-14S	72
116th Av. N&S	14W-14S	72
116th Ct. E&W	10W-13S	73
118th St. E&W	10W-13S	73
119th Pl. E&W	10W-14S	73
119th St. E&W	11W-13S,10W-13S	72
120th St. E&W	11W-14S,9W-14S	72
121st St. E&W	11W-14S,10W-14S	72
122nd St. E&W	11W-14S,10W-14S	72
123rd Pl. E&W	14W-14S	72
123rd Pl. E&W	9W-14S	73
123rd St. E&W	11W-14S,10W-14S	72
124th St. E&W	11W-14S,10W-14S	72
125th St. E&W	11W-14S,10W-14S	72
126th St. E&W	11W-14S,10W-14S	72
127th St. E&W	14W-14S,9W-14S	72
131st St. E&W	10W-15S	73
Algoma Dr. E&W	10W-14S	73

Palos Township

Street	Grid	Page
Aberdeen Ct. E&W	10W-15S	73
Aberdeen Dr. E&W	10W-15S	73
Adsit Rd. N&S	13W-15S	72
Aquilla Dr. E&W	9W-15S	73
Archer Av. NE&SW	14W-11S,14W-10S	64
Arlington Ct. E&W	13W-15S	72
Arlington Ln. E&W	13W-15S	72
Barberry Ct. N&S	13W-14S	72
Bernice Dr. E&W	13W-15S	72
Birchwood Dr. E&W	13W-14S	72
Bishop Ct. NE&SW	12W-15S	72
Black Walnut Tr. NS&EW	11W-15S,11W-14S	72
Bloomfield Dr. E&W	13W-15S	72
Blue Heron Dr. N&S	13W-15S	72
Brian Pl. N&S	12W-15S	72
Brittany Dr. N&S	13W-15S	72
Brodt Dr. NW&SE	12W-15S	72
Carriage House Ln. NS&EW	13W-15S	72
Charlotte Ct. NE&SW	13W-15S	72
Cherrywood Dr. E&W	13W-14S	72
Circle Pkwy. NS&EW	12W-15S	72
Cornell Ln. N&S	10W-15S	73
Country Club Ct., North N&S	9W-15S	73
Country Club Ct., South N&S	9W-15S	73
Creek Rd. E&W	12W-15S	72
Deerwood Dr. E&W	9W-15S	73
E. Tanglewood Cir. N&S	12W-15S	72
Edgewater Ln. N&S	12W-15S	72
Edinburgh Dr. N&S	10W-15S	73
Elm Av. N&S	14W-10S	64
Farm Hill Av. E&W	10W-15S	73
Farm Hill Ln. N&S	10W-15S	73
Fitzsimmons Dr. E&W	13W-15S	72
Flavin Rd. N&S	13W-11S	64
Ford Rd. NW&SE	14W-14S,14W-13S	72
Forest Hill Ct. E&W	9W-15S	73
Forest Hill Ln. E&W	9W-15S	73
Forest Woods Dr. NE&SW	13W-10S	64
Frances Ct. E&W	13W-15S	72
Georgetown Dr. N&S	13W-15S	72
Glen Rd. N&S	14W-13S	72
Golf Dr. E&W	9W-15S	73
Grandview Dr. E&W	13W-14S	72
Haas Dr. NE&SW	13W-14S	72
Hampton Ct. N&S	13W-15S	72
Hampton Rd. E&W	13W-15S	72
Hillcrest Dr. E&W	13W-15S,10W-15S	72
Indian Trail Dr. NS&EW	12W-14S	72
Joseph Dr., South N&S	12W-15S	72
Joseph Dr., West E&W	12W-15S	72
Kean Av. N&S	11W-13S,11W-10S	72
Kindling Ct. E&W	13W-15S	72
Kinvarra Dr. N&S	12W-14S	72
Kirkcaldy Ct. E&W	10W-15S	73
Knightbridge Rd. NW&SE	11W-12S	64
Laflin Ct. E&W	12W-15S	72
Lake Trail Dr. N&S	12W-14S	72
Lakeland Dr. E&W	12W-14S	72
Lakeview Ct. NW&SE	12W-15S	72
Longwood Ct. N&S	10W-15S	73
Lucille Ln. N&S	12W-14S	72
Maple Av. N&S	14W-10S	64
McCarthy Rd. E&W	14W-14S,12W-14S	72
McIntosh Dr. E&W	9W-15S	73
Meadowlark Dr. N&S	10W-15S	73
Mill Rd. N&S	12W-15S	72
Mission Dr. NE&SW	12W-15S	72
Misty Meadow Dr. N&S	10W-15S	73
Newport Dr. N&S	12W-15S	72
Northbridge Ct. E&W	13W-15S	72
Northbridge Ln. E&W	13W-15S	72
Oak Hills Ct. E&W	9W-15S	73
Oak Hills Pkwy. N&S	9W-15S	73
Oak Ln. N&S	9W-15S	73
Oak Ridge Ct. E&W	9W-15S	73
Oak Ridge Tr., North N&S	9W-15S	73
Oakgrove Av. N&S	14W-10S	64
Oakview Ct. N&S	9W-15S	73
Oakwood Dr. E&W	13W-14S	72
Old George's Way NW&SE	9W-15S	73
Orchard St. NS&EW	14W-10S	64
Paloma Dr. E&W	10W-15S	73
Palos West Dr. NS&EW	13W-14S	72
Park Lane Dr. E&W	12W-14S	72
Parkside Dr. N&S	12W-15S	72
Parkview Dr. E&W	12W-15S	72
Parkview Rd. N&S	14W-10S	64
Pauline Ct. N&S	10W-15S	73
Peace Memorial Dr. N&S	12W-15S	72
Pebble Dr. NS&EW	12W-15S	72
Powell Rd. E&W	12W-14S	72
Prarie View Dr. E&W	12W-14S	72
Public St. E&W	14W-10S	64
Railroad Av. NE&SW	14W-10S	64
Regent Dr. E&W	11W-12S	64
Richmond Ct. E&W	13W-15S	72
Ridgewood Dr. NS&EW	13W-15S	72
Rt. 171 NE&SW	14W-11S,13W-10S	64
Rt. 45 N&S	12W-15S,11W-10S	72
Sandburg Ct. N&S	12W-15S	72
Sedgewick Dr. NS&EW	13W-14S	72
Sheffield Ct. N&S	13W-14S	72
Somerset Ln. E&W	12W-14S	72
Stark Dr. N&S	14W-10S	64
Stephen Dr. N&S	13W-15S	72
Suffield Dr. N&S	13W-15S	72
Surrey Ct. N&S	12W-15S	72
Tanglewood Cir., West N&S	12W-15S	72
Tanglewood Ct. NW&SE	12W-15S	72
Timber Edge Ln. E&W	12W-14S	72
Turtle Pond Ln. NW&SE	9W-15S	73
Walter Dr. E&W	13W-15S	72
Westminster Ct. E&W	11W-12S	64
Westport Dr. N&S	12W-15S	72
Westview Dr. N&S	9W-15S	73
Whispering Lake Dr. E&W	12W-14S	72
Wild Cherry Ln. E&W	12W-14S	72
Wildwood Dr. NS&EW	13W-14S	72
Will Cook Rd. N&S	14W-15S,14W-13S	72
Willow St. NW&SE	13W-10S	64
Windsor Dr. N&S	12W-14S	72
Windward Trail N&S	10W-15S	73
Wisteria Ct. N&S	13W-14S	72
Wolf Rd. N&S	14W-15S	72
76th Av. N&S	9W-15S	73
77th Pl. N&S	9W-15S	73
80th Av. N&S	10W-15S	73
81st Av. N&S	10W-15S	73
82nd Av. N&S	10W-15S	73
82nd Ct. N&S	10W-15S	73
83rd Av. N&S	10W-15S	73
83rd Ct. N&S	10W-15S	73
84th Av. N&S	10W-15S	73
84th Ct. N&S	10W-15S	73
85th Av. N&S	10W-13S	73
85th Ct. N&S	10W-15S	73
86th Av. N&S	10W-15S	73
87th Av. N&S	10W-15S	73
87th St. E&W	14W-10S,11W-10S	64
91st St. E&W	14W-10S	64
95th Av. N&S	12W-15S,12W-14S	72
95th St. E&W	13W-11S,11W-11S	64
100th Av. N&S	12W-12S,12W-10S	64
104th Av. N&S	13W-15S,13W-10S	72
107th St. E&W	14W-12S,11W-12S	64
108th Av. N&S	13W-15S	72
111th St. E&W	14W-12S,11W-12S	64
118th St. E&W	14W-13S	72
126th St. E&W	12W-14S	72
127th St. E&W	10W-15S,9W-15S	73
128th Pl. E&W	10W-15S	73
129th Pl. E&W	10W-15S	73
129th St. E&W	10W-15S	73
130th St. E&W	12W-15S,10W-15S	72
131st Pl. E&W	11W-15S	72
131st St. E&W	14W-15S,10W-15S	72
132nd St. E&W	10W-15S	73
133rd St. E&W	13W-15S,10W-15S	72
134th St. E&W	13W-15S	72
135th St. E&W	14W-15S,10W-15S	72

Park Forest*

(Also see Will County)

Street	Grid	Page
Algonquin St. E&W	2W-26S	93
Allegheny Ct. E&W	2W-26S	93
Allegheny St. E&W	2W-26S	93
Antietam St. E&W	4W-25S	92
Antioch Pl. N&S	2W-26S	93
Apache St. E&W	2W-26S	93
Apple Ct. NE&SW	3W-27S	92
Apple Ln. NW&SE	3W-27S	92
Arcadia St. E&W	2W-26S	93
Arrowhead Ct. E&W	2W-26S	93
Arrowhead St. E&W	2W-26S	93
Ash St. E&W	2W-26S	93
Bailey Rd. E&W	2W-26S	93
Bender Rd. E&W	2W-26S	93
Berry St. E&W	3W-25S	92
Bertoldo Rd. NW&SE	2W-26S	93
Bigelow Rd. NW&SE	2W-26S	93
Birch St. N&S	2W-26S	93
Blackhawk Dr. NW&SE	4W-27S,3W-27S	92
Blair St. E&W	4W-25S	92
Cedar St. N&S	2W-27S	93
Central Park Av. N&S	4W-27S	92
Chase St. E&W	4W-25S	92
Cherry St. N&S	3W-27S	92
Chestnut Ct. NW&SE	2W-27S	93
Chestnut St. NS&EW	2W-27S	93
Choate Rd. NW&SE	2W-26S	93
Cromwell Rd. E&W	2W-27S	93
Davis St. E&W	4W-25S	92
Dogwood St. N&S	2W-27S	93
Douglas St. N&S	3W-25S	92
Dunham Rd. N&S	2W-27S	93
Dunlap Rd. E&W	2W-27S	93
Early St. E&W	3W-25S	92
Elm St. N&S	3W-26S	92
Farragut St. E&W	3W-25S	92
Fir St. E&W	3W-26S	92
Forest Bl. N&S	3W-27S,3W-26S	92
Garman Rd. NW&SE	2W-27S	93
Gentry St. E&W	3W-25S	92
Gerstung Rd. N&S	3W-26S	92
Gettysburg St. N&S	4W-25S	92
Gibson Rd. N&S	3W-26S	92
Gold St. E&W	3W-27S	92
Grant St. E&W	3W-25S	92
Green St. E&W	3W-27S	92
Hamlin St. N&S	3W-25S	92
Hay St. E&W	3W-25S	92
Hemlock St. N&S	3W-27S	92
Herndon St. E&W	3W-25S	92
Hickory Ct. N&S	4W-25S	92
Hickory St. E&W	4W-25S	92
Holly St. N&S	3W-26S	92
Homan Av. NS&EW	4W-25S	92
Homan Ct. E&W	4W-25S	92
Horwood Blvd. E&W	3W-26S	92
Huron St. NE&SW	4W-27S	92
Illinois Ct. NE&SW	3W-25S	92
Illinois St. N&S	4W-25S,3W-25S	92
Indiana St. NS&EW	4W-25S,3W-25S	92
Indianwood Blvd. NE&SW	3W-27S	92
Iroquois St. NW&SE	3W-27S	92
Jackson St. N&S	4W-25S	92
Juniper St. N&S	3W-26S	92
Kentucky Ct. E&W	3W-25S	92
Kentucky St. E&W	3W-25S	92
Krotiak Rd. E&W	3W-26S	92
Lakewood Blvd. E&W	4W-26S,3W-26S	92
Lakewood Ct. N&S	4W-27S	92
Larue St. E&W	3W-25S	92
Lee St. E&W	3W-25S	92
Leims Rd. E&W	3W-27S	92
Lester Rd. N&S	3W-27S	92
Main St. N&S	4W-26S	92
Manitowac St. NE&SW	3W-27S	92
Mantua Ct. NE&SW	4W-27S	92
Mantua St. NW&SE	4W-27S,3W-27S	92
Marquette Pl. NE&SW	3W-27S	92
Marquette St. NW&SE	4W-27S,3W-27S	92
McCarthy Rd. E&W	3W-27S	92
McGarity Rd. NE&SW	3W-27S	92
Meota St. NE&SW	3W-27S	92
Merrimac St. NE&SW	3W-27S	92
Miami St. NW&SE	4W-27S	92
Michael Rd. E&W	2W-27S	93
Minocqua Ct. E&W	3W-27S	92
Minocqua St. N&S	4W-27S	92
Mohawk St. NW&SE	4W-27S	92
Monitor St. N&S	3W-25S	92
Nashua St. E&W	4W-27S,3W-27S	92
Nassau St. NE&SW	3W-27S	92
Neola St. NE&SW	3W-27S	92
New Salem St. N&S	3W-25S	92
Nokomis St. NE&SW	3W-27S	92
North St. E&W	3W-26S	92
Norwood Bl. E&W	3W-26S	92
Oak Ln. E&W	3W-27S	92
Oakwood St. NE&SW	3W-27S	92
Onarga St. NE&SW	3W-27S	92
Orchard Dr., North N&S	3W-25S	92
Orchard Dr., South N&S	3W-27S,3W-26S	92
Osage St. NE&SW	3W-27S	92
Oswego St. E&W	3W-27S	92
Ottawa St. NW&SE	3W-27S	92
Park St. N&S	3W-26S	92
Peach St. NE&SW	3W-27S	92
Rich Ct. E&W	4W-27S	92
Rich Rd. E&W	3W-27S	92
Rocket Cir., East N&S	3W-27S	92
Rocket Cir., West N&S	3W-27S	92
Rockwell St. N&S	3W-25S	92
Rutledge St. N&S	3W-25S	92
Saginaw Pl. N&S	4W-27S	92
Sandburg St. N&S	3W-25S	92
Sangamon Ct. N&S	4W-27S	92
Sangamon St. NW&SE	4W-27S	92
Sassabee St. NE&SW	4W-27S	92
Sauganash St. E&W	4W-27S	92
Saugatuck St. N&S	4W-27S	92
Sauk Ct. NE&SW	3W-27S	92
Sauk Trail NW&SE	3W-27S	92
Seminole St. E&W	4W-27S	92
Seneca St. E&W	4W-27S	92
Seward St. E&W	3W-25S	92
Shabbona Dr. N&S	4W-27S	92
Shawnee St. E&W	4W-27S	92
Sheridan St. E&W	3W-25S	92
Sherman St. E&W	3W-25S	92
Sioux St. E&W	4W-27S	92
Somonauk Ct. N&S	4W-27S	92
Somonauk St. E&W	4W-27S	92
South St. E&W	3W-26S	92
Springfield St. NS&EW	3W-25S	92
Stanton St. E&W	3W-25S	92
Suwanee St. N&S	4W-27S	92
Thomas St. E&W	3W-25S	92
Todd St. N&S	3W-25S	92
U.S. 30 E&W	3W-25S	92
Victory Bl. E&W	3W-27S	92
Waldmann Dr. E&W	4W-26S	92
Walnut St. NE&SW	4W-26S	92
Warwick Ct. E&W	3W-26S	92
Warwick St. NE&SW	3W-26S	92
Washington Ct. E&W	4W-26S	92
Washington St. N&S	4W-26S	92
Water St. N&S	3W-26S	92
Watseka St. E&W	4W-26S	92
Waverly Ct. N&S	4W-27S	92
Waverly St. N&S	4W-26S	92
Wayne Ct. N&S	4W-26S,3W-26S	92
Wayne St. E&W	4W-26S	92
Well Ct. E&W	3W-26S	92
Well St. E&W	3W-26S	92
Western Av. N&S	3W-27S,3W-25S	92
Westgate Dr. E&W	4W-26S	92
Westwood Ct. E&W	3W-26S	92
Westwood Dr. NS&EW	4W-27S,3W-26S	92
Wildwood Dr. NE&SW	4W-26S	92
Willow St. NE&SW	3W-26S	92
Wilshire St. E&W	4W-26S	92
Wilson Ct. NE&SW	4W-26S	92
Wilson St. NE&SW	4W-26S	92
Windsor St. N&S	4W-27S	92
Winnebago St. NS&EW	4W-26S	92
Winnemac St. E&W	4W-26S	92
Winona St. E&W	4W-26S	92
Winslow St. E&W	3W-26S	92
26th St. E&W	2W-26S	93

Park Ridge

Street	Grid	Page
Albion Av. E&W	11W-8N,9W-8N	36
Aldine Av. N&S	11W-8N,11W-9N	36
Archbury Ln. NW&SE	11W-8N	36
Arthur St. E&W	11W-8N,9W-8N	36
Ascot Dr. NS&EW	11W-8N	36
Ashbury Cir.* E&W	11W-8N	36
Ashland Av. N&S	10W-7N,10W-9N	36
Astoria Way* NE&SW	11W-8N	36
Austin Av. E&W	10W-9N	36
Avondale Av. NW&SE	11W-9N,9W-8N	36
Babetta Ln. N&S	11W-9N	36
Ballard Rd. E&W	11W-11N	26
Beau Dr. NS&EW	10W-10N	36
Belle Plaine Av. E&W	11W-8N,10W-8N	36
Berry Pkwy. N&S	9W-8N	37
Birch St. NE&SW	11W-10N	36
Boardwalk Pl.* E&W	11W-8N	36
Bonita Dr. E&W	10W-7N	36
Bonnie Dr. E&W	10W-7N	36
Bouterse St. E&W	11W-9N	36
Broadway Av. N&S	11W-8N,11W-9N	36
Brookline Av. NW&SE	10W-9N	36
Brophy Av. N&S	10W-7N	36
Burton Ln. E&W	11W-10N	36
Busse Hwy. NW&SE	11W-9N,11W-10N	36
Canfield Av. N&S	9W-7N	37
Carol St. E&W	10W-10N	36
Carolyn Ln. N&S	11W-9N	36
Castle E&W	10W-7N	36
Cedar St. E&W	11W-9N,10W-9N	36
Cherry St. E&W	11W-9N,10W-9N	36
Chester Av. N&S	10W-7N,10W-9N	36
Cleveland Av. N&S	9W-8N	37
Clifton Av. N&S	10W-7N,10W-10N	36
Clinton Av. N&S	10W-9N	36
Columbia Av. E&W	9W-8N	37
Courtland Av. N&S	10W-7N,10W-8N	36
Crain St. E&W	10W-10N	36
Crescent Av. N&S	11W-8N,10W-8N	36
Cumberland Av. N&S	10W-8N	36
Cuttriss St. E&W	10W-9N	36
Cynthia Av. NW&SE	10W-9N	36
Davis St. E&W	11W-11N	26
DeCook Av. E&W	11W-10N	36
Dee Rd. N&S	11W-8N,11W-10N	36
Delphia Av. N&S	10W-8N,10W-9N	36
Des Plaines Av. E&W	11W-8N	36
Devon Av. E&W	10W-8N	36
East Av. N&S	9W-9N	37
Edgemont Ln. E&W	11W-9N,9W-9N	36
Edna Av. E&W	11W-10N	36
Elliot Av. N&S	11W-10N	36
Elm St. E&W	11W-9N,10W-9N	36
Elmore St. N&S	9W-9N	37
Engel Blvd. NE&SW	11W-8N	36
Evergreen Ln. E&W	11W-10N	36
Fairview Av. N&S	10W-7N,10W-9N	36
Farrell Av. E&W	11W-10N	36
Fenton Ln. E&W	11W-10N	36
Florence Dr. N&S	11W-9N	36
Forestview Av. N&S	11W-9N	36
Fortuna Av. NE&SW	11W-10N	36
Frances Pkwy. E&W	10W-7N	36
Gables Av. N&S	11W-10N	36
Garden Av. NW&SE	10W-8N	36
Garden St. E&W	11W-8N,10W-8N	36
Gillick St. E&W	11W-8N,10W-8N	36
Glenlake Av. E&W	10W-7N,9W-7N	36
Glenview Av. E&W	11W-10N	36
Good Av. N&S	11W-10N	36
Goodwin Dr. N&S	11W-9N	36
Grace Av. N&S	10W-7N,10W-9N	36
Grand Blvd. N&S	10W-9N	36
Grant Pl. E&W	10W-9N	36
Granville Av. E&W	10W-7N,9W-7N	36

	Grid	Page
Greendale Av. E&W	11W-10N	36
Greenwood Av. N&S	10W-7N,10W-8N	36
Greenwood Rd. N&S	10W-9N	36
Grove Av. NE&SW	9W-7N	37
Habberton St. N&S	11W-10N	36
Halien Ter.* N&S	11W-8N	36
Hamlin Av. N&S	11W-8N,11W-10N	36
Hamlin Ct. NE&SW	11W-8N	36
Hansen Pl. E&W	10W-9N	36
Harrison St. N&S	9W-8N	37
Hastings St. E&W	10W-9N	36
Helen St. N&S	11W-10N	36
Higgins Rd. NW&SE	10W-7N	36
Hoffman Av. N&S	11W-10N	36
Home Av. N&S	11W-8N,11W-11N	36
Imperial St. E&W	9W-8N	37
Interstate 294 N&S	12W-10N	36
Irwin Av. E&W	11W-9N	36
Irwin St. E&W	11W-9N	36
Joyce Pl. E&W	9W-9N	37
Kathleen Dr. E&W	9W-9N	37
Kent Av. E&W	10W-7N	36
Knight Av. N&S	10W-7N,10W-10N	36
Lahon St. E&W	11W-9N,9W-9N	36
Lake Av. NE&SW	10W-8N	36
Laverne Av. E&W	11W-9N,10W-9N	36
Leonard St. N&S	10W-9N	36
Lincoln Av. N&S	10W-7N,10W-10N	36
Linden Av. N&S	9W-7N	37
Lois Av. E&W	10W-7N	36
Lois Ct. N&S	10W-7N	36
Lundergan Av. N&S	12W-10N	36
Luther Ln. N&S	10W-10N	36
Main St. NW&SE	10W-8N	36
Manor Ln. E&W	11W-10N	36
Marguerite St. E&W	11W-10N	36
Marlowe E&W	11W-8N	36
Marvin Pkwy. NW&SE	11W-9N	36
Mary Jane Ln. E&W	11W-9N	36
Mayfield Dr. E&W	11W-10N	36
Meacham Av. N&S	10W-9N	36
Merrill St. N&S	9W-8N,9W-9N	37
Michael John Dr. E&W	10W-9N,9W-9N	36
Milton Av. E&W	11W-9N	36
Morris St. N&S	10W-9N	36
Murphy Lake Ln.* E&W	11W-8N	36
Murphy Lake Rd. E&W	11W-9N	36
Newton Av. N&S	9W-7N	37
Norman Blvd. E&W	11W-8N	36
Northwest Hwy. NW&SE	10W-9N	36
Oak St. NE&SW	10W-8N	36
Oakton St. E&W	11W-10N	36
Oaktree Ln. E&W	11W-10N	36
Oriole Av. N&S	9W-9N	37
Ottawa Av. N&S	9W-9N	37
Park Ln.* E&W	11W-8N	36
Park Pl. NE&SW	10W-8N	36
Park Plaine Av. N&S	11W-9N	36
Park Ridge Blvd. NE&SW	10W-7N	36
Parkside Av. N&S	11W-10N	36
Parkside Dr. N&S	11W-11N	26
Parkwood Av. N&S	11W-9N	36
Peale Av. N&S	9W-8N	37
Peterson Av. E&W	10W-7N	36
Poplar St. NW&SE	11W-10N	36
Potter Rd. N&S	11W-10N	36
Prairie Av. NS&EW	11W-8N,10W-8N	36
Prospect Av. N&S	10W-7N,10W-9N	36
Rabe Ct. NE&SW	10W-9N	36
Redfield Ct. N&S	11W-8N	36
Redfield Rd. N&S	11W-9N	36
Renaissance Dr. NW&SE	11W-10N	36
Rene Ct. E&W	10W-10N	36
Ridge Ter. NE&SW	10W-8N	36
Riverside Dr. N&S	11W-9N	36
Root St. N&S	10W-9N	36
Rose Av. N&S	11W-8N,11W-9N	36
Rosemont Av. E&W	10W-7N	36
Rowe Av. NE&SW	11W-9N	36
Ruth Av. E&W	11W-10N	36
Scottlynne Dr. NW&SE	11W-9N	36
Seeley Av. NE&SW	11W-9N	36
Seminary Av. N&S	11W-8N,11W-9N	36
Shibley Av. NE&SW	11W-9N	36
Shore Line Dr. N&S	11W-9N	36
Sibley St. E&W	11W-9N,9W-9N	36
Spring St. NE&SW	10W-8N	36
St. James Pl. E&W	10W-8N	36
Stanley Av. E&W	11W-8N,9W-8N	36
Stewart Av. E&W	11W-8N,10W-8N	36
Summit Av. N&S	10W-8N,10W-9N	36
Sylviawood Av. N&S	11W-9N	36
Talcott Pl. NW&SE	10W-7N	36
Talcott Rd. NW&SE	9W-7N	37
Thames E&W	11W-8N	36
Thames Pkwy. N&S	11W-8N	36
Thorndale Av. E&W	10W-7N	36
Tomawadee Dr. N&S	10W-9N	36
Touhy Av. E&W	11W-9N	36
Tri-State Tollway N&S	12W-10N	36
Tyrell Av. N&S	11W-10N	36
U.S. Route 14 E&W	11W-10N	36
Vernon Av. N&S	11W-10N	36
Vine Av. NS&EW	10W-7N,10W-8N	36
Virginia St. E&W	11W-9N	36
Walnut St. NE&SW	11W-10N	36
Warren NE&SW	11W-8N	36
Washington Av. N&S	10W-7N,10W-9N	36
Washington Pl. N&S	9W-7N	37
Weeg Way E&W	11W-10N	36
Wesley Dr. N&S	11W-9N	36
Western Av. N&S	11W-9N,11W-11N	36
Wilkinson Pkwy N&S	11W-9N	36
Wilma Pl. E&W	9W-9N	37
Windsor Mall E&W	11W-8N	36
Wisnen St. N&S	9W-9N	37
Woodland Av. E&W	11W-10N	36
Woodview Ln. N&S	11W-11N	26
Yost Av. E&W	9W-7N	37
1st St. N&S	10W-10N	36
3rd St. NE&SW	10W-8N	36

Phoenix

	Grid	Page
Belle Ct. E&W	0W-18S	82
Frances St. E&W	0W-18S	82
Halsted St. N&S	1W-18S	82
Miller Pl. E&W	0W-18S	82
Vincennes Av. NW&SE	0W-18S	82
Wallace Av. N&S	0W-17S	82
Walton Av. NW&SE	0W-18S	82
1st Av. N&S	0W-18S,0W-17S	82
2nd Av. N&S	0W-18S,0W-17S	82
3rd Av. N&S	0W-18S,0W-17S	82
4th Av. N&S	0W-18S	82
5th Av. N&S	0W-18S	82
6th Av. N&S	0W-18S	82
7th Av. N&S	0W-18S	82
8th Av. N&S	0W-18S	82
9th Av. N&S	0W-18S	82
149th St. E&W	0W-17S	82
151st Pl. E&W	0W-18S	82
151st St. E&W	0W-17S	82
152nd Pl. E&W	0W-18S	82
152nd St. E&W	0W-18S	82
153rd St. E&W	0W-18S	82
154th Pl. E&W	0W-18S	82
154th St. E&W	0W-18S	82
155th Ct. E&W	0W-18S	82
155th Pl. E&W	0W-18S	82
155th St. E&W	0W-18S	82

Posen

	Grid	Page
Albany Av. N&S	3W-17S	81
Albert Av. N&S	3W-17S	81
Blaine Av. N&S	3W-17S,3W-16S	81
California Av. N&S	3W-17S,3W-16S	81
Campbell Av. N&S	3W-17S	81
Cleveland Av. N&S	3W-17S,3W-16S	81
Division St. N&S	3W-17S	81
Elm St. E&W	3W-16S	75
Fairfield Av. NW&SE	3W-17S	81
Francisco Av. N&S	3W-17S	81
Grove St. E&W	2W-17S	81
Harrison Av. N&S	3W-17S,3W-16S	81
Kedzie Av. N&S	3W-17S	81
Maple St. E&W	3W-16S	75
Mc Kinley Av. N&S	3W-17S	81
Mozart Av. N&S	3W-17S	81
Palmer Av. N&S	3W-17S,3W-16S	81
Richmond Av. N&S	3W-17S	81
Sacramento Av. N&S	3W-17S	81
Sherman Av. N&S	3W-17S	81
Short St. N&S	2W-17S	81
Tri State Tollway SE&NW	3W-17S	81
Troy St. N&S	3W-17S	81
Utica Av. N&S	3W-17S	81
Western Av. N&S	2W-16S	75
Whipple St. N&S	3W-17S	81
139th Pl. E&W	3W-16S	75
139th St. E&W	3W-16S	75
140th St. E&W	3W-16S	75
141st St. E&W	3W-16S	75
142nd St. E&W	3W-16S	75
143rd St. E&W	3W-16S	75
144th St. E&W	3W-17S	81
145th St. E&W	3W-17S	81
146th St. E&W	3W-17S	81
147th St. E&W	3W-17S	81
148th St. E&W	3W-17S	81
149th St. E&W	4W-17S	81

Prospect Heights

	Grid	Page
Aberdeen Ln.* E&W	14W-15N	15
Alderman Av. E&W	13W-15N	16
Alton Rd. N&S	14W-14N	25
Andover Dr. E&W	15W-16N	15
Anne Ct. E&W	15W-16N	15
Apple Dr. E&W	13W-15N	16
Blossom Ln. N&S	13W-15N	16
Bonniebrook Ct. E&W	14W-15N	15
Break St. E&W	14W-14N	25
Brian Ln. N&S	14W-15N	15
Brook Rd. NW&SE	15W-14N	25
Burning Bush Ln. N&S	13W-15N	16
Burr Oak Ln.* E&W	14W-15N	15
Calway E&W	14W-15N	15
Camp McDonald Rd. E&W	16W-15N,14W-15N	15
Carl Ct. NW&SE	15W-15N	15
Carrbridge Ln.* NE&SW	14W-15N	15
Center Ln. NE&SW	16W-15N	15
Chester Ln. N&S	15W-14N	25
Cider Ln. N&S	13W-15N	16
Circle Av., East N&S	15W-15N	15
Circle Av., West N&S	15W-15N	15
Claire Ln. E&W	14W-15N	15
Clarendon St. E&W	16W-15N,15W-15N	15
Coldren Dr. N&S	14W-15N	15
Compton Ln. E&W	15W-14N	25
Country Club Dr. NS&EW	14W-14N	25
Countryside Ln. E&W	15W-16N	15
Court St. E&W	15W-15N	15
Cove Dr. NS&EW	13W-15N	16
Crabapple Dr. E&W	13W-15N	16
Creek Ct. E&W	15W-15N	15
Crest Hill Dr. E&W	16W-16N	15
Crimson Ct. N&S	13W-15N	16
Dale Av. N&S	16W-15N	15
DeCook Ln. E&W	11W-10N	36
Delicious Ct. N&S	13W-14N,13W-15N	26
Derbyshire Ct. NE&SW	15W-16N	15
DesPlaines River Rd. N&S	13W-15N	16
Dorset St. E&W	16W-15N	15
Drake Av. N&S	15W-16N	15
Drake Ter. E&W	15W-16N	15
Drury Ln. N&S	16W-15N	15
Duchess Ct. N&S	13W-15N	16
East Kenilworth Av. E&W	15W-15N	15
Edinburgh Ln. NE&SW	15W-14N	25
Edward Cul De Sac N&S	14W-14N	25
Edward Rd. E&W	14W-14N	25
Elaine Cir., East N&S	14W-15N	15
Elaine Cir., West N&S	14W-15N	15
Eleanor Dr. E&W	15W-16N	15
Elm St. N&S	15W-15N	15
Elmhurst Rd. N&S	15W-14N,15W-16N	25
Essex St. N&S	15W-16N	15
Etowah Av. N&S	16W-15N	15
Euclid Av. E&W	14W-14N	25
Fairway Ct. E&W	14W-14N	25
Fairway Dr. N&S	14W-14N	25
Ferndale Ct. NW&SE	14W-14N	25
Ferndale Ln. NE&SW	14W-14N	25
Frankie Ct. E&W	16W-15N	15
Gail Ct., North N&S	16W-15N	15
Gail Ct., South N&S	16W-15N	15
Garden Ct. N&S	15W-16N	15
Garden Ln. E&W	15W-16N	15
Glasgow Ln.* N&S	14W-15N	15
Glenbrook Dr. E&W	15W-16N	15
Glendale Dr. N&S	15W-15N	15
Golfview Cir. E&W	14W-14N	25
Green Bridge Ln. E&W	14W-14N	25
Greening Ct. N&S	13W-15N	16
Grego Ct. E&W	14W-15N	15
Greystone Ln. NW&SE	14W-14N	25
Grove Pl. NE&SW	15W-14N	25
Hawthorne Pl. E&W	15W-14N	25
Highland Dr. N&S	15W-16N	15
Hill Ct. NW&SE	15W-15N	15
Hillcrest Dr. N&S	15W-15N	15
Hillside Av. NE&SW	15W-15N	15
Jonathan Ct. E&W	13W-15N	16
Kenilworth Av. E&W	15W-15N	15
Kenilworth Ct. E&W	15W-15N	15
Kenneth Av. E&W	15W-16N	15
Kerry Ct. E&W	16W-16N	15
Kewaunee Ct. N&S	16W-15N	15
Kingsmill Ln.* NW&SE	14W-15N	15
Lancaster Av. N&S	16W-15N	15
Lanford Ln. E&W	15W-14N	25
Leon Ln., East N&S	14W-15N	15
Leon Ln., West N&S	14W-15N	15
Lewis Isle Ln.* NW&SE	14W-15N	15
Linden N. N&S	16W-15N	15
Linden S. N&S	16W-15N	15
Loch Lomond Ln.* NE&SW	14W-15N	15
Lonsdale Rd. E&W	15W-14N	25
Love Dr. E&W	13W-15N	16
Lynnbrook Dr. E&W	15W-16N	15
Mandel Ln. N&S	13W-15N	16
Manor Av. N&S	15W-15N	15
Maple Av. N&S	15W-15N	15
Maple Ln. N&S	15W-14N,15W-16N	25
Maple St. N&S	15W-15N,15W-16N	15
Marberry Dr. E&W	15W-14N	25
Margate Ln.* E&W	14W-15N	15
Marion Av. E&W	15W-15N	15
Marion St. E&W	16W-15N,15W-15N	15
Mars Pl. E&W	15W-14N	25
McIntosch Ct. E&W	13W-15N	16
Meadow Ridge Ln. E&W	16W-16N	15
Minnaqua Dr. E&W	16W-15N	15
Natawa Pl. E&W	16W-15N	15
Newcastle E&W	14W-15N	15
Newgate Ln.* N&S	14W-15N	15
North Pkwy E&W	15W-15N	15
Oak Av. N&S	15W-16N	15
Oakwood Dr. E&W	15W-16N	15
Olive Av. E&W	15W-15N	15
Olive St. E&W	16W-15N	15
Owen Ct. E&W	15W-15N	15
Owen Pl. N&S	15W-14N	25
Owen St. N&S	15W-15N	15
Palatine Rd. E&W	13W-15N	16
Parkview West NE&SW	13W-15N	16
Patricia Ln. NW&SE	14W-15N	15
Pembridge Ln.* N&S	14W-15N	15
Phelps Av. N&S	16W-15N	15
Pin Oak Dr. N&S	16W-16N	15
Pin Oak Ln. N&S	16W-16N	15
Pine Forest Ln. E&W	14W-14N	25
Pine St. N&S	15W-15N	15
Pinecrest Dr. N&S	13W-15N	16
Piper Cir. N&S	13W-15N	16
Piper Ln. E&W	13W-15N	16
Plaza Dr. N&S	13W-15N	16
Prospect Ct. N&S	15W-15N	15
Prospect Dr. E&W	15W-16N	15
Quaker Ln. N&S	13W-15N	16
Rand Rd. NW&SE	16W-14N	25
Regent Ln.* N&S	14W-15N	15
Ridge Av. E&W	15W-15N	15
Riley Av. N&S	14W-15N	15
Roberts Ln. E&W	13W-15N	16
Robyn Ct. E&W	14W-15N	15
Rose Av. E&W	14W-15N	15
Royal Ct. N&S	16W-15N	15
Rt. 83 N&S	15W-16N	15
Saunders Rd. E&W	12W-15N	16
School Ln. NS&EW	15W-14N	25
Shannon Dr. N&S	14W-15N	15
Shawn Ln. E&W	14W-15N	15
Sherwood Dr. N&S	15W-16N	15
South Pkwy. E&W	15W-15N	15
Spruce Dr. N&S	15W-15N,15W-16N	15
Stirling Ln.* N&S	14W-15N	15
Stonegate Dr., East NE&SW	15W-16N	15
Stonegate Dr., West NW&SE	15W-16N	15
Sussex Corner Ln. N&S	14W-14N	25
Sutherland Ln.* E&W	14W-15N	15
Thierry Ln. N&S	14W-15N	15
Thistle Ln. NW&SE	14W-14N	25
Tomah Av. N&S	16W-15N	15
Tree Ln. N&S	13W-15N	16
Tully Pl. NW&SE	15W-14N	25
Viola Ln. E&W	15W-15N	15
Cir.	13W-15N	16
Walden Ln. E&W	16W-16N	15
Waterford Dr. N&S	14W-15N	15
Waterman Av. N&S	16W-15N	15
Wheeling Rd. N&S	15W-15N	15
Wildwood Dr., North E&W	15W-16N	15
Wildwood Dr., South N&S	15W-16N	15
Wildwood Dr., West N&S	15W-16N	15
Wildwood Rd., East E&W	15W-16N	15
Williamsburg Ln.* E&W	14W-15N	15
Willow Hills Ln. E&W	14W-14N	25
Willow Rd. E&W	15W-15N	15
Wimbledon Cir. E&W	13W-15N	16
Windsor Dr. N&S	16W-15N	15
Winesap Ct. E&W	13W-15N	16
Winkelman Rd. NW&SE	12W-15N	16
Wolf Rd. N&S	14W-15N	15
Woodview Dr. N&S	16W-16N	15
Wurtz Ct. E&W	15W-15N	15

Prospect Hts.

	Grid	Page
Break St. E&W	14W-14N	25
Court St. E&W	15W-15N	15
Crest Hill Dr. E&W	16W-16N	15
Kenilworth Ct. E&W	15W-15N	15
Kerry Ct. E&W	16W-16N	15
Meadow Ridge Ln. E&W	16W-16N	15
Pin Oak Dr. N&S	16W-16N	15
Pin Oak Ln. N&S	16W-16N	15
Walden Ln. E&W	16W-16N	15
Woodview Dr. N&S	16W-16N	15

Proviso Township

	Grid	Page
Buck Rd. N&S	14W-0S	46
Concord St. E&W	14W-0S	46
DesPlaines N&S	9W-1S	54
Harvard St. E&W	14W-0S	46
18th E&W	9W-1S	54
19th E&W	9W-1S	54

Rich Township

	Grid	Page
Blackstone Av. N&S	5W-24S	92
Blackthorn Rd. NS&EW	8W-26S	90
Briar Ln. E&W	5W-24S	92
Briarbranch Terr. NS&EW	8W-26S	90
Brushwood Dr. E&W	8W-26S	90
Burlwood Ct. E&W	8W-26S	90
Candlegate Ct. NS&EW	8W-26S	90
Central Av. N&S	7W-26S,7W-22S	91
Central Park Av. N&S	4W-23S	87
Chapparal Terr. NS&EW	8W-26S	90
Cicero Av. N&S	6W-23S	86
Davis Av. E&W	5W-27S	92
Dewey Av. E&W	5W-27S	92
Elmwood Ln. E&W	5W-25S	92
Flossmorr Rd. E&W	7W-22S,4W-22S	86
Green Sward Way N&S	8W-26S	90
Hearthemeade Rd. NW&SE	8W-26S	90
Heartside Rd. NS&EW	8W-26S	90
Hedgewick Ct. E&W	8W-26S	90
Hickory Glen NW&SE	8W-26S	90
Homan Av. N&S	4W-23S	87
Homeland Rd. N&S	5W-25S	92
Huntsbridge Rd. NS&EW	8W-26S	90
Interstate 57 N&S	6W-23S,6W-22S	86
Ivvlog Terr. NS&EW	8W-26S	90
Keeler Av. N&S	5W-24S	92
Keystone Av. N&S	5W-24S	92
Knollwood Cir. NS&EW	8W-26S	90
Kostner Av. N&S	5W-27S,5W-24S	92
Leclaire Av. N&S	6W-22S	86
Lincoln Hwy. E&W	7W-25S	91
Maple Av. E&W	5W-26S	92
Maplewood Ln. E&W	5W-25S	92
Moorfield Rd. NS&EW	8W-26S	90
N. Windmere Cir. NS&EW	8W-26S	90
Orchard Ln. E&W	5W-24S	92
Pleasant Terr. NS&EW	8W-26S	90
Polk Av. E&W	5W-27S	92
Poplar Av. E&W	5W-26S	92
Prairie Rd. NW&SE	8W-26S	90

	Grid	Page
Ridgeland Av. N&S	8W-27S,8W-22S	90
Rt. 43 N&S	8W-25S,8W-23S	90
Sauk Trail E&W	8W-26S	90
Sprucewood Ln. E&W	5W-25S	92
Steger Rd. E&W	8W-27S,4W-28S	90
Sunset Rd. E&W	5W-24S	92
Tanglewood Rd. E&W	8W-26S	90
Thistle Ct. E&W	8W-26S	90
Thorntree Terr. NS&EW	8W-26S	90
Thornwood Av. E&W	5W-24S	92
Timber Ridge Rd. NW&SE	8W-26S	90
Tullamore Terr. NS&EW	8W-26S	90
Vollmer Rd. E&W	8W-23S,5W-23S	85
Wildrock Terr. NS&EW	8W-26S	90
Woodbine Terr. NS&EW	8W-26S	90
66th Av. N&S	8W-22S	85
192nd St. E&W	4W-23S	87
194th St. E&W	4W-23S	87
196th St. E&W	4W-23S	87
197th St. E&W	4W-23S	87
198th St. E&W	4W-23S	87
204th St. E&W	5W-24S	92
205th St. E&W	5W-24S	92
206th Pl. E&W	5W-24S	92
206th St. E&W	5W-24S	92

Richton Park

	Grid	Page
Adams Dr. N&S	5W-27S	92
Amy Dr. N&S	6W-27S	91
Andover Dr. NW&SE	5W-26S	92
Appleby Ln. E&W	4W-26S	92
Arlington Dr. E&W	5W-26S	92
Arquilla Dr. NS&EW	6W-27S	91
Ascot Ct. E&W	5W-26S	92
Balmoral Dr. N&S	5W-26S	92
Belmont Rd. NE&SW	5W-26S	92
Birchwood Rd. E&W	5W-26S,4W-26S	92
Bretz Dr. E&W	5W-27S	92
Brighton Ln. E&W	4W-26S	92
Bruce Dr. N&S	6W-27S	91
Butterfield Rd. N&S	5W-27S	92
Camden Ct. E&W	5W-26S	92
Carlborg Ct. E&W	6W-27S	91
Cedar Rd. E&W	4W-26S	92
Central Av. N&S	7W-26S	91
Central Park Av. N&S	4W-27S	92
Cherie Ct. E&W	6W-26S	91
Churchill Dr. NS&EW	5W-26S	92
Cicero Av. N&S	6W-27S	91
Clarendon Av. N&S	5W-27S	92
Clark Dr. NW&SE	5W-26S	92
Crawford Av. N&S	5W-26S	92
Cresent Way E&W	5W-27S	92
Cypress Ct. E&W	6W-27S	91
Davis Av. E&W	4W-27S	92
Dewey Av. E&W	4W-27S	92
East Dr. N&S	6W-27S	91
Elm Rd. E&W	4W-26S	92
Euclid Ln. NS&EW	4W-26S	92
Farmington Av. NW&SE	5W-27S	92
Franklin Dr. N&S	5W-27S	92
Governors Hwy. N&S	5W-27S	92
Grant Av. E&W	4W-27S	92
Greenbriar Ln. E&W	5W-26S	92
Hamilton Dr. N&S	5W-27S	92
Hawthorne Way N&S	5W-26S	92
Heartland Dr. N&S	5W-27S	92
Hillside Dr. N&S	6W-26S	91
Imperial Ct. NE&SW	6W-27S	91
Imperial Dr. NS&EW	6W-27S	91
Jackson Av. E&W	4W-27S	92
Jefferson Dr. E&W	5W-27S	92
Kara Ln. E&W	6W-27S	91
Karlov Av. N&S	5W-26S	92
Keenehand Ct. E&W	5W-26S	92
Keith Dr. E&W	6W-27S	91
Kings Ct. E&W	6W-27S	91
Kolin Av. N&S	5W-27S	92
Kostner Av. N&S	5W-26S	92
Lake Shore Dr. N&S	6W-27S	91
Latonia Ct. NW&SE	5W-27S	92
Latonia Ln. N&S	5W-27S	92
Laurel Dr. E&W	5W-26S	92
Lawndale Av. N&S	4W-27S	92
Lee Av. E&W	4W-27S	92
Lee Ct. E&W	5W-26S	92
Linden Rd. E&W	4W-26S	92
Lioncrest Dr. E&W	4W-26S	92
Lorraine Ct. N&S	6W-27S	91
Main St. N&S	4W-26S	92
Maple Av. E&W	5W-26S	92
Millard Av. N&S	4W-27S,4W-26S	92
Mission Dr. N&S	6W-27S	91
Monterey Dr. E&W	6W-27S	91
Parkview Dr. NE&SW	5W-27S	92
Patterson Ct. N&S	4W-26S	92
Pleasant Dr. N&S	6W-27S	91
Polk Av. E&W	4W-27S	92
Poplar Av. E&W	5W-26S	92
Poplar St. E&W	5W-26S	92
Redwood Dr. N&S	6W-27S	91
Richton Pl. N&S	4W-26S	92
Richton Rd. N&S	4W-26S	92
Richton Square N&S	4W-27S	92
Ridgeway Av. N&S	4W-27S,4W-26S	92
Riverside Dr. NS&EW	6W-27S	91
Roberta Ln. E&W	6W-27S	91
Rockingham Rd. NE&SW	5W-26S	92
Salem Ct. E&W	5W-26S	92
Saratoga Rd. E&W	5W-26S	92
Sauk Trail E&W	4W-26S	92
Schaff Ct. E&W	6W-26S	91
Scott Dr. N&S	6W-26S	91
St. Ives Ln. E&W	4W-26S	92
Steger Rd. E&W	6W-27S	91
Taylor Av. E&W	4W-27S	92
Thomas Ct. E&W	4W-26S	92
Thomas Dr. E&W	6W-27S	91
Tyler Dr. E&W	5W-27S	92
Valley Dr. N&S	6W-27S	91
Washington Dr. E&W	5W-27S	92
Windsor Ct. E&W	5W-26S	92
219th St. E&W	4W-26S	92

River Forest

	Grid	Page
Ashland N&S	9W-0N,9W-1N	48
Augusta E&W	9W-1N	48
Auvergne N&S	10W-0N	47
Berkshire E&W	9W-1N	48
Bonnie Brae N&S	9W-0N,9W-1N	48
Central E&W	10W-0N,9W-0N	47
Chicago E&W	9W-1N	48
Clinton Pl. N&S	9W-0N,9W-1N	48
Division E&W	9W-1N	48
Edgewood Pl. N&S	10W-0N	47
Forest N&S	9W-0N,9W-1N	48
Franklin N&S	9W-0N,9W-1N	48
Gale N&S	10W-0N	47
Garden E&W	9W-0N	48
Greenfield E&W	9W-1N	48
Hawthorne E&W	9W-0N	48
Holly E&W	9W-0N	48
Iowa E&W	9W-1N	48
Jackson N&S	9W-0N,9W-1N	48
Keystone N&S	9W-0N,9W-1N	48
Lake E&W	9W-0N	48
Lathrop N&S	9W-0N,9W-1N	48
Le Moyne E&W	9W-1N	48
Linden E&W	9W-0N	48
Monroe N&S	9W-0N,9W-1N	48
Oak E&W	9W-0N	48
Park N&S	9W-0N,9W-1N	48
Park Dr. E&W	9W-0N	48
Quick E&W	9W-0N	48
Rt. 64 E&W	9W-1N	48
Thatcher N&S	10W-0N,10W-1N	47
Thomas E&W	9W-1N	48
Vine E&W	9W-0N	48
Vinson N&S	10W-1N,9W-1N	47
William N&S	9W-0N,9W-1N	48

River Grove

	Grid	Page
Arnold E&W	10W-3N	41
AuxPlaines N&S	10W-3N	41
Belden E&W	10W-2N	47
Beulah N&S	11W-3N	41
Boyle St. N&S	10W-3N	41
Budd N&S	10W-3N	41
Carey E&W	10W-3N	41
Center E&W	10W-3N	41
Clark N&S	11W-3N	41
Clinton N&S	10W-3N	41
Davisson N&S	10W-3N	41
Des Plaines River Rd. NW&SE	11W-3N,10W-2N	41
Elm N&S	11W-2N	47
Enger St. E&W	10W-3N	41
Erie N&S	10W-3N	41
Forest N&S	10W-3N	41
Forest View N&S	10W-3N	41
Fullerton E&W	11W-3N	41
Greenwood Ter. E&W	10W-3N	41
Grove N&S	10W-3N	41
Haymond N&S	10W-3N	41
Herrick E&W	11W-3N,10W-3N	41
Hessing N&S	10W-3N	41
Indian Rd. NE&SW	10W-3N	41
Julian Ter. N&S	10W-3N	41
Leyden N&S	11W-3N	41
Lyndale E&W	10W-2N	47
Maple N&S	10W-3N	41
Marwood N&S	10W-3N	41
O'Conner E&W	10W-3N	41
Oak N&S	10W-3N	41
Pacific E&W	11W-3N	41
Palmer E&W	10W-2N	47
Paris N&S	10W-3N	41
Park N&S	10W-3N	41
Rhodes N&S	11W-3N	41
Richard E&W	11W-3N	41
Ridge E&W	10W-3N	41
River N&S	10W-3N	41
River Grove E&W	10W-3N	41
Roosevelt Ln. E&W	10W-3N	41
Smith St. E&W	10W-3N	41
Spruce N&S	11W-3N	41
Struckmann N&S	10W-3N	41
Thatcher Rd. N&S	10W-3N,10W-2N	41
Trumbull N&S	10W-2N	47
Walsh St. E&W	10W-3N	41
West N&S	11W-2N,11W-3N	47
Wood N&S	10W-3N	41
Wrightwood E&W	11W-3N,10W-3N	41
1st N&S	10W-2N	47
9th N&S	11W-2N	47
80th N&S	10W-3N	41

Riverdale

	Grid	Page
Acme Av. E&W	0W-15S	76
Ashland Av. N&S	1W-16S,1W-15S	82
Atlantic Av. N&S	0W-16S,0W-15S	82
Basic Steel Dr. N&S	1W-16S	82
Blue Island-Riverdale Rd. E&W	1W-16S,0W-15S	82
Charles Dr. E&W	1W-15S	76
Clark St. N&S	0W-16S	82
Dearborn St. N&S	0W-16S	82
Edbrook Av. N&S	0W-16S	82
Eggleston N&S	0W-16S,0W-15S	82
Emerald Av. N&S	1W-17S,1W-16S	82
Forest View Av. E&W	0W-15S	76
Fritz Blvd. N&S	1W-15S	76
Glen Ln. E&W	0W-16S	82
Halsted St. N&W	1W-16S,1W-15S	82
Highlawn Av. NE&SW	0W-16S	82
Illinois St. NE&SW	0W-16S,0W-15S	82
Indiana Av. N&S	0W-16S	82
Industrial Dr. N&S	1W-15S	76
Ivanhoe Ct. E&W	0W-16S	82
Jackson St. E&W	1W-15S	76
La Salle St. N&S	0W-16S	82
Lotz Dr. N&S	1W-16S	82
Lowe Av. N&S	1W-16S	82
Michigan Av. N&S	0W-16S,0W-15S	82
Normal Av. N&S	0W-16S,0W-15S	82
Pacesetter Pkwy. E&W	1W-15S,0W-15S	76
Parnell Av. N&S	0W-16S,0W-15S	82
Perry Av. N&S	0W-15S	76
Rt. 1 N&S	1W-16S	82
School St. N&S	0W-16S,0W-15S	82
State St. N&S	0W-16S	82
Stewart Av. N&S	0W-16S,0W-15S	82
Tracy Av. N&S	0W-16S,0W-15S	82
Union Av. N&S	1W-17S,1W-16S	82
Wabash Av. N&S	0W-16S,0W-15S	82
Wallace St. N&S	0W-16S,0W-15S	82
Wentworth Av. N&S	0W-16S,0W-15S	82
135th Pl. E&W	0W-15S	76
136th Pl. E&W	0W-15S	76
136th St. E&W	0W-15S	76
137th Pl. E&W	0W-15S	76
137th St. E&W	0W-15S	76
138th St. E&W	1W-15S,0W-15S	76
139th St. E&W	0W-16S	82
140th Ct. E&W	0W-16S	82
140th Pl. E&W	0W-16S	82
140th St. E&W	0W-16S	82
141st St. E&W	0W-16S	82
142nd St. E&W	1W-16S,0W-16S	82
143rd St. E&W	1W-16S,0W-16S	82
144th St. E&W	1W-16S,0W-16S	82
145th Pl. E&W	0W-16S	82
145th St. E&W	1W-16S,0W-16S	82
146th St. E&W	1W-16S,0W-16S	82

Riverside

	Grid	Page
Addison E&W	9W-3S	54
Akenside N&S	9W-3S	54
Arlington N&S	9W-2S	54
Audubon N&S	9W-2S	54
Barry Point N&S	9W-3S	54
Bartram N&S	9W-2S	54
Berkeley E&W	9W-2S	54
Blackhawk E&W	9W-3S	54
Blooming Bank N&S	10W-3S	53
Blythe E&W	9W-3S	54
Burling E&W	9W-3S	54
Burlington E&W	10W-3S,9W-3S	53
Byrd N&S	9W-2S	54
Cowley N&S	9W-3S	54
Cir.	9W-3S	54
Downing N&S	9W-2S	54
E. Longcommon E&W	9W-3S	54
Cir.	9W-3S	54
Evelyn E&W	9W-2S	54
Fairbank N&S	10W-3S	53
Forbes N&S	10W-3S	53
Forest E&W	10W-3S	53
Forest N&S	10W-3S	53
Gage E&W	9W-3S	54
Gatesby E&W	9W-2S	54
Groveland N&S	10W-3S	53
Haas N&S	9W-3S	54
Harris St. N&S	9W-3S	53
Herbert N&S	9W-3S	54
Herrick E&W	9W-3S	54
Kent E&W	9W-2S	54
Kimbark N&S	10W-3S	53
Lawton E&W	9W-3S	54
Leesley N&S	9W-2S	54
Lincoln N&S	10W-3S	53
Lindberg E&W	9W-2S	54
Lionel E&W	9W-3S	54
Longcommon E&W	9W-2S	54
Loudoun E&W	9W-2S	54
Maplewood N&S	10W-3S	53
Michavy N&S	9W-3S	54
Millbridge E&W	10W-3S,9W-3S	53
Northgate Ct. N&S	9W-2S	54
Northgate Rd. E&W	9W-2S	54
Northwood N&S	9W-3S,9W-2S	54
Nuttail N&S	9W-3S,9W-2S	54
Ogden E&W	9W-3S	54
Olmstead E&W	9W-3S	54
Park E&W	10W-3S	53
Parkway N&S	9W-2S	54
Pine E&W	10W-3S	53
Quincy E&W	10W-3S,9W-3S	53
Repton N&S	9W-2S	54
Ridgewood E&W	10W-3S	53
Riverside N&S	9W-3S	54
Robinson E&W	9W-3S	54
Scottswood Dr. N&S	10W-3S	53
Scottswood Rd. N&S	10W-3S	53
Selbourne E&W	9W-2S	54
Shenstone E&W	9W-3S	54
Southcote E&W	9W-2S	54
Uvedale E&W	9W-2S	54
Uvedale Ct. NW&SE	9W-2S	53
Waubansee E&W	10W-3S	53
West N&S	10W-3S	53
Wood Side N&S	9W-3S	54
York E&W	9W-2S	54
1st Av. N&S	10W-3S	53

Riverside Township

	Grid	Page
Bismarck E&W	9W-3S	54
Columbus N&S	9W-3S	54
Gladstone N&S	9W-3S	54
Shakespeare N&S	9W-3S	54
Stanley N&S	9W-3S	54
Washington N&S	9W-3S	54
39th E&W	9W-3S	54

Robbins

	Grid	Page
Albany Av. N&S	3W-16S	75
Avers Av. N&S	4W-16S,4W-15S	75
Broadway NE&SW	3W-15S	75
Central Park Av. N&S	4W-16S,4W-15S	75
Clifton Park Av. N&S	4W-16S,4W-15S	75
Emery Ln. E&W	4W-16S	75
Finley Av. NE&SW	4W-16S	75
Francisco Av. N&S	3W-15S	75
Glenn Av. NW&SE	4W-16S	75
Grace Av. NE&SW	4W-16S	75
Hamlin Av. N&S	4W-16S,4W-15S	75
Harding Av. N&S	4W-15S	75
Hendricks Rd. NW&SE	4W-16S	75
Homan Av. N&S	4W-16S,4W-15S	75
Karlov Av. N&S	5W-15S	74
Kedvale Av. N&S	5W-16S,5W-15S	74
Kedzie Av. N&S	4W-16S,4W-15S	75
Keystone Av. N&S	5W-16S	74
Kildare Av. N&S	5W-15S	74
Kinne Rd. E&W	4W-16S	75
Kolin Av. N&S	5W-15S	74
Komensky Av. N&S	5W-15S	74
Kostner Av. N&S	5W-15S	74
Lawndale Av. N&S	4W-16S,4W-15S	75
Lincoln Ln., East E&W	4W-15S	75
Lincoln Ln., West E&W	5W-15S,4W-15S	74
Lydia Av. NE&SW	4W-16S	75
Maxey Ct. E&W	4W-16S	75
Mcbreen Av. NW&SE	4W-16S	75
Midlothian Turnpike NE&SW	5W-16S,4W-16S	74
Monticello Av. N&S	4W-16S,4W-15S	75
Nelson Av. E&W	4W-16S	75
Pulaski Rd. N&S	5W-16S	74
Reeves Rd. NE&SW	4W-16S	75
Rexford Rd. NE&SW	4W-16S	75
Richardson Av. NW&SE	4W-16S	75
Richmond St. N&S	3W-15S	75
Ridgeway Av. N&S	4W-16S,4W-15S	75
Sacramento Av. N&S	3W-16S	75
Sawyer Av. N&S	4W-16S,4W-15S	75
Spaulding Av. N&S	4W-16S	75
Springfield Av. N&S	4W-15S	75
St. Louis Av. N&S	4W-16S,4W-15S	75
Tripp Av. N&S	5W-15S	74
Troy St. N&S	3W-16S	75
Trumbull Av. N&S	4W-16S	75
Turner Av. N&S	4W-16S,4W-15S	75
Utica Av. N&S	3W-16S	75
Woodlawn Av. N&S	3W-15S	75
134th Pl. E&W	4W-15S	75
134th St. E&W	5W-15S,4W-15S	74
135th Pl. E&W	5W-16S,4W-16S	74
135th St. E&W	5W-15S,3W-15S	74
136th Pl. E&W	5W-16S,4W-16S	74
136th St. E&W	5W-16S,3W-16S	74
137th Pl. E&W	5W-16S	74
137th St. E&W	5W-16S,3W-16S	74
138th St. E&W	5W-16S,3W-16S	74
139th Pl. E&W	4W-16S,3W-16S	75
139th St. E&W	4W-16S,3W-16S	75
140th Pl. E&W	4W-16S	75
140th St. E&W	4W-16S	75
141st St. E&W	4W-16S	75
142nd St. NE&SW	4W-16S	75

Rolling Meadows

	Grid	Page
Adams St. NW&SE	20W-14N	23
Adler Ct.* N&S	21W-14N	23
Algonquin Pkwy. NS&EW	19W-12N	24
Algonquin Rd. NW&SE	20W-12N,19W-12N	23
Alton E&W	19W-11N	33
Amanda Ct. NS&EW	21W-13N	23
Anseline Ct. E&W	20W-14N	23
Apollo Dr. N&S	20W-12N	23
Apple Jack Rd. NS&EW	20W-13N	23
Arbor Dr. N&S	21W-12N	23
Arlingdale Ct. NE&SW	22W-14N	23
Arlingdale Dr. NW&SE	22W-14N	23
Arrowood Ln. N&S	21W-14N	23
Asbury E&W	20W-13N	23
Ashland Av. N&S	20W-14N	23
Astor Ln. N&S	19W-12N	24
Attleboro E&W	19W-10N	33
Auburn E&W	20W-13N	23
Barker Av. N&S	20W-12N,19W-13N	23
Barrow NW&SE	20W-11N	33
Belford Ln. E&W	20W-11N	33
Bent Creek Ct. NS&EW	21W-13N	23
Benton St. N&S	20W-14N	23
Berdnick St. E&W	20W-15N	23
Bethel N&S	19W-11N	33
Birch Ln. E&W	19W-12N	24
Black Twig N&S	20W-13N	23
Blackhawk Dr. E&W	20W-15N	23
Bluebird Ln. NS&EW	19W-13N,19W-14N	24
Bob White Ln. N&S	20W-13N	23
Bob-o-Link Ln. NS&EW	20W-13N	23
Bobwhite Ln. E&W	20W-13N	23
Bridgeton N&S	19W-11N	33
Brookmeade Dr. NW&SE	20W-12N,20W-13N	23
Brookton NW&SE	19W-11N	33
Brookwood Way Dr. N&S	20W-14N	23
Bryant Av. NS&EW	20W-14N	23
Burning Trees N&S	20W-13N	23
Butterfield Ct. N&S	21W-14N	23
Calchester E&W	19W-11N	33
California Av. N&S	20W-13N	23
California Ct. E&W	20W-13N	23
California St. N&S	20W-14N	23
Calvert Dr. N&S	19W-12N	24

	Grid	Page
Campbell St. E&W		
	20W-14N,19W-14N	23
Cardinal Dr. NS&EW		
	19W-13N,19W-14N	24
Carlisle NW&SE	20W-11N	33
Carnegie St. N&S	20W-15N	23
Carr St. N&S	19W-15N	14
Carriageway Dr. NW&SE		
	19W-12N	24
Cedar Glen Dr. N&S		
	20W-13N	23
Cedar St. E&W	20W-12N	24
Central Rd. E&W		
	20W-13N,19W-13N	23
Chapman Ct. N&S	20W-14N	23
Chas Dr. N&S	20W-14N	23
Chelsea N&S	20W-11N	33
Cheshire Dr. E&W	19W-12N	24
Cheviot NW&SE	20W-11N	33
Chichester N&S	20W-11N	33
Chicory Ct. N&S	21W-14N	23
Cold Springs Ct. E&W		
	21W-13N	23
Coleen Ct. N&S	20W-13N	23
College Crossing NW&SE		
	21W-13N	23
Comstock N&S	20W-11N	33
Coventry NW&SE	20W-11N	33
Crane Ct. E&W	19W-14N	24
Creekside Dr. E&W		
	20W-13N	23
Creekside Dr. N&S	21W-13N	23
Crestwood Dr. N&S		
	20W-13N	23
Croftwood Ct. N&S		
	21W-14N	23
Cross Rds. Of Commerce Dr		
E&W	20W-12N	23
Crossing Ct. NE&SW		
	21W-13N	23
Croydon NW&SE	20W-11N	33
Cyndi Ct. N&S	18W-13N	24
Dahlia Ct. N&S	21W-14N	23
Davis Ct. NW&SE	20W-14N	23
Dawngate Ct. N&S	20W-13N	23
Dawngate Ln. N&S		
	20W-13N	23
Debra Ct. N&S	20W-13N	23
Deepwood Ln. E&W		
	21W-13N	23
Deerfield Ln. N&S		
	21W-13N,21W-14N	23
Denny Ct. E&W	20W-14N	23
Dogwood Rd. E&W		
	20W-13N	23
Donegal E&W	20W-11N	33
Dove Ct. NW&SE	19W-13N	24
Dove St. N&S		
	19W-13N,19W-14N	24
Dunbar E&W	20W-11N	33
Duxbury N&S	20W-11N	33
Eagle Ln. N&S	19W-13N	24
Ealing NW&SE	20W-11N	33
East Frontage Rd. E&W		
	20W-13N	23
Eastman St. E&W	20W-14N	24
Eaton NW&SE	20W-11N	33
Edgar St. N&S	20W-13N	23
Edgewood Ct. N&S		
	21W-14N	23
Edison Pl. NW&SE	20W-15N	23
Eleanore Ct. N&S	20W-13N	23
Elk Grove Rd. NE&SW		
	18W-11N	24
Emerson Av. E&W	19W-12N	24
Essex Way N&S	19W-12N	24
Euclid Av. E&W		
	20W-14N,19W-14N	23
Exeter NW&SE	20W-11N	33
Fairfax E&W	20W-14N	23
Falcon Ct., North NS&EW		
	20W-13N	23
Falcon Ct., South E&W		
	20W-13N	23
Falkirk N&S	20W-11N	33
Falmouth NW&SE	20W-11N	33
Farmington Ct. N&S		
	21W-14N	23
Fieldstone N&S	20W-13N	23
Finch Ct. E&W	20W-14N	23
Flicker Ln. N&S	20W-14N	23
Forest Av. N&S	20W-14N	23
Fox Ln. N&S	21W-13N	23
Foxcroft E&W	19W-11N	33
Fremont St. E&W		
	20W-14N,19W-14N	23
Fulle St. E&W	19W-13N	24
George Ct. NW&SE		
	19W-14N	24
George St. E&W	19W-13N	24
Gettysburg Dr. E&W		
	20W-14N	23
Ginger Ct. E&W	21W-14N	23
Golf Place E&W	19W-12N	24
Golf Rd. E&W		
	20W-12N,18W-12N	23
Green Meadow Ct. N&S		
	21W-13N	23
Greenwich Ct. E&W		
	21W-13N	23
Grimsby NW&SE	20W-11N	33
Grouse Ct. N&S	19W-13N	24
Grouse Ln. E&W	19W-13N	24
Grove St. E&W	19W-13N	24
Groveside Ln. E&W		
	21W-13N	23
Gull Ct. NE&SW	19W-13N	24
Hampton E&W	19W-10N	33

	Grid	Page
Hartung Rd. E&W	21W-13N	23
Haverhill E&W	19W-11N	33
Hawk Ct. NE&SW	19W-13N	24
Hawk Ln. N&S	19W-13N	24
Hawthorne Ln. N&S		
	20W-12N	23
Heather Ct. N&S	21W-14N	23
Heron Ct. N&S	19W-14N	23
Hicks Rd. N&S	20W-14N	23
Highland Dr. NS&EW		
	22W-14N	23
Holliston NE&SW	19W-11N	33
Holly Ln. E&W	20W-13N	23
Holyoke E&W	19W-10N	33
Honeysuckle Ln. NE&SW		
	21W-14N	23
Hoover Ct. NE&SW		
	20W-12N	23
Hoover St. NW&SE		
	20W-12N	23
Huron Ct. N&S	19W-14N	24
Industrial Av. N&S		
	20W-15N,20W-14N	23
Interstate 290 N&S		
	20W-12N,20W-14N	23
Interstate 90 NW&SE		
	20W-12N	23
Ironwood Ct. E&W	21W-14N	23
Jasmine Ln. E&W	21W-14N	23
Jay Ln. N&S		
	19W-13N,19W-14N	24
Jessica Ct. E&W	21W-14N	23
Jill Ct. N&S	20W-14N	23
Jody Ct. NS&EW	20W-13N	23
Jonquil Ct. E&W	21W-14N	23
Juniper Rd. NE&SW		
	20W-13N	23
Kendal N&S	20W-11N	33
Kenilworth Dr. N&S		
	19W-12N	24
Kettering N&S	20W-11N	33
Kevin Ln. N&S	21W-14N	23
Keystone Ct. N&S	19W-12N	24
Kieth Ct. NW&SE	20W-14N	23
Killarney Ct. N&S	19W-13N	24
Kimball Hill Dr. N&S		
	19W-12N	24
Kimberry Ct. E&W	21W-14N	23
King's Walk Dr. E&W		
	20W-14N	23
Kingfisher Ct. E&W	20W-13N	23
Kingfisher Ct. NE&SW		
	19W-13N	24
Kingfisher Ln., East NW&SE		
	19W-13N	24
Kirchoff Rd. E&W		
	20W-14N,19W-13N	23
Kittery E&W	19W-11N	33
Knoll Ridge Rd. N&S		
	20W-13N	23
Lark Ct. NW&SE	19W-14N	24
Lilac Ct. N&S	21W-14N	23
Lincoln Av. E&W	20W-14N	23
Lincoln Ct. N&S	20W-14N	23
Linden Ln. N&S		
	19W-12N,19W-13N	23
Lisa Ct. E&W	21W-14N	23
Lois Dr. NS&EW	19W-12N	24
Magnolia Dr. N&S	20W-12N	23
Mallard Ct. E&W	20W-13N	23
Mallory Ct. N&S	21W-14N	23
Manomet E&W	19W-10N	33
Maple Ct. E&W	19W-12N	24
Maple Ln. E&W	19W-12N	24
Margate NW&SE	20W-11N	33
Martin Ct. N&S	19W-13N	24
Martin Ln. E&W	19W-13N	24
Mayberry Ct. E&W	21W-14N	23
McKone Ct. E&W	20W-13N	23
Meacham Rd. N&S		
	21W-13N	23
Meadow Dr. NE&SW		
	19W-13N,19W-14N	24
Meadow Edge Dr. N&S		
	21W-14N	23
Meadowbrook Ct. E&W		
	19W-12N	24
Meadowbrook-Industrial Ct		
N&S	19W-12N	24
Michael Ct. E&W	21W-14N	23
Mill Creek Ln. N&S		
	21W-13N	23
Millstone Ln. NW&SE		
	21W-13N	23
New Castle Ct. E&W		
	21W-14N	23
Newport Dr. NW&SE		
	19W-12N	24
North Carona Dr. NS&EW		
	22W-13N,22W-14N	23
North Hampton Dr. E&W		
	19W-12N	24
Northwest Tollway NW&SE		
	20W-15N,19W-15N	23
Norwood Ct. N&S	21W-14N	23
Oak Ln. E&W	19W-12N	24
Oaksbury Ct. E&W	20W-13N	23
Oaksbury Ln. E&W		
	20W-13N	23
Old Creek Rd. E&W		
	21W-13N	23
Old Hickory Rd. N&S		
	20W-13N	23
Old Mill Ln. E&W		
	21W-13N,21W-14N	23
Old Saybrook E&W		
	19W-10N	33
Old Valley N&S	20W-13N	23

	Grid	Page
Oriole Ln. N&S		
	19W-13N,19W-14N	24
Owl Dr. N&S		
	20W-13N,19W-13N	23
Owl Ln. NE&SW		
	20W-13N,19W-14N	23
Oxford NS&EW	20W-14N	23
Park Ct. N&S	19W-13N	24
Park St. NW&SE	19W-13N	24
Peacock Ct. NE&SW		
	19W-13N	24
Peacock Dr. NW&SE		
	20W-12N,19W-13N	23
Peacock Ln. NW&SE		
	19W-13N	24
Pebblebrook Ln. N&S		
	21W-13N,21W-14N	23
Pembroke E&W	20W-11N	33
Pepperell N&S	19W-11N	33
Pheasant Dr. N&S	20W-14N	23
Pine Valley N&S	20W-13N	23
Piper Ct. E&W	20W-13N	23
Plum Grove Dr. NW&SE		
	20W-12N	23
Plum Grove Rd. N&S		
	21W-14N	23
Plum Tree Ln. N&S		
	21W-13N,21W-14N	23
Pocassett N&S	19W-11N	33
Polk Ave. N&S	20W-14N	23
Prarie Ln. E&W	22W-14N	23
Prestwick E&W	20W-11N	33
Pride Ct. E&W		
	21W-14N,20W-14N	23
Quail Ct. NE&SW	19W-13N	24
Quail Ln. E&W	19W-13N	24
Quentin Rd. N&S		
	21W-13N,21W-18N	23
Raven Ln. N&S	19W-13N	24
Red Haw N&S	20W-13N	23
Redbud Rd. NW&SE		
	20W-13N	23
Redwing Ct. NE&SW		
	20W-13N	23
Rhiannon Ct. E&W	21W-14N	23
Richnee Ln. N&S	20W-13N	23
Robin Ln. N&S	19W-14N	24
Rohlwing Rd. N&S		
	20W-14N,20W-15N	23
Rose Wood Dr. N&S		
	20W-13N	23
Rt. 53 N&S		
	20W-13N,20W-14N	23
Rt. 58 E&W		
	20W-12N,18W-12N	23
Rt. 62 NW&SE		
	20W-14N,19W-14N	23
Rumford N&S	19W-11N	33
Rywick Dr. E&W	20W-13N	23
Saint George Dr. NS&EW		
	20W-14N	23
Salt Creek Ln. NS&EW		
	20W-14N,19W-14N	23
Scarborough N&S	20W-11N	33
School Dr. NE&SW		
	19W-14N	24
School St. NW&SE		
	19W-13N,19W-14N	24
Shady St. NE&SW	20W-13N	23
Shagbark N&S	20W-13N	23
Shannon E&W	20W-11N	33
Sigwalt St. E&W		
	19W-14N,19W-13N	24
Silent Brook Ln. NS&EW		
	21W-14N	23
Smith St. N&S	21W-14N	23
South Corona Dr. E&W		
	22W-13N	23
South Ct. N&S	19W-13N	24
South Hampton Dr. E&W		
	19W-12N	24
South St. E&W	19W-13N	24
Spruce Ct. NW&SE		
	20W-14N	23
Squibb Av. N&S	20W-12N	23
St. Alban's N&S	20W-11N	33
St. James St. E&W		
	19W-14N	24
Starling Ln. E&W	19W-14N	24
Stirling St. E&W	20W-11N	33
Stoneridge Rd. N&S		
	20W-13N	23
Stork Ct. E&W	19W-14N	24
Stuart N&S	20W-11N	33
Swallow Ln. E&W	19W-13N	24
Swan Ln. N&S	19W-14N	24
Sycamore Av. N&S		
	20W-12N	23
Taft Av. N&S	20W-14N	23
Tall Oaks Ln. N&S	20W-13N	23
Teal Ct. E&W	20W-13N	23
Teonia Woods Rd. E&W		
	20W-13N	23
Tern Ct. N&S	19W-14N	24
Theda Ln. N&S	20W-13N	23
Thorntree N&S	20W-12N	23
Thrush Ct. N&S	19W-13N	24
Thrush Ln. E&W		
	19W-13N,19W-13N	23
Tollview Dr. NS&EW		
	20W-12N,19W-12N	23
Troy N&S	20W-11N	33
Upton N&S	19W-11N	33
Vermont Ct. N&S	20W-14N	23
Vermont St. NS&EW		
	20W-14N	23
Villa Cir. NS&EW	21W-13N	23
Viola Ct. NS&EW	21W-13N	23

	Grid	Page
Walnut Ct. E&W	20W-13N	23
Weber Dr. N&S	19W-12N	24
West Frontage Rd. N&S		
	20W-13N	23
Wildwood Dr. N&S	20W-13N	23
Wilke Rd. N&S		
	20W-12N,19W-14N	24
Williamsburg Ct. E&W		
	20W-14N	23
Willow Ln. E&W	20W-12N	24
Wilmette Av. N&S	20W-14N	23
Wilson Av. NW&SE		
	20W-14N	23
Windemere N&S	20W-11N	33
Wing St. E&W	19W-14N	24
Winnetka Av. NE&SW		
	20W-14N,20W-15N	23
Winnetka Cir. N&S	20W-14N	23
Winterport NW&SE		
	19W-11N	33
Wiscassett N&S	19W-11N	33
Woodbine N&S	20W-13N	23
Woodcliff E&W	21W-13N	23
Woodcliff Ln. E&W	20W-13N	23
Woodland Ct. E&W		
	20W-13N	23
Woods Chapel N&S		
	20W-13N	23
Wren Ct. NE&SW	19W-13N	24
Wren Ln. NS&EW		
	19W-13N,19W-14N	24
Yarrow Ct. N&S	21W-14N	23
Yarrow Ln. N&S	21W-14N	23

Roselle*
(Also see Du Page County)

	Grid	Page
Acadia Bay * E&W	23W-8N	32
Acadia Ct. N&S	22W-8N	33
Acadia Tr. E&W	22W-8N	33
Bryce Tr. E&W	22W-8N	33
Canterbury Tr.* NE&SW		
	23W-8N	32
Carlsbad Tr. E&W	22W-8N	33
Chisolm Tr. E&W	22W-8N	33
Conway Bay* NE&SW		
	23W-8N	32
Cross Creek Ct. E&W		
	22W-8N	33
Cross Creek Dr. E&W		
	22W-8N	33
Cumberland Rd. N&S		
	22W-8N	33
Glacier Bay* N&S	23W-8N	32
Glacier Ct.* NW&SE	23W-8N	32
Glacier Tr. E&W	22W-8N	33
Isle Royal Bay E&W	22W-8N	33
Nerge Rd. E&W	22W-8N	33
Oregon Tr. E&W	22W-8N	33
Overland Ct. E&W	22W-8N	33
Sequoia Bay N&S	22W-8N	33
Sequoia Tr. NW&SE	22W-8N	33
Shadow Lake Bay* NE&SW		
	23W-8N	32
Shawnee Tr. E&W	22W-8N	33
White Sands Bay N&S		
	22W-8N	33
Woodfield Tr. NS&EW		
	22W-8N	33
Yosemite Ct. N&S	22W-8N	33
Yosemite Tr. N&S	22W-8N	33

Rosemont

	Grid	Page
Allen St. Ext. E&W	11W-6N	41
Anne Ct. E&W	13W-8N	35
Ardmore St. E&W	12W-7N	36
Balmoral Av. E&W	12W-6N	40
Barry St. N&S	13W-8N	35
Betty Ct. E&W	13W-8N	35
Bryn Mawr Av. E&W		
	12W-7N	36
Byron St. E&W	12W-7N	36
Carol St. E&W	13W-8N	35
Chestnut Av. N&S	13W-8N	35
Devon Av. E&W	12W-8N	36
Devon Av. E&W	12W-7N	36
Doris Ct. E&W	13W-8N	35
Emerson St. E&W	12W-7N	36
Estelle Dr. E&W	12W-8N	35
Ethel Ct. E&W	13W-8N	35
Evenhouse Av. E&W		
	11W-7N	36
Foster Av. E&W		
	12W-6N,11W-6N	40
Gage St. E&W	11W-7N	36
Graff Dr. E&W	12W-7N	36
Granville St. E&W	12W-7N	36
Hawthorne St. N&S	12W-7N	36
Higgins Rd. NW&SE	13W-8N	35
Hirschman St. E&W	11W-7N	36
Hope Dr. E&W	12W-7N	36
Kirschoff St. N&S	12W-7N	36
Lunt St. E&W	13W-8N	35
Lyman St. N&S	11W-7N	41
Lyndon St. N&S	13W-8N	35
Mannheim Rd. N&S	13W-8N	35
Michigan Av. N&S	12W-6N	40
Milton Ct. N&S	12W-6N	41
Milton Parkway N&S		
	12W-6N	40
Morse St. E&W	12W-7N	36
Northwest Tollway NW&SE		
	12W-7N,11W-7N	36
Norwood Av. N&S	12W-7N	36
Otto Av. N&S	12W-6N	40
Pearl St. N&S	12W-6N	40
Pine St. N&S	13W-8N	35

	Grid	Page
River Rd. N&S	11W-6N	41
River Rose St. E&W		
	11W-7N	36
Rose St. N&S	11W-6N	41
Rosemont St. E&W	12W-7N	36
Ruby St. N&S	12W-7N	36
Scott St. N&S	12W-7N	36
Thorndale Av. E&W	12W-7N	36
Wesley Ter. N&S	11W-6N	41
William St. E&W	11W-6N	41
Willow Creek Dr. N&S		
	11W-7N	36
60th St. E&W	12W-6N	40
61st St. E&W	12W-6N	40
62nd St. E&W	12W-6N	40

Sauk Village

	Grid	Page
Apache Dr. E&W	2E-26S	95
Arlene Ct. N&S	2E-26S	95
Astor St. E&W	2E-25S	95
Barry Ln. NE&SW	2E-25S	95
Brookwood Av. N&S	3E-27S	95
Calumet Expwy. N&S		
	1E-27S,1E-26S	94
Carol Av. N&S	2E-26S	95
Carole Ln. E&W	2E-26S	95
Cedar Ln. NE&SW	2E-25S	95
Chappel Av. N&S		
	2E-27S,2E-26S	95
Charlotte Ct. NW&SE		
	2E-26S	95
Chestnut Av. N&S	2E-25S	95
Clyde Av. N&S	2E-26S	95
Constance Av. E&W	2E-26S	95
Cornell Av. N&S	2E-27S	95
Cynthia Av. N&S	2E-26S	95
Eastbrook Ct. NW&SE		
	2E-27S	95
Eastbrook Dr. N&S	2E-27S	95
Evergreen Ln. E&W	2E-25S	95
Galine Av. N&S	2E-25S	95
Grand St. E&W	2E-25S	95
Jeffery Av. N&S		
	2E-27S,2E-26S	95
Kalvelage Dr. E&W	3E-26S	95
Kathy Dr. N&S	2E-27S	95
Katz-Corner Rd. E&W		
	3E-25S	95
Lincoln Hwy. E&W	2E-25S	95
Locust Av. N&S	2E-25S	95
Luella Ct. N&S	2E-26S	95
Maple Ct. N&S	2E-25S	95
Maple Ln. NE&SW	2E-25S	95
Merrill Av. N&S		
	2E-27S,2E-26S	95
Merrill Ct. N&S	2E-26S	95
Murphy Av. N&S	3E-27S	95
Navaho Av. N&S	2E-26S	95
Nichols Av. N&S	3E-27S	95
Nichols Dr. E&W	3E-27S	95
Oak St. E&W	2E-25S	95
Oakbrook Ct. E&W	2E-27S	95
Oakbrook Ln. E&W	2E-27S	95
Olivia Av. N&S	2E-26S	95
Orion Av. N&S	2E-26S	95
Paxton Av. N&S	2E-26S	95
Peach Tree Av. N&S	3E-26S	95
Penrose Av. N&S	2E-25S	95
Peterson Av. N&S	2E-25S	95
Pine Av. N&S	2E-25S	95
Pomo Ct. E&W	2E-25S	95
Poplar Ln. E&W	2E-25S	95
Prairie Av. N&S	2E-26S	95
Reichert Av. N&S	2E-27S	95
Ross Ct. N&S	2E-27S	95
Rt. 394 N&S	2E-26S	95
Sauk Trail E&W	2E-26S	95
Scott Dr. NW&SE	2E-26S	95
Shirley Av. N&S	2E-26S	95
Southbrook Dr. NE&SW		
	2E-27S	95
Spencer Av. N&S	3E-27S	95
Steger Rd. E&W	2E-27S	95
Strassburg Av. NE&SW		
	2E-27S	95
Talandis Dr. N&S	2E-26S	95
Theisen Av. N&S	2E-26S	95
Theodore Av. N&S	3E-27S	95
Torrence Av. N&S		
	2E-27S,2E-26S	95
U.S. Rt. 30 E&W	2E-26S	95
Walton Pl. NW&SE	2E-25S	95
Westbrook Ct. NW&SE		
	2E-27S	95
Westbrook Dr. N&S	2E-27S	95
Westbrook Tr. N&S	2E-27S	95
Willow Tree Ln. N&S	3E-26S	95
Yates Av. NE&SW		
	2E-27S,2E-26S	95
30th St. E&W	3E-27S	95
215th Pl. E&W	2E-26S	95
216th Ct. E&W	2E-26S	95
216th Pl. E&W	2E-26S	95
216th St. E&W	2E-26S	95
217th Pl. E&W	2E-26S	95
217th St. E&W	2E-26S	95
218th Pl. E&W	2E-26S	95
218th St. E&W	2E-26S	95
219th Pl. E&W	2E-26S	95
219th St. E&W	2E-26S	95
220th St. NE&SW	2E-26S	95
221st St. E&W	2E-26S	95
222nd Pl. E&W	2E-26S	95
222nd St. E&W	2E-26S	95
223rd Pl. E&W		
	2E-27S,3E-27S	95
223rd St. E&W	2E-26S	95

Grid Page
224th Pl. E&W 3E-27S 95
224th St. E&W 2E-27S,3E-27S 95
225th Pl. E&W 2E-27S,3E-27S 95
225th St. E&W 2E-27S,3E-27S 95
226th St. E&W 3E-27S 95

Schaumburg*

(Also see Du Page County)

Abbington Pl. N&S . . 24W-11N 32
Aberdeen Ct. N&S . . 26W-10N 31
Academy Ct.* N&S 24W-13N 22
Acorn Ct. N&S 23W-10N 32
Adams Ct. NW&SE 27W-10N 31
Adler Ct. N&S 26W-11N 31
Aegaean Dr. N&S 24W-9N 32
Aimee Ln. E&W 26W-10N 31
Aimee Way N&S 26W-10N 31
Aimtree Pl. E&W 24W-10N 32
Akron Ct. E&W 25W-9N 32
Albany Ct. NS&EW . . 25W-8N 32
Albion Av. E&W 24W-8N,23W-8N 32
Alcott Ct. E&W 25W-9N 32
Alden Ln. N&S 24W-10N 32
Alder Ct. N&S 24W-10N 32
Alexandra Ct. E&W . . 24W-9N 32
Algonquin Rd. NW&SE 22W-13N 23
Allcott Ct. E&W 25W-9N 32
Allison Ln. E&W 25W-11N 32
Allonby Dr. NW&SE 24W-11N 32
Alpine Dr. N&S 25W-10N 32
Altoona Ct. E&W . . 25W-11N 32
Amanda Ln. E&W . . 23W-12N 22
Amboy Ln. N&S 25W-10N 32
American Ln. E&W 22W-11N,21W-11N 33
Amherst Dr. NE&SW 25W-10N 32
Andover Ct.* E&W . . 24W-14N 22
Andrew Ct. E&W 24W-9N 32
Andrew Ln. N&S 24W-8N,24W-9N 32
Andria Ct. E&W 26W-11N 31
Angus Ct.* E&W 23W-13N 22
Apple Dr. N&S 25W-11N 32
Arbor Dr. NS&EW 21W-12N,20W-12N 23
Arbor Sq. N&S 21W-12N 23
Ardmore Ct. E&W . . 24W-9N 32
Argyll Ln. N&S 24W-11N 32
Arklow Pl. E&W 26W-10N 31
Arleen Ct. NW&SE . . 26W-10N 31
Arlington Ln. E&W . . 25W-10N 32
Arrowwood Ct.* NE&SW 21W-9N 33
Arthur Av. E&W 23W-8N 32
Asbury Ln. N&S 23W-9N 32
Ascot Cir. NS&EW . . 23W-10N 32
Ash Ct. N&S 24W-9N 32
Ashcroft Ln. N&S . . 23W-10N 32
Ashley Ln. N&S 25W-11N 32
Ashton Ct. N&S 23W-9N 32
Ashwood Ct. N&S . . 24W-9N 32
Ashwood Dr. NE&SW 24W-9N,23W-9N 32
Aspen Ct. E&W 25W-10N 32
Aspen Dr. N&S 25W-10N 32
Aster Dr. N&S 21W-10N 33
Athena Ct. E&W 24W-9N 32
Attleboro Ct. N&S . . 24W-9N 32
Auburn Cir. E&W 25W-9N 32
Auburn Ln. N&S 25W-9N 32
Austin Ln. NS&EW 23W-12N 22
Avon Ct. E&W 25W-8N 32
Azalea Dr.* NS&EW 20W-10N 33
Bahama Ct. N&S 24W-9N 32
Bahama Ln. N&S 24W-9N 32
Balboa Ct. NE&SW . . 24W-9N 32
Baldwin Ct. E&W . . 26W-10N 31
Balsam Ct. E&W 24W-10N,23W-10N 32
Banbury Ct. NW&SE 26W-11N 31
Bank Dr. NS&EW . . 21W-11N 33
Bar Harbour Rd. N&S 21W-10N 33
Barcliffe Ln. N&S . . 24W-10N 32
Bardsey Dr. N&S . . 25W-10N 32
Bark Wood Rd. NE&SW 21W-13N 23
Barrett Ln. E&W 23W-9N 32
Barrington Rd. N&S 27W-10N 31
Barton Cir. NS&EW 23W-10N 32
Basswood Rd. N&S 22W-12N 23
Bates Ct. N&S 25W-9N 32
Bates Ln. E&W 25W-9N 32
Bayshore Dr. NE&SW 24W-11N 32
Bayview Pt. E&W . . 22W-10N 33
Beach Comber Dr. N&S 24W-9N 32
Beacon Dr. N&S 24W-9N 32
Beckett Ln. E&W . . 21W-10N 33
Bedford Ct. N&S 24W-9N 32
Beech Ct. NE&SW . . 23W-9N 32
Beech Dr. E&W 23W-10N,22W-10N 32
Beechwood Ct. SW&NE 21W-10N 33
Belinder Ln. E&W . . 21W-13N 23
Belmont Ct.* E&W . . 23W-14N 22
Bent Tree Ln. E&W 23W-12N 22
Bentley Ln. E&W 23W-9N 32
Benwick Ct. NW&SE 23W-9N 32
Berkley Ct. E&W . . 22W-10N 33
Berkshire Ct. N&S . . 22W-9N 33
Berkshire Ln. E&W . . 22W-9N 33
Berry Ct. E&W 26W-11N 31
Beverly Ct. E&W 23W-9N 32
Birch Pl. E&W 21W-12N 23
Birdsong Ct. E&W . . 26W-11N 31
Bishop Ct.* N&S 23W-13N 22
Bittersweet Ct. NW&SE 23W-10N 32
Blackhawk Ct. N&S . . 22W-9N 33
Bladon Rd. N&S 24W-12N 22
Blaine Ct. NW&SE . . 21W-13N 23
Blandford Ct. E&W 21W-10N 33
Blenheim Dr. NW&SE 24W-12N 22
Bode Rd. E&W 26W-11N,25W-11N 31
Bonded Pkwy. N&S 27W-10N 31
Cir. 22W-9N 33
Bourne Ln. NW&SE . . 24W-9N 32
Boxwood Dr. NS&EW 24W-9N 32
Bradford Ct. N&S 24W-9N 32
Bradford Ln. E&W . . 24W-9N 32
Braintree Ct. NE&SW 25W-10N 32
Braintree Dr. N&S 25W-9N,25W-10N 32
Bramble Ct. NW&SE 24W-10N 32
Bramble Ln. NE&SW 24W-10N 32
Branchwood Ct. E&W 23W-10N 32
Branchwood Dr. NE&SW 23W-10N 32
Breakers Pt. E&W . . 22W-10N 33
Brendon Dr. N&S . . 26W-11N 31
Brent Ct. E&W 26W-10N 31
Brentwood Ct. N&S . . 24W-9N 32
Brewster Ln. N&S . . 25W-9N 32
Brian Av. N&S 26W-11N 31
Briar Hill Dr. E&W . . 26W-11N 31
Briar Trail N&S 21W-13N 23
Briarwood Ct. N&S 24W-10N 32
Bridgeport Dr. N&S . . 22W-8N 33
Bridgeview Pt. E&W 22W-10N 33
Bridle Ln. N&S 21W-11N 33
Bridlewood Ct.* NW&SE 20W-10N 33
Bright Ridge Dr. NS&EW 22W-10N 33
Brighton Ct. NW&SE 24W-9N 32
Brightridge Dr. E&W 22W-10N 33
Bristol Ln. N&S 24W-11N 32
Brittany Ct.* E&W . . 24W-13N 22
Brixham Pl. N&S 25W-10N 32
Broadway Ln. NS&EW 26W-10N 31
Brockton Ln. NS&EW 25W-9N 32
Bromley Ct.* E&W . . 23W-13N 22
Brookhill Ct. E&W . . 23W-9N 32
Brookside Ln. N&S 21W-13N 23
Brookston Ct.* N&S 21W-9N 33
Brookston Dr. N&S 21W-10N 33
Brown Ct. N&S 25W-9N 32
Brunswick Cir. NW&SE 23W-9N 32
Brunswick Harbor* N&S 22W-9N 33
Brush Rd. N&S 21W-13N 23
Bryn Mawr Ct.* E&W 24W-13N 22
Buccaneer Dr. E&W 21W-13N 23
Buckingham Ct.* N&S 21W-9N 33
Burberry Cir.* NW&SE 20W-10N 33
Burberry Ln.* E&W 20W-10N 33
Burgess Ct.* N&S . . 24W-13N 22
Burke Ct. N&S 22W-10N 33
Burnley Cir. E&W . . 22W-10N 33
Burr Oak Ln. NW&SE 23W-10N 32
Buttercup Ln. N&S 21W-11N 33
Buttonwood Cir.* NS&EW 20W-9N 33
Cable Ct. NW&SE . . 25W-9N 32
Cabot Ln. E&W 25W-9N 32
Cambia Dr. E&W 23W-8N 32
Cambourne Ln. E&W 25W-10N 32
Cambridge Dr. N&S 25W-9N,24W-9N 32
Camden Ct. NW&SE 26W-10N 31
Camellia Ln. E&W . . 21W-11N 33
Cannon Ct. NW&SE 25W-9N 32
Canterbury Dr. NW&SE 24W-12N 22
Cape Ln. N&S 25W-9N 32
Capital Ct. NW&SE . . 25W-9N 32
Capri Ln. NW&SE 25W-9N 32
Cardiff Ct.* E&W . . 23W-13N 22
Cardinal Ct.* E&W . . 24W-13N 22
Carlisle Ct.* E&W . . 23W-13N 22
Carlton Ct. N&S 24W-10N 32
Carlton Ln. NW&SE 24W-10N 32
Carmelhead Ln. E&W 23W-10N 32
Carnaby Ct.* N&S . . 23W-13N 22
Carolina Ct. N&S 25W-8N 32
Caron Ct. N&S 25W-9N 32
Carr Ct. N&S 25W-9N 32
Carriage Ct. E&W 22W-9N 33
Carriage Ln. N&S 22W-9N 33
Carson Ct. E&W 25W-9N 32
Carver Ln. N&S 25W-10N 32
Caryville Ln. N&S . . 26W-10N 31
Casa Dr. E&W 22W-11N 33
Case Ct. NE&SW 25W-9N 32
Casey Ct. E&W 21W-13N 23
Cedar Ct. N&S 24W-10N 32
Cedarcrest Dr. NS&EW 24W-9N,24W-10N 32
Center Ct. NS&EW 23W-12N 22
Century Ct. E&W . . 23W-10N 32
Chalfront Dr. E&W . . 25W-11N 32
Champlaine Ct. N&S 22W-8N 33
Charlene Ln. NW&SE 24W-10N 32
Charleston Ct.* NE&SW 27W-10N 31
Charleston Dr. E&W 26W-10N 31
Charlotte Ct.* NE&SW 27W-10N 31
Chartwell Rd. N&S . . 24W-12N 22
Chatham Ct. N&S . . 25W-10N 32
Chatham Ln. E&W . . 25W-10N 32
Chatsworth Cir. NS&EW 23W-10N,22W-10N 32
Chaucer Ct. E&W 24W-9N 32
Chelsea Ln. E&W 24W-9N 32
Cheltenham Pl. E&W 25W-11N 32
Cherry Ct. N&S 25W-11N 32
Cherry Dr. N&S 25W-11N 32
Cherrywood Dr. N&S 23W-10N 32
Chesapeake Ln. E&W 25W-8N 32
Chestnut Ct. NE&SW 24W-9N 32
Cheswick Ct. NW&SE 26W-10N 31
Chilmark Ln. N&S . . 26W-10N 31
Chopin Ct. E&W 24W-9N 32
Churchill Rd. N&S . . 24W-12N 22
Civic Dr. N&S 25W-10N 32
Claredon Springs Ct.* N&S 23W-13N 22
Claridge Ct. NE&SW 26W-11N 31
Classic Rd. E&W 23W-9N 32
Clayton Cir.* NE&SW 24W-13N 22
Clearbrook Ct. E&W 23W-9N 32
Clearwater Ln. N&S 22W-10N 33
Cleveland Ct. N&S . . 23W-10N 32
Clifton Ct. E&W 24W-9N 32
Clipper Dr. N&S 21W-13N 23
Cloud Ct. N&S 25W-9N 32
Clover Ln. N&S 25W-9N 32
Cloverdale Ln. NE&SW 25W-10N 32
Cobbler Ln. NS&EW 22W-10N 33
Cobblestone Ct.* SE&NW 20W-10N 33
Colby Ln. NW&SE . . 25W-9N 32
Cole Ct. NW&SE . . 25W-9N 32
College Dr. NE&SW 22W-13N 23
College Hill Cir. NS&EW 22W-13N 23
Colony Lake Dr. NS&EW 24W-11N 32
Columbia Ct.* N&S 27W-10N 31
Columbine Cir. NW&SE 21W-11N 33
Columbine Rd. NS&EW 21W-11N 33
Colwyn Dr. E&W 25W-11N,25W-10N 32
Commerce Dr. NS&EW 23W-12N,22W-12N 22
Commodore Ct. E&W 25W-8N 32
Commodore Ln. N&S 25W-8N 32
Commons Dr. N&S 22W-11N 33
Compton Ln. NW&SE 25W-10N 32
Concord Ln. E&W . . 25W-9N 32
Concordia Ln. NW&SE 25W-9N 32
Continental Ln. NS&EW 26W-10N 31
Cooper Ct. E&W . . 22W-12N 23
Copperfield Ln. N&S 23W-9N 32
Coral Ct. N&S 25W-9N 32
Cornell Ln. N&S 25W-9N 32
Corporate Crossing NE&SW 21W-10N 33
Cottington Dr. E&W 25W-11N 32
Cottington.Ct. NE&SW 25W-11N 32
Cottonwood Ct. E&W 23W-9N 32
Cottonwood Ln. N&S 23W-9N 32
Cotuit Ct. N&S 25W-10N 32
Cougar Trail N&S . . 22W-10N 33
Country Club Ln. N&S 25W-9N 32
County Farm Ln. NS&EW 26W-10N 31
Courtenay Ln. E&W . . 23W-9N 32
Courtland Ct. N&S . . 24W-9N 32
Cove Pt. E&W 22W-10N 33
Covehill Ct. NE&SW 24W-10N 32
Coventry Rd. N&S . . 24W-12N 22
Coveside Ln. E&W . . 24W-9N 32
Covington Pl. N&S . . 25W-10N 32
Crabtree Ln. N&S 23W-8N 32
Cranbrook Ct. NS&EW 24W-8N,24W-9N 32
Crandell Ln. N&S 23W-9N 32
Crandon Ln. E&W . . 25W-9N 32
Creek Bend Rd. N&S 21W-13N 23
Creighton Ln. N&S . . 23W-9N 32
Crescent Ln. NE&SW 27W-10N 31
Crest E&W 25W-8N 32
Crest Ct. N&S 23W-8N,23W-9N 32
Crest St. N&S 23W-8N 32
Crest Wood Ct. NE&SW 23W-12N 22
Crooked Creek Rd. N&S 21W-13N 23
Crown Ct. N&S 23W-9N 32
Crystal Ct. N&S 22W-9N 33
Crystal Ct. N. NW&SE 22W-9N 33
Crystal Ct. S. NW&SE 22W-9N 33
Cumberland Dr. N&S 25W-10N 32
Cutters Mill Ct. N&S 25W-11N 32
Cutters Mill Ln. N&S 25W-11N 32
Cypress Ct. NW&SE 23W-9N 32
Dana Ct. E&W 23W-9N 32
Danbury Ct. E&W 24W-9N 32
Daniels Ct. E&W 24W-11N 32
Dansforth Dr. E&W . . 23W-9N 32
Dante Ct. N&S 24W-9N 32
Danvers Ct.* E&W . . 23W-14N 22
Dartford Ln. N&S . . 25W-10N 32
Dartmouth Ln. NS&EW 25W-9N 32
Debbie Ln. NW&SE 27W-10N 31
Deborah Ct. N&S 23W-9N 32
Dedham Ln. E&W 25W-9N 32
Deerfield Ct. E&W . . 24W-13N 22
Deerpath Ct.* N&S . . 21W-9N 33
Del Lago Dr. N&S . . 22W-11N 33
Delaware Ct. NE&SW 27W-10N 31
Denham Pl. E&W . . 25W-11N 32
Dennis Ct. N&S 25W-10N 32
Denton Ct.* E&W . . 24W-14N 22
Derby Ct. N&S 23W-9N 32
Derry Ct. E&W 23W-9N 32
Desmond Dr. E&W 24W-9N,23W-9N 32
Devon Av. E&W 23W-8N 32
Devonshire Ct.* NE&SW 20W-10N 33
Dickens Way N&S . . 23W-9N 32
Dighton Ln. N&S . . 22W-11N 33
Discovery Dr. E&W 26W-11N 31
Dixie Ct. NW&SE 25W-8N 32
Dogwood Ct.* E&W 24W-13N 22
Donna Ct. N&S 23W-9N 32
Dorchester Ct.* N&S 24W-13N 22
Dorset Ln. E&W 23W-9N 32
Dover Ct. E&W 25W-9N 32
Downing Dr. E&W . . 24W-12N 22
Dracut Ln. NS&EW 22W-11N 33
Drake Cir. E&W 22W-9N 33
Drake Ct. NW&SE . . 22W-9N 33
Driftwood Ln.* N&S . . 21W-9N 33
Drummer Dr. E&W . . 21W-12N 23
Dublin Ln. NW&SE . . 26W-10N 31
Dumont Ln. E&W . . 25W-11N 32
Dunbar Ct.* N&S . . 23W-13N 22
Dunlap Pl. N&S 25W-10N 32
Dunsford Ct.* E&W 23W-13N 22
Durham Ct. N&S 22W-9N 33
Duxbury Ct. NW&SE 24W-9N 32
Duxbury Ln. E&W . . 24W-9N 32
Eagle Ct.* N&S 24W-13N 22
East Point Rd. N&S . . 22W-9N 33
East Wood Ct. SW&NE 23W-12N 22
Eastham Ct. N&S . . 25W-10N 32
Eastham Ln. E&W . . 25W-10N 32
Eastview Ct. E&W . . 22W-10N 33
Eaton Ct. NW&SE . . 23W-9N 32
Ebbtide Pt. E&W . . 22W-10N 33
Eden Dr. E&W 24W-12N 22
Edgelake Pt. E&W . . 22W-10N 33
Effingham Ct. NE&SW 27W-10N 31
Elgin Ln. NE&SW . . 25W-10N 32
Ellington Dr. N&S . . 25W-10N 32
Ellisville Ln. E&W . . 25W-10N 32
Elm Dr. NE&SW 25W-10N 32
Elmhurst Ln. NW&SE 25W-11N 32
Elmont Ct. NE&SW 22W-10N 33
Elmwood Ln.* N&S 21W-10N 33
Emerald Dr. E&W . . 22W-11N 33
Emerson Dr. N&S . . 25W-10N 32
Epping Pl. E&W 25W-10N 32
Essex Ct. N&S 25W-10N 32
Estes Av. E&W 24W-8N 32
Estes Ct. E&W 24W-8N 32
Evening Song Ct. N&S 26W-11N 31
Evergreen Ct. E&W 23W-10N 32
Exeter Ct. N&S 23W-9N 32
Exmore Dr. E&W . . 25W-10N 32
Fabish Ct. NW&SE 26W-10N 31
Fairbanks Ct. E&W 22W-10N 33
Fairfield Ct. N&S 23W-9N 32
Fairfield Ln. E&W . . 25W-10N 32
Fairhaven Ln. E&W 25W-11N 32
Fairlane Ct. E&W 25W-8N 32
Fairlane Dr. N&S 25W-8N,25W-9N 32
Fairview Ln. N&S 24W-9N 32
Falbrook Ct. NW&SE 27W-10N 31
Falmouth Ln. NW&SE 25W-9N 32
Farmgate Dr. E&W . . 23W-9N 32
Farnham Ct. E&W . . 25W-11N 32
Featherstone Ct. E&W 26W-10N 31
Fennel Ct. N&S 22W-9N 33
Fennel Ln. E&W 22W-9N 33
Fenwick Ct.* E&W . . 23W-13N 22
Fermi Ct. NW&SE . . 24W-10N 32
Fern Ct. N&S 23W-9N 32
Ferndale Ct.* N&S . . 21W-9N 33
Fieldstone Ct. E&W 24W-11N 32
Finchley Ct.* E&W . . 23W-13N 22
Fiskeville Ln.* E&W 23W-14N 22
Fleming Ln. E&W 23W-9N 32
Flintshire Dr. N&S . . 25W-11N 32
Flower Ct.* E&W . . 24W-13N 22
Forest Ct. N&S 24W-10N 32
Forest Ln. E&W 24W-10N,23W-10N 32
Forestdale Ct. E&W 25W-10N 32
Foxboro Ct. N&S 24W-9N 32
Foxboro Ln. N&S 24W-9N 32
Foxwood Ct. NW&SE 24W-11N 32
Fremont Ct. NW&SE 27W-10N 31
Gareth Ln. N&S 22W-10N 33
Garnet Cir. NW&SE . . 23W-9N 32
Gayton Ln. N&S 22W-9N 33
Geneva Ct. E&W 23W-9N 32
Georgean Ln. E&W 24W-10N 32
Georgia Ct. E&W 25W-8N 32
Gilcrest Ct. E&W . . 27W-10N 31
Glasgow Ct. N&S . . 26W-10N 31
Glasgow Ln. N&S . . 26W-10N 31
Glastonbury Ln.* E&W 23W-13N 22
Glen Bryn Ct. NE&SW 27W-10N 31
Glengary Ct. E&W . . 26W-10N 31
Glenn Ct. N&S 23W-9N 32
Glenridge Ln. N&S . . 26W-10N 31
Glenview Ct. E&W . . 24W-13N 22
Glouchester Cir. N&S 23W-9N 32
Glouchester Harbor* N&S 22W-9N 33
Golf Rd. E&W 26W-11N,21W-11N 31
Good Speed Ln. NE&SW 26W-11N 31
Governor's Dr. E&W 23W-9N 32
Grace Ct. E&W 24W-9N 32
Grace Ln. N&S 24W-9N 32
Grand Central Ln. NW&SE 24W-10N 32
Green Briar Dr. E&W 21W-10N 33
Green Bridge Ct. NE&SW 26W-10N 31
Green River Ct. NW&SE 25W-11N 32

Grid Page

Green River Dr. NS&EW
.... 25W-11N 32
Greenhill Ln. NS&EW
.... 22W-10N 33
Greenview Ct. N&S . 23W-8N 32
Gregory Ct. E&W .. 23W-9N 32
Gregory Ln.
.... 24W-9N,23W-9N 32
Greystone Ct.* N&S..21W-9N 33
Groen Ct. E&W 22W-9N 33
Groen Ln. E&W 22W-9N 33
Groton Ct.* E&W .. 23W-13N 22
Grovenor Dr. E&W .. 22W-9N 32
Hackberry Ct. N&S
.... 23W-10N 32
Hadley Run Ln. E&W
.... 22W-10N 33
Halifax Ct. N&S 23W-9N 32
Hamilton Cir.* N&S 24W-13N 22
Hamilton Pl.* E&W ..24W-13N 22
Hammond Dr. NE&SW
.... 22W-12N,21W-13N 23
Hampshire Ct. N&S ..22W-9N 33
Hampton Cir. N&S .. 22W-9N 33
Hampton Ct. E&W .. 25W-9N 32
Hanley Ct.* E&W ...23W-13N 22
Hanover Ct. E&W .. 22W-10N 33
Hanscom Ct. E&W..26W-10N 31
Harbour Ct. NW&SE
.... 25W-8N 32
Hartford Ct. NE&SW
.... 23W-9N 32
Hartford Dr.
.... 23W-9N,22W-9N 32
Hartley Rd. E&W .. 21W-12N 23
Hartmann Dr. E&W ..25W-9N 32
Harvard Ct. E&W....25W-9N 32
Harvard Ln. NS&EW
.... 25W-9N 32
Harwich Ln. E&W . 24W-10N 32
Hastings Ct.* E&W 23W-14N 22
Haverford Ct.* NW&SE
.... 20W-10N 33
Haverhill Ln. N&S .. 26W-10N 31
Hawk Ln. N&S 27W-10N 31
Hawthorne Ct.* N&S
.... 21W-9N 33
Hazel Dr. N&S
.... 23W-9N,23W-10N 32
Heather Ct.* N&S...21W-9N 33
Heatherstone Dr. NS&EW
.... 22W-11N 33
Helene Ln. E&W ...24W-10N 32
Hemlock Pl. N&S . 21W-12N 23
Hickory Ct. N&S .. 24W-10N 32
Hickory Ln. E&W
.... 24W-10N,23W-10N 32
Higgins Rd. NW&SE
.... 22W-11N,21W-11N 33
Highland Ct. N&S
.... 26W-10N 31
Highpoint Ct. N&S .. 22W-9N 33
Hillside Ct. E&W ... 23W-9N 32
Hilltop Ct. E&W ... 24W-10N 32
Hilltop Dr. N&S 24W-10N 32
Hillwick Ln. N&S...22W-10N 33
Hingham Ct. E&W . 24W-9N 32
Hingham Ln. NE&SW
.... 24W-9N 32
Hinkle Ct. E&W 24W-10N 32
Hinkle Ln. N&S 24W-10N 32
Hitching Post Ln. NS&EW
.... 26W-10N,26W-11N 31
Holliston Ct.* N&S..23W-13N 22
Holmes Way E&W ..26W-10N 31
Holmes Way NE&SW
.... 27W-10N 31
Holyoke Ct. NE&SW
.... 23W-9N 32
Houlton Ct. NW&SE 23W-9N 32
Howard Dr. N&S ... 23W-9N 32
Hunt Wyck Ct.* E&W
.... 20W-10N 33
Hunterdon Ct. NE&SW
.... 27W-10N 31
Hunters Cir. E&W .. 23W-9N 32
Huntington Ln. N&S..23W-9N 32
Huntly Ct.* N&S .. 24W-14N 22
Hyannis Ct. E&W .. 24W-10N 32
Hyde Ct.* E&W ... 24W-13N 22
Idle Wild Ct. E&W . 23W-12N 22
Idlestone Ln. N&S .. 25W-9N 32
Illinois Av. E&W ... 23W-10N 32
Indian Hill Dr. N&S .. 23W-8N 32
Interstate 290 N&S
.... 21W-10N,20W-12N 33
Interstate 90 E&W
.... 23W-12N,21W-11N 22
Ironwood Ct.* E&W ..21W-9N 33
Irving Park Rd. NW&SE
.... 25W-8N,24W-8N 32
Island Ct. N&S 27W-10N 31
Islington Ln. N&S .. 24W-10N 32
Iverson Ln. N&S....25W-11N 32
Ivy Ct. E&W.........24W-9N 32
James Ct. NE&SW 25W-11N 32
Jamestown Ct. N&S 23W-9N 32
Janine Ln. E&W ... 24W-9N 32
Jeffrey Ct. NW&SE
.... 24W-10N 32
Jeffrey Ln. NE&SW
.... 24W-10N 32
Jessica Ct. N&S .. 21W-13N 23
John Ct. N&S 23W-9N 32
John Rolfe Dr. E&W
.... 26W-11N 31
John Smith Dr. E&W
.... 26W-11N 31

Grid Page

Jones Rd. N&S
.... 24W-11N,24W-12N 32
Juli Ct. N&S 23W-9N 32
Juli Dr. N&S 23W-9N 32
Juniper Ln. E&W ... 24W-9N 32
Kavalier Ct.* E&W .. 24W-13N 22
Keats Ct. N&S 24W-9N 32
Kemah Ct. NE&SW
.... 24W-10N 32
Kemah Ln. NS&EW
.... 24W-10N 32
Kendall Ct.* N&S .. 24W-13N 22
Kendrick Ct. N&S .. 24W-10N 32
Kenilworth Dr. E&W
.... 22W-10N 33
Kenmore Ct. E&W ..26W-10N 31
Kensington Dr. E&W
.... 26W-11N 31
Kent Ct. E&W 24W-9N 32
Kern Dr. E&W 21W-11N 33
Killarney Ct. N&S .. 23W-9N 32
Kimberly Dr. E&W .. 21W-11N 33
Kings Mill Ct. NW&SE
.... 23W-10N 32
Kingsbury Dr. N&S 26W-10N 31
Kingsley Dr. E&W .. 25W-11N 32
Kingsport Dr. E&W 23W-9N 32
Kingston Ct. N&S .. 25W-9N 32
Kingston Ln. E&W .. 22W-10N 32
Kirkwall Ln. E&W .. 22W-10N 33
Knightsbridge Ln. N&S
.... 24W-12N 22
Knollwood Ct.* E&W
.... 24W-13N 22
Knollwood Dr. N&S
.... 26W-10N,26W-11N 31
Knottingham Ct. N&S
.... 22W-8N 33
Knottingham Dr. NS&EW
.... 22W-8N 33
Kristin Cir. E&W .. 23W-12N 22
Kristin Dr. E&W ... 23W-12N 22
La Ron Ct. NW&SE
.... 25W-11N 32
Laconia Ln. E&W .. 24W-9N 32
Lakeland Ct. NE&SW
.... 22W-10N 33
Lakeland Dr. E&W.. 22W-10N 33
Lakeview Pt. E&W .. 22W-10N 33
Lambert Dr. E&W .. 23W-10N 32
Lamorak Ct. N&S .. 22W-10N 33
Lamorak Dr. N&S .. 22W-10N 33
Lamorak Ln. E&W .. 22W-10N 33
Lancashire Ct.* E&W
.... 23W-13N 22
Langley Ct. N&S 23W-9N 32
Langley Dr. E&W .. 23W-9N 32
Largo Ct.* E&W .. 23W-13N 22
Laurel Ln. N&S 23W-8N 32
Laurette Ct. N&S .. 22W-9N 33
Lawn Ct. E&W ... 26W-10N 31
Lawson Rd. E&W
.... 26W-11N,25W-11N 31
Lear Ct.* E&W ... 24W-13N 22
Leeds Pass NW&SE
.... 25W-11N 32
Lela Ct. NW&SE...24W-10N 32
Lela Ln. NE&SW ... 24W-10N 32
Lengl Dr. E&W 23W-10N 32
Lenox Ct. N&S ... 26W-10N 31
Leonard Dr. NE&SW
.... 23W-8N 32
Leslie Ln. N&S ... 26W-11N 31
Lewis Ct. N&S ... 26W-11N 31
Lexington Cir. E&W
.... 21W-10N 33
Lexington Dr. N&S..21W-11N 33
Liberty Ct.* E&W . 24W-13N 22
Library Ct. N&S ... 23W-10N 32
Library Ln. E&W ... 23W-10N 32
Lighthouse Dr. N&S ..24W-9N 32
Lily Ct. N&S 23W-9N 32
Limerick Ln. N&S.. 23W-9N 32
Lincoln Meadows Ct. NS&EW
.... 20W-10N 33
Lincoln Meadows Dr.*
NE&SW.......... 20W-10N 33
Lincoln St. N&S
.... 23W-8N,23W-10N 32
Lincolnshire Ct. NW&SE
.... 23W-10N 32
Linsey Av. N&S ... 26W-11N 31
Linton Ct. E&W ... 25W-8N 32
Linwood Ln. E&W .. 24W-10N 32
Lippitt St. NE&SW . 24W-9N 32
Locust Pl. E&W .. 21W-12N 23
Long Boat Ct. N&S
.... 24W-11N 32
Long Boat Ln. E&W
.... 24W-11N 32
Long Meadow Dr. NS&EW
.... 23W-9N 32
Longtree Ct. E&W ..21W-12N 23
Lorelei Dr. N&S ... 24W-9N 32
Lowell Ct. N&S ... 24W-9N 32
Lowell Dr. N&S ... 24W-9N 32
Lundy Ln. N&S ... 24W-9N 32
Lunt Av. E&W
.... 24W-8N,23W-8N 32
Luther Dr. N&S ... 23W-10N 32
Lynnfield Ln. N&S .. 26W-10N 31
Magnolia Ct. N&S .. 23W-10N 32
Maidstone Ct. N&S
.... 27W-10N 31
Malden Ct. N&S .. 23W-9N 32
Mall Ct. N&S 23W-9N 32
Mallard Cir. E&W . 22W-9N 33
Mallard Ct. NW&SE..22W-9N 33

Grid Page

Mallow Ct. E&W....26W-10N 31
Manchester Rd. N&S
.... 23W-9N 32
Manomet Ct. E&W .. 24W-10N 32
Manomet Ln. NS&EW..22W-11N 33
Manor Cir.* N&S....24W-13N 22
Mansfield Ct. NW&SE
.... 25W-11N 32
Maple Pl. N&S......21W-12N 23
Maplebrook Ct. E&W
.... 24W-10N 33
Maplewood Ct.* N&S
.... 21W-9N 33
Marble Hill Ct. N&S
.... 23W-10N 32
Margate Ct. E&W .. 23W-10N 32
Marie Ln. N&S
.... 24W-9N,24W-10N 32
Mariner Pt.* N&S..22W-10N 33
Martingale Rd. N&S
.... 21W-10N,21W-11N 33
Mashpee Ln. N&S .. 24W-10N 32
Matfield Ct.* E&W .. 23W-13N 22
Mayfair Ln.* N&S....21W-9N 33
Maywood Ln. N&S..26W-11N 31
McConnor Pkwy. E&W
.... 21W-12N 23
Meacham Rd. N&S
.... 21W-9N,21W-13N 33
Medford Cir. N&S .. 23W-9N 32
Melody Ct. E&W...26W-11N 31
Mendon Ln. N&S .. 26W-10N 31
Mercury Ct. N&S .. 25W-9N 32
Mercury Dr. N&S
.... 25W-8N,25W-9N 32
Merlin Dr. NS&EW .. 22W-9N 33
Mermaid Ct. NW&SE
.... 24W-10N 32
Middlebury Ct.* N&S
.... 21W-9N 33
Milan Ct. E&W 23W-9N 32
Milford Ct. E&W .. 24W-10N 32
Millbrook Ct. E&W ..25W-9N 32
Millbrook Ln. N&S .. 25W-10N 32
Millis Ln. N&S 26W-10N 31
Milton Ln. E&W ... 25W-9N 32
Mistic Harbour Ln. E&W
.... 24W-10N 32
Mitchell Blvd. N&S .. 24W-8N 32
Mohawk Ct. N&S .. 22W-8N 33
Mohegan Ln.* N&S
.... 23W-13N 22
Monet Ct. N&S 24W-9N 32
Monson Ct. N&S .. 21W-10N 33
Monterey Av. E&W .. 23W-8N 32
Mora Ct. E&W 25W-8N 32
Morning Song Ct. N&S
.... 26W-11N 31
Morse Av. E&W ... 24W-8N 32
Mullingar Ct. E&W .. 23W-9N 32
N. Brookdale Dr. E&W
.... 25W-11N 32
Nantucket Harbor E&W
.... 23W-9N 32
National Pkwy. NS&EW
.... 22W-11N,23W-12N 33
Natwick Ln. N&S .. 26W-10N 31
Nauset Ln. N&S ... 24W-10N 32
Neal Ct. N&S 22W-9N 33
New Mills Ct. E&W
.... 23W-10N 32
Newbury Ct. NW&SE
.... 22W-11N 33
Newbury Ln. NS&EW
.... 22W-11N 33
Newgate Ct. E&W .. 23W-10N 32
Newport Harbor* NS&EW
.... 22W-9N 33
Newton Ct.* E&W .. 24W-13N 22
Niagara Av. E&W
.... 23W-8N,22W-8N 32
Night Song Ct. NS&EW
.... 26W-11N 31
Norman Ct. E&W .. 23W-9N 32
North Brookdale Rd. E&W
.... 25W-11N 32
North Point Dr. E&W
.... 22W-9N 33
Northbury Ct. E&W
.... 21W-10N 33
Northway Dr. N&S..26W-10N 31
Northwest Tollway NW&SE
.... 23W-12N,21W-12N 22
Northwind Cir. E&W
.... 26W-11N 31
Norwell Ct. N&S ... 25W-9N 32
Norwell Ln. E&W .. 25W-9N 32
Norwood Ln. NE&SW
.... 24W-9N 32
Notis Ct. N&S 24W-9N 32
Oak Creek Cir. NW&SE
.... 23W-10N 32
Oak Knoll Ct.* N&S..21W-9N 33
Oak Lawn Ct. NS&EW
.... 23W-12N 22
Oak Meadow Ct. E&W
.... 21W-10N 33
Oakmont Ln. N&S .. 22W-11N 33
Oakview Ct. E&W .. 24W-10N 32
Oakwood Ct. NW&SE
.... 22W-9N 33
Odlum Dr. NS&EW 27W-10N 31
Office Sq. N&S .. 21W-12N 23
Old Farm Rd. E&W
.... 21W-13N 23
Old Kings Ct.* N&S
.... 24W-13N 22
Old Mills Dr. N&S....23W-9N 32

Grid Page

Old Schaumburg Rd. E&W
.... 21W-10N 33
Olde Oaks Rd. E&W
.... 21W-13N 23
Oleander Dr. NS&EW
.... 21W-11N 33
Oliver Ct. E&W 23W-9N 32
Omni Dr. E&W 23W-9N 32
Onyx Ct.* N&S 24W-13N 22
Orleans Ln. N&S
.... 25W-9N,25W-10N 32
Orrington Ct. N&S..22W-10N 33
Osage Ln. E&W ... 24W-10N 32
Oxford Ct. NW&SE
.... 26W-11N,25W-11N 31
Oxhill Ct.* E&W .. 24W-13N 22
Palace Ct.* E&W .. 24W-13N 22
Palisades Pt. N&S..22W-10N 33
Palmer Dr. NE&SW
.... 21W-13N 23
Park Dr. NW&SE . 25W-10N 32
Park Trail Ct.* N&S
.... 20W-10N 33
Parker Dr. E&W .. 24W-11N 32
Parkville Rd. E&W . 26W-11N 31
Partridge Dr. E&W .. 22W-8N 33
Patricia Ct. NW&SE
.... 24W-10N 32
Patricia Dr. N&S .. 24W-10N 32
Patuxet Ct. N&S...24W-10N 32
Paxton Ln. N&S .. 24W-10N 32
Payne Rd. N&S .. 22W-12N 23
Peach Ln. E&W ... 25W-11N 32
Pebble Ct. E&W .. 22W-9N 33
Pembridge Ln. NE&SW
.... 24W-10N 32
Pembroke Ct.* NE&SW
.... 27W-10N 31
Pembroke Dr. N&S 26W-10N 31
Pennsbury Ct. E&W
.... 26W-11N 31
Pennview Ln. NW&SE
.... 26W-11N 31
Penny Ln. NS&EW..22W-12N 23
Penrith Pl. E&W .. 25W-10N 32
Penwood Ct. NE&SW
.... 21W-10N 33
Perimeter Dr. NS&EW
.... 21W-11N 33
Perth Dr. E&W ... 24W-10N 32
Petersham Ln. E&W
.... 21W-10N 33
Pheasant Walk Dr. N&S
.... 23W-9N 32
Pickwick Dr. NW&SE
.... 23W-9N 32
Pine Valley Dr. N&S
.... 21W-13N 23
Pinehurst Ln. N&S .. 24W-9N 32
Pinetree Ln.* N&S .. 21W-9N 33
Pirates Cove Ct. . 21W-13N 23
Plaza Dr. N&S.....21W-11N 33
Pleasant Dr. NS&EW
.... 23W-10N 32
Plum Grove Rd. N&S
.... 21W-13N 23
Plum Grove Rd.
.... 22W-8N,22W-12N 33
Plum Tree Ct.* E&W
.... 21W-9N 33
Plumrose Ct. N&S .. 23W-10N 32
Plumrose Ln. NS&EW
.... 23W-10N 32
Plumwood Ct. N&S
.... 22W-11N 33
Plumwood Dr. E&W
.... 22W-11N 33
Plymouth Ln. E&W .. 25W-9N 32
Pocasset Ct. N&S .. 25W-10N 32
Pochet Ct. NE&SW
.... 24W-10N 32
Pochet Ln. NW&SE
.... 24W-10N 32
Pond View Ct. NE&SW
.... 24W-10N 32
Popular Pl. N&S .. 21W-12N 23
Portland Dr. N&S .. 25W-11N 32
Portsmouth Ct. N&S
.... 25W-10N 32
Portsmouth Ln. E&W
.... 25W-10N 32
Post Oak Pl. N&S .. 21W-12N 23
Prairie Sq. N&S .. 21W-12N 23
Prairie Wind Ln. E&W
.... 21W-12N 23
Pratt Av., North E&W
.... 24W-8N 32
Pratt Av., South E&W
.... 24W-8N 32
Presidio Ct. E&W .. 23W-12N 22
Preston Ct. E&W .. 22W-9N 33
Primrose Ln. N&S .. 26W-11N 31
Prince Charles Ct. E&W
.... 24W-12N 22
Prince Charles Dr. NW&SE
.... 24W-12N 22
Prince Edward Cir. NE&SW
.... 22W-9N 33
Prince Edward Dr. E&W
.... 22W-9N 33
Princeton Ct. N&S .. 25W-9N 32
Princeton Ln. NS&EW
.... 25W-9N 32
Putnam Ct. E&W .. 24W-9N 32
Pvt. Rd. NW&SE .. 23W-12N 22
Quanset Ct. . 24W-10N 32
Queen's Ct. N&S .. 23W-9N 32

Grid Page

Quentin Rd. N&S
.... 22W-13N,21W-13N 23
Quincy Ct. E&W ... 23W-9N 32
Quindel Av. E&W .. 23W-10N 32
Radcliffe Ln. NE&SW
.... 25W-9N 32
Raleigh Ct.* NW&SE
.... 27W-10N 31
Ramsey Cir.* NS&EW
.... 24W-13N 22
Raymond Ct. N&S .. 22W-9N 33
Rebecca Ct. E&W .. 24W-9N 32
Redwood Ln. N&S .. 24W-9N 32
Reedham Pass NE&SW
.... 25W-10N 32
Regal Ct. E&W ... 25W-11N 32
Regatta Pt. NW&SE
.... 22W-10N 33
Regency Ct. N&S....24W-8N 33
Regency Dr. NS&EW
.... 24W-8N 32
Regent Cir. N&S . 23W-10N 32
Remington Cir. N&S
.... 23W-12N 22
Remington Rd. N&S
.... 22W-12N,21W-12N 23
Republic Ct. N&S .. 23W-9N 32
Revere Cir. NS&EW
.... 25W-10N 32
Richmond Ct.* E&W
.... 23W-13N 22
Ridge Ct. N&S 22W-9N 33
Ridgeway Ct. NE&SW
.... 24W-11N 32
Ripplebrook Ct.* N&S
.... 20W-10N 33
Rob Roy Ct. NE&SW
.... 26W-10N 31
Rochester Ct. E&W
.... 26W-11N 31
Rockne Ct. NW&SE
.... 26W-10N 31
Rodenburg Rd. N&S
.... 24W-8N 32
Romm Ct. E&W ... 26W-10N 31
Rose Ct. E&W ... 26W-11N 31
Roselle Rd. N&S
.... 23W-8N,23W-10N 32
Rosewood Ct. E&W
.... 21W-10N 33
Roslyn Ln. E&W .. 24W-11N 32
Rothbury Ct. N&S .. 22W-9N 33
Routh Ct. NS&EW.. 23W-12N 22
Roxbury Ln. NW&SE
.... 24W-11N 32
Royal Ct. E&W ... 24W-9N 32
Rt. 19 NW&SE ... 24W-8N 32
Rt. 58 E&W
.... 25W-12N,21W-12N 22
Rt. 72 E&W
.... 26W-12N,21W-11N 21
Rugby Pl. N&S ... 24W-10N 32
Ruskin Ct.* N&S...23W-13N 22
Russelwood Ct. NE&SW
.... 26W-10N 31
Russett Ct. N&S ... 23W-9N 32
Rutland Ct.* N&S .. 20W-10N 33
Rutland Ln. E&W .. 21W-10N 33
Sagamore Ct. N&S
.... 24W-10N 32
Sagamore Dr. NS&EW
.... 24W-10N 32
Salado Ct. NS&EW
.... 23W-12N 22
Salem Ct. N&S 24W-10N 32
Salem Dr. N&S
.... 24W-9N,24W-10N 32
Salford Dr. N&S ... 22W-8N 33
Samoset Ct. E&W .. 24W-10N 32
Samoset Ln. N&S .. 24W-10N 32
Sandalwood Ln. NW&SE
.... 21W-10N 33
Sandburg Dr.* E&W
.... 20W-10N 33
Sandhurst Ct.* NW&SE
.... 21W-9N 33
Sandpebble Dr. NE&SW
.... 24W-9N 32
Sandy Ct. N&S ... 24W-9N 32
Santuit Ct. E&W....24W-10N 32
Sarah Constant Ln. N&S
.... 26W-11N 31
Sarahs Grove Ln. N&S
.... 23W-10N 32
Saugus Ln. N&S....21W-10N 33
Savannah Ln. NS&EW
.... 25W-8N 32
Savoy Ct. N&S 23W-9N 32
Saylesville Ln.* E&W
.... 23W-13N 22
Scarsdale Ct.* N&S ..21W-9N 33
Schaumburg Ct. NW&SE
.... 22W-10N 33
Schaumburg Rd.
.... 26W-10N,22W-10N 31
Schreiber Av. E&W .. 23W-8N 32
Scott Dr. N&S 22W-8N 33
Scully Ct. N&S ... 23W-10N 32
Scully Dr. E&W ... 23W-10N 32
Seafarer Dr. E&W...24W-9N 32
Seaside Ct. NW&SE 24W-9N 32
Seaton Ln. E&W .. 25W-10N 32
Seaview Ct. N&S .. 25W-9N 32
Sedgfield Ct. E&W ..26W-10N 31
Selkirk Dr. N&S ... 25W-11N 32
Sequoia Ct. N&S ... 23W-9N 32
Serenade Ct. E&W .. 23W-9N 32

	Grid	Page
Seven Pines Rd. E&W		
.... 21W-10N		33
Seville Ct. NW&SE .. 25W-8N		32
Shady Ln. E&W 21W-10N		33
Shagbark Ct. N&S ..23W-10N		32
Shakespeare Ct. E&W		
.... 25W-10N		32
Shannock Ln.* N&S		
.... 23W-13N		22
Sharon N&S		
.... 25W-10N,24W-10N		32
Shattuck Ct. N&S .. 24W-10N		32
Shattuck Ln. E&W ..24W-10N		32
Shaw Ct.* N&S 23W-13N		22
Shell Ct. NE&SW 24W-9N		32
Sherborn Ln. E&W ..26W-10N		31
Sheridan Ln. N&S....23W-9N		32
Sherwood Ln. E&W		
.... 22W-10N		33
Shore Ct. N&S 23W-9N		32
Shore Dr. N&S&E&W		
.... 22W-10N		33
Shoreline Cir. NW&SE		
.... 22W-10N		33
Sienna Ct. NW&SE .. 23W-9N		32
Sienna Dr. NS&EW ..23W-9N		32
Silverwood Ct. N&S ..21W-9N		33
Sky Water Dr. NW&SE		
.... 21W-12N		23
Skyvue Ln. N&S....26W-10N		31
Sleepy Hollow Ct. N&S		
.... 23W-12N		22
Slingerland Dr. SW&NE		
.... 25W-9N		32
Small Dr. N&S 27W-10N		31
Somerset Ct. NW&SE		
.... 25W-10N		32
Somerset Ln. E&W		
.... 25W-10N		32
Song Sparrow Ct. E&W		
.... 21W-12N		23
South Point Ct. N&S		
.... 22W-8N		33
South Point Dr. E&W		
.... 22W-8N		33
Southbridge Ct. E&W		
.... 25W-11N		32
Southbridge Ln. NS&EW		
.... 25W-11N		32
Southbury Ct.* E&W		
.... 21W-9N		33
Southwick Ln. N&S		
.... 21W-10N		33
Southwind Cir E&W		
.... 26W-11N		31
Spinnaker Pt. E&W		
.... 22W-10N		33
Spring Cove Dr. N&S		
.... 24W-9N		32
Spring Creek Cir.* NE&SW		
.... 20W-10N		33
Spring Valley Ct. NS&EW		
.... 22W-10N		33
Springinsguth Rd. N&S		
.... 25W-8N,25W-10N		32
Springwood Ct. E&W		
.... 23W-8N		32
Springwood Dr. NW&SE		
.... 23W-8N		32
Springwood Dr. NW&SE		
.... 21W-10N		33
Spruce Ln. N&S .. 24W-10N		32
Spruce Dr. NW&SE		
.... 24W-10N		32
Squanto Ct. N&S .. 24W-10N		32
Staffire Dr. N&S....26W-10N		31
Staffmark Ln. N&S 23W-10N		32
Stamford Ct. N&S .. 23W-9N		32
Standish Ln. .. 25W-10N		32
Stanley Ct. NS&EW		
.... 26W-11N		31
Stanton Ct. N&S .. 23W-10N		32
Starboard Pt. E&W		
.... 22W-10N		33
State Parkway NS&EW		
.... 22W-11N,23W-12N		33
Steeplechase Ct.* NW&SE		
.... 20W-10N		33
Stevens Dr. N&S .. 21W-10N		33
Stirling Ln. NS&EW		
.... 26W-11N		31
Stock Port Ln. NE&SW		
.... 23W-10N		32
Stockbridge Ct.* E&W		
.... 23W-14N		22
Stone Circle Ct. E&W		
.... 22W-10N		33
Stone Gate Cir. NS&EW		
.... 23W-9N		32
Stonefield Ct.* NE&SW		
.... 20W-10N		33
Stonehedge Dr. N&S		
.... 24W-11N		32
Stonehill Ln. E&W .. 23W-10N		32
Stonington Ct.* NW&SE		
.... 23W-14N		22
Stormy Ct. N&S .. 25W-8N		32
Stoughton Ct. E&W		
.... 24W-10N		32
Stratford Dr. E&W .. 22W-9N		33
Stratham Ct. NW&SE		
.... 24W-9N		32
Stratton Pond Ln. E&W		
.... 25W-11N		32
Sturnbridge Ct. N&S		
.... 22W-11N		33
Sturnbridge Ln. E&W		
.... 22W-11N		33
Sudbury Ct.* E&W ..23W-13N		22

	Grid	Page
Suffield Terr. N&S .. 23W-9N		32
Sumac Ct. NE&SW		
.... 23W-10N		32
Sumac Ln. N&S .. 23W-10N		32
Summer Song Ct. E&W		
.... 26W-11N		31
Summit Ct. E&W .. 23W-9N		32
Summit Dr. N&S		
.... 23W-8N,23W-9N		32
Sunfish Pt. E&W....22W-10N		33
Superior Ct. E&W .. 23W-10N		32
Surf Dr. E&W 22W-11N		33
Surfside Pt. E&W .. 22W-10N		33
Surrey Ln. NS&EW .. 22W-8N		33
Susan Ct. N&S 22W-9N		33
Sussex Cir. N&S .. 23W-10N		32
Sutton Ln. E&W .. 25W-10N		32
Swansea Ct.* E&W		
.... 23W-14N		22
Swartmore Ct. N&S.. 25W-9N		32
Sycamore Pl. E&W .. 21W-12N		23
Syracuse Ln. NW&SE		
.... 25W-9N		32
Tadmore Ct. NW&SE		
.... 25W-11N		32
Tall Timbers Rd. N&S		
.... 21W-13N		23
Tamworth Pl. N&S .. 25W-10N		32
Tarpon Ct. NW&SE .. 24W-9N		32
Taunton Ct.* N&S .. 23W-14N		22
Teal Ct. E&W........23W-9N		32
Tebay Pl. N&S		
.... 26W-11N,25W-11N		31
Terrace Ln. N&S .. 22W-9N		33
Thacker Dr. E&W ... 23W-9N		33
Thacker St. N&S .. 22W-11N		33
Thacker St. E. E&W		
.... 22W-11N		33
Thames Ln. N&S .. 22W-9N		33
Thames Dr. NW&SE		
.... 22W-9N		33
Thistle Ct. N&S .. 26W-10N		31
Thoreau Ct. E&W .. 25W-10N		32
Thoreau Dr., North NS&EW		
.... 21W-12N		23
Thorney Lea Terr. E&W		
.... 24W-9N		32
Thornhill Ct.* N&S .. 21W-9N		33
Thornton Ct.* N&S .. 22W-8N		33
Thornwood Dr. N&S 23W-8N		32
Ticknor Ct. E&W .. 25W-8N		32
Tiffany Dr. E&W 26W-10N		31
Tilipi Ct. NE&SW .. 24W-10N		32
Tilipi Ln. NW&SE .. 24W-10N		32
Timbercrest Dr. N&S		
.... 22W-9N		33
Timberwood Ct. NS&EW		
.... 21W-10N		33
Timothy Ln. N&S .. 22W-9N		33
Tipperary Ct. N&S .. 23W-9N		32
Tisbury Ln. E&W 24W-9N		32
Tiverton Ct. N&S .. 24W-9N		32
Tobey Ct. E&W 24W-10N		32
Tonset Ct. E&W .. 25W-10N		32
Tonset Ln. E&W .. 25W-10N		32
Tower Rd. E&W		
........22W-12N,21W-12N		23
Tracy Ct. N&S 23W-9N		32
Trails Dr. NE&SW .. 22W-10N		33
Tralee Ct. N&S 23W-9N		32
Travis Ct. E&W .. 23W-12N		22
Treebark Ct. E&W .. 24W-10N		32
Treebark Dr. NW&SE		
.... 24W-10N		32
Trent Ln. NS&EW 22W-9N		33
Trenton Ct. N&S....22W-10N		33
Truro Ct. N&S 24W-10N		32
Tudor Ln. N&S 23W-10N		32
Tulip Ct. N&S 26W-10N		31
Tullamore Ct. N&S .. 23W-9N		32
Tyburn Ct. E&W .. 25W-10N		32
Tyler Ct. N&S 23W-9N		32
Vada Ct. E&W 23W-8N		32
Valley Lake Dr. NS&EW		
.... 23W-12N		22
Valley View Dr. E&W		
.... 25W-8N		32
Vasser Ct. NE&SW .. 25W-9N		32
Vasser Ln. NW&SE .. 25W-9N		32
Venice Ct. N&S....24W-9N		32
Verde Ct. N&S 22W-11N		33
Verona Ct. N&S .. 24W-9N		32
Victoria Ct. N&S .. 24W-10N		32
Viola Ct. N&S 26W-11N		31
Vista Ct. N&S 25W-8N		32
W. Remington Cir. N&S		
.... 23W-12N		22
W. Remington Ln. E&W		
.... 23W-12N		22
Waban Ct. N&S .. 25W-10N		32
Waban Ln. NE&SW		
.... 25W-10N		32
Wakeby Ln. E&W .. 25W-10N		32
Wakefield Ln. NS&EW		
.... 24W-10N		32
Walden Ct. N&S 25W-9N		32
Walden Office Square N&S		
.... 21W-12N		23
Walnut Ln. N&S		
........26W-10N,26W-11N		31
Walpole Ln.* N&S .. 23W-13N		22
Wapoos Ct. N&S .. 24W-10N		32
Wareham Ln. N&S .. 25W-10N		32
Warwick Ct. NE&SW		
.... 26W-11N		31
Warwick Ln. E&W		
........26W-10N,25W-10N		31

	Grid	Page
Washington St. N&S		
.... 23W-8N		32
Waterbury Ln. N&S		
.... 23W-10N		32
Waterford Dr. E&W 22W-10N		33
Waterford Dr. NS&EW		
.... 23W-9N		32
Waterfront Dr. N&S		
.... 22W-11N		33
Waterville Ln. E&W		
.... 25W-11N		32
Waverly Ln. N&S .. 25W-10N		32
Wax Wing Ct. E&W		
.... 21W-12N		23
Wayland Ct. E&W .. 25W-10N		32
Wayland Ln. E&W .. 25W-10N		32
Weathersfield Way N&S		
.... 26W-10N,22W-9N		31
Webley Ct. E&W .. 23W-9N		33
Webley Ln. N&S .. 23W-9N		33
Webster Ct. N&S .. 25W-10N		32
Webster Ln. N&S		
.... 25W-9N,25W-10N		32
Wedgewood Ln. N&S		
.... 25W-10N		32
Wellesly Ln. NE&SW		
.... 25W-10N		32
Wellington Rd. E&W		
.... 23W-12N		22
West Brookdale Rd. NE&SW		
.... 25W-11N		32
West Frontage Rd. N&S		
.... 21W-11N		33
West Point Dr. N&S .. 22W-9N		33
West Remington Cir. NS&EW		
.... 23W-12N		22
Westbridge Ct. N&S		
.... 25W-11N		32
Westbridge Ln. E&W		
.... 25W-11N		32
Westchester Cir. N&S		
.... 22W-9N		33
Westchester Rd. E&W		
.... 22W-9N		33
Westfield Ln. E&W .. 24W-9N		32
Weston Ln. E&W .. 25W-10N		32
Westover Ct. N&S .. 25W-10N		32
Westover Ln. NS&EW		
.... 25W-9N,25W-10N		32
Weyers Ct. NW&SE ..25W-8N		32
Weymouth Ct. E&W		
.... 25W-10N		32
Weymouth Ln. E&W		
.... 25W-10N		32
Whalom Ln. N&S .. 22W-11N		33
Whidah Ct. NE&SW		
.... 24W-10N		32
White Branch Ln., North		
NE&SW..........26W-10N		31
White Branch Ct., South		
NE&SW..........26W-10N		31
White Oak Ct. E&W		
.... 23W-12N		22
White Oak Ln. N&S		
.... 23W-12N		22
White Pine Dr. NS&EW		
.... 23W-10N		32
Whitehall Ct. N&S .. 25W-11N		32
Whitesail E&W 22W-10N		33
Whitheger Dr. N&S .. 23W-12N		22
Whitman Ct.* E&W .. 20W-10N		33
Whitman Dr. NE&SW		
.... 21W-10N		33
Whittier Ct. N&S....25W-10N		32
Whittier Ln. NS&EW		
.... 25W-10N		32
Wianno Ln. N&S....24W-10N		32
Wickham Dr. N&S .. 25W-10N		32
Wildberry Ct.* N&S .. 21W-9N		33
Wildflower Ln. NE&SW		
.... 21W-10N		33
Wiley Rd. E&W		
........22W-12N,21W-12N		23
Wilkening Ct. N&S .. 23W-12N		22
Wilkening Rd. N&S 23W-12N		22
Williams Ct. E&W .. 24W-10N		32
Williams Dr. N&S .. 24W-10N		32
Williamsburg Dr. E&W		
.... 21W-10N		33
Willington Rd. E&W		
.... 25W-11N		32
Willoby Ct.* E&W .. 20W-10N		33
Willow Brook Cir. SE&NW		
.... 23W-12N		22
Willow Rd. E&W .. 21W-10N		33
Wilmette Ct. NW&SE		
.... 21W-9N		33
Wilmslow Ln. N&S.. 25W-10N		32
Wilton Ln. N&S 25W-10N		32
Wiltshire Ct. N&S .. 23W-9N		32
Winchester Ct. NW&SE		
.... 25W-10N		32
Winchester Ln. E&W		
.... 25W-10N		32
Windemere Cir. N&S		
.... 26W-11N		31
Windsong Ct. E&W		
.... 26W-11N		31
Windsong Dr. NE&SW		
.... 26W-11N		31
Windsor Dr. N&S .. 23W-10N		32
Windward Dr. N&S .. 22W-10N		33
Winfall Ct. NE&SW 22W-10N		33
Wingate Dr. N&S		
.... 22W-9N,22W-10N		33
Winston Ln. NS&EW		
.... 25W-10N		32
Winthrop Ct. N&S .. 25W-10N		32

	Grid	Page
Winthrop Ln. E&W ..25W-10N		32
Wise Rd. E&W		
.... 25W-9N,22W-9N		32
Wisteria Ct.* NW&SE		
.... 20W-9N		33
Wisteria Ln.* NW&SE		
.... 20W-9N		33
Woburn Ln. N&S .. 21W-10N		33
Wolcott Ct. E&W .. 23W-10N		32
Wood Creek Rd. N&S		
.... 22W-13N		23
Woodbury Ct. N&S		
.... 21W-10N		33
Woodcroft Ct. N&S		
.... 22W-11N		33
Woodcroft Ln. N&S		
.... 22W-11N		33
Woodfield Office Ct. E&W		
.... 22W-11N		33
Woodfield Rd. E&W		
........22W-11N,21W-11N		33
Woodland Dr. E&W		
.... 21W-12N		23
Woodside Ct. NE&SW		
.... 21W-10N		33
Woodville Ln. N&S..23W-10N		32
Woonsocket Ct.* N&S		
.... 23W-14N		22
Worcester Ct. N&S .. 24W-9N		32
Wright Blvd. N&S .. 24W-8N		32
Wyoma Ln. N&S....26W-10N		31
Yale Ln. E&W 25W-9N		32
Yardley Ln. NW&SE		
.... 26W-10N		31
Yarmouth Ct. E&W		
.... 25W-10N		32
Yarmouth Ln. E&W .. 24W-9N		32
Yorkshire Ln. E&W 25W-10N		32

Schaumburg Township

	Grid	Page
Algonquin Dr. N&S .. 22W-8N		33
Blackhawk Dr. E&W 23W-8N		32
Bulrush Dr. E&W .. 25W-11N		32
Circle Ct. N&S 21W-12N		23
Circle Dr. NE&SW .. 21W-12N		23
Council Ct. E&W .. 23W-8N		32
Crest Av. E&W 25W-8N		32
Crest St. N&S 23W-8N		32
Crowfoot Cir, N. E&W		
.... 25W-11N		32
Crowfoot Cir, S. E&W		
.... 25W-11N		32
Elmwood Ln. NE&SW		
.... 21W-12N		23
Fenz Rd. E&W 25W-8N		32
Grant St. N&S 23W-8N		32
Greenbriar Ln. N&S.. 22W-8N		33
Greenview St. N&S .. 23W-8N		32
Half Ln. E&W.......23W-8N		32
Harmon Rd. N&S .. 25W-11N		32
Hillcrest St. E&W .. 22W-8N		33
Home Av. N&S 21W-8N		33
Indian Hill Dr. N&S .. 23W-8N		32
Irving Park Rd. NW&SE		
.... 25W-8N		32
Larson Ln. N&S .. 23W-8N		32
Lincoln St. N&S .. 23W-8N		32
Lincoln St. N&S .. 20W-8N		33
Logan St. N&S 23W-8N		32
Long Av. N&S 25W-8N		32
Marion St. N&S .. 23W-8N		32
Martha St. N&S .. 20W-8N		33
May St. N&S 23W-8N		32
Meadow Ln. NS&EW		
.... 25W-8N		32
Mitchell St. N&S .. 23W-8N		32
Mohawk Dr. N&S .. 23W-8N		32
Monterey Av. NS&EW		
.... 22W-8N		33
Morse Av. E&W .. 23W-8N		32
Myrtle St. N&S .. 23W-8N		32
Nerge Rd. E&W		
.... 25W-8N,21W-8N		32
Northshore Av. E&W		
.... 23W-8N		32
Northwest Tollway N&S		
.... 22W-12N		23
Omaha Cir. N&S .. 21W-8N		33
Oneida Ct. E&W .. 22W-8N		33
Park Av. N&S 25W-8N		32
Park St. N&S 23W-8N		32
Parkview St. N&S .. 25W-8N		32
Pleasant Dr. N&S .. 25W-8N		32
Plum Grove Rd. N&S		
.... 22W-8N		33
Pratt Av. E&W 23W-8N		32
Rodenburg Rd. E&W		
.... 25W-8N		32
Roselle Rd. N&S .. 23W-8N		32
Roslyn Ln. N&S .. 23W-8N		32
Roslyn Rd. N&S .. 23W-8N		32
Rucks St. N&S 21W-8N		33
Seward St. N&S .. 23W-8N		32
Sunset Dr. N&S .. 22W-8N		33
Valleyview Dr. E&W .. 25W-8N		32
Wise Rd. E&W 25W-8N		32

Schiller Park

	Grid	Page
Agatite Ct. E&W .. 12W-5N		40
Ainsiee E&W 11W-6N		41
Ashland St. N&S .. 12W-6N		40
Atlantic N&S 12W-5N		40
Belle Plaine E&W .. 12W-5N		40
Bernice E&W........11W-4N		41
Berteau E&W........12W-5N		40

	Grid	Page
Bobby Ln. N&S 11W-5N		41
Byron E&W 11W-4N		41
Cary Av. E&W 12W-4N		40
Cora Av. E&W 12W-6N		40
Cullom E&W 11W-5N		41
Denley N&S 12W-4N		40
Des Plaines River Rd. N&S		
.... 11W-6N,11W-4N		41
Eastwood E&W 12W-5N		40
Eastwood Ct. E&W .. 12W-5N		40
Eden Av. E&W 12W-4N		40
Emerson N&S		
.... 12W-4N,12W-5N		40
Florence Av. E&W .. 11W-4N		41
Forence E&W 12W-6N		40
Foster N&S 12W-4N		40
Garden Ct. E&W .. 12W-5N		40
Grace N&S 11W-5N		41
Gremley N&S........11W-5N		41
Gross N&S 12W-5N		40
Harold E&W 12W-6N		40
Harold N&S 12W-6N		40
Hartford E&W 12W-4N		40
Interstate 294 N&S .. 12W-4N		40
Ivanhoe Ct. N&S .. 12W-4N		40
Ivanhoe Dr. N&S .. 12W-4N		40
Ivanhoe St. E&W .. 11W-4N		41
Jill Ln. E&W 11W-5N		41
Judd E&W 12W-5N		40
Kelvin E&W 11W-6N		41
Kolze N&S 11W-5N		41
Lawrence E&W 11W-5N		41
Lawrence Ct. E&W .. 12W-5N		40
Leland E&W 12W-6N		40
Linn E&W 12W-6N		40
Lucy E&W 11W-5N		41
Marie E&W 11W-5N		41
Michigan N&S 12W-6N		40
Montrose E&W		
.... 12W-5N,11W-5N		40
Moore Ln. E&W 12W-5N		40
Pace N&S 12W-4N		40
Panoramic Dr. N&S .. 12W-4N		40
Prairie N&S........12W-5N		40
River E&W..........12W-6N		40
River St. E&W 11W-6N		41
Rose N&S .. 12W-4N,12W-5N		40
Ruby N&S 12W-5N		40
Sally Ln. E&W 11W-5N		41
Sarah E&W 12W-5N		40
Scott Av. N&S		
.... 12W-4N,12W-6N		40
Scott St. N&S 12W-5N		40
Seymour E&W 12W-5N		40
Sod Line Ter. N&S .. 11W-5N		41
Soreng E&W 12W-5N		40
South E&W 12W-4N		40
Sterling Ln. E&W .. 12W-4N		40
Sunnyside E&W		
.... 12W-5N,11W-5N		40
Wagner N&S........12W-5N		40
Waveland E&W 12W-4N		40
Webster E&W 12W-6N		40
Wehrman Pl. E&W .. 12W-4N		40
Wesley N&S		
.... 11W-6N,11W-5N		41
Wesley Ter. N&S .. 11W-5N		41
Willow N&S 11W-4N		41
Wilson E&W 12W-5N		40
Winona E&W 11W-6N		41
5th N&S 12W-6N		40
9th N&S 12W-6N		40
25th N&S .. 12W-4N,12W-5N		40

Skokie

	Grid	Page
Arcadia St. E&W		
.... 6W-11N,4W-11N		28
Austin Av. NW&SE....7W-9N		37
Avers Av. N&S		
.... 4W-10N,4W-11N		38
Babb Av. NE&SW .. 6W-9N		38
Bennett Av. N&S		
.... 4W-10N,4W-11N		38
Beverly Dr. N&S .. 5W-12N		28
Birchwood Av. E&W		
.... 6W-9N,4W-9N		38
Bobolink Ter. E&W .. 6W-11N		28
Brandess Ter. E&W .. 6W-11N		28
Bronx Av. N&S		
.... 6W-10N,6W-12N		38
Bronx Pl. (Pvt.) N&S		
.... 6W-11N		28
Brown St. E&W 6W-10N		38
Brummel St. E&W		
.... 6W-9N,4W-9N		38
Capitol St. E&W .. 4W-10N		38
Carol St. E&W		
.... 6W-10N,4W-10N		38
Central Av. E&W 7W-12N		28
Central Av. N&S .. 6W-9N		38
Central Park Av. N&S		
.... 4W-9N,4W-11N		38
Chase Av. E&W		
.... 6W-9N,4W-9N		38
Cherry Pkwy. (Pvt.) N&S		
.... 5W-12N		28
Christiana Av. N&S .. 4W-10N		38
Church St. E&W 5W-11N		28
Clark St. E&W 5W-11N		28
Cleveland St. E&W		
.... 6W-10N,4W-10N		38
Clifford Ter. N&S .. 4W-10N		38
Colfax N&S 6W-12N		28
Colfax Pl. N&S 6W-12N		28
Colfax St. E&W		
.... 6W-12N,5W-12N		28
Concord Ln. N&S .. 5W-10N		38
Conrad St. E&W .. 6W-10N		38

	Grid	Page
Coyle Av. E&W	6W-8N	38
Crain St. N&S		
....	6W-10N,4W-10N	38
Crawford Av. N&S ..	5W-11N	28
Culver St. E&W.....	5W-12N	28
Dahlin Dr. (Pvt.) E&W	7W-9N	37
Dartmoth Pl. E&W ..	6W-12N	28
Davis St. E&W		
....	6W-11N,4W-11N	28
Dobson St. N&S		
....	6W-9N,4W-9N	38
Drake Av. N&S		
....	4W-10N,4W-11N	38
East Prairie Rd. NE&SW		
....	4W-9N,4W-11N	38
Edens Expwy. N&S .	6W-12N	28
Elgin St. NW&SE ...	4W-11N	29
Elm St. E&W	6W-10N	38
Elm Ter. E&W	5W-10N	38
Elmwood St. NE&SW		
....	6W-10N	38
Emerson St. E&W		
....	6W-11N,4W-11N	28
Enfield St. E&W		
....	6W-11N,4W-11N	28
Estes Av. E&W	6W-8N	38
Ewing Av. N&S	4W-11N	29
Fairview Ln. E&W ..	6W-8N	38
Fargo Av. E&W		
....	6W-9N,4W-9N	38
Farwell Av. E&W ...	6W-8N	38
Fitch Av. E&W	6W-8N	38
Floral Av. N&S	6W-10N	38
Forestview Rd. N&S	4W-11N	29
Foster St. E&W		
....	6W-11N,4W-11N	28
Four Winds Way N&S		
....	4W-11N	29
Galitz St. E&W	6W-9N	38
George St. E&W	6W-9N	38
Golf Rd. E&W		
....	7W-12N,5W-12N	28
Grant St. E&W		
....	6W-12N,5W-12N	28
Greenleaf St. E&W		
....	6W-10N,4W-10N	38
Greenwood St. E&W		
....	6W-11N,4W-11N	28
Gross Point Rd. NE&SW		
....	6W-11N,5W-12N	28
Grove St. E&W		
....	6W-11N,4W-11N	28
Hamlin Av. N&S		
....	4W-9N,4W-11N	38
Harding Av. N&S		
....	4W-10N,4W-11N	38
Harms Rd. N&S		
....	7W-12N,6W-®0N	28
Harmswood Ter. E&W		
....	6W-11N	28
Harvard Ter. E&W		
....	6W-9N,4W-9N	38
Harvest Ln. E&W ...	5W-11N	28
Hipple St. E&W	5W-11N	28
Hoffman St. E&W ...	6W-10N	38
Howard St. E&W		
....	6W-9N,5W-9N	38
Hull St. E&W ..	6W-9N,5W-9N	38
Interstate 94 E&W ..	6W-12N	28
Isabel St. E&W	4W-10N	38
James Dr. E&W	5W-10N	38
Jarlath St. E&W	6W-9N	38
Jarvis Av. E&W		
....	6W-9N,4W-9N	38
Jerome Av. E&W		
....	6W-9N,4W-9N	38
Jonathan Ln. (pvt) E&W		
....	5W-12N	28
Karlov Av. N&S		
....	5W-9N,5W-12N	38
Keating Av. N&S		
....	5W-10N,5W-11N	38
Kedvale Av. N&S		
....	5W-9N,5W-12N	38
Kedzie Av. N&S	4W-9N	38
Keeler Av. N&S		
....	5W-9N,5W-12N	38
Keeney St. E&W		
....	5W-10N,4W-10N	38
Kellen Av. E&W ...	6W-9N	38
Kenneth Av. N&S		
....	5W-9N,5W-11N	38
Kenneth Ter. N&S ..	5W-11N	28
Kenton Av. N&S		
....	5W-9N,5W-12N	38
Keystone Av. N&S		
....	5W-9N,5W-12N	38
Kilbourn Av. N&S		
....	5W-9N,5W-11N	38
Kildare Av. N&S		
....	5W-9N,5W-11N	38
Kilpatrick Av. NW&SE		
....	5W-9N,5W-11N	38
Kimball Av. N&S ...	4W-10N	38
Kirk St. E&W ..	6W-9N,4W-9N	38
Klehm Ct. E&W.....	6W-10N	38
Knox Av. N&S		
....	5W-10N,5W-12N	38
Kolmar Av. N&S		
....	5W-9N,5W-12N	38
Kostner Av. N&S		
....	5W-9N,5W-12N	38
Kostner Ter. N&S....	5W-11N	38
LaCrosse Av. N&S		
....	6W-9N,6W-12N	38
Lake St. E&W	4W-11N	29
Lamon Av. N&S		
....	6W-8N,6W-12N	38
Lamon Av. (Pvt.) N&S		
....	6W-11N	28
Laramie Av. N&S		
....	6W-8N,6W-12N	38
Latrobe Av. N&S		
....	6W-8N,6W-11N	38
Lavergne Av. N&S		
....	6W-8N,6W-12N	38
Lawler Av. N&S ...	6W-11N	28
Lawndale Av. N&S		
....	4W-9N,4W-11N	38
LeClaire Av. N&S		
....	6W-8N,6W-12N	38
Lee St. E&W		
....	6W-10N,4W-10N	38
Lemington Av. N&S .	6W-12N	28
Lewis Ter. E&W	5W-11N	28
Lincolnwood Dr. N&S		
....	4W-11N	29
Linder Av. N&S ...	6W-9N	38
Little Elm Bend N&S		
....	4W-11N	29
Little One Path N&S	4W-11N	29
Lockwood Av. N&S .	6W-12N	28
Lockwood Av. N&S		
....	6W-8N,6W-12N	38
Long Av. N&S		
....	6W-9N,6W-11N	38
Lorel Av. N&S		
....	6W-8N,6W-12N	38
Lotus Av. N&S		
....	6W-10N,6W-11N	38
Louise St. E&W		
....	6W-9N,4W-9N	38
Lowell Av. N&S		
....	5W-9N,5W-11N	38
Lowell Ter. N&S	5W-11N	28
Lunt Av. E&W	6W-8N	38
Lyons Av. E&W		
....	6W-11N,4W-11N	38
Madison St. E&W		
....	6W-10N,4W-10N	38
Main St. E&W	6W-10N	38
Major Av. N&S		
....	7W-9N,7W-12N	37
Mango Av. N&S	7W-9N	37
Manor Ct. E&W	4W-11N	29
Marmora Av. N&S ...	7W-9N	37
McCormick Blvd. N&S		
....	4W-9N,4W-10N	38
Menard Av. N&S ...	7W-9N	37
Monroe Av. N&S ...	6W-10N	38
Monticello Av. N&S		
....	4W-9N,4W-11N	38
Morgan N&S	4W-9N	29
Morse Av. E&W	6W-8N	38
Mulford St. E&W		
....	6W-9N,5W-7N	38
Murphy Ln. N&S ...	4W-11N	29
New Gross Point Rd. N&S		
....	7W-9N	37
Niles Av. N&S	6W-10N	38
Niles Center Rd. NE&SW		
....	6W-9N,6W-11N	38
Oakton St. E&W ..	6W-10N	38
Old Orchard Rd. E&W		
....	6W-12N,5W-12N	28
Palma Pl. E&W		
....	6W-11N,5W-11N	28
Park Av. NE&SW	6W-9N	38
Parkside Av. N&S		
....	7W-9N,7W-12N	37
Payne St. E&W		
....	6W-12N,5W-12N	28
Peach Pkwy. (Pvt.) N&S		
....	5W-12N	28
Pottawattami Dr. N&S		
....	4W-11N	29
Pratt Av. E&W	6W-8N	38
Railroad Av. N&S ..	6W-11N	28
Ridgeway Av. N&S		
....	4W-9N,4W-11N	38
Ridgeway Ter. N&S		
....	4W-10N,4W-11N	38
Roth Ter. E&W	4W-10N	38
Russet Ln. (Pvt.) E&W		
....	5W-12N	28
Salem Ln. NS&EW .	4W-11N	29
Samoset Blvd. N&S .	4W-11N	29
Searle Pkwy. .	6W-10N	38
Sherwin Av. E&W ...	6W-9N	38
Skokiana Ter. E&W .	5W-11N	28
Skokie Blvd. N&S		
....	6W-9N,6W-11N	38
Sleeping Bear Rd. N&S		
....	4W-11N	29
Springfield Av. N&S		
....	4W-10N,4W-11N	38
St. Louis Av. N&S		
....	4W-9N,4W-10N	38
Suffield Ct. E&W		
....	6W-11N,5W-11N	28
Suffield St. E&W.....	6W-11N	28
Sugarloaf Ln. E&W .	4W-11N	29
Tamora Ter. N&S....	4W-11N	29
Terminal Av. NW&SE		
....	6W-11N,6W-12N	28
Tierney Dr. N&S ...	7W-9N	37
Touhy Av. E&W	4W-9N	38
Tower Cir. E&W	5W-11N	38
Tripp Av. N&S		
....	5W-9N,5W-11N	38
Trumbull Av. N&S ..	4W-11N	38
Turner Av. N&S	4W-11N	29
U.S. Route 41 N&S .	6W-12N	28
Warren St. E&W		
....	6W-10N,4W-10N	38
Washington St. E&W		
....	6W-10N,5W-10N	38
Weber Ln. E&W ...	6W-12N	28
Whitecloud Dr. E&W		
....	4W-11N	29
Wilder St. E&W	4W-10N	38
Williamsburg Cir. N&S		
....	4W-11N	29
Williamsburg Ct. N&S		
....	4W-11N	29
Williamsburg Ln. N&S		
....	4W-11N	29
Williamsburg Ter. N&S		
....	4W-11N	29
Wilson St. E&W	5W-11N	28
Wright Ter. E&W		
....	6W-10N,4W-10N	38

South Barrington

	Grid	Page
Algonquin Rd. NW&SE		
....	27W-15N,25W-14N	21
Ambrose Ln. N&S ..	26W-13N	21
Ashford Ct. NW&SE		
....	25W-15N	22
Avon Ln. N&S......	24W-14N	22
Back Bay NE&SW .	26W-14N	21
Barrington Rd. N&S		
....	26W-13N,25W-15N	21
Bartlett Rd. N&S		
....	27W-13N,27W-15N	21
Bay-Reef Dr. N&S .	27W-13N	21
Beacon Hill Dr. E&W		
....	27W-15N	21
Beechnut Dr. NS&EW		
....	27W-15N	21
Berkshire Ct. N&S .	27W-13N	21
Blanchard Cir. N&S		
....	26W-13N	21
Brackley Ln. NW&SE		
....	25W-15N	22
Brandywine Rd. NS&EW		
....	27W-14N	21
Bridlewood Trail NS&EW		
....	26W-14N,26W-15N	21
Brookhaven Cir. N&S		
....	27W-15N	21
Buckthorn Rd. NS&EW		
....	26W-15N	21
Butera Ln. NW&SE		
....	27W-14N	21
Campden Dr. N&S .	25W-15N	22
Canterbury Ct. NW&SE		
....	24W-14N	22
Champlain Rd. NS&EW		
....	27W-14N	21
Chipping Dr. E&W .	25W-15N	22
Concord Ct. NE&SW		
....	27W-14N	21
Corey Dr. N&S	26W-13N	21
Coverd Bridge Rd. E&W		
....	26W-14N	21
Creet Ln. N&S	26W-13N	21
Cutter's Run NS&EW		
....	27W-13N	21
Deveaux NW&SE ..	26W-15N	21
East Cove N&S	26W-14N	21
Easting Way NE&SW		
....	27W-15N	21
Eton Ct. NE&SW ..	24W-14N	22
Evergreen Pkwy. E&W		
....	26W-14N	21
Forest Ln. NS&EW		
....	25W-14N,25W-15N	22
Foxshire Ct. NE&SW		
....	26W-14N	21
Gibbs Dr. E&W	26W-13N	21
Gregory Ln. N&S ..	26W-13N	21
Hackney Cir. NE&SW		
....	27W-13N	21
Headwater Cove N&S		
....	27W-14N	21
Heath Way NS&EW		
....	26W-15N	21
Higgins Rd. NW&SE		
....	27W-13N	21
Kitson Dr. E&W ..	26W-13N	21
Knoll Ct. E&W......	25W-15N	22
Lake Adalyn Dr. NS&EW		
....	26W-13N	21
Lakeside Ct. N&S .	27W-15N	21
Lakeside Dr. NS&EW		
....	27W-15N	21
Lexington Rd. NW&SE		
....	27W-14N	21
Liberty Dr. NS&EW		
....	27W-14N	21
Loch Ln. E&W......	26W-15N	21
McGlashen Dr. N&S		
....	26W-13N	21
Midlands Dr. E&W .	26W-13N	21
Mohawk Ct. N&S ..	25W-13N	22
Mohawk Dr. N&S ..	25W-13N	22
Morgan Ln. E&W ..	27W-14N	21
Mundhank Rd. NS&EW		
....	27W-13N,25W-13N	21
North Meadow Ct. N&S		
....	27W-13N	21
Old Barrington Rd. N&S		
....	26W-13N	21
Old Coach Dr. N&S		
....	26W-14N	21
Overbrook Rd. NS&EW		
....	26W-14N	21
Pacer Tr. N&S	27W-14N	21
Palatine Rd. E&W		
....	27W-15N,25W-15N	21
Pembury Way NW&SE		
....	25W-15N	22
Penny Rd. E&W		
....	27W-15N,26W-15N	21
Pentwater Dr. NS&EW		
....	27W-13N	21
Polo Tr. NS&EW ..	26W-14N	21
Quincy Cir. SE&NW		
....	27W-15N	21
Red Ridge Cir. E&W		
....	27W-15N	21
Revere Dr. N&S	27W-14N	21
Rose Blvd. E&W ...	26W-13N	21
Rt. 59 N&S	28W-13N	21
Rt. 62 NW&SE		
....	27W-15N,25W-14N	21
Rt. 72 NW&SE	27W-13N	21
Shire Tr. N&S	27W-13N	21
Shirebrook Way E&W		
....	25W-15N	22
Shoreside Dr. NS&EW		
....	27W-15N	21
Somerset Ct. N&S ..	24W-14N	22
South Cove E&W ..	26W-14N	21
South Meadow Ct. NE&SW		
....	27W-13N	21
Spring Creek Dr. N&S		
....	27W-15N	21
Squire Ln. N&S ...	26W-14N	21
Stannington Way N&S		
....	25W-15N	22
Stonebrook Ct. N&S		
....	27W-13N	21
Stoneridge Dr. NS&EW		
....	26W-15N	21
Taynton Ln. E&W ..	25W-15N	22
Tennis Club Ln. N&S		
....	26W-13N	21
Tewkesbury Ln. E&W		
....	25W-15N	22
Tiffany Cir. N&S ..	27W-13N	21
Trenton Ct. NE&SW		
....	27W-14N	21
Turning Shore N&S		
....	27W-14N,26W-14N	21
Upper Pond Rd. NS&EW		
....	27W-14N	21
Vandenbergh Dr. E&W		
....	26W-13N	21
Walnut Ln. N&S		
....	27W-13N	21
Watercrest Ct. NE&SW		
....	27W-13N	21
Watergate Rd. NS&EW		
....	27W-14N	21
Wellingborough Ct. E&W		
....	25W-15N	22
Wescott Dr. NS&EW		
....	26W-13N	21
Westlake Dr. NS&EW		
....	27W-13N	21
Willow Bay Dr. N&S		
....	27W-13N	21
Willowmere Dr. N&S		
....	25W-14N,25W-15N	22
Windemere Ln. NE&SW		
....	24W-14N	22
Windridge Dr. NS&EW		
....	27W-15N	21
Windsor Ct. N&S		
....	26W-14N,26W-15N	21
Witt Rd. N&S	26W-14N	21
Woodbury Ct. NE&SW		
....	27W-15N	21
Woodhaven Dr. NS&EW		
....	27W-13N	21
Wynchwood Ln. N&S		
....	25W-15N	22
Yorktown Ct. E&W	27W-14N	21

South Chicago Heights

	Grid	Page
Aberdeen St. N&S .	1W-26S	93
Benton Av. N&S	1W-27S	93
Butler Av. N&S......	0W-27S	94
Campbell Av. N&S .	1W-27S	93
Cappelletti Ln. N&S .	1W-27S	93
Cherry Ln. E&W	1W-27S	93
Chestnut Av. E&W .	1W-27S	93
Chicago & Vincennes Rd.		
N&S............	1W-26S	93
Chicago Pl. N&S ..	1W-27S	93
Commercial Av. N&S		
....	1W-26S	93
Courtney Ln. N&S ..	1W-27S	94
Crescenzo Dr. N&S .	1W-27S	93
Deer Path N&S	1W-27S	94
Dornell Dr. N&S...	1W-27S	93
East End Av. N&S ..	0W-26S	94
Enterprise Park Av. N&S		
....	1W-27S	93
Euclid Av. N&S	1W-27S	93
Euclid Dr. N&S	1W-27S	93
Fairview Av. N&S ..	1W-27S	93
Forest Dr. N&S	1W-27S	93
Helfred Av. N&S ...	1W-27S	93
Holeman Av. N&S ..	1W-27S	94
Interocean Av. E&W .	1W-26S	93
Jackson Av. N&S ..	1W-26S	93
Lawrence Av. N&S .	1W-27S	93
Lynnwood Dr. N&S .	1W-27S	93
Magnolia Plaza NS&EW		
....	1W-27S	93
Maple Av. N&S	1W-27S	93
Miller Av. N&S	1W-27S	93
Park Terr. E&W......	1W-27S	93
Paulsen St. E&W		
....	1W-26S,0W-26S	93
Rennine-Smith Dr. N&S		
....	0W-27S	94
Rosiclaire Ct. N&S ..	1W-27S	94
Rt. 1 N&S	1W-27S	93
Sauk Trial N&S	0W-27S	94
State St. N&S	0W-27S	94
Vincennes Rd. N&S		
....	1W-27S,0W-27S	93
Willow Rd. N&S	1W-27S	93
26th St. E&W	1W-26S	93
27th Pl. E&W	1W-26S	93
27th St. E&W	1W-26S	93
28th Pl. E&W	1W-26S	93
28th St. E&W	1W-26S	93
29th Pl. E&W	1W-26S	93
29th St. E&W	1W-26S	93
30th Pl. E&W	1W-27S	93
30th St. E&W	1W-26S	93
31st St. E&W	1W-27S	93
33rd St. E&W		
....	1W-27S,0W-27S	93
34th St. E&W	1W-27S	93

South Holland

	Grid	Page
Armory Dr. E&W	0W-19S	82
Avalon Av. N&S		
....	1E-19S,1E-18S	82
Bennett Av. NE&SW .	2E-20S	89
Bernice Rd. E&W ...	2E-20S	89
Betty Ln. N&S.......	0E-19S	82
Calumet Expwy. N&S		
....	1E-18S	82
Canal St. N&S		
....	0W-20S,0W-19S	88
Champlain St. N&S .	0E-18S	82
Cherry N&S	0E-18S	82
Church Dr. NW&SE ..	0E-18S	82
Claire Ln. N&S	0E-19S	82
Clark St. N&S	0W-17S	82
Clyde Av. N&S		
....	2E-20S,20E-19S	89
Constance Av. NE&SW		
....	2E-20S	89
Cornell Av. N&S	1E-20S	89
Cottage Grove Av. N&S		
....	0E-19S,0E-18S	82
Cregier Av. N&S...	0E-18S	89
Dante Dr. N&S	1E-18S	82
Dearborn St. N&S		
....	0W-18S,0W-17S	82
Debbie Ln. N&S ...	1E-18S	82
Dobson Av. N&S		
....	1E-20S,1E-17S	89
Dobson Ct. N&S ...	1E-18S	82
Dorchester Av. N&S .	1E-18S	82
Drexel Av. N&S		
....	1E-20S,1E-17S	89
Ellis Av. N&S	1E-19S,1E-17S	82
Ellis Ct. N&S..	1E-20S,1E-19S	89
Elm Ct. N&S	0E-19S	82
Elm St. N&S ..	0E-19S,0E-18S	82
Evans Av. N&S		
....	0E-20S,0E-18S	88
Evans Ct. NW&SE		
....	0E-20S,0E-19S	88
Evans Dr. NW&SE ..	0E-20S	88
Everett Av. N&S......	1E-20S	89
Gouwens Ln. NE&SW		
....	0E-18S	82
Greenwood Av. N&S		
....	1E-20S,1E-19S	89
Holland Rd. NW&SE .	0E-19S	82
Indiana Av. N&S	0W-19S	82
Ingleside Av. N&S		
....	1E-20S,1E-17S	89
Ingleside Ct. NE&SW		
....	1E-19S	82
Ingleside Dr. NW&SE		
....	1E-19S	82
Interstate 294-80 E&W		
....	0E-20S	88
Interstate 94 N&S		
....	1E-20S,1E-19S	89
Jeffrey St. N&S	2E-20S	89
Joyce Cir. E&W	0E-19S	82
Joyce Ct. N&S	0E-19S	82
Kenwood Av. N&S		
....	1E-20S,1E-18S	89
Kenwood Ct. NE&SW		
....	1E-19S	82
Kenwood Dr. NW&SE		
....	1E-19S	82
Kimbark Av. N&S		
....	1E-20S,1E-18S	89
Kimbark Ct. N&S ..	1E-19S	82
King Av. NW&SE ...	1E-18S	82
King Cir. NE&SW ...	1E-18S	82
La Salle St. N&S		
....	0W-18S,0W-17S	82
Langley Av. N&S		
....	0E-20S,0E-18S	88
Louis Av. N&S		
....	0E-20S,0E-18S	88
Louis Ct. N&S		
....	0E-20S,0E-18S	88
Lowell Av. N&S	0E-19S	82
Luella Av. N&S	2E-19S	89
Maple Ct. N&S	0E-19S	82
Maple St. N&S		
....	0E-19S,0E-18S	82
Marie Dr. E&W	0E-19S	82
Marion Dr. N&S	1E-18S	82
Maryland Av. N&S		
....	1E-20S,1E-19S	89
Merrill Av. N&S	2E-19S	83

	Grid	Page
Michigan Av. N&S	0E-19S,0E-18S	82
Michigan Av. N&S	0W-17S	82
Minerva Av. N&S	1E-18S	82
Mutual Terr. NW&SE	0E-18S	82
Naughton Av. N&S	1E-17S	82
Orchid Dr. NE&SW	0E-18S	82
Park Ln. NE&SW	0E-18S	82
Park Plaza E&W	0E-18S	82
Parkside Av. N&S	1E-19S,1E-18S	82
Parkside St. N&S	0E-20S	88
Paxton Av. N&S	2E-20S	89
Perry Av. N&S	0W-17S	82
Prairie Av. N&S	0E-19S,0E-18S	82
Prince Dr. NW&SE	1E-18S	82
Reitveldt Rd. N&S	2E-19S	83
Riverside Dr. E&W	1E-18S	82
Riverside Dr. NW&SE	0E-17S	82
Riverside St. E&W	0E-18S	82
Rose Dr. NE&SW	0E-18S	82
Rt. 83 E&W	0W-17S	82
School St. N&S	0E-20S,0E-19S	88
Cir.	1E-18S	82
Shirley Ct. N&S	0E-19S	82
Sibley Blvd. E&W	0E-17S	82
South Park Av. N&S	0E-20S,0E-18S	88
State St. E&W	0W-18S,0W-17S	82
Sun Tone Dr. N&S	0W-18S	82
Taft Dr. E&W	0W-19S	82
Thornton Av. NW&SE	0E-19S	82
Thornwood Ct. NW&SE	0E-20S	88
Thornwood Dr. NW&SE	0E-20S	88
Union Av. N&S	0W-20S	88
Univeristy Av. N&S	1E-20S,1E-18S	89
Univeristy Ct. N&S	1E-19S	82
Van Drunen Rd. N&S	0W-18S	82
Vincennes Av. NW&SE	0W-18S,0E-20S	82
Volbrecht Ct. E&W	2E-20S	89
Volbrecht Dr. N&S	2E-20S	89
Volbrecht Rd. N&S	1E-20S	89
Wabash Av. N&S	0E-19S,0E-17S	82
Wabash Ct. NE&SW	0E-18S	82
Wallace Av. N&S	0W-20S	88
Waterman Ct. NE&SW	1E-17S	82
Waterman Dr. NW&SE	1E-17S	82
Wausau Av. N&S	0E-20S,0E-18S	88
Wausau Ct. N&S	0E-20S,0E-19S	88
Wentworth Av. N&S	0W-18S	82
Westview Av. N&S	0W-20S	88
Woodlawn Av. N&S	0E-19S,0E-18S	82
Woodlawn Av. East N&S	1E-19S,1E-18S	82
Woodlawn Ct. N&S	1E-19S	82
Woodlawn Dr. East E&W	1E-18S	82
Woodlawn Rd. West N&S	0E-17S	82
Yates Av. N&S	2E-20S	89
148th St. E&W	0W-17S,0E-17S	82
149th St. E&W	0W-17S,0E-17S	82
150th Pl. E&W	0W-17S,1E-17S	82
152nd St. E&W	1E-17S	82
153rd Pl. E&W	1E-17S	82
153rd St. E&W	1E-17S	82
154th Pl. E&W	0E-18S	82
154th St. E&W	0W-17S,0E-17S	82
155th Pl. E&W	0E-18S,1E-18S	82
155th St. E&W	0W-18S,0E-18S	82
156th Pl. E&W	0E-18S	82
156th St. E&W	0E-18S	82
157th Pl. E&W	0E-18S	82
157th St. E&W	0E-18S	82
158th Pl. E&W	0E-18S	82
158th St. E&W	0E-18S	82
159th Ct. E&W	0E-18S,1E-18S	82
159th Pl. E&W	0E-18S,1E-18S	82
159th St. E&W	0E-18S,1E-18S	82
160th Ct. E&W	0E-18S	82
160th Pl. E&W	0E-18S,1E-18S	82
160th St. E&W	0E-18S,1E-18S	82
161st Pl. E&W	0E-18S,1E-18S	82
161st St. E&W	1E-18S	82
162nd Pl. E&W	0E-19S,1E-19S	82
162nd St. E&W	0W-18S,1E-18S	82
163rd Ct. E&W	0E-19S	82
163rd Pl. E&W	0E-19S,1E-19S	82
163rd St. E&W	0E-19S,1E-19S	82
164th Cir. E&W	0E-19S	82
164th Ct. E&W	1E-19S	82
164th Pl. E&W	0E-19S,1E-19S	82
164th St. E&W	0E-19S,1E-19S	82
165th Pl. E&W	0E-19S,1E-19S	82
165th St. E&W	0E-19S,1E-19S	82
166th Pl. E&W	0E-19S,2E-19S	82
166th St. E&W	0W-19S,2E-19S	82
167th Pl. E&W	0E-19S,1E-19S	82
167th St. E&W	0W-19S,1E-19S	82
168th Ct. E&W	0E-19S	82
168th Pl. E&W	0E-19S,1E-19S	82
168th St. E&W	0W-19S,2E-19S	82
169th Pl. E&W	1E-19S,2E-19S	82
169th St. E&W	0W-19S,1E-19S	82
170th Pl. E&W	0E-20S,2E-20S	88
170th St. E&W	0W-19S,2E-19S	82
171st Pl. E&W	0E-20S,2E-20S	88
171st St. E&W	0E-20S,2E-20S	88
172nd St. E&W	0W-20S,2E-20S	88
173rd Pl. E&W	0E-20S,2E-20S	88
173rd St. E&W	0E-20S,2E-20S	88

Steger*

(Also see Will County)

	Grid	Page
Adair Rd. N&S	0E-27S	94
Butler Av. N&S	0W-27S	51
Carpenter St. N&S	1W-27S	51
Cottage Grove Av. N&S	1E-27S	94
Dorchester Av. N&S	1E-27S	94
Emerald Av. N&S	1W-27S	51
Florence Av. N&S	0W-27S	51
Frederick Rd. N&S	0E-27S	94
George St. NW&SE	0E-27S	94
Green St. N&S	1W-27S	51
Halsted St. N&S	1W-27S	51
Holeman Av. N&S	0W-27S	51
Hopkins Av. N&S	0W-27S	51
Keeney Av. N&S	0W-27S	94
Lahon Rd. N&S	0E-27S	94
Lewis Av. N&S	0W-27S	51
Lisa Ln. E&W	0E-27S	94
Loverock Av. N&S	0W-27S	51
Mach Dr. N&S	0E-27S	94
Michigan Av. N&S	0E-27S	94
Miller Rd. N&S	0E-27S	94
Morgan St. N&S	1W-27S	51
Oakland Dr. N&S	0E-27S	94
Patricia Rd. N&S	0E-27S	94
Peoria St. N&S	1W-27S	51
Phillips Av. N&S	0W-27S	51
Route 1 N&S	1W-27S	51
Sangamon St. N&S	1W-27S	51
Sherman Rd. N&S	0E-27S	94
State St. N&S	0W-27S	94
Steger Rd. E&W	0E-27S	94
Stewart Av. N&S	0W-27S	94
Union Av. N&S	0W-27S	51
Wallace Av. N&S	0W-27S	51
Wentworth Av. N&S	0W-27S	94
Woodlawn Av. N&S	1E-27S	94
30th Pl. E&W	1W-27S,0W-27S	93
30th St. E&W	1W-26S,0W-26S	93
30th St. E&W	1E-27S	94
31st Pl. E&W	1W-27S,0W-27S	93
31st St. E&W	1W-27S,0W-27S	93
32nd Pl. E&W	1W-27S	93
32nd St. E&W	1W-27S,0W-27S	93
32nd St. E&W	1E-27S	94
33rd Pl. E&W	1W-27S,0W-27S	93
33rd St. E&W	1W-27S,0W-27S	93
33rd St. E&W	0E-27S	94
34th Pl. E&W	0E-27S	94
34th St. E&W	1W-27S,0W-27S	93
34th St. E&W	0E-27S	94

Stickney

	Grid	Page
Clarence N&S	8W-4S	54
East N&S	8W-4S	54
Elmwood N&S	8W-4S	54
Euclid N&S	8W-4S	54
Gunderson N&S	8W-4S	54
Kenilworth N&S	8W-4S	54
Lorraine E&W	8W-4S	54
Pershing E&W	8W-4S,7W-4S	54
Scoville N&S	8W-4S	54
Wenonah N&S	8W-4S	54
Wesley N&S	8W-4S	54
Wisconsin N&S	8W-4S	54
40th E&W	8W-4S	54
40th Pl. E&W	8W-4S	53
41st E&W	8W-4S	54
42nd E&W	8W-4S	54
43rd E&W	8W-4S	54
44th E&W	8W-4S	54
45th E&W	8W-4S	54

Stickney Township

	Grid	Page
Central N&S	6W-5S	60
Laramie N&S	6W-5S	60
Latrobe N&S	6W-5S	60
48th E&W	6W-5S	60
49th E&W	6W-5S	60
50th E&W	6W-5S	60
51st E&W	6W-5S	60

Stone Park

	Grid	Page
Soffel E&W	12W-1N	46
37th N&S	12W-1N	46
38th N&S	12W-1N	46
39th N&S	12W-1N	46
40th N&S	12W-1N	46
43rd N&S	13W-1N	46

Streamwood

	Grid	Page
Abbeywood Cir. NE&SW	28W-11N	31
Abington Ct. N&S	29W-9N,29W-10N	30
Acorn Dr. E&W	27W-10N	31
Adams Ct. NW&SE	29W-10N	30
Alexander Av. N&S	28W-9N	31
Alexander Ct. NW&SE	28W-9N	31
Alexander Ln. N&S	28W-9N	31
Alexander Pl. E&W	28W-9N	31
Andover Ct. NW&SE	28W-9N	31
Apple Hill Ln. NE&SW	28W-11N	31
Arabian Ct. E&W	30W-10N	30
Arnold Av. E&W	28W-9N	31
Arrowood Ct.* E&W	24W-13N	22
Arthur Ct. N&S	28W-10N	31
Ascot Ln. NS&EW	29W-11N	30
Ash Ct. E&W	27W-10N	31
Ashton Ct. NS&EW	29W-9N,28W-9N	30
Aspen Ct. E&W	27W-10N	31
Attleboro Ln. E&W	29W-9N	30
Audubon Rd. E&W	28W-10N	31
Autumn Ln. NS&EW	29W-9N	30
Azalea Cir. NS&EW	27W-10N	31
Barrington Rd. N&S	27W-9N,27W-11N	31
Bartlett Rd. N&S	29W-9N,28W-11N	30
Bayberry Ct. NE&SW	27W-10N	31
Beaver Dr. N&S	28W-10N	31
Beebe Ct. E&W	28W-9N	31
Berkely Pl. E&W	28W-10N	31
Berkshire Ct. N&S	29W-9N	30
Beverley Ct. E&W	29W-9N	30
Beverley Ln. N&S	29W-9N	30
Big Oaks Ct. E&W	28W-10N	31
Big Oaks Rd. NW&SE	28W-10N	31
Birchwood Ct.* N&S	24W-13N	22
Bittersweet Ln. N&S	30W-10N	30
Blackberry Ct. E&W	30W-10N	30
Bluff Ct. E&W	29W-10N	30
Bode Rd. E&W	29W-11N,26W-11N	30
Bonded Pkwy. N&S	27W-10N	31
Borris Cir. NS&EW	28W-11N	31
Bourbon Pkwy. N&S	27W-10N	31
Boxwood Ct.* E&W	24W-13N	22
Brandy Pkwy. E&W	27W-10N	31
Briarwood Dr. E&W	28W-9N	31
Bristol Ct. NE&SW	29W-9N	30
Brittany Dr. NS&EW	29W-9N	30
Brooks Dr. NE&SW	27W-11N	31
Brookstone Ct. N&S	29W-10N	30
Brookstone Dr. NS&EW	29W-10N	30
Brunswick Ct. NW&SE	28W-9N	31
Buchanan Ln. E&W	29W-10N	30
Burgundy Pkwy. E&W	27W-10N	31
Bussey Ct. E&W	27W-9N	31
Butternut Ln. E&W	27W-10N	31
Buttitta Dr. E&W	27W-9N	31
Cahill Rd. E&W	29W-10N	30
Cambridge Av. E&W	28W-9N	31
Canterbury Ct. E&W	28W-11N	31
Canton Ln. NS&EW	29W-11N	30
Carey Ln. NE&SW	28W-11N	31
Carlson Dr. NS&EW	28W-9N	31
Carol Ann Dr. E&W	28W-11N	31
Cedar Cir. NS&EW	28W-10N	31
Cedar Ct. NE&SW	28W-10N	31
Cedarcrest Dr. N&S	28W-10N	31
Center Rd. N&S	29W-10N	30
Chase Ct. NW&SE	28W-10N	31
Chase Ter. N&S	28W-10N	31
Chaucer Ln. N&S	28W-11N	31
Cherry Ln. NW&SE	28W-10N	31
Chestnut Dr. NS&EW	28W-10N	31
Chrisman Dr. E&W	28W-11N	31
Claridge Ct.* NW&SE	24W-13N	22
Clearwater Ct. N&S	29W-10N	30
Clematis Dr. E&W	27W-10N	31
Club Tree Dr. NS&EW	28W-9N	31
Colony Ct. E&W	29W-9N,28W-9N	30
Concord Dr. N&S	28W-10N	31
Coolidge Ct. NW&SE	29W-10N	30
Corrington Ct. E&W	27W-11N	31
Country Ln. NW&SE	28W-10N	31
Creekside Ct. NW&SE	29W-10N	30
Crestwood Ct. NE&SW	28W-9N	31
Crestwood Dr. NW&SE	28W-9N	31
Cypress Dr. NE&SW	28W-10N	31
Dana Ln. E&W	28W-11N	31
Dartmouth Ct. E&W	29W-9N	30
Dato Ct. E&W	27W-11N	31
Dato Dr. N&S	27W-11N	31
David Dr. N&S	27W-11N	31
Debbie Ln. N&S	27W-11N	31
Deerfield Dr. E&W	29W-10N	30
Diane Dr. E&W	29W-10N	30
Dogwood Ct.* E&W	24W-13N	22
Dorchester Ct. E&W	29W-9N	30
Dorman Dr. N&S	28W-9N,28W-10N	31
Dover Ct. NW&SE	29W-11N	30
Driftwood Ct.* E&W	24W-13N	22
Dunbar Ct. E&W	28W-9N	31
Duxbury Ct. E&W	28W-9N	31
E. Bartlett Rd. E&W	28W-9N	31
E. Oltendorf Rd. E&W	28W-10N	31
E. Pine St. E&W	28W-10N	31
E. Shagbark Lane E&W	28W-11N	31
E. Streamwood Blvd. NS&EW	28W-10N	31
E. Valley Ln. E&W	28W-9N	31
East Av. N&S	27W-9N,27W-10N	31
Edgewood Dr. NW&SE	28W-9N	31
Egan Ct. E&W	27W-11N	31
Egan Dr. N&S	27W-11N	31
Eliasek Ct. NW&SE	27W-11N	31
Elm Ln. NE&SW	28W-10N	31
Essex Ct. E&W	28W-9N	31
Evans Ct. NE&SW	28W-9N	31
Evergreen Dr. NS&EW	29W-9N	30
Exmoor Dr. E&W	28W-11N	31
Fairview Ln. E&W	28W-9N	31
Fallstone Dr. NW&SE	28W-11N	31
Falmouth Ct. E&W	28W-10N	31
Fernwood Ct.* E&W	24W-13N	22
Field Ln. NW&SE	28W-10N	31
Filbert Dr. NS&EW	27W-10N	31
Fillmore Ln. E&W	29W-10N	30
Fir Ct. E&W	27W-10N	31
Flowers Av. E&W	27W-8N	31
Forest Dr. NW&SE	28W-9N	31
Foxboro Ct. N&S	29W-9N	30
Frances Dr. N&S	27W-9N	31
Franklin Ct. E&W	28W-11N	31
Frederick Av. NE&SW	28W-9N	31
Freeman Av. NE&SW	28W-9N	31
Fulton Dr. N&S	29W-9N	30
Fulton St. N&S	29W-9N	30
Gant Cir. NS&EW	28W-11N	31
Garden Cir. NS&EW	27W-11N	31
Garfield Ln. E&W	29W-10N	30
Gayle Ct. E&W	27W-11N	31
Genauldi Av. N&S	28W-8N	31
Glendale Ct. N&S	27W-11N	31
Golf Keys Rd. (Pvt.) NS&EW	30W-9N	30
Green Ct. N&S	27W-11N	31
Green Knoll Ln. N&S	26W-11N	31
Greenbriar Ln. E&W	29W-10N	30
Greenwood Ct. E&W	28W-9N	31
Gregg Ct. N&S	27W-11N	31
Greystone Ct.* NW&SE	24W-13N	22
Grow Ln. NW&SE	28W-10N	31
Hackberry Dr. NS&EW	27W-10N	31
Halick Dr. N&S	29W-11N	30
Hampton Ct. E&W	28W-9N	31
Harrison Ln. NE&SW	29W-10N	30
Hartwood Dr. NS&EW	27W-9N	31
Harvest Dr. E&W	29W-9N	30
Hastings Mill Rd. N&S	28W-11N	31
Hawthorne Ln NW&SE	28W-10N	31
Hayward Av. NW&SE	28W-8N,27W-8N	31
Hazelnut Dr. N&S	27W-10N	31
Heath Ct. NE&SW	28W-9N	31
Heather Ct. E&W	27W-11N	31
Heather Ln. NE&SW	27W-11N	31
Hecht Rd. N&S	27W-9N	31
Heine Ct. E&W	27W-11N	31
Heine Dr. N&S	27W-11N	31
Helen Ct. N&S	29W-9N	30
Hickory Av. NE&SW	28W-10N	31
Hillside Ct. NW&SE	28W-10N	31
Hillside Dr. NE&SW	28W-10N,27W-10N	31
Hise Ct. N&S	27W-11N	31
Holly Ct. N&S	28W-10N	31
Holly Dr. N&S	28W-10N	31
Hoover Ct. NW&SE	29W-10N	30
Huntington Dr. N&S	28W-11N	31
Innsbrook Dr. E&W	29W-9N	30
Iris Dr. NS&EW	27W-10N	31
Ironwood Ct.* N&S	24W-13N	22
Irving Park Rd. NW&SE	29W-10N,27W-9N	30
Ivy Ct. E&W	27W-10N	31
Jackson Ln. E&W	29W-10N	30
Jamestown Ct. N&S	28W-9N	31
Janet Av. E&W	29W-10N	30
Jefferson Ct. NW&SE	29W-10N	30
Jefferson Ln. NS&EW	29W-10N,28W-10N	30
Jill Ln. NW&SE	27W-11N	31
Jonquil Ct. E&W	30W-10N	30
Joyce Ln. N&S	29W-10N	30
Judy Ln. NE&SW	27W-11N	31
Juniper Ct. NS&EW	27W-10N	31
Kennedy Dr. NW&SE	27W-11N	31
Kensington Ct. NW&SE	29W-11N	30
Kensington Dr. NS&EW	29W-11N	30
Kevin Morris Ct. N&S	27W-11N	31
Kimberley Ln. E&W	27W-10N	31
King Dr. NW&SE	27W-9N	31
Kingston Ct. NW&SE	28W-9N	31
Klafter Ct. N&S	27W-11N	31
Klein Dr. E&W	27W-11N	31
Kosan Cir. NS&EW	27W-11N	31
Krause Av. E&W	28W-9N,27W-9N	31
LaSalle Ct. N&S	29W-9N	30
LaSalle Rd. E&W	29W-9N	30
Lacy Av. E&W	28W-9N,27W-9N	31
Lake St. NW&SE	29W-9N,28W-8N	30
Lancaster Ct. E&W	29W-9N	30
Larkspur Ln. NS&EW	30W-10N	30
Larsen Av. E&W	28W-9N,27W-9N	31
Laurel Ct. NW&SE	28W-10N	31
Laurel Ln. E&W	28W-10N	31
Lee Ct. N&S	28W-9N	31
Lexington Ct.* E&W	24W-13N	22
Library Ln. E&W	28W-9N	31
Lincoln Av. NE&SW	27W-9N	31
Lincoln Ct. NW&SE	27W-9N	31
Lincolnwood Dr. N&S	28W-10N	31
Linda Ln. N&S	29W-10N	30
Lisa Ln. N&S	29W-10N	30
Little Creek Dr. E&W	30W-10N	30
Longboat Key Ln. E&W	30W-9N	30
Madison Dr. N&S	29W-10N	30

	Grid	Page
Magnolia Ct., West E&W		
	30W-10N	30
Magnolia Dr. NS&EW		
	30W-10N	30
Manchester Ct. E&W		
	29W-11N	30
Manor Dr. N&S	28W-9N	31
Marion Ln. NS&EW		
	28W-11N	31
Mark Ln. N&S	27W-11N	31
Marrayat Pl. NW&SE		
	29W-10N	30
Maxon Ln. NS&EW		
	28W-10N	31
Mayfield Dr. N&S	28W-10N	31
Mc Cabe Dr. NS&EW		
	28W-10N	31
Mc Kool Av. N&S	27W-9N	31
McKinley Ln. NS&EW		
	29W-10N	30
Meadow Ct. N&S	28W-9N	31
Meadow Ln. E&W		
	28W-9N,27W-9N	31
Medford Ct. N&S	28W-10N	31
Merryoaks Rd. N&S	29W-9N	30
Meyer Ct. N&S	27W-11N	31
Middlebury Ct.* NS&EW		
	24W-14N	22
Miller Av. NW&SE		
	28W-9N,27W-9N	31
Monroe Ct. N&S	28W-9N	31
Moore Av. N&S	28W-9N	31
Mulberry Ln. NW&SE		
	28W-9N	31
Mustang Ct. E&W	30W-10N	30
Myrtle Ln. N&S	29W-10N	30
N. Green Meadow Blvd. E&W		
	27W-11N	31
N. Kennedy Dr. NS&EW		
	27W-11N	31
N. Oltendorf Rd. N&S		
	28W-11N	31
N. Victoria Ln. NS&EW		
	28W-11N	31
N. Walnut Ct. NS&EW		
	29W-11N	30
Newberry Dr. N&S	27W-10N	31
Nippert Ct. NW&SE	28W-9N	31
Nippert Dr. N&S	28W-9N	31
Norwood Ct. N&S	28W-9N	31
Oak Knoll Ct.* E&W		
	24W-14N	22
Oak Meadow Ct.* N&S		
	24W-13N	22
Oak Ridge Dr. E&W		
	29W-9N	30
Oakland Dr. NW&SE		
	28W-9N	31
Oakmont Ct. N&S	24W-13N	22
Old Church Rd. E&W		
	27W-11N	31
Old Oak Ln. E&W	28W-9N	31
Oltendorf Rd. N&S		
	28W-10N,28W-11N	31
Oriole Dr. NS&EW		
	28W-9N,27W-9N	31
Oxford Ct. N&S	28W-9N	31
Park Blvd. N&S		
	28W-9N,28W-11N	31
Parkside Dr. N&S		
	28W-9N,28W-10N	31
Parkside Ct. N&S	28W-9N	31
Parkwood Ct. E&W	27W-9N	31
Parkwood Dr. N&S	27W-9N	31
Patricia Pkwy. E&W		
	29W-10N	30
Pembroke Ct. N&S	29W-11N	30
Pepperidge Cir. NS&EW		
	27W-10N	31
Petrie Cir. NS&EW	27W-11N	31
Phillippi Creek Dr. NS&EW		
	30W-9N	30
Pinto Ct. E&W	30W-10N	30
Pleasant Pl. E&W	28W-9N	31
Plum Tree Ct.* N&S		
	24W-13N	22
Plymouth Dr. N&S	28W-11N	31
Polk Ct. N&S	29W-10N	30
Poplar Creek Ct. N&S		
	28W-10N	31
Post Ln. NE&SW	27W-10N	31
Post Pl. NW&SE	27W-10N	31
Princeton Ct. N&S	28W-9N	31
Quacker Hollow Ln. E&W		
	26W-11N	31
Quincy Ct. NS&EW	28W-9N	31
Rambler Ct. N&S	28W-10N	31
Rambler Ln. N&S	28W-10N	31
Rambler Pl. N&S	28W-10N	31
Ramblewood E&W	27W-9N	31
Red Cedar Dr. NS&EW		
	27W-10N	31
Redwood Ct. E&W	27W-10N	31
Regency Ct. E&W	29W-9N	30
Ridge Cir. NS&EW		
	28W-10N,27W-10N	31
Ridge Ct. N&S	27W-10N	31
Ridge Ct. N&S	27W-11N	31
Ridge Pl. N&S	27W-10N	31
Ridgewood Rd. NE&SW		
	29W-10N	30
Robinhood Ct. N&S		
	28W-10N	31
Robinhood Dr. N&S		
	28W-10N	31
Robinson Av. NS&EW		
	27W-9N	31
Roder Ct. N&S	28W-9N	31

	Grid	Page
Romajean Pkwy. NS&EW		
	28W-9N	31
Rowley Ct. NS&EW	28W-9N	31
Rt. 19 E&W		
	29W-10N,27W-9N	30
Rt. 59 N&S		
	29W-10N,29W-11N	30
Russet Ln. E&W	28W-10N	31
S. Green Meadow Blvd. E&W		
	27W-11N	31
S. Oltendorf Rd. N&S		
	28W-10N	31
S. Pine St. N&S	28W-10N	31
Sandalwood Ct.* NW&SE		
	24W-13N	22
Sandhurst Ct.* E&W		
	24W-14N	22
Sarasota Dr. E&W	28W-11N	31
Saybrook Ct. E&W	28W-9N	31
Schaumburg Rd. E&W		
	29W-11N,27W-11N	30
Seneca Ct. NE&SW		
	29W-10N	30
Seton Ct. E&W	27W-10N	31
Seton Pl. N&S	27W-10N	31
Shadywood Ln. N&S		
	29W-10N	30
Sheffield Ct. E&W	29W-11N	30
Sherwood Dr. E&W		
	28W-10N,27W-10N	31
Shirley Av. N&S	28W-9N	31
Short Av. N&S	27W-9N	31
Short St. E&W	27W-9N	31
Siesta Key Ln. NW&SE		
	30W-9N	30
Sieverwood Ct.* NW&SE		
	24W-13N	22
Smith Ct. N&S	27W-11N	31
Somerset Dr. NS&EW		
	28W-11N	31
Southbury Ct.* N&S		
	24W-13N	22
Southwood Cir. NS&EW		
	27W-9N	31
Spring Valley Ln. NE&SW		
	28W-10N	31
Spruce Dr. E&W		
	28W-8N,27W-9N	31
Spur Ct. E&W	30W-10N	30
Stowell Av. E&W		
	28W-9N,27W-9N	31
Stowell Pl. NE&SW	28W-9N	31
Stratford Cir. NS&EW		
	28W-9N	31
Stratford Ct. NW&SE		
	28W-9N	31
Stratford Ln. NE&SW		
	28W-9N	31
Streamwood Blvd. EW&NS		
	29W-10N,27W-10N	30
Suffolk Ct. N&S	28W-10N	31
Suffolk Pl. N&S		
	28W-10N,27W-10N	31
Summac Dr. NS&EW		
	28W-9N	31
Sunnydale Blvd. N&S		
	27W-9N	31
Sunset Cir. NS&EW		
	27W-9N	31
Surrey Ct. E&W	28W-9N	31
Surrey Dr. NW&SE		
	28W-9N,27W-9N	31
Sutton Rd. N&S		
	29W-10N,29W-11N	30
Taft Ct. N&S	29W-10N	30
Tall Tree Rd. N&S	28W-9N	31
Tanglewood Dr. NS&EW		
	29W-10N	30
Taylor Ct. E&W	29W-10N	30
Teak Ln. N&S	27W-10N	31
Thorndale Ct. NW&SE		
	29W-10N	30
Timber Ct. NE&SW		
	28W-10N	31
Timber Tr. N&S	28W-10N	31
Tinnerella Av. NW&SE		
	27W-9N	31
Troy Ct. E&W	28W-9N	31
Truman Ct. NW&SE		
	29W-10N	30
Truman Ln. E&W	29W-10N	30
Twilight Tr. E&W	28W-9N	31
Tyler Ct. E&W	29W-10N	30
Uxbridge Ct. E&W	28W-9N	31
Valley Ln. N&S	28W-9N	31
Victoria Ln. N&S	28W-10N	31
Villa Rd. N&S		
	28W-10N,27W-10N	31
Vine St. N&S	28W-9N	31
Vinton Ct. NW&SE	28W-9N	31
Virginia St. NS&EW	29W-9N	30
W. Green Meadow Blvd. NE&SW	27W-11N	31
W. Streamwood Blvd. E&W		
	29W-10N,28W-10N	30
Walden Tr. E&W	28W-11N	31
Walker Av. N&S	28W-9N	31
Walnut Dr. N&S		
	29W-10N,28W-10N	30
Warwick Ct.* NW&SE		
	24W-13N	22
Washington Av. NE&SW		
	29W-10N	30
Washington Ct. NW&SE		
	29W-10N	30
Waverley Av. NE&SW		
	28W-10N,27W-10N	31

	Grid	Page
Wellington Dr. E&W		
	28W-11N	31
Westgate Ct. E&W	27W-10N	31
Westgate Ter. NS&EW		
	27W-10N	31
Weston Ct. E&W	29W-11N	30
Whispering Ct. NW&SE		
	29W-10N	30
Whispering Dr. N&S		
	29W-10N	30
White Fence Tr. E&W		
	30W-10N	30
Whitehall Ct.* NW&SE		
	24W-13N	22
Whitewood Dr. NS&EW		
	27W-10N	31
Wicker Av. N&S		
	28W-9N,27W-9N	31
Wildrose Dr. E&W	29W-9N	30
Wildwood Ct. E&W	27W-9N	31
Wildwood Ln. N&S		
	28W-9N	31
Williamsburg Dr. NS&EW		
	28W-11N	31
Willow Ct. NE&SW	27W-10N	31
Willow Rd. N&S	27W-10N	31
Wilshire Ct. NE&SW		
	29W-10N	30
Winchester Dr. NS&EW		
	29W-9N	30
Winding Run Ln. E&W		
	26W-11N	31
Windsor Ct. E&W	27W-11N	31
Wingate Ct. E&W	27W-11N	31
Wisteria Dr. E&W	27W-10N	31
Woodbury Ct.* NW&SE		
	24W-13N	22
Woodcrest Cir. N&S		
	28W-10N	31
Woodland Hts. Blvd. E&W		
	28W-10N,27W-10N	31
Woodmar Ct. E&W		
	27W-11N	31
Woodridge Ln. NS&EW		
	28W-10N	31
Woodview Ct. N&S	27W-9N	31
Woodview Dr. N&S	27W-9N	31
Yorkshire Dr. E&W	27W-9N	31

Summit

	Grid	Page
Archer N&S	9W-6S	59
Center N&S	9W-5S	59
Douglas E&W	9W-5S	59
Hanover E&W	9W-5S	59
Harlem N&S	9W-6S	59
Hunt N&S	9W-5S	59
Lawndale Rd. NW&SE		
	10W-5S	59
Park E&W	9W-5S	59
Rt. 171 E&W	9W-5S	59
Rt. 43 N&S	9W-6S,9W-5S	59
55th E&W	9W-5S	59
55th Pl. E&W	9W-5S	59
56th Pl. E&W	9W-6S	59
57th Pl. E&W	9W-6S	59
58th Pl. E&W	9W-6S	59
59th E&W	9W-6S	59
60th E&W	9W-6S	59
61st E&W	9W-6S	59
61st Pl. E&W	9W-6S	59
62nd E&W	9W-6S	59
62nd Pl. E&W	9W-6S	59
63rd E&W	9W-6S	59
63rd Pl. E&W	9W-7S	59
64th E&W	9W-7S	59
64th Pl. E&W	9W-7S	59
72nd Ct. N&S	9W-6S	59
73rd N&S	9W-5S	59
73rd Ct. N&S	9W-5S	59
74th N&S	9W-6S,9W-5S	59
75th N&S	9W-6S	59
76th N&S	9W-6S	59

Thornton

	Grid	Page
Anne Ct. E&W	0E-20S	88
Apache Dr. E&W	0E-20S	88
Arapaho Dr. N&S	0E-20S	88
Armory Dr. E&W	0W-19S	82
Arrowhead Dr. NS&EW		
	0E-20S	88
Blackhawk Dr. NS&EW		
	0E-20S	88
Blackstone St. N&S		
	0E-21S,0E-20S	88
Bonnie Ct. E&W	0E-20S	88
Brown Derby Rd. N&S		
	0W-21S	88
Brownell Av. N&S	0E-21S	88
Canal St. N&S	0E-21S	88
Cherry Ln. E&W	0E-21S	88
Chicago Rd. N&S	0E-20S	88
Chippewa Dr. NE&SW		
	0E-20S	88
Clark St. N&S	0W-20S	88
Cora Ct. N&S	0E-20S	88
Dorie Ct. E&W	0E-20S	88
Eleanor St. E&W	0E-20S	88
Evans Av. N&S	0W-17S	82
Forest Ln. E&W	0E-20S	88
Francis St. E&W	0E-21S	88
Harriet St. E&W	0E-21S	88
Highland Av. N&S	0E-20S	88
Hubbard St. N&S	0E-21S	88
Hunter St. N&S	0E-21S	88
Indianwood Dr. NS&EW		
	0E-20S	88

	Grid	Page
Interstate 294-80 E&W		
	0W-20S,0E-20S	88
Julian St. N&S		
	0E-21S,0E-20S	88
Juliette St. E&W	0E-21S	88
Kinzie Pl. E&W	0E-20S	88
Kinzie St. N&S	0E-20S	88
La Salle St. N&S	0W-20S	88
Laura Ln. E&W	0E-20S	88
Leavenworth St. N&S		
	0W-21S	88
Leverette St. N&S	0E-21S	88
Malette St. N&S	0E-21S	88
Maple Ln. E&W	0E-21S	88
Margaret St. E&W	0W-21S	88
Maria St. E&W	0E-21S	88
Marion St. E&W	0E-21S	88
Michigan Av. N&S	0W-20S	88
Mohawk Dr. E&W	0E-20S	88
Park Av. N&S	0E-21S	88
Parnell Av. N&S	0W-20S	88
Pawnee Dr. E&W	0E-20S	88
Queens Ln. E&W	0E-20S	88
Ridge Rd. E&W	0W-21S	88
Rt. 294-80 E&W		
	0W-20S,0E-20S	88
Schwab St. E&W	0E-21S	88
Sunnyside Av. N&S	0E-20S	88
Tahoe St. E&W	0E-20S	88
Thorncreek Rd. NW&SE		
	0E-21S	88
Thornton Rd. E&W	0E-21S	88
Vincennes Rd. N&S		
	0W-20S,0E-20S	88
Wabash Av. N&S	0W-20S	88
Water St. N&S	0E-20S	88
Wentworth Av. N&S	0W-20S	88
Williams St. N&S	0E-20S	88
Wolcott St. N&S	0E-21S	88
168th St. E&W	0W-20S	88
169th St. E&W	0W-20S	88
170th St. E&W	0W-20S	88
171st St. E&W	0E-20S	88
172nd St. E&W	0E-20S	88
175th St. E&W	0E-20S	88

Thornton Township

	Grid	Page
Beaubien St. E&W	1E-21S	89
Blackstone Av. N&S	1E-21S	89
Blitz St. N&S	1E-21S	89
Bock Av. N&S	2E-21S	89
Calumet Expwy. N&S		
	1E-21S,1E-20S	89
Chappel Av. N&S	2E-18S	83
Chappell Av. N&S	2E-21S	89
Claremont Av. N&S	2W-19S	81
Clyde Av. N&S		
	2E-21S,2E-18S	89
Cornell Av. N&S	2E-21S	89
Cottage Grove Av. N&S		
	0E-21S	88
Country Ln. N&S	2E-21S	89
Dante Av. N&S	1E-21S	89
East End Av. N&S	2E-21S	89
Elizabeth Av. N&S	2E-21S	89
Eustace St. E&W	1E-21S	89
Fern St. NW&SE	2E-21S	89
Greenwood Rd. NW&SE		
	2E-18S	83
Hamilton Av. N&S	2W-19S	81
Harper Av. N&S	1E-21S	89
Hoyne Av. N&S	2W-19S	81
Indiana Av. NW&SE	2E-21S	89
Irving Av. N&S	2W-19S	81
Jason Ln. N&S	2E-21S	89
Jeffrey Av. N&S	2E-21S	89
Keegan St. E&W	1E-20S	89
Lansing Rd. E&W	2E-21S	89
Leavitt Av. N&S	2W-19S	81
Locust Av. N&S	2E-21S	89
Merrill Av. N&S	2E-18S	83
Oakley Av. N&S	2W-19S	81
Olde Farm Rd. N&S	2E-21S	89
Paxton Av. N&S		
	2E-21S,2E-18S	89
Ridge Rd. E&W	2E-21S	89
Ridgeland Av. N&S	2E-21S	89
St. Andrews St. E&W		
	1E-21S	89
Stony Island Av. N&S		
	1E-21S	89
Whitman Ln. N&S	2E-21S	89
158th St. E&W	2E-18S	83
160th St. E&W	2W-19S	81
161st St. E&W	2W-19S	81
177th St. E&W	1E-21S	89
178th St. E&W	1E-21S	89
180th Pl. E&W	2E-21S	89
181st St. E&W	2E-21S	89
183rd St. E&W	2E-21S	89
185th St. E&W	2E-21S	89

Tinley Park

	Grid	Page
Alexandria Dr. N&S		
	10W-19S	79
Almond Ln. E&W	11W-21S	85
Amherst St. N&S	9W-20S	85
Andres Av. E&W	7W-20S	86
Anne-Marie Av. N&S		
	7W-20S	86
Anvil Pl. E&W	10W-19S	79
Apache Dr. N&S	11W-21S	85
Apple Ln. N&S	10W-19S	79
Arcadia Dr. N&S	7W-20S	86
Arlington St. E&W	9W-19S	79
Ash St. N&S	11W-21S	85

	Grid	Page
Ashley Ln. E&W	10W-19S	79
Aspen Ln. E&W	10W-19S	79
Aster Ln. N&S	11W-21S	85
Autumn Dr. N&S	7W-20S	86
Avon Ln. N&S	10W-20S	85
Barbara E&W	7W-20S	86
Barbara Ln. N&S	7W-20S	86
Bayberry Ln. NS&EW		
	11W-21S	85
Bedford Ln. N&S	9W-20S	85
Bethany Ln. E&W	10W-20S	85
Birchwoode Ct. E&W		
	8W-19S	79
Blossom Ln. N&S	10W-19S	79
Blue Spruce Ct. E&W		
	10W-19S	79
Bluebird E&W	11W-21S	85
Bormet Dr. N&S	10W-19S	79
Brementowne Dr. NS&EW		
	8W-19S	79
Brementowne Rd. NS&EW		
	8W-19S	79
Briar Dr. N&S	10W-20S	84
Briarwood Dr. N&S	11W-20S	84
Bridalwood Ln. N&S		
	11W-20S	85
Bristol Ln. N&S	9W-19S	79
Brittney Ln. E&W	7W-20S	86
Budingen Ln. E&W	10W-20S	85
Cambridge Pl. N&S		
	11W-21S	85
Cardinal Dr. N&S	11W-21S	85
Carlsbad Dr. E&W	7W-20S	86
Carriage Ln. E&W	10W-20S	85
Castle Ln. NE&SW	10W-20S	85
Catalpa Ct. NW&SE		
	11W-20S	85
Cedar Ln. N&S	9W-21S	85
Centennial Cir. NS&EW		
	8W-19S	79
Centennial Dr. N&S	8W-19S	79
Charnswood Ct.* N&S		
	9W-19S	79
Charnswood Dr.* E&W		
	9W-19S	79
Chelsea Rd. E&W	8W-19S	79
Cherokee Tr. E&W	10W-21S	85
Cherry Creek Av. N&S		
	10W-20S	85
Cherry Hill Av. N&S		
	10W-19S	79
Cherry Stone Pl. NW&SE		
	10W-19S	79
Chestnut Dr. NE&SW		
	11W-20S	85
Chippewa Tr. N&S	11W-21S	85
Christopher Ct. N&S		
	7W-20S	86
Circle Dr. N&S	10W-20S	85
Claremont Dr. E&W	9W-20S	85
Clark Ln. N&S	8W-19S	79
Clover Ln. N&S	10W-20S	85
Cloverview Dr. NS&EW		
	10W-21S	85
Coachwood Tr. E&W		
	8W-19S	79
Cobbler Ct. NE&SW		
	10W-20S	85
Cobblestone Ct. NS&EW		
	8W-19S	79
Cottage Ct. N&S	10W-20S	85
Cranberry Ct. N&S	10W-19S	79
Creekside Av. N&S		
	10W-20S	85
Crescent Av. N&S	9W-19S	79
Crown Ln. E&W	8W-19S	79
Currant Av. N&S	10W-19S	79
Cynthia Ct.* N&S	10W-20S	85
Cypress Ct. NE&SW		
	11W-21S	85
Debra Av. E&W	7W-20S	86
Dee Ct. E&W	10W-20S	85
Derwent Ln. N&S	10W-20S	85
Desiree Dr. N&S	10W-19S	79
Doneen Av. NS&EW	10W-20S	85
Dooneen Av. NS&EW		
	10W-20S	85
Dorothy Ln. E&W	9W-20S	85
Drummond Dr. NS&EW		
	11W-21S	85
Duvan Dr. N&S	9W-21S	85
Eagle Ridge Dr. N&S		
	10W-19S	79
Edgewood Ct. NE&SW		
	11W-21S	85
Elm Lane Dr. N&S	8W-20S	85
Elmwood Dr. N&S	11W-21S	85
Everdon Dr. N&S	9W-19S	79
Evergreen Dr. N&S	10W-19S	79
Farm View Ct. N&S		
	10W-20S	85
Flamingo Dr. N&S	11W-21S	85
Forest View Dr. N&S		
	8W-19S	79
Fox Grove Ln. NW&SE		
	10W-20S	85
Gaynelle Rd. N&S	7W-20S	86
Gentry Ln. N&S	7W-20S	86
Glenview Dr. E&W	8W-20S	85
Greenwood Dr. E&W		
	11W-21S	85
Grissom Dr. N&S	10W-20S	85
Hamilton Av. N&S	9W-19S	79
Hanover Dr. E&W	9W-20S	85
Harlem N&S		
	9W-20S,8W-23S	85
Heather Ct. E&W	10W-21S	85

	Grid	Page
Heather Ln. NW&SE		
	10W-21S	85
Helen Sandidge Ct. N&S		
	10W-20S	85
Henry Ct. N&S	10W-19S	79
Hickory St. NE&SW . . 9W-21S		85
Highland Av. 7W-21S		86
Hillcrest Dr. N&S . . . 10W-19S		79
Hillcrest Ln. E&W . . . 10W-19S		79
Hillside Pl. N&S 7W-20S		86
Holly Ct. NW&SE . . . 11W-21S		85
Honey Lane Dr. E&W		
	8W-20S,7W-20S	85
Horsehoe Dr. N&S . . 10W-19S		79
Hubbard Lane Dr. E&W		
	8W-20S	85
Humber Ln. N&S . . . 10W-20S		85
Hummingbird Dr. N&S		
	11W-21S	85
Hunter Tr. N&S 8W-19S		79
Interstate 80 E&W . . 8W-22S		85
Inverness Dr. N&S . . 11W-20S		85
Ironwood Dr. N&S . . 10W-19S		79
Iroquois Trace NS&EW		
	10W-21S	85
Jacquelin Ct. E&W . . 7W-20S		86
Jean Ln. N&S 8W-19S		79
Jeanette Ct. E&W . . . 8W-21S		85
Jennifer Av. E&W 7W-20S		86
Jessica Ln. E&W . . . 10W-19S		79
Juniper Ct. NE&SW		
	11W-20S	85
Justin Ct. E&W 10W-20S		85
Kathleen Ln. E&W . . 10W-20S		85
Kensington Av. N&S . 9W-19S		79
Kingston Ct. NW&SE		
	8W-19S	79
Kingston Rd. E&W . . 8W-19S		79
Lake Bluff Dr. NE&SW		
	7W-20S	86
Lake Villa Av. N&S . . 9W-19S		79
Lakeside Pl. NE&SW		
	7W-20S	86
Lakeview Ct. N&S . . . 8W-22S		86
Lakewood Dr. N&S . . 7W-20S		86
Laura Ln. NW&SE . . . 7W-20S		86
Laverne Ln. E&W . . . 8W-19S		79
Lilac Ln. N&S 10W-21S		85
Linden Dr. NS&EW 11W-21S		85
Lindsay St. E&W . . . 12W-20S		84
Live Oak Ct. N&S . . . 8W-21S		85
Locust Dr. N&S 10W-21S		85
Magnolia Ln. E&W . .11W-21S		85
Mallard Rd. E&W . . . 11W-21S		85
Manchester St. N&S 8W-19S		79
Maple Creek Dr. E&W		
	9W-22S	85
Maple Ln. E&W 8W-20S		85
Maplewoode Ct. N&S		
	8W-19S	79
Meadow Ct. NW&SE		
	10W-19S	79
Meadow Ln. N&S . . 10W-19S		79
Michelle Ln. E&W . . 10W-19S		79
Milford Av. N&S 9W-20S		85
Mockingbird Ln. N&S		
	11W-21S	85
Mulberry Av. N&S		
	11W-21S,11W-20S	85
Mulberry St. N&S . . 11W-21S		85
Navajo Trace N&S . . 10W-21S		85
New England Av. N&S		
	8W-20S	85
Normandy Ln. NE&SW		
	8W-20S	85
North Creek Dr. N&S		
	9W-22S	85
North St. NE&SW 8W-20S		85
Nottingham Ct. NW&SE		
	9W-19S	79
Nottingham Dr. N&S		
	10W-19S,9W-19S	79
Oak Forest Av. NE&SW		
	8W-20S	85
Oak Park Av. N&S		
	8W-22S,8W-19S	85
Oakwood Dr. N&S . .11W-21S		85
Oconto Av. N&S 9W-20S		85
Odell Av. N&S 9W-20S		85
Oketo Av. N&S 9W-20S		85
Olcott Av. N&S		
	9W-20S,9W-19S	85
Olcott Ct. NW&SE		
	9W-20S,9W-19S	85
Old Barn Ct. N&S . . 10W-20S		85
Olde Gatehouse Rd. NS&EW		
	8W-19S	79
Oleander Av. N&S . . . 9W-20S		85
Orchard Ct. E&W . . 10W-19S		79
Oriole Av. N&S 9W-20S		85
Osceola Av. NE&SW . 9W-20S		85
Ottawa Av. N&S 9W-20S		85
Overhill Av. N&S 9W-20S		85
Oxford Dr. N&S 8W-19S		79
Ozark Av. N&S		
	9W-20S,9W-19S	85
Parkside Dr. E&W . . . 9W-20S		85
Parkview Dr. N&S . . . 9W-19S		79
Parliament Av. N&S . . 9W-19S		79
Paxton Av. N&S		
	9W-20S,9W-19S	85
Peacock Ln. N&S . . 11W-21S		85
Pecan Ln. N&S 11W-21S		85
Pelican Dr. N&S . . . 11W-21S		85
Pheasant Lake Dr. E&W		
	11W-21S	85
Pheasant Ln. N&S . .11W-21S		85
Pin Oak Ct. N&S 8W-21S		85

	Grid	Page
Pine Dr. N&S 10W-21S		79
Pine Dr. NW&SE 8W-22S		86
Pine Lake Dr. N&S . . 8W-22S		86
Pine Point Dr. E&W . 8W-20S		85
Pinewood Av. N&S . . 8W-22S		86
Piute Tr. E&W 10W-21S		85
Plum Ct. E&W 10W-21S		85
Poplar St. E&W 11W-21S		85
Pottawattomi Tr. N&S		
	10W-21S	85
Prairie Dr. E&W . . . 10W-19S		79
Prairie Dr. N&S 10W-19S		79
Prestwick Dr. E&W . . 9W-20S		85
Princess Elizabeth Ct. N&S		
	11W-21S	85
Princeton Av. N&S . . 9W-19S		79
Quail Cross E&W . . 11W-20S		85
Queen Ann Ln. N&S		
	10W-20S	85
Queen Elizabeth Ln. N&S		
	10W-20S	85
Queen Mary Ln. N&S		
	10W-20S	85
Queen Victoria Ln. N&S		
	10W-20S	85
Raintree Ln. E&W . . 10W-20S		85
Ravinia Dr. E&W 8W-20S		85
Redwood Ln. N&S		
	11W-21S	85
Richards Ct. NW&SE		
	10W-20S	85
Richards Dr. N&S		
	10W-20S,10W-19S	85
Ridgeland Av. N&S		
	8W-24S,8W-19S	90
Riverside Dr. NS&EW		
	8W-20S	85
Rosewood Ln. NW&SE		
	11W-21S	85
Royal Oak Ct. N&S . . 8W-21S		85
Rt. 43 N&S . .9W-22S,8W-22S		85
Salem Ct. N&S 9W-20S		85
Sandalwood Dr. E&W		
	9W-21S	85
Sandy Ln. NW&SE . . 9W-20S		85
Sayre Av. N&S 8W-20S		85
Scott Ct. E&W 7W-20S		86
Shannon Ct. N&S . . 10W-20S		85
Shetland Dr. N&S . . 11W-20S		85
Shoshone Tr. E&W 10W-21S		85
Skyline Dr. N&S 8W-19S		79
South St. NE&SW		
	8W-20S,7W-20S	85
Spring Creek Dr. N&S		
	9W-22S	85
Spruce Ln. E&W . . . 11W-20S		85
Steeple Dr. NE&SW		
	10W-20S	85
Steven Pl. NW&SE . .10W-19S		79
Stone Bridge Walk* NE&SW		
	9W-19S	79
Surrey Dr. N&S 10W-19S		79
Sussex Rd. E&W . . . 8W-19S		79
Sycamore Ct. E&W		
	11W-21S	85
Tamar Ln. N&S 10W-20S		85
Tanbark Dr. NS&EW		
	10W-19S	79
Tayside Ln. E&W . . . 10W-20S		85
Teakwood Dr. N&S		
	11W-20S	85
Terrace Dr. E&W . . . 8W-19S		79
Theresa Ln. N&S . . . 8W-19S		79
Thornwood Dr. E&W		
	11W-20S	85
Timber Ct. E&W . . . 11W-20S		85
Tinley Park Dr. NE&SW		
	8W-20S	85
Tower Ct. N&S 10W-19S		79
Trent Ct. E&W 10W-21S		85
Tudor Ln. NS&EW . . 10W-20S		85
Tulip Ln. N&S 10W-21S		85
Turnberry Ln. E&W		
	11W-20S	85
Valley Forge Dr. N&S		
	8W-20S	85
Valleyview Ct. NW&SE		
	10W-20S	85
Valleyview Dr. NS&EW		
	10W-20S	85
Vogt St. E&W 8W-20S		85
Waterford Dr. E&W		
	10W-20S	85
West Creek Dr. E&W		
	9W-22S	85
Westberry Ln. E&W		
	8W-20S	79
Westwind Ct. NW&SE		
	8W-20S	85
Whittington Dr. E&W		
	9W-19S	79
Willow Ln. N&S 8W-20S		85
Windsor Pkwy. E&W		
	11W-21S	85
Winston Dr. E&W . . . 8W-19S		79
Witham Ct. N&S 8W-19S		79
Wittington Ct. N&S . . 9W-16S		79
Woodland Dr. E&W		
	10W-19S	79
Woodstock Av. E&W		
	9W-20S	85
64th Ct. N&S		
	8W-20S,8W-19S	85
65th Av. N&S		
	8W-21S,8W-19S	85
65th Ct. N&S		
	8W-21S,8W-19S	85

	Grid	Page
66th Av. N&S		
	8W-21S,8W-19S	85
66th Ct. N&S		
	8W-21S,8W-20S	85
67th Av. N&S		
	8W-21S,8W-20S	85
67th Ct. N&S		
	8W-21S,8W-19S	85
68th Ct. N&S		
	8W-21S,8W-19S	85
69th Av. N&S 8W-20S		85
70th Av. N&S 8W-20S		85
70th Ct. N&S 8W-21S		85
71st Av. N&S		
	8W-21S,8W-19S	85
71st Ct. N&S		
	8W-21S,8W-19S	85
75th Av. N&S 9W-19S		79
76th Ct. N&S 9W-19S		79
78th Ave. N&S 9W-19S		79
80th Av. N&S 10W-20S		85
81st Av. N&S 10W-20S		85
81st Ct. N&S 10W-20S		85
82nd Av. N&S 10W-20S		85
82nd Ct. N&S 10W-20S		85
83rd Ct. N&S 10W-20S		85
84th Av. N&S		
	10W-20S,10W-19S	85
84th Ct. N&S 10W-19S		79
84th Pl. N&S 10W-19S		79
85th Av. N&S 10W-19S		79
85th Ct. N&S 10W-19S		79
87th Ct. N&S 10W-20S		85
88th Av. N&S 11W-20S		85
92nd Av. N&S		
	11W-20S,11W-21S	85
94th Ct. N&S 11W-20S		84
159th Pl. E&W 9W-19S		79
159th St. E&W		
	10W-19S,9W-19S	79
160th Pl. E&W		
	10W-19S,9W-19S	79
160th St. E&W 9W-19S		79
161st Pl. E&W		
	10W-19S,9W-19S	79
161st St. E&W		
	10W-19S,9W-19S	79
162nd Pl. E&W		
	10W-19S,9W-19S	79
162nd St. E&W 9W-19S		79
163rd Pl. E&W		
	9W-19S,8W-19S	79
163rd St. E&W		
	10W-19S,9W-19S	79
164th Pl. N&S 10W-19S		79
164th Pl. E&W		
	10W-19S,9W-19S	79
164th St. E&W		
	10W-19S,9W-19S	79
165th Pl. E&W		
	9W-19S,8W-19S	79
165th St. E&W		
	10W-19S,9W-19S	79
166th Pl. E&W 9W-19S		79
166th St. E&W		
	9W-19S,8W-19S	79
167th Pl. E&W 10W-20S		85
167th St. E&W		
	10W-19S,8W-19S	79
168th Pl. E&W		
	10W-20S,9W-20S	85
168th St. E&W 10W-20S		85
169th Pl. E&W 10W-20S		85
169th St. E&W		
	10W-20S,8W-20S	85
170th Pl. E&W		
	10W-20S,9W-20S	85
170th St. E&W		
	10W-20S,8W-20S	85
171st Pl. E&W 10W-20S		85
171st St. E&W		
	11W-20S,8W-20S	85
172nd Pl. E&W 9W-20S		85
172nd St. E&W		
	11W-20S,8W-20S	85
173rd Pl. E&W		
	9W-20S,8W-20S	85
173rd Rd. NW&SE . .11W-20S		85
173rd St. E&W		
	9W-20S,8W-20S	85
174th Pl. E&W		
	10W-20S,8W-20S	85
174th St. E&W		
	11W-20S,8W-20S	85
175th St. E&W 8W-21S		85
175th St. E&W		
	11W-20S,7W-20S	85
176th Pl. E&W 8W-21S		85
176th St. E&W		
	11W-21S,8W-21S	85
177th Pl. E&W 8W-21S		85
177th St. E&W 8W-21S		85
178th St. E&W		
	11W-21S,8W-21S	85
179th Pl. E&W 8W-21S		85
179th St. E&W 8W-21S		85
180th Pl. E&W 8W-21S		85
180th St. E&W 8W-21S		85
181st Pl. E&W 8W-21S		85
181st St. E&W 8W-21S		85
182nd Pl. E&W 8W-21S		85
182nd St. E&W 8W-21S		85
183rd St. E&W		
	11W-21S,8W-21S	85

Westchester

	Grid	Page
Alexandria St. NS&EW		
	14W-2S	52
Ashley Woods Ln. NE&SW		
	14W-2S	52
Balmoral N&S		
	13W-2S,12W-1S	52
Becket N&S 13W-2S		52
Bedford E&W 13W-2S		52
Belleview N&S 13W-2S		52
Berkshire E&W 13W-2S		52
Boeger N&S 13W-1S		52
Boeger Ct. SW&NE		
	13W-1S	52
Bond E&W 12W-1S		52
Bristol N&S . .12W-1S,12W-0S		52
Buckingham N&S . . . 13W-2S		52
Cir. 13W-1S		52
Burton Dr. E&W 14W-2S		52
Cambridge E&W . . . 12W-1S		52
Camelot E&W 13W-1S		52
Canterbury E&W		
	13W-1S,12W-1S	52
Carlisle E&W 13W-2S		52
Charles E&W 13W-2S		52
Chaucer E&W 12W-1S		52
Chesapeake St. NS&EW		
	14W-2S	52
Claridge E&W 13W-2S		52
Concord N&S 13W-2S		52
Crestwood St. E&W . 12W-1S		52
Cromwell N&S 12W-0S		46
Derby Ln. NW&SE . . 12W-1S		52
Devonshire E&W . . . 12W-0S		46
Devonshire Ln. NW&SE		
	12W-0S	46
Dickens E&W		
	13W-1S,12W-1S	52
Dorchester E&W . . . 13W-1S		52
Dover N&S 12W-1S		52
Downing N&S		
	13W-2S,13W-1S	52
Drake N&S 14W-2S		52
Drury Ln. NW&SE . . 12W-1S		52
Elizabeth E&W 12W-1S		52
Enterprise St. N&S . . 14W-2S		52
Essex E&W 13W-1S		52
Evers N&S 13W-1S		52
Fairfield E&W 13W-2S		52
Fielding E&W		
	14W-1S,13W-1S	52
Fleet E&W 14W-2S		52
Forest St. N&S 14W-2S		52
Gardner N&S 12W-1S		52
Gladstone E&W 12W-0S		46
Haas N&S 13W-1S		52
Halifax N&S 13W-2S		52
Hastings E&W 13W-2S		52
Hawthorne N&S 13W-2S		52
Heidorn N&S 13W-1S		52
Highridge N&S 13W-1S		52
Hull N&S 12W-1S,12W-0S		52
Interstate 290 E&W . 12W-0S		46
Kensington		
	13W-2S,13W-1S	52
Kent E&W 12W-1S		52
Kingston E&W 13W-1S		52
Kipling E&W 12W-1S		52
Kitchener E&W 12W-0S		46
LaGrange Rd. N&S . . 12W-1S		52
Lancaster E&W 13W-2S		52
Manchester N&S		
	12W-1S,12W-0S	52
Mandel N&S 13W-1S		52
Mannheim Rd. N&S . 13W-1S		52
Martindale E&W 13W-2S		52
Mayfair N&S		
	13W-2S,13W-1S	52
Milford E&W 12W-1S		52
Nelson E&W 13W-2S		52
Nelson Sq. N&S . . . 13W-2S		52
Newberry E&W 13W-2S		52
Newcastle N&S		
	12W-1S,12S-0S	52
Norfolk N&S		
	12W-1S,12W-0S	52
Oxford E&W 13W-1S		52
Park St. N&S 14W-2S		52
Pelham E&W 12W-1S		52
Pell N&S 13W-2S		52
Plymouth N&S 13W-2S		52
Portsmouth N&S		
	12W-1S,12W-0S	52
Prescott Ln. NS&EW		
	14W-2S	52
Preston E&W 13W-2S		52
Raliegh E&W 13W-2S		52
Robin Hood E&W . . . 13W-1S		52
Rt. 38 E&W 13W-1S		52
Shakespeare E&W . . 13W-1S		52
Shaw SW&NE 13W-1S		52
Shelly E&W 13W-1S		52
Sherwood N&S 13W-2S		52
Somerset N&S 13W-2S		52
Stratford N&S 13W-2S		52
Suffolk N&S . .12W-1S,12S-0S		52
Sunnyside N&S 13W-1S		52
The Strand N&S 12W-0S		46
U.S. Rt. 12 N&S 13W-1S		52
U.S. Rt. 20 E&W 13W-1S		52
U.S. Rt. 45 N&S 13W-1S		52
Wakefield E&W 13W-1S		52
Waterford Dr. E&W . . 13W-1S		52
Waverly N&S 13W-1S		52
Wellington N&S 13W-2S		52
Westchester Blvd. N&S		
	12W-1S,12W-0S	52

	Grid	Page
Westminster E&W . . 13W-2S		52
Windsor E&W 13W-2S		52
Woodland St. N&S . . 14W-2S		52
Worcester N&S 12W-0S		46
Wright E&W 12W-1S		52
26th St. E&W 14W-2S		52
28th St. E&W 14W-2S		52
29th E&W 13W-2S		52
30th St. E&W 14W-2S		52

Western Springs

	Grid	Page
Birch E&W 14W-4S		58
Burlington SW&NE . . 14W-4S		52
Caroline N&S 13W-5S		58
Central N&S		
	14W-5S,14W-4S	58
Chestnut E&W 14W-4S		52
Clausen N&S		
	13W-5S,13W-4S	58
Commonwealth N&S		
	14W-5S	58
Courtland E&W 13W-5S		58
Crest E&W 14W-4S		52
Crestview Dr. N&S . . 14W-6S		58
Crestview Pl. E&W . . 14W-6S		58
Dierks E&W 13W-5S		58
Ellington N&S		
	13W-5S,13W-4S	58
Elm SW&NE 14W-4S		52
Ernst N&S 14W-4S		58
Fairels N&S 13W-5S		58
Forest N&S 14W-4S		52
Franklin N&S		
	13W-5S,13W-4S	58
Garden N&S 14W-4S		52
Gilbert N&S 14W-4S		52
Grand N&S . .14W-6S,14W-4S		58
Grove N&S 14W-4S		52
Hampton N&S 14W-4S		52
Harvey N&S		
	13W-5S,13W-4S	58
Hill Grove SW&NE . . 14W-4S		52
Howard N&S		
	13W-5S,13W-4S	58
Johnson N&S		
	13W-5S,13W-4S	58
Kirkland E&W 13W-5S		58
Lawn N&S . . 14W-5S,15W-4S		58
Lawn Cir. E&W 14W-6S		58
Lawn Ct. E&W 14W-6S		58
Lawn Dr. N&S 14W-6S		58
Linden N&S 13W-4S		52
Linden Cir. N&S 14W-6S		58
Linden Cir. NS&EW . . 14W-6S		58
Linden Ct. NW&SE . . 14W-6S		58
Longmeadow E&W . . 14W-4S		58
Maple E&W 14W-4S		52
Maple E&W 14W-6S		58
Oak E&W 14W-4S		52
Oak E&W 14W-6S		58
Ogden E&W 14W-4S		52
Park Ln. E&W 14W-6S		58
Park Pl. E&W 14W-6S		58
Prospect N&S 14W-4S		52
Race E&W 14W-4S		52
Reed E&W 14W-4S		52
Ridge E&W 14W-4S		58
Ridgewood N&S 14W-6S		58
Rose N&S 14W-4S		52
Rugeley E&W 13W-5S		58
Sping N&S 14W-4S		52
Sunset E&W 14W-4S		52
U.S. Rt. 34 E&W . . . 13W-3S		52
Walker E&W 14W-4S		52
Walnut E&W 14W-4S		52
Western N&S 14W-4S		52
Willow E&W 14W-4S		52
Wolf N&S 14W-5S		58
Woodland N&S		
	14W-5S,14W-4S	58
39th E&W 13W-3S		52
40th E&W 13W-4S		52
41st E&W 13W-4S		52
42nd E&W 13W-4S		52
43rd E&W 13W-4S		52
45th E&W 13W-4S		52
46th E&W 13W-4S		52
47th E&W 14W-4S		58
48th E&W . . 14W-5S,13W-5S		58
49th E&W 14W-5S		58
50th E&W 14W-5S		58
50th Pl. E&W 13W-5S		58
51st E&W 13W-5S		58
51st Pl. E&W 13W-5S		58
52nd E&W 14W-5S		58
52nd Pl. E&W 13W-5S		58
53rd E&W . .14W-5S,13W-5S		58
54th E&W 14W-5S		58
Cir. 13W-5S		58
55th E&W 14W-5S		58

Wheeling

	Grid	Page
Abbey Ct.* NW&SE		
	13W-18N	16
Abbott Dr. E&W . . . 14W-16N		15
Abourndale Ct. E&W		
	16W-16N	15
Acco Plaza Dr. N&S		
	14W-17N	15
Acorn Ct.* E&W . . . 13W-17N		16
Albert Ter. NE&SW		
	15W-17N	15
Alder Dr. N&S 13W-17N		16
Alderman Av. E&W		
	14W-15N	15
Aldgate Ct.* E&W . 13W-17N		16

	Grid	Page
Alice St. NW&SE	14W-16N	15
Allen Ct. N&S	14W-17N	15
Allendale Dr. E&W	15W-18N	15
Alpine Ct.* E&W	16W-18N	15
Amy Ct.* NW&SE	16W-18N	15
Ande Av. N&S	14W-16N	15
Ander Ct. E&W	15W-16N	15
Anita Pl. N&S	14W-17N	15
Anne Ter. N&S	14W-17N	15
Anthony Rd. E&W	15W-17N	15
Apache Tr.* N&S	16W-17N	15
Aptakisic Rd. NE&SW	16W-18N	15
Arbour Ct. E&W	14W-18N	15
Arlene Ct. NE&SW	14W-17N	15
Arlington Dr.* NS&EW	13W-18N	16
Armand Ln. N&S	15W-18N	15
Armand St. E&W	15W-18N	15
Arrow Tr. N&S	16W-17N	15
Ash Ln.* N&S	16W-16N	15
Ashton Ct.* N&S	13W-17N	16
Audrey Ct. NE&SW	15W-17N	15
Baldwin Ct.* N&S	13W-17N	16
Barberry Ln.* N&S	16W-16N	15
Barnaby Pl.* NE&SW	15W-18N	15
Bayside Ct. * N&S	16W-16N	15
Bayside Dr.* E&W	16W-16N	15
Bayside Ln.* E&W	16W-16N	15
Beech Dr.* NW&SE	13W-16N	16
Berkshire Dr. N&S	16W-18N	15
Bernice Ct. NE&SW	15W-17N	15
Beverly Dr. E&W	15W-18N	15
Bina Ct. N&S	14W-17N	15
Birch Tr. N&S	16W-17N	15
Blackfoot Tr.* N&S	16W-17N	15
Blackhawk Tr.* E&W	16W-17N	15
Blaze Tr. E&W	15W-16N	15
Boehmer Av. N&S	14W-17N	15
Bow Tr.* E&W	16W-17N	15
Boxwood Ct.* E&W	15W-18N	15
Braeburn Ct.* N&S	13W-17N	16
Brandon Pl. E&W	15W-18N	15
Braver Ct.* E&W	16W-18N	15
Brian Ln. NS&EW	14W-17N	15
Briarwood Dr. N&S	15W-17N	15
Bridgeport Pl.* E&W	15W-18N	15
Bridget Pl. E&W	15W-17N	15
Bridgeview Ct.* NS&EW	16W-16N	15
Bridle Tr. E&W	15W-16N	15
Brighton Pl.* N&S	15W-18N	15
Brittany Ct.* N&S	13W-17N	16
Broadway Ct.* E&W	13W-17N	16
Brougham Dr. E&W	15W-18N	15
Buckboard Dr. E&W	15W-18N	15
Buckeye Dr. N&S	13W-17N	16
Buckingham Ct.* E&W	15W-18N	15
Buffalo Grove Rd. N&S	16W-17N	15
Buffalo Tr. N&S	16W-17N	15
Buxton Ct.* E&W	16W-17N	15
Cambridge Pl., West N&S	15W-18N	15
Camden Ct.* NW&SE	13W-17N	16
Camp McDonald Rd. E&W	14W-15N	15
Canbury Ct.* N&S	13W-17N	16
Candlewood Ct.* E&W	16W-16N	15
Capitol Dr. NS&EW	14W-16N	15
Capri Ter. E&W	14W-18N	15
Captains Ln. E&W	15W-17N	15
Cardinal Ct.* NW&SE	13W-18N	16
Carpenter Av. E&W	14W-17N	15
Carriage Hill Rd. N&S	15W-18N	15
Catherine Ct. E&W	14W-17N	15
Cedar Dr. NW&SE	15W-18N	15
Cedar Run Dr. N&S	16W-18N	15
Cedarwood Ct.* N&S	16W-16N	15
Cedarwood Ln.* N&S	16W-16N	15
Center Av. E&W	14W-17N	15
Century Dr. E&W	14W-17N	15
Chaddick Dr. NS&EW	14W-17N	15
Charabanc Ln. E&W	15W-18N	15
Chariot Ct.* N&S	14W-18N	15
Chariot Rd.* N&S	14W-18N	15
Chelsea Dr.* N&S	15W-18N	15
Cherrywood Dr. N&S	15W-17N	15
Chestnut Lane N&S	15W-18N	15
Cheswick Ct.* E&W	13W-17N	16
Chippewa Tr. NS&EW	16W-17N	15
Chukker Ct. E&W	15W-16N	15
Cindy Ln. NS&EW	15W-17N	15
Clearwater Ct.* N&S	16W-16N	15
Clearwater Dr.* E&W	16W-16N	15
Clearwater Ln.* E&W	16W-16N	15
Cleo Ct.* NW&SE	16W-18N	15
Coach Rd.* E&W	14W-18N	15
Cobbler Ln.* E&W	13W-18N	16
Colonial Dr. NW&SE	15W-18N	15
Commanche Tr. NS&EW	16W-17N	15
Commercial Dr. E&W	14W-17N	15
Coral Ln. N&S	16W-18N	15
Corey Ln. E&W	16W-16N	15
Cornell Av. N&S	16W-16N	15
Cottonwood Ct. N&S	16W-16N	15
Coventry Pl.* E&W	15W-18N	15
Creekside Ct. NE&SW	16W-16N	15
Crescent Dr. NW&SE	14W-17N	15
Crimson Dr. N&S	15W-16N	15
Crow Tr.* N&S	16W-17N	15
Curricle Rd.* N&S	14W-18N	15
Custer Tr.* N&S	16W-17N	15
Cypress Dr.* E&W	16W-18N	15
Cyrille St. E&W	14W-18N	15
Dakota Tr. NE&SW	16W-17N	15
Deborah Ln. E&W	14W-18N	15
Deerpath Ct.* N&S	14W-18N	15
Delaware Tr.* E&W	16W-17N	15
Denniston Ct. E&W	15W-17N	15
Denoyer Tr. E&W	15W-18N	15
Denoyer Trail N&S	15W-18N	15
Derby St. N&S	15W-16N	15
Diens Dr. N&S	15W-18N	15
Donna Ct. NE&SW	14W-17N	15
Dorset Cir.* E&W	15W-18N	15
Dorset Ct. E&W	15W-18N	15
Dover Pl.* N&S	15W-18N	15
Drae Ct.* NW&SE	16W-18N	15
Driftwood Ct.* N&S	15W-18N	15
Dundee Rd. E&W	15W-18N,14W-18N	15
Eagle Grove Ct.* N&S	13W-18N	16
East Dennis Rd. N&S	15W-17N	15
East Dr. E&W	14W-16N	15
East Norman Ln. N&S	15W-17N	15
Eastchester Rd. N&S	15W-17N	15
Easton Ct.* E&W	13W-17N	16
Edgewood Dr. E&W	14W-18N	15
Egidi Dr. E&W	14W-18N	15
Elden Ct.* N&S	16W-18N	15
Elizabeth Ln. N&S	16W-16N	15
Elm Dr. E&W	13W-17N	16
Elmwood Ln.* N&S	15W-18N	15
Equestrian Dr. E&W	15W-16N	15
Exchange Ct. NS&EW	14W-17N	15
Exeter Ct.* E&W	16W-18N	15
Fairfield Ct.* E&W	13W-17N	16
Fairway View Dr. NS&EW	16W-18N	15
Fall Ct. NW&SE	16W-18N	15
Fernak Ct. E&W	15W-18N	15
Ferndale Ct.* N&S	15W-18N	15
Ferne Dr. E&W	13W-17N	16
Fletcher Dr. N&S	15W-17N	15
Fore Ct.* E&W	16W-18N	15
Forestway Ln.* N&S	13W-18N	16
Forums Ct. N&S	15W-16N	15
Foster Av. N&S	14W-15N	15
Fox Ln. E&W	14W-16N	15
Foxboro Dr. NW&SE	14W-17N	15
Garden E&W	15W-16N	15
Garth Rd. N&S	15W-17N	15
Gayle Ct. NE&SW	14W-17N	15
Gee Ct.* NW&SE	16W-18N	15
George Rd. N&S	15W-17N	15
Gilman Av. E&W	14W-16N	15
Glendale St. E&W	14W-18N	15
Glengary Ct.* NW&SE	13W-17N	16
Glengary Ln.* NE&SW	13W-17N	16
Glenn Av. N&S	14W-17N	15
Gray Ct.* E&W	16W-18N	15
Greenview Ln. E&W	15W-18N	15
Gregor Ln. E&W	15W-17N	15
Greystone Ln.* N&S	15W-18N	15
Haben Ln. N&S	15W-18N	15
Hadley Ct.* NE&SW	13W-18N	16
Hansom Ct. N&S	15W-18N	15
Hansom Dr. E&W	15W-18N	15
Harbour Ct. N&S	16W-16N	15
Harbour Dr. E&W	16W-16N	15
Harmony Dr. N&S	13W-17N	16
Harms Ct.* N&S	16W-18N	15
Harvester Ct. E&W	15W-17N	15
Hastings Ct. E&W	15W-17N	15
Hastings Rd. E&W	15W-17N	15
Hawthorne Ct.* E&W	15W-18N	15
Heather Ct. E&W	16W-17N	15
Henley Ct.* N&S	13W-17N	16
Hickory Dr. E&W	13W-17N	16
High Goal Dr. NS&EW	15W-16N	15
Highland Av. E&W	14W-17N,13W-17N	15
Hintz Ln. E&W	16W-17N	15
Hintz Rd. E&W	14W-17N	15
Holbrook Dr. NS&EW	15W-16N,14W-16N	15
Holly Ct. E&W	16W-18N	15
Honey Locust Dr. N&S	13W-17N	16
Honey Suckle Dr. NW&SE	15W-16N	15
Hopi Ter.* E&W	16W-17N	15
Hunter Dr. E&W	16W-17N	15
Huntington Ln.* N&S	16W-18N	15
Inwood Dr. E&W	14W-15N	15
Iota Ct.* E&W	16W-18N	15
Ironwood Ct.* E&W	15W-18N	15
Irvine Ct. N&S	16W-17N	15
Isa Dr. NE&SW	15W-17N	15
Ivy Ct.* N&S	16W-16N	15
Jackson N&S	16W-17N	15
Janice Ct. E&W	15W-16N	15
Jaspen Ct.* E&W	16W-18N	15
Jeanne Ter. N&S	14W-17N	15
Jefferson Ct. E&W	15W-16N	15
Jeffrey Av. E&W	13W-17N	16
Jeffrey Ln. N&S	14W-17N	15
Jenkins Ct. E&W	15W-17N	15
Jerome Pl. N&S	13W-17N	16
Kenilworth E&W	16W-18N,15W-18N	15
King Ct. N&S	15W-17N	15
Kingsport Dr. E&W	15W-17N	15
Kingswood Ln.* N&S	13W-18N	16
Kiowa Tr.* N&S	16W-17N	15
Kranse Ln. E&W	14W-18N	15
Kristy Ln. E&W	15W-17N	15
Laguna Ct.* NW&SE	16W-17N	15
Lakeland Ct. N&S	16W-16N	15
Lakeside Dr. N&S	16W-17N	15
Lakeview N&S	16W-17N	15
Lakeview Dr. E&W	16W-17N	15
Landau Ln. E&W	15W-18N	15
Larkin Dr. E&W	14W-16N	15
Laurel Ct. NW&SE	15W-17N	15
Laurel Tr. N&S	16W-17N	15
Lee St. E&W	15W-17N	15
Lenox Ct.* NS&EW	13W-17N	16
Leslie Ln. N&S	14W-17N	15
Lexington Dr. N&S	15W-18N	15
Lilac Ln. N&S	14W-17N	15
Lincoln Rd. N&S	14W-17N	15
Linda Ter. NE&SW	15W-17N	15
Linden Ln.* N&S	13W-18N	16
Locust Dr. NS&EW	16W-18N	15
London Pl.* E&W	15W-18N	15
Longacre Ln. NE&SW	15W-16N	15
Longbow Ct.* N&S	13W-17N	16
Longtree Dr. N&S	15W-17N	15
Longtree Ln. E&W	15W-17N	15
Lotus Ct.* NE&SW	16W-18N	15
Lynn Ln. E&W	14W-16N	15
Mae Ct.* E&W	16W-18N	15
Mallard Ln. N&S	16W-17N	15
Manchester Dr. E&W	14W-17N,13W-17N	15
Manda Ln. NS&EW	16W-18N	15
Maple Dr. N&S	13W-17N	16
Marcy Ln. E&W	16W-17N,15W-17N	15
Marion Ct. NE&SW	15W-17N	15
Mark Ln. E&W	16W-16N	15
Marquardt Dr. E&W	14W-16N	15
Marvin Pl. N&S	13W-17N	16
Maureen Dr. NE&SW	15W-17N	15
Mayer Av. E&W	14W-18N	15
McHenry Rd. NW&SE	15W-18N	15
Meadow Brook Ln. N&S	14W-17N	15
Meadow Ln. E&W	14W-18N	15
Melvin Pl. N&S	14W-17N	15
Mercantile Ct. NS&EW	15W-17N	15
Merle Ln. N&S	15W-17N	15
Merle Ln. N. E&W	15W-17N	15
Messner Dr. E&W	14W-16N	15
Middlebury Ln.* NE&SW	15W-18N	15
Milwaukee Av. NW&SE	14W-18N,13W-17N	15
Mockingbird Ln. N&S	16W-18N	15
Mohawk Tr.* N&S	16W-17N	15
Molly Ct. NE&SW	16W-18N	15
Mors Av. E&W	14W-17N,13W-17N	15
Muriel Ct. E&W	15W-16N	15
Nancy Ln. N&S	14W-17N	15
Navajo Tr.* N&S	16W-17N	15
Newburn Ct.* E&W	13W-18N	16
Noel Av. N&S	14W-16N	15
North Dennis Rd. E&W	15W-17N	15
North Green Dr. E&W	15W-17N	15
North Norman Ln. E&W	15W-17N	15
North Wayne Pl. E&W	15W-17N	15
Northbury Ln.* E&W	15W-18N	15
Northgate Pkwy. N&S	15W-18N,14W-18N	15
Nottingham Ct.* NW&SE	13W-17N	16
Nova Ct.* E&W	16W-18N	15
Oak Dr. NW&SE	16W-18N	15
Oak St. N&S	14W-17N	15
Oakmeadow Ct.* NW&SE	13W-18N	16
Oakmont Ln.* N&S	15W-18N	15
Oakwood Ln.* N&S	13W-18N	16
Oboe Ct.* E&W	16W-18N	15
Old McHenry Rd. NW&SE	16W-18N	15
Orrington Ct.* E&W	13W-17N	16
Osage Tr. NE&SW	16W-17N	15
Ottawa Tr.* N&S	16W-17N	15
Oxford Pl. N&S	15W-18N	15
Oxley Ct.* N&S	13W-17N	16
Pacific Ct. N&S	16W-17N	15
Paddock Dr. NS&EW	15W-16N	15
Palatine Rd. E&W	14W-16N	15
Palm Dr. E&W	16W-17N,15W-17N	15
Palwaukee Dr. N&S	14W-17N	15
Pam Ct.* N&S	16W-18N	15
Park Av. NW&SE	13W-17N	16
Partridge Ln.* E&W	13W-18N	16
Peace Dr. N&S	15W-16N	15
Pear Tree Ln. NE&SW	15W-16N	15
Pebble Dr. E&W	15W-18N	15
Pennsbury Ct.* NW&SE	13W-18N	16
Peterson Dr. N&S	14W-16N	15
Phaeton Dr. E&W	15W-18N	15
Pleasant Run Dr. E&W	15W-16N	15
Plumtree Ct.* N&S	15W-18N	15
Poplar Dr.* N&S	13W-16N	16
Portsmouth Pl.* E&W	15W-18N	15
Pueblo Tr.* N&S	16W-17N	15
Quad Ct.* N&S	16W-18N	15
Quail Hollow Dr. E&W	15W-18N	15
Queens Ct.* E&W	13W-18N	16
Quincy Ct.* E&W	16W-18N	15
Railroad Av. N&S	15W-17N	15
Ransom Ct.* E&W	14W-18N	15
Redwood Tr. N&S	15W-17N	15
Reef Ct. N&S	15W-16N	15
Regent Ln. N&S	14W-18N	15
Regent Ln.* N&S	13W-18N	16
Renee Ter. NS&EW	15W-17N	15
Ridgefield Ln. N&S	15W-16N	15
Robert Av. N&S	15W-16N	15
Rose Ln. E&W	15W-16N	15
Roth Ct.* E&W	16W-18N	15
Rt. 68 E&W	16W-18N,13W-18N	15
Rt. 83 N&S	16W-17N	15
Russettwood Ct. E&W	15W-17N	15
Rustic Dr.* E&W	13W-17N	16
Salvington Pl. E&W	16W-17N	15
Salvington Pl. E&W	16W-18N	15
Sandalwood Ln. E&W	14W-18N	15
Sandpebble Dr. N&S	15W-16N	15
Sandra Ln. E&W	15W-17N	15
Sandstone Dr. N&S	15W-16N	15
Sarah Ct. NW&SE	15W-17N	15
Sarasota Dr. E&W	15W-18N	15
Schoenbeck Rd. N&S	15W-17N	15
Scott St. N&S	16W-17N	15
Seton Ct. N&S	15W-17N	15
Seville Ct.* N&S	13W-17N	16
Shawn Ct. E&W	15W-16N	15
Shay Ct.* N&S	14W-18N	15
Sheldrake Dr. E&W	16W-17N	15
Shelly Ct. NW&SE	14W-17N	15
Sheppard Av. E&W	14W-17N	15
Sheridan Ct.* NE&SW	13W-17N	16
Shore Ct.* N&S	13W-17N	16
Shoshone Tr.* E&W	16W-17N	15
Silverwood Ct.* E&W	15W-18N	15
South Dennis Rd. E&W	15W-17N	15
South Dr. N&S	14W-15N	15
South Fletcher Dr. E&W	15W-17N	15
South Merle Ln. E&W	15W-17N	15
South Pine St. N&S	15W-17N	15
South Wayne Pl. E&W	15W-17N	15
Southbury Ln.* E&W	15W-18N	15
Springview Ct.* E&W	16W-17N	15
Spruce Dr. N&S	16W-18N	15
Spur Ct.* E&W	16W-18N	15
Stafford Ln.* N&S	13W-18N	16
Stone Pl. N&S	14W-17N	15
Stonehedge Ct.* E&W	13W-18N	16
Stratford Ct.* N&S	13W-18N	16
Strong Av. E&W	14W-18N	15
Summerhill Ln.* N&S	13W-18N	16
Sunrise Dr. E&W	14W-17N	15
Sunset Ln. N&S	14W-17N	15
Surf Ct. N&S	16W-17N	15
Surrey Rd.* N&S	14W-18N	15
Sutton Ct.* E&W	15W-18N	15
Sycamore Ln. N&S	15W-17N	15
Tahoe Cir. Dr. E&W	16W-17N	15
Tanglewood Dr. N&S	15W-16N	15
Teal Ln. E&W	16W-17N	15
Thelma Ct. E&W	15W-17N	15
Thorndale Ct.* E&W	15W-18N	15
Thornhill Ln.* NS&EW	15W-18N	15
Thyne Ct.* E&W	16W-18N	15
Tide Ct. N&S	15W-16N	15
Tilbury Ln. NW&SE	15W-18N	15
Tulip Ct.* N&S	13W-17N	16
Twilight Ln. E&W	15W-16N	15
Union Ct.* E&W	16W-18N	15
Vail Ct.* N&S	16W-18N	15
Valley Stream Dr. E&W	16W-18N,15W-18N	15
Virginia Pl. N&S	14W-17N	15
Vita Dr.* NE&W	13W-16N	16
Walnut Av. E&W	14W-17N,13W-17N	15
Waltz Dr. E&W	14W-17N	15
Warwick Ct.* E&W	13W-17N	16
Wayne Pl. N&S	15W-17N	15
Weeping Willow Dr. NE&SW	15W-16N	15
West Green Dr. N&S	15W-17N	15
West Lodge Ter.* N&S	16W-17N	15
West Pine St. E&W	15W-17N	15
West Wayne Pl. N&S	15W-17N	15
Westwood Ct.* NE&SW	16W-16N	15
Wetumka Tr.* N&S	16W-17N	15
Wheeling Av. N&S	14W-17N	15
Wheeling Rd. NS&EW	15W-16N,14W-17N	15
Whippletree Rd. N&S	15W-18N	15
White Hall Ct.* NW&SE	13W-18N	16
Widgeon Dr. N&S	16W-17N	15
Wildberry Ct.* N&S	15W-18N	15
Wille Av. N&S	14W-17N	15
Williamsburg Ct.* N&S	15W-18N	15
Willis Av. N&S	14W-16N	15
Willow Rd. E&W	14W-15N	15
Willow Tr. N&S	15W-17N	15
Willowbrook Dr. NW&SE	15W-16N	15
Wilshire Dr. E&W	15W-18N	15
Wolf Rd. N&S	14W-15N,13W-18N	15
Wood Creek Dr. E&W	14W-15N	15
Wood Duck Dr. E&W	16W-17N	15
Woodbury Ln.* E&W	14W-18N	15
Woodlake Rd. NS&EW	14W-15N	15
Woodland Dr. E&W	15W-18N	15
Woodmere Ln.* N&S	13W-18N	16
Wye Ct.* E&W	16W-18N	15
Wynn Ct.* N&S	16W-18N	15
Yorkshire Pl.* E&W	15W-18N	15
Zee Ct.* E&W	16W-18N	15
1st St. N&S	14W-18N	15
2nd St. N&S	14W-18N	15
3rd St. N&S	14W-18N	15
4th St. N&S	14W-18N	15

	Grid	Page
5th St. N&S	14W-18N	15
6th St. N&S	14W-18N	15
7th St. N&S	14W-18N	15
8th St. N&S	14W-18N	15
9th St. N&S	14W-18N	15
10th St. N&S	14W-18N	15
11th St. N&S	14W-18N	15
12th St. N&S	14W-18N	15

Wheeling Township

	Grid	Page
Ande Av. N&S	14W-16N	15
Anita Av. E&W	13W-13N	26
Betty Dr. N&S	16W-17N	15
Carol Av. N&S	14W-16N	15
Chestnut Av. N&S	18W-16N	14
Chicago Av. N&S	18W-15N	14
Cindy Ln. E&W	14W-16N	15
Darryl Dr. E&W	16W-17N	15
Dean Av. N&S	14W-16N	15
Debra Ln. E&W	14W-16N	15
Des Plaines River Rd. N&S	13W-13N,13W-14N	26
Dryden Av. N&S	17W-15N	14
Dun-lo Av. E&W	16W-17N	15
Edward St. E&W	15W-17N	15
Ellen Dr. N&S	16W-17N	15
Elm Ln. N&S	17W-17N	14
Fernandez Av. N&S	18W-16N	14
Foundry Rd. E&W	14W-13N	25
Graylynn Dr. N&S	13W-13N	26
Gregory St. E&W	13W-13N	26
Harold St. E&W	18W-16N	14
Hill St. E&W	13W-13N	26
Industrial Ln. E&W	13W-16N	16
Island Dr. E&W	13W-17N	16
Jane Av. E&W	17W-15N	14
LaSalle St. E&W	18W-16N	14
Lee St. N&S	13W-13N	26
Lillian Av. E&W	18W-15N,17W-15N	14
Lynwood Av. E&W	18W-15N	14
Maude Av. E&W	18W-15N	14
Morrison Av. E&W	13W-13N	26
Norman St. E&W	18W-16N	14
Park Pl. E&W	16W-17N	15
Phelps Av. N&S	16W-15N	15
Plant Rd. N&S	13W-16N	16
Plum Creek Dr. E&W	13W-17N	16
Ridge Av. N&S	18W-16N	14
Russell St. N&S	15W-17N	15
Salk Rd. N&S	16W-17N	15
South St. E&W	15W-17N	15
Stavros Rd. N&S	13W-16N	16
Stratford Rd. N&S	16W-15N	15
Vera Ln. E&W	15W-17N	15
Verde Av. N&S	18W-15N	14
Vista Rd. N&S	17W-17N	14
Wilson Ln. NS&EW	18W-15N	14
Woodland Dr. N&S	13W-13N	26
Yale Av. N&S	18W-15N	14
Yale Ct. E&W	18W-15N	14

Willow Springs

	Grid	Page
Abbot St. N&S	12W-9S	64
Archer Av. NE&SW	12W-9S,11W-9S	64
Arizona Trail N&S	14W-9S	64
Beech St. N&S	12W-9S	64
Beverly Ln. N&S	11W-9S	64
Blackstone Av. NW&SE	12W-9S	64
Buck Ln. N&S	13W-9S	64
Buckingham Ct. NW&SE	13W-9S	64
Candlelight Dr. NS&EW	13W-9S	64
Cedar St. NE&SW	12W-9S	64
Charleston Av. NW&SE	12W-9S	64
Chaucer Dr. NS&EW	13W-9S	64
Chelsea Ln. N&S	13W-9S	64
Cliff St. N&S	12W-9S	64
Colonel Av. NE&SW	13W-9S	64
Columbia Woods Dr. NW&SE	13W-9S	64
Crescent Ct. NS&EW	14W-9S	64
Crestview Dr. N&S	14W-9S	64
Crown Ct. N&S	13W-9S	64
Doogan Av. E&W	13W-9S	64
Dunbar St. N&S	13W-9S	64
Elm St. N&S	12W-9S	64
Fieldcrest Av. N&S	14W-9S	64
Flag Av. N&S	14W-9S	64
Forest Av. N&S	12W-9S	64
German Church Rd. E&W	13W-9S	64
Glenwood St. NW&SE	12W-9S	64
Grove St. NW&SE	12W-9S	64
Hess Av. N&S	13W-9S	64
High St. E&W	12W-9S	64
Hill St. NW&SE	12W-9S	64
Hilton St. NW&SE	12W-9S	64
Independence Dr. N&S	14W-9S	64
Interstate 294 NW&SE	11W-9S	64
Joyce Av. E&W	14W-9S	64
La Grange Rd. NW&SE	12W-8S	58
Lake Av. NE&SW	12W-9S	64
Liberty Grove Dr. NE&SW	13W-9S	64
Louis Dr. NW&SE	11W-9S	64
Maple Av. N&S	12W-9S	64
Mona St. NS&EW	12W-9S	64
Mound St. NW&SE	12W-9S	64
Nolton Av. NW&SE	12W-9S	64
Nueport Ct. NE&SW	14W-9S	64
Nueport Dr. N&S	14W-9S	64
Oakwood Av. NW&SE	12W-9S	64
Orlak Dr. N&S	14W-9S	64
Park St. NW&SE	12W-9S	64
Pearl St. NW&SE	12W-9S	64
Pine St. N&S	12W-9S	64
Pleasant View Av. N&S	14W-9S	64
Poston Rd. NW&SE	12W-9S	64
Prospect Av. E&W	12W-9S	64
Providence Ct. N&S	13W-9S	64
Rainford Ln. E&W	13W-9S	64
Ravine Av. East N&S	12W-9S	64
Ravine Av. West N&S	12W-9S	64
Regency Ct. N&S	13W-9S	64
Ridge Ct. N&S	12W-9S	64
Rosemere Ct. E&W	14W-9S	64
Rt. 171 NE&SW	13W-10S,12W-9S	64
Rust St. N&S	12W-9S	64
Rust Trail NE&SW	12W-9S,11W-9S	64
Scenic Dr. N&S	14W-9S	64
School Av. N&S	13W-9S	64
Sherwood St. NE&SW	12W-9S	64
Spring St. NE&SW	13W-9S	64
Spring St. NW&SE	12W-9S	64
Stradford Dr. E&W	13W-9S	64
Tec-Air Rd. NW&SE	12W-9S	64
Tower Rd. N&S	13W-9S	64
Tri-State Tollway E&W	12W-8S,11W-9S	58
Tudor Cir. N&S	13W-9S	64
U.S. Route 45 N&S	11W-9S	64
Union St. NW&SE	12W-9S	64
Ursula Dr. NE&SW	13W-9S	64
Vana Dr. N&S	11W-9S	64
Vinewood Av. NE&SW	12W-9S	64
Wadsworth Rd. N&S	13W-9S	64
Walnut St. E&W	12W-9S	64
Wentworth Av. NW&SE	13W-9S	64
Willow Springs Rd. N&S	13W-9S	64
Willow West Dr. N&S	13W-9S	64
Willows Edge Ct. NW&SE	13W-10S	64
79th St. E&W	13W-8S	58
83rd St. E&W	14W-9S	64
84th Pl. E&W	14W-9S,13W-9S	64
85th Pl. E&W	14W-9S	64
85th St. E&W	14W-9S,13W-9S	64
86th St. E&W	14W-9S,13W-9S	64
87th St. E&W	14W-9S,12W-9S	64

Wilmette

	Grid	Page
Alison Ln. E&W	5W-13N	28
Alpine Ln. N&S	4W-13N	29
Amhurst Ln. N&S	4W-14N	29
Appletree Ln. N&S	5W-13N	28
Ashland Av. E&W	3W-14N	29
Ashland Ln. E&W	4W-14N	29
Avondale Ct. E&W	4W-13N	29
Avondale Ln. E&W	4W-13N	29
Beechwood Av. E&W	5W-14N	28
Beverly Dr. E&W	5W-13N	28
Big Tree Ln. E&W	6W-13N	28
Birchwood Av. E&W	5W-13N,4W-13N	28
Birchwood Ct. E&W	5W-13N	28
Birchwood Ln. E&W	5W-13N	28
Blackhawk Rd. E&W	5W-14N	28
Briar Dr. N&S	6W-13N	28
Broadway N&S	3W-13N	29
Brookside Dr. N&S	5W-13N	28
Bunker Ln. N&S	7W-14N	28
Cambridge Ln. N&S	4W-14N	29
Canterbury Ct. E&W	2W-13N	29
Cardinal Ln. E&W	5W-13N	28
Carriage Way N&S	5W-12N	28
Catalpa Pl. N&S	3W-13N	29
Cedar Ln. N&S	4W-13N	29
Central Av. E&W	6W-13N,3W-13N	28
Central Park Av. N&S	4W-13N	29
Cherokee Rd. N&S	5W-14N	28
Chestnut Av. E&W	4W-14N,3W-14N	29
Chilton Ln. N&S	5W-13N	28
Chippewa N&S	5W-14N	28
Cleveland St. N&S	5W-14N	28
Colgate St. N&S	5W-14N	28
Columbus Av. N&S	4W-14N	29
Concord Ln. E&W	5W-13N	28
Cornell St. N&S	4W-14N	29
Country Ln. E&W	6W-14N	28
Cove Ln. N&S	5W-13N	28
Crabtree Ln. E&W	5W-13N	28
Cranston Ct. E&W	6W-14N	28
Crescent Pl. E&W	3W-13N	29
Crestview Ln. NE&SW	4W-13N	29
Dartmouth St. N&S	5W-14N	28
Dupee Pl. N&S	2W-13N	29
Edgewood Ct. NW&SE	6W-14N	28
Electric Pl. NW&SE	3W-13N	29
Elmwood Av. E&W	6W-14N,3W-14N	28
Fairway Dr. N&S	7W-14N	28
Forest Av. E&W	6W-14N,3W-14N	28
Garrison Av. N&S	2W-13N	29
Girard Av. N&S	2W-13N	29
Glenview Rd. E&W	4W-13N	29
Golf Ter. N&S	2W-13N	29
Grant St. N&S	5W-14N	28
Greenbay Rd. NW&SE	4W-14N	29
Greenleaf Av. E&W	6W-13N,3W-13N	28
Greenwood Av. E&W	6W-14N,3W-14N	28
Gregory Av. E&W	5W-13N,3W-13N	28
Hartzell St. E&W	5W-13N	28
Harvard Ln. N&S	4W-13N,4W-14N	29
Harvard St. N&S	4W-13N	29
Hawthorne Ln. E&W	5W-13N	28
Heather Ln. N&S	6W-13N	28
Hibbard Rd. N&S	6W-13N	28
Highcrest E&W	4W-13N	29
Highland Av. E&W	5W-13N,4W-13N	28
Hill Ln. E&W	6W-13N	28
Hill St. E&W	6W-13N	28
Hollywood Ct. NS&EW	4W-13N	29
Hunter Ct. E&W	5W-13N	28
Hunter Rd. N&S	5W-13N	28
Illinois Rd. NW&SE	5W-14N	28
Indian Hill Rd. E&W	5W-14N	28
Indianwood Rd. E&W	5W-14N	28
Iroquois Rd. E&W	5W-14N,4W-14N	28
Karey Ct. N&S	5W-13N	28
Kenilworth Av. E&W	5W-14N,4W-14N	28
Kilpatrick Av. N&S	5W-13N	28
Kin Ct. N&S	5W-13N	28
Knoll Ln. E&W	4W-14N	29
Knox N&S	5W-13N,5W-14N	28
Koerper Ct. N&S	5W-13N	28
LaCrosse Ct. N&S	6W-13N	28
LaPorte Av. N&S	6W-13N	28
Lacrosse Av. N&S	6W-13N	28
Lake Av. E&W	5W-14N,3W-14N	28
Lamon Av. N&S	6W-13N	28
Laramie Av. N&S	6W-13N,6W-14N	28
Laurel Av. E&W	3W-13N,2W-13N	29
Lavergne Av. N&S	6W-13N	28
Lawler Av. N&S	6W-13N	28
Lawndale Av. N&S	4W-13N	29
LeClaire Av. N&S	6W-13N	28
Leamington Av. N&S	6W-13N	28
Leydon Ln. N&S	5W-13N	28
Lilac Ln. E&W	6W-13N	28
Lincoln Ln. E&W	5W-13N	28
Linden Av. E&W	3W-13N,2W-13N	29
Lockerbie Ln. N&S	6W-13N	28
Locust Rd. N&S	5W-13N	28
Manor Dr. N&S	6W-14N	28
Maple Av. E&W	4W-13N,3W-13N	29
Marian Ln. E&W	5W-13N	28
Meadow Dr., East N&S	5W-13N	28
Meadow Dr., South N&S	5W-13N	28
Meadow Dr., West N&S	5W-13N	28
Meadows Dr., North NS&EW	5W-13N	28
Melrose Ct. E&W	6W-14N	28
Miami Rd. N&S	5W-14N	28
Michigan Av. NW&SE	3W-14N	29
Middlebury Ln. N&S	5W-14N	28
Millbrook Ln. N&S	6W-13N	28
Mohawk Rd. NS&EW	5W-14N	28
New Trier Ct. N&S	6W-14N	28
Nina Av. E&W	6W-13N	28
North Branch Rd. N&S	6W-14N	28
Oak Cir. Dr. N&S	3W-13N	29
Oakwood Av. E&W	3W-13N	29
Old Glenview Rd. NE&SW	5W-13N	28
Osage Ln. N&S	5W-14N	28
Ottawa N&S	5W-14N	28
Oxford Ln. N&S	4W-14N	29
Park Av. NW&SE	3W-13N	29
Parkview Ct. NW&SE	4W-13N	29
Pawnee Rd. N&S	5W-14N	28
Pin Oak Dr. E&W	4W-13N	29
Pine Manor Dr. N&S	4W-13N	29
Pine St. N&S	5W-13N	28
Pinecrest Ln. N&S	4W-13N	29
Pioneer Ln. E&W	4W-13N	29
Plum Tree Ln. N&S	5W-12N	28
Pomona Ln. E&W	5W-13N	28
Pontiac Rd. NW&SE	5W-14N	28
Poplar Dr. NW&SE	3W-13N	29
Prairie Av. NW&SE	4W-13N	29
Princeton St. N&S	5W-14N	28
Red Bud Ln. N&S	5W-13N	28
Richmond Ln. NW&SE	4W-14N	29
Ridge Rd. N&S	4W-13N	29
Riverside Dr. E&W	6W-14N	28
Romona Ct. E&W	5W-14N	28
Romona Rd. N&S	5W-13N	28
Sandy Ln. N&S	4W-13N	29
Schiller Av. E&W	4W-13N	29
Seger St. E&W	5W-13N	28
Seminole Rd. N&S	5W-14N	28
Seneca Rd. N&S	5W-14N	28
Shabona Ln. N&S	5W-14N	28
Sheridan Rd. NW&SE	3W-14N	29
Skokie Ct. E&W	5W-13N	28
Skokie Rd. NW&SE	6W-13N	28
Spencer Av. E&W	4W-13N	29
Sprucewood Ln. NW&SE	6W-14N	28
Sprucewood Rd. E&W	6W-14N	28
Sterling Ln. N&S	4W-13N	29
Sunset Dr. N&S	5W-13N	28
Temple Ln. E&W	6W-13N	28
Thelin Ct. N&S	5W-13N	28
Thornwood Av. E&W	6W-14N,5W-14N	28
Timber Ln. N&S	4W-14N	29
Valley View Ct. E&W	5W-13N	28
Valley View Dr. N&S	5W-13N	28
Vine Ct. E&W	5W-13N	28
Vine St. N&S	5W-13N	28
Virginia Ln. N&S	5W-14N	28
Vista Ct. E&W	5W-13N	28
Vista Dr. N&S	5W-13N	28
Walden Ln. E&W	6W-14N	28
Walnut Av. E&W	6W-14N,4W-14N	28
Washington Av. E&W	6W-13N,3W-13N	28
Washington Ct. E&W	3W-13N	29
Westerfield Dr. E&W	3W-14N	29
Westmoreland Dr. N&S	5W-13N	28
Westwood Ln. NS&EW	5W-13N	28
Wilmette Av. E&W	5W-13N,3W-13N	28
Wilshire Dr., East N&S	5W-13N	28
Wilshire Dr., West N&S	5W-13N	28
Wood Ct. N&S	3W-13N	29
Woodbine Av. N&S	3W-13N	29
Yale St. N&S	4W-14N	29
3rd St. N&S	2W-13N	29
4th St. N&S	3W-13N	29
5th St. N&S	3W-13N	29
6th St. N&S	3W-13N	29
7th St. N&S	3W-13N,3W-14N	29
8th St. N&S	3W-13N,3W-14N	29
9th St. N&S	3W-13N,3W-14N	29
10th St. N&S	3W-13N,3W-14N	29
11th St. N&S	3W-13N,3W-14N	29
12th St. N&S	3W-13N	29
13th St. N&S	3W-13N,3W-14N	29
14th St. N&S	4W-13N	29
15th St. N&S	4W-13N,4W-14N	29
16th St. N&S	4W-13N,4W-14N	29
17th St. N&S	4W-13N,4W-14N	29
18th St. N&S	4W-14N	29
20th St. N&S	4W-13N	29
21st St. N&S	4W-14N	29
21st St. N&S	4W-13N,4W-14N	29
25th St. E&W	5W-14N	28
36th Pl. N&S	6W-14N	28

Winnetka

	Grid	Page
Alles Rd. E&W	5W-15N	19
Appletree Rd. N&S	5W-15N	19
Arbor Vitae N&S	4W-15N	19
Ardsley Rd. N&S	5W-16N	19
Asbury Av. E&W	6W-16N	18
Asbury Ct. N&S	6W-16N	18
Asbury Ln. N&S	6W-16N	18
Ash St. E&W	5W-15N,4W-15N	19
Auburn Av. N&S	5W-15N	19
Auburn Rd. N&S	5W-16N	19
Bell Ln. N&S	6W-16N	18
Berkeley Av. N&S	5W-15N	19
Bertling Ln. N&S	4W-15N	19
Beverly Av. E&W	5W-15N	19
Birch St. N&S	5W-15N	19
Blackthorn Rd. N&S	5W-16N	19
Boal Pkwy. N&S	6W-16N	18
Briar Ln. NE&SW	4W-15N	19
Broadmeadow Rd. E&W	5W-15N	19
Bryant Av. N&S	5W-16N	19
Burr Av. N&S	5W-16N	19
Cedar St. N&S	4W-15N	19
Chatfield Rd. E&W	5W-16N	19
Cherry St. E&W	5W-15N,4W-15N	19
Chestnut Ct. E&W	5W-15N	19
Chestnut St. N&S	5W-15N	19
Church Rd. N&S	4W-15N	19
Compton St. E&W	6W-15N	19
DeWindt Rd. E&W	5W-15N	19
Dinsmore Rd. E&W	5W-16N	19
Edgewood Ln. E&W	6W-16N	18
Elder Ln. E&W	4W-15N	19
Eldorado St. NE&SW	5W-16N	19
Elm St. E&W	5W-15N	19
Essex Rd. N&S	4W-15N	19
Euclid N&S	5W-15N	19
Fairview Av. N&S	4W-15N	19
Fisher Cres. N&S	5W-16N	19
Fisher Ln. E&W	5W-16N	19
Forest Glen Dr., East N&S	5W-16N	19
Forest Glen Dr., North E&W	6W-16N,5W-16N	18
Forest Glen Dr., South E&W	6W-16N,5W-16N	18
Forest Glen Dr., West N&S	6W-16N	18
Forest St. N&S	4W-15N	19
Fox Ln. N&S	5W-14N	28
Foxdale Av. NW&SE	5W-16N	19
Fuller Ln. NW&SE	4W-15N	19
Gage St. E&W	5W-16N	19
Garland Av. E&W	4W-15N	19
Glen Oak Dr. E&W	5W-16N	19
Glendale Av. N&S	5W-15N	19
Golf Rd. N&S	5W-14N	28
Gordon Ter. N&S	5W-16N	19
Green Bay Rd. NW&SE	5W-16N	19
Hackberry Ln. E&W	6W-16N	18
Hamptondale Av. NE&SW	5W-16N	19
Hawthorn Ln. E&W	4W-15N	19
Hazel Ln. E&W	6W-16N	18
Heather Ln. N&S	6W-16N	18
Hibbard Rd. N&S	6W-15N	18
Hickory Ln. E&W	6W-16N	18
Higginson Ln. N&S	5W-15N	19
High St. N&S	4W-14N	29
Hill Rd. E&W	5W-15N,4W-15N	19
Hill Ter. E&W	4W-15N	19
Holly Ln. E&W	6W-16N	18
Hoyt Ln. NW&SE	4W-15N	19
Hubbard Pl. E&W	5W-16N	19
Humboldt Av. E&W	5W-16N	19
Indian Hill Rd. NE&SW	5W-14N	28
Indian Mill Rd. E&W	5W-15N	19
Indian Way Rd. NS&EW	5W-14N	28
Kent E&W	5W-16N	19
Lake Ln. E&W	4W-15N	19
Lamson Dr. E&W	5W-16N	19
Laurel Av. E&W	5W-16N	19
Lincoln Av. NW&SE	6W-15N,5W-15N	18
Linden St. N&S	5W-15N	19
Lindenwood Dr. E&W	5W-15N	19
Lloyd Pl. E&W	5W-16N	19
Locust Rd. N&S	5W-15N	19
Locust St. N&S	5W-15N,5W-16N	19
Maple St. N&S	4W-15N	19
Meadow Ln. E&W	5W-15N	19
Meadow Rd. E&W	4W-14N	29
Merrill St. E&W	5W-16N	19
Mt. Pleasant Ln. E&W	5W-15N	19
Myrtle St. N&S	4W-15N	19
Oak St. E&W	5W-15N,4W-15N	19
Oak St. E&W	6W-15N,4W-15N	19
Oakley Av. E&W	5W-16N	19
Old Green Bay Rd. N&S	5W-16N	19
Orchard Ln. NS&EW	4W-15N	19
Park Ln. N&S	4W-16N	19
Pelham Rd. (Pvt.) E&W	5W-16N	19
Pine Ln. (Pvt.) N&S	5W-16N	19
Pine St. E&W	6W-16N,4W-16N	19
Pine St. E&W	5W-16N,4W-16N	19
Pine Tree Ln. NS&EW	6W-16N	18

	Grid	Page
Plum Tree Ln. E&W	6W-16N	18
Poplar St. N&S	4W-15N	19
Private Rd., South E&W		
	5W-16N	19
Prospect Av. NW&SE		
	5W-16N	19
Provident Av. N&S	5W-16N	19
Ravine Rd. E&W	5W-16N	19
Ridge Av. N&S	5W-16N	19
Rosewood Av. N&S		
	5W-15N,5W-16N	19
Scott Av. E&W	6W-16N	18
Sheridan Rd. NW&SE		
	4W-16N	19
Smith St. N&S	6W-15N	18
Spring Ln. N&S	5W-15N	19
Spruce Ln. NE&SW	4W-15N	19
Spruce St. E&W		
	6W-15N,4W-15N	19
Spruce St. E&W		
	5W-15N,4W-15N	19
Starr Rd. E&W	5W-16N	19
Stockton St. N&S	6W-15N	19
Sumac Ln. N&S	6W-16N	18
Summit St. NE&SW	5W-16N	19
Sunset Rd. E&W		
	5W-15N,4W-15N	19
Sunview Ln. E&W	6W-16N	18
Swinton St. E&W	6W-15N	19
Taylorsport Ln. N&S		
	5W-17N	19
Thorn Tree Ln. (Pvt.) N&S		
	5W-15N	19
Tower Manor N&S	5W-16N	19
Tower Rd. E&W	6W-16N	18
Trapp Ln. E&W	6W-16N	18
Vernon St. N&S	6W-16N	18
Vine St. E&W	5W-16N	19
Walden Rd. N&S	5W-16N	19
Walnut St. N&S	4W-15N	19
Warwick Rd. E&W	4W-14N	29
Westmoor Rd. E&W	5W-16N	19
Westmoor Trail E&W		
	6W-16N	18
Westview Rd. N&S	6W-16N	18
White Oak Ln. N&S	5W-15N	19
Whitebridge Hill St. E&W		
	5W-16N	19
Whitebridge Ln. NW&SE		
	5W-17N	19
Willow Rd. E&W		
	5W-15N,4W-15N	19
Wilson St. NW&SE	4W-15N	19
Winnetka Rd. E&W	4W-15N	19
Woodland Av. N&S	4W-15N	19

Worth

	Grid	Page
Autullo Dr. E&W	9W-12S	65
Beloit Av. N&S		
	9W-13S,9W-12S	73
Columbus Dr. NE&SW		
	9W-13S	73
Crandall Av. E&W	8W-12S	65
Depot St. N&S		
	8W-13S,8W-12S	73
Harlem Av. N&S		
	9W-13S,9W-12S	73
Highland Rd. N&S	8W-12S	65
Home Av. NW&SE	8W-13S	73
Lloyd Dr. N&S	8W-12S	65
Lode Dr. NE&SW	8W-12S	65
Nagle Av. N&S		
	8W-13S,8W-12S	73
Nashville Av. N&S		
	8W-13S,8W-12S	73
Natchez Av. N&S		
	8W-13S,8W-12S	73
Natoma Av. N&S	8W-13S	73
Neenah Av. N&S		
	8W-13S,8W-12S	73
New England Av. N&S		
	8W-12S	65
New England Av. N&S		
	8W-13S,8W-12S	73
Nordica Av. N&S	8W-12S	65
Normandy Av. N&S		
	8W-13S,8W-12S	73
Oak Park Av. N&S		
	8W-13S,8W-12S	73
Oak Tree Dr. N&S	9W-12S	65
Oconto Ct. N&S	9W-12S	65
Octavia Av. N&S		
	9W-13S,9W-12S	73
Odell Ct. N&S	9W-12S	65
Oketo Av. N&S		
	9W-13S,9S-12S	73
Pacific Dr. NE&SW	9W-13S	73
Plahm Ct. NE&SW	8W-12S	65
Preller Av. N&S	8W-13S	73
Ridgeland Av. N&S	8W-12S	65
Rt. 43 N&S	8W-12S	65
Rutherford Av. N&S	8W-12S	65
Southwest Hwy. NE&SW		
	8W-12S	65
Tri-State Tollway SE&NW		
	8W-13S,8W-12S	73
Wabash Dr. N&S	8W-12S	65
Wood Av. E&W	8W-13S	73
Worth Av. N&S		
	8W-13S,8W-12S	73
71st Ct. N&S	8W-12S	65
105th Pl. E&W	8W-12S	65
106th Pl. E&W	8W-12S	65
107th Pl. E&W	8W-12S	65
107th St. E&W		
	9W-12S,8W-12S	65
108th Pl. E&W	9W-12S	65

	Grid	Page
108th St. E&W		
	9W-12S,8W-12S	65
109th Pl. E&W		
	9W-12S,8W-12S	65
109th St. E&W		
	9W-12S,8W-12S	65
110th Pl. E&W	9W-12S	65
110th St. E&W		
	9W-12S,8W-12S	65
111th Pl. E&W		
	9W-12S,8W-12S	65
111th St. E&W		
	9W-12S,8W-12S	65
112th Pl. E&W		
	9W-13S,8W-13S	73
112th St. E&W		
	9W-13S,8W-13S	73
113th Pl. E&W		
	9W-13S,8W-13S	73
113th St. E&W		
	9W-13S,8W-13S	73
114th Pl. E&W		
	9W-13S,8W-13S	73
114th St. E&W		
	9W-13S,8W-13S	73
115th Pl. E&W	8W-13S	73
115th St. E&W		
	9W-13S,8W-13S	73
116th St. E&W	8W-13S	73

Worth Township

	Grid	Page
Austin Av. N&S		
	7W-15S,7W-14S	74
Brayton St. N&S	3W-15S	75
Francisco Av. N&S	3W-15S	75
Major Av. N&S	7W-14S	74
Mason Av. N&S	7W-15S	74
Massasoit Av. N&S	7W-14S	74
McVicker Av. N&S		
	7W-15S,7W-14S	74
Meade Av. N&S		
	7W-15S,7W-14S	74
Melvina Av. N&S	7W-15S	74
Moblie Av. N&S	7W-15S	74
Monitor Av. N&S	7W-15S	74
Moody Av. N&S		
	7W-15S,7W-14S	74
Parkside Av. N&S	7W-14S	74
Ridgeland Av. N&S		
	8W-17S,8W-15S	79
Sacramento Av. N&S		
	3W-15S	75
Sheridan Av. N&S	3W-15S	75
123rd St. E&W	7W-14S	74
124th St. E&W	7W-14S	74
125th St. E&W	7W-14S	74
127th Pl. E&W	7W-15S	74
127th St. E&W	7W-15S	74
128th Pl. E&W	7W-15S	74
128th St. E&W	7W-15S	74
129th Pl. E&W	7W-15S	74
129th St. E&W	7W-15S	74
130th Pl. E&W	7W-15S	74
131st Pl. E&W	3W-15S	75
131st St. E&W	7W-15S	74
132nd St. E&W	3W-15S	75
135th St. E&W	7W-15S	74

RAND McNALLY

DuPage County

StreetFinder®

Contents

page	
A	Using the StreetFinder
B	PageFinder Map
C	Vicinity Map
1	Contents
2	Quick Reference Guide
4	StreetFinder Maps
37	StreetFinder Index

Photo credit: Amoco Building, DuPage County Development Department

Reproducing or recording maps, tables, or any other material in this publication by photocopying, by electronic storage and retrieval, or by any other means is prohibited.

PageFinder™ Map patent pending.

Information included in this publication has been checked for accuracy prior to publication. Since changes do occur, the publisher cannot be responsible for any variations from the information printed.

DuPage County Municipal Offices

	Location	Page
Addison Village Hall	18W-3N	14
130 Army Trail Road; 543-4100		
Bensenville Village Hall	16W-5N	19
700 W Irving Park Rd; 766-8200		
Bloomingdale Village Hall	23W-5N	7
201 S Bloomingdale Rd; 893-7000		
Carol Stream Village Hall	25W-2N	12
500 N Gary Ave; 665-7050		
Clarendon Hills Village Hall	16W-5S	27
1 N Prospect Ave; 323-3500		
Darien City Hall	18W-8S	32,33
1702 Plainfield Rd; 852-5000		
Downers Grove Village Hall	19W-5S	26
801 Burlington Ave; 964-0300		
Elmhurst City Hall	15W-1N	15
119 Schiller; 530-3000		
Glendale Heights Village Hall	23W-3N	13
300 E Fullerton Ave; 260-6000		
Glen Ellyn Village Hall	22W-0N	13
535 Duane; 469-5000		
Hinsdale Village Hall	15W-4S	27
19 E Chicago Ave; 789-7000		
Itasca Village Hall	19W-6N	8
100 N Walnut; 773-0835		
Lisle Village Hall	23W-5S	24,25
1040 Burlington Avenue; 968-1200		
Lombard Civic Center	19W-0S	20
255 E Wilson Ave; 620-5700		
Naperville City Hall	27W-6S	23
175 W Jackson Ave; 420-6111		
Oak Brook Village Hall	16W-2S	21
1200 Oak Brook Rd; 624-2220		
Roselle Village Hall	23W-7N	6,7
31 S Prospect Ave; 980-2000		
Villa Park Village Hall	18W-0N	14,15
20 S Ardmore Ave; 834-8500		
Warrenville City Hall	28W-3S	17
28 W 630 Stafford Pl; 393-9427		
Wayne Village Hall	32W-5N	4
5 N 430 Railroad; 584-3031		
Westmont Village Hall	18W-5S	26
31 W Quincy; 968-0560		
West Chicago Village Hall	29W-0N	10,11
475 Main St; 231-3322		
Wheaton City Hall	24W-0S	18
214 N Wheaton; 260-2000		
Winfield Village Hall	27W-0S	17
27 W 465 Jewell Rd; 665-1778		
Wood Dale City Hall	18W-6N	8
404 N Wood Dale Rd; 766-4900		
Woodridge Village Hall	20W-8S	32
1900 W 75th; 852-7000		

Cemeteries

	Location	Page
Allerton Ridge	20W-2S	19
Arlington Cem	15W-2N	15
Assumption Cem	28W-1S	17
Blodgetts Cem	21W-5S	25
Bloomingdale Cem	22W-5N	7
Bronswood Cem	16W-3S	21
Chapel Hills Garden Cem	17W-1S	21
Clarendon Hills Cem	18W-7S	26
Elm Lawn Cem	15W-2N	15
Forest Hill Cem	22W-1N	13
Fullersburg Cem	15W-4S	27
Glen Oak Cem	30W-1N	10
Hinsdale Animal Cem	17W-7S	27
Illinois Pet Cem	27W-6N	5
Mt Emblem Cem	15W-3N	15
Naperville Cem	27W-7S	23
Oakwood Cem	29W-1N	10
Pet Haven Cem	20W-7N	7,8
Resthaven Cem	15W-5N	9
Roselle Cem	23W-7N	6,7
SS Peter and Paul Cem	26W-6S	24
St Bernard Cem	21W-6S	25
St Johannes Cem	15W-6N	9
St John Cem	15W-3S	21
St Johns Cem	17W-7S	27
St Marys Cem	16W-1N	15
St Michaels Cem	25W-1S	18
St Stevens Cem	24W-1N	12
Wayne Center Cem	29W-4N	5
Wheaton Cem	25W-1S	18
York Twp Cem	15W-1S	21
Zion Cem	15W-3S	21

Colleges & Universities

	Location	Page
Col of DuPage	22W-2S	19
DeVry Inst of Tech	20W-1S	19,20
Illinis Benedictine Col	24W-2N	12
Midwest Col of Engineering	20W-0N	13,14
National Col of Education	20W-2S	20
National Col of Chiropractic	20W-0S	19
North Central Col	26W-6S	24
Northern Ill Univ Col of Law	21W-1S	19
Wheaton Col	24W-0S	18

Forest Preserves

	Location	Page
Belleau Woods FP	26W-1S	18
Churchill Woods FP	21W-0N	13
Cricket Creek FP	17W-3N	15
Danada FP	25W-3S	18
East Branch FP	21,22W-2,3N	13
Egerman Woods	24W-7S	30
Fischer Woods FP	17W-4N	8,9

	Location	Page
Fullersburg FP	15W-3S	21
Fullerton FP	19W-2N	14
Green Valley FP	23W-9S	31
Herrick Lake FP	26W-3S	17,18
Hidden Lake FP	17W-3N	15
Mallard Lake FP	26W-5, 6N	6
Maple Grove FP	20W-5S	26
McDowell Grove FP	28W-4S	23
Meachem Grove FP	23W-6N	6,7
Pratts Wayne Woods FP	31W-6N	4
Roy C. Blackwell FP	28,29W-2,3S	17
Salt Creek FP	19W-5S	8
Salt Creek Park FP	18W-5N	8
Spring Brook FP	28W-8S	29
Timber Ridge FP	28W-1, 2N	11
Warrenville Grove FP	28W-3S	17
Waterfall Glen FP	18W-10,12S	34
Wayne Grove FP	29W-7N	5
West Branch Reservoir FP	28W-4,6N	5
West Chicago Prairie FP	30,31W-1N	10
West DuPage FP	28,29W-0S	17
Willowbrook FP	22W-2S	19
Winfield Mounds FP	27N-0N	11
Wood Dale Grove FP	17W-4N	8,9
York Woods FP	15W-1S	21

Golf Courses & Country Clubs

	Location	Page
Addison CC	19W-3N	14
Apple Orchard GC	29W-7N	5
Arrowhead GC	26W-3S	18
Black Horse GC	18W-7S	26
Brookwood GC	18W-5N	8
Butler International Sports Core	16W-2S	21
Butterfield CC	18W-2S	21
Carrieage Greens GC	19W-10S	32
Chicago GC	25W-1S	18
Country Lakes GC	30W-5S	22
Cress Creek CC	28W-5S	23
Downers Grove GC	21W-4S	25
Elmhurst GC	18W-4N	8
Glendale G&CC	22W-5N	7
Glen Briar GC	21W-2S	19
Glen Ellyn Village Links	22W-1S	19
Glen Oak CC	21W-0N	13,14
Hinsdale GC	17W-4S	27
Indian Lakes CC	24W-5N	6
Itasca CC	19W-6N	8
Ken Loch GL	20W-1S	20
Medinah CC	21W-6N	7
Naperville CC	25W-6S	24
Nordic Hills CC	21W-5N	7
Old Wayne GC	29W-3N	10,11

	Location	Page
Pheasant Run GC	32W-3N	10
River Forest CC	15W-4N	9
Ruth Lake CC	16W-6S	27
Salt Creek CC	18W-7N	8
St Andrews GC	29,30W-3N	10,11
Springbrook GC	29W-9S	29
Sugar Creek GC	17W-0S	20,21
The Village Green	23W-1S	18,19
Twin Lakes GC	18W-6S	26
Village Greens of Woodbridge CC	20W-8S	32
White Eagle GC	30W-9S	28
White Pines GC	16W-4N	9
Woodbridge GC	23W-7S	24,25

Shopping Centers

	Location	Page
Bloomingdale Plaza	23W-4N	6, 7
Brush Hill Mall	28W-6S	23
Courtyard SC	18W-0S	20
Cress Creek SC	27W-5S	23
Downers Park SC	19W-8S	29
Eastgate SC	18W-0S	20
Finley Sq Mall	20W-2S	19
Fox Run Square SC	25W-8S	30
Fox Valley SC	30W-7S	28
Geneva Plaza SC	24W-1N	12
Gina's Plaza SC	24W-1N	12
Georgetown Sq SC	18W-6N	18
Glendale Heights SC	22W-1, 2N	13
Greenbrook Shopping Plaza	26W-7N	6
Greenknolls SC	19W-6S	26
Green Meadow SC	18W-3N	14
Hawthorne Square SC	26W-9S	30
Hobson West Commons SC	27W-8S	29
Lombard Pines SC	20W-0S	20
Main Plaza SC	24W-0N	12
Market Meadows SC	25W-8S	30
Meadowbrook SC	21W-7S	25
Naper Plaza SC	26W-8S	30
Naper West Plaza SC	29W-7S	23
North Park Shopping Mall	18W-2N	14
Oakbrook SC	16W-1S	21
Ogden Mall SC	25W-5S	24
Ogden Place SC	26W-5S	24
Plaza Westlake	23W-4N	6, 7
Springbrook SC	23W-5N	7
Stratford Square SC	25W-5N	6
Twin Ctrs SC	26W-5S	24
Villa DuPage SC	18W-1N	14
Villa Oaks SC	18W-0S	20
Woodgrove Festival SC	20W-8S	32
Yorktown SC	19W-2S	20

SEE COOK COUNTY PAGES 30-31

SEE KANE COUNTY PAGES 26-27

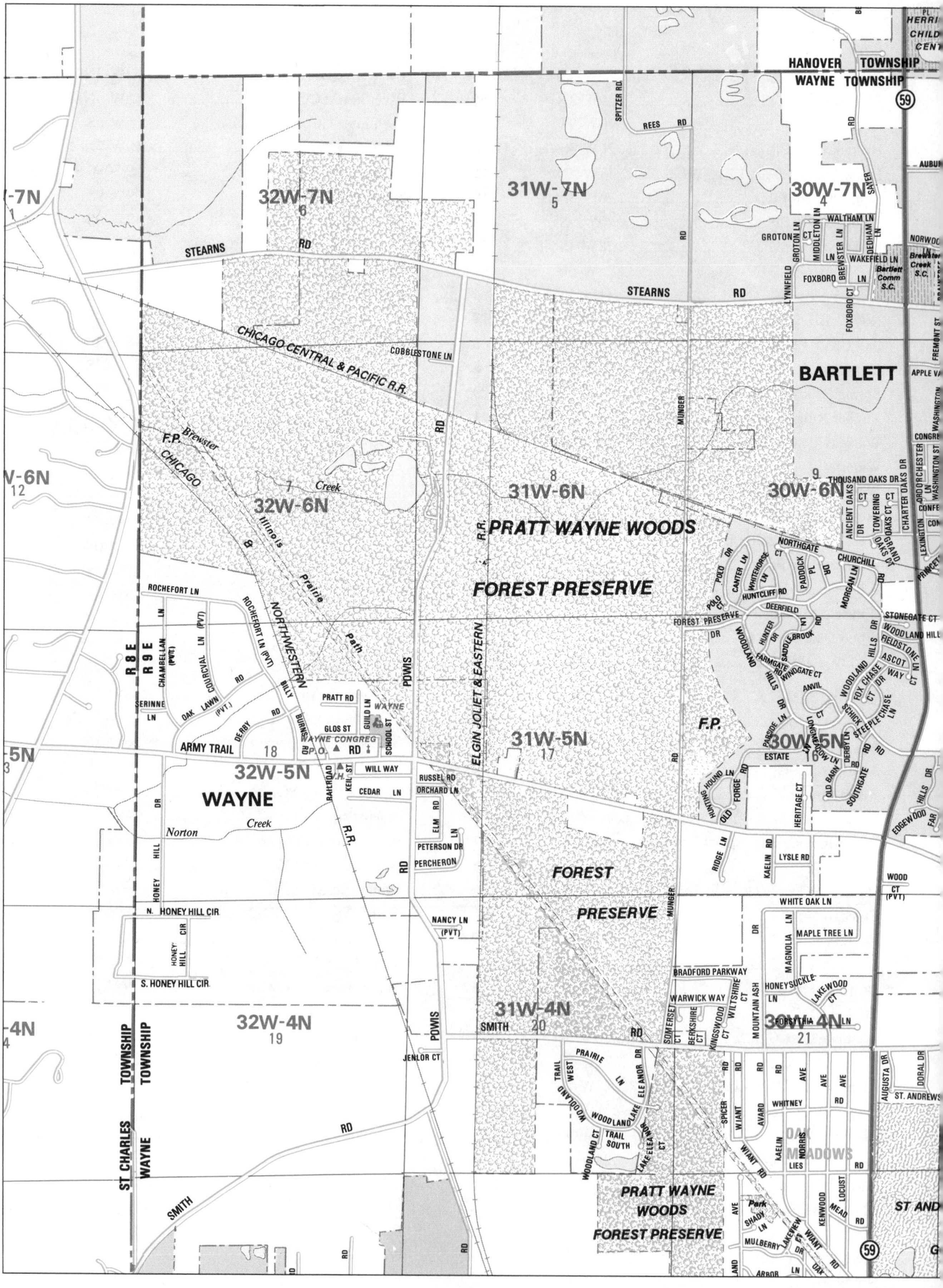

HANOVER TOWNSHIP
WAYNE TOWNSHIP

59

-7N

32W-7N
6

31W-7N
5

30W-7N
4

STEARNS   RD

SPITZER RD
REES   RD
RD

GROTON LN
MIDDLETON LN
BREWSTER LN
WALTHAM LN
WAKEFIELD LN
BEDHAM LN
NORWOOD
Brewster Creek S.C.
Bartlett Comm S.C.

LYNNFIELD
FOXBORO
BREWSTER CT
FOXBORO CT

FREMONT ST
APPLE VA

CHICAGO CENTRAL & PACIFIC R.R.

COBBLESTONE LN

STEARNS   RD

BARTLETT

CONGRE

-6N
12

F.P.
CHICAGO
Brewster
Creek
7
Illinois
Prairie
Path

32W-6N

31W-6N
8

R.R.

PRATT WAYNE WOODS

FOREST PRESERVE

MUNGER

30W-6N
9

THOUSAND OAKS DR
ANCIENT OAKS
CHARTER OAKS DR
DORCHESTER
WASHINGTON
LEXINGTON
CONFE

NORTHGATE
POLO CT
POLO DR
CANTER LN
WHITEHORSE LN
HUNTCLIFF RD
DEERFIELD
HUNTER
SADDLE BROOK
MORGAN LN
PADDOCK
CHURCHILL
FOREST PRESERVE DR
STONEGATE CT
WOODLAND HILLS
WOODLAND HILLS DR
ASCOT CT
FIELDSTONE

ROCHEFORT LN
CHAMBELLAN (PVT)
COURCVAL LN (PVT)
OAK LAWN (PVT)
SERINNE LN
ROCHEFORT LN (PVT)
DERBY RD
BILLY BURNS RD
NORTHWESTERN

WINDGATE CT
WOODLAND HILLS
FOX CHASE CT
FARMGATE CT
ANVIL
SCHICK
STEEPLE CHASE LN

PRATT RD
WAYNE
GUILD LN
GLOS ST
SCHOOL ST
WAYNE CONGREG P.O.
ARMY TRAIL   RD
18

F.P.
ESTATE
30W-5N

32W-5N

WAYNE

KEHL ST
BAER ROAD
WILL WAY
CEDAR   LN
RUSSEL RD
ORCHARD LN
ELM RD
PETERSON DR
PERCHERON

Norton   Creek

R.R.

ELGIN JOLIET & EASTERN

31W-5N
17

RD

HUNTING HOUND LN
OLD FORGE
OLD BARN
DERBY CT
SOUTHGATE
HERITAGE CT

HONEY HILL
DR

-5N
3

N. HONEY HILL CIR.
HONEY HILL CIR
S. HONEY HILL CIR.

NANCY LN (PVT)

FOREST

PRESERVE

MUNGER

WHITE OAK LN
RIDGE LN
KAELIN RD
LYSLE RD
WOOD CT (PVT)

-4N
4

ST CHARLES TOWNSHIP
WAYNE TOWNSHIP

32W-4N
19

POWIS
SMITH

JENLOR CT

31W-4N
20

SMITH   RD

TRAIL WEST
WOODLAND CT
WOODLAND TRAIL SOUTH
PRAIRIE LN
LAKE ELEANOR DR
LAKE ELEANOR CT
SPICER
WIANT RD
AVARD
WHITNEY
KAELIN
LIES

MAGNOLIA LN
MAPLE TREE LN
BRADFORD PARKWAY
HONEYSUCKLE LN
LAKEWOOD CT
LAKEWOOD
WARWICK WAY
SOMERSET
BERKSHIRE CT
WILTSHIRE CT
KINGSWOOD CT
MOUNTAIN ASH

30W-4N
21

AUGUSTA DR
DORAL CT
ST. ANDREWS
OAK MEADOWS
ST AND

59

RD
SMITH
RD

PRATT WAYNE
WOODS
FOREST PRESERVE

Park
SHADY
AVE
LAKEVIEW
KENWOOD RD
WIANT RD
LOCUST
MULBERRY LN
LAND
ARBOR LN

R 8 E   R 9 E

SEE PAGES 10-11

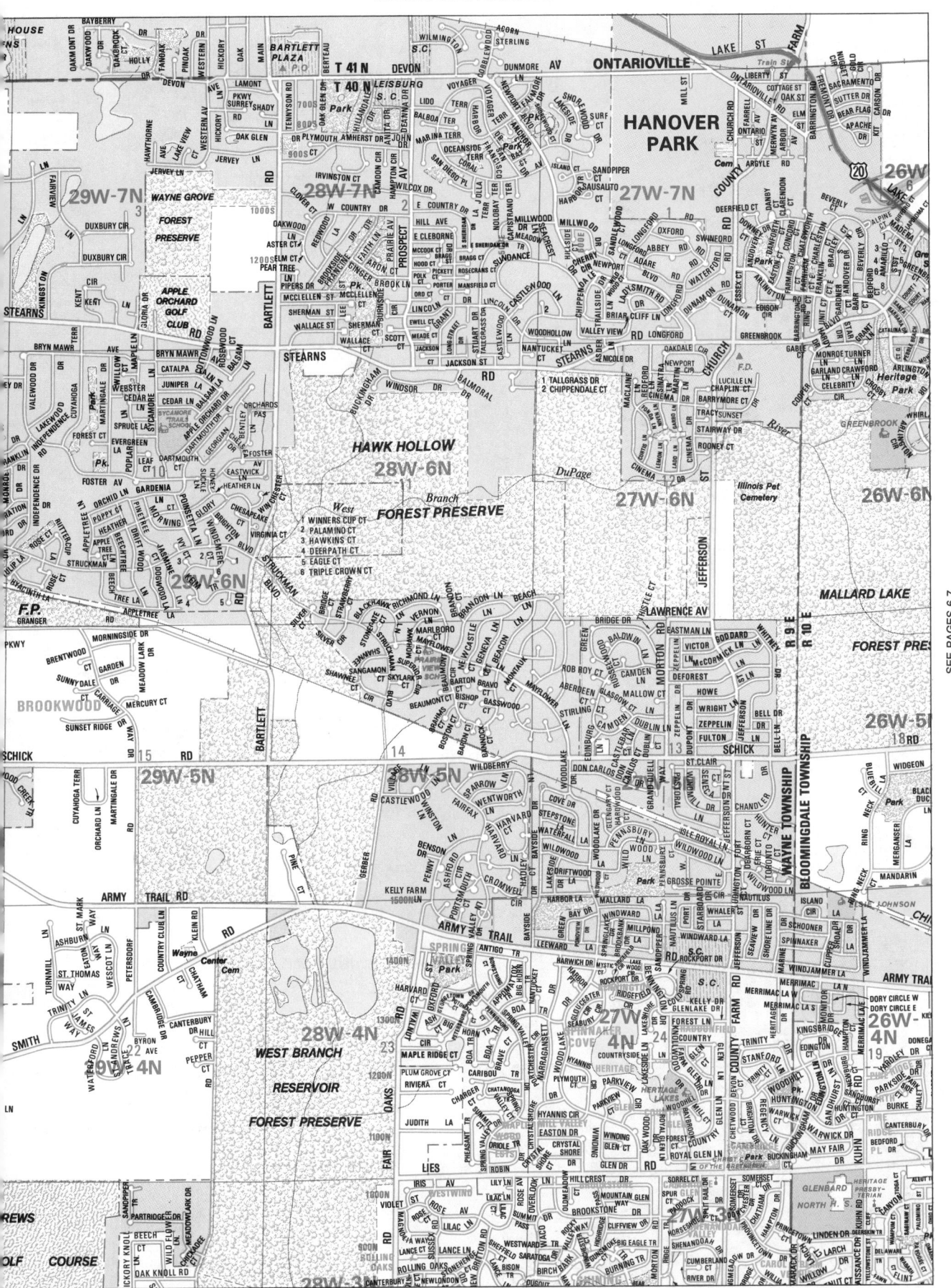

SEE PAGES 6-7

6 DU PAGE COUNTY

SEE COOK COUNTY PAGES 32-33

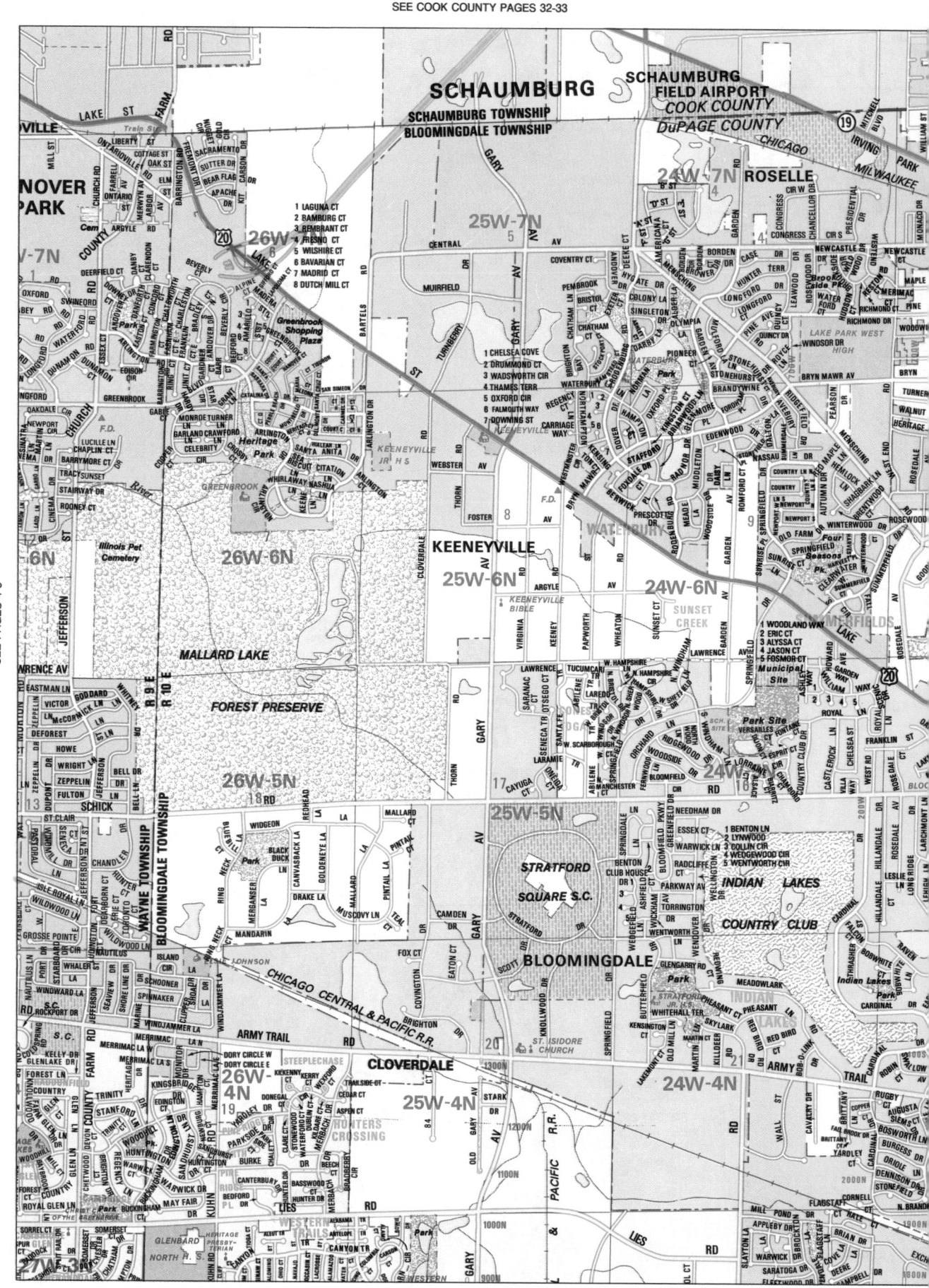

SEE PAGES 4-5

SEE PAGES 12-13

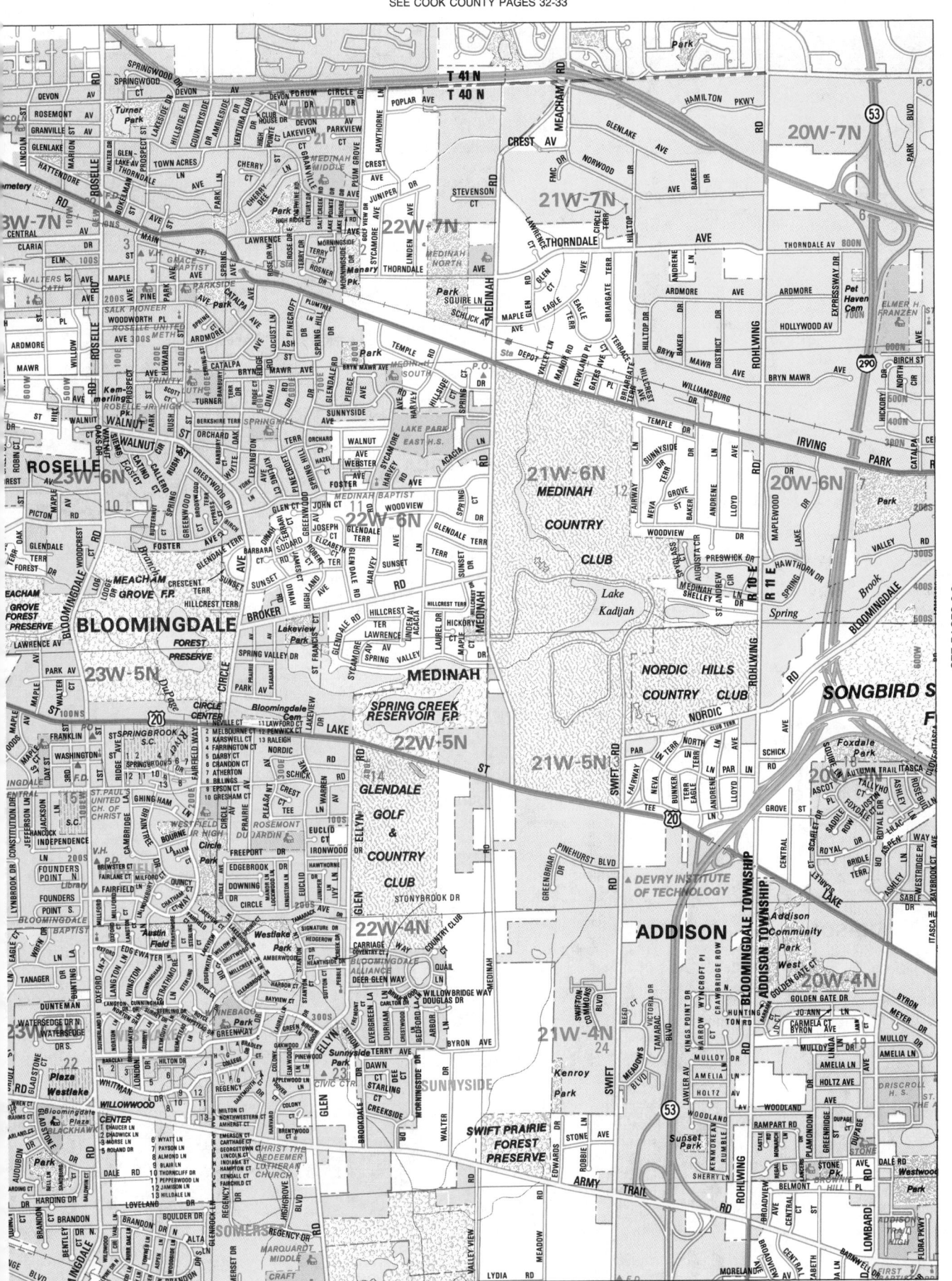

SEE PAGES 8-9

8 DU PAGE COUNTY

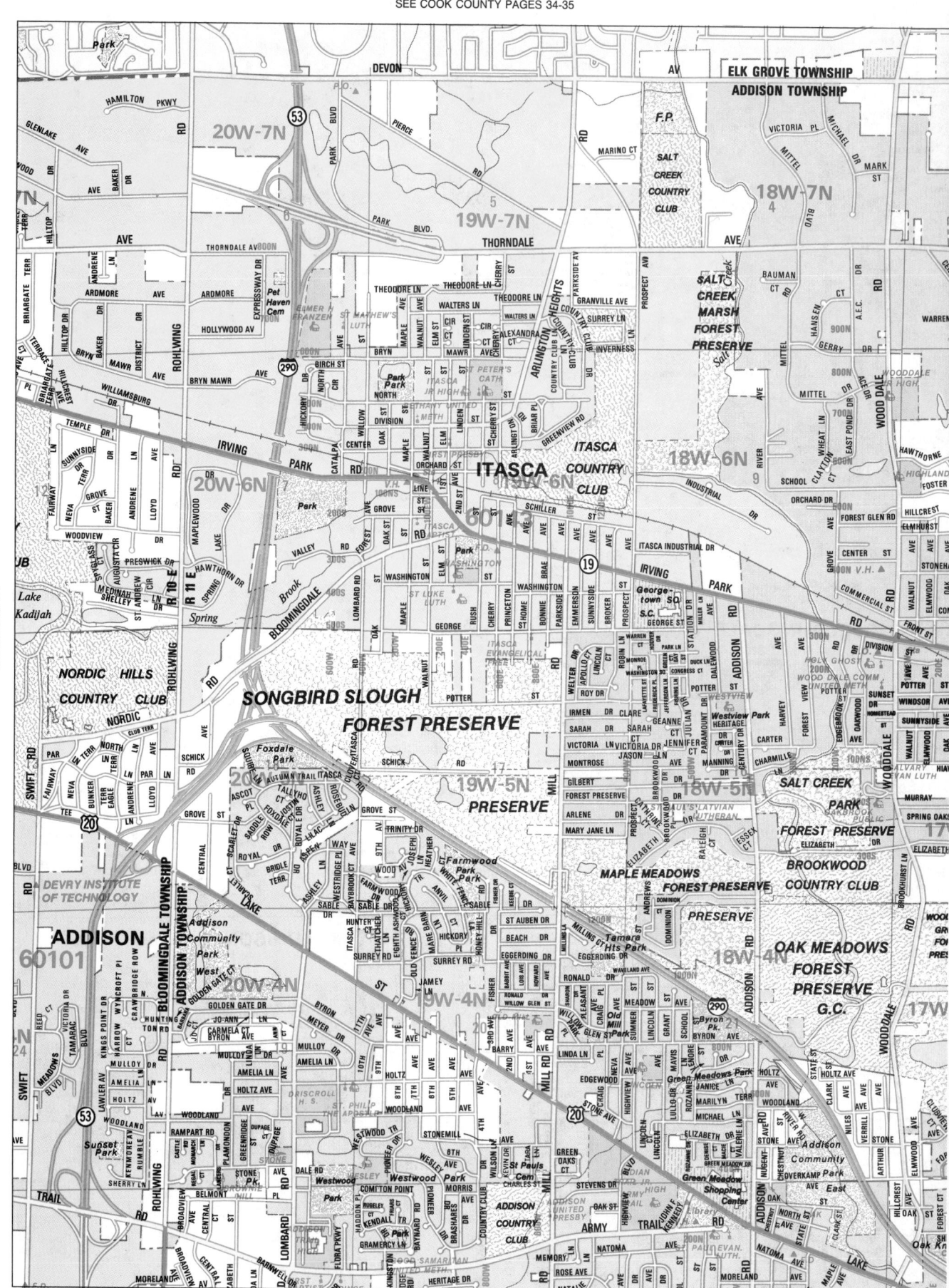

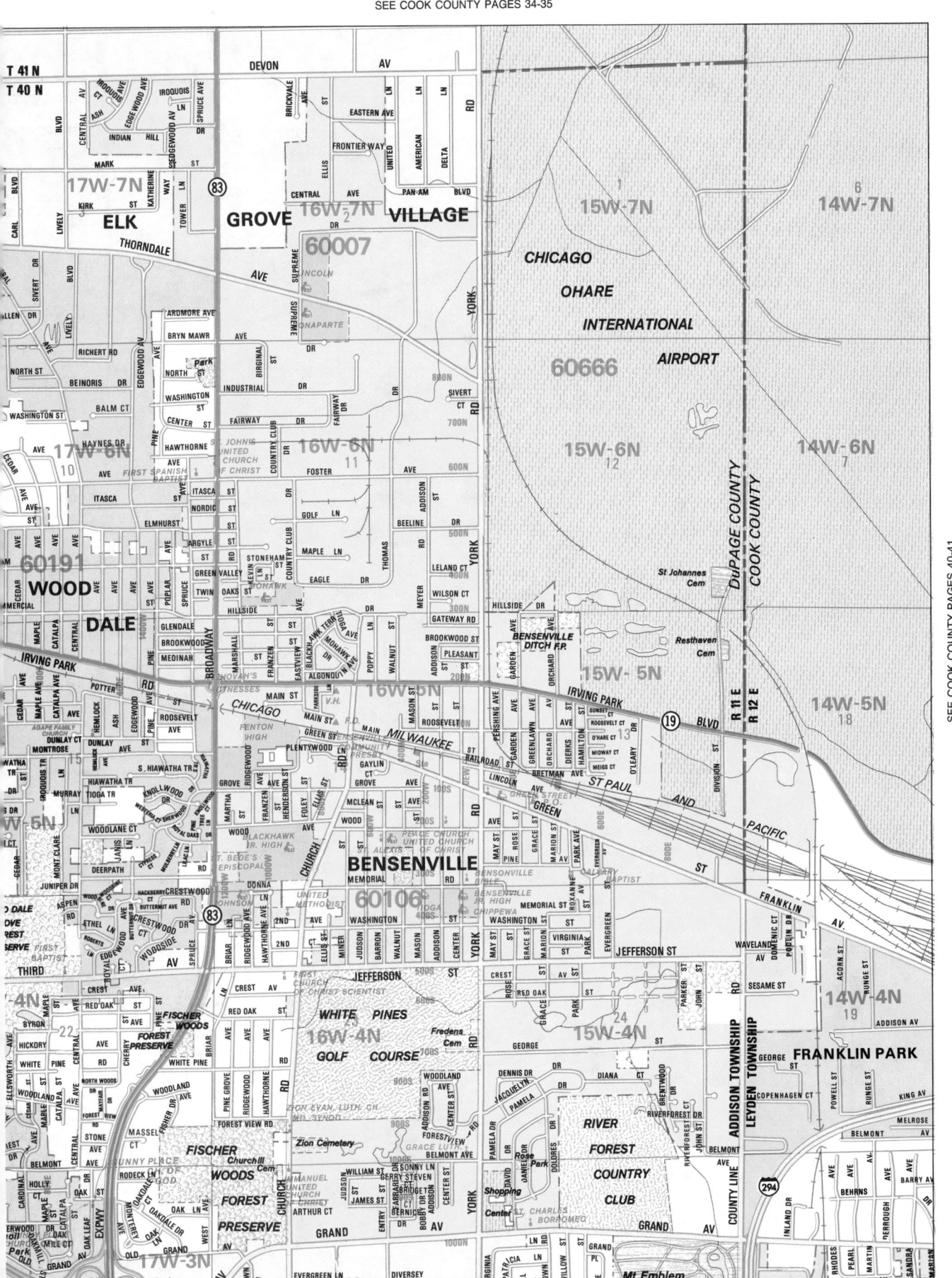

SEE PAGES 4-5

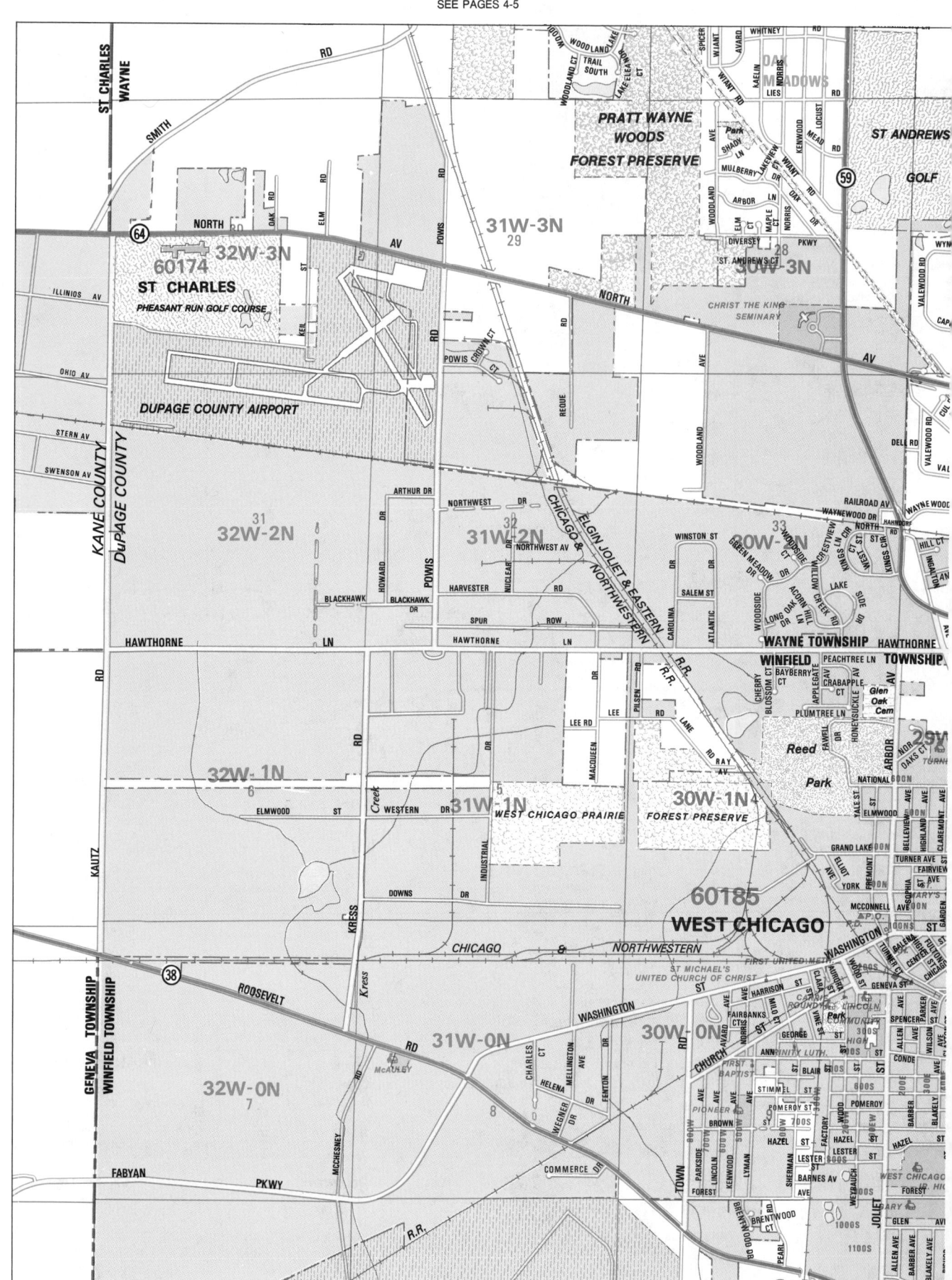

SEE KANE COUNTY PAGES 32-33

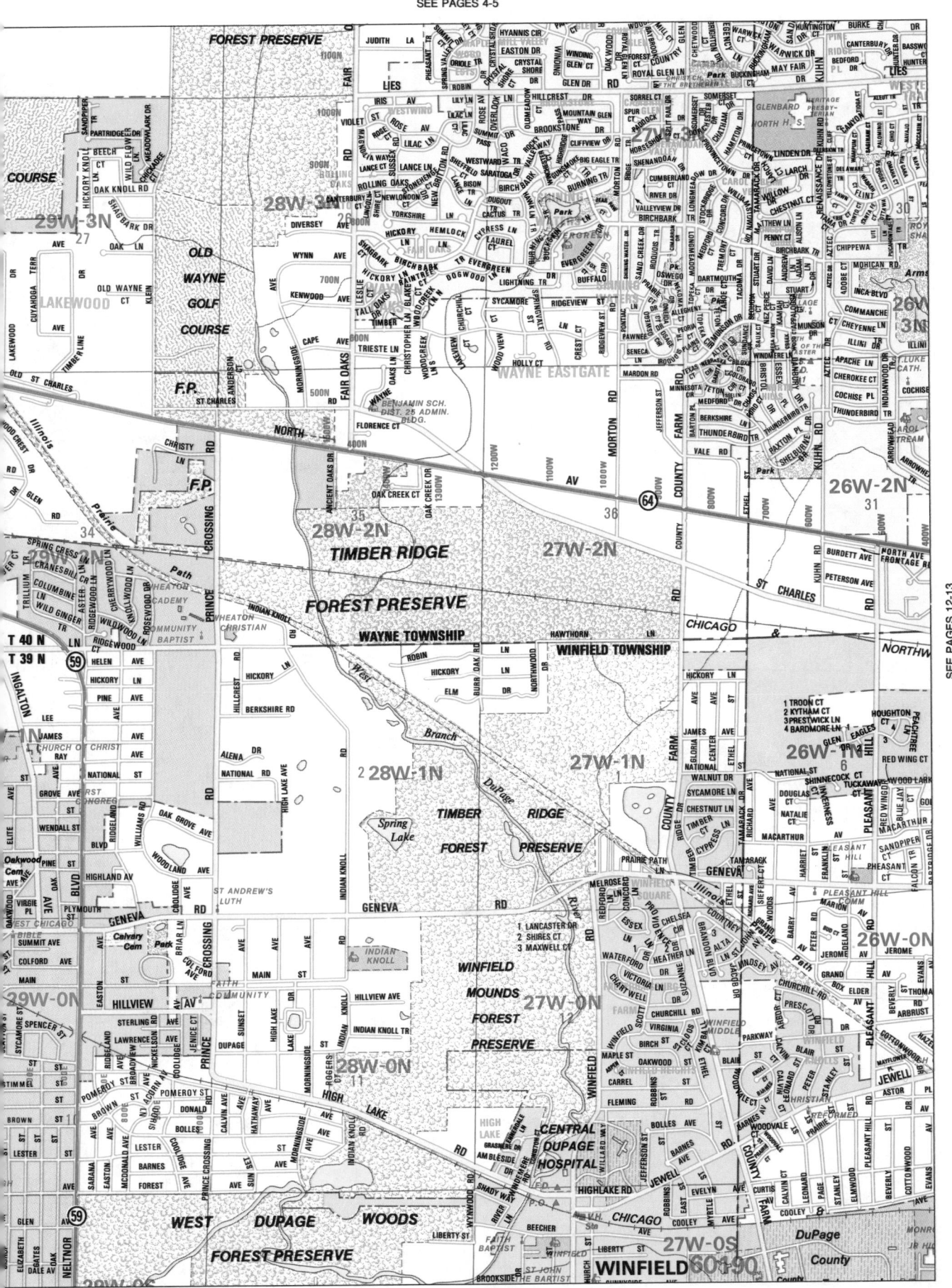

SEE PAGES 12-13

SEE PAGES 6-7

SEE PAGES 10-11

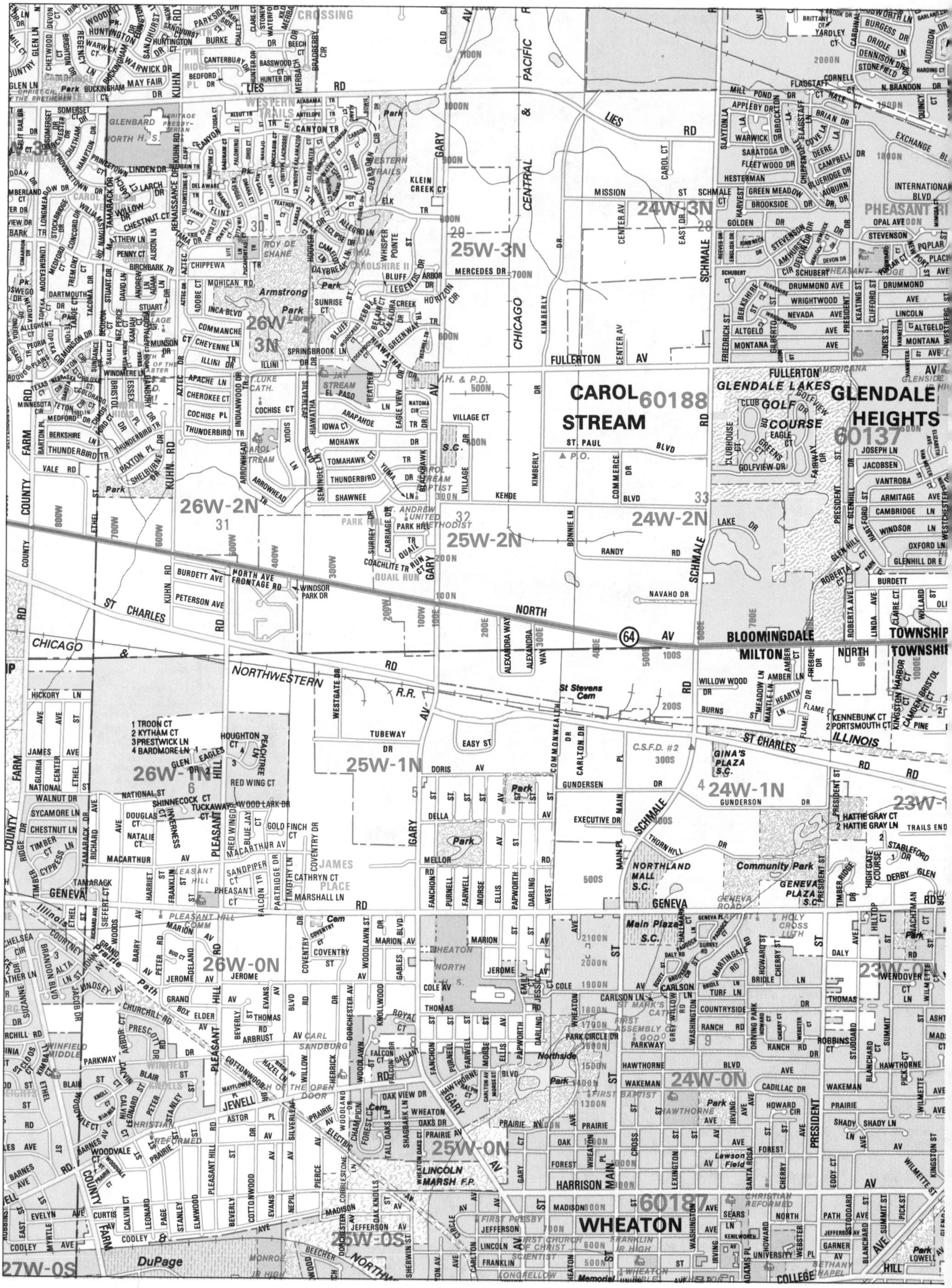

CAROL STREAM 60188

GLENDALE LAKES GOLF COURSE

GLENDALE HEIGHTS 60137

WHEATON 60187

24W-3N   25W-3N   26W-3N

26W-2N   25W-2N   24W-2N

26W-1N   25W-1N   24W-1N   23W

26W-0N   25W-0N   24W-0N   23W-0N

25W-0S   27W-0S

BLOOMINGDALE TOWNSHIP

MILTON TOWNSHIP

NORTH AV (64)

ST CHARLES RD

NORTHWESTERN RR

CHICAGO

ILLINOIS

1 TROON CT
2 KYTHAM CT
3 PRESTWICK LN
4 BARDMORE LN

1 KENNEBUNK CT
2 PORTSMOUTH CT

1 TROON CT

GINA'S PLAZA S.C.

GENEVA PLAZA S.C.

NORTHLAND MALL S.C.

Community Park

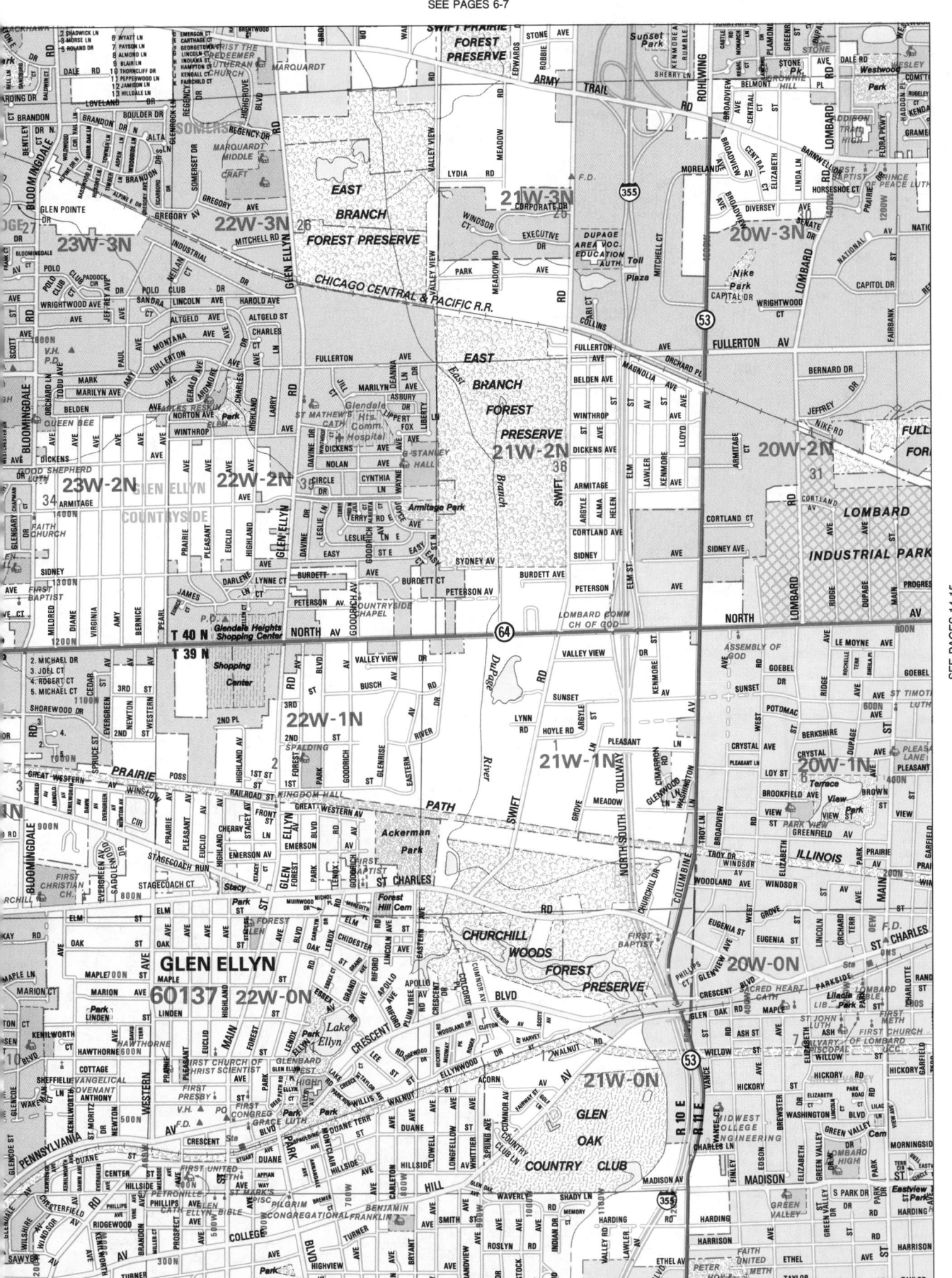

14 DU PAGE COUNTY

SEE PAGES 12-13

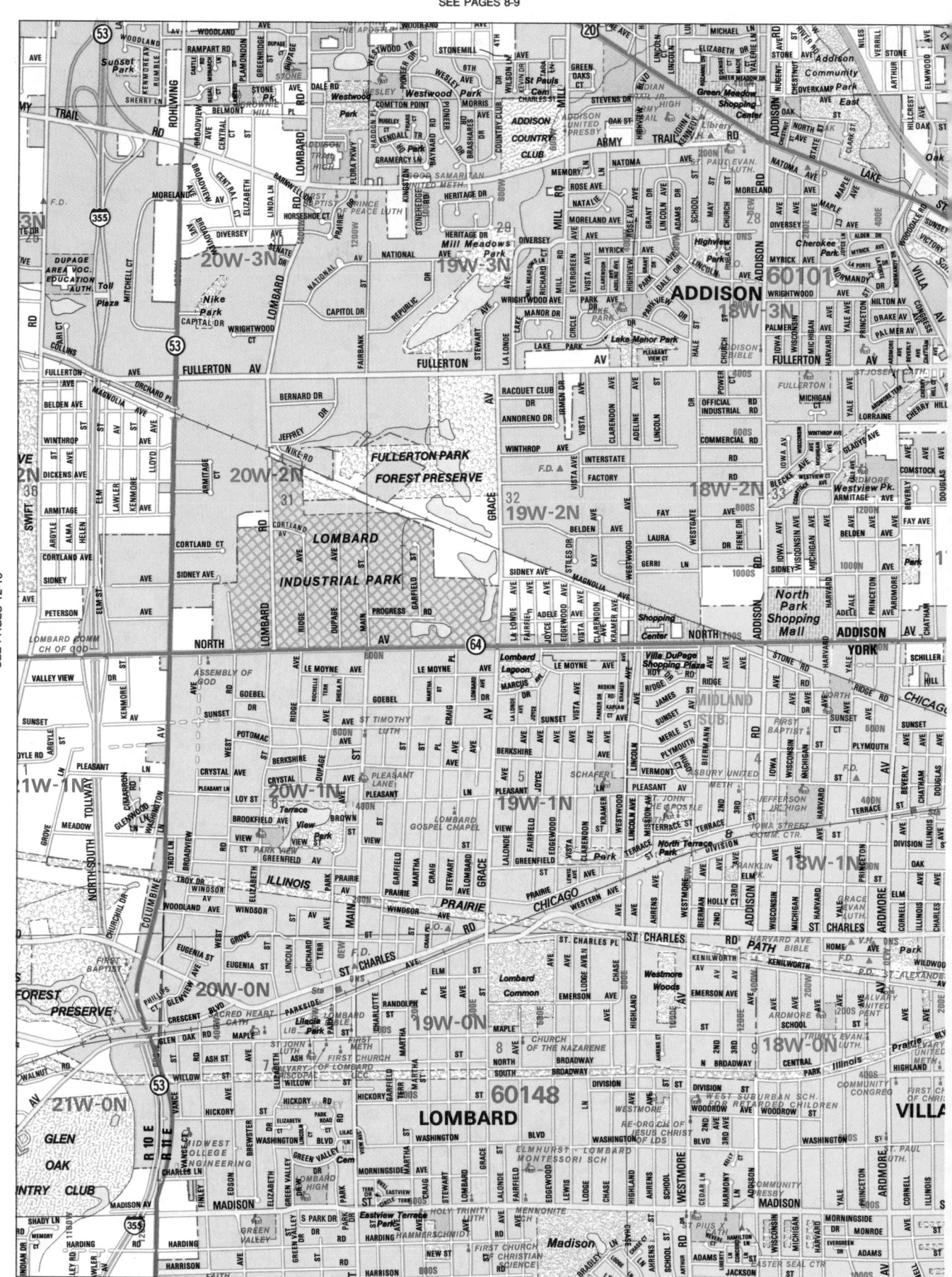

SEE PAGES 8-9

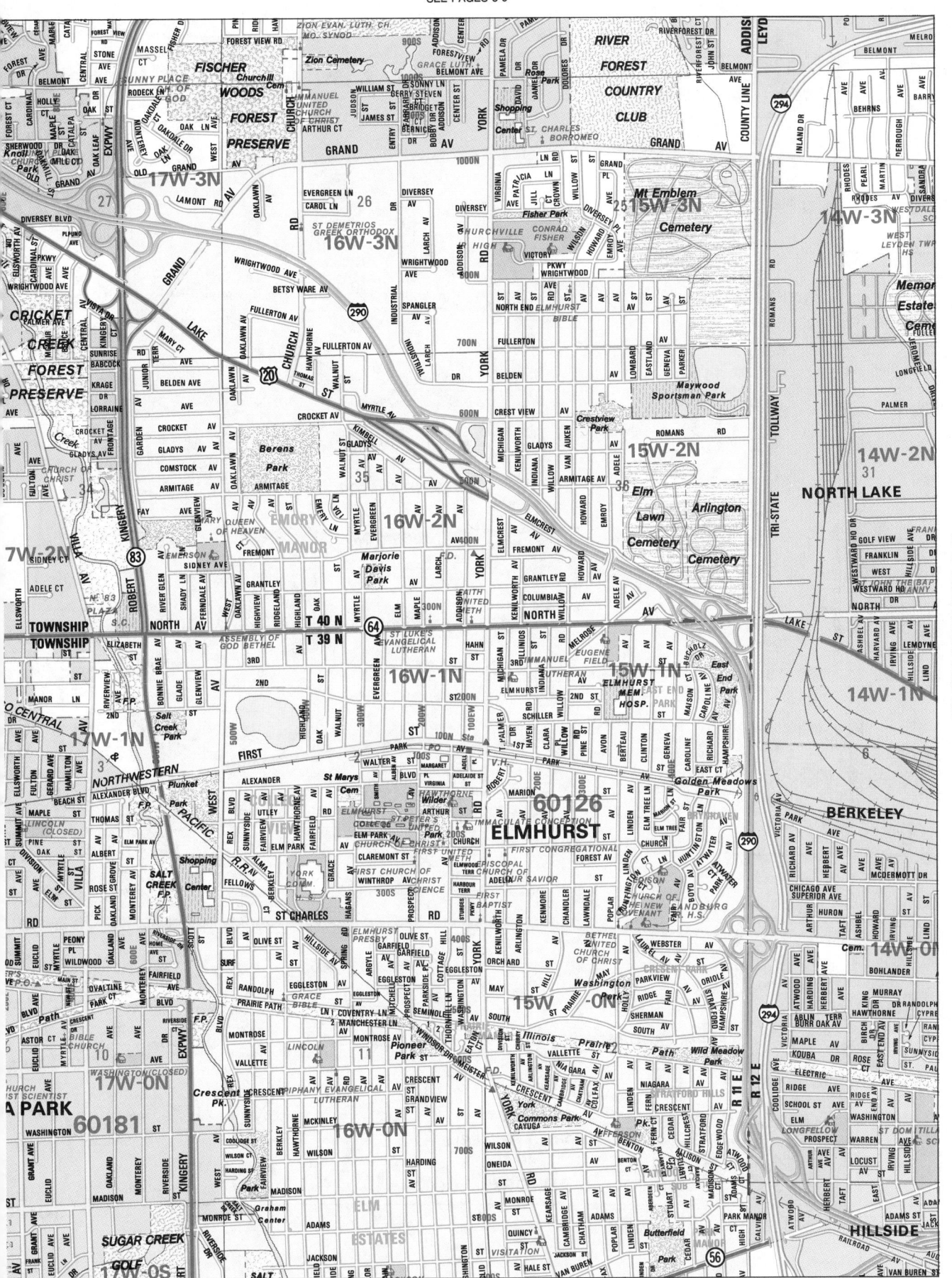

SEE COOK COUNTY PAGES 46-47

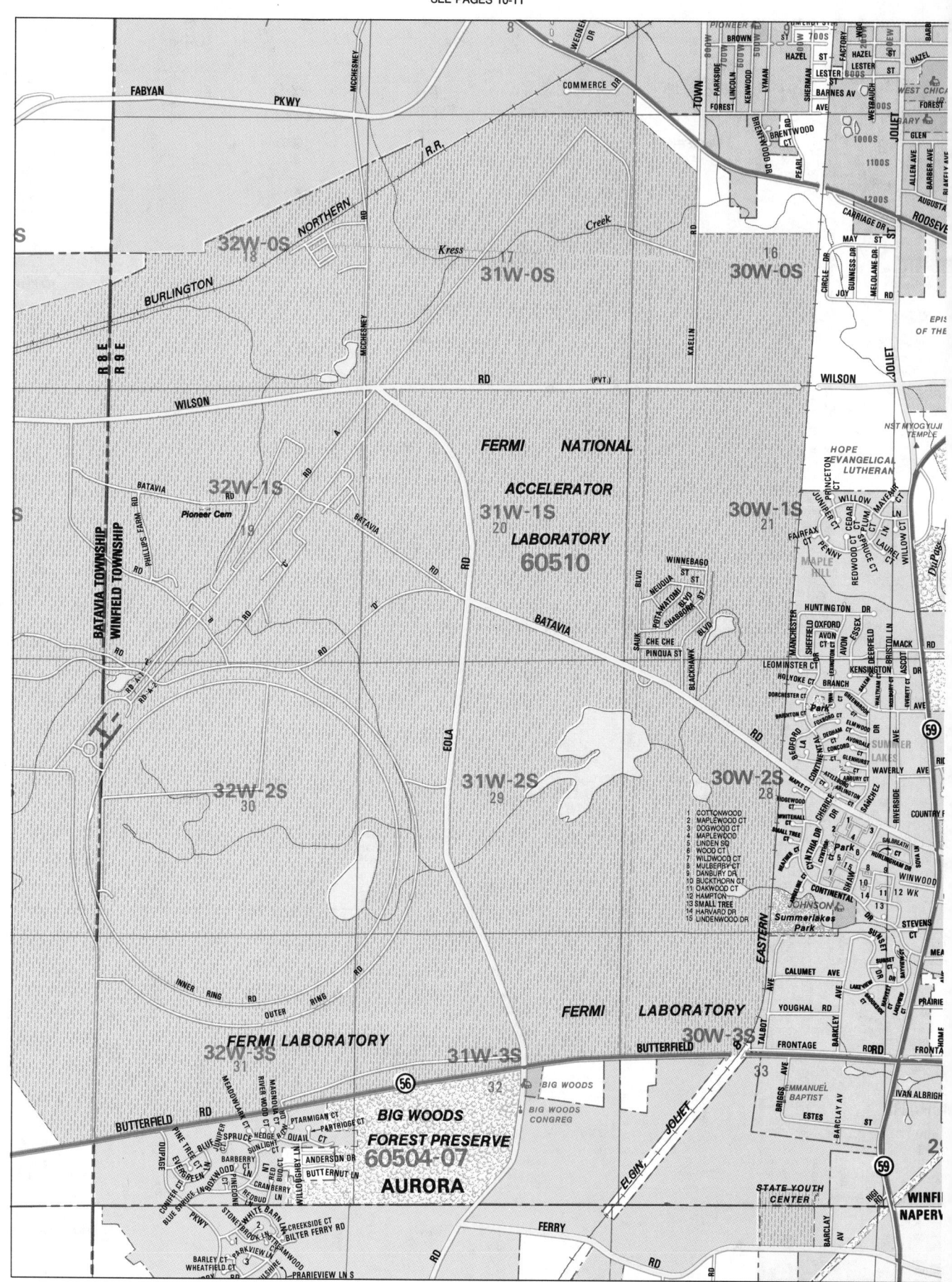

SEE KANE COUNTY PAGES 38-39

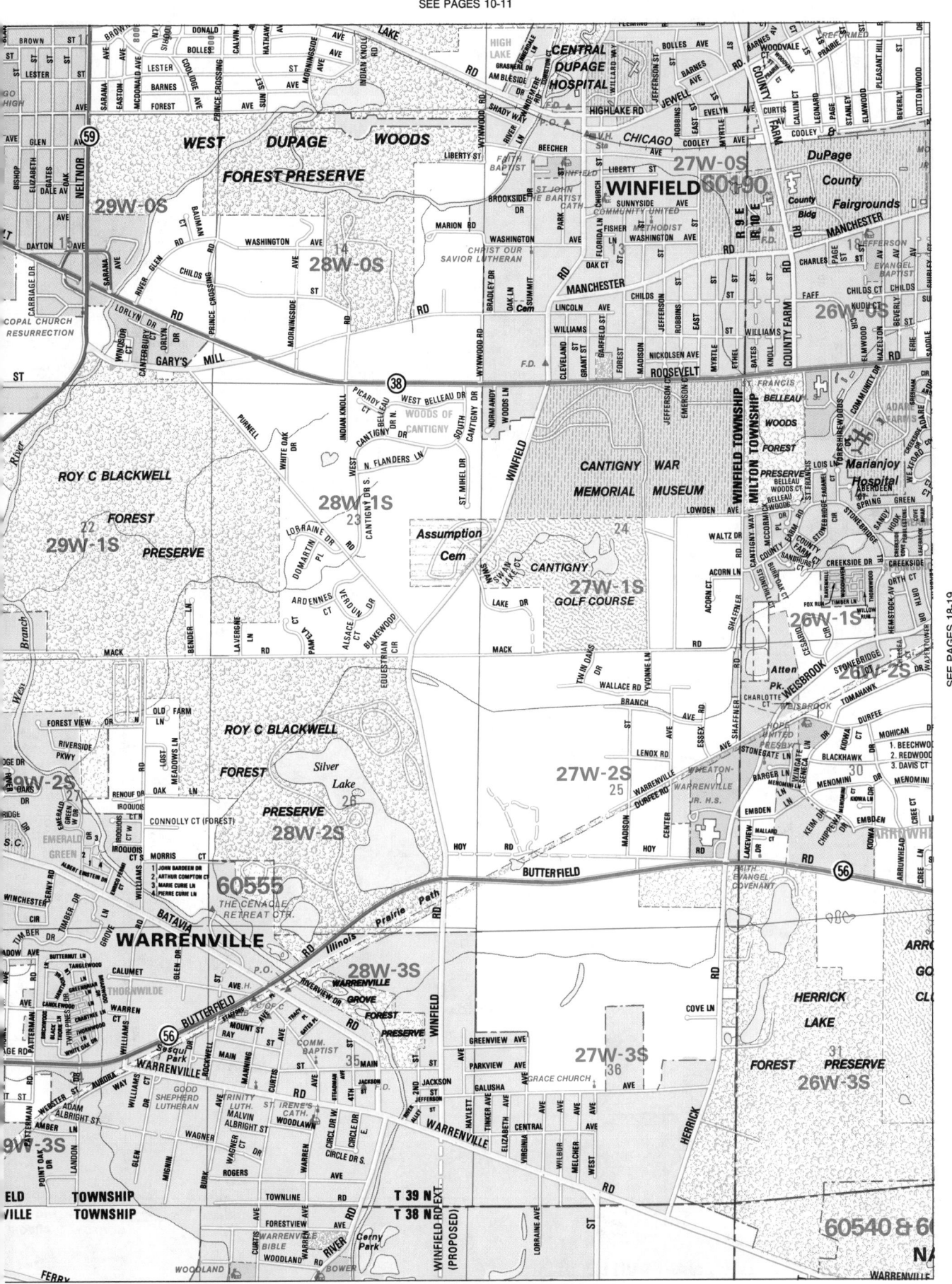

SEE PAGES 12-13

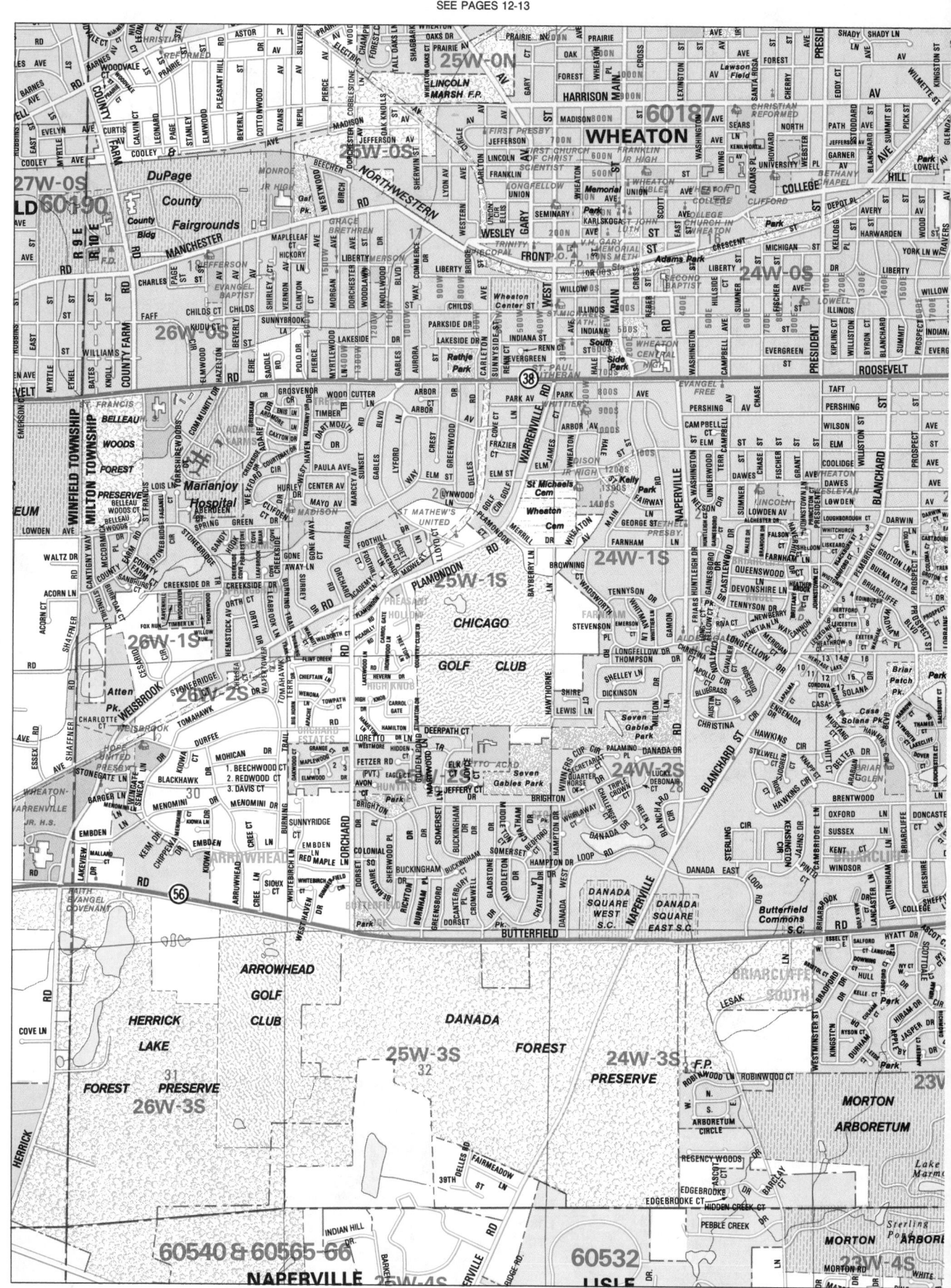

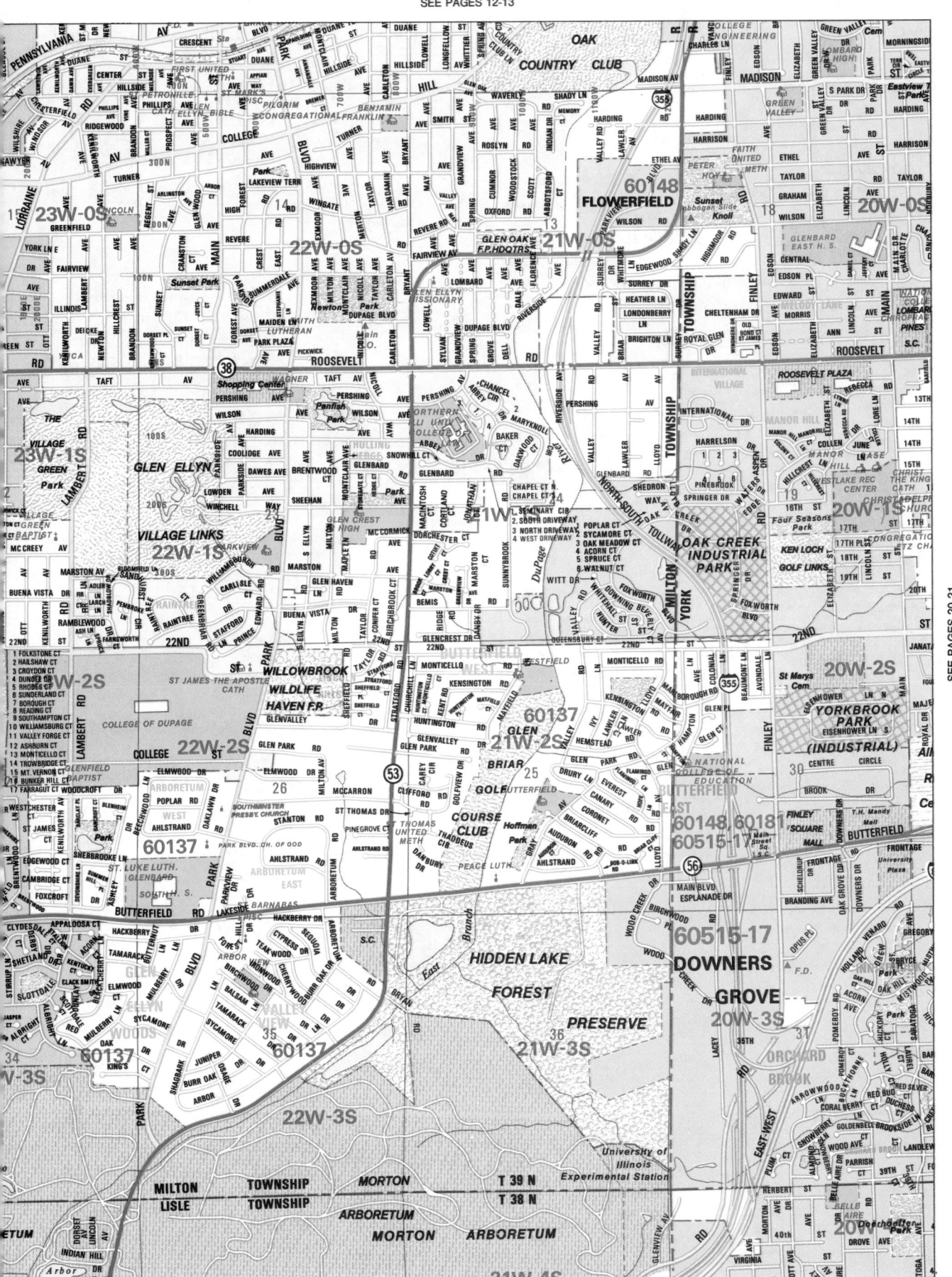

SEE PAGES 14-15

SEE PAGES 18-19

SEE PAGES 26-27

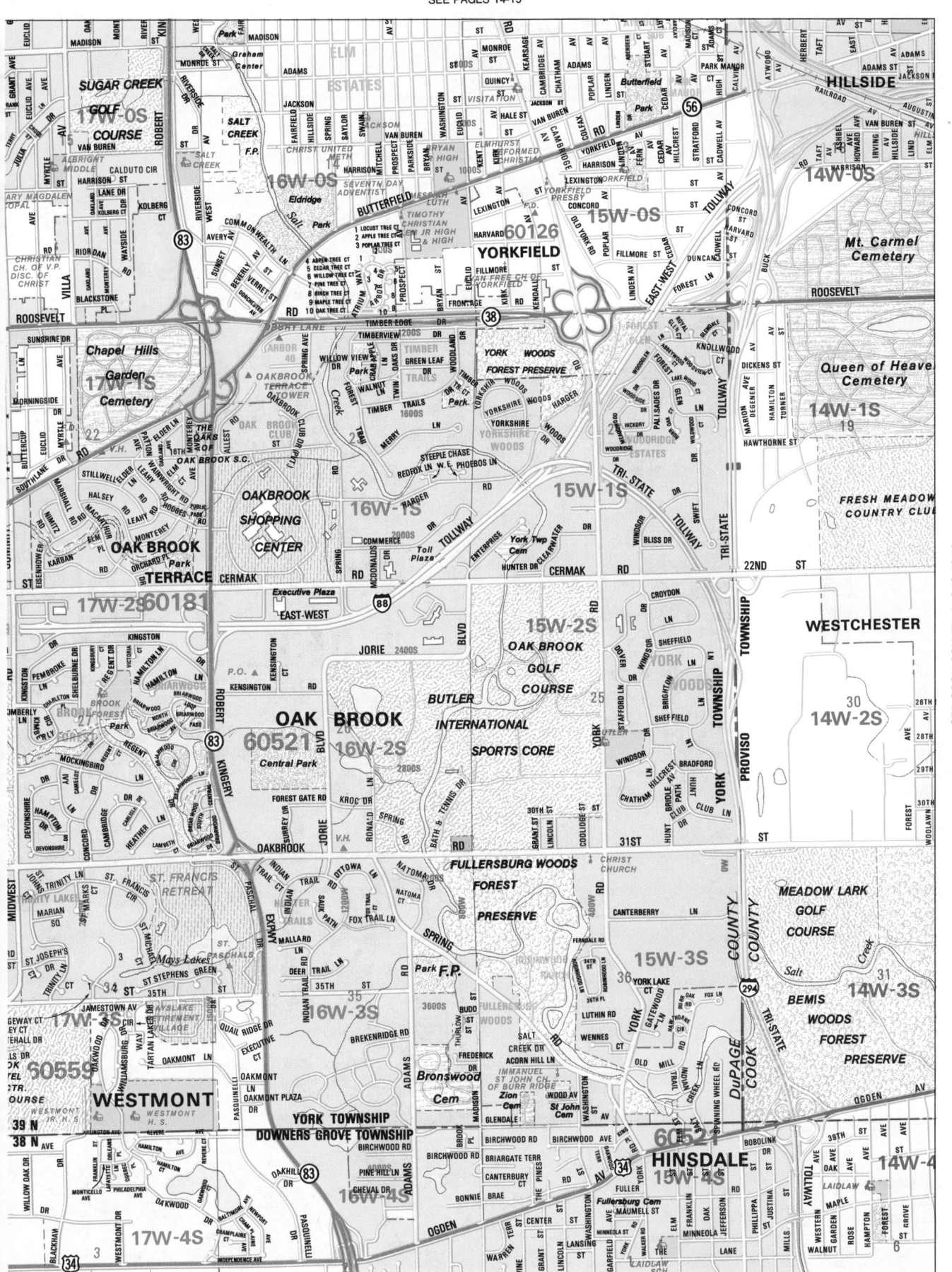

SEE PAGES 16-17

BUTTERFIELD

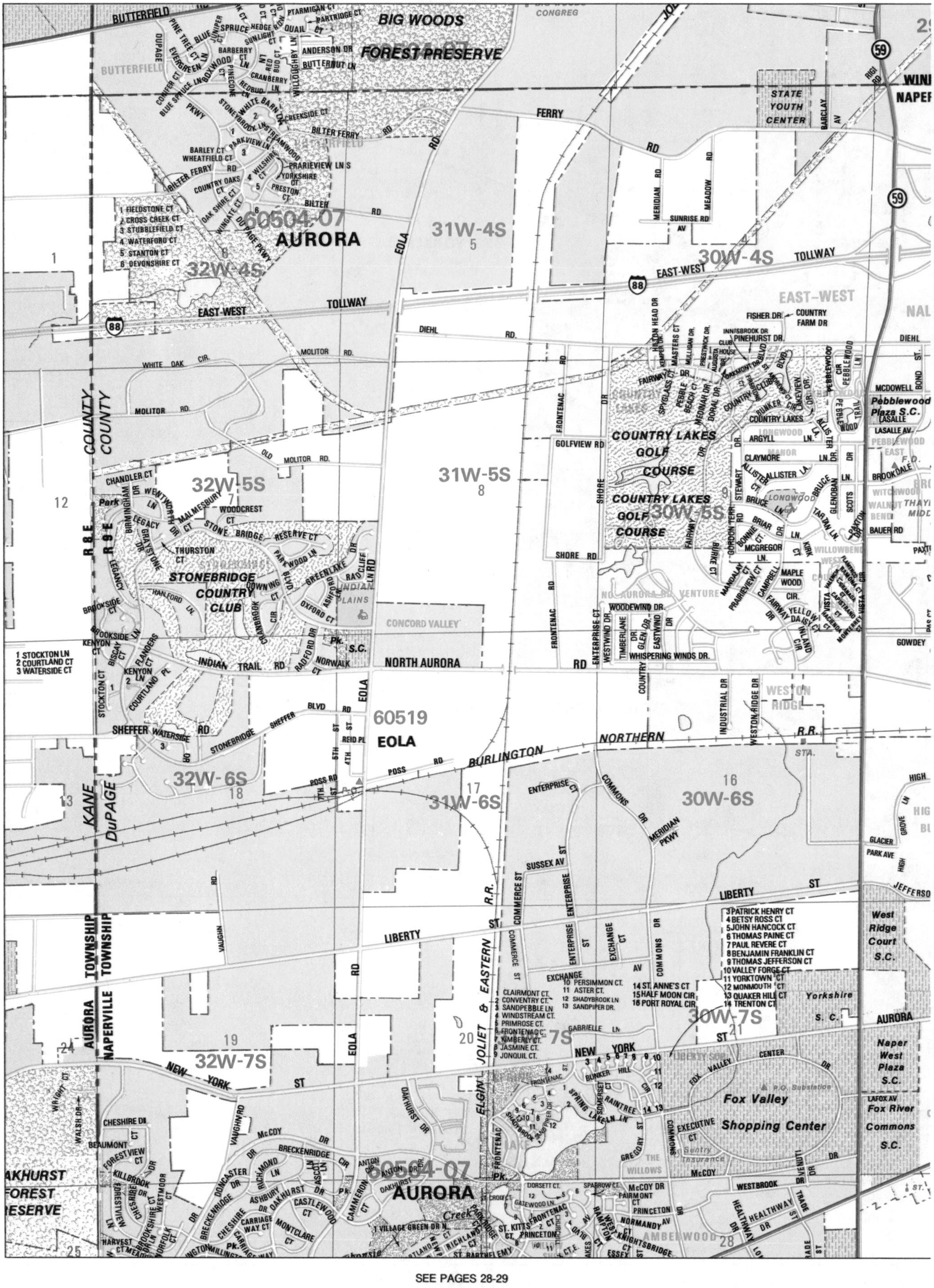

BIG WOODS FOREST PRESERVE

CONGREG

59

STATE YOUTH CENTER

WIN NAPER

FERRY

BUTTERFIELD

60504-07

AURORA

31W-4S
5

30W-4S TOLLWAY

32W-4S

EAST-WEST TOLLWAY

88 EAST-WEST TOLLWAY

EAST-WEST

88

EAST-WEST NAL

DIEHL RD.

DIEHL

COUNTRY LAKES GOLF COURSE

COUNTRY LAKES GOLF COURSE

Pebblewood Plaza S.C.

MOLITOR RD.

WHITE OAK CIR.

MOLITOR RD.

OLD MOLITOR RD.

31W-5S
8

GOLFVIEW RD

30W-5S

Pebblewood East

COUNTY COUNTY

R 8 E R 9 E

32W-5S
7

Park

WOODCREST

STONEBRIDGE COUNTRY CLUB

CONCORD VALLEY

S.C.

NORTH AURORA

12

60519

EOLA

1 STOCKTON LN
2 COURTLAND CT
3 WATERSIDE CT

SHEFFER WATERSIDE

32W-6S
18

31W-6S
17

30W-6S
16

WESTON RIDGE

R.R.

SHEFFER

BURLINGTON

NORTHERN

STA.

KANE DuPAGE

13

LIBERTY

LIBERTY

West Ridge Court S.C.

3 PATRICK HENRY CT
4 BETSY ROSS CT
5 JOHN HANCOCK CT
6 THOMAS PAINE CT
7 PAUL REVERE CT
8 BENJAMIN FRANKLIN CT
9 THOMAS JEFFERSON CT
10 VALLEY FORGE CT
11 YORKTOWN CT
12 MONMOUTH CT
13 QUAKER HILL CT
14 TRENTON CT

10 PERSIMMON CT.
11 ASTER CT.
12 SHADYBROOK LN.
13 SANDPIPER DR.

14 ST. ANNE'S CT.
15 HALF MOON CIR.
16 PORT ROYAL CIR.

30W-7S
21

Yorkshire S.C.

AURORA

AURORA NAPERVILLE TOWNSHIP TOWNSHIP

19

32W-7S

20

7S

NEW YORK

Fox Valley Shopping Center

Naper West Plaza S.C.

Fox River Commons S.C.

60504-07

AURORA

OAKHURST FOREST RESERVE

25

28

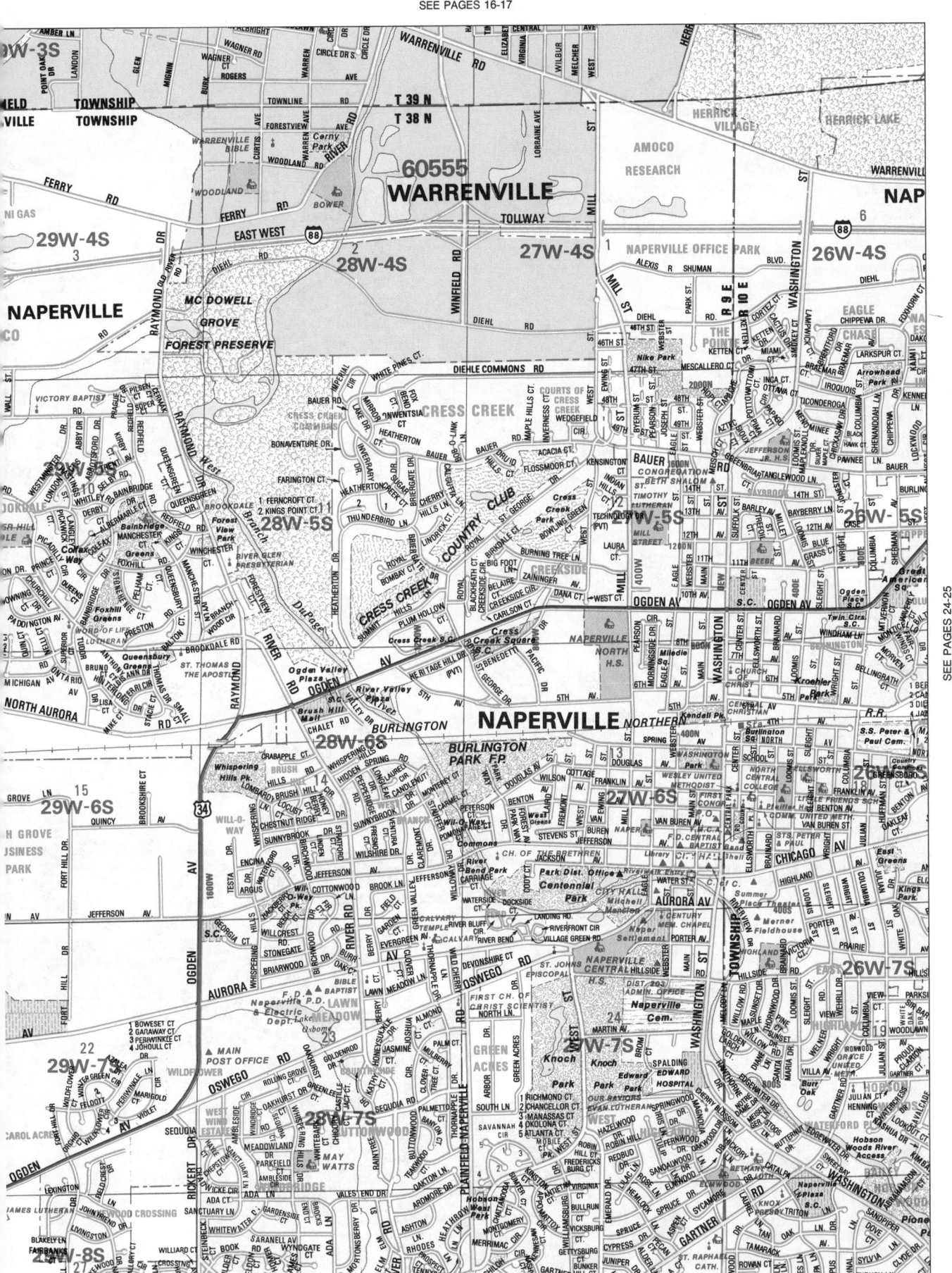

SEE PAGES 18-19

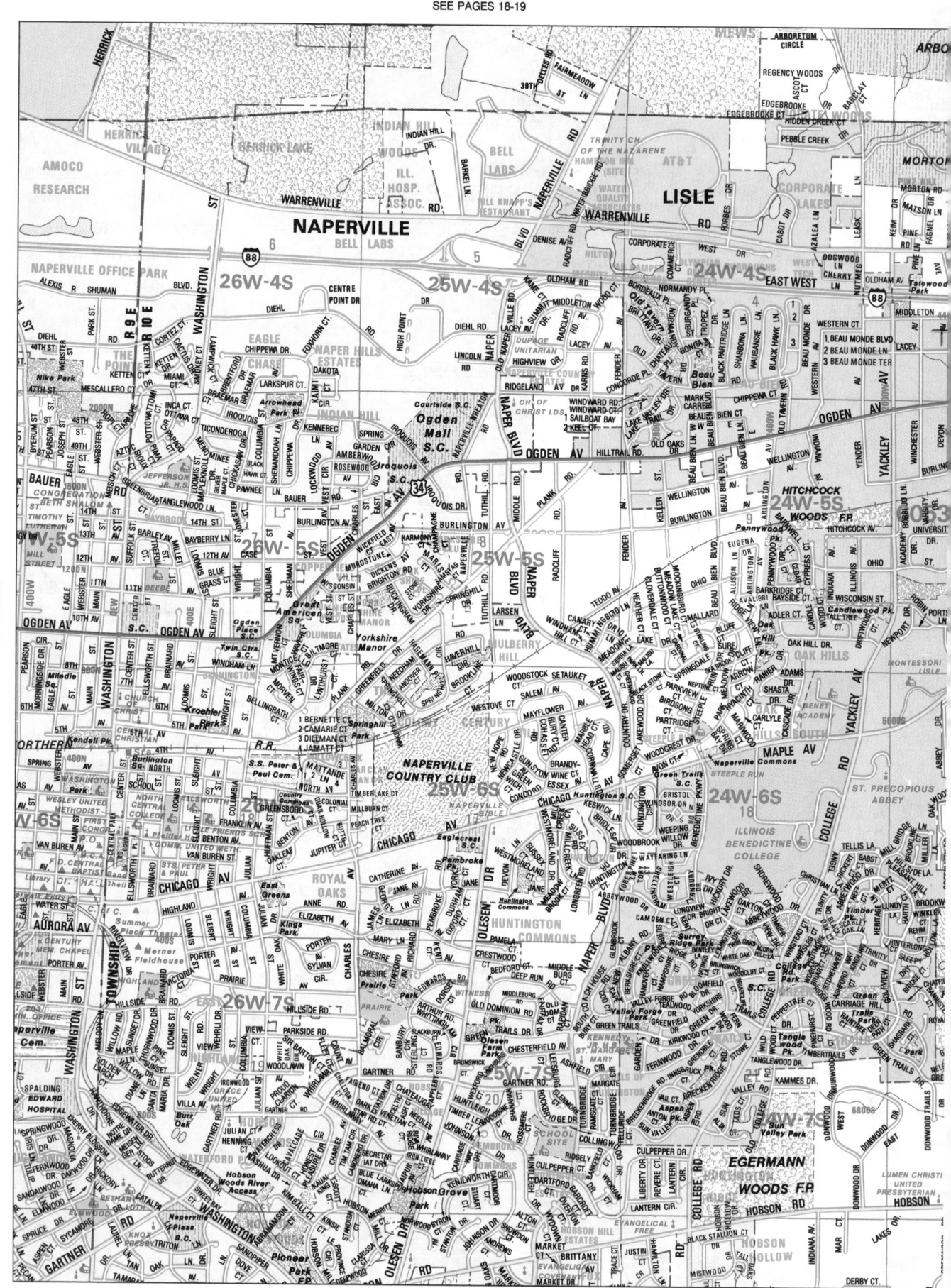

SEE PAGES 30-31

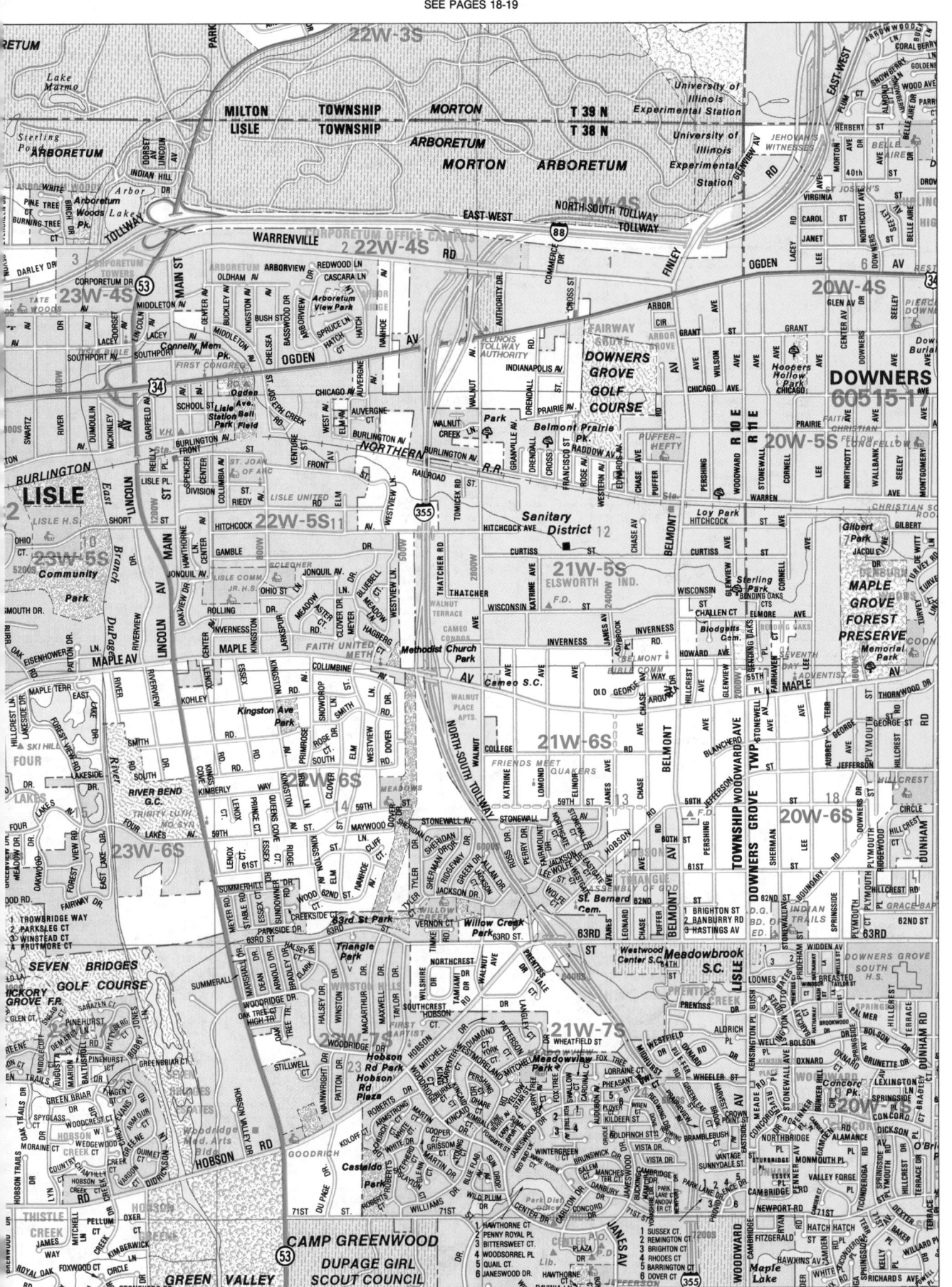

SEE PAGES 26-27

SEE PAGES 20-21

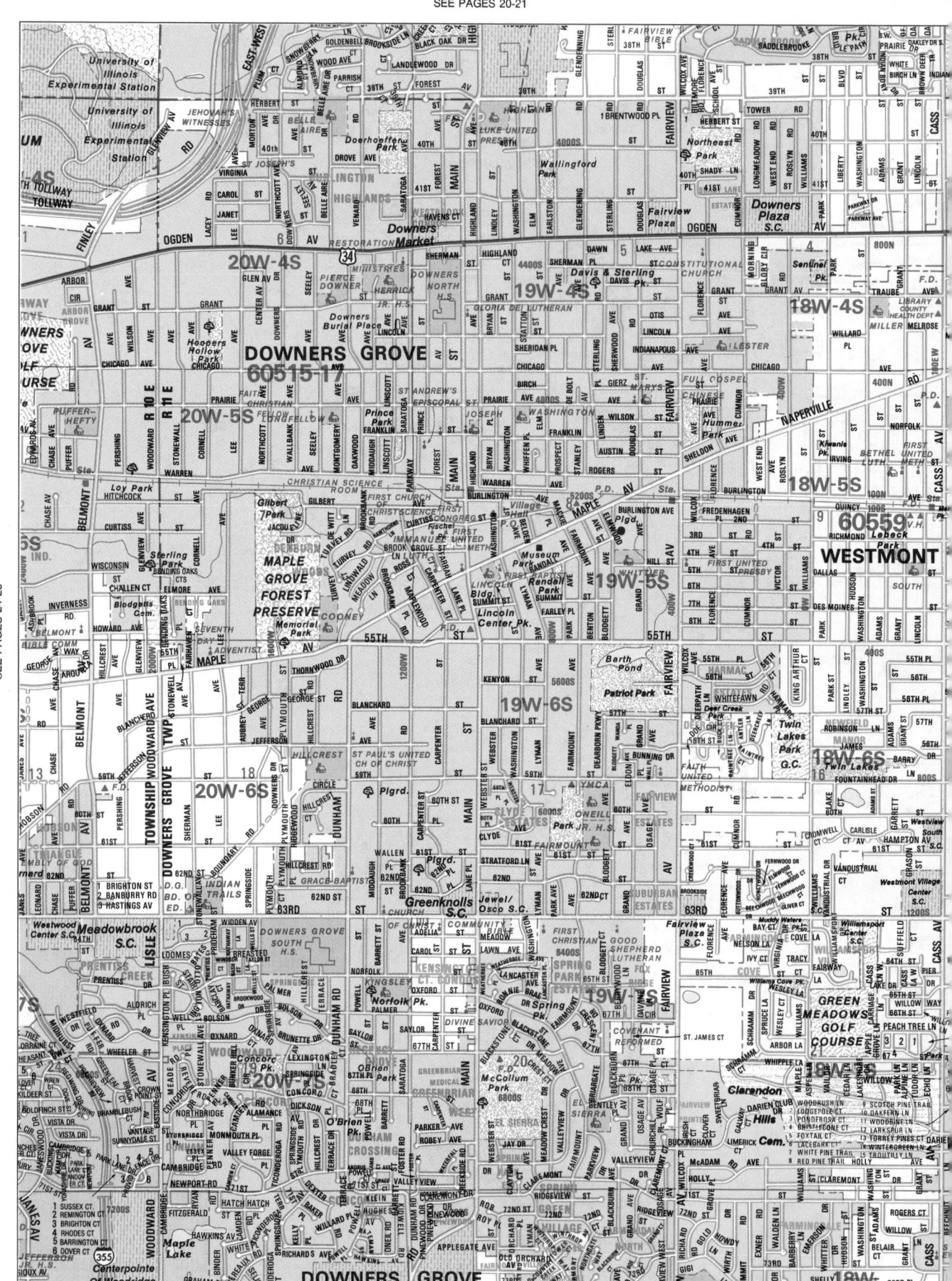

SEE PAGES 20-21

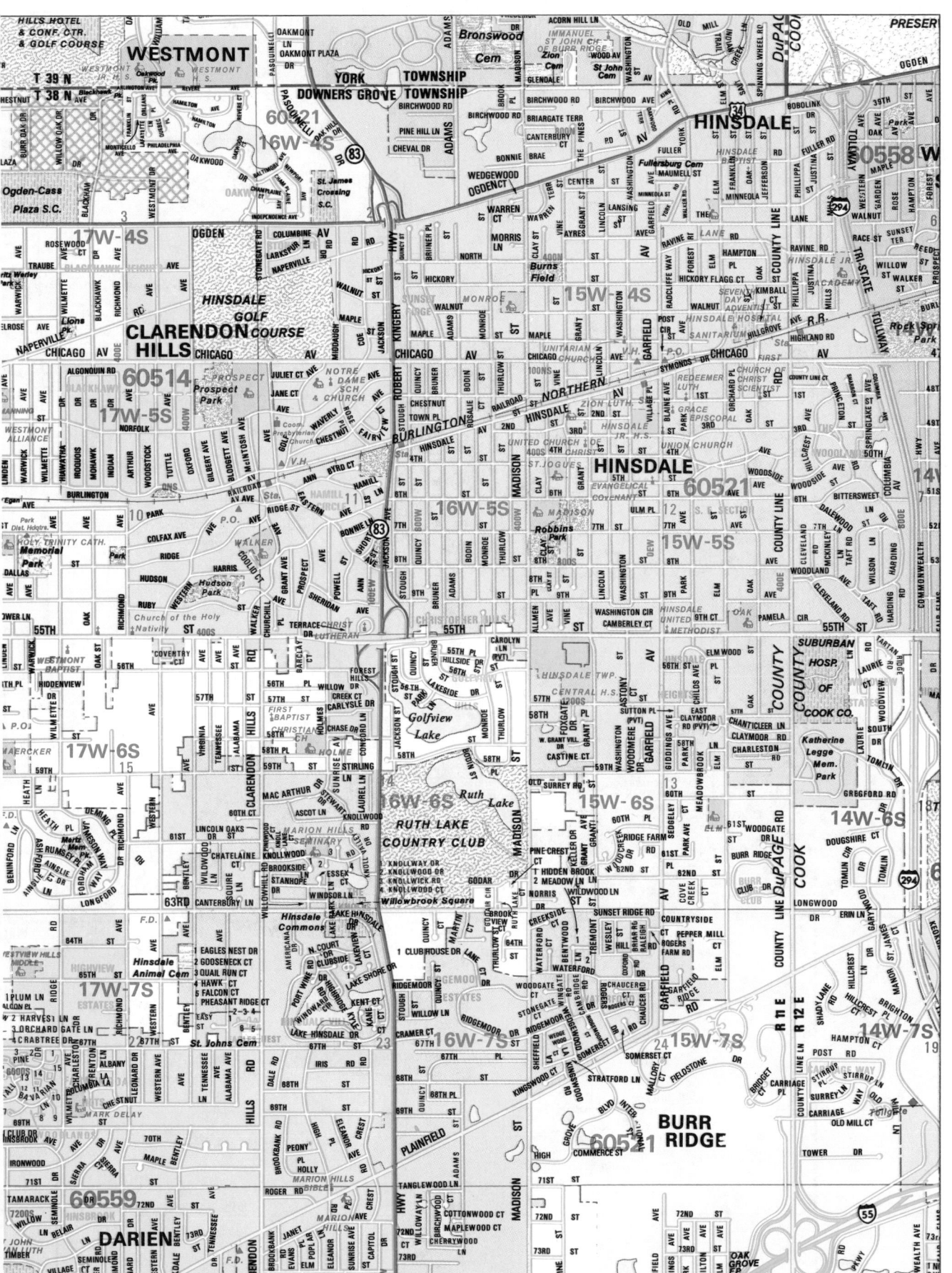

SEE PAGES 32-33

SEE COOK COUNTY PAGES 58-59

SEE PAGES 22-23

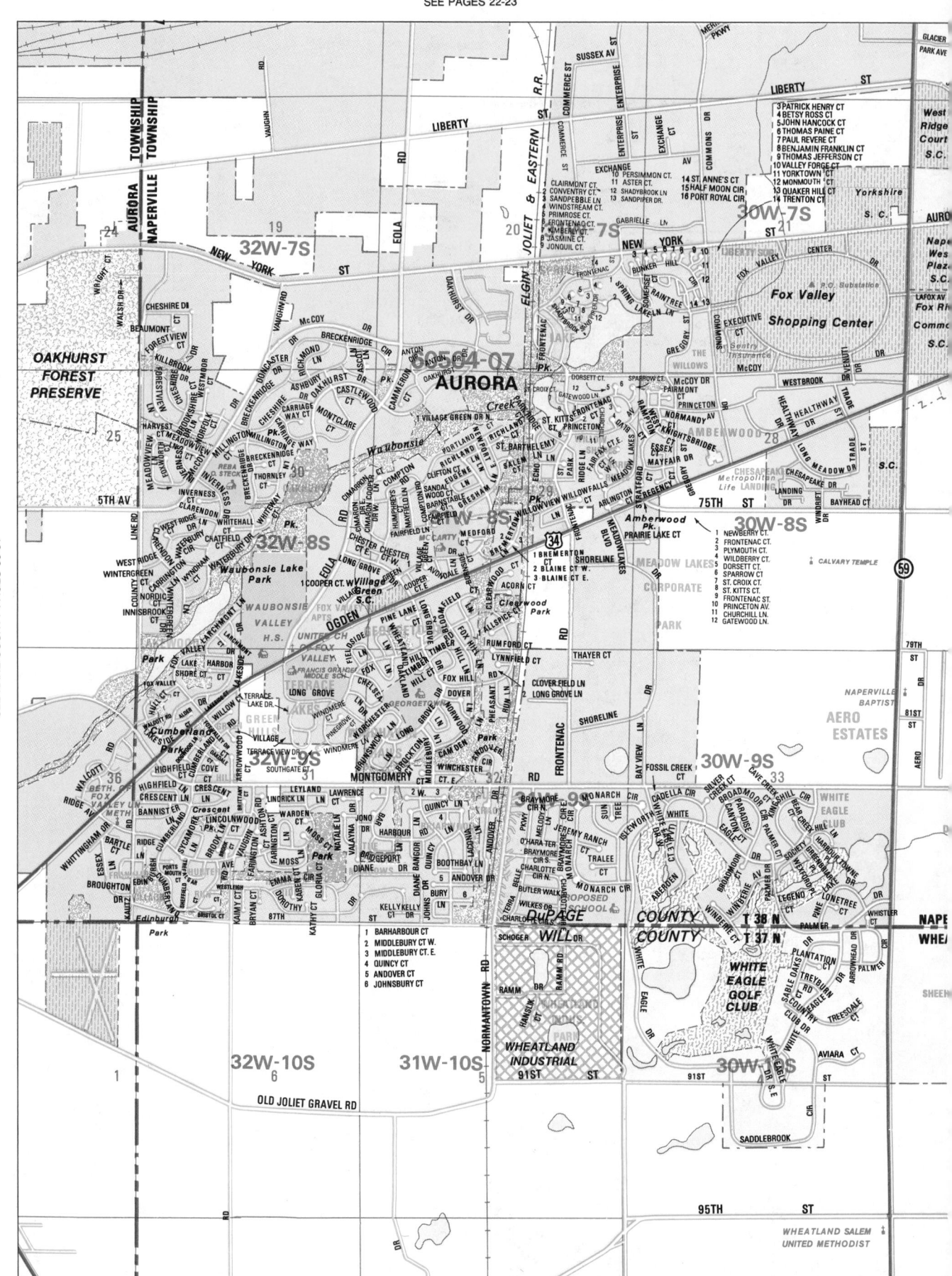

SEE KANE COUNTY PAGES 50-51

SEE PAGES 22-23

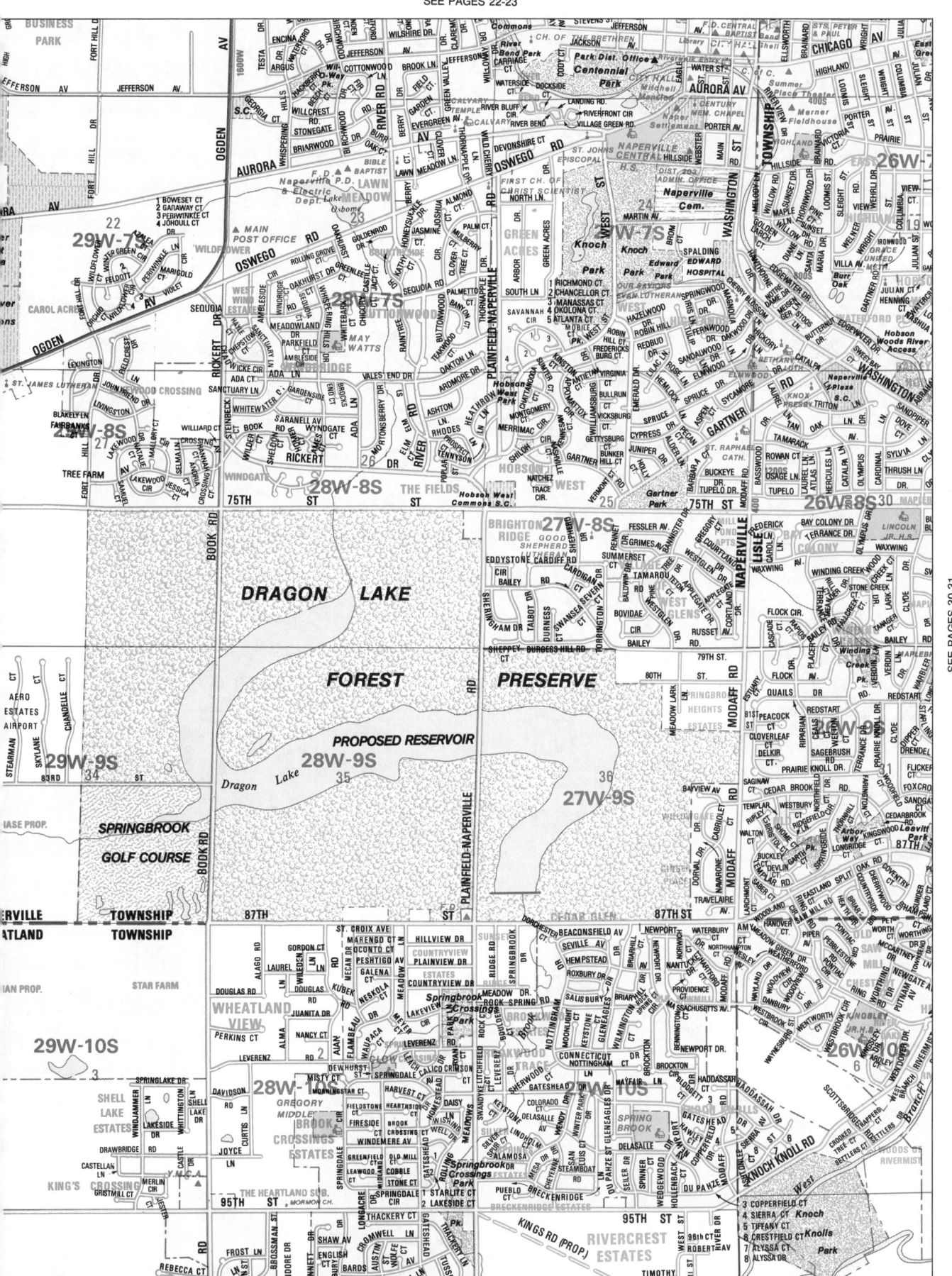

SEE PAGES 30-31

SEE PAGES 24-25

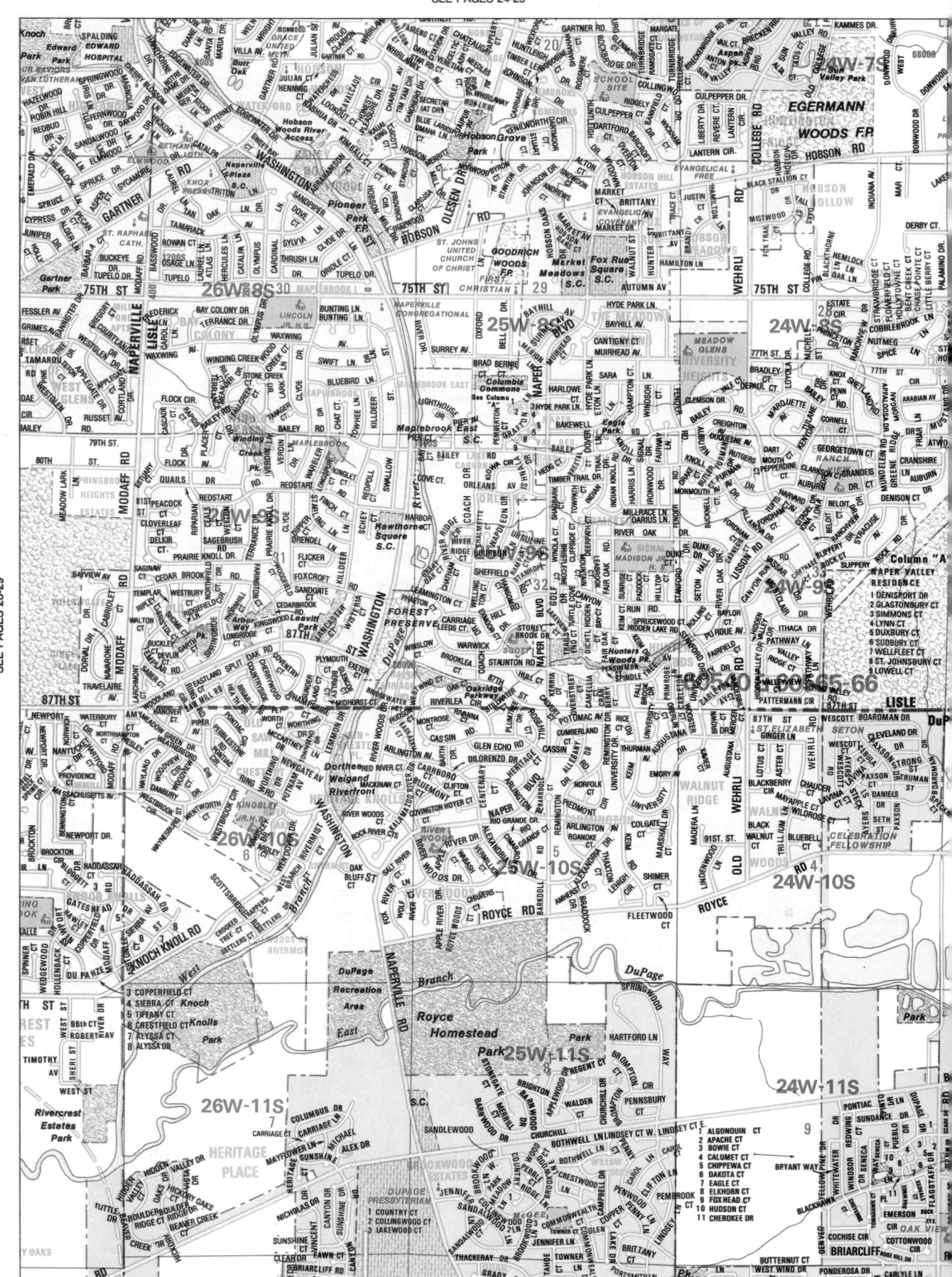

SEE WILL COUNTY PAGES 6-7

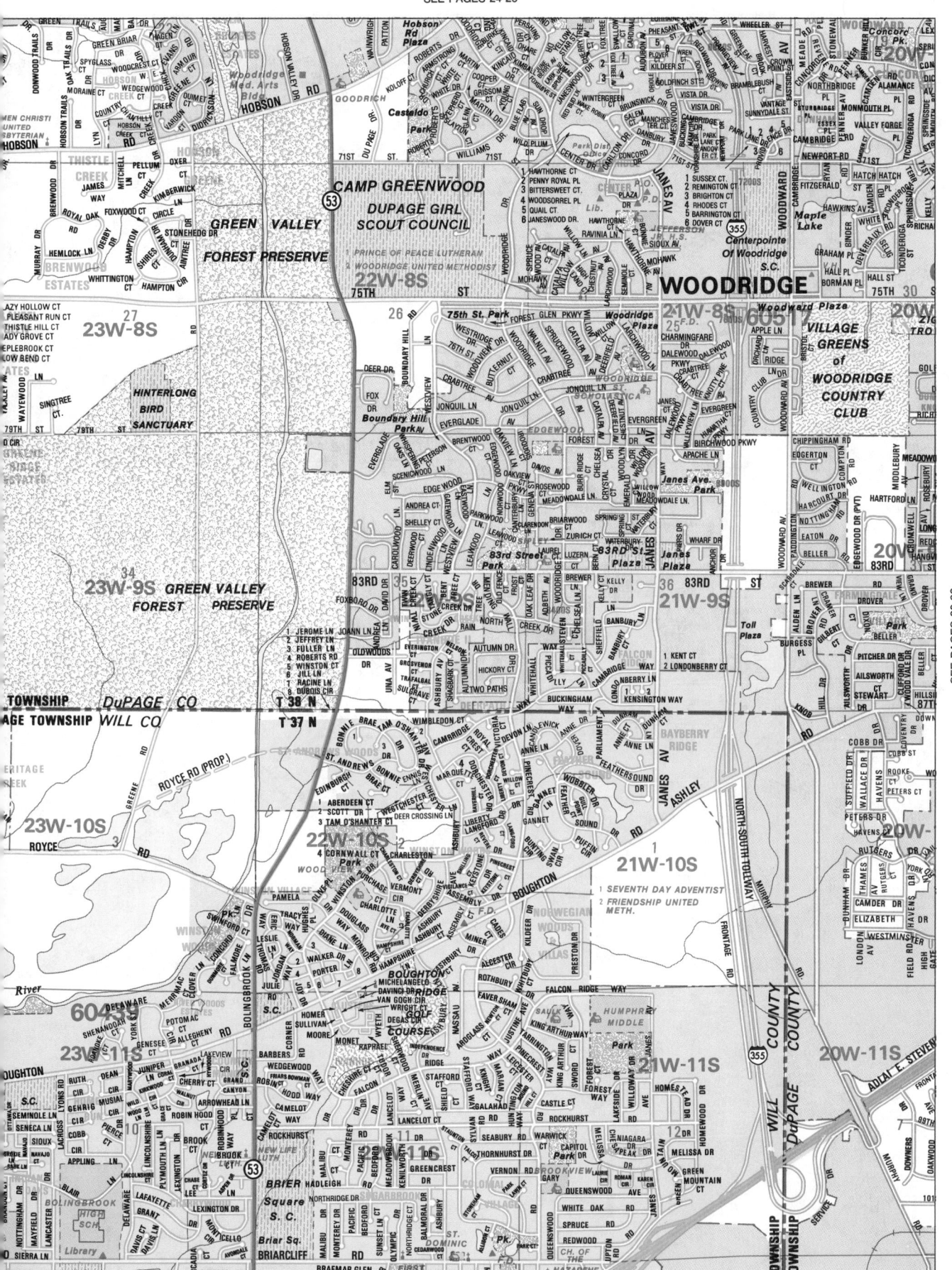

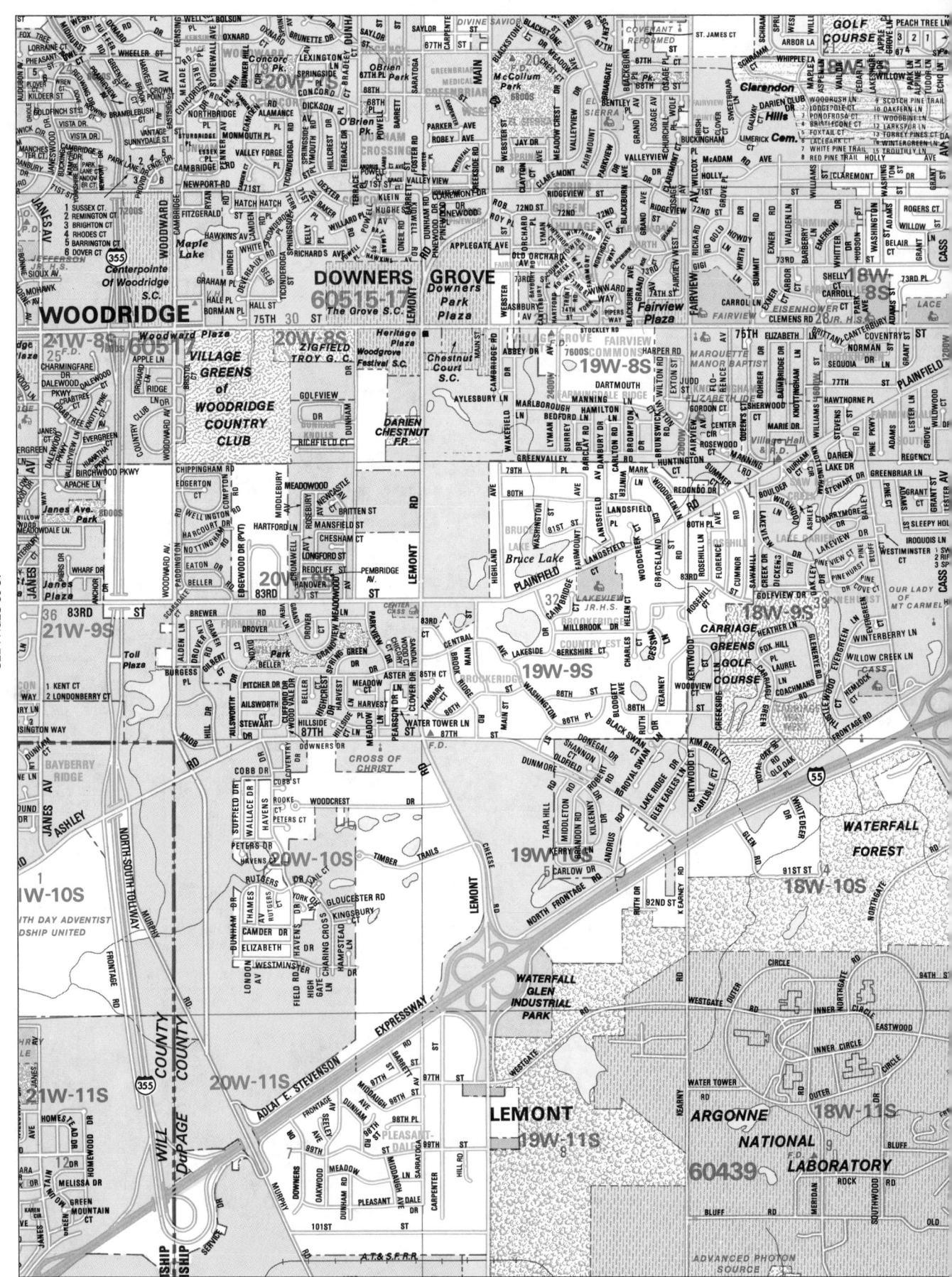

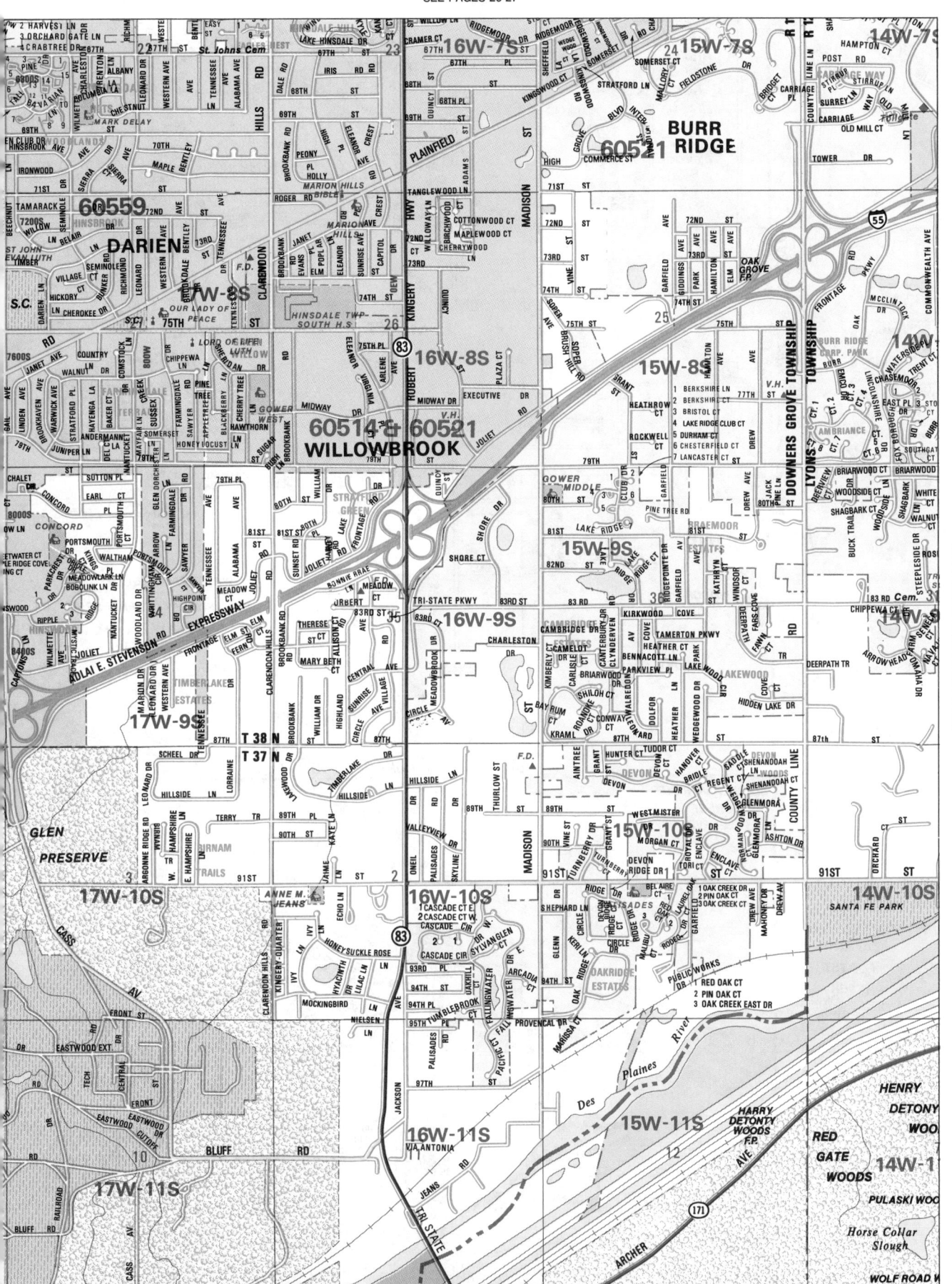

SEE COOK COUNTY PAGES 64-65

WATERFALL FOREST

21W-10S

20W-10S

19W-10S

18W-10S

SEVENTH DAY ADVENTIST
FRIENDSHIP UNITED
CH.

WATERFALL
GLEN
INDUSTRIAL
PARK

NORTH-SOUTH TOLLWAY

ADLAI E. STEVENSON EXPRESSWAY

21W-11S

20W-11S

LEMONT
19W-11S

ARGONNE
NATIONAL
LABORATORY
60439

18W-11S

WATER TOWER

ADVANCED PHOTON
SOURCE

ST. PATRICK

St. Patricks
Cem.

WOODRIDGE

WILL COUNTY / DuPAGE COUNTY

DOWNERS GROVE TOWNSHIP

PLEASANT-
DALE

A.T.& S.F.R.R.

DEERPATH LN

INTERNATIONALE
CENTRE

WOODRIDGE
F.P.

BROMBEREK

20W-12S

19W-12S

WATERFALL GLEN
FOREST PRESERVE

18W-12S

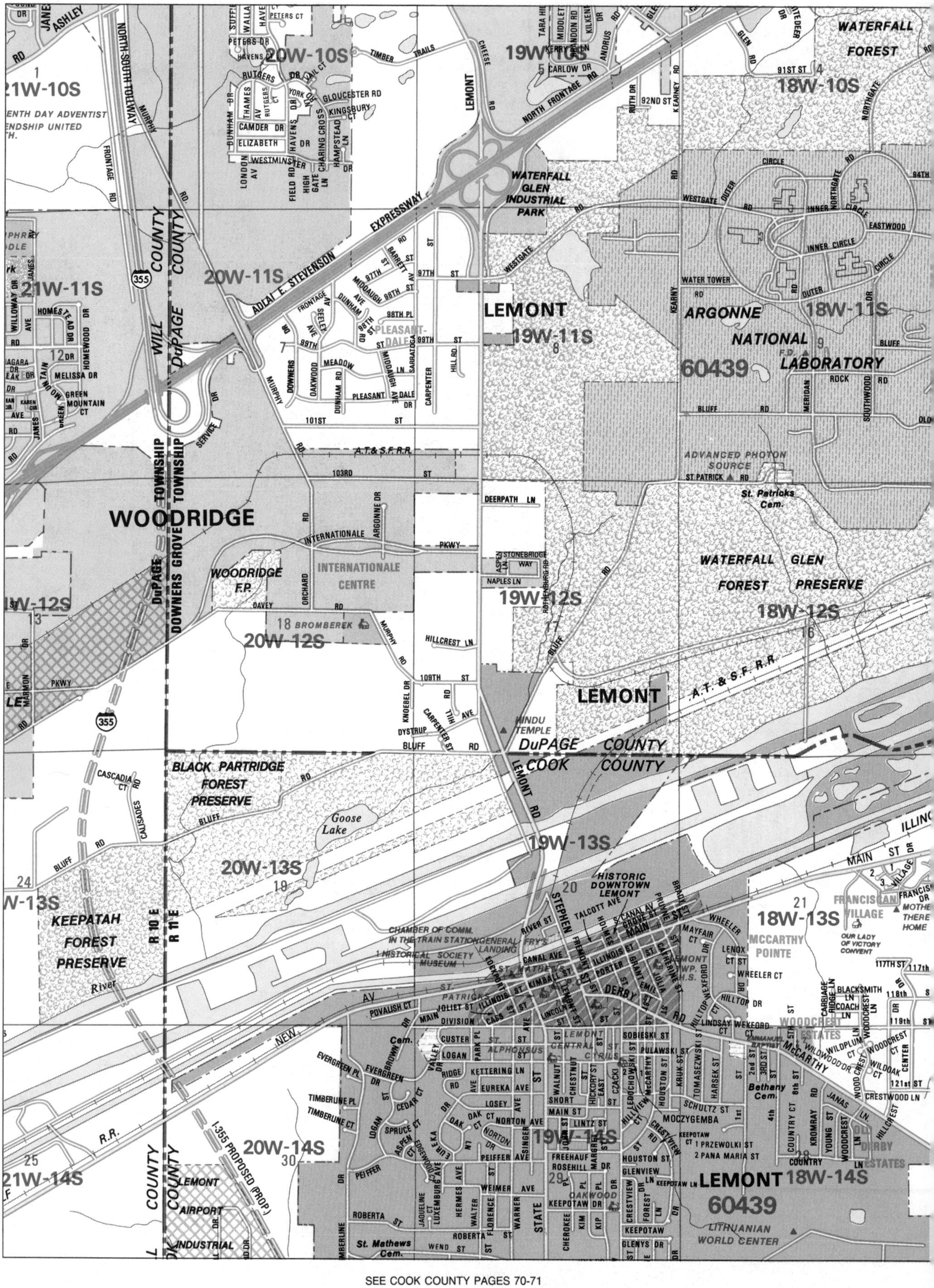

SEE WILL COUNTY PAGES 6-7

355

HILLCREST LN

KNOEBEL DR

DYSTRUP
BLUFF RD

LEMONT

HINDU
TEMPLE

DuPAGE COUNTY
COOK COUNTY

A.T. & S.F.R.R.

BLACK PARTRIDGE
FOREST
PRESERVE

Goose
Lake

20W-13S

19W-13S

ILLINOIS

MAIN ST

FRANCISCAN
VILLAGE

W-13S

KEEPATAH
FOREST
PRESERVE

River

R 10 E R 11 E

HISTORIC
DOWNTOWN
LEMONT

18W-13S

MOTHER
THERE
HOME

McCARTHY
POINTE

OUR LADY
OF VICTORY
CONVENT

CHAMBER OF COMM.
IN THE TRAIN STATION

GENERAL
LANDING

HISTORICAL SOCIETY
MUSEUM

WOODCREST
ESTATES

BLACKSMITH

R.R.

20W-14S

19W-14S

18W-14S

21W-14S

LEMONT
AIRPORT

INDUSTRIAL

St. Mathews
Cem.

LEMONT
60439

LITHUANIAN
WORLD CENTER

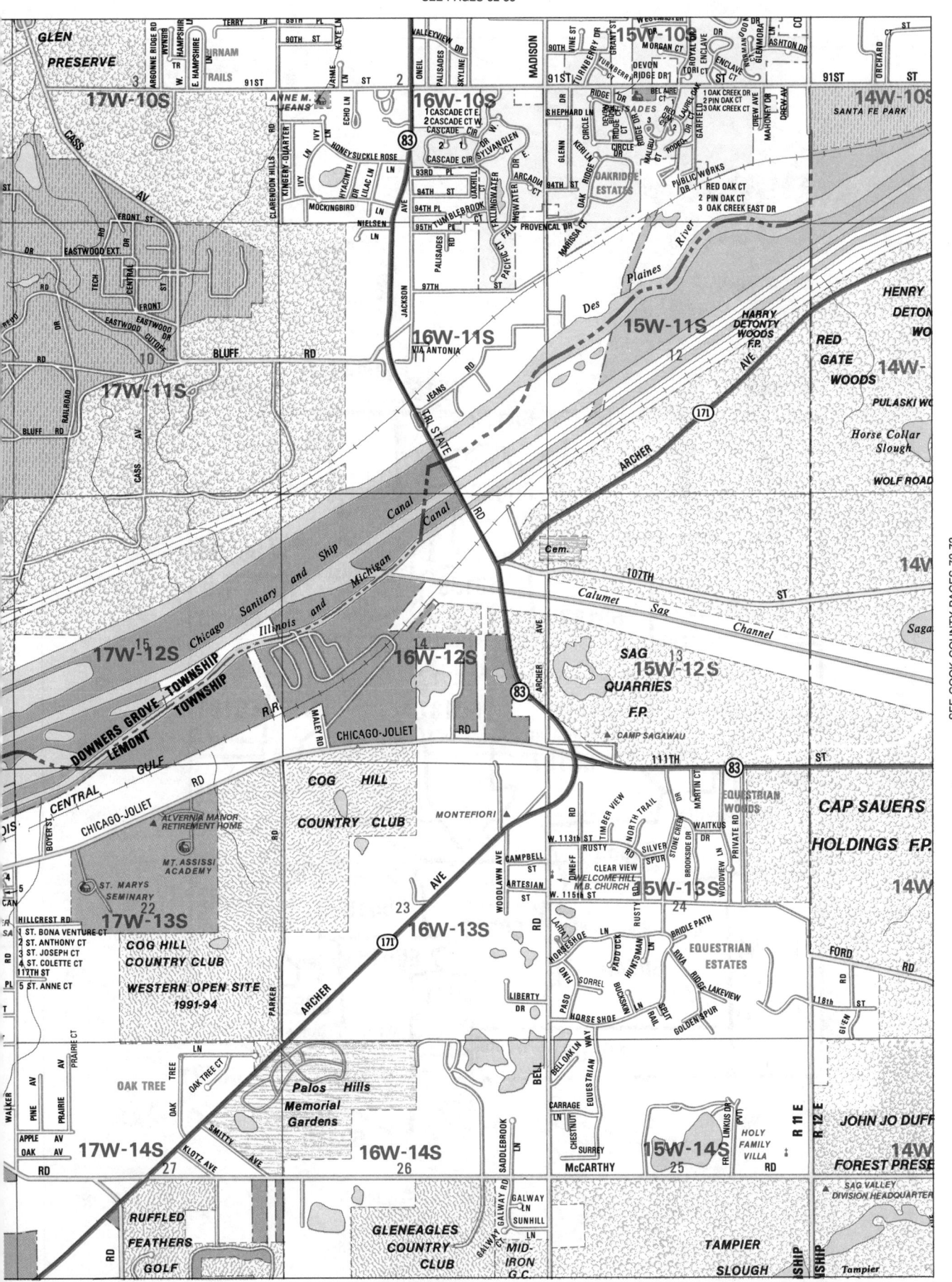

SEE COOK COUNTY PAGES 72-73

DuPage
County

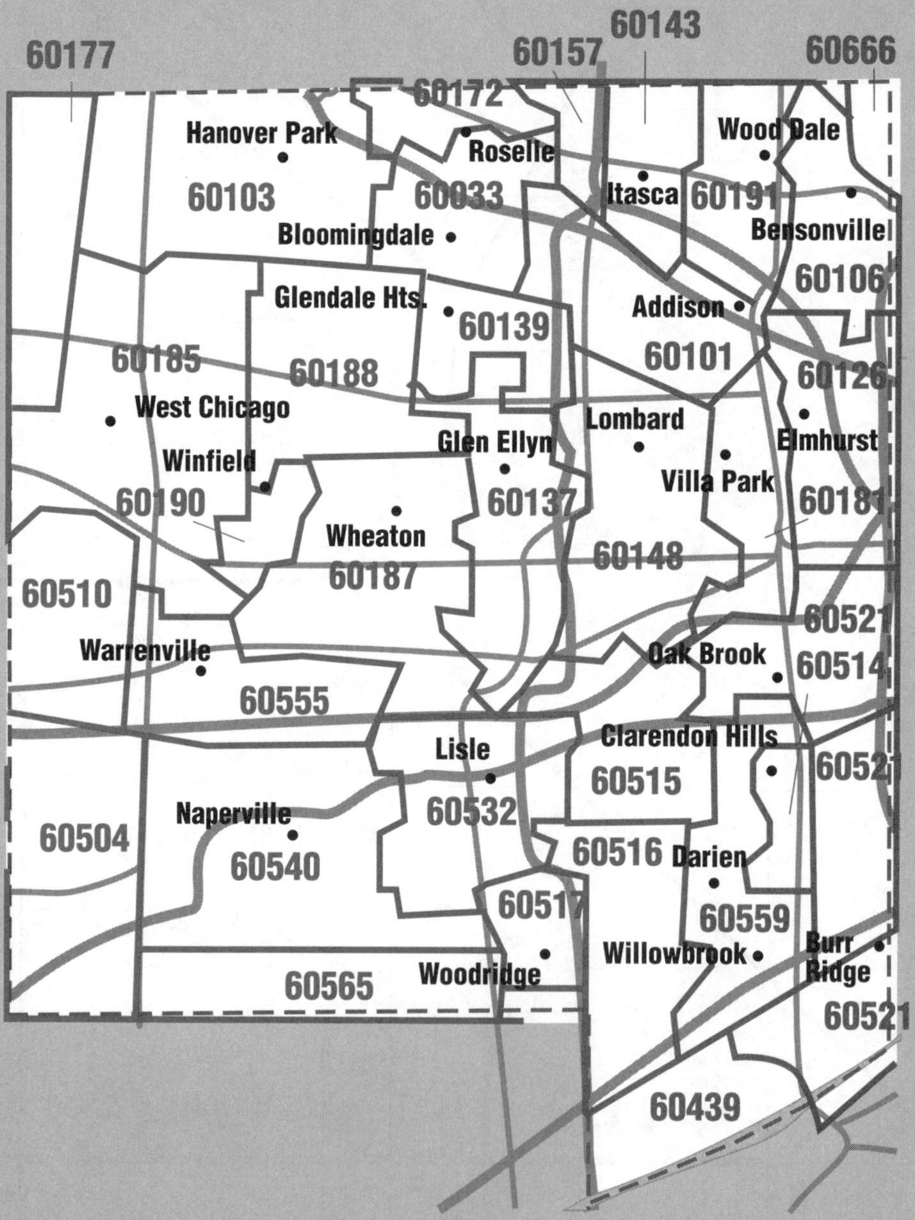

N

60177

60143

60157

60666

60172

Hanover Park

Roselle

Wood Dale

60103

60033

Itasca   60191

Bloomingdale •

Bensonville

60106

Glendale Hts.

Addison •

60139

60185

60188

60101

West Chicago •

Lombard •

60126

Winfield

Glen Ellyn •

Villa Park •

Elmhurst

60190

60137

60181

Wheaton •

60187

60148

60510

60521

Warrenville •

Oak Brook •

60514

60555

Lisle

Clarendon Hills

60515

60525

Naperville •

60532

60504

60516   Darien •

60540

60517

60559

Burr
Ridge

60565

Woodridge

Willowbrook •

60521

60439

Street Index

Addison

	Grid	Page
Adams Dr. N&S 18W-3N		14
Addison Rd. N&S		
.... 18W-2N,18W-4N		14
Adeline Av. E&W		
.... 18W-2N,19W-3N		14
Alden Dr. E&W 18W-3N		14
Amelia Ln. E&W		
.... 21W-4N,20W-4N		7
Annoreno Dr. E&W .. 19W-2N		14
Anvil Ct. NW&SE		
.... 19W-4N,19W-5N		8
Ardmore Av. N&S 18W-3N		14
Ardmore Terr. NE&SW		
.... 18W-2N		14
Ari Ct. N&S 21W-3N		13
Armitage Av. .. 18W-2N		14
Armitage Ct. N&S ... 20W-2N		14
Army Trail Rd. E&W		
.... 21W-3N,18W-3N		13
Ascot Pl. E&W 20W-5N		8
Ashley Ln. NW&SE .. 20W-5N		8
Ashwood Ct. N&S		
.... 19W-4N,19W-5N		8
Aspen Way E&W 20W-5N		8
Autumn Tr. E&W 20W-5N		8
Babbit Av. N&S 19W-4N		8
Barbara Ct. NE&SW		
.... 20W-4N		8
Barnwell Dr. NW&SE		
.... 20W-3N		14
Baybrook Ct. N&S		
.... 20W-5N,19W-5N		8
Baynard Av. N&S 19W-3N		14
Beach Dr. E&W 19W-4N		8
Belden Av. E&W ... 19W-2N		14
Belmont Av. E&W .. 17W-4N		9
Bernard Dr. E&W .. 20N-2N		14
Beverly Av. N&S		
.... 17W-2N,17W-3N		14
Blecke Av. N&S .. 18W-2N		14
Brashares Dr. N&S .. 19W-3N		14
Byron Av. E&W		
.... 20W-4N,18W-4N		8
Capitol Dr. E&W .. 20W-3N		14
Cardinal Av. N&S		
.... 17W-3N,17W-4N		14
Carmela Ct. N&S 20W-4N		8
Castle Rd. N&S...... 20W-4N		8
Catalpa Ct. N&S 17W-3N		14
Chatham Av. N&S		
.... 17W-2N,17W-3N		14
Cherry Hill Ct. N&S .. 17W-2N		14
Cherry Hill Rd. N&S .. 17W-2N		14
Chestnut St. N&S... 18W-3N		14
Church St. N&S 18W-3N		14
Circle Dr. N&S 19W-3N		14
Clarendon Av. N&S		
.... 19W-2N,19W-3N		14
Clover Ct. N&S 20W-5N		8
Clubview Ct. NW&SE		
.... 17W-4N		9
Collins Av. NW&SE .. 21W-3N		13
Commercial Rd. E&W		
.... 18W-2N		14
Compton Point E&W		
.... 19W-3N		14
Comstock Av. E&W .. 18W-2N		14
Congress Av. NW&SE		
.... 17W-3N		14
Corporate Dr. E&W .. 21W-3N		13
Cortland Ct. E&W....20W-2N		14
Country Club Dr. N&S		
.... 19W-3N,19W-4N		14
Craig Pl. N&S 19W-4N		8
Crawbridge Row N&S		
.... 21W-4N		7
Crocket Av. E&W ... 17W-2N		14
Dale Rd. E&W 20W-4N		8
Denise Ct. N&S......18W-4N		8
Diversey Av. E&W		
.... 19W-3N,18W-3N		14
Diversey Blvd. N&S.. 17W-3N		14
Douglas Av. N&S ... 17W-2N		14
Drake Av. E&W 18W-3N		14
DuPage Av. N&S ... 20W-4N		8
DuPage St. E&W 20W-4N		8
Edgewood Av. E&W		
.... 19W-4N		8
Eggerding Dr. E&W .. 18W-4N		8
Eighth Av. N&S..... 19W-4N		8
Elizabeth Dr. E&W .. 19W-4N		8
Ellsworth Av. N&S .. 17W-2N		14
Evergreen Av. N&S .. 19W-3N		14
Executive Dr. E&W .. 21W-3N		13
Factory Rd. N&S		
.... 19W-2N,18W-2N		14
Fairbank St. N&S.... 20W-4N		8
Farmwood Dr. E&W..19W-5N		8
Fay Av. E&W........18W-2N		14
Fiene Dr. N&S 18W-4N		8
Fisher Dr. N&S		
.... 19W-4N,19W-5N		8
Flora Pkwy N&S 20W-3N		14

	Grid	Page
Forest Ct. N&S 17W-3N		14
Forest Dr. E&W 17W-4N		9
Friars Ct. N&S 19W-3N		14
Fullerton Av. N&S		
.... 21W-3N,17W-3N		13
Gerri Ln. E&W 18W-2N		14
Gladys Av. NE&SW		
.... 20W-2N,17W-2N		14
Golden Gate Dr. E&W		
.... 20W-4N		8
Grace Av. N&S 19W-2N		14
Gramercy Ln. N&S .. 19W-3N		14
Grant Dr. N&S 18W-3N		14
Grant St. N&S 18W-4N		8
Green Meadow Dr. E&W		
.... 18W-4N		8
Green Oaks Ct. E&W		
.... 19W-4N		8
Greenbriar Dr. N&S .. 21W-5N		7
Greenridge St. N&S .. 20W-4N		8
Haddon Pl. N&S 19W-3N		14
Hale St. N&S........19W-3N		14
Harrow Ct. N&S 21W-4N		7
Harvard Av. N&S		
.... 18W-2N,18W-3N		14
Heather Ct. N&S ... 19W-5N		8
Heritage Dr. Cir. ... 19W-3N		14
Hickory Pl. E&W 19W-4N		8
Hickory Tr. N&S 19W-5N		8
Highview Av. N&S		
.... 19W-3N,18W-4N		14
Hilton Av. E&W 18W-3N		14
Holly Ct. E&W 17W-3N		14
Holtz Av. E&W		
.... 21W-4N,18W-4N		7
Honey Hill Rd. N&S		
.... 19W-4N,19W-5N		8
Horseshoe Ct. E&W 20W-3N		14
Howard Av. N&S 19W-4N		8
Hunter Ct. E&W 19W-4N		8
Huntington Rd. E&W		
.... 21W-4N		7
Industrial Rd. E&W .. 18W-2N		14
Interstate Rd. E&W		
.... 20W-2N,18W-2N		14
Iowa Av. N&S		
.... 18W-2N,18W-3N		14
Irmen Dr. N&S 19W-2N		14
Itasca Rd. N&S 20W-5N		8
Jamey Ln. NS&EW .. 19W-4N		8
Jefferey Dr. NE&SW		
................\.....20W-2N		14
John F. Kennedy Dr. N&S		
.... 18W-3N		14
Joseph Ln. N&S 19W-5N		8
Joyce Ln. N&S 18W-3N		14
Kay Av. N&S........19W-2N		14
Keebie Ct. N&S 19W-5N		8
Kendall Tr. E&W 19W-3N		14
Kenmore Av. N&S .. 21W-4N		7
Kings Point Dr. N&S 21W-4N		7
Kingston Dr. N&S ... 19W-3N		14
La Londe Av. N&S .. 19W-3N		14
Lake Manor Dr. E&W		
.... 19W-3N		14
Lake Park Dr. E&W		
.... 19W-3N,18W-3N		14
Lake St. NW&SE		
.... 21W-5N,17W-3N		7
Lancers Dr. N&S 20W-4N		8
Laporte Ct. N&S 18W-3N		14
Laporte Dr. N&S 18W-3N		14
Laura Dr. E&W 18W-2N		14
Lawler Av. N&S 21W-4N		7
Lenmore St. N&S ... 18W-4N		8
Lilac Ln. NE&SW ... 20W-5N		8
Lincoln Av. E&W		
.... 18W-3N,18W-4N		14
Lincoln Ct. N&S 18W-4N		8
Lincoln St. N&S		
.... 18W-2N,18W-4N		14
Linda Ln. E&W 19W-4N		8
Lois Av. N&S........19W-4N		8
Lombard Rd. N&S		
.... 20W-2N,20W-4N		14
Lorraine Av. E&W		
.... 19W-2N,17W-2N		14
Macie St. N&S...... 18W-4N		8
Maple Av. NE&SW .. 18W-3N		14
Maple Ct. N&S...... 18W-3N		14
Mare Barn Ln. N&S .. 19W-4N		8
Marilyn Terr. E&W .. 18W-4N		8
Mavis Ln. N&S...... 18W-4N		8
May St. N&S........18W-3N		14
Meadow Av. E&W .. 18W-4N		8
Meadows Blvd. N&S		
.... 21W-4N		7
Memory Ln. E&W ... 19W-3N		14
Meyer Dr. NW&SE .. 20W-4N		8
Michael Ln. E&W ... 18W-4N		8
Michigan Av. N&S		
.... 18W-2N,18W-3N		14
Michigan Ct. E&W .. 18W-2N		14
Mill Meadows Ln. N&S		
.... 19W-3N		14

	Grid	Page
Mill Rd. N&S		
.... 19W-3N,19W-4N		14
Millins St. NW&SE .. 19W-4N		8
Millins Ln. N&S...... 19W-4N		8
Mitchell Ct. N&S 21W-3N		13
Monarch Ln. N&S....20W-4N		8
Moreland Av. E&W		
.... 19W-3N,18W-3N		14
Morris Av. N&S 19W-3N		14
Mulloy Dr. E&W		
.... 21W-4N,20W-4N		7
Myrick Av. E&W		
.... 19W-3N,18W-3N		14
Natalie Ln. NS&EW .. 19W-3N		14
National Av. N&S		
.... 20W-3N,19W-3N		14
Natoma Av. N&S		
.... 19W-3N,19W-4N		14
Neva Av. N&S 19W-4N		8
Nike Rd. NW&SE....20W-2N		14
Normandy Dr. NS&EW		
.... 19W-4N		8
North Av. NW&SE .. 18W-3N		14
North-South Tollway N&S		
.... 21W-2N,21W-3N		13
Oak Leaf Dr. N&S .. 17W-3N		14
Oak Mill Ct. E&W 17W-3N		14
Oak Mill St. N&S.... 17W-3N		14
Oak St. E&W		
.... 18W-3N,17W-3N		14
Official Rd. E&W 18W-4N		8
Old Fence Rd. N&S .. 19W-4N		8
Old Grand Av. E&W .. 18W-3N		15
Palmer Av. E&W 19W-3N		14
Park Av. E&W		
.... 19W-3N,18W-3N		14
Park Pl. NW&SE ... 19W-4N		8
Parkview Dr. NW&SE		
.... 18W-3N		14
Pinehurst Blvd. E&W		
.... 21W-5N		7
Pioneer Dr. N&S		
.... 19W-3N,19W-4N		14
Plamondon Dr. N&S..20W-4N		8
Pleasant Av. N&S.... 19W-4N		8
Pleasant View Ct. E&W		
.... 18W-3N		14
Power Ct. N&S...... 18W-2N		14
Prairie Dr. N&S...... 19W-4N		8
Princeton St. N&S .. 18W-3N		14
Racquet Club Dr. E&W		
.... 19W-2N		14
Rampart Rd. E&W .. 20W-4N		8
Reed Ct. N&S...... 21W-4N		7
Regal Ct. N&S...... 20W-4N		8
Republic Dr. NE&SW		
.... 19W-3N		14
Richard Ct. N&S 19W-3N		14
Rohlwing Rd. N&S .. 21W-4N		7
Ronald Dr. E&W 19W-4N		8
Rose Av. N&S...... 19W-3N		14
Rosebud NW&SE ... 20W-5N		8
Royal Dr. E&W...... 20W-5N		8
Rozanne Dr. N&S.... 18W-4N		8
Rugeley Ct. N&S 19W-3N		14
Rumble Ln. N&S 21W-4N		7
Sable Dr. E&W		
.... 20W-5N,19W-5N		8
School St. N&S		
.... 18W-3N,18W-4N		14
Senate Dr. NW&SE .. 20W-3N		14
Sharon Dr. N&S 19W-4N		8
Sherry Ln. E&W 21W-4N		7
Sidney Av. N&S		
.... 21W-2N,18W-2N		13
Squire Ln. N&S 20W-5N		8
St. Auben Dr. E&W .. 19W-4N		8
State St. NE&SW ... 18W-3N		14
Stevens Dr. N&S 19W-4N		8
Stewart Av. N&S 19W-3N		14
Stiles Dr. N&S...... 19W-2N		14
Stone Av. E&W		
.... 20W-4N,19W-4N		8
Stonehedge Rd. N&S		
.... 19W-3N		14
Sumner St. N&S 18W-4N		8
Surrey Rd. E&W 19W-4N		8
Swift Rd. N&S		
.... 21W-3N,21W-5N		13
Swifton Commons Blvd. N&S		
.... 21W-4N		7
Tallyho Ct. N&S 20W-5N		8
Tamarac Blvd. N&S .. 21W-4N		7
Trinity Dr. E&W......19W-5N		8
Valerie Ln. N&S 18W-4N		8
Victoria Dr. E&W 21W-4N		7
Villa Av. NW&SE		
.... 17W-2N,17W-3N		14
Vista Av. N&S		
.... 19W-2N,19W-3N		14
Waveland Av. E&W .. 18W-4N		8
Wesley Dr. NS&EW		
.... 19W-3N,19W-4N		14
Westgate Dr. N&S .. 18W-2N		14
Westridge Pl. N&S .. 20W-5N		8

	Grid	Page
Westview Ct. E&W .. 18W-2N		14
Westwood Av. N&S .. 19W-2N		14
Westwood Tr. E&W ..19W-4N		8
White Fence Ln. NW&SE		
.... 19W-5N		8
Willow Glen St. E&W		
.... 19W-4N		8
Wilson Ln. N&S...... 19W-4N		8
Windsor Ct. E&W ... 21W-3N		13
Winthrop Ct. N&S		
.... 19W-2N,18W-2N		14
Wisconsin N&S		
.... 18W-2N,18W-3N		14
Wood Av. E&W......19W-5N		8
Woodland Av. E&W		
.... 21W-4N,18W-4N		7
Wrightwood Ct. E&W		
.... 20W-3N		14
Wyncroft Pl. N&S ...21W-4N		7
Yale Av. N&S		
.... 18W-2N,18W-3N		14
9th Av. N&S		
.... 19W-4N,19W-5N		8
10th Av. N&S...... 19W-4N		8
11th Av. NE&SW ... 19W-4N		8

Addison Township

	Grid	Page
Addison Rd. N&S ..18W-4N		8
Adele Av. E&W 19W-2N		14
Ann Ct. N&S...... 15W-4N		15
Ardmore Av. E&W .. 17W-7N		9
Arlington Heights Rd. N&S		
.... 19W-7N		8
Armitage Av. E&W .. 17W-4N		14
Arthur Av. N&S...... 18W-4N		8
Arthur Ct. E&W 16W-3N		15
Belmont Av. E&W		
.... 17W-4N,15W-4N		9
Belmont Pl. E&W ... 20W-3N		14
Bloomingdale Rd. NE&SW		
.... 20W-5N		8
Briar Ln. N&S		
.... 17W-4N,16W-4N		9
Broadview Av. N&S ..20W-3N		14
Bryn Mawr Av. E&W		
.... 17W-7N		9
Byron Av. E&W		
.... 20W-4N,18W-4N		8
Catalpa Av. N&S 17W-6N		9
Catalpa St. N&S 17W-4N		9
Cedar Av. N&S...... 17W-6N		9
Cedar St. N&S...... 17W-4N		9
Center St. E&W 18W-4N		8
Central Av. N&S		
.... 17W-2N,17W-6N		14
Chatham Av. N&S .. 17W-2N		14
Cherry St. N&S...... 17W-4N		9
Chestnut St. N&S....18W-4N		8
Church Rd. N&S 16W-4N		9
Clarendon Av. N&S .. 19W-2N		14
Clark St. N&S 18W-4N		8
County Line Rd. N&S		
.... 15W-3N		15
Crest Av. N&S		
.... 17W-4N,16W-4N		9
Crown Rd. N&S 15W-3N		15
Diversey Av. E&W		
.... 20W-3N,15W-3N		14
Diversey Pl. NW&SE		
.... 15W-3N		15
Donna Ln. E&W 16W-4N		9
Edgewood Av. N&S..17W-2N		14
Elizabeth St. N&S....20W-3N		14
Ellis St. N&S 16W-4N		9
Ellsworth Av. N&S		
.... 17W-2N,17W-4N		14
Elmhurst St. E&W .. 17W-6N		9
Elmwood Av. N&S .. 17W-4N		9
Emroy Av. N&S...... 15W-3N		15
Fairfield Av. N&S 19W-2N		14
Fay Av. E&W........17W-2N		14
Fisher Dr. N&S...... 17W-4N		9
Forest View Rd. E&W		
.... 17W-4N,16W-4N		9
Foster Av. E&W 17W-6N		9
George St. E&W 15W-4N		9
Grand Av. E&W		
.... 17W-3N,15W-3N		14
Grand Pl. E&W 15W-3N		15
Granville Av. E&W .. 19W-7N		8
Grove St. E&W		
.... 20W-5N,19W-5N		8
Hawthorne Av. N&S..16W-4N		9
Hickory Av. N&S 17W-4N		9
Hillcrest Av. N&S 17W-3N		14
Holtz Av. E&W		
.... 19W-4N,18W-4N		8
Howard Av. N&S 15W-3N		15
Itasca Rd. N&S		
.... 20W-4N,20W-5N		8
Itasca St. N&S...... 17W-6N		9
Jill Ct. N&S.......... 15W-3N		15
Jo-Ann Ln. E&W 19W-2N		14
John St. N&S........15W-4N		9

	Grid	Page
Joyce Av. N&S 19W-2N		14
Kramer Av. N&S 19W-2N		14
La Londe Av. N&S .. 19W-2N		14
Linda Ln. N&S		
.... 20W-3N,20W-4N		14
Lombard Rd. N&S .. 20W-3N		14
Lorraine Av. E&W ...17W-2N		14
Magnolia Av. NW&SE		
.... 19W-2N		14
Maple Av. N&S...... 17W-6N		9
Maple St. N&S...... 17W-4N		9
Marino Ct. E&W 19W-7N		8
Massel Ct. E&W 17W-4N		9
Mayfair Dr. N&S ... 17W-4N		9
Mill Rd. N&S		
.... 19W-4N,19W-5N		8
Monterey Av. N&S .. 19W-2N		14
Niles Av. N&S 18W-4N		8
North Av. E&W		
.... 20W-2N,18W-2N		14
North St. E&W...... 17W-6N		9
North Woods Dr. E&W		
.... 17W-4N		9
Nugent Av. N&S 18W-4N		8
Oak Leaf Dr. N&S .. 17W-3N		14
Oak Ln. E&W 17W-3N		14
Oak St. E&W		
.... 19W-3N,17W-3N		14
Oakdale Av. NW&SE		
.... 17W-3N		14
Parker St. N&S...... 15W-4N		9
Parkside Av. N&S....19W-7N		8
Patricia Ln. E&W 15W-3N		15
Pierce Rd. NW&SE .. 19W-7N		8
Pine Av. N&S........17W-6N		9
Pine Grove Av. N&S 16W-4N		9
Pine St. N&S........17W-4N		9
Pfund Av. E&W 17W-3N		14
Potter St. E&W 19W-5N		8
Prospect Av. N&S .. 19W-7N		8
Red Oak St. E&W		
.... 17W-4N,16W-4N		9
Ridgewood Av. N&S		
.... 16W-4N		9
Rodeck Ln. E&W 17W-3N		14
Schick Rd. E&W		
.... 20W-5N,19W-5N		8
Sherwood Dr. E&W .. 17W-3N		14
Sidney Av. N&S 19W-2N		14
Spruce Av. N&S 17W-6N		9
State St. N&S...... 18W-4N		8
Stone Av. E&W		
.... 18W-4N,17W-4N		8
Stonemill Ct. E&W .. 19W-4N		8
Third Av. N&S...... 17W-4N		9
Thorndale Av. E&W		
.... 19W-7N,18W-7N		8
Verrill Av. N&S...... 18W-4N		8
Victory Pkwy E&W		
.... 17W-3N,15W-3N		14
Virginia Ln. E&W 15W-3N		15
Vista Av. N&S 19W-2N		14
Walnut St. N&S...... 19W-5N		8
Washington St. E&W		
.... 17W-6N		9
West Av. N&S...... 17W-3N		14
White Pine Rd. E&W		
.... 17W-4N,16W-4N		9
Willow St. N&S...... 15W-3N		15
Wilson St. N&S...... 15W-3N		15
Wood Dale Rd. N&S		
.... 18W-4N,17W-4N		8
Woodland Av. E&W		
.... 19W-4N,17W-4N		8
Wrightwood Av. E&W		
.... 17W-3N,15W-3N		14
1st Av. N&S........19W-4N		8
2nd Av. N&S........16W-4N		9
2nd Av. N&S........19W-4N		8
2nd Ct. E&W 16W-4N		9
4th Av. N&S........19W-4N		8
5th Av. N&S........19W-4N		8
6th Av. N&S........19W-4N		8
7th Av. N&S........19W-4N		8
8th Av. N&S........19W-4N		8
9th Av. N&S........19W-4N		8

Aurora*
(Also see Kane County)

	Grid	Page
Aberdeen Ct. (Pvt) N&S		
.... 30W-9S		28
Acorn Ct E&W 31W-8S		28
Alder Ct E&W 32W-9S		28
Allspice Ct SW&NE .. 31W-8S		28
Andover Cir NS&EW		
.... 31W-9S		28
Andover Ct.* E&W .. 31W-10S		28
Andover Dr. NS&EW		
.... 31W-9S		28
Anton Dr. NE&SW .. 31W-8S		28
Arlington Ct. E&W .. 31W-8S		28
Arrowwood Ct N&S .. 32W-9S		28
Asbury Ln. NE&SW		
.... 32W-8S,32W-7S		28

	Grid	Page
Ashford Ln. NE&SW		
.................32W-5S	22
Ashton Ct N&S......32W-9S	28
Bangor Ln. N&S 31W-9S	28
Bannister Ln E&W . 32W-9S	28
Bar Harbour Ct.* E&W		
.................31W-10S	28
Bar Harbour Rd. NS&EW		
.................31W-9S	28
Barberry Ct N&S 32W-3S	16
Barley Ct. NW&SE ... 32W-4S	22
Bay View Ln. N&S . 30W-9S	28
Bayhead Ct. E&W ... 30W-8S	28
Bear Creek Ct. NW&SE		
.................30W-9S	28
Benjamin Franklin Ct. E&W		
.................30W-7S	22
Betsy Ross Ct. E&W		
.................30W-7S	22
Bilter Ferry Rd. E&W		
...........32W-4S,31W-4S	22
Bilter Rd. E&W		
...........32W-4S,31W-4S	22
Birmingham Dr. N&S		
.................32W-5S	28
Blaine Ct. E* NW&SE		
.................31W-8S	28
Blomfield Ct E&W ...31W-8S	28
Blomfield Ln E&W . 31W-8S	28
Blue Spruce Ln. E&W		
.................32W-3S	16
Boothbay Ln. E&W .. 31W-9S	28
Boxwood Ln. E&W .. 32W-3S	16
Brechenridge Dr. N&S		
.................32W-8S	28
Breckenridge Cir. E&W		
.................32W-7S	28
Breckenridge Ct. E&W		
.................32W-8S	28
Breckenridge Dr. NE&SW		
.........32W-8S,32W-7S	28
Breckenridge Ln. N&S		
.................32W-8S	28
Bremerton Ct.* E&W		
.................31W-8S	28
Bremerton Ln N&S .. 31W-8S	28
Bridgeport Ln. NS&EW		
.................31W-9S	28
Bristol Ct E&W 32W-9S	28
Brittany Ct E&W 32W-9S	28
Broadmoor Cir. NS&EW		
.................30W-9S	28
Brockton Ct. NE&SW		
.................31W-9S	28
Brook Ct. NE&SW .. 32W-9S	28
Brook Ln. NE&SW .. 32W-9S	28
Brookshire Ct. N&S ..32W-8S	28
Brookshire Ln. NE&SW		
.................32W-8S	28
Brookside Ct. NE&SW		
.................32W-5S	22
Brunswick Ln. NE&SW		
.................31W-9S	28
Bryan Ct. N&S 32W-9S	28
Bunker Hill Cir. E&W		
.........31W-7S,30W-7S	22
Butterfield Rd. E&W		
...........32W-3S,30W-3S	16
Camden Ln. NE&SW		
.................31W-9S	28
Cameron Ct. NE&SW		
.................31W-8S	28
Carriage Way NS&EW		
.................32W-8S	28
Carriage Way Ct. E&W		
.................32W-8S	28
Casper Ct. NE&SW . 32W-8S	28
Casper Ln. N N&S .. 32W-8S	28
Casper Ln. S N&S .. 32W-8S	28
Castlewood Ct. E&W		
.................32W-8S	28
Cave Creek Ct. NW&SE		
.................30W-9S	28
Chandler Ct. NE&SW		
.................32W-5S	22
Chatfield Ct. E&W .. 32W-8S	28
Chesapeake Dr. E&W		
.................30W-8S	28
Cheshire Dr. E&W .. 32W-8S	28
Churchill Ln. N&S ... 32W-8S	28
Cimarron Ct. NE&SW		
.................32W-8S	28
Clairmont Ct. NE&SW		
.................31W-7S	22
Clarendon Ln. E&W .32W-8S	28
Clearwood Ct. N&S .. 31W-8S	28
Clifton Ct. E&W.....31W-8S	28
Clover Field Ln. NW&SE		
.................31W-9S	28
Commerce St. N&S		
.........31W-7S,31W-6S	22
Commons Dr. N&S		
.........30W-8S,30W-7S	28
Commons Rd. N&S		
.........30W-7S,30W-6S	22
Compton Rd. NS&EW		
.................31W-8S	28
Cooper Ct. E. N&S		
.................31W-8S	28
Cooper Ct. W.* E&W		
.................32W-8S	28

	Grid	Page
Cooper Ln. N&S		
.........32W-8S,31W-8S	28
Corvalis Ct., South N&S		
.................31W-8S	28
Country Oaks Ct. E&W		
.................32W-4S	22
County Line Rd. N&S		
.................32W-8S	28
Courtland Ct.* E&W . 33W-6S	22
Cove Ct. E&W 32W-9S	28
Coventry Ct. E&W .. 31W-7S	22
Cranberry Ln. E&W . 32W-3S	16
Cranbrook Cir. NS&EW		
.................32W-5S	22
Creekside Ct. E&W .. 32W-4S	22
Crescent Ln., North N&S		
.................32W-5S	22
Crescent Ln., South N&S		
.................32W-5S	22
Cross Creek Ct.* NE&SW		
.................32W-4S	22
Cumberland Rd. N&S		
.................31W-9S	28
Devonshire Ct.* N&S		
.................32W-4S	22
Diane Dr. E&W		
.........32W-9S,31W-9S	28
Diane Dr. NS&EW . 31W-9S	28
Dogwood Ln. NE&SW		
.................32W-9S	28
Doncaster Dr. NS&EW		
.........32W-8S,32W-7S	28
Dorothy Dr. E&W .. 32W-9S	28
Dover Ln. E&W 31W-9S	28
Downing Ct. E&W .. 32W-5S	22
Driftwood Ln. NS&EW		
.................30W-7S	22
East West Tollway E&W		
.........32W-4S,31W-4S	22
Echo Ln. N&S 31W-8S	28
Emma Ct. EW&NS .. 32W-9S	28
Enterprise St. N&S		
.........31W-7S,31W-6S	22
Eola Rd. N&S		
.........32W-9S,31W-3S	28
Essex Ln. E&W 30W-8S	28
Eugene Ln. NE&SW . 31W-9S	28
Exchange Av. E&W		
.........31W-7S,30W-7S	22
Exchange Ct. N&S . 30W-7S	22
Executive Ct. NS&EW		
.................30W-7S	22
Fairfax Ct., East E&W		
.................31W-8S	28
Fairfax Ct., West NS&EW		
.................31W-8S	28
Fairmount Ct. E&W . 30W-8S	28
Farington Ct. N&S .. 32W-9S	28
Farington Ln. N&S . 32W-9S	28
Ferry Rd. E&W		
.........31W-4S,30W-4S	22
Fieldside Ln. N&S		
.........32W-9S,32W-8S	28
Flanders Ct. NE&SW		
.................32W-6S	22
Forest Dr. E&W......32W-7S	28
Forestview Ct. NE&SW		
.................32W-7S	28
Forestview Ln. NE&SW		
.................32W-8S	28
Fossil Creek Ct. NW&SE		
.................30W-9S	28
Fox Hill Ln. N&S		
.........31W-8S,31W-9S	28
Fox Hill Rd. E&W		
.........31W-9S,31W-8S	28
Fox Valley Ct. E&W .32W-9S	28
Fox Valley Ctr. Dr. NS&EW		
.................30W-7S	22
Fox Valley Dr. NE&SW		
.................32W-9S	28
Foxworth Ct. N&S .. 32W-8S	28
Frontenac Ct. N&S		
.........31W-9S,31W-8S	28
Frontenac St E&W		
.........31W-8S,31W-7S	28
Gloria Ct. N&S 32W-9S	28
Graystone Dr. N&S . 32W-5S	22
Greenbriar Pl. NW&SE		
.................30W-9S	28
Greenlake Dr. NE&SW		
.........32W-5S,31W-5S	22
Gregory St. NS&EW		
.........30W-8S,30W-7S	22
Gresham Ln. E. N&S		
.................31W-8S	28
Harbor Ct. E&W 32W-9S	28
Harbour Towne Pl. NW&SE		
.................30W-9S	28
Harvest Ct. NW&SE . 32W-8S	28
Hedgerow Dr. E&W . 32W-3S	16
Highfield Ct. E&W ... 32W-9S	28
Highfield Ln. E&W .. 32W-9S	28
Hull Ct N&S 32W-9S	28
Humphreys Ct. N&S .31W-8S	28
Interstate 88 E&W		
.........32W-4S,31W-4S	22
Inverness Ct. N&S .. 31W-8S	28
Inverness Dr. N&S .. 32W-8S	28
Isleworth NE&SW		
.........31W-9S,30W-9S	28
Jeremy Ranch Ct. E&W		
.................31W-9S	28

	Grid	Page
John Hancock Ct. E&W		
.................30W-7S	22
Johnsbury Ct.* E&W		
.................31W-10S	28
Johnsbury Ln. NS&EW		
.................31W-9S	28
Jono Dr. E&W		
.........32W-9S,31W-9S	28
Juniper Ct. N&S 32W-3S	16
Kaimy Ct. N&S 32W-9S	28
Karen Ct. N&S 32W-9S	28
Kathy Ct. N&S 32W-9S	28
Kenyon Ct. E&W ... 32W-6S	22
Killbrook Dr. NW&SE		
.................32W-7S	22
Kimberly Ct. NE&SW		
.................31W-7S	22
Kingshill Cir. NS&EW		
.................30W-9S	28
Kirkwood Ln. N&S . 31W-9S	28
Knightsbridge Ct. NW&SE		
.................30W-9S	28
Laconia Ln. N&S ... 31W-9S	28
Lake Dr. N&S 32W-9S	28
Lake Shore Ct. E&W		
.................32W-9S	28
Lakeside Dr. E&W		
.........32W-9S,32W-8S	28
Landing Dr. E&W 30W-8S	28
Larchmont Ct. NW&SE		
.................32W-8S	28
Lawrence Ct. E&W .. 31W-9S	28
Legacy Dr. NW&SE . 32W-5S	22
Legacy Dr. NS&EW . 32W-5S	22
Legend Ct. E&W ... 30W-9S	28
Leyland Ln. E&W ... 32W-9S	28
Liberty St. E&W 30W-6S	22
Lindrick Ln. E&W ... 32W-9S	28
Lonetree Ct. E&W .. 30W-9S	28
Long Grove Dr. N&S		
.........32W-9S,32W-8S	28
Long Grove Ln. N&S		
.................31W-9S	28
Longmeadow Dr. N&S		
.................30W-8S	28
Lynnfield Ct. E&W .. 31W-9S	28
Magnolia Ct. N&S . 32W-3S	16
Malmesbury Ct. NE&SW		
.................32W-5S	22
Mayfair Ct. E&W ... 30W-8S	28
Mayfair Ln. E&W ... 30W-8S	28
McCoy Dr. E&W		
.........32W-8S,30W-7S	28
Meadow Lakes Blvd. N&S		
.................31W-8S	28
Meadowlark Ct. NW&SE		
.................32W-3S	16
Meadowview Ln. NS&EW		
.................32W-8S	28
Medford Ct. N&S ...31W-8S	28
Middlebury Ct., East* E&W		
.................31W-10S	28
Middlebury Ct., West* E&W		
.................31W-10S	28
Middlebury Ln. N&S . 31W-9S	28
Millington Ct. E&W .. 32W-8S	28
Millington Ln. NS&EW		
.................32W-8S	28
Molitor Rd. N&S ... 32W-5S	22
Monarch Cir. NS&EW		
.........31W-9S,30W-9S	28
Montclare Ct. NW&SE		
.................32W-8S	28
Montgomery Rd. E&W		
.........32W-9S,30W-9S	28
Moss Ct. NW&SE ...32W-9S	28
Moss Ln. E&W 32W-9S	28
Natalie Ln. N&S ... 32W-9S	28
Newport Ln. NW&SE		
.................31W-9S	28
Normandy Ct. E&W . 30W-8S	28
Normantown Rd. N&S		
.................31W-9S	28
Norwalk Ct. E&W... 32W-6S	22
Norwood Ln. N&S		
.................31W-9S	28
Oak Shire Ct. SW&NE		
.................32W-4S	22
Oakdale Ct. NW&SE		
.................32W-9S	28
Oakhurst Dr. NS&EW		
.........32W-8S,31W-7S	28
Oakland Ln. N&S .. 31W-9S	28
Oxford Ct. NW&SE .. 32W-5S	22
Palmer Ct. N&S.....30W-9S	28
Paradise Canyon Ct. E&W		
.................30W-9S	28
Park Ridge Ct. NW&SE		
.................31W-8S	28
Parkwood Ln. E&W . 32W-5S	22
Partridge Ct. N&S .. 32W-3S	16
Patrick Henry Ct.* E&W		
.................30W-7S	22
Paul Revere Ct. E&W		
.................30W-7S	22
Peppertree N&S 30W-7S	22
Pheasant Run Ln. N&S		
.................31W-9S	28
Pine Lake Dr. NS&EW		
.................30W-9S	28
Pine Tree Ct. N&S . 32W-3S	16
Pinecone Ct. N&S...32W-3S	16
Pinegrove Ln. N&S . 32W-9S	28

	Grid	Page
Portland Ct. NE&SW		
.................31W-8S	28
Ports Mouth Ct. E&W		
.................32W-9S	28
Prairie Lake Ct. E&W		
.........31W-8S,30W-8S	28
Preston Ct. E&W.... 32W-4S	22
Primrose Ct. E&W . 31W-7S	22
Princeton Av. E&W		
.........31W-8S,30W-8S	28
Ptarmigan Ct. E&W . 32W-3S	16
Quail Ct. E&W 32W-3S	16
Quaker Hill Ct. E&W 30W-7S	22
Quincy Ct.* E&W . 31W-10S	28
Quincy Ln. NS&EW . 31W-9S	28
Radcliff Ln. N&S .32W-5S	22
Radford Dr. NE&SW		
.........32W-6S,32W-5S	22
Raintree Ln. E&W .. 30W-7S	22
Redbud Ct. N&S ... 32W-3S	16
Regency Ct. NE&SW		
.................30W-8S	28
Reserve Ct. N&S ... 32W-5S	22
Richland Ct. NE&SW		
.................31W-9S	28
Richmond Ln. NS&EW		
.........32W-8S,32W-7S	28
Ridge Av. E&W 32W-9S	28
Ridge Ln. N&S 32W-9S	28
River Wood Ct. N&S		
.................32W-3S	16
Rumford Ct. N&S . 31W-8S	28
Salem Ct. E&W.... 31W-8S	28
Sandpebble Ln. NW&SE		
.................31W-7S	22
Sandpiper Ct. N&S . 30W-7S	22
Sandpiper Dr. NW&SE		
.................31W-7S	22
Scheffer Rd. E&W . 32W-6S	22
Shadybrook Ln. N&S		
.................31W-7S	22
Sheffer Rd. NE&SW		
.........32W-6S,31W-6S	22
Sheffield Ct. N&S .. 32W-9S	28
Shoreline Dr. NS&EW		
.........31W-9S,31W-8S	28
Silver Creek Ct. NE&SW		
.................30W-9S	28
Society Hill Ln. NE&SW		
.................30W-9S	28
Society Hill Way NE&SW		
.................30W-9S	28
Southgate Ct. NW&SE		
.................30W-8S	28
St. Barthelemy Ln. NS&EW		
.................31W-8S	28
St. Croix Ct. NS&EW		
.................31W-8S	28
St. Kitts Ct. E&W .. 31W-8S	28
Stanton Ct.* NW&SE		
.................32W-4S	22
Stockton Ln.* E&W . 33W-6S	22
Stone Bridge Blvd. E&W		
.........32W-5S,31W-5S	22
Stratford Ct. N&S .. 30W-8S	28
Streamwood Ct. NW&SE		
.................32W-4S	22
Stubblefield Ct.* NE&SW		
.................32W-4S	22
Sun Tree N&S 31W-9S	28
Sunlight Ct. NE&SW		
.................32W-3S	16
Sussex Av. E&W ... 31W-6S	22
Sycamore Ln. N&S . 32W-9S	28
Terraceview Dr. N&S		
.................32W-9S	28
Thayer Ct. E&W ... 32W-9S	28
Thomas Jefferson Ct. E&W		
.................30W-7S	22
Thomas Paine Ct. E&W		
.................30W-7S	22
Thornley Ct. E&W ...32W-8S	28
Thurston Ct. N&S .. 32W-5S	22
Timber Hill Ct. NE&SW		
.................31W-9S	28
Timber Hill Ln. E&W 31W-9S	28
Trade St. N&S 30W-8S	28
Tralee Ct. E&W 31W-9S	28
Trefalgar NS&EW .. 32W-9S	28
Trenton Ct. E&W .. 30W-7S	22
Valayna Dr. N&S ... 32W-9S	28
Valley Forge Ct. E&W		
.................30W-7S	22
Vaughan Rd. N&S		
.........32W-8S,32W-6S	28
Village Ct. E&W ... 32W-9S	28
Village Green Ct. N&S		
.................31W-8S	28
Village Green Dr. E&W		
.........32W-8S,31W-8S	28
Village Green Dr., North N&S		
.................31W-8S	28
Walcott Rd. E&W .. 32W-9S	28
Warden Ct. E&W ... 32W-9S	28
Waterbury Cir. NE&SW		
.................32W-8S	28
Waterbury Dr. N&S . 32W-8S	28
Waterside Ct.* E&W . 33W-6S	22
Wentworth Ln. NW&SE		
.................32W-5S	22
West Hampton Ct. N&S		
.................30W-8S	28
West Ridge Ct. NE&SW		
.................32W-8S	28

	Grid	Page
Westbrook Dr. E&W 30W-8S	28
Westleigh Ct. E&W .. 32W-9S	28
Westmoor Ct. N&S . 32W-8S	28
Wexford Pl. NW&SE		
.................30W-9S	28
Wheatfield Ct. E&W . 32W-4S	22
Wheatland Ln.		
.........31W-8S,31W-9S	28
White Barn Ln. E&W		
.................32W-4S	22
White Eagle Dr. NS&EW		
.................30W-9S	28
White Eagle Dr., West N&S		
.................30W-9S	28
Whitehall Ct. E&W .. 32W-8S	28
Whiteway Ct. E&W .. 32W-8S	28
Willow Ct. NE&SW . 32W-9S	28
Willowfalls Ct. E&W . 32W-8S	28
Willowview Ln. E&W		
.................31W-8S	28
Wilshire Ct. NE&SW. 32W-4S	22
Winberie Ct. NW&SE		
.................30W-9S	28
Winchester Ct., East E&W		
.................31W-9S	28
Winchester Ct., West E&W		
.................31W-9S	28
Windemere Ln. N&S . 32W-9S	28
Windrift Dr. N&S ... 30W-8S	28
Wingate Ln. N&S .. 32W-4S	22
Woodcrest Ct. NS&EW		
.................32W-5S	22
Worchester Ln. NE&SW		
.........32W-9S,31W-9S	28
Wyndham Ct. NE&SW		
.................32W-8S	28
Yorkshire Ct. E&W . 32W-4S	22
Yorktown Ct. E&W . 30W-7S	22

Bartlett*
(Also see Cook County)

	Grid	Page
Abbey Rd. E&W 27W-7N	5
Adare Rd. E&W 27W-7N	5
Amherst Dr. E&W .. 28W-7N	5
Anchor Ct. NW&SE . 28W-7N	5
Ancient Oaks Ct. E&W		
.................30W-6N	4
Ancient Oaks Dr. N&S		
.................30W-6N	4
Anita Dr. N&S 28W-7N	5
Apple Orchard Dr. NE&SW		
.................29W-6N	5
Apple Valley Dr. E&W		
.................30W-6N	4
Appletree Ct. E&W . 29W-6N	5
Appletree Ln. N&S . 29W-6N	5
Arbor St. E&W 29W-6N	5
Army Trail Rd. E&W		
.........29W-5N,28W-4N	5
Aron Ct. NW&SE ... 29W-6N	5
Ascot Way NW&SE . 30W-5N	4
Asder Ln. N&S 27W-6N	5
Ashford Cir. NS&EW		
.................28W-5N	5
Aster Ct. N&S 28W-7N	5
Auburn Ln. E&W		
.........30W-7N,29W-7N	5
Balboa Terr. E&W . 28W-7N	5
Balmoral Dr. NW&SE		
.................28W-6N	5
Balsam Ln. N&S.....29W-6N	5
Bartlett Rd. N&S		
.........29W-6N,29W-7N	5
Bay Ct. NE&SW ... 28W-7N	5
Beach Ln. NE&SW . 28W-6N	5
Beechtree Ln. N&S . 29W-6N	5
Bragg Ct. E&W.....28W-7N	5
Bragg St. E&W.....28W-7N	5
Braintree Ln. N&S		
.........30W-7N,29W-7N	4
Branden Ln. NE&SW		
.................28W-6N	5
Briar Cliff Ln. E&W . 27W-7N	5
Brighton Ct. N&S .. 29W-6N	5
Brookside Dr. NE&SW		
.................28W-7N	5
Bryn Mawr Av. E&W		
.................29W-6N	5
Buckingham Dr. NE&SW		
.................28W-6N	5
Burnside Cir. N&S .. 28W-7N	5
Buttercup Ln. N&S . 28W-7N	5
Camdon Cir. N&S...28W-7N	5
Capistrano Terr. N&S		
.................28W-7N	5
Castlewood Ln. NW&SE		
.................27W-7N	5
Catalpa Ln. E&W ... 29W-6N	5
Cedar Ln. E&W.....29W-6N	5
Charter Oaks Ct. E&W		
.................30W-6N	4
Charter Oaks Dr. N&S		
.................30W-6N	4
Cherry Cir. E&W ... 27W-7N	5
Chesapeake Ct. NW&SE		
.................29W-6N	5
Chippendale Ct.* E&W		
.................27W-6N	5
Chippendale Dr. N&S		
.................27W-7N	5
Church Rd. N&S 27W-7N	5
Circle Dr. NW&SE .. 29W-6N	5
Clover Ct. NW&SE .. 28W-7N	5

	Grid	Page
Cobblestone Ln. E&W		
	31W-6N	4
Concord Dr. E&W		
	30W-6N,29W-6N	4
Confederation Dr. E&W		
	30W-6N,29W-6N	4
Congress Dr. E&W	30W-6N	4
Coral Av. E&W	28W-7N	5
Cottonwood Ln. N&S		
	29W-6N	5
Cromwell Cir. E&W	28W-7N	5
Cuyahoga Terr. N&S		
	29W-6N	5
Deanna Dr. N&S	28W-7N	5
Deerpath Ct. E&W	28W-6N	5
Devon Av. E&W		
	29W-7N,28W-7N	5
Dogwood Ln. NW&SE		
	29W-6N	5
Dorchester Ln. N&S	30W-6N	4
Driftwood Ln. N&S	29W-6N	5
Dunamon Ct. NW&SE		
	27W-7N	5
Dunamon Rd. E&W	27W-7N	5
Duxbury Cir. Cir.	29W-7N	5
E. Cleborne Dr. E&W		
	28W-7N	5
E. Country Dr. E&W	28W-7N	5
E. Sheridan Dr. E&W		
	28W-7N	5
Eagle Ct. N&S	28W-7N	5
Edgewood E&W	30W-5N	4
Elm Ct. E&W	28W-7N	5
Evergreen Av. E&W	29W-6N	5
Evergreen Ln. E&W	29W-6N	5
Ewell Ct. E&W	28W-7N	5
Fairfax Ln. NW&SE	28W-5N	5
Fairview Ln. N&S	29W-7N	5
Faith Ln. NE&SW	28W-7N	5
Far Hills Dr. N&S	30W-5N	4
Fieldstone Ct. N&S	30W-5N	4
Fieldstone Ln. NW&SE		
	30W-5N	4
Forest Ct. E&W	29W-6N	5
Forest Preserve Dr. E&W		
	30W-5N	4
Foster Av. E&W	29W-6N	5
Fox Chase Ct. NE&SW		
	30W-5N	4
Fox Chase Dr. NE&SW		
	30W-5N	4
Francine Dr. NE&SW		
	28W-7N	5
Franklin Dr. E&W		
	30W-6N,29W-6N	4
Fremont St. N&S	30W-6N	4
Garden St. NE&SW	29W-6N	5
Gardenia Ln. E&W	29W-6N	5
Ginger Brook Ln. E&W		
	28W-7N	5
Gloria Dr. N&S	29W-7N	5
Grand Oak Ct. NW&SE		
	30W-6N	4
Grandview Av. E&W	29W-6N	5
Granger Rd. E&W	29W-6N	5
Grant N&S	28W-6N,28W-7N	5
Groton Ct. E&W	30W-7N	4
Hadley Ct. N&S	28W-5N	5
Hampton Cir. E&W	28W-7N	5
Harbor Terr. N&S	28W-7N	5
Harbor Tr. NE&SW	27W-7N	5
Harvard Ct. N&S	28W-5N	5
Harvard Ln. NW&SE		
	28W-7N	5
Hawkins Ct. NW&SE		
	28W-6N	5
Hawthorne Av. N&S	29W-7N	5
Heather Ct. E&W	29W-6N	5
Hickory Ln. N&S	29W-7N	5
Hill Av. E&W	28W-7N	5
Hillandale Dr. N&S	29W-7N	5
Hillside N&S	27W-7N	5
Honeysuckle Ln. N&S		
	29W-6N	5
Hood Ct. E&W	28W-7N	5
Hyacinth Ln. E&W	29W-6N	5
Independence Dr. N&S		
	29W-6N	5
Independence Rd. NE&SW		
	29W-6N	5
Irvington Ct. E&W	28W-7N	5
Island Ct. E&W	27W-7N	5
Ivy Ct. NW&SE	29W-6N	5
Jackson Ct. E&W	28W-7N	5
Jackson St. N&S	28W-7N	5
Jasmine Ln. NW&SE		
	29W-6N	5
Jervey Ln. E&W	29W-7N	5
John Dr. E&W	28W-7N	5
Juniper Ln. E&W	29W-6N	5
Kelly Farm Ln. N&S	28W-4N	5
Kent Cir. E&W	29W-7N	5
Kingston Ln. N&S	29W-7N	5
La Jolla Terr. N&S	28W-7N	5
Ladysmith Rd. E&W	27W-7N	5
Lakeside Dr. NW&SE		
	27W-7N	5
Lakeview Ct. NE&SW		
	29W-7N	5
Lakewood Dr. N&S	29W-6N	5
Lamont Pkwy. E&W	29W-7N	5
Leaf Ct. N&S	29W-6N	5
Lee St. N&S	28W-7N	5
Lexington Dr. N&S	30W-6N	4

	Grid	Page
Lido Terr. E&W	28W-7N	5
Lincoln Dr. E&W	28W-7N	5
Longford Ct. E&W	27W-7N	5
Longford Rd. NE&SW		
	27W-7N	5
Longstreet Dr. N&S	28W-7N	5
Mansfield Ct. E&W	28W-7N	5
Maple Ct. N&S		
	29W-6N,29W-7N	5
Maple Ln. N&S	29W-6N	5
Marlboro Ct. E&W	28W-5N	5
Martingale Dr. N&S	29W-6N	5
McClellen Ct. E&W	28W-7N	5
McClellen St. E&W	28W-7N	5
McCook Ct. E&W	28W-7N	5
Meade Ct. E&W	28W-7N	5
Meadow Trail E&W	28W-7N	5
Millwood Dr. E&W	27W-7N	5
Monroe Dr. N&S	29W-6N	5
Morning Glory Ln. E&W		
	29W-6N	5
Munger Rd. N&S		
	31W-5N,31W-7N	4
Nantucket Ct. E&W	28W-6N	5
Newcastle Ln. NE&SW		
	28W-5N	5
Newport Blvd. NW&SE		
	28W-7N,27W-7N	5
Nicole Dr. NS&EW	27W-6N	5
Nolobay Terr. N&S	28W-7N	5
Northbridge Ct. NS&EW		
	30W-6N	4
Northgate Ln. N&S	30W-6N	4
Northgate Dr. NW&SE		
	30W-6N	4
Norwood Ln. E&W	30W-7N	4
Oak Glen Dr. E&W	29W-7N	5
Oakwood Ln. E&W	28W-7N	5
Oceanside Terr. E&W		
	28W-7N	5
Orchid Ln. E&W	29W-6N	5
Ord Ct. E&W	28W-7N	5
Oxford Rd. E&W	27W-7N	5
Paddock Pl. N&S	30W-6N	4
Palamino Ct. E&W	28W-6N	5
Pear Tree Ln. NS&EW		
	28W-7N	5
Penny Ln. NE&SW	28W-4N	5
Pickett Ct. E&W	28W-7N	5
Pinetree Ln. N&S	29W-6N	5
Pipers Dr. E&W	29W-7N	5
Plymouth Ct. E&W	28W-7N	5
Poinsetta Ln. NW&SE		
	29W-6N	5
Polk Ct. E&W	28W-7N	5
Poplar Ln. N&S	29W-6N	5
Poppy Ct. NE&SW	29W-6N	5
Porter Ct. E&W	28W-7N	5
Portsmouth Ct. N&S		
	28W-4N	5
Powis Rd. N&S		
	31W-6N,31W-7N	4
Prairie Av. N&S	28W-7N	5
Princeton Dr. N&S	30W-6N	4
Prospect Av. N&S		
	28W-7N,28W-8N	5
Redwood Ln. NE&SW		
	28W-7N	5
Rees Rd. E&W	31W-7N	4
Richmond Ln. E&W	28W-6N	5
Ridgefield Ln. N&S	30W-6N	4
Rose Ct. NE&SW	29W-6N	5
Rosecrans Ct. E&W	28W-7N	5
Rosewood Ct. N&S	29W-6N	5
S. Sheridan Dr. N&S		
	28W-7N	5
San Diego Pl. NW&SE		
	28W-7N	5
San Francisco Ct. N&S	28W-7N	5
Sandlewood Ct. N&S		
	27W-7N	5
Sandpiper Ct. E&W	27W-7N	5
Sausalito Ct. E&W	27W-7N	5
Schick Rd. E&W		
	30W-5N,27W-5N	4
Schick Rd. NW&SE	30W-5N	4
Scott Ct. E&W	28W-7N	5
Seacrest Ln. N&S	27W-7N	5
Shady Ln. E&W	29W-7N	5
Sherman Ct. E&W	28W-7N	5
Sherman St. E&W	28W-7N	5
Shorewood Ln. N&S		
	27W-7N	5
Shorewood Dr. NW&SE		
	27W-7N	5
Sparrow Ln. E&W	28W-5N	5
Spitzer Rd. N&S	31W-7N	4
Spruce Ln. E&W	29W-6N	5
Stearns Rd. E&W		
	32W-7N,27W-7N	4
Steeplechase Ln. NE&SW		
	30W-5N	4
Stonegate Ct. E&W	30W-5N	4
Struckman Blvd. E&W		
	29W-6N	5
Stuart Dr. N&S	28W-7N	5
Sundance Dr. NW&SE		
	27W-7N	5
Sundance Dr. E&W	28W-7N	5
Surf Ct. E&W	27W-7N	5
Surrey Rd. E&W	29W-7N	5
Swinford Rd. E&W	27W-7N	5
Sycamore Ln. E&W	29W-6N	5
Tallgrass Dr. N&S	28W-7N	5

	Grid	Page
Tennyson Rd. N&S	28W-7N	5
Thousand Oaks Dr. E&W		
	30W-6N	4
Towering Oaks Ct. N&S		
	30W-6N	4
Trailside Dr. N&S	27W-7N	5
Triple Crown Ct. N&S		
	28W-6N	5
Tulip Ln. NE&SW	29W-6N	5
Valewood Dr. N&S	29W-6N	5
Valley View Dr. E&W		
	28W-7N	5
Vernon Ln. NE&SW	28W-7N	5
Village Ln. NS&EW	28W-5N	5
Virginia Ct. E&W	29W-6N	5
Voyager Dr. NW&SE		
	28W-7N	5
W. Country Dr. E&W		
	28W-7N	5
Wallace Ct. E&W	28W-7N	5
Wallace St. E&W	28W-7N	5
Washington Ct. N&S		
	30W-6N	4
Washington St. N&S		
	30W-6N	4
Waterford Rd. N&S	27W-7N	5
Webster Av. E&W	29W-6N	5
Wentworth Ln. E&W		
	28W-5N	5
Wilcox Dr. E&W	28W-7N	5
Wildberry Ln. E&W	28W-7N	5
Wildwood Ct. NE&SW		
	27W-7N	5
Willow Ct. E&W	29W-6N	5
Winchester Ct. NE&SW		
	29W-6N	5
Windemere Ct. N&S	29W-6N	5
Windsor Dr. E&W	28W-6N	5
Winner's Cup Ct. NW&SE		
	28W-6N	5
Winston Ln. NW&SE		
	28W-6N	5
Wood Creek Tr. NW&SE		
	30W-5N	4
Woodhollow Ln. E&W		
	28W-7N,27W-6N	5
Woodland Hills Dr. NS&EW		
	30W-5N,30W-6N	4
Woodland Hills Pkwy. E&W		
	30W-5N	4

Bensenville

	Grid	Page
Addison Rd. N&S		
	16W-3N,16W-4N	15
Addison St. N&S		
	16W-4N,16W-6N	9
Algonquin Av. E&W	16W-5N	9
Argyle St. E&W		
	17W-6N,16W-6N	9
Ash Av. NE&SW	17W-7N	9
Barron St. N&S		
	16W-4N,16W-5N	9
Beeline Dr. E&W	16W-6N	9
Belmont Av. E&W	16W-4N	9
Bernice Dr. E&W	16W-3N	15
Birginal St. N&S	16W-6N	9
Blackhawk Terr. NE&SW		
	16W-5N	9
Bobby Dr. N&S	16W-3N	15
Brentwood Dr. N&S	15W-5N	9
Bretman Av. E&W	15W-5N	9
Bridget Ct. E&W	16W-3N	15
Broadway N&S	17W-5N	9
Brookwood St. N&S		
	17W-5N,16W-5N	9
Bryn Mawr Av. E&W		
	16W-7N	9
Center St. N&S		
	16W-3N,16W-5N	15
Central Av. E&W	16W-7N	9
Central Av. N&S	16W-7N	9
Church Rd. N&S		
	16W-4N,16W-5N	9
Country Club Dr. N&S		
	16W-6N	9
Crest Av. E&W	15W-4N	9
Daniel Dr. N&S		
	16W-3N,15W-4N	15
David Dr. NE&SW		
	15W-3N,15W-4N	15
Dennis Dr. E&W	15W-4N	9
Devon Av. E&W		
	17W-7N,16W-7N	9
Diana Ct. E&W	15W-4N	9
Dierks St. N&S	15W-5N	9
Division St. N&S	15W-5N	9
Dolores Dr. N&S		
	15W-3N,15W-4N	15
Donna Ln. E&W	15W-4N	9
Eagle Dr. E&W	16W-6N	9
Eastview Av. N&S	16W-6N	9
Edgewood Av. NE&SW		
	17W-7N	9
Ellis St. N&S		
	16W-4N,16W-7N	9
Elmhurst St. N&S		
	17W-6N,16W-6N	9
Entry St. N&S	16W-3N	15
Evergreen Av. N&S	15W-4N	9
Evergreen St. N&S	15W-4N	9
Farrari N&S	16W-3N	15
Foley St. N&S	16W-5N	9

	Grid	Page
Forestview Rd. E&W		
	16W-4N	9
Foster Av. E&W	16W-6N	9
Franzen St. N&S	16W-5N	9
Frontier Way E&W	16W-7N	9
Garden Av. N&S	15W-5N	9
Gateway Rd. E&W	16W-5N	9
Gaylin Ct. E&W	16W-5N	9
Glendale St. E&W		
	17W-5N,16W-5N	9
Golf Ln. E&W	16W-6N	9
Grace St. N&S		
	15W-4N,15W-5N	9
Grand Av. E&W		
	16W-3N,15W-3N	15
Green St. NW&SE		
	16W-5N,15W-5N	9
Greenlawn Av. N&S	15W-5N	9
Grove Av. E&W	16W-5N	9
Hamilton St. N&S	15W-5N	9
Henderson St. N&S	16W-5N	9
Hillside Dr. E&W		
	16W-6N,15W-6N	9
Indian Hill Dr. E&W	17W-7N	9
Industrial Dr. E&W	16W-6N	9
Iroquois Ct. NW&SE		
	17W-7N	9
Iroquois Ln. E&W	17W-7N	9
Irving Park Blvd. E&W		
	16W-5N,15W-5N	9
Itasca St. E&W		
	17W-6N,16W-6N	9
Jacquelyn Dr. E&W	15W-4N	9
James St. E&W	16W-3N	15
Jefferson St. E&W		
	16W-4N,15W-4N	9
John St. N&S	15W-4N	9
Judson St. N&S		
	16W-3N,15W-3N	15
Kevin Ln. N&S	16W-6N	9
Leland Ct. E&W	16W-6N	9
Lincoln Av. NW&SE	15W-5N	9
Main St. NW&SE	16W-5N	9
Maple Ln. E&W	16W-6N	9
Marion St. N&S		
	15W-4N,15W-5N	9
Marshall Rd. N&S		
	16W-5N,16W-6N	9
Martha St. N&S	16W-5N	9
Mason St. N&S		
	16W-4N,16W-5N	9
May St. N&S		
	15W-4N,15W-5N	9
McClean Av. E&W	16W-6N	9
Medinah St. E&W		
	17W-5N,15W-5N	9
Meigs Ct. E&W	15W-5N	9
Memorial Rd. E&W	15W-5N	9
Meyer Rd. N&S	16W-6N	9
Midway St. E&W	15W-5N	9
Miner St. N&S		
	16W-4N,16W-5N	9
Mohawk Dr. NW&SE		
	16W-5N	9
Nordic St. E&W		
	17W-6N,16W-6N	9
O'hare Ct. E&W	15W-5N	9
O'leary Dr. N&S	15W-5N	9
Orchard Av. N&S	15W-5N	9
Pamela Dr. N&S	15W-4N	9
Park St. N&S	15W-4N	9
Parkside Ln. N&S	15W-5N	9
Pershing Av. N&S	15W-5N	9
Pine Av. N&S	15W-5N	9
Pleasant St. E&W	16W-5N	9
Plentywood Ln. E&W		
	16W-5N	9
Poppy Ln. N&S	16W-5N	9
Railroad St. NW&SE		
	16W-5N,15W-5N	9
Ridgewood Av. N&S		
	16W-5N	9
Riverforest N&S	15W-4N	9
Roosevelt Av. E&W		
	16W-5N,15W-5N	9
Roosevelt St. N&S	16W-5N	9
Rose St. N&S		
	15W-4N,15W-5N	9
Roxanne Av. E&W	15W-4N	9
Rt. 19 E&W		
	16W-5N,15W-5N	9
Rt. 83 N&S	16W-5N,16W-7N	9
Sivert Ct. E&W	16W-6N	9
Sonny Ln. E&W	16W-3N	15
Spruce Av. N&S		
	17W-5N,17W-7N	9
Stoneham St. E&W		
	17W-6N,16W-6N	9
Sunset Ct. E&W	15W-5N	9
Supreme Dr. NS&EW		
	16W-6N	9
Thomas St. N&S	16W-6N	9
Tioga Av. N&S	16W-5N	9
Tower Ln. N&S	17W-7N	9
Twin Oaks St. N&S		
	17W-6N,16W-6N	9
Virginia Av. E&W	15W-4N	9
Walnut St. N&S		
	16W-4N,16W-5N	9
Washington St. E&W		
	16W-4N,15W-4N	9
William St. N&S	16W-3N	15
Wilson Ct. E&W	16W-6N	9

	Grid	Page
Wood Av. E&W		
	16W-5N,15W-5N	9
Woodland Av. N&S	16W-4N	9
York Rd. N&S		
	16W-3N,16W-7N	15

Bloomingdale

	Grid	Page
Alsace Ct. N&S	24W-5N	6
Alyssa Ct.* N&S	24W-6N	6
Amberwood Ct. N&S		
	22W-4N	7
Amherst Ct. E&W	22W-4N	7
Applewood Ln. E&W		
	22W-4N	7
Arbor Ln. N&S	22W-4N	7
Army Trail Rd. N&S		
	25W-4N,22W-4N	6
Ashfield Ct. NS&EW		
	24W-5N	6
Atherton Ct. E&W	23W-5N	6
Bayview Ct. E&W	22W-4N	7
Bedford Ln. N&S	22W-4N	7
Benton Ln. NS&EW	24W-4N	6
Benton Ln. N&S	24W-5N	6
Biarritz Ct. N&S	24W-5N	6
Billings E&W	23W-5N	6
Birchwood Ln. NW&SE		
	22W-4N	7
Bloomfield Cir. E&W	24W-5N	6
Bloomfield Pkwy. N&S		
	24W-5N	6
Bloomingdale Rd. N&S		
	23W-4N,23W-6N	6
Bob-o-link Dr. NE&SW		
	24W-4N	6
Bobwhite Ct. E&W	24W-4N	6
Bobwhite Ln. N&S	23W-4N	6
Bourne Ln. NE&SW	23W-5N	6
Bradley Ct. E&W	22W-4N	7
Braintree Dr. N&S	23W-5N	6
Brentwood Ct. E&W		
	22W-4N	7
Brewster Ln. E&W	23W-5N	6
Briarwood Ln. E&W	22W-4N	7
Bridgewater Ln. NE&SW		
	23W-4N	6
Brighton Dr. E&W	25W-4N	6
Broker Rd. NE&SW	22W-6N	7
Brookdale Dr. N&S	22W-4N	7
Bunting Ln. N&S	23W-4N	6
Butterfield Dr. N&S	24W-4N	6
Byron Av. E&W	22W-4N	7
Cambridge Ln. N&S	23W-5N	6
Camden Dr. E&W	25W-5N	6
Cardinal Dr. N&S		
	24W-5N,23W-4N	6
Carlton Dr. E&W	22W-4N	7
Carriage Way E&W	22W-4N	7
Carthage Ct. NE&SW		
	22W-4N	7
Castlerock Ln. N&S	24W-5N	6
Cayuga St. NE&SW	25W-5N	6
Chambord Ct. NW&SE		
	24W-5N	6
Chatham Ct. NW&SE		
	23W-5N	6
Chelsea St. N&S	24W-5N	6
Circle Av. N&S	22W-4N	7
Clearbrook Ln. NE&SW		
	22W-4N	7
Club House Dr. E&W		
	24W-5N	6
College Dr. N&S	22W-4N	7
Colony Ct. E&W	22W-4N	7
Colony Green Dr. NS&EW		
	22W-4N	7
Constitution Dr. N&S		
	23W-5N	6
Country Club Ct. NE&SW		
	22W-4N	7
Country Club Dr. N&S		
	24W-5N	6
Coventry Ct. N&S	22W-4N	7
Covington Dr. N&S		
	25W-4N,25W-5N	6
Crandon Ct. N&S	23W-5N	6
Creekside Dr. NW&SE		
	22W-4N	7
Crest Ct. E&W	22W-4N	7
Crestwood Ln. N&S	22W-4N	7
Cunningham Dr. E&W		
	23W-4N	6
Cunningham Ln. N&S		
	23W-4N	6
Darby St. N&S	23W-5N	6
Dartmouth Ct. NE&SW		
	22W-4N	7
Dawn Ct. E&W	22W-4N	7
Day St. N&S	23W-5N	6
Dee Ct. N&S	22W-4N	7
Deeke Ct. N&S	25W-7N	6
Deer Glen Way E&W		
	22W-4N	7
Dijon Ct. N&S	24W-5N	6
Douglas Dr. E&W	22W-4N	7
Dover Ct. NW&SE	22W-4N	7
Downing Dr. E&W	22W-4N	7
Driftwood Ln. NE&SW		
	22W-4N	7
Durham Ln. N&S	22W-4N	7
Duxbury Ct. NW&SE		
	23W-5N	6
Eagle Ct. N&S	23W-4N	6

Street	Grid	Page
Eagle Ln. NE&SW	23W-4N	6
Eaton Ct. N&S	25W-4N	6
Edgebrook Dr. E&W	22W-5N	7
Edgewater Ct. N&S	23W-4N	6
Edgewater Dr. E&W	23W-4N,22W-4N	6
Elmwood Ln. N&S	22W-4N	7
Emerson Ct. NE&SW	22W-4N	7
Epson Ct. N&S	23W-5N	6
Eric Ct.* E&W	24W-6N	6
Esprit Ct. E&W	24W-5N	6
Essex Ct. E&W	24W-5N	6
Euclid Av. N&S	22W-5N	7
Euclid Ct. E&W	22W-5N	7
Evergreen Ln. N&S	22W-4N	7
Fairchild Ct. NW&SE	22W-4N	7
Fairfield Ct. NW&SE	23W-5N	6
Fairfield Way NS&EW	23W-5N	6
Fairlane Ct. E&W	23W-5N	6
Falcon Ct. NW&SE	24W-4N	6
Farrington Ct. N&S	23W-5N	6
Fernwood Ln. N&S	24W-5N	6
Fontaine Ct. E&W	24W-5N	6
Forest Dr. E&W	23W-6N	6
Fosmor Ct. E&W	24W-5N	6
Foster Av. E&W	23W-6N	6
Founders Point N. Cir.	23W-5N	6
Founders Point S. Cir.	23W-5N	6
Fox Ct. E&W	25W-4N	6
Franklin St. E&W	24W-5N,23W-5N	6
Freeport Dr. E&W	22W-5N	7
Fremont Ct. N&S	22W-4N	7
Garden Way NW&SE	24W-5N	6
Gary Av. N&S	25W-4N,25W-6N	6
Georgetown Ct. NW&SE	22W-4N	7
Ghingham Ln. E&W	23W-5N	6
Glen Ellyn Rd. N&S	22W-4N,22W-5N	7
Glengarry Rd. E&W	24W-4N	6
Greenfield Dr. N&S	24W-5N	6
Greenway Dr. E&W	23W-4N,22W-4N	6
Gresham Ct. N&S	23W-5N	6
Hampton Ct. N&S	22W-4N	7
Hancock E&W	23W-5N	6
Harbor Ct. E&W	22W-4N	7
Harvard Ln. N&S	22W-4N	7
Hawthorne Dr. E&W	22W-5N	7
Hearthside Dr. E&W	22W-4N	7
Hedgegrow Dr. E&W	22W-4N	7
Hempstead Dr. NW&SE	23W-4N	6
Hempstead Ln. NE&SW	23W-4N	6
Hillandale Ct. N&S	23W-5N	6
Hillandale Dr. N&S	23W-5N	6
Hillcrest Terr. E&W	23W-6N	6
Independence Ln. E&W	23W-5N	6
Indiana Ct. NW&SE	22W-4N	7
Ironwood Dr. E&W	22W-5N	7
Ivy Ln. N&S	22W-5N	7
Jackson Ln. N&S	23W-5N	6
Jason Ct.* E&W	24W-6N	6
Jefferson Ln. N&S	23W-5N	6
Juniper Ln. N&S	22W-5N	7
Karswell Ct. N&S	23W-5N	6
Kendall Ct. E&W	22W-4N	7
Kensington Ct. E&W	24W-4N	6
Killdeer Rd. N&S	24W-4N	6
Kingston Ln. N&S	22W-5N	7
Knollwood Dr. N&S	25W-4N	6
Lake St. NW&SE	24W-6N,22W-5N	6
Lakemont Ct. NE&SW	24W-4N	6
Lakeshore Ln. NW&SE	22W-4N	7
Lakeview Ct. NW&SE	22W-4N	7
Lakeview Dr. N&S	22W-5N	7
Lakeview Ln. NE&SW	23W-4N,22W-4N	6
Lakewoods Ct. NE&SW	23W-5N	6
Langton Ct. N&S	23W-4N	6
Langton Ln. N&S	23W-4N	6
Larchmont Ln. N&S	23W-5N	6
Laurel Ln. NE&SW	22W-4N	7
Lawford Ct.* N&S	22W-5N	7
Lawrence Av. E&W	25W-5N,23W-6N	6
Lehigh Ln. N&S	23W-5N	6
Leslie Ln. E&W	23W-5N	6
Lincoln Ct. N&S	22W-4N	7
Lockwood Ln N&S	22W-5N	7
Log Lodge Dr. NE&SW	23W-6N	6
Long Ridge Dr. N&S	23W-5N	6
Lorraine Cir. NS&EW	24W-5N	6
Lynbrook Dr. E&W	23W-5N	6
Manor Ln. N&S	22W-5N	7
Maple Av. N&S	23W-5N	6
Maple Ct. NE&SW	23W-5N	6
Martin Ct. E&W	24W-4N	6
Martin Ln. N&S	24W-4N	6
Meadowlark Rd. NW&SE	24W-4N	6
Melbourne Ct. N&S	23W-5N	6
Millcreek Ln. NE&SW	22W-4N	7
Millford Ct. N&S	23W-5N	6
Millford Ln. N&S	23W-5N	6
Milton Ct. N&S	22W-4N	7
Morningside Dr. N&S	22W-4N	7
N. Bristol Dr. N&S	25W-5N	6
N. Hampshire Cir. NE&SW	24W-5N	6
N. Manchester Ct. E&W	25W-5N	6
N. Sherwood NE&SW	24W-5N	6
N. Windham Ln. N&S	24W-5N	6
N. Windsor NE&SW	24W-5N	6
Needham Dr. E&W	24W-5N	6
Neville Ct. N&S	23W-5N	6
Nordic Rd. E&W	22W-5N	7
Northwestern Ct. N&S	22W-4N	7
Northwood Ln. N&S	24W-5N	6
Norton Dr. NW&SE	23W-4N	6
Norton Ln. N&S	23W-4N	6
Oak St. NW&SE	23W-5N	6
Oakwood Ln. E&W	22W-4N	7
Old Mill Ln. N&S	24W-4N	6
Oneida Ct. NW&SE	25W-5N	6
Orchard Ln. NE&SW	24W-5N	6
Oriole St. N&S	23W-4N	6
Otsego Ct. N&S	25W-5N	6
Oxford Ct. N&S	23W-4N	6
Oxford Ln. N&S	23W-4N	6
Oxford Pl. NW&SE	23W-4N	6
Park Av. E&W	23W-5N,22W-5N	6
Parkway Av. E&W	24W-5N	6
Pebble Creek Dr. N&S	22W-4N	7
Penwick Ct.* N&S	22W-5N	7
Pheasant Ct. NW&SE	24W-4N	6
Pheasant Ln. E&W	24W-4N	6
Picton Rd. E&W	23W-6N	6
Pinewood Ln. E&W	22W-4N	7
Pleasant Av. N&S	22W-5N	7
Plymouth Ln. NE&SW	23W-4N	6
Prairie Av. N&S	22W-5N	7
Quail Ln. E&W	22W-4N	7
Quincy Ct. E&W	23W-5N	6
Radcliffe Ct. E&W	24W-5N	6
Raleigh* E&W	22W-5N	7
Raven Ln. E&W	23W-4N	6
Red Bird Ct. NE&SW	24W-4N	6
Red Bird Rd. NW&SE	24W-4N	6
Redwing Ct. N&S	24W-4N	6
Regency Dr. NS&EW	22W-4N	7
Ridge Av. N&S	23W-5N	6
Ridgewood Dr. NW&SE	24W-5N	6
Robin Ct. NE&SW	23W-4N	6
Rosedale Av. N&S	23W-5N	6
Rosedale Ct. N&S	23W-5N	6
Royal Ln. E&W	24W-5N	6
Royce Ct. NW&SE	23W-4N	6
Royce Dr. E&W	23W-4N	6
Salem Ct. NE&SW	23W-5N	6
Saranac Ct. N&S	25W-5N	6
Schick Rd. E&W	25W-5N,22W-5N	6
Scott Dr. E&W	25W-4N	6
Seneca Tr. N&S	25W-5N	6
Signature Dr. E&W	22W-4N	7
Skylark Dr. NW&SE	24W-4N	6
Spring Ct. NE&SW	22W-4N	7
Spring Valley Dr. E&W	22W-5N	7
Springbrook Dr. E&W	23W-5N	6
Springdale Ln. NS&EW	24W-5N	6
Springfield Dr. NE&SW	25W-4N,24W-6N	6
St. Francis Ct. N&S	22W-5N	7
Stanton Ln. N&S	22W-4N	7
Stanyon Ct. N&S	22W-4N	7
Starling Ct. NW&SE	22W-4N	7
Sterling Ct. N&S	23W-4N	6
Sterling Dr. E&W	23W-4N	6
Sterling Ln. N&S	23W-4N	6
Stonybrook Dr. NS&EW	22W-5N	7
Stratford Dr. Cir.	25W-5N	6
Strathmore Ct. N&S	23W-4N	6
Strathmore Ln. N&S	23W-4N	6
Surrey Dr. NW&SE	23W-4N	6
Surrey Ln. NE&SW	23W-4N	6
Sutherland Ln. N&S	23W-4N	6
Sutton Ct. N&S	22W-4N	7
Swallow Av. E&W	23W-4N	6
Tamarack Dr. E&W	22W-4N	7
Tanager Dr. E&W	23W-4N	6
Tee Ln. E&W	22W-5N	7
Terry Av. E&W	22W-4N	7
Thrasher St. N&S	24W-4N	6
Torrington Dr. E&W	24W-5N	6
U.S. Route 20 NW&SE	24W-6N,22W-5N	6
Versailles Ct. N&S	24W-5N	6
Villa Way N&S	24W-5N	6
W. Bristol NE&SW	25W-5N	6
W. Hampshire Dr. NW&SE	24W-5N	6
W. Hampshire Ln. E&W	25W-5N	6
W. Scarborough Ct. E&W	25W-5N	6
W. Sheffield Dr. E&W	24W-5N	6
W. Windsor Dr. NS&EW	25W-5N,24W-5N	6
Walter Ct. N&S	23W-5N	6
Warren Av. N&S	22W-5N	7
Warwick Ln. E&W	24W-5N	6
Washington St. E&W	23W-5N	6
Wedgefield Ln. N&S	24W-5N	6
Wellington Dr. N&S	24W-5N	6
Wendover Dr. N&S	24W-5N	6
Wentworth Ln. E&W	24W-5N	6
West Rd. N&S	24W-5N	6
Whitehall Ter. E&W	24W-4N	6
Wickham Av. N&S	24W-5N	6
Willow Ln. NE&SW	22W-4N	7
Willowbridge Way E&W	22W-4N	7
Winston Ln. N&S	23W-4N	6
Woodcrest Ct. N&S	23W-6N	6
Woodland Way* E&W	24W-6N	6
Woodside Dr. NW&SE	24W-5N	6
Wren Dr. NE&SW	23W-4N	6
1st St. N&S	23W-5N	6
3rd St. N&S	23W-5N	6

Bloomingdale Township

Street	Grid	Page
Abilene Tr. N&S	25W-5N	6
Acacia Ln. NE&SW	22W-6N	7
Alma Av. N&S	21W-2N	13
Amy Av. N&S	23W-2N	12
Andrene Ln. N&S	21W-5N,22W-7N	7
Ardmore Av. E&W	23W-7N	6
Argyle Av. E&W	25W-6N	6
Argyle St. N&S	21W-2N	13
Armitage Av. E&W	23W-2N,21W-2N	12
Army Trail Rd. NW&SE	26W-4N,21W-3N	6
Baker Dr. N&S	21W-6N	7
Barbara Ct. E&W	22W-6N	7
Bartels Rd. N&S	26W-7N	6
Belden Av. E&W	21W-2N	13
Bernice Av. N&S	23W-2N	12
Blackduck Ln. E&W	26W-5N	6
Bluebill Ct. NW&SE	26W-5N	6
Briargate Terr. N&S	21W-6N,21W-7N	7
Broker Rd. NE&SW	22W-6N	7
Bryn Mawr Av. E&W	23W-7N	6
Bunker Terr. N&S	21W-5N	7
Burdett Av. E&W	26W-2N,21W-2N	12
Byron Av. E&W	22W-4N	7
Canvasback Ln. N&S	26W-5N	6
Cavalry Dr. N&S	24W-4N	6
Central Av. E&W	25W-7N	6
Circle Av. N&S	22W-6N	7
Circle Terr. N&S	21W-7N	7
Claire Ct. N&S	23W-2N	12
Cloverdale Rd. N&S	25W-6N	6
Club Terr. NE&SW	21W-5N	7
Cortland Av. E&W	21W-2N	13
Crescent Terr. N&S	22W-6N	7
Crest Av. E&W	22W-7N,21W-7N	7
Depot Pl. NW&SE	21W-6N	7
Diane Av. N&S	23W-2N	12
Dickens Av. E&W	23W-2N,21W-2N	12
Dinah Ct. NE&SW	22W-6N	7
Dinah Rd. N&S	22W-6N	7
Drake Ln. E&W	26W-5N	6
Eagle Av. NE&SW	21W-7N	7
Eagle Terr. NS&EW	21W-5N,21W-7N	7
Edwards Dr. N&S	21W-4N	7
Elizabeth Ct. NW&SE	22W-6N	7
Elm St. N&S	21W-2N	13
Euclid Av. N&S	22W-2N	13
Fairway Ln. N&S	21W-5N,21W-6N	7
Ferrari Ct. N&S	22W-6N	7
Foster Av. E&W	25W-6N,22W-6N	6
Fullerton Av. E&W	24W-3N,21W-3N	12
Garden Av. N&S	24W-6N	6
Gary Av. N&S	25W-3N,25W-6N	12
Gates Av. NE&SW	21W-6N	7
Glen Ct. NE&SW	22W-6N,21W-7N	7
Glen Ellyn Rd. N&S	22W-4N	7
Glen Rd. N&S	21W-7N	7
Glendale Rd. N&S	22W-6N	7
Glendale Terr. NE&SW	23W-6N,22W-6N	6
Goldeneye Ln. N&S	26W-5N	6
Goodridge Terr. NE&SW	23W-6N	6
Greenwood Av. N&S	22W-6N	7
Grove St. E&W	21W-6N	7
Harvey Rd. N&S	22W-6N	7
Hawthorne Ln. N&S	22W-7N	7
Helen St. N&S	21W-2N	13
Hickory Ct. E&W	22W-6N	7
Highland Av. N&S	22W-2N,22W-6N	13
Hill St. N&S	23W-6N,23W-7N	6
Hillcrest Dr. N&S	22W-6N	7
Hillcrest Terr. E&W	22W-6N	7
Hillside Ct. NE&SW	22W-6N	7
Hilltop Dr. N&S	21W-7N	7
James Ct. N&S	22W-6N	7
John Ct. E&W	22W-6N	7
Joseph Ct. E&W	22W-6N	7
Juniper Dr. NE&SW	22W-7N	7
Keeney Rd. N&S	25W-6N	6
Kenmore St. N&S	21W-2N	13
Kuhn Rd. N&S	26W-2N,26W-4N	12
Lake St. NW&SE	26W-7N,21W-5N	6
Laramie Tr. E&W	25W-5N	6
Laredo Tr. E&W	25W-5N	6
Laurel Dr. N&S	22W-5N,22W-6N	7
Lawler Av. N&S	21W-2N	13
Lawrence Av. E&W	25W-6N,22W-6N	6
Lawrence Ct. NW&SE	21W-7N	7
Lies Rd. E&W	28W-4N,25W-4N	5
Lincoln St. N&S	23W-7N	6
Linda Av. N&S	23W-2N	12
Linden Av. N&S	22W-6N,22W-7N	7
Lloyd Av. N&S	21W-2N,21W-6N	13
Lydia Rd. E&W	21W-3N	13
Magnolia Av. NW&SE	21W-2N	13
Mallard Ct. E&W	25W-5N	6
Mallard Ln. N&S	26W-5N,25W-5N	6
Mandarin Ln. E&W	26W-5N	6
Manor Rd. NE&SW	21W-6N	7
Maple Av. E&W	23W-7N,21W-7N	6
Maple Ct. N&S	22W-5N	7
Meacham Rd. N&S	21W-7N	7
Meadow Rd. N&S	21W-3N	13
Medinah Rd. N&S	22W-7N,21W-4N	7
Mensching Rd. NW&SE	24W-6N	6
Merganser Ln. N&S	26W-5N	6
Mildred Av. N&S	23W-2N	12
Muscovy Ln. E&W	26W-5N,25W-5N	6
Neva Terr. N&S	21W-5N,21W-6N	7
Newland Pl. NE&SW	21W-6N	7
Nordic Rd. NE&SW	21W-5N	7
North Av. E&W	26W-2N,21W-2N	12
North Ln. E&W	21W-5N	7
North-South Tollway N&S	21W-2N	13
Oak St. N&S	23W-6N	6
Old Gary Av. N&S	25W-4N	6
Olive Ct. E&W	23W-2N	12
Papworth St. N&S	25W-6N	6
Par Ln. E&W	21W-5N	7
Park Av. E&W	21W-3N	13
Pearl Av. N&S	23W-2N	12
Pearson Dr. N&S	24W-6N	6
Peterson Av. E&W	26W-2N,21W-2N	12
Pine Av. E&W	23W-7N	6
Pintail Ct. NE&SW	25W-5N	6
Pleasant Av. N&S	22W-2N	13
Poplar Av. E&W	22W-7N	7
Prairie Av. N&S	22W-2N	13
Redhead Ln. N&S	26W-5N	6
Ring Neck Ct. NE&SW	26W-5N	6
Ring Neck Ln. N&S	26W-5N	6
Robbie Ln. N&S	21W-4N	7
Robert Ct. N&S	22W-6N	7
Roberta Av. N&S	23W-2N	12
Rohlwing Rd. N&S	21W-3N,21W-7N	13
Rosedale Av. N&S	23W-6N	6
Santa Fe Tr. N&S	25W-5N	6
Schick Rd. E&W	26W-5N,25W-5N	6
Schlick Av. E&W	22W-7N	7
Schmale Rd. N&S	24W-3N,24W-4N	12
Shelley Dr. E&W	21W-6N	7
Sidney Av. E&W	23W-2N,21W-2N	12
Sodard Rd. E&W	22W-6N	7
Spring Ct. N&S	22W-6N	7
Squire Ln. E&W	22W-7N	7
St. Charles Rd. NW&SE	26W-2N	12
Stone Av. E&W	21W-4N	7
Sunnyside Dr. NE&SW	21W-6N	7
Sunset Ct. N&S	24W-6N	6
Sunset Terr. E&W	22W-6N	7
Swift Rd. N&S	21W-2N,21W-5N	13
Sycamore Av. N&S	22W-5N,22W-7N	7
Teal Ct. NW&SE	25W-5N	6
Tee Ln. E&W	21W-5N	7
Temple Dr. NW&SE	22W-7N,21W-6N	7
Terrace Ct. NW&SE	21W-7N	7
Thorn Rd. N&S	25W-5N	6
Tucumcari Tr. E&W	25W-5N	6
Turner Av. E&W	23W-6N	6
Valley Ln. NE&SW	21W-6N	7
Valley Rd. N&S	23W-6N,23W-7N	6
Valley View Rd. N&S	22W-3N	13
Virginia Av. N&S	23W-2N	12
Virginia Rd. N&S	25W-6N	6
Walnut Av. E&W	22W-6N	7
Walnut St. NW&SE	23W-6N	6
Walter Dr. N&S	22W-4N	7
Webster Av. E&W	22W-6N	7
West End Rd. N&S	24W-6N	6
Wheaton Rd. N&S	25W-6N	6
Widgeon Ln. E&W	26W-5N	6
Willard St. N&S	23W-2N	12
William St. N&S	23W-7N	6
Willow St. N&S	23W-7N	6
Winthrop Av. E&W	21W-2N	13
Woodview Dr. E&W	22W-6N,21W-6N	7
Woodworth Pl. E&W	23W-7N	6
84th Ct. N&S	25W-4N	6

Bolingbrook*
(Also see Will County)

Street	Grid	Page
Banbury Ct. NE&SW	21W-9S	31
Bent Tree Ct. N&S	22W-9S	31
Buckingham Way E&W	21W-9S	31
Cambridge Way E&W	22W-10S,21W-9S	13
Everington Ct. E&W	22W-9S	31
Grosvernor Ct. E&W	22W-9S	31
Janes Av. N&S	21W-9S	31
Kensington Way E&W	21W-9S	31
Kent Ct. N&S	21W-9S	31
Londonberry Ct. N&S	21W-9S	31
Londonberry Ln. NS&EW	21W-9S	31
Nelson Ct. NW&SE	22W-9S	31
Piccadilly Ln. NS&EW	21W-9S	31
Rain Tree Dr. NS&EW	22W-9S	31
Sheffield Ln. NS&EW	21W-9S	31
Stone Creek Dr. E&W	22W-9S	31
Sulgrave Ct. E&W	22W-9S	31
Tangly Ct. N&S	22W-9S	31
Trafalgal Ct. E&W	22W-9S	31
Twin Creek Ct. N&S	22W-9S	31
Twin Creek Dr. NS&EW	22W-9S	31
Whitehall Ct. N&S	21W-9S	31

Burr Ridge*
(Also see Cook County)

Street	Grid	Page
Arcadia Ct. E&W	16W-10S	33
Ashton Dr. E&W	15W-10S	33
Bel Aire Ct. E&W	15W-10S	33
Bennacott Ln. E&W	15W-9S	33
Berkshire Ct.* E&W	15W-8S	33
Berkshire Ln.* E&W	15W-8S	33
Briarwood Dr. E&W	15W-9S	33
Bridle Ct. NE&SW	15W-10S	33
Bristol Ct.* N&S	15W-8S	33
Burr Ridge Club Dr. NS&EW	15W-6S	27
Cambridge Dr. E&W	15W-9S	33
Camelot Dr. E&W	15W-9S	33
Canterbury Dr. N&S	15W-9S	33
Carlisle Ct. N&S	15W-9S	33
Cascade Cir. NS&EW	16W-10S	33
Cascade Ct. E. E&W	16W-10S	33

	Grid	Page
Cascade Ct. W. E&W	16W-10S	33
Charleston Dr. E&W	16W-9S	33
Chesterfield Ct.* E&W	15W-8S	33
Circle Ridge Ct. N&S	15W-10S	33
Circle Ridge Dr. NS&EW	15W-10S	33
Clynderven Rd. N&S	15W-9S	33
Conway Ct. E&W	15W-9S	33
Countryside Ct. E&W	15W-7S	27
County Line Creek Dr. E&W	15W-10S	33
County Line Rd. N&S	15W-10S,15W-6S	33
Cove Creek Ct. E&W	15W-6S	27
Deerpath Tr. NS&EW	15W-9S	33
Devon Ct. N&S	15W-10S	33
Devon Dr. E&W	15W-10S	33
Devon Ridge Dr. N&S	15W-10S	33
Dolfor Cove N&S	15W-9S,15W-8S	33
Drew Av. N&S	15W-9S,15W-8S	33
Durham Ct.* E&W	15W-8S	33
Elm Av. N&S	15W-8S	33
Enclave Ct. NW&SE	15W-10S	33
Fallingwater Dr. E. N&S	16W-10S	33
Fallingwater Dr. W N&S	16W-10S	33
Fawn Ct. NE&SW	15W-9S	33
Fieldstone Dr. NE&SW	15W-7S	27
Garfield Av. N&S	15W-9S,15W-8S	33
Glenmora Ln. N&S	15W-10S	33
Glenn Dr. N&S	15W-10S	33
Grant St. NW&SE	15W-10S,15W-8S	33
Hanover Ct. NE&SW	15W-10S	33
Heather Ct. N&S	15W-9S	33
Heather Ln. N&S	15W-10S	33
Hunter Ct. E&W	15W-10S	33
Jack Pine Ln. N&S	15W-9S	33
Joy Ct. E&W	15W-10S	33
Kathryn Ct. NS&EW	15W-9S	33
Keller Dr. N&S	15W-6S	27
Keri Ln. NW&SE	15W-10S	33
Kimberly Ct. N&S	15W-9S	33
Kirkwood Cove E&W	15W-9S	33
Lake Ridge Club Ct.* E&W	15W-8S	33
Lake Ridge Club Dr. NS&EW	15W-9S	33
Lake Ridge Ct. NE&SW	15W-9S	33
Lake Ridge Dr. N&S	15W-9S	33
Lakewood Cir. NS&EW	15W-9S	33
Lancaster Ct.* E&W	15W-8S	33
Laurel Oak Ct. N&S	15W-10S	33
Laurel Oak Ct. NE&SW	15W-10S	33
Leonard Ln. NW&SE	15W-9S	33
Linden Ct. NE&SW	15W-10S	33
Madison St. N&S	15W-9S,15W-8S	33
Marissa Ct. NE&SW	15W-11S	15
Morgan Ct. E&W	15W-10S	33
Norman Ct. N&S	15W-10S	33
Norris Dr. E&W	15W-6S	27
Oak Creek Ct. NS&EW	15W-10S	33
Oak Creek Ct. E&W	15W-10S	33
Oak Creek Dr. N&S	15W-10S	33
Oak Creek E&W	15W-10S	33
Oak Creek East Dr. E&W	15W-10S	33
Oak Ct. N&S	15W-10S	33
Oak Ridge Dr. NE&SW	15W-10S	33
Old Surrey Rd. E&W	15W-6S	27
Pacific St. N&S	16W-11S	35
Park Av. N&S	15W-9S,15W-6S	33
Parkview Pl. E&W	15W-9S	33
Pepper Mill Ct. E&W	15W-7S	27
Pin Oak Ln. N&S	15W-9S	33
Pine Tree Rd. E&W	15W-9S	33
Prairie Ln. E&W	15W-10S	33
Provencal Dr. E&W	15W-10S	33
Red Oak Ct. NW&SE	15W-10S	33
Regent Ct. E&W	15W-10S	33
Ridge Farm Rd. NS&EW	15W-6S	27
Ridgepointe Dr. N&S	15W-9S	33
Rodeo Dr. NE&SW	15W-10S	33
Royal Dr. N&S	15W-10S	33
Saddle Ct. NE&SW	15W-10S	33
Sedgeley Ct. N&S	15W-6S	27
Shenandoah Ct. E&W	15W-10S	33
Shenandoah Ln. N&S	15W-10S	33
Shepherd Ln. E&W	15W-10S	33
Shore Ct. E&W	16W-9S	33
Shore Dr. N&S	16W-9S	33
Soper Av. NW&SE	15W-8S	33
Sylvan Glen Ct. NE&SW	16W-10S	33
Tamarack Dr. N&S	15W-10S	33
Tamerton Pkwy. E&W	15W-10S	33
Tori Ct. E&W	15W-10S	33
Tudor Ct. E&W	15W-10S	33
Tumblebrook Ct. E&W	16W-10S	33
Turnberry Ct. NW&SE	15W-9S	33
Turnberry Dr. NE&SW	15W-9S	33
Walredon Av. N&S	15W-9S	33
Wedgewood Dr. N&S	15W-10S,15W-9S	33
Westmister Dr. E&W	15W-10S	33
White Oak Ct. E&W	15W-10S	33
Windsor Ct. N&S	15W-9S	33
Woodcreek Dr. NS&EW	15W-6S	27
60th Pl. E&W	15W-6S	27
60th St. E&W	15W-6S	27
61st Pl. E&W	15W-6S	27
61st St. E&W	15W-6S	27
62nd St. E&W	15W-6S	27
71st St. E&W	15W-7S	27
72nd St. E&W	15W-8S	33
73rd St. E&W	15W-8S	33
74th St. E&W	15W-8S	33
75th St. E&W	15W-8S	33
77th St. E&W	15W-8S	33
79th St. E&W	15W-8S	33
80th St. E&W	15W-8S	33
83rd St. E&W	15W-9S,14W-9S	33
87th St. E&W	15W-9S	33
91st St. E&W	15W-10S	33
94th St. E&W	15W-10S	33

Carol Stream

	Grid	Page
Adam Ln. NS&EW	26W-3N	12
Adobe Ct. E&W	26W-3N	12
Alabama Tr. E&W	26W-3N,25W-3N	12
Alamo Ct. E&W	25W-3N	12
Aleut Tr. E&W	26W-3N	12
Alexandra Way NS&EW	25W-1N,25W-2N	12
Alison Ln. N&S	26W-3N	12
Allegheny Ct. E&W	27W-3N	11
Allegro Ln. NW&SE	25W-3N	12
Alton Ct. NE&SW	27W-3N	11
Amber Ct. N&S	24W-1N	12
Amber Ln. E&W	24W-1N	12
Andrew Ln. NE&SW	26W-3N	12
Antelope Tr. E&W	26W-3N,25W-3N	12
Antigo Tr. E&W	28W-4N	5
Apache Ln. E&W	26W-2N	12
Appaloosa Ct. N&S	26W-3N	12
Appomattox Tr. NE&SW	28W-4N	5
Arapahoe Tr. E&W	25W-2N	12
Arbor Dr. E&W	25W-3N	12
Army Trail Rd. E&W	28W-4N,25W-4N	5
Arrowhead Tr. N&S	26W-2N	12
Ash Ct. N&S	28W-4N	5
Aspen Ct. E&W	26W-4N	6
Aztec Dr. N&S	26W-2N,26W-3N	12
Barton Pl. N&S	27W-2N	11
Basswood Ct. E&W	26W-4N	6
Bear Paw Ct. E&W	27W-3N	11
Bedford Dr. E&W	26W-4N	6
Beech Ct. E&W	26W-4N	6
Belair St. NE&SW	25W-3N	12
Bennington Dr. N&S	27W-3N	11
Berkshire Dr. E&W	27W-2N	11
Big Eagle Tr. E&W	27W-3N	11
Big Horn Tr. E&W	28W-4N	5
Biloxie Ct. E&W	27W-2N	11
Birchbark Tr. E&W	28W-3N,26W-3N	11
Bison Tr. E&W	25W-2N	11
Blackhawk Dr. N&S	25W-2N	12
Blake Ct. N&S	25W-3N	12
Bluff St. N&S	25W-3N	12
Boa Tr. NE&SW	28W-4N	5
Bonnie Ln. N&S	25W-3N	12
Boone Dr. N&S	25W-3N	12
Bowie Dr. N&S	25W-3N	12
Bowstring Ct. NW&SE	28W-4N	5
Bradbury Cir. N&S	26W-4N	6
Brave Ct. N&S	28W-4N	5
Brighton Dr. N&S	27W-4N	5
Bristol Dr. N&S	26W-2N	12
Brookstone Ct. E&W	27W-3N	11
Buckskin Ln. NE&SW	27W-3N	11
Buffalo Cir. N&S	27W-3N	11
Burke Dr. E&W	26W-4N	6
Burning Tr. N&S	27W-3N	11
Burns St. E&W	24W-1N	12
Cactus Tr. N&S	27W-3N	11
Camelot Ln. E&W	25W-3N	12
Canterbury Ct. E&W	28W-3N	11
Canterbury Dr. E&W	26W-4N	6
Canyon Tr. E&W	26W-3N,25W-3N	12
Caribou Tr. N&S	28W-4N	5
Carleton Dr. N&S	24W-1N	12
Carol Ct. N&S	24W-3N	12
Carriage Dr. N&S	25W-2N	12
Carson Ct. NE&SW	25W-3N	12
Cedar Ct. E&W	26W-4N	6
Center Av. N&S	24W-3N	12
Chadsford Ct. N&S	27W-2N	11
Chalet Dr. N&S	26W-4N	6
Charger Ct. E&W	28W-4N	5
Chattanooga Tr. E&W	28W-4N	5
Cherokee Ct. E&W	26W-2N	12
Chestnut Ct. E&W	26W-3N	12
Cheyenne Tr. N&S	26W-3N	12
Chippewa Tr. E&W	26W-3N	12
Christopher Ln. N&S	28W-3N	11
Cimarron Dr. N&S	28W-3N	11
Clare Ct. E&W	26W-4N	6
Clearwater Ct. N&S	26W-3N,25W-3N	12
Cliff Ct. N&S	26W-3N	12
Cliffview Dr. E&W	27W-3N	11
Coachlite Ct. E&W	25W-3N	12
Cochise Ct. E&W	26W-2N	12
Cochise Pl. E&W	26W-2N	12
Colorado Ct. N&S	27W-2N	11
Columbia Ct. NE&SW	26W-3N	12
Commanche Ct. E&W	26W-3N	12
Commanche Ln. E&W	26W-3N	12
Commerce Dr. N&S	24W-2N	12
Commonwealth Dr. N&S	25W-1N	12
Concord Dr. N&S	27W-3N	11
Country Glen Ln. N&S	27W-4N	5
Countryside Ln. E&W	27W-4N	5
County Farm Rd. N&S	27W-2N,27W-4N	11
Creekwood Ct. N&S	26W-3N	12
Crystal Shore Ct. NE&SW	25W-2N	12
Crystal Shore Dr. NS&EW	27W-4N	5
Cumberland Ct. E&W	27W-3N	11
Cypress Ln. N&S	28W-3N,27W-3N	11
Dakota Tr. N&S	27W-2N	11
Danbury Dr. N&S	26W-2N	12
Dancing Water Ct. NE&SW	25W-3N	12
Dartmouth Dr. E&W	27W-3N	11
David Ln. N&S	26W-3N	12
Daybreak Ln. N&S	25W-3N	12
Dearborn Cir. N&S	25W-3N	12
Deerskin Tr. N&S	26W-3N	12
Delaware Tr. E&W	26W-3N	12
Dodge Ct. N&S	25W-3N	12
Dogwood Ln. N&S	28W-3N	11
Donegal Ln. NS&EW	26W-4N	6
Dorchester Dr. N&S	27W-3N	11
Doris Av. N&S	25W-1N	12
Dublin Ct. NW&SE	26W-4N	6
Dugout Tr. N&S	27W-3N	11
Eagle View Dr. N&S	25W-2N	12
East Dr. N&S	24W-3N	12
Easton Dr. NS&EW	27W-4N	5
Easy St. E&W	25W-1N	12
Eclipse Dr. NW&SE	25W-3N	12
Edgebrook Ct. NE&SW	25W-3N	12
Edington Ct. N&S	26W-4N	6
Edington Ln. N&S	26W-4N	6
El Paso Ln. N&S	25W-2N	12
Elk Tr. E&W	25W-3N	12
Erie Ct. E&W	26W-3N	12
Esselen Tr. E&W	27W-2N	11
Essex Pl. N&S	26W-2N	12
Evergreen Dr. E&W	28W-3N,27W-3N	11
Executive Dr. E&W	24W-1N	12
Fair Oaks Rd. N&S	28W-3N,28W-4N	11
Farm Glen Ln. NW&SE	27W-4N	5
Fawn Ct. E&W	26W-3N	12
Feather Ct. E&W	26W-3N	12
Fireside Dr. N&S	24W-1N	12
Flame Ct. E&W	24W-1N	12
Flame Dr. E&W	24W-1N	12
Flint Tr. E&W	26W-3N	12
Forest Ct. E&W	27W-4N	5
Forest Ln. E&W	27W-4N	5
Fox Ct. NW&SE	27W-3N	11
Frontage Rd. E&W	26W-2N	12
Fullerton Av. E&W	25W-4N,25W-3N	6
Gary Av. N&S	25W-1N,25W-4N	12
Geneva Rd. E&W	24W-1N	12
Georgetown Dr. N&S	28W-4N	5
Glen Ct. NW&SE	27W-4N	5
Glen Flora Dr. NE&SW	25W-3N	12
Glenlake Dr. E&W	27W-4N	5
Gloucester Cir. N&S	27W-4N	5
Greenway Tr. NE&SW	25W-3N	12
Gundersen Dr. E&W	24W-1N	12
Gunsmoke Ct. NE&SW	27W-3N	11
Hampton Ct. N&S	26W-4N	6
Hampton Ln. N&S	27W-3N	11
Harbor Pt. NW&SE	27W-4N	5
Harwich Ct. E&W	27W-4N	5
Hawk Ln. N&S	27W-3N	11
Hearth Ln. NE&SW	24W-1N	12
Heather Ln. N&S	25W-2N	12
Hemlock Ln. E&W	28W-3N	11
Heritage Dr. N&S	27W-4N	5
Hiawatha Tr. E&W	25W-2N,25W-3N	12
Hickory Ln. E&W	28W-3N	11
Highridge Pass N&S	27W-3N	11
Hoover Dr. N&S	25W-3N	12
Hopi Ct. NW&SE	25W-3N	12
Horizon Cir. NS&EW	25W-3N	12
Horseshoe Ct. NE&SW	27W-3N	11
Hunter Dr. NS&EW	26W-4N	6
Huntington Ct. N&S	26W-4N	6
Huron Ct. N&S	25W-3N	12
Hyannis Cir. NS&EW	27W-4N	5
Idaho St. N&S	25W-3N	12
Illini Dr. E&W	26W-3N	12
Inca Blvd. E&W	26W-3N	12
Indianwood Dr. N&S	26W-2N	12
Iowa Ct. E&W	25W-3N	12
Iris Av. E&W	28N-3N	11
Ironhawk Ct. NW&SE	27W-3N	11
Iroquois Tr. N&S	27W-3N	11
Juniper Ct. NW&SE	26W-3N	12
Kalamazoo Ct. N&S	26W-3N	12
Kamiah Ct. N&S	26W-3N	12
Kansas St. N&S	26W-3N	12
Kehoe Blvd. E&W	25W-2N,24W-2N	12
Kelly Dr. E&W	27W-4N	5
Kerry Ct. N&S	26W-4N	6
Kildare Ct. NS&EW	26W-4N	6
Kilkenny Ct. NW&SE	26W-4N	6
Kimberly Dr. N&S	25W-2N,25W-3N	12
Klein Creek Ct. E&W	25W-3N	12
Knollwood Dr. N&S	27W-4N	5
Kuhn Rd. N&S	26W-2N,26W-4N	12
Lacrosse St. N&S	26W-3N	12
Laguna Ct. N&S	26W-3N	12
Lakeside Ln. N&S	27W-4N	5
Laurel Ct. E&W	27W-3N	11
Legends Dr. N&S	25W-3N	12
Leslie Ct. N&S	28W-3N	11
Lies Rd. E&W	28W-4N,24W-3N	11
Lightning Ct. N&S	27W-3N	11
Lilac Ct. N&S	27W-3N	11
Lilac Ln. N&S	27W-3N	11
Lincolnshire Ct. N&S	28W-3N	11
Longmeadow Dr. NE&SW	27W-3N	11
Ludington Ct. N&S	27W-5N	5
Main Pl. N&S	24W-1N	12
Malibu Ct. E&W	26W-3N	12
Mantle Ct. N&S	24W-1N	12
Maple Ridge Dr. E&W	28W-4N	5
Matthew Ln. NS&EW	26W-3N	12
Meadow Ln. N&S	24W-1N	12
Medford Dr. E&W	27W-2N	11
Merbach Dr. N&S	26W-4N	6
Mercedes Dr. N&S	26W-3N	12
Mesa Verde Ct. N&S	27W-3N	11
Minnesota Cir. E&W	27W-2N	11
Mission St. E&W	26W-3N	12
Moccasin Ct. N&S	26W-3N	12
Mohawk Tr. N&S	26W-3N	12
Mohican Rd. E&W	26W-3N	12
Morton Rd. E&W	27W-3N	11
Mountain Glen Way E&W	27W-3N	11
Munson Dr. E&W	27W-3N,26W-3N	11
Nantucket Ct. N&S	27W-4N	5
Napa St. N&S	26W-3N	12
Narragansett Dr. N&S	27W-4N	5
Natoma Cir. E&W	25W-2N	12
Navajo St. N&S	26W-3N	12
Nebraska Cir. NE&SW	27W-2N	11
Nekoma St. N&S	27W-3N	11
New Britton Rd. N&S	28W-3N	11
New London Ct. NE&SW	27W-3N	11
Nez Perce Ct. N&S	26W-3N	12
Niagara St. N&S	26W-3N	12
North Av. E&W	26W-2N,24W-2N	12
Oak Wood Dr. NS&EW	27W-4N	5
Ohio Ct. N&S	26W-3N	12
Old Gary Ave. N&S	25W-4N	6
Old Meadow Ct. N&S	27W-3N	11
Omaha Ct. N&S	26W-3N	11
Oneida Ct. N&S	27W-3N	11
Oriole Tr. N&S	28W-4N	5
Osage Ct. N&S	27W-3N	11
Oswego Dr. NS&EW	27W-3N	11
Ottawa Ct. N&S	26W-3N	12
Overlook Ln. NS&EW	27W-3N	11
Oxford St. N&S	28W-4N	5
Paddock Dr. E&W	27W-3N	11
Palomino St. N&S	26W-3N	12
Papoose Ct. N&S	26W-3N	12
Park Hill Tr. E&W	25W-2N	12
Parkside Ct. NW&SE	26W-4N	6
Parkside Dr. E&W	26W-4N	6
Parkview Cir. NS&EW	27W-4N	5
Parkview Ct. NE&SW	27W-4N	5
Pawnee Dr. E&W	27W-3N	11
Paxton St. N&S	26W-2N	12
Pebble Creek Tr. E&W	25W-2N	12
Pennsboro Ct. NW&SE	28W-4N	5
Penny Ct. E&W	27W-3N	11
Peoria St. NE&SW	27W-3N	11
Petersburg Ct. N&S	28W-4N	5
Pheasant Tr. NS&EW	28W-4N	5
Plains Ct. NE&SW	27W-3N	11
Pleasant Hill Rd. N&S	26W-2N	12
Plymouth Ct. E&W	27W-4N	5
Pocahontas Tr. N&S	26W-3N	12
Pontiac Ln. N&S	27W-3N	11
Portchester Cir. N&S	28W-4N	5
Portsmouth Ct. NE&SW	28W-4N	5
Potomac Ct. E&W	28W-4N	5
Prairie Ct. NW&SE	27W-3N	11
President St. N&S	24W-1N	12
Princetown Ct. NW&SE	27W-3N	11
Provincetown Dr. NW&SE	27W-3N	11
Quail Run Ct. NE&SW	25W-2N	12
Raintree Ct. N&S	28W-3N	11
Randy Rd. E&W	24W-2N	12
Redhill Tr. N&S	25W-3N	12
Ridge Tr. N&S	27W-3N	11
River Dr. NE&SW	27W-3N	11
Rockport Ct. E&W	27W-4N	5
Rocky Valley Way NS&EW	27W-3N	11
Rolling Oaks Dr. E&W	28W-3N	11
Rose Av. E&W	28W-3N	11
Rose Ct. NE&SW	28W-3N	11
Royal Glen Ln. NS&EW	27W-4N	5
Rt. 64 NW&SE	26W-2N,24W-2N	12
Saginaw Dr. N&S	26W-3N	12
Sand Creek Dr. N&S	27W-3N	11
Santa Fe Ct. N&S	26W-3N	12
Saratoga Dr. NS&EW	27W-3N	11
Sauk Ct. N&S	26W-3N	12
Schmale Rd. N&S	24W-1N,24W-3N	12
Seabury Cir. N&S	27W-4N	5
Seminole Ct. N&S	26W-3N	12
Seneca Dr. N&S	27W-5N	5
Seneca Ln. N&S	27W-3N	11
Sequoia Ct. N&S	26W-3N	12
Shagbark Ct. NW&SE	28W-3N	11
Shawnee Dr. E&W	25W-2N	12

	Grid	Page
Sheffield Ct. E&W	28W-3N	11
Shelburne Dr. NE&SW	26W-2N	12
Shenandoah Dr. E&W	27W-3N	11
Shining Water Dr. N&S	27W-4N	5
Silverleaf Blvd. N&S	26W-2N	12
Sioux Ln. N&S	26W-2N	12
Somerset Dr. NS&EW	27W-3N	11
Sorrel Ct. E&W	27W-3N	11
Split Rail Dr. N&S	27W-3N	11
Spring Valley Ct. NW&SE	28W-4N	5
Spring Valley Dr. N&S	28W-4N	5
Springbrook Ln. E&W	25W-3N	12
Spur Ct. NW&SE	27W-3N	11
St. Charles Rd. E&W	25W-1N,24W-1N	12
St. Paul Blvd. E&W	25W-2N,24W-2N	12
Stark Dr. E&W	25W-4N	6
Stockbridge Dr. N&S	27W-3N	11
Stonehenge Ct. E&W	28W-3N	11
Stonewood Cir. N&S	26W-4N	6
Stuart Dr. N&S	26W-3N	12
Summit Pass E&W	27W-3N	11
Sundance Ct. N&S	27W-3N	11
Sunrise Ct. N&S	25W-3N	12
Surrey Dr. N&S	25W-2N	12
Sussex Rd. N&S	28W-3N	11
Tacoma Ct. N&S	27W-3N	11
Tahoe Ct. N&S	27W-3N	11
Tall Oaks Dr. E&W	28W-3N	11
Tama Ct. NW&SE	26W-3N	12
Teton Cir. NE&SW	27W-2N,27W-3N	11
Texas Ct. NE&SW	27W-3N	11
Thornhill Dr. E&W	24W-1N	12
Thunderbird Tr. E&W	27W-2N,25W-2N	11
Timber Ridge Dr. N&S	23W-1N	12
Tioga Ct. N&S	26W-2N	12
Tomahawk Ct. E&W	25W-2N	12
Tonto Ct. N&S	26W-3N	12
Topeka Cir. N&S	27W-3N	11
Topeka Ct. N&S	27W-3N	11
Trailside Ct. E&W	26W-4N	6
Tremont Ct. N&S	27W-3N	11
Tubeway Dr. E&W	25W-1N	12
Ute Ln. N&S	26W-2N	12
Vale Rd. E&W	27W-2N	11
Valleyview Dr. NS&EW		11
Village Ct. E&W	25W-2N	12
Village Dr. N&S	25W-2N	12
Violet St. E&W	28W-3N	11
Wabash Ct. N&S	26W-3N	12
Waco Dr. NS&EW	27W-3N	11
Walnut Cir. E&W	28W-4N	5
Wampum Ct. N&S	26W-3N	12
Waterford Ct. E&W	26W-4N	6
Westgate Dr. N&S	25W-1N	12
Westward Tr. NS&EW	26W-4N	6
Wexford Ct. N&S	26W-4N	6
Whisper Pointe N&S	25W-3N	12
Williamstown Dr. N&S	27W-3N	11
Willow Dr. NE&SW	26W-3N	12
Willow Wood Dr. E&W	24W-1N	12
Winchester Ct. NW&SE	27W-3N	11
Winding Glen Ct. E&W	27W-4N	5
Winding Glen Dr. NS&EW	27W-4N	5
Windmere Ln. E&W	26W-3N	12
Windsor Park Dr. Cir.	26W-2N	12
Woodcrest Ct. NE&SW	25W-3N	12
Woodhill Dr. E&W	27W-4N,26W-4N	5
Yardley Dr. E&W	26W-4N	6
Yellowstone St. N&S	26W-3N	12
Yorkshire Ln. N&S	27W-3N	11
Yuma Ln. NW&SE	25W-2N	12

Clarendon Hills

	Grid	Page
Algonquin Rd. E&W	17W-5S	27
Ann Byrd Ct. E&W	16W-5S	27
Arthur Av. E&W	17W-5S	27
Barclay Ct. N&S	16W-6S	27
Blodgett Av. N&S	17W-5S	27
Bonnie Ln. E&W	16W-5S	27
Burlington Av. E&W	17W-5S	27
Chestnut Av. NE&SW	16W-5S	27
Churchill Pl. N&S	16W-5S	27
Coe Rd. N&S	16W-4S	27
Colfax Av. E&W	17W-5S	27
Columbine Dr. NS&EW	16W-4S	27
Coolid Ct. NW&SE	17W-5S	27
Coventry Ct. E&W	17W-6S	27
Eastern Av. E&W	16W-5S	27
Fairview Ct. E&W	16W-5S	27
Forest Hills Dr. E&W	16W-6S	27
Gilbert Av. N&S	17W-5S	27
Golf Av. N&S	16W-5S	27
Grant Av. N&S	16W-5S	27
Hamill Ln. E&W	16W-5S	27
Harris Av. E&W	17W-5S,16W-5S	27
Hiawatha Av. N&S	17W-5S	27
Hickory St. E&W	16W-4S	27
Hudson Av. E&W	17W-5S	27
Indian Dr. N&S	17W-5S	27
Iroquois Dr. N&S	17W-5S	27
Jackson St. N&S	16W-4S	27
Jane Ct. E&W	17W-5S	27
Jane Rd. E&W	17W-5S	27
Juliet Ct. E&W	17W-5S	27
Larkspur Ln. E&W	16W-4S	27
Maple St. E&W	16W-4S	27
McIntosh Av. N&S	17W-5S	27
Middaugh Rd. N&S	16W-4S	27
Mohawk Dr. N&S	17W-5S	27
Naperville Rd. NE&SW	17W-4S	27
Norfolk Av. E&W	17W-5S,16W-5S	27
Oxford Av. N&S	17W-5S,17W-4S	27
Park Av. E&W	17W-5S	27
Powell St. N&S	16W-5S	27
Prospect Av. N&S	16W-5S	27
Railroad Av. E&W	17W-5S	27
Ridge Av. N&S	16W-5S	27
Ridge St. E&W	16W-5S	27
Rose Pl. NW&SE	16W-5S	27
Ruby St. E&W	17W-5S	27
Sheridan Av. NW&SE	16W-5S	27
Short St. NE&SW	16W-5S	27
Stonegate Rd. N&S	16W-4S	27
Terrace Dr. E&W	16W-5S	27
Tuttle Av. N&S	17W-5S	27
Walker Av. N&S	17W-5S,16W-5S	27
Walnut St. E&W	16W-4S	27
Waverly Av. NE&SW	16W-5S	27
Western Av. NE&SW	17W-5S,16W-5S	27
Willow Creek Ct. E&W	16W-6S	27
Woodstock Av. N&S	17W-5S,17W-4S	27

Darien

	Grid	Page
Adams St. N&S	18W-9S,18W-8S	32
Adlai Stevenson Expres. NE&SW	18W-9S,17W-9S	32
Alabama Av. N&S	17W-7S	27
Albany La. E&W	17W-7S	27
Andermann La. E&W	17W-8S	33
Arbor Ct. N&S	18W-8S	32
Arrow Ln. N&S	17W-9S	33
Ashley St. NE&SW	18W-9S	32
Aylesbury Ln. E&W	19W-8S	32
Bailey Rd. N&S	18W-9S,18W-8S	32
Baker Ct. N&S	17W-8S	33
Barclay Rd. N&S	19W-8S	32
Barrymore Ln. E&W	18W-9S	32
Bavarian Ln. E&W	17W-7S	27
Bedford Ln. E&W	19W-8S	32
Beechnut Ln. N&S	17W-8S,17W-7S	33
Belair Dr. N&S	17W-8S	33
Bentley Av. N&S	17W-8S,17W-7S	33
Black Swan Ct. NW&SE	19W-10S	32
Bobolink Ct. N&S	17W-9S	33
Boulder Dr. E&W	18W-9S	32
Bristlecone Ct.* NS&EW	18W-7S	26
Brompton Dr. N&S	19W-8S	32
Brookbank Rd. N&S	16W-8S,16W-7S	33
Brookhaven Av. N&S	17W-8S	33
Brunswick Rd. N&S	19W-8S	32
Bunker Rd. N&S	17W-8S	33
Cambridge Rd. N&S	19W-8S	32
Capitol Dr. N&S	18W-8S	33
Carriage Green Dr. N&S	18W-9S	32
Carrol Ct. E&W	18W-8S	32
Carrol Ln. E&W	18W-8S	32
Cass Av. N&S	18W-9S,17W-8S	32
Center Dr. E&W	18W-8S	32
Chalet Dr. N&S	17W-9S	33
Cherokee Dr. E&W	18W-8S	32
Chestnut Ln. E&W	17W-7S	27
Chippewa Ln. E&W	17W-8S	33
Clarendon Hills Rd. N&S	17W-8S,17W-7S	33
Clemens Rd. E&W	18W-8S	32
Coachmans Rd. NE&SW	18W-9S	32
Columbia La. NE&SW	17W-7S	27
Comstock Ln. N&S	18W-8S	33
Country Ln. E&W	17W-8S	33
Creekside Ln. N&S	18W-8S	32
Crest Rd. N&S	16W-8S,16W-7S	33
Dale Rd. N&S	16W-7S	27
Danbury Rd. N&S	19W-8S	32
Darien Lake Dr. NS&EW	18W-8S,18W-9S	32
Darien Ln. N&S	18W-8S	33
Del Ct. N&S	17W-8S	33
Dickens Ct. E&W	18W-9S	32
Dorchester Ln. N&S	17W-8S	33
Durham Ct. N&S	18W-9S	32
Eleanor Pl. NW&SE	16W-8S,16S-7S	33
Eleanor Rd. N&S	16W-8S	33
Elm St. E&W	16W-8S	33
Emerson Dr. N&S	18W-8S	32
Evans Pl. N&S	16W-8S	33
Evergreen Ct. N&S	18W-9S	32
Evergreen Ln. NE&SW	18W-9S	32
Exner Rd. N&S	18W-8S	32
Fairview Av. N&S	18W-8S	32
Farmingdale Rd. N&S	17W-9S,17W-8S	33
Fox Hill Pl. E&W	18W-9S	32
Foxtail Ct.* NS&EW	18W-7S	26
Frontage Rd. NE&SW	18W-10S,18W-9S	32
Gail Av. N&S	18W-8S	33
Glen Eagles Ln. NE&SW	19W-10S	32
Glen Ln. NS&EW	19W-9S	33
Glenerye Rd. N&S	18W-9S	32
Golfview Dr. E&W	18W-9S	32
Gordon Ct. E&W	18W-9S	32
Grant Ct. E&W	18W-9S	32
Grant St. N&S	18W-9S,18W-8S	32
Greenbriar Ln. E&W	18W-9S	32
Greenvalley Rd. E&W	19W-8S	32
Grove Dr. NS&EW	18W-8S	32
Hayenga La. N&S	17W-8S	33
Heather Ln. E&W	18W-9S	32
Hemlock Ct. NE&SW	18W-9S	32
Hickory Ln. E&W	17W-8S	33
High Rd. N&S	16W-7S,16W-8S	27
Highpoint Cir. NS&EW	17W-9S	33
Hinsbrook Av. N&S	17W-7S	27
Hinswood Dr. NE&SW	17W-9S	33
Holly Av. E&W	18W-7S,16W-7S	26
Huntington Ct. NS&EW	19W-8S	32
Iris Rd. N&S	17W-7S	27
Ironwood Av. E&W	17W-7S	27
Iroquois Ln. E&W	18W-9S	32
Janet Av. NE&SW	17W-8S,16W-8S	33
Juniper Ln. E&W	17W-8S	33
Kentwood Ct.. N&S	18W-10S,18W-9S	32
Kimberly Ct. NW&SE	18W-10S	32
Knottingham Cir. NW&SE	18W-9S,18W-8S	32
Laceback Ct.* NE&SW	18W-7S	26
Lake Ridge Dr. NE&SW	19W-10S	32
Lakeview Dr. NS&EW	18W-9S	32
Larkspur Ln.* NE&SW	18W-7S	26
Laurel Ln. E&W	18W-9S	32
Leonard Dr. N&S	17W-8S,17W-7S	33
Lester Ln. E&W	18W-9S	32
Lester Ln. N&S	18W-8S	33
Linden Av. N&S	17W-8S	33
Lodgepole Ct.* E&W	18W-7S	26
Lyman Av. N&S	19W-8S	32
Maple Ln. E&W	17W-7S	27
Marlborough Ln. E&W	19W-8S	32
Mayfair Ln. N&S	19W-8S	33
McAdam Rd. E&W	18W-7S	26
Meadowlark Ln. E&W	17W-9S	33
Nantucket Dr. N&S	18W-8S	33
Norman Dr. NS&EW	18W-9S	33
Oakfern Ln.* NE&SW	18W-7S	26
Oakley Dr. N&S	18W-9S	32
Old Oak Pl. NE&SW	18W-10S	32
Oriole Ln. E&W	18W-9S	33
Park Av. N&S	18W-8S	33
Parkcrest NE&SW	17W-9S	33
Peony Pl. E&W	16W-7S	27
Pine Bluff Ct. N&S	18W-9S	32
Pine Cove Ct. E&W	18W-9S	32
Pine Ct. N&S	18W-9S	32
Pine Pkwy. N&S	18W-9S	32
Pine View Ct. E&W	18W-9S	32
Pinehurst Dr. NS&EW	18W-9S	32
Plainfield Rd. NE&SW	18W-9S,16W-8S	32
Ponderosa Ct.* N&S	18W-7S	26
Poplar Ln. N&S	16W-8S	33
Portsmouth Dr. NW&SE	17W-9S	33
Red Pine Trail* E&W	18W-7S	26
Redondo Dr. E&W	18W-8S	32
Regency St. E&W	18W-8S	32
Richmond Av. N&S	17W-8S,17W-7S	33
Ridge Rd. N&S	17W-7S	27
Ripple Ridge Cove E&W	18W-9S,17W-9S	32
Robert Kingery Hwy. N&S	16W-8S,16W-7S	33
Roger Rd. E&W	16W-8S	33
Rosewood Ct. E&W	18W-8S	32
Royal Oak Rd. NE&SW	18W-10S	32
Royal Swan Ln. NE&SW	19W-10S	32
Sawmill Creek Dr. N&S	18W-9S	32
Sawyer Ct. NW&SE	17W-9S	33
Sawyer Rd. N&S	18W-9S,17W-9S	33
Scotch Pine Trail* N&S	18W-7S	26
Seminole Dr. NS&EW	17W-8S	33
Sequoia Ln. N&S	18W-8S	32
Shelley Ct. E&W	18W-8S	32
Sierra Ct. N&S	17W-7S	27
Sleepy Hollow Ln. E&W	18W-9S,17W-9S	32
Somerset Ln. E&W	17W-8S	33
Spring Ct. E&W	18W-9S,17W-9S	32
Stevens St. N&S	18W-8S	32
Stewart Dr. E&W	18W-9S	32
Stratford Pl. N&S	17W-8S	33
Summit Rd. N&S	18W-8S	32
Sunrise Av. N&S	16W-8S	33
Surrey Dr. N&S	19W-8S	32
Sussex Creek Ln.	17W-8S	33
Sweetbriar Ln. N&S	18W-7S	26
Sweetwater Ct. E&W	18W-9S,17W-9S	32
Tamarack Dr. E&W	17W-8S	33
Tennessee Dr. N&S	17W-8S,17W-7S	33
Thistlewood Ct. NS&EW	18W-9S	32
Timber Ln. E&W	18W-9S	32
Torrey Pines Ct.* NE&SW	18W-7S	26
Troutilly Ln.* NS&EW	18W-7S	26
Village Ct. E&W	17W-8S	33
Wakefield Dr. N&S	19W-8S	32
Walden Ln. N&S	18W-8S	32
Walnut Dr. E&W	17W-8S	33
Westminister Ct. NS&EW	18W-9S	32
White Pine Trail* NE&SW	18W-7S	26
Whitter Dr. N&S	18W-8S	32
Wilcox Av. N&S	18W-7S	26
Wildwood Ct. N&S	18W-8S	32
Wildwood Ln. NW&SE	18W-9S	32
Willow Creek Ln. E&W	18W-9S	32
Willow Ln. E&W	17W-8S	33
Wilmette Av. N&S	17W-9S,17W-7S	33
Winterberry Ln. E&W	18W-9S	32
Wintergreen Ln.* N&S	18W-7S	26
Woodbine Ln.* N&S	18W-7S	26
Woodrush Ln.* E&W	18W-7S	26
Woodview Ct. E&W	18W-9S	32
63rd St. E&W	17W-7S	27
64th St. E&W	17W-7S	27
65th St. E&W	17W-7S	27
67th St. E&W	17W-7S,17W-6S	27
68th St. E&W	16W-7S	27
69th St. E&W	17W-7S,16W-7S	27
71st St. E&W	17W-7S	27
72nd St. E&W	17W-7S	33
73rd St. E&W	18W-8S,17W-8S	33
74th St. E&W	16W-8S	33
75th St. E&W	18W-8S,17W-8S	33
77th St. E&W	18W-8S	33
79th St. E&W	17W-8S,17W-7S	33

Downers Grove

	Grid	Page
Abbey Dr. E&W	19W-8S	32
Acorn Av. E&W	20W-3S	20
Acorn Dr. NE&SW	19W-3S	20
Acorn Dr. NE&SW	20W-3S	20
Alamance Pl. E&W	20W-7S	26
Aldrich Pl. E&W	21W-7S	25
Almond Ct. N&S	20W-3S	20
Andrus Av. E&W	20W-7S	26
Applegate Av. E&W	19W-8S	32
Arbor Circle NS&EW	21W-4S	25
Ashbury Ave. E&W	19W-8S	32
Aubrey Ter. N&S	20W-6S	26
Austin St. E&W	19W-5S	26
Baimbridge Dr. N&S	18W-8S	32
Baker Pl. NW&SE	20W-8S	32
Banbury Rd. N&S	21W-6S	25
Barclay Ln. N&S	20W-7S	26
Barrett St. N&S	20W-7S	26
Bateman St. NS&EW	20W-8S	32
Bates Pl. NE&SW	20W-7S	26
Baybury Rd. N&S	19W-8S	32
Belden Av. NW&SE	19W-5S	26
Bell Aire Dr. N&S	20W-4S,20W-3S	26
Belmont Av. N&S	21W-6S,21W-5S	25
Bending Oaks Cts. E&W	20W-5S	26
Bending Oaks Pl. N&S	20W-5S	26
Bentley Pl. E&W	19W-7S	26
Benton Av. N&S	19W-5S	26
Binder Rd. N&S	20W-8S	32
Birch Av. E&W	19W-5S	26
Birchwood Pl. E&W	21W-3S	19
Black Oak Dr. N&S	19W-3S	20
Blackburn Av. N&S	19W-8S,19W-7S	32
Blackburn Av. N&S	19W-8S	32
Blackburn Pl. N&S	19W-7S	26
Blackstone Ct. NE&SW	19W-7S	26
Blackstone Dr. NW&SE	19W-7S	26
Blanchard St. E&W	20W-6S,19W-6S	26
Blodgett Av. N&S	19W-7S,19W-5S	26
Bolson Dr. E&W	20W-7S	26
Bonnie Brae Dr. E&W	19W-7S	26
Borman Pl. E&W	20W-8S	32
Bradley Ct. NS&EW	20W-7S	26
Braemoor Dr. E&W	19W-3S	20
Breasted Av. E&W	20W-7S	26
Brentwood Pl.* E&W	19W-4S	26
Briargate Dr. NS&EW	19W-7S	26
Brighton St. NS&EW	21W-6S	25
Brook Dr. E&W	20W-2S	20
Brookbank Rd. N&S	20W-6S,20W-5S	26
Brookwood Dr. E&W	20W-7S	26
Brunette Dr. E&W	20W-7S	26
Bryan Pl. N&S	19W-5S	26
Buckingham Pl. E&W	19W-7S	26
Buckthorn Ln. N&S	20W-3S	20
Bunker Hill Cir. N&S	20W-7S	26
Bunning Dr. E&W	19W-6S	26
Burlington Av. E&W	20W-5S,18W-5S	26
Bush N&S	20W-7S	26
Butterfield Rd. NE&SW	20W-2S	20
Camden Ct. N&S	20W-7S	26
Camden Pl. NW&SE	20W-7S	26
Camden Rd. N&S	20W-8S,20W-7S	32
Candlewood Dr. E&W	20W-3S,19W-3S	20
Carol St. E&W	20W-4S	26
Carpenter St. N&S	19W-7S,19W-5S	26
Center Av. N&S	20W-4S	26
Centre Circle E&W	20W-2S	20
Challen Ct. E&W	21W-5S	25
Chase Av. N&S	21W-6S,21W-5S	25
Chicago Av. N&S	21W-4S,19W-4S	25
Churchill Pl. N&S	19W-7S	26
Claremont Ct. NE&SW	19W-7S	26
Claremont Dr. E&W	19W-8S,19W-7S	32
Clayton St. N&S	19W-7S	26
Clyde Av. E&W	19W-6S	26
Commerce Dr. N&S	21W-4S	25
Concord Ct. NE&SW	20W-7S	26
Concord Dr. E&W	20W-7S	26
Concord Pl. N&S	20W-7S	26
Coralberry Ln. E&W	20W-3S	20
Cornell Av. N&S	20W-5S	26
Country Creek Way NE&SW	19W-8S	32

	Grid	Page
Creekside Rd. N&S ... 19W-7S		26
Creekwood Ct. NE&SW		
.... 19W-3S		20
Crescent Dr. NW&SE		
.... 19W-7S		26
Cumnor Rd. N&S		
.... 18W-5S,18W-4S		26
Curtiss St. E&W		
.... 21W-5S,19W-5S		25
Davis St. N&S 19W-4S		26
Dawn Pl. E&W 19W-4S		26
De Bolt Av. N&S 19W-5S		26
De Witt Ln. N&S 20W-5S		26
Dearborn Pkwy. N&S		
.... 19W-6S		26
Deerpath Ln. N&S .. 18W-6S		26
Devereaux Dr. NE&SW		
.... 20W-8S		32
Dexter Rd. N&S		
.... 20W-8S,20W-7S		32
Dickson Av. E&W .. 20W-7S		26
Douglas Rd. N&S		
.... 19W-5S,19W-4S		26
Downers Dr. N&S		
.... 20W-4S,20W-3S		26
Drew St. N&S 20W-3S		20
Drove Av. E&W 20W-4S		26
Duchess Ct. N&S .. 20W-3S		20
Dunham Rd. N&S		
.... 20W-8S,19W-8S		32
Earlston Rd. N&S .. 19W-4S		26
East-West Tollway NE&SW		
.... 21W-4S,20W-3S		25
Eldon Pl. N&S 19W-6S		26
Elizabeth Ln. A. N&S .. 18W-8S		32
Elm St. N&S		
.... 19W-5S,19W-4S		26
Elmore Av. E&W		
.... 21W-5S,20W-5S		25
Elmwood Av. NW&SE		
.... 19W-5S		26
Esplanade Dr. E&W . 20W-2S		20
Essex Pl. E&W 20W-7S		26
Fairhaven Ct. N&S .. 20W-6S		26
Fairmount Av. N&S		
.... 19W-8S,19W-4S		32
Fairview Av. N&S		
.... 19W-8S,19W-4S		32
Fairview West N&S .. 19W-8S		32
Farley Pl. N&S 19W-5S		26
Farrah Ct. NW&SE .. 20W-5S		26
Finley Rd. N&S		
.... 21W-4S,20W-4S		25
Florence Av. N&S		
.... 18W-8S,18W-4S		32
Forest Av. N&S		
.... 19W-5S,19W-4S		26
Foster Pl. N&S 20W-7S		26
Foster Rd. N&S 20W-7S		26
Franklin St. E&W		
.... 20W-5S,19W-5S		26
Frontage Rd. N&S		
.... 20W-2S,19W-2S		20
George St. E&W ... 20W-6S		26
Gierz St. E&W		
.... 19W-5S,18W-5S		26
Gilbert Av. E&W ... 20W-5S		26
Glen Av. E&W 20W-4S		26
Glendenning Rd. N&S		
.... 19W-4S		26
Glenview Av. N&S .. 21W-5S		25
Golden Bell Ct. E&W		
.... 20W-3S		20
Goldenbell Ct. E&W . 20W-3S		20
Grace Ct. E&W 20W-7S		26
Graham Pl. E&W ... 20W-8S		32
Grand Av. N&S		
.... 19W-8S,19W-5S		32
Grand Ct. NE&SW .. 19W-8S		32
Grant St. E&W		
.... 20W-4S,18W-4S		26
Granville Av. N&S .. 21W-5S		25
Gregory Pl. E&W ... 20W-3S		20
Grove St. E&W 19W-5S		26
Haddow Av. E&W .. 21W-5S		25
Hall Pl. E&W 20W-8S		32
Harmac Pl. NW&SE . 18W-6S		26
Hartford Rd. N&S .. 19W-8S		32
Hastings Av. E&W .. 21W-6S		25
Hatch Pl. E&W 20W-8S		32
Hatch St. E&W 20W-8S		32
Hathaway Ln. N&S . 20W-7S		26
Havens Ct. E&W ... 19W-4S		26
Hawkins Av. E&W .. 20W-8S		32
Hawthorne Ln. NE&SW		
.... 20W-5S		26
Herbert St. E&W		
.... 20W-4S,18W-4S		26
Hickory Ct. N&S ... 20W-3S		20
Hickory Tr. NW&SE . 20W-3S		20
Hickory Trail E&W .. 19W-3S		20
Highland Av. N&S		
.... 19W-5S,19W-3S		26
Highland Ct. E&W .. 19W-4S		26
Hill St. E&W 19W-5S		26
Hillcrest Ct. NS&EW 20W-6S		26
Hillcrest Dr. E&W .. 20W-7S		26
Hillcrest Rd. NS&EW		
.... 20W-5S		26
Hitchcock Av. E&W . 21W-5S		25
Hitchcock St. N&S		
.... 21W-5S,20W-5S		25
Hobart Av. E&W ... 19W-8S		32
Holland Pl. NE&SW . 20W-3S		20

	Grid	Page
Holly Ct. N&S 20W-3S		20
Hughes Av. E&W ... 20W-8S		32
Indianapolis Av. E&W		
.... 19W-4S		26
Interstate 88 E&W .. 20W-3S		20
Inverness Rd. E&W . 21W-5S		25
Jacqueline Dr. NS&EW		
.... 20W-5S		26
Janes Av. N&S		
.... 21W-6S,21W-5S		25
Janet St. E&W 20W-4S		26
Jay Dr. E&W 19W-7S		26
Jefferson Av. E&W . 20W-6S		26
Katrine Av. N&S ... 21W-5S		25
Kelly Pl. N&S 20W-8S		32
Kensington Pl. NS&EW		
.... 20W-7S		26
Kenyon St. E&W ... 19W-6S		26
Kidwell Pl. E&W ... 20W-8S		32
Knottingham Ln. N&S		
.... 18W-8S		32
Lacey Rd. N&S		
.... 20W-4S,20W-3S		26
Lake Av. E&W 19W-4S		26
Lamb Ct. N&S 20W-7S		26
Lancaster Av. E&W . 19W-7S		26
Lancaster Pl. N&S .. 19W-7S		26
Lane Pl. N&S		
.... 19W-6S,19W-5S		26
Laurel Ct. N&S 20W-3S		20
Lee Av. N&S		
.... 20W-6S,20W-4S		26
Lee St. N&S 20W-4S		26
Lemont Rd. N&S		
.... 20W-8S,19W-8S		32
Leonard Av. N&S .. 21W-6S		25
Lexington Ln. E&W . 20W-7S		26
Lincoln Ln. N&S		
.... 19W-4S,18W-4S		26
Lincoln St. N&S		
.... 20W-4S,19W-4S		26
Linden Av. N&S ... 19W-5S		26
Lindley St. N&S 19W-5S		26
Lindwald NE&SW ... 20W-5S		26
Linscott Av. N&S		
.... 20W-5S,20W-4S		26
Longmeadow N&S		
.... 18W-4S		26
Loomes NW&SE ... 20W-7S		26
Lyman Av. N&S		
.... 19W-8S,19W-5S		32
Mackie Pl. NW&SE . 19W-5S		26
Main Blvd. E&W ... 20W-2S		20
Main St. N&S		
.... 19W-7S,19W-4S		26
Maplewood Pl. NW&SE		
.... 20W-5S,19W-5S		26
Marie Dr. E&W 18W-8S		32
Meade Rd. N&S ... 20W-7S		26
Meadow Crest Ct. NW&SE		
.... 19W-7S		26
Meadow Crest Dr. N&S		
.... 19W-7S		26
Meadow Ln. NE&SW		
.... 20W-5S		26
Middaugh Av. N&S		
.... 20W-6S,20W-4S		26
Midhurst Rd. NW&SE		
.... 21W-7S		25
Mistwood Ct. N&S . 19W-3S		20
Mistwood Ln. E&W . 19W-3S		20
Mistwood Pl. NE&SW		
.... 20W-3S		20
Monmouth Pl. E&W . 20W-7S		26
Montgomery Av. N&S		
.... 20W-5S		26
Morton Av. N&S ... 20W-4S		26
Nash St. N&S 20W-7S		26
Newport Rd. E&W . 20W-7S		26
Norfolk St. E&W ... 20W-7S		26
Northcott Av. N&S		
.... 20W-5S,20W-4S		26
Northgate Way N&S 19W-8S		32
Oak Grove Dr. N&S . 20W-2S		20
Oak Grove Rd. 20W-2S		20
Oak Hill Ct. NW&SE 20W-3S		20
Oak Hill Ct. E&W .. 20W-3S		20
Oak Hill Rd. E&W		
.... 20W-3S,19W-3S		20
Oakwood Av. . 20W-5S		26
Ogden Av. E&W		
.... 19W-8S		32
Oneil Pl. N&S 19W-8S		32
Opus Pl. N&S 20W-3S		20
Orchard Pl. N&S ... 19W-8S		32
Osage Av. N&S		
.... 19W-8S,19W-6S		32
Otis Av. E&W 19W-4S		26
Otto N&S 20W-7S		26
Oxford St. NW&SE . 19W-7S		26
Oxnard Ct. NW&SE . 20W-7S		26
Oxnard Dr. N&S		
.... 21W-7S,20W-7S		25
Palmer St. NE&SW . 20W-7S		26
Park Av. N&S		
.... 19W-6S,19W-5S		26
Parker Av. N&S 19W-7S		26
Parkview Dr. NE&SW		
.... 19W-7S		26
Parrish Ct. N&S 20W-3S		20
Penner Ave. N&S .. 20W-7S		26

	Grid	Page
Pershing Av. N&S		
.... 21W-5S,21W-4S		25
Pipers Way NE&SW . 19W-8S		32
Plum Ct. NS&EW ... 20W-3S		20
Plum Ct. E&W 20W-3S		20
Plymouth Ct. N&S .. 20W-6S		26
Plymouth Pl. N&S .. 20W-6S		26
Plymouth Rd. N&S . 20W-7S		26
Plymouth St. N&S .. 20W-6S		26
Pomeroy Ct. N&S .. 20W-3S		20
Pomeroy Rd. N&S .. 20W-3S		20
Powell Ct. N&S 20W-7S		26
Powell Pl. NE&SW . 20W-8S		32
Powell St. N&S		
.... 20W-8S,20W-7S		32
Prairie Av. N&S		
.... 21W-5S,18W-5S		25
Prentiss Ct. N&S ... 20W-7S		26
Prentiss Dr. E&W		
.... 21W-7S,20W-7S		25
Prideham St. N&S .. 19W-7S		26
Prince St. N&S		
.... 19W-5S,19W-4S		26
Prospect Av. N&S .. 19W-5S		26
Puffer Rd. N&S		
.... 21W-7S,21W-5S		25
Queens Ct. N&S ... 18W-8S		32
Quince Ct. NW&SE . 19W-3S		20
Railroad St. NW&SE		
.... 22W-5S,21W-5S		25
Randall St. NE&SW . 19W-5S		26
Red Bud Ct. N&S .. 20W-3S		20
Red Silver Ct. N&S . 20W-3S		20
Revere Rd. NW&SE . 20W-7S		26
Richards Av. E&W .. 20W-8S		32
Ridgeview St. E&W . 19W-8S		32
Ridgewood Circle NS&EW		
.... 20W-6S		26
Rob Roy Pl. NW&SE		
.... 19W-8S		32
Robey Av. E&W ... 19W-7S		26
Roe Ct. NE&SW ... 19W-8S		32
Rogers St. E&W ... 19W-5S		26
Rohrer Dr. N&S 18W-8S		32
Rosewood Ln. N&S . 19W-3S		20
Roslyn Rd. N&S ... 18W-4S		26
Ross Ct. NE&SW ... 20W-5S		26
Saratoga Av. N&S		
.... 20W-7S,20W-3S		26
Saylor St. N&S 20W-3S		20
Scheldrup Dr. N&S . 20W-2S		20
Seeley N&S		
.... 20W-5S,20W-4S		26
Selig NW&SE 20W-8S		32
Shady Ln. E&W 18W-4S		26
Sheldon Av. NE&SW		
.... 18W-5S		26
Sheridan Pl. E&W .. 20W-5S		26
Sherman St. E&W .. 20W-5S		26
Sherwood Av. N&S . 19W-4S		26
Sherwood Ct. E&W . 18W-8S		32
Snowberry Ct. NW&SE		
.... 20W-3S		20
Snowberry Ln. N&S		
.... 20W-3S		20
Springside Av. N&S		
.... 20W-8S,20W-7S		32
Springside Pl. E&W . 20W-7S		26
Stair St. NW&SE ... 20W-7S		26
Stanley Av. N&S ... 19W-5S		26
Sterling Rd. N&S ... 19W-5S		26
Stockley Rd. E&W . 19W-8S		32
Stonewall Av. N&S		
.... 20W-7S,20W-4S		26
Stratford Ln. E&W .. 19W-6S		26
Sturbridge Pl. E&W . 18W-8S		32
Summit St. E&W ... 19W-5S		26
Taylor St. E&W 20W-7S		26
Terrace Dr. N&S ... 20W-7S		26
Terrace Pl. N&S ... 20W-7S		26
Thatcher Rd. NS&EW		
.... 22W-5S		25
Thornwood Dr. E&W		
.... 20W-6S		26
Ticonderoga Pl. N&S		
.... 20W-8S		32
Ticonderoga Rd. N&S		
.... 20W-7S		26
Ticonderoga St. NS&EW		
.... 20W-8S		32
Tomicek Rd. N&S ... 20W-5S		25
Tower Ct. NE&SW .. 19W-8S		32
Tower Rd. E&W 18W-4S		26
Trent Rd. N&S 19W-8S		32
Turvey Ct. N&S 20W-5S		26
Turvey Rd. NE&SW . 20W-5S		26
Valley Forge Pl. E&W		
.... 20W-7S		26
Valleyview Dr. NS&EW		
.... 20W-7S,19W-7S		26
Venard Rd. NE&SW		
.... 20W-4S,20W-3S		26
Victor St. N&S 18W-5S		26
Virginia St. E&W ... 20W-4S		26
Wall Pl. N&S 19W-6S		26
Wallbank Av. N&S .. 20W-5S		26
Wallen Pl. E&W 20W-6S		26
Walnut Av. N&S		
.... 20W-6S,22W-5S		25
Wanda Pl. N&S 19W-6S		26
Warren Av. N&S		
.... 20W-5S,19W-5S		26
Washington St. N&S		
.... 19W-7S,19W-4S		26

	Grid	Page
Waterfall Pl. NE&SW		
.... 19W-7S		26
Weatherbee Av. E&W		
.... 19W-7S		26
Weatherbee Pl. N&S		
.... 19W-7S		26
Webster Pl. NW&SE 19W-6S		26
Webster St. N&S		
.... 19W-7S,19W-6S		26
Webster St. N&S ... 19W-6S		26
Wellington Pl. NS&EW		
.... 20W-7S		26
Wells St. N&S 20W-7S		26
Westend Rd. N&S .. 18W-4S		26
Westfield Dr. NE&SW		
.... 21W-7S		25
Whiffen Pl. N&S 19W-5S		26
White Pl. E&W 20W-8S		32
Whitefawn Rd. N&S		
.... 18W-6S		26
Widden Av. E&W ... 19W-7S		26
Wilcox Av. N&S		
.... 19W-6S,18W-5S		26
Willard Pl. NE&SW . 20W-8S		32
Williams St. N&S		
.... 18W-8S,18W-4S		32
Wilson Av. N&S 19W-4S		25
Wilson St. E&W 19W-5S		26
Windsor Ct. E&W .. 20W-7S		26
Winthrop Ct. NW&SE		
.... 19W-8S		32
Winward Way NE&SW		
.... 19W-8S		32
Wisconsin St. N&S .. 21W-5S		25
Wolf Pl. N&S 19W-7S		26
Wood Av. E&W 20W-3S		20
Wood Creek Dr. N&S		
.... 21W-3S,21W-2S		19
Woodcreek Dr. NS&EW		
.... 21W-2S,20W-3S		19
Woodward Av. N&S		
.... 21W-5S,21W-4S		25
York Rd. NE&SW ... 19W-8S		32
2nd St. E&W 18W-5S		26
4th St. E&W 18W-5S		26
5th St. E&W 18W-5S		26
6th St. E&W 18W-5S		26
7th St. E&W 18W-5S		26
8th St. E&W 18W-5S		26
31st St. E&W 19W-2S		20
35th St. E&W		
.... 20W-3S,19W-3S		20
36th St. E&W 19W-3S		20
39th St. NW&SE ... 20W-4S		26
39th St. E&W		
.... 20W-3S,18W-3S		20
39th St. E&W 20W-3S		20
40th Pl. E&W 18W-4S		26
40th St. E&W		
.... 20W-4S,19W-4S		26
41st St. E&W		
.... 20W-4S,18W-4S		26
55th Pl. E&W 18W-6S		26
55th St. E&W		
.... 20W-5S,19W-5S		26
56th St. NE&SW ... 18W-6S		26
56th St. E&W 18W-6S		26
59th Pl. E&W 19W-6S		26
59th St. E&W		
.... 20W-6S,19W-6S		26
60th Pl. E&W		
.... 20W-6S,19W-6S		26
60th St. E&W		
.... 21W-6S,19W-6S		25
61st St. E&W 19W-6S		26
62nd Ct. E&W 19W-6S		26
62nd Pl. NE&SW ... 19W-6S		26
62nd St. E&W		
.... 21W-6S,19W-6S		25
63rd St. E&W		
.... 20W-7S,19W-7S		26
64th St. E&W 21W-7S		25
65th St. E&W 19W-7S		26
67th Pl. E&W 20W-7S		26
67th St. E&W		
.... 20W-7S,20W-7S		26
68th Pl. E&W 20W-7S		26
68th St. E&W		
.... 20W-7S,19W-7S		26
71st St. E&W		
.... 20W-8S,19W-7S		32
71st Ter. NW&SE .. 20W-8S		32
72nd Ct. E&W 19W-8S		32
72nd St. E&W 19W-8S		32
73rd St. E&W 19W-8S		32
74th St. E&W 19W-8S		32
75th St. E&W		
.... 20W-8S,19W-8S		32

Downers Grove Township

	Grid	Page
Adams St. N&S		
.... 18W-8S,18W-7S		32
Adelia St. E&W 19W-7S		26
Ailsworth Ct. E&W . 20W-9S		32
Ailsworth Dr. N&S .. 20W-9S		32
Alabama Av. N&S		
.... 17W-9S,17W-6S		33
Alden Ln. N&S 20W-9S		32
Allison Ct. N&S 16W-9S		33
Andrus Rd. N&S ... 19W-10S		32
Argonne Ridge Rd. N&S		
.... 17W-10S		33

	Grid	Page
Barrett St. NW&SE		
.... 20W-11S		34
Bay Rum Ct. NW&SE		
.... 15W-9S		33
Belair Ct. E&W 18W-8S		32
Beller Ct. N&S 20W-9S		32
Beller Dr. E&W 20W-9S		32
Berkshire Ct. E&W . 19W-9S		32
Birnam Tr. NS&EW		
.... 17W-10S		33
Blancherd St. E&W . 20W-6S		26
Blodgett Av. N&S .. 19W-9S		32
Bluff Rd. E&W		
.... 19W-12S,16W-11S		34
Bonnie Brae Ln. NE&SW		
.... 16W-9S		33
Boundary Rd. NE&SW		
.... 20W-9S		26
Brewer Rd. E&W ... 20W-9S		32
Brook Ridge Rd. N&S		
.... 19W-9S		32
Brook View Ct. NS&EW		
.... 19W-6S		27
Brookbank Rd. N&S 16W-9S		33
Burgess Pl. E&W ... 20W-9S		32
Caimbridge Ct. NE&SW		
.... 19W-9S		32
Camder Dr. E&W .. 20W-10S		32
Carl Ct. N&S 20W-9S		32
Carol St. E&W 19W-7S		26
Carpenter St. N&S		
.... 19W-12S,19W-7S		34
Cass Av. N&S		
.... 17W-11S,17W-8S		35
Central Av. N&S		
.... 19W-9S,16W-9S		32
Central Ct. N&S ... 17W-11S		35
Charles St. N&S ... 19W-9S		32
Circle Av. NE&SW . 16W-9S		33
Claremont Dr. E&W . 18W-7S		26
Clarendon Hills Rd. N&S		
.... 17W-10S,17W-6S		33
Cobb Dr. E&W ... 20W-10S		32
Cobb St. E&W 20W-10S		32
Concord Pl. NW&SE		
.... 17W-9S		33
Cramer Ln. NW&SE . 20W-9S		32
Cumnor Rd. N&S		
.... 18W-9S,18W-6S		32
Darien Club Dr. E&W		
.... 18W-7S		26
Deerpath Ln. E&W . 19W-12S		34
Dixon Ct. N&S 20W-9S		32
Don Ct. E&W 19W-9S		32
Downers Dr. N&S		
.... 20W-11S,20W-6S		34
Drew Av. N&S 15W-10S		33
Drover Ct. N&S ... 20W-9S		32
Drover Ln. N&S ... 20W-9S		32
Drover Rd. N&S ... 20W-9S		32
Dunham Dr. N&S .. 20W-10S		32
Dunham Rd. N&S		
.... 20W-11S,20W-9S		34
Dystrup Av. E&W .. 19W-12S		34
E. Hampshire Ln. N&S		
.... 17W-10S		33
Earl Ct. E&W 17W-9S		33
Eastwood Cutoff E&W		
.... 17W-11S		35
Eastwood Dr. E&W		
.... 18W-11S,17W-11S		34
Echo Ln. N&S 16W-10S		33
Edgewood Dr. (Pvt.) N&S		
.... 20W-9S		32
Elizabeth Dr. E&W . 20W-10S		32
Elm Ct. NE&SW ... 17W-9S		33
Elm St. NE&SW ... 17W-9S		33
Fairmount Av. N&S . 19W-9S		32
Fern St. NE&SW		
.... 17W-9S		33
Florence Av. N&S		
.... 18W-9S,18W-7S		32
Freud Rd. N&S		
.... 18W-11S,17W-11S		34
Front St. NS&EW		
.... 17W-11S,17W-10S		35
Frontage Rd. NE&SW		
.... 18W-9S,16W-9S		33
Gilbert Ct. NE&SW . 20W-9S		32
Godair Ct. N&S ... 16W-9S		27
Graceland St. N&S . 19W-9S		32
Grandview Ln. NW&SE		
.... 20W-9S		32
Grandview Pl. N&S . 20W-9S		32
Grant Av. E&W ... 18W-4S		26
Grant St. N&S		
.... 18W-8S,18W-4S		32
Harvest Ln. N&S .. 20W-9S		32
Harvest Pl. E&W .. 20W-9S		32
Harvey Rd. N&S .. 16W-9S		33
Havens Ct. E&W .. 20W-10S		32
Helen Ct. N&S 19W-9S		32
Helix Rd. E&W ... 19W-12S		34
Highland Av. N&S .. 19W-9S		32
Highland Rd. N&S . 16W-9S		33
Hill Rd. N&S 19W-12S		34
Hillcrest Dr. N&S .. 20W-9S		32
Hillcrest Ln. E&W .. 19W-12S		34
Hillside Ct. NE&SW . 20W-9S		32
Hillside Ln. E&W		
.... 17W-10S,16W-10S		33
Honeysuckle Rose Ln. E&W		
.... 16W-10S		33

	Grid	Page
Hyacinth Dr. NE&SW		
	16W-10S	33
Inner Circle E&W	18W-11S	34
Ivy Ln. N&S	16W-10S	33
Jackson St. N&S		
	16W-11S,16W-10S	35
Jackson St. N&S	16W-6S	27
Jaime Ln. N&S	16W-10S	33
Jeans Rd. NE&SW	16W-11S	35
Jefferson St. E&W	20W-6S	26
Joliet Rd. NE&SW		
	17W-9S,16W-9S	33
Kearney Rd. N&S		
	19W-11S,19W-10S	34
Kingery Quarter N&S		
	16W-10S	33
Kings Ct. NW&SE	17W-9S	33
Kraml Dr. N&S	15W-9S	33
Lake Dr. NE&SW	16W-9S	33
Lakeside Dr. E&W		
	17W-9S,16W-6S	32
Landsfield Av. N&S	19W-9S	32
Landsfield Ct. NE&SW		
	19W-9S	32
Landsfield Pl. E&W	19W-9S	32
Lemont Rd. N&S		
	19W-12S,19W-9S	34
Leonard Dr. N&S		
	17W-10S,17W-9S	33
Lilac Ln. NE&SW	16W-10S	33
Lincoln St. N&S	18W-4S	26
Linda Ln. E&W	19W-9S	32
London Av. N&S	20W-10S	32
Lorraine Dr. N&S		
	17W-10S,17W-9S	33
Main St. N&S		
	19W-9S,19W-7S	32
Mark Ln. E&W	19W-9S	32
Mary Beth Ct. E&W	16W-9S	33
Meadow Ct. E&W		
	20W-9S,16W-9S	32
Meadow Lawn Av.		
	19W-7S	26
Meadow Ln. E&W		
	20W-11S,20W-9S	34
Meridan Rd. N&S	18W-11S	34
Middaugh Av. NW&SE		
	20W-11S	34
Millbrook Dr. E&W	19W-9S	32
Mockingbird Ln. E&W		
	16W-10S	33
Nantucket Dr. N&S	17W-9S	33
Nielsen Ln. E&W	17W-10S	33
Northgate Rd. N&S		
	18W-10S	32
Oakhill Ct. N&S	16W-10S	33
Oakwood Av. N&S	20W-11S	34
Old Bluff Rd. N&S		
	18W-11S,17W-11S	34
Oneil Dr. N&S	16W-10S	33
Orchard Rd. N&S		
	20W-12S,20W-11S	34
Oxford St. E&W	19W-7S	26
Palisades Rd. N&S		
	16W-11S,16W-10S	35
Parkview Dr. N&S	20W-9S	32
Peters Ct. E&W	20W-10S	32
Peters Dr. E&W	20W-10S	32
Pine Crest Ct. N&S	16W-6S	27
Pitcher Dr. E&W	20W-9S	32
Pleasant Dale Dr. E&W		
	20W-11S	34
Plymouth St. N&S	20W-6S	26
Portsmouth Ct. N&S		
	17W-9S	33
Portsmouth Dr. E&W		
	17W-9S	33
Quincy Ct. N&S	16W-7S	27
Quincy St. N&S		
	16W-9S,16W-6S	33
Railroad Dr. N&S	17W-11S	35
Roanoke Ct. NE&SW		
	15W-9S	33
Rock Rd. E&W	18W-11S	34
Rogers Ct. E&W	18W-8S	32
Rooke Ct. E&W	20W-10S	32
Rosehill Av. NE&SW	18W-9S	32
Rosehill Ln. N&S	20W-10S	32
Rutgers Ct. N&S	20W-10S	32
Rutgers Dr. E&W	20W-10S	32
Ruth Dr. N&S		
	19W-10S,19W-9S	32
Ruth Lake Ct. N&S	16W-6S	27
Scheel Dr. N&S	17W-10S	33
Seeley Av. N&S	20W-11S	34
Shiloh Dr. N&S	15W-9S	33
Skyline Dr. N&S	16W-10S	33
Spring Green Dr. E&W		
	20W-9S	32
Springside Av. N&S	20W-6S	26
St. Patrick Rd. E&W		
	18W-11S	34
Stewart St. E&W	20W-9S	32
Stonewall Av. N&S	20W-6S	26
Stough St. N&S	16W-6S	27
Suffield Dr. N&S	20W-10S	32
Summer Ct. NW&SE		
	20W-9S	32
Sunrise Av. N&S	16W-6S	27
Sunset Rd. N&S	19W-9S	33
Sutton Pl. E&W	17W-9S	33
Tanbark Ct. NE&SW		
	19W-9S	32

	Grid	Page
Tennessee Av. N&S		
	17W-9S,17W-6S	33
Terry Tr. E&W	17W-10S	33
Thames Av. N&S	20W-10S	32
Therese Ct. E&W	16W-9S	33
Thurlow St. N&S		
	16W-10S,16W-6S	33
Timber Trails E&W	20W-10S	32
Tolios Dr. N&S	19W-12S	34
Vail Ct. NE&SW	20W-10S	32
Valleyview Dr. E&W		
	16W-10S	33
Village Rd. N&S	16W-9S	33
Vine St. N&S	19W-10S	33
Virginia Av. N&S	17W-6S	27
Wallace Dr. N&S	20W-10S	32
Waltham Pl. E&W	17W-9S	33
Washington St. N&S		
	19W-9S,18W-4S	32
Water Tower Rd. E&W		
	18W-11S	34
Webster St. N&S	19W-8S	32
Western Av. N&S		
	17W-9S,17W-6S	33
Westgate Rd. E&W		
	19W-10S,18W-11S	32
Westminister Dr. E&W		
	20W-10S	32
White Deer Dr. N&S		
	18W-10S	32
William Dr. N&S	16W-9S	33
Willow Ln. E&W	18W-8S	32
Winter Cir. NS&EW	19W-9S	32
Wood Vale Dr. N&S	20W-9S	32
Woodcreek Ct. N&S	19W-9S	32
Woodcrest Dr. E&W		
	20W-10S,19W-10S	32
Woodglen Ln. NW&SE		
	19W-9S	32
York Ln. NW&SE	20W-10S	32
1st St. E&W		
	20W-11S,19W-11S	34
3rd St. E&W	20W-11S	34
9th St. E&W	19W-12S	34
40th St. E&W	18W-4S	26
41st St. E&W	18W-4S	26
55th Pl. E&W	20W-6S	26
56th Pl. E&W	16W-6S	27
56th St. E&W		
	17W-6S,16W-6S	27
57th St. E&W		
	17W-6S,16W-6S	27
58th Pl. E&W	16W-6S	27
58th St. E&W		
	17W-6S,16W-6S	27
59th St. E&W		
	20W-6S,17W-6S	26
61st St. E&W		
	20W-6S,17W-6S	26
62nd St. E&W	20W-6S	26
64th St. E&W	16W-7S	27
67th Ct. E&W	18W-7S	26
72nd St. E&W		
	18W-8S,15W-8S	32
73rd Pl. E&W	18W-8S	32
73rd St. E&W	15W-8S	33
74th St. E&W	15W-8S	33
75th St. E&W	15W-8S	33
79th Pl. E&W		
	19W-9S,17W-9S	32
80th St. E&W		
	19W-9S,16W-9S	32
81st St. E&W		
	19W-9S,15W-9S	32
82nd St. E&W	15W-9S	32
83rd Ct. E&W	19W-9S	32
83rd St. E&W		
	20W-9S,15W-9S	32
85th Ct. E&W	19W-9S	32
86th Pl. E&W	19W-9S	32
86th St. E&W	19W-9S	32
87th St. E&W		
	20W-9S,16W-9S	32
89th Pl. E&W	16W-10S	33
89th St. E&W		
	16W-10S,15W-10S	33
90th St. E&W		
	16W-10S,15W-10S	33
91st St. E&W		
	18W-10S,14W-10S	32
92nd St. E&W	19W-10S	32
93rd Pl. E&W	16W-10S	33
94th Pl. E&W	16W-10S	33
94th St. E&W		
	18W-10S,16W-10S	33
95th Pl. E&W	16W-11S	35
97th St. E&W		
	20W-11S,16W-11S	34
98th Pl. NE&SW	20W-11S	34
99th St. E&W		
	20W-11S,19W-11S	34

Elk Grove Village*

(Also see Cook County)

	Grid	Page
American Ln. N&S	16W-7N	9
Brickvale Dr. N&S	16W-7N	9
Carl Blvd. N&S	17W-7N	9
Central Av. N&S	17W-7N	9
Delta Ln. N&S	16W-7N	9
Devon Av. E&W		
	17W-7N,16W-7N	9
Eastern Av. E&W	16W-7N	9
Katherine Way N&S	17W-7N	9
Kirk St. E&W	17W-7N	9
Lively Blvd. N&S	17W-7N	9
Mark St. E&W	17W-7N	9
Pan Am Blvd. E&W	16W-7N	9
Thorndale Av. E&W	17W-7N	9
United Ln. N&S	16W-7N	9
York Rd. N&S	16W-7N	9

Elmhurst

	Grid	Page
Aberdeen Ct. N&S	15W-0S	21
Adams Ct. N&S	15W-0S	21
Adams St. E&W		
	16W-0S,15W-0S	21
Addison Av. N&S		
	16W-2N,16W-3N	15
Adelaide St. E&W	16W-1N	15
Adele Av. N&S		
	15W-2N,15W-3N	15
Adelia St. E&W	15W-1N	15
Adell Pl. N&S	16W-1N	15
Alben Av. N&S	15W-1N	15
Albert St. E&W	17W-1N	14
Alexander Blvd. E&W		
	17W-1N,16W-1N	14
Allison St. NW&SE	15W-0N	15
Alma Av. N&S	16W-1N	15
Apple Tree Ct. N&S	16W-0S	21
Arbor Dr. NE&SW	16W-0S	21
Argyle N&S	16W-0N	15
Arlington Av. N&S		
	15W-0N,15W-1N	15
Armitage Av. E&W		
	17W-2N,15W-2N	14
Arthur St. E&W	16W-1N	15
Aspen Tree Ct. E&W		
	16W-0S	21
Atwater Av. NE&SW		
	15W-1N	15
Atwood St. E&W	15W-1N	15
Avery St. NW&SE	16W-0S	21
Avon Rd. N&S	15W-1N	15
Babcock Av. E&W	16W-1N	14
Barclay Ct. N&S	15W-0N	15
Belden Av. E&W		
	17W-2N,15W-2N	14
Benton Av. NW&SE	15W-1N	15
Benton Ct. E&W	15W-1N	15
Berkley Av. N&S		
	16W-0S,16W-1N	21
Berteau Av. N&S	15W-1N	15
Betsy Ware Av. E&W		
	16W-3N	15
Beverly Av. NE&SW	16W-0S	21
Birch Tree Ct. E&W	16W-0S	21
Bonnie Brae Av. N&S		
	17W-1N	14
Boyd Av. N&S	15W-1N	15
Bryan St. N&S		
	16W-0S,16W-0N	21
Bucholz Dr. NE&SW		
	15W-1N	15
Butterfield Rd. NE&SW		
	16W-0S,15W-0S	21
Cadwell Av. N&S	15W-0S	21
Cambridge Av. N&S		
	15W-0S,15W-0N	21
Carol Ln. E&W	16W-3N	15
Caroline Av. N&S	15W-1N	15
Cayuga Av. E&W	15W-0N	15
Cedar Av. N&S		
	15W-0S,15W-0N	21
Cedar Tree Ct. E&W		
	16W-0S	21
Chandler Av. N&S	15W-1N	15
Chatham Av. N&S		
	15W-0S,15W-0N	21
Cherry St. N&S	15W-0N	15
Church St. E&W		
	16W-1N,15W-1N	15
Clara Pl. N&S	15W-1N	15
Claremont St. E&W	16W-1N	15
Clinton Av. N&S	15W-1N	15
Colfax Av. N&S		
	15W-0S,15W-0N	21
Colombia Av. N&S	15W-0N	15
Commonwealth Ln. NW&SE		
	16W-0S	21
Comstock Av. E&W	17W-2N	14
Coolidge Av. E&W	16W-0N	15
Cottage Hill Av. N&S		
	16W-0N,16W-1N	15
Coventry Ln. N&S	16W-0N	15
Crescent Av. E&W		
	16W-0N,15W-0N	15
Crescent St. E&W	16W-0N	15
Crest View Av. E&W		
	15W-2N	15
Crockett Av. N&S		
	17W-2N,16W-2N	14
Diversey Av. N&S	16W-3N	15
Division St. N&S	15W-0N	15
Dorchester Av. NW&SE		
	16W-0S	21
East Ct. E&W	15W-1N	15
Eastland St. N&S		
	15W-2N,15W-3N	15
Eaton St. N&S	15W-1N	15
Edgewood Av. N&S	15W-0N	15
Eggleston Av. E&W	16W-0N	15
Elizabeth St. N&S	17W-1N	14
Elm Av. N&S		
	16W-1N,16W-2N	15

	Grid	Page
Elm Park Av. E&W		
	17W-1N,16W-1N	14
Elm Tree Ln. N&S	15W-1N	15
Elmcrest Av. NW&SE		
	15W-2N	15
Elmhurst Av. E&W	15W-1N	15
Elmwood Terr. E&W		
	16W-1N	15
Emery Ln. N&S	16W-2N	15
Emroy Av. N&S		
	15W-2N,15W-3N	15
Euclid Av. N&S		
	16W-0S,16W-0N	21
Evergreen Av. N&S		
	16W-1N,16W-2N	15
Evergreen Ln. E&W	16W-3N	15
Fair Av. N&S		
	15W-0N,15W-1N	15
Fairfield Av. N&S		
	16W-0S,16W-1N	21
Fay Av. N&S	15W-1N	15
Fellows Ct. NW&SE	16W-1N	15
Fern Av. N&S		
	15W-0S,15W-0N	21
Fern Ct. N&S	15W-0N	15
Ferndale Av. N&S	17W-2N	14
First St. E&W		
	17W-1N,15W-1N	15
Forest Av. E&W	15W-1N	15
Fremont Av. E&W		
	16W-2N,15W-2N	15
Fullerton Av. E&W		
	16W-3N,15W-3N	15
Garfield Av. N&S	16W-0N	15
Geneva Av. N&S		
	15W-1N,15W-3N	15
Glade Av. N&S	17W-1N	14
Gladys Av. E&W		
	17W-2N,15W-2N	14
Glenview Av. N&S		
	17W-1N,17W-2N	14
Grace Av. N&S	16W-1N	15
Grand Av. NE&SW		
	17W-3N,15W-3N	14
Grandview St. E&W	16W-0N	15
Grantley Av. E&W		
	16W-2N,15W-2N	15
Hagans Av. N&S	16W-1N	15
Hahn St. E&W	16W-1N	15
Hale St. E&W	15W-0S	21
Hampshire Av. N&S		
	15W-0N,15W-1N	15
Harbour Terr. E&W	16W-1N	15
Harding St. E&W	16W-0N	15
Harrison St. E&W		
	16W-0S,15W-0S	21
Haven Rd. N&S	15W-1N	15
Hawthorne Av. N&S		
	16W-0N,16W-2N	15
High St. N&S	15W-0S	21
Highland Av. N&S		
	16W-1N,16W-2N	15
Highview Av. N&S	16W-2N	15
Hill Av. N&S	15W-0N	15
Hillcrest Av. N&S		
	15W-0S,15W-0N	21
Hillside Av. N&S		
	16W-0S,16W-0N	21
Howard Av. N&S		
	15W-2N,15W-3N	15
Huntington Ln. NE&SW		
	15W-1N	15
Ida Ln. N&S	16W-2N	15
Illinois St. N&S	15W-1N	15
Indiana St. N&S		
	15W-1N,15W-3N	15
Industrial Dr. NS&EW		
	16W-2N,16W-3N	15
Interstate 290 NW&SE		
	17W-3N,15W-0N	14
Jackson St. E&W		
	16W-0S,15W-0S	21
Junior Terr. N&S	17W-2N	14
Kearsage Av. N&S		
	15W-0S,15W-0N	21
Kenilworth Av. N&S		
	15W-0N,15W-3N	15
Kenmore Av. N&S	15W-1N	15
Kent Av. N&S	15W-0S	21
Killarny Ct. N&S	15W-0N	15
Kimbell Av. N&S	16W-1N	15
Kirk Av. N&S	15W-0S	21
Lake St. NW&SE		
	17W-3N,16W-2N	14
Lamont Rd. E&W	17W-3N	14
Larch Av. N&S		
	16W-1N,16W-3N	15
Laurel Av. NW&SE	15W-0N	15
Lawndale Av. N&S	15W-1N	15
Linden Av. N&S		
	15W-0S,15W-1N	21
Linden Dr. N&S	15W-0S	21
Locust Tree Ct. NW&SE		
	16W-0S	21
Lombard St. N&S		
	15W-2N,15W-3N	15
Lorraine Av. E&W	17W-2N	14
Madison Ct. N&S	15W-0S	21
Madison St. E&W		
	16W-0N,15W-0N	15
Manchester Ln. N&S		
	16W-0N	15
Maple Av. N&S		
	16W-1N,16W-2N	15

	Grid	Page
Maple Tree Ct. NW&SE		
	16W-0S	21
Margaret Pl. E&W	16W-1N	15
Marion St. E&W	15W-1N	15
Mary Ct. NW&SE	17W-3N	14
May St. E&W	15W-0N	15
Mckinley Av. E&W	16W-0N	15
Meister Av. NW&SE	16W-0N	15
Melrose Av. NE&SW		
	15W-1N,15W-2N	15
Michigan St. N&S		
	15W-1N,15W-3N	15
Mitchell Av. N&S		
	16W-0S,16W-0N	21
Monroe St. E&W		
	16W-0S,15W-0S	21
Monterey Av. N&S	17W-1N	14
Montrose Av. E&W	15W-1N	15
Myrtle Av. E&W		
	16W-1N,16W-2N	15
Niagara Av. E&W	15W-0N	15
North Av. E&W		
	17W-2N,15W-2N	14
North End Av. E&W	15W-3N	15
Oak St. N&S		
	16W-1N,16W-2N	15
Oak Tree Ct. N&S	16W-0S	21
Oakland Grove Av. N&S		
	17W-1N	14
Oaklawn Av. N&S		
	16W-1N,16W-3N	15
Olive St. E&W	16W-0N	15
Oneida Av. N&S	15W-0N	15
Orchard St. E&W	15W-0N	15
Oriole Av. N&S	15W-0N	15
Park Av. E&W		
	16W-1N,15W-1N	15
Park Manor Ct. E&W		
	15W-0S	21
Parker St. N&S		
	15W-2N,15W-3N	15
Parkside Av. N&S		
	16W-0S,16W-0N	21
Parkview Av. NS&EW		
	15W-0N	15
Pick Av. N&S	17W-1N	14
Pine St. N&S	15W-1N	15
Pine Tree Ct. E&W	16W-0S	21
Poplar Av. N&S		
	15W-0S,15W-1N	21
Poplar Tree Ct. NW&SE		
	16W-0S	21
Prairie Av. N&S	15W-0N	15
Prairie Path Ln. E&W		
	16W-0N	15
Prospect Av. N&S		
	16W-0S,16W-1N	21
Quincy St. E&W	15W-0S	21
Randolph St. NW&SE		
	16W-0N	15
Rex Blvd. N&S		
	16W-0N,16W-1N	15
Richard Av. N&S	15W-0N	15
Ridge Av. E&W	15W-0N	15
Ridgeland Av. N&S	15W-0N	15
River Glen Av. N&S	17W-2N	14
Riverside Dr. N&S		
	17W-0S,17W-0N	21
Robert Kingery Expwy. N&S		
	17W-0S,17W-3N	21
Robert T. Palmer Dr. N&S		
	15W-1N	15
Roosevelt Rd. E&W	16W-0S	21
Rose St. E&W	17W-1N	14
Rt. 38 E&W	16W-0S	21
Rt. 64 E&W		
	17W-2N,15W-2N	14
Rt. 83 N&S	17W-0S,17W-3N	21
Saylor Av. N&S		
	16W-0S,16W-0N	21
Schiller St. E&W	15W-1N	15
Scott St. N&S	17W-0N	14
Seminole Av. NW&SE		
	16W-0N	15
Shady Ln. N&S	17W-2N	14
Sherman Av. E&W	15W-0N	15
Sidney Av. E&W	17W-2N	14
Smith Av. N&S	16W-1N	15
South St. E&W	15W-0N	15
Spangler Av. N&S	16W-3N	15
Spring Rd. N&S		
	16W-0S,16W-0N	21
St. Charles Rd. E&W		
	17W-1N,15W-1N	14
Stratford Av. N&S		
	15W-0S,15W-0N	21
Stuart Av. N&S	15W-0S	21
Stuart Ct. N&S	15W-0N	15
Sturges Pkwy. N&S	16W-1N	15
Sunnyside Av. N&S		
	16W-0N,16W-1N	15
Sunrise Rd. E&W	17W-3N	14
Sunset Av. NE&SW	16W-0S	21
Surf St. E&W	17W-0N	14
Swain Av. N&S		
	16W-1N,16W-2N	15
Thomas St. NW&SE	16W-2N	15
Thornhill Ln. N&S	16W-0N	15
U.S. Route 20 NW&SE		
	17W-3N,16W-2N	14
Utley Rd. E&W	16W-1N	15
Vallette St. E&W		
	16W-0N,15W-0N	15

	Grid	Page
Van Auken St. N&S	15W-2N,15W-3N	15
Van Buren St. E&W	16W-0S,15W-OS	21
Verret St. NW&SE	16W-0S	21
Virginia Ln. N&S	15W-3N	15
Virginia St. E&W	16W-1N	15
Walnut St. N&S	16W-1N,16W-2N	15
Walter St. E&W	16W-1N	15
Washington St. N&S	16W-0S,16W-0N	21
Webster Av. E&W	15W-0N	15
West Av. N&S	17W-0N,17W-2N	14
West Av. N&S	16W-0S	21
Willow Rd. N&S	15W-1N,15W-3N	15
Willow Tree Ct. E&W	16W-0S	21
Wilson St. E&W	16W-0N,15W-0N	15
Windsor Dr. NW&SE	16W-0N	15
Winthrop Av. E&W	16W-1N	15
Wrightwood Av. E&W	16W-3N,15W-3N	15
York Rd. N&S	15W-0S,16W-3N	21
Yorkfield Av. NE&SW	15W-0S	21
2nd St. E&W	17W-1N,15W-1N	14
3rd St. E&W	17W-1N,15W-1N	14

Glen Ellyn

	Grid	Page
Abbey Dr. E&W	21W-1S	19
Abbotsford Ct. N&S	21W-0S	19
Adler Ln. E&W	23W-1S	18
Annandale Av. NW&SE	22W-0N	13
Anthony St. E&W	23W-0N,22W-0N	12
Apollo Av. NE&SW	22W-0N	13
Appian Way E&W	22W-0N	13
Arbor Ct. E&W	22W-0S	19
Arlington Av. E&W	22W-0S	19
Ash Ln. E&W	23W-1S	18
Baker Ct. E&W	21W-1S	19
Bloomfield Ln. E&W	23W-1S	18
Brandon Av. N&S	23W-0S	18
Bremer Ct. E&W	22W-0S	19
Briar St. N&S	21W-0S	19
Brighton Ln. E&W	21W-0S	19
Bryant Av. N&S	22W-0S,22W-ON	19
Carleton Av. N&S	22W-0S,22W-0N	19
Carlisle Ct. E&W	22W-1S	19
Carolyn Dr. N&S	22W-0N	13
Center St. E&W	23W-0N	12
Chancel Cir. NW&SE	21W-1S	19
Chapel Ct. N. E&W	21W-1S	19
Chapel Ct. S. E&W	21W-1S	19
Cheltenham Dr. E&W	20W-0S	20
Chesterfield Av. E&W	23W-0S	18
Chidester Av. E&W	22W-0N	13
Clifton Av. E&W	21W-0N	13
Colcord Pl. N&S	21W-0N	13
College Hill Av. NE&SW	23W-0S,21W-0N	19
College St. E&W	22W-2S	19
Concord Pl. N&S	21W-0N	13
Coolidge Av. E&W	21W-1S	19
Cottage Av. E&W	23W-0N,22W-0N	12
Country Club Ln. NW&SE	21W-ON	13
Cranston Ct. N&S	22W-0S	19
Crescent Blvd. NE&SW	22W-0N,21W-0N	13
Crescent Ct. E&W	22W-0N	13
Crescent Dr. N&S	21W-0N	13
Crest Rd. N&S	22W-0S	19
Cumnor Av. N&S	21W-0S	19
Dale Av. N&S	21W-0S	19
Davis Terr. N&S	23W-0N	12
Dawes Av. E&W	22W-1S	19
Dawn Av. N&S	23W-0N	12
Deer Ct. NE&SW	22W-0N	13
Deer Path Rd. N&S	22W-0N	13
Deicke Dr. E&W	23W-0S	18
Dell Av. N&S	21W-0S	19
Dorset Av. E&W	22W-0S	19
Dorset Ct. N&S	22W-0S	19
Dorset Pl. E&W	22W-0S	19
DuPage Blvd. E&W	22W-0S,21W-0S	19
Duane St. E&W	23W-0N,21W-0N	12
Duane Terr. NE&SW	22W-0N	13
East Rd. N&S	22W-0S	19
Edgewood Dr. E&W	22W-0N	13
Ellyn Av. NE&SW	22W-0N	13
Ellyn Ct. E&W	22W-0N	13
Ellynwood Dr. NE&SW	21W-0N	13

	Grid	Page
Elm St. E&W	23W-1N,22W-1N	12
Essex Ct. NE&SW	22W-0N	13
Essex Rd. NW&SE	22W-0N	13
Euclid Av. N&S	22W-0N	13
Evergreen Av. N&S	23W-0N,23W-1N	12
Exmoor Av. N&S	22W-0S	19
Fairview Av. E&W	23W-0S,22W-0S	18
Fairview Av. E&W	23W-0S,21W-0S	19
Farnsworth Ct. E&W	23W-1S	18
Finley Rd. N&S	20W-0S	20
Fir Ct. E&W	23W-1S	18
Fir Ln. N&S	23W-1S	18
Florence Av. N&S	21W-0S	19
Forest Av. N&S	22W-0S,22W-0N	19
Geneva Rd. E&W	23W-0N,23W-1N	12
Glen Ellyn Pl. E&W	22W-0N	13
Glen Haven Ln. E&W	22W-1S	19
Glen Oak Av. E&W	21W-0N	13
Glenbard Rd. E&W	22W-1S,21W-1S	19
Glenwood Av. N&S	22W-0S	19
Grand Av. NE&SW	22W-0N	13
Grandview Av. N&S	21W-0S	19
Greenbriar Rd. N&S	22W-1S	19
Greenfield Av. E&W	23W-0S,22W-0S	18
Greenwood Ct. N&S	22W-0S	19
Grove Av. N&S	21W-0S	19
Harding Av. E&W	22W-1S	19
Hawthorne St. E&W	23W-0N,22W-0N	12
Heather Ln. E&W	21W-0S	19
Hedge Ct. N&S	22W-1S	19
Hickory Rd. N&S	21W-0N	13
High Rd. E&W	22W-0S	19
Highland Av. N&S	22W-0N	13
Highview Av. E&W	22W-0S,21W-0S	19
Hill Av. E&W	23W-0S,21W-0N	18
Hillcrest Av. N&S	23W-0S	18
Hillside Av. E&W	23W-0N,21W-0N	12
Illinois St. E&W	23W-0S	18
Indian Dr. N&S	21W-0S	19
Joyce Ct. N&S	22W-0S	19
Kenilworth Av. N&S	23W-0S,23W-0N	18
Kenilworth Ct. E&W	23W-0N	12
Lake Rd. NW&SE	22W-0N	13
Lakeview Terr. E&W	22W-0S	19
Lambert Av. N&S	23W-0S	18
Lambert Rd. N&S	23W-1S	18
Larch Ln. E&W	23W-1S	18
Lawrence Av. N&S	23W-0N	12
Lee St. NW&SE	22W-0N	13
Lenox Rd. NE&SW	22W-0N,22W-1N	13
Lincoln Av. N&S	22W-0N	13
Linden St. E&W	23W-0N,22W-0N	12
Lombard Av. E&W	21W-0S	19
Londonberry Ln. E&W	21W-0S	19
Longfellow Av. N&S	21W-0N	13
Lorraine Rd. NE&SW	23W-0S	18
Lowden Av. E&W	22W-1S	19
Lowell Av. N&S	21W-0S,21W-0N	19
Maiden Ln. E&W	22W-0S	19
Main St. N&S	22W-0S,22W-0N	19
Maple St. E&W	23W-0N,22W-0N	12
Marion Av. E&W	23W-0N	12
Marston Av. E&W	23W-1S,22W-1S	18
Maryknoll Cir. Cir.	21W-1S	19
May Av. N&S	21W-0S	19
Melrose Av. N&S	22W-0N	13
Memory Ct. E&W	21W-0S	19
Meredith Pl. NW&SE	22W-1N	13
Merton Av. N&S	22W-0S	19
Midway Pk. N&S	21W-0N	13
Miller Ct. N&S	23W-0S	18
Milton Av. N&S	22W-1S,22W-0S	19
Montclair Ave. N&S	22W-1S,22W-0N	19
Muirwood Dr. N&S	22W-1N	13
Newton Av. N&S	23W-0S,23W-0N	18
Nichol Pl. E&W	22W-1N	13
Nicoll Av. N&S	22W-1S,22W-0S	19
North Driveway NW&SE	21W-1S	19
Oak St. E&W	23W-0N,22W-0N	12
Oakwood Ct. N&S	21W-1S	19
Old Bond Ct. E&W	20W-0S	20
Ott Av. N&S	23W-0S	18

	Grid	Page
Oxford Rd. E&W	21W-0S	19
Park Blvd. N&S	22W-2S,22W-0N	19
Park Plaza Av. E&W	22W-0S	19
Park Row NW&SE	22W-0N	13
Parkside Av. N&S	22W-1S,22W-0S	19
Pembroke Ln. E&W	23W-1S	18
Pennsylvania Av. E&W	23W-0N,22W-0N	12
Pershing Av. E&W	22W-1S,21W-1S	19
Phillips Av. E&W	23W-0S,22W-0S	18
Pickwick Pl. E&W	22W-0S	19
Pleasant Av. N&S	22W-0N	13
Plum Tree Rd. NE&SW	22W-0N	13
Prairie Av. N&S	22W-0N	13
Prince Edward Rd. NE&SW	22W-1S	19
Prospect Av. N&S	22W-0S,22W-0N	19
Raintree Ct. NS&EW	22W-1S	19
Raintree Dr. E&W	22W-1S	19
Ramblewood Dr. E&W	23W-1S	18
Regent St. N&S	22W-0S	19
Revere Rd. E&W	22W-0S,21W-0S	19
Ridgewood Av. E&W	23W-0S,22W-0S	18
Riford Rd. N&S	22W-0N	13
Riverside Av. NE&SW	21W-0S	19
Roger Rd. N&S	21W-0N	13
Roslyn Rd. E&W	21W-0S	19
Royal Glen Dr. E&W	20W-0S	20
Rt. 38 E&W	23W-0S,20W-0S	18
Rt. 53 N&S	21W-1S,21W-0S	19
S. Ellyn Av. N&S	22W-1S	19
Saddlewood Dr. Cir.	23W-1N	12
Sandhurst Cir. N&S	23W-1S	18
Scott Av. N&S	21W-0S	19
Seminary Cir. NW&SE	21W-1S	19
Shadblow Dr. N&S	23W-1S	18
Sheehan Av. E&W	21W-1S	19
Sheffield Ln. E&W	23W-0N	12
Smith St. E&W	21W-0S	19
Snowhill Ct. E&W	21W-1S	19
South Driveway E&W	21W-1S	19
Spaulding Av. NE&SW	22W-0N	13
Spring Av. N&S	21W-0S	19
Spruce Ln. N&S	23W-1S	18
St. Charles Rd. E&W	23W-1N,22W-1N	12
St. James Pl. E&W	20W-0S	20
St. Moritz Dr. N&S	23W-0N	12
Stacy Ct. N&S	22W-0N	13
Stafford Ln. NE&SW	22W-1S	19
Stagecoach Ct. E&W	22W-1N	13
Stagecoach Run E&W	22W-1N	13
Stephenie Ln. N&S	22W-0S	19
Stonegate Ct. N&S	22W-1S	19
Stuart Av. NE&SW	22W-0N	13
Summerdale Av. NE&SW	22W-0S	19
Sunset Av. N&S	22W-0S	19
Sunset Ct. E&W	22W-0S	19
Surrey Dr. N&S	21W-0S	19
Sylvan Av. N&S	21W-0S	19
Taft Av. E&W	23W-1S,22W-1S	18
Taylor Av. N&S	22W-0S,22W-0N	19
Turner Av. E&W	23W-0S,22W-0S	18
Valley Av. E&W	21W-0S	19
Vandamin Av. N&S	22W-0S	19
Vine Av. N&S	23W-0S	18
Walnut St. E&W	22W-0N	13
Waverly Rd. E&W	21W-0S	19
West Driveway N&S	21W-1S	19
Western Av. N&S	23W-0N	12
Whittier Av. N&S	21W-0N	13
Williamsburgh Rd. E&W	22W-1S	19
Willis St. NE&SW	22W-0N	13
Wilson Av. E&W	22W-1S	19
Winchell Way E&W	22W-1S	19
Windemere Dr. N&S	20W-0S	20
Windsor Av. NE&SW	23W-0S	18
Wingate Rd. E&W	22W-0S	19
Woodland Dr. NE&SW	21W-0N	13
Woodstock Av. N&S	21W-0S	19
22nd St. E&W	22W-2S,22W-1S	19

Glendale Heights

	Grid	Page
Alberta Ct. N&S	22W-2N	13
Almond Ln. NW&SE	23W-4N	6
Alpine Dr. N. NE&SW	23W-3N	12
Alpine E. Dr. E&W	23W-3N	12
Alta Ln. E&W	23W-3N	12
Altgeld Av. E&W	24W-3N,22W-3N	12
Altgeld St. E&W	22W-3N	13
Amhurst Cir. E&W	24W-3N	12
Amy Av. N&S	23W-2N,23W-3N	12
Appleby Dr. E&W	24W-3N	12
Ardmore Av. NE&SW	22W-2N,22W-3N	13
Armitage Av. E&W	23W-2N,22W-2N	12
Army Trail Rd. E&W	24W-4N,22W-4N	6
Asbury Dr. E&W	22W-2N	13
Aspen Ln. N&S	23W-3N	12
Auburn Dr. NE&SW	24W-3N	12
Audubon Dr. N&S	23W-4N	6
Baldwin Ct. N&S	23W-4N	6
Barclay Dr. E&W	23W-4N	6
Basswood Ln. N&S	23W-3N	12
Belden Av. E&W	23W-2N	12
Bell Ln. N&S	23W-4N	6
Bentley Ct. N&S	23W-3N	12
Berkshire Ct. NW&SE	24W-3N	12
Berkshire St. NE&SW	24W-3N	12
Blair Ln. N&S	23W-4N	6
Bloomingdale Rd. N&S	23W-1N,23W-4N	12
Blueridge Dr. NE&SW	24W-3N	12
Boulder Dr. E&W	23W-3N	12
Brahms Ct. E&W	23W-4N	6
Brandon Ct. N&S	23W-3N	12
Brandon Dr. N. E&W	23W-3N	12
Brandon Dr. S. E&W	23W-3N	12
Brian Dr. NW&SE	24W-3N	12
Brittany Ct. E&W	24W-4N	6
Brittany Ln. N&S	24W-4N	6
Brockton Ln. N&S	24W-3N	12
Brookside Dr. E&W	24W-3N	12
Burdett Ct. NE&SW	22W-2N	13
Burgess Dr. E&W	23W-4N	6
Burr Oak Ln. N&S	23W-3N	12
Cambridge Ln. E&W	23W-2N	12
Camden Ct. NE&SW	23W-1N	12
Campbell Dr. NE&SW	24W-3N	12
Cardinal Dr. N&S	24W-4N	6
Cedar St. N&S	23W-1N	12
Chadwick Ln. N&S	23W-4N	6
Chapman Ct. N&S	23W-2N	12
Charles Ct. E&W	23W-3N	12
Charles Dr. N&S	22W-2N	13
Chaucer Ln. N&S	23W-4N	6
Chippendale Ln. N&S	24W-3N	12
Circle Dr. E&W	22W-2N	13
Clifford St. N&S	23W-3N	12
Club Dr. NS&EW	24W-2N	12
Clubhouse Ct. N&S	24W-2N	12
Copper Ct. E&W	24W-4N	6
Cornell Ct. E&W	24W-4N	6
Cove Ln. NE&SW	24W-3N	12
Cynthia Ln. E&W	22W-2N	13
Dale Rd. E&W	23W-4N	6
Darlene Ln. E&W	22W-2N	13
Davine Dr. N&S	22W-2N	13
Deanna Ln. N&S	22W-2N	13
Deere Ln. NE&SW	24W-3N	12
Dennison Dr. E&W	23W-4N	6
Devon Dr. N&S	24W-3N	12
Dickens Av. E&W	22W-2N	13
Drummond Av. E&W	24W-3N,23W-3N	12
Dunteman Dr. E&W	23W-4N	6
Eagle Ct. NW&SE	24W-2N	12
Easy Ct. NW&SE	22W-2N	13
Easy St. E. E&W	22W-2N	13
Easy St. N. N&S	22W-2N	13
Elizabeth Av. N&S	23W-2N	12
English Dr. N&S	24W-3N	12
Eunice Ct. NE&SW	22W-2N	13
Evergreen Av. N&S	23W-1N	12
Exchange Blvd. NW&SE	23W-3N	12
Fairway Dr. N&S	24W-2N	12
Fallbrook Dr. E&W	24W-4N	6
Flagstaff Ct. E&W	24W-3N	12
Flagstaff Ln. N&S	24W-3N	12
Fleetwood Dr. E&W	24W-3N	12
Fox Av. E&W	22W-2N	13
Frank Ct. N&S	23W-3N	12
Friedrich St. N&S	24W-3N	12
Fullerton Av. E&W	24W-2N,22W-3N	12
Garland Ct. NE&SW	23W-4N	6
Gerald Av. NE&SW	22W-2N	13
Gilberto St. N&S	24W-3N	12
Gladstone Ct. N&S	23W-4N	6

	Grid	Page
Glen Ellyn Rd. N&S	22W-1N,22W-4N	13
Glen Point Dr. E&W	23W-3N	12
Glengary Dr. N&S	23W-2N	12
Glenhill Ct. NE&SW	23W-2N	12
Glenhill Dr. E. E&W	23W-2N	12
Glenrock Ln. N&S	23W-3N	12
Golfview Dr. NS&EW	24W-2N	12
Goodrich Av. N&S	22W-2N	13
Green Meadow Dr. E&W	24W-3N	12
Greens Ct. NW&SE	24W-2N	12
Gregory Av. NS&EW	23W-3N,22W-3N	12
Hale Ct. E&W	24W-3N	12
Harding Ct. E&W	23W-4N	6
Harding Dr. E&W	23W-4N	6
Harold Av. E&W	22W-3N	13
Hartford St. N&S	23W-2N	12
Harvest Ln. N&S	24W-3N	12
Hemlock Ct. N&S	24W-3N	12
Hemlock Dr. NE&SW	24W-3N	12
Hesterman Dr. E&W	24W-3N	12
Hickory Ln. N&S	23W-3N	12
Highgrove Blvd. N&S	22W-3N,22W-4N	13
Highland Av. N&S	22W-2N,22W-3N	13
Hilldale Ln. N&S	23W-4N	6
Hilton Dr. E&W	23W-4N	6
Industrial Dr. NW&SE	22W-3N	13
International Blvd. E&W	23W-3N	12
Jacobsen Av. E&W	23W-2N	12
James Ct. E&W	22W-2N	13
Jamison Ln. E&W	23W-4N	6
Jeffrey Av. N&S	23W-3N	12
Jill Ct. NW&SE	22W-2N	13
Joel Ct. E&W	23W-1N	12
John St. N&S	24W-3N	12
Jones St. N&S	23W-3N	12
Joseph Ln. E&W	23W-2N	12
Joyce Av. NW&SE	22W-2N	13
Keating St. N&S	23W-3N	12
Kennebunk Ct. E&W	23W-1N	12
Lake Dr. E&W	24W-2N	12
Larry Ln. N&S	22W-2N,22W-3N	13
Leslie Ln. N&S	22W-2N	13
Leslie Ln. E. E&W	22W-2N	13
Liberty Dr. N&S	22W-2N	13
Lincoln Av. E&W	23W-3N,22W-3N	12
Lippert Ln. E&W	22W-2N	13
Lombard Ct. N&S	23W-3N	12
London Dr. N&S	23W-4N	6
Loveland Dr. E&W	23W-3N	12
Lynne Ct. E&W	22W-2N	13
Marilyn Av. E&W	23W-2N,22W-2N	12
Mark Av. E&W	23W-2N	12
Michael Ct. N&S	23W-1N	12
Michael Dr. N&S	23W-1N	12
Mill Pond Dr. E&W	24W-3N	12
Mimosa Ct. N&S	23W-3N	12
Mitchell Rd. E&W	22W-3N	13
Montana Av. E&W	24W-3N,22W-3N	12
Morse Ln. N&S	23W-4N	6
N. Brandon Dr. E&W	23W-3N	12
Neilan Ct. NE&SW	22W-3N	13
Nevada Av. E&W	24W-3N	12
Newport St. NE&SW	23W-3N	12
Niess Ct. NW&SE	23W-2N	12
Nolan Av. E&W	22W-2N	13
North Av. E&W	24W-2N,22W-2N	12
Norton Av. E&W	22W-2N	13
Opal Av. E&W	23W-3N	12
Orchard Ln. N&S	23W-2N	12
Oriole Ln. NS&EW	23W-4N	6
Oxford Ln. E&W	23W-2N	12
Paddock Cir. E&W	23W-3N	12
Park Ct. NE&SW	23W-2N	12
Paul Av. N&S	23W-3N	12
Payson Ln. NE&SW	23W-4N	6
Pepperwood Ln. N&S	23W-4N	6
Peterson Av. E&W	21W-2N	13
Placid Av. E&W	23W-3N	12
Polo Club Ct. NE&SW	23W-3N	12
Polo Club Dr. E&W	23W-3N,22W-3N	12
Poplar St. E&W	23W-3N	12
Portsmouth Ct. E&W	23W-1N	12
President Dr. N&S	24W-3N	12
President St. N&S	24W-2N	12
Quincy Ct. N&S	23W-3N	12
Reeves Dr. N&S	24W-3N	12
Regency Dr. NS&EW	22W-3N,22W-4N	13
Ringneck Dr. E&W	24W-3N	12
Robert Ct. E&W	23W-1N	12
Roberta Ct. E&W	24W-2N,23W-2N	12
Roland Dr. E&W	23W-4N	6

	Grid	Page
Rugby Ct. E&W	23W-4N	6
Sandburg Ct. N&S	23W-4N	6
Sandra Ct. E&W	23W-3N	12
Saratoga Dr. E&W	24W-3N	12
Scarboro Dr. N&S	23W-3N	12
Schmale Ct. E&W	24W-3N	12
Schubert Av. E&W 24W-3N,23W-3N		12
Schubert St. N&S	24W-3N	12
Scott St. N&S	23W-3N	12
Shorewood Dr. E&W	23W-1N	12
Sidney Av. E&W	22W-2N	13
Siems Dr. E&W	23W-4W	25
Slayton Ln. N&S	24W-3N	12
Somerset Dr. N&S	22W-3N	13
Spruce St. N&S	23W-1N	12
Stevenson Dr. E&W 24W-3N,23W-3N		12
Sydney Av. E&W	21W-2N	13
Terry Rd. E. E&W	22W-2N	13
Terry Rd. N. N&S	22W-2N	13
Thorncliff Dr. E&W	23W-4N	6
Timber Ln. E&W	23W-3N	12
Todd Av. N&S	23W-3N	12
Towner Ln. E&W	23W-3N	12
Vail Ln. N&S	23W-3N	12
Vanmeter Ct. E&W	23W-3N	12
Vanmeter St. N&S 23W-2N,23W-3N		12
Vantroba Dr. E&W	23W-3N	12
Vista Ct. E&W	23W-4N	6
W. Glenhill Dr. N&S	23W-3N	12
Wall St. N&S	24W-4N	6
Warwick Dr. E&W	24W-3N	12
Watersedge Dr. N. E&W	23W-4N	6
Wayne Av. N&S	22W-2N	13
Westberg St. N&S	23W-3N	12
Westchester Dr. N&S	23W-2N	12
Whitman Dr. E&W	23W-4N	6
Wildwood Cir. N&S	23W-3N	12
Windsor Ln. E&W	23W-2N	12
Winthrop Av. E&W	22W-2N	13
Winthrop Ct. N&S	22W-2N	13
Woodside Ln. N&S	23W-3N	12
Wren Ct. E&W	23W-4N	6
Wrightwood Av. E&W 24W-3N,23W-3N		12
Wyatt Ln. N&S	23W-4N	6
Yardley Ct. E&W	24W-4N	6

Hanover Park*
(Also see Cook County)

	Grid	Page
Aberdeen Ct. E&W	27W-5N	5
Alpine Ct. N&S	26W-7N	6
Amarillo St. N&S	26W-7N	6
Andover Dr. N&S 27W-7N,26W-7N		5
Apache Ct. E&W	26W-7N	6
Arbor Av. N&S	27W-7N	5
Argyle Rd. E&W	27W-7N	5
Arlington Cir. N&S	26W-6N	6
Arlington Ct. NW&SE	26W-6N	6
Arlington Dr. N&S	26W-6N	6
Arlington Rd. E&W	27W-7N	5
Army Trail Rd. E&W	27W-4N	5
Baldwin Ct. NS&SW	27W-5N	5
Bamburg Ct. E&W	26W-7N	6
Barr Ct. N&S	26W-7N	6
Barrington Rd. N&S 27W-7N,26W-7N		5
Barrymore Ct. E&W	27W-6N	5
Bavarian Ct. E&W	26W-7N	6
Bayside Dr. N&S 27W-4N,27W-5N		5
Bear Flag Dr. E&W	27W-5N	5
Bedford Ct. N&S	26W-7N	6
Bell Dr. E&W	27W-5N	5
Bell Ln. N&S	27W-5N	5
Beverly Cir. N&S	26W-7N	6
Beverly Ct. E&W	26W-7N	6
Bradley Ct. N. & S. N&S	26W-7N	6
Brookbank Dr. N&S	27W-4N	5
Camden Ln. E&W	27W-5N	5
Carmel Dr. N&S	26W-6N	6
Castlebar Ct. E&W	27W-5N	5
Castlebar Ln. N&S	26W-6N	6
Catalina Dr. E&W 26W-6N,26W-6N		6
Celebrity Cir. Cir.	26W-6N	6
Chandler Dr. NS&EW	27W-5N	5
Chaplin Ct. E&W	27W-5N	5
Charleston Ct. N&S	26W-7N	6
Chatsworth Ct. N&S	26W-7N	6
Church Rd. N&S 27W-6N,27W-7N		5
Cinema Dr. E&W	27W-6N	5
Citation Ct. N&S	26W-6N	6
Clarendon Ct. E&W	27W-5N	5
Clipper Dr. N&S	26W-4N	6
Concord Ct. N&S	27W-5N	5
Cooper Ct. NE&SW	26W-6N	6
Cottage St. N&S	27W-7N	5
County Farm Rd. NE&SW	27W-7N	5
Cove Dr. NW&SE	27W-5N	5
Crawford Ln. E&W	26W-6N	6
Crosby Ct. NW&SE	26W-6N	6
Ct. 'A' E&W	26W-6N	6
Ct. 'B' E&W	26W-6N	6
Ct. 'C' NE&SW	26W-7N	6
Ct. 'D' E&W	26W-6N	6
Ct. 'G' E&W	26W-6N	6
Ct. 'H' E&W	26W-6N	6
Ct. 'I' NE&SW	26W-7N	6
Ct. 'J' NE&SW	26W-7N	6
Ct. 'Q' N&S	26W-6N	6
Ct. Leona E&W	26W-7N	6
Ct. Maria N&S	26W-7N	6
Ct. Tiburon NE&SW	26W-7N	6
Curtis Ln. N&S	27W-6N	5
Danby Ct. N&S	27W-7N	5
Danforth Ct. N&S	27W-7N	5
Deerfield Ct. E&W	27W-7N	5
Deforest Ln. N&S	27W-7N	5
Del Monte Dr. N&S	26W-6N	6
Devon Av. E&W	27W-7N	5
Don Carlos Ct. N&S	27W-5N	5
Don Carlos Dr. E&W	27W-5N	5
Dory Cir. E. E&W	26W-4N	6
Downey Ct. NW&SE	27W-7N	5
Driftwood Ln. E&W	27W-5N	5
Dublin Ct. N&S	27W-7N	5
Dublin Ln. E&W	27W-5N	5
Dupont St. N&S	27W-5N	5
Dutch Mill Ct. N&S	26W-7N	6
Eastman Ln. N&S	27W-7N	5
Easton Ct. N&S	27W-7N	5
Edinburg Ln. N&S	27W-7N	5
Edison Cir. E&W	27W-7N	5
Elm St. E&W	27W-7N	5
Erie Ct. NE&SW	27W-5N	5
Essex Ct. N&S	27W-7N	5
Farmington Ct. N&S	27W-7N	5
Farnham Ct. E. N&S	26W-7N	6
Farrell Av. N&S	27W-7N	5
Fonda Ln. N&S	27W-6N	5
Franklin Ct. E. N&S	26W-7N	6
Fremont Dr. NW&SE	26W-7N	6
Fresno Ct. E&W	27W-7N	5
Fulton Ln. E&W	27W-5N	5
Gable Ct. NW&SE	26W-6N	6
Garbo Ln. N&S	27W-5N	5
Gardner Ct. N&S	26W-6N	6
Garland Ln. E&W	26W-6N	6
Gifford Ct. NE&SW	26W-7N	6
Glasgow Ct. NW&SE	27W-5N	5
Glengary Ct. N&S	27W-5N	5
Goddard Ln. E&W	27W-5N	5
Gold Cir. N&S	26W-7N	6
Grand Duell Way N&S	27W-5N	5
Grant Ln. NE&SW	26W-6N	6
Green Bay Dr. E&W	27W-4N	5
Greenbridge Dr. NS&EW	27W-7N	5
Greenbrook Ct. NW&SE	27W-7N	6
Grosse Pointe Cir.,W. & E NE&SW	27W-5N	5
Hanover Ct. NE&SW	26W-7N	6
Harbor Ln. E&W	27W-4N	5
Hardwood Ct. N&S	27W-5N	5
Hardy Dr. NW&SE	26W-7N	6
Heritage Ct. NE&SW	26W-7N	6
Hialeah Ct. N&S	26W-6N	6
Howe Ln. E&W	27W-5N	5
Hunter Ct. E&W	27W-5N	5
Island Cir. E&W	26W-4N	6
Isle Royal Ln. NW&SE	27W-5N	5
Jefferson St. N&S 27W-4N,27W-6N		5
Keene Ln. E&W	26W-6N	6
Kit Carson Dr. N&S	26W-7N	6
Kuhn Rd. N&S	26W-4N	6
Ladd Ct. NE&SW	27W-6N	5
Laguna Ct. NE&SW	26W-7N	6
Lake St. NW&SE 26W-7N,25W-6N		5
Lakeside Dr. N&S	27W-5N	5
Lapoloma Ct. NE&SW	26W-7N	6
Leeward Ln. E&W	27W-4N	5
Lemon Ln. E&W	27W-6N	5
Liberty St. E&W	27W-7N	5
Maclaine Ln. N&S	27W-6N	5
Madera St. NW&SE	26W-7N	6
Madrid Ct. E&W	26W-7N	6
Mallard Ln. E&W	27W-4N	5
Mallow Ct. E&W	27W-5N	5
Marine Dr. N&S	27W-4N	5
McCormick Ln. E&W	27W-5N	5
Merrimac Ln. N. E&W	26W-4N	6
Merwyn Av. N&S	27W-7N	5
Mill Ln. N&S	27W-5N	5
Millpond Ln. E&W	27W-4N	5
Monitor Dr. N&S	26W-4N	6
Monroe Ln. E&W	26W-6N	6
Montebello Dr. NE&SW	26W-6N	6
Morton Rd. N&S	27W-5N	5
Muirfield Dr. E&W	25W-7N	6
Nashua Ln. E&W	27W-4N	5
Nautilus St. N&S	27W-4N	5
Niven Ln. N&S	27W-6N	5
Nugget Cir. NW&SE	26W-7N	6
Oak St. N&S	27W-7N	5
Ontario St. N&S	27W-7N	5
Ontarioville Rd. NW&SE 27W-7N,26W-7N		5
Pastoral Ln. NS&EW	27W-5N	5
Pebble Beach Dr. N&S	26W-6N	6
Pennsbury Ct. N&S	27W-5N	5
Pennsbury Ct. E&W	27W-5N	5
Pondview Dr. N&S	27W-4N	5
Port Dr. N&S	27W-4N	5
Redford Ln. N&S	27W-5N	5
Rembrant Ct. E&W	26W-7N	6
Ring Ct. N&S	26W-7N	6
Rob Roy Ct. E&W	27W-5N	5
Rooney Ct. E&W	27W-6N	5
Russelwood Ct. N&S	27W-5N	5
Sacramento Ct. E&W	26W-7N	6
San Simeon Dr. E&W	26W-7N	6
Sandpiper Ln. N&S	27W-4N	5
Santa Anita Dr. E&W	26W-6N	6
Santa Barbara Dr. NW&SE 26W-6N,26W-7N		6
Santa Cruz N&S	26W-7N	6
Schick Rd. E&W	27W-5N	5
Schooner Ln. E&W 27W-4N,26W-4N		5
Sea Biscuit Ln. E&W	26W-6N	6
Seaview Dr. N&S	27W-4N	5
Shoal Dr. N&S	26W-4N	6
Shoreline Dr. N&S	27W-4N	5
Sinatra Ln. N&S	27W-6N	5
Spinnaker Ln. E&W 27W-4N,26W-4N		5
Springlake Dr. N&S	27W-4N	5
St. Claire Ct. N&S	27W-5N	5
Stairway Dr. E&W	27W-6N	5
Star Dr. N&S	26W-7N	6
Starboard Dr. N&S	27W-4N	5
Stearns Rd. E&W	27W-6N	5
Stepstone Ln. NW&SE	27W-5N	5
Stirling Ct. E&W	27W-5N	5
Sutter Dr. E&W	27W-7N	5
Thistle St. NE&SW	27W-5N	5
Toronto Ct. N&S	27W-5N	5
Tracy Ct. E&W	27W-6N	5
Turnberry Dr. N&S	25W-7N	6
Turner Ln. E&W	26W-6N	6
U.S. Route 20 NW&SE 26W-7N,25W-6N		6
Unit Ct. N&S	26W-7N	6
Victor Ln. E&W	27W-5N	5
Waterfall Ln. NW&SE	27W-5N	5
Whaler Ln. E&W	27W-4N	5
Whirlaway Ln. E&W	26W-7N	6
Whitney Dr. NW&SE	27W-5N	5
Wildwood Ct. N&S	27W-5N	5
Wildwood Ln. NW&SE	27W-5N	5
Wilshire Ct. E&W	27W-7N	5
Windjammer Ln. NS&EW 27W-4N,26W-4N		5
Windmill Dr. NS&EW	27W-5N	5
Windward Ct. E&W	27W-4N	5
Woodlake Dr. N&S	27W-5N	5
Wright Ln. E&W	27W-5N	5
Zeppelin Dr. NS&EW	27W-5N	5

Hinsdale*
(Also see Cook County)

	Grid	Page
Adams St. N&S 16W-5S,16W-4S		27
Allmen Av. N&S	15W-5S	27
Astony Ct. N&S	15W-6S	27
Ayres Av. N&S	15W-4S	27
Birchwood Av. E&W	15W-4S	27
Birchwood Rd. N&S 16W-4S,15W-4S		27
Blaine Av. N&S	15W-5S	27
Bodin St. N&S	16W-5S	27
Bonnie Brae Rd. E&W	15W-4S	27
Briargate Terr. E&W	15W-4S	27
Brook Pl. N&S	16W-4S	27
Bruner St. N&S 16W-5S,16W-4S		27
Burr Oak Ln. E&W	15W-3S	21
Burr Oak Rd. Cir.	15W-3S	21
Canterbury Ct. E&W	15W-4S	27
Carolyn Ln. (Pvt.) E&W	16W-6S	27
Castine Ct. N&S	15W-6S	27
Center St. N&S	15W-4S	27
Chanticleer Ln. E&W	15W-6S	27
Chestnut St. E&W	16W-4S	27
Chicago Av. E&W 16W-4S,14W-4S		27
Childs Av. N&S	15W-6S	27
Clay St. N&S 15W-5S,15W-4S		27
Claymoor Rd. E&W	15W-6S	27
County Line Rd. N&S 15W-5S,15W-4S		27
East Claymoor Rd. (Pvt.) E&W	15W-4S	27
Elm St. N&S 15W-6S,15W-4S		27
Elmwood Pl. E&W	15W-5S	27
Flagg Ct. E&W	15W-5S	27
Forest Rd. N&S	15W-4S	27
Fox Ln. E&W	15W-3S	21
Foxgate Dr. N&S	15W-6S	27
Franklin St. N&S	15W-4S	27
Fuller Rd. E&W 15W-4S,14W-4S		27
Garfield Av. N&S 15W-6S,15W-4S		27
Giddings Av. N&S	15W-6S	27
Glendale Av. N&S	15W-4S	27
Grant St. N&S	15W-4S	27
Grason St. N&S	18W-6S	26
Hampton Pl. E&W	15W-4S	27
Hawthorne Cir. E&W	15W-3S	21
Hawthorne Cir. E&W	15W-3S	21
Hickory St. E&W	16W-4S	27
Hillgrove Av. NE&SW	15W-4S	27
Hinsdale Av. NE&SW 16W-5S,15W-5S		27
Indian Trail N&S	15W-3S	21
Jackson Av. N&S	16W-5S	27
Jefferson St. N&S	15W-4S	27
Kimball Ct. E&W	15W-4S	27
King Pl. NE&SW	15W-4S	27
Lansing St. E&W	15W-4S	27
Lincoln St. N&S 15W-5S,15W-4S		27
Madison St. N&S 16W-5S,16W-4S		27
Maple St. E&W	15W-4S	27
Maumell St. E&W	15W-4S	27
Meadowbrook Ln. NS&EW	15W-6S	27
Merrill Woods Rd. N&S	15W-4S	27
Mills St. N&S	15W-4S	27
Minneola St. E&W 15W-4S,14W-4S		27
Monroe St. N&S 16W-5S,16W-4S		27
Morris Ln. E&W	16W-4S	27
North St. E&W 16W-4S,15W-4S		27
Oak St. N&S 15W-6S,15W-4S		27
Oakwood Ter. N&S	15W-4S	27
Ogden Av. E&W 16W-4S,14W-4S		27
Old Mill Rd. E&W	15W-3S	21
Orchard Pl. N&S	15W-3S	27
Pamela Cir. E&W 15W-5S,14W-5S		27
Park Av. N&S 15W-5S,15W-4S		27
Post Cir. NS&EW	15W-4S	27
Quincy St. N&S 16W-5S,16W-4S		27
Radcliffe Way N&S	15W-4S	27
Railroad St. N&S	16W-5S	27
Ravine Rd. E&W 15W-4S,14W-4S		27
Rosalie Ct. E&W	16W-5S	27
Salt Creek Ln. N&S 15W-4S,15W-3S		27
Salt Creek Ln. N&S	15W-3S	21
Spinning Wheel Rd. N&S	15W-3S	21
Spinning Wheel Rd. N&S 15W-4S,15W-3S		27
Stough St. N&S 16W-5S,16W-4S		27
Sutton Pl. (Pvt.)	15W-6S	27
Symonds Dr. E&W	15W-5S	27
The Lane E&W 15W-4S,14W-4S		27
Thurlow St. N&S	16W-5S	27
Town Pl. E&W	16W-5S	27
Tri State Tollway N&S 15W-5S,15W-4S		27
U.S. Route 34 NE&SW 15W-4S,15W-4S		27
Ulm Pl. N&S	15W-5S	27
Village Pl. N&S	16W-5S	27
Vine St. N&S 15W-5S,15W-4S		27
W. Grant Vill. Dr. E&W	15W-6S	27
Walker Rd. N&S	15W-4S	27
Walnut St. E&W 16W-4S,15W-4S		27
Warren Ct. E&W	16W-4S	27
Warren Terr. NS&EW	15W-4S	27
Washington St. N&S 15W-6S,15W-4S		27
Wedgewood Ct. E&W	16W-4S	27
Wildwood Ln. NS&EW	17W-5S	27
Woodmere Dr. NS&EW		27
Woodside Av. E&W	15W-5S	27
York Rd. N&S	15W-4S	27
1st St E&W 16W-5S,14W-5S		27
2nd St. E&W 16W-5S,15W-5S		27
3rd St. E&W	15W-5S	27
4th Av. E&W	15W-5S	27
5th St. E&W	15W-5S	27
6th Av. E&W	15W-5S	27
6th St. E&W 16W-5S,14W-5S		27
7th Av. E&W	15W-5S	27
7th St. E&W 16W-5S,15W-5S		27
8th Av. E&W	15W-5S	27
8th Pl. E&W	15W-5S	27
8th St. E&W	16W-4S	27
9th Av. E&W	15W-5S	27
9th St. E&W 16W-5S,15W-5S		27
55th St. E&W 15W-5S,14W-5S		27
56th St. E&W	15W-6S	27
57th St. E&W	15W-6S	27
58th Pl. E&W	15W-6S	27
58th St. E&W	15W-6S	27

Itasca

	Grid	Page
Alexandra Ct. E&W	19W-7N	8
Ardmore Av. E&W 21W-7N,20W-7N		7
Arlington Heights Rd. NE&SW 19W-6N,19W-7N		8
Arlington Rd. N&S	19W-6N	8
Augusta Cir. N&S	21W-7N	7
Baker Dr. N&S	21W-7N	7
Birch St. E&W	20W-6N	8
Bloomingdale Rd. NE&SW 20W-6N,19W-6N		8
Bonnie Brae Av. N&S	19W-6N	8
Briar Pl. E&W	19W-6N	8
Broker Av. N&S	19W-6N	8
Bryn Mawr Av. E&W 21W-6N,19W-6N		7
Catalpa Av. N&S	20W-6N	8
Center St. E&W 20W-6N,19W-6N		8
Cherry Cir. E&W	19W-7N	8
Cherry Ct. E&W	19W-7N	8
Cherry St. N&S 19W-6N,19W-7N		8
Country Club Dr. N&S 19W-6N,19W-7N		8
Country Club Ln. NW&SE 19W-6N,19W-7N		8
Crest Av. E&W	21W-7N	7
Devon Av. E&W 20W-7N,19W-6N		8
District Dr. N&S 21W-6N,21W-7N		7
Division St. E&W 20W-6N,19W-6N		8
Elm Ct. E&W	19W-7N	8
Elm St. N&S 19W-6N,19W-7N		8
Emmerson Av. N&S	19W-6N	8
Expressway Dr. N&S	20W-7N	8
Fmc Dr. E&W	21W-7N	7
Forest Av. N&S	19W-6N	8
George St. E&W 19W-6N,18W-6N		8
Glenlake Av. NW&SE	21W-7N	7
Greenview Rd. NE&SW	19W-6N	8
Grove St. E&W	19W-6N	8
Hamilton Pkwy. E&W	21W-7N	7
Hawthorn Dr. E&W	20W-6N	8
Hickory Dr. N&S	20W-6N	8
Hillcrest Av. NW&SE	21W-6N	7
Hilltop Dr. N&S	21W-7N	7
Hollywood Av. E&W	20W-7N	8
Home Av. N&S	19W-6N	8
Industrial Dr. NW&SE	18W-6N	8
Interstate 290 N&S 20W-6N,20W-7N		8
Inverness Ln. NS&EW	19W-7N	8
Irving Park Rd. E&W 21W-6N,18W-6N		7
Itasca Industrial Dr. E&W	18W-6N	8
Linden Cir. E&W	19W-7N	8
Linden Ct. E&W	19W-7N	8
Linden St. N&S 19W-6N,19W-7N		8
Line St. N&S	19W-6N	8
Lombard Rd. N&S	20W-6N	8

Grid / Page

Maple Av. N&S
..........19W-6N,19W-7N   8
Maple St. N&S ... 19W-6N   8
Maplewood Dr. NS&EW
..........20W-6N
Meacham Rd. N&S .. 21W-7N   7
Medinah Ln. E&W .. 21W-6N   7
Mill Rd. N&S 19W-5N   8
North Cir. N&S 20W-6N   8
Norwood Av. NW&SE
..........21W-7N   7
Oak St. N&S 19W-6N   8
Orchard St. E&W .. 19W-6N   8
Park Blvd. N&S....20W-7N   8
Parkside Av. N&S .. 19W-7N   8
Pierce Rd. NW&SE .. 19W-7N   8
Preswick Dr. E&W .. 19W-6N   7
Princeton Av. N&S .. 19W-6N   8
Prospect Av. E&W
..........19W-6N,18W-7N   8
Rohlwing Rd. N&S
..........21W-6N,21W-7N   7
Rt. 19 NW&SE
..........21W-6N,18W-6N   7
Rush St. N&S 19W-6N   8
Schiller St. N&S .. 19W-6N   8
Spring Lake Dr. NS&EW
..........20W-6N   8
Spyglass Ct. N&S....21W-6N   8
St. Andrew's Cir. N&S
..........21W-6N   7
Sunnyside Av. N&S .. 19W-6N   8
Surrey Ln. E&W .. 19W-7N   8
Theodore Ln. E&W .. 19W-7N   8
Thorndale Av. N&S
..........21W-7N,18W-7N   7
Walnut Av. N&S 19W-7N   8
Walnut St. N&S .. 19W-6N   8
Walters Ln. E&W .. 19W-7N   8
Washington St. E&W
..........19W-6N   8
Willow St. N&S 20W-6N   8
1st Av. N&S 19W-6N   8
2nd St. N&S 19W-6N   8

Lisle

Abbey Dr. N&S .. 23W-6S   25
Abbeywood Dr. NW&SE
..........24W-6S   24
Academy Dr. N&S .. 23W-5S   25
Acorn Hill La. E&W .. 24W-7S   24
Adams Dr. E&W .. 24W-6S   24
Adler Ct. E&W 24W-5S   24
Alta Ct. NW&SE .. 24W-7S   24
Angel La. N&S 24W-6S   24
Anton Ct. NW&SE .. 24W-7S   24
Arborview Dr. N&S .. 22W-4S   25
Ascot Ct. N&S 24W-3S   18
Ashford Ct. NE&SW 24W-7S   24
Aspen Rd. N&S....24W-7S   24
Aster Ct. NW&SE .. 22W-5S   25
Auvergne Av. N&S .. 22W-4S   25
Auvergne Ct. E&W .. 22W-5S   25
Azalea Ln. N&S .. 24W-4S   24
Babst Ct. E&W 24W-4S   24
Bannister Ct. NW&SE
..........23W-7S   25
Barclay Ct. NE&SW .. 24W-3S   18
Barnwell Ct. NW&SE
..........24W-5S   24
Basswood Ct. N&S .. 22W-4S   25
Bayside Ct. E&W .. 24W-5S   24
Beau Bien Blvd. N&S
..........25W-5S,25W-4S   24
Beau Bien Ct. E&W .. 24W-5S   24
Beau Monde Blvd. N&S
..........24W-4S   24
Beau Monde Dr. N&S
..........24W-4S   24
Beau Monde Ln. E&W
..........24W-4S   24
Beau Monde Terr. E&W
..........24W-4S   24
Benedictine Pkwy. N&S
..........24W-6S   24
Bentley La. E&W .. 24W-7S   24
Berkshire Ct. N&S .. 24W-7S   24
Black Hawk Ln. N&S
..........24W-4S   24
Black Oak Ct. E&W .. 24W-6S   24
Black Partridge Ln. N&S
..........24W-4S   24
Bloomfield Dr. NS&EW
..........24W-7S   24
Bluebell Ct. E&W 22W-5S   25
Bobbi Jo Ln. N&S....23W-5S   25
Bonita Ct. NE&SW .. 24W-4S   24
Bordeaux Pl. NW&SE
..........24W-4S   24
Breckenridge Rd. NE&SW
..........24W-7S   24
Briarcliff Ct. N&S .. 24W-7S   24
Brighton Ct. N&S
..........24W-6S   24
Bristol Dr. N&S 24W-6S   24
Brittany Ct. NW&SE..24W-4S   24
Brookline Ct. N&S .. 23W-6S   25
Brookwood Rd. N&S
..........23W-6S   25
Brunswick Ct. N&S .. 24W-4S   25
Buckley Ct. N&S 22W-4S   25
Burgandy Pl. N&S ..24W-4S   24

Grid / Page

Burlington Av. E&W
..........23W-5S,22W-5S   25
Burning Tree Ct. E&W
..........23W-4S   25
Burr Oak Rd. NW&SE
..........23W-5S   25
Bush St. E&W 22W-4S   25
Cabot Ct. N&S 24W-4S   24
Camden Ct. E&W .. 24W-6S   24
Candle Wood Ct. N&S
..........24W-5S   24
Cape Cod Ct. N&S .. 25W-7S   24
Carriage Hill Rd. E&W
..........24W-7S,23W-7S   24
Cascade Dr. N&S
..........24W-6S,24W-5S   24
Cascara Ln. E&W....22W-4S   25
Cedar Ct. NW&SE .. 24W-5S   24
Center Av. N&S
..........23W-5S,22W-4S   25
Chateau Pl. NE&SW 24W-4S   24
Chatfield Ln. E&W .. 23W-7S   25
Chelsea Av. N&S 24W-4S   24
Cherry Ln. E&W 24W-4S   24
Chicago Av. E&W .. 22W-5S   25
Chippewa Ct. E&W .. 24W-4S   24
Christian Ln. NW&SE
..........24W-6S   24
Clover Dr. N&S 22W-5S   25
Coach House Rd. NS&EW
..........25W-7S   24
College Rd. NE&SW
..........24W-7S,24W-6S   24
Columbia Av. N&S .. 22W-5S   25
Commerce Ct. .. 24W-4S   24
Concorde Pl. NE&SW
..........24W-4S   24
Corporate West Dr. E&W
..........24W-4S   24
Corporetum Dr. E&W
..........23W-4S   25
Cypress Ct. N&S 24W-4S   24
Darley Dr. E&W......23W-4S   25
Deer Path Ct. NE&SW
..........24W-7S   24
Devon Av. N&S
..........23W-5S,23W-4S   25
Division St. E&W
..........23W-5S,22W-5S   25
Dogwood Ln. E&W .. 24W-4S   24
Dorset Av. N&S....23W-4S   25
Driftwood Ct. N&S .. 24W-5S   24
Dry Bridge Ct. NE&SW
..........23W-7S   25
Dumoulin Av. N&S .. 23W-5S   25
Edgebrook Ct. N&S .. 24W-3S   18
Edgebrook Dr. E&W 24W-3S   18
Eisenhower Ln. E&W
..........23W-5S   25
Elm Av. N&S 22W-5S   25
Fernwood Dr. NW&SE
..........24W-7S   24
Forbes Dr. N&S....24W-4S   24
Forest View Rd. N&S
..........23W-6S   25
Front St. E&W
..........23W-5S,22W-5S   25
Gamble Dr. N&S 22W-5S   25
Garden Dr. N&S 24W-7S   24
Garfield Ct. E&W .. 22W-5S   25
Glenbrook Ct. N&S .. 25W-7S   24
Golfview Dr. N&S
..........24W-7S,23W-6S   24
Green Trails Dr. E&W
..........25W-7S,23W-7S   24
Greenfield Ct. E&W .. 24W-7S   24
Grenoble Ct. NE&SW
..........24W-7S   24
Hagberg Ct. NW&SE
..........22W-5S   25
Hampshire Ct. NE&SW
..........24W-7S   24
Hanover Ct. N&S 24W-7S   24
Harth Ct. N&S 23W-6S   25
Hastings La. N&S..24W-7S   24
Hatch Ct. NE&SW .. 22W-4S   25
Hatch Ln. N&S 22W-4S   25
Hawthorne Ln. N&S .. 23W-5S   25
Heritage La. N&S .. 24W-6S   24
Hickory Dr. N&S .. 24W-6S   24
Hidden Creek Ct. E&W
..........24W-4S   24
Hillside Ln. NW&SE .. 23W-4S   25
Hilltrail Rd. E&W .. 24W-5S   24
Hinterlong Ct. N&S .. 24W-5S   24
Hitchcock Av. E&W
..........24W-5S,24W-4S   24
Holly Ct. NW&SE .. 24W-7S   24
Illinois Av. N&S 24W-5S   24
Indiana Av. N&S .. 24W-5S   24
Innsbruck Ln. NE&SW
..........24W-7S   24
Inverness Rd. E&W .. 22W-5S   25
Ivanhoe Av. N&S .. 22W-4S   25
Ivy Dr. NS&EW 24W-6S   24
Jonquil Av. N&S
..........23W-5S,22W-5S   25
Keel Ct.* N&S 24W-5S   24
Keller St. N&S 24W-5S   24
Kent Ct. NW&SE .. 24W-6S   24
Kindling Ct. N&S 24W-7S   24
Kingston Av. N&S
..........22W-5S,22W-4S   25

Grid / Page

Kirkwood Ct. NE&SW
..........24W-7S   24
Knollwood Ct. NE&SW
..........24W-7S   24
Lacey Av. N&S
..........23W-4S,22W-4S   25
Lake Trail Dr. N&S
..........24W-5S   24
Lake Valley Dr. N&S
..........24W-5S   24
Lakewood Dr. N&S
..........24W-6S   24
Larkspur Ln. N&S....22W-5S   25
Lexington Dr. N&S
..........24W-7S,24W-6S   24
Lincoln Av. N&S
..........23W-5S,23W-4S   25
Lisle Pl. E&W........23W-5S   25
Longview Dr. E&W .. 24W-6S   24
Lundy Ct. E&W
..........24W-6S,23W-6S   25
Main St. N&S
..........23W-5S,23W-4S   25
Maple Av. N&S
..........24W-6S,22W-5S   25
Mark Carre E&W 24W-4S   24
Mckinley Av. N&S....23W-5S   25
Meadow Ln. NE&SW
..........22W-5S   24
Mertz Ct. NE&SW .. 24W-4S   24
Meyer Dr. N&S 22W-5S   25
Middleton Av. N&S
..........23W-4S,22W-4S   25
Mill Bridge Ln. NE&SW
..........23W-6S   25
Miller Ln. N&S 23W-6S   25
Muirwood Ct. N&S .. 24W-7S   24
Neff Ct. NE&SW .. 24W-6S   24
New Albany Rd. NE&SW
..........24W-7S,24W-6S   24
Newport Dr. NE&SW
..........23W-5S   25
Normandy Dr. N&S .. 24W-4S   24
Normandy Pl. NE&SW
..........24W-4S   24
North-South Tollway N&S
..........22W-4S,22W-5S   25
Nutmeg Ln. N&S .. 24W-4S   24
Oak Hill Dr. N&S .. 24W-5S   24
Oakton Ct. E&W .. 24W-6S   24
Oakview St. N&S .. 23W-5S   25
Oakwood Dr. NS&EW
..........23W-6S   25
Ogden Av. NE&SW
..........24W-5S,22W-5S   25
Ohio St. E&W
..........22W-5S,22W-5S   25
Old College Rd. N&S
..........24W-7S   24
Old Oaks Dr. N&S
..........24W-5S   24
Old Tavern Rd. E&W
..........24W-4S,24W-5S   24
Oldham Av. N&S
..........23W-4S,22W-4S   25
Parksleg Ct.* N&S
..........24W-7S,24W-6S   24
Patton Dr. N&S......23W-5S   25
Pebble Creek Dr. E&W
..........24W-4S   24
Pennywood Dr. N&S
..........24W-5S   24
Peppertree Ct. NE&SW
..........24W-7S   24
Pine Tree Ct. E&W .. 23W-4S   25
Pinehurst Ct. N&S .. 24W-7S   24
Pleasant Hill Ln. NW&SE
..........23W-6S   25
Portsmouth Dr. E&W
..........23W-5S   25
Prutmore Ct.* NE&SW
..........24W-7S   24
Raintree Ct. NW&SE
..........24W-7S   24
Ranier Dr. NE&SW
..........24W-6S,24W-5S   24
Redwood Ln. E&W .. 22W-4S   25
Regency Woods Dr. E&W
..........24W-3S   18
Rehm Ct. E&W 23W-5S   25
Reilly Pl. N&S......23W-5S   25
Rickert Ct. N&S 24W-4S   24
Ridgewood Rd. N&S
..........24W-7S   24
Riedy Rd. E&W
..........23W-5S,22W-5S   25
River Dr. N&S
..........23W-5S,23W-4S   25
Riverview Dr. N&S .. 23W-5S   25
Robin Ln. N&S 23W-4S   25
Rolling Dr. E&W
..........23W-5S,22W-5S   25
Royal Glen Ct. N&S .. 23W-7S   25
Sailboat Bay* NW&SE
..........24W-5S   24
Scarlet Oak Ln. N&S
..........24W-6S   24
School St. E&W 23W-5S   25
Shabbona Ln. N&S .. 24W-4S   24
Shagbark Ct. NE&SW
..........23W-7S   25
Shasta Dr. E&W 24W-7S   24
Shellingham Dr. NS&EW
..........24W-6S   24

Grid / Page

Shorewood Ct. NW&SE
..........24W-6S   24
Short St. E&W 23W-5S   25
Sleepy Hollow Ln. N&S
..........23W-7S,23W-6S   25
Southbury Ct. NE&SW
..........25W-7S   24
Southport Av. E&W .. 23W-4S   25
Spencer Av. N&S 23W-5S   25
Spruce Ln. N&S&SE .. 22W-4S   25
St. Joseph Creek Rd.
.....NW&SE .. 22W-4S,22W-5S   25
Stonehaven Way NE&SW
..........24W-7S   24
Stowe Ct. NE&SW .. 24W-7S   24
Sun Valley Rd. NE&SW
..........24W-7S   24
Surrey Ridge Rd. NW&SE
..........24W-6S,24W-5S   24
Tall Tree Ct. E&W .. 24W-5S   24
Tanglewood Dr. E&W
..........24W-7S   24
Taos Ct. N&S 24W-7S   24
Tealwood Dr. E&W .. 24W-6S   24
Tellis Ln. E&W 24W-6S   24
Tellurude Ct. N&S .. 24W-7S   24
Timbertrails Dr. E&W
..........24W-7S   24
Timberview Dr. NW&SE
..........24W-6S   24
Torey Pine Ln. N&S .. 24W-6S   24
Tory Ct. N&S...... 24W-6S   24
Trinity Dr. NS&EW .. 24W-6S   24
Tropez Dr. N&S 24W-4S   24
Trowbridge Way* NE&SW
..........23W-5S   25
Twin Oaks Ln. NS&EW
..........24W-7S   24
Tyrnbury Dr. N&S....24W-6S   24
University Dr. E&W .. 23W-5S   25
Vail Ct. E&W 24W-7S   24
Valley Forge Rd. E&W
..........24W-7S   24
Varsity Dr. N&S......23W-5S   25
Venture St. N&S 22W-5S   25
Walnut Creek Ln. N&S
..........22W-5S   25
Warrenville Rd. E&W
..........22W-4S   25
Waterbridge Rd. N&S
..........25W-4S   24
Waubansie Ln. N&S .. 24W-4S   24
Wayfaring Ln. E&W .. 24W-4S   24
Weeping Willow Dr. E&W
..........24W-6S   24
Wellington Av. E&W .. 24W-5S   24
West Av. N&S 22W-5S   25
Western Av. N&S 24W-4S   24
Western Ct. E&W .. 24W-4S   24
Westleigh Ct. N&S .. 24W-4S   24
Westview Ln. N&S .. 24W-5S   24
White Birch Dr. NS&EW
..........23W-4S   25
White Oak La. N&S .. 24W-7S   24
Wildwood Ct. N&S .. 24W-7S   24
Winchester Av. N&S
..........23W-5S,23W-4S   25
Windsor Dr., North E&W
..........24W-6S   24
Windsor Dr., South E&W
..........24W-6S   24
Windward Ct.* NE&SW
..........25W-5S   24
Windward Rd.* NE&SW
..........25W-5S   24
Winkler Ct. E&W 23W-6S   25
Winstead Ct.* NW&SE
..........24W-6S   24
Wisconsin St. E&W .. 24W-5S   24
Woodbriar Dr. NE&SW
..........24W-7S   24
Woodcliff Ct. N&S .. 24W-7S   24
Yackley Av. N&S
..........24W-6S,24W-4S   24
Yender Av. N&S
..........24W-5S,24W-4S   24
Yorkshire Dr. N&S .. 24W-7S   24

Lisle Township

Aintree Dr. N&S .. 23W-8S   31
Allison Ln. N&S .. 24W-5S   24
Arguilla Dr. E&W .. 21W-6S   25
Arlington Av. N&S .. 24W-5S   24
Arrow Ct. NE&SW .. 24W-6S   24
Augusta N&S 23W-7S   25
Avalieri Ln. NE&SW .. 24W-5S   24
Avon Ct. N&S 24W-6S   24
Baltusroll N&S 23W-7S   25
Beau Bien Blvd. N&S
..........24W-5S   24
Bell Dr. N&S 25W-8S   30
Birdsong Ct. E&W .. 24W-6S   24
Black Oak Dr. N&S .. 23W-4S   25
Black Stone N&S
..........24W-6S   24
Blackthorne N&S
..........24W-8S   30
Blanchard St. NE&SW
..........21W-6S   25
Bluff Ct. NE&SW .. 24W-5S   24
Bluff Ln. E&W 24W-5S   24

Grid / Page

Boundary Hill Rd. N&S
..........22W-8S   31
Brandywine Dr. E&W
..........25W-6S   24
Brenwood Dr. N&S .. 23W-8S   31
Bridlespur Dr. NS&EW
..........25W-6S   24
Burlington Av. E&W
..........25W-5S,24W-5S   24
Buttonwood Dr. N&S
..........24W-5S   24
Canary Ct. E&W .. 25W-5S   24
Canterbury Ct. NW&SE
..........25W-6S   24
Cape Rd. NS&EW .. 25W-6S   24
Cardinal Ct. NW&SE
..........24W-5S   24
Carlyle Ct. NS&EW .. 24W-6S   24
Case St. E&W 26W-5S   24
Chantilly Ct. E&W....23W-7S   25
Charles St. N&S 21W-6S   25
Chase Av. N&S 21W-6S   25
Chicago Av. N&S
..........25W-6S,24W-6S   24
Cliff Ct. E&W
..........24W-5S,22W-6S   24
Clover Dr. N&S 22W-6S   25
Cloverdale Ct. N&S .. 24W-5S   24
Cohasset Rd. N&S .. 25W-6S   24
College Rd. N&S
..........24W-8S,21W-6S   30
Columbine Av. E&W 22W-6S   25
Concord Rd. NE&SW
..........25W-6S   24
Cornwall Rd. NW&SE
..........25W-6S   24
Country Ct. E&W .. 23W-7S   25
Country Dr. N&S 24W-6S   24
Creek Ct. E&W 23W-7S   25
Creek Dr. N&S
..........23W-8S,23W-7S   31
Cross St. N&S
..........21W-5S,21W-6S   25
Derby Ct. E&W 24W-8S   30
Derby Dr. NE&SW .. 23W-8S   31
Devon Av. N&S 23W-4S   25
Donwood Dr. N&S .. 24W-7S   24
Donwood Dr. East NE&SW
..........24W-7S   24
Donwood Dr. West N&S
..........24W-7S   24
Donwood Trails Dr. N&S
..........23W-7S   25
Dorset Av. N&S......23W-4S   25
Dover Dr. N&S 22W-6S   25
Drendall Rd. N&S
..........21W-5S,21W-4S   25
Dunham Dr. NW&SE
..........23W-8S   31
Dupage Dr. N&S .. 22W-5S   25
East Lake Dr. NE&SW
..........23W-6S   25
Elinor Av. N&S 21W-6S   25
Elm St. N&S 22W-6S   25
Essex Rd. N&S......22W-6S   25
Eugena Dr. NE&SW .. 23W-4S   25
Evergreen Ct. E&W .. 23W-4S   25
Fairway Dr. E&W 23W-7S   25
Fender Av. N&S
..........24W-5S,24W-4S   24
Forest View Rd. N&S
..........23W-6S   25
Four Lakes Av. E&W
..........23W-6S   25
Foxwood Ct. NW&SE
..........23W-8S   31
Francisco St. N&S .. 21W-5S   25
Glenview Av. N&S
..........21W-6S,21W-5S   25
Green Briar Ct. E&W
..........23W-7S   25
Green Briar Dr. E&W
..........23W-7S   25
Green Trails Dr. E&W
..........23W-7S   25
Greene Rd. N&S
..........23W-9S,23W-7S   31
Greenview Dr. N&S .. 23W-6S   25
Greenwich Ct. NW&SE
..........25W-6S   24
Greenwood Ct. N&S
..........24W-6S   24
Gunston Av. NW&SE
..........25W-6S   24
Hampton Cir. NS&EW
..........23W-8S   31
Heather Ct. N&S .. 24W-5S   24
Hemlock Ct. N&S
..........24W-8S,23W-8S   30
Highview Dr. N&S .. 25W-4S   24
Hillcrest Av. N&S .. 21W-6S   25
Hillcrest Ln. N&S .. 21W-6S   25
Hillside Rd. E&W .. 26W-7S   31
Hobson Rd. E&W
..........25W-8S,23W-7S   30
Hobson Trails Dr. N&S
..........23W-7S   25
Howard Av. E&W .. 21W-5S   25
Hummingbird Ct. N&S
..........25W-5S,24W-5S   24
Hyacinth Ct. NE&SW
..........24W-6S   24
Indian Hill Dr. E&W .. 23W-4S   25
Iris Ln. N&S 22W-6S   25

Ivanhoe Av. N&S 22W-6S 25
James Way E&W ... 23W-8S 31
Janes Av. E&W 21W-6S 25
Jefferson St. NE&SW
.... 21W-6S 25
Julian St. N&S ... 26W-7S 24
Kame Ct. NW&SE 25W-4S 24
Kammes Dr. E&W ... 24W-7S 24
Karns Rd. N&S ... 25W-4S 24
Katrine Av. N&S 21W-6S 25
Keim Dr. N&S 23W-4S 25
Kimberly Way E&W .. 22W-6S 25
Kimberwick Ln. E&W
.... 23W-8S 31
Kings Cove N&S ... 23W-6S 25
Kingston Av. N&S ...22W-6S 25
Kohley Rd. E&W
.... 23W-6S,22W-6S 25
Lacey Av. E&W......25W-4S 24
Lakeside Dr. N&S .. 23W-6S 25
Lakewood Dr. N&S .. 24W-6S 24
Leask Ln. N&S 24W-4S 24
Lenox Ct. N&S ... 22W-6S 25
Lenox Rd. N&S
.... 23W-6S,22W-6S 25
Lincoln Av N&S
.... 23W-6S,23W-4S 25
Lomond Av. N&S
.... 21W-6S,21W-5S 25
Longreen Rd. N&S
................25W-6S5S 24
Lyn Dr. N&S 23W-7S 25
Malibu Ct. E&W
.... 24W-6S,24W-4S 24
Malibu Ln. NE&SW .. 25W-6S 24
Mallard Ct. E&W ... 24W-5S 24
Maple Terr. E&W ... 24W-4S 24
Marble Head Ct. NE&SW
.... 25W-6S 24
Marion N&S 23W-7S 25
Mariwood Ct. NW&SE
.... 24W-6S 24
Matson Ln. E&W ... 23W-4S 25
Mayflower Av. NE&SW
.... 25W-6S 24
Maywood Ln. E&W .. 23W-6S 25
Mcintyre Rd. N&S...21W-4S 25
Meadow Dr. N&S ... 24W-6S 25
Meadow Lake Dr. E&W
.... 25W-5S,24W-5S 24
Meadowbrook Ct. NE&SW
.... 25W-5S 24
Middle Rd. N&S ... 25W-5S 24
Middleton Av. E&W
.... 25W-4S,23W-4S 24
Millcreek Ct. E&W ... 24W-6S 24
Millcreek Ln. N&S ... 25W-6S 24
Mitchell Ln. N&S
.... 23W-8S,23W-7S 31
Mockingbird Ct. N&S
.... 24W-5S 24
Moraine Ct. E&W ... 23W-7S 25
Morton Rd. E&W ... 23W-4S 25
Murray Dr. N&S ... 23W-8S 31
Neptune Ct. E&W...24W-6S 24
New Hope Rd. NW&SE
.... 25W-6S 24
Newcastle Dr. NE&SW
.... 25W-6S 24
Northcrest Dr. E&W..22W-7S 25
Oak Trails Dr. N&S .. 23W-7S 25
Ohio St. E&W ... 24W-5S 24
Oldham Rd. E&W ... 24W-4S 24
Oldwoods Dr. E&W .. 22W-9S 31
Oriole Ln. NW&SE
.... 25W-5S,24W-5S 24
Oxer Ct. E&W ... 23W-8S 31
Oxford Ln. N&S ... 24W-8S 30
Palamino Dr. N&S...24W-8S 30
Park Meadow Dr. N&S
.... 24W-6S,24W-5S 24
Parkside Dr. E&W .. 26W-7S 24
Parkview Ct. E&W ... 24W-6S 24
Partridge Ct. E&W .. 24W-6S 24
Pellum Ct. E&W ... 23W-8S 31
Pershing Av. N&S...21W-6S 25
Pin Oak Ln. E&W ... 24W-8S 30
Pine Ln. N&S........ 23W-4S 25
Pine Rd. E&W 23W-4S 25
Pinehurst Ct. E&W .. 23W-7S 25
Pinehurst Ln. E&W .. 23W-7S 25
Plank Rd. NE&SW
.... 26W-6S,25W-5S 24
Prairie Av. E&W ... 21W-5S 25
Prentiss Dr. NW&SE
.... 21W-7S 25
Primrose Av. N&S...22W-6S 25
Prince Ct. N&S ... 22W-6S 25
Queens Cove N&S . 22W-6S 25
Radcliff Rd. N&S
.... 25W-5S,25W-4S 24
Ridge View Ln. NW&SE
.... 24W-5S 24
Ridgeland Av. E&W .. 25W-4S 24
River Dr. N&S 23W-8S 30
River Rd. N&S 23W-6S 25
Riverview Dr. N&S .. 23W-6S 25
Rose Av. N&S 21W-5S 25
Rose Ct. E&W 22W-6S 25
Royal Oak Rd. E&W
.... 23W-8S 31
Salem Av. N&S .. 25W-6S 24
Setauket Av. E&W
.... 25W-6S,25W-5S 24

Shires Ct. NE&SW .. 23W-8S 31
Sierra Ln. NE&SW .. 24W-5S 24
Singtree Ct. N&S ... 23W-8S 31
Smith Rd. E&W
.... 23W-6S,22W-6S 25
Snowdrop Ln. N&S .. 22W-6S 25
Somerset Ct. NE&SW
.... 24W-6S 24
South Rd. E&W
.... 23W-6S,22W-6S 25
Southcrest Dr. E&W
.... 22W-7S,21W-7S 25
Spring Bay Dr. NW&SE
.... 24W-6S 24
Springdale Dr. NW&SE
.... 24W-6S,24W-5S 24
Spyglass Ct. E&W .. 23W-7S 25
Starling Ct. NW&SE .. 24W-5S 24
Steeple Run Dr. N&S
.... 24W-6S,24W-5S 24
Stonehedg Ct. E&W .. 23W-8S 31
Stonehedg Pl. E&W .. 23W-8S 31
Summit Dr. NE&SW .. 25W-4S 24
Surf Ct. E&W 24W-5S 24
Surrey Av. E&W 23W-8S 30
Tamiami Dr. N&S ... 22W-7S 25
Teddo Rd. NE&SW
.... 25W-5S,24W-5S 24
Tuthill Rd. N&S ... 25W-5S 24
Una Av. N&S 22W-9S 31
Vale Ct. NW&SE ... 21W-7S 25
Vest St. N&S 26W-5S 24
Walnut Av. N&S ... 22W-7S 25
Wayewood Ln. NS&EW
.... 23W-8S 31
Wedgewood Ct. N&S
.... 23W-7S 25
Wellington Av. NE&SW
.... 24W-5S 24
Westove Ct. E&W .. 25W-6S 24
Westview Ln. N&S
.... 22W-6S,22W-5S 25
White Oak Dr. N&S .. 26W-7S 24
Whittington Ct. NE&SW
.... 23W-8S 31
Wilshire Ct. E&W ... 23W-8S 31
Windham Hill Ct. NW&SE
.... 25W-5S 24
Wood Ct. NE&SW
.... 25W-4S,22W-6S 24
Woodbrook Dr. E&W
.... 24W-6S 24
Woodcrest Ct., East E&W
.... 23W-7S 25
Woodcrest Ct., West E&W
.... 23W-7S 25
Woodlawn Av. E&W..26W-7S 24
Woodstock Ct.
.... 25W-6S 24
Woodward Av. N&S .. 21W-6S 25
Yackley Av. N&S ... 24W-8S 30
59th St. E&W
.... 22W-6S,21W-6S 25
61st St. E&W
.... 22W-6S,21W-6S 25
62nd St. E&W
.... 22W-6S,21W-6S 25
63rd St. E&W
.... 22W-7S,21W-7S 25
75th St. E&W
.... 24W-8S,22W-8S 30
79th St. E&W.......23W-8S 31

Lombard

Acorn Ct.* N&S......21W-1S 19
Adams St. E&W ... 18W-0S 20
Addison Av. N&S.....18W-0N 14
Ahrens Av. N&S
.... 18W-0S,18W-1N 20
Ahrens Ct. N&S 18W-0N 14
Ainsley Ct. N&S ... 19W-1N 20
Ann St. E&W
.... 20W-0S,19W-0S 20
Apple Ln. N&S 19W-0S 20
Arthur Dr. N&S ... 18W-0S 20
Ash St. E&W 20W-0N 14
Berkshire Av. E&W
.... 20W-1N,19W-1N 14
Beverly Ct. N&S ... 21W-1S 19
Bradley Ln. NE&SW..19W-0S 20
Brewster Av. N&S .. 20W-0N 14
Broadview Av. N&S...20W-1N 14
Brookfield Av. E&W..20W-1N 14
Brown St. E&W ... 20W-1N 14
Cedar Ln. N&S 18W-0N 14
Central Av. E&W
.... 20W-0S,19W-0S 20
Charing Cross Rd. N&S
.... 19W-0S 20
Charles Ln. E&W .. 20W-0N 14
Charlotte St. N&S
.... 19W-1S,19W-1N 20
Chase Av. N&S
.... 19W-0N,19W-1N 14
Chase Ct. N&S 18W-0S 20
Chase Ln. N&S 19W-0S 20
Cherry Ln. N&S ... 19W-0S 20
Circle Av. E&W ... 19W-0S 20
Circle Terr. N&S ... 19W-0S 20
Clarendon Av. N&S .. 19W-1N 14
Collen Ct. N&S 20W-1S 20
Collen Dr. N&S 20W-1S 20
Colony Ct. NW&SE .. 20W-1S 20

Columbine Av. N&S .. 21W-1N 13
Concord Ln. N&S .. 18W-0S 20
Cortland Ln. N&S .. 20W-2N 14
Craig Pl. N&S
.... 19W-0N,19W-1N 14
Crescent Blvd. E&W
.... 20W-0N 14
Crystal Av. N&S 20W-0N 14
Daniel Ct. N&S 20W-0S 20
Division St. N&S
.... 19W-0N,18W-0N 14
Downing St. NW&SE
.... 21W-1S 19
DuPage Av. N&S
.... 20W-1N,20W-2N 14
East West Tollway E&W
.... 19W-2S 20
Eastview Terr. E&W 19W-0N 14
Edgewood Av. N&S
.... 19W-0S,19W-1N 20
Edson Av. N&S
.... 20W-0S,20W-0N 20
Edson Pl. E&W 20W-0S 20
Edward St. N&S
.... 20W-0S,19W-0S 20
Eisenhower Ln. N. E&W
.... 20W-2S 20
Eisenhower Ln. S. E&W
.... 20W-2S 20
Elizabeth Ct. E&W . 20W-0N 14
Elizabeth Dr. N&S .. 20W-0N 14
Elizabeth St. N&S
.... 20W-1S,20W-1N 20
Elm St. E&W 19W-0N 14
Emerson Av. N&S
.... 19W-0N,18W-0N 14
Ethel Av. E&W 20W-0S 20
Eugenia St. E&W 20W-0N 14
Fairfield Av. N&S
.... 19W-2S,19W-1N 20
Fairview Av. N&S .. 18W-0S 20
Finley Rd. N&S
.... 20W-2S,20W-0N 20
Fountainview Ln. E&W
.... 20W-2S 20
Foxworth Blvd. NW&SE
.... 21W-1S,20W-1S 19
Garfield St. N&S
.... 19W-1S,19W-2N 20
Garfield Terr. N&S . 19W-0N 14
Glen Oak Rd. E&W .. 20W-0N 14
Glenview Av. N&S
.... 20W-0N 14
Glenwood Ln. NW&SE
.... 21W-1N 13
Goebel Dr. E&W
.... 20W-1N,19W-1N 14
Grace Av. N&S 19W-1N 14
Grace St. N&S
.... 19W-2S,19W-0N 20
Graham Av. E&W....20W-0S 20
Graywood Dr. N&S . 19W-1S 20
Green Valley Dr. NS&EW
.... 20W-0S,20W-0N 20
Green Valley Dr. W. N&S
.... 20W-0N 14
Greenfield Av. E&W . 20W-1N 14
Grove St. E&W 20W-0N 14
Hamilton Ln. E&W .. 18W-0S 20
Hammerschmidt Av. N&S
.... 19W-0S 20
Harding Rd. E&W
.... 20W-0S,19W-0S 20
Harmony Ln. N&S .. 18W-0N 14
Harrelson Dr. E&W . 20W-1S 20
Harrison Rd. E&W
.... 20W-0S,19W-0S 20
Hawthorn Cir. E&W
.... 19W-0S,18W-0S 20
Hickory Rd. E&W .. 20W-0N 14
Hickory St. E&W
.... 20W-0N,19W-0N 14
High Ridge Rd. E&W
.... 19W-0S,18W-0S 20
Highland Av. N&S
.... 19W-2S,19W-0N 20
Highlands Lake Dr. N&S
.... 20W-2S 20
Hillcrest Ct. NW&SE 20W-1S 20
Hillcrest Ln. N&S 20W-1S 20
Hunter St. N&S 21W-1S 19
International Dr. E&W
.... 20W-1S 20
Interstate 88 NE&SW
.... 19W-1S 20
Jackson St. E&W .. 18W-0S 20
Janata Blvd. E&W .. 20W-2S 20
Jeffrey Ct. N&S......20W-0S 20
Joyce Av. N&S 19W-1N 14
June Ln. N&S 20W-1S 20
Kaplan Ct. N&S 19W-1N 14
Kelly Ct. N&S........19W-1N 14
Kenilworth Av. E&W 18W-0N 14
Kramer Av. N&S 19W-1N 14
Lalonde Av. N&S
.... 19W-0S,19W-1N 20
Lalonde St. N&S ... 19W-0N 14
LeMoyne Av. E&W
.... 19W-0N,19W-1N 14
Lewis Av. N&S
.... 19W-0S,19W-0N 20
Liberty Ln. N&S......18W-0S 20
Lilac Ln. E&W 20W-0N 14

Lilac Way E&W
.... 19W-0S,18W-0S 20
Lincoln Ct. N&S ... 20W-0N 14
Lincoln St. N&S
.... 20W-1S,20W-0N 20
Lloyd Av. N&S 19W-0N 14
Lodge Av. N&S ... 19W-0N 14
Lodge Ln. N&S......19W-0S 20
Lombard Av. N&S
.... 20W-1S,19W-1N 20
Lombard Cir. NW&SE
.... 18W-0S 20
Lombard Rd. N&S .. 20W-2N 14
Lore Ln. N&S.......20W-1S 20
Loy St. E&W 20W-1N 14
Lynne Ln. E&W 20W-1S 20
Madison St. N&S
.... 20W-0S,18W-0S 20
Magnolia Ct. E&W .. 19W-0S 20
Main Dr. N&S......19W-0S 20
Main St. N&S
.... 20W-2S,20W-2N 20
Majestic Dr. E&W ... 19W-2S 20
Manor Hill Ct. E&W . 20W-1S 20
Manor Hill Ln. E&W . 20W-1S 20
Maple St. N&S
.... 20W-0N,18W-0N 14
Marcus Dr. E&W ... 19W-1N 14
Martha St. N&S
.... 19W-0N,19W-1N 14
Meadow Ln. E&W .. 19W-1S 20
Meadow Ln. E&W .. 20W-1S 20
Michele Ln. NE&SW..19W-0S 20
Morningside Av. E&W
.... 19W-0N 14
Morris Av. E&W
.... 19W-0S,19W-0N 20
Morris Ct. N&S 19W-0S 20
New St. E&W 19W-0S 20
Norbury Av. N&S ... 19W-0S 20
North Av. E&W 20W-2N 14
North Broadway E&W
.... 19W-0N,18W-0N 14
North Lake Rd. NS&EW
.... 19W-1S 20
North-South Tollway N&S
.... 21W-1S,20W-1S 13
Norton St. E&W
.... 19W-0S,18W-0S 20
Oak Creek Dr. E&W
.... 21W-0S,20W-0S 19
Oak Meadow Ct.* N&S
.... 21W-1S 19
Orchard Terr. N&S .. 20W-0N 14
Park Av. N&S
.... 20W-0N,20W-1N 14
Park Dr. E&W 20W-0S 20
Park Rd. N&S 20W-0N 14
Park Road Ct. E&W .. 20W-0N 14
Parker Dr. N&S 20W-0N 14
Parkside Av. NE&SW
.... 20W-0N,19W-0N 14
Parkview Blvd. N&S..21W-0S 19
Phillips Ct. N&S 21W-0N 13
Pinebrook Dr. E&W . 20W-1S 20
Pleasant Ln. E&W
.... 20W-1N,19W-1N 14
Poplar Ct.* N&S ... 21W-1S 19
Potomac Av. E&W .. 20W-1N 14
Prairie Av. E&W
.... 20W-1N,19W-1N 14
Progress Rd. E&W .. 19W-2N 14
Queensbury Ct. E&W
.... 21W-1S 19
Randolph St. E&W .. 19W-0N 14
Rebecca Rd. NS&EW
.... 20W-1S 20
Regency Dr. E&W .. 19W-2S 20
Reskin Rd. E&W ... 19W-1N 14
Revere Ct. E&W ... 18W-0S 20
Ridge Av. N&S
.... 20W-1N,20W-2N 14
Rochelle Terr. N&S .. 20W-1N 14
Roosevelt Rd. E&W
.... 20W-0S,19W-0S 20
Royal Dr. N&S 19W-2S 20
Rt. 38 E&W
.... 20W-0S,19W-0S 20
Rt. 64 E&W
.... 20W-2N,19W-2N 14
School Ct. E&W 18W-0S 20
School St. N&S
.... 18W-0S,18W-0N 20
Shedron Way E&W .. 19W-1S 20
Sheila Pl. N&S ... 20W-1N 14
South Broadway E&W
.... 19W-0N 14
Springer Dr. NS&EW
.... 20W-1S 20
Spruce Ct.* N&S ... 21W-1S 19
St. Charles Pl. E&W..19W-0N 14
St. Charles Rd. NE&SW
.... 20W-0N,18W-0N 14
St. Regis Dr. N&S . 19W-1S 20
Stewart Av. N&S
.... 19W-0S,19W-1N 20
Stewart Ln. N&S ... 19W-2S 20
Sunset Av. N&S
.... 20W-1N,19W-1N 14
Sycamore St.* N&S..21W-1S 19
Taylor Rd. E&W
.... 20W-1N,19W-1N 14
Troy Dr. E&W 20W-1N 14
Troy Ln. N&S.......20W-1N 14

Valley Rd. NS&EW .. 21W-1S 19
Vance Ct. N&S ... 20W-0N 14
Vance St. N&S 20W-0N 14
View Av. N&S 19W-0N 14
View St. E&W
.... 20W-1N,19W-1N 14
Vista Av. N&S 19W-1N 14
Walnut St.* N&S ... 21W-1S 19
Washington Blvd. E&W
.... 20W-0N,18W-0N 14
Waters Edge NE&SW
.... 20W-1S 20
West Rd. N&S
.... 20W-0N,20W-1N 14
Western Av. NE&SW
.... 19W-1N 14
Westmore Av. N&S
.... 18W-0N,18W-1N 14
Westwood Av. N&S..19W-1N 14
Whitehall St. NW&SE
.... 21W-1S 19
Willow St. E&W 20W-0N 14
Wilson Av. E&W
.... 19W-0S,18W-0S 20
Wilson Ct. N&S......18W-0S 20
Windsor Av. N&S
.... 20W-1N,19W-1N 14
Witt Dr. NW&SE 21W-1S 19
Woodland Av..20W-1N 14
Woodridge Way N&S
.... 20W-2S 20
Woodrow Av. E&W .. 18W-0N 14
2nd Av. N&S 18W-0N 14
3rd Av. N&S 18W-0N 14
13th St. E&W......19W-1S 20
16th St. E&W
.... 20W-1S,19W-1S 20
17th Pl. E&W 20W-1S 20
17th St. E&W......20W-1S 20
18th St. E&W......20W-1S 20
19th St. E&W 20W-1S 20
22nd St. E&W
.... 20W-1S,19W-1S 20

Milton Township

Acorn Av. E&W......21W-0N 13
Ahlstrand Rd. E&W
.... 22W-2S,21W-2S 19
Apache Dr. N&S 25W-2S 18
Arbor Ln. N&S 22W-3S 19
Arboretum Cir. E&W
.... 24W-3S 18
Arboretum Rd. N&S
.... 22W-3S,22W-2S 19
Arbrust Av. E&W ... 26W-0N 12
Argyle St. N&S ... 21W-1N 13
Arnold Av. N&S......23W-1N 12
Arrowhead Dr. N&S . 26W-2S 18
Ashley Dr. NE&SW .. 23W-2S 18
Astor Pl. N&S 26W-0N 12
Audubon Ct. NE&SW
.... 21W-2S 19
Audubon Rd. NW&SE
.... 21W-2S 19
Balsam Dr. NW&SE .. 22W-3S 19
Bancroft Dr. N&S .. 23W-2S 18
Barclay Pl. N&S.....23W-2S 18
Barry Av. N&S 26W-0N 12
Bayberry Ln. N&S...25W-1S 18
Beechwood Ln. N&S
.... 22W-2S 19
Bemis Rd. E&W ... 21W-1S 19
Beverly St. N&S
.... 26W-0S,26W-0N 18
Big Horn Dr. N&S ... 25W-2S 18
Birch Av. NW&SE...26W-0N 12
Birchwood Dr. NW&SE
.... 22W-3S 19
Birchwood Pl. E&W .. 21W-3S 19
Blackcherry Ln. N&S
.... 23W-3S 18
Blackhawk Dr. N&S . 26W-2S 18
Blenheim Ct. E&W .. 23W-2S 18
Bloomingdale Rd. N&S
.... 23W-1N 12
Blue Jay Ct. N&S...26W-1N 12
Bob-o-link Rd. N&S . 21W-2S 19
Box Elder Av. E&W . 26W-0N 12
Briarcliff Ct. NE&SW
.... 21W-2S 19
Briarcliff Rd. NW&SE
.... 21W-2S 19
Brook Ct. NE&SW .. 21W-1S 19
Bryan Rd. N&S 22W-3S 19
Bud Ct. NW&SE 26W-0N 12
Buena Vista Dr. E&W
.... 23W-1S,22W-1S 18
Burning Trail N&S .. 26W-2S 18
Burr Oak Dr. NE&SW
.... 22W-3S 19
Busch Rd. E&W ... 22W-1N 13
Butternut Ln. N&S .. 22W-3S 19
Calvin Ct. N&S 26W-0S 18
Cambridge Ct. E&W 23W-2S 18
Canary Rd. NW&SE .. 21W-3S 19
Carey Cir. N&S ... 21W-2S 19
Carrol Gate NS&EW
.... 25W-2S,25W-1S 18
Cathryn Ct. N&S ... 26W-1N 12
Cesario Cir. NS&EW
.... 26W-2S,25W-2S 18
Cheiftain E&W 25W-2S 18
Cherry Ln. E&W ... 22W-1N 13

Street	Grid	Page
Cherrywood Dr. NW&SE	22W-3S	19
Chippewa Dr. NE&SW	26W-2S	18
Churchill Ln. N&S	21W-2S	19
Clifford Rd. E&W	21W-2S	19
Cole Av. E&W	25W-0N	12
Conifer Ct. N&S	22W-1S	19
Cooley Av. E&W	26W-0S	18
Coronet Rd. NW&SE	21W-2S	19
Cottonwood Dr. N&S	26W-0S,26W-0N	18
Cotuit Ct. NE&SW	21W-1S	19
Country Club Ln. N&S	25W-2S,25W-1S	18
Coventry Ct. NE&SW	25W-0N	12
Coventry Dr. N&S	25W-0N,25W-1N	12
Coventry St. E&W	25W-0N	12
Cree Ct. N&S	26W-2S	18
Crest Ct. N&S	21W-1S	19
Cumnor Av. NW&SE	21W-0N	13
Curtis Av. E&W	26W-0S	18
Cypress Dr. NW&SE	22W-3S	19
Daly Rd. E&W	23W-0N	12
Danbury Dr. N&S	21W-2S	19
Danby Dr. N&S	21W-1S	19
Darling St. N&S	25W-0N,25W-1N	12
Dawn Av. N&S	23W-1N	12
Delano St. N&S	26W-0N	12
Della Av. E&W	25W-1N	12
Delles Rd. N&S	25W-3S	18
Derby Glen Dr. E&W	23W-1N	12
Devonshire Ln. N&S	23W-2S	18
Dorchester Av. N&S	25W-0N	12
Dorchester Ct. E&W	21W-1S	19
Dorchester Rd. N&S	25W-0N	12
Douglas Ct. E&W	26W-1N	12
Drury Ln E&W	21W-2S	19
Durfee Rd. E&W	26W-2S,25W-2S	18
Eastern Av. N&S	22W-1N	13
Eddy Rd. N&S	22W-1S	19
Edgebrook Pl. E&W	21W-3S	19
Edgewood Ct. E&W	23W-2S	18
Edgewood Rd. E&W	21W-0S	19
Electric Av. NW&SE	25W-0N	12
Ellis Av. N&S	25W-0N,25W-1N	12
Ellynwood Dr. NE&SW	21W-0N	13
Elmwood Ct. E&W	23W-3S	18
Elmwood Dr. E&W	25W-2S,22W-2S	18
Elmwood St. N&S	26W-0S,26W-0N	18
Embden Ln. E&W	26W-2S,25W-2S	18
Emerson Av. E&W	22W-1N	13
Ethel Av. E&W	21W-0S	19
Euclid Av. N&S	22W-1N	13
Evans Av. N&S	26W-0S,26W-0N	18
Everest Rd. NW&SE	21W-2S	19
Evergreen Av. N&S	23W-1N	12
Fairmeadow Ln. NW&SE	25W-3S	18
Fairway St. NE&SW	21W-0N	13
Falcon Tr. N&S	26W-1N	12
Fanchon St. N&S	25W-0N,25W-1N	12
Farwell St. N&S	25W-0N,25W-1N	12
Fetzer Rd. E&W	25W-2S	18
Flamingo Ct. NE&SW	21W-2S	19
Flamingo Ln. NW&SE	21W-2S	19
Flint Creek Rd. E&W	25W-2S	18
Forest Av. N&S	22W-1N	13
Forest Hill Dr. NS&EW	22W-3S	19
Foxcroft Dr. E&W	23W-2S	18
Franklin St. N&S	26W-1N	12
Front St. E&W	22W-1N	13
Gables Blvd. N&S	25W-0N	12
Glen Av. NE&SW	21W-2S	19
Glen Park Rd. E&W	22W-2S,21W-2S	19
Glenbard Rd. E&W	21W-1S	19
Glencrest Dr. E&W	22W-1S	19
Glenrise Av. N&S	22W-1N	13
Glenvalley Dr. E&W	22W-2S,21W-2S	19
Gold Finch Ct. E&W	26W-1N	12
Golf Ln. NW&SE	21W-0N	13
Golfview Dr. N&S	21W-2S	19
Goodrich Av. N&S	22W-1N	13
Grand Av. NW&SE	26W-0N	12
Grange Ct. E&W	25W-2S	18
Gray Av. NE&SW	21W-2S	19
Great Western Av. NW&SE	23W-1N,22W-1N	12
Greenview Dr. N&S	21W-1S	19
Grove Ln. NE&SW	21W-1N	13
Hackberry Dr. E&W	23W-3S,22W-3S	18
Hamilton Dr. E&W	25W-2S	18
Hamilton Ln. NW&SE	25W-2S	18
Hampton Ct. NW&SE	21W-2S	19
Harding Rd. E&W	21W-0S	19
Harriet St. N&S	26W-1N	12
Harvey St. E&W	21W-0N	13
Hattie Gray Ct. E&W	23W-1N	12
Hattie Gray Ln. E&W	23W-1N	12
Hawthorne Ln. N&S	25W-2S,25W-1S	18
Hazel Ln. NW&SE	26W-0N	12
Hemstead Rd. E&W	21W-2S	19
Herrick Dr. N&S	25W-0N	12
Hevern Dr. E&W	25W-2S	18
High Knob Dr. E&W	25W-2S	18
Highland Av. N&S	22W-1N	13
Hill Av. NE&SW	21W-0N	13
Hoyle Rd. E&W	21W-1N	13
Huntington Ct. N&S	21W-2S	19
Huntington Pl. N&S	21W-2S	19
Huntington Rd. E&W	21W-2S	19
Ironwood Ln. N&S	25W-2S,25W-1S	18
Ivy Ln. N&S	21W-2S	19
Jerome Av. E&W	26W-0N,25W-0N	12
Jewell Rd. NE&SW	26W-0N,25W-0N	12
Jewell Rd. NE&SW	26W-0N	12
Juniper Ln. NE&SW	22W-3S	19
Keim Dr. NE&SW	26W-2S	18
Kenilworth Av. N&S	23W-2S,23W-1N	18
Kenmore St. N&S	21W-1N	13
Kensington Rd. E&W	21W-2S	19
Kent Rd. N&S	21W-2S	19
King's Ct. E&W	23W-3S,22W-3S	18
Kiowa Ct. N&S	26W-2S	18
Kiowa Dr. N&S	26W-2S	18
Kiowa Ln. E&W	26W-2S	18
Knollwood Dr. N&S	25W-0N	12
Lakeview Dr. N&S	26W-2S	18
Lakewood Ln. N&S	25W-2S	18
Lambert Rd. N&S	23W-2S	18
Lawler Av. N&S	21W-1S,21W-0S	19
Lawler Ln. NE&SW	21W-2S	19
Lenox Rd. N&S	22W-1N	13
Leonard St. N&S	26W-0S,26W-0N	18
Lloyd Av. N&S	21W-2S,21W-1S	19
Lorry Ct. NE&SW	22W-1S	19
Lynn Rd. E&W	21W-1N	13
Macarthur Av. E&W	26W-1N	12
Madison Av. E&W	21W-0N	13
Mallard Ct. E&W	26W-2S	18
Maple Ln. N&S	22W-1S	19
Marion Av. NW&SE	26W-0N,25W-0N	12
Marlborough Rd. E&W	21W-2S	19
Marshall Ln. E&W	25W-1N	12
Marston Ct. N&S	21W-1S	19
Mayfair Rd. E&W	21W-2S	19
Mayfield Ct. E&W	21W-2S	19
Mayfield Ln. NE&SW	21W-2S	19
Mayflower Pl. NE&SW	26W-0N	12
Mc Carron Rd. E&W	22W-2S	19
Mc Cormick Av. E&W	22W-1S	19
Mc Creey Av. E&W	23W-1S	18
Meadow Ln. E&W	21W-1N	13
Mellor Rd. E&W	25W-1N	12
Menomini Ct. N&S	26W-2S	18
Menomini Dr. E&W	26W-2S	18
Menomini Ln. E&W	26W-2S	18
Mildred Av. N&S	23W-1N	12
Milton Av. N&S	22W-2S	19
Milton Rd. N&S	22W-1S	19
Mohican Dr. E&W	26W-2S	18
Monticello Ct. N&S	21W-2S	19
Monticello Rd. E&W	21W-2S	19
Morse St. N&S	25W-0N,25W-1N	12
Mulberry Ln. NE&SW	23W-3S,22W-3S	18
Naperville Rd. NE&SW	25W-3S,24W-3S	18
Natalie Ct. E&W	26W-1N	12
National St. E&W	26W-1N	12
Nepil Av. N&S	25W-0S,25W-0N	18
Newton Av. N&S	23W-1N	12
North-South Tollway N&S	21W-1S,21W-1N	19
Oaklawn Dr. NE&SW	22W-2S	19
Oakwood Ct. N&S	25W-2S	18
Orchard Rd. N&S	25W-2S	18
Osage Dr. NW&SE	22W-3S	19
Ott Av. N&S	23W-1S	18
Page St. N&S	26W-0S,26W-0N	18
Papworth St. N&S	25W-0N,25W-1N	12
Park Blvd. NE&SW	22W-3S,22W-1N	19
Parkway Dr. E&W	26W-0N	12
Partridge Dr. N&S	26W-1N	12
Pershing Av. E&W	21W-1S	19
Peter Rd. N&S	26W-0N	12
Pheasant Ct. E&W	26W-1N	12
Picadilly Rd. NE&SW	25W-1S	18
Pierce Av. N&S	25W-0N	12
Pine Dr. E&W	23W-1N	12
Plamondon Rd. NE&SW	25W-1S	18
Pleasant Av. N&S	22W-1N	13
Pleasant Hill Rd. N&S	26W-0N,26W-1N	12
Pleasant Ln. E&W	21W-1N	13
Poplar Rd. E&W	22W-2S	19
Poss St. E&W	22W-1N	13
Prairie Av. N&S	22W-1N	13
Prairie Av. E&W	26W-0N,25W-0N	12
Prairie Ct. NE&SW	26W-0N	12
President St. N&S	23W-1N	12
Purnell St. N&S	25W-0N,25W-1N	12
Railroad St. NW&SE	22W-1N	13
Red Maple Ln. E&W	25W-2N	12
Red Oak Dr. E&W	23W-3S,22W-3S	18
Red Wing Ct. E&W	26W-1N	12
Red Wing Dr. N&S	26W-1N	12
Richard Av. N&S	26W-1N	12
Ridge Rd. N&S	22W-1S	19
River Dr. NE&SW	22W-1N,21W-1N	13
Riverside Av. N&S	21W-1S	19
Robinwood Ln. E&W	24W-3S	18
S. Ellyn Rd. N&S	22W-1S	19
Sandpiper Ct. E&W	26W-1N	12
Scott Av. N&S	21W-0N	13
Seneca Dr. N&S	26W-2S	18
Sequoia Dr. NW&SE	22W-3S	19
Shagbark Ln. NE&SW	22W-3S	19
Sheffield Ct. E&W	22W-2S	19
Sheffield Pl. E&W	22W-2S	19
Sherbrooke Ln. E&W	23W-2S	18
Silverleaf Blvd. N&S	25W-0N	12
Sioux Ct. E&W	26W-2S	18
Spring Av. N&S	21W-0N	13
St. Charles Rd. E&W	24W-1N,21W-1N	12
St. James Ct. E&W	23W-2S	18
St. Thomas Dr. N&S	22W-2S	19
Stableford Dr. E&W	23W-1N	12
Stacey Av. N&S	22W-1N	13
Stacey Ct. N&S	22W-1N	13
Stanley St. N&S	26W-0S,26W-0N	18
Stanton Rd. E&W	22W-2S	19
Stoddard Av. N&S	23W-0N	12
Stratford Ct. E&W	22W-2S	19
Stratford Rd. N&S	22W-2S	19
Stuarton Dr. N&S	25W-2S	18
Summerfield Cir. E&W	25W-2S	18
Summerhill Pl. NW&SE	23W-2S	18
Sunnybrook Rd. N&S	21W-1S	19
Sunset Av. E&W	21W-1N	13
Swift Rd. NE&SW	21W-1N	13
Sycamore Dr. NW&SE	23W-3S,22W-3S	18
Tamarack Dr. NW&SE	23W-3S,22W-3S	18
Taylor Rd. N&S	22W-2S,22W-1S	19
Teakwood Dr. E&W	22W-3S	19
Thaddeus Cir. E&W	21W-2S	19
Thomas Rd. E&W	26W-0N,25W-0N	12
Timothy Ln. N&S	26W-1N	12
Tomahawk Dr. NE&SW	26W-2S	18
Tomahawk Terr. N&S	26W-2S	18
Towpath Ct. E&W	25W-2S	18
Trails End Rd. E&W	23W-1N	12
Tree Tops Ln. N&S	25W-2S,25W-1S	18
Valley View Dr. E&W	22W-1N,21W-1N	13
Waldorth Ct. E&W	25W-1S	18
Walnut Rd. NE&SW	21W-0N	13
Washington Ln. NE&SW	21W-1N	13
Weisbrook Rd. NE&SW	26W-2S	18
Wenona Ln. E&W	25W-2S	18
West Haven Dr. N&S	25W-2S	18
West St. N&S	25W-1N	12
Westchester Ct. E&W	23W-2N	12
Western Av. N&S	23W-1N	12
Whitebirch Ct. E&W	25W-2S	18
Whitebirch Ln. N&S	25W-2N	12
Whitmore Ln. N&S	21W-0S	19
Willow Rd. N&S	25W-0N	12
Wilson Rd. E&W	21W-0S	19
Woodcreek Dr. N&S	21W-3S,21W-2S	19
Woodcroft Dr. E&W	23W-2S	18
Woodland Dr. N&S	25W-0N	12
Woodlark Dr. E&W	26W-1N	12
Woodlawn St. N&S	25W-0N	12
Woods Av. N&S	26W-0N,26W-1N	12
Woodvale Ct. E&W	26W-0N	12
Woodvale St. NW&SE	26W-0N	12
1st St. E&W	22W-1N	13
2nd Pl. E&W	22W-1N	13
2nd St. E&W	23W-1N,22W-1N	12
3rd St. E&W	23W-1N,22W-1N	12
22nd St. E&W	23W-1S,21W-2S	18
39th St. E&W	25W-3S	18

Naperville*

(Also see Will County)

Street	Grid	Page
Abbotsford Dr. NS&EW	29W-5S	23
Abby Dr. N&S	29W-5S	23
Ableside Cir. N&S	28W-7S	23
Abrahamson Ct. NE&SW	26W-8S	30
Acacia Ct. N&S	27W-5S	23
Ada Ct. E&W	28W-8S	29
Ada Ln. E&W	28W-8S	29
Albermarle Ct. N&S	29W-5S	23
Alder Ln. NW&SE	27W-8S	29
Alexis R Shuman Blvd. E&W	27W-4S,26W-4S	23
Almond Ct. NE&SW	28W-7S	23
Alton Ct. E&W	25W-8S	30
Amberwood Cir. NS&EW	26W-5S,25W-5S	24
Ambleside Cir. NS&EW	28W-7S	23
Andover Ct. NE&SW	25W-6S	24
Andrews Ct. NW&SE	25W-8S	30
Anne Rd. E&W	26W-6S	24
Anthony Ct. NW&SE	29W-6S	23
Antietam Ct. E&W	27W-8S	29
Apache Dr. NE&SW	27W-5S,26W-4S	23
Appaloosa Dr. NS&EW	24W-8S	30
Appleby Ct. N&S	25W-7S	24
Applegate Ct. NE&SW	27W-8S	29
Applegate Dr. NW&SE	27W-8S	29
Appomattox Cir. N&S	27W-8S	29
Arabian Av. E&W	24W-8S	30
Ardmore Dr. NE&SW	28W-8S,28W-7S	29
Argus E&W	28W-6S	23
Ashfield Rd. NW&SE	25W-7S	24
Aspen Ct. NE&SW	27W-8S	29
Atlanta Ct. NW&SE	27W-7S	23
Atlas Ln. N&S	26W-8S	30
Atwood Cir. NE&SW	24W-9S	30
Auburn Dr. NS&EW	24W-9S	30
Augusta Dr. N&S	30W-5S	22
Aurora Av. E&W	29W-7S,29W-6S	23
Autumn Av. E&W	25W-8S	30
Aztec Cir. NE&SW	27W-5S,26W-5S	23
Bailey Rd. E&W	27W-8S,24W-8S	29
Bainbridge Dr. N&S	29W-5S	23
Bakewell Ln. E&W	25W-8S	30
Baldwin Dr. N&S	27W-8S	29
Balmoral Cir. NS&EW	25W-7S	24
Balton Ct. N&S	29W-5S	23
Banbury Cir. NW&SE	25W-7S	24
Bankfield Ct. NE&SW	25W-7S	24
Bannister Dr. NE&SW	27W-8S	29
Banyon Ct. NS&EW	28W-7S	23
Barbara Ct. NE&SW	27W-8S	29
Barcroft Ct. NW&SE	25W-7S	24
Barkei Rd. N&S	25W-4S	24
Barley Av. E&W	26W-5S	24
Basswood Dr. N&S	26W-8S	30
Bauer Rd. E&W	28W-5S,26W-5S	23
Bay Colony Dr. E&W	26W-8S	30
Bay Ct. N&S	26W-9S	30
Bayberry Ln. E&W	26W-5S	24
Bayhill Av. E&W	25W-8S	30
Baylor Ct. E&W	24W-9S	30
Bayview Ct. E&W	27W-9S	29
Bedford Ct. NW&SE	25W-7S	24
Beech Ct. NE&SW	28W-6S	23
Belaire Ct. NE&SW	27W-5S	23
Bellingrath Ct. NW&SE	26W-6S	24
Beloit Ct. N&S	24W-9S	30
Beloit Dr. E&W	24W-9S	30
Benedetti Dr. NE&SW	27W-6S,27W-5S	23
Bent Creek Ct.* N&S	23W-9S	31
Benton Av. E&W	27W-6S,26W-6S	23
Berkley Ct. N&S	26W-9S	30
Bernie Ct. E&W	25W-8S	30
Berry Dr. N&S	28W-7S,28W-6S	23
Big Foot Ln. N&S	27W-5S	23
Biltmore Ct. N&S	26W-6S	24
Birchwood Dr. N&S	28W-7S,28W-6S	23
Birkdale Ct. NE&SW	27W-5S	23
Black Hawk Ct. E&W	26W-5S	24
Black Stallion Ct. E&W	24W-8S	30
Black Stallion Dr. NS&EW	24W-8S	30
Blackburn Ct. NE&SW	25W-7S	24
Blue Grass Ct. E&W	26W-5S	24
Blue Larkspur Ln. E&W	25W-7S	24
Bluebird Ln. E&W	26W-8S	30
Bob-o-Link Ln. N&S	28W-5S	23
Bonaventure Dr. N&S	28W-5S	23
Bond St. N&S	29W-5S,29W-4S	23
Book Rd. E&W	28W-8S	29
Bourbon Ct. N&S	25W-9S	30
Bovidae Cir. NS&EW	27W-8S	29
Boweset Ct. E&W	29W-7S	29
Bowling Green Ct. NE&SW	27W-5S	23
Brad Ct. E&W	25W-8S	30
Bradley Ct. E&W	24W-8S	30
Braemar Av. NS&EW	26W-5S,26W-4S	24
Braemar Ct. E&W	26W-5S	24
Brainard N&S	26W-7S	24
Brainard St. N&S	26W-6S,26W-5S	24
Brandeis Ct. E&W	24W-9S	30
Brandy Cir. NS&EW	25W-8S,24W-8S	30
Brentford Dr. NE&SW	26W-4S	24
Brestal Ct. N&S	25W-9S	30
Briarheath Cir. NW&SE	26W-9S	30
Briarwood Dr. E&W	28W-7S	23
Bridgewater Ct. E&W	26W-9S	30
Briergate Dr. N&S	28W-5S	23
Brighton Rd. E&W	25W-5S	24
Bristlecone Ct. N&S	25W-9S	30
Bristol Ct. NW&SE	26W-9S	30
Brittany Av. E&W	25W-8S,24W-8S	30
Brom Ct. N&S	27W-7S	23
Brook Ln. E&W	28W-6S	23
Brookdale Rd. NS&EW	29W-5S,28W-5S	23
Brooklea Ct. E&W	25W-9S	30
Brookline Ct. NE&SW	25W-6S	24
Brookshire Ct. N&S	29W-6S	23
Bruno Ct. N&S	29W-6S	23
Brunswick Ct. E&W	25W-7S	24
Brush Hill Cir. NS&EW	28W-6S	23
Buckeye Dr. E&W	27W-8S	29
Buckingham Dr. N&S	25W-5S	24
Buckley Ct. N&S	26W-9S	30
Bucknell Ct. N&S	24W-9S	30
Bullrun Ct. E&W	27W-8S	29
Burke Ct. N&S	30W-5S	22
Burning Tree Ln. E&W	27W-5S	23
Burr Oak Ct. NW&SE	28W-7S	23
Butler Ct. NE&SW	24W-9S	30
Butte Ct. N&S	26W-6S	24
Butternut Dr. NE&SW	26W-7S	24
Buttonwood Cir. N&S	28W-7S	23
Byron Ct. NE&SW	25W-8S	30
Cabriolet Ct. N&S	27W-9S	29
Cactus Dr. NW&SE	26W-4S	24
Calcutta Ln. N&S	28W-5S	23
Camellia Ct. N&S	25W-9S	30
Candlenut Dr. E&W	28W-6S	23

Street	Grid	Page
Canonero Dr. NE&SW	26W-7S,25W-7S	24
Cantigny Ct. NS&EW	25W-8S	30
Canyon Run Rd. E&W	25W-9S,24W-9S	30
Cardiff Rd. E&W	27W-8S	29
Cardigan Ct. NW&SE	27W-8S	29
Cardinal Ln. N&S	26W-8S	30
Carleton Av. E&W	24W-9S	30
Carleton Cir. NS&EW	24W-9S	30
Carlson Ct. N&S	27W-5S	23
Carmel Ct. NE&SW	28W-5S	23
Carnege Ct. NE&SW	24W-9S	30
Carol Ln. N&S	26W-8S	30
Carriage Ct. E&W	27W-6S	23
Carriage Hill Rd. E&W	25W-9S	30
Carriage Way Dr. NE&SW	25W-7S	24
Carthage Ct. E&W	24W-9S	30
Castilleject Ct. N&S	28W-7S	23
Catalpa Ln. NW&SE	26W-8S,26W-7S	23
Catherine Av. E&W	25W-6S	24
Ceals Ct. N&S	26W-9S	30
Cedarbrook Rd. NS&EW	26W-9S	30
Celtic Ash Ct. N&S	25W-7S	24
Center St. N&S	26W-6S,26W-5S	24
Central Park Rd. N&S	26W-6S	24
Centre Point Dr. NS&EW	26W-4S	24
Chadwicke Cir. EW&NS	28W-7S	23
Chalet Rd. E&W	28W-6S	23
Challdon Ct. NW&SE	26W-7S	24
Chalmette Cir. NS&EW	26W-9S	30
Chancellor Ct. NW&SE	27W-5S	23
Charles St. N&S	26W-7S,26W-5S	24
Chat Ct. N&S	26W-8S	30
Chateaugay Av. NS&EW	25W-7S	24
Chatham Ct. N&S	26W-9S	30
Chattanooga Ct. NE&SW	27W-8S	29
Chepstow Ct. NE&SW	28W-7S	23
Cherry Blossom Ln. NW&SE	27W-7S,26W-7S	23
Cherry Hills Ln. E&W	28W-5S	23
Cherrywood Cir. NW&SE	26W-9S	30
Chesire Av. E&W	25W-7S	24
Chesterfield Av. E&W	25W-7S	24
Chestnut Ridge Dr. E&W	28W-6S	23
Chicago Av. E&W	26W-6S,25W-6S	24
Chickasaw Dr. N&S	26W-5S	24
Chicory Ct. N&S	28W-7S	23
Chippewa Dr. NS&EW	26W-5S,26W-4S	24
Churchill Dr. NW&SE	29W-5S	23
Citadel Dr. NE&SW	24W-9S	30
Claremont Dr. N&S	28W-6S	23
Clarissa Ct. NE&SW	28W-8S	30
Clarkson Ct. E&W	24W-9S	30
Clemson Dr. NS&EW	24W-8S	30
Cliffside Ct. N&S	25W-9S	30
Cloverleaf Ct. N&S	26W-9S	30
Clovertree Ct. N&S	28W-7S	23
Clyde Dr. NE&SW	26W-9S,26W-8S	30
Coach Dr. N&S	25W-9S	30
Cody Ct. N&S	27W-6S	23
Colfax Ct. E&W	29W-5S	23
Collingwood Dr. NS&EW	25W-7S	24
Colonial Ct. E&W	26W-6S	24
Columbia St. N&S	26W-7S,26W-5S	24
Cornell Ct. E&W	24W-8S	30
Cortez Ct. NE&SW	26W-4S	24
Cottage Av. E&W	27W-6S	23
Cottonwood Ln. E&W	28W-6S	23
Count Fleet Ct. N&S	26W-7S	24
Country Farm Dr. N&S	30W-4S	22
Countryside Cir. NW&SE	26W-9S	30
Court Pl. N&S	26W-9S	24
Courtland Dr. N&S	28W-7S	29
Cove Ct. E&W	26W-9S	30
Coventry Ct. E&W	26W-8S	30
Crabapple Ct. N&S	28W-6S	23
Cranberry Ct. N&S	25W-9S	30
Cranshire Ln. E&W	24W-9S	30
Creekside Cir. NS&EW	27W-5S	23
Creighton Av. E&W	24W-8S	30
Crestwood Ct. E&W	25W-7S	24
Culpepper Dr. NS&EW	25W-7S,24W-7S	24
Cypress Dr. E&W	27W-8S	29
Dakota Cir. NS&EW	26W-4S	24
Dana Ct. E&W	27W-5S	23
Darius Ln. N&S	25W-9S	30
Dark Star Rd. NW&SE	25W-7S	24
Dartford Ct. E&W	25W-7S	24
Dartmoor Ct. E&W	25W-7S	24
Dartmouth Ct. E&W	24W-9S	30
DePaul Ct. E&W	24W-8S	30
Deep Run Rd. E&W	25W-7S	24
Deepwood Ct. NW&SE	26W-8S	30
Deerpath Ct. N&S	25W-9S	30
Delkir Ct. E&W	26W-9S	30
Denison Ct. E&W	24W-9S	30
Denisport Dr. E&W	25W-8S	30
Derby Ct. E&W	29W-5S	23
Devlin Ct. NW&SE	26W-9S	30
Devon Ct. E&W	25W-6S	24
Devonshire Ct. E&W	27W-7S	23
Diane Ln. NE&SW	26W-7S	24
Dichtl Ct. N&S	25W-9S	30
Dickens Dr. NW&SE	26W-5S,25W-5S	24
Diehl Rd. E&W	29W-4S,25W-4S	24
Dockside Ct. N&S	27W-5S	23
Dorset Ct. E&W	25W-7S	24
Dorval Dr. N&S	27W-9S	29
Douglas Av. NE&SW	27W-6S	23
Dove Ct. NW&SE	26W-8S	30
Dover Ln. N&S	25W-9S	30
Downing Ct. NE&SW	29W-5S	23
Drake Ct. N&S	24W-9S	30
Drendel Ln. E&W	26W-9S	30
Druid Hills Ct. NW&SE	27W-5S	23
Duke Ct. E&W	24W-9S	30
Duke Dr. E&W	24W-9S	30
Duquesne Av. E&W	24W-9S	30
Durham Ct. N&S	25W-6S	24
Durness Ct. N&S	27W-8S	29
Eagle St. N&S	27W-6S,27W-5S	23
East Av. N&S	25W-5S	24
East West Tollway E&W	30W-4S,25W-4S	22
Eastland Ct. N&S	26W-9S	30
Eddystone Cir. NS&EW	27W-9S	29
Edgewater Dr. NW&SE	26W-7S	24
Edwards Rd. NE&SW	25W-7S	24
Elizabeth Av. E&W	26W-6S,25W-5S	24
Ellsworth St. N&S	26W-6S,25W-5S	24
Elm Ct. NE&SW	28W-8S	29
Elm St. N&S	28W-8S	29
Elmwood Dr. NW&SE	27W-8S,26W-7S	29
Emerald Dr. N&S	27W-8S,27W-7S	29
Encina Ct. E&W	28W-6S	23
Estate Cir. E&W	24W-8S	30
Estuary Ct. N&S	26W-9S	30
Eton Ln. N&S	25W-8S	30
Evergreen Av. E&W	28W-7S	23
Ewing St. N&S	27W-6S	23
Fair Oak Rd. N&S	25W-9S	30
Fair Oaks Ct. NE&SW	30W-5S	22
Faireno Ct. E&W	25W-7S	24
Fairfax Rd. N&S	25W-7S	24
Fairfield Ct. E&W	24W-9S	30
Fairway Dr. N&S	30W-6S,30W-5S	22
Fairway Ln. N&S	25W-7S	24
Fairwinds Ct. NS&SE	26W-6S	24
Fall Brook Ct. N&S	25W-9S	30
Fallcreek Ct. N&S	26W-8S	30
Farington Ct. E&W	26W-5S	24
Feldott Dr. N&S	29W-7S	29
Fender Rd. N&S	25W-9S,25W-8S	30
Ferncroft Ct. N&S	28W-5S	23
Fernwood Dr. E&W	27W-7S	23
Ferry Rd. E&W	30W-4S,29W-4S	22
Fessler Av. N&S	27W-8S	29
Field Ct. NE&SW	28W-6S	23
Finch Ct. NW&SE	26W-9S	30
Fir Ct. NW&SE	28W-6S	23
Fisher Dr. E&W	30W-4S	22
Flicker Ct. E&W	26W-9S	30
Flock Av. E&W	26W-9S	30
Flock Cir. NS&EW	26W-9S	30
Floret Ct. N&S	24W-9S	30
Flossmoor Ct. NW&SE	27W-5S	23
Flowerfield Ct.* N&S	23W-9S	31
Fordham Ct. NE&SW	24W-9S	30
Forest Av. N&S	27W-6S	23
Forestview Ct. N&S	28W-5S	23
Fort Hill Dr. N&S	29W-7S,29W-6S	23
Fox Bend Ct. N&S	28W-5S	23
Fox Trail Ct. N&S	24W-8S	30
Foxcroft Rd. E&W	26W-9S	30
Foxhill Rd. E&W	29W-5S	23
Foxhorn Ct. NE&SW	26W-4S	24
Franklin Av. E&W	27W-6S,26W-6S	23
Frederick Ln. E&W	26W-8S	30
Fredericksburg Ct. E&W	27W-7S	23
Fresno Ct. N&S	28W-6S	23
Friar Ln. N&S	24W-9S,24W-8S	30
Furman Av. E&W	24W-9S	30
Garaway Ct. E&W	29W-7S	29
Garden Ct. NE&SW	28W-7S	23
Gardenside Ct. N&S	28W-8S	29
Garth Ct. NE&SW	26W-9S	30
Gartner Rd. E&W	28W-7S,26W-7S	29
George Ln. NS&EW	25W-6S	24
Georgetown Ct. E&W	24W-9S	30
Georgia Ct. NE&SW	28W-7S	23
Gettysburg Ct. E&W	27W-8S	29
Glencoe Ct. E&W	26W-9S	30
Glenmore Ct. NW&SE	25W-7S	24
Goldenrod Ct. NE&SW	28W-7S	23
Golf Ct. N&S	25W-9S	30
Golfview Rd. E&W	31W-5S	22
Gowdey Rd. E&W	29W-6S,29W-5S	23
Great Oak Ct. NE&SW	25W-9S	30
Green Ridge Dr. NS&EW	24W-8S	30
Green Trails Dr. E&W	25W-7S	24
Greenlee Ct. E&W	28W-7S	23
Greensboro Ct. E&W	26W-6S	24
Greenfield Dr. E&W	26W-6S,25W-6S	24
Gregory Ct. NW&SE	27W-8S	29
Grimes Av. E&W	27W-8S	29
Hackberry Ln. E&W	28W-7S,28W-6S	23
Hagemann Pl. NW&SE	25W-6S,25W-5S	24
Hamilton Ln. NS&EW	25W-8S,24W-8S	30
Hampton Ct. N&S	25W-8S	30
Hansom Ct. N&S	25W-9S	30
Harbor Ct. N&S	26W-9S	30
Harlowe Ct. E&W	25W-8S	30
Harris Ln. N&S	25W-9S	30
Harvard Av. NW&SE	24W-9S	30
Hastings St. N&S	29W-5S	23
Haverhill Cir. NE&SW	25W-5S	24
Hawthorne Dr. NW&SE	27W-7S,26W-7S	23
Hayle St. NW&SE	28W-7S	23
Hazelwood Dr. E&W	27W-7S	23
Heatherton Ct. NW&SE	28W-6S	23
Heatherton Dr. NS&EW	28W-5S	23
Heathrow Ln. NE&SW	28W-8S,27W-8S	29
Hedg Ct. NE&SW	25W-9S	30
Hemlock Ln. N&S	27W-8S	29
Henning Ct. NE&SW	26W-7S	24
Hercules Ln. N&S	26W-8S	30
Heritage Hill Dr. (Pvt) E&W	28W-6S	23
Hickory Ln. NW&SE	26W-7S	23
Hidden Bay Ct. N&S	25W-9S	30
Hidden Lake Rd. E&W	26W-9S	30
Hidden Spring Rd. E&W	28W-6S	23
Hidden Valley Dr. E&W	24W-9S	30
High Grove Ln. E&W	29W-6S	23
High Point Dr. NS&EW	25W-4S	24
Highland Av. E&W	26W-6S	24
Hillside Rd. E&W	27W-7S,26W-6S	23
Hilltop Ct. N&S	25W-9S	30
Hilton Head Dr. N&S	30W-5S	22
Hinterlong Dr. E&W	29W-6S	23
Hobson Hollow Dr. N&S	24W-8S	30
Hobson Mill Dr. E&W	26W-8S,25W-8S	30
Hobson Oaks Ct. NW&SE	25W-8S	30
Hobson Oaks Dr. N&S	25W-8S	30
Hobson Rd. E&W	26W-8S,24W-8S	30
Holly Ct. NW&SE	27W-8S	29
Hollytowne Ct.* N&S	23W-9S	31
Honest Pleasure Dr. NE&SW	26W-7S	24
Honey Berry Ct. N&S	28W-6S	23
Honeysuckle Dr. N&S	28W-7S	23
Hopi Ct. NW&SE	27W-5S	23
Huffman St. N&S	26W-6S	24
Hull Ct. NW&SE	25W-9S	30
Hunter St. N&S	25W-8S	30
Huntington Cir. NS&EW	24W-6S	24
Huntleigh Dr. NW&SE	25W-7S	24
Hyde Park Ln. NS&EW	25W-9S	30
I-88 Tollway E&W	30W-4S,25W-4S	22
Imperial Cir. NE&SW	28W-5S	23
Inca Ct. E&W	26W-5S	24
Indian Hills Ct. NW&SE	27W-5S	23
Indian Knoll Rd. N&S	25W-9S	30
Indian Trail Dr. N&S	25W-9S	30
Industrial Dr. N&S	30W-6S	22
Inland Cir. N&S	30W-6S,30W-5S	22
Inverness Ct. N&S	27W-5S	23
Inverrary Dr. N&S	25W-7S	24
Iona Ct. E&W	24W-9S	30
Iona Dr. N&S	24W-9S	30
Iris Ln. N&S	27W-9S	23
Iron Liege Ct. N&S	26W-7S	24
Ironwood N&S	26W-7S	24
Iroquois Av. E&W	26W-5S,25W-5S	24
Ishnala Dr. N&S	25W-9S	30
Jackson Av. E&W	27W-6S	23
Jaipur Av. N&S	25W-7S	24
James Ln. NE&SW	25W-7S	24
Jane Av. E&W	25W-6S	24
Jasmine Ct. NE&SW	28W-7S	23
Jefferson Av. E&W	29W-6S,27W-6S	23
John Friend Dr. NW&SE	29W-9S	29
Johnson Dr. NE&SW	25W-8S,25W-7S	30
Johoull Ct. E&W	29W-7S	29
Joshua Ct. N&S	28W-7S	23
Julian St. N&S	26W-7S,26W-6S	24
Juniper Ct. E&W	27W-8S	29
Jupiter Ct. E&W	26W-5S	24
Justin Ct. E&W	24W-8S	30
Kaimi Ct. N&S	26W-4S	24
Kathy Ct. N&S	25W-7S	24
Kauai King Ct. E&W	26W-7S	24
Keim Ct. N&S	25W-9S	30
Keim Dr. NW&SE	25W-9S	30
Kemper Dr. N&S	30W-5S	22
Kennebec Ln. E&W	26W-5S	24
Kennesaw Ct. N&S	27W-8S	29
Kensington Ct. E&W	27W-5S	23
Kent Ct. N&S	25W-7S	24
Kenyon Dr. NE&SW	24W-9S,24W-8S	30
Kerria Ct. N&S	25W-6S	24
Kesick Ln. E&W	25W-6S	24
Ketten Ct. E&W	26W-4S	24
Ketten Dr. N&S	26W-4S	24
Killdeer Ln. N&S	26W-9S,26W-8S	30
Kimball Ct. NW&SE	26W-8S,26W-7S	30
Kinglet Ct. N&S	26W-9S	30
Kings Ct. NE&SW	29W-5S	23
Kings Point Ct. E&W	28W-5S	23
Kingswood Ct. E&W	26W-9S	30
Kinsie St. NW&SE	28W-6S	30
Kinston Ct. NW&SE	27W-5S	23
Kiowa Dr. N&S	25W-9S	30
Kirby Ct. N&S	29W-5S	23
Knoll Dr. N&S	25W-10S,25W-9S	12
Knox Ct. E&W	24W-8S	30
Laird St. N&S	27W-6S	23
Lakeview Ct. NW&SE	25W-9S	30
Lambsgate Dr. N&S	28W-9S	29
Lampwick Ct. N&S	26W-4S	24
Lancaster Cir. NS&EW	26W-9S	30
Lancaster Ct. NE&SW	26W-9S	30
Langley Cir. NS&EW	29W-5S	23
Lantern Cir. NS&EW	24W-7S	24
Larchmont Ct. N&S	26W-9S	30
Lark Ln. N&S	26W-8S	30
Larkspur Ct. E&W	26W-4S	24
Larsen Ln. NE&SW	25W-7S	24
Laura Ct. E&W	27W-5S	23
Laurel Ln. N&S	26W-8S	30
Laureli Ct. E&W	28W-6S	23
Le Province Av. NS&EW	26W-9S	30
Leabrook Ct. N&S	25W-9S	30
Leamington Ct. NW&SE	25W-9S	30
Leeds Ct. E&W	25W-9S	30
Leesburg Ct. N&S	25W-9S	30
Lexington Ln. E&W	29W-7S	29
Liberty Dr. N&S	24W-7S	24
Lighthouse Dr. N&S	25W-8S	30
Lilac Ln. NE&SW	27W-8S,27W-7S	29
Lincoln Rd. E&W	25W-4S	24
Linden Ct. E&W	28W-6S	23
Lindrick Ct. NE&SW	28W-5S	23
Lisa Ct. E&W	29W-6S	23
Lisson Rd. NE&SW	24W-9S	30
Lockwood Ct. NS&EW	26W-5S	24
Locust Ct. N&S	28W-6S	23
Lois Ann Dr. E&W	29W-6S	23
Lombardy Ln. NW&SE	28W-6S	23
London Ct. S. N&S	29W-5S	23
Longleaf Dr. N&S	28W-6S	23
Longridge Ct. E&W	26W-9S	30
Longspur Ct. NE&SW	26W-7S	30
Lookout Ct. NW&SE	26W-7S	24
Loomis St. N&S	26W-7S,26W-5S	24
Loyola Dr. N&S	24W-8S	30
Lynchburg Ct. NE&SW	25W-7S	24
Lyndhurst Ct. NS&EW	26W-9S	30
Magnolia Ln. N&S	27W-7S	23
Main St. N&S	27W-7S,27W-5S	23
Manassas Ct. NE&SW	27W-7S	23
Manchester Ct. NW&SE	29W-5S	23
Manchester St. N&S	29W-5S	23
Maple Hills Ct. N&S	25W-7S	23
Maple Ln. E&W	26W-7S	24
Mapleknoll Dr. N&S	26W-5S	24
Maplewood Cir. E&W	30W-5S	22
Margate Ct. N&S	25W-7S	24
Market Av. NE&SW	25W-8S	30
Market Ct. N&S	25W-8S	30
Market Dr. E&W	25W-8S	30
Marls Ct. NE&SW	25W-8S	24
Marquette Av. NE&SW	24W-8S	30
Martin Av. E&W	27W-7S	23
Mary Ln. E&W	25W-6S	24
Masters Ct. N&S	30W-5S,30W-4S	22
Mattande Dr. N&S	26W-6S	24
Mcdowell Rd. E&W	29W-5S	23
Meadowland Dr. E&W	28W-7S	23
Meander Dr. N&S	26W-8S	30
Meisch Ct. N&S	27W-5S	23
Meisenger Ln. NE&SW	26W-7S	24
Melody Ln. N&S	26W-7S	24
Menominee Ln. E&W	26W-5S	24
Merrimac Cir. NS&EW	27W-8S	29
Merritt Ct. NE&SW	25W-8S	30
Mescallero Ct. E&W	26W-4S	24
Miami Ct. N&S	26W-4S	24
Michele St. N&S	24W-8S	30
Michigan Av. E&W	29W-6S	23
Middleburg Ct. NE&SW	25W-7S	24
Middleburg Rd. NE&SW	25W-7S	24
Mike Ct. NE&SW	29W-6S	23
Mill St. N&S	27W-6S,27W-4S	23
Millburn Ct. E&W	26W-6S,25W-6S	24
Millet St. N&S	26W-5S	24
Millrace Ln. N&S	25W-9S	30
Milton Dr. E&W	25W-6S	24
Mirror Lake Dr. E&W	28W-5S	23
Mistwood Dr. NS&EW	24W-8S	30
Misty Ct. N&S	28W-10S	29
Mobile Ct. NW&SE	27W-5S	23
Monmouth Av. E&W	24W-9S	30

Street	Grid	Page
Montclair Dr. NW&SE	24W-9S	30
Monterey Ct. NE&SW	28W-6S	23
Montgomery Ct. NE&SW	27W-8S	29
Monticello Dr. NE&SW	26W-6S	24
Morgan Cir. NS&EW	24W-8S	30
Morningside Dr. N&S	27W-6S,27W-5S	23
Mortonsberry Dr. N&S	28W-8S	29
Morven Ct. NW&SE	26W-6S	24
Mountain Ct. N&S	25W-9S	30
Mt. Vernon Ct. N&S	26W-6S,26W-5S	24
Muirhead Ct. E&W	25W-8S	30
Muirhead St. N&S	25W-8S	30
Mulberry Ct. NW&SE	28W-7S	23
Mulligan Dr. N&S	30W-5S,30W-4S	22
Mundelein Rd. NS&EW	24W-9S	30
Mundstone Ct. NE&SW	25W-5S	24
Mustang Dr. NS&EW	24W-8S	30
Naper Blvd. N&S	25W-9S,25W-4S	30
Naperville Rd. NE&SW	25W-4S	24
Naperville-Wheaton Rd. NE&SW	25W-5S	24
Napoleon Dr. N&S	25W-9S	30
Nashua Dr. NE&SW	26W-7S	24
Nashville Ct. NW&SE	27W-8S	29
Natchez Trace Cir. E&W	27W-8S	29
Navarone Dr. N&S	27W-9S	29
Needham Rd. NE&SW	25W-6S,25W-5S	24
Needles Ct. NW&SE	25W-7S	24
North Av. E&W	26W-6S	24
Northfield Ct. N&S	26W-9S	30
Northfield Dr. N&S	26W-9S	30
Norwood Av. NE&SW	25W-8S	30
Notre Dame Dr. NE&SW	26W-7S	24
Oak Ct. NE&SW	25W-9S	30
Oakhurst Dr. NS&EW	28W-7S	23
Oakleaf Ct. E&W	26W-6S	24
Oakton Ln. NE&SW	28W-8S,28W-7S	29
Oakwood Dr. NE&SW	27W-7S	23
Ogden Av. E&W	28W-6S,25W-5S	23
Okolona Ct. NE&SW	27W-7S	23
Old Dominion Ct. E&W	25W-7S	24
Old Dominion Rd. E&W	25W-7S	24
Old Naperville Rd. N&S	25W-4S	24
Olesen Dr. NE&SW	25W-8S,25W-6S	30
Olympus Dr. N&S	26W-8S	30
Omaha Ct. NW&SE	25W-7S	24
Ontario Av. E&W	29W-6S	23
Oriole Ct. NE&SW	26W-8S	30
Orleans Av. E&W	25W-8S	30
Osage Ln. E&W	26W-8S	30
Ottawa Ct. E&W	26W-5S	24
Overstreet Ct. N&S	25W-9S	30
Overton Ct. NW&SE	25W-7S	24
Oxford Ct. NW&SE	25W-6S	24
Pacific Dr. N&S	27W-6S	23
Paddington Av. E&W	29W-5S	23
Paddock Ct. N&S	25W-9S	30
Page Ct. E&W	27W-6S	23
Palm Ct. E&W	28W-7S	23
Palmetto Ct. E&W	28W-7S	23
Pamela Ct. E&W	25W-7S	24
Papago Ct. NW&SE	26W-5S	24
Park St. N&S	27W-4S	23
Park Way Dr. N&S	27W-6S	23
Parkfield Ct. E&W	28W-7S	23
Pas Ct. N&S	29W-6S,29W-5S	23
Pathway Ln. N&S	24W-9S	30
Pattermann Cir. NS&EW	24W-9S	30
Pawnee Ln. E&W	26W-5S	24
Peachtree Ct. N&S	25W-6S,25W-5S	24
Peacock Ct. E&W	26W-9S	30
Pearson Cir. NE&SW	27W-6S,27W-5S	23
Pecan Ct. E&W	27W-8S	29
Pelham Ct. N&S	29W-5S	23
Pemberton Ct. N&S	25W-8S	30
Pembroke Rd. N&S	25W-6S	24
Penn Ct. E&W	24W-8S	30
Pepperdine Ct. E&W	24W-9S	30
Pepperidge Rd. N&S	28W-6S	23
Periwinkle Ct. E&W	29W-7S	29
Persimmon Ct. N&S	25W-9S	30
Peterson Ct. E&W	27W-6S	23
Phaeton Ct. NW&SE	25W-9S	30
Picadilly Cir. N&S	29W-5S	23
Pickwick Ct. NW&SE	29W-5S	23
Pier Av. E&W	25W-8S	30
Pier Ct. E&W	25W-8S	30
Pima Ct. NW&SE	26W-5S	24
Pine Ct. NW&SE	26W-7S	24
Pinetree Dr. N&S	27W-8S	29
Placer Ct. N&S	26W-9S,26W-8S	30
Plainfield-Naperville Rd. N&S	27W-8S,27W-7S	29
Plank Rd. N&S	26W-6S,25W-5S	24
Plum Hollow Ct. NE&SW	28W-5S	23
Pomona Ct. E&W	28W-6S	23
Poplar St. N&S	28W-8S	29
Porter Av. E&W	26W-7S	24
Pottowattomi Ct. N&S	26W-5S	24
Prague Ct. N&S	29W-5S	23
Prairie Knoll Dr. NS&EW	26W-9S	30
Prairie St. E&W	26W-7S	24
Prairieview Ct. NE&SW	30W-5S	22
Preston Rd. E&W	29W-5S	23
Prestwick Dr. N&S	30W-5S	22
Primrose Ln. N&S	29W-5S	30
Prince Ct. E&W	29W-5S	23
Princeton Cir. NE&SW	24W-8S	30
Proud Clarion Ct. NE&SW	26W-7S	24
Purdue Av. E&W	24W-9S	30
Quail Hollow Ct. N&S	26W-6S	24
Quails Dr. E&W	26W-9S	30
Queen Ct. NW&SE	29W-5S	23
Queensbury Ct. N&S	29W-5S	23
Queensgreen Cir. E&W	29W-5S	23
Queensgreen Ct. N&S	29W-5S	23
Quin Ct. N&S	29W-6S,29W-5S	23
Quincy Ave. E&W	29W-6S	23
Raintree Ct. N&S	28W-7S	23
Rampart Cir. N&S	25W-9S	30
Ranchview Dr. N&S	24W-9S,24W-8S	30
Rapidan Ct. N&S	25W-7S	24
Rapids Ct. NW&SE	26W-8S	30
Raymond Dr. N&S	28W-6S,29W-4S	23
Redbud Ct. E&W	27W-7S	23
Redfield Rd. NS&EW	29W-5S	23
Redpoll Ct. N&S	26W-9S	30
Redstart Rd. NW&SE	26W-9S	30
Rennet Dr. N&S	27W-8S	29
Revere Ct. N&S	24W-7S	24
Rhodes Dr. E&W	28W-8S	29
Richard Rd. N&S	25W-7S,25W-6S	24
Richmond Ct. N&S	27W-7S	23
Rickert Dr. NS&EW	29W-7S,28W-8S	23
Ridgefield Ln. E&W	26W-9S	30
Ridgely Ct. E&W	25W-7S	24
Rill Ct. N&S	26W-8S	30
Ring Rd. N&S	26W-10S,26W-9S	30
Riparian Dr. N&S	26W-9S	30
Ripley Ct. NE&SW	26W-9S	30
Riva Ridge Ct. N&S	25W-7S	24
River Bluff Cir. NE&SW	27W-7S	23
River Oak Dr. NE&SW	25W-9S,24W-9S	30
River Rd. N&S	28W-8S,28W-5S	29
River Ridge Cir. NS&EW	25W-9S	30
Riverfront Cir. NS&EW	27W-7S	23
Riverlea Cir. E&W	25W-9S	30
Riveridge Ct. E&W	25W-9S	30
Riverview Ln. NW&SE	26W-7S	24
Robin Hill Ct. E&W	27W-7S	23
Robin Hill Dr. N&S	27W-7S	23
Rockbridge Dr. NE&SW	25W-7S	24
Rolling Grove Ct. N&S	28W-7S	23
Rollins Ct. N&S	24W-9S	30
Rose Ln. NW&SE	27W-7S,27W-8S	23
Rosewood Ln. N&S	26W-5S,25W-5S	24
Rossmere Ct. N&S	25W-5S	24
Rowan Ct. E&W	26W-8S	30
Royal Blackheath Ct. N&S	28W-5S,27W-5S	23
Royal Bombay Ct. NE&SW	28W-5S	23
Royal St. George Dr. NE&SW	28W-5S,27W-5S	23
Russet Av. E&W	27W-8S	29
Rutgers Ct. NS&EW	24W-9S	30
Saber Ct. N&S	26W-9S	30
Sagebrush Rd. E&W	26W-9S	30
Saginaw Ct. E&W	26W-9S	30
Samstag Ct. NE&SW	25W-5S	24
Sanctuary Ln. EW&NS	28W-8S,28W-7S	29
Sandalwood Dr. NE&SW	27W-7S	23
Sandgate Ct. NW&SE	26W-9S	30
Sandpiper Ln. NW&SE	26W-8S	30
Sara Ln. E&W	25W-8S	30
Saranell Av. E&W	28W-8S	29
Savannah Cir. NS&EW	27W-7S	23
Schey Ct. N&S	26W-9S	30
School St. E&W	26W-6S	24
Scott Ct. N&S	29W-5S	23
Secretariat Dr. N&S	25W-7S	24
Selby Rd. E&W	29W-5S	23
Sequoia Ct. N&S	28W-7S	23
Sequoia Rd. N&S	28W-7S	23
Seton Hall Dr. N&S	24W-9S	30
Severn Ct. NE&SW	27W-8S	29
Shanahan Ct. N&S	25W-7S	24
Sheffield Rd. E&W	25W-9S	30
Shenandoah Ln. N&S	26W-5S	24
Shepherd Dr. N&S	27W-8S	29
Sheppey Ct. N&S	27W-8S	29
Sheringham Dr. NS&EW	27W-8S	29
Sherman Av. N&S	26W-5S	24
Shetland Rd. NW&SE	24W-8S	30
Shiloh Cir. E&W	27W-8S	29
Shome Ct. NW&SE	26W-9S	30
Signal Dr. N&S	25W-9S	30
Silver Maple Ct. N&S	26W-5S	24
Sir Barton Ct. N&S	26W-7S	24
Sleepy Hollow Ct. E&W	24W-9S	30
Sleight St. N&S	26W-7S,26W-5S	24
Slippery Rock Ct. NE&SW	24W-9S	30
Slippery Rock Dr. NE&SW	24W-9S	30
Small Ct. E&W	29W-6S	23
Smokey Ct. N&S	26W-4S	24
Snowdon Ct. NW&SE	25W-8S	30
Spalding Dr. E&W	27W-7S	23
Spindle Tree Av. E&W	25W-9S	30
Split Oak Rd. NE&SW	26W-9S	30
Spring Garden Cir. E&W	25W-5S	24
Spring St. E&W	27W-6S	23
Spring Valley Rd. E&W	24W-9S	30
Springhill Ct. N&S	25W-6S,25W-5S	24
Springside Cir. NE&SW	26W-9S	30
Springwood Dr. NE&SW	27W-7S	23
Spruce Dr. E&W	27W-8S	29
Sprucewood Ct. E&W	25W-9S	30
Stacie Ct. N&S	29W-6S	23
Stafford Av. N&S	25W-7S	24
Standish Ct. NE&SW	26W-8S	30
Stanford Dr. N&S	24W-9S	30
Stanhope Ct. NE&SW	25W-9S	30
Stanton Dr. NE&SW	25W-8S	30
Starling Ln. NW&SE	26W-9S	30
Stauffer Dr. N&S	28W-6S	23
Staunton Rd. E&W	25W-9S	30
Steinbeck St. N&S	28W-8S	29
Sterling Ct. N&S	25W-7S	24
Stevens St. E&W	27W-6S	23
Stone Creek Ct. E&W	26W-8S	30
Stone Ct. N&S	25W-5S	24
Stonegate Rd. E&W	28W-7S	23
Stonehenge Ct. NW&SE	29W-5S	23
Stoney Brook Dr. E&W	25W-9S	30
Stoos Ln. NW&SE	26W-7S	24
Stratford Ct. N&S	28W-6S	23
Strawbridge Ct. N&S	23W-9S	31
Stuart Ct. N&S	25W-7S	24
Suffolk St. N&S	26W-5S	24
Sugar Creek Ct. NW&SE	28W-5S	23
Summerset Ct. N&S	27W-8S	29
Summit Hills Dr. NE&SW	28W-5S	23
Sumter Ct. N&S	27W-8S	29
Sunderland Ct. N&S	26W-9S	30
Sunrise Ct. N&S	25W-9S	30
Sunset Dr. NS&EW	26W-7S	24
Surrey Av. N&S	25W-8S	30
Sussex Ct. NW&SE	25W-6S	24
Sussex Rd. NW&SE	25W-6S	24
Swallow St. N&S	26W-9S,26W-8S	30
Swansea Ct. NE&SW	27W-8S	29
Sweetbay Ct. NW&SE	26W-7S	24
Swift Ln. E&W	26W-8S	30
Sycamore Dr. NE&SW	27W-8S,26W-7S	29
Sylvan Cir. NW&SE	26W-7S	24
Sylvia Ln. E&W	26W-8S	30
Syracuse Av. NE&SW	24W-9S	30
Talbot Dr. E&W	27W-8S	29
Tall Oaks Ct. NW&SE	24W-8S	30
Tamarack Av. E&W	26W-8S	30
Tan Oak Ln. E&W	26W-8S	30
Tanager Ct. NE&SW	26W-8S	30
Tanglewood Ln. E&W	26W-7S	24
Teakwood Ct. NE&SW	26W-7S	23
Technology Dr. (Pvt) E&W	27W-5S	23
Templar Rd. N&S	26W-9S	30
Tennyson Ln. EW&NS	28W-8S,27W-8S	29
Terrance Dr. NS&EW	26W-9S,26W-8S	30
Terri Cir. E&W	29W-6S	23
Testa Dr. N&S	28W-6S	23
Teton Ct. NW&SE	27W-8S	29
Thomas Dr. N&S	29W-6S	23
Thornwood Dr. NE&SW	26W-7S	24
Thrush Ln. E&W	26W-8S	30
Thunderbird Ln. E&W	28W-5S	23
Ticonderoga Ln. E&W	26W-5S	24
Tim Tam Cir. NS&EW	26W-7S	24
Timber Lee Ct. N&S	25W-7S	24
Timber Trail Dr. E&W	25W-9S	30
Timberlake Ct. E&W	26W-6S,25W-5S	24
Torrington Dr. N&S	27W-8S	29
Towhee Ln. N&S	26W-8S	30
Towpath Ct. N&S	25W-9S	30
Trace Ct. N&S	27W-8S	29
Trails End Ct. N&S	25W-9S	30
Travelaire Av. E&W	27W-9S	29
Trefoil Ct. N&S	26W-9S	30
Trent Av. E&W	29W-5S	23
Triton Ln. E&W	26W-8S	30
Tudor Dr. N&S	29W-5S,29W-6S	23
Tufts Ct. NE&SW	24W-9S	30
Tulane Dr. N&S	24W-8S	30
Tulip Ln. N&S	27W-7S	23
Tupelo Dr. E&W	27W-8S,26W-8S	29
Turnbridge Cir. NS&EW	25W-7S	24
Turtle Cove Ct. N&S	25W-9S	30
Union Ct. NE&SW	30W-5S	22
University Ct. N&S	25W,10S	30
University Dr. NS&EW	25W-10S,24W-9S	30
Vales End Dr. NE&SW	28W-8S	29
Valley Dr. N&S	28W-6S	23
Valley Ridge Ct. E&W	24W-9S	30
Van Buren Av. E&W	27W-6S	23
Van Buren St. E&W	26W-6S	24
Vanderbilt Ct. N&S	24W-9S	30
Vassar Dr. N&S	24W-9S	30
Venetian Ct. E&W	25W-7S	24
Ventura Ct. E&W	28W-6S	23
Vermont Ct. NE&SW	27W-8S	29
Vest Dr. N&S	26W-5S	24
Vest St. N&S	26W-5S	24
Vicksburg Ct. E&W	27W-8S	29
Victoria Ct. NE&SW	26W-7S	24
View Ct. N&S	26W-7S	24
Villa Av. E&W	26W-5S	24
Village Green Rd. E&W	27W-7S	23
Villanova Dr. NW&SE	24W-9S	30
Violet Ct. NE&SW	29W-7S	29
Virginia Ct. E&W	27W-9S	23
Wakefield Ct. N&S	25W-6S	24
Wall St. N&S	29W-5S,29W-4S	23
Walnut St. N&S	25W-8S	30
Walton Ct. E&W	26W-9S	30
Warbler Ln. N&S	26W-9S,26W-8S	30
Warrenville Rd. E&W	27W-4S,25W-4S	23
Warwick Dr. E&W	25W-9S	30
Washington St. N&S	26W-10S,26W-9S	30
Water St. E&W	27W-6S	23
WaterCress Dr. NE&SW	26W-7S	24
Waterford Ct. N&S	28W-6S	23
Waterside Ct. NW&SE	27W-6S	23
Waverly Ct. N&S	26W-6S	24
Waxwing Av. E&W	26W-8S	30
Webster St. N&S	27W-7S,27W-5S	23
Wedgefield Cir. NS&EW	27W-5S	23
Wehrli Dr. N&S	26W-7S	24
Wehrli Rd. N&S	24W-9S	30
Wellington Ct. N&S	25W-9S	30
Welner Rd. N&S	24W-9S	30
Welton Ct. N&S	26W-9S	30
West Ct. N&S	27W-5S	23
West St. N&S	27W-8S,27W-4S	29
Westbury Ct. E&W	26W-9S	30
Westglen Dr. NS&EW	27W-8S	29
Westminster Dr. N&S	29W-5S	23
Westmoreland Ct. NW&SE	25W-6S	24
Westmoreland Dr. N&S	25W-7S	24
Weston Ridge Dr. N&S	30W-6S	22
Westwind Ct. NE&SW	25W-9S	30
Wexford Ct. NE&SW	25W-7S	24
Whirlaway Av. NW&SE	26W-7S,25W-7S	24
Whispering Hills Ct. E&W	28W-6S	23
Whispering Hills Rd. NS&EW	28W-8S,28W-6S	29
White Oak Dr. N&S	26W-7S,26W-6S	24
White Pines Ct. E&W	28W-5S	23
Whitebark Ct. N&S	28W-7S	23
Whitewater Ln. E&W	28W-8S	29
Whitley Rd. E&W	29W-5S	23
Whittingham Cir. Cir.	25W-7S	24
Wickfield Ct. NE&SW	25W-5S	24
Wickham Ct. NW&SE	25W-7S	24
Wildflower Cir. NE&SW	29W-7S	29
Wildflower Ct. E&W	29W-7S	29
Willcrest Rd. E&W	28W-7S	23
Williamsburg Dr. N&S	27W-8S,27W-7S	29
Williard Ct. E&W	29W-8S	29
Willow Rd. NW&SE	26W-7S	24
Willoway Dr. N&S	28W-6S	23
Wilshire Dr. E&W	28W-6S	23
Wilson Av. E&W	27W-6S	23
Winchester Ct. E&W	28W-5S	23
Windham Ln. E&W	26W-5S	24
Winding Creek Dr. NS&EW	26W-8S	30
Windridge Ct. N&S	28W-7S	23
Windsor Ct. N&S	25W-8S	30
Winola Ct. N&S	25W-9S	30
Winslow Ct. E&W	25W-9S	30
Wisconsin St. E&W	26W-5S	24
Wisteria Ct. NE&SW	26W-9S	30
Wood Creek Ct. NE&SW	26W-8S	30
Woodfield Ct. NW&SE	26W-9S	30
Wright St. N&S	26W-7S,26W-5S	24
Wydown Ct. NW&SE	25W-8S	30
Wyndgate Ct. E&W	28W-8S	29
Yale Ct. NE&SW	24W-8S	30
Yellow Daisy Ct. E&W	30W-5S	22
York Ct. N&S	25W-6S	24
Yorkshire Dr. E&W	25W-5S	24
Zaininger Av. NW&SE	27W-5S	23
4th Av. E&W	26W-6S	24
5th Av. E&W	28W-6S,26W-6S	23
6th Av. E&W	27W-6S,26W-6S	23
7th Av. E&W	26W-6S	23
8th Av. E&W	27W-6S,26W-6S	23
10th Av. E&W	27W-5S	23
11th Av. E&W	27W-5S,26W-5S	23
12th Av. E&W	27W-5S,26W-5S	23
13th Av. E&W	27W-5S	23

	Grid	Page
14th St. E&W	27W-5S,26W-5S	23
75th St. E&W	27W-8S,25W-8S	29
77th St. E&W	24W-8S	30
79th St. E&W	24W-8S	30
81st St. E&W	26W-9S	30
87th St. E&W	28W-9S,24W-9S	29

Naperville Township

	Grid	Page
Ada Av. E&W	28W-8S	29
Aero Dr. N&S	29W-9S	29
Allister Ct. NW&SE	30W-5S	22
Allister La. E&W	30W-5S	22
Arbor Dr. N&S	27W-7S	23
Argyll Ln. E&W	30W-5S	22
Aurora Av. E&W	29W-7S,28W-7S	23
Bailey Rd. E&W	27W-9S	29
Bauer Rd. E&W	27W-5S	23
Berry Ct. N&S	28W-7S	23
Bonnie Ct. NE&SW	30W-5S	22
Book Rd. N&S	29W-9S,29W-8S	29
Briar Ln. E&W	30W-5S	22
Bruce Ln. E&W	30W-5S	22
Bunker Cir. E&W	30W-5S	22
Byerum St. N&S	27W-5S	23
Campbell Dr. N&S	30W-5S	22
Chandelle Ct. N&S	29W-9S	29
Claymore Ln. E&W	30W-5S	22
Clover Ct. N&S	28W-7S	23
Country Club Blvd. N&S	30W-5S	22
Country Glen Dr. N&S	30W-6S	22
Country Lakes Dr. E&W	30W-5S	22
Diehl Rd. E&W	31W-4S,28W-4S	22
Doral Dr. E&W	30W-5S	22
Eagle St. N&S	27W-5S	23
Eola Rd. N&S	31W-7S,31W-5S	22
Estuary Ct. NW&SE	26W-9S	30
Ewing St. N&S	27W-5S	23
Fairway Dr. N&S	30W-6S,30W-5S	22
Ferry Rd. E&W	31W-4S,30W-4S	22
Firestone Ct. N&S	30W-5S	22
Flamenco Ct. NW&SE	30W-5S	22
Frontenac Rd. N&S	31W-5S	22
Glenoban Dr. N&S	30W-5S	22
Gordon Terr. Rd. N&S	30W-5S	22
Granada Ct. NW&SE	30W-5S	22
Green Acres Dr. N&S	27W-7S	23
Hacienda Ct. NW&SE	30W-5S	22
Huntington Ct. NE&SW	25W-6S	24
Innisbrook Dr. E&W	30W-5S	22
Innsbrook N&S	30W-5S,30W-4S	22
Joseph Rd. N&S	27W-5S	23
Joseph St. N&S	27W-5S	23
Kautz Rd. N&S	32W-5S	22
Kirk Ct. N&S	30W-5S	22
Lawn Meadow Ln. E&W	28W-7S	23
Liberty St. E&W	32W-7S,31W-7S	22
McGregor Ln. E&W	30W-5S	22
Meadow Lark Ln. N&S	27W-9S	29
Meadow Rd. N&S	30W-4S	22
Medinah Ct. N&S	30W-5S	22
Medinah Dr. N&S	30W-5S	22
Mill St. N&S	27W-5S	23
Modaff Rd. N&S	27W-9S	29
Molitor Rd. E&W	32W-5S	22
Monterrey Ct. NE&SW	30W-5S	22
New York St. E&W	32W-7S,30W-7S	22
North Aurora Rd. E&W	31W-6S,29W-6S	22
North Ln. E&W	27W-7S	23
Northfield Ct. N&S	26W-9S	30
Ogden Av. NE&SW	29W-8S	29
Oswego Rd. NE&SW	28W-7S	23
Paxton Dr. E&W	30W-5S	22
Pearson St. N&S	27W-5S	23
Pebble Beach Ln. N&S	30W-5S	22
Pebblewood Ln. N&S	30W-5S,30W-4S	22
Pebblewood Trail NS&EW	30W-5S,30W-4S	22
Pinehurst Dr. E&W	30W-5S	22
Pioneer Ct. E&W	26W-9S	30
Poss Rd. E&W	32W-6S,31W-6S	22
Ramona Ct. NW&SE	30W-5S	22
Reid Pl. E&W	31W-6S	22
River Rd. NW&SE	28W-6S	23
Rt. 59 N&S	29W-9S,29W-6S	29
Scheffer Rd. E&W	32W-6S	22
Schooner Ct. E&W	26W-9S	30
Scotts Dr. N&S	30W-5S	22
Skylane Ct. N&S	29W-9S	29
South Ln. E&W	27W-7S	23
Spyglass Ct. N&S	30W-5S	22
St. Andrews Ct. N&S	30W-5S	22
Stearman Ct. N&S	29W-9S	29
Stewart Dr. N&S	30W-5S	22
Stone Creek Ct. E&W	26W-8S	30
Stonebridge Blvd. NE&SW	32W-6S	22
Tartan Ln. N&S	30W-5S	22
Thornapple Dr. N&S	28W-7S	23
Timberlane Dr. N&S	30W-6S,30W-5S	22
Valencia Ct. NE&SW	30W-5S	22
Vista Cir. N&S	30W-5S	22
Wagon Wheel Ct. E&W	26W-9S	30
Webster St. N&S	27W-5S	23
Westwind Dr. N&S	30W-6S,30W-5S	22
Whispering Winds Dr. E&W	30W-6S	22
Wild Cherry Ln. N&S	28W-7S	23
Woodewind Dr. E&W	30W-5S	22
4th St. N&S	31W-6S	22
5th St. N&S	32W-6S	22
7th St. N&S	32W-6S	22
46th St. E&W	27W-4S	23
47th St. E&W	27W-4S	23
48th St. E&W	27W-5S	23
49th St. E&W	27W-5S	23
75th St. E&W	30W-8S,28W-8S	28
79th St. E&W	27W-9S	29
80th St. E&W	27W-9S	29
81st St. E&W	29W-9S	29
82nd Ct. E&W	27W-9S	29
82nd Pl. E&W	27W-9S	29
83rd St. E&W	27W-9S	29
87th St. E&W	28W-9S,24W-9S	29

Oak Brook

	Grid	Page
Abbeywood Ct. NW&SE	15W-1S	21
Acorn Hill Ln. E&W	15W-3S	21
Adams Rd. N&S	16W-3S	21
Arden Ct. N&S	18W-2S	20
Ascot Ln. N&S	18W-3S	20
Avenue Loire N&S	18W-2S	20
Bath & Tennis Dr. N&S	16W-2S,15W-2S	21
Baybrook Ct. NW&SE	18W-2S	20
Baybrook Ln. E&W	18W-2S,17W-2S	20
Berseem Ct. NE&SW	18W-2S	20
Birchwood Rd. E&W	16W-4S	27
Bliss Dr. E&W	15W-1S	21
Bluegrass Ct. N&S	18W-3S	20
Bradford Ln. NS&EW	15W-2S	21
Breckenridge Rd. E&W	16W-3S	21
Briarwood Av. E&W	17W-2S	21
Briarwood Central N&S	17W-2S	21
Briarwood Cir. Cir.	17W-2S	21
Briarwood Dr. E&W	17W-2S	21
Briarwood Ln. E&W	17W-2S	21
Briarwood Loop E&W	17W-2S	21
Briarwood North E&W	17W-2S	21
Briarwood Pass E&W	17W-2S	21
Briarwood South N&S	17W-2S	21
Bridle Path Cir. NS&EW	18W-3S	20
Brighton Ln. N&S	15W-2S	21
Brougham Ln. N&S	18W-3S	20
Burr Oak Ct. N&S	15W-1S	21
Butterfield Rd. NE&SW	18W-2S	20
Cambridge Dr. N&S	17W-2S	21
Camden Ct. N&S	18W-2S	20
Camelot Dr. NE&SW	17W-2S	21
Canterberry Ln. E&W	15W-3S	21
Carlisle Dr. NE&SW	17W-2S	21
Carriage Ct. NW&SE	18W-3S	20
Cass Ct. N&S	18W-3S,17W-3S	20
Cermak Rd. E&W	16W-1S,15W-1S	21
Charleton Pl. E&W	17W-2S	21
Cheval Dr. E&W	16W-4S	27
Cochise Ct. NE&SW	18W-2S	20
Commerce Dr. E&W	16W-1S	21
Concord Dr. N&S	17W-2S	21
Coolidge St. N&S	15W-2S	21
Court 1 N&S	18W-3S	20
Court 2 N&S	18W-3S	20
Court 3 N&S	18W-3S	20
Court 4 N&S	18W-3S	20
Court 5 N&S	18W-3S	20
Court 6 N&S	18W-3S	20
Court 7 E&W	18W-3S	20
Court 8 E&W	18W-3S	20
Court 9 N&S	18W-3S	20
Court 10 N&S	18W-3S	20
Court 11 N&S	18W-3S	20
Court 12 N&S	18W-3S	20
Court 13 N&S	18W-3S	20
Court 14 N&S	18W-3S	20
Court 15 N&S	18W-3S	20
Court 16 N&S	18W-1S	20
Court 17 N&S	17W-3S	21
Court 18 E&W	17W-3S	21
Court 19 E&W	18W-3S	20
Court 20 E&W	18W-3S	20
Crab Apple Ln. N&S	16W-1S	21
Croydon Ln. E&W	15W-2S	21
Deer Trail Ln. E&W	16W-3S	21
Derby Ct. N&S	18W-3S	20
Devonshire Dr. NS&EW	17W-2S	21
Dover Dr. E&W	15W-2S	21
East-West Tollway E&W	18W-2S,15W-1S	20
Enterprise NE&SW	15W-1S	21
Fairview Av. N&S	18W-3S	20
Ferndale Rd. E&W	15W-3S	21
Forest Gate Rd. E&W	16W-2S	21
Forest Glen Ln. NW&SE	15W-1S	21
Forest Mews Dr. E&W	18W-3S	20
Forest Trail N&S	16W-1S	21
Fox Trail Ct. N&S	16W-3S	21
Fox Trail Ln. E&W	16W-3S	21
Fox Trail Rd. E&W	16W-3S	21
Foxiana Ct. NW&SE	18W-2S	20
Frederick Dr. E&W	16W-3S	21
Gatewood Ln. E&W	15W-3S	21
Glendale Av. E&W	15W-3S	21
Green Leaf Dr. E&W	16W-1S	21
Hambletonian Dr. E&W	18W-3S	20
Hamilton Ln. E&W	17W-2S	21
Hampton Dr. E&W	17W-2S	21
Harger Rd. E&W	16W-1S,15W-1S	21
Heather Ln. NE&SW	17W-2S	21
Heritage Oaks Ct. N&S	18W-3S	20
Heritage Oaks Ln. (pvt.) E&W	18W-3S	20
Hickory Dr. E&W	15W-1S	21
Hunt Club Dr. N&S	15W-2S	21
Hunt Club Ln. E&W	15W-2S	21
Hunter Dr. E&W	15W-1S	21
Indian Trail Ct. N&S	16W-3S	21
Indian Trail Dr. N&S	16W-3S	21
Indian Trail Rd. NS&EW	16W-3S	21
Interstate 88 E&W	18W-2S,15W-1S	20
Interstate 294 N&S	15W-2S,15W-1S	21
Ivy Ln. N&S	17W-2S	21
Jorie Blvd. N&S	16W-2S	21
Kensington Ct. N&S	16W-2S	21
Kensington Rd. E&W	16W-2S	21
Kimberly Cir. Cir.	17W-2S	21
Kimberly Ln. E&W	17W-2S	21
Kingsbury Ct. N&S	17W-2S	21
Kingston Dr. E&W	17W-2S	21
Lakewood Ct. E&W	15W-1S	21
Lambeth Ct. E&W	17W-2S	21
Liberty Blvd. N&S	18W-2S	20
Lincoln St. N&S	15W-2S	21
Livery Circle N&S	18W-3S	20
Lockinvar Ln. N&S	18W-2S	20
Luthin Rd. E&W	15W-3S	21
Madison St. N&S	16W-3S	21
Mallard Ln. E&W	16W-3S	21
Marian Sq. E&W	17W-3S	21
McDonalds Dr. N&S	16W-1S	21
Meadowood Dr. N&S	15W-1S	21
Merry Ln. E&W	16W-1S	21
Meyers Rd. N&S	18W-3S,18W-2S	20
Midwest Club Pkwy. Cir.	18W-3S	20
Midwest Rd. N&S	17W-3S,17W-2S	21
Mockingbird Ln. E&W	17W-2S	21
Monterey Av. N&S	17W-1S	21
Mulberry Ct. N&S	18W-3S	20
Oak Brook Dr. N&S	18W-2S	20
Oak Brook Hills Dr. N&S	18W-2S	20
Oak Brook Rd. E&W	18W-2S,16W-2S	20
Olympia Ct. E&W	18W-2S	20
Ottawa Ln. E&W	16W-3S	21
Palisades Dr. N&S	15W-1S	21
Pembroke Ln. E&W	17W-2S	21
Phoebos Ln. E&W	16W-1S	21
Pine Hill Ln. E&W	16W-4S	27
Polo Ln. N&S	18W-3S	20
Red Stable Way E&W	18W-3S	20
Redfox Ln. E&W	16W-1S	21
Regent Ct. NE&SW	17W-2S	21
Regent Dr. N&S	17W-2S	21
Ridgewood Ct. E&W	18W-3S	20
Robert Kingery Expwy. N&S	16W-3S,16W-1S	21
Robinhood Ln. N&S	15W-3S	21
Roslyn Rd. N&S	18W-3S	20
Royal Glen Ct. N&S	15W-1S	21
Royal Vale Dr. NE&SW	18W-2S	20
Saddlebrooke Dr. NS&EW	18W-3S	20
Salt Creek Cir. E&W	15W-3S	21
Salt Creek Dr. E&W	15W-3S	21
Sauk Path N&S	16W-3S	21
Sheffield Ln. NS&EW	15W-2S	21
Shelburne Dr. N&S	17W-2S	21
Spring Rd. N&S	16W-1S,15W-3S	21
St. Francis Cir. Cir.	17W-3S	21
St. Johns Ct. NW&SE	17W-3S	21
St. Joseph's Ct. NS&EW	17W-3S	21
St. Joseph's Dr. E&W	17W-3S	21
St. Marks Ct. N&S	17W-3S	21
St. Michael Ct. N&S	17W-3S	21
St. Stephen's Green E&W	17W-3S	21
Stafford Ln. N&S	15W-2S	21
Steeplechase West E&W	16W-1S	21
Steepleridge Ct. N&S	18W-3S	20
Suffolk Ln. N&S	18W-3S	20
Surrey Dr. N&S	16W-2S	21
Swaps Ct. NW&SE	18W-2S	20
Swift Dr. N&S	15W-1S	21
Thurlow St. N&S	16W-3S	21
Timber Edge Dr. E&W	16W-1S	21
Timber Trails Ct. NW&SE	16W-1S	21
Timber Trails Dr. E&W	16W-1S	21
Timberview Dr. E&W	16W-1S	21
Tower Dr. N&S	18W-2S	20
Tri-state Tollway N&S	15W-2S,15W-1S	21
Trinity Ct. E&W	17W-3S	21
Trinity Ln. NS&EW	17W-3S	21
Twin Oaks Dr. N&S	16W-1S	21
Victoria Ct. N&S	17W-2S	21
Walnut Ln. E&W	16W-1S	21
Wennes Ct. E&W	15W-3S	21
White Oak Ln. E&W	18W-3S	20
Wildwood Ct. N&S	15W-1S	21
Willow View Dr. E&W	16W-1S	21
Windsor Dr. N&S	15W-2S,15W-1S	21
Wood Av. E&W	15W-3S	21
Woodglen Ln. NE&SW	15W-1S	21
Woodland Dr. N&S	16W-1S	21
Woodridge Dr. E&W	15W-1S	21
Woodside Dr. E&W	15W-1S	21
Woodview Ct. E&W	15W-1S	21
Wyndham Ct. E&W	18W-2S	20
York Rd. N&S	15W-3S,15W-1S	21
Yorkshire Woods Dr. Cir.	15W-1S	21
22nd St. E&W	18W-1S,17W-1S	20
30th St. E&W	15W-2S	21
31st St. E&W	15W-2S	21
32nd St. E&W	18W-3S	20
33rd St. E&W	17W-3S	21
34th St. E&W	15W-3S	21
35th Pl. E&W	15W-3S	21
35th St. E&W	18W-3S,16W-3S	20
38th St. E&W	18W-3S	20

Oakbrook Terrace

	Grid	Page
Ardmore Av. N&S	18W-1S	20
Butterfield Rd. NE&SW	18W-1S,17W-1S	20
Eisenhower Rd. N&S	17W-1S	21
Elder Ln. NE&SW	17W-1S	21
Elm Ct. NE&SW	17W-1S	21
Elm Pl. E&W	17W-1S	21
Halsey Rd. E&W	17W-1S	21
Hodges Rd. E&W	17W-1S	21
Karban Rd. E&W	17W-1S	21
Leahy Rd. N&S	17W-1S	21
Lincoln Av. E&W	18W-1S	20
MacArthur Dr. N&S	17W-1S	21
Marshall Rd. NW&SE	17W-1S	21
Monterey Av. NE&SW	17W-1S	21
Nimitz Rd. NW&SE	17W-1S	21
Oakland Av. N&S	17W-1S	21
Orchard Pl. E&W	17W-1S	21
Patton Av. N&S	17W-1S	21
Renaissance Blvd. N&S	18W-1S	20
Robert Kingery Expwy. N&S	17W-1S	21
Roosevelt Rd. E&W	17W-0S	21
Stillwell Rd. E&W	17W-1S	21
Summit Av. N&S	17W-1S	21
Terrace Blvd. E&W	18W-1S	20
Wainwright Rd. NW&SE	17W-1S	21
Washington Av. N&S	18W-1S	20
16th St. E&W	17W-1S	21

Roselle*

(Also see Cook County)

	Grid	Page
Acacia Ln. N&S	22W-6N	7
Alder Ln. N&S	24W-7N	6
Ambleside Dr. NE&SW	23W-7N	6
Americana Ct. N&S	24W-7N	6
Andover Dr. N&S	25W-7N	6
Ardmore Av. E&W	23W-7N	6
Ash St. E&W	22W-7N	7
Ashley Ct. NW&SE	24W-7N	6
Autumn Dr. N&S	24W-6N	6
Avebury Ct. N&S	24W-6N	6
Avebury Ln. NW&SE	24W-6N	6
Banbury Ct. N&S	23W-6N	6
Banbury Terr. N&S	23W-6N	6
Berkshire Terr. E&W	23W-6N	6
Berwick Pl. E&W	25W-6N,24W-6N	6
Birch Ct. NW&SE	22W-6N	7
Bokelman St. NE&SW	23W-7N	6
Borden Ct. N&S	24W-7N	6
Borden Dr. E&W	24W-7N	6
Brandywine Dr. E&W	24W-6N	6
Briarwood Ln. N&S	24W-6N	6
Brighton Bay N&S	25W-7N	6
Bristol Ct. E&W	25W-7N	6
Brookside Dr. N&S	24W-7N	6
Brookwood Terr. N&S	23W-6N	6
Brower Dr. E&W	24W-7N	6
Bryn Mawr Av. E&W	25W-6N,22W-6N	6
Butternut Ct. N&S	23W-6N	6
Callero Ct. NW&SE	23W-6N	6
Carriage Way E&W	25W-6N	6
Case Dr. E&W	24W-7N	6
Catalpa Av. NW&SE	23W-7N,22W-7N	6
Catino Ct. NW&SE	23W-6N	6
Central Av. E&W	24W-7N,23W-7N	6
Century Dr. N&S	22W-7N	7
Chancellor Dr. N&S	24W-7N	6
Chatham Ct. E&W	25W-7N	6
Chatham Ln. N&S	25W-7N	6
Chelsea Cove* E&W	25W-7N	6
Cherry Ct. E&W	22W-7N	7
Cherry St. NE&SW	22W-7N	7
Circle Dr. E&W	22W-7N	7
Claria Dr. E&W	23W-7N	6
Clearwater N. E&W	24W-6N	6
Clearwater S. NW&SE	24W-6N	6
Club House Dr. NE&SW	22W-7N	7
Colony Ln. E&W	24W-7N	6
Congress Cir. S. E&W	24W-7N	6
Congress Cir. W. NS&EW	24W-7N	6
Country Ln. E. N&S	24W-6N	6
Country Ln. N. E&W	24W-6N	6
Country Ln. S. E&W	24W-6N	6
Countryside Dr. NE&SW	23W-7N	6
Coventry Ct. E&W	25W-7N	6
Crestwood Dr. NW&SE	23W-6N	6
Cypress Ct. N&S	23W-6N	6
Daisy Ln. N&S	24W-6N	6
Dalton Ln. N&S	24W-6N	6
Darby Ln. NE&SW	25W-7N	6
Dee Ln. NE&SW	22W-7N	7
Devon Av. E&W	23W-7N,22W-7N	6
Dinah Rd. N&S	22W-6N	7
Dover Dr. N&S	25W-6N	6
Downing St. E&W	25W-6N	6
Drummond Ct.* E&W	25W-7N	6
Edenwood Dr. E&W	24W-6N	6
Elm St. E&W	23W-7N	6
Exeter Ct. N&S	25W-7N	6

Street	Grid	Page
Fall Cir. NW&SE	24W-6N	6
Falmouth Way NW&SE	25W-6N	6
Fordham Pl. NE&SW	24W-6N	6
Forest Av. E&W	23W-6N	6
Forum Dr. E&W	22W-7N	7
Foster Av. E&W	23W-6N,22W-6N	6
Foxdale Ct. NE&SW	25W-6N,24W-6N	6
Garden Av. N&S	24W-7N	6
Garden Rd. N&S	24W-7N	6
Gary Av. N&S	25W-7N	6
Glendale Terr. E&W	23W-6N	6
Glenlake Av. E&W	23W-7N	6
Glenmore Pl. NE&SW	24W-6N	6
Golf View Dr. E&W	22W-7N	7
Granville Av. E&W	23W-7N,22W-7N	6
Greenwood Ct. N&S	23W-6N	6
Hampton Ln. NW&SE	25W-6N	6
Hamstead Ct. NE&SW	25W-7N	6
Harvest Ln. N&S	24W-6N	6
Harvest Pl. NE&SW	24W-6N	6
Hattendorf Av. NW&SE	23W-7N	6
Hazel Ct. E&W	22W-6N	7
Hemlock Ln. NW&SE	24W-6N	6
Heritage Dr. E&W	23W-6N	6
High Pointe Ct. NE&SW	22W-7N	7
High Ridge Rd. E&W	22W-7N	7
Hill St. N&S	23W-6N,23W-7N	6
Hillside Dr. NE&SW	23W-7N	6
Howard Av. N&S	23W-7N	6
Hudson Ct. N&S	24W-7N	6
Hunter Terr. E&W	24W-7N	6
Hygate Dr. E&W	24W-7N	6
Irving Park Rd. NW&SE	24W-7N,22W-7N	6
Kensington Ct. NW&SE	25W-6N	6
Kingston Ct. N&S	24W-6N	6
Kipling Ct. NW&SE	22W-6N	7
Lake Pointe Dr. N&S	22W-7N	7
Lake Shore Dr. N&S	22W-7N	7
Lake St. NW&SE	24W-6N	6
Lakeside Dr. NE&SW	23W-7N	6
Lakeview Ct. E&W	22W-7N	7
Lawrence Av. E&W	22W-7N	7
Leawood Dr. N&S	24W-7N	6
Lexington Av. N&S	22W-6N	7
Lincoln St. N&S	23W-7N	6
Locust Ln. N&S	22W-7N	7
Longford Ct. N&S	24W-7N	6
Longford Dr. E&W	24W-7N	6
Main St. NW&SE	23W-7N	6
Maple Av. E&W	23W-7N	6
Marion St. N&S	23W-7N	6
Meade Ln. N&S	24W-6N	6
Mensching Rd. NW&SE	24W-7N,23W-6N	6
Merimac Ct. E&W	24W-7N	6
Middleton Dr. N&S	24W-6N	6
Monaco Dr. N&S	23W-7N	6
Morningside Ct. E&W	22W-7N	7
Morningside Dr. N&S	22W-7N	7
Mulford Ln. N&S	24W-7N	6
Nassau Dr. E&W	24W-6N	6
Newcastle Ct. E&W	24W-7N	6
Newcastle Dr. E&W	24W-7N	6
Newport N. E&W	24W-6N	6
Newport S. E&W	24W-6N	6
Newport W. N&S	24W-6N	6
Norman Ln. NE&SW	24W-6N	6
Northhampton Ln. N&S	25W-6N	6
Oak St. N&S	23W-6N	6
Old Farm Dr. E&W	24W-6N	6
Olympia Ln. E&W	24W-7N	6
Orchard Ct. E&W	22W-6N	7
Orchard Terr. E&W	23W-6N,22W-6N	6
Oxford Cir. E&W	25W-6N	6
Oxford Pl. NE&SW	24W-6N	6
Park Ln. N&S	23W-7N	6
Park St. N&S	23W-6N	6
Parkview Ct. E&W	22W-7N	7
Pembrook Ct. E&W	25W-7N	6
Picton Rd. E&W	23W-6N	6
Pierce Av. N&S	22W-6N	7
Pine Av. E&W	24W-7N,23W-7N	6
Pinecroft Dr. N&S	22W-6N,22W-7N	7
Pioneer Ct. NW&SE	24W-7N	6
Plum Grove Rd. N&S	22W-7N	7
Plumtree Ln. NW&SE	22W-7N	7
Portwine Rd. N&S	22W-7N	7
Prescott Dr. N&S	24W-6N	6
Presidential Dr. N&S	24W-7N	6
Prospect St. N&S	23W-7N	6
Quincy Ct. N&S	24W-7N	6
Quincy Dr. E&W	24W-7N	6
Radnor Dr. N&S	24W-6N	6
Red Maple Ln. N&S	24W-6N	6
Regency Ct. E&W	25W-6N	6
Reston Ct. NE&SW	24W-7N	6
Richmond Ct. N&S	24W-7N	6
Richmond Dr. E&W	24W-7N	6
Ridge Ct. N&S	22W-6N	7
Ridge Rd. N&S	22W-6N,22W-7N	7
Ridgefield Dr. N&S	24W-6N	6
Robin Ct. N&S	23W-6N	6
Rodenburg Rd. N&S	25W-7N,24W-6N	6
Romford Ct. N&S	24W-6N	6
Rose Dr. E. N&S	22W-7N	7
Rose Dr. W. N&S	22W-7N	7
Rosedale Av. N&S	23W-6N	6
Roselle Rd. N&S	23W-6N,23W-7N	6
Rosemont Av. E&W	23W-7N	6
Rosewood Ct. NE&SW	24W-6N	6
Rosewood Dr. N&S	24W-7N	6
Rosner Dr. E&W	22W-7N	7
Royce Ln. NE&SW	24W-7N	6
Rt. 19 NW&SE	23W-7N,22W-7N	6
Rush St. N&S	23W-6N	6
Sally Ct. N&S	23W-6N	6
Salt Creek Rd. N&S	22W-7N	7
Scott Ct. NW&SE	23W-6N	6
Shagbark Ln. NE&SW	24W-6N	6
Siems Cir. E&W	23W-6N	6
Singleton Dr. E&W	24W-7N	6
Spring Av. N&S	23W-7N	6
Spring Ct. NE&SW	23W-7N	6
Spring Hill Dr. N&S	22W-6N,22W-7N	7
Spring St. N&S	23W-6N,23W-7N	6
Springfield Ct. E&W	24W-6N	6
Springfield Dr. N&S	24W-6N	6
Stafford Dr. NE&SW	25W-6N,24W-6N	6
Stevenson Ct. E&W	22W-7N	7
Stockport Ct. N&S	25W-7N	6
Stonefield Pl. E&W	24W-6N	6
Stonehurst Dr. E&W	24W-7N	6
Stonehurst Ln. E&W	24W-7N	6
Summerdale Ln. N&S	24W-6N	6
Summerfield Ct. NW&SE	24W-6N	6
Summerfield Dr. NE&SW	24W-6N	6
Sunnyside Rd. E&W	22W-6N	7
Sunrise Ln. E&W	24W-6N	6
Sunrise Pl. N&S	24W-6N	6
Terry Ct. E&W	22W-7N	7
Terry Dr. N&S	22W-7N	7
Thames Terr.* E&W	25W-7N	6
Thorndale Av. E&W	23W-7N	6
Town Acres Ln. E&W	23W-7N	6
Turner Av. E&W	23W-6N,22W-6N	6
Ventura Club Dr. NE&SW	22W-7N	7
Wadsworth Cir. E&W	25W-6N	6
Wadsworth Cir.* E&W	25W-7N	6
Walnut Ct. E&W	23W-6N	6
Walnut St. E&W	23W-6N	6
Walter Dr. N&S	23W-7N	6
Waterbury Ln. NW&SE	25W-7N,24W-6N	6
Waterford Ct. E&W	24W-7N	6
West End Rd. N&S	24W-7N,23W-6N	6
Westminster Cir. N&S	25W-6N	6
White Oak Dr. N&S	22W-6N	7
Wildwood Dr. N&S	24W-7N	6
Willow St. N&S	23W-7N	6
Windsor Dr. N&S	24W-7N	6
Winterwood Ct. N&S	24W-6N	6
Winterwood Dr. E&W	24W-6N	6
Woodside Dr. N&S	24W-6N	6
Woodworth Pl. E&W	23W-7N	6
York Ln. NE&SW	22W-6N	7

Schaumburg*
(Also see Cook County)

Street	Grid	Page
Mitchell Blvd. NE&SW	24W-7N	6

St. Charles*
(Also see Kane County)

Street	Grid	Page
North Av. E&W	32W-3N	10
Oak Rd. N&S	32W-3N	10

Villa Park

Street	Grid	Page
Adams St. E&W	18W-0S,18W-0N	20
Addison Rd. N&S	18W-1N	14
Adele Av. E&W	18W-2N	14
Adele Ct. E&W	17W-2N	14
Ardmore Av. N&S	18W-0S,18W-1N	20
Armitage Av. E&W	18W-2N	14
Astor Ct. E&W	17W-0N	14
Beach St. E&W	17W-1N	14
Belden Way E&W	18W-2N	14
Beverly Av. N&S	17W-1N,17W-2N	14
Bierman Av. N&S	18W-1N	14
Blackstone Pl. E&W	17W-0S	21
Calduto Cir. E&W	17W-0S	21
Central Blvd. NE&SW	18W-0N,17W-0N	14
Charles Av. N&S	17W-1N	14
Chatham Av. N&S	17W-1N,17W-2N	14
Congress St. E&W	17W-0S	21
Cornell Av. N&S	17W-0S,17W-1N	21
Crescent Dr. NE&SW	17W-0N	14
Cross St. NW&SE	17W-0S	21
Division St. E&W	18W-1N,17W-1N	14
Douglas Av. N&S	17W-1N,17W-2N	14
Ellsworth Av. N&S	17W-1N,17W-2N	14
Elm St. E&W	18W-1N,17W-1N	14
Euclid Av. N&S	17W-0S,17W-1N	21
Evergreen Dr. E&W	18W-0S	20
Fairfield Av. E&W	17W-0N	14
Frank St. E&W	17W-0S	21
Fulton Av. N&S	17W-1N,17W-2N	14
Gerald Av. N&S	17W-1N	14
Grant Av. N&S	17W-0S,17W-0N	21
Hamilton Av. N&S	17W-1N	14
Harrison St. E&W	17W-0S	21
Harvard Av. N&S	18W-0S,18W-2N	20
Highland Av. E&W	17W-0N	14
Holly Ct. E&W	18W-1N	14
Home Av. E&W	18W-0N,17W-0N	14
Hugo Ct. NW&SE	18W-1N	14
Illinois Av. N&S	17W-0N,17W-1N	14
Iowa Av. N&S	18W-1N	14
Jackson St. E&W	18W-0S,17W-0S	20
James St. E&W	18W-1N	14
Julia Dr. NE&SW	17W-0S	21
Kenilworth Av. E&W	18W-0N,17W-0N	14
Kolberg Ct. E&W	17W-0S	21
Lane Dr. E&W	17W-0S	21
Leslie Ln. N&S	18W-0S	20
Lincoln Av. N&S	18W-1N	14
Madison St. E&W	18W-0N,17W-0N	14
Main St. E&W	17W-0N	14
Maple St. E&W	17W-1N	14
Merle St. E&W	18W-1N	14
Michigan Av. N&S	18W-0S,18W-1N	20
Michigan St. E&W	18W-1S	20
Mission Av. N&S	18W-1N	14
Monroe St. E&W	18W-0S,17W-0S	20
Monterey Av. N&S	17W-0S,17W-0N	21
Morningside Dr. E&W	18W-0S	20
Myrtle Av. N&S	17W-0S,17W-0N	21
North Av. E&W	18W-2N,17W-2N	14
Oak Ct. E&W	17W-1N	14
Oak St. E&W	17W-1N	14
Oakland Av. N&S	17W-0S,17W-0N	21
Orchard Hill Ct. N&S	17W-0S	21
Ovaltine Ct. E&W	17W-0N	14
Park Blvd. E&W	18W-0N,17W-0N	14
Peony Pl. N&S	17W-0N	14
Pine St. E&W	17W-1N	14
Pleasant Av. E&W	18W-1N	14
Plymouth St. E&W	18W-1N,17W-1N	14
Princeton Av. N&S	18W-0N,18W-2N	14
Rand Rd. NS&EW	18W-0S	20
Ridge Rd. E&W	18W-1N	14
Riordan Rd. NW&SE	17W-0S	21
Riverside Ct. E&W	17W-0N	14
Riverside Dr. N&S	17W-0N	14
Robert Kingery Expressway N&S	17W-0S,17W-0N	21
Roosevelt Rd. E&W	18W-0S,17W-0S	20
Roy Dr. E&W	18W-1N	14
Rt. 38 E&W	18W-0S,17W-0S	20
Rt. 64 E&W	18W-2N,17W-2N	14
Rt. 83 N&S	17W-0S,17W-0N	21
Salt Creek Dr. NW&SE	16W-0S	21
School St. E&W	18W-0N	14
Sidney Av. E&W	18W-2N	14
Sidney Ct. E&W	17W-2N	14
St. Charles Rd. E&W	18W-1N,17W-1N	14
Stone Rd. NW&SE	18W-1N	14
Summit Av. N&S	17W-0S,17W-1N	21
Sunset Ct. E&W	18W-1N	14
Sunset Dr. E&W	18W-1N,17W-1N	14
Terrace St. E&W	18W-1N	14
Terry Ln. NE&SW	17W-0S	21
Van Buren St. E&W	17W-0S	21
Vermont St. E&W	18W-1N,17W-1N	14
Villa Av. N&S	17W-0S,17W-2N	21
Washington St. E&W	18W-0N,17W-0N	14
Wayside Dr. N&S	17W-0S	21
Westmore Av. N&S	18W-1N	14
Westwood Av. N&S	19W-1N	14
Wildwood St. E&W	17W-0N	14
Willowcrest Dr. NE&SW	18W-0S	20
Wisconsin Av. N&S	18W-1N,18W-0S	14
Woodrow St. E&W	18W-0N	14
Yale Av. N&S	18W-0N,18W-2N	14
2nd Av. N&S	18W-1N	14
3rd Av. N&S	18W-1N	14

Warrenville

Street	Grid	Page
Albert Einstein Dr. E&W	29W-2S	16
Albright St. NW&SE	29W-3S,28W-3S	16
Amber Ln. E&W	29W-3S	16
Angeline Ct. N&S	30W-2S	16
Arbury Ct. E&W	30W-2S	16
Arlington Ct. NW&SE	30W-2S	16
Arthur Compton Ct. N&S	29W-2S	16
Ascot N&S	29W-1S,29W-2S	16
Attleboro Ct. NW&SE	30W-2S	16
Aurora Way NE&SW	29W-3S	16
Avon Ct. E&W	30W-1S	16
Avondale Ct. E&W	30W-2S	16
Barclay Av. N&S	30W-3S,30W-4S	16
Barkley Av. N&S	30W-3S	16
Batavia Rd. NW&SE	30W-2S,28W-3S	16
Bayview Ct. N&S	29W-3S	16
Bedford Ln. N&S	30W-2S	16
Birchwood Ln. N&S	29W-3S	16
Blackthorn Ln. N&S	29W-3S	16
Branch Av. E&W	30W-2S	16
Briarwood Dr. N&S	29W-3S	16
Briggs Av. N&S	30W-3S	16
Brighton Ct. E&W	30W-2S	16
Bristol Ln. N&S	30W-1S,30W-2S	16
Brookside Ct. NW&SE	30W-3S	16
Buckthorn Ct.* E&W	30W-2S	16
Burk Av. NE&SW	29W-3S,28W-3S	16
Butterfield Rd. E&W	30W-3S,28W-3S	16
Calumet Av. E&W	30W-3S,28W-3S	16
Candlewood Ln. E&W	29W-3S	16
Cedar Ct. N&S	30W-1S	16
Central Av. E&W	27W-3S	17
Cerny Rd. N&S	29W-2S	16
Cherice Dr. N&S	30W-2S	16
Circle Dr., E. N&S	28W-3S	17
Circle Dr., S. E&W	28W-3S	17
Circle Dr., W. N&S	28W-3S	17
Concord Ct. E&W	30W-2S	16
Continental Dr. N&S	30W-2S	16
Cottonwood* NE&SW	30W-2S	16
Country Ridge Dr. NW&SE	29W-2S	16
Crabtree Ln. NE&SW	29W-3S	16
Curtis Av. N&S	28W-4S,28W-3S	23
Cynthia Ct. N&S	30W-2S	16
Cynthia Dr. N&S	30W-2S	16
Danbury Dr.* E&W	30W-2S	16
Dedham Ct. E&W	30W-2S	16
Deerfield N&S	30W-1S,30W-2S	16
Diehl Rd. E&W	28W-4S,27W-4S	23
Dogwood Ct.* NE&SW	30W-2S	16
Dorchester Ct. E&W	30W-2S	16
East-West Tollway E&W	29W-4S,28W-4S	23
Elizabeth Av. N&S	27W-3S	17
Elmwood Ct. E&W	30W-2S	16
Emerald Green Dr. N&S	29W-2S	16
Essex N&S	30W-1S	16
Estes St. E&W	30W-3S	16
Everett Ct. N&S	29W-2S	16
Fairfax Ct. E&W	30W-1S	16
Ferry Rd. E&W	29W-4S,28W-4S	23
Forestview Av. E&W	28W-4S	23
Forestview Dr. N. E&W	29W-2S	16
Foxboro Ct. E&W	30W-2S	16
Frontage Rd. E&W	30W-3S,29W-3S	16
Galbreath E&W	30W-2S,29W-2S	16
Galusha Av. E&W	27W-3S	17
Gates Pl. NE&SW	28W-3S	17
Glen Dr. NE&SW	29W-3S	16
Glenhurst Ct. E&W	30W-2S	16
Greenbriar Ln. E&W	29W-3S	16
Greenbrook Ct. NW&SE	30W-2S	16
Greenview Av. E&W	27W-3S	17
Grove Ln. NS&EW	29W-2S,29W-3S	16
Hampton* N&S	30W-2S	16
Harvard Dr.* N&S	30W-2S	16
Hawthorne Ln. N&S	29W-3S	16
Haylett Av. N&S	28W-3S	17
Heather Ct. NE&SW	30W-2S	16
Herrick Rd. N&S	27W-3S	17
Home Av. N&S	29W-3S	16
Huntington Dr. E&W	30W-1S	16
Hurlingham Ct. N&S	29W-2S	16
Hurlingham Dr. E&W	30W-2S,29W-2S	16
Interstate 88 E&W	28W-4S,27W-4S	23
Iroquois Ct. N. E&W	29W-2S	16
Jackson St. E&W	28W-3S	17
Jefferson St. E&W	28W-3S	17
John Bardeen Dr. N&S	29W-2S	17
Juniper Ct. N&S	30W-1S	16
Kensington Dr. E&W	30W-2S	16
Lakeview Ct. N&S	29W-3S	16
Lakeview Dr. E&W	30W-3S,29W-3S	16
Landon Av. N&S	29W-3S	16
Laurel Ct. NW&SE	30W-1S	16
Leominster Ct. E&W	30W-2S	16
Lexington Ct. N&S	30W-2S	16
Linden Sq.* NS&EW	30W-2S	16
Lindenwood Dr.* NW&SE	30W-2S	16
Lorraine Av. N&S	27W-4S,27W-3S	23
Lynn Ct. N&S	30W-2S	16
Mack Rd. E&W	30W-1S,29W-1S	16
Main St. E&W	28W-3S	17
Manchester N&S	30W-1S	16
Manning Av. NE&SW	28W-3S	17
Maple Ct. NW&SE	30W-2S	16
Maplewood* NW&SE	30W-2S	16
Maplewood Ct.* NE&SW	30W-2S	16
Marie Curie Ln. E&W	29W-2S	16
Mayfair Ct. NE&SW	30W-1S	16
Meadow Av. E&W	29W-3S	16
Melcher Av. N&S	27W-3S	17
Melvin Albright St. E&W	28W-3S	17
Middleton Av. E&W	27W-4S	23
Mignin Dr. N&S	29W-3S	16
Mill St. N&S	27W-4S	23
Mount St. NW&SE	28W-3S	17
Mulberry Ct.* E&W	30W-2S	16
Oakwood Ct.* E&W	30W-2S	16
Oxford E&W	30W-1S	16
Parkview Av. E&W	27W-3S	17
Patterman Rd. NE&SW	29W-3S	16
Penny Ln. E&W	30W-1S	16
Pierre Curie Ln. NE&SW	29W-2S	16
Plum Ct. NE&SW	30W-1S	16
Point Oak Dr. N&S	29W-3S	16
Prairie Av. E&W	29W-3S	16
Princeton Ct. N&S	30W-1S	16
Ray St. NW&SE	28W-3S	17
Raymond Dr. N&S	29W-4S	23
Redwood Ct. N&S	30W-1S	16
Ridge Dr. E&W	29W-2S	16

Grid | Page

Ridgewood Ct. E&W 30W-2S 16
Rigi Rd. (pvt.) NE&SW 30W-3S 16
River Alley NE&SW .. 28W-3S 17
River Oaks Dr. NS&EW 30W-2S 16
River Rd. NE&SW 30W-3S,28W-4S 17
Riverside Av. N&S .. 29W-2S 16
Riverside Pkwy. NS&EW 29W-2S 16
Riverview Dr. NW&SE 28W-3S 17
Rockwell St. NE&SW 29W-3S,28W-3S 16
Rogers Av. E&W .. 28W-3S 17
Route 56 E&W 30W-3S,28W-3S 16
Route 59 N&S 29W-1S,29W-3S 16
Roxbury Ct. N&S 30W-2S,29W-2S 16
Salem Ct. E&W . 30W-2S 16
Sanchez Dr. N&S .. 30W-2S 16
Shaw Dr. N&S 30W-2S 16
Sheffield N&S 30W-1S 16
Small Trees Ct. NW&SE 30W-2S 16
Sova Ln. N&S .. 29W-2S 16
Spruce Ct. NW&SE . 30W-1S 16
Stafford Pl. NE&SW .. 28W-3S 17
Steadman Av. N&S .. 28W-3S 17
Stevens Ct. E&W .. 29W-2S 16
Sunset Ct. E&W . 30W-3S 16
Sunset Dr. N&S 30W-3S 16
Talbot Av. N&S 30W-3S 16
Tanglewood Ln. E&W 29W-3S 16
Thornwood Ln. NE&SW 29W-3S 16
Timber Dr. NE&SW 29W-3S,28W-3S 16
Tinker Av. N&S 27W-3S 17
Townline Rd. N&S 29W-4S,28W-4S 23
Townline Rd. N&S 29W-3S,28W-3S 16
Tracy Pl. NE&SW 28W-3S 17
Virginia Av. N&S 27W-3S 17
Wagner Ct. N&S 28W-3S 17
Wagner Dr. E&W 29W-3S,28W-3S 16
Wagner Rd. N&S .. 28W-3S 17
Waltham Ct. N&S .. 30W-2S 16
Warren Av. N&S 28W-4S,28W-3S 23
Warren Ct. E&W 29W-3S 16
Warrenville Rd. E&W 28W-3S,27W-3S 17
Waverly Av. E&W 30W-2S,29W-2S 16
Webster St. NE&SW 29W-3S 16
Wembly Dr.* NS&EW 30W-2S 16
West Av. N&S 27W-3S 17
White Oak Dr. NE&SW 29W-3S 16
Whitehall Ct. E&W . 30W-3S 16
Wilbur Av. N&S 28W-3S,27W-3S 17
Wildwood Ct.* NE&SW 30W-2S 16
Williams Ct. NE&SW 29W-3S 16
Williams Rd. N&S 29W-3S,29W-2S 16
Willow Ln. E&W 30W-1S,29W-1S 16
Winchester Cir. E&W 29W-2S 16
Winfield Rd. N&S .. 28W-3S 17
Winwood Wk. E&W .. 29W-2S 16
Wood Ct.* NW&SE . 30W-2S 16
Woodland Rd. E&W .. 28W-4S 23
Woodlawn Av. E&W .. 28W-3S 17
Youghal Rd. E&W .. 30W-3S 16
2nd St. N&S 28W-3S 17
4th St. N&S 28W-3S 17

Wayne*
(Also see Kane County)
Army Trail Rd. E&W 32W-5N,30W-5N 4
Berkshire Ct. N&S .. 30W-4N 4
Billy Burns Rd. NW&SE 32W-5N 4
Bradford Pkwy. E&W 30W-4N 4
Cedar Ln. E&W....32W-5N 4
Chambellan Ln. N&S 32W-5N 4
Courcval Ln. NW&SE 32W-5N 4
Derby Rd. NE&SW .. 32W-5N 4
Elm Rd. N&S .. 31W-5N 4
Forsynthia Ln. E&W .. 32W-5N 4
Glos St. E&W 32W-5N 4
Guild Ln. N&S .. 32W-5N 4
Heritage Ct. N&S . 30W-5N 4
Honey Hill Cir. NS&EW 32W-4N 4

Honey Hill Dr. N&S .. 32W-5N 4
Honeysuckle Ln. 30W-4N 4
Kaelin Rd. N&S.....30W-5N 4
Keil St. N&S 32W-5N 4
Kingswood Ct. N&S .. 30W-4N 4
Lakewood Ct. NE&SW 30W-4N 4
Lysle Rd. N&S 30W-5N 4
Magnolia Ln. N&S . 30W-4N 4
Maple Tree Ln. E&W 30W-4N 4
Mountain Ash Dr. N&S 30W-4N 4
Oak Lawn Rd. NE&SW 32W-5N 4
Orchard Ln. E&W ...31W-5N 4
Percheron Ln. NS&EW 31W-5N 4
Peterson Dr. E&W . 31W-5N 4
Powis Rd. N&S 32W-5N,31W-4N 4
Pratt Rd. E&W 32W-5N 4
Railroad N&S...... 32W-5N 4
Ridge Ln. N&S 30W-5N 4
Rochefort Ln. E&W . 32W-6N 4
Russel Rd. E&W 31W-5N 4
School St. N&S 32W-5N 4
Serinne Ln. N&S .. 30W-4N 4
Somerset Ct. N&S . 30W-4N 4
Warwick Way E&W . 30W-4N 4
White Oak Ln. E&W 30W-4N 4
Will Way E&W .. 32W-5N 4
Wiltshire Ln. N&S .30W-4N 4

Wayne Township
Ancient Oaks Dr. N&S 28W-2N 11
Arbor Ln. E&W 30W-3N 10
Army Trail Rd. E&W 29W-5N 5
Ashburn Ln. E&W ...29W-4N 5
Augusta Dr. N&S ...30W-4N 4
Avard Rd. N&S 30W-4N 4
Bartlett Rd. N&S 29W-5N,29W-7N 5
Beech Ct. E&W 29W-5N 11
Brentwood Ct. E&W 29W-5N 5
Byron Av. E&W 29W-5N 5
Cambridge Dr. NW&SE 29W-4N 5
Canterbury Dr. E&W 29W-4N 5
Cape Av. E&W 29W-3N,29W-4N 11
Carriage Way N&S .. 29W-5N 5
Chatham Ct. NW&SE 29W-4N 5
Christy Ln. E&W .. 29W-2N 11
Churchill Ct. N&S . 28W-3N 11
County Club Ln. N&S 29W-4N 5
County Farm Rd. N&S 27W-2N 11
Crest Ct. N&S 27W-3N 11
Cul De Sac NE&SW . 29W-2N 11
Cuyahoga Terr. N&S 29W-2N,29W-5N 11
Dell Rd. N&S......29W-2N 11
Diversey Av. E&W .. 28W-3N 11
Diversey Pkwy. 30W-3N 10
Doral Dr. N&S 30W-4N 4
Eaton Way N&S .. 29W-4N 5
Elm Ct. N&S 30W-3N 10
Elm Rd. N&S 32W-3N 10
Ethel St. N&S 27W-2N 11
Fair Oaks Rd. N&S 28W-2N,28W-4N 11
Florence Ct. E&W....28W-2N 11
Garden Dr. E&W .. 29W-5N 5
Gerber Rd. NE&SW...28W-5N 5
Glen Rd. NW&SE .. 29W-5N 11
Hawthorn Ln. E&W .. 27W-2N 11
Hickory Knoll Ln. N&S 29W-3N 11
Hill Ct. E&W 29W-5N 5
Holly Ct. E&W 27W-3N 11
Indian Knoll Rd. NW&SE 28W-2N 11
Ingalton Av. NW&SE 29W-2N 11
Jefferson St. N&S .. 27W-2N 11
Jenlor Ct. E&W 31W-4N 4
Judith Ln. E&W 28W-4N 5
Keil St. N&S 32W-3N 10
Kenwood Av. E&W .. 28W-2N 11
Klein Rd. N&S 29W-3N,29W-4N 11
Lake Eleanor Ct. N&S 31W-4N 4
Lake Eleanor Dr. N&S 31W-4N 4
Lakeview Ct. N&S .. 28W-3N 11
Lakewood Dr. N&S .. 29W-3N 11
Lawrence Av. E&W .. 27W-6N 5
Lies Rd. E&W 30W-4N 4
Locust Av. N&S 30W-3N,30W-4N 10
Maple Ct. N&S 30W-3N 10
Mardon Rd. N&S 27W-2N 11
Martingale Av. E&W 29W-5N 5

Mead Rd. NW&SE .. 30W-3N 10
Meadow Lark Dr. N&S 29W-5N 5
Mercury Ct. E&W 29W-5N 5
Morningside Av. N&S 28W-2N,28W-3N 11
Morningside Dr. E&W 29W-5N 5
Morton Rd. N&S 27W-2N 11
Mulberry Dr. E&W . 30W-3N 10
Munger Rd. N&S 31W-4N,31W-7N 4
Norris Av. N&S 30W-3N,30W-4N 10
North Av. E&W 30W-3N,27W-2N 10
Oak Creek Ct. E&W 28W-2N 11
Oak Creek Dr. N&S .. 28W-2N 11
Oak Dr. NW&SE 30W-3N 10
Oak Knoll Rd. E&W . 29W-3N 11
Oak Ln. E&W 29W-3N 11
Oak Rd. N&S........32W-3N 10
Old St. Charles Rd. NW&SE 29W-2N 11
Old Wayne Ct. E&W 29W-3N 11
Orchard Ln. E&W .. 29W-5N 5
Pepper Ct. E&W .. 29W-4N 5
Petersdorf Rd. N&S 29W-4N,29W-5N 5
Pine Ct. NW&SE .. 28W-5N 5
Powis Rd. N&S 31W-3N,31W-7N 10
Prairie Ln. NW&SE .. 31W-4N 4
Prince Crossing Rd. N&S 29W-2N 11
Reque Rd. N&S 31W-2N,31W-3N 10
Ridgeview St. NS&EW 27W-3N 11
Riviera Ct. E&W 28W-4N 5
Sayer Rd. N&S......30W-7N 4
Schick Rd. E&W 30W-5N 4
Shady Ln. E&W 30W-3N 10
Shagbark Dr. .. 29W-3N 11
Smith Rd. E&W 31W-4N,29W-4N 4
Spicer Rd. N&S....30W-4N 4
Springvale Rd. N&S 27W-3N 11
St. Andrews Ln. E&W 30W-4N 4
St. Andrews Trace N&S 29W-4N 5
St. Charles Rd. E&W 29W-2N,27W-2N 11
St. James Way NW&SE 29W-4N 5
St. Mark Way N&S .. 29W-4N 5
St. Thomas Way E&W 29W-4N 5
Stearns Rd. E&W 32W-7N,28W-6N 4
Sunnydale Ct. ..29W-5N 5
Sunset Ridge Dr. E&W 29W-5N 5
Sycamore St. E&W .. 27W-3N 11
Timber Ln. E&W 28W-3N,27W-3N 11
Timberline R. N&S .. 29W-3N 11
Trieste Ln. E&W .. 28W-3N 11
Trinity Ln. NS&EW .. 29W-4N 5
Turnmill Ln. N&S 29W-4N 5
Vale Rd. E&W 29W-2N 11
Valewood Rd. N&S 29W-2N,29W-3N 11
Waterford Ln. E&W .. 29W-4N 5
Wayne Oaks Ln. N&S 28W-2N 11
Waynewood Dr. E&W 29W-2N 11
Wescot Ln. E&W .. 29W-4N 5
Whitney Rd. E&W .. 30W-4N 4
Wiant Rd. N&S 30W-3N,30W-4N 4
Wild Flower Ln. N&S 29W-3N 11
Wood Ct. E&W 30W-5N 4
Woodcreek Ln. N. N&S 28W-3N 11
Woodcreek Ln. S. N&S 28W-3N 11
Woodcrest Dr. NW&SE 29W-2N 11
Woodland Av. N&S 30W-3N,30W-3N 10
Woodland Ln. N&S .. 31W-4N 4
Woodland Trail South E&W 31W-4N 4
Woodland Trail West N&S 31W-4N 4
Woodview Ct. N&S .. 27W-3N 11
Wynn Av. E&W 29W-3N,28W-3N 11

West Chicago
Acorn Hill Ln. NW&SE 30W-2N 10
Allen Av. N&S 29W-0S,29W-0N 16
Amber Ct. E&W .. 29W-0N 11
Ann St. E&W 30W-0N 10
Applegate Av. N&S . 30W-1N 10
Arbor Av. N&S 30W-1N 10

Arthur Dr. E&W 31W-2N 10
Aster Ln. N&S 29W-2N 11
Atlantic Dr. N&S 30W-2N 10
Augusta Ct. E&W ...29W-0S 16
Aurora St. NW&SE .. 30W-0N 10
Avard Av. N&S 30W-0N 10
Barber Av. N&S 29W-0S,29W-0N 16
Bayberry Ct. E&W .. 30W-1N 10
Belleview Av. N&S .. 29W-1N 11
Bishop St. N&S 29W-0S,29W-0N 16
Blackhawk Dr. E&W 32W-2N,31W-2N 10
Blair St. E&W 30W-0N,29W-0N 10
Blakeley St. N&S .. 29W-0N 10
Brentwood Ct. E&W 30W-0S 16
Brentwood Dr. N&S .30W-0S 16
Brown St. E&W 30W-0N,29W-0N 10
Canterbury Ct. N&S .. 30W-2N 10
Carolina Dr. N&S .. 30W-2N 10
Carriage Dr. N&S .. 29W-0S 16
Carriage Dr. NW&SE 30W-0S 16
Center St. NE&SW .. 29W-0N 11
Charles Ct. N&S .. 31W-0N 10
Cherry Blossom Ct. N&S 30W-1N 10
Cherrywood Ln. N&S 29W-2N 11
Chicago St. NE&SW 29W-0N 11
Church St. N&S...... 30W-0N 10
Clara St. NW&SE .. 30W-0N 10
Claremont Av. N&S .. 29W-2N 11
Clayton St. N&S .. 29W-0N 11
Colford Av. E&W .. 29W-0N 11
Columbine Ln. NW&SE 29W-2N 11
Commerce Dr. E&W 31W-0N 10
Conde St. NW&SE .. 29W-0N 11
Coolidge Av. N&S .. 29W-0N 11
Crabapple Ct. E&W .. 30W-1N 10
Cranesbill Dr. NW&SE 29W-2N 11
Crestview Ln. N&S . 30W-2N 10
Crown Ct. NE&SW .. 31W-3N 10
Dale Av. E&W 29W-0S 16
Dayton Av. N&S 30W-0N 10
Downs Dr. E&W .. 31W-1N 10
Easton Av. E&W .. 29W-0N 11
Elite Av. N&S........29W-1N 11
Elizabeth St. N&S 29W-0S,29W-0N 16
Elliot Av. NW&SE .. 30W-1N 10
Elmwood St. E&W 30W-1N,29W-1N 10
Fabyan Pkwy E&W .. 32W-0N 10
Factory St. N&S 30W-0N 10
Fairbanks Ct. E&W . 30W-0N 10
Fairview Av. E&W .. 29W-0N 11
Fawell Dr. N&S...... 30W-1N 10
Fenton St. N&S 31W-0N 10
Forest Av. E&W 30W-0N,29W-0N 10
Fremont St. N&S .. 30W-1N 10
Fulton St. NW&SE .. 29W-0N 11
Galena NE&SW 29W-0N 11
Garden St. N&S .. 29W-1N 11
Garys Mill Rd. E&W . 29W-0S 16
Gates St. N&S 29W-0S,29W-0N 16
Geneva Rd. E&W .. 29W-0N 11
Geneva St. E&W 30W-0N,29W-0N 10
George St. E&W .. 30W-0N 10
Glen Av. E&W 29W-0S 16
Grand Lake Blvd. E&W 31W-1N,29W-1N 10
Green Meadow Dr. NW&SE 30W-2N 10
Grove Av. E&W....29W-1N 11
Hahndorf Rd. E&W . 30W-2N 10
Harrison St. NE&SW 30W-0N 10
Harvester Rd. E&W . 31W-2N 10
Hawthorne Ln. E&W 28W-2N,30W-2N 10
Hazel St. N&S 30W-0N,29W-0N 10
Helena Dr. NW&SE . 31W-2N 10
High St. NW&SE 29W-0N 11
Highland Av. N&S .. 29W-1N 11
Highland Av. E&W .. 29W-1N 11
Hill Ct. E&W 29W-0N 11
Hillview Av. E&W .. 29W-0N 11
Honeysuckle Av. N&S 30W-1N 10
Howard St. N&S .. 31W-1N 10
Industrial Dr. N&S .. 31W-1N 10
Ingalton St. N&S .. 29W-1N 11
Joliet St. N&S 29W-0S,30W-0N 16
Kenwood Av. N&S .. 30W-1N 10
Kings Cir. Cir. 30W-2N 10
Knollwood Ln. N&S .. 29W-2N 11
Kress Rd. N&S 30W-2N 10
Lake Side Dr. NS&EW 30W-2N 10
Lawrence Av. E&W .. 29W-0N 11

Lester St. E&W 30W-0N,29W-0N 10
Lincoln Av. N&S 30W-0N 10
Lorlyn Dr. NW&SE .. 29W-0S 16
Lyman Av. N&S 30W-0N 10
Main St. E&W 29W-0N 11
McChesney Rd. N&S 32W-0N 10
McConnell Av. E&W 30W-1N,29W-1N 10
Mellington Av. N&S .. 31W-0N 10
Milo Ct. N&S 30W-0N 10
National St. N&S 30W-1N,29W-1N 10
Neltnor Blvd. N&S 29W-0S,29W-1N 16
Nickelson Rd. N&S .. 29W-0N 11
Norris Av. N&S 30W-0N 10
North St. E&W 30W-2N 10
Northwest Av. E&W .31W-2N 10
Northwest Dr. N&S . 31W-2N 10
Nuclear Dr. N&S 31W-2N 10
Oak Av. N&S 29W-1N 11
Oak St. N&S 29W-0S,29W-0N 16
Oakwood Av. N&S 30W-0N,29W-1N 11
Orlyn Dr. N&S 29W-0S 16
Parker Av. N&S 29W-0N 11
Parkside Av. N&S .. 30W-0N 10
Peachtree Ln. E&W . 30W-1N 10
Pine St. E&W 29W-1N 11
Plumtree Ln. E&W . 30W-1N 10
Plymouth St. E&W . 29W-1N 11
Pomeroy St. E&W 30W-0N,29W-0N 10
Powis Ct. NW&SE 31W-2N,31W-3N 10
Powis Rd. N&S 31W-2N,31W-3N 10
Railroad Av. E&W .. 30W-0N 10
Ridgewood Ct. E&W 29W-2N 11
Ridgewood Ln. N&S 29W-2N 11
Roosevelt Rd. NW&SE 32W-0N,29W-0S 10
Rosewood Dr. NS&EW 29W-2N 11
Salem St. E&W .. 30W-2N 10
Sarana Av. N&S .. 29W-0S 16
Sherman St. N&S....30W-0N 10
Sophia St. N&S .. 29W-1N 11
Spencer St. E&W .. 29W-1N 11
Spring Cress Ln. NW&SE 29W-2N 11
Spur Row E&W 31W-2N 10
St. Andrew's Ct. E&W 30W-3N 10
Sterling Av. E&W .. 29W-0N 11
Stimmel St. E&W 30W-0N,29W-0N 10
Summit Av. E&W .. 29W-0N 11
Sycamore St. N&S . 29W-0N 11
Town Rd. N&S 30W-0N 10
Trillium Tr. N&S....29W-2N 11
Turner Av. N&S .. 29W-1N 11
Turner Ct. NW&SE 30W-0N,29W-0N 10
Vine St. N&S...... 30W-0N 10
Virgie Pl. E&W 29W-1N 11
Washington St. NE&SW 30W-0N,31W-0N 10
Waynewood Dr. N&S 30W-2N 10
Wegner Dr. NE&SW 31W-0N 10
Wendall St. E&W .. 29W-1N 11
West St. N&S 30W-2N 10
Western Dr. E&W...31W-1N 10
Weyrauch St. N&S .. 30W-0N 10
Wild Ginger Tr. NW&SE 29W-2N 11
Wildwood Ln. E&W . 29W-2N 11
Willow Creek Rd. N&S 30W-2N 10
Wilson Av. N&S 29W-0N 11
Windsor Dr. E&W .. 29W-0S 16
Winston St. E&W .. 30W-2N 10
Wood St. N&S 30W-0N 10
Woodland Av. N&S 30W-2N,30W-2N 10
Woodside Ct. NW&SE 30W-2N 10
Woodside Dr. NS&EW 30W-2N 10
Yale St. N&S...... 30W-1N 10
York Av. E&W 30W-1N,29W-1N 10

Westmont
Adams St. E&W 18W-6S,18W-5S 26
Ainslie Dr. NW&SE .. 17W-5S 27
Alpine Ln. N&S 18W-7S 26
Antler Ln. N&S 18W-6S 26
Arbor La. E&W 18W-7S 26
Arlington Av. E&W .. 17W-4S 27
Ashford Ln. NS&EW 17W-6S 27
Aspen Ln. S. N&S .. 18W-7S 26
Baltimore Av. NE&SW 16W-4S 27
Barry Ln. E&W 18W-6S 26

DU PAGE COUNTY 55

Street	Grid	Page
Bay Ct. E&W	18W-7S	26
Beechwood Ct. NE&SW	18W-6S	26
Beechwood Dr. E&W	18W-6S	26
Beninford Ln. N&S	17W-6S	27
Blackhawk Dr. N&S	17W-4S,17W-3S	27
Brookside Dr. E&W	18W-6S	26
Brown Deer Dr. N&S	18W-3S	20
Brown Deer Tr. N&S	18W-3S	20
Buck Ct. N&S	18W-6S	26
Burr Oak Dr. N&S	17W-4S	27
Buttonwood Dr. N&S	18W-6S	26
Carlisle Av. NS&EW	18W-6S	26
Carriage Ln. N&S	18W-7S	26
Cass Av. N&S	18W-7S,18W-4S	26
Cass Ln. W. N&S	18W-7S	26
Cedar Ln. N&S	18W-7S	26
Champlaine Av. N&S	16W-4S	27
Chestnut Av. E&W	17W-4S	27
Chicago Av. E&W	18W-5S,17W-5S	26
Cove Ln. E&W	18W-7S	26
Creekwood Ct. N&S	18W-6S	26
Cromwell Ct. NE&SW	18W-6S	26
Cumnor Rd. N&S	18W-6S	26
Dallas St. E&W	18W-5S,17W-5S	26
Deercreek Ln. N&S	18W-6S	26
Deming Pl. NW&SE	17W-6S	27
Des Moines St. E&W	18W-5S,17W-5S	26
Doe Cir. NS&EW	18W-6S	26
Echo Ln. N&S	18W-7S	26
Elmwood Ct. NE&SW	18W-6S	26
Executive Ct. NE&SW	16W-3S	21
Fairway Ln. E&W	18W-7S	26
Fernwood Ct. NE&SW	18W-6S	26
Fernwood Dr. NS&EW	18W-6S	26
Fordham Way NE&SW	17W-6S	27
Fountainhead Dr. E&W	18W-6S	26
Franklin St. N&S	17W-4S	27
Garrett St. NS&EW	18W-6S	26
Grant St. N&S	18W-6S,18W-5S	26
Hamilton Av. E&W	17W-4S	27
Hamilton Ct. E&W	17W-4S	27
Hampton Av. E&W	18W-6S	26
Health Ln. N&S	17W-6S	27
Heath Pl. NW&SE	17W-6S	27
Hiddenview Dr. E&W	17W-6S	27
Honeywood Dr. N&S	18W-6S	26
Hudson St. N&S	18W-5S	26
Independence Av. E&W	16W-4S	27
Indian Bdry Dr. N&S	18W-3S	20
Indian Tr. E&W	18W-3S	20
Irving St. E&W	18W-5S,17W-5S	26
Ivy Ct. E&W	18W-7S	26
James Dr. E&W	18W-6S	26
Jameson Way NW&SE	17W-6S	27
Jamestown Av. E&W	17W-3S	21
King Arthur Ct. N&S	18W-6S	26
Lafayette Ln. E&W	17W-4S	27
Lake Ct. N&S	18W-6S	26
Lake Dr. E&W	18W-7S	26
Lincoln St. N&S	18W-5S,18W-4S	26
Linden Av. N&S	17W-6S,17W-5S	27
Lindley St. N&S	18W-6S	26
Longford Dr. NE&SW	17W-6S	27
Maple N&S	18W-7S	26
Maple Av. NE&SW	18W-5S	26
Melrose Av. E&W	18W-4S,17W-4S	26
Monticello Av. E&W	17W-4S	27
Morning Glory Cir. N&S	18W-4S	26
Naperville Rd. NE&SW	18W-5S,17W-4S	26
Nelson Ln. E&W	18W-7S	26
New Port Av. N&S	16W-4S	27
Norfolk St. E&W	18W-5S,17W-5S	26
Oak Av. N&S	17W-5S	27
Oak Hill Dr. E&W	16W-4S	27
Oak St. N&S	17W-6S	27
Oakbrook Hills Dr. E&W	17W-3S	21
Oakley Ct. N&S	18W-3S	20
Oakley Dr. E&W	18W-3S	20
Oakley Ln. N&S	18W-3S	20
Oakmont Ln. E&W	17W-3S,16W-3S	21
Oakmont Plaza Dr. E&W	16W-3S	21
Oakwood Ct. NE&SW	17W-3S	21
Oakwood Dr. N&S	17W-4S,17W-3S	27
Office Dr. N&S	18W-3S	20
Oliver Ct. E&W	18W-6S	26
Orleans Pl. N&S	17W-4S	27
Park Ln. N&S	18W-7S	26
Park St. N&S	18W-6S,18W-4S	26
Parkway Av. E&W	18W-4S	26
Parkway Dr. NE&SW	18W-4S	26
Pasquinelli Dr. NW&SE	16W-4S,16W-3S	27
Philadelphia Av. E&W	17W-4S	27
Pier Dr. E&W	18W-7S	26
Plaza Dr. E&W	17W-4S	27
Prairie Dr. E&W	18W-3S	20
Quail Ridge Dr. E&W	16W-3S	21
Quebec Pl. N&S	17W-4S	27
Quincy Av. E&W	18W-5S,17W-5S	26
Raintree Ct. NW&SE	18W-6S	26
Raintree Ln. N&S	18W-6S	26
Revere Av. E&W	17W-4S,16W-4S	27
Revere Ct. N&S	17W-4S	27
Richmond Av. N&S	17W-6S,17W-4S	27
Richmond St. E&W	18W-5S,17W-5S	26
Robinson Ln. E&W	18W-6S	26
Rosewood Ct. E&W	17W-4S	27
Roslyn Rd. N&S	18W-5S,18W-4S	26
Rumsey Pl. NW&SE	17W-6S	27
Schramm Ct. NE&SW	18W-7S	26
Schramm Dr. N&S	18W-7S	26
Spruce Ln. N&S	18W-7S	26
Suffield Ct. N&S	18W-7S	26
Tartan Lakes Cir. E&W	17W-3S	21
Tartan Lakes Ct. E&W	17W-3S	21
Tartan Lakes Dr. N&S	17W-3S	21
Tartan Lakes Way N&S	17W-3S	21
Tower Ln. E&W	17W-5S	27
Tracy Ct. E&W	18W-7S	26
Traube Av. E&W	18W-4S,17W-4S	26
Tudor Ln. N&S	18W-7S	26
Vail Dr. N&S	18W-7S	26
Vandustrial Ct. E&W	18W-6S	26
Vandustrial Dr. N&S	18W-6S	26
Virginia Dr. N&S	18W-7S	26
Warwick Av. N&S	17W-6S,17W-4S	27
Washington St. N&S	18W-6S,18W-4S	26
Wesley Ct. N&S	18W-7S	26
Wesley Ln. NW&SE	18W-7S	26
West End Av. N&S	18W-5S	26
Westmont Dr. N&S	17W-4S	27
Whipple La. E&W	18W-7S	26
White Birch Ln. E&W	18W-3S	20
Willard Pl. E&W	18W-4S	26
Williams St. N&S	18W-7S,18W-5S	26
Williamsburg Dr. N&S	17W-3S	21
Williamsport Dr. NE&SW	18W-7S	26
Willow La. N&S	18W-7S	26
Willow Oak Dr. N&S	17W-4S	27
Willow Way E&W	18W-7S	26
Wilmette Av. N&S	17W-6S,17W-4S	27
35th St. N&S	17W-3S	21
55th Pl. E&W	18W-6S	26
55th St. E&W	18W-6S,17W-6S	26
56th Pl. E&W	18W-6S,17W-6S	26
56th St. E&W	18W-6S,17W-6S	26
57th St. E&W	18W-6S,17W-6S	26
58th St. E&W	18W-6S,17W-6S	26
60th St. E&W	18W-6S,17W-6S	26
61st St. E&W	18W-6S	26
64th St. E&W	18W-7S	26
65th St. E&W	18W-7S	26
66th St. E&W	18W-7S	26

Wheaton

Street	Grid	Page
Aberdeen Ct. E&W	26W-1S	18
Academy Ln. NE&SW	25W-1S	18
Acorn Ct. NE&SW	23W-3S	18
Adare Dr. N&S	26W-1S	18
Albright Ct. NE&SW	23W-3S	18
Albright Ln. NW&SE	23W-3S	18
Alchester Dr. E&W	24W-1S	18
Appaloosa Ct. E. E&W	23W-3S	18
Appaloosa Ct. W. E&W	23W-3S	18
Appleby Ct. N&S	23W-3S	18
Appleby Dr. NS&EW	23W-3S	18
Arbor Av. E&W	25W-1S,24W-1S	18
Arbor Ct. E&W	25W-1S	18
Ardmore Ln. E&W	26W-1S	18
Ascot Ct. NW&SE	23W-3S	18
Ashburn Ct. N&S	23W-2S	18
Ashford Ct. NE&SW	23W-1S	18
Ashton Ct. E&W	23W-0N	12
Aurora Way N&S	25W-1S,25W-0S	18
Avery Av. E&W	23W-0S	18
Barger Ln. E&W	26W-2S	18
Bates St. N&S	26W-0S	18
Beecher NW&SE	25W-0S	18
Beechwood Ct.* E&W	26W-2S	18
Belleau Woods Dr. E&W	26W-1S	18
Berkshire Pl. N&S	25W-2S	18
Birch N&S	25W-0S	18
Blackburn St. NE&SW	23W-1S	18
Blacksmith Dr. NS&EW	23W-3S	18
Blanchard Ct. N&S	23W-0N	12
Blanchard St. N&S	24W-2S,23W-0S	18
Borough Ct.* E&W	23W-2S	18
Bradford Dr. NE&SW	23W-3S	18
Brandon Dr. N&S	24W-1S	18
Brentwood Ct. NW&SE	23W-2S	18
Brentwood Ln. NS&EW	23W-2S	18
Briar Cove N&S	26W-1S	18
Briarbrook Dr. NE&SW	23W-2S	18
Briarcliffe Blvd. N&S	23W-2S,23W-1S	18
Bridge St. N&S	25W-0S	18
Bridle Ln. E&W	24W-0N	12
Brighton Dr. E&W	25W-2S,24W-2S	18
Bristol Ct. NW&SE	23W-3S	18
Brittany Ct. N&S	24W-1S	18
Brookside Cir. NE&SW	24W-0N	12
Browning Ct. E&W	24W-1S	18
Buckingham Dr. NS&EW	25W-2S	18
Buena Vista Dr. NW&SE	23W-1S	18
Bunker Hill Ct. NW&SE	23W-2S	18
Burnham Pl. N&S	25W-2S	18
Burning Trail N&S	26W-1S	18
Burning Trail Ct. N&S	25W-1S	18
Burr Oak Ct. NE&SW	26W-1S	18
Butterfield Rd. E&W	26W-2S,23W-2S	18
Byron Ct. N&S	23W-0S	18
Cadillac Dr. E&W	24W-0N	12
Cambridge Ln. N&S	24W-2S,23W-2S	18
Campbell Av. N&S	24W-1S,24W-0S	18
Campbell Ct. E&W	24W-1S	18
Canterbury Pl. N&S	25W-2S	18
Cantigny Way N&S	26W-1S	18
Carlson Ln. E&W	24W-0N	12
Carlton Av. N&S	25W-0S,25W-0N	18
Casa Solana Dr. NE&SW	23W-2S	18
Castbourne Ct. E&W	23W-1S	18
Castlewood Dr. N&S	24W-1S	18
Caxton Dr. E&W	26W-1S	18
Center Av. E&W	25W-1S	18
Charles St. E&W	26W-0S	18
Charlotte Ct. NW&SE	26W-2S	18
Chase St. N&S	24W-1S,24W-0S	18
Chatham Dr. N&S	25W-2S	18
Chelsea Ct. N&S	26W-2S	18
Cherry Ct. N&S	24W-0N	12
Cherry St. N&S	24W-0N	12
Cheshire Ct. NW&SE	23W-2S	18
Cheshire Ln. N&S	23W-2S	18
Chesterfield Av. NW&SE	23W-0S	18
Childs Ct. E&W	26W-0S	18
Childs St. E&W	26W-0S,25W-0S	18
Circle Av. N&S	25W-0S	18
Clifden Ct. NW&SE	26W-1S	18
Clinton Ct. N&S	25W-0S	18
Clydesdale Dr. NS&EW	23W-3S	18
Cole Av. E&W	24W-0N	12
College Av. NE&SW	24W-0S,23W-0S	18
College Hill Av. E&W	24W-0S,23W-0S	18
College Ln. E&W	23W-2S	18
Coloma Ct. N&S	23W-1S	18
Coloma Pl. N&S	23W-1S	18
Colonial Sq. E&W	25W-2S	18
Community Dr. NE&SW	26W-1S	18
Coolidge Av. E&W	23W-1S	18
Cordova Ct. E&W	24W-2S,23W-2S	18
Countryside Dr. E&W	24W-0N	12
County Farm Rd. N&S	26W-1S,26W-0S	18
Courtenay Dr. NE&SW	25W-1S	18
Cove Ct. N&S	25W-1S	18
Creekside Cove N&S	26W-1S	18
Creekside Dr. NS&EW	26W-1S	18
Crescent St. E&W	24W-0S	18
Crest St. N&S	25W-1S,25W-0S	18
Cromwell Dr. N&S	25W-2S	18
Cross St. N&S	24W-0S,24W-0N	18
Croydon Ct.* E&W	23W-2S	18
Culham Ct. NE&SW	23W-3S	18
Daly Rd. E&W	24W-0N,23W-0N	12
Darling St. N&S	25W-0N	12
Dartmouth Dr. Cir.	25W-1S	18
Darwin Ct. NE&SW	23W-1S	18
Darwin Ln. E&W	23W-1S	18
Davis Ct.* E&W	26W-2S	18
Dawes Av. E&W	24W-1S,23W-1S	18
Delles Rd. N&S	25W-1S	18
Depot Pl. NE&SW	23W-0S	18
Derby Ct. N&S	23W-3S	18
Devonshire Ln. E&W	24W-1S	18
Dickinson Dr. NS&EW	24W-2S	18
Doncaster Ct. E&W	23W-2S	18
Dorchester Av. N&S	25W-0S	18
Dorset Dr. NS&EW	25W-2S	18
Downing Ct. E&W	23W-3S	18
Driving Park Rd. N&S	24W-0N	12
Dundee Dr. N&S	23W-2S	18
Dunlay N&S	23W-3S	18
Durham Dr. NS&EW	23W-3S	18
Eagle Ct. E&W	25W-2S	18
East St. N&S	24W-0S	18
Eaton Ct. E&W	23W-1S	18
Eddy Ct. N&S	23W-0N	12
Edgewood Av. N&S	26W-0S	18
Edinburgh Ct. E&W	23W-1S	18
Ellis Av. N&S	25W-0S,25W-0N	18
Elm St. E&W	25W-1S,23W-1S	18
Elmwood St. N&S	26W-0S	18
Emerson Ct. E&W	24W-1S	18
Enis Ln. E&W	26W-1S	18
Ensenada Dr. NW&SE	24W-2S	18
Erie St. N&S	26W-0S	18
Essel Ct. E. E&W	23W-3S	18
Essel Ct. W. E&W	23W-3S	18
Evergreen St. E&W	25W-0S,23W-0S	18
Exeter Ct. E&W	23W-1S	18
Faff Cir. NS&EW	26W-0S	18
Faganel Ct. N&S	26W-1S	18
Fairway Ln. E&W	24W-1S	18
Falson Ct. E&W	24W-1S	18
Fanchon St. N&S	25W-0N	12
Farnham Ln. E&W	24W-1S	18
Farragut Ct. N&S	23W-2S	18
Farragut Ct. E&W	23W-2S	19
Farwell St. N&S	25W-0N	12
Fischer St. N&S	24W-1S,24W-0S	18
Folkstone Ct.* E&W	23W-2S	18
Foothill Ct. NW&SE	25W-1S	18
Foothill Dr. NE&SW	25W-1S	18
Forest Av. E&W	24W-0N,23W-0N	12
Fox Run E&W	26W-1S	18
Franklin St. E&W	25W-0S,24W-0S	18
Frazier Ct. E&W	25W-1S	18
Friars Ct. N&S	24W-1S	18
Front St. E&W	25W-0S,24W-0S	18
Gables Blvd. N&S	25W-1S,25W-0S	18
Gainesboro Ct. N&S	24W-1S	18
Gainesboro Dr. N&S	24W-1S	18
Gamon Rd. N&S	24W-1S	18
Garner Av. E&W	23W-0S	18
Gary Av. N&S	25W-0S,25W-0N	18
Gary Ct. N&S	25W-0N	12
Geneva Pl. E&W	24W-0N	12
Geneva Rd. E&W	24W-1N,23W-1N	12
George St. E&W	24W-1S	18
Gladstone Dr. N&S	25W-2S	18
Glencoe St. N&S	23W-0N	12
Glendale Av. NE&SW	23W-0S	18
Golden Pond Ln. N&S	25W-2S	18
Golf Cir. N&S	25W-1S	18
Golf Ln. N&S	25W-1S	18
Golf View Ct. N&S	23W-2S	18
Gone Away Ct. N&S	25W-1S	18
Gone Away Ln. E&W	26W-1S,25W-1S	18
Grant St. N&S	24W-1S	18
Greensboro Dr. N&S	25W-2S	18
Greenwood Dr. N&S	25W-1S	18
Gresham Cir. N&S	26W-1S	18
Grey Willow Rd. N&S	24W-0N	12
Grosvenor Cr. E&W	26W-1S,25W-1S	18
Groton Ct. E&W	23W-1S	18
Groton Ln. NW&SE	23W-1S	18
Hailshaw Ct.* E&W	23W-2S	18
Hale St. N&S	24W-1S,24W-0S	18
Harrison Av. E&W	25W-0N,23W-0N	12
Harwarden St. E&W	23W-0S	18
Haverhill Dr. N&S	24W-1S	18
Hawthorne Blvd. E&W	25W-0N,23W-0N	12
Hazelton Av N&S	26W-0S	18
Heatherbrook Ct. E&W	24W-1S	18
Heathrow Ct. E&W	23W-1S	18
Hemstock Av. N&S	26W-1S	18
Heritage Lake Dr. E&W	24W-2S,23W-2S	18
Hertford Ct. E&W	23W-1S	18
Hickory Ln. E&W	25W-0S	18
Hidden Ct. E&W	25W-2S	18
Hill Av. E&W	23W-0S	18
Hillside Ct. N&S	24W-0S	18
Hiram Ct. N&S	23W-3S	18
Hiram Dr. NE&SW	23W-3S	18
Howard Cir. E&W	24W-0N	12
Howard Ct. N&S	24W-0N	12
Howard St. N&S	24W-0S,24W-0N	18
Hull Dr. E&W	23W-3S	18
Huntleigh Ct. N&S	24W-1S	18
Huntleigh Dr. N&S	24W-1S	18
Hurley Dr. E&W	26W-1S,25W-1S	18
Hyatt Dr. E&W	23W-3S	18
Illinois St. E&W	24W-0S,23W-0S	18
Indiana St. E&W	25W-0S,23W-0S	18
Irving Av. N&S	24W-0S,24W-0N	18
Ivy Ct. E. E&W	23W-3S	18
Ivy Ct. W. E&W	23W-3S	18
James Ct. N&S	25W-1S,24W-1S	18
Jasper Ct. E&W	23W-3S	18
Jasper Dr. NE&SW	23W-3S	18
Jefferson Av. E&W	25W-0S,23W-0S	18
Jewell Rd. E&W	25W-0N	12
Johnstown Ct. N&S	24W-1S	18
Joyce Ct. N&S	25W-2S	18
Karlskoga Av. E&W	24W-0S	18
Kay Rd. E&W	23W-0N	12
Kelle Ct. E&W	23W-3S	18
Kellogg Pl. N&S	23W-0S	18
Kenilworth Av. E&W	24W-0S	18
Kensington Cir. NS&EW	24W-2S	18
Kensington Cir. N&S	24W-2S	18
Kent Ct. E&W	23W-2S	18
Kentucky Ct. E&W	23W-3S	18
Kilkenny Dr. N&S	25W-1S	18
Kingston Dr. N&S	23W-3S	18
Kingston St. N&S	23W-0N	12
Kipling Ct. N&S	23W-0S	18
Knoll St. N&S	26W-0S	18
Knollwood Dr. N&S	25W-0S	18
Kudu Ct. E&W	26W-0S	18
LaPalma Ct. NE&SW	24W-2S	18
Lakeside Dr. E&W	25W-0S	18
Lancaster Ln. N&S	23W-2S	18
Langford Ct. NW&SE	23W-3S	18
Langford Ln. N&S	23W-3S	18
Leabrook Ln. N&S	26W-1S	18
Leafbrook Cove N&S	26W-1S	18
Leask Ln. NW&SE	24W-3S	18
Leeds Ct. NW&SE	23W-3S	18
Leicester Ct. E&W	23W-1S	18
Lewis Ln. E&W	24W-2S	18
Lexington St. N&S	24W-0N	12
Leystone Ln. NE&SW	23W-1S	18

56 DU PAGE COUNTY

	Grid	Page
Liberty Dr. E&W	25W-0S,23W-0S	18
Lincoln Av. E&W	25W-0S,24W-0S	18
Liskeard Ct. E&W	23W-1S	18
Lloyd Ct. N&S	25W-1S	18
Lois Ln. E&W	26W-1S	18
Lorraine St N&S	23W-1S,23W-0S	18
Loughborough Ct. E&W	23W-1S	18
Lowden Av. E&W	24W-1S,23W-1S	18
Lowell Av. E&W	23W-0S	18
Lyford Ln. N&S	25W-1S	18
Lynwood Ln. E&W	25W-1S	18
Lyon Av. N&S	25W-0S	18
Madera Dr. NW&SE	23W-2S	18
Madison Av. E&W	25W-0S,24W-0S	18
Madsen Ct. E&W	23W-0N	12
Main St. N&S	24W-1S,24W-0N	18
Manchester Rd. E&W	26W-0S,25W-0S	18
Maple Ln. E&W	23W-0N	12
Mapleleaf Ct. E&W	25W-0S	18
Maplewood Dr. E&W	25W-2S	18
Marcey Av. N&S	25W-1S	18
Martingale Rd. NE&SW	24W-0N	12
Mayo Av. E&W	25W-1S	18
Merrill Dr. E&W	25W-1S	18
Michigan St. E&W	24W-0S	18
Middleton Ct. N&S	25W-2S	18
Middleton Dr. N&S	25W-2S	18
Milton Ln. N&S	24W-2S	18
Monticello Ct. N&S	23W-2S	18
Morgan Av. N&S	25W-0S	18
Morse St. N&S	25W-0N	12
Mt. Vernon Ct. E&W	23W-2S	18
Myrtlewood Ln. N&S	25W-0S	18
Nachtman Ct. N&S	23W-0N	12
Naperville Rd. N&S	24W-2S,24W-0S	18
Nelson Cir. E&W	24W-1S	18
Newberry Ln. N&S	24W-1S	18
North Path Av. E&W	24W-0S,23W-0S	18
Nottingham Ln. N&S	23W-2S	18
Oak Av. E&W	24W-0N	12
Oak Knolls St. N&S	25W-0S	18
Oak View Dr. E&W	25W-0N	12
Orchard Rd. NW&SE	25W-1S	18
Orchard Rd. N&S	25W-2S	18
Orth Ct. E&W	26W-1S	18
Orth Dr. NW&SE	26W-1S	18
Ott Av. N&S	23W-0S	18
Oxford Ln. E&W	23W-2S	18
Paddock Ct. NE&SW	24W-0N	12
Paddock Ln. N&S	24W-0N	12
Page St. N&S	26W-0S	18
Papworth St. N&S	25W-0N	12
Park Av. E&W	25W-1S,24W-1S	18
Park Circle Dr. E&W	24W-0N	12
Parkside Dr. E&W	25W-0S	18
Parkway Dr. E&W	24W-0N	12
Paula Av. E&W	25W-1S	18
Pebblestone Cove N&S	26W-1S	18
Pembroke Ln. NE&SW	23W-1S	18
Pershing Av. E&W	24W-1S,23W-1S	18
Pick Ct. N&S	23W-0N	12
Pick St. N&S	23W-0S	18
Pierce Av. N&S	25W-0S	18
Pinto Ct. NW&SE	24W-2S	18
Plamondon Ct. NW&SE	25W-1S	18
Plamondon Rd. NE&SW	25W-1S	18
Prairie Av. E&W	25W-0N,23W-0N	12
President St. N&S	24W-1S,24W-0N	18
Princeton Ct. N&S	24W-1S	18
Prospect Ct. NE&SW	23W-1S	18
Prospect St. N&S	23W-1S,23W-0S	18
Purnell St. N&S	25W-0N	12
Queenswood Ct. E&W	24W-1S	18
Ralph Ct. E&W	25W-0N	12
Ranch Rd. E&W	24W-0N	12
Ravenhill N&S	26W-1S	18
Reading Ct.* E&W	23W-2S	18
Reber St. N&S	24W-0S	18
Redwood Ct.* E&W	26W-2S	18
Renn Ct. N&S	25W-0S	18
Rhodes Ct.* NW&SE	23W-2S	18
Richmond Dr. N&S	23W-3S	18
Richton Dr. N&S	25W-2S	18
Robinwood Ct. E&W	24W-3S	18
Robinwood Ln. E&W	24W-3S	18
Roosevelt Rd E&W	26W-0S,23W-0S	18
Royal Ct. E&W	25W-0N	12
Ryson Ct. E&W	23W-3S	18
Salford Ct. E&W	23W-3S	18
Sandy Hook N&S	26W-1S	18
Santa Rosa Av. N&S	24W-0N	12
Scott St. N&S	24W-0S,24W-0N	18
Scottdale Cir. NS&EW	23W-3S	18
Scottdale Ct. NW&SE	23W-3S	18
Sears Ln. E&W	24W-0S	18
Seminary Av. E&W	25W-0S,24W-0S	18
Shady Ln. E&W	23W-0N	12
Shaffner Rd. N&S	26W-2S,26W-1S	18
Shagbark Ln. N&S	25W-0N	12
Sheffield Ln. N&S	23W-2S	18
Sheldon Ct. E&W	24W-1S	18
Shelley Ln. E&W	24W-2S	18
Sherwin St. N&S	25W-0S	18
Sherwood Pl. N&S	25W-2S	18
Shetland Dr. E&W	23W-3S	18
Shire Ct. E&W	24W-2S	18
Shirley Ct. N&S	26W-0S	18
Somerset Ln. NS&EW	25W-2S	18
Southampton Ct.* N&S	23W-2S	18
Spring Green Dr E&W	26W-1S	18
St. Francis Ct. N&S	26W-1S	18
Stallion Ct. NW&SE	23W-3S	18
Stanley St. N&S	26W-0S	18
Stevenson Pl. E&W	24W-1S	18
Stirrup Ln. NS&EW	23W-3S	18
Stoddard Av. N&S	23W-0S,23W-0N	18
Stonebridge Cir. N&S	26W-1S	18
Stonebridge Ct. NE&SW	26W-2S	18
Stonebridge Tr. NW&SE	26W-1S	18
Stonegate Ln. E&W	26W-2S	18
Stonehill Ct. N&S	26W-1S	18
Summit St. N&S	23W-0S,23W-0N	18
Sumner St N&S	24W-1S,24W-0S	18
Sunderland Ct.* NW&SE	23W-2S	18
Sunnybrook Ln. E&W	26W-0S,25W-0S	18
Sunnyside St. N&S	25W-0S	18
Sunset Rd. N&S	25W-1S	18
Surrey Dr. N&S	25W-1S	18
Surrey Rd. E&W	24W-0N	12
Sussex Ln. E&W	23W-2S	18
Taft Av. E&W	23W-1S	18
Tall Oaks Ln. N&S	25W-0N	12
Tennyson Dr. E&W	24W-1S	18
Thomas Ln. E&W	23W-0N	12
Thompson Dr. E&W	24W-2S	18
Thornwood N&S	26W-1S	18
Timber Ln. E&W	26W-1S	18
Timber Tr. NS&EW	25W-1S	18
Trent Ct. E&W	23W-1S	18
Trowbridge Ct. NW&SE	23W-2S	18
Turf Ln. E&W	24W-0N	12
Underwood Terr N&S	24W-1S	18
Union Av. E&W	25W-0S,24W-0S	18
University Pl. E&W	24W-0S	18
Valley Forge Ct.* NE&SW	23W-2S	18
Wadham Ct. E&W	23W-1S	18
Wadham Pl. N&S	23W-1S	18
Wadsworth Rd. NW&SE	24W-1S	18
Wakeman Av. E&W	24W-0N,23W-0N	12
Wakeman Ct. E&W	23W-0N	12
Wales Dr. N&S	24W-1S	18
Walnut Ct. NW&SE	25W-1S	18
Warrenville Rd. N&S	25W-1S	18
Warwick Ct. E&W	23W-1S	18
Washington St. N&S	24W-1S,24W-0N	18
Watertower Ct. N&S	26W-1S	18
Webster Av. N&S	24W-0S,24W-0N	18
Weisbrook Rd. NE&SW	26W-2S,25W-1S	18
Wesley St. E&W	25W-0S,24W-0S	18
West St. N&S	25W-0S	18
Western Av. N&S	25W-0S	18
Westhaven Dr. N&S	25W-1S	18
Westminister St. N&S	23W-3S	18
Westwood N&S	25W-0S	18
Wexford Cir. NE&SW	26W-1S	18
Wheaton Av. N&S	24W-1S,24W-0N	18
Wheaton Oaks Ct. N&S	25W-0N	12
Wheaton Oaks Dr. E&W	25W-0N	12
Wheaton Pl. N&S	24W-0N	12
Whitchurch Ct. E&W	23W-1S	18
Whitman Ln. NW&SE	24W-1S	18
Whitt Ct. N&S	23W-2S	18
Whittier Ln. N&S	24W-1S	18
Wigtown Ct. NW&SE	23W-1S	18
Williams St. E&W	26W-0S	18
Williamsburg Ct.* NE&SW	23W-2S	18
Williston St. N&S	23W-1S,23W-0S	18
Willow Av. E&W	24W-0S,23W-0S	18
Willow Run E&W	26W-1S	18
Wilmette St. N&S	23W-0N	12
Wilshire Av. NE&SW	23W-0S	18
Wilson Av. E&W	23W-1S	18
Windsor Dr. E&W	23W-2S	18
Wingate Ln. N&S	26W-2S	18
Wood St. N&S	23W-0S	18
Woodcutter Ln. E&W	25W-1S	18
Woodhaven N&S	26W-1S	18
Woodlawn St. N&S	25W-0S,25W-0N	18
York Ln. E. E&W	23W-0S	18
York Ln. W. E&W	23W-0S	18
Yorkshire Woods Ct. N&S	26W-1S	18

Willowbrook

	Grid	Page
Adams St. N&S	16W-8S,16W-7S	33
Americana Dr. N&S	16W-7S	27
Appletree Ln. N&S	17W-8S	33
Arlene Av. N&S	16W-8S	33
Ascot Ln. E&W	16W-6S	27
Birchwood Ct. N&S	16W-8S	33
Blackberry Ln. N&S	17W-8S	33
Briar Rd. N&S	15W-7S	27
Brookbank Rd. N&S	16W-8S	33
Brookside Ln. E&W	16W-6S	27
Cambridge Rd. N&S	15W-7S	27
Canterbury Ln. N&S	16W-7S	27
Chatelaine Ct. E&W	17W-6S	27
Chaucer Ct. E&W	15W-7S	27
Chaucer Rd. N&S	15W-7S	27
Cherry Tree Ln. N&S	17W-8S	33
Cherrywood Ln. E&W	16W-8S	33
Clarendon Hills Rd. N&S	16W-7S,16W-6S	27
Clubside Dr. E&W	16W-7S	27
Cottonwood Ct. E&W	16W-8S	33
Cramer Ct. E&W	16W-7S	27
Eagles Nest Dr. E&W	17W-7S	27
Easy St. E&W	17W-7S	27
Eleanor Pl. N&S	16W-8S	33
Essex Ct. N&S	16W-6S	27
Executive Dr. E&W	16W-8S	33
Falcon Ct. N&S	17W-7S	27
Garfield Ridge NS&EW	15W-7S	27
Gooseneck Ct. N&S	17W-7S	27
Hawk Ct. N&S	17W-7S	27
Hawthorn Ln. E&W	17W-8S	33
Hidden Brook Ln. E&W	15W-6S	27
Highridge Rd. NW&SE	16W-7S	27
Hill Rd. E&W	15W-7S	27
Honey Locust Ln. E&W	17W-8S	33
Joliet Rd. NE&SW	16W-8S	33
Kane Ct. N&S	16W-7S	27
Kent Ct. E&W	16W-7S	27
Kingswood Ct. NE&SW	15W-7S	27
Kingswood Rd. NW&SE	15W-7S	27
Knoll Lane Ct. N&S	16W-6S	27
Knoll Valley Dr. N&S	16W-6S	27
Knollwick Rd. N&S	16W-6S	27
Knollwood Dr. N&S	16W-6S	27
Knollwood Rd. E&W	16W-6S	27
Kyle Ct. E&W	16W-7S	27
Lake Hinsdale Dr. E&W	16W-7S	27
Lake Park Ln. N&S	16W-7S,16W-6S	27
Lake Shore Dr. NE&SW	16W-7S	27
Lakeview Ct. N&S	16W-7S	27
Lane Ct. NW&SE	16W-7S	27
Laurel Ln. N&S	16W-6S	27
Lincoln Oaks Dr. E&W	17W-6S	27
Mac Arthur Dr. NS&EW	16W-6S	27
Madison St. N&S	16W-8S,16W-7S	33
Maplewood Ct. E&W	16W-8S	33
Martin Ct. NE&SW	16W-7S	27
Meadow Ln. N&S	15W-6S	27
Midway Dr. E&W	16W-8S	33
N. Court Dr. NE&SW	16W-7S	27
Oxford Rd. N&S	15W-7S	27
Pine Tree Ln. E&W	17W-8S	33
Pinewood Ct. N&S	16W-6S	27
Plainfield Rd. NE&SW	16W-7S	27
Plaza Ct. N&S	16W-8S	33
Port Wine Rd. N&S	16W-7S	27
Quail Run Ct. N&S	17W-7S	27
Quincy St. N&S	16W-8S,16W-7S	33
Raleigh Rd. N&S	15W-7S	27
Ridgemoor Ct. NW&SE	15W-7S	27
Ridgemoor Dr. E&W	16W-7S,15W-7S	27
Rogers Ct. E&W	15W-7S	27
Rogers Dr. N&S	15W-7S	27
Rogers Farm Rd. E&W	15W-7S	27
Sheffield La. N&S	15W-7S	27
Sheridan Dr. NW&SE	17W-8S	33
Somerset Rd. NE&SW	15W-7S	27
Squire Ln. N&S	17W-6S	27
Stanhope Dr. E&W	16W-6S	27
Stewart Dr. NW&SE	16W-6S	27
Stirling Ln. E&W	16W-6S	27
Stonegate Ct. NE&SW	15W-7S	27
Stough St. N&S	16W-7S	27
Stratford Ln. E&W	15W-7S	27
Sugar Bush Ln. NE&SW	17W-8S	33
Sunset Ridge Rd. E&W	15W-7S	27
Tanglewood Ln. E&W	16W-8S	33
Virginia Ct. N&S	16W-8S	33
Waterford Dr. E&W	15W-7S	27
Wedgewood Ct. NE&SW	15W-7S	27
Wedgewood La. NW&SE	15W-7S	27
Willow Ln. E&W	16W-7S	27
Willoway Ln. N&S	16W-8S	33
Windsor Ln. E&W	16W-6S	27
Windward Cir. NS&EW	16W-7S	27
Wingate Rd. N&S	15W-7S	27
Woodgate Ct. NS&EW	15W-7S	27
67th Pl. E&W	16W-7S	27
67th St. E&W	16W-7S	27
68th Pl. E&W	16W-7S	27
68th St. E&W	16W-7S	27
69th St. E&W	16W-7S	27
72nd Ct. E&W	16W-8S	33
73rd Ct. E&W	16W-8S	33

Winfield

	Grid	Page
Alta Ln. E&W	27W-0N	11
Ambleside Dr. E&W	27W-0N	11
Arbor Ct. N&S	26W-0N	12
Aspen Ct. NE&SW	27W-0N	11
Bardmore Ln. E&W	26W-1N	12
Barnes Av. NE&SW	27W-0N,26W-0N	11
Barnes Ct. NE&SW	26W-0N	12
Beecher Av. E&W	27W-0S	17
Birch St. E&W	27W-0N	11
Blair St. E&W	27W-0N,26W-0N	11
Bolles Av. E&W	27W-0N	11
Bradley Dr. N&S	27W-0S	17
Brandon Blvd. N&S	27W-0N	11
Brookside Dr. E&W	27W-0S	17
Calvin Ct. N&S	26W-0N	12
Carrel St. E&W	27W-0N	11
Chartwell Dr. E&W	27W-0N	11
Chelsea Cir. E&W	27W-0N	11
Chestnut Ln. E&W	27W-1N	11
Childs St. E&W	27W-0S	17
Church St. N&S	27W-0S	17
Churchill Rd. E&W	27W-0N,26W-0N	11
Cleveland St. N&S	27W-0S	17
Cloos Ct. N&S	27W-0N	11
Concord Ln. N&S	27W-0N	11
Conniston Ct. N&S	27W-0N	11
Cooley Av. E&W	27W-0S	17
County Farm Rd. NW&SE	27W-0N,27W-1N	11
Courtney NW&SE	27W-0N	11
Cypress Ln. NE&SW	27W-1N	11
East St. N&S	27W-0S	17
Emerson Ct. N&S	27W-1S	17
Ennerdale Ln. N&S	27W-0N	11
Essex Ln. E&W	27W-0N	11
Ethel St. N&S	27W-0S,27W-0N	17
Evelyn Av. E&W	27W-0S	17
Fisher Ln. E&W	27W-0S	17
Fleming Rd. E&W	27W-0N	11
Florida Ln. N&S	27W-0S	17
Forest St. N&S	27W-0S	17
Garfield St. N&S	27W-0S	17
Grant St. N&S	27W-0S	17
Grasnere Dr. E&W	27W-0N	11
Heather Ln. E&W	27W-0N	11
Highlake Rd. E&W	27W-0S	11
Jacob Dr. N&S	27W-0N	11
Jefferson Ct. N&S	27W-1S,27W-0S	17
Jefferson St. N&S	27W-0S,27W-0N	17
Jewell Rd. NE&SW	27W-0S	11
Kimball Ct. N&S	27W-0N	11
Knoll Ct. NE&SW	26W-0N	12
Kytham Ct. E&W	26W-1N	12
Lancaster Dr. NS&EW	27W-0N	11
Leonard St. N&S	26W-0N	12
Liberty St. E&W	27W-0S	17
Lincoln Av. E&W	27W-0S	17
Lindsey Av. E&W	27W-0N,26W-0N	11
Madison St. N&S	27W-0S	17
Manchester Rd. E&W	27W-0S	17
Maple St. E&W	27W-0N	11
Melrose Ln. E&W	27W-1N	11
Myrtle St. N&S	27W-0S	17
Nickolsen Av. E&W	27W-0S	17
Oak Ct. E&W	27W-0S	17
Oak Ln. N&S	27W-0S	17
Oakwood St. E&W	27W-0N	11
Park St. N&S	27W-0S	17
Peter Rd. N&S	26W-0N	12
Pleasant Hill Rd. N&S	26W-1N	12
Prairie Path Ln. N&S	27W-1N	11
Prescott Dr. NS&EW	26W-0N	12
Providence Ln. E&W	27W-0N	11
Redford Ln. N&S	27W-0N,27W-1N	11
River Ln. NE&SW	27W-0S	17
Robbins St. N&S	27W-0S,27W-0N	17
Scott Dr. NE&SW	27W-0N	11
Shady Way NW&SE	27W-0S	17
Shires Ct. N&S	27W-0N	11
Siefert Ct. N&S	26W-1N	12
St. John Av. E&W	27W-0N,26W-0N	11
Summit Dr. N&S	27W-0S	17
Sunnyside Av. E&W	27W-0S	17
Suzanne Dr. N&S	27W-0N	11
Sycamore Ln. E&W	27W-1N	11
Tamarack Ct. NW&SE	27W-1N	11
Tamarack Dr. NS&EW	27W-1N	11
Timber Ct. NE&SW	27W-1N	11
Timber Ridge Dr. N&S	27W-1N	11
Troon Ct. E&W	26W-1N	12
Victoria Ln. NS&EW	27W-0N	11
Virginia St. E&W	27W-0N	11
Walnut Dr. E&W	27W-1N	11
Washington Av. E&W	27W-0S	17
Waterford Dr. E&W	27W-0N	11
Williams St. E&W	27W-0S	17
Windemere Rd. NE&SW	27W-0N	11
Winfield Rd. N&S	27W-0S,27W-0N	17
Woodvale Ct. N&S	27W-0N,26W-0N	11
Wynwood Rd. N&S	28W-0S	17

Winfield Township

	Grid	Page
Acorn Av. NE&SW	29W-0N	11
Acorn Ct. N&S	27W-1S	17
Acorn Ln. E&W	27W-1S	17
Alena Dr. E&W	28W-1N	11
Alsace Ct. N&S	28W-1S	17
Anderson Dr. E&W	32W-3S	16
Ardennes Ct. E&W	28W-1S	17
Barnes Av. E&W	30W-0N,28W-0N	10
Batavia Rd. NW&SE	32W-1S,30W-2S	16
Bauman Ct. NW&SE	29W-0S	16
Belleau Dr,N. N&S	28W-1S	17
Bender Ln. N&S	29W-1S	16
Berkshire Rd. E&W	28W-1N	11
Blackhawk Blvd. N&S	30W-2S,30W-1S	16
Bolles Av. E&W	29W-0N,28W-0N	11
Branch Av. E&W	27W-2S	17
Briar Ln. N&S	29W-0N	11
Broadview Av. N&S	29W-0N	11
Brown St. E&W	29W-0N	11
Burr Oak Rd. N&S	28W-1N	11
Butterfield Rd. E&W	32W-3S,27W-2S	16
Butternut Ln. E&W	32W-3S	16

DU PAGE COUNTY 57

	Grid	Page
Calvin Av. N&S	28W-0N	11
Cantigny Dr,S. N&S	28W-1S	17
Center Av. N&S	27W-2N,27W-1N	11
Che Che Pinqua St. E&W	30W-1S	16
Childs St. E&W	29W-0S,28W-0S	16
Circle Dr. N&S	30W-0S	16
Colford Av. NW&SE	29W-0N	11
Connolly Ct. E&W	29W-2S	16
Coolidge Av. N&S	29W-0N,29W-1N	11
County Farm Rd. N&S	27W-1N	11
Cove Ln. E&W	27W-3S	17
Domartin Pl. NE&SW	28W-1S	17
Donald Av. E&W	29W-0N,28W-0N	11
Doris Ln. N&S	29W-0N	11
DuPage St. E&W	28W-0N	11
Easton Av. N&S	29W-0N	11
Edgewood Walk N&S	29W-1S	16
Elm Dr. E&W	28W-1N,27W-1N	11
Elmwood St. E&W	32W-1N	10
Eola Rd. N&S	31W-3S,31W-1S	16
Essex Rd. N&S	27W-2S	17
Ethel St. N&S	27W-1N	11
Fabyan Pkwy. E&W	32W-0N	10
Forest Av. E&W	29W-0N,28W-0N	11
Gary's Mill Rd. E&W	29W-0S,28W-0S	16
Geneva Rd. E&W	29W-1N,27W-1N	11
Gloria Av. N&S	27W-1N	11
Guiness Dr. N&S	30W-0S	16
Hathaway Av. N&S	28W-0N	11
Helen Av. E&W	29W-1N	11
Herrick Rd. N&S	27W-3S	17
Hickory Ln. E&W	29W-1N,27W-1N	11
High Lake Av. N&S	28W-0N,28W-1N	11
High Lake Rd. NW&SE	28W-0N	11
Hillcrest Rd. N&S	28W-1N	11
Hillview Av. E&W	28W-0N	11
Home Av. N&S	29W-3S	16
Hoy Rd. E&W	28W-2S,27W-2S	17
Indian Knoll Rd. N&S	28W-1S,28W-1N	17
Indian Knoll Tr. E&W	28W-0N	11
Ingalton Av. NW&SE	29W-1N,29W-2N	11
Inner Ring Rd. Cir.	32W-3S	16
James Av. E&W	29W-1N,27W-1N	11
Joliet St. N&S	30W-0S	16
Joy Rd. E&W	30W-0S	16
Kaelin Rd. N&S	30W-1S,30W-0S	16
Kautz Rd. N&S	32W-1N	10
Kress Rd. N&S	32W-1N	10
Lake Dr. N&S	28W-0N	11
Lane Rd. NW&SE	30W-1N	10
Lavergne Ln. N&S	28W-1S	17
Lee Rd. E&W	31W-1N,29W-1N	10
Lenox Rd. E&W	27W-2S	17
Lester St. E&W	30W-0N,28W-0N	10
Liberty St. E&W	28W-0S	17
Lorraine Dr. E&W	28W-1S	17
Lost Meadows Ln. N&S	29W-2S	16
Lowden Av. E&W	27W-1S	17
Mack Rd. E&W	29W-1S,27W-1S	16
Macqueen Dr. N&S	31W-1N	10
Madison St. N&S	27W-2S	17
Main St. E&W	28W-0N	11
Marion Rd. E&W	28W-0S	17
May St. E&W	30W-0S	16
McChesney Rd. N&S	32W-0S	16
McDonald Av. N&S	29W-0N	11
Melolane Dr. N&S	30W-0S	16
Morningside Av. N&S	28W-0S,28W-0N	17
Morris Ct. E&W	29W-2S	16
N. Flanders Ln. E&W	28W-1S	17
National Rd. E&W	28W-1N	11
National St. E&W	29W-1N,27W-1N	11
Neuqua St. E&W	30W-1S	16
Northwood Dr. N&S	27W-1N	11
Oak Grove Av. E&W	29W-1N	11
Oak Ln. E&W	29W-2S	16
Old Farm Ln. E&W	29W-2S	16
Outer Ring Rd. Cir.	32W-3S	16
Pamela Ct. N&S	28W-1S	17
Pearl Rd. N&S	30W-0S	16
Phillips Farm Rd. N&S	32W-1S	16
Picardy Ct. NW&SE	28W-1S	17

	Grid	Page
Pilsen Rd. N&S	30W-1N	10
Pine Av. E&W	29W-1N	11
Pomeroy St. E&W	30W-0N,29W-0N	10
Potawatomi Blvd. NE&SW	30W-1S	16
Prairie Av. E&W	29W-3S	16
Prince Crossing Rd. N&S	29W-0S,29W-1N	16
Purnell Rd. NW&SE	28W-1S	17
Ray Av. E&W	30W-1N,29W-1N	10
Rd. A NE&SW	32W-1S,31W-0S	16
Rd. B NW&SE	32W-1S	16
Rd. C NE&SW	32W-1S	16
Rd. D NE&SW	32W-1S,31W-1S	16
Rd. L NW&SE	32W-1S	16
Renouf Dr. E&W	29W-2S	16
Ridgeland Av. N&S	29W-0N,29W-1N	11
River Glen Rd. NE&SW	29W-0S	16
Robin Ln. E&W	28W-1N,27W-1N	11
Rogers Ct. N&S	28W-0N	11
Roosevelt Rd. NW&SE	32W-0N,28W-0S	10
Sarana Av. N&S	29W-0N	11
Sauk Blvd. N&S	30W-1S	16
Shabbona St. NE&SW	30W-1S	16
Shaffner Rd. N&S	27W-2S,27W-1S	17
Sherman St. N&S	30W-0N	10
South Cantigny Dr. N&S	28W-1S	17
St. Mihel Dr. N&S	28W-1S	17
Sunset Av. N&S	28W-0N	11
Swan Lake Ct. NE&SW	27W-1S	17
Swan Lake Dr. N&S	27W-1S	17
Twin Oaks Dr. N&S	27W-2S	17
Verdun Dr. NS&EW	28W-1S	17
Wallace Rd. E&W	27W-2S	17
Waltz Dr. E&W	27W-1S	17
Warrenville Av. NE&SW	27W-2S	17
Washington Av. E&W	28W-0S	17
West Belleau Dr. E&W	28W-1S	17
West Cantigny Dr. E&W	28W-1S	17
White Oak Dr. N&S	28W-1S	17
Williams Rd. N&S	29W-1N	11
Willoughby Ln. N&S	32W-3S	16
Wilson Rd. E&W	32W-1S,31W-0S	16
Wilson St. E&W	30W-0S,29W-0S	16
Winfield Rd. N&S	28W-2S,27W-1S	17
Winnebago St. E&W	30W-1S	16
Woodland Av. E&W	29W-1N	11
Wynwood Rd. N&S	28W-0S	17
Yvonne Ln. N&S	27W-2S	17

Wood Dale

	Grid	Page
A.E.C. Dr. N&S	18W-7N	8
Abbotsford Rd. E&W	18W-5N	8
Ace Dr. N&S	18W-6N	8
Addison Rd. N&S	18W-4N,18W-6N	8
Apollo Ct. N&S	19W-5N	8
Arlene Dr. E&W	19W-5N	8
Ash Av. N&S	17W-5N,17W-6N	9
Aspen Rd. E&W	17W-4N	9
Balm Ct. E&W	17W-6N	9
Bauman Ct. E&W	18W-7N	8
Beinoris Dr. E&W	17W-6N	9
Brookhurst Ln. N&S	17W-5N	9
Brookwood Dr. N&S	18W-5N	8
Brookwood Pl. N&S	18W-5N	8
Butternut Av. E&W	17W-4N	9
Butternut Dr. N&S	17W-4N	9
Carter Av. E&W	18W-5N	8
Carter Dr. E&W	18W-5N	8
Catalpa Av. N&S	17W-5N,17W-6N	9
Catherine Ct. NW&SE	18W-5N	8
Cedar Av. N&S	17W-5N,17W-6N	9
Cedar St. N&S	17W-5N	9
Center St. E&W	18W-6N	8
Central Av. N&S	17W-5N,17W-7N	9
Century Dr. N&S	18W-5N	8
Charmille Ln. E&W	18W-5N	8
Clare Ct. NS&EW	18W-5N	8
Clayton Ct. NE&SW	18W-6N	8
Commercial St. E&W	18W-6N,17W-6N	8
Crestwood Ct. E&W	17W-4N	9
Crestwood Rd. E&W	17W-4N	9
Cypress Ct. NW&SE	17W-5N	9

	Grid	Page
Dalewood Av. N&S	18W-5N,18W-6N	8
Deerpath Rd. E&W	17W-5N	9
Division St. E&W	18W-5N,17W-5N	8
Dominion Ct. E&W	18W-5N	8
Dominion Dr. N&S	18W-4N,18W-5N	8
Duck Ln. E&W	18W-5N	8
Dunlay Ct. E&W	17W-5N	9
Dunlay St. E&W	17W-5N	9
East Pond Dr. N&S	18W-6N	8
Edgebrook Rd. N&S	18W-5N	8
Edgewood Av. N&S	17W-4N,17W-7N	9
Elizabeth Ct. E&W	17W-5N	9
Elizabeth Dr. E&W	18W-5N	8
Elmhurst St. E&W	17W-6N	9
Elmwood Av. N&S	17W-5N,17W-6N	9
Essex Ct. NE&SW	18W-5N	8
Ethel Ln. E&W	17W-4N	9
Fishing Ln. N&S	18W-5N	8
Forest Glen Rd. E&W	18W-6N	8
Forest Preserve Dr. E&W	19W-5N,18W-5N	8
Forest View Av. N&S	18W-5N	8
Frederick Pl. N&S	18W-5N	8
Front St. NW&SE	17W-5N	9
Gay St. N&S	18W-5N	8
Geanne Ct. E&W	18W-5N	8
George St. E&W	18W-6N	8
Gerry Dr. E&W	18W-7N	8
Gilbert Dr. E&W	19W-5N,18W-5N	8
Green Ct. N&S	18W-5N	8
Grove Av. N&S	18W-6N	8
Hackberry Ct. E&W	17W-4N	9
Hansen Ct. NE&SW	18W-7N	8
Harvey Av. N&S	18W-5N	8
Haynes Dr. E&W	17W-6N	9
Hemlock Av. N&S	17W-5N,17W-6N	9
Heritage Dr. E&W	18W-5N	8
Hiawatha Tr. E&W	17W-5N	9
Homestead St. E&W	18W-5N	8
Hoover Dr. N&S	18W-5N	8
Irmen Dr. E&W	19W-5N	8
Iroquois Tr. N&S	17W-5N	9
Irving Park Rd. E&W	18W-6N,17W-5N	8
Jason Ln. E&W	18W-5N	8
Jefferson Ln. N&S	18W-5N	8
Jennifer Ct. E&W	18W-5N	8
Julian Dr. N&S	18W-5N	8
Juniper Dr. E&W	17W-5N	9
Knollwood Ct. E&W	17W-5N	9
Knollwood Dr. E&W	17W-5N	9
Lafayette St. N&S	18W-5N	8
Lilac Ln. N&S	17W-5N	9
Lincoln Ct. N&S	19W-5N	8
Lively Blvd. N&S	17W-7N	9
Manning Dr. E&W	18W-5N	8
Maple Av. N&S	17W-5N,17W-6N	9
Mark St. E&W	18W-7N,17W-7N	8
Mary Jane Ln. E&W	19W-5N	8
Michael Dr. N&S	18W-7N	8
Miller Ln. N&S	18W-6N	8
Mittel Blvd. N&S	18W-7N	8
Mittel Dr. E&W	18W-6N	8
Mittel Rd. N&S	18W-6N,18W-7N	8
Monroe Pl. E&W	18W-5N	8
Mont Clare Ln. N&S	17W-5N	9
Montrose Av. E&W	19W-5N,17W-5N	8
Mulberry Ln. NE&SW	17W-5N	9
Murray Dr. E&W	17W-5N	9
Murray Ln. E&W	17W-5N	9
North St. E&W	17W-6N	9
Oak Av. N&S	17W-5N,17W-6N	9
Oakwood Dr. N&S	18W-5N	8
Orchard Dr. E&W	18W-6N	8
Paramount Dr. N&S	18W-5N	8
Park Ln. E&W	18W-5N	8
Pine Av. N&S	17W-5N,17W-6N	9
Pine Tree Ln. N&S	17W-5N	9
Poplar Av. N&S	17W-6N	9
Potter St. E&W	19W-5N,17W-5N	8
Prospect Av. N&S	18W-5N	8
Raleigh Ct. N&S	18W-5N	8
Richert Rd. E&W	17W-6N	9
River Av. N&S	18W-6N	8
Roberts Ln. NW&SE	17W-4N	9
Robin Ln. N&S	19W-5N	8
Roosevelt Av. E&W	17W-5N	9
Roy Dr. E&W	19W-5N	8
Royal Oaks Dr. E&W	17W-5N	9
Royal St. N&S	17W-4N	9
S. Hiawatha Tr. E&W	17W-5N	9
Sarah Ct. E&W	18W-5N	8
Sarah Dr. E&W	19W-5N	8

	Grid	Page
School E&W	18W-6N	8
Sherwood Dr. NE&SW	17W-5N	9
Sivert Ct. N&S	16W-6N	9
Sivert Dr. N&S	17W-7N	9
Spring Oaks Dr. E&W	17W-5N	9
St. Andrews Dr. N&S	18W-4N,18W-5N	8
Station Dr. N&S	18W-5N,18W-6N	8
Stoneham St. E&W	17W-6N	9
Sunnyside Av. E&W	17W-5N	9
Sunset Dr. E&W	18W-5N	8
Thorndale Av. E&W	18W-7N,17W-7N	8
Tioga Tr. E&W	17W-5N	9
Victoria Dr. E&W	18W-5N	8
Victoria Ln. E&W	19W-5N	8
Victoria Pl. E&W	18W-7N	8
Walnut Av. N&S	17W-5N,17W-6N	9
Warren Ct. E&W	18W-5N	8
Warrenallen Dr. E&W	17W-7N	9
Washington Sq. E&W	18W-5N	8
Welter Dr. N&S	19W-5N	8
Westeria Ct. NW&SE	17W-5N	9
Wheat Ln. N&S	18W-6N	8
Windsor Av. E&W	17W-5N	9
Woodbine Ct. NE&SW	17W-4N	9
Woodbine Dr. E&W	17W-4N	9
Wooddale Rd. N&S	18W-7N,17W-5N	8
Woodside Dr. NE&SW	17W-4N	9

Woodridge*

(Also see Will County)

	Grid	Page
Adbeth Av. N&S	21W-9S	31
Allan Dr. NW&SE	22W-6S	25
Anchor Dr. N&S	21W-9S	31
Andover Ct. E&W	21W-7S	25
Andrea Ct. E&W	22W-9S	31
Andrea Ln. N&S	22W-9S	31
Apache Ln. NW&SE	21W-9S	31
Apple Ln. E&W	21W-8S	31
Armour Ct. E&W	23W-7S	25
Armstrong Ct. NE&SW	22W-7S	25
Arnold Dr. N&S	22W-7S	25
Audubon Av. N&S	21W-7S	25
Autumn Dr. NS&EW	22W-9S	31
Barrington Ct.* N&S	21W-8S	31
Beller Rd. NS&EW	20W-9S	32
Bern Ct. N&S	21W-9S	31
Birchwood Pkwy. E&W	21W-9S	31
Bittersweet Ct.* E&W	21W-8S	31
Blue Flag Av. N&S	22W-7S	25
Bobby Jones Ln. N&S	23W-7S	25
Bonnie Ct. N&S	21W-8S	31
Bradley Dr. N&S	22W-7S	25
Bramblebush Ct. E&W	21W-7S	25
Brentwood Ct. E&W	22W-9S	31
Brewer Ln. E&W	22W-9S	31
Briarwood Ct. E&W	21W-9S	31
Brighton Ct.* NE&SW	21W-8S	31
Bristol Ct. N&S	20W-8S	32
Britten St. E&W	20W-9S	32
Brook Ct. N&S	21W-7S	25
Brunswick Cir. E&W	21W-7S	25
Buckingham Cir. NS&EW	21W-7S	25
Burke Ct. NE&SW	22W-7S	25
Burr Ridge Ct. N&S	21W-9S	31
Butternut Ct. NE&SW	22W-8S	31
Cambridge N&S	20W-8S	32
Cambridge Dr. E&W	21W-7S	25
Canterbury Ln. N&S	21W-9S	31
Cardinal Ct. N&S	21W-7S	25
Carlton Dr. NE&SW	21W-8S,21W-7S	31
Carolwood Ln. N&S	22W-9S	31
Carpenter Ct. N&S	22W-7S	25
Catalpa Av. NW&SE	21W-8S	31
Catalpa Ct. NW&SE	21W-8S	31
Charmingfare Dr. E&W	21W-8S	31
Chelsea Dr. N&S	21W-9S	31
Chelsea Ln. N&S	21W-9S	31
Cherry Tree Av. N&S	21W-7S	25
Cherry Tree Ct. E&W	21W-7S	25
Chesham Ct. E&W	20W-9S	32
Chestnut Av. N&S	21W-8S	31
Chick Evans Ln. N&S	23W-7S	25
Chippingham Rd. E&W	20W-9S	32
Church Ct. NE&SW	22W-7S	25
Claredon Ln. E&W	21W-9S	31
Clark Dr. NE&SW	22W-7S	25
Compton Rd. N&S	20W-9S	32

	Grid	Page
Concord Dr. E&W	21W-7S	25
Cooper Ct. E&W	22W-7S	25
Country Club Dr. NS&EW	21W-8S	31
Crabtree Av. E&W	22W-8S,21W-8S	31
Crabtree Ct. NE&SW	21W-8S	31
Creekside Ct. E&W	22W-6S	25
Cromwell Av. N&S	20W-9S	32
Crown Point St. E&W	21W-7S	25
Crystal Ct. N&S	21W-9S	31
Dalewood Ct. NE&SW	21W-8S	31
Dalewood Pkwy. NS&EW	21W-9S	31
Danbury Dr. NW&SE	21W-7S	25
David Dr. N&S	22W-9S	31
Davos Av. NW&SE	21W-9S	31
Dean Dr. N&S	22W-7S	25
Deer Dr. NS&EW	22W-8S	31
Deerfield Av. N&S	21W-8S	31
Deerwood Ct. N&S	22W-9S	31
Demaret Ct. NE&SW	23W-7S	25
Diamond Ct. NE&SW	22W-7S	25
Didrickson Ln. N&S	23W-7S	25
Dove St. N&S	21W-7S	25
Dover Ct.* NS&EW	21W-8S	31
Dunham Dr. N&S	20W-9S,20W-8S	32
Eastgate Ct. NW&SE	21W-6S	25
Eastside Av. N&S	21W-7S	25
Eastwood Ln. N&S	22W-9S	31
Eaton Dr. NS&EW	20W-9S	32
Edgerton Ct. E&W	20W-9S	32
Edgewood Ct. N&S	22W-9S	31
Edgewood Pkwy. E&W	22W-9S	31
Elm St. N&S	22W-9S	31
Emerald Ct. N&S	21W-9S	31
Essex Ct. N&S	22W-6S	25
Everglade Av. NE&SW	22W-9S,21W-8S	31
Evergreen Ct. E&W	21W-8S	31
Evergreen Ln. E&W	21W-8S	31
Fairmount Dr. N&S	21W-6S	25
Field Rd. N&S	20W-10S	32
Fitzgerald E&W	20W-8S	32
Forest Dr. E&W	21W-9S	31
Forest Glen Pkwy. E&W	22W-8S,21W-8S	31
Fox Dr. E&W	22W-8S	31
Fox Tree Av. NS&EW	21W-7S	25
Foxboro Dr. NW&SE	22W-9S	31
Foxglove St. NW&SE	21W-7S	25
Frost Ct. NE&SW	21W-9S	31
Gatewood Ln. N&S	22W-9S	31
Geneva St. N&S	21W-9S	31
Glenn Ct. N&S	22W-7S	25
Goldfinch St. E&W	21W-7S	25
Golfview Dr. NS&EW	20W-8S	32
Green Dr. NE&SW	22W-6S	25
Greene Rd. NS&EW	23W-7S	25
Greenleaf St. NW&SE	21W-7S	25
Grissom Ct. E&W	22W-7S	25
Hagen Ct. E&W	23W-7S	25
Halsey Ct. NW&SE	22W-7S	25
Halsey Dr. N&S	22W-7S	25
Harcourt Dr. E&W	20W-9S	32
Hartford Ln. E&W	20W-9S	32
Harvest Av. N&S	21W-7S	25
Hawthorne Av. N&S	21W-8S	31
Hiawatha Pkwy. NE&SW	21W-9S	31
Hickory Ct. E&W	22W-9S	31
High Gate Ln. N&S	20W-10S	32
High Tr. NE&SW	22W-7S	25
Highland Ct. NW&SE	21W-8S	31
Hobson Ct. E&W	22W-7S	25
Hobson Rd. NE&SW	23W-7S,22W-7S	25
Hobson Valley Dr. N&S	22W-7S	25
Iroquois Ct. N&S	22W-9S	31
Jackson Ct. N&S	22W-6S	25
Jackson Dr. NE&SW	22W-6S,21W-6S	25
Janes Ct. E&W	21W-8S	31
Janeswood Dr.* NS&EW	21W-7S	25
Jo Ann Ln. E&W	22W-9S	31
Jonquil Ln. E&W	22W-8S,21W-8S	31
Juneberry Av. NE&SW	21W-7S	25
Kelly Ct. N&S	21W-9S	31
Kelly Dr. E&W	21W-9S	31
Kildeer St. E&W	21W-7S	25
Kimball Ct. NW&SE	22W-7S	25
Kincaid Ct. NW&SE	22W-7S	25
Kincaid Dr. NE&SW	22W-7S	25
Knob Hill Dr. N&S	20W-10S,20W-9S	32

	Grid	Page
Knotty Pine Ct. NE&SW		
....	21W-8S	31
Koloff Ct. NE&SW ..	22W-7S	25
Langley Ct. NW&SE..	21W-7S	25
Larchwood Av. N&S	21W-8S	31
Larchwood Ln. NW&SE		
....	21W-8S	31
Larkspur Ln. NE&SW		
....	21W-7S	25
Laurel Ct. E&W ...	21W-9S	31
Leawood Ct. NW&SE		
....	22W-9S	31
Leawood Ln. N&S ..	22W-9S	31
Lee St. NW&SE ..	21W-6S	25
Lindenwood Ln. N&S		
....	22W-9S	31
Longford St. N&S ..	20W-9S	32
Lorraine Ct. E&W	21W-7S	25
Luzern Ct. E&W	21W-7S	25
Macarthur Dr. N&S ..	22W-7S	25
Manchester Ct. E&W		
....	21W-7S	25
Mansfield St. E&W ..	20W-9S	32
Marshall Dr. N&S ..	22W-7S	25
Martin Ct. NW&SE ..	22W-7S	25
Martin Dr. NW&SE ..	22W-7S	25
Meadowdale Ln. E&W		
....	21W-9S	31
Meadowood Ave. NW&SE		
....	20W-9S	32
Mendingwall Dr. N&S		
....	22W-9S	31
Meyer Rd. N&S	22W-6S	25
Middlebury Av. N&S..	20W-9S	32
Middlecoff Ct. N&S ..	23W-7S	25
Mitchell Dr. NE&SW		
....	22W-7S,21W-7S	25
Mohawk Av. E&W ..	21W-8S	31
Nelson Ct. N&S ..	23W-7S	25
Newcastle Ct. NE&SW		
....	20W-9S	32
Newport Ct. E&W...	21W-7S	25
Newport Dr. N&S ..	21W-7S	25
North Creek Dr. E&W		
....	22W-9S,21W-9S	31
North-South Tollway N&S		
....	21W-9S,21W-6S	31
Northgate Ct. NW&SE		
....	21W-6S	25
Norwood Ct. N&S....	22W-9S	31
Nottingham Rd. E&W		
....	20W-9S	32
O'Hare Ct. NW&SE ..	22W-7S	25
Oak Leaf Ct. N&S ..	21W-9S	31
Oak Tree Ct. E&W ..	22W-7S	25
Oak Tree Tr. N&S ..	22W-7S	25
Oakview Ct. E&W....	22W-9S	31
Oakview Ln. N&S ..	22W-9S	31
Old Fence Ct. N&S ..	22W-9S	31
Orchard Ln. N&S ..	21W-9S	31
Orchard Rd. N&S		
....	20W-12S,20W-11S	34
Oriole Ct. N&S ..	21W-7S	25
Ouimet Ct. E&W ..	23W-7S	25
Paddington Rd. N&S		
....	20W-9S	32
Park Lane Ct. E&W ..	21W-7S	25
Park Lane Dr. NW&SE		
....	21W-7S	25
Parkside Dr. E&W....	22W-6S	25
Parkwood Ln. NW&SE		
....	22W-9S	31
Patterson Ct. NW&SE		
....	21W-7S	25
Patton Dr. N&S ..	22W-7S	25
Patty Berg Ct. NE&SW		
....	23W-7S	25
Pembridge Av. N&S..	20W-9S	32
Penny Royal Pl.* E&W		
....	21W-8S	31
Perry Dr. N&S ..	21W-6S	25
Pershing Ct. NW&SE		
....	22W-7S	25
Peterson Ct. NE&SW		
....	22W-9S	31
Pheasant St. E&W ..	21W-7S	25
Piers Dr. N&S ..	21W-7S	25
Plover Ct. E&W...	21W-7S	25
Prairieview Av. N&S..	21W-7S	25
Providence Dr. NE&SW		
....	22W-7S	25
Quail Ct.* N&S ..	21W-8S	31
Ravinia Ln. E&W ..	21W-8S	31
Red Bud Ln. NE&SW		
....	21W-7S	25
Redcliff St. E&W ..	20W-9S	32
Redwing Ct. NW&SE		
....	21W-7S	25
Redwing Dr. NW&SE		
....	21W-7S	25
Remington Ct.* N&S		
....	21W-8S	31
Rhodes Ct.* N&S ..	21W-8S	31
Richfield Ct. E&W....	20W-8S	32
Ridge Ln. E&W ..	21W-8S	31
Ridgeway Dr. NE&SW		
....	22W-6S	25
Roberts Ct. NE&SW	22W-7S	25
Roberts Dr. N&S ..	22W-7S	25
Rosebury Ave. N&S..	20W-9S	32
Rosewood Ct. E&W..	21W-9S	31
Ross Dr. N&S ..	21W-8S	25
Ryan N&S	20W-8S	32
Salem Ct. NE&SW ..	21W-7S	25
Sarazen Ct. E&W....	23W-7S	25
Scarsdale Ct. NE&SW		
....	20W-9S	32
Scenicwood Ln. E&W		
....	22W-9S	31
Seminole Ct. N&S..	21W-8S	31
Shagbark N&S ..	22W-9S	31
Shelley Ct. E&W ..	22W-9S	31
Shepherd Ct. E&W ..	22W-7S	25
Sheridan Ct. NW&SE		
....	22W-6S	25
Sheridan Dr. NE&SW		
....	22W-6S	25
Sherman Dr. NE&SW		
....	22W-6S	25
Sioux Av. NE&SW ..	21W-8S	31
Slayton Ct. N&S ..	22W-7S	25
Snead Ct. NS&SW ..	23W-7S	25
Spring St. E&W...	21W-9S	31
Sprucewood Av. NW&SE		
....	21W-8S	31
Stable Rd. N&S....	22W-6S	25
Steven Ct. N&S....	21W-9S	31
Stevenson Expressway		
NE&SW..........	20W-11S	34
Stillwell Ct. E&W ..	22W-7S	25
Stonewall Av. E&W		
....	22W-6S,21W-6S	25
Stonewall Ct. NW&SE		
....	21W-6S	25
Sumac St. NW&SE ..	21W-7S	25
Summerall St. E&W ..	22W-7S	25
Summerhill Dr. E&W	22W-6S	25
Sun Drop Av. N&S ..	21W-7S	25
Sundowner Rd. N&S		
....	22W-6S	25
Sunnydale St. E&W ..	21W-7S	25
Sussex Ct.* E&W ..	21W-8S	31
Swallow Ct. N&S ..	21W-7S	25
Taylor Dr. N&S ..	22W-7S	25
Timke Rd. N&S....	22W-6S	25
Two Paths E&W ..	22W-9S	31
Tyler Ct. NE&SW ..	22W-6S	25
Tyler Dr. N&S ..	22W-6S	25
Valleyview Ln. N&S		
....	21W-9S,21W-8S	31
Vantage St. E&W....	21W-7S	25
Vardon Ct. N&S ..	23W-7S	25
Vernon Ct. E&W ..	22W-6S	25
Vista Dr. E&W ..	21W-7S	25
Wainwright Dr. N&S..	22W-7S	25
Wake Robin Ct. NW&SE		
....	21W-7S	25
Walnut Av. NW&SE..	21W-8S	31
Waterbury Ct. NE&SW		
....	21W-9S	31
Waterbury Dr. N&S..	21W-9S	31
Wellington Rd. E&W	20W-9S	32
Westgate Ct. NW&SE		
....	21W-6S	25
Westmoreland Dr. NW&SE		
....	22W-7S,21W-7S	25
Westridge Dr. NW&SE		
....	22W-8S	31
Westview Ln. N&S		
....	22W-9S,22W-8S	31
Wharf Dr. E&W....	21W-9S	31
Wheatfield St. E&W..	21W-7S	25
Wheeler St. E&W ...	21W-7S	25
Whispering Oaks Ln. NW&SE		
....	22W-9S	31
White Ct. NE&SW ..	22W-7S	25
White Dr. E&W ..	22W-7S	25
Wild Plum Dr. E&W		
....	22W-7S,21W-7S	25
Williams Dr. NE&SW		
....	22W-7S	25
Willow Av. N&S......	21W-8S	31
Willow Ct. NE&SW ..	21W-8S	31
Willow Ln. NW&SE ..	21W-8S	31
Willow Wood Dr. NE&SW		
....	21W-9S	31
Winston Dr. N&S	22W-7S	25
Wintergreen Ct. E&W		
....	21W-7S	25
Wolfe Ct. NE&SW ..	21W-6S	25
Wolfe Dr. E&W ..	21W-6S	25
Woodlyn Dr. N&S....	21W-9S	31
Woodridge Dr.* NW&SE		
....	21W-9S,22W-7S	31
Woodsorrel Pl.* E&W		
....	21W-8S	31
Woodview Dr. NE&SW		
....	22W-8S	31
Woodward Av. N&S		
....	21W-9S,21W-7S	31
Wren Ct. E&W ..	21W-7S	25
Yellow Star St. N&S	21W-7S	25
York Ct. E&W ..	21W-7S	25
Young Ct. N&S....	22W-7S	25
Zurich St. E&W......	21W-9S	31
63rd St. E&W		
....	22W-7S,21W-6S	25
71st St. E&W		
....	22W-7S,21W-7S	25
76th St. NW&SE	22W-8S	31
83rd St. E&W		
....	22W-9S,21W-9S	31

York Township

	Grid	Page
Addison Av. N&S....	18W-1S	20
Ahrens Av. N&S	18W-0S	20
Apley Ln.* E&W	17W-1S	21
Ardmore Av. N&S....	18W-1S	20
Ashby Ln.* NW&SE ..	17W-1S	21
Ave. Barbizon* E&W		
....	19W-3S	20
Ave. Chateau E. N&S		
....	18W-2S	20
Ave. Chateau N. E&W		
....	19W-2S	20
Ave. Cherburg* NE&SW		
....	19W-3S	20
Ave. LaTours N&S ..	19W-2S	20
Ave. Normande* E&W		
....	19W-3S	20
Ave. Orleans* NW&SE		
....	19W-3S	20
Ave. Royal* NW&SE		
....	19W-3S	20
Avondale Ln. N&S ..	20W-2S	20
Barbara Ln.* E&W ..	20W-2S	20
Beaumont Ln. ..	20W-2S	20
Bryan St. N&S	16W-0S	21
Buckingham Ln.* E&W		
....	17W-1S	21
Buttercup Ln.* E&W..	17W-1S	21
Cadwell Av. N&S ..	15W-0S	21
Cedar Av. N&S......	15W-0S	21
Charlotte Ln.* E&W ..	19W-3S	20
Chase Av. N&S......	19W-1S	20
Church St. N&S ..	18W-1S	20
Clara Ct.* E&W......	19W-3S	20
Colonial St. N&S ..	20W-2S	20
Concord St. E&W ..	15W-0S	21
Cornell Av. N&S ..	15W-0S	21
Danby St. N&S......	18W-1S	20
Dillon Ln. N&S......	18W-1S	20
Douglas Rd. N&S ..	19W-3S	20
Duncan St. N&S ..	15W-0S	21
Eastgate Rd. E&W ..	19W-1S	20
Edgewood Av. N&S ..	18W-0S	20
Edgewood Rd. E&W		
....	20W-0S	20
Edward St. E&W ..	18W-0S	20
Elliot Ln.* N&S ..	18W-1S	20
Euclid Av. N&S		
....	17W-1S,17W-0S	21
Fairfield Av. N&S		
....	19W-0S,19W-1S	20
Fairfield Ln.* N&S ..	19W-2S	20
Fillmore St. E&W ..	15W-0S	21
Florence Av. N&S ..	18W-3S	20
Forest Grove Ln. N&S		
....	19W-3S	20
Forest Ln. NE&SW ..	15W-0S	21
Formento Ct.* E&W ..	19W-3S	20
Frontage Rd. E&W		
....	16W-0S,15W-0S	21
Ginger Brook Dr.* E&W		
....	19W-3S	20
Ginny Ln.* E&W ..	19W-3S	20
Glen Av. N&S	20W-2S	20
Glen Ct. E&W	20W-2S	20
Glen Pl. N&S	20W-2S	20
Glendenning Rd. N&S		
....	19W-3S	20
Gloria Ln.* E&W ..	19W-3S	20
Grace St. N&S	19W-1S	20
Gramercy Park Ln. E&W		
....	19W-1S	20
Hampton Ln. N&S ..	20W-2S	20
Harding Rd. E&W ..	20W-0S	20
Harvard St. E&W ..	15W-0S	21
Highland Av. N&S...	19W-1S	20
Highmoor Rd. NE&SW		
....	20W-0S	20
Hill St. E&W	17W-1N	14
Holyoke Ln.* N&S...	17W-1S	21
Ingersoll Ln.* N&S .	17W-1S	21
Ingrid Ln.* E&W ..	20W-2S	20
Jamestowne Ln. E&W		
....	18W-1S	20
Karlin Dr.* N&S ..	19W-3S	20
Kendall Av. N&S ..	15W-0S	21
Kirk Av. N&S	15W-0S	21
Kirkland Ln.* E&W .	17W-1S	21
Knollwood Ln.* E&W		
....	17W-1S	21
Lathrop Ln.* N&S ..	17W-1S	21
Lewis Av. N&S	19W-1S	20
Lexington St. NE&SW		
....	15W-0S	21
Liberty Blvd. N&S....	18W-3S	20
Linden Av. N&S ..	15W-0S	21
Lombard Av. N&S ..	19W-1S	20
Lowell Ln.* E&W ..	18W-1S	20
Luther Av. N&S......	18W-1S	20
Manor Ln. E&W ..	17W-1N	14
Margaret Ln.* E&W ..	19W-3S	20
Marion Ln.* E&W ..	19W-3S	20
Marlborough Rd. E&W		
....	20W-2S	20
Mary's Ln.* N&S ..	19W-3S	20
Mayfair Rd. E&W	20W-2S	20
Meadow Ln. N&S ..	19W-3S	20
Michigan Av. N&S..	18W-1S	20
Morningside Dr. E&W		
....	17W-1S	21
Morris Av. N&S ..	18W-0S	20
Myrtle Av. N&S......	17W-1S	21
Oakbrook Club Dr. N&S		
....	16W-1S	21
Old York Rd. N&S .	15W-0S	21
Park St. N&S ..	18W-3S	20
Pine Ln. N&S........	19W-1S	20
Pine View Ct. N&S ..	19W-1S	20
Pine View St. N&S ..	19W-1S	20
Poplar Av. N&S......	15W-0S	21
Prospect Av. N&S ..	16W-0S	21
Radford Ln. N&S ..	18W-1S	20
Riordan Rd. E&W	17W-0S	21
Riverview Av. N&S ..	17W-1N	14
Rochdale Cir. NS&EW		
....	19W-1S	20
Rosebud Dr. N. NS&EW		
....	19W-1S	20
Rosebud Dr. S. N&S		
Schiller St. E&W	17W-1N	14
School Av. N&S		
....	18W-0S,18W-1S	20
School St. N&S......	18W-3S	20
Shady Ln. NE&SW ..	20W-0S	20
Southlane Dr. NE&SW		
....	17W-1S	21
Standish Ln. E&W ..	17W-1S	21
Sterling Rd. N&S ..	19W-3S	20
Stewart Av. N&S ..	19W-1S	20
Stratford Ln.* N&S ..	17W-1S	21
Summit Av. N&S	17W-0S	21
Sunshine Dr. E&W ..	17W-1S	21
Theresa Ln.* E&W ..	19W-3S	20
Tower Rd. N. E&W ..	18W-2S	20
Tower Rd. S. E&W ..	18W-2S	20
Villa Av. N&S	17W-0S	21
Vista St. NE&SW ..	19W-1S	20
Vivian Ln.* E&W ..	19W-3S	20
Washington St. N&S		
....	18W-3S	20
Westview Av. N&S ..	18W-1S	20
Wilcox Av. N&S......	18W-1S	20
Williams St. N&S	18W-3S	20
Williamsburg Ln. E&W		
....	18W-1S	20
Windsor Ln.* N&S ..	17W-1S	21
Winthrop Ln.* N&S ..	18W-1S	20
Wisconsin Av. N&S..	18W-1S	20
Woodland Ln. N&S ..	19W-3S	20
Woodland Rd. N&S ..	19W-3S	20
2nd St. E&W	17W-1N	14
3rd St. N&S	18W-1S	20
13th Pl. E&W	19W-1S	20
13th St. E&W		
....	19W-1S,18W-1S	20
14th Pl. E&W	19W-1S	20
14th St. E&W		
....	19W-1S,17W-1S	20
15th Pl. E&W	19W-1S	20
15th St. E&W		
....	19W-1S,18W-1S	20
16th St. E&W		
....	19W-1S,18W-1S	20
17th Pl. E&W	19W-1S	20
17th St. E&W	19W-1S	20
18th St. E&W		
....	19W-1S,18W-1S	20
20th Pl. E&W	18W-1S	20
20th St. E&W	19W-1S	20
21st Pl. E&W	18W-1S	20
22nd St. E&W		
....	20W-2S,18W-1S	20
31st St. E&W........	19W-2S	20
36th St. E&W........	19W-3S	20
37th St. E&W........	19W-3S	20
38th St. E&W........	19W-3S	20
39th St. E&W		
....	19W-3S,18W-3S	20

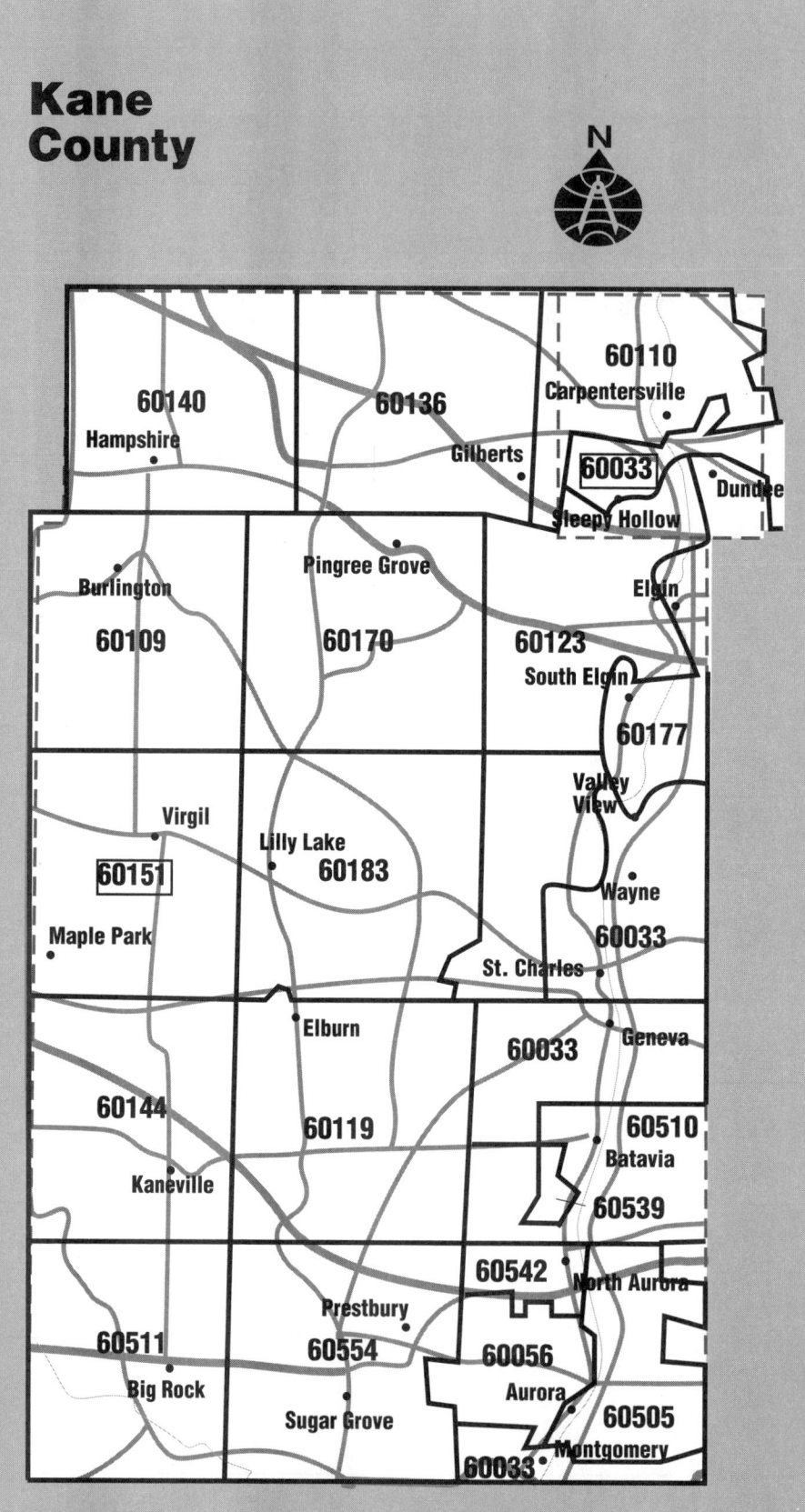

Kane County

N

60140
Hampshire

60136
Gilberts

60110
Carpentersville

60033

Dundee

Sleepy Hollow

Burlington

Pingree Grove

Elgin

60109

60170

60123

South Elgin

60177

Valley
View

Virgil

Lilly Lake

60151

60183

Wayne

60033

Maple Park

St. Charles

Elburn

Geneva

60033

60144

60510
Batavia

60119

Kaneville

60539

60542

North Aurora

Prestbury

60511

60554

60056

Big Rock

Aurora

60505

Sugar Grove

Montgomery

60033

Kane
County

StreetFinder®

page	**Contents**
A	Using the StreetFinder
B	PageFinder Map
C	Vicinity Map
1	Contents
2	Quick Reference Guide
4	StreetFinder Maps
52	StreetFinder Index

Photo credit: Geneva Dam, Kane County Development Department

Kane County Municipal Offices

	Location	Page
Aurora City Hall	35W-7S	51
44 E Downer Pl; 892-8811		
Batavia City Hall	35W-1S	39
101 N Island Ave; 879-1424		
Carpentersville Village Hall	32W-16N	51
1200 L W Besinger Dr; 426-3439		
East Dundee Village Hall	32W-15N	15
120 Barrington Ave; 426-2822		
Elgin City Hall	34W-11N	21
150 Dexter Ct; 695-6500		
Geneva City Hall	35W-1N	33
21 S 1st; 232-0854		
Montgomery Village Hall	36W-9S	50,51
1300 S Broadway Ave; 896-8080		
North Aurora Village Hall	36W-4S	44
25 E State; 897-8228		
St. Charles City Hall	35W-3N	33
2 E Main; 377-4446		
Sleepy Hollow Village Hall	34W-14N	15
Thoroughbred Lane & Sleepy Hollow Rd; 426-6700		
South Elgin Village Hall	34W-8N	27
10 N Water; 742-5780		
West Dundee Village Hall	33W-15N	15
102 S Second; 426-6161		

Cemeteries

	Location	Page
Baker Cem	44W-8N	24
Blackberry Cem	43W-1N	30
Bluff City Cem	32W-10N	21
Burlington Union Cem	48W-11W	16
Calvary Cem	35W-6S	44,45
Dundee Twp Cem	31W-14N	17
East Aurora Cem	35W-7S	51
Eastside Cem	35W-0S	39
Fowler Grove Cem	48W-9S	46
French Cem	36W-9S	50,51
Gardner Cem	49W-1N	28
Hampshire Center Cem	46W-16N	11
Jericho Cem	43W-9S	48
Lily Lake Cem	44W-5N	24
Little Woods Cem	33W-7N	27
Mt. Hope Cem	32W-10N	21
Mt. Olivet Cem	36W-8S	50
New Hampshire Cem	44W-4N	30
North Cem	35W-4N	33
Oak Hill Cem	34W-1N	33
Old St. Mary's Cem	48W-0N	34
Resurrection Cem	37W-0S	38
Riverside Cem	36W-9S	50,51
River Valley Mem Gardens	33W-13N	21
Spring Lake Cem	36W-8S	50
St. Joseph Cem	35W-5S	45
St. Michael Cem	35W-6S	45

	Location	Page
St. Nicholas Cem	34W-7S	51
St. Paul's Cem	35W-9S	51
Sts. Peter & Paul Cem	47W-4N	28,29
South Burlington Cem	49W-8N	22
Sugar Grove Cem	42W-4S	42
Thatcher Cem	48W-3N	28
Union Cem	35W-4N	33
Welch Cem	47W-8S	46,47
West Aurora Cem	35W-6S	44,45
West Big Rock Cem	50W-6S	40
Westside Cem	35W-2S	39

Colleges & Universities

	Location	Page
Aurora Univ	37W-7S	50
Elgin Comm Col	36W-10N	20
Judson Col	34W-12N	21
Waubonsie Comm Col	43W-4S	42
Waubonsie Comm Col D T Campus	35W-7S	51

Forest Preserves

	Location	Page
Aurora West	39, 40W-5S	43
Binnie	37W-17N	8
Blackhawk	34W-7N	27
Bliss Woods	42W-5S	42
Burnidge	37W-13N	20
Campton	41W-4N	31
Elburn	45W-2N	30
Fabyan	35W-0N	39
Fox River Shores	33W-17N	9
Glenwood Park	35W-2S	39
Gunar Anderson	35W-0N	39
Hampshire	43W-16N	12
Helm Woods	31W-17N	11
Johnson's Mound	41W-0N	37
Leroy Oaks	37W-4N	32
Les Arends	36W-3S	44
Lone Grove	48W-1S	34
Nelson Lake	39W-2S	38
Oakhurst	33W-8S	51
Rutland	40W-15N	12,13
Tyler Creek	34W-12N	21
Voyageurs Landing	32W-13N	15

Golf & Country Clubs

	Location	Page
Aurora CC	37W-8S	50
Bonnie Dundee CC	31W-16N	11
Elgin CC	37W-11N	19
Fox Valley CC	35W-3S	44,45
Geneva GC	35W-1N	33
Phillips Park & GC	34W-9S	51
Prestbury GC	41W-6S	49

	Location	Page
Randall Oaks GC	36W-16N	14
St. Charles CC	35W-4N	33
Spartan Meadows GC	36W-10N	20
Valley Green GC	36W-4S	44

Hospitals

	Location	Page
Copley Mem Hosp	35W-7S	51
Copley Hosp	35W-8S	51
Delnor Hosp	35W-3N	32,33
Mercy Center for Health Care Services	36W-5S	44
St. Josephs Hosp	36W-11N	20
Sherman Hosp	33W-12N	21

Shopping Centers

	Location	Page
Ashland Plaza SC	35W-9S	51
Clock Tower Plaza SC	33W-10N	21
Eastgate Plaza SC	35W-6S	44,45
Fox River Business Plaza SC	33W-13N	21
Indian Trail SC	35W-6S	44,45
Indian Trail Court SC	35W-5S	45
Meadowdale SC	32W-16N	15
Northgate SC	35W-6S	44,45
South Ctr SC	36W-9S	50,51
Spring Hill Mall SC	33W-15N	15
Town & Country Plaza SC	35W-10N	20,21
Tyler Creek Plaza SC	35W-12N	20
Village Mart SC	34W-6S	45
West Aurora Plaza SC	38W-7S	50
Wing Park Manor SC	35W-12N	20

Kane County Close-up

Elgin Public Museum
225 Grand Blvd., Elgin
This natural history museum features rotating exhibits in the fields of geology, Native American history and endangered species studies. Discovery Room features hands-on exhibits for children. Closed during January.

Beith House
8 Indiana St., St. Charles
Constructed in 1850 by stonemason William Beith, this preservation study house remains one of the few riverstone homes with its original design features. Those who appreciate fine architecture will appreciate its masonry, woodwork and unusual decorative details. This fully restored dwelling includes a research library and exhibits. Touch-It Table displays fabrics, tools, maps and other artifacts. Listed on National Register of Hisoric Places. Open June-Nov. or by appointment.

Aurora Historical Society
corner of Cedar St. and Oak Ave., Aurora
Dedicated to preserving Aurora's history, this site features 2 exhibition facilities: the 18-room William Tanner house, built in 1857, is a fully furnished Victorian home open to the public; the History Center is a multi-use exhibition space that houses artifacts relating to the history of this once big "railroad town": furniture, musical instruments, motor cycles, a railway hand car and even locally found Mastodon bones. Open mid-April to mid-December.

Blackberry Farm Village
corner of Barnes Rd. and Galena Blvd., Aurora
This 54-acre historical farm village features a carriage house, a 19th century farm museum, Huntoon House and a one-room schoolhouse. Blacksmiths, potters, weavers and other artisans give frequent craft demonstrations. Visitors can enjoy pony rides, hay rides, fishing and train rides. Open from mid-April to Labor Day.

SciTech
18 W. Benton, Aurora
This interactive museum, with its hands-on exhibits, lets visitors experience the wonders of science. Changing exhibits feature experiments in light, sound and electricity. Comprehensive programs for children. Auditorium, activity room, concessions and gift shop. Open daily.

Kane County Flea Market
Kane County Fairgrounds, Randall Rd., between Rtes. 64 & 38, St. Charles
Not just another flea market! Since its founding in 1967, this flea market has grown into a major event with fans the world over. Unique merchandise includes antiques and other collectibles. Country breakfasts available. Open the first Sunday of every month and the preceding Saturday, all year.

Garfield Farm Museum
5 miles W of Geneva, IL, off Rt. 38 on Garfield Rd., LaFox
This historically intact farm features a homestead and teamster's inn constructed in the 1840's. Resting on 250 acres, this unique site has undergone substantial prairie restoration and includes an 1842 hay barn, an 1849 horse barn and heirloom-variety vegetable and flower gardens. Rare animal breeds include Devon oxen and Marino sheep. Open by appointment all year.

Schingoethe Center for Native American Cultures
347 S. Gladstone Ave., Aurora
Located on the campus of Aurora University, Schingoethe Center displays artifacts relating to America's first inhabitants. Its collection of 10,000 objects include Kachinas, pottery, musical instruments, pipes and textiles. Many of its artifacts are touchable objects. Also, workshops, lectures and demonstrations.

Red Oak Nature Center
Rte. 25, 1 mile N of Rte. 56, N. Aurora
This site covers 40 acres of oak and maple forest on the east bank of the Fox River. This beautiful region includes a visitor's center, an observation deck, hiking trails and separate biking trails. Also, one of the few limestone caves in northeast Illinois is on the premises. Abundant wildlife.

4 KANE COUNTY

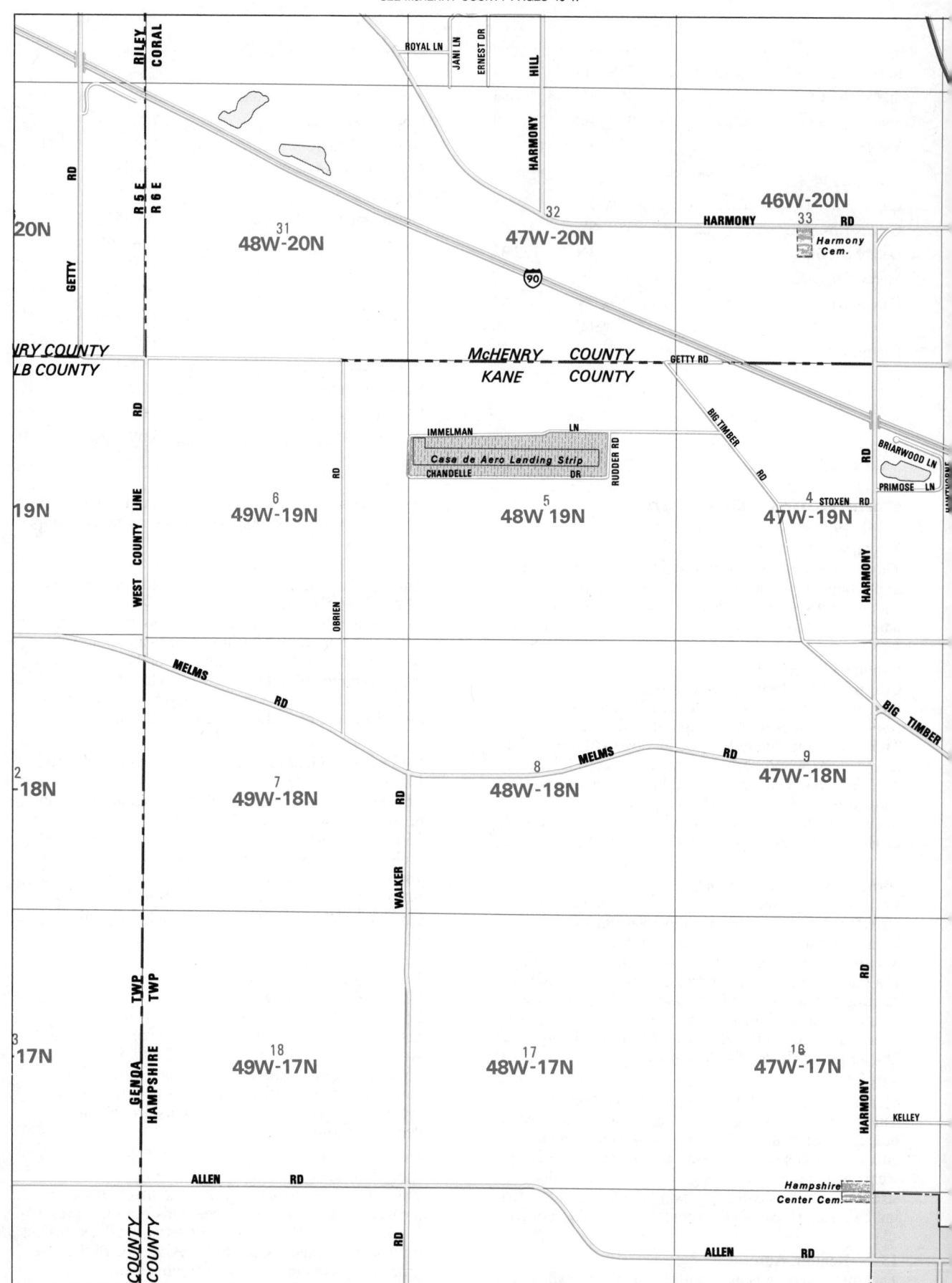

RILEY
CORAL

ROYAL LN

JANI LN
ERNEST DR
HARMONY
HILL

GETTY

R 5 E
R 6 E

RD

20N

48W-20N
31

47W-20N

46W-20N

HARMONY
33
RD

Harmony
Cem.

I-90

NRY COUNTY
LB COUNTY

McHENRY   COUNTY
KANE   COUNTY

GETTY RD

BIG TIMBER
RD

BRIARWOOD LN

RD

PRIMOSE LN

RD

WEST COUNTY LINE

RD

IMMELMAN
LN

Casa de Aero Landing Strip

CHANDELLE
DR

RUDDER RD

STOXEN RD

4

19N

49W-19N
6

48W 19N
5

47W-19N

HARMONY

OBRIEN

MELMS
RD

BIG   TIMBER

18N

49W-18N
7

RD

WALKER

8
MELMS
RD
48W-18N

RD
9
47W-18N

TWP
TWP

17N

GENOA
HAMPSHIRE

49W-17N
18

48W-17N
17

47W-17N
16

HARMONY
RD

KELLEY

ALLEN   RD

Hampshire
Center Cem

COUNTY
COUNTY

RD

ALLEN   RD

CORAL GRAFTON

PEBBLE DR
ARBOR LN
BURR OAK
OVERLOOK LN
HILLSBORO DR
HILLSBORO COVE

SEEMAN RD

RD

HARMONY

HARMONY

34

45W-20N

35

44W-20N

36

HARMONY

43W- 20N

R 6 E

R 7 E

31

42W-2

PEREGRINE
MANDA DR
DR
TR
RD
CONDOR LN
HERON DR
CAMP CT
TRAIL
PINOAK
BEALL LN

T 43 N
T 42 N

CORAL   TWP

HAMPSHIRE   TWP

CLANYARD   RD

NORTHWEST

TOLLWAY

3

DIETRICH

2

RD

1

46W-19N

45W-19N

44W-19N

HENNIG   RD

43W-1

6

WOODVIEW LN
GREEN MEADOW LN
WOODVIEW PKWY
HILLCREST DR
FELSMITH RD
RD
WOODVIEW PKWY
RD

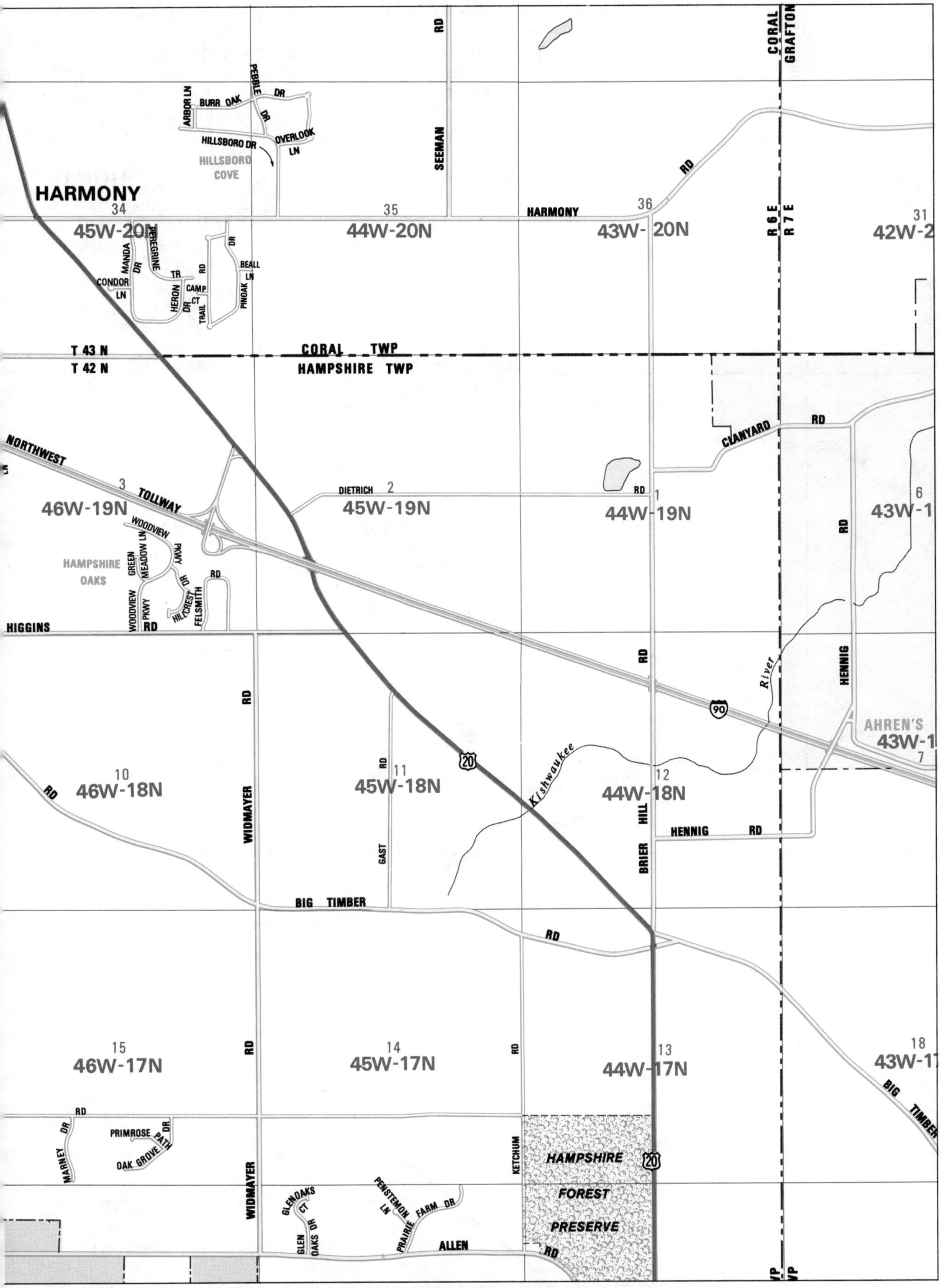

HAMPSHIRE OAKS

HIGGINS

River

SEE PAGES 6-7

AHREN'S

43W-1

RD

WIDMAYER RD

RD

RD

Kishwaukee

90

20

10

46W-18N

11

45W-18N

GAST RD

12

44W-18N

BRIER HILL

HENNIG   RD

BIG   TIMBER

RD

15

46W-17N

WIDMAYER RD

14

45W-17N

RD

13

44W-17N

18

43W-1

BIG   TIMBER

MARNEY DR
RD
PRIMROSE DR
OAK GROVE PATH

KETCHUM

HAMPSHIRE

20

GLEN OAKS CT
GLEN OAKS DR
PENSTEMON LN
PRAIRIE FARM DR

WIDMAYER

FOREST

PRESERVE

ALLEN   RD

TWP

6 KANE COUNTY

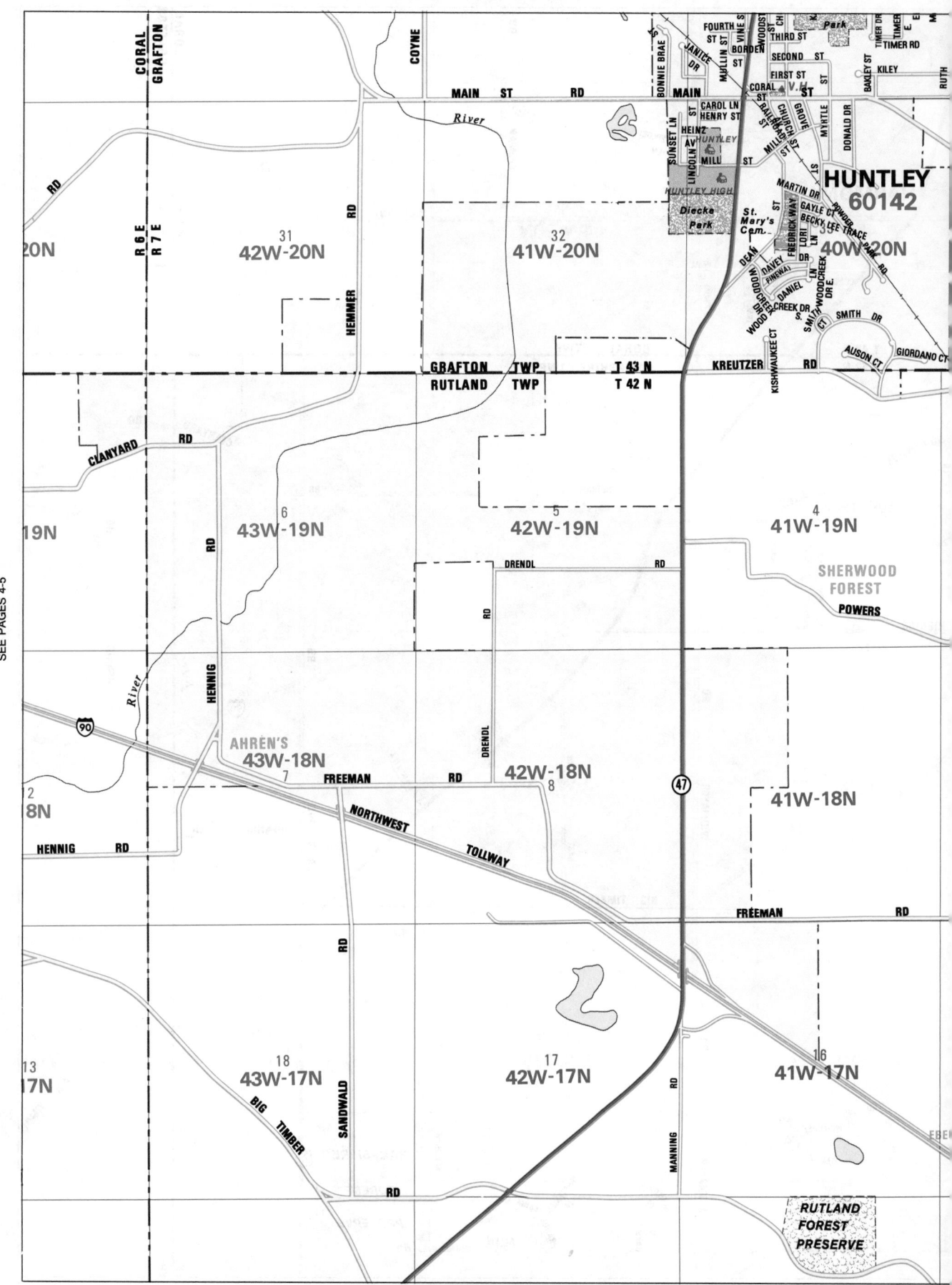

SEE McHENRY COUNTY PAGES 48-49

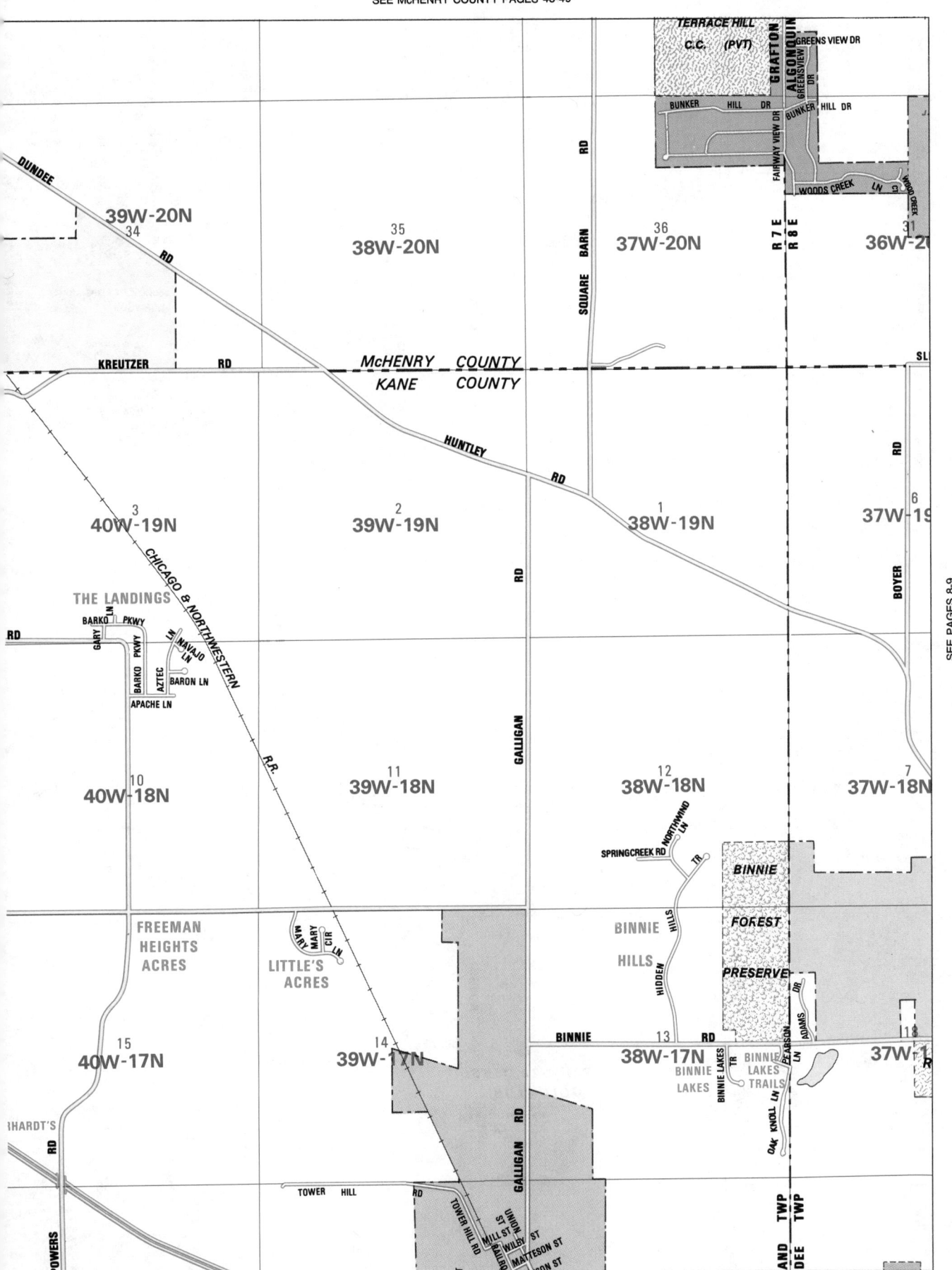

DUNDEE

TERRACE HILL
C.C. (PVT)

GRAFTON

ALGONQUIN

GREENS VIEW DR

GREENSVIEW DR

BUNKER HILL DR

BUNKER HILL DR

FAIRWAY VIEW DR

WOODS CREEK LN

WOOD CREEK CT

39W-20N
34

35
38W-20N

SQUARE BARN RD

36
37W-20N

R 7 E
R 8 E

31
36W-20

KREUTZER RD

SL

McHENRY COUNTY
KANE COUNTY

HUNTLEY RD

3
40W-19N

CHICAGO & NORTHWESTERN

THE LANDINGS

BARKO LN PKWY

GARY LN

NAVAJO LN

AZTEC LN

BARON LN

BARKO PKWY

APACHE LN

RD

2
39W-19N

RD

1
38W-19N

BOYER RD

6
37W-19

SEE PAGES 8-9

10
40W-18N

R.R.

11
39W-18N

GALLIGAN RD

12
38W-18N

SPRINGCREEK RD

NORTHWIND LN

TR

7
37W-18N

BINNIE

FOREST

PRESERVE

BINNIE
HILLS

HILLS

HIDDEN

ADAMS DR

FREEMAN
HEIGHTS
ACRES

MARY LN

MARY CIR

LITTLE'S
ACRES

BINNIE RD

13
38W-17N

BINNIE LAKES

18

37W-1

15
40W-17N

RHARDT'S RD

POWERS

14
39W-17N

GALLIGAN RD

TOWER HILL RD

TOWER HILL RD

MILL ST

UNION ST

RAILROAD ST

WILBY ST

MATTESON ST

SON ST

BINNIE
LAKES

BINNIE LAKES TR

BINNIE
LAKES
TRAILS

PEARSON LN

OAK KNOLL LN

AND TWP
DEE TWP

SEE PAGES 12-13

SEE McHENRY COUNTY PAGES 50-51

SEE PAGES 6-7

TERRACE HILL C.C. (PVT)

GRAFTON

ALGONQUIN

GREENS VIEW DR
GREENSVIEW DR

KAPER DR

HUNTINGTON DR
HUNTINGTON
PARTRIDGE CT
CRESTWOOD
GRANDVIEW CT
BRIAR
FIELDCREST
AMBERWOOD
SOMERSET CT
GRASS
NEUBERT ELEM
HOLLY
SANDPEBBLE
CT

BUNKER HILL DR
BUNKER HILL DR

N. HUNTINGTON DR.

Park

HUNTINGTON DR. S
CHERRYWOOD CIR
BRENTWOOD
BUTTERFIELD
CLEARVIEW

WOODS CREEK LN
WOOD CREEK LN

JACOBS H.S.

RD

SADDLEBROOK CT
PARKVIEW TERR. S
FAIRMONT
SHAGBARK
CANDLEWOOD
CHESTNUT
ORCHARD

HUNTINGTON DR.
BROOK LN
FAIRFIELD
EAGLE

FAIRWAY VIEW DR

R 7 E
R 8 E

36

37W-20N

31

36W-20N

RANDALL

32

35W-20N GOLF CLUB OF ILLINOIS

HANSON RD

WESTBURY DR
ZANGE
MAYFIELD LN
WINDY KNOLL DR
LEXINGTON LN W.
YORKTOWN
OAKVIEW CT

33

34W-20N

WINDY KNOLL DR
FOREST CIR
MULBERRY CT
JUNIPER CIR
HARPER
EVERGREEN
CT
TWISTED OAK DR
WEST END DR

SURREY CT
SURREY LN
SOUTH. Park Surrey LN
HARPER DR

EDGEWOOD

ST. ANDREWS CT
CARLISLE ST
SOMERSET ST
DARLINGTON CT
EDGEWOOD RD
INTERLOCH CT
ST. ANDREWS CT

KINGSTON PL

GREEN PASTURE
CARDINAL
BRIGHTON LN
DEVONSHIRE RD

SLEEPY HOLLOW RD

COUNTY LINE RD. N.

GASLIGHT DR
BRANDYWINE DR
SPRING HILL RD
SPRUCE TREE
GASLIGHT LN
RUSTIC LN

BRAEWOOD DR
HAMPTON
Park

1

38W-19N

BOYER RD

6

37W-19N

5

36W-19N

RD

4

35W-19N
ALPINE DR
CRESCENT

SLEEPY HOLLOW RD

WOODLAND SPRINGS

HOLLOWSIDE DR
FIELD CT
SPRINGBLUFF
WOODHAVEN LN
BURR OAK LN

RIDGE RD
WEST HILL RD
COUNTRY LN
WOODCREST LN

12

38W-18N

NORTHWIND LN
SPRINGCREEK RD
TR

7

37W-18N

HUNTLEY

RD

8

36W-18N

SAWYER RD

VALLEY VIEW RD

OLD BARN RD
PARSONS RD

SCHOOL

9

RD

35W-18N

MILLER

BINNIE

FOREST

PRESERVE

BINNIE HILLS

HIDDEN HILLS

ADAMS DR
PEARSON LN

RANDALL RD

RD

BINNIE

WINDING TR
SHADY LN

STURGIS CT

BURNING OAK DR

SLEEPY HOLLOW RD

13

38W-17N

BINNIE LAKES

BINNIE LAKES LAKES
BINNIE LAKES TRAILS
TR

OAK KNOLL LN

18

37W-17N

RANDALL OAKS GOLF CLUB

Randall Oaks Park

17

36W-17N

RD

HICKORY AV

OAK AV
ELM AV

16

35W-17N

RIVER

SPRING CT

HUNTLEY RD

CHADWICK LN
WESTLEY LN
PRESTON LN
HAMPTON CT
THATCHER LN
HARBOR
INLET
COVE
INLET CT

FIDDLER CT
GREEN

KITTRIDGE

BARNES
SPAIN
BREWER
AHOCHE
CIR
WATERBURY
SPY GLASS COVE

ND TWP
EE TWP

MERRIWEATHER LN

WAT

SEE PAGES 14-15

SEE McHENRY COUNTY PAGES 50-51

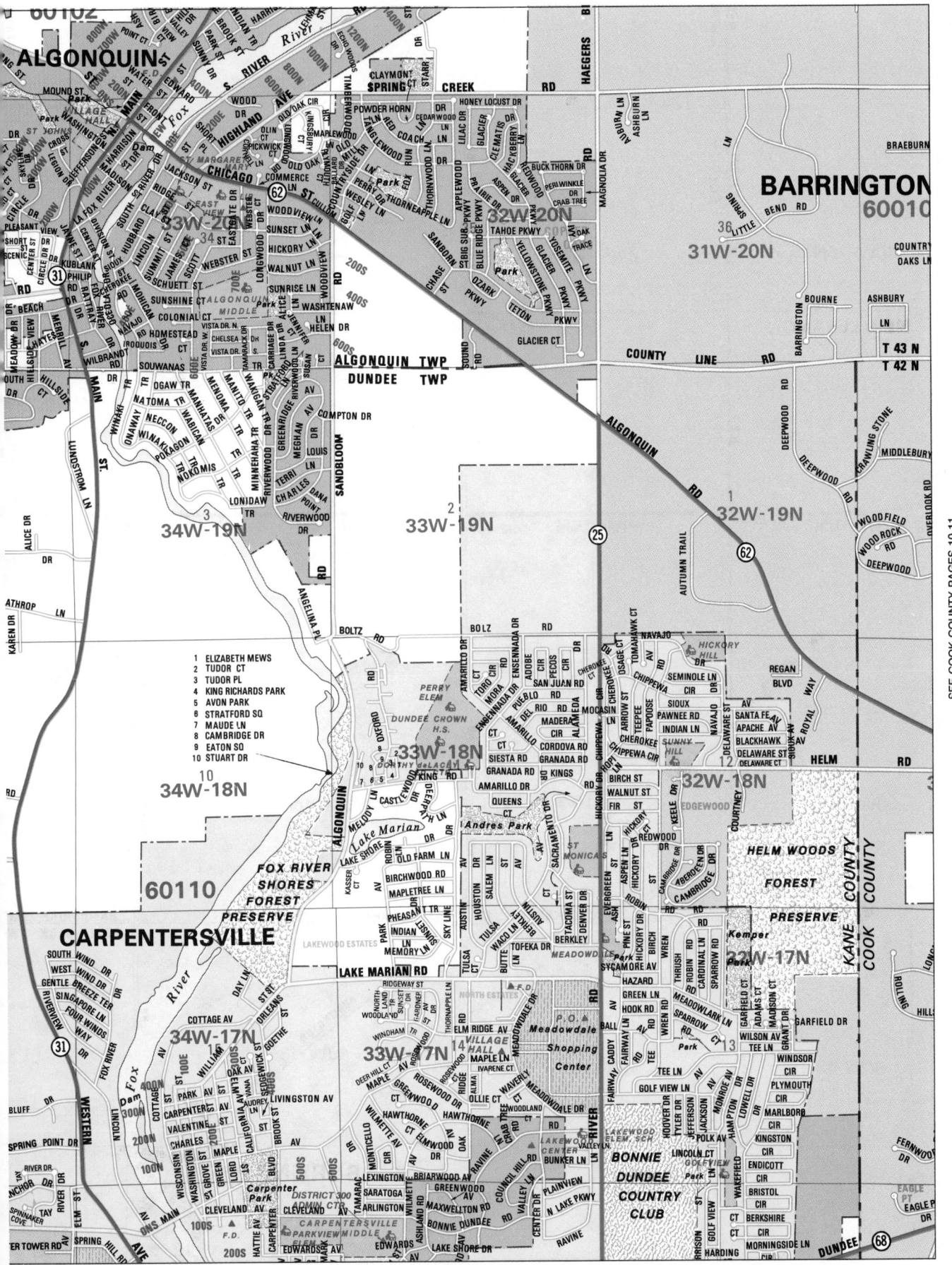

SEE COOK COUNTY PAGES 10-11

SEE PAGES 4-5

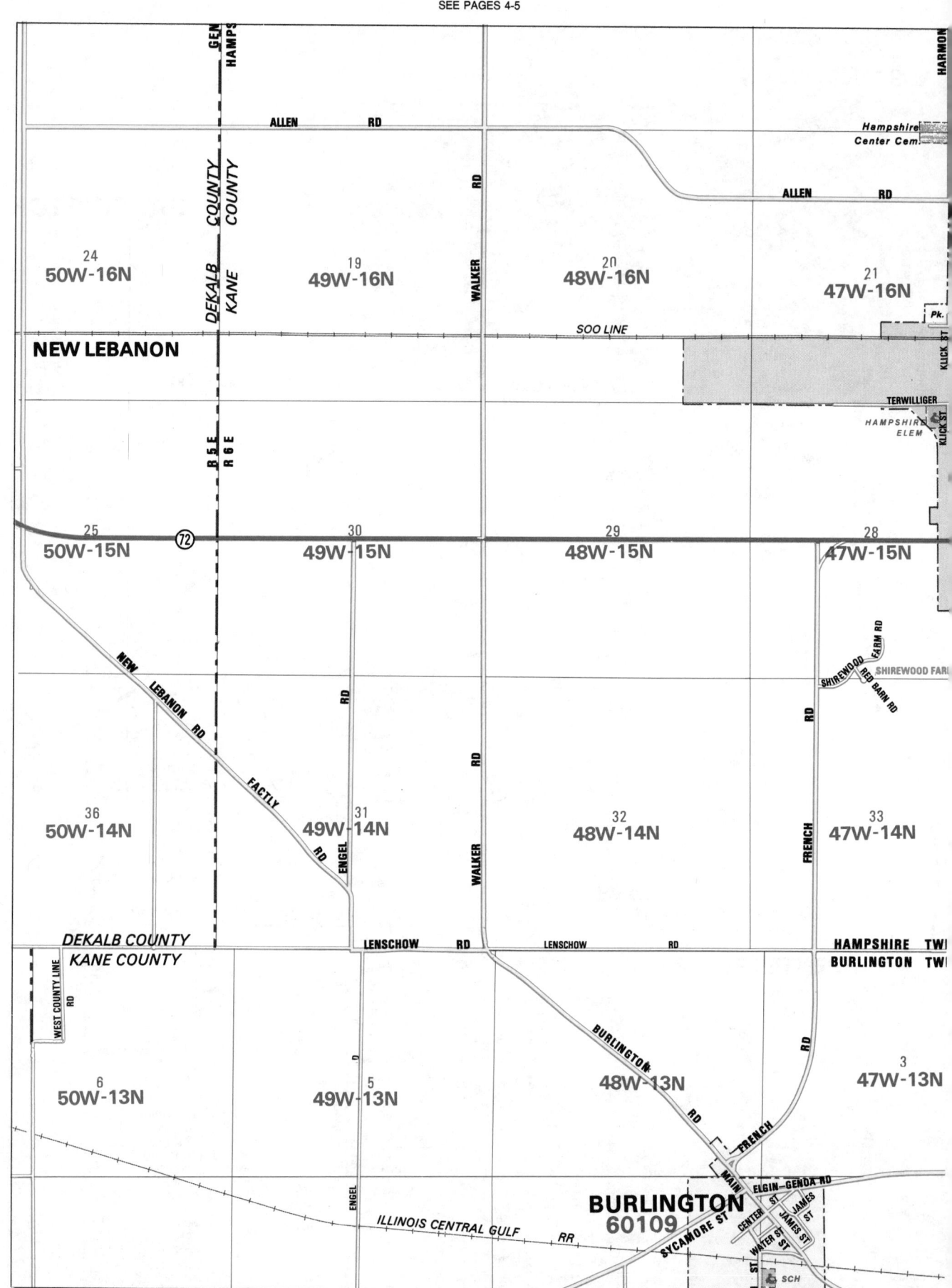

SEE PAGES 16-17

SEE PAGES 4-5

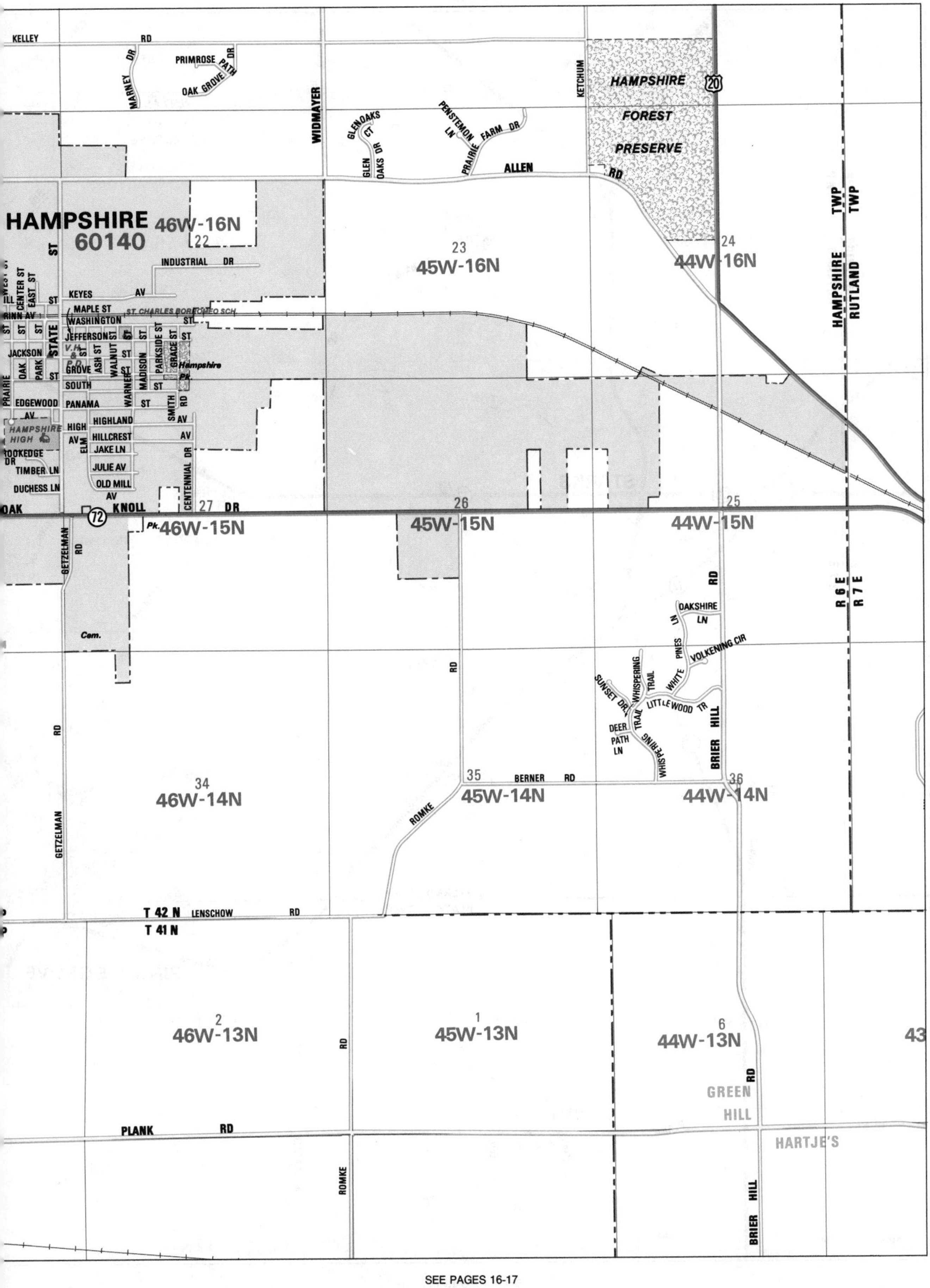

SEE PAGES 12-13

KELLEY RD

MARNEY DR

PRIMROSE PATH

OAK GROVE DR

WIDMAYER

GLENOAKS CT

GLEN OAKS DR

PENSTEMON LN

PRAIRIE FARM DR

ALLEN RD

KETCHUM

HAMPSHIRE

FOREST

PRESERVE

20

HAMPSHIRE TWP

RUTLAND TWP

HAMPSHIRE 46W-16N
60140

22

23
45W-16N

24
44W-16N

INDUSTRIAL DR

WEST ST
CENTER ST
EAST ST
STATE ST

KEYES AV

MAPLE ST

ST. CHARLES BORROMEO SCH

WASHINGTON ST

JEFFERSON ST

V.H. & P.O.

ASH ST

WALNUT ST

PARKSIDE ST

GRACE ST

JACKSON ST

OAK ST

PARK ST

PRAIRIE ST

GROVE

MADISON ST

WARNER ST

SMITH RD

Hampshire Pk.

SOUTH

PANAMA

RINN AV

EDGEWOOD AV

HAMPSHIRE HIGH

HIGH AV

ELM

HIGHLAND

HILLCREST AV

JAKE LN

CENTENNIAL DR

BROOKEDGE DR

TIMBER LN

JULIE AV

OLD MILL AV

DUCHESS LN

OAK

72

KNOLL

Pk. 46W-15N

27 DR

26
45W-15N

25
44W-15N

GETZELMAN RD

Cem.

RD

OAKSHIRE LN

LN

WHITE PINES LN

RD

VOLKENING CIR

SUNSET DR

WHISPERING TRAIL

LITTLEWOOD TR

BRIER HILL

DEER PATH LN

WHISPERING

R 6 E

R 7 E

34
46W-14N

GETZELMAN

RD

35

BERNER RD

45W-14N

36
44W-14N

ROMKE

T 42 N LENSCHOW RD

T 41 N

RD

2
46W-13N

1
45W-13N

6
44W-13N

43

BRIER HILL RD

GREEN HILL

PLANK RD

ROMKE

HARTJE'S

BRIER HILL

SEE PAGES 16-17

SEE PAGES 6-7

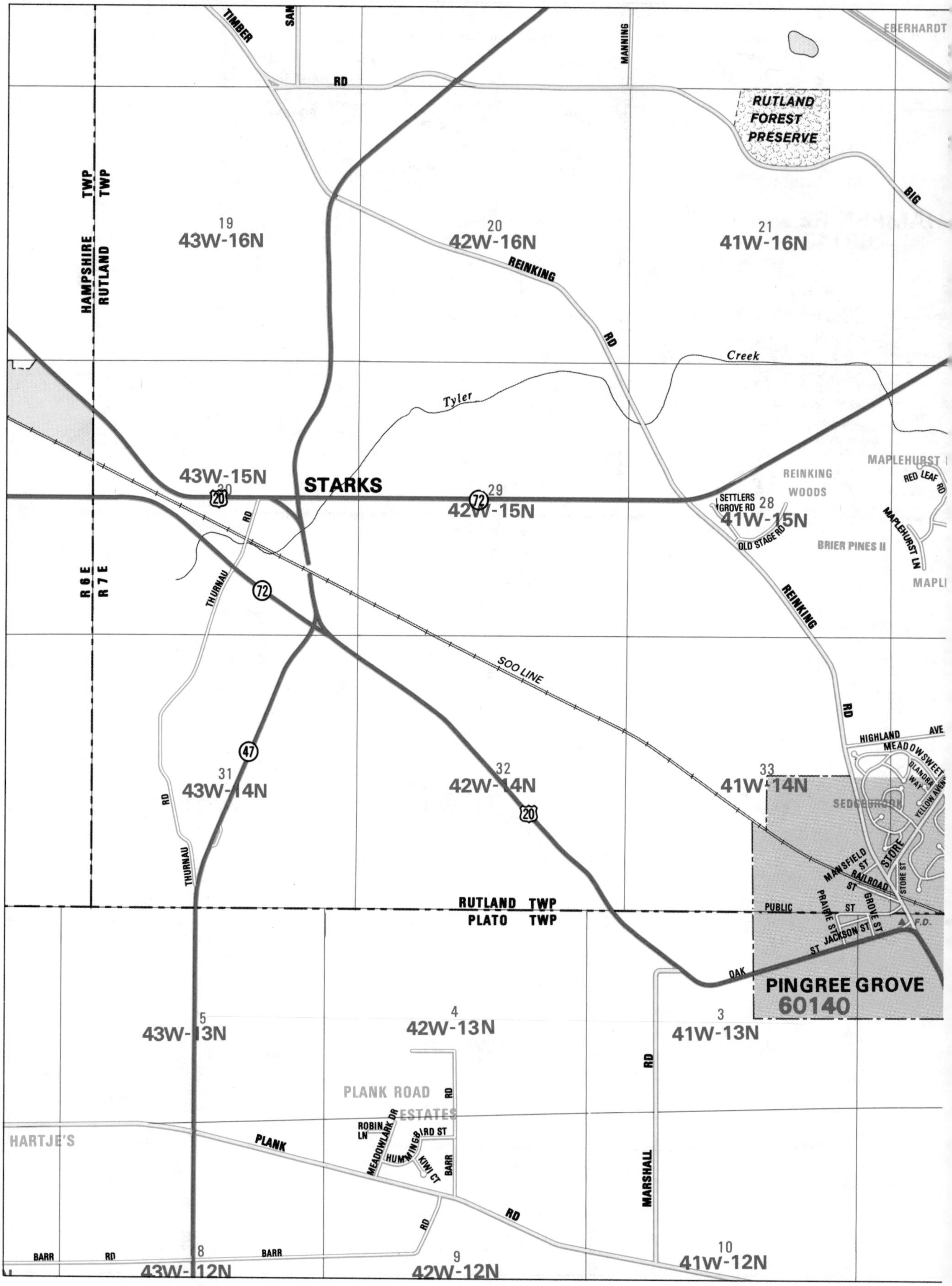

EBERHARDT

RUTLAND
FOREST
PRESERVE

BIG

TIMBER

SAN

MANNING

RD

HAMPSHIRE TWP

RUTLAND TWP

19
43W-16N

20
42W-16N

21
41W-16N

REINKING

RD

Creek

Tyler

SEE PAGES 10-11

R 6 E

R 7 E

43W-15N

20

STARKS

72 29
42W-15N

REINKING
WOODS

MAPLEHURST

RED LEAF RD

MAPLEHURST LN

SETTLERS
GROVE RD

28
41W-15N

OLD STAGE RD

BRIER PINES II

MAPLE

THURNAU

RD

72

REINKING

RD

SOO LINE

HIGHLAND
AVE

MEADOWSWEET

GLANDRA
WAY

47

31
43W-14N

32
42W-14N

20

33
41W-14N

SEDGEGROVE

YELLOW AVE

THURNAU
RD

MANSFIELD
ST

STORE

RAILROAD
ST

GROVE ST

STORE ST

PUBLIC

PRAIRIE ST

ST

F.D.

RUTLAND TWP
PLATO TWP

ST JACKSON ST

OAK

PINGREE GROVE
60140

5
43W-13N

4
42W-13N

RD

3
41W-13N

MARSHALL

RD

PLANK ROAD ESTATES

RD

HARTJE'S

PLANK

ROBIN
LN

WINGBIRD ST

BARR

MEADOWLARK DR

HUM

KIWI CT

RD

RD

BARR

RD

BARR

8
43W-12N

9
42W-12N

10
41W-12N

SEE PAGES 18-19

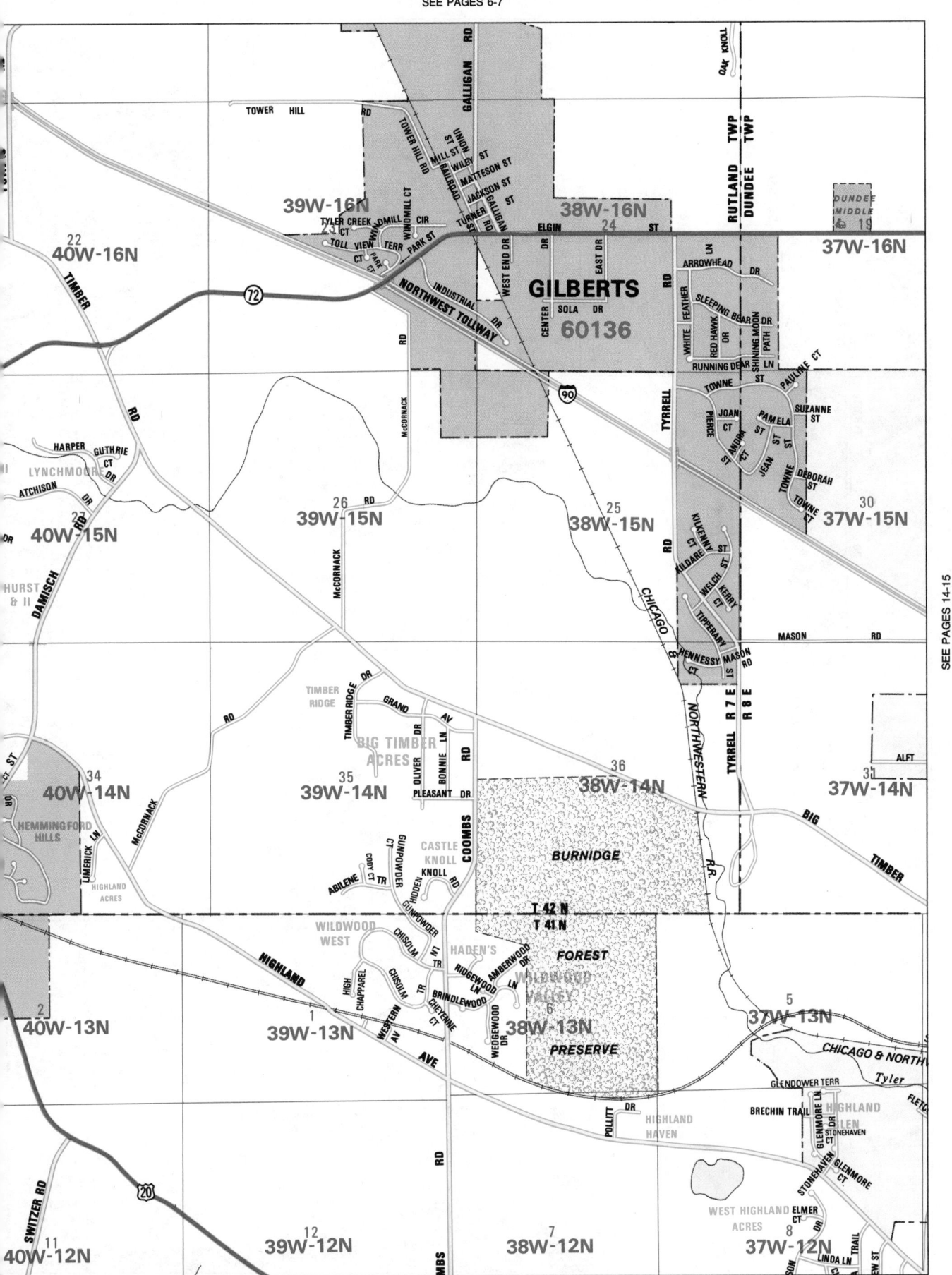

SEE PAGES 14-15

GILBERTS
60136

SEE PAGES 8-9

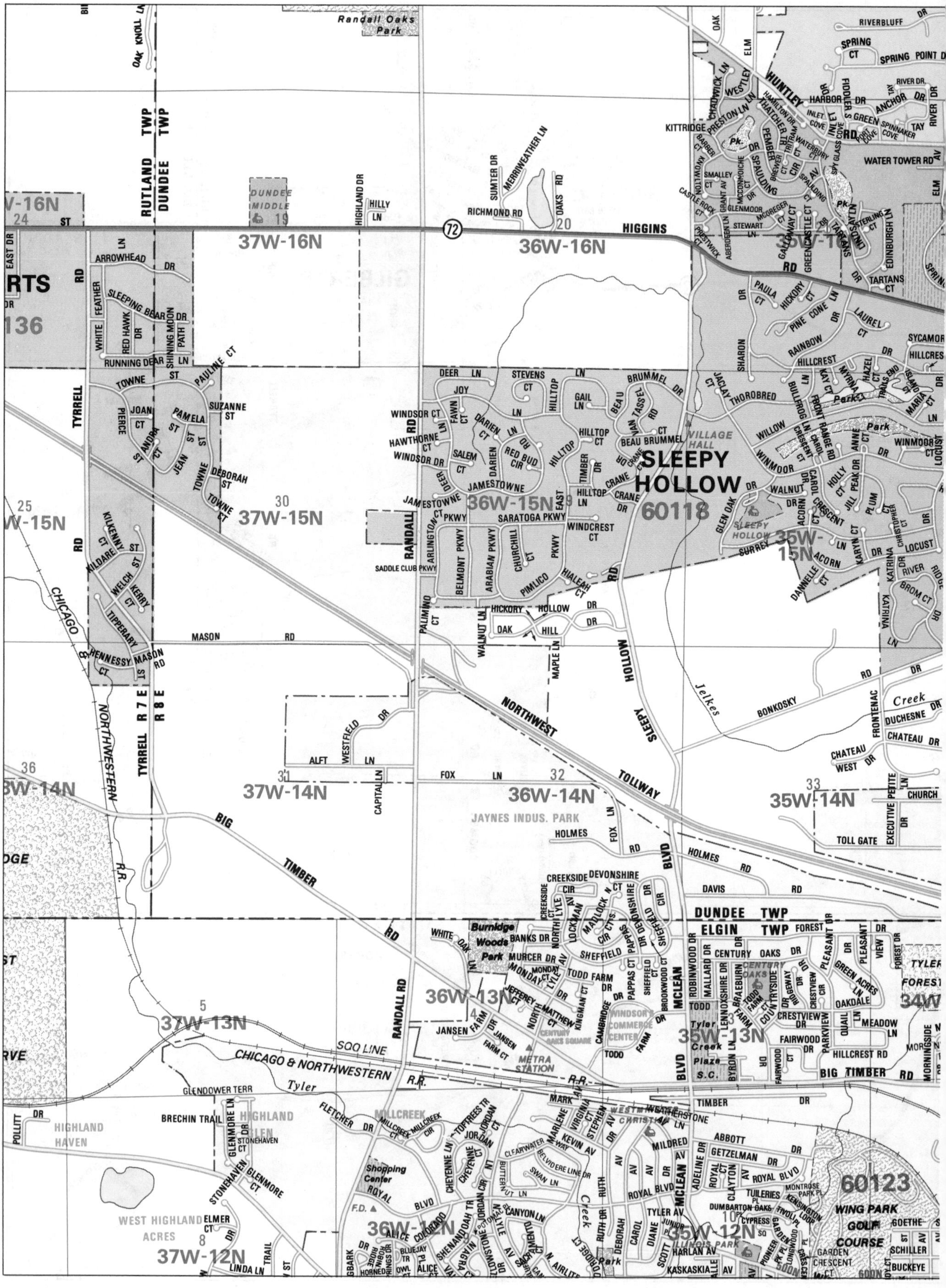

SEE PAGES 12-13

SLEEPY
HOLLOW
60118

60123

WING PARK
GOLF
COURSE

DUNDEE TWP
ELGIN TWP

RUTLAND TWP
DUNDEE TWP

Randall Oaks
Park

Dundee Middle

VILLAGE HALL

JAYNES INDUS. PARK

Burnidge Woods Park

WEST HIGHLAND ACRES

HIGHLAND HAVEN

37W-16N 36W-16N

37W-15N 36W-15N 35W-15N

37W-14N 36W-14N 35W-14N

37W-13N 36W-13N 35W-13N

37W-12N 36W-12N 35W-12N

SEE PAGES 20-21

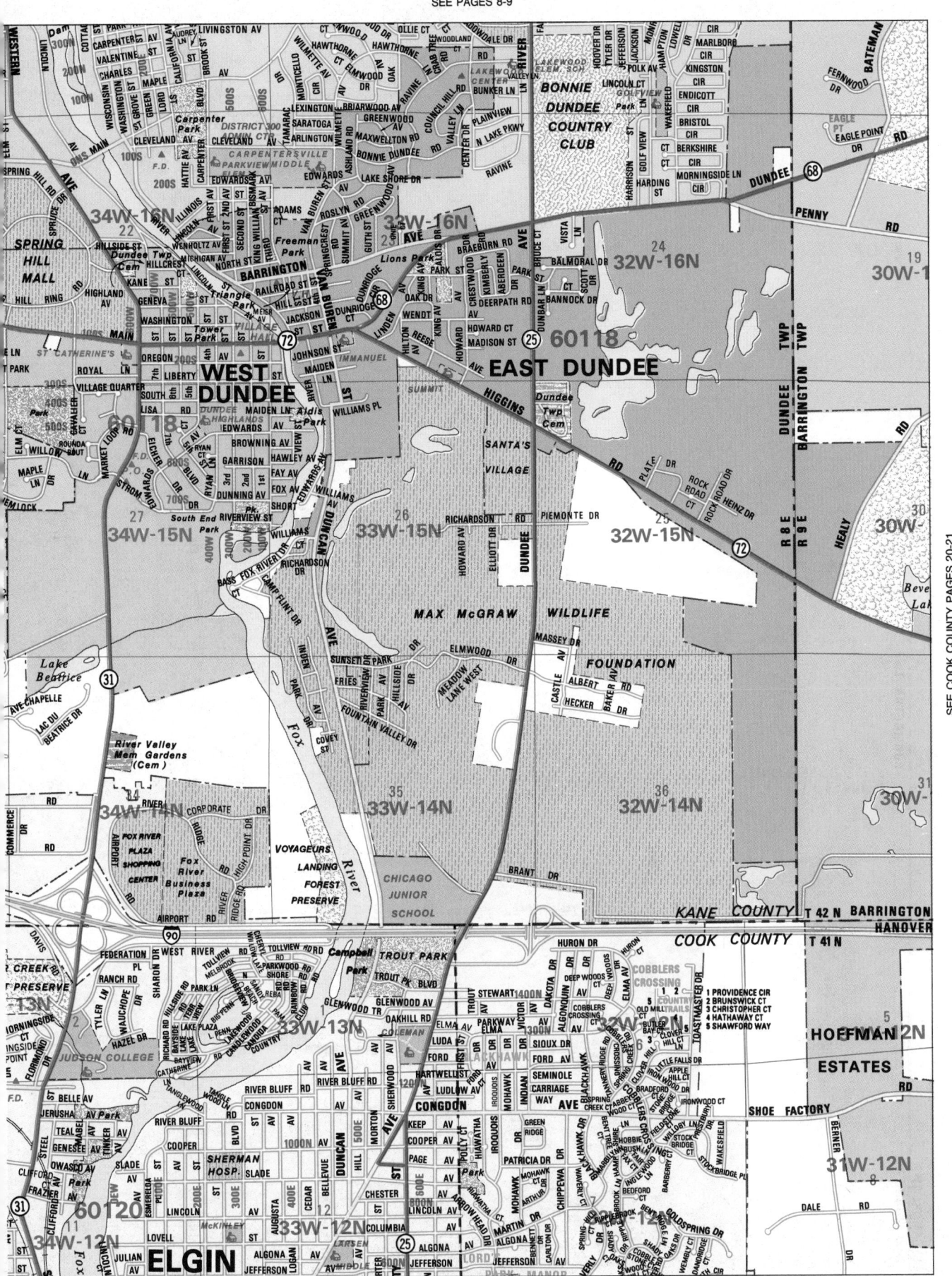

SEE COOK COUNTY PAGES 20-21

SEE PAGES 10-11

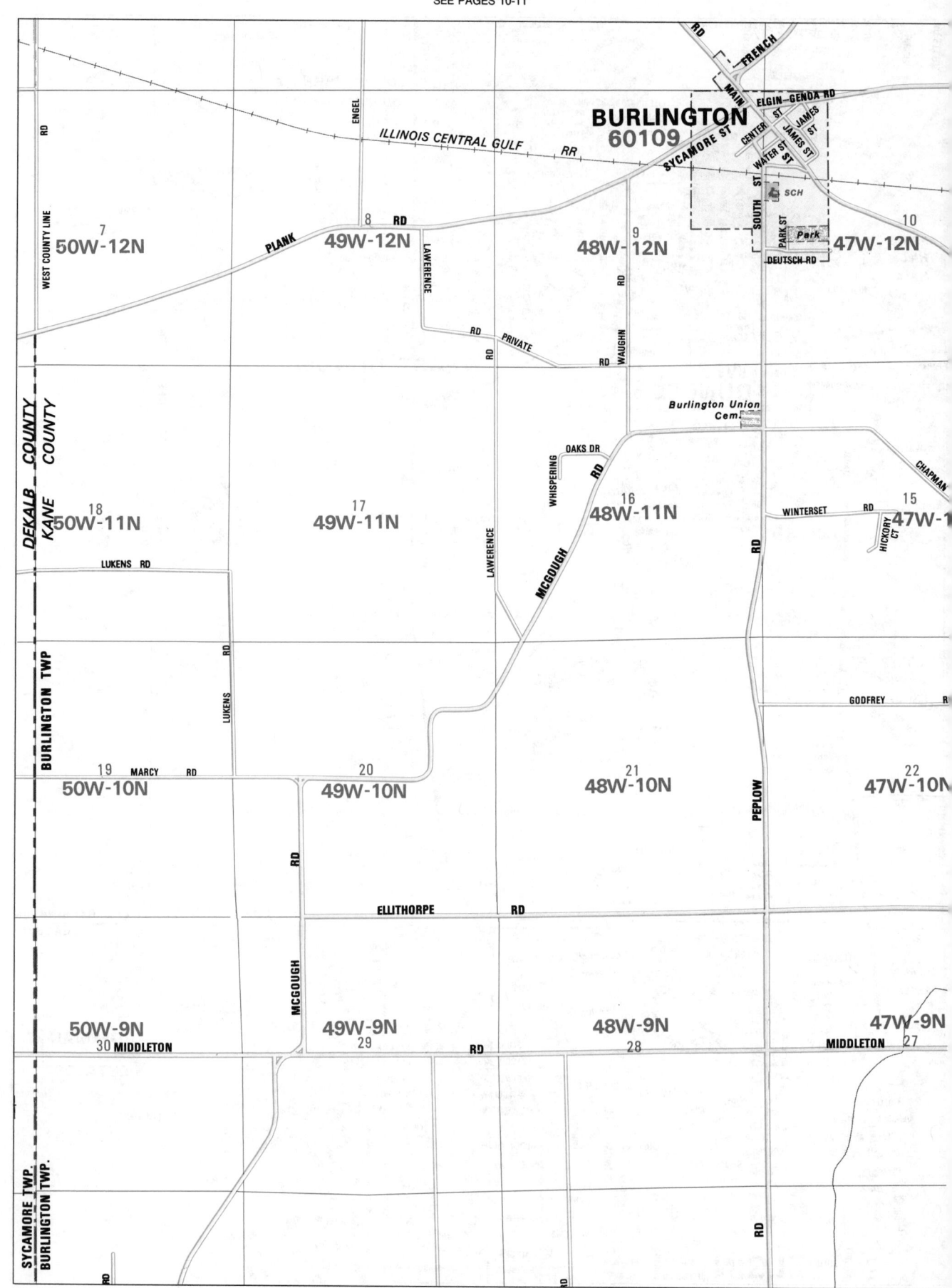

BURLINGTON
60109

ILLINOIS CENTRAL GULF RR

ENGEL

FRENCH

ELGIN—GENOA RD

MAIN ST

SYCAMORE ST

CENTER ST

JAMES ST

WATER ST

JAMES ST

SOUTH ST

PARK ST

Park

SCH

DEUTSCH RD

RD

7
50W-12N

WEST COUNTY LINE

PLANK

8 RD
49W-12N

LAWERENCE

RD

PRIVATE

RD

9
48W-12N

WAUGHN

RD

10
47W-12N

DEKALB COUNTY

KANE COUNTY

BURLINGTON TWP

18
50W-11N

LUKENS RD

RD

LUKENS

17
49W-11N

LAWERENCE

MCGOUGH

WHISPERING

OAKS DR

RD

Burlington Union
Cem.

16
48W-11N

WINTERSET RD

CHAPMAN

15
47W-1

HICKORY CT

RD

GODFREY R

19 MARCY RD
50W-10N

20
49W-10N

RD

21
48W-10N

PEPLOW

22
47W-10N

ELLITHORPE RD

MCGOUGH

50W-9N
30 MIDDLETON

49W-9N
29

RD

48W-9N
28

MIDDLETON
47W-9N
27

SYCAMORE TWP.
BURLINGTON TWP.

RD

RD

SEE PAGES 22-23

SEE PAGES 10-11

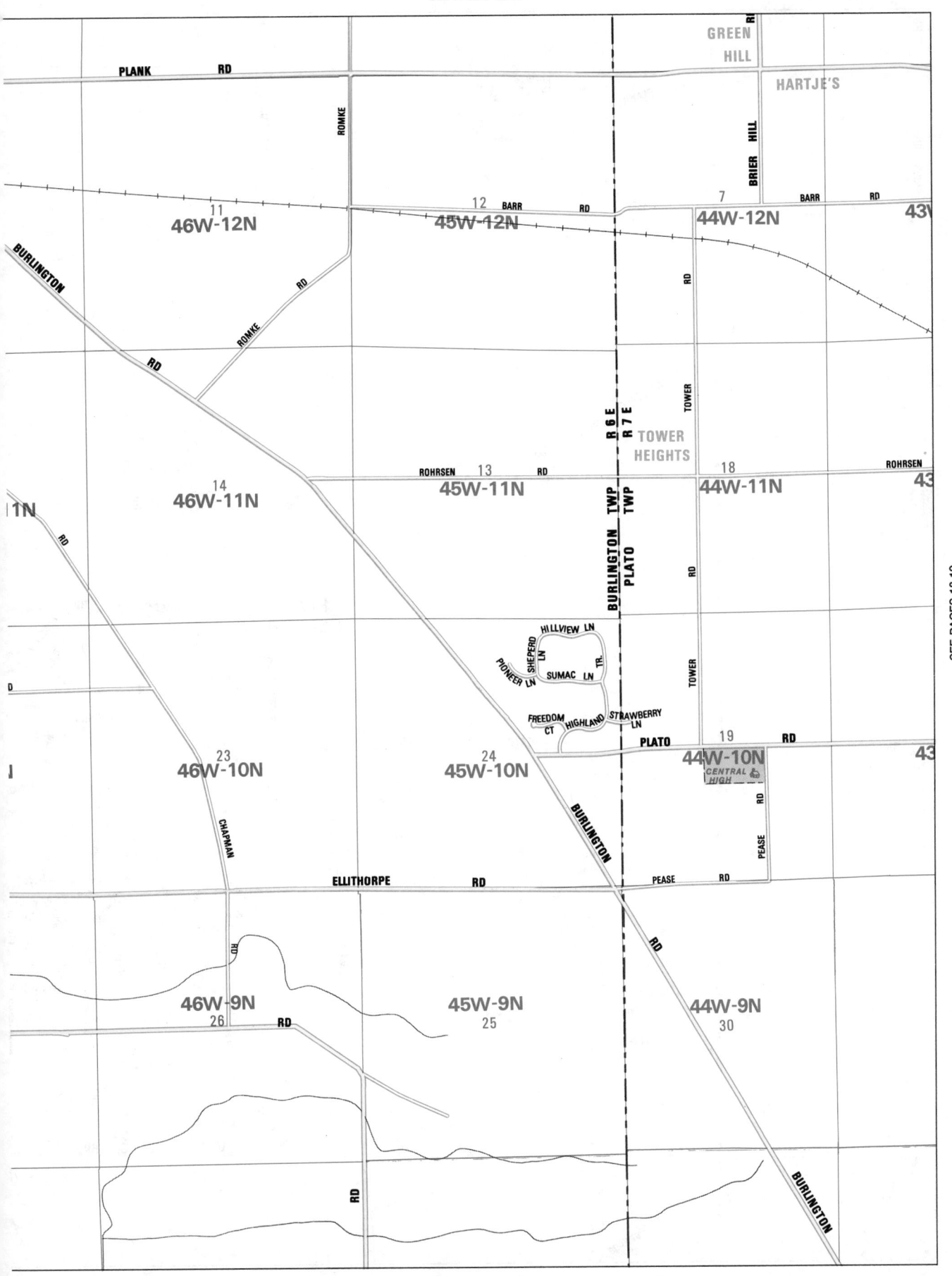

SEE PAGES 18-19

SEE PAGES 12-13

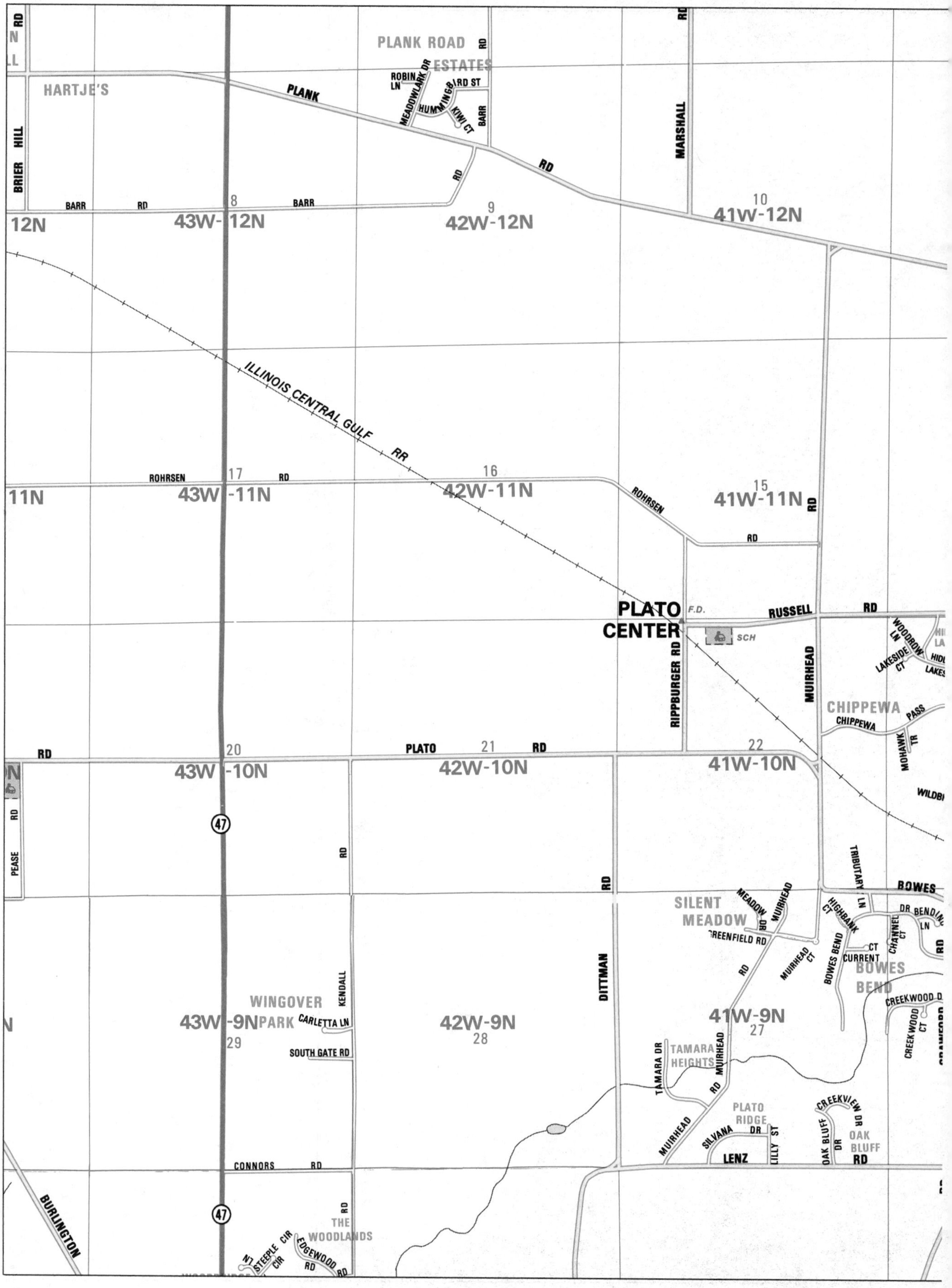

PLANK ROAD ESTATES

HARTJE'S

PLANK

ROBIN LN
MEADOWLARK DR
HUMMINGBIRD ST
KIWI CT
BIRD ST
BARR

RD

MARSHALL

BRIER HILL RD

BARR RD
BARR

8
43W-12N

RD

9
42W-12N

10
41W-12N

12N

ILLINOIS CENTRAL GULF RR

SEE PAGES 16-17

ROHRSEN RD
17
43W-11N

16
42W-11N

ROHRSEN

RD

15
41W-11N

RD

11N

RD

PLATO CENTER
F.D.
SCH

RUSSELL RD

WOODROW LN
HILL LA
LAKESIDE CT
HIDD
LAKES

RIPPBURGER RD

MUIRHEAD

CHIPPEWA
CHIPPEWA PASS
MOHAWK TR

RD

20
43W-10N

PLATO
21
RD
42W-10N

22
41W-10N

WILDBI

RD

PEASE RD

(47)

RD

RD

DITTMAN

TRIBUTARY LN

BOWES

HIGHBANK LN
DR BENDING
CHANNEL CT

SILENT MEADOW
MEADOW CT
MUIRHEAD
Greenfield RD
RD
MUIRHEAD CT
BOWES BEND
CT
CURRENT

BOWES BEND
CREEKWOOD

WINGOVER PARK
KENDALL RD
CARLETTA LN
43W-9N
29

42W-9N
28

RD

41W-9N
27

CREEKWOOD D

SOUTH GATE RD

TAMARA DR
TAMARA HEIGHTS
MUIRHEAD

CREEKVIEW DR

CRAWFORD

PLATO RIDGE DR
SILVANA
LENZ
LILLY ST
OAK BLUFF DR
OAK BLUFF RD

CONNORS RD

RD

BURLINGTON

(47)

N1
STEEPLE CIR
CIR
THE WOODLANDS
EDGEWOOD RD
WOODFORD RD
RD

SEE PAGES 24-25

SEE PAGES 12-13

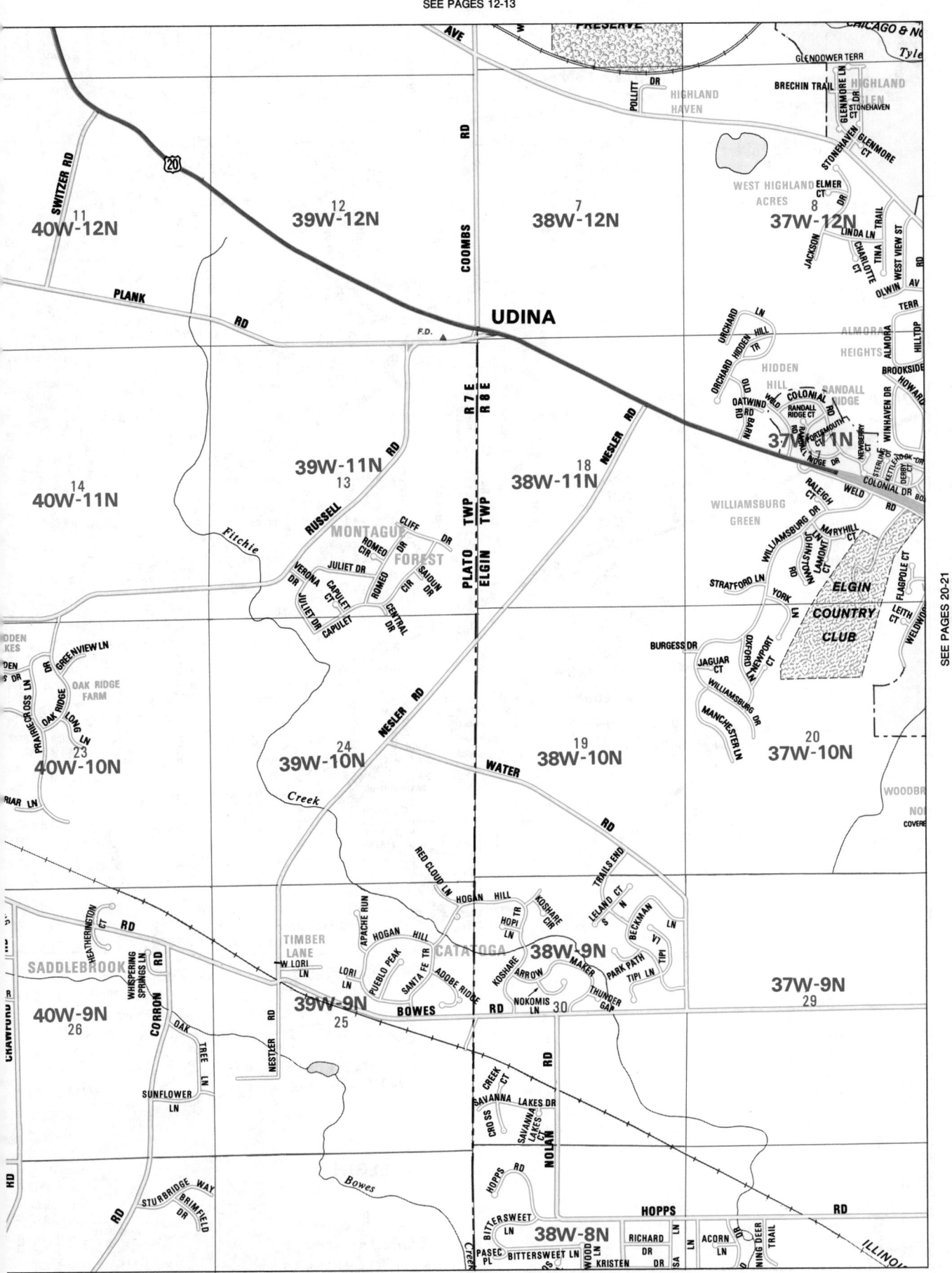

UDINA

40W-12N 11
39W-12N 12
38W-12N 7
37W-12N 8

40W-11N 14
39W-11N 13
38W-11N 18
37W-11N

40W-10N 23
39W-10N 24
38W-10N 19
37W-10N 20

40W-9N 26
39W-9N 25
38W-9N
37W-9N 29

38W-8N

SEE PAGES 20-21

SEE PAGES 14-15

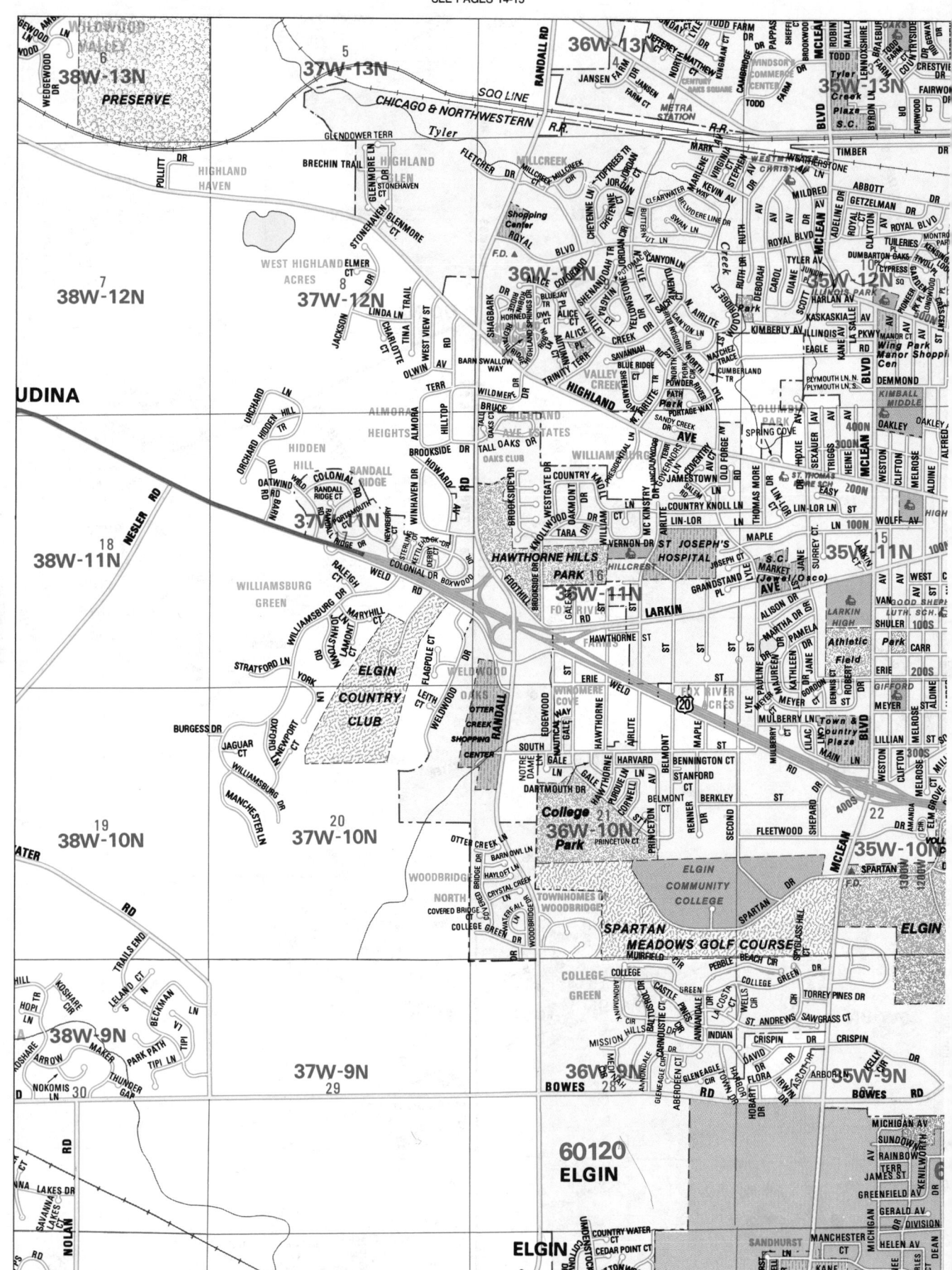

SEE PAGES 18-19

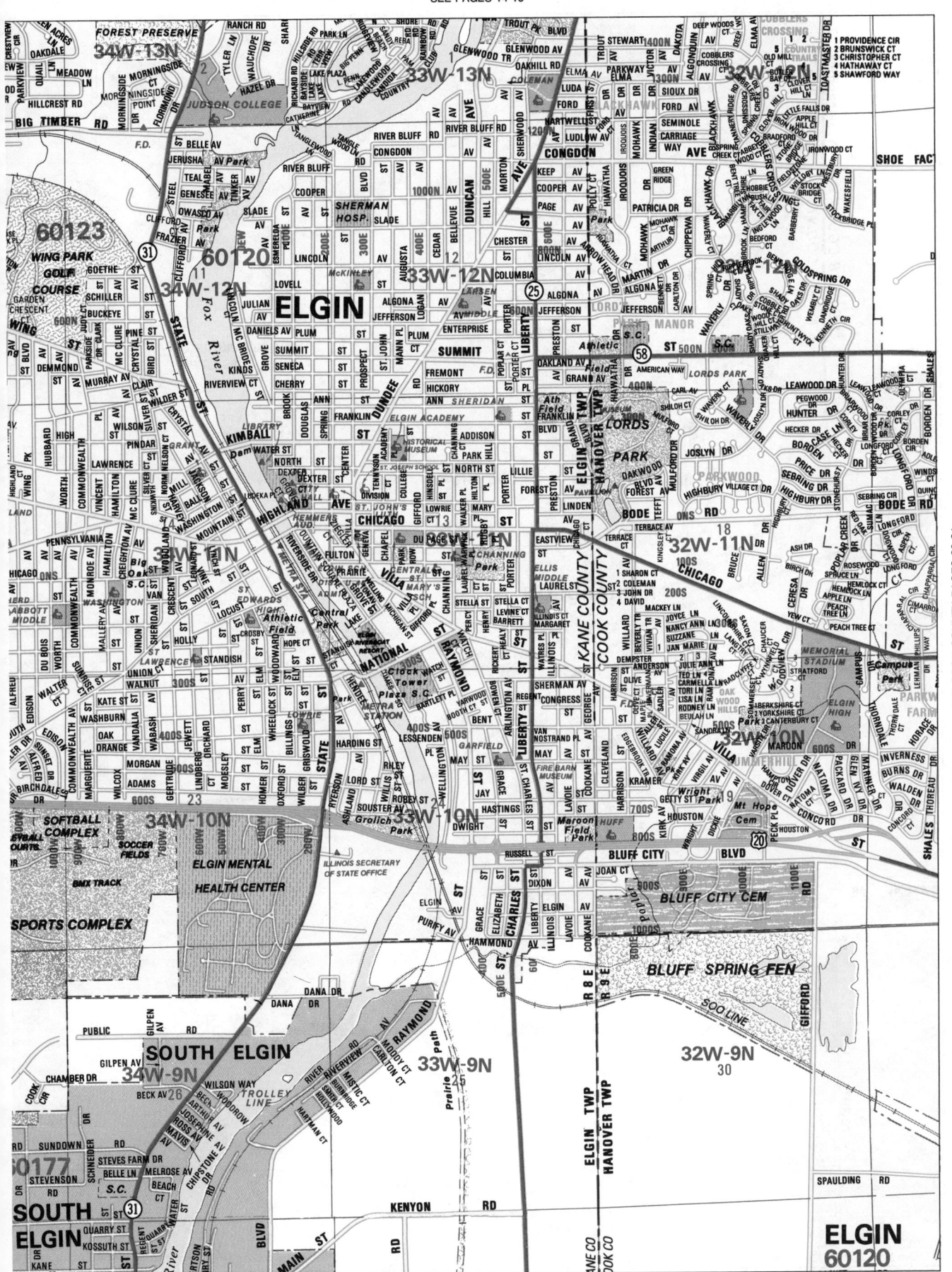

SEE PAGES 16-17

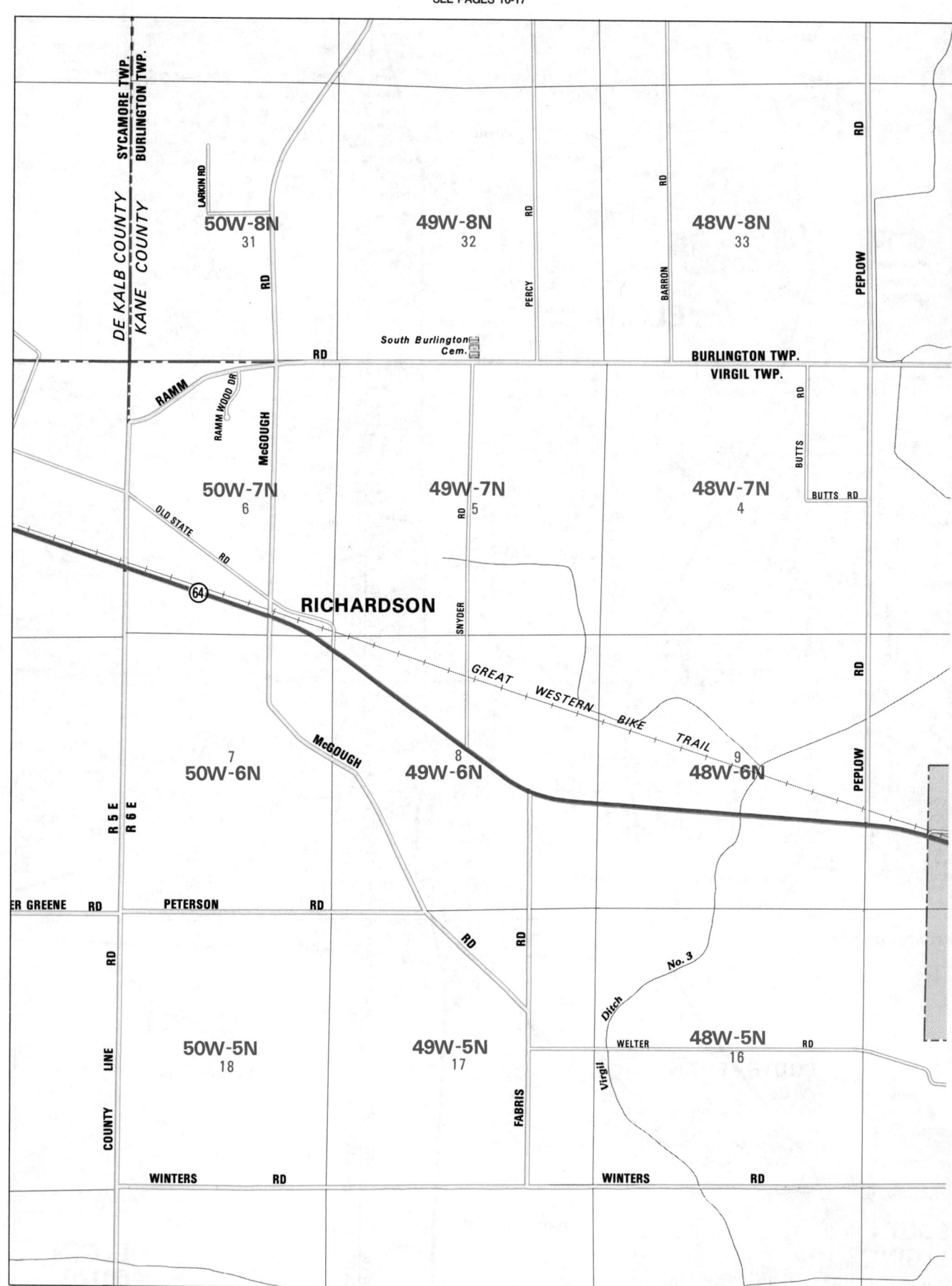

SEE PAGES 28-29

SEE PAGES 16-17

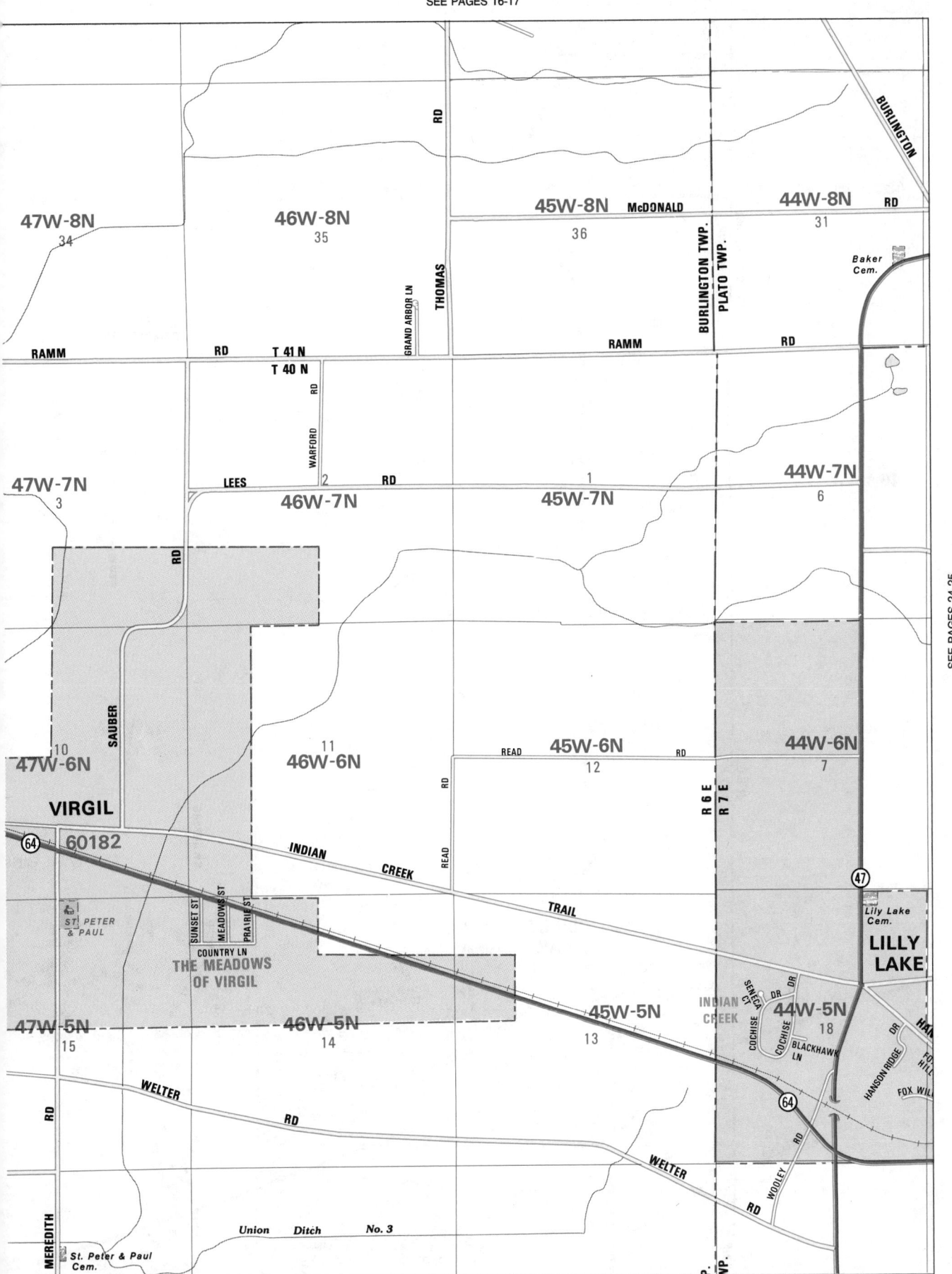

RD

BURLINGTON

47W-8N
34

46W-8N
35

45W-8N McDONALD
36

44W-8N
31

RD

THOMAS

GRAND ARBOR LN

BURLINGTON TWP.
PLATO TWP.

Baker
Cem.

RAMM RD T 41 N
T 40 N

RAMM RD

47W-7N
3

WARFORD RD

LEES RD
2

46W-7N

45W-7N
1

44W-7N
6

RD

SAUBER

47W-6N
10

46W-6N
11

READ RD

READ RD
45W-6N
12

RD

44W-6N
7

R 6 E
R 7 E

VIRGIL

64 60182

INDIAN CREEK

READ RD

47

ST. PETER
& PAUL

SUNSET ST
MEADOWS ST
PRAIRIE ST

TRAIL

Lily Lake
Cem.

**LILLY
LAKE**

COUNTRY LN

**THE MEADOWS
OF VIRGIL**

SENECA CT
COCHISE DR

DR

INDIAN
CREEK

47W-5N
15

46W-5N
14

45W-5N
13

COCHISE DR
BLACKHAWK
LN

44W-5N
18

HANSON RIDGE DR

FOX
HILL

RD

WELTER RD

64

FOX WIL

WOOLEY RD

MEREDITH

St. Peter & Paul
Cem.

Union Ditch No. 3

WELTER

RD

SEE PAGES 24-25

SEE PAGES 28-29

SEE PAGES 18-19

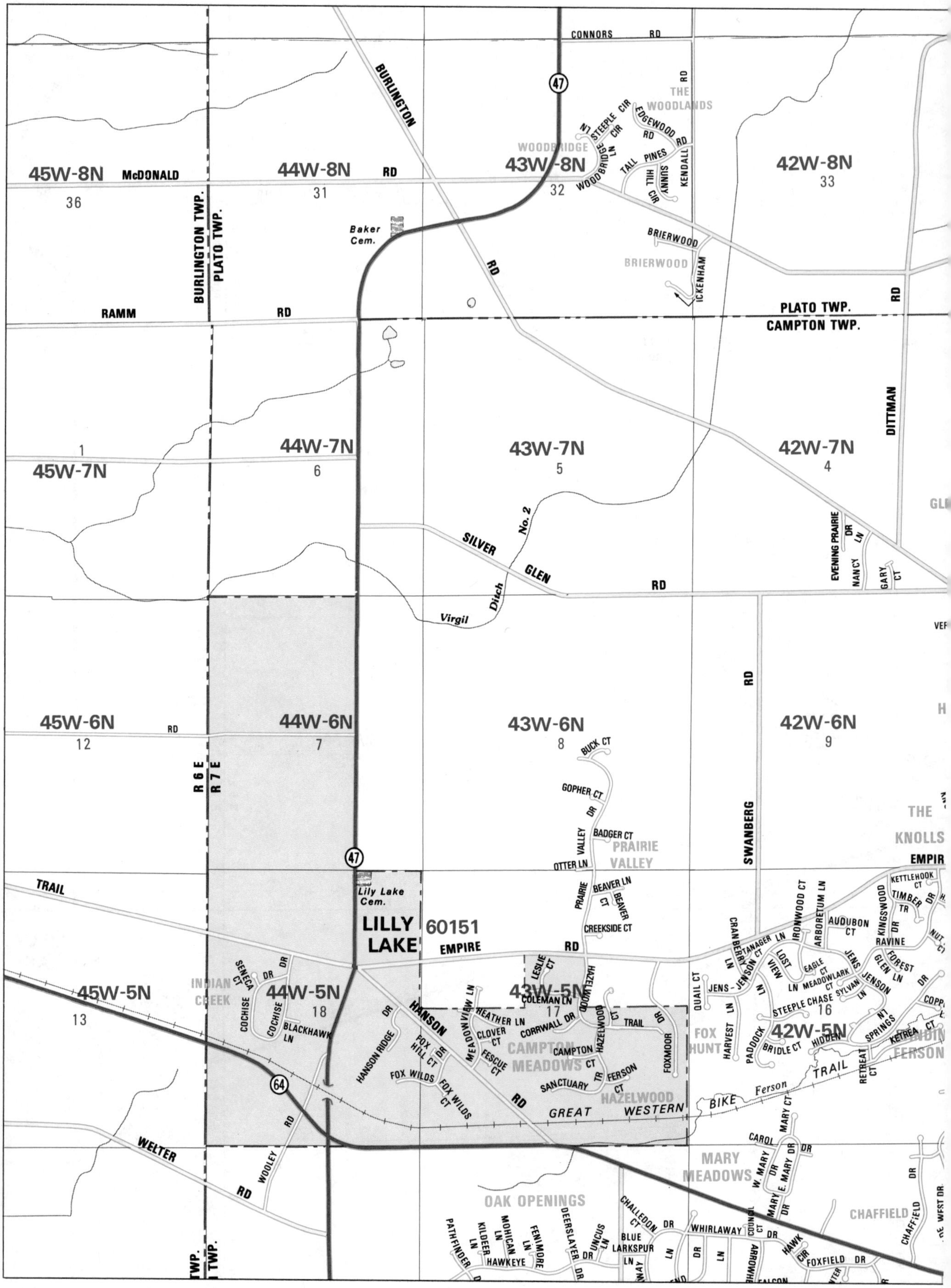

SEE PAGES 22-23

CONNORS RD

BURLINGTON

47

THE WOODLANDS

STEEPLE CIR
N
CIR
EDGEWOOD RD
RD
TALL PINES
WOODBRIDGE LN
HILL CIR
SUNNY
KENDALL
RD

45W-8N McDONALD
36

44W-8N RD
31

WOODBRIDGE
43W-8N
32

42W-8N
33

Baker Cem.

BRIERWOOD

BRIERWOOD
ICKENHAM

PLATO TWP.
CAMPTON TWP.

RAMM RD

BURLINGTON TWP.
PLATO TWP.

RD

DITTMAN

1
45W-7N

44W-7N
6

43W-7N
5

42W-7N
4

SILVER
No. 2
GLEN

EVENING PRAIRIE
DR
LN
NANCY
GARY
CT

Virgil
Ditch
GLEN RD

GLE

R 6 E
R 7 E

12
45W-6N RD

44W-6N
7

43W-6N
8

BUCK CT

GOPHER CT

DR

BADGER CT

OTTER LN

PRAIRIE VALLEY

SWANBERG RD

42W-6N
9

THE KNOLLS

EMPIR

TRAIL

47

Lily Lake Cem.

LILLY LAKE 60151

EMPIRE RD

PRAIRIE
BEAVER LN
BEAVER

CREEKSIDE CT

KETTLEHOOK CT
TIMBER TR
DR
AUDUBON CT
KINGSWOOD
RAVINE
NUT CT

45W-5N
13

INDIAN CREEK

SENECA
CT
DR
COCHISE
CT
BLACKHAWK
LN

44W-5N
18

HANSON
DR
HANSON RIDGE
FOX HILL CT
FOX WILDS
FOX WILDS CT

MEADOW VIEW LN
HEATHER LN
CLOVER CT
CORNWALL DR
FESCUE CT

LESLIE CT
COLEMAN LN

43W-5N
17

HAZELWOOD DR
CAMPTON CT
SANCTUARY
TR
FERSON
FOXMOOR DR

TRAIL

CAMPTON MEADOWS

HAZELWOOD

QUAIL CT

CRANBERRY LN
JENS-JENSON CT
TANAGER LN
VIEW
LOST
STEEPLE CHASE LN
PADDOCK
BRIDLE CT

IRONWOOD CT
EAGLE LN
MEADOWLARK
JENS JENSON
GLEN LN
SYLVAN
HIDDEN CT

ARBORETUM LN

FOREST DR
COPP

42W-5N
16

SPRINGS
RETREAT CT
KETREA CT

FOX HUNT

NDI
FERSON

64

WELTER

WOOLEY RD

RD

GREAT WESTERN

BIKE

Ferson

MARY MEADOWS

CAROL CT
W MARY DR
MARY E MARY DR

TRAIL

CHAFFIELD

CHAFFIELD DR

OAK OPENINGS

PATHFINDER
KILDEER LN
MOHICAN LN
FENIMORE LN
HAWKEYE

DEERSLAYER DR
UNCUS CT
BLUE LARKSPUR LN
WAY

CHALLEDON CT
DR

COUNCIL
WHIRLAWAY
DR

HAWK CIR
FOXFIELD DR
FALCON

WEST DR

TWP.
TWP.

SEE PAGES 30-31

SEE PAGES 18-19

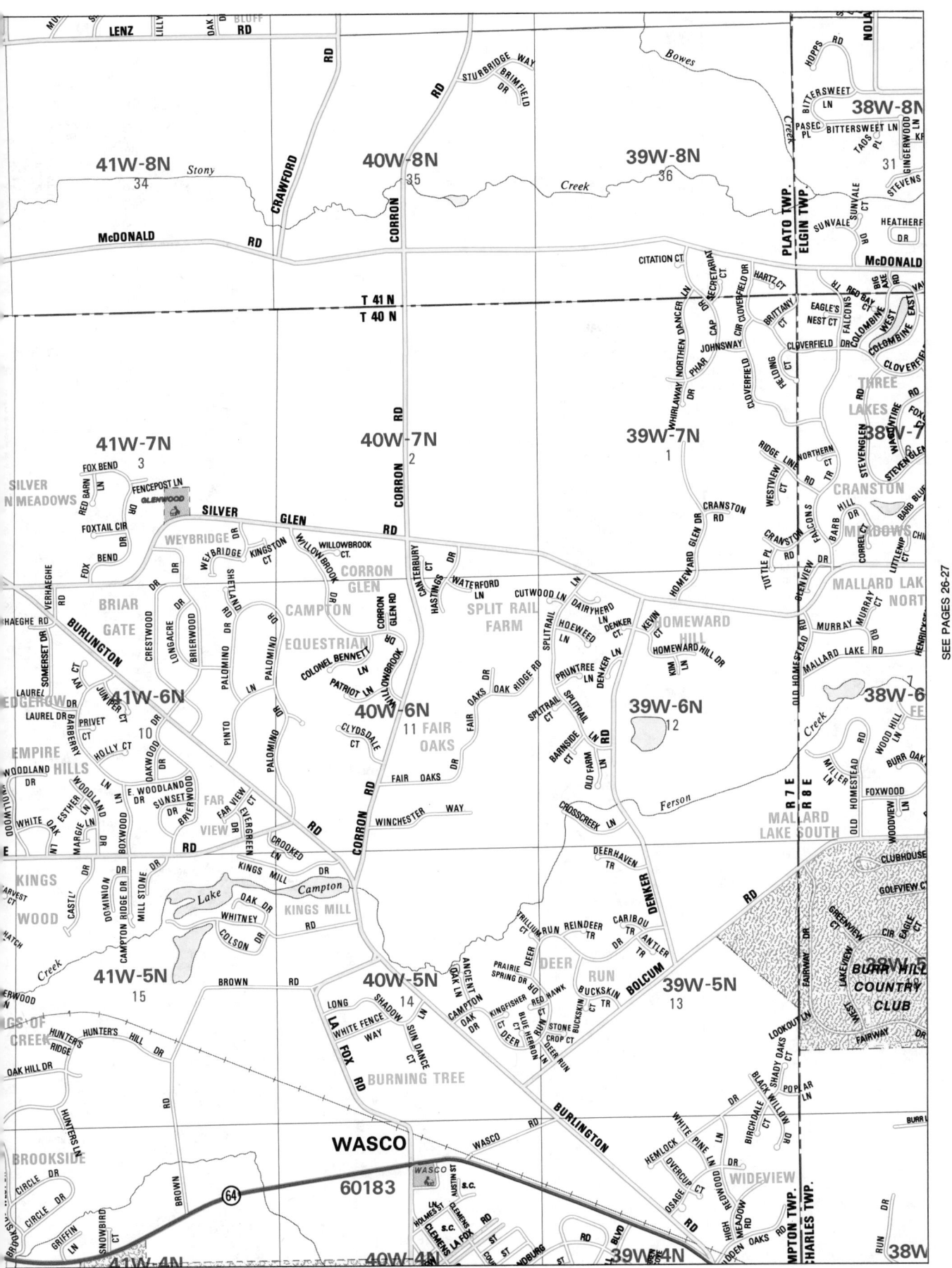

SEE PAGES 26-27

SEE PAGES 20-21

ELGIN

ELGIN

SAVANNA LAKES DR
SAVANNA LAKES CT
CROSS
NOLAN

COUNTRY WATER CT
UMDENSTOCK RD
CEDAR POINT CT
COTTONWOOD RD
COTTONWOOD DR

Bowes
Creek

HOPPS RD
HOPPS RD
HOPPS

ILLINOIS

HOPPS

AMBER LN
KNOTTY
PINE DR
MISTHAVEN
ASHWOOD DR
DEERPATH DR
SWEETBRIAR
FOXMOOR
CREEK DR
ASHWOOD CT
WALNUT
WESTON
PEMBROKE
EXETER
PEM BR
ANDOVER

BITTERSWEET LN
38W-8N
RICHARD DR
ACORN LN
RUNNING DEER TRAIL
37W-8N
36W-8N

39W-8N
36
PASEC PL
BITTERSWEET LN
GINGERWOOD
KRISTEN
TAOS PL
31
LISA
WILLOW LN
WILDWOOD
SHADY
32
33

WOODBRIDGE SOUTH

CENTRAL

RANDALL RD

STEVENS
SUNVALE CT
SUNVALE RD
HEATHERFIELD DR
HEATHERFIELD EAST
PEPPERTREE CT
PEPPERTREE LN
PLATO TWP.
ELGIN TWP.

McDONALD
RD

ELGIN TWP.
ST. CHARLES TWP.

CITATION CT
SECRETARIAT CT
CLOVERFIELD DR
HARTZ CT
BIG RED BAY
AXE
VAUPON CT
MARGATE & OTTER CREEK

NORTHERN DANCER LN
WHIRLAWAY
PHAR
CAP
CIR
BRITTANY
EAGLE'S NEST CT
FALCONS
COLOMBINE
WEST
OAK
SYCAMORE
DOGWOOD
CREEK

JOHNSWAY
CLOVERFIELD
FIELDING
CLOVERFIELD CT
CLOVERFIELD
COLOMBINE EAST
STEVENS
LILAC
PINE
RD

THREE LAKES
WAGONTIRE
FOXGLOVE

39W-7N
1
RIDGE LINE RD
WESTVIEW
NORTHERN CT
TR
STEVENGLEN
STEVEN GLEN RD
38W-7N
ADELE LN
LONGRIDGE
SYDNEY READ SUBDIVISION
37W-7N
5
36W-7N
4

HOMEWARD GLEN DR
CRANSTON RD
FALCONS
CRANSTON
TUTTLE PL
CRANSTON MEADOWS
BARB HILL DR
BARB BLUESTEM
HILL
LITTLELEAF CT
CHICKASAW CT
CORRELL
GLENVIEW DR
SIDNEY

SILVER GLEN RD

WHISPERING TRAIL
HAMPTON DR
HICKORY CT
WINDSOR
ALDREA
LANCASTER CT
RD
MICHAEL
SILVER GLEN ESTATES

WHISPERING
WHISPERING TRAIL CT
PLYMOUTH
BRISTOL CT
BRISTOL RD
RIVER VIEW
SILVER GLEN

MALLARD LAKE NORTH
MURRAY
MURRAY
HENDRICKSEN RD
MALLARD LAKE RD
MCK POINT CT
Otter

FOXBOROUGH
SILVER RIDGE DR
McKAY

MYRA HYERD
DENKER
KEVIN CT
DENKER CT
HOMEWARD HILL
HOMEWARD HILL DR
KIM LN
MALLARD LAKE
HERITAGE CT
BURR
MALLARD LAKE RD
PROMONTORY CT

36W-6N

REED LN
KIM
38W-6N
FERSON
WOODSIDE
CREEKSIDE CT
37W-6N
8
TWIN SILOS DR
RIVER
GRANGE RD
FOLEY DR
BARTON DR
RIVER

39W-6N
12
WOOD HILL RD
MILLER
BURR OAK LN
WOODS DR
WOODSIDE DR
BURSIDE DR
CREEK

RIVER GRANGE LN

OLD FARM LN RD
Ferson Creek
OLD HOMESTEAD RD
R7E R8E
FOXWOOD
WOODVIEW
OLD HOMESTEAD DR
FERSON
WOODS
PERSON CREEK
BOLCUM

WOODVIEW CT
WOODVIEW
RIDGEWOOD
E. RIDGEWOOD DR
TIMBER RIDGE

MALLARD LAKE SOUTH
RD
WESTWOOD DR
OAK RUN

DEERHAVEN TR
CLUBHOUSE
GOLFVIEW CT
RED GATE RD
RED GATE

DEER TR
CARIBOU DR
ANTLER TR
DENKER RD
GREENVIEW CT
EAGLE CT
CREEKVIEW LN
LONGVIEW DR
MYLES
FENCE RAIL
LONGVIEW ESTATES
36W-5N
16

BOLCUM
BUCKSKIN TR
39W-5N
13
FAIRWAY DR
LAKEVIEW
38W-5N
BURR HILL COUNTRY CLUB
GLENOAK LN
HAWKINS LN
HAWKINS GLEN
37W-5N
17
LEDLA
RONSU
BAKER
BAKERS ACRES
WOODLAND CT
DEERPATH
MEADOW WAY
WILTON CRAFT RD
DORCHESTER RD
FIELDCR

RUN
LOOKOUT LN
FAIRWAY DR

OVERCUP CT
HEMLOCK
WHITE PINE LN
BIRCHDALE CT
SHADY OAKS CT
BLACK WILLOW
POPLAR RD
BURR LN
WOODGATE RD
GREY BARN CT
MAPLE
CRANE
NORTHFIELD RD
GREY BARN RD
OAK OAK
STONEBRIDGE LN
FIELDSTONE
MIDDLECREEK LN
HUNTER'S GATE
OLD FARM RD
ABBEY

WIDEVIEW
ARRINGTON TWP.
GRANDMA'S LN
MIDDLECREEK
STONELEAT
FIELDCREST DR
FIELDSTONE CT
RANDALL RD

CRANE WOODS ESTATES

SEE PAGES 20-21

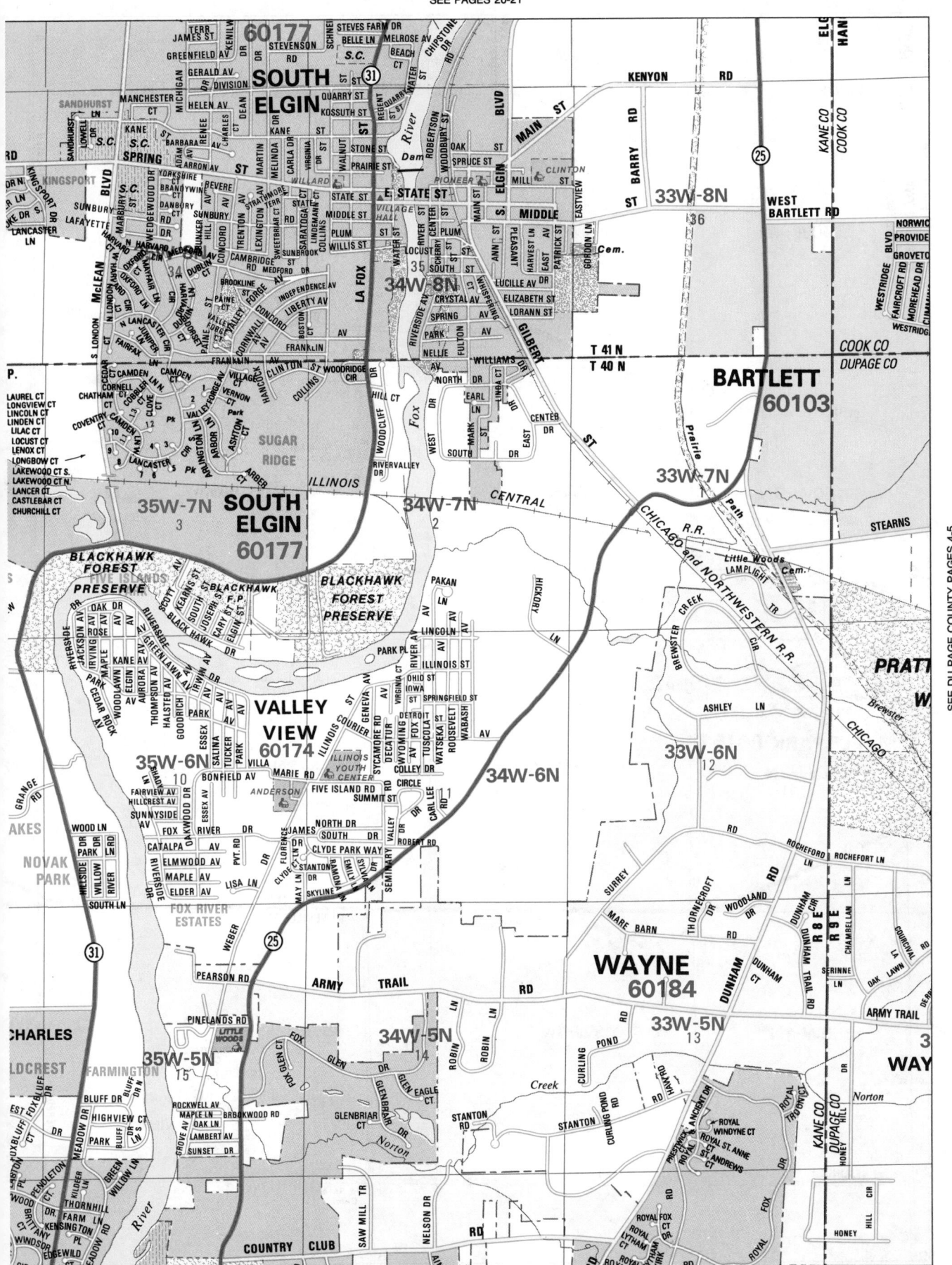

SEE DU PAGE COUNTY PAGES 4-5

SEE PAGES 22-23

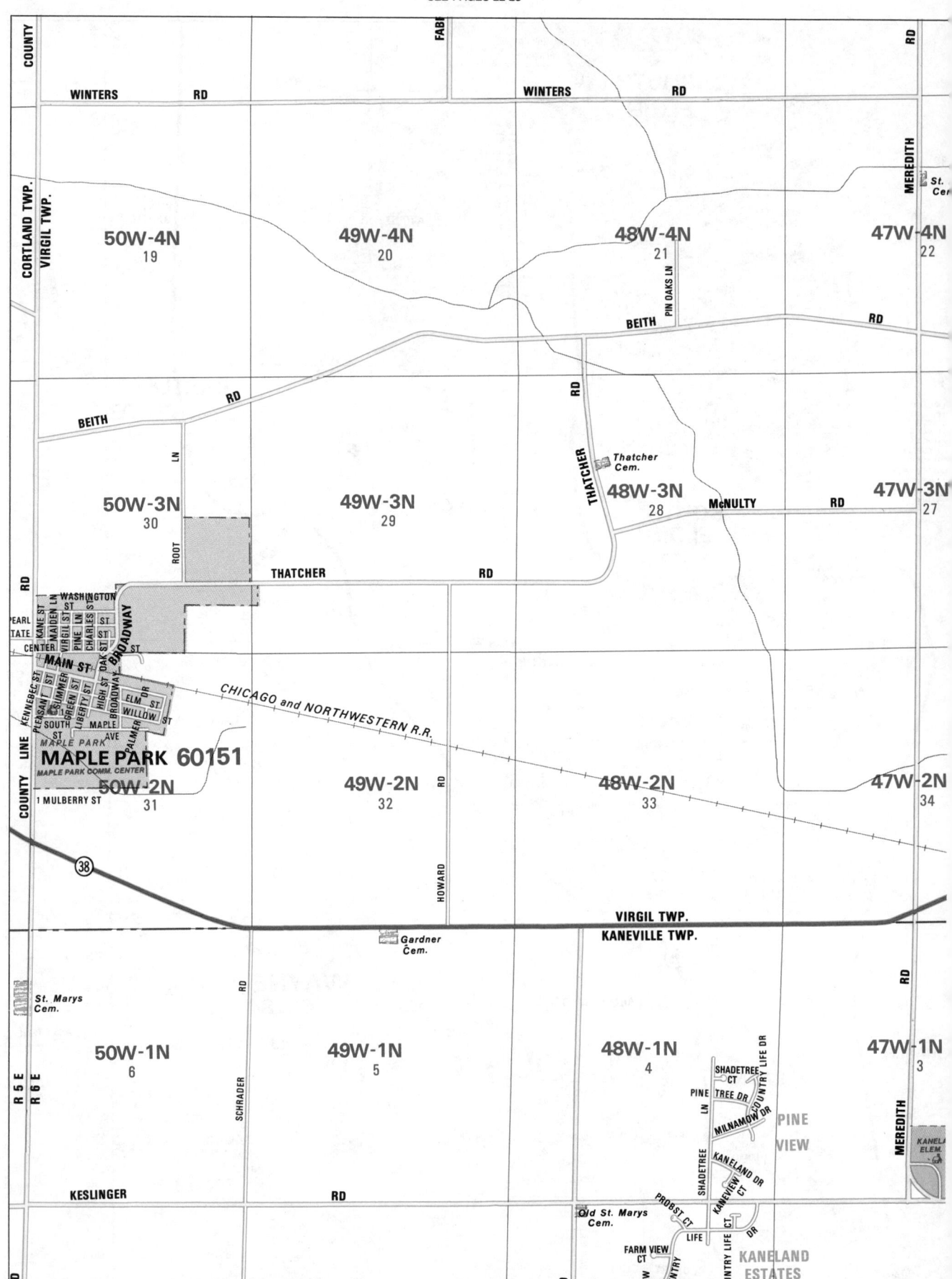

COUNTY

FABE

WINTERS RD

WINTERS RD

MEREDITH

St. Cer

CORTLAND TWP.

VIRGIL TWP.

50W-4N
19

49W-4N
20

48W-4N
21

PIN OAKS LN

47W-4N
22

BEITH RD

RD

BEITH RD

LN

50W-3N
30

ROOT

49W-3N
29

THATCHER

Thatcher
Cem.

48W-3N
28

McNULTY RD

47W-3N
27

THATCHER RD

THATCHER RD

WASHINGTON ST

PEARL
TATE

KANE ST

MAIDEN LN

VIRGIL ST

PINE LN

CHARLES ST

ST

CENTER

BROADWAY

MAIN ST

KENNEBEC ST

PLEASANT ST

SUMMER ST

GREEN ST

LIBERTY ST

HIGH ST

OAK ST

BROADWAY

ELM ST

DR

WILLOW ST

SOUTH ST

MAPLE AVE

PALMER

MAPLE PARK 60151

MAPLE PARK COMM. CENTER

MAPLE PARK

50W-2N
31

1 MULBERRY ST

CHICAGO and NORTHWESTERN R.R.

COUNTY LINE

49W-2N
32

RD

48W-2N
33

47W-2N
34

HOWARD

38

VIRGIL TWP.

KANEVILLE TWP.

Gardner
Cem.

RD

St. Marys
Cem.

R 5 E
R 6 E

50W-1N
6

SCHRADER

49W-1N
5

48W-1N
4

SHADETREE CT

PINE TREE DR

COUNTRY LIFE DR

MILNAMOW DR

SHADETREE LN

KANELAND DR

KANEVIEW CT

PINE VIEW

RD

MEREDITH

47W-1N
3

KANELA
ELEM.

PROBST CT

Old St. Marys
Cem.

FARM VIEW CT

COUNTRY LIFE

COUNTRY LIFE DR

KESLINGER RD

KANELAND ESTATES

SEE PAGES 34-35

SEE PAGES 22-23

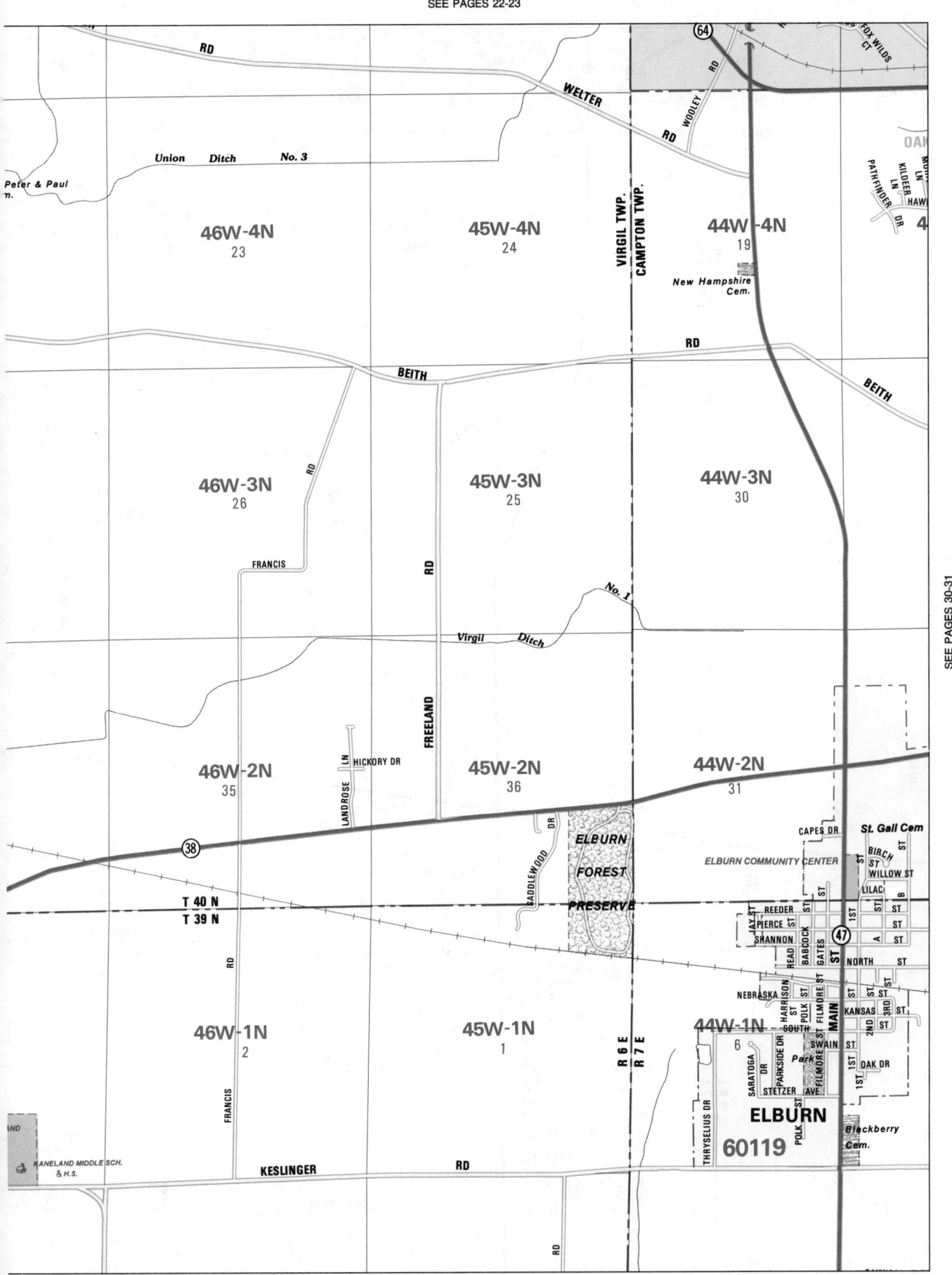

SEE PAGES 30-31

SEE PAGES 34-35

SEE PAGES 24-25

CAMPTON MEADOWS CT
HANSON RIDGE
FOX HILL CT DR
MEADOW FESCUE CT
CAMPTON
FERSON CT
FOXWOOD
HUNT HAR PADD BRIDLE CT
TRAIL FERS
RETREAT CT
64
FOX WILDS
FOX WILDS CT
SANCTUARY TR
HAZELWOOD RD
GREAT WESTERN BIKE
Ferson
WELTER RD
WOOLEY RD
CAROL
MARY MEADOWS
W. MARY DR
E. MARY DR
CHAFFIELD
CHAFFIELD
DR

OAK OPENINGS
PATHFINDER DR
MOHICAN LN
KILDEER LN
FENIMORE LN
DEERSLAYER DR
UNCUS DR
HAWKEYE
BLUE LARKSPUR LN
CHALLEDON CT DR
WHIRLAWAY
FAIREND
WHIRLAWAY DR LN
CHATEAUGAY
CHATEAUGAY
CHATEAUGAY CT
CITATION
VENETIAN WAY
CAMPTON
WHIRLAWAY
COLUMBINE DR
ARROWHEAD DR
HAWK CIR
FALCON LN
FOXFIELD DR
HUNTER DR
CT
DERBY CT
WINDSOR CT EAST RD
FOXFIELD
CHEVEL de SALLE
CAMPTON HILLS FARM
COUNTRYVIEW LN
HILLS
FARM VIEW
FOX

45W-4N 24
VIRGIL TWP. CAMPTON TWP.
44W-4N 19
43W-4N 20
42W-4N 21

New Hampshire Cem.

RD
BEITH RD
ANDERSON RD

45W-3N 25
44W-3N 30
43W-3N 29
42W-3N 28

WOODCREST CT
SPRING WOOD LN

FREELAND RD
No. 1
Virgil Ditch

38

45W-2N 36
44W-2N 31
43W-2N 32
42W-2N 33
RD
POULEY RD

SADDLEWOOD DR
ELBURN FOREST PRESERVE

CAPES DR
St. Gall Cem
BIRCH ST
WILLOW ST
LILAC
ELBURN COMMUNITY CENTER
1ST ST
B
HICKS DR
DEMPSEY DR
PAUL ST
47
CAMPTON TWP.
BLACKBERRY TWP.

JAY ST
REEDER ST
PIERCE ST
SHANNON ST
READ
BABCOCK
GATES ST
MAIN ST
NORTH ST
A ST

R 6 E R 7 E
NEBRASKA
HARRISON ST
POLK
FILMORE ST
SOUTH
SWAIN ST
KANSAS
2ND ST
3RD ST
MEADOWSWEET DR

45W-1N 1
44W-1N 6
SARATOGA DR
PARKSIDE DR
Park
STETZER AVE
FILMORE ST
1ST ST
OAK DR
43W-1N 5
42W-1N 4

THRYSELIUS DR
POLK
ELBURN 60119
Blackberry Cem.
STILL MEADOWS
STILL MEADOWS LN
STARGRASS CT
PENNYCRESS
RD

RD
KESLINGER RD
RD

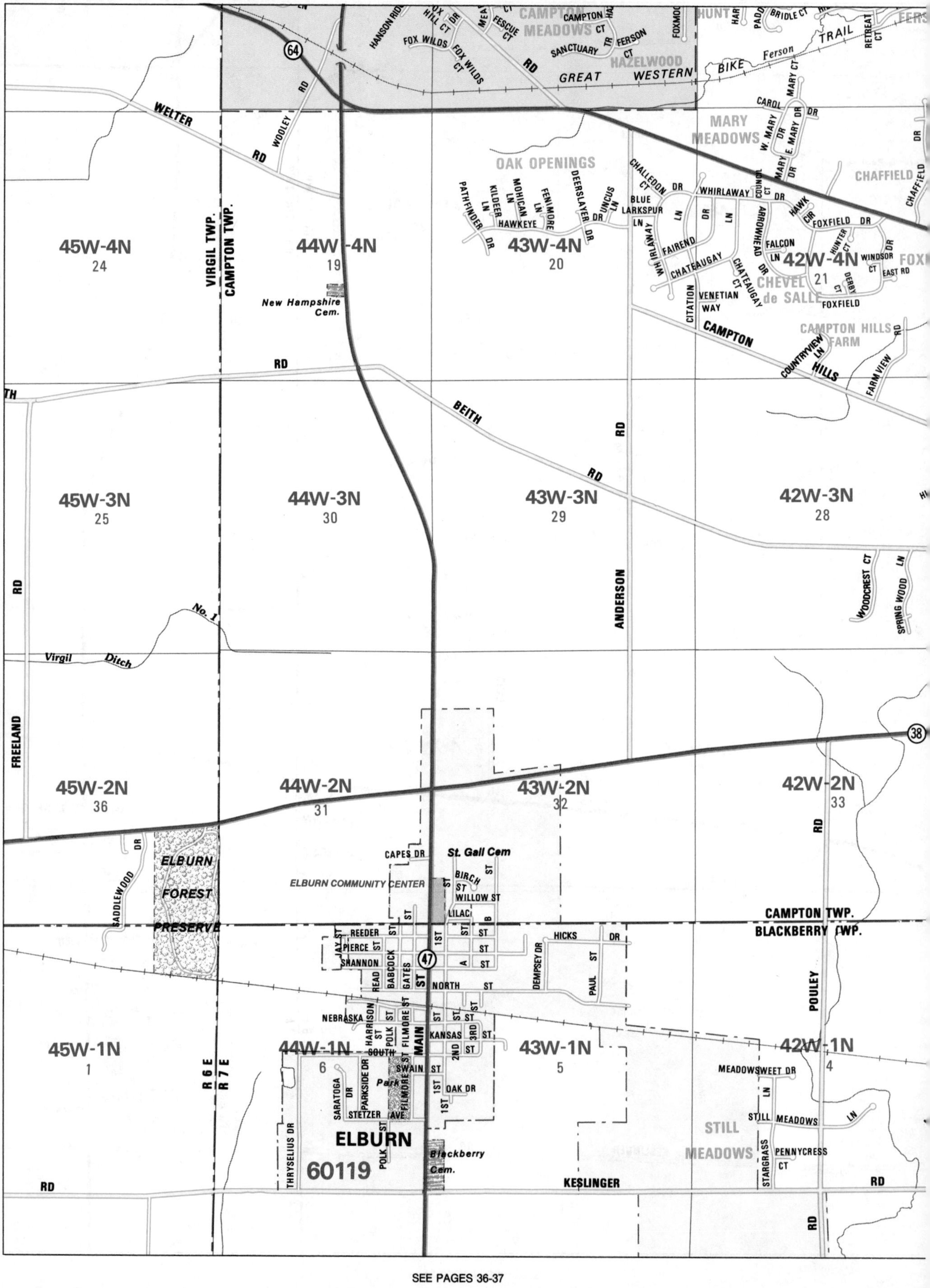

SEE PAGES 36-37

SEE PAGES 28-29

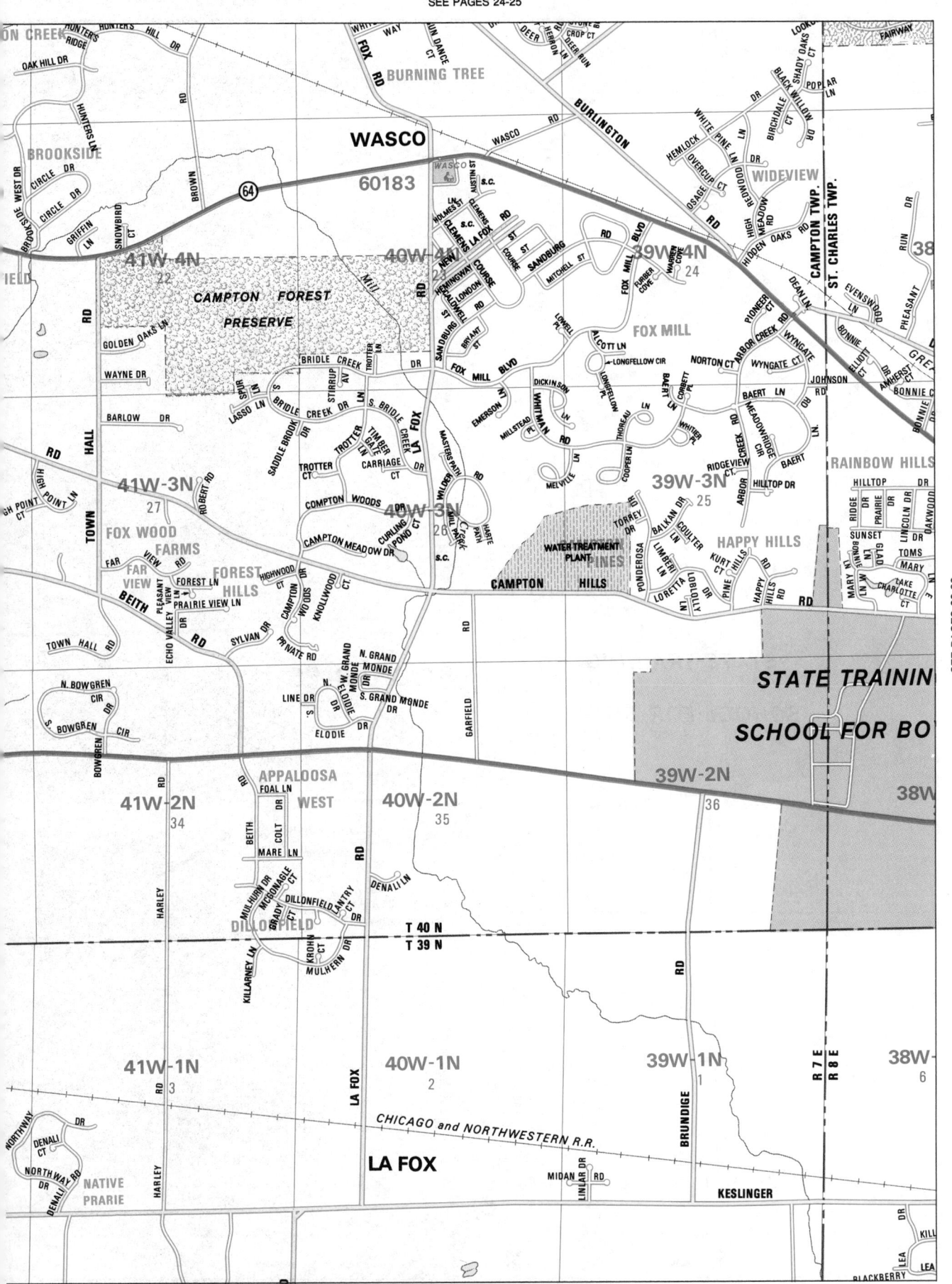

SEE PAGES 26-27

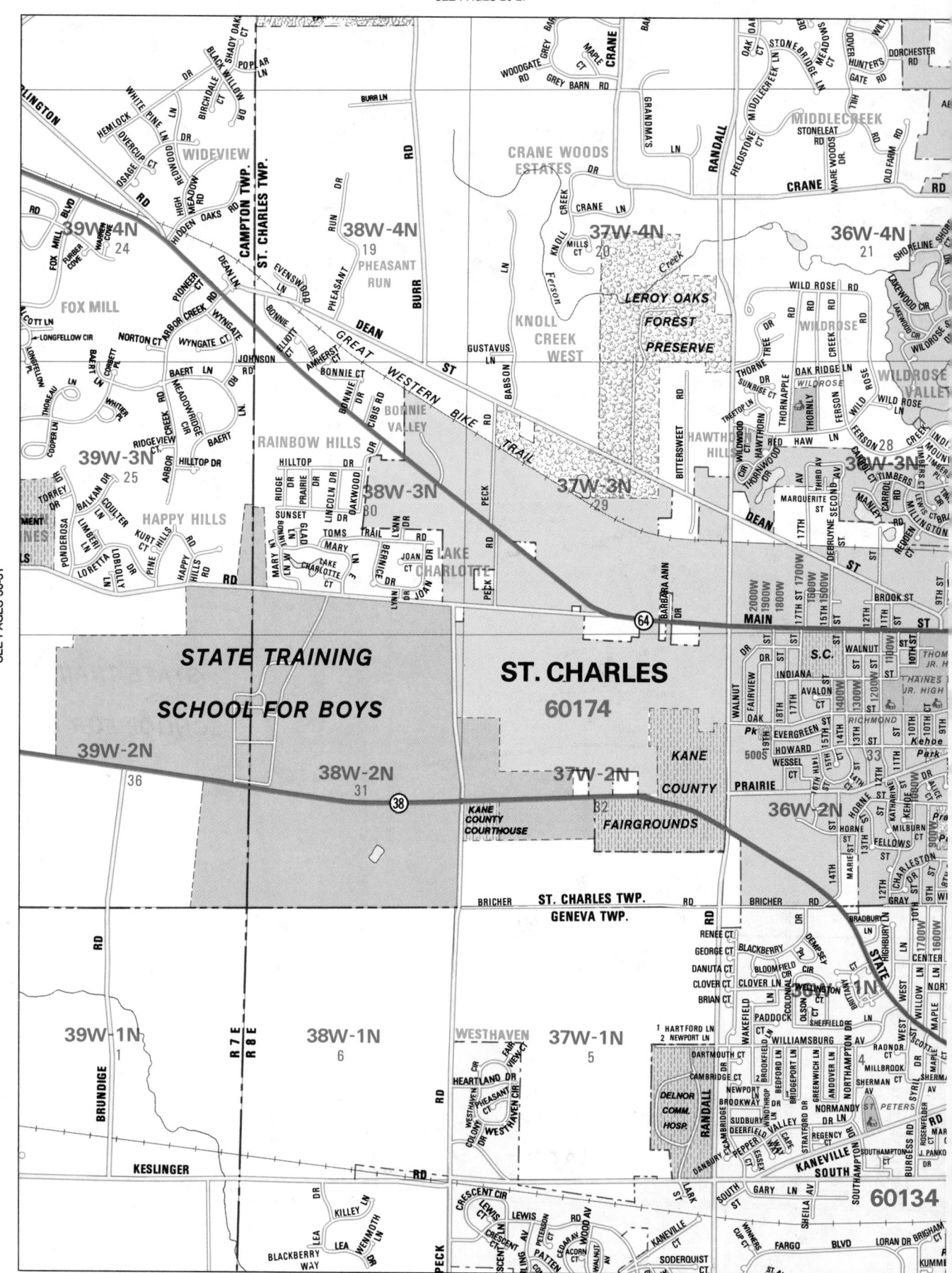

STATE TRAINING

SCHOOL FOR BOYS

ST. CHARLES

60174

60134

SEE PAGES 30-31

SEE PAGES 38-39

SEE PAGES 26-27

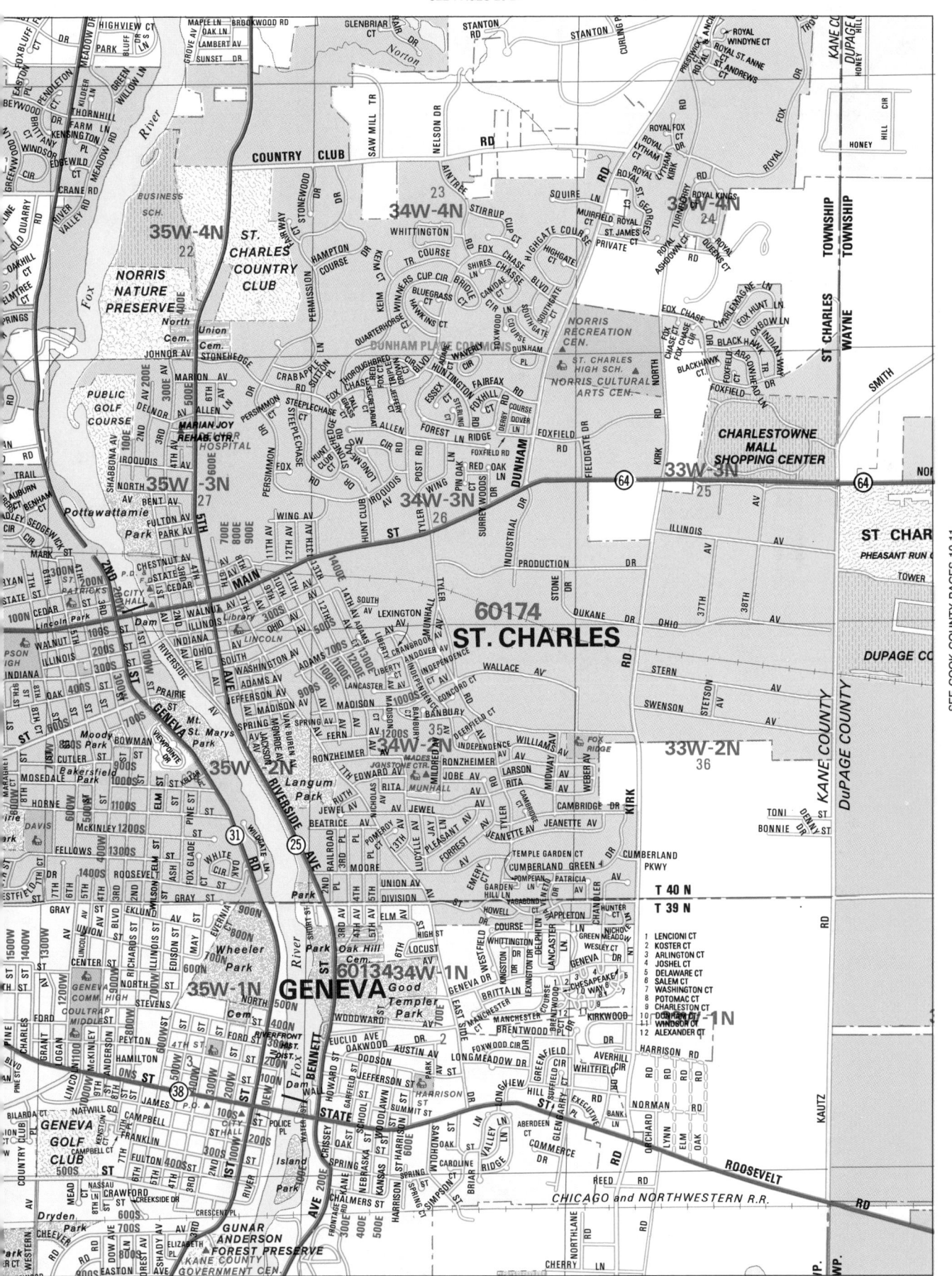

SEE PAGES 38-39

SEE PAGES 28-29

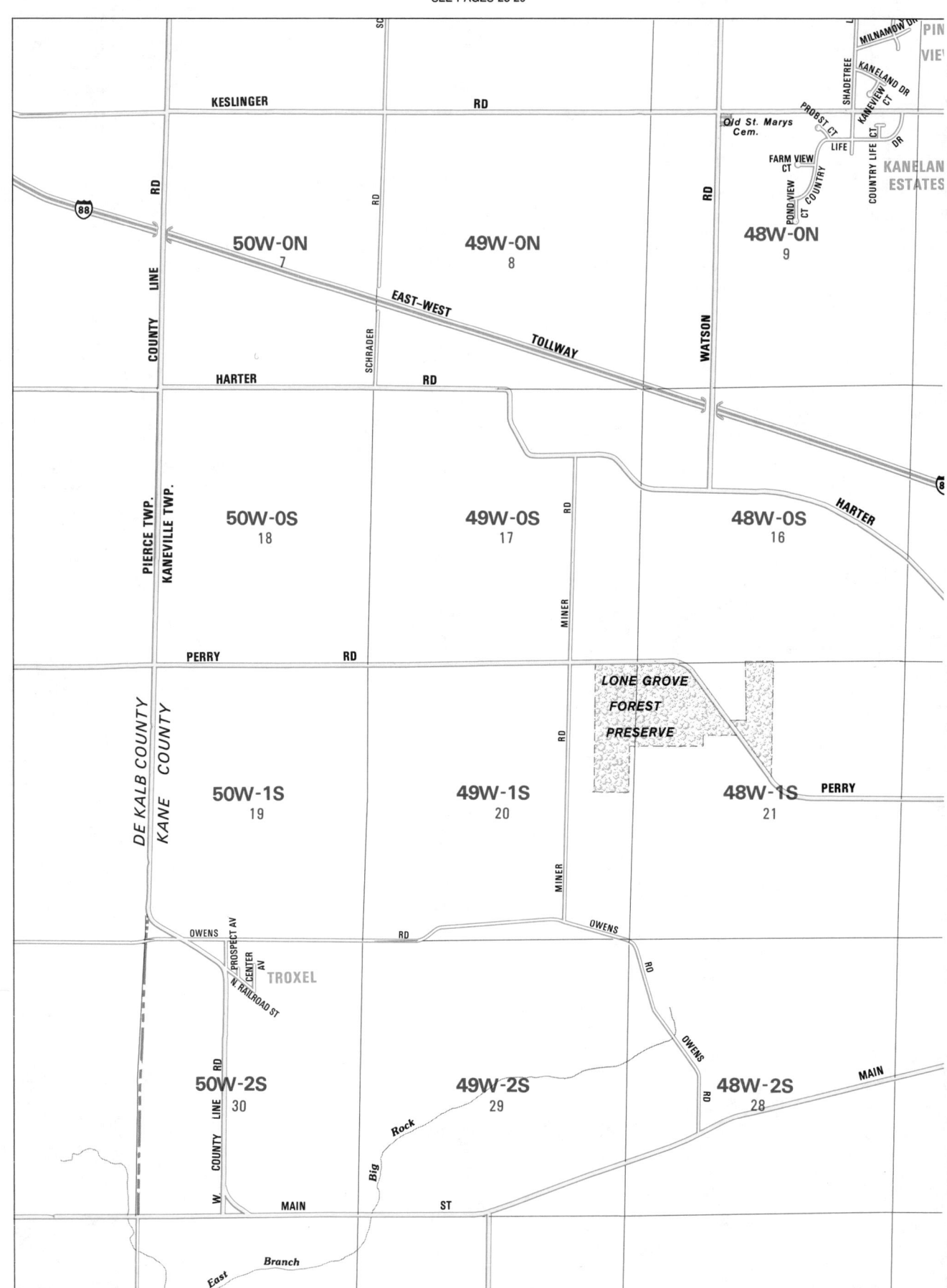

SEE PAGES 40-41

SEE PAGES 28-29

MEREDIT

KANELAND ELEM.

KANELAND MIDDLE SCH. & H.S.

FRANC

STETZER

THRYSELIUS DR

ELBU

60119

KESLINGER RD

RD

47W-0N
10

46W-0N
11

45W-0N
12

44W-0N
7

SCHNEIDER

47W-0S
15

46W-0S
14

45W-0S
13

RD

ROWE

RD

44W-0S
18

Creek

ROWE

KANEVILLE TWP.
BLACKBERRY TWP.

BATEMAN

RD

BATEMAN
RD

WEST

TOLLWAY

RD

45W-1S
24

LORANG

CONCORD
CT

GREENBRIER DR

44W-1S
19

BLACKBERRY
ACRES

BATEMAN

ST

MAIN

SOUTH LORANG RD

RD

LORANG

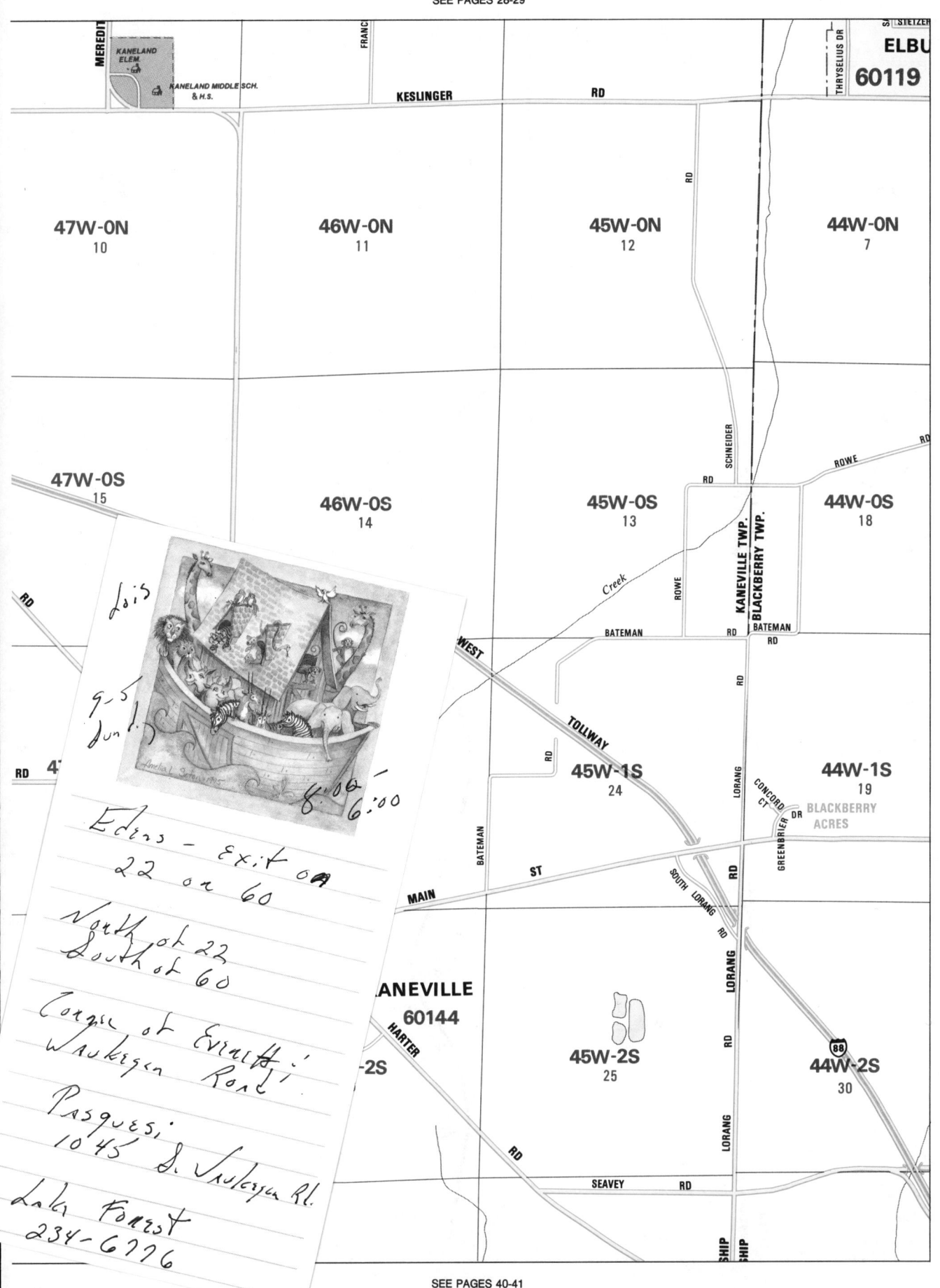

ANEVILLE
60144

HARTER

-2S

45W-2S
25

RD

LORANG

RD

88

44W-2S
30

LORANG

RD

SEAVEY RD

SHIP
HIP

SEE PAGES 40-41

SEE PAGES 36-37

SEE PAGES 30-31

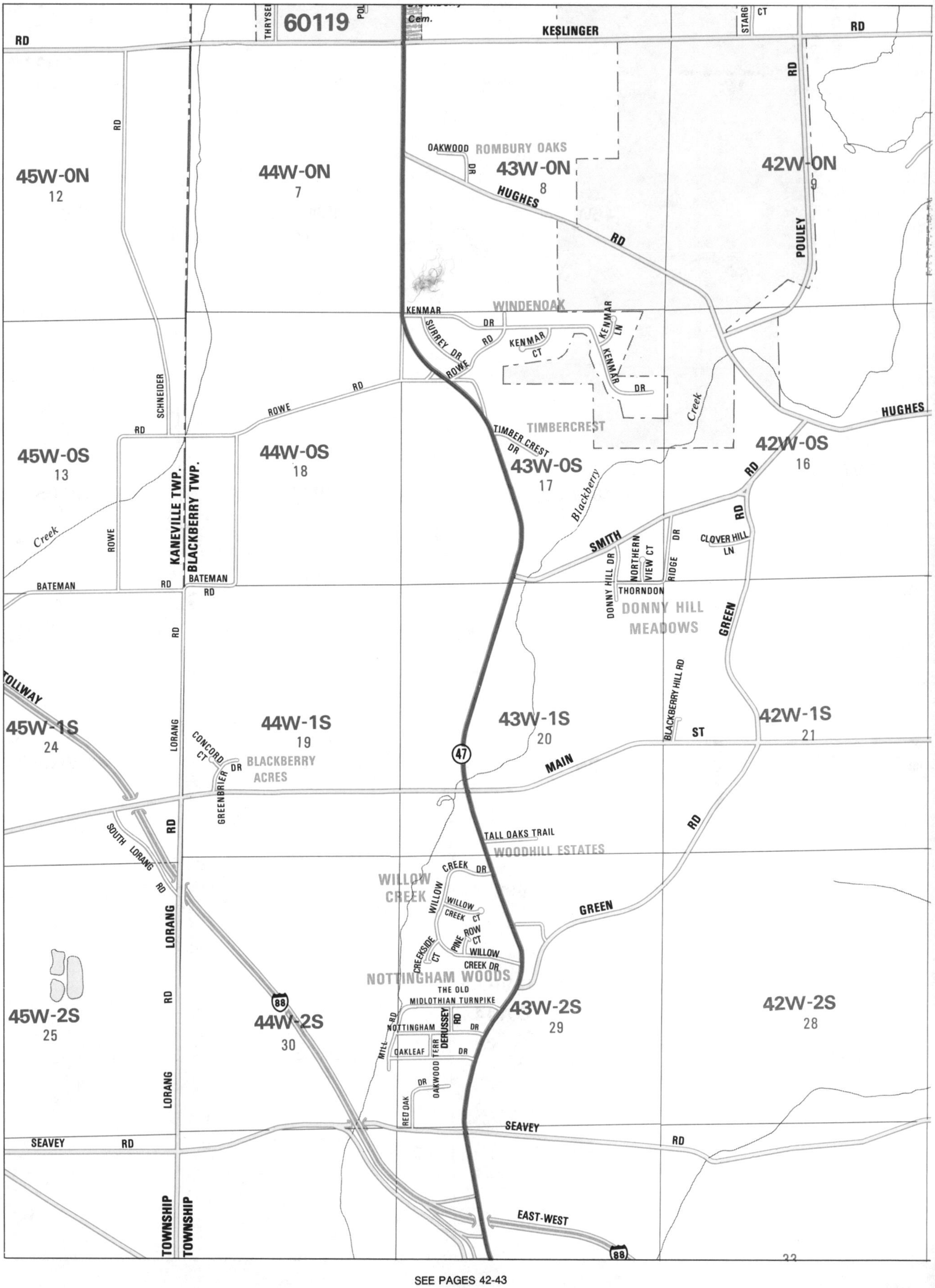

RD

THRYSE

60119

POL

Cem.

STARG

CT

KESLINGER

RD

RD

RD

RD

SCHNEIDER RD

45W-0N
12

44W-0N
7

OAKWOOD DR

ROMBURY OAKS

43W-0N
8

HUGHES

RD

POULEY

42W-0N
9

SEE PAGES 34-35

KENMAR

SURREY DR

WINDENOAK

DR

ROWE

RD

KENMAR
CT

KENMAR
CT

KENMAR

KENMAR
DR

Creek

HUGHES

ROWE

RD

ROWE

RD

45W-0S
13

SCHNEIDER RD

KANEVILLE TWP.
BLACKBERRY TWP.

44W-0S
18

TIMBER CREST DR

TIMBERCREST

43W-0S
17

Blackberry

RD

42W-0S
16

Creek

ROWE

SMITH

DONNY HILL DR

NORTHERN VIEW CT

THORNDON

RIDGE DR

CLOVER HILL
LN

GREEN

BATEMAN

RD

BATEMAN
RD

DONNY HILL
MEADOWS

RD

LORANG RD

BLACKBERRY HILL RD

TOLLWAY

45W-1S
24

CONCORD
CT

GREENBRIER DR

44W-1S
19

BLACKBERRY
ACRES

(47)

MAIN

ST

RD

43W-1S
20

42W-1S
21

SOUTH LORANG RD

LORANG RD

(88)

TALL OAKS TRAIL

WOODHILL ESTATES

WILLOW
CREEK

CREEK DR

WILLOW

WILLOW
CREEK CT

CREEKSIDE CT

PINE ROW
CT

WILLOW
CREEK DR

GREEN

NOTTINGHAM WOODS

THE OLD
MIDLOTHIAN TURNPIKE

45W-2S
25

44W-2S
30

NOTTINGHAM RD

MILL RD

OAKLEAF

OAKWOOD

DERUSSEY RD

TERR

RD

DR

43W-2S
29

42W-2S
28

RED OAK

DR

SEAVEY

RD

RD

SEAVEY

RD

SEAVEY RD

TOWNSHIP

TOWNSHIP

EAST-WEST

(88)

SEE PAGES 30-31

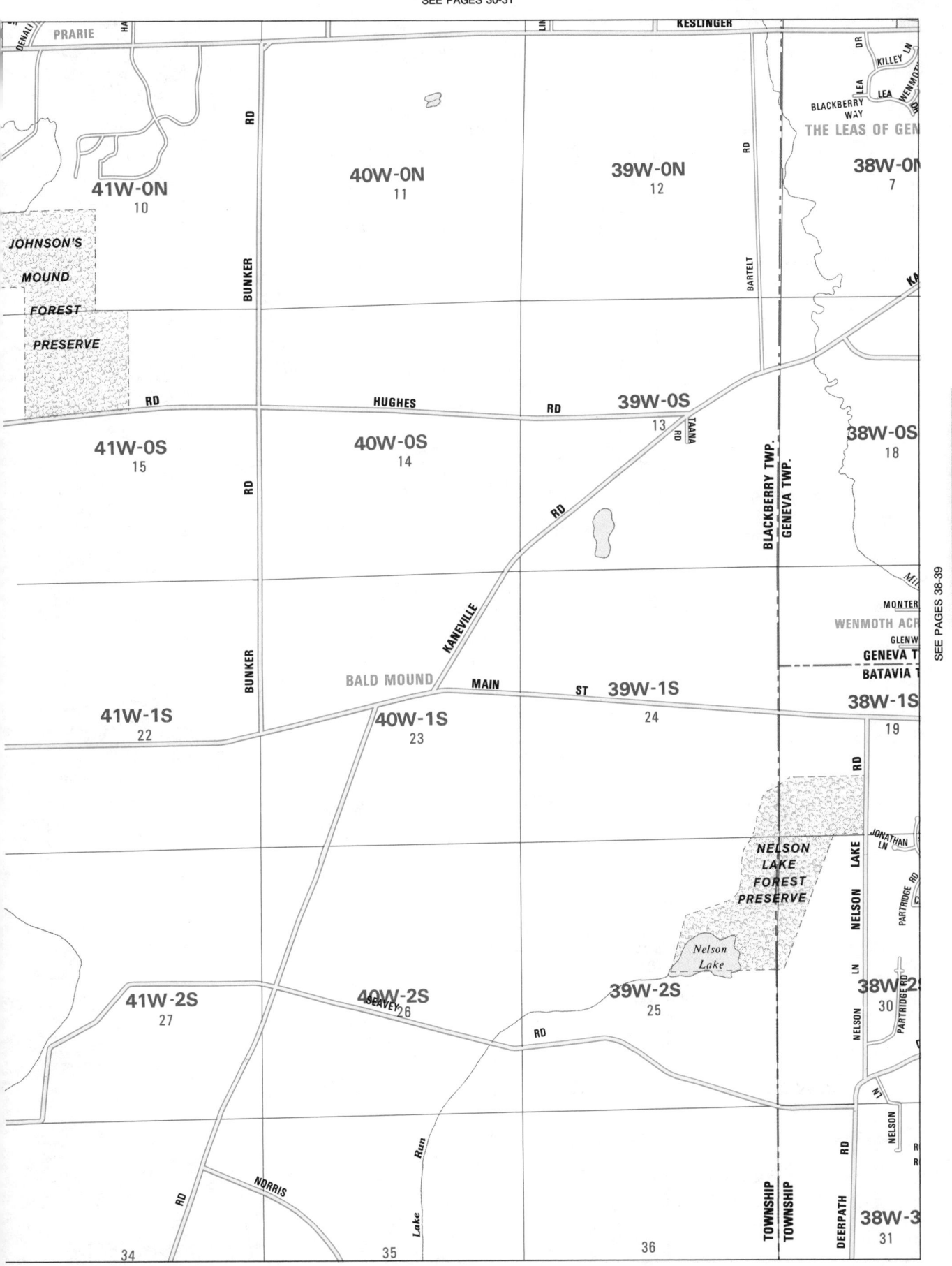

PRARIE

KESLINGER

KILLEY LN

BLACKBERRY WAY

THE LEAS OF GEN

41W-0N
10

40W-0N
11

39W-0N
12

38W-0N
7

RD

BUNKER

JOHNSON'S
MOUND
FOREST
PRESERVE

RD

HUGHES

RD

39W-0S
13

38W-0S
18

41W-0S
15

40W-0S
14

TAANA

RD

RD

BUNKER

KANEVILLE

BLACKBERRY TWP.
GENEVA TWP.

MONTER

WENMOTH ACR
GLENW

GENEVA T
BATAVIA T

BALD MOUND

MAIN

ST

39W-1S
24

38W-1S
19

41W-1S
22

40W-1S
23

RD

JONATHAN
LN

NELSON

LN

PARTRIDGE RD

NELSON
LAKE
FOREST
PRESERVE

Nelson
Lake

41W-2S
27

40W-2S
26

SEAVEY

39W-2S
25

38W-2S
30

NELSON

PARTRIDGE RD

RD

Run

Lake

NORRIS

TOWNSHIP

TOWNSHIP

DEERPATH

NELSON

R
R

38W-3
31

34

35

36

SEE PAGES 42-43

SEE PAGES 38-39

SEE PAGES 32-33

SEE PAGES 36-37

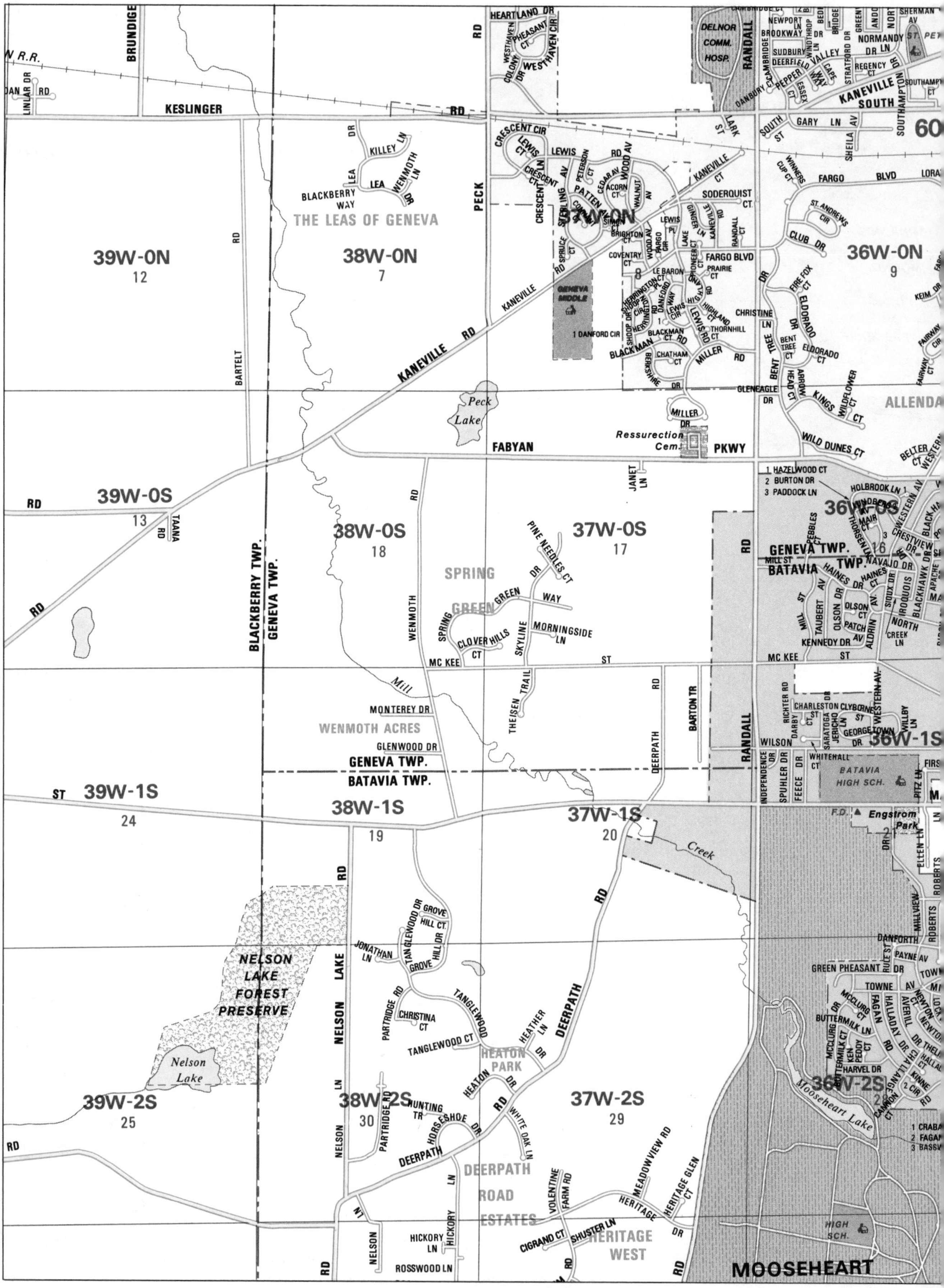

KESLINGER RD

BRUNDIGE

N.R.R.

LINLAR DR

DELNOR COMM. HOSP.

RANDALL

KANEVILLE SOUTH

60

HEARTLAND DR
PHEASANT CT
WESTHAVEN CIR
COLONY CT
WESTHAM CT

CRESCENT CIR
LEWIS CT
LEWIS
CRESCENT CT
STERLING
PATTEN
CRESCENT CT
ACORN CT
CEDAR AV
WOOD AV
WALNUT AV
PETERSON

KILLEY
LEA DR
WENMOTH LN
LEA
BLACKBERRY WAY

THE LEAS OF GENEVA

39W-0N
12

38W-0N
7

37W-0N

PECK RD

KANEVILLE RD

KANEVILLE RD

BARTELT

GENEVA MIDDLE
1 DANFORD CIR

SODERQUIST CT
KANEVILLE CT
LEWIS
SIMON
BRIGHTON
WOOD AV
HIGH
RANDALL
FARGO BLVD
PRAIRIE CT
COVENTRY CT
LE BARON DR
HERRINGTON CT
LEWIS RD
LEWIS CIR
BLACKMAN CT
THORNHILL
CHATHAM CT
BLACKMAN RD
BERKSHIRE DR
MILLER RD

CAMBRIDGE CT
NEWPORT
BROOKWAY
SUDBURY
DEERFIELD WAY
PEPPER
ESSEX
DANBURY
WINDTHROP
STRATFORD DR
REGENCY DR
SOUTHAMPTON CT

LARK ST
SOUTH ST
GARY LN
SHEILA AV
WINNERS CIR CT
FARGO BLVD
LORA

ST. ANDREWS CIR
CLUB DR
FIRE FOX DR
ELDORADO
BENT TREE DR
BENT TREE CT
HEAD CT
ARROW CT
KINGS DR
WILDFLOWER CT
FAIRWAY
KEIM DR

36W-0N
9

CHRISTINE
ELDORADO
GLENEAGLE DR
WILD DUNES CT
BELTER CT
WESTERN

ALLENDA

36W-0S
16

Peck Lake

FABYAN

Ressurection Cem.
PKWY

MILLER DR

JANET LN

1 HAZELWOOD CT
2 BURTON DR
3 PADDOCK LN

HOLBROOK LN
WINDBERG
THORSEN LN
MAIN
WESTERN AV
PEBBLES
CRESTVIEW

GENEVA TWP.
BATAVIA TWP.

39W-0S
13

38W-0S
18

37W-0S
17

RD

TAANA RD

RD

SPRING GREEN

PINE NEEDLES CT
SPRING GREEN DR
GREEN WAY
MORNINGSIDE LN
SKYLINE
CLOVER HILLS CT
MC KEE
THEISEN TRAIL
ST

WENMOTH

MILL ST
BATAVIA
HAINES DR
HAINES
TAUBERT AV
OLSON
OLSON CT
PATCH
ALDRIN DR
KENNEDY DR
MC KEE ST
SIOUX DR
NAVAJO DR
IROQUOIS
BLACKHAWK DR
APACHE
NORTH CREEK LN

RANDALL RD

BARTON TR

DEERPATH RD

MILL

MONTEREY DR

WENMOTH ACRES

GLENWOOD DR

GENEVA TWP.
BATAVIA TWP.

39W-1S
24

38W-1S
19

37W-1S
20

RICHTER RD
DARBY
SARATOGA DR
JERICHO
GEORGETOWN
WILLBY
CHARLESTON
CLYBORNE ST
INDEPENDENCE DR
SPUHLER DR
FEECE DR
WILSON
WHITEHALL CT

36W-1S

BLACKBERRY TWP.
GENEVA TWP.

Creek

BATAVIA HIGH SCH.

FIRST

F.D.
Engstrom Park

NELSON LAKE RD

NELSON LAKE

TANGLEWOOD DR
GROVE HILL CT
GROVE HILL DR
JONATHAN LN
PARTRIDGE RD
CHRISTINA CT
TANGLEWOOD CT
TANGLEWOOD
HEATHER
HEATON DR
HEATON
HEATON PARK

DEERPATH RD

ELLEN LN
DANFORTH
PAYNE AV
ROBERTS
ROBERTS
HILLVIEW
GREEN PHEASANT DR
TOWNE AV
MCCLURG
HALLADAY DR
FAGAN
AVERILL
BUTTERMILK LN
MCCLURG DR
KEN NEWTON
FEDDY
HARVEL DR
CANNON

NELSON LAKE FOREST PRESERVE

Nelson Lake

39W-2S
25

38W-2S
30

37W-2S
29

36W-2S

Mooseheart Lake

NELSON LN

PARTRIDGE RD
HUNTING TR
HORSESHOE DR
WHITE OAK LN
DEERPATH RD

DEERPATH ROAD ESTATES

HICKORY
HICKORY LN
ROSSWOOD LN

VOLENTINE
FARM RD
CIGRAND CT
SHUSTER LN
MEADOWVIEW RD
HERITAGE GLEN CT

HERITAGE WEST

HERITAGE DR

HIGH SCH.

1 CRABA
2 FAGAN
3 BASS

MOOSEHEART

SEE PAGES 44-45

SEE PAGES 32-33

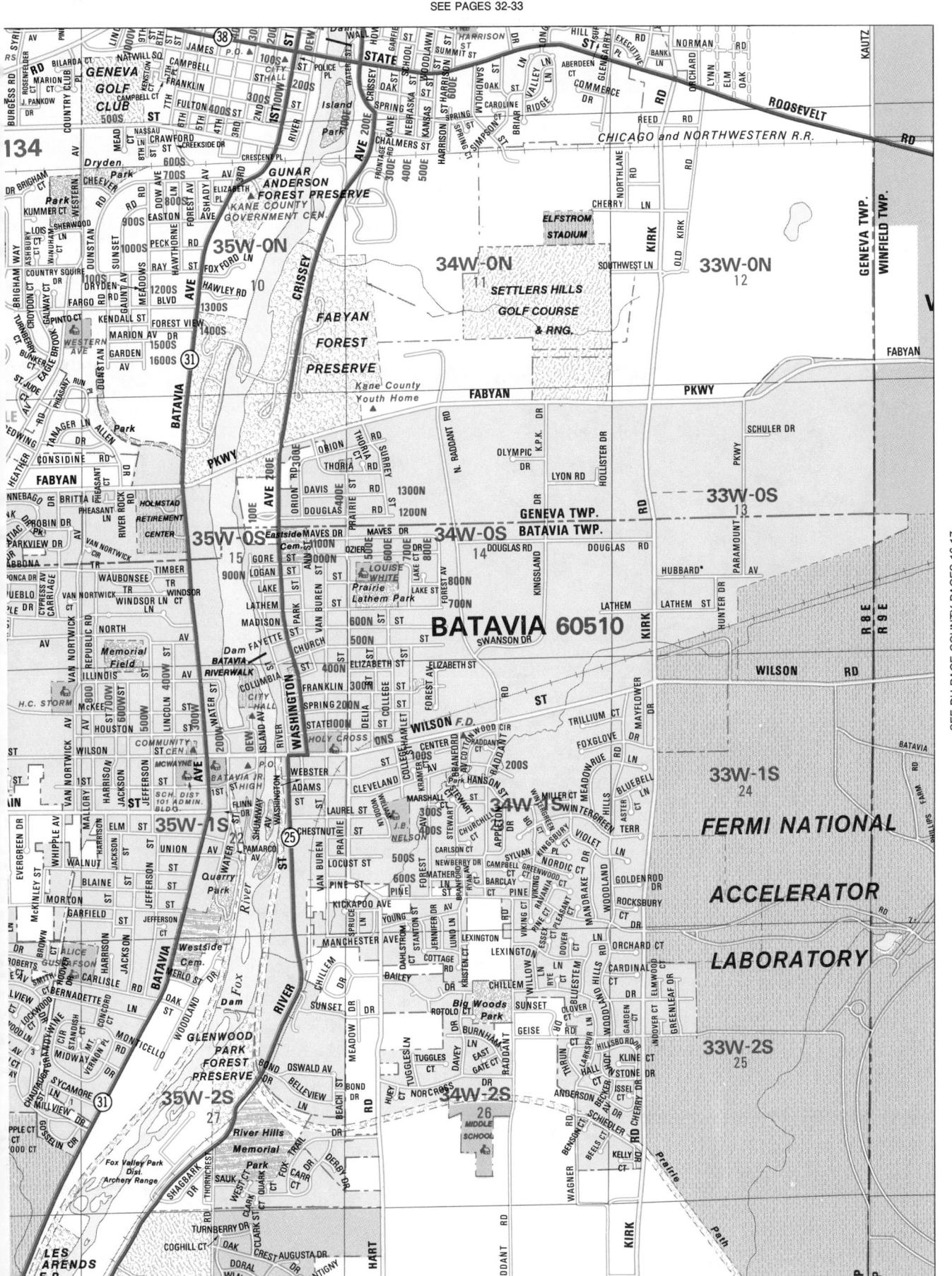

BATAVIA 60510

FERMI NATIONAL

ACCELERATOR

LABORATORY

SEE DU PAGE COUNTY PAGES 16-17

SEE PAGES 34-35

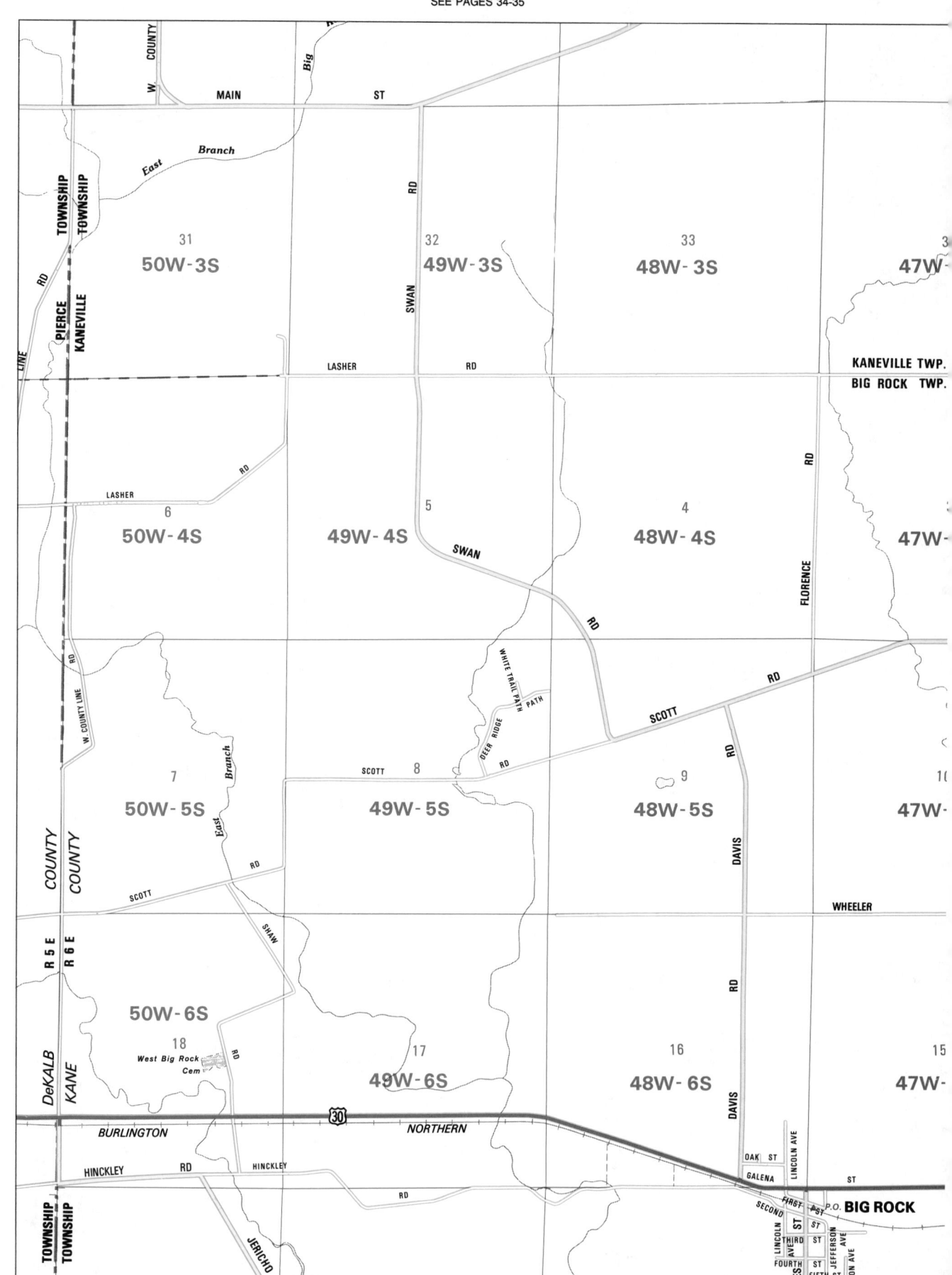

SEE PAGES 46-47

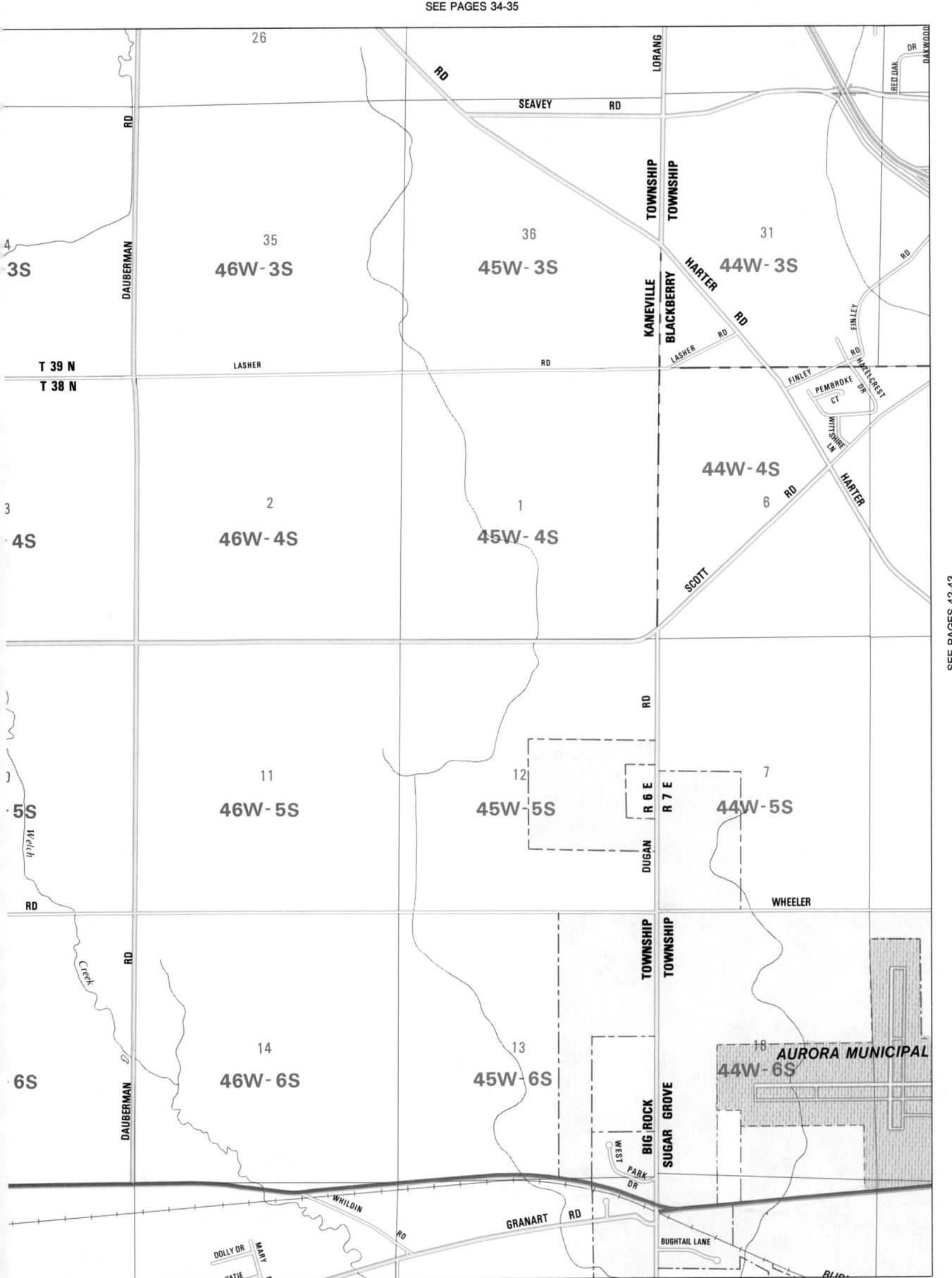

26

LORANG

RED OAK DR OAKWOOD

SEAVEY RD

RD

TOWNSHIP
TOWNSHIP

35
46W-3S

36
45W-3S

31
44W-3S

DAUBERMAN RD

KANEVILLE
BLACKBERRY

HARTER RD

FINLEY RD

T 39 N
T 38 N

LASHER RD

LASHER RD

RD

FINLEY

PEMBROKE
CT

HAZELCREST DR

WILTSHIRE LN

44W-4S

2
46W-4S

1
45W-4S

6 RD

HARTER

SCOTT

SEE PAGES 42-43

RD

11
46W-5S

12
45W-5S

R 6 E
R 7 E

7
44W-5S

Welch

DUGAN

Creek

RD

RD

WHEELER

TOWNSHIP
TOWNSHIP

14
46W-6S

13
45W-6S

18
44W-6S

AURORA MUNICIPAL

DAUBERMAN

BIG ROCK
SUGAR GROVE

WEST
PARK
DR

WHILDIN
RD

GRANART RD

BUGHTAIL LANE

DOLLY DR

MARY

KATIE

BUR

SEE PAGES 36-37

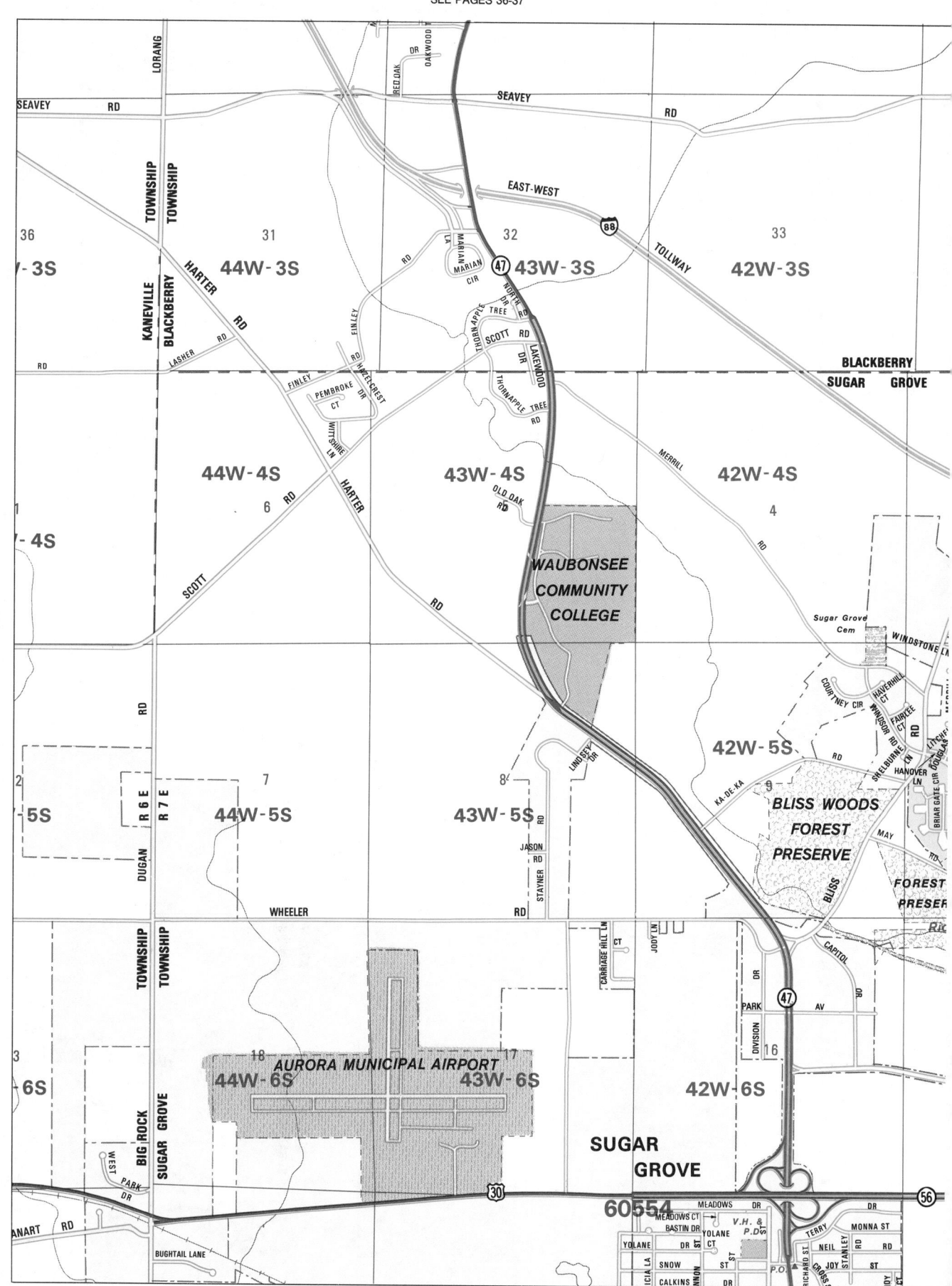

SEE PAGES 40-41

SEE PAGES 48-49

SEE PAGES 36-37

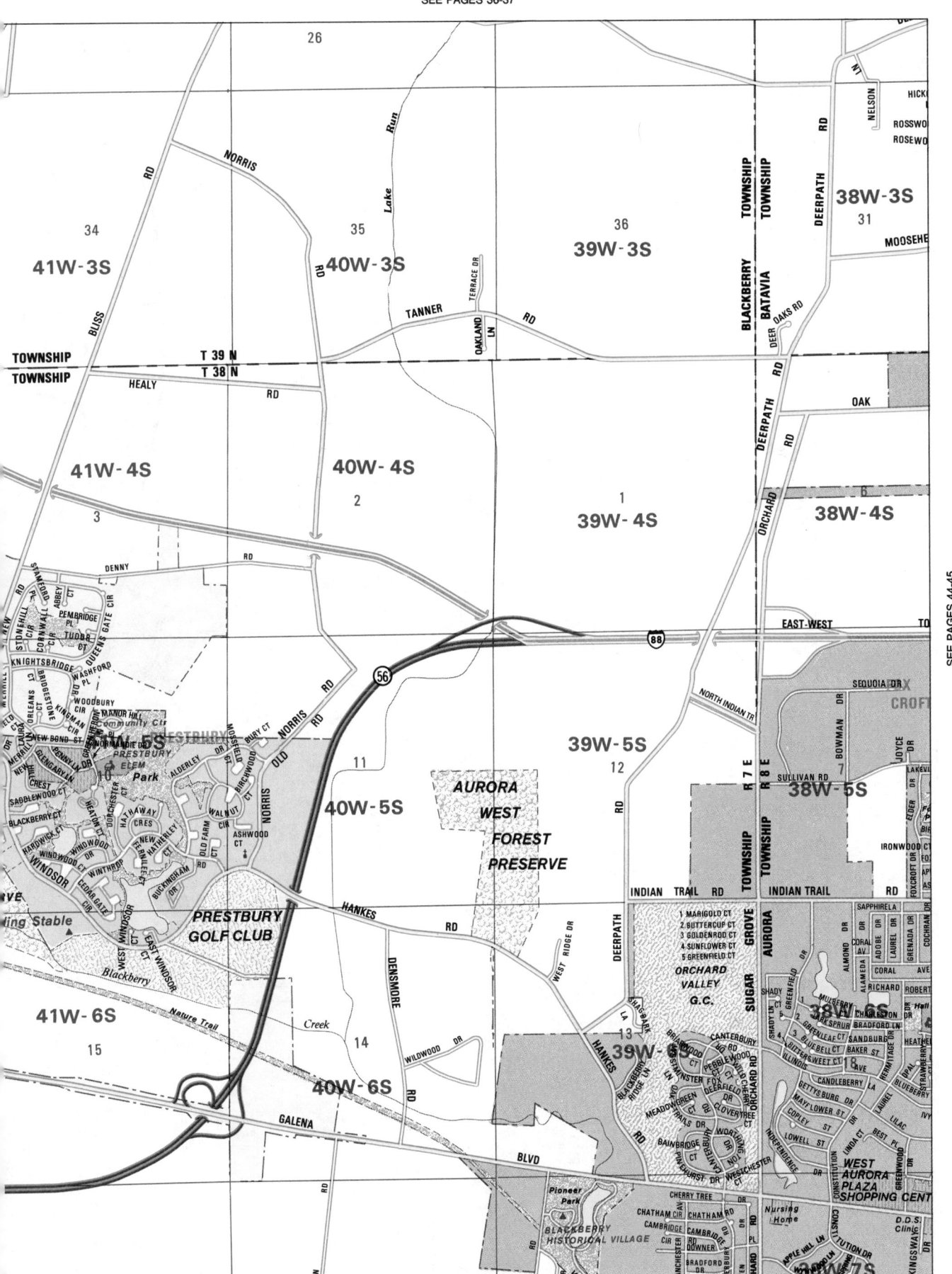

SEE PAGES 44-45

SEE PAGES 48-49

SEE PAGES 38-39

MOOSEHEART
Moosehart

LES ARENDS F.P.

FOX COUNT

Fox Valley Dist Archery Re

Giles HIGH SCH

Red Oak Nature Center

North Aurora Shoreline Park

DEERPARK LN

HERITAGE GLEN CT

HERITAGE DR

MEADOW

VOLENTINE FARM RD

CIGRAND CT   SHUSTER LN

HICKORY LN   HICKORY

NELSON

ROSSWOOD LN
ROSEWOOD LN

VOLENTINE FARM

ELFSTROM

DEERPATH RD

BLACKBERRY TOWNSHIP
BATAVIA TOWNSHIP

38W-3S
31

MOOSEHEART RD 32

MOOSEHEART

37W-3S

MOOSEHEART RD

VALLEY RD

33 RD

36W-3S

NORTH AURORA

HILLTOP
TIMBER EASTVIEW OAKS

ELM AVE
MAPLE AVE

31

25

BATAVIA AURORA

DEER OAKS RD

DEERPATH RD

OAK ST

NAPA
SUNSHINE DR
BULL CT

GERONIMO

REDWING

SUMAC DR

38W-4S
6

37W-4S

RANDALL RD

MAGNOLIA DR
MAGNOLIA CT
DOGWOOD CT
DOGWOOD DR

ACORN DR   JUNIPER LN

BUTTERNUT CT
COTTONWOOD DR

PINEWOOD DR
PINE OAK DR
CHERRYWOOD DR
BIRCHWOOD

REDWOOD
LARCHWOOD

BASSWOOD DR   MAPLEWOOD DR

OAK DR

ACORN DR

MAGNOLIA DR

BUCKTHORN

SHAGBARK

WALNUT DR

GOODWIN ELEM
Goodwin Park

STAGHORN

NORTH AURORA
60542

JUNIPER DR

CYPRESS

POPLAR

SYCAMORE

PRINCETON

PRINCETON

CEDAR DR

STATE
VIEW CT

HAWTHORNE

CANDLEWICK

CHANTILLY LN
ERICKSON Park

FAR VIEW DR

HARMONY

STATE ST

HARMONY DR

JOHN ST
KINGSWOOD

FARVIEW DR

WILLOW WAY

ROBERT

ADAMS

CHERRY

JOHN ST
B.O.

HILLSIDE PL

REC. CTR.

VILLAGE HALL & POLICE DEPT

F.D.

OAK DR

BOWMAN

LINCOLN WAY

North Aurora
Island Park

Library

56

BUTTER

STATE ST

PIERCE ST

HETTINGER LN

VALLEY GREEN GOLF COURSE

36W-4S

BURLINGTON

4

POPLAR

ADLER

FOXFIELDS
Overland DR
RACETRACK

AIRPORT RD

RIVERVIEW AVE

GRANT

CONKEY ST

River

SOUTH ST

SEE PAGES 42-43

88

EAST-WEST TOLLWAY

AIRPORT RD

NORTH INDIAN TR

SEQUOIA DR   FOX CROFT

BOWMAN DR

SEQUOIA DR

SEQUOIA RD

Randall Plaza

MAPLE ST
GILMORE ST
INDIANA ST
DAWES ST
BRYANT ST
PERSHING ST
DOUGLAS ST

HINDU TEMPLE

LANDMARK RD

INDUSTRIAL ST

1 HOLIDAY DR
2 BLACK OAK TR
3 GREEN OAK TR

1600N

Smoke Tree Plaza

36W-5S

TINLEY DR
CALICO DR
SANDY

CAMBRIDGE
SPRINGBROOK DR

EASTWOOD
LINCOLNSHIRE AVE
CLOVERDALE PL

LAMBRIDGE
EVERGREEN DR

HOLLY FOREST AVE

FAIRVIEW DR

U.S. Army Reserves

BURLINGTON

OFFUTT LA

LOVEDALE

LA

JOYCE DR

SULLIVAN

R 7 E   R 8 E

SULLIVAN RD   7

38W-5S

LAKEVIEW DR   BRIGHTON

ELDER

Foxcroft Park   CIR   WALDEN
BIRCH   CIR   LA

IRONWOOD CT

FOXCROFT CT

APPLETREE LA

ASHWOOD

FOXCROFT DR

INDIAN TRAIL TOWNSHIP

INDIAN TRAIL RD

ILLINOIS MATHMATICAL & SCIENCE ACD.

FRANKLIN MIDDLE

1 CARNATION CT
2 CROFTON CT

Randall Park

8

BURR OAK CIR

LONE OAKS

BIG OAK TR

SULLIVAN

HIDDEN OAKS LA

GOLDEN OAK

PIN OAK CT

SCARLET AV

RED OAK CT

OAK CIR

RED OAK TR

SMITH ELEM
UPPER BRANDON DR

FORAL

HUNTINGTON
DR

HUNTINGTON PL

ARLON RD

FALLEN OAK

ELMWOOD

EASTWOOD
CRESTWOOD
ROBINHOOD

BEAU RIDGE DR

NANTUCKET DR

9

BEAU
RIDGE PL

CEDARDALE PL
Beau Ridge Park
F.A.A.
Admin Office

RUSSELL AVE

RUSSELL AVE

MERCY DR

Mercy Center For Health Care Services

MERCY DR

NEW INDIAN CT

HOPE WALL

INDIAN TRAIL

RD

INDIAN TRAIL RD

OLD INDIAN

BRIDGE

NEW INDIAN TRAIL

Convenient Ctr.

1 MARIGOLD CT
2 BUTTERCUP CT
3 GOLDENROD CT
4 SUNFLOWER CT
5 GREENFIELD CT

ORCHARD VALLEY G.C.

SUGAR GROVE AURORA TOWNSHIP

SAPPHIRE LA

ALMOND DR

CORAL
LA

LAUREL LA

GRENADA DR

COCHRAN

EMERALD DR

CORAL AV
RICHARD
ROBERT

CARRIAGE LN

CASCADE DR

NEW CASTLE

COLORADO

RAINWOOD

BRANDON LN

AMBERGATE

COLORADO CT

COLORADO

BEACON
LA

BEACON LN

CALUMET AVE

CIRRUS AV

LEGRANDE AVE

COLORADO

CALIFORNIA

FORDHAM AVE

SHELDON AVE

ELMWOOD

St RITA OF CASCIA SCH

RUSSELL AVE

RICHARD

ROBERT

LANCASTER DR

MORTON AVE

NORTHERN

RUSSELL AVE

LAKELAWN

HIGHLAND

NORTH PARK

IOWA AVE

GRAND

MICHIGAN AVE

MANOR AVE

CULVARY

BLVD

NORTHGATE Shopping CENT

38W-6S

CANTERBURY

BRIARWOOD

WESTMINSTER FOX

PEBBLEWOOD CT

DEERFIELD RD

CLOVERTREE

MEADOWGREEN DR

BAINBRIDGE DR

CANTERBURY CIR

WORTHINGHURST

WESTCHESTER

CHATHAM CIR

CAMBRIDGE CIR

CHATHAM RD

CAMBRIDGE RD

BRADFORD

SHADY LA
GREENFIELD
MULBERRY

OAK SPUR
GREENLEAF CT
SANDBURG

BLUEBELL CT
SWEET CT

CHARLESTON
BRADFORD LN

BAKER ST

CYPRESS

CANDLEBERRY LA

BETTYS BURG CT

COPLEY ST

LOWELL ST

INDEPENDENCE

CONSTITUTION

Hall Park
HALL ELEM

HEATHER

HERMITAGE DR

LAUREL
IVY

REDWOOD
LA

LILAC LA

EDGELAWN

STRAWBERRY

BLUEBERRY CT

CYPRESS

SHERWOOD

DUNCAN DR

HAT CT

Rob Roy Park

ROSESTONE

CRESTWOOD

CT   17

GLADSTONE TERR

MONOMOY

WESTLAWN

WEST GATE

WESTGATE

HOYT

SEVENTH DAY ADVENTIST

38W-6S

37W-6S

ILLINOIS

MCCLEERY

BOON

NEW HAVEN

GLADSTONE

MCCLEERY ELEM

MONONA AV

MERRIMAC AV

WINONA AV

CHEYENNE AV

OTTAWA AVE

SENECA AV

IROQUOIS

PLEASURE CT

W PLEASURE CT

McCleary Park

RANDALL RD

ELMWOOD
FORDHAM AVE

GLEN CIR

FORAN AVE

TAYLOR

MORTON AVE

AURORA CHRISTIAN SCH   600N

Greene Playground

RUSSELL AVE

BURLINGTON AVE

GILLETTE AVE

Gillette Park

MAY

VIEW

HAMMOND

HILL

ILLINOIS

400W

SUNSET AVE

LAWNDALE

AURORA ELEM

PENNSYLVANIA

36W-6S

16

COLORADO AVE

CALIFORNIA AVE

FLORIDA AVE

PLUM

300W

GALENA

WEST AURORA PLAZA SHOPPING CENTER

Nursing Home

D.D.S. Clinic

Medical Clinic

CHERRY TREE

APPLE HILL LN

CONSTITUTION DR

KINGSWAY

GREENWOOD

LINDA CT

PLUM
GREENE

REDWOOD

ROSEDALE

WEST GATE

GALENA

DOWNER

BUELL

WEST GATE

RANDALL RD

Jefferson Park

JEFFERSON JR. H.S.

WEST AURORA H.S.

CHARLES

SPRUCE

NEW YORK

NEW HAVEN

GREENMAN ELEM

HOWARD

PERRY

WEST PARK

F.D.

VIEW

VIEW Street Playground

GRAND

PALACE

AURORA CHRISTIAN SCH

HIGHLAND

100N

AURORA

BLACKHAWK

BLVD

ALTERNATE

TODD

Edna Children's

37W-7S

Nursing Home

SPRUCE

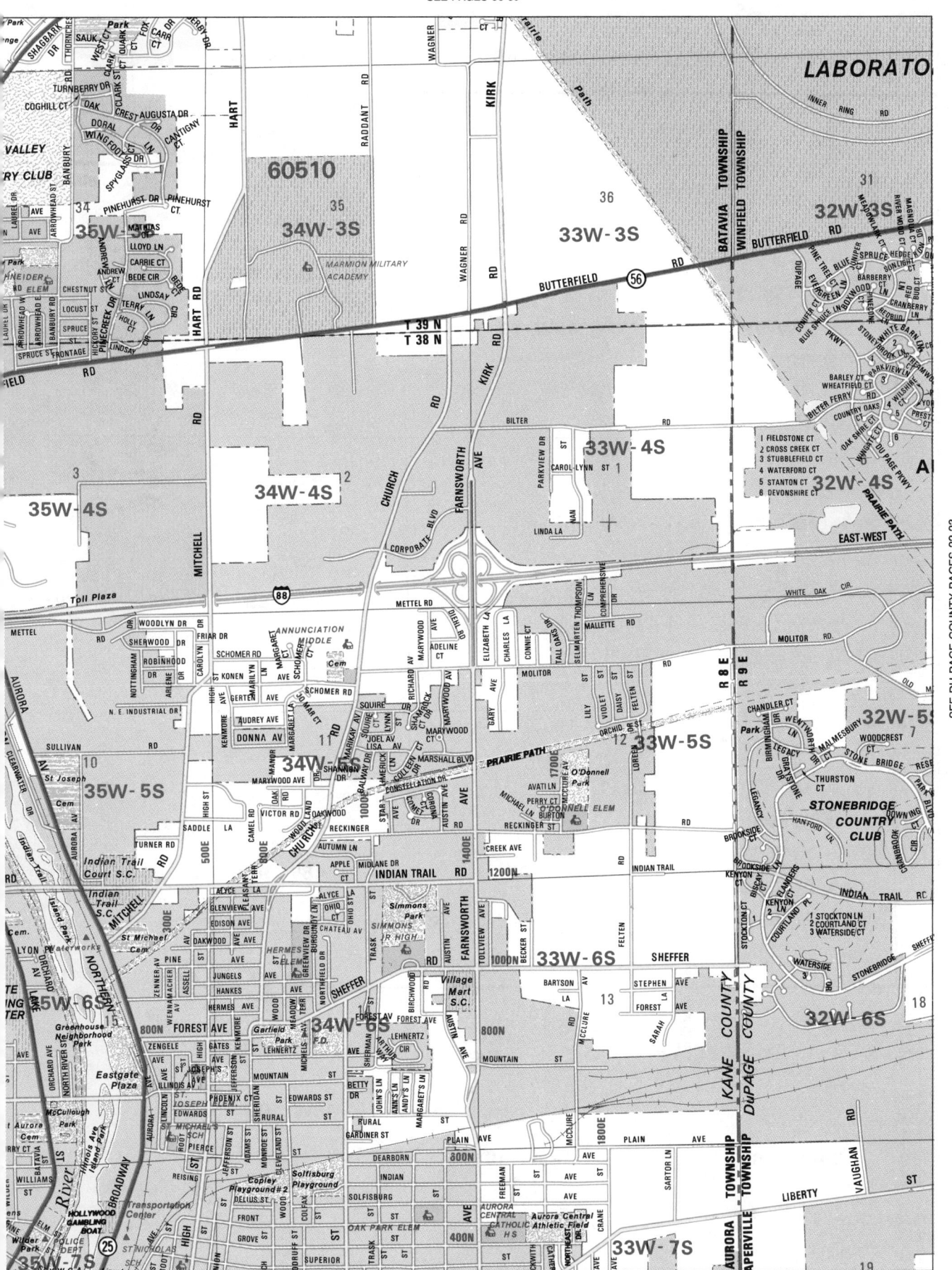

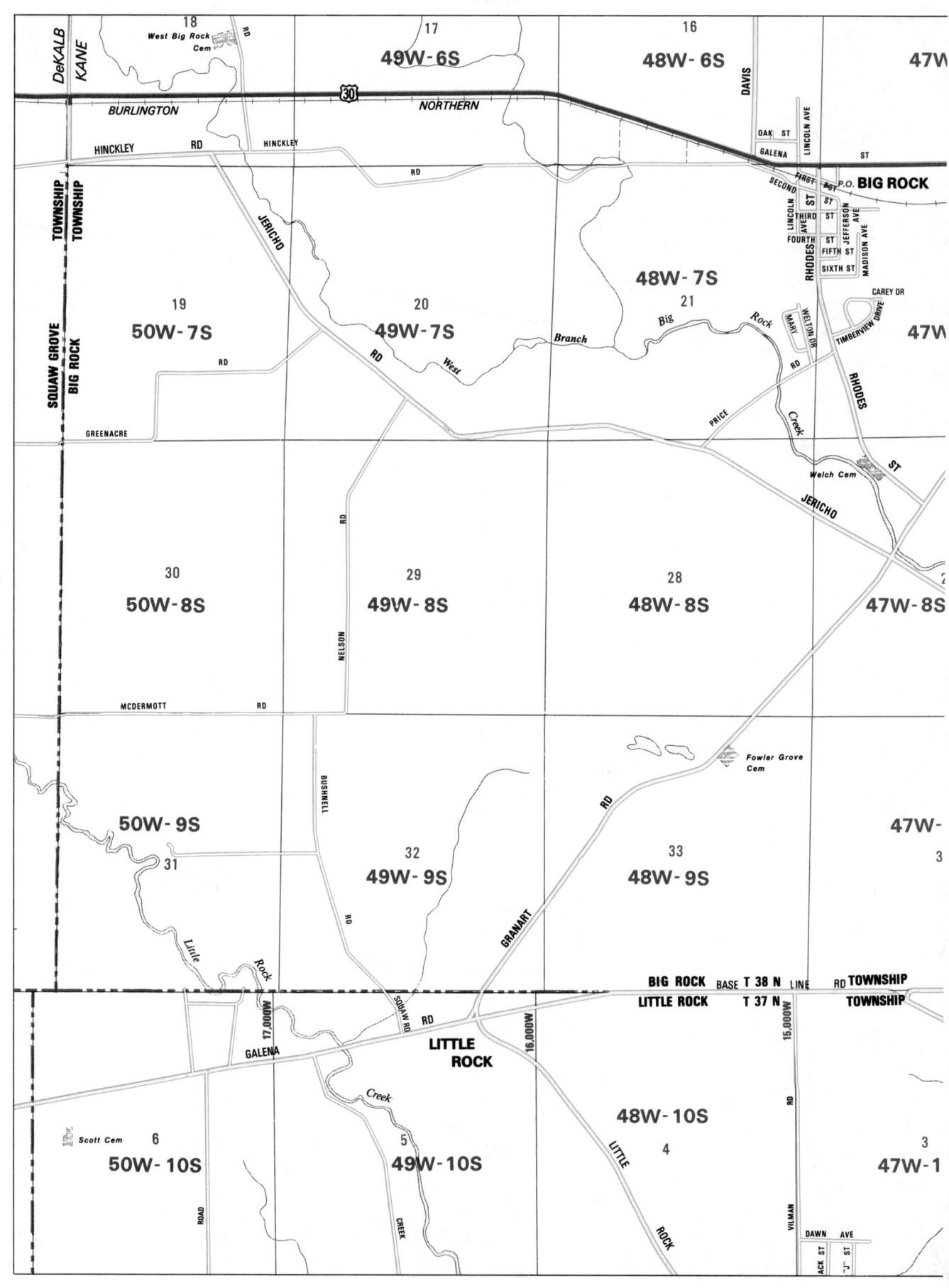

DeKALB
KANE

18
West Big Rock Cem
RD

17
49W-6S

16
48W-6S

DAVIS

47W

BURLINGTON   NORTHERN

HINCKLEY   RD   HINCKLEY

RD

OAK ST
GALENA

LINCOLN AVE

P.O. BIG ROCK

ST
FIRST
SECOND

JERICHO

TOWNSHIP   TOWNSHIP

LINCOLN AVE
THIRD   ST
FOURTH   ST
FIFTH   ST
SIXTH ST

ST
ST
ST

JEFFERSON AVE

MADISON AVE

RHODES

CAREY DR

48W-7S

21

WELTON DR

TIMBERVIEW DRIVE

19
50W-7S

20
49W-7S

Big   Rock

MARY

47W

SQUAW GROVE
BIG ROCK

RD

RD

Branch

West

PRICE

RD

Creek

RHODES   ST

GREENACRE

Welch Cem

JERICHO

30
50W-8S

29
49W-8S

28
48W-8S

RD

47W-8S

NELSON

MCDERMOTT   RD

Fowler Grove
Cem

BUSHNELL

RD

50W-9S

31

32
49W-9S

33
48W-9S

47W-

RD

GRANART

Little   Rock

17,000W

SQUAW RD

16,000W

BIG ROCK BASE **T 38 N** LINE RD **TOWNSHIP**
LITTLE ROCK   **T 37 N**   **TOWNSHIP**

15,000W

GALENA

**LITTLE
ROCK**

Creek

Scott Cem   6
50W-10S

5
49W-10S

4
48W-10S

LITTLE

3
47W-1

ROAD

CREEK

ROCK

VILMAN
RD

DAWN   AVE

ACK ST
"J" ST

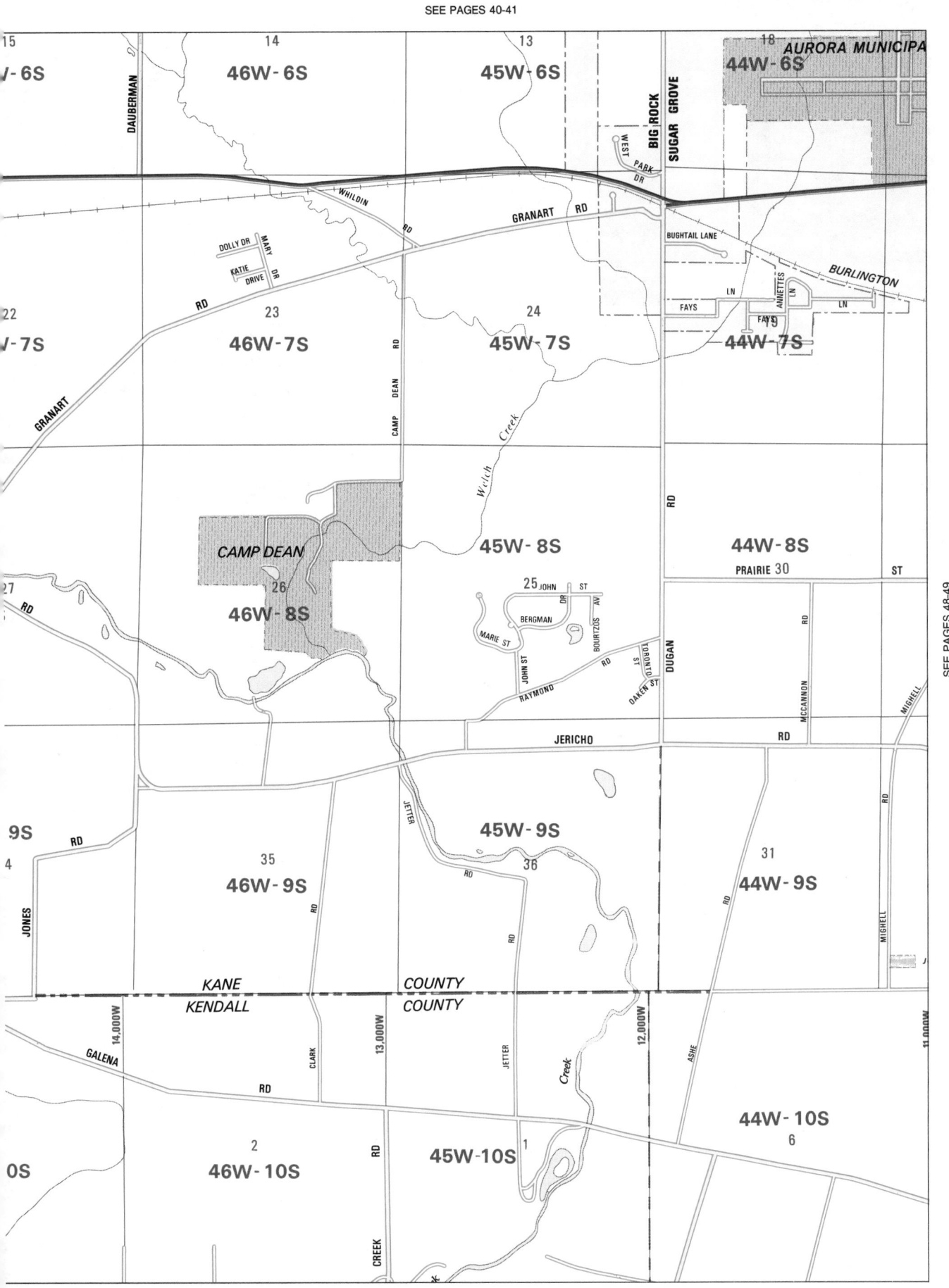

15

14

13

46W-6S

45W-6S

44W-6S

18 **AURORA MUNICIPA**

-6S

V-6S

DAUBERMAN

BIG ROCK

SUGAR GROVE

WEST PARK DR

WHILDIN RD

GRANART RD

BUGHTAIL LANE

BURLINGTON

DOLLY DR

MARY DR

KATIE DRIVE

RD

22

23

24

19

FAYS LN

ANNETTES LN

LN

FAYS

V-7S

46W-7S

45W-7S

44W-7S

GRANART

CAMP DEAN RD

Welch Creek

RD

27

CAMP DEAN

26

45W-8S

44W-8S

PRAIRIE 30

ST

RD

46W-8S

25 JOHN ST

BERGMAN

DR

BOURTZOS AV

DUGAN

RD

MCCANNON ST

MIGHELL

ST

MARIE ST

JOHN ST

RAYMOND RD

TORONTO ST

OAKEN ST

JERICHO

RD

9S

RD

JETTER

45W-9S

31

44W-9S

RD

35

46W-9S

36

RD

4

JONES

RD

MIGHELL RD

KANE COUNTY

KENDALL COUNTY

14,000W

13,000W

12,000W

11,000W

GALENA

CLARK

JETTER

ASHE

Creek

RD

2

46W-10S

RD

1

45W-10S

44W-10S

6

OS

CREEK

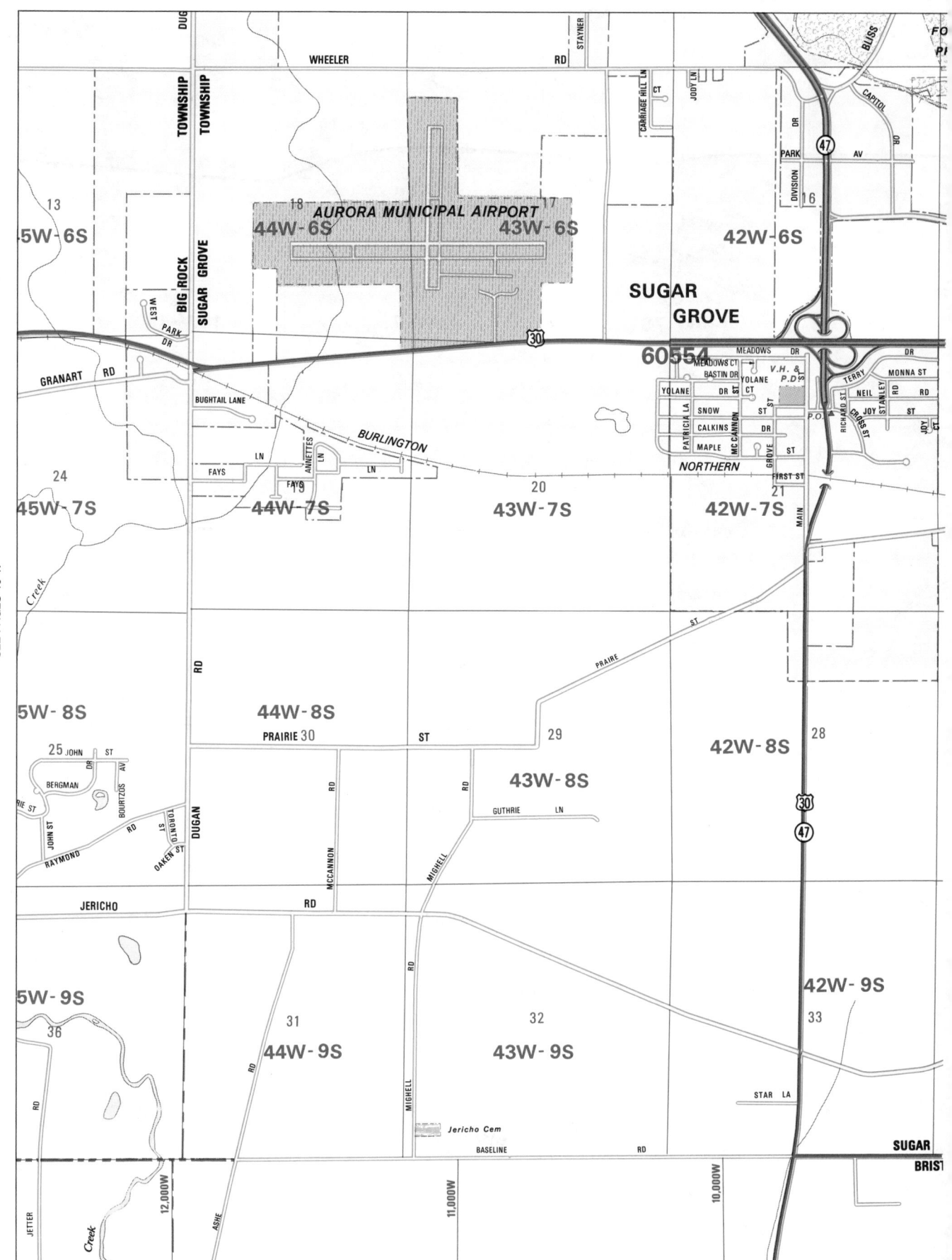

AURORA MUNICIPAL AIRPORT

44W-6S   43W-6S

42W-6S

SUGAR
GROVE

60554

45W-7S   44W-7S   43W-7S   42W-7S

NORTHERN

5W-6S   13

TOWNSHIP TOWNSHIP

BIG ROCK SUGAR GROVE

WHEELER   RD

GRANART   RD

BUGHTAIL LANE

BURLINGTON

FAYS   FAYS

44W-8S

PRAIRIE 30   ST

43W-8S

GUTHRIE   LN

42W-8S

5W-8S   24

JERICHO   RD

44W-9S   43W-9S   42W-9S

5W-9S   36   31   32   33

Jericho Cem

BASELINE   RD

SUGAR
BRIST

SEE PAGES 46-47

SEE PAGES 42-43

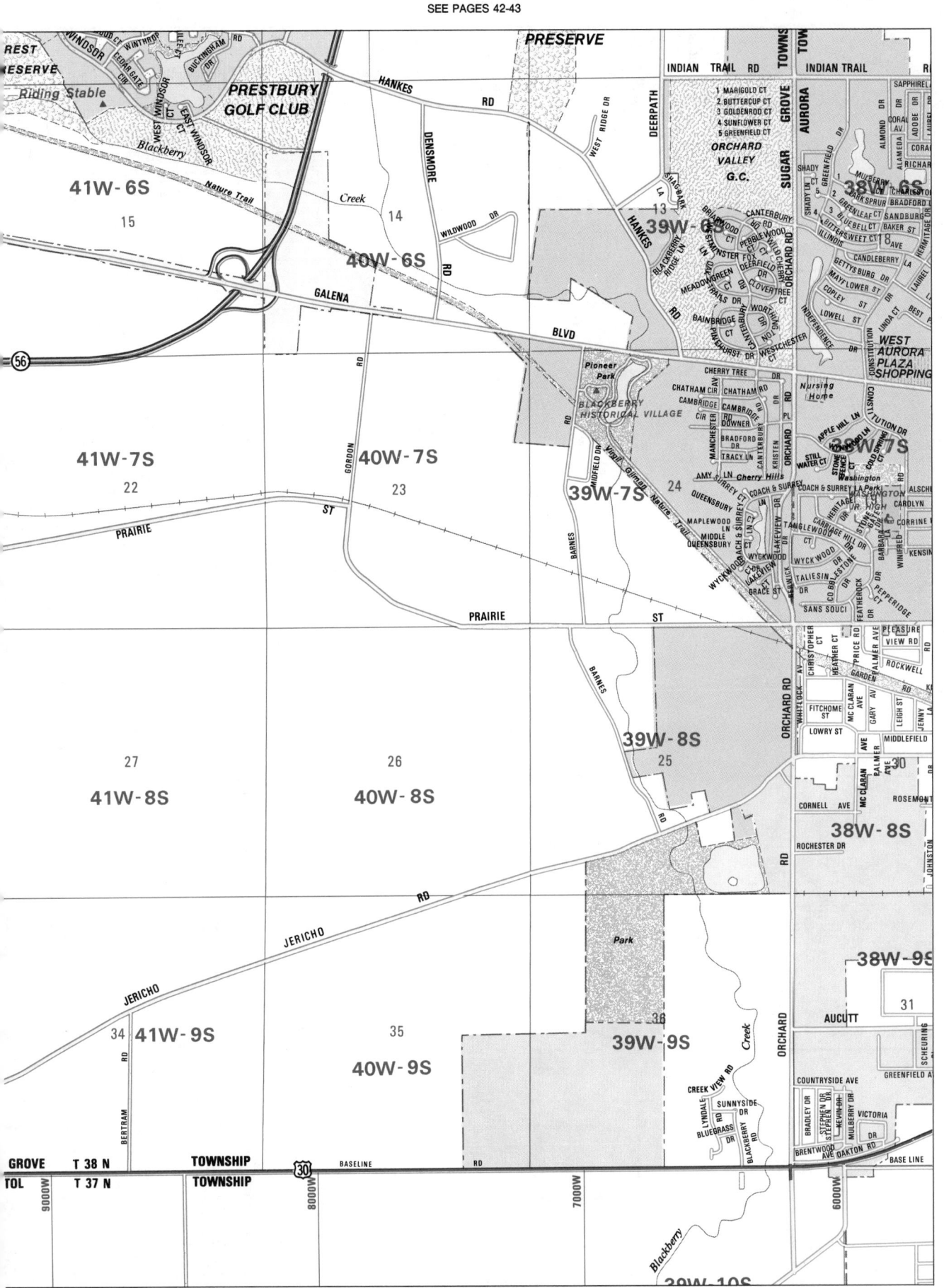

SEE PAGES 50-51

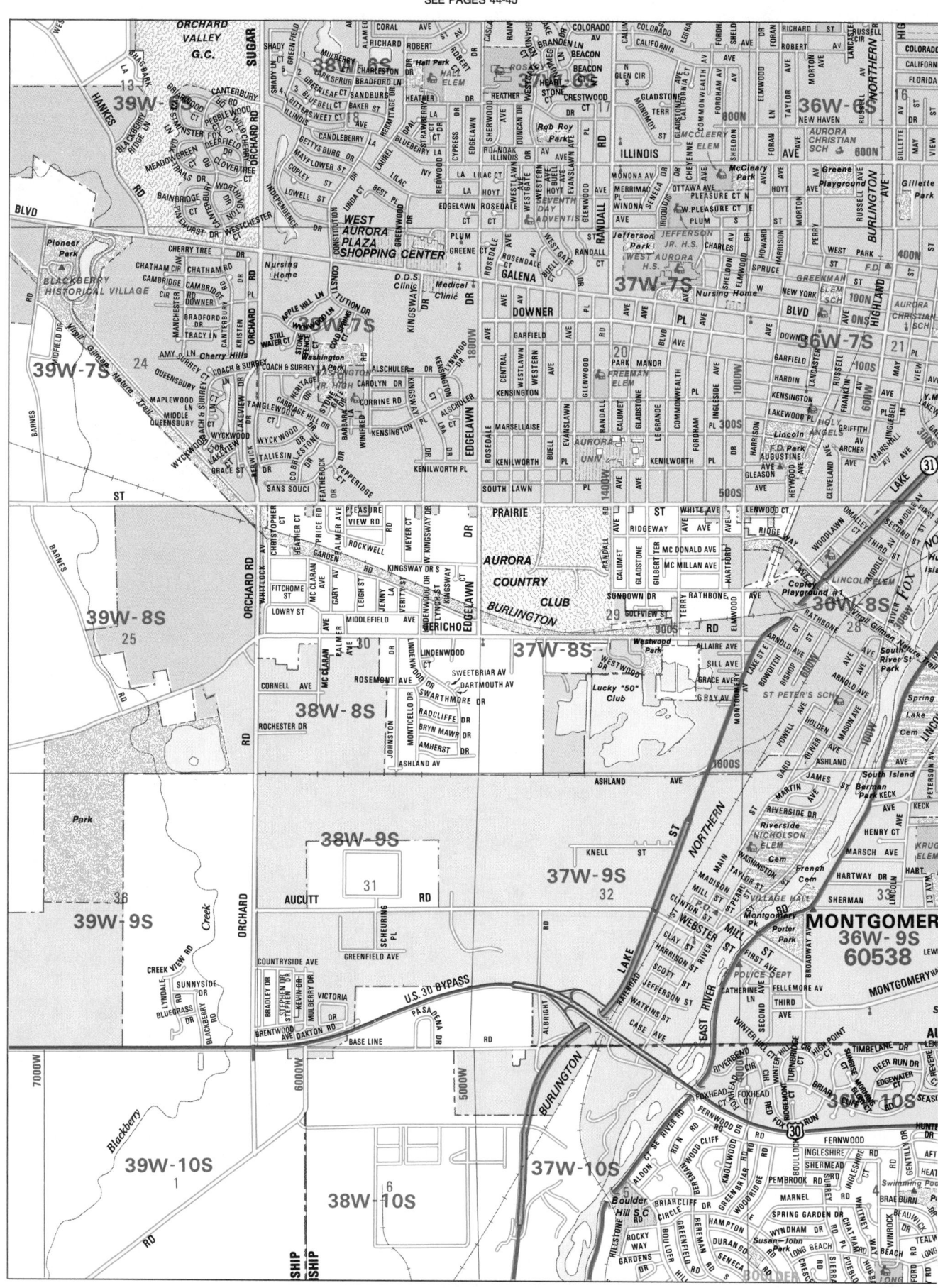

SEE PAGES 48-49

SEE PAGES 44-45

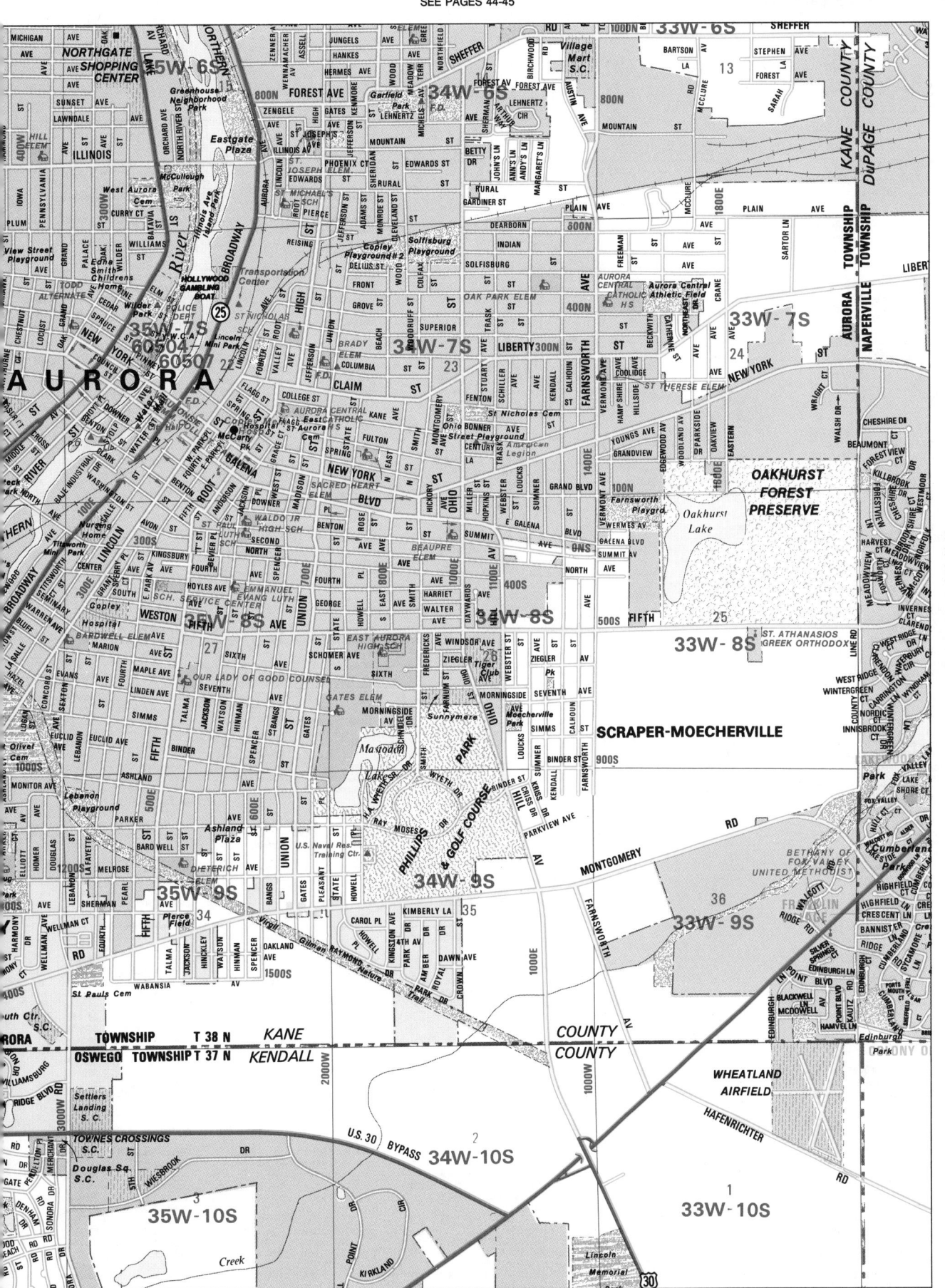

SEE DU PAGE COUNTY PAGES 28-29

Street Index

Algonquin*
(Also see McHenry County)

	Grid	Page
Braewood Dr. N&S	35W-19N	8
Brandywine Cir. E&W	35W-19N	8
Charles Dr. N&S	34W-19N	9
Compton Dr. E&W	34W-19N	9
Dana Point NW&SE	34W-19N	9
Gaslight Dr. N&S	35W-19N	8
Greenridge Av. N&S	34W-19N	9
Hampton Ct. NE&SW	35W-19N	8
Hanson Rd. E&W	35W-19N	8
Hillside Ct. NW&SE	35W-19N	8
Louis Av. E&W	34W-19N	9
Meghan Av. N&S	34W-19N	9
Minnehaha Tr. N&S	34W-19N	9
Riverwood Dr. N&S	34W-19N	9
Riverwood Ln. N&S	34W-19N	9
Rustic Ln. N&S	35W-19N	8
Spring Hill Dr. NS&EW	35W-19N	8
Spruce Tree Ln. N&S	35W-19N	8
Stratford Ln. NE&SW	34W-19N	9
Terri Ln. E&W	34W-19N	9

Aurora*
(Also see DuPage County)

	Grid	Page
Adams St. N&S	34W-6S	45
Adobe Dr. N&S	38W-6S	44
Alameda Dr. N&S	38W-6S	44
Allaire Av. E&W	38W-7S	50
Almond Dr. N&S	38W-6S	44
Alschuler Ct. E&W	38W-7S	50
Alschuler Dr. NS&EW	38W-7S	50
Alyce Ln. E&W	34W-6S	45
Amy Ln. E&W	39W-7S	49
Anderson St. NE&SW	35W-8S,35W-7S	51
Apple Ct. E&W	34W-5S	45
Apple Hill Ln. SW&NE	38W-7S	49
Appletree Ln. E&W	38W-5S	44
Archer Av. N&S	36W-7S	50
Arlene Dr. N&S	35W-5S	45
Arlon Rd. E&W	36W-5S	44
Arnold Av. N&S	36W-5S	44
Arthur Way NW&SE	34W-6S	45
Ashland Av. E&W	36W-9S,35W-9S	50
Ashland Ct. N&S	36W-9S	50
Ashwood Ct. NW&SE	38W-5S	44
Ashwood Ln. NS&EW	38W-5S	44
Assel Av. N&S	35W-6S	45
Augustine Av. E&W	36W-7S	50
Aurora Av. N&S	36W-6S,35W-9S	45
Austin Av. N&S	34W-6S,34W-5S	45
Autumn Ln. E&W	34W-5S	45
Avati Ln. E&W	33W-5S	45
Avon St. E&W	35W-8S	51
Baje Industrial Dr. NE&SW	36W-7S	50
Baker St. E&W	38W-6S	44
Bangs St. N&S	35W-9S,35W-8S	51
Barbara Ln. N&S	38W-7S	50
Bardwell St. E&W	35W-9S	51
Barnes Rd. N&S	39W-7S	49
Bartle Ct. E&W	33W-9S	51
Batavia St. N&S	35W-6S	45
Beach St. N&S	34W-7S	51
Beau Ridge Ct. NW&SE	36W-3S	44
Beau Ridge Dr. N&S	36W-5S	44
Beckwith St. N&S	33W-7S	51
Benton St. NW&SE	35W-7S,34W-8S	51
Berwick St. N&S	39W-7S	49
Best Pl. NE&SW	38W-6S	44
Bevier Pl. N&S	35W-8S	51
Bilter Rd. E&W	34W-4S,33W-4S	45
Binder St. NW&SE	35W-9S,35W-9S	51
Birch Ln. E&W	38W-5S	44
Bishop St. NE&SW	36W-8S	50

	Grid	Page
Bittersweet Ct. NW&SE	38W-6S	44
Blackhawk St. N&S	36W-7S	50
Blackwell Ln. NS&EW	33W-9S	28
Bluebell Ct. E&W	38W-6S	44
Blueberry Ln. NW&SE	38W-6S	44
Bluff St. NW&SE	36W-8S	50
Bonner Av. N&S	34W-7S	51
Bowditch St. NE&SW	36W-8S	50
Bowman Dr. N&S	38W-6S	44
Bradford Dr. E&W	39W-7S	49
Bradford Ln. E&W	38W-6S	44
Brighton Cir. NS&EW	38W-5S	44
Broadway NE&SW	36W-8S,35W-6S	50
Broughton Dr. E&W	33W-9S	51
Buell Av. N&S	37W-7S	50
Buell Ct. NE&SW	37W-7S	50
Burgundy N&S	34W-6S	45
Burton St. E&W	33W-5S	45
Buttercup Ct.* NW&SE	39W-6S	43
Calhoun St. N&S	34W-8S,34W-7S	51
Calico Dr. E&W	36W-5S	44
California Av. E&W	37W-6S,35W-6S	44
California St. N&S	37W-6S	44
Calumet Av. N&S	37W-7S,37W-6S	44
Cambridge Av. N&S	36W-5S	44
Cambridge Cir. E&W	39W-7S	49
Cambridge Rd. E&W	39W-7S	49
Candleberry Ln. N&S	38W-6S	44
Canterbury Rd. N&S	39W-7S	49
Carnation Ct.* N&S	37W-5S	44
Carolyn Dr. E&W	38W-7S	50
Carriage Hill Dr. NW&SE	38W-7S	50
Carriage Ln. E&W	37W-6S	44
Cascade Dr. N&S	37W-6S	44
Cedar St. NW&SE	36W-7S,35W-7S	50
Cedardale Pl. E&W	36W-5S	44
Center Av. E&W	35W-8S	51
Central Av. N&S	37W-7S	50
Century Ln. E&W	34W-7S	51
Charles St. NW&SE	37W-7S,37W-6S	50
Charleston Dr. E&W	38W-6S	44
Chateau Av. E&W	34W-6S	45
Chatham Cir. E&W	39W-7S	49
Chatham Rd. E&W	39W-7S	49
Cherry Tree Dr. E&W	39W-7S	49
Chestnut St. NE&SW	36W-7S	50
Cheyenne Av. E&W	37W-6S	44
Church Rd. NE&SW	34W-6S,34W-4S	45
Claim St. E&W	35W-7S,34W-7S	51
Clark St. NW&SE	35W-8S,35W-7S	51
Clearwater Dr. NW&SE	35W-5S	45
Cleveland Av. N&S	37W-5S	50
Cleveland St. N&S	34W-6S	45
Cloverdale Pl. E&W	36W-5S	44
Coach & Surrey Ln. E&W	39W-7S,38W-7S	49
Cobblestone Dr. NE&SW	38W-7S	50
Cochran St. N&S	38W-6S	44
Cold Spring Ct. N&S	38W-7S	49
Colfax St. N&S	34W-7S	51
College St. E&W	35W-7S	51
Colorado Av. E&W	37W-6S,35W-6S	44
Columbia St. N&S	35W-7S,34W-7S	51
Comet Dr. NE&SW	34W-5S	45
Commonwealth Av. N&S	37W-7S,37W-6S	50
Comprehensive Dr. E&W	33W-5S,33W-4S	45
Concord St. N&S	36W-8S	50
Constellation Dr. NE&SW	34W-5S	45
Constitution Dr. N&S	38W-7S,38W-6S	50
Constitution Dr. N&S	38W-7S	49
Coolidge Av. E&W	33W-7S	51

	Grid	Page
Copley St. NW&SE	38W-6S	44
Coral Av. E&W	38W-6S	44
Corona Ct. N&S	34W-5S	45
Corporate Blvd. NE&SW	34W-4S	45
Corrine Rd. E&W	38W-7S	50
Council St. NW&SE	35W-7S	51
Coventry Pl. NS&EW	36W-5S	44
Crestwood Ct. E&W	37W-6S	44
Crestwood Dr. N&S	36W-5S	44
Crofton Ct.* NE&SW	37W-5S	44
Cross St. NW&SE	36W-7S	50
Curry Ct. E&W	35W-6S	45
Cypress Dr. N&S	38W-6S	44
Daisy St. N&S	33W-5S	45
Daywards Ct. N&S	34W-8S	51
Dearborn Av. E&W	34W-7S	51
Deerpath Rd. N&S	39W-6S	43
Delius St. E&W	34W-7S	51
Donna Av. E&W	34W-5S	45
Douglas Av. N&S	36W-9S,36W-8S	50
Downer Pl. E&W	39W-7S,34W-8S	49
Duncan Av. E&W	38W-6S	44
Durham Dr. N&S	36W-5S	44
E. Galena Blvd. NW&SE	34W-8S	51
E. Park Av. E&W	35W-8S	51
E. Park Pl. NE&SW	35W-8S,35W-7S	51
E. Pleasure Ct. E&W	36W-6S	44
East-West Tollway E&W	38W-5S,34W-4S	44
Eastern Av. N&S	33W-7S	51
Eastwood Dr. N&S	36W-5S	44
Eastwood Dr. N&S	37W-5S	44
Edgelawn Ct. E&W	38W-6S	44
Edgelawn Av. N&S	38W-7S,38W-6S	50
Edgewood Av. N&S	33W-7S	51
Edinburgh Ln. E&W	33W-9S	28
Edison Av. E&W	35W-6S,34W-6S	45
Edwards St. E&W	35W-6S,34W-6S	45
Elder Dr. N&S	34W-5S	44
Elliot Av. N&S	36W-9S	50
Elm St. NW&SE	35W-7S	51
Elmwood Dr. N&S	37W-6S,36W-5S	44
Emerald Dr. N&S	38W-6S	44
Essex Ln. N&S	33W-9S	51
Euclid Av. E&W	36W-8S,35W-8S	50
Evans Av. E&W	36W-8S,35W-8S	50
Evanslawn Av. N&S	37W-7S,37W-6S	50
Evergreen Dr. N&S	36W-5S	44
Farnsworth Av. N&S	34W-7S,34W-4S	51
Farnum St. N&S	34W-8S	51
Featherock Dr. N&S	38W-7S	50
Felten St. N&S	33W-5S	45
Fenton St. E&W	34W-7S	51
Fifth Av. E&W	35W-8S,33W-8S	51
Fifth St. N&S	35W-9S,35W-8S	51
First St. NW&SE	36W-8S	50
Flagg St. NW&SE	35W-7S	51
Florida Av. N&S	36W-6S,35W-6S	44
Foran Ln. N&S	36W-6S,36W-5S	44
Fordham Av. N&S	37W-7S,37W-6S	50
Forest Av. E&W	35W-6S,34W-6S	45
Fourth Av. E&W	35W-8S,34W-8S	51
Fourth St. N&S	35W-9S,35W-7S	51
Foxcroft Dr. NS&EW	38W-5S	44
Franklin Av. N&S	36W-7S	50
Frazier Ct. NW&SE	36W-7S	50
Fredericks Av. N&S	34W-8S	51
Front St. E&W	35W-7S,34W-7S	51
Fulton St. E&W	34W-7S	51
Gale St. NW&SE	36W-7S	50
Galena Blvd. NW&SE	39W-6S,34W-6S	44
Garfield Av. E&W	37W-7S,36W-7S	50
Gates Av. E&W	35W-6S,34W-6S	45

	Grid	Page
Gates St. N&S	35W-8S	51
George Av. E&W	35W-8S,34W-8S	51
Gettysburg Dr. NW&SE	38W-6S	44
Gillette Av. N&S	36W-6S	44
Gladstone Av. N&S	37W-7S,37W-6S	50
Gladstone Ter. E&W	37W-6S	44
Gleason Av. E&W	36W-7S	50
Glenview Av. E&W	34W-6S	45
Glenwood Pl. N&S	37W-7S,37W-6S	50
Golden Oaks Pkwy. N&S	37W-5S	44
Goldenrod Ct.* NW&SE	39W-6S	43
Grace Av. E&W	37W-8S	50
Grace Ct. N&S	35W-7S	51
Grace St. E&W	39W-7S	49
Grand Av. N&S	35W-7S,35W-6S	51
Grand Blvd. E&W	34W-7S,33W-7S	51
Grandview Dr. E&W	33W-7S	51
Grant Pl. NE&SW	35W-8S	51
Gray Av. E&W	37W-8S	50
Greene Ct. E&W	37W-7S	50
Greenfield Ct.* E&W	39W-6S	43
Greenfield Dr. N&S	38W-6S	44
Greenleaf Ct. E&W	38W-6S	44
Greenview Av. N&S	34W-6S	45
Greenwood Dr. N&S	38W-6S	44
Grenada Dr. N&S	38W-6S	44
Griffith Av. E&W	36W-7S	50
Grove St. E&W	35W-7S,33W-7S	51
Hamilton St. NE&SW	36W-8S	50
Hammond Av. N&S	36W-6S	44
Hampshire Av. N&S	33W-7S	51
Hamuel Ln. E&W	33W-9S	28
Hankes Av. N&S	35W-6S,34W-6S	45
Hankes Rd. NW&SE	39W-6S	43
Hardin Av. E&W	36W-5S	50
Harriet Av. E&W	34W-8S	51
Harrison Av. N&S	36W-7S,36W-6S	50
Hartford Av. N&S	37W-8S	50
Hawthorne Ct. N&S	36W-7S	50
Hazel Av. N&S	36W-8S	50
Heartstone Ct. NS&EW	37W-6S	44
Heather Dr. E&W	36W-6S,37W-6S	44
Heritage Dr. NE&SW	38W-7S	50
Hermes Av. E&W	34W-6S	45
Hermitage Dr. N&S	38W-6S	44
Heywood Av. N&S	36W-7S	50
Hickory Av. N&S	34W-8S	51
High St. NE&SW	35W-7S,35W-6S	51
Highland Av. N&S	36W-7S,36W-5S	50
Hill Av. N&S	34W-8S	51
Hillside Av. N&S	33W-7S	51
Hinman St. NE&SW	35W-9S,35W-8S	51
Holden Av. NW&SE	36W-8S	50
Holiday Dr. E&W	36W-5S	44
Hollycrest Av. N&S	36W-5S	44
Homer Av. N&S	36W-9S	50
Hopkins St. N&S	34W-8S	51
Howard Av. N&S	36W-7S,36W-6S	50
Howell Pl. N&S	34W-8S	51
Hoyles Av. E&W	35W-8S	51
Hoyt Av. E&W	37W-6S,36W-6S	44
Hoyt Pl. E&W	35W-7S	51
Huntington Dr. E&W	36W-5S	44
Huntington Dr. E&W	37W-6S	44
Illinois Av. E&W	38W-6S,35W-6S	44
Independence Dr. NS&EW	38W-6S	44
Indian Av. E&W	34W-7S	51
Indian Trail Rd. E&W	38W-5S,34W-5S	44
Industrial Dr. E&W	36W-5S	44
Ingleside Av. N&S	37W-5S	50
Iowa St. N&S	36W-7S,36W-6S	50
Ironwood Ct. NE&SW	38W-5S	44
Iroquois Dr. N&S	37W-6S	44
Ivy Ln. E&W	38W-6S	44

	Grid	Page
Jackson Pl. NE&SW	35W-8S,35W-7S	51
Jackson St. N&S	35W-9S,35W-8S	51
Jefferson St. NE&SW	35W-7S,34W-6S	51
Jericho Rd. E&W	39W-8S,36W-8S	49
Jinglebell Ln. N&S	36W-5S	50
Joel Av. E&W	34W-5S	45
Joyce Dr. N&S	38W-5S	44
Jungels Av. E&W	35W-6S,34W-6S	45
Kane St. E&W	34W-7S	51
Kautz Rd. N&S	33W-9S	51
Keck Av. E&W	36W-9S	50
Kendall St. N&S	34W-8S,34W-7S	51
Kenilworth Pl. E&W	38W-7S,37W-7S	50
Kenmore Av. N&S	34W-6S,35W-4S	45
Kensington Ct. NW&SE	38W-7S	50
Kensington Pl. E&W	38W-7S,36W-7S	50
Kingsbury Av. E&W	35W-8S	51
Kingsway Dr. N&S	38W-7S	50
Kirk Rd. N&S	33W-4S	45
Kristen Dr. N&S	39W-7S	49
Lafayette St. NE&SW	35W-9S,35W-8S	51
Lake St. N&S	36W-8S,35W-6S	50
Lake St. E. NE&SW	36W-8S	50
Lakelawn Blvd. E&W	36W-6S,35W-6S	44
Lakeview Ct. E&W	39W-7S	49
Lakeview Dr. NE&SW	37W-6S	44
Lakeview Dr. E&W	38W-5S	44
Lakeview Dr. N&S	39W-7S	49
Lakewood Ct. NW&SE	36W-7S	50
Lakewood Pl. E&W	36W-7S	50
Lancaster Av. N&S	37W-5S,36W-6S	50
Landmark Rd. N&S	37W-5S	44
Larkspur Ct. E&W	38W-6S	44
Lasalle St. N&S	36W-8S,35W-8S	50
Laurel Dr. N&S	38W-6S	44
Lawndale Av. N&S	35W-6S	45
Le Grande Blvd. N&S	37W-7S	50
Lebanon St. N&S	35W-9S,35W-8S	51
Legrande Av. N&S	37W-6S	44
Lehnertz Av. NS&EW	34W-6S	45
Lehnertz Cir. E&W	34W-6S	45
Lenwood Ct. E&W	36W-8S	50
Liberty St. E&W	34W-7S	51
Lilac Ct. E&W	37W-6S	44
Lilac Ln. NW&SE	38W-6S	44
Lily St. N&S	33W-5S	45
Lincoln Av. NE&SW	36W-8S,35W-6S	50
Lincolnshire N&S	36W-5S	44
Linda Ct. NE&SW	38W-6S	44
Linda Ln. E&W	33W-4S	45
Linden Av. E&W	35W-8S	51
Locust St. NE&SW	36W-7S	50
Logan St. N&S	36W-8S	50
Loreen Dr. N&S	33W-5S	45
Loucks St. N&S	34W-8S,34W-7S	51
Lowell St. E&W	38W-6S	44
Lra Dr. N&S	38W-7S	50
Lynn St. N&S	34W-5S	45
Lynwood Dr. NE&SW	38W-7S	50
Lyon Pl. E&W	35W-6S	45
Madison St. NE&SW	38W-5S	51
Mallette Rd. E&W	33W-5S	45
Manchester Av. N&S	39W-7S	49
Manor Pl. E&W	36W-6S,35W-6S	44
Maple Av. N&S	35W-8S	51
Maplewood Ln. E&W	39W-7S	49
Margaret Ln. N&S	34W-5S	45
Marigold Ct.* NW&SE	39W-6S	43
Marikay Dr. N&S	34W-5S	45
Marion Av. E&W	35W-8S	51
Marsellaise Pl. E&W	37W-7S	50
Marshall Av. E&W	36W-7S	50

Name	Grid	Page
Marshall Blvd. E&W . 34W-5S		45
Marywood Av. N&S . 34W-5S		45
Mason Av. NE&SW . 36W-8S		50
May St. N&S		
.... 36W-7S,36W-6S		50
Mayflower St. NW&SE		
.... 38W-6S		44
McClure Av. N&S 33W-5S		45
McDowell Av. E&W .. 33W-9S		28
Meadow Ter. N&S . 34W-6S		45
Melrose Av. E&W 35W-9S		51
Merrimac Pl. E&W . 37W-7S		50
Mettle Dr. E&W 35W-5S		45
Michael Ln. NW&SE .. 33W-9S		45
Michels Av. N&S 34W-6S		45
Michigan Av. E&W		
.... 36W-6S,35W-6S		44
Middle Av. NE&SW		
.... 36W-8S,35W-7S		50
Middle Queensbury Ct. E&W		
.... 39W-7S		49
Midlane Dr. E&W .. 34W-5S		45
Miller St. N&S 34W-8S		51
Mitchell Rd. NE&SW		
.... 35W-6S,34W-4S		45
Molitor Rd. E&W .. 33W-5S		45
Monitor Av. E&W ... 36W-9S		50
Monomoy St. NW&SE		
.... 37W-6S		44
Monona Av. E&W ... 37W-6S		44
Monroe St. N&S ... 34W-6S		45
Montgomery Av. N&S		
.... 37W-8S		50
Montgomery Rd. NE&SW		
.... 33W-9S		51
Montgomery St. N&S		
.... 34W-7S		51
Morningside Av. E&W		
.... 34W-8S		51
Morton Av. N&S		
.... 36W-7S,36W-6S		50
Mountain St. E&W		
.... 34W-6S,33W-6S		45
Mulberry Ct. E&W .. 34W-8S		44
N. East Av. N&S .. 34W-7S		51
N. Glen Cir. E&W ... 37W-6S		44
N. Kingsway Dr. N&S		
.... 38W-7S		50
N. Lake St. NW&SE		
.... 35W-5S,36W-5S		45
N. Smith St. N&S ... 34W-7S		51
N.E. Industrial Dr. E&W		
.... 35W-5S		45
Nantucket Dr. N&S		
.... 36W-6S,36W-5S		44
New Haven St. E&W		
.... 37W-6S,36W-6S		44
New Indian Trail Ct. NS&SW		
.... 36W-5S		44
New York St. E&W		
.... 36W-7S,33W-7S		50
North Av. E&W		
.... 36W-7S,34W-8S		50
North Park Av. E&W		
.... 36W-6S,35W-6S		44
Northfield Dr. N&S .. 34W-6S		45
Nottingham Dr. N&S. 35W-5S		45
O'Malley Ct. NE&SW		
.... 36W-8S		50
Oak Av. N&S		
.... 35W-7S,35W-6S		51
Oak St. E&W 38W-4S		44
Oakview Av. N&S ... 33W-7S		51
Oakwood Av. E&W		
.... 35W-6S,34W-6S		45
Ohio Ct. E&W 34W-6S		45
Ohio St. N&S		
.... 34W-8S,33W-8S		51
Old Indian Trail E&W		
.... 36W-6S,35W-6S		44
Oliver Av. NE&SW .. 36W-8S		50
Opal Ln. NE&SW .. 38W-6S		44
Orchard Av. N&S ... 35W-6S		45
Orchard Rd. N&S .. 38W-7S		50
Orchid St. E&W ... 33W-5S		45
Ottawa Av. E&W ... 37W-6S		44
Palace St. N&S		
.... 37W-7S,35W-7S		51
Park Manor E&W ... 37W-7S		50
Parker Av. E&W		
.... 36W-9S,35W-9S		50
Parkside Av. N&S ... 33W-7S		51
Pearl St. N&S		
.... 35W-9S,35W-8S		51
Pennsylvania Av. E&W		
.... 36W-7S,36W-6S		50
Pepperidge Ct. NW&SE		
.... 38W-7S		50
Perry Av. N&S		
.... 36W-7S,36W-6S		50
Perry Ct. E&W 33W-5S		45
Peterson Av. N&S .. 36W-9S		50
Phoenix St. E&W .. 34W-6S		45
Pierce St. N&S		
.... 36W-6S,34W-6S		45
Pine Av. E&W		
.... 36W-6S,34W-6S		45
Pinney St. NW&SE .. 35W-7S		51
Pleasant Ter. N&S . 34W-6S		45
Pleasure Ct. N. E&W		
.... 37W-6S		44
Plum St. E&W		
.... 38W-7S,35W-7S		50
Post Rd. N&S 36W-5S		44
Powell Av. NE&SW .. 36W-8S		50
Prairie St. E&W		
.... 39W-8S,36W-8S		49
Queensbury Ct. NW&SE		
.... 39W-7S		49
Rainwood Dr. N&S . 37W-6S		44
Randall Ct. E&W . 37W-7S		50
Randall Rd. N&S		
.... 37W-7S,37W-5S		50
Rathbone Av. NW&SE		
.... 36W-8S		50
Ray Moses Dr. E&W		
.... 34W-9S		51
Reckinger Rd. E&W		
.... 34W-5S,33W-5S		45
Redwood Dr. N&S .. 38W-6S		44
Reising St. E&W		
.... 35W-7S,34W-7S		51
Richard St. E&W		
.... 38W-6S,36W-6S		44
Ridge Rd. NW&SE .. 33W-9S		28
Ridgeway Av. E&W		
.... 36W-8S		50
River St. NE&SW		
.... 36W-8S,35W-5S		50
Roanoak Av. E&W . 37W-6S		44
Robert Av. E&W ... 36W-6S		44
Robert Ct. NW&SE . 38W-6S		44
Robinhood Dr. E&W		
.... 36W-5S,35W-5S		44
Root St. N&S		
.... 35W-7S,35W-6S		51
Rose St. N&S 34W-8S		51
Rosedale Av. N&S		
.... 37W-7S,37W-6S		50
Rosedale Ct. E&W . 37W-6S		44
Rosendale Ct. E&W . 37W-7S		50
Rosewood Av. NE&SW		
.... 36W-8S		50
Rt. 25 NW&SE .. 36W-8S,35W-5S		50
Rt. 31 N&S .. 36W-8S,35W-6S		50
Rural St. E&W		
.... 35W-6S,34W-6S		45
Russell Av. N&S		
.... 35W-7S,36W-6S		50
Russell Cir. E&W 36W-6S		44
Russell Ct. E&W 36W-6S		44
Russell Dr. E&W 36W-6S		44
S. Bridge Dr. E&W . 36W-5S		44
S. East Av. N&S ... 34W-8S		51
S. Glen Cir. E&W ... 37W-6S		44
S. Pleasure Ct. E&W		
.... 37W-6S		44
Sandburg St. E&W . 38W-6S		44
Sandy Ln. N&S ... 36W-5S		44
Sans Souci Dr. NS&EW		
.... 38W-7S		50
Sapphire Ln. E&W . 38W-6S		44
Saro Av. NE&SW .. 36W-8S		50
Scheffer Rd. E&W . 34W-6S		45
Schiller Av. N&S ... 34W-7S		51
Schnidel Dr. N&S ... 34W-8S		51
Schomer Av. E&W		
.... 35W-8S,34W-8S		51
Second Av. E&W		
.... 35W-8S,34W-8S		51
Second St. NW&SE . 36W-8S		50
Selmarten Rd. N&S .. 33W-5S		45
Seminary Av. NW&SE		
.... 36W-8S,35W-8S		51
Seneca Dr. NE&SW . 37W-6S		44
Sequoia Dr. E&W		
.... 38W-5S,37W-5S		44
Seventh Av. NW&SE		
.... 35W-8S		51
Sexton St. N&S 35W-8S		51
Shady Ct. E&W 38W-6S		44
Shady Ln. NS&EW .. 38W-6S		44
Shamrock Ct. N&S . 34W-5S		45
Shannon Dr. N&S . 34W-8S		51
Sheldon Av. N&S		
.... 37W-7S,37W-6S		50
Sheridan St. N&S .. 34W-6S		45
Sherman Av. E&W .. 35W-9S		51
Sherman St. E&W .. 34W-6S		45
Sherwood Av. N&S .. 36W-6S		44
Sherwood Dr. E&W . 35W-5S		45
Sill Av. E&W 37W-8S		50
Silver Springs Ct. E&W		
.... 33W-9S		28
Simms St. E&W		
.... 36W-8S,35W-8S		50
Sirus Av. N&S 37W-6S		44
Sivas Rd. E&W 36W-5S		44
Sixth Av. E&W		
.... 34W-8S,34W-7S		51
Smith St. N&S		
.... 34W-8S,34W-7S		51
Solfisburg Av. E&W . 34W-7S		51
South Av. E&W ... 35W-8S		51
South Lawn Pl. E&W		
.... 37W-7S		50
Spencer St. N&S		
.... 35W-8S,35W-7S		51
Sperry Ct. N&S 35W-8S		51
Spring St. NW&SE		
.... 35W-7S,34W-7S		51
Springbrook Dr. E&W		
.... 36W-5S		44
Spruce St. NW&SE		
.... 37W-7S,35W-7S		50
Squire St. N&S 34W-5S		45
Squire Dr. E&W 34W-5S		45
St. Joseph's Av. E&W		
.... 35W-5S		45
Star Av. N&S 34W-5S		45
State St. N&S		
.... 34W-8S,34W-7S		51
Stillwater Ct. N&S 38W-7S		49
Stolp Av. NE&SW 35W-7S		51
Stone Fence Ct. E&W		
.... 38W-7S		49
Stone Gate Dr. NE&SW		
.... 38W-7S		50
Strawberry Ct. N&S . 38W-6S		44
Stuart Av. N&S 34W-7S		51
Sullivan Rd. N&S		
.... 38W-5S,35W-5S		44
Summit Av. N&S		
.... 34W-8S,33W-8S		51
Sumner Av. N&S		
.... 34W-8S,34W-7S		51
Sunflower Ct. NW&SE		
.... 38W-7S		44
Sunset Av. E&W		
.... 36W-5S,35W-6S		44
Superior St. E&W		
.... 37W-5S,33W-7S		51
Surrey Ct. NW&SE . 39W-7S		49
T St. N&S 35W-8S		51
Taliesin Dr. E&W ... 38W-7S		50
Tall Oaks Dr. N&S . 33W-5S		45
Talma St. NE&SW		
.... 35W-9S,35W-8S		51
Tanglewood Ct. N&S		
.... 38W-7S		50
Tanglewood Dr. NW&SE		
.... 38W-7S		50
Taylor Av. N&S 36W-6S		44
Third St. NW&SE .. 36W-8S		50
Thompson Ln. N&S		
.... 33W-5S,33W-4S		45
Timber Lake Ct. N&S		
.... 37W-6S		44
Tinley Dr. E&W 36W-5S		44
Titsworth Ct. NW&SE		
.... 36W-8S		50
Tollview Av. N&S .. 33W-6S		45
Tracy Ln. E&W 39W-7S		49
Trask St. N&S		
.... 34W-7S,34W-6S		51
Union St. N&S		
.... 35W-9S,34W-7S		51
Upper Brandon Dr. E&W		
.... 36W-5S		44
Valley Av. N&S 35W-7S		51
Vermont Av. N&S		
.... 33W-7S,33W-8S		51
View St. N&S		
.... 36W-7S,36W-6S		50
Vine St. NW&SE ... 35W-7S		51
Violet St. N&S 33W-5S		45
W. New York St. E&W		
.... 36W-7S		50
W. Park Pl. NE&SW . 35W-7S		51
W. Pleasure Ct. E&W		
.... 37W-6S		44
Walden Cir. NW&SE 38W-5S		44
Walter Av. E&W 34W-8S		51
Warren Av. NW&SE . 36W-8S		50
Washington St. NW&SE		
.... 35W-8S		51
Water St. NE&SW . 35W-7S		51
Watson St. NE&SW		
.... 35W-9S,35W-8S		51
Webster St. N&S .. 34W-8S		51
Welsh Dr. NS&EW . 33W-7S		51
Wennamacher Av. N&S		
.... 35W-6S		45
West Park Av. N&S		
.... 36W-7S,35W-7S		50
West St. N&S		
.... 35W-8S,35W-7S		51
Western Av. N&S		
.... 37W-7S,37W-6S		50
Westgate Dr. N&S		
.... 37W-7S,37W-6S		50
Westlawn Av. N&S		
.... 37W-7S,35W-8S		51
Weston Av. E&W ... 35W-8S		51
Westwood Dr. NS&EW		
.... 37W-8S		50
White Av. E&W ... 37W-8S		50
Whitingham Dr. NE&SW		
.... 33W-9S		51
Wilder St. N&S		
.... 35W-7S,35W-6S		51
Williams St. E&W .. 35W-7S		51
Windsor Av. E&W .. 34W-8S		51
Winifred St. N&S .. 38W-7S		50
Winona Av. E&W .. 37W-6S		44
Wood St. N&S		
.... 34W-7S,34W-6S		51
Woodlawn Av. N&S. 33W-7S		51
Woodlawn Av. NE&SW		
.... 36W-8S,36W-7S		50
Woodruff St. N&S . 34W-8S		51
Wright St. NE&SW . 37W-8S		50
Wyckwood Ct. NE&SW		
.... 39W-7S		49
Wyckwood Dr. E&W		
.... 39W-7S,38W-7S		49
Wyeth Dr. NW&SE .. 34W-9S		51
Wynwood Ln. SW&NE		
.... 38W-7S		49
Youngs Av. NE&SW 33W-9S		51
Zengele Av. E&W .. 35W-6S		45
Zenner Av. N&S ... 35W-6S		45
Ziegler Av. E&W ... 34W-8S		51

Aurora Township

Name	Grid	Page
Adeline Ct. E&W .. 34W-5S		45
Amherst Dr. E&W .. 38W-8S		50
Andy's Ln. N&S 34W-6S		45
Ann's Ln. N&S 34W-6S		45
Ashland Av. E&W .. 38W-8S		50
Aucutt Rd. E&W .. 38W-9S		50
Audrey Av. E&W ... 34W-5S		45
Austin Av. NW&SE . 34W-5S		45
Bangs St. N&S 35W-9S		51
Bartson Ln. E&W .. 34W-5S		45
Base Line Rd. E&W		
.... 38W-9S,37W-9S		50
Becker St. N&S 33W-6S		45
Beckwith St. N&S ... 33W-7S		51
Bilter Rd. E&W 33W-4S		45
Binder St. E&W		
.... 34W-9S,34W-8S		51
Birchwood Rd. N&S . 34W-6S		45
Bradley Dr. N&S .. 38W-9S		50
Brentwood Av. E&W		
.... 38W-9S		50
Bryant St. E&W 37W-5S		44
Bryn Mawr Dr. E&W		
.... 38W-8S		50
Calhoun St. N&S		
.... 34W-9S,34W-8S		51
Calumet Av. N&S .. 37W-8S		50
Camel Rd. E&W 34W-5S		45
Carol-Lynn St. E&W . 33W-4S		45
Carolyn Dr. E&W .. 35W-5S		45
Catherine St. N&S . 33W-7S		51
Charles St. N&S ... 33W-5S		45
Christopher Ct. N&S		
.... 38W-8S		50
Church Rd. NE&SW		
.... 34W-5S,34W-4S		45
Connie Ct. N&S 33W-5S		45
Coolidge Av. E&W .. 33W-7S		51
County Line Rd. N&S		
.... 33W-8S		51
Crane St. N&S 33W-7S		51
Creek Av. N&S 33W-5S		45
Criss Dr. NW&SE .. 34W-9S		51
Dartmouth Av. NE&SW		
.... 38W-8S		50
Dawes St. E&W ... 37W-5S		44
Dearborn Av. E&W . 33W-7S		51
Deerpath Rd. NE&SW		
.... 38W-4S		44
Diehl Rd. E&W 34W-5S		45
Douglas St. E&W ... 37W-5S		44
East-West Tollway E&W		
.... 38W-4S,33W-4S		44
Eastern Av. N&S ... 33W-7S		51
Edgelawn Dr. N&S . 38W-8S		50
Elizabeth Ln. N&S .. 33W-5S		45
Elmwood Dr. N&S .. 37W-8S		50
Exposition Av. N&S . 37W-5S		44
Farnsworth Av. N&S		
.... 34W-8S,33W-9S		51
Felten Rd. N&S		
.... 33W-6S,33W-5S		45
Fifth Av. E&W		
.... 34W-8S,33W-8S		51
Fifth St. N&S 35W-8S		51
Fitchome St. E&W . 38W-8S		50
Forest Av. E&W		
.... 34W-6S,33W-6S		45
Freeman St. N&S ... 33W-7S		51
Friar Dr. E&W 35W-5S		45
Galena Blvd. E&W . 33W-8S		51
Garden Rd. NW&SE 38W-8S		50
Gardiner St. E&W .. 34W-6S		45
Gary Av. N&S 33W-5S		45
Gary Av. N&S 38W-8S		50
Gates St. N&S 35W-9S		51
Gerten Av. E&W ... 34W-5S		45
Gilbert Ter. N&S .. 37W-8S		50
Gilmore St. E&W .. 37W-5S		44
Gladstone Av. N&S . 37W-8S		50
Golfview St. E&W .. 37W-8S		50
Grand Blvd. E&W .. 33W-8S		51
Hartford Av. N&S .. 37W-8S		50
Heather Ct. N&S .. 38W-8S		50
High St. N&S 35W-5S		45
Hinman St. N&S ... 35W-9S		51
Howell Pl. N&S 34W-9S		51
Indian Av. E&W ... 33W-7S		51
Indian Trail E&W .. 33W-5S		45
Indiana St. E&W ... 37W-5S		44
Jenny Ln. N&S 38W-9S		50
Jericho Rd. E&W		
.... 38W-8S,37W-9S		50
Jo Mar Ct. NW&SE . 34W-5S		45
John's Ln. N&S 34W-6S		45
Johnston Dr. N&S .. 38W-8S		50
Kendall St. N&S		
.... 34W-9S,34W-8S		51
Kevin Dr. N&S 38W-9S		50
Kingsway Ct. N&S . 38W-8S		50
Kingsway Dr. S. E&W		
.... 38W-8S		50
Konen Av. E&W		
.... 35W-5S,34W-5S		45
Kriss Dr. NW&SE 34W-9S		51
Leigh St. N&S 38W-8S		50
Liberty St. E&W ... 33W-7S		51
Lindenwood Ct. E&W		
.... 38W-8S		50
Lindenwood Dr. N&S		
.... 38W-8S		50
Loucks St. N&S 34W-8S		51
Lowry St. E&W 38W-8S		50
Lynch St. N&S 38W-8S		50
Maple St. E&W 37W-5S		44
Margaret Ln. N&S . 34W-5S		45
Margaret's Ln. N&S. 34W-6S		45
Marilyn Ln. N&S ... 34W-5S		45
Marywood Av. N&S . 34W-5S		45
McClaran Av. N&S . 38W-6S		50
McClure Av. N&S .. 33W-6S		50
McClure Rd. N&S		
.... 33W-7S,33W-6S		51
McDonald Av. E&W . 37W-8S		50
McMillan Av. E&W . 38W-8S		50
Melrose Av. E&W .. 35W-9S		51
Mettel Rd. E&W		
.... 35W-4S,34W-5S		45
Meyer Ct. E&W ... 38W-8S		50
Middlefield Av. E&W 38W-8S		50
Montgomery Rd. NE&SW		
.... 35W-9S,33W-9S		51
Monticello Dr. E&W. 33W-9S		51
Morningside Av. E&W		
.... 34W-8S		51
Nan St. N&S 33W-4S		45
North Av. E&W		
.... 34W-8S,33W-8S		51
Oak Rd. N&S 34W-5S		45
Oakland Av. E&W .. 35W-9S		51
Oakview Av. N&S .. 33W-7S		51
Oakwood St. E&W . 34W-5S		45
Ohio St. N&S 34W-9S		51
Palmer Av. N&S .. 38W-8S		50
Parkside Av. N&S ... 33W-7S		51
Parkview Av. NE&SW		
.... 34W-9S		51
Parkview Dr. N&S ... 33W-4S		45
Pasadena Dr. NS&EW		
.... 38W-9S		50
Pershing St. E&W ... 37W-5S		44
Plain Av. E&W		
.... 34W-6S,33W-6S		45
Pleasant Pl. N&S .. 35W-9S		51
Pleasure View Rd. E&W		
.... 38W-8S		50
Price Rd. N&S 38W-8S		50
Radcliffe Dr. E&W . 38W-8S		50
Randall Rd. N&S .. 37W-8S		50
Rathbone Av. E&W . 37W-8S		50
Reckinger Rd. E&W. 34W-5S		45
Richard Av. N&S .. 34W-5S		45
Ridgeway Av. E&W . 37W-8S		50
Rockwell Rd. NS&EW		
.... 38W-8S		50
Rosemont Av. E&W. 38W-8S		50
Rural St. E&W 34W-6S		45
Saddle Ln. E&W		
.... 35W-5S,34W-5S		45
Sarah Ln. N&S 33W-6S		45
Sartor Ln. N&S 33W-7S		51
Schomer Rd. E&W . 34W-5S		45
Seventh Av. E&W .. 34W-8S		51
Sheffer Rd. E&W .. 33W-5S		45
Sherman Av. E&W . 35W-9S		51
Simms St. E&W ... 34W-8S		51
Spencer St. N&S .. 35W-9S		51
State St. N&S 34W-9S		51
Stephen St. N&S .. 33W-6S		45
Stephen Dr. N&S .. 38W-9S		50
Summit Av. E&W		
.... 34W-8S,33W-8S		51
Sumner Av. N&S		
.... 34W-7S,34W-8S		51
Sundown Dr. E&W . 37W-8S		50
Swarthmore Dr. E&W		
.... 38W-8S		50
Sweetbriar Av. NE&SW		
.... 38W-8S		50
Talma St. N&S 35W-9S		51
Terry Av. N&S 37W-8S		50
Turner Rd. E&W ... 35W-5S		45
U. S. 30 Bypass NE&SW		
.... 38W-9S,37W-9S		50
Union St. N&S 35W-9S		51
Verity St. N&S 38W-8S		50
Victor Rd. E&W 34W-5S		45
W. Kingsway Dr. N&S		
.... 38W-8S		50
Wabansia Av. N&S . 35W-9S		51
Watson St. N&S ... 35W-9S		51
Webster St. N&S .. 34W-8S		51
Wermes Av. N&S .. 33W-8S		50
White Av. E&W 37W-8S		50
Whitlock Av. N&S .. 38W-8S		50
Woodland Dr. NE&SW		
.... 34W-5S		45
Woodlyn Dr. E&W . 35W-5S		45
Ziegler Av. E&W ... 34W-8S		51

Barrington Hills*
(Also see Cook, Lake and McHenry Counties)

	Grid	Page
Aberdeen Dr. NE&SW	32W-18N	9
Algonquin Rd. (Rt. 62) NW&SE	32W-19N	9
Autumn Trail NS&EW	32W-19N	9
Cambridge Dr. N&S	32W-18N	9
Courtney St. N&S	32W-18N	9
Deepwood Rd. NW&SE	32W-19N	9
Helm Rd. E&W	32W-18N	9
Keele Dr. N&S	32W-18N	9
Regan Blvd. NS&EW	32W-18N	9
Royal Way N&S	32W-18N	9

Batavia*
(Also see DuPage County)

	Grid	Page
Adams St. E&W	35W-1S	39
Anderson Dr. NW&SE	34W-2S,33W-2S	39
Andover Ct. N&S	33W-2S	39
Ann St. N&S	35W-0S	39
Apache Dr. N&S	36W-0S	38
Appleton Dr. N&S	34W-1S	39
Aster Ct. N&S	33W-1S	39
Averill Dr. NW&SE	36W-2S	38
Barclay Ct. E&W	34W-1S	39
Basswood Ct. NW&SE	36W-2S	38
Batavia Av. N&S	35W-2S,35W-0S	39
Beach St. N&S	35W-2S	39
Becker Av. N&S	34W-2S	39
Belleview Ln. NW&SE	35W-2S	39
Bernadette Ln. NW&SE	36W-2S,35W-2S	38
Birch St. N&S	36W-0S	38
Blackhawk Dr. NS&EW	36W-0S	38
Blaine St. E&W	36W-1S,35W-1S	38
Bluebell Ln. E&W	33W-1S	39
Bluestem Ln. NS&EW	34W-2S	39
Bond Dr. NW&SE	35W-2S	39
Brandywine Cir. NS&EW	36W-2S,35W-2S	38
Branford Av. N&S	34W-1S	39
Branford Av.. N&S	34W-1S	39
Britta Ln. E&W	36W-0S	38
Brown Ct. N&S	36W-2S	38
Burton Dr. E&W	36W-0S	38
Buttermilk Ct. N&S	36W-2S	38
Campbell Ct. E&W	34W-1S	39
Cannon Ct. N&S	34W-1S	39
Cardinal Ct. E&W	34W-2S,33W-2S	39
Carlisle Rd. NW&SE	36W-2S,35W-2S	38
Carlson Ct. E&W	34W-1S	39
Carriage Dr. N&S	36W-0S	38
Center Av NE&SW	34W-1S	39
Challange Rd. NE&SW	36W-2S	38
Charleston St. E&W	36W-1S	38
Cherry Dr. N&S	33W-2S	39
Chestnut St. E&W	35W-1S,34W-1S	39
Chillem Dr. E&W	34W-2S,33W-2S	39
Church St. E&W	35W-0S,34W-0S	39
Churchill Ct. NE&SW	34W-1S	39
Cleveland Av. NE&SW	34W-1S	39
Clover Ct. E&W	34W-2S	39
College St. N&S	34W-1S,34W-0S	39
Columbia St. NE&SW	35W-1S	39
Concord Ct. NE&SW	35W-2S	39
Cottage Rd. E&W	34W-2S	39
Cottonwood Cir. NE&SW	34W-1S	39
Creek Ln. E&W	36W-0S	38
Crestview Dr. NW&SE	36W-0S	38
Cypress Av. N&S	36W-0S	38
Danforth Dr. E&W	36W-2S	38
Darby Ct. N&S	36W-1S	38
Davis Rd. E&W	35W-0S,34W-0S	39
Delia St. N&S	34W-1S,34W-0S	39
Douglas Rd. E&W	35W-0S,33W-0S	39
Dover Ct. N&S	34W-2S	39
Elizabeth St. E&W	34W-1S	39
Elm St. E&W	35W-1S	39
Elmwood Ct. N&S	33W-2S	39
Essex Ct. N&S	34W-2S	39
Fabyan Pkwy. E&W	36W-0S,33W-0S	38
Fagan Ct. NW&SE	36W-2S	38
Fagan Rd. NW&SE	36W-2S	38
Fayette St. NE&SW	35W-0S	39
Feece Dr. N&S	36W-1S	38
First St. E&W	36W-1S	38
Flinn Dr. E&W	35W-1S	39
Forest Av. N&S	34W-1S,34W-0S	39
Foxglove Dr. E&W	34W-1S,33W-1S	39
Franklin St. E&W	35W-1S,34W-1S	39
Garden Ct. N&S	33W-2S	39
Garfield St. E&W	36W-1S,35W-1S	38
Geise Rd. E&W	34W-2S	39
Goldenrod Dr. E&W	33W-1S	39
Gore St. E&W	35W-0S	39
Green Pheasant DR. E&W	36W-2S	38
Greenleaf Dr. N&S	33W-2S	39
Greenwood Ct. NW&SE	34W-1S	39
Hall Ct. E&W	34W-2S	39
Halladay Ct. NW&SE	36W-2S	38
Halladay Dr. N&S	36W-2S	38
Hamlet St. N&S	34W-1S,34W-0S	39
Hanson St. NW&SE	34W-1S	39
Harrison St. N&S	35W-2S,35W-1S	39
Hazelwood Ct. E&W	36W-0S	38
Hillsboro Dr. NS&EW	34W-2S,33W-2S	39
Holbrook Ln. E&W	36W-0S	38
Hollister Dr. N&S	34W-0S	39
Houston St. E&W	36W-1S,35W-1S	38
Hubbard Av. E&W	33W-0S	39
Hunter Dr. N&S	33W-0S	39
Illinois Av. E&W	35W-1S	39
Independence Dr. NE&SW	36W-1S	38
Iroquois Dr. NE&SW	36W-0S	38
Island Av. N&S	35W-1S	39
Issel Ct. E&W	33W-2S	39
Jackson St. N&S	35W-2S,35W-1S	39
Jefferson Ct. E&W	35W-1S	39
Jefferson St. N&S	35W-2S,35W-0S	39
Jennifer Dr. N&S	34W-2S,34W-1S	39
Johnston Dr. NS&EW	34W-2S,33W-2S	39
K.P.K. Dr. N&S	34W-0S	39
Ken Peddy Ct. N&S	36W-2S	38
Kickapoo Ave. E&W	34W-1S	39
Kingsbury Ct. NE&SW	34W-1S	39
Kingsland Dr. N&S	34W-0S	39
Kirk Rd. N&S	33W-2S,33W-0S	39
Kline Ct. E&W	34W-2S	39
Kramer Ct. N&S	34W-1S	39
Kristen Ct. N&S	34W-2S	39
Lake Ct. N&S	34W-0S	39
Lake St. E&W	35W-0S,34W-0S	39
Larkspur Ln. N&S	34W-2S	39
Lathem St. E&W	35W-0S,33W-0S	39
Laurel St. E&W	35W-1S,34W-1S	39
Lexington E&W	34W-2S	39
Lexington Ln. E&W	34W-2S	39
Lincoln St. N&S	35W-1S,35W-0S	39
Lockwood Ct. NE&SW	36W-2S	38
Lockwood Ln. NW&SE	36W-2S	38
Locust St. E&W	35W-1S,34W-1S	39
Logan St. E&W	35W-0S	39
Lund Ln. N&S	34W-2S,34W-1S	39
Lundberg Av. E&W	36W-0S	38
Lyon Rd. E&W	34W-0S	39
Madison St. E&W	35W-0S,34W-0S	39
Main St. E&W	36W-1S,35W-1S	38
Mair Ct. E&W	36W-0S	38
Mallory Av. N&S	35W-1S	39
Manchester Av. NS&EW	35W-2S,34W-1S	39
Mandrake Dr. N&S	34W-1S	39
Maple Ln. E&W	36W-0S,35W-0S	38
Marshall Ct. E&W	34W-1S	39
Mather Ln. E&W	34W-1S	39
Maves Dr. E&W	35W-0S	39
Mayflower Dr. N&S	33W-1S	39
McClurg Ct. NW&SE	36W-2S	38
McClurg Dr. N&S	36W-2S	38
McKee St. E&W	36W-1S,35W-1S	38
Meadowrue Ln. NS&EW	34W-1S,33W-1S	39
Midway Dr. NW&SE	36W-2S,35W-2S	38
Mill St. NS&EW	36W-0S	38
Millview Ct. NE&SW	36W-2S	38
Millview Dr. NS&EW	36W-2S,36W-1S	38
Monticello Rd. NW&SE	35W-2S	39
Morton St. E&W	36W-1S,35W-1S	38
Mt. Vernon Pl. NE&SW	35W-2S	39
N. Raddant Rd. NW&SE	34W-0S	39
Navajo Dr. E&W	36W-0S	38
Newton Av. NW&SE	36W-2S	38
Nordic Ct. E&W	34W-1S	39
North Av. E&W	36W-0S,35W-0S	38
Olympic Dr. E&W	34W-0S	39
Orchard Ct. E&W	33W-2S	39
Orion Rd. NS&EW	35W-0S,34W-0S	39
Oswald Av. E&W	35W-2S	39
Ozier Dr. E&W	35W-0S	39
Paddock Ln. E&W	36W-0S	38
Pamarco Av. N&S	35W-1S	39
Paramount Pkwy. N&S	33W-0S	39
Park St. N&S	35W-0S	39
Parkview Dr. E&W	36W-0S	38
Payne Av. E&W	36W-2S	38
Peebles Ct. N&S	36W-0S	38
Pheasant Ct. N&S	35W-0S	39
Pheasant Ln. E&W	35W-0S	39
Pine Ct. NE&SW	34W-1S	39
Pine Dr. E&W	34W-1S	39
Pine St. E&W	35W-1S,34W-1S	39
Pitz Ln. N&S	36W-1S	38
Pleasant Ct. NE&SW	34W-1S	39
Ponca Dr. E&W	36W-0S	38
Pontiac Dr. NE&SW	36W-0S	38
Prairie St. N&S	35W-1S,34W-0S	39
Pueblo Dr. E&W	36W-0S	38
Raddant Rd. N&S	34W-1S	39
Randall Rd. N&S	37W-1S,36W-1S	38
Ravania Ct. NS&EW	34W-1S	39
Republic Rd. N&S	35W-1S,35W-0S	39
Richter Rd. N&S	36W-1S	38
River Rock Rd. N&S	35W-0S	39
River St. N&S	35W-2S,35W-0S	39
Roberts Ct. E&W	36W-2S	38
Roberts Ln. N&S	36W-2S,36W-1S	38
Robin Dr. E&W	36W-0S	38
Rocksbury Ct. E&W	33W-1S	39
Rt. 25 N&S	35W-2S,35W-0S	39
Rt. 31 N&S	35W-2S,35W-0S	39
Rule St. N&S	36W-2S	38
Ryan Ct. N&S	34W-1S	39
Rye Ct. N&S	34W-2S	39
Saratoga Dr. N&S	36W-1S	38
Schuler Dr. E&W	33W-0S	39
Schumway Av. N&S	35W-1S	39
Shabbona Tr. E&W	36W-0S,35W-0S	38
Sioux Dr. NE&SW	36W-0S	38
Smith Ct. NE&SW	36W-2S	38
Spring St. E&W	35W-1S,34W-1S	39
Spruce Ln. N&S	34W-1S	39
Spuhler Dr. N&S	36W-1S	38
Standish Ct. N&S	36W-2S	38
Stanton St. N&S	34W-2S,34W-1S	39
State St. E&W	35W-1S,34W-1S	39
Stewart Ct. NW&SE	34W-1S	39
Stewart St. NS&EW	34W-1S	39
Sunset Dr. E&W	35W-2S,34W-2S	39
Surrey St. NW&SE	34W-0S	39
Sylvan Pl. E&W	34W-1S	39
Thoria Ct. NW&SE	34W-0S	39
Thoria Rd. NS&EW	35W-0S,34W-0S	39
Thorsen Ln. NW&SE	36W-0S	38
Thrun Ct. NS&EW	34W-2S	39
Timber Tr. E&W	35W-0S	39
Towne Av. E&W	36W-2S	38
Trillium Ct. E&W	34W-1S,33W-1S	39
Union Av. E&W	35W-1S	39
Van Buren St. N&S	35W-1S,35W-0S	39
Van Nortwick Av. N&S	36W-1S,36W-0S	38
Van Nortwick Cir. NW&SE	35W-0S	39
Van Nortwick Ct. E&W	36W-0S	38
Viking Ct. N&S	34W-1S	39
Viking Dr. N&S	34W-1S	39
Violet Ln. NW&SE	34W-1S	39
Wagner Rd. N&S	34W-2S	39
Walnut St. E&W	36W-1S,35W-1S	38
Washington Av. N&S	35W-1S,35W-0S	39
Water St. NE&SW	35W-1S	39
Waubonsee Tr. E&W	35W-0S	39
Webster St. E&W	35W-1S,34W-1S	39
Western Av. N&S	36W-0S	38
Whitehall Ct. E&W	36W-1S	38
Willby Ln. N&S	36W-1S	38
Willow Ln. N&S	34W-2S	39
Wilson Rd. E&W	33W-1N	33
Wilson St. E&W	36W-1S,33W-1S	38
Windsor Ln. E&W	35W-0S	39
Winnebago Tr. NS&EW	36W-0S	38
Wintergreen Ct. NW&SE	34W-1S	39
Wintergreen Ter. E&W	34W-1N	33
Woodland Hills Rd. N&S	34W-2S,33W-1S	39
Young Av. NE&SW	34W-1S	39
1st St. NE&SW	35W-1S	39

Batavia Township

	Grid	Page
Arrowhead St. N&S	35W-3S	45
Banbury Rd. N&S	35W-3S,35W-2S	45
Butterfield Rd. E&W	34W-3S,33W-3S	45
Chillem Dr. NE&SW	35W-2S,34W-2S	39
Christina Ct. E&W	38W-2S	38
Cigrand Ct. NE&SW	37W-3S	44
Deer Oaks Rd. N&S	38W-3S	44
Deerpath Rd. NE&SW	38W-3S,37W-1S	44
Eastview Rd. E&W	36W-3S	44
Elfstrom Tr. N&S	37W-3S	44
Ellen Ln. N&S	36W-1S	38
Evergreen Dr. N&S	36W-1S	38
Geise Rd. E&W	34W-2S	39
Grove Hill Dr. NS&EW	38W-2S	38
Hart Rd. N&S	34W-3S,34W-2S	45
Heather Ln. NE&SW	37W-2S	38
Heaton Dr. E&W	38W-2S,37W-2S	38
Heritage Dr. NW&SE	37W-3S,37W-2S	44
Heritage Glen Ct. NE&SW	37W-2S	38
Hickory Ln. NS&EW	38W-3S	44
Hilltop Dr. N&S	36W-3S	44
Horseshoe Dr. NS&EW	38W-2S	38
Hunting Tr. E&W	38W-2S	38
Jonathan Ln. E&W	38W-2S	38
Kirk Rd. N&S	33W-3S,33W-2S	45
Long Av. E&W	36W-3S,35W-3S	44
Main St. E&W	38W-1S,36W-1S	38
McKee St. E&W	36W-0S	38
McKinley St. N&S	36W-1S	38
Meadow Dr. N&S	34W-2S	39
Meadowview Rd. NE&SW	37W-2S	38
Mooseheart Rd. E&W	38W-3S,36W-3S	44
Morton St. E&W	36W-1S	38
Nelson Lake Rd. N&S	38W-2S,38W-1S	38
Nelson Ln. N&S	38W-3S,38W-2S	44
Oak St. NW&SE	35W-2S	39
Partridge Rd. NS&EW	38W-2S	38
Raddant Rd. N&S	34W-3S,34W-2S	45
Randall Rd. NE&SW	37W-3S,36W-1S	44
Roberts Ln. N&S	36W-1S	38
Rosewood Ln. E&W	38W-3S	44
Rt. 25 NE&SW	36W-3S,35W-2S	44
Rt. 31 NE&SW	36W-3S,35W-2S	44
Shagbark Dr. NE&SW	35W-2S	39
Shuster Ln. NE&SW	37W-3S	44
Sunset Dr. E&W	34W-2S	39
Tanglewood Ct. E&W	38W-2S	38
Tanglewood Dr. NW&SE	38W-1S,37W-2S	38
Thorncrest N&S	35W-2S	39
Valley Rd. E&W	36W-3S	44
Volintine Farm Rd. N&S	37W-3S	44
Wagner Rd. N&S	34W-3S,34W-2S	45
Western Av. NE&SW	36W-0S	38
Whipple Av. N&S	36W-1S	38
White Oak Ln. NW&SE	37W-2S	38
Wilson Rd. E&W	33W-1S	39
Woodland Dr. NE&SW	35W-2S	39

Big Rock Township

	Grid	Page
Base Line Rd. E&W	48W-9S	46
Bergman Dr. NS&EW	45W-8S	47
Bourtzos Av. N&S	45W-8S	47
Bushnell Rd. NW&SE	49W-9S	46
Camp Dean Rd. N&S	46W-8S,46W-7S	47
Carey Dr. E&W	47W-7S	47
Clark Rd. NE&SW	46W-9S	47
Dauberman Rd. N&S	47W-6S,47W-4S	41
Davis Rd. N&S	48W-6S,48W-5S	40
Deer Ridge Path NE&SW	49W-5S	40
Dolly Dr. NE&SW	46W-7S	47
Dugan Rd. N&S	45W-5S	41
Fifth St. E&W	47W-7S	47
First St. NW&SE	48W-7S,47W-7S	46
Florence Rd. N&S	48W-5S,48W-4S	40
Fourth St. E&W	48W-7S,47W-7S	46
Galena St. E&W	48W-6S,47W-6S	40
Granart Rd. NE&SW	49W-9S,45W-7S	46
Greenacre Rd. NS&EW	50W-7S,49W-7S	46
Hincley Rd. E&W	50W-6S,48W-6S	40
Jefferson Av. N&S	47W-7S	47
Jericho Rd. NW&SE	50W-7S,45W-9S	46
Jetter Rd. NS&EW	45W-9S	47
John St. NS&EW	45W-8S	47
Jones Rd. NS&EW	47W-9S	47
Katie Dr. NE&SW	46W-7S	47
Lasher Rd. NS&EW	50W-4S,45W-4S	40
Lincoln Av. N&S	48W-7S,48W-6S	46
Madison Av. N&S	47W-7S	47
Marie St. NS&EW	45W-8S	47
Mary Dr. NW&SE	46W-7S	47
Mary St. NW&SE	48W-7S	46
McDermott Rd. E&W	50W-8S	46
Nelson Rd. N&S	49W-8S,49W-7S	46
Oak St. E&W	48W-6S	40
Oaken St. NE&SW	45W-8S	47
Price Rd. NE&SW	48W-7S,47W-7S	46
Raymond Rd. NE&SW	45W-9S,45W-8S	47
Rhodes St. NW&SE	47W-8S,48W-7S	47
Scott Rd. NE&SW	50W-5S,47W-5S	40
Second St. NW&SE	48W-7S,47W-7S	46
Shaw Rd. NS&EW	50W-6S,50W-5S	40
Sixth St. E&W	47W-7S	47
Swan Rd. NS&EW	48W-5S,49W-4S	40
Third St. E&W	48W-7S,47W-7S	46
Timberview Dr. NE&SW	47W-7S	47
Toronto St. N&S	45W-8S	47
U.S. Route 30 E&W	50W-6S,45W-6S	40
W. County Line Rd. N&S	50W-5S,50W-4S	40
Welton Dr. NW&SE	48W-7S	46
Wheeler Rd. E&W	48W-5S,45W-5S	40
Whildin Rd. NW&SE	46W-7S,45W-7S	47
White Trail Path NW&SE	49W-5S	40

Blackberry Township

	Grid	Page
Bartlett Rd. N&S	39W-0S,39W-0N	37
Bateman Rd. E&W	44W-0S	36
Blackberry Hill Rd. N&S	42W-1S	36
Bliss Rd. NE&SW	41W-3S,40W-1S	43
Brundige Rd. N&S	39W-1N	31
Bunker Rd. N&S	41W-1S,41W-0N	37

	Grid	Page

Concord Ct. NW&SE
..........44W-1S 36
Creekside Ct. NE&SW
..........43W-2S 36
Denali Ct. E&W 41W-1N 31
Denali Rd. NE&SW . 41W-1N 31
Derussey Rd. N&S .. 43W-2S 36
Donny Hill Dr. N&S
..........43W-1S,43W-0S 36
East West Tollway NW&SE
..........44W-2S,42W-3S 36
Finley Rd. NE&SW
..........44W-3S,43W-3S 42
Green Rd. NE&SW
..........43W-2S,42W-0S 36
Greenbriar Dr. NE&SW
..........44W-1S 36
Harley Rd. N&S 41W-1N 31
Hughes Rd. E&W
..........43W-0N,39W-0S 36
Interstate 88 NW&SE
..........44W-2S,43W-3S 36
Kaneville Rd. NE&SW
..........40W-1S,39W-0S 37
Kenmar Ct. NE&SW . 43W-0S 36
Kenmar Dr. NW&SE.. 43W-0S 36
Kenmar Ln. N&S 43W-0S 36
Keslinger Rd. E&W
..........44W-1N,39W-1N 30
Krohn Ct. N&S 40W-1N 37
La Fox Rd. N&S .. 40W-1N 37
Lakewood Dr. NW&SE
..........43W-3S 42
Lasher Rd. N&S..44W-3S 42
Linlar Dr. N&S .. 39W-1N 31
Lorang Rd. N&S
..........44W-3S,44W-1S 42
Main St. E&W
..........44W-1S,39W-1S 36
Marian Cir. NS&EW . 43W-3S 42
Marian Ln. N&S 43W-3S 42
Meadowsweet Dr. E&W
..........42W-1N 30
Midan Rd. E&W 39W-1N 31
Mill Rd. NE&SW .. 44W-2S 36
Mulhern Dr. NE&SW
..........41W-1N 31
Norris Rd. NW&SE
..........41W-3S,40W-3S 43
North Dr. NW&SE....43W-3S 42
Northern View Ct. N&S
..........43W-0S 36
Northway Dr. NE&SW
..........42W-1N,41W-1N 30
Nottingham Dr. E&W
..........43W-2S 36
Oakland Ln. N&S 40W-3S 43
Oakleaf Dr. E&W
..........44W-2S,43W-2S 36
Oakwood Dr. NS&EW
..........43W-0N 36
Oakwood Ter. N&S .. 43W-2S 36
Pennycress Ct. E&W
..........42W-1N 30
Pine Row Ct. N&S .. 43W-2S 36
Pouley Rd. N&S
..........42W-0N,42W-1N 36
Red Oak Dr. NS&EW
..........43W-2S 36
Ridge Dr. NE&SW .. 42W-0S 36
Rowe Rd. NE&SW
..........44W-0S,43W-0S 36
Rt. 47 N&S 43W-3S,43W-0N 42
Scott Rd. NE&SW .. 43W-3S 42
Seavey Rd. E&W
..........44W-3S,39W-2S 42
Smith Rd. NE&SW
..........43W-0S,42W-0S 36
Stargrass Ln. N&S .. 42W-1N 30
Still Meadows Ln. E&W
..........42W-1N 30
Surrey Dr. NW&SE .. 43W-0S 36
Taana Rd. N&S......39W-0S 37
Tall Oaks Trail E&W .43W-1S 36
Tanner Rd. E&W
..........40W-3S,39W-3S 43
Terrace Dr. N&S 40W-3S 43
The Old Midlothian Turnpike
E&W.............43W-2S 36
Thorn Apple Tree Rd.
NS&EW 43W-3S 42
Thorndon Ridge Dr. NS&EW
..........43W-1S,42W-0S 36
Timber Crest Dr. NW&SE
..........43W-0S 36
Willow Creek Ct. NW&SE
..........43W-2S 36
Willow Creek Dr. NS&EW
..........43W-2S 36

Burlington

Burlington Rd. NW&SE
..........48W-13N,47W-12N 10
Center St. NE&SW 48W-12N 16
Deutsch Rd. E&W ...47W-12N 16
Elgin-Genoa Rd. E&W
..........47W-12N 16
French Rd. NE&SW
..........48W-13N,47W-13N 10
James St. N&S 47W-12N 16
James St. NW&SE..47W-12N 16

Main St. NW&SE
..........48W-12N,47W-12N 16
Park St. N&S 47W-12N 16
South St. N&S 48W-12N 16
Sycamore St. NE&SW
..........48W-12N 16
Water St. NE&SW .. 47W-12N 16

Burlington Township

Barr Rd. E&W.....45W-12N 17
Barron Rd. N&S
..........48W-8N,48W-9N 22
Burlington Rd. NW&SE
..........48W-13N,45W-10N 10
Chapman Rd. NW&SE
..........47W-11N,46W-10N 16
Ellithorpe Rd. E&W
..........49W-10N,45W-10N 16
Engel Rd. N&S
..........49W-12N,49W-13N 16
Freedom Ct. E&W . 45W-10N 17
Godfrey Rd. E&W .. 47W-10N 16
Grand Arbor Ln. N&S
..........46W-8N 23
Hickory Ct. N&S....47W-11N 16
Highland Tr. N&S .. 45W-10N 17
Hillview Ln. E&W .. 45W-10N 17
Indian Creek Tr. NW&SE
..........46W-6N 23
Larkin Rd. NE&SW .. 50W-8N 22
Lawrence Rd. N&S
..........49W-11N,49W-12N 16
Lukens Rd. NS&EW
..........50W-10N,50W-11N 16
Marcy Rd. E&W .. 50W-10N 16
Mc Gough Rd. E&W
..........49W-9N,48W-11N 16
McDonald Rd. E&W ..45W-8N 23
McGough Rd. N&S
..........50W-8N,49W-9N 22
Middleton Rd. E&W
..........50W-9N,47W-9N 16
Peplow Rd. N&S
..........48W-8N,48W-10N 22
Percy Rd. N&S
..........49W-8N,49W-9N 22
Pioneer Ln. NW&SE
..........45W-10N 17
Plank Rd. E&W 49W-12N 16
Private Rd. E&W . 48W-12N 16
Ramm Rd. E&W
..........50W-8N,45W-8N 22
Read Rd. E&W 45W-6N 23
Rohrsen Rd. E&W .. 45W-11N 17
Romke Rd. E&W .. 46W-12N 17
Shepherds Ln. N&S
..........45W-10N 17
Strawberry Ln. E&W
..........45W-10N 17
Sumac Ln. E&W 45W-10N 17
Thomas Rd. N&S
..........46W-8N,46W-9N 23
Waughn Rd. N&S .. 48W-12N 16
West County Line Rd. N&S
..........50W-12N,50W-13N 16
Whispering Oaks Dr. E&W
..........48W-11N 16
Winterset Rd. E&W
..........47W-11N 16

Campton Township

Ancient Oak Ln. NW&SE
..........40W-5N 25
Anderson Rd. N&S
..........43W-2N,43W-4N 30
Antler Tr. NW&SE . 39W-5N 25
Arbor Creek Rd. N&S
..........39W-3N 31
Arboretum Ln. N&S . 42W-5N 24
Arrowhead Dr. NS&EW
..........42W-4N 30
Audubon Ct. E&W .. 42W-5N 24
Badger Ct. E&W .. 43W-6N 24
Baert Ln. E&W 39W-3N 31
Balkan Dr. NE&SW . 39W-3N 31
Barberry Ln. NW&SE
..........41W-6N 25
Barlow Dr. E&W .. 41W-3N 31
Barnside Ct. NE&SW
..........39W-6N 25
Beaver Ct. NE&SW .. 43W-5N 24
Beaver Ln. E&W .. 43W-5N 24
Beith Rd. NW&SE
..........44W-4N,41W-2N 30
Birchdale Ct. E&W
..........39W-4N,39W-5N 31
Black Willow Dr. NW&SE
..........39W-3N,39W-5N 31
Blue Heron Ct. NW&SE
..........40W-5N 25
Blue Larkspur Ln. E&W
..........43W-4N 30
Bolcum Rd. NE&SW
..........40W-5N,39W-5N 31
Bowgren Dr. N&S...41W-2N 31
Boxwood Ln. N&S .. 41W-6N 25
Bridle Creek Dr. E&W
..........40W-4N 31
Bridle Ct. E&W .. 42W-5N 24
Brierwood Dr. N&S
..........41W-6N,41W-7N 25

Brittany Ct. NE&SW 39W-7N 25
Brookside West Dr. N&S
..........42W-4N,41W-4N 30
Brown Rd. NS&EW
..........41W-4N,40W-5N 31
Buck Ct. NW&SE .. 39W-6N 24
Buckskin Ct. N&S . 39W-5N 25
Buckskin Tr. NE&SW
..........39W-5N 25
Burlington Rd. NW&SE
..........43W-7N,39W-4N 24
Campton Hills Rd. NW&SE
..........43W-4N,39W-3N 30
Campton Oak Dr. NE&SW
..........40W-5N 25
Campton Ridge Dr. N&S
..........41W-5N 25
Campton Woods Dr. N&S
..........41W-3N 31
Canterbury Ct. N&S
..........40W-6N,40W-7N 25
Caribou Tr. E&W .. 39W-5N 25
Carol Dr. NW&SE ..42W-4N 30
Carriage Ct. E&W ..40W-3N 31
Castle Dr. NE&SW .. 41W-5N 25
Chaffield Ct. N&S .. 42W-4N 30
Challedon Ct. N&S
..........43W-4N 30
Chateaugay Ct. NW&SE
..........42W-4N 30
Chateaugay Ln. NS&EW
..........43W-4N,42W-4N 30
Circle Dr. NE&SW . 41W-4N 31
Citation Ln. N&S
..........43W-4N,42W-4N 30
Cloverfield Cir. NS&EW
..........39W-7N 25
Cloverfield Dr. N&S
..........39W-7N 25
Clydsdale Ct. N&S
..........40W-6N 25
Colonel Bennett Ln. NE&SW
..........40W-6N 25
Colson Dr. NS&EW . 41W-5N 25
Colt Dr. N&S 41W-2N 31
Copperwood Ln. NW&SE
..........42W-5N 24
Corron Glen Rd. N&S
..........40W-6N 25
Corron Rd. N&S
..........40W-5N,40W-7N 25
Coulter Ln. NW&SE .39W-3N 31
Council Ct. N&S .. 42W-4N 30
Country View Ln. NE&SW
..........42W-4N 30
Cranberry Ln. NW&SE
..........42W-5N 24
Cranston Rd. E&W . 39W-7N 25
Creekside Ct. E&W .. 43W-5N 24
Crestwood Dr. N&S .. 41W-5N 25
Crooked Ln. NW&SE
..........41W-6N,40W-5N 25
Crosscreek Ln. E&W
..........39W-6N 25
Curling Pond Ct. NE&SW
..........40W-3N 31
Cutwood Ln. E&W . 39W-6N 25
Dairyherd Ln. N&S .. 39W-6N 25
Deer Run Dr. NS&EW
..........40W-5N,39W-5N 25
Deer Run Ln. NE&SW
..........39W-5N 25
Deerhaven Tr. NW&SE
..........39W-5N 25
Deerslayer Dr. NW&SE
..........43W-4N 30
Denali Ln. E&W....40W-2N 31
Denker Rd. N&S
..........39W-5N,39W-6N 25
Derby Ct. NW&SE . 42W-4N 30
Dillonfield Dr. NW&SE
..........41W-2N,40W-2N 31
Dittman Rd. N&S .. 42W-7N 24
Dominion Dr. NS&EW
..........41W-5N 25
E. Mary Dr. NE&SW 42W-4N 30
E. Woodland Dr. E&W
..........41W-6N 25
Eagle Ct. NE&SW .. 42W-5N 24
East Rd. E&W 42W-4N 30
Echo Valley Dr. N&S
..........41W-3N 31
Empire Rd. NE&SW
..........44W-5N,40W-6N 24
Esther Ln. NE&SW . 41W-6N 25
Evening Prairie Dr. N&S
..........42W-7N 24
Evergreen Dr. N&S . 41W-6N 25
Fair Oaks Dr. NE&SW
..........40W-6N 25
Fairend Dr. NW&SE
..........43W-4N,42W-4N 30
Falcon Ln. E&W .. 42W-4N 30
Far View Ct. NE&SW
..........41W-6N 25
Far View Rd. E&W . 41W-3N 31
Farmview Rd. NE&SW
..........40W-3N,42W-4N 30
Fencepost Ln. E&W . 41W-7N 25
Fenimore Ln. N&S .. 43W-4N 30
Fielding Ct. NW&SE .39W-7N 25
Foal Ln. E&W 41W-2N 31

Forest Glen Ln. NW&SE
..........42W-5N 24
Forest Ln. E&W ... 41W-3N 31
Fox Bend Dr. NS&EW
..........41W-7N 25
Foxfield Dr. NS&EW 42W-4N 30
Foxtail Cir. E&W .. 41W-7N 25
Garfield Rd. N&S
..........40W-2N,40W-3N 31
Gary Ct. N&S 42W-7N 24
Golden Oaks Ln. E&W
..........41W-4N 25
Gopher Ct. NE&SW . 43W-6N 24
Grady Ct. NE&SW .. 41W-2N 31
Griffin Ln. NE&SW .. 41W-4N 31
Hanson Rd. NW&SE
..........41W-5N,43W-5N 24
Happy Hills Rd. NE&SW
..........39W-3N 31
Harley Rd. N&S....41W-2N 31
Harvest Ct. NE&SW 42W-5N 24
Harvest Ln. N&S .. 42W-5N 24
Hastings Dr. N&S
..........40W-6N,40W-7N 25
Hawk Cir. NE&SW . 42W-4N 30
Hawkeye Dr. E&W .. 43W-4N 30
Hazelwood Ct. N&S . 42W-5N 24
Hazelwood Trail NS&EW
..........42W-5N 24
Hemlock Dr. NE&SW
..........39W-4N,39W-5N 31
Hidden Oaks Rd. NE&SW
..........39W-4N 41
Hidden Springs Dr. N&S
..........42W-5N 24
High Meadow Rd. N&S
..........39W-4N 31
High Point Ct. E&W
..........42W-3N,41W-3N 30
High Point Ln. NW&SE
..........41W-3N 31
Highwood Ct. E&W . 41W-3N 31
Hilltop Dr. E&W 39W-3N 31
Hoeweed Ln. NW&SE
..........39W-6N 25
Holly Ct. NE&SW .. 41W-6N 25
Homeward Glen Dr. N&S
..........39W-6N,39W-7N 25
Homeward Hill Dr. E&W
..........39W-6N 25
Hunter Ct. N&S....42W-4N 30
Hunter's Hill Dr. E&W
..........41W-5N 25
Hunter's Ridge NW&SE
..........41W-5N 25
Indian Creek Tr. NW&SE
..........41W-5N 24
Ironwood Ct. N&S . 42W-5N 24
Ivy Ct. N&S....41W-6N 25
Jens-Jenson Ln. NS&EW
..........42W-5N 24
Johnson Rd. E&W . 39W-3N 31
Johnsway E&W......39W-7N 25
Juniper Ct. NE&SW . 41W-6N 25
Kettlehook Ct. E&W .42W-5N 24
Kevin Ct. N&S....39W-6N 25
Kildeer Ln. NE&SW .43W-4N 30
Killarney Ln. N&S .. 41W-1N 31
Kim Ln. N&S 39W-6N 25
Kingfisher Ct. NE&SW
..........40W-5N 25
Kings Mill Dr. E&W
..........41W-5N,40W-5N 25
Kingston Ct. NE&SW
..........41W-7N 25
Kingswood Dr. N&S..42W-5N 24
Knollwood Dr. NW&SE
..........42W-6N,41W-6N 24
Kurt Ct. NW&SE ... 39W-3N 31
La Fox Rd. N&S
..........40W-2N,40W-5N 31
Lantry Ct. NE&SW . 41W-2N 31
Lasso Ct. NE&SW .. 41W-3N 31
Laurel Dr. E&W....41W-6N 25
Limberi Ln. NW&SE . 39W-3N 31
Line Dr. E&W........40W-2N 31
Loblolly Ln. NW&SE .39W-3N 31
Long Shadow Ln. E&W
..........40W-5N 25
Longacre Dr. NS&EW
..........41W-6N 25
Loretta Dr. E&W .. 39W-3N 31
Lost View Ln. NW&SE
..........42W-5N 24
Mare Ln. E&W 41W-2N 31
Margie Ln. N&S .. 41W-6N 25
Mary Ct. N&S 42W-5N 24
Mary Dr. NE&SW .. 42W-4N 30
McGonagle Ct. NE&SW
..........41W-2N 31
Meadowlark Ct. NE&SW
..........42W-5N 24
Meadowridge Cir. N&S
..........39W-3N 31
Meadowview Ln. N&S
..........43W-5N 24
Millstone Dr. NE&SW
..........41W-5N 25
Mohican Ln. NW&SE
..........43W-4N 30
Mulhurn Dr. NE&SW
..........41W-2N 31

N. Bowgren Cir. E&W
..........41W-2N 31
N. Elodie Dr. NW&SE
..........40W-2N 31
N. Grand Monde Dr. E&W
..........40W-2N 31
Nancy Ln. N&S .. 42W-7N 24
Norton Ct. E&W .. 39W-4N 31
Nut Hatch Ct. NW&SE
..........42W-5N 24
Oak Dr. NS&EW 41W-5N 25
Oak Ridge Rd. NE&SW
..........40W-6N 25
Oakwood Dr. N&S . 41W-6N 25
Old Farm Ln. N&S .. 39W-6N 25
Osage Dr. NE&SW . 39W-4N 31
Otter Ln. E&W 43W-6N 24
Overcup Ct. NW&SE
..........39W-4N 31
Paddock Ln. N&S...42W-5N 24
Palomino Dr. NS&EW
..........41W-6N,40W-6N 25
Pathfinder Ct. E&W
..........43W-4N 30
Patriot Hill Ln. E&W .40W-6N 25
Pine Hills Rd. NS&EW
..........39W-3N 31
Pinto Ln. NE&SW....41W-6N 25
Pioneer Ln. N&S .. 39W-4N 31
Pleasant View Ln. NW&SE
..........41W-3N 31
Ponderosa Dr. N&S . 39W-3N 31
Poplar Ln. N&S .. 39W-5N 25
Pouley Rd. N&S .. 42W-3N 30
Praire View Ln. E&W
..........41W-3N 31
Prairie Spring Dr. NW&SE
..........40W-5N 25
Prairie Valley Dr. N&S
..........43W-5N,43W-6N 24
Private Rd. NW&SE
..........41W-3N,40W-3N 31
Privey Ct. E&W 41W-6N 25
Pruntree Ln. NE&SW
..........39W-6N 25
Quail Ct. N&S 42W-5N 24
Ravine Dr. NS&EW . 42W-5N 24
Red Barn Ln. N&S . 41W-7N 25
Red Hawk Ct. E&W
..........40W-5N,39W-5N 25
Red Line Rd. NW&SE
..........39W-7N 25
Redwood Ln. N&S . 39W-4N 31
Reindeer Tr. E&W . 39W-5N 25
Retreat Ct. N&S .. 42W-5N 24
Ridge Line Rd. NW&SE
..........39W-7N 25
Ridgeview Ct. E&W . 39W-3N 31
Robert Rd. N&S .. 41W-3N 31
Rt. 38 E&W
..........44W-2N,39W-2N 30
Rt. 47 N&S 44W-2N,44W-7N 30
Rt. 64 E&W
..........44W-5N,39W-4N 24
S. Bowgren Cir. NW&SE
..........41W-2N 31
S. Bridle Creek Dr. NW&SE
..........41W-4N,40W-3N 31
S. Elodie Dr. NS&EW
..........40W-2N 31
S. Grand Monde Dr. E&W
..........40W-2N 31
Saddle Brook Dr. NW&SE
..........41W-3N 31
Shady Oaks Ct. NE&SW
..........39W-5N 25
Shetland Rd. N&S . 41W-6N 25
Silver Glen Rd. E&W
..........44W-7N,39W-6N 24
Snowbird Ct. N&S . 41W-4N 31
Somerset Dr. N&S . 42W-6N 25
Somerset Dr. N&S . 42W-6N 25
Splitrail Ct. NE&SW . 39W-6N 25
Splitrail Ln. NE&SW . 39W-6N 25
Springwood Ln. N&S
..........42W-3N 30
Spur Ln. NW&SE
..........41W-3N,41W-4N 31
Steeple Chase E&W 42W-5N 24
Stirrup Av. NE&SW .. 40W-4N 31
Stone Crop Ct. E&W
..........39W-5N 25
Sun Dance Ct. NW&SE
..........40W-5N 25
Sunset Dr. NS&EW . 41W-6N 25
Swanberg Rd. N&S
..........42W-5N,42W-6N 24
Sylvan Dr. NE&SW .. 42W-3N 31
Sylvan Ln. NE&SW . 42W-5N 24
Tanager Ct. NW&SE
..........42W-5N 24
Timber Tr. NW&SE . 42W-5N 24
Torrey Dr. NE&SW . 39W-3N 31
Town Hall Rd. NS&EW
..........41W-3N,41W-4N 31
Trillium Ct. NE&SW . 40W-5N 25
Trotter Ct. E&W .. 40W-3N 31
Trotter Ln. N&S
..........40W-3N,40W-4N 31
Tuttle Pl. N&S 39W-7N 25
Uncus Ln. NE&SW . 43W-4N 30
Venetian Way E&W .. 42W-4N 30

	Grid	Page
Verhaeghe Rd. N&S	41W-6N	25
W. Grand Monde Dr. N&S		
....	40W-2N	31
W. Mary Dr. NE&SW		
....	42W-4N	30
Wasco Rd. NE&SW		
....	40W-4N,39W-4N	31
Waterford Ln. E&W	40W-6N	25
Wayne Dr. E&W	41W-4N	31
Welter Rd. NW&SE	44W-4N	30
Westview Ct. N&S	39W-7N	25
Weybridge Dr. NS&EW		
....	41W-7N	25
Whirlaway Dr. NS&EW		
....	43W-4N,42W-4N	30
White Fence Way NE&SW		
....	40W-5N	25
White Oak Ln. NS&EW		
....	41W-6N	25
White Pine Ln. NW&SE		
....	39W-4N	31
Whitney Rd. E&W		
....	41W-5N,40W-5N	25
Willowbrook Ct. E&W		
....	40W-6N	25
Willowbrook Dr. NW&SE		
....	40W-6N	25
Winchester Way E&W		
....	40W-6N	25
Windsor Ct. E&W	42W-4N	30
Woodcrest Ct. N&S	42W-3N	30
Woodland Dr. NS&EW		
....	42W-6N,41W-6N	24
Wooley Rd. NE&SW		
....	44W-4N,44W-5N	30
Wyngate Ct. E&W	39W-4N	31
Wyngate Rd. NW&SE		
....	39W-4N	31

Carpentersville

	Grid	Page
Adams Ct. N&S	32W-17N	9
Adobe Cir. N&S	33W-18N	9
Alameda Dr. N&S	33W-18N	9
Algonquin Rd. N&S		
....	33W-18N	9
Alma N&S	33W-17N	9
Amarillo Ct. E&W	33W-18N	9
Amarillo Dr. NS&EW		
....	33W-18N	9
Anchor Dr. E&W	35W-16N	14
Apache Av. E&W	32W-18N	9
Arlington E&W	33W-16N	15
Arrow St. N&S	32W-18N	9
Ash St. N&S	32W-17N	9
Aspen Ln. N&S	32W-18N	9
Audrey Ln. E&W	34W-17N	9
Austin Av. NS&EW	33W-17N	9
Avon Park* E&W	34W-18N	9
Ball Av. E&W	32W-17N	9
Berkley St. NS&EW		
....	33W-17N	9
Berkshire Cir. NS&EW		
....	32W-16N	15
Berkshire Ct. E&W	32W-15N	15
Birch St. NS&EW		
....	32W-17N,32W-18N	9
Bismark Av. N&S	34W-16N	15
Blackhawk Av. E&W		
....	32W-18N	9
Briarwood Av. E&W		
....	33W-17N	9
Bristol Cir. NS&EW		
....	32W-16N	15
Bristol Ct. E&W	32W-16N	15
Bristol Ln. E&W	32W-15N	15
Brook St. N&S	34W-17N	9
Bunker Ln. N&S	33W-17N	9
Butte St. N&S	33W-18N	9
Caddy Av. N&S	32W-17N	9
California Av. N&S	34W-17N	9
Cambridge Dr.* N&S		
....	34W-18N	9
Cardinal Ln. E&W	34W-17N	9
Carpenter Blvd. N&S		
....	34W-16N	15
Carpenter St. E&W		
....	34W-17N	9
Center Dr. N&S	33W-16N	15
Charles St. E&W	34W-17N	9
Cherokee Ct. NE&SW		
....	32W-18N	9
Cherokee Rd. NS&EW		
....	32W-18N	9
Cir.	32W-18N	9
Cleveland Av. E&W		
....	34W-16N	15
Cordova Rd. E&W	33W-18N	9
Cottage Av. NS&EW		
....	32W-17N	9
Day Ln. NE&SW	34W-17N	9
Deer Hill Ct. NE&SW		
....	33W-17N	9
Del Rio Rd. N&S	33W-18N	9
Delaware Ct. NE&SW		
....	32W-17N	9
Delaware St. NS&EW		
....	32W-18N	9
Deleware Ct. N&S	32W-18N	9
Denver Dr. N&S	32W-18N	9
Eaton Sq.* N&S	34W-18N	9
Edwards Av. E&W	34W-16N	15

	Grid	Page
Elizabeth Mews* E&W		
....	34W-18N	9
Elm Ridge Av. E&W		
....	33W-17N	9
Elm St. N&S	35W-16N	14
Elmwood Dr. E&W	33W-17N	9
Endicott Cir. NS&EW		
....	32W-17N	9
Ensennada Ct. E&W		
....	34W-18N	9
Ensennada Dr. NE&SW		
....	34W-18N	9
Evergreen Ln. N&S		
....	32W-17N,32W-18N	9
Fairway Ln. N&S	32W-17N	9
Fairway Rd. NE&SW		
....	32W-17N	9
Fiddler's Green NS&EW		
....	35W-16N	14
Fir St. E&W	32W-18N	9
First Av. N&S	34W-16N	15
Four Winds Way NW&SE		
....	34W-17N	9
Fox River Dr. N&S	34W-17N	9
Gardner Av. N&S	33W-17N	9
Garfield Ct. N&S	32W-17N	9
Garfield Dr. E&W	32W-17N	9
Gentle Breeze Ter. E&W		
....	34W-17N	9
Goethe St. NE&SW		
....	34W-17N	9
Golf View Ln. E&W		
....	32W-17N	9
Golf View Ln. N&S	32W-16N	15
Granada Rd. E&W	33W-18N	9
Grant Dr. N&S	32W-17N	9
Green Ln. E&W	32W-17N	9
Green St. N&S	34W-17N	9
Greenwood Av. NW&SE		
....	33W-17N	9
Greenwood Ct. E&W		
....	33W-17N	9
Grove St. N&S	33W-17N	9
Hampton Dr. N&S	32W-17N	9
Harbor St. NS&EW		
....	35W-16N	14
Harding St. E&W	32W-16N	15
Harrison St. N&S	32W-16N	15
Hattie Av. N&S	34W-16N	15
Hawthorne Ct. E&W		
....	33W-17N	9
Hawthorne Ln. E&W		
....	33W-17N	9
Hazard Rd. E&W	32W-17N	9
Helm Rd. N&S	32W-17N	9
Hickory Ct. NE&SW		
....	32W-18N	9
Hickory Dr. N&S		
....	32W-17N,32W-18N	9
Hillside St. E&W	34W-16N	15
Hook Rd. E&W	32W-17N	9
Hoover Dr. N&S	32W-17N	9
Hopi Ln. NE&SW	32W-18N	9
Houston Dr. N&S	33W-17N	9
Huntley Rd. E&W		
....	35W-17N,35W-16N	14
Illinois St. E&W	34W-16N	15
Indian Ln. E&W	32W-18N	9
Inlet Cove E&W	35W-16N	14
Inlet Dr. N&S	35W-16N	14
Ivarene Ct. E&W	33W-17N	9
Jackson Av. N&S	32W-17N	9
Jefferson Av. NS&EW		
....	32W-17N	9
King Rd. E&W	33W-18N	9
King Richards Park* E&W		
....	34W-18N	9
Kingston Cir. 32W-17N		9
Kingston Cir. NS&EW		
....	32W-16N	15
Lake Marian Rd. E&W		
....	33W-17N	9
Lexington Av. E&W	33W-17N	9
Lincoln Av. N&S	34W-17N	9
Lincoln Ct. E&W	34W-17N	9
Livingston Av. NS&EW		
....	34W-17N	9
Lord St. N&S	34W-17N	9
Lowell Dr. N&S	32W-17N	9
Madera Cir. E&W	33W-18N	9
Madison Ct. N&S	32W-17N	9
Main St. E&W	34W-16N	15
Maple Av. E&W		
....	34W-17N,33W-17N	9
Maple Ln. NE&SW	33W-17N	9
Marlboro Cir. NS&EW		
....	33W-18N	9
Maude Ln.* N&S	34W-18N	9
Meadowdale Dr. NS&EW		
....	33W-17N	9
Meadowlark Ln. NW&SE		
....	33W-17N	9
Miller Rd. NE&SW	36W-18N	8
Mocasin Ln. N&S	32W-18N	9
Monroe Av. N&S	32W-17N	9
Monticello Cir. N&S		
....	32W-17N	9
Mora Rd. NS&EW	33W-18N	9
Cir.	32W-18N	9
Morningside Ln. E&W		
....	32W-16N	15

	Grid	Page
Navajo Dr. NS&EW		
....	32W-18N	9
North Lake Pkwy. E&W		
....	33W-16N	15
Northland Tr. N&S	33W-17N	9
Oak Av. E&W	34W-17N	9
Oak Ridge Rd. N&S		
....	33W-17N	9
Ollie Ct. E&W	33W-17N	9
Orleans St. NE&SW		
....	34W-17N	9
Osage Ct. N&S	32W-18N	9
Oxford Rd. N&S	33W-18N	9
Papoose Rd. N&S	32W-18N	9
Park Av. N&S	34W-17N	9
Pawnee Rd. E&W	32W-18N	9
Pecos Cir. N&S	33W-18N	9
Pine St. N&S	32W-17N	9
Plainview E&W	33W-16N	15
Plymouth Dr. NS&EW		
....	32W-17N	9
Polk Av. E&W	32W-17N	9
Port Cove NE&SW	35W-16N	14
Pueblo Rd. E&W	32W-18N	9
Queens Ct. E&W	33W-18N	9
Randall Rd. N&S	37W-17N	8
Ravine Ln. NS&EW		
....	33W-16N	15
Redwood Dr. E&W	32W-18N	9
Ridgeway St. E&W		
....	33W-17N	9
River Rd. N&S	33W-17N	9
River St. NW&SE	34W-16N	15
Riverbluff Dr. E&W	35W-17N	9
Riverview Dr. N&S		
....	35W-17N,34W-17N	8
Robin Rd. NS&EW		
....	32W-17N,32W-18N	9
Rosewood Ct. NE&SW		
....	33W-17N	9
Rosewood Dr. NW&SE		
....	33W-17N	9
Rt. 25 N&S	33W-16N,33W-19N	15
Rt. 31 N&S	35W-17N	8
Rt. 68 E&W		
....	33W-16N,32W-16N	15
Sacramento Ct. E&W		
....	33W-18N	9
Sacramento Dr. N&S		
....	33W-18N	9
Salem Ln. N&S	33W-18N	9
San Juan Rd. E&W		
....	33W-18N	9
Santa Fe Av. E&W	32W-18N	9
Saratoga E&W	33W-16N	15
Second St. N&S	34W-16N	15
Sedgewick St. N&S		
....	34W-18N	9
Seminole Av. N&S	32W-18N	9
Siesta Ct. N&S	33W-18N	9
Siesta Rd. N&S	33W-18N	9
Singapore Ln. NW&SE		
....	35W-17N,34W-17N	8
Sioux Av. NS&EW	32W-18N	9
Sleepy Hollow Rd. N&S		
....	36W-17N	8
South Wind Dr. NW&SE		
....	35W-17N,34W-17N	8
Sparrow Ct. NW&SE		
....	32W-17N	9
Sparrow Rd. NS&EW		
....	32W-17N	9
Spinnare Cove E&W		
....	35W-16N	14
Spring Ct. E&W	35W-17N	8
Spring Hill Rd. E&W		
....	34W-16N	15
Spring Point Dr. E&W		
....	32W-17N	9
Spruce Dr. NE&SW		
....	34W-16N	15
Spy Glass Cove N&S		
....	35W-16N	14
Stratford Sq.* N&S	34W-18N	9
Stuart Dr.* N&S	34W-18N	9
Sunset Dr. N&S	33W-17N	9
Sycamore Av. E&W		
....	32W-17N	9
Tacoma St. N&S	33W-18N	9
Tamarac Dr. N&S	33W-16N	15
Tay River Dr. NS&EW		
....	35W-16N,35W-17N	14
Tee Ln. E&W	32W-17N	9
Tee Rd. N&S	32W-17N	9
Tepee Av. N&S	32W-18N	9
Thornapple Ln. N&S		
....	32W-17N	9
Thrush Rd. NS&EW		
....	32W-17N	9
Thrush St. N&S	33W-17N	9
Tomahawk Ct. N&S		
....	32W-18N	9
Topeka Dr. E&W	33W-18N	9
Toro Cir. N&S	33W-18N	9
Tudor Dr.* N&S	34W-18N	9
Tudor Pl.* N&S	34W-18N	9
Tulsa Av. NE&SW	33W-17N	9
Tulsa Ct. E&W	33W-17N	9
Tyler Dr. N&S	32W-17N	9
Valentine Av E&W	34W-17N	9
Valley Ln. E&W	33W-17N	9

	Grid	Page
Van Buren St. N&S		
....	33W-16N	15
Vana Ct. N&S	34W-17N	9
Vista Ln. N&S	32W-16N	15
Waco Ln. NE&SW	33W-17N	9
Wakefield Dr. N&S		
....	32W-16N,32W-17N	15
Walnut St. N&S	32W-18N	9
Washington St. N&S		
....	34W-17N	9
Waverly Ct. N&S		
....	33W-17N	9
West Wind St. N&S		
....	35W-17N,34W-17N	8
Western Av. N&S	34W-17N	9
William St. N&S		
....	34W-17N	9
Wilmette Av. NW&SE		
....	33W-17N	9
Wilson Av. N&S	32W-17N	9
Windham Tr. NS&EW		
....	32W-17N	9
Windsor Cir. NS&EW		
....	32W-17N	9
Wisconsin St. N&S	34W-17N	9
Woodland Ct. E&W		
....	33W-17N	9
Woodland St. E&W		
....	33W-17N	9
Wren Rd. NS&EW	32W-17N	9
2nd Av. N&S	34W-16N	15

Dundee Township

	Grid	Page
Adams Dr. N&S	37W-17N	8
Airport Rd. E&W	34W-14N	15
Albert E&W	32W-14N	15
Alice Dr. N&S	35W-19N	8
Alpine Dr. N&S	35W-19N	8
Angelina Pl. NW&SE		
....	34W-14N	9
Ave Chapelle E&W	34W-14N	15
Baker Av. N&S	33W-14N	15
Bass Ct. NE&SW	34W-15N	15
Big Timber Rd. NW&SE		
....	37W-14N	14
Binnie Rd. E&W	36W-17N	8
Birchwood Rd. E&W		
....	33W-18N	9
Bolz Rd. E&W	33W-19N	9
Bonkosky Rd. NE&SW		
....	35W-14N	14
Boyer Rd. N&S	37W-19N	8
Brandt Dr. N&S	33W-14N	15
Burning Oak Dr. N&S		
....	36W-17N	8
Burr Oak Ln. NW&SE		
....	35W-18N	8
Camp Flint Dr. N&S		
....	33W-15N	15
Castle Av. N&S	34W-15N	15
Castlewood Dr. NE&SW		
....	33W-18N	9
Chateau Dr. N&S	34W-14N	15
Chateau West Dr. E&W		
....	35W-14N	14
Commerce Dr. E&W		
....	34W-15N	15
Country School Rd. NW&SE		
....	35W-18N	8
Cover St. N&S	34W-15N	15
Crescent Dr. E&W	35W-19N	8
Deerpath Ln. NS&EW		
....	33W-18N	9
Duchesne Dr. E&W		
....	35W-14N	14
Duncan Av. N&S	33W-15N	15
Dundee Av. N&S	33W-15N	15
Edwards Av. NE&SW		
....	33W-15N	15
Elliot Dr. N&S	33W-15N	15
Elm Av. N&S	35W-16N	14
Elm Av. N&S	34W-16N	8
Elmwood Dr. E&W	33W-15N	15
Field Ct. N&S	36W-18N	8
Fountain Valley Dr. NW&SE		
....	34W-14N	15
Fox River Dr. NE&SW		
....	34W-15N	15
Fries Av. E&W	33W-14N	15
Frontenac Dr. NS&EW		
....	35W-14N	14
Hecker Dr. E&W	32W-14N	15
Hickory Av. NS&EW		
....	35W-17N	8
Hickory Hollow Dr. E&W		
....	36W-15N	14
Higgins Rd. E&W	36W-16N	14
High Point Dr. N&S		
....	34W-14N	15
Highland Dr. N&S	34W-14N	15
Hillside Dr. N&S	33W-14N	15
Hilly Ln. E&W	37W-16N	14
Hollowside Dr. E&W		
....	36W-18N	8
Howard Av. N&S	33W-15N	15
Huntley Rd. NW&SE		
....	37W-18N,35W-17N	8
Indian Ln. E&W	33W-17N	9
Karen Dr. N&S		
....	35W-18N,35W-19N	8
Kasser Ct. N&S	33W-18N	9

	Grid	Page
Lac Du Beatrice Dr. NE&SW		
....	34W-14N	15
Lake Shore Dr. E&W		
....	33W-18N	9
Lathrop Ln. E&W	35W-19N	8
Linden Av. N&S		
....	33W-15N,33W-14N	15
Lonidaw Tr. E&W	34W-19N	9
Lundstrom Ln. NW&SE		
....	34W-19N	9
Manhatas Tr. NW&SE		
....	34W-19N	9
Manito Tr. NW&SE	34W-19N	9
Maple Ln. N&S	36W-14N	14
Mapletree Ln. E&W		
....	33W-18N	9
Mason Rd. E&W	37W-15N	14
Massey Dr. E&W	32W-15N	15
Meadow Lane West NW&SE		
....	34W-14N	15
Melody Ln. NE&SW		
....	33W-18N	9
Memory Ln. E&W	33W-17N	9
Menoma Tr. NW&SE		
....	34W-19N	9
Merriweather Ln. NE&SW		
....	36W-16N	14
Miller Rd. E&W	36W-18N	8
Minnehaha Tr. N&S		
....	34W-19N	9
Natoma Tr. E&W	34W-19N	9
Neccon Tr. NW&SE		
....	34W-19N	9
Nokomis Tr. NE&SW		
....	34W-19N	9
Northwest Tollway NW&SE		
....	36W-14N	14
Oak Av. N&S	34W-16N	8
Oak Hill Dr. E&W	36W-15N	14
Oaks Rd. N&S	36W-16N	14
Ogaw Tr. E&W	34W-19N	9
Old Barn Rd. NW&SE		
....	35W-18N	8
Old Farm Ln. E&W	33W-18N	9
Onaway Tr. N&S	34W-19N	9
Park Av. N&S	33W-14N	15
Park Av. N&S	33W-17N	9
Park Dr. N&S	33W-14N	15
Parsons Rd. NS&EW		
....	35W-18N	8
Petite Ln. N&S	35W-14N	14
Pheasant Tr. E&W	33W-17N	9
Piemonte Dr. E&W	32W-15N	15
Pokagon Dr. NE&SW		
....	34W-19N	9
Randall Rd. N&S		
....	37W-15N,37W-17N	14
Richardson Dr. E&W		
....	33W-15N	15
Richardson Rd. E&W		
....	33W-15N	15
Richmond Rd. E&W		
....	36W-16N	14
Ridge Rd. E&W	35W-18N	8
River Ridge Rd. NW&SE		
....	34W-14N	15
Riverview Dr. N&S	33W-14N	15
Robin Ln. N&S	33W-18N	9
Rock Ridge Ct. NW&SE		
....	32W-15N	15
Rock Ridge Dr. NE&SW		
....	32W-15N	15
Rt. 31 N&S		
....	34W-14N,35W-17N	15
Rt. 72 E&W	36W-16N,32W-15N	14
Sawyer Rd. NW&SE		
....	35W-18N	8
Shady Ln. N&S	36W-17N	8
Skyline Dr. N&S	33W-17N	9
Sleepy Hollow Rd. N&S		
....	36W-14N,36W-19N	14
Cir.	35W-16N,34W-16N	14
Springbluff Dr. N&S		
....	36W-18N	8
Sturgis Ct. N&S	36W-17N	8
Sumter St. N&S	36W-16N	14
Sunset Dr. N&S	33W-17N	9
Sunset Park Dr. E&W		
....	33W-14N	15
Tyrrell Rd. N&S	38W-14N	13
Valley View Rd. E&W		
....	35W-18N	8
Wabican Tr. NW&SE		
....	34W-19N	9
Wakigan Tr. NW&SE		
....	34W-19N	9
Walnut Ln. N&S	36W-15N	14
West Hill Rd. NS&EW		
....	35W-18N	8
Williams Av. E&W	33W-15N	15
Williams Ct. E&W	33W-15N	15
Winaki Tr. NS&EW	34W-19N	9
Winding Tr. NS&EW		
....	36W-17N	8
Woodcrest Ln. N&S		
....	35W-18N	8
Woodhaven Ln. E&W		
....	36W-18N,35W-18N	8

East Dundee

Street	Grid	Page
Aberdeen Dr. N&S	33W-16N	15
Adams Ct. E&W	33W-16N	15
Alois Dr. N&S	33W-16N	15
Ashland Rd. N&S	33W-16N	15
Balmoral Dr. E&W	32W-16N	15
Bannock Ct. N&S	32W-16N	15
Bannock Dr. E&W	32W-16N	15
Bannock Rd. E&W	32W-15N	15
Barrington Av. E&W	34W-16N,33W-16N	15
Bonnie Dundee Rd. E&W	33W-16N	15
Braeburn Rd. E&W	33W-16N	15
Bruce Ct. E&W	32W-16N	15
Council Hill Rd. NE&SW	33W-16N,33W-17N	15
Crabtree Rd. N&S	33W-17N	9
Crestwood Dr. N&S	33W-16N	15
Dawn Ct. NE&SW	32W-15N	15
Deerpath Rd. E&W	33W-16N	15
Dunbar Ln. N&S	32W-16N	15
Dundee Av. N&S	33W-15N	15
Dunridge Cir. NS&EW	33W-16N	15
Dunridge Ct. E&W	33W-16N	15
Edwards Av. E&W	33W-16N	15
First St. N&S	34W-16N	15
Fourth St. N&S	33W-16N	15
Fox River Dr. NE&SW	33W-15N	15
Greenwood Av. N&S	33W-16N	15
Guth St. N&S	33W-16N	15
Hawthorne Ln. E&W	33W-17N	9
Heinz Dr. NW&SE	32W-15N	15
Higgins Rd. NW&SE	33W-15N	15
Hill St. E&W	33W-16N	15
Hilton Av. N&S	33W-16N	15
Howard Av. N&S	33W-16N	15
Howard Ct. E&W	33W-16N	15
Jackson St. E&W	33W-16N	15
Johnson St. E&W	33W-16N	15
Kimberly Rd. N&S	33W-16N	15
King Av. N&S	33W-16N	15
King William St. N&S	34W-16N	15
Lake Shore Dr. E&W	33W-16N	15
Lincoln Av. E&W	34W-16N	15
Linden Av. NE&SW	33W-16N	15
Madison St. E&W	33W-16N	15
Maiden Ln. E&W	33W-16N	15
Main St. E&W	33W-16N	15
Maxwellton Rd. E&W	33W-16N	15
Meier Av. N&S	34W-16N	15
Michigan Av. E&W	34W-16N	15
North St. E&W	34W-16N	15
Oak Dr. E&W	33W-16N	15
Onie Ct. N&S	33W-16N	15
Park St. E&W	33W-16N	15
Plate Dr. NS&EW	32W-15N	15
Railroad St. E&W	33W-16N	15
Ravine Rd. NE&SW	33W-17N	9
Reese Av. NW&SE	33W-16N	15
River St. N&S	33W-15N,34W-16N	15
Rock Ridge Ct. NW&SE	32W-15N	15
Rock Ridge Dr. NE&SW	32W-15N	15
Roslyn Rd. NE&SW	33W-16N	15
Rt. 25 N&S	32W-16N	15
Rt. 68 E&W	33W-16N	15
Rt. 72 NW&SE	33W-16N,32W-15N	15
Scott Dr. N&S	33W-16N	15
Second St. N&S	34W-16N	15
Springcrest Rd. N&S	33W-16N	15
Summit Av. N&S	33W-16N	15
Third St. N&S	34W-16N	15
Valley Ln. N&S	33W-16N	15
Van Buren St. N&S	33W-16N	15
Wendt Av. E&W	33W-16N	15
Wenholtz Av. E&W	34W-16N	15
Williams Ct. E&W	33W-15N	15
Williams Pl. E&W	33W-15N	15
Wilmette Av. N&S	33W-16N	15

Elburn

Street	Grid	Page
Babcock St. N&S	44W-1N	30
Birch St. E&W	43W-2N	30
Capes Dr. E&W	44W-2N	30
Dempsey Dr. N&S	43W-1N	30
Filmore St. N&S	44N-1N	30
Gates St. N&S	44W-1N,44W-2N	30
Harrison St. N&S	44W-1N	30
Jay St. N&S	44W-1N	30
Kansas St. E&W	43W-1N	30
Keslinger Rd. E&W	44W-1N	30
Lilac E&W	43W-2N	30
Main St. N&S	44W-1N,44W-2N	30
Nebraska St. E&W	44W-1N,43W-1N	30
North St. E&W	44W-1N,43W-1N	30
Oak Dr. E&W	43W-1N	30
Pierce St. E&W	44W-1N,43W-1N	30
Polk St. N&S	44W-1N	30
Read St. N&S	44W-1N	30
Reeder St. E&W	44W-1N,43W-1N	30
Rt. 47 N&S	44W-1N,44W-2N	30
Saratoga Dr. N&S	44W-1N	30
Shannon St. E&W	44W-1N,43W-1N	30
South St. E&W	44W-1N,43W-1N	30
Stetzer Av. E&W	44W-1N	30
Swain St. E&W	44W-1N,43W-1N	30
Thryselius Dr. N&S	44W-1N	30
Willow St. E&W	43W-2N	30
1st St. N&S	43W-1N	30
2nd St. N&S	43W-1N	30
3rd St. N&S	43W-1N	30

Elgin*

(Also see Cook County)

Street	Grid	Page
Abbot Dr. E&W	35W-12N	20
Aberdeen Ct. N&S	36W-9N	20
Academy Pl. N&S	33W-11N	21
Adams St. E&W	34W-10N	21
Addison St. E&W	33W-11N	21
Adelaide Av. E&W	34W-12N	21
Adeline Dr. N&S	35W-12N	20
Cir.	36W-9N	20
Airlite St. N&S	36W-10N,36W-11N	20
Airport Rd. N&S	34W-14N	15
Aldine St. N&S	35W-10N,35W-11N	20
Alfred Av. N&S	35W-10N,35W-11N	20
Alft Ln. E&W	37W-14N	14
Algona Av. E&W	33W-12N	21
Algonquin Dr. N&S	32W-13N	15
Alice Ct. E&W	36W-12N	20
Alice Pl. NS&EW	36W-12N	20
Alison Dr. NE&SW	35W-11N	20
Cir.	35W-10N	20
Amber Ln. E&W	36W-8N	26
Ann St. E&W	34W-11N,33W-11N	21
Annandale Dr. N&S	36W-9N	20
Arbor Ln. E&W	35W-9N	20
Arlington Av. N&S	33W-10N	21
Aronomink Cir. N&S	36W-9N	20
Ascot Dr. N&S	35W-9N	20
Ashland Av. N&S	33W-10N	21
Augusta Av. N&S	33W-12N	21
Autumn Ct. N&S	36W-12N	20
Ball St. NE&SW	34W-11N	21
Baltusrol Dr. N&S	36W-9N	20
Banks Dr. E&W	36W-13N	14
Barn Owl Ln. E&W	36W-10N	20
Barn Swallow Way N&S	36W-12N	20
Barrett St. E&W	33W-11N	21
Bartlett Pl. NE&SW	33W-10N	21
Bayside Rd. N&S	34W-13N	15
Bayview Rd. E&W	33W-13N	15
Bel-Aire Rd. E&W	33W-13N	15
Belle Av. E&W	34W-12N	21
Bellevue Av. N&S	33W-12N	21
Belmont E&W	36W-10N	20
Belmont St. N&S	36W-10N	20
Belvidere Line Dr. NW&SE	36W-12N	20
Bennington Ct. E&W	36W-10N	20
Bent St. E&W	33W-10N	21
Berkley St. E&W	36W-10N	20
Big Penn E&W	33W-13N	15
Big Timber Rd. E&W	35W-13N,34W-13N	14
Billings St. N&S	34W-10N	21
Birchdale Dr. E&W	35W-10N	20
Bird St. N&S	34W-12N	21
Blue Ridge Ct. E&W	36W-12N	20
Cir.	36W-12N	20
Booth Ct. NE&SW	33W-10N	21
Bowen Ct. N&S	33W-10N	21
Bowes Rd. E&W	36W-9N	20
Cir.	37W-11N	20
Braeburn Dr. N&S	35W-13N	14
Brandt Dr. E&W	33W-14N	15
Brechin Tr. E&W	37W-12N	20
Bridgeview NW&SE	33W-13N	15
Brook St. N&S	34W-11N	21
Brookside Dr. N&S	36W-11N	20
Brookwood Ct. N&S	35W-13N	14
Buckeye St. E&W	34W-12N	21
Butternut Ln. NS&EW	36W-12N	20
Byron Ln. N&S	35W-13N	14
Cambridge Dr. NE&SW	36W-13N	14
Candida Rd. NE&SW	33W-13N	15
Candlewood Rd. NE&SW	33W-13N	15
Canyon Ln. NS&EW	36W-12N	20
Capital Ln. N&S	37W-14N	14
Carnoustie Ct. NE&SW	36W-9N	20
Carol Av. N&S	35W-12N	20
Carr St. E&W	35W-11N	20
Castle Pines Cir. NW&SE	36W-9N	20
Catherine Ln. NE&SW	34W-13N	15
Cedar Av. N&S	33W-12N	21
Cedar Point Ct. E&W	36W-8N	26
Center St. N&S	33W-11N	21
Century Oaks Dr. E&W	35W-13N	14
Channing Ct. N&S	33W-11N	21
Channing St. N&S	33W-11N	21
Chapel N&S	33W-11N	21
Charles St. N&S	33W-10N	21
Cherry St. E&W	34W-12N	21
Cheryl N&S	33W-13N	15
Chester Av. E&W	33W-12N	21
Cheyenne Ct. N&S	36W-12N	20
Cheyenne Ln. N&S	36W-12N	20
Chicago Ct. N&S	33W-11N	21
Chicago St. E&W	33W-11N,32W-11N	21
Church Dr. N&S	36W-10N	20
Church Rd. E&W	35W-14N	14
Clair St. E&W	34W-12N	21
Clark St. N&S	34W-12N	21
Clayton Av. N&S	35W-12N	20
Clearwater Way E&W	36W-12N	20
Clifford Av. N&S	34W-12N	21
Clifford Ct. E&W	34W-12N	21
Clifton Av. N&S	35W-10N,35W-11N	20
College N&S	33W-11N	21
College Green E&W	36W-9N	20
Colonial Rd. NS&EW	37W-11N	20
Colorado NE&SW	36W-12N	20
Columbia Av. E&W	33W-12N	21
Commerce Dr. N&S	34W-14N	15
Commonwealth Av. N&S	34W-10N,34W-11N	21
Congdon Av. E&W	33W-12N	21
Congress St. E&W	33W-10N	21
Cookane Av. N&S	33W-10N	21
Cooper Av. E&W	34W-12N,33W-12N	21
Cornell N&S	36W-10N	20
Corporate Dr. E&W	34W-14N	15
Costa Ct. N&S	36W-9N	20
Cottonwood Dr. NS&EW	36W-8N	26
Country Club Rd. NE&SW	33W-13N	15
Country Knoll Ct. N&S	36W-11N	20
Country Knoll Ln. NS&EW	36W-11N	20
Country Water Ct. E&W	36W-8N	26
Countryside Dr. N&S	35W-13N	14
Coventry Ct. NE&SW	36W-11N	20
Covered Bridge Ct. E&W	36W-10N	20
Covered Bridge Dr. N&S	36W-10N	20
Creekside Cir. E&W	36W-14N	14
Creekside Ct. N&S	36W-14N	14
Creighton Av. N&S	34W-11N	21
Crescent St. N&S	34W-11N	21
Crestview Cir. N&S	35W-13N	14
Crestview Dr. E&W	35W-13N	14
Crispin Dr. E&W	35W-9N	20
Crosby St. E&W	34W-11N	21
Crystal Av. N&S	34W-11N,34W-12N	21
Crystal Creek Ln. NE&SW	36W-10N	20
Cumberland Tr. N&S	36W-12N	20
Cypress Ct. E&W	35W-12N	20
Dana Dr. E&W	34W-9N	21
Daniels Av. E&W	34W-12N	21
Dartmonth Dr. NW&SE	36W-10N	20
Davis Rd. E&W	35W-14N	14
Deborah Av. N&S	35W-12N	20
Deerpath Dr. E&W	36W-8N	26
Demmond St. E&W	35W-12N	20
Dennis Ct. N&S	35W-10N	20
Derby Ct. N&S	37W-11N	20
Cir.	36W-14N	14
Devonshire Ct. E&W	36W-14N	14
Dexter Av. E&W	34W-11N	21
Dexter Ct. E&W	34W-11N	21
Diane Av. N&S	35W-12N	20
Division St. E&W	33W-11N	21
Dixon Av. E&W	33W-10N	21
Douglas Av. N&S	34W-11N	21
Du Bois Av. N&S	35W-11N	20
Dumbarton Oaks E&W	35W-12N	20
Duncan Av. N&S	33W-12N	21
Dundee Av. NE&SW	33W-11N	21
Dupage St. E&W	33W-11N	21
Dwight St. E&W	33W-10N	21
Eagle Rd. E&W	35W-12N	20
Eastview St. E&W	33W-11N	21
Easy St. E&W	35W-11N	20
Edgewood N&S	36W-10N	20
Edison Av. N&S	35W-10N	20
Edison St. N&S	35W-10N	20
Elgin Av. E&W	33W-10N	21
Elizabeth St. N&S	33W-10N	21
Elm Grove Dr. N&S	35W-10N	20
Elm St. N&S	34W-10N	21
Ely St. N&S	34W-10N	21
Enterprise St. E&W	33W-12N	21
Erie St. E&W	36W-11N,35W-11N	20
Esmerelda Pl. N&S	34W-12N	21
Executive Dr. N&S	35W-14N	14
Fairwood Ct. N&S	35W-13N	14
Fairwood Dr. E&W	35W-13N	14
Federation Pl. E&W	34W-13N	15
Fleetwood Dr. E&W	35W-10N	20
Fletcher Dr. NW&SE	36W-12N	20
Florimond Dr. NE&SW	34W-13N	15
Foothill Rd. NW&SE	36W-11N	20
Ford Av. E&W	33W-13N	15
Forest Av. E&W	33W-11N,32W-11N	21
Forest Dr. NS&EW	35W-13N	14
Fountain Square Plaza NW&SE	34W-11N	21
Fox Ln. NS&EW	36W-14N	14
Franklin Blvd. E&W	33W-11N	21
Franklin St. E&W	33W-11N	21
Frazier Av. E&W	34W-12N	21
Fremont St. E&W	33W-12N	21
Fulton St. E&W	33W-11N	21
Gale Ln. E&W	36W-10N	20
Gale St. NW&SE	36W-10N	20
Garden Crescent N&S	35W-12N	20
Genesee Av. E&W	34W-12N	21
Geneva N&S	33W-11N	21
George St. N&S	33W-10N	21
Gertrude St. N&S	34W-10N,34W-11N	21
Getzelman Dr. E&W	35W-12N	20
Gifford Pl. NE&SW	33W-11N	21
Glen Eagle Cir. NS&EW	36W-9N	20
Glendower Terr. E&W	37W-13N	14
Glenmore Ct. NW&SE	37W-12N	20
Glenmore Ln. N&S	37W-12N	20
Glenwood Av. E&W	33W-13N	15
Glenwood Tr. E&W	33W-13N	15
Goethe St. E&W	34W-12N	21
Gordon Ct. NE&SW	35W-10N	20
Governors Ln. NE&SW	36W-11N	20
Grace St. N&S	33W-10N	21
Grand Av. E&W	33W-12N	21
Grand Blvd. N&S	33W-11N	21
Grandstand Pl. E&W	36W-11N	20
Grant NW&SE	34W-12N	21
Green Acres Ln. NW&SE	35W-13N	14
Griswold St. N&S	34W-10N	21
Grove Av. NW&SE	33W-11N,34W-12N	21
Hamilton Av. N&S	34W-11N	21
Hamlin Ct. NW&SE	35W-11N	20
Hammond Av. E&W	33W-9N	21
Harbor Town Dr. N&S	36W-9N	20
Harding St. E&W	33W-10N	21
Harlan Av. E&W	35W-12N	20
Hartwell St. E&W	33W-12N	21
Harvard Ln. E&W	36W-10N	20
Harvey St. NW&SE	34W-11N	21
Hastings St. E&W	33W-10N	21
Hawkins St. NW&SE	34W-11N	21
Hawthorne Ln. E&W	36W-11N	20
Hawthorne St. N&S	36W-10N	20
Hayloft Ln. E&W	36W-10N	20
Hazel Dr. E&W	34W-13N	15
Healy St. N&S	33W-11N	21
Heine Av. N&S	35W-11N	20
Hendee St. NW&SE	33W-10N	21
Henry St. N&S	33W-11N	21
Hickory Pl. E&W	33W-12N	21
High St. E&W	34W-11N	21
Highland Av. NW&SE	36W-12N,34W-11N	20
Highland Ct. N&S	35W-11N	20
Highland Springs Dr. N&S	36W-12N	20
Hill Av. N&S	33W-11N,33W-12N	21
Hillcrest Rd. E&W	35W-13N	14
Hillside Rd. NE&SW	34W-13N	15
Hilton Pl. N&S	33W-11N	21
Hinsdell Pl. N&S	33W-11N	21
Holly St. E&W	34W-11N	21
Holmes Rd. E&W	36W-14N,35W-14N	14
Homer St. N&S	34W-10N	21
Hope Ct. E&W	34W-11N	21
Horned Owl Ct. E&W	36W-12N	20
Hoxie Av. N&S	35W-11N	20
Hubbard Av. N&S	35W-11N	20
Hudson Bluff Dr. N&S	36W-12N	20
I-90 E&W	34W-13N	15
Illinois Pkwy. E&W	35W-12N	20
Indian Wells Cir. NS&EW	36W-9N	20
Jackson St. NW&SE	34W-11N	21
Jamestown Ln. E&W	36W-11N	20
Jane Dr. N&S	35W-11N	20
Jansen Farm Ct. NW&SE	36W-13N	14
Jansen Farm Dr. NE&SW	36W-13N	14
Jay St. E&W	33W-10N	21
Jefferey Ct. NE&SW	36W-13N	14
Jefferson Av. E&W	33W-12N	21
Jerusha Av. E&W	34W-12N	21
Jewett St. N&S	34W-10N	21
Jordan Cir. N&S	36W-12N	20
Jordan Ct. NE&SW	36W-12N	20
Jordan Ln. NS&EW	36W-12N	20
Joseph Ct. E&W	36W-11N	20
Judy Ct. N&S	34W-12N	21
Julian Av. E&W	34W-12N	21
Junior Pl. E&W	35W-12N	20
Kane Av. N&S	35W-12N	20
Kaskaskia Av. E&W	35W-12N	20
Kate St. E&W	34W-10N	21
Kathleen Dr. N&S	35W-11N	20
Keep Av. E&W	33W-12N	21
Kelly Cir. NE&SW	35W-9N	20
Kensington Loop N&S	35W-12N	20
Kettlehook Dr. NS&EW	37W-11N	20
Kevin Dr. NW&SE	36W-12N	20
Kimball St. E&W	34W-11N	21
Kimberly Av. E&W	35W-12N	20
Kinds Riverview Ct. E&W	34W-12N	21
Kingman Ct. N&S	36W-13N	14
Knollwood Dr. E&W	36W-11N	20
Knotty Pine Dr. NE&SW	36W-8N	26
L. Penn E&W	33W-13N	15
Lake Plaza E&W	34W-13N	15
Lake St. NE&SW	33W-11N	21
Lake Terr. N&S	34W-13N	15
Lake View N&S	34W-13N	15
Lakewood NE&SW	33W-13N	15
Larkin Av. NE&SW	36W-11N	20
Lasalle N&S	35W-12N	20
Laurel Ct. N&S	33W-11N	21
Laurel St. E&W	33W-11N	21
Lavoie Av. N&S	33W-10N	21
Lawrence Av. E&W	34W-11N	21
Lennoxshire Dr. N&S	35W-13N	14
Leonard St. N&S	34W-11N	21
Lessenden Pl. E&W	33W-10N	21
Levine Ct. E&W	33W-11N	21
Liberty St. N&S	33W-10N,33W-12N	21
Lilac Ln. N&S	35W-10N	20
Lillian St. E&W	35W-10N	20

Street	Grid	Page
Lillie St. E&W	33W-11N	21
Lin-Lor Ct. NW&SE	35W-11N	20
Lin-Lor Ln. E&W	36W-11N,35W-11N	20
Lincoln Av. N&S	34W-12N,33W-12N	21
Lincoln Mc Bride St. N&S	34W-12N	21
Lincolnwood Terr. N&S	36W-11N	20
Lindbergh Ct. N&S	34W-10N	21
Linden Av. E&W	33W-11N	21
Cir.	36W-13N	14
Locust St. NE&SW	34W-11N	21
Logan Av. N&S	34W-12N	21
Longwood Pl. N&S	35W-12N	20
Lord St. E&W	33W-10N	21
Lovell St. E&W	34W-11N	21
Lowrie Ct. E&W	33W-11N	21
Luda St. E&W	33W-13N	15
Ludeka Pl. N&S	34W-12N	21
Ludlow Av. E&W	33W-12N	21
Lyle Av. N&S	35W-11N	20
Lyle St. N&S	35W-10N	20
Lynch St. N&S	34W-11N	21
Mable Av. N&S	34W-12N	21
Madlock Ct. N&S	36W-13N,36W-14N	14
Main Ln. E&W	35W-10N	20
Mallard Dr. N&S	35W-13N	14
Mallery Av. N&S	34W-11N	21
Mann Pl. N&S	33W-12N	21
Manor Ct. E&W	35W-12N	20
Maple Ln. E&W	35W-11N	20
Maple St. N&S	36W-10N	20
Margaret Pl. E&W	34W-11N	21
Marguerite St. N&S	34W-10N	21
Mark Av. E&W	35W-12N	20
Market St. E&W	35W-11N	20
Marlene Av. NE&SW	36W-12N	20
Martha Dr. N&S	35W-11N	20
Mary Pl. E&W	33W-11N	21
Matthew Ct. NW&SE	36W-13N	14
Maureen Dr. N&S	35W-11N	20
May St. E&W	33W-10N	21
McClure Av. N&S	34W-11N,34W-12N	21
McKinstry Dr. N&S	36W-11N	20
McLean Blvd. N&S	35W-10N,35W-11N	20
Meadow Ln. E&W	34W-13N	15
Medinah Cir. NS&EW	36W-9N	20
Melrose Av. N&S	35W-10N,35W-11N	20
Melrose Ct. N&S	35W-10N	20
Meyer Ct. N&S	35W-10N	20
Meyer St. E&W	35W-10N	20
Michigan St. NW&SE	33W-11N	21
Mildred Av. E&W	35W-12N	20
Mill St. E&W	34W-11N	21
Millbrook Rd. E&W	33W-13N	15
Cir.	36W-12N	20
Millcreek Ct. NW&SE	36W-12N	20
Miller Dr. NE&SW	35W-10N	20
Mission Hills Dr. E&W	36W-9N	20
Misthaven Ln. NS&EW	36W-8N	26
Moesley St. N&S	34W-10N	21
Monday Ct. NE&SW	36W-13N	14
Monday Dr. NW&SE	36W-13N	14
Monroe Av. N&S	34W-11N	21
Montrose Pl. E&W	35W-12N	20
Morgan St. E&W	34W-10N	21
Morningside Ct. E&W	34W-13N	15
Morningside Dr. N&S	34W-13N	15
Morningside Point E&W	34W-13N	14
Morton Av. N&S	33W-12N	21
Mountain St. NE&SW	34W-11N	21
Muirfield Cir. E&W	36W-9N	20
Mulberry Ct. N&S	35W-10N	20
Mulberry Ln. E&W	35W-10N	20
Murcer Dr. N&S	36W-13N	14
Murray Av. E&W	34W-11N	21
N. Airlite St. N&S	36W-12N	20
Natchez Trace E&W	36W-12N	20
National St. NE&SW	33W-10N,33W-11N	21
Nautical Way N&S	36W-12N	20
Newberry Ct. N&S	37W-11N	20
Niagra Ct. N&S	36W-12N	20
Norm Nelson Ct. N&S	34W-11N	21
North Lyle Av. N&S	34W-12N,36W-14N	21
North Shore Rd. E&W	33W-13N	15
North St. E&W	34W-11N,33W-11N	21
Northfork Cir. N&S	36W-12N	20
Notre Dame Ln. N&S	36W-10N	20
Oak St. E&W	34W-10N	21
Oakdale Dr. NS&EW	35W-13N	14
Oakhill Rd. E&W	33W-13N	15
Oakland Av. E&W	33W-12N	20
Oakley Av. E&W	35W-11N	20
Oakmont Dr. N&S	36W-11N	20
Old Forge Rd. N&S	36W-11N	20
Orange St. E&W	34W-10N	21
Orchard St. N&S	34W-10N	21
Otter Creek Ln. E&W	36W-11N	20
Owasco Av. E&W	34W-12N	21
Oxford St. N&S	34W-10N	21
Page Av. E&W	33W-12N	21
Pam NW&SE	33W-13N	15
Pamela Dr. NE&SW	35W-11N	20
Pappas Ct. N&S	35W-13N	14
Pappas Dr. N&S	35W-13N	14
Park Ln. E&W	33W-13N	15
Park Row N&S	33W-11N	21
Park St. N&S	33W-11N	21
Parkside Dr. N&S	34W-12N	21
Parkview Dr. N&S	35W-13N	14
Parkwood Rd. E&W	33W-13N	15
Pauline St. N&S	35W-11N	20
Pebble Beach Cir. E&W	36W-9N,35W-9N	20
Pennsylvania Av. E&W	34W-11N	21
Percy St. N&S	33W-11N	21
Perry St. N&S	34W-10N	21
Pindar St. E&W	34W-11N	21
Pine St. E&W	34W-11N	21
Pioneer Pk. Pl. NE&SW	35W-12N	20
Pleasant Dr. N&S	35W-13N	14
Pleasant View N&S	35W-13N	14
Plum Ct. E&W	33W-12N	21
Plum St. E&W	34W-12N	21
Plymouth Ct. N. E&W	35W-12N	20
Plymouth Ct. S. E&W	35W-12N	20
Poplar Ct. N&S	33W-12N	21
Portage Way E&W	33W-12N	20
Porter Ct. N&S	33W-12N	21
Porter St. N&S	33W-11N,33W-12N	21
Potomac Pl. E&W	36W-12N	20
Powder River Path E&W	36W-12N	20
Prairie St. E&W	33W-11N	21
Preston N&S	32W-11N,32W-12N	21
Princeton Av. N&S	36W-10N	20
Princeton Ct. E&W	36W-10N	20
Prospect Blvd. N&S	33W-12N	21
Purdue Ln. N&S	36W-10N	20
Purify Av. NW&SE	33W-12N	21
Quail Ln. N&S	35W-13N	14
Rainbow Rd. N&S	33W-13N	15
Ranch Rd. E&W	34W-13N	15
Randall Rd. N&S	36W-10N	20
Randall Ridge Ct. NW&SE	37W-11N	20
Cir.	37W-11N	20
Raymond St. N&S	33W-10N,33W-11N	21
Reba E&W	33W-13N	15
Red Tail Ridge N&S	36W-12N	20
Regent St. E&W	33W-10N	21
Renner Dr. N&S	34W-13N	15
Richard Rd. N&S	34W-13N	15
Rickert Ct. N&S	33W-11N	21
Ridgeway Dr. N&S	35W-13N	14
Riley St. E&W	35W-10N	21
River Bluff Rd. E&W	34W-12N,33W-12N	21
River Ridge Rd. E&W	34W-14N	15
Riverside Dr. NW&SE	35W-11N	21
Robert Dr. N&S	35W-10N	20
Robey St. E&W	33W-10N	21
Cir.	36W-12N	20
Robin Ridge N&S	36W-12N	20
Robinwood Dr. N&S	35W-13N	14
Royal Blvd. E&W	36W-12N,35W-12N	20
Royal Ct. N&S	35W-12N	20
Rt. 25 N&S	33W-12N	21
Rt. 31 N&S	34W-12N	21
Rugby Pl. N&S	33W-11N	21
Russel St. E&W	35W-12N	20
Ruth Dr. N&S	35W-12N	20
Ryerson Av. N&S	33W-10N	21
Sacramento Ct. N&S	36W-12N	20
Salem Rd. NW&SE	36W-11N	20
Sandy Beach NW&SE	33W-13N	15
Savannah Rd. E&W	36W-12N	20
Sawgrass Ct. E&W	35W-9N	20
Schiller St. E&W	34W-12N	21
Schuler St. N&S	33W-12N	20
Scott Dr. N&S	35W-12N	20
Second St. N&S	36W-10N	20
Seneca St. E&W	34W-12N	21
Sexauer Av. N&S	35W-11N	20
Shagbark Dr. N&S	36W-12N	20
Sharon Dr. N&S	34W-13N	15
Sheffield Ct. N&S	33W-13N	14
Sheffield Dr. NS&EW	36W-13N,33W-13N	14
Shenandoah Tr. N&S	36W-12N	20
Shepard Dr. N&S	35W-10N	20
Sheridan St. N&S	34W-11N	21
Sherman Av. N&S	34W-10N	21
Sherwood Av. N&S	34W-11N	21
Silver Ct. N&S	34W-11N	21
Silver St. N&S	34W-11N	21
Slade Av. E&W	34W-12N,33W-12N	21
Souster Av. E&W	33W-10N	21
South St. NE&SW	35W-10N,34W-11N	20
South St. E&W	36W-10N	20
Spartan Dr. E&W	35W-12N	20
Spring Cove E&W	35W-12N	20
Spring Creek Ct. E&W	32W-12N	21
Spring St. N&S	34W-11N	21
Spyglass Hill Ct. N&S	35W-10N	20
St. Andrews Cir. NS&EW	35W-9N	20
St. Charles St. N&S	33W-10N	21
St. John St. N&S	33W-12N	21
Standish Ct. E&W	33W-11N	21
Standish St. E&W	34W-11N	21
Stanford Ct. E&W	36W-10N	20
State St. N&S	34W-10N,34W-12N	21
Steel St. N&S	34W-12N	21
Stella Ct. E&W	33W-11N	21
Stella St. N&S	33W-11N	21
Stephen Av. NE&SW	36W-12N	20
Sterling Ct. E&W	37W-11N	20
Stonehaven Ct. N&S	37W-12N	20
Stonehaven Dr. NE&SW	37W-12N	20
Summit St. E&W	34W-12N,33W-12N	20
Sunrise Ct. NW&SE	34W-10N	21
Sunset Dr. N&S	35W-10N	20
Surrey Dr. N&S	35W-11N	20
Swan Ln. N&S	36W-12N	20
Tall Oaks Ct. N&S	36W-11N	20
Cir.	36W-11N	20
Tanglewood Ln. NW&SE	33W-12N	21
Tannery Ridge Rd. N&S	32W-12N	21
Tara Dr. N&S	36W-12N	20
Teal Av. E&W	34W-12N	21
Tenneyson Ct. N&S	33W-11N	21
Thomas More Dr. N&S	35W-11N	20
Timber Dr. E&W	35W-12N	20
Times Sq. NW&SE	33W-10N	21
Tinker Av. N&S	34W-12N	21
Tivoli Pl. E&W	35W-12N	20
Todd Farm Ct. NE&SW	35W-13N	14
Todd Farm Dr. E&W	36W-13N,35W-13N	14
Toftrees Tr. NW&SE	36W-12N	20
Toll Gate Rd. E&W	34W-13N	15
Tollview Rd. E&W	33W-13N	15
Torrey Pines Dr. E&W	35W-9N	20
Tower Ct. NW&SE	33W-10N	21
Triggs Av. N&S	35W-11N	20
Trinity Ter. NE&SW	35W-9N	20
Trout Pk. Blvd. E&W	33W-13N	15
Tuileries E&W	35W-12N	20
Tyler Av. E&W	35W-12N	20
Tyler Ln. N&S	34W-13N	15
U.S. Route 20 E&W	35W-10N,34W-10N	20
Union N&S	34W-11N	21
Union Ct. E&W	34W-10N	21
Valley Creek Dr. E&W	36W-12N	20
Van Nostrand Pl. E&W	33W-10N	21
Van St. E&W	35W-11N,34W-11N	20
Vandalia St. N&S	34W-10N	21
Vernon Dr. E&W	36W-11N	20
Villa Pl. NE&SW	33W-11N	21
Villa St. NW&SE	33W-11N,32W-10N	21
Vincent Pl. N&S	34W-11N	21
Vine St. N&S	34W-11N	21
Virginia Ct. NE&SW	36W-12N	20
Wabash Av. N&S	34W-10N	21
Walker Pl. N&S	33W-11N	21
Walnut Av. E&W	34W-10N	21
Walnut Creek Dr. NW&SE	36W-8N	26
Walter Ct. NE&SW	35W-10N	20
Warwick Pl. N&S	33W-11N	21
Washburn St. E&W	34W-10N	21
Washington St. NE&SW	34W-11N	21
Watch St. N&S	33W-10N	21
Water St. E&W	34W-11N	21
Waterfall Ln. NS&EW	36W-10N	20
Watres Dr. E&W	34W-11N	21
Wauchope Dr. N&S	34W-13N	15
Weatherstone Av. N&S	35W-12N	20
Weld Rd. NW&SE	36W-10N	20
Wellington Av. N&S	35W-10N	21
West Chicago St. E&W	35W-11N	20
West River Rd. N&S	34W-13N,33W-13N	15
Westfield Dr. NE&SW	37W-14N	14
Westgate Dr. N&S	36W-11N	20
Weston Av. N&S	35W-10N,35W-11N	20
Wheelock St. N&S	34W-10N	21
White Oak Ln. N&S	36W-13N	14
Wilber St. N&S	34W-10N	21
Wilcox Av. N&S	34W-10N	21
Wilder St. E&W	34W-12N	21
William N&S	36W-10N	20
Willis St. N&S	33W-10N	21
Willow Lake Rd. NW&SE	33W-13N	15
Wilson St. E&W	34W-11N	21
Wing Pk. Blvd. N&S	35W-11N	20
Wing St. NW&SE	35W-12N	20
Wolff Av. E&W	35W-11N	20
Woodbridge Dr. N&S	36W-10N	20
Woodland Av. N&S	34W-11N	21
Woodridge Ct. N&S	36W-12N	20
Woodward St. N&S	34W-11N	21
Worth Av. E&W	33W-10N	21
Yardwood St. NE&SW	33W-10N	21
Yellowstone Dr. N&S	36W-12N	20
Ziegler Ct. NW&SE	33W-11N	21

Elgin Township

Street	Grid	Page
Acorn Ln. E&W	37W-8N	26
Airlite St. N&S	36W-10N	20
Almora Terr. N&S	37W-11N	20
Amberwood Dr. NE&SW	38W-13N	13
Arrow Maker NS&EW	38W-9N	19
Beckman Ln. NS&EW	38W-9N	19
Belmont St. N&S	36W-10N	20
Big Axe Rd. N&S	38W-8N	26
Bittersweet Ln. E&W	38W-8N	26
Bowes Rd. E&W	36W-9N	20
Brindlewood Ln. E&W	38W-13N	13
Brookside Dr. E&W	37W-11N	20
Bruce Dr. NS&EW	36W-12N	20
Burgess Dr. NE&SW	37W-11N	20
Charlotte Ct. N&S	37W-12N	20
Crispin Dr. E&W	35W-9N	20
Cross Creek Ct. N&S	38W-9N	19
David Dr. N&S	35W-9N	20
Elgin C.C NE&SW	37W-11N	20
Elmer Ct. E&W	36W-12N	20
Erie St. E&W	36W-12N	20
Flagpole Ln. E&W	38W-13N	13
Fletcher Dr. E&W	36W-12N,37W-12N	38
Flora Dr. NE&SW	35W-9N	20
Foothill Rd. E&W	36W-11N	20
Gale St. N&S	36W-12N	20
Gingerwood Ln. N&S	38W-8N	26
Gordon Ln. N&S	33W-8N	27
Hawthorne Ln. E&W	36W-11N	20
Hawthorne St. N&S	36W-10N	20
Heatherfield Dr. NS&EW	38W-8N	26
Heatherfield East Dr. E&W	38W-8N,37W-8N	26
Hidden Hill Tr. NE&SW	37W-11N	20
Highland Av. NW&SE	36W-12N	20
Hilltop Rd. N&S	37W-11N	20
Hobart Dr. N&S	35W-9N	20
Hogan Hill NS&EW	39W-9N,38W-9N	19
Hopi Ln. N&S	38W-9N	19
Hopps Rd. E&W	38W-8N,36W-8N	26
Howard Av. N&S	37W-11N	20
Hoxie Av. N&S	35W-9N	20
Irwin Dr. NW&SE	35W-9N	20
Jackson Dr. N&S	37W-12N	20
Jaguar Ct. E&W	37W-10N	20
Johnstown Rd. N&S	37W-11N	20
Koshare Cir. NW&SE	38W-9N	19
Koshare Tr. N&S	38W-9N	19
Kristen Dr. N&S	38W-8N	26
Lamont Ct. N&S	37W-11N	20
Leith Ct. NW&SE	37W-10N	20
Leland Ct. N. NE&SW	38W-9N	19
Leland Ct. S. NE&SW	38W-9N	19
Linda Ln. E&W	37W-12N	20
Lisa Ln. N&S	38W-8N	26
Manchester Ln. NW&SE	37W-10N	20
Maple St. N&S	36W-10N	20
Maryhill Ln. NW&SE	37W-11N	20
McDonald St. N&S	38W-8N,37W-8N	26
McLean Blvd. N&S	35W-8N,35W-9N	27
Middle St. E&W	33W-8N	27
Newport Ct. NE&SW	37W-10N	20
Nokomis Ln. NS&EW	38W-9N	19
Nolan Rd. N&S	38W-8N,39W-7N	26
Oatwind Rd. N&S	37W-11N	20
Old Barn Rd. N&S	37W-11N	20
Olwin Av. E&W	37W-12N	20
Orchard Ln. NS&EW	37W-11N,37W-12N	20
Oxford Ln. N&S	37W-10N	20
Park Path E&W	38W-9N	19
Pasec Pl. N&S	38W-9N	19
Peppertree Ct. N&S	37W-8N	26
Peppertree Ln. N&S	37W-8N	26
Pollitt Dr. N&S	38W-12N	19
Presidential Ln. NE&SW	36W-11N	20
Public Rd. E&W	34W-9N	21
Raleigh Ct. NW&SE	37W-11N	20
Randall Rd. N&S	36W-10N	20
Raymond St. NE&SW	33W-9N	21
Richard Dr. N&S	38W-8N	26
Ridgewood Ln. NW&SE	38W-13N	13
Rt. 25 N&S	33W-9N,33W-12N	21
Running Deer Trail N&S	37W-8N	26
Savanna Lakes Ct. N&S	38W-9N	19
Savanna Lakes Dr. E&W	38W-9N	19
Second St. N&S	36W-10N	20
Sexauer Av. N&S	35W-11N	20
Shady Ln. N&S	37W-8N	26
South St. E&W	36W-10N	20
Stevens Ln. NE&SW	38W-8N	26
Stevens Rd. N&S	38W-8N	26
Stratford Ln. E&W	37W-11N	20
Sunvale Ct. N&S	38W-8N	26
Sunvale Dr. NS&EW	38W-8N	26
Taos Pl. N&S	38W-8N	26
Thunder Gap NW&SE	38W-9N	19
Tina Trail N&S	37W-12N	20
Tipi Ln. NS&EW	38W-9N	19
Trails NW&SE	38W-9N	19
Trails End N&S	38W-9N,38W-10N	19
U.S. Route 20 NW&SE	36W-10N	20
Umdenstock Rd. N&S	36W-8N	26
Water Rd. NW&SE	38W-10N	19
Wedgewood Dr. N&S	38W-13N	13
Weld Rd. NW&SE	37W-11N,36W-10N	20

	Grid	Page
Weldwood Dr. N&S	37W-10N	20
West View St. N&S	37W-12N	20
Wildwood Dr. NS&EW	37W-8N	26
Williamsburg Dr. NS&EW	37W-10N,37W-11N	20
Willow Ln. E&W	37W-8N	26
Winhaven Dr. NS&EW	37W-11N	20
York Ln. NS&EW	37W-11N	20

Geneva

	Grid	Page
Aberdeen Ct. NS&EW	34W-1N	33
Acorn Ct. E&W	37W-0N	38
Allen Pl. NW&SE	36W-0S,35W-0S	38
Anderson Blvd. N&S	35W-1N	33
Andover Ln. N&S	36W-1N	32
Appleton Ln. NS&EW	34W-1N,33W-1N	33
Arrowhead Ct. N&S	36W-0S	38
Ashbury Ct. N&S	36W-0N	38
Austin Av. NW&SE	34W-1N	33
Averhill Cir. E&W	33W-1N	33
Bank Ln. E&W	33W-1N	33
Batavia Ave. NE&SW	35W-0S,35W-0N	39
Bedford Ln. N&S	36W-1N	32
Bennett St. N&S	34W-1N	33
Bent Tree Dr. N&S	36W-0N	38
Berkshire Dr. NS&EW	37W-0N	38
Bilarda Ct. E&W	36W-1N	32
Blackberry Dr. NS&EW	36W-1N	32
Blackman Ct. N&S	37W-0N	38
Blackman Rd. E&W	37W-0N	38
Bloomfield Cir. NS&EW	36W-1N	32
Brentwood Ct. N&S	34W-1N	33
Brentwood Pl. E&W	34W-1N	33
Brian Ct. E&W	36W-1N	32
Briar Ln. N&S	34W-1N	33
Bridgeport Ln. N&S	36W-1N	32
Brigham Ct. NE&SW	36W-0N	38
Brigham Way N&S	36W-0N	38
Brighton Ct. E&W	37W-0N	38
Britta Ln. E&W	34W-1N	33
Brittany Ct. NS&EW	36W-1N	32
Brookfield Ln. N&S	36W-1N	32
Brookway Dr. E&W	36W-1N	32
Burgess Rd. N&S	36W-1N	32
Cambridge Ct. E&W	36W-1N	32
Cambridge Dr. N&S	36W-1N	32
Campbell Ct. E&W	35W-1N	33
Campbell St. NW&SE	35W-1N	33
Cape Way NW&SE	36W-1N	32
Caroline Ct. E&W	34W-1N	33
Cedar Av. NE&SW	37W-0N	38
Center St. E&W	36W-1N,35W-1N	32
Chalmers St. NW&SE	34W-0N	39
Chandler Av. N&S	33W-1N	33
Charles St. N&S	36W-1N	32
Chatham Ct. E&W	37W-0N	38
Cheever Av. E&W	35W-0N	39
Chesapeake Way E&W	33W-1N	33
Christina Ln. E&W	36W-0N	38
Clover Ct. E&W	36W-1N	32
Clover Ln. E&W	36W-1N	32
Club Dr. E&W	36W-0N	38
Colony Dr. N&S	37W-1N	32
Commerce Dr. NW&SE	34W-1N	33
Connant Ct. NW&SE	37W-0N	38
Country Club Pl. N&S	36W-1N	32
Country Squire Dr. E&W	36W-0N	38
Coventry Ct. E&W	37W-0N	38
Crawford St. E&W	35W-0N	39
Creekside Dr. N&S	35W-1N	33
Crescent Cir. E&W	37W-0N	38
Crescent Ct. NW&SE	37W-0N	38
Crescent Ln. N&S	37W-0N	38
Crescent Pl. E&W	35W-0N	39
Crissey N&S	34W-1N	33
Crissey Av. NE&SW	35W-0N,34W-1N	39
Croydon Ct. N&S	36W-0N	38
Danbury Ct. NE&SW	36W-1N	32
Danford Cir. E&W	37W-0N	38
Danford Way NS&EW	37W-0N	38
Danuta Ct. E&W	36W-1N	32
Dartmouth Ct. E&W	36W-1N	32
Deerfield Way NW&SE	36W-1N	32
Delphi Ln. N&S	34W-1N	33
Dodson St. NW&SE	34W-1N	33
Dow Av. N&S	35W-0N	39
Dryden Rd. E&W	35W-0N	39
Dunstan Rd. N&S	35W-0N	39
East Side Dr. NW&SE	34W-1N	33
Easton Av. E&W	35W-0N	39
Edison St. N&S	35W-1N	33
Eklund Av. NW&SE	35W-1N	33
Eldorado Ct. E&W	36W-0N	38
Eldorado Dr. N&S	36W-0N	38
Elizabeth Pl. E&W	35W-0N	39
Elm St. NS&EW	34W-1N	33
Essex Ct. NW&SE	36W-1N	32
Euclid Av. NE&SW	34W-1N	33
Evernia Ct. NE&SW	35W-1N	33
Executive Pl. NW&SE	33W-1N	33
Fabyan Pkwy E&W	34W-0S	39
Fairview Ct. NE&SW	37W-1N	32
Fargo Blvd. E&W	37W-0N,35W-0N	38
Fargo Cir. N&S	37W-0N	38
Fire Fox Ct. NE&SW	36W-0N	38
Ford St. E&W	36W-1N,35W-1N	32
Forest Av. N&S	35W-0N	39
Forest View Dr. E&W	35W-0N	39
Foxford Ln. E&W	35W-0N	39
Foxwood Cir. E&W	34W-1N	33
Franklin St. NW&SE	35W-1N	33
Frontage Rd. NE&SW	34W-0N	39
Fulton St. NW&SE	35W-1N,35W-0N	33
Galway Ct. N&S	36W-0N	39
Garden Av. E&W	35W-0N	39
Garfield NE&SW	34W-1N	33
Gary Ln. E&W	36W-1N	32
Gaunt Av. N&S	35W-0N	39
Geneva Dr. E&W	34W-1N,33W-1N	33
George Ct. E&W	36W-1N	32
Ginger Ln. NW&SE	37W-0N	38
Gleneagle Dr. E&W	36W-0N	38
Glengarry Dr. NE&SW	34W-1N,33W-1N	33
Grant Av. N&S	35W-1N	33
Gray St. E&W	36W-1N,35W-1N	32
Greenfield Cir. NS&EW	34W-1N	33
Greenmeadow Ln. E&W	33W-1N	33
Greenwich Ln. N&S	36W-1N	32
Hamilton St. E&W	35W-1N	33
Hampton Cir. NW&SE	36W-1N	32
Harrison St. N&S	34W-0N,34W-1N	39
Hartford Ln. E&W	36W-1N	32
Hawthorne Ln. N&S	35W-0N	39
Heartland Dr. E&W	37W-1N	32
Heather Rd. NE&SW	36W-0S	38
Herrington Pl. NE&SW	37W-0N	38
Herrington Rd. NS&EW	37W-0N	38
High St. E&W	34W-1N	33
Highbury Ln. N&S	36W-1N	32
Highland Ct. NW&SE	37W-0N	38
Cir.	37W-0N	38
Hill Rd. NW&SE	34W-1N,33W-1N	33
Howard St. NE&SW	34W-1N	33
Howell Dr. E&W	34W-1N	33
Hunter Ct. E&W	33W-1N	33
Illinois St. N&S	35W-1N	33
J. Pankow Dr. E&W	36W-1N	32
James St. NW&SE	35W-1N	33
Jefferson St. NW&SE	34W-1N	33
Kane St. NE&SW	34W-0N,34W-1N	39
Kaneville Rd. NE&SW	37W-1N,35W-1N	32
Kansas St. NE&SW	34W-0N,34W-1N	39
Kendall St. E&W	35W-0N	39
Kenston Ct. NS&EW	35W-1N	33
Kings Ct. NW&SE	36W-0S	38
Kingston Dr. N&S	34W-1N	33
Kirkwood Cir. NS&EW	33W-1N	33
Kummer Ct. E&W	36W-0N	38
Lake Ct. N&S	37W-0N	38
Lark St. NW&SE	37W-0N	38
Le Baron Ct. NW&SE	37W-0N	38
Lewis Cir. E&W	37W-0N	38
Lewis Ct. NW&SE	37W-0N	38
Lewis Pl. E&W	37W-0N	38
Lewis Rd. N&S	37W-0N	38
Lexington Dr. N&S	34W-1N	33
Lincoln Av. N&S	35W-1N	33
Locust Av. E&W	34W-1N	33
Logan Av. N&S	35W-1N	33
Lois Ct. E&W	36W-0N	38
Long Meadow Dr. NW&SE	34W-1N	33
Longview Dr. NS&EW	34W-1N	33
Loran Dr. E&W	36W-0N	38
Manchester Course NS&EW	34W-1N	33
Manchester Ct. NE&SW	34W-1N	33
Maple Ct. N&S	36W-1N	32
Maple Ln. N&S	36W-1N	32
Marion Av. E&W	35W-0N	39
Marion Ct. E&W	36W-1N	32
May St. N&S	35W-1N	33
McKinley Av. N&S	35W-1N	33
Mead Ct. N&S	35W-0N	39
Meadows Rd. N&S	35W-0N	39
Millbrook Ct. E&W	36W-1N	32
Miller Rd. NE&SW	37W-0N	38
Nassau Ln. E&W	35W-0N	39
Natwill Sq. E&W	35W-1N	33
Nebraska St. NE&SW	34W-0N,34W-1N	39
Newport Ln. E&W	36W-1N	32
Nichole Ln. E&W	33W-1N	33
Normandy Ln. E&W	36W-1N	32
North St. E&W	36W-1N,35W-1N	32
Northampton Dr. N&S	36W-1N	32
Oak St. NW&SE	34W-1N	33
Oakwood Av. NE&SW	34W-1N	33
Paddock Ct. E&W	36W-1N	32
Park Av. N&S	34W-1N	33
Patten Av. NW&SE	37W-0N	38
Peck Rd. E&W	35W-0N	39
Pepper Valley Dr. E&W	36W-1N	32
Peterson Ct. N&S	37W-0N	38
Peyton St. NW&SE	35W-1N	33
Pheasant Ct. E&W	37W-1N	32
Pheasant Run Pl. NS&EW	36W-0S,35W-0N	38
Pine St. N&S	36W-1N	32
Pioneer Ct. N&S	37W-0N	38
Police Pl. NE&SW	35W-1N	33
Prairie Ct. NE&SW	37W-0N	38
Radnor Ct. E&W	36W-1N	32
Randall Ct. N&S	37W-0N	38
Randall Rd. N&S	37W-0N	38
Ray St. E&W	35W-0N	39
Redwing Dr. E&W	36W-0S,35W-0S	38
Regency Ct. E&W	36W-1N	32
Renee Ct. E&W	36W-1N	32
Richards St. N&S	35W-1N	33
Ridge Ln. NS&EW	34W-1N	33
River Ln. N&S	35W-0N,35W-1N	39
Rosenfelder Ct. N&S	36W-1N	32
Sandholm St. N&S	34W-0N,34W-1N	39
School St. NE&SW	34W-1N	33
Scott Blvd. NW&SE	36W-1N	32
Shady Av. N&S	35W-0N	39
Sheffield Ln. E&W	36W-1N	32
Sheila Av. N&S	36W-0N	38
Sherman Av. E&W	36W-1N	32
Sherwood Ln. E&W	36W-0N	38
Shoop Dr. N&S	37W-0N	38
Short St. NW&SE	34W-1N	33
Simon Ct. E&W	37W-0N	38
Simpson St. NE&SW	34W-0N,34W-1N	39
South St. E&W	36W-1N,35W-0N	32
Southampton Ct. NS&EW	36W-0N	38
Southhampton Dr. N&S	36W-1N	32
Spring Ct. NS&EW	34W-0N	39
Spring St. NW&SE	34W-1N,34W-0N	33
Spruce Ct. N&S	37W-0N	38
Cir.	36W-0N	38
State St. NW&SE	36W-1N,34W-1N	32
Sterling Av. N&S	37W-0N	38
Stevens St. NW&SE	35W-1N	33
Stratford Dr. N&S	36W-1N	32
Sudbury Ln. E&W	36W-1N	32
Suffield Ct. N&S	34W-1N	33
Summit St. NW&SE	34W-1N	33
Sunset Rd. N&S	35W-0N	39
Syril Dr. N&S	36W-1N	32
Tanager Ln. NE&SW	36W-0S	38
Thornhill Ct. E&W	37W-0N	38
Union St. NW&SE	35W-1N	33
Valley Ln. NE&SW	34W-1N	33
Wakefield Ln. N&S	36W-1N	32
Wall St. E&W	34W-1N	33
Walnut Av. N&S	37W-0N	38
Water St. N&S	34W-1N	33
Wesley Ct. E&W	33W-1N	33
West Ln. N&S	36W-1N	32
West St. N&S	36W-1N	32
Western Av. N&S	36W-0N	38
Westfield Course NS&EW	34W-1N	33
Westhaven Cir. N&S	37W-1N	32
Whitfield Dr. NS&EW	33W-1N	33
Whittington Dr. E&W	34W-1N	33
Wild Dunes Ct. E&W	36W-0S	38
Wildflower Ct. N&S	36W-0S	38
Williamsburg Av. E&W	36W-1N	32
Willow Ln. N&S	36W-1N	32
Wilson St. N&S	35W-1N	33
Windham Ct. N&S	36W-0N	38
Windthrop Ln. N&S	36W-1N	32
Winners Cup Ct. NW&SE	36W-0N	38
Wood Av. N&S	37W-0N	38
Woodlawn St. NE&SW	34W-1N	33
Woodward Av. NW&SE	34W-1N	33
1st St. NE&SW	35W-0N,35W-1N	39
2nd St. NE&SW	35W-1N	33
3rd Av. N&S	34W-1N	33
3rd St. NE&SW	35W-0N,35W-1N	39
4th Av. N&S	34W-1N	33
4th St. NE&SW	35W-0N,35W-1N	39
5th Av. N&S	34W-1N	33
5th St. NE&SW	35W-0N,35W-1N	39
6th Av. N&S	34W-1N	33
6th St. NE&SW	35W-1N	33
7th Pl. N&S	35W-1N	33
7th St. NE&SW	35W-1N	33
8th St. N&S	35W-0N,35W-1N	39
9th St. N&S	35W-1N	33

Geneva Township

	Grid	Page
Blackberry Way E&W	38W-0N	38
Bricker Rd. E&W	37W-1N	32
Cherry Ln. E&W	34W-0N,33W-0N	39
Clover Hills Ct. E&W	38W-0S,37W-0S	38
Considine Rd. E&W	36W-0S,35W-0S	38
Deerpath Rd. N&S	37W-1S	38
Elm Rd. N&S	33W-1N	33
Fabyan Pkwy. E&W	38W-0S,36W-0S	38
Glenwood Dr. E&W	38W-1S	38
Harrison Rd. NW&SE	33W-1N	33
Heather Rd. NE&SW	36W-0S	38
Janet Ln. N&S	37W-0S	38
Kaneville Rd. NE&SW	38W-0S,38W-0N	38
Kautz Rd. N&S	33W-0N,33W-1N	39
Keslinger Rd. E&W	38W-1N,37W-1N	32
Killey Ln. NE&SW	38W-0N	38
Kirk Rd. N&S	33W-0N,33W-1N	39
Lea Dr. N&S	38W-0N	38
Lynn Rd. N&S	33W-1N	33
McKee St. E&W	38W-0S,37W-0S	38
Miller Dr. E&W	37W-0S	38
Miller Rd. NW&SE	37W-0S	38
Minor Rd. NE&SW	37W-0N	38
Monterey Dr. E&W	38W-1S	38
Morningside Ln. NW&SE	37W-0S	38
Norman Rd. E&W	33W-1N	33
Northlane Rd. N&S	33W-0N	39
Oak Rd. N&S	33W-1N	33
Old Kirk Rd. N&S	33W-0N,33W-1N	39
Orchard Rd. N&S	33W-1N	33
Peck Av. N&S	38W-0N,38W-1N	38
Pine Needles Ct. NW&SE	37W-0S	38
Randall Rd. N&S	37W-0S,37W-1N	38
Reed Rd. E&W	33W-0N	39
River Ln. NE&SW	35W-0N,35W-1N	39
Roosevelt Rd. NW&SE	33W-0N,33W-1N	39
Rt. 31 NE&SW	35W-0N,35W-1N	39
Rt. 38 NW&SE	36W-1N,34W-1N	32
Skyline Dr. NE&SW	37W-0S	38
Southwest Ln. E&W	33W-0N	39
Spring Green Way NE&SW	38W-0S,37W-0S	38
Theisen Trail N&S	37W-1S	38
Wenmoth Ln. NE&SW	38W-0N	38
Wenmoth Rd. N&S	38W-1S,38W-0S	38
Western Av. NE&SW	36W-0S,36W-0N	38

Gilberts

	Grid	Page
Andra Ct. NE&SW	38W-15N	13
Arrowhead Dr. E&W	38W-16N,37W-16N	13
Center Dr. N&S	38W-16N	13
Deborah St. E&W	37W-15N	13
East Dr. N&S	38W-16N	13
Elgin St. E&W	38W-16N	13
Galligan Rd. N&S	38W-16N,39W-17N	13
Hennessy Ct. E&W	38W-14N	13
Industrial Dr. NW&SE	39W-16N	13
Jackson St. NE&SW	38W-16N	13
Jean St. N&S	37W-15N	13
Joan Ct. E&W	38W-15N	13
Kerry Ct. NW&SE	38W-15N	13
Kildare St. NE&SW	38W-15N	13
Kilkenny Ct. N&S	38W-15N	13
Mason Rd. E&W	38W-14N,37W-15N	13
Matteson St. NE&SW	39W-16N	13
McCornack Rd. N&S	39W-15N	13
Mill St. NE&SW	39W-16N	13
Northwest Tollway NW&SE	39W-16N	13
Pamela St. E&W	37W-15N	13
Park Ct. NW&SE	39W-16N	13
Park St. NE&SW	39W-16N	13
Pauline Ct. NE&SW	37W-15N	13
Pierce St. NW&SE	38W-15N	13
Railroad St. NW&SE	39W-16N	13
Red Hawk Dr. N&S	38W-16N	13
Rt. 72 E&W	39W-16N	13
Running Dear Ln. E&W	38W-16N,37W-16N	13
Shining Moon Path N&S	37W-16N	13
Sleeping Bear Tr. NW&SE	38W-16N,37W-16N	13
Sola Dr. E&W	38W-16N	13
Suzanne St. E&W	37W-15N	13
Tipperary St. NW&SE	38W-15N	13
Tollview Ct. N&S	39W-16N	13
Tollview Ter. E&W	39W-16N	13
Tower Hill Rd. NW&SE	39W-16N	13
Towne Ct. E&W	37W-15N	13
Towne St. NE&SW	38W-15N,37W-15N	13
Turner St. NE&SW	38W-16N	13
Tyler Creek Ct. E&W	39W-16N	13
Tyrrell Rd. N&S	38W-15N	13
Union St. NW&SE	39W-16N	13
Welch St. NE&SW	38W-15N	13
West End Dr. N&S	38W-16N	13
White Feather Ln. N&S	38W-16N	13
Wiley St. NE&SW	39W-16N	13
Wind Mill St. E&W	39W-16N	13
Windmill Cir. N&S	39W-16N	13
Windmill Ct. N&S	39W-16N	13

Hampshire

	Grid	Page
Allen Rd. E&W	47W-16N	11
Ash St. N&S	46W-16N	11
Brookedge Dr. E&W	47W-15N	10
Centennial Dr. N&S	46W-15N	11
Center St. N&S	47W-16N	10
Duchess Ln. E&W	47W-15N	10
East St. N&S	47W-16N	10
Edgewood Av. E&W	47W-15N	10
Elm St. N&S	46W-15N	11
Getzelman Rd. E&W	46W-15N	11
Grace St. N&S	46W-16N	11
Grove St. E&W	46W-16N	11
High Av. E&W	46W-15N	11
Highland Av. E&W	46W-15N	11
Hillcrest Av. E&W	46W-15N	11
Industrial Dr. E&W	46W-16N	11
Jackson St. E&W	47W-16N	10
Jake Ln. E&W	46W-15N	11
Jefferson St. E&W	46W-16N	11
Keys Av. E&W	46W-16N	11
Klick St. N&S	47W-15N,47W-16N	10
Madison St. N&S	46W-16N	11
Maple St. E&W	46W-16N	11
Mill St. E&W	47W-16N	10
Oak Knoll Dr. E&W	47W-15N,46W-15N	10
Oak St. N&S	47W-16N	10

	Grid	Page
Panama St. E&W	46W-15N	11
Park St. N&S	47W-16N	10
Parkside St. N&S	46W-16N	11
Prairie St. N&S	47W-15N	10
Rinn Av. E&W	47W-16N	10
Smith Rd. N&S	46W-15N	11
South St. E&W	46W-15N	11
State St. N&S	47W-16N	10
Terwilliger St. E&W	47W-16N	10
Timber Ln. E&W	47W-16N	10
Walnut St. N&S	46W-16N	11
Warner St. N&S	46W-15N	11
Washington St. E&W	46W-16N	11
West St. N&S	47W-16N	10

Hampshire Township

	Grid	Page
Aileron Av. N&S	49W-19N	4
Allen Rd. E&W	49W-17N,45W-16N	4
Berner Rd. E&W	45W-14N	11
Big Timber Rd. NW&SE	47W-19N,45W-18N	4
Briarwood Ln. NW&SE	47W-19N	4
Brier Hill Rd. N&S	44W-14N,44W-18N	11
Chandelle Dr. E&W	48W-19N	4
Clanyard Rd. E&W	44W-19N	5
Deer Path Ln. E&W	44W-14N	11
Dietrich Rd. E&W	45W-19N	5
Engle Rd. N&S	49W-14N	10
Factly Rd. NW&SE	49W-14N	10
Cir.	46W-19N	5
Gast Rd. N&S	45W-18N	5
Getty Rd. E&W	47W-19N	4
Getzelman Rd. N&S	47W-14N	10
Glen Oaks Ct. E&W	45W-16N	11
Glen Oaks Dr. N&S	45W-16N	11
Green Meadow Ln. N&S	46W-19N	5
Harmony Rd. N&S	47W-17N,47W-19N	4
Hawthorne Ln. N&S	46W-19N	5
Hennig Rd. E&W	44W-18N	5
Higgins Rd. E&W	46W-19N	5
Hillcrest Dr. N&S	46W-19N	5
I-90 NW&SE	44W-18N	5
Immelman Ln. E&W	48W-19N	4
Kelly Rd. E&W	47W-17N	4
Ketchum Rd. N&S	45W-17N	5
Lenschow Rd. E&W	49W-14N,46W-14N	10
Littlewood Trail E&W	44W-14N	11
Marney Dr. N&S	46W-17N	5
Melms Rd. E&W	49W-18N,48W-18N	4
New Lebanon Rd. NW&SE	50W-14N	10
Northwest Tollway NW&SE	46W-19N	5
Oak Grove Dr. NE&SW	46W-17N	5
Oakshire Ln. E&W	44W-15N	11
Obrien Rd. N&S	49W-19N	4
Penstemon Ln. NW&SE	45W-16N	11
Prairie Farm Dr. NE&SW	45W-16N	11
Primrose Ln. E&W	47W-19N	4
Primrose Path E&W	46W-17N	5
Red Barn Rd. NW&SE	47W-14N	10
Romke Rd. N&S	45W-14N	11
Rt. 72 E&W	46W-15N	11
Rudder Rd. N&S	48W-19N	4
Shirewood Farm Rd. NE&SW	47W-15N	10
Stoxen Rd. E&W	47W-19N	4
Sunset Dr. NW&SE	44W-14N	11
U.S. Rt. 20 NW&SE	44W-17N,45W-18N	5
Volkening Cir. E&W	44W-14N	11
Walker Rd. N&S	49W-14N,49W-18N	10
West County Line Rd. N&S	49W-19N	4
Whispering Tr. N&S	44W-14N	11
White Pines Ln. N&S	44W-14N	11
Widmayer Rd. N&S	46W-16N,46W-18N	11
Woodview Pkwy. NS&EW	46W-19N	5

Kaneville Township

	Grid	Page
Bateman Rd. NS&EW	46W-1S,45W-0S	35
Cedar Ct. NW&SE	46W-2S	35
Center Av. N&S	50W-2S	34
Country Life Ct. N&S	48W-0N	34
Country Life Dr. NS&EW	48W-0N,48W-1N	34
County Line Rd. N&S	50W-1S,50W-1N	34
Dauberman Rd. N&S	47W-3S,47W-0N	41
East West Tollway NW&SE	50W-0N,45W-1S	34
Elm Ct. N&S	46W-2S	35
Elm St. E&W	46W-2S	35
Farm View Ct. E&W	48W-0N	34
Francis Rd. N&S	46W-1N	29
Harter Rd. NW&SE	50W-0N,45W-3S	34
Interstate 88 NW&SE	50W-0N,45W-1S	34
Kaneland Dr. NW&SE	48W-1N	28
Kaneview Ct. NE&SW	48W-1N	28
Keslinger Rd. E&W	50W-1N,45W-1N	28
Lasher Rd. E&W	46W-3S	41
Locust Ct. NW&SE	46W-2S	35
Locust St. NS&EW	46W-2S	35
Lorang Rd. N&S	45W-2S,45W-1S	35
Lovell St. NE&SW	46W-2S	35
Main St. NE&SW	50W-2S,45W-1S	34
Maple St. N&S	46W-2S	35
Meredith Rd. N&S	47W-1N	28
Merril Av. N&S	46W-2S	35
Milnamow Dr. NE&SW	48W-1N	28
Miner Rd. N&S	49W-1S,49W-0S	34
N. Railroad St. NW&SE	50W-2S	34
Owens Rd. NS&EW	50W-1S,48W-2S	34
Perry Rd. E&W	50W-0S,47W-1S	34
Pine St. N&S	46W-2S	35
Pine Tree Dr. E&W	48W-1N	28
Pond View Ct. N&S	48W-0N	34
Probst Ct. NW&SE	48W-0N	34
Prospect Av. N&S	50W-2S	34
Rowe Rd. NS&EW	45W-0S	35
Rt. 38 E&W	49W-1N,47W-1N	28
Schneider Rd. N&S	45W-0S,45W-0N	35
School St. NE&SW	46W-2S	35
Schrader Rd. N&S	50W-0N,50W-1N	34
Seavey Rd. E&W	45W-3S	41
Shadetree Ct. E&W	48W-1N	28
Shadetree Ln. N&S	48W-0N,48W-1N	34
South Lorang Rd. NW&SE	45W-2S,45W-1S	35
Swan Rd. N&S	49W-3S	40
W. County Line Rd. N&S	50W-2S	34
Watson Rd. N&S	48W-0S,48W-1N	34

Lilly Lake

	Grid	Page
Blackhawk Ln. E&W	44W-5N	24
Campton Ct. E&W	43W-5N	24
Clover Ct. E&W	43W-5N	24
Cir.	44W-5N	24
Coleman Ln. E&W	43W-5N	24
Cornwall Dr. E&W	43W-5N	24
Empire Rd. E&W	43W-5N	24
Ferson Ct. E&W	43W-5N	24
Fescue Ct. NW&SE	43W-5N	24
Fox Hill Ct. NW&SE	44W-5N	24
Fox Wilds Ct. NW&SE	43W-5N	24
Fox Wilds Dr. EW&NS	44W-5N	24
Foxmoor Dr. N&S	43W-5N	24
Hanson Rd. NW&SE	44W-5N	24
Hanson Ridge Dr. N&S	44W-5N	24
Hazelwood Ct. N&S	43W-5N	24
Hazelwood Tr. NS&EW	43W-5N	24
Heather Ln. NW&SE	43W-5N	24
Leslie Ct. N&S	43W-5N	24
Meadowview Ln. N&S	43W-5N	24
Route 47 (State) N&S	44W-6N	24
Route 64 (State) NW&SE	44W-5N	24
Sanctuary Tr. E&W	43W-5N	24
Seneca Ct. NW&SE	44W-5N	24
Wooley Rd. N&S	44W-4N	30

Maple Park

	Grid	Page
Broadway St. N&S	50W-2N,50W-3N	28
Center St. E&W	50W-2N	28
Charles St. N&S	50W-3N	28
County Line Rd. N&S	50W-2N,50W-3N	28
Elm St. NW&SE	50W-2N	28
Green St. NE&SW	50W-2N	28
High St. N&S	50W-2N	28
Kane St. N&S	50W-3N	28
Kennberg St. NE&SW	50W-2N	28
Liberty St. NE&SW	50W-2N	28
Maiden Ln. N&S	50W-3N	28
Main St. NW&SE	50W-2N	28
Maple Av. E&W	50W-2N	28
Mulberry St. NE&SW	50W-2N	28
Oak St. N&S	50W-3N	28
Palmer Dr. N&S	50W-2N	28
Pearl St. E&W	50W-3N	28
Pine Ln. N&S	50W-3N	28
Pleasant St. NE&SW	50W-2N	28
Root Ln. N&S	50W-3N	28
South St. E&W	50W-2N	28
State St. E&W	50W-3N	28
Summer St. NE&SW	50W-2N	28
Thatcher Rd. E&W	50W-3N	28
Virgil St. N&S	50W-3N	28
Washington St. E&W	50W-3N	28
Willow St. NW&SE	50W-2N	28

Montgomery

	Grid	Page
Albright Rd. N&S	37W-9S	50
Amber Dr. N&S	34W-9S	51
Ashland Av. E&W	37W-9S	50
Aucutt Rd. E&W	38W-9S	50
Broadway Av. N&S	36W-9S	50
Carol Pl. E&W	34W-9S	51
Case Av. NW&SE	37W-9S	50
Catherine Ln. E&W	36W-9S	50
Clay St. NW&SE	37W-9S	50
Clinton St. NW&SE	37W-9S	50
Cornell Av. E&W	38W-8S	50
Countryside Av. E&W	38W-9S	50
Crown St. N&S	34W-9S	51
Dawn Av. E&W	34W-9S	51
Fellemore Av. E&W	36W-9S	50
First Av. NW&SE	36W-9S	50
Greenfield Av. E&W	38W-9S	50
Harmony Ct. NE&SW	36W-9S	50
Harmony Dr. N&S	36W-9S	50
Harrison St. NW&SE	37W-9S	50
Hartway Ct. NS&EW	36W-9S	50
Hartway Dr. E&W	36W-9S	50
Henry Ct. E&W	36W-9S	50
Hinkley St. N&S	35W-9S	51
Howell Pl. NW&SE	34W-9S	51
Jackson St. N&S	35W-9S	51
James St. E&W	36W-9S	50
Jefferson St. NW&SE	37W-9S	50
Keck Av. E&W	36W-9S	50
Kimberly Ln. E&W	34W-9S	51
Kingston Av. N&S	34W-9S	51
Knell St. E&W	37W-9S	50
Lake St. NE&SW	37W-9S	50
Lebanon St. N&S	35W-9S	51
Lewis St. E&W	36W-9S	50
Lincoln Av. N&S	36W-9S	50
Madison St. NW&SE	37W-9S	50
Main St. NE&SW	37W-9S,36W-9S	50
Marsch Av. E&W	36W-9S	50
Martin Av. NE&SW	36W-9S	50
Mill St. NW&SE	37W-9S,36W-9S	50
Montgomery Rd. NE&SW	36W-9S,35W-9S	50
Mulberry Dr. N&S	38W-9S	50
Oakton Rd. E&W	38W-9S	50
Orchard Rd. N&S	38W-9S	50
Park Dr. N&S	34W-9S	51
Parker Ct. N&S	36W-9S	50
Pearl St. NE&SW	37W-9S	50
Railroad St. NE&SW	37W-9S	50
Raymond Dr. NW&SE	34W-9S	51
River St. NE&SW	37W-9S,36W-9S	50
Riverside Dr. NS&EW	36W-9S	50
Rochester Dr. E&W	38W-8S	50
Royal Dr. N&S	34W-9S	51
Sard Av. NE&SW	36W-9S,36W-8S	50
Scheuring Pl. N&S	38W-9S	50
Scott St. NW&SE	37W-9S	50
Second Av. N&S	36W-9S	50
Sherman Av. E&W	36W-9S	50
Taylor St. NW&SE	36W-9S	50
Third Av. E&W	36W-9S	50
Victoria Dr. NS&EW	38W-9S	50
Washington St. NE&SW	36W-9S	50
Watkins St. NW&SE	37W-9S	50
Webster St. NW&SE	37W-9S	50
Wellman Av. N&S	36W-9S	50
Wellman Ct. E&W	36W-9S	50
14th Av. E&W	34W-9S	51

North Aurora

	Grid	Page
Abbeywood Ln. NS&EW	36W-4S,36W-3S	44
Acorn Dr. NE&SW	37W-4S,37W-3S	44
Adams St. NW&SE	36W-4S	44
Adler St. N&S	37W-4S	44
Airport Rd. NE&SW	37W-4S,36W-4S	44
Andrew Ct. E&W	35W-3S	45
Andrew Ln. N&S	35W-3S	45
Anna St. N&S	36W-3S	44
April Ln. NS&EW	36W-3S,35W-3S	44
Arrowhead E. N&S	35W-3S	45
Arrowhead St. N&S	35W-3S	45
Arrowhead W. N&S	35W-3S	45
Banbury Rd. N&S	35W-3S	45
Basswood Dr. N&S	37W-3S	44
Bede Cir. E&W	35W-3S	45
Bede Ct. NW&SE	35W-3S	45
Birchwood Dr. E&W	37W-4S,36W-4S	44
Briar Ln. E&W	36W-3S,35W-3S	44
Buckthorn N&S	37W-4S	44
Bull Ct. N&S	37W-4S	44
Butterfield Rd. NE&SW	36W-4S,35W-4S	44
Butternut Dr. E&W	37W-3S	44
Candlewick Ct. NE&SW	37W-4S	44
Carrie Ct. E&W	35W-3S	45
Cedar Dr. N&S	36W-4S	44
Chantilly Ln. N&S	36W-4S	44
Cherry Tree Ct. N&S	36W-4S	44
Cherrywood Dr. E&W	37W-4S,36W-4S	44
Chestnut St. E&W	35W-3S	45
Conco St. NE&SW	35W-4S	45
Cottonwood Dr. E&W	37W-3S	44
Cypress Ln. N&S	37W-4S	44
Dee Rd. E&W	36W-3S,35W-3S	44
East-West Tollway E&W	36W-5S	44
Eastview Rd. E&W	36W-3S	44
Elm Av. E&W	36W-3S	44
Fairview Dr. NS&EW	36W-5S	44
Farview Ct. E&W	36W-4S	44
Farview Dr. E&W	36W-4S	44
Frontage Rd. NE&SW	35W-4S	45
Geronimo Dr. NW&SE	37W-4S	44
Grace St. NW&SE	36W-4S	44
Grant St. NW&SE	35W-4S,36W-4S	45
Harmony Dr. E&W	37W-4S,36W-4S	44
Hawthorne Dr. N&S	37W-4S	44
Hettinger Ln. E&W	35W-4S	45
Hickory St. N&S	35W-4S,35W-3S	45
Hill Av. N&S	35W-3S	45
Hillside Pl. E&W	36W-4S	44
Holly Ct. E&W	35W-3S	45
John St. E&W	36W-4S	44
Juniper Dr. N&S	37W-4S	44
Kingswood Dr. E&W	36W-4S	44
Larchwood Ln. N&S	36W-4S,36W-3S	44
Laurel Dr. N&S	35W-3S	45
Lilac Ln. N&S	35W-3S	45
Lincoln St. N&S	36W-4S	44
Lincoln Way N&S	36W-5S,36W-3S	44
Lindsay Cir. E&W	35W-3S	45
Linn Ct. N&S	35W-3S	45
Lloyd Ln. E&W	35W-3S	45
Lovedale Ln. E&W	36W-4S	44
Magnolia Dr. E&W	37W-4S	44
Maple Av. E&W	36W-3S	44
Matthias Ct. E&W	35W-3S	45
Mistwood Ln. NS&EW	36W-4S,36W-3S	44
Monroe St. N&S	36W-4S	44
Napa Ct. E&W	37W-4S	44
Oak Av. E&W	36W-4S	44
Oak Dr. E&W	36W-4S	44
Oak St. E&W	38W-4S,36W-4S	44
Oakwood NW&SE	36W-4S	44
Offutt Ln. N&S	36W-4S	44
Pierce St. E&W	36W-4S,35W-4S	44
Pin Oak Ct. E&W	36W-3S	44
Pine Oak Dr. N&S	36W-4S	44
Pinecreek Dr. N&S	35W-3S	45
Pinewood Dr. E&W	36W-3S	44
Poplar Pl. N&S	37W-4S	44
Poplar St. N&S	37W-4S	44
Princeton Ct. N&S	36W-4S	44
Princeton Dr. E&W	37W-4S	44
Randall Rd. N&S	37W-4S	44
Redwing Dr. N&S	38W-4S	44
Redwood Ct. E&W	36W-3S	44
Riverview St. NE&SW	36W-4S	44
Robert St. NW&SE	36W-4S	44
Rt. 25 NW&SE	36W-4S	44
Rt. 31 N&S	36W-4S,36W-3S	44
Rt. 56 NW&SE	36W-4S	44
Shagbark E&W	37W-4S	44
Sharon Ln. E&W	36W-3S,35W-3S	44
South St. E&W	35W-5S	45
Spruce St. NE&SW	35W-4S	45
Staghorn E&W	37W-4S	44
State St. E&W	37W-4S,36W-4S	44
Stone St. N&S	35W-4S	45
Sumac Dr. E&W	38W-4S,37W-4S	44
Sunshine Dr. NE&SW	37W-4S	44
Sycamore Ln. N&S	36W-4S,36W-3S	44
Terry Ln. E&W	35W-3S	45
Timber Oaks Dr. N&S	36W-4S,36W-3S	44
Walnut Dr. N&S	37W-4S	44
Wildwood Dr. N&S	36W-4S,36W-3S	44
Willow Way N&S	36W-4S	44

Pingree Grove

	Grid	Page
Grove St. N&S	40W-13N	13
Jackson St. E&W	41W-13N	12
Mansfield St. NE&SW	41W-14N	12
Meadowsweet Dr. NW&SE	40W-14N	13
Oak St. E&W	41W-13N	12
Olandra Way NW&SE	40W-14N	13
Prairie St. N&S	41W-13N	12
Public St. E&W	41W-14N	12
Railroad St. NW&SE	41W-14N	12
Reinking Rd. N&S	41W-15N	12
Store St. NE&SW	40W-14N	13
U.S. Rt. 20 E&W	42W-14N,40W-12N	12
Yellow Avens Ct. NE&SW	40W-14N	13

Plato Township

	Grid	Page
Adobe Ridge NW&SE	39W-9N	19
Apache Run N&S	39W-9N	19
Barr Rd. NS&EW	44W-12N,42W-12N	17
Bending Ln. E&W	40W-9N	19
Bowes Bend Dr. NS&EW	41W-9N,40W-9N	18
Bowes Rd. NW&SE	40W-10N,39W-9N	19
Brier Hill Rd. N&S	44W-12N	17
Brierwood NW&SE	43W-8N	24
Brimfield Dr. NW&SE	40W-8N	25
Burlington Rd. NW&SE	43W-8N,44W-9N	24
Burlington Rd. NW&SE	45W-10N	17
Capulet Cir NE&SW	39W-10N	19
Capulet Ct. NW&SE	39W-11N	19
Carletta Ln. E&W	43W-9N	18
Central Dr. NW&SE	39W-11N	19
Channel Ct. N&S	40W-9N	19
Channel Dr. N&S	40W-9N	19
Cheyenne Ct. NW&SE	39W-13N	13
Chippewa Pass E&W	41W-10N	18
Cir.	39W-13N	13
Cliff Dr. NW&SE	39W-11N	19
Cloverfield Dr. N&S	39W-8N	25
Connors Rd. E&W	43W-9N	18
Coombs Rd. N&S	39W-12N	19
Corron Rd. N&S	40W-8N,40W-9N	25
Crawford Rd. N&S	40W-8N,40W-9N	25
Creekview Dr. E&W	41W-9N	18
Creekwood Ct. N&S	40W-9N	19
Creekwood Dr. E&W	40W-9N	19
Current Ct. E&W	41W-9N	18
Dittman Rd. N&S	42W-9N	18
Greenfield Rd. E&W	41W-9N	18

Greenview Ln. NE&SW
.........40W-10N 19
Gunpowder Ln. N&S
.........39W-13N 13
Hartz Ct. E&W 39W-8N 25
Heatherington Ct. NS&EW
.........40W-9N 19
Hidden Lakes Dr. E&W
.........40W-10N 19
High Chapparel N&S
.........39W-13N 13
Highbank Ct. NW&SE
.........41W-9N 18
Highland Av. NW&SE
.........39W-13N 13
Hogan Hill E&W 39W-9N 19
Hummingbird St. E&W
.........42W-12N 18
Ickenham Ln. NS&EW
.........43W-8N,42W-8N 24
Juliet Dr. E&W . 39W-11N 19
Kendall Rd. N&S
.........43W-8N,43W-9N 24
Kiwi Ct. NW&SE ...42W-12N 18
Lakeside Ct. NE&SW
.........40W-10N 19
Lenz Rd. E&W
.........41W-9N,40W-9N 18
Lilly St. N&S 41W-9N 18
Long Ln. NW&SE . 40W-10N 19
Lori Ln. E&W........39W-9N 19
Marshall Rd. N&S . 41W-12N 18
McDonald Rd. E&W
.........44W-8N,39W-8N 24
Meadow Dr. NW&SE
.........41W-9N 18
Meadowlark Dr. NE&SW
.........42W-12N 18
Mohawk Tr. N&S ...41W-9N 19
Muirhead Ct. E&W . 41W-9N 18
Muirhead Ln. NE&SW
.........41W-9N 18
Muirhead Rd. NE&SW
.........41W-9N,41W-10N 18
Nesler Rd. NE&SW
.........39W-9N,39W-10N 19
Oak Bluff Dr. N&S . 41W-9N 18
Oak Ridge Dr. N&S
.........40W-10N 19
Oak Tree Ln. NW&SE
.........40W-9N 19
Pease Rd. NS&EW
.........44W-10N 17
Plank Rd. E&W
.........43W-12N,40W-12N 18
Plato Rd. E&W
.........44W-10N,42W-10N 17
Prairie Crossing N&S
.........40W-10N 19
Pueblo Peak NE&SW
.........39W-9N 19
Red Cloud Ln. NW&SE
.........39W-10N 19
Rippburger Rd. N&S
.........41W-10N 18
Robin Ln. E&W .. 42W-12N 18
Rohrsen Rd. E&W
.........43W-11N,41W-11N 18
Romeo Cir. NW&SE
.........39W-11N 19
Romeo Dr. NE&SW
.........39W-11N 19
Route 47 NS&EW
.........43W-8N,43W-10N 24
Russell Rd. E&W
.........41W-11N,39W-11N 18
Saidon Dr. NW&SE
.........39W-11N 19
Santa Fe Tr. NE&SW
.........39W-9N 19
Silvana Dr. NS&EW..41W-9N 18
South Gate Rd. E&W
.........43W-9N 18
Steeple Cir. NE&SW
.........43W-8N 24
Sturbridge Way E&W
.........40W-8N 25
Sunflower Ln. E&W .39W-9N 19
Switzer Rd. N&S .. 40W-12N 19
Tamara Dr. NS&EW..41W-9N 18
Tower Rd. N&S
.........44W-10N,44W-11N 17
Tributary Ln. N&S . 41W-10N 18
U.S. Route 20 NW&SE
.........40W-12N 19
Verona Dr. NW&SE
.........39W-11N 19
W. Lori Ln. E&W ... 39W-9N 19
Western Av. NE&SW
.........39W-13N 13
Whispering Springs Ln.
NS&EW 40W-13N 19
Wildbriar Ln. N&S .. 40W-13N 19
Woodbridge Ln. NS&EW
.........43W-8N 24
Woodrow Ln. N&S
.........40W-10N 19

Prestbury

Alderley Dr. NE&SW
.........41W-5S,40W-5S 43
Ashwood Ct. E&W .. 40W-5S 43
Birchwood Ct. NE&SW
.........40W-5S 43
Bliss Rd. NE&SW 41W-5S 43
Briar Gate Cir. N&S . 41W-5S 43
Buckingham Dr. NW&SE
.........41W-5S 43
Bury Ct. NE&SW 40W-5S 43
Cedar Gate Cir. NW&SE
.........41W-5S 43
Dorchester Ct. N&S .. 41W-5S 43
Fernilee Ct. NW&SE . .41W-5S 43
Hardwick Ct. NE&SW
.........41W-5S 43
Hathaway Cres. NS&EW
.........41W-5S 43
Hatherley Ct. NE&SW
.........41W-5S 43
Heaton Ct. N&S 41W-5S 43
Merrill New Rd. NE&SW
.........41W-5S 43
Mossfield Ct. NW&SE
.........40W-5S 43
N. Windsor St. NE&SW
.........41W-5S 43
New Bond St. E&W .41W-5S 43
Normandie Dr. NW&SE
.........41W-5S 43
Norris Old Rd. NE&SW
.........40W-5S 43
Old Farm Ct. N&S .. 41W-5S 43
Rt. 56 NE&SW 40W-5S 43
Saddlewood Ct. E&W
.........41W-5S 43
Walnut Cir. NS&EW
.........41W-5S,40W-5S 43
Windsor Rd. NW&SE
.........41W-6S,41W-5S 43
Windwood Ct. NW&SE
.........41W-5S 43
Windwood Dr. E&W .41W-5S 43
Winthrop New Rd. E&W
.........41W-4S 43

Rutland Township

Abilene Tr. E&W....39W-14N 13
Apache Ln. E&W .. 40W-18N 7
Atchison Dr. E&W .. 40W-15N 13
Aztec Ln. N&S 40W-18N 7
Barko Pkwy. NS&EW
.........40W-18N,40W-19N 7
Baron Ln. E&W ... 40W-18N 7
Big Timber Rd. NW&SE
.........43W-17N,40W-16N 6
Binnie Lakes Tr. N&S
.........38W-17N 7
Binnie Rd. E&W ... 38W-17N 7
Bonnie Ln. N&S ... 39W-14N 13
Cody Ct. N&S 39W-14N 13
Coombs Rd. N&S .. 39W-14N 13
Damisch Rd. NE&SW
.........40W-14N 13
Drendl Rd. NS&EW
.........42W-18N,42W-19N 6
Freeman Rd. E&W
.........43W-18N,41W-18N 6
Galligan Rd. N&S . 39W-18N 7
Gary Ln. N&S 40W-19N 7
Grand Av. E&W ... 39W-14N 13
Gunpowder Ct. N&S
.........39W-14N 13
Guthrie Ct. E&W .. 40W-15N 13
Harper Dr. E&W .. 40W-15N 13
Hennig Rd. N&S ...43W-18N 6
Hidden Hills Tr. N&S
.........38W-17N 7
Hidden Knoll Rd. E&W
.........39W-14N 13
Highland Av. E&W .. 41W-14N 12
Homestead Dr. NW&SE
.........40W-15N 12
Huntley Rd. NW&SE
.........39W-19N 7
Limerick Ln. N&S .. 40W-14N 13
Manning Rd. N&S .. 42W-17N 6
Maplehurst Ln. NW&SE
.........40W-15N 12
Mary Ln. N&S.....39W-17N 7
Mary Ln. NS&EW .. 39W-17N 7
McCornack Rd. NW&SE
.........40W-14N,39W-15N 13
Navajo Ln. NW&SE
.........40W-18N 7
Northwest Tollway NW&SE
.........43W-18N,39W-16N 6
Northwind Ln. N&S
.........38W-18N 7
Oak Knoll Ln. N&S..38W-17N 8
Old Stage Rd. NE&SW
.........41W-15N 12
Oliver Dr. N&S ... 39W-14N 13
Pleasant Dr. E&W .. 39W-14N 13
Powers Rd. NS&EW
.........40W-16N,41W-19N 6
Red Leaf Ln. NW&SE
.........40W-15N 13
Reinking Rd. N&S
.........41W-15N,42W-16N 12

Route 47 N&S
.........43W-14N,42W-18N 12
Route 72 E&W
.........43W-15N,39W-16N 12
Sandwald Rd. N&S
.........43W-17N 6
Settlers Grove Rd. NW&SE
.........41W-15N 12
Springcreek Rd. E&W
.........38W-18N 8
Thurnau Rd. E&W
.........43W-14N,43W-15N 12
Timber Ridge Dr. N&S
.........39W-14N 13
Tower Hill Rd. E&W
.........39W-16N 13
U.S. Route 20 NW&SE
.........43W-15N,42W-14N 12

Sleepy Hollow

Acorn Ct. N&S 35W-15N 14
Acorn Dr. E&W 35W-15N 14
Cir................35W-15N 14
Arabian Pkwy. N&S
.........36W-15N 14
Arlington Pkwy. NS&EW
.........36W-15N 14
Beau Brummel Ct. E&W
.........36W-15N 14
Beau Brummel Dr. E&W
.........36W-15N 14
Belmont Pkwy. N&S
.........36W-15N 14
Brom Ct. NW&SE .. 35W-15N 14
Bullfrog Ln. N&S .. 35W-15N 14
Cir...............35W-15N 14
Carol Crescent Dr. N&S
.........35W-15N 14
Christopher Ct. N&S
.........35W-15N 14
Churchill Ct. N&S . 36W-15N 14
Crane Ct. NE&SW . 36W-15N 14
Crane Dr. NS&EW . 36W-15N 14
Dannelle Ct. NE&SW
.........35W-15N 14
Darien Ct. NW&SE..36W-15N 14
Darien Ln. N&S ... 35W-15N 14
Deer Ln. NS&EW . 36W-15N 14
East Hilltop Ln. NS&EW
.........36W-15N 14
Elm Ct. N&S 34W-15N 15
Fawn Ct. N&S 35W-15N 14
Front Range Rd. N&S
.........35W-15N 14
Gail Ln. E&W 36W-15N 14
Glen Oak Dr. NE&SW
.........35W-15N 14
Hawthorne Ct. E&W
.........36W-15N 14
Hazel Ct. N&S......35W-16N 14
Hemlock Dr. NS&EW
.........34W-15N 15
Hialeah Ct. NW&SE
.........36W-15N 14
Hickory Ct. NE&SW
.........35W-16N 14
Higgins Rd. E&W .. 36W-16N 14
Hillcrest Dr. E&W .. 35W-16N 14
Cir................35W-16N 14
Hilltop Ct. E&W 36W-15N 14
Hilltop Ln. E&W ... 36W-15N 14
Holly Ct. NE&SW . 35W-15N 14
Island Ct. N&S 35W-15N 14
Jaclay Ct. NW&SE..35W-15N 14
Jamestowne Ct. E&W
.........36W-15N 14
Jamestowne Rd. NS&EW
.........36W-15N 14
Jill Peak Dr. N&S . 35W-15N 14
Joy Ln. E&W 35W-15N 14
Karyn Ct. N&S ... 35W-15N 14
Katrina Dr. N&S ... 35W-15N 14
Katrina Ln. NS&EW
.........35W-15N 14
Kay Ct. N&S 35W-15N 14
Laurel Ct. E&W .. 35W-16N 14
Locust Dr. NS&EW
.........35W-15N 14
Maple Ln. E&W .. 34W-15N 15
Maria Ct. NE&SW .. 35W-15N 14
Cir................35W-15N 14
Palimino Ct. NE&SW
.........36W-15N 14
Paula Ct. NW&SE . 35W-16N 14
Pimlico Pkwy. NS&EW
.........36W-15N 14
Pine Cone Ln. NE&SW
.........35W-15N 14
Plum Ct. N&S 35W-15N 14
Rainbow Dr. N&S
.........35W-16N 14
Randall Rd. N&S .. 37W-15N 14
Red Bud Cir. E&W..36W-15N 14
River Ridge Dr. NS&EW
.........35W-15N 14
Saddle Club Pkwy. E&W
.........35W-15N 14
Salem Ct. E&W .. 36W-15N 14
Saratoga Pkwy. E&W
.........36W-15N 14
Sharon Dr. N&S .. 35W-16N 14

Sleepy Hollow Rd. N&S
.........36W-14N 14
Cir................36W-15N 14
Surrey Ln. E&W .. 35W-15N 14
Sycamore Ln. E&W
.........36W-16N 14
Thorobred Ln. E&W
.........35W-15N 14
Timber Dr. N&S .. 36W-15N 14
Trails End NE&SW..35W-15N 14
Van Tassel Rd. N&S
.........36W-15N 14
Walnut Dr. E&W .. 35W-15N 14
Willow Ln. E&W
.........35W-15N,34W-15N 14
Windcrest Ct. N&S
.........36W-15N 14
Windsor Ct. E&W .. 35W-15N 14
Windsor Dr. E&W . 36W-15N 14
Winmoor Ct. E&W . 35W-15N 14
Winmoor Dr. NS&EW
.........35W-15N 14

South Elgin

Adam Av. N&S 35W-8N 27
Andover Av. NW&SE
.........36W-8N 26
Ann St. N&S 34W-8N 27
Arron Av. E&W 35W-8N 27
Arthur Av. NW&SE . 34W-9N 21
Barbara Av. E&W .. 34W-8N 27
Beach Ct. E&W 34W-8N 27
Beck Av. NW&SE .. 34W-9N 21
Belle Ln. E&W 34W-8N 27
Boston Ct. N&S 34W-8N 27
Brandywine Ct. E&W
.........35W-8N 27
Brookline St. E&W .. 35W-8N 27
Bunker Hill Av. N&S..35W-8N 27
Burnridge Av. N&S . 33W-9N 21
Cambridge Rd. E&W
.........35W-8N 27
Camden Ln. N. E&W
.........35W-7N 27
Camden Ln. W. N&S
.........35W-7N 27
Carla Dr. N&S 35W-8N 27
Carlton Ct. NW&SE..33W-9N 21
Castlebar Ct.* E&W..36W-7N 27
Cedar Ct. N&S 35W-7N 27
Center St. N&S 34W-8N 27
Charles Ct. E&W .. 35W-8N 27
Chatham Ct. E&W .. 35W-7N 27
Cherry St. N&S..... 34W-8N 27
Chipstone Dr. NE&SW
.........34W-9N 21
Churchill Ct.* E&W . 36W-7N 27
Clinton St. E&W
.........35W-7N,34W-7N 27
Clove Ct. N&S 35W-7N 27
Cobbler St. E&W .. 35W-8N 27
Collins St. N&S
.........35W-7N,35W-8N 27
Concord Av. NS&EW
.........35W-8N,34W-8N 27
Cornell Ct. E&W 35W-7N 27
Cornwell Av. NE&SW
.........35W-8N,34W-8N 27
Coventry Ct. E&W .. 35W-7N 27
Crystal Av. E&W .. 34W-8N 27
Dana Dr. E&W 34W-9N 21
Danbury Ct. E&W .. 35W-8N 27
Dean Dr. N&S
.........35W-8N,35W-9N 27
Division St. N&S
.........35W-8N,34W-8N 27
Dorset Ct. NW&SE . 35W-8N 27
Dublin Ct. NE&SW . 35W-8N 27
Dublin Ln. N&S..... 35W-8N 27
E. Harvard Cir. N&S 35W-8N 27
Earl Ln. E&W........34W-7N 27
East Av. N&S 34W-8N 27
Eastview N&S 34W-8N 27
Elizabeth St. E&W . 34W-8N 27
Exeter Ln. NE&SW . 36W-8N 26
Fairfax Ln. NW&SE..35W-8N 27
Franklin Av. E&W .. 34W-8N 27
Fulton St. N&S
.........34W-7N,34W-8N 27
Gerald Av. E&W ... 35W-9N 20
Gilbert St. NW&SE..34W-8N 27
Gilpen Av. N&S.... 34W-9N 21
Greenfield Av. E&W .35W-9N 20
Hancock Av. NE&SW
.........35W-7N,35W-8N 27
Hartman Ct. NW&SE
.........34W-9N 21
Harvard Ct. N&S .. 35W-8N 27
Harvest Ln. N&S .. 35W-8N 27
Helen Av. E&W 35W-8N 27
Hollywood NW&SE . 34W-9N 21
Hollywood Ct. NW&SE
.........33W-9N 21
James St. E&W 35W-9N 20
Josephine Av. NW&SE
.........34W-9N 21
Juniper Ln. E&W .. 34W-8N 27
Kane St. E&W
.........34W-8N 27
Kenilworth Av. N&S..35W-9N 20
Kingsport Dr. N&S .. 36W-8N 27

Kossuth St. E&W ... 34W-8N 27
La Fox St. N&S
.........34W-8N,34W-9N 27
Lafayette Dr. E&W . 35W-8N 27
Lakewood Ct. N.* E&W
.........36W-7N 27
Lakewood Ct. S.* E&W
.........36W-7N 27
Lancaster Cir. S. NS&EW
.........35W-7N 27
Lancaster Ln. N&S . 36W-8N 27
Lancer Ct.* E&W .. 36W-7N 27
Laurel Ct.* E&W ... 36W-7N 27
Lenox Ct.* E&W ... 36W-7N 27
Lexington Av. N&S . 35W-8N 27
Lilac Ct.* E&W 36W-7N 27
Lincoln Ct.* E&W .. 36W-7N 27
Linda Ct. N&S 35W-8N 27
Lindemann Ct. N&S..34W-8N 27
Linden Ct.* E&W ... 36W-7N 27
Locust Ct.* E&W .. 36W-7N 27
Locust St. N&S 34W-8N 27
Longbow Ct.* E&W . 36W-7N 27
Longview Ct.* E&W . 36W-7N 27
Lorann St. E&W 35W-8N 27
Lowell Dr. N&S 35W-8N 27
Lucille Av. E&W ... 34W-8N 27
Main St. NE&SW .. 34W-8N 27
Manchester Ct. E&W
.........35W-8N 27
Marbury St. N&S .. 35W-8N 27
Mark St. N&S 35W-8N 27
Martin Dr. N&S
.........34W-8N,34W-9N 27
Mavis Av. N&S .. 34W-9N 21
Mayfair Ln. N&S .. 35W-8N 27
McLean Blvd. N&S . 35W-9N 27
Medford Av. E&W .. 35W-8N 27
Melinda Dr. N&S .. 35W-8N 27
Melrose Av. E&W .. 34W-9N 21
Michigan Av. N&S
.........35W-8N,35W-9N 27
Middle St. E&W 34W-8N 27
Mill St. E&W 34W-8N 27
Mistic Ct. NW&SE .. 33W-9N 21
Moody Ct. NW&SE . 33W-9N 21
N. Harvard Cir. E&W
.........35W-8N 27
N. Lancaster Cir. NS&EW
.........35W-8N 27
N. London Ct. N&S . 35W-8N 27
Nellie Av. E&W 34W-8N 27
North Av. E&W 34W-7N 27
Oak St. E&W 34W-8N 27
Oxford Ct. NW&SE . 35W-8N 27
Oxford Ln. NW&SE . 35W-8N 27
Paine Ct. E&W 35W-8N 27
Paine St. NE&SW .. 35W-8N 27
Park Av. E&W 34W-8N 27
Patrick St. N&S.....34W-8N 27
Pembroke Dr. N. NE&SW
.........36W-8N 26
Pembroke Dr. S. NE&SW
.........36W-8N 26
Pleasant Dr. NS&EW
.........34W-8N 27
Plum St. E&W 34W-8N 27
Prairie St. E&W 34W-8N 27
Quarry St. E&W ... 34W-8N 27
Rainbow Ter. E&W . 35W-9N 20
Raymond St. NE&SW
.........33W-9N 21
Regent St. N&S ... 34W-8N 27
Renee Dr. N&S
.........35W-8N,35W-9N 27
Revere St. E&W ... 35W-8N 27
River Rd. NE&SW
.........34W-9N,33W-9N 21
River St. N&S 34W-8N 27
Riverside Av. NE&SW
.........34W-8N 27
Riverview Av. NE&SW
.........33W-9N 21
Robertson Rd. N&S
.........34W-8N 27
Ross Av. NW&SE .. 34W-9N 21
Rt. 25 N&S 33W-7N,33W-8N 27
Rt. 31 NE&SW ... 34W-9N 21
S. Elgin Blvd. E&W
.........34W-8N,34W-9N 27
S. London Ct. N&S . 35W-8N 27
Sandhurst Ln. NS&EW
.........35W-8N 27
Saratoga Ct. N&S . 35W-8N 27
Schneider Ct. N&S . 34W-9N 21
Smith Ct. NW&SE . 34W-9N 21
South St. E&W 34W-8N 27
Spring Av. E&W .. 34W-8N 27
Spring St. E&W 35W-8N 27
Spruce St. E&W .. 34W-8N 27
State St. E&W
.........35W-8N,34W-8N 27
Stearns Rd. E&W .. 32W-7N 27
Stevenson Rd. E&W
.........35W-9N,34W-9N 20
Steves Farm Rd. E&W
.........34W-9N 21
Stone St. E&W 35W-8N 27
Strathmore Ter. N&S
.........34W-8N 27
Sunbrook St. E&W . 35W-8N 27
Sunbury Rd. E&W .. 35W-8N 27

	Grid	Page
Sundown Rd. E&W 35W-9N,34W-9N		20
Sweetbriar Ct. N&S..35W-8N		27
Trenton Av. N&S ... 35W-8N		27
Valley Forge Av. NE&SW 35W-8N,34W-8N		27
Valley Forge Ct. NW&SE 35W-8N		27
Vernon Ct. NW&SE..35W-7N		27
Village Ct. E&W ... 35W-7N		27
Virginia Dr. N&S ... 34W-8N		27
W. Harvard Cir. N&S 35W-8N		27
Walnut St. N&S......34W-8N		27
Water St. N&S 34W-8N,34W-9N		27
Wedgewood Dr. N&S 35W-8N		27
Weston Ct. N&S . 36W-8N		26
Whispering Ct. NW&SE 34W-8N		27
Williams Ct. E&W ...34W-7N		27
Willis St. E&W 34W-8N		27
Wilson Way E&W ...34W-9N		21
Woodbridge Cir. E&W 34W-7N		27
Woodbury St. N&S..34W-8N		27
Woodrow NW&SE .. 34W-9N		21
Yorkshire Ct. E&W .. 35W-8N		27

St. Charles

	Grid	Page
Adams Av. NE&SW 35W-2N,34W-2N		33
Adams Ct. NW&SE....34W-2N		33
Adare Ct. N&S.......34W-3N		33
Aintree Rd. NW&SE..34W-4N		33
Alice Ct. NW&SE 36W-2N		32
Allen Ln. E&W 35W-3N,34W-3N		33
Andover Av. NE&SW 34W-2N		33
Arrowhead Ln. N&S..35W-3N		33
Ash St. N&S 35W-2N		33
Auburn Ct. E&W .. 36W-3N		33
Avalon Ct. E&W 36W-2N		32
Banbury Av. NS&EW 34W-2N		33
Banbury Ct. N&S...34W-2N		33
Beatrice Av. E&W....34W-2N		33
Benham Ct. E&W....36W-3N		33
Bent Av. E&W ... 35W-3N		33
Blackhawk Ct. E&W..35W-3N		33
Blackhawk Tr. E&W..33W-4N		33
Bowman St. E&W .. 35W-2N		33
Bradley Cir. E&W...36W-3N		32
Brook St. E&W 36W-3N		32
Canidae Ct. NE&SW 34W-4N		33
Carrol Ct. N&S .. 34W-3N		32
Carrol Rd. N&S....36W-3N		32
Cedar Av. N&S . 35W-3N		33
Cedar St. NE&SW 36W-3N,35W-3N		32
Chandler Av. NS&EW 33W-2N		33
Charlemagne Ln. NE&SW 33W-4N		33
Charleston Dr. NE&SW 36W-2N		32
Chasse Cir. NS&EW 34W-4N		33
Chestnut Av. NE&SW 35W-3N		33
Concord Ct. NE&SW 34W-2N		33
Country Club Rd. E&W 35W-4N,33W-4N		33
Crabapple Ln. NE&SW 35W-3N,34W-4N		33
Cranbrook Av. NE&SW 34W-2N		33
Crane Rd. E&W 35W-4N		33
Cumberland Green Dr. NE&SW .. 34W-2N,33W-2N		33
Cumberland Pkwy. E&W 33W-2N		33
Cutler St. E&W 36W-2N,35W-2N		32
De Bruyne St. N&S .. 36W-3N		32
Dean St. NW&SE ..36W-3N		32
Deerfield Ct. NE&SW 34W-2N		33
Delnor Av. E&W 35W-3N		33
Derby Course NS&EW 34W-3N		33
Division St. E&W 34W-3N		33
Dover Ln. E&W 34W-3N		33
Dukane Dr. NW&SE..33W-3N		33
Dunham Pl. E&W 34W-3N,34W-4N		33
Dunham Rd. E&W 34W-3N,33W-4N		33
Edgewild Ct. E&W 34W-2N		33
Edward Av. E&W 34W-2N		33
Elm St. NE&SW 35W-2N		33
Elmtree Ct. NE&SW..34W-2N		32
Emery Ct. E&W 34W-2N		33
Essex Ct. E&W 34W-2N		32
Evergreen E&W 36W-2N		32
Fairfax Rd. E&W 34W-3N		33

	Grid	Page
Fairview Dr. NS&EW 36W-2N		32
Fairway Ct. NW&SE 35W-4N		33
Fellows St. E&W 36W-2N,35W-2N		32
Fern Av. E&W 34W-2N		33
Ferson Creek Rd. NE&SW 36W-3N		33
Fieldgate Dr. N&S .. 33W-3N		33
Forest Av. NE&SW .. 34W-2N		33
Forest Ridge Rd. NS&EW 34W-3N		33
Fox Chase Blvd. NS&EW 34W-3N,34W-4N		33
Fox Chase Cir. NE&SW 33W-4N		33
Fox Chase Ct. N&S..33W-4N		33
Fox Chase Dr. E&W 33W-4N		33
Fox Ct. E&W........35W-3N		33
Fox Hunt Ln. NE&SW 33W-4N		33
Foxfield Ct. N&S .. 33W-3N		33
Foxfield Dr. E&W .. 33W-3N		33
Foxfield Rd. E&W....33W-3N		33
Foxglade Ct. N&S .. 35W-2N		33
Foxhill Ct. NS&EW .. 34W-3N		33
Foxwood Ln. NS&EW 34W-4N		33
Fulton Av. E&W 34W-3N		33
Garden Hill Ln. E&W 34W-3N		33
Geneva Rd. NW&SE 35W-2N		33
Gray St. E&W 36W-2N,35W-2N		32
Green Willow Ln. NE&SW 35W-4N,35W-5N		33
Hampton Course Dr. NE&SW 34W-4N		33
Highgate Course NE&SW 34W-4N,33W-4N		33
Highgate Ct. E&W .. 34W-4N		33
Horne St. E&W 36W-2N,35W-2N		32
Howard St. E&W .. 36W-2N		32
Huntington Rd. NW&SE 34W-3N		33
Illinois Av. E&W 35W-3N,33W-3N		33
Illinois St. NE&SW .. 33W-3N		33
Independence Av. NS&EW 34W-2N		33
Independence Ct. NE&SW 34W-2N		33
Indian Way NW&SE..33W-4N		33
Indiana Av. NE&SW 35W-2N,34W-2N		33
Indiana St. E&W 36W-2N,35W-2N		32
Industrial Dr. NE&SW 34W-2N		33
Iroquois Av. E&W 35W-3N,34W-3N		33
Jackson Av. NW&SE 35W-2N		33
Jay Ln. N&S ... 34W-2N		33
Jeanette Av. E&W .. 34W-2N		33
Jefferson Av. NS&EW 35W-2N		33
Jeffery Ct. N&S....34W-3N		33
Jewel Av. E&W 34W-2N		33
Jobe Av. E&W ... 34W-2N		33
Johnor Av. E&W 34W-2N		33
Katharine St. N&S . 36W-2N		32
Kautz Rd. N&S 33W-2N,33W-3N		33
Kehoe Dr. NS&EW .. 36W-3N		32
Kellar Pl. NE&SW ... 35W-2N		33
Kensington Pl. E&W..35W-3N		33
Kildeer Ln. N&S 35W-4N		33
Kirk Ln. N&S 33W-2N		33
Kirk Rd. North N&S 33W-3N,33W-4N		33
Lakewood Cir. NS&EW 36W-4N		32
Lancaster Av. E&W .. 34W-2N		33
Larson Av. E&W 34W-2N,33W-2N		33
Lewis St. NW&SE .. 36W-3N		32
Liberty Av. N&S 36W-3N		32
Liberty Ct. NE&SW .. 34W-2N		33
Long Meadow Cir. NS&EW 34W-4N		33
Lucylle Av. N&S 34W-2N		33
Madison Av. E&W 35W-2N,34W-2N		33
Madison St. N&S....34W-2N		33
Main St. E&W 36W-3N,33W-3N		32
Manley Rd. NW&SE..36W-3N		32
Maple Av. NE&SW .. 35W-2N		33
Margaret Ct. N&S...36W-2N		32
Marie St. E&W 36W-2N		32
Marion Av. E&W 35W-3N		33
Mark St. E&W 36W-3N,35W-3N		32
McKinley St. E&W .. 35W-2N		33
Meadows Rd. NE&SW 35W-4N		33
Midway Ln. N&S .. 34W-2N		33
Milburn Ct. E&W .. 36W-2N		32
Mildred Av. N&S .. 34W-2N		33

	Grid	Page
Millington Way NS&EW 36W-3N		32
Monroe Av. NW&SE 35W-2N		33
Moore Av. E&W 34W-2N		33
Mosedale St. E&W 36W-2N,35W-2N		32
Muirfield Ct. E&W....33W-4N		33
Munhall Av. N&S 34W-2N,34W-3N		33
Nicholas Av. N&S .. 36W-3N		33
North Av. E&W 35W-3N,33W-3N		33
Oak St. NE&SW 34W-2N,33W-2N		33
Oakhill Ct. NE&SW .. 36W-4N		32
Ohio Av. E&W 35W-2N,33W-2N		33
Oxbow Ln. E&W 33W-4N		33
Park Av. NE&SW .. 35W-3N		33
Patricia Av. NS&EW .34W-2N		33
Permission Ct. NE&SW 35W-3N		33
Permission Dr. NE&SW 35W-3N,34W-4N		33
Pine St. N&S 35W-2N		33
Pleasant Av. NE&SW 34W-2N		33
Pomeroy Ct. NE&SW 34W-2N		33
Post Rd. N&S .. 34W-3N		33
Prairie St. NE&SW 36W-2N,35W-2N		32
Prestwick Ct. N&S .. 33W-5N		27
Private Rd. NW&SE .33W-3N		33
Production Dr. E&W 34W-3N,33W-3N		33
Railroad Pl. N&S 34W-2N		33
Redden Ct. NE&SW 36W-3N		32
Rita Av. E&W 34W-4N,33W-4N		33
River Valley Rd. NE&SW 35W-3N		33
Riverside Av. NW&SE 35W-3N		33
Ronzheimer Av. E&W 34W-2N		33
Roosevelt St. E&W .. 35W-2N		33
Royal & Ancient Dr. N&S 33W-5N		27
Royal Ashdown Ct. NE&SW 33W-4N		33
Royal Fox Ct. E&W .33W-4N		33
Royal Fox Dr. N&S .. 33W-4N		33
Royal Kings Ct. E&W 33W-4N		33
Royal Lytham Ct. NW&SE 33W-4N		33
Royal Lytham Dr. NE&SW 33W-4N		33
Royal Queens Ct. NW&SE 33W-4N		33
Royal St.Anne Ct. E&W 33W-5N		27
Royal St.Georges Ct. NW&SE 33W-4N		33
Royal St.James Ct. E&W 33W-4N		33
Royal Troon Ct. N&S 33W-5N		27
Royal Windyne Ct. E&W 33W-5N		27
Rt. 25 N&S 34W-2N,35W-4N		33
Rt. 31 N&S 35W-2N,35W-4N		33
Rt. 64 NW&SE 36W-3N,33W-3N		32
Ruth Av. NE&SW 34W-2N		33
Ryan St. N&S 36W-3N		32
Secretariat Ct. N&S..34W-3N		33
Sedgewick Cir. NW&SE 36W-3N		32
Shabbona Av. N&S . 35W-3N		33
Shires Ln. E&W 34W-4N		33
Shoreline Ct. NE&SW 36W-4N		32
Shoreline Dr. NS&EW 36W-4N		32
Smith Rd. E&W 33W-3N		33
South Av. NE&SW 35W-2N,34W-2N		33
Southgate Course NW&SE 34W-4N		33
Southgate Ct. NE&SW 34W-4N		33
Spring Av. NE&SW 35W-2N,34W-2N		33
Squire Ln. E&W 34W-4N,33W-4N		33
St. Andrews Ct. E&W 33W-5N		27
State Av. E&W 35W-3N		33
State St. E&W 35W-3N,34W-3N		32
Sterling Ct. N&S .. 34W-3N		33
Stern Av. NW&SE .. 34W-2N		33
Stetson Av. NE&SW 33W-2N		33
Stirrup Cup Ct. NW&SE 34W-4N		33
Stone Dr. N&S .. 35W-3N		33
Stonehedge Rd. NW&SE 35W-3N,34W-3N		33

	Grid	Page
Stonewood Dr. N&S 34W-4N		33
Swenson Av. NW&SE 33W-2N		33
Tall Grass Ct. N&S .. 33W-3N		33
Temple Garden Ct. N&S 34W-2N		33
Thornhill Farm Ln. E&W 35W-4N		33
Timbers Cir. NW&SE 36W-3N		32
Timbers Ct. N&S 36W-3N		32
Timbers Pl. NW&SE .36W-3N		32
Timbers Trail E&W .. 36W-3N		32
Turnberry Rd. N&S . 33W-4N		33
Tyler Rd. N&S 34W-2N,34W-3N		33
Union Av. E&W......34W-2N		33
Van Buren Av. NW&SE 35W-2N		33
Via Veneto Dr. N&S .34W-3N		33
Walnut Av. NE&SW . 35W-3N		33
Walnut Dr. N&S .. 36W-3N		32
Walnut St. N&S 36W-2N,35W-2N		33
Washington Av. NE&SW 35W-2N		33
Waverly Ct. E&W....34W-3N		33
Weber Av. N&S....33W-2N		33
Wessel Ct. E&W .. 36W-2N		32
Westfield Dr. NE&SW 36W-2N		32
White Oak Cir. NS&EW 35W-2N		33
Whittington Course NS&EW 34W-4N		33
Wildrose Springs Dr. NE&SW 36W-4N		32
Williams Av. NW&SE 34W-2N		33
Wing Av. E&W 35W-3N,34W-3NW		33
Wing Ln. NS&EW .. 34W-3N		33
1st Av. NW&SE 35W-2N,35W-3N		33
1st St. NW&SE 35W-2N,35W-3N		33
2nd Av. NW&SE 35W-2N,35W-3N		33
2nd Pl. N&S 34W-2N		33
2nd St. NW&SE 35W-2N,35W-3N		33
3rd Av. NW&SE 35W-2N,35W-3N		33
3rd Pl. N&S .. 34W-2N		33
3rd St. NW&SE 35W-2N,35W-3N		33
4th Av. NW&SE 35W-2N,35W-3N		33
4th Pl. N&S 34W-2N		33
4th St. N&S 35W-2N,35W-3N		33
5th Av. NW&SE 35W-2N,35W-3N		33
5th Pl. N&S 34W-2N		33
5th St. NW&SE 35W-2N,35W-3N		33
6th Av. NW&SE 35W-2N,35W-3N		33
6th St. NW&SE 36W-2N,35W-2N		32
7th Av. NW&SE 34W-2N,35W-3N		33
7th Ct. NW&SE .. 35W-3N		33
7th St. NW&SE 36W-3N,35W-2N		32
8th Av. NW&SE ... 35W-3N		33
8th Ct. N&S 36W-2N		32
8th St. N&S 36W-2N		32
9th Av. NW&SE 34W-2N,35W-2N		33
9th St. N&S 36W-2N,36W-3N		32
10th Av. NW&SE 35W-2N,35W-3N		33
10th Ct. N&S 36W-2N		32
10th St. N&S........36W-2N		32
11th Av. NW&SE 34W-2N,35W-2N		33
11th St. N&S 36W-2N,36W-3N		32
12th Av. NW&SE 34W-2N		33
12th St. N&S 36W-2N,36W-3N		32
13th Av. NW&SE 34W-3N,34W-2N		33
13th St. N&S .. 36W-2N		32
14th Av. NW&SE .. 34W-2N		33
14th Ct. N&S 36W-2N		32
14th St. NS&EW .. 36W-2N		32
15th Ct. N&S .. 36W-2N		32
15th St. N&S 36W-2N,36W-3N		32
16th St. N&S .. 36W-2N		32
17th St. N&S 36W-2N,36W-3N		32
18th St. N&S .. 36W-2N		32
19th St. N&S .. 36W-2N		32
37th Av. NE&SW .. 33W-3N		33
38th Av. NE&SW....33W-3N		33

St. Charles Township

	Grid	Page
Adele Ln. E&W 38W-7N		26
Aldrea Ct. NE&SW .. 36W-7N		26
Amherst Ct. NE&SW 38W-4N		32
Aurora Av. N&S . 35W-6N		27
Babson Ln. N&S 37W-3N,37W-4N		32
Baker Ln. N&S . 37W-5N		26
Barb Hill Dr. NS&EW 38W-7N		26
Barbara Ann Dr. N&S 37W-3N		32
Barton Dr. E&W .. 36W-6N		26
Bernice Dr. NS&EW .38W-3N		32
Bittersweet Rd. N&S 37W-3N		32
Blackhawk Dr. NW&SE 35W-7N		27
Bluestem Ct. N&S .. 38W-7N		26
Bluff Dr. E&W 35W-5N		27
Bluff Dr. N. NE&SW 35W-5N		27
Bluff Dr. S. N&S .. 35W-5N		27
Bolcum Rd. E&W 38W-6N,37W-6N		26
Bonfield Av. E&W....35W-6N		27
Bonnie Ct. E&W .. 38W-3N		32
Bonnie Dr. NW&SE 38W-3N,38W-4N		32
Bonnie Ln. N&S .. 38W-3N		32
Bonnie St. E&W .. 33W-2N		33
Brewster Creek Ct. NW&SE 34W-7N		27
Bricker Rd. E&W .. 37W-2N		32
Bristol Ct. N&S 36W-6N		26
Bristol Rd. E&W .. 36W-6N		26
Brookwood Rd. E&W 35W-5N		27
Burr Oak Ln. NW&SE 38W-6N		26
Burr Rd. N&S 38W-4N,38W-6N		32
Burrside Ln. SW&NE 38W-6N		26
Carl Lee Rd. N&S....34W-6N		27
Cary St. NE&SW 35W-6N,35W-7N		27
Catalpa Av. E&W .. 35W-6N		27
Cedar Rock Av. NW&SE 35W-6N		27
Center Dr. E&W .. 34W-7N		27
Chickasaw Ct. E&W 38W-7N		26
Cibis Rd. NE&SW .. 38W-3N		32
Circle Dr. NE&SW .. 34W-6N		27
Cloverfield Dr. E&W .38W-7N		26
Cloverfield Dr. E&W .38W-7N		26
Clubhouse Dr. E&W .38W-5N		26
Clyde Ct. NE&SW 35W-6N,34W-6N		27
Clyde Park Way E&W 34W-6N		27
Colley Dr. E&W 34W-6N		27
Colombine East NE&SW 38W-7N		26
Colombine West NE&SW 38W-7N		26
Correl Ct. N&S 38W-7N		26
Courier Av. E&W 34W-6N		27
Crane Ln. E&W .. 37W-4N		32
Crane Rd. NE&SW 37W-7N,36W-4N		26
Creekside Dr. NE&SW 38W-6N,37W-6N		26
Dean St. NW&SE 38W-4N,36W-3N		32
Decatur Av. N&S .. 34W-6N		27
Deerpath Way NW&SE 36W-5N		26
Denny Dr. NW&SE .. 33W-2N		33
Detroit St. E&W .. 34W-6N		27
Dogwood Ln. NW&SE 38W-7N		26
Dorchester Rd. E&W 36W-5N		26
Dover Hill Rd. NW&SE 38W-4N,36W-5N		32
E. Ridgewood Dr. N&S 36W-5N,36W-6N		26
Eagle Ct. NE&SW .. 38W-5N		26
Eagle's Nest Ct. E&W 38W-7N		26
East Dr. N&S........34W-7N		27
Elder Av. E&W .. 35W-6N		27
Elgin Av. N&S 35W-6N,35W-7N		27
Elgin St. NE&SW 35W-6N,35W-7N		27
Elliott Ct. N&S 38W-4N		32
Elmwood Av. E&W .. 35W-6N		27
Emily Ln. NW&SE....34W-6N		27
Essex Av. N&S 35W-6N		27
Evenswood Ln. NW&SE 38W-4N		32
Fairview Av. E&W .. 35W-6N		27
Fairway Dr. NS&EW .38W-5N		26
Falcons Tr. N&S 38W-7N,38W-8N		26
Fence Rail Ct. NW&SE 35W-5N		26
Ferson Creek Rd. NS&EW 36W-3N,36W-4N		32

	Grid	Page
Ferson Woods Dr. NS&EW	38W-6N	26
Fieldcrest Dr. N&S	36W-5N,35W-5N	26
Fieldstone Ct. NE&SW	36W-4N	32
Five Island Rd. E&W	34W-6N	27
Florence Ln. N&S	35W-6N	27
Foley Ct. N&S	36W-6N	26
Fox River Av. N&S	34W-6N	27
Fox River Dr. E&W	35W-6N	27
Foxbluff Ct. NE&SW	36W-5N	26
Foxbluff Dr. E&W	36W-5N	26
Foxglove Ct. N&S	38W-5N	26
Foxwood Ln. E&W	38W-6N	26
Geneva Av. N&S	34W-6N	27
Glad Ln. N&S	38W-3N	32
Glenoak Ln. NE&SW	38W-5N	26
Glenview Dr. NS&EW	38W-6N	26
Golfview Ct. E&W	38W-5N	26
Goodrich Av. N&S	35W-6N	27
Grandma's Ln. NS&EW	37W-4N	32
Greenlawn Av. NW&SE	35W-6N	27
Greenview Ct. NW&SE	38W-5N	26
Grey Barn Rd. NS&EW	37W-5N	26
Grove Av. NE&SW	35W-5N	27
Gustavus Ln. E&W	37W-4N	32
Halsted Av. N&S	35W-6N	27
Hampton Dr. N&S	36W-7N	26
Hawkins Ln. E&W	36W-5N	26
Hawthorn Dr. N&S	36W-3N	32
Henricksen Rd. N&S	38W-6N	26
Heritage Ct. N&S	38W-6N	26
Hickory Ct. E&W	36W-7N	26
Hickory Ln. NW&SE	34W-6N,34W-7N	27
Highview Ct. E&W	35W-5N	27
Hill Ct. NW&SE	34W-7N	27
Hillcrest Av. E&W	35W-6N	27
Hillside Dr. N&S	35W-6N	27
Hilltop Dr. E&W	38W-3N	32
Hunter's Gate Rd. NS&EW	36W-5N	26
Illinois St. NE&SW	34W-6N	27
Indian Mound Rd. E&W	36W-3N	32
Iowa St. E&W	34W-6N	27
Irving Av. N&S	35W-6N,35W-7N	27
Irwin Av. N&S	35W-6N	27
Jackson Av. N&S	35W-6N,35W-7N	27
James Dr. E&W	35W-6N,34W-6N	27
Joan Ct. E&W	38W-3N	32
Joan Dr. NS&EW	38W-3N	32
Joseph St. N&S	35W-7N	27
Kane Av. E&W	35W-6N	27
Kautz Rd. N&S	33W-2N	33
Kearns St. N&S	35W-7N	27
Kimberly Dr. E&W	36W-7N	26
Knoll Creek Dr. NS&EW	37W-4N	32
Lake Charlotte Ct. NW&SE	38W-3N	32
Lambert Av. E&W	35W-5N	27
Lancaster Rd. NW&SE	36W-7N	26
Landing Dr. N&S	34W-6N,34W-7N	27
Leola Ln. NE&SW	37W-5N	26
Lilac Ln. NS&EW	38W-7N	26
Lincoln Dr. N&S	38W-3N	32
Lincoln St. NW&SE	34W-6N	27
Lisa Ln. E&W	35W-6N	27
Littlenell Ct. N&S	38W-7N	26
Longridge Rd. N&S	38W-7N	26
Longview Dr. NW&SE	36W-5N	26
Lookout Ln. NE&SW	38W-5N	26
Lynn Dr. N&S	38W-3N	32
Mallard Lake Rd. N&S	38W-6N	26
Maple Av. N&S	36W-6N	26
Maple Ct. NW&SE	37W-5N	26
Maple Ln. N&S	35W-6N	27
Marquerite St. E&W	36W-3N	32
Marre Dr. NW&SE	34W-7N	27
Mary Ln. E. E&SW	38W-3N	32
Mary Ln. W. N&S	38W-3N	32
May Ln. N&S	34W-6N	27
McKay Dr. NW&SE	36W-6N	26
McLean Blvd. N&S	35W-7N	26
Mck Point Ct. E&W	37W-6N	26
Meadow Dr. NE&SW	35W-5N	27
Meadows Ct. N&S	36W-5N	26
Michael Ct. NE&SW	36W-7N	26
Middlecreek Ln. NS&EW	36W-4N,36W-5N	32
Miller Ln. NW&SE	38W-6N	26
Mills Ct. E&W	37W-4N	32
Murray Ct. NE&SW	38W-6N	26
Murray Rd. NS&EW	38W-6N	26
Myles Rd. N&S	36W-5N	26
North Dr. E&W	34W-6N	27
Northern Ct. NE&SW	38W-7N	26
Northfield Rd. N&S	37W-5N	26
Oak Ct. N&S	36W-5N	26
Oak Dr. E&W	35W-7N	27
Oak Dr. NE&SW	38W-7N	26
Oak Ln. E&W	35W-5N	27
Oak Rd. NS&EW	36W-5N	26
Oak Ridge Ln. E&W	36W-3N	32
Oak Run Ct. NE&SW	36W-6N	26
Oakwood Dr. N&S	35W-7N	27
Oakwood Dr. NE&SW	38W-3N	32
Ohio St. E&W	34W-6N	27
Old Farm Rd. N&S	36W-4N	32
Old Homestead Rd. N&S	38W-6N	26
Old Quarry Rd. NE&SW	36W-4N	32
Pakan Ln. E&W	34W-7N	27
Park Av. NW&SE	35W-6N	27
Park Ln. E&W	35W-5N,35W-6N	27
Park Pl. E&W	34W-6N	27
Pearson Rd. E&W	35W-5N	27
Peck Rd. N&S	37W-3N	32
Pheasant Run Dr. N&S	38W-4N	32
Pine Rd. E&W	38W-7N	26
Pinelands Rd. E&W	35W-5N	26
Plymouth Ct. N&S	36W-7N	26
Prairie Dr. N&S	38W-3N	32
Private Rd. E&W	33W-4N	33
Promontory Ct. NE&SW	38W-6N,37W-6N	26
Pvt. Rd. N&S	35W-6N	27
Ramona Ln. NW&SE	34W-6N	27
Randall Rd. N&S	37W-2N,37W-7N	32
Red Bay Ct. NW&SE	38W-7N	26
Red Gate Rd. NW&SE	37W-5N,35W-5N	26
Red Haw Ln. E&W	36W-3N	32
Ridge Dr. N&S	38W-3N	32
Ridgewood Dr. NW&SE	36W-6N	26
River Grange Rd. E&W	36W-6N,35W-6N	26
River Rd. N&S	35W-6N	26
River Valley Dr. E&W	34W-7N	27
River View Ct. E&W	36W-7N	26
Riverside Dr. NW&SE	35W-6N,35W-7N	27
Robert Rd. E&W	36W-6N	27
Rockwell Av. E&W	35W-5N	27
Ronsu Ln. NS&EW	37W-5N	26
Roosevelt Av. N&S	34W-6N,34W-7N	27
Rose Av. E&W	35W-7N	27
Rt. 25 NE&SW	35W-5N,33W-7N	27
Rt. 31 N&S	36W-3N,34W-7N	32
Rt. 38 E&W	38W-2N,37W-2N	32
Salina Av. N&S	35W-6N	27
Scott Av. N&S	35W-7N	27
Second Av. N&S	36W-3N	32
Seminary Rd. N&S	34W-6N	27
Shady Ln. NW&SE	35W-6N	27
Sidney Ct. E&W	38W-7N	26
Silver Glen Ct. N&S	36W-6N	26
Silver Glen Rd. E&W	37W-6N,36W-6N	26
Silver Ridge Dr. E&W	36W-6N	26
Skyline Dr. NE&SW	34W-6N	27
Smith Rd. N&S	33W-3N	33
South Dr. E&W	34W-6N,34W-7N	27
South Ln. E&W	35W-6N	26
South St. NE&SW	35W-7N	27
Springfield St. E&W	34W-6N	27
Stanton Dr. E&W	35W-6N,33W-5N	27
Steven Glen Rd. N&S	38W-7N	26
Stevens Rd. E&W	38W-7N	26
Stonebridge Ln. NW&SE	36W-5N	26
Stoneleat Rd. E&W	36W-4N	32
Summit St. E&W	35W-6N	27
Sunnyside Av. E&W	35W-6N	27
Sunrise Ct. NW&SE	36W-3N	32
Sunset Dr. E&W	35W-6N	27
Sunset Ct. E&W	38W-3N	32
Sycamore Rd. N&S	38W-7N,34W-6N	26
Sylvia Ln. N&S	36W-3N	32
Third Av. N&S	36W-3N	32
Thompson Av. N&S	35W-6N	27
Thornapple Rd. N&S	36W-3N,36W-4N	32
Thorne Tree Dr. NE&SW	36W-3N,36W-4N	32
Thornly Rd. N&S	36W-3N,36W-4N	32
Timber Ridge Dr. NW&SE	36W-5N	26
Toms Trail Rd. E&W	38W-3N	32
Toni St. E&W	33W-2N	33
Treetop Ln. NE&SW	36W-3N	32
Tucker Av. N&S	35W-6N	27
Tuscola Rd. N&S	34W-6N,34W-7N	27
Twin Silos Dr. NW&SE	37W-6N	26
Valley Dr. N&S	34W-6N	27
Vaupon Ct. E&W	38W-7N	26
Villa Marie Rd. E&W	36W-5N,34W-6N	27
Virginia Ct. N&S	34W-6N	27
W. Ridgewood Ln. NS&EW	36W-6N	26
Wabash Av. N&S	34W-6N	27
Wagon Tire Rd. N&S	38W-7N	26
Watseka Av. N&S	34W-6N	27
Weber Dr. NE&SW	35W-5N,35W-6N	27
West Dr. N&S	34W-7N	27
Cir.	38W-5N	26
Westwood Ln. N&S	36W-5N,36W-6N	26
Whispering Tr. NS&EW	36W-7N	26
Whispering Tr. Ct. NE&SW	36W-7N	26
Wild Rose Ln. NW&SE	36W-3N	32
Wild Rose Rd. NS&EW	36W-4N	32
Willow Dr. N&S	35W-6N	27
Wilton Craft Rd. NE&SW	36W-5N	26
Windsor Dr. NE&SW	36W-7N	26
Wood Hill Ln. NE&SW	36W-6N	26
Wood Ln. E&W	35W-6N	27
Woodcliff Dr. N&S	34W-7N	27
Woodgate Rd. NE&SW	37W-5N	26
Woodland Ct. E&W	36W-5N	26
Woodlawn Av. N&S	35W-6N,35W-7N	27
Woodside Ln. E&W	38W-6N	26
Woodview Ct. NW&SE	36W-6N	26
Woodview Ln. N&S	38W-6N	26
Wyoming Av. N&S	36W-6N	27
17th Av. N&S	36W-3N	32

Sugar Grove

	Grid	Page
Annettes Ln. NS&EW	44W-7S	48
Bastian Dr. E&W	43W-7S,42W-7S	48
Bliss Rd. N&S	41W-5S	43
Bridgestone Ct. N&S	41W-5S	43
Brock Dr. N&S	45W-7S	47
Bughtail Lane E&W	44W-7S	48
Calkins Dr. E&W	43W-7S,42W-7S	48
Cardinal Crest Dr. NW&SE	42W-5S	42
Carriage Hill Ct. E&W	43W-6S	42
Carriage Hill Ln. N&S	43W-6S	42
Cornwall Cir. N&S	41W-5S	43
Cross St. NW&SE	42W-7S	48
Cyndi Dr. E&W	45W-7S	47
Fays Ln. E&W	44W-7S	48
First St. E&W	42W-7S	48
Gordon Sq. NW&SE	41W-4S	43
Granart Rd. NE&SW	45W-7S	47
Grove St. N&S	42W-7S	48
Hampshire Ct. NE&SW	41W-4S	43
Jody Ln. N&S	42W-6S	42
Jonathan Dr. N&S	45W-7S	47
Joy Ct. N&S	42W-7S	48
Joy St. E&W	42W-7S	48
Kimbery Ct. E&W	45W-7S	47
Kingman Cir. NW&SE	41W-5S	43
Knightsbridge Dr. NS&EW	41W-5S	43
Litchfield Ct. NE&SW	41W-5S	43
Main St. N&S	42W-7S	48
Maple St. E&W	42W-7S	48
Marybeth Ln. E&W	45W-7S	47
McCannon St. N&S	44W-7S	48
Meadowlark Run NE&SW	42W-5S	42
Meadows Ct. N&S	42W-7S	48
Meadows Dr. E&W	42W-7S	48
Merrill New Rd. N&S	41W-5S,41W-4S	43
Molly Dr. N&S	45W-7S	47
Monna St. E&W	42W-7S	48
Neil Rd. E&W	42W-7S	48
Orleans Ct. N&S	41W-5S	43
Patricia Ln. N&S	45W-7S	47
Queens Gate Cir. NS&EW	41W-5S,41W-4S	43
Richard St. N&S	42W-7S	48
Rt. 30 E&W	45W-7S	47
Rt. 47 N&S	42W-7S	48
Snow St. E&W	43W-7S,42W-7S	48
Stamford Ct. NW&SE	41W-4S	43
Stanley Rd. N&S	42W-7S	48
Stonehill Cir. N&S	41W-5S	43
Terry Dr. E&W	42W-7S	48
Tom St. E&W	45W-7S	47
Tudor Ct. E&W	41W-4S	43
Washford Pl. NE&SW	41W-4S	43
West Park Dr. NW&SE	45W-6S	41
Wheeler Rd. N&S	43W-6S	42
Whipporwill Ln. E&W	42W-5S	42
Windstone Ln. NW&SE	41W-5S	43
Woodbury Cir. E&W	41W-5S	43
Woodthrush Ln. NE&SW	42W-5S	42
Yolane Ct. E&W	42W-7S	48
Yolane Dr. E&W	43W-7S,42W-7S	48

Sugar Grove Township

	Grid	Page
Ashe Rd. NE&SW	44W-9S	48
Barnes Rd. N&S	39W-8S,39W-7S	49
Baseline Rd. E&W	44W-9S,39W-9S	48
Bertram Rd. N&S	41W-9S	49
Blackberry Rd. N&S	39W-9S	49
Bliss Rd. NE&SW	42W-6S,41W-4S	42
Bluegrass Dr. E&W	39W-9S	49
Creekview Rd. NE&SW	39W-9S	49
Deerpath Rd. NE&SW	39W-6S,39W-4S	43
Denny Rd. N&S	41W-4S,40W-4S	43
Densmore Rd. N&S	40W-6S	43
Dugan Rd. N&S	44W-9S,44W-5S	48
East-West Tollway NW&SE	41W-4S	43
Finley Rd. NE&SW	44W-4S	42
Galena Blvd. NW&SE	42W-6S,39W-6S	42
Gordon Rd. N&S	40W-7S,40W-6S	49
Guthrie Ln. E&W	43W-8S	48
Hankes Rd. NW&SE	40W-6S,39W-6S	43
Harter Rd. NW&SE	44W-4S,42W-5S	42
Hazelcrest Dr. NW&SE	44W-4S	42
Healy Rd. NW&SE	41W-4S,40W-4S	43
Jericho Rd. NE&SW	44W-9S,39W-8S	48
Ka-de-ka Rd. E&W	42W-5S	42
Lyndale Rd. N&S	39W-9S	49
May Rd. NW&SE	42W-5S,41W-4S	42
McCannon Rd. N&S	44W-8S	48
Merrill Rd. NW&SE	44W-4S,41W-5S	42
Midfield Dr. N&S	39W-7S	49
Mighell Rd. N&S	43W-9S,43W-8S	48
Norris Rd. N&S	40W-5S,40W-4S	43
Old Oak Rd. NW&SE	43W-4S	42
Orchard Rd. N&S	39W-9S,39W-9S	49
Pembroke Ct. NE&SW	44W-4S	42
Prairie St. NE&SW	44W-8S,39W-7S	48
Rt. 47 N&S	42W-9S	48
Rt. 56 NE&SW	42W-6S	42
Scott Rd. E&W	44W-4S,43W-4S	42
Shagbark Ln. NW&SE	39W-6S	43
Star Ln. E&W	42W-9S	48
Sunnyside Dr. E&W	39W-9S	49
Thornapple Tree Rd. NW&SE	43W-4S	42
U.S. Rt. 30 N&S	42W-8S	48
West Ridge Dr. N&S	39W-6S	43
Wheeler Rd. E&W	44W-5S,42W-5S	42
Wildwood Dr. NS&EW	40W-6S	43
Wittshire Ln. N&S	44W-4S	42

Virgil Township

	Grid	Page
Beith Rd. E&W	50W-3N,45W-3N	28
Butts Rd. NS&EW	48W-7N	22
Country Ln. E&W	46W-5N	23
County Line Rd. N&S	50W-2N,50W-7N	28
Fabris Rd. N&S	49W-5N,49W-6N	22
Francis Rd. N&S	46W-2N,46W-3N	29
Freeland Rd. N&S	45W-2N,45W-3N	29
Hickory Dr. E&W	46W-2N	29
Howard Rd. N&S	49W-2N,49W-3N	28
Indian Creek Trail NW&SE	47W-6N,45W-5N	23
Landrose Ln. N&S	46W-2N	29
Lees Rd. E&W	46W-7N,45W-7N	23
McGough Rd. NW&SE	50W-7N,49W-5N	22
McNulty Rd. N&S	48W-3N,47W-3N	28
Meadows St. N&S	46W-5N	23
Meredith Rd. N&S	47W-2N,47W-6N	28
Old State Rd. NW&SE	50W-7N	22
Peplow Rd. N&S	48W-6N,47W-7N	22
Peterson Rd. N&S	50W-6N,49W-6N	22
Pin Oaks Ln. N&S	48W-4N	28
Prairie St. N&S	46W-5N	23
Ramm Rd. E&W	50W-7N,49W-7N	22
Ramm Wood Dr. N&S	50W-7N	22
Read Rd. NS&EW	45W-6N	23
Root Rd. N&S	50W-3N	28
Rt. 38 E&W	50W-2N,46W-2N	28
Rt. 64 NE&SW	50W-7N,45W-5N	22
Saddlewood Dr. NE&SW	45W-2N	29
Sauber Rd. N&S	47W-6N,47W-7N	23
Snyder Rd. N&S	49W-6N,49W-7N	22
Sunset St. N&S	46W-5N	23
Thatcher Rd. N&S	49W-3N,48W-4N	28
Warford Rd. N&S	46W-7N	23
Welter Rd. NW&SE	49W-5N,45W-4N	22
Winters Rd. N&S	50W-5N,47W-5N	22

Wayne

	Grid	Page
Army Trail Rd. E&W	35W-5N,33W-5N	27
Ashely Ln. E&W	33W-6N	27
Brewster Creek Cir. NE&SW	33W-6N,33W-7N	27
Country Club Rd. E&W	34W-4N,33W-4N	33
Curling Pond Rd. NS&EW	33W-5N	27
Dunham Cir. NE&SW	33W-5N	27
Dunham Ct. NW&SE	33W-5N	27
Dunham Rd. NE&SW	33W-4N,33W-6N	33
Dunham Trail Rd. NW&SE	33W-5N	27
Lamplight Tr. NW&SE	33W-7N	27
Mare Barn Rd. E&W	33W-5N	27
Nelson Dr. N&S	34W-4N	33
Robin Ln. NS&EW	34W-5N	27
Rocheford Ln. NW&SE	33W-6N	27
Saw Mill Tr. N&S	34W-4N	33
Surrey Rd. NE&SW	33W-5N,33W-6N	27
Thorncroft Dr. NW&SE	33W-5N	27
White Thorn Rd. E&W	34W-5N,33W-5N	27
Woodland Dr. N&S	33W-5N	27

West Dundee

	Grid	Page
Aberdeen Ln. N&S	35W-16N	14
Barber Ct. NW&SE	35W-16N	14
Brewer Ct. N&S	35W-16N	14
Browning Av. E&W	34W-15N	15
Castle Rock Ct. NW&SE	35W-16N	14
Cavalier Ct. N&S	34W-15N	15

	Grid	Page
Chadwick Ln. N&S		
..........35W-16N,35W-17N		14
Dunning Av. E&W .. 34W-15N		15
Edinburgh Ln. NS&EW		
.... 35W-16N		14
Edwards Av. NS&EW		
.... 34W-15N		15
Eicher Dr. NW&SE..34W-15N		15
Fay Av. E&W 33W-15N		15
Fox Av. E&W 33W-15N		15
Galloway Ct. N&S .. 35W-16N		14
Garrison E&W......34W-15N		15
Geneva St. E&W .. 34W-16N		15
Glenmoor Dr. E&W		
.... 35W-16N		14
Grant Av. N&S 35W-16N		14
Greencastle Ct. N&S		
.... 35W-16N		14
Hamilton Dr. NW&SE		
.... 35W-16N		14
Hawley Av. E&W .. 33W-15N		15
Highland Av. N&S .. 34W-16N		15
Hillcrest Ct. E&W .. 34W-16N		15
Kane St. E&W......34W-16N		15
Kitridge Dr. NW&SE		
.... 35W-16N		14
Knowlton Dr. NS&EW		
.... 35W-16N		14
Liberty St. E&W 34W-16N		15
Lincoln Av. NW&SE		
.... 34W-16N		15
Lindsay Ln. N&S .. 35W-16N		14
Lisa Rd. E&W......34W-15N		15
Maiden Ln. E&W .. 34W-15N		15
Main St. E&W......34W-16N		15
Market Loop Rd. N&S		
.... 34W-15N		15
McConnoiche Ct. N&S		
.... 35W-16N		14
McGregor Ct. NE&SW		
.... 35W-16N		14
Oregon Av. E&W .. 34W-16N		15
Pember Cir. NS&EW		
.... 35W-16N		14
Preston Ln. NE&SW		
.... 35W-16N		14
Prestwick Ct. NW&SE		
.... 35W-16N		14
Riverview St. E&W 34W-15N		15
Cir................34W-15N		15
Royal Ln. E&W 34W-16N		15
Ryan Ct. E&W 34W-15N		15
Ryan Ln. N&S......34W-15N		15
Short E&W 33W-15N		15
Smalley Ct. E&W .. 35W-16N		14
South E&W 34W-15N		15
Spaulding Av. NS&EW		
.... 35W-16N		14
Spaulding Ct. NW&SE		
.... 35W-16N		14
Cir................35W-16N		14
Sterling Ct. NE&SW		
.... 35W-16N		14
Stewart Ln. E&W .. 35W-16N		14
Strom Dr. E&W 34W-15N		15
Tartans Ct. E&W .. 35W-16N		14
Tartans Dr. NW&SE		
.... 35W-16N		14
Thatcher Tr. NW&SE		
.... 35W-16N		14
Tritram Ct. NE&SW		
.... 35W-16N		14
View St. N&S 33W-15N		15
Village Quarter E&W		
.... 34W-15N		15
Washington St. E&W		
.... 34W-16N		15
Water Tower Rd. E&W		
.... 35W-16N		14
Waterbury Ct. NW&SE		
.... 35W-16N		14
Westley Ln. NS&EW		
.... 35W-17N		8
Willow Ln. NW&SE		
.... 34W-15N		15
1st St. N&S........34W-15N		15
2nd St. N&S .. 34W-15N		15
3rd St. N&S.......34W-15N		15
4th St. N&S........34W-16N		15
5th Ct. NW&SE 34W-15N		15
5th St. N&S........34W-15N		15
6th St. N&S........34W-15N		15
7th Blvd. NW&SE .. 34W-15N		15
7th St. N&S........34W-16N		15

Lake
County

StreetFinder®

page	**Contents**

A	Using the StreetFinder
B	PageFinder Map
C	Vicinity Map
1	Contents
2	Quick Reference Guide
4	StreetFinder Maps
53	StreetFinder Index

Photo credit: Shores of Lake Michigan / Lake County Chamber of Commerce

PageFinder™ Map patent pending.

Information included in this publication has been checked for accuracy prior to publication. Since changes do occur, the publisher cannot be responsible for any variations from the information printed.

Lake County Municipal Offices

	Location	Page
Antioch Village Hall	23W-42N	6
874 Main St; 395-1000		
Bannockburn Village Hall	11W-22N	50
2165 Telegraph Rd; 945-6080		
Deerfield Town Hall	10W-21N	50
850 Waukegan Rd; 945-5000		
Fox Lake Village Hall	26W-36N	13
305 Rt 59; 587-2151		
Grayslake Village Hall	20W-33N	23
164 Hawley St; 223-8515		
Gurnee Village Hall	14W-35N	25
4573 Grand Ave; 673-7650		
Hainesville Village Hall	22W-33N	22
221 E Pine View; 223-2032		
Highland Park City Hall	8W-22N	51
1707 St Johns; 432-0800		
Highwood City Hall	9W-23N	51
17 Highwood Ave; 432-1924		
Island Lake Village Hall	29W-28N	28
Rt 176, PO Box 41; 526-8764		
Lake Bluff City Hall	10W-28N	35
40 E Center; 234-0774		
Lake Forest City Hall	10W-26N	43
220 E Deerpath; 234-2600		
Lake Villa Village Hall	22W-38N	14
65 Cedar Ave; 356-6100		
Lake Zurich Village Hall	22W-22N	46
61 W Main St; 438-5141		
Libertyville City Hall	16W-29N	33
200 E Cook; 362-2430		
Lindenhurst Village Hall	20W-37N	15
2301 Sand Lake Rd; 356-8252		
Mundelein Village Hall	18W-28N	32
440 E Hawley; 566-7070		
North Chicago City Hall	11W-31N	35
1850 S Lewis Ave; 578-7750		
Riverwoods Village Hall	13W-20N	44
2300 Portwine Rd; 945-3990		
Round Lake Village Hall	23W-33N	22
322 W Railroad Ave; 546-5400		
Round Lake Beach Village Hall	23W-34N	22
1212 N Cedar Lake Rd; 546-3466		
Round Lake Heights Village Hall	23W-35N	22
629 W Pontiac Ct; 546-1206		
Tower Lakes Village Hall	25W-25N	37
115 South Dr; 526-2226		
Vernon Hills Village Hall	17W-24N	40
290 Evergreen Dr; 367-3700		
Wauconda Hills Village Hall	26W-27N	37
101 N. Main St; 526-8786		
Waukegan Mun Bldg	10W-34N	27
106 N Utica St; 360-9000		
Winthrop Harbor City Hall	9W42N	11
830 Sheridan Rd; 872-3846		
Zion City Hall	10W-40N	11
2828 Sheridan Rd; 872-4546		

Cemeteries

	Location	Page
Angolian Cem	22W-37N	14
Ascension Cem	14W-30N	33
Avon Center Cem	21W-35N	22,23
Benton Greenwood Cem	12W-39N	18
Cemetery	25W-20N	45
Cemetery	25,26W-27N	37
Diamond Lake Cem	18W-25N	40
Druce Cem	19W-34N	23
Fort Hill Cem	24W-32N	22
Fox Lake Cem	24W-36N	13,14
Grant Cem	26W-33N	21
Grass Lake Cem	26W-39N	13
Home Oaks Cem	21W-39N	14,15
Knopf Cem	18W-20N	47,48
Lakeside Cem	16W-29N	33
Lake Forest Cem	10W-27N	43
Lake Mound Cem	9W-39N	19
Lake Zurich Cem	23W-23N	46
Mooney Cem	9W-21N	50,51
Mount Oliver Mem Cem	12W-41N	10
Mount Rest Cem	16W-42N	8
Naval Cem	10W-30N	35
Nicholas Dowden Mem Cem	17W-28N	32
North Shore Garden of Mem	12W-32N	26
Oakdale Cem	13W-42N	9
Oakwood Cem	10W-33N	27
Orvis Cem	29W-41N	4
Pineview Cem	10W-38N	19
St Marys Cem	10W-27N	43
St Marys Cem	10W-33N	27
St Patrick Cem	12W-24N	42
St Patricks Cem	16W-40N	8,9
Sand Lake Cem	21W-37N	14,15
Vernon Cem	15W-23N	49
Warren Cem	16W-35N	25
White Cem	26W-21N	45

Colleges & Universities

	Location	Page
Barat Col	10W-25N	43
Benedictine Col	17W-28N	32
College of Lake County	19W-34N	23
Lake Forest Univ	10W-26N	43
St Marys of the Lake Sem	18W-28N	32
Trinity Sem	12W-22N	50

Forest Preserves

	Location	Page
Capt Daniel Wright FP	15,16W-24N	41
Countryside FP	20W-27N	39
Cuba Marsh FP	24W-21N	45,46
Forest Preserve	16W-30,31N	33
	15W-39,41N	9
Greenbelt FP	12W-32N	26
Gurnee Woods	14W-36N	17

	Location	Page
Lake County FP	13W-20N	49
Lake Forest Nature Pres	9W-25N	43
Lakewood FP	24W-26,27N	37,38
MacDonald Woods FP	19W-38N	15,16
Nature And Conservation Area	9W-38N	19
Old School FP	14W-27N	41,42
Riverhill FP	15W-33N	25
Ryerson Conservation Area	14W-21N	49
Spring Bluff FP	9W-43N	11
Van Patten Woods FP	15W-42N	9
Wilmont FP	15W-30N	34

Golf & Country Clubs

	Location	Page
Antioch CC	24W-40N	5,6
Arboretum GC	17W-22N	48
Barrington Hills GC	27W-20N	44,45
Biltmore CC	25W-23N	44
Bob O Link GC	8W-21N	51
Bonnie Brook Mun GC	11W-36,37N	18
Brae Loch CC	19W-33N	24
Briarwood CC	10W-20N	50
Buffalo Grove GC	17W-20N	48
Chevy Chase CC	15W-20N	49
Countryside GC	20W-27N	39
Deerfield Park Dist GC	12W-21N	50
Deerpath Park & GC	11W-26N	42,43
Exmoor CC	9W-22N	51
Four Winds GC	23W-27N	38
Foss Park GC	12W-31N	35
Fox Lake CC	28W-39N	12
Glen Flora CC	10W-36N	19
Great Lakes GC	10W-30N	35
Greenshire GC	11W-38N	18
Highland Park CC	9W-22N	51
Hillcrest CC	19W-21N	47
Indian Valley CC	18W-25N	40

	Location	Page
Knollwood CC	13W-28N	34,35
Lake Barrington Shores GC	26W-24N	37
Lake Bluff GC	11W-29N	35
Lake Zurich GC	23W-23N	46
Municipal GC	17W-24N	40
Northmoor CC	8W-21N	51
Old Elm GC	10W-24N	43
Orchard Hills GC & CC	12,13W-38N	18
Owentsia GC	11W-26N	43
Ravinia Green CC	12W-21N	50
Renwood CC	22W-35N	22
Seminary GC	18W-28N	32
Shiloh Park GC	10W-40N	11
Shore Acres CC	10W-30N	35
Stonehenge GC	27W-22N	45
Sunset Valley GC	9W-21N	50,51
Tally Ho CC	17W-24N	40
Thorngate CC	13W-20N	49
Twin Orchard CC	18W-22N	47,48
USN GC	12W-30N	35
Village Green GC	19W-29N	32

Hospitals

	Location	Page
American International Hosp	10W-40N	11
Condell Mem Hosp	16W-28N	33
County Hosp	12W-35N	26
Good Shepherd Hosp	28W-22N	44
Great Lakes Training Center Hosps	10W-30N	35
Highland Park Hosp	8W-22N	51
Lake Forest Hosp	11W-26N	43
Round Lake Medical Clinic	23W-33N	22
St Theresa Medical Center	12W-34N	26
VA Medical Center	11W-30N	35
Victory Mem Hosp	10W-35N	27

Lake County Close-up

Ravinia Festival

Lake Cook & Green Bay Rds., Highland Park
Known to music-lovers the world over, Ravinia is the nation's oldest performance arts festival. Its busy summer schedule includes performances by world-class musicians and dancers. Ravinia is also the summer home of Chicago Symphony Orchestra. Rising Star, an indoor chamber music series, runs during the fall and winter months. Also, programs for children.

Raupp Memorial Museum

901 Dunham Lane, Buffalo Grove
This charming museum honors local history with a number of exhibits and artifacts. Its collection features clothing, farm equipment and household goods dating back to the mid-1800's. Of special interest is a life-sized re-creation of the area's early business district. Open all year.

Lake County Museum

Lakewood Forest Preserve, 27277 Forest Preserve Dr., Wauconda
This comprehensive museum, with its more than 18,000 objects, celebrates Lake County history. Its exhibits include costumes, paintings, tools and several Native American artifacts. The museum sponsors Civil War Days, a battle reenactment of the war between the states.

The Power House

100 Shiloh Blvd., Zion
This unusual museum and resource center features more than 50 hands-on exhibits relating to energy. Educational programs explain alternative energy sources, energy principles and energy conservation. Visitors learn about everything from solar panels to wind turbines. Closed Sundays.

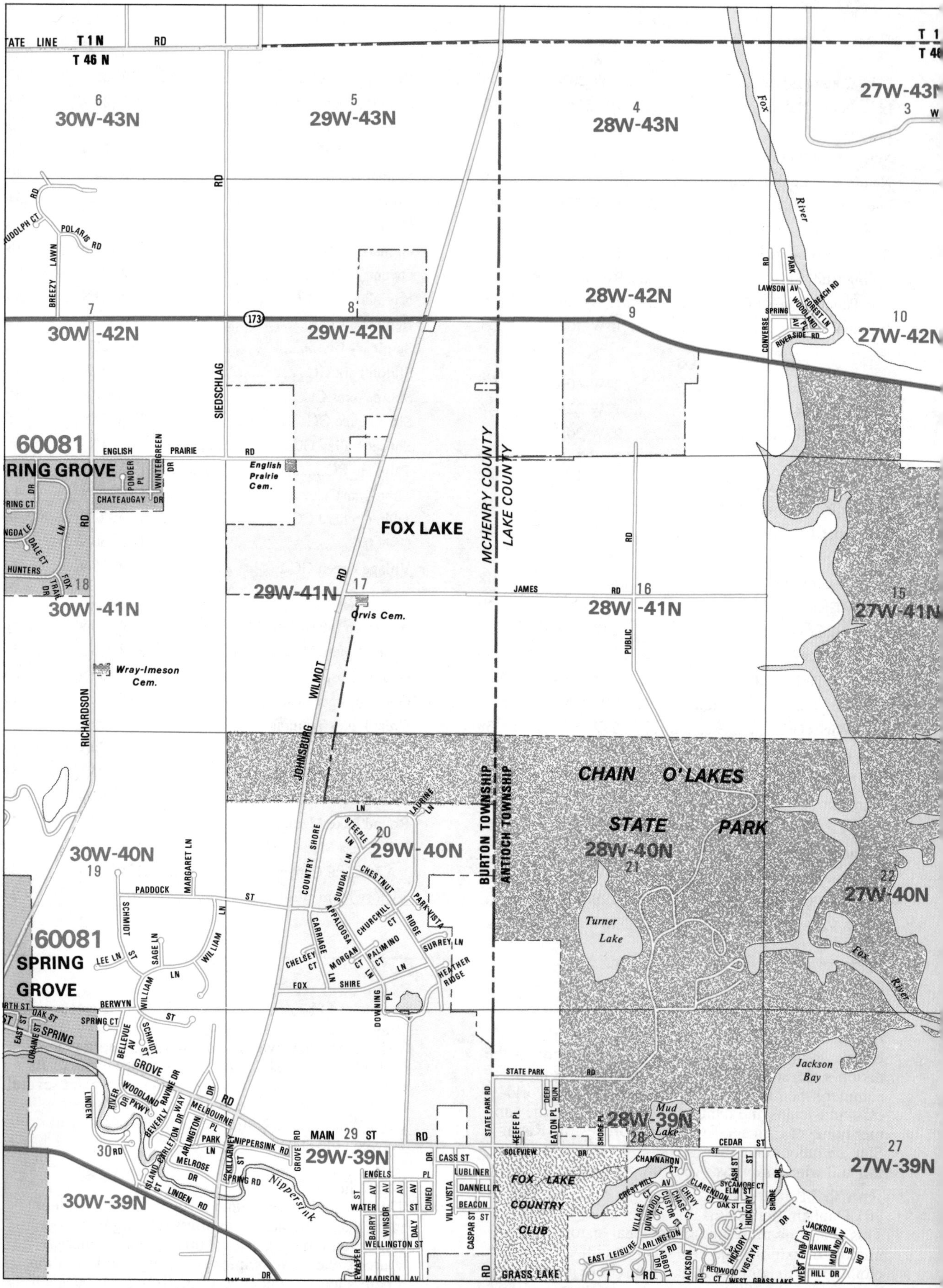

STATE LINE T 1 N RD
T 46 N

T 1
T 4

6
30W-43N

5
29W-43N

4
28W-43N

27W-43N
3

Fox

River

RD

RUDOLPH CT
POLARIS RD
BREEZY LAWN

7
30W -42N

173

8
29W-42N

28W-42N

9

PARK
LAWSON AV
FOREST LN
FORBEACH RD
SPRING
WOODLAND
CONVERSE
AV
RIVER-SIDE
RD

10
27W-42N

SIEDSCHLAG

60081
RING GROVE

ENGLISH
WINTERGREEN
DR
PONDER
PL
PRAIRIE

RD
English
Prairie
Cem.

SPRING DR
RING CT
NGDALE
DALE CT
LN
HUNTERS
RD
FOX
TRAIL
DR
18

CHATEAUGAY DR

RD

FOX LAKE

McHENRY COUNTY
LAKE COUNTY

RD

30W -41N

29W-41N 17

JAMES RD 16

28W -41N

15
27W-41N

Orvis Cem.

WILMOT RD

PUBLIC

RICHARDSON

Wray-Imeson
Cem.

JOHNSBURG

CHAIN   O'LAKES

STATE   PARK
28W-40N
21

22
27W-40N

30W-40N
19

MARGARET LN

PADDOCK ST

LN
STEEPLE
SUNDIAL LN
COUNTRY SHORE
CHESTNUT
PARKVISTA
RIDGE
20
29W-40N

LAUPINE
LN

BURTON TOWNSHIP
ANTIOCH TOWNSHIP

Turner
Lake

SCHMIDT ST
SAGE LN
WILLIAM LN
LEE LN
LN
CARRIAGE CT
APPALOSSA
CHURCHILL
CT
MORGAN
PALIMINO
CT
CT
LN
SURREY LN
HEATHER RIDGE

CHELSEY
CT
FOX
SHIRE
DOWNING
PL

60081
SPRING
GROVE

WILLIAM ST
BERWYN

Fox
River

Jackson
Bay

BELLEVUE
AV
SCHMIDT ST
SPRING CT
ST
GROVE

Mud
Lake

RTH ST
OAK ST
EAST ST
LORAINE ST
SPRING

RIVER DR
WOODLAND
DR
BEVERLY RAVINE DR
CARLETON DR WAY
MELBOURNE
DR
PARK
ARLINGTON
PL
LN
KILLARNEY CT
NIPPERSINK RD
GROVE
RD
RD

STATE PARK RD
STATE PARK RD
KEEFE PL
EATON PL
DEER
RUN
SHORE PL

CHANNAHON
CT
CEDAR
RD

28
28W-39N

27
27W-39N

N 30EN
MELROSE
DR
LINDEN RD
30 RD
ISLAND
SPRING RD
MAIN 29 ST
RD
29W-39N
Nippersink

CASS ST
CUNEO DR
VILLA VISTA
BEACON
ST
LUBLINER
DR
GOLFVIEW

CREST HILL
VILLAGE
DUNWOOD
CHEVY
CHASE CT
CUSTOR
OAK ST
CLARENDON
CT
SYCAMORE CT
ELM ST
HICKORY
SHORE DR

JACKSON
RAVINE DR
MON DAY
HILL DR

30W-39N

ENGELS ST
WATER
BARRY AV
WINSOR AV
DALY AV
DANNELL PL
CASPAR ST
ST

FOX LAKE
COUNTRY
CLUB

CHEST HILL CT
CHASE CT
AV
ARLINGTON DR
ABBOTT DR
A RD
REDWOOD
VISCAYA
WEST END DR
HICKORY

EWAPER
WELLINGTON ST
MADISON AV

EAST LEISURE
JACKSON
WEST GRASS LAKE
GRASS LAKE

SEE McHENRY COUNTY PAGES 10-11

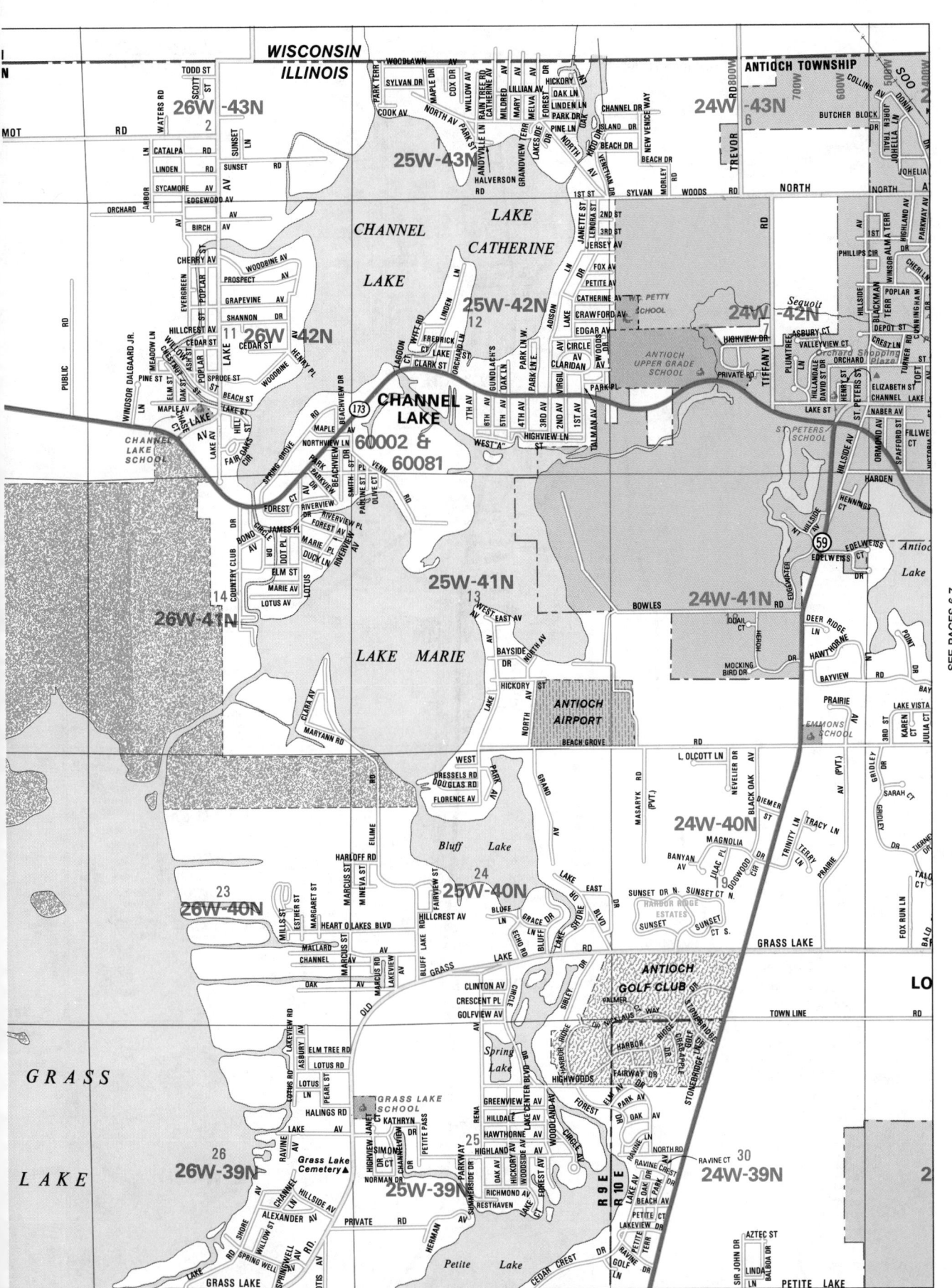

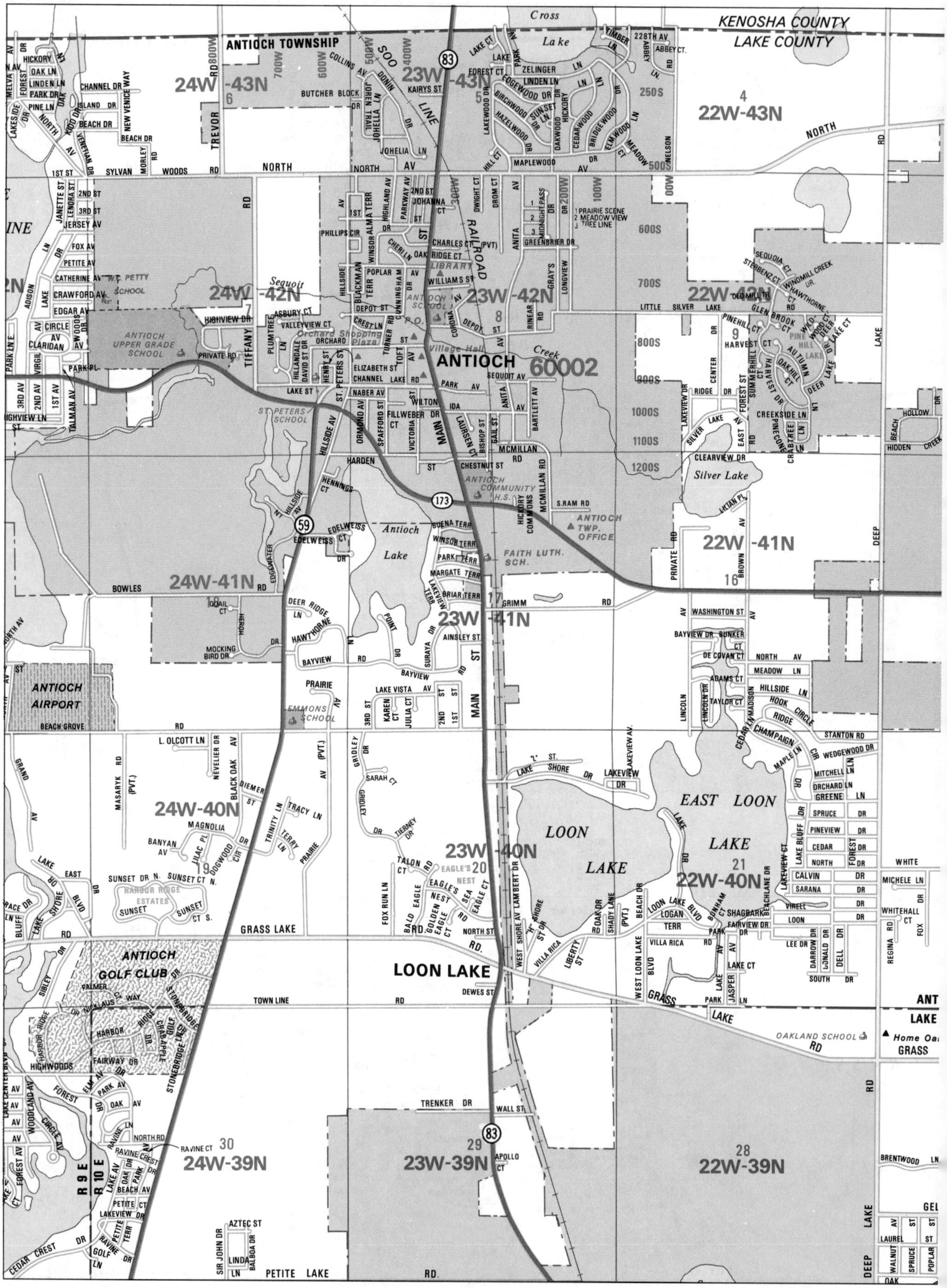

KENOSHA COUNTY
LAKE COUNTY

Cross Lake

ANTIOCH TOWNSHIP

24W -43N
6

23W -43N
5

22W -43N
4

ANTIOCH

24W -42N
7

23W -42N
8

22W -42N
9

60002

SEE PAGES 4-5

ANTIOCH UPPER GRADE SCHOOL

ANTIOCH AIRPORT

ANTIOCH GOLF CLUB

HARBOR RIDGE ESTATES

24W -41N

23W -41N
17

22W -41N
16

ANTIOCH TWP. OFFICE

24W -40N
19

23W -40N
20

22W -40N
21

LOON LAKE

EAST LOON LAKE

Silver Lake

Little Silver Lake

Antioch Lake

LOON LAKE

GRASS LAKE

24W -39N
30

23W -39N
29

22W -39N
28

OAKLAND SCHOOL

PETITE LAKE

DEEP LAKE

R 9 E   R 10 E

Home

GRASS
ANT
LAKE

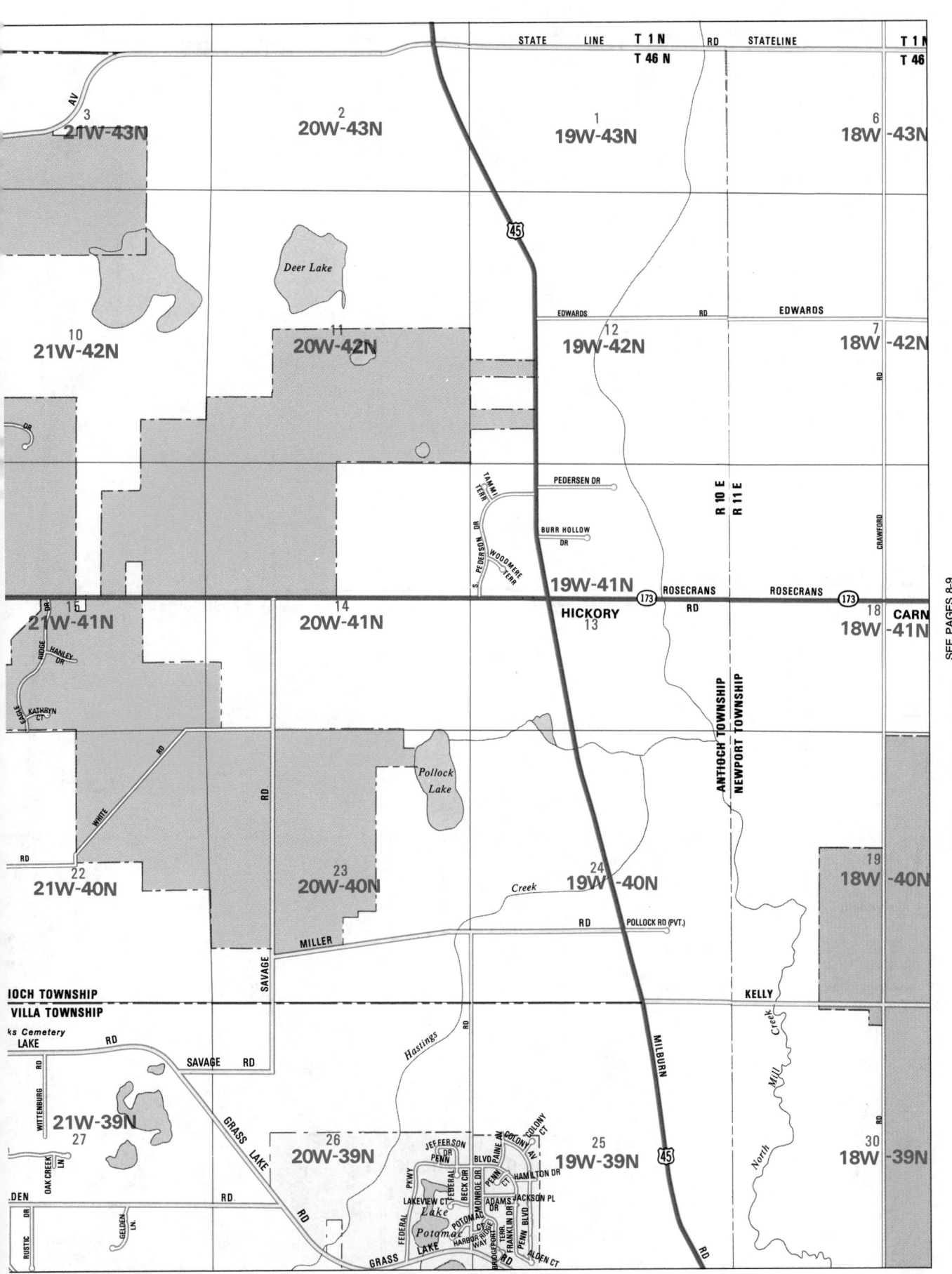

AV
3
21W-43N

2
20W-43N

1
19W-43N

6
18W -43N

45

Deer Lake

10
21W-42N

11
20W-42N

EDWARDS RD EDWARDS

12
19W-42N

7
18W -42N

RD

PEDERSEN DR

R 10 E
R 11 E

TAMMI TERR

S. PEDERSON DR

WOODMERE TERR

BURR HOLLOW DR

CRAWFORD

19W-41N

ROSECRANS 173 ROSECRANS RD 173

15
21W-41N

RIDGE DR
HANLEY DR

EAGLE CT
KATHRYN CT

14
20W-41N

HICKORY
13

18
CARN
18W -41N

RD

WHITE

RD

Pollock Lake

ANTIOCH TOWNSHIP
NEWPORT TOWNSHIP

RD

22
21W-40N

23
20W-40N

Creek

24
19W -40N

19
18W -40N

MILLER

SAVAGE

RD POLLOCK RD (PVT.)

KELLY

IOCH TOWNSHIP
VILLA TOWNSHIP

ks Cemetery
LAKE RD

SAVAGE RD

Hastings

RD

MILBURN

North Mill Creek

WITTENBURG RD

21W-39N
27

GRASS LAKE RD

26
20W-39N

JEFFERSON DR
PENN
FEDERAL
BECK CIR
BLVD
PENN PAINE AV
COLONY CT
COLONY AV
HAMILTON DR

25
19W-39N

45

30
18W -39N

OAK CREEK LN

DEN
RUSTIC DR
GELDEN LN

RD

LAKEVIEW CT
FEDERAL PKWY
Lake Potomac
POTOMAC CT
MONROE DR
ADAMS DR
JACKSON PL
FRANKLIN DR
HARBOR RIDGE TERR
BRIDGEPORT RD
PENN BLVD
ALDEN CT

GRASS LAKE

RD

SEE PAGES 8-9

SEE PAGES 14-15

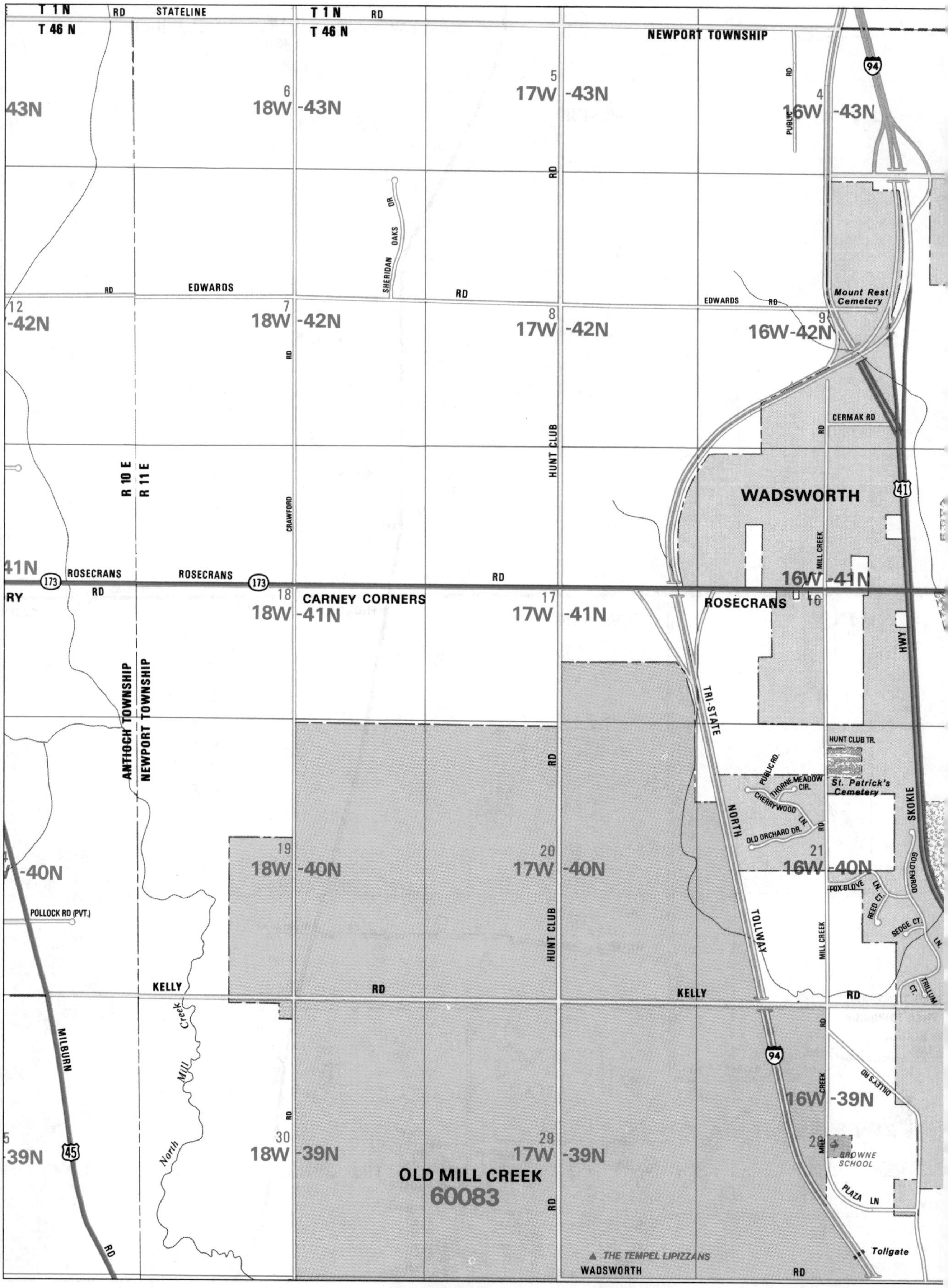

T 1 N RD STATELINE T 1 N RD

T 46 N T 46 N NEWPORT TOWNSHIP

I-94

6
43N 18W -43N 5 17W -43N 4 16W -43N

PUBLIC RD

Mount Rest
Cemetery

SHERIDAN OAKS DR

RD EDWARDS RD EDWARDS RD

12
-42N 7 18W -42N 8 17W -42N 9 16W-42N

CRAWFORD HUNT CLUB RD

CERMAK RD

R 10 E
R 11 E

WADSWORTH

41N 173 ROSECRANS RD ROSECRANS 173 RD 16W -41N 16

RD 18 17 ROSECRANS

RY RD 18W -41N CARNEY CORNERS 17W -41N HWY

SKOKIE

ANTIOCH TOWNSHIP
NEWPORT TOWNSHIP

TRI-STATE

HUNT CLUB TR.

PUBLIC RD

THORNE MEADOW CIR.

CHERRYWOOD

OLD ORCHARD DR.

St. Patrick's
Cemetery

40N 19 18W -40N 20 17W -40N 21 16W -40N

POLLOCK RD (PVT.)

NORTH

TOLLWAY

HUNT CLUB RD

MILL CREEK

FOX GLOVE LN

REED CT.

GOLDENROD LN

SEDGE CT.

TRILLIUM CT.

KELLY RD KELLY RD

MILBURN

Mill Creek

North

RD

I-94

MILL CREEK RD

16W -39N

DILEY'S RD

5
39N 45 30 18W -39N 29 17W -39N 28

BROWNE
SCHOOL

PLAZA LN

OLD MILL CREEK
60083

▲ THE TEMPEL LIPIZZANS

Tollgate

WADSWORTH RD

SEE PAGES 6-7

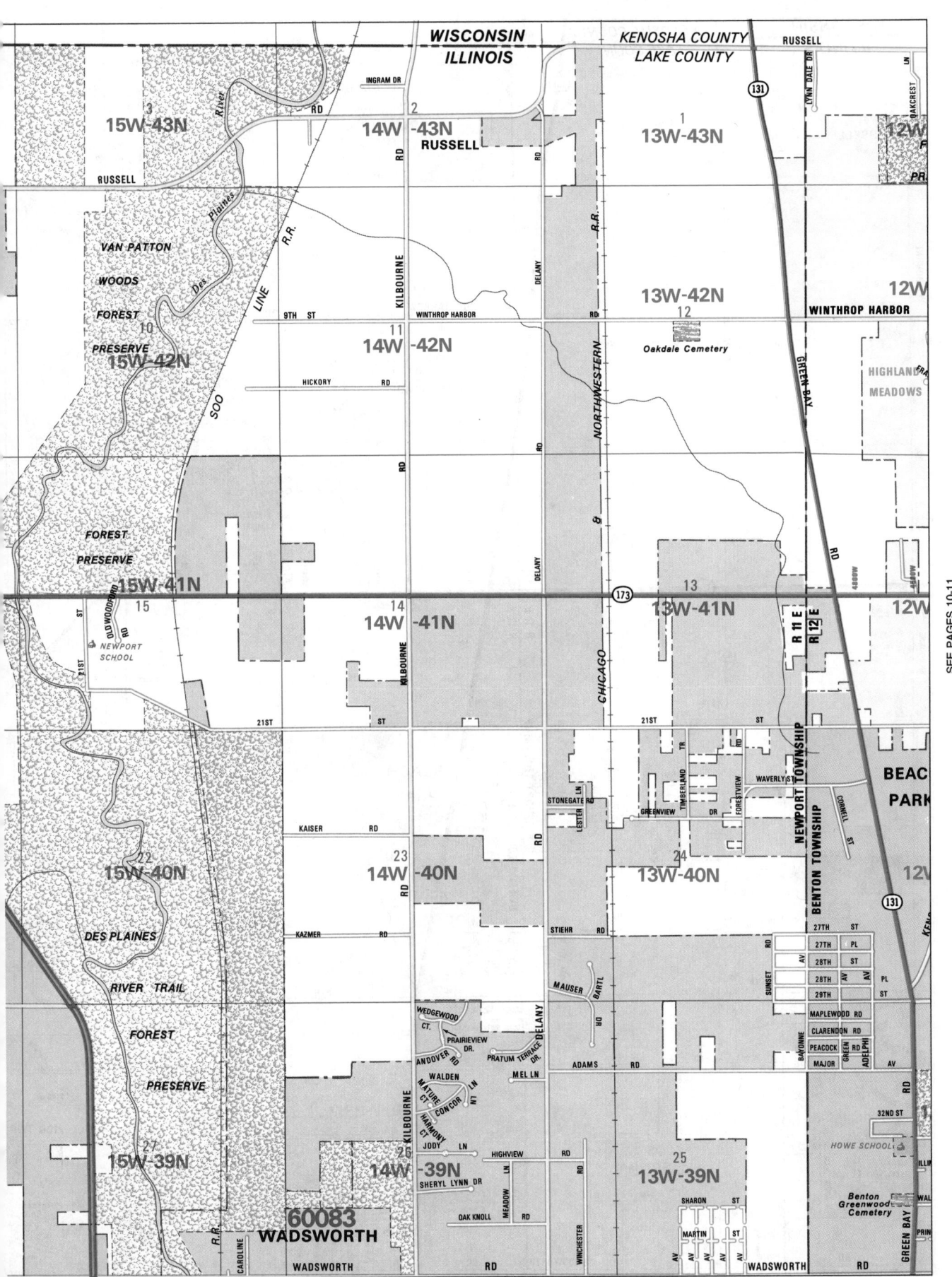

SEE PAGES 10-11

SEE PAGES 16-17

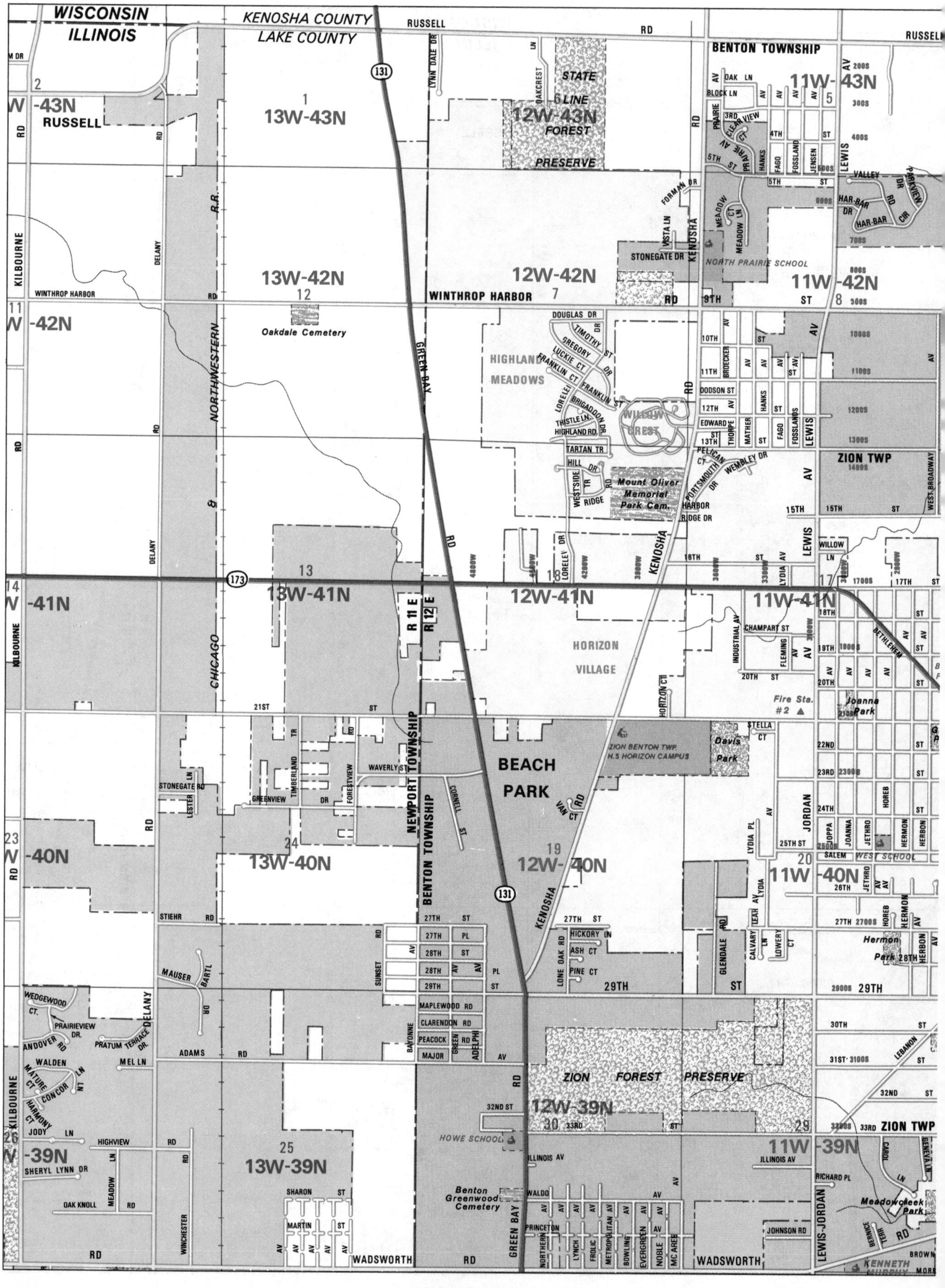

WISCONSIN
ILLINOIS

KENOSHA COUNTY
LAKE COUNTY

RUSSELL   RD   RUSSELL

131

BENTON TOWNSHIP

11W- 43N

OAK LN
BLOCK LN

STATE
LINE

12W-43N

FOREST

PRESERVE

1
13W-43N

2

W -43N
RUSSELL

RD

DELANY

KILBOURNE

R.R.

NORTHWESTERN

PRAIRIE AV
3RD
4TH
5TH

FABO
JENSEN

CLEAR VIEW AV
PINE AV
HANKS

LEWIS AV

VALLEY DR
PARKVIEW DR

HAR-BAR DR
HAR-BAR CIR

FORMAN DR

VISTA LN

STONEGATE DR

MEADOW LN
KENOSHA

NORTH PRAIRIE SCHOOL

13W-42N

12

Oakdale Cemetery

WINTHROP HARBOR RD

12W-42N

7

9TH ST

11W-42N

W -42N

Winthrop Harbor

DELANY

RD

CHICAGO

&

DOUGLAS DR
TIMOTHY ST
GREGORY DR
LUCKIE ST
FRANKLIN CT
FRANKLIN ST

HIGHLAND
MEADOWS

LORELEI
BRIGADOON DR

TWISTLE LN
HIGHLAND RD.
TARTAN TR

WESTSIDE
TR
HILL DR.
RIDGE

WILLOW
CREST

Mount Oliver
Memorial
Park Cem.

PELICAN
CT.

PORTSMOUTH DR
HARBOR
RIDGE DR

WEMBLEY DR

KENOSHA RD

10TH ST
11TH ST
DODSON ST
12TH ST
EDWARD ST
13TH ST

BROECKER AV
HANKS
THORPE
MATHER

FABO
FOSSLAND

LEWIS AV

ZION TWP

1000S
1100S
1200S
1300S
1400S

15TH ST

15TH ST

WEST BROADWAY

LEWIS AV

WILLOW

16TH ST

LYDIA ST

17TH ST

173

R 11 E
R 12 E

13
13W-41N

18
12W-41N

17
11W-41N

W -41N

KILBOURNE

RD

21ST ST

HORIZON
VILLAGE

4800W

4200W

3800W

3600W

3300W

3000W

2900W

INDUSTRIAL AV

CHAMPART ST

FLEMING AV

HORIZON CT

18TH ST
19TH ST
20TH ST

Fire Sta.
#2

STELLA
CT.

ZION BENTON TWP.
H.S HORIZON CAMPUS

DAVIS
Park

Joanna
Park

BETHLEHEM

1000S
1700S

21ST
22ND
23RD 2300S

24
13W-40N

NEWPORT TOWNSHIP

BENTON TOWNSHIP

BEACH
PARK

19
12W-40N

131

TIMBERLAND
TR

STONEGATE LN
LESTER RD

GREENVIEW

FORESTVIEW
DR

WAVERLY ST

CORNELL ST

VAN CT

KENOSHA RD

JORDAN

HOREB

SALEM WEST SCHOOL

HERMON
HERMON

LYDIA PL

LYDIA
LEAH

CALVARY
CT.
LOWERY

2600PPA
2600N

JETHRO
JOANNA

HERMON
HOREB

HERMON AV

Hermon
Park 28TH

2800S

24TH
25TH ST

26TH ST

27TH 27000S

20
11W -40N

25
13W-40N

STIEHR RD

27TH ST
27TH PL
28TH ST
28TH PL
29TH ST

SUNSET RD

LONE OAK RD

27TH ST

HICKORY LN
ASH CT
PINE CT

29TH ST

GLENDALE RD

29TH ST

2900S

28TH

29TH

MAUSER
BARTL

WEDGEWOOD
CT.

PRAIRIEVIEW
DR.

ANDOVER RD

PRATUM TERRACE DR.

WALDEN

MEL LN

ADAMS RD

MAPLEWOOD RD
CLARENDON RD
PEACOCK RD
MAJOR GREEN ADELPHI

BAYONNE DR

GREEN BAY RD

32ND ST

ZION FOREST PRESERVE

12W-39N
30

33RD

31ST- 3100S

32ND ST

33000S

30TH ST

31ST 31000S

32ND ST

29

33RD ZION TWP

LEBANON

11W -39N
29

WEDGEWOOD

MATURE CT
CONCORD

HARMONY
CT.
JODY LN

HIGHVIEW

SHERYL LYNN DR

OAK KNOLL

MEADOW
LN

RD

RD

25
13W-39N

SHARON ST

MARTIN AV

HOWE SCHOOL

Benton
Greenwood
Cemetery

GREEN BAY RD

Illinois Av

WALDO

PRINCETON AV

NORTHERN
LYNCH
FROLIC

METROPOLITAN AV
BOTTLING
EVERGREEN
NOBLE
MC AREE AV

WADSWORTH RD

ILLINOIS AV

JOHNSON RD

RICHARD PL

BERNICE AV
FERRIE RD

LEWIS-JORDAN

CAROL

Meadowcreek
Park

KENNETH
MURPHY

BROWN
MORR

11W -39N

WADSWORTH

W -40N

23

W -39N

26

KILBOURNE

WINCHESTER RD

SEE PAGES 8-9

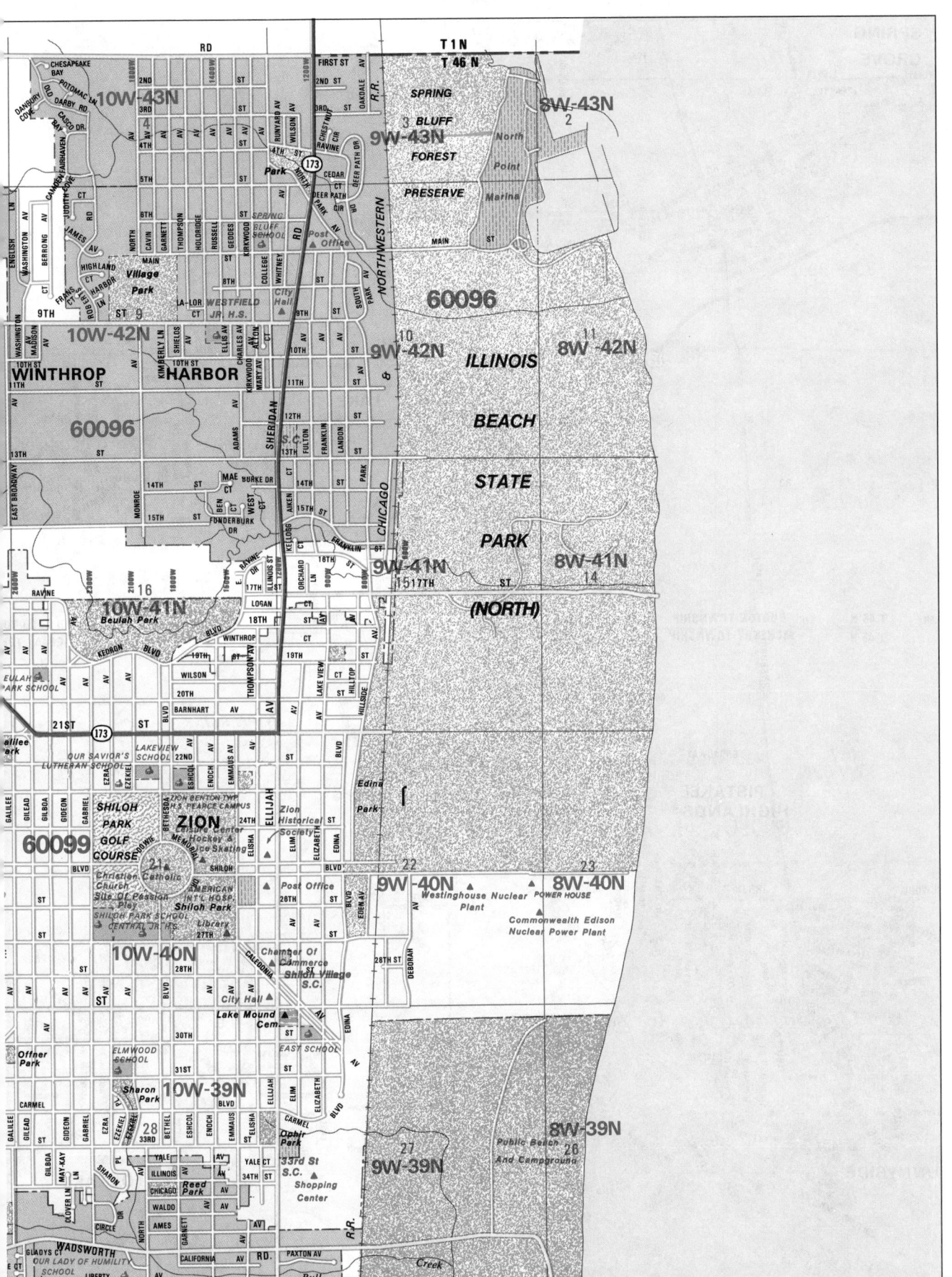

SEE PAGES 4-5

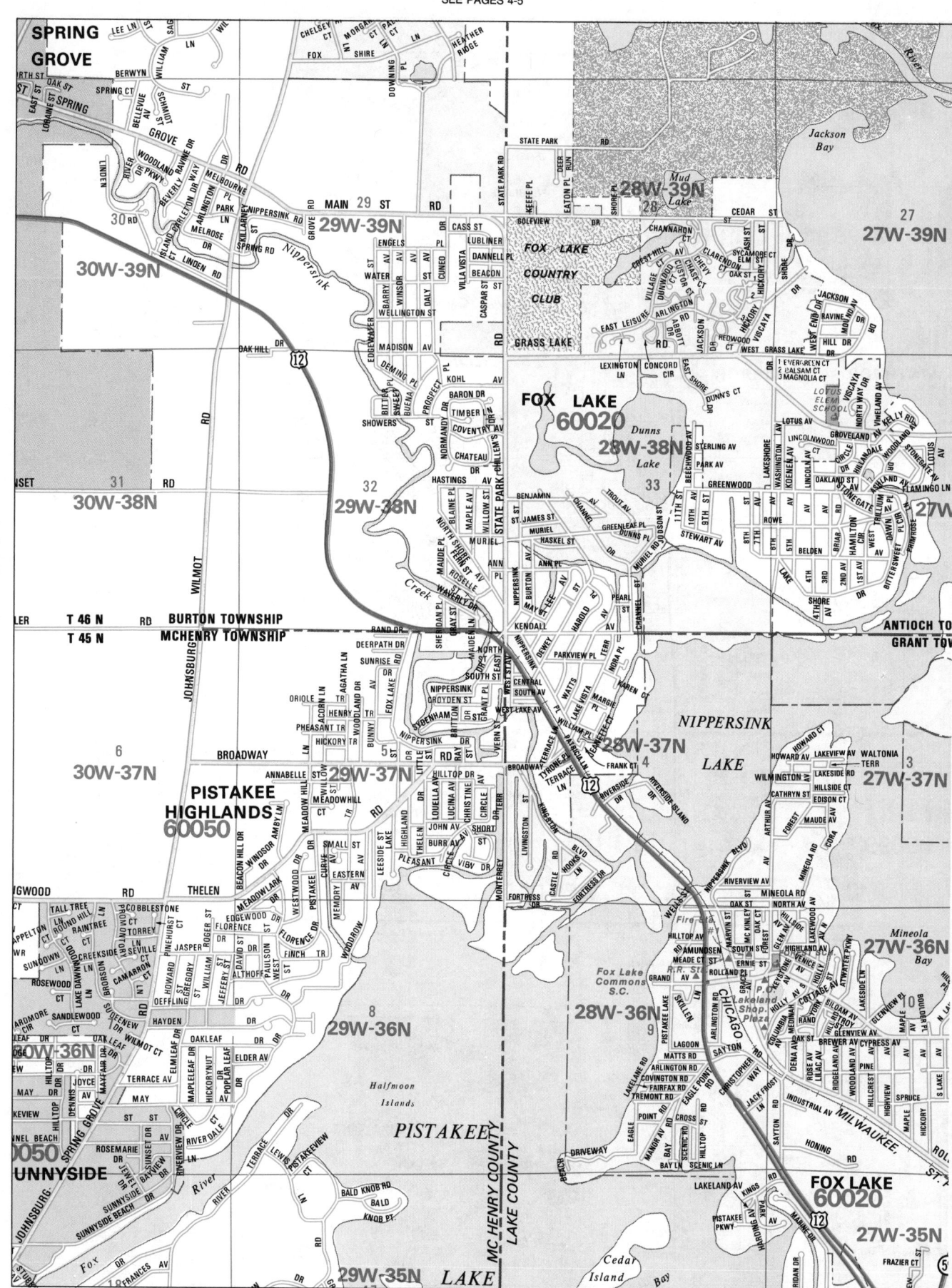

SEE McHENRY COUNTY PAGES 18-19

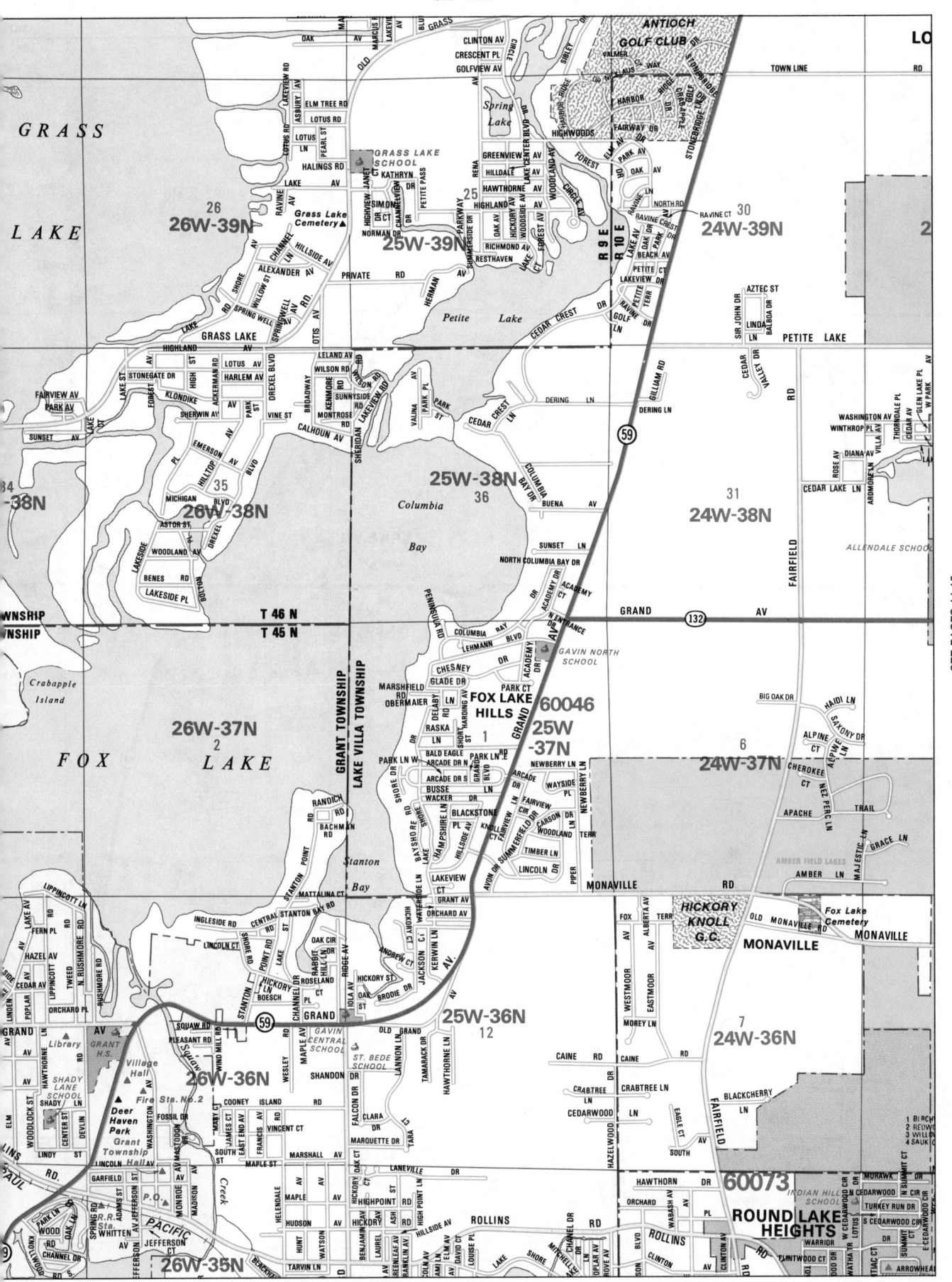

SEE PAGES 14-15

SEE PAGES 6-7

SEE PAGES 12-13

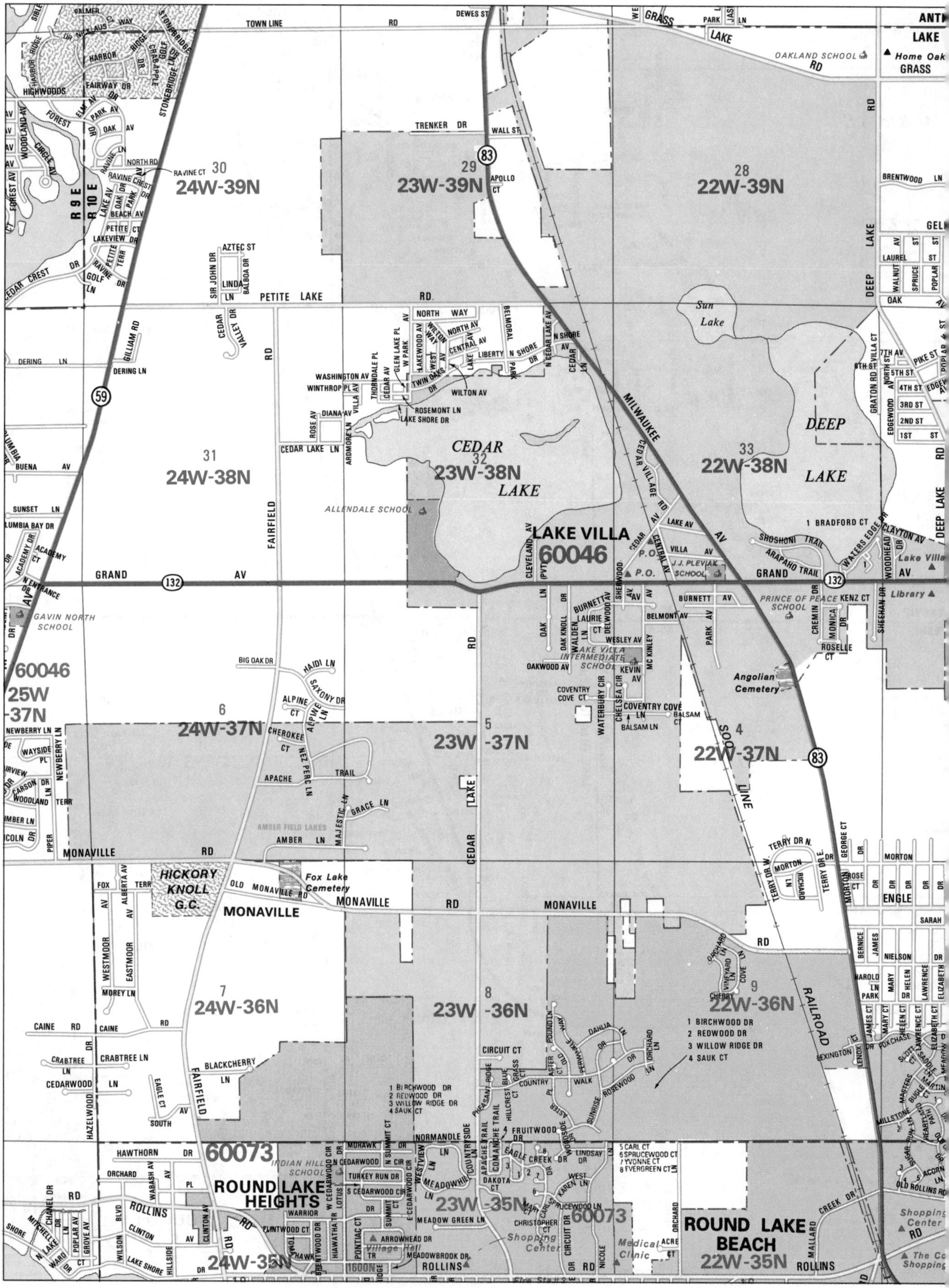

SEE PAGES 22-23

LAKE VILLA
60046

ROUND LAKE HEIGHTS

ROUND LAKE BEACH

60073

60046

30
24W-39N

29
23W-39N

28
22W-39N

31
24W-38N

32
23W-38N

33
22W-38N

6
24W-37N

5
23W-37N

4
22W-37N

25W-37N

7
24W-36N

8
23W-36N

9
22W-36N

24W-35N

23W-35N

22W-35N

60073

CEDAR LAKE

DEEP LAKE

Sun Lake

GRASS LAKE

ANTH LAKE

1 BIRCHWOOD DR
2 REDWOOD DR
3 WILLOW RIDGE DR
4 SAUK CT

1 BIRCHWOOD DR
2 REDWOOD DR
3 WILLOW RIDGE DR
4 SAUK CT

5 CARL CT
6 SPRUCEWOOD CT
7 YVONNE CT
8 EVERGREEN LN

HICKORY KNOLL G.C.

MONAVILLE

GRAND AV

TOWN LINE RD

PETITE LAKE RD.

MONAVILLE RD

ROLLINS RD

SEE PAGES 6-7

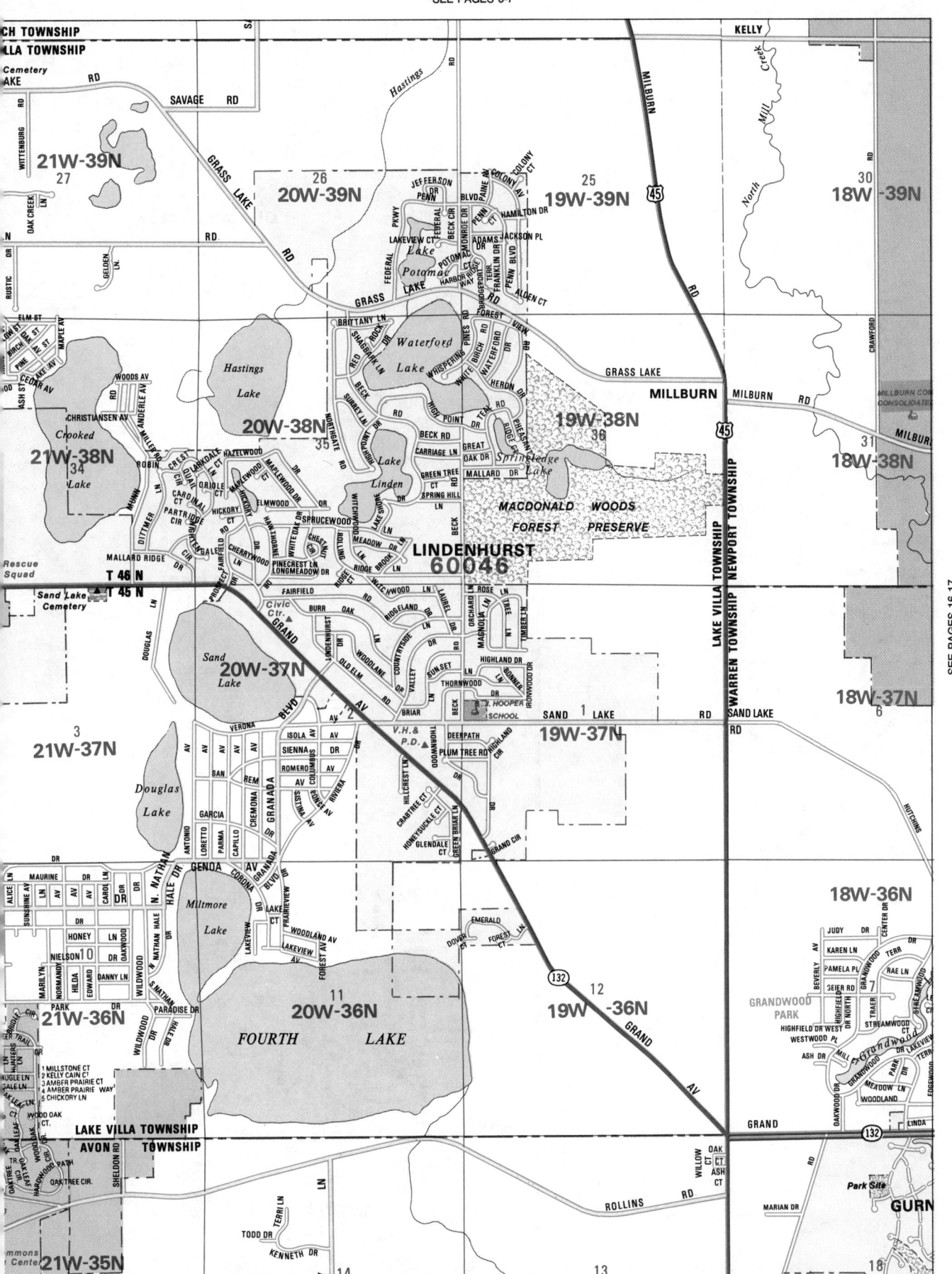

SEE PAGES 16-17

SEE PAGES 14-15

KELLY RD

MILBURN RD

North Mill Creek

Hastings RD

JEFFERSON DR
PENN
PKWY
FEDERAL
BECK CIR
MONROE DR
PENN PAINE AV
COLONY AV
COLONY CT
LAKEVIEW CT
HAMILTON DR
JACKSON PL
POTOMAC
ADAMS TERR
FRANKLIN DR
PENN BLVD
HARBOR RIDGE WAY
ALDEN CT

Lake Potomac
Lake

19W-39N   25

18W-39N   30

17W-39N   29

OLD MILL CREEK
60083

△ THE TEMPEL LIP

WADSWORTH

GRASS LAKE

BRITTANY LN
RED ROCK LN
SHAGBARK LN
FOREST RD
FOREST VIEW RD
WATERFORD RD
WHITE BIRCH
PHEASANT RIDGE DR
WHISPERING PINES RD
HERON DR
TEAL DR
HIGH POINT DR
Waterford Lake
GRASS LAKE

Springledge Lake

MILLBURN   MILBURN   RD

MILLBURN COMMUNITY CONSOLIDATED SCHOOL

19W-38N   36

18W-38N   31   MILBURN   RD   32

17W-38N

CRAWFORD RD

HUNT CLUB RD

BECK
SURRY LN
NORTHGATE RD
HIGH POINT DR
BECK RD
CARRIAGE LN
GREAT OAK DR
GREEN TREE CT
SPRING HILL LN
MALLARD DR
Lake Linden

MACDONALD WOODS FOREST PRESERVE

LINDENHURST
60046

WITCHWOOD
MEADOW
ROLLING RIDGE
RIDGE BROOK CT
LAKE SHORE LN

Mill Creek   T 46 N
T 45 N

OAK RD
LINDENHURST
WOODLANE
OLD ELM AV
RIDGELAND LN
COURTSIDE DR
VALLEY DR
SUNSET DR
THORNWOOD DR
HIGHLAND DR
BONNIE LN
BRONWOOD DR
TIMBER LN
MAGNOLIA LN
ORCHARD LN
ROSE LN
TREE LN
BRIAR
HOOPER SCHOOL

LAKE VILLA TOWNSHIP
NEWPORT TOWNSHIP

SAND   1   LAKE   RD

SAND LAKE RD   6

18W-37N

5   SAND   LAKE   RD

V.H. & P.D.
DEERPATH
PLUM TREE RD
HIGHLAND CIR
CRABTREE CT
HILLCREST LN
HONEYSUCKLE CT
GLENDALE CT
GREEN BRIAR LN
GRAND CIR
THORNWOOD

19W-37N   WARREN TOWNSHIP

17W-37N

RIVERA AV

EMERALD
FOREST LN
DOVER CT

HUTCHINS

VELVET LN
BLACK DR
THOROUGHBRED
YEARLING DR
HUNT CLUB FARMS I

STEARNS SCHOOL

Nature Pres.
POND RIDGE
MILL POND
MILL CREEK CROSSING
MILL CREEK CIR
FIELD VIEW DR
BROOKSIDE DR
BRIDLE TRAIL
HUNT CLUB FARMS
Field & Fence Equestrian
WOODS

132   12   19W-36N

18W-36N

JUDY DR
KAREN LN
BEVERLY AV
PAMELA PL
GEIER RD
CENTER DR
GRANDWOOD TERR
RAE LN
STRAER DR
STREAMWOOD CT
LEE CIR
DOUGLAS TERR
BANBURY DR
HAMPSHIRE DR
BEECHWOOD DR
WALDEN WAY
17W-36N   9

GRANDWOOD PARK
HIGHFIELD DR WEST
HIGHFIELD DR NORTH
WESTWOOD PL
ASH LN
MILL DR
GRANDWOOD DR
PARK LAKEVIEW DR
MEADOW LN
WOODLAND
EDGEWOOD DR
BRIDLEWOOD
Grandwood Lake

Nature Pres.
WEST LN
MEADOW RIDGE DR
Brookside Green
BROOKSIDE

LAKE
11
W-36N

GRAND   AV

GRAND   AV   132

LINDA LN

OAK CT
WILLOW CT
ASH CT

ROLLINS   RD

MARIAN DR

Park Site

GURNEE

ALMOND RD

1 OVERBROOK CT
2 GOLDSPRING CT
3 OLD CREEK CT
4 STILLWATER CT
5 BACK BAY CT
6 RUNNING CREEK CT
7 EDGEWATER CT
8 SPRINGBROOK LN
9 NEW BRIDGE CT
10 WALDEN WAY
11 SALISBURY DR
12 YORKSHIRE DR
13 BEECHWOOD CT
14 HAMPSHIRE CT
15 WALDEN WAY
16 BANBURY CT

FIRE DEPARTME
SITE

ROSEWOOD

SEE PAGES 8-9

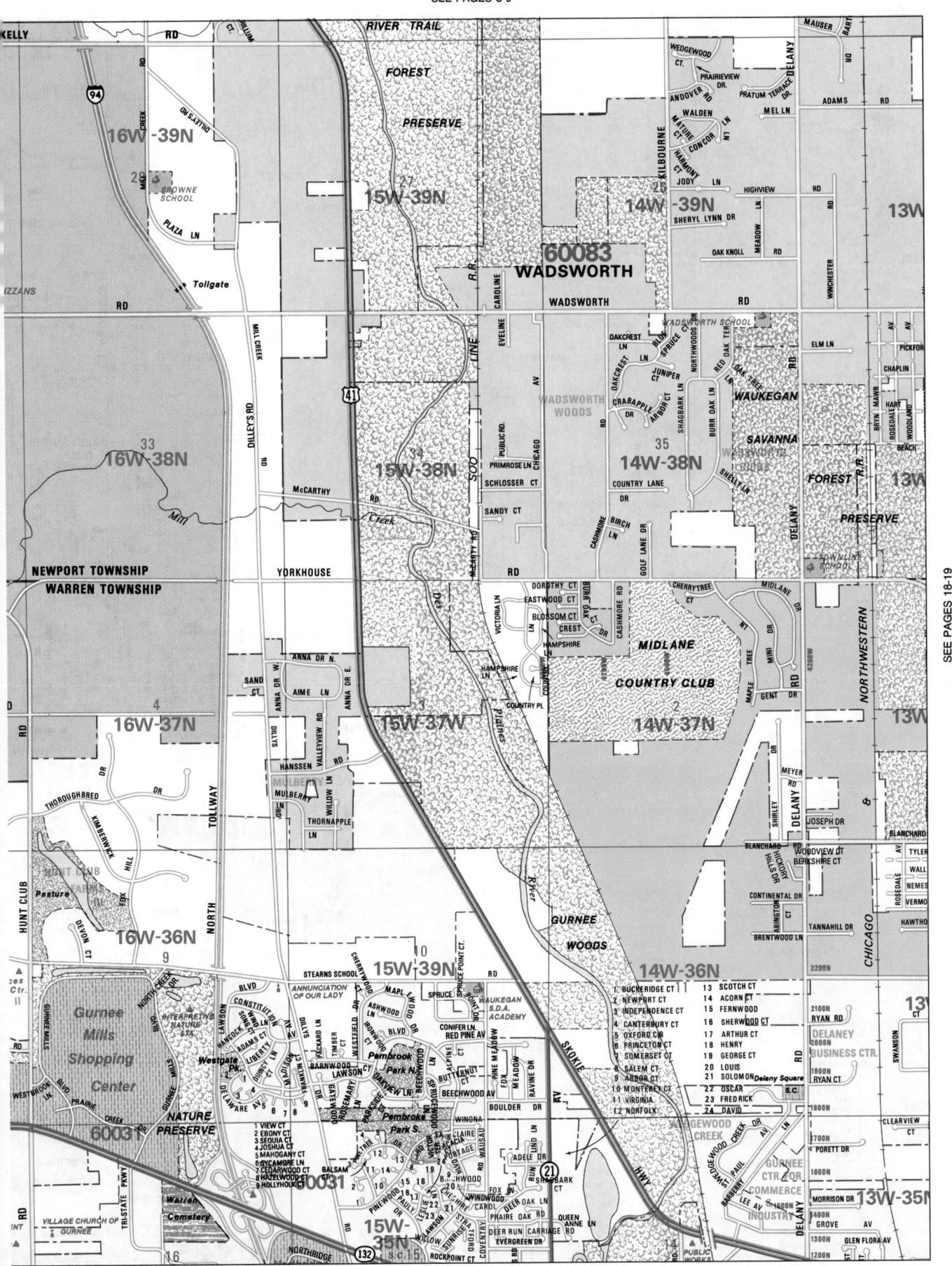

SEE PAGES 18-19

SEE PAGES 10-11

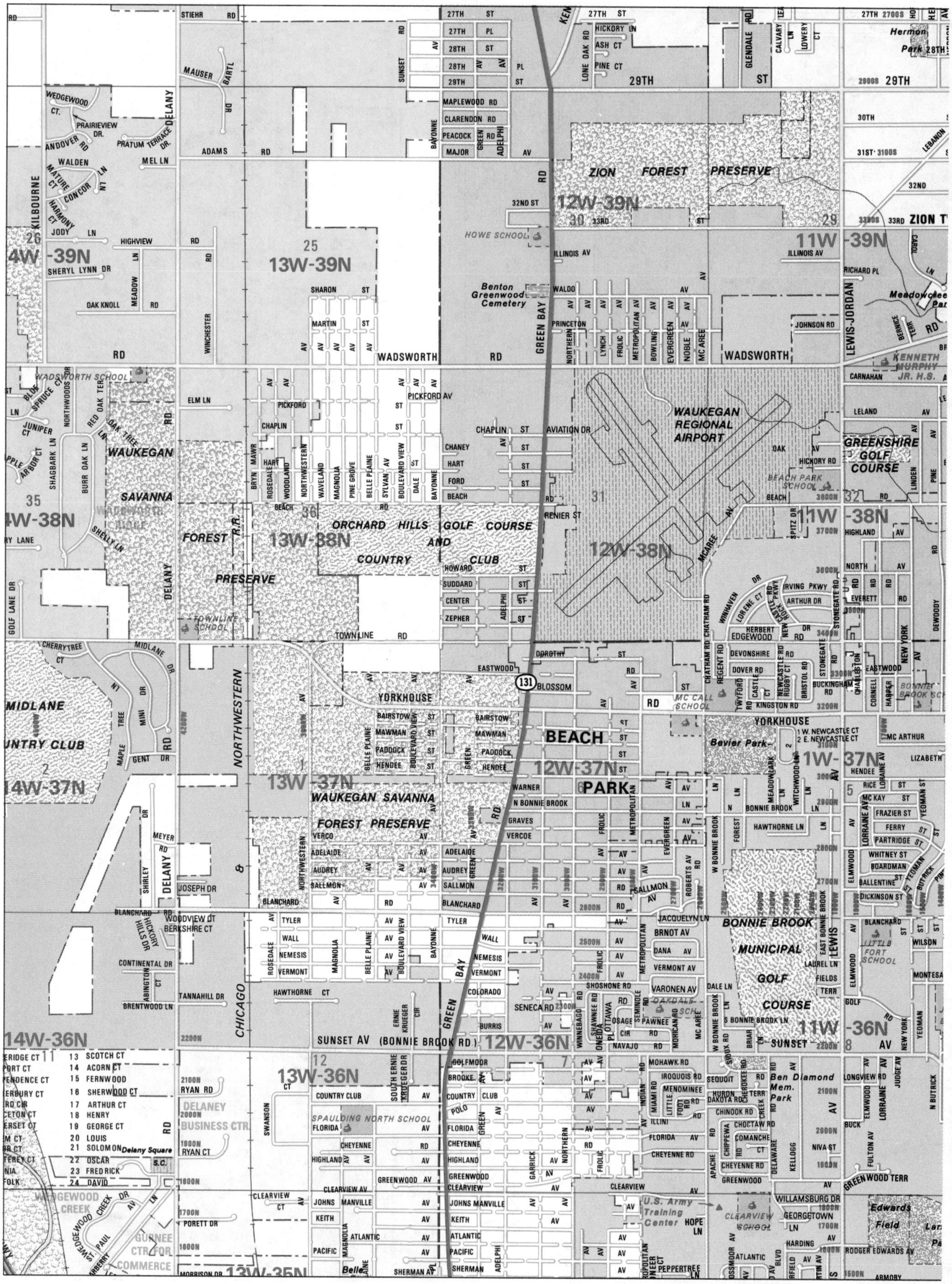

ZION FOREST PRESERVE

12W-39N

11W-39N

ZION T

13W-39N

4W-39N

WAUKEGAN REGIONAL AIRPORT

GREENSHIRE GOLF COURSE

WAUKEGAN SAVANNA FOREST PRESERVE

ORCHARD HILLS GOLF COURSE AND COUNTRY CLUB

13W-38N

12W-38N

11W-38N

4W-38N

MIDLANE COUNTRY CLUB

WAUKEGAN SAVANNA FOREST PRESERVE

BEACH PARK

BONNIE BROOK MUNICIPAL GOLF COURSE

14W-37N

13W-37N

12W-37N

11W-37N

14W-36N

13W-36N

12W-36N

11W-36N

SPAULDING NORTH SCHOOL

Ben Diamond Mem. Park

U.S. Army Training Center

13W-35N

DELANEY BUSINESS CTR.

GURNEE CTR FOR COMMERCE

SEE PAGES 16-17

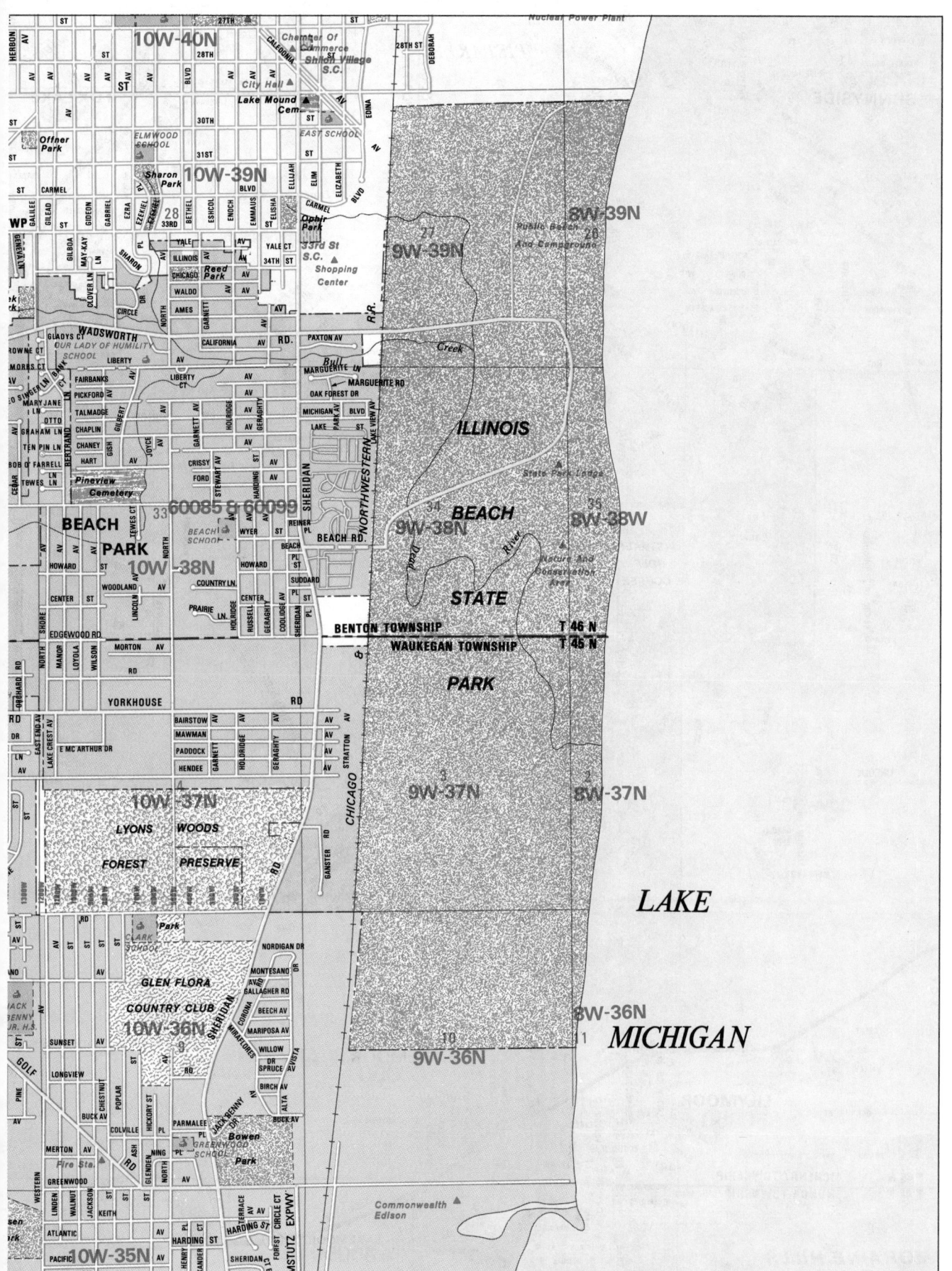

SEE PAGES 12-13

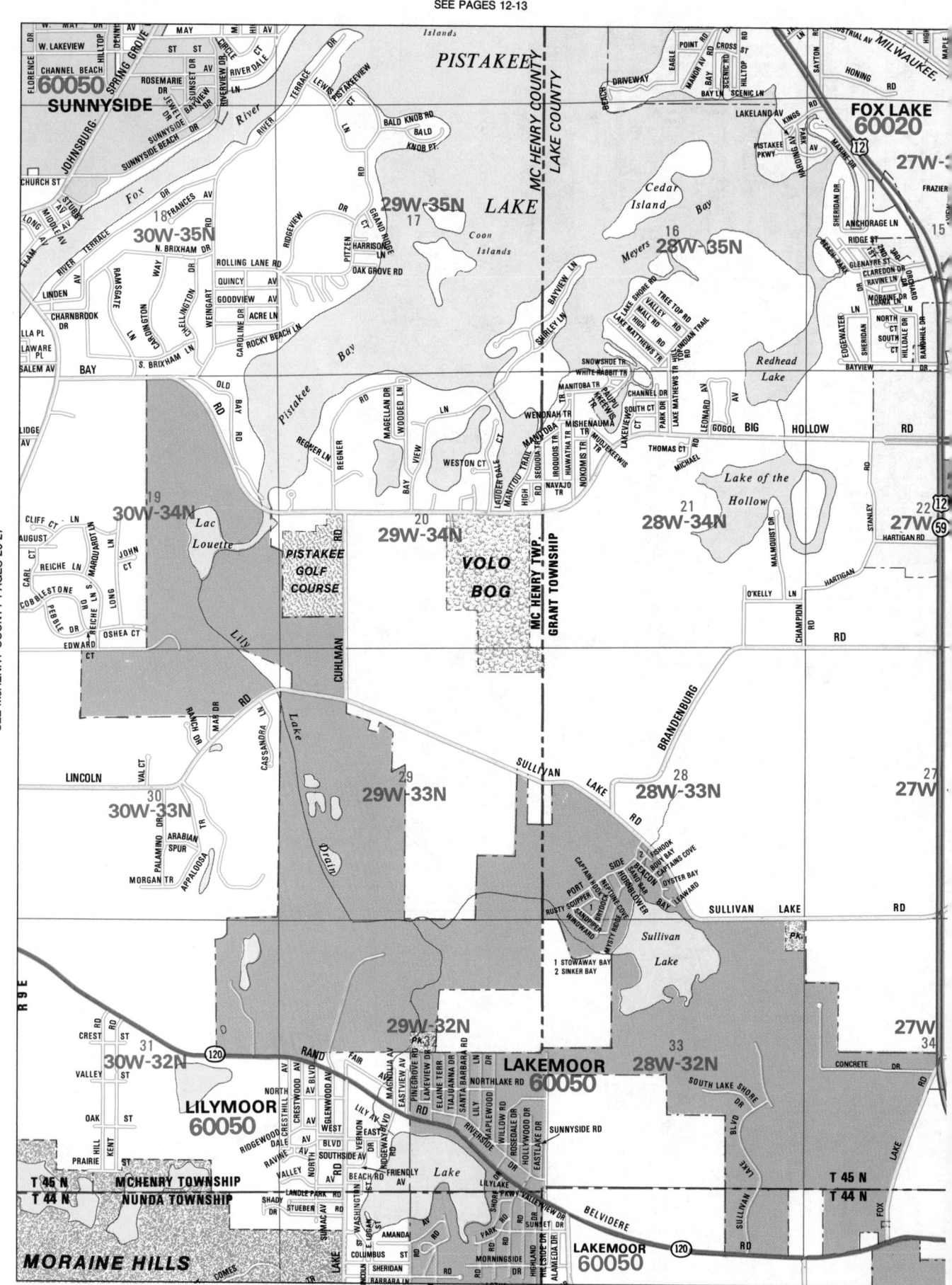

SEE PAGES 28-29

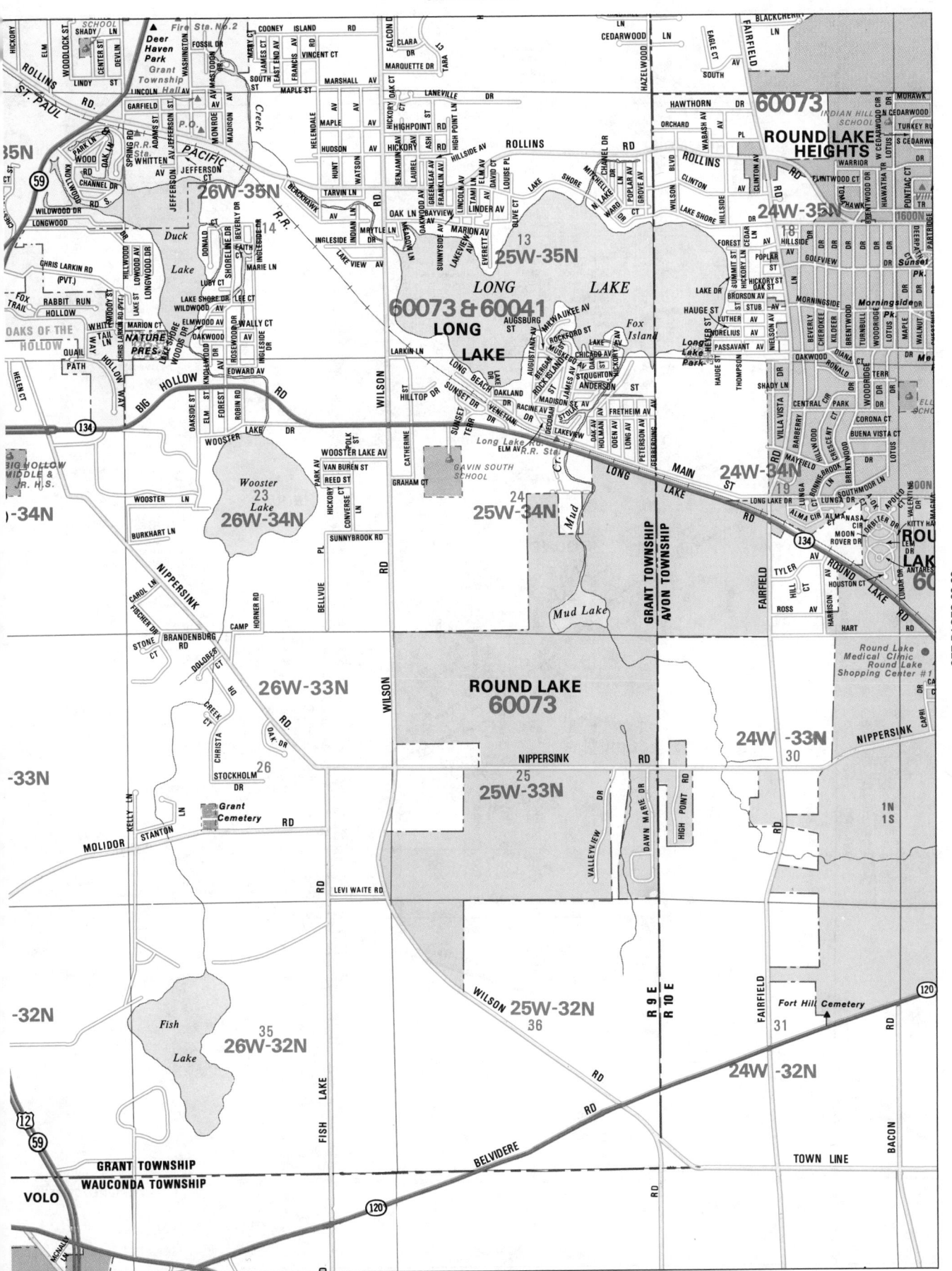

22 LAKE COUNTY

SEE PAGES 20-21

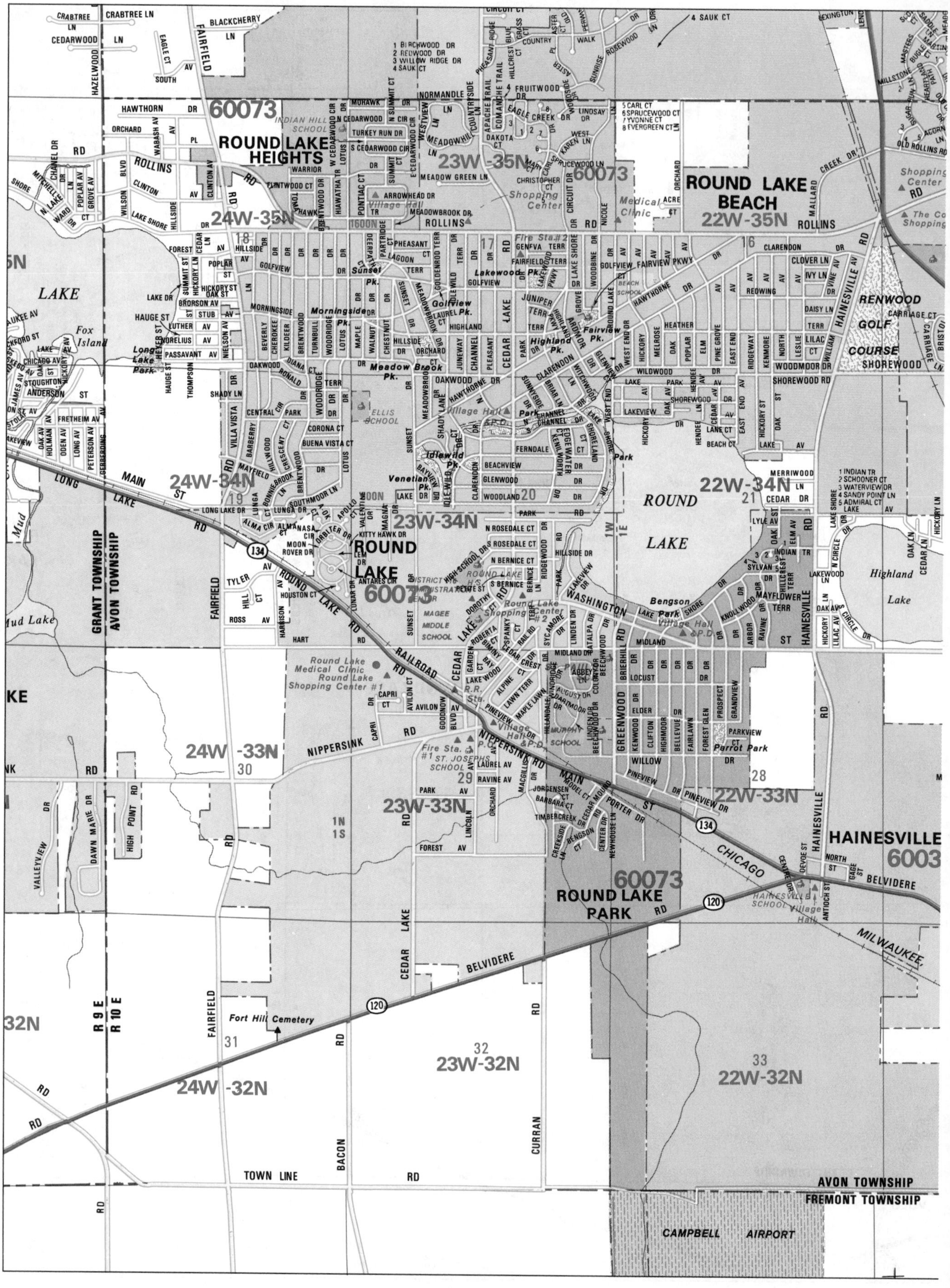

ROUND LAKE HEIGHTS 60073

23W-35N

24W-35N

ROUND LAKE BEACH 22W-35N

RENWOOD GOLF COURSE

SHOREWOOD

24W-34N

23W-34N

22W-34N

ROUND LAKE 60073

ROUND LAKE

Highland Lake

24W-33N

23W-33N

22W-33N

ROUND LAKE PARK 60073

HAINESVILLE 6003

GRANT TOWNSHIP

AVON TOWNSHIP

24W-32N

23W-32N

22W-32N

R 9 E   R 10 E

Fort Hill Cemetery

CAMPBELL AIRPORT

AVON TOWNSHIP

FREMONT TOWNSHIP

SEE PAGES 14-15

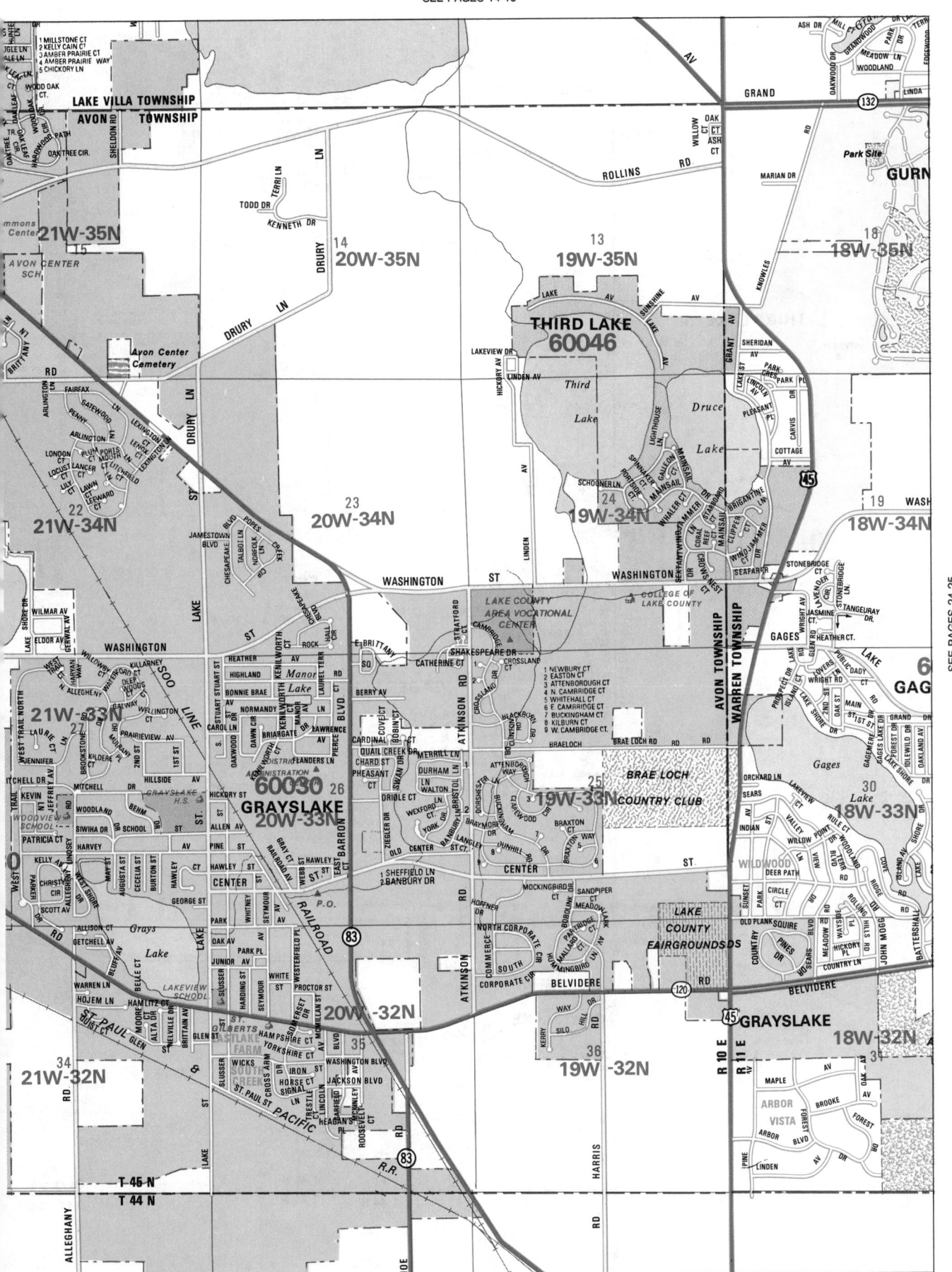

SEE PAGES 24-25

SEE PAGES 30-31

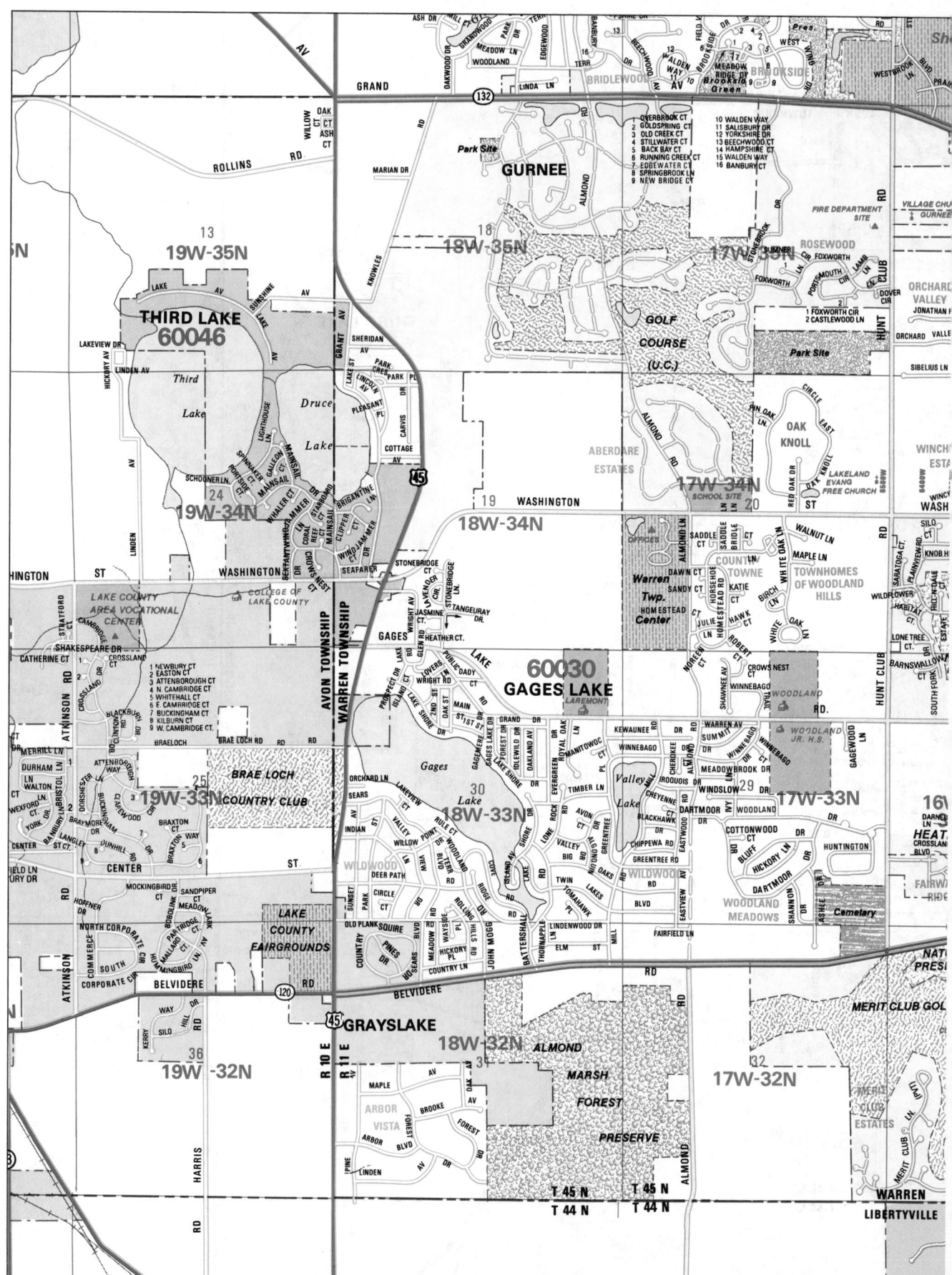

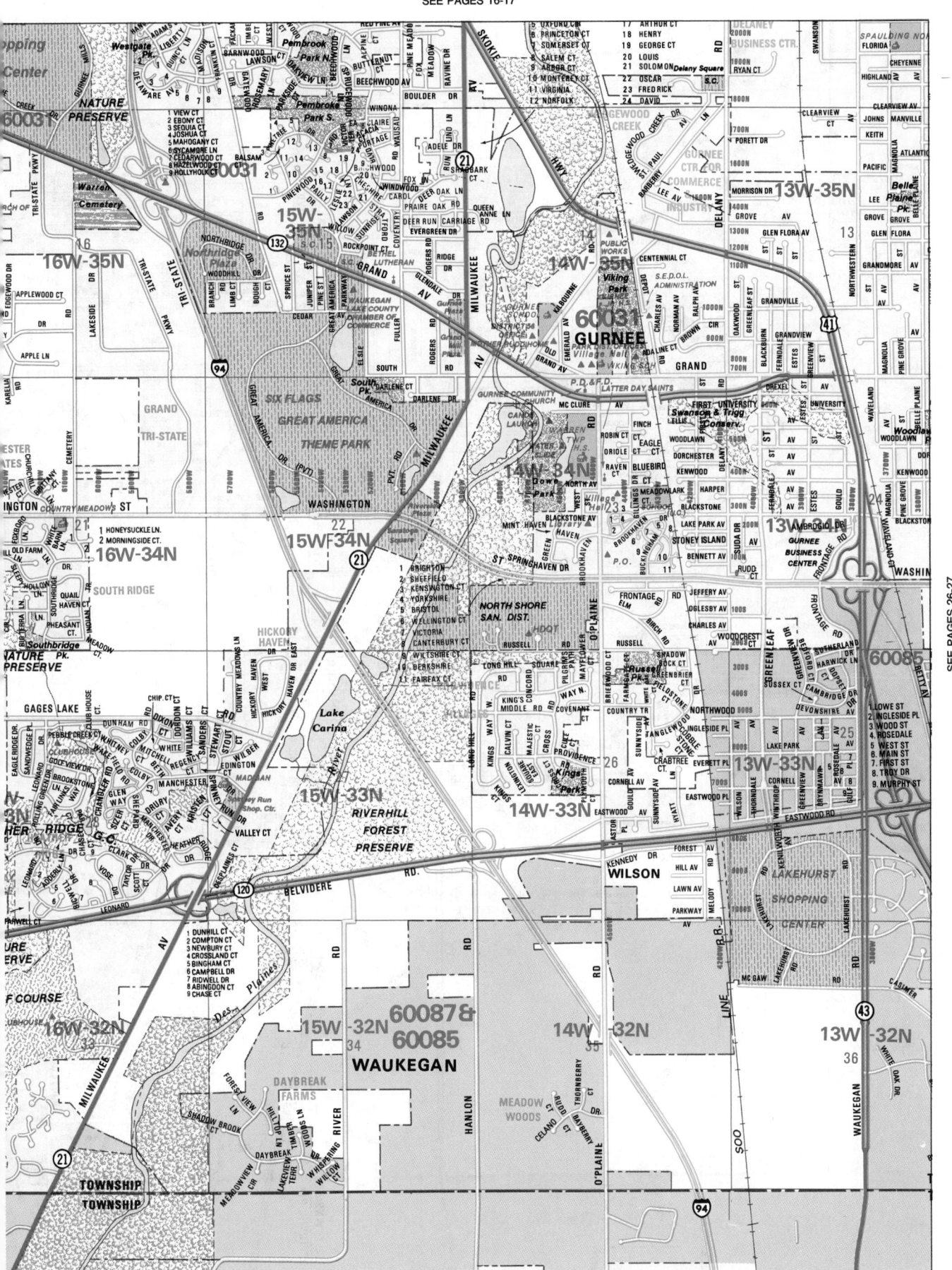

SEE PAGES 26-27

SEE PAGES 18-19

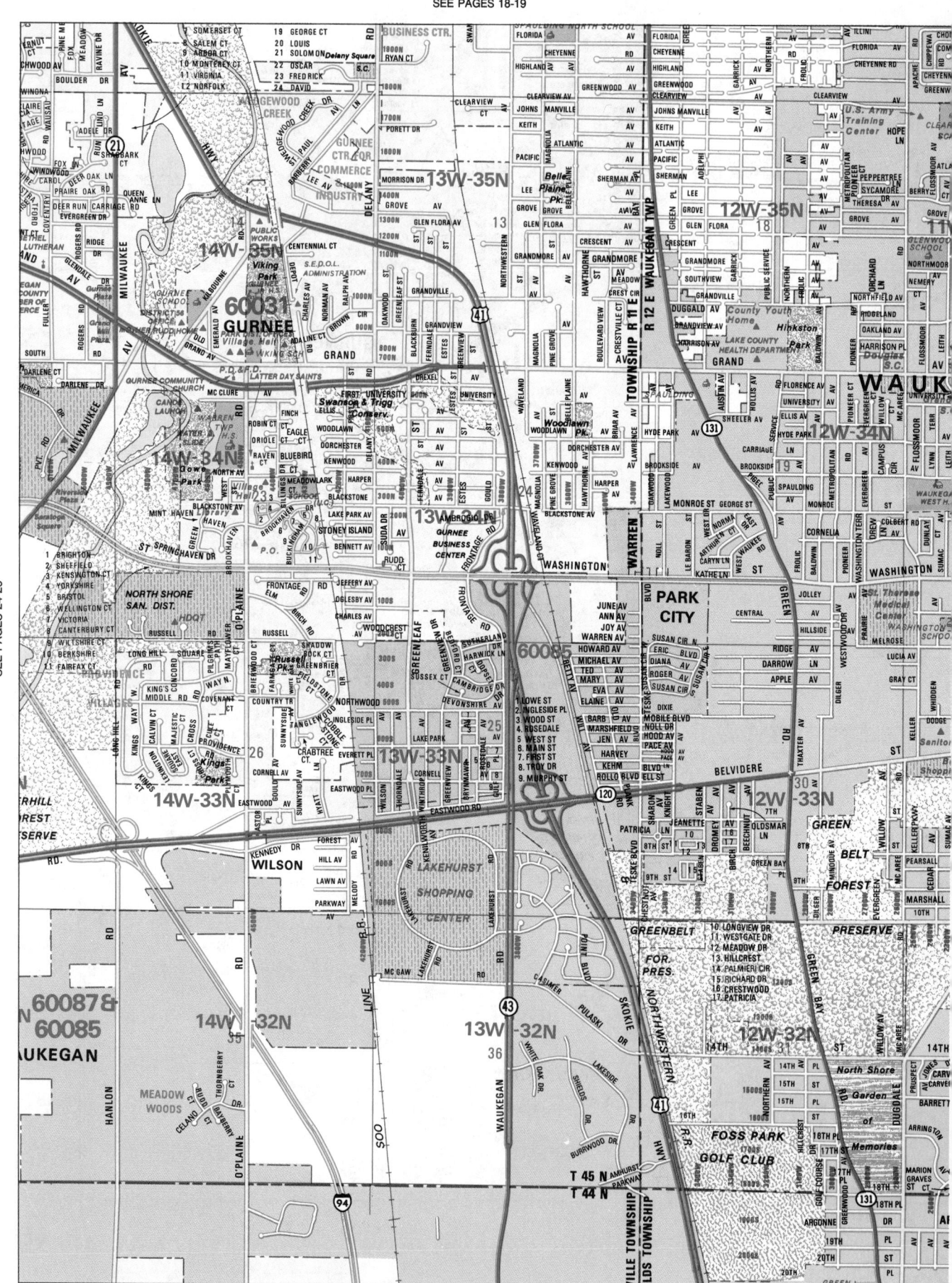

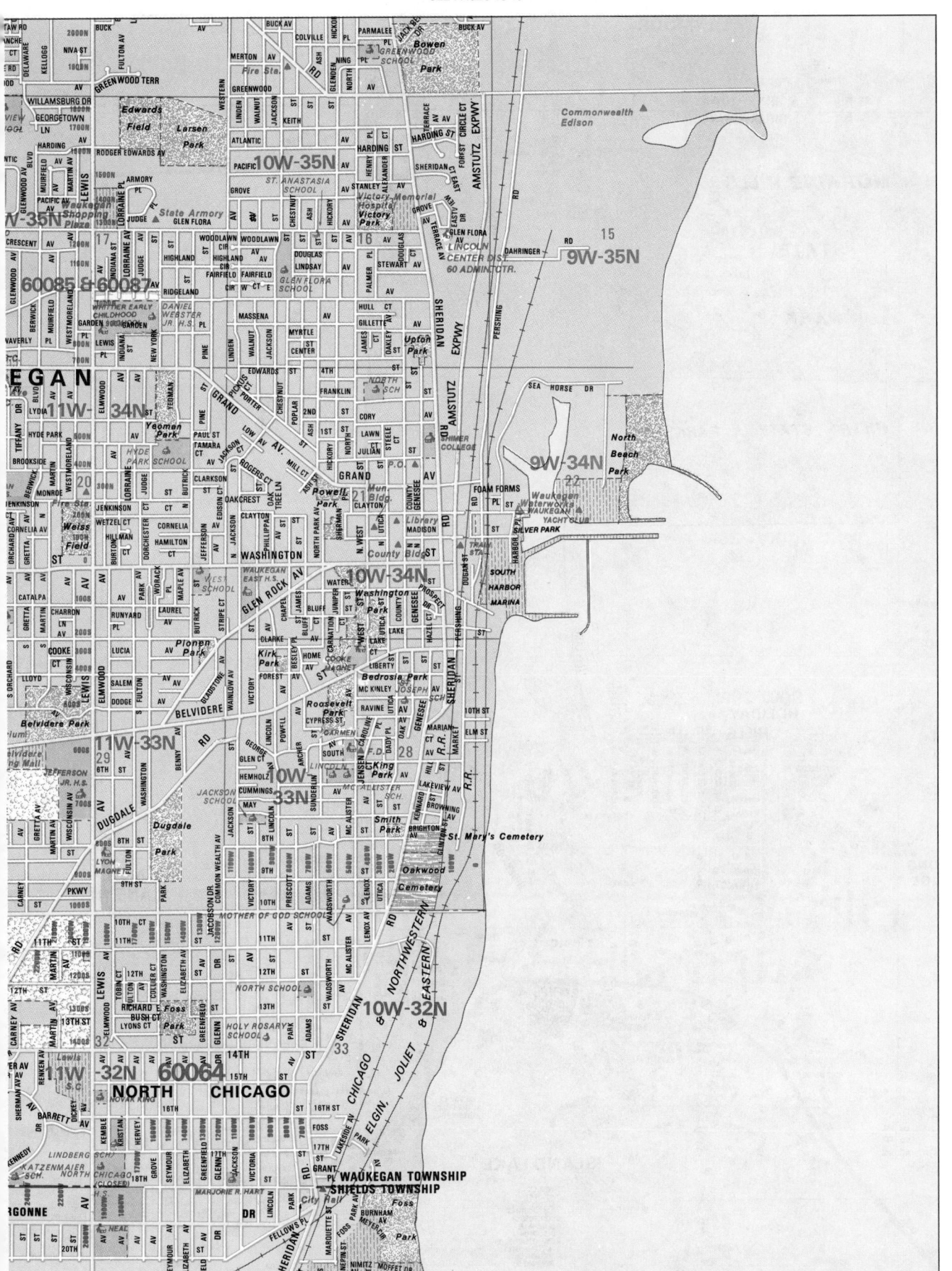

LILYMOOR
60050

T 45 N
T 44 N

MCHENRY TOWNSHIP
NUNDA TOWNSHIP

MORAINE HILLS

30W-31N

STATE

Lake
Defiance

PARK

DEFIANCE RD

29W-31N

LAKEMOOR
60050

4
28W-31N

120

T 45 N
T 44 N

FOX LAKE

DARRELL RD

FISHER RD

DARRELL RD

FISHER RD

HILLS STATE PARK

30W-30N

8
29W-30N

9
28W-30N

NEVILLE RD

DOWELL RD

60050 **30W-29N**
HOLIDAY
HILLS

29W-29N

MCHENRY

16
28W-29N

DARRELL RD

NUNDA TOWNSHIP
WAUCONDA TOWNSHIP

Highwood
Lake

**ISLAND
LAKE**

Griswold
Lake

BURNETT RD

21
28W-28N

176

30W-28N

20
29W-28N

ISLAND LAKE

BONNER RD

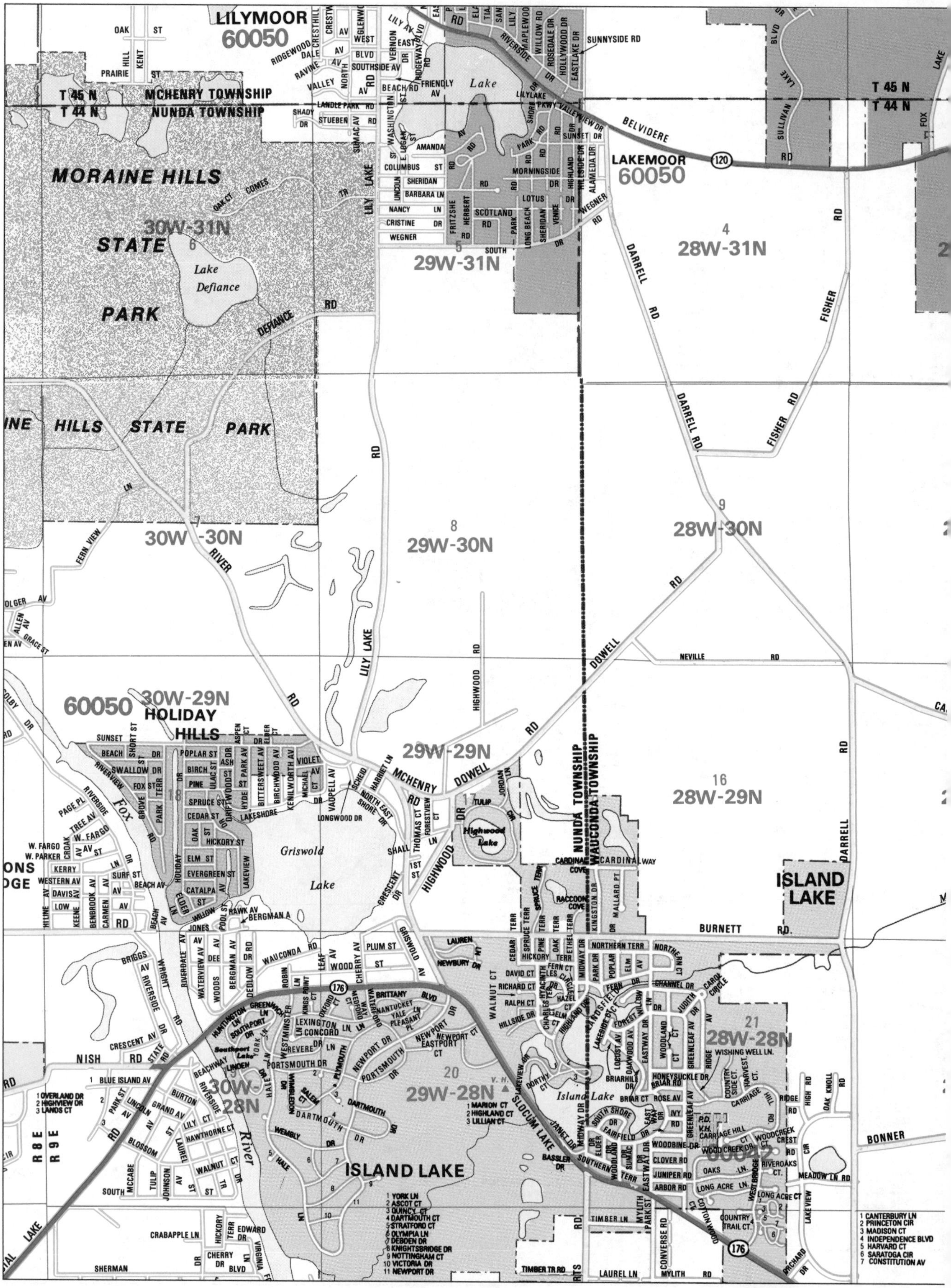

SEE McHENRY COUNTY PAGES 34-35

R 8 E
R 9 E

176

SEE PAGES 20-21

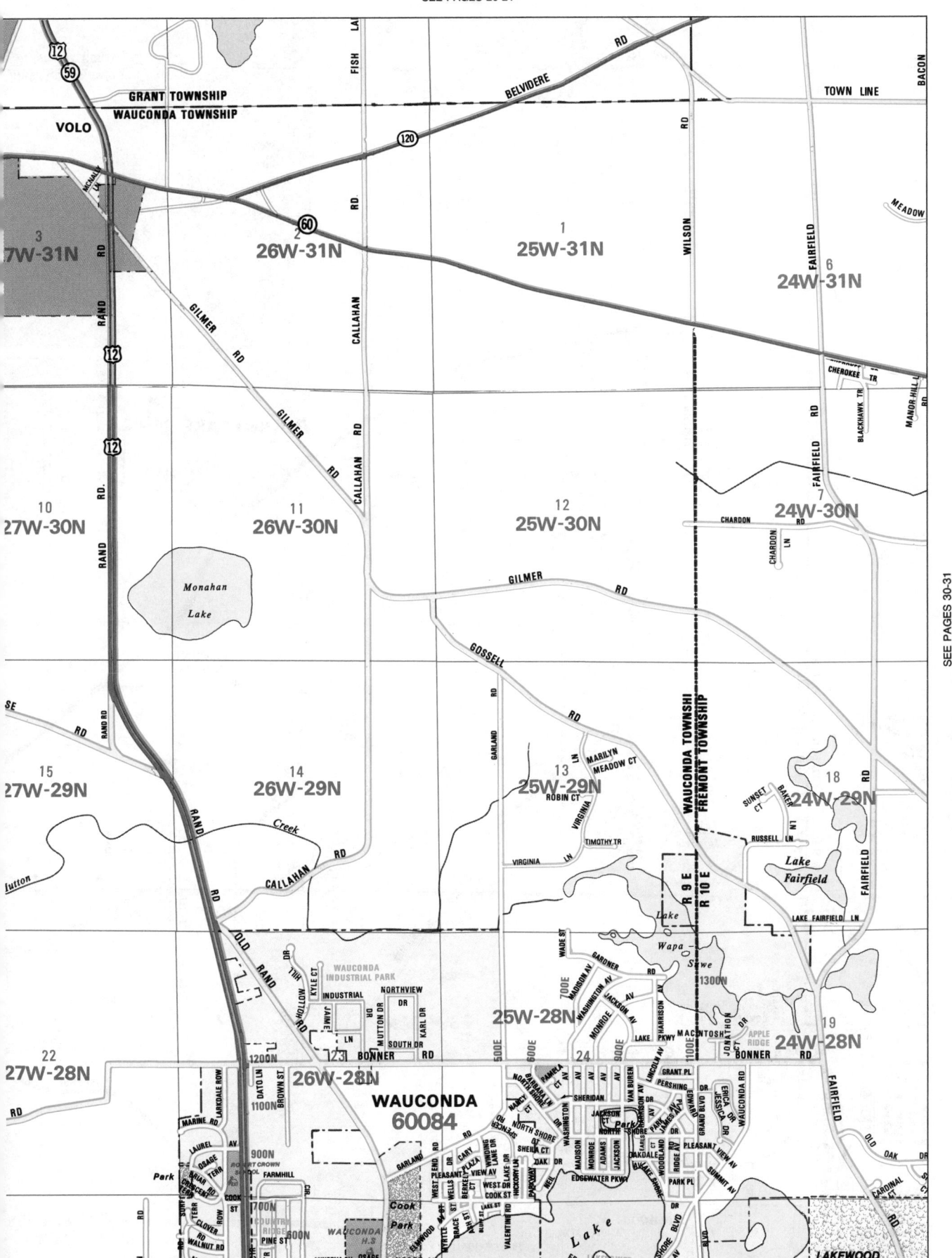

SEE PAGES 30-31

SEE PAGES 36-37

SEE PAGES 22-23

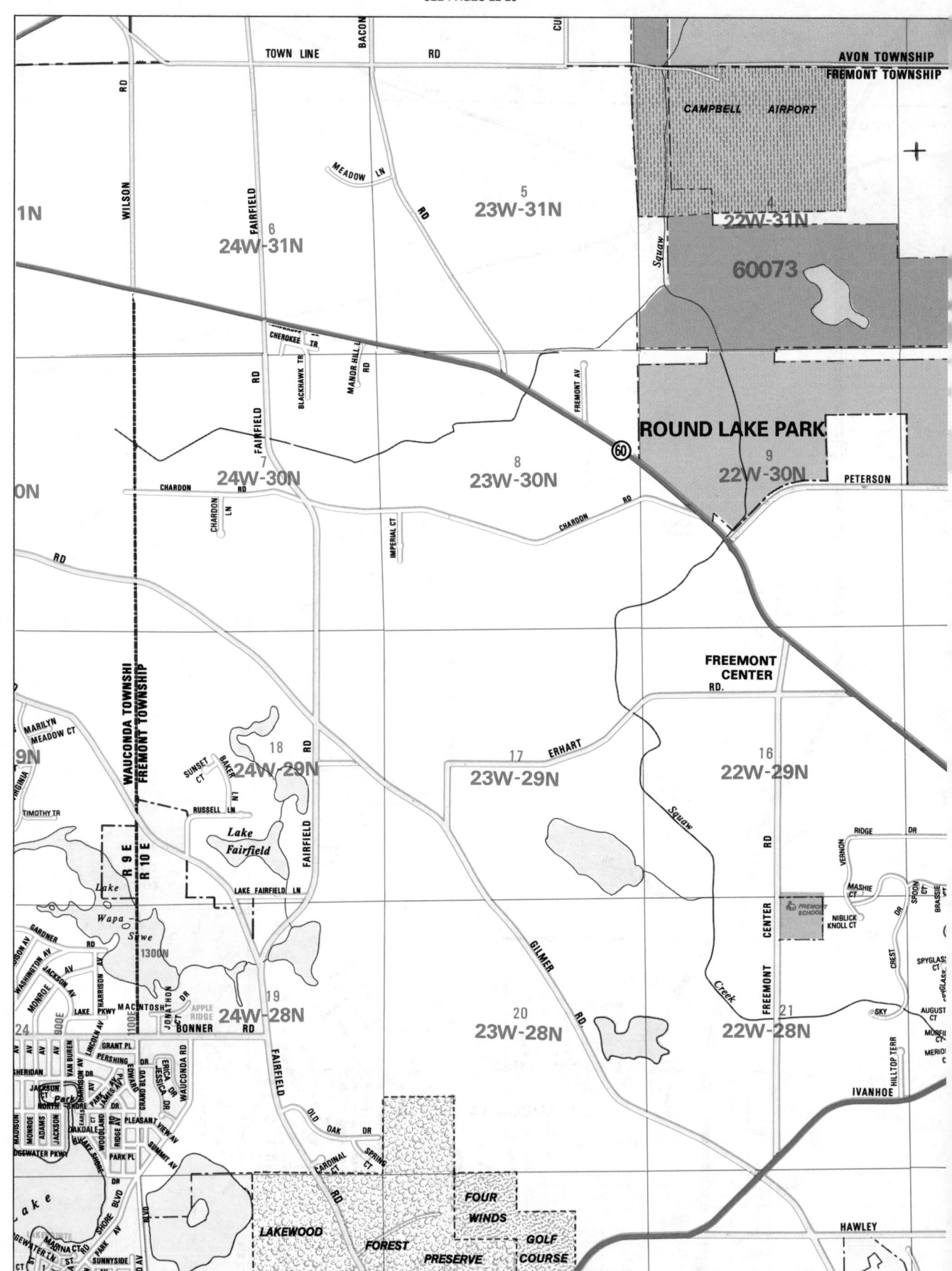

SEE PAGES 38-39

SEE PAGES 22-23

SEE PAGES 32-33

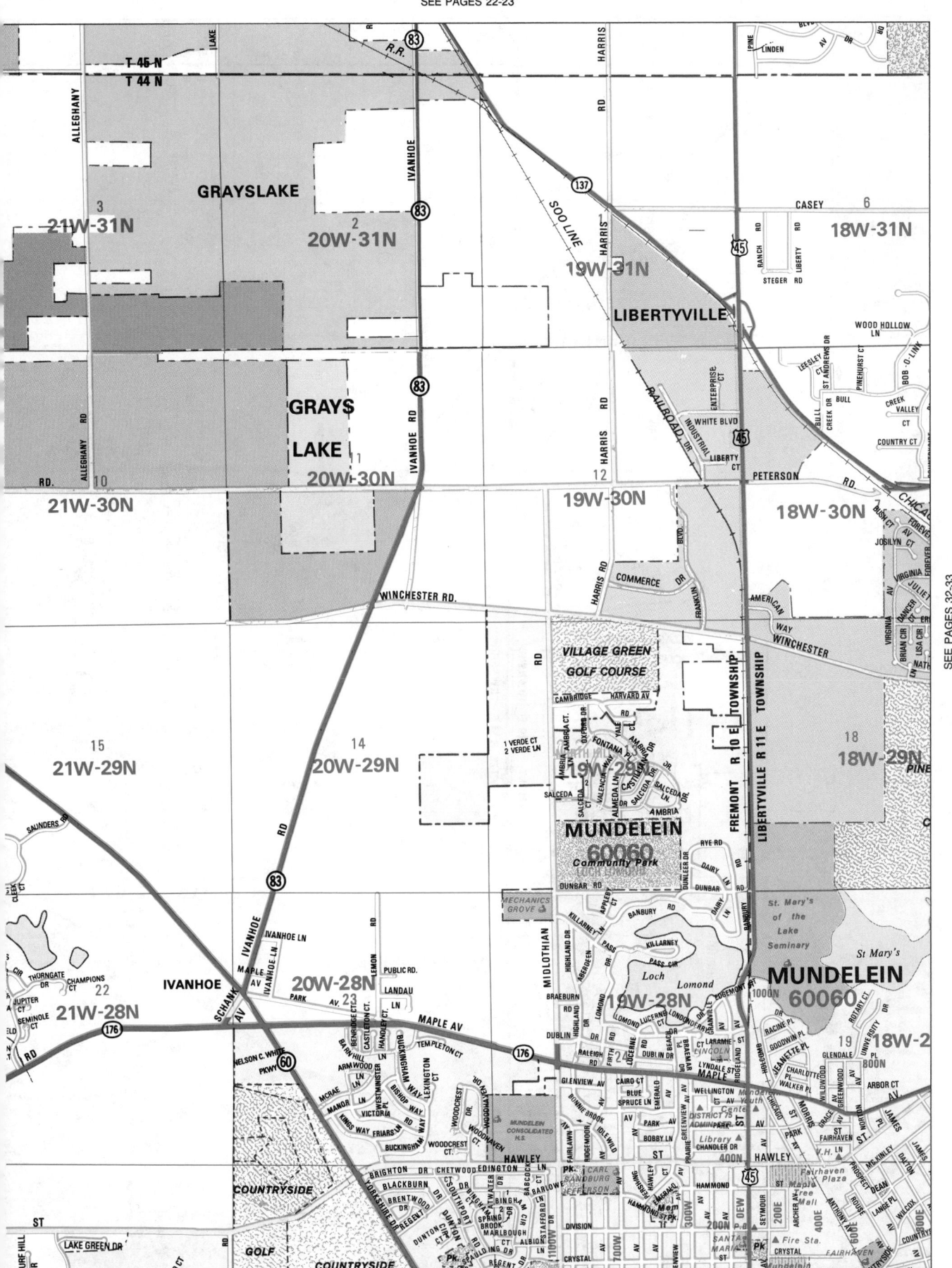

SEE PAGES 38-39

SEE PAGES 24-25

SEE PAGES 30-31

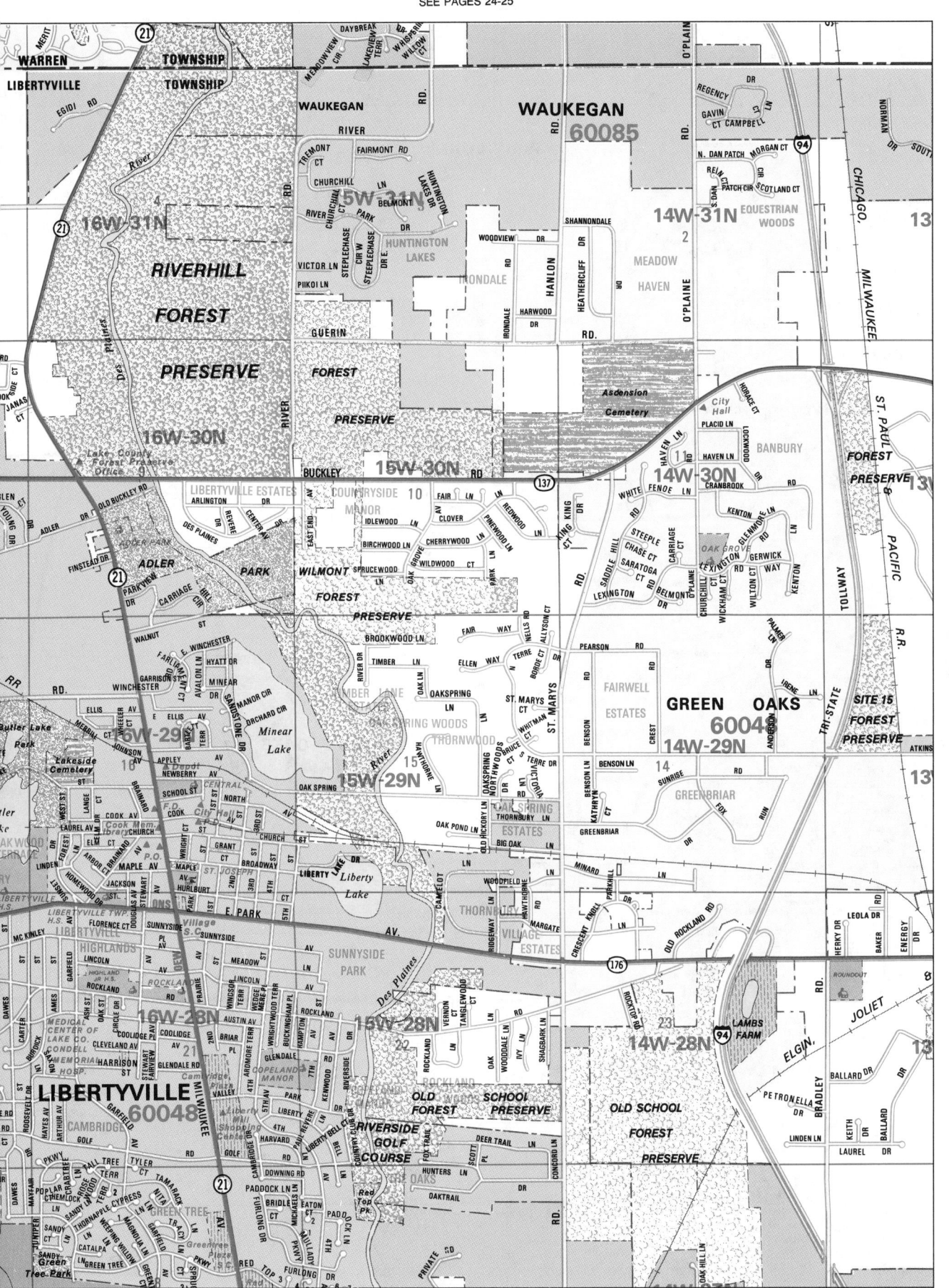

SEE PAGES 34-35

SEE PAGES 26-27

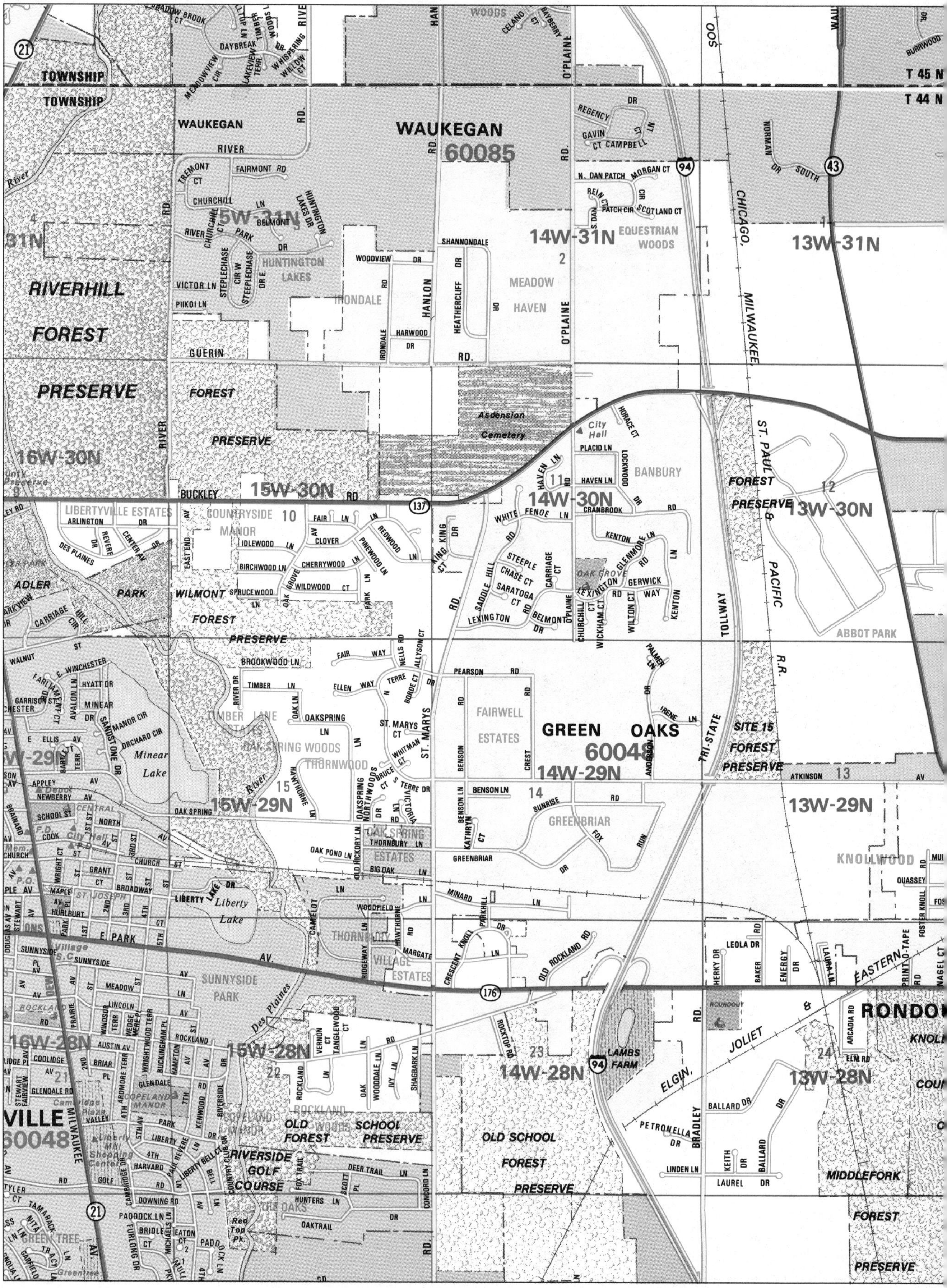

TOWNSHIP

TOWNSHIP

WAUKEGAN

WAUKEGAN
60085

RIVERHILL

FOREST

PRESERVE

16W-30N

15W-31N

14W-31N

13W-31N

EQUESTRIAN WOODS

MEADOW HAVEN

RIVERHILL FOREST PRESERVE

GUERIN

FOREST

PRESERVE

BUCKLEY

15W-30N

14W-30N

13W-30N

ST. PAUL FOREST PRESERVE

BANBURY

Ascension Cemetery

City Hall

LIBERTYVILLE ESTATES

COUNTRYSIDE MANOR

ADLER PARK

WILMONT FOREST PRESERVE

Minear Lake

TIMBER LANE ESTATES

OAK SPRING WOODS

THORNWOOD

FAIRWELL ESTATES

GREEN OAKS
60048

14W-29N

13W-29N

FOREST PRESERVE

SITE 15

15W-29N

OAK SPRING ESTATES

Liberty Lake

THORNBURY

SUNNYSIDE PARK

THORNBURY VILLAGE ESTATES

GREENBRIAR

KNOLLWOOD

16W-28N

15W-28N

14W-28N

13W-28N

ROCKLAND

OLD WOODS SCHOOL PRESERVE FOREST

OLD SCHOOL FOREST PRESERVE

RIVERSIDE GOLF COURSE

VILLE
60048

LAMBS FARM

RONDO

MIDDLEFORK

FOREST PRESERVE

SEE PAGES 32-33

SEE PAGES 26-27

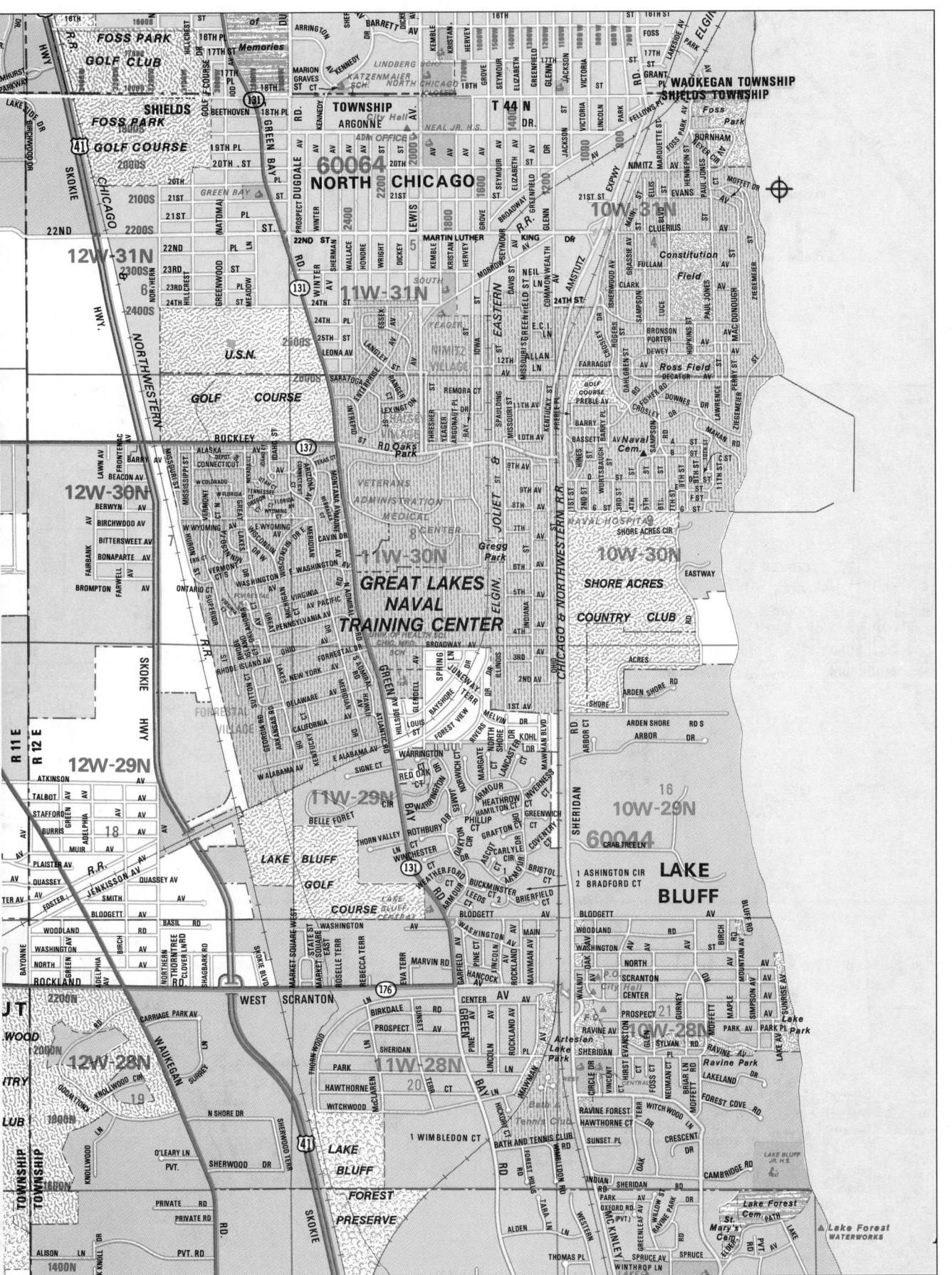

SEE PAGES 42-43

SEE PAGES 28-29

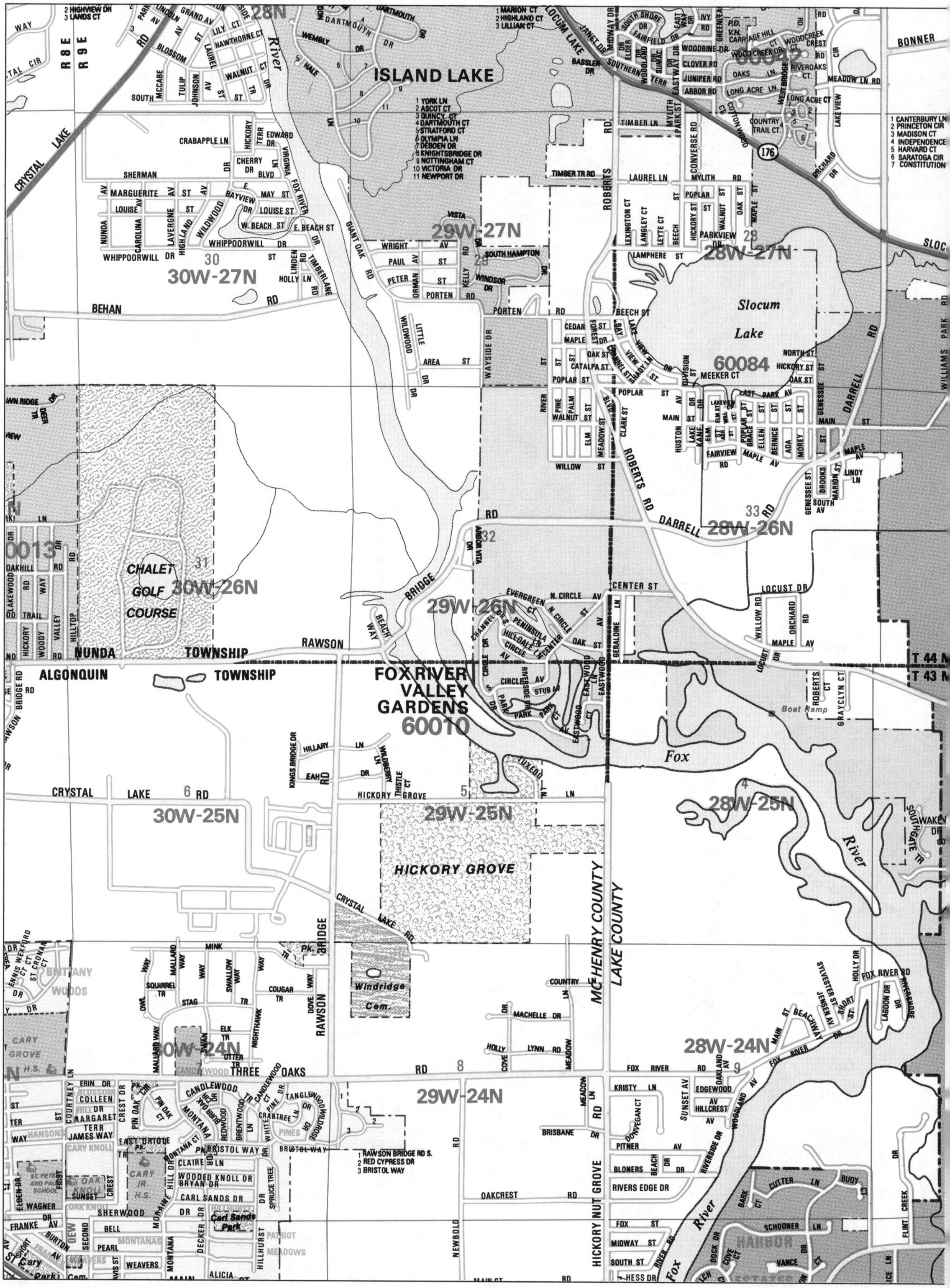

ISLAND LAKE

28N

29W-27N

28W-27N

30W-27N

30W-26N

29W-26N

28W-26N

30W-25N

29W-25N

28W-25N

30W-24N

29W-24N

28W-24N

FOX RIVER VALLEY GARDENS
60010

HICKORY GROVE

Windridge Cem.

CHALET GOLF COURSE

Slocum Lake

60084

Fox River

NUNDA TOWNSHIP

ALGONQUIN TOWNSHIP

THREE OAKS

60013

CRYSTAL LAKE

MC HENRY COUNTY
LAKE COUNTY

SEE McHENRY COUNTY PAGES 42-43

1 YORK LN
2 ASCOT CT
3 QUINCY CT
4 DARTMOUTH CT
5 STRATFORD CT
6 OLYMPIA LN
7 DEBDEN DR
8 KNIGHTSBRIDGE DR
9 NOTTINGHAM CT
10 VICTORIA DR
11 NEWPORT DR

1 MARION CT
2 HIGHLAND CT
3 LILLIAN CT

1 CANTERBURY LN
2 PRINCETON CIR
3 MADISON CT
4 INDEPENDENCE
5 HARVARD CT
6 SARATOGA CIR
7 CONSTITUTION

1 RAWSON BRIDGE RD S.
2 RED CYPRESS DR
3 BRISTOL WAY

176

T 44 N
T 43 N

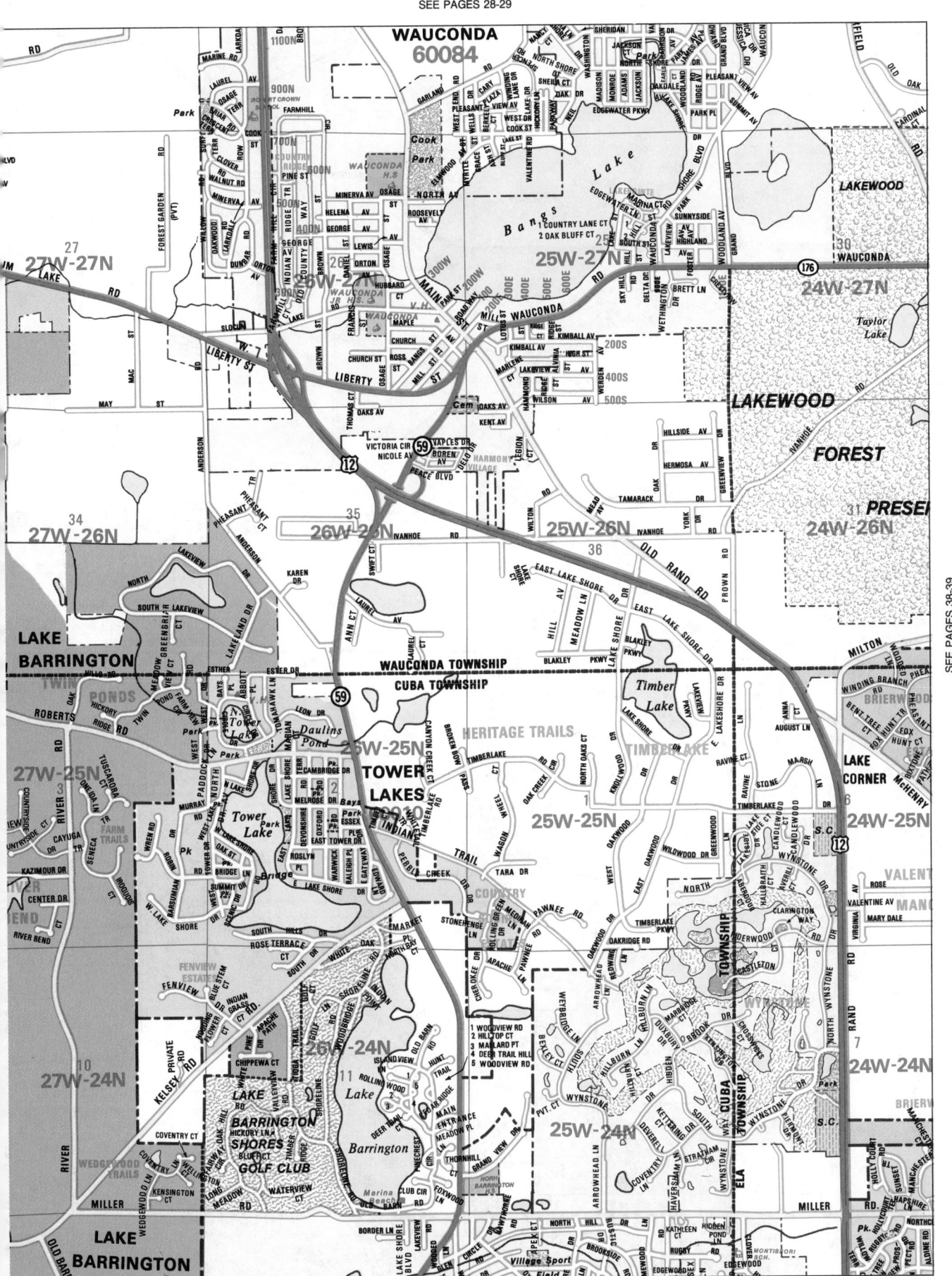

SEE PAGES 38-39

SEE PAGES 30-31

SEE PAGES 36-37

IVANHOE

LAKEWOOD

FOUR WINDS

FOREST PRESERVE

GOLF COURSE

HAWLEY

N. TURF HILL
LA VISTA
PRIMROSE
PRIMROSE CT
HILL DR

WAUCONDA RD

24W-27N

23W-27N

Davis Lake

28 22W-27N

GILMER RD

OWENS

Taylor Lake

Schreiber Lake

LAKEWOOD

FOREST

PRESERVE

24W-26N 31

23W-26N 32

OWENS RD

SCHWERMAN RD.

22W-26N 33

ROSEWO

FAIRFIELD RD

MILTON RD

PHEASANT RUN

SHEARWATER CT
MEADOWLARK
ORIOLE CT
LAKE LORRAINE ESTATES
BLUEJAY LN
GOLDFINCH CT
BLUEBIRD CT
CARDINAL
FALCON CT
CAMDEN TRAC
PEREGRINE

FREEMONT TOWNSHIP
ELA TOWNSHIP

T 44 N
T 43 N

HAWTHORN WOO

MILTON RD

BRIERWOODS
WINDING BRANCH RD
BENT TREE CT
HUNT TR
EDX HUNT CT
BOTTOM TRAIL
MOOREGATE
THORNFIELD
PADDOCK DR
EQUESTRIAN WAY
STEEPLECHASE
BRIDLEWOODS CT
WESTWIND
FUR LONG
NORTH TRAIL
NORTH TR

SOMERSET CT
STONEY KIRK
N. KYLE CT
KYLE S
KYLE CT

22W-25N 4

KNUCKENBERG

LAKE CORNER
McHENRY

24W-25N 6

HAWTHORN WOODS

23W-25N 5
60047

LOCHANDRA
BONNY RIGG CT
BAGPIPE CT
GENTRY DR
GOVERNORS WAY

KREUSER WHITE BIRCH LAKES
OVERLOOK DEER
POINT DR
HIGH POINT CIR
HIGH POINT

COOPER CT
BYRON CT
DAVID
NANCY CT
MARLYN
KATHY

WHITMAN TERR
KATHY
NORMESS LN
MARLES

EGEN
KNOLLS

KNOTT

VALENTINE MANOR
ROSE AV
VALENTINE AV
VIRGINIA AV
MARY DALE AV
ELLRIE
LAKEWOOD BOXWOOD ESTATES

SYCAMORE
REDWOOD
COBBLEWOOD LN
NEWHAVEN
ARROWWOOD
MIDDLETREE
ELDERBERT DR
WALNUT CREEK

Copperfield Park
COPPERFIELD DR
CHANTILLY
COPPERFIELD DR
WEDGEWOOD LN
BURNETT WHITE BIRCH LAKES
MULBERRY DR
COPPERFIELD
CHANCELLOR DR
L. Leo
BIRCH LAKE RD
L. Naomi
WHITE BIRCH LAKES

60047 9
22W-24N

WYNSTONE

24W-24N 7

CUBA TOWNSHIP

N. BRUCE CIR
BRUCE CIR
WAYNE LN
BRUCE CIR LN
ACORN DR
LYDIA CT
ACORN ACRES
ECHO CT

STONE CREEK CT
MEADOW
HARVEST GLEN CT
DAN VERA
MARCH ST

HEATHER HIGHLANDS

ROBIN CREST RD
23W-24N 8

MILLER RD

SHAGBARK
ELMWOOD

ECHO LAKE RD

MID LOTHIAN RD

GABRIEL DR
SACANAND CT

DAKWOOD

McGREGOR CT
QUENTIN
HIGHLAND TER
PIPER
Pk.

BRIERWOOD
BRIERWOODS
MANCHESTER
TANGLEWOOD LN

SETH PAINE

HOLLY COURT
SUNSET
LANCASTER
HAPSHIRE
MILLER RD.

NORTHCREST RD
NORTH LAKEWOOD LN
LONETREE
CRESTWOOD

Echo Lake
ECHO LAKE

SOUTH LAKEWOOD LN
HILLCREST

SEE PAGES 30-31

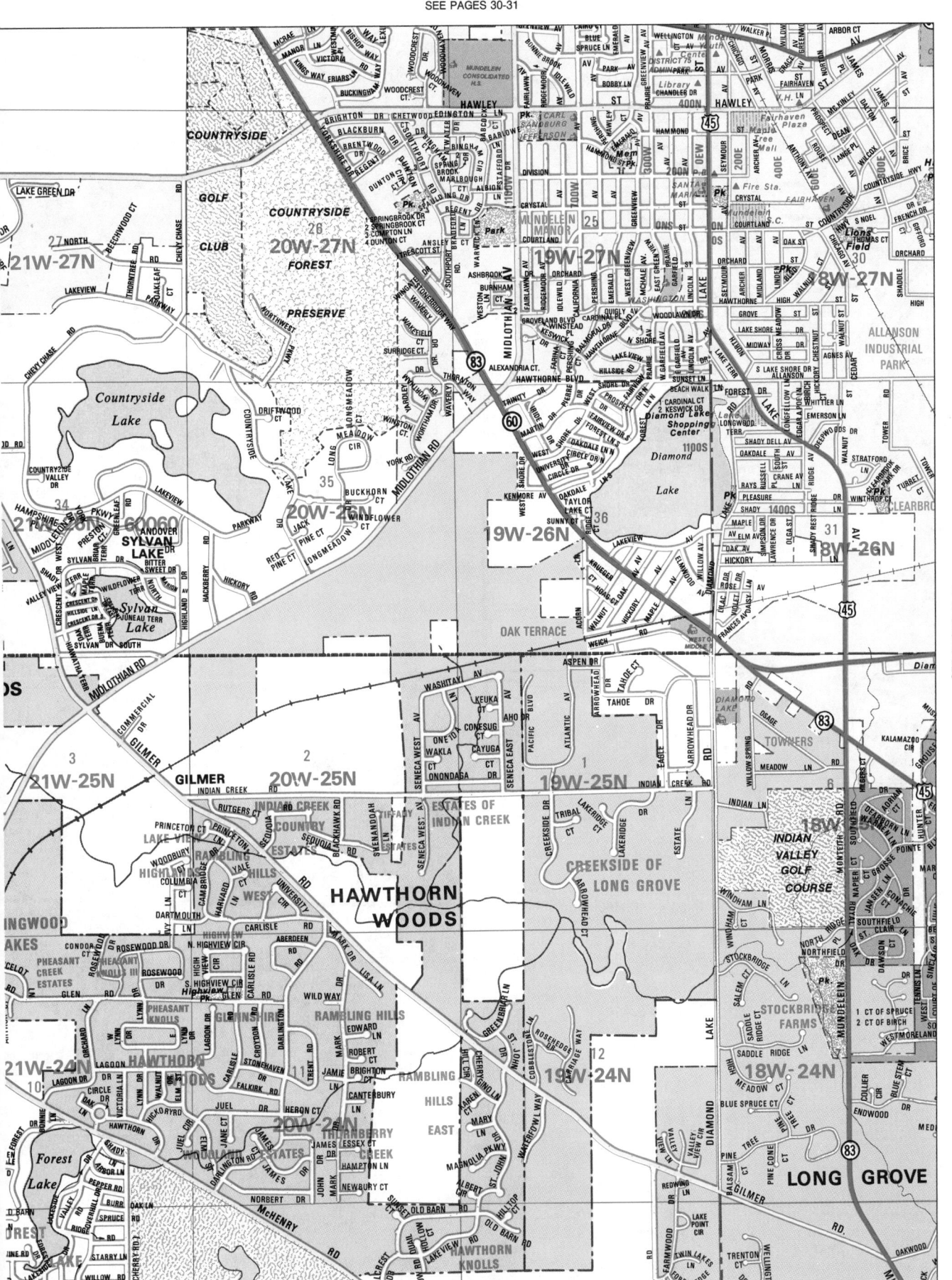

SEE PAGES 40-41

SEE PAGES 46-47

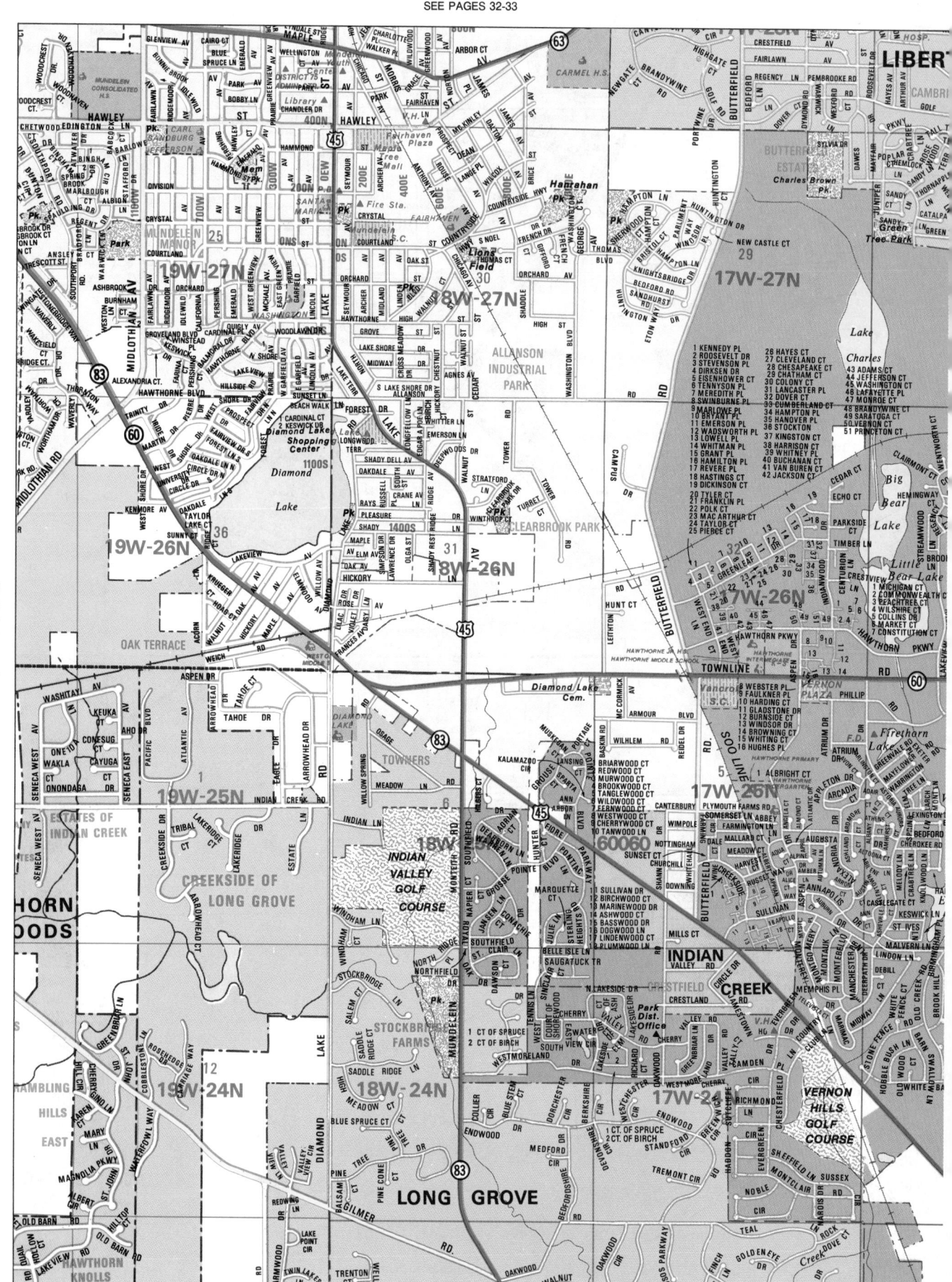

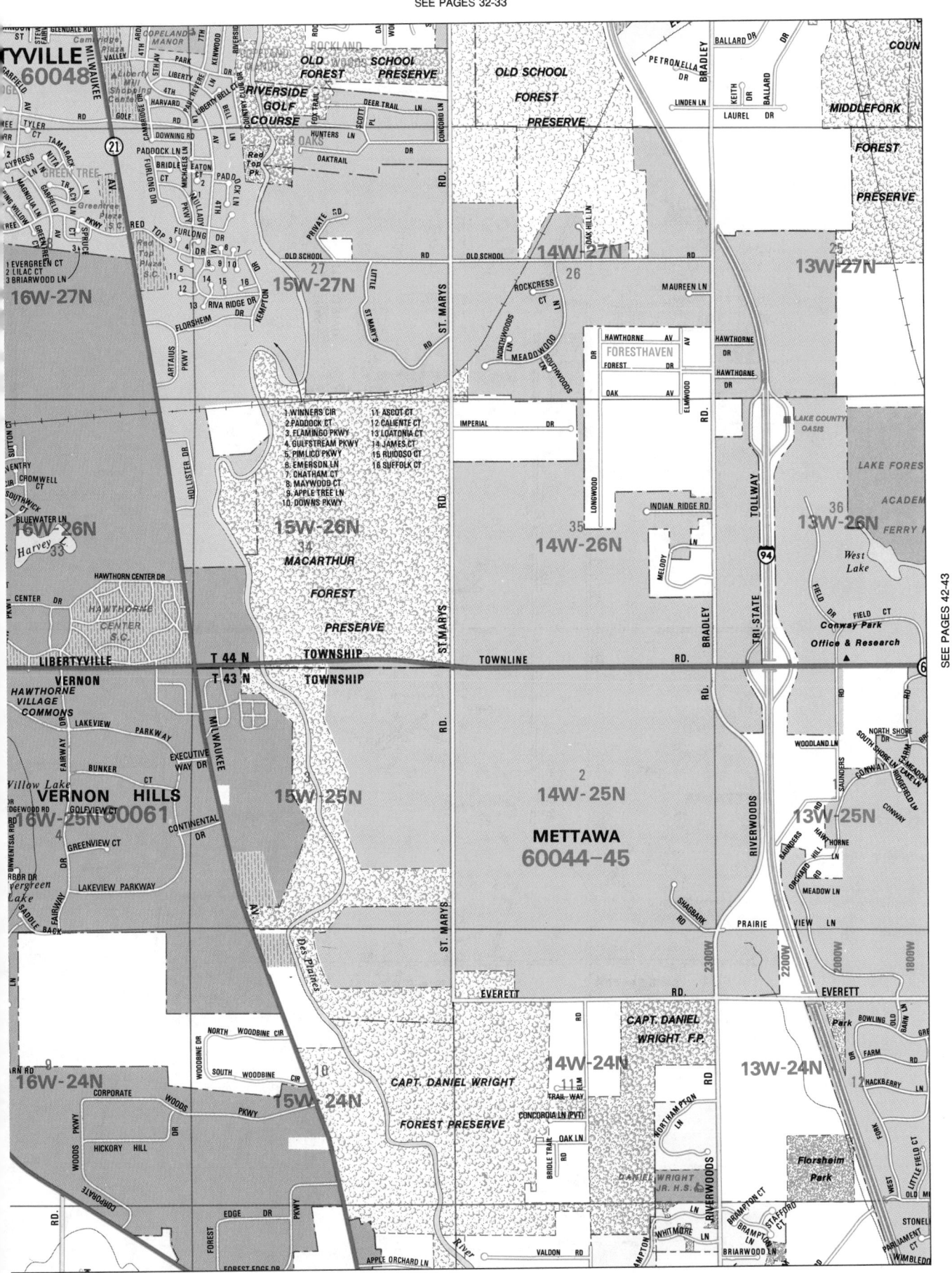

SEE PAGES 32-33
SEE PAGES 42-43
SEE PAGES 48-49

SEE PAGES 34-35

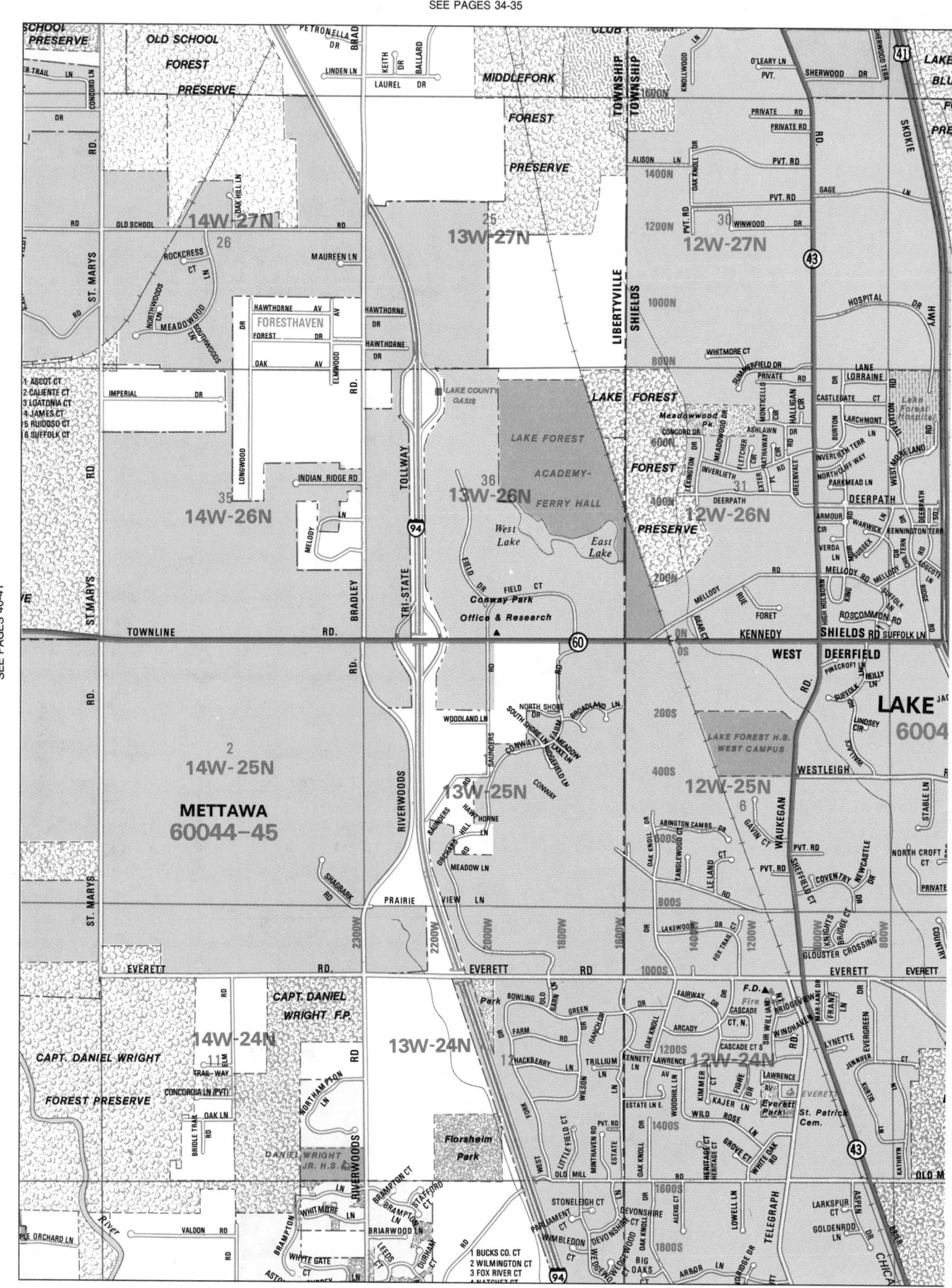

SEE PAGES 40-41

SEE PAGES 50-51

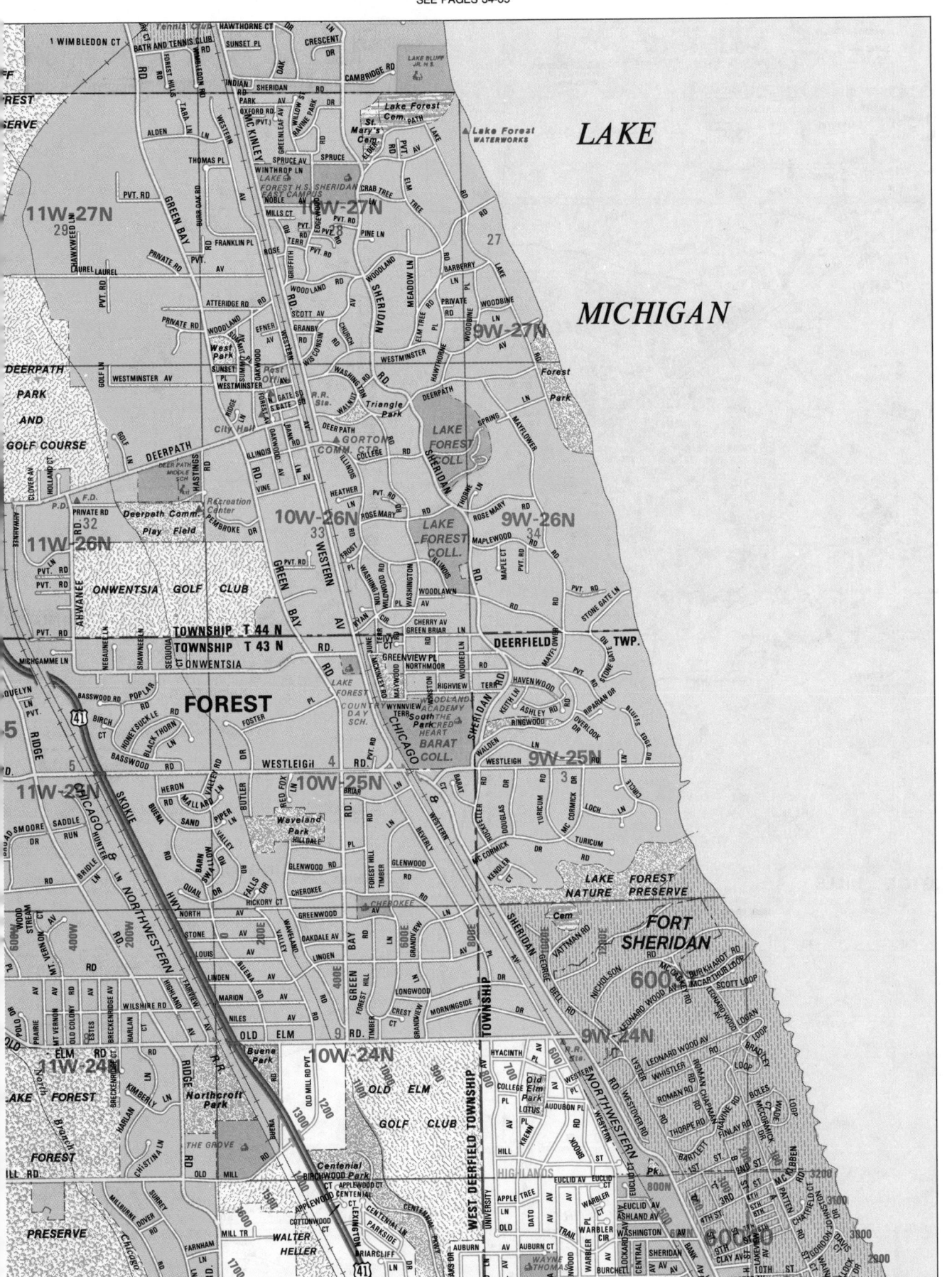

LAKE

MICHIGAN

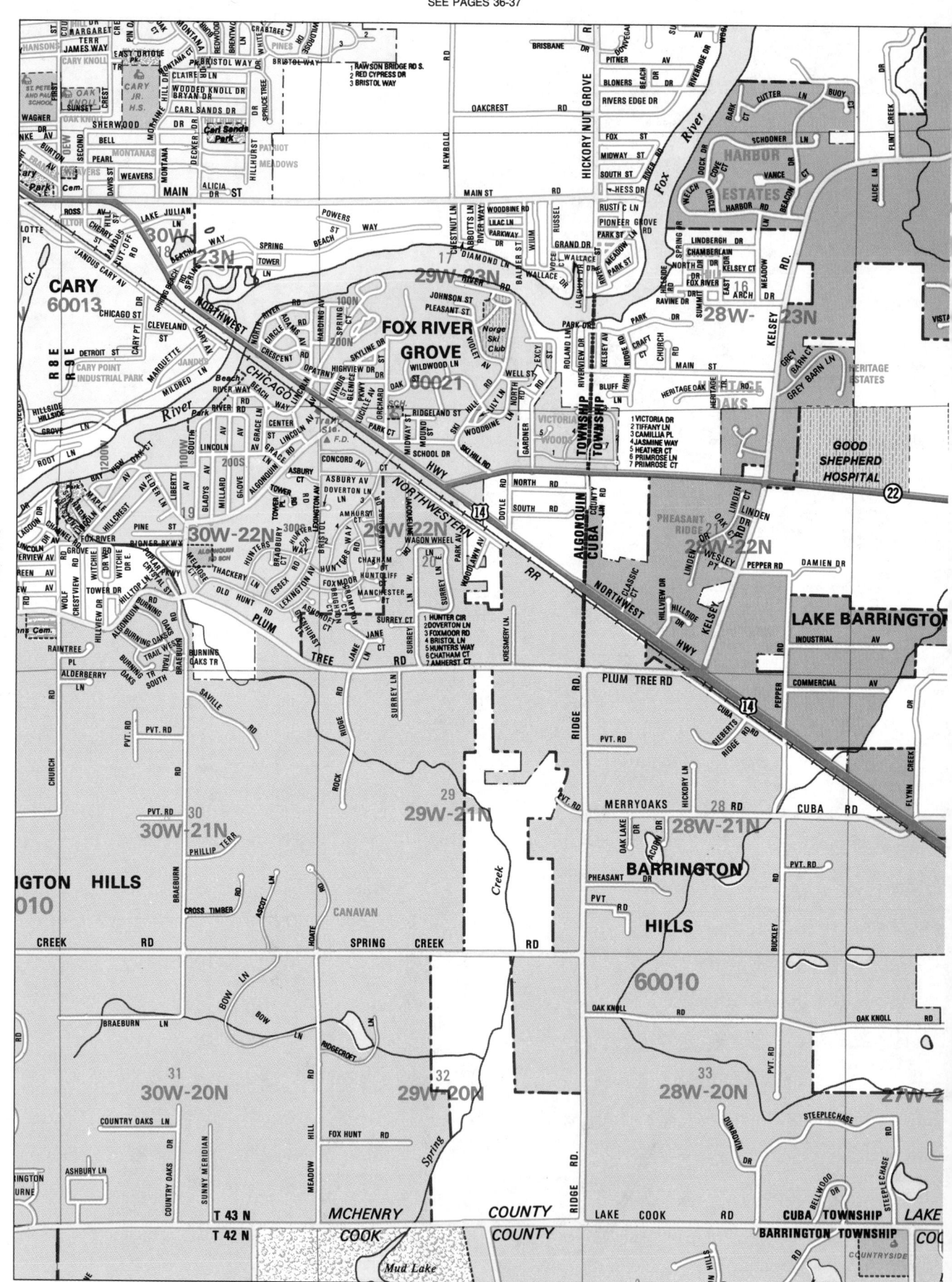

SEE PAGES 36-37

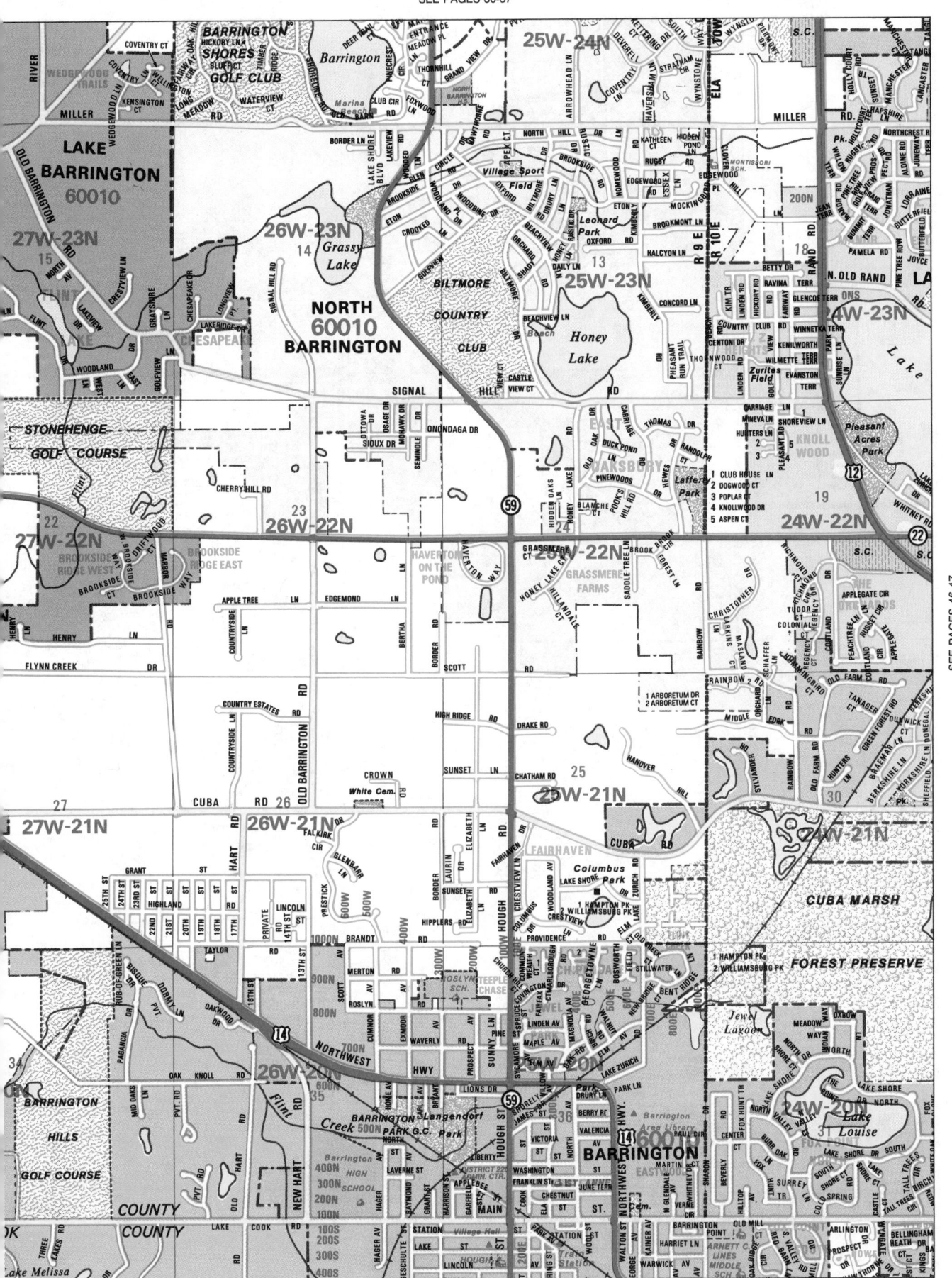

SEE PAGES 46-47

46 LAKE COUNTY

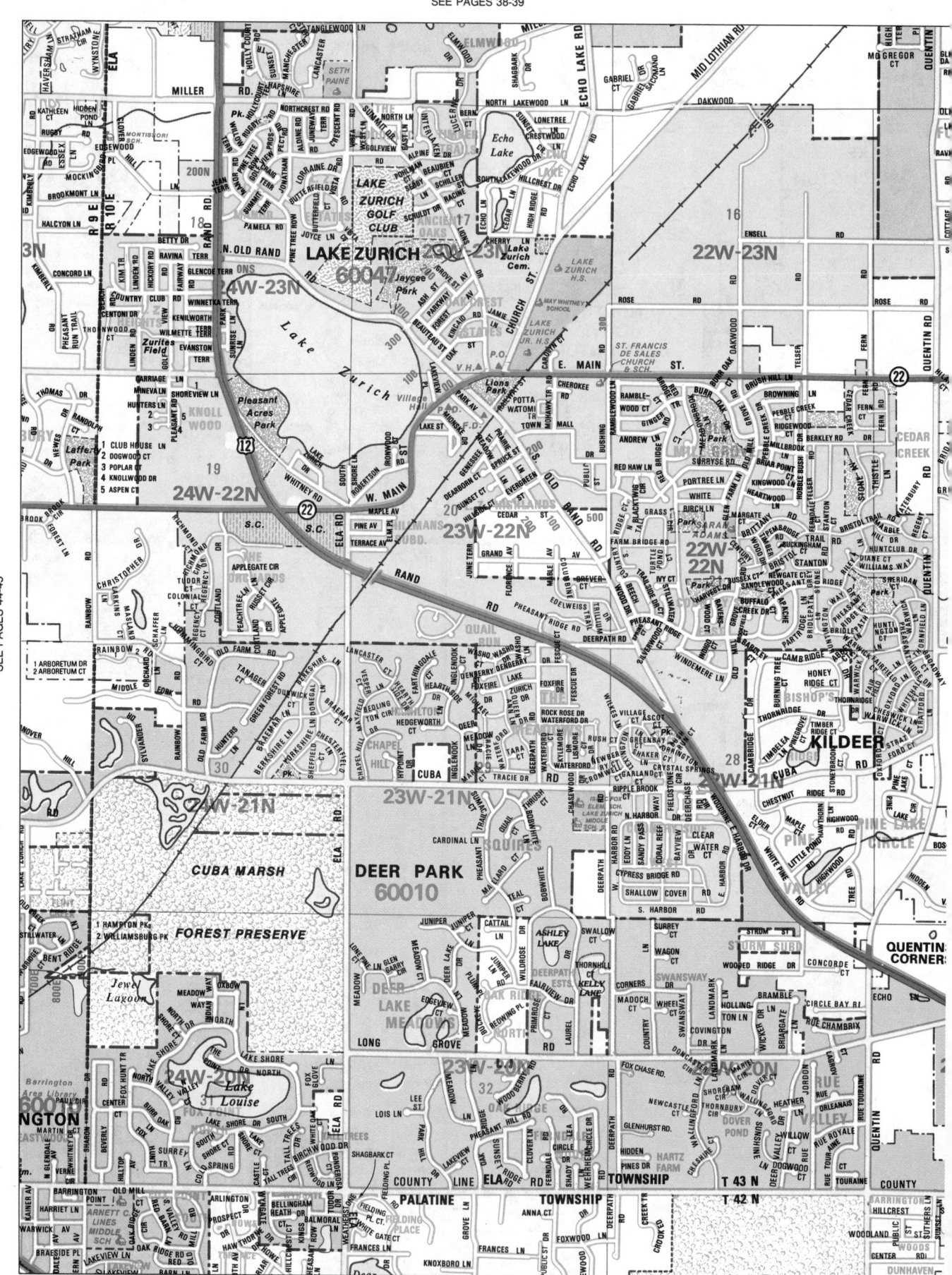

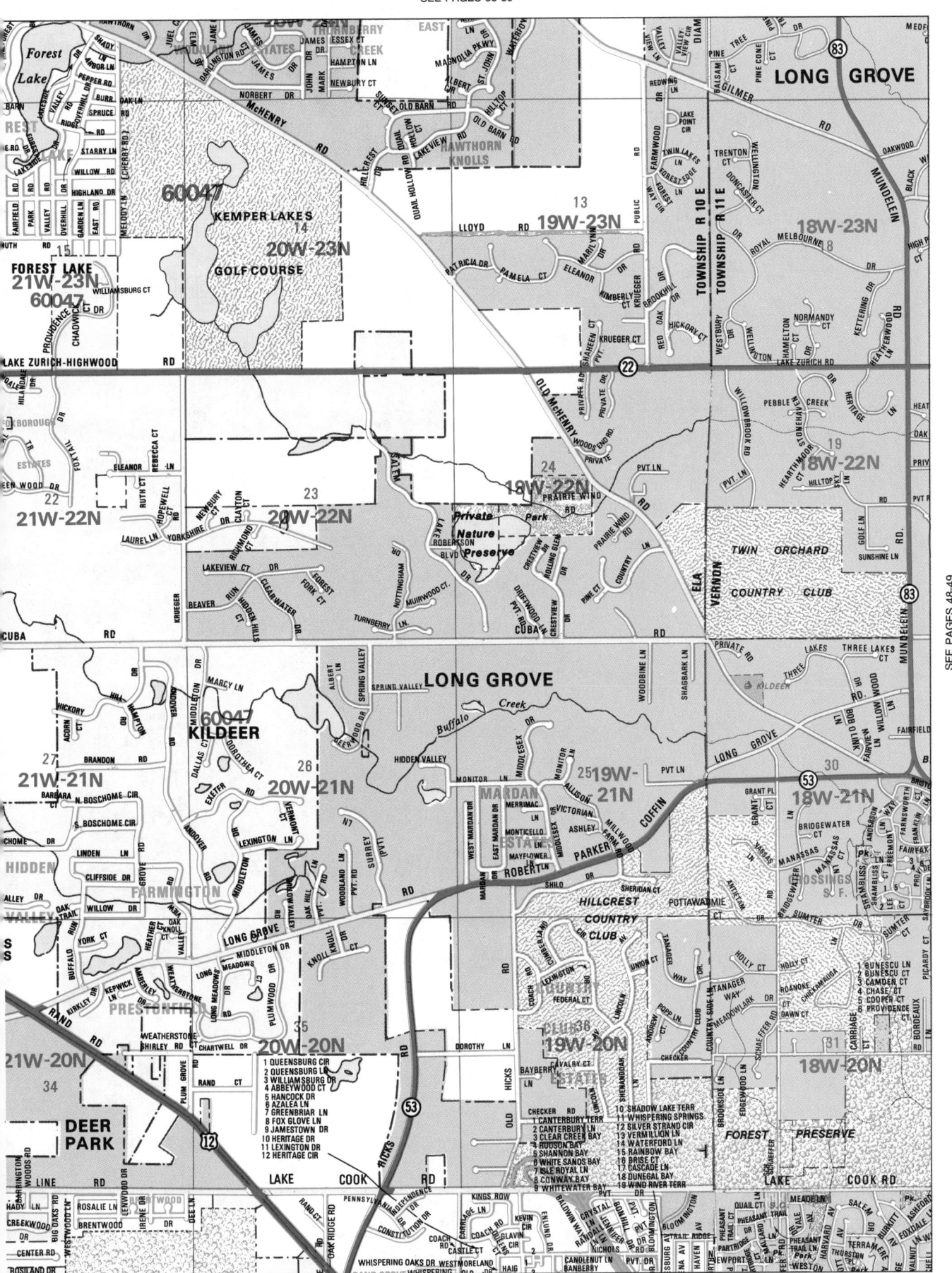

SEE PAGES 40-41

SEE PAGES 46-47

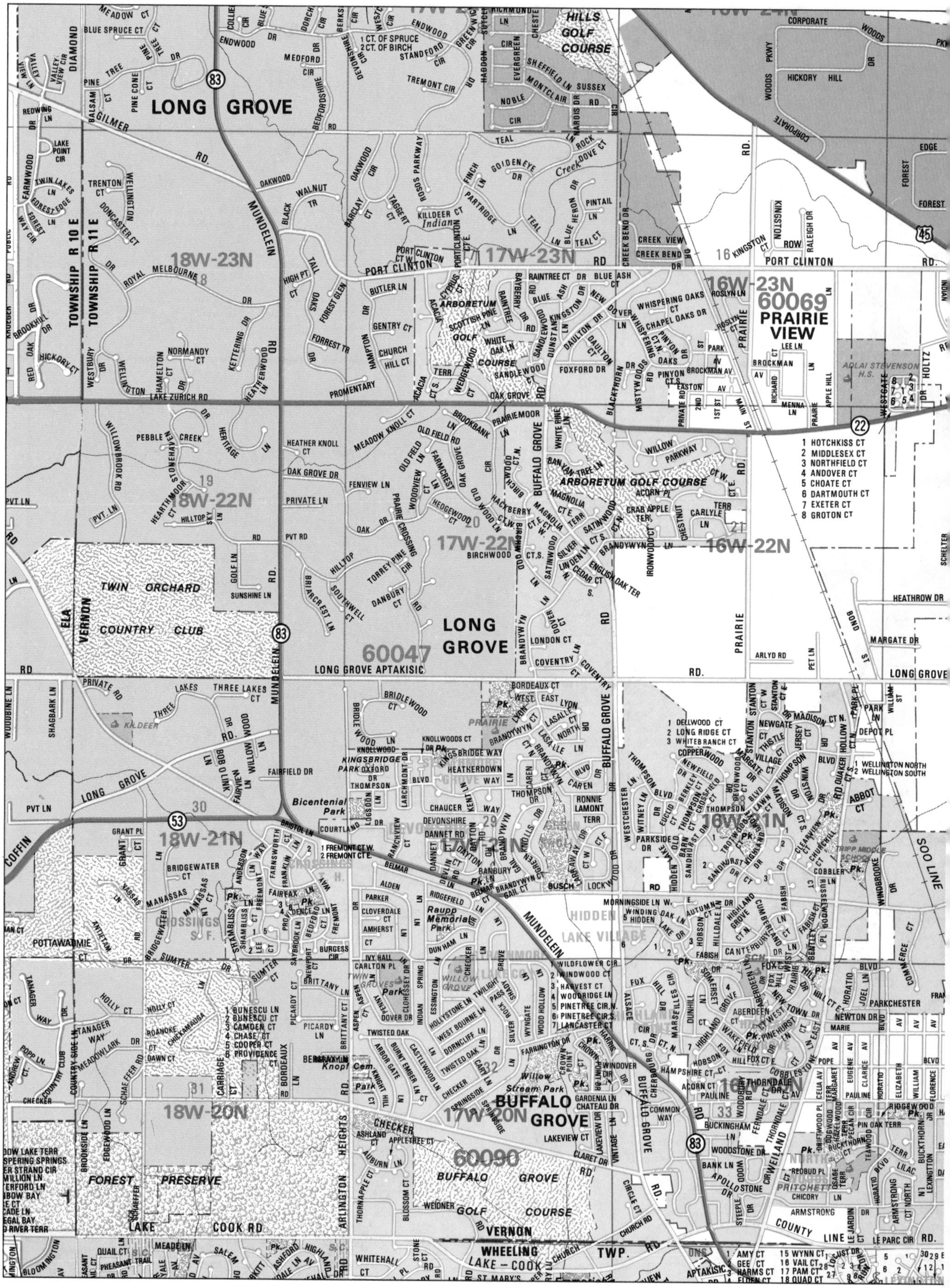

LONG GROVE

PORT CLINTON

TOWNSHIP R 10 E

TOWNSHIP R 11 E

18W-23N

17W-23N

16W-23N

60069 PRAIRIE VIEW

HILLS GOLF COURSE

ARBORETUM GOLF COURSE

18W-22N

17W-22N

16W-22N

TWIN ORCHARD COUNTRY CLUB

LONG GROVE

60047

LONG GROVE APTAKISIC

BUFFALO GROVE

ADLAI STEVENSON H.S.

18W-21N

17W-21N

16W-21N

18W-20N

17W-20N

16W-20N

BUFFALO GROVE

60090

BUFFALO GROVE GOLF COURSE

FOREST PRESERVE

MUNDELEIN

VERNON TWP.

WHEELING TWP.

LAKE-COOK RD.

VERNON

1 HOTCHKISS CT
2 MIDDLESEX CT
3 NORTHFIELD CT
4 ANDOVER CT
5 CHOATE CT
6 DARTMOUTH CT
7 EXETER CT
8 GROTON CT

1 DELLWOOD CT
2 LONG RIDGE CT
3 WHITE BRANCH CT

1 WELLINGTON NORTH
2 WELLINGTON SOUTH

1 FREMONT CT W.
2 FREMONT CT E.

1 BUNESCU LN
2 BUNESCU CT
3 CAMDEN CT
4 CHASE CT
5 COOPER CT
6 PROVIDENCE CT

1 WILDFLOWER CIR
2 WINDWOOD CT
3 HARVEST CT
4 WOODRIDGE LN
5 PINETREE CIR N.
6 PINETREE CIR S.
7 LANCASTER CT

SEE COOK COUNTY PAGES 14-15

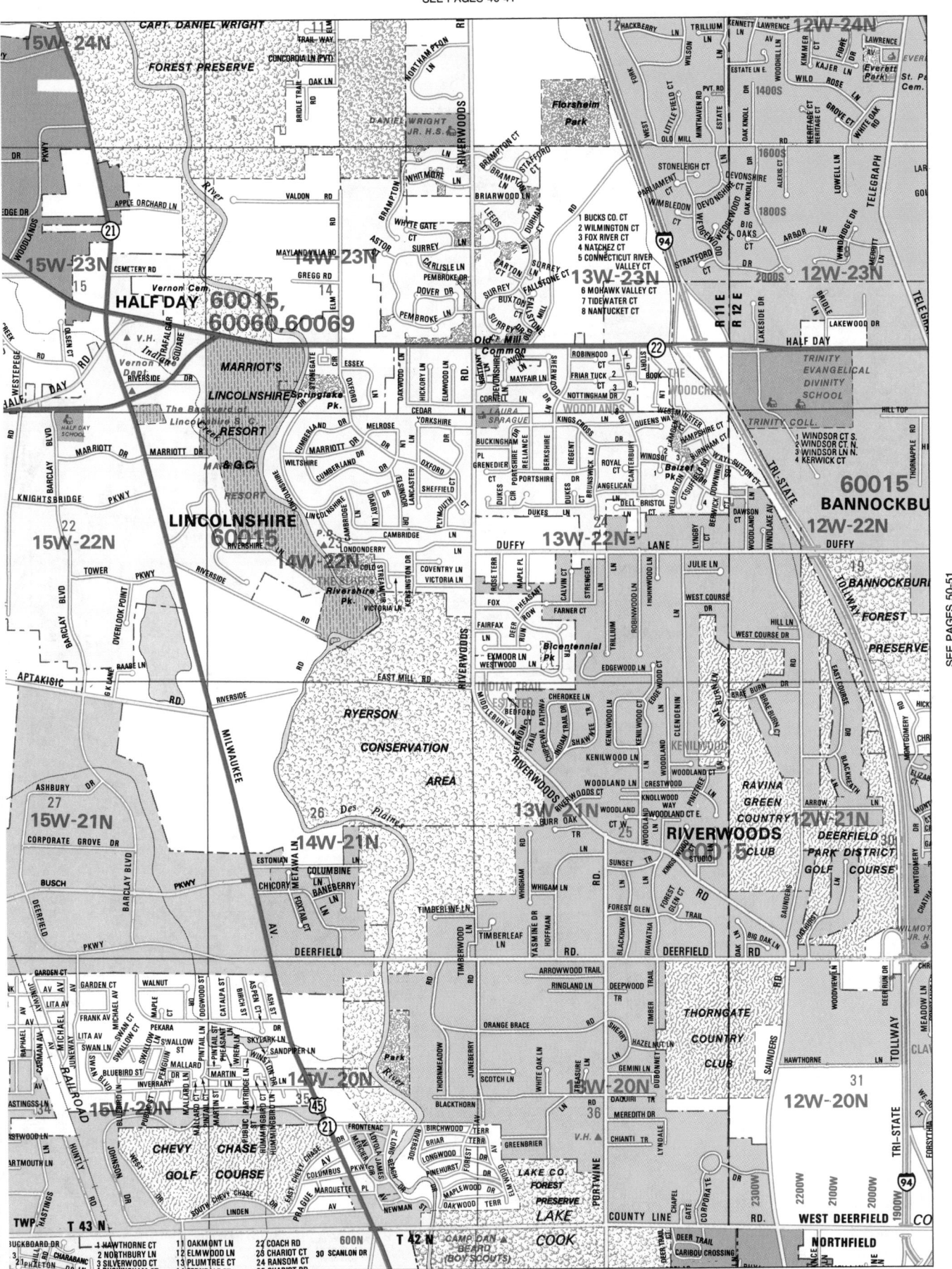

SEE PAGES 50-51

SEE PAGES 42-43

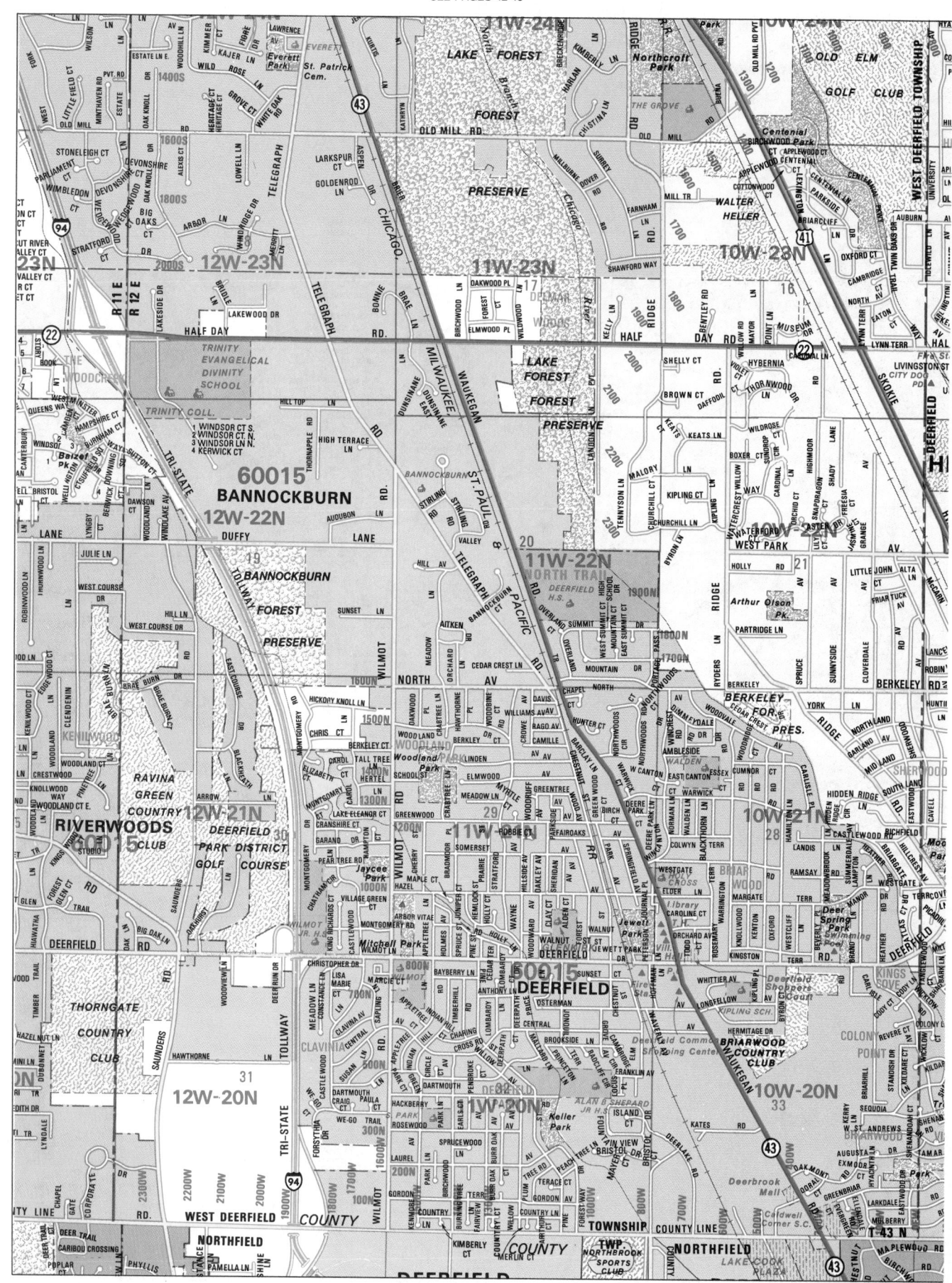

SEE PAGES 48-49

SEE PAGES 42-43

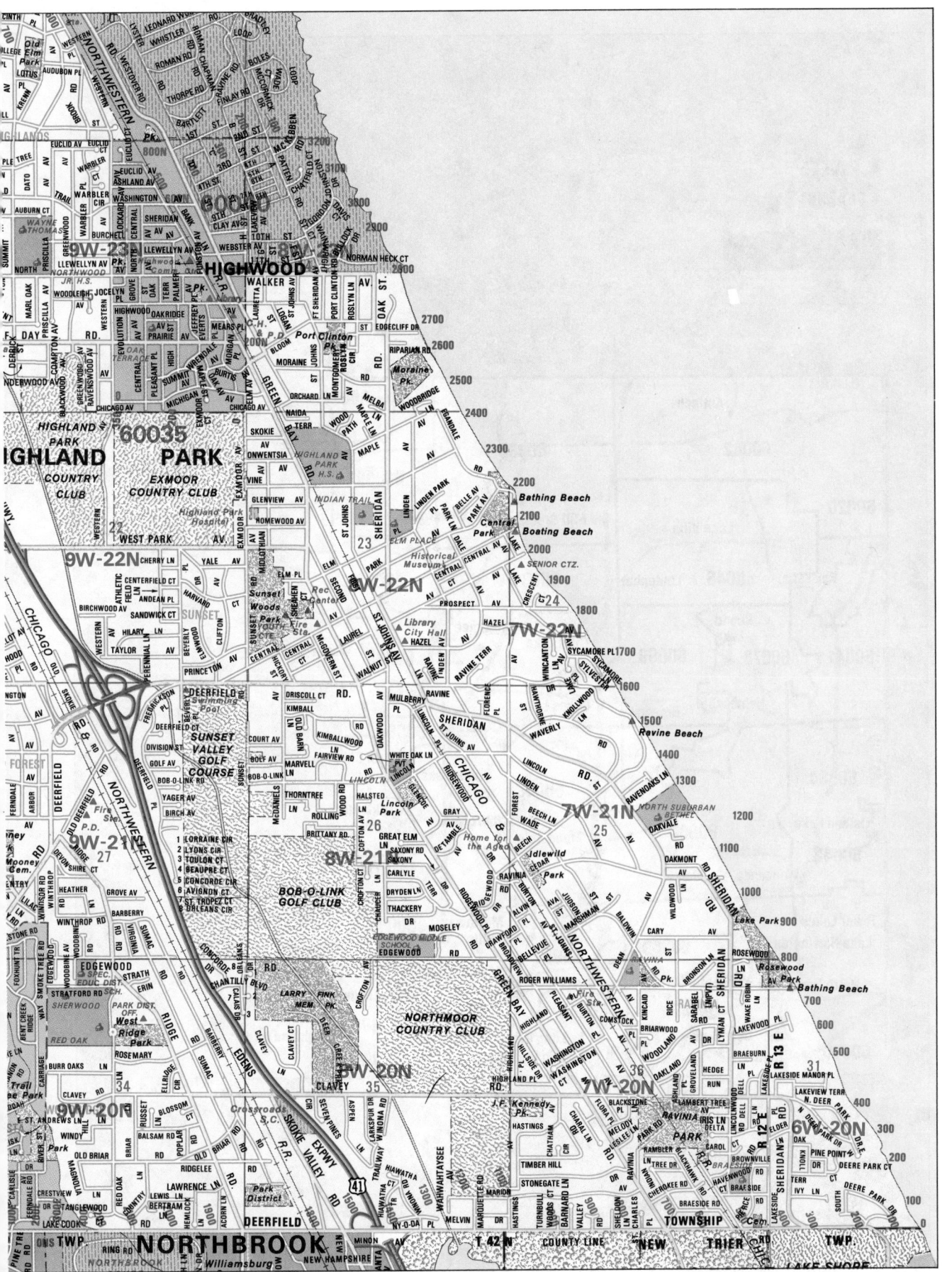

SEE COOK COUNTY PAGES 18-19

Lake County

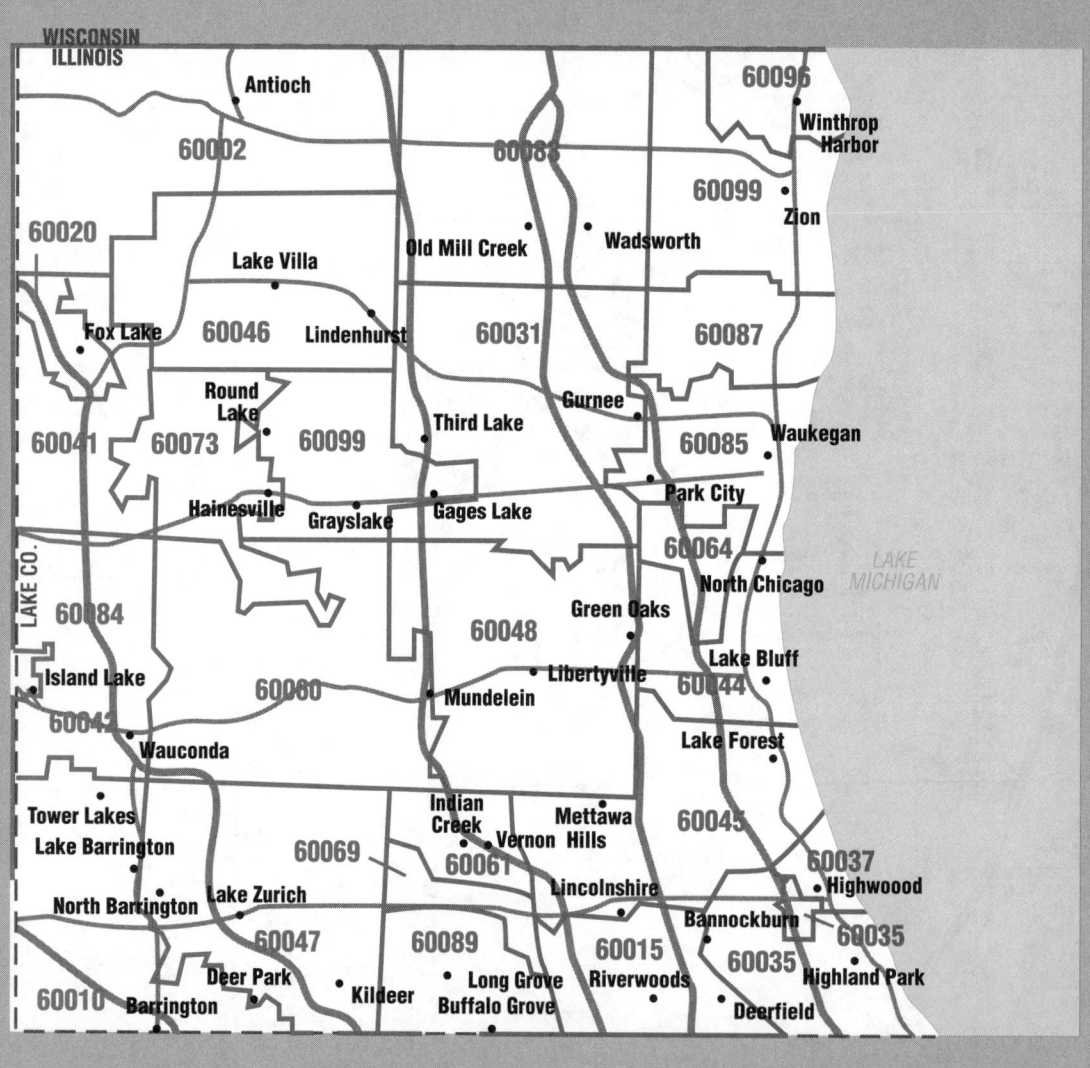

WISCONSIN
ILLINOIS

Antioch

60002

60086

60096

Winthrop Harbor

60099

Zion

60020

Old Mill Creek   Wadsworth

Lake Villa

Fox Lake   60046   Lindenhurst   60031   60087

Round Lake

Gurnee

60041   60073   60099   Third Lake   60085   Waukegan

Hainesville   Grayslake   Gages Lake   Park City

LAKE CO.

60064   North Chicago   LAKE MICHIGAN

60084   Green Oaks

60048   Lake Bluff

Island Lake   60060   Libertyville   60044

60042   Mundelein   Lake Forest

Wauconda

Tower Lakes   Indian Creek   Mettawa Hills   60045

Lake Barrington   60069   Vernon Hills   60037

60061   Highwoood

North Barrington   Lake Zurich   Lincolnshire   Bannockburn   60035

60047   60089   60015   60035

60010   Deer Park   Long Grove   Riverwoods   Highland Park

Barrington   Kildeer   Buffalo Grove   Deerfield

Street Index

Antioch

Street	Grid	Page
Abbey Ct. E&W	22W-43N	6
Abbey Ln. NS&EW	22W-43N	6
Alma Ter. N&S	23W-42N	6
Anita Av. N&S	23W-42N	6
Asbury Ct. E&W	24W-42N	5
Autumn Ln. N&S	22W-42N	6
Bartlett Av. N&S	23W-42N	6
Beach Hollow Dr. NS&EW	21W-42N	6
Birchwood Dr. NW&SE	23W-43N	6
Bishop St. N&S	23W-42N	6
Blackman Ter. N&S	23W-42N	6
Bowles Rd. E&W	24W-41N	5
Bridgewood Dr. N&S	23W-43N	6
Buena Ter. E&W	23W-41N	6
Butcher Block Dr. E&W	24W-43N	5
Cedarwood Ln. N&S	23W-43N	6
Channel Lake Rd. E&W	23W-42N	6
Charles Ct. (Pvt.) E&W	23W-42N	6
Cheri Ln. NW&SE	23W-42N	6
Chestnut St. N&S	23W-41N	6
Collins Av. NW&SE	23W-43N	6
Corona Av. N&S	23W-42N	6
Crabtree Ln. N&S	22W-42N	6
Creekside Ln. E&W	22W-42N	6
Crest Ln. E&W	23W-42N	6
Cunningham Dr. N&S	23W-42N	6
David St. N&S	24W-42N	5
Deep Lake Rd. N&S	22W-41N,22W-42N	6
Deer Lake Ct. NE&SW	22W-42N	6
Deer Lake Dr. N&S	22W-42N	6
Depot St. E&W	23W-42N	6
Donin Dr. NW&SE	23W-43N	6
Drom Ct. N&S	23W-42N	6
Dwight Ct. N&S	23W-42N	6
East Rd. N&S	22W-42N	6
Edelweiss Ct. E&W	24W-41N	5
Edelweiss Dr. E&W	24W-41N	5
Edgewood Dr. E&W	23W-43N	6
Elizabeth St. E&W	23W-42N	6
Elmwood Ln. NE&SW	23W-43N,22W-43N	6
Fillweber Dr. E&W	23W-42N	6
Forest Ct. N&S	23W-43N	6
Gail St. N&S	23W-42N	6
Glen Brook Ct. E&W	22W-42N	6
Gray's Dr. N&S	23W-42N	6
Greenbrier Dr. E&W	23W-42N	6
Harden St. E&W	23W-41N	6
Harvest Ct. N&S	22W-42N	6
Harvest Dr. N&S	22W-42N	6
Hawthorne Ct. NW&SE	22W-42N	6
Hazelwood Dr. NW&SE	23W-43N	6
Hennings Ct. NW&SE	24W-41N	5
Henry St. N&S	24W-42N	5
Hickory Commons N&S	23W-41N	6
Hickory Ln. N&S	23W-42N	6
Hidden Creek Dr. E&W	21W-42N	6
Highland Av. N&S	23W-42N	6
Highview Dr. E&W	24W-42N	5
Hill Ct. NE&SW	23W-43N	6
Hillandale Dr.	24W-42N	5
Hillside Av. NE&SW	24W-42N	6
Ida Av. E&W	23W-42N	6
Johanna Ct. E&W	23W-42N	6
Johella Ln. N&S	23W-43N	6
Joren Tr. N&S	23N-43N	6
Lake E&W	23W-43N	6
Lake Ct. NE&SW	23W-43N	6
Lake St. N&S	24W-42N	5
Lakewood Dr. N&S	23W-43N	6
Laursen Ct. NW&SE	23W-42N	6
Linden Ln. E&W	23W-42N	6
Little Silver Lake Rd. E&W	22W-42N	6
Longview Dr. N&S	23W-42N	6
Main St. N&S	23W-42N	6
Maplewood Dr. E&W	23W-43N	6
McMillan Rd. NS&EW	23W-41N,23W-42N	6
Meadow Ct. NW&SE	22W-43N	6
Miller Rd. E&W	20W-40N	7
Naber Av. E&W	23W-42N	6
Nelson Rd. N&S	22W-43N	6
North Av. E&W	24W-43N,22W-43N	5
Oak Ridge Ct. E&W	23W-42N	6
Oakhill Ct. NW&SE	22W-42N	6
Oakwood Dr. N&S	23W-43N	6
Old Mill Tr. E&W	22W-42N	6
Orchard St. E&W	24W-42N	5
Ormond Av. N&S	23W-42N	6
Park Av. N&S	23W-43N	6
Park Av. E&W	23W-42N	6
Parkway Av. N&S	23W-42N	6
Phillips Cir. NS&EW	24W-42N,23W-42N	5
Pine Cone Ln. NS&EW	22W-42N	6
Pinehill Ct. NW&SE	22W-42N	6
Plumtree Ln. N&S	24W-42N	5
Poplar Av. E&W	23W-42N	6
Private Rd. E&W	24W-42N	5
Ram Rd., S. E&W	23W-41N	6
Rinear Rd. N&S	23W-42N	6
Rt. 59 N&S	24W-41N	5
Rt. 83 N&S	23W-41N,23W-43N	6
Rt. 173 E&W	24W-42N,20W-41N	5
Savage Rd. N&S	20W-40N	7
Sequoia Ct. NW&SE	22W-42N	6
Sequoit Av. E&W	23W-42N	6
Spafford St. N&S	23W-42N	6
St. Peters St. N&S	24W-42N	5
Sterbenz Ct. NW&SE	22W-42N	6
Summerhill Ct. N&S	22W-42N	6
Sunset Ln. NE&SW	23W-43N	6
Tiffany Rd. N&S	24W-42N	5
Timber Ln. E&W	23W-43N,22W-43N	6
Toft Av. N&S	23W-42N	6
Turner Rd. E&W	23W-42N	6
U.S. Route 45 N&S	19W-42N	7
Valleyview Ct. E&W	24W-42N	5
Victoria St. N&S	23W-42N	6
White Rd. NE&SW	21W-40N	7
Wildwood Ct. NE&SW	22W-42N	6
Williams St. E&W	23W-42N	6
Wilton Dr. E&W	23W-42N	6
Windmill Creek Dr. NW&SW	22W-42N	6
Winsor Dr. E&W	23W-42N	6
Winsor Ter. E&W	23W-41N	6
Zelinger Ln. E&W	23W-43N	6
1st St. E&W	23W-42N	6
2nd St. E&W	23W-42N	6
228th Av. E&W	22W-43N	6

Antioch Township

Street	Grid	Page
Ackerman Rd. N&S	26W-38N	13
Adams Ct. E&W	22W-41N	6
Adison Ln. N&S	25W-40N	5
Ainsley St. E&W	23W-41N	6
Alexander Av. E&W	26W-39N	13
Andyville Ln. N&S	25W-39N	5
Ann Pl. E&W	28W-38N	12
Arbor Ln. N&S	26W-40N	5
Asbury Av. N&S	26W-39N	13
Ash St. N&S	26W-42N	5
Ashland Av. E&W	27W-38N	12
Astor St. E&W	26W-38N	13
Bald Eagle Rd. N&S	23W-40N	6
Banyan Av. E&W	24W-40N	5
Bayside Dr. E&W	25W-41N	5
Bayview Av. N&S	22W-41N	6
Bayview Rd. E&W	24W-41N,23W-41N	5
Beach Dr. E&W	24W-40N	6
Beach Dr. E&W	24W-43N	5
Beach Grove Rd. E&W	25W-41N	5
Beach Rd. NE&SW	27W-42N	4
Beach St. E&W	26W-42N	5
Beachlane Dr. N&S	22W-40N	6
Beachview Dr. N&S	26W-41N,26W-42N	5
Beechwood Av. N&S	28W-38N	12
Belden Av. E&W	27W-38N	12
Benes Rd. E&W	26W-38N	13
Birch Av. E&W	26W-42N	5
Bittersweet Pl. N&S	27W-38N	12
Black Oak Av. N&S	24W-40N	5
Bluff Dr. N&S	25W-40N	5
Bluff Lake Rd. N&S	25W-40N	5
Bluff Ln.	25W-40N	5
Bolton Pl. N&S	26W-38N	13
Bond Av. NE&SW	24W-40N	5
Briar Rd. N&S	27W-38N	12
Briar Ter. E&W	23W-41N	6
Broadway Av. N&S	26W-38N	13
Brown Av. N&S	22W-41N	6
Bunham Ct. NE&SW	22W-40N	6
Bunker Ct. E&W	22W-41N	6
Burr Hollow Dr. E&W	19W-41N	7
Burton Av. E&W	28W-38N	12
Calhoun Av. E&W	26W-38N	13
Calvin Dr. E&W	22W-40N	6
Catalpa Rd. E&W	26W-43N	5
Catherine Av. N&S	25W-43N	5
Catherine Av. E&W	25W-42N	5
Cedar Crest Dr. NE&SW	25W-39N	13
Cedar Dr. N&S	22W-40N	6
Cedar Ln. NE&SW	22W-40N,22W-41N	6
Cedar St. E&W	26W-42N	5
Cedar E&W	28W-39N	12
Center Dr. N&S	22W-42N	6
Champaign Dr. NW&SE	22W-40N	6
Channel Av. E&W	26W-40N	5
Channel Dr. E&W	24W-43N	5
Channel Ln. NE&SW	26W-39N	13
Channel St. N&S	28W-38N	12
Channelview Dr. N&S	25W-39N	13
Chase Ct. N&S	26W-42N	5
Cherry Av. E&W	26W-42N	5
Chestnut St. NW&SE	26W-42N	5
Circle Av. NW&SE	25W-39N	13
Circle Av. Cir.	25W-42N	5
Circle Dr. N&S	25W-40N	5
Circle Dr. N&S	26W-41N	5
Circle Dr. NE&SW	27W-38N	12
Clara Av. N&S	26W-41N	5
Claridan Av. E&W	25W-40N	5
Clark St. E&W	25W-42N	5
Clearview Dr. E&W	22W-41N	6
Clinton Av. E&W	25W-40N	5
Converse Rd. N&S	28W-42N	4
Cook Av. E&W	25W-43N	5
Country Club Dr. N&S	26W-41N	5
Cox Av. N&S	25W-43N	5
Crawford Av. E&W	25W-42N	5
Crescent Pl. N&S	25W-40N	5
Darrow Dr. N&S	26W-40N	5
Dawn Cir. SE&NW	27W-38N	12
De Coven Ct. E&W	22W-41N	6
Deep Lake Rd. N&S	22W-40N,22W-43N	6
Deer Ridge Ln. E&W	24W-41N	5
Dell Dr. N&S	22W-40N	6
Dewes St. E&W	25W-40N	5
Diemer St. E&W	24W-40N	5
Dogwood Cir. NE&SW	24W-40N	5
Donald Dr. N&S	25W-40N	5
Dot Pl. N&S	26W-41N	5
Douglas Rd. E&W	25W-40N	5
Dressels Rd. E&W	25W-40N	5
Drexel Blvd. N&S	26W-38N	13
Duck Ln. NW&SE	26W-41N	5
Eagle's Nest Rd. E&W	24W-40N	5
East Av. N&S	22W-40N	6
East Av. N&S	25W-41N	5
East Dr. NS&EW	22W-40N	6
Echo Rd. NW&SE	25W-40N	5
Edgar Av. E&W	25W-40N	5
Edgewood Av. E&W	26W-43N	5
Edwards Rd. E&W	19W-42N	7
Elime Rd. E&W	25W-40N	5
Elm St. N&S	26W-42N	5
Elm St. E&W	26W-41N	5
Elm St. E&W	28W-39N	12
Elm Tree Rd. E&W	26W-39N	13
Emerson Av. NW&SE	26W-38N	13
Esther St. N&S	26W-40N	5
Evergreen Av. N&S	26W-42N	5
Fair Oaks Cir. NE&SW	26W-42N	5
Fairview Av. E&W	26W-38N	12
Fairview Dr. E&W	22W-40N	6
Fairview St. N&S	26W-42N	5
Flamingo Ln. E&W	27W-38N	12
Florence Av. E&W	25W-40N	5
Forest Av. N&S	25W-39N	13
Forest Av. NW&SE	26W-41N	5
Forest Av. N&S	26W-38N	13
Forest Ct. E&W	26W-41N	5
Forest Dr. N&S	25W-43N	5
Forest Dr. NW&SE	25W-39N	13
Forest Ln. N&S	22W-40N	6
Forest Ln. NW&SE	27W-42N	4
Forest St. N&S	22W-42N	6
Fox Av. E&W	25W-42N	5
Fox Dr. N&S	21W-40N	7
Fox Run Ln. N&S	23W-40N	6
Fredrick Ct. E&W	25W-42N	5
Golden Eagle Ct. NE&SW	23W-40N	6
Golf Ln. NW&SE	24W-39N	13
Golfview Av. N&S	25W-40N	5
Golfview Dr. E&W	28W-39N	12
Grace Ln. E&W	25W-40N	5
Grand Av. N&S	25W-40N	5
Grandview Ter. NE&SW	25W-43N	5
Grapevine Av. E&W	26W-42N	5
Grass Lake Rd. E&W	27W-38N,22W-40N	12
Greene Ln. E&W	22W-40N	6
Greenleaf Pl. E&W	28W-38N	12
Greenview Av. E&W	25W-39N	13
Gridley Av. N&S	23W-40N	6
Gridley Dr. N&S	23W-40N	6
Grimm Rd. E&W	23W-40N	6
Groveland Av. E&W	27W-38N	12
Gundlach's Ln. E&W	25W-42N	5
'H' St. NE&SW	23W-40N	6
Halings Rd. E&W	26W-39N	13
Halverson Rd. E&W	25W-43N	5
Hamilton Cir., West N&S	27W-38N	12
Harlem Av. E&W	26W-39N	13
Harloff Rd. E&W	26W-40N,25W-40N	5
Harold Pl. NE&SW	28W-38N	12
Hawthorne Av. E&W	25W-39N	13
Hawthorne Ln. E&W	24W-41N	5
Heart O Lakes Blvd. E&W	26W-40N,25W-40N	5
Henry Pl. NW&SE	26W-42N	5
Herman Av. E&W	25W-39N	13
Hickory Av. N&S	25W-39N	13
Hickory Ln. N&S	25W-43N	5
Hickory St. N&S	25W-41N	5
Hickory St. N&S	28W-39N	12
High St. N&S	26W-38N	13
Highland Av. E&W	25W-39N	13
Highland Av. E&W	26W-38N	13
Highview Dr. N&S	25W-39N	13
Highview Ln. E&W	25W-42N	5
Highwoods Dr. E&W	25W-39N	13
Hill Dr. E&W	27W-39N	12
Hill St. E&W	26W-42N	5
Hillandale Dr. NS&EW	27W-38N	12
Hillcrest Av. E&W	25W-40N	5
Hillcrest Av. E&W	26W-42N	5
Hilldale Av. E&W	25W-39N	13
Hillside Av. NW&SE	26W-39N	13
Hillside Ln. E&W	22W-41N	6
Hilltop Av. NE&SW	26W-39N	13
Hook Cir. NW&SE	22W-41N	6
Island Dr. E&W	24W-43N	5
Jackson Dr. E&W	27W-39N	12
James Pl. E&W	26W-41N	5
James Rd. E&W	28W-41N	4
Janet Ct. N&S	25W-39N	13
Janette St. E&W	25W-42N	5
Jasper Av. N&S	22W-40N	6
Jersey Av. E&W	25W-42N	5
Julia Ct. N&S	23W-41N	6
Kairys St. N&S	23W-43N	6
Karen Ct. N&S	23W-41N	6
Kathryn Dr. E&W	25W-39N	13
Kelly Rd. E&W	27W-38N	12
Kendall Av. E&W	28W-38N	12
Kenmore Rd. N&S	26W-38N	13
Kidd Ct. N&S	25W-43N	5
Klondike Av. E&W	26W-38N	13
Koenen Av. N&S	27W-38N	12
'L' St. E&W	23W-40N	6
Lagoon Ct. N&S	25W-42N	5
Lake Av. N&S	22W-40N	6
Lake Av. N&S	25W-41N	5
Lake Av. E&W	26W-39N	13
Lake Av. N&S	26W-42N,26W-43N	5
Lake Bluff Dr. N&S	22W-40N	6
Lake Blvd. N&S	25W-40N	6
Lake Ct. N&S	22W-40N	6
Lake Ct. N&S	25W-39N	13
Lake Ct. N&S	26W-38N	13
Lake Dr. N&S	22W-40N	6
Lake Dr. N&S	25W-42N	5
Lake Rd. NE&SW	26W-39N	13
Lake Shore Dr. E&W	23W-40N	6
Lake Shore Dr. E&W	27W-38N	12
Lake Shore Dr. N&S	25W-40N	5
Lake St. N&S	25W-42N	5
Lake St. N&S	26W-42N	5
Lake St. N&S	26W-38N	13
Lake Vista Av. E&W	23W-41N	6
Lakes Center Blvd. N&S	25W-39N	13
Lakeside Dr. N&S	25W-43N	5
Lakeside Pl. E&W	26W-38N	13
Lakeview Av. N&S	22W-40N	6
Lakeview Av. N&S	25W-40N	5
Lakeview Ct. N&S	22W-40N	6
Lakeview Dr. E&W	22W-40N	6
Lakeview Dr. N&S	22W-42N	6
Lakeview Rd. N&S	26W-39N	13
Lakeview Ter. N&S	23W-41N	6
Lambert Dr. N&S	23W-40N	6
Lawson Av. E&W	27W-42N	4
Lee Av. NE&SW	28W-38N	12
Lee Dr. E&W	22W-40N	6
Leland Av. E&W	26W-38N	13
Lenora St. N&S	25W-42N	5
Liberty St. NE&SW	23W-40N	6
Lilac Pl. N&S	24W-40N	5
Lillian Av. E&W	25W-43N	5
Lillian Pl. N&S	22W-41N	6
Lincoln Av. N&S	22W-41N	6
Lincoln Av. N&S	27W-38N	12
Lincoln Dr. N&S	22W-41N	6
Linden Ln. N&S	25W-43N	5
Linden Ln. N&S	25W-42N	5
Linden Rd. E&W	26W-43N	5
Logan Ter. E&W	22W-40N	6
Loon Dr. E&W	22W-40N	6
Loon Lake Blvd. E&W	22W-40N	6
Lotus Av. NS&EW	26W-41N	5
Lotus Av. E&W	26W-38N	13
Lotus Av. E&W	27W-38N	12
Lotus Ln. E&W	26W-39N	13
Lotus Rd. NE&SW	26W-39N	13
Madison Av. N&S	22W-41N	6
Magnolia Av. E&W	24W-40N	5
Main St. N&S	23W-41N	6
Mallard Av. N&S	26W-40N	5
Maple Av. E&W	26W-42N	5
Maple Dr. N&S	25W-43N	5
Maple Ln. NE&SW	22W-40N	6
Marcus Rd. N&S	25W-40N	5
Marcus St. N&S	26W-40N	5
Margaret St. N&S	26W-40N	5
Margate Ter.	23W-41N	6
Marie Av. E&W	26W-41N	5
Marie Pl. NW&SE	26W-41N	5
Mary Av. N&S	25W-43N	5
Maryann Rd. E&W	26W-41N	5
Masaryk Rd. (Pvt) N&S	24W-40N	5
May St. NW&SE	28W-38N	12
Meadow Ln. E&W	22W-41N	6
Meadow Ln. N&S	26W-42N	5
Melva St. N&S	25W-43N	5
Michele Ln. E&W	21W-40N	7
Michigan Blvd. E&W	26W-38N	13
Mildred Av. N&S	25W-43N	5
Miller Rd. E&W	20W-40N,19W-40N	7
Mills St. N&S	25W-40N	5
Mineva St. N&S	25W-40N	5
Mitchell Ln. E&W	22W-40N	6
Montrose Rd. E&W	26W-38N	13
Morley Rd. N&S	24W-43N	5

	Grid	Page
Mound Av. N&S	27W-39N	12
Nevelier Dr. N&S	24W-40N	5
New Venice Way N&S		
	24W-43N	5
Nicklaus Way E&W		
	25W-39N,24W-40N	5
Nippersink St. N&S		
	28W-38N	12
Norman Dr. E&W	25W-39N	13
North Av. E&W	22W-41N	6
North Av. NE&SW		
	22W-43N,21W-43N	6
North Av. NW&SE	25W-43N	5
North Av. N&S	25W-41N	5
North Dr. E&W	22W-40N	6
North St. E&W	23W-40N	6
Northview Ln. E&W		
	26W-42N	5
Oak Av. N&S	25W-39N	13
Oak Av. E&W	26W-40N	5
Oak Dr. N&S	23W-40N	6
Oak Ln. NS&EW	25W-43N	5
Oak Ln. N&S	25W-42N	5
Oak St. N&S	26W-42N	5
Oakland St. E&W	27W-38N	12
Olcott Ln. E&W	24W-40N	5
Old Grass Lake Rd. E&W		
	25W-40N	5
Olive Ct. N&S	25W-41N	5
Orchard Av. E&W	26W-42N	5
Orchard Ln. E&W	22W-40N	6
Orchard Ln. N&S	25W-42N	5
Otis Av. N&S	26W-39N	13
Palmer Ct. E&W		
	25W-40N,24W-40N	5
Park Av. N&S	25W-40N	5
Park Av. N&S	27W-42N	4
Park Av. E&W	27W-38N	12
Park Av. E&W	28W-38N	12
Park Dr. E&W	22W-40N	6
Park Dr. E&W	25W-43N	5
Park Ln. E&W	22W-40N	6
Park Ln. E. N&S	25W-42N	5
Park Ln. W. N&S	25W-42N	5
Park Pl. E&W		
	26W-42N,25W-42N	5
Park St. N&S	26W-42N	5
Park St. N&S	26W-38N	13
Park Ter. E&W	23W-41N	6
Park Ter. N&S	25W-43N	5
Parkview Dr. NW&SE		
	26W-41N	5
Parkway N&S	25W-39N	13
Pauline St. N&S	25W-41N	5
Pearl St. N&S	26W-39N	13
Pearl St. E&W	28W-38N	12
Pedersen Dr. E&W	19W-41N	7
Petite Av. E&W	25W-42N	5
Petite Pass N&S	25W-42N	5
Pine Ln. E&W	25W-43N	5
Pine St. E&W	26W-42N	5
Pineview Dr. E&W	22W-40N	6
Point Dr. N&S	23W-41N	6
Pollock Rd. (Pvt) E&W		
	19W-40N	7
Poplar St. E&W	26W-42N	5
Prairie Av. (Pvt) N&S		
	24W-40N,24W-41N	5
Primrose Ln. E&W	27W-38N	12
Private Rd. N&S	22W-41N	6
Private Rd. E&W		
	26W-39N,25W-39N	13
Prospect Av. E&W	26W-42N	5
Public Rd. N&S	27W-42N	4
Public Rd. N&S	28W-41N	4
Rain Tree Rd. N&S		
	25W-43N	5
Ravine Av. N&S	26W-39N	13
Ravine Dr. NW&SE		
	24W-39N	13
Ravine Dr. E&W	27W-39N	12
Regina Rd. E&W	21W-40N	7
Rena Av. N&S	25W-39N	13
Resthaven E&W	25W-39N	13
Richmond Av. E&W		
	25W-39N	13
Ridge Cir. NW&SE		
	22W-40N,22W-41N	6
Ridge Dr. E&W	22W-42N	6
Riverside Rd. E&W		
	27W-42N	4
Riverview Av. NE&SW		
	26W-41N	5
Riverview Dr. E&W	26W-41N	5
Riverview Pl. NW&SE		
	26W-41N	5
Rosecrans Rd. E&W		
	19W-41N	7
Rowe Av. E&W	27W-38N	12
Rt. 173 E&W		
	28W-42N,19W-41N	4
S. Pedersen Dr. N&S		
	19W-41N	7
Sarah St. E&W	23W-40N	6
Sarana Dr. E&W	22W-40N	6
Savage Rd. N&S	20W-40N	7
Scott St. N&S	26W-43N	5
Sea Eagle Ct. NE&SW		
	26W-42N	6
Shady Lane (Pvt) N&S		
	26W-42N	6
Shagbark E&W	22W-40N	6
Shannon Dr. E&W	26W-42N	5
Sherwin Av. N&S	26W-38N	13

	Grid	Page
Shore Av. NE&SW	26W-39N	13
Shore Dr. NE&SW	23W-40N	6
Shore Pl. N&S	28W-39N	12
Sibley Dr. N&S	25W-40N	5
Silver Lake Av. E&W		
	22W-42N	6
Simon Ct. N&S	25W-39N	13
Smith St. N&S	25W-41N	6
South Dr. E&W	22W-40N	6
Spring E&W	27W-42N	4
Spring Grove Rd. NE&SW		
	27W-42N	4
Spring Well Av. NS&EW		
	26W-39N	13
Spruce Dr. E&W	22W-40N	6
Spruce St. E&W	26W-42N	5
Stanton Rd. E&W	22W-40N	6
State Line Rd. E&W		
	19W-43N	7
State Park Rd. E&W		
	28W-39N	12
Sterling Av. E&W	28W-38N	12
Stewart Av. NW&SE		
	28W-38N	12
Stonegate Av. NW&SE		
	27W-38N	12
Stonegate Dr. E&W		
	26W-38N	13
Summerside Dr. N&S		
	25W-39N	13
Sunset E&W	24W-40N	5
Sunset Av. E&W	27W-38N	12
Sunset Ct., N. E&W		
	24W-40N	5
Sunset Ct., S. E&W		
	24W-40N	5
Sunset Dr., N. E&W		
	24W-40N	5
Sunset Ln. N&S	26W-43N	5
Sunset Rd. E&W	26W-43N	5
Suraya Dr. N&S	23W-41N	6
Sycamore Av. E&W		
	26W-43N	5
Sylvan Dr. E&W	25W-43N	5
Sylvan Woods Rd. E&W		
	24W-43N	5
Talman Av. N&S	25W-42N	5
Talon Ct. E&W	23W-40N	6
Tammi Ter. NW&SE		
	19W-41N	7
Taylor Ct. E&W	22W-41N	6
Terry Ln. NW&SE	24W-40N	5
Tierney Dr. NE&SW		
	23W-40N	6
Tiffany St. N&S	24W-42N	5
Todd St. E&W	26W-43N	5
Town Line Rd. E&W		
	24W-40N	5
Tracy Ln. NW&SE	24W-40N	5
Trevor Rd. N&S	24W-43N	5
Trillium Pl. N&S	27W-38N	12
Trinity Ln. N&S	24W-40N	5
U.S. Route 45 E&W		
	19W-40N,20W-43N	7
Venetian Dr. N&S	25W-43N	5
Venn Rd. NW&SE	25W-42N	5
Villa Rica Rd. E&W		
	23W-40N,22W-40N	6
Vine St. E&W	26W-38N	13
Virell Dr. E&W	22W-40N	6
Virgil Av. E&W	25W-42N	5
Washington Av. N&S		
	27W-38N	12
Washington St. E&W		
	22W-41N	6
Waters Rd. N&S	26W-43N	5
Wedgewood Dr. E&W		
	22W-40N	6
West E&W	25W-40N	5
West 'A' St. E&W	25W-42N	5
West Av. NW&SE	25W-41N	5
West End Dr. N&S	27W-39N	12
West Loon Lake Blvd. N&S		
	22W-40N	6
West Shore Av. N&S		
	23W-40N	6
White Rd. N&S	21W-40N	7
Whitehall Ct. E&W	21W-40N	7
Willow Av. N&S	25W-43N	5
Willow St. NW&SE	26W-42N	5
Willow St. N&S	26W-39N	13
Wilmot Rd. E&W		
	27W-43N,26W-43N	4
Wilson Rd. E&W	26W-38N	13
Windsor Dalgaard Jr. Ln. N&S		
	26W-42N	5
Winsor Ter. E&W	23W-41N	6
Witt Rd. N&S	25W-42N	5
Woodbine Av. NS&EW		
	26W-42N	5
Woodland Av. N&S		
	25W-39N	13
Woodland Av. E&W		
	26W-38N	13
Woodland Av. NE&SW		
	27W-38N	12
Woodland Pl. NW&SE		
	27W-42N	4
Woodlawn Av. N&S		
	26W-43N	5
Woodmere Ter. NW&SE		
	19W-41N	7
Woods Dr. N&S	25W-42N	5

	Grid	Page
Woodside Av. N&S		
	25W-39N	13
1st Av. N&S	25W-42N	5
1st Av. N&S	27W-38N	12
1st St. N&S	23W-41N	6
1st St. N&S	25W-43N	5
2nd Av. N&S	25W-42N	5
2nd Av. N&S	27W-38N	12
2nd St. N&S	23W-41N	6
2nd St. E&W	25W-42N	5
3rd Av. N&S	25W-42N	5
3rd Av. N&S	27W-38N	12
3rd St. N&S	23W-41N	6
3rd St. E&W	25W-42N	5
4th Av. N&S	25W-42N	5
4th Av. N&S	27W-38N	12
5th Av. N&S	25W-42N	5
5th Av. N&S	27W-38N	12
6th Av. N&S	25W-42N	5
6th Av. N&S	27W-38N	12
7th Av. N&S	25W-42N	5
7th Av. N&S	28W-38N	12
8th St. N&S	28W-38N	12
9th St. N&S	28W-38N	12
10th Av. N&S	28W-38N	12
11th St. N&S	28W-38N	12

Avon Township

	Grid	Page
Acre Ct. E&W	22W-35N	22
Alleghany Rd. N&S		
	21W-33N,21W-33N	23
Ash Ct. E&W	19W-35N	23
Bacon Rd. N&S	24W-32N	22
Beach Ct. E&W	22W-34N	22
Belvidere Rd. E&W		
	24W-32N,19W-32N	22
Brorson Av. E&W	24W-35N	22
Cedar Av. N&S	22W-34N	22
Cedar Dr. E&W	22W-34N	22
Cedar Ln. N&S	21W-34N	23
Cedar Ln. N&S	24W-35N	22
Center St. E&W	19W-33N	23
Circle Dr., S. NW&SE		
	22W-34N	22
Clinton Av. NS&EW		
	24W-35N	22
Curran Rd. N&S	23W-32N	22
Drury Ln. NS&EW		
	21W-34N,20W-35N	23
East End Av. N&S	22W-34N	22
Eldor Av. E&W	21W-34N	23
Fairfield Rd. N&S		
	24W-32N,24W-35N	22
Forest Av. E&W	24W-35N	22
Garfield Blvd. N&S	20W-32N	23
Gerwal Av. N&S	21W-34N	23
Harris Rd. N&S	19W-32N	23
Harrison Av. N&S	24W-34N	22
Hart Rd. E&W	24W-34N	22
Hauge St. N&S		
	24W-34N,24W-35N	22
Hawthorn Dr. E&W		
	24W-35N	22
Hendee Av. N&S	22W-34N	22
Hendee Ln. N&S	22W-34N	22
Heyer St. N&S	24W-35N	22
Hickory Av. N&S		
	19W-34N,19W-35N	23
Hickory Av. N&S	22W-34N	22
Hickory Ln. N&S	21W-34N	23
Hickory Ln. N&S	22W-34N	22
Hickory Ln. N&S	24W-35N	22
Hickory St. E&W	24W-35N	22
Hill Ct. N&S	24W-34N	22
Hillside Av. N&S	24W-35N	22
Indian Trail E&W	22W-34N	22
Ivanhoe Rd. N&S	20W-32N	23
Kenneth Dr. E&W	20W-35N	23
Lake Av. E&W	24W-35N	22
Lake Dr. E&W	24W-35N	22
Lake Park Av. E&W		
	22W-34N	22
Lake Shore Dr. N&S		
	21W-33N	23
Lake Shore Dr. E&W		
	22W-34N	22
Lake Shore Dr. E&W		
	24W-35N	22
Lake St. N&S		
	21W-32N,21W-34N	23
Lakeview Av. E&W		
	22W-34N	22
Lakeview Dr. E&W	19W-35N	23
Lakeview Av. N&S	22W-34N	22
Lakewood Ln. E&W		
	22W-34N	22
Lane Ct. E&W	22W-34N	22
Lilac Av. N&S	22W-34N	22
Linden Av. N&S		
	19W-34N,19W-35N	23
Luther Av. E&W	24W-35N	22
Main St. NW&SE	24W-34N	22
McKinley Av. N&S	20W-32N	23
Merriwood Ln. E&W		
	22W-34N	22
Nielson Av. N&S	24W-35N	22
Nippersink Rd. E&W		
	24W-33N	22
Norelius Av. E&W	24W-35N	22
Oak Av. N&S	22W-34N	22
Oak Av. E&W	24W-34N	22
Oak Ct. E&W	19W-35N	23
Oak Ln. N&S	21W-34N	23

	Grid	Page
Oak St. E&W	24W-35N	22
Orchard Ln. N&S	22W-35N	22
Orchard Pl. E&W	24W-35N	22
Passavant Av. E&W		
	24W-35N	22
Poplar St. E&W	24W-35N	22
Rollins Rd. E&W		
	24W-35N,19W-35N	22
Ross Av. E&W	24W-34N	22
Round Lake Rd. E&W		
	24W-34N	22
Rt. 45 N&S		
	19W-32N,19W-33N	23
Rt. 83 NW&SE		
	21W-35N,20W-32N	23
Rt. 120 E&W		
	24W-32N,19W-32N	22
Rt. 134 NW&SE	24W-34N	22
Sheldon Rd. N&S	21W-35N	23
Shorewood Dr. E&W		
	22W-34N	22
Shorewood Rd. E&W		
	22W-35N,21W-35N	22
Stub Av. E&W	24W-35N	22
Summit St. N&S	24W-35N	22
Sunshine Av. E&W	19W-35N	23
Terri St. N&S	20W-35N	23
Thompson St. N&S		
	24W-34N	22
Todd Ct. E&W	20W-35N	23
Town Line Rd. E&W		
	24W-32N,23W-32N	22
Tyler Av. E&W	24W-34N	22
Wabash Av. N&S	24W-35N	22
Washington St. E&W		
	21W-34N,19W-34N	23
Willow Ct. N&S	19W-35N	23
Wilmar Av. E&W	21W-34N	23
Wilson Blvd. N&S	24W-35N	22

Bannockburn

	Grid	Page
Aitken Dr. E&W	11W-22N	50
Audubon Ln. E&W	12W-23N	50
Bannockburn Ct. NW&SW		
	11W-22N	50
Bridle Ln. NW&SE	12W-23N	50
Cedar Crest Ln. E&W		
	11W-22N	50
Duffy Ln. E&W	11W-22N	50
Dunsinane East NW&SE		
	11W-23N	50
Dunsinane Ln. N&S		
	11W-23N	50
Half Day Rd. E&W	12W-23N	50
High Terrace Ln. E&W		
	12W-23N	50
Hill Av. E&W	11W-22N	50
Hilltop Ln. E&W	11W-22N	50
Lakeside Dr. N&S	12W-23N	50
Lakewood Dr. E&W		
	12W-23N	50
Meadow Ln. N&S	11W-22N	50
North Av. E&W	11W-22N	50
Orchard Ln. N&S	11W-22N	50
Rt. 22 E&W		
	12W-23N,11W-23N	50
Rt. 43 N&S		
	11W-22N,11W-23N	50
Stirling Rd. NS&EW		
	11W-22N	50
Sunset Ln. E&W	12W-22N	50
Telegraph Rd. N&S		
	12W-23N,11W-23N	50
Thornapple Rd. N&S		
	12W-22N	50
Tri-State Tollway N&S		
	12W-22N	50
Valley Rd. NS&EW	11W-22N	50
Waukegan Rd. N&S		
	11W-22N,11W-23N	50
Wilmot Rd. N&S	12W-22N	50

Barrington*
(Also see Cook County)

	Grid	Page
Applebee St. NW&SE		
	25W-20N	45
Bellwood Dr. NS&EW		
	28W-20N	11
Bent Ridge Ln. NS&EW		
	25W-20N	45
Berry Rd. E&W	25W-20N	45
Beverly Rd. N&S	24W-20N	46
Birch Rd. N&S	25W-20N	45
Birchwood Dr. E&W		
	24W-20N	46
Bosworth Field N&S		
	25W-20N	45
Bryant Av. N&S	26W-20N	45
Burr Oak Ln. NW&SE		
	24W-20N	46
Carl Av. N&S	26W-20N	45
Castle Ct. N&S	24W-20N	46
Center Ct. E&W	24W-20N	46
Chestnut St. E&W	25W-20N	45
Church Hill Ct. NW&SE		
	25W-20N	45
Cold Spring Rd. NS&EW		
	24W-20N	46
Commonwealth Ct. N&S		
	25W-20N	45
Covington Dr. E&W		
	25W-20N	45

	Grid	Page
Cumnor Av. N&S	26W-20N	45
Drury Ln. E&W	25W-20N	45
Ela St. N&S	25W-20N	45
Elm Av. NE&SW	25W-20N	45
Exmoor Av. N&S	26W-20N	45
Fairfax Ct. N&S	25W-20N	45
Fox Hunt Rd. NS&EW		
	24W-20N	46
Foxglove Ln. NS&EW		
	24W-20N	46
Franklin St. E&W	25W-20N	45
Garfield St. N&S	25W-20N	45
Georgetowne Ln. N&S		
	25W-20N	45
Grant St. N&S	26W-20N	45
Hampton Pk.* E&W		
	24W-20N	46
Harrison St. N&S	25W-20N	45
Hilltop Av. N&S	24W-20N	46
Home Av. N&S	26W-20N	45
Hough St. N&S	25W-20N	45
Indian Way N&S	24W-20N	46
James St. N&S	25W-20N	45
June Ter. E&W	25W-20N	45
Lake Shore Ct. NW&SE		
	24W-20N	46
Lake Shore Dr., North NS&EW		
	24W-20N	46
Lake Shore Dr., South E&W		
	24W-20N	46
Lake Zurich Rd. NE&SW		
	25W-20N	45
Laverne St. E&W	26W-20N	45
Liberty St. E&W	25W-20N	45
Linden Av. N&S	25W-20N	45
Lions Dr. E&W	25W-20N	45
Magnolia Av. N&S	25W-20N	45
Main St. E&W		
	26W-20N,24W-20N	45
Maple Av. E&W	25W-20N	45
Marlborough Rd. N&S		
	25W-20N	45
Martin Ct. E&W	25W-20N	45
Meadow Way E&W		
	24W-20N	46
N. Glendale Av. N&S		
	25W-20N	45
N. Valley Ct. NE&SW		
	24W-20N	46
New Bridge Ct. NE&SW		
	25W-20N	45
New Hart Rd. N&S	26W-20N	45
North Av. N&S	25W-20N	45
North Shore Ct. NW&SE		
	24W-20N	46
North St. E&W	25W-20N	45
North Valley Rd. NW&SE		
	24W-20N	46
Northwest Hwy. NW&SE		
	26W-20N,25W-20N	45
Oak Rd. NE&SW	25W-20N	45
Old Creek Ct. SE&NW		
	25W-20N	45
Oxbow Ln. NS&EW		
	24W-20N	46
Park Ln. NW&SE	25W-20N	45
Paul Cir. E&W	25W-20N	45
Pine St. E&W	25W-20N	45
Ponderosa Ct. N&S		
	24W-20N	46
Prospect Av. N&S	25W-20N	45
Raymond Av. N&S	26W-20N	45
Redwood Ln. NS&EW		
	24W-20N	46
Roslyn Rd. N&S		
	26W-20N,25W-20N	45
Scott Av. N&S	26W-20N	45
Sharon Dr. N&S	24W-20N	46
Shorely Dr. NE&SW		
	25W-20N	45
South Shore Ct. NE&SW		
	24W-20N	46
Spruce St. N&S	25W-20N	45
Stillwater Ln. SE&NW		
	25W-20N	45
Sunny Ln. N&S	25W-20N	45
Surrey Ln. NS&EW		
	24W-20N	46
Sycamore Av. N&S		
	25W-20N	45
Tall Trees Cir. E&W		
	24W-20N	46
Tall Trees Dr. NS&EW		
	24W-20N	46
Taylor Rd. E&W	26W-20N	45
The Point E&W	24W-20N	46
U.S. Route 14 NW&SE		
	26W-20N,25W-20N	45
Valencia Av. E&W	25W-20N	45
Verne Cir. E&W	25W-20N	45
Victoria St. E&W	25W-20N	45
Walnut Rd. NW&SE		
	25W-20N	45
Washington St. E&W		
	25W-20N	45
Waverly Rd. E&W	26W-20N	45
White Oak Ct. N&S		
	24W-20N	46
Whitney Rd. N&S	25W-20N	45
Williamsburg Pk.* E&W		
	25W-20N	45
Willow Av. N&S	25W-20N	45
16th St. N&S	26W-20N	45
20th St. N&S	26W-20N	45

LAKE COUNTY 55

Barrington Hills*
(Also see Cook, Kane and McHenry Counties)

Street	Grid	Page
Acorn Dr. N&S	28W-21N	44
Bellwood Dr. NS&EW	28W-20N	44
Bisque Dr. NW&SE	27W-20N	44
Buckley Rd. N&S	28W-21N	44
Cuba Rd. NW&SE	28W-21N	44
Dormy Ln. (Pvt.) NW&SE	27W-20N	44
Dunrovin Dr. NW&SE	28W-20N	44
Hickory Ln. N&S	28W-21N	44
Lake Cook Rd. E&W	26W-20N	45
Merryoaks Rd. E&W	28W-21N	44
Mid Oaks Ln. N&S	27W-20N	44
New Hart Rd. N&S	26W-20N	45
Northwest Highway NW&SE	28W-22N,26W-20N	44
Oak Knoll Rd. E&W	28W-20N,26W-20N	44
Oak Lake Dr. N&S	28W-21N	44
Oakwood Dr. NW&SE	26W-20N	45
Old Hart Rd. N&S	26W-20N	45
Pagancia Dr. N&S	27W-20N	44
Pheasant Dr. E&W	28W-21N	44
Plum Tree Rd. E&W	28W-21N	44
Private Rd. E&W	28W-21N	44
Pvt. Rd. N&S	28W-20N	44
Pvt. Rd. N&S	26W-20N	45
Rub-of-Green Ln. N&S	27W-20N	44
Sieberts Ridge Rd. NE&SW	28W-21N	44
Steeplechase Rd. NS&EW	28W-20N,27W-20N	44

Beach Park

Street	Grid	Page
Adelaide Av. E&W	12W-37N	18
Adelphi Av. N&S	12W-37N,12W-40N	18
Ames Av. E&W	10W-39N	19
Ash Ct. E&W	12W-40N	10
Audrey Av. E&W	12W-37N	18
Bairstow Av. E&W	10W-37N,9W-37N	19
Bairstow St. E&W	12W-37N	18
Beach Pl. E&W	10W-38N	19
Beach Rd. E&W	11W-38N,9W-38N	18
Bernice Ter. N&S	11W-39N	18
Blanchard Rd. E&W	12W-37N	18
Blossom St. E&W	12W-37N	18
Bonnie Brook Ln., North E&W	12W-37N	18
Bowling Av. N&S	12W-39N	18
California Av. E&W	10W-39N	19
Carnahan Av. E&W	11W-38N	18
Carol Ln. NS&EW	11W-39N	18
Cedar Av. N&S	11W-38N	18
Center St. E&W	12W-38N,10W-38N	18
Chaney Av. E&W	10W-38N	19
Chaney St. E&W	12W-38N	18
Chaplin Av. E&W	10W-38N	19
Chaplin St. E&W	12W-38N	18
Charleston Rd. N&S	11W-38N	18
Chicago Av. E&W	10W-39N	19
Clarendon Rd. E&W	12W-39N	18
Coolidge Av. N&S	10W-38N	19
Cornell Rd. N&S	11W-37N,11W-38N	18
Cornell St. NW&SE	12W-40N	10
Country Ln. E&W	10W-38N	19
Crissy Av. E&W	10W-38N	19
DeWoody Rd. N&S	11W-37N,11W-38N	18
Delores Av. E&W	12W-38N	18
Dorothy St. E&W	12W-37N	18
E. McArthur Dr. E&W	10W-37N	19
Eastwood Rd. E&W	12W-37N,10W-37N	18
Emanuel Av. N&S	12W-38N	18
Everett Rd. E&W	11W-38N	18
Evergreen Av. N&S	12W-38N	18
Evergreen Av. N&S	12W-37N	18
Fairbanks Av. E&W	10W-38N	19
Ford Av. E&W	10W-38N	19
Ford St. E&W	12W-38N	18
Frolic Av. N&S	12W-39N	18
Ganster Rd. NS&EW	10W-37N	19
Garnett Av. N&S	10W-37N,10W-39N	19
Geraghty Av. N&S	10W-37N,10W-39N	19
Gilbert Av. N&S	10W-38N	19
Gish Av. N&S	10W-38N	19
Glendale Rd. N&S	11W-40N	10
Graves Av. E&W	12W-37N	18
Green Av. N&S	12W-39N,12W-40N	18
Green Bay Rd. N&S	12W-37N,12W-40N	18
Harding St. N&S	10W-38N	19
Harper Rd. N&S	11W-37N,11W-38N	18
Hart Av. E&W	10W-38N	19
Hart St. E&W	12W-38N	18
Hendee Av. E&W	10W-37N,9W-37N	19
Hendee St. E&W	12W-37N	18
Hickory Ln. E&W	12W-40N	10
Hickory Rd. E&W	11W-38N	18
Holdridge Av. N&S	10W-37N	19
Holridge Av. N&S	10W-38N,10W-39N	19
Howard St. E&W	12W-38N,10W-38N	18
Illinois Av. E&W	12W-39N,10W-39N	18
Johnson Rd. E&W	11W-39N	18
Joyce Av. N&S	10W-38N	19
Kenosha Rd. NE&SW	12W-40N	10
Lake Crest Av. N&S	10W-37N	19
Lake St. E&W	9W-38N	19
Lake View Av. N&S	9W-38N	19
Lakeside Av. N&S	9W-40N,9W-41N	11
Leland Av. E&W	11W-38N	18
Lewis-Jordan Rd. N&S	11W-38N,11W-39N	18
Liberty Av. E&W	10W-38N,10W-39N	19
Liberty Ct. E&W	10W-38N	19
Lincoln Av. N&S	10W-38N	19
Linden Av. N&S	11W-38N	18
Lone Oak Rd. N&S	12W-40N	10
Loyola Av. N&S	10W-37N	19
Lynch Av. N&S	12W-39N	18
Major Av. E&W	12W-39N	18
Manor Av. N&S	10W-37N,10W-38N	19
Maplewood Rd. E&W	12W-39N	18
Marguerite Ln. E&W	9W-38N	19
Marguerite Rd. N&S	9W-38N	19
Mawman Av. E&W	10W-37N,9W-37N	19
Mawman St. E&W	12W-37N	18
McAree Av. N&S	12W-39N	18
Metropolitan Av. N&S	12W-39N	18
Michigan Blvd. E&W	9W-38N	19
Minturn Av. E&W	12W-38N	18
Morse Av. N&S	12W-38N	18
Morton Av. E&W	10W-37N	19
New York Av. N&S	11W-37N,11W-38N	18
Noble Av. N&S	12W-38N	18
North Av. N&S	11W-38N,11W-39N	18
North Shore Av. N&S	10W-37N,10W-38N	19
Oak Av. E&W	11W-38N	18
Oak Forest Dr. E&W	9W-38N	19
Oberg Av. N&S	12W-38N	18
Orchard Rd. N&S	11W-37N	18
Paddock Av. E&W	10W-37N,9W-37N	19
Paxton Av. E&W	9W-39N	19
Peacock Rd. E&W	12W-39N	18
Pickford Av. E&W	12W-38N	18
Pickford St. E&W	12W-38N	18
Pine Av. N&S	11W-38N	18
Pine Ct. E&W	12W-40N	10
Prairie Ln. E&W	10W-38N	19
Princeton Av. E&W	12W-39N	18
Reiner Pl. E&W	10W-38N	19
Richard Pl. E&W	11W-39N	18
Russell Av. N&S	10W-38N	19
Ruth Av. N&S	12W-38N	18
Sallmon Rd. E&W	12W-37N	18
Sheridan Pl. N&S	9W-38N	19
Sheridan Rd. N&S	10W-37N,10W-39N	19
Spitz Dr. N&S	11W-38N	18
Stewart Av. N&S	10W-38N	19
Stratton Av. N&S	9W-37N	19
Suddard Pl. E&W	9W-38N,10W-38N	19
Talmadge Av. E&W	10W-38N	19
Tewes Ct. N&S	10W-38N	19
Van Ct. NS&EW	12W-40N	10
Vercoe Av. E&W	12W-37N	18
Wadsworth Rd. E&W	12W-39N,10W-39N	18
Waldo Av. E&W	12W-39N,10W-39N	18
Warner Av. E&W	12W-37N	18
Wilson Av. N&S	10W-37N,10W-38N	19
Woodland Av. E&W	10W-38N	19
Wyer St. E&W	10W-38N	19
Yale Av. E&W	10W-39N	19
Yorkhouse Rd. E&W	12W-37N,10W-37N	18
27th Pl. E&W	12W-40N	10
27th St. E&W	12W-40N	10
28th Pl. E&W	12W-40N	10
28th St. E&W	12W-40N	10
29th St. E&W	12W-40N,11W-40N	10
32nd St. E&W	12W-39N	18
33rd St. E&W	12W-39N,11W-39N	18

Benton Township

Street	Grid	Page
Barnhart Ct. E&W	9W-41N,8W-41N	11
Berrong Av. N&S	10W-42N	11
Berrong Ct. N&S	10W-42N	11
Block Ln. E&W	11W-43N	10
Brockway Av. N&S	9W-40N	11
Broecker Av. N&S	11W-42N	10
Burnett Av. N&S	8W-40N,8W-41N	11
Dodson St. E&W	11W-42N	10
E. Ravine Dr. NE&SW	10W-41N	11
Edward St. E&W	11W-42N	10
English Ln. N&S	10W-42N	11
Fago Av. N&S	11W-42N,11W-43N	10
Forman Dr. E&W	12W-42N,11W-42N	10
Fosslands Av. N&S	11W-42N,11W-43N	10
Fulton Av. N&S	8W-40N,8W-41N	11
Geneva Ln. N&S	11W-39N	18
Green Bay Rd. NW&SE	12W-41N	10
Hanks Av. N&S	11W-42N,11W-43N	10
Highland Av. E&W	11W-38N	18
Howland Av. N&S	9W-40N	11
Illinois St. N&S	10W-41N	11
Jensen Av. N&S	11W-43N	10
Johnson Rd. E&W	11W-39N	18
Kellogg Ct. N&S	9W-41N	11
Kenosha Rd. N&S	12W-41N,11W-43N	10
Lake Front Dr. N&S	8W-40N,8W-41N	11
Lewis Av. N&S	11W-42N,11W-43N	10
Logan Ct. E&W	10W-41N,8W-41N	11
Lynn Dale Dr. N&S	12W-43N	10
Mather Av. N&S	11W-42N	10
Mills Av. N&S	8W-40N	11
Morse Av. N&S	9W-40N,9W-41N	11
Oak Ln. E&W	11W-43N	10
Oakcrest Ln. N&S	12W-43N	10
Orchard Ln. N&S	9W-41N	11
Parkview Av. N&S	9W-40N,9W-41N	11
Pearsal Av. N&S	9W-40N,9W-41N	11
Prairie Av. N&S	11W-43N	10
Prospect Av. E&W	9W-40N,8W-40N	11
Rita Av. E&W	9W-40N,8W-40N	11
Russell Rd. E&W	13W-43N,10W-43N	10
Short Av. N&S	8W-40N	11
Spitz Dr. N&S	11W-38N	18
Strong Av. N&S	9W-40N	11
Suddard St. E&W	12W-38N	18
Thorpe Av. N&S	11W-42N	10
Transportation Av. N&S	9W-40N,9W-41N	11
Vista Ln. N&S	12W-42N	10
Washington Av. N&S	10W-42N	11
Wilmont Av. N&S	9W-40N	11
Wilson Ct. E&W	9W-41N,8W-41N	11
Winthrop Ct. E&W	10W-41N,8W-41N	11
Winthrop Harbor Rd. E&W	12W-42N	10
Wintrhop Ct. E&W	10W-41N,8W-41N	11
Zepher St. E&W	12W-38N	18
3rd St. E&W	11W-43N	10
4th St. E&W	11W-43N	10
5th St. E&W	11W-42N	10
9th St. E&W	11W-42N,10W-42N	10
10th St. E&W	11W-42N	10
11th St. E&W	11W-42N	10
12th St. E&W	11W-42N	10
13th St. E&W	11W-42N	10
16th St. E&W	9W-41N	11
17th St. E&W	11W-41N,8W-41N	10
18th St. E&W	10W-41N,8W-41N	11
19th St. E&W	10W-41N,8W-41N	11
20th St. E&W	9W-41N,8W-41N	11
22nd St. E&W	9W-40N,8W-40N	11
23rd St. E&W	9W-40N,8W-40N	11

Buffalo Grove*
(Also see Cook County)

Street	Grid	Page
Aberdeen Ct. E&W	16W-20N	48
Aberdeen Ln. N&S	16W-20N	48
Acacia Ct. S. NE&SW	17W-23N	48
Acacia Ter. N&S	17W-23N	48
Acorn Ct. E&W	16W-20N	48
Acorn Pl. E&W	16W-22N	48
Alden Ln. NS&EW	17W-21N	48
Alsace Cir. NS&EW	16W-20N	48
Amherst Ct. E&W	17W-21N	48
Anderson Ln. N&S	18W-21N	48
Apollo Dr. NW&SE	16W-20N	48
Appletree Ct. N&S	17W-20N	48
Arbor Gate Ln. NW&SE	17W-20N	48
Arlington Heights Rd. N&S	17W-20N,18W-20N	48
Armstrong Ct., North N&S	15W-20N	49
Armstrong Dr. E&W	16W-20N	48
Ashbury Dr. E&W	15W-21N	49
Ashland Ct. NE&SW	17W-20N	48
Aspen Ct. N&S	17W-20N	48
Aspen Dr. N&S	17W-20N	48
Auburn Ln. E&W	17W-20N	48
Autumn Ct. E&W	16W-21N	48
Banbury Ln. E&W	17W-21N	48
Bank Ln. E&W	16W-20N	48
Banyan Tree Ln. E&W	17W-22N	48
Barclay Blvd. N&S	15W-21N	49
Bayberry Dr. N&S	17W-23N	48
Bedford Ct. N&S	18W-21N	48
Belmar Ct. NE&SW	17W-21N	48
Belmar Ln. NW&SE	17W-21N	48
Bentley Pl. NS&EW	16W-21N	48
Berkley Ct. N&S	16W-21N	48
Birchwood Ct. N. NW&SE	17W-22N	48
Birchwood Ct. S. E&W	17W-22N	48
Birchwood Ln. N&S	17W-22N	48
Blackthorn Dr. NE&SW	16W-23N	48
Blossom Ct. N&S	17W-20N	48
Blue Ash Ct. E&W	17W-23N	48
Blue Ash Dr. E&W	17W-23N	48
Bordeaux Ct. West E&W	17W-21N	48
Bordeaux Ct., East E&W	17W-21N	48
Brandwyn Ct. E&W	17W-21N	48
Brandywyn Ct., N. NW&SE	17W-21N	48
Brandywyn Ln. NS&EW	17W-22N,16W-22N	48
Bristol Ln. NW&SE	18W-21N	48
Buckingham Ln. E&W	16W-20N	48
Buckthorn Ct. E&W	16W-20N	48
Buckthorn Ter. NS&EW	15W-20N	49
Buffalo Grove Rd. N&S	17W-22N,16W-20N	48
Bunescu Ct.* N&S	18W-20N	48
Bunescu Ln.* E&W	18W-20N	48
Burgess Cir. E&W	18W-21N	48
Burnt Ember Ct. E&W	17W-20N	48
Burnt Ember Ln. NW&SE	17W-20N	48
Busch Pkwy. E&W	15W-21N	49
Busch Rd. E&W	17W-21N,16W-21N	48
Camden Ct.* E&W	18W-20N	48
Canterbury Ln. E&W	16W-21N	48
Caren Ct. N&S	17W-21N	48
Caren Dr. E&W	17W-21N	48
Carlton Pl. E&W	17W-20N	48
Carlyle Ln. E&W	16W-22N	48
Carman Av. N&S	15W-20N	49
Cary Ln. NW&SE	17W-20N	48
Castlewood Ln. NW&SE	17W-20N	48
Cedar Ct. N. NW&SE	17W-22N	48
Cedar Ct. S. E&W	17W-22N	48
Chapel Oaks Dr. E&W	16W-23N	48
Chase Ct.* E&W	18W-20N	48
Chateau Dr. E&W	17W-20N	48
Chaucer Way E&W	17W-21N	48
Checker Dr. N&S	17W-20N,17W-21N	48
Checker Rd. E&W	17W-20N	48
Cherbourg Ct. N E&W	16W-20N	48
Cherbourg Ct. S. E&W	16W-20N	48
Cherbourg Dr. N&S	16W-20N	48
Chestnut Ct., E. N&S	16W-22N	48
Chestnut Ct., W. N&S	16W-22N	48
Chestnut Ter. NS&EW	16W-22N	48
Chicory Ln. E&W	16W-20N	48
Church Rd. NE&SW	16W-20N	48
Churchill Ct. N&S	16W-21N	48
Circle Ct. NW&SE	16W-20N	48
Circle Dr. NS&EW	10W-39N	19
Claret Dr. E&W	17W-20N	48
Clearview Ct. N&S	16W-21N	48
Clohessey Dr. N&S	17W-20N	48
Cloverdale Ct. E&W	17W-21N	48
Cobbler Ln. E&W	16W-21N	48
Cobblestone Ct. N&S	16W-20N	48
Cobblestone Ln. N&S	16W-20N	48
Commerce Ct. N&S	15W-20N,15W-21N	49
Common Way E&W	16W-20N	48
Cooper Ct.* E&W	18W-20N	48
Copperwood Dr. NS&EW	16W-21N	48
Corporate Grove Dr. E&W	15W-21N	49
County Line Rd. E&W	16W-20N	48
Courtland Dr. E&W	17W-21N	48
Coventry Ct. NW&SE	17W-22N	48
Coventry Ln. E&W	17W-22N	48
Crab Apple Ter. E&W	16W-22N	48
Crossfield Ct. N&S	16W-21N	48
Crown Point Ct. N&S	17W-20N	48
Crown Point Dr. N&S	17W-20N	48
Cumberland Ct. NE&SW	16W-21N	48
Cumberland Ln. NW&SE	16W-21N	48
Cyprus Ct. NE&SW	17W-23N	48
Dannet Ct. N&S	17W-21N	48
Dannet Rd. E&W	17W-21N	48
Dartmouth Ln. E&W	15W-20N	49
Daulton Ct. NW&SE	17W-23N	48
Daulton Dr. NE&SW	17W-23N	48
Dayton Ct. E&W	17W-21N	48
Dayton Rd. N&S	17W-21N	48
Deerfield Pkwy. NS&EW	15W-21N	49
Dellwood Ct. E&W	16W-21N	48
Devlin Rd. N&S	17W-21N	48
Devonshire Rd. Cir.	17W-21N	48
Devonwood Ct. N&S	16W-21N	48
Dogwood Ter. N&S	16W-20N	48
Dorncliff Ln. NE&SW	17W-20N	48
Dover Ct. NW&SE	17W-22N	48
Dover Dr. E&W	17W-20N	48
Driftwood Pl. N&S	16W-20N	48
Dunham Ln. NE&SW	17W-21N	48
Dunhill Dr. NS&EW	16W-20N	48
Dunstan Ln. N&S	17W-23N	48
East Chevy Chase Dr. NE&SW	14W-20N	49
Eastwood Ln. E&W	15W-20N	49
English Oak Ter. E&W	17W-22N	48
Essington Ln. N&S	17W-20N	48
Euclid Ct. N&S	16W-21N	48
Fabish Ct. N&S	16W-21N	48
Fabish Dr. NS&EW	16W-21N	48
Fairfax Ln. E&W	18W-21N	48
Farnsworth Ct. N&S	18W-21N	48
Farrington Dr. E&W	17W-20N	48
Ferndale Ct. N&S	16W-20N	48
Fox Hill Ct. E. NW&SE	16W-20N	48
Fox Hill Ct. W. N&S	16W-20N	48
Fox Hill Dr. NS&EW	16W-20N	48
Foxford Dr. NS&EW	17W-23N	48
Franklin Ln. N&S	18W-21N	48

	Grid	Page
Fremont Ct. E. E&W	18W-21N	48
Fremont Ct. W. E&W	18W-21N	48
Fremont Way Cir. . E&W	18W-21N	48
Gail Ct. NE&SW	17W-21N	48
Gail Dr. N&S	17W-21N	48
Gardenia Ln. E&W	17W-20N	48
Green Knolls Dr. N&S		
	17W-21N	48
Hackberry Ct. E E&W	17W-22N	48
Hackberry Ct. W. E&W	17W-22N	48
Hampshire Ct. E&W	16W-20N	48
Harris Dr. N&S	17W-20N	48
Harvest Ct.* NE&SW	17W-20N	48
Hastings Ln. NS&EW	15W-20N	49
Hazelwood Ter. N&S	16W-20N	48
Heatherdown Way E&W	17W-21N	48
Hidden Lake Dr. NS&EW	16W-21N	48
Highland Grove Ct. N. NE&SW	16W-21N	48
Highland Grove Dr. N&S	16W-20N,16W-21N	48
Hill Ct. E. E&W	16W-20N	48
Hilldale Ln. N&S	16W-20N	48
Hobson Ct. NE&SW	16W-20N	48
Hobson Dr. N&S	16W-21N	48
Hollystone Ln. NE&SW	17W-20N	48
Horatio Blvd. E&W	16W-20N,15W-20N	48
Huntley Rd. NW&SE	15W-20N	49
Indian Spring Ln. N&S	17W-20N	48
Ironwood Ct. N&S	16W-22N	48
Ironwood Rd. E&S	16W-22N	48
Ivy Hall Ln. E&W	17W-21N	48
Jersey Ct. N&S	16W-21N	48
Joel Ln. N&S	16W-20N	48
Johnson Dr. N&S	15W-20N	49
Kent Ln. N&S	17W-21N	48
Kingsbridge Way E&W	17W-21N	48
Kingston Dr. NE&SW	17W-23N	48
Knight Hill Ct. NW&SE	17W-20N	48
Knollwood Dr. E&W	17W-21N	48
Knollwoods Ct. N&S	17W-21N	48
LaSalle Ct. N. E&W	17W-21N	48
LaSalle E&W	17W-21N	48
Lake Cook Rd. E&W	16W-20N	48
Lakeview Ln. N&S	17W-20N	48
Lakeview Dr. N&S	17W-20N	48
Lamont Ter. E&W	17W-21N	48
Lancaster Ct.* E&W	17W-20N	48
Larchmont Dr. N&S	17W-21N	48
Larraway Dr. N&S	17W-21N	48
Lawn Ct. NE&SW	16W-21N	48
Le Parc Cir. E&W	16W-20N	48
LeJardin Ct. N&S	16W-20N	48
Lee Ct. N&S	18W-21N	48
Lexington Dr. N&S	15W-20N	49
Lilac Ln. NS&EW	15W-20N	49
Linden Av. E&W	14W-20N	49
Lockwood Ct. E. N&S	17W-21N	48
Lockwood Ct. W. N&S	17W-21N	48
Lockwood Dr. NS&EW	17W-21N	48
Logsdon Ln. N&S	17W-21N	48
London Ct. E&W	17W-22N	48
Long Grove Aptakisic Rd. E&W	17W-22N	48
Long Ridge Ct. NW&SE	16W-21N	48
Lyon Ct. E&W	17W-21N	48
Lyon Dr. E&W	17W-21N	48
Madison Ct. N. E&W	16W-21N	48
Madison Ct. S. N&S	16W-21N	48
Madison Dr. NS&EW	16W-21N	48
Magnolia Ct. E. E&W	17W-22N	48
Magnolia Ct. W. NW&SE	17W-22N	48
Margate Dr. NW&SE	16W-21N	48
Marseilles Cir. N&S	16W-20N	48
Milwaukee Av. NW&SE	14W-21N	49
Mistywoods Rd. N&S	16W-23N	48

	Grid	Page
Morningside Ln. W. E&W	18W-21N	48
Morningside Ln. E. E&W	16W-21N	48
Mundelein Rd. NW&SE	17W-21N	48
New Dover Ln. E&W	17W-23N,16W-23N	48
New Town Ct., East N&S	16W-20N	48
New Town Ct., West NE&SW	16W-20N	48
New Town Dr. NW&SE	16W-20N	48
Newfield Dr. E&W	16W-21N	48
Newgate Ct. E&W	16W-21N	48
Newman St. E&W	14W-20N	49
Newport Ct. N&S	18W-21N	48
Newton Dr. E&W	16W-20N	48
Old Barn N&S	16W-21N	48
Osage Ter. N&S	16W-20N	48
Oxford Dr. E&W	17W-21N	48
Park Av. E&W	16W-20N	48
Parkchester E&W	16W-20N	48
Parker Ln. NS&EW	17W-21N	48
Parkside Dr. N&S	16W-21N	48
Pauline Av. E&W	16W-20N	48
Pecan Cir. N&S	16W-20N	48
Penny Ln. N&S	17W-20N	48
Pin Oak Ter. E&W	16W-20N	48
Pinehurst Ln. N&S	16W-21N,15W-21N	48
Pinetree Cir. N.* NS&EW	17W-20N	48
Pinetree Cir. S.* NS&EW	17W-20N	48
Pinyon Ct. N. N&S	16W-23N	48
Pinyon Ct. S. N&S	16W-23N	48
Port Clinton Ct. E. N&S	17W-23N	48
Port Clinton Ct. W. N&S	17W-23N	48
Prairie Ln. N&S	16W-20N	48
Providence Ct.* N&S	18W-20N	48
Providence Ln. NS&EW	18W-21N	48
Quaker Hollow Ct. N. N&S	16W-21N	48
Quaker Hollow Ct. S. N&S	16W-21N	48
Raintree Ln. N&S	17W-23N	48
Raintree Rd. N&S	17W-23N	48
Ranchview Ct. N&S	17W-21N	48
Raphael Av. N&S	15W-20N	49
Redbud Pl. E&W	16W-20N	48
Ridgefield Ln. NW&SE	17W-21N	48
Ridgewood N&S	15W-20N	49
Ronnie Dr. NS&EW	17W-21N	48
Roslyn Ct. NE&SW	16W-23N	48
Roslyn Ln. E&W	16W-23N	48
Rt. 21 N&S	14W-21N	49
Rt. 83 NW&SE	18W-21N,16W-20N	48
Russelwood Ct. E&W	16W-21N	48
Sandalwood Ct. E&W	17W-23N	48
Sandalwood Rd. N&S	17W-23N	48
Sandhurst Ct. NW&SE	16W-21N	48
Sandhurst Dr. NS&EW	16W-21N	48
Satinwood Ct. N. NE&SW	17W-22N	48
Satinwood Ct. S. E&W	17W-22N	48
Satinwood Ter. NS&EW	17W-22N	48
Saybrook Ln. N&S	18W-21N	48
Scottish Pine Ln. E&W	17W-23N	48
Shady Grove Ln. N&S	17W-20N	48
Shambliss St. N&S	18W-21N	48
Shambliss Ln. N&S	18W-21N	48
Silver Linden Ln. N&S	17W-22N	48
Silver Rock Ln. N&S	17W-20N	48
Somerset Ln. NW&SE	16W-20N	48
South Chevy Chase Dr. E&W	14W-20N	49
Springside Ct. NW&SE	17W-20N	48
Springside Ln. NE&SW	17W-20N	48
Stanton Ct. E. N&S	16W-21N	48
Stanton Ct. W. N&S	16W-21N	48
Stanton Dr. NS&EW	16W-21N	48
Steeple Dr. N&S	16W-20N	48
Teawood Cir. NS&EW	16W-20N	48

	Grid	Page
Thistle Ct. SW&NE	16W-21N	48
Thompson Blvd. E&W	17W-21N,16W-21N	48
Thompson Ct. N&S	16W-21N	48
Thornapple Ct. NE&SW	17W-20N	48
Thorndale Ct. N&S	16W-20N	48
Thorndale Dr. E&W	16W-20N	48
Trotwood Ct. N&S	16W-21N	48
Twilight Pass NE&SW	17W-20N	48
Twisted Oak Ln. E&W	17W-20N	48
Village Ct. N&S	16W-21N	48
Vintage Ln. N&S	17W-20N	48
Wakefield Ct. NW&SE	16W-20N	48
Wakefield Ln. NW&SE	16W-20N	48
Wedgewood Ct. SW&NE	17W-23N	48
Weidner Rd. E&W	17W-20N	48
Weiland Rd. N&S	16W-20N,16W-21N	48
Wellington Ct. N&S	16W-21N	48
Wellington North E&W	16W-21N,15W-21N	48
Wellington South E&W	16W-21N,15W-21N	48
West Bourne Ln. NE&SW	17W-20N	48
West Dr. NW&SE	14W-20N	49
Westchester Ln. N&S	16W-21N	48
Whispering Oaks Ct. E&W	16W-23N	48
Whispering Oaks Dr. Cir.	16W-23N	48
White Oak Ln. E&W	17W-23N	48
White Pine Ln. N&S	17W-23N	48
Whitebranch Ct. N&S	16W-20N	48
Wildflower Cir. N&S	17W-20N	48
Willow Parkway E&W	16W-22N	48
Windbrooke Dr. N&S	15W-21N	49
Winding Oak Ln. N&S	16W-21N	48
Windover Cir. N&S	17W-20N,16W-20N	48
Windwood Ct.* N&S	17W-21N	48
Winston Dr. N&S	16W-21N	48
Witney Ln. N&S	16W-21N	48
Wood Hollow Ln. N&S	17W-20N	48
Woodbury Rd. N&S	16W-20N	48
Woodridge Ln.* N&S	17W-20N	48
Woodstone Cir. Cir.	16W-20N	48
Woodstone Dr. E&W	16W-20N	48
Wyngate Ln. N&S	17W-20N	48

Cuba Township

	Grid	Page
Apple Tree Ln. E&W	26W-22N	44
Arch Dr. E&W	28W-23N	44
Beach Dr. N&S	28W-24N	36
Beachway NW&SE	28W-24N	36
Bertha Ln. N&S	26W-22N	45
Bloners Dr. E&W	28W-24N	36
Bluff Ct. NS&EW	26W-24N	37
Bluff Ln. E&W	28W-23N	44
Border Rd. N&S	28W-21N,25W-22N	45
Brandt Rd. E&W	28W-21N,25W-21N	45
Broken Bow Pass NW&SE	28W-25N,25W-25N	37
Canyon Creek Ct. N&S	26W-25N	37
Chamberlain Dr. E&W	28W-23N	44
Chatham Rd. E&W	25W-21N	45
Church Rd. N&S	28W-23N	44
Columbus Dr. NE&SW	25W-21N	45
Country Estates Rd. E&W	26W-22N	
Countryside Ln. N&S	26W-22N	45
County Line Rd. N&S	28W-22N	44
Craft Ct. NE&SW	28W-23N	44
Crestview Ln. NS&EW	25W-21N	45
Crown Rd. NS&EW	26W-21N	45
Cuba Rd. E&W	28W-21N,25W-21N	44
Damien Dr. E&W	28W-22N	44
Drake E&W	25W-21N	45

	Grid	Page
E. Lake Shore Dr. N&S	25W-25N	37
East Dr. N&S	28W-23N	44
East Oakwood Dr. NE&SW	25W-25N	37
Edgemond Ln. E&W	26W-22N	45
Edgewood Av. E&W	28W-24N	36
Elizabeth Ln. N&S	25W-21N	45
Elm Ct. NE&SW	28W-21N	45
Exmoor Av. N&S	26W-20N	45
Fairhaven Dr. NE&SW	25W-21N	45
Fairway Cir. NE&SW	26W-24N	37
Falkirk Ct. N&S	25W-21N	45
Flint Creek Dr. N&S	27W-23N,27W-24N	44
Flynn Creek Dr. NS&EW	27W-21N	44
Fox River Dr. NE&SW	28W-23N,28W-24N	44
Fox River Rd. E&W	28W-24N,27W-24N	36
Fox St. E&W	28W-23N	44
Glenbarr Ln. NW&SE	26W-21N	45
Golf Ct. N&S	26W-24N	37
Golf Ln. N&S	26W-24N	37
Grand View Dr. NE&SW	25W-24N	37
Grant St. E&W	27W-21N,26W-21N	44
Grayclyn Ct. N&S	28W-25N	36
Greenwood Ln. N&S	25W-25N	37
Hanover Hill NW&SE	25W-21N	45
Hart Rd. N&S	26W-20N,26W-21N	45
Heritage Oak Dr. E&W	28W-23N	44
Heritage Tr. N&S	28W-23N	44
Hess Dr. N&S	28W-23N	44
Hickory Ln. N&S	26W-24N	37
Hickory Nut Grove Rd. N&S	28W-23N,28W-25N	44
High Ridge Dr. E&W	25W-21N	45
High Ridge Rd. N&S	28W-23N	44
Highland Dr. E&W	27W-21N,26W-21N	44
Hillcrest Av. E&W	28W-24N	36
Hillside Rd. N&S	28W-23N	44
Hipplers Rd. E&W	25W-21N	45
Holly Dr. N&S	28W-24N	36
Hough Rd. N&S	25W-20N,25W-21N	45
Indian Tr. NW&SE	25W-25N,25W-25N	37
Jensen Av. NW&SE	28W-24N	36
Kelsey Av. N&S	28W-23N	44
Kelsey Ct. N&S	28W-23N	44
Kelsey Rd. N&S	28W-22N,28W-23N	44
Knollwood Dr. N&S	25W-25N	37
Kristy Ln. NS&EW	28W-24N	36
Lagoon Dr. N&S	27W-24N	37
Lake Shore Dr. N&S	25W-25N,25W-25N	37
Lake Shore Dr. E&W	25W-21N	45
Lake Zurich Rd. N&S	25W-21N	45
Lakeview Pkwy. NS&EW	25W-25N	37
Laurin Dr. N&S	25W-21N	45
Lincoln St. E&W	26W-21N	45
Lindbergh Dr. E&W	28W-23N	44
Long Meadow NE&SW	26W-24N	37
Main St. NE&SW	28W-24N	36
Main St. E&W	28W-23N	44
Meadow Ln. NE&SW	28W-23N	44
Merton Rd. N&S	26W-20N	45
Midway St. E&W	28W-23N	44
North Dr. E&W	28W-23N	44
North Oaks Ct. N&S	25W-25N	37
Oak Creek Cir. NE&SW	25W-25N	37
Oakland Av. N&S	28W-24N	36
Oakridge Rd. E&W	25W-25N	37
Oakwood Dr. N&S	25W-24N,25W-25N	37
Old Barrington Rd. N&S	26W-21N	45
Park Dr. E&W	28W-23N	44
Park St. NS&EW	28W-23N	44
Pawnee Rd. NS&EW	25W-21N	45
Pepper Rd. E&W	28W-23N	44
Pioneer Grove Rd. N&S	28W-23N	44
Pitner Av. E&W	28W-24N	36
Prestick Dr. NS&EW	26W-21N	45
Private Rd. N&S	26W-21N	45

	Grid	Page
Providence Rd. E&W	25W-21N	45
Ravine Dr. E&W	28W-23N	44
River Rd. NE&SW	28W-23N	44
River St. N&S	28W-23N	44
Rivers Edge Dr. NS&EW	28W-24N	36
Rivershore Dr. N&S	27W-24N	37
Riverside Dr. NE&SW	28W-24N	36
Roberts Ct. N&S	28W-25N	36
Roslyn Rd. E&W	26W-20N	45
Rustic Dr. N&S	25W-23N	45
Rustic Ln. E&W	28W-23N	44
Scott Rd. E&W	26W-22N,25W-22N	45
Shoreline Rd. E&W	26W-24N	37
Short St. NS&EW	28W-24N	36
South St. E&W	28W-23N	44
Spring Dr. N&S	28W-23N	44
Summit Dr. N&S	28W-23N	44
Sunset Av. N&S	28W-24N	36
Sunset Ln. N&S	25W-21N	45
Sunset Rd. E&W	25W-21N	45
Sylvester St. NW&SE	28W-24N	36
Tara Dr. E&W	25W-25N	37
Taylor Rd. E&W	26W-20N	45
Timberlake Pkwy. E&W	25W-25N	37
Timberlake Rd. NS&EW	26W-25N,25W-25N	37
Wagon Wheel Ct. N&S	25W-25N	37
Waterview Ct. E&W	26W-24N	37
Waveland Av. NE&SW	28W-24N	36
West Oakwood Dr. N&S	25W-25N	37
White Oak NE&SW	26W-24N,26W-25N	37
Wildwood Dr. E&W	25W-25N	37
Woodbridge N&S	26W-24N	37
Woodland Av. N&S	25W-21N	45
13th St. N&S	26W-20N,26W-21N	45
14th St. N&S	26W-21N	45
17th St. N&S	26W-21N	45
18th St. N&S	26W-21N	45
19th St. N&S	26W-21N	45
20th St. N&S	26W-21N	45
21st St. N&S	26W-21N	45
22nd St. N&S	27W-21N	44
23rd St. N&S	27W-21N	44
24th St. N&S	27W-21N	44
25th St. N&S	27W-21N	44

Deer Park

	Grid	Page
Bedlington Cir. E&W	23W-21N	46
Bobwhite Ln. N&S	23W-21N	46
Bramble Ln. E&W	22W-20N	46
Briargate Ln. N&S	22W-20N	46
Cardinal Ln. E&W	23W-21N	46
Chapel Hill Dr. NS&EW	23W-21N	46
Cheshire Ct. NE&SW	22W-20N	46
Circle Dr. NS&EW	23W-20N	46
Clover Ln. N&S	23W-20N	46
Corners Dr. E&W	22W-20N	46
Covington Dr. E&W	22W-20N	46
Cuba Rd. E&W	25W-21N,23W-21N	45
Deer Lake Dr. N&S	23W-20N	46
Deer Valley Dr. N&S	22W-20N	46
Deerpath Rd. N&S	23W-20N,23W-21N	46
Dogwood Ct. E&W	22W-20N	46
Doncaster Cir. NW&SE	22W-20N	46
Dover St. NS&EW	22W-20N	46
Echo Ln. E&W	21W-20N	47
Edgeview Ct. NW&SE	23W-20N	46
Ela Rd. N&S	24W-21N	46
Fairview Dr. E&W	22W-20N	46
Farthingdale Ct. NE&SW	23W-21N	46
Ferndale Rd. N&S	23W-20N	46
Foxchase Rd. NW&SE	22W-20N	46
Glen Garry Cir. E&W	23W-20N	46
Glenhurst Rd. E&W	22W-20N	46
Green Forest Rd. NE&SW	24W-21N	46
Hearthside Ct. N&S	23W-21N	46
Heather Ln. E&W	22W-20N	46
Hedgeworth Ct. E&W	23W-21N	46
Hidden Pines Dr. E&W	22W-20N	46

	Grid	Page
Hollington Ln. E&W	22W-20N	46
Hummingbird Ct. NW&SE	24W-21N	46
Hunters Ln. NE&SW	24W-21N	46
Hypoint Dr. N&S	23W-21N	46
Inglenook Ct. N&S	23W-21N	46
Inglenook Ln. N&S	23W-21N	46
Juniper Ct. E&W	23W-20N	46
Juniper Ln. NW&SE	23W-20N	46
Lagrov Ct. N&S	22W-20N	46
Lakeview Ct. NE&SW	23W-20N	46
Lancaster Ct. NW&SE	23W-21N	46
Landmark Ln. N&S	22W-20N	46
Laurel Dr. N&S	23W-20N	46
Lea Rd. N&S	23W-20N	46
Lee St. E&W	23W-20N	46
Lois Ln. E&W	23W-20N	46
Lone Pine Ct. NW&SE	23W-20N	46
Long Grove Rd. E&W	23W-20N,22W-20N	46
Madoch Ct. E&W	22W-20N	46
Mallard Ct. NE&SW	23W-21N	46
Mariel Cir. NE&SW	22W-20N	46
Mayfield Ln. NE&SW	23W-21N	46
Meadow Ct. N&S	23W-20N	46
Meadow Ln. NS&EW	23W-20N	46
Newcastle Ct. E&W	22W-20N	46
Oak Ridge Ln. N&S	23W-20N	46
Old Farm Rd. NS&EW	24W-21N	46
Park Hill Dr. N&S	23W-20N	46
Pheasant Hill Rd. NE&SW	23W-20N	46
Pheasant Trail N&S	23W-21N	46
Primrose Ct. N&S	23W-20N	46
Quail Ct. NS&EW	23W-21N	46
Quentin Rd. N&S	22W-20N	46
Rainbow Rd. N&S	24W-21N	46
Rand Rd. NW&SE	22W-20N,21W-20N	46
Regency Ct. N&S	24W-22N	46
Rt. 12 NW&SE	21W-20N,20W-20N	47
Rue Chambrix E&W	22W-20N	46
Rue Jordon N&S	22W-20N	46
Rue Orleanais E&W	22W-20N	46
Rue Royale E&W	22W-20N	46
Rue Tourain N&S	22W-20N	46
Shady Ln. N&S	23W-20N	46
Shoreham Cir. E&W	22W-20N	46
Sumac Ct. NW&SE	23W-21N	46
Sunset Ridge Rd. NS&EW	23W-20N	46
Sunshine Ln. NS&EW	22W-20N	46
Surrey Ct. E&W	22W-20N	46
Swallow Ct. E&W	23W-20N,22W-20N	46
Swansway N&S	22W-20N	46
Sylvander Dr. N&S	24W-21N	46
Tanager Ct. NW&SE	24W-21N	46
Teal Ct. E&W	23W-21N	46
Thornbury Cir. E&W	22W-20N	46
Thornhill Ct. E&W	23W-20N,22W-20N	46
Thrush Ct. NE&SW	23W-21N	46
Touraine Ct. E&W	22W-20N	46
Vesper Ct. NW&SE	23W-21N	46
Wagon Ct. E&W	22W-20N	46
Wallingford Ln. NS&EW	22W-20N	46
Wehrheim Rd. N&S	22W-20N	46
Wheel Ct. E&W	22W-20N	46
Wicker Dr. N&S	22W-20N	46
Willow Ct. E&W	22W-20N	46
Woodberry Rd. NE&SW	23W-20N	46

Deerfield*

(Also see Cook County)

	Grid	Page
Alden Ct. N&S	11W-21N	50
Ambleside Dr. E&W	10W-21N	50
Anthony Ln. E&W	11W-20N	50
Appletree Ct. NW&SE	11W-20N	50
Appletree Ln. N&S	11W-20N,11W-21N	50
Arbor Vitae Rd. E&W	11W-21N	50
Arthur Ct. N&S	11W-20N	50
Augusta Dr. E&W	10W-20N	50
Barclay Ln. NW&SE	11W-21N	50
Bayberry Ln. E&W	11W-20N	50
Bent Creek Ridge N&S	9W-20N	51
Berkeley Ct. E&W	12W-21N,11W-21N	50
Beverly Pl. N&S	10W-21N	50
Birch Ct. E&W	11W-21N	50
Birchwood Av. N&S	11W-20N	50
Blackthorn Ln. N&S	10W-21N	50
Bradmoor Pl. N&S	11W-21N	50
Brand Ln. N&S	10W-21N	50
Briarhill Rd. N&S	10W-20N	50
Brookside Ln. E&W	11W-20N	50
Burning Tree Ln. N&S	11W-20N	50
Burr Oak Av. N&S	11W-20N	50
Burr Oak Ct. N&S	11W-20N	50
Byron Ct. N&S	10W-20N	50
Cambridge Cir. NW&SE	11W-20N	50
Camille Av. E&W	11W-21N	50
Carlisle Av. N&S	10W-20N,9W-20N	50
Carlisle Pl. N&S	10W-21N	50
Carol Ct. E&W	12W-21N	50
Carol Ln. N&S	12W-21N	50
Caroline Ct. E&W	10W-21N	50
Carriage Way N&S	9W-20N	51
Castlewood Ln. N&S	12W-20N,12W-21N	50
Cedar Crest Dr. NW&SE	10W-21N	50
Cedar Ter. N&S	11W-20N,11W-21N	50
Central Av. E&W	12W-20N,11W-20N	50
Chapel Ct. E&W	11W-21N	50
Charing Cross Rd. E&W	11W-21N	50
Chatham Cir. NE&SE	11W-21N	50
Cherry St. N&S	11W-21N	50
Chestnut St. NW&SE	11W-20N,11W-21N	50
Christopher Dr. E&W	12W-20N	50
Circle Ct. N&S	11W-20N	50
Clavina Av. NE&SW	12W-20N	50
Clay Ct. N&S	11W-21N	50
Cody Ct. NE&SW	10W-20N	50
Cody Ln. NE&SW	10W-20N	50
Colony Ln. E&W	9W-20N	51
Colwyn Ter. E&W	10W-21N	50
Constance Ln. N&S	12W-20N	50
Country Ct. N&S	11W-20N	50
Country Ln. E&W	11W-20N	50
County Line Rd. E&W	10W-20N	50
Crabtree Ln. N&S	11W-21N	50
Craig Ct. N&S	12W-20N	50
Cranshire Ct. E&W	12W-21N	50
Crestview Dr. E&W	9W-20N	51
Crowe Av. N&S	11W-21N	50
Cumnor Ct. E&W	10W-21N	50
Dartmouth Ln. E&W	12W-20N,11W-20N	50
Davis Av. E&W	11W-21N	50
Deere Park Ct. E&W	11W-21N	50
Deere Park Ln. N&S	11W-21N	50
Deerfield Rd. E&W	11W-21N,11W-20N	50
Deerlake Rd. N&S	10W-20N	50
Deerpath Ct. N&S	11W-20N	50
Deerpath Dr. N&S	11W-20N	50
Dimmeydale Dr. E&W	10W-21N	50
Doral Ct. E&W	10W-20N	50
Earls Ct. N&S	11W-20N	50
East Canton Ct. E&W	10W-21N	50
East Summit Ct. N&S	11W-22N	50
Eastwood Dr. N&S	10W-20N	50
Elder Ln. E&W	10W-21N	50
Elizabeth Ct. NW&SE	12W-21N	50
Ellendale Rd. N&S	10W-20N	50
Elm St. N&S	11W-20N	50
Elmwood Av. E&W	11W-21N	50
Essex Ct. E&W	10W-21N	50
Evergreen Ct. NW&SE	10W-20N	50
Exmoor Ct. E&W	10W-20N	50
Fairoaks Av. E&W	11W-21N	50
Fairview Av. N&S	11W-20N	50
Ferndale Rd. N&S	9W-20N	51
Forest Av. N&S	11W-21N	50
Forest Way Dr. N&S	11W-20N	50
Forsythia Dr. N&S	12W-20N	50
Fountain View Dr. NS&EW	11W-20N	50
Foxhunt Tr. N&S	9W-20N	51
Franklin Av. E&W	11W-20N	50
Garand Dr. E&W	12W-21N	50
Gordon Av. E&W	11W-20N	50
Gordon Ter. E&W	11W-20N	50
Green Park Ct. NW&SE	11W-20N	50
Greenbriar Dr. E&W	10W-20N	50
Greentree Av. E&W	11W-21N	50
Greenwood Av. E&W	11W-21N	50
Greenwood Ct. N&S	11W-21N	50
Grove Av. N&S	11W-20N	50
Hackberry Rd. E&W	11W-20N	50
Hamilton Ln. N&S	10W-21N	50
Hampton Ct. N&S	12W-21N	50
Hawthorne Pl. N&S	11W-21N	50
Hazel Av. E&W	11W-21N	50
Heather Rd. N&S	10W-21N	50
Hemlock St. N&S	11W-21N	50
Hermitage Dr. E&W	10W-20N	50
Hertel Ln. E&W	12W-21N	50
High School Dr. N&S	11W-20N	50
Hillside Av. N&S	11W-21N	50
Hoffman Ln. N&S	10W-20N	50
Holly Ct. N&S	11W-21N	50
Holly Ln. NW&SE	11W-21N	50
Holmes Av. N&S	11W-21N	50
Hunter Ct. E&W	11W-21N	50
Hyacinth Ln. N&S	10W-20N	50
Indian Hill Ct. NW&SE	11W-20N	50
Indian Hill Rd. N&S	11W-20N	50
Interstate 94 N&S	12W-20N,12W-21N	50
Island Ct. E&W	11W-20N	50
Jewett Park Dr. E&W	11W-21N	50
Jonquil Ter. N&S	11W-20N	50
Journal Pl. N&S	11W-21N	50
Juniper Ct. N&S	11W-21N	50
Kates Rd. E&W	10W-20N	50
Kenmore Av. N&S	11W-20N	50
Kenton Rd. N&S	10W-21N	50
Kerry Ln. N&S	10W-20N	50
Kildare Ct. E&W	10W-20N	50
Kildare Ln. E&W	9W-20N	51
Kimberly Ct. E&W	11W-20N	50
King Richards Ct. N&S	12W-21N	50
Kingston Ter. E&W	10W-21N	50
Kipling Pl. N&S	10W-20N	50
Knollwood Rd. N&S	10W-20N	50
Lake Cook Rd. E&W	10W-20N,9W-20N	50
Lake Eleanor Ct. E&W	12W-21N	50
Lampton Ln. N&S	10W-21N	50
Landis Ln. E&W	10W-21N	50
Larkdale Rd. E&W	10W-20N	50
Laurel Av. E&W	11W-20N	50
Linden Av. E&W	11W-21N	50
Lisa Marie Ct. E&W	12W-20N	50
Locus Pl. N&S	11W-20N	50
Lombardy Ct. N&S	11W-20N	50
Lombardy Ln. N&S	11W-20N	50
Longfellow Av. E&W	10W-20N	50
Mallard Ln. NW&SE	11W-20N	50
Manor Dr. NE&SW	10W-21N	50
Maple Ct. NW&SE	11W-21N	50
Marcie Ct. E&W	12W-20N	50
Margate Ter. E&W	10W-21N	50
Meadow Ln. E&W	11W-21N	50
Meadow Ln. N&S	12W-20N	50
Meadowbrook Ln. N&S	10W-21N	50
Merlin Ct. N&S	11W-20N	50
Millstone Rd. E&W	9W-21N	51
Montgomery Ct. NE&SW	12W-21N	50
Montgomery Dr. N&S	12W-21N	50
Montgomery Rd. E&W	12W-21N	50
Mountain Ct. N&S	11W-22N	50
Mountain Dr. E&W	11W-22N	50
Mulberry Rd. E&W	10W-20N	50
Myrtle Ct. N&S	11W-21N	50
Norman Ln. N&S	10W-21N	50
North Av. E&W	11W-22N	50
Northwoods Cir. N&S	11W-21N	50
Northwoods Ct. NE&SW	11W-21N	50
Northwoods Rd. N&S	11W-21N	50
Oakley Av. N&S	11W-21N	50
Oakmont Rd. E&W	10W-20N	50
Oakwood Pl. N&S	11W-21N	50
Orchard Av. E&W	10W-21N	50
Osterman Av. E&W	11W-20N	50
Overland Ct. NW&SE	11W-22N	50
Overland Tr. N&S	11W-22N	50
Oxford Rd. N&S	10W-21N	50
Park Av. N&S	11W-21N	50
Park Ln. N&S	11W-20N	50
Parkside Ln. N&S	11W-21N	50
Paula Ct. N&S	12W-20N	50
Peachtree Ln. NE&SW	11W-20N	50
Pear Tree Rd. E&W	12W-21N	50
Penbroke Ct. N&S	11W-20N	50
Peterson Ln. N&S	11W-21N	50
Pine St. N&S	11W-20N,11W-21N	50
Plum Tree Rd. N&S	11W-20N	50
Portage Pass N&S	11W-22N	50
Prairie St. N&S	11W-21N	50
Price Ln. N&S	11W-20N	50
Princeton Ln. NW&SE	11W-20N	50
Radcliff Cir. NW&SE	11W-20N	50
Rago Av. E&W	11W-21N	50
Ramsay Rd. E&W	10W-21N	50
Revere Ct. NE&SW	10W-20N	50
River Rd. N&S	9W-20N	51
Robbie Ct. E&W	11W-21N	50
Rosemary Ter. N&S	10W-21N	50
Rosewood Av. E&W	11W-20N	50
Rt. 43 NW&SE	11W-21N,10W-20N	50
Sapling Ln. N&S	12W-20N	50
School St. E&W	11W-21N	50
Sequoia Ln. E&W	10W-20N	50
Shagbark Ln. N&S	9W-20N	51
Shanon Rd. NE&SW	9W-20N	51
Shenandoah Ct. N&S	10W-20N	50
Shenandoah Rd. E&W	9W-20N	51
Sheridan Av. N&S	11W-21N	50
Smoke Tree Rd. N&S	9W-20N	51
Somerset Av. E&W	11W-21N	50
Springfield Av. NW&SE	11W-21N	50
Spruce St. N&S	11W-21N	50
Sprucewood Ln. E&W	11W-20N	50
St. Andrews Ln. E&W	9W-20N	51
Standish Dr. N&S	10W-20N	50
Stratford Rd. N&S	11W-21N	50
Summit Dr. E&W	11W-22N	50
Sunset Ct. E&W	12W-20N	50
Susan Ln. NE&SW	12W-20N	50
Tall Tree Ln. E&W	12W-21N	50
Tamarisk Ln. E&W	9W-20N	51
Tanglewood Ct. N&S	9W-20N	51
Terrace Ct. E&W	11W-20N	50
Timberhill Rd. N&S	11W-20N	50
Todd Ct. N&S	10W-21N	50
Tri-State Tollway N&S	12W-20N,12W-21N	50
Village Green Ct. E&W	12W-21N	50
W. Canton Ct. E&W	11W-21N	50
W. St. Andrews Ln. E&W	10W-20N	50
Walden Ln. N&S	10W-21N	50
Walnut St. E&W	11W-21N	50
Warrington Rd. N&S	10W-21N	50
Warwick Ct. NW&SE	11W-21N	50
Warwick Rd. E&W	10W-21N	50
Waukegan Rd. NW&SE	11W-22N,10W-20N	50
Waverly Av. NW&SE	10W-20N	50
Wayne Av. N&S	11W-21N	50
We-go Ct. NW&SE	12W-20N	50
We-go Trail E&W	12W-20N	50
West Summit Ct. N&S	11W-22N	50
Westcliff Ln. N&S	10W-21N	50
Westgate Rd. E&W	10W-21N	50
Whittier Av. E&W	10W-20N	50
Wicklow Ct. N&S	9W-20N	51
Wicklow Rd. NW&SE	9W-20N	51
Williams Av. E&W	11W-21N	50
Willow Av N&S	11W-20N	50
Wilmot Ln. E&W	12W-21N	50
Wilmot Rd. N&S	12W-20N,11W-21N	50
Wincanton Dr. N&S	10W-21N	50
Wincrest Rd. N&S	10W-21N	50
Wood Av. N&S	11W-21N	50
Woodbine Ct. N&S	11W-21N	50
Woodland Dr. E&W	11W-21N	50
Woodridge Ct. N&S	10W-21N	50
Woodruff Av. N&S	11W-21N	50
Woodvale Av. NW&SE	10W-21N	50
Woodward Av. N&S	11W-21N	50

Ela Township

	Grid	Page
Anna Ct. N&S	24W-25N	38
Arbor Ln. E&W	21W-24N	38
Arrowhead Dr. N&S	19W-25N	39
August Ln. E&W	24W-25N	38
Bagpipe Ct. E&W	23W-25N	38
Berkley Rd. E&W	22W-22N	46
Bonnie Ln. N&S	21W-24N	38
Bonnyrigg Ct. E&W	23W-25N	38
Buckeye Rd. NS&EW	23W-20N	46
Bur Oak Ln. E&W	21W-23N	47
Cattail Ln. E&W	23W-20N	46
Cedar Ln. N&S	23W-23N	46
Cedar St. E&W	23W-22N	46
Cherry Ln. E&W	23W-23N	46
Clover Hill Ln. NS&EW	24W-23N	46
Columbia Ct. E&W	21W-25N	38
Commercial Dr. NE&SW	21W-25N	38
Cottage Rd. N&S	21W-23N	47
Crestwood Ln. E&W	23W-23N	46
Dan Vera Ln. NE&SW	22W-24N	38
Dartmouth Ln. E&W	21W-25N	38
Deer Meadow Ln. E&W	23W-21N	46
Deerpath Rd. N&S	23W-21N	46
Diamond Lake Rd. N&S	19W-24N,19W-25N	39
Dorothy Ln. E&W	20W-20N,19W-20N	47
Eagle Dr. N&S	19W-25N	39
East Ln. N&S	23W-23N	46
East Rd. N&S	21W-23N	47
Echo Lake Rd. N&S	23W-23N	46
Echo Ln. N&S	23W-23N	46
Edgewood Pl. E&W	24W-23N	46
Ellrie Ter. N&S	24W-25N	38
Fairfield Rd. NW&SE	21W-23N,22W-25N	47
Florence Av. N&S	23W-22N	46
Forest Dr. NS&EW	21W-23N,21W-24N	47
Gabriel Ct. E&W	22W-24N	38
Gabriel Dr. N&S	22W-24N	38
Garden Ln. N&S	21W-23N	47
Gilmer Rd. NW&SE	21W-25N,19W-24N	38
Glendale Rd. E&W	21W-24N	38
Grand Av. E&W	23W-22N	46
High Ridge Rd. N&S	23W-23N	46
Highland Dr. E&W	21W-23N	47
Hillcrest Dr. E&W	23W-23N	46
Ivy Ln. N&S	21W-25N	38
June Ter. N&S	23W-22N	46
Juniper Ln. NW&SE	23W-20N	46
Kruckenberg Rd. NS&EW	22W-25N,21W-25N	38
Lake Zurich-Highwood Rd. E&W	21W-23N,19W-23N	47
Lake-Cook Rd. E&W	20W-20S,19W-20S	47
Lakeside Dr. NE&SW	21W-23N,21W-24N	47
Lloyd Rd. E&W	20W-23N,19W-23N	47
Lochanora Dr. NS&EW	23W-25N,22W-25N	38
Lonetree Ln. E&W	23W-23N	46
Mable Av. N&S	23W-22N	46
March St. NW&SE	22W-24N	38
Marilyn Ln. NS&EW	21W-25N	38
Mary Dale Av. E&W	24W-25N	38
Middle Fork Rd. E&W	24W-21N	46
Midlothian Rd. NE&SW	22W-24N,21W-25N	38
Millcrest Dr. E&W	23W-23N	46
Miller Rd. NE&SW	24W-24N,23W-24N	38
N. Kyle Ct. NW&SE	22W-25N	38
Nancy Ct. E&W	21W-25N	38
North Lakewood Ln. E&W	23W-23N	46
North Ln. N&S	23W-23N	46
Old Barn Ln. E&W	21W-23N	47
Orchard Ln. N&S	24W-21N	46
Overhill Dr. NE&SW	21W-23N,21W-24N	47
Park Rd. N&S	21W-23N	47
Pepper Rd. NW&SE	21W-24N	38

	Grid	Page
Plum Pl. NW&SE .. 23W-20N		46
Public Rd. N&S		
.........19W-23N,19W-24N		47
Quentin Rd. N&S		
.......21W-20N,21W-24N		47
Rand Ct. E&W 20W-20N		47
Rand Rd. N&S		
.......24W-24N,24W-25N		38
Ravine Ct. NE&SW		
24W-25N		38
Ravine Ln. N&S .. 24W-25N		38
Ravine Rd. E&W .. 21W-23N		47
Redwing Pl. NE&SW		
23W-20N		46
Ridge Rd. E&W .. 21W-23N		47
Rose Av. E&W 24W-25N		38
Rose Rd. E&W .. 21W-23N		47
S. Kyle Ct. SE&NW		
22W-25N		38
Sacomand Ln. NS&EW		
21W-24N		38
Shady Ln. NW&SE 21W-24N		38
Shagbark Dr. N&S 23W-24N		38
Shirley Rd. E&W .. 21W-20N		47
Somerset Ct. N&S ..23W-25N		38
South Lakewood Dr. E&W		
23W-23N		46
South Rd. E&W .. 21W-23N		47
Spruce Rd. E&W .. 21W-23N		47
Starry Ln. E&W .. 21W-23N		47
Stone Marsh Ln. E&W		
24W-25N		38
Stoney Kirk Ct. N&S		
23W-25N		38
Strum St. E&W 22W-20N		46
Summit Dr. NW&SE		
23W-23N		46
Sunset Dr. NW&SE		
23W-23N		46
Tahoe Ct. NE&SW ..19W-25N		39
Tahoe Dr. E&W .. 19W-25N		39
Telser Rd. N&S		
.......22W-22N,22W-23N		46
Timberlake Dr. E&W		
24W-25N		38
Valentine Av. E&W ..24W-25N		38
Valley Rd. N&S		
.......21W-23N,21W-24N		47
Virginia Av. N&S .. 24W-25N		38
West Ln. N&S 23W-23N		46
Wildrose Dr. N&S .. 23W-20N		46
Willow Rd. E&W .. 21W-23N		47
Woodbury Ct. E&W		
21W-25N		38

Fox Lake*
(Also see McHenry County)

	Grid	Page
Abbott Dr. N&S .. 28W-39N		12
Adam St. N&S 26W-35N		21
Amundsen St. E&W		
28W-36N		12
Arlington Rd. E&W .. 28W-39N		12
Arlington Rd. NE&SW		
28W-36N		12
Arthur Av. N&S .. 28W-37N		12
Ash St. N&S 28W-39N		12
Atwater Pkwy. N&S		
27W-36N		12
Balsam Ct.* E&W .. 27W-38N		12
Bay Ln. E&W 28W-36N		12
Bay Rd. N&S 28W-36N		12
Beach Driveway NS&EW		
28W-36N		12
Benjamin Av. E&W 28W-38N		12
Boulder Pl. N&S ...27W-36N		12
Brewer E&W 27W-36N		12
Castle Rd. N&S .. 28W-37N		12
Cathryn St. E&W .. 27W-37N		12
Cedar Av. E&W .. 27W-36N		12
Cedar St. E&W .. 27W-36N		12
Center St. N&S .. 27W-36N		12
Channahon Ct. N&S		
28W-39N		12
Channel Dr. E&W .. 27W-35N		20
Channel Dr. NW&SE		
28W-38N		12
Chevy Chase Ct. NW&SE		
28W-39N		12
Christopher Way SW&NE		
28W-36N		12
Clarendon Ct. NW&SE		
28W-39N		12
Columbia Av. NE&SW		
27W-36N		12
Concord Cir. Cir. .. 28W-38N		12
Cora Av. N&S 27W-37N		12
Cottage Av. NE&SW		
27W-36N		12
Covington Rd. E&W		
28W-36N		12
Crest Hill Ct. NE&SW		
28W-39N		12
Cross St. E&W 28W-39N		12
Custor Ct. N&S .. 28W-39N		12
Cypress Av. E&W .. 27W-36N		12
Dena Av. N&S 27W-36N		12
Devlin Rd. N&S .. 27W-36N		12
Dunns Pl. E&W .. 28W-38N		12
Dunwood Ct. E&W .. 28W-39N		12
Eagle Point Rd. NE&SW		
27W-36N		12
East Leisure Village Av.		
NE&SW..........28W-39N		12
East Shore Dr. N&S		
28W-38N		12
Eaton St. N&S 28W-39N		12
Edison Ct. E&W .. 27W-37N		12
Elm Av. N&S 27W-36N		12
Ernie St. E&W 28W-36N		12
Evergreen Ct.* E&W		
27W-38N		12
Fairfax Rd. E&W .. 28W-36N		12
Fern Pl. E&W 27W-36N		12
Forest Av. N&S		
.......28W-36N,27W-37N		12
Fortress Dr. E&W		
28W-37N		12
Fossil Dr. E&W 28W-38N		13
Fox Trail NW&SE .. 27W-35N		20
Frazier Ct. E&W 27W-35N		20
Garfield Av. E&W .. 26W-35N		21
Glen Av. N&S 27W-36N		12
Glenview Av. E&W .. 27W-36N		12
Glenview Pl. NE&SW		
27W-36N		12
Golfview Dr. E&W .. 28W-39N		12
Grace Av. N&S 28W-36N		12
Grand Av. E&W		
28W-36N,27W-36N		12
Grass Lake Rd. E&W		
28W-39N		12
Greenwood Av. E&W		
28W-38N		12
Harding Av. N&S .. 28W-35N		20
Hartigan Rd. E&W .. 27W-34N		20
Haskel St. E&W .. 28W-38N		12
Hawthorne Ln. N&S		
27W-36N		12
Hazel Av E&W .. 27W-36N		12
Hickory Av. N&S .. 27W-36N		12
Hickory St. N&S .. 28W-39N		12
Highland Av. E&W .. 27W-36N		12
Highview Av. N&S .. 27W-36N		12
Hill Rd. NE&SW .. 27W-36N		12
Hillcrest Av. N&S .. 27W-36N		12
Hillcrest Dr. N&S .. 27W-35N		12
Hillside Av. NW&SE		
27W-36N		12
Hillside Ct. E&W ...27W-37N		12
Hillside Pl. NE&SW 28W-36N		12
Hilltop Av. E&W .. 28W-36N		12
Hilltop Rd. N&S .. 28W-36N		12
Hollow Way NS&EW		
27W-34N,27W-35N		20
Holly Av., N. N&S .. 27W-36N		12
Holly Av., S. NE&SW		
27W-36N		12
Honing Rd NW&SE		
27W-36N		12
Hooks Ln. NE&SW 28W-37N		12
Howard Av. N&S .. 27W-37N		12
Howard Ct. NE&SW		
27W-37N		12
Industrial Av. NW&SE		
28W-36N		12
Jack Frost Ln. NE&SW		
28W-36N		12
Jackson Dr. N&S .. 28W-39N		12
Jefferson Av. N&S ..26W-35N		21
Jefferson Ct. E&W ..26W-35N		21
Jefferson St. N&S .. 26W-35N		21
Judson St. N&S .. 28W-38N		12
Keefe Pl. N&S 28W-39N		12
Keystone Av. NE&SW		
27W-36N		12
Kings Rd. NE&SW .. 28W-35N		20
Kingston Blvd. NW&SE		
28W-37N		12
Knollwood Rd., S. NW&SE		
27W-35N		20
Lagoon Rd. E&W .. 28W-36N		12
Lake Av. N&S 28W-39N		12
Lakeland Av. E&W .. 28W-35N		20
Lakelane Rd. NE&SW		
28W-36N		12
Lakeshore Dr. N&S		
28W-38N		12
Lakeside Ln. N&S .. 27W-36N		12
Lakeside Rd. E&W .. 27W-37N		12
Lakeview Av. E&W		
27W-37N		12
Lakewood Av. N&S		
27W-36N		12
Lexington Ln. E&W		
28W-38N		12
Lilac Av. N&S 27W-36N		12
Lincoln Av. N&S .. 26W-36N		13
Linden N&S 27W-36N		12
Lindy St. E&W 27W-36N		12
Lippencott Ln. NW&SE		
27W-36N		12
Lippencott Rd. N&S		
27W-36N		12
Livingston St. N&S 28W-37N		12
Longwood Dr. E&W		
27W-36N		12
Lotus Av. E&W 27W-38N		12
Madison Av. N&S .. 26W-35N		21
Magnolia Ct.* E&W		
27W-38N		12
Manor Av. NE&SW		
28W-36N		12
Maple Av. N&S .. 27W-36N		12
Marvin St. N&S .. 27W-36N		12
Mastodon Dr. N&S 26W-36N		13
Matts Rd. E&W .. 28W-36N		12
Maude Av. E&W ...27W-37N		12
McKinley St. N&S .. 28W-36N		12
Meade Ct. E&W .. 28W-36N		12
Medinah N&S 28W-36N		12
Mineola Rd. NS&EW		
27W-37N		12
Monroe Av. N&S .. 26W-35N		21
Monterrey Ter. N&S		
29W-37N		12
Muriel Rd. E&W .. 28W-38N		12
N. Lake Av. N&S .. 27W-36N		12
N. Rushmore Rd. N&S		
27W-36N		12
Nippersink Blvd. NE&SW		
28W-37N		12
Nippersink St. N&S		
28W-38N		12
North Av. E&W 27W-37N		12
Northway N&S 28W-36N		12
Oak Ct. N&S 28W-36N		12
Oak Ln. N&S 27W-35N		20
Oak St. N&S 28W-36N		12
Oak St. E&W 28W-37N		12
Orchard Pl. E&W .. 27W-36N		12
Park Av. N&S 28W-35N		20
Park Ln. NE&SW .. 27W-35N		20
Park Rd. E&W 28W-39N		12
Pine Av. E&W 27W-36N		12
Pistakee Lake Rd. N&S		
28W-36N		12
Pistakee Pkwy E&W		
28W-35N		20
Poplar Av. N&S .. 27W-36N		12
Quail Path E&W		
27W-34N,27W-35N		20
Rabbit Run E&W .. 27W-35N		20
Rand E&W 27W-36N		12
Ridgeland Av. N&S		
27W-36N		12
Riverside Dr. NE&SW		
28W-37N		12
Riverside Island Dr. NW&SE		
28W-37N		12
Riverview Av. E&W		
28W-37N		12
Rolland Pl. E&W .. 28W-36N		12
Rollins Rd. N&S		
.......27W-36N,26W-35N		12
Rose Av. N&S......27W-36N		12
Rt. 59 NE&SW		
27W-35N,26W-36N		20
Rushmore Rd. N&S		
26W-36N		13
S. Lake Av. N&S .. 27W-36N		12
Sayton Rd. N&S .. 27W-36N		12
Sayton Rd. NE&SW		
27W-36N		12
Scenic Ln. E&W .. 28W-36N		12
Scenic Rd. N&S .. 28W-36N		12
Shady Ln. NW&SE .. 27W-36N		12
Shore Dr. N&S 27W-39N		12
Siloam Av. NW&SE		
27W-36N		12
Skillen St. N&S .. 28W-36N		12
South St. E&W 28W-36N		12
Spring Rd. N&S .. 27W-35N		20
Spruce Av. E&W .. 27W-36N		12
Squaw Rd. E&W .. 26W-36N		13
St. James St. E&W		
28W-38N		12
Stanton Point Rd. NE&SW		
26W-36N		13
State Park Rd. N&S		
.......28W-38N,28W-39N		12
Tremont St. E&W .. 28W-36N		12
Trout Av. NW&SE .. 28W-38N		12
Troy St. NW&SE .. 27W-36N		12
Tweed Rd. N&S .. 27W-36N		12
U.S. Route 12 N&S		
.......28W-37N,27W-35N		12
Venice Av. NW&SE		
28W-36N		12
Viscaya Dr. NE&SW		
.......28W-39N,27W-39N		12
Waltonia Ter. N&S .. 27W-37N		12
Washington Av. N&S		
26W-36N		13
Wells St. NE&SW .. 28W-37N		12
West Grass Lake Dr. E&W		
.......28W-39N,27W-39N		12
White Tail Ln. E&W		
27W-35N		20
Whitten Av. E&W .. 26W-35N		21
Wildwood Dr. E&W		
27W-35N		20
Wilmington E&W .. 27W-37N		12
Wood Rd. E&W .. 27W-35N		20
Woodland Av. N&S		
27W-36N		12
Woodlock St. N&S..27W-36N		12
York St. N&S 27W-36N		12

Fox River Grove*
(Also see McHenry County)

	Grid	Page
Camilla Pl. E&W .. 28W-22N		44
Primrose Ln. E&W .. 28W-22N		44
Tiffany Ln. N&S .. 28W-22N		44
Victoria Dr.* E&W .. 28W-22N		44

Fremont Township

	Grid	Page
Acorn Ln. N&S .. 19W-26N		39
Alleghany Rd. N&S		
.......21W-31N,21W-30N		31
Bacon Rd. SE&NW		
23W-31N		30
Baker St. N&S 24W-29N		30
Barnswallow Ln. NE&SW		
.......23W-26N,23W-27N		38
Beechwood Ct. NE&SW		
21W-27N		38
Bitter Sweet Dr. E&W		
21W-27N		38
Blackhawk Tr. N&S		
.......24W-30N,24W-31N		30
Brassie Ct. N&S		
.......21W-28N,21W-29N		31
Briar Ter. N&S 21W-26N		38
Buckthorn Ct. E&W		
20W-26N		39
Buena Ter. N&S ...21W-26N		38
Cambridge Rd. E&W		
19W-29N		31
Cardinal Ct. NE&SW		
24W-28N		30
Casey Rd. N&S .. 24W-28N		30
Chardon Ln. N&S .. 24W-30N		30
Chardon Rd. N&S		
.......24W-30N,22W-30N		30
Cherokee Tr. N&S		
.......24W-30N,24W-31N		30
Chevy Chase Rd. NS&EW		
21W-27N		38
Circle Dr. N. NW&SE		
19W-26N,18W-26N		39
Circle Dr. S. NE&SW		
19W-26N,18W-26N		39
Countryside Lake Dr. N&S		
.......20W-26N,20W-27N		39
Countryside Valley Dr. E&W		
21W-26N		38
Crescent Dr. N. E&W		
21W-26N		38
Crescent Dr. S. E&W		
21W-26N		38
Crescent Dr., East N&S		
21W-26N		38
Crescent Dr., West E&W		
21W-26N		38
Driftwood Ct. E&W		
20W-26N		39
Elmwood A. NW&SE		
19W-26N,18W-26N		39
Erhart Rd. E&W		
.......23W-29N,22W-29N		30
Fairfield Rd. N&S		
.......24W-31N,23W-26N		30
Fairview Dr., N. NE&SW		
19W-26N,18W-26N		39
Fairview Dr., S. NW&SE		
19W-26N,18W-26N		39
Forest Ln., N. NW&SE		
19W-26N		39
Forest Ln., S. NW&SE		
19W-26N,18W-26N		39
Freemont Center Rd. N&S		
.......22W-28N,22W-29N		30
Fremont Av. N&S .. 23W-30N		30
Gilmer Rd. NW&SE		
.......24W-30N,21W-26N		30
Greeleaf Rd. N&S .. 21W-26N		38
Hackberry Rd. N&S		
20W-26N		39
Harris Rd. N&S		
.......19W-30N,19W-31N		31
Harvard Av. N&S .. 19W-29N		31
Hawley St. E&W		
.......22W-27N,21W-27N		38
Hawley St. 22W-27N		38
Hawthorne Blvd. E&W		
19W-26N		39
Hiawatha Ter. N&S		
21W-26N		38
Hickory Av. NE&SW		
19W-26N,18W-26N		39
Hickory Rd. NW&SE		
20W-26N		39
Highland Dr. E&W .. 21W-26N		38
Hillside E&W....21W-26N		38
Hilltop Ter. N&S .. 22W-28N		30
Horizon Spur E&W 23W-26N		38
Imperial Ct. N&S...23W-30N		30
Iride Dr. NW&SE		
.......19W-26N,18W-26N		39
Ivanhoe Ct. N&S .. 23W-27N		38
Ivanhoe Ln. NS&EW		
20W-26N		39
Ivanhoe Rd. N&S .. 20W-31N		31
Ivanhoe Rd. NE&SW		
24W-26N		38
Jack Pine Ct. NE&SW		
20W-26N		39
Kenmore Av. E&W		
19W-26N,18W-26N		39
Krueger Ct. NW&SE		
19W-26N		39
Lake Fairfield Ln. E&W		
24W-29N		30
Lakeview Av. E&W		
.......19W-26N,18W-26N		39
Lakeview Pkwy. E&W		
.......21W-27N,20W-26N		38
Landau Ln. NS&EW		
20W-28N		31
Lemon Rd. E&W .. 20W-26N		39
Longmeadow Cir. NS&EW		
19W-26N		39
Longmeadow Ct. N&S		
20W-26N		39
Longmeadow Dr. N&S		
20W-26N		39
Manor Hill Rd. N&S		
.......24W-30N,24W-31N		30
Maple Av. N&S		
.......19W-26N,18W-26N		39
Maple E&W 20W-28N		31
Maple Ter. N&S .. 21W-26N		38
Marion Av. NW&SE		
21W-26N		38
Martin Dr. NE&SW		
.......19W-26N,18W-26N		39
Meadow Ln. E&W		
.......24W-31N,23W-31N		30
Midlothian Rd. NE&SW		
.......20W-26N,19W-30N		30
Milton Rd. E&W		
.......24W-26N,22W-26N		38
Niblick Knoll Ct. E&W		
22W-28N		30
North Rd. E&W 21W-27N		38
Northwest Pkwy. NW&SE		
20W-27N		39
Oak Av. NE&SW		
.......19W-26N,18W-26N		39
Oak Ter. N&S 21W-26N		38
Oakdale Ln. N. NE&SW		
19W-26N,18W-26N		39
Oakdale Ln. S. NE&SW		
19W-26N,18W-26N		39
Oakleaf Ct. N&S....21W-27N		38
Old Oak Dr. NW&SE		
24W-28N		30
Owens Rd. N&S		
.......23W-26N,22W-27N		38
Oxford Dr. N&S .. 19W-29N		31
Park Av. NW&SE . 20W-28N		31
Peterson Rd. E&W		
.......22W-30N,19W-30N		30
Pheasant Run NW&SE		
22W-26N		38
Pierre Dr. NE&SW . 19W-26N		39
Pond Shore Dr. NE&SW		
23W-26N		38
Prospect Dr. NW&SE		
19W-26N		39
Public Rd. E&W .. 20W-28N		31
Red Pine Ct. NE&SW		
20W-26N		39
Ridge Ct. N&S 19W-26N		39
Rosewood Rd. E&W		
21W-26N		38
Rt. 45 N&S 19W-31N		31
Rt. 60 NW&SE		
.......24W-31N,19W-26N		30
Rt. 83 N&S		
.......20W-28N,20W-31N		31
Rt. 137 NW&SE 19W-31N		31
Rt. 176 N&S		
.......24W-27N,19W-26N		38
Russell Ln. E&W . 24W-29N		30
Schank Rd. N&S		
.......21W-28N,20W-28N		31
Schwerman Rd. NW&SE		
.......23W-26N,21W-26N		38
Seminole Ct. E&W . 21W-28N		31
Shady Ln. NW&SE . 21W-26N		38
Sky Crest Dr. NS&EW		
22W-28N		30
Spring Ct. NW&SE 24W-28N		30
Spyglass Cir. N&S . 19W-28N		31
Spyglass Ct. E&W . 19W-28N		31
Sunny Ct. E&W .. 19W-26N		39
Sunset Ct. NE&SW		
24W-29N		30
Sylvan Dr., North NW&SE		
19W-26N		39
Sylvan Dr., South E&W		
19W-26N		39
Taylor Lake Ct. E&W		
19W-26N		39
Thorngate Dr. E&W		
21W-28N		31
Thorntree Rd. N&S		
21W-27N		38
Trinity Dr. E&W		
.......19W-26N,18W-26N		39
University Dr. NE&SW		
19W-26N		39
Valley View Ter. N&S		
19W-26N		39
Vernon Ridge Dr. NS&EW		
22W-29N		30
Walnut Av. NE&SW		
19W-26N		39
Wauconda Rd. E&W		
.......24W-27N,23W-27N		38
Weich Rd. NE&SW		
19W-26N		39
West Shore Dr. N&S		
.......19W-26N,18W-26N		39
Wildflower Ter. E&W		
21W-26N		38
Willow Av. N&S		
.......19W-26N,18W-26N		39
Wilson Rd. N&S .. 24W-31N		30
Winchester Rd. E&W		
.......20W-30N,19W-30N		31
Windflower Ct. E&W		
20W-26N		39
Yale Ct. N&S 19W-29N		31

Grant Township

Street	Grid	Page
Anchorage Ln. E&W	27W-35N	20
Anderson St. E&W	25W-34N	21
Andrew Ct. NW&SE	25W-36N	13
Ash St. N&S	25W-35N	21
Augsburg St. E&W	25W-35N	21
Augustana Av. N&S	25W-34N,25W-35N	21
Bachman Rd. E&W	26W-37N	13
Bayview Av. E&W	25W-35N	21
Bayview Dr. E&W	27W-35N	20
Bayview Ln. NE&SW	28W-35N	20
Beach Park NW&SE	27W-35N	20
Bellvue Pl. N&S	26W-34N	21
Belvidere Rd. NE&SW	25W-32N	21
Benjamin Av. N&S	25W-35N	21
Bergan St. NE&SW	25W-34N	21
Beverly Dr. N&S	26W-35N	21
Big Hollow Rd. E&W	28W-34N,26W-34N	20
Blackhawk Av. E&W	26W-35N	21
Boesch Pl. E&W	26W-36N	13
Brandenburg Rd. E&W	28W-33N,26W-33N	20
Broadway E&W	28W-27N	36
Brodie Dr. NE&SW	25W-36N	13
Burkhart Ln. E&W	26W-34N	21
Caine Rd. E&W	25W-36N	13
Camp Horner Rd. N&S	26W-34N	21
Carol Ln. NE&SW	26W-34N	21
Catherine St. N&S	25W-34N	21
Cedarwood Ln. E&W	25W-36N	13
Central Av. E&W	28W-37N	12
Central Rd. NE&SW	26W-36N	13
Champion Rd. N&S	28W-34N	20
Chanel Dr. N&S	25W-35N	21
Channel Dr. N&S	26W-36N	13
Channel Dr. E&W	28W-34N	20
Chicago Av. E&W	25W-35N	21
Chris Larkin Rd. (pvt) NS&EW	27W-35N	20
Christa Dr. N&S	26W-33N	21
Clara Dr. E&W	25W-36N	13
Claredon Dr. E&W	27W-35N	20
Converse Ln. N&S	26W-34N	21
Cooney Island Rd. E&W	26W-36N	13
Crabtree Ln. E&W	25W-36N	13
Creek Ct. NW&SE	26W-33N	21
David Ct. N&S	25W-35N	21
Decorah Av. N&S	25W-34N	21
Dewey St. NE&SW	28W-37N	12
Dolores Ct. NE&SW	26W-33N	21
Donald Ct. N&S	26W-35N	21
East End Av. N&S	26W-36N	13
Edgewater Ln. N&S	27W-35N	20
Edward Av. E&W	26W-34N	21
Elm Av. N&S	25W-35N	21
Elm Av. E&W	25W-34N	21
Elm St. N&S	26W-34N	21
Elmwood Av. E&W	26W-35N	21
Everett Av. N&S	25W-35N	21
Faith Ct. E&W	26W-35N	21
Falcon Dr. N&S	25W-36N	13
Fischer Dr. NW&SE	26W-34N	21
Fish Lake Rd. N&S	26W-32N	21
Forest Av. N&S	26W-34N	21
Fox Lake Rd. NE&SW	27W-32N	20
Francis Av. N&S	26W-36N	13
Frank Ct. E&W	28W-37N	12
Franklin Av. N&S	25W-35N	21
Fretheim Av. E&W	25W-34N	21
Gerberding Av. N&S	25W-34N	21
Glenayre St. E&W	27W-35N	20
Gogol Av. NS&EW	28W-34N	20
Graham Ct. E&W	25W-34N	21
Grand Av. E&W	26W-34N,25W-34N	21
Grant Av. E&W	25W-36N	13
Greenleaf Av. N&S	25W-35N	21
Grove Av. N&S	25W-35N	21
Hartigan Rd. NE&SW	27W-34N	20
Hawthorne Ln. N&S	25W-36N	13
Hazelwood Dr. N&S	25W-36N	13
Helen Ct. N&S	27W-34N	20
Helendale Rd. N&S	26W-35N	21
Hiawatha Tr. N&S	28W-34N	20
Hickory Av. N&S	25W-34N,25W-35N	21
Hickory Ct. N&S	25W-35N	21
Hickory Ct. N&S	25W-36N	13
Hickory Ct. N&S	26W-34N	21
Hickory Ln. E&W	26W-36N	13
Hickory Rd. E&W	25W-35N	21
Hickory St. E&W	25W-36N	13
High Point Ln. N&S	25W-35N	21
High Rd. NW&SE	28W-35N	20
Highpoint Rd. E&W	25W-35N	21
Hill Top Rd. N&S	28W-35N	20
Hilldale Dr. N&S	27W-35N	20
Hillside Av. E&W	25W-35N	21
Hilltop Dr. E&W	25W-34N	21
Hillwood N&S	26W-35N	21
Holman Av. N&S	25W-34N	21
Hudson Av. E&W	26W-35N	21
Hunt Av. N&S	26W-35N	21
Indian Ln. N&S	26W-35N	21
Indian Trail NE&SW	28W-35N	20
Ingleside Dr. E&W	26W-35N	21
Ingleside Dr. N&S	26W-35N	21
Ingleside Rd. E&W	26W-36N	13
Iola Av. N&S	25W-36N	13
Iroquois Tr. N&S	28W-34N	20
Jackson Ct. N&S	25W-36N	13
James Av. N&S	25W-34N	21
James Ct. N&S	26W-36N	13
Jeanette Ct. NE&SW	28W-37N	12
Karen Ct. NW&SE	28W-37N	12
Kelly Ln. N&S	26W-33N	21
Kerwin Ln. N&S	25W-36N	13
Knollwood Dr. N&S	26W-34N,26W-35N	21
Lake Av. E&W	25W-35N	21
Lake Dr. N&S	25W-34N	21
Lake Mathews Tr. NW&SE	28W-34N,28W-35N	20
Lake Shore Dr. E&W	25W-35N	21
Lake Shore Dr. E&W	26W-35N	21
Lake Shore Rd. NE&SW	28W-35N	20
Lake Shore Woods Dr. N&S	26W-35N	21
Lake St. N&S	26W-35N	21
Lake St. N&S	26W-36N	13
Lake Vista Ter. NE&SW	28W-37N	12
Lakeview E&W	25W-34N	21
Lakeview Av. E&W	25W-35N	21
Lakeview Av. E&W	26W-35N	21
Lakeview Ct. N&S	28W-34N	20
Laneville Dr. E&W	25W-35N	21
Lannon Ln. N&S	25W-36N	13
Larkin Ln. E&W	25W-35N	21
Laurel Av. N&S	25W-35N	21
Lee Ct. E&W	26W-35N	21
Leonard Av. N&S	28W-34N	20
Levi Waite Rd. E&W	26W-33N	21
Lincoln Av. N&S	25W-35N	21
Lincoln Ct. E&W	26W-36N	13
Linder Av. E&W	25W-35N	21
Long Av. N&S	25W-34N	21
Long Beach Dr. NW&SE	25W-34N	21
Longwood Dr. N&S	26W-35N	21
Louise Pl. N&S	25W-35N	21
Lowood Av. N&S	26W-35N	21
Luana Ln. E&W	27W-35N	20
Luby Ct. E&W	26W-35N	21
Madison St. E&W	25W-34N	21
Mall Rd. NW&SE	28W-35N	20
Malmquist Dr. N&S	28W-34N	20
Manitoba Tr. NE&SW	28W-34N	20
Maple Av. N&S	26W-36N	13
Maple Av. E&W	26W-35N	21
Maple St. E&W	26W-36N	13
Margie Pl. NW&SE	28W-37N	12
Marie Ln. E&W	26W-35N	21
Marine Dr. N&S	27W-35N	20
Marion Av. E&W	25W-35N	21
Marion Ct. E&W	26W-35N	21
Marquette Dr. E&W	25W-36N	13
Marshall Av. E&W	26W-36N	13
Mary Ct. N&S	26W-36N	13
Mattalina Ct. E&W	26W-37N	13
Meadow Ln. N&S	25W-35N	21
Michael Rd. N&S	28W-34N	20
Milwaukee Av. NE&SW	25W-35N	21
Mishenauma Tr. E&W	28W-34N	20
Mitchell Ct. NW&SE	25W-35N	21
Molidor Rd. E&W	27W-33N,26W-33N	20
Moody St. N&S	27W-35N	20
Moraine Dr. E&W	27W-35N	20
Myrtle Ln. E&W	26W-35N	21
Mudjekeewis Tr. NW&SE	28W-34N	20
Muskego Av. NW&SE	25W-34N,25W-35N	21
N. Lake Dr. NE&SW	25W-35N	21
Navajo Tr. E&W	28W-34N	20
Nippersink Pl. NW&SE	28W-37N	12
Nippersink Rd. NW&SE	26W-34N	21
Nokomis Tr. N&S	28W-34N	20
Nora Pl. NE&SW	28W-37N	12
North Ct. E&W	27W-35N	20
O'Kelly Ln. E&W	28W-34N	20
Oak Av. N&S	25W-34N	21
Oak Cir. Dr. E&W	26W-36N	13
Oak Ct. N&S	25W-35N	21
Oak Dr. N&S	26W-33N	21
Oak Ln. E&W	25W-35N	21
Oak St. E&W	25W-36N	13
Oak St. N&S	25W-34N	21
Oakland Dr. E&W	25W-34N	21
Oakside St. N&S	26W-34N	21
Oakwood Av. N&S	25W-35N	21
Oakwood Av. E&W	26W-35N	21
Oden Av. N&S	25W-34N	21
Old Grand Av. E&W	25W-36N	13
Olive Ct. N&S	25W-35N	21
Orchard Av. E&W	25W-36N	13
Orchard Dr. N&S	27W-35N	20
Park Av. N&S	26W-34N	21
Park Dr. N&S	28W-34N	20
Parkview Pl. E&W	28W-37N	12
Patricia Ln. NW&SE	28W-37N	12
Paupukkeewis Tr. NW&SE	28W-34N	20
Peterson Av. N&S	25W-34N	21
Pleasant Rd. E&W	26W-36N	13
Polk St. N&S	26W-34N	21
Poplar Av. N&S	25W-35N	21
Rabbit Hill Ln. N&S	26W-36N	13
Racine Av. E&W	25W-34N	21
Randhill Dr. N&S	27W-35N	20
Randich Rd. E&W	26W-37N	13
Ravine Ln. E&W	27W-35N	20
Reed St. E&W	26W-34N	21
Ridge Av. N&S	26W-36N	13
Ridge St. E&W	27W-35N	20
Robin Rd. N&S	26W-34N	21
Rock Island St. NE&SW	25W-34N	21
Rockford St. E&W	25W-35N	21
Rollins Rd. E&W	25W-35N	21
Roseland Ct. E&W	26W-36N	13
Rosewood Dr. N&S	26W-35N	21
Rt. 12 N&S	27W-32N,28W-37N	20
Rt. 59 N&S	27W-32N,25W-36N	20
Rt. 134 E&W	27W-34N,25W-34N	20
Sequoia Tr. N&S	28W-34N	20
Shandon Dr. E&W	26W-36N,25W-36N	13
Sheridan Dr. N&S	27W-35N	20
Shirley Ln. NE&SW	28W-35N	20
Shore Rd. N&S	26W-36N	13
Shoreline Dr. N&S	26W-35N	21
Snowshoe Tr. E&W	28W-34N	20
South Av. E&W	28W-37N	12
South Ct. E&W	27W-35N	20
South Ct. E&W	28W-34N	20
South St. E&W	26W-36N	13
Squaw Rd. E&W	26W-36N	13
Stanley Rd. N&S	27W-34N	20
Stanton Bay Rd. E&W	26W-36N	13
Stanton Ln. NS&EW	26W-33N	21
Stanton Point Rd. NE&SW	26W-36N,26W-37N	13
Stockholm Dr. E&W	26W-33N	21
Stolaf Av. NE&SW	25W-34N	21
Stoughton Av. E&W	25W-34N	21
Sullivan Lake Rd. E&W	28W-33N	20
Sunnybrook Rd. E&W	26W-34N	21
Sunnyside Av. N&S	25W-35N	21
Sunset Dr. NW&SE	25W-34N	21
Sunset Ter. NE&SW	25W-34N	21
Tamarack Dr. N&S	25W-36N	13
Tami Ln. N&S	25W-35N	21
Tara Ct. N&S	25W-36N	13
Tarvin Ln. E&W	26W-35N	21
Terrace Ln. NE&SW	28W-37N	12
Thomas Ct. E&W	28W-34N	20
Tree Top Rd. NW&SE	28W-35N	20
Tyrone Pl. NE&SW	28W-37N	12
Valley Rd. NW&SE	28W-35N	20
Van Buren St. E&W	26W-34N	21
Venetian Dr. NW&SE	25W-34N	21
Vincent Ct. E&W	26W-36N	13
Wally Ct. E&W	26W-35N	21
Ward Ln. N&S	25W-35N	21
Watson Av. N&S	26W-35N	21
Watts Av. NE&SW	28W-37N	12
Wenonah Tr. E&W	28W-34N	20
Wesley Rd. N&S	26W-36N	13
West Lake Av. E&W	28W-37N	12
West St. N&S	28W-37N	12
White Rabbit Tr. E&W	28W-34N	20
Wildwood Av. E&W	26W-35N	21
William Pl. NW&SE	28W-37N	12
Wilson Rd. N&S	26W-36N,25W-32N	13
Wind Mill Rd. N&S	26W-36N	13
Wooster Lake Av. E&W	26W-34N	21
Wooster Lake Dr. E&W	26W-34N	21
Wooster Ln. E&W	26W-34N	21
1st Av. NW&SE	27W-35N	20
2nd Av. NW&SE	27W-35N	20
3rd Av. NW&SE	27W-35N	20

Grayslake

Street	Grid	Page
Allegheny Rd. N&S	21W-31N,21W-32N	31
Allen Av. E&W	20W-33N	23
Allison Ct. E&W	21W-32N,21W-33N	23
Alta Dr. N&S	21W-32N	23
Arlington Ln. NW&SE	21W-34N	23
Atkinson Rd. N&S	20W-32N,20W-33N	23
Attenborough Ct. E&W	19W-33N	23
Attenborough Way NS&EW	19W-33N	23
Augusta St. N&S	21W-33N	23
Banbury Dr. E&W	20W-33N	23
Banbury Ln. NE&SW	20W-33N	23
Barron Blvd. N&S	20W-33N	23
Behm Dr. NW&SE	21W-33N	23
Belle Ct. N&S	21W-32N	23
Belvidere Rd. E&W	21W-32N,18W-32N	23
Berry Av. E&W	20W-33N	23
Bluff Av. NE&SW	21W-32N	23
Bob-o-link Dr. N&S	19W-32N	23
Bonnie Brae Av. E&W	20W-33N	23
Brae Loch Rd. E&W	19W-33N	23
Braxton Ct. E&W	19W-33N	23
Braxton Way NE&SW	19W-33N	23
Briargate Dr. E&W	20W-33N	23
Bristol Ln. N&S	20W-33N	23
Brittain Av. N&S	21W-32N	23
Buckingham Ct. E&W	19W-33N	23
Burton St. N&S	21W-33N	23
Cambridge Dr. NW&SE	19W-34N	23
Cardinal Ct. E&W	20W-33N	23
Carol Ln. E&W	20W-33N	23
Catherine Ct. E&W	20W-33N	23
Cecelia St. N&S	21W-33N	23
Center St. E&W	20W-33N,19W-33N	23
Chard St. E&W	20W-33N	23
Chesapeake Blvd. N&S	20W-34N	23
Christy Cir. E&W	21W-33N	23
Clarewood Cir. NS&EW	19W-33N	23
Commerce N&S	19W-32N	23
Country Squire Dr. NS&EW	18W-32N	23
Cove Ct. N&S	20W-33N	23
Cross Arm Dr. N&S	20W-32N	23
Dawn Cir. N&S	20W-33N	23
Deep Woods Ct. E&W	21W-33N	23
Durham Ln. E&W	20W-33N	23
E. Brittany Sq. Cir.	20W-33N,20W-34N	23
E. Cambridge Ct. E&W	19W-33N	23
East Ct. N&S	20W-33N	23
Fairfax Ln. E&W	21W-34N	23
Flanders Ln. E&W	20W-33N	23
Garfield Blvd. N&S	20W-32N	23
George St. E&W	21W-33N	23
Getchell Av. E&W	21W-32N	23
Glen St. E&W	21W-32N	23
Gray Ct. E&W	20W-33N	23
Greenwood Av. E&W	21W-32N	23
Hamlitz Ct. E&W	21W-32N	23
Hampshire Ct. E&W	20W-32N	23
Harding St. N&S	20W-32N	23
Harris Rd. N&S	19W-32N	23
Harvey Av. E&W	21W-33N	23
Hawley Ct. NS&EW	21W-33N	23
Hawley St. E&W	20W-33N	23
Heather Av. E&W	20W-33N	23
Hickory St. E&W	20W-33N	23
Highland Rd. E&W	20W-33N	23
Hillside Av. E&W	21W-33N	23
Hojem Ln. E&W	21W-32N	23
Hummingbird Ln. NS&EW	19W-32N	23
Iron Horse Ct. E&W	20W-32N	23
Ivanhoe Rd. N&S	20W-31N,20W-32N	31
Jackson Blvd. E&W	20W-32N	23
Jeffrey Av. N&S	21W-33N	23
Jennifer Ln. E&W	21W-33N	23
Junior Av. E&W	20W-32N	23
Kelly Av. NW&SE	21W-33N	23
Kenilworth Ct. N&S	20W-33N,20W-34N	23
Kerry Way NS&EW	19W-32N	23
Kevin Ln. NS&EW	21W-33N	23
Kilburn Ct. E&W	19W-33N	33
Lake St. N&S	21W-32N,21W-34N	23
Langley Ct. NW&SE	20W-33N	23
Laurel Ter. N&S	20W-33N	23
Laurie Ct. NW&SE	21W-33N	23
Lawrence Av. E&W	20W-33N	23
Lexington Ln. NE&SW	21W-34N	23
Lincoln Av. N&S	20W-32N	23
Lindsey N&S	21W-33N	23
Mallard Ct. NE&SW	19W-32N	23
Manor Av. N&S	20W-33N	23
May St. N&S	21W-33N	23
McMillan St. N&S	20W-33N	23
Meadowlark Av. NS&EW	19W-32N,19W-33N	23
Merrill Ln. E&W	20W-33N	23
Mitchell Dr. E&W	21W-33N	23
Moore Ct. N&S	21W-32N	23
N. Allegheny Rd. E&W	21W-33N	23
N. Cambridge Ct. E&W	19W-33N	23
Neville Dr. N&S	21W-32N	23
Normandy Ln. E&W	20W-33N	23
North Corporate Cir. E&W	19W-32N	23
Oak Av. E&W	20W-32N	23
Oakwood Dr. N&S	20W-33N	23
Old Center St. E&W	20W-33N	23
Old Plank Rd. E&W	18W-32N	23
Oriole Ct. E&W	20W-33N	23
Park Av. E&W	20W-33N	23
Park Pl. E&W	20W-32N	23
Parker Dr. N&S	21W-33N	23
Partridge Ct. NE&SW	19W-32N	23
Patricia Ct. E&W	21W-33N	23
Peterson Rd. E&W	20W-30N	31
Pheasant Ct. E&W	20W-33N	23
Pierce Ct. N&S	20W-33N	23
Pine St. E&W	20W-33N	23
Pines Dr. NW&SE	18W-32N	24
Popes Creek Cir. NW&SE	20W-34N	23
Prairieview Av. E&W	21W-33N	23
Proctor St. E&W	20W-33N	23
Quail Creek Dr. E&W	20W-33N	23
Quist Ct. E&W	21W-32N	23
Railroad Av. NW&SE	20W-33N	23
Reagan's Pl. E&W	20W-32N	23
Robin Ct. N&S	20W-33N	23
Rock Hall Cir. NW&SE	20W-34N	23
Rollins Rd. NE&SW	21W-35N	23
Roosevelt Ct. N&S	20W-32N	23
Rt. 83 N&S	20W-31N,21W-34N	31
Rt. 120 E&W	21W-32N,18W-32N	23
S. Stuart St. N&S	20W-33N	23
School St. E&W	21W-33N	23
Scott Av. E&W	21W-33N	23
Sears Blvd. N&S	18W-32N	23
Seymour Av. N&S	20W-32N,20W-33N	23
Shakespeare Dr. E&W	20W-34N,19W-34N	23
Sheffield Ln. E&W	19W-33N	23
Sheldon Rd. N&S	21W-35N	23
Shorewood Rd. E&W	21W-35N	23
Signal Ln. E&W	20W-32N	23
Silo Hill Dr. NE&SW	19W-32N	23
Silo Rill Ct. E&W	19W-32N	23
Siwiha Dr. E&W	21W-33N	23
Slusser St. N&S	20W-32N	23

Street	Grid	Page
Somerset Dr. N&S	20W-32N	23
South Corporte Cir. E&W	19W-32N	23
St. Paul St. NW&SE	20W-32N	23
Stratford Ct. N&S	20W-34N	23
Stuart St. N&S	20W-33N	23
Swan Dr. N&S	20W-33N	23
Talbot Ln. N&S	20W-34N	23
Trestle Ct. N&S	20W-32N	23
Vista Ct. N&S	19W-32N	23
W. Cambridge Ct. E&W	19W-33N	23
Walton Ln. E&W	20W-33N	23
Warren Ln. E&W	21W-32N	23
Washington Blvd. E&W	20W-32N	23
Washington St. E&W	21W-34N,19W-34N	23
Waterford Dr. NS&EW	21W-33N	23
Webb St. N&S	20W-33N	23
West Shore Dr. NW&SE	21W-33N	23
West Trail N&S	21W-33N	23
West Trail North N&S	21W-33N	23
Westerfield Pl. N&S	20W-32N	23
Wexford Ct. E&W	20W-33N	23
White St. E&W	20W-33N	23
Whitney St. N&S	20W-33N	23
Wicks St. E&W	20W-32N	23
Woodland Dr. E&W	21W-33N	23
York Dr. NE&SW	20W-33N	23
Yorkshire Ct. E&W	20W-32N	23
Ziegler Dr. N&S	20W-33N	23
1st St. N&S	21W-33N	23
2nd St. N&S	21W-33N	23

Great Lakes

Street	Grid	Page
'A' St. E&W	10W-30N	35
Alaska Av. E&W	12W-30N	35
Allan Ln. E&W	11W-31N	35
Argonaut Pl. N&S	11W-30N	35
Arizona Av. N&S	11W-30N	35
Arkansas Rd. N&S	12W-29N	35
Atlantic Rd. N&S	11W-29N	35
'B' St. E&W	10W-30N	35
Barry Pl. N&S	10W-30N	35
Barry Rd. N&S	10W-30N	35
Bassett Av. E&W	10W-30N	35
Broadway Av. E&W	11W-30N	35
Bronson Av. N&S	10W-31N	35
Buckley Rd. E&W	12W-30N	35
Burnham Av. E&W	11W-31N	35
'C' St. E&W	10W-30N	35
California Av. E&W	11W-29N	35
Cavin Dr. N&S	11W-30N	35
Clark Av. E&W	10W-31N	35
Cluerius Av. N&S	10W-31N	35
Connecticut Av. E&W	12W-30N	35
Connecticut Ct. E&W	11W-30N	35
Crosley Dr. N&S	10W-30N,10W-31N	35
'D' St. E&W	10W-30N	35
Dahlgren St. N&S	10W-31N	35
Decatur Av. E&W	11W-30N	35
Delaware Av. E&W	11W-29N	35
Depot Dr. E&W	12W-30N	35
Dewey Av. N&S	10W-31N	35
Downes Dr. N&S	10W-30N	35
'E' St. E&W	10W-30N	35
E. Alabama Av. E&W	11W-30N	35
E. Washington Av. E&W	11W-30N	35
E. Wyoming Av. E&W	12W-30N	35
E.C.I. Ln. E&W	11W-30N	35
Ellis St. N&S	10W-31N	35
Enterprise Av. NE&SW	11W-30N	35
Erie Ct. E&W	12W-30N	35
Essex Av. N&S	11W-30N	35
Evans Av. E&W	10W-31N	35
'F' St. E&W	10W-30N	35
Farragut Av. E&W	10W-31N	35
Fisher Rd. NE&SW	10W-30N	35
Florida Av. E&W	11W-29N	35
Forrestal Dr. E&W	11W-30N	35
Fullam Av. E&W	10W-31N	35
'G' St. E&W	10W-30N	35
Georgia Rd. N&S	12W-29N	35
Grassie Av. N&S	10W-31N	35
Great Lakes Dr. N&S	12W-30N	35
Greenfield St. N&S	11W-30N	35
Hawaii Av. N&S	11W-29N	35
Hennepin St. N&S	11W-31N	35
Hines St. N&S	10W-30N	35
Hopkins St. N&S	11W-30N	35
Huron St. N&S	10W-30N	35
Idaho Ct. N&S	12W-30N	35
Idaho St. N&S	12W-30N	35
Illinois St. N&S	11W-29N,11W-30N	35
Indiana St. N&S	11W-30N	35
Intrepid St. N&S	11W-30N	35
Iowa St. N&S	11W-31N	35
Isherwood Av. N&S	10W-31N	35
Kentucky Ct. N&S	11W-29N	35
Kentucky St. N&S	11W-30N	35
Langley St. NW&SE	11W-31N	35
Lawrence St. N&S	10W-31N	35
Lexington Av. N&S	11W-30N	35
Luce Blvd. N&S	10W-31N	35
Mac Dough St. N&S	10W-31N	35
Mahan Rd. NW&SE	10W-30N	35
Main St. N&S	10W-31N	35
Maine Av. N&S	11W-30N	35
Meridian St. N&S	11W-30N,11W-29N	35
Michigan Ct. N&S	12W-30N	35
Minnesota Av. N&S	12W-30N	35
Mississippi St. N&S	12W-30N	35
Missouri St. N&S	11W-30N,11W-31N	35
Missouri St. N&S	11W-30N	35
Moffet Dr. E&W	10W-31N	35
Montana Av. N&S	11W-30N	35
N. Admiral Rd. N&S	11W-30N	35
Nebarska St. N&S	11W-30N	35
Nevada Ct. N&S	12W-30N	35
New York Av. N&S	11W-29N	35
Neyer Cir. NW&SE	10W-31N	35
Nimitz St. N&S	10W-31N	35
Ohio Av. E&W	11W-29N	35
Ohio St. N&S	11W-29N,11W-30N	35
Oklahoma Av. N&S	12W-30N	35
Oklahoma Ct. N&S	12W-30N	35
Ontario Ct. N&S	12W-30N	35
Oregon Ct. E&W	12W-30N	35
Pacific Av. N&S	11W-30N	35
Paul Jones Ct. N&S	10W-31N	35
Paul Jones St. N&S	10W-31N	35
Pennsylvania Av. E&W	11W-30N	35
Perry St. N&S	10W-31N	35
Porter Av. E&W	10W-31N	35
Preble Av. E&W	10W-31N	35
Preble Pl. N&S	10W-30N	35
Ranger Ct. N&S	11W-30N	35
Ray St. N&S	11W-30N	35
Remora Ct. E&W	11W-30N	35
Rhode Island Av. E&W	12W-29N	35
Rhode Island Ct. E&W	12W-30N	35
Rogers St. N&S	10W-31N	35
S. Admiral Rd. N&S	11W-29N	35
Sampson Rd. N&S	10W-30N	35
Sampson St. N&S	10W-31N	35
Saratoga St. NW&SE	11W-30N	35
Spaulding St. N&S	11W-30N	35
Superior St. N&S	12W-30N	35
Sutton St. N&S	12W-29N	35
Tennessee Ct. E&W	12W-30N	35
Texas Ct. E&W	11W-30N	35
Thresher St. N&S	11W-30N	35
Utah Ct. E&W	12W-30N	35
Vermont Ct. N. N&S	12W-30N	35
Vermont Ct. S. N&S	12W-30N	35
Virginia Ct. NW&SE	12W-30N	35
Virginia Rd. E&W	11W-30N	35
W. Alabama Av. E&W	12W-29N	35
W. Colorado Ct. E&W	12W-30N	35
W. Florida Ct. E&W	12W-30N	35
W. Wyoming Av. E&W	12W-30N	35
Washington Av. E&W	12W-30N	35
Wisconsin Dr. E. NE&SW	12W-30N	35
Wisconsin Dr. W. NW&SE	12W-30N	35
Wurtsbaugh St. N&S	10W-30N	35
Wyoming Ct. E&W	12W-30N	35
Yeager St. N&S	11W-30N	35
Ziegemeier St. N&S	10W-30N,10W-31N	35
1st Av. E&W	11W-29N	35
1st St. N&S	10W-30N	35
2nd Av. E&W	11W-29N	35
2nd St. N&S	10W-30N	35
3rd Av. E&W	11W-30N	35
3rd St. N&S	10W-30N	35
4th Av. E&W	10W-30N	35
4th St. N&S	10W-30N	35
5th Av. E&W	10W-30N	35
5th St. N&S	10W-30N	35
6th Av. E&W	11W-30N	35
6th St. N&S	10W-30N	35
7th Av. E&W	11W-30N	35
7th St. N&S	10W-30N	35
8th Av. E&W	11W-30N	35
8th St. N&S	10W-30N	35
9th Av. E&W	11W-30N	35
9th St. N&S	10W-30N	35
10th Av. E&W	11W-30N	35
10th St. N&S	10W-30N	35
11th Av. E&W	10W-30N	35
11th St. N&S	10W-30N	35
12th Av. E&W	11W-30N	35
24th E&W	11W-31N	35

Green Oaks

Street	Grid	Page
Atkinson Av. E&W	14W-29N,13W-19N	33
Belmont Dr. E&W	14W-30N	33
Benson Ln. N&S	14W-29N	33
Benson Rd. N&S	14W-29N	33
Brentwood Dr. E&W	14W-30N	33
Buckley St. N&S	14W-30N	33
Carriage Ct. N&S	14W-30N	33
Churchill Ct. N&S	14W-30N	33
Cranbrook Rd. E&W	14W-30N	33
Crescent Knoll Dr. NE&SW	14W-28N	33
Crest Rd. N&S	14W-29N	33
Dan Patch Cir. E&W	14W-31N	33
Fox Run NS&EW	14W-29N	33
Gerwick Way E&W	14W-30N	33
Glenmore Rd. N&S	14W-30N	33
Greenbriar Dr. E&W	14W-29N	33
Greenbriar Ln. E&W	14W-29N	33
Gurin Rd. E&W	15W-31N,14W-31N	33
Hanlon Rd. N&S	15W-31N	33
Harwood Dr. E&W	14W-30N	33
Haven Ln. NS&EW	14W-30N	33
Hawthorne Rd. N&S	15W-28N	33
Heathercliff Dr. N&S	14W-31N	33
Horace Ct. NW&SE	14W-30N	33
Interstate 94 N&S	14W-28N,14W-31N	33
Irene Ln. E&W	14W-29N	33
Irondale Rd. N&S	15W-31N	33
Kathryn Ct. N&S	14W-29N	33
Kenton Ln. E&W	14W-30N	33
King Ct. E&W	14W-30N	33
King Dr. N&S	14W-30N	33
Lexington Rd. NE&SW	14W-30N	33
Lockwood Dr. N&S	14W-30N	33
Margate Ln. E&W	15W-28N	33
Minard Ln. E&W	14W-29N	33
Morgan Ct. E&W	14W-31N	33
N. Dan Patch Cir. E&W	14W-31N	33
N. Dan Patch Cir. E&W	14W-31N	33
O'Plaine Rd. N&S	14W-30N,14W-31N	33
Old Rockland Rd. NE&SW	14W-28N	33
Parkhill N&S	14W-29W	33
Pearson Rd. E&W	14W-29N	33
Placid Ln. E&W	14W-30N	33
Rein Ct. NW&SE	14W-31N	33
Rockland Rd. NE&SW	14W-28N	33
Rocktop Rd. N&S	14W-28N	33
Rt. 137 E&W	14W-30N,15W-30N	33
S. Dan Patch Cir. NS&EW	14W-31N	33
Saddle Hill Rd. N&S	14W-30N	33
Saratoga Ct. E&W	14W-30N	33
Scotland Ct. E&W	14W-30N	33
Shagbark Ln. N&S	15W-28N	33
Shannondale Dr. NS&EW	14W-31N	33
St. Mary's Rd. N&S	15W-29N,14W-30N	33
Steeplechase Ct. E&W	14W-30N	33
Sunrise Rd. E&W	14W-29N	33
Thornhill Rd. N&S	14W-30N	33
Tri-State Tollway N&S	14W-28N,14W-31N	33
White Fenoe Ln. E&W	14W-30N	33
Wickham St. N&S	14W-30N	33
Wickham Dr. N&S	14W-30N	33
Wilton Ct. N&S	14W-30N	33
Wilton Ln. N&S	14W-30N	33
Woodview Dr. E&W	15W-31N	33

Gurnee

Street	Grid	Page
Abingdon Ct. N&S	16W-33N	24
Acorn Ct.* NW&SE	14W-36N	17
Adaline Ct. E&W	14W-35N	25
Adams Ct. NE&SW	16W-36N	17
Adele Dr. E&W	15W-35N	25
Albingdon Ct.* E&W	16W-32N	25
Arbor Ct.* N&S	14W-36N	17
Arthur* E&W	14W-36N	17
Ashwood Ln. E&W	15W-36N	17
Astor Pl. N&S	14W-33N	25
Atlantic Av. E&W	13W-35N	26
Avery Ct. NE&SW	16W-33N	24
Barberry Ln. NE&SW	14W-35N	25
Barnwood Ct. E&W	15W-36N	17
Bay Pl. N&S	13W-35N	26
Beechwood Av. E&W	15W-36N	17
Beechwood Ln. N&S	15W-36N	17
Belleplaine Av. N&S	13W-34N,13W-36N	26
Belvidere Dr. E&W	16W-33N,16W-32N	24
Bennett Av. E&W	14W-34N	25
Berkshire* NE&SW	15W-34N	25
Beth Ct. NW&SE	14W-33N	24
Bicwell Dr. NW&SE	16W-33N	24
Big Terra Ln. N&S	16W-34N	24
Bingham Ct.* E&W	16W-32N	25
Birch Rd. NW&SE	14W-34N	25
Birchwood Dr. E&W	15W-35N	25
Blackburn St. N&S	13W-35N	26
Blackstone Dr. E&W	14W-35N,13W-35N	25
Bluebird Ct. E&W	14W-34N	25
Boulder Dr. E&W	15W-36N	17
Boulevard View Av. N&S	13W-35N,13W-36N	26
Briar Av. N&S	13W-34N	26
Brierwood Ct. N&S	14W-33N	25
Brighton* E&W	15W-34N	25
Bristol* NW&SE	15W-34N	25
Brittany Ct. NE&SW	16W-34N	24
Brookhaven Dr. NE&SW	14W-34N	25
Brookstone Ln. NW&SE	16W-33N	24
Brown Cir. E&W	14W-35N	25
Brown Pl. N&S	14W-34N	25
Buckeridge Ct.* NW&SE	14W-36N	17
Buckingham N&S	14W-34N	25
Butternut Ct. E&W	15W-36N	17
Calvin Ct. N&S	14W-33N	25
Campbell Dr. NS&EW	16W-33N	24
Campbell Dr.* E&W	16W-32N	25
Canterbury N&S	15W-34N	25
Canterbury Ct.* E&W	14W-36N	17
Carol Ln. E&W	15W-35N	25
Carriage Rd. E&W	14W-35N	25
Castlewood Ln. E&W	17W-35N	24
Cedar Av. E&W	15W-35N	25
Cedarwood Ct.* E&W	16W-35N	25
Cemetary Rd. N&S	16W-34N,16W-35N	24
Centennial Ct. E&W	14W-35N	25
Chandler Rd. NE&SW	16W-33N	24
Charles Av. E&W	14W-34N	25
Charles Av. N&S	14W-35N	25
Chase Ct. N&S	16W-33N	24
Chase Ct.* E&W	16W-32N	25
Cherrywood Ct. NW&SE	15W-36N	17
Cheshire Rd. E&W	16W-33N	24
Cheyenne Rd. E&W	13W-36N	18
Chip Ct. NE&SW	16W-33N	24
Churchill Ln. NW&SE	16W-34N	24
Clark Dr. E&W	16W-33N	24
Clearview Av. N&S	13W-35N	26
Clearview Ct. E&W	12W-35N	26
Club House Ct. N&S	16W-33N	24
Colby Ct. N&S	16W-33N	24
Colby Rd. N&S	16W-33N	24
Compton Ct.* E&W	16W-32N	24
Concord Sq. NS&EW	14W-33N	25
Concord Square NS&EW	14W-33N	25
Constitution Av. NW&SE	16W-36N	17
Cornell Av. E&W	14W-33N	25
Country Club Av. N&S	13W-36N	18
Country Tr. E&W	14W-33N	25
Covenant Ct. NS&EW	14W-33N	25
Coventry Rd. N&S	15W-35N	25
Crescent Av. E&W	13W-35N	26
Cross Ct. NW&SE	14W-33N	25
Cross Rd. N&S	14W-33N	25
Crossland Blvd. E&W		24
Crossland Ct.* E&W	16W-32N	25
Cule Ct. N&S	14W-33N	25
Darlene Ct. E&W	15W-34N	25
Darlene Dr. E&W	15W-34N	25
Darnell Ln. N&S	16W-33N	24
David* NW&SE	14W-36N	17
Deer Run NE&SW	15W-35N	25
Delany Rd. N&S	14W-34N,14W-36N	25
Delaware Av. N&S	16W-36N,15W-36N	17
Depot Rd. N&S	14W-35N	25
Desplaines Ct. NE&SW	15W-33N	25
Dillys Rd. N&S	15W-35N,16W-37N	25
Dixon Ct. N&S	16W-33N	24
Dorchester Av. E&W	14W-34N,13W-34N	25
Dordan Ct. N&S	16W-33N	24
Dover Cir. E&W	17W-35N	24
Drexel Av. E&W	13W-34N	26
Drury Ct. NE&SW	14W-33N	24
Dunham Rd. E&W	14W-33N	24
Dunhill Ct.* E&W	16W-32N	25
Eagle Ridge Rd. N&S	16W-33N	24
Eastwood Av. E&W	14W-33N	25
Eau Claire E&W	15W-35N	25
Ebony Ct.* E&W	14W-33N	25
Edington Ct. N&S	16W-33N	24
Ellis Av. E&W	14W-34N,13W-34N	25
Elm Rd. N&S	14W-34N	25
Elsie Av. N&S	15W-35N	25
Emerald Av. N&S	14W-35N	25
Estate Cir. NW&SE	16W-34N	24
Estes Rd. N&S	13W-34N,13W-35N	26
Evergreen Dr. E&W	15W-35N	25
Fair Links Way NE&SW	16W-33N	24
Fairfax Ct.* N&S	14W-34N	25
Farmgate Ct. N&S	14W-33N	25
Farwell Ct. N&S	16W-33N	24
Farwell Ct.	16W-32N,16W-33N	25
Ferndale St. N&S	13W-34N,13W-35N	26
Fernwood* NW&SE	14W-36N	17
Fieldstone Ct. NE&SW	14W-33N	25
Finch Ct. E&W	14W-34N	25
First Pl. E&W	14W-34N	25
First St. N&S	14W-34N	25
Florida Av. E&W	13W-36N	18
Fox Ln. NW&SE	15W-35N	25
Fox Meadow N&S	15W-36N	17
Foxboro Ln. NS&EW	16W-34N	24
Franklin Ct. N&S	15W-36N	17
Fredrick* N&S	14W-36N	17
Fuller Rd. N&S	15W-35N,15W-36N	25
Gages Lake Rd. E&W	16W-33N	24
Gatewood Dr. NS&EW	15W-36N	17
George Ct.* N&S	14W-36N	17
Gillings Dr. N&S	14W-34N	25
Glen Flora Av. E&W	13W-35N	26
Glen Way E&W	16W-33N	24
Glendale Dr. NW&SE	15W-35N	25
Golf View Dr. NW&SE	16W-33N	24
Gould St. N&S	13W-34N	26
Grace St. N&S	13W-35N	26
Grand Av. NW&SE	18W-35N,13W-35N	24
Grandmore Av. E&W	13W-35N	26
Grandview Av. E&W	13W-35N	26
Grandville Av. E&W	13W-35N	26
Grass Lake Rd. E&W	16W-33N,15W-33N	24
Great America Parkway N&S	15W-35N	25
Green Haven NS&EW	14W-34N	25
Greenbrier Ct. E&W	13W-35N	26
Greenleaf St. N&S	13W-34N,13W-35N	26
Greenview St. N&S	13W-35N	26
Greenwood Av. E&W	13W-36N	18

	Grid	Page
Grove Av. E&W	14W-35N,13W-35N	25
Gurnee Mills Blvd. NS&EW	16W-35N,16W-36N	24
Habitat Ct. NW&SE	16W-34N	24
Hancock Ln. E&W	16W-36N	17
Harper Av. E&W	14W-34N,13W-34N	25
Hawthorne Av. N&S	13W-34N,13W-35N	26
Hawthorne St. N&S	13W-35N	26
Hazlewood Ct.* E&W	16W-35N	25
Heather Ridge Dr. E&W	16W-33N	24
Henry* NW&SE	14W-36N	17
Hickory Haven Dr., East N&S	15W-33N	25
Hickory Haven Dr., West NS&EW	15W-33N	25
Highland Av. E&W	13W-36N	18
Hill-n-Dale Ct. E&W	16W-34N	24
Hollyholk Ct.* E&W	16W-35N	25
Honey Suckle Ln. E&W	16W-34N	24
Hunt Club Rd. N&S	17W-34N,17W-36N	24
Independence Ct.* NE&SW	14W-36N	17
Indian Tr. E&W	14W-34N	24
Interstate 94 N&S	15W-35N	25
Ironwood Ct. NW&SE	15W-36N	17
Jeffery Av. N&S	14W-34N,14W-34N	25
Johns Manville Av. E&W	13W-35N	26
Joshua Ct.* E&W	16W-35N	25
Keith Av. E&W	13W-35N	26
Kensington Ct.* NS&EW	15W-34N	25
Kenwood Av. N&S	14W-34N,13W-34N	25
Kilbourne Rd. NE&SW	14W-35N	25
King's Ct. NW&SE	14W-33N	25
King's Way N. E&W	14W-33N	25
King's Way W. N&S	14W-33N	25
Kisten Ct. NW&SE	16W-33N	24
Knob Hill Ln. E&W	16W-33N	24
Lake Park Av. E&W	14W-34N	25
Lakeside Dr. NS&EW	16W-35N	24
Lawrence Av. N&S	13W-34N	26
Lawson Blvd. NS&EW	16W-36N,15W-35N	17
Lee Av. E&W	14W-35N,13W-35N	25
Leonard Dr. NE&SW	16W-33N	24
Leonard Dr. N&S	16W-33N	24
Leslie Ln. NS&EW	15W-35N	25
Lexington Square, East NS&EW	14W-33N	25
Liberty Ln. NE&SW	16W-36N	17
Lind Ln. N&S	15W-35N	25
Lone Tree Ct. E&W	16W-34N	24
Long Hill Rd. NS&EW	14W-33N	25
Louis* NE&SW	14W-36N	17
Madison Av. N&S	14W-36N	17
Magnolia Av. N&S	13W-34N,13W-35N	26
Mahogany Ct.* E&W	16W-35N	25
Majestic Ct. N&S	14W-33N	25
Manchester Dr. E&W	16W-36N,15W-36N	24
Maplewood Dr. NS&EW	15W-36N	17
Mayflower Ct. N&S	14W-33N	25
McClure Av. E&W	14W-34N	25
Meadow Crest Cir. E&W	13W-35N	26
Meadow Ct. E&W	16W-34N	24
Meadow Lark Ct. E&W	14W-34N	25
Middle Rd. E&W	14W-33N	25
Milwaukee Av. NE&SW	16W-33N,16W-35N	
Mint Haven Ct. E&W	14W-34N	25
Mitchell Ct. NW&SE	16W-33N	24
Monterey Ct.* NW&SE	14W-36N	17
Morningside Ct. E&W	16W-34N	24
Morrison Dr. E&W	13W-36N	26
Newbury Ct.* E&W	16W-32N	25
Newport Ct.* E&W	14W-36N	17
Norfolk* N&S	14W-36N	17
Norman Av. N&S	14W-35N	25
North Av. E&W	14W-34N	25
Northwestern Av. N&S	13W-35N,13W-36N	26

	Grid	Page
O'Plaine Rd. N&S	14W-33N,14W-34N	25
Oak Ln. E&W	15W-35N	25
Oakview Ln. NE&SW	15W-36N	17
Oakwood St. N&S	13W-35N	26
Oglesby Av. N&S	14W-34N	25
Old Farm Ln. NW&SE	16W-34N	24
Old Grand Av. NW&SE	14W-35N	25
Oriole Ct. E&W	14W-34N	25
Oscar* E&W	14W-36N	17
Oxford Cir.* NS&EW	14W-36N	17
Pacific Av. E&W	13W-35N	26
Packard Ln. N&S	15W-35N	17
Par Ct. NW&SE	16W-33N	24
Parkside Ct. N&S	16W-36N	17
Pauly Dr. NW&SE	15W-35N	25
Pebble Creek Ct. E&W	16W-33N	24
Pheasant Ct. E&W	16W-34N	24
Pilgrim's Path NW&SE	14W-33N	25
Pine Ct. N&S	15W-34N	17
Pine Grove St. N&S	13W-34N,13W-35N	26
Pine Meadow N&S	15W-36N	17
Pine Tree Dr. NS&EW	15W-35N	25
Pinewood Rd. NE&SW	15W-35N	25
Plainview Rd. N&S	16W-34N	24
Plymouth Ct. N&S	14W-33N	25
Ponderosa NW&SE	15W-35N	25
Porett Dr. E&W	13W-35N	26
Portage Ln. NE&SW	15W-35N	25
Portsmouth Cir. NE&SW	17W-35N	24
Prairie Creek Dr. NW&SE	16W-36N,16W-35N	17
Prairie Oak Rd. E&W	15W-35N	25
Princeton Ct.* NS&EW	14W-36N	17
Providence Rd. E&W	14W-33N	25
Quail Haven Ct. E&W	16W-34N	24
Queen Anne Ln. E&W	16W-34N	25
Quincy Ct. NE&SW	16W-36N	17
Ralph Av. N&S	14W-35N	25
Raven Ct. E&W	14W-34N	25
Ravine Dr. N&S	15W-36N	17
Red Pine Av. E&W	15W-36N	17
Regency Ct. E&W	14W-33N	24
Ridge Dr. E&W	15W-35N	25
Ridwell Dr.* E&W	16W-32N	25
Robin Ct. E&W	14W-34N	25
Rockpoint Ct. E&W	15W-35N	25
Rogers Rd. N&S	15W-35N	25
Rolling Green Dr. NE&SW	16W-33N	24
Rosemary Ln. E&W	15W-36N	17
Rt. 21 NE&SW	15W-34N,14W-35N	25
Rt. 120 E&W	15W-33N	25
Rt. 132 NW&SE	18W-35N,13W-34N	24
Russell Av. E&W	14W-33N,14W-34N	25
Salem Ct.* N&S	14W-36N	17
Sanders Ct. N&S	16W-33N	24
Sandwedge Pl. N&S	16W-33N	24
Saratoga Ct. N&S	16W-33N	24
Scot Ct. N&S	16W-33N	24
Scotch Ct.* NE&SW	14W-36N	17
Scott Ct. N&S	16W-33N	25
Sequia Ct.* N&S	15W-36N	25
Shadow Rock Ct. E&W	14W-33N	25
Shagbark Ct. E&W	15W-33N	25
Sheffield* E&W	14W-33N	25
Shepard Ct. N&S	16W-33N	24
Sherman Av. E&W	13W-35N	26
Sherwood Ct.* NW&SE	14W-36N	17
Sizer Ct. NW&SE	16W-33N	24
Skokie Hwy. N&S	13W-34N,14W-36N	26
Sleepy Hollow Ln. NS&EW	16W-34N	24
Solomon* E&W	14W-36N	17
Somerset Ct.* N&S	16W-34N	17
South Rd. E&W	15W-34N,15W-35N	25
Southridge Dr. E&W	14W-34N	24
Spinney Run Dr. NE&SW	16W-33N	24
Springhaven Dr. E&W	14W-34N	25
Spruce St. N&S	15W-35N,15W-35N	25
Sprucewood Ln. NE&SW	15W-36N	17

	Grid	Page
St. Paul Av. NE&SW	14W-35N	25
Stearns School Rd. E&W	16W-36N,15W-36N	17
Stewart Ct. N&S	15W-33N	25
Stony Island Av. E&W	14W-34N	25
Stout Ct. N&S	15W-35N	25
Stratford Dr. N&S	15W-35N	25
Suda Dr. N&S	13W-34N	26
Sumner Cir. E&W	17W-35N	24
Sunnyside Av. N&S	14W-33N	25
Sunrise NE&SW	15W-35N	25
Swanson Ct. NS&EW	13W-36N	18
Sycamore Ln.* E&W	16W-35N	25
Tanglewood Dr. NS&EW	14W-33N	25
Taylor Dr. N&S	16W-33N	24
Timber Ct. N&S	15W-36N	17
Tri State North Tollway N&S	15W-33N,16W-36N	25
Tri-State Pkwy. NS&EW	16W-35N	25
U.S. Route 41 N&S	13W-35N	26
U.S. Route 45 N&S	18W-35N,18W-35N	24
University Av. E&W	14W-34N,13W-34N	25
Valley Ct. NE&SW	16W-33N	24
Victoria* NW&SE	15W-34N	25
View Ct.* E&W	16W-35N	25
Virginia* NE&SW	14W-36N	17
Vose Dr. N&S	16W-33N	24
Wakefield Rd. NW&SE	16W-33N	24
Washington St. E&W	18W-34N,13W-34N	24
Wausau Ln. N&S	15W-35N,15W-36N	25
Waveland Av. NW&SE	13W-34N	26
Waveland St. N&S	13W-34N,13W-35N	26
Wellington Ct.* N&S	15W-34N	25
West St N&S	14W-34N	25
Westbrook Ln. NE&SW	15W-36N	17
Westfield Dr. N&S	15W-36N	17
White Barn Ln. N&S	16W-34N	24
White Ct. E&W	16W-33N	24
White Oak Ct. N&S	14W-33N	25
Whitney Ct. NW&SE	16W-33N	24
Wilber Ct. NE&SW	15W-33N	25
Wild Flower Ln. N&S	16W-34N	24
Williams Ct. N&S	16W-33N	24
Willow Ln. NW&SE	14W-35N	25
Wiltshire Ct.* NE&SW	15W-34N	25
Wind Song Ct. NW&SE	16W-36N	17
Windwood NW&SE	15W-35N	25
Winona Ln. E&W	15W-35N	25
Woodhill Dr. E&W	15W-35N	25
Woodlawn Av. E&W	14W-34N,13W-34N	25
Yorkshire* N&S	15W-34N	25

Hainesville

	Grid	Page
Antioch St. N&S	22W-33N	22
Belvidere Rd. E&W	22W-33N,21W-33N	22
Center Dr. NW&SE	22W-33N	22
Gage St. N&S	22W-33N	22
Hainesville Rd. N&S	22W-33N	22
North St. E&W	22W-33N	22
Pineview Dr. NW&SE	22W-33N	22
Rt. 120 E&W	22W-33N	22
Rt. 134 E&W	22W-33N	22
Washington St. E&W	22W-33N	22

Half Day

	Grid	Page
Andover Ct. E&W	16W-23N	48
Cemetery Rd. E&W	15W-23N	49
Choate Ct. N&S	16W-23N	48
Dartmouth Ct. E&W	16W-23N	48
Exeter Ct. E&W	16W-23N	48
Groton Ct. E&W	16W-23N	48
Hotchkiss Ct. N&S	16W-23N	48
Middlesex Ct. E&W	16W-23N	48
Northfield Ct. E&W	16W-23N	48
Trafalgar Square N&S	15W-23N	49
Westgate Dr. NS&EW	15W-23N	49

Hawthorn Woods

	Grid	Page
Aberdeen Rd. E&W	20W-24N	39
Acorn Dr. NE&SW	23W-24N	38
Aho Dr. E&W	19W-25N	39
Albert Cir. NW&SE	19W-24N	39
Arrowwood Dr. N&S	23W-25N	38
Arthur Ct. N&S	21W-24N	38
Bent Tree Ct. NW&SE	24W-25N	38
Birch Lake Rd. E&W	22W-25N	38
Bixtone Path NE&SW	24W-25N	38
Blackhawk Rd. N&S	20W-25N	39
Bluebird Ct. N&S	21W-26N	38
Bluejay Ln. E&W	22W-26N	38
Boxwood Dr. E&W	23W-25N	38
Bridle Path NW&SE	23W-26N	38
Brierwoods Ln. NS&EW	23W-24N	38
Brighton Ct. E&W	20W-24N	39
Bruce Cir. N. NS&EW	23W-24N	38
Byron Ct. N&S	22W-25N	38
Cambridge Dr. N&S	20W-25N	39
Canterbury Ln. E&W	20W-24N	39
Cardinal Dr. N&S	21W-26N	38
Carlisle Rd. NS&EW	20W-24N,20W-25N	39
Cayuga Ct. E&W	19W-25N	39
Chancellor Ct. NW&SE	22W-24N	38
Chantilly Ln. N&S	23W-24N,22W-24N	38
Cherry Hill Cir. N&S	19W-24N	39
Circle Dr. N&S	22W-25N	38
Cobblewood Ln. NW&SE	23W-25N	38
Condor Ct. E&W	21W-24N	38
Conesug Ct. E&W	19W-25N	39
Cooper Ct. N&S	22W-25N	38
Copperfield Dr. NS&EW	23W-24N,22W-24N	38
Croydon Rd. N&S	20W-25N	39
Darlington Rd. N&S	20W-24N	39
David Ln. N&S	22W-25N	38
Deer Point Dr. NS&EW	22W-24N,22W-25N	38
Derby Ln. N&S	23W-25N	38
Dorset Ct. NW&SE	23W-25N	38
E. Lynn Dr. N&S	21W-24N	38
Echo Ct. NW&SE	23W-24N	38
Edward Ln. E&W	20W-24N	39
Elm St. N&S	21W-24N,20W-24N	38
Elmwood Dr. NW&SE	23W-24N	38
Equestrian Way N&S	23W-25N	38
Essex Ct. E&W	20W-24N	39
Fairfield Rd. N&S	22W-25N	38
Falcon Ct. E&W	22W-26N,21W-26N	38
Falkirk Rd. E&S	20W-24N	39
Fox Hunt Ct. NW&SE	24W-25N	38
Fox Hunt Tr. NE&SW	24W-25N	38
Furlong Dr. E&W	23W-25N	38
Gentry Dr. N&S	23W-24N,23W-25N	38
Gilmer Rd. E&W	19W-24N,21W-26N	39
Gino Ln. NW&SE	19W-24N	39
Glen Rd. E&W	23W-25N	38
Goldfinch Ct. E&W	22W-26N	38
Govenors Way E&W	23W-25N	38
Greenbriar Ln. NS&EW	19W-24N	39
Hampton Ln. E&W	20W-24N	39
Harvard Ln. N&S	20W-25N	39
Harvest Glen Ct. N&S	23W-24N	38
Hawthorne Dr. E&W	21W-24N	38
Heather Ln. E&W	21W-24N	38
Heron Ct. E&W	21W-24N	38
Hickory Rd. E&W	21W-24N	38
Highland Ter. N&S	21W-24N	38
Highpoint Cir. NW&SE	22W-25N	38
Highpoint Dr. NE&SW	22W-25N	38
Highview Ct. E&W	20W-23N	47
Hillcrest Dr. N&S	20W-23N	47
Hilltop Ct. NE&SW	19W-23N	47
Holmes Av. N&S	22W-25N	38
Indian Creek Rd. E&W	20W-25N,19W-25N	39
James Ct. NW&SE	20W-24N	39

	Grid	Page
James Dr. E&W	20W-24N	39
Jamie Ln. E&W	20W-24N	39
Jane Ct. N&S	20W-24N	39
John Dr. NS&EW	20W-24N	39
Juel Cir. NE&SW	21W-24N,20W-24N	38
Juel Dr. E&W	20W-24N	39
Karen Ct. NE&SW	19W-24N	39
Kathy Ln. E&W	22W-25N,21W-25N	38
Keuka Ct. E&W	19W-25N	39
Krueger Rd. E&W	22W-25N	38
Lagoon Dr. NS&EW	21W-24N,20W-24N	38
Lakeview Rd. NS&EW	20W-23N,19W-23N	47
Lancelot Ln. NW&SE	21W-24N	38
Lisa Ln. NW&SE	20W-24N	39
Lydia Ct. E&W	23W-25N	38
Lynn Dr. N&S	21W-24N	38
Magnolia Pkwy. E&W	19W-24N	39
Mark Dr. N&S	20W-24N	39
Mary Ln. E&W	19W-24N	39
McGregor Ct. E&W	21W-24N	38
McHenry Rd. NW&SE	24W-25N,20W-23N	38
Meadow Ln. N&S	22W-24N	38
Meadowlark Dr. NS&EW	22W-26N	38
Melody Ln. (Cherry Rd.) N&S	21W-23N,21W-24N	47
Midlothian Rd. NE&SW	22W-24N,21W-25N	38
Moorgate Trail NW&SE	24W-25N	38
Mulberry Dr. NE&SW	22W-24N	38
N. Bruce Cir. E&W	23W-24N	38
N. Highview Circle E&W	20W-24N	39
Newbury Ct. E&W	20W-24N	39
Newhaven Dr. E&W	23W-25N	38
Norbert Dr. E&W	20W-24N	39
North Trail NS&EW	23W-25N,23W-26N	38
Oak Ln. NW&SE	21W-24N	38
Old Barn Rd. NS&EW	20W-23N,19W-23N	47
Old Lake Rd. NS&EW	22W-24N	38
Oneida Ln. NS&EW	20W-25N	39
Onondaga Dr. E&W	20W-25N,19W-25N	39
Orchard Ln. N&S	21W-24N	38
Oriole Ct. NW&SE	22W-26N	38
Overlook Dr. SE&NW	22W-24N	38
Paddock Ct. N&S	23W-25N	38
Peregrine Ln. NS&EW	21W-26N	38
Pheasant Run Rd. NS&EW	24W-25N	38
Piper Ln. N&S	21W-24N	38
Princeton Ct. N&S	21W-25N	38
Princeton Ln. NW&SE	20W-25N	39
Quail Hollow Ct. N&S	20W-23N	47
Quail Hollow Rd. N&S	20W-23N	47
Redwood Ln. NE&SW	23W-25N	38
Robert Ct. E&W	20W-24N	39
Robin Crest Rd. N&S	23W-25N	38
Rosewood Dr. NS&EW	21W-24N	38
Rutgers Ct. E&W	20W-25N	39
S. Bruce Cir. E&W	23W-24N	38
S. Higgins Circle E&W	20W-24N	39
Sandpiper Ln. E&W	22W-26N	38
Schrewman Rd. E&W	22W-26N,21W-26N	38
Seneca East Av. N&S	19W-24N	39
Seneca West Av. N&S	20W-24N	39
Sequoia Rd. NS&EW	20W-25N	39
Shearwater Ct. E&W	22W-26N	38
Shenandoah Ln. NS&EW	20W-25N	39
Squire Rd. NE&SW	21W-24N	38
St. John Dr. N&S	19W-24N	39
Steeplchase Dr. NE&SW	23W-25N	38
Stone Creek Ct. N&S	23W-24N,22W-24N	38
Stonehaven Dr. E&W	20W-24N	39
Sunset Ct. NW&SE	20W-23N	47
Sycamore Dr. NW&SE	23W-25N	38

	Grid	Page
Tanglewood Ct. N&S	24W-24N	38
Tanglewood Ln. E&W	24W-24N	38
Thornfield Ln. NW&SE	24W-25N	38
Trent Rd. N&S	20W-24N	39
University Cir. NE&SW	20W-25N	39
Victoria Ln. N&S	21W-24N	38
W. Lynn Dr. N&S	21W-24N	38
Wakla Ct. E&W	20W-25N	39
Walnut Dr. N&S	21W-24N	38
Washitay Av. NE&SW	20W-25N,19W-25N	39
Wayne Ln. N&S	23W-24N	38
Wedgewood Dr. NS&EW	23W-24N,22W-24N	38
Westwind Ct. NW&SE	23W-25N	38
Whitman Ter. E&W	22W-25N	38
Wild Way E&W	20W-24N	39
Winding Branch Rd. E&W	24W-25N	38
Wooded Ln. N&S	24W-25N,24W-26N	38
Yale Ct. E&W	20W-25N	39

Highland Park

	Grid	Page
'A' St. NW&SE	8W-23N	51
Acorn Ln. N&S	9W-20N	51
Alta Ln. E&W	10W-22N	50
Alvin Pl. NE&SW	7W-21N	51
Andean Pl. E&W	9W-22N	51
Apple Tree Ln. E&W	9W-23N	51
Applewood Ct. E&W	10W-23N	50
Arbor Av. N&S	9W-21N	51
Arlington Av. N&S	9W-23N	51
Ashland Pl. N&S	7W-20N	51
Aspen Ln. N&S	8W-20N	51
Aster Dr. E&W	10W-22N	50
Athletic Field Ln. N&S	9W-22N	51
Atlas Ct. N&S	10W-21N	50
Auburn Av. E&W	10W-23N	50
Auburn Ct. E&W	9W-23N	51
Audubon Pl. E&W	9W-24N	43
Ava St. NE&SW	7W-21N	51
Avignon Ct.* E&W	9W-21N	51
'B' St. NW&SE	9W-24N,8W-23N	43
Baldwin Rd. NW&SE	7W-21N	51
Balsam Rd. E&W	9W-20N	51
Barberry Rd. NS&EW	9W-20N,9W-21N	51
Barnard Ln. N&S	7W-20N	51
Beaupre Ct.* E&W	9W-21N	51
Beech Ln. NW&SE	7W-21N	51
Beech St. NE&SW	7W-21N	51
Belle Av. NE&SW	8W-22N	51
Bellevue Pl. NE&SW	7W-21N	51
Bentley Rd. N&S	10W-23N	50
Berkeley Rd. E&W	10W-22N	50
Bertram Ln. E&W	9W-20N	51
Beverly Pl. N&S	9W-21N,9W-22N	51
Birch Av. E&W	9W-21N	51
Birchwood Av. E&W	9W-22N	51
Birchwood Ct. E&W	10W-24N	43
Blackhawk Rd. NW&SE	7W-20N	51
Blackstone Pl. E&W	7W-20N	51
Bloom St. NE&SW	8W-23N	51
Blossom Ct. NE&SW	9W-20N	51
Bob-O-Link Rd. E&W	9W-21N,8W-21N	51
Boxer Ct. E&W	10W-22N	50
Braeburn Ln. E&W	7W-20N	51
Braeside Rd. E&W	7W-20N	51
Briar Ln. N&S	9W-20N	51
Briarcliff Ln. E&W	10W-23N	50
Briargate Dr. NW&SE	10W-21N	50
Briarwood Pl. E&W	7W-20N	51
Brittany Rd. E&W	8W-21N	51
Broadview Av. NW&SE	8W-21N,7W-20N	51
Bronson Ln. NE&SW	7W-20N	51
Brook Rd. N&S	9W-24N	43
Brown Ct. E&W	10W-23N	50
Brownville Rd. E&W	7W-20N	51
Burchell Av. E&W	9W-23N	51
Burr Oaks Ln. E&W	9W-20N	51
Burton Av. NW&SE	7W-20N,7W-21N	51
Byron Ln. NE&SW	10W-22N	50
Calais Dr. N&S	9W-20N	51
Cambridge Ct. E&W	10W-23N	50
Cardinal Ln. N&S	10W-22N	50
Cardinal Ln. N&S	10W-23N	50
Carlyle Ter. E&W	8W-21N	51
Carol Ct. E&W	7W-20N	51
Cary Av. E&W	7W-21N	51
Castlewood Rd. E&W	10W-21N	50
Cavell Av. N&S	9W-21N	51
Cedar Av. E&W	7W-21N	51
Centenial Ct. N&S	10W-23N	50
Centenial Ln. NW&SE	10W-23N	50
Centennial Pkwy NW&SE	10W-23N	50
Centerfield Ct. E&W	9W-22N	51
Central Av. NE&SW	8W-22N	51
Central Ct. NE&SW	8W-22N	51
Chantilly Blvd. E&W	9W-20N	51
Charal Ln. NW&SE	7W-20N	51
Chatham Cir. N&S	7W-20N	51
Chaucer Ln. N&S	8W-21N	51
Cherokee Rd. E&W	7W-20N	51
Cherry Ln. E&W	9W-22N	51
Chicago Av. E&W	8W-23N	51
Churchill Ct. N&S	11W-22N	50
Churchill Ln. E&W	10W-22N	50
Clavey Ct. N&S	8W-20N	51
Clavey Ln. NW&SE	8W-20N	51
Clavey Rd. E&W	9W-20N,8W-20N	51
Clifton Av. N&S	9W-22N	51
Cloverdale Av. N&S	10W-22N	50
Cofton Av. N&S	8W-21N	51
College Pl. E&W	9W-24N	43
Compton Av. N&S	9W-23N	51
Comstock Pl. E&W	7W-20N	51
Concorde Cir.* E&W	9W-21N	51
Concorde Dr. NW&SE	9W-20N	51
Conventry Ln. E&W	9W-21N	51
Cottonwood Ct. NW&SE	10W-23N	50
Country Ln. NE&SW	9W-20N	51
County Line Rd. E&W	8W-20N,6W-20N	51
Court Av. E&W	8W-21N	51
Crawford Pl. NE&SW	8W-21N,7W-21N	51
Crescent Ct. NE&SW	7W-22N	51
Crofton Av. N&S	8W-21N	51
Crofton Ct. N&S	8W-21N	51
Daffodil Ct. E&W	10W-23N	50
Dale Av. NW&SE	8W-22N	51
Dato Av. N&S	9W-23N	51
Dean Av. NE&SW	7W-20N,7W-21N	51
Deer Creek Pkwy. N&S	8W-20N	51
Deere Park Ct. N&S	6W-20N	51
Deerfield Ct. E&W	9W-21N	51
Deerfield Pl. N&S	9W-21N	51
Deerfield Rd. NS&EW	9W-21N,8W-21N	51
Dell Ln. N&S	7W-20N	51
Dell Pl. N&S	7W-20N	51
Delta Rd. E&W	7W-20N	51
Derrick St. N&S	9W-23N	51
Detamble Av. NE&SW	8W-21N	51
Devonshire Ct. NW&SE	9W-21N	51
Division St. E&W	9W-21N	51
Driscoll Ct. E&W	8W-21N	51
Dryden Ln. E&W	8W-21N	51
Eastwood Av. N&S	10W-21N	50
Eaton Ct. NE&SW	10W-23N	50
Edens Expwy. NW&SE	8W-20N	51
Edgecliff Dr. E&W	8W-23N	51
Edgewood Ct. N&S	9W-20N,9W-21N	51
Edgewood Rd. E&W	9W-20N,8W-21N	51
Egandale Rd. NW&SE	8W-22N	51
Elder Ln. NE&SW	6W-20N	51
Ellridge Cir. N&S	9W-20N	51
Elm Av. N&S	8W-23N	51
Elm Pl. E&W	8W-22N	51
Elmwood Dr. N&S	9W-22N	51
Euclid Av. E&W	9W-23N	51
Euclid Ct E&W	9W-23N	51
Exmoor Av. N&S	9W-22N	51
'F' St. NW&SE	8W-23N	51
Fairview Rd. E&W	8W-21N	51
Ferndale Av. N&S	9W-21N	51
First St. NW&SE	8W-22N	51
Flora Pl. NW&SE	7W-20N	51
Florence Pl. N&S	8W-21N	51
Forest Av. N&S	7W-21N	51
Fredrickson Pl. NE&SW	9W-21N	51
Freesia Cir. N&S	10W-22N	50
Friar Tuck Av. E&W	10W-22N	50
Ft. Sheridan Av. N&S	8W-23N	51
'G' St. N&S	8W-23N	51
Garland Av. E&W	10W-21N	50
Glencoe Av. NW&SE	8W-21N	51
Glenview Av. E&W	8W-22N	51
Golf Av. E&W	9W-21N,8W-21N	51
Grange Av. N&S	10W-22N	50
Gray Av. E&W	8W-21N	51
Great Elm Ln. E&W	8W-21N	51
Green Bay Rd. NW&SE	8W-23N,7W-20N	51
Greenwood Av. N&S	9W-23N	51
Grove Av. E&W	9W-21N	51
Groveland Av. N&S	7W-20N	51
'H' St. N&S	8W-23N	51
Half Day Rd. E&W	11W-23N,9W-23N	50
Halsted Ln. E&W	8W-21N	51
Harvard Ct. E&W	9W-22N	51
Hastings Av. N&S	7W-20N	51
Havenwood Ct. NE&SW	7W-20N	51
Hawthorne St. N&S	7W-21N	51
Hazel Av. E&W	8W-22N	51
Heather Ln. E&W	9W-21N	51
Hedge Run E&W	7W-20N	51
Hemlock Ln. N&S	9W-20N	51
Hiawatha Ct. E&W	8W-20N	51
Hiawatha Tr. N&S	8W-20N	51
Hickory St. NW&SE	8W-22N	51
Hidden Ridge Cir. N&S	10W-21N	50
Hidden Ridge Ln. E&W	10W-21N	50
Highland Pl. NS&EW	7W-20N	51
Highmoor Rd. N&S	10W-22N	50
Hill St. E&W	9W-24N	43
Hillcrest Av. NW&SE	10W-21N	50
Hillside Dr. NW&SE	7W-20N	51
Holly Rd. E&W	10W-22N	50
Homewood Av. E&W	8W-22N	51
Huntington Ln. E&W	9W-21N	51
Hyacinth Pl. E&W	9W-24N	43
Hybernia Dr. NS&EW	10W-22N,10W-23N	50
Idlewild Ln. N&S	9W-23N	51
Indian Ln. NW&SE	7W-20N	51
Iris Ln. E&W	7W-20N	51
Ivy Ln. E&W	6W-20N	51
Jasmine Ct. NW&SE	10W-22N	50
Johnston Loop NS&EW	8W-23N	51
Judson Av. NW&SE	7W-21N	51
Keats Ct. NW&SE	10W-22N	50
Keats Ln. E&W	10W-22N	50
Kelly Ln. N&S	11W-23N	50
Kent Av. E&W	9W-23N	51
Kimball Rd. E&W	8W-21N	51
Kimballwood Ln. E&W	8W-21N	51
Kincaid Av. N&S	7W-20N	51
Kipling Ct. E&W	10W-22N	50
Kipling Ln. N&S	10W-22N	50
Knollwood Ln. NE&SW	7W-21N	51
Krenn Av. NE&SW	9W-24N	43
Lake Av. NW&SE	7W-22N	51
Lake Cook Rd. E&W	9W-20N	51
Lakeside Manor Pl. E&W	6W-20N	51
Lakeside Pl. N&S	7W-20N,6W-20N	51
Lakeview Av. N&S	8W-23N	51
Lakeview Ter. E&W	6W-20N	51
Lakewood Pl. E&W	7W-20N	51
Lambert Tree Av. E&W	7W-20N	51
Lancelot Av. E&W	9W-22N	51
Larkspur Dr. N&S	8W-20N	51
Laurel Av. NE&SW	8W-22N	51
Lauretta Av. N&S	8W-23N	51
Lawrence Ln. E&W	9W-20N	51
Leslee Ln. NE&SW	7W-20N	51
Lewis Ln. E&W	9W-20N	51
Lexington Ln. NW&SE	10W-23N	50
Lilac Ln. NW&SE	9W-21N	51
Lily Ct. NW&SE	10W-22N	50
Lincoln Av. E&W	8W-21N,7W-21N	51
Lincoln Pl. NW&SE	8W-21N	51
Lincolnwood Rd. N&S	7W-20N	51
Linden Av. N&S	8W-22N,7W-21N	51
Linden Park Pl. NE&SW	8W-22N	51
Little John Ct. E&W	10W-22N	50
Livingston St. E&W	10W-23N,9W-23N	50
Llewellyn Av. E&W	9W-23N	51
Logan St. NW&SE	8W-23N	51
Lorraine Cir.* E&W	8W-21N	51
Lotus Pl. E&W	9W-24N	43
Lyman Ct. N&S	7W-20N	51
Lynn Ter. NS&EW	10W-23N	50
Lyons Cir.* E&W	8W-21N	51
Magnolia Ln. NW&SE	9W-20N	51
Malory Ln. E&W	11W-22N	50
Manor Rd. N&S	10W-22N	50
Maple Av. NE&SW	8W-22N	51
Maple Ln. NW&SE	8W-22N	51
Marion Av. E&W	8W-20N,7W-20N	51
Marl Oak Dr. N&S	9W-23N	51
Marquette Rd. N&S	8W-20N	51
Marshman St. NE&SW	7W-21N	51
Marvell Ln. E&W	8W-21N	51
Mayor Ln. N&S	10W-23N	50
McCarn Rd. NW&SE	9W-22N	51
McDaniels Av. N&S	8W-21N	51
McGovern St. NW&SE	8W-22N	51
McKibben Rd NE&SW	8W-23N	51
Meadow Ln. E&W	10W-23N	50
Melba Ln. NW&SE	8W-23N	51
Melody Ln. NE&SW	7W-20N	51
Melvin Dr. E&W	8W-20N	51
Midland Av. E&W	10W-21N	50
Midlothian Av. N&S	8W-22N	51
Mill Tr. E&W	10W-23N	50
Montgomery Av. N&S	8W-23N	51
Moraine Rd. E&W	8W-23N	51
Moseley Rd. E&W	8W-21N	51
Mulberry Pl. E&W	8W-21N	51
Museum Dr. E&W	10W-23N	50
N. Deer Park Dr., W. NW&SE	6W-20N	51
N. Deer Park Dr., E. NW&SE	6W-20N	51
Naida Ter. E&W	8W-22N	51
North Av. E&W	10W-23N,9W-23N	50
Northland Av. E&W	10W-21N	50
Ny-O-Da Pl. E&W	8W-20N	51
Oak Knoll Ter. NS&EW	6W-20N	51
Oak St. N&S	8W-23N	51
Oakland Dr. NE&SW	7W-20N	51
Oakmont Rd. E&W	7W-21N	51
Oakvale Rd. E&W	7W-21N	51
Oakwood Av. N&S	8W-21N	51
Old Barn Ln. N&S	8W-21N	51
Old Briar Rd. E&W	9W-20N	51
Old Deerfield Rd. NE&SW	9W-21N	51
Old Mill Rd.(pvt) N&S	10W-24N	43
Old Skokie Rd. NW&SE	9W-21N,9W-22N	51
Old Trail E&W	9W-23N	51
Onwentsia Av. E&W	8W-22N	51
Orchard Ln. E&W	8W-23N	51
Orchid Ct. N&S	10W-22N	50
Orleans Cir.* N&S	9W-21N	51
Orleans Dr. N&S	9W-20N	51
Oxford Ct. E&W	10W-23N	50
Park Av. NE&SW	8W-22N	51
Park Ln. NW&SE	8W-22N	51
Parkside Dr. NW&SE	10W-23N	50
Partridge Ln. E&W	10W-22N	50
Patten St. NW&SE	8W-23N	51
Perennial Ln. N&S	9W-22N	51
Picadilly Rd. NW&SE	9W-21N	51
Pierce Rd. NE&SW	7W-20N	51
Pine Point Ct. E&W	6W-20N	51
Pine Point Dr. E&W	6W-20N	51
Pleasant Av. NW&SE	7W-20N	51
Point Ln. N&S	10W-23N	50
Poplar Rd. N&S	9W-20N	51
Port Clinton Rd. N&S	8W-23N	51
Pricilla Av. N&S	9W-23N	51
Princeton Av. E&W	9W-22N	51
Prospect Av. E&W	8W-22N	51
Rambler Ln. E&W	7W-20N	51
Ravenoaks Ln. NE&SW	7W-21N	51
Ravenswood Av. N&S	9W-23N	51
Ravine Dr. E&W	8W-21N	51
Ravine Ln. NW&SE	8W-22N	51
Ravine Ter. NE&SW	8W-22N	51
Ravinia Park Rd. N&S	7W-20N	51
Ravinia Rd. E&W	8W-21N,7W-21N	51
Red Oak Ln. N&S	9W-20N	51
Rice St. N&S	7W-20N	51
Richfield Av. E&W	10W-21N	50
Ridge Rd. NW&SE	11W-23N,9W-20N	50
Ridgelee Rd. E&W	9W-20N	51
Ridgewood Dr. NW&SE	8W-21N	51
Ridgewood Pl. NW&SE	8W-21N	51
Riparian Rd. E&W	8W-23N	51
Robin Hood Pl. E&W	9W-22N	51
Roger Williams Av. E&W	7W-20N	51
Rolling Wood Rd. NS&EW	8W-21N	51
Rosemary Rd. E&W	9W-20N	51
Rosewood Rd. E&W	7W-21N	51
Roslyn Cir. N&S	8W-23N	51
Roslyn Ln. N&S	8W-23N	51
Rt. 22 E&W	10W-23N	50
Rt. 41 NW&SE	10W-23N,8W-20N	50
Russet Ln. N&S	9W-20N	51
Ryders Ln. N&S	10W-22N	50
Sandwick Ct. E&W	9W-22N	51
Sarabell Ln. (pvt) N&S	7W-20N	51
Saxony Dr. NW&SE	8W-21N	51
Saxony Rd. NE&SW	8W-21N	51
Second St. NW&SE	8W-22N	51
Seven Pines Cir. NW&SE	8W-20N	51
Shady Ln. N&S	10W-22N	50
Sheahen Ct. N&S	8W-22N	51
Sheldon Ln. N&S	7W-20N	51
Shelly Ct. E&W	10W-23N	50
Sheridan Rd. N&S	8W-22W,6W-20N	51
Sherwood Rd. N&S	10W-21N	50
Skokie Av. E&W	8W-22N	51
Skokie Hwy. NW&SE	10W-23N,8W-20N	50
Skokie Valley Rd. NE&SW	8W-20N	51
Snapdragon Ct. NW&SE	10W-22N	50
South Deer Park Dr. Cir.	6W-20N	51
Southland Av. E&W	10W-21N	50
Spruce Av. N&S	10W-22N	50
St. Charles Pl. N&S	7W-20N	51
St. Johns Av. N&S	8W-23N,7W-20N	51
St. Tropez Ct.* E&W	9W-21N	51
Stonegate Dr. E&W	7W-21N	51
Stratford Rd. E&W	9W-20N	51
Strath Erin E&W	9W-20N	51
Sumac Rd. NW&SE	9W-20N,9W-21N	51
Summerdale Av. N&S	10W-21N	50
Summit Av. N&S	9W-23N	51
Sundrop Cir. N&S	10W-22N	50
Sunnyside Av. N&S	10W-22N	50
Sunset Rd. N&S	9W-21N,8W-22N	51
Sycamore Ln. N&S	7W-22N	51
Sycamore Pl. E&W	7W-22N	51
Sylvester Pl. NW&SE	7W-22N	51
Tanglewood Ct. E&W	9W-20N	51
Taylor Av. E&W	9W-22N	51
Tennyson Ln. N&S	11W-22N	50
Thackery Dr. E&W	8W-21N	51
Thorntree Ln. E&W	8W-21N	51
Thornwood Ln. E&W	10W-23N	50
Timber Hill Rd. E&W	7W-20N	51
Toulon Ct.* N&S	8W-21N	51
Trail Way N&S	10W-23N	50
Trailway N&S	8W-20N	51
Tree Dr. E&W	7W-20N	51
Turnbull Woods Ct. N&S	7W-20N	51
Twin Oaks Dr. N&S	10W-23N	50
U.S. Route 41 NW&SE	10W-24N,8W-20N	43
Underwood Av. E&W	9W-23N	51
University Av. N&S	9W-23N	51
Valley Rd. N&S	7W-20N	51
Vine Av. E&W	8W-22N	51
Virginia Rd. NW&SE	9W-21N	51
Wade St. NW&SE	7W-21N	51
Wahwahtaysee Av. N&S	8W-20N	51
Wake Robin Ln. N&S	7W-20N	51
Walker Av. E&W	8W-23N	51
Walnut St. NE&SW	8W-22N	51
Warbler Cir. E&W	9W-23N	51
Warbler Ct. NE&SW	9W-23N	51
Warbler Pl. N&S	9W-23N	51
Washington Ct. NE&SW	7W-20N	51
Washington Pl. NE&SW	7W-20N	51
Watercrest Way NS&EW	10W-22N	50
Waterford Ct. E&W	10W-22N	50
Waverly Rd. NE&SW	7W-21N	51
West Park Av. E&W	10W-22N,9W-22N	50
Wester Pl. NE&SW	9W-24N	43
Western Av. N&S	9W-22N,9W-24N	51
Westgate Ter. E&W	10W-21N	50

LAKE COUNTY 63

	Grid	Page
White Oak Ln. (Pvt.) E&W	8W-21N	51
Wildrose Ct. NE&SW	10W-22N	50
Wildwood Ln. N&S	7W-21N	51
Willow Pl. E&W	10W-23N	50
Willow Rd. N&S	10W-22N	50
Willow Rd. N&S	10W-23N	50
Wincanton Ln. N&S	7W-22N	51
Windsor Rd. N&S	9W-21N	51
Windy Hill Ln. E&W	9W-20N	51
Winona Rd. N&S	8W-20N	51
Winthrop Rd. E&W	9W-21N	51
Wood Path NE&SW	8W-22N	51
Woodbine Av. N&S	9W-20N	51
Woodbine Rd. N&S	9W-21N	51
Woodbridge Ln. NE&SW	8W-23N	51
Woodland Rd. NE&SW	7W-20N	51
Woodleigh Av. E&W	9W-23N	51
Yager Av. E&W	9W-21N	51
Yale Av. E&W	9W-22N	51
York Ln. E&W	10W-21N	50
2nd St. NE&SW	9W-24N	43
3rd St. NE&SW	8W-23N	51
4th St. NE&SW	8W-23N	51
5th St. NE&SW	8W-23N	51
6th St. NE&SW	8W-23N	51
7th St. NE&SW	8W-23N	51
8th St. NE&SW	8W-23N	51
9th St. NE&SW	8W-23N	51
10th St. E&W	8W-23N	51
11th St. E&W	8W-23N	51

Highwood

	Grid	Page
Ashland Av. E&W	9W-23N	51
Bank Ln. N&S	9W-23N	51
Burchell Av. E&W	9W-23N	51
Burtis Av. E&W	9W-23N	51
Central Av. N&S	9W-23N	51
Clay Av. E&W	9W-23N	51
Euclid Av. E&W	9W-23N	51
Euclid Ct. N&S	9W-23N	51
Everts Pl. N&S	9W-23N	51
Evolution Av. N&S	9W-23N	51
Exmoor Ct. N&S	9W-23N	51
Funston Av. N&S	9W-23N	51
Green Bay Rd. NW&SE	9W-23N	51
Grove St. N&S	9W-23N	51
H St. N&S	9W-23N	51
High St. N&S	9W-23N	51
Highwood Av. E&W	9W-23N	51
Jeffrey Pl. N&S	9W-23N	51
Jocelyn Pl. E&W	9W-23N	51
Llewellyn Av. E&W	9W-23N	51
Lockard Av. N&S	9W-23N	51
Maple St. N&S	9W-23N	51
Mears Pl. E&W	9W-23N	51
Michigan Av. E&W	9W-23N	51
Morgan Pl. N&S	9W-23N	51
North Av. N&S	9W-23N	51
Oak Av. N&S	9W-23N	51
Oak Ter. N&S	9W-23N	51
Oakridge Av. E&W	9W-23N	51
Palmer Av. N&S	9W-23N	51
Pleasant Pl. N&S	9W-23N	51
Prairie Av. E&W	9W-23N	51
Sheridan Av. E&W	9W-23N	51
Sheridan Rd. NW&SE	9W-23N	51
Summit Av. E&W	9W-23N	51
Washington Av. E&W	9W-23N	51
Webster Av. E&W	9W-23N	51
Western Av. N&S	9W-23N,9W-24N	51
Wrendale Av. E&W	9W-23N	51
3rd St. NE&SW	9W-23N	51
9th St. NE&SW	9W-23N	51

Indian Creek

	Grid	Page
Circle Dr. NE&SW	17W-24N	40
Crestland Rd. E&W	17W-24N	40
Mills Ct. E&W	17W-25N	40
Shann Dr. N&S	17W-25N	40
U.S. Route 45 NW&SE	17W-24N,17W-25N	40
Valley Rd. E&W	17W-24N	40

Island Lake*

(Also see McHenry County)

	Grid	Page
Arbor Rd. E&W	28W-28N	28
Briar Ct. E&W	28W-28N	28
Briar Hill Dr. NW&SE	28W-28N	28
Briar Rd. E&W	28W-28N	28
Canterbury Ln.* E&W	27W-27N	36
Cardinal E&W	28W-29N	28
Carol Cir. NE&SW	28W-28N	28
Carriage Hill Ct. E&W	28W-28N	28
Carriage Hill Rd. NS&EW	28W-28N	28
Channel Dr. E&W	28W-28N	28
Clover Rd. E&W	28W-28N	28
Constitution Av.* E&W	27W-27N	36
Cottonwood Ct. NW&SE	28W-28N	28
Country Trail Ct. E&W	28W-27N	36
Countryside Ct. N&S	28W-28N	28
Eastway Dr. N&S	28W-28N	28
Elder Dr. N&S	28W-28N	28
Elm Av. N&S	28W-28N	28
Fairfield Dr. E&W	28W-28N	28
Fern Dr. E&W	28W-28N	28
Forest Dr. E&W	28W-28N	28
Greenleaf Av. N&S	28W-28N	28
Harvard Ct.* E&W	27W-27N	36
Harvest Ct. N&S	28W-28N	28
Independence Blvd.* E&W	27W-27N	36
Ivy Rd. E&W	28W-28N	28
Judith Dr. NE&SW	28W-28N	28
Juniper Rd. E&W	28W-28N	28
Kingston Dr. N&S	28W-29N	28
Lakeside Ct. NE&SW	28W-28N	28
Landsfield Ct. NE&SW	28W-28N	28
Locust Av. N&S	28W-28N	28
Longacre Ln. E&W	28W-28N	28
Madison Ct.* E&W	27W-27N	36
Mallard Pt. N&S	28W-29N	28
Mylith Park St. N&S	28W-28N	28
Northern Ct. NS&EW	28W-28N	28
Northern Ter. E&W	28W-28N	28
Oaks Ln. E&W	28W-28N	28
Oakwood Av. N&S	28W-28N	28
Park Dr. N&S	28W-28N	28
Poplar Dr. N&S	28W-28N	28
Princeton Cir.* E&W	27W-27N	36
Ridge Av. N&S	28W-28N	28
River Oaks Ct. E&W	28W-28N	28
Rose Av. E&W	28W-28N	28
Saratoga Cir.* E&W	27W-27N	36
Slocum Lake Rd. NW&SE	28W-28N	28
South Shore Dr. NS&EW	28W-28N	28
Southern Ter. NW&SE	28W-28N	28
Sumac Dr. N&S	28W-28N	28
Timber Ln. E&W	28W-27N	36
Westridge Dr. N&S	28W-27N,28W-28N	36
Willow Ln. N&S	28W-28N	28
Wishing Well Ln. NW&SE	28W-28N	28
Woodbine Dr. E&W	28W-28N	28
Woodcreek Ct. E&W	28W-28N	28
Woodcreek Dr. E&W	28W-28N	28
Woodland Ct. N. N&S	28W-28N	28
Woodland Ct. S. N&S	28W-28N	28
Woodland Dr. N&S	28W-28N	28

Kildeer

	Grid	Page
Abbey Ct. NW&SE	22W-21N	46
Acorn Ct. N&S	21W-21N	47
Amberley Dr. NW&SE	21W-20N	47
Andover Rd. NW&SE	21W-21N,20W-21N	47
Barbara Ct. E&W	21W-21N	47
Barcley Ct. NW&SE	22W-21N	46
Boschome Dr. E&W	21W-21N	47
Brandon Rd. E&W	21W-21N	47
Bridal Tr. NS&EW	21W-22N	47
Buffalo Run N&S	21W-20N,21W-21N	47
Burning Tree Ct. N&S	22W-21N	46
Cambridge Dr. NS&EW	22W-21N	46
Chartwell Dr. E&W	20W-20N	47
Chestnut Ridge Rd. E&W	22W-21N	46
Circle Bay Rd. E&W	22W-20N	46
Clayton Ct. N&S	20W-22N	47
Cliffside Dr. E&W	21W-21N	47
Concorde Ct. E&W	22W-20N	46
Cuba Rd. E&W	22W-21N,20W-21N	46
Dallas Ct. N&S	20W-21N	47
Dorothea Ct. NW&SE	20W-21N	47
Elder Ct. NW&SE	22W-21N	46
Eleanor Ln. E&W	21W-22N	47
Exeter Rd. E&W	20W-21N	47
Foxtail Dr. N&S	21W-22N	47
Green Wood Dr. E&W	21W-22N	47
Grove Rd. N&S	21W-21N	47
Hampton Rd. NW&SE	21W-21N	47
Hawthorn Ln. N&S	22W-21N	46
Heather Ct. N&S	21W-20N	47
Hickory Hill Dr. E&W	21W-21N	47
Hidden Valley Dr. E&W	21W-21N	47
Highwood Rd. NS&EW	22W-21N	46
Hilandale Ct. E&W	21W-22N	47
Hilandale Dr. N&S	21W-22N	47
Honey Ridge Ct. E&W	22W-21N	46
Hopewell Ct. NE&SW	21W-22N	47
Kepwick Ln. NE&SW	21W-20N	47
Kirkley Dr. NE&SW	21W-20N	47
Krueger Rd. N&S	21W-22N	47
Laurel Ln. E&W	21W-22N	47
Lexington Ln. E&W	20W-21N	47
Linden Ln. E&W	21W-21N	47
Little Pond Rd. NE&SW	22W-21N	46
Long Grove Rd. NE&SW	21W-20N,20W-20N	47
Long Meadows Ct. NS&EW	20W-20N	47
Long Meadows Dr. N&S	20W-20N	47
Maple Ct. NW&SE	22W-21N	46
Marcy Ln. NW&SE	20W-21N	47
Middleton Dr. N&S	20W-20N,20W-21N	47
N. Boschome Cir. E&W	21W-21N	47
Newbury Ct. N&S	20W-22N	47
Oak Knoll Ct. E&W	21W-20N	47
Oak Trail E&W	21W-21N	47
Pine Lake Cir. NS&EW	21W-21N	47
Pine Lake Ct. N&S	21W-21N	47
Pinegrove Ct. N&S	22W-21N	46
Plumwood Dr. N&S	20W-20N	47
Rebecca Ct. N&S	21W-22N	47
Richmond Ct. NE&SW	20W-22N	47
Ruth Ct. N&S	21W-22N	47
S. Boschome Dr. E&W	21W-21N	47
Stoneybrook Ct. N&S	22W-21N	46
Thornridge Dr. E&W	22W-21N	46
Timber Ridge Ct. N&S	22W-21N	46
Timberlea Ct. NE&SW	22W-21N	46
Tree Rd. N&S	22W-21N	46
Valley View Rd. NS&EW	21W-21N	47
Vermont Ct. N&S	20W-21N	47
Weatherstone Rd. NW&SE	21W-20N	47
White Pine Rd. NW&SE	22W-21N	46
Willow Dr. E&W	21W-21N	47
Wooded Ridge Dr. E&W	22W-20N	46
York Ct. E&W	21W-20N	47
Yorkshire Dr. E&W	21W-22N,20W-22N	47

Lake Barrington

	Grid	Page
Alice Ln. N&S	27W-23N	44
Apache Ln. NW&SE	25W-24N	37
Apache Path NE&SW	26W-24N	37
Bark Ct. N&S	28W-24N	36
Beacon Dr. N&S	28W-23N	44
Brookside Ct. E&W	27W-22N	44
Brookside Way NS&EW	27W-22N,26W-22N	44
Buoy Ct. NS&EW	28W-24N	36
Cayuga Tr. E&W	27W-25N	37
Cedar Ridge NE&SW	26W-24N	37
Center Dr. E&W	27W-25N	37
Cherokee Dr. N&S	25W-24N	37
Chesapeake Dr. NS&EW	26W-23N	45
Chippewa Ct. E&W	26W-24N	37
Classic Ct. NE&SW	28W-22N	44
Club Cir. E&W	26W-24N	37
Commercial Av. E&W	28W-21N,27W-21N	44
Countryside Ct. NE&SW	27W-25N	37
Countryside Dr. NS&EW	27W-25N	37
Cove Ct. N&S	28W-23N	44
Coventry Ct. E&W	27W-24N	37
Coventry Ln. E&W	27W-24N	37
Crestview Ln. NE&SW	27W-23N	44
Cutter Ln. E&W	28W-24N	36
Darrell Rd. NE&SW	28W-26N,28W-27N	36
Deer Trail Ct. NE&SW	26W-24N	37
Deer Trail Hill* N&S	25W-24N	37
Dock Dr. N&S	28W-23N	44
Driftwood Ct. NE&SW	27W-22N	44
East Ln. NW&SE	27W-23N	44
Farm View Cir. NW&SE	27W-25N	37
Flint Dr. NW&SE	27W-23N	44
Flynn Creek Dr. N&S	27W-21N	44
Foxwood Ln. NW&SE	26W-24N	37
Golfview Ln. NS&EW	27W-23N	44
Grayshire Ln. N&S	27W-23N	44
Grey Barn Ln. NE&SW	28W-23N	44
Harbor Rd. NS&EW	28W-23N,28W-22N	44
Henry Ln. NS&EW	27W-22N	44
Henry Rd. E&W	27W-22N	44
Hickory Ln. N&S	26W-24N	37
Hickory Ridge NW&SE	27W-25N	37
Hillside Dr. E&W	28W-22N	44
Hilltop Ct.* NW&SE	25W-24N	37
Hillview Dr. N&S	28W-22N	44
Hunt Trail NE&SW	26W-24N	37
Indian Point NW&SE	26W-24N	37
Industrial Av. E&W	28W-22N,27W-22N	44
Iroquois Ct. NW&SE	27W-25N	37
Island View Ln. E&W	26W-24N	37
Kazimour Dr. E&W	27W-25N	37
Kelsey Rd. NE&SW	28W-23N,27W-24N	44
Kensington Ct. E&W	27W-24N	37
Lakeridge Dr. E&W	26W-23N	45
Lakeview Dr. NW&SE	27W-23N	44
Linden Dr. NE&SW	28W-22N	44
Longview Pt. NE&SW	26W-23N	45
Main Entrance E&W	26W-24N	37
Main St. E&W	28W-26N	36
Mallard Pt.* E&W	25W-24N	37
Market Pl. E&W	26W-25N	37
Meadow Pl. E&W	26W-24N	37
Meadow View Ct. N&S	27W-25N	37
Miller Rd. E&W	27W-24N,26W-24N	37
North Av. NE&SW	27W-23N	44
North Bay Ct. NE&SW	26W-24N,26W-25N	37
Northwest Highway NW&SE	28W-22N,27W-21N	44
Oak Ct. NW&SE	28W-22N	44
Oak Hill Rd. NS&EW	26W-24N	37
Oak Hills Rd. NS&EW	27W-25N	37
Old Barn Rd. NE&SW	26W-24N	37
Old Barrington Rd. NW&SE	27W-23N	44
Oneida Ln. N&S	27W-25N	37
Pawnee Rd. NE&SW	25W-24N,25W-25N	37
Pepper Rd. N&S	28W-21N,28W-22N	44
Pinecrest Cir. NS&EW	26W-24N	37
River Rd. N&S	27W-24N,27W-25N	37
Riverbend Ct. NS&EW	27W-25N	37
Roberts Rd. E&W	27W-25N	37
Rolling Wood E&W	26W-24N	37
Schooner Ln. E&W	28W-23N	44
Seneca Tr. NS&EW	27W-25N	37
Shoreline Rd. N&S	26W-24N	37
Thornhill Ct. E&W	26W-24N	37
Thornhill Ln. E&W	26W-24N	37
Timber Ridge N&S	26W-24N	37
Tioga Trail N&S	26W-24N	37
Tuscarora Ct. N&S	27W-25N	37
Twin Pond Rd. NE&SW	27W-25N,27W-16N	37
Valley View Rd. N&S	26W-24N	37
Vance Ct. E&W	28W-23N	44
Vista Ln. E&W	27W-23N	44
W. Brookside Way NS&EW	27W-22N	44
Wedgewood Ln. N&S	27W-24N	37
Welch Cir. NS&EW	28W-23N	44
Wellington Ct. NW&SE	27W-24N	37
West Ln. N&S	27W-23N	44
White Pine Dr. N&S	26W-24N	37
Woodland Dr. NS&EW	27W-23N	44
Woodview Rd.* N&S	25W-24N	37

Lake Bluff

	Grid	Page
Arbor Ct. N&S	10W-29N	35
Arbor Dr. E&W	10W-29N	35
Arden Shore Rd. E&W	10W-29N	35
Arden Shore Rd. S. E&W	10W-29N	35
Armour Dr. Cir.	11W-29N	35
Ascot Ct. NE&SW	11W-29N	35
Ashington Cir.* E&W	10W-29N	35
Bath And Tennis Club Rd. E&W	11W-28N	35
Belle Foret Cir. Cir.	11W-29N	35
Birch Rd. N&S	10W-28N	35
Birkdale Rd. E&W	11W-28N	35
Blodgett Av. E&W	11W-29N,10W-29N	35
Bluff Rd. N&S	10W-28N,10W-29N	35
Bradford Ct.* E&W	10W-29N	35
Briar Ln. N&S	10W-28N	35
Brierfield Ct. E&W	11W-29N	35
Bristol Ct. E&W	11W-29N	35
Buckminster Ct. E&W	11W-29N	35
Cambridge Rd. E&W	10W-28N	35
Carlyle Cir. N&S	11W-29N	35
Carriage Park Av. E&W	12W-28N	35
Center Av. E&W	11W-28N,10W-28N	35
Circle Dr. N&S	10W-28N	35
Coventry Ct. E&W	11W-29N	35
Crescent Dr. E&W	10W-28N	35
Eastway E&W	10W-30N	35
Eva Ter. N&S	11W-28N	35
Evanston Av. N&S	10W-28N	35
Forest Cove Rd. E&W	10W-28N	35
Forest Hills Rd. N&S	11W-28N	35
Foss Ct. N&S	10W-28N	35
Garfield Av. N&S	11W-28N	35
Glen Av. N&S	10W-28N	35
Grafton Ct. E&W	11W-29N	35
Green Bay Rd. N&S	11W-28N,11W-29N	35
Greenwich Ct. E&W	11W-29N	35
Gurney Av. N&S	10W-28N	35
Hamilton Ct. E&W	11W-29N	35
Hancock Av. NW&SE	11W-28N	35
Hawthorne Ct. E&W	11W-28N,10W-28N	35
Heathrow Ct. E&W	11W-29N	35
Hickory Ct. N&S	11W-28N	35
Hirst Ct. N&S	10W-28N	35
Indian Rd. E&W	10W-28N	35
Inverness Ct. E&W	11W-29N	35
James N&S	11W-29N	35
Kohl Dr. E&W	11W-29N	35
Lake Av. N&S	10W-28N	35
Lakeland Dr. E&W	10W-28N	35
Lancaster Ct. N&S	11W-29N	35
Leeds Ct. NW&SE	11W-29N	35
Lincoln Av. N&S	11W-28N	35
Main St. E&W	11W-28N	35
Maple Av. N&S	10W-28N	35
Margate Ct. N&S	11W-29N	35
Market Sq. East E&W	11W-28N	35
Market Sq. West E&W	11W-28N	35
Marvin Rd. E&W	11W-28N	35
Mawman Av. NE&SW	11W-28N	35
McClaren Ln. N&S	11W-28N	35
Moffet Rd. N&S	10W-28N	35
Mountain Av. N&S	10W-28N	35
N. Shore Dr. E&W	12W-28N	35
Neuman Ct. N&S	11W-28N	35
North Av. E&W	10W-28N	35
Norwich Ct. N&S	11W-29N	35
Oak Av. N&S	10W-28N	35
Oak Ter. N&S	10W-28N	35
Oakton Cir. N&S	11W-29N	35
Park Av. E&W	10W-28N	35
Park Ln. E&W	11W-28N	35
Park Pl. E&W	10W-28N	35
Phillip Ct. E&W	11W-29N	35
Pine Av. N&S	11W-28N	35
Pine Ct. N&S	11W-28N	35
Prospect Av. E&W	11W-29N,10W-29N	35
Ravine Av. E&W	10W-28N	35

	Grid	Page
Ravine Forest Dr. E&W	10W-28N	35
Rebecca Ter. N&S	11W-28N	35
Rockland Av. N&S	11W-28N	35
Rockland Rd. E&W	12W-28N	35
Roselle Ter. N&S	11W-28N	35
Rothbury Ct. E&W	11W-29N	35
Rt. 43 N&S	11W-28N	35
Rt. 131 N&S	11W-28N	35
Rt. 176 E&W	11W-28N	35
Scranton Av. N&S	10W-28N	35
Sheridan Pl. E&W	11W-28N,10W-28N	35
Sheridan Rd. NS&EW	11W-28N,10W-29N	35
Sherwood Dr. E&W	12W-28N	35
Sherwood Ter. N&S	12W-28N	35
Shore Acres Cir. E&W	10W-30N	35
Shore Acres Rd. E&W	10W-29N	35
Signe Ct. E&W	11W-29N	35
Simpson Av. N&S	10W-28N	35
Skokie Hwy. N&S	12W-29N,11W-28N	35
State St. N&S	11W-28N	35
Sunrise Av. N&S	10W-28N	35
Sunset Pl. E&W	10W-28N	35
Sunset Ter. N&S	11W-28N	35
Surrey Ln. NS&EW	11W-28N	35
Sylvan Rd. E&W	10W-28N	35
Thorn Valley Ln. E&W	11W-29N	35
Thornwood Ln. NE&SW	11W-28N	35
U.S. Route 41 N&S	12W-28N,11W-28N	35
Vincent Ct. N&S	10W-28N	35
Walnut Av. N&S	10W-28N	35
Warrington Ct. E&W	11W-29N	35
Warrington Dr. N&S	11W-29N	35
Washington Av. E&W	11W-28N	35
Washington St. E&W	10W-28N	35
Waukegan Rd. NW&SE	12W-28N	35
Weatherford Ct. E&W	11W-29N	35
West Scranton Av. E&W	12W-28N,11W-28N	35
Wimbledon Ct. N&S	11W-28N	35
Wimbleton Rd. N&S	11W-28N	35
Winchester Ct. E&W	11W-29N	35
Witchwood Ln. N&S	11W-28N,10W-28N	35
Woodland Rd. E&W	10W-28N	35

Lake Forest

	Grid	Page
Abington Cambs Dr. E&W	12W-25N	42
Ahwahnee Ln. N&S	11W-26N	43
Ahwanee Rd. N&S	11W-26N	43
Alden Ln. E&W	11W-26N	43
Alexis Ct. N&S	12W-23N	50
Alison Ln. E&W	12W-27N	42
Arbor Ln. E&W	12W-23N	50
Arcady Dr. E&W	12W-26N	42
Armour Cir. E&W	12W-26N	42
Ashlawn Dr. E&W	12W-26N	42
Ashley Rd. E&W	9W-25N	43
Aspen Dr. NW&SE	12W-23N	50
Atteridge Rd. E&W	10W-27N	43
Bank Ln. N&S	10W-26N	43
Barat Ct. N&S	10W-26N	43
Barberry Ln. E&W	10W-27N,10W-26N	43
Barn Swallow Rd. N&S	11W-25N	43
Basswood Rd. E&W	11W-25N	43
Bear Ct. NW&SE	12W-25N	42
Beverly Pl. N&S	10W-25N	43
Big Oaks Ct. E&W	12W-23N	50
Birch Ct. E&W	11W-25N	43
Blackthorn Ln. NE&SW	11W-25N	43
Bluffs Edge Dr. NW&SE	9W-25N	43
Bowling Green Dr. E&W	13W-24N	42
Breckenridge Av. N&S	11W-23N	50
Breckenridge Ct. N&S	12W-24N	42
Bridgeview E&W	12W-24N	42
Bridle Ln. E&W	11W-25N	43
Brir Ln. E&W	10W-25N	43
Broadsmore Dr. NS&EW	11W-25N	43
Buena Rd. N&S	11W-25N,10W-24N	43

	Grid	Page
Burr Oak Rd. N&S	10W-27N	43
Burton Dr. N&S	12W-26N	42
Butler Dr. N&S	10W-25N	43
Cascade Ct. N. E&W	12W-24N	42
Cascade Ct. S. E&W	12W-24N	42
Castlegate Ct. E&W	12W-26N	42
Cherokee Rd. E&W	10W-25N	43
Cherry Av. E&W	10W-26N	43
Chiltern Dr. N&S	11W-26N	43
Chistina Ln. NE&SW	11W-24N	43
Church Rd. NW&SE	10W-27N	43
Circle Ln. N&S	9W-25N	43
Clover Av. N&S	11W-26N	43
College Rd. E&W	11W-26N	43
Conway Farm Rd. NE&SW	13W-25N	42
Conway Rd. NW&SE	13W-26N,12W-26N	42
Country Pl. N&S	11W-24N	43
Coventry Dr. E&W	12W-25N	42
Crab Tree Ln. E&W	10W-27N	43
Crest Ct. E&W	10W-24N	43
Deerpath E&W	12W-26N,10W-26N	42
Deerpath Sq. N&S	11W-26N	43
Devonshire Ct. NW&SE	12W-23N	50
Devonshire Ln. NE&SW	13W-23N	49
Douglas Dr. N&S	9W-25N	43
Dover Rd. E&W	11W-23N	50
Edgecote Ln. NW&SE	11W-26N	43
Edgewood Rd. N&S	10W-27N	43
Efner Av. E&W	10W-27N	43
Elder Path NE&SW	10W-27N	43
Elm Tree Rd. N&S	10W-27N	43
Estate Ln. N&S	13W-24N	42
Estate Ln.E. E&W	12W-24N	42
Estes Av. N&S	11W-24N	43
Everett Rd. E&W	13W-24N,11W-24N	42
Evergreen Dr. N&S	12W-24N	42
Exter Pl. N&S	12W-25N	42
Fairview Av. N&S	11W-24N	43
Fairway Dr. E&W	12W-25N	42
Falls Cir. SW&NE	12W-25N	42
Farm Rd. E&W	13W-24N	42
Farnham Ln. E&W	11W-23N	50
Field Ct. NS&EW	13W-24N	42
Field Dr. N&S	13W-24N	42
Fiore Dr. N&S	12W-24N	42
Fletcher Ct. N&S	12W-24N	42
Forest Av. N&S	10W-26N	43
Forest Hill Rd. N&S	10W-24N,10W-25N	43
Foster Pl. N&S	10W-25N	43
Fox Trail Ct. N&S	12W-24N	42
Franklin Pl. E&W	10W-27N	43
Franz Ln. N&S	12W-24N	42
Frost Pl. NE&SW	12W-27N	42
Gage Ln. E&W	12W-27N	42
Gavin Ct. NS&EW	12W-24N	42
Glenwood Rd. E&W	10W-25N	43
Glouchester Crossing NE&SW	12W-24N	42
Goldenrod Ln. E&W	12W-23N	50
Golf Ln. N&S	11W-26N	43
Granby Rd. E&W	10W-27N	43
Grandview Ln. N&S	10W-24N	43
Green Bay Rd. N&S	10W-24N,11W-27N	43
Green Briar Ln. E&W	10W-26N	43
Greenleaf Rd. N&S	10W-27N	43
Greenvale Rd. N&S	12W-26N	42
Greenview Pl. E&W	10W-25N	43
Greenwood Av. E&W	10W-24N	43
Griffith Rd. N&S	10W-27N	43
Grove Ct. NW&SE	12W-24N	42
Hackberry Ln. E&W	13W-24N	42
Halligan Cir. N&S	12W-26N	43
Harlan Ln. N&S	11W-24N	43
Harlan Ln. N&S	11W-24N	43
Hastings Ct. N&S	11W-26N	43
Hathaway Cir. N&S	12W-26N	42
Havenwood Ln. E&W	9W-25N	43
Hawkweed Ln. E&W	12W-27N	42
Hawthorne Ln. E&W	13W-24N	42
Hawthorne Pl. N&S	10W-26N,10W-27N	43
Heather Ct. E&W	11W-25N	43
Heritage Ct. N&S	12W-24N	42
Heron Rd. E&W	11W-25N	43

	Grid	Page
Hickory Ct. E&W	10W-25N	43
High Holborn N&S	12W-26N	42
Highland Av. E&W	11W-24N	43
Highview Ter. E&W	10W-25N	43
Hilldale Pl. E&W	10W-25N	43
Holland Ct. N&S	11W-26N	43
Honeysuckle Rd. NE&SW	11W-25N	43
Hospital Dr. NS&EW	12W-27N,11W-27N	42
Hunter Ln. NW&SE	11W-25N	43
Illinois Rd. N&S	10W-26N	43
Inverlieth Rd. E&W	12W-26N	42
Inverlieth Ter. E&W	12W-26N	42
Ivy Ct. E&W	10W-25N	43
Jacquelyn Ln. (Pvt.) E&W	11W-25N	43
Jennifer Ct. E&W	12W-24N	42
June Ter. N&S	12W-24N	42
Kajer Ln. E&W	12W-24N	42
Kathryn Ln. N&S	11W-24N	43
Keith Ln. NE&SW	9W-25N	43
Kendler Ct. NE&SW	9W-25N	43
Kennedy Rd. E&W	12W-26N	42
Kennett Ln. E&W	12W-24N	42
Kennington Ter. E&W	11W-26N	43
Kimberly Ln. E&W	11W-24N	43
Kimmer Ct. N&S	12W-26N	42
King Muir Rd. N&S	12W-26N	42
Knight's Bridge Ct. N&S	12W-24N	42
Knollwood Cir. NE&SW	12W-28N	35
Knollwood Ln. N&S	12W-28N	35
Knollwood Rd. NS&EW	12W-28N	35
Kurtis Ln. NW&SE	12W-24N	42
Lake Rd. NW&SE	10W-27N,9W-27N	43
Lakewood Dr. NS&EW	12W-26N	42
Lane Lorraine E&W	12W-26N	42
Larchmont Ln. E&W	12W-26N	42
Larkspur Ct. E&W	12W-23N	50
Laurel Av. E&W	11W-27N	43
Lawrence Av. E&W	12W-24N	42
Leilly Ln. E&W	12W-25N	42
Leland Ct. NS&EW	12W-25N	42
Lexington Dr. N&S	12W-26N	42
Linden Av. E&W	10W-24N	43
Littlefield Ct. N&S	13W-24N	42
Loch Ln. E&W	9W-25N	43
Longwood Dr. E&W	9W-26N	43
Louis Av. E&W	11W-24N	43
Lowell Ln. N&S	12W-23N	50
Lynette NE&SW	12W-24N	42
Mallard Ln. N&S	11W-25N	43
Maple Ct. N&S	9W-26N	43
Maplewood Rd. E&W	9W-26N	43
Mar-Lane Dr. N&S	12W-24N	42
Marion Av. E&W	10W-24N	43
Mayflower Rd. N&S	9W-25N,9W-26N	43
Maywood Rd. N&S	10W-25N	43
McCormick Dr. NS&EW	9W-25N	43
McKinley Rd. N&S	10W-25N,10W-27N	43
Meadow Lake Ln. NW&SE	13W-25N	42
Meadow Ln. N&S	10W-25N	43
Meadow Ln. E&W	13W-25N	42
Meadowood Dr. N&S	12W-26N	42
Mellody Rd. E&W	12W-26N	42
Memory Ln. N&S	12W-23N	50
Michgamme Ln. E&W	11W-25N	43
Millburne Rd. N&S	11W-23N	50
Mills Ct. N&S	10W-27N	43
Minthaven Rd. N&S	13W-24N	42
Monticello Cir. N&S	12W-26N	42
Morningside Dr. E&W	10W-24N	43
Mt. Vernon Av. N&S	11W-24N	43
Negaunee Ln. N&S	11W-25N	43
Newcastle Ln. NE&SW	12W-25N	42
Niles Av. E&W	10W-24N	43
Noble Av. N&S	10W-24N	43
North Av. E&W	11W-24N	43
North Cliff Way E&W	12W-24N	42
North Croft Ct. E&W	11W-25N	43
North Gate Sq. E&W	10W-25N	43
Northmoor Rd. E&W	10W-25N	43

	Grid	Page
O'Leary Ln. (Pvt.) E&W	12W-28N	35
Oak Knoll Dr. N&S	12W-23N,12W-27N	50
Oakdale Av. E&W	10W-24N	43
Oakwood Av. N&S	10W-26N,10W-25N	43
Old Barn Ln. N&S	13W-24N	42
Old Colony Rd. N&S	11W-24N	43
Old Elm Rd. E&W	11W-24N,10W-24N	43
Old Mill Rd. E&W	12W-26N	42
Onwentsia Rd. E&W	11W-25N,10W-25N	43
Orchard Hill Rd. NS&EW	13W-25N	42
Overlook Dr. NW&SE	9W-25N	43
Oxford Rd, (pvt.) E&W	10W-27N	43
Park Av. E&W	10W-27N	43
Parkmead Ln. E&W	12W-26N	42
Parliament Ct. E&W	13W-23N	49
Pembroke Dr. E&W	10W-26N	43
Pine Ln. E&W	10W-27N	43
Pinecroft Ln. NE&SW	12W-25N	42
Polo Dr. N&S	12W-24N	43
Poplar Rd. E&W	11W-25N	43
Prairie Av. N&S	11W-25N	43
Prairie View Ln. E&W	13W-25N	42
Private Rd. N&S	12W-25N	42
Private Rd. E&W	12W-27N	42
Pvt. Rd E&W	11W-26N	43
Pvt. Rd. E&W	11W-26N	43
Pvt. Rd. N&S	12W-25N	42
Quail Dr. E&W	11W-25N	43
Ranch Rd. N&S	13W-24N	42
Ravine Park Dr. E&W	10W-27N	43
Red Fox Ln. N&S	10W-25N	43
Ridge Ln. NE&SW	10W-26N	43
Ridge Rd. N&S	11W-23N,11W-26N	50
Ridgefield Ln. NW&SE	13W-25N	42
Ringwood Rd. NS&EW	9W-25N	43
Riparian Dr. NE&SW	9W-25N	43
Rockfeller Rd. N&S	9W-25N	43
Rose Ter. E&W	10W-27N	43
Rosemary Rd. E&W	10W-26N,9W-26N	43
Rt. 43 N&S	12W-24N,12W-26N	42
Rt. 60 E&W	13W-26N	42
Rue Foret NW&SE	12W-26N	42
Ryan Pl. NE&SW	10W-26N	43
S. Gate Sq. E&W	10W-26N	43
Saddle Run E&W	11W-25N	43
Sand Piper Ln. E&W	11W-25N,10W-25N	43
Saunders Rd. N&S	13W-25N	42
Scott Av. E&W	10W-27N	43
Sequoia Ct. N&S	11W-26N	43
Shawford Way E&W	11W-23N	50
Shawnee Ln. N&S	11W-25N	43
Sheffield Ct. NW&SE	13W-25N	42
Sheridan Rd. N&S	10W-27N,9W-24N	43
Sir William Ln. N&S	12W-24N	42
Skokie Hwy. N&S	11W-25N,11W-27N	43
South Shore Ln. NW&SE	13W-25N	42
Spring Ln. E&W	9W-26N	43
Spruce Av. E&W	10W-27N	43
Stable Ln. N&S	11W-24N	43
Stone Av. E&W	11W-24N	43
Stone Gate Ln. NE&SW	9W-26N	43
Stone Gate Rd. N&S	9W-26N	43
Stoneleigh Ct. E&W	13W-23N	49
Stratford Ct. E&W	13W-23N	49
Suffolk Ct. N&S	12W-26N,11W-26N	42
Suffolk Ln. E&W	12W-25N,12W-26N	50
Summerfield Dr. NE&SW	12W-26N,12W-27N	42
Summit Av. N&S	10W-25N,10W-27N	43
Summit Pl. NW&SE	10W-27N	43
Sunset Pl. E&W	10W-26N	43
Surrey Ln. N&S	11W-23N	50
Sussex Ln. NE&SW	12W-26N	42
Tanglewood Ct. N&S	12W-25N	42
Tara Ln. N&S	11W-27N	43

	Grid	Page
Telegraph Rd. N&S	12W-23N	50
Thomas Pl. E&W	10W-27N	43
Thorne Ln. NE&SW	10W-26N	43
Timber Ln. N&S	10W-24N,10W-25N	43
Tiverton Rd. N&S	11W-26N	43
Trillium Ln. E&W	13W-24N	42
Turicum Rd. NS&EW	9W-25N	43
U.S. Route 41 N&S	11W-24N,11W-27N	43
Valley Rd. N&S	10W-24N,10W-25N	43
Verda Ln. E&W	12W-26N	42
Vine Av. E&W	10W-26N	43
Walden Ln. E&W	9W-25N	43
Walnut Rd. NE&SW	10W-26N	43
Warwick Rd. E&W	12W-26N	42
Washington Cir. NW&SE	10W-26N	43
Washington Rd. NW&SE	10W-26N	43
Waukegan Rd. N&S	12W-25N	42
Waveland Rd. N&S	10W-24N	43
Wedgewood Ct. NE&SW	13W-23N	49
Wedgewood Dr. NW&SE	13W-23N	49
West Fork Dr. N&S	13W-24N	42
Western Av. N&S	10W-25N,10W-27N	43
Westleigh Rd. E&W	12W-25N,9W-25N	42
Westminster Av. E&W	10W-27N	43
Westmoreland Rd. N&S	11W-26N	43
White Oak Rd. N&S	12W-24N	42
Wild Rose Ln. N&S	12W-24N	42
Wildwood Rd. N&S	10W-26N	43
Willow St. N&S	10W-27N	43
Wilshire Rd. E&W	10W-24N	43
Wilson Dr. N&S	13W-24N	42
Wimbledon Ct. E&W	13W-23N	49
Windhaven E&W	12W-24N	42
Windridge Dr. NS&EW	12W-23N	50
Winston Rd. N&S	10W-25N	43
Winwood Dr. E&W	12W-27N	42
Wisconsin Av. NE&SW	9W-27N	43
Woodbine Ln. N&S	9W-27N	43
Woodbine Pl. N&S	9W-27N	43
Wooded Ln. N&S	10W-25N	43
Woodhill Ln. N&S	12W-24N	42
Woodland Rd. E&W	10W-27N	43
Woodlawn Av. E&W	10W-26N	43
Woodstream Ct. N&S	11W-24N	43
Wynnfield Ter. E&W	10W-25N	43

Lake Villa

	Grid	Page
Amber Ln. E&W	24W-37N,23W-27N	14
Apache Trail NS&EW	24W-37N,23W-37N	14
Arapaho Trail E&W	22W-37N	14
Balsam Ct. E&W	22W-37N	14
Balsam Ln. E&W	22W-37N	14
Belmont Av. E&W	22W-37N	14
Bradford Ct. NW&SE	22W-38N	14
Burnett Av. E&W	22W-37N,22W-37N	14
Cedar Av. NE&SW	22W-38N	14
Cedar Lake Rd. N&S	23W-36N,23W-37N	14
Cedar Ln. N&S	23W-38N	14
Cedar Village Rd. NW&SE	22W-38N	14
Central Av. NW&SE	22W-38N	14
Chelsea Cir. N&S	22W-37N	14
Cherokee Ct. N&S	24W-37N	14
Circuit Ct. E&W	23W-36N	14
Clayton Av. NW&SE	21W-38N	14
Cleveland Av. (Pvt.) N&S	23W-38N	14
Coventry Cove Ct. E&W	23W-37N	14
Coventry Cove Ln. E&W	22W-37N	14
Cremin Dr. N&S	22W-37N	14
Deep Lake Rd. E&W	22W-39N,21W-38N	14
Delwood Av. N&S	23W-37N	14
Fairfield Rd. N&S	24W-37N	14
Grace Ln. E&W	23W-37N	14

	Grid	Page
Grand Av. E&W	22W-38N	14
Grass Lake Rd. NW&SE		
....	22W-39N	14
Kenz Ct. E&W......	22W-37N	14
Kevin Av. E&W	22W-37N	14
Lake Av. NW&SE ..	22W-38N	14
Lake Shore Dr. E&W		
....	23W-38N	14
Laurie Ct. E&W	23W-37N	14
Majestic Ln. N&S ..	24W-37N	14
McKinley Av. N&S ..	22W-37N	14
Milwaukee Av. E&W		
........22W-38N,23W-38N		14
Monaville Rd. E&W		
....	23W-36N	14
Monica Dr. N&S	23W-37N	14
N. Cedar Lake Ave. N&S		
....	23W-38N	14
N. Shore Av. E&W ..	23W-38N	14
Nez Perc Ln. N&S ..	24W-37N	14
Oak Knoll Dr. N&S..	23W-37N	14
Oak Ln. N&S	23W-37N	14
Oakwood Av. E&W		
....	23W-37N	14
Park Av. N&S	23W-37N	14
Petite Lake Rd. E&W		
....	23W-39N	14
Roselle Ct E&W	22W-37N	14
Rosemont Ln. E&W		
....	23W-38N	14
Rt. 83 NW&SE		
........22W-37N,23W-38N		14
Rt. 132 E&W		
........23W-38N,22W-38N		14
Sheehan Dr. N&S ..	22W-37N	14
Sherwood Av. N&S		
....	22W-37N	14
Shoshoni Trail E&W		
....	22W-38N	14
Trenker Dr. E&W ..	23W-39N	14
Twin Oaks Dr. E&W		
....	23W-38N	14
Villa Av. E&W	22W-38N	14
Walden Ln. N&S	23W-37N	14
Waterbury Cir. N&S		
....	22W-38N	14
Waters Edge Dr. NE&SW		
....	22W-38N	14
Wesley Av. E&W		
........23W-37N,22W-37N		14
Woodhead Dr. N&S		
....	21W-38N	14

Lake Villa Township

	Grid	Page
Academy Ct. NW&SE		
....	25W-38N	13
Academy Dr. NE&SW		
........25W-37N,25W-38N		13
Alberta Av. N&S	24W-36N	13
Alice Ln. N&S	21W-36N	15
Alpine Ct. E&W	24W-37N	13
Alpine Ln. NE&SW ..	24W-37N	13
Anderle Av. N&S ..	21W-38N	15
Antonio Av. N&S ..	21W-37N	15
Apollo Ct. E&W	23W-39N	14
Arcade Dr. E&W	24W-37N	13
Arcade Dr., N. E&W		
....	25W-37N	13
Arcade Dr., S. E&W		
....	25W-37N	13
Ardmore Ln. N&S ..	23W-38N	14
Ash St. N&S	21W-38N	15
Avon Dr. N&S	25W-37N	13
Aztec St. E&W	24W-39N	13
Balboa Dr. N&S	24W-39N	13
Bald Eagle Rd. E&W		
....	25W-37N	13
Bayshore Rd. N&S	25W-37N	13
Beach Av. E&W	24W-39N	13
Beck Rd. N&S		
........20W-38N,20W-39N		15
Belmoral Park N&S		
....	23W-38N	14
Bernice Av. E&W ..	22W-36N	14
Big Oak Dr. E&W ..	24W-37N	13
Birch St. NE&SW ..	21W-38N	15
Blackcherry Ln. E&W		
....	24W-36N	13
Blackstone Pl. E&W		
....	25W-37N	13
Brentwood Ln. E&W		
....	21W-39N	15
Buena Av. N&S	25W-38N	13
Busse Ln. E&W	25W-37N	13
Caine Dr. E&W	24W-36N	13
Capillo Av. N&S	20W-37N	15
Carol Ln. N&S......	21W-36N	15
Carson Dr. E&W ..	25W-37N	13
Cedar Av. E&W	21W-38N	15
Cedar Av. N&S	23W-38N	14
Cedar Crest Ln. E&W		
....	25W-38N	13
Cedar Lake Ln. E&W		
....	24W-38N	13
Cedar Lake Rd. N&S		
....	24W-38N	13
Cedar Valley Dr. N&S		
....	24W-38N	13
Cedarwood Ln. N&S		
....	24W-36N	13
Central Av. NE&SW		
....	23W-38N	14
Chesney Dr. E&W ..	25W-37N	13

	Grid	Page
Christiansen Av. E&W		
....	21W-38N	15
Columbia Bay Dr. N&S		
........25W-37N,25W-38N		13
Columbus Av. N&S		
....	20W-37N	15
Corona Dr. E&W		
....	20W-36N	15
Crabapple Dr. N&S		
....	24W-39N	13
Crabtree Ln. E&W .	24W-36N	13
Cremona Av. N&S ..	20W-37N	15
Danny Ln. E&W	21W-36N	15
Deep Lake Rd. N&S		
........22W-39N,21W-38N		14
Delaby Ln. N&S	25W-37N	13
Dering Ln. N&S		
........25W-38N,24W-38N		13
Diana Av. E&W	24W-38N	13
Douglas Ln. N&S ..	21W-37N	15
Eagle Ct. NW&SE ..	24W-36N	13
Eastmore Av. N&S .	24W-36N	13
Edgewood Av. N&S		
....	21W-38N	15
Edgewood Av. E&W		
....	21W-38N	15
Edward Av. N&S ..	21W-36N	15
Elizabeth Ct. N&S ..	21W-36N	15
Elizabeth Dr. N&S ..	21W-36N	15
Elm Av. NE&SW	24W-39N	13
Elm St. E&W	21W-38N	15
Engle Dr. E&W	21W-36N	15
Entrance Dr. NW&SE		
....	25W-38N	13
Fairfield Rd. N&S		
........24W-36N,24W-38N		13
Fairview Cir. E&W ..	25W-37N	13
Fairview Ln. NE&SW		
....	25W-37N	13
Fairway Dr. NS&EW		
....	24W-39N	13
Forest Av. N&S	20W-36N	15
Fox Ter. E&W	24W-36N	13
Garcia Dr. E&W ..	20W-37N	15
Gelden Rd. E&W ..	21W-39N	15
Genoa Av. E&W ..	20W-36N	15
George Ct. N&S ..	22W-37N	14
Gilliam Rd. NE&SW		
....	24W-38N	13
Glade Dr. E&W ..	25W-37N	13
Glen Lake Pl. N&S..	23W-38N	14
Golf Ln. N&S	24W-39N	13
Granada Blvd. NE&SW		
....	20W-36N,20W-37N	15
Grand Av. NS&EW		
........25W-37N,19W-36N		13
Grand Blvd. N&S ..	25W-37N	13
Grass Lake Rd. NW&SE		
........22W-39N,19W-38N		14
Graton Rd. N&S....	22W-38N	14
Haidi Ln. E&W 24W-37N		13
Hampshire Ln. N&S		
....	25W-37N	13
Harbor Ridge Dr. NS&EW		
....	23W-38N	14
Harding Av. N&S ..	25W-37N	13
Harold Ln. E&W ..	22W-36N	14
Hazelwood Dr. N&S		
....	24W-36N	13
Helen Ct. N&S	21W-36N	15
Helen Dr. N&S	21W-36N	15
Hickory St. N&S ..	21W-38N	15
Highwood Av. E&W		
....	24W-39N	13
Hilda Av. N&S 21W-36N		15
Hillside Av. N&S 25W-37N		13
Honey Ln. E&W ..	21W-36N	15
Ironwood Dr. N&S ..	19W-37N	15
Isola Av. N&S 20W-37N		15
James Ct. N&S ..	22W-36N	14
James St. N&S ..	22W-36N	14
Knoll Ct. E&W	25W-37N	13
Lake Av. NE&SW ..	21W-38N	15
Lake Av. N&S	23W-38N	14
Cremona Av. N&S ..	24W-39N	13
Lake Ct. E&W 20W-36N		15
Lake Shore Dr. E&W		
....	23W-38N	14
Lake Shore Dr. N&S		
....	25W-37N	13
Lakeview Av. NW&SE		
....	20W-36N	15
Lakeview Av. E&W ..	20W-36N	15
Lakeview Ct. E&W .25W-37N		13
Lakeview Ct. E&W .24W-39N		13
Lakeview Rd. NE&SW		
....	25W-38N	13
Lakewood Av. N&S		
....	23W-38N	14
Laurel St. N&S 21W-38N		15
Lawrence Ct. N&S .21W-36N		15
Lawrence Dr. N&S ..21W-36N		15
Lehmann Blvd. E&W		
....	25W-37N	13
Liberty Av. E&W 23W-38N		14
Lincoln Dr. N&S ..	25W-37N	13
Linda Ln. E&W	21W-36N	15
Loretto Av. N&S ..	20W-37N	15
Maple Av. N&S ..	21W-38N	15
Marilyn Ln. N&S ..	21W-36N	15
Marshfield Rd. E&W		
....	25W-37N	13
Mary Ct. N&S 21W-36N		15
Mary Dr. N&S ..	21W-36N	15

	Grid	Page
Maurine Dr. E&W ..	21W-36N	15
Milburn Rd. NW&SE		
....	19W-39N	15
Miller Rd. NW&SE ..	21W-38N	15
Monaville Rd. E&W		
........24W-37N,22W-36N		13
Morey Ln. E&W ..	24W-36N	13
Morton N&S 22W-36N		14
Morton Dr. E&W		
........22W-36N,21W-37N		14
Munn Rd. N&S	21W-38N	15
N. Shore Dr. E&W ..	23W-38N	14
Nathan Hale Dr. N&S		
....	21W-36N	15
Newberry Ln. NS&EW		
....	25W-37N	13
Nielson Dr. E&W ..	21W-36N	15
Normandy Av. N&S		
....	21W-36N	15
North Av. NE&SW ..	23W-38N	14
North Columbia Bay Dr. E&W		
....	25W-38N	13
North Dr. E&W ..	24W-39N	13
North Rd. E&W ..	24W-39N	13
North St. N&S 21W-38N		15
North Way E&W		
....	23W-38N	14
Oak Av. NW&SE		
........21W-38N,21W-39N		15
Oak Av. E&W 24W-39N		13
Oak Creek Ln. N&S		
....	21W-36N	15
Oak Dr. N&S	24W-39N	13
Oakwood Dr. N&S ..21W-36N		15
Obermaier Ln. E&W		
....	25W-37N	13
Old Monaville Rd. E&W		
....	24W-36N	13
Paradise Ln. E&W ..	24W-36N	13
Park Av. E&W 24W-39N		13
Park Av. N&S 24W-39N		13
Park Ct. E&W 25W-37N		13
Park Dr. E&W		
........22W-36N,21W-36N		14
Park Ln., E. N&S ..	25W-37N	13
Park Ln., W. N&S ..	25W-37N	13
Park Pl. N&S	25W-38N	13
Park St. NW&SE ..	25W-38N	13
Parma Av. N&S ..	20W-37N	15
Peninsula Ln. N&S		
........25W-37N,25W-38N		13
Petite Ct. E&W	24W-39N	13
Petite Lake Rd. E&W		
........24W-39N,23W-39N		13
Petite Ter. N&S ..	24W-39N	13
Pike St. NE&SW ..	21W-39N	15
Pine St. NE&SW ..	21W-38N	15
Piper Ln. N&S......25W-37N		13
Ponce Av. NW&SE 20W-37N		15
Poplar St. N&S		
........21W-38N,21W-39N		15
Prairieview Dr. N&S		
....	20W-36N	15
Raska Ln. N&S ..	24W-36N	13
Ravine Crest Dr. NW&SE		
....	24W-39N	13
Ravine Ct. NW&SE		
....	24W-39N	13
Ravine Dr. NW&SE		
....	24W-39N	13
Ravine Ln. NE&SW		
....	24W-39N	13
Richard Ln. E&W ..	22W-36N	14
Riviera Dr. NE&SW		
....	24W-39N	13
Romero Av. E&W ..	20W-37N	15
Rose Av. N&S......24W-38N		13
Rose Ct. E&W	22W-36N	14
Rosemont Ln. E&W		
....	23W-38N	14
Rt. 83 N&S		
........23W-39N,22W-36N		14
Rt. 132 E&W		
........25W-38N,19W-36N		13
Rustic Dr. N&S ..	21W-39N	15
S. Nathan Hale Dr. NE&SW		
....	21W-36N	15
San Rem Av. E&W 20W-37N		15
Sand Lake Rd. E&W		
....	19W-37N	15
Sarah Dr. E&W ..	21W-36N	15
Savage Rd. NS&EW		
........21W-39N,20W-39N		15
Saxony Dr. NW&SE		
....	24W-37N	13
Sheridan Rd. N&S ..	25W-38N	13
Shore Dr. N&S	25W-37N	13
Short St. N&S	25W-37N	13
Sienna Dr. N&S ..	20W-37N	15
Sir John Dr. N&S ..	24W-39N	13
Sistina Av. NW&SE		
....	20W-37N	15
South Av. E&W ..	24W-38N	13
Spruce St. N&S ..	21W-39N	15
Stonebridge Dr. NW&SE		
....	24W-39N	13
Summerfield Dr. NE&SW		
....	25W-37N	13
Sunnyside Rd. E&W		
....	25W-38N	13
Sunset Ln. E&W ..25W-38N		13
Sunshine Av. N&S ..21W-36N		15
Terry Dr., E. NW&SE		
....	22W-36N	14

	Grid	Page
Terry Dr., N. E&W ..22W-37N		14
Terry Dr., W. N&S ..	22W-36N	14
Thorndale Pl. N&S ..	23W-38N	14
Timber Ln. E&W ..	25W-37N	13
Trenker Dr. E&W ..	23W-39N	14
U.S. Route 45 N&S		
........19W-39N,19W-36N		15
Valina Av. N&S ..	25W-38N	13
Verona Av. E&W ..	20W-37N	15
Villa Av. N&S	23W-38N	14
Villa Ct. N&S	22W-38N	14
W. Park Av. N&S ..	23W-38N	14
Wacker Dr. E&W ..	25W-37N	13
Wall St. E&W	23W-38N	14
Walnut Av. N&S ..	21W-39N	15
Washington Av. E&W		
........24W-38N,23W-38N		13
Wayside Pl. E&W ..	25W-37N	13
West Av. N&S......23W-38N		14
Westmoor Av. N&S		
....	24W-36N	13
Wildwood Dr. N&S ..21W-38N		15
Willow St. N&S......21W-38N		15
Wilson Rd. E&W ..25W-38N		13
Wilton Av. N&S	23W-38N	14
Wilton Way NW&SE		
....	23W-38N	14
Winthrop Pl. E&W		
........24W-38N,23W-38N		13
Wittenburg Rd. N&S		
....	21W-39N	15
Woodland Av. NW&SE		
....	20W-36N	15
Woodland Ter. E&W		
....	25W-37N	13
Woods Av. N&S ..	21W-38N	15
1st St. E&W	21W-38N	15
2nd St. E&W	21W-38N	15
3rd St. E&W	21W-38N	15
4th St. E&W	21W-38N	15
5th St. E&W	21W-38N	15
6th St. E&W	22W-38N	14
7th Av. E&W	21W-38N	15

Lake Zurich

	Grid	Page
Aldine Rd. N&S	24W-23N	46
Alpine Dr. E&W	23W-23N	46
Andrew Ln. E&W ..	22W-22N	46
Apache Ct. N&S ..	22W-22N	46
Applegate Cir. E&W		
....	24W-22N	46
Applegate Ln. NW&SE		
....	24W-22N	46
Ascot Ct. E&W	22W-21N	46
Ash St. NE&SW	23W-23N	46
Aspen Ct. E&W	24W-22N	46
Bayview Dr. N&S ..	24W-23N	46
Beach Rd. N&S	24W-23N	46
Beaubien Ct. E&W ..23W-23N		46
Beauteau St. NW&SE		
....	23W-23N	46
Beechwood Dr. NW&SE		
....	22W-22N	46
Berkshire Ln. NS&EW		
....	24W-21N	46
Bern Ct. E&W 23W-23N		46
Betty Dr. E&W	24W-23N	46
Blacktwig Cir. N&S 22W-22N		46
Braemar Ct. NW&SE		
....	24W-21N	46
Braemar Ln. NE&SW		
....	24W-21N	46
Briar Point Ct. E&W		
....	22W-22N	46
Bridlepath Ln. NS&EW		
....	22W-22N	46
Bristol Trail Rd. E&W		
........22W-22N,21W-22N		46
Broadway Ct. NW&SE		
....	21W-21N	47
Brookfield St. N&S 22W-22N		46
Browning Ln. E&W 22W-22N		46
Brush Hill Ln. E&W		
....	22W-22N	46
Buckingham Ct. NW&SE		
....	22W-22N	46
Buffalo Creek Dr. E&W		
....	22W-22N	46
Burr Oak Ct. NE&SW		
....	22W-22N	46
Burr Oak Dr. N&S		
....	22W-22N	46
Bushing Rd. E&W		
........23W-22N,23W-23N		46
Bushrun Ct. NE&SW		
....	22W-21N	46
Butterfield Ct. N&S		
....	23W-23N	46
Butterfield Rd. E&W		
....	24W-23N	46
Carolyn Ct. N&S....23W-23N		46
Carriage Ln. E&W 24W-22N		46
Cedar Creek Dr. NS&EW		
........22W-22N,21W-22N		46
Centoni Ct. E&W ..	24W-23N	46
Century Ct. N&S ..	22W-22N	46
Chasewood Dr. N&S		
....	23W-21N	46
Cherokee Rd. E&W		
....	23W-23N	46
Cherry Ln. E&W ..	23W-23N	46

	Grid	Page
Chesterfield Ln. NS&EW		
....	24W-21N	46
Cheswick Dr. E&W		
....	21W-21N	47
Church St. NE&SW		
....	23W-23N	46
Clearwater Ct. E&W		
....	22W-21N	46
Clubhouse Ln. NW&SE		
....	24W-22N	46
Colonial Ct. E&W ..	24W-22N	46
Columbine Ct. E&W	23W-22N	46
Coral Reef Way E&W		
....	22W-21N	46
Cortland Cir. N&S ..	24W-22N	46
Cortland Dr. N&S ..	24W-22N	46
Country Club Rd. E&W		
....	24W-23N	46
Country Ridge Ct., N. N&S		
....	22W-22N	46
Country Ridge Ct., S. N&S		
....	22W-22N	46
Craig Ter. NW&SE ..24W-23N		46
Crescent Rd. NS&EW		
....	24W-23N	46
Cromwell Ct. E&W ..22W-21N		46
Cypress Bridge Rd. E&W		
....	22W-21N	46
Dearborn Ct. NE&SW		
....	23W-22N	46
Deerchase Rd. N&S		
....	22W-21N	46
Deerpath Rd. NS&EW		
........23W-21N,22W-21N		46
Denberry Ct. E&W ..	22W-21N	46
Denberry Dr. E&W ..22W-21N		46
Diane Ct. NW&SE .	22W-22N	46
Dogwood Ct. E&W 24W-22N		46
Donegal Ln. N&S ..	24W-21N	46
Dunwick Ct. E&W ..	24W-21N	46
E. Main St. E&W		
........23W-23N,22W-23N		46
Eddy Ln. N&S......22W-21N		46
Edelweiss Dr. E&W		
....	23W-22N	46
Ela Rd. N&S......24W-22N		46
Elm Pl. N&S......22W-22N		46
Ensell Rd. N&S		
........22W-23N,21W-23N		46
Evanston Ter. E&W		
....	24W-23N	46
Evergreen St. NE&SW		
....	23W-22N	46
Fairfield Ct. NW&SE		
....	21W-21N	47
Fairfield Ln. NE&SW		
........21W-21N,21W-22N		47
Fairway Rd. N&S ..	24W-23N	46
Farm Bridge Rd. E&W		
....	22W-22N	46
Fern Ct. E&W		
........22W-22N,21W-22N		46
Fern Rd. N&S		
........21W-22N,22W-22N		47
Ferndale Ct. N&S ..	22W-22N	46
Fescue Ct. N&S....23W-21N		46
Fescue Dr. N&S ..23W-21N		46
Fieldstone Cir. N&S		
....	22W-21N	46
Forest Av. NE&SW		
....	23W-23N	46
Foxfire Dr. E&W....23W-21N		46
Garland Ct. E&W ..	22W-21N	46
Genesee St. NE&SW		
....	23W-23N	46
Ginger Tr. NE&SW ..22W-22N		46
Glen Farm Ln. N&S		
....	22W-22N	46
Glencoe Ter. E&W ..24W-23N		46
Golf View Rd. NE&SW		
........24W-23N,23W-23N		46
Greenbay Ct. E&W 22W-21N		46
Grever Ct. E&W ..	23W-22N	46
Greystone N&S ..	22W-22N	46
Grove St. NW&SE ..23W-23N		46
Hapshire Ln. N&S ..	24W-24N	38
Harvest Dr. E&W ..	22W-22N	46
Heartwood Ln. E&W		
....	22W-22N	46
Hickory Rd. N&S ..	24W-23N	46
Hillside Ct. NE&SW		
....	23W-22N	46
Hobble Bush Dr. N&S		
....	22W-22N	46
Holly Court Rd. E&W		
....	24W-24N	38
Hollycourt Ter. NE&SW		
....	24W-24N	38
Huntclub Dr. E&W		
........22W-22N,21W-22N		46
Hunters Ln. E&W ..	24W-22N	46
Huntington Ct. E&W		
....	21W-22N	47
Huntington Way NS&EW		
....	22W-22N	46
Indigo Ct. NE&SW ..22W-22N		46
Interlaken Dr. N&S ..22W-22N		46
Ironwood Ct. N&S ..23W-22N		46
Ivy Ct. E&W	22W-22N	46
Jamie Ln. NW&SE ..23W-23N		46
Jean Ter. E&W	24W-23N	46
Jonathan Rd. N&S ..24W-23N		46
Joyce Ln. E&W	24W-23N	46
June Ter. N&S	23W-23N	46

	Grid	Page
Juneway Ter. N&S	24W-23N	46
Kenilworth Ter. E&W		
	24W-23N	46
Kim Tr. N&S	24W-23N	46
Kincaid Rd. NS&EW		
	23W-23N	46
Kingwood Ln. E&W		
	22W-23N	46
Knollwood Dr. E&W		
	24W-22N	46
Kylemore Dr. N&S	23W-21N	46
Lake St. E&W	23W-21N	46
Lake Zurich Dr. E&W		
	23W-21N	46
Lake Zurich Dr. NW&SE		
	24W-22N	46
Lakeview Pl. N&S	23W-22N	46
Lancaster N&S	24W-24N	38
Lexington Ln. NS&EW		
	22W-21N	46
Linden Rd. N&S	24W-23N	46
Lions Dr. NW&SE	23W-23N	46
Lismore Ct. N&S	23W-21N	46
Lorraine Dr. NS&EW		
	24W-22N	46
Lucerne Dr. N&S		
	23N-23N,23W-24N	46
Manchester NE&SW		
	24W-24N	38
Manor Rd. N&S	24W-23N	46
Maple Av. E&W	23W-22N	46
Marble Hill Dr. E&W	21W-22N	47
Margate Ct. E&W	22W-22N	46
Maria Ct. E&W	23W-21N	46
Meadow Ln. NW&SE		
	22W-22N	46
Meadowbrook Ct. E&W		
	22W-22N	46
Meadowbrook Ln. N&S		
	22W-22N	46
Micheal Ct. NW&SE		
	23W-21N	46
Millbrook Ln. E&W	22W-22N	46
Miller Rd. E&W		
	24W-23N,23W-23N	46
Minerva Ln. E&W	24W-22N	46
Mionske Dr. NW&SE		
	23W-22N	46
Mohawk Rd. N&S	23W-22N	46
Mohawk Tr. N&S	23W-22N	46
N. Harbor Dr. E&W		
	22W-21N	46
N. Old Rand Rd. N&S		
	24W-23N	46
Newberry Ct. E&W	22W-21N	46
Newgate Ct. NW&SE		
	24W-23N	46
Northcrest Rd. E&W		
	22W-23N	46
Oak St. NE&SW	23W-23N	46
Oakwood Rd. N&S	22W-23N	46
Old Mill Grove Rd. N&S		
	22W-22N,22W-23N	46
Orrington Ct. N&S	24W-21N	46
Oxford Ct. N&S	21W-21N	47
Oxford Ln. N&S	21W-21N	47
Pamela Rd. E&W	24W-23N	46
Park Av. NW&SE	22W-22N	46
Park Ln N&S	24W-23N	46
Parkway Av. NE&SW		
	23W-23N	46
Partridge Ln. NS&EW		
	22W-22N	46
Peachtree Ln. N&S		
	24W-23N	46
Pebblecreek Ct. E&W		
	22W-22N	46
Pebblecreek Dr. N&S		
	22W-22N	46
Pembridge Rd. NS&EW		
	22W-22N	46
Pheasant Ridge Ct. NE&SW		
	22W-22N	46
Pheasant Ridge Dr. E&W		
	22W-22N,21W-21N	46
Pheasant Ridge Rd. E&W		
	23W-22N,21W-21N	46
Pine Av E&W	23W-22N	46
Pine Tree Row NE&SW		
	24W-23N	46
Pleasant Rd. N&S	24W-22N	46
Pohlman Rd. E&W	23W-23N	46
Poplar Ct. E&W	24W-22N	46
Portree Ln. E&W	24W-22N	46
Pottawatomi Tr. NS&EW		
	22W-22N	46
Prairie Ln. NW&SE	23W-22N	46
Prospect Rd. N&S	24W-23N	46
Public St. N&S	23W-23N	46
Queen Ann Ln. N&S		
	21W-21N,21W-23N	47
Quentin Rd. N&S	23W-23N	46
Railway St. N&S		
	23W-22N	46
Ramblewood Ct. E&W		
	22W-22N	46
Ramblewood Ln. N&S		
	22W-22N	46
Rand Rd. N&S		
	24W-23N,22W-21N	46

	Grid	Page
Ravenswood Ct. N&S		
	22W-22N	46
Ravinia Ter. E&W	24W-23N	46
Red Bridge Ct. NW&SE		
	22W-22N	46
Red Bridge Rd. N&S		
	22W-22N	46
Red Haw Ln. E&W	22W-22N	46
Regency Dr. N&S	24W-22N	46
Regent Ct. NE&SW		
	21W-22N	47
Richmond Cir. NE&SW		
	24W-22N	46
Richmond Ct. NW&SE		
	24W-22N	46
Ridgewood Ct. E&W		
	22W-22N	46
Riley Ln. N&S	22W-22N	46
Ripple Brook Ct. E&W		
	22W-21N	46
Robertson Rd. NE&SW		
	23W-22N	46
Rock Rose Dr. N&S		
	23W-21N	46
Rose Rd. E&W	22W-23N	46
Rugby Rd. NE&SW		
	24W-23N	46
Rush Ct. E&W	22W-21N	46
Russet Cir. N&S		
	22W-21N	46
S. Harbor Rd. E&W		
	22W-21N	46
S. Old Rand Rd. NW&SE		
	23W-22N	46
Sandlewood Ct. E&W		
	22W-22N	46
Sandy Pass N&S	22W-21N	46
Schiller St. E&W	23W-23N	46
Schuldt Dr. E&W	23W-23N	46
Sebby Ln. NE&SW	23W-23N	46
Shaker Ln. NS&EW		
	22W-21N	46
Shallow Cover Rd. E&W		
	22W-21N	46
Sheffield Ct. N&S	24W-21N	46
Sheridan Ct. NW&SE		
	21W-22N	47
Shoreview Ln. E&W		
	24W-22N	46
Silverwood Ct. NE&SW		
	22W-22N	46
South Shore Ln. N&S		
	23W-22N	46
Spring Ct. NE&SW	22W-22N	46
Spruce St. N&S		
	23W-22N	46
Stanton Ct. N&S	22W-22N	46
Stanton Rd. NS&EW		
	22W-22N	46
Stillwater Ct. N&S	22W-22N	46
Stone Av. N&S	22W-22N	46
Stratford Ln. N&S	21W-21N	47
Summit Ter. NW&SE		
	23W-23N	46
Sunrise Ln. N&S	24W-23N	46
Sunset Ct. NE&SW		
	23W-22N	46
Sunset Tr. NS&EW		
	24W-24N	38
Surryse Rd. E&W	22W-22N	46
Sussex Ct. N&S	22W-22N	46
Tall Grass Cir. NS&EW		
	24W-23N	46
Tara Ct. E&W	23W-21N	46
Telser Rd. N&S		
	22W-22N,22W-23N	46
Terrace Av. E&W	23W-22N	46
Thistle Ln. NS&EW		
	22W-22N,21W-22N	46
Thornfield Ln. NS&EW		
	21W-22N	47
Thornwood Ct. E&W		
	24W-23N	46
Timberwood Dr. N&S		
	22W-22N	46
Town Mall Dr. N&S		
	23W-22N	46
Tracie Dr. NS&EW	23W-21N	46
Trailside Dr. NW&SE		
	22W-22N	46
Trillium Trail N&S	23W-22N	46
Tudor Ct. E&W	24W-22N	46
Turtle Pond Ct. NW&SE		
	22W-22N	46
Vista Ct. NW&SE	24W-23N	46
Vista Rd. N&S	24W-23N	46
W. Harbor Rd. N&S		
	22W-21N	46
W. Main St. NE&SW		
	24W-22N,23W-22N	46
Warwick Ct. NW&SE		
	22W-21N	46
Warwick Ln. NS&EW		
	21W-21N,21W-22N	47
Washo Ct. E&W	23W-21N	46
Washo Dr. E&W	23W-21N	46
Washo Ln. N&S	23W-21N	46
Waterbury Ct. N&S		
	21W-22N	47
Waterford Ct. N&S	23W-21N	46
Waterford Dr. NS&EW		
	23W-21N	46
White Birch Ln. E&W		
	22W-22N	46

	Grid	Page
Whitehall Ct. N&S	21W-21N	47
Whitney Rd. NW&SE		
	24W-22N	46
Wilkes Ln. NW&SE		
	22W-21N	46
Williams Way E&W		
	22W-22N,21W-22N	46
Willow Ter. E&W		
	24W-23N	46
Wilmette Ter. E&W		
	24W-23N	46
Windermere Ln. NW&SE		
	22W-21N	46
Winnetka Ter. E&W		
	24W-23N	46
Woodbine Cir. NW&SE		
	22W-22N	46
Yorkshire Ln. NE&SW		
	24W-21N	46

Lakemoor*
(Also see McHenry County)

	Grid	Page
Beacon Bay NW&SE		
	28W-33N	20
Belvidere Rd. E&W		
	28W-31N,27W-31N	28
Bouy Bay NE&SW	28W-33N	20
Captain Hook NW&SE		
	28W-33N	20
Captains Cove NE&SW		
	28W-33N	20
Concrete Dr. NS&EW		
	27W-32N	20
Drydock NE&SW	28W-33N	20
Fishook NE&SW	28W-33N	20
Fox Lake Rd. NE&SW		
	27W-32N	20
Fox Lake Rd. N&S		
	27W-31N,27W-32N	20
Gilmer Rd. NW&SE		
	27W-31N	29
Hornblower NW&SE		
	28W-33N	20
Leaward NE&SW	28W-33N	20
Mysty Ridge N&S	28W-32N	20
Neptune Cove NW&SE		
	28W-33N	20
Oyster Bay E&W	28W-33N	20
Port Side NE&SW	28W-33N	20
Rand Rd. N&S	27W-31N	29
Route 59 N&S	27W-32N	20
Rusty Scupper NE&SW		
	28W-33N	20
Sand Bar NW&SE	28W-33N	20
Sandpiper NE&SW	28W-33N	20
Sinker Bay* NW&SE		
	28W-32N	20
South Lake Shore Dr. E&W		
	28W-32N	20
Stowaway Bay* NE&SW		
	28W-32N	20
Sullivan Lake Blvd. N&S		
	28W-31N,28W-32N	28
Sullivan Lake Rd. E&W		
	28W-33N	20
Sunnyside Rd. E&W		
	28W-32N	20
U.S. Route 12 N&S		
	27W-32N	20
Windward NW&SE	28W-32N	20

Libertyville

	Grid	Page
Abby Ct. N&S	17W-29N	32
Acorn Ln. N&S	17W-28N	32
Adler Dr. E&W	16W-30N	33
American Way NS&EW		
	18W-29N,18W-30N	32
Ames St. N&S	16W-28N	33
Amy St. N&S	18W-29N	32
Amy Ln. N&S		
	18W-29N,18W-30N	32
Annie Ln. N&S	18W-30N	32
Apple Tree Ln. N&S		
	15W-26N	41
Appley Av. E&W	16W-29N	33
Arbor Ct. NW&SE	16W-29N	33
Ardmore Ter. N&S	16W-28N	33
Artaius Pkwy. N&S	16W-27N	41
Arthur Av. N&S	16W-28N	33
Ascot Ct.* N&S	15W-26N	41
Ash St. N&S	16W-28N	33
Ashley Ln. N&S	17W-29N	32
Austin Av. N&S	16W-28N	33
Avalon Ln. N&S	16W-29N	33
Bartlett Ter. N&S	16W-28N	33
Bedford Ln. N&S	17W-28N	32
Big Oak Ln. N&S	15W-29N	33
Blackberry Ct. NW&SE		
	17W-30N	32
Blackburn Cir. E&W		
	17W-29N	32
Blueberry Rd. N&S	17W-30N	32
Brainard Av. N&S	16W-29N	33
Brandywine Rd. NW&SE		
	17W-28N	32
Brian Dr. N&S	18W-29N	32
Briar Pl. E&W	16W-28N	33
Briarwood Dr. E&W		
	16W-27N	41
Bridle Ct. E&W	16W-27N	41
Broadway E&W	16W-29N	33

	Grid	Page
Buckingham Pl. N&S		
	16W-28N	33
Buckley Rd. E&W	17W-30N	32
Burdick St. N&S	16W-28N	33
Burridge Ct. E&W	17W-28N	32
Bush Ct. NW&SE	18W-30N	32
Butterfield Rd. N&S		
	17W-28N,18W-30N	32
Caliente Ct.* E&W	15W-26N	41
Cambridge Dr. N&S		
	16W-28N	33
Camelot Ln. NS&EW		
	15W-28N,15W-29N	33
Canterbury Cir. N&S		
	17W-28N	32
Carriage Hill Cir. NE&SW		
	16W-30N	33
Carter St. N&S	17W-28N	32
Cass Ct. E&W	17W-30N	32
Cass Dr. N&S	17W-30N	32
Catalpa Ln. E&W	16W-27N	41
Cedar Glen Ct. NE&SW		
	17W-30N	32
Cedar Glen Dr. E&W		
	17W-30N	32
Chatham Ln.* N&S	15W-26N	41
Checkberry Ct. E&W		
	17W-30N	32
Church St. E&W	16W-29N	33
Circle Dr. N&S	16W-28N	33
Claridge Dr. N&S	17W-29N	32
Cleveland Av. E&W		
	16W-28N	33
Concord Ln. N&S		
	15W-27N,15W-28N	41
Cook Av. N&S	16W-29N	33
Coolidge Av. N&S	16W-28N	33
Coolidge Pl. E&W	16W-28N	33
Country Club Dr. N&S		
	15W-28N	33
Crabtree Ln. N&S	16W-27N	41
Crane Blvd. E&W	17W-28N	32
Crestfield Av. E&W		
	17W-28N	32
Cypress Ln. E&W	16W-27N	41
Dancer Ct. NE&SW		
	18W-30N	32
Dawes St. N&S		
	17W-27N,17W-28N	40
Deborah Cir. N&S	18W-29N	32
Deborah Ct. N&S	18W-30N	32
Deer Trail Ln. E&W		
	15W-28N	33
Douglas Av. N&S	16W-28N	33
Dover Ct. N&S	17W-28N	32
Downing Rd. E&W	16W-27N	41
Downs Pkwy.* N&S		
	15W-26N	41
Drake St. N&S	17W-28N	32
Dymond Rd. N&S	17W-28N	32
E. Ellis Av. E&W	16W-29N	33
E. Park Av. E&W	16W-28N	33
E. Winchester E&W		
	16W-29N	33
Eaton Ct. E&W	15W-27N	41
Eldeberry Dr. N&S	17W-30N	32
Ellis Av. E&W	16W-29N	33
Elm Ct. E&W	16W-29N	33
Elm Dr. N&S	16W-29N	33
Emerson Ln.* N&S	15W-26N	41
Enterprise Ct. N&S	19W-30N	31
Eric Ln. E&W	18W-29N	32
Evergreen Ct. E&W		
	16W-27N	41
Fairlawn Av. E&W	17W-28N	32
Fairview Av. N&S	16W-28N	33
Finstead Dr. E&W	16W-30N	33
Flamingo Pkwy.* E&W		
	15W-26N	41
Florence Ct. E&W	16W-28N	33
Florsheim Dr. E&W		
	16W-27N,15W-27N	41
Forest Creek Ln. N&S		
	17W-30N	32
Forest Ln. N&S	16W-29N	33
Forestview Ln. NW&SE		
	17W-30N	32
Forever Av. NW&SE		
	18W-30N	32
Fox Trail N&S	15W-28N	33
Furlong Dr. NS&EW		
	16W-27N	41
Garfield Av. NW&SE		
	16W-27N,16W-28N	41
Garrison St. E&W	16W-29N	33
Glendale Rd. E&W	16W-28N	33
Golf Rd. E&W		
	17W-28N,16W-28N	32
Gracewood Dr. E&W		
	17W-28N	32
Grant Ct. E&W	16W-29N	33
Green Tree Pkwy. E&W		
	16W-27N	41
Greentree Ct. NW&SE		
	16W-27N	41
Gulfstream Pkwy.* E&W		
	15W-26N	41
Hackberry Ct. E&W		
	17W-30N	32
Hampton Av. N&S	15W-28N	33
Harding Av. N&S	17W-28N	32
Harms Av. E&W	17W-28N	32
Harris Rd. N&S		
	19W-30N,19W-31N	31

	Grid	Page
Harrison St E&W	16W-28N	33
Harvard Ln. E&W	16W-28N	33
Havenwood Ct. N&S		
	17W-28N	32
Havenwood Dr. E&W		
	17W-28N	32
Hayes Av. N&S	16W-28N	33
Hemlock Ln. E&W	16W-27N	41
Highgate Ct. NW&SE		
	17W-28N	32
Hillcrest Dr. N&S	17W-29N	32
Hollister Dr. N&S	16W-26N	41
Homewood Dr. NW&SE		
	16W-29N	33
Hunters Ln. E&W	15W-27N	41
Hurlburt Ln. E&W	16W-29N	33
Hyatt Dr. E&W	16W-29N	33
Industrial Dr. NW&SE		
	19W-30N	31
Innsbruck Ct. N&S	17W-29N	32
Interlaken Ct. E&W		
	17W-29N	32
Interlaken Dr. E&W		
	17W-29N	32
Jackson St. E&W	16W-29N	33
James Ct.* N&S	15W-26N	41
Jeremy Ln. E&W	17W-29N	32
Johnson Av. NW&SE		
	16W-29N	33
Josilyn Ct. E&W	18W-30N	32
Julie Ln. N&S	18W-30N	32
Juliet Cir. NW&SE	18W-30N	32
Juniper Pkwy. N&S		
	16W-27N	41
Kempton Ln. N&S	15W-27N	41
Kenlock Av. N&S	17W-28N	32
Kenwood Av. N&S	15W-28N	33
Kristen Dr. N&S	17W-29N	32
Lake St. E&W		
	17W-29N,16W-29N	32
Lange Ct. E&W	16W-29N	33
Laurel Av. E&W	16W-29N	33
Liberty Bell Ct. NE&SW		
	15W-28N	33
Liberty Bell Ln. E&W		
	16W-28N,15W-28N	33
Liberty Ct. E&W	19W-30N	31
Liberty Lake Dr. E&W		
	15W-29N	33
Lilac Ct. N&S	16W-27N	41
Lincoln Av. E&W	16W-28N	33
Linden Ln. E&W	16W-29N	33
Lingonberry Ct. NW&SE		
	17W-30N	32
Lisa Cir. N&S	18W-29N	32
Loatonia Ct.* N&S	15W-26N	41
Lothair Av. E&W	17W-28N	32
Loyola Dr. N&S	17W-29N	32
Magnolia Ln. NW&SE		
	16W-27N	41
Manor Ct. E&W	16W-29N	33
Maple Av. E&W	16W-29N	33
Margate Ln. E&W	15W-28N	33
Mayfair Ct. N&S	17W-27N	40
Maywood Ct.* E&W		
	15W-26N	41
McKinley Av. E&W	16W-28N	33
Meadow Ln. E&W	16W-28N	33
Merril Ct. E&W	16W-29N	33
Michaels Ln. N&S	16W-27N	41
Miller Ct. N&S	16W-30N	33
Milwaukee Av. NW&SE		
	16W-28N	33
Minear Dr. E&W	16W-29N	33
Mulberry Dr. NW&SE		
	17W-30N	32
Mullady Pkwy.* N&S		
	15W-27N	41
Nathan Ln. E&W	18W-29N	32
New Castle Dr. NW&SE		
	17W-30N	32
Newberry Av. E&W		
	16W-29N	33
Newgate Ct. NE&SW		
	17W-28N	32
Nita Ln. E&W	16W-27N	41
Nordic Ct. E&W	17W-28N	32
North Av. E&W	16W-29N	33
Oak Spring Rd. E&W		
	15W-29N	33
Oak St. N&S	16W-28N	33
Oaktrail Dr. E&W	15W-27N	41
Old Barn Cir. E&W	17W-30N	32
Old Buckley Rd. NE&SW		
	16W-30N	33
Orchard Cir. NE&SW		
	16W-29N	33
Oxford Ct. E&W	17W-29N	32
Paddock Ct.* N&S	15W-26N	41
Paddock Ln. N&S		
	16W-27N,15W-27N	41
Paradise Ln. N&S	17W-29N	32
Park Crest Ct. E&W		
	18W-30N	32
Park Pl. N&S	16W-28N	33
Parkview Dr. NE&SW		
	16W-30N	33
Parliament Ct. NW&SE		
	16W-29N	33
Paul Revere Ln. NE&SW		
	16W-28N	33
Pembrooke Rd. E&W		
	17W-28N	32

Name	Grid	Page
Peterson Rd. E&W	19W-30N,18W-30N	31
Pimlico Pkwy.* E&W	15W-26N	41
Pine Tree Ln. NE&SW	17W-28N	32
Pleasant Ct. E&W	18W-30N	32
Pond Ridge Rd. NE&SW	17W-28N	32
Poplar Ct. E&W	16W-27N	41
Portwine Dr. N&S	17W-28N	32
Prairie Av. N&S	16W-28N	33
Quaker Hill N&S	17W-29N	32
Red Top Dr. E&W	16W-27N	41
Regency Ln. E&W	17W-28N	32
Ridgeway Ln. N&S	15W-28N	33
Ridgewood Ln. N&S	17W-29N	32
Riva Ridge Dr. E&W	15W-27N	41
Riverside Dr. N&S	15W-28N	33
Rockland Ln. N&S	16W-28N,15W-28N	33
Romeo Cir. N&S	18W-30N	32
Ronan Ct. NE&SW	17W-30N	32
Roosevelt Dr. N&S	17W-28N	32
Rosewood Ter. N&S	16W-27N	41
Rt. 21 N&S	16W-27N,16W-28N	41
Rt. 45 N&S	18W-29N,19W-30N	32
Rt. 137 E&W	17W-30N	32
Rt. 176 E&W	17W-28N,15W-28N	32
Ruidoso Ct.* N&S	15W-26N	41
Sandstone Dr. N&S	16W-29N	33
Sandy Ct. E&W	16W-27N	41
Sandy Ln. E&W	16W-27N	41
Saxon Ln. N&S	16W-28N	33
School St. N&S	16W-29N	33
Scott Pl. N&S	15W-28N	33
Sedgwick Dr. NE&SW	17W-29N	32
Shari Ln. E&W	17W-28N	32
Springhaven Dr. N&S	17W-29N	32
Spruce Ct. N&S	16W-27N	41
St. James Pl. E&W	17W-29N	32
St. William Dr. NS&EW	17W-29N,17W-30N	32
Stevenson Dr. NW&SE	16W-28N	33
Stewart Av. N&S	16W-28N,16W-29N	33
Stonegate Ct. E&W	17W-29N	32
Stonegate Rd. N&S	17W-29N	32
Suffolk Ct.* E&W	15W-26N	41
Sunnyside Av. E&W	16W-28N	33
Sunnyside Pl. E&W	16W-28N	33
Sunset Dr. N&S	16W-29N	33
Sussex Ln.	17W-29N	32
Sylvia Dr. E&W	17W-27N	40
Tall Tree Ter. E&W	16W-28N	33
Tamarack Ln. NW&SE	16W-27N	41
Tempel Dr. E&W	19W-30N	31
Thomas Ct. E&W	17W-28N	32
Thornapple Ln. NE&SW	16W-27N	41
Thornbury Ln. E&W	15W-29N	33
Tiffany Ln. N&S	18W-29N	32
Tracy Ln. NW&SE	16W-27N	41
Trevor Cir. N&S	18W-29N	32
Trinity Pl. N&S	17W-29N	32
Tyler Ct. E&W	16W-28N	33
U.S. Rt. 45 N&S	18W-29N,19W-30N	32
Valley Park Dr. E&W	16W-28N	33
Victory Dr. E&W	17W-28N	32
Vineyard Ln. E&W	17W-30N	32
Virginia Av. N&S	18W-29N,18W-30N	32
Virginia Av. N&S	18W-29N,18W-30N	32
W. Park Av. E&W	17W-28N	32
Walnut St. E&W	16W-29N	33
Warwick Ln. E&W	17W-28N	32
Wedgemere Pl. N&S	16W-28N	33
Weeping Willow Ln. NW&SE	16W-27N	41
Wellington Av. E&W	17W-29N	32
West St. N&S	16W-29N	33
Wexford Ct. N&S	17W-28N	32
Wheeler Ct. N&S	16W-29N	33
White Blvd. E&W	19W-30N	31
Willow Av. E&W	16W-29N	33
Wilshire Ct. N&S	17W-29N	32
Wilshire Dr. E&W	17W-29N	32
Winchester Rd. E&W	18W-29N,16W-28N	32
Windhaven Rd. N&S	17W-29N	32
Windsor Ter. N&S	16W-28N	33
Winners Cir.* N&S	15W-26N	41
Woodfield Ln. E&W	15W-29N	33
Woodland Rd. N&S	17W-28N	32
Wright Ct. N&S	16W-29N	33
Wrightwood Ter. N&S	16W-28N	33
Young Dr. N&S	17W-30N	32
1st St. N&S	16W-28N,16W-29N	33
2nd St. N&S	16W-28N,16W-29N	33
3rd St. N&S	16W-29N	33
4th Av. N&S	16W-28N,15W-27N	33
4th St. N&S	16W-28N,16W-29N	33
5th Av. N&S	16W-28N	33
5th St. N&S	16W-28N	33
7th St. N&S	15W-28N	33

Libertyville Township

Name	Grid	Page
Allyson Ct. N&S	15W-29N	33
Almond Rd. N&S	17W-31N	32
Arcadia Rd. N&S	13W-28N	34
Arlington Dr. E&W	16W-30N	33
Atkinson Av. E&W	13W-29N	34
Baker Rd. N&S	13W-28N	34
Ballard Dr. NS&EW	13W-28N	34
Bayonne Av. N&S	13W-28N	34
Bell Ln. N&S	17W-30N	32
Birchwood Ln. E&W	15W-30N	33
Bob-O-Link Ln. N&S	18W-30N,18W-31N	32
Borde Ct. NE&SW	15W-29N	33
Bradley Rd. N&S	14W-28N	33
Brookhill Rd. E&W	17W-30N	32
Brookside Ct. NS&EW	17W-30N	32
Brookwood Ln. E&W	15W-29N	33
Bruce Ct. NE&SW	15W-29N	33
Buckley Rd. N&S	15W-29N	33
Bull Creek Dr. NS&EW	18W-30N	32
Casey Rd. E&W	18W-31N	32
Center Av. NW&SE	16W-30N	33
Cherrywood Ln. E&W	15W-30N	33
Clover Ln. E&W	15W-30N	33
Country Ct. E&W	18W-30N	32
Countryside Rd. N&S	18W-30N	32
Daisy Ln. N&S	18W-26N	40
Des Plaines Dr. E&W	16W-30N	33
Diamond Lake Rd. N&S	18W-26N	40
East End Av. N&S	15W-30N	33
Ellen Way E&W	15W-29N	33
Elm Rd. E&W	13W-28N	34
Elmwood Ln. E&W	14W-26N,14W-27N	41
Energy Dr. N&S	13W-28N	34
Fair Ln. E&W	15W-30N	33
Fair Way E&W	15W-29N	33
Fairhill Rd. E&W	17W-30N	32
Forest Dr. E&W	14W-27N	41
Foster Av. E&W	13W-29N	34
Foster Knoll Rd. N&S	13W-29N	34
Frances Av. NE&SW	18W-26N	40
Greenacre Dr. E&W	17W-30N	32
Gurin Rd. E&W	15W-31N	33
Hawthorne Av. E&W	14W-27N	41
Hawthorne Ln. NW&SE	15W-29N	33
Herky Dr. N&S	13W-28N	34
Hickory Ln. E&W	18W-26N	40
Idlewood Ln. E&W	15W-30N	33
Imperial Dr. E&W	14W-26N	41
Ivy Ln. N&S	15W-28N	33
Janas Ct. NE&SW	17W-30N	32
Keith Dr. N&S	13W-28N	34
Kildare Av. E&W	17W-30N	32
Laura Ln. NS&EW	13W-28N	34
Laurel Dr. E&W	13W-28N	34
Leesley Ct. E&W	18W-30N	32
Leola Dr. E&W	13W-28N	34
Liberty Rd. N&S	18W-31N	32
Lilac Dr. N&S	18W-26N	40
Linden Ln. E&W	14W-28N	33
Longwood Dr. N&S	14W-26N	41
Melody Ln. NS&EW	14W-26N	41
Muir Av. E&W	13W-29N	34
N. Terre Dr. E&W	15W-29N	33
Nagel Ct. N&S	13W-28N	34
Nells Rd. N&S	15W-29N	33
Northwoods Dr. N&S	15W-29N	33
Northwoods Dr. NS&EW	15W-29N	33
Oak Av. E&W	14W-27N	41
Oak Av. E&W	18W-26N	40
Oak Grove Av. N&S	15W-30N	33
Oak Ln. N&S	15W-28N,15W-29N	33
Oak Pond Ln. E&W	15W-29N	33
Oaksprings Ln. N&S	15W-29N	33
Old Hickory Ln. N&S	15W-29N	33
Park Ln. N&S	15W-30N	33
Peterson Rd. E&W	18W-30N	32
Petronella Dr. E&W	14W-28N	33
Pinehurst Dr. E&W	18W-30N	32
Pinewood Ln. NW&SE	15W-30N	33
Print-O-Tade Rd. N&S	13W-28N	34
Providence Dr. N&S	17W-30N	32
Quassey Av. E&W	15W-30N	34
Ranch Rd. N&S	18W-31N	32
Redwood Ln. N&S	15W-30N	33
Revere Dr. N&S	16W-30N	33
River Dr. N&S	15W-29N	33
River Rd. N&S	16W-30N	33
Rockland Ln. N&S	16W-30N	33
Rose Av. E&W	18W-26N	40
Rt. 21 N&S	16W-31N	33
Rt. 137 E&W	18W-30N,17W-30N	32
Rt. 176 E&W	14W-28N,13W-28N	33
Shagbark Ln. N&S	15W-28N	33
Sprucewood Ln. E&W	15W-30N	33
St. Andrew's Dr. N&S	18W-30N	32
St. Mary's Ct. E&W	15W-29N	33
St. Mary's Rd. N&S	15W-29N	33
Steger Rd. E&W	18W-31N	32
Sunnyview Rd. N&S	17W-30N	32
Tanglewood Ct. N&S	15W-28N	33
Terre Dr., S. E&W	15W-29N	33
Timber Ln. E&W	15W-29N	33
U.S. Rt. 45 N&S	18W-30N,18W-31N	32
Valley Ct. E&W	18W-30N	32
Vernon Ct. N&S	15W-28N	33
Victor Ln. E&W	15W-31N	33
Victoria Ln. NW&SE	15W-29N	33
Violet Dr. N&S	18W-26N	40
Whitman Ct. NE&SW	15W-29N	33
Wildwood Ct. E&W	15W-30N	33
Wood Hollow Ln. E&W	18W-31N	32
Wooddale Ln. N&S	15W-28N	33

Lincolnshire

Name	Grid	Page
Angelican Ln. E&W	13W-22N	49
Astor Ct. E&W	14W-23N	49
Avon Ln. E&W	13W-23N	49
Barclay Blvd. N&S	15W-22N	49
Bedford Ct. N&S	13W-21N	49
Berkshire Ln. N&S	13W-22N	49
Berwick Ct. N&S	13W-22N	49
Bond St. N&S	16W-22N	48
Brampton Ct. NE&SW	13W-23N	49
Brampton Ln. E&W	14W-23N,13W-23N	49
Briarwood Ln. E&W	13W-23N	49
Bristol Ct. E&W	13W-22N	49
Brunswick Ln. N&S	13W-22N	49
Buckingham Pl. E&W	13W-22N	49
Bucks Co. Ct.* E&W	13W-23N	49
Burnham Ct. E&W	13W-22N	49
Buxton Ct. E&W	13W-22N	49
Cambridge Ln. NS&EW	14W-22N	49
Camden Ct. E&W	13W-22N	49
Canterbury Rd. N&S	13W-22N	49
Carlisle Ln. N&S	14W-23N	49
Cedar Ln. E&W	14W-22N	49
Coldstream Cir. NS&EW	14W-22N	49
Connecticut River Val. Ct E&W	14W-22N	49
Cornell Ln. E&W	13W-23N	49
Coventry Ln. E&W	14W-22N	49
Cumberland Dr. E&W	13W-23N	49
Darby Ln. N&S	14W-22N	49
Deer Run N&S	13W-22N	49
Devonshire Dr. N&S	13W-23N	49
Dover Dr. E&W	14W-23N	49
Downing Sq. N&S	12W-22N	50
Dukes Cir. N&S	13W-22N	49
Dukes Ct. N&S	13W-22N	49
Dukes E&W	13W-22N	49
Durham Ct. NE&SW	13W-23N	49
Elmwood Ln. E&W	14W-23N	49
Elsinoor Dr. N&S	14W-23N	49
Essex Ln. N&S	14W-23N	49
Exmoor Ln. E&W	13W-22N	49
Fairfax Ln. N&S	13W-22N	49
Fallstone Ct. E&W	13W-23N	49
Fallstone Dr. N&S	13W-23N	49
Fox River Ct. E&W	13W-23N	49
Fox Tr. E&W	13W-22N	49
Friar Truck Ct. E&W	13W-23N	49
Grenadier Ct. E&W	13W-22N	49
Half Day Rd. E&W	15W-23N	49
Hampshire Ct. E&W	13W-23N	49
Heathrow Dr. E&W	15W-22N	49
Hickory Ln. N&S	14W-23N	49
Interstate 94 N&S	13W-23N	49
Kensington Dr. N&S	14W-22N	49
Kent NE&SW	14W-22N	49
Kerwick Ct.* N&S	12W-22N	49
Kings Cross Dr. E&W	13W-22N	49
Knightsbridge Pkwy. E&W	13W-23N	49
Lancaster Ln.	14W-22N	49
Leeds Ct. NW&SE	13W-23N	49
Lincolnshire Dr. NS&EW	13W-22N	49
Londonderry Ln. E&W	14W-22N	49
Margate Dr. E&W	16W-22N,15W-22N	48
Marriott Dr. E&W	15W-22N,14W-22N	49
Mayfair Ln. N&S	13W-23N	49
Melrose Ln. NS&EW	13W-22N	49
Middlebury Ln. N&S	13W-21N	49
Mohawk Valley Ct. E&W	13W-23N	49
Nantucket Ct. N&S	13W-23N	49
Natchez Ct. E&W	14W-23N	49
Northampton Ln. NS&EW	14W-24N	41
Nottingham Dr. E&W	13W-23N	49
Oakwood Ln. N&S	14W-23N	49
Overlook Point N&S	15W-22N	49
Oxford Dr. NW&SE	14W-22N,14W-23N	49
Park Ln. E&W	16W-21N	48
Park Pl. E&W	16W-21N	48
Parton Ct. NE&SW	13W-23N	49
Pembroke Dr. E&W	14W-23N	49
Pembroke E&W	14W-23N	49
Pheasant Row NE&SW	13W-22N	49
Plymouth Ct. N&S	14W-22N	49
Portshire Dr. NS&EW	13W-22N	49
Queens Way E&W	13W-22N	49
Regent Ln. N&S	13W-22N	49
Reliance N&S	13W-22N	49
Rivershire Ln. E&W	14W-22N	49
Riverside Dr. E&W	14W-22N	49
Riverside Rd. NW&SE	13W-22N	49
Riverwoods Rd. N&S	14W-22N,14W-24N	49
Robinhood Ct. E&W	13W-22N	49
Royal Ct. E&W	13W-22N	49
Rt. 21 N&S	15W-23N	49
Rt. 22 E&W	15W-23N,13W-23N	49
Schelter Rd. N&S	15W-22N	49
Sheffield Ct. E&W	14W-22N	49
Sherwood Dr. NW&SE	13W-23N	49
Stafford Ct. N&S	13W-23N	49
Stonegate Cir. N&S	14W-22N	49
Suffield Sq. N&S	13W-22N	49
Surrey Ct. E&W	13W-23N	49
Surrey Ln. N&S	14W-23N,13W-23N	49
Sutton Ct. NW&SE	12W-22N	49
Tidewater Ct. E&W	13W-23N	49
Tower Pkwy. E&W	15W-22N	49
Trafalgar Square N&S	15W-23N	49
Tri-state Tollway N&S	13W-23N	49
U.S. Route 45 E&W	15W-23N	49
Victoria Ln. N&S	14W-22N	49
Wellington Ct. N&S	13W-23N	49
Westminster Way NW&SE	13W-22N	49
Westwood Ln. E&W	13W-22N	49
Whitby Ct. N&S	13W-22N	49
Whitmore E&W	14W-23N	49
Whyte Gate Ln. E&W	14W-23N	49
William St. N&S	15W-21N	49
Wilmington Ct. E&W	13W-23N	49
Wiltshire Dr. E&W	14W-22N	49
Windsor Ct. S. N&S	13W-22N	49
Windsor Dr. E&W	13W-22N	49
Yorkshire Dr. E&W	14W-22N	49

Lindenhurst

Name	Grid	Page
Adams Dr. E&W	19W-39N	15
Alden Ct. NW&SE	19W-39N	15
Beck Cir. N&S	19W-39N	15
Beck Rd. NS&EW	20W-37N,20W-38N	15
Bonner Ln. NW&SE	19W-37N	15
Briar Ln. E&W	20W-37N	15
Bridgeport Ter. N&S	19W-37N	15
Brittany Ln. E&W	20W-38N	15
Brook Ln. N&S	20W-38N	15
Burr Oak Ln. NW&SE	20W-37N	15
Cardinal Ct. NW&SE	21W-38N,20W-38N	15
Carriage Ln. E&W	20W-38N	15
Cherrywood Ln. NW&SE	20W-38N	15
Chestnut Cir. N&S	20W-38N	15
Colony Av. NW&SE	19W-39N	15
Colony Ct. NE&SW	19W-39N	15
Columbus Av. NW&SE	19W-37N	15
Countryside Ln. NE&SW	20W-37N	15
Crabtree Ct. NE&SW	19W-37N	15
Deep Lake Rd. N&S	21W-38N	15
Deerpath Dr. NS&EW	20W-37N,19W-37N	15
Dittmer Ln. N&S	21W-38N	15
Dover Ct. E&W	19W-36N,20W-36N	15
Elmwood Dr.	20W-38N	15
Fairfield Rd. N&S	20W-37N,20W-38N	15
Federal Ct. N&S	20W-39N	15
Federal Pkwy N&S	20W-39N	15
Forest View Rd. NS&EW	19W-38N	15
Franklin Dr. N&S	19W-39N	15
Glendale Ct. E&W	20W-37N	15
Grand Av. NW&SE	20W-37N	15
Grand Cir. NE&SW	19W-37N	15
Grass Lake Rd. N&S	20W-39N	15
Great Oak Dr. N&S	19W-38N	15
Green Briar Ln. N&S	20W-37N	15
Green Tree Ct. E&W	20W-38N	15
Hamilton Dr. E&W	19W-39N	15
Harbor Ridge Way NE&SW	19W-39N	15
Hawthorne Dr. N&S	20W-38N	15
Hazelwood Dr. E&W	20W-38N	15
Heron Dr. NW&SE	19W-39N	15
Hickory Ct. E&W	20W-38N	15
Hickory Dr. N&S	20W-38N	15
Highland Cir. NE&SW	19W-37N	15
Highland Dr. E&W	19W-37N	15
Highpoint Dr. NS&EW	19W-37N	15
Hillcrest Ln. N&S	20W-37N	15
Honeysuckle Ct. NE&SW	20W-38N	15
Ironwood Dr. N&S	19W-37N	15
Jackson Pl. E&W	20W-39N	15
Jefferson Dr. E&W	20W-39N	15
Lakeshore Dr. N&S	20W-38N	15
Lakeview Ct. E&W	20W-39N	15
Larkdale Ln. NE&SW	20W-38N	15
Laurel Dr. N&S	20W-37N	15
Lindenhurst Dr. N&S	20W-38N	15
Longmeadow Dr. E&W	20W-38N	15
Magnolia Ln. N&S	19W-37N	15
Mallard Dr. E&W	19W-38N	15

Street	Grid	Page
Mallard Ridge Dr. NS&EW	21W-38N	15
Maplewood Ct. NE&SW	20W-38N	15
Maplewood Dr. NW&SE	20W-38N	15
Meadow Dr. E&W	20W-38N	15
Monroe Dr. N&S	19W-39N	15
Munn Rd. N&S	21W-38N	15
Nightengale Cir. NS&EW	21W-38N,20W-38N	15
Northgate Rd. N&S	20W-38N	15
Old Elm Rd. NW&SE	20W-37N	15
Orchard Ln. N&S	19W-37N	15
Oriole Ct. E&W	20W-38N	15
Paine Av. N&S	19W-39N	15
Partridge Cir. E&W	21W-38N	15
Penn Blvd. NS&EW	20W-39N,19W-39N	15
Penn Ct. NE&SW	19W-39N	15
Pheasant Ridge Ct. NW&SE	19W-38N	15
Pinecrest Ln. E&W	20W-38N	15
Plum Tree Rd. E&W	20W-37N,19W-37N	15
Potomac Ct. NE&SW	20W-39N	15
Prospect Dr. NE&SW	20W-37N,20W-38N	15
Quail Cir. N&S	21W-38N	15
Red Rock Dr. NE&SW	20W-38N	15
Ridge Ct. NE&SW	20W-38N	15
Ridgeland Dr. E&W	20W-37N	15
Robin Crest Ln. NS&EW	21W-38N,20W-38N	15
Rolling Ridge Ln. NS&EW	20W-38N	15
Rose Tree Ln. NS&EW	19W-37N	15
Rt. 132 NW&SE	21W-38N,20W-37N	15
Sand Lake Rd. E&W	20W-37N,19W-37N	15
Shagbark Ln. NW&SE	20W-38N	15
Spring Hill Ln. E&W	20W-38N	15
Sprucewood Ln. E&W	20W-38N	15
Sunset Ln. E&W	20W-37N	15
Surrey Ln. NW&SE	20W-38N	15
Teal Rd. E&W	19W-38N	15
Thornwood Dr. NS&EW	20W-37N,19W-37N	15
Timber Ln. N&S	19W-37N	15
Valley Dr. N&S	20W-37N	15
Waterford Dr. NE&SW	19W-38N	15
Whispering Pines Rd. NE&SW	20W-38N,19W-38N	15
White Birch Rd. NE&SW	19W-38N	15
White Oak Dr. N&S	20W-38N	15
Witchwood Ln. NS&EW	20W-37N,20W-38N	15
Woodlane Dr. NW&SE	20W-37N	15

Long Grove

Street	Grid	Page
Albert Ln. NW&SE	20W-21N	47
Allison NW&SE	19W-21N	47
Andrew Ct. NE&SW	19W-20N	47
Antietam Dr. NW&SE	18W-21N	48
Arlington Heights Rd. N&S	18W-20N	48
Arrowhead Ct. N&S	19W-25N	39
Ashley E&W	19W-21N	47
Aspen Dr. E&W	19W-25N	39
Atlantic Av. N&S	19W-25N	39
Barclay Ct. SW&NE	17W-23N	48
Bayberry Ln. E&W	19W-20N	47
Beaver Run Dr. NS&EW	20W-22N	47
Bedfordshire Dr. N&S	18W-24N	40
Berkshire Cir. N&S	18W-24N	40
Bernay Ln. E&W	18W-20N	48
Black Walnut Tr. E&W	18W-23N	48
Blue Heron Dr. N&S	17W-23N	48
Blue Stem Ct. N&S	18W-24N	40
Bob-O-Link Ln. N&S	18W-21N	48
Bordeaux Ln. N&S	18W-20N	48
Briarcrest Ln. NW&SE	18W-22N	48
Bridgewater Ct. E&W	18W-21N	48
Bridgewater Ln. N&S	18W-21N	48
Bridlewood Ct. NW&SE	17W-21N	48
Bridlewood Ln. NS&EW	17W-21N	48
Brittany Ct. N&S	18W-20N	48
Brittany Ln. E&W	18W-20N	48
Brookbank Ln. E&W	17W-22N	48
Brookhill Dr. NE&SW	19W-23N	47
Brookside Ln. N&S	18W-20N	48
Butler Ln. E&W	17W-23N	48
Carriage Ct. N&S	18W-20N	48
Carriage Way N&S	19W-24N	39
Cavalry Ct. E&W	19W-20N	47
Cedar Ct. N&S	18W-24N	40
Checker Rd. E&W	19W-20N	47
Chickamauga Ln. N&S	18W-20N	48
Chickamuga Ln. NS&EW	18W-20N	48
Church Hill Ct. E&W	17W-23N	48
Clearwater Dr. NW&SE	20W-22N	47
Coach Rd. N&S	19W-20N	47
Cobblestone Ln. N&S	19W-24N	39
Collier Cir. N&S	18W-24N	40
Country Club Dr. NE&SW	19W-20N	47
Country Ln. NE&SW	19W-22N	47
Country Ln. N&S	19W-22N	47
Country Side Ln. N&S	18W-20N	48
Countryside Ln. N&S	18W-20N	48
Creekside Dr. N&S	19W-25N	39
Crestview Dr. N&S	19W-22N	47
Cuba Rd. E&W	20W-22N,19W-22N	47
Cumberland Cir. NS&EW	19W-20N	47
Danbury Ct. NE&SW	17W-22N	48
Dawn Ct. E&W	18W-20N	48
Deerwood Dr. NS&EW	20W-21N	47
Devonshire Cir. N&S	17W-24N	40
Diamond Lake Rd. N&S	19W-24N,19W-25N	39
Doncaster Ct. NW&SE	18W-23N	48
Dorchester Cir. N&S	18W-24N	40
Driftwood Ln. (Pvt.) NW&SE	19W-22N	47
Driftwood Rd. (Pvt.) NW&SE	19W-22N	47
East Mardan Dr. N&S	19W-21N	47
Edgewood Ln. N&S	18W-20N	48
Eleanor Dr. E&W	19W-23N	47
Endwood Dr. NS&EW	18W-24N,17W-24N	40
Estate Ln. N&S	19W-25N	39
Evergreen Ct. E&W	18W-24N	40
Fairfield Dr. E&W	18W-21N	48
Fairview Ln. N&S	18W-21N	48
Farmcrest Ln. NW&SE	17W-22N	48
Farmwood Dr. N&S	19W-23N,19W-24N	47
Federal Ct. E&W	19W-20N	47
Fenview Ln. E&W	17W-22N	48
Finch Ln. NE&SW	17W-23N	48
Forest Edge Ln. E&W	19W-23N	47
Forest Fork Ct. NW&SE	20W-22N	47
Forest Glen NE&SW	18W-23N	48
Forest Way Cir. NW&SE	19W-23N	47
Forrest Tr. E&W	18W-23N	48
Gentry Ct. E&W	17W-23N	48
Gilmer Rd. NW&SE	18W-24N	40
Goldeneye Dr. E&W	17W-23N	48
Golf Ln. N&S	18W-22N	48
Grant Ct. N&S	18W-21N	48
Grant Pl. E&W	18W-21N	48
Greenwich Cir. NE&SW	17W-24N	40
Hamelton Ct. N&S	18W-23N	48
Hampton Dr. N&S	17W-23N	48
Heather Knoll Ct. NW&SE	18W-22N	48
Heatherwood Ln. N&S	18W-23N	48
Heathmoor Ct. NE&SW	18W-22N	48
Hedgewood Ct. NW&SE	17W-22N	48
Heritage Ln. NS&EW	18W-22N	48
Hickory Ct. E&W	19W-23N	47
Hidden Hills Ct. NW&SE	20W-22N	47
Hidden Valley NS&EW	20W-21N	47
High Meadow Ct. NW&SE	18W-24N	40
High Point Ct. NE&SW	18W-23N	48
Highland Dr. E&W	18W-24N	40
Hilltop Rd. NW&SE	18W-22N,17W-22N	48
Holly Ct. E&W	18W-20N	48
Holly Ct. NW&SE	18W-20N	48
Indian Creek Rd. E&W	19W-25N	39
Indian Ln. E&W	18W-25N	40
Juniper Ln. E&W	19W-20N	47
Kettering Dr. NE&SW	18W-23N	48
Killdeer Ct. E&W	17W-23N	48
Kimberly Ct. NS&EW	19W-23N	47
Kimberly Ct. E&W	19W-23N	47
Knoll Ct. E&W	20W-20N	47
Knoll Dr. N&S	20W-20N	47
Krueger Ct. E&W	19W-23N	47
Krueger Ln. E&W	19W-23N	47
Krueger Rd. N&S	19W-23N	47
Lake Point Cir. NS&EW	19W-23N	47
Lake Zurich Rd. E&W	18W-23N	48
Lakeridge Ct. NW&SE	19W-25N	39
Lakeridge Dr. NS&EW	19W-25N	39
Lakeridge Dr. N&S	19W-25N	39
Lakeview Ct. E&W	20W-22N	47
Lexington Dr. NS&EW	19W-20N	47
Lincoln Av. N&S	19W-20N	47
Long Grove Aptakisic Rd. E&W	18W-22N,17W-22N	48
Long Grove Rd. NE&SW	20W-20N,18W-21N	47
Manassas Ct. N&S	18W-21N	48
Manassas Ln. NW&SE	18W-21N	48
Mardan Dr. N&S	19W-21N	47
Marilynn Dr. NE&SW	19W-23N	47
Mayflower Ln. E&W	19W-21N	47
Meadow Knoll Ct. NE&SW	17W-22N	48
Meadow Lark Dr. E&W	18W-20N	48
Meadow Ln. E&W	18W-25N	40
Meadowlark Dr. E&W	18W-20N	48
Medford Cir. E&W	18W-24N	40
Merrimac Ln. E&W	19W-21N	47
Middlesex Dr. N&S	19W-21N	47
Millwood Farm Rd. NW&SE	19W-21N	47
Moniter Ln. E&W	19W-21N	47
Monitor Ln. NS&EW	19W-21N	47
Monticello Ln. E&W	19W-21N	47
Muirwood Ct. N&S	20W-22N	47
Mundelein Rd. N&S	18W-21N,18W-24N	48
Normandy Ct. E&W	18W-23N	48
North Ridge Pl. E&W	18W-24N	40
Nottinghamn Dr. N&S	20W-22N	47
Oak Dr. E&W	17W-22N	48
Oak Grove Cir. Cir.	17W-22N	48
Oak Grove Dr. E&W	18W-22N	48
Oak Hill Ln. N&S	20W-21N	47
Oakwood Cir. NS&EW	17W-23N	48
Oakwood Rd. E&W	18W-23N	48
Old Field Ln. SW&NE	17W-22N	48
Old Field Rd. NW&SE	17W-22N	48
Old Hicks Rd. N&S	19W-20N,19W-21N	47
Old McHenry Rd. NW&SE	19W-22N,18W-21N	47
Old Wood Ln. NW&SE	17W-22N	48
Osage Rd. NW&SE	18W-25N	40
Pacific Blvd. N&S	19W-25N	39
Pamela Ct. NE&SW	19W-23N	47
Pamela Ct. E&W	19W-23N	47
Partridge Ln. NW&SE	17W-23N	48
Patricia Dr. E&W	19W-23N	47
Pebble Creek E&W	18W-22N	48
Picardy Ct. N&S	18W-20N	48
Picardy Ln. E&W	18W-20N	48
Pine Cone Ct. N&S	18W-24N	40
Pine Ct. NE&SW	19W-22N	47
Pine Tree Ct. N&S	18W-24N	40
Pintail Ln. E&W	17W-23N	48
Popp Ln. NW&SE	19W-20N	47
Port Clinton Rd. E&W	17W-23N	48
Pottawatomie Ct. E&W	19W-21N,19W-20N	47
Prairie Crossing NW&SE	17W-22N	48
Prairie Wind Rd. NE&SW	19W-22N	47
Prairiemoor Ln. E&W	17W-22N	48
Private Dr. N&S	19W-22N	47
Private Ln. NS&EW	19W-21N	47
Private Ln. NS&EW	19W-22N	47
Private Rd. E&W	18W-21N	48
Private Rd. N&S	19W-22N	47
Promentary E&W	17W-23N	48
Pvt. Ln. NE&SW	18W-22N	48
Red Oak Dr. NS&EW	19W-23N	47
Redwing Ln. E&W	19W-24N	39
Roanoke Ct. E&W	18W-20N	48
Robert Parker Coffin Rd. E&W	19W-21N,18W-21N	47
Robertson Blvd. E&W	20W-22N	47
Rock Dove Ct. NE&SW	17W-23N	48
Rolling Glen Dr. N&S	19W-22N	47
Rosehedge Dr. NW&SE	19W-24N	39
Rosos Parkway N&S	17W-23N	48
Royal Melbourne Dr. E&W	18W-23N	48
Rt. 22 E&W	19W-23N,17W-23N	47
Rt. 53 E&W	20W-20N,18W-21N	47
Rt. 83 N&S	18W-21N,18W-25N	48
Saddle Ridge Ct. N&S	18W-24N	40
Saddle Ridge Ln. E&W	18W-24N	40
Salem Ct. NE&SW	18W-24N	40
Salem Lake Dr. NW&SE	20W-22N,19W-22N	47
Schaeffer Rd. N&S	18W-20N,18W-21N	48
Shagbark Ln. N&S	19W-21N	47
Shaheen Ct. (Pvt.) N&S	19W-23N	47
Shenandoah Ln. N&S	19W-20N	47
Sheridan Ct. E&W	19W-21N	47
Shilo Dr. E&W	19W-21N	47
Sky Ln. N&S	18W-22N	48
Southwell Ct. NW&SE	18W-22N	48
Spring Valley NS&EW	20W-21N	47
Standford Cir. NE&SW	17W-24N	40
Stockbridge Ln. NW&SE	18W-24N	40
Stonehaven Dr. NS&EW	18W-22N	48
Sumter Ct. NE&SW	18W-20N	48
Sumter Dr. E&W	18W-20N	48
Sunshine Ln. E&W	18W-22N	48
Surrey Ln.(Pvt.) NW&SE	20W-21N	47
Taggert Ct. NW&SE	17W-23N	48
Tall Oaks Dr. N&S	18W-23N	48
Tanager Way NS&EW	19W-20N,18W-20N	47
Teal Ct. E&W	17W-23N	48
Teal Ln. NS&EW	17W-23N	48
Three Lakes Ct. E&W	18W-21N	48
Three Lakes Dr. NS&EW	18W-21N	48
Tremont Cir. E&W	17W-24N	40
Trenton Ct. E&W	18W-23N	48
Tribal Ct. E&W	19W-25N	39
Turnberry Ln. E&W	20W-22N	47
Twin Lakes Ln. E&W	19W-23N	47
Union Ct. E&W	19W-20N	47
Valley View Cir. NS&EW	19W-24N	39
Valley View Ln. NW&SE	19W-24N	39
Victorian E&W	19W-21N	47
Walnut Ln. E&W	19W-20N	47
Waterfowl Way N&S	19W-24N	39
Wellington Dr. NS&EW	18W-23N	48
West Mardan Dr. N&S	19W-21N	47
Westbury Dr. N&S	18W-23N	48
Westchester Cir. N&S	17W-24N	40
Westmoreland Dr. E&W	18W-24N	40
Wildwood Ln. N&W	18W-21N	48
Wildwood Ln. N&S	18W-21N	48
Willow Brook Rd. N&S	18W-22N	48
Willow Spring Rd. N&S	18W-25N	40
Willow Valley Rd. N&S	20W-20N,20W-21N	47
Willowbrook Rd. N&S	18W-22N	48
Windham Ct. N&S	18W-25N	40
Windham Ln. NW&SE	18W-25N	40
Woodbine Ln. N&S	19W-21N	47
Woodland Ln. N&S	20W-21N	47
Woods End Rd. E&W	19W-22N	47
Woodview Ct. N&S	17W-22N	48

Mettawa

Street	Grid	Page
Bradley Rd. N&S	14W-26N	41
Everett Rd. E&W	14W-24N	41
Hawthorne Dr. E&W	13W-27N	42
Indian Ridge Rd. E&W	14W-26N	41
Interstate 94 N&S	13W-27N	42
Little St. Mary's Rd. NS&EW	15W-27N	41
Maureen Ln. E&W	14W-27N	41
Meadowood Ct. NW&SE	14W-27N	41
Meadowood Ln. N&S	14W-27N	41
Oak Hill Ln. N&S	14W-27N	41
Old School Rd. E&W	15W-27N,14W-27N	41
Prairie View Ln. E&W	14W-25N	41
Private Rd. NS&EW	15W-27N	41
Riverwoods Rd. N&S	13W-25N	42
Rockcress Ct. NS&EW	14W-27N	41
Rt. 60 E&W	15W-26N,14W-26N	41
Shagbark Rd. NE&SW	14W-25N	41
St. Mary's Rd. N&S	15W-25N,15W-27N	41
Townline Rd. E&W	14W-26N	41
Tri-State Tollway N&S	13W-26N	42

Mundelein

Street	Grid	Page
Aberdeen Ln. NE&SW	19W-28N	31
Agnes Av. E&W	18W-27N	40
Albion Ln. E&W	19W-27N	39
Alexandra Ct. E&W	19W-27N	39
Allanson Rd. E&W	18W-27N	40
Ambria Ct. N&S	19W-29N	31
Ambria Dr. NW&SE	19W-29N	31
Anthony Av. NW&SE	18W-27N	40
Appleby Cir. N&S	19W-28N	31
Appleby Ct. N&S	19W-28N	31
Arbor Ct. E&W	18W-28N	32
Archer Av. N&S	18W-27N	40
Armour Blvd. E&W	17W-25N	40
Armwood Ln. E&W	20W-28N	31
Ashbrook Dr. E&W	19W-27N	39
Atwater Dr. N&S	19W-27N	39
Babcock Ct. N&S	19W-27N	39
Balmoral Dr. NE&SW	19W-27N	39
Banbury Rd. NS&EW	19W-28N	31
Barlowe Ln. E&W	19W-27N	39
Barnhill Ln. NS&EW	20W-28N	31
Baskin Rd. N&S	17W-25N	40
Beach Pl. N&S	19W-28N	31
Bedford Rd. E&W	17W-27N	40
Benridge Ct. N&S	20W-28N	31
Bingham Cir. NS&EW	20W-27N,19W-27N	39
Bingham Ct. NW&SE	20W-27N	39
Bishop Way NW&SE	20W-28N	31
Blackburn Dr. E&W	20W-27N	39
Blue Spruce Ln. E&W	19W-28N	31
Bobby Ln. E&W	19W-28N	31
Bonnie Brook Av. E&W	19W-28N	31
Bradford Ln. N&S	19W-27N	39
Braeburn Rd. E&W	19W-28N	31
Braemar Dr. N&S	19W-28N	31

	Grid	Page
Brentwood Dr. E&W	20W-27N	39
Brice Av. N&S	18W-27N	40
Brighton Dr. E&W	20W-27N	39
Bristol Ct. NE&SW	17W-27N	40
Buckingham Way NS&EW	20W-28N	31
Burnham Ct. E&W	19W-27N	39
Butterfield Rd. N&S	17W-25N,17W-27N	40
Cairo Ct. E&W	19W-28N	31
California Av. N&S	19W-27N	39
Campus Dr. NS&EW	17W-26N	40
Canterbury E&W	17W-25N	40
Cardinal Pl. E&W	19W-27N	39
Castleton Ct. N&S	20W-28N	31
Cedar St. N&S	18W-27N	40
Chandler Dr. E&W	19W-28N	31
Charlotte Pl. E&W	18W-28N	32
Chestnut St. N&S	18W-27N	40
Chetwood Ct. E&W	20W-27N	39
Chicago Av. NW&SE	18W-27N,18W-28N	40
Churchill E&W	17W-25N	40
Clearbrook Park Dr. NE&SW	18W-26N	40
Courtland St. E&W	19W-27N,18W-27N	39
Crane Av. E&W	18W-26N	40
Cross Meadow Dr. N&S	18W-27N	40
Crystal St. E&W	19W-27N,18W-27N	39
Dairy Ln. N&S	19W-28N,19W-29N	31
Dalton Av. NW&SE	18W-27N	40
Dean Pl. E&W	18W-27N	40
Deepwoods Dr. NE&SW	18W-26N	40
Diamond Lake Rd. N&S	18W-26W	40
Division St. E&W	19W-27N	39
Downing E&W	17W-25N	40
Dublin Dr. E&W	19W-28N	31
Dunbar Rd. E&W	19W-28N,19W-29N	31
Dunleer Dr. N&S	19W-29N	31
Dunton Cir. NS&EW	20W-27N	39
Dunton Ct. NE&SW	20W-27N	39
E. Garfield Av. N&S	19W-27N	39
East Greenview N&S	19W-27N	39
Edgar A. Poe Ln. N&S	18W-26N	40
Edgemont St. E&W	19W-28N	31
Edington Ln. E&W	20W-27N,19W-27N	39
Elm Av. E&W	18W-26N	40
Emerald Av. N&S	19W-27N,19W-28N	39
Emerson Ln. E&W	18W-26N	40
Eton Way N&S	17W-27N	40
Fair Lawn Av. N&S	19W-27N,19W-28N	39
Fairhaven Ln. E&W	18W-28N	32
Farina Dr. N&S	19W-27N	39
Firth Rd. N&S	19W-28N	31
Forest Dr. E&W	18W-26N	40
French Ct. N&S	18W-27N	40
French Dr. E&W	18W-27N	40
Friars Ln. E&W	20W-28N	31
Garfield Av. N&S	19W-27N	39
George Av. N&S	18W-27N	40
Gifford Ct. NW&SE	18W-27N	40
Glendale Pl. E&W	18W-28N	32
Glenview Av. E&W	19W-28N	31
Goodwin Pl. NE&SW	18W-28N	32
Grace Av. NE&SW	18W-28N	32
Granville Av. N&S	19W-28N	31
Greenview Av. N&S	19W-27N,19W-28N	39
Greenwood Av. N&S	18W-28N	32
Grove St. E&W	18W-27N	40
Groveland Blvd. E&W	19W-27N	39
Hammond St. E&W	19W-27N	39
Hampton Ct. N&S	17W-27N	40
Hampton Ln. E&W	17W-27N	40
Handley Ct. N&S	20W-28N	31
Hawley Ct. N&S	19W-27N	39
Hawley St. E&W	19W-28N,18W-28N	31
Hawthorne Blvd. E&W	19W-27N,18W-27N	39
Hickory St. N&S	18W-27N	40
High St. E&W	18W-27N	40
Highland Dr. N&S	19W-28N	31
Highland Rd. N&S	19W-28N	31
Hilgers Ct. N&S	18W-25N	40
Hillside Rd. E&W	19W-27N	39
Holcomb Dr. N&S	18W-28N	32
Hunt Ct. E&W	17W-26N	40

	Grid	Page
Huntington Ct. NE&SW	17W-27N	40
Huntington Dr. NS&EW	17W-27N	40
Idlewild Av. N&S	17W-27N,17W-28N	40
James Av. N&S	18W-27N,18W-28N	40
James Pl. NW&SE	18W-28N	32
Jeanette Pl. NE&SW	18W-28N	32
Karl Ct. N&S	18W-26N	40
Killarney Cir. E&W	19W-28N	31
Killarney Pass NW&SE	19W-28N	31
Killarney Pass Cir. E&W	19W-28N	31
Kings Way NW&SE	20W-28N	31
Knightsbridge Dr. E&W	17W-27N	40
Kreswick Dr. NW&SE	19W-27N	39
La Vista Dr. E&W	21W-27N	38
Lake Av. N&S	18W-26N,19W-28N	40
Lake Shore Dr. E&W	18W-27N	40
Lake St. N&S	18W-26N,19W-27N	40
Lake Ter. NW&SE	18W-27N	40
Lake View Dr. E&W	19W-27N	39
Lange St. NE&SW	18W-27N	40
Laramie St. E&W	19W-28N	31
Lawrence Dr. N&S	18W-26N	40
Leithton Rd. NS&EW	17W-26N	40
Lexington Ct. N&S	20W-28N	31
Lincoln Av. N&S	19W-27N	39
Linden Av. N&S	18W-27N	40
Lomond Dr. NS&EW	19W-28N	31
Londonderry Ct. E&W	19W-28N	31
Longfellow Ln. N&S	18W-26N	40
Longwood Ter. E&W	18W-26N	40
Lucerine Rd. N&S	19W-28N	31
Lucerne Ct. E&W	19W-28N	31
Lucerne Rd. N&S	19W-28N	31
Lyndale St. E&W	19W-28N	31
Manor Ln. NE&SW	20W-28N	31
Maple Av. NW&SE	19W-28N,18W-28N	31
Maplewood Av. E&W	18W-26N	40
Marlbough Ct. E&W	19W-27N	39
McCormick Av. N&S	17W-25N	40
McKinley Av. NE&SW	18W-27N,18W-28N	40
McRae Ln. NE&SW	20W-28N	31
Michale Av. N&S	19W-27N	39
Midland Av. N&S	18W-27N	40
Midlothian Av. N&S	19W-27N	39
Midlothian Rd. N&S	19W-28N,19W-29N	31
Midway Dr. E&W	18W-27N	40
Morris St. NW&SE	18W-28N	32
N. Shore Dr. E&W	19W-27N	39
New Castle Ct. N&S	17W-27N	40
North Av. N&S	18W-27N	40
North Shore Dr. E&W	19W-27N	39
Norton Av. N&S	18W-28N	32
Nottingham E&W	17W-25N	40
Oak St. E&W	18W-27N	40
Oakdale Av. E&W	18W-26N	40
Olga St. N&S	18W-26N	40
Orchard Av. E&W	18W-27N	40
Orchard St. E&W	19W-27N,18W-27N	39
Park Av. E&W	19W-28N	31
Park St. E&W	18W-28N	32
Parliment Way N&S	17W-27N	40
Pershing Av. N&S	19W-27N	39
Pershing Dr. NE&SW	19W-27N	39
Pleasure Dr. E&W	18W-26N	40
Prairie Av. N&S	19W-27N,19W-28N	39
Primrose Ln. NW&SE	22W-27N	38
Prospect Av. NW&SE	18W-27N	40
Pymouth Farms Rd. E&W	17W-25N	40
Quigley Av. E&W	19W-27N	39
Racine Pl. NE&SW	18W-28N	32
Raleigh Rd. E&W	19W-28N	31
Rays Ln. E&W	18W-26N	40
Regent Dr. NS&EW	20W-27N,19W-27N	39
Reidel Dr. N&S	17W-25N	40
Ridge Av. N&S	18W-26N	40
Ridgeland Av. N&S	19W-28N	31

	Grid	Page
Ridgemoor Av. N&S	19W-27N,19W-28N	39
Rotary Ct. N&S	18W-28N	32
Rouse Av. NW&SE	18W-27N	40
Rt. 60 NW&SE	20W-28N	31
Rt. 63 N&S	19W-27N,19W-29N	39
Rt. 176 E&W	19W-28N,18W-28N	31
Russell Pl. N&S	18W-26N	40
Rye Rd. E&W	19W-29N	31
S. Lake Shore Dr. E&W	18W-27N	40
S. Noel Dr. E&W	18W-27N	40
Salceda Dr. NE&SW	19W-29N	31
Salceda Ln. NW&SE	19W-29N	31
Sandhurst Rd. E&W	17W-27N	40
Seymour Av. N&S	18W-27N	40
Shaddle Av. N&S	18W-27N	40
Shady Dell Av. E&W	18W-26N	40
Shady Ln. E&W	18W-26N	40
Shady Rest Ridge N&S	18W-26N	40
Sherwood Ct. NE&SW	17W-27N	40
Simpson Dr. N&S	18W-26N	40
South St. N&S	18W-26N	40
Southport Rd. NW&SE	20W-27N	39
Spaulding Dr. E&W	20W-27N,19W-27N	39
Springbrook Ct.* E&W	20W-27N	39
Springbrook Dr.* E&W	20W-27N	39
Stafford Dr. N&S	19W-27N	39
Stratford Ln. E&W	18W-26N	40
Sunset Ln. E&W	19W-27N	39
Surridge Ct. E&W	20W-27N	39
Templeton Ct. E&W	20W-28N	31
Thomas Blvd. E&W	18W-27N	40
Thomas Ct. N&S	18W-27N	40
Thornton Way NW&SE	20W-27N,19W-26N	39
Tower Rd. N&S	18W-26N	40
Townline Rd. E&W	18W-25N,17W-25N	40
Trescott St. E&W	20W-27N	39
Turret Ct. NE&SW	18W-26N	40
U.S. Rt. 45 N&S	18W-26N,19W-27N	40
University Dr. N&S	18W-28N	32
Verde Ln. E&W	19W-29N	31
Victoria Rd. E&W	20W-28N	31
W. Garfield Av. N&S	19W-27N	39
Wakefield Ct. NW&SE	20W-27N	39
Walker Pl. E&W	18W-28N	32
Walnut Ct. NE&SW	18W-27N	40
Walnut St. N&S	18W-26N,18W-27N	40
Warwick Ln. N&S	19W-27N	39
Washington Blvd. N&S	18W-27N	40
Washington Ct. N&S	18W-27N	40
Waverly Dr. NW&SE	20W-26N,20W-27N	39
Wellington Av. E&W	19W-28N	31
Wellington Ct. N&S	19W-28N	31
West Greenview N&S	19W-27N	39
Westminister Pl. N&S	20W-28N	31
Weston Ln. N&S	19W-27N	39
Whitehall N&S	17W-25N	40
Whittier Ln. E&W	18W-26N	40
Wilcox St. E&W	18W-27N	40
Wildwood Av. N&S	18W-28N	32
Wilhelm Rd. E&W	17W-25N	40
Wimpole E&W	17W-25N	40
Windsor Pl. NE&SW	17W-27N	40
Wingate Dr. NE&SW	20W-27N	39
Winstead Pl. NE&SW	19W-27N	39
Winston Ct. NW&SE	20W-26N	39
Winthrop Ct. E&W	18W-26N	40
Woodcrest Cir. E&W	20W-28N	31
Woodcrest Ct. E&W	20W-28N	31
Woodcrest Dr. N&S	20W-28N	31
Woodhaven Ct. NE&SW	20W-28N	31
Woodhaven Dr. NS&EW	20W-28N	31
Woodlawn Dr. E&W	19W-27N	39

	Grid	Page
Wortham Cir. NW&SE	20W-26N	39
Wortham Dr. NE&SW	20W-26N	39
York Rd. NE&SW	20W-26N	39
Yorkshire Dr. NW&SE	20W-27N	39

Newport Township

	Grid	Page
Bayonne Av. N&S	13W-38N,3W-39N	18
Beach Rd. E&W	13W-38N	18
Belle Plaine Av. N&S	13W-38N,13W-39N	18
Boulevard View Av. N&S	13W-38N	18
Browne School Rd. N&S	16W-39N	17
Chaplin St. E&W	13W-38N	18
Crawford Rd. N&S	18W-38N,18W-43N	16
Dale Av. N&S	13W-38N	18
Delany Rd. N&S	14W-40N,14W-43N	9
Edwards Rd. E&W	18W-42N,16W-42N	8
Forestview Rd. N&S	13W-40N	10
Green Bay Rd. N&S	13W-42N,13W-43N	10
Greenview Dr. E&W	13W-40N	10
Hart St. E&W	13W-38N	18
Hickory Rd. E&W	15W-42N,14W-42N	9
Hunt Club Rd. N&S	17W-41N,17W-43N	8
Ingram Dr. E&W	14W-43N	9
Kaiser Rd. E&W	14W-40N	9
Kazmer Rd. E&W	14W-40N	9
Kelly Rd. E&W	18W-40N,16W-40N	8
Kilbourne Rd. N&S	14W-39N,14W-43N	17
Lester Ln. N&S	13W-40N	10
Magnolia Av. N&S	13W-38N,13W-39N	18
Martin St. E&W	13W-39N	18
McCarthy Rd. E&W	16W-38N,15W-38N	17
Milbourne Rd. E&W	18W-38N	16
Mill Creek Rd. N&S	16W-38N,16W-41N	17
Northwestern Av. N&S	13W-38N,13W-39N	18
Old Woodford Rd. N&S	15W-41N	9
Pickford St. E&W	13W-38N	18
Pine Grove Av. N&S	13W-38N,13W-39N	18
Plaza Ln. NW&SE	16W-39N	17
Rosecrans Rd. E&W	18W-41N,13W-41N	8
Rosedale Av. N&S	13W-38N	18
Rt. 131 N&S	13W-42N,13W-43N	10
Rt. 173 E&W	18W-41N,13W-41N	8
Russell Rd. E&W	16W-43N,14W-43N	8
Sharon St. E&W	13W-39N	18
Sheridan Oaks Dr. N&S	18W-43N	8
Stateline Rd. E&W	18W-43N,16W-43N	8
Stonegate Rd. E&W	13W-40N	10
Sunset Rd. N&S	13W-40N	10
Sylvan Av. N&S	13W-38N	18
Timberland Tr. N&S	13W-40N	10
Town Line Rd. E&W	13W-38N	18
Tri State Tollway N&S	16W-38N,16W-43N	17
Wadsworth Rd. E&W	13W-39N	18
Waveland Av. N&S	13W-38N	18
Waverly St. E&W	13W-40N	10
Winthrop Harbor Rd. E&W	14W-42N,13W-42N	9
Woodland Av. N&S	13W-38N	18
Yorkhouse Rd. E&W	15W-38N	17
9th St. E&W	15W-42N,14W-42N	9
27th Pl. E&W	13W-40N	10
27th St. E&W	13W-40N	10
28th Pl. E&W	13W-40N	10
28th St. E&W	13W-40N	10
29th St. E&W	13W-40N	10

North Barrington

	Grid	Page
Aberdour Ct. N&S	24W-25N	38
Apex Ct. N&S	25W-23N	45
Arboretum Ct.* E&W	25W-21N	45
Arboretum Dr.* E&W	25W-21N	45
Arrowhead Ln. N&S	25W-24N	37
Averill Ct. N&S	24W-25N	38
Beachview Ln. NS&EW	25W-23N	45
Bexley Ct. NW&SE	25W-24N	37
Biltmore Dr. N&S	25W-23N	45
Blanche Ct. E&W	25W-22N	45
Border Ln. E&W	26W-23N	45
Brook Cir. NE&SW	25W-22N	45
Brook Forest Ln. NW&SE	25W-22N	45
Brookmont Ln. E&W	25W-23N	45
Brookside Rd. NE&SW	26W-23N,25W-23N	45
Candlewood Ct. NS&EW	24W-25N	38
Candlewood Dr. N&S	24W-25N	38
Carriage Rd. N&S	25W-24N	37
Castle View Ct. E&W	25W-23N	45
Castleton Ct. NS&EW	25W-24N,24W-25N	37
Cherry Hill Rd. E&W	26W-23N	45
Christopher Dr. NS&EW	24W-22N	46
Clarington Way NS&EW	24W-25N	38
Clover Hill Ln. NS&EW	24W-23N	46
Concord Ln. E&W	25W-23N	45
Coventry Ln. NE&SW	25W-24N	37
Crooked Ln. E&W	26W-23N,25W-23N	45
Crosswicks Ct. NW&SE	24W-24N	38
Daily Ln. E&W	25W-23N	45
Deverell Dr. NS&EW	25W-24N	37
Drury Ln. N&S	25W-23N	45
Duck Pond Ln. E&W	25W-22N	45
Duxbury Dr. NW&SE	25W-24N	37
Edgewood Rd. E&W	25W-23N	45
Essex Ln. N&S	25W-23N	45
Eton Dr. NE&SW	26W-23N,25W-23N	45
Glen Circle Dr. NE&SW	26W-23N,25W-23N	45
Golfview Dr. NE&SW	26W-23N,25W-23N	45
Grassmere Ct. E&W	26W-22N	45
Halcyon Ln. E&W	25W-23N	45
Hallbraith Ct. N&S	24W-25N	38
Haversham Ln. N&S	25W-24N	37
Haverton Way NS&EW	25W-22N	45
Hawthorne Rd. NE&SW	265W-23N	45
Hewes Dr. N&S	25W-22N	45
Hidden Brook Dr. NS&EW	25W-24N,24W-24N	37
Hidden Oaks Ln. NS&EW	25W-22N	45
Hidden Pond Ln. E&W	25W-23N	45
Hillandale Ct. NW&SE	25W-22N	45
Hillburn Ct. N&S	25W-24N	37
Hillburn Ln. N&S	25W-24N	37
Homewood Ln. N&S	25W-23N	45
Honey Lake Ct. NE&SW	25W-22N	45
Honey Lake Rd. N&S	25W-22N	45
Honey Ln. NE&SW	25W-23N	45
Kathleen Ct. E&W	25W-23N	45
Kensington Dr. NW&SE	25W-24N	37
Kettering Ct. NW&SE	25W-24N	37
Kimberly Rd. N&S	25W-23N	45
Lake Shore Blvd. N&S	26W-23N	45
Lakeside Ct. NE&SW	24W-25N	38
Lakeside Dr. NS&EW	24W-25N	38
Lakeview Rd. N&S	26W-23N	45
Larkins Ln. N&S	24W-22N	46
Marbridge Ct. NE&SW	25W-24N	37
Masland Ct. N&S	24W-22N	46
Miller Rd. E&W	26W-23N,25W-23N	45
Mockingbird Ln. NS&EW	25W-23N	45
Mohawk Dr. N&S	26W-22N	45
North Hill Dr. E&W	25W-23N	45
North Wynstone Dr. NS&EW	25W-25N,24W-24N	37
Old Oak Dr. N&S	25W-22N	45
Onondaga Dr. E&W	26W-22N	45

	Grid	Page
Orchard Rd. NW&SE 25W-23N		45
Osage Dr. N&S	26W-22N	45
Ottawa Dr. N&S	26W-22N	45
Oxford Rd. NW&SE 25W-23N		45
Pheasant Run Trail N&S 25W-23N		45
Piermont Dr. N&S	24W-24N	38
Pinewoods Dr. E&W 25W-22N		45
Pook's Hill Rd. NE&SW 25W-22N		45
Private Ct. NS&EW 25W-24N		37
Rainbow Rd. NS&EW 25W-22N,24W-21N		45
Randolph Ct. NW&SE 25W-22N		45
Redwing Ln. N&S	25W-24N	37
Riderwood Rd. E&W 24W-25N		38
Rugby Rd. E&W	25W-23N	45
Rustic Dr. N&S	25W-23N	45
Saddle Tree Ln. N&S 25W-22N		45
Schaffer Ln. N&S 24W-21N,24W-22N		46
Seminole Dr. N&S	25W-23N	45
Shady Ln. NE&SW	25W-23N	45
Signal Hill Rd. NS&EW 26W-23N,25W-23N		45
Sioux Dr. E&W	26W-22N	45
South Wynstone Dr. NS&EW 25W-24N,24W-24N		37
Stratham Cir. E&W	25W-24N	37
Thomas Dr.	25W-22N	45
U.S. Route 59 N&S 25W-22N,26W-23N		45
View Ct. N&S	25W-23N	45
Weybridge Ln. NW&SE 25W-24N		37
Williams Park Rd. N&S 27W-27N		36
Woodbine Pl. NW&SE 25W-23N		45
Wooded Ln. N&S	26W-23N	45
Woodland Dr. NW&SE 26W-23N,25W-23N		45
Wynstone Way N&S 25W-24N		37

North Chicago

	Grid	Page
Adams St. N&S	10W-32N	27
Argonne Dr. E&W	11W-31N	35
Arlington Av. NS&EW 11W-31N		26
Barrett Av. E&W	11W-32N	26
Barry Av. E&W	12W-30N	35
Beacon Av. E&W	12W-30N	35
Berwyn Av. E&W	12W-30N	35
Broadway Av. NE&SW 11W-31N		35
Brompton Rd. E&W 12W-30N		35
Buckley Rd. E&W	12W-30N	35
Carver Av. E&W	11W-32N	26
Commonwealth Av. N&S 11W-31N		35
Davis St. N&S	11W-31N	35
Dickey Av. N&S	11W-31N	26
Dickey St. N&S	11W-31N	35
Dugdale Rd. N&S 11W-31N,11W-32N		35
Elizabeth Av. N&S 11W-31N,11W-32N		35
Fairbank Av. N&S	12W-30N	35
Farwell Av. N&S	12W-30N	35
Fellows Pl. NE&SW 10W-31N		35
Foss Park Av. NS&EW 11W-31N,10W-32N		35
Frontenac Av. N&S 12W-30N		35
Glenn Dr. N&S 11W-31N,11W-32N		35
Golf Course Dr. N&S 12W-31N,12W-32N		35
Grant Pl. E&W	10W-32N	27
Green Bay Rd. NW&SE 12W-31N		35
Greenfield Av. N&S 11W-31N,11W-32N		35
Greenwood (Natoma) Av. N&S	12W-31N	35
Grove Av. N&S 11W-31N,11W-32N		35
Hervey Av. N&S 11W-31N,11W-32N		35
Hillcrest Av. N&S 12W-31N,12W-32N		35
Honore Av. N&S	11W-31N	35
Honore St. N&S	11W-31N	35
Jackson St. N&S 10W-31N,10W-32N		35
Jones Dr. E&W	12W-32N	26
Kemble Av. N&S 11W-31N,11W-32N		35
Kennedy Dr. NE&SW 11W-32N		26
Kristan Av. N&S 11W-31N,11W-32N		35

	Grid	Page
Lakeside Av. NE&SW 10W-32N		27
Lawn Av. N&S	12W-30N	35
Lenox Av. N&S	10W-32N	27
Leona Av. E&W	11W-31N	35
Lewis Av. N&S	11W-31N	35
Lincoln St. N&S	10W-31N	35
Main St. NE&SW	10W-31N	35
Marquette St. N&S	10W-31N	35
Mc Alister Av. N&S 10W-32N		27
Meadow Ln. N&S	12W-31N	35
Morrow Av. NE&SW 11W-31N		35
Natoma (Greenwood) Av. N&S	12W-31N	35
Neil Ln. E&W	11W-31N	35
Northern Av. N&S 12W-31N,12W-32N		35
Park Av. N&S 10W-31N,10W-32N		35
Prospect Av. N&S 11W-31N,11W-32N		35
Renken Av. N&S	11W-32N	26
Rt. 131 N&S 11W-31N,11W-31N		35
Rt. 137 N&S	11W-30N	35
Seymour Av. N&S 11W-31N,11W-32N		35
Sheridan Rd. NE&SW 10W-31N,10W-32N		35
Sherman Av. N&S 11W-31N,11W-32N		35
U.S. Route 41 N&S 12W-32N		26
Victoria N&S 10W-31N,10W-32N		35
Wadsworth Av. N&S 10W-32N		27
Wallace Av. N&S	11W-31N	35
Wallace St. N&S	11W-31N	35
Willow Av. N&S	12W-32N	26
Winter Av. N&S	11W-31N	35
Wright St. N&S	11W-31N	35
11th St. E&W	10W-32N	27
12th St. E&W	10W-32N	27
13th St. E&W	10W-32N	27
14th Pl. E&W	10W-32N	26
14th St. E&W 12W-32N,10W-32N		26
15th Pl. E&W	10W-32N	26
15th St. E&W 12W-32N,10W-32N		26
16th Pl. E&W	10W-32N	26
16th St. E&W 12W-32N,10W-32N		26
17th Pl. E&W	12W-32N	26
17th St. E&W 12W-32N,10W-32N		26
18th Pl. E&W	12W-32N	35
18th St. E&W 12W-32N,10W-32N		26
19th Pl. E&W	12W-31N	35
20th Pl. E&W	12W-31N	35
20th St. E&W 12W-31N,11W-31N		35
21st Pl. E&W	12W-31N	35
21st St. E&W 12W-31N,10W-31N		35
22nd Pl. E&W	12W-31N	35
22nd St. E&W	11W-31N	35
23rd Pl. E&W	12W-31N	35
23rd St. E&W	12W-31N	35
24th Pl. E&W	11W-31N	35
24th St. E&W 12W-31N,11W-31N		35
25th St. E&W	11W-31N	35

Old Mill Creek

	Grid	Page
Brownie Sch. Rd. N&S 16W-39N		17
Crawford Rd. N&S 18W-38N,18W-40N		16
Hunt Club Rd. N&S 17W-37N,17W-41N		16
Interstate 94 N&S	16W-39N	17
Kelly Rd. E&W 18W-40N,16W-40N		8
Milburn Rd. E&W 18W-38N,17W-38N		16
Sand Lake Rd. E&W 17W-37N,16W-37N		16
Tri State North Tollway N&S 16W-37N,16W-40N		17
Wadsworth Rd. E&W 17W-39N,16W-39N		16

Park City

	Grid	Page
Ann Av. E&W	13W-34N	26
Barb Av. E&W	13W-33N	26
Bedford Ct. N&S	13W-33N	26
Belvidere Rd. E&W 13W-33N		26
Betty Av. NW&SE	13W-33N	26
Birch Av. N&S	13W-33N	26
Bryn Mawr Av. N&S 13W-33N		26
Cambridge Dr. E&W 13W-33N		26
Cornell Av. N&S 14W-33N,13W-33N		26
Crestwood* E&W	12W-32N	26

	Grid	Page
Devonshire Av. E&W 13W-33N		26
Diana Av. E&W	13W-33N	26
Dixie Av. E&W 13W-33N,12W-33N		26
Dorset Ct. NW&SE 13W-33N		26
Dromey Av. N&S	12W-33N	26
Eastwood Pl. E&W	14W-33N	26
Elaine Av. E&W	13W-33N	26
Eli St. E&W	12W-33N	26
Eric Blvd. E&W	13W-33N	26
Eva Av. E&W	13W-33N	26
Everett Pl. E&W	14W-33N	26
First St. E&W	13W-33N	26
Greenleaf St. N&S	13W-33N	26
Greenview Av. N&S 13W-33N		26
Greenview Dr. N&S 13W-33N		26
Gulf Av. E&W	13W-33N	26
Harvey E&W	13W-33N	26
Harwick Ln. E&W	13W-33N	26
Hillcrest* N&S	12W-33N	26
Hood Av. N&S	12W-33N	26
Howard Av. E&W	13W-33N	26
Ingleside Pl. N&S 14W-33N,13W-33N		26
Jeanette Av. E&W	13W-33N	26
Jen Av. E&W	13W-33N	26
Joy Av. E&W	13W-34N	26
June Av. E&W	13W-34N	26
Kehm Blvd. E&W 13W-33N,12W-33N		26
Knight Av. N&S	12W-33N	26
Lake Park Av. E&W 14W-33N,13W-33N		26
Longview Dr.* E&W 12W-32N		26
Lowe St. E&W	13W-33N	26
Main St. E&W	13W-33N	26
Marshfield Av. E&W 13W-33N		26
Mary Av. E&W	13W-33N	26
Meadow Dr.* E&W	13W-33N	26
Michael Av. E&W	13W-33N	26
Mobile Blvd. E&W	12W-33N	26
Murphy St. E&W	13W-33N	26
Noll Dr. E&W	12W-33N	26
Northwood Av. E&W 14W-33N,13W-33N		26
Old Plank Rd. NW&SE 13W-33N		26
Oldsmar Ln. N&S	12W-33N	26
Pace Av. E&W	13W-33N	26
Palmori Cir.* N&S	12W-32N	26
Patricia Ln. E&W	12W-33N	26
Richard Av.* N&S	12W-32N	26
Roger Av. E&W	13W-33N	26
Rollo Blvd. E&W	13W-33N	26
Rosedale Av. N&S	13W-33N	26
Rt. 120 E&W	13W-33N	26
Sharon Av. N&S	12W-33N	26
Staben Av. N&S	12W-33N	26
Susan Circle, East N&S 13W-33N		26
Susan Circle, North E&W 13W-33N		26
Susan Circle, South E&W 12W-33N		26
Susan Circle, West N&S 13W-33N		26
Sussex Ct. NE&SW 13W-33N		26
Sutherland Dr. E&W 13W-33N		26
Ted Av. E&W	13W-33N	26
Teske Blvd. N&S 13W-33N,12W-33N		26
Thorndale Av. N&S 13W-33N		26
Troy Dr. E&W	13W-33N	26
Warren Av. E&W	13W-34N	26
West St. N&S	13W-33N	26
Westgate* N&S	12W-32N	26
Will Av. NW&SE	13W-33N	26
Wilson Av. N&S	13W-33N	26
Winthrop Av. N&S	13W-33N	26
Wood St. N&S	13W-33N	26
7th St. E&W	13W-33N	26
8th St. E&W	12W-33N	26
9th St. E&W	12W-33N	26

Riverwoods

	Grid	Page
Arrow Ln. E&W	12W-21N	49
Arrowwood Tr. E&W 13W-20N		49
Baneberry Ln. E&W 14W-21N		49
Big Oak Ln. E&W	12W-21N	49
Blackhawk Ln. N&S 13W-21N		49
Blackheath Ln. N&S 14W-21N		49
Blackhorn Rd. E&W 14W-20N		49
Brae Burn Ln. N&S 14W-21N		49
Braeburn Dr. NS&EW 12W-21N		49
Burr Oak Ln. E&W	12W-21N	49
Calvin Ct. N&S	13W-22N	49
Chapel Gate N&S	13W-20N	49

	Grid	Page
Cherokee Ln. E&W 13W-21N		49
Chianti Tr. E&W	13W-20N	49
Chicory Ln. NS&EW 14W-21N		49
Chippewa Pathway N&S 13W-21N		49
Clendenin Ln. N&S	13W-21N	49
Columbine Ln. NS&EW 14W-21N		49
Corporate Dr. NS&EW 13W-20N		49
County Line Rd. E&W 13W-20N,12W-20N		49
Crestwood Ln. E&W 13W-21N		49
Daquiri Tr. E&W	13W-20N	49
Dawson Ct. N&S	12W-22N	49
Deepwood Tr. E&W 13W-20N		49
Deerfield Rd. E&W 14W-21N,12W-21N		49
Dubonnet N&S	13W-20N	49
East Course Dr. N&S 12W-21N,12W-22N		49
Edgewood Dr. N&S 13W-21N,13W-22N		49
Edgewood Ln. E&W 13W-22N		49
Farner Ct. E&W	12W-22N	49
Forest Glen Ct. NE&SW 13W-21N		49
Forest Glen Trail E&W 13W-21N		49
Foxtail Ct. N&S	14W-21N	49
Gemini Ln. E&W	13W-20N	49
Greenbrier Ln. E&W 13W-20N		49
Hazel Nut Ln. E&W 13W-20N		49
Hiawatha Ln. N&S	13W-21N	49
Hill Ln. E&W	12W-22N	49
Hoffman Ln. N&S	13W-21N	49
Indian Trail Dr. N&S 13W-21N		49
Julie Ln. E&W	13W-22N	49
Juneberry Rd. N&S 14W-20N		49
Kenilwood Ct. N&S 13W-21N		49
Kenilwood Ln. NS&EW 13W-21N		49
Kingswood Ct. NE&SW 13W-21N		49
Knollwood Way E&W 13W-21N		49
Lyndale Ln. N&S	13W-21N	49
Lyngby Ct. N&S	13W-22N	49
Meredith Dr. E&W	13W-20N	49
Metawa Ln. N&S	14W-21N	49
Oak Ln N&S	12W-21N	49
Oakhurst Ln. NE&SW 12W-21N		49
Orange Brace Rd. E&W 13W-20N		49
Pinetree Ln. NE&SW 13W-21N		49
Portwine Rd. N&S	13W-20N	49
Ringland Ln. N&S	13W-20N	49
Riverwoods Ct. E&W 13W-21N		49
Riverwoods Rd. NW&SE 13W-21N		49
Robinhood Ln. N&S 13W-22N		49
Saunders Rd. N&S	12W-21N	49
Scotch Ln. E&W	13W-20N	49
Shawnee Tr. N&S	13W-21N	49
Sherry Ln. NW&SE 13W-21N		49
Strenger Ln. N&S	13W-22N	49
Studio Ln. NS&EW	13W-21N	49
Sunset Tr. E&W	13W-21N	49
Thornmeadow Rd. N&S 14W-20N		49
Thornwood Ln. N&S 13W-22N		49
Timber Trail N&S	13W-20N	49
Timberleaf Ln. E&W 13W-21N		49
Timberline Ln. E&W 14W-21N		49
Timberwood Ln. N&S 14W-21N		49
Treasure Ln. N&S	13W-20N	49
Trillium Ln. N&S	13W-20N	49
Vernon Trail N&S	13W-21N	49
West Course Dr. N&S 13W-22N,12W-22N		49
Whigam Ln. E&W	13W-21N	49
Whigham Rd. N&S	13W-21N	49
White Oak Ln. N&S 13W-20N		49
Windlake Av. N&S	12W-22N	49
Woodland Ct. N&S 13W-21N		49
Woodland Ct., E. E&W 13W-21N		49
Woodland Ct., W. E&W 13W-21N		49
Woodland Ln. N&S 12W-22N		49
Woodland Ln. NS&EW 13W-21N		49

	Grid	Page
Yasmine Dr. N&S	13W-21N	49

Round Lake

	Grid	Page
A'ok Ct. NW&SE	24W-34N	22
Alma Cir. E&W	24W-34N	22
Alma Ct. E&W	24W-34N	22
Alpine Dr. NE&SW	23W-33N	22
Antares Cir. E&W	24W-34N	22
Apollo Ct. N&S	24W-34N	22
Avilon Av. E&W	23W-33N	22
Avilon Ct. N&S	23W-33N	22
Beechwood Dr. N&S 23W-33N,23W-34N		22
Belvidere Rd. NE&SW 23W-32N		22
Bernice Ln. N&S	23W-34N	22
Biminy Bay NW&SE 23W-33N		22
Capri Ct. E&W	23W-33N	22
Capri Dr. N&S	23W-33N	22
Catalpa Dr. N&S	23W-34N	22
Cedar Crest Ct. NW&SE 23W-33N		22
Cedar Lake Rd. N&S 23W-32N,23W-34N		22
Dawn Marie Dr. N&S 25W-33N		21
Dorothy Ct. NE&SW 23W-34N		22
Fairfield Rd. N&S 24W-33N,24W-34N		22
Forest Av. E&W	23W-33N	22
Garden Ct. N&S	23W-33N	22
Goodnow Blvd. N&S 23W-33N		22
Hart Rd. E&W 24W-34N,23W-34N		22
High Point Rd. N&S 24W-33N		22
High School Dr. NE&SW 23W-34N		22
Hillandale Dr. N&S	23W-33N	22
Hillside Dr. E&W	23W-34N	22
Houston Ct. N&S	24W-34N	22
Kitty Hawk Dr. N&S 23W-34N		22
Lake Dr. E&W	23W-34N	22
Lakeview Dr. NE&SW 23W-34N		22
Lakewood Ter. N&S 23W-33N,23W-34N		22
Laurel Av. E&W	23W-33N	22
Lawn Ter. NE&SW	23W-34N	22
Lem Dr. Cir.	23W-34N	22
Lincoln Av. N&S	23W-33N	22
Linden Dr. N&S	23W-34N	22
Long Lake Dr. E&W 24W-34N		22
Long Lake Rd. E&W 25W-34N,24W-34N		21
Lunar Dr. N&S	24W-34N	22
Lunga Ct. N&S	24W-34N	22
Lunga Dr. N&S	24W-34N	22
MacGillis Dr. N&S	23W-33N	22
Magna Ct. N&S	23W-34N	22
Maple Lawn NE&SW 23W-33N		22
Midland Dr. E&W	23W-33N	22
Moon Rover Dr. N&S 24W-34N		22
N. Bernice Ct. E&W 23W-34N		22
N. Rosedale Ct. E&W 23W-34N		22
Nasa Cir. Cir.	24W-34N	22
Nippersink Rd. E&W 25W-33N,23W-33N		21
Orbiter Dr. E&W	24W-34N	22
Orchard Av. N&S	23W-33N	22
Park Av. N&S	23W-34N	22
Park Rd. NS&EW	23W-34N	22
Petite St. N&S	23W-34N	22
Pineview Dr. NW&SE 23W-33N		22
Rail Rd. N&S	23W-34N	22
Railroad Rd. NW&SE 23W-33N		22
Ravine Av. E&W	23W-33N	22
Ridgewood Dr. N&S 23W-33N		22
Roberta Ct. NE&SW 23W-34N		22
Round Lake Rd. NW&SE 24W-34N		22
Rt. 120 E&W	23W-32N	22
Rt. 134 NW&SE 25W-34N,23W-33N		21
S. Bernice Ct. E&W 23W-34N		22
S. Rosedale Ct. E&W 23W-33N		22
Spanky Ct. N&S	23W-33N	22
Sunset Dr. E&W	23W-33N	22
Sycamore Dr. NE&SW 23W-34N		22
Valleyview Dr. N&S 25W-33N		21
Washington St. E&W 23W-34N		22
Wilson Rd. N&S	25W-33N	21

	Grid	Page

Round Lake Beach

Acorn Ln. NE&SW . 21W-35N — 23
Amber Prairie Ct.* E&W
.... 21W-36N — 15
Amber Prairie Way* E&W
.... 21W-36N — 15
Apache Trail N&S .. 23W-35N — 22
Ardmor Dr. N&S .. 23W-35N — 22
Aster Pl. NW&SE .. 23W-36N — 14
Barberry Ln. N&S .. 24W-34N — 22
Bayview Dr. NW&SE
.... 23W-34N — 22
Beachview Dr. E&W
.... 23W-34N — 22
Beverly Dr. N&S .... 24W-35N — 22
Birchwood Dr.* E&W
.... 22W-36N — 14
Bonnie Brook Ln. N&S
.... 24W-34N — 22
Brentwood Dr. N&S
.... 24W-34N,24W-35N — 22
Briar Ln. NW&SE . 23W-34N — 22
Bristol Ln. N&S .... 23W-35N — 23
Brittany Ln. N&S .. 21W-35N — 23
Buena Vista Ct. E&W
.... 24W-34N — 22
Carl Ct.* E&W ...... 22W-35N — 22
Carriage Ct. N&S .. 21W-35N — 23
Carriage Ln. N&S .. 21W-35N — 23
Cedar Lake Rd. N&S
.... 23W-35N — 22
Central Park Dr. E&W
.... 24W-34N — 22
Cherokee Dr. N&S.. 24W-35N — 22
Cherry Cove Ln. NS&SW
.... 22W-36N — 14
Chestnut Dr. N&S .. 24W-34N — 22
Chickory Ln.* E&W
.... 21W-36N — 15
Circuit Dr. N&S .. 23W-35N — 22
Clarendon Dr. NE&SW
.... 23W-34N,22W-35N — 22
Clover Ln. E&W .... 22W-35N — 22
Comanche Trail N&S
.... 23W-35N — 22
Corona Ct. E&W .. 24W-34N — 22
Country Walk Dr. E&W
.... 23W-36N — 14
Countryside Ln. N&S
.... 23W-35N — 22
Crescent Ct. N&S .. 24W-34N — 22
Dahlia Ln. NE&SW .. 23W-36N — 14
Daisy Ln. E&W .... 22W-35N — 22
Dakota Ct. E&W .. 23W-35N — 22
Deerpath Ct. N&S .. 24W-35N — 22
Diana Ct. NW&SE .. 24W-35N — 22
Eagle Creek Dr. E&W
.... 23W-35N — 22
East End Av. N&S .. 23W-35N — 22
Edgewater Ct. N&S
.... 23W-34N — 22
Elm Av. N&S ...... 23W-35N — 22
Evergreen Ct.* E&W
.... 22W-35N — 22
Fairfield Rd. N&S
.... 24W-34N,24W-35N — 22
Fairfield Ter. E&W .. 23W-35N — 22
Fairview Pkwy. E&W
.... 22W-35N — 22
Ferndale Ct. E&W .. 23W-34N — 22
Foxchase Dr. E&W 21W-36N — 14
Fruitwood Dr. E&W
.... 23W-36N — 14
Geneva Ter. E&W .. 23W-35N — 22
Glenview Ct. NW&SE
.... 23W-35N — 22
Glenwood Dr. E&W
.... 23W-34N — 22
Goldenwood Ter. N&S
.... 23W-35N — 22
Golfview Ct. E&W
.... 23W-35N,22W-35N — 22
Golfview Dr. E&W
.... 24W-35N,23W-35N — 22
Grove Dr. N&S .... 23W-35N — 22
Hainesville Rd. NE&SW
.... 23W-35N — 22
Hardwood Path NE&SW
.... 21W-35N — 23
Hawthorne Dr. NE&SW
.... 23W-34N,22W-35N — 22
Heather Ter. E&W .. 22W-35N — 22
Hickory Av. N&S .. 22W-35N — 22
Hickory St. N&S .... 22W-34N — 22
Highland Pkwy. N&S
.... 23W-35N — 22
Highland Ter. E&W 23W-35N — 22
Hillside Av. N&S .... 24W-35N — 22
Hillside Dr. E&W .. 22W-35N — 22
Hillwood Ct. N&S .. 24W-34N — 22
Idlewild Dr. N&S
.... 23W-34N,23W-35N — 22
Ivy Ln. E&W ...... 23W-35N — 22
Juneway Ter. N&S . 23W-35N — 22
Juniper Ter. E&W .. 23W-35N — 22
Kelly Cain Ct. E&W
.... 21W-36N — 15
Kenilworth Dr. N&S
.... 23W-34N — 22
Kenmore Av. N&S .. 22W-35N — 22
Kildeer Dr. N&S .... 24W-35N — 22
Lagoon Ter. N&S .. 23W-35N — 22
Lake Av. E&W .... 22W-34N — 22

Lake Shore Dr. N&S
.... 23W-34N,23W-35N — 22
Lakewood Pkwy. NE&SW
.... 23W-35N — 22
Laurel Ct. E&W .. 23W-35N — 22
Leslie Av. N&S .... 22W-35N — 22
Lilac Ct. E&W .. 22W-35N — 22
Lotus Dr. N&S
.... 24W-34N,24W-35N — 22
Mallard Creek Dr. NS&EW
.... 23W-35N — 22
Maple Dr. N&S .. 23W-35N — 22
Mayfield Dr. N&S
.... 24W-34N — 22
Meadow Green Ln. E&W
.... 23W-34N — 22
Meadowbrook Dr. NW&SE
.... 23W-35N — 22
Meadowhill Ln. E&W
.... 23W-35N — 22
Melrose Av. N&S .. 22W-35N — 22
Millstone Ct. E&W .. 21W-36N — 15
Morningside Dr. E&W
.... 24W-35N — 22
N. Channel Dr. NS&EW
.... 23W-34N,23W-35N — 22
Normandle Ln. E&W
.... 23W-36N — 14
North Av. N&S .. 22W-35N — 22
Oak Av. N&S .... 22W-35N — 22
Oak Leaf Cir. NW&SE
.... 21W-35N — 23
Oak Leaf Ct. NW&SE
.... 21W-36N — 15
Oak Leaf Ln. N&S .. 21W-35N — 23
Oak St. N&S .... 22W-34N — 22
Oak Tree Cir. E&W
.... 21W-35N — 23
Oaktree Tr. Cir. .... 21W-35N — 23
Oakwood Dr. E&W
.... 24W-35N,23W-34N — 22
Old Oak Dr. NW&SE
.... 21W-35N — 23
Old Rollins Rd. E&W
.... 21W-35N — 23
Orchard Dr. N&S .. 23W-35N — 22
Park Dr. N&S .... 23W-35N — 22
Partridge Ct. N&S .. 23W-35N — 22
Pheasant Ct. E&W.. 23W-35N — 22
Pine Grove Av. N&S
.... 23W-35N — 22
Pleasant Dr. N&S .. 23W-35N — 22
Poplar Av. N&S .... 22W-35N — 22
Redwing Dr. E&W .. 22W-35N — 22
Redwood Dr.* N&S
.... 22W-36N — 14
Ridgeway Av. N&S 22W-35N — 22
Rollins Rd. E&W
.... 24W-35N,21W-35N — 22
Ronald Ter. NW&SE
.... 24W-34N — 22
Rosewood Ln. NE&SW
.... 23W-36N,22W-36N — 14
Round Lake Dr. N&S
.... 23W-35N — 22
Rt. 83 N&S .... 21W-35N — 23
S. Channel Dr. E&W
.... 23W-34N — 22
Sauk Ct.* E&W .. 23W-36N — 14
Shady Lane Ct. N&S
.... 23W-34N — 22
Shady Ln. E&W .. 24W-34N — 22
Shoreland Ct. N&S 23W-34N — 22
Shorewood Rd. E&W
.... 24W-34N — 22
Southmoor Ln. E&W
.... 24W-34N — 22
Sprucewood Ct.* E&W
.... 22W-35N — 22
Sunnyside Dr. NW&SE
.... 23W-34N — 22
Sunrise Dr. N&S .... 23W-36N — 14
Sunset Dr. N&S
.... 23W-34N,23W-35N — 22
Turnbull Dr. N&S .. 23W-35N — 22
Villa Vista Dr. N&S..24W-34N — 22
Vine Av. NE&SW .. 22W-35N — 22
Walnut Dr. N&S .. 23W-35N — 22
West End Dr. N&S
.... 23W-34N,22W-35N — 22
West End Ln. N&S..23W-34N — 22
Westview Ln. N&S .. 23W-35N — 22
Wildwood Dr. E&W
.... 22W-35N — 22
William Av. NE&SW
.... 22W-35N — 22
Willow Ridge Dr.* N&S
.... 22W-36N — 14
Witchwood Ln. NW&SE
.... 23W-34N — 22
Wood Oak Cir. N&S
.... 21W-35N — 23
Wood Oak Ct. N&S
.... 21W-36N — 15
Wood Oak Dr. N&S
.... 21W-35N,21W-36N — 23
Woodbine N&S 23W-35N — 22
Woodland Dr. E&W
.... 23W-34N — 22
Woodmoor Dr. N&S
.... 22W-35N — 22
Woodridge Dr. E&W
.... 24W-34N,24W-35N — 22
Yvonne Ct.* E&W . 22W-35N — 22

Round Lake Heights

Arrowhead Dr. E&W
.... 23W-35N — 22
Brentwood Dr. N&S
.... 24W-35N — 22
Clinton Av. N&S .... 24W-35N — 22
E. Cedarwood Cir. N&S
.... 23W-35N — 22
Fairfield Rd. N&S .. 24W-35N — 22
Flintwood Ct. E&W 24W-35N — 22
Hiawatha Tr. N&S .. 24W-35N — 22
Lotus Dr. N&S .... 24W-35N — 22
Meadowbrook Dr. N&S
.... 23W-35N — 22
Mohawk Dr. E&W . 24W-35N — 22
N. Cedarwood Cir. E&W
.... 23W-35N — 22
N. Summit Ct. N&S
.... 23W-35N — 22
Pontiac Ct. N&S .... 24W-35N — 22
Rollins Rd. E&W ..24W-35N — 22
S. Cedarwood Cir. E&W
.... 23W-35N — 22
Summit Ct. N&S.... 23W-35N — 22
Tomahawk Tr. NS&EW
.... 24W-35N — 22
Turkey Run Dr. E&W
.... 23W-35N — 22
W. Cedarwood Cir. N&S
.... 24W-35N — 22
Warrior Dr. E&W .. 24W-35N — 22

Round Lake Park

Alleghany Rd. N&S
.... 21W-30N,21W-31N — 31
Arbor Dr. N&S......22W-34N — 22
Barbara Ct. NW&SE
.... 23W-33N — 22
Beechwood Dr. N&S
.... 23W-33N — 22
Bellevue Dr. N&S .. 22W-33N — 22
Belvidere Rd. E&W
.... 23W-33N,22W-33N — 22
Bengson Ct. E&W .. 22W-33N — 22
Brierhill Dr. N&S ... 22W-33N — 22
Cedar Mound Rd. NE&SW
.... 23W-33N — 22
Center Dr. N&S .... 23W-33N — 22
Clifton Dr. N&S .... 22W-33N — 22
Creekside Ln. N&S 23W-33N — 22
Elder Dr. E&W......22W-33N — 22
Elm Av. N&S ...... 22W-34N — 22
Fairlawn Dr. N&S .. 22W-33N — 22
Forest Glen Dr. N&S
.... 22W-33N — 22
Grandview Dr. N&S
.... 23W-33N — 22
Greenwood Rd. N&S
.... 23W-33N — 22
Hainesville Rd. N&S
.... 22W-34N — 22
Highmoor Dr. N&S..22W-34N — 22
Hillcrest Ter. N&S . 22W-34N — 22
Indian Tr. E&W .... 22W-34N — 22
Jorgensen Ct. E&W
.... 22W-34N — 22
Kenwood Dr. N&S .. 22W-33N — 22
Knollwood Dr. NE&SW
.... 22W-34N — 22
Lake Shore Dr. N&S
.... 22W-34N — 22
Locust Dr. E&W .... 22W-33N — 22
Lyle Av. E&W ...... 22W-34N — 22
Main St. NW&SE .. 23W-33N — 22
Mayflower Ter. E&W
.... 22W-34N — 22
Midland E&W ...... 22W-34N — 22
Model Ct. NW&SE .. 23W-33N — 22
Newhouse Ln. N&S
.... 23W-33N — 22
Oak St. N&S ...... 22W-34N — 22
Parkview Ct. E&W .. 22W-33N — 22
Peterson Rd. E&W
.... 22W-30N,20W-30N — 30
Pineview Dr. NW&SE
.... 22W-33N — 22
Porter Dr. NW&SE ..22W-33N — 22
Prospect Dr. N&S .. 22W-33N — 22
Ravine Dr. E&W .... 22W-34N — 22
Rt. 83 N&S
.... 20W-30N,20W-31N — 31
Rt. 120 E&W
.... 23W-33N,22W-33N — 22
Rt. 134 E&W
.... 23W-33N,22W-33N — 22
Sandy Point Ln. E&W
.... 22W-34N — 22
Schooner Ct. E&W .. 22W-34N — 22
Sylvan Dr. E&W .. 22W-34N — 22
Timbercreek Dr. E&W
.... 23W-33N — 22
Washington St. E&W
.... 22W-34N — 22
Waterview Dr. E&W
.... 22W-34N — 22
Willow Dr. E&W .. 22W-34N — 22

Shields Township

Adelphia Av. N&S
.... 12W-28N,12W-29N — 35
Atkinson Av. E&W .. 12W-29N — 35
Basil Rd. E&W .... 12W-28N — 35

Bayshore Dr. NE&SW
.... 11W-29N — 35
Birch Av. E&W .... 12W-28N — 35
Birchwood Av. E&W
.... 12W-30N — 35
Bittersweet Av. E&W
.... 12W-30N — 35
Blodgett Av. E&W .. 12W-29N — 35
Bonaparte Av. E&W
.... 12W-30N — 35
Brompton Av. E&W
.... 12W-30N — 35
Burris Av. E&W .... 12W-29N — 35
Clover Ln. N&S .... 12W-28N — 35
Fairbank Av. N&S .. 12W-30N — 35
Farwell Av. N&S .... 12W-30N — 35
Forest View Dr. NE&SW
.... 11W-29N — 35
Foster Av. NE&SW
.... 12W-29N — 35
Glendell Av. N&S .. 11W-29N — 35
Green Av. N&S
.... 12W-28N,12W-29N — 35
Hillside N&S ....11W-29N — 35
Jenkisson Av. NE&SW
.... 12W-29N — 35
Juneway Ter. E&W
.... 11W-29N — 35
Kohl Dr. E&W .... 11W-29N — 35
Louis St. E&W .... 11W-29N — 35
Mawman Blvd. N&S
.... 11W-29N — 35
Melvin Dr. E&W .... 12W-29N — 35
Muir Av. E&W .... 12W-29N — 35
North Av. E&W .... 12W-28N — 35
North Shore Dr. N&S
.... 11W-29N — 35
Northern Av. N&S .. 12W-28N — 35
Plaister Av. N&S .. 12W-29N — 35
Quassey Av. E&W .. 12W-29N — 35
Red Oak Ct. NE&SW
.... 11W-29N — 35
Rivers Dr. N&S .... 11W-29N — 35
Rockland Rd. N&S
.... 12W-28N — 35
Shagbark Ln. E&W 12W-28N — 35
Skokie Blvd. N&S .. 12W-28N — 35
Skokie Hwy. N&S
.... 12W-29N,12W-31N — 35
Smith Av. E&W .... 12W-29N — 35
Spring Ln. N&S .... 11W-29N — 35
Stafford Av. E&W .. 12W-29N — 35
Talbot Av. E&W .... 12W-29N — 35
Thorntree Rd. N&S
.... 12W-28N — 35
Washington Av. E&W
.... 12W-28N — 35
Woodland Rd. E&W
.... 12W-28N — 35

Third Lake

Brigantine Ln. E&W
.... 18W-34N — 24
Clipper Ct. NE&SW
.... 18W-34N — 24
Coral Reef Ct. N&S
.... 19W-34N — 23
Crows Nest Ct. N&S
.... 19W-34N — 23
Galleon Ct. NW&SE
.... 19W-34N — 23
Grant Av. N&S .... 18W-35N — 24
Lake Av. E&W ...... 18W-34N — 24
Lighthouse Ln. NE&SW
.... 19W-34N — 23
Mainsail N&S ...... 18W-34N — 24
Mainsail Dr. N&S
.... 19W-34N — 23
Portside Ct. NW&SE
.... 19W-34N — 23
Schooner Ln. E&W
.... 19W-34N — 23
Seafarer Dr. NS&EW
.... 18W-34N — 24
Sextant Dr. N&S....19W-34N — 23
Sextant Windjammer NE&SW
.... 18W-34N — 24
Spinnaker Ct. NW&SE
.... 19W-34N — 23
Starboard Ct. NW&SE
.... 19W-34N — 23
Sunshine Av. E&W .19W-35N — 23
U.S. Route 45 N&S
.... 18W-34N — 24
Whaler Ct. NE&SW
.... 19W-34N — 23
Windjammer Ct. NW&SE
.... 18W-34N — 24
Windjammer Ln. NS&EW
.... 19W-34N — 23

Tower Lakes

Abbott Pl. N&S .. 26W-25N — 37
Barsumian Dr. N&S
.... 27W-25N — 37
Bays Pl. N&S .... 26W-25N — 37
Blue Stem Cr. N&S
.... 26W-24N — 37
Bridge Ln. E&W .... 26W-25N — 37
Cambridge Ct. E&W
.... 26W-25N — 37
Circle Dr. NS&EW . 26W-25N — 37

Devonshire Rd. N&S
.... 26W-25N — 37
E. Gateway Dr. N&S
.... 26W-25N — 37
East Lake Shore Dr. NS&EW
.... 26W-25N — 37
East Tower Dr. E&W
.... 26W-25N — 37
Edward Ln. N&S .. 26W-25N — 37
Essex Pl. E&W .... 26W-25N — 37
Esther Ct. E&W .... 26W-25N — 37
Fenview Dr. NW&SE
.... 27W-24N,26W-24N — 37
Indian Grass Ct. NE&SW
.... 26W-24N — 37
Indian Trail N&S
.... 26W-25N,25W-25N — 37
Kelsey Rd. NE&SW
.... 27W-24N,26W-25N — 37
Lake Shore Ter. N&S
.... 26W-25N — 37
Leon Dr. E&W......26W-25N — 37
Marian Ct. N&S .. 26W-25N — 37
Medinah Ln. NW&SE
.... 25W-25N — 37
Melrose Dr. E&W .. 26W-25N — 37
Murray Rd. E&W
.... 27W-25N,26W-25N — 37
N. Lake Shore Dr. E&W
.... 26W-25N — 37
Nooding Flower Ct. NE&SW
.... 26W-24N — 37
North Av. N&S .. 26W-25N — 37
Oak St. NW&SE .. 26W-25N — 37
Oxford Rd. N&S .. 26W-25N — 37
Paddock Ln. N&S .. 26W-25N — 37
Pebble Creek Dr. N&S
.... 26W-25N,25W-25N — 37
Raleigh Pl. N&S .. 26W-25N — 37
Roberts Rd. E&W
.... 27W-25N,26W-25N — 37
Robin Rd. NS&EW . 27W-25N — 37
Rolling Green Dr. N&S
.... 25W-25N — 37
Rose Terrace Ct. E&W
.... 26W-24N — 37
Roslyn Pl. E&W .. 26W-25N — 37
Scenic Dr. N&S .. 26W-25N — 37
South Dr. NE&SW . 26W-24N — 37
South Hills Dr. N&S
.... 26W-25N — 37
Stonehenge Ln. E&W
.... 25W-25N — 37
Summit Dr. E&W .. 25W-25N — 37
Timber Ct. N&S .. 26W-25N — 37
Tomahawk Ln. N&S
.... 26W-25N — 37
Warwick Rd. N&S . 26W-25N — 37
West Dr. N&S .. 26W-25N — 37
West Lake Dr. N&S
.... 26W-25N — 37
West Lake Shore Dr. NS&EW
.... 27W-25N,26W-25N — 37
West Tower Dr. N&S
.... 26W-25N — 37
Wren Rd. N&S .. 27W-25N — 37

Vernon Hills

Abbey Ln. E&W .... 17W-25N — 40
Abilene Ln. NS&EW
.... 16W-25N — 41
Adair Ct. E&W .... 16W-25N — 41
Adams Ct.* E&W .. 17W-27N — 40
Adrian Ct. NE&SW ..18W-25N — 40
Albany Ln. N&S .... 17W-25N — 40
Albert Dr. NS&EW .. 17W-25N — 40
Albright Ct. NW&SE
.... 17W-25N — 40
Alexandria Dr. N&S
.... 17W-25N — 40
Alice Ct. E&W .... 17W-25N — 40
Allentown Ct. N&S.. 16W-25N — 41
Almond Ct. N&S.... 17W-25N — 40
Alpine Dr. E&W .... 17W-25N — 40
Altoona Ct. N&S .. 16W-25N — 41
Amber Ln. E&W .... 17W-25N — 40
Amherst Ct. N&S .. 16W-25N — 41
Angela Ct. N&S .... 17W-25N — 40
Ann Arbor Ln. E&W
.... 18W-25N — 40
Ann Ct. E&W .... 16W-25N — 41
Annapolis Dr. NW&SE
.... 17W-25N — 40
Apollo Ct. E&W .... 17W-25N — 40
Appian Way NW&SE
.... 17W-25N — 40
Appleton Dr. NS&EW
.... 17W-25N — 40
April Av. N&S .. 17W-25N — 40
Aqua Ct. NE&SW . 17W-25N — 40
Arcadia Ct. E&W .. 17W-25N — 40
Ardmore Ct. N&S .. 17W-25N — 40
Arlington Ct. NW&SE
.... 17W-25N — 40
Arron Ct. E&W .... 17W-25N — 40
Ashland Ct. E&W .. 17W-25N — 40
Ashville Ct. N&S....16W-25N — 41
Ashwood Ct. NE&SW
.... 17W-25N — 40
Aspen Dr. N&S
.... 17W-25N,17W-26N — 40
Astoria Ct. NW&SE
.... 16W-25N — 41

	Grid	Page
Athens Ct. E&W	16W-25N	41
Atlantic Dr. E&W	17W-25N	40
Atrium Dr. NS&EW	17W-25N	40
Auburn Ct. N&S	17W-25N	40
Augusta Dr. E&W	17W-25N	40
Aurora Ct. NW&SE	16W-25N	41
Austin Ct. NE&SW	17W-25N	40
Autumn Ln. N&S	17W-25N	40
Avon Ct. NW&SE	17W-25N	40
Barn Swallow Ln. N&S	16W-24N	41
Basswood Dr. NW&SE	17W-25N	40
Bedford Rd. E&W	16W-25N	41
Belle Isle Ln. E&W	18W-25N	40
Birchwood Ct. NE&SW	17W-25N	40
Birmingham Pl. N&S	16W-25N	41
Bluewater Ln. N&S	16W-26N	41
Brandywine Ct. NE&SW	17W-26N	40
Briarwood Ct. NE&SW	17W-25N	40
Brook Hill Ln. N&S	16W-24N	41
Brook Ln. E&W	16W-26N	41
Brookwood Ct. NW&SE	17W-25N	40
Browning Ct.* N&S	17W-25N	40
Bryant Pl. NE&SW	17W-26N	40
Buchanan Ct. E&W	17W-26N	40
Bunker Ct. E&W	16W-25N	41
Burnside Ct.* E&W	17W-25N	40
Butterfield Rd. N&S	17W-25N	40
Camden Pl. E&W	17W-24N	40
Castlegate Ct. E&W	16W-25N	41
Cedar Ct. NE&SW	17W-26N	40
Center Dr. N&S	16W-26N	41
Centurion Ln. N&S	17W-26N	40
Chatham Ct.* N&S	17W-27N	40
Cherokee Rd. E&W	16W-25N	41
Cherry Valley Rd. E&W	18W-24N,17W-24N	40
Cherrywood Ct. NE&SW	17W-25N	40
Chesapeake Ct. E&W	17W-27N	40
Chesterfield Ln. N&S	17W-24N	40
Clairmont Ct. NW&SE	16W-26N	41
Cleveland Ct.* NW&SE	17W-27N	40
Collins Dr.* E&W	17W-26N	41
Colony Ct.* E&W	17W-27N	40
Commonwealth Ct.* N&S	16W-26N	41
Conachie Ct. NW&SE	18W-25N	40
Constitution Ct.* N&S	16W-26N	41
Continental Dr. NE&SW	16W-25N,15W-25N	41
Corporate Woods Pkwy. NS&EW	16W-24N,15W-24N	41
Country Club Ln. NE&SW	17W-24N	40
Court Of Ash E&W	17W-24N	40
Court Of Birch* E&W	18W-24N	40
Court Of Elm NE&SW	17W-24N	40
Court Of Shorewood NS&EW	18W-24N	
Court Of Spruce* E&W	18W-24N	40
Coventry Cir. NS&EW	16W-26N	41
Crabtree Ln. NE&SW	16W-26N	41
Creek Bend Dr. NS&EW	16W-23N	48
Creek View Dr. E&W	16W-23N	48
Creekside Dr. NW&SE	17W-25N	40
Crestview Ln. E&W	17W-26N	40
Cromwell Ct.* E&W	16W-26N	41
Cumberland Ct. N&S	17W-26N	40
Dawson Ct. N&S	18W-24N	40
Dearborn Ln. NW&SE	18W-25N	40
Debill Ct. E&W	16W-24N	41
Deerpath Dr. N&S	16W-24N,16W-25N	41
Dickinson Ct. NE&SW	17W-26N	40
Dirksen Dr.* NW&SE	17W-27N	40
Dogwood Ln. NW&SE	17W-25N	40
Dover Ct.* N&S	17W-26N	40
Echo Ct.* E&W	17W-26N	40
Edgewood Rd. E&W	16W-25N	41

	Grid	Page
Eisenhower Ct.* NW&SE	17W-27N	40
Elm Tree Ln. NE&SW	16W-25N	41
Emerson Pl. NW&SE	17W-26N	40
Evergreen Dr. N&S	17W-24N	40
Executive Way Dr. E&W	16W-25N,15W-25N	41
Exeter Pl. N&S	16W-25N	41
Fairway Dr. N&S	16W-25N	41
Farmingdale Cir. N&S	17W-25N	40
Farmington Ln. E&W	17W-25N	40
Faulkner Pl.* N&S	17W-25N	40
Fernwood Ct. NE&SW	17W-25N	40
Fiore Parkway N&S	18W-25N	40
Forest Edge Dr. NW&SE	15W-23N	49
Franklin Pl. NW&SE	17W-26N	40
Gladstone Dr.* N&S	17W-25N	40
Grant Pl. NW&SE	17W-26N	40
Greenbriar Ln. N&S	17W-24N	40
Greenleaf Dr. NE&SW	17W-26N	40
Greenvale Rd. NE&SW	16W-25N	41
Grosse Pointe Blvd. NE&SW	18W-25N	40
Haddon Cir. NS&EW	17W-24N	40
Hamilton Pl. NE&SW	17W-26N	40
Hampton Pl. E&W	17W-26N	40
Hanover Pl. E&W	17W-26N	40
Harbor Dr. E&W	16W-25N	41
Harding Ct.* E&W	17W-25N	40
Harrison Ct. E&W	17W-26N	40
Harvest Ct. NE&SW	17W-25N	40
Hastings Ct. E&W	17W-26N	40
Hawthorne Pkwy. E&W	17W-26N,16W-26N	40
Hayes Ct.* NW&SE	17W-27N	40
Hemingway Ct. E&W	17W-26N	40
Hickory Hill Dr. E&W	16W-24N	41
Hobble Bush Ln. NE&SW	16W-24N	41
Hughes Pl.* N&S	17W-25N	40
Hunter Ct. N&S	18W-25N	40
Indianwood Dr. N&S	17W-26N	40
Jackson Ct. N&S	17W-26N	40
Jansen LN. NE&SW	18W-25N	40
Jefferson Ct. E&W	17W-27N	40
Julie Ln. N&S	18W-25N	40
Kalamazoo Cir. NW&SE	18W-25N	40
Kennedy Pl.* NE&SW	17W-27N	40
Keswick Ln. E&W	16W-25N	41
Kingston Ct. N&S	17W-26N	40
Knollwood Ln. N&S	16W-25N	41
Lafayette Pl.* N&S	17W-26N	40
Lakeside Dr. N&S	17W-24N	40
Lakeside Dr., E. N&S	17W-24N	40
Lakeside Dr., N. E&W	17W-24N	40
Lakeview Parkway NS&EW	16W-25N	41
Lancaster Pl.* E&W	17W-26N	40
Lansing Ct. E&W	18W-25N	40
Larchmont Ln. N&S	16W-25N	41
Lexington Dr. E&W	16W-25N	41
Lindenwood Ct. NW&SE	17W-25N	40
Lindon Ln. E&W	16W-24N	41
Lowell Pl. NE&SW	17W-26N	40
Mac Arthur Ct. NW&SE	17W-26N	40
Mallard Ct.* E&W	17W-25N	40
Malvern Ln. E&W	16W-25N	41
Manchester Ln. N&S	17W-24N	40
Marimac Ln. N&S	17W-24N	40
Marinewood Dr. NW&SE	17W-25N	40
Market Ct.* E&W	16W-26N	41
Marlowe Pl. NW&SE	17W-26N	40
Marquette Ct. N&S	18W-25N	40
Mayflower Rd. NE&SW	16W-25N	41
Meadow Ct. E&W	16W-25N	41
Melody Ln. N&S	16W-25N	41
Memphis Pl. E&W	17W-24N	40
Meredith Pl.* NE&SW	17W-27N	40

	Grid	Page
Michigan Ct.* E&W	16W-26N	41
Midway Ln. NE&SW	16W-24N	41
Milwaukee Av. NW&SE	15W-25N	41
Monroe Ct.* E&W	17W-27N	40
Montauk Ln. N&S	17W-24N	40
Montclair Rd. E&W	17W-24N	40
Montebello Dr. N&S	17W-24N	40
Monterrey Dr. N&S	17W-24N	40
Montgomery Ln. N&S	17W-24N	40
Murwood Ct. NE&SW	17W-25N	40
Muskegon Ct. NW&SE	18W-25N	40
Mystic Pl. N&S	17W-25N	40
Napier Ct. N&S	18W-25N	40
Narois Dr. N&S	17W-24N	40
Noble Cir. NS&EW	17W-24N	40
Northfield Dr. E&W	18W-25N	40
Oakwood Rd. N&S	17W-24N	40
Old Creek Rd. N&S	16W-24N	41
Old Wood St. N&S	16W-24N	41
Onwentsia Rd. N&S	16W-25N	41
Parkside Ct. E&W	17W-26N	40
Peachtree Ct.* E&W	16W-26N	41
Phillip Rd. E&W	17W-25N	40
Pierce Ct. E&W	17W-26N	40
Plumwood Ln. N&S	17W-24N	40
Plymouth Farms Rd. E&W	17W-25N	40
Polk Ct. NW&SE	17W-26N	40
Pontiac Ln. NW&SE	18W-25N	40
Portage Ct. NE&SW	18W-25N	40
Princeton Ct. N&S	17W-26N	40
Redwood Ct. NW&SE	17W-25N	40
Regency Ct. N&S	16W-26N	41
Revere Pl. NW&SE	17W-26N	40
Richard Ct. N&S	17W-24N	40
Richmond Ln. E&W	17W-24N	40
Roosevelt Dr.* NW&SE	17W-27N	40
Royal Oak Rd. NS&EW	18W-24N,18W-25N	40
Rt. 21 N&S	16W-27N,15W-24N	41
Rt. 60 E&W	17W-26N,16W-26N	40
Russet Way NE&SW	17W-25N	40
Saddle Back NW&SE	16W-25N	41
Saratoga Ct. E&W	17W-26N	40
Saugatuck Tr. NS&EW	18W-24N	40
Sheffield Ln. E&W	17W-25N	40
Sinclair Ct. N&S	18W-24N	40
Somerset Ln. E&W	17W-25N	40
Southfield Dr. NS&EW	18W-25N	40
Southwick Ct. E&W	16W-26N	41
Sparta Ct. NE&SW	18W-25N	40
Spring Ln. N&S	16W-25N	41
St. Clair Ln. NS&EW	18W-24N	40
St. Ives Ln. E&W	16W-25N	41
Sterling Heights Rd. N&S	18W-25N	40
Stevenson Pl.* N&S	17W-27N	40
Stockton N&S	17W-26N	40
Stone Fence Rd. NE&SW	16W-25N	41
Streamwood Ln. N&S	16W-25N	41
Sullivan Dr. E&W	17W-25N	40
Sunset Ct. N&S	17W-25N	40
Sussex Cir. NS&EW	17W-24N	40
Sutcliffe Cir. NS&EW	17W-24N	40
Sutton Ct. N&S	16W-26N	41
Swinburne Pl.* NW&SE	17W-27N	40
Tally Ho Dr. E&W	17W-24N	40
Tanglewood Ct. NE&SW	17W-25N	40
Tanwood Ln. NW&SE	17W-25N	40
Taylor Ct. NE&SW	17W-26N	40
Televista Ct. NW&SE	17W-24N	40
Tennis Ln. N&S	18W-24N	40
Tennyson Pl.* NW&SE	17W-27N	40
Timber Ln. E&W	17W-26N	40
Townline Rd. E&W	17W-26N	40
Tyler Ct. N&S	16W-26N	41
U.S. Route 45 NW&SE	18W-25N,15W-23N	40

	Grid	Page
Van Buren Ct. N&S	17W-26N	40
Vernon Ct. N&S	17W-26N	40
Wadsworth Pl. NW&SE	17W-26N	40
Warren Ln. N&S	18W-25N	40
Warrington Rd. NE&SW	16W-26N	41
Washington Ct.* E&W	17W-27N	40
Waterview Cir. E&W	18W-24N	40
Webster Pl.* E&W	17W-26N	40
Wentworth Ct. N&S	16W-26N	41
West End Ct. N&S	17W-26N	40
West End Ln. NW&SE	17W-26N	40
Westmoreland Dr. E&W	18W-24N,17W-24N	40
Westwood Ct. NE&SW	17W-25N	40
White Barn Rd. E&W	16W-24N	41
White Fence Ct. NW&SE	16W-24N	41
Whiting Ct.* E&W	17W-25N	40
Whitman Pl. NW&SE	17W-26N	40
Whitney Pl. N&S	17W-26N	40
Wildwood Ct. NE&SW	17W-25N	40
Wilshire Ct.* E&W	16W-26N	41
Windsor Dr.* E&W	17W-25N	40
Wine Ct. N&S	17W-25N	40
Woodlands Pkwy. N&S	15W-23N	49

Vernon Township

	Grid	Page
Apple Hill Ln. N&S	16W-22N	48
Apple Orchard Ln. E&W	15W-23N	49
Arlyd Rd. E&W	16W-22N	48
Ash St. N&S	14W-20N	49
Aspen St. NW&SE	14W-20N	49
Birch St. N&S	14W-20N	49
Birchwood Ter. E&W	14W-20N	49
Bluebird Ln. N&S	15W-20N	49
Bluebird St. N&S	15W-20N	49
Brae Burn Ln. N&S	13W-21N	49
Briar Ter. E&W	14W-20N	49
Bridle Trail Rd. N&S	14W-24N	41
Brockman Av. E&W	16W-23N	48
Busch Rd. E&W	17W-21N	48
Carman Av. N&S	15W-20N	49
Catalpa St. N&S	14W-20N	49
Celia Av. N&S	16W-20N	48
Cemetery Rd. E&W	15W-23N	49
Clarice Av. N&S	16W-20N	48
Columbus Pkwy E&W	14W-20N	49
Concordia Ln. (pvt)	14W-24N	41
Dell Ln. E&W	13W-22N	49
Depot Pl. E&W	16W-21N	48
Diamond Lake Rd. N&S	19W-24N	39
Dogwood St. N&S	15W-20N	49
Duffy Ln. E&W	13W-22N	49
East Chevy Chase Dr. NE&SW	14W-20N	49
East Mill Rd. E&W	14W-21N	49
Easton Av. N&S	16W-23N	48
Elizabeth Av. N&S	15W-20N	49
Elm Rd. N&S	14W-23N,14W-24N	49
Elmwood Av. N&S	13W-20N	49
Estonian Ln. E&W	14W-21N	49
Eugene Av. N&S	16W-20N	48
Florence Av. N&S	15W-20N	49
Forest Av. N&S	14W-20N	49
Frank Av. N&S	15W-20N	49
Frontenac Pl. E&W	14W-20N	49
G K Lane N&S	15W-21N,15W-22N	49
Garden Ct. E&W	15W-20N	49
Gregg Rd. E&W	14W-23N	49
Hawthorne Ln. E&W	13W-23N	42
Holtz Rd. N&S	15W-23N	49
Horatio Blvd. N&S	15W-20N	49
Hummingbird Ct. E&W	14W-20N	49
Hummingbird Ln. E&W	14W-20N	49
Indian Creek Rd. N&S	15W-23N	49
Inverrary Ln. N&S	15W-20N	49
James Cir. N&S	14W-20N	49
Juneway Av. NW&SE	15W-20N	49
Kingston Ct. NE&SW	14W-23N	49
Kingston Row NS&EW	16W-23N	48
Lee Ln. E&W	16W-23N	48
Linden Av. E&W	14W-20N	49

	Grid	Page
Lita Av. E&W	15W-20N	49
Long Beach Dr. N&S	14W-20N	49
Long Grove Aptakistic Rd. E&W	17W-22N,15W-22N	48
Longwood Dr. E&W	14W-20N	49
Loyola Av. NW&SE	14W-20N	49
Main St. NW&SE	16W-22N,16W-23N	48
Mallard Ct. E&W	15W-20N	49
Mallard Dr. E&W	15W-20N	49
Mallard Ln. N&S	15W-20N	49
Maple Ct. N&S	15W-20N	49
Maple Pl. N&S	13W-22N	49
Maplewood Dr. E&W	14W-20N	49
Margaret Av. N&S	16W-20N	48
Marie Av. E&W	16W-20N	48
Marquette Pl. E&W	14W-20N	49
Martin Ln. E&W	14W-20N	49
Martin St. N&S	14W-20N	49
Maryland Villa Rd. E&W	14W-23N	49
Menna Ln. E&W	16W-23N	48
Mercier Av. NW&SE	14W-20N	49
Michael Av. N&S	15W-20N	49
Milwaukee Av. E&W	15W-25N,14W-20N	41
Newman St. E&W	14W-20N	49
North Woodbine Cir. E&W	15W-24N	41
Oak Ln. E&W	14W-24N	41
Oakwood Ter. E&W	14W-20N	49
Old Mill Rd. NE&SW	13W-23N	49
Olsen N&S	15W-23N	49
Orchard Hill Rd. E&W	13W-25N	42
Osage Rd. NW&SE	18W-25N	40
Park Av. E&W	16W-23N	48
Park Pl. NE&SW	16W-21N	48
Partridge Ln. N&S	14W-20N	49
Pauline Av. E&W	16W-20N	48
Pekara Dr. E&W	15W-20N,14W-20N	49
Penguin Ln. N&S	15W-20N	49
Pet Ln. N&S	16W-22N	48
Pheasant Ln. E&W	14W-20N	49
Pinehurst Dr. E&W	14W-20N	49
Pintail Ct. N&S	15W-20N	49
Pintail Ln. N&S	15W-20N	49
Pintail St. E&W	14W-20N	49
Pope Blvd. E&W	16W-20N	48
Port Clinton Rd. E&W	16W-23N	48
Prague Av. N&S	14W-20N	49
Prairie Ln. N&S	16W-22N,16W-23N	48
Prairie Rd. N&S	16W-22N,16W-23N	48
Prairie View Ln. E&W	13W-25N	42
Private Rd. N&S	16W-22N,16W-23N	48
Public St. N&S	14W-20N	49
Public St. N&S	15W-20N	49
Rabbe Ln. E&W	15W-22N	49
Raleigh Dr. N&S	16W-23N	48
Raphael Av. N&S	15W-20N	49
Richard Ct. N&S	16W-23N	48
Riverside Dr. N&S	14W-20N	49
Riverside Rd. E&W	14W-21N	49
Rose Ter. N&S	13W-20N	49
Sandpiper Ln. E&W	14W-20N	49
Saunders Rd. N&S	13W-25N	42
Schaffer Rd. N&S	18W-20N	48
Skylark Ln. NE&SW	14W-20N	49
South Woodbine Cir. E&W	15W-24N	41
Story Book Ln. N&S	13W-23N	49
Strenger Ln. N&S	13W-22N	49
Swallow Ct. NE&SW	15W-20N	49
Swallow Ln. NE&SW	15W-20N	49
Swallow St. N&S	15W-20N	49
Swan Blvd. N&S	15W-20N	49
Swan Ct. NE&SW	15W-20N	49
Swan Ln. E&W	15W-20N	49
Trail Way E&W	14W-24N	41
Valdon Rd. E&W	14W-23N	49
Walnut Dr. E&W	15W-20N	49
Weiland Rd. NE&SW	16W-23N	48
Westedge N&S	15W-23N	49
William Av. N&S	15W-20N	49
Winston Dr. NW&SE	14W-20N	49
Woodbine Dr. N&S	15W-24N	41
Woodland Ln. E&W	13W-25N	42
Wren Ln. N&S	14W-20N	49
1st St. N&S	16W-22N	48
2nd St. N&S	16W-23N	48

	Grid	Page

Wadsworth

Street	Grid	Page
Adams Rd. E&W		
........14W-39N,13W-39N		17
Aime Ln. E&W 15W-37N		17
Andover Rd. E&W .. 14W-39N		17
Anna Dr., East N&S		
.... 16W-37N		17
Anna Dr., North N&S		
.... 15W-37N		17
Anna Dr., West N&S		
.... 16W-37N		17
Arbor Ct. NE&SW .. 14W-38N		17
August Zupec Dr. E&W		
........14W-39N,13W-39N		17
Barth Dr. N&S 13W-40N		10
Bartlett Ln. N&S.... 13W-40N		10
Birch Ln. E&W 14W-38N		17
Blue Spruce Ct. NE&SW		
.... 14W-38N		17
Bryn Mawr N&S 13W-38N		18
Burr Oak Ln. N&S .. 14W-38N		17
Caroline N&S 15W-39N		17
Cashmore Rd. N&S		
.... 14W-38N		17
Cermak Rd. E&W .. 16W-42N		8
Chaplin St. E&W .. 13W-38N		18
Cherrywood Ln. NW&SE		
.... 16W-40N		8
Chicago Av. N&S .. 15W-38N		17
Concord Ln. NE&SW		
.... 14W-39N		17
Country Lane Dr. E&W		
.... 14W-38N		17
Crabapple E&W ... 14W-38N		17
Delany Rd. N&S		
........14W-38N,14W-40N		17
Delany Rd. N&S 14W-39N		9
Dilleys Rd. N&S		
........16W-37N,16W-39N		17
Elm Ln. E&W 13W-38N		17
Eveline N&S........ 15W-38N		17
Fox Glove Ln. NW&SE		
.... 16W-40N		8
Goldenrod Ln. N&S		
.... 16W-40N		8
Golf Lane Dr. N&S .. 14W-38N		17
Hanssen Rd. E&W		
........16W-37N,15W-37N		17
Harmony Ct. NW&SE		
.... 14W-39N		17
Hart E&W.......... 13W-38N		18
Highview Rd. E&W . 14W-39N		9
Hunt Club Tr. E&W		
.... 16W-40N		8
Jody Ln. E&W 14W-39N		17
Juniper E&W 14W-38N		17
Kilbourne Rd. N&S		
........14W-39N,14W-40N		17
Mature Ct. NW&SE		
.... 14W-39N		17
Mauser Dr. NS&EW		
........13W-39N,13W-40N		18
McCarthy Rd. E&W		
.... 15W-38N		17
Meadow Ln. N&S .. 14W-39N		17
Mel Ln. E&W 14W-39N		17
Mill Creek Rd. N&S		
........16W-40N,16W-41N		8
Mulberry Ln. E&W		
........16W-37N,15W-37N		17
Nature Ct. NW&SE		
.... 14W-39N		17
Northwoods Dr. N&S		
.... 14W-38N		17
Oak Knoll Rd. N&S		
.... 14W-39N		17
Oak Tree Ln. NW&SE		
.... 14W-39N		17
Oakcrest Ln. NS&EW		
.... 14W-38N		17
Old Orchard Dr. NE&SW		
.... 16W-40N		8
Pickford St. E&W .. 13W-38N		18
Prairie View Dr. N&S		
.... 14W-39N		17
Pratum Ter. Dr. E&W		
.... 14W-39N		17
Primrose Ln. E&W . 15W-38N		17
Public Rd. E&W .. 15W-38N		17
Public Rd. N&S .. 16W-40N		8
Red Oak Ter. N&S 14W-38N		17
Reed St. NE&SW .. 16W-40N		8
Rosedale Av. N&S .. 13W-38N		18
Rt. 173 E&W		
........17W-41N,15W-41N		10
Sand Ct. E&W ... 16W-37N		17
Sandy Ct. E&W .. 15W-38N		17
Schlosser Ct. E&W		
.... 15W-38N		17
Sedge Ct. E&W .. 16W-40N		8
Shagbark Ln. N&S.. 14W-38N		17
Shelley Ln. NW&SE		
.... 14W-38N		17
Shelly Ln. NW&SE .. 14W-38N		17
Sheryl Lynn Dr. E&W		
.... 14W-39N		17
Skokie Hwy. N&S		
........15W-37N,16W-42N		17
Stiehr Rd. E&W ... 13W-40N		10
Sunset Rd. N&S.... 13W-40N		10
Thorn Meadow Cir. NE&SW		
.... 16W-40N		8

Street	Grid	Page
Thornapple Ln. E&W		
.... 15W-37N		17
Tri-State North Tollway N&S		
........16W-40N,16W-41N		8
Trillum Ct. NW&SE 16W-40N		8
U.S. Route 41 N&S		
........15W-37N,16W-42N		17
Valleyview Rd. N&S		
.... 15W-37N		17
Wadsworth Rd. E&W		
........15W-39N,13W-39N		17
Walden Ln. NS&EW		
.... 14W-39N		17
Wedgewood Ct. NW&SE		
.... 14W-39N		17
Willow Ln. N&S 15W-37N		17
Winchester Rd. N&S		
.... 13W-39N		18
Woodland Av. N&S		
.... 13W-38N		18

Warren Township

Street	Grid	Page
Adelaide Av. E&W .. 12W-37N		18
Algonquin Dr. N&S .. 18W-33N		24
Almond Ln. N&S.... 17W-34N		24
Almond Rd. N&S		
........17W-32N,17W-33N		24
Apple Ln. E&W 16W-35N		24
Applewood Ct. E&W		
.... 16W-35N		24
Arbor Blvd. E&W .. 18W-32N		24
Ash Dr. E&W 18W-36N		16
Ashley Dr. N&S ... 17W-33N		24
Audrey Av. E&W .. 13W-37N		18
Avon Ct. NW&SE .. 18W-33N		24
Back Bay Ct.* E&W		
.... 17W-35N		16
Banbury Ct.* E&W .. 17W-35N		16
Banbury Dr. N&S .. 18W-36N		16
Battershall Rd. N&S		
........18W-32N,18W-33N		24
Bayonne Av. N&S		
........13W-36N,13W-37N		18
Beechwood Ct.* E&W		
.... 17W-35N		16
Belle Plaine Av. N&S		
........13W-36N,13W-37N		18
Belvidere Rd. E&W		
........18W-32N,16W-32N		24
Beverly Av. N&S....18W-36N		16
Big Oaks Rd. E&W		
........18W-33N,17W-33N		24
Black Velvet Ln. NS&EW		
.... 17W-37N		16
Blackhawk Dr. E&W		
.... 17W-33N		24
Blanchard Rd. E&W		
.... 13W-37N		18
Blossom St. E&W . 13W-37N		18
Bluff Dr. NE&SW .. 17W-33N		24
Bough Ct. N&S 15W-35N		25
Boulevard View Av. N&S		
........13W-36N,13W-37N		18
Branch Rd. N&S..15W-35N		25
Bridle Ct. N&S 17W-34N		24
Bridle Ln. N&S 17W-34N		24
Bridle Trail Rd. NW&SE		
.... 17W-36N		16
Brooke Av. E&W .. 18W-32N		24
Brookside Dr. NS&EW		
.... 17W-36N		16
Carvis Dr. E&W .. 18W-34N		24
Cemetery Rd. N&S		
........16W-34N,16W-35N		24
Center Dr. N&S .. 18W-36N		16
Cherokee Ct. N&S.. 17W-33N		24
Cheyenne Ct. E&W		
.... 17W-33N		24
Chippewa Rd. E&W		
.... 17W-33N		24
Circle Ct. N&S 18W-33N		24
Circle East SE&NW		
.... 17W-34N		24
Conifer Ln. NS&EW		
.... 15W-36N		17
Cottage Av. E&W . 18W-34N		24
Cottonwood Ct. E&W		
.... 17W-33N		24
Country Ln. N&S .. 18W-33N		24
Country Ln. N&S .. 15W-37N		17
Country Pl. E&W .. 15W-37N		17
Cove Rd. NW&SE .. 18W-33N		24
Crows Nest Ct. N&S		
.... 17W-33N		24
Dady Ln. E&W 18W-33N		24
Dartmoor Dr. E&W .. 17W-33N		24
Dawn Ct. E&W .. 17W-34N		24
Deer Path Rd. E&W		
.... 18W-33N		24
Delany Rd. N&S		
........14W-35N,14W-37N		25
Devon Ct. N&S .. 18W-36N		17
Dillys Rd. N&S		
........16W-37N,15W-36N		17
Douglas Ter. N&S .. 18W-36N		16
Eastview Av. N&S .. 17W-33N		24
Eastwood Av. E&W		
.... 14W-33N		25
Eastwood Rd. N&S		
.... 17W-33N		24
Edgewater Ct.* N&S		
.... 17W-35N		24

Street	Grid	Page
Edgewood Dr. N&S		
.... 16W-35N		24
Edgewood Dr. N&S		
.... 18W-36N		16
Elizabeth St. E&W .. 13W-37N		18
Elm St. E&W 18W-32N		24
Evergreen Dr. N&S 18W-33N		24
Fairfield Ln. E&W .. 17W-32N		24
Field View Dr. N&S		
.... 17W-36N		16
Forest Av. E&W .. 14W-33N		25
Forest Dr. N&S		
........18W-32N,18W-33N		24
Fox Hill Dr. N&S		
........16W-36N,16W-37N		17
Gagemere Dr. N&S		
.... 18W-33N		24
Gages Lake Dr. N&S		
.... 18W-33N		24
Gages Lake Rd. E&W		
........18W-33N,15W-33N		24
Gagewood Ln. N&S		
.... 17W-33N		24
Geier Rd. E&W ... 18W-36N		16
Glen Rd. N&S...... 18W-33N		24
Goldspring Ct.* E&W		
.... 17W-35N		16
Grand Av. E&W ... 17W-36N		16
Grand Dr. E&W ... 17W-33N		24
Grandwood Dr. NE&SW		
.... 18W-36N		16
Greentree Pl. N&S .. 18W-33N		24
Greentree Rd. E&W		
.... 17W-33N		24
Hampshire Ct.* E&W		
.... 17W-35N		16
Hampshire Ln. E&W		
.... 15W-37N		17
Hawk Ct. E&W ... 17W-34N		24
Heather Ct. E&W .. 17W-34N		24
Hendee St. E&W .. 13W-37N		18
Hickory Ln. NE&SW		
.... 17W-33N		24
Hickory Pl. E&W ... 18W-32N		24
Highfield Dr. North N&S		
.... 18W-36N		16
Highfield Dr. West N&S		
.... 18W-36N		16
Hill Av. E&W 14W-33N		25
Homestead Ct. N&S		
.... 17W-34N		24
Homestead Rd. NS&EW		
.... 17W-34N		24
Horseshoe Ln. NS&EW		
.... 17W-34N		24
Hunt Club Rd. N&S		
........17W-33N,17W-36N		24
Huntington Cir. NS&EW		
.... 17W-33N		24
Hutchins Rd. N&S		
........18W-36N,18W-37N		16
Hyatt Ln. N&S.....14W-33N		25
Idlewild Ln. E&W .. 18W-33N		24
Indian Ln. NW&SE .. 18W-33N		24
Iroquois Dr. E&W .. 18W-33N		24
Island Av. N&S ... 18W-33N		24
Island Ct. N&S 18W-33N		24
Ivy Ln. N&S........ 17W-33N		24
Jasmine Ct. E&W .. 18W-34N		24
John Mogg Rd. N&S		
........18W-32N,18W-33N		24
Jonathan Rd. E&W		
.... 16W-35N		24
Judy Ln. E&W 18W-36N		16
Julie Ln. E&W 17W-34N		24
Juniper St. N&S ... 15W-35N		25
Karelia Rd. N&S		
........16W-34N,16W-35N		24
Karen Ln. E&W 18W-36N		16
Kennedy Dr. E&W .. 14W-33N		25
Kewaunee Dr. E&W		
........18W-33N,17W-33N		24
Kimberwick Ln. N&S		
........16W-36N,16W-37N		17
Knowles Rd. NE&SW		
.... 18W-35N		24
Lake Rd. N&S....18W-33N		24
Lake Shore Dr. NW&SE		
.... 18W-33N		24
Lake St. N&S 18W-34N		24
Lakeview Ct. NW&SE		
.... 18W-33N		24
Lakeview Ter. NE&SW		
.... 18W-36N		16
Lavender Cir. NS&EW		
.... 18W-34N		24
Lawn Av. E&W 14W-33N		25
Lee Cir. E&W 18W-36N		16
Limb Ct. N&S 15W-35N		25
Lincoln Av. NW&SE		
.... 18W-34N		24
Linda Ln. E&W ... 18W-36N		16
Linden Av. E&W .. 18W-32N		24
Lindenwood Dr. E&W		
.... 18W-33N		24
Lone Rock Rd. N&S		
.... 18W-33N		24
Lovers Ln. NE&SW		
.... 18W-33N		24
Magnolia Ln. N&S		
........13W-36N,13W-37N		18
Main St. E&W ... 18W-33N		24
Manitowoc Ct. NE&SW		
.... 18W-33N		24

Street	Grid	Page
Maple Av. E&W ... 18W-32N		24
Maple Ln. E&W ... 17W-34N		24
Marian Dr. E&W .. 18W-35N		24
Mawman St. E&W .. 13W-37N		18
Meadow Ln. E&W .. 18W-36N		16
Meadow Rd. N&S.. 18W-33N		24
Meadow Ridge Dr. E&W		
.... 17W-36N		16
Meadowbrook Dr. E&W		
.... 17W-33N		24
Melody Rd. N&S....14W-33N		25
Meyer Rd. E&W 14W-37N		17
Mill Creek Dr. NE&SW		
.... 17W-36N		16
Mill Crossing Dr. E&W		
.... 17W-36N		16
Mill Ct. NW&SE ... 18W-36N		16
Mill Rd. N&S		
........18W-32N,17W-33N		24
Nemesis Av. E&W .. 13W-36N		18
New Bridge Ct.* NW&SE		
.... 17W-35N		24
Noreen Ct. NE&SW		
.... 17W-34N		24
Northwestern Av. N&S		
........13W-36N,13W-37N		18
O'Plaine Rd. N&S .. 14W-32N		25
Oak Av. N&S 18W-32N		24
Oak Knoll NE&SW .. 17W-34N		24
Oak St. N&S 18W-33N		24
Oakland Av. N&S .. 18W-33N		24
Oakwood Dr. N&S.. 18W-36N		16
Old Creek Ct.* E&W		
.... 17W-35N		16
Old Woods Trail NE&SW		
.... 17W-36N		16
Orchard Ln. E&W .. 18W-33N		24
Orchard Valley Dr. E&W		
.... 16W-35N		24
Overbrook Ct.* E&W		
.... 17W-35N		16
Paddock St. E&W .. 13W-37N		18
Pamela Dr. E&W .. 18W-36N		16
Park Cres. E&W ... 18W-34N		24
Park Dr. N&S 18W-36N		16
Park Pl. E&W 18W-34N		24
Park St. N&S 18W-33N		24
Parkway Av. E&W .. 14W-33N		25
Pin Oak Ln. E&W .. 17W-34N		24
Pine Av. N&S 18W-32N		24
Pine St. N&S 15W-35N		25
Pleasant Pl. E&W .. 18W-34N		24
Pond Ridge Cir. NE&SW		
........18W-36N,17W-36N		16
Prospect Dr. NE&SW		
.... 18W-33N		24
Public Rd. NW&SE 18W-33N		24
Rae Ln. E&W 18W-36N		16
Red Oak Dr. N&S .. 17W-34N		24
Ridge Rd. NW&SE .. 18W-33N		24
River Rd. N&S 15W-32N		25
Robert Ct. NW&SE		
.... 17W-34N		24
Rolling Hills Rd. NW&SE		
........18W-32N,18W-33N		24
Roosevelt Rd. E&W		
.... 18W-35N		24
Rosedale Av. N&S .. 13W-36N		18
Royal Oak Ln. N&S		
.... 18W-33N		24
Rt. 120 E&W ... 15W-33N		25
Rt. 132 E&W ... 18W-36N		16
Rule Ct. NW&SE .. 18W-33N		24
Running Creek Ct.* E&W		
.... 17W-35N		16
Saddle Ct. E&W ... 17W-34N		24
Saddle Ln. N&S ... 17W-34N		24
Salisbury Dr.* E&W		
.... 17W-35N		16
Salmon Av. E&W .. 13W-37N		18
Sand Lake Rd. E&W		
........18W-37N,17W-37N		16
Sandy Ct. E&W ... 17W-34N		24
Sears Blvd. NS&EW		
.... 18W-33N		24
Shannon Dr. N&S .. 17W-33N		24
Sheridan Av. E&W .. 18W-34N		24
Shirley Dr. N&S ... 14W-37N		17
Sibelius Ln. E&W .. 16W-35N		24
Skokie Hwy. N&S		
........14W-35N,14W-35N		25
Springbrook Ln.* E&W		
.... 17W-35N		16
Spruce Point Ct. N&S		
.... 15W-36N		17
Spruce Point Dr. NW&SE		
.... 15W-36N		17
Stearns School Rd. E&W		
........18W-34N,15W-34N		24
Stillwater Ct.* E&W		
.... 17W-35N		16
Stonebridge Ct. E&W		
.... 18W-34N		24
Streamwood Ct. N&S		
.... 18W-36N		16
Streamwood Dr. N&S		
.... 18W-36N		16
Summit Dr. NE&SW		
.... 17W-33N		24
Sunnyside Av. N&S		
.... 14W-33N		25
Sunset Av. N&S ... 18W-33N		24

Street	Grid	Page
Tangueray Dr. NS&EW		
.... 18W-34N		24
Telegraph Rd. N&S		
.... 13W-32N		26
Thornapple Ln. N&S		
.... 18W-32N		24
Thoroughbred Dr. NS&EW		
........17W-36N,16W-37N		16
Timber Ln. E&W .. 18W-33N		24
Tomahawk Pl. NW&SE		
.... 18W-33N		24
Treaer Ter. NE&SW		
.... 18W-36N		16
Twin Lakes Blvd. E&W		
........18W-33N,17W-33N		24
Tyler Av. E&W 13W-36N		18
U.S. Route 41 N&S		
........14W-35N,15W-37N		25
U.S. Route 45 N&S		
........18W-32N,18W-37N		24
Valley Dr. E&W .. 18W-33N		24
Valley View Dr. NW&SE		
.... 18W-33N		24
Verco Av. E&W .. 13W-37N		18
Vermont Av. E&W .. 13W-36N		18
Victoria Ln. N&S.... 15W-37N		17
Walden Way* E&W		
.... 17W-35N		16
Wall Av. E&W 13W-36N		18
Walnut Ln. E&W ... 17W-34N		24
Warren Av. E&W .. 17W-33N		24
Washington St. E&W		
........18W-34N,14W-34N		24
Wayside Pl. N&S .. 18W-33N		24
West Av. N&S 14W-33N		25
West Wind Dr. NS&EW		
.... 17W-36N		16
Westwood Pl. E&W		
.... 18W-36N		16
White Oak Ln. N&S		
.... 17W-34N		24
Willow Point Dr. E&W		
.... 18W-33N		24
Windslow Dr. N&S.. 17W-33N		24
Winnebago Ct. NW&SE		
.... 17W-33N		24
Winnebago Dr. E&W		
.... 17W-33N		24
Winnebago Trail NS&EW		
.... 17W-33N		24
Woodland Dr. E&W		
.... 17W-33N		24
Woodland Ter. E&W		
.... 18W-36N		16
Woodland Ter. N&S		
.... 18W-33N		24
Wright Av. N&S ... 18W-33N		24
Wright Rd. E&W .. 18W-33N		24
Yearling Dr. E&W .. 17W-36N		16
Yorkhouse Rd. E&W		
........15W-37N,13W-37N		17
Yorkshire Dr.* E&W		
.... 17W-35N		16
1st St. E&W 18W-33N		24
2nd St. N&S 18W-33N		24

Wauconda

Street	Grid	Page
Adams Av. N&S 25W-28N		29
Anderson Rd. N&S		
........27W-26N,27W-27N		37
Bangs St. NE&SW .. 26W-27N		37
Barbara Ln. NW&SE		
.... 25W-28N		29
Bonner Rd. E&W		
........26W-28N,24W-28N		29
Brett Ln. E&W.....25W-27N		37
Briar Rd. NW&SE .. 26W-28N		29
Brown St. N&S		
........26W-27N,26W-28N		37
Buren Av. E&W ... 26W-26N		37
Church St. E&W ... 26W-27N		37
Clover Rd. E&W ... 26W-27N		37
Cook St. E&W ... 26W-27N		37
Country Lane Ct. N&S		
.... 25W-27N		37
Crescent Ter. NW&SE		
.... 26W-28N		29
Daniel St. N&S ... 26W-27N		37
Dato Ln. N&S 26W-28N		29
Delo Dr. NE&SW ... 26W-26N		37
Delta Dr. N&S 25W-27N		37
Dunbar Rd. NS&EW		
.... 26W-27N		37
Earls Ct. N&S 25W-28N		29
Edgewater Ln. NW&SE		
.... 25W-27N		37
Edgewater Pkwy. E&W		
.... 25W-28N		29
Edward Pl. N&S ... 25W-28N		29
Farm Hill Cir. N&S .. 26W-27N		37
Foster Av. N&S ... 25W-28N		37
Francis St. N&S ... 26W-27N		37
George Av. E&W .. 26W-27N		37
Grand Blvd. N&S		
........25W-27N,25W-28N		37
Grant St. N&S..... 25W-28N		29
Hammond St. N&S 25W-27N		37
Harrison Av. N&S .. 26W-28N		29
Helena Av. E&W .. 26W-27N		37
High St. N&S 25W-27N		37
Highland Av. E&W .. 25W-27N		37
Hill St. N&S 25W-27N		37

	Grid	Page
Hollow Hill Rd. N&S	26W-28N	29
Hubbard Ct. E&W	26W-27N	37
Indian Ridge Trail N&S	26W-27N	37
Industrial Dr. NS&EW	25W-28N	29
Jackson Av. N&S	25W-28N	29
Jackson Ct. E&W	25W-28N	29
Jaime Ln. NS&EW	25W-28N	29
James Av. NE&SW	25W-28N	29
Jonathon Ct. N&S	24W-28N	30
Karl Dr. N&S	25W-28N	29
Kent Av. E&W	25W-26N	37
Kimball Av. E&W	25W-27N	37
Kyle Ct. N&S	26W-28N	29
Lake Shore Dr. NW&SE	25W-28N	29
Lake St. N&S	25W-27N	37
Lakeview Av. N&S	25W-27N	37
Larkdale Row N&S	26W-27N,26W-28N	37
Laurel Av. N&S	26W-28N	29
Legion Ct. E&W	25W-26N	37
Lewis Av. E&W	26W-28N	37
Liberty St. E&W	26W-27N	37
Lincoln Av. NE&SW	25W-28N	29
Lotus St. N&S	25W-27N	37
Macintosh Dr. NE&SW	24W-28N	30
Madison Av. N&S	25W-28N	29
Main St. N&S	26W-27N,25W-26N	37
Maple Av. E&W	26W-27N	37
Marine Rd. E&W	26W-28N	29
Marlene Ct. NE&SW	25W-27N	37
Mill St. NE&SW	26W-27N,25W-27N	37
Minerva Av. E&W	26W-27N	37
Monroe Av. N&S	25W-28N	29
Mutton Dr. N&S	26W-28N	29
Nancy Ct. NE&SW	25W-28N	29
Naples Dr. E&W	26W-28N	29
Neil Ct. N&S	25W-28N	29
North Av. E&W	26W-27N	37
North Shore Ct. NW&SE	25W-28N	29
North Shore Dr. NW&SE	25W-28N	29
Northview Dr. E&W	26W-28N	29
Oak Bluff Ct. E&W	25W-27N	37
Oak Dr. E&W	25W-28N	29
Oakdale Av. E&W	25W-28N	29
Oaks Av. E&W	26W-26N,25W-26N	37
Oakwood Rd. N&S	26W-27N	37
Old Country Way N&S	26W-27N	37
Old Rand Rd. NW&SE	26W-28N	29
Orton Av. NW&SE	26W-27N	37
Osage St. N&S	26W-27N	37
Osage Ter. NE&SW	26W-28N	29
Pamela Ct. NE&SW	25W-28N	29
Park Av. N&S	25W-27N,25W-28N	37
Park Pl. E&W	25W-28N	29
Park St. E&W	26W-27N	37
Peace Blvd. E&W	26W-26N	37
Pershing Dr. E&W	25W-28N	29
Pine St. E&W	26W-27N	37
Pleasant View Av. N&S	25W-28N,24W-28N	29
Ridge Av. N&S	25W-28N	29
Ridge Ct. E&W	25W-27N	37
Ridge St. N&S	25W-27N	37
Road Way NE&SW	26W-27N,25W-27N	37
Roosevelt Av. E&W	26W-27N	37
Ross St. E&W	26W-27N	37
Sheila St. E&W	25W-28N	29
Sheridan Dr. E&W	25W-28N	29
Sky Hill Rd. N&S	25W-27N	37
Slocum Lake Rd. E&W	26W-27N	37
South Dr. E&W	25W-28N	29
South St. E&W	25W-27N	37
Summit Av. NW&SE	25W-28N,24W-28N	29
Sunnyside Av. E&W	25W-27N	37
Surf Ter. N&S	26W-28N	29
Thomas St. N&S	26W-27N	37
Van Buren Av. N&S	25W-28N	29
W. Liberty St. NW&SE	26W-27N	37
Walnut Rd. E&W	26W-28N	29
Washington Av. N&S	25W-28N	29
Wauconda Rd. NE&SW	25W-27N,24W-28N	37
Wethington St. N&S	25W-27N	37
Willow Rd. N&S	26W-27N	37
Woodland Av. N&S	25W-27N	37

	Grid	Page
Woodland Rd. N&S	25W-28N	29

Wauconda Township

	Grid	Page
Ada St. N&S	28W-26N	36
Alameda Dr. N&S	28W-31N	28
Alvinia St. N&S	25W-27N	37
Ann Ct. N&S	26W-26N	37
Ash St. N&S	28W-26N,25W-27N	36
Bayview St. NW&SE	28W-27N	36
Beech St. NS&EW	28W-27N	36
Belvidere Rd. E&W	28W-31N	28
Berkely Ct. N&S	25W-28N	29
Bernice St. N&S	28W-26N	36
Blakley Pkwy. E&W	25W-26N	37
Bluff St. N&S	25W-27N	37
Bonner Rd. E&W	27W-28N,26W-28N	28
Brooks St. N&S	28W-26N	36
Brown Rd. N&S	25W-26N	37
Burnett Rd. E&W	28W-29N	28
Callahan Rd. NS&EW	26W-29N,26W-31N	29
Callahan Rd. N&S	26W-31N	29
Cary Plaza NE&SW	25W-28N	29
Case Rd. NW&SE	27W-29N	28
Channel St. NW&SE	28W-27N	36
Clark St. N&S	28W-26N	36
Converse Rd. N&S	28W-27N	36
Cook St. E&W	25W-28N	29
Crest Rd. E&W	28W-28N	28
Division St. N&S	28W-27N	36
Dowell Rd. N&S	28W-29N,28W-30N	28
East Lake Shore Dr. NW&SE	25W-26N	37
East Park Av. E&W	28W-26N	36
Ellen St. N&S	28W-26N	36
Elm St. N&S	28W-26N	36
Elmwood Av. NE&SW	26W-27N	37
Fairview Av. E&W	28W-27N	36
Fisher Rd. NE&SW	28W-30N,28W-31N	28
Fisher Rd. N&S	28W-31N	28
Forest Garden Rd. (Pvt.) N&S	27W-27N	37
Gardner Rd. NW&SE	25W-28N	29
Garland Rd. NS&EW	26W-28N,25W-29N	29
Genessee St. N&S	28W-26N,28W-27N	36
Geraldine Ln. N&S	28W-26N	36
Gilmer Rd. NW&SE	26W-31N,25W-30N	29
Gossell Rd. N&S	25W-30N,25W-29N	29
Grace St. N&S	28W-26N,25W-27N	36
Greenbriar Ct. N&S	27W-26N	37
Greenview Dr. N&S	25W-26N	37
Harrison Av. N&S	25W-29N	29
Hermosa Av. E&W	25W-28N	29
Hickory Av. N&S	28W-28N	36
Hickory St. E&W	28W-27N	36
Hickory St. N&S	28W-27N	36
High Rd. N&S	28W-28N	28
Hill Av. N&S	25W-26N	37
Hillside Av. E&W	25W-26N	37
Hillside Dr. N&S	28W-31N	28
Huston Av. N&S	28W-26N	36
Ivanhoe Rd. E&W	26W-26N,25W-26N	37
Kane Dr. N&S	28W-26N	36
Karen Dr. E&W	26W-26N	37
Lake Dr. N&S	28W-28N	28
Lake Dr. N&S	28W-26N	36
Lake Pkwy. E&W	28W-28N	36
Lake Shore Ct. N&S	25W-26N	37
Lake Shore Dr. N&S	25W-26N	37
Lake St. E&W	25W-27N	37
Lakeland Dr. NE&SW	26W-26N	37
Lakeview Av. E&W	25W-27N	37
Lakeview Cir. NS&EW	28W-27N,28W-28N	36
Lakeview Ct. E&W	28W-26N	36
Lakeview Dr. NW&SE	28W-27N	36
Lamphere St. E&W	28W-27N	36
Langley Ct. N&S	28W-27N	36
Laurel Av. N&S	28W-27N	36
Laurel Ct. N&S	26W-26N	37
Laurel Ln. E&W	28W-27N	36
Lexington St. N&S	28W-27N	36
Leyte Ct. N&S	28W-27N	36
Lindy Ln. E&W	28W-26N	36
Locust Dr. NS&EW	28W-26N	36

	Grid	Page
Mac St. N&S	27W-27N	37
Madison Av. NE&SW	25W-29N	29
Maple Av. E&W	28W-26N	36
Maple St. N&S	28W-27N	36
Marilyn Meadow Ct. E&W	25W-29N	29
Marion St. N&S	28W-26N	36
May St. E&W	27W-27N	37
McNally Ln. NE&SW	27W-31N	29
Mead Av. N&S	28W-26N	37
Meadow Lane Rd. E&W	28W-28N	28
Meadow Ln. N&S	25W-26N	37
Meeker Ct. E&W	28W-26N	36
Monroe Av. N&S	25W-29N	29
Morey St. N&S	28W-26N	36
Mylith Rd. E&W	28W-27N	36
Myrtle St. N&S	25W-27N	37
Neville Rd. E&W	28W-29N	28
North Lakeview Dr. E&W	27W-26N,26W-26N	37
North St. E&W	28W-27N	36
Oak Dr. N&S	25W-26N	37
Oak Knoll Rd. N&S	28W-28N	28
Oak St. NS&EW	28W-26N,28W-27N	36
Old Rand Rd. NW&SE	25W-26N	37
Orchard Dr. NE&SW	28W-26N	36
Orchard Rd. N&S	28W-26N	36
Parkview Dr. E&W	28W-27N	36
Pheasant Ct. NW&SE	26W-26N	37
Pheasant Trail NS&EW	26W-26N	37
Pleasant View Av. E&W	25W-28N	29
Poplar St. NS&EW	28W-26N,28W-27N	36
Rand Rd. NW&SE	25W-26N,27W-31N	29
Ridge Rd. E&W	28W-28N	28
Ridge St. N&S	25W-27N	37
Roberts Rd. NS&EW	28W-26N,28W-27N	36
Robin Ct. E&W	25W-29N	29
Shady Ln. NE&SW	28W-27N	36
Slocum Lake Rd. E&W	28W-27N,27W-27N	36
South Av. E&W	28W-26N	36
South Lakeview Dr. E&W	27W-26N,26W-26N	37
State Hwy. 60 NW&SE	26W-31N,25W-31N	29
Sunset Dr. E&W	28W-31N	28
Swift Ct. N&S	26W-26N	37
Tamarack Dr. E&W	25W-26N	37
Valentine Rd. N&S	25W-27N	37
Virginia Ln. N&S	25W-29N	29
Wade St. N&S	25W-29N	29
Walnut St. N&S	28W-27N	36
Washington Av. NE&SW	25W-29N	29
Wegner Rd. NE&SW	28W-31N	28
Wells Dr. N&S	25W-28N	29
Werden Av. N&S	25W-27N	37
West Dr. E&W	25W-28N	29
West End Rd. N&S	28W-28N	29
Will St. N&S	28W-26N	36
Willow Rd. N&S	28W-26N	36
Wilson Av. E&W	25W-27N	37
Wilson Rd. N&S	25W-31N	29
Wilton Rd. N&S	25W-27N	37
Winding Lane Dr. N&S	25W-28N	29
York Dr. N&S	25W-26N	37

Waukegan

	Grid	Page
Abington Ct. N&S	14W-36N	17
Adams St. N&S	10W-33N	27
Alexander Ct. N&S	10W-35N	27
Alta Vista Dr. N&S	10W-36N	19
Amhurst Pkwy. NE&SW	13W-32N	26
Apache Rd. N&S	11W-36N	18
Apple Av. E&W	12W-33N	26
Archer Av. N&S	10W-33N	27
Arizona Av. E&W	12W-36N	18
Armory Pl. E&W	11W-35N	26
Arthur Ct. NE&SW	12W-34N	26
Arthur Dr. E&W	11W-38N	18
Ash St. N&S	10W-34N,10W-36N	27
Atlantic Av. E&W	11W-35N,11W-35N	26
Austin Av. N&S	12W-34N	26
Avoneida Pl. N&S	12W-36N	18
Baldwin Av. E&W	12W-34N,12W-35N	26
Ballentine St. E&W	11W-37N	18
Bank Ct. N&S	10W-38N	19
Batberry Ct. NE&SW	14W-32N	25
Beach Rd. E&W	12W-38N,10W-38N	18

	Grid	Page
Beech Av. E&W	10W-36N	19
Beechnut Av. N&S	12W-33N	26
Belvidere St. E&W	12W-33N,10W-33N	26
Benny Av. N&S	11W-33N	26
Berry Ct. E&W	11W-35N	26
Bertrand Ln. N&S	10W-38N	19
Berwick Blvd. N&S	11W-34N,11W-35N	26
Besley Pl. N&S	10W-33N	27
Birch Av. E&W	10W-36N	19
Birch Av. N&S	12W-33N	26
Birchwood Dr. N&S	12W-31N	35
Blanchard Rd. E&W	14W-37N,10W-37N	17
Blossom Av. E&W	14W-37N	17
Bluff Ct. N&S	10W-34N	27
Bluff St. N&S	10W-34N	27
Boardman St. E&W	11W-37N	18
Bob O' Farrell Ln. E&W	10W-38N	19
Bonnie Brook Ln. N&S	11W-37N	18
Bonnie Brook Rd. E&W	13W-36N,11W-36N	18
Brentwood Ln. E&W	14W-36N	17
Briar Ln. N&S	11W-36N	18
Brighton Av. E&W	11W-33N	27
Bristol Rd. N&S	11W-37N	18
Brookside Av. E&W	12W-34N,11W-34N	26
Browne Ct. E&W	10W-39N	19
Browning Av. E&W	10W-33N	27
Buck Av. E&W	11W-36N,10W-36N	18
Buckingham Rd. E&W	11W-37N	18
Burr Oak Ter. N&S	14W-37N	17
Burris Av. E&W	12W-36N	18
Burrwood Dr. NE&SW	13W-32N	26
Burton Av. N&S	11W-34N	26
Bush Ct. E&W	11W-32N	26
Butrick St. N&S	11W-34N	26
Campbell Ln. NS&EW	14W-31N	33
Carnation Ct. N&S	10W-34N	27
Carney Av. N&S	11W-32N,11W-33N	26
Caroline Pl. NE&SW	10W-33N	27
Carriage Ln. E&W	12W-34N	26
Caryn Ln. E&W	12W-34N	26
Cashmore Rd. N&S	14W-37N	17
Casimer Pulaski Dr. NW&SE	13W-32N	26
Castle Ct. N&S	11W-37N	18
Catalpa St. E&W	11W-34N	26
Cedar Av. N&S	11W-33N,11W-38N	26
Celand Dr. NE&SW	14W-32N	25
Center St. E&W	10W-35N	27
Central Av. E&W	12W-34N	26
Chaney St. E&W	12W-38N	18
Chapel St. N&S	10W-34N	27
Chaplin St. E&W	12W-38N	18
Charleston Rd. N&S	11W-37N,11W-38N	19
Charron Ln. E&W	11W-34N	26
Chatham Rd. N&S	11W-37N	18
Cherokee Rd. N&S	11W-36N	18
Cherry Tree Ct. E&W	14W-37N	17
Chestnut St. N&S	10W-34N,10W-36N	27
Cheyenne Rd. E&W	12W-36N,11W-36N	18
Chinook Rd. E&W	11W-36N	18
Chippewa Rd. N&S	11W-36N	18
Choctaw Rd. E&W	11W-36N	18
Churchill Ct. N&S	15W-31N	33
Churchill St. N&S	15W-31N	33
Circle Ct. N&S	10W-35N	27
Clarke Av. E&W	10W-33N	27
Clarkson St. E&W	11W-34N	26
Clayton St. E&W	10W-34N,9W-34N	27
Clinton St. N&S	10W-33N	27
Colbert Rd. N&S	12W-34N	26
Coller Ct. N&S	11W-32N	26
Colorado Av. E&W	12W-36N	18
Colville Ct. E&W	10W-36N	19
Commanche Ct. E&W	11W-36N	18
Commonwealth Av. N&S	11W-33N,10W-33N	26
Continental Dr. E&W	14W-31N	17
Cooke Ct. E&W	11W-33N	26
Cornelia Av. E&W	12W-34N,11W-34N	26
Cornell Rd. N&S	11W-37N	18
Corona Rd. NE&SW	10W-36N	19
Cory Av. E&W	10W-34N	27
County St. N&S	10W-34N,10W-35N	27

	Grid	Page
Creek Rd. N&S	11W-36N	18
Crescent Av. N&S	11W-35N	26
Crest Ct. E&W	14W-37N	17
Cummings Av. E&W	10W-33N	27
Cypress St. E&W	10W-33N	27
Dady Pl. N&S	10W-33N	27
Dahringer Rd. E&W	9W-35N	27
Dakota Rd. E&W	11W-36N	18
Dale Ln. E&W	11W-36N	18
Dana Av. E&W	12W-36N	18
Darrow Ln. E&W	12W-33N	26
Daybreak Dr. NS&EW	15W-32N	25
Delany Rd. N&S	14W-36N,14W-37N	17
Delaware Rd. E&W	11W-35N,11W-36N	26
Devonshire Rd. E&W	11W-37N	18
Dickinson St. E&W	11W-37N	18
Dilger Av. N&S	12W-33N,12W-34N	26
Dixie Av. E&W	12W-33N	26
Dodge Av. E&W	11W-33N	26
Dorchester Av. E&W	12W-34N	26
Dorchester Ct. N&S	11W-34N	26
Dorothy Ct. E&W	14W-37N	17
Douglas Av. E&W	10W-35N	27
Douglas Cir. N&S	12W-34N	26
Douglas Ct. N&S	10W-35N	27
Douglas St. N&S	12W-38N	18
Dover Rd. E&W	11W-37N	18
Dugan St. N&S	10W-34N	27
Dugdale Rd. NE&SW	11W-32N,11W-33N	26
Duggald Av. E&W	12W-35N	26
Dunlay Ct. NS&EW	11W-34N	26
E. Newcastle Ct. E&W	11W-37N	18
East Bonnie Brook Ln. N&S	11W-36N,11W-37N	18
East Dr. NW&SE	12W-34N	26
East End Av. N&S	11W-37N	18
Eastview Dr. N&S	10W-35N	27
Eastwood Ct. E&W	14W-37N	17
Eastwood Rd. E&W	11W-37N	18
Edison Ct. N&S	11W-34N,10W-34N	26
Edwards St. E&W	10W-35N	27
Egidi Rd. NW&SE	16W-31N	33
Elizabeth Av. N&S	11W-32N	26
Ellis Av. E&W	12W-34N	26
Elm St. E&W	10W-33N	27
Elmwood Av. N&S	11W-32N,11W-37N	26
Emanuel Av. N&S	12W-38N	18
Ernie Krueger Cir. N&S	13W-36N	18
Evergreen Av. N&S	12W-38N,12W-38N	26
Evergreen Ct. N&S	12W-34N	26
Fairfield Av. E&W	10W-35N	27
Fairfield Ct., East E&W	10W-35N	27
Fairfield Ct., West E&W	10W-35N	27
Ferry St. E&W	11W-37N	18
Fields Ter. E&W	11W-36N	18
Florence Av. E&W	12W-34N	26
Florida Av. E&W	12W-36N	18
Flossmoor Av. N&S	11W-34N,11W-35N	26
Foam Forms Pl. E&W	9W-34N	27
Ford St. E&W	12W-38N	18
Forest Av. E&W	10W-33N	27
Forest Ct. N&S	10W-36N	19
Forest Ln. N&S	11W-37N	18
Forest View Ln. NW&SE	15W-32N	25
Franklin St. E&W	10W-34N	27
Frazier St. E&W	11W-37N	18
Frolic Av. E&W	12W-34N,12W-38N	26
Fulton Av. E&W	11W-32N,11W-34N	26
Gallagher Rd. E&W	10W-36N	19
Garden Pl. E&W	11W-35N	26
Gavin Ct. E&W	14W-31N	33
Genesee St. N&S	10W-33N,10W-34N	27
Gent Dr. E&W	14W-37N	17
George Av. NW&SE	10W-33N	27
George St. E&W	12W-34N	26
Georgetown Av. E&W	11W-35N	26
Gillette Av. E&W	10W-35N	27
Gladstone Av. NS&EW	11W-35N	26
Gladys Ct. E&W	10W-39N	19
Glen Ct. E&W	10W-33N	27
Glen Flora Av. E&W	12W-35N,10W-35N	26
Glen Rock Av. NE&SW	11W-33N,10W-34N	26

	Grid	Page
Glendening Pl. NS&EW		
	10W-36N	19
Glenn Dr. N&S 11W-32N		26
Glenwood Av. N&S		
	11W-35N	26
Golf Rd. NW&SE		
	11W-36N,10W-36N	18
Golfmoor Av. E&W 12W-36N		18
Grand Av. E&W		
	12W-35N,10W-34N	26
Gray Ct. E&W		
	12W-33N,11W-33N	26
Green Bay Pl. E&W		
	12W-33N	26
Green Bay Rd. N&S		
	12W-33N,12W-38N	26
Greenfield Av. N&S		
	11W-32N	26
Greenwood Av. E&W		
	12W-36N,9W-36N	18
Greenwood Ter. E&W		
	11W-36N	18
Gretta Av. N&S		
	11W-35N,11W-34N	26
Grove Av. E&W		
	12W-35N,10W-35N	26
Grove St. E&W		27
Hamilton Ct. E&W .. 11W-34N		26
Hanlon Rd. N&S		
	15W-31N,15W-32N	33
Harbor Pl. N&S 9W-34N		27
Harding Ct. E&W		
	11W-35N,10W-35N	26
Harper Rd. N&S 11W-37N		18
Harrison Av. E&W .. 12W-35N		26
Hawthorne Ct. E&W		
	13W-36N	18
Hawthorne Ln. E&W		
	11W-37N	18
Hazel Ct. N&S...... 10W-34N		27
Hemholz Rd. E&W .. 10W-33N		27
Hendee Av. N&S		
	12W-37N,11W-37N	18
Henry Pl. N&S...... 10W-35N		27
Herbert Dr. E&W .. 11W-38N		18
Hickory Hills Rd. N&S		
	14W-36N	17
Hickory St. N&S		
	10W-34N,10W-36N	27
Highland Av. E&W		
	11W-35N,10W-35N	26
Highland Av. E&W .. 11W-36N		18
Highland Cir. E&W .. 10W-35N		27
Hill St. N&S........ 10W-33N		27
Hillside Av. E&W...12W-34N		26
Hilltop Ln. NW&SE .. 15W-32N		25
Hilman Ct. E&W 12W-34N		26
Hollis Av. N&S 12W-34N		26
Home Av. E&W 10W-33N		27
Hope Ln. E&W 11W-36N		18
Howard St. E&W .. 11W-38N		18
Hull Ct. E&W 10W-35N		27
Huntington Lakes Dr. N&S		
	15W-31N	33
Huron Ter. N&S 11W-36N		18
Hyde Park Av. E&W		
	12W-34N,11W-34N	26
Illini Rd. E&W 12W-36N		18
Indian Rd. N&S 12W-36N		18
Indiana St. N&S 11W-36N		26
Interstate 94 N&S .. 14W-31N		33
Iroquois Rd. E&W .. 12W-36N		18
Irving Pkwy. E&W .. 11W-38N		18
Jackson Ct. NE&SW		
	10W-34N	27
Jackson St. N&S		
	10W-34N,10W-36N	27
James Ct. N&S 10W-35N		27
James St. N&S 10W-34N		27
Jaquelyn Ln. E&W .. 11W-34N		18
Jefferson Av. N&S ..11W-34N		26
Jenkinson Ct. E&W		
	11W-34N	26
Jensen Ct. N&S .. 10W-33N		27
Jolley Av. E&W .. 12W-34N		26
Judge Av. N&S		
	11W-34N,11W-36N	26
Judge Pl. NS&EW .. 11W-35N		26
Julian St. N&S 10W-34N		27
Juniper St. N&S 10W-54N		27
Kathy Ln. E&W 11W-34N		26
Keith Av. E&W .. 10W-35N		27
Keller Av. N&S		
	11W-33N,11W-34N	26
Kellogg Av. N&S...11W-36N		18
Kennard St. N&S .. 10W-33N		27
Kingston Rd. E&W..11W-37N		18
Lake Ct. E&W 10W-33N		27
Lake St. E&W		
	10W-34N,9W-34N	27
Lakehurst Rd. NS&EW		
	13W-32N,13W-33N	26
Lakeside Dr. NW&SE		
	13W-32N	26
Lakeview Av. E&W		
	10W-33N	27
Lakeview Ter. N&S		
	15W-32N	25
Lakewood Av. N&S		
	12W-34N	26
Laurel Av. E&W ... 11W-34N		26
Laurel Ln. E&W 11W-36N		18
Lawn Ct. E&W ... 10W-34N		27
Le Baron St. N&S . 12W-34N		26

	Grid	Page
Leith Av. N&S		
	11W-34N,11W-35N	26
Lenox Av. N&S 10W-33N		27
Leo Singer Ln. NE&SW		
	10W-38N	19
Lewis Av. N&S		
	11W-32N,11W-38N	26
Lewis Pl. E&W 11W-35N		26
Liberty St. E&W 10W-33N		27
Lincoln Av. N&S		
	10W-33N,10W-34N	27
Linden Av. N&S		
	10W-34N,10W-36N	27
Lindsay Av. E&W .. 10W-35N		27
Little Foot Rd. N&S		
	12W-36N	18
Lizabeth Ln. E&W .. 11W-37N		18
Lloyd Av. E&W 11W-33N		27
Longview Rd. E&W		
	11W-36N,10W-36N	18
Lorene Ct. N&S		
	11W-38N	18
Lorraine Av. N&S		
	11W-34N,11W-37N	26
Lorraine Pl. N&S...11W-35N		26
Low Av. NW&SE .. 10W-34N		27
Lucia Av. E&W		
	12W-33N,11W-33N	26
Lydia St. E&W 11W-34N		26
Lynn Ter. N&S 11W-34N		26
Lyons Ct. E&W 11W-32N		26
Madison St. E&W		
	10W-34N,9W-34N	27
Main St. E&W		
	12W-38N,11W-34N	18
Maple Av. N&S .. 11W-34N		26
Maple Tree Ln. N&S		
	14W-37N	17
Marian Ct. E&W 10W-33N		27
Mariposa Av. E&W 10W-36N		19
Market St. N&S		
	10W-33N,10W-34N	27
Marshall Pkwy. E&W		
	11W-33N	26
Martin Av. N&S		
	11W-32N,11W-35N	26
Mary Jane Ln. N&S		
	10W-38N	19
Massena Av. N&S ..10W-35N		27
May St. E&W 10W-33N		27
McAlister Av. N&S .. 10W-33N		27
McAree Av. N&S .. 11W-38N		18
McAree Rd. N&S		
	12W-33N,12W-36N	26
McArthur Dr. E&W ..11W-37N		18
McGaw Rd. E&W .. 13W-32N		26
McKay St. E&W 11W-37N		18
McKinley Av. N&S .. 10W-33N		27
Meadowlark Ln. N&S		
	11W-37N	18
Meadowview Cir. N&S		
	15W-32N	25
Melrose Av. N&S		
	12W-34N,11W-34N	26
Menominee Rd. E&W		
	12W-36N	18
Merton Av. N&S .. 10W-36N		19
Metropolitan Av. N&S		
	12W-34N,12W-36N	26
Miami Rd. N&S 12W-36N		18
Midlane Dr. NS&EW		
	14W-37N	17
Mill Ct. NW&SE 10W-34N		27
Milwaukee Av. NE&SW		
	16W-32N	24
Mini Dr. N&S 14W-37N		17
Minoque Av. N&S .. 12W-33N		26
Miraflores Av. NW&SE		
	10W-36N	19
Mohawk Rd. E&W ..12W-36N		18
Mohican Rd. E&W .. 12W-36N		18
Monroe St. E&W		
	12W-34N,11W-34N	26
Montesano Av. E&W		
	11W-36N,10W-36N	18
Morrs Ct. E&W 10W-38N		19
Muirfield Ln. E&W .. 11W-35N		26
Myrtle St. E&W 11W-35N		27
N. Bonnie Brook Ln. E&W		
	11W-37N	18
N. Butrick St. N&S		
	11W-36N	26
N. Orchard Av. N&S		
	11W-34N	26
Navajo Rd. E&W .. 12W-36N		18
Nemery Ct. E&W .. 11W-35N		26
Nemesis Av. E&W .. 12W-36N		18
New York St. N&S		
	11W-35N,11W-36N	26
Newcastle Rd. N&S		
	11W-37N,11W-38N	18
Niva St. E&W 11W-36N		18
Noble Av. N&S .. 12W-38N		18
Noll Ct. N&S 12W-34N		26
Nordigan Dr. E&W .. 10W-36N		19
Norma Ln. NE&SW		
	12W-36N	26
Norman Dr., South NS&EW		
	13W-31N	34
North Av. N&S		
	10W-34N,10W-37N	27
North Av. E&W 11W-38N		18
North Park Av. N&S		
	10W-34N	27

	Grid	Page
Northern Av. N&S		
	12W-32N,12W-38N	26
Northfield Av. E&W		
	12W-35N	26
Northmoor Av. E&W		
	11W-35N	26
Northwestern Av. N&S		
	13W-37N	18
O'Plaine Rd. N&S		
	14W-32N,14W-33N	25
O'Plaine Rd. N&S .. 14W-31N		33
Oak Av. E&W 11W-38N		18
Oak Crest St. NW&SE		
	10W-34N	27
Oak St. N&S 10W-34N		27
Oak Tree Ln. E&W 10W-34N		27
Oakland Av. E&W .. 12W-35N		26
Oakley Av. N&S 10W-35N		27
Oakwood Av. N&S.. 12W-34N		26
Oneida Pl. N&S 12W-36N		18
Orchard Ln. N&S .. 12W-35N		26
Osage Cir. E&W 12W-36N		18
Otto Graham Ln. E&W		
	10W-38N	19
Ottowa Rd. NS&EW		
	12W-36N	18
Pacific Av. E&W .. 10W-35N		27
Palmer Pl. N&S 10W-35N		27
Parkway Av. N&S .. 14W-33N		25
Parmalee Pl. E&W .. 10W-36N		19
Partridge St. N&S .. 11W-37N		18
Paul St. E&W 11W-34N		26
Pawnee Rd. E&W . 12W-36N		18
Pearsall St. E&W		
	12W-33N,11W-33N	26
Pershing Rd. N&S		
	10W-34N,9W-35N	27
Phillippa Av. N&S .. 10W-34N		27
Piccadilly Ct. N&S .. 10W-34N		27
Pickford St. E&W .. 12W-38N		18
Pickus Ct. NE&SW 10W-34N		27
Pigeon Av. NW&SE		
	12W-34N	26
Piikoi Ln. N&S 15W-31N		33
Pine St. N&S		
	11W-34N,11W-36N	26
Pioneer Ct. N&S .. 12W-34N		26
Pioneer Rd. N&S		
	12W-34N,12W-35N	26
Poplar St. N&S		
	10W-34N,10W-36N	27
Porter St. NW&SE .. 10W-34N		27
Powell Av. N&S 10W-33N		27
Prairie Av. N&S 12W-34N		26
Prescott St. N&S .. 10W-33N		27
Prospect Ct. NW&SE		
	10W-34N	27
Public Service Rd. N&S		
	12W-34N	26
Ravine Av. E&W ... 10W-33N		27
Regency Dr. NS&EW		
	14W-31N	33
Reiner St. N&S 12W-38N		18
Rice St. E&W 11W-37N		18
Richard E. Buch Ct. E&W		
	11W-32N	26
Ridge Av. E&W . 12W-33N		26
Ridgeland Av. E&W		
	12W-35N,10W-35N	26
River Park Dr. N&S		
	15W-31N	33
River Rd. N&S		
	15W-31N,15W-32N	33
Roberts Av. N&S .. 12W-37N		18
Rock Pkwy. NW&SE		
	11W-38N	18
Roger Edwards Av. E&W		
	11W-35N	26
Rt. 21 NE&SW 16W-31N		33
Rt. 43 N&S		
	13W-31N,13W-32N	34
Rt. 120 E&W 12W-33N		26
Rt. 131 N&S		
	12W-33N,12W-38N	26
Rudd St. NW&SE .. 14W-32N		25
Rugby Ct. N&S 11W-37N		18
Runyard Pl. E&W .. 11W-34N		26
S. Bonnie Brook Ln. E&W		
	11W-36N	18
S. Orchard Av. N&S		
	11W-33N,11W-34N	26
Salem Av. E&W ... 11W-33N		26
Sea Horse Dr. NS&EW		
	9W-34N	27
Seminole Rd. N&S .. 12W-36N		18
Sequoit Ct. E&W .. 11W-36N		18
Shadow Brook Ct. E&W		
	15W-32N	25
Shawnee Rd. N&S.. 12W-36N		18
Sheeler Av. E&W .. 12W-34N		26
Sheridan Ct., East E&W		
	10W-35N	27
Sheridan Rd. N&S		
	10W-33N,10W-34N	27
Sherman Pl. N&S .. 10W-34N		27
Shields St. E&W .. 13W-32N		26
Shoshone Rd. E&W		
	12W-36N	18
Sioux Rd. N&S 11W-36N		18
Skokie Hwy. N&S .. 13W-31N		34
South Av. E&W 10W-33N		27
Spaulding Av. N&S		
	12W-36N	26
Spring St. N&S 10W-34N		27

	Grid	Page
Spruce Av. E&W .. 10W-36N		19
St. James St. N&S 10W-34N		27
Stanley Av. N&S .. 10W-35N		27
Steele Ct. N&S 10W-34N		27
Steeplechase Cir. W. N&S		
	15W-31N	33
Steeplechase Dr. E. N&S		
	15W-31N	33
Stewart Av. N&S .. 10W-35N		27
Stonegate Rd. N&S		
	11W-37N,11W-38N	18
Stripe Ct. N&S 10W-34N		27
Sumac Av. N&S		
	11W-33N,11W-34N	26
Sunderlin St. N&S .. 10W-33N		27
Sunset Av. E&W		
	13W-36N,10W-36N	18
Tamara Ct. E&W .. 11W-34N		26
Tanahill Dr. E&W .. 13W-36N		18
Telegraph Rd. N&S		
	13W-31N,13W-33N	34
Ten Pin Ln. N&S .. 10W-38N		19
Terrace Av. N&S .. 10W-35N		27
Teske Blvd. N&S .. 13W-33N		26
Tewes Ln. E&W .. 10W-38N		19
Thaxter Av. N&S .. 12W-33N		26
Theresa Av. E&W .. 12W-36N		26
Thornberry Ct. N&S		
	14W-32N	25
Tiffany Dr. N&S .. 12W-34N		26
Timber Woods Ln. N&S		
	15W-32N	25
Tobin Ct. N&S...... 11W-37N		18
Tremont Ct. NS&EW		
	15W-31N	33
Twyford Rd. N&S .. 11W-37N		18
U.S. Rt. 41 NW&SE		
	12W-31N,12W-32N	35
University Av. E&W		
	12W-34N,11W-34N	26
Utica St. N&S		
	10W-33N,10W-34N	27
Vermont Av. E&W .. 13W-36N		18
Vernon Av. E&W .. 12W-36N		18
Victory St. N&S		
	10W-33N,10W-34N	27
W. Newcastle Ct. E&W		
	11W-37N	18
Wadsworth Av. N&S		
	10W-33N	27
Wadsworth Rd. E&W		
	11W-39N	18
Wainlow Av. N&S		
	10W-33N,11W-33N	27
Wall Av. E&W 12W-36N		18
Walnut St. N&S		
	10W-34N,10W-36N	27
Washington Park N&S		
	11W-32N,11W-34N	26
Washington St. E&W		
	12W-34N,10W-34N	26
Washington Ter. N&S		
	12W-34N	26
Water St. E&W 10W-34N		27
Waverly St. E&W .. 11W-35N		26
West Dr. N&S ...12W-34N		26
West St. N&S		
	10W-33N,10W-34N	27
West Waukee Rd. NE&SW		
	12W-34N	26
Western Av. N&S		
	10W-35N,10W-36N	27
Westmoreland Av. N&S		
	11W-34N,11W-35N	26
Wetzel Ct. E&W....11W-34N		26
Whidden Av. N&S		
	11W-33N,11W-34N	26
Whispering Willow Ct.		
NE&SW........15W-32N		25
White Oak Dr. NW&SE		
	13W-32N	26
Whitney St. E&W .. 11W-37N		18
Willamsburg Av. E&W		
	11W-35N	26
Willow Av. N&S .. 12W-33N		26
Willow Ct. N&S .. 12W-34N		26
Willow Dr. E&W .. 10W-36N		19
Wilson Av. E&W...11W-36N		18
Winhaven Dr. N&S		
	11W-37N,11W-38N	18
Winnebago Rd. N&S		
	12W-36N	18
Wisconsin Av. N&S		
	11W-33N,11W-34N	26
Witchford Ln. N&S .. 11W-37N		18
Witchwood Ln. N&S		
	11W-37N	18
Wood Av. E&W .. 12W-33N		26
Woodlawn Av. E&W		
	10W-35N	27
Woodlawn Cir. E&W		
	10W-35N	27
Worack Pl. N&S.... 11W-34N		26
Wyer St. E&W ... 12W-38N		18
Yeoman St. N&S		
	11W-34N,11W-37N	26
Yorkhouse Rd. E&W		
	14W-37N,11W-37N	17
Zepher St. E&W .. 12W-38N		18
1st St. N&S 10W-34N		27
2nd St. E&W 10W-34N		27
4th St. E&W 10W-34N		27
6th St. E&W		
	12W-33N,11W-33N	26

	Grid	Page
7th St. E&W 12W-33N		26
8th St. E&W		
	12W-33N,10W-33N	26
9th St. E&W		
	12W-33N,10W-33N	26
10th Ct. E&W 11W-32N		26
10th St. E&W		
	11W-33N,10W-33N	26
11th Ct. E&W 11W-32N		26
11th St. E&W 11W-32N		26
12th St. E&W 11W-32N		26
13th St. E&W 11W-32N		26
14th St. E&W 11W-32N		26

Waukegan Township

	Grid	Page
Adelphi Av. N&S		
	12W-35N,12W-36N	26
Arizona Av. E&W .. 12W-36N		18
Atlantic Av. E&W .. 12W-35N		26
Austin Av. N&S 12W-34N		26
Brooke Av. E&W .. 12W-36N		18
Cheyenne Rd. E&W		
	12W-36N	18
Clearview Av E&W .. 12W-35N		26
Colorado Av. E&W .. 12W-36N		18
Country Club Av. E&W		
	12W-36N	18
Crescent Av. E&W .. 12W-35N		26
Duggald Av. E&W .. 12W-35N		26
Florida Av. E&W .. 12W-36N		18
Frolic Av. N&S		
	12W-35N,12W-36N	26
Garrick Av. N&S		
	12W-35N,12W-36N	26
Glen Flora Av. E&W		
	12W-35N	26
Golfmoor Av. E&W 12W-36N		18
Grandmore Av. E&W		
	12W-35N	26
Grandview Av. E&W		
	12W-35N	26
Grandville Av. E&W		
	12W-35N	26
Green Av. N&S		
	12W-35N,12W-36N	26
Green Bay Rd. NE&SW		
	12W-35N,12W-36N	26
Greenwood Av. E&W		
	12W-36N	18
Grove Av. E&W 12W-35N		26
Harrison Av. E&W .. 12W-35N		26
Highland Av. E&W .. 12W-36N		18
Johns Manville Av. E&W		
	12W-36N	18
Keith Av. E&W 12W-35N		26
Lakewood Av. N&S		
	12W-34N	26
Lee Av. E&W 12W-35N		26
Metropolitan Av. N&S		
	12W-35N,12W-36N	26
Nemesis E&W 12W-36N		18
North Av. N&S 10W-37N		19
Northern Av. N&S		
	12W-35N,12W-36N	26
Northfield Av. E&W		
	12W-35N	26
Pacific Av. E&W 12W-35N		26
Paddock St. N&S .. 12W-37N		18
Polo Ct. E&W 12W-36N		18
Public Service Av. N&S		
	12W-35N	26
Sherman Av. E&W .. 12W-35N		26
Southview Av. E&W		
	12W-35N	26
Tyler Av. E&W 12W-36N		18
University Av. E&W		
	12W-34N	26
Vermont Av. E&W .. 12W-36N		18
Wall Av. E&W ... 12W-36N		18

West Deerfield Township

	Grid	Page
Birchwood Ln. N&S		
	11W-23N	50
Carol Ln. N&S....12W-21N		50
Chris Ct. E&W .. 12W-21N		50
Deer Run Dr. N&S .. 12W-20N		50
Elmwood Pl. E&W .. 11W-23N		50
Forest Ct. N&S .. 11W-23N		50
Hawthorne Ln. E&W		
	12W-20N	50
Hickory Knoll Ln. E&W		
	12W-21N	50
London Ln. (Pvt.) N&S		
	11W-23N	50
Montgomery Rd. N&S		
	12W-21N	50
Oakwood Pl. E&W..11W-23N		50
Sannders Rd. N&S .. 12W-20N		50
Wildwood Ln. N&S 11W-23N		50
Woodview Ln. N&S		
	12W-20N	50

Winthrop Harbor

	Grid	Page
Adams Av. N&S 10W-42N		11
Aiken Ct. N&S 9W-41N		11
Alton Ct. N&S 10W-41N		11
Ben Ct. N&S 10W-41N		11
Burke Av. E&W 10W-41N		11
Camden Cove NE&SW		
	10W-43N	11

	Grid	Page
Casco Bay NW&SE	10W-43N	11
Cavin Av. N&S	10W-42N,10W-43N	11
Cedar Ct. E&W	9W-43N	11
Charles Av. N&S	10W-42N	11
Chestnut Cir. N&S	9W-43N	11
Clearview Ct. NE&SW	11W-43N	10
College Av. N&S	10W-42N,10W-43N	11
Deer Path Cir. E&W	9W-43N	11
Deer Path Dr. N&S	9W-42N,9W-43N	11
East Broadway Av. N&S	11W-41N,11W-42N	10
Ellis Av. N&S	10W-42N	11
English Ln. N&S	10W-42N	11
Fairhaven Dr. NS&EW	10W-43N	11
Franklin Av. N&S	9W-41N,9W-42N	11
Franklin St. E&W	9W-41N	11
Frans Ct. NE&SW	10W-42N	11
Fulton Av. N&S	9W-41N,9W-42N	11
Funderburk Dr. E&W	10W-41N	11
Garnett Av N&S	10W-42N,10W-43N	11
Geddes Av. N&S	10W-42N,10W-43N	11
Har-Bar Cir. E&W	11W-42N	10
Har-Bar Dr. E&W	11W-42N	10
Harbor Ln. NE&SW	10W-42N	11
Highland Ct. E&W	10W-42N	11
Holdridge Av. N&S	10W-42N,10W-43N	11
James Av. NW&SE	10W-42N	11
Judith Ct. NS&EW	10W-42N	11
Kenosha Rd. N&S	11W-42N,11W-43N	10
Kimberly Ln. N&S	10W-42N	11
Kirkwood Av. N&S	10W-42N,10W-43N	11
La Lor Ct. E&W	10W-42N	11
Landon Av. N&S	9W-41N,9W-42N	11
Lewis Av. N&S	11W-41N,11W-42N	10
Madison Av. N&S	10W-42N	11
Mae Ct. E&W	10W-41N	11
Main St. E&W	10W-42N,9W-42N	11
Mary Av. N&S	10W-42N	11
Meadow Ct. N&S	11W-42N	10
Meadow Ln. N&S	11W-42N	10
Monroe Av. N&S	10W-41N,10W-42N	11
North Ave. N&S	10W-42N,10W-43N	11
North Park Av. NW&SE	9W-42N,9W-43N	11
Oakdale Av. N&S	9W-43N	11
Old Darby Rd. NW&SE	10W-43N	11
Park Av. N&S	9W-41N,9W-42N	11
Parkview Dr. NW&SE	11W-42N	10
Potomac Ln. N&S	10W-43N	11
Prairie Av. NW&SE	11W-43N	10
Ravine Dr. NS&EW	9W-42N,9W-43N	11
Roberts Rd. N&S	10W-42N	11
Rt. 173 N&S	9W-41N,9W-43N	11
Runyard Av. N&S	9W-43N	11
Russell Av. N&S	10W-42N,10W-43N	11
Russell Rd. E&W	10W-43N,9W-43N	11
Sheridan Rd. E&W	9W-41N,9W-43N	11
Shields Av. N&S	10W-42N	11
South Park Av. N&S	9W-42N	11
Stonegate Dr. E&W	12W-42N	10
Thompson Av. N&S	10W-42N,10W-43N	11
Valley Rd. NS&EW	11W-42N	10
Vista Ln. N&S	12W-42N	10
Washington Av. N&S	10W-42N	11
West Broadway Av. N&S	11W-41N,11W-42N	10
West Ct. N&S	10W-41N	11
Whitney Av. N&S	9W-42N	11
Wilson Av. N&S	9W-43N	11
2nd St. E&W	10W-43N,9W-43N	11
3rd St. E&W	10W-43N,9W-43N	11
4th St. E&W	10W-43N,9W-43N	11
5th St. E&W	11W-43N,9W-43N	10
6th St. E&W	10W-42N,9W-42N	11
8th St. E&W	10W-42N,9W-42N	11
9th St. E&W	11W-42N,9W-42N	10
10th St. E&W	10W-42N,9W-42N	11
11th St. E&W	11W-42N,9W-42N	10
12th St. E&W	11W-42N,9W-42N	10
13th St. E&W	11W-42N,9W-42N	10
14th St. E&W	10W-41N,9W-41N	11
15th St. E&W	11W-41N,9W-41N	10

Zion

	Grid	Page
Barnhart Av. E&W	10W-41N,9W-41N	11
Bethel Blvd. N&S	10W-39N,10W-41N	19
Bethesda Blvd. N&S	12W-40N,10W-41N	11
Bethlehem Av. NW&SE	11W-41N	10
Big Highlands Rd. NS&EW	12W-41N	10
Brigadoon Dr. NW&SE	12W-42N	10
Caledonia Av. NW&SE	10W-40N,9W-39N	11
Calvary Ln. N&S	11W-40N	10
Carmel Blvd. E&W	10W-39N,9W-39N	19
Champart Ct. E&W	11W-41N	10
Clover Ln. N&S	10W-39N	19
Deborah Av. N&S	9W-40N	11
Douglas Dr. E&W	12W-42N	10
Dowie Memorial Dr. Cir.	10W-40N	11
Eden Av. N&S	9W-40N	11
Edina Blvd. N&S	9W-39N,9W-40N	19
Elim Av. N&S	9W-39N,9W-40N	19
Elisha Av. N&S	10W-39N,10W-40N	19
Elizabeth Av. N&S	9W-39N,9W-40N	19
Ellijah Av. N&S	10W-39N,10W-41N	19
Emmaus Av. N&S	10W-39N,10W-40N	19
Enoch Av. N&S	10W-39N,10W-41N	19
Eshcol Av. N&S	10W-39N,10W-40N	19
Ezekiel Av. N&S	10W-39N,10W-41N	19
Ezekiel Pl. N&S	10W-39N	19
Ezra Av. N&S	10W-39N,10W-41N	19
Fleming Av. N&S	11W-41N	10
Franklin Ct. NW&SE	12W-42N	10
Franklin St. NW&SE	12W-42N	10
Gabriel Av. N&S	10W-39N,10W-41N	19
Galilee Av. N&S	11W-39N,11W-41N	18
Gideon Av. N&S	10W-39N,10W-41N	19
Gilboa Av. N&S	10W-39N,10W-41N	19
Gilead Av. N&S	10W-39N,10W-41N	19
Gregory Dr. NW&SE	12W-42N	10
Hebron Av. N&S	11W-40N,11W-41N	10
Hermon Av. N&S	11W-40N,11W-41N	10
Hillside Av. NE&SW	9W-41N	11
Hilltop Av. N&S	9W-41N	11
Horeb Av. N&S	11W-40N	10
Industrial Av. N&S	11W-41N	10
Jethro Av. N&S	11W-40N,11W-41N	10
Joanna Av. N&S	11W-40N,11W-41N	10
Joppa Av. N&S	11W-40N,11W-41N	10
Jordan Av. N&S	11W-39N,11W-40N	18
Kebron Blvd. E&W	10W-41N	11
Kenosha Rd. N&S	12W-40N,11W-42N	10
Lake View Av. N&S	9W-41N	11
Leah Av. N&S	11W-40N	10
Lebanon Av. NE&SW	11W-39N	18
Lewis Av. N&S	11W-41N	10
Lorlei Dr. NE&SW	12W-42N	10
Lowery Ct. N&S	11W-40N	10
Luckie Ct. NW&SE	12W-42N	10
Lydia Av. N&S	11W-40N,11W-41N	10
Lydia Pl. N&S	11W-40N	10
Mary-Kay Ln. N&S	10W-39N	19
Ravine Av. NS&EW	10W-41N	11

	Grid	Page
Rt. 173 E&W	12W-41N,10W-41N	10
Salem Blvd. E&W	11W-40N,9W-40N	10
Sharon Pl. NS&EW	10W-39N	19
Shiloh Blvd. E&W	10W-40N,9W-40N	11
Stella Ct. E&W	11W-40N	10
Thistle Ln. E&W	12W-42N	10
Thompson Av. N&S	10W-41N	11
Timothy St. NW&SE	12W-42N	10
Willow Ln. E&W	11W-41N	10
Wilson Ct. E&W	10W-41N,9W-41N	11
Winthrop Ct. E&W	9W-41N	11
Winthrop Harbor Rd. E&W	12W-42N	10
Yale Ct. E&W	10W-39N	19
15th St. E&W	11W-41N	10
16th St. E&W	12W-41N,11W-41N	10
17th St. E&W	11W-41N,9W-41N	10
18th St. E&W	11W-41N,9W-41N	10
19th St. E&W	11W-41N,9W-41N	10
20th St. E&W	11W-41N,9W-41N	10
21st St. E&W	11W-41N,9W-41N	10
22nd St. E&W	11W-40N,9W-40N	10
23rd St. E&W	11W-40N,9W-40N	10
24th St. E&W	11W-40N,9W-40N	10
25th St. E&W	11W-40N	10
26th St. E&W	11W-40N,9W-40N	10
27th St. E&W	12W-40N,9W-40N	10
28th St. E&W	10W-40N,9W-40N	11
29th St. E&W	11W-40N,9W-40N	10
30th St. E&W	11W-39N,9W-39N	18
31st St. E&W	11W-39N,9W-39N	18
32nd St. E&W	11W-39N	18
33rd St. E&W	11W-39N,10W-39N	18
34th St. E&W	10W-39N	19

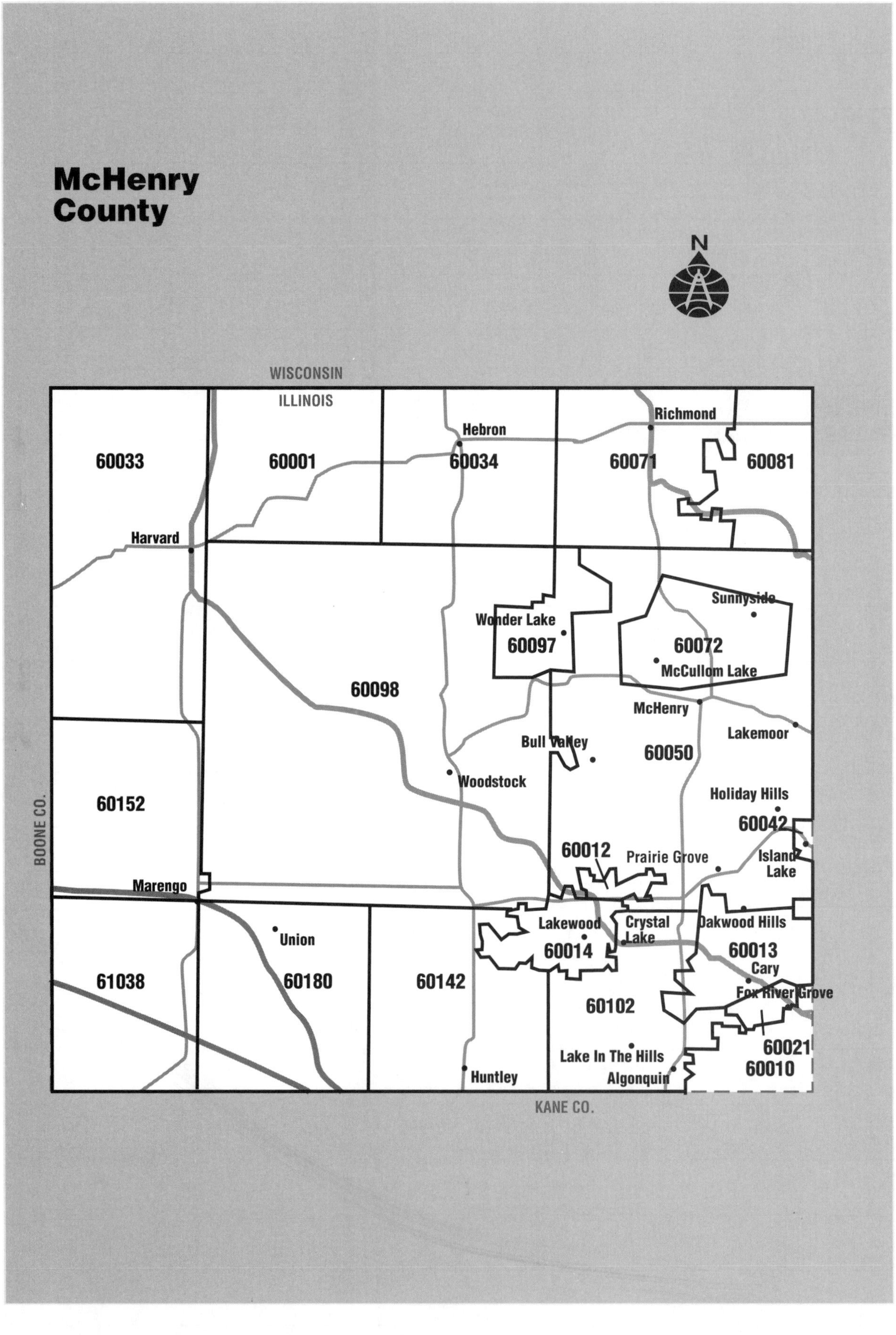

McHenry County

N

WISCONSIN

ILLINOIS

60033

60001

Hebron

60034

Richmond

60071

60081

Harvard

Sunnyside

Wonder Lake

60097

60072

McCullom Lake

60098

McHenry

Lakemoor

BOONE CO.

Bull Valley

60050

60152

Woodstock

Holiday Hills

60042

60012

Prairie Grove

Island Lake

Marengo

Lakewood

Crystal Lake

Oakwood Hills

60013

Union

60014

Cary

61038

60180

60142

60102

Fox River Grove

60021

Lake In The Hills

60010

Huntley

Algonquin

KANE CO.

McHenry County

StreetFinder®

page **Contents**

A Using the StreetFinder

B PageFinder Map

C Vicinity Map

1 Contents

2 Quick Reference Guide

4 StreetFinder Maps

52 StreetFinder Index

Photo credit: by Don Peasley 388 Lincoln Ave. Woodstock IL

PageFinder™ Map patent pending.

Information included in this publication has been checked for accuracy prior to publication. Since changes do occur, the publisher cannot be responsible for any variations from the information printed.

McHenry County Municipal Offices

	Location	Page
Algonquin Village Hall	33W-20N	50
2 S Main St; 658-4322		
Lake In The Hills Village Hall	35W-22N	50
1111 Crystal Lake Rd; 658-4213		
Marengo Village Hall	49W-26N	37
132 E Prairie St; 568-7112		
McHenry Twp City Hall	32W-33N	26,27
1111 N Green St; 385-8500		
Woodstock County Couthouse	41W-32N	24
2200 N Seminary Ave; 338-2040		

Cemeteries

	Location	Page
Alden Cem	45W-41N	7
Big Foot Cem	49W-43N	6
Brandon Cem	44W-41N	7
Calvary Cem	42W-31N	24
Cedarvale Cem	32W-39N	10,11
Chemung Cem	52W-38N	12
Chemung-Dunham Cem	53W-37N	12
Christ The King Cem	36W-36N	17
Crystal Lake Mem Park Cem	36W-27N	33
Eckert Cem	40W-34N	24
English Prairie Cem	29W-41N	11
First Street Cem	30W-23N	43
Greenwood Cem	38W-35N	17
Harmony Cem	46W-20N	46
Hebron Cem	40W-41N	8
Holcombville Cem	35W-29N	34
Jerome Cem	49W-37N	13
Lawrence Cem	52W-39N	4
Linn-Hebron Cem	42W-43N	8
Marengo City Cem	49W-26N	37
McHenry County Cem	45W-34N	23
McMillian Cem	33W-29N	34
Mosgrove Cem	32W-30N	34,35
Mt Auburn Cem	49W-37N	13
Mt Thabor Cem	39W-26N	40,41
North Solon Cem	32W-40N	10,11
Oakland Cem	42W-31N	24
Opfergelt Cem	43W-32N	23
Orvis Cem	29W-41N	11
Ostend Cem	35W-34N	26
Prairie Grove Cem	31W-27N	34
Richmond Cem	34W-42N	10
Ridgefield Cem	37W-27N	33
Ringwood Cem	34W-36N	18
Sacred Heart Cem	50W-27N	29
Scandinavian Cem	41W-34N	24
Scottish Cem	49W-28N	30
South Dunham Cem	50W-33N	21
St John's Cem	31W-35N	19
St John's Cem	32W-22N	50,51
St Joseph's Cem	49W-37N	13

	Location	Page
St Joseph's Cem	34W-43N	10
St Mary's Cem	40W-20N	48
St Mary's Cem	32W-33N	26,27
St Patrick's Cem	43,44W-35N	15
St Patrick's Countryside Cem	32W-31N	26,27
Stewart Cem	39W-39N	8,9
Union Cem	47W-25N	38
Union Cem	35W-26N	42
Unity Church Cem	54W-34N	20
Washington Cem	32W-42N	10,11
Windridge Cem	29W-24N	43
Woodland Cem	32W-33N	26,27
Woodstock Mem Cem	40W-29N	32

Colleges & Universities

	Location	Page
McHenry County Col	37W-27N	33

Forest Preserves

	Location	Page
Beck Woods FP	53W-38N	12
Borrow Woods FP	49W-36N	14
Butternut Preserve	38W-25N	41
Chain O'Lakes State Park	29W-40N	11
Coral Woods FP	48W-24N	38
Deep Cut Marsh	46W-33N	22,23
Farm Hills Acres	36W-24N	41
Harrison Benwell	35W-35N	18
Hickory Grove FP	29W-25N	43
Indian Ridge FP	31,32W-29N	34,35
Marengo Ridge	49W-29N	30
Marie Gundstrom Nat Area	37W-38N	17
Moraine Hills State Park	31,30W-31N	27
Ryders Woods FP	41W-30N	32
The Hollows FP	32W-24,25N	42,43
Volo Bog FP	29W-34N	27
Weers Nature Area	42,41W-27N	32

Golf Courses & Country Clubs

	Location	Page
Cary CC	31W-22N	51
Chalet GC	30W-26N	43
Crystal Lake CC	36W-24N	41
Crystal Woods GC	40W-26,25N	40
Hunter CC	34W-42N	10
McHenry CC	32W-32N	26,27
Pinecrest GC	40W-21N	48
Pistakee GC	29W-34N	27
Turnberry CC	38W-24N	41
Twin Ponds GC	33W24N	42
Woodstock CC	39W-30N	32,33

Shopping Centers

	Location	Page
Country Corner Plaza	35W-25N	42
Crystal Lake Plaza	35W-25N	42
Crystal Point Mall	34W-25N	42
Manchester Mall	33W-33N	26
McHenry Market Place	33W-33N	26
Somerset Mall	33W-33N	26

McHenry County Close-up

Woodstock Opera Company
121 Van Buren Street, Woodstock
Constructed in 1890, this facility offers a mixed program of dance, theater, opera, lectures and art exhibits. It is also home to two resident performance companies, Town Square Players and Woodstock Musical Theater Company. Special events include an annual Mozart Festival. Designed in the architectural style of Steamboat Gothic, this structure is listed with the National Register of Historic Places. Open all year.

Antique Village Museum & Wild West Town
8512 S. Union Road, Union
This unique site covers 20 acres and consists of a museum and a re-created western town. The museum displays war relics and numerous antique music boxes and phonographs. Wild west town features a replica of a turn-of-the-century street, replete with saloon, blacksmith, sheriff's office and print shop. An old time movie theater shows vintage films. Pony rides and train tours available. Children encouraged to pan for gold.

McHenry County Historical Museum
6422 Main Street, Union
This comprehensive museum honors the history of McHenry County. On the premises are a one-room schoolhouse built in 1895, a log cabin constructed in 1847, and a tourist cabin built in 1948. Collections include clothing, looms, quilts, toys and farm tools. There is also a research library. Open to the public May through October.

Illinois Railway Museum
7000 Olson Rd., Union
Visit one of the country's major museums for railway history. Set on 56 acres, this site features a depot constructed in 1851, a restored elevated station, a 1910 signal tower and 300 pieces of historic railway equipment. Vintage trains include several steam and diesel-powered locomotives and a 1936 stainless steel Nebraska Zephyr. Visitors can ride trains and tour the seven large storage barns that house many of the trains. Seasonal.

Colonel Palmer House
5516 S. Terra Cotta Rd., Crystal Lake
This charming two-story house was built in 1858 for Colonel Gustavus Parker. A fine example of Greek Revival architecture, this dwelling features original furnishings including a vintage piano. Open by appointment.

Dole Mansion
401 Country Club Rd., Crystal Lake
This handsome dwelling was built for grain merchant Charles Dole in the 1860's. Constructed from imported lumber, this distinctive home includes parquet floors and hand-carved banisters. Open by appointment.

Moraine Hills State Park
914 S. River Rd., McHenry
This beautiful outdoor getaway covers 1,700 acres of rolling hills and wetlands. The area includes picnic shelters, two nature preserves and several miles of multi-purpose trails. Fox River and three lakes are available for fishing and boating. McHenry Dam is on the premises. No overnight camping. Open all year.

Chester Gould-Dick Tracy Memorial Museum
101 N. Johnson St., Woodstock
This site honors Chester Gould, creator of one of America's most cherished cartoon characters, Dick Tracy. A Woodstock resident for some fifty years, Gould's syndicated cartoon strip entertained millions of readers. Exhibits includes sketches, storyboards, life-sized figures and other memorabilia. The artist's desk and chair are displayed. Also, the museum holds an annual Dick Tracy Days Parade. Located in historic courthouse building.

Landmark School
3614 W. Waukegan St., McHenry
Revisit the past at this school building built in 1894. Constructed from local materials, this historic structure remains the area's oldest active school. A registered city landmark.

Glacial Park
6512 Harts Rd., Ringwood
Nature at its best! Hundreds of acres of glacial sand, wetlands, prairies and marshes. A boardwalk provides visitors with a breathtaking view of the Nippersink valley. Intrepretive staff available for lectures, workshops and nature programs. Abundant wildlife.

The Hollows
off Rte. 14, between Crystal Lake and Cary,
This unusual terrain, with its large sand and gravel content, features trails for both hiking and cross-country skiing. The area also includes campsites, shelters and play areas for children. Lake Atwood is available for non-motorized boating.

Rush Creek Conservation Area
Rte. 14 to Harvard, E on McGuire Rd., Harvard
Not just another conservation area, this unique site consists of flood plain, meadows and upland forest. It is available for a wide range of activities including hiking, fishing, cross-country skiing, skating, ice fishing and horseback riding. Campsites and picnic shelters are on the premises.

America's Cardboard Cup Regatta
Main Beach, 300 Lake Shore Drive, Crystal Lake
This whimsical one-day event delights visitors every year. Crews of up to twelve individuals race in boats made from corrugated cardboard. Proceeds go to charity. June.

Crystal Lake Gala & Lakeside Festival
401 Country Club Rd., Crystal Lake
This two-weekend festival features a parade, fireworks, live entertainment, a golf touroment, beer garden and a carnival. The Cardboard Cup Regatta is the centerpiece event. Proceeds go to charity. June.

Illinois Storytelling Festival
Spring Grove Village Park, Spring Grove
Experience an art form rich in folk lore. This major one-day event hosts some of America's most accomplished storytellers. Other highlights include old time music. Last weekend in July.

McHenry County Fair
County Fairgrounds, Woodstock
Fun for all ages! This major event includes carnival rides, special programs for children and various livestock shows. Also, the fair holds a Miss McHenry County Contest. Food and concessions. August.

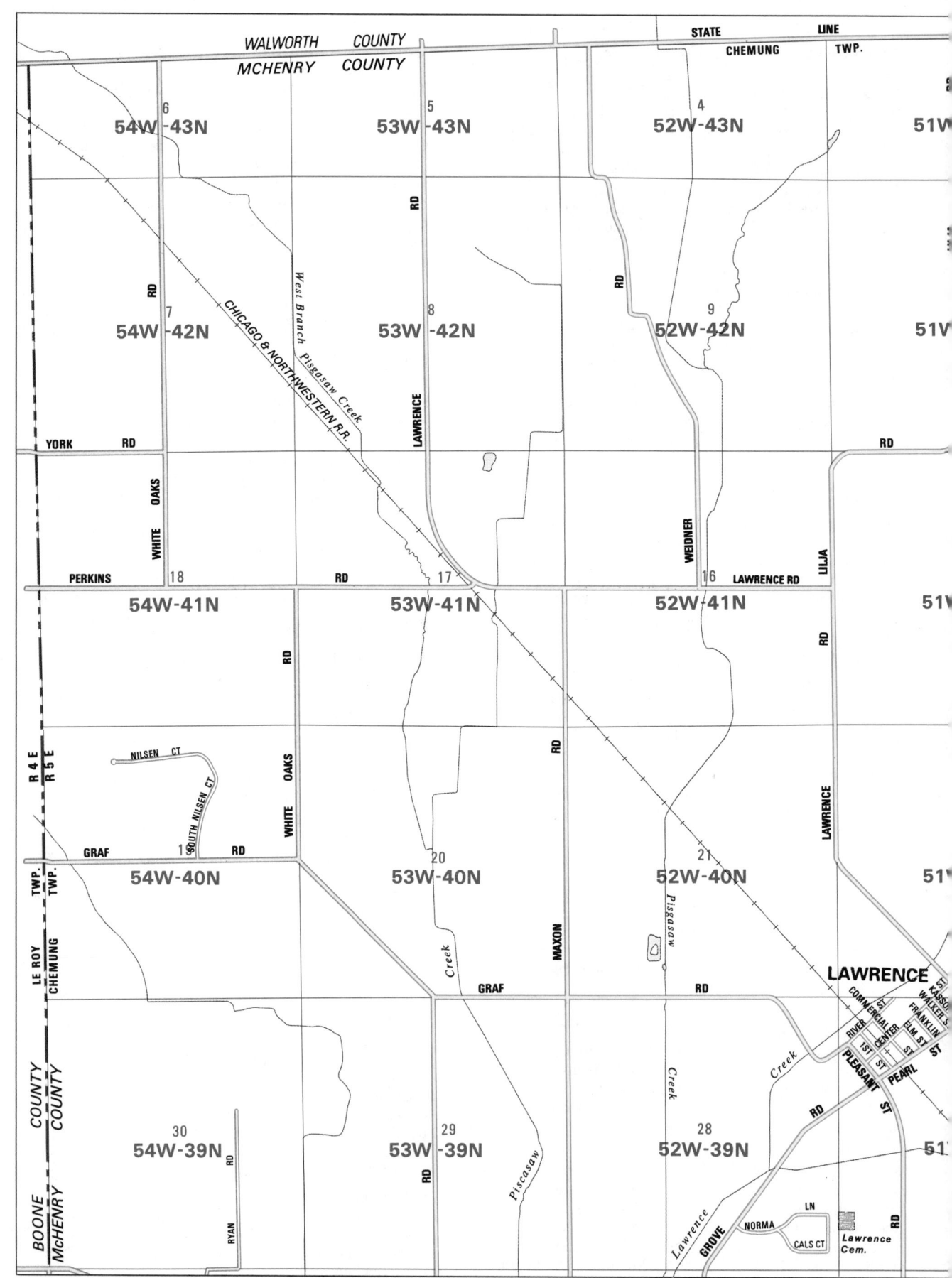

WALWORTH COUNTY

MCHENRY COUNTY

STATE LINE

CHEMUNG TWP.

6
54W-43N

5
53W-43N

4
52W-43N

51W

7
54W-42N

8
53W-42N

9
52W-42N

51W

CHICAGO & NORTHWESTERN R.R.

West Branch Pisgasaw Creek

RD

RD

LAWRENCE

YORK RD

RD

WHITE OAKS

WEIDNER

LILJA

PERKINS 18

RD

17

16 LAWRENCE RD

54W-41N

53W-41N

52W-41N

51W

RD

RD

RD

WHITE OAKS

NILSEN CT

SOUTH NILSEN CT

LAWRENCE

R 4 E
R 5 E

GRAF 1 RD

20

21

54W-40N

53W-40N

52W-40N

51W

Creek

MAXON

Pisgasaw

LAWRENCE

GRAF RD

RD

KASSON ST

WALKER S.

FRANKLIN ST

COMMERCIAL ST

RIVER ST

1ST ST

CENTER ST

ELM ST

LE ROY TWP.
CHEMUNG TWP.

Creek

PEARL

PLEASANT ST

BOONE COUNTY
McHENRY COUNTY

30
54W-39N

29
53W-39N

Piscasaw

28
52W-39N

51

RD

RYAN

RD

Piscasaw

LN

NORMA

Lawrence Cem.

CALS CT

Lawrence GROVE

RD

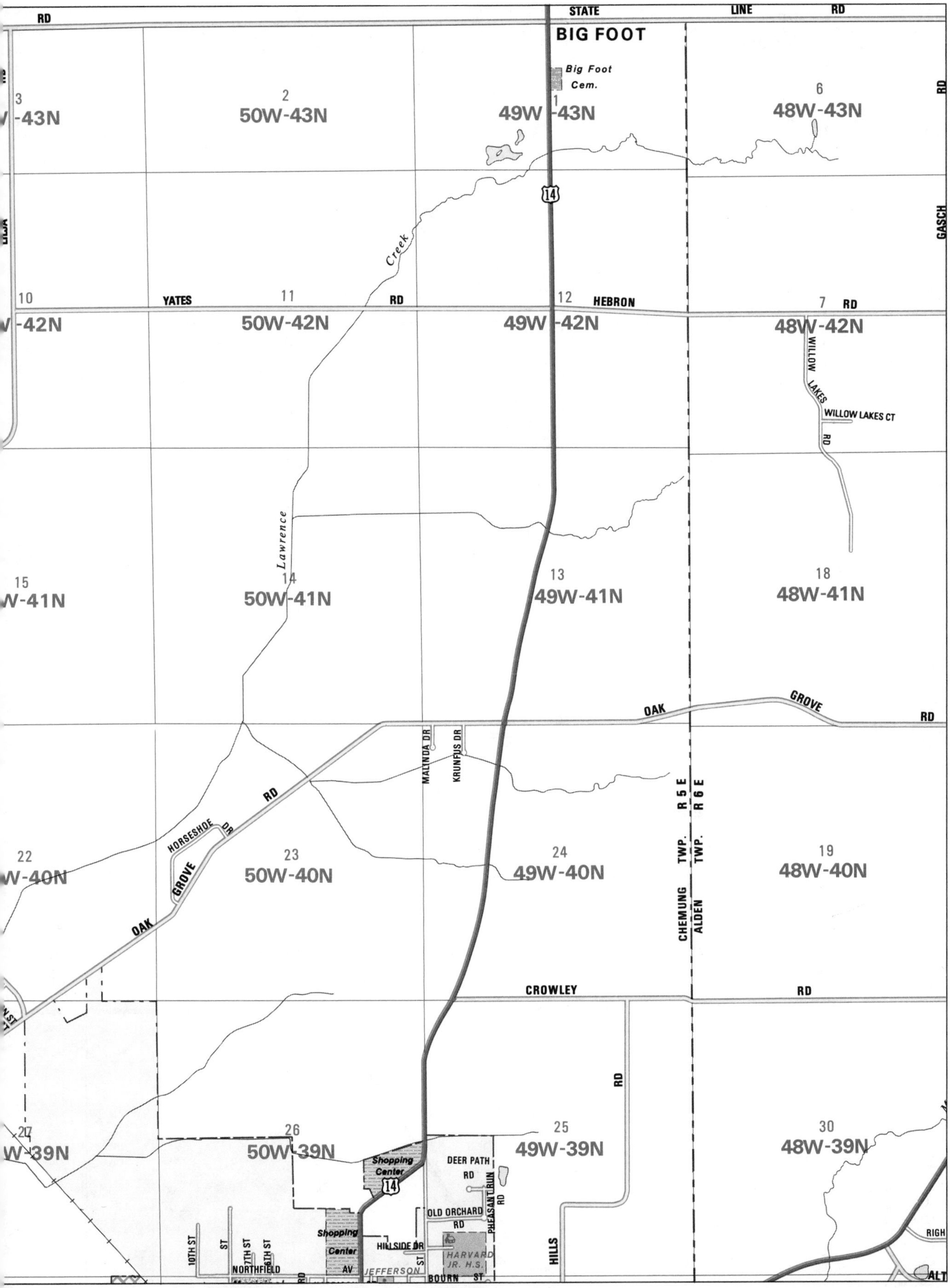

RD

STATE LINE RD

BIG FOOT

Big Foot Cem.

3

-43N

2

50W-43N

49W **-43N**

6

48W-43N

RD

GASCH

14

10

-42N

YATES

11

50W-42N

RD

12

HEBRON

49W **-42N**

7 RD

48W-42N

WILLOW LAKES

WILLOW LAKES CT

RD

15

-41N

Lawrence

14

50W-41N

13

49W-41N

18

48W-41N

OAK GROVE RD

MALINDA DR

KRUNFUS DR

22

-40N

HORSESHOE DR

RD

GROVE

OAK

23

50W-40N

24

49W-40N

R 5 E

R 6 E

CHEMUNG TWP.

ALDEN TWP.

19

48W-40N

CROWLEY RD

RD

27

-39N

26

50W-39N

Shopping Center

14

DEER PATH RD

PHEASANT RUN RD

25

49W-39N

30

48W-39N

10TH ST

7TH ST

5TH ST

Shopping Center

NORTHFIELD

AV

OLD ORCHARD RD

HILLSIDE DR

ST

JEFFERSON

BOURN ST

HARVARD JR. H.S.

HILLS

ST

RIGH

ALT

SEE PAGES 6-7

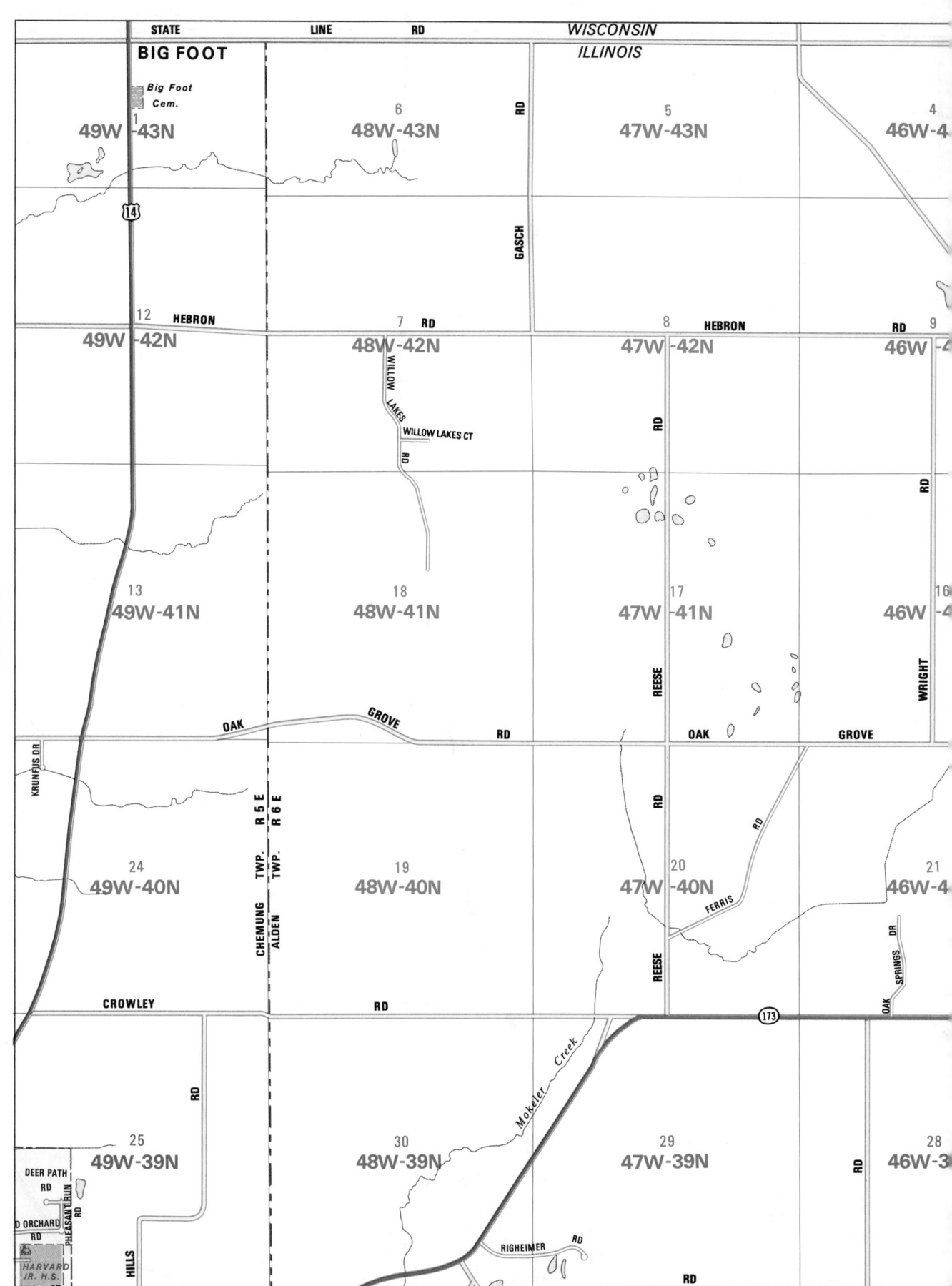

STATE LINE RD WISCONSIN
ILLINOIS

BIG FOOT

Big Foot Cem.

49W-43N 1
48W-43N 6
47W-43N 5
46W-4 4

14

GASCH RD

12 HEBRON
49W-42N
7 RD
48W-42N
8 HEBRON
47W-42N
RD 9
46W-4

WILLOW LAKES
WILLOW LAKES CT
RD

RD

SEE PAGES 4-5

13
49W-41N
18
48W-41N
17
47W-41N
16
46W-4

REESE

WRIGHT RD

OAK GROVE RD
OAK GROVE

KRUNFLS DR

RD

24
49W-40N
19
48W-40N
20
47W-40N
21
46W-4

CHEMUNG TWP. R 5 E
ALDEN TWP. R 6 E

FERRIS

REESE

OAK SPRINGS DR

CROWLEY RD
173

Mokeler Creek

RD

25
49W-39N
30
48W-39N
29
47W-39N
28
46W-3

DEER PATH RD
D ORCHARD RD
RD

PHEASANT RUN RD

HARVARD JR. H.S.

HILLS

RIGHEIMER RD
RD

RD

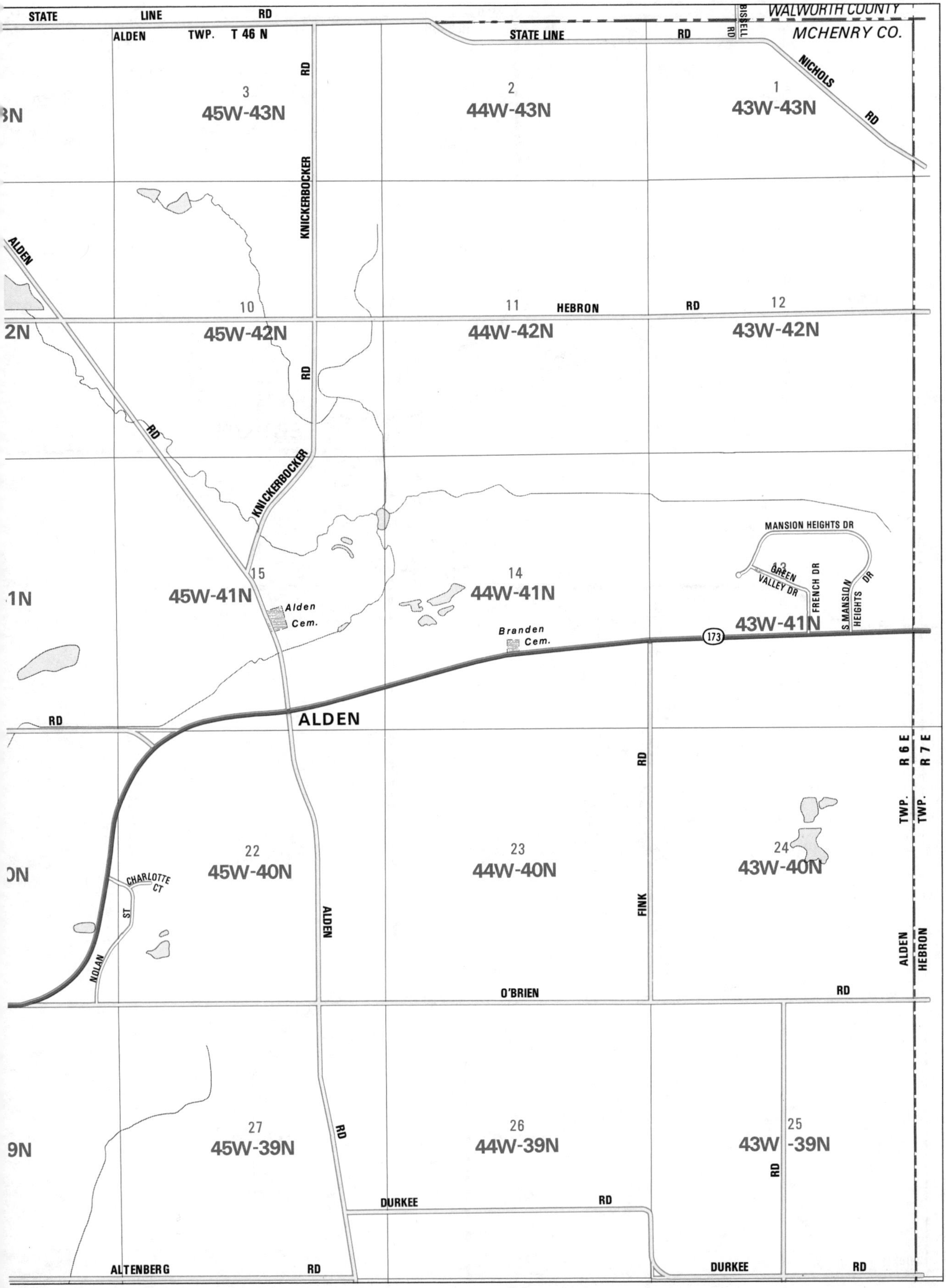

STATE LINE RD

WALWORTH COUNTY

ALDEN TWP. T 46 N

BISSELL RD

STATE LINE RD

MCHENRY CO.

NICHOLS RD

BN

3
45W-43N

2
44W-43N

1
43W-43N

KNICKERBOCKER RD

ALDEN

2N

10
45W-42N

11   HEBRON   RD

44W-42N

12
43W-42N

RD

RD

KNICKERBOCKER RD

MANSION HEIGHTS DR

GREEN VALLEY DR

FRENCH DR

S. MANSION HEIGHTS DR

SEE PAGES 8-9

1N

15

45W-41N

Alden Cem.

14
44W-41N

Branden Cem.

173

43W-41N

RD

ALDEN

RD

FINK RD

R 6 E TWP.

R 7 E TWP.

ON

22
45W-40N

CHARLOTTE CT

NOLAN ST

23
44W-40N

24
43W-40N

ALDEN

ALDEN

HEBRON

O'BRIEN

RD

9N

27
45W-39N

RD

26
44W-39N

25
43W -39N

RD

DURKEE   RD

ALTENBERG   RD

DURKEE   RD

SEE PAGES 14-15

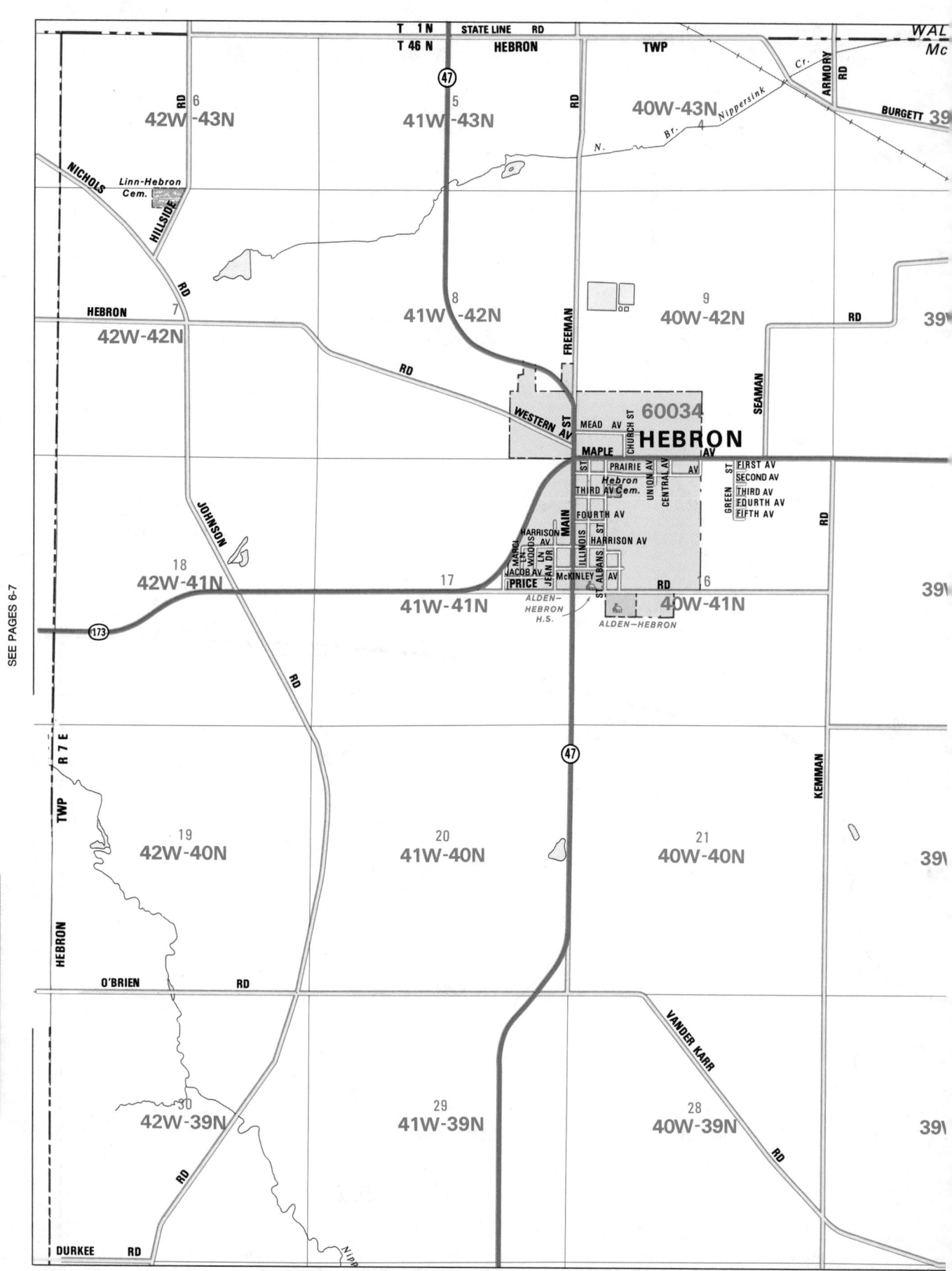

WAL
Mc

T 1N
T 46 N
STATE LINE RD
HEBRON TWP

BURGETT 39

42W-43N 6

41W-43N 5

40W-43N 4

Nippersink
N. Br. Cr.
ARMORY RD

NICHOLS
HILLSIDE RD

Linn-Hebron Cem.

41W-42N 8

40W-42N 9

RD 39

HEBRON RD
42W-42N 7

SEAMAN

FREEMAN ST

RD

WESTERN AV ST

60034

MEAD AV
CHURCH ST
HEBRON

MAPLE ST
PRAIRIE AV

FIRST AV
SECOND AV
THIRD AV
FOURTH AV
FIFTH AV

GREEN ST

JOHNSON RD

Hebron Cem.
THIRD AV
UNION AV
CENTRAL AV
AV

MAIN ST
FOURTH AV
ILLINOIS ST

HARRISON AV
MAPLE LN
WOODS LN
JEAN DR
HARRISON AV
ST. ALBANS ST

RD

42W-41N 18

JACOB AV
PRICE
McKINLEY ST

41W-41N 17

40W-41N 16

39

ALDEN-HEBRON H.S.
ALDEN-HEBRON

173

SEE PAGES 6-7

47

42W-40N 19

41W-40N 20

40W-40N 21

39

TWP R 7 E

KEMMAN

HEBRON

O'BRIEN RD

VANDER KARR

42W-39N 30

41W-39N 29

40W-39N 28

RD

39

DURKEE RD

NIPP

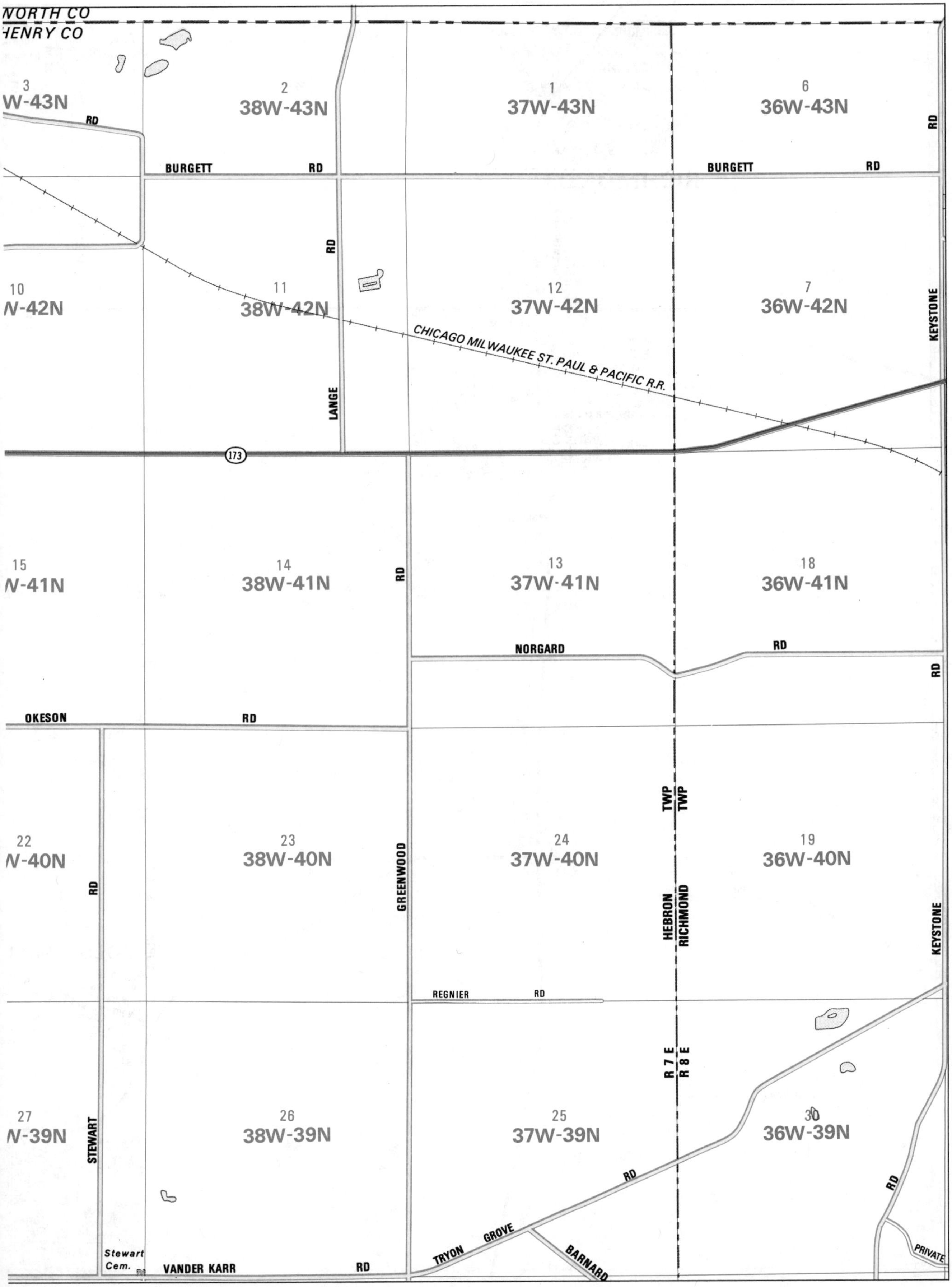

NORTH CO
HENRY CO

3
W-43N

2
38W-43N

1
37W-43N

6
36W-43N

RD

RD

BURGETT RD

BURGETT RD

RD

10
N-42N

11
38W-42N

12
37W-42N

7
36W-42N

LANGE RD

KEYSTONE

CHICAGO MILWAUKEE ST. PAUL & PACIFIC R.R.

173

15
N-41N

14
38W-41N

RD

13
37W-41N

18
36W-41N

SEE PAGES 10-11

NORGARD RD

RD

OKESON RD

22
N-40N

23
38W-40N

GREENWOOD RD

24
37W-40N

TWP
TWP

19
36W-40N

RD

HEBRON
RICHMOND

KEYSTONE

REGNIER RD

R 7 E
R 8 E

27
N-39N

STEWART

26
38W-39N

25
37W-39N

RD

30
36W-39N

RD

TRYON GROVE

BARNARD

PRIVATE

Stewart
Cem.

VANDER KARR RD

SEE PAGES 16-17

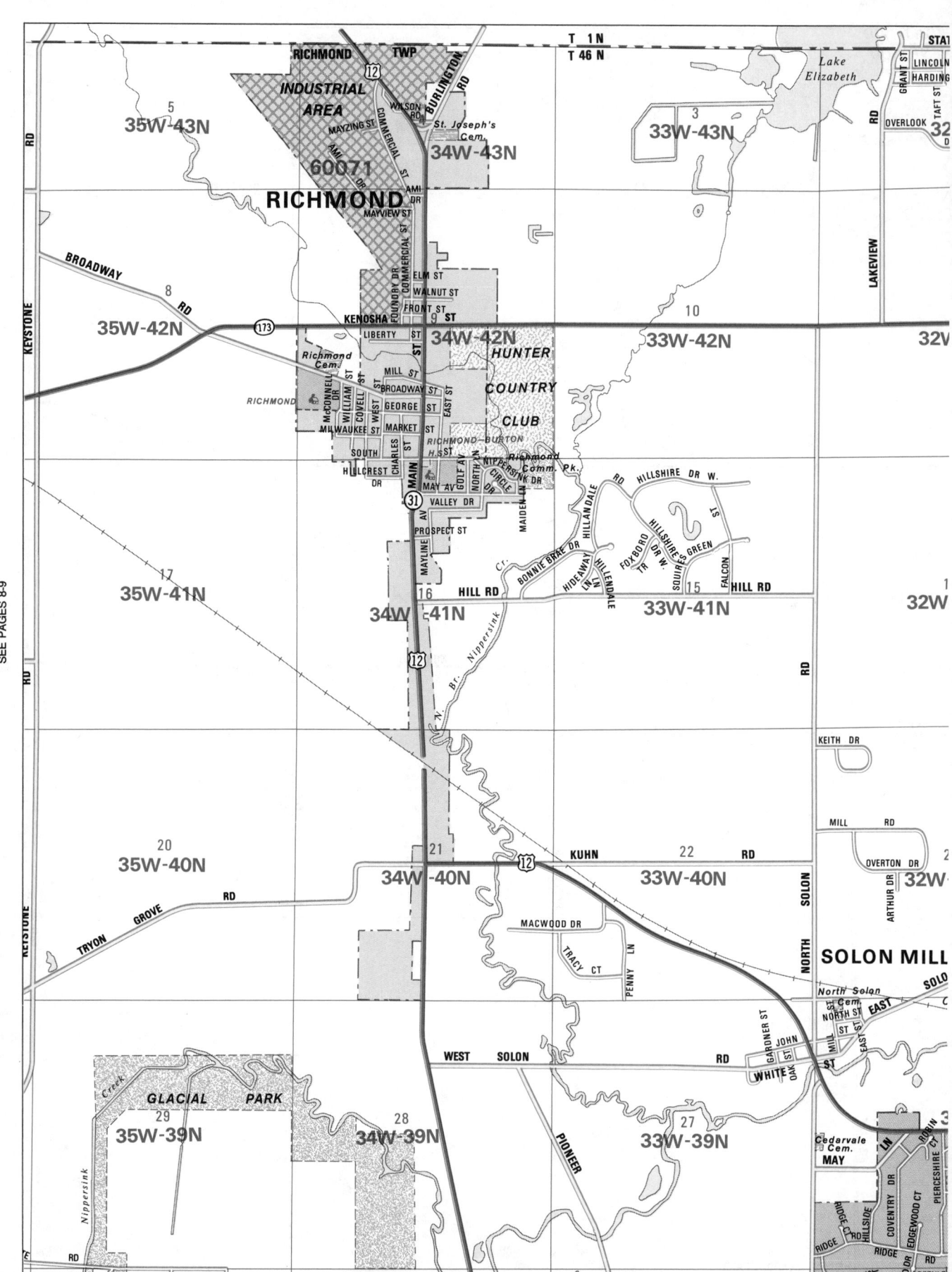

SEE PAGES 8-9

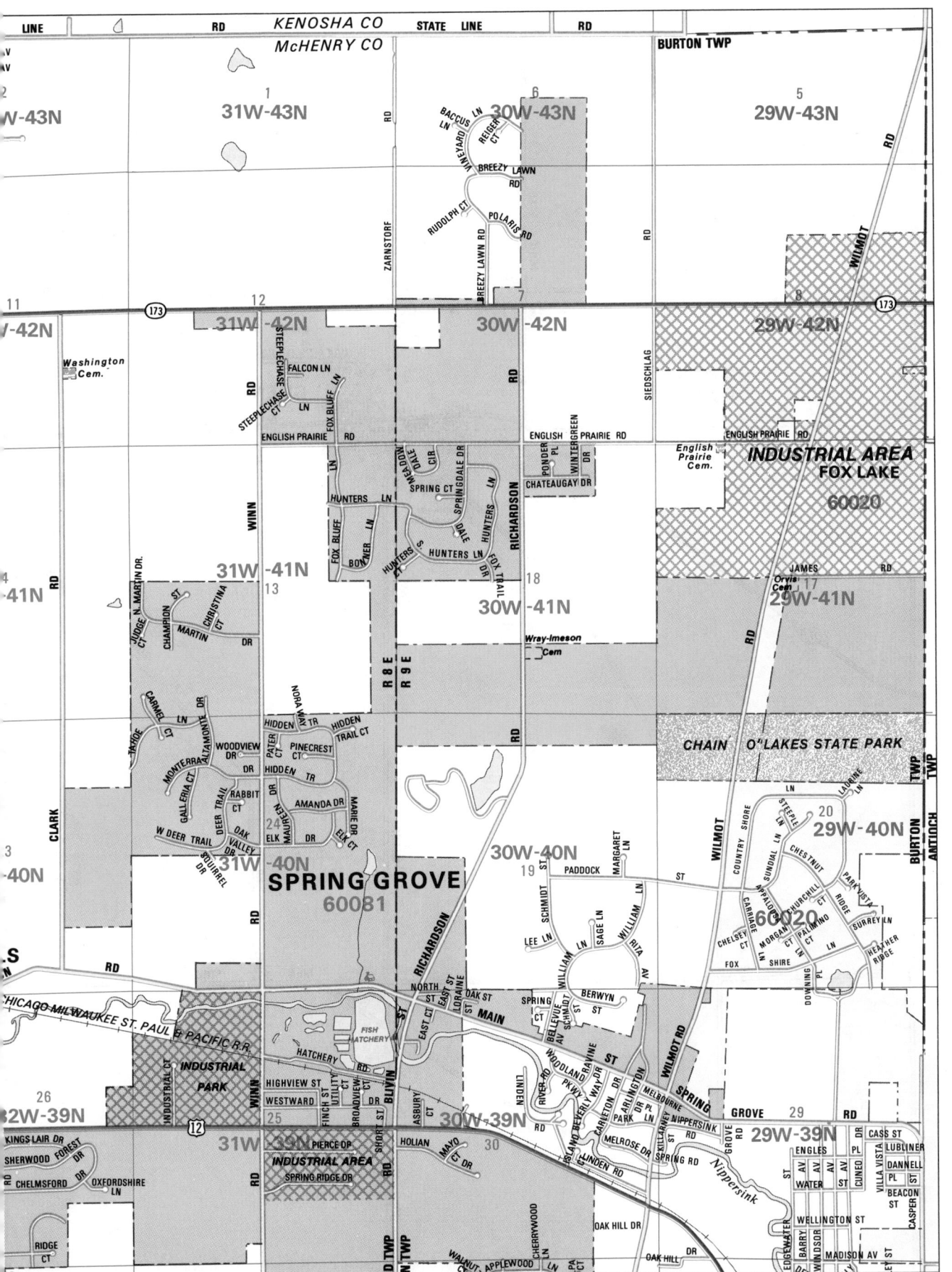

SPRING GROVE
60081

INDUSTRIAL AREA
FOX LAKE
60020

CHAIN O' LAKES STATE PARK

SEE PAGES 18-19

SEE PAGES 4-5

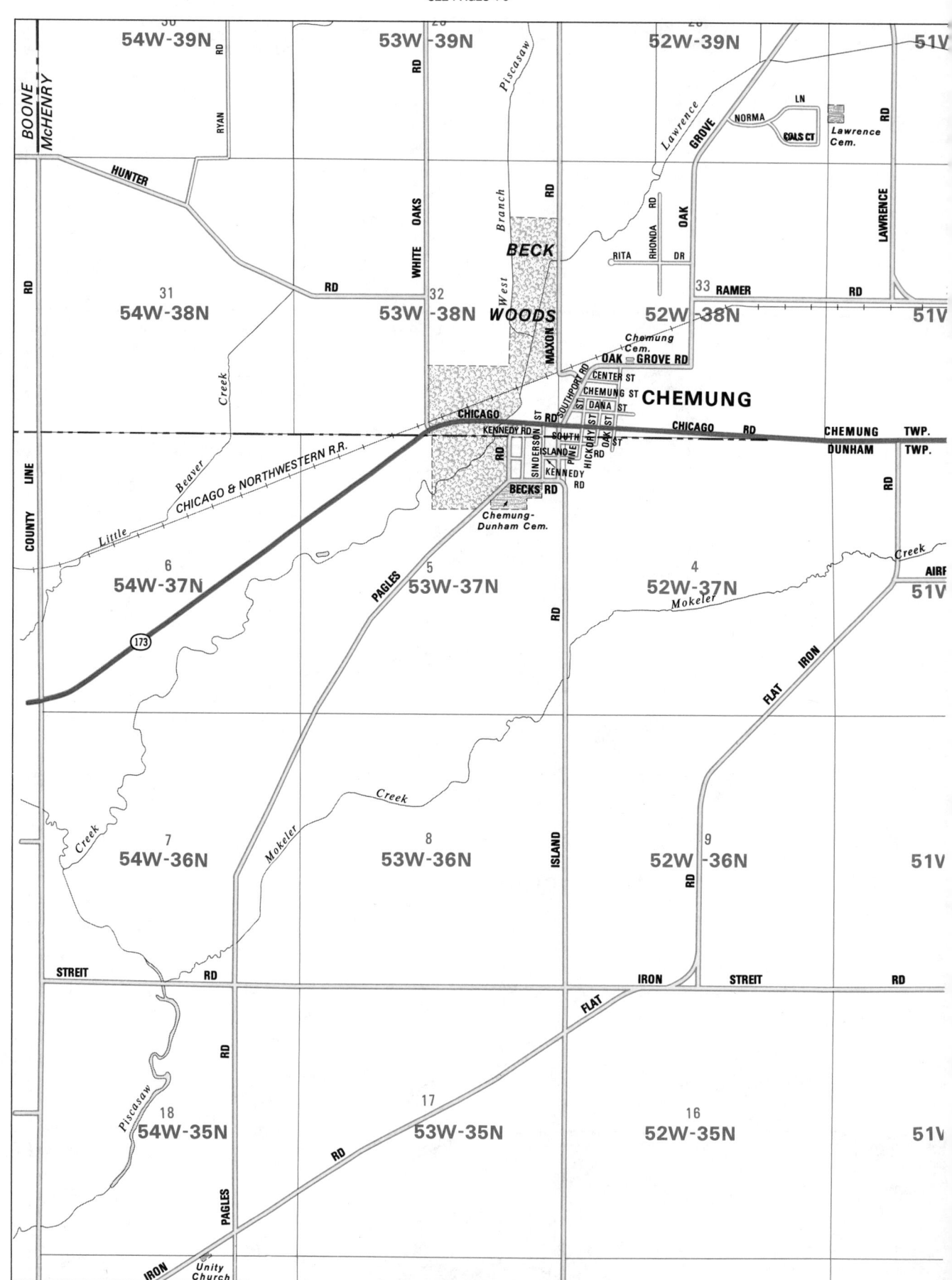

SEE PAGES 4-5

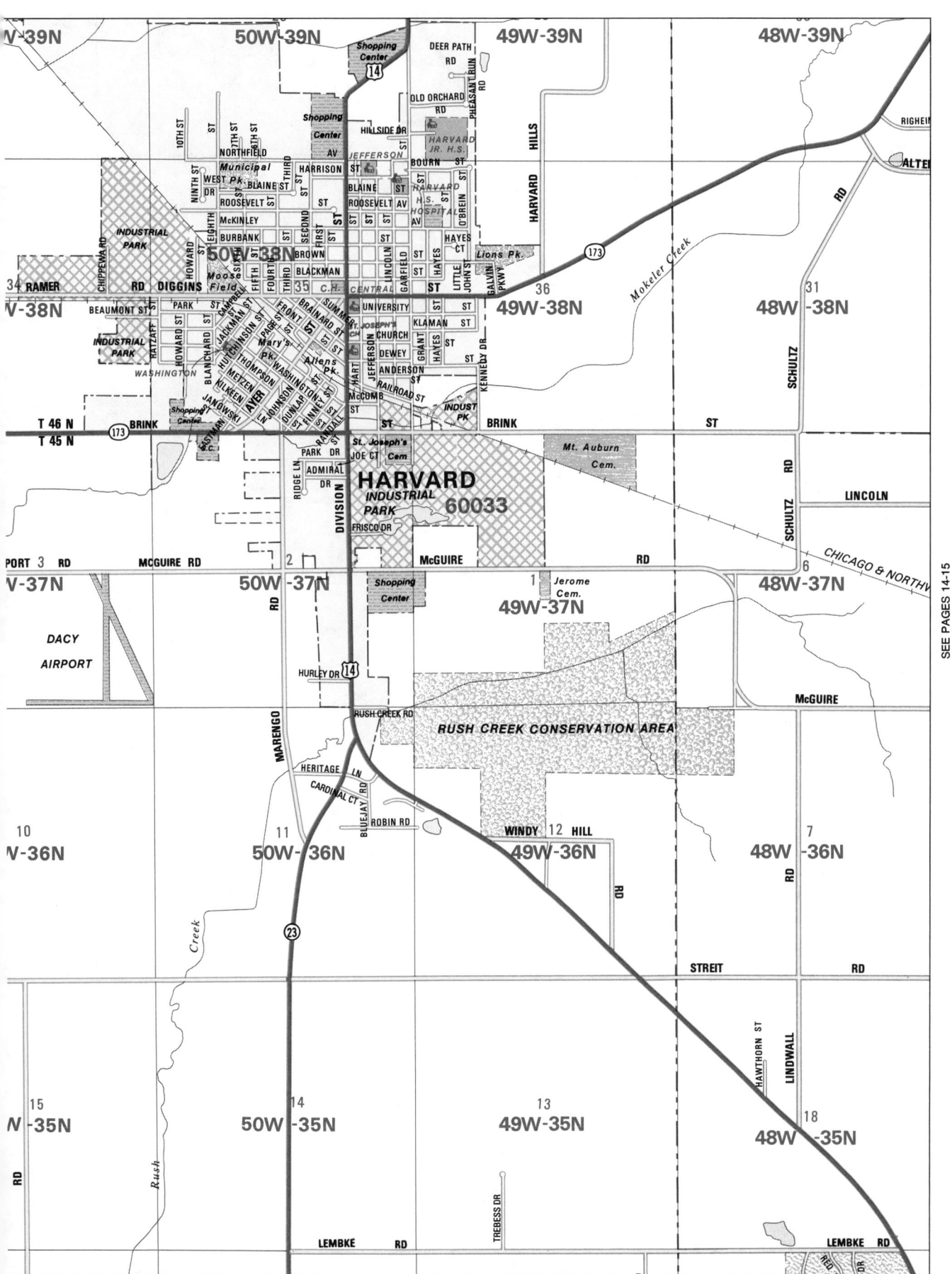

SEE PAGES 14-15

SEE PAGES 20-21

SEE PAGES 6-7

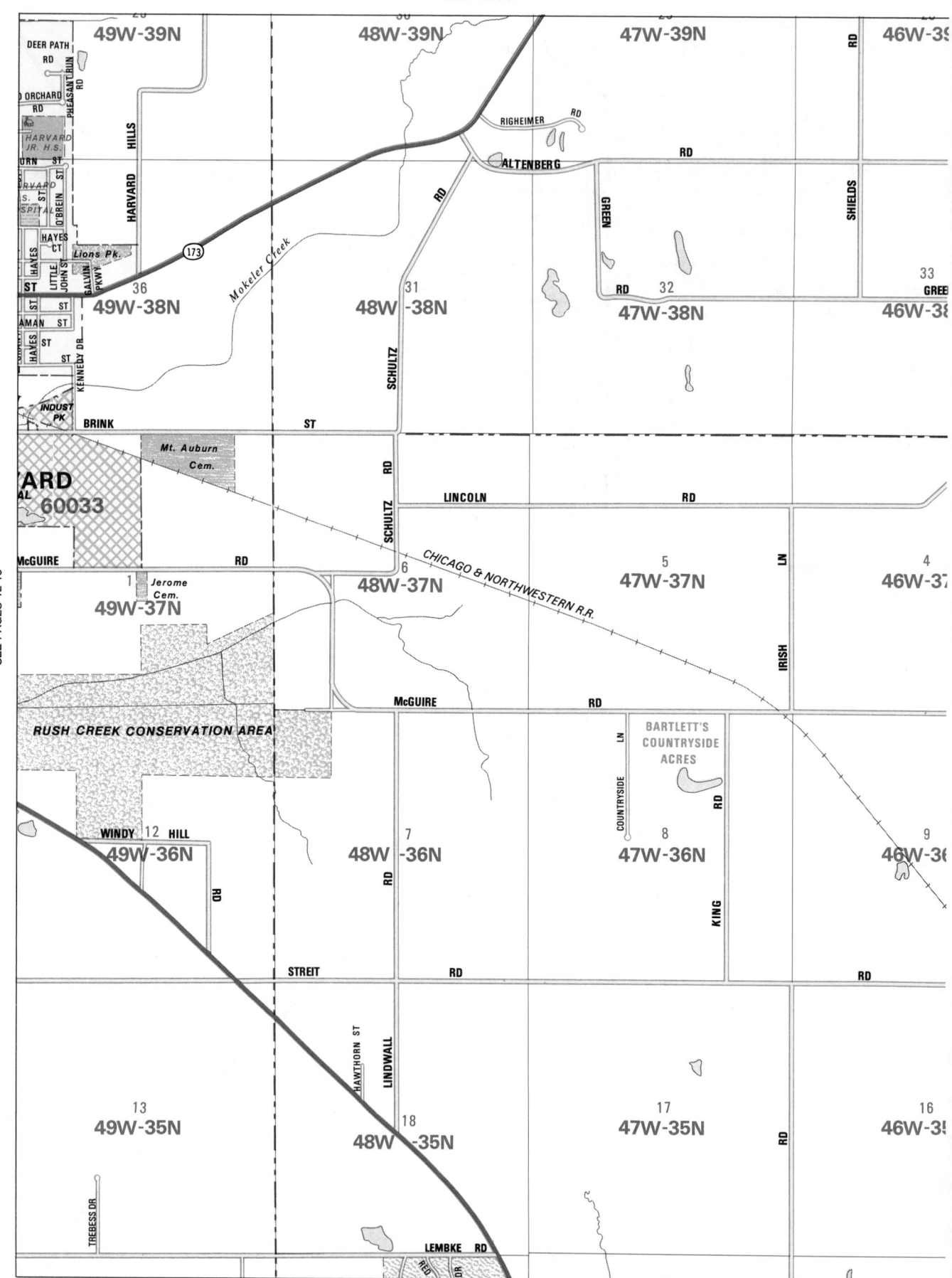

SEE PAGES 12-13

SEE PAGES 6-7

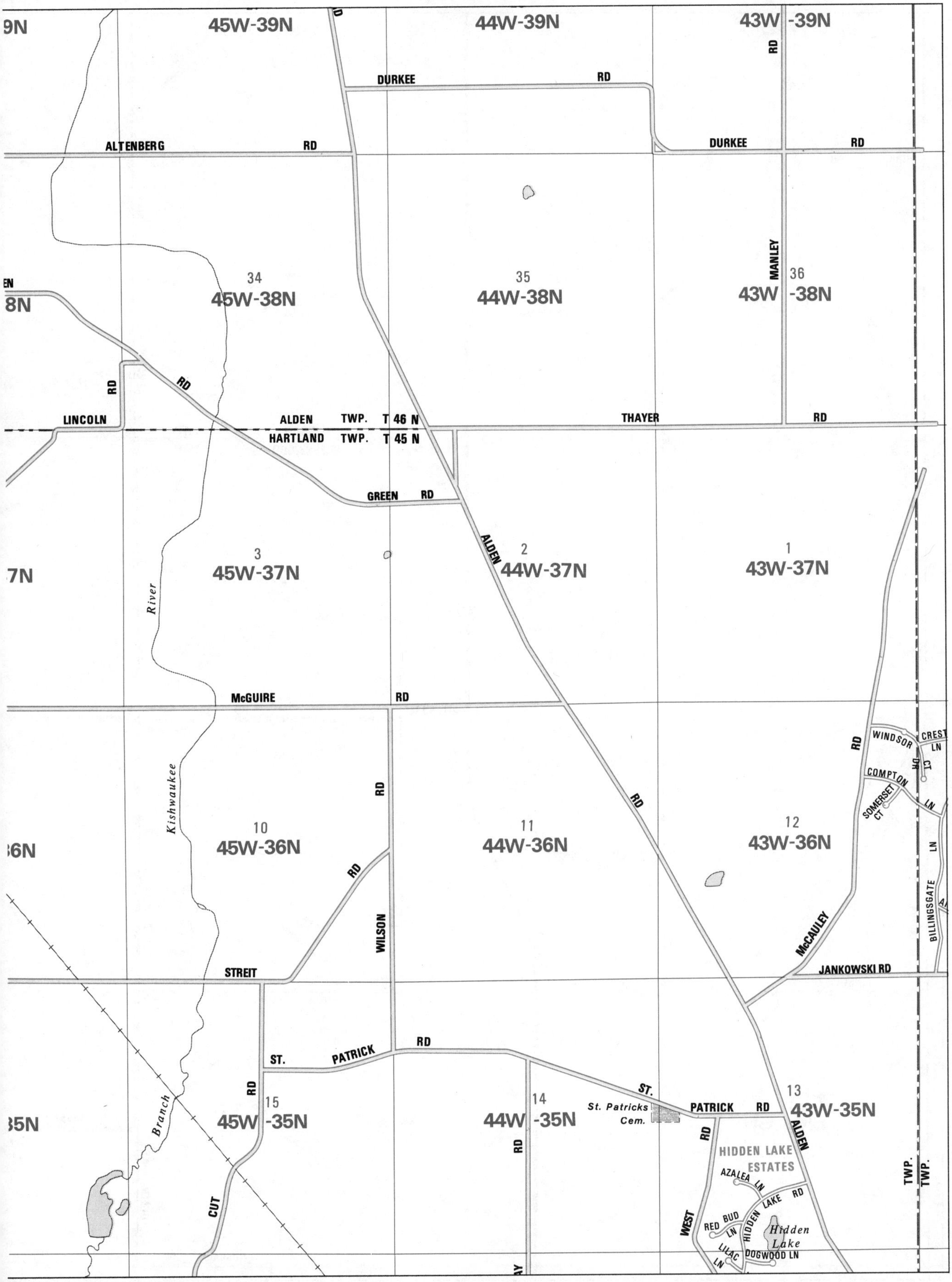

45W-39N

44W-39N

43W -39N

DURKEE RD

DURKEE RD

ALTENBERG RD

34
45W-38N

35
44W-38N

36
43W -38N

MANLEY

RD

RD

LINCOLN

ALDEN TWP. T 46 N
HARTLAND TWP. T 45 N

THAYER RD

GREEN RD

River

ALDEN

3
45W-37N

2
44W-37N

1
43W-37N

SEE PAGES 16-17

McGUIRE RD

Kishwaukee

RD

WINDSOR CREST LN

COMPTON LN

SOMERSET CT

RD

RD

10
45W-36N

11
44W-36N

12
43W-36N

LN

BILLINGSGATE

WILSON

RD

McCAULEY

STREIT

JANKOWSKI RD

ST. PATRICK RD

RD

Branch

CUT

RD

15
45W -35N

14
44W -35N

St. Patricks
Cem.

ST.
PATRICK RD

13
43W-35N

ALDEN

HIDDEN LAKE
ESTATES

AZALEA LN

WEST

RED BUD LN

HIDDEN LAKE RD

Hidden
Lake

LILAC LN

DOGWOOD LN

TWP.

TWP.

SEE PAGES 22-23

SEE PAGES 8-9

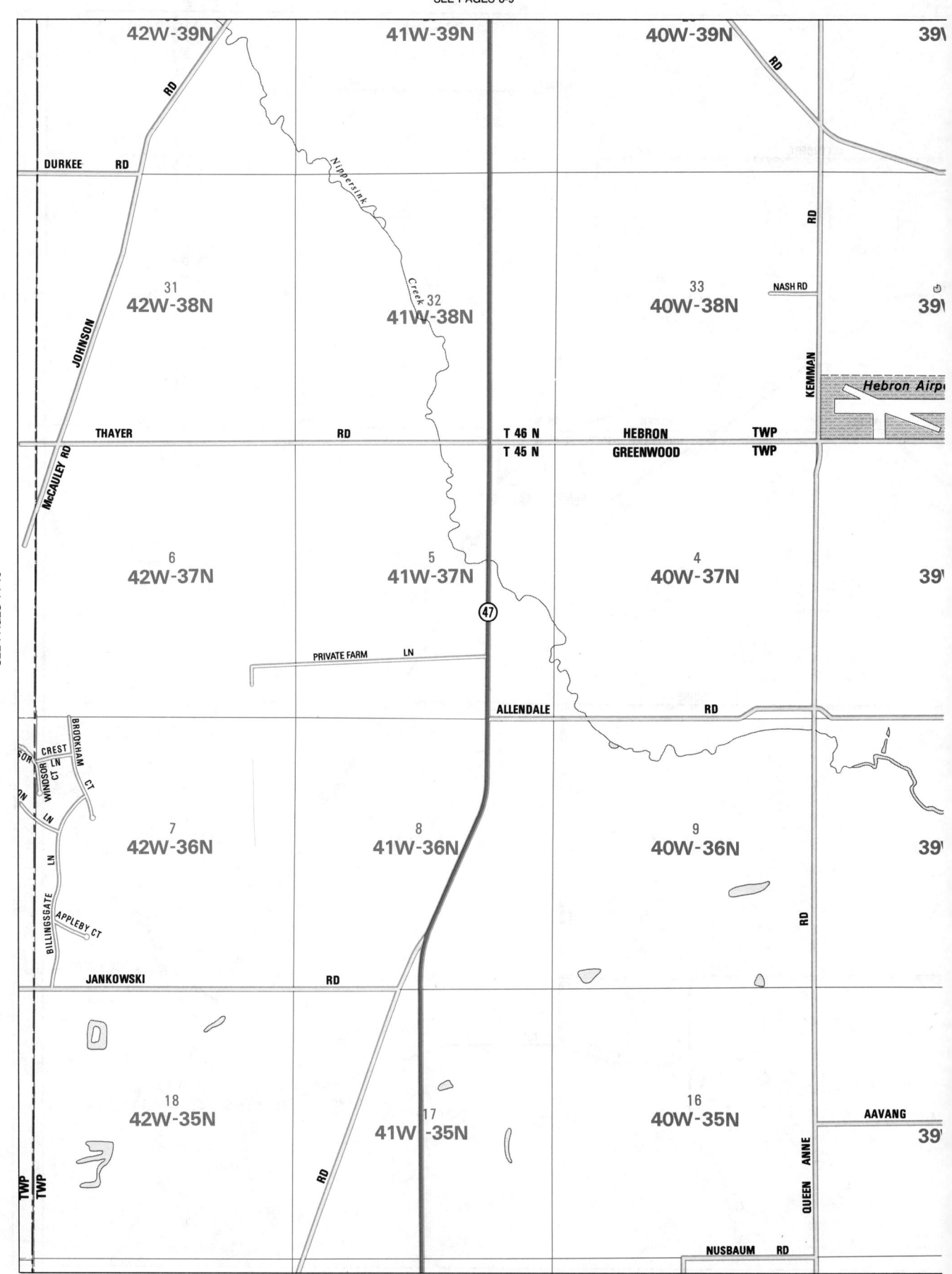

42W-39N

41W-39N

40W-39N

39W

RD

Nippersink

DURKEE RD

Creek

31
42W-38N

32
41W-38N

33
40W-38N

NASH RD

39W

JOHNSON

KEMMAN

RD

Hebron Airp

THAYER RD

T 46 N
T 45 N

HEBRON TWP
GREENWOOD TWP

McCAULEY RD

6
42W-37N

5
41W-37N

4
40W-37N

39W

47

PRIVATE FARM LN

ALLENDALE RD

BROOKHAM CT

CREST

WINDSOR CT

SOR

LN

7
42W-36N

8
41W-36N

9
40W-36N

39W

RD

LN

BILLINGSGATE

APPLEBY CT

JANKOWSKI RD

18
42W-35N

17
41W-35N

16
40W-35N

AAVANG

39W

RD

QUEEN ANNE RD

TWP
TWP

NUSBAUM RD

SEE PAGES 14-15

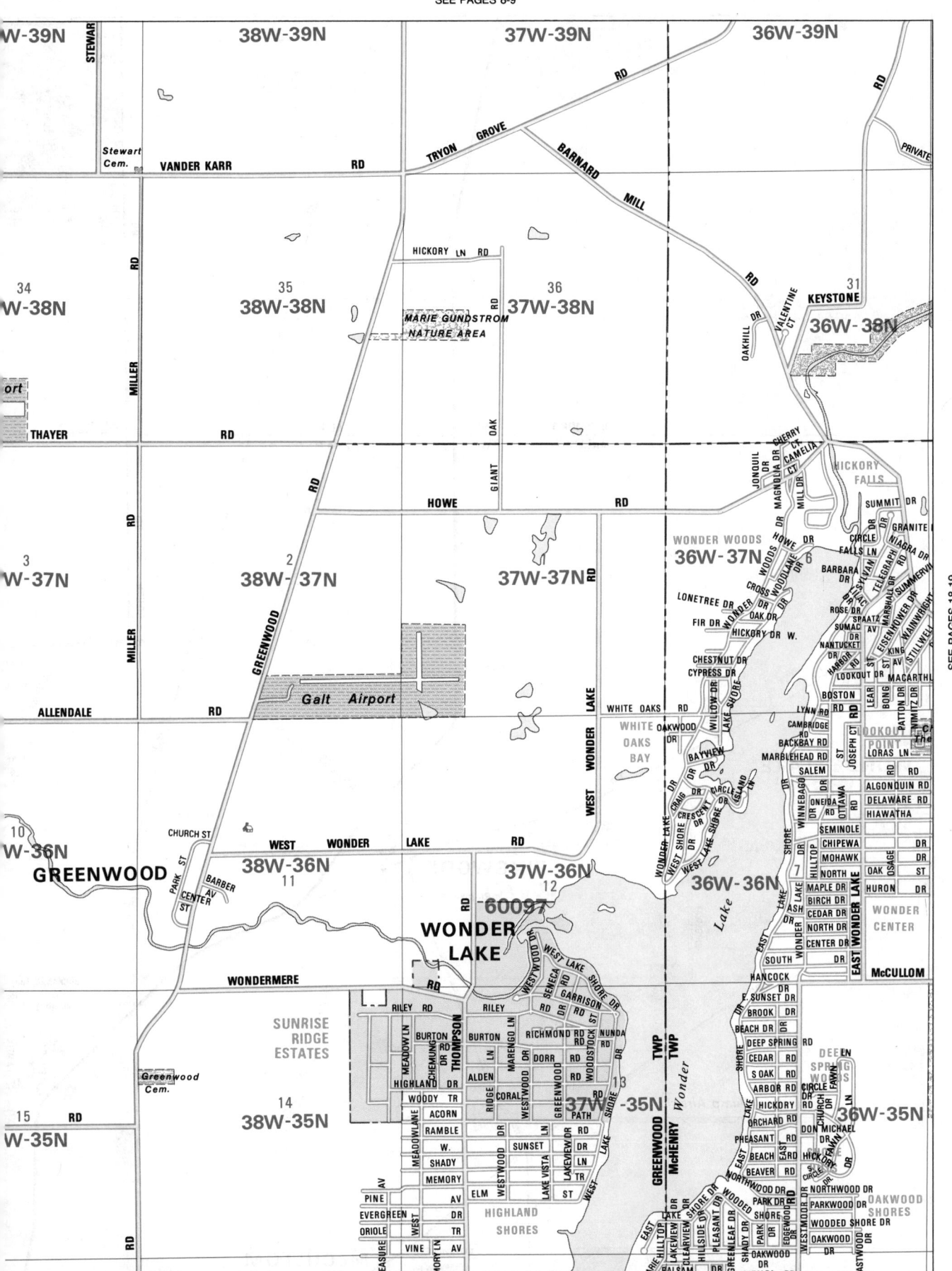

SEE PAGES 18-19

SEE PAGES 10-11

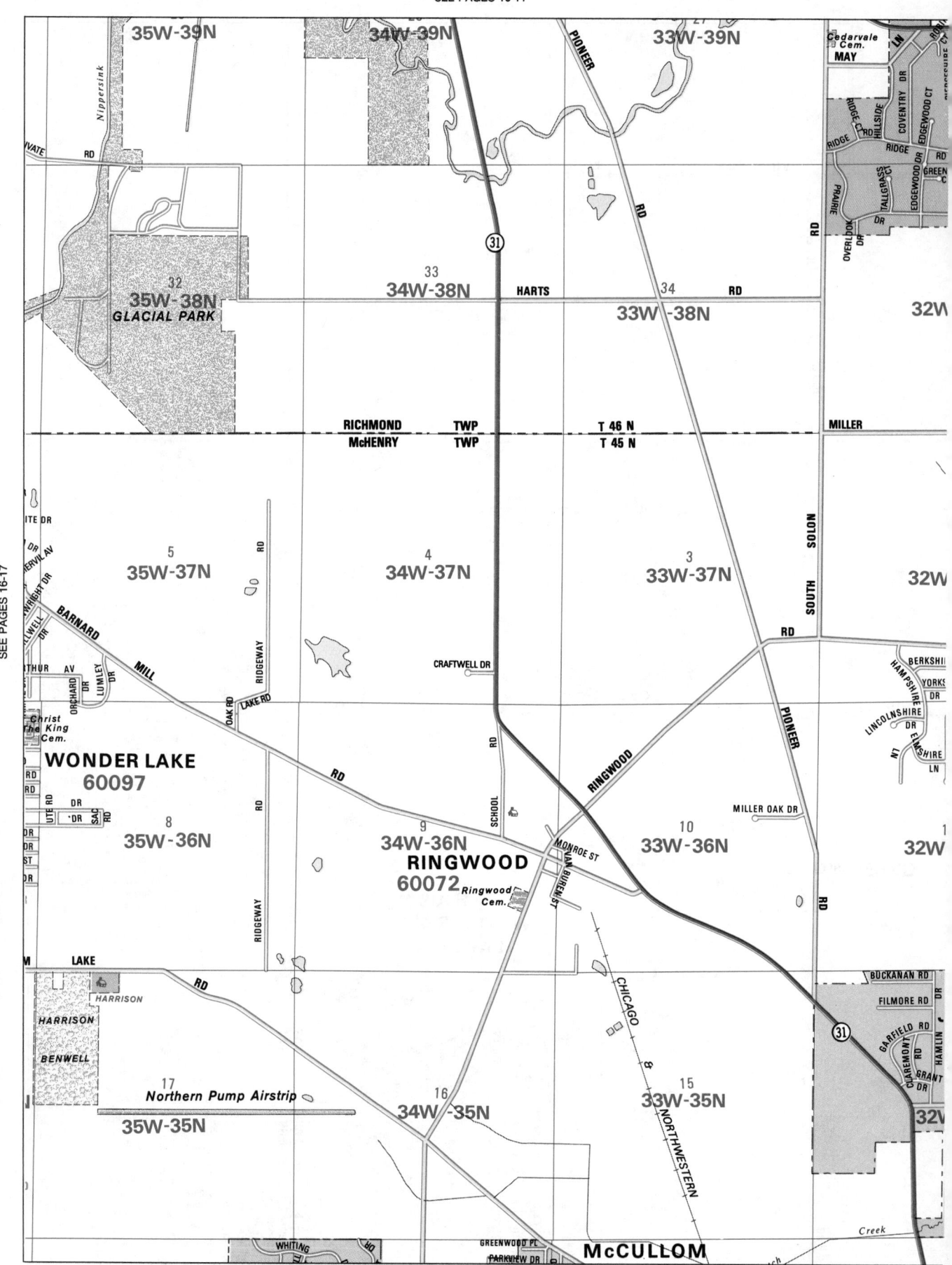

35W-39N

34W-39N

33W-39N

Cedarvale Cem.

MAY

Nippersink

PRIVATE   RD

PIONEER

RD

31

33

34W-38N   HARTS   34   RD

32   35W-38N
GLACIAL PARK

33W-38N

32W

RICHMOND   TWP   T 46 N   MILLER
McHENRY   TWP   T 45 N

5

35W-37N

RD

RIDGEWAY

4

34W-37N

3

33W-37N

SOUTH SOLON

32W

BARNARD   MILL

OAK RD   LAKE RD

CRAFTWELL DR

RD

RD

BERKSHI
HAMPSHIRE   YORKS
DR
LINCOLNSHIRE   DR   ELMSHIRE
DR   LN
LN

Christ
The King
Cem.

WONDER LAKE
60097

RIDGEWAY

RD

RD

RINGWOOD

SCHOOL

RD

PIONEER

MILLER OAK DR

8
35W-36N

9
34W-36N

RINGWOOD
60072

MONROE ST
VAN BUREN ST

10
33W-36N

RD

32W

Ringwood
Cem.

RIDGEWAY

LAKE

RD

HARRISON

HARRISON

BENWELL

CHICAGO

BUCKANAN RD

FILMORE RD

31

GARFIELD RD

CLAREMONT   HAMLIN
GRANT
DR

17
Northern Pump Airstrip

16
34W-35N

15
33W-35N

NORTHWESTERN

35W-35N

32W

Creek

WHITING

GREENWOOD PL
PARKVIEW DR

McCULLOM

SEE PAGES 16-17

SEE PAGES 10-11

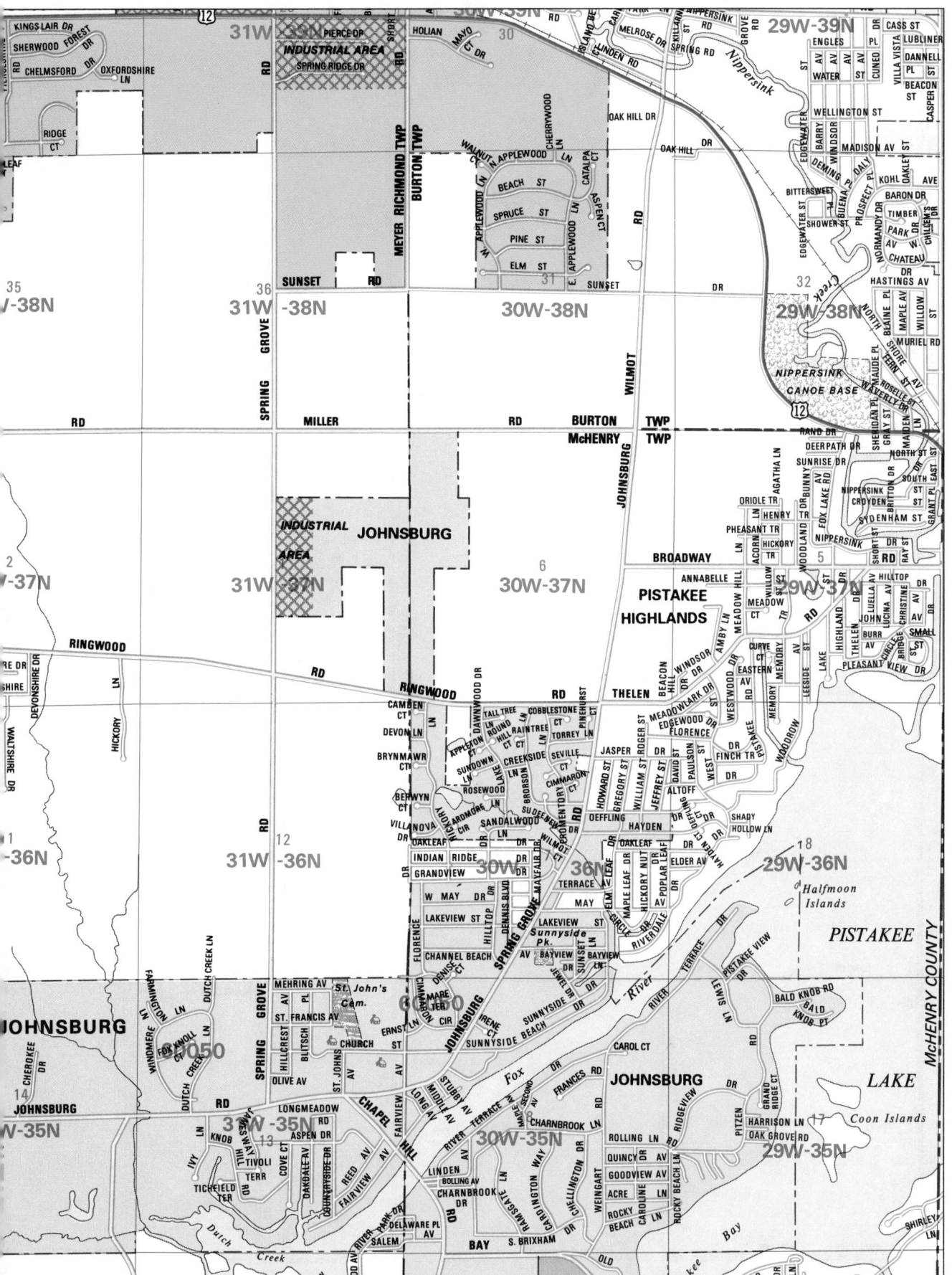

SEE PAGES 12-13

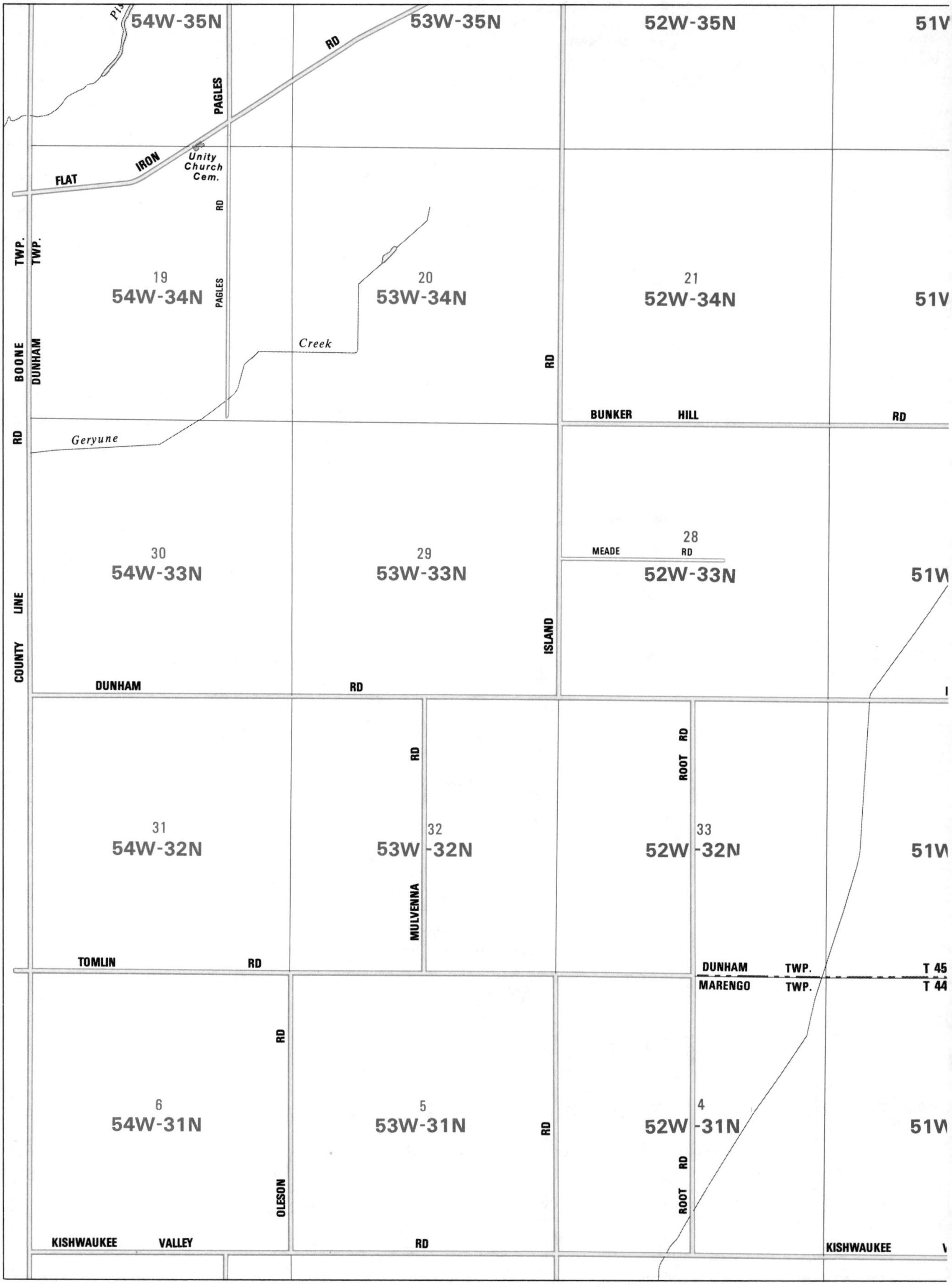

54W-35N

53W-35N

52W-35N

51V

PAGLES

RD

IRON

FLAT

Unity
Church
Cem.

RD

TWP.

TWP.

BOONE

DUNHAM

PAGLES

19
54W-34N

20
53W-34N

21
52W-34N

51V

Creek

RD

BUNKER   HILL   RD

RD

COUNTY   LINE

Geryune

30
54W-33N

29
53W-33N

MEADE   RD

28

52W-33N

51W

ISLAND

DUNHAM   RD

RD

ROOT   RD

31
54W-32N

32
53W-32N

33
52W-32N

51W

MULVENNA

TOMLIN   RD

DUNHAM   TWP.   T 45
MARENGO   TWP.   T 44

RD

6
54W-31N

5
53W-31N

RD

4
52W-31N

51W

OLESON

ROOT   RD

KISHWAUKEE   VALLEY   RD

KISHWAUKEE

SEE PAGES 28-29

SEE PAGES 12-13

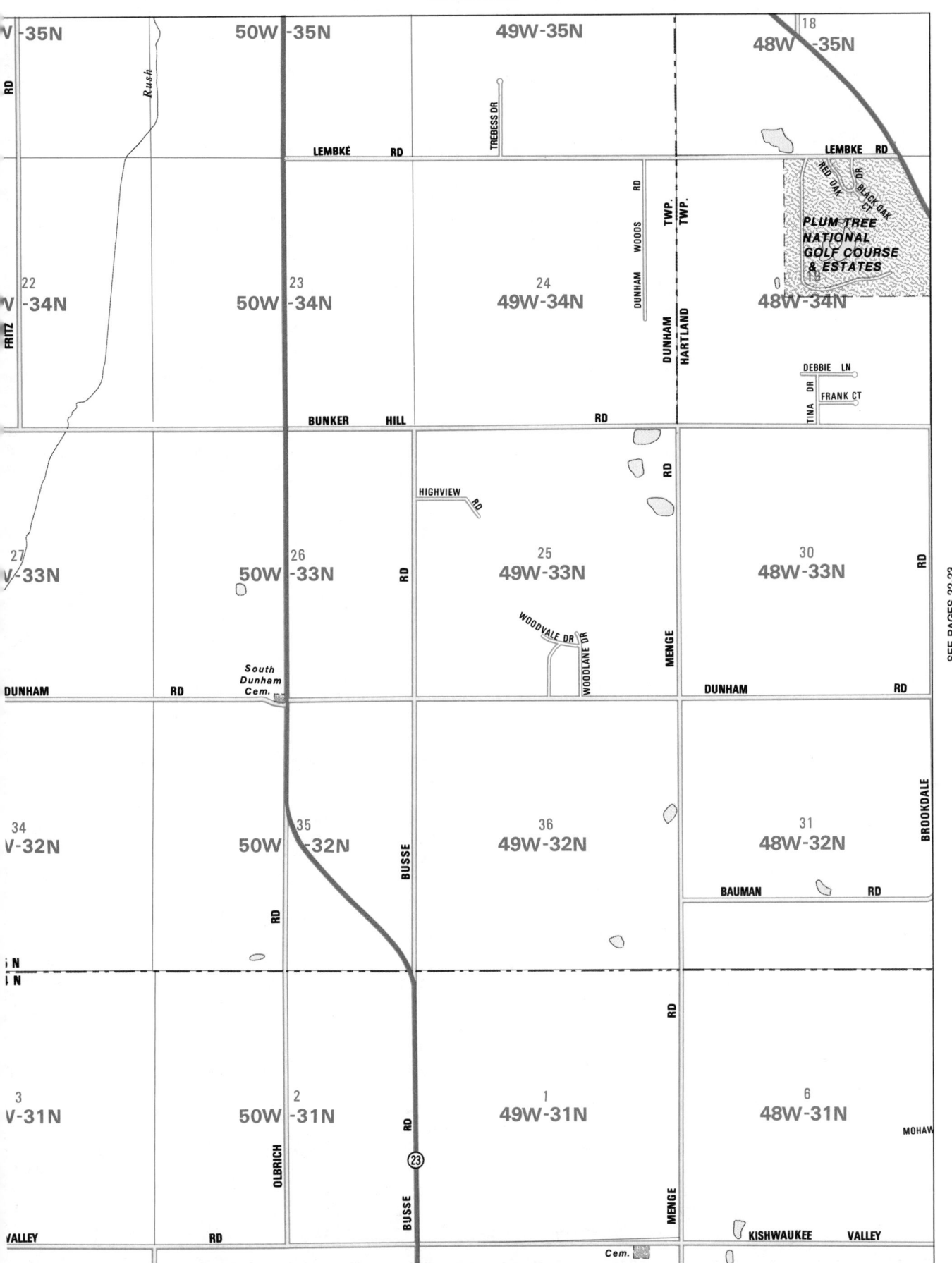

V -35N

50W -35N

49W-35N

48W -35N

RD

Rush

TREBESS DR

LEMBKE   RD

LEMBKE   RD

RD

DUNHAM WOODS   RD

TWP.
TWP.

PLUM TREE
NATIONAL
GOLF COURSE
& ESTATES

RED OAK   BLACK OAK CT

22
V -34N

FRITZ

23
50W -34N

24
49W-34N

DUNHAM
HARTLAND

48W -34N

18

19

DEBBIE   LN

TINA DR

FRANK CT

BUNKER   HILL

RD

27
V -33N

26
50W -33N

RD

HIGHVIEW   RD

25
49W-33N

RD

MENGE

30
48W-33N

RD

SEE PAGES 22-23

WOODVALE DR

WOODLANE DR

South
Dunham
Cem.

DUNHAM   RD

DUNHAM   RD

34
V -32N

35
50W -32N

BUSSE

RD

36
49W-32N

RD

31
48W-32N

BROOKDALE

BAUMAN   RD

N
N

3
V -31N

OLBRICH

2
50W -31N

BUSSE   RD

23

1
49W-31N

RD

MENGE

6
48W-31N

MOHAW

VALLEY   RD

BUSSE

Cem.

KISHWAUKEE   VALLEY

SEE PAGES 28-29

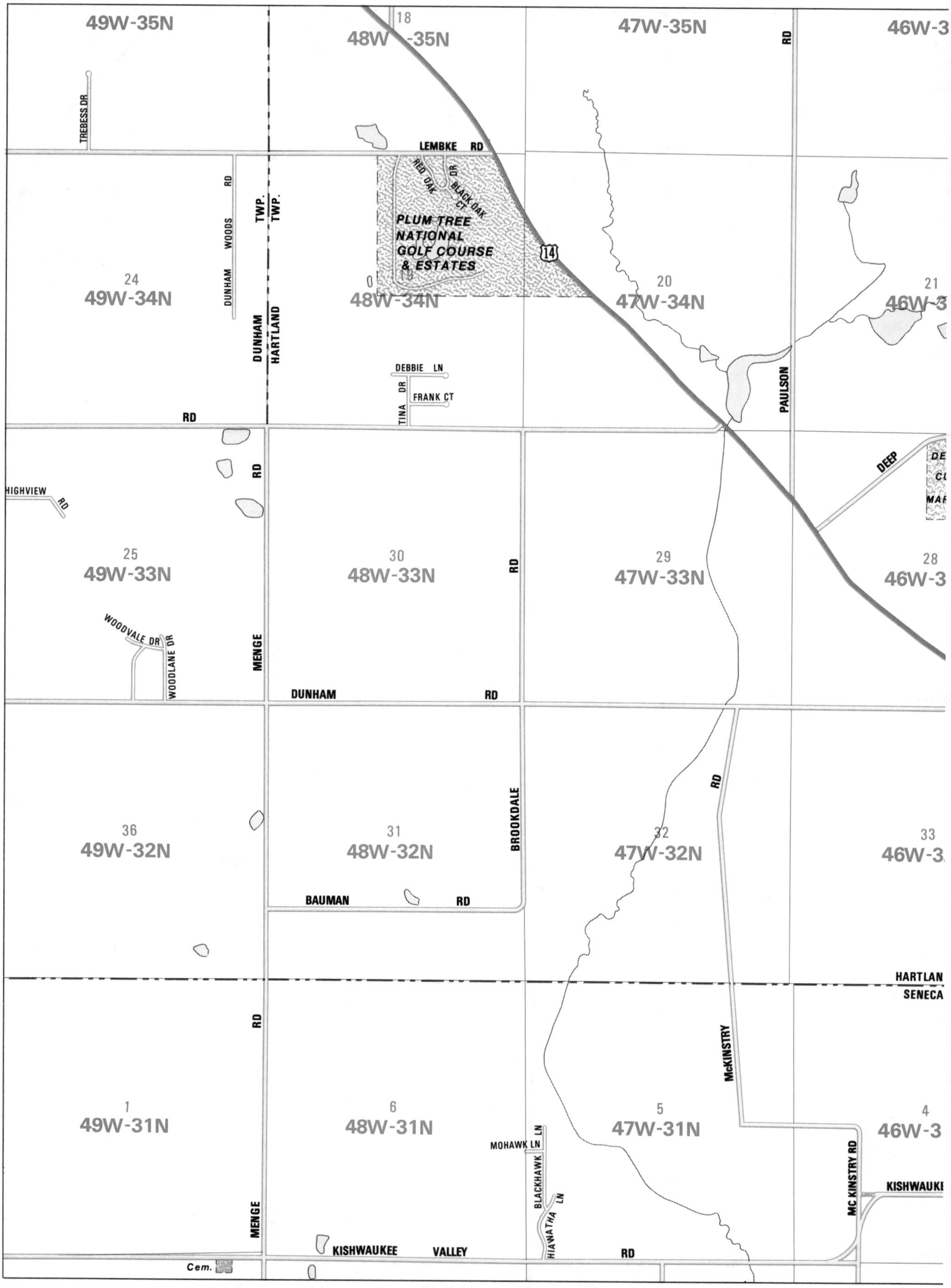

49W-35N

18

48W-35N

47W-35N

46W-3

TREBESS DR

LEMBKE RD

RED OAK DR

BLACK OAK CT

PLUM TREE NATIONAL GOLF COURSE & ESTATES

14

DUNHAM WOODS RD

TWP. TWP.

24

49W-34N

DUNHAM HARTLAND

48W-34N

20

47W-34N

21

46W-3

PAULSON

DEBBIE LN

TINA DR

FRANK CT

RD

RD

HIGHVIEW RD

DEEP

DE C MAF

25

49W-33N

30

48W-33N

RD

29

47W-33N

28

46W-3

WOODVALE DR

WOODLANE DR

MENGE

DUNHAM RD

36

49W-32N

31

48W-32N

BROOKDALE

32

47W-32N

33

46W-3

RD

BAUMAN RD

HARTLAN

SENECA

RD

MCKINSTRY

1

49W-31N

6

48W-31N

MOHAWK LN

BLACKHAWK LN

HIAWATHA LN

5

47W-31N

MCKINSTRY RD

4

46W-3

KISHWAUKE

MENGE

KISHWAUKEE VALLEY

RD

Cem.

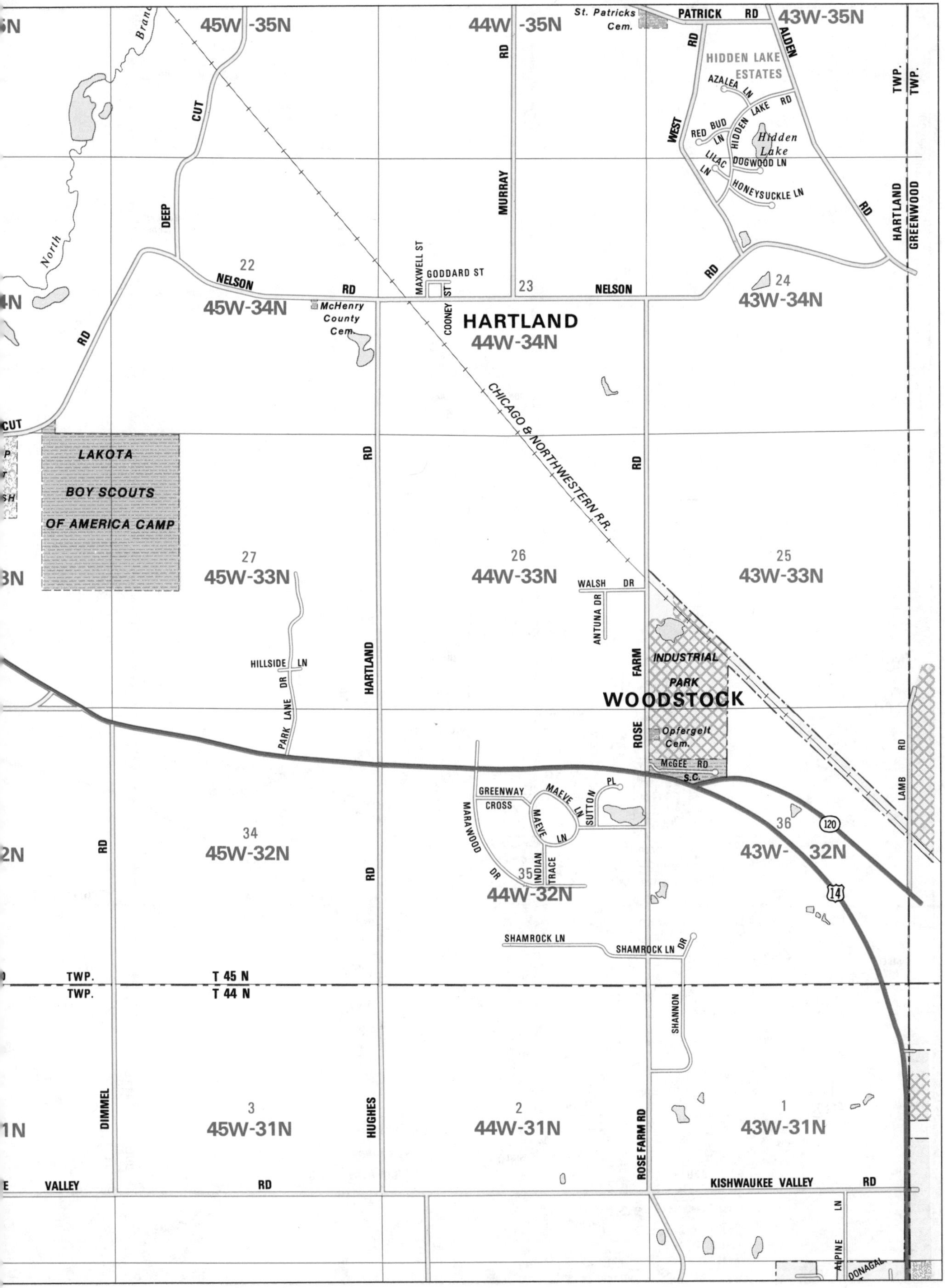

45W -35N

44W -35N

St. Patricks Cem.

PATRICK   RD   43W-35N

RD

MURRAY

ALDEN

PATRICK RD

HIDDEN LAKE ESTATES

AZALEA LN

WEST

RED BUD LN

HIDDEN LAKE RD

Hidden Lake

LILAC LN

DOGWOOD LN

HONEYSUCKLE LN

TWP.

TWP.

RD

HARTLAND

GREENWOOD

DEEP CUT

NELSON   RD

22

Branch

North

CUT

RD

45W-34N

McHenry County Cem.

MAXWELL ST

GODDARD ST

COONEY ST

23

NELSON

HARTLAND
44W-34N

NELSON

RD

24

43W-34N

CHICAGO & NORTHWESTERN R.R.

RD

RD

LAKOTA
BOY SCOUTS
OF AMERICA CAMP

27
45W-33N

26
44W-33N

WALSH   DR

ANTUNA DR

25
43W-33N

SEE PAGES 24-25

HILLSIDE LN

HARTLAND

PARK LANE DR

INDUSTRIAL PARK

WOODSTOCK

ROSE FARM

Opfergelt Cem.

McGEE RD

S.C.

LAMB RD

RD

RD

34
45W-32N

GREENWAY CROSS

MARAWOOD DR

MAEVE LN

MAEVE LN

SUTTON PL

PL

35
44W-32N

INDIAN TRACE

36

43W- 32N

120

14

SHAMROCK LN

SHAMROCK LN DR

SHANNON DR

TWP.

TWP.

T 45 N

T 44 N

DIMMEL RD

3
45W-31N

HUGHES RD

2
44W-31N

ROSE FARM RD

1
43W-31N

VALLEY RD

KISHWAUKEE VALLEY   RD

ALPINE LN

DONAGAL

SEE PAGES 22-23

QUEEN

NUSBAUM RD

RD

RD

HARTLAND
GREENWOOD

RAYCRAFT

47

CHARLES 19

42W-34N

41W-34N
20

RAFFEL

40W-34N
21

Eckert
Cem.

PHEASANT LN

39

CHARLES

RL

Scandinavian
Cem.

R 6 E
R 7 E

LAMB
RD

PARKER
CT

GLEASON
CT

COONEY DR

BRODY
LN

BAKER TERR

30
42W-33N

29
41W-33N

28
40W-33N

39

NORTHWOOD LN

WARE RD

RD

39

INDUSTRIAL

AREA

31
42W-32N

Woodstock

NORTHWOOD
JR. HIGH

McHENRY CO.
GOV CTR.

NORTHWOOD

RUSSELL
CT

MILLER

POWERS RD

CENTRAL

PKWY

McCANNON RD

BANFORD 33
RD

40W-32N

RD

39

CHICAGO

120

Airport

NORTHWESTERN

PATRICIA CT

MELODY LN

KEATING DR
DIANE DR
JOSEF

BELAIR LN
HILLTOP CT

SUNSHINE
LN SHELIA

PEACH
TREE LN

BIRCH
RD

CHARLES

JULIE

HICKORY

BROADWAY CT

WOODSIDE DR

REDWING DR
BARNSWALLOW DR

TANAGER DR

SPARROW DR

WICKER ST

QUAIL
WICKER DR

FARM TR

TERRY
FOX LN

ORCHARD

CHERRY CT
ASH DR
CLAY
AV

LOCUST
ST. JOHN
RD

BIRCH

ROGER
HICKORY

POPLAR

TWELVE
OAKS PKWY

41W-32N

ASHLEY
CT

RAFFEL

120

RD

WILLOW

MEADOW
AV

SUMMIT
AV

BEECH
AV

GREENWOOD
AV

VICTORIA DR
REGINA DR
THOMAS DR

CAIRNS

WALNUT

CLAY ST

MADISON

CLAY

MAPLE
AV

SILVER
Creek Pk.

BROADWAY RD

GREENWOOD

AV

T 45 N
T 44 N

GREENWOOD
DORR

BAGLEY

DONOVAN

OLIVE

Olson
Pk.

ST

TAPPAN

SEMINARY

GREENWOOD PL

SUMNER ST

NORTHAMPTON ST

MARVEL AV

HUFFMAN
ST

MARIAN CENTRAL
CATHOLIC
H.S.

RD

ROSE McHENRY RD

TIMOTHY
CT

4
40W-31N

39W

SUNSET

HILLCREST ST

CLAUSSEN
DR

WASHINGTON

RIDGE RD

CAROL AV

OAK

WEST AV

CAROL
ST

SUZANNE ST

DONNA
ST

MARY ANN ST
PLEASANT

LISA
ST
ANNE
ST

ARTHUR
CT

ST

DANE

BORDEN

42W-31N

6

MARGARET ST

Oakland
Cem.

OLSON

JACKSON

OAKLAND
HILL

Calvary
Cem.

City
Park

RHODES ST

AMSTERDAM

WICKER

JEWETT

ST

QUEEN
ANNE

THIRD
ST

SECOND
ST

FIRST

WHEELER

CONWAY
ST

DACY ST

BECKING
ST

LINCOLN

QUINLAN LN

TODD

St.
Mary's

MADISON

McHENRY

CLAY

CASS

JACKSON

JUDD ST

DOUGLAS ST

FLAGG

FLAGG
LN

ELLEN LINDA

SHARON

DAISY
CT

ESTHER
CT

Tourist
Park

SOUTH
ST

City
Sq.

JUDD

City Ct

VAN BUREN
ST

CALHOUN

NEBRASKA

ST

CENTER ST
CENTER ST

GRACY
ST

IRVING

NEWELL

HUTCHINS
ST

CHURCH ST

HICKMAN

SHARON
RD

ESTHER
CT

RD

McHENRY CO.
FAIRGROUNDS

COUNTRY CLUB

QUEEN
ANNE
RD

BULL
VALLEY

39W

41W-31N
5

14

Woodstock
City
Park

HILL ST

WOODSTOCK
H.S.

FOREST AV

STEWART
ST

RIDGELAND
ST

PLEASANT ST

HERRINGTON
PL

TRYON ST

DEAN ST

TARA DR

THROOP ST

JOHNSON ST
MAIN

CLAY
ST

P.O. ST

BENTON ST

LAWRENCE

GREENLEY

DEAN
ST
HOY
AV

GRIFFING

FREMONT

DAVIS

LAWNDALE AV

VINEY

BROWN ST
LAKE
ST

RIDINGS ST

BRINK ST
SMITH ST
KING

FAIR
RD

Reintree Pk.

CENTERVILLE
CT

WASHBURN ST

S.C.

DORHAM LN

NIDDRIE DR

JOHNSTON
ST

GALLOWAY DR

AYRSHIRE LN

ABRAHAM
WHITE

KEATS CT

BYRON

SHELLEY

BULL VALLEY
DR

BULL VALLEY
DR

BRABBIN
CIR

PINE AV
EVERGREEN DR
ORIOLE WEST TR
VINE AV

ELM ST

HIGHLAND
SHORES

PLEASANT
MEMORY LN

WONDER VIEW

WONDER LA
60097

NORTHWOOD DR
OAKWOOD
SHORES
PARKWOOD DR
WOOD PARK DR
SHORE DR
WOODED SHORE DR
EASTWOOD DR
WESTWOOD DR
WOODED SHORE DR
EDGEWOOD RD
OAKWOOD DR

EAST LAKE SHORE DR
LAKEVIEW DR
HILLSIDE DR
HILLTOP
GREENLEAF DR
PLEASANT DR
SHADY DR
CATALPA DR
PAMELA DR
PRESTON DR
WIDOFF DR
GENE DR
LUCY DR
MAPLEWOOD DR

ARABIAN TR

22
W-34N

THOROUGHBRED TR
MUSTANG TR
SADDLEBRED TR

GREENWOOD RD

23
38W-34N

LONGWOOD CT
PINE NEEDLE PASS
SHADOWOOD DR
NORTHWOODS CT
WINDHILL CT
SHADOW LN

37W-34N
24

ROSE MARIE RD
LAKE SHORE DR
BALSAM DR
ELMWOOD DR
GERSON DR
DEER DR
WONDERVIEW DR
BIRCHWOOD DR
ASHWOOD DR
BASSWOOD DR
HIGHVIEW DR
MEADOW LN
PINDAK
MAPLEWOOD DR
OAKWOOD DR

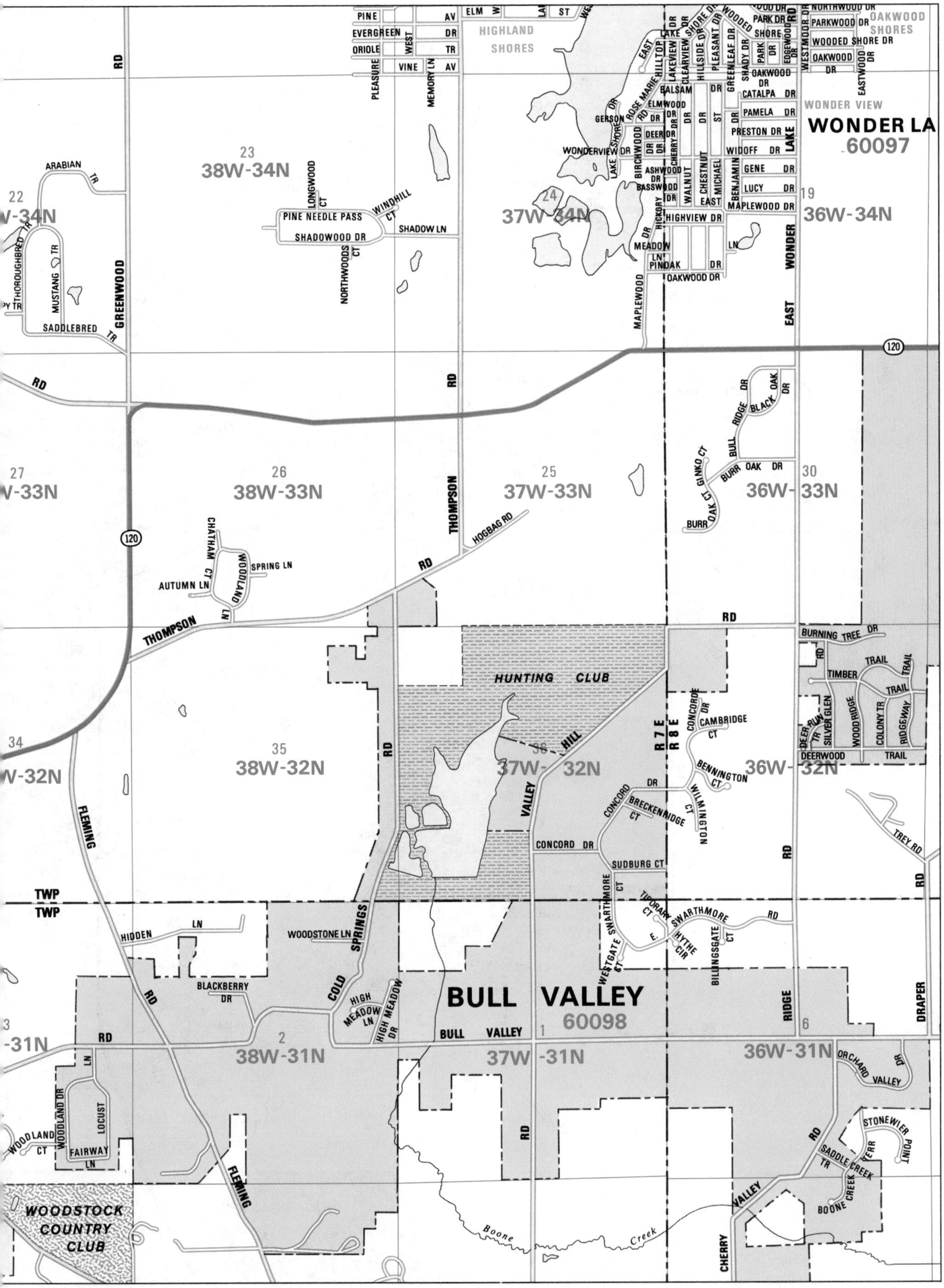

CHESTNUT ST
WALNUT ST
HICKORY ST
CHERRY ST

BENJAMIN
MICHAEL
EAST
LN

19
36W-34N

WONDER LAKE

EAST WONDER

120

RD

RD

THOMPSON RD

27
W-33N

26
38W-33N

25
37W-33N

HOGBAG RD

30
36W-33N

BULL RIDGE DR
BLACK OAK DR
OAK DR
GINKO CT
BURR OAK DR
BURR OAK CT

120

CHATHAM CT
SPRING LN
WOODLAND LN
AUTUMN LN

THOMPSON RD

RD

SEE PAGES 26-27

34
W-32N

FLEMING RD

35
38W-32N

RD

HUNTING CLUB

VALLEY HILL RD

37W- 32N

R 7 E
R 8 E

CONCORDE DR
CAMBRIDGE CT
BENNINGTON CT
CONCORD DR
BRECKENRIDGE CT
WILMINGTON CT
CONCORD DR

36W-32N

RD

BURNING TREE DR
TIMBER TRAIL
TRAIL
DEER RUN TR
SILVER GLEN
WOODRIDGE
COLONY TR
RIDGEWAY
DEERWOOD TRAIL

TREY RD

RD

TWP
TWP

HIDDEN LN

WOODSTONE LN

COLD SPRINGS RD

BLACKBERRY DR

HIGH MEADOW LN
HIGH MEADOW DR

BULL VALLEY
60098

SUDBURG CT
WESTGATE CT
SWARTHMORE CT
TUDOR LN
E.
SWARTHMORE
HYTHE CIR
BILLINGSGATE CT
RD

RIDGE RD

DRAPER RD

3
-31N

RD

WOODLAND DR
WOODLAND CT
LOCUST LN
FAIRWAY LN

FLEMING RD

2
38W-31N

BULL VALLEY RD

1
37W -31N

RD

36W-31N

ORCHARD VALLEY DR
STONEWIER POINT
SADDLE CREEK TR
BOONE CREEK TERR
CHERRY VALLEY RD
BOONE

WOODSTOCK
COUNTRY
CLUB

Boone Creek

SEE PAGES 18-19

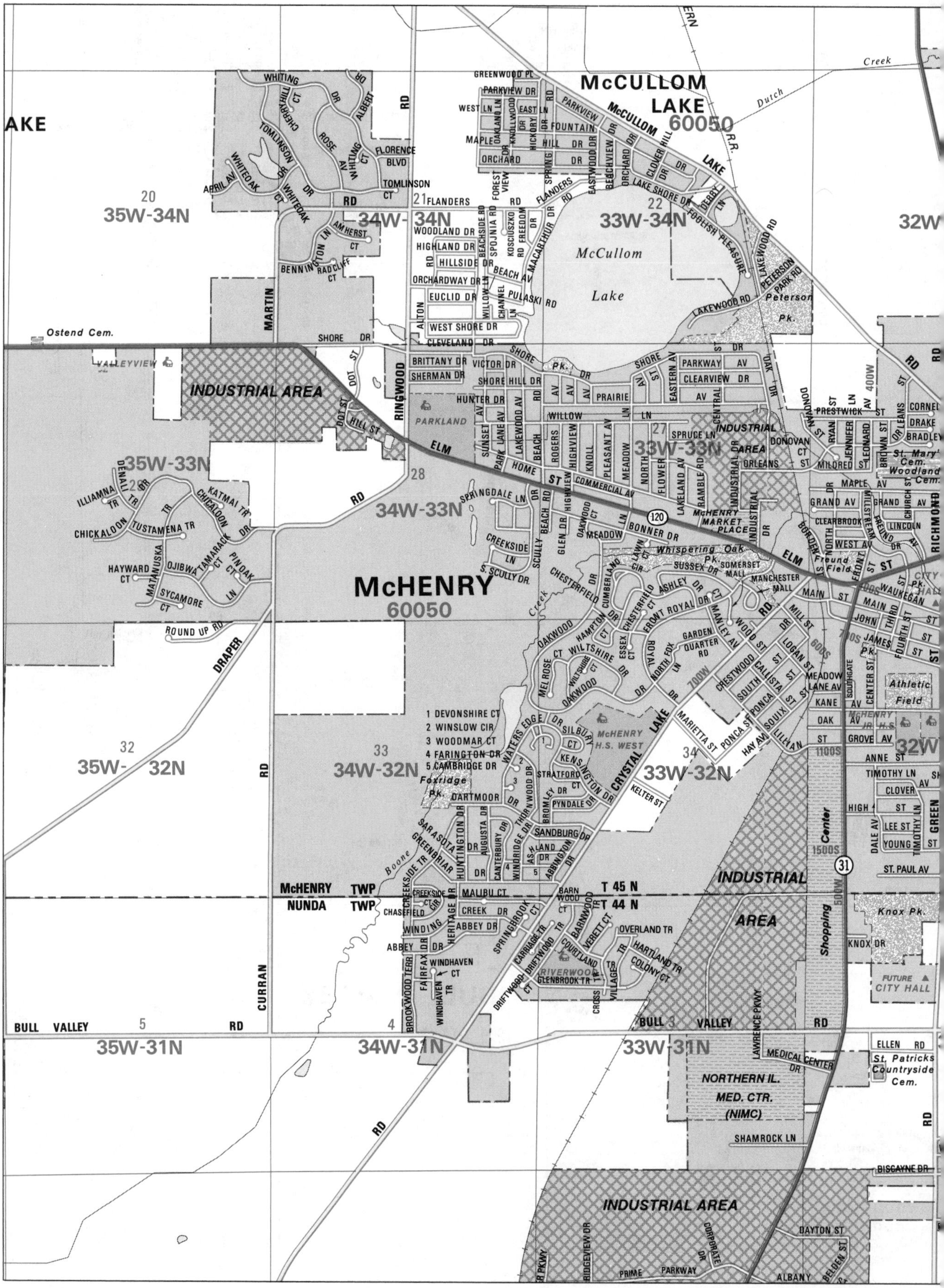

SEE PAGES 24-25

McCULLOM LAKE 60050

McHENRY 60050

LAKE

McCULLOM LAKE

Creek

Dutch

R.R.

20
35W-34N

21
34W-34N

22
33W-34N

32W

McCullom Lake

26
35W-33N

28
34W-33N

27
33W-33N

INDUSTRIAL AREA

Ostend Cem.

VALLEYVIEW

Parkland

McHenry Market Place

Whispering Oak Pk

Somerset Mall

Manchester Mall

St. Mary Cem.
Woodland Cem.

32
35W-32N

33
34W-32N

34
33W-32N

32W

McHenry H.S. West

1 DEVONSHIRE CT
2 WINSLOW CIR
3 WOODMAR CT
4 FARINGTON DR
5 CAMBRIDGE DR

Foxridge Pk.

Athletic Field

McHENRY TWP
NUNDA TWP

T 45 N
T 44 N

INDUSTRIAL AREA

Knox Pk.

FUTURE CITY HALL

5
35W-31N

4
34W-31N

3
33W-31N

BULL VALLEY RD

Bull Valley RD

NORTHERN IL. MED. CTR. (NIMC)

SHAMROCK LN

St. Patricks Countryside Cem.

INDUSTRIAL AREA

BISCAYNE DR

(120)
(31)

SEE PAGES 34-35

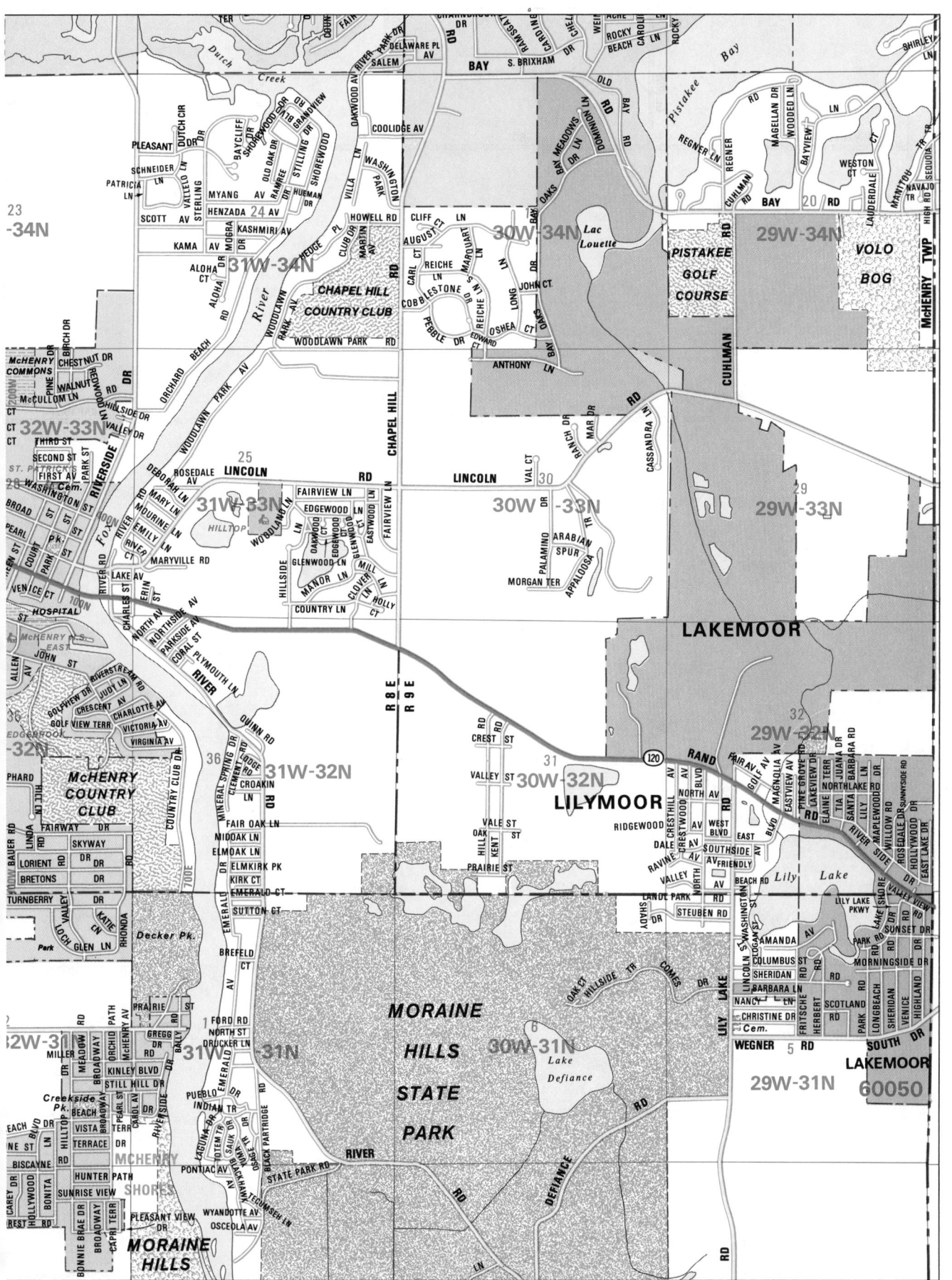

SEE PAGES 20-21

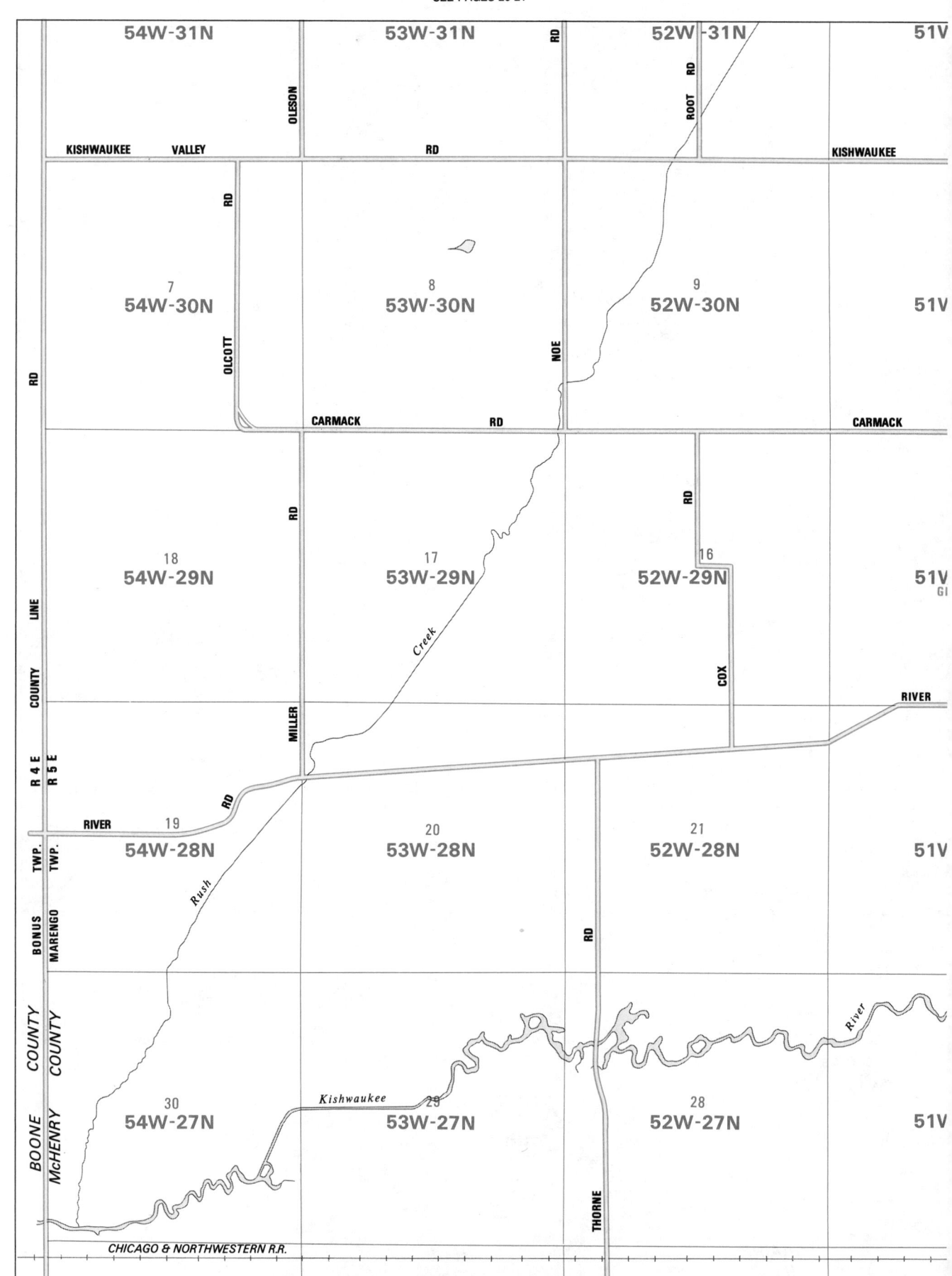

SEE PAGES 36-37

SEE PAGES 20-21

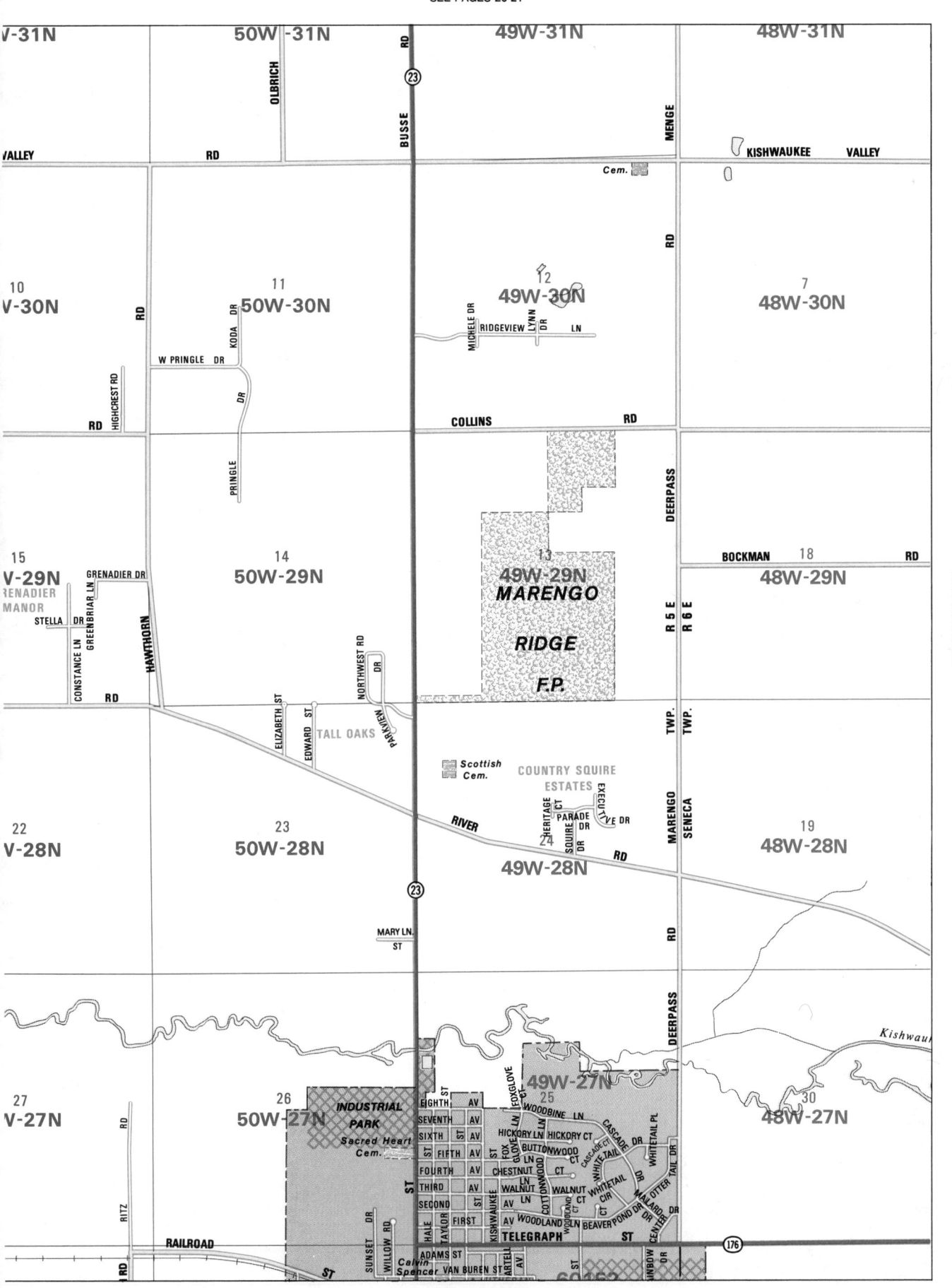

SEE PAGES 30-31

SEE PAGES 22-23

SEE PAGES 28-29

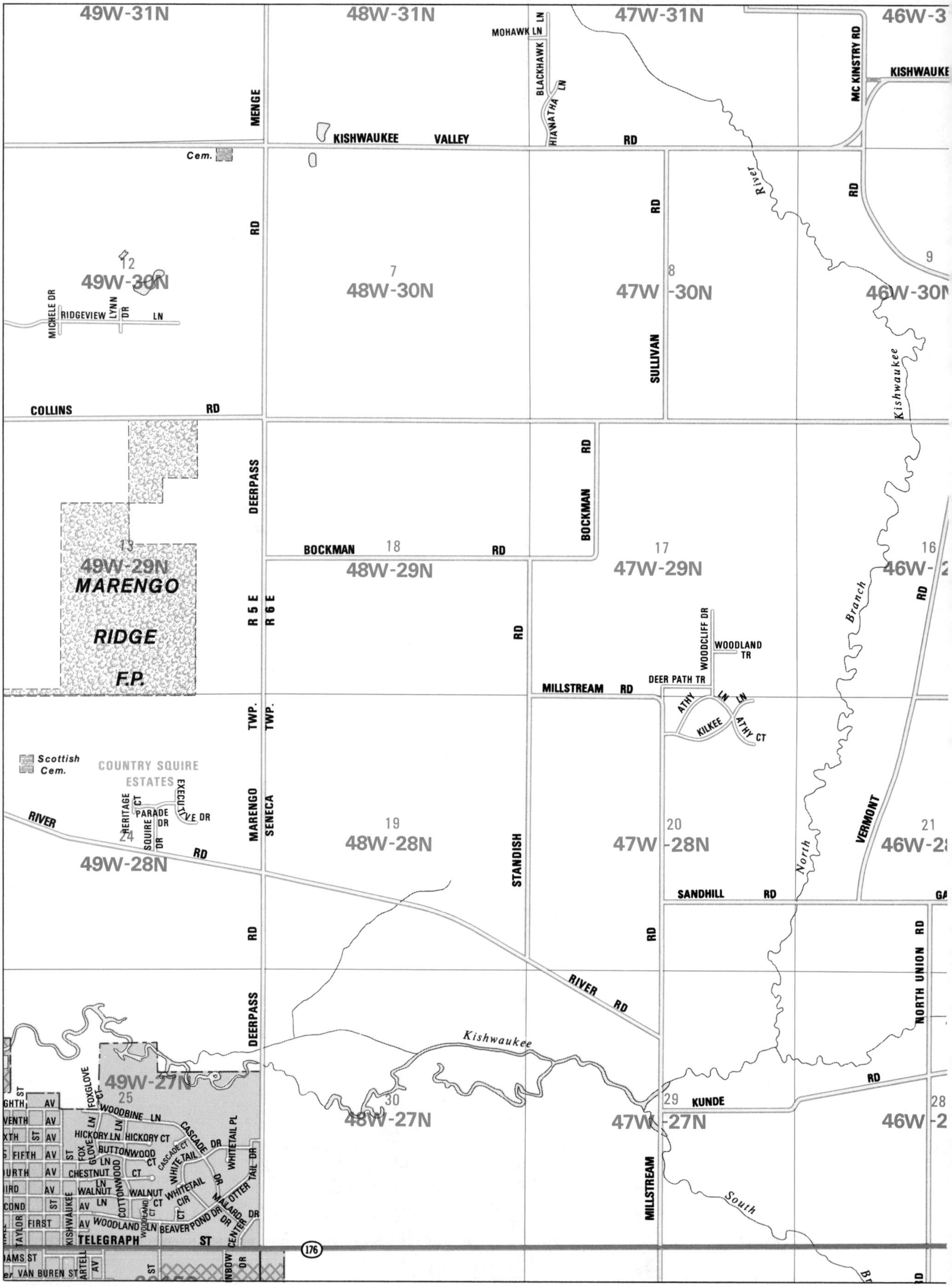

49W-31N

48W-31N

47W-31N

46W-3

MOHAWK LN

BLACKHAWK LN

HIAWATHA LN

MC KINSTRY RD

KISHWAUKE

KISHWAUKEE   VALLEY   RD

MENGE

RD

RD

River

RD

Cem.

7

8

9

49W-30N

48W-30N

47W -30N

46W-30N

12

MICHELE DR

RIDGEVIEW LYNN DR LN

Kishwaukee

SULLIVAN

COLLINS   RD

DEERPASS

BOCKMAN   RD

13

18

17

16

49W-29N

BOCKMAN   RD

48W-29N

47W-29N

46W-2

MARENGO

R 5 E   R 6 E

RD

RD

RIDGE

Branch

F.P.

WOODCLIFF DR

WOODLAND TR

MILLSTREAM   RD

DEER PATH TR

Scottish Cem.

ATHY LN LN

COUNTRY SQUIRE ESTATES

KILKEE   ATHY CT

TWP. TWP.

EXECUTIVE DR

RIVER

MARENGO SENECA

19

20

21

HERITAGE CT

PARADE DR

48W-28N

STANDISH

47W -28N

VERMONT

46W-28

24

SQUIRE DR

RD

North

49W-28N

RD

SANDHILL   RD

GA

RD

RD

NORTH UNION RD

DEERPASS

RIVER   RD

Kishwaukee

FOXGLOVE LN

49W-27N

25

RD

WOODBINE LN

30

29

KUNDE

28

HICKORY LN HICKORY CT

CASCADE CT

WHITETAIL PL

48W-27N

47W-27N

46W-2

FOX GLOVE LN

BUTTONWOOD LN

CASCADE DR

GHTH AV

ENTH AV

XTH ST AV

FIFTH AV

CHESTNUT LN CT

WHITE TAIL DR

OTTER TAIL DR

URTH AV

WALNUT WALNUT CT

WHITETAIL CIR

MILLARD DR

MALLARD DR

MILLSTREAM

IRD ST

COTTONWOOD LN

CENTER DR

South

COND AV

WOODLAND LN

BEAVER POND DR

KISHWAUKEE

FIRST AV

WOODLAND ST

TAYLOR ST

TELEGRAPH   ST

176

DAMS ST

VAN BUREN ST

ARTELL AV

RINBOW DR

ST

B

RD

SEE PAGES 38-39

SEE PAGES 22-23

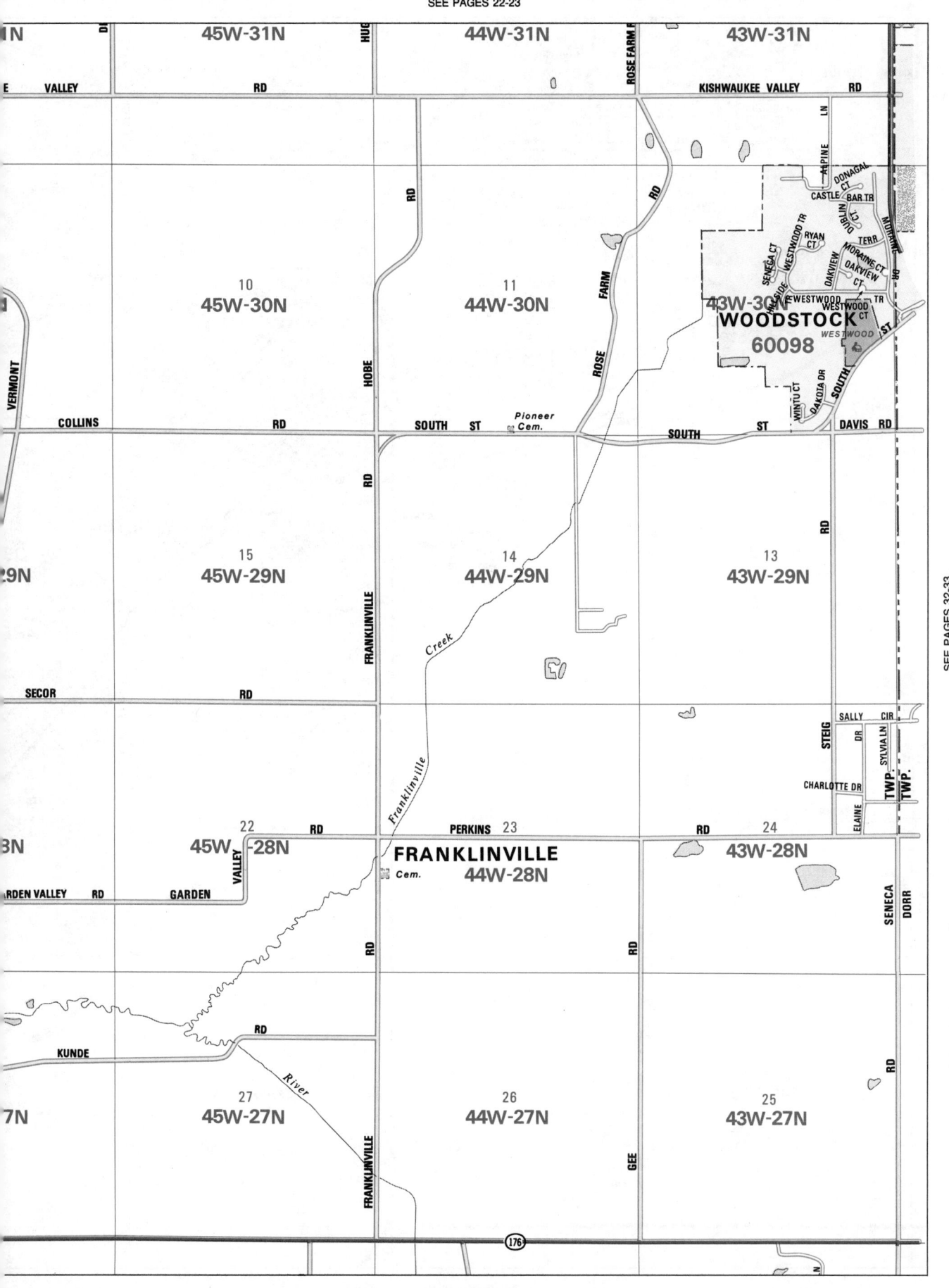

SEE PAGES 32-33

SEE PAGES 38-39

SEE PAGES 24-25

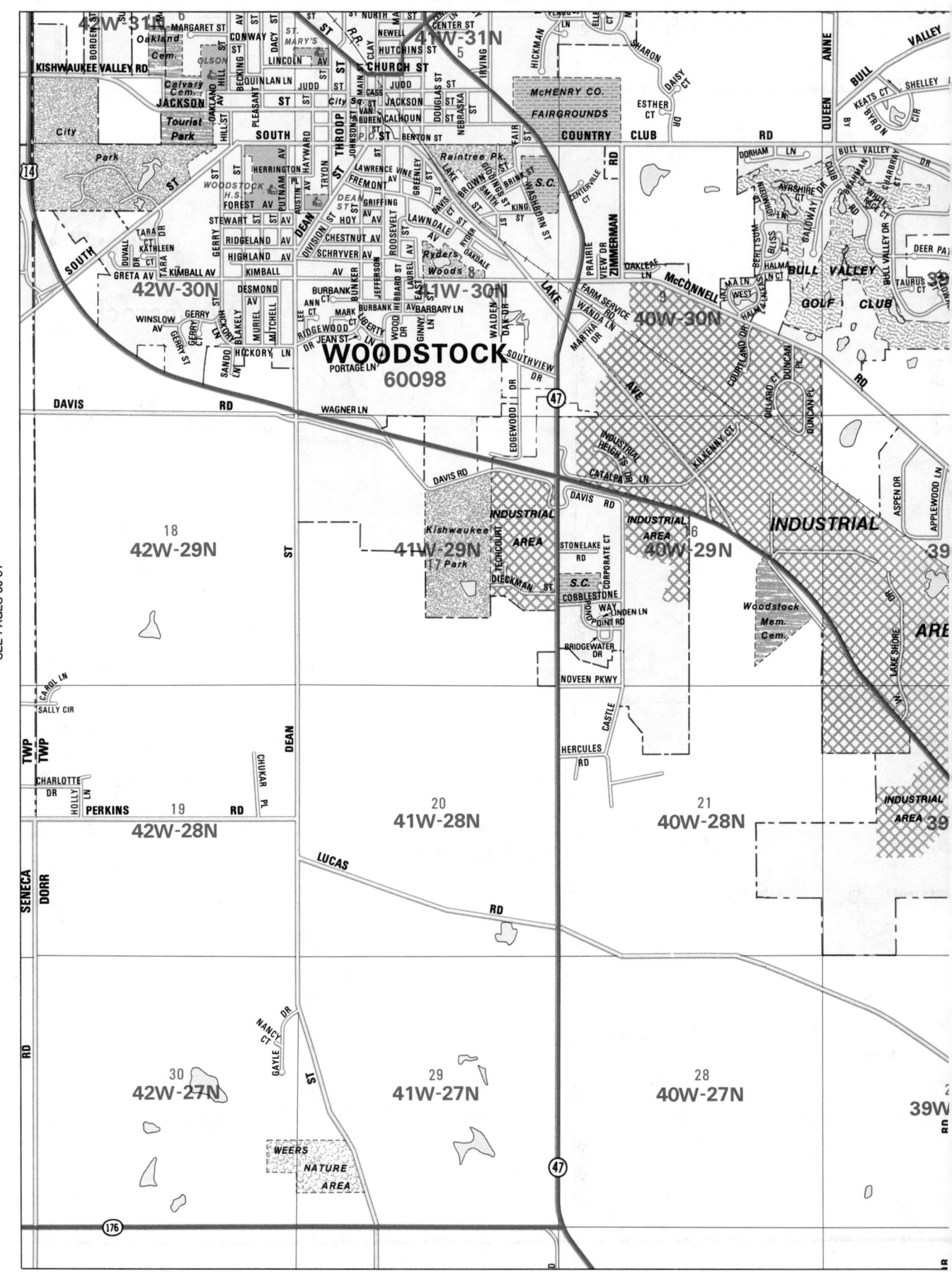

WOODSTOCK
60098

SEE PAGES 30-31

38W-31N   37W-31N   36W-31N

ORCHARD VALLEY DR

WOODLAND DR
LOCUST LN
WOODLAND CT
FAIRWAY LN

FLEMING RD

RD

STONEWIER POINT
SADDLE CREEK TR
CREEK TERR
BOONE

WOODSTOCK COUNTRY CLUB

Boone Creek

VALLEY

N. CHERRY

COUNTRY CLUB RD

RD

38W-30N   37W-30N   36W-30N

W-30N

11

12

7   MASON   HILL

BOB 'O' LINK CIR
BULL VALLEY LN

HILL

VALLEY

RD

MASON   HILL

RD

N. CHERRY VALLEY RD

BULL RUN TR

HUNTING CLUB SPRINGS

CRYSTAL

McCONNELL

15

14

13

18

RD

W-29N

38W-29N   37W-29N   36W-29N   RD

RD

RD

PHEASANT RUN

DR

PHEASANT

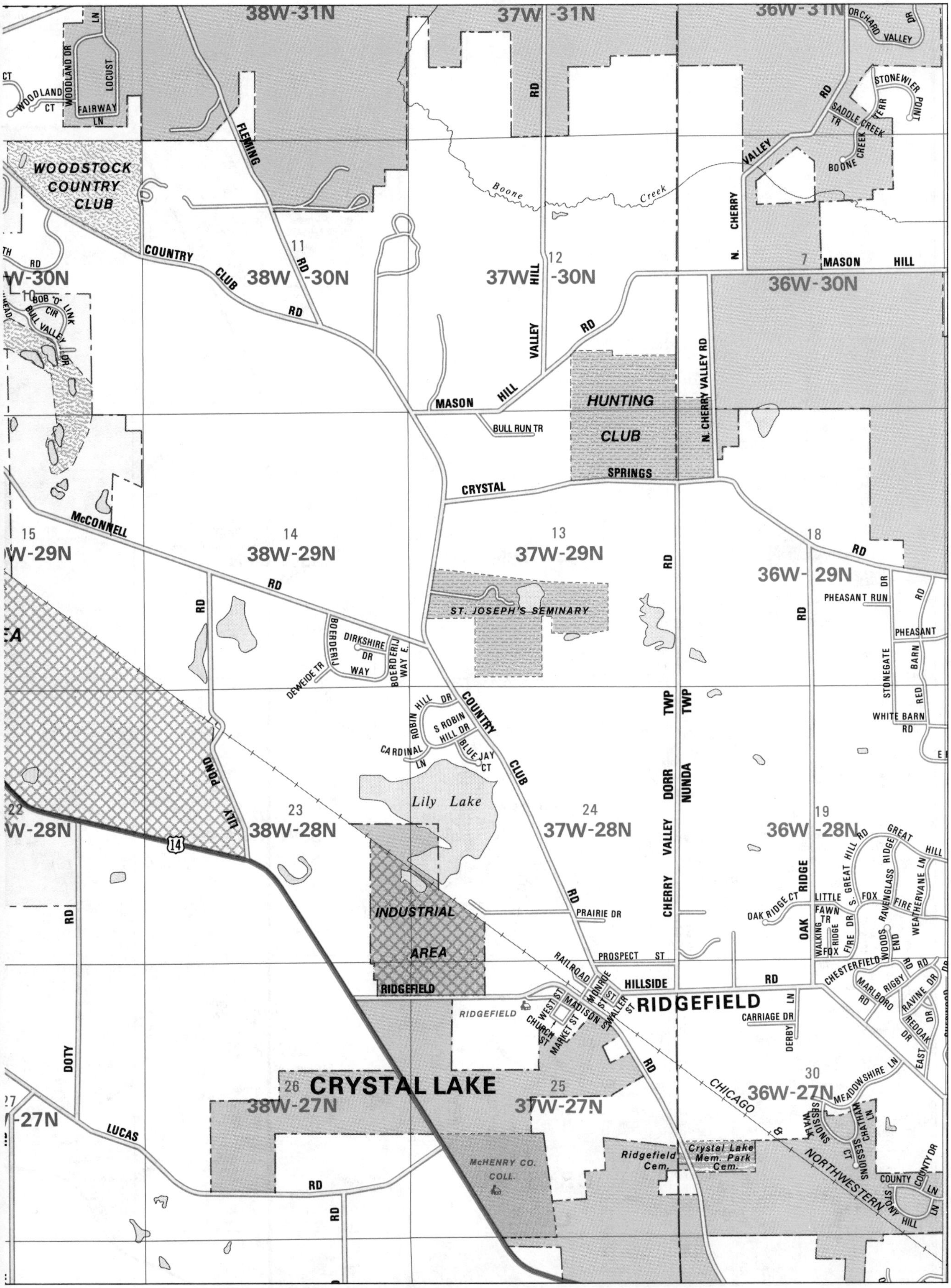

BOERDERIJ

DIRKSHIRE DR WAY

BOERDERIJ WAY E.

ST. JOSEPH'S SEMINARY

RED BARN

STONEGATE

DEWEIDE TR

TWP

TWP

WHITE BARN RD

RED RD

ROBIN HILL DR
S ROBIN HILL DR
CARDINAL LN
BLUE JAY CT

COUNTRY

CLUB

Lily Lake

24

37W-28N

19

22

23

38W-28N

DORR

NUNDA

36W-28N

GREAT

HILL

OAK RIDGE CT
LITTLE
GREAT HILL
S. FOX
FAWN TR
WALKING
FIRE DR
FOX
RIDGE
WOODS

RAVENGLASS RIDGE
FIRE LN
WEATHERVANE LN
RIDGE END

GREAT HILL

W-28N

LILY POND RD

14

CHERRY

VALLEY

OAK RIDGE

CHESTERFIELD

RIGBY RD

MARLBORO RD

RAVINE DR

REDOAK DR

EAST RD

RD

PRAIRIE DR

RD

RAILROAD

PROSPECT ST

HILLSIDE RD

INDUSTRIAL AREA

RIDGEFIELD

RIDGEFIELD

RIDGEFIELD

WEST ST
MADISON ST
MARKET ST
CHURCH
MONROE ST
WALLER ST

CARRIAGE DR

DERBY LN

DOTY RD

26   CRYSTAL LAKE

38W-27N

25

37W-27N

30

36W-27N

CHICAGO &

MEADOWSHIRE LN

CHATHAM CT

LUCAS

RD

McHENRY CO. COLL.

Ridgefield Cem.

Crystal Lake Mem. Park Cem.

NORTHWESTERN

SESSIONS

WALK

COUNTY LN

COUNTY DR

STONY HILL

27

27N

SEE PAGES 34-35

SEE PAGES 26-27

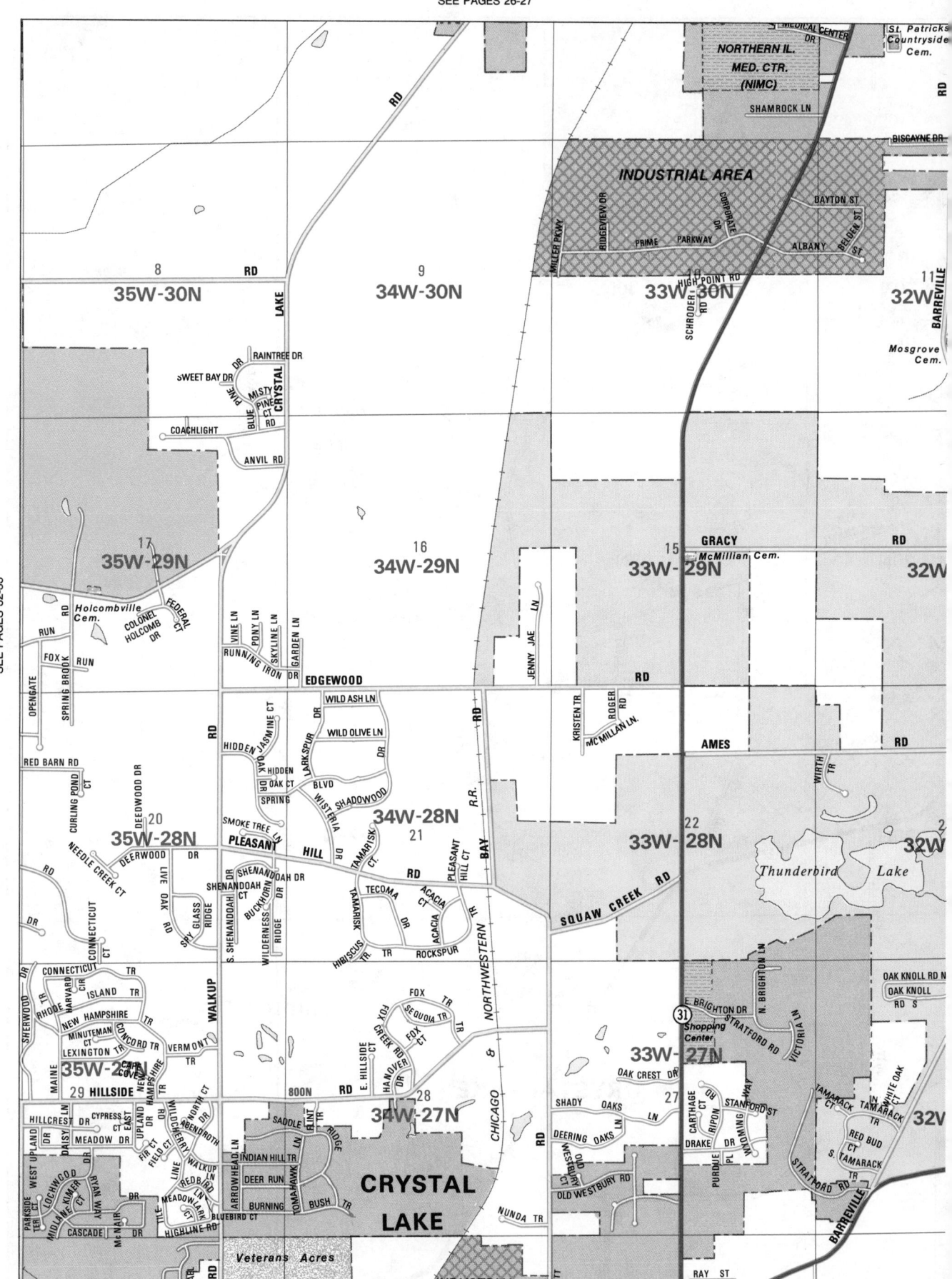

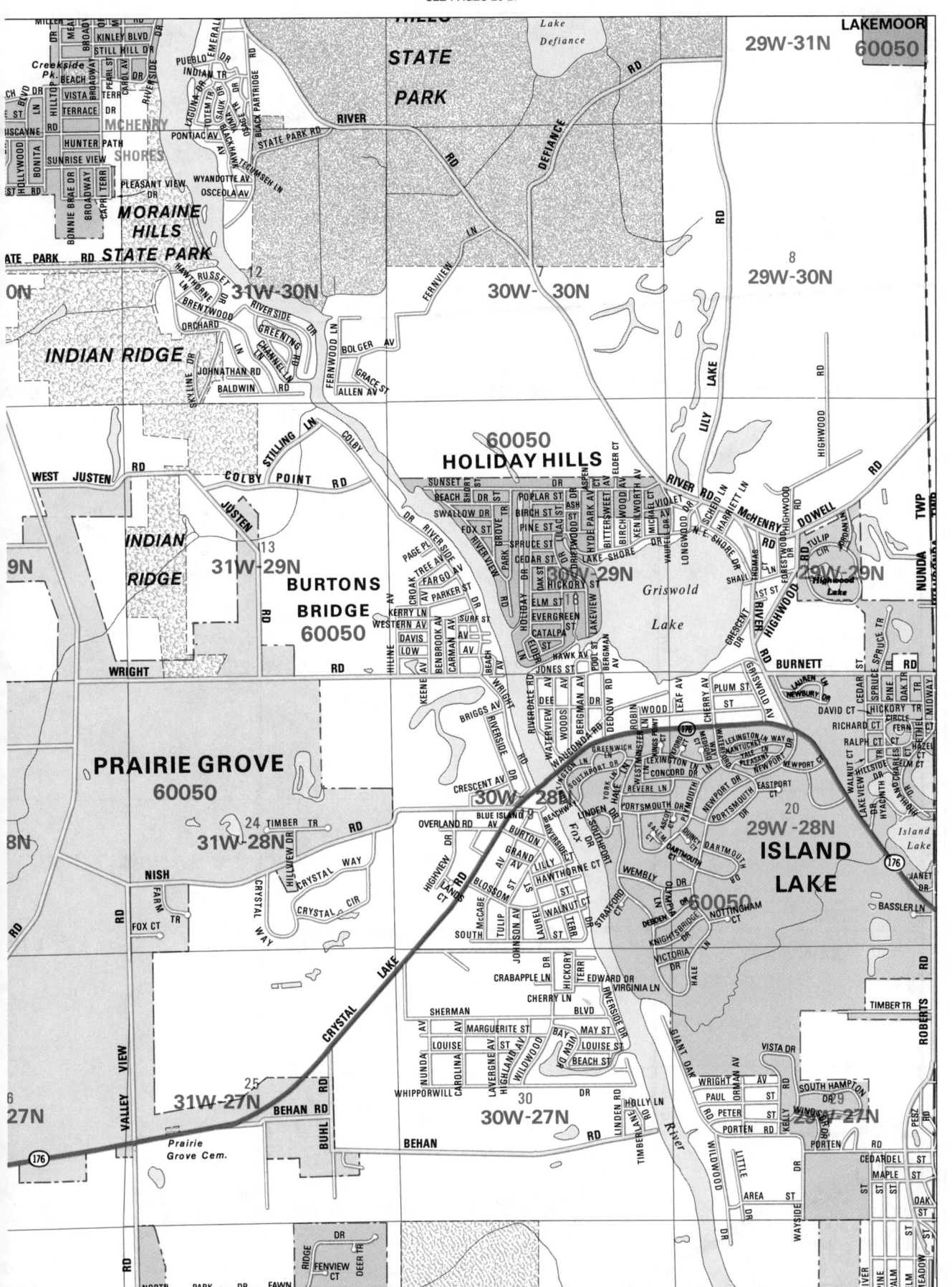

SEE PAGES 28-29

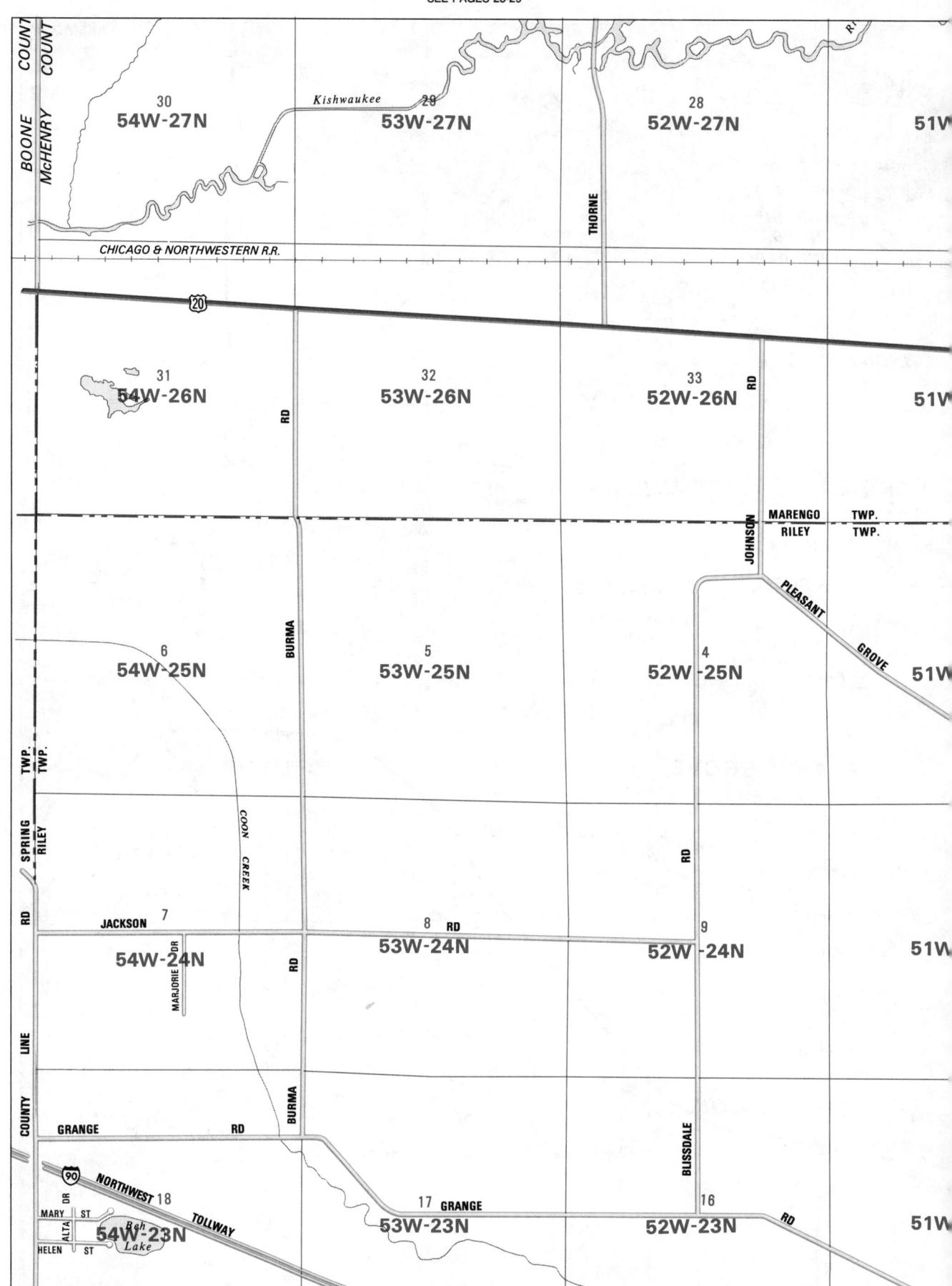

SEE PAGES 28-29

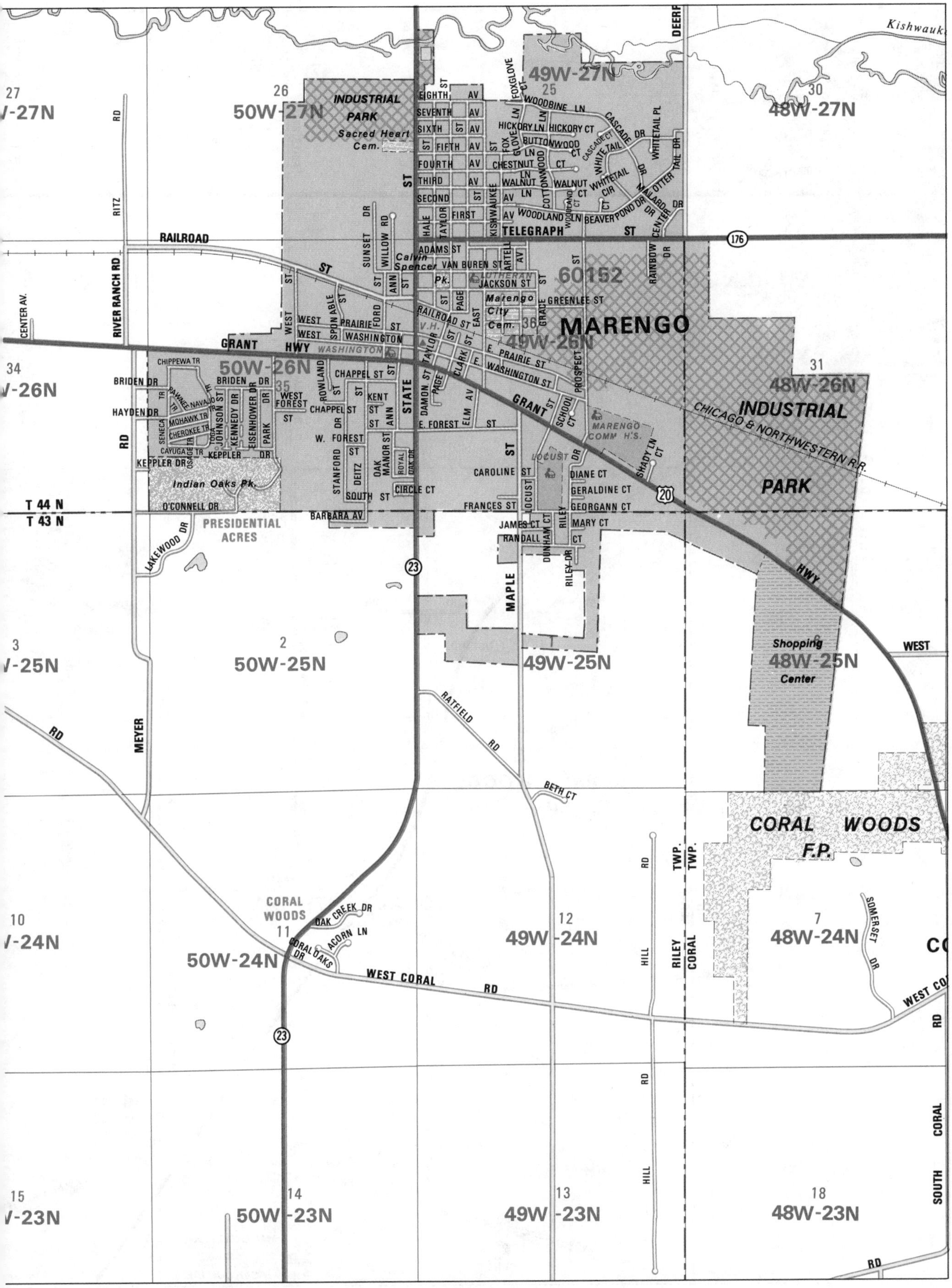

Kishwauk

27
W-27N

50W-27N

INDUSTRIAL
PARK

Sacred Heart
Cem.

26

RITZ RD

49W-27N
25

FOXGLOVE
WOODBINE LN

EIGHTH ST AV
SEVENTH AV
SIXTH AV
FIFTH AV
FOURTH AV
THIRD AV
SECOND AV
FIRST AV

ST

HICKORY LN HICKORY CT
FOX BUTTONWOOD CT
GLOVE LN
CHESTNUT LN
WALNUT LN
WALNUT CT
WOODLAND LN

CASCADE DR
CASCADE CT
WHITE TAIL CT
WHITETAIL CIR
WHITETAIL DR

WHITETAIL PL

TAIL DR
OTTER
NATARO DR
BEAVER POND DR

30

48W-27N

DEER

SUNSET DR
WILLOW RD
HALE
TAYLOR ST
KISHWAUKE AV

RAILROAD

RAILROAD ST

ADAMS ST
Calvin SPENCER VAN BUREN ST
Pk. JACKSON ST
Marengo
City
Cem.

ARTELL
GREENLEE ST

BRADIE DR

RAINBOW DR

TELEGRAPH ST

60152

MARENGO

176

49W-26N

CENTER AV

RIVER RANCH RD

GRANT HWY
WASHINGTON

34
W-26N

50W-26N

CHIPPEWA TR

BRIDEN DR
HAYDEN DR

PAWNEE TR
NAVAJO
MOHAWK TR
SENECA TR
CHEROKEE TR
CAYUGA TR

BRIDEN DR
JOHNSON ST
KENNEDY DR
EISENHOWER DR
PARK
KEPPLER DR

WEST
WEST
WEST WASHINGTON
WASHINGTON

SPINNABLE
PRAIRIE
FORD ST

35

ROWLAND ST

CHAPPEL ST
WEST
FOREST
ST
CHAPPEL ST
W. FOREST ST

KENT ST
ST ANN ST

STATE

DAMON ST
PAGE
CLARK ST

E. PRAIRIE ST
E. WASHINGTON ST

RAILROAD ST
TAYLOR ST
EAST

PROSPECT

31

48W-26N

INDUSTRIAL

CHICAGO & NORTHWESTERN R.R.

KEPPLER DR

Indian Oaks Pk.

O'CONNELL DR

STANFORD DR
DEITZ ST
SOUTH
OAK MANOR ST
ROYAL
OAK DR

CIRCLE CT

CAROLINE ST

E. FOREST ST
ELM AV

GRANT ST

LOCUST

SCHOOL CT
SHADY LN CT

Marengo
COMM. H.S.

DIANE CT
GERALDINE CT
GEORGANN CT

20

PARK

T 44 N
T 43 N

PRESIDENTIAL
ACRES

BARBARA AV

FRANCES ST

JAMES CT
RANDALL ST

LOCUST

DUNHAM CT
RILEY DR
RILEY DR

MARY CT

HWY

LAKEWOOD DR

23

MAPLE ST

Shopping

Center

3
W-25N

50W-25N

2

RATFIELD RD

BETH CT

49W-25N

WEST

48W-25N

MEYER

RD

SEE PAGES 38-39

10
W-24N

50W-24N

11

CORAL
WOODS

OAK CREEK DR
ACORN LN

CORAL OAKS
DR

WEST CORAL RD

12

RD

TWP.
TWP.

HILL

RILEY
CORAL

49W-24N

CORAL WOODS
F.P.

7

SOMERSET DR

48W-24N

C

WEST CO

RD

23

RD

HILL

CORAL

SOUTH CORAL

15
W-23N

50W-23N

14

13

49W-23N

18

48W-23N

RD

SEE PAGES 30-31

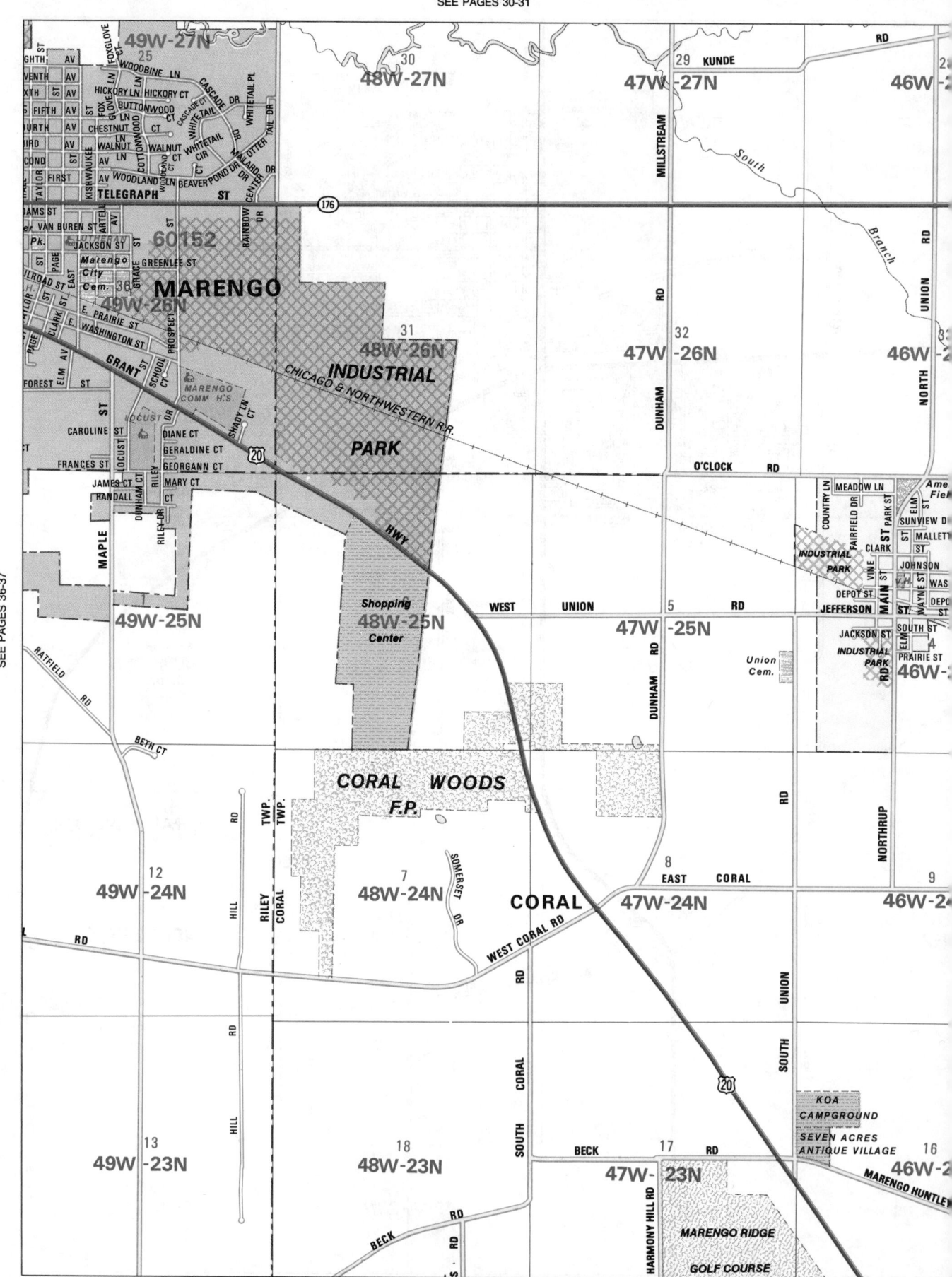

SEE PAGES 36-37

SEE PAGES 30-31

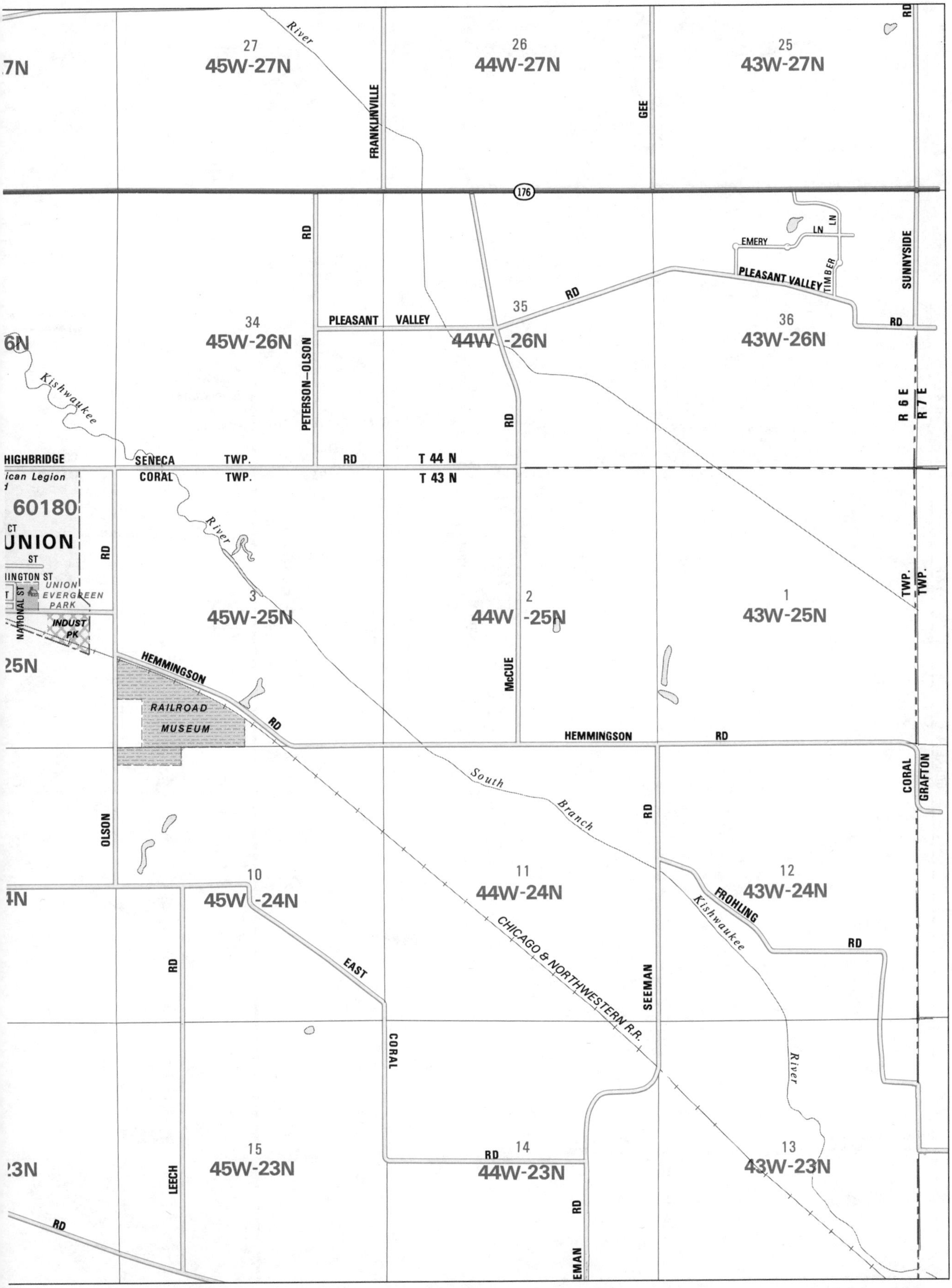

27
45W-27N

26
44W-27N

25
43W-27N

River

FRANKLINVILLE

GEE

RD

176

EMERY

LN

LN

PLEASANT VALLEY

TIMBER

SUNNYSIDE

RD

RD

PLEASANT   VALLEY

34
45W-26N

35
44W -26N

36
43W-26N

PETERSON-OLSON

RD

Kishwaukee

R 6 E

R 7 E

HIGHBRIDGE

SENECA   TWP.

CORAL   TWP.

RD

T 44 N

T 43 N

River

ican Legion

60180

CT

UNION

ST

INGTON ST

UNION
EVERGREEN
PARK

NATIONAL ST

25N

INDUST
PK

RD

River

3
45W-25N

2
44W -25N

1
43W-25N

McCUE

TWP.

TWP.

SEE PAGES 40-41

HEMMINGSON

RAILROAD
MUSEUM

RD

HEMMINGSON

RD

South

Branch

RD

OLSON

CORAL

GRAFTON

10
45W -24N

11
44W-24N

12
43W-24N

FROHLING

Kishwaukee

RD

RD

EAST

CHICAGO & NORTHWESTERN R.R.

SEEMAN

River

CORAL

15
45W-23N

14
44W-23N

13
43W-23N

LEECH

RD

RD

RD

EMAN

23N

RD

SEE PAGES 46-47

SEE PAGES 32-33

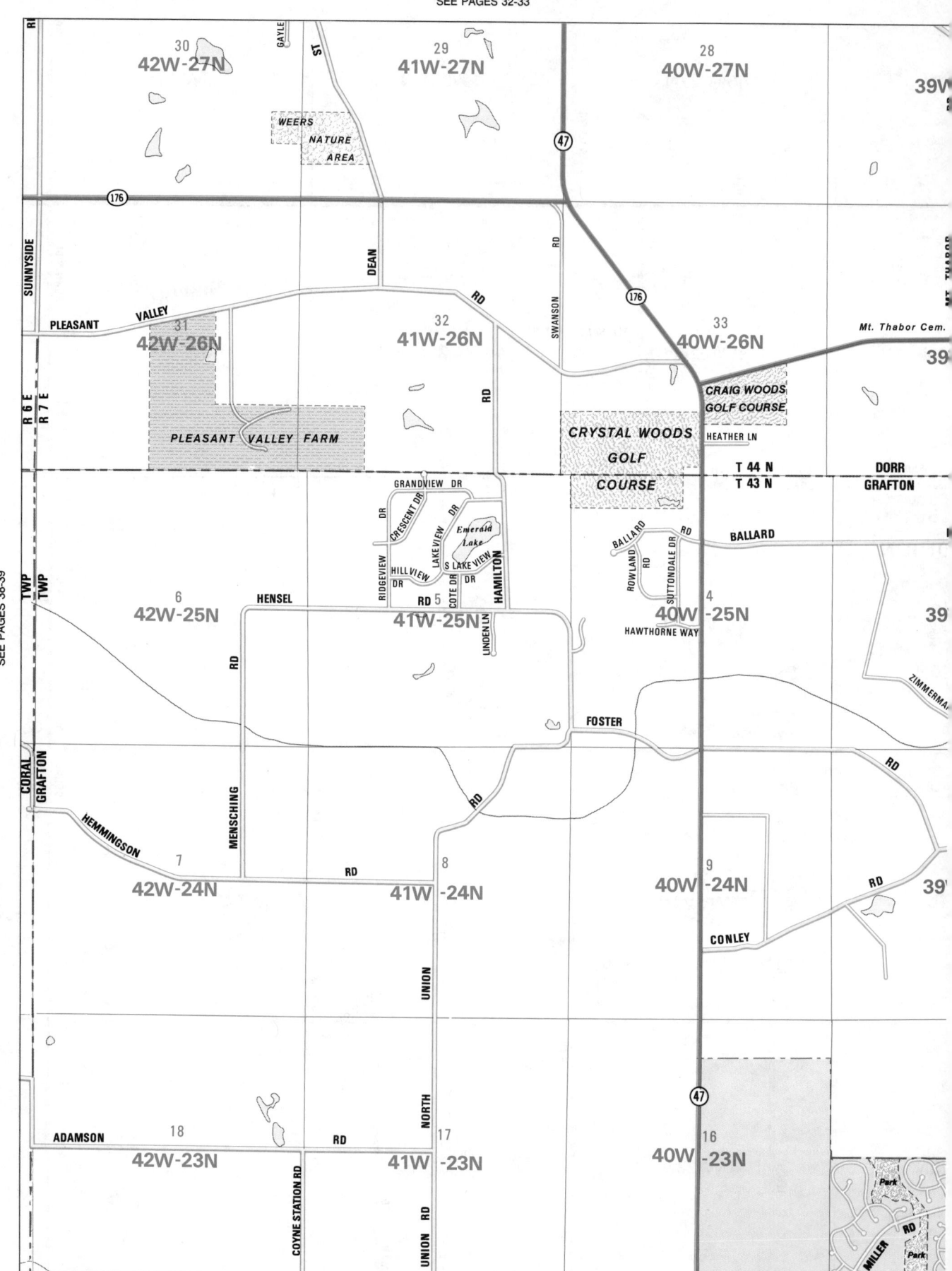

SEE PAGES 48-49

SEE PAGES 38-39

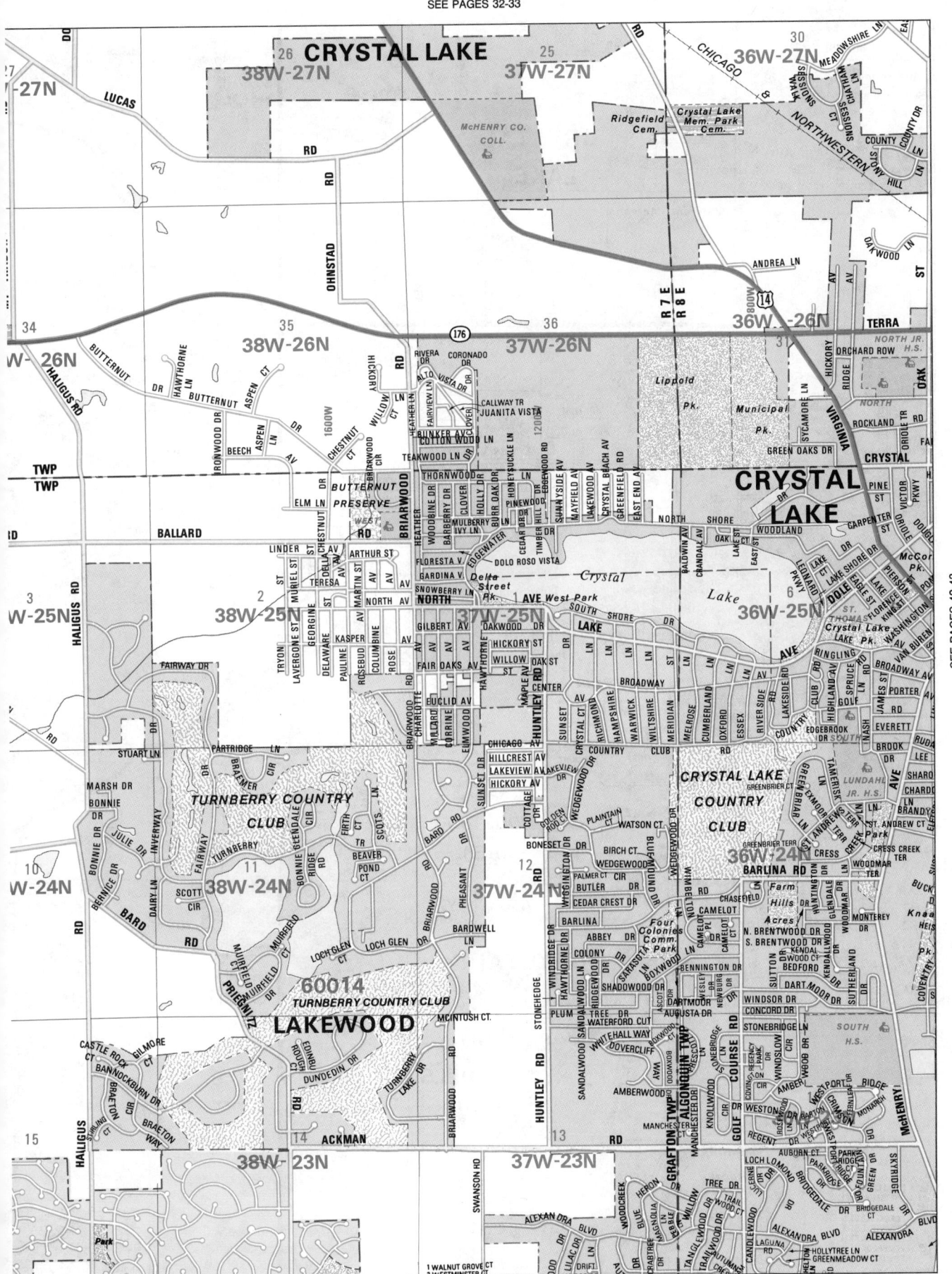

SEE PAGES 42-43

SEE PAGES 40-41

CRYSTAL LAKE

34W-27N

35W-27N

32W

HILLSIDE

Veterans Acres
Park

NATURE
CENTER

INDUSTRIAL
AREA

PINE PARADISE
NORTH

PINE PARADISE

35W-26N

34W-26N

33W-26N

32W

CRYSTAL   LAKE

STEEPLE RUN

WYNDWOOD

35W-25N

34W-25N
60014

INDUSTRIAL AREA

33W-25N

INDUSTRIAL
AREA

32W

Crystal Point
Mall

Shopping Center

ILL. INSTITUTE
OF TECHNOLOGY

CRYSTAL
COURT
MALL

TWIN
PONDS
G.C.

OAKBROOK
ESTATES

TOWNSHIP
OFFICE

INDUSTRIAL
AREA

34W-24N
(QUARRY)

35W-24N

33W-24N

32W
60013

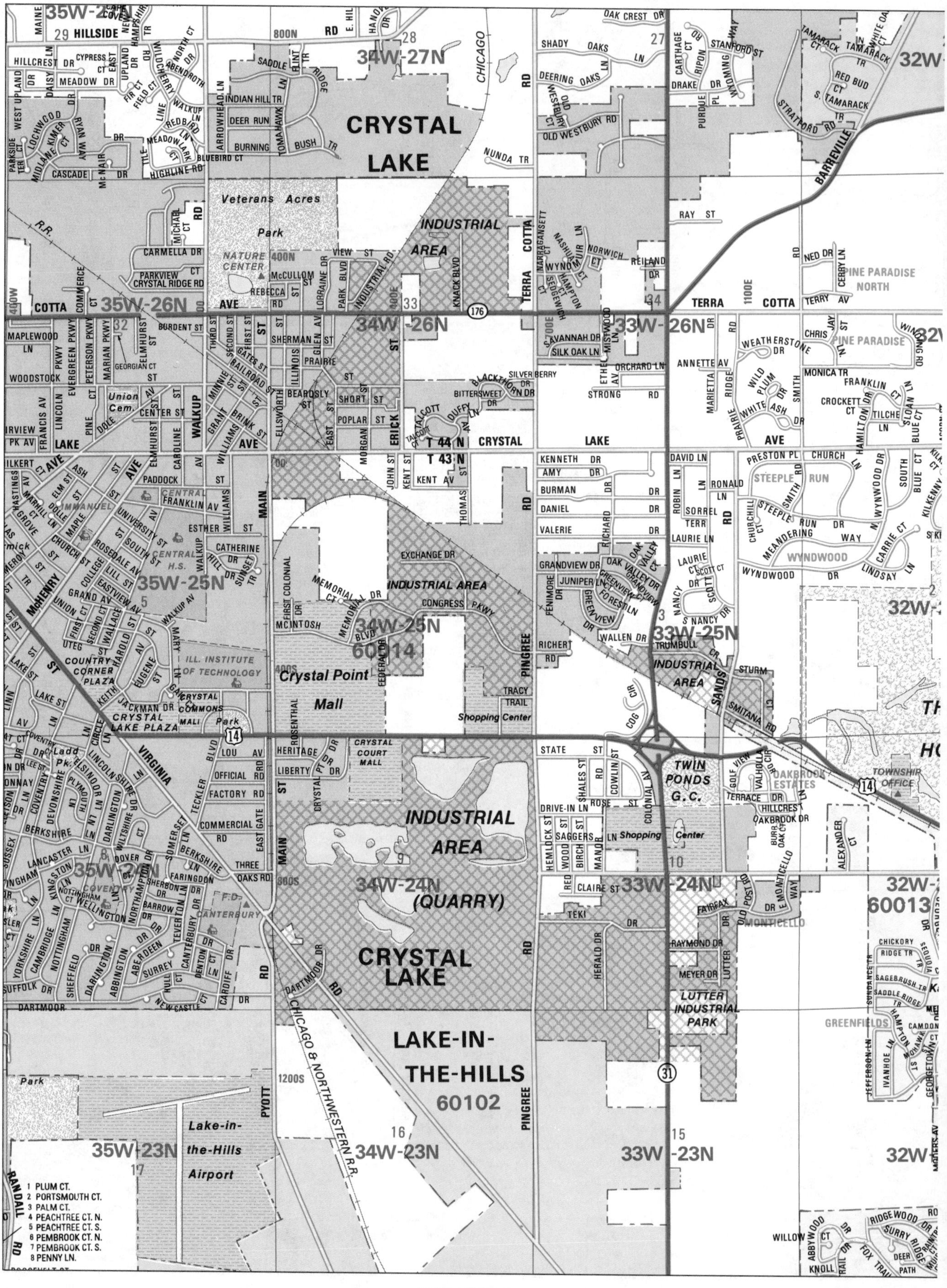

CRYSTAL
LAKE

LUTTER
INDUSTRIAL
PARK

GREENFIELDS

LAKE-IN-
THE-HILLS
60102

Lake-in-
the-Hills
Airport

35W-23N

34W-23N

33W-23N

32W

1 PLUM CT.
2 PORTSMOUTH CT.
3 PALM CT.
4 PEACHTREE CT. N.
5 PEACHTREE CT. S.
6 PEMBROOK CT. N.
7 PEMBROOK CT. S.
8 PENNY LN.

SEE PAGES 34-35

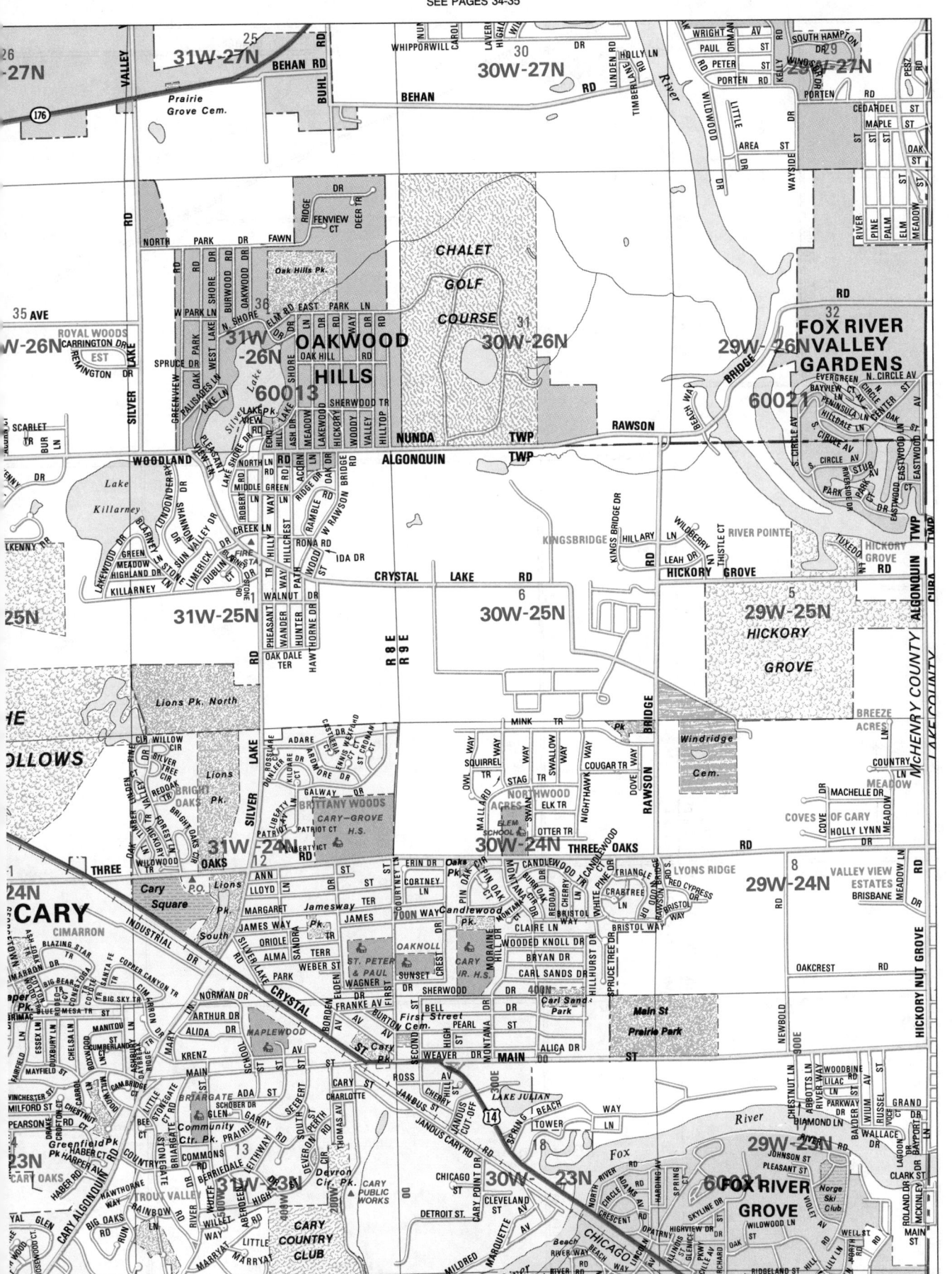

SEE PAGES 50-51

13

MARY ST
ALTA DR
HELEN ST

Bch Lake

54W-23N

NORTHWEST 18 TOLLWAY

17 GRANGE

53W-23N

52W-2

Coon Creek

ANTHONY RD

GARDEN PRAIRIE RD

COUNTY LINE RD

24

Williamson Creek

McKEOWIN RD

19

54W-22N

ANTHONY

I-90

20

53W-22N

21

52W-2

HARMONY RD

25

PINEGER RD

30

54W-21N

29

53W-21N

28

52W-

HARTMAN

6

R 4 E
R 5 E
BOONE COUNTY
McHENRY COUNTY

GENOA RD

RUDOLPH RD

31

54W-20N

32

53W-20N

BURROWS RD

33

52W-

POPLAR RD

DEKALB COUNTY

RILEY TWP   T 4

GENOA TWP   T 4

NORTH STATE RD

1

6

55W-19N

5

54W-19N

EISENHOWER RD

Cem.

4

53W-

MELMS RD

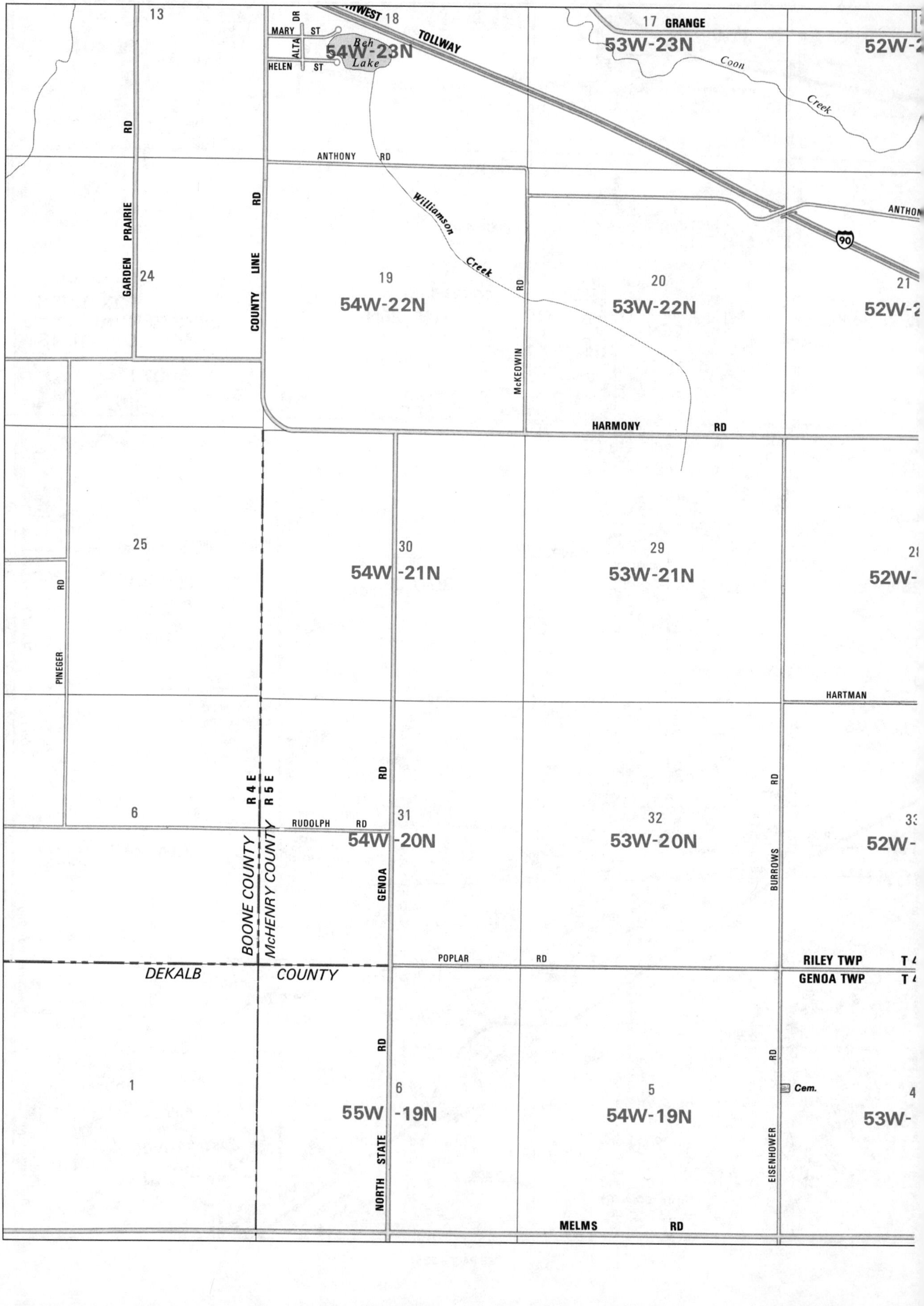

6

23N RD

15
51W-23N

14
50W-23N

FAR FIELD

13
49W-23N

GRANGE HALL ▲ **GRANGE** RD

Riley *Creek*

BECK

RD

Cem. ▲ RILEY

22
RILEY

22N

51W-22N

RILEY RD

23
ANTHONY RD
50W-22N

24
49W-22N

MAPLE STREET

LANDING STRIP

LANDING STRIP

HARMONY RD

NORTHWEST

GROSSEN RD

TOLLWAY

21N

27
51W-21N

RD

26
50W-21N

90

25
49W-21N

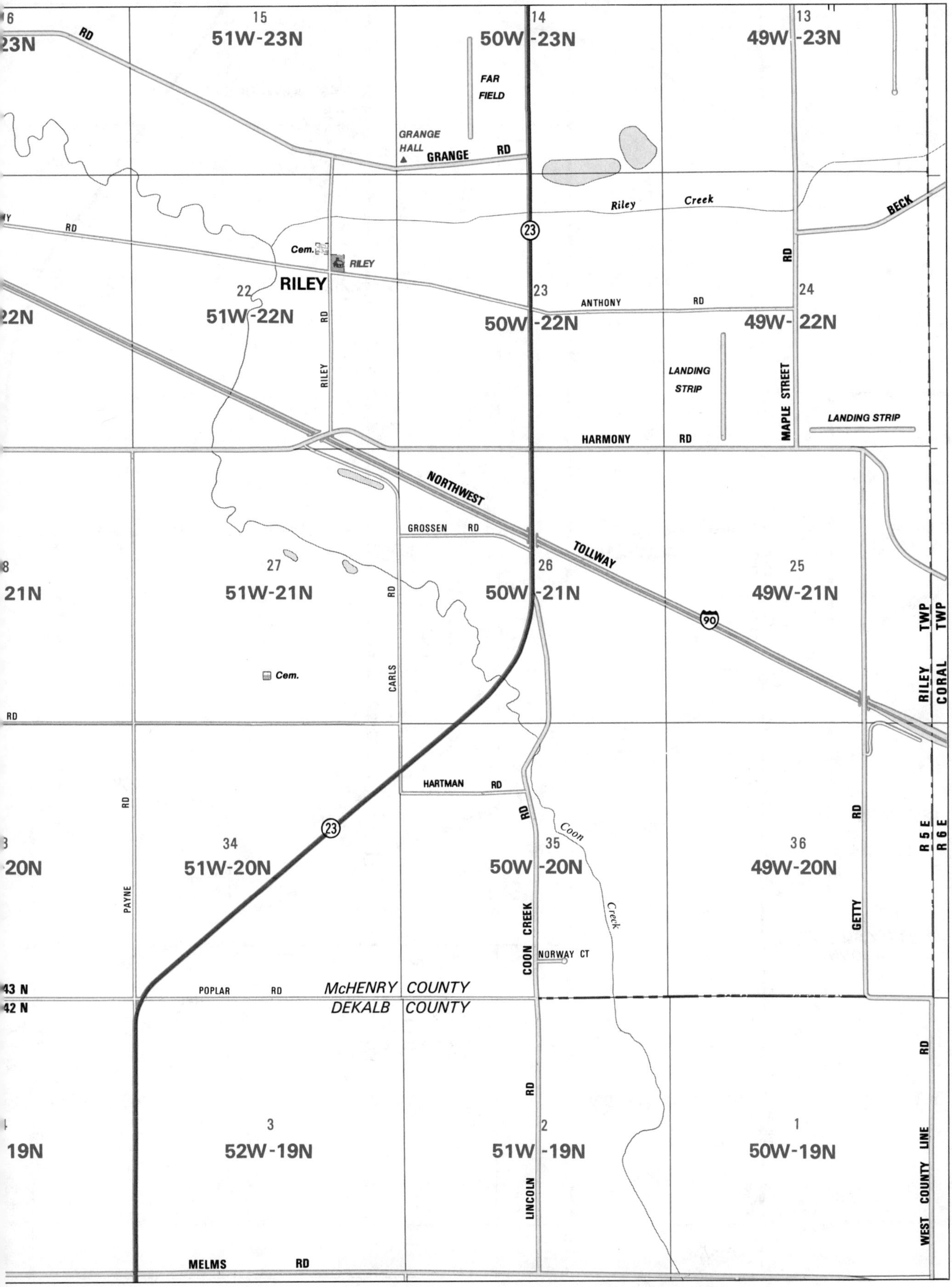

SEE PAGES 46-47

RILEY TWP
CORAL TWP

Cem.

CARLS

RD

HARTMAN RD

RD

Coon

R 5 E
R 6 E

20N

PAYNE RD

34
51W-20N

23

35
50W-20N

Creek

36
49W-20N

GETTY RD

COON CREEK

NORWAY CT

43 N
42 N

POPLAR RD

McHENRY COUNTY
DEKALB COUNTY

RD

19N

3
52W-19N

2
51W-19N

LINCOLN

1
50W-19N

RD

WEST COUNTY LINE

MELMS RD

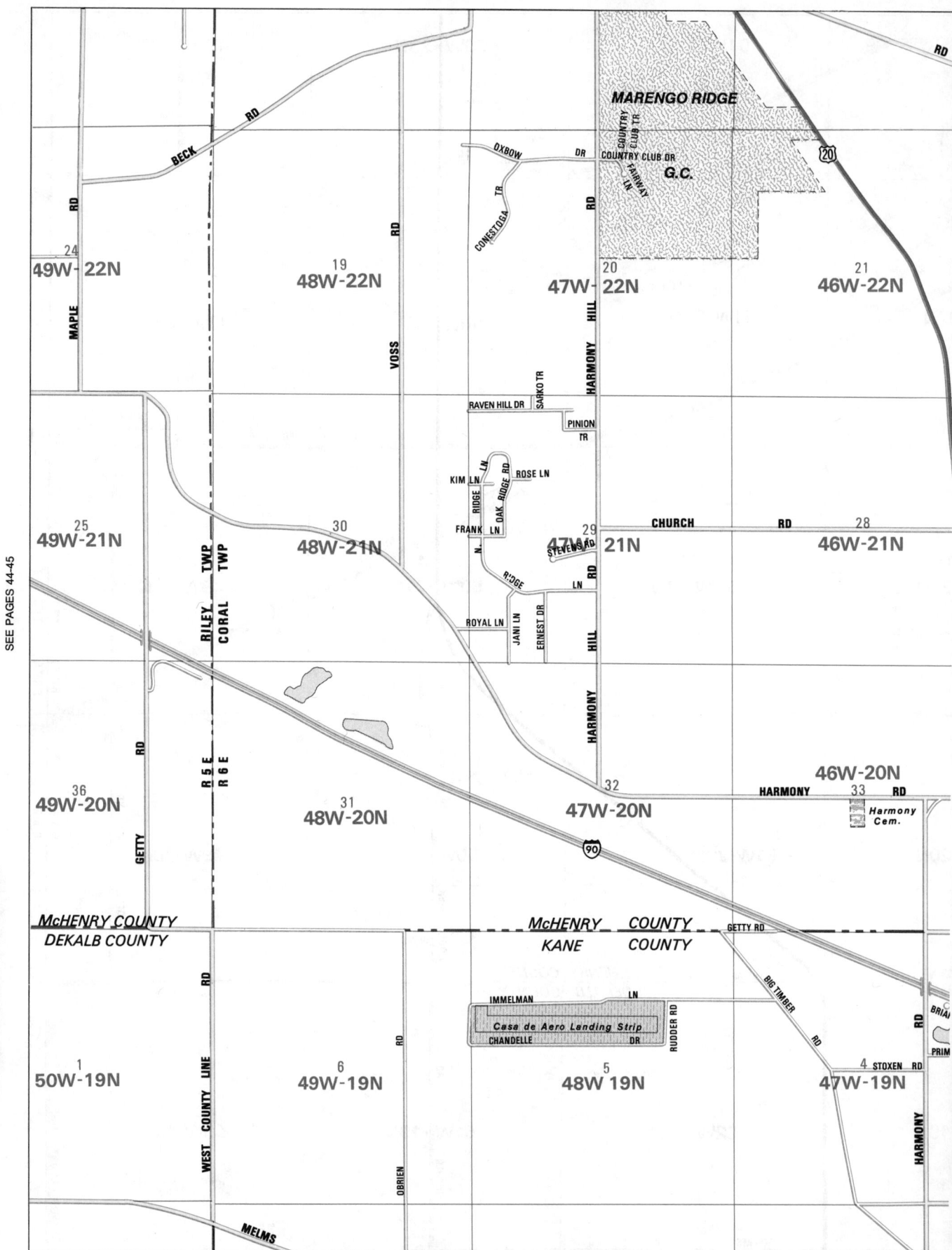

SEE PAGES 44-45

BECK RD

MAPLE RD

24
49W-22N

19
48W-22N

VOSS RD

OXBOW DR

CONESTOGA TR

COUNTRY CLUB TR
COUNTRY CLUB DR
FAIRWAY LN

MARENGO RIDGE

G.C.

20
47W-22N

HARMONY HILL RD

20

21
46W-22N

RAVEN HILL DR

SABKO TR

PINION TR

KIM LN
RIDGE LN
ROSE LN
OAK RIDGE RD
FRANK LN
N RIDGE RD

STEVENS RD

RIDGE LN

ROYAL LN
JANI LN
ERNEST DR

CHURCH RD

28

29
47W-21N

46W-21N

25
49W-21N

RILEY TWP
CORAL TWP

30
48W-21N

HARMONY HILL RD

R 5 E
R 6 E

GETTY RD

36
49W-20N

31
48W-20N

HARMONY HILL RD

32

47W-20N

HARMONY RD

33

46W-20N

Harmony Cem.

I-90

McHENRY COUNTY
DEKALB COUNTY

McHENRY COUNTY
KANE COUNTY

GETTY RD

RD

BIG TIMBER RD

BRIA

IMMELMAN LN

RUDDER RD

Casa de Aero Landing Strip

CHANDELLE DR

PRIM

1
50W-19N

WEST COUNTY LINE RD

OBRIEN RD

6
49W-19N

5
48W 19N

4 STOXEN RD

47W-19N

HARMONY

MELMS

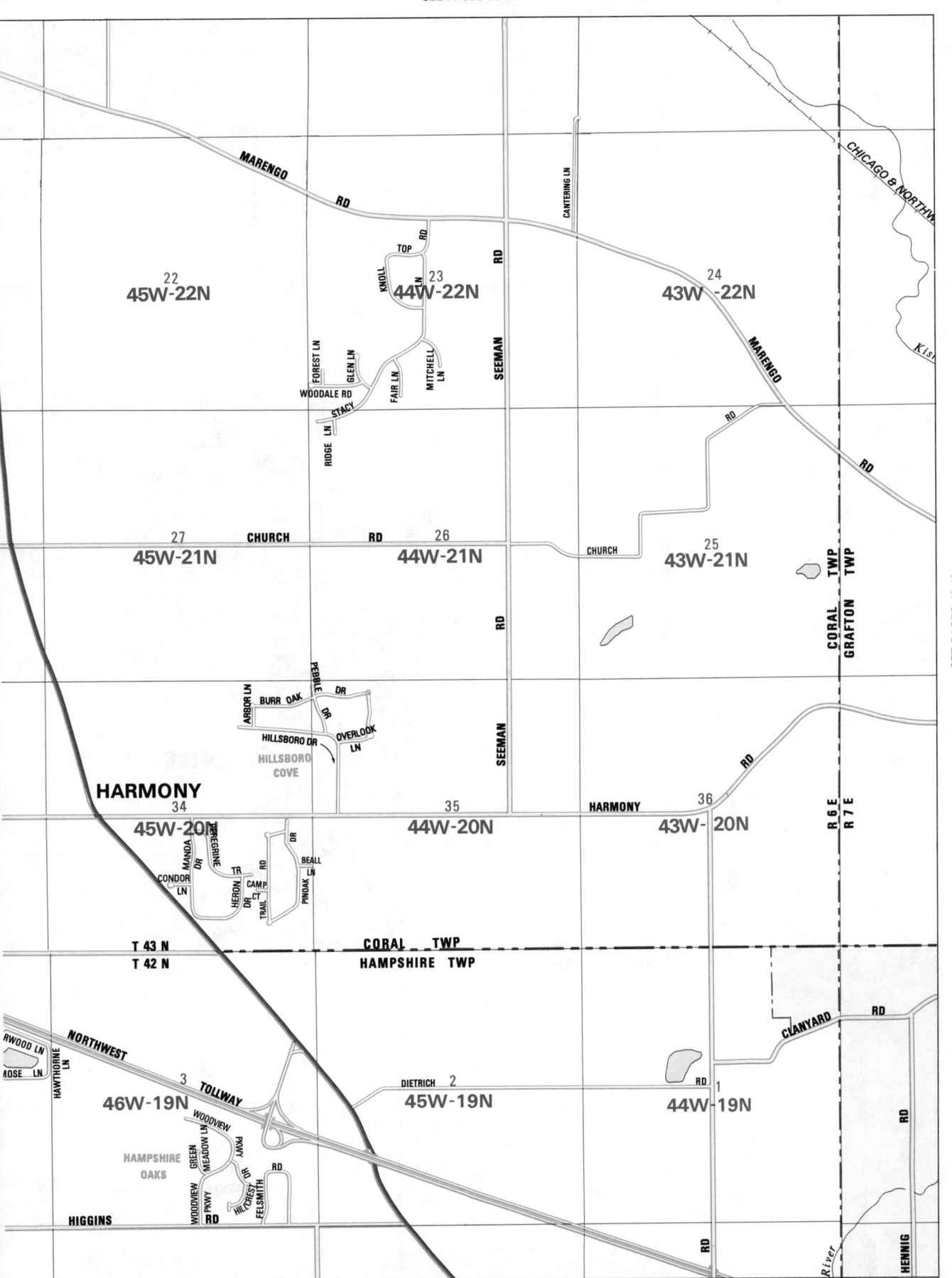

SEE PAGES 48-49

MARENGO RD

CHICAGO & NORTHW

22
45W-22N

TOP KNOLL RD

23
44W-22N

CANTERING LN

24
43W-22N

MARENGO RD

Kisi

FOREST LN
GLEN LN
WOODALE RD
FAIR LN
MITCHELL LN
SEEMAN RD

RIDGE LN
STACY

RD
RD

27
45W-21N
CHURCH RD
26
44W-21N

CHURCH

25
43W-21N

CORAL TWP
GRAFTON TWP

RD

SEEMAN RD

ARBOR LN
BURR OAK
PEBBLE DR
DR
OVERLOOK LN
HILLSBORO DR
HILLSBORO COVE

SEEMAN RD

RD

HARMONY
34

PEREGRINE DR
MANDA DR
CONDOR LN
HERON DR
CAMP CT
TRAIL
TR
RD
DR
PINOAK
BEALL LN

35
44W-20N

HARMONY
36
43W-20N

R 6 E
R 7 E

45W-20N

T 43 N
T 42 N

CORAL TWP
HAMPSHIRE TWP

CLANYARD RD

RWOOD LN
NOSE LN
HAWTHORNE LN
NORTHWEST
3
TOLLWAY
46W-19N

DIETRICH 2
45W-19N

RD 1
44W-19N

RD

WOODVIEW
GREEN MEADOW LN
PKWY
WOODVIEW PKWY
HILLCREST DR
FELSMITH
RD
HAMPSHIRE OAKS

HIGGINS RD

River

HENNIG

SEE PAGES 40-41

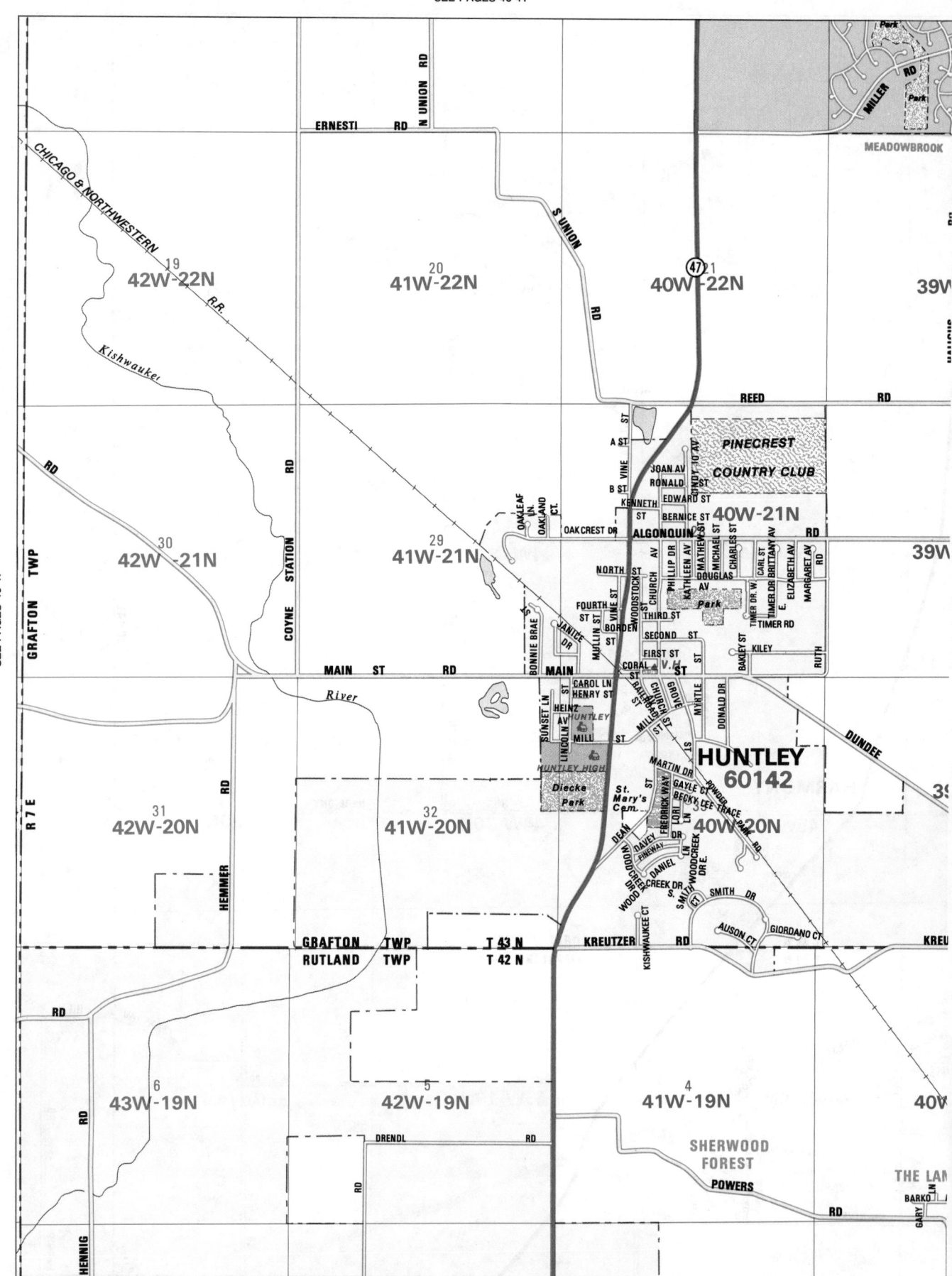

SEE PAGES 46-47

MEADOWBROOK

CHICAGO & NORTHWESTERN R.R.

Kishwaukee

N UNION RD

ERNESTI RD

S UNION RD

42W-22N 19

41W-22N 20

40W-22N 21
47

39W

REED RD

PINECREST COUNTRY CLUB

GRAFTON TWP

R 7 E

RD

COYNE STATION RD

A ST
VINE ST
B ST
JOAN AV
RONALD
KENNETH ST
EDWARD ST
BERNICE ST
CINDY JO AV

42W-21N 30

41W-21N 29

ALGONQUIN 40W-21N RD

OAKLEAF LN
OAKLAND CT.
OAKCREST DR

39W

NORTH ST
FOURTH ST
VINE ST
WOODSTOCK
CHURCH AV
PHILLIP DR
KATHLEEN AV
MATHEW ST
MICHAEL ST
DOUGLAS AV
CHARLES ST
CARL ST
TIMER DR W.
TIMER DR E.
BRITANY AV
ELIZABETH AV
MARGARET AV
RD

Park

THIRD ST
MULLIN ST
VINE ST
BORDEN ST
SECOND ST
FIRST ST
BAGLEY ST
KILEY
RUTH
TIMER RD

BONNIE BRAE
JANICE DR

MAIN ST RD

MAIN

CORAL ST
V.H.
DONALD DR

DUNDEE

River

SUNSET LN
HEINZ AV
LINCOLN AV

CAROL LN
HENRY ST
RAILROAD ST
CHURCH ST
MILL ST
MYRTLE ST
GROVE ST

HUNTLEY MILL

HUNTLEY 60142

HUNTLEY HIGH

Diecke Park

St. Mary's Cem.

MARTIN DR
GAYLE CT
LORI LN
BECKY LN
GARY LEE TRACE
SPINDLER RD
PARK RD
40W-20N

42W-20N 31

41W-20N 32

HEMMER RD

DEAN
FREDRICK WAY
PINEWAY
DAVEY
DANIEL
WOOD CT
CREEK DR
S. SMITH
WOODCREEK DR E.
SMITH DR
CT
AUSON CT.
GIORDANO CT.

KISHWAUKEE CT

GRAFTON TWP T 43 N
RUTLAND TWP T 42 N

KREUTZER RD

KREU

RD

43W-19N 6

42W-19N 5

41W-19N 4

40W

SHERWOOD FOREST

DRENDL RD

RD

POWERS

THE LAM

HENNIG

BARKO LN

GARY LN RD

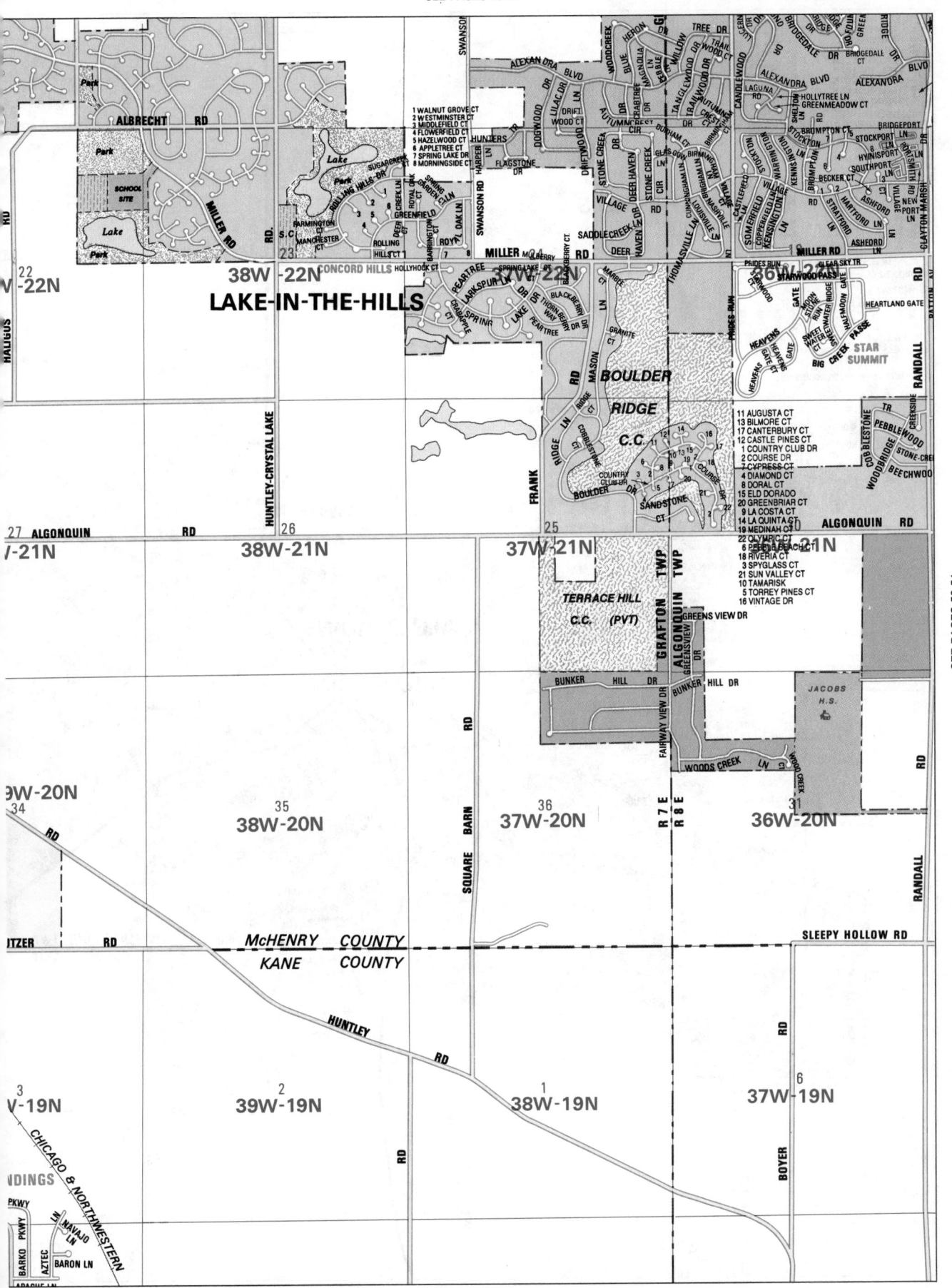

SEE PAGES 48-49

Airport

LAKE-IN-THE-HILLS **60102**

1 PLUM CT.
2 PORTSMOUTH CT.
3 PALM CT.
4 PEACHTREE CT. N.
5 PEACHTREE CT. S.
6 PEMBROOK CT. N.
7 PEMBROOK CT. S.
8 PENNY LN.

35W-22N
34W-22N
33W-22N
32W

35W-21N
34W-21N
33W-21N
32W

ALGONQUIN **60102**

35W-20N
34W-20N
33W-20

GOLF CLUB OF ILLINOIS

CARY

ALGONQUIN TWP
DUNDEE TWP

36W-19N
35W-19N
34W-19N
33W-19

WOODLAND SPRINGS

Fox River

VILLAGE HALL

PINGREE RD
VIRGINIA RD
PYOTT RD
KLASEN RD
COUNTY LINE RD. N.
EDGEWOOD
HANSON RD
SLEEPY HOLLOW
LINDSTROM LN
MAIN ST
BOLTZ RD

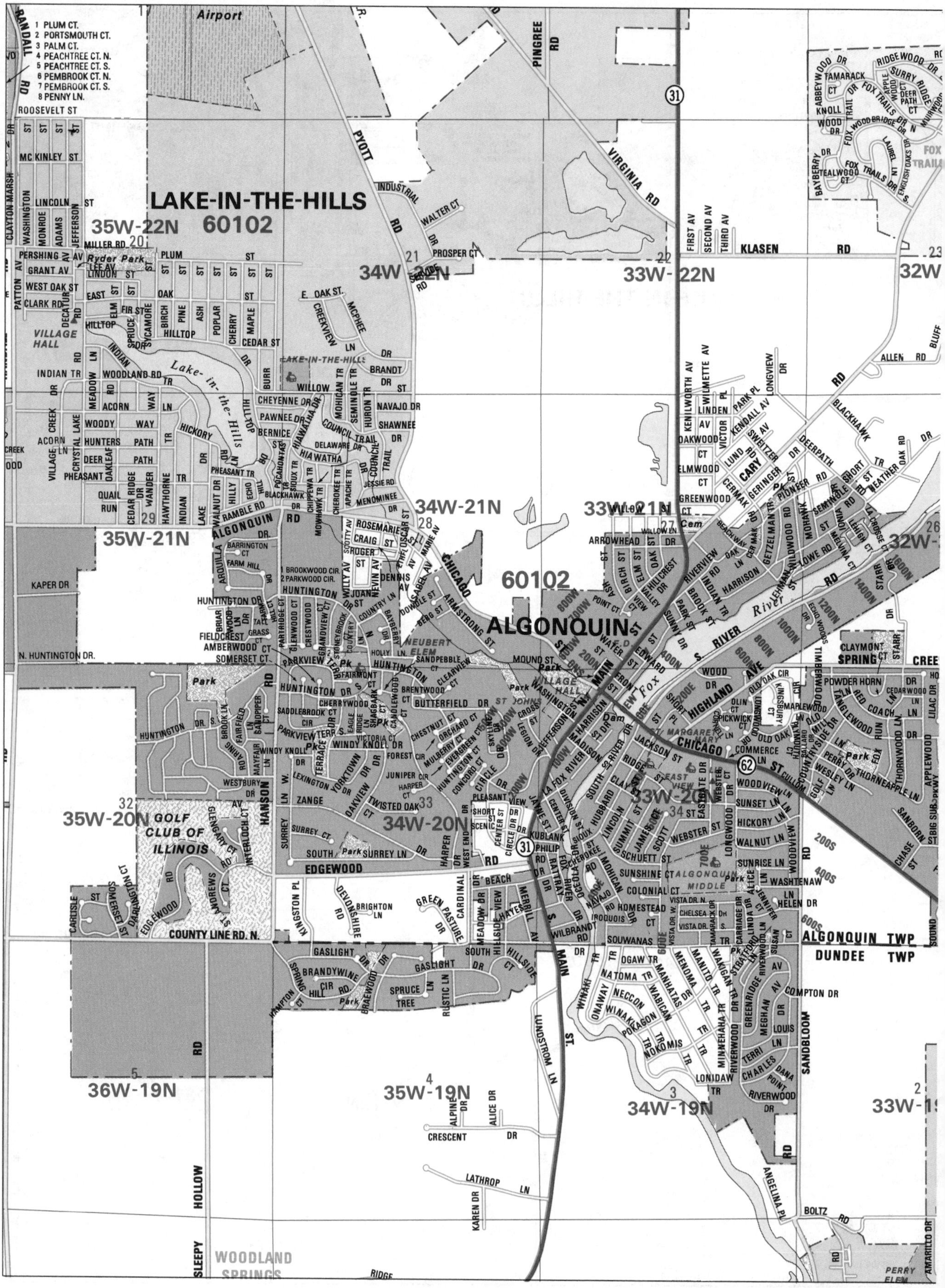

SEE PAGES 42-43

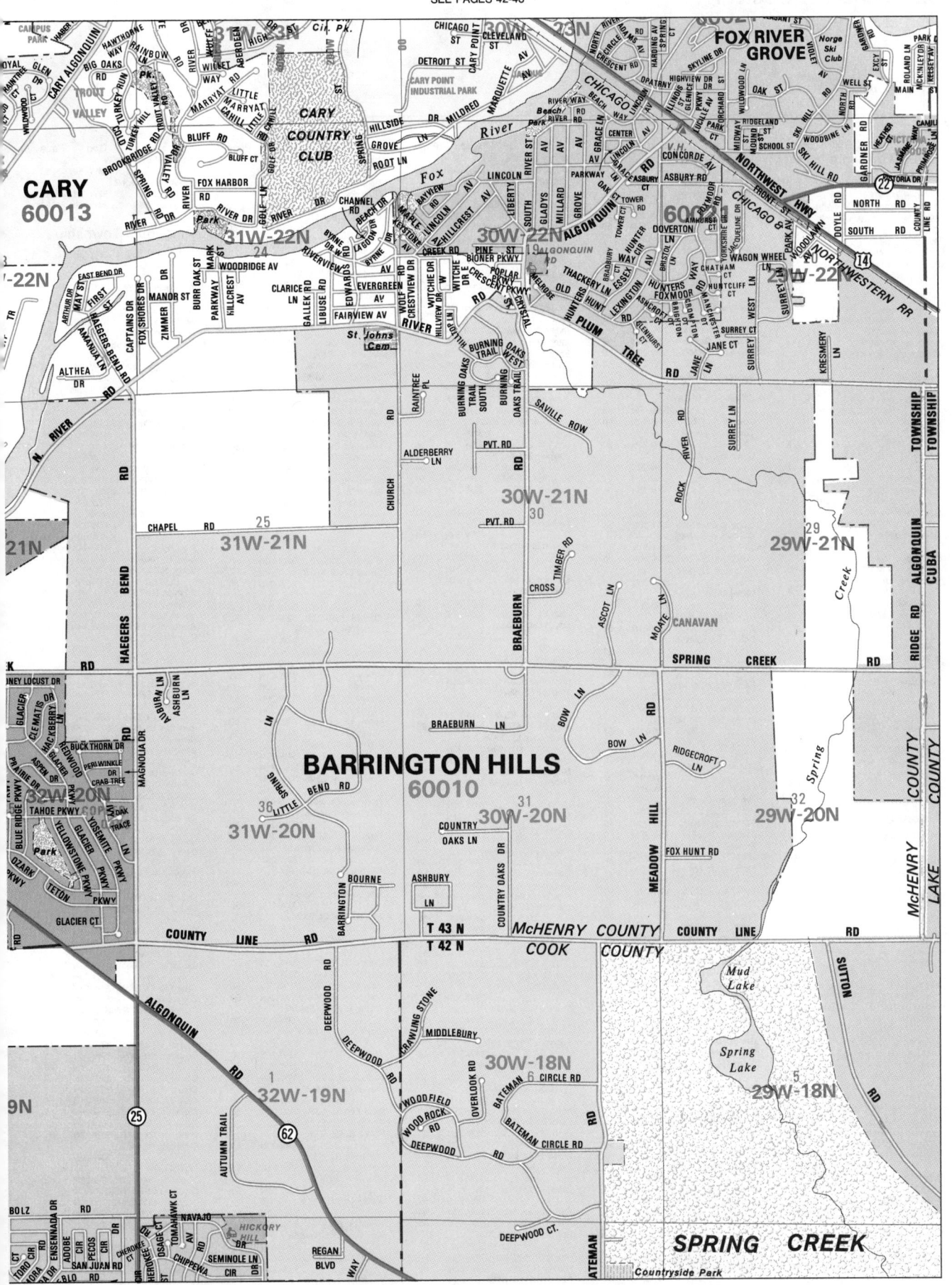

Street Index

Alden Township

	Grid	Page
Alden Rd. NW&SE		
........46W-43N,45W-38N	6
Altenberg Rd. E&W		
........48W-39N,45W-39N	6
Bissell Rd. N&S ... 43W-43N	7
Crowley Rd. E&W .. 48W-40N	6
Durkee Rd. E&W		
........45W-39N,43W-39N	7
Ferris Rd. NE&SW ..47W-40N	6
Fink Rd. N&S		
........44W-40N,44W-41N	7
French Dr. N&S 43W-41N	7
Gasch Rd. N&S		
........48W-42N,48W-43N	6
Green Rd. N&S		
........47W-38N,45W-38N	14
Green Valley Dr. NW&SE		
.... 43W-41N	7
Hebron Rd. E&W		
........48W-42N,43W-42N	6
Knickerbocker Rd. N&S		
........45W-41N,45W-43N	7
Lincoln Rd. N&S... 46W-38N	14
Manley Rd. N&S		
........43W-38N,43W-39N	15
Mansion Heights Dr. E&W		
.... 43W-41N	7
Nolan St. N&S......46W-40N	6
O'Brien Rd. E&W		
........46W-40N,43W-40N	6
Oak Grove Rd. E&W		
........46W-41N,46W-41N	6
Oak Springs Dr. N&S		
.... 46W-40N	6
Reese Rd. N&S		
........47W-40N,47W-42N	6
Righeimer Rd. E&W		
........48W-39N,47W-39N	6
Rt. 173 NE&SW		
........48W-38N,43W-41N	14
S. Mansion Heights Dr. N&S		
.... 43W-41N	7
Schultz Rd. N&S .. 48W-38N	14
Shields Rd. N&S		
........46W-38N,46W-39N	14
State Line Rd. E&W		
........49W-43N,44W-43N	5
Thayer Rd. E&W		
........44W-38N,43W-38N	15
Willow Lakes Ct. E&W		
.... 48W-42N	6
Willow Lakes Rd. N&S		
........48W-41N,48W-42N	6
Wright Rd. N&S		
........46W-41N,46W-42N	6

Algonquin*

(Also see Kane County)

	Grid	Page
Alice Ln. N&S 33W-20N	50
Amberwood Ct. N&S		
.... 35W-21N	50
Applewood Ln. N&S		
.... 32W-20N	50
Arguilla Dr. NS&EW		
.... 35W-21N	50
Armstrong St. NW&SE		
.... 34W-21N	50
Arrowhead Dr. E&W		
.... 33W-21N	50
Ash St. N&S 33W-21N	50
Aspen Dr. E&W ... 32W-20N	50
Ballard Dr. N&S .. 32W-20N	50
Barrington Ct. E&W		
.... 35W-21N	50
Bayberry Dr. N. N&S		
.... 34W-21N	50
Beach Dr. E&W .. 34W-20N	50
Beach Way NW&SE		
.... 34W-21N	50
Berg St. NW&SE .. 34W-21N	50
Big Sur Pkwy N&S 32W-20N	50
Birch St. N&S 33W-21N	50
Blue Ridge Pkwy N&S		
.... 32W-20N	50
Brentwood Ct. E&W		
.... 34W-20N	50
Briarwood Ln. NW&SE		
.... 35W-21N	49
Brook St. NW&SE . 33W-21N	50
Brookwood Cir. E&W		
.... 32W-20N	50
Buck Thorn Dr. E&W		
.... 34W-20N	50
Bunker Hill Dr. E&W		
........37W-20N,36W-20N	49
Butterfield Dr. E&W		
.... 34W-20N	50
Candlewood Ct. N&S		
.... 34W-20N	50

	Grid	Page
Carlisle St. NS&EW		
.... 35W-20N	50
Carriage Dr. N&S .. 33W-20N	50
Cedarwood Ln. E&W		
.... 32W-20N	50
Center St. NW&SE 33W-20N	50
Cermak Rd. N&S .. 33W-21N	50
Chase St. NE&SW..32W-20N	50
Chelsea Dr. E&W .. 33W-20N	50
Cherokee Rd. NE&SW		
.... 33W-20N	50
Cherrywood Ct. E&W		
.... 34W-20N	50
Chestnut Ct. NE&SW		
.... 34W-20N	50
Chicago St. NW&SE		
........34W-21N,33W-20N	50
Clay St. NW&SE...33W-20N	50
Cir. 32W-21N	50
Clearview Ct. NE&SW		
.... 33W-21N	50
Clematis Dr. NE&SW		
.... 32W-20N	50
Colonial Ct. E&W .. 33W-20N	50
Commerce Ln. N&S		
.... 34W-20N	50
Concord Ct. NE&SW		
.... 34W-20N	50
Cornell Ln. N&S 33W-20N	50
Country Ln. N&S .. 34W-21N	50
Countryside Dr. NE&SW		
.... 32W-20N	50
Crabtree Ln. E&W . 32W-20N	50
Crestwood Ct. N&S		
.... 34W-21N	50
Cross St. NE&SW .. 34W-20N	50
Cullom Ln. NW&SE		
.... 33W-20N	50
Darlington Ct. NE&SW		
.... 35W-20N	50
Deer Path Rd. NW&SE		
.... 32W-21N	50
Division St. N&S		
.... 33W-20N	50
Dowdle St. NW&SE		
.... 34W-21N	50
Eagle Ridge Ln. N&S		
.... 34W-20N	50
Eastgate Dr. N&S .. 33W-20N	50
Echo Woods Dr. NS&EW		
.... 32W-21N	50
Edgewood Rd. E&W		
........35W-20N,34W-20N	50
Edward St. NW&SE		
.... 33W-21N	50
Elm St. N&S 33W-21N	50
Evergreen Ct. NE&SW		
.... 34W-20N	50
Fairfield Ln. N&S....35W-20N	50
Fairmont Ct. E&W . 34W-20N	50
Fairway View Dr. N&S		
.... 37W-21N	49
Farm Hill Ct. NW&SE		
.... 35W-21N	50
Farm Hill Dr. E&W . 35W-21N	50
Fieldcrest Dr. N&S..35W-21N	50
Filip Rd. E&W 33W-20N	50
Forest Ct. NW&SE..34W-20N	50
Fox River Dr. N&S .. 33W-20N	50
Fox Run Ln. N&S .. 32W-20N	50
Front St. NW&SE .. 33W-20N	50
Getzelman Tr. N&S		
.... 33W-21N	50
Glacier Ct. E&W ... 32W-20N	50
Glacier Pkwy N&S . 32W-20N	50
Glengary Ct. NW&SE		
.... 35W-20N	50
Glenwood Ct. E&W		
.... 34W-21N	50
Golf Ln. NE&SW . 32W-20N	50
Grandview Ct. N&S		
.... 34W-21N	50
Greens View Dr. N&S		
........36W-20N,36W-21N	49
Hackberry Ln N&S..32W-20N	50
Hanson Rd. N&S .. 35W-20N	50
Harper Ct. NW&SE		
.... 34W-20N	50
Harper Dr. N&S 34W-20N	50
Harrison St. NW&SE		
........33W-20N,33W-21N	50
Hayes E&W 34W-20N	50
Helen Dr. E&W ... 34W-20N	50
Hickory Ln. E&W .. 33W-20N	50
Highland Av. NE&SW		
.... 33W-20N	50
Hillcrest Dr. NE&SW		
.... 33W-21N	50
Hillside Ln. N&S		
........35W-19N,34W-19N	50
Holly Ln. E&W...34W-21N	50
Homestead St. E&W		
.... 33W-20N	50

	Grid	Page
Honey Locust Dr. E&W		
.... 32W-20N	50
Hubbard St. NE&SW		
.... 34W-20N	50
Huntington Ct. NE&SW		
.... 34W-20N	50
Huntington Dr. NW&SE		
.... 34W-20N	50
Huntington Dr. E&W		
.... 35W-21N	50
Huntington Dr. N. NW&SE		
.... 34W-21N	50
Huntington Dr. S. E&W		
........35W-20N,34W-20N	50
Inca St. N&S 32W-21N	50
Indian Tr. NW&SE .. 33W-20N	50
Interloch Ct. N&S .. 35W-20N	50
Ionia Tr. N&S 32W-21N	50
Ivy Ln. NW&SE ... 32W-20N	50
Jackson St. NW&SE		
.... 33W-20N	50
James Ct. NE&SW 33W-20N	50
Jayne St. N&S 33W-20N	50
Jefferson St. NE&SW		
.... 33W-20N	50
Jennifer Ct. N&S .. 33W-20N	50
Juniper Ct. NW&SE		
.... 34W-20N	50
Kingsbury Ct N&S .. 33W-20N	50
Kublank E&W 33W-20N	50
La Crosse Ct. N&S		
.... 32W-21N	50
La Fox River Dr. NE&SW		
.... 33W-20N	50
Legion Dr. NW&SE		
.... 34W-20N	50
Lehigh Ct. N&S ... 32W-21N	50
Lehman St. N&S .. 33W-21N	50
Lexington Dr. E&W		
.... 34W-20N	50
Lilac Dr. N&S ... 32W-20N	50
Lincoln St. N&S		
.... 33W-20N	50
Linda Dr. N&S ... 33W-20N	50
Longwood Dr. N&S		
.... 33W-20N	50
Lowe Rd. NE&SW . 32W-21N	50
Madison St. NW&SE		
.... 33W-20N	50
Magnolia Dr. E&W . 32W-20N	50
Main St. NE&SW .. 34W-20N	50
Maplewood Ln. E&W		
........33W-20N,32W-20N	50
Mayfair Ln. N&S .. 35W-20N	50
Meadow Dr. N&S .. 34W-20N	50
Melvina Ct. N&S		
.... 32W-21N	50
Merrill Av. N&S .. 34W-20N	50
Mohawk St. N&S .. 32W-21N	50
Mound St. NE&SW		
........33W-20N,33W-21N	50
Mulberry Ct. NE&SW		
.... 34W-20N	50
Oak Ln. NE&SW .. 33W-21N	50
Oak St. N&S 33W-21N	50
Oak Trace E&W 32W-20N	50
Oakview Dr. NE&SW		
.... 34W-20N	50
Oceola Dr. N&S 33W-20N	50
Old Mill Ln. NE&SW		
.... 32W-20N	50
Cir. 33W-20N	50
Olin Ct. E&W ... 33W-20N	50
Orchard Ct. NE&SW		
.... 34W-20N	50
Ozark Pkwy. NW&SE		
.... 32W-20N	50
Park St. NW&SE .. 33W-21N	50
Parkview Terr. N&S		
.... 34W-20N	50
Parkview Terr. S. E&W		
.... 34W-20N	50
Parkwood Cir. E&W		
.... 34W-21N	50
Partridge Ct. N&S .. 34W-21N	50
Perry Dr. NW&SE . 32W-20N	50
Perwinkle Dr. E&W 32W-20N	50
Philip Rd. E&W ... 33W-20N	50
Pickwick Ct. E&W . 33W-20N	50
Pioneer Rd. NE&SW		
.... 33W-21N	50
Pleasant View Dr. E&W		
.... 34W-20N	50
Plymouth Ct. N&S . 33W-20N	50
Point Ct. N&S ... 33W-20N	50
Powder Horn Dr. E&W		
.... 32W-20N	50
Prairie Dr. NW&SE .32W-20N	50
Rattray Dr. N&S....33W-20N	50
Red Coach Ln. E&W		
.... 32W-20N	50
Redwood Dr. NW&SE		
.... 32W-20N	50

	Grid	Page
Ridge Ln. N&S 34W-20N	50
Ridge St. NW&SE .. 33W-20N	50
River Rd. NE&SW .. 33W-21N	50
Riverview Rd. NE&SW		
.... 33W-21N	50
Roaring Brook Ln. N&S		
.... 35W-20N	50
Rt. 62 NW&SE ... 33W-20N	50
Saddlebrook Cir. N&S		
.... 34W-20N	50
Sanborn St. NW&SE		
.... 32W-20N	50
Sandbloom Rd. N&S		
.... 32W-20N	50
Sandpebble Ct. NW&SE		
.... 34W-21N	50
Sandpiper Ct. N&S 35W-20N	50
Schuett St. E&W .. 33W-20N	50
Scott St. NE&SW .. 34W-20N	50
Seminole Rd. NE&SW		
.... 33W-20N	50
Shagbark Ct. N&S.. 34W-20N	50
Short Pl. E&W 33W-20N	50
Sioux St. NE&SW .. 33W-20N	50
Skyline Dr. N&S .. 34W-20N	50
Somerset Ct. N&S .. 35W-21N	50
Somerset St. N&S		
.... 35W-20N	50
Sound Rd. N&S ... 32W-20N	50
South Dr. E&W ... 35W-20N	50
South River Dr. NE&SW		
.... 33W-20N	50
South Surrey Ln. E&W		
.... 34W-20N	50
Souwanas Tr. E&W		
.... 33W-20N	50
Spring Creek Rd. E&W		
.... 32W-21N	50
St. Andrews Ct. N&S		
.... 35W-20N	50
Starr Dr. N&S 32W-21N	50
Stoneybrook Ct. N&S		
.... 34W-21N	50
Summit St. NE&SW		
.... 33W-20N	50
Sunny Dr. NW&SE..33W-21N	50
Sunrise Ln. E&W .. 33W-20N	50
Sunset Ln. E&W ... 33W-20N	50
Sunshine Ct. E&W . 33W-20N	50
Surrey Ct. E&W ... 34W-20N	50
Surrey Ln. E&W....34W-20N	50
Surrey Ln. W. N&S		
.... 34W-20N	50
Susan Ct. N&S 33W-20N	50
Tahoe Pkwy. E&W.. 32W-20N	50
Tallgrass Ct. N&S .. 35W-21N	50
Tamarack Dr. N&S.. 33W-20N	50
Tanglewood Dr. E&W		
.... 32W-20N	50
Terrace Dr. N&S....34W-20N	50
Thorneapple Ln. NW&SE		
.... 32W-20N	50
Thornwood Dr. N&S		
.... 32W-20N	50
Timberwood Ln. N&S		
.... 33W-20N	50
Twisted Oak Ct. E&W		
.... 34W-20N	50
Valley View Ct. NW&SE		
.... 33W-21N	50
Victoria Ct. E&W .. 33W-20N	50
Vista Dr. N. E&W .. 33W-20N	50
Vista Dr. W E&W .. 33W-20N	50
Walnut Ln. E&W ... 33W-20N	50
Washington St. NW&SE		
.... 33W-20N	50
Washtenaw Ln. E&W		
.... 33W-20N	50
Water St. NW&SE . 33W-21N	50
Webster Ct. N&S .. 33W-20N	50
Webster St. N&S .. 33W-20N	50
Wesley Ln. NW&SE		
.... 32W-20N	50
West End Dr. N&S.. 34W-20N	50
Westbury Dr. E&W 35W-20N	50
Wilbrandt Rd. E&W		
.... 33W-20N	50
Wildwood Rd. N&S		
.... 33W-21N	50
Willow Ln. NW&SE		
.... 33W-21N	50
Windy Knoll Dr. E&W		
.... 34W-20N	50
Wood Dr. E&W ... 33W-20N	50
Woodcreek Ct. NW&SE		
.... 36W-20N	49
Woods Creek Ln. E&W		
.... 36W-20N	49
Woodview Ln. E&W		
.... 33W-20N	50
Yellowstone Pkwy. NW&SE		
.... 32W-20N	50

	Grid	Page
Yorktown Dr. NE&SW		
.... 34W-20N	50
Yosemite Pkwy. N&S		
.... 32W-20N	50
Zange Dr. E&W .. 34W-20N	50

Algonquin Township

	Grid	Page
Abbotts Ln. N&S .. 29W-23N	43
Algonquin Rd. N&S		
.... 30W-22N	51
Algonquin Rd. E&W		
........36W-21N,35W-21N	49
Allen Rd. E&W ... 32W-22N	50
Amy Dr. E&W ... 33W-25N	42
Arthur Dr. NS&EW .. 32W-22N	50
Ashford Ln. NW&SE		
.... 36W-22N	49
Autumn Ln. E&W .. 32W-22N	50
Balder St. N&S ... 29W-23N	43
Bayport Ln. N&S .. 29W-23N	43
Beach Dr. NE&SW ..31W-22N	51
Becker Ct. N&S 36W-22N	49
Big Creek Passe NE&SW		
.... 36W-22N	49
Big Oaks Rd. NS&EW		
.... 32W-23N	42
Bioner Pkwy. E&W 30W-22N	51
Birch St. N&S 33W-24N	42
Blackhawk Tr. NW&SE		
.... 32W-21N	50
Blarney Stone Rd. NW&SE		
.... 31W-25N	43
Bluff Ct. E&W ... 31W-22N	51
Bluff Rd. E&W ... 31W-22N	51
Bluff Tr. N&S ... 32W-22N	50
Bridgeport Ln. E&W		
.... 36W-22N	49
Brighton Ln. E&W . 34W-20N	50
Brisbane Dr. E&W . 29W-24N	43
Bristol Way E&W .. 30W-24N	43
Bristol Way E&W .. 29W-24N	43
Brompton Ct. E&W		
.... 36W-22N	49
Brompton Rd. N&S		
.... 36W-22N	49
Brookbridge Rd. NE&SW		
........32W-22N,31W-22N	50
Burman Dr. E&W .. 33W-25N	42
Burr Oak Ct. N&S .. 33W-24N	42
Burr Oak St. N&S .. 31W-22N	51
Byrne Dr. NE&SW .. 31W-22N	51
Byrne Dr. W. E&W..31W-22N	51
Cahill Rd. NW&SE . 31W-23N	43
Captains Dr. N&S .. 32W-22N	50
Cardinal Dr. N&S .. 34W-20N	50
Carrie Ct. NE&SW .. 33W-25N	42
Cary Rd. NE&SW .. 33W-21N	50
Cary-Algonquin Rd. NE&SW		
........32W-22N,31W-22N	50
Center St. N&S 34W-20N	50
Cermak Rd. NW&SE		
.... 33W-21N	50
Channel Rd. NE&SW		
.... 31W-22N	51
Chestnut Ln. N&S . 29W-23N	43
Church Ln. NS&EW		
.... 32W-25N	42
Churchill Ct. N&S . 33W-25N	42
Circle Dr. N&S......34W-20N	50
Claire St. E&W ... 33W-24N	42
Clarice Ln. E&W 31W-22N	51
Clark St. E&W.....29W-23N	43
Clear Sky Tr. E&W . 36W-22N	49
Cog Circle N&S ... 33W-25N	42
Cold Spring Rd. NW&SE		
........32W-22N,31W-22N	50
Colonial Av. N&S .. 33W-24N	42
Cougar Tr. E&W ... 30W-24N	43
Country Ln. E&W .. 29W-24N	43
Cove Dr. N&S.....29W-24N	43
Cowlin St. N&S ... 33W-24N	42
Crabtree Ln. E&W .. 30W-24N	43
Craig St. E&W ... 34W-21N	50
Creek Ln. E&W ... 31W-25N	43
Creek Rd. E&W ... 31W-25N	43
Crestview Dr. N&S..30W-22N	51
Crestview Rd. N&S		
.... 30W-22N	51
Crystal Lake Av. E&W		
........34W-25N,32W-25N	42
Crystal Lake Rd. NS&EW		
........31W-25N,30W-25N	43
Crystal St. N&S ... 30W-22N	51
Daniel Dr. E&W ... 33W-25N	42
David Ln. E&W ... 33W-25N	42
Deerpath Rd. NW&SE		
.... 32W-21N	50
Dennis Av. N&S .. 34W-21N	50
Devonshire Rd. N&S		
.... 34W-20N	50
Diamond Ln. E&W.. 29W-23N	43
Dove Way N&S 30W-24N	43

Street	Grid	Page
Doyle Rd. N&S	29W-22N	51
Drive-In Ln. E&W	33W-24N	42
Dublin Ct. NE&SW	31W-25N	43
Dublin Dr. NW&SE	31W-25N	43
E. Monticello Way N&S	33W-24N	42
Edwards Rd. N&S	31W-22N	51
Elk Tr. E&W	30W-24N	43
Elmwood Ct. E&W	33W-21N	50
Ethel Av. N&S	34W-21N	50
Evergreen Av. E&W	31W-22N	51
Excy St. N&S	29W-23N	43
Fairfax Dr. E&W	33W-24N	42
Fairview Av. E&W	31W-22N	51
First Av. N&S	33W-22N	50
First St. NE&SW	32W-22N	50
Fox Harbor Rd. E&W	31W-22N	51
Fox River Grove Rd. E&W	30W-22N	51
Fox Shores Dr. N&S	31W-22N	51
Gallek Rd. N&S	31W-22N	51
Garden Ln. E&W	32W-22N	50
Gardner Rd. N&S	29W-22N	51
Geringer Dr. NE&SW	33W-21N	50
Golf Course Rd. N&S	36W-22N	49
Golf Dr. N&S	31W-22N	51
Golf Ln. N&S	31W-22N	51
Golf View Dr. NS&EW	33W-24N	42
Grand Dr. E&W	29W-23N	43
Green Ln. E&W	31W-25N	43
Green Meadow Ln. N&S	32W-25N,31W-25N	42
Green Meadow Ln. E&W	32W-25N,31W-25N	42
Green Pasture Dr. NW&SE	34W-20N	50
Green Ridge Rd. E&W	31W-25N	43
Greensview Dr. N&S	36W-21N	49
Greenwood Ct. E&W	33W-21N	50
Grove Ln. E&W	31W-22N	51
Haegers Bend Rd. N&S	32W-20N,32W-22N	50
Halfmoon Gate N&S	36W-22N	49
Hartford Ln. NW&SE	36W-22N	49
Hawthorne Dr. NE&SW	32W-23N,31W-25N	42
Hawthorne Way NE&SW	32W-23N	42
Heather Dr. NE&SW	32W-21N	50
Heatland Gate E&W	36w-22N	49
Heavens Gate NE&SW	36W-22N	49
Heavens Gate Ct. NW&SE	36W-22N	49
Hemlock St. N&S	33W-24N	42
Hickory Grove Ln. E&W	29W-25N	43
Hickory Grove Rd. E&W	29W-25N	43
Highland Dr. E&W	32W-25N,31W-25N	42
Hillary Ln. E&W	30W-25N	43
Hillcrest Av. N&S	31W-22N	51
Hillcrest Ln. E&W	33W-24N	42
Hillcrest Rd. N&S	31W-25N	43
Hillside Dr. E&W	31W-22N	51
Hilltop Ln. N&S	30W-22N	51
Hillview Dr. N&S	30W-22N	51
Holly Lynn Dr. E&W	29W-24N	43
Holly Lynn Rd. E&W	29W-24N	43
Hunter Path N&S	31W-25N	43
Hyinisport Ln. E&W	36W-22N	49
Ida Dr. NW&SE	31W-25N	43
Iroquois E&W	33W-20N	50
Isabel Av. NE&SW	34W-21N	50
Joan St. E&W	34W-21N	50
John St. N&S	34W-25N	42
Kendall Av. NE&SW	33W-21N	50
Kenilworth Av. N&S	33W-21N	50
Kenneth Dr. E&W	33W-25N	42
Kensington Ln. N&S	36W-22N	49
Kent Ave. E&W	34W-25N	42
Kent St. N&S	34W-25N	42
Kilkenny Ct. NW&SE	32W-25N	42
Killarney Dr. E&W	32W-25N,31W-25N	42
Kings Bridge Dr. N&S	30W-25N	43
Kingsbury Ct. N&S	33W-20N	50
Kingston Pl. N&S	34W-20N	50
Klasen Rd. E&W	33W-22N	50
Knoll Dr. N&S	30W-24N	43
Lagoon Dr. N&S	29W-23N	43
Lagoon Dr. NE&SW	31W-22N	51
Lake Shore Dr. N&S	31W-25N	43
Lakewood Dr. NS&EW	32W-25N,31W-25N	42
Laurie Ct. E&W	33W-25N	42
Laurie Ln. E&W	33W-25N	42
Leah Dr. E&W	29W-25N	43
Libuse Rd. N&S	31W-22N	51
Lilac Ln. E&W	29W-23N	43
Limerick Dr. NE&SW	31W-25N	43
Linden Av. E&W	33W-21N	50
Lindsay Ln. NS&EW	32W-25N	42
Little Cahill N&S	31W-22N,31W-23N	51
Little Marayat E&W	31W-23N	43
Little Marryat E&W	31W-23N	43
Little Marryat Ct. E&W	31W-23N	43
Londonderry Dr. N&S	31W-25N	43
Longview Dr. N&S	33W-22N	50
Lund Av. NE&SW	33W-21N	50
Machelle Dr. E&W	29W-24N	43
Main St. E&W	29W-23N	43
Mallard Stag Tr. NS&EW	30W-24N	43
Manor Rd. N&S	33W-24N	42
Manor St. E&W	31W-22N	51
Marayat Rd. NE&SW	31W-23N	43
Marie Av. NE&SW	34W-21N	50
Mark St. N&S	31W-22N	51
Marryat Rd. NE&SW	31W-23N	43
May Av. N&S	32W-22N	50
May St. N&S	32W-22N	50
McKinley Dr. N&S	29W-23N	43
Meadow Ln. N&S	29W-24N	43
Meandering Way E&W	33W-25N,32W-25N	42
Middle Ln. E&W	31W-25N	43
Miller Rd. E&W	36W-22N	49
Mink Tr. E&W	30W-24N	43
Mohican Dr. NW&SE	33W-20N	50
Moonstone Run NE&SW	36W-22N	49
N. Wynwood Dr. N&S	32W-25N	42
Nancy Dr. NS&EW	33W-25N	42
Navajo Rd. NE&SW	33W-20N	50
Nevin Av. N&S	34W-21N	50
Newbold Rd. N&S	29W-23N,29W-24N	43
Newport Ln. E&W	36W-22N	49
Nighthawk Way N&S	30W-24N	43
North Ln. E&W	31W-25N	43
North Rd. E&W	29W-22N	51
Northlane Rd. E&W	31W-25N	43
Oak Rd. N&S	32W-21N	50
Oakbrook Dr. E&W	33W-24N	42
Oakcrest Rd. E&W	29W-24N	43
Oakdale Ter. E&W	31W-25N	43
Oakwood Ct. E&W	33W-21N	50
Old Post Rd. N&S	33W-24N	42
Oscar St. N&S	34W-21N	50
Otter Tr. E&W	30W-24N	43
Owl Way N&S	30W-24N	43
Park Dr. E&W	29W-23N	43
Park Pl. NE&SW	33W-21N	50
Parkway N&S	31W-22N	51
Parkway Dr. E&W	29W-23N	43
Pheasant Dr. N&S	31W-25N	43
Pleasant View Ln. NW&SE	31W-25N	43
Poplar Pkwy. NW&SE	30W-22N	51
Poplar Pkwy. E&W	30W-22N	51
Portsmith Rd. N&S	36W-22N	49
Powers St. E&W	29W-23N	43
Preston Pl. NS&EW	33W-25N	42
Prides Run N&S	36W-22N	49
Pringee Rd. N&S	33W-24N,33W-25N	42
Rainbow Ln. E&W	31W-23N	43
Ramble Rd. NS&EW	31W-25N	43
Randall Rd. N&S	36W-20N,36W-22N	49
Rawson Bridge Rd. N&S	30W-24N,30W-25N	43
Rawson Bridge Rd. S. N&S	29W-24N	43
Raymond Dr. E&W	33W-24N	42
Red Cypress Dr. E&W	29W-24N	43
Redwood St. N&S	33W-24N	42
Richard Dr. N&S	33W-25N	42
Ridge Rd. N&S	29W-20N	51
River Dr. N&S	31W-23N	43
River Dr. E&W	32W-22N,31W-22N	50
River Rd. E&W	30W-22N	51
River Way N&S	29W-23N	43
Riverview Av. E&W	31W-22N	51
Riverview Dr. N&S	29W-23N	43
Robert Rd. N&S	31W-25N	43
Robin Ln. N&S	33W-25N	42
Rodger St. E&W	34W-21N	50
Roland Dr. N&S	29W-23N	43
Roland Ln. N&S	29W-23N	43
Rona Rd. E&W	31W-25N	43
Ronald Ln. E&W	33W-25N	42
Roosevelt St. E&W	35W-23N	50
Root Ln. E&W	31W-22N	51
Rose St. E&W	33W-24N	42
Rosemarie St. E&W	34W-21N	50
Rt. 31 N&S	34W-20N,33W-23N	50
Russel St. N&S	29W-23N	43
S. Kilkenny Dr. NS&EW	32W-25N	42
S. Nancy Dr. E&W	33W-25N	42
Saggers Ln. E&W	33W-24N	42
Sands Rd. N&S	33W-24N,33W-25N	42
Sauwanas Tr. E&W	33W-20N	50
Scenic Dr. E&W	34W-20N	50
Scott Ct. E&W	33W-25N	42
Scott Ln. N&S	33W-25N	42
Scotty Av. N&S	34W-21N	50
Second Av. N&S	33W-22N	50
Shales St. N&S	33W-24N	42
Shannon Dr. N&S	31W-25N	43
Short St. E&W	34W-20N	50
Short St. NE&SW	32W-21N	50
Sleepy Hollow Rd. E&W	36W-20N,35W-20N	49
Smith Rd. N&S	33W-25N	42
Sorrel Ter. E&W	33W-25N	42
South Blue Ct. N&S	32W-25N	42
South Rd. E&W	29W-22N	51
Southlane E&W	31W-25N	43
Southport Ln. E&W	36W-22N	49
Spring Beach Way E&W	30W-23N	43
Spring Creek Rd. E&W	32W-21N	50
Spring St. N&S	31W-22N	51
Squirrel Tr. E&W	30W-24N	43
Starr Dr. N&S	32W-21N	50
Starwood Pass E&W	36W-22N	49
State St. E&W	33W-24N	42
Steeple Run Dr. E&W	33W-25N,32W-25N	42
Stockport Ln. E&W	36W-22N	49
Stockton Ln. NS&EW	36W-22N	49
Stonegate Rd. N&S	31W-23N	43
Stratford Ln. NW&SE	36W-22N	49
Summer Dr. NE&SW	32W-22N	50
Sun Valley Dr. NE&SW	31W-25N	43
Swan Way N&S	30W-24N	43
Sweetwater Ct. NE&SW	36W-22N	49
Sweetwater Ridge N&S	36W-22N	49
Sweitzer NW&SE	33W-21N	50
Teki Dr. E&W	33W-24N	42
Terrace Dr. NE&SW	33W-24N	42
Third Av. N&S	33W-22N	50
Thistle Ct. N&S	29W-25N	43
Thomas St. N&S	34W-25N	42
Three Oaks Rd. E&W	33W-24N,29W-24N	42
Topview Rd. NW&SE	31W-25N	43
Tower Dr. E&W	30W-22N	51
Tower Ln. E&W	30W-23N	43
Triangle Wood Dr. EW&NS	30W-24N	43
Trout Valley Rd. N&S	31W-23N	43
Turkey Hill NE&SW	32W-22N,31W-23N	50
Turkey Run N&S	32W-23N	42
Tuxedo Ln. NW&SE	29W-25N	43
U.S. Route 14 E&W	32W-24N,29W-22N	42
Valerie Dr. E&W	33W-25N	42
Valholla Circle NW&SE	33W-24N	42
Valley Rd. N&S	31W-22N	51
Victor Pl. N&S	33W-21N	50
Village Ct. N&S	36W-22N	49
Village Rd. E&W	36W-22N	49
Virginia Rd. NW&SE	33W-22N	50
Voce Ct. N&S	29W-23N	43
W. Rawson Bridge Rd. N&S	31W-25N	43
Wallace Dr. E&W	29W-23N	43
Wallen Dr. E&W	33W-25N	42
Walnut Dr. E&W	31W-25N	43
Wander Way N&S	31W-25N	43
Warrington Ln. N&S	36W-22N	49
White Pine Dr. N&S	30W-24N	43
Wichie Dr., East N&S	30W-22N	51
Wichie Dr., West N&S	30W-22N	51
Wildberry Ln. N&S	29W-25N	43
Willet Way E&W	31W-23N	43
Willow St. E&W	33W-21N	50
Willy Av. N&S	34W-21N	50
Wilmette Av. N&S	33W-21N	50
Winding Rd. N&S	32W-26N	42
Witchie Dr. N&S	30W-22N	51
Witchie Dr. E. N&S	30W-22N	51
Witchie Dr. W. N&S	30W-22N	51
Wium Av. N&S	29W-23N	43
Wolf Rd. N&S	31W-22N	51
Wood St. N&S	31W-25N	43
Woodbine Rd. E&W	29W-23N	43
Woodlawn Av. NE&SW	29W-22N	51
Woodridge Av. E&W	31W-22N	51
Wyndwood Dr. NS&EW	33W-25N,32W-25N	42
Zimmer Dr. N&S	31W-22N	51
1st St. NE&SW	32W-22N	50

Barrington Hills*
(Also see Cook, Kane and Lake Counties)

Street	Grid	Page
Alderberry Ln. E&W	30W-21N	51
Cir.	30W-20N	51
Auburn Ln. N&S	31W-20N	51
Barrington Bourne NS&EW	31W-20N	51
Bow Ln. NW&SE	30W-20N	51
Braeburn Ln. E&W	30W-20N	51
Braeburn Rd. N&S	30W-21N	51
Burning Oaks Trail NW&SE	30W-22N	51
Burning Oaks Trail South NW&SE	30W-22N	51
Burning Oaks Trail West E&W	30W-22N	51
Chapel Rd. E&W	31W-21N	51
Church Rd. N&S	31W-21N	51
Country Oaks Dr. N&S	30W-20N	51
Country Oaks Ln. E&W	30W-20N	51
County Line Rd. E&W	31W-20N,29W-20N	51
Crescent Pkwy. NW&SE	30W-22N	51
Cross Timber Rd. NS&EW	30W-21N	51
Fox Hunt Rd. E&W	29W-20N	51
Front St. NW&SE	29W-22N	51
Haegers Bend Rd. N&S	32W-21N	51
Jane Ct. E&W	29W-22N	51
Jane Ln. NE&SW	29W-22N	51
Kresmery Ln. N&S	29W-22N	51
Little Bend Rd. NS&EW	31W-20N	51
Meadow Hill Rd. N&S	30W-20N	51
N. County Line Rd. E&W	31W-20N	51
Park Av. N&S	29W-22N	51
Phillip Ter. E&W	30W-21N	51
Plum Tree Rd. E&W	30W-22N	51
Pvt. Rd. NW&SE	29W-21N	51
Pvt. Rd. E&W	30W-21N,26W-18N	51
Pvt. Rd. N&S	30W-21N	51
Raintree Dr. EW&NS	30W-22N	51
Raintree Pl. E&W	30W-22N	51
Ridge Rd. N&S	29W-21N	51
Ridgecroft Ln. NS&EW	29W-20N	51
Rock Ridge Rd. N&S	29W-21N	51
Saville Row NW&SE	30W-21N	51
Spring Creek Rd. E&W	30W-21N	51
Spring Ln. N&S	31W-20N	51
Sunny Meridian N&S	30W-20N	51
Surrey Ct. E&W	29W-22N	51
Surrey Ln. N&S	29W-21N	51
Surrey Ln. NS&EW	29W-22N	51
Surrey West N&S	29W-22N	51
Wagon Wheel Ln. E&W	29W-22N	51

Bull Valley

Street	Grid	Page
Blackberry Dr. NS&EW	38W-31N	25
Boone Creek Ter. NS&EW	36W-30N,36W-31N	33
Breckenridge Ct. NW&SE	37W-32N	25
Bull Valley Rd. E&W	39W-31N,36W-31N	24
Cold Springs Rd. N&S	38W-31N,38W-32N	25
Concord Dr. E&W	37W-32N	25
Crystal Springs Rd. E&W	36W-29N,35W-29N	33
Fairway Ln. E&W	39W-31N	25
Fleming Rd. NW&SE	39W-31N,38W-31N	25
High Meadow Dr. NE&SW	38W-31N	25
High Meadow Ln. E&W	38W-31N	25
Locust Ln. N&S	39W-31N	25
N. Cherry Valley Rd. N&S	36W-30N,36W-31N	25
Orchard Valley Dr. E&W	36W-31N	25
Ridge Rd. N&S	36W-31N	25
Saddle Creek Tr. NW&SE	36W-31N	25
Sudberg Ct. E&W	37W-32N	25
Swarthmore Rd. N&S	37W-31N,37W-32N	25
Valley Hill Rd. N&S	37W-31N,37W-32N	25
Valley Rd. NE&SW	36W-30N,36W-31N	33
Woodland Dr. N&S	39W-31N	25
Woodstone Ln. E&W	38W-31N	25

Burton Township

Street	Grid	Page
Ann Pl. E&W	29W-38N	19
Arlington Dr. NE&SW	30W-39N	11
Baccus Ln. NW&SE	30W-43N	11
Baron Dr. E&W	29W-38N	19
Barry Av. N&S	29W-39N	11
Beacon St. E&W	29W-39N	11
Bellevue Av. N&S	30W-39N	11
Berwyn St. E&W	30W-39N	11
Beverly Way NE&SW	30W-39N	11
Bittersweet Pl. NS&EW	29W-38N	19
Blaine Pl. N&S	29W-38N	19
Breezy Lawn Rd. NS&EW	30W-42N	11
Buena Av. N&S	29W-38N,29W-39N	19
Carleton Dr. N&S	30W-39N	11
Casper St. N&S	29W-39N	11
Cass St. E&W	29W-39N	11
Chateau Dr. E&W	29W-38N	19
Coventry Av. E&W	29W-38N	19
Cuneo Dr. N&S	29W-39N	11
Daly Av. N&S	29W-39N	11
Dannell Pl. E&W	29W-39N	11
Deming Pl. NW&SE	29W-38N	19
Edgewater St. N&S	29W-39N	11
Engels Pl. E&W	29W-39N	11
English Prairie Rd. E&W	30W-42N	11
Fern St. NW&SE	29W-38N	19
Gray St. N&S	29W-38N	19
Grove Rd. N&S	29W-39N	11
Hastings Av. E&W	29W-38N	19
Island Ct. NE&SW	30W-39N	11
James Rd. E&W	29W-41N	11
Johnsburg Wilmot Rd. N&S	30W-38N,29W-43N	19
Killarney St. N&S	29W-39N	11
Kohl Av. E&W	29W-38N	19
Lee Ln. E&W	30W-40N	11
Linden Rd. NS&EW	30W-39N	11
Lubliner St. E&W	29W-39N	11
Madison Av. E&W	29W-39N	11
Maiden Ln. N&S	29W-37N,29W-38N	19
Main St. E&W	29W-39N	11
Maple Av. N&S	29W-38N	19
Margaret Ln. N&S	30W-40N	11
Maude Pl. N&S	29W-38N	19
Melbourne Pl. NW&SE	30W-39N	11
Melrose Dr. NW&SE	30W-39N	11
Miller Rd. E&W	30W-38N	19
Muriel Rd. E&W	29W-38N	19
Nippersink Rd. E&W	29W-39N	11
Normandy Dr. N&S	29W-38N	19
North Shore Ave. NW&SE	29W-38N	19

Street	Grid	Page
Oak Hill Dr. E&W	29W-38N,29W-39N	19
Paddock St. E&W	30W-40N	11
Park Ln. E&W	30W-39N	11
Polaris Rd. NW&SE	30W-42N	11
Prospect Pl. E&W	29W-38N	19
Ravine Dr. NE&SW	30W-39N	11
Reiger Ct. NE&SW	30W-43N	11
Richardson Rd. N&S	30W-40N	11
Rita Av. N&S	30W-40N	11
River Rd. N&S	30W-39N	11
Roselle St. NW&SE	29W-38N	19
Rt. 173 E&W	30W-42N,29W-42N	11
Rudolph Ct. NE&SW	30W-42N	11
Sage Ln. N&S	30W-40N	11
Schmidt St. N&S	29W-39N,30W-40N	11
Sheridan Pl. N&S	29W-38N	19
Showers St. N&S	29W-38N	19
Siedschlag Rd. N&S	30W-40N	11
Spring Ct. E&W	30W-39N	11
Spring Grove Rd. NW&SE	30W-39N,29W-39N	11
Spring Rd. E&W	29W-39N	11
State Park Rd. N&S	29W-38N,29W-39N	19
Timber Ln. E&W	29W-38N	19
U.S. Route 12 E&W	30W-39N,29W-38N	11
Valley Rd. NE&SW	36W-29N,36W-31N	33
Villa Vista Rd. N&S	29W-39N	11
Vineyard Ln. NS&EW	30W-42N,30W-43N	11
Water St. E&W	29W-39N	11
Waverly Dr. NW&SE	29W-38N	19
Wellington St. E&W	29W-39N	11
William Ln. NE&SW	30W-40N	11
Willow St. N&S	29W-38N	19
Wilmot Rd. N&S	30W-38N,29W-39W	11
Winsor Av. N&S	29W-39N	11
Woodland Pkwy. NW&SE	30W-39N	11

Cary

Street	Grid	Page
Abbywood Dr. NS&EW	32W-23N	42
Aberdeen Rd. N&S	31W-23N	43
Ada St. E&W	31W-23N	43
Adare Rd. E&W	31W-24N	43
Alexander Ct. N&S	32W-24N	42
Alica Dr. E&W	30W-23N	43
Alida Dr. E&W	31W-23N	43
Alma Ter. E&W	31W-24N	43
Ann St. E&W	31W-24N	43
Applewood Ct. N&S	32W-22N	50
Ardmore Dr. NW&SE	31W-24N	43
Arthur Dr. E&W	31W-23N	43
Asbury Ln. N&S	32W-23N	42
Ash Fork Tr. N&S	32W-24N	42
Bayberry Dr. NS&EW	32W-22N	50
Bee Ct. E&W	31W-23N	43
Bell Dr. E&W	30W-23N	43
Berriedale Dr. E&W	31W-23N	43
Big Bear Tr. E&W	32W-23N	42
Big Sky Tr. E&W	32W-24N,31W-24N	42
Blazing Star Tr. NW&SE	32W-24N	42
Blue Mesa Tr. E&W	32W-24N	42
Borden Av. N&S	31W-24N	43
Boxwood Ln. N&S	32W-23N	42
Brentwood Ln. N&S	30W-24N	43
Briargate Rd. N&S	31W-23N	43
Bright Oaks Cir. NW&SE	31W-24N	43
Bristol Way E&W	30W-24N	43
Bryan Dr. E&W	30W-24N	43
Burr Oak Cir. NW&SE	30W-23N	43
Burton Av. NW&SE	31W-23N	43
Cambridge Ct. E&W	32W-23N	42
Camden Ct. E&W	32W-23N	42
Candlewood Ct. NE&SW	30W-24N	43
Candlewood Tr. E&W	30W-24N	43
Carl Sands Dr. E&W	30W-24N	43
Carlisle Ct. NE&SW	32W-23N	42
Carrol Ln. N&S	32W-23N	42

Street	Grid	Page
Cary Algonquin Rd. NE&SW	32W-22N	50
Cary Av. NW&SE	30W-23N	43
Cary St. E&W	31W-23N	43
Cary-Algonquin Rd. NE&SW	32W-23N	42
Castleria Ct. N&S	31W-24N	43
Charlotte Pl. E&W	31W-23N	43
Chelsea Ln. N&S	32W-23N	42
Cherry Ln. N&S	30W-23N	43
Cherry St. NW&SE	30W-23N	43
Chestnut Ct. E&W	32W-23N	42
Chicago St. E&W	30W-23N	43
Chickory Ridge Tr. E&W	32W-24N	42
Cimarron Dr. N&S	32W-22N	50
Claire Ln. E&W	32W-24N,31W-24N	42
Cleveland St. E&W	30W-24N	43
Colleen Dr. E&W	30W-23N	43
Conestoga Tr. NE&SW	32W-24N	42
Copper Canyon Tr. NW&SE	32W-24N,31W-24N	42
Cottonwood Tr. NW&SE	32W-24N	42
Country Commons Rd. E&W	32W-23N,31W-23N	42
Courtney Ln. NS&EW	30W-24N	43
Coyote Tr. N&S	32W-24N	42
Crabtree Ln. N&S	30W-24N	43
Crest Dr. N&S	30W-24N	43
Crofton Ct. N&S	32W-23N	42
Crystal St. NW&SE	32W-24N,31W-23N	42
Cumberland Ct. E&W	32W-22N	50
Daybreak Ridge Tr. NS&EW	31W-23N	43
Decker Dr. N&S	30W-23N,30W-24N	43
Deerpath Ct. N&S	32W-23N	42
Detroit Dr. E&W	30W-23N	43
Deveron Cir. N&S	31W-23N	43
Drake Ln. N&S	32W-24N	42
Dunleer Dr. N&S	31W-24N	43
Duxbury Ln. N&S	32W-23N	42
East Oriole Tr. E&W	30W-24N	43
Elden Dr. N&S	30W-24N	43
English Oaks Dr. N&S	32W-22N	50
Ennis St. NE&SW	31W-24N	43
Erin Dr. E&W	30W-24N	43
Essex Ln. N&S	32W-23N	42
Fairfield Ln. N&S	32W-23N	42
First St. N&S	31W-24N	43
Forest Ln. NW&SE	31W-24N	43
Fox Trails Dr. N&S	32W-22N,32W-23N	50
Fox Trails Dr. N. NW&SE	32W-23N,32W-22N	42
Fox Trails Dr. S. E&W	32W-22N	50
Franke Av. E&W	31W-23N	43
Galway Dr. E&W	31W-24N	43
Georgetown N&S	32W-23N,32W-24N	42
Glen Garry Rd. E&W	31W-23N	43
Haber Ct. E&W	32W-23N	42
Haber Rd. NE&SW	32W-23N	42
Hampton St. NW&SE	32W-23N	42
Harper Av. NW&SE	32W-23N	42
Hickory Ln. N&S	31W-24N	43
High Av. N&S	30W-23N	43
High Rd. NS&EW	30W-23N	43
High St. N&S	30W-23N	43
Hill St. N&S	30W-23N	43
Hillhurst Dr. N&S	30W-23N,30W-24N	43
Hillside Dr. E&W	30W-22N,30W-22N	51
Industrial Dr. NW&SE	32W-24N,31W-24N	42
Ivanhoe Ln. N&S	32W-24N	42
James Way E&W	31W-24N,30W-24N	43
Jandus Cary Av. NW&SE	30W-23N	43
Jandus Cary Rd. NW&SE	30W-23N	43
Jandus Cut-Off N&S	30W-23N	43
Jefferson Ln. NS&EW	32W-23N	42
Keith Way NE&SW	31W-23N	43
Kildare Ct. N&S	31W-24N	43
Knollwood Dr. NS&EW	32W-23N,32W-23N	50
Krenz Av. E&W	31W-23N	43
Lake Julian Ln. NE&SW	30W-23N	43
Laurel Ln. N&S	32W-22N	50
Leith Way NE&SW	31W-23N	43

Street	Grid	Page
Liberty Av. NS&EW	31W-24N	43
Liberty Ct. N&S	31W-24N	43
Linden St. E&W	31W-24N	43
Little Stonegate Ct. NE&SW	31W-23N	43
Lloyd St. E&W	31W-24N	43
Main St. E&W	30W-23N	43
Manchester St. E&W	32W-23N	42
Manitou Ct. N&S	32W-23N	42
Manitou St. E&W	32W-23N	42
Margaret Ter. E&W	31W-24N,30W-24N	43
Marquette Av. NE&SW	30W-23N	43
Mary Ln. NE&SW	31W-23N	43
Mayfield St. E&W	32W-23N	42
Merrimac St. E&W	32W-23N	42
Mildred Av. NE&SW	30W-23N	43
Milford St. E&W	32W-23N	42
Millwood St. NW&SE	32W-23N	42
Moders Ave. N&S	32W-23N	42
Mohawk St. N&S	32W-23N	42
Montana Ct. NE&SW	30W-24N	43
Montana Dr. N&S	30W-23N,30W-24N	43
Moraine Hill Dr. N&S	30W-23N,30W-24N	43
Muir Wood Ct. NE&SW	32W-23N	42
Norman Dr. E&W	31W-24N	43
Oak Valley Dr. N&S	31W-24N	43
Oriole Tr. E&W	31W-24N	43
Park Av. NW&SE	31W-24N,31W-23N	43
Patriot Ct. N&S	31W-24N	43
Patriot Ln. NS&EW	31W-24N	43
Pearl St. E&W	30W-23N	43
Pearson Rd. NS&EW	32W-23N	42
Perth Ln. NE&SW	31W-23N	43
Pin Oak Cir. NS&EW	30W-24N	43
Pin Oak Ct. NW&SE	30W-24N	43
Pine Cir. NS&EW	31W-24N	43
Prairie Av. NE&SW	31W-23N	43
Raintree Ct. NE&SW	32W-23N	42
Red Oak Dr. N&S	30W-24N	43
Red Oak Tr. E&W	31W-24N	43
Ridgewood Dr. N&S	32W-23N	42
Rodeo Ct. N&S	32W-24N	42
Rosewood Ct. N&S	32W-22N,32W-23N	50
Ross Av. N&S	30W-23N	43
Rosslare Ct. N&S	31W-24N	43
Royal Glen Dr. NS&EW	32W-22N	50
Saddle Ridge Tr. E&W	32W-24N	42
Sagebrush Tr. E&W	32W-24N	42
Sandra Ln. N&S	31W-24N	43
Santa Fe Tr. N&S	32W-24N	42
Schober Dr. NS&EW	31W-23N	43
School St. N&S	31W-23N	43
Second St. N&S	30W-23N	43
Seebert St. NS&EW	31W-23N	43
Sequoia Tr. N&S	32W-24N	42
Sherwood Dr. E&W	30W-24N	43
Silver Lake Rd. N&S	31W-24N	43
Silver Tree Cir. NS&EW	31W-24N	43
South St. N&S	31W-23N	43
Spring Beach Rd. NE&SW	30W-24N	43
Spring St. N&S	31W-23N	43
Spruce Tree Dr. N&S	30W-24N	43
St. Cronan St. N&S	31W-23N	43
Stonegate Rd. N&S	31W-23N	43
Sundance Tr. N&S	32W-24N	42
Sunset Dr. E&W	30W-24N	43
Surry Ridge Dr. NW&SE	32W-23N	42
Tamarack Ct. E&W	32W-24N	42
Tealwood Ct. N&S	32W-22N	50
Thomas Av. N&S	31W-23N	43
Three Oaks Rd. N&S	32W-24N,30W-24N	42
Timber Ter. N&S	31W-24N	43
Trianglewood Dr. N&S	30W-24N	43
W. Main St. E&W	32W-23N,31W-23N	42

Street	Grid	Page
Wagner Dr. E&W	31W-24N	43
Weavers Dr. N&S	30W-23N	43
Weber Dr. E&W	31W-24N	43
Wexford Ct. N&S	31W-24N	43
Whitepine Dr. N&S	30W-24N	43
Wildrose Dr. N&S	30W-24N	43
Wildwood Ct. N&S	32W-23N	42
Wildwood Tr. E&W	31W-24N	43
Willow Cir. NS&EW	31W-24N	43
Willow Ct. E&W	32W-23N	42
Winchester St. E&W	32W-23N	42
Woodbridge Dr. E&W	32W-22N	50
Wooded Knoll Dr. E&W	30W-23N	43
Wulff St. N&S	31W-23N	43

Chemung Township

Street	Grid	Page
Brink St. E&W	49W-38N	13
Cals Ct. E&W	51W-39N,51W-40N	4
Center St. NE&SW	52W-38N	12
Chemung St. E&W	52W-38N	12
Chicago Rd. E&W	52W-38N	12
Chippewa Rd. N&S	51W-38N	12
Commercial St. NW&SE	51W-39N	4
County Line Rd. N&S	54W-38N	12
Crowley Rd. E&W	49W-40N	5
Dana St. E&W	52W-38N	12
Elm St. NW&SE	51W-39N	4
Franklin NW&SE	51W-39N	4
Graf Rd. E&W	54W-40N,52W-40N	4
Grove Rd. E&W	52W-38N	12
Harvard Hills Rd. N&S	49W-36N,49W-39N	13
Hebron Rd. E&W	49W-42N	5
Hickory St. N&S	52W-38N	12
Horseshoe Dr. NE&SW	50W-40N	5
Hunter Rd. E&W	54W-38N,53W-38N	12
Kasson St. NW&SE	51W-39N	4
Kennedy Dr. N&S	49W-38N	13
Kennedy Rd. N&S	53W-38N	12
Krunfus Rd. N&S	49W-40N	5
Lawrence Rd. E&W	53W-43N,51W-38N	4
Lilja Rd. N&S	52W-41N,51W-43N	4
Malinda Dr. N&S	49W-40N	5
Maxon Rd. N&S	53W-38N,58W-40N	12
Nilsen Ct. E&W	54W-40N	4
Norma Ln. E&W	52W-39N	4
Oak Grove Rd. NE&SW	52W-38N,49W-41N	12
Oak St. N&S	52W-38N	12
Pearl St. NE&SW	51W-39N	4
Perkins Rd. E&W	54W-41N,53W-41N	12
Pine St. N&S	52W-38N	12
Pleasant St. NW&SE	51W-39N	4
Ramer Rd. E&W	52W-38N,51W-38N	12
Rhonda Dr. N&S	52W-38N	12
Rita Dr. E&W	52W-38N	12
River St. NE&SW	51W-39N	4
Rt. 173 NE&SW	49W-38N	13
Ruhff St. NW&SE	51W-39N	4
Ryan Rd. NS&EW	54W-38N,54W-39N	12
South Nilsen Ct. N&S	54W-40N	4
Southport Rd. NE&SW	52W-38N	12
State Line Rd. E&W	54W-43N,49W-43N	4
U.S. Route 14 N&S	50W-39N,49W-43N	5
Walker St. NW&SE	51W-39N	4
Weidner Rd. N&S	52W-41N,52W-43N	4
White Oaks Rd. N&S	53W-38N,54W-43N	12
Yates Rd. E&W	51W-42N,49W-42N	4
York Rd. E&W	54W-42N	4

Coral Township

Street	Grid	Page
Arbor Ln. N&S	45W-20N	47
Beall Ln. E&W	45W-20N	47
Beck Rd. E&W	48W-23N,47W-23N	38
Burr Oak Rd. N&S	45W-20N	47
Camp Ct. E&W	45W-20N	47
Cantering Ln. N&S	44W-22N	47
Church Rd. E&W	47W-21N,43W-21N	46
Condor Ln. E&W	45W-20N	47
Conestoga Tr. N&S	47W-22N	46

Street	Grid	Page
Country Club Dr. E&W	47W-22N	46
Country Club Tr. N&S	47W-22N	46
Dunham Rd. N&S	47W-25N,47W-26N	38
East Coral Rd. N&S	47W-24N,44W-23N	38
Ernest Dr. N&S	47W-21N	46
Fair Ln. N&S	44W-22N	47
Fairway Ln. N&S	47W-22N	46
Forest Ln. N&S	44W-22N	47
Frank Ln. E&W	47W-21N	46
Frohling Rd. NW&SE	43W-24N,43W-23N	39
Glen Ln. N&S	44W-22N	47
Harmony Hill Rd. N&S	47W-20N,47W-22N	46
Harmony Hill Rd. N&S	47W-23N	38
Harmony Rd. E&W	46W-20N,43W-20N	46
Hemmington Rd. E&W	45W-25N,43W-25N	39
Heron Dr. N&S	45W-20N	47
Highbridge Rd. E&W	46W-25N,44W-25N	38
Hill Rd. N&S	47W-23N	38
Hillsboro Dr. NS&EW	44W-20N	47
I-90 NW&SE	47W-20N	46
Jani Ln. N&S	47W-21N	46
Kim Ln. E&W	47W-21N	46
Knoll Top Ln. NS&EW	44W-22N	47
Leech Rd. N&S	45W-23N,45W-24N	39
Manda Dr. N&S	45W-20N	47
Marengo Rd. NW&SE	45W-22N,43W-22N	47
Marengo-Huntley Rd. NW&SE	46W-23N,46W-25N	38
McCue Rd. N&S	44W-25N	39
Mitchell Ln. N&S	44W-22N	47
N. Ridge Ln. NS&EW	47W-21N	46
Northrup Rd. N&S	46W-24N,46W-25N	38
O'Cock Rd. N&S	47W-25N,46W-25N	38
Oak Ridge Rd. N&S	47W-21N	46
Olson Rd. N&S	45W-24N,45W-25N	39
Overlook Ln. N&S	44W-20N	47
Oxbow Dr. E&W	47W-22N	46
Pebble Dr. N&S	44W-20N	47
Peregrine Tr. NS&EW	45W-20N	47
Pinion Tr. N&S	47W-21N	47
Pinoak Tr. N&S	45W-20N	47
Raven Hill Dr. E&W	47W-21N	47
Ridge Ln. N&S	44W-21N	47
Rose Ln. E&W	47W-21N	46
Royal Ln. N&S	47W-21N	46
Sarko Tr. N&S	47W-21N	47
Seeman Rd. N&S	44W-20N,44W-22N	47
Seeman Rd. N&S	44W-23N,44W-24N	39
Somerset Dr. N&S	48W-24N	38
South Coral Rd. N&S	48W-23N,48W-24N	38
South Union Rd. N&S	47W-24N,47W-25N	38
Stacy Ln. NE&SW	44W-22N	47
Stevens Rd. E&W	47W-21N	46
Trail Rd. N&S	45W-20N	47
U.S. Route 20 NW&SE	46W-22N,45W-20N	46
U.S. Route 20 NW&SE	48W-25N,46W-23N	38
Voss Rd. N&S	48W-22N	46
Voss Rd. N&S	48W-23N	38
West Coral Rd. N&S	48W-24N,47W-24N	38
West Union Rd. E&W	48W-25N	38
Woodale Rd. E&W	44W-22N	47
Woodwind Dr. N&S	44W-20N	47

Crystal Lake

Street	Grid	Page
Abbey Dr. E&W	37W-24N	41
Abbington Dr. NE&SW	35W-24N	42
Aberdeen Dr. NS&EW	35W-24N	42
Ackman Rd. E&W	37W-23N,36W-23N	41
Alexandra Blvd. E&W	37W-23N,36W-23N	41
Amberwood Dr. EW&NS	37W-23N,36W-23N	41
Arrowhead Ln. N&S	35W-27N	34
Ascot Ln. E&W	37W-24N	41
Ash St. NW&SE	35W-25N	42
Ashford Ln. NS&EW	36W-22N	49

	Grid	Page
Auburn Ct. E&W	36W-23N	41
Augusta Dr. E&W	37W-24N,36W-24N	41
Autumn Crest Ct. NW&SE	36W-23N	41
Autumn Crest Dr. E&W	37W-23N	49
Baldwin Av. N&S	36W-25N	41
Barberry Dr. N&S	37W-25N	41
Barlina Rd. E&W	37W-24N,36W-24N	41
Barrow Dr. E&W	35W-24N	42
Barton Ln. NE&SW	36W-23N	41
Beardsly St. E&W	34W-26N	42
Becker Ct. N&S	36W-22N	49
Bedford Dr. E&W	36W-24N	41
Bennington Dr. E&W	36W-24N	41
Berkshire Ln. E&W	36W-25N	42
Birch Ct. NW&SE	37W-24N	41
Birmingham Ct. NE&SW	36W-22N	49
Birmingham Ln. NS&EW	36W-22N	49
Bittersweet Dr. E&W	34W-26N	42
Blackthorn Dr. NW&SE	34W-26N	42
Blue Heron Dr. NW&SE	37W-23N	41
Blue Mound Dr. N&S	37W-24N	41
Boneset Dr. E&W	37W-24N	41
Boxwood Ct. NE&SW	37W-23N	41
Boxwood Ln. N&S	37W-23N,36W-24N	41
Brandy Dr. E&W	36W-24N,35W-24N	41
Briarwood Cir. NS&EW	38W-25N	41
Briarwood Rd. N&S	37W-25N,37W-26N	41
Bridgedale Ct. E&W	36W-23N	41
Bridgedale Dr. NW&SE	36W-23N	41
Bridgeport Ln. NS&EW	36W-22N	49
Brigham Ln. NW&SE	36W-22N	49
Brink St. E&W	35W-26N	42
Broadway Av. NW&SE	36W-25N	41
Brompton Ct. E&W	36W-22N	49
Brompton Rd. N&S	36W-22N	49
Brook Dr. E&W	36W-24N	41
Buckingham Dr. E&W	35W-24N	42
Burdent St. E&W	35W-24N	42
Burning Bush Tr. E&W	35W-27N,34W-27N	34
Burr Oak Dr. N&S	37W-25N	41
Butler Dr. E&W	37W-24N	41
Cambridge Ln. NE&SW	35W-24N	42
Camelot Ct. N&S	36W-24N	41
Camelot Dr. NS&EW	36W-24N	41
Camelot Pl. N&S	36W-24N	41
Candlewood Dr. N&S	36W-23N	41
Canterbury Dr. N&S	35W-24N	42
Cardiff Dr. N&S	35W-24N	42
Carmella Dr. E&W	35W-26N	42
Caroline St. N&S	35W-25N,35W-26N	42
Carpenter St. NE&SW	36W-25N	41
Cascade Dr. E&W	35W-27N	34
Castlefield Ln. E&W	36W-24N	49
Catherine Dr. E&W	35W-25N	42
Cedar Crest Dr. N&S	37W-24N	41
Cedar Dr. N&S	37W-25N	41
Center Av. E&W	35W-25N	41
Center St. E&W	35W-26N	42
Chardonnay Ln. E&W	36W-24N,35W-24N	41
Charlotte Av. N&S	37W-25N	41
Chasefield Ln. NS&EW	36W-24N	41
Church St. NW&SE	35W-25N	42
Circle Ln. N&S	35W-24N,35W-25N	42
Clover Dr. N&S	37W-25N,37W-26N	41
College St. NE&SW	35W-25N	42
Colony Dr. E&W	37W-24N	41
Commerce Ct. N&S	35W-26N	42
Commercial Rd. E&W	35W-24N	42
Concord Dr. E&W	36W-23N	41
Congress Pkwy. E&W	34W-25N	42
Copperfield Ln. NE&SW	36W-22N	49
Corrine Av. N&S	37W-25N	41
Cotton Wood Ln. N&S	37W-26N	41
Country Club Rd. NS&EW	37W-24N,36W-25N	41
Country Dr. NE&SW	36W-27N	33
Coventry Ct. NW&SE	35W-24N	42
Coventry Ln. NE&SW	36W-24N,35W-25N	41
Covington Cir. NS&EW	36W-23N	41
Crabtree Dr. N&S	37W-23N	49
Crandall Av. N&S	36W-25N	41
Cress Creek Ln. E&W	36W-24N	41
Cress Creek Ter. NE&SW	36W-24N	41
Crystal Lake Av. E&W	36W-26N,34W-26N	41
Crystal Pt.Dr. N&S	34W-24N	42
Crystal Ridge Rd. N&S	35W-26N	42
Cumberland Ln. N&S	36W-25N	41
Cunningham Ln. N&S	36W-22N	49
Darlington Ln. NE&SW	36W-22N	42
Dartmoor Dr. E&W	36W-24N,34W-24N	41
Deer Haven Dr. N&S	37W-22N	49
Deer Run Dr. E&W	35W-27N,34W-27N	34
Denton Ct. N&S	34W-24N	42
Devonshire Ln. NE&SW	35W-24N	42
Dogwood Dr. NS&EW	37W-23N	41
Dole Av. E&W	36W-25N,35W-26N	41
Dolle Ln. NW&SE	35W-26N	42
Dolo Roso Vista E&W	37W-23N	41
Douglas St. NW&SE	36W-25N,35W-25N	41
Dover Ct. NS&EW	35W-24N	42
Dovercliff Way NS&EW	37W-23N	41
Driftwood Ct. E&W	37W-22N	49
Driftwood Ln. N&S	37W-22N	49
Duffy Ln. NE&SW	34W-26N	42
Durham Ct. NW&SE	36W-22N	49
E. Brighton Dr. E&W	33W-27N	34
Eagle St. NW&SE	36W-25N	41
East Gate Rd. N&S	35W-24N	42
East St. N&S	34W-26N	41
East St. N&S	36W-25N	41
Eastview Av. NW&SE	35W-25N	42
Edgebrook Dr. E&W	36W-25N	41
Edgewater Dr. NS&SW	37W-25N	41
Edgewood Rd. N&S	37W-25N	41
Eletson Dr. NE&SW	35W-24N	42
Ellsworth St. N&S	34W-26N	42
Elm St. NE&SW	35W-25N	42
Elmhurst St. N&S	35W-26N	42
Elmwood Av. N&S	35W-25N	41
Elsinoor Ln. NW&SE	35W-24N	42
Erick St. N&S	34W-26N	42
Essex Ln. N&S	36W-25N	41
Esther St. E&W	35W-25N	42
Euclid Av. E&W	37W-25N	42
Eugene St. N&S	35W-25N	42
Everett Av. E&W	36W-25N,35W-25N	41
Evergreen Pkwy. N&S	35W-26N	42
Exchange Dr. E&W	34W-25N	42
Exmoor Ter. NW&SE	36W-24N	41
Factory Rd. E&W	35W-24N	42
Fair Oaks Av. E&W	35W-25N	41
Fairview Pk. Av. E&W	36W-26N	42
Faringdon Dr. E&W	35W-25N	42
Federal Dr. N&S	33W-25N	42
Fenimore Dr. N&S	33W-25N	42
Fernleaf Dr. N&S	36W-23N	41
First Colonial Dr. N&S	34W-25N	42
First Ct. NE&SW	35W-25N	42
First St. N&S	35W-26N	42
Flagstone Dr. E&W	37W-22N	49
Flint Tr. N&S	34W-27N	34
Florence St. NE&SW	36W-25N	41
Floresta Vista E&W	37W-25N	41
Fountain Green Dr. N&S	36W-23N	41
Francis Av. N&S	35W-26N	42
Franklin Av. N&S	35W-25N	42
Gail Ct. NW&SE	35W-25N	42
Gardina Vista E&W	37W-25N	41
Gates St. NE&SW	35W-26N	42
Georgian Ct. E&W	35W-26N	42
Gilbert Av. E&W	37W-25N	41
Glasgow Ct. E&W	37W-22N,36W-22N	49
Glen Av. N&S	34W-26N	42
Glendale Dr. N&S	36W-24N	41
Golden Rod Ct. SW&NE	37W-24N	41
Golf Course Rd. N&S	36W-26N,36W-24N	49
Golf Rd. E&W	36W-25N	41
Grand Av. NW&SE	35W-25N	42
Grandview Ct. NE&SE	33W-25N	42
Grant St. NE&SW	35W-26N	42
Green Meadow Ct. N&S	36W-24N	41
Green Oaks Dr. E&W	36W-26N	41
Greenbriar Ln. NS&EW	36W-24N	41
Greenbrier Ln. E&W	36W-24N	41
Greenbrier Ter. NW&SE	36W-24N	41
Greenview Ct. NE&SE	33W-25N	42
Grove St. NW&SE	35W-25N	42
Hampton Ct. N&S	33W-26N	42
Harold St. NE&SW	35W-25N	42
Harper Ln. N&S	37W-22N	49
Hartford Ln. NW&SE	33W-26N	49
Hastings Av. N&S	35W-25N	42
Hawthorne Dr. N&S	37W-24N	41
Heather Ln. N&S	37W-26N,37W-26N	41
Heisler Ct. E&W	35W-24N	42
Herald Dr. N&S	33W-26N	42
Heritage Dr. E&W	34W-24N	42
Hickory Av. N&S	36W-26N	41
Highland Av. N&S	36W-26N	41
Hilkert Ct. E&W	35W-25N	42
Hill Dr. NW&SE	35W-25N	42
Holly Dr. N&S	37W-25N	41
Holly Tree Ln. N&S	36W-23N	41
Honeysuckle. N&S	37W-25N	41
Hull Ct. N&S	35W-24N	42
Hunters Tr. NS&EW	37W-23N	49
Huntington Dr. N&S	36W-24N	41
Huntley Rd. N&S	37W-23N,37W-24N	41
Hyinisport Ln. E&W	36W-23N	49
Illinois St. N&S	34W-26N	42
Indian Hill Tr. E&W	35W-27N,34W-27N	34
Industrial Rd. NE&SW	34W-26N	42
Ivy Ln. E&W	37W-25N	41
Jackman Dr. N&S	35W-25N	42
James St. N&S	36W-25N	41
Juniper Dr. E&W	33W-25N	42
Keith Av. N&S	35W-25N	42
Kendallwood Ct. E&W	36W-24N	41
Kendallwood Dr. N&S	36W-24N	41
Kensington Ln. N&S	36W-22N	49
Kimer Ct. N&S	35W-27N	34
King St. NE&SW	36W-25N	41
Kingston Ln. NS&EW	35W-24N	42
Knack Blvd. N&S	34W-26N	42
Knollwood Cir. NS&EW	36W-23N	41
Laguna Rd. E&W	36W-25N	41
Lake Av. E&W	36W-25N	41
Lake Ct. NW&SE	36W-25N	41
Lake Shore Dr. NE&SW	36W-25N	41
Lake St. NW&SE	36W-25N,35W-25N	41
Lakeside Rd. N&S	36W-25N	41
Lakeview Dr. E&W	37W-25N	41
Lancaster Ln. E&W	35W-24N	42
Lee Dr. E&W	36W-24N,35W-24N	41
Lee St. NW&SE	35W-24N	42
Leonard Pkwy. NW&SE	36W-25N	41
Liberty Dr. E&W	35W-26N	42
Lilac Dr. NE&SE	37W-23N	41
Lill St. NW&SE	35W-25N	42
Lincoln Pkwy. N&S	35W-26N	42
Lincolnshire Dr. NW&SE	35W-24N	42
Linn St. NE&SW	35W-25N	42
Lochwood Dr. NW&SE	35W-27N	34
Lock Lomond Dr. NS&EW	36W-23N	41
Lorraine Dr. N&S	34W-26N	42
Lou Av. E&W	35W-24N	42
Louisville Ln. NW&SE	36W-22N	49
Lucerne Dr. N&S	36W-23N	41
Lutter Dr. N&S	33W-24N	42
Magnolia Ln. N&S	37W-23N	41
Main St. N&S	34W-24N,34W-26N	42
Manchester Ct. E&W	36W-23N	41
Manchester Dr. N&S	36W-23N	41
Maple St. NE&SW	35W-25N	42
Maplewood Ln. E&W	35W-25N	42
Marhill Ct. NW&SE	35W-25N	42
Marian Pkwy. N&S	35W-26N	42
Mary Ln. N&S	35W-25N	42
McCullom St. E&W	34W-26N	42
McHenry Av. N&S	36W-23N,35W-25N	41
McIntosh Blvd. E&W	34W-25N	42
McNair St. N&S	35W-27N	34
Melrose Ln. N&S	36W-25N	41
Memorial Ct. NW&SE	34W-25N	42
Memorial Dr. NS&EW	34W-25N	42
Meyer Dr. N&S	33W-24N	42
Michael Ct. N&S	35W-26N	42
Midlane Dr.	35W-27N	34
Millard Av. N&S	37W-25N	41
Miller Rd. E&W	36W-22N	49
Minnie St. NE&SW	35W-26N	42
Mistwood Ln. N&S	33W-26N	42
Monarch NE&SW	36W-23N	41
Monterey Dr. E&W	36W-24N	41
Morgan St. N&S	34W-26N	42
Mulberry Ln. E&W	37W-25N	41
N.Brentwood Dr. E&W	36W-24N	41
Narragansett Ct. N&S	33W-26N	42
Nash Rd. N&S	36W-25N	41
Nashua Ct. NE&SW	33W-26N	42
Nashville Ln. NW&SE	36W-22N	49
Newburg Dr. N&S	36W-25N	41
Newcastle Ct. E&W	35W-24N	42
Newport Ln. NS&EW	36W-22N	49
North Ave. E&W	37W-25N	41
North Shore Dr. E&W	37W-25N,36W-25N	41
Northampton Dr. NE&SW	35W-24N	42
Norwich Ct. E&W	33W-24N	42
Nottingham Ct. E&W	35W-24N	42
Nottingham Ln. N&S	35W-24N	42
Oak Ct. E&W	35W-24N	42
Oak Valley Ct. NE&SW	33W-25N	42
Oak Valley Dr. NW&SE	33W-25N	42
Official Rd. E&W	35W-24N	42
Old Westbury Ct. NW&SE	33W-27N	34
Old Westbury Rd. NE&SW	33W-27N	34
Orchard Row E&W	36W-26N	41
Oriole Tr. NW&SE	36W-25N,35W-25N	41
Oxford Ln. N&S	36W-25N	41
Paddock St. E&W	35W-25N	42
Palm Ct. N&S	36W-22N	49
Palmer Ct. E&W	37W-24N	41
Park Blvd. N&S	34W-26N	42
Parkridge Dr. N&S	36W-23N	41
Parkridge Dr. NE&SE	36W-23N	41
Parkside Ct. N&S	35W-27N	34
Parkside Dr. N&S	36W-23N	41
Parkside Ter. N&S	35W-27N	34
Parkview Ct. E&W	35W-26N	42
Peachtree Ct. N. N&S	36W-22N	49
Peachtree Ct. S. N&S	36W-22N	49
Pebble Ln. N&S	37W-23N	41
Pembroke Ct. N. N&S	36W-22N	49
Pembroke Ct. S. N&S	36W-22N	49
Penny Ln. NW&SE	36W-22N	49
Peterson Pkwy. N&S	35W-26N	42
Pierson St. NW&SE	36W-25N,35W-25N	41
Pine Ct N&S	35W-26N	42
Pine St. E&W	36W-25N	41
Pinewood Dr. E&W	37W-25N	41
Pingree Rd. N&S	34W-24N,34W-25N	41
Plaintain Ct. E&W	37W-24N	41
Plum Ct. N&S	36W-22N	49
Plum Tree Dr. E&W	37W-24N	41
Plymouth Ln. NS&EW	35W-24N	42
Pomeroy St. NE&SW	35W-25N	42
Poplar St. E&W	34W-26N	42
Porter Av. E&W	36W-25N	41
Portsmouth Ct. N&S	36W-22N	49
Portsmouth Rd. N&S	36W-22N	49
Prairie St. E&W	34W-26N	42
Prescott Ln. N&S	36W-23N	41
Pyott Rd. N&S	35W-24N	42
Railroad St. NW&SE	35W-25N	42
Rebecca Rd. E&W	34W-26N	42
Regency Park Dr. NS&EW	36W-23N	41
Regent Dr. E&W	36W-23N	41
Richert Rd. E&W	33W-25N	42
Ridge Av. N&S	36W-26N	41
Ridgefield Rd. NW&SE	36W-26N,36W-27N	41
Ridgewood Dr. N&S	37W-24N	41
Ringling Rd. E&W	36W-25N	41
Riverside Rd. N&S	36W-25N	41
Rockland Rd. E&W	36W-26N	41
Rosedale Av. NW&SE	35W-25N	42
Rosenthal Dr. N&S	34W-24N	42
Rosewood Ln. N&S	36W-23N	41
Rt. 176 (Terra Cotta) E&W	37W-26N,33W-26N	41
Rt. 31 N&S	33W-25N,33W-27N	42
Rudat Ct. NW&SE	36W-24N,25W-24N	41
Ryan Way N&S	35W-27N	34
S.Brentwood Dr. E&W	36W-24N	41
Saddle Creek Ln. NE&SW	37W-22N	49
Saddle Ridge NS&EW	34W-27N	34
Sandlewood Ln. N&S	37W-23N,37W-24N	41
Sarasota Ln. NE&SW	37W-24N	41
Savannah Dr. E&W	33W-26N	42
Second Ct. NE&SW	35W-26N	42
Second St. N&S	35W-26N	42
Sedgewich Ct. N&S	33W-26N	42
Shadowwood Dr. E&W	37W-24N	41
Sharon Dr. E&W	36W-24N,35W-24N	41
Sheffield Dr. NS&EW	35W-24N	42
Shelton Ln. N&S	36W-23N	41
Sherbon Dr. N&S	35W-24N	42
Sherman St. E&W	35W-26N,34W-26N	42
Short St. E&W	34W-26N	42
Silk Oak Ln. E&W	33W-26N	42
Silver Berry Dr. E&W	34W-26N	42
Smitana Rd. NW&SE	33W-25N	42
Snowberry Ln. E&W	37W-25N	41
Somerfield Ln. NE&SW	36W-22N	49
Somerset Ln. NE&SW	35W-24N	42
South St. NW&SE	35W-25N	42
Southport Ln. E&W	36W-23N	49
Spruce Ln. N&S	36W-25N	41
St.Andrew Ct. NW&SE	36W-24N	41
St.Andrews NS&EW	36W-24N	41
Stockport Ln. NS&EW	36W-22N	49
Stockton Ln. NW&SE	36W-22N	49

	Grid	Page
Stone Creek Cir. NS&EW	37W-22N	49
Stonebridge Ln. E&W	36W-23N	41
Stonehedge Dr. N&S	37W-24N	41
Stony Hill Ln. NS&EW	36W-27N	33
Stratford Ln. NW&SE	37W-22N	49
Sturm Ct. NS&EW	33W-25N	42
Suffolk Dr. E&W	35W-24N	42
Summercrest Ct. NW&SE	36W-23N	49
Summercrest Dr. W&W	36W-23N	49
Sunset Tr. NE&SW	35W-25N	42
Surrey Ln. E&W	35W-24N	42
Sussex Ln. NE&SW	35W-24N	42
Sutherland Dr. N&S	36W-24N	41
Sutton Dr. N&S	36W-24N	41
Sycamore Ln. N&S	36W-26N	41
Talcott Av. NS&EW	34W-26N	42
Talcott Ct. NE&SW	34W-26N	42
Tamerisk Ln. N&S	36W-26N	41
Tamerisk Ter. NW&SE	36W-26N	41
Tanglewood Dr. NE&SW	36W-23N	41
Teakwood Ln. E&W	37W-26N	41
Teckler Blvd. NE&SW	35W-24N	42
Teki Dr. E&W	33W-24N	42
Terra Cotta Av. E&W	37W-26N,33W-26N	41
Teverton Ln. N&S	35W-24N	42
Third St. N&S	35W-26N	42
Thomasville Ln. NE&SW	36W-22N	49
Thornwood Ln. E&W	37W-25N	41
Three Oaks Rd. E&W	35W-24N	42
Timber Hill Dr. N&S	37W-25N	41
Tomahawk Ln. N&S	34W-27N	34
Tracy Trail NS&EW	34W-25N	42
Trailwood Ct. NW&SE	36W-23N	41
Trailwood Dr. NE&SW	36W-23N	41
Trumbull Cir. E&W	33W-25N	42
U.S. Route 14 NW&SE	36W-26N,36W-24N	41
Union St. NW&SE	36W-24N	42
University St. NW&SE	35W-25N	42
Uteg St. E&W	35W-25N	42
Van Buren St. NW&SE	36W-25N,35W-25N	41
Victor Pkwy. N&S	36W-25N	41
View St. E&W	34W-26N	42
Village Ct. NW&SE	36W-22N	49
Village Rd. E&W	37W-22N,36W-22N	49
Virginia Rd. E&W	36W-26N,34W-24N	41
Walkup Av. N&S	35W-25N	42
Walkup Rd. N&S	35W-26N,35W-27N	42
Wallace Av. NE&SW	35W-25N	42
Warrington Ln. NW&SE	36W-22N	49
Washington St. NE&SW	36W-25N	41
Waterford Cut E&W	37W-23N	41
Waterford Dr. E&W	37W-23N	41
Watson Ct. NS&EW	37W-24N	41
Wedgewood Cir. NS&EW	37W-24N	41
Wedgewood Ct. N&S	37W-24N	41
Wedgewood Dr. NS&EW	37W-24N	41
Wellington Dr. E&W	35W-24N	42
Wesley Dr. N&S	36W-24N	41
West Porridge Dr. N&S	36W-23N	41
Weston Dr. E&W	36W-23N	41
Westport Ct. NE&SW	36W-23N	41
Westport Ridge Dr. N&S	36W-23N	41
Whitehall Way E&W	37W-23N	41
Wiggington Dr. N&S	37W-24N	41
Williams St. N&S	35W-25N	42

	Grid	Page
Willow Tree Dr. NS&EW	36W-23N	41
Wiltshire Ct. NE&SW	35W-24N	42
Wimbelton Ln. N&S	36W-24N	41
Windridge Dr. N&S	37W-24N	41
Windsor Dr. E&W	36W-24N	41
Winslow Cir. NS&EW	36W-23N	41
Woodbine Dr. N&S	37W-25N	41
Woodcreek Cir. NS&EW	37W-23N	41
Woodhaven Dr. NE&SW	36W-22N	49
Woodland Dr. E&W	36W-25N	41
Woodmar Ln. N&S	36W-24N	41
Woodmar Ter. E&W	36W-24N	41
Woodstock St. E&W	36W-26N,35W-26N	41
Wyndmuir Ln. NS&EW	33W-25N	42
Yorkshire Ln. NE&SW	35W-24N	42

Dorr Township

	Grid	Page
Aspen Ct. NE&SW	38W-26N	41
Aspen Ln. N&S	38W-26N	41
Beech Av. NS&EW	38W-25N,38W-26N	41
Blue Jay Ct. NW&SE	37W-28N	33
Boerderij Way NS&EW	38W-29N	33
Boerderij Way E. N&S	38W-29N	33
Bryon Cir. NS&EW	39W-31N	24
Bull Valley Rd. NE&SW	39W-31N	24
Bunker Av. E&W	37W-26N	41
Butternut Dr. NW&SE	39W-26N,38W-26N	40
Callway Tr. N&S	37W-26N	41
Cardinal Ln. E&W	38W-28N,37W-28N	33
Carol Ln. NE&SW	42W-28N	32
Castle Rd. N&S	40W-28N,40W-29N	32
Charlotte Dr. E&W	42W-28N	32
Cherry Valley Rd. N&S	37W-28N,37W-29N	33
Chestnut Ct. N&S	38W-26N	41
Chukar Pl. N&S	42W-28N	32
Church St. NW&SE	37W-27N	33
Country Club Rd. NW&SE	40W-31N,37W-28N	24
Crystal Springs Rd. E&W	37W-29N	33
Daisy Ct. NE&SW	40W-31N	24
Davis Rd. E&W	42W-29N	32
Dean St. N&S	41W-26N,42W-29N	40
Deer Path Rd. NS&EW	39W-30N	32
Deweio Tr. NE&SW	38W-29N	33
Dirkshire Dr. E&W	38W-29N	33
Doty Rd. N&S	39W-27N,39W-28N	32
Esther Ct. E&W	40W-31N	24
Flemming Rd. NW&SE	39W-31N,38W-30N	24
Gayle Dr. N&S	42W-27N	32
Haligus Rd. NW&SE	39W-26N	40
Hamilton Rd. N&S	41W-26N	40
Hawthorne Ln. N&S	38W-26N	41
Heather Ln. E&W	40W-26N	40
Hercules Rd. E&W	40W-28N	32
Hickory Ln. NS&EW	38W-26N	41
Hidden Ln. E&W	38W-31N	25
Hidden Ln. E&W	38W-31N	25
Hillside Rd. E&W	37W-27N	33
Holly Ln. N&S	42W-28N	32
Ironwood Dr. N&S	38W-26N	41
Juanita Vista NW&SE		
Keats Ct. E&W	39W-31N	24
Lily Pond Rd. N&S	38W-28N,38W-29N	33
Lucas Rd. NW&SE	41W-28N,37W-27N	32
Madison St. NW&SE	37W-27N	33
Market St. NE&SW	37W-27N	33
Mason Hill Rd. NE&SW	37W-30N	33
McConnell Rd. E&W	40W-30N,38W-29N	32
Monroe St. NE&SW	37W-27N	33
Mt. Thabor Rd. E&W	39W-26N,39W-27N	40
Nancy Ct. NW&SE	42W-27N	32

	Grid	Page
Noveen Pkwy. E&W	40W-28N	32
Oakleaf Ln. N&S	40W-30N	32
Ohnstad Rd. N&S	38W-26N,38W-27N	41
Perkins Rd. E&W	42W-28N	32
Pleasant Valley Rd. E&W	42W-26N,40W-26N	40
Prairie Dr. E&W	37W-28N	33
Prospect St. E&W	37W-27N	33
Queen Anne Rd. N&S	39W-31N	24
Railroad St. NW&SE	37W-27N	33
Ridgefield Rd. NS&EW	38W-27N,37W-27N	33
Robin Hill Dr. NS&EW	37W-28N	33
Rt. 176 E&W	42W-26N,40W-26N	40
Rt. 47 N&S	41W-27N,41W-29N	32
S. Robin Hill Rd. NE&SW	37W-28N	33
Sally Circle E&W	42W-28N	32
Sharon Dr. NW&SE	40W-31N	24
Shelly St. E&W	39W-31N	24
Sunnyside Rd. N&S	42W-26N,42W-28N	40
Swanson Rd. N&S	41W-26N	40
Tiporary Ct. NW&SE	37W-31N	25
U.S. Route 14 NW&SE	40W-28N,38W-28N	32
Valley Hill Rd. N&S	37W-30N	33
Waller St. NS&EW	37W-27N	33
West St. NE&SW	37W-27N	33
Westgate St. NE&SW	37W-31N	25
Willow Ct. N&S	38W-26N	41
Woodland Ct. E&W	39W-31N	24

Dunham Township

	Grid	Page
Airport Rd. E&W	51W-37N,50W-37N	12
Becks Rd. E&W	53W-37N	12
Bluejay Rd. N&S	50W-36N	13
Bunker Hill Rd. E&W	52W-34N,49W-34N	20
Busse Rd. N&S	50W-32N,50W-33N	21
Cardinal Ct. NW&SE	50W-36N	13
County Line Rd. N&S	54W-32N,54W-37N	20
Dunham Rd. N&S	54W-33N,49W-33N	20
Dunham Woods Rd. N&S	49W-34N	21
Flat Iron Rd. NE&SW	54W-34N,51W-37N	20
Fritz Rd. N&S	51W-34N,51W-35N	20
Heritage Ln. E&W	50W-36N	13
Hickory St. N&S	52W-37N	12
Highview Rd. E&W	49W-33N	21
Hurley Dr. E&W	50W-37N	13
Island Rd. N&S	53W-37N	12
Island Rd. E&W	53W-37N,52W-37N	12
Kennedy Rd. E&W	53W-37N	12
Lembke Rd. E&W	50W-35N,49W-35N	13
Marengo Rd. N&S	50W-36N,50W-37N	13
McGuire Rd. E&W	51W-37N,49W-37N	13
Meade Rd. E&W	52W-33N	21
Menge Rd. N&S	54W-32N,49W-33N	21
Mulvenna Rd. N&S	53W-32N	20
Olbrich Rd. N&S	50W-32N	21
Pagles Rd. N&S	54W-34N,53W-37N	20
Pine St. N&S	52W-37N	12
Ramer Rd. E&W	52W-38N	12
Robin Rd. E&W	50W-36N	13
Root Rd. N&S	52W-32N	20
Rt. 173 E&W	54W-37N,51W-37N	12
Rt. 23 N&S	50W-32N,50W-36N	21
Rush Creek Rd. E&W	50W-36N	13
Sinderson St. N&S	53W-37N	12
South St. E&W	52W-37N	12
Streit Rd. E&W	54W-36N,49W-36N	12
Tomlin Rd. N&S	54W-32N	20
Trebess Dr. N&S	49W-35N	13
Windy Hill Rd. NS&EW	49W-36N	13
Woodland Dr. N&S	49W-33N	21
Woodvale Dr. E&W	49W-33N	21

Fox Lake*
(Also see Lake County)

	Grid	Page
Appaloosa Ln. NW&SE	29W-40N	11
Carriage Ln. N&S	29W-40N	11
Chelsey Ct. NE&SW	29W-40N	11
Chestnut Ridge NW&SE	29W-40N	11
Churchill Ct. NE&SW	29W-40N	11
Country Shore Ln. NS&EW	29W-40N	11
Downing Pl. N&S	29W-40N	11
English Prairie Rd. E&W	29W-42N	11
Fox Shire Ln. E&W	29W-40N	11
Heather Ridge NE&SW	29W-40N	11
Laurine Ln. N&S	29W-40N	11
Main St. E&W	29W-39N	11
Morgan Ct. NE&SW	29W-40N	11
Palimino Ct. N&S	29W-40N	11
Park Vista NW&SE	29W-40N	11
Steeple Ct. NE&SW	29W-40N	11
Sundial Ln. NS&EW	29W-40N	11
Surrey Ln. N&S	29W-40N	11
Wilmot Rd. N&S	29W-40N,29W-41N	11

Fox River Grove

	Grid	Page
Adams Av. NW&SE	30W-23N	43
Algonquin Ln. E&W	30W-22N,29W-22N	51
Amhurst Ct. E&W	29W-22N	51
Asbury Av. E&W	29W-22N	51
Asbury Ct. E&W	30W-22N,29W-22N	51
Asbury Rd. E&W	29W-22N	51
Ashburn Av. E&W	29W-22N	51
Ashbury Ct. E&W	30W-22N	51
Ashcroft Ct. NW&SE	30W-22N	51
Bay View Av. NE&SW	30W-22N	51
Beach Way NW&SE	30W-22N,30W-23N	51
Bradbury Rd. N&S	30W-22N	51
Brighton Ct. NE&SW	29W-22N	51
Bristol Ln. NE&SW	29W-22N	51
Cary Rd. NW&SE	29W-22N	51
Center St. E&W	29W-22N	51
Chatham Ct. E&W	29W-22N	51
Circle Rd. NE&SW	30W-23N	43
Concorde Av. E&W	29W-22N	51
Crescent Rd. E&W	30W-23N	43
Crompton Ct. NW&SE	30W-22N	51
Doverton Ln. E&W	29W-22N	51
Elder Ln. N&S	30W-22N	51
Essex Rd. NE&SW	30W-22N	51
Foxmoor Rd. NE&SW	30W-22N,29W-22N	51
Gladys Av. N&S	30W-22N	51
Glenhurst Ct. NE&SW	30W-22N	51
Glenice Pkwy. N&S	29W-23N	43
Grace Ln. N&S	30W-22N	51
Grace Ln. NW&SE	30W-22N	51
Grove Av. N&S	30W-22N	51
Harding Av. N&S	29W-23N	43
Highview Dr. E&W	29W-23N	43
Hillcrest Av. NE&SW	30W-22N	51
Huntcliff Ct. NW&SE	29W-22N	51
Hunter Cir. NE&SW	30W-22N	51
Hunters Way NS&EW	30W-22N,29W-22N	51
Illinois St. N&S	29W-23N	43
Johnson St. N&S	29W-23N	43
Keystone Ln. NW&SE	30W-22N	51
Lexington Av. NE&SW	30W-22N	51
Liberty Av. N&S	30W-22N	51
Lily Ln. NE&SW	29W-22N	51
Lincoln Av. N&S	30W-22N,30W-23N	51
Lucille Av. N&S	29W-22N,29W-23N	51
Manchester Ct. NW&SE	30W-22N	51
Maple Av. NW&SE	30W-22N	51
Melrose Ct. NW&SE	30W-22N	51
Midway St. N&S	30W-22N	51
Millard Av. N&S	30W-22N	51
Mound St. N&S	29W-22N	51

	Grid	Page
North Rd. N&S	29W-23N	43
North River Rd. NS&EW	30W-23N	43
Northwest Highway NW&SE	30W-23N,29W-22N	43
Oak Ct. NW&SE	30W-22N	51
Oak St. E&W	29W-23N	43
Old Hunt Rd. NW&SE	30W-22N	51
Opatmy Dr. E&W	30W-23N,29W-23N	43
Orchard St. N&S	29W-22N,29W-23N	51
Park Ct. NW&SE	29W-22N	51
Pine St. E&W	30W-22N	51
Pleasant St. E&W	29W-23N	43
Ridgeland St. E&W	29W-22N	51
River Rd. E&W	29W-23N	43
River Way Rd. NS&EW	30W-23N	43
School Dr. N&S	29W-22N	51
School St. E&W	29W-22N	51
Shannon Dr. NW&SE	30W-22N	51
Ski Hill Rd. NE&SW	29W-22N,29W-23N	51
Skyline Dr. NE&SW	29W-23N	43
South River Rd. NS&EW	30W-23N	43
Spring Ct. N&S	29W-23N	43
Thackery Ln. NW&SE	30W-22N	51
Tower Ct. N&S	30W-22N	51
Tower Pl. N&S	30W-22N	51
Tower Rd. NE&SW	30W-22N	51
Violet Av. NS&EW	29W-23N	43
Well St. E&W	29W-23N	43
Wildwood Ln. E&W	29W-23N	43
Woodbine Ln. NE&SW	29W-22N	51
Yorkshire Dr. N&S	29W-22N	51

Fox River Valley Gardens*
(Also see Lake County)

	Grid	Page
Center St. NE&SW	29W-26N,28W-26N	43
Channel Dr., S. NE&SW	29W-26N	43
Circle Av. NW&SE	29W-26N	43
Eastwood Av. N&S	29W-25N	43
Eastwood Ln. N&S	29W-25N,29W-26N	43
Evergreen Ct. E&W	29W-26N	43
Hilldale Ln. NW&SE	29W-26N	43
N. Circle Av. NW&SE	29W-26N	43
Oak St. NW&SE	29W-26N	43
Park Av. E&W	29W-25N	43
Park Ct. NE&SW	29W-25N	43
Park Dr. NW&SE	29W-25N	43
Peninsula Ln. NW&SE	29W-26N	43
Riverside Dr. N&S	29W-25N	43
S. Circle Av. NS&EW	29W-25N,29W-26N	43
Stub Av. NS&EW	29W-25N	43

Grafton Township

	Grid	Page
Adamson Rd. E&W	42W-23N,41W-23N	40
Albrecht Rd. E&W	39W-23N	40
Algonquin Rd. E&W	39W-21N	48
Arthur St. E&W	38W-25N	41
Ballard Rd. E&W	40W-25N,38W-25N	40
Bard Rd. E&W	39W-24N,37W-24N	40
Charlotte Av. N&S	37W-25N	41
Chestnut Dr. N&S	38W-25N,38W-26N	41
Chicago Av. E&W	37W-24N	41
Columbine Av. N&S	38W-25N	41
Conley Rd. NE&SW	40W-24N,39W-24N	40
Cote Dr. N&S	41W-25N	40
Cottage Dr. N&S	37W-24N	41
Coyne Station Rd. N&S	42W-21N,42W-23N	48
Crescent Dr. NE&SW	41W-25N	40
Crystal Beach Av. N&S	37W-25N	41
Crystal Lake Rd. N&S	38W-23N	49
Delaware St. N&S	38W-25N	41
Della Av. N&S	38W-25N	41
Dundee Rd. NW&SE	39W-20N	48
East End Av. N&S	37W-25N	41

Grid Page

Elm Ln. E&W 38W-25N 41
Ernesti Rd. E&W .. 41W-23N 40
Foster Rd. E&W
.... 40W-25N,39W-24N 40
Frank Rd. N&S 37W-21N 49
Georgine St. N&S .. 38W-25N 41
Grandview Dr. E&W
.... 41W-25N 40
Greenfield Rd. N&S
.... 37W-25N 41
Haligus Rd. N&S
.... 39W-22N,39W-25N 48
Hamilton Rd. N&S .. 41W-25N 40
Hawthorne Way E&W
.... 40W-25N 40
Hemmer Rd. N&S .. 42W-20N 48
Hemmingsen Rd. E&W
.... 42W-24N,41W-24N 40
Hensel Rd. E&W
.... 42W-25N,41W-25N 40
Hickory Av. E&W .. 37W-24N 41
Hillcrest Av. E&W .. 37W-24N 41
Hillview Dr. E&W .. 41W-25N 40
Huntley Rd. N&S
.... 37W-23N,37W-24N 41
Kasper Av. E&W .. 38W-25N 41
Kruetzer Rd. E&W .. 39W-20N 48
Lakeview Av. E&W
.... 37W-24N 41
Lakeview Dr. N&S .. 41W-25N 40
Lakewood Av. N&S
.... 37W-25N 41
Lavergone St. N&S
.... 38W-25N 41
Linder Av. E&W 38W-25N 41
Main St. Rd. E&W .. 41W-21N 48
Marengo Rd. NW&SE
.... 43W-23N 39
Martin St. N&S 38W-25N 41
Mayfield Av. N&S .. 37W-25N 41
Mensching Rd. N&S
.... 42W-24N,42W-25N 40
Miller Rd. E&W 38W-22N 49
Muriel St. N&S 38W-25N 41
N. Union Rd. N&S
.... 41W-23N,41W-24N 40
North Av. E&W 38W-25N 41
North Union Rd. N&S
.... 41W-23N,41W-24N 40
Pauline Av. N&S 38W-25N 41
Priegnitz Rd. N&S
.... 38W-22N,38W-24N 49
Reed Rd. E&W 40W-22N 48
Ridgeview Dr. N&S
.... 41W-25N 40
Rose Av. N&S 38W-25N 41
Rosebud Av. N&S .. 38W-25N 41
Rowland Rd. NS&EW
.... 40W-25N 40
Rt. 47 N&S 40W-22N 48
Ruth Rd. N&S 40W-21N 48
S. Lakeview Dr. E&W
.... 41W-25N 40
S. Union Rd. NS&EW
.... 40W-22N,41W-23N 48
Square Barn Rd. N&S
.... 37W-20N 49
Sunnyside Av. N&S
.... 37W-25N 41
Sunset Dr. N&S 37W-24N 41
Suttondale Dr. N&S
.... 40W-25N 40
Swanson Rd. N&S
.... 37W-23N,37W-22N 41
Teresa Av. E&W 38W-25N 41
Tryon St. N&S 38W-25N 41
Zimmerman Rd. NS&EW
.... 39W-24N,39W-25N 40

Greenwood Township

Aavang Rd. E&W .. 39W-35N 17
Acorn Path E&W .. 37W-35N 17
Allendale Rd. E&W
.... 41W-37N,38W-37N 16
Appleby Ct. NW&SE
.... 42W-36N 16
Arabian Tr. E&W .. 39W-34N 24
Autumn Ln. E&W .. 38W-33N 25
Baker Ter. E&W 41W-33N 24
Banford Rd. E&W .. 40W-32N 24
Barber Av. NW&SE
.... 38W-36N 17
Bennington Ct. NW&SE
.... 36W-32N 25
Billingsgate Ln. N&S
.... 42W-36N 16
Birchwood Dr. N&S
.... 37W-34N 25
Brody Ln. N&S 41W-33N 24
Brookham Ct. N&S
.... 42W-36N 16
Cambridge Ct. E&W
.... 36W-32N 25
Center St. NW&SE 38W-36N 17
Charles Rd. NW&SE
.... 42W-34N,39W-33N 24
Chatham Ct. N&S .. 38W-33N 25
Church St. E&W 38W-36N 17
Compton Ln. NW&SE
.... 42W-36N 16
Concorde Dr. N&S .. 36W-32N 25

Grid Page

Cooney Dr. E&W .. 41W-33N 24
Crest Ln. E&W 42W-36N 16
Deer Dr. E&W 37W-34N 25
East Lake Shore Dr. NE&SW
.... 37W-35N 17
Elm St. E&W 37W-35N 17
Elmwood Dr. E&W .. 33W-34N 26
Evergreen Dr. E&W
.... 38W-35N,37W-35N 17
Fleming Rd. N&S .. 39W-32N 24
Gerson Dr. E&W 37W-34N 25
Giant Oak Dr. N&S 37W-37N 17
Gleason Ct. N&S .. 41W-33N 24
Greenwood Rd. NE&SW
.... 39W-33N,38W-37N 24
Happy Tr. E&W 39W-34N 25
Hickory Dr. N&S 37W-34N 25
Hilltop Dr. N&S 37W-35N 17
Hogbag Rd. NE&SW
.... 37W-33N 25
Howe Rd. E&W
.... 38W-37N,36W-37N 17
Jankowski Rd. E&W
.... 42W-36N,41W-36N 16
Johnson Rd. NE&SW
.... 42W-37N 16
Lake Shore Dr. N&S
.... 37W-34N 25
Lake Vista Ln. N&S
.... 37W-35N 17
Lamb Rd. NE&SW
.... 42W-33N,42W-34N 24
Lathrop Dr. NE&SW
.... 38W-36N 17
Longwood Ct. N&S
.... 38W-34N 25
Maplewood Dr. N&S
.... 37W-34N 25
Marie Dr. NE&SW .. 37W-34N 25
McCannon Rd. E&W
.... 41W-32N 24
McCauley Rd. N&S
.... 42W-37N 16
Meadow Ln. E&W .. 37W-34N 25
Memory Ln. N&S .. 37W-34N 25
Memory Tr. E&W .. 37W-35N 17
Miller Rd. N&S 39W-37N 17
Miller Rd. E&W 41W-32N 24
Mustang Tr. N&S .. 39W-34N 25
Northwoods Ct. N&S
.... 38W-34N 25
Nusbaum Rd. E&W
.... 40W-34N 24
Oriole Tr. E&W
.... 38W-35N,37W-35N 17
Park St. SW&NE .. 38W-36N 17
Parker Ct. NW&SE 41W-33N 24
Pheasant Ln. E&W 39W-34N 24
Pine Av. E&W
.... 38W-35N,37W-35N 17
Pine Needle Pass E&W
.... 38W-34N 25
Pinoak Ln. E&W 37W-34N 25
Pleasure Av. N&S .. 38W-35N 17
Powers Rd. N&S .. 41W-32N 24
Private Farm Ln. E&W
.... 42W-37N,41W-37N 16
Queen Anne Rd. N&S
.... 40W-32N,40W-37N 24
Raffel Rd. NS&EW
.... 40W-32N,40W-34N 24
Ramble Rd. E&W .. 37W-35N 17
Raycraft Rd. NE&SW
.... 42W-34N,41W-36N 24
Rose Marie Rd. SW&NE
.... 37W-34N 25
Rose St. N&S 37W-34N 25
Rt. 120 NW&SE
.... 40W-32N,37W-33N 24
Rt. 47 N&S
.... 41W-33N,41W-37N 24
Saddlebred Tr. E&W
.... 39W-34N 24
Shadow Ln. E&W .. 37W-34N 25
Shadowood Dr. E&W
.... 38W-34N 25
Shady Ln. E&W 37W-35N 17
Spring Ln. E&W 38W-33N 25
Thayer Rd. E&W
.... 42W-37N,38W-37N 16
Thompson Rd. NS&EW
.... 38W-32N,37W-36N 25
Thoroughbred Tr. N&S
.... 39W-34N 25
Trey Rd. NW&SE .. 36W-32N 25
Valley Hill Rd. E&W
.... 36W-33N 25
Vine Av. E&W
.... 38W-35N,37W-35N 17
W. Sunset Dr. E&W
.... 37W-35N 17
Ware Rd. E&W 41W-33N 24
West Lake Shore Dr. NE&SW
.... 37W-35N 17
West Meadowlane N&S
.... 37W-35N 17
West Wonder Lake Rd.
NS&EW 38W-36N,37W-37N 17
Westwood Dr. N&S
.... 37W-35N 17
White Oaks Rd. E&W
.... 37W-37N 17

Grid Page

Wilmington Ct. N&S
.... 36W-32N 25
Windhill Ct. NE&SW
.... 38W-34N 25
Wondermere Rd. E&W
.... 38W-36N,37W-36N 17
Wonderview Dr. E&W
.... 37W-34N 25
Woodland Ln. N&S
.... 38W-33N 25
Woody Tr. E&W 37W-35N 17

Hartland Township

Alden Rd. NW&SE
.... 44W-37N,43W-34N 15
Antuna Dr. N&S 44W-33N 23
Azalea Ln. NW&SE
.... 43W-35N 15
Bauman Rd. E&W .. 48W-32N 22
Black Oak Ct. NW&SE
.... 48W-34N 22
Brink St. E&W 48W-37N 14
Brookdale Rd. N&S
.... 48W-32N,48W-33N 22
Bunker Hill Rd. E&W
.... 48W-34N,47W-34N 22
Compton Ln. NW&SE
.... 43W-36N 15
Cooney St. N&S 44W-34N 23
Countryside Ln. N&S
.... 47W-36N 14
Debbie Ln. E&W 48W-34N 22
Deep Cut Rd. NE&SW
.... 46W-33N,45W-35N 22
Dimmel Rd. N&S .. 46W-32N 22
Dogwood Ln. E&W
.... 43W-34N 23
Dunham Rd. E&W
.... 48W-33N,46W-33N 22
Frank Ct. E&W 48W-34N 22
Goddard St. E&W .. 44W-34N 23
Green Rd. E&W
.... 45W-37N,44W-37N 15
Green Rd. NW&SE 45W-37N 15
Greenway Cross E&W
.... 44W-32N 23
Hartland Rd. N&S
.... 45W-32N,45W-34N 23
Hawthorn St. N&S .. 48W-35N 14
Hidden Lake Rd. NE&SW
.... 43W-34N,43W-35N 23
Hillside Ln. E&W 45W-33N 23
Honeysuckle Ln. E&W
.... 43W-34N 23
Indian Trace N&S .. 44W-32N 23
Irish Ln. N&S 47W-37N 14
Jankowski Rd. E&W
.... 43W-36N 15
King Rd. N&S 47W-36N 14
Lamb Rd. N&S 43W-32N 23
Lembke Rd. E&W .. 48W-35N 14
Lilac Ln. NW&SE .. 43W-34N 23
Lincoln Rd. E&W
.... 48W-37N,46W-37N 14
Lindwall Rd. N&S
.... 48W-35N,48W-36N 14
Maeve Ln. NS&EW
.... 44W-32N 23
Marawood Dr. NW&SE
.... 44W-32N 23
Maxwell St. N&S .. 44W-34N 23
McCauley Rd. NE&SW
.... 43W-36N,43W-37N 15
McGee Rd. E&W .. 43W-32N 23
McGuire Rd. NS&EW
.... 48W-37N,44W-37N 14
Menge Rd. N&S
.... 48W-32N,48W-33N 22
Murray Rd. N&S
.... 44W-34N,44W-35N 23
Nelson Rd. E&W
.... 45W-34N,43W-34N 23
Park Lane Dr. N&S
.... 45W-32N,45W-33N 23
Paulson Rd. N&S
.... 47W-33N,47W-35N 22
Red Bud Ln. E&W .. 43W-35N 15
Red Oak Dr. N&S .. 48W-34N 22
Rose Farm Rd. N&S
.... 44W-32N,44W-34N 23
Rt. 120 E&W 43W-32N 23
Schultz Rd. N&S .. 48W-37N 14
Shamrock Ln. E&W
.... 44W-32N,43W-32N 23
Shannon Dr. N&S .. 43W-32N 23
Somerset Ct. NE&SW
.... 43W-36N 15
St.Patrick Rd. E&W
.... 45W-35N,43W-35N 15
Streit Rd. E&W
.... 48W-36N,45W-36N 14
Sutton Pl. NS&EW .. 44W-32N 23
Thayer Rd. E&W
.... 44W-37N,43W-37N 15
Tina Dr. N&S 48W-34N 22
U.S. Route 14 NW&SE
.... 48W-35N,43W-31N 14
Walsh Dr. E&W 44W-33N 23
West Rd. N&S
.... 43W-34N,43W-35N 23

Grid Page

Wilson Rd. N&S
.... 45W-35N,45W-36N 15
Windsor Dr. NS&EW
.... 43W-36N 15

Harvard

Admiral Dr. E&W .. 50W-37N 13
Anderson St. E&W .. 50W-38N 13
Ayer St. NE&SW .. 50W-38N 13
Beaumont St. E&W
.... 51W-38N,50W-38N 13
Blackman St. E&W
.... 50W-38N,49W-38N 13
Blaine St. E&W
.... 50W-38N,49W-38N 13
Blanchard St. N&S .. 50W-38N 13
Bourn St. E&W 49W-38N 13
Brainard St. NW&SE
.... 50W-38N 13
Brink St. E&W
.... 50W-37N,49W-37N 13
Brown St. E&W
.... 50W-38N,49W-38N 13
Burbank St. E&W .. 50W-38N 13
Campell St. NE&SW
.... 50W-38N 13
Church St. E&W
.... 50W-38N,49W-38N 13
Deer Path Rd. E&W
.... 49W-39N 13
Dewey St. E&W
.... 50W-38N,49W-38N 13
Diggins St. E&W
.... 50W-38N,49W-38N 13
Division St. N&S
.... 50W-37N,50W-38N 13
Dunlap St. NE&SW
.... 50W-38N 13
Eastman St. NE&SW
.... 50W-38N 13
Eighth St. N&S
.... 50W-38N,50W-39N 13
Fifth St. N&S 50W-38N 13
Finney St. NE&SW .. 50W-38N 13
First St. N&S 50W-38N 13
Fourth St. N&S 50W-38N 13
Frisco Dr. E&W 49W-38N 13
Front St. NW&SE .. 50W-38N 13
Galvin Pkwy. N&S .. 49W-38N 13
Garfield St. N&S
.... 50W-38N,50W-39N 13
Grant St. N&S 49W-38N 13
Harrison St. E&W .. 50W-38N 13
Hart St. N&S 50W-38N 13
Harvard Hills Rd. N&S
.... 49W-38N 13
Hayes Ct. E&W 49W-38N 13
Hayes St. N&S 49W-38N 13
Hillside Dr. E&W 50W-39N 13
Howard St. N&S 50W-38N 13
Hutchinson St. NE&SW
.... 50W-38N 13
Jackman St. NE&SW
.... 50W-38N 13
Janowski St. NW&SE
.... 50W-38N 13
Jefferson St. N&S .. 50W-38N 13
Joe Ct. N&S 49W-38N 13
Johnson St. NE&SW
.... 50W-38N 13
Kennedy Dr. N&S .. 49W-38N 13
Kilkeen Ln. NW&SE
.... 50W-38N 13
Klaman St. E&W 49W-38N 13
Lincoln St. N&S 50W-38N 13
Little John St. N&S
.... 49W-38N 13
Marengo Rd. N&S .. 50W-37N 13
McComb St. E&W .. 50W-38N 13
McKinley Av. E&W
.... 50W-38N,49W-38N 13
Mcguire Rd E&W .. 49W-37N 13
Metzen St. NW&SE
.... 50W-38N 13
Ninth St. N&S 50W-38N 13
Northfield Av. E&W
.... 50W-39N 13
O'Brien St. N&S 49W-38N 13
Old Orchard Rd. E&W
.... 49W-39N 13
Page St. NE&SW .. 50W-38N 13
Park Dr. E&W 50W-37N 13
Park St. NW&SE .. 50W-38N 13
Pheasant Run Rd. N&S
.... 49W-39N 13
Railroad St. NE&SE
.... 50W-38N 13
Ramer Rd. E&W 52W-38N 13
Randall St. NE&SW
.... 50W-38N 13
Ratzaff St. N&S 50W-38N 13
Ridge Ln. N&S 50W-37N 13
Roosevelt St. E&W
.... 50W-38N 13
Rt. 173 NE&SW 49W-38N 13
Second St. N&S 50W-38N 13
Seventh St. N&S .. 50W-39N 13
Sixth St. N&S
.... 50W-38N,50W-39N 13
Summer NW&SE .. 50W-38N 13
Tenth St. N&S 50W-39N 13

Grid Page

Third St. N&S 50W-38N 13
Thompson St. NW&SE
.... 50W-38N 13
U.S. Rt. 14 N&S
.... 50W-38N,50W-39N 13
University St. E&W
.... 50W-38N,49W-38N 13
Washington St. NW&SE
.... 50W-38N 13
West Dr. E&W 50W-38N 13

Hebron

Central Av. N&S 40W-41N 8
Church St. N&S 40W-42N 8
Fourth Av. E&W 40W-41N 8
Harrison Av. E&W
.... 40W-41N,41W-41N 8
Illinois St. N&S 40W-41N 8
Jacob Av. E&W 41W-41N 8
Jean Dr. N&S 41W-41N 8
Main St. N&S
.... 41W-41N,41W-42N 8
Maple Av. E&W 40W-42N 8
Marci Ln. N&S 41W-41N 8
McKinley Av. E&W
.... 41W-41N,40W-41N 8
Mead Av. E&W 40W-42N 8
Prairie Av. E&W 40W-41N 8
Price Rd. E&W
.... 41W-41N,40W-41N 8
Rt. 173 (Maple Av.) NS&EW
.... 41W-41N,40W-42N 8
St. Albans St. N&S
.... 40W-41N 8
Third Av. E&W 40W-41N 8
Union Av. N&S 40W-41N 8
Western Av. NW&SE
.... 41W-42N 8
Woods Ln. N&S 41W-41N 8

Hebron Township

Armory Rd. N&S .. 39W-43N 8
Barnard Mill Rd. NW&SE
.... 37W-39N,37W-38N 9
Burgett Rd. E&W
.... 38W-43N,37W-43N 9
Durkee Rd. E&W .. 42W-39N 8
Fifth Av. E&W 40W-41N 8
First Av. E&W 40W-41N 8
Fourth Av. E&W 40W-41N 8
Freeman Rd. N&S
.... 41W-42N,41W-43N 8
Giant Oak Dr. N&S 37W-38N 17
Green St. N&S 40W-41N 8
Greenwood Rd. N&S
.... 38W-38N,38W-41N 17
Hebron Rd. E&W
.... 42W-42N,41W-42N 8
Hickory Ln. Rd. E&W
.... 38W-38N,37W-38N 17
Hillside Rd. N&S
.... 42W-42N,42W-43N 8
Johnson Rd. N&S
.... 42W-38N,42W-41N 16
Kemman Rd. N&S
.... 40W-38N,40W-41N 16
Lange Rd. N&S
.... 38W-42N,38W-43N 9
Miller Rd. N&S 39W-38N 17
Nash Rd. E&W 40W-38N 16
Nichols Rd. NW&SE
.... 42W-43N,42W-42N 8
Norgard Rd. E&W .. 37W-41N 9
O'Brien Rd. E&W
.... 42W-40N,40W-40N 8
Okeson Rd. E&W
.... 39W-41N,38W-41N 8
Price Rd. E&W
.... 41W-41N,40W-41N 8
Regnier Rd. E&W .. 37W-40N 9
Route 173 E&W
.... 42W-41N,37W-41N 8
Route 47 N&S
.... 41W-38N,41W-43N 16
Seaman Rd. NS&EW
.... 40W-42N,39W-42N 8
Second Av. E&W .. 40W-41N 8
State Line Rd. E&W
.... 41W-43N 8
Stewart Rd. N&S
.... 39W-39N,39W-40N 8
Thayer Rd. E&W
.... 42W-38N,38W-38N 16
Third Av. E&W 40W-41N 8
Tryon Grove Rd. NE&SW
.... 37W-39N 9
Vander Karr Rd. NW&SE
.... 40W-39N,38W-39N 8

Holiday Hills

Ash St. E&W 30W-29N 35
Aspen Ct. N&S 30W-29N 35
Beach Dr. E&W 30W-29N 35
Birch St. E&W 30W-29N 35
Birchwood Av. N&S
.... 30W-29N 35
Bittersweet Av. N&S
.... 30W-29N 35
Catalpa St. E&W .. 30W-29N 35
Cedar St. E&W 30W-29N 35

	Grid	Page
Driftwood Dr. N&S	30W-29N	35
Elder Ct. N&S	30W-29N	35
Elder Ln. NS&EW	30W-29N	35
Elm St. E&W	30W-29N	35
Evergreen St. E&W	30W-29N	35
Fox St. E&W	30W-29N	35
Grove St. N&S	30W-29N	35
Hickory St. E&W	30W-29N	35
Holiday Dr. N&S	30W-29N	35
Hyde Park Av. N&S	30W-29N	35
Kenilworth Av. N&S	30W-29N	35
Lakeshore Dr. E&W	30W-29N	35
Lakeview Dr. N&S	30W-29N	35
Lilac St. N&S	30W-29N	35
Michael Ct. N&S	30W-29N	35
Oak St. N&S	30W-29N	35
Park Ter. N&S	30W-29N	35
Pine St. E&W	30W-29N	35
Poplar St. E&W	30W-29N	35
Riverview Rd. NW&SE	30W-29N	35
Short St. N&S	30W-29N	35
Spruce St. E&W	30W-29N	35
Sunset Dr. E&W	30W-29N	35
Swallow Dr. E&W	30W-29N	35
Violet Av. E&W	30W-29N	35
Willow Av. NS&EW	30W-29N	35

Huntley

	Grid	Page
Algonquin Rd. E&W	40W-21N	48
Auson Ct. NW&SE	40W-20N	48
Bakley St. N&S	40W-21N	48
Becky Lee Trace E&W	40W-20N	48
Bernice St. E&W	40W-21N	48
Bonnie Brae St. N&S	41W-21N	48
Borden St. E&W	40W-21N	48
Brittany Av. N&S	40W-21N	48
Carl St. N&S	40W-21N	48
Carol Ln. E&W	40W-20N	48
Charles St. N&S	40W-21N	48
Church Av. N&S	40W-21N	48
Church St. N&S	40W-20N	48
Coral St. E&W	40W-21N	48
Daniel Ln. E&W	40W-20N	48
Davey Dr. E&W	40W-20N	48
Dean St. N&S	40W-20N	48
Donald Dr. N&S	40W-20N	48
Douglas Av. E&W	40W-21N	48
Edward St. E&W	40W-21N	48
Elizabeth Av. N&S	40W-21N	48
First St. E&W	40W-21N	48
Fourth St. E&W	40W-21N	48
Fredrick Way N&S	40W-20N	48
Gayle Ct. NW&SE	40W-20N	48
Giordano Pl. E&W	40W-20N	48
Grove St. N&S	40W-20N	48
Heinz Av. E&W	40W-20N	48
Henry St. E&W	40W-20N	48
Janice Dr. NW&SE	40W-21N	48
Joan Av. E&W	40W-21N	48
Kathleen Av. N&S	40W-21N	48
Kenneth St. E&W	40W-21N	48
Kiley E&W	40W-21N	48
Kishwaukee Ct. N&S	40W-20N	48
Kruetzer Rd. E&W	40W-20N	48
Lincoln St. N&S	40W-20N	48
Lori Ln. N&S	40W-20N	48
Main St. E&W	40W-21N	48
Margaret Av. N&S	40W-21N	48
Martin Dr. NW&SE	40W-20N	48
Mathews St. N&S	40W-21N	48
Michael St. N&S	40W-20N	48
Mill St. E&W	40W-20N	48
Mullin St. N&S	40W-21N	48
Myrtle St. N&S	40W-20N	48
North St. E&W	40W-21N	48
Phillip Dr. N&S	40W-21N	48
Pineway E&W	40W-20N	48
Powder Park Rd. NW&SE	40W-20N	48
Ronald St. E&W	40W-21N	48
Rt. 47 N&S	40W-22N	48
Second St. E&W	40W-21N	48
Cir.	40W-20N	48
Smith Dr. NS&EW	40W-20N	48
Sunset Ln. N&S	41W-20N	48
Third St. E&W	40W-21N	48
Timber Dr. E&W	40W-21N	48
Timber Dr. E. N&S	40W-21N	48
Timer Dr. W. N&S	40W-21N	48
Vine St. N&S	40W-21N	48
Woodcreek Dr. N&S	40W-20N	48
Woodstock St. N&S	40W-21N	48

Island Lake*
(Also see Lake County)

	Grid	Page
Ascot Ct. N&S	29W-28N	35
Bassler Dr. E&W	29W-28N	35
Bassler Ln. E&W	29W-28N	35
Burnett Rd. E&W	29W-28N,28W-28N	35
Cedar Ter. N&S	29W-28N	35
Charles Ct. E&W	29W-28N	35
Cir.	29W-28N	35
Concord Dr. E&W	30W-28N,29W-28N	35
Dartmouth Ct. NW&SE	29W-28N	35
Dartmouth Dr. NS&EW	30W-28N,29W-28N	35
David Ct. E&W	29W-28N	35
Debden Dr. SW&NE	30W-28N	35
Dorthy Ct. E&W	29W-28N	35
Eastport Ct. E&W	29W-28N	35
Elm Ct. E&W	29W-28N	35
Ethel Ter. N&S	29W-28N	35
Fern Ct. E&W	29W-28N	35
Greenwich Ln. E&W	30W-28N	35
Hale Ln. N&S	30W-28N,29W-27N	35
Hale Ln. NW&SE	30W-28N	35
Hazel Ct. E&W	29W-28N	35
Hickory Ter. E&W	29W-28N	35
Highland Dr. NE&SW	29W-28N	35
Hillside Dr. NE&SW	29W-28N	35
Honeysuckle Dr. E&W	29W-28N	35
Huntington Ln. NS&EW	30W-28N	35
Hyacinth Ter. N&S	29W-28N	35
Island Dr. NS&EW	29W-28N	35
Janet Dr. E&W	29W-28N	35
Jordan Ln. N&S	29W-29N	35
Kelly Rd. N&S	29W-27N	35
Kings Point Ct. N&S	30W-28N	35
Knights Bridge Dr. SW&NE	30W-28N	35
Lakeview Dr. N&S	29W-28N	35
Lauren Ln. NW&SE	29W-29N	35
Lexington Ln. E&W	30W-28N,29W-28N	35
Lillian Ct. NW&SE	29W-28N	35
Linden Dr. NS&EW	30W-28N	35
Marion Ct. NW&SE	29W-28N	35
Medford Ct. N&S	29W-28N	35
Midway Dr. N&S	29W-28N	35
Nantucket Way E&W	29W-28N	35
Newbury Dr. NW&E	29W-29N	35
Newport Ct. E&W	29W-28N	35
Newport Dr. NE&SW	29W-28N	35
Nottingham Ct. E&W	29W-28N	35
Oak Ter. N&S	29W-28N	35
Olympia Ln. N&SE	30W-28N	35
Oxford Ct. SW&NE	29W-28N	35
Pine Ter. N&S	29W-28N	35
Pleasant Pl. E&W	29W-28N	35
Plymouth Ln. SW&NE	29W-28N	35
Porten Rd. E&W	29W-27N	35
Portsmouth Dr. E&W	30W-28N,29W-28N	35
Quiney Ct. E&W	29W-28N	35
Ralph Ct. E&W	29W-28N	35
Revere Ln. E&W	30W-28N,29W-28N	35
Richard Ct. E&W	29W-28N	35
River St. N&S	29W-26N,29W-27N	43
Salem Ct. NW&SE	30W-28N,29W-28N	35
South Hampton Dr. E&W	29W-27N	35
Southport Dr. NE&SW	30W-28N	35
Spruce Ter. N&S	29W-28N	35
Spruce Tr. N&S	29W-29N	35
Stafford Ct. NE&SW	30W-28N	35
Tulip Cir. SW&NE	29W-29N	35
Victoria Dr. E&W	29W-27N	35
Vista Dr. E&W	29W-27N	35
Walnut Ct. N&S	29W-28N	35
Waterford Way NS&EW	29W-28N	35
Wayside Dr. N&S	29W-27N	35
Wembly Dr. NW&SE	30W-28N	35
Westminister Ln. N&S	30W-28N	35
Windsor Dr. NE&SE	29W-27N	35
Yale Ln. E&W	29W-28N	35
York Ln. N&S	30W-28N	35

Johnsburg

	Grid	Page
Acre Ln. E&W	30W-35N	19
Appleton Ct. NE&SW	30W-36N	19
Ardmore Cir. NE&SW	30W-36N	19
Aspen Dr. E&W	31W-35N	19
Bald Knob Pt. E&W	29W-35N	19
Bald Knob Rd. E&W	29W-35N	19
Bayview Dr. E&W	30W-36N	19
Bayview Ln. NE&SW	30W-36N	19
Berwyn Ct. E&W	30W-36N	19
Blitsch Pl. N&S	31W-35N	19
Bolling Av. E&W	30W-35N	19
Brorson Ln. NS&EW	30W-36N	19
Brynmawr Ct. E&W	30W-36N	19
Buckanan Rd. E&W	32W-35N	18
Camden Ct. E&W	30W-36N	19
Cardington Way N&S	30W-35N	19
Carol Ct. E&W	30W-35N	19
Caroline Dr. N&S	30W-35N	19
Channel Beach Av. E&W	30W-36N	19
Chapel Hill Rd. NW&SE	31W-35N	19
Charnbrook Dr. E&W	30W-35N	19
Charnbrook Ln. E&W	30W-35N	19
Chellington Dr. N&S	30W-35N	19
Cherokee Dr. N&S	32W-35N	19
Church St. E&W	31W-35N	19
Cimmaron Cir. NS&EW	30W-35N	19
Cimmaron Ct. E&W	30W-36N	19
Claremont Rd. N&S	32W-35N	18
Cobblestone Ct. E&W	30W-36N	19
Countryside Dr. N&S	31W-35N	19
Cove Ct. N&S	31W-35N	19
Creekside Ln. E&W	30W-36N	19
Dawnwood Dr. N&S	30W-36N	19
Delaware Pl. E&W	31W-35N,30W-35N	19
Denise Ct. SW&NE	30W-36N	19
Dennis Blvd. N&S	30W-36N	19
Devon Ln. E&W	30W-36N	19
Dutch Creek Ln. N&S	31W-35N	19
Ernst Ln. SW&NE	30W-35N	19
Fairview Av. N&S	31W-35N	19
Farmington Ln. N&S	31W-35N	19
Filmore Rd. E&W	32W-35N	18
Florence Dr. N&S	30W-36N	19
Fox Knoll Ct. E&W	31W-35N	19
Frances Rd. NE&SW	30W-35N	19
Garfield Rd. NE&SW	32W-35N	18
Goodview Av. E&W	30W-35N	19
Grand Ridge Ct. N&S	29W-35N	19
Grant Dr. E&W	32W-35N	18
Hamblin Dr. N&S	32W-35N	18
Harrison Ln. E&W	29W-35N	19
Hayden Dr. E&W	30W-36N	19
Hickory Ln. N&S	30W-36N	19
Hickory Way N&S	30W-36N	19
Hillcrest Av. N&S	31W-35N	19
Hilltop Dr. N&S	30W-36N	19
Irene Ct. NW&SE	30W-35N	19
Ivy Ln. N&S	31W-35N	19
Jamesway N&S	31W-35N	19
Jewel Dr. NW&SE	30W-35N,30W-36N	19
Johnsburg Rd. E&W	32W-35N	19
Johnsburg Spring Grove Rd NE&SW	30W-35N,30W-36N	19
Knob Hill Rd. NS&EW	31W-35N	19
Lake Dawnwood Ln. N&S	30W-36N	19
Lake St. N&S	30W-36N	19
Lakeview St. E&W	30W-36N	19
Lewis Ln. NW&SE	29W-35N	19
Linden Av. NS&EW	30W-35N	19
Long Av. NW&SE	30W-35N	19
Longmeadow Dr. E&W	31W-35N	19
Maple Av. NW&SE	30W-35N	19
Maple Ter. E&W	30W-35N	19
Mayfair Dr. N&S	30W-36N	19
Mehring Av. E&W	31W-35N	19
Middle Av. NW&SE	30W-35N	19
Oak Grove Rd. E&W	29W-35N	19
Oak Leaf Dr. NS&EW	30W-36N	19
Oakdale Av. N&S	31W-35N	19
Oeffling Dr. E&W	30W-36N	19
Old Bay Rd. NW&SE	30W-34N	27
Olive Av. E&W	31W-35N	19
Pinehurst Ct. N&S	30W-36N	19
Pistakee View Dr. NE&SW	29W-36N	19
Pitzen Rd. N&S	29W-35N	19
Promentory Ln. N&S	30W-36N	19
Quincy Av. E&W	30W-35N	19
Raintree Ct. E&W	30W-36N	19
Ramsgate Ln. N&S	30W-35N	19
Reed Av. NE&SW	31W-35N	19
Ridgeview Dr. NS&EW	29W-35N	19
Ringwood Rd. E&W	30W-36N	19
River Park Dr. NE&SW	31W-35N	19
River Terrace Dr. NE&SW	30W-35N,29W-36N	19
Riverview Dr. N&S	30W-36N	19
Rocky Beach Ln. NS&EW	30W-35N,29W-35N	19
Rolling Lane Rd. E&W	30W-35N	19
Rosemarie Dr. E&W	30W-36N	19
Rosewood Ct. E&W	30W-36N	19
Rosewood Dr. E&W	30W-36N	19
Round Hill Ct. NE&SW	30W-36N	19
S. Brixham Dr. E&W	30W-35N	19
Salem Av. E&W	31W-35N,30W-35N	19
Sandalwood Ct. E&W	30W-36N	19
Sandalwood Dr. NW&SE	30W-36N	19
Sandalwood Ln. E&W	30W-36N	19
Second Av. NE&SW	30W-35N	19
Seville Ct. E&W	30W-36N	19
Spring Grove Rd. N&S	31W-35N	19
St. Francis Av. E&W	31W-35N	19
St. Johns Av. N&S	31W-35N	19
Stubby Av. NW&SE	30W-35N	19
Sudeenew Dr. E&W	30W-36N	19
Sundown Ln. E&W	30W-36N	19
Sunnyside Beach Dr. NE&SW	30W-35N	19
Sunnyside Dr. NE&SW	30W-35N	19
Sunset Dr. N&S	30W-36N	19
Sunset Ln. N&S	30W-36N	19
Tall Tree Ln. E&W	30W-36N	19
Tichfield Ter. E&W	31W-35N	19
Tivoli Ter. E&W	31W-35N	19
Torrey Ln. E&W	30W-36N	19
Villanova Dr. E&W	30W-36N	19
W. Lakeview St. E&W	30W-36N	19
W. May Dr. E&W	30W-36N	19
Weingart Rd. N&S	30W-35N	19
Wilmot Ct. NW&SE	30W-36N	19
Windmere Ln. NS&EW	31W-35N	19

Lake In The Hills

	Grid	Page
Acorn Ln. E&W	35W-21N	50
Adams St. N&S	35W-22N	50
Albrecht Rd. E&W	39W-23N	49
Algonquin Rd. E&W	36W-21N	49
Apache Tr. N&S	34W-21N	50
Appletree Ct. NW&SE	38W-22N	49
Ash St. N&S	35W-22N	50
Barrington Ct. N&S	37W-22N	49
Beechwood E&W	36W-21N	49
Bernice St. E&W	35W-21N	50
Birch St. N&S	35W-22N	50
Blackberry Ct. E&W	37W-22N	49
Blackberry Dr. NS&EW	37W-22N	49
Blackhawk Dr. E&W	34W-21N	50
Boulder Dr. NS&EW	37W-21N	49
Brandt Dr. E&W	34W-22N	50
Burr St. N&S	35W-22N	50
Cedar Ridge Dr. N&S	35W-21N	50
Cedar St. E&W	35W-22N	50
Cherokee Tr. N&S	34W-21N	50
Cherry St. N&S	35W-22N	50
Cheyenne St. E&W	35W-21N	50
Chippewa Tr. N&S	34W-21N	50
Clark Rd. E&W	36W-22N	49
Clayton Marsh Dr. N&S	36W-22N	49
Cobblestone Ct. N&S	37W-21N	49
Cobblestone Tr. NS&EW	36W-21N	49
Council Tr. NS&EW	34W-21N	50
Country Club Dr. NE&SW	37W-21N	49
Crabapple Ct. NW&SE	37W-22N	49
Creekside N&S	36W-21N	49
Creekview Ln. NW&SE	34W-22N	50
Crystal Lake Rd. N&S	35W-21N,35W-22N	50
Decatur Av. N&S	35W-22N	50
Deer Creek Ct. N&S	37W-22N	49
Deer Creek Ln. N&S	37W-22N	49
Deer Path E&W	35W-21N	50
Delaware Dr. E&W	34W-21N	50
E. Oak St. E&W	34W-22N	50
Echo Hill N&S	35W-21N	50
Elm St. N&S	35W-22N	50
Farmington Ct. E&W	38W-22N	49
Fir St. E&W	35W-22N	50
Flowerfield Ct. NW&SE	38W-22N	49
Granite Ct. E&W	37W-22N	49
Grant Av. E&W	35W-22N	50
Greenfield Ln. E&W	38W-22N	49
Hawthorne Tr. N&S	35W-21N	50
Hazelwood Ct. NW&SE	38W-22N	49
Hiawatha Dr. NS&EW	34W-21N	50
Hickory Rd. NW&SE	35W-21N	50
Hilltop Dr. NS&EW	35W-21N,35W-22N	50
Hilly Ln. N&S	35W-21N	50
Hollyhock Ct. E&W	37W-22N	49
Hunters Path E&W	35W-21N	50
Huron Tr. N&S	34W-21N	50
Indian Tr. NS&EW	35W-21N,35W-22N	50
Industrial Dr. NW&SE	34W-22N	50
Jefferson St. N&S	35W-22N	50
Jessie Rd. E&W	34W-21N	50
Lake Tr. N&S	35W-21N	50
Larkspur Ln. N&S	37W-22N	49
Lee Av. E&W	35W-22N	50
Lincoln St. E&W	35W-22N	50
Lindon St. E&W	35W-22N	50
Manchester Ct. E&W	38W-22N	49
Maple St. N&S	35W-22N	50
Marble Ct. NW&SE	37W-22N	49
Mason Ln. NS&EW	37W-22N	49
McKinley St. E&W	35W-22N	50
McPhee Dr. NW&SE	34W-22N	50
Meadow Ln. N&S	35W-22N	50
Menominee Dr. E&W	34W-21N	50
Middlefield Ct. NW&SE	38W-22N	49
Miller Rd. E&W	35W-22N	50
Miller Rd. E&W	39W-23N	40
Miller Rd. NS&EW	38W-23N	49
Mohawk Tr. N&S	34W-21N	50
Mohican Tr. N&S	34W-21N	50
Monroe St. N&S	35W-22N	50
Morningside Ct. NW&SE	37W-22N	49
Mulberry Ct. E&W	37W-22N	49
Navajo Dr. E&W	34W-21N	50
Oakleaf Rd. N&S	35W-21N	50
Patton Av. N&S	35W-22N	50
Pawnee St. E&W	35W-21N	50
Peartree Dr. NW&SE	37W-22N	49
Pebblewood E&W	36W-21N	49
Pershing Av. E&W	35W-22N	50
Pheasant Tr. E&W	35W-21N	50
Pine St. N&S	35W-22N	50
Pingree Rd. N&S	34W-23N	42
Plum St. E&W	35W-22N	50
Pocahontas Tr. N&S	34W-21N	50
Poplar St. N&S	35W-22N	50
Prosper Ct. E&W	34W-22N	50
Pyott Rd. NW&SE	35W-23N,34W-22N	42

Street	Grid	Page
Quail Run E&W	35W-21N	50
Ramble Rd. E&W	35W-21N	50
Randall Rd. N&S	36W-22N	49
Ridge Ct. NE&SW	37W-21N	49
Ridge Ln. NE&SW	37W-21N	49
Rolling Hills Ct. E&W	38W-22N	49
Rolling Hills Dr. SW&NE	38W-22N	49
Roosevelt St. E&W	35W-23N	42
Royal Oak Ct. N&S	38W-22N	49
Royal Oak Ln. NS&EW	37W-22N	49
Rt. 31 N&S	33W-23N	42
Sandstone Ct. E&W	37W-21N,36W-21N	49
Seminole Tr. N&S	34W-21N	50
Service Rd. NE&SW	34W-22N	50
Shawnee Dr. E&W	34W-21N	50
Sioux Tr. N&S	34W-21N	50
Spring Garden Ct. NW&SE	37W-22N	49
Springlake Ct. E&W	37W-22N	49
Springlake Dr. NS&EW	37W-22N	49
Spruce St. N&S	35W-22N	50
Stone Creek NW&SE	36W-21N	49
Sugarcreek Ct. SW&NE	38W-22N	49
Sycamore St. N&S	35W-22N	50
Thornberry Way NW&SE	37W-22N	49
Village Creek Dr. N&S	35W-21N,35W-22N	50
Virginia Rd. NW&SE	34W-23N,33W-22N	42
Walnut Dr. N&S	35W-21N	50
Walnut Grove Ct. NW&SE	38W-22N	49
Walter Ct. NE&SW	34W-22N	50
Wander Way N&S	35W-21N	50
Washington St. N&S	35W-22N	50
West Oak St. E&W	35W-22N	50
Westminster Ct. NW&SE	38W-22N	49
Willow St. E&W	34W-22N	50
Woodbridge NE&SW	36W-21N	49
Woodland Rd. E&W	35W-22N	50
Woody Way E&W	35W-21N	50

Lakemoor*

(Also see Lake County)

Street	Grid	Page
Amanda Av. NE&SW	29W-31N	27
Anthony Ln. E&W	30W-33N	27
Bay Meadows Ln. NE&SW	30W-34N	27
Bay Oaks Dr. N&S	30W-33N,30W-34N	27
Bay Rd. E&W	30W-34N	27
Cuhlman Rd. N&S	29W-33N,29W-34N	27
Dominion Ln. NW&SE	30W-34N	27
East Lake Dr. N&S	29W-32N	27
Edward Ct. N&S	30W-33N	27
Elaine Ter. N&S	29W-32N	27
Fritsche Rd. N&S	29W-31N	27
Herbert Rd. N&S	29W-31N	27
Highland Dr. N&S	29W-31N	27
Hollywood Dr. N&S	29W-32N	27
Lake Shore Dr. N&S	29W-31N,29W-32N	27
Lakeview Dr. N&S	29W-32N	27
Lily Lake Pkwy. E&W	29W-31N	27
Lily Ln. N&S	29W-32N	27
Lincoln Rd. E&W	30W-33N	27
Longbeach Rd. N&S	29W-31N	27
Lotus Dr. E&W	29W-31N	27
Maplewood Dr. N&S	29W-32N	27
Morningside Dr. E&W	29W-31N	27
Northlake Rd. E&W	29W-32N	27
Oshea Ct. E&W	30W-34N	27
Park Rd. NS&EW	29W-31N	27
Pine Grove Rd. N&S	29W-32N	27
Rand Rd. NW&SE	29W-32N	27
River Side Dr. NW&SE	29W-32N	27
Rosedale Dr. N&S	29W-32N	27
Route 120 NW&SE	30W-32N	27
Santa Barbara Rd. N&S	29W-32N	27
Scotland Rd. E&W	29W-31N	27
Sheridan Rd. NS&EW	29W-31N	27
Shore Dr. N&S	29W-31N,29W-32N	27
South Dr. E&W	29W-31N	27
Sunnyside Rd. E&W	29W-32N	27
Sunset Dr. E&W	29W-31N	27
Tiajuana Dr. N&S	29W-32N	27
Valley View Dr. NW&SE	29W-31N	27
Venice Rd. N&S	29W-31N	27
Wegner Rd. E&W	29W-31N	27
Willow Rd. N&S	29W-32N	27

Lakewood

Street	Grid	Page
Ackman Rd. E&W	38W-23N,37W-23N	41
Bannockburn Dr. NW&SE	39W-23N	40
Bard Rd. NE&SW	39W-24N,37W-24N	40
Bardwell Ln. E&W	37W-24N	41
Beaver Pond Ct. E&W	38W-24N	41
Bernice Dr. NE&SW	39W-24N	40
Bonnie Dr. NS&EW	39W-24N	40
Bonnie Ridge Rd. N&S	38W-24N	41
Braemer Cir. NS&EW	38W-24N	41
Braeton Cir. NS&EW	39W-23N	40
Braeton Way NW&SE	39W-23N	40
Briarwood Rd. N&S	37W-24N,37W-23N	41
Broadway Av. E&W	37W-25N,36W-25N	41
Castle Rock Ct. NW&SE	39W-23N	40
Center Av. E&W	37W-25N	41
Crystal Ct. N&S	37W-25N	41
Cumberland Ln. N&S	36W-25N	41
Dairy Ln. NS&EW	38W-24N	41
Dundedin Dr. SW&NE	38W-23N	41
Edinburough Ct. N&S	38W-23N	41
Essex Ln. N&S	36W-25N	41
Fairway Dr. N&S	38W-24N,38W-25N	41
Firth Ct. NS&EW	38W-24N	41
Gilmore Ct. NE&SW	39W-23N	40
Glendale Cir. NS&EW	38W-24N	41
Haligus Rd. N&S	39W-23N,39W-25N	40
Hampshire Ln. N&S	37W-25N	41
Hawthorn St. N&S	37W-25N	41
Hickory St. E&W	37W-25N	41
Huntley Rd. N&S	37W-25N	41
Inverway Dr. NS&EW	39W-25N,38W-25N	40
Julie Dr. SE&NW	39W-24N,38W-24N	40
Lake Av. NW&SE	37W-25N,36W-25N	41
Loch Glen Ct. NE&SW	38W-24N	41
Loch Glen Dr. NS&EW	38W-24N,37W-24N	41
Marsh Dr. NS&EW	39W-24N	40
McIntosh Ct. E&W	37W-23N	41
Melrose Ln. N&S	36W-25N	41
Meridian St N&S	37W-25N	41
Muirfield Ct. N&S	38W-24N	41
Muirfield Dr. NE&SW	38W-24N	41
North Av. E&W	37W-25N	41
Oak St. E&W	37W-25N	41
Oakwood Dr. E&W	37W-25N	41
Oxford Ln. N&S	36W-25N	41
Partridge Ln. E&W	38W-24N	41
Pheasant Dr. N&S	37W-24N	41
Priegnitz Rd. N&S	38W-23N,38W-24N	41
Richmond Ln. N&S	37W-25N	41
Scots Ln. N&S	38W-24N	41
Scott Cir. E&W	38W-24N	41
South Shore Dr. E&W	37W-25N,36W-25N	41
Stirling Ct. NE&SW	39W-23N	40
Stuart Ln. E&W	39W-24N,38W-24N	40
Sunset Dr. N&S	37W-25N	41
Turnberry Lake Dr. NE&SW	38W-23N,37W-23N	41
Turnberry Tr. E&W	38W-24N	41
Warwick Ln. N&S	37W-25N	41
Willow St. E&W	37W-25N	41
Wiltshire Ln. N&S	37W-25N	41

Marengo

Street	Grid	Page
Adams St. E&W	49W-26N	37
Ann St. N&S	50W-26N	37
Artell Av. N&S	49W-26N	37
Barbara Av. E&W	50W-25N	37
Beaverpond Ct. N&S	49W-27N	29
Beaverpond Dr. NE&SW	49W-27N	29
Briden Dr. E&W	50W-26N	37
Buttonwood Ct. E&W	49W-27N	29
Buttonwood Ln. E&W	49W-27N	29
Caroline St. E&W	49W-26N	37
Cascade Ct. NE&SW	49W-27N	29
Cascade Dr. NW&SE	49W-27N	29
Cayuga Tr. E&W	50W-26N	37
Center Dr. NE&SW	49W-27N	29
Chappel St. E&W	50W-26N	37
Cherokee Tr. E&W	50W-26N	37
Chestnut Ct. E&W	49W-27N	29
Chestnut Ln. E&W	49W-27N	29
Chippewater E&W	50W-26N	37
Circle Ct. E&W	50W-26N	37
Clark St. N&S	49W-26N	37
Cottonwood Ln. N&S	49W-27N	29
Damon St. N&S	49W-26N	37
Deerpass Rd. N&S	49W-27N	29
Deitz St. N&S	50W-26N	37
Diane Ct. E&W	49W-26N	37
Dunham Ct. N&S	49W-25N	37
East Forest St. E&W	50W-26N,49W-26N	37
East Prairie St. E&W	49W-26N	37
East St. N&S	49W-26N,49W-27N	37
East Washington St. E&W	49W-26N	37
Eighth Av. E&W	49W-27N	29
Eisenhower Dr. N&S	50W-26N	37
Elm Av. N&S	49W-26N	37
Fifth Av. E&W	49W-27N	29
First Av. E&W	49W-27N	29
Ford St. N&S	50W-26N	37
Fourth Av. E&W	49W-27N	29
Foxglove Ct. N&S	49W-27N	29
Foxglove Ln. N&S	49W-27N	29
Frances St. E&W	49W-26N	37
Georgeann Ct. E&W	49W-26N	37
Geraldine Ct. E&W	49W-26N	37
Grace St. N&S	49W-26N	37
Grant Hwy. E&W	50W-26N,49W-26N	37
Greenlee St. E&W	49W-26N	37
Hale St. N&S	49W-27N	37
Hickory Ct. E&W	49W-27N	29
Hickory Ln. E&W	49W-27N	29
Jackson St. E&W	49W-26N	37
James Ct. E&W	49W-25N	37
Johnson St. N&S	50W-26N	37
Kennedy Dr. N&S	50W-26N	37
Kent St. E&W	50W-26N	37
Keppler Dr. E&W	50W-26N	37
Kishwaukee St. N&S	49W-27N	29
Locust St. N&S	49W-26N	37
Mallard Dr. NW&SE	49W-27N	29
Maple St. N&S	49W-25N,49W-26N	37
Mary Ct. E&W	49W-25N	37
Mohawk Tr. E&W	50W-26N	37
Navajo Tr. E&W	50W-26N	37
Oak Manor St. N&S	50W-26N	37
Osage Tr. N&S	50W-26N	37
Otter Tail Dr. NE&SW	49W-27N	29
Page St. N&S	49W-26N,49W-27N	37
Park Dr. N&S	50W-26N	37
Pawnee Tr. NW&SE	50W-26N	37
Prospect St. N&S	49W-26N	37
Railroad St. NW&SE	49W-26N	37
Rainbow Dr. N&S	49W-26N	37
Randall Ct. E&W	49W-25N	37
Riley Dr. N&S	49W-25N,49W-26N	37
Rowland St. N&S	50W-26N	37
Royal Oak Dr. N&S	50W-26N	37
Rt. 176 E&W	49W-26N	37
Rt. 23 N&S	50W-26N,50W-27N	37
School Ct. NE&SW	49W-26N	37
Second Av. E&W	49W-27N	29
Seneca Tr. N&S	50W-26N	37
Seventh Av. E&W	49W-27N	29
Shady Ln. Ct. NE&SW	49W-26N	37
Sixth Av. E&W	49W-27N	29
South St. E&W	50W-26N	37
Sponable St. N&S	50W-26N	37
Stanford Dr. N&S	50W-26N	37
State St. N&S	50W-26N,50W-27N	37
Sunset Dr. N&S	50W-26N,50W-27N	37
Taylor St. N&S	49W-26N,49W-27N	37
Telegraph St. E&W	49W-27N	29
Third Av. E&W	49W-27N	29
Toga Tr. N&S	50W-26N	37
U.S. Route 20 E&W	49W-26N	37
Van Buren St. E&W	50W-26N,49W-26N	37
Walnut Ct. E&W	49W-27N	29
Walnut Ln. E&W	49W-27N	29
West Forest St. E&W	50W-26N	37
West Prairie St. E&W	50W-26N	37
West St. N&S	50W-26N	37
West Washington St. E&W	50W-26N	37
Whitetail Cir. NE&SW	49W-27N	29
Whitetail Dr. NS&EW	49W-27N	29
Whitetail Pl. N&S	49W-27N	29
Willow Rd. N&S	50W-26N,50W-27N	37
Woodbine Ln. E&W	49W-27N	29
Woodland Ct. N&S	49W-27N	29
Woodland Ln. E&W	49W-27N	29

Marengo Township

Street	Grid	Page
Briden Dr. E&W	51W-26N	36
Busse Rd. N&S	50W-31N	29
Busse Rd. N&S	49W-31N	21
Carmack Rd. E&W	53W-30N,51W-30N	28
Center Av. N&S	51W-27N	28
Collins Rd. E&W	49W-30N	29
Constance Ln. N&S	51W-29N	28
County Line Rd. N&S	54W-26N,54W-31N	36
Cox Rd. N&S	52W-28N,52W-29N	28
Deerpass Rd. N&S	49W-27N,49W-30N	29
Edward St. N&S	50W-28N	29
Elizabeth St. N&S	50W-28N	29
Executive Dr. N&S	49W-28N	29
Grant Hwy. E&W	54W-26N,50W-26N	36
Greenbriar Ln. N&S	51W-29N	28
Grenadier Dr. E&W	51W-29N	28
Hawthorn Rd. N&S	50W-29N,51W-30N	29
Hayden Dr. E&W	51W-26N	36
Heritage Ct. N&S	49W-28N	29
Highcrest Rd. N&S	51W-30N	28
Johnson Rd. N&S	52W-26N	36
Keppler Dr. E&W	51W-26N	36
Kishwaukee Valley Rd. E&W	54W-31N,49W-31N	20
Koda Dr. N&S	50W-30N	29
Lynn Dr. N&S	49W-30N	29
Mary Ln. St. E&W	50W-28N	29
Menge Rd. N&S	49W-31N	21
Meyer Rd. N&S	51W-26N	36
Michele Dr. N&S	49W-30N	29
Miller Rd. N&S	54W-28N,54W-29N	28
Noe Rd. N&S	53W-30N,53W-31N	28
Northwest Rd. NS&EW	50W-28N,50W-29N	29
O'Connell Dr. E&W	50W-26N	37
Olbrich Rd. N&S	50W-31N	21
Olcott Rd. N&S	54W-30N	28
Oleson Rd. N&S	54W-31N	20
Parade Dr. E&W	49W-28N	29
Parkview Dr. N&S	50W-28N,50W-29N	29
Pringle Dr. N&S	50W-29N,50W-30N	29
Railroad St. E&W	51W-26N,50W-26N	36
Ridgeview Ln. E&W	49W-30N	29
Ritz Rd. N&S	51W-27N	37
River Ranch Rd. N&S	51W-26N	37
River Rd. E&W	54W-28N,49W-28N	28
Root Rd. N&S	52W-31N	20
Rt. 23 N&S	50W-27N,50W-31N	29
Squire Dr. N&S	49W-28N	29
Stella Dr. E&W	51W-29N	28
Thorne Rd. N&S	52W-26N,52W-28N	36
Tomlin Rd. E&W	54W-31N,52W-31N	20
W. Pringle Dr. E&W	50W-30N	29

McCullom Lake

Street	Grid	Page
Beachview Dr. N&S	33W-34N	26
Clover Hill Dr. NE&SW	33W-34N	26
East Ln. E&W	34W-34N	26
Eastwood Dr. N&S	33W-34N	26
Flanders Rd. SW&NE	33W-34N	26
Forest View Dr. N&S	34W-34N	26
Fountain E&W	34W-34N,33W-34N	26
Greenwood Pl. E&W	34W-34N	26
Hickory Dr. N&S	34W-34N	26
Knollwood Dr. N&S	34W-34N	26
Lake Shore Dr. NW&SE	33W-34N	26
Maple Hill Dr. E&W	34W-34N,33W-34N	26
Oakland Ln. N&S	34W-34N	26
Orchard Dr. NS&EW	34W-34N,33W-34N	26
Parkview Dr. NW&SE	33W-34N	26
Parkview Dr. E&W	34W-34N	26
Spring Rd. N&S	33W-34N	26
West Ln. E&W	34W-34N	26

McHenry

Street	Grid	Page
Abbey Dr. E&W	34W-31N	26
Abbington Dr. NE&SW	33W-32N	26
Albany St. E&W	33W-30N,32W-30N	34
Albert Dr. N&S	34W-34N	26
Allen Av. NS&EW	32W-32N	26
Amherst Ct. E&W	34W-34N	26
Anne St. E&W	32W-32N	26
April Av. NE&SW	35W-34N	26
Ashland Dr. E&W	34W-32N,33W-32N	26
Ashley Dr. E&W	33W-33N	26
Augusta Dr. N&S	34W-32N	26
Bally Rd. N&S	31W-31N	27
Barn Wood Ct. E&W	33W-31N	26
Barnwood Tr. NE&SW	33W-31N	26
Bauer Rd. N&S	32W-32N	26
Beach Dr. E&W	32W-31N,31W-31N	26
Beach Rd. N&S	33W-33N,34W-33N	26
Belden St. NE&SW	32W-30N	34
Bennington Ln. NS&EW	34W-34N	26
Birch Dr. N&S	32W-33N	26
Biscayne Dr. E&W	32W-30N	34
Biscayne Rd. E&W	32W-30N	34
Bonita Ln. N&S	32W-30N,32W-31N	34
Bonner Dr. NW&SE	33W-33N	26
Bonnie Brae Dr. N&S	32W-30N	34
Borden St. NW&SE	32W-33N	26
Bradley Ct. E&W	32W-33N	26
Bretons Dr. E&W	32W-32N	26
Brittany Dr. E&W	34W-33N	26
Broad St. NW&SE	32W-33N	26
Broadway Dr. N&S	32W-30N,32W-31N	34
Bromley Dr. NS&EW	33W-32N	26
Brookwood Tr. N&S	34W-31N	26
Brown St. N&S	32W-33N	26
Bull Valley Rd. E&W	34W-31N,32W-31N	26
Burning Tree Dr. E&W	36W-32N	25
Callista St. NW&SE	33W-32N	26
Cambridge Dr. E&W	34W-32N,33W-32N	26
Canterbury Dr. N&S	34W-32N	26
Capri Ter. N&S	32W-30N	34
Carey Dr. N&S	32W-30N	34
Carol Av. N&S	31W-31N	27
Carriage Tr. NE&SW	34W-31N	26
Center St. N&S	32W-32N	26
Central St. N&S	33W-33N	26
Charlotte Av. NE&SW	32W-32N,31W-32N	26
Chasefield Circle NS&EW	34W-31N,34W-32N	26
Cherryhill Ct. N&S	34W-34N	26

	Grid	Page
Chesterfield Ct. NE&SW	33W-32N,33W-33N	26
Chesterfield Dr. N&S	33W-32N,33W-33N	26
Chestnut Dr. E&W	32W-33N	26
Chickaloon Dr. NS&EW	35W-33N	26
Church St. N&S	32W-33N	26
Clearbrook Av. E&W	32W-33N	26
Clearview Dr. E&W	33W-33N	26
Clover Av. E&W	32W-32N	26
Colony Ct. NW&SE	33W-31N	26
Colony Tr. N&S	36W-32N	25
Commecial Av. NW&SE	33W-33N	26
Cornell Ct. E&W	32W-33N	26
Corporate Dr. N&S	33W-30N	34
Court St. NE&SW	32W-33N	26
Courtland Tr. NW&SE	33W-31N	26
Creek Dr. E&W	34W-31N	26
Creekside Ct. E&W	34W-31N	26
Creekside Ln. E&W	33W-33N	26
Creekside Tr. NE&SW	34W-31N,34W-32N	26
Crescent Av. E&W	32W-32N	26
Crestwood Dr. NE&SW	33W-32N	26
Cross Tr. N&S	33W-32N	26
Crystal Lake Rd. NE&SW	34W-31N,33W-33N	26
Cumberland Circle NS&EW	33W-33N	26
Curran Rd. N&S	34W-32N,34W-33N	26
Dale Av. N&S	32W-32N	26
Dartmoor Dr. E&W	34W-32N,33W-32N	26
Dayton St. E&W	33W-30N,32W-30N	34
Deer Run Tr. NE&SW	36W-32N	25
Deerwood Tr. E&W	36W-32N	25
Denali Tr. NW&SE	35W-33N	26
Derby Ln. N&S	34W-34N	26
Devonshire Ct.* E&W	34W-32N,33W-32N	26
Donovan Ct. NW&SE	32W-33N	26
Donovan St. NW&SE	33W-33N,32W-33N	26
Dot St. N&S	33W-33N	26
Drake Ct. E&W	32W-33N	26
Draper Rd. N&S	35W-33N,34W-33N	26
Driftwood Ct. NE&SW	34W-31N	26
Driftwood Tr. NE&SW	33W-33N,33W-31N	26
Eastern Av. N&S	33W-33N	26
Elm St. NW&SE	34W-33N,32W-33N	26
Essex Ct. N&S	33W-32N	26
Fairfax Dr. N&S	34W-31N	26
Fairway Dr. E&W	32W-32N	26
Farington Dr. E&W	34W-32N	26
Florence Blvd. E&W	34W-34N	26
Flower St. N&S	33W-33N	26
Foolish Pleasure Dr. NW&SE	33W-34N	26
Forest Rd. E&W	32W-30N	34
Fourth St. NE&SW	32W-32N	26
Freund Av. NS&EW	33W-33N	26
Front Royal Dr. NE&SW	33W-32N,33W-33N	26
Front St. NE&SW	32W-32N	26
Garden Quarter Rd. E&W	33W-32N	26
Glen Dr. N&S	33W-33N	26
Glenbrook Dr. E&W	34W-31N,33W-31N	26
Golf View Ter. E&W	32W-32N	26
Golfview Dr. SW&NE	32W-32N	26
Grand Av. E&W	32W-33N	26
Green St. N&S	32W-32N,32W-33N	26
Greenbriar Dr. NW&SE	34W-32N	26
Gregg Dr. E&W	31W-31N	27
Grove Av. E&W	32W-32N	26
Hampton Ct. NE&SW	33W-32N	26
Hartland Tr. NW&SE	33W-31N	26
Hayward St. E&W	35W-33N	26
Heritage Dr. N&S	34W-31N,33W-31N	26
High St. E&W	32W-33N	26
Highview Av. N&S	33W-33N	26
Hill St. NW&SE	34W-33N	26
Hillside Dr. E&W	32W-33N	27
Hilltop Dr. N&S	32W-30N,32W-31N	34
Hollywood Blvd. N&S	32W-30N,32W-31N	34
Home Av. E&W	34W-33N,33W-33N	26
Hunter Dr. E&W	34W-33N	26
Hunter Path E&W	32W-30N	34
Huntington Dr. N&S	34W-32N	26
Illiamna Tr. NE&SW	35W-33N	26
Industrial Dr. N&S	33W-33N	26
James St. NW&SE	32W-32N	26
Jennifer Ln. N&S	32W-33N	26
John St. NW&SE	32W-32N	26
Judy Ln. NE&SW	32W-32N	26
Kane Av. E&W	32W-32N	26
Katie Ln. NW&SE	32W-31N	26
Katmai Tr. NE&SW	35W-33N	26
Kensington Dr. NW&SE	33W-32N	26
Kinley Blvd. E&W	32W-31N,31W-31N	26
Knoll Av. N&S	33W-33N	26
Knox Dr. E&W	33W-33N	26
Lakeland Av. N&S	33W-33N	26
Lakewood Av. N&S	33W-33N	26
Lakewood Rd. N&S	33W-34N	26
Lawn Ct. N&S	33W-33N	26
Lawrence Pkwy. N&S	33W-31N	26
Lee St. N&S	32W-32N	26
Leonard Av. N&S	32W-33N	26
Lillian St. NW&SE	33W-32N,32W-32N	26
Lincoln St. E&W	33W-33N	26
Linda Rd. N&S	32W-32N	26
Loch Glen Ln. NS&EW	32W-31N	26
Logan St. NW&SE	33W-33N	26
Lorient Dr. E&W	32W-32N	26
Main St. NW&SE	33W-33N,32W-32N	26
Malibu Ct. E&W	34W-31N	26
Manchester Mall NS&EW	33W-33N	26
Manley Av. NW&SE	33W-32N,33W-33N	26
Maple Av. E&W	32W-33N	26
Martin Rd. NS&EW	34W-34N	26
Matanuska Tr. N&S	35W-33N	26
McCullom Lake Rd. NW&SE	32W-33N	26
McCullom Rd. E&W	32W-33N	26
McHenry Av. N&S	32W-31N	26
Meadow Lake Av. E&W	32W-32N	26
Meadow Ln. NS&EW	33W-33N	26
Meadow Rd. N&S	32W-31N	26
Medical Center Dr. NS&EW	33W-31N,32W-31N	26
Melrose Ct. NS&EW	32W-32N	26
Mildred St. E&W	32W-33N	26
Mill St. NW&SE	33W-32N,33W-33N	26
Miller Pkwy. N&S	33W-30N	34
Miller Rd. E&W	32W-31N,31W-31N	26
Millstream Dr. NS&EW	32W-32N	26
North Av. N&S	33W-33N	26
North Dr. N&S	32W-33N	26
North Fox Ln. NE&SW	32W-31N	26
Oak Av. E&W	32W-32N	26
Oak Dr. NW&SE	33W-33N	26
Oakwood Ct. N&S	33W-33N	26
Oakwood Dr. N&S	33W-32N,33W-33N	26
Ojibwa Ln. NW&SE	35W-33N	26
Orchid Path N&S	32W-31N	26
Orleans St. NE&SW	33W-33N,32W-33N	26
Overland Tr. E&W	33W-31N	26
Park Lane Av. N&S	34W-32N	26
Park St. N&S	32W-32N	26
Parkway Av. E&W	33W-33N	26
Pearl St. N&S	32W-31N,32W-32N	26
Peterson Park Rd. NS&EW	33W-34N	26
Pine Dr. N&S	32W-33N	26
Pinoak Ct. NW&SE	35W-33N	26
Pleasant Av. N&S	33W-33N	26
Pleasant View Dr. E&W	32W-30N	34
Ponca St. NE&SW	33W-32N	26
Prairie Av. E&W	33W-33N	26
Prairie Ln. N&S	31W-31N	27
Prestwick St. E&W	32W-33N	26
Prime Parkway E&W	33W-30N	34
Pyndale Dr. E&W	33W-32N	26
Radcliff Ct. E&W	34W-34N	26
Ramble Rd. N&S	33W-33N	26
Redwood Ln. N&S	32W-33N	26
Rhonda Rd. N&S	32W-31N,32W-32N	26
Richmond Rd. N&S	33W-33N	26
Ridgeview Dr. N&S	33W-30N	34
Ridgeway Tr. N&S	36W-32N	25
Ringwood Rd. N&S	34W-33N,34W-34N	26
Riverside Dr. N&S	31W-31N,32W-33N	27
Riverstream Rd. NE&SW	32W-32N	26
Rogers Av. N&S	33W-33N	26
Rose Av. NW&SE	34W-34N	26
Royal Dr. NW&SE	33W-32N	26
Rt. 120 NW&SE	35W-33N,32W-33N	26
Rt. 31 E&W	33W-30N,32W-34N	34
Ryan St. N&S	32W-33N	26
S. Scully Dr. E&W	34W-33N	26
Sandburg Dr. E&W	33W-32N	26
Sarasota Dr. NW&SE	34W-32N	26
Scully Dr. N&S	34W-33N	26
Shamrock Ln. E&W	33W-31N	26
Shephard Hill NS&EW	32W-32N	26
Sherman Dr. E&W	34W-33N	26
Shore Dr. E&W	34W-33N,33W-33N	26
Shore Hill Dr. E&W	34W-33N	26
Silbury Ct. NS&EW	33W-32N	26
Silver Glen Rd. N&S	36W-32N	25
Sioux St. NE&SW	33W-32N	26
Skyway Dr. E&W	32W-32N	26
Somerset Mall N&S	33W-33N	26
South St. NE&SW	33W-32N	26
Southgate N&S	32W-32N	26
Springbrook Ct. NE&SW	34W-31N	26
Springdale Ln. E&W	34W-33N	26
Spruce Ln. E&W	33W-33N	26
St. Pauls Av. E&W	32W-32N	26
Still Hill Dr. E&W	32W-31N,31W-31N	26
Stratford Dr. E&W	33W-32N	26
Sunrise View E&W	32W-30N	34
Sunset Av. N&S	34W-33N	26
Sussex Dr. E&W	33W-33N	26
Sycamore Ct. E&W	35W-33N	26
Tamarack Ct. NE&SW	35W-33N	26
Terrace Dr. E&W	32W-31N	26
Third St. NE&SW	33W-32N,32W-33N	26
Thornwood Dr. N&S	34W-32N	26
Timber Trail E&W	36W-32N	25
Timothy Ln. N&S	32W-32N	26
Tomlinson Ct. E&W	34W-32N	26
Tomlinson Dr. NW&SE	35W-34N,34W-34N	26
Turnberry Dr. E&W	32W-31N	26
Tustamena Tr. E&W	35W-33N	26
Valley Dr. E&W	32W-33N	26
Valley Rd. N&S	32W-32N	27
Venice Ct. E&W	32W-32N	26
Veretta Ct. NE&SW	33W-31N	26
Victor Dr. E&W	33W-33N	26
Victoria Av. E&W	32W-32N,31W-32N	26
Village Tr. N&S	33W-31N	26
Vine St. E&W	32W-33N	26
Virginia Av. E&W	32W-32N,32W-33N	26
Vista Ter. E&W	32W-31N	26
Walnut Ln. E&W	32W-33N	26
Washington St. NW&SE	32W-32N	26
Waters Edge Dr. NE&SW	34W-33N,33W-32N	26
Waukegan St. NW&SE	32W-32N	26
West Av. E&W	32W-32N	26
Whiteoak Ct. NW&SE	34W-34N	26
Whiteoak Dr. NS&EW	35W-34N,34W-34N	26
Whiting Ct. N&S	34W-34N	26
Whiting Dr. NW&SE	35W-34N,34W-34N	26
Willow Ln. E&W	34W-33N,33W-33N	26
Wiltshire Ct. NE&SW	33W-32N	26
Wiltshire Dr. NW&SE	33W-32N	26
Windhaven Ct. N&S	34W-31N	26
Windhaven Tr. NS&EW	34W-31N	26
Winding Creek Dr. E&W	34W-31N	26
Windridge Dr. N&S	34W-32N	26
Winslow Circle NW&SE	34W-32N	26
Wood St. NW&SE	33W-32N	26
Woodmar Ct. E&W	34W-31N	26
Woodridge Tr. NS&EW	36W-32N	25
Young St. E&W	32W-32N	26

McHenry Township

	Grid	Page
Acorn Ln. N&S	29W-37N	19
Agatha Ln. N&S	29W-37N	19
Algonquin Rd. E&W	36W-36N	17
Aloha Ct. E&W	31W-34N	27
Aloha Dr. N&S	31W-34N	27
Althoff Dr. E&W	29W-36N	19
Altoff Dr. E&W	29W-36N	19
Alton Rd. N&S	34W-34N	26
Amanda Av. E&W	29W-31N	27
Amby Ln. N&S	29W-37N	19
Annabelle St. E&W	29W-37N	19
Appaloosa Tr. N&S	30W-33N	27
Arabian Spur E&W	30W-33N	27
Arbor Rd. E&W	36W-35N	17
Ash Dr. E&W	36W-36N	17
Ashwood Dr. E&W	36W-34N	25
August Ln. N&S	30W-34N	27
Backbay Rd. E&W	36W-36N	17
Balsam Dr. E&W	36W-34N	25
Barbara Dr. E&W	36W-36N	17
Barnard Mill Rd. NW&SE	35W-37N,33W-36N	18
Basswood Dr. E&W	36W-34N	25
Bay Rd. NW&SE	29W-34N	27
Bay View Ln. N&S	29W-34N	27
Baycliff Dr. N&S	31W-34N	27
Bayview Dr. E&W	36W-36N	17
Bayview Ln. NS&EW	29W-34N	27
Beach Av. E&W	34W-34N	26
Beach Dr. E&W	36W-36N	17
Beach Rd. E&W	29W-35N,34W-35N	19
Beach Rd. E&W	29W-32N	27
Beachside Rd. N&S	34W-34N	26
Beacon Hill Dr. N&S	29W-37N	19
Beaver Rd. E&W	36W-35N	17
Benjamin Dr. N&S	36W-34N	25
Bennington Ct. E&W	36W-32N	25
Berkshire Dr. E&W	32W-37N	18
Birch Dr. E&W	36W-36N	17
Black Oak Dr. NE&SW	36W-33N	25
Bong St. N&S	36W-37N	17
Boston Rd. E&W	36W-37N	17
Bridge St. N&S	29W-37N	19
Britton Dr. N&S	29W-37N	19
Broadway Rd. E&W	30W-37N,29W-37N	19
Brook Dr. E&W	36W-35N	17
Bull Ridge Dr. NS&EW	36W-33N	25
Bunny Av. E&W	29W-37N	19
Burr Av. E&W	29W-37N	19
Burr Oak Ct. NS&EW	36W-33N	25
Burr Oak Dr.	36W-33N	25
Cambridge Ct. E&W	36W-32N	25
Cambridge Rd. E&W	36W-36N	17
Camelia Ct. NE&SW	36W-37N	17
Carl Ct. N&S	30W-34N	27
Cassandra Ln. N&S	30W-33N	27
Catalpa Dr. E&W	36W-34N	25
Cedar Dr. E&W	36W-36N	17
Cedar Rd. E&W	36W-35N	17
Center Dr. E&W	36W-36N	17
Channel St. N&S	34W-34N	26
Chapel Hill Rd. N&S	31W-32N	27
Charles St. N&S	32W-33N	26
Cherry Ct. NE&SW	36W-37N	17
Cherry Dr. N&S	36W-34N	25
Chestnut Dr. E&W	36W-34N,36W-36N	25
Chipewa Dr. E&W	29W-37N	19
Christine Av. N&S	29W-37N	19
Church Dr. N&S	36W-35N	17
Circle Av. NE&SW	29W-37N	19
Circle Ct. E&W	30W-36N	19
Circle Dr. NE&SW	29W-37N	19
Circle Dr. NS&EW	30W-36N	19
Circle Dr. NE&SW	36W-37N	17
Circle Dr. E&W	36W-36N	17
Clearview Dr. N&S	36W-35N	17
Clement Rd. N&S	31W-32N	27
Cleveland Dr. E&W	34W-34N	26
Cliff Ct. E&W	30W-34N	27
Clover Ln. NE&SW	31W-33N	27
Club Dr. N&S	31W-34N	27
Cobblestone Dr. NS&EW	30W-34N	27
Concorde Dr. N&S	36W-32N	25
Coolidge Av. E&W	31W-34N,30W-34N	27
Coral St. NE&SW	31W-32N	27
Country Club Dr. N&S	31W-32N	27
Country Ln. E&W	31W-33N	27
Craftwell Dr. E&W	34W-37N	18
Craig Dr. E&W	36W-36N	17
Crescent Dr. E&W	36W-36N	17
Crest St. E&W	30W-32N	27
Cresthill Av. N&S	29W-32N	27
Crestwood Av. N&S	29W-32N	27
Croakin Ln. E&W	31W-32N	27
Cross Dr. NW&SE	36W-37N	17
Croyden St. E&W	29W-37N	19
Cuhlman Rd. N&S	29W-33N,29W-34N	27
Curve Dr. E&W	29W-37N	19
Curve Ct. NS&EW	29W-37N	19
Cushman Rd. NE&SW	29W-34N	27
Cypress Dr. E&W	36W-37N	17
Dale Av. E&W	29W-32N	27
David St. N&S	29W-36N	19
Deborah Ln . NW&SE	31W-33N	27
Deep Spring Rd. E&W	36W-35N	17
Deerpath Dr. E&W	29W-37N	19
Delaware Rd. E&W	36W-36N	17
Della Pl. E&W	30W-35N	19
Dennis Dr. N&S	30W-36N	19
Devonshire Dr. N&S	32W-37N	18
Don Michael Dr. E&W	36W-35N	17
Dot St. N&S	34W-33N	26
Draper Rd. NE&SW	29W-32N	27
Dutch Cir. N&S	31W-34N	27
E. Sunset Dr. E&W	36W-35N	17
East Av. N&S	29W-37N	19
East Blvd. NE&SW	29W-32N	27
East Dr. N&S	36W-35N	17
East Lake Shore Dr. N&S	36W-35N,36W-37N	17
East Maplewood Dr. E&W	36W-34N	25
East St. N&S	29W-37N	19
East Wonder Lake Rd. N&S	36W-32N,36W-37N	25
Eastern Av. E&W	29W-37N	19
Eastview Av. N&S	29W-32N	27
Eastwood Dr. N&S	36W-35N	17
Eastwood Ln. N&S	31W-33N	27
Edgewood Ct. N&S	31W-33N	27
Edgewood Dr. E&W	29W-36N	19
Edgewood Dr. N&S	36W-35N	17
Edgewood Ln. N&S	31W-33N	27
Edward Ct. N&S	30W-34N	27
Eisenhower Dr. NE&SW	36W-37N	17
Elam Av. NE&SW	30W-35N	19
Elder Av. E&W	29W-36N	19
Elm Leaf Dr. N&S	30W-36N	19
Elmkirk Pk. E&W	31W-32N	27
Elmoak Ln. E&W	31W-32N	27
Elmshire Ln. NS&EW	32W-36N	18
Emerald Dr. N&S	31W-32N	27
Emily Ln. NW&SE	31W-33N	27
Erin St. N&S	31W-33N	27
Euclid Dr. E&W	34W-34N	26
Fair Av. NW&SE	29W-32N	27
Fair Oak Ln. E&W	31W-32N	27
Fairview Av. NW&SE	29W-32N	27
Fairview Ln. NS&EW	31W-33N	27
Falls Ln. E&W	36W-35N	17
Fawn Ln. N&S	36W-35N	17
Finch Tr. E&W	29W-36N	19
Fir Dr. E&W	36W-36N	17
First Av. E&W	32W-33N	26
Flanders Rd. E&W	34W-34N,33W-34N	26
Florence Dr. N&S	30W-36N	19
Florence Dr. E&W	36W-36N	17
Fox Lake Rd. N&S	29W-37N	19

	Grid	Page
Freedom Dr. N&S ... 34W-34N		26
Friendly Av. NS&EW		
.... 29W-32N		27
Gene Dr. E&W .. 36W-34N		25
Ginko Ct. N&S ... 36W-33N		25
Glenwood Av. N&S		
.... 29W-32N		27
Glenwood Ct. NW&SE		
.... 31W-33N		27
Glenwood Ln. E&W		
.... 31W-33N		27
Grandview Dr. E&W		
.... 30W-36N		19
Grandview Dr. NE&SW		
.... 31W-34N		27
Granite Dr. E&W ..36W-37N		17
Grant Pl. N&S 29W-32N		19
Greenleaf Dr. N&S .36W-35N		17
Gregory St. N&S .. 30W-36N		19
Hampshire Ln. N&S		
........32W-36N,32W-37N		18
Hancock Dr. E&W .. 36W-35N		17
Harbor Rd. NE&SW		
.... 36W-37N		17
Hay Av. NE&SW... 33W-32N		26
Hayden Ct. NW&SE		
.... 29W-36N		19
Hayden Dr. E&W .. 29W-36N		19
Hedge Pl. SW&NE .. 31W-34N		27
Henry Tr. E&W 29W-37N		19
Henzada Av. N&S ..31W-34N		27
Hiawatha Dr. E&W		
........36W,35W-36N		17
Hickman Ln. NE&SW		
........41W-30N,40W-30N		32
Hickory Dr. NS&EW		
.... 36W-35N		17
Hickory Dr. W E&W		
.... 36W-37N		17
Hickory Ln. N&S		
........32W-36N,32W-37N		18
Hickory Rd. E&W .. 36W-35N		17
Hickory Tr. E&W .. 29W-37N		19
Hickorynut Dr. N&S		
.... 30W-36N		19
High Rd. N&S 29W-34N		27
Highland Dr. N&S .. 29W-37N		19
Highland Dr. E&W ..34W-34N		26
Highview Dr. E&W ..36W-34N		25
Hill Rd. N&S 30W-32N		27
Hillside Dr. N&S .. 36W-35N		17
Hillside Ln. E&W .. 31W-33N		27
Hilltop Dr. E&W .. 29W-37N		19
Hilltop Dr. N&S .. 30W-36N		19
Hilltop Dr. N&S		
........36W-36N,36W-37N		17
Holly Ct. E&W 31W-33N		27
Howard St. N&S ..30W-36N		19
Howe Dr. E&W 30W-36N		17
Howell Rd. E&W ...31W-34N		27
Huron Dr. E&W .. 36W-36N		17
Indian Ridge Dr. E&W		
.... 30W-36N		19
Island Ln. N&S 30W-36N		17
Jasper Dr. E&W .. 30W-36N		19
Jeffery St. N&S ... 30W-36N		19
John Av. E&W 29W-36N		19
John Ct. E&W 30W-34N		27
Johnsburg Wilmot Rd. N&S		
.... 30W-37N		19
Jonquil Dr. N&S .. 36W-36N		17
Joseph Ct. N&S .. 36W-36N		17
Joyce Av. E&W ... 30W-36N		19
Kama Av. E&W ... 31W-34N		27
Kashmiri Av. E&W ..31W-34N		27
Kelter St. NW&SE ..33W-32N		26
Kent Rd. N&S.....30W-32N		27
King Av. E&W 36W-37N		17
Kirk Ct. E&W 31W-32N		27
Kosciuszko Rd. N&S		
.... 34W-34N		26
Lake Av. E&W ... 32W-33N		26
Lake Rd. NE&SW .. 35W-37N		18
Lake Shore Dr. NE&SW		
........36W-36N,36W-37N		17
Lake St. N&S 29W-37N		19
Lakeview Dr. N&S .36W-35N		17
Lauderdale Ct. N&S		
.... 29W-34N		27
Lear St. N&S 36W-37N		17
Leeside St. N&S ...29W-37N		19
Lilac Dr. NW&SE .. 36W-37N		17
Lily Av. N&S 29W-32N		27
Lincoln Rd. E&W		
........31W-33N,30W-33N		27
Lincolnshire Dr. E&W		
.... 32W-36N		18
Little St. N&S ... 29W-37N		19
Lodge Rd. NW&SE		
.... 31W-32N		27
Lonetree Dr. E&W ..36W-37N		17
Long Ln. N&S..... 30W-34N		27
Lookout Dr. E&W .. 36W-37N		17
Loras Ln. N&S ... 36W-36N		17
Louella Av. N&S .. 29W-36N		19
Lucina Av. N&S .. 29W-36N		19
Lucy Dr. E&W ... 36W-34N		25
Luella Av. N&S .. 29W-36N		19
Lumley Dr. N&S .. 35W-36N		18
Lynn Dr. E&W .. 36W-36N		17
Macarthur Av. E&W		
........36W-37N,35W-37N		17

	Grid	Page
Macarthur Dr. NE&SW		
.... 34W-34N		26
Magellan Dr. N&S .29W-34N		27
Magnolia Av. N&S .29W-32N		27
Magnolia Dr. N&S .. 36W-37N		17
Manitou Tr. NE&SW		
.... 29W-34N		27
Manor Ln. NE&SW 31W-33N		27
Maple Dr. E&W 36W-36N		17
Maple Leaf Dr. N&S		
.... 30W-36N		19
Mar Dr. N&S 30W-33N		27
Marblehead Rd. E&W		
.... 36W-36N		17
Marietta St. NW&SE		
.... 33W-32N		26
Marquardt Ln. N&S		
.... 30W-34N		27
Marquart Ln. N&S .. 30W-34N		27
Marshall Dr. N&S .. 36W-37N		17
Martin Av. N&S 31W-34N		27
Mary Ln. NW&SE .. 31W-33N		27
Maryville Rd. E&W .31W-33N		27
May Av. E&W 30W-36N		19
Mayfair Dr. N&S .. 30W-36N		19
McCullom Lake Rd. NW&SE		
........36W-36N,32W-34N		17
Meadow Ct. N&S .. 29W-37N		19
Meadow Hill Ct. E&W		
.... 29W-37N		19
Meadow Hill Ln. N&S		
.... 29W-37N		19
Meadow Ln. E&W .. 29W-34N		25
Meadowlark Dr. NE&SW		
........29W-36N,29W-37N		19
Memory Tr. N&S		
........29W-36N,29W-37N		19
Michael St. N&S ...36W-34N		25
Midoak Ln. E&W .. 31W-32N		27
Mill Dr. N&S....... 36W-37N		17
Mill Ln. NW&SE .. 31W-33N		27
Miller Oak Dr. E&W		
.... 33W-36N		18
Miller Rd. E&W		
........32W-37N,30W-37N		18
Mineral Spring Dr. N&S		
.... 31W-32N		27
Mogra Dr. N&S .. 31W-34N		27
Mohawk Dr. E&W .. 36W-37N		17
Monroe St. NW&SE		
........34W-36N,33W-36N		18
Morgan Ter. E&W . 30W-33N		27
Mourine Ln. NW&SE		
.... 31W-33N		27
Myang Av. E&W ..31W-34N		27
N. Brixham Dr. E&W		
.... 30W-35N		19
Nantucket Dr. E&W		
.... 36W-37N		17
Navajo Tr. E&W 29W-34N		27
Niagra Dr. NW&SE 36W-37N		17
Nimitz Dr. N&S .. 36W-37N		17
Nippersink Dr. E&W		
.... 29W-37N		19
North Av. E&W .. 29W-32N		27
North Av. NW&SE .. 31W-32N		27
North Blvd. N&S ..29W-32N		27
North Dr. E&W .. 36W-36N		17
North Oak St. E&W		
.... 36W-36N		17
North St. N&S ... 29W-37N		19
Northside Av. NE&SW		
.... 31W-32N		27
Northwood Dr. E&W		
.... 36W-35N		17
O'Shea Ct. E&W .. 30W-34N		27
Oak Dr. E&W 36W-37N		17
Oak Rd. N&S 35W-36N		18
Oak St. E&W 30W-34N		27
Oakleaf Dr. E&W		
........30W-36N,29W-36N		19
Oakwood Av. N&S .31W-34N		27
Oakwood Ct. N&S ..31W-33N		27
Oakwood Dr. E&W 36W-36N		17
Oakwood Dr. E&W .. 36W-34N		25
Oeffling Ct. NW&SE		
.... 29W-36N		19
Oeffling Dr. E&W .. 29W-36N		19
Old Oak Dr. N&S .. 31W-34N		27
Oneida Rd. E&W .. 36W-36N		17
Orchard Beach Rd. NE&SW		
........31W-33N,31W-34N		27
Orchard Dr. N&S .. 35W-37N		18
Orchard Rd. E&W .36W-35N		17
Orchardway Dr. E&W		
.... 34W-34N		26
Oriole Tr. E&W 29W-37N		19
Osage Rd. N&S ... 36W-36N		17
Oshea Ct. E&W .. 30W-34N		27
Ottawa Rd. N&S .. 36W-36N		17
Palamino Dr. N&S .. 30W-33N		27
Pamela Dr. N&S .. 36W-34N		25
Park Dr. NS&EW .. 36W-35N		17
Parkside Av. NE&SW		
.... 30W-34N		27
Parkwood Dr. E&W		
.... 30W-35N		19
Patricia Ln. N&S .. 31W-34N		27
Patton St. N&S 36W-37N		17
Paulson St. N&S .. 36W-37N		17
Pebble Dr. NW&SE		
.... 30W-34N		27

	Grid	Page
Pheasant Rd. E&W		
.... 36W-35N		17
Pheasant Tr. E&W .. 29W-37N		19
Pinoak Dr. N&S .. 36W-34N		25
Pioneer Rd. NW&SE		
........32W-35N,33W-37N		18
Pistakee Dr. N&S .. 29W-37N		19
Pistakee Rd. N&S		
........29W-36N,29W-37N		19
Pleasant Dr. N&S .. 31W-34N		27
Pleasant Dr. N&S .. 36W-35N		17
Pleasant View Dr. NS&EW		
.... 29W-37N		19
Plymouth Ln. NW&SE		
.... 31W-32N		27
Ponca St. NE&SW .. 33W-32N		26
Poplar Leaf Dr. N&S		
.... 30W-36N		19
Prairie St. E&W ... 30W-32N		27
Preston Dr. E&W .. 36W-34N		25
Pulaski Rd. E&W .. 34W-34N		26
Quinn Rd. N&S .. 30W-32N		27
Ramree N&S . 31W-34N		27
Ranch Dr. NW&SE . 30W-33N		27
Rand Dr. E&W .. 29W-37N		19
Ravine Av. E&W		
........30W-32N,29W-32N		27
Ray St. N&S 29W-32N		27
Regner Ln. NW&SE		
.... 29W-34N		27
Regner Rd. NS&EW		
.... 29W-34N		27
Reiche Ln. E&W .. 30W-34N		27
Reiche Ln., S. N&S		
.... 30W-34N		27
Ridgeway Rd. N&S		
.... 29W-32N		27
Ridgeway Rd. N&S		
........35W-36N,35W-37N		18
Ridgewood Av. E&W		
........30W-32N,29W-32N		27
Ringwood Rd. NE&SW		
........34W-33N,30W-34N		26
River Rd. NW&SE		
........31W-32N,31W-33N		27
Riverdale Dr. NE&SW		
.... 30W-34N		27
Roger St. N&S ..30W-36N		19
Rose Dr. E&W 36W-37N		17
Rosedale Av. E&W 31W-33N		27
Round Up Rd. E&W		
.... 30W-33N		27
Rt. 120 E&W 35W-32N		26
Rt. 120 E&W 30W-32N		27
........36W-34N,30W-32N		25
Rt. 31 N&S		
........32W-34N,34W-37N		26
S. Brixham Ln. E&W		
.... 30W-35N		19
S. Circle Dr. NE&SW		
.... 36W-35N		17
S.Brixham Dr. E&W		
.... 30W-35N		19
S.Oak Rd. E&W .. 36W-35N		17
Sac Rd. N&S 35W-36N		18
Salem Rd. E&W .. 36W-36N		17
Schneider Ln. E&W		
.... 31W-34N		27
School Rd. N&S .. 34W-36N		18
Scott Av. E&W 31W-34N		27
Second St. E&W .. 32W-33N		26
Seminole Dr. N&S		
........36W-36N,35W-36N		17
Sequoia Tr. N&S .. 29W-34N		27
Shady Dr. NS&EW .30W-31N		27
Shady Dr. N&S .. 36W-35N		17
Shady Hollow Ln. E&W		
.... 29W-36N		19
Shirley Ln. N&S		
.... 29W-35N		19
Shorewood Dr. SW&NE		
.... 31W-34N		27
Shorewood Rd. NS&EW		
.... 31W-34N		27
Short St. N&S..... 29W-37N		19
Small St. E&W .. 29W-37N		19
South Dr. E&W .. 36W-36N		17
South Solon Rd. N&S		
.... 33W-37N		18
South St. E&W .. 29W-37N		19
Southside Av. E&W		
.... 29W-32N		27
Spaatz Av. E&W .. 36W-37N		17
Spojnia St. N&S .. 34W-34N		26
Spring Grove Rd. N&S		
........31W-34N,31W-37N		19
Sterling Dr. NS&EW		
.... 31W-34N		27
Stilling Blvd. NS&EW		
.... 31W-34N		27
Stillwell Dr. NE&SW		
........36W-37N,35W-37N		17
Sumac Dr. N&S .. 36W-37N		17
Summervil Av. NE&SW		
.... 36W-37N		17
Summit Dr. E&W .. 36W-36N		17
Sunrise Dr. E&W .. 29W-37N		19
Sydenham St. E&W		
.... 29W-37N		19
Sylvan St. NE&SW 36W-37N		17
Telegraph Rd. NE&SW		
.... 36W-37N		17

	Grid	Page
Terrace Av. E&W .. 30W-36N		19
Thelen Dr. E&W .. 29W-37N		19
Thelen Rd. NE&SW		
........30W-37N,29W-37N		19
Third St. E&W 32W-33N		26
Tichfield Ter. E&W .31W-35N		19
Trey Rd. NW&SE .. 36W-32N		25
Ute Rd. N&S 35W-36N		18
Val Ct. N&S....... 30W-33N		27
Vale St. E&W 30W-32N		27
Vallelo Ln. N&S ... 31W-34N		27
Valley Av. E&W .. 29W-32N		27
Valley St. E&W .. 30W-32N		27
Van Buren St. N&S		
........34W-36N,33W-36N		18
Vern St. N&S 29W-37N		19
Vernon Dr. N&S .. 29W-32N		27
Villa Ln. NE&SW ...31W-34N		27
Wainwright Dr. NE&SW		
.... 36W-37N		17
Walnut Dr. N&S .. 29W-34N		25
Waltshire Dr. N&S ..32W-36N		18
Washington Park NW&SE		
.... 31W-34N		27
Washington St. N&S		
.... 29W-32N		27
West Blvd. E&W ...29W-32N		27
West Lake Shore Dr. NE&SW		
.... 36W-36N		17
West Shore Dr. E&W		
.... 34W-34N		26
West Shore Dr. N&S		
.... 36W-36N		17
West St. N&S ... 29W-36N		19
Westmoor Dr. N&S		
.... 36W-35N		17
Weston Ct. E&W .. 29W-34N		27
Westwood Dr. N&S		
........29W-36N,29W-37N		19
White Oaks Rd. E&W		
........37W-37N,36W-37N		17
Widoff Dr. E&W ... 36W-34N		25
William St. N&S .. 30W-36N		19
Willow Dr. N&S		
........36W-36N,36W-37N		17
Willow Ln. N&S .. 34W-34N		26
Willow St. N&S .. 29W-37N		19
Wilmington Ct. N&S		
.... 36W-32N		25
Windsor Dr. NE&SW		
.... 36W-37N		17
Winnebago Dr. N&S		
.... 36W-36N		17
Wonder Lake Dr. N&S		
.... 36W-36N		17
Wonder Woods Dr. N&S		
.... 36W-37N		17
Wooded Ln. N&S .. 29W-34N		27
Wooded Shore Dr. E&W		
.... 36W-35N		17
Woodland Dr. N&S 29W-37N		19
Woodland Dr. E&W		
.... 34W-34N		26
Woodland Ln. NE&SW		
.... 31W-33N		27
Woodlane Dr. NE&SW		
.... 36W-37N		17
Woodlawn Park Av. NE&SW		
........31W-33N,31W-34N		27
Woodrow Av. N&S		
........29W-36N,29W-37N		19
Yorkshire Dr. E&W 32W-37N		18

Nunda Township

	Grid	Page
Abendroth Dr. NW&SE		
.... 35W-27N		34
Acacia NS&EW ... 34W-28N		34
Acacia NW&SE .. 34W-28N		34
Acorn Ct. N&S .. 32W-26N		42
Allen Av. NS&EW ..31W-30N		35
Amanda Av. E&W .. 29W-31N		27
Ames Rd. E&W .. 33W-28N		34
Andrea Ln. E&W .. 36W-26N		41
Annette Av. E&W .. 33W-26N		42
Anvil Rd. E&W 35W-29N		34
Area St. N&S 29W-27N		35
Baldwin Rd. E&W .. 31W-30N		35
Barbara Ln. E&W .. 31W-27N		34
Barreville Rd. N&S		
........32W-27N,32W-31N		34
Bay Rd. N&S 34W-28N		34
Beach Av. NS&EW 30W-29N		35
Beach Dr. E&W .. 32W-31N		26
Beach St. E&W .. 30W-29N		35
Beach Way N&S		
........29W-26N,30W-28N		43
Behan Rd. E&W .. 30W-27N		35
Benbrook Av. N&S 30W-29N		35
Bergman Av. N&S		
........30W-28N,30W-29N		35
Billingsgate Ct. N&S		
.... 36W-31N		25
Black Partridge Rd. N&S		
........31W-30N,31W-31N		35
Blackhawk Av. N&S		
.... 30W-30N		35
Blossom St. NW&SE		
.... 30W-28N		35
Blue Ct. N&S 32W-26N		42
Blue Island Av. N&S		
.... 30W-28N		35

	Grid	Page
Blue Pine Dr. N&S		
........35W-29N,35W-30N		34
Bluebird Ct. NE&SW		
.... 35W-27N		34
Bolger Av. E&W ... 31W-30N		35
Brefeld Ct. E&W .. 31W-31N		27
Brentwood Ln. NW&SE		
.... 31W-30N		35
Briggs Av. NE&SW		
.... 30W-28N		35
Buckhorn Dr. N&S		
.... 35W-28N		34
Bull Valley Rd. E&W		
........35W-31N,32W-31N		25
Bur Ln. N&S 32W-26N		42
Burton Terr. NW&SE		
.... 30W-28N		35
Cape Cove NW&SE		
.... 35W-27N		34
Carman Av. N&S .. 30W-29N		35
Carolina Av. N&S .. 30W-27N		35
Carraige Dr. E&W .36W-27N		33
Carrington Dr. EW&NS		
.... 32W-26N		42
Carthage Ct. N&S .33W-27N		34
Catalpa St. E&W .. 29W-27N		35
Cedar St. E&W .. 29W-27N		35
Cedardel St. E&W .. 29W-27N		35
Channel Ln. NW&SE		
.... 31W-30N		35
Chatham Ln. N&S .. 36W-27N		33
Cherry Av. N&S .. 29W-28N		35
Cherry Dr. E&W .. 30W-27N		35
Cherry Ln. E&W .. 30W-27N		35
Cherry Ln. N&S .. 32W-26N		42
Cherry Valley Rd. N&S		
........36W-27N,36W-29N		33
Chesterfield Rd. E&W		
.... 36W-27N		33
Chris Ln. NS&EW .. 32W-26N		42
Christine Dr. E&W .. 29W-31N		27
Coachlight Rd. E&W		
.... 35W-29N		34
Colby Dr. NW&SE		
........31W-29N,30W-29N		35
Colby Point Rd. E&W		
.... 31W-29N		35
Colonel Holcomb Dr. NE&SW		
.... 35W-29N		34
Columbus St. E&W		
.... 29W-31N		27
Comes Tr. E&W		
........30W-31N,29W-31N		27
Concord Tr. NW&SE		
.... 35W-27N		34
Connecticut Ct. N&S		
.... 35W-28N		34
Connecticut Tr. N&S		
.... 35W-27N		34
Crabapple Ln. E&W		
.... 30W-27N		35
Crescent Av. NE&SW		
.... 30W-28N		35
Crescent Dr. NE&SW		
.... 29W-29N		35
Cristine Dr. E&W .. 29W-31N		27
Croak Av. N&S .. 30W-29N		35
Crockett Ct. E&W .. 32W-26N		42
Crystal Cir. NS&EW		
.... 35W-28N		35
Crystal Lake Av. E&W		
.... 34W-26N		42
Crystal Lake Rd. NE&SW		
.... 31W-27N		34
Crystal Way NS&EW		
.... 31W-28N		35
Curling Pond Ct. N&S		
.... 35W-28N		34
Curran Rd. N&S .. 35W-31N		26
Cypress Ct. E&W .. 35W-27N		34
Daisy Ln. N&S 35W-27N		34
Davis Av. E&W ... 30W-29N		35
Dedlow Rd. N&S .. 30W-28N		35
Dee Dr. E&W 30W-28N		35
Deedwood Dr. N&S		
.... 35W-28N		34
Deering Oaks Ln. E&W		
.... 33W-27N		34
Deerwood Dr. E&W		
.... 35W-28N		34
Defiance St. N&S		
........30W-30N,30W-31N		35
Derby Ln. N&S ... 36W-27N		33
Drake Dr. E&W .. 33W-27N		34
Draper Rd. N&S .. 35W-31N		26
Drucker Ln. E&W .. 31W-31N		27
E. Bayview Dr. E&W		
.... 30W-27N		35
E. Beach St. E&W .. 30W-27N		35
E. Hillside Ln. E&W .34W-27N		34
E. Logan St. N&S .. 29W-31N		27
E. Red Barn Rd. E&W		
........36W-28N,35W-28N		33
E. Swarthmore Rd. E&W		
.... 36W-27N		33
East Dr. N&S 36W-27N		33
East Upland Dr. N&S		
.... 35W-27N		34
Edgewood Rd. E&W		
........35W-29N,33W-29N		34
Edward Dr. E&W .. 30W-27N		35
Ellen Rd. E&W .. 32W-31N		26

	Grid	Page
Elm St. N&S	29W-26N	43
Emerald Av. N&S	31W-31N	27
Emerald Ct. E&W	31W-31N	27
Emerald Dr. N&S	31W-31N	27
Ethel Av. N&S	33W-26N	42
Fargo Av. NE&SW	30W-29N	35
Federal Ct. NW&SE	35W-29N	34
Femwood Ln. N&S	31W-30N	35
Fernview Ln. NE&SW	30W-30N	35
Field Ct. NE&SW	35W-27N	34
Fir Ct. N&S	35W-27N	34
Ford Rd. E&W	31W-31N	27
Forest Blvd.	29W-26N,29W-27N	43
Forestview Ct. N&S	29W-29N	35
Forestwood Dr. N&S	29W-29N	35
Fox Creek Rd. NW&SE	34W-27N	34
Fox Ct. NE&SW	34W-27N	34
Fox Fire Dr. N&S	36W-28N,35W-28N	33
Fox River Dr. NW&SE	30W-27N	35
Fox Run E&W	35W-29N	34
Fox Tr. NS&EW	34W-27N	34
Franklin Ct. E&W	32W-26N	42
Garden Ln. N&S	34W-29N	34
Giant Oak Rd. N&S	29W-27N	35
Grace St. NW&SE	31W-30N	35
Gracy Rd. E&W	33W-29N,32W-29N	34
Grand Av. NW&SE	30W-28N	35
Great Hill Rd.	36W-28N,35W-28N	33
Greening Rd. NW&SE	31W-30N	35
Griswold Av. NW&SE	29W-28N	35
Hamilton Dr. N&S	32W-26N	42
Hanover Dr. NS&EW	34W-27N	34
Harriett Ln. NE&SW	29W-29N	35
Harvard Circle N&S	35W-27N	34
Hawk Av. E&W	30W-29N	35
Hawthorne Ct. NE&SW	30W-28N	35
Hawthorne Ln. NW&SE	31W-30N	35
Hibiscus Tr. NE&SW	34W-28N	34
Hickory Ter. N&S	30W-27N	35
Hidden Oak Ct. E&W	35W-28N	34
Hidden Oak Dr. NS&EW	35W-28N	34
High Point Rd. E&W	33W-30N	34
Highland Av. NE&SW	30W-27N	35
Highline Rd. E&W	35W-27N	34
Highview Dr. NE&SW	30W-28N	35
Highwood Rd. N&S	29W-29N	35
Hiline Av. N&S	30W-29N	35
Hillcrest Dr. N&S	35W-27N	34
Hillside Rd. E&W	36W-27N,34W-27N	33
Hillside Tr. N&S	30W-31N,29W-31N	27
Holly Ln. E&W	30W-27N	35
Hythe Cir. NW&SE	36W-31N	25
Indian Tr. E&W	31W-31N	27
Jasmine Ct. N&S	34W-28N	34
Jenny Jae Ln. N&S	34W-29N	34
Johnathan Rd. E&W	31W-30N	35
Johnson Av. N&S	30W-28N	35
Jones Av. E&W	30W-28N	35
Jones St. E&W	30W-29N	35
Justen Rd. N&S	32W-29N,31W-29N	34
Keene Av. N&S	30W-28N,30W-29N	35
Keene Av. N&S	30W-29N	35
Kenilworth Av. N&S	30W-29N	35
Kerry Ln. E&W	30W-29N	35
Kirsten Tr. N&S	33W-28N	34
Laguna Dr. NE&SW	31W-31N	27
Lake View Rd. NS&EW	31W-26N	43
Lakeshore Dr. NE&SW	29W-29N	35
Landl Park Rd. E&W	29W-31N	27
Lands Ct. NW&SE	30W-28N	35
Larkspur Dr. N&S	34W-28N	34
Laurel St. NW&SE	30W-27N	35
Lavergne Av. N&S	30W-27N	35
Leaf Av. N&S	30W-29N	35
Lexington Tr. E&W	35W-27N	34
Lilly Ct. NE&SW	30W-28N	35

	Grid	Page
Lily Lake Rd. N&S	29W-29N,29W-32N	35
Lincoln Av. NW&SE	30W-28N	35
Lincoln St. N&S	29W-31N	27
Linden Rd. N&S	30W-27N	35
Little Dr. N&S	29W-27N	35
Little Fawn Tr. E&W	36W-28N	33
Live Oak Rd. NS&EW	35W-28N	34
Logan St. N&S	29W-31N	27
Longwood Dr. N&S	29W-29N	35
Louise St. E&W	30W-27N	35
Low Av. E&W	30W-29N	35
Maine Tr. N&S	35W-27N	34
Maple Dr. E&W	29W-27N	35
Marguerite St. E&W	30W-27N	35
Marietta Dr. N&S	33W-26N	42
Marlboro Rd. NW&SE	36W-27N	33
Mason Hill Rd. E&W	36W-30N,35W-30N	33
May St. E&W	30W-27N	35
McCabe Av. N&S	30W-28N	35
McHenry Dowell Rd. NE&SW	29W-29N	35
McMillan Ln. NE&SW	33W-28N	34
Meadow Dr. N&S	35W-27N	34
Meadow St. N&S	29W-26N	43
Meadowlark Ct. NW&SE	35W-27N	34
Meadowshire Ln. N&S	36W-27N	33
Minuteman Ct. E&W	35W-27N	34
Misty Pine Ct. E&W	35W-30N	34
Monica Tr. E&W	32W-26N	42
N.E. Shore Dr. NW&SE	29W-29N	35
Nancy Ln. E&W	29W-31N	27
Ned Dr. E&W	32W-26N	42
Needle Creek Ct. NW&SE	35W-28N	34
New Hampshire Tr. E&W	35W-27N	34
Nish Rd. E&W	30W-28N	35
North Ct. NE&SW	35W-27N	34
North East Shore Dr. NW&SE	29W-29N	35
North St. E&W	31W-31N	27
Nunda Av. N&S	30W-27N	35
Nunda Tr. N&S	34W-27N	34
Oak Crest Dr. E&W	33W-27N	34
Oak Ct. NE&SW	30W-31N	27
Oak Ridge Ct. E&W	36W-28N	33
Oak Ridge Rd. N&S	36W-27N,36W-29N	33
Oak St. E&W	29W-27N	35
Oak St. E&W	36W-26N,36W-29N	41
Oakwood Ln. E&W	36W-26N	41
Opengate Rd. N&S	35W-28N,35W-29N	34
Orchard Ln. NS&EW	31W-30N	35
Orchard Ln. E&W	33W-26N	42
Orman Av. N&S	29W-27N	35
Osage Dr. N&S	31W-31N	27
Osceola Av. E&W	31W-30N	35
Overland Rd. E&W	30W-28N	35
Page Pl. NE&SW	30W-29N	35
Palm St. N&S	29W-26N	43
Park St. NE&SW	30W-28N	35
Parker St. E&W	30W-29N	35
Paul St. E&W	29W-27N	35
Pesz Rd. N&S	29W-27N	35
Peter St. E&W	29W-27N	35
Pheasant Run E&W	36W-29N,35W-29N	33
Pine St. N&S	29W-26N	43
Pleasant Hill Ct. N&S	34W-28N	34
Pleasant Hill Rd. N&S	35W-28N,34W-28N	34
Plum St. E&W	29W-28N	35
Pontiac Av. E&W	31W-30N	35
Pony Ln. N&S	35W-29N	34
Pool St. N&S	30W-29N	35
Poplar St. E&W	29W-27N	35
Porten Rd. E&W	29W-27N	35
Prairie Ridge Rd. N&S	33W-26N	42
Prairie St. E&W	31W-30N	35
Pueblo Dr. NW&SE	31W-31N	27
Pueblo Dr. E&W	31W-31N	27
Purdue Pl. N&S	33W-27N	34
Raintree Dr. E&W	35W-30N	34
Ravenglass Ridge N&S	36W-28N	33
Ravine Dr. NE&SW	36W-27N	33
Rawson Bridge Rd. NE&SW	30W-26N,29W-26N	43

	Grid	Page
Ray St. E&W	33W-26N	42
Red Barn Rd. N&S	36W-28N,36W-29N	33
Redbird Ln. E&W	35W-27N	34
Redoak Dr. NW&SE	36W-27N	33
Reiland Dr. E&W	33W-26N	42
Remington Dr. N&S	32W-26N,31W-26N	42
Rhode Island Tr. E&W	35W-27N	34
Rigby Rd. N&S	36W-27N	33
Ripon Rd. N&S	33W-27N	34
River Rd. E&W	31W-30N,29W-30N	35
River St. N&S	29W-26N,29W-27N	43
Riverdale Rd. N&S	30W-28N	35
Riverside Dr. NW&SE	31W-30N,30W-30N	35
Roberts Rd. N&S	29W-27N	35
Robin Ln. N&S	30W-28N	35
Rockspur Tr. NS&EW	34W-28N	34
Roger Rd. N&S	33W-28N	34
Rona Rd. N&S	31W-25N	43
Rt. 176 NE&SW	34N-26N,29W-28N	42
Rt. 31 N&S	33W-26N,32W-31N	42
Running Iron Dr. NW&SE	35W-29N,34W-29N	34
Russet Dr. NS&EW	31W-30N	35
S. Great Hill Rd. NS&EW	36W-28N,35W-28N	33
S. Shenandoah Dr. N&S	35W-28N	34
Sauk Dr. N&S	31W-31N	27
Sauk Tr. N&S	31W-31N	27
Scarlet Tr. E&W	32W-26N	42
Scheid Ln. NE&SW	29W-29N	35
Schroder Rd. N&S	33W-30N	34
Sequoia Tr. N&S	34W-27N	34
Sessions Ct. N&S	36W-27N	33
Sessions Walk NW&SE	36W-27N	33
Shadowood Dr. NS&EW	34W-28N	34
Shady Dr. NS&EW	30W-31N	27
Shady Oaks Ln. E&W	33W-27N	34
Shall Ln. E&W	29W-29N	35
Shenandoah Ct. E&W	35W-28N	34
Shenandoah Dr. E&W	35W-28N	34
Sheridan Rd. E&W	29W-31N	27
Sherman Blvd. E&W	30W-27N	35
Sherwood Dr. N&S	35W-27N	34
Silver Lake Rd. N&S	31W-26N	43
Skyline Dr. N&S	31W-30N	35
Skyline Ln. N&S	35W-29N	34
Sloan Ln. N&S	32W-26N	42
Smith Rd. N&S	33W-26N	42
Smoke Tree Ln. NW&SE	35W-28N	34
South St. E&W	30W-28N	35
Spring Blvd. E&W	35W-28N,34W-28N	34
Spring Brook Rd. N&S	31W-29N	34
Spy Glass Ridge N&S	35W-28N	34
Stanford St. N&S	33W-27N	34
State Park Rd. E&W	32W-30N	34
State Rd. NE&SW	30W-28N	35
Steuben Rd. N&S	29W-31N	27
Stilling Ln. NE&SW	31W-29N	35
Stonegate Dr. N&S	36W-28N,36W-29N	33
Stonewier Pt. NS&EW	36W-31N	25
Strong Rd. E&W	33W-26N	42
Sumac Av. N&S	29W-31N	27
Surf St. E&W	30W-29N	35
Sutton Ct. E&W	31W-31N	27
Sweet Bay Dr. E&W	35W-30N	34
Tamarisk Ct. NE&SW	34W-28N	34
Tamarisk Tr. NS&EW	34W-28N	34
Tecoma Dr. NS&EW	34W-28N	34
Tecumseh Av. NW&SE	31W-30N	35
Tecumseh Ln. NW&SE	31W-30N	35
Terra Cotta Av. E&W	32W-26N,31W-26N	42
Terra Cotta Rd. N&S	34W-26N,34W-28N	42
Terry Av. E&W	32W-26N	42
Thomas Ct. N&S	29W-29N	35

	Grid	Page
Tilche Ln. E&W	32W-26N	42
Tile Line Dr. N&S	35W-27N	34
Timber Tr. E&W	29W-27N	35
Timberlane Rd. N&S	30W-27N	35
Totem Tr. N&S	31W-31N	27
Tree Av. NE&SW	30W-29N	35
Tulip St. N&S	30W-28N	35
U.S. Route 14 NW&SE	36W-26N	41
Valley Rd. N&S	36W-29N,36W-30N	33
Vaupell Rd. N&S	30W-29N,29W-29N	35
Vermont Tr. E&W	35W-27N	34
Vine Ln. N&S	35W-29N	34
Virginia Ln. N&S	35W-29N	35
W. Beach St. E&W	30W-27N	35
W. Fargo Av. NE&SW	30W-29N	35
W. Parker St. E&W	30W-29N	35
Walking Ridge N&S	36W-28N	33
Walkup Av. N&S	35W-27N	34
Walkup Rd. N&S	35W-27N,35W-29N	34
Walnut Ct. NE&SW	30W-28N	35
Walnut St. E&W	29W-26N	43
Waterview Av. N&S	30W-28N	35
Wauconda Rd. NE&SW	30W-28N	35
Wayside Dr. N&S	29W-27N	35
Weatherstone Dr. E&W	33W-26N	42
Weathervane Ln. N&S	36W-28N	33
Wegner Rd. E&W	29W-31N	27
West Justen Rd. E&W	32W-29N	34
West Upland Dr. N&S	35W-27N	34
Western Av. E&W	30W-29N	35
Whipporwill Dr. E&W	30W-27N	35
White Ash Dr. E&W	36W-26N	42
White Barn Rd. E&W	36W-28N	33
Wild Ash Ln. N&S	34W-28N	34
Wild Olive Ln. E&W	34W-28N	34
Wild Plum Dr. NE&SW	33W-26N	42
Wildcherry Rd. N&S	35W-27N	34
Wilderness Ridge N&S	35W-28N	34
Wildwood Dr. N&S	29W-27N	35
Wildwood Rd. NE&SW	30W-27N	35
Wisteria Dr. NW&SE	34W-28N	34
Wood St. E&W	30W-28N,29W-28N	35
Woods Av. N&S	35W-28N	34
Woods End N&S	36W-28N	33
Wright Av. E&W	29W-27N	35
Wright Rd. NW&SE	29W-29N,30W-28N	35
Wyandotte Av. E&W	31W-30N	35
Wyoming Way N&S	33W-27N	34
Yuma Tr. N&S	31W-31N	27
1st St. E&W	29W-29N	35

Oakwood Hills

	Grid	Page
Acorn Ln. N&S	31W-25N	43
Ash Dr. N&S	31W-26N	43
Birch Ln. NS&EW	31W-26N	43
Burwood Rd. N&S	31W-26N	43
Deer Tr. N&S	31W-26N	43
East Park Ln. E&W	31W-26N	43
Echo Hill Dr. N&S	31W-26N	43
Elm Rd. NE&SW	31W-26N	43
Fawn Ridge Dr. NS&EW	31W-26N	43
Fenview Ct. E&W	31W-26N	43
Greenview Rd. N&S	31W-26N	43
Hickory Rd. N&S	31W-26N	43
Hilltop Rd. N&S	31W-26N	43
Lake Ln. NE&SW	31W-26N	43
Lake Shore Dr. N&S	31W-26N	43
Lakeview NW&SE	31W-26N	43
Lakewood Dr. N&S	31W-26N	43
Meadow Ln. N&S	31W-26N	43
North Park Dr. E&W	31W-26N	43
North Shore Dr. E&W	31W-26N	43
Oak Dr. N&S	31W-25N	43
Oak Hill Rd. E&W	31W-26N	43
Oak Park Rd. N&S	31W-26N	43

	Grid	Page
Oakwood Dr. N&S	31W-26N	43
Palisades Ln. NE&SW	31W-26N	43
Sherwood Trail E&W	31W-26N	43
Spruce Rd. E&W	31W-26N	43
Valley Dr. N&S	31W-26N	43
West Lake Shore Dr. N&S	31W-26N	43
West Park Ln. N&S	31W-26N	43
Woodland Rd. E&W	31W-25N	43
Woodland Rd. W. E&W	31W-26N	43
Woody Way N&S	31W-26N	43

Prairie Grove

	Grid	Page
Ames Rd. E&W	33W-28N,32W-28N	34
Barreville Rd. N&S	32W-27N,32W-29N	34
Behan Rd. E&W	31W-27N	35
Buhl Rd. N&S	31W-27N	35
Crystal Lake Rd. E&W	32W-27N,31W-27N	34
Fawn Tr. NS&EW	31W-28N	35
Fox Ct. E&W	31W-28N	35
Gracy Rd. E&W	32W-29N	34
Hillview Dr. N&S	31W-28N	35
N. Brighton Ln. N&S	33W-27N	34
N. Tamarack Ln. NW&SE	32W-27N	34
Nish Rd. E&W	31W-28N	35
Oak Knoll North Rd. N&S	32W-27N	34
Oak Knoll South Rd. N&S	32W-27N	34
Red Bud Ct. NW&SE	32W-27N	34
Rt. 176 E&W	32W-27N	34
S. Tamarack Tr. N&S	32W-27N	34
Squaw Creek Rd. NE&SW	33W-28N	34
Stratford Rd. NW&SE	33W-27N,32W-27N	34
Tamarack Ct. NW&SE	32W-27N	34
Timber Trail Dr. E&W	31W-28N	35
Valley View Rd. N&S	32W-27N,32W-28N	35
Victoria N&S	33W-27N	34
West Justen Rd. E&W	32W-29N,31W-29N	34
White Oak Ct. NW&SE	32W-27N	34
Wirth Tr. NS&EW	32W-28N	34
Wright Rd. E&W	32W-29N,31W-29N	34

Richmond

	Grid	Page
Ami Dr. NW&SE	34W-42N,34W-43N	10
Broadway St. E&W	34W-42N	10
Charles St. N&S	34W-41N,34W-42N	10
Circle Dr. NW&SE	34W-41N	10
Commercial St. NW&SE	34W-42N,34W-41N	10
Covell St. N&S	34W-42N	10
East St. N&S	34W-42N	10
Elm St. E&W	34W-42N	10
Foundry Dr. N&S	34W-42N	10
Front St. E&W	34W-42N	10
George St. E&W	34W-42N	10
Golf Av. N&S	34W-41N	10
Hillcrest Dr. N&S	34W-41N	10
Kenosha St. E&W	34W-42N	10
Liberty St. E&W	34W-42N	10
Maiden Ln. N&S	34W-41N	10
Main St. N&S	34W-41N,34W-42N	10
Market St. N&S	34W-42N	10
May Av. E&W	34W-41N	10
Mayview St. E&W	34W-41N	10
Mayzing St. E&W	34W-43N	10
McDonnell Dr. N&S	34W-42N	10
Mill St. E&W	34W-42N	10
Milwaukee St. E&W	34W-42N	10
Nippersink Dr. NW&SE	34W-41N	10
North Ln. N&S	34W-41N	10
Rt. 31 N&S	34W-41N,34W-43N	10
Rt.173 E&W	35W-42N,31W-42N	10
South St. E&W	34W-42N	10
Valley Dr. N&S	34W-41N	10
Walnut St. E&W	34W-42N	10
West St. N&S	34W-42N	10
William St. N&S	34W-42N	10
Wilson Rd. E&W	34W-43N	10

Grid / Page

Richmond Township
Arthur Dr. N&S .... 32W-40N — 10
Barnard Mill Rd. NW&SE
.... 36W-38N — 17
Bonnie Brae Dr. NE&SW
........34W-41N,33W-41N — 10
Broadway Rd. NW&SE
........35W-42N,34W-42N — 10
Burgett Rd. E&W . 36W-43N — 9
Burlington Rd. N&S
........33W-43N,33W-44N — 10
Clark Rd. N&S
........32W-40N,32W-42N — 10
East Solon Rd. E&W
........32W-39N,31W-39N — 10
East St. N&S ...... 32W-39N — 10
Falcon St. N&S&EW 33W-41N — 10
Foxboro Tr. NE&SW
.... 33W-41N — 10
Gardner St. N&S . 33W-39N — 10
Grant St. N&S .... 32W-43N — 10
Harding Av. E&W . 32W-43N — 10
Harts Rd. E&W
........34W-38N,33W-38N — 18
Hideaway Ln. NE&SW
.... 33W-41N — 10
Hill Rd. E&W
........34W-41N,33W-41N — 10
Hillandale Rd. NS&EW
.... 33W-41N — 10
Hillendale Ln. NW&SE
.... 33W-41N — 10
Hillshire Dr. W. NS&EW
.... 33W-41N — 10
John St. NE&SW
........33W-39N,32W-39N — 10
Keith Rd. E&W . 32W-40N — 10
Keystone Rd. N&S
........36W-38N,36W-43N — 17
Kuhn Rd. E&W
........34W-40N,33W-40N — 10
Lakeview Rd. N&S
........32W-42N,32W-43N — 10
Lincoln Av. E&W . 32W-43N — 10
Macwood Dr. E&W
........34W-40N,33W-40N — 10
May Ln. E&W ...... 32W-39N — 10
Mayline Av. N&S . 34W-41N — 10
Meyer Rd. N&S
........31W-38N,31W-39N — 19
Mill Rd. E&W ...... 32W-40N — 10
Mill St. N&S ...... 32W-39N — 10
Miller Rd. E&W
........32W-38N,31W-38N — 18
Norgard Rd. E&W . 36W-41N — 9
North Solon Rd. N&S
........33W-40N,33W-42N — 10
North St. N&S . 32W-39N — 10
Oak St. N&S ...... 33W-39N — 10
Oakhill Dr. NS&EW
.... 36W-38N — 17
Overlook Dr. E&W . 32W-43N — 10
Overton St. E&W . 32W-43N — 10
Penny Ln. N&S .. 33W-40N — 10
Pioneer Rd. NW&SE
........33W-38N,34W-39N — 18
Private Rd. N&S
........36W-39N,35W-39N — 9
Prospect St. E&W . 34W-41N — 10
Rt. 12 NS&EW
........34W-41N,32W-39N — 10
Rt. 31 N&S
........34W-38N,34W-41N — 18
South Solon Rd. N&S
........33W-38N,33W-39N — 18
Spring Grove Rd. N&S
.... 31W-38N — 19
Squires Green NS&EW
.... 33W-41N — 10
State Line Rd. E&W
........32W-43N,31W-43N — 10
Sunset Rd. E&W . 31W-38N — 19
Taft St. N&S ...... 32W-43N — 10
Tracy Ct. NS&EW . 33W-40N — 10
Tryon Grove Rd. NE&SW
........36W-39N,34W-40N — 9
Valentine Ct. NE&SW
.... 36W-38N — 17
West Solon Rd. E&W
........34W-39N,33W-39N — 10
White St. N&S
........39W-34N,32W-39N — 10
Winn Rd. N&S
........31W-40N,31W-42N — 11
Zarnstorf Rd. N&S
........31W-42N,31W-43N — 11

Riley Township
Acorn Ln. E&W ... 50W-24N — 37
Alta Dr. N&S ... 54W-23N — 44
Anthony Rd. E&W . 50W-22N — 45
Anthony Rd. E&W . 52W-22N — 44
Beck Rd. E&W ... 49W-22N — 45
Beth Ct. E&W ... 49W-25N — 37
Blisdale Rd. N&S
........52W-23N,52W-25N — 45
Burma Rd. N&S
........54W-23N,54W-26N — 44
Burrows Rd. N&S . 53W-20N — 44
Carls Rd. N&S ... 51W-21N — 45

Coon Creek Rd. N&S
.... 50W-20N — 45
Coral Oaks Dr. NW&SE
.... 50W-24N — 37
County Line Rd. N&S
........54W-20N,54W-25N — 44
Genoa Rd. N&S ... 54W-20N — 44
Getty Rd. N&S ... 49W-20N — 45
Grange Rd. E&W
........54W-23N,50W-23N — 44
Grange Rd. E&W . 50W-23N — 45
Grossen Rd. E&W . 50W-21N — 45
Hartman Rd. E&W . 50W-20N — 45
Hartman Rd. E&W . 52W-21N — 44
Helen St. E&W ... 54W-23N — 44
Hill Rd. N&S
........49W-23N,49W-24N — 45
Johnson Rd. N&S . 52W-25N — 36
Lakewood Dr. NE&SW
........51W-25N,50W-25N — 36
Maple Rd. N&S
........49W-22N,49W-25N — 45
Maple St. N&S ... 49W-25N — 37
Marjorie Dr. N&S . 54W-24N — 36
Mary St. E&W ... 54W-23N — 44
McKeowin Rd. N&S
.... 54W-22N — 44
Meyer Rd. N&S ... 51W-25N — 36
Northwest Tollway NW&SE
.... 54W-23N — 44
Northwest Tollway NW&SE
.... 50W-21N — 45
Norway Ct. E&W . 50W-20N — 45
O'Connell Dr. E&W
.... 50W-25N — 37
Oak Creek Dr. E&W
.... 50W-24N — 37
Payne Rd. N&S ... 51W-20N — 45
Pleasant Grove Rd. NW&SE
........52W-25N,50W-24N — 36
Poplar Rd. E&W ... 51W-20N — 45
Poplar Rd. E&W ... 54W-20N — 44
Ratfield Rd. NW&SE
.... 49W-25N — 37
Riley Dr. N&S ... 49W-25N — 37
Riley Harmony Rd. E&W
.... 50W-25N — 45
Riley Harmony Rd. E&W
.... 53W-22N — 44
Riley Rd. N&S..... 51W-22N — 45
Rt. 23 N&S
........50W-23N,50W-25N — 45
Rudolph Rd. E&W . 54W-20N — 44
West Coral Rd. E&W
........50W-24N,49W-24N — 37

Seneca Township
Alpine Ln. NS&EW
........43W-30N,43W-31N — 31
Athy Ct. NW&SE .. 47W-28N — 30
Athy Ln. NS&EW .. 47W-28N — 30
Blackhawk Ln. N&S
.... 47W-31N — 22
Bockman Rd. NS&EW
........48W-29N,43W-28N — 30
Charlotte Dr. E&W . 43W-28N — 31
Collins Rd. E&W
........48W-30N,45W-30N — 30
Davis Rd. E&W ... 43W-30N — 31
Deer Path Tr. E&W
.... 47W-29N — 30
Dimmel Rd. N&S .. 46W-31N — 22
Dunham Rd. N&S .. 47W-26N — 38
Elaine Dr. N&S ... 43W-28N — 31
Emery Ln. EW&NS . 43W-26N — 39
Franklinville Rd. N&S
........45W-27N,45W-29N — 31
Garden Valley Rd. E&W
........46W-28N,45W-28N — 31
Gee Rd. N&S
........44W-27N,44W-28N — 31
Hiawatha Ln. N&S .. 47W-31N — 22
Highbridge Rd. E&W
.... 46W-26N — 38
Hobe Rd. N&S
........45W-30N,44W-31N — 31
Hughes Rd. N&S .. 45W-31N — 23
Kilkee Ln. NE&SW ..47W-28N — 30
Kishwaukee Valley Rd. E&W
........48W-31N,43W-31N — 22
Kunde Rd. E&W
........47W-27N,45W-27N — 30
McKinstry Rd. NS&EW
........47W-31N,46W-31N — 22
Millstream Rd. N&S
........47W-28N,47W-28N — 38
Millstream Rd. E&W
.... 47W-29N — 30
Mohawk Ln. E&W .. 47W-31N — 22
North Union Rd. N&S
........46W-26N,46W-28N — 38
O'Clock Rd. E&W
........47W-26N,46W-26N — 38
Perkins Rd. E&W
........44W-28N,43W-28N — 31
Peterson-Olson Rd. N&S
.... 45W-26N — 39
Pleasant Valley Rd. E&W
........45W-26N,43W-26N — 39
River Rd. NW&SE
........48W-28N,47W-27N — 30

Rose Farm Rd. NE&SW
........44W-30N,43W-31N — 31
Rt. 176 E&W
........48W-27N,44W-27N — 30
Sally Cir. E&W .... 43W-28N — 31
Sandhill Rd. E&W .. 47W-28N — 30
Secor Rd. E&W
........46W-29N,45W-29N — 30
Shannon Dr. N&S . 43W-31N — 23
South St. E&W
........44W-30N,43W-30N — 31
Standish Rd. N&S
........48W-28N,48W-29N — 30
Steig Rd. N&S
........43W-28N,43W-29N — 31
Sullivan Rd. N&S . 47W-30N — 30
Sunnyside Rd. N&S
........43W-26N,43W-28N — 39
Sylvia Ln. N&S .... 43W-29N — 31
Timber Ln. NS&EW
.... 43W-26N — 39
Vermont Rd. N&S
........46W-28N,46W-30N — 30
Woodcliff Dr. N&S .. 47W-29N — 30
Woodland Tr. E&W
.... 47W-29N — 30

Spring Grove
Altamonte Dr. N&S 31W-40N — 11
Amanda Dr. E&W .. 31W-40N — 11
Asbury Ct. N&S ... 30W-39N — 11
Aspen Ct. N&S .... 30W-38N — 19
Beach St. E&W ... 30W-38N — 19
Belvin St. N&S ... 30W-39N — 11
Blivin St. N&S
........31W-39N,30W-39N — 11
Broadview Ct. N&S
.... 31W-39N — 11
Carmel Ct. NW&SE
........31W-40N,31W-41N — 11
Catalpa Ct. N&S.... 30W-38N — 19
Champion St. NE&SW
.... 31W-41N — 11
Chateaugay Dr. E&W
.... 30W-41N — 11
Chelmsford Dr. E&W
.... 32W-39N — 11
Cherrywood Ln. N&S
.... 30W-38N — 19
Christina Ct. NE&SW
.... 31W-41N — 11
Coventry Dr. N&S . 32W-39N — 10
Dale Ct. NW&SE .. 30W-41N — 11
Deer Trail Dr. N&S..31W-40N — 11
E. Applewood Ln. N&S
.... 30W-38N — 19
East Ct. N&S ...... 30W-39N — 11
East St. N&S ...... 30W-39N — 11
Edgewood Ct. NE&SW
.... 31W-39N — 11
Edgewood Dr. N&S
.... 32W-38N — 18
Elk Ct. NW&SE ... 31W-40N — 11
Elk St. E&W ...... 31W-40N — 11
Elm St. E&W ...... 30W-38N — 19
English Prairie Rd. E&W
.... 31W-42N — 11
Falcon Ln. E&W ... 31W-42N — 11
Finch St. N&S .... 31W-40N — 11
Fox Bluff Ln. N&S..31W-42N — 11
Fox Trail Dr. N&S .. 30W-41N — 11
Galleria Ct. N&S ... 31W-40N — 11
Greenleaf Ct. E&W 32W-38N — 10
Hatchery Rd. NW&SE
.... 31W-39N — 11
Hidden Tr. N&S ... 31W-40N — 11
Hidden Trail Ct. SW&NE
.... 31W-39N — 11
Highview St. E&W . 31W-39N — 11
Hillside Ridge Rd. NS&EW
.... 30W-39N — 10
Hunters Ct. NE&SW
.... 30W-41N — 11
Hunters Ln. EW&NS
........31W-41N,30W-41N — 11
Industrial Ln. E&W . 31W-39N — 11
Judge Ct. NE&SW . 31W-41N — 11
Kings Lair Dr. E&W
.... 32W-39N — 11
Lair St. N&S ...... 32W-39N — 10
Loraine St. N&S ... 30W-39N — 11
Main St. NW&SE .. 30W-39N — 11
Marie Dr. N&S..... 31W-40N — 11
Martin Dr. N&S .... 31W-41N — 11
Maureen Dr. N&S . 31W-40N — 11
May Ln. NE&SW .. 32W-39N — 10
Mayo Ct. NE&SW . 31W-40N — 11
Meadowdale Cir. NS&EW
.... 30W-41N — 11
Meyer Rd. N&S ... 31W-39N — 11
Monterra Dr. E&W . 31W-40N — 11
N. Applewood Ln. N&S
.... 30W-38N — 19
N. Martin Dr. N&S . 31W-40N — 11
Nora Way N&S .... 31W-40N — 11
North St. E&W .... 30W-39N — 11
Oak St. E&W ...... 30W-39N — 11
Oak Valley Dr. N&S
.... 31W-40N — 11
Overlook Dr. N&S . 32W-38N — 18

Oxfordshire Ln. N&S
.... 32W-39N — 10
Pater St. N&S..... 31W-40N — 11
Pierce Dr. E&W ... 31W-40N — 11
Pierceshire Rd. N&S
.... 32W-39N — 10
Pine St. E&W ..... 30W-38N — 19
Pinecrest Ct. E&W .. 31W-40N — 11
Ponder Pl. E&W .. 30W-41N — 11
Rabbit Ct. E&W ... 31W-40N — 11
Richardson Rd. N&S
........30W-40N,30W-41N — 11
Ridge Ct. N&S .... 32W-39N — 10
Ridge Rd. E&W ... 32W-39N — 10
Robin Ct. NE&SW . 32W-39N — 10
S. Hunters Ln. E&W
.... 30W-41N — 11
S. Springdale Dr. E&W
........31W-41N,30W-41N — 11
Sherwood Forest Dr. E&W
.... 32W-39N — 10
Short St. N&S ..... 30W-38N — 19
Spring Ct. E&W ... 30W-41N — 11
Spring Grove Rd. NW&SE
.... 30W-39N — 11
Springdale Dr. N&S
.... 30W-41N — 11
Spruce St. E&W .. 30W-38N — 19
Squirrel Dr. NW&SE
.... 31W-40N — 11
Steeplechase Ct. SW&NE
.... 31W-40N — 11
Steeplechase Ln. NS&EW
.... 31W-40N — 11
Sunset Dr. N&S ... 30W-38N — 19
Sunset Rd. E&W .. 30W-38N — 19
Tahoe Ln. NE&SW 31W-40N — 11
Tallgrass Ct. N&S . 32W-38N — 18
U.S. Route 12 E&W
.... 31W-39N — 11
Utility Ct. N&S..... 31W-39N — 11
W. Applewood Ln. N&S
.... 30W-38N — 19
W. Deer Trail. NW&SE
.... 30W-40N — 11
Walnut Ct. NW&SE
.... 30W-38N — 19
Westward Dr. E&W
.... 31W-39N — 11
Winn Rd. N&S
........31W-39N,31W-41N — 11
Wintergreen Ct. N&S
.... 30W-41N — 11
Woodview Dr. E&W
.... 31W-40N — 11

Union
Clark St. E&W .. 46W-25N — 38
Country Ln. N&S . 46W-25N — 38
Depot St. N&S ... 46W-25N — 38
Elm St. N&S ..... 46W-25N — 38
Fairfield Rd. E&W . 46W-25N — 38
Highbridge Rd. E&W
.... 46W-26N — 38
Jackson St. N&S .. 46W-25N — 38
Jefferson St. E&W .46W-25N — 38
Johnson St. N&S . 46W-25N — 38
Main St. N&S ..... 46W-25N — 38
Mallett Ct. E&W ... 46W-25N — 38
Meadow Ln. E&W . 46W-25N — 38
Northrup Rd. N&S . 46W-25N — 38
O'Clock Rd. E&W . 46W-26N — 38
Park St. N&S ..... 46W-25N — 38
Prairie St. E&W ... 46W-25N — 38
South St. E&W ... 46W-25N — 38
Sunview Dr. E&W . 46W-25N — 38
Vine St. N&S ..... 46W-25N — 38
Washington St. E&W
.... 46W-25N — 38
Wayne St. N&S ... 46W-25N — 38

Wonder Lake
Acorn Path E&W
........38W-35N,37W-35N — 17
Alden Rd. E&W ... 37W-35N — 17
Burton Rd. E&W...37W-35N — 17
Chemung Dr. N&S . 37W-35N — 17
Coral Rd. E&W .... 37W-35N — 17
Dorr Rd. E&W ..... 37W-35N — 17
Garrison Rd. NW&SE
.... 37W-35N — 17
Greenwood Dr. N&S
.... 37W-35N — 17
Highland Dr. E&W
........38W-35N,37W-35N — 17
Marengo Rd. N&S . 37W-35N — 17
Meadow Ln. N&S .. 37W-35N — 17
Nunda Rd. E&W ... 37W-35N — 17
Richmond Rd. E&W
.... 37W-35N — 17
Ridge Ln. N&S ... 37W-35N — 17
Riley Rd. N&S
........38W-35N,37W-35N — 17
Seneca Rd. NE&SW
.... 37W-35N — 17
Thompson Rd. N&S
.... 37W-35N — 17
W. Lake Shore Dr. NW&SE
.... 37W-36N — 17
Westwood Dr. N&S
........37W-35N,37W-36N — 17

Wondermere Rd. E&W
........38W-35N,37W-35N — 17
Woodstock St. N&S
.... 37W-35N — 17

Woodstock
Amsterdam St. NE&SW
.... 42W-31N — 24
Ann Ct. E&W ..... 41W-30N — 32
Anne St. E&W .... 42W-31N — 24
Applewood Ln. N&S
.... 39W-29N — 32
Arthur Dr. NE&SW . 42W-31N — 24
Ash Av. NW&SE ... 41W-32N — 24
Ashley Ct. N&S ... 41W-32N — 24
Aspen Dr. N&S ... 39W-29N — 32
Austin Av. N&S ... 42W-30N — 32
Ayrshire Ct. E&W . 40W-30N — 32
Bagley St. E&W ... 41W-31N — 24
Banford Rd. E&W .. 41W-30N — 32
Barbary Ln. E&W .. 41W-30N — 32
Barnswallow Dr. N&S
.... 40W-30N — 32
Becking Av. N&S . 42W-31N — 24
Beech Av. N&S ... 41W-32N — 24
Bellair Ln. N&S .... 41W-32N — 24
Benton St. N&S ... 41W-31N — 24
Beritsum Ln. NS&EW
.... 40W-30N — 32
Birch Rd. E&W ... 41W-32N — 24
Blakely St. N&S ... 42W-30N — 32
Bliss Ct. N&S .... 40W-30N — 32
Bob 'O' Link Cir. NS&EW
.... 39W-30N — 32
Borden St. N&S ... 41W-30N — 24
Brahman Ct. N&S .. 39W-30N — 32
Bridgewater Dr. NW&NE
.... 40W-29N — 32
Brink St. NE&SW . 41W-30N — 32
Broadway Ct. N&S . 41W-30N — 24
Broadway Rd. N&S
.... 41W-32N — 24
Brown St. NE&SW .41W-30N — 32
Bull Valley Dr. NS&EW
.... 39W-30N — 32
Bunker Av. N&S ... 41W-30N — 32
Burbank Av. E&W . 41W-30N — 32
Burbank St. E&W .. 41W-30N — 32
Cairns Ct. NE&SW .42W-32N — 24
Calhoun St. E&W . 41W-31N — 24
Carol Av. NE&SW . 42W-31N — 24
Carol St. E&W .... 42W-31N — 24
Cass St. E&W .... 41W-31N — 24
Castle Bar Tr. N&S
.... 43W-30N — 31
Castle Rd. N&S
........40W-28N,40W-29N — 32
Catalpa Ln. NE&SW . 40W-29N — 32
Center Ln. N&S
.... 41W-31N — 24
Center St. E&W ... 41W-31N — 24
Centerville Ct. NE&SW
.... 40W-30N — 32
Central Pkwy. E&W
.... 41W-32N — 24
Charbray Ct. N&S . 39W-30N — 32
Charles St. N&S .. 41W-30N — 24
Cherry Ct. E&W ... 41W-32N — 24
Chestnut Av. E&W .41W-30N — 32
Church St. N&S ... 41W-31N — 24
Claussen Dr. E&W . 42W-31N — 24
Clay Ct. NW&SE .. 41W-32N — 24
Clay St. N&S
........41W-31N,41W-32N — 24
Club Rd. NW&SE . 39W-30N — 32
Cobblestone Way E&W
.... 40W-29N — 32
Conway St. E&W . 42W-31N — 24
Corporate Ct. E&W 40W-29N — 32
Country Club Rd. N&S
.... 40W-30N — 24
Courtland St. N&S .. 40W-30N — 32
Dacy St. N&S .... 42W-31N — 24
Dakota Dr. NS&EW
.... 43W-30N — 31
Davis St. N&S ... 41W-30N — 32
Davis Rd. NW&SE . 40W-29N — 32
Dean St. NE&SW
........41W-29N,41W-30N — 32
Desmond Av. E&W
.... 42W-30N — 32
Diane Dr. NW&SE . 41W-32N — 24
Dieckman St. E&W 41W-30N — 24
Dillard Ct. NS&EW
........40W-29N,40W-30N — 32
Division St. N&S
.... 41W-30N — 32
Donagal Ct. NE&SW
.... 43W-30N — 31
Donna Dr. E&W .. 42W-31N — 24
Donovan Av. E&W
........42W-31N,41W-31N — 24
Dorham E&W ..... 40W-30N — 32
Douglas St. N&S .. 41W-30N — 24
Dublin Ct. NS&EW ..43W-30N — 31
Duncan Pl. N&S .. 40W-30N — 32
Duvall St. E&W ... 41W-30N — 32
East St. N&S ..... 41W-30N — 32
Edgewood Dr. E&W
........41W-29N,41W-30N — 32
Ellen Ct. N&S ..... 40W-31N — 24

	Grid	Page
Fair St. N&S	41W-31N	24
Farm Service Rd. NW&SE		
	40W-30N	32
Farm Tr. NW&SE	41W-32N	24
First St. E&W	41W-31N	24
Flagg Ln. NS&EW		
	41W-31N,40W-31N	24
Forest Av. E&W		
	42W-30N,41W-30N	32
Fox Ln. E&W	41W-32N	24
Fremont Av. NW&SE		
	41W-30N	32
Friesian Ct. NS&EW		
	40W-30N	32
Galoway Dr. NE&SW		
	40W-30N,39W-30N	32
Gerry Ct. N&S	42W-30N	32
Gerry St. E&W	42W-30N	32
Giddings St. NW&SE		
	41W-30N	32
Ginny Ln. N&S	41W-30N	32
Gracy St. NW&SE	41W-31N	24
Greenley St. N&S	41W-30N	32
Greenwood Av. E&W		
	41W-32N	24
Greenwood Pl. NE&SW		
	41W-31N	24
Greta Av. E&W	42W-30N	32
Griffing Av. E&W	41W-30N	32
Grove St. E&W	41W-31N	24
Halma Ln. Ct. E&W		
	40W-30N	32
Halma Ln. East E&W		
	40W-30N	32
Halma Ln. West E&W		
	40W-30N	32
Hayward St. N&S		
	41W-30N,41W-31N	32
Herrington Pl. E&W		
	42W-30N,41W-30N	32
Hibbard St. N&S	41W-30N	32
Hickman Ln. N&S	41W-31N	24
Hickory Ln. N&S	42W-30N	32
Hickory Rd. N&S	41W-32N	24
Highland Av. E&W	42W-30N	32
Hill St. N&S	42W-31N	24
Hillcrest St. N&S	42W-31N	24
Hillside Tr. NE&SW		
	43W-30N	31
Hilltop Ct. N&S	41W-32N	24
Hoy Av. E&W	41W-30N	32
Hucthins St. E&W	41W-31N	24
Huffman St. NW&SE		
	41W-31N	24
Industrial Heights Dr. NW&SE		
	40W-29N	32
Irving Av. N&S	41W-31N	24
Island Ct. N&S	41W-32N	24
Jackson St. E&W		
	42W-31N,41W-31N	24
Jean St. NW&SE	41W-30N	32
Jefferson St. N&S		
	41W-30N,41W-31N	32
Jewett St. N&S		
	41W-31N,41W-32N	24
Johnson St. N&S	41W-31N	24
Joseph St. NS&EW		
	41W-32N	24
Judd St. E&W		
	42W-31N,41W-31N	24
Julie St. N&S	41W-32N	24
Kathleen Ct. E&W	42W-30N	32
Keating Dr. N&S	41W-32N	24
Kilkenny Ct. NW&SE		
	40W-29N	32
Kimball Av. E&W		
	42W-30N,41W-30N	32
King St. E&W	41W-30N	32
Lake Av. NW&SE		
	41W-30N,40W-29N	32
Lake St. NW&SE	41W-30N	32
Lamb Rd. N&S	43W-32N	23
Laurel Av. N&S	41W-30N	32
Lawndale Av. E&W		
	41W-30N	32
Lawrence Av. E&W		
	41W-30N	32
Lee St. N&S	41W-30N	32
Liberty Ln. NW&SE		
	41W-30N	32
Lincoln Av. E&W		
	42W-31N,41W-31N	24
Linda Ct. NE&SW	40W-31N	24
Linden Ln. E&W	40W-29N	32
Lisa St. E&W	42W-31N	24
Locust Av. E&W	41W-32N	24
Madison St. N&S		
	41W-31N,41W-32N	24
Main St. N&S	41W-31N	24
Mansfield St. N&S	41W-31N	24
Maple Av. E&W	41W-32N	24
Margaret St. E&W	42W-31N	24
Mark Ct. E&W	41W-30N	32
Martha Dr. NE&SW		
	40W-30N	32
Marvel Av. N&S	41W-31N	24
Mary Ann St. N&S	42W-31N	24
Mary Anne St. NE&SW		
	42W-31N	24
McConnell Rd. NW&SE		
	40W-30N,39W-29N	32
McGee Rd. E&W	43W-32N	23

	Grid	Page
McHenry Av. NE&SW		
	41W-31N,40W-31N	24
McHenry Rd. NE&SW		
	40W-31N	24
Meadow Av. E&W	41W-32N	24
Melody Ln. E&W	41W-32N	24
Mitchell St. N&S	42W-30N	32
Moraine Ct. E&W	43W-30N	31
Moraine Dr. N&S	43W-30N	31
Muriel St. N&S	42W-30N	32
Nebraska St. N&S	41W-31N	24
Newell St. E&W	41W-31N	24
Nieuwhof Ct. N&S	40W-30N	32
North St. E&W	41W-31N	24
Northampton St. N&S		
	41W-31N	24
Oak St. NW&SE	42W-31N	24
Oakdale St. NW&SE		
	41W-30N	32
Oakland Av. E&W	42W-31N	24
Oakview Ct. NW&SE		
	43W-30N	31
Oakview Ter. NS&EW		
	43W-30N	31
Olive Av. E&W	42W-31N	24
Olson St. NE&SW	41W-31N	24
Orchard Ct. NW&SE		
	41W-32N	24
Park St. E&W	41W-31N	24
Patricia Ct. E&W	41W-32N	24
Peach Tree Ln. E&W		
	41W-32N	24
Pine Ct. NW&SE	41W-31N	24
Pinecrest Ct. E&W	43W-30N	31
Pleasant St. NW&SE		
	42W-31N	24
Pond Point Rd. N&S		
	40W-29N	32
Poplar Ln. N&S	41W-32N	24
Portage Ln. N&S	41W-30N	32
Prairie View Dr. N&S		
	40W-30N	32
Putnam Av. N&S	42W-30N	32
Quail Ct. N&S	41W-32N	24
Queen Anne St. N&S		
	41W-31N,41W-32N	24
Quinlan Ln. E&W	42W-31N	24
Raffel Rd. N&S	40W-32N	24
Redwing Dr. N&S	40W-32N	24
Regina St. E&W	42W-32N	24
Rhodes St. N&S	42W-31N	24
Ridgeland Av. E&W		
	42W-30N	32
Ridgewood Dr. E&W		
	41W-31N	24
Roger Rd. E&W	41W-32N	24
Roosevelt St. N&S	41W-30N	32
Rose Ct. E&W	40W-31N	24
Rose Farm Rd. N&S		
	44W-30N,44W-33N	31
Rt. 120 E&W		
	42W-32N,40W-32N	24
Rt. 47 N&S		
	41W-29N,41W-32N	32
Russell Ct. E&W	41W-32N	24
Ryan Ct. N&S	43W-30N	31
Ryder Dr. NE&SW	41W-30N	32
Sando Ln. N&S	42W-30N	32
Schryver Av. E&W	41W-30N	32
Second St. E&W	41W-31N	24
Seminary Av. N&S		
	41W-31N,41W-32N	24
Seneca Ct. N&S	43W-30N	31
Sharon Dr. N&S	40W-31N	24
Shelia St. NS&EW	41W-32N	24
Short St. E&W	41W-31N	24
Smith St. NW&SE	41W-30N	32
South St. SW&NE	43W-30N	31
Southview Dr. NW&SE		
	41W-30N	32
Sparrow Dr. E&W	40W-32N	24
St. John Rd. E&W		
	41W-32N,40W-32N	24
Stewart Av. E&W	42W-30N	32
Stonelake Rd. E&W		
	40W-29N	32
Summit Av. E&W		
	42W-32N,41W-32N	24
Sunset Ridge Dr. N&S		
	42W-31N	24
Sunshine Ln. E&W	41W-32N	24
Suzanne St. N&S	42W-31N	24
Tanager Dr. N&S	40W-32N	24
Tappan St. N&S		
	41W-31N,41W-32N	24
Tara Ct. E&W	42W-30N	32
Tara Dr. N&S	42W-30N	32
Taurus Ct. E&W	39W-30N	32
Tech Court N&S	41W-29N	32
Terry Ct. E&W	41W-32N	24
Third St. E&W	41W-31N	24
Thomas Dr. N&S	42W-30N	32
Throop St. N&S	41W-31N	24
Timothy Ln. NS&EW		
	40W-31N	24
Todd Av. E&W	41W-31N	24
Tryon St. N&S		
	41W-30N,41W-31N	32
Twelve Oaks Pkwy. SW&NE		
	40W-32N	24
U.S. Route 14 NW&SE		
	42W-30N,39W-28N	32

	Grid	Page
Van Buren St. E&W		
	41W-31N	24
Victoria Dr. NW&SE		
	42W-32N	24
Vine St. NS&EW	41W-30N	32
W. Lake Shore Dr. N&S		
	39W-28N,39W-29N	32
Wagner Ln. E&W	41W-30N	32
Walden Oak Dr. N&S		
	41W-30N	32
Walnut Dr. NS&EW		
	41W-32N	24
Wanda Ln. NW&SE		
	40W-30N	32
Ware Rd. E&W		
	41W-33N,40W-33N	24
Washburn St. NW&SE		
	41W-30N	32
Washington St. NW&SE		
	42W-31N,41W-31N	24
West Av. E&W	42W-31N	24
Westwood Ct. N&S		
	43W-30N	31
Westwood Tr. E&W		
	43W-30N	31
Wheeler St. N&S		
	41W-31N,41W-32N	24
White Face Ct. NW&SE		
	39W-30N	32
Wicker St. N&S	41W-32N	24
Wicker St. N&S	42W-31N	24
Willow Av. E&W	41W-32N	24
Winslow Av. E&W	42W-30N	32
Wintu Ct. N&S	43W-30N	31
Wood Dr. N&S	41W-30N	32
Woodside Ct. E&W		
	40W-32N	24
Yellowhead Ct. NW&SE		
	39W-30N	32
Zimmerman Rd. N&S		
	40W-30N	32

Will County

StreetFinder®

page	Contents
A	Using the StreetFinder
B	PageFinder Map
C	Vicinity Map
1	Contents
2	Quick Reference Guide
4	StreetFinder Maps
79	StreetFinder Index

Photo credit: Downtown Joliet, Joliet Region Chamber of Commerce

Will County Municipal Offices

	Location	Page
Crete Village Hall	1W-29S	44
524 Exchange St; 672-5431		
Frankfort Twp Village Hall	12W-25S	32
123 W Kansas St; 469-2177		
Joilet City Hall	24W-23S	20
150 W Jefferson St; 740-2220		
Monee Village Hall	6W-31S	42
500 E Court St; 534-8301		

Cemeteries

	Location	Page
Adem Cem	1W-29S	44
Barrett Cem	20W-20S	22
Brooks Cem	17W-17S	17
County Farm Cem	27W-24S	27
Eagle Lake Cem	3E-32S	57
Elmhurst Cem	22W-24S	29
Frankfort Cem	11W-25S	32,33
Hadley Cem	17W-21S	23
Hickory Hill Cem	11W-24S	24,25
Hills of Rest Cem	28W-23S	19
Holy Cross Cem	26W-21S	19
Lockport Cem	22W-19S	15
Marshall Cem	15W-23S	23
Mt Calvary Cem	22W-21S	21
Mt Monah Cem	24W-22S	20
Mt Olivet Cem	22W-23S	21
Pleasant Hill Cem	12W-25S	32
Rose Hill Cem	12W-28S	40
St Cyril's Cem	21W-22S	22
St John's Cem	24W-22S	20
St John's Cem	13W-24S	32
St Joseph Cem	24W-22S	20
St Mary's Cem	24W-21S	20
St Patrick's Cem	24W-23S	20
St Peter's Cem	9W-29S	41
Trinity Cem	0W-30S	44
Twining Cem	13W-30S	40

	Location	Page
Union Cem	10W-30S	41
Woodland Mem Park Cem	28W-23S	19
Zion Cem	0W-30S	44

Colleges & Universities

	Location	Page
Col of St Francis	24W-23S	20
Governors State Univ	5W-29S	42
Joilet Jr Col	23W-24S	29
Lewis Univ of Science	23W-18S	14,15

Forest Preserves

	Location	Page
Hammel Woods FP	29W-23S	18,19
Higinbotham Woods FP	19W-23S	22
Messenger Woods FP	17W-20S	23
Van Horn Woods FP	13W-25S	32

Golf Courses & Country Clubs

	Location	Page
Big Run GC	21W-16S	15
Boughton Ridge GC	22W-10S	7
Green Garden CC	11W-31S	40,41
Inwood GC	27,28W-24S	27
Joilet CC	22W-25S	29
Old Oak CC	17W-16S	17
Prestwick CC	9W-26S	33
Tamarack Fairways GC	30W-12S	8
Wedgewood GC	30W-20S	18
Willow Run CC	15W-22S	23
Woodruff GC	20W-23S	22

Hospitals

	Location	Page
St Joseph Hosp	26W-23S	20
Silver Cross Hosp	22W-23S	21
Sunny Acres Sanitarium	23W-25S	28,29

Shopping Centers

	Location	Page
Hillcrest SC	26W-21S	20
Jefferson Square Mall	27W-24S	27
Louis Joilet Mall	28W-20S	19

Will County Close-up

Rialto Square Theatre
102 N. Chicago St., Joliet
Constructed in 1926, this designated landmark combines several different architectural styles. Its ornate interior boasts gold and silver inlay, a main foyer patterned after the Hall of Versailles and a rotunda inspired by the Parthenon. Known as "the jewel of Joliet," this 1,900-seat theater has hosted Indianapolis Dance Company, Ruth Page Ballet Company, Hubbard Street Dance Company and Joffree Ballet Company.

Jacob Henry Mansion
20 S. Eastern Ave., Joliet
This magnificent 44-room dwelling, built in 1873 for railroad magnet Jacob Henry, remains the largest and best example of Renaissance architecture in Illinois. Its interior of hand-rubbed woods was designed by some of the era's finest German artisans. This one-of-a-kind house features an elaborate staircase with 119 handcarved octagon spindles of burled walnut. Open by appointment.

Joliet Union Station
50 E. Jefferson, Joliet
Relive Illinois history with a visit to this fully-restored train station. A designated landmark, this structure was

first made possible by Abraham Lincoln who was then a corporate lawyer for Rock Island Railroad. Vintage architecture features a 50-ft. dome ceiling and cast-iron window frames. Buiding materials include Italian marble.

Isle A La Cache Museum
501 E. Romeo Rd., Romeoville
This unique museum highlights the fur trade of the late 1700's. Museum interpreters, costumed as French Voyageurs, provide demonstrations. Exhibits feature a birch bark canoe and an actual wigwam. Artifacts include pottery and bone tools. Hands-on exhibits. Closed Mondays.

Will County Historical Society
803 S. State St., Lockport
This site is the location of the Illinois & Michigan Canal Museum and Pioneer Settlement. The I. & M. Canal Museum displays artifacts relating to the canal's history including tools, photographs, shipping rosters and Native American objects. Pioneer Settlement is a living museum staffed by costumed interpreters. It features a smoke house, blacksmith, tinsmith, log cabin, schoolhouse and a mid-19th-century farmhouse.

Plum Creek Nature Center
27064 S. Dutton Rd., Beecher
Enjoy the outdoors on 689-acres of oak and hickory forest. This region features campsites, picnic shelters and three miles of multi-purpose trails. Available for ice skating and sledding during the winter months.

Monee Reservoir
27341 Ridgeland Ave., Monee
This public reservoir, surrounded by 195 acres of fields and woods, is an ideal fishing lake. Fish population includes small-mouth bass, large-mouth bass, catfish and bluegill. Free to the public.

Plainfield Historical Society Museum
217 Main St., Plainfield
Dedicated to the preservation of local history, this museum features an artifact collection that includes tools, farm equipment, toys and several military relics. Also, archives and a research library are on the premises. Open Saturdays.

Lake Renwick Heron Rookery
23144 W. Renwick Rd., Plainfield
This breeding area for endangered birds rests on a small island. Site features an interpretive staff, a bird viewing area and brief presentations. Spotting scopes provided. Open Wednesdays and Saturdays.

Illinois State Museum Lockport Gallery
200 W. 8th St., Lockport
This center for visual arts showcases the work of Illinois artists, past and present. Its changing exhibits include paintings, sculpture, photographs and weavings. It also sponsors lectures and educational programs. Open all year.

Gladys Fox Museum
231 9th St., Lockport
This museum, with its focus on the community of Lockport, is housed in a church constructed from native limestone in 1840. Closed Fridays, Saturdays and Sundays.

Round Barn Farm Museum and Recreation Area
Rte. 52 South of I-80 (take Briggs St. Exit South), Joliet
Highlights include pony rides, a petting zoo and horse carriage rides. Artifacts include household, agricultural antiques and a barn constructed in 1898.

Slovenian Union Women's Heritage Museum
431 N. Chicago St., Joliet
This ethnic museum honors Slovenian culture. Exhibits feature clothing, dolls, books and jewelry. Also, video presentations.

The Herbert Trackman Planetarium
1216 Houbolt Ave., Joliet
Astronomy-lovers enjoy multi-media planetary shows. Several programs for children. Located on the campus of Joliet Junior College. Free admission.

Bolingbrook Aquatic Center
180 S. Canterbury Lane, Bolingbrook
Enjoy swimming, scuba diving and water aerobics at this indoor wave pool. Swimming and scuba lessons available. Ideal for private parties. Open all year.

Riegel Mini Farm
580 Farmview Rd., University Park
This small petting zoo covers 7 acres and has a wide variety of farm animals. Highlights include pony rides and hay rides. Special events. Open May thru October.

Thorn Creek Nature Preserve
247 Monee Rd., Park Forest
Hike some of the most beautiful trails in the region. Experience 180 acres of oak and hickory forest. Variable terrain includes creeks, marshes, flood plains and upland forest. Ideal for bird watching. Interpretive staff available for lectures, workshops and various educational programs.

Channahon State Park
2 W. Story St., Channahon
Twenty acres of woodlands, ideal for outdoor recreation. This picturesque site includes multi-purpose trails, picnic shelters, horseshoe pits and 75 campsites. Play areas for children. Open all year.

I & M Canal State Trail
Channahon to LaSalle
This trail of limestone screenings extends 55.5 miles and is available for hiking, jogging, biking and cross-country skiing. Camping and water sports nearby.

Dellwood Park
State Street, Lockport
Centrally located, this 150-acre park boasts six pavilions, a performance arts center, band shell and an enclosed field house. Play areas for children.

Gaylord Building
200 W. 8th St., Lockport
Winner of the 1988 President's Award for Historic Preservation. The first of its two wings was constructed in 1838 by the I & M Canal Commission. Currently, it houses a restaurant, art gallery and a visitors center.

4 WILL COUNTY

SEE DU PAGE COUNTY PAGES 28-29

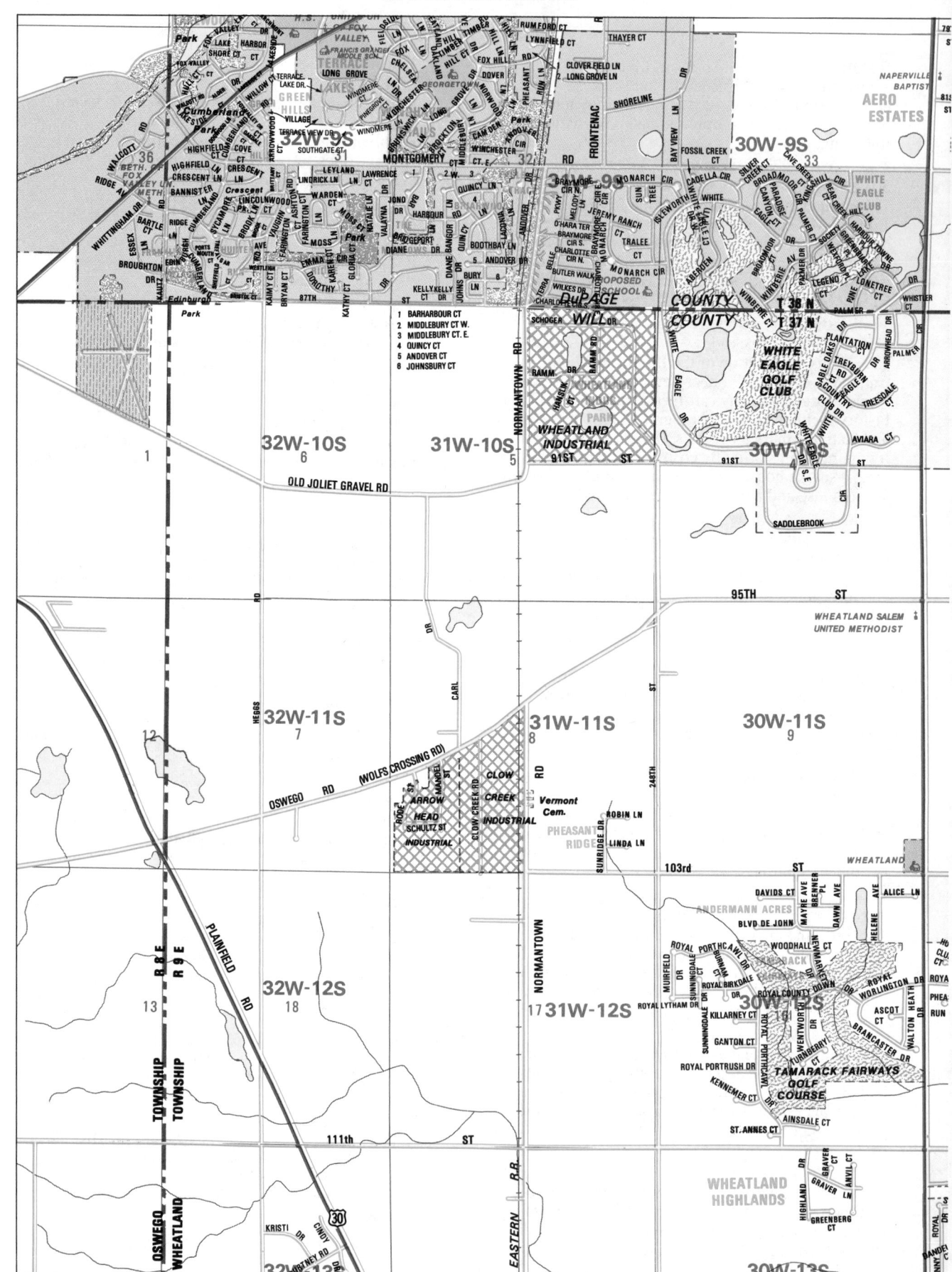

SEE PAGES 8-9

SEE DU PAGE COUNTY PAGES 28-29

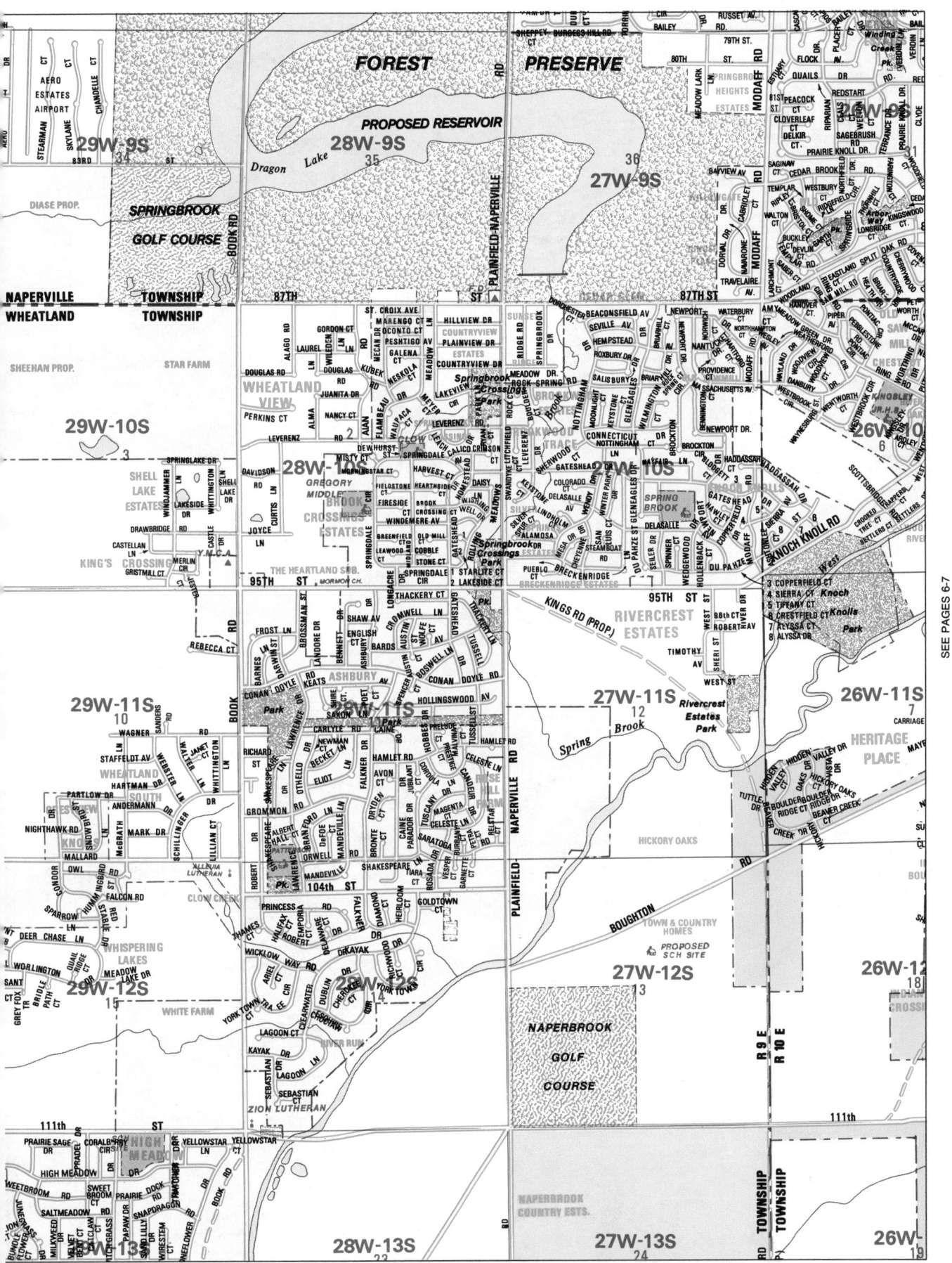

SEE PAGES 6-7

SEE PAGES 8-9

SEE DU PAGE COUNTY PAGES 30-31

BOLINGBROOK

Grid references: 26W-9S, 25W-9S, 24W-9S, 26W-10S, 25W-10S, 24W-10S, 26W-11S, 25W-11S, 24W-11S, 26W-12S, 25W-12S, 24W-12S, 26W-13S, 25W-13S, 24W-13S

Column "A"
NAPER VALLEY
RESIDENCE
1 DENISPORT DR
2 GLASTONBURY CT
3 SIMMONS CT
4 LYNN CT
5 DUXBURY CT
6 SUDBURY CT
7 WELLFLEET CT
8 ST. JOHNSBURY CT
9 LOWELL CT

1 ALGONQUIN CT
2 APACHE CT
3 BOWIE CT
4 CALUMET CT
5 CHIPPEWA CT
6 DAKOTA CT
7 EAGLE CT
8 ELKHORN CT
9 FOX HEAD CT
10 HUDSON CT
11 CHEROKEE DR

DuPAGE PRESBYTERIAN
1 COUNTRY CT
2 COLLINGWOOD CT
3 LAKEWOOD CT

3 COPPERFIELD CT
4 SIERRA CT
5 TIFFANY CT
6 CRESTFIELD CT
7 ALYSSA CT
8 ALYSSA DR

Other labels: FOREST PRESERVE, WALNUT RIDGE, WALNUT WOODS, CELEBRATION FELLOWSHIP, DuPage Recreation Area, Royce Homestead, HERITAGE PLACE, Clow International Airport, JANE ADDAMS MIDDLE, REMINGTON LAKE, BUS. CENTER, Bolingbrook Medical Center, DuPage River, LISLE TOWNSHIP, DuPAGE TOWNSHIP

SEE PAGES 4-5

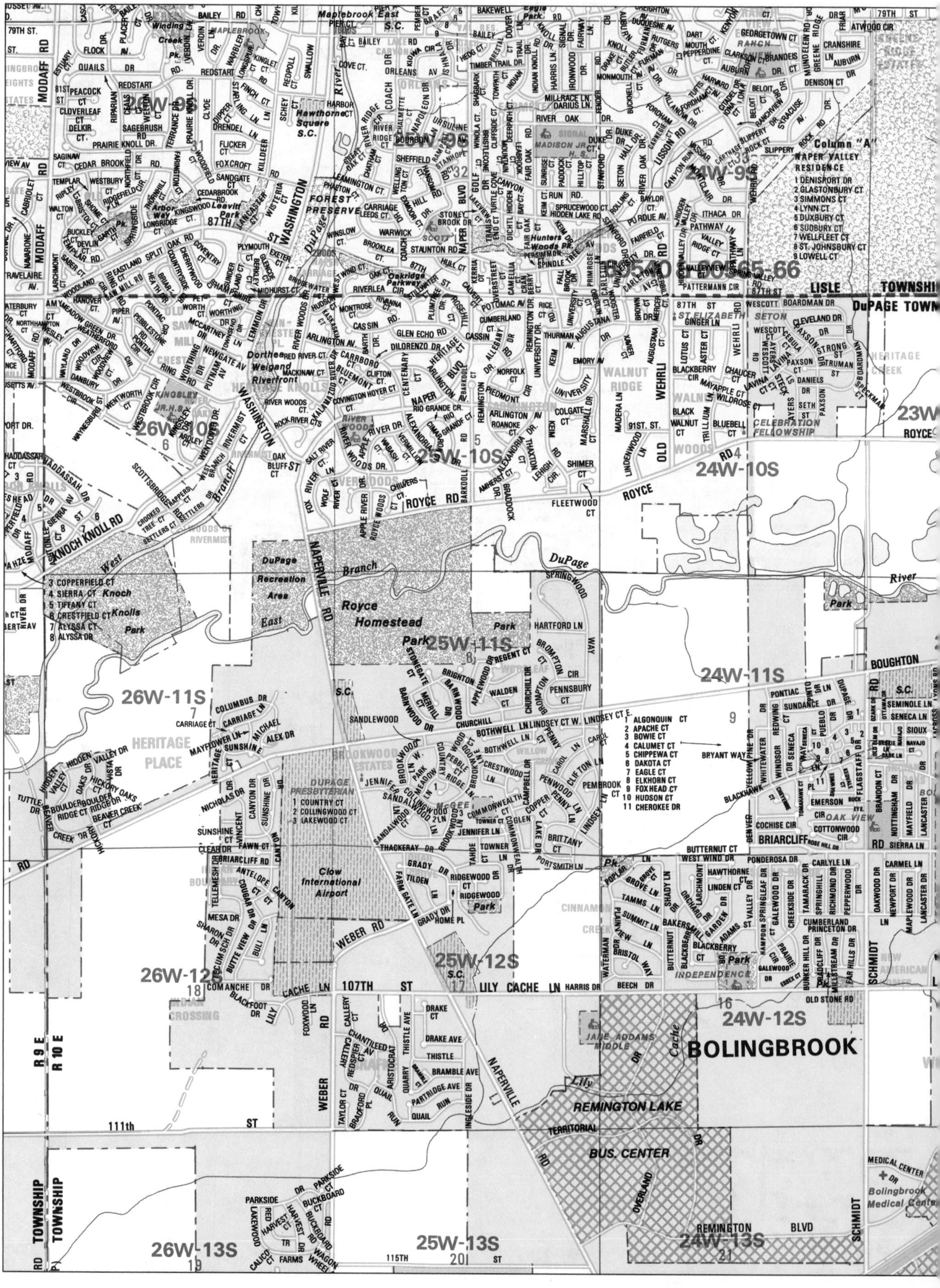

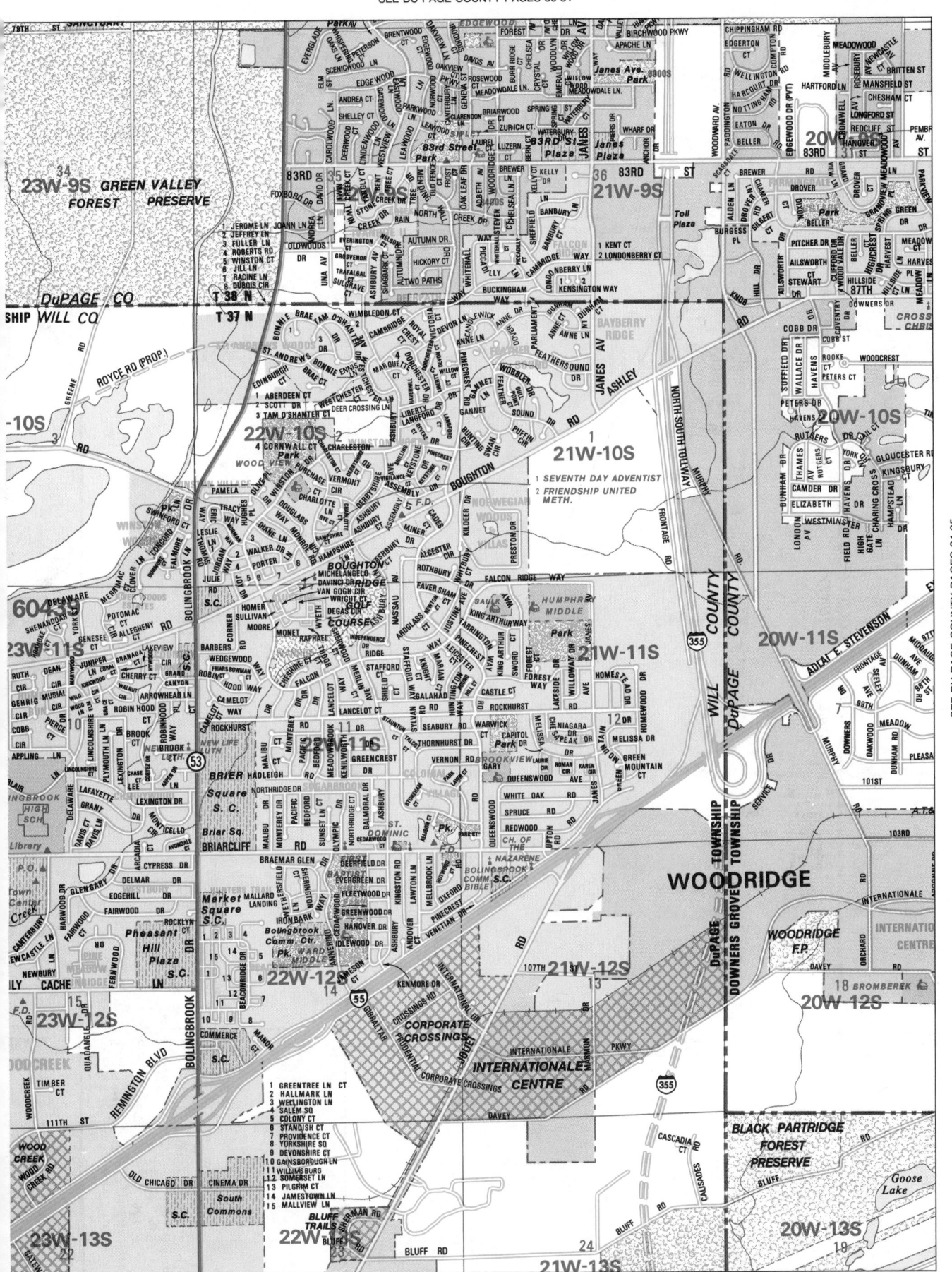

SEE DU PAGE COUNTY PAGES 34-35

SEE PAGES 4-5

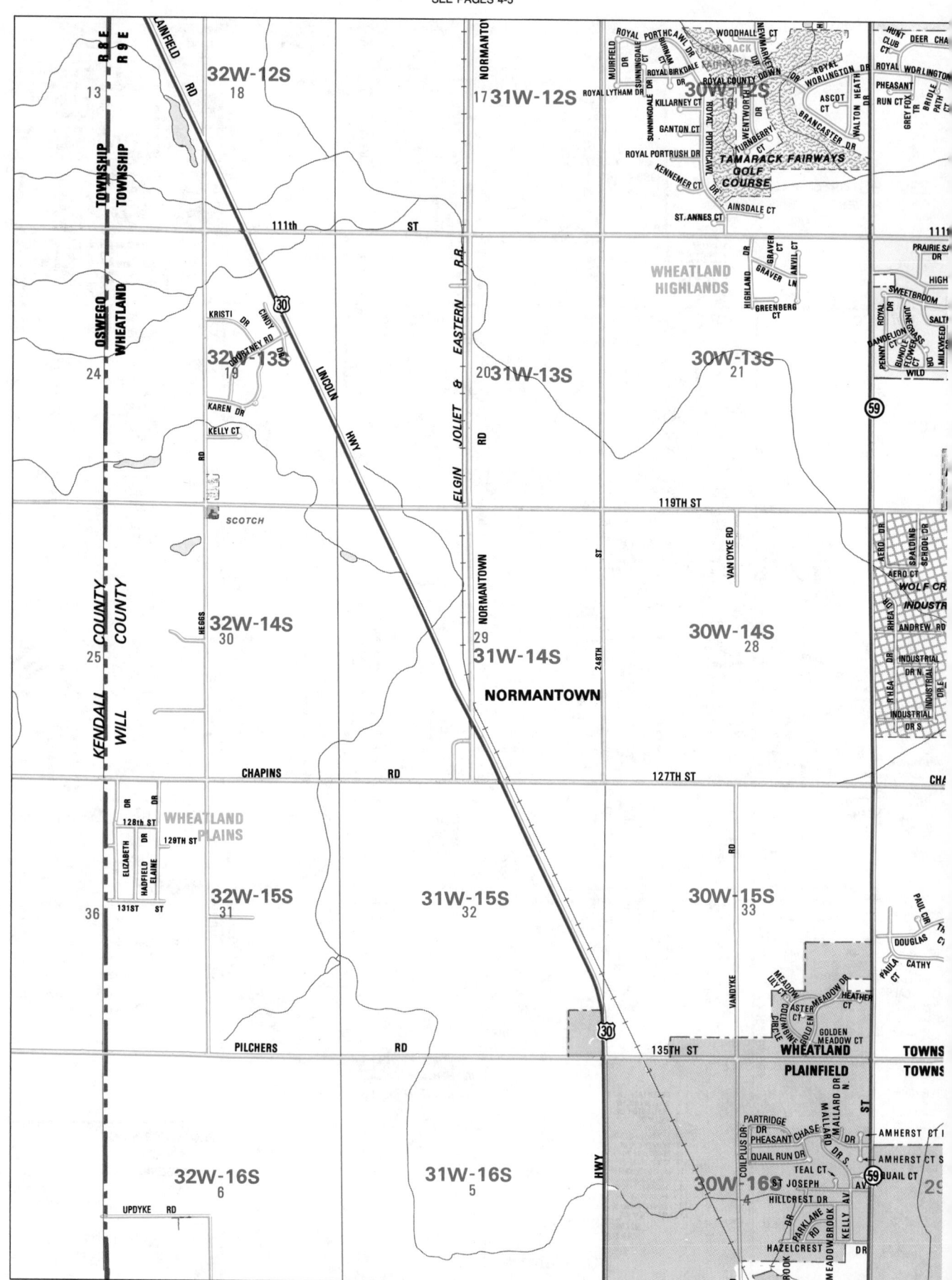

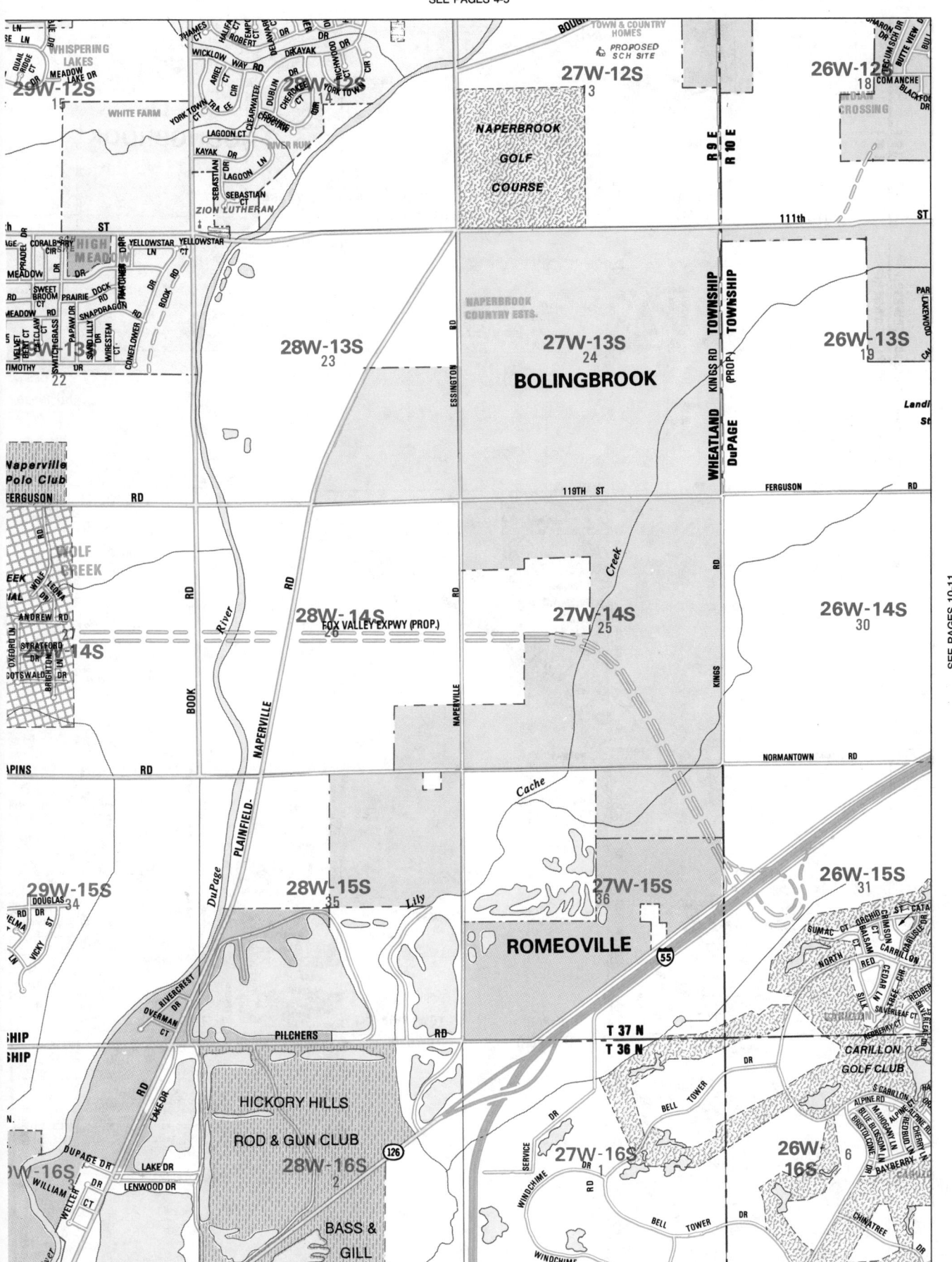

SEE PAGES 10-11

SEE PAGES 6-7

SEE PAGES 8-9

26W-12S
18
INDIAN CROSSING

SHARON DR
BUTTE VIEW DR
BULL RUN
OSCUMAN SCH DR
COMANCHE DR
BLACKFOOT DR
LILY
FOXWOOD
CACHE LN
WEBER RD
WEBER RD
107TH ST
CALLERY CT
THISTLE AVE
ABSTHY
REDSPUR AV
CHANTILEED AV
ARISTOCRAT
QUARRY CT
DRAKE CT
DRAKE AVE
THISTLE
BRAMBLE AVE
PARTRIDGE AVE
QUAIL RUN
QUAIL RUN
TAYLOR CT
BRACFORD PL
QUAIL

25W-12S
S.C.
17   LILY CACHE LN   HARRIS DR   BEECH DR

WATERMAN
BROR
BRISTOL
LN
CRE...
VIEW
LN

BUTTERNUT
BLACKBERRY
CT
BLACKBERRY
INDEPENDENCE
Park
KAMPOO
CIR
PRAIRIE
GALEWOOD
DR
PRINCETON DR
BUNKER HILL DR
RADCLIFF DR
MILLSTREAM RD
FAR HILLS DR
ESSEX CT
NEW AMERICAN...

16
OLD STONE RD
SCHMIDT

24W-12S
BOLINGBROOK

JANE ADDAMS MIDDLE
Lily
NAPERVILLE
REMINGTON LAKE
CACHE DR

TERRITORIAL
BUS. CENTER
OVERLAND
MEDICAL CENTER
Bolingbrook Medical Center

REMINGTON   BLVD
24W-13S
21
SCHMIDT

WHEATLAND TOWNSHIP
DuPAGE TOWNSHIP
KINGS RD
R 9 E
R 10 E
111th ST

26W-13S (PROP)
19

PARKSIDE DR
PARKSIDE
LAKEWOOD
RED HARVEST CT
HARVEST DR
BUCKBOARD
BUCKBOARD RD
WAGON
CALICO CT
FARMS CT
WHEEL CT

25W-13S
20
115TH ST

CROSS RO...
BUSINESS P...

Landing Strip

FERGUSON RD
ARBOR

26W-14S
30

RD
LAKEVIEW CT
LAKEVIEW
WINDHAM PKWY
WINDHAM LAKES BUSINESS
LAKEVIEW DR
WILLIAMS

25W-14S
29
FRONTAGE RD
ADLAI E. STEVENSON
EXPRESSWAY
NAPERVILLE RD

24W-14S
28

1 DUNBRIDGE LN
2 HARRIS DR
3 WINDSOR LN
4 DEVON LN
5 CRANBROOK CT
6 FARMBROOK CT
7 BEECHWOOD RD
8 CEDARBEND DR
9 ROBBIN DR
10 HONEYBEAR LN
11 BIRCHWOOD DR
12 WALDEN CT
13 WILDWOOD CT
14 WILLOW CT
15 GARDINER CT
16 SHERWOOD
17 ARBURY CT
18 SAYBROOK CT
19 SIX PINES DR
20 PINETREE CT
21 CEDAR CT
22 ALLENHURST C...
23 ELMWOOD RD

HONEYTREE TOWNHOMES
JEHOVAH HALL
SCHMIDT RD

FARRAGUT
GENEVA
HILLCREST AVE
COPONA
ABBEYWOOD
ROCKLEDGE DR
ABBEY...
VALLEY VIEW DIST. ADMIN. CT.
BEACON AVE
ARCADIA
GOOD SHEPHARD EVAN. LUTH.
ROGERS RD
HUDSON RD
NORMANTOWN RD
NORMANTOWN RD

KINGS RD

26W-15S
31

NORMANTOWN RD
NORMANTOWN RD
WEBER RD

WINNEBAGO CT
MICHIGAN DR
HURON
ONTARIO LN
MENDOTA LN
SUPERIOR DR
NEWMAN
ST. CLAIRE DR
ERIE DR
RAMBEAU CT
LAKEWOOD ESTATES

BIBLE BAPTIST
ROBERT C. HILL
ARLINGTON Park
BEVERLY J. GRIFFIN DR
ST. ANDREWS
BELMONT
HAMPTON DR
NELSON AVE
GLEN AVE
HUDSON AVE
KINGSTON AVE
GARLAND AVE
FENTON AVE
EVERETTE
CONCORD AVE
AMHERST AVE
DOVER AVE
CAMDEN
ELGIN AV
DALHART AV
S.C.
PHELPS
ALEXANDER CIR
BRISTOL
SPANGLER
P.D.

24W-15S
33

ORCHARD DR
CATALINA DR
CRIMSON DR
SUMAC
NORTH CT
CARLISLE DR
BALSAM CT
RED CT
CARILLON
CEDAR CT
REDBERRY
SILVERLEAF CT
SILK
TALL PINES LN
CHERRY LN
REDBERRY
MAGNOLIA DR
CARRILON

FOREST PRESERVE
Rec. Center
RECREATION DR
WOODLAWN AV
YATES AV
UNION AVE
NEWLAND AVE
A. VITO MARTINEZ MIDDLE
BERKSHIRE
PALMER AVE
HALSTEAD CT
BELMONTON
BELMONT
CLIFTON
MONTCLAIR AV
LAUREL AVE
MONTROSE
GLEN AV
KENYON
HUDSON
KINGSTON
GARLAND
EVERETTE
DALHART
CIVIC CENTER DR
Resurrection Cem.
Park

25W-15S
32

DuPAGE   TOWNSHIP
LOCKPORT   TOWNSHIP
JOLIET JR. COLLEGE

CARILLON GOLF CLUB

TOWER DR
S CARILLON
HAZELNUT LN
COTTONWOOD LN
ALPINE RD
ORANGE DR
ALPINE RD
REDBUD LN
MAGNOLIA DR
REDBUD
BLUE BLOSSOM DR
BAYBERRY
GINSON DR
CARILLON
CHINATREE DR
TOWER DR

26W 16S
6

135th ST
MONTROSE
ROOF AV
BRIARWOOD AV
AVALON AVE
POPLAR AVE
POPLAR BLVD
HOMER AVE
Atchley Park
MAGON AVE
HALE AVE
EMERY AVE
EATON AVE
HAYES
KAREN AV
GORDON AVE
ALINANZA CHRISTIAN CHURCH
IRENE AVE
H. KING AV
PELL AVE
MURPHY DR
SPANGLER DR
P.O.
McKOOL

25W-16S
5

ASHTON AV
DRIFTWOOD AV
EVERGREEN
POPLAR CT
POPLAR AV
MURPHY
GAVIN
BELMONT
RIPPERT AV
FREMONT AV
HARVEY AV
KENT AV
LINDEN AVE
HICKORY AVE
HEMLOCK
HALLER AVE
HEALY AVE
TALLMAN AVE
TRO EL AV
KIRMAN AVE
CHURCH OF GOD

OAKTON AV
HAMERICK AVE
24W-16S

INDEPENDENCE BLVD
ROMEOVILLE HIGH

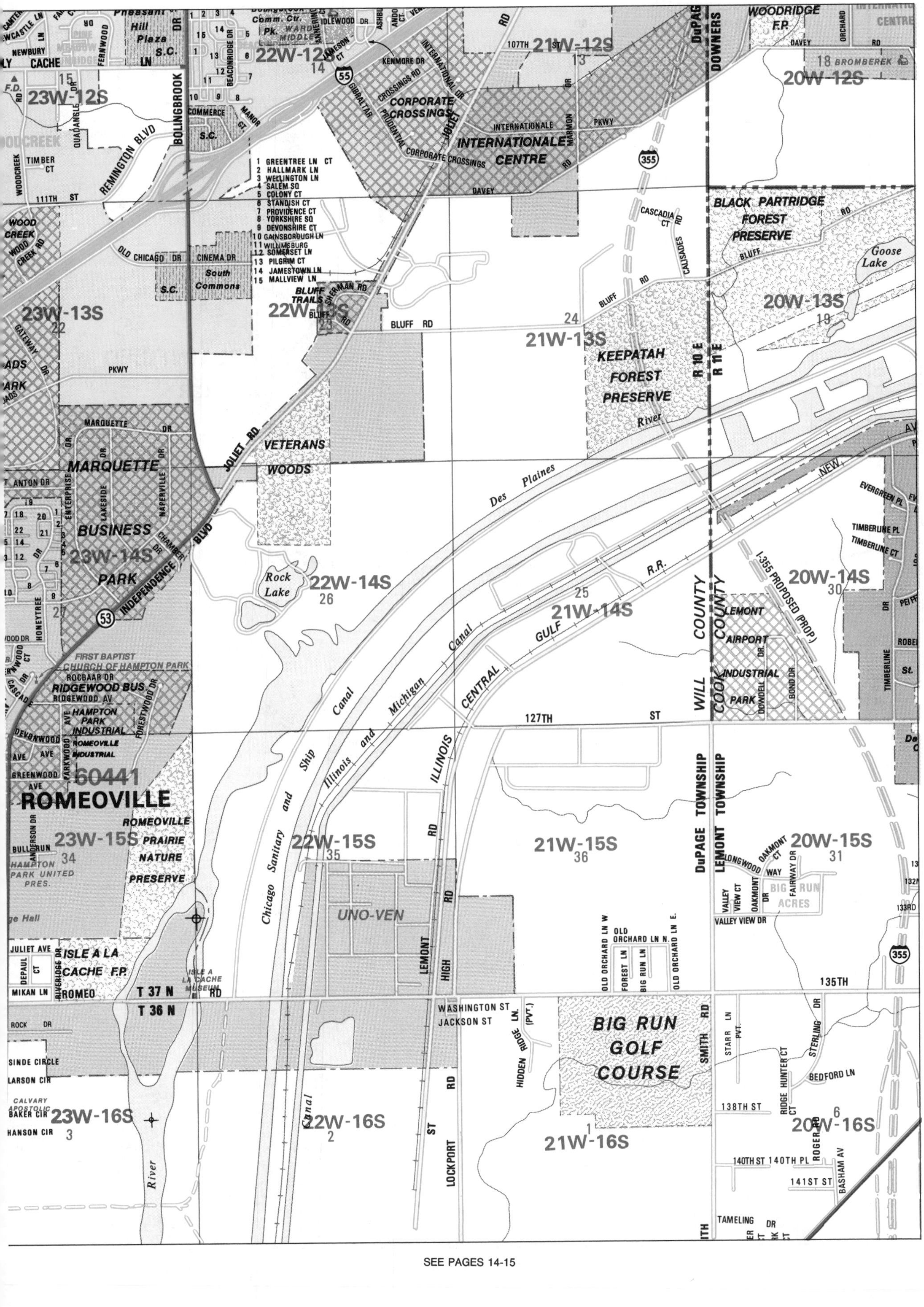

SEE COOK COUNTY PAGES 70-71

23W-12S
22W-12S
21W-12S
20W-12S

WOODRIDGE F.P.

18 BROMBEREK

BOLINGBROOK

Pheasant Hill Plaza S.C.

CORPORATE CROSSINGS

INTERNATIONALE CENTRE

I-355

BLACK PARTRIDGE FOREST PRESERVE

Goose Lake

23W-13S
22W-13S
21W-13S
20W-13S

KEEPATAH FOREST PRESERVE

BLUFF RD

1 GREENTREE LN CT
2 HALLMARK LN
3 WELLINGTON LN
4 SALEM SQ
5 COLONY CT
6 STANDISH CT
7 PROVIDENCE CT
8 YORKSHIRE SQ
9 DEVONSHIRE CT
10 GAINSBOROUGH LN
11 WILLIAMSBURG
12 SOMERSET LN
13 PILGRIM CT
14 JAMESTOWN LN
15 MALLVIEW LN

BLUFF TRAILS

MARQUETTE

VETERANS WOODS

BUSINESS PARK

23W-14S
22W-14S
21W-14S
20W-14S

Des Plaines

Rock Lake

LEMONT AIRPORT

INDUSTRIAL PARK

FIRST BAPTIST CHURCH OF HAMPTON PARK

RIDGEWOOD BUS.

HAMPTON PARK INDUSTRIAL

ROMEOVILLE INDUSTRIAL

60441

ROMEOVILLE

23W-15S
22W-15S
21W-15S
20W-15S

PRAIRIE NATURE PRESERVE

HAMPTON PARK UNITED PRES.

UNO-VEN

BIG RUN ACRES

DuPAGE TOWNSHIP
LEMONT TOWNSHIP

WILL COUNTY
COOK COUNTY

ISLE A LA CACHE F.P.

ISLE A LA CACHE MUSEUM

T 37 N
T 36 N
ROMEO

WASHINGTON ST
JACKSON ST

BIG RUN GOLF COURSE

CALVARY APOSTOLIC

23W-16S
22W-16S
21W-16S
20W-16S

135TH

138TH ST

140TH ST 140TH PL
141ST ST

TAMELING

SEE PAGES 8-9

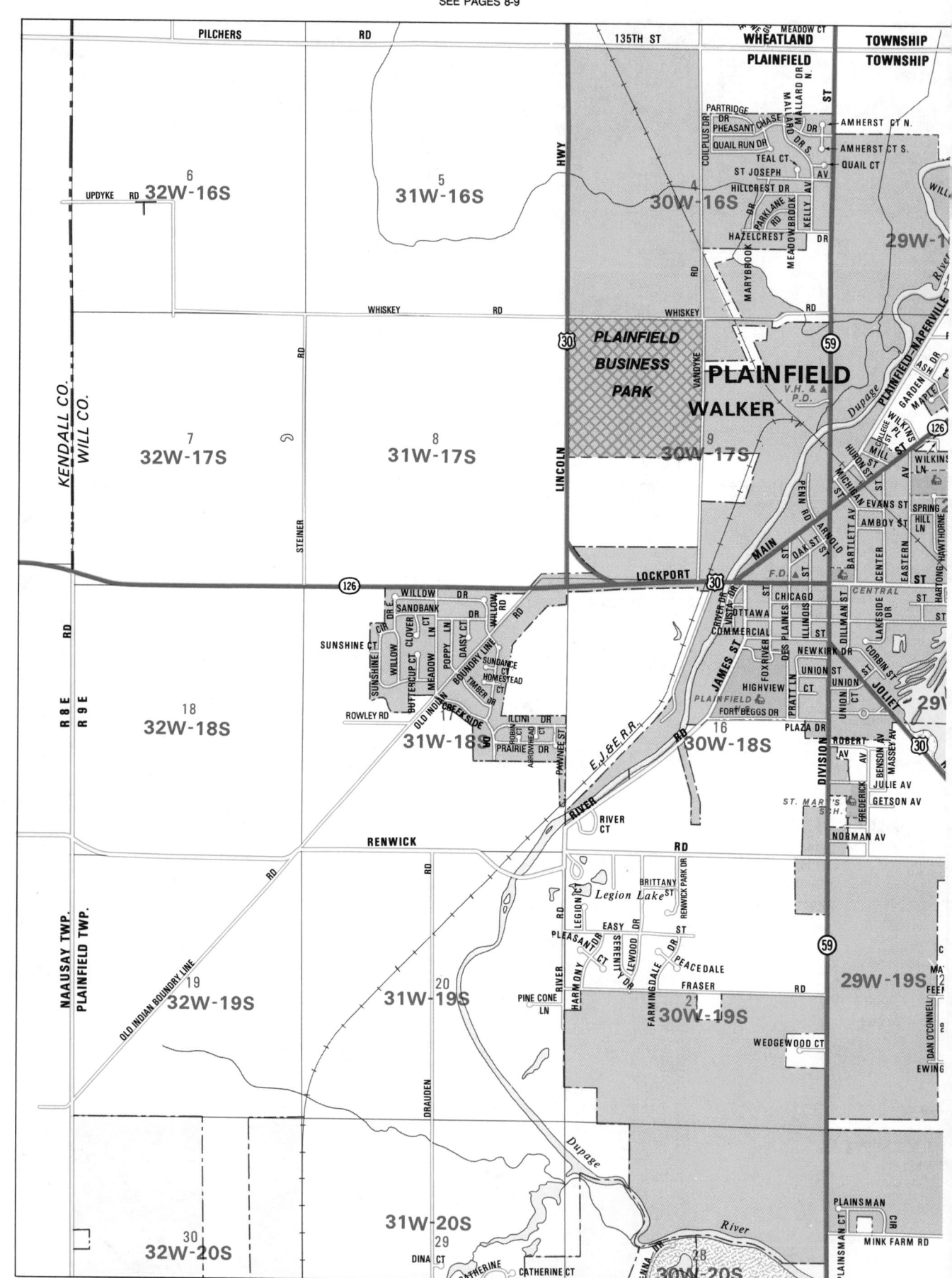

SEE PAGES 18-19

PILCHERS RD

T 37 N
T 36 N

HICKORY HILLS
ROD & GUN CLUB

28W-16S
2

BASS &
GILL
CLUB

143RD ST

CARILLON
GOLF CLUB

27W-16S
1

26W-16S

S CARILLON
ALPINE RD
BLUE BLOSSOM LN
BRISTOLCONE LN
MARCEAU LN
ALPINE DR
PINE RD
CHERRY LN
REDBUD LN
ORANGE BLOSSOM
HAZELNUT LN
BAYBERRY
COTT
CHINATREE DR

WINDCHIME DR
BELL TOWER DR
SERVICE RD
BELL TOWER DR
WINDCHIME DR
BELL TOWER ST
143RD ST

I-55

COPPER
PENNINGTON LN
GOLDEN LN
DR
SILVER LN
PENNY
KERNS DR

Lake
Plainfield

INDIAN TRAIL
JR. P.S.

29W-17S

PURE
60544
ROD & GUN CLUB

28W-17S
1

OAK PARK
SPORTSMEN CLUB

FOUR SEASONS PARK

27W-17S
12

INDUSTRIAL   R.R.

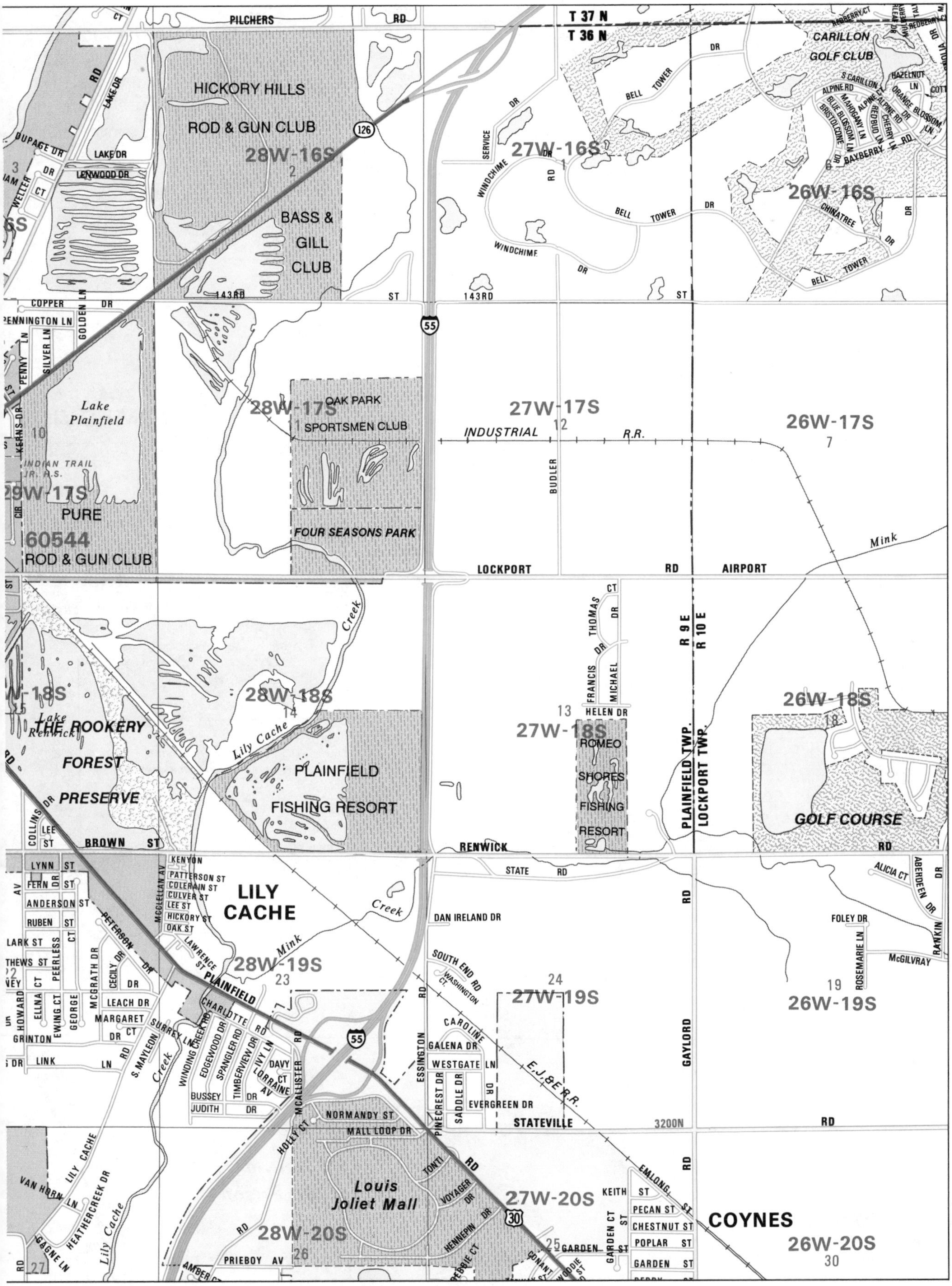

BUDLER

LOCKPORT   RD   AIRPORT

26W-17S
7

Mink

Creek

W-18S

Lake
Renwick

THE ROOKERY
FOREST
PRESERVE

COLLINS DR
LEE ST
BROWN ST

28W-18S
14

Lily Cache

PLAINFIELD
FISHING RESORT

RENWICK

THOMAS CT
DR
FRANCIS DR
MICHAEL DR
HELEN DR

27W-18S
13

ROMEO
SHORES
FISHING
RESORT

R 9 E
R 10 E
PLAINFIELD TWP.
LOCKPORT TWP.

26W-18S
18

GOLF COURSE

RD

STATE RD

LYNN ST
FERN ST
ANDERSON ST
RUBEN ST
AV
PEERLESS CT

KENYON
PATTERSON ST
COLERAIN ST
CULVER ST
LEE ST
HICKORY ST
OAK ST

MCGELLAN AV

LILY
CACHE

Creek

DAN IRELAND DR

ALICIA CT
ABERDEEN DR
RANKIN

FOLEY DR
ROSEMARIE LN
McGILVRAY

LARK ST
THEWS ST
NEY
HOWARD
ELNA CT
EWING CT
GRINTON
LINK
DR

PETERSON DR
McGRATH DR
CECILY DR
LEACH DR
MARGARET DR
GEORGE CT
LN

S. MAYLEON
SURREY LN

Creek
WINDING CREEK RD
EDGEWOOD DR
SPANGLER RD
TIMBERVIEW DR
IVY LN
CHARLOTTE RD
LAWRENCE ST
PLAINFIELD
DAVY CT
LORRAINE AV
BUSSEY DR
JUDITH DR

28W-19S
23

Mink

SOUTH END RD
WASHINGTON CT

ESSINGTON RD
CAROLINE
GALENA RD
WESTGATE LN
PINECREST DR
SADDLE DR
EVERGREEN DR

27W-19S
24

GAYLORD RD

26W-19S
19

I-55

E.J. & E.R.R.

3200N   RD

STATEVILLE

McALLISTER RD
HOLLY CT
NORMANDY ST
MALL LOOP DR

Louis
Joliet Mall

28W-20S
26

VAN HORN
LILY CACHE
HEATHERCREEK DR
GAGNE LN
Lily Cache
PRIEBOY AV
AMBER RD

27

TONTI RD
VOYAGER DR
HENNEPIN CT
PEBBLE CT

US 30

27W-20S
25

GARDEN

KEITH ST
PECAN ST
CHESTNUT ST
POPLAR ST
GARDEN ST
EMLONG RD
GARDEN CT

COYNES

26W-20S
30

SEE PAGES 14-15

SEE PAGES 10-11

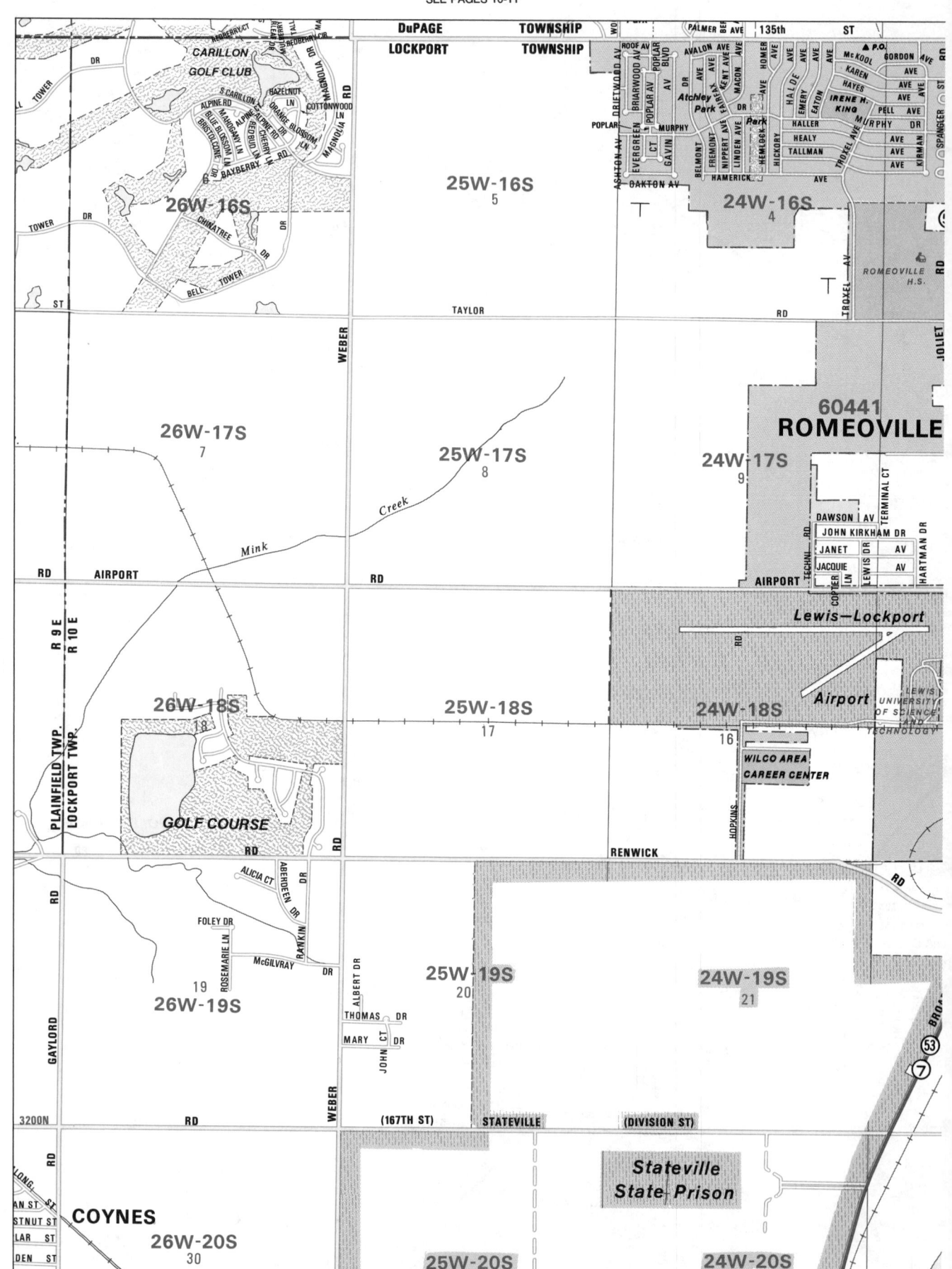

SEE PAGES 12-13

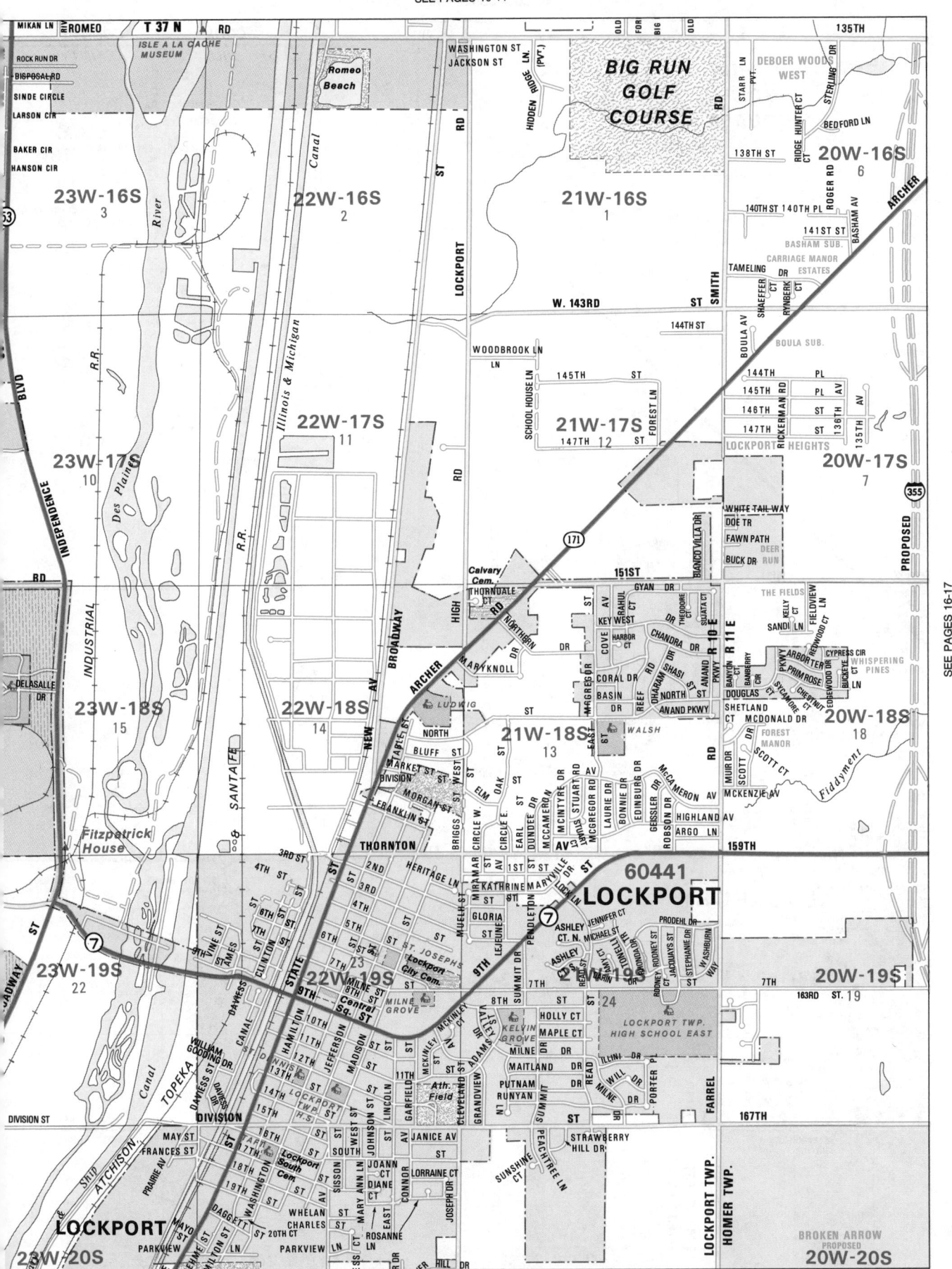

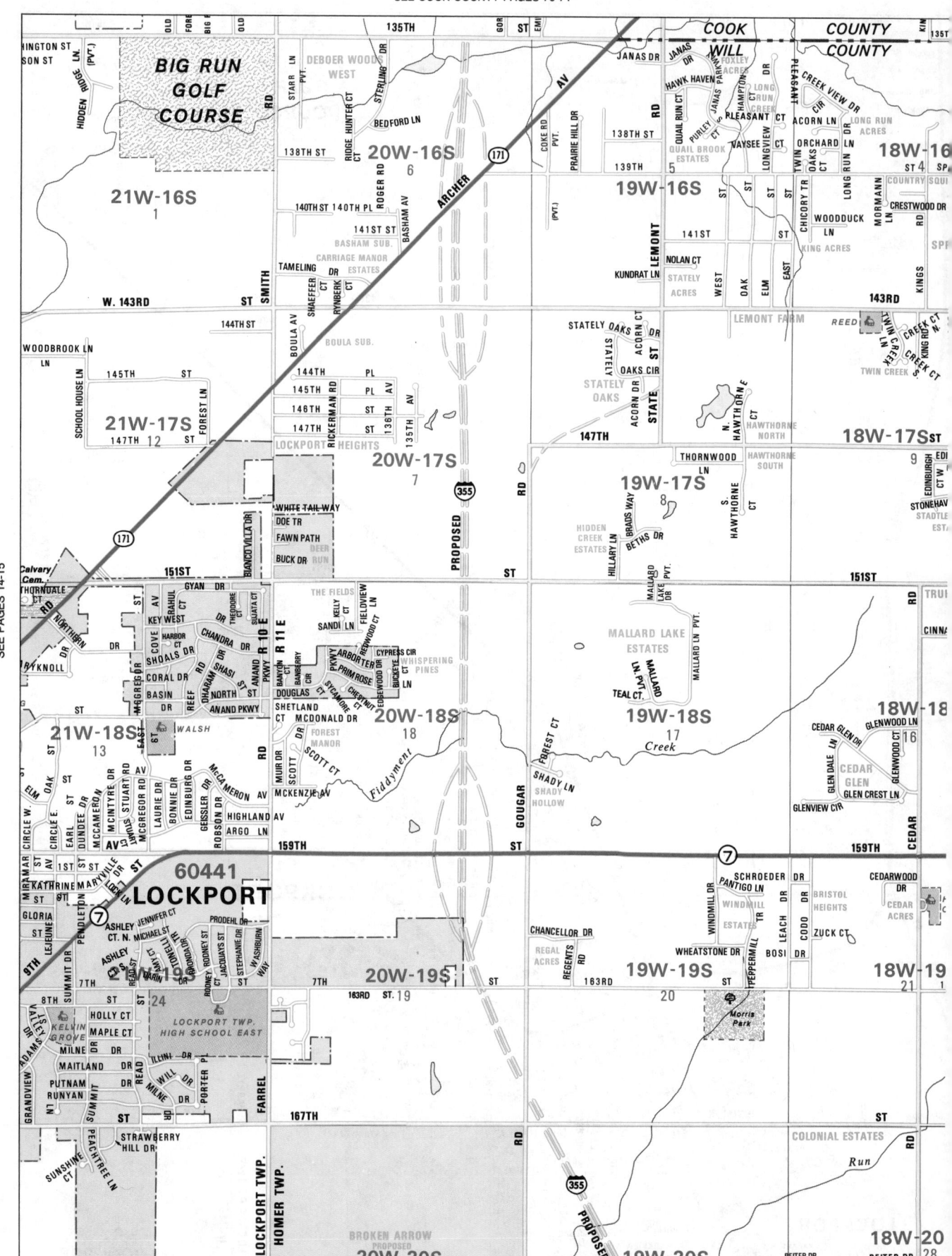

SEE COOK COUNTY PAGES 70-71

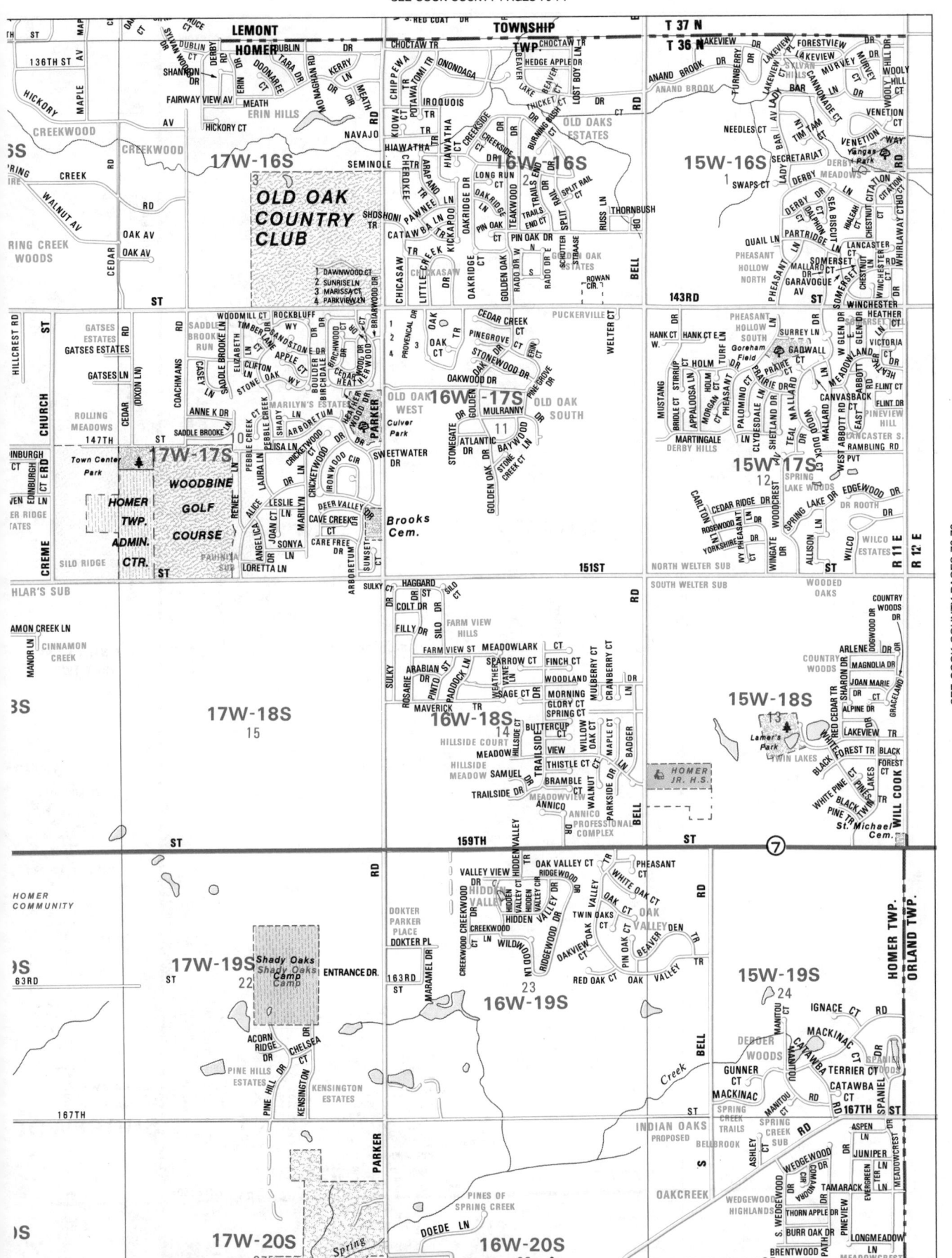

SEE COOK COUNTY PAGES 78-79

SEE PAGES 12-13

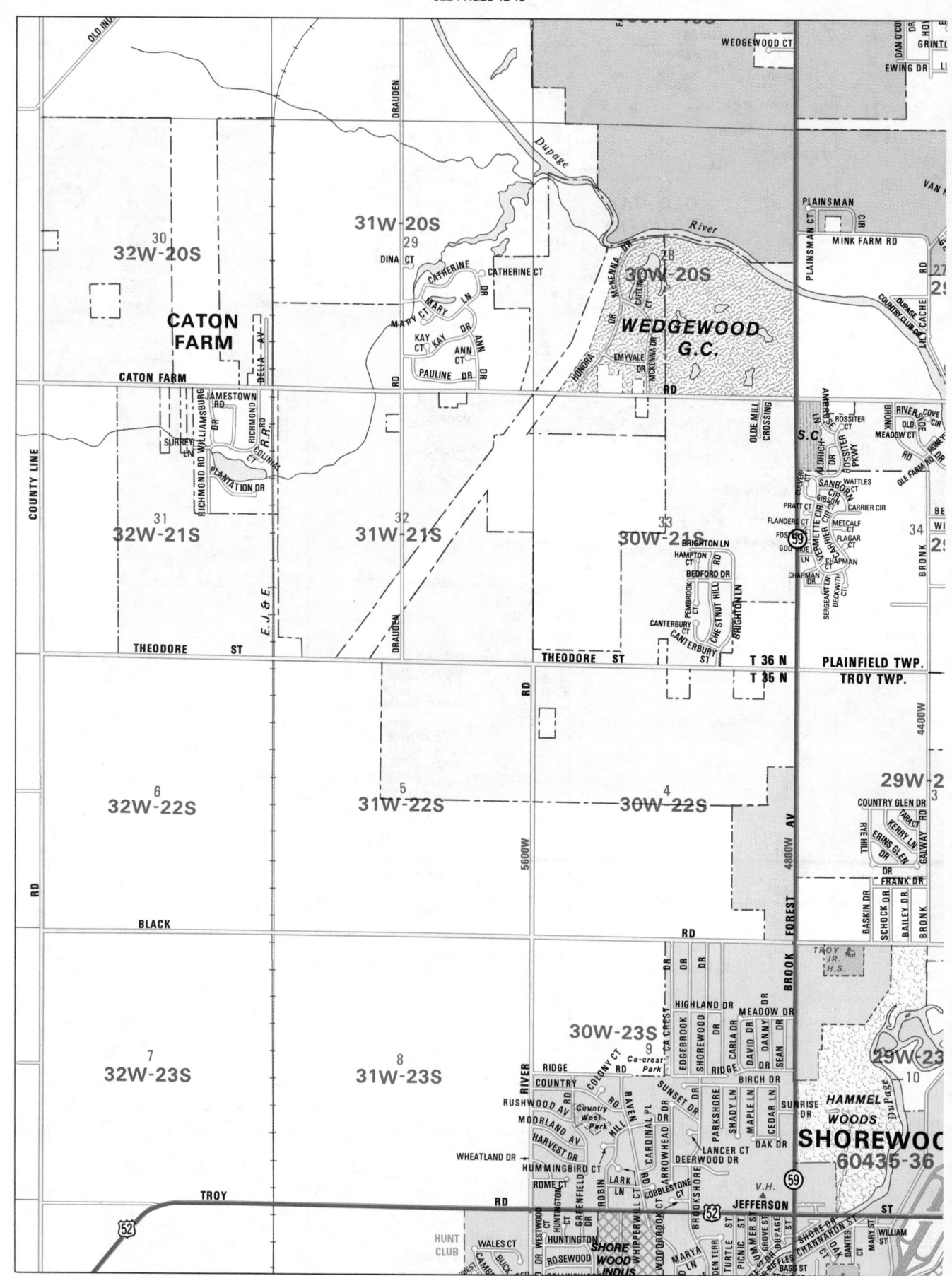

SEE PAGES 26-27

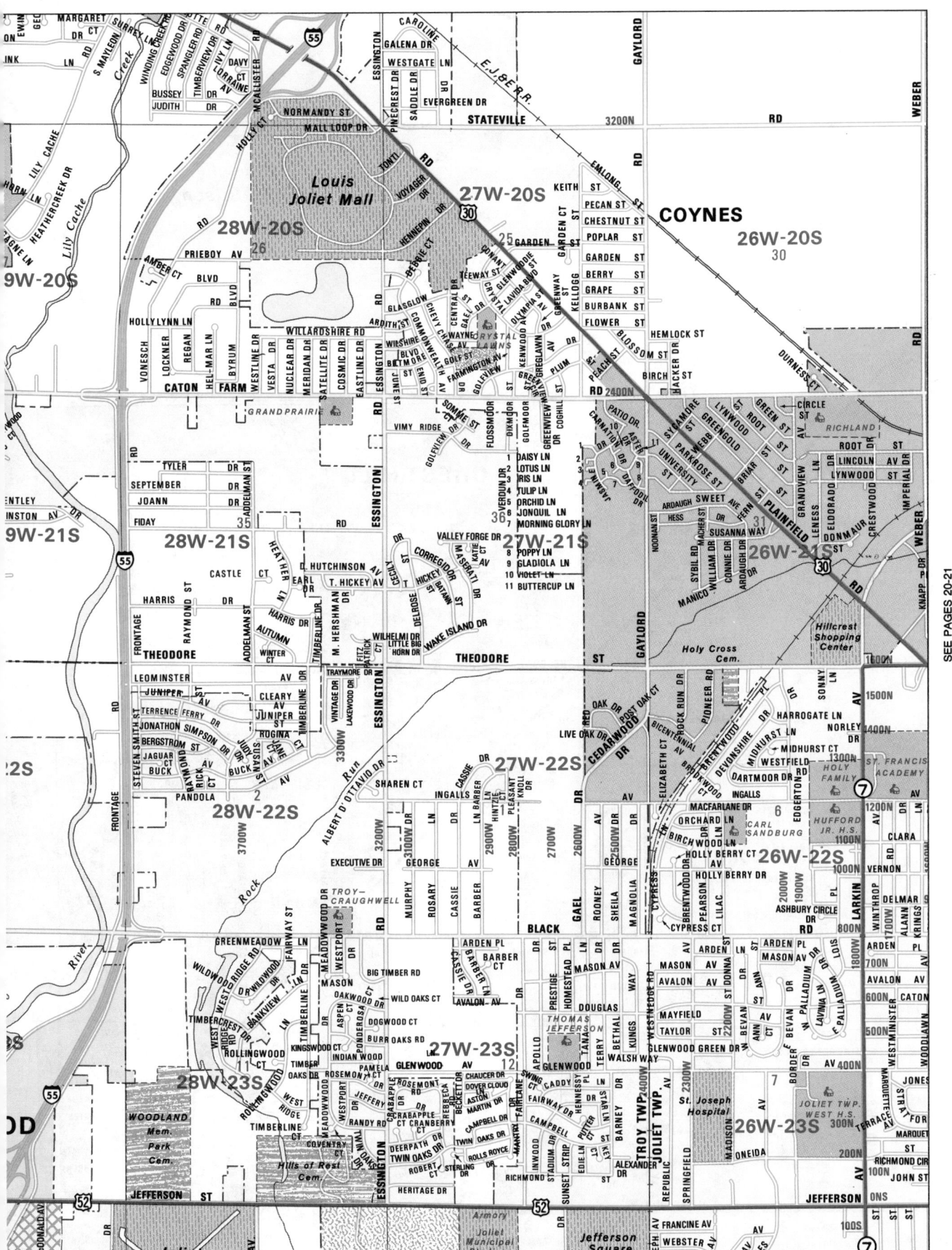

SEE PAGES 14-15

SEE PAGES 18-19

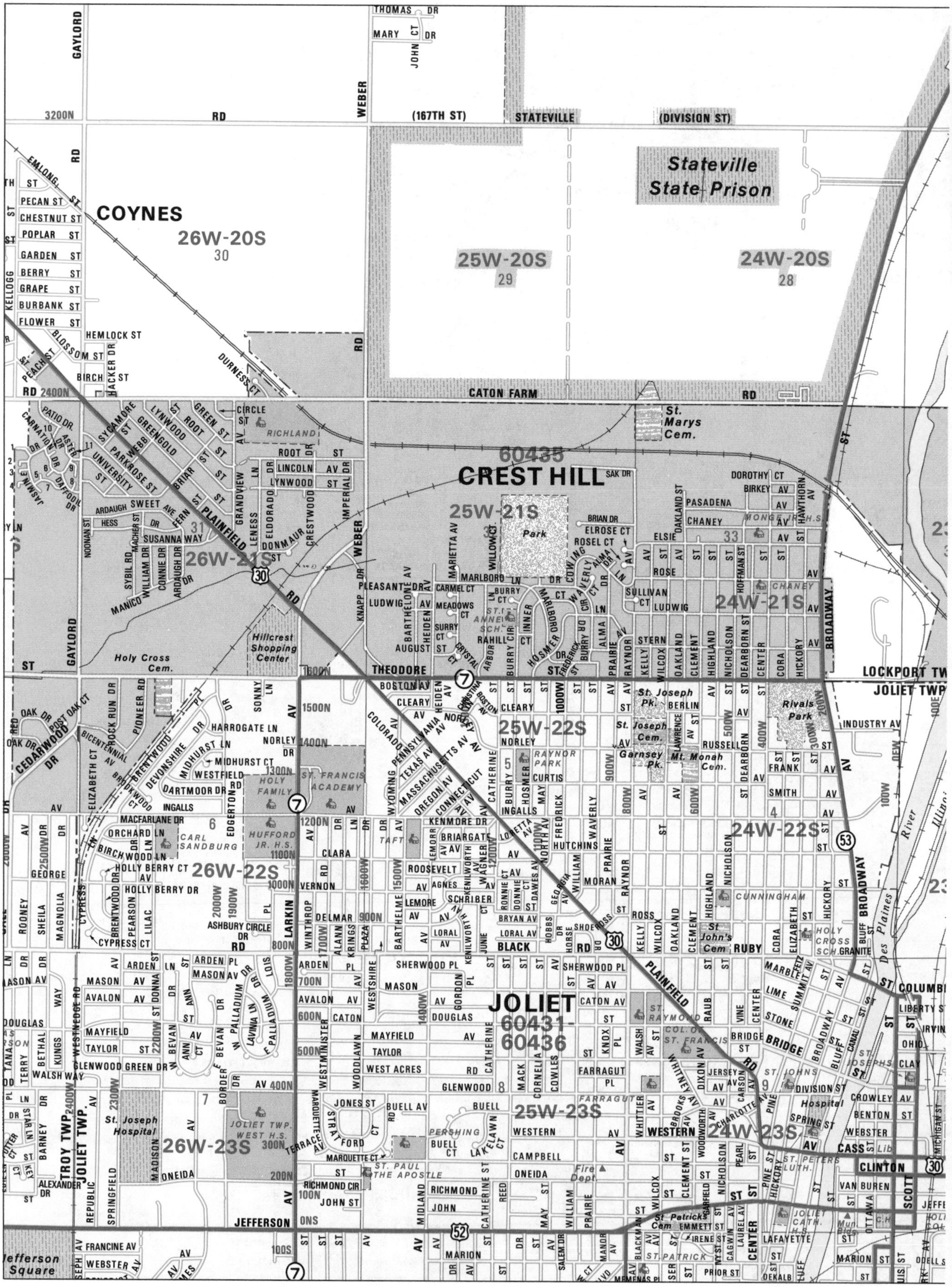

COYNES
26W-20S
30

Stateville State Prison

25W-20S
29

24W-20S
28

STATEVILLE

(167TH ST)

(DIVISION ST)

GAYLORD RD

3200N

WEBER RD

THOMAS DR

MARY CT

JOHN DR

CATON FARM RD

2400N

2400N RD

St. Marys Cem.

60435

CREST HILL
60435

25W-21S
32

26W-21S

24W-21S

Holy Cross Cem.

Hillcrest Shopping Center

Richland

THEODORE

25W-22S

24W-22S

26W-22S

St. Francis Academy

Holy Family

Hufford Jr. H.S.

Carl Sandburg

Raynor Park

St. Joseph Cem.

Mt. Monah Cem.

Garnsey Pk.

St. Joseph Pk.

Rivals Park

JOLIET
60431-60436

25W-23S

24W-23S

26W-23S

St. Joseph Hospital

Joliet Twp. West H.S.

St. Paul The Apostle

Fire Dept.

St. Raymond

St. Francis

Holy Cross Sch

St. John's Cem.

Farragut

LOCKPORT TW
JOLIET TWP

JOLIET TWP.
TROY TWP.

Jefferson Square

JEFFERSON

St. Patricks Cem

St. Peters Luth.

Joliet Cath.

CLINTON

Des Plaines River

Illinois & Michigan Canal

PLAINFIELD RD

BROADWAY

CEDARWOOD DR

LARKIN AV

SEE PAGES 28-29

SEE PAGES 14-15

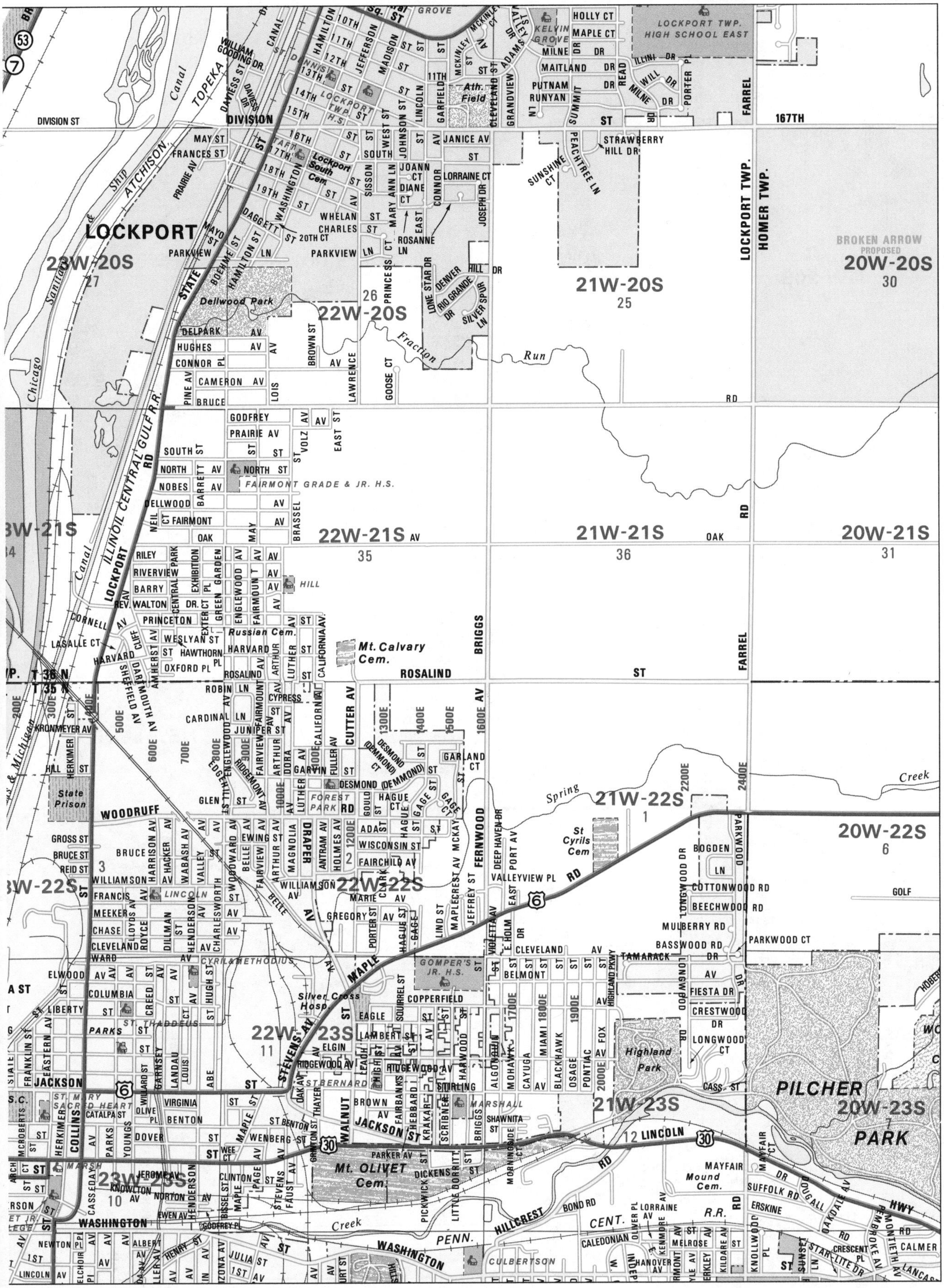

LOCKPORT TWP. HIGH SCHOOL EAST

DIVISION ST

167TH

LOCKPORT TWP.

HOMER TWP.

LOCKPORT

23W-20S
27

21W-20S
25

20W-20S
30

BROKEN ARROW
PROPOSED

22W-20S
26

Dellwood Park

Lockport South Cem.

Run

Fraction

RD

ILLINOIL CENTRAL GULF R.R.

3W-21S
34

22W-21S
35

21W-21S
36

20W-21S
31

FAIRMONT GRADE & JR. H.S.

Mt. Calvary Cem.

ROSALIND

Russian Cem.

BRIGGS

ST

FARREL

OAK

Creek

Spring

21W-22S
1

20W-22S
6

State Prison

WOODRUFF

Forest Park

St Cyrils Cem.

GOLF

3W-22S

22W-22S

Lincoln

Silver Cross Hosp

GOMPER'S JR. H.S.

Highland Park

PILCHER
PARK

22W-23S
11

21W-23S

20W-23S
6

23W-23S
10

Mt. OLIVET Cem.

ST. MARY SACRED HEART

JACKSON

LINCOLN

30

WASHINGTON

PENN.

Creek

Mt. OLIVET Cem.

CENT.

R.R.

HWY

SEE PAGES 22-23

SEE PAGES 16-17

SEE PAGES 20-21

Morris Park

KELVIN GROVE
HOLLY CT
MAPLE CT
MILNE DR
MAITLAND DR
PUTNAM DR
RUNYAN
GRANDVIEW
ILLINI DR
WILL DR
PORTER PL
MILNE
READ
SUMMIT
ST
FARREL

LOCKPORT TWP. HIGH SCHOOL EAST

167TH

COLONIAL ESTATES
ST

STRAWBERRY HILL DR
SUNSHINE CT
PEACHTREE LN

LOCKPORT TWP.
HOMER TWP.

BROKEN ARROW
PROPOSED
20W-20S
30

Run
PROPOSED
355
Fraction

19W-20S
29

18W-20
28

REITER DR
REITER DR
RD
MCCARRON
RON CT
RON CT
HERITAGE ESTATES
BRUCE DR
CEDAR CT
REITER SUB CIR
CEDAR

COLONIAL ACRES

21W-20S
25

Run
RD

Barrett Cem.

LITTLE ACRES
JAMES CT
MCCARRON RD
PARNELL CIR
BRUCE CIR
CEDAR CIR
CEDAR CT
O'CONNELL DR
KYLEMORE
ROBERT EMMETT DR
GOMBIS DR
GOMBIS SUB

LINSEY LN
CIMARRON RIDGE

21W-21S
36

OAK
20W-21S
31

AV

19W-21S
32

18W-2
33

RD

ARROWHEAD
BLACKHAWK ESTATES
EAGLE BEND
LACROSSE DR
DR

FARREL
ST
ST
GOUGAR

HOM
NEW L

Spring
2200E
2400E
Creek
3200E

21W-22S
1

20W-22S
6

19W-22S
SOUTHWEST

NEW LENOX

6

18W-22
4

St Cyrils Cem
VALLEYVIEW PL
EAST PORT AV
E. HOLM DR
RD
6

BOGDEN
LN
PARKWOOD
LONGWOOD DR
COTTONWOOD RD
BEECHWOOD RD
MULBERRY RD
BASSWOOD RD
TAMARACK AV
FIESTA DR
CRESTWOOD DR
LONGWOOD CT
CASS ST

WIMBORNE AV
VAIL CT
ALAMOSA ST
GOLF
CRICKET
SOMERSET ST
SPRINGCREEK AV
BLANDFORD AVE
BLANDFORD AV

GOLF RD

PARKWOOD CT

DR
ROBERG

EDGECREEK
HICKORY STEFANITH
LEWIS LN
WINSOR LN
COVENTRY
PINE GROVE
LAMBETH LN
CREEK RD
LAMBETH DR
IONE DR

CLEVELAND AV
BELMONT ST
MOHAWK
MIAMI
BLACKHAWK
OSAGE
PONTIAC
FOX
CAYUGA AV
1700E
1800E
1900E
2000E
HIGHLAND PKWY
LONGWOOD RD

Highland Park

21W-23S

PILCHER
20W-23S
PARK

WOODRUFF GOLF COURSE

HIGINBOTHAM
FRANCIS
WOODS

19W-23S
8

18W-23
9
BRA

ABRAHAM DR
MARLA LN
CLINTON
SPECTOR RD
CAROL RD
MENNO
MENNO DR
RD

12 LINCOLN
30
MAYFAIR MOUND CEM.

MAYFAIR CT
DOUGALL RD
SUFFOLK RD
ERSKINE

TRINITY SCH.

HOLLY HILL CIR
BOORK DR
CLINTON ST
GORDON ST
SYCAMORE LN
GREEN
WILLOW ST

ALL
NVITA
LLCREST
BOND RD
CENT.
MORNINGSIDE
RD
CALEDONIAN AV
KENMORE AV
LORRAINE AV
MELROSE AV
HANOVER AV
ERKLEY LA
KILDARE AV
KNOLLWOOD PL
SUNSET ST
STAR LITE DR
CRESCENT PL
MONTIETH AV
PEMBROKE AV
LANCAST
CALMER RD
HWY
HMSTEAD

HOUSE CRESCENT
GRANDHOUSE CRESCENT
BURE

CH

16W-19S

ACORN RIDGE DR
PINE HILLS ESTATES
PINE HILL DR
CHELSEA CT
KENSINGTON CT
KENSINGTON ESTATES

IGNACE CT RD
MACKINAC
CATAWBA
MANITOU
MANITOU CT
DEBOER WOODS
BELL
Creek
ST
TERRIER CT
SPANIEL WOODS
GUNNER CT
MACKINAC
MANITOU RD
CATAWBA CT
SPANIEL
167TH RD
167TH ST

167TH

PARKER

INDIAN OAKS PROPOSED
BELLBROOK
SPRING CREEK
S
SPRING CREEK SUB
RD
ASHLEY CT
ASPEN LN
JUNIPER LN
MEADOWCREST

OAKCREEK
WEDGEWOOD HIGHLANDS
S. WEDGEWOOD
WEDGEWOOD COMMON
DR
COMADOMA
THORN APPLE DR
BURR OAK DR
BRENTWOOD DR
DEER PATH DR
TAMARACK
EVERGREEN TER
PINEVIEW
LONGMEADOW LN
MEADOWCREST

17W-20S
27
MESSENGER WOODS
Spring
DEER CREEK

DOEDE LN
PINES OF SPRING CREEK

16W-20S
26
OAK MEADOW CT
OAK MEADOW
HADLEY
LAUFFER RD
RD

15W-20S
25

WILL CO.
COOK CO.

175TH ST
RD
RD
RD
LARSEN'S SUB
PARKER WOODS PROPOSED
RYCON DR
CRYSTAL LAKE
CRYSTAL LAKES ESTATES
CRYSTAL LAKE CT
RD
BRUCE RD
175TH ST.
MARTI
179TH ST
36
15W-21S

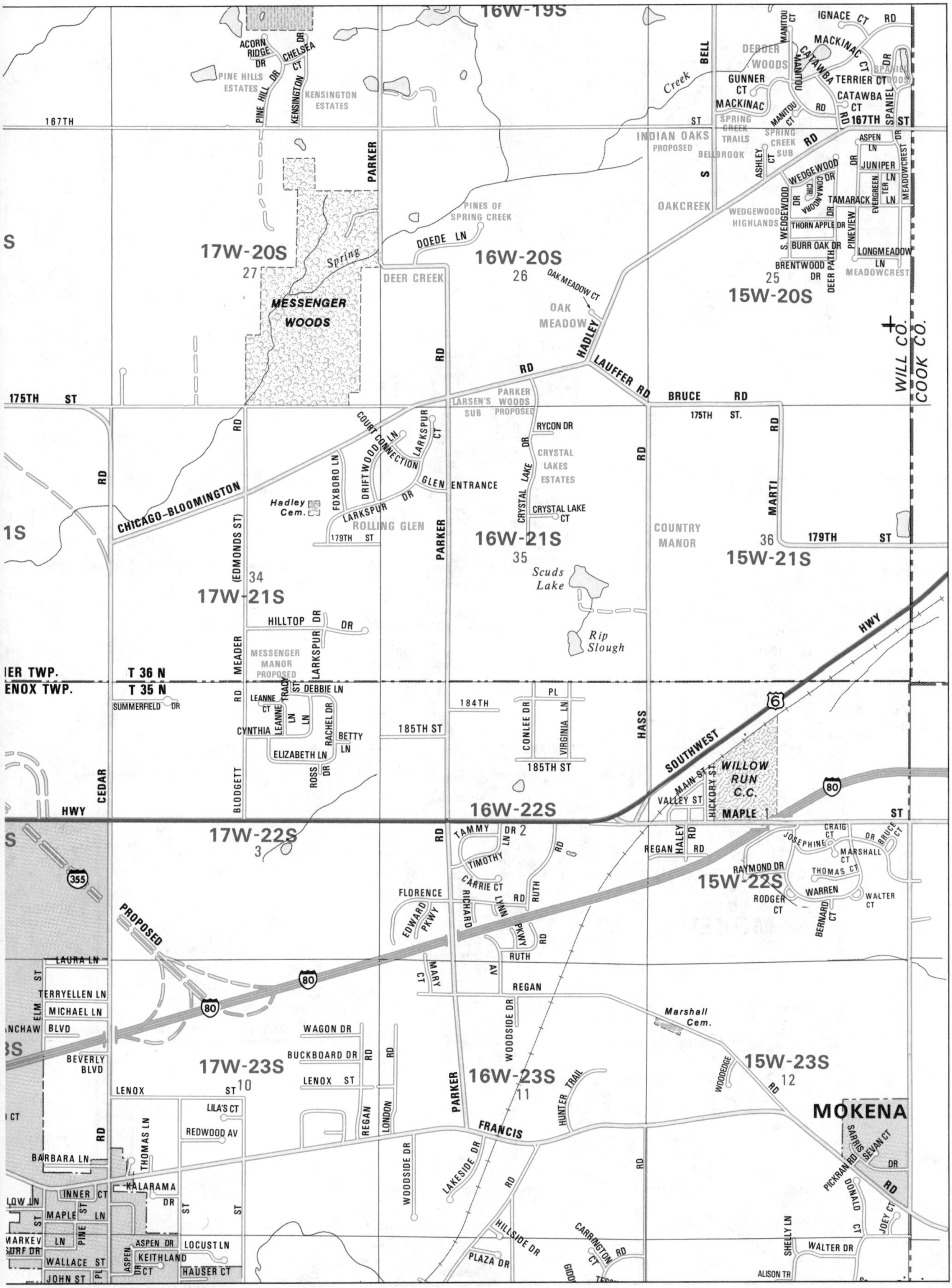

COURT CONNECTION
DRIFTWOOD LN
LARKSPUR CT
GLEN ENTRANCE
FOXBORO LN
LARKSPUR DR
LARKSPUR ST
ROLLING GLEN
179TH ST
Hadley Cem.
CHICAGO-BLOOMINGTON
(EDMONDS ST)
34
17W-21S
PARKER
16W-21S
35
COUNTRY MANOR
Scuds Lake
Rip Slough

HILLTOP DR
MEADER RD
MESSENGER MANOR PROPOSED
LARKSPUR DR
HWY
6
SOUTHWEST

IER TWP.
ENOX TWP.
T 36 N
T 35 N
SUMMERFIELD DR
LEANNE LN
LEANNE TRAIL
DEBBIE LN
CYNTHIA
RACHEL LN
BETTY LN
ELIZABETH LN
ROSS DR
BLODGETT RD
184TH
CONLEE DR
PL
VIRGINIA LN
185TH ST
185TH ST
HASS
MAIN ST
HICKORY ST
WILLOW RUN C.C.
MAPLE
CRAIG CT
DR
BRUCE CT
JOSEPHINE
MARSHALL
THOMAS CT
RAYMOND DR
WARREN
WALTER CT
RODGER CT
BERNARD CT

CEDAR HWY
17W-22S
3
16W-22S
TAMMY DR
2
TIMOTHY
CARRIE CT
RICHARD
FLORENCE
LYNN PKWY
RUTH
RD
VALLEY ST
REGAN
HALEY RD
15W-22S

355
PROPOSED
80
EDWARD PKWY
RUTH
MARY CT
RUTH AV
REGAN
Marshall Cem.
WOODSIDE DR
WOODEDGE
15W-23S
12

LAURA LN
ELM ST
TERRYELLEN LN
MICHAEL LN
BLVD
ANCHAW
BEVERLY BLVD
WAGON DR
BUCKBOARD DR
LENOX ST
17W-23S
10
16W-23S
11
HUNTER TRAIL
MOKENA

LENOX ST
THOMAS LN
LILA'S CT
REDWOOD AV
REGAN
LONDON
PARKER RD
FRANCIS
WOODSIDE DR
LAKESIDE DR
SARRIS RD
SEVAN CT
DONALD
PICKRAN RD
JOEY CT
BARBARA LN
RD
INNER CT
KALARAMA DR
WOODSIDE DR
HILLSIDE DR
CARRINGTON RD
SHELLY LN
WALTER DR

LOW LN
MAPLE LN
PINE LN
ASPEN DR
LOCUST LN
KEITHLAND
HAUSER CT
PLAZA DR
ALISON TR
MARKEV LN
SURF DR
WALLACE ST
JOHN ST

SEE PAGES 24-25

SEE COOK COUNTY PAGES 84-85

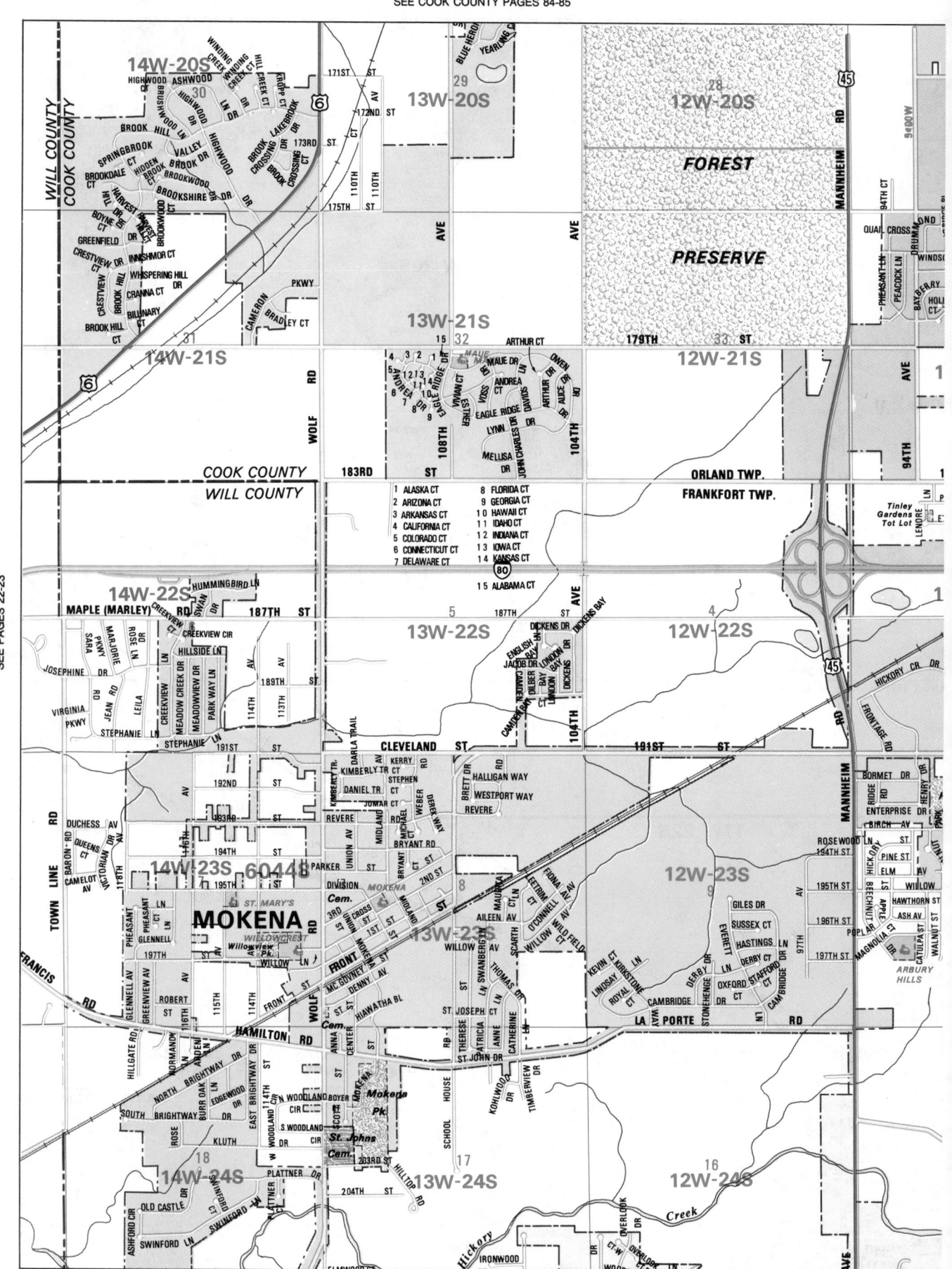

14W-20S

13W-20S

12W-20S

FOREST

PRESERVE

14W-21S

13W-21S

12W-21S

14W-22S

13W-22S

12W-22S

1	ALASKA CT	8	FLORIDA CT
2	ARIZONA CT	9	GEORGIA CT
3	ARKANSAS CT	10	HAWAII CT
4	CALIFORNIA CT	11	IDAHO CT
5	COLORADO CT	12	INDIANA CT
6	CONNECTICUT CT	13	IOWA CT
7	DELAWARE CT	14	KANSAS CT
		15	ALABAMA CT

COOK COUNTY
WILL COUNTY

ORLAND TWP.
FRANKFORT TWP.

14W-23S

13W-23S

12W-23S

MOKENA

14W-24S

13W-24S

12W-24S

SEE PAGES 32-33

SEE PAGES 22-23

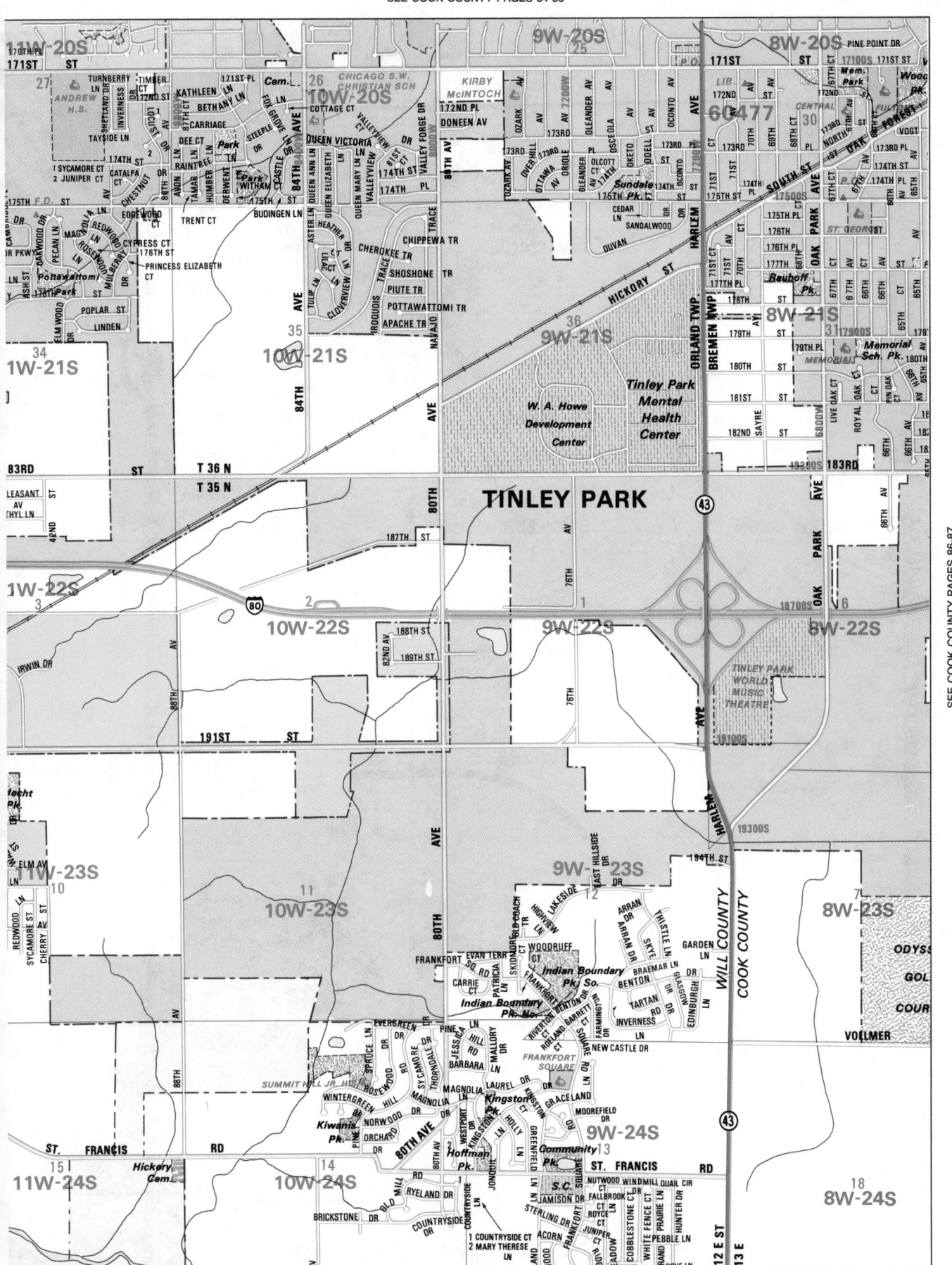

TINLEY PARK

Tinley Park Mental Health Center

W. A. Howe Development Center

Tinley Park World Music Theatre

SEE COOK COUNTY PAGES 86-87

SEE PAGES 18-19

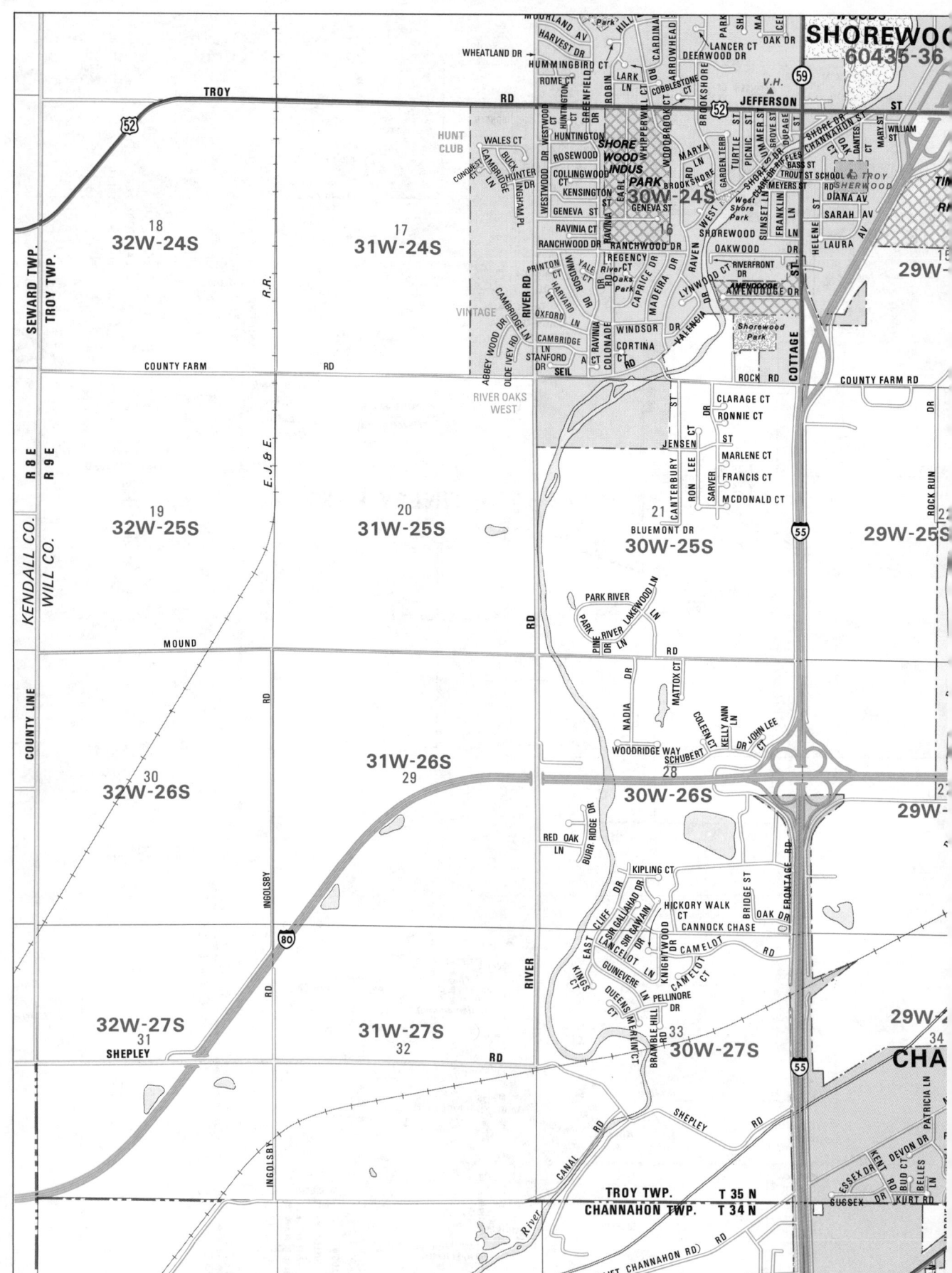

SEE PAGES 34-35

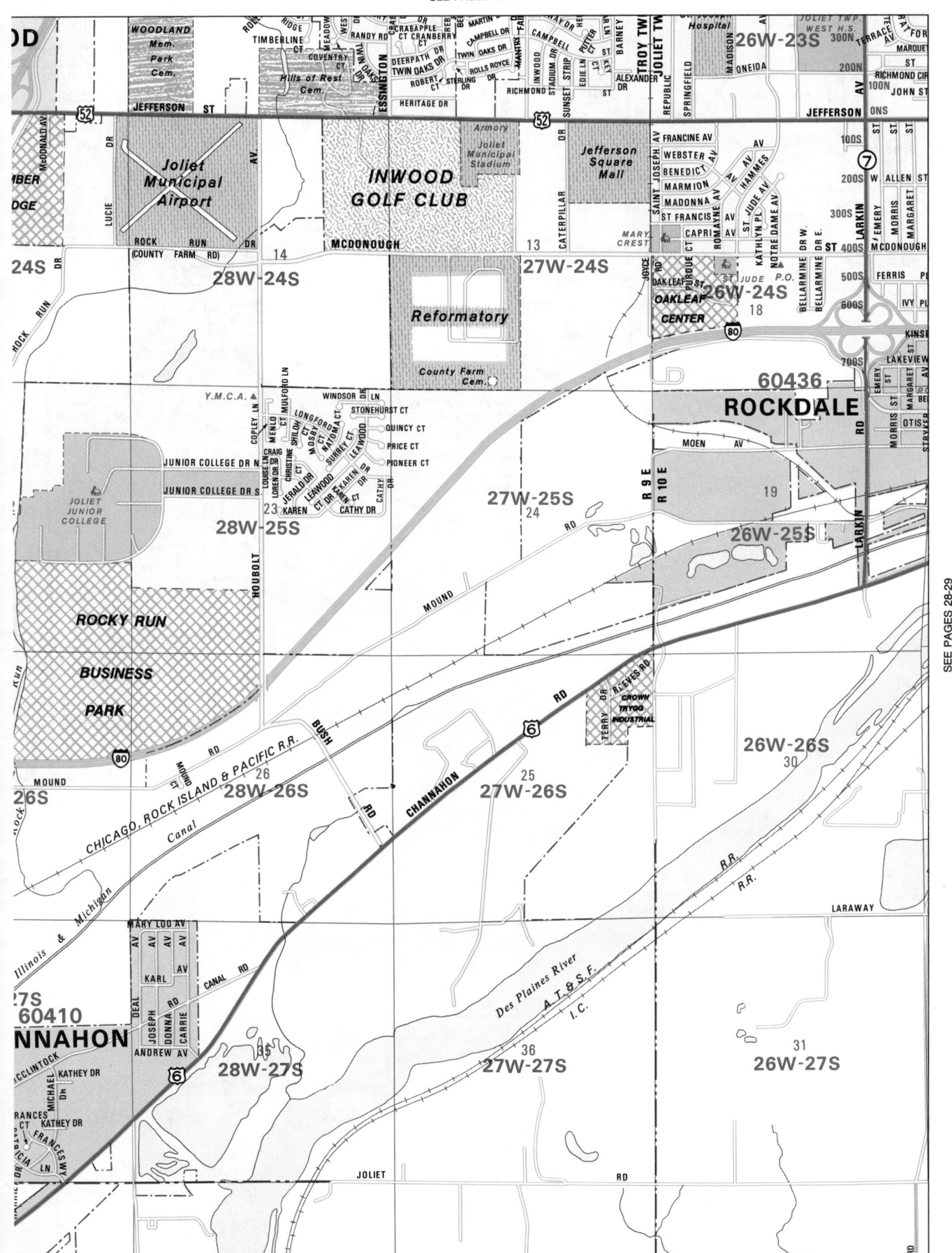

WOODLAND Mem. Park Cem.

Hills of Rest Cem.

JEFFERSON ST

52

Joliet Municipal Airport

ROCK RUN (COUNTY FARM RD) 14

28W-24S

McDONOUGH 13

27W-24S

INWOOD GOLF CLUB

Armory Joliet Municipal Stadium

Reformatory

County Farm Cem.

Jefferson Square Mall

MARY CREST

52

FRANCINE AV
WEBSTER AV
BENEDICT AV
MARMION AV
MADONNA AV
ST FRANCIS AV
CAPRI AV
ROMANE AV

HAMMES
ST JUDE AV
NOTRE DAME AV
KATHLYN PL

BELLARMINE DR W.
BELLARMINE DR E.

ST JUDE P.O.

OAKLEAF CENTER

26W-24S 18

7

100S
200S W ALLEN ST
300S
400S McDONOUGH
500S FERRIS
600S IVY PL

RICHMOND CIR
JOHN ST

MORRIS
MARGARET

80

60436

ROCKDALE

MOEN AV

R 9 E R 10 E

19

26W-25S

700S KINS
LAKEVIEW

LARKIN RD
EMERY ST
MORRIS ST
MARGARET ST
OTIS
STRYKER

Y.M.C.A.

WINDSOR LN
STONEHURST CT
QUINCY CT
PRICE CT
PIONEER CT

COPLEY LN
MENLO
CRAIG
SHILOH CT
CHRISTINE
JERALD DR
LEAWOOD CT
KAREN CT
KAREN DR
CATHY DR
CATHY DR

LOUISE LN
LOREN DR
MOSBY
NATOMA CT
SURREY CT
LEAWOOD DR

LONGFORD CT
MULFORD LN

23

JUNIOR COLLEGE DR N
JUNIOR COLLEGE DR S

JOLIET JUNIOR COLLEGE

28W-25S

27W-25S 24

MOUND RD

HOUBOLT

ROCKY RUN

BUSINESS

PARK

80

MOUND RD

CHICAGO, ROCK ISLAND & PACIFIC R.R. 26

26S

BUSH RD

CHANNAHON RD

6

25

27W-26S

TERRY DR
REEVES RD

CROWN TRYOG INDUSTRIAL

26W-26S 30

MOUND

28W-26S

R.R.
R.R.

LARAWAY

Illinois & Michigan Canal

MARY LOU AV
AV AV AV
KARL
DEAL
KARL RD
JOSEPH
DONNA RD
CARRIE
ANDREW AV

CANAL RD

27S
60410

CHANNAHON

McCLINTOCK
MICHAEL
KATHEY DR
KATHEY DR
FRANCES CT
FRANCES WAY
PATRICIA LN

6

28W-27S 35

Des Plaines River A.T.&S.F. I.C.

36

27W-27S

26W-27S 31

JOLIET RD

RD

SEE PAGES 28-29

SEE PAGES 20-21

26W-23S

25W-24S

24W-23S

24W-24S

26W-24S

60436
ROCKDALE

26W-25S

25W-25S

24W-25S

26W-26S

25W-26S

24W-26S

26W-27S

25W-27S

24W-27S

Jefferson
Square
Mall

St. Joseph
Hospital

Mary
Crest

West
Park

McKinley
Park

SEE PAGES 26-27

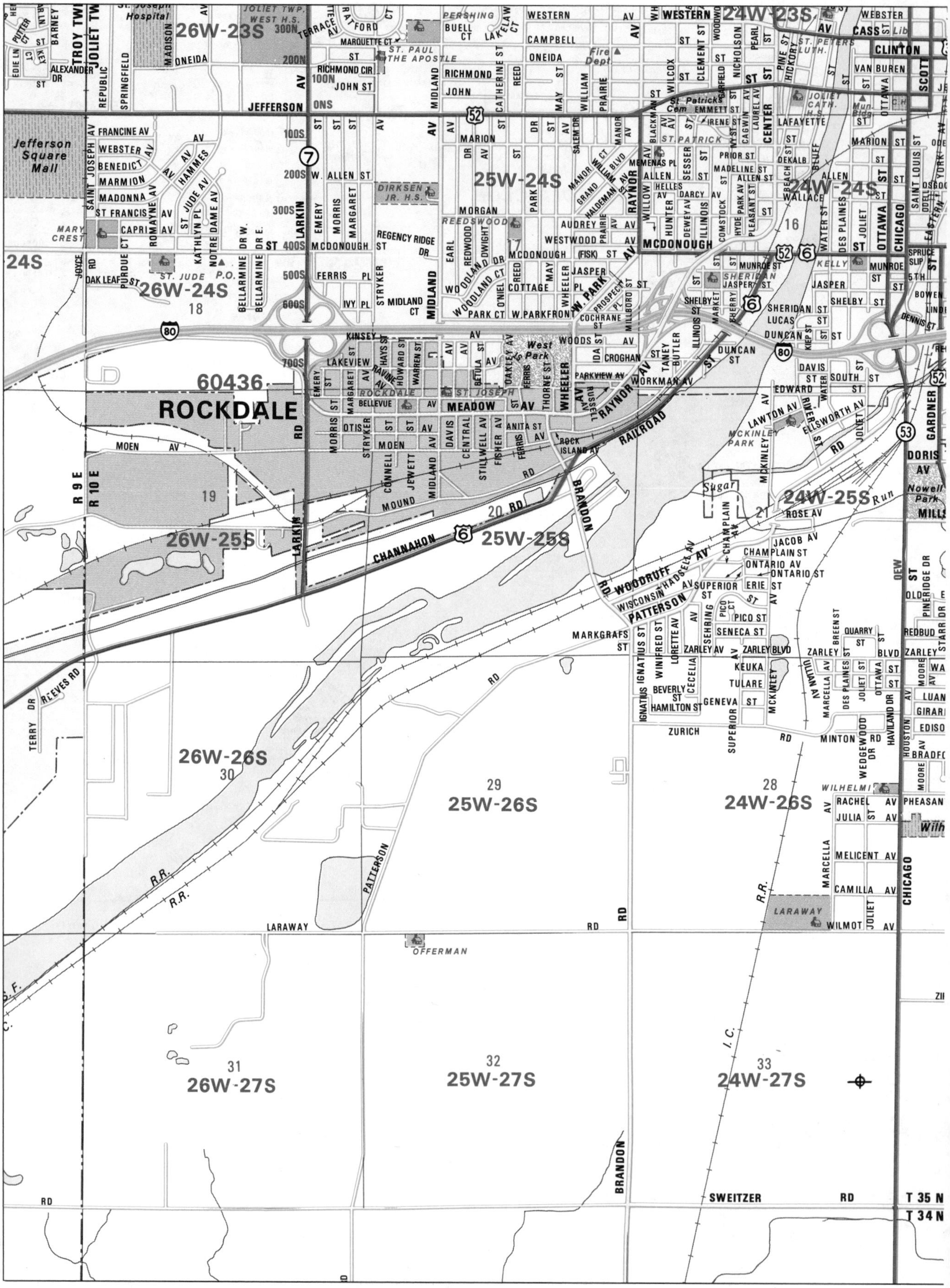

SEE PAGES 36-37

SEE PAGES 20-21

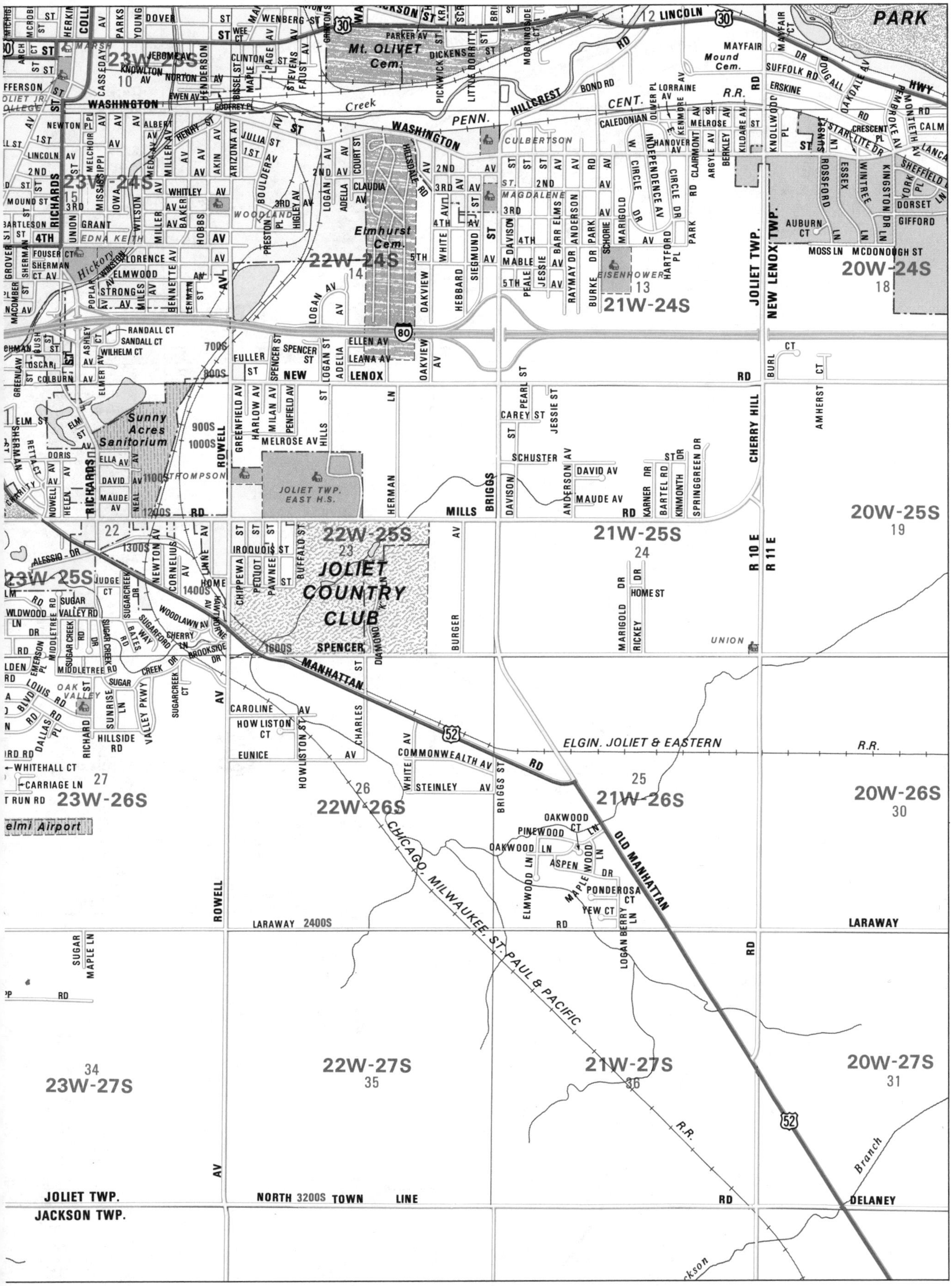

PARK

Mt. OLIVET Cem.

Elmhurst Cem.

Woodland

JOLIET TWP. EAST H.S.

Sunny Acres Sanitorium

JOLIET COUNTRY CLUB

23W-24S

22W-24S

21W-24S

20W-24S

23W-25S

22W-25S

21W-25S

20W-25S

23W-26S

22W-26S

21W-26S

20W-26S

23W-27S

22W-27S

21W-27S

20W-27S

ELGIN. JOLIET & EASTERN R.R.

CHICAGO. MILWAUKEE. ST. PAUL & PACIFIC

JOLIET TWP.
JACKSON TWP.

NORTH 3200S TOWN LINE

SEE PAGES 30-31

SEE PAGES 22-23

FRANCIS
WOODS
TRINITY SCH.

MOUND Cem.
MAYFAIR
MONROE
BOND RD
ERSKINE
SUFFOLK RD
DOUGALL
R.R.
CENT.
CALEDONIAN
HILLCREST
OLIVER PL
LORRAINE AV
KENMORE
MELROSE AV
INDEPENDENCE AV
HANOVER AV
CLAIRMONT
ARGYLE AV
KILDARE AV
BERKLEY AV
KNOLLWOOD PL
SUNSET ST
STARLITE DR
CRESCENT PL
PEMBROKE AV
MONTIETH AV
OAKDALE AV
DR
CALMER RD
HWY
LANCASTER
CHICAGO, ROCK ISLAND & PACIFIC R.R.
19W-24S
17
HOLLY HILL CIR
BJORK DR
CLINTON
GORDON ST
SYCAMORE LN
THORNHOUSE
BAMBURKE
CRESCENT
GREEN
WILLOW ST
ALBERTSON
ST
W. CIRCLE DR
CIRCLE DR
ANDERSON
RAYMAY DR
RD
BARR ELMS AV
SCHORIE
MARIGOLD
HARTFORD PL
PARK AV
CLAIRMONT RD
ROSSFORD LN
ESSEX LN
WINTREE LN
KINGSTON DR LN
DORSET LN
GIFFORD PL
SHEFFIELD DR
PARK PL
DOXBURY LN
HEMPSTEAD PL
DR
PROVIDENCE H.S.
WILLOW RD WEST
OLD NEW LENOX
Hickory
18W-24
16
VINE
MAGDALENE
2ND ST
4TH ST
PEALE AV
JESSIE AV
BURKE DR
PARK AV
EISENHOWER
13
AUBURN CT
MOSS LN
MCDONOUGH ST
MCDONOUGH ST
20W-24S
18
STONE CT
FERRO DR
GARNET DR
I-80
60451
NEW LENOX
GREENBRIAR DR
PARK LN
MANOR
GREENBRIAR DR
VINE ST
OLD NEW LENOX RD
HILLSIDE RD
30
21W-24S
JOLIET TWP.
NEW LENOX TWP.
NEW LENOX
LIVINGSTON DR
HONEY
KRIS
NELSON RD
DR
PEARL ST
JESSIE ST
CHUSTER AV
ANDERSON AV
DAVID AV
MAUDE AV
RD
KARNER DR
BARTEL RD
KINMONTH DR
SPRINGGREEN DR
CHERRY HILL
BURL CT
AMHERST CT
RD
RD
CT
RD
20W-25S
19
19W-25S
20
MUSTANG LN
PIPER DR
LEAR LN
CENTRAL RD
AERONCA CT
MISTY CREEK DR
BEECHCRAFT DR
BEECH LN
CORSAIR CT
CESSNA CT
GRUMMAN CT
JOLIET STAFFORD ARMSTRONG
STAFFORD DR
CT
GLENN DR
MCDIVITT DR
HWY
GEAR DR
BOEING DR
WARREN AV
WILDWOOD DR
18W-25
21
JOLIET
AV
WOOD
21W-25S
24
MARIGOLD DR
DR
HOME ST
RICKEY
R 10 E
R 11 E
WHITE LN
YOUNG DR
GRISSOM LN
NELSON RD
WESTERN
WISCONSIN
UNION
RD
SPENCER
RD
RD
SEE PAGES 28-29
ELGIN, JOLIET & EASTERN
R.R.
19W-26S
29
CIMARRON
18W-2
28
RD
25
20W-26S
30
ANDREA DR
COUNTRY CREEK DR
21W-26S
OAKWOOD CT
PINEWOOD LN
WOOD LN
ASPEN
MAPLEWOOD DR
PONDEROSA CT
YEW CT
ELMWOOD LN
LOGAN BERRY LN
OLD MANHATTAN
RD
STONEBRIDGE
CHERRYWOOD LN
Park
EAGLESVISTA DR
DELMAR DR
BRIARCREST LN
ANDREA LN
STONEBRIDGE DR
GRANDVIEW Park
SHAGBARK CT
SHAGBARK
FERNWOOD TER
RD
WINTER PARK DR
MEADOW RIDGE LN
TIMBER PL
FOXWOOD
RD
LARAWAY
GOUGAR
TIMBER CT
PL
HEATHERWAY LN
NELSON
Park
REITER
RD
CARLTON CT
SANFORD AV
JACKSON
BRANCH DR
DIGBY DR
ARGYLE LN
Park
DANIEL LEWIS DR
GINGER LN
KERRY WINDE DR
19W-27S
32
18W-27
33
PAUL & PACIFIC
21W-27S
36
20W-27S
31
52
R.R.
Branch
RD
DELANEY
RD
NEW LE
MANHAT
MALIBU DR
AV
Jackson

SEE PAGES 32-33

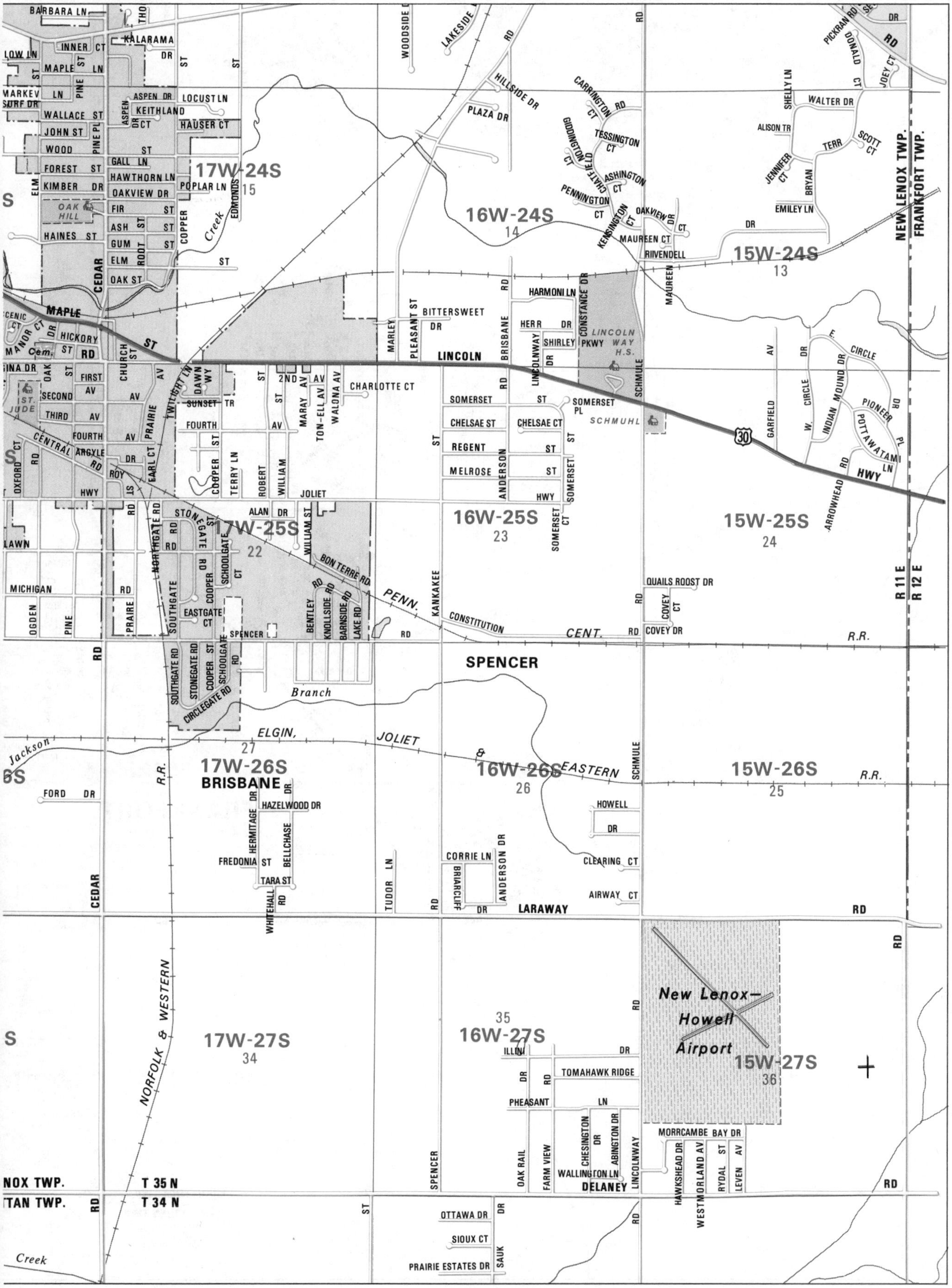

17W-24S 15

16W-24S 14

15W-24S 13

NEW LENOX TWP.
FRANKFORT TWP.

Lincoln Way H.S.

LINCOLN

HWY 30

17W-25S 22

16W-25S 23

15W-25S 24

R 11 E
R 12 E

SPENCER

17W-26S
BRISBANE

16W-26S 26

15W-26S 25

LARAWAY

New Lenox—
Howell
Airport

17W-27S 34

16W-27S 35

15W-27S 36

DELANEY

NOX TWP.   T 35 N
TAN TWP.   T 34 N

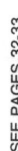

SEE PAGES 30-31

R 11 E
R 12 E

14W-24S 18

13W-24S 17

12W-24S 16

14W-25S

13W-25S 20

12W-25S

VAN HORNE WOODS

MAPLE RIDGE DR

LINCOLN HWY 19

WOLF RD

OSAGE RD

Old Plank Road Trail
(Under development)

13W-26S 29
E. J. & E. RR

14W-26S 30

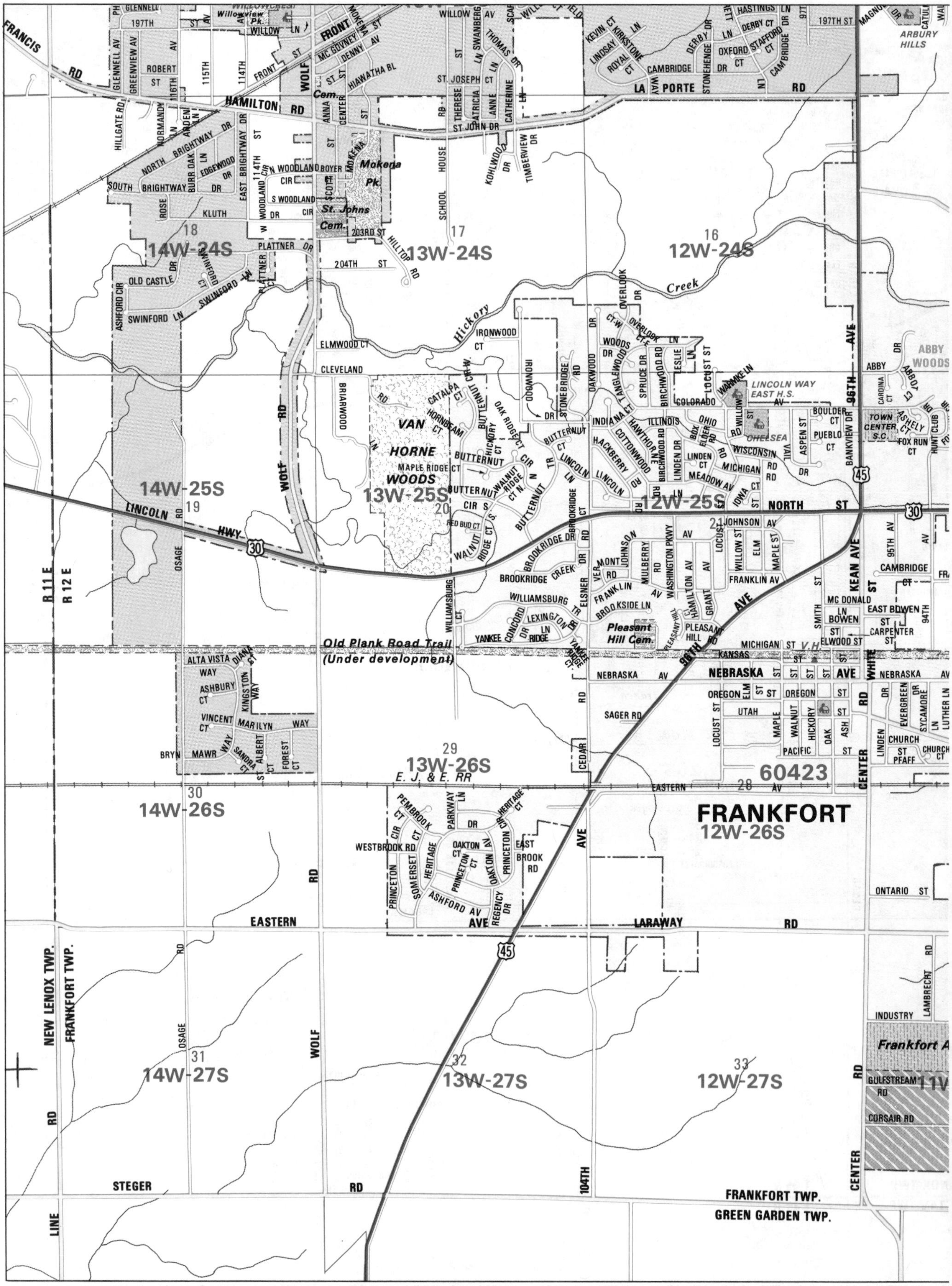

12W-26S

FRANKFORT

60423 28

EASTERN AVE

LARAWAY RD

NEW LENOX TWP.
FRANKFORT TWP.

14W-27S 31

13W-27S 32

12W-27S 33

STEGER RD

WOLF RD

104TH

FRANKFORT TWP.
GREEN GARDEN TWP.

Frankfort A
GULFSTREAM RD
CORSAIR RD
INDUSTRY
LAMBRECHT RD

CENTER RD

ARBURY HILLS

ABBY WOODS

Mokena Pk.
St. Johns Cem.

Pleasant Hill Cem.

Lincoln Way East H.S.

HAMILTON RD
FRONT ST
WOLF RD
LA PORTE RD
96TH AVE

NORTH ST
KEAN AVE
96TH AVE

NEBRASKA AVE

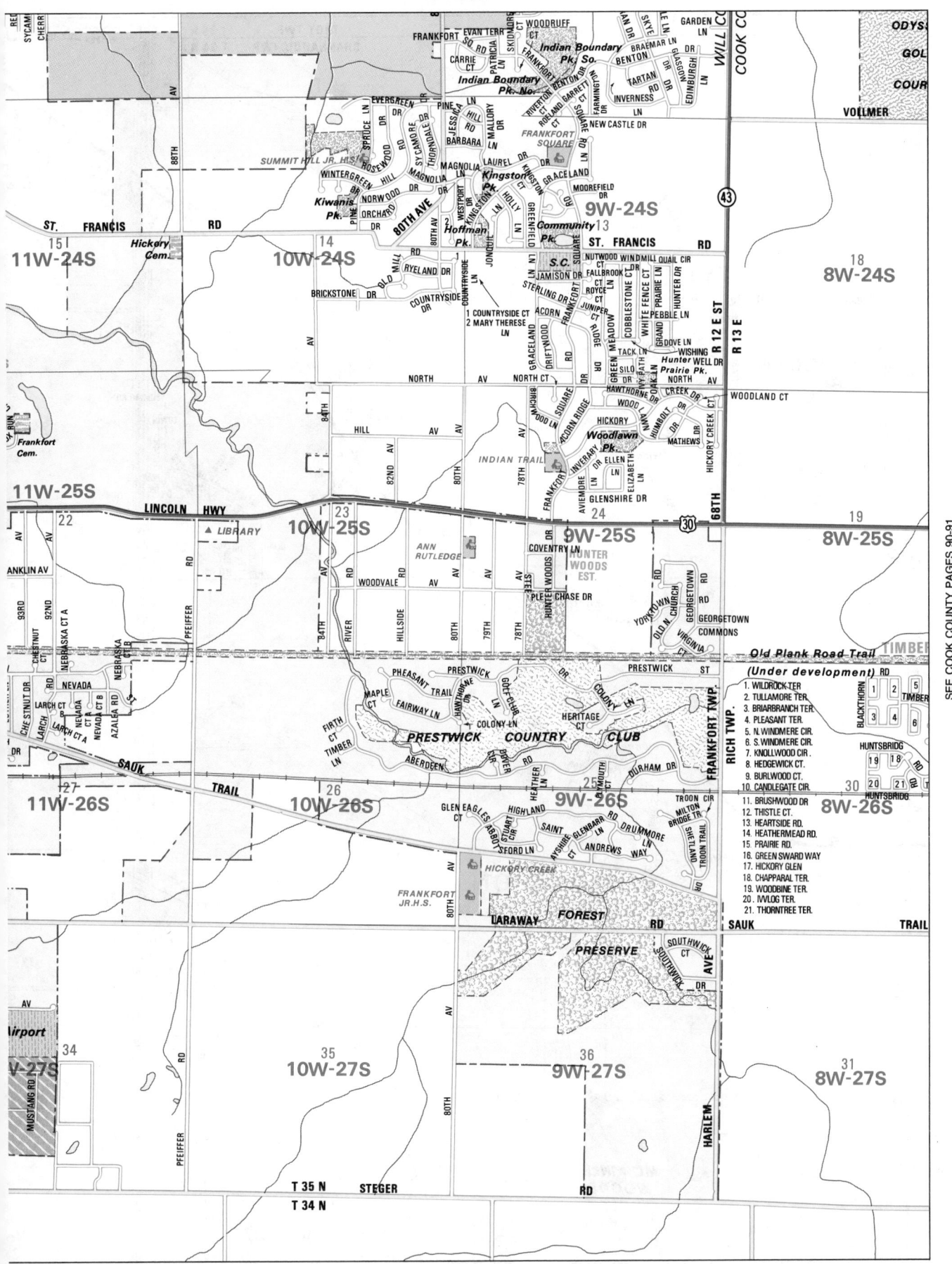

ODYS.

GOL

COUR

VOLLMER

WILL CO

COOK CO

Indian Boundary Pk. So.

Indian Boundary Pk. Nor.

FRANKFORT

EVAN TERR

FRANKFORT SQ

CARRIE CT

PATRICIA LN

SKIDMORE

WOODRUFF CT

AN DR

SKYE

GARDEN LN

BRAEMAR LN

BENTON

GLASGOW LN

TARTAN

EDINBURGH

INVERNESS RD

FARMINGTON LN

NEW CASTLE DR

43

EVERGREEN

PINE LN

HILL RD

JESSICA LN

SPRUCE LN

ROSEWOOD DR

SYCAMORE DR

THORNDALE DR

MALLORY

RIVERTON BENTON DR

ROLAND GARRETT SQUARE

SUMMIT HILL JR. H.S.

WINTERGREEN

HILL

MAGNOLIA

NORWOOD

ORCHARD

Kingston Pk.

LAUREL DR

KINGSTON DR

KINGSTON LN

HOLLY LN

GRACELAND

Frankfort Square

MOOREFIELD DR

9W-24S

13

Kiwanis Pk.

PINE

80TH AVE

WESTPORT

GREENFIELD

15

St. Francis Cem.

Hickory Cem.

11W-24S

ST. FRANCIS RD

14

10W-24S

80TH AV

Hoffman Pk.

JONQUIL

Community Pk.

S.C.

ST. FRANCIS RD

NUTWOOD

WINDMILL QUAIL CIR

18

8W-24S

OLD MILL RD

RYELAND DR

JAMISON DR

FALLBROOK CT

WHITE FENCE CT

PEBBLE LN

GRAND PRAIRIE LN

HUNTER DR

R 12 E ST

R 13 E

BRICKSTONE DR

COUNTRYSIDE DR

COUNTRYSIDE LN

STERLING DR

ACORN

ROYCE CT

JUNIPER

390TH

GREEN MEADOW

TACK LN

DOVE LN

WISHING WELL DR

1 COUNTRYSIDE CT
2 MARY THERESE LN

GRACELAND

DRIFTWOOD

SILO DR

PATH

Hunter Prairie Pk. North

NORTH AV

NORTH CT

NORTH AV

82ND

80TH

78TH

MICHIGAN

FRANKFORT SQUARE

ACORN RIDGE

Woodlawn Pk.

Hickory Pk.

WOOD

OAK LN

HUMBOLT DR

MATHEWS

WOODLAND CT

WOODLAND CT

68TH

HILL AV

84TH

82ND

80TH

78TH

INDIAN TRAIL

INVERARY

DR ELLEN

ELIZABETH LN

AVEMORE

GLENSHIRE DR

HICKORY CREEK DR

11W-25S

LINCOLN HWY

22

LIBRARY

23

10W-25S

84TH

RIVER RD

WOODVALE AV

HILLSIDE

80TH

79TH

78TH

Ann Rutledge

COVENTRY LN

HUNTER WOODS EST.

HUNTER WOODS

30

24

9W-25S

STEG

HUNTER CHASE DR

YORKTOWN

OLD N CHURCH

GEORGETOWN

Georgetown Commons

VIRGINIA

19

8W-25S

FRANKLIN AV

93RD

92ND

CHESTNUT CT

NEBRASKA

PFEIFFER RD

NEVADA

NEBRASKA CT A

NEBRASKA CT B

AZALEA RD

LARCH CT

CHESTNUT DR

LARCH RD

LARCH CT A

Old Plank Road Trail

(Under development)

TIMBER

RD

SEE COOK COUNTY PAGES 90-91

PHEASANT TRAIL LN

PRESTWICK

HAWTHORNE LN

MAPLE CT

FAIRWAY LN

GOLF CLUB DR

PRESTWICK

COLONY LN

ST

1. WILDROCK TER.
2. TULLAMORE TER.
3. BRIARBRANCH TER.
4. PLEASANT TER.
5. N. WINDMERE CIR.
6. S. WINDMERE CIR.
7. KNOLLWOOD CIR.
8. HEDGEWICK CT.
9. BURLWOOD CT.
10. CANDLEGATE CIR.

BLACKTHORN

TIMBER

1 **2** **5**

3 **4** **6**

HUNTSBRIDGE

19 **18**

FIRTH CT

TIMBER LN

PRESTWICK COUNTRY CLUB

HERITAGE CT

COLONY LN

BOYER DR

HEATHER LN

DURHAM DR

FRANKFORT TWP.

RICH TWP.

20 **21**

ABERDEEN

HUNTSBRIDGE RD

30

27

SAUK TRAIL

26

10W-26S

GLEN EAGLES CT

GLEN EAGLES DR

HIGHLAND

STUART

SAINT

ASHTON

GLENBARR LN

DRUMMORE LN

ANDREWS WAY

ABBOTS

SFORD LN

AYSHIRE

25

PLYMOUTH

TROON CIR

MILTON BRIDGE TER

SHETLAND

TROON TRAIL

9W-26S

11. BRUSHWOOD DR
12. THISTLE CT.
13. HEARTSIDE RD.
14. HEATHERMEAD RD.
15. PRAIRIE RD.
16. GREEN SWARD WAY
17. HICKORY GLEN
18. CHAPPARAL TER.
19. WOODBINE TER.
20. IVYLOG TER.
21. THORNTREE TER.

8W-26S

11W-26S

80TH AV

Hickory Creek

FRANKFORT JR. H.S.

LARAWAY

FOREST RD

SAUK TRAIL

PRESERVE

SOUTHWICK CT

SOUTHWICK

SOUTHWICK AVE

Airport

34

V-27S

MUSTANG RD

PFEIFFER RD

35

10W-27S

80TH AV

36

9W-27S

HARLEM AV

31

8W-27S

T 35 N

STEGER RD

T 34 N

Frankfort Cem.

SEE PAGES 26-27

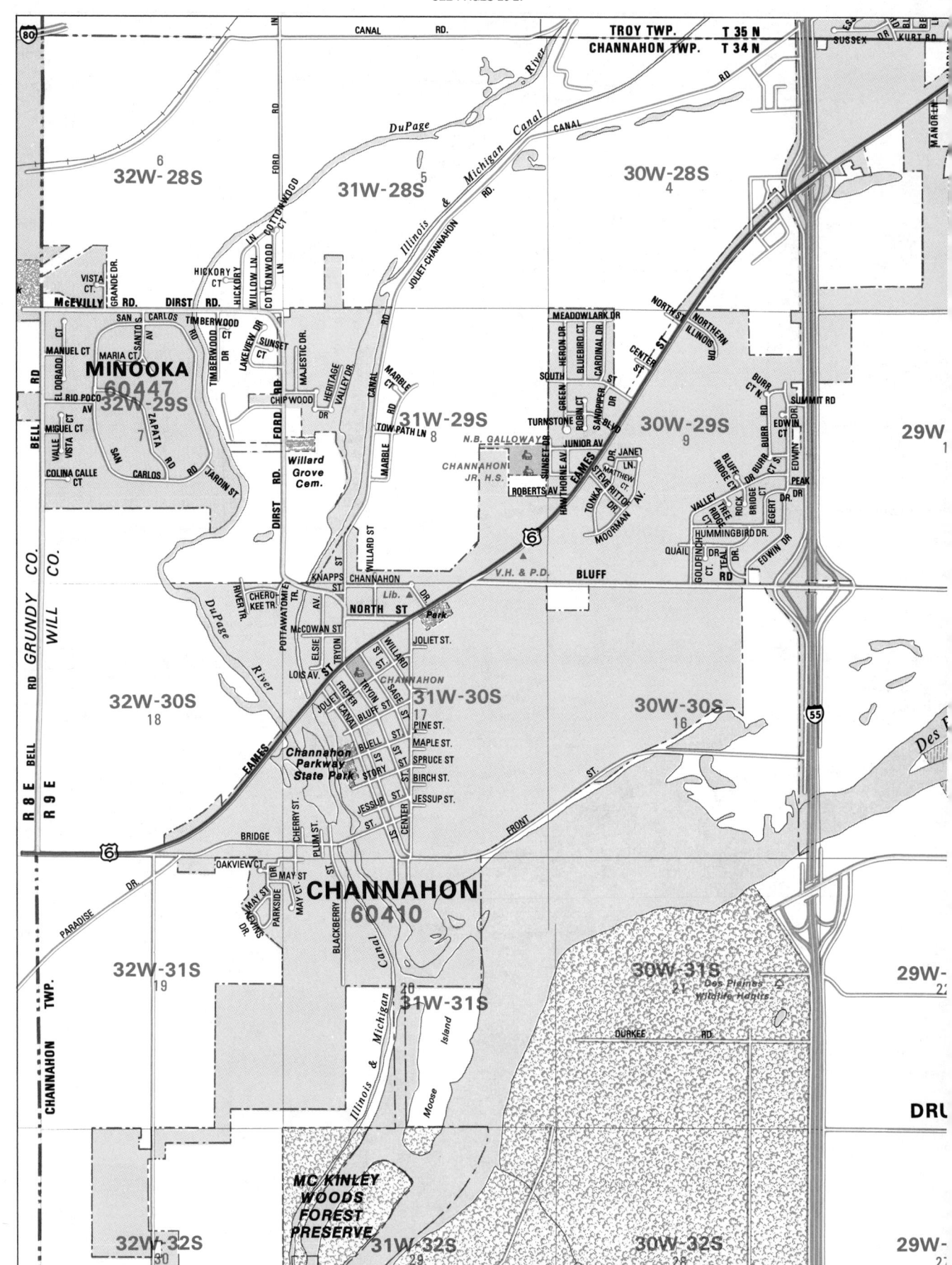

JOLIET
60436

29W-28S

28W-28S
2

27W-28S
1

26W-28S
6

Troutman Grove
Cem.

WILMINGTON

JOLIET RD

28W-29S
11

29S

OLIN RD

MILLSDALE 12 RD

27W-29S

26W-29S
7

CRAIG RD.
MILLSDALE RD.

PATTERSON RD.

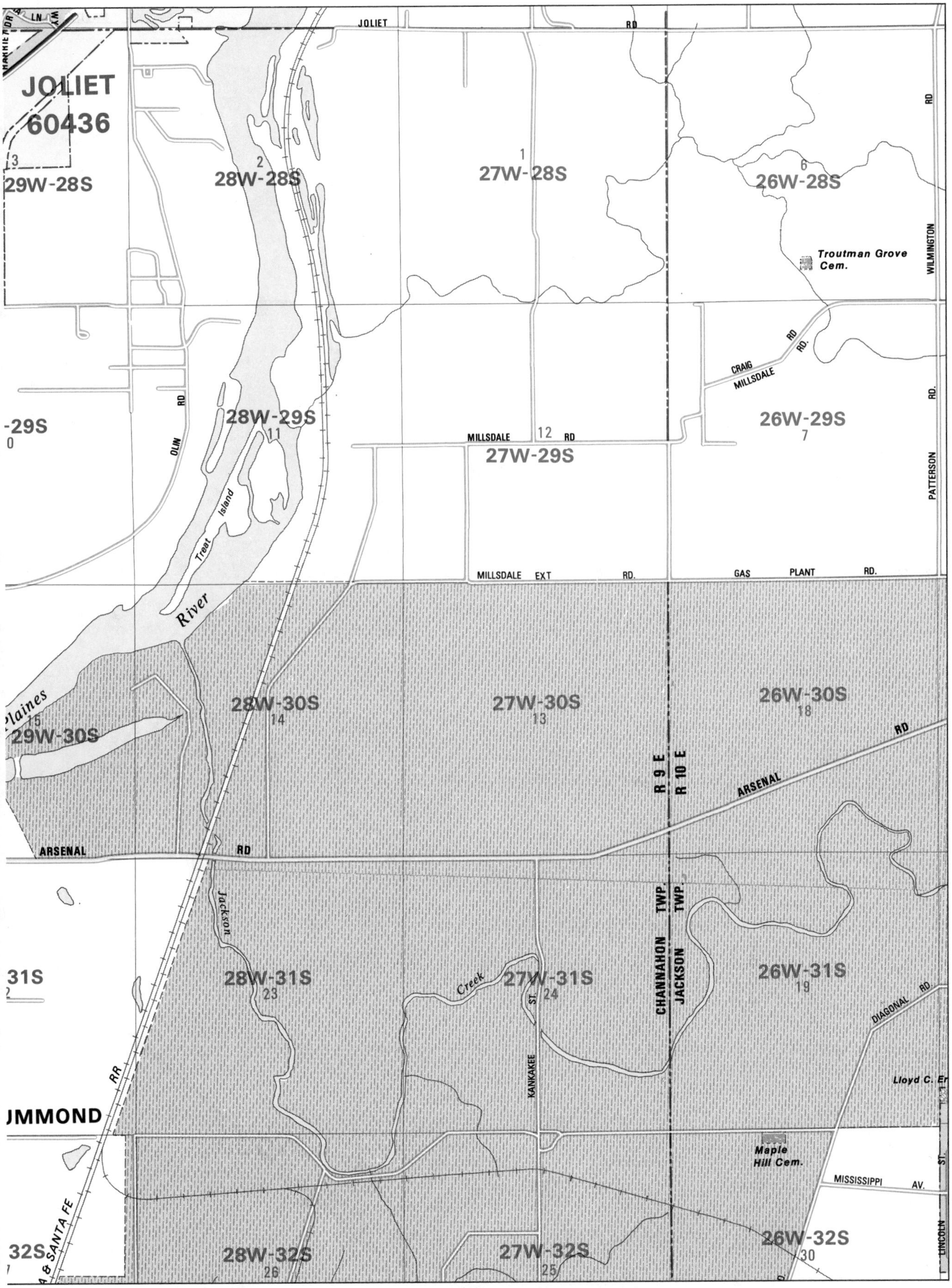

Treat Island

River

MILLSDALE EXT RD.

GAS PLANT RD.

SEE PAGES 36-37

Plaines
15

29W-30S

28W-30S
14

27W-30S
13

26W-30S
18

R 9 E
R 10 E

ARSENAL RD

ARSENAL RD

31S
2

Jackson

28W-31S
23

Creek

27W-31S
24

KANKAKEE ST.

CHANNAHON TWP.
JACKSON TWP.

26W-31S
19

DIAGONAL RD.

Lloyd C. Er

JMMOND

R.R.

Maple
Hill Cem.

MISSISSIPPI AV.

LINCOLN ST.

A & SANTA FE

32S

28W-32S
26

27W-32S
25

26W-32S
30

SEE PAGES 28-29

SEE PAGES 34-35

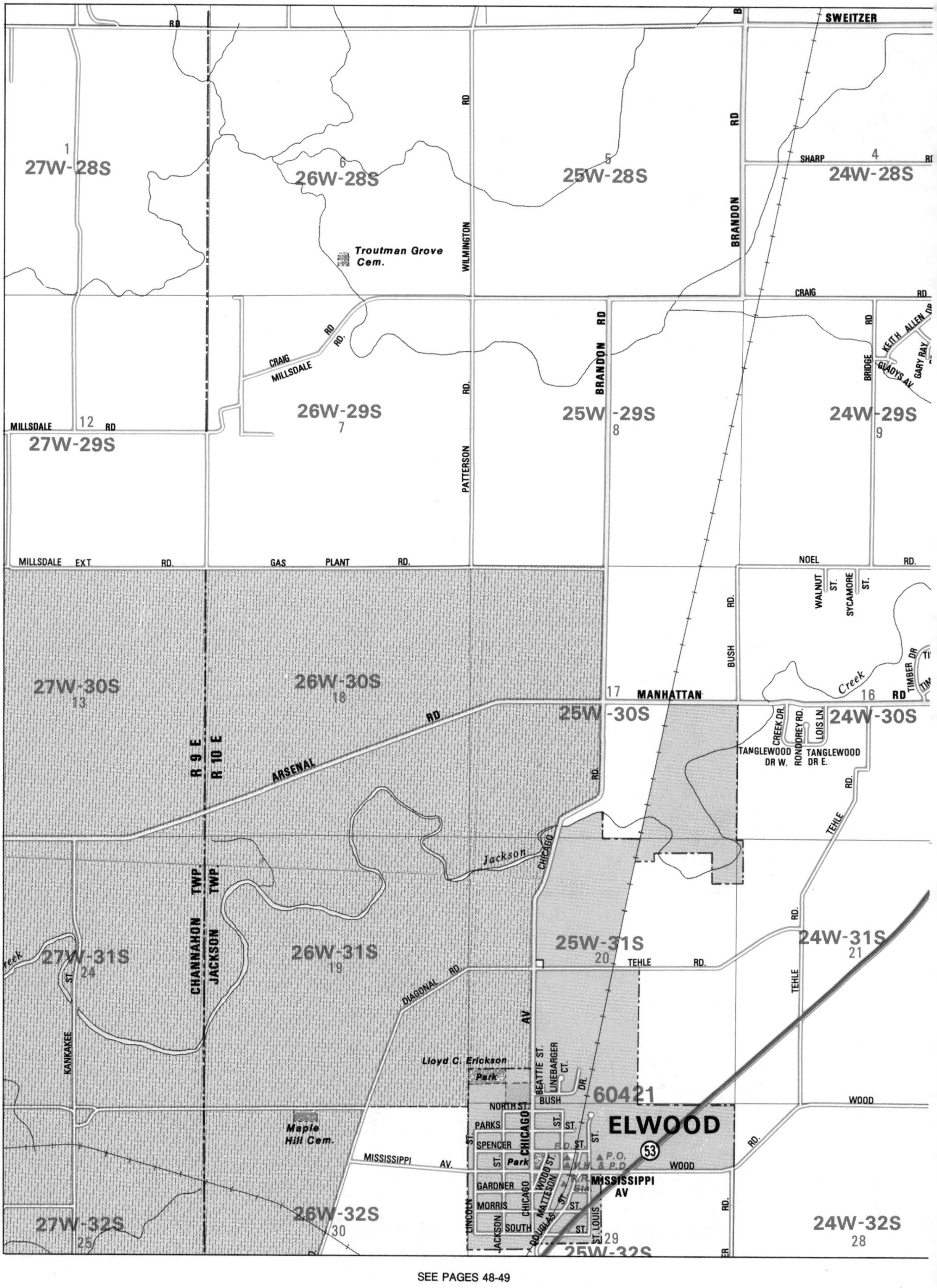

SWEITZER

27W-28S 1

26W-28S 6

25W-28S 5

24W-28S 4

SHARP RD

BRANDON RD

Troutman Grove Cem.

WILMINGTON

CRAIG RD

CRAIG RD

MILLSDALE RD

26W-29S 7

25W-29S 8

24W-29S 9

BRANDON RD

BRIDGE RD

KEITH ALLEN DR

GLADYS AV

GARY RAY

MILLSDALE RD 12

27W-29S

PATTERSON RD

NOEL RD

WALNUT ST

SYCAMORE ST

MILLSDALE EXT. RD.

GAS PLANT RD.

27W-30S 13

26W-30S 16

25W-30S 17 MANHATTAN

BUSH RD

Creek

TIMBER DR

24W-30S 16

R 9 E

R 10 E

ARSENAL RD

CREEK DR

RONDOREY RD

LOIS LN

TANGLEWOOD DR W.

TANGLEWOOD DR E.

Jackson

CHICAGO

CHANNAHON TWP.

JACKSON TWP.

TEHLE RD

Creek

27W-31S 24

26W-31S 19

25W-31S 20

24W-31S 21

KANKAKEE ST

DIAGONAL RD.

AV

TEHLE RD.

TEHLE RD

Lloyd C. Erickson Park

BEATTIE ST.

LINEBARGER CT.

BUSH DR.

60421

WOOD

Maple Hill Cem.

NORTH ST.

PARKS ST.

CHICAGO

ST.

ST.

ELWOOD

53

SPENCER ST.

Park

F.D. ST.

ST.

P.O.

MISSISSIPPI AV.

GARDNER ST.

WOOD ST.

CHICAGO

MATTESON ST.

ST. LOUIS ST.

MISSISSIPPI AV

WOOD RD

MORRIS ST.

LINCOLN ST.

JACKSON ST.

SOUTH ST.

DOUGLAS ST.

27W-32S 25

26W-32S 30

25W-32S 29

24W-32S 28

SEE PAGES 48-49

RD T 35 N JOLIET TWP. NORTH TOWN LINE RD

T 34 N JACKSON TWP.

Jackson

(53)

RD

3 2 BERNHARD RD. STONE 1 RD

23W-28S **22W-28S** **21W-28S**

RIDGE

ROWELL

CHERRY HILL

RAYMOND

DR

EATON AV

Cem.

23W-29S **22W-29S** **21W-29S**

10 BREEN RD 11 12

(53)

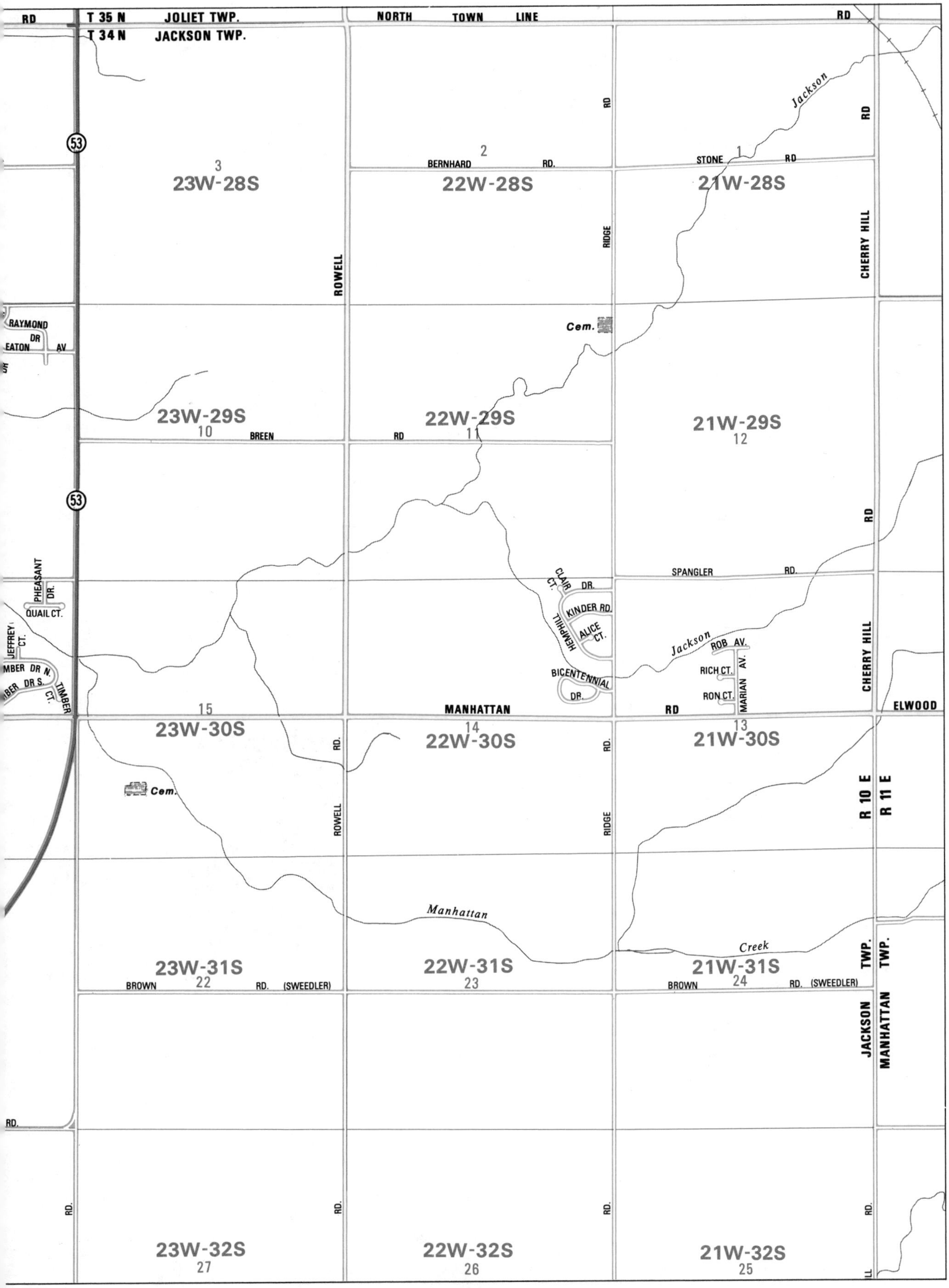

RD

PHEASANT

DR

QUAIL CT.

SPANGLER RD

CLAIR CT.

JEFFREY

CT.

KINDER RD.

DR.

MBER DR N

ALICE

TIMBER

CT.

TIMBER DR S.

CT.

Jackson ROB AV.

BICENTENNIAL

RICH CT.

DR.

RON CT.

MARIAN AV.

15 MANHATTAN RD ELWOOD

CHERRY HILL

23W-30S 14 **22W-30S** 13 **21W-30S**

RD.

RD.

RIDGE

ROWELL

R 10 E

R 11 E

Cem.

Manhattan

Creek

23W-31S **22W-31S** **21W-31S**

BROWN 22 RD. (SWEEDLER) 23 BROWN 24 RD. (SWEEDLER)

JACKSON TWP.

MANHATTAN TWP.

RD.

RD.

RD.

RD.

RD.

23W-32S **22W-32S** **21W-32S**

27 26 25

SEE PAGES 30-31

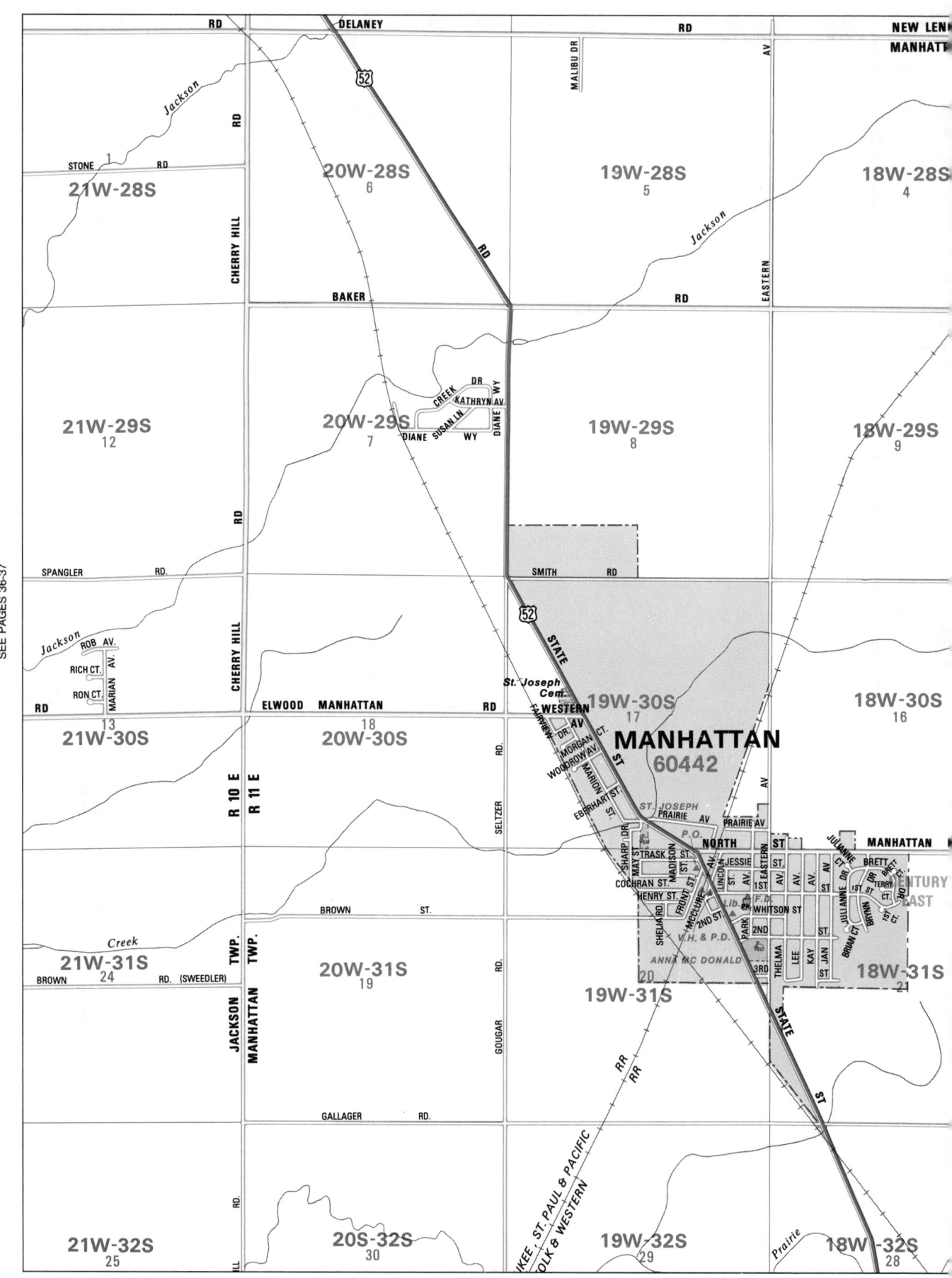

RD   DELANEY   RD   NEW LEN

MALIBU DR   AV   MANHATT

Jackson

52

21W-28S   20W-28S   19W-28S   18W-28S
STONE   1   RD   6   5   4

CHERRY HILL RD

Jackson

BAKER   RD

EASTERN

21W-29S   20W-29S   19W-29S   18W-29S
12   7   8   9
CREEK   DR
KATHRYN AV
DIANE SUSAN LN WY
DIANE WY

SPANGLER   RD.   SMITH   RD

SEE PAGES 36-37

52   STATE

Jackson   ROB AV.   St. Joseph
RICH CT. MARIAN AV.   Cem.   19W-30S   18W-30S
RON CT.   WESTERN AV   17   16
RD   ELWOOD MANHATTAN RD   FAIRVIEW ST
21W-30S   13   20W-30S 18   MANHATTAN
21W-30S   20W-30S   60442
MORGAN CT
WOODROW AV MARION ST

R 10 E R 11 E   SELTZER RD   EBERHART ST   PRAIRIE AV
St. Joseph
Prairie AV
P.O.
PRAIRIE AV
NORTH   ST   MANHATTAN
SHARP DR MAY ST TRASK JESSIE   JULIANNE DR BRETT
COCHRAN ST MADISON ST LINCOLN EASTERN AV BRETT CT
HENRY ST FRONT ST ST 1ST AV AV TERRY CT
SHELIA RD McCLURE AV BRIAN CT
2ND ST Lib. F.D. WHITSON ST JULIANNE DR ENTURY
PARK 2ND BRYAN CT EAST
BROWN   ST   V.H. & P.D. 3RD THELMA 1ST CT
Creek ANNA MC DONALD LEE KAY JAN ST
21W-31S   20W-31S 20   18W-31S
24   19   21
BROWN   RD. (SWEEDLER)   19W-31S

JACKSON TWP. MANHATTAN TWP. GOUGAR RD

BROWN   ST.   STATE ST

GALLAGER   RD.

RR
RR

21W-32S   20S-32S   IKEE ST. PAUL & PACIFIC 19W-32S Prairie 18W-32S
25   30   OLK & WESTERN 29   28

RD

SEE PAGES 50-51

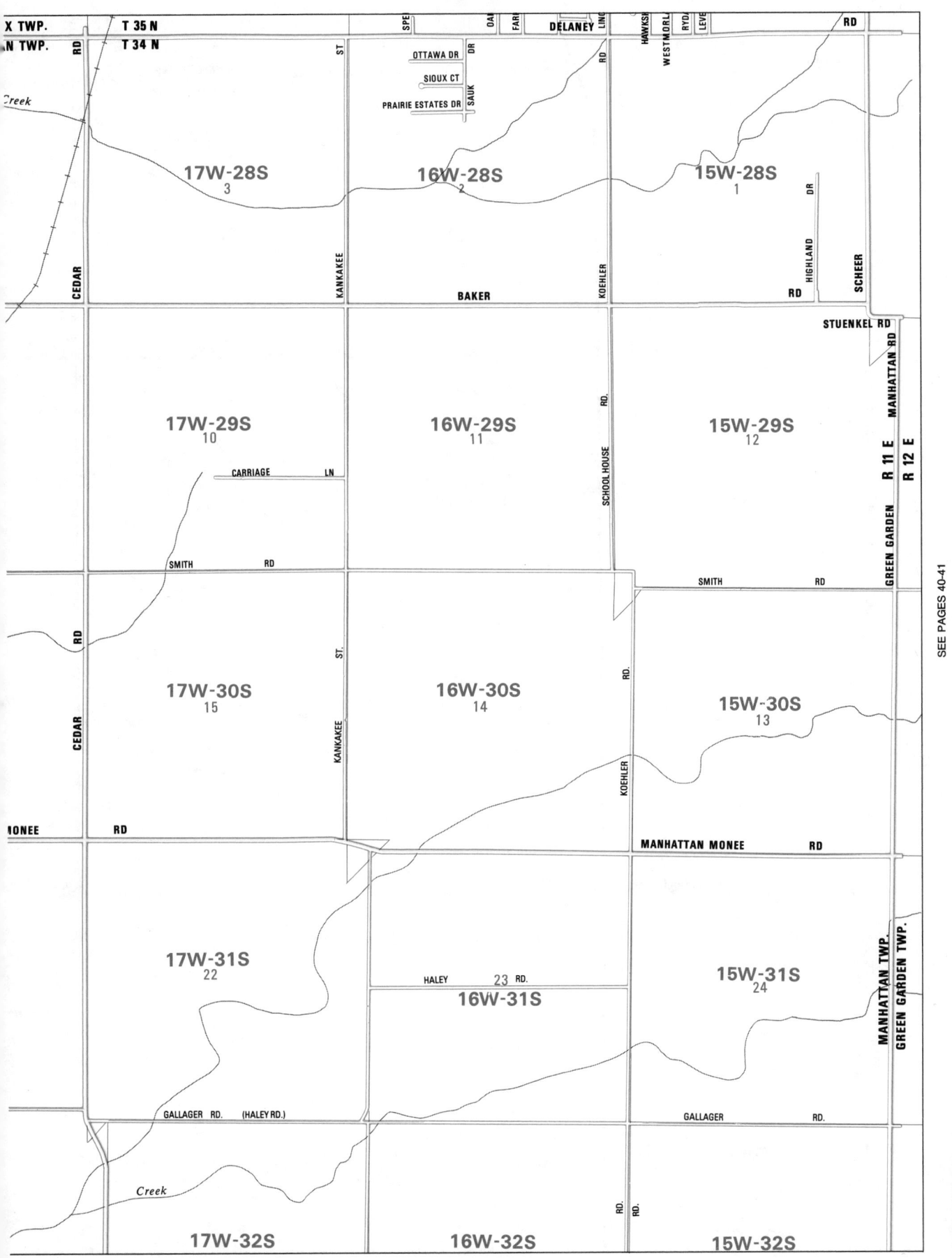

SEE PAGES 32-33

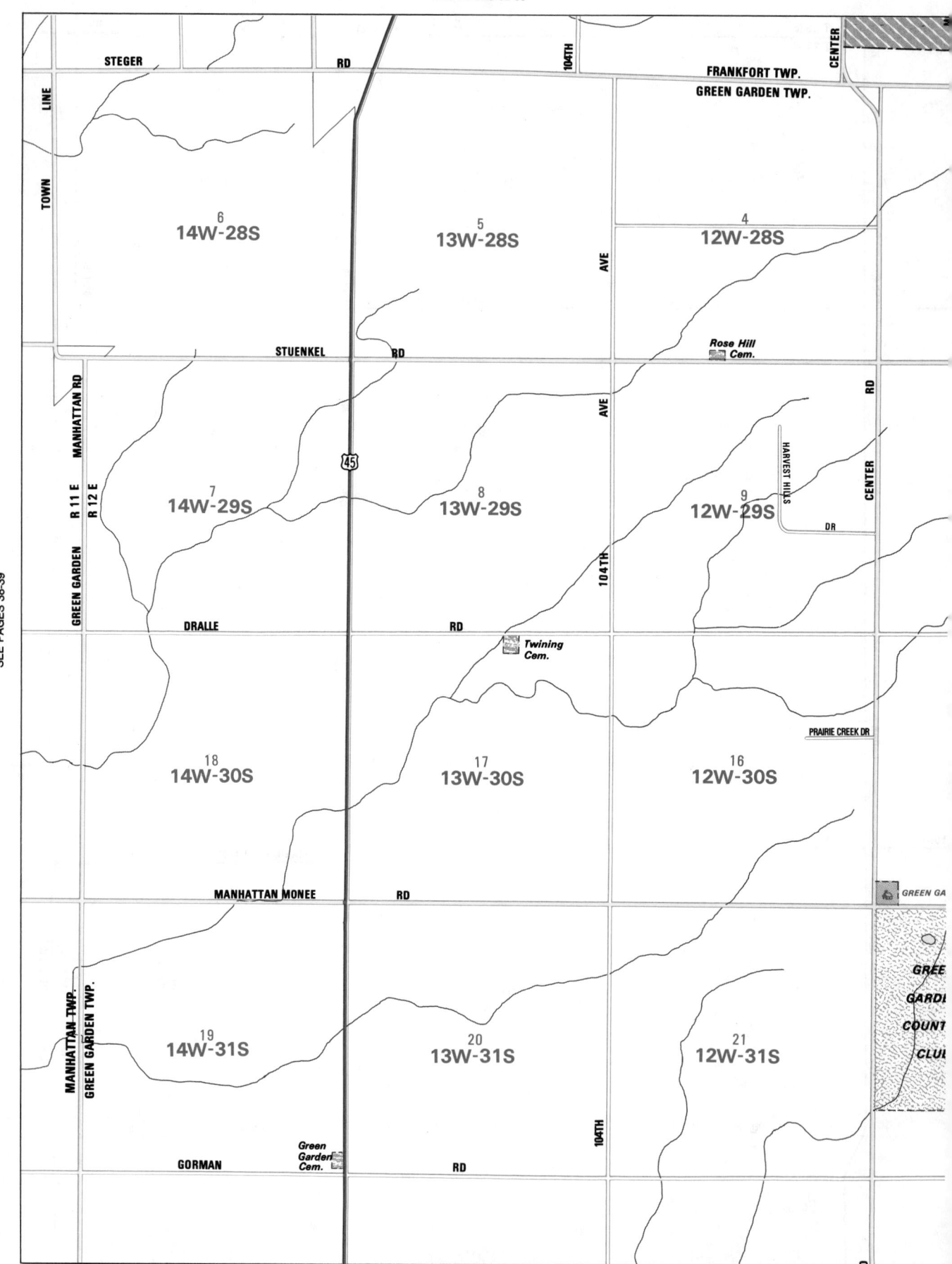

STEGER   RD

104TH

CENTER

FRANKFORT TWP.
GREEN GARDEN TWP.

TOWN LINE

6
14W-28S

5
13W-28S

AVE

4
12W-28S

STUENKEL   RD

AVE

Rose Hill
Cem.

MANHATTAN RD

GREEN GARDEN

R 11 E

R 12 E

45

7
14W-29S

8
13W-29S

9
12W-29S

HARVEST HILLS

CENTER   RD

DR

104TH

SEE PAGES 38-39

DRALLE   RD

Twining
Cem.

PRAIRIE CREEK DR

18
14W-30S

17
13W-30S

16
12W-30S

MANHATTAN MONEE   RD

GREEN GA

MANHATTAN TWP.
GREEN GARDEN TWP.

19
14W-31S

20
13W-31S

21
12W-31S

GREE
GARDE
COUNT
CLU

104TH

GORMAN

Green
Garden
Cem.

RD

SEE PAGES 52-53

SEE PAGES 32-33

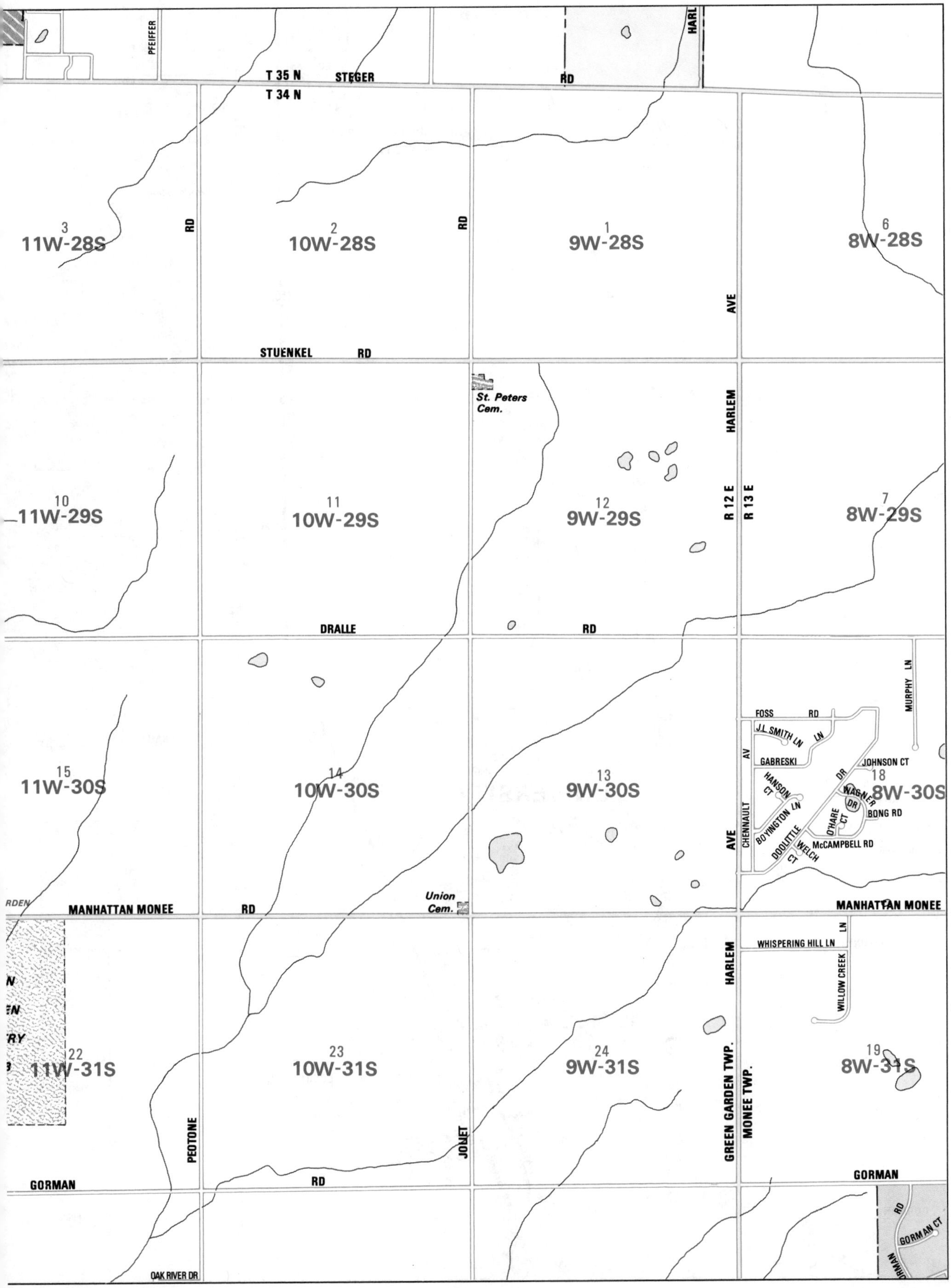

PEIFFER

HARL

T 35 N   STEGER
T 34 N   RD

RD

RD

^{3}
11W-28S

^{2}
10W-28S

^{1}
9W-28S

AVE

^{6}
8W-28S

STUENKEL   RD

St. Peters
Cem.

HARLEM

^{10}
11W-29S

^{11}
10W-29S

^{12}
9W-29S

R 12 E
R 13 E

^{7}
8W-29S

SEE PAGES 42-43

DRALLE   RD

MURPHY LN

FOSS   RD
J.L. SMITH LN
GABRESKI   LN
JOHNSON CT

^{15}
11W-30S

^{14}
10W-30S

^{13}
9W-30S

CHENNAULT   AV
HANSON CT
BOYINGTON LN
DOOLITTLE
WELCH CT

WAGNER DR
DR
O'HARE CT
BONG RD
McCAMPBELL RD

^{18}
8W-30S

AVE

RDEN

MANHATTAN MONEE   RD

Union
Cem.

MANHATTAN MONEE

N
EN
RY

^{22}
11W-31S

^{23}
10W-31S

JOLIET

^{24}
9W-31S

HARLEM
GREEN GARDEN TWP.
MONEE TWP.

WHISPERING HILL LN

WILLOW CREEK LN

^{19}
8W-31S

PEOTONE

GORMAN   RD

GORMAN

RD
GORMAN CT
NMAN

OAK RIVER DR

SEE PAGES 52-53

SEE COOK COUNTY PAGES 92-93

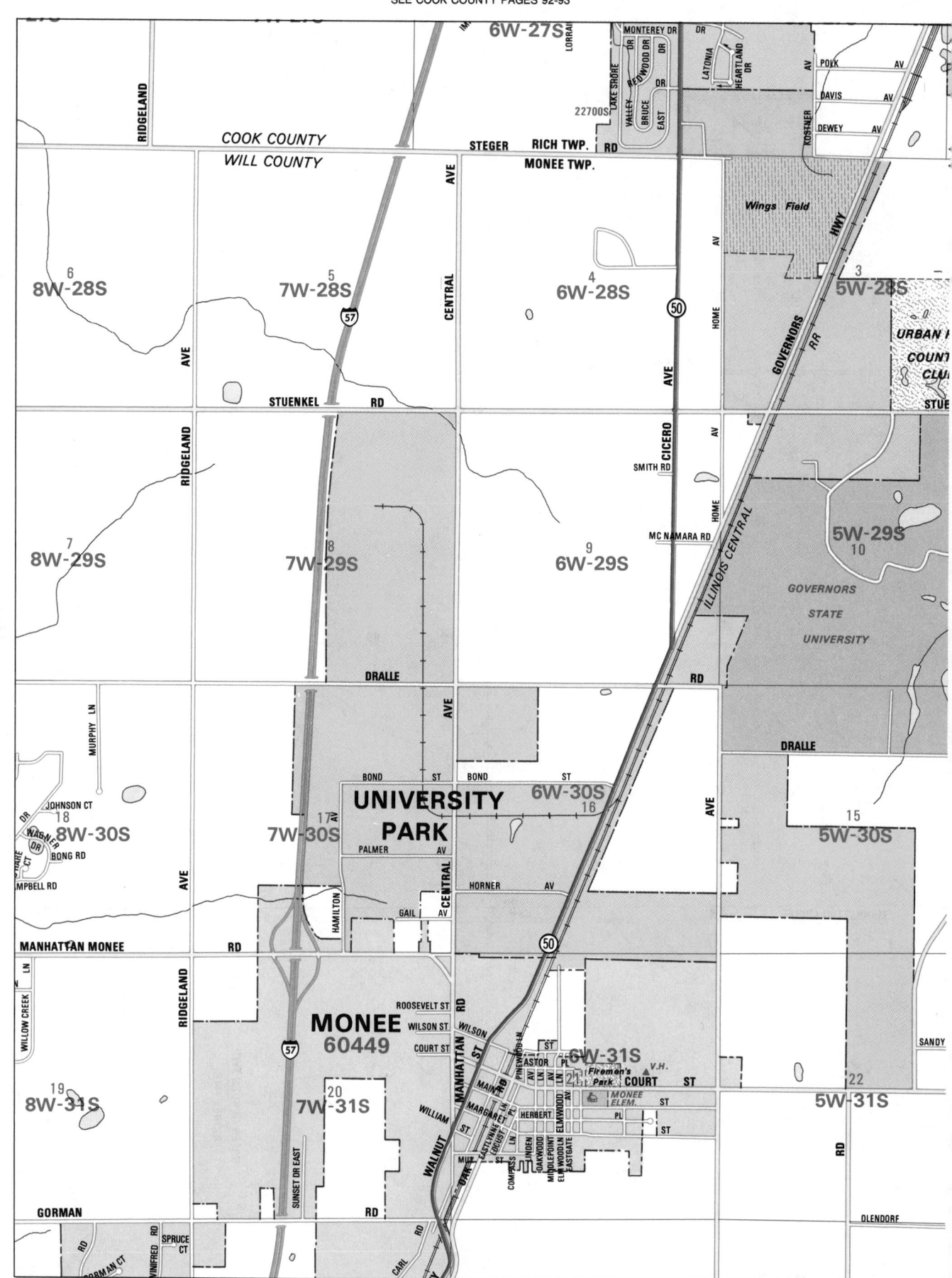

6W-27S

MONTEREY DR

LATONIA

HEARTLAND DR

POLK   AV

DAVIS   AV

DEWEY   AV

22700S

LAKE SHORE DR

VALLEY DR

REDWOOD DR

BRUCE DR

EAST

Wings Field

STEGER   RICH TWP.   RD

MONEE TWP.

COOK COUNTY

WILL COUNTY

8W-28S   7W-28S   6W-28S   5W-28S

6   5   4   3

RIDGELAND

CENTRAL   AVE

HOME   AV

50

GOVERNORS   R.R.

KOSTNER AV

HWY

URBAN
COUNTY
CLUB
STUE

SEE PAGES 40-41

STUENKEL   RD

RIDGELAND

AVE

ILLINOIS CENTRAL

CICERO   AV

HOME   AV

SMITH RD

MC NAMARA RD

GOVERNORS
STATE
UNIVERSITY

8W-29S   7W-29S   6W-29S   5W-29S

7   8   9   10

DRALLE

AVE

DRALLE

BOND   ST   BOND   ST

UNIVERSITY
PARK

6W-30S

16

AVE

5W-30S

15

8W-30S   7W-30S   5W-30S

18   17

JOHNSON CT

WAGNER DR

DR

SHORE CT

BONG RD

MPBELL RD

MURPHY LN

PALMER   AV

HAMILTON

GAIL

CENTRAL   AV

HORNER   AV

50

MANHATTAN MONEE   RD

RIDGELAND   AVE

WILLOW CREEK   LN

LN

57

MONEE
60449

ROOSEVELT ST

WILSON ST

COURT ST

MANHATTAN   RD

WILSON   ST

PINEWOOD LN

ASTOR   PL

MAIN   RD

6W-31S

21

Firemen's
Park

V.H.

MONEE
ELEM.

COURT   ST

8W-31S   7W-31S   5W-31S

19   20   22

SANDY

WILLIAM

MARGARET   ST

HERBERT   ELMWOOD

ST   ST

PL   ST

RD

WALNUT

SUNSET DR EAST

MUX

EASTLYNN LN

LOCUST   LN

LINDEN

OAKWOOD LN

MIDDLEPOINT

ELM WOOD LN

ELMWOOD LN

EASTGATE

COMPASS

PL

GORMAN   RD   OLENDORF

RD

WINIFRED RD

SPRUCE
CT

GORMAN CT

CARL

SEE PAGES 54-55

1W-27S

35 N
34 N   STEGER   RD

STEGER   RD

INDIANWOOD

BLACKHAWK
CENTER

Calvary
Cem.

235TH ST
235TH PL

4W-28S   3W-28S   2W-28S

St. Mary's

Thorn Creekwoods
Nature Center

PARK FOREST GOLF CENTER   RD

THORN   CREEK

WOODS

OAK HILL DR

PINE TRACE CT

Park Forest Comm. Gardens

NORFOLK

WOLPERS   RD

FOREST   LN
BROADVIEW

4W-29S   3W-29S   2W-29S
              12   7

Pine Lake Pk

EXCHANGE ST

UNIVERSITY PARK
60466

EXCHANGE   ST

CRETE MONEE H.S.

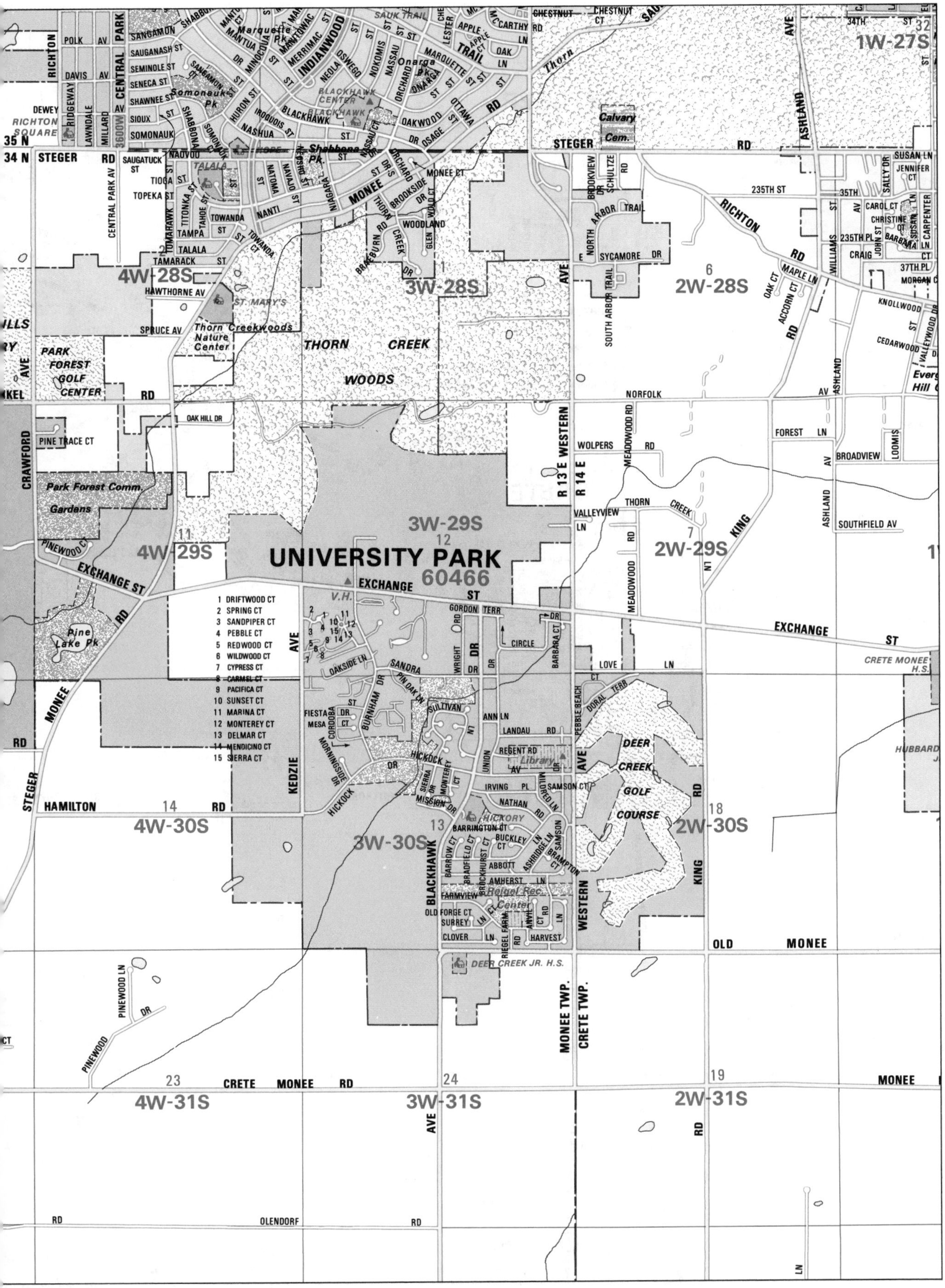

1 DRIFTWOOD CT
2 SPRING CT
3 SANDPIPER CT
4 PEBBLE CT
5 REDWOOD CT
6 WILDWOOD CT
7 CYPRESS CT
8 CARMEL CT
9 PACIFICA CT
10 SUNSET CT
11 MARINA CT
12 MONTEREY CT
13 DELMAR CT
14 MENDICINO CT
15 SIERRA CT

V.H.

Library

DEER CREEK GOLF COURSE

HAMILTON   14   RD

4W-30S   3W-30S   2W-30S
              13   18

Riegel Rec. Center

OLD   MONEE

Deer Creek Jr. H.S.

MONEE TWP.
CRETE TWP.

PINEWOOD LN

23   CRETE   MONEE   RD   24   19   MONEE

4W-31S   3W-31S   2W-31S

RD   OLENDORF   RD

SEE PAGES 44-45

SEE COOK COUNTY-PAGES 94-95

SEE PAGES 42-43

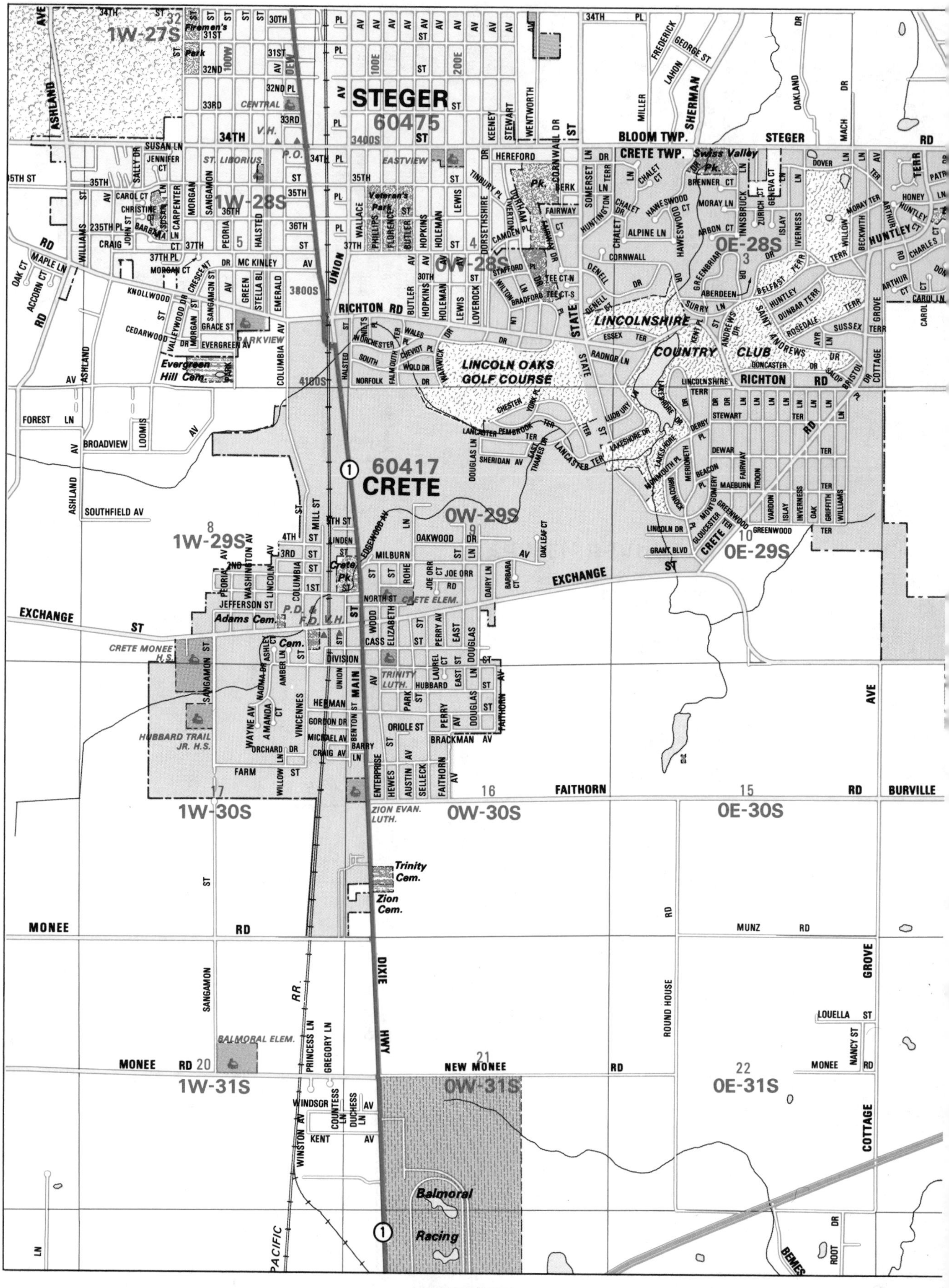

STEGER
60475

CRETE
60417

BLOOM TWP.
CRETE TWP.
STEGER

LINCOLNSHIRE
COUNTRY CLUB

LINCOLN OAKS
GOLF COURSE

Swiss Valley Pk.

Veteran's Park
Firemen's Park
CENTRAL
St. Liborius
Evergreen Hill Cem.
Parkview
Eastview

Crete Pk.
Adams Cem.
Cem.
CRETE MONEE H.S.
Hubbard Trail Jr. H.S.
Crete Elem.
Trinity Luth.
P.D. & F.D. V.H.

Zion Evan. Luth.
Trinity Cem.
Zion Cem.

Balmoral Elem.
NEW MONEE
Balmoral Racing

1W-27S
1W-28S   0W-28S   OE-28S
1W-29S   0W-29S   OE-29S
1W-30S   0W-30S   OE-30S
1W-31S   0W-31S   OE-31S

MONEE   RD
DIXIE   HWY
PACIFIC   RR
EXCHANGE   ST
RICHTON   RD
STATE   ST
CRETE   RD
FAITHORN   RD
BURVILLE
COTTAGE   GROVE   AVE
MUNZ   RD
ASHLAND   AVE
HUNTLEY

SEE PAGES 56-57

SEE COOK COUNTY PAGES 94-95

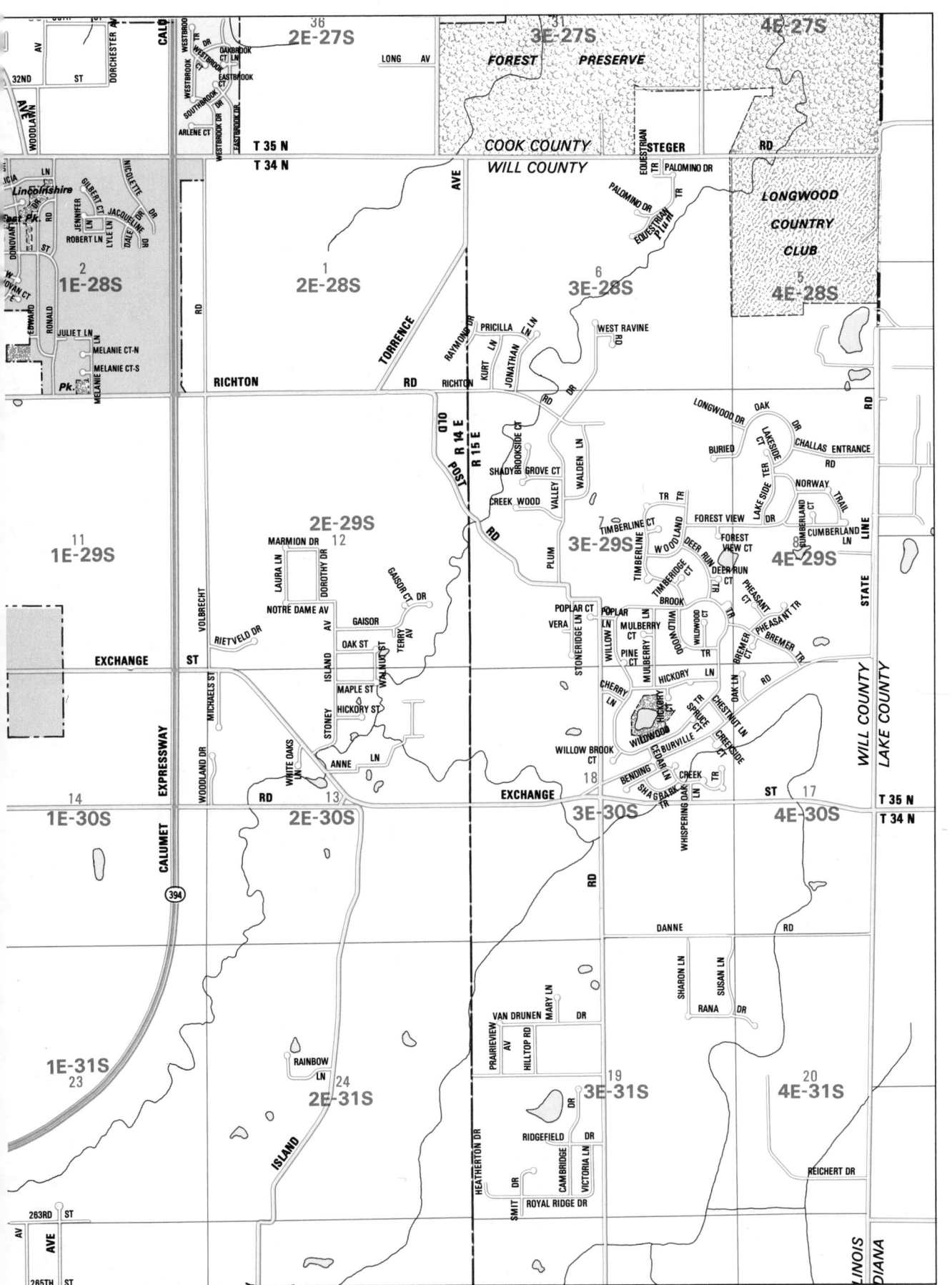

SEE PAGES 56-57

SEE PAGES 34-35

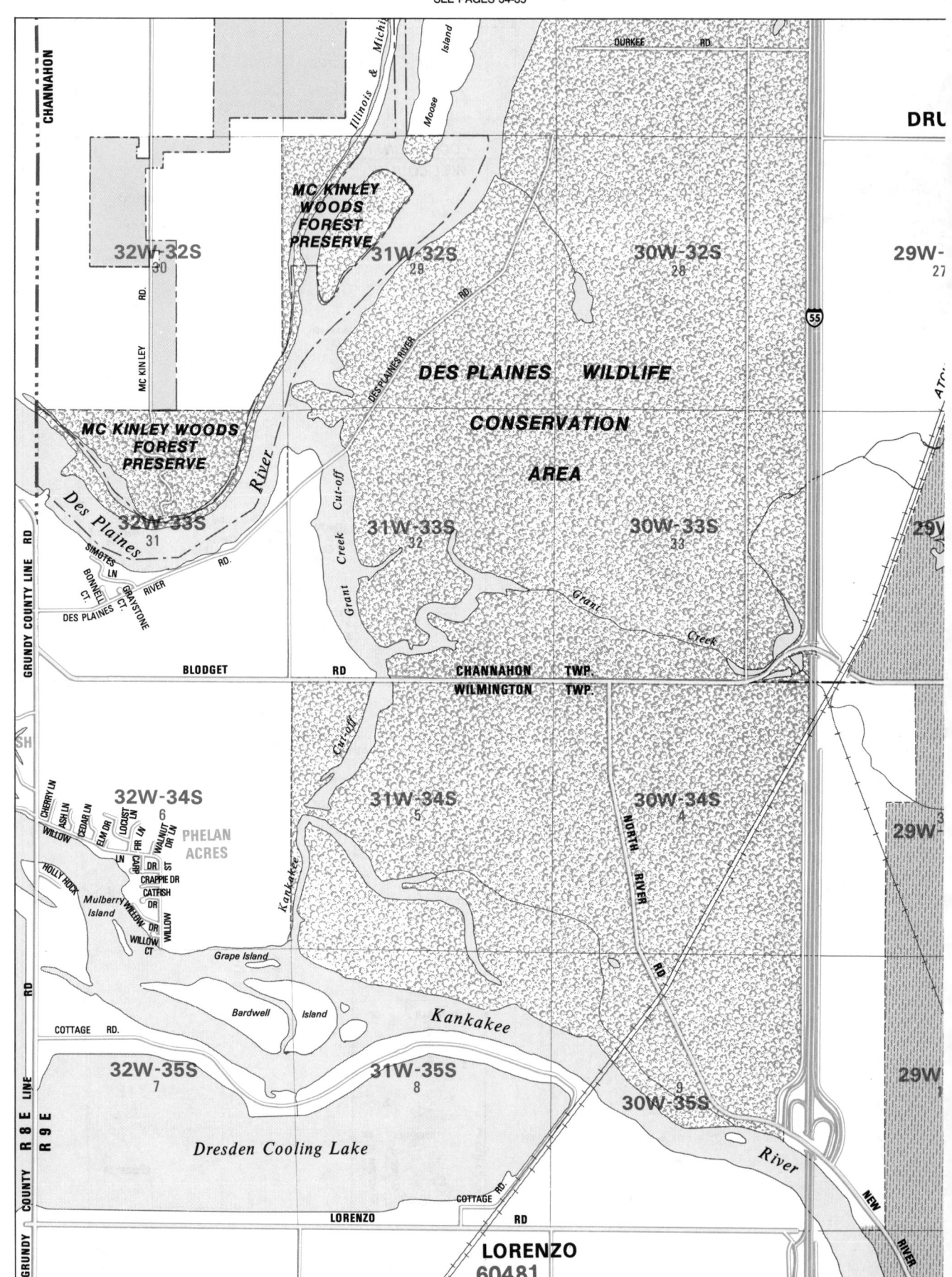

SEE PAGES 58-59

UMMOND

RR

ATCHISON, TOPEKA & SANTA FE

KANKAKEE

Lloyd C. Er

Maple
Hill Cem.

MISSISSIPPI AV.

LINCOLN ST.

32S

28W-32S
26

27W-32S
25

26W-32S
30

OLD WILMINGTON RD.

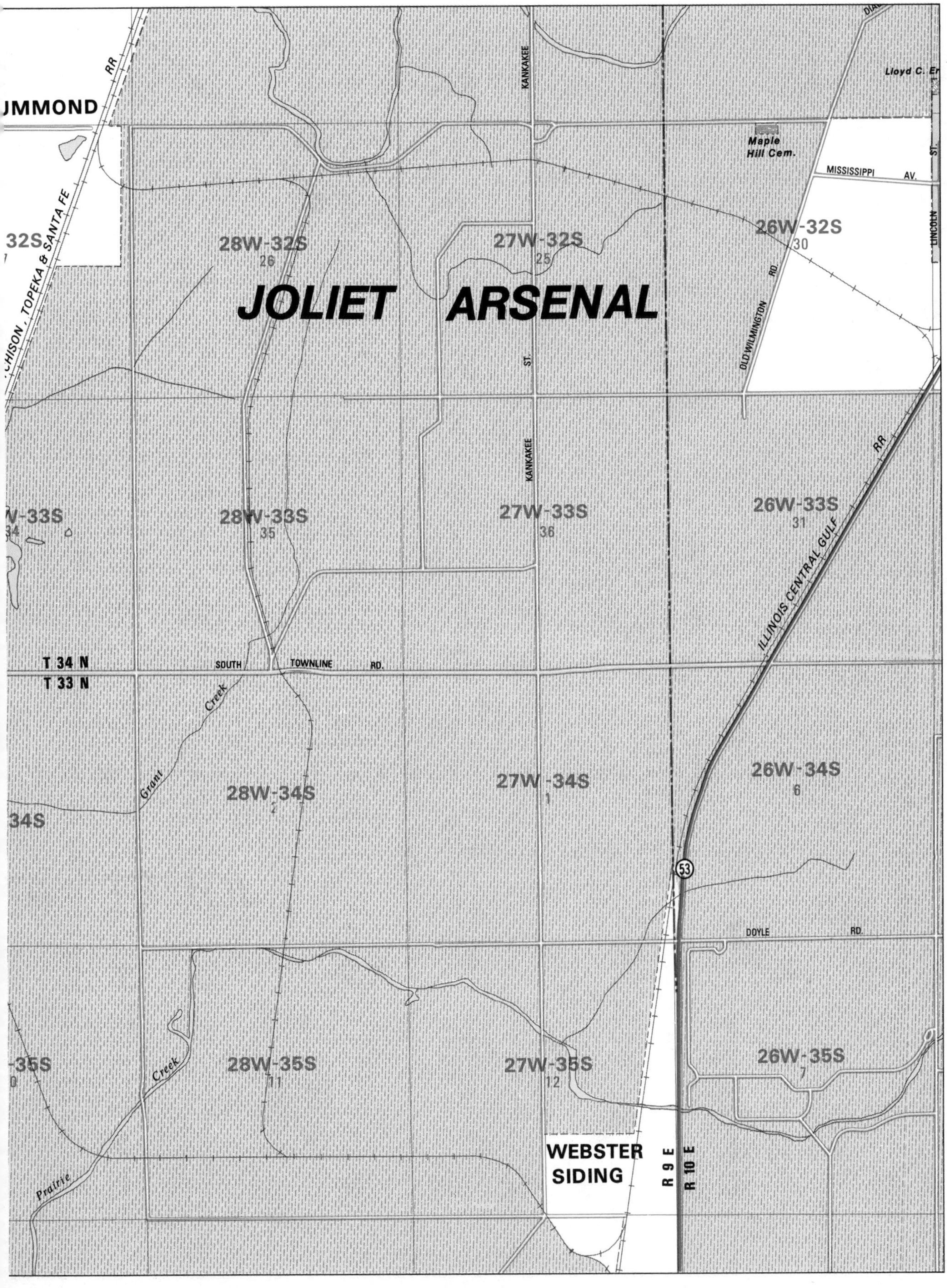

JOLIET   ARSENAL

KANKAKEE ST.

W-33S
34

28W-33S
35

27W-33S
36

26W-33S
31

ILLINOIS CENTRAL GULF RR

T 34 N
T 33 N

SOUTH   TOWNLINE   RD.

Creek

Grant

28W-34S
2

27W-34S
1

26W-34S
6

34S

(53)

DOYLE   RD.

Creek

28W-35S
11

27W-35S
12

26W-35S
7

35S

Prairie

WEBSTER
SIDING

R 9 E

R 10 E

SEE PAGES 36-37

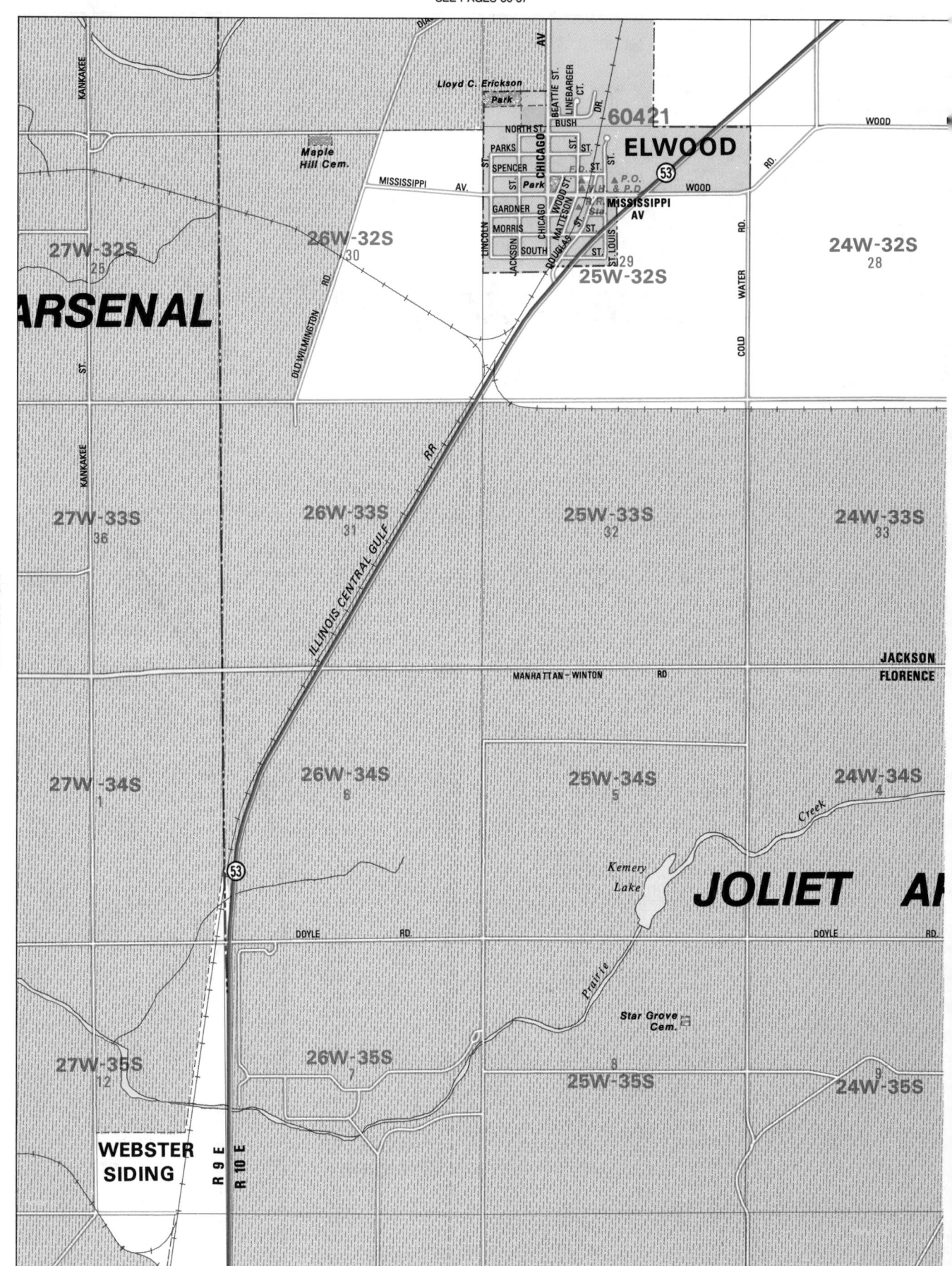

SEE PAGES 46-47

SEE PAGES 60-61

SEE PAGES 36-37

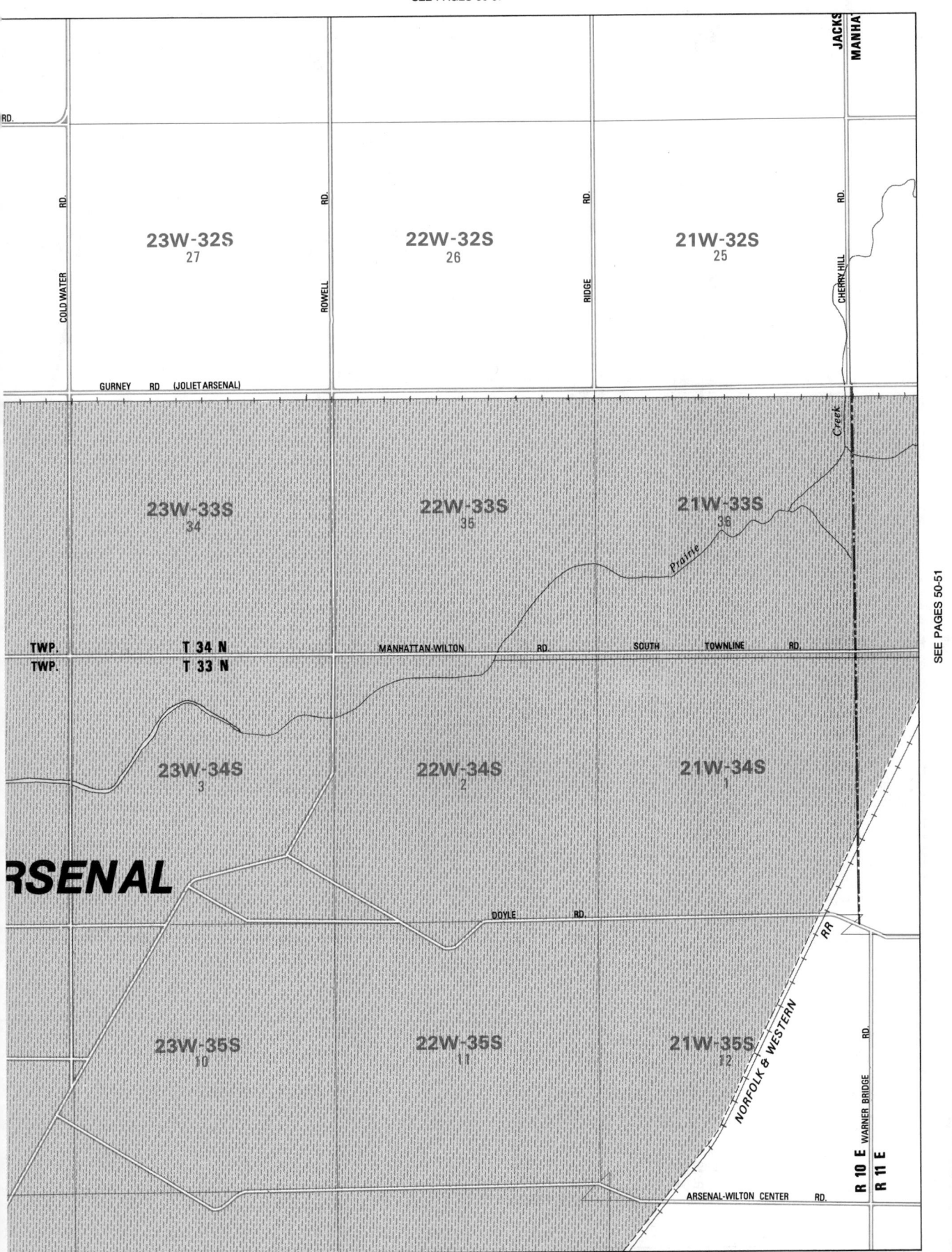

SEE PAGES 50-51

SEE PAGES 60-61

SEE PAGES 38-39

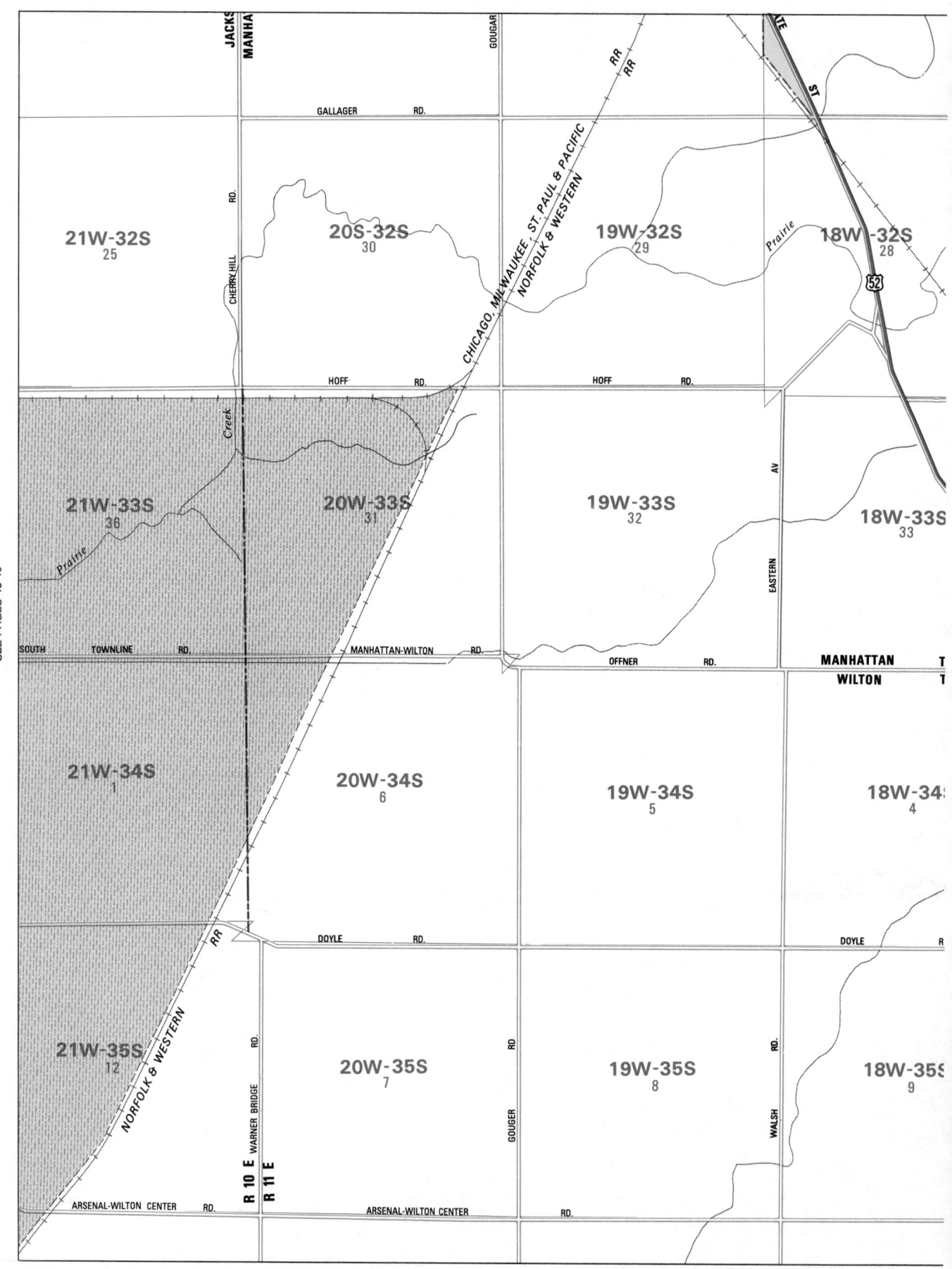

SEE PAGES 48-49

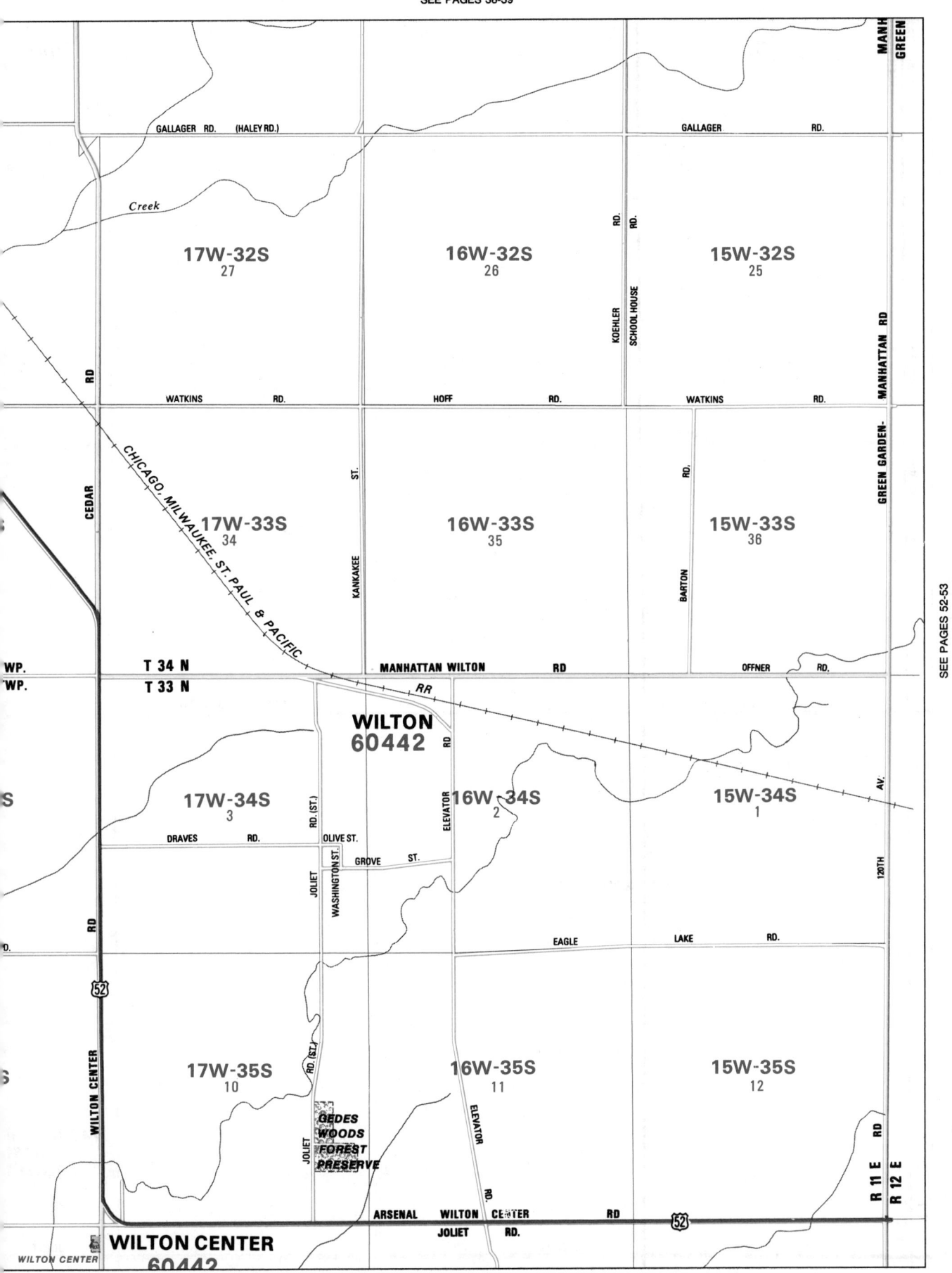

GALLAGER RD. (HALEY RD.)

GALLAGER   RD.

Creek

17W-32S
27

16W-32S
26

15W-32S
25

KOEHLER   RD.

SCHOOL HOUSE   RD.

MANHATTAN   RD.

WATKINS   RD.

HOFF   RD.

WATKINS   RD.

CEDAR

CHICAGO, MILWAUKEE, ST. PAUL & PACIFIC

17W-33S
34

ST.

KANKAKEE

16W-33S
35

15W-33S
36

RD.

BARTON   RD.

GREEN GARDEN-

WP.

WP.

T 34 N

T 33 N

MANHATTAN WILTON   RD

OFFNER   RD.

RR

WILTON
60442

RD.

17W-34S
3

RD. (ST.)

ELEVATOR

16W-34S
2

15W-34S
1

AV.

DRAVES   RD.

OLIVE ST.

JOLIET

WASHINGTON ST.

GROVE   ST.

120TH

EAGLE   LAKE   RD.

RD.

RD. (ST.)

17W-35S
10

JOLIET

16W-35S
11

ELEVATOR

15W-35S
12

R 11 E   RD

R 12 E

GEDES
WOODS
FOREST
PRESERVE

ARSENAL   WILTON CENTER   RD

JOLIET   RD.

52

WILTON CENTER

WILTON CENTER
60442

WILTON CENTER

SEE PAGES 52-53

SEE PAGES 40-41

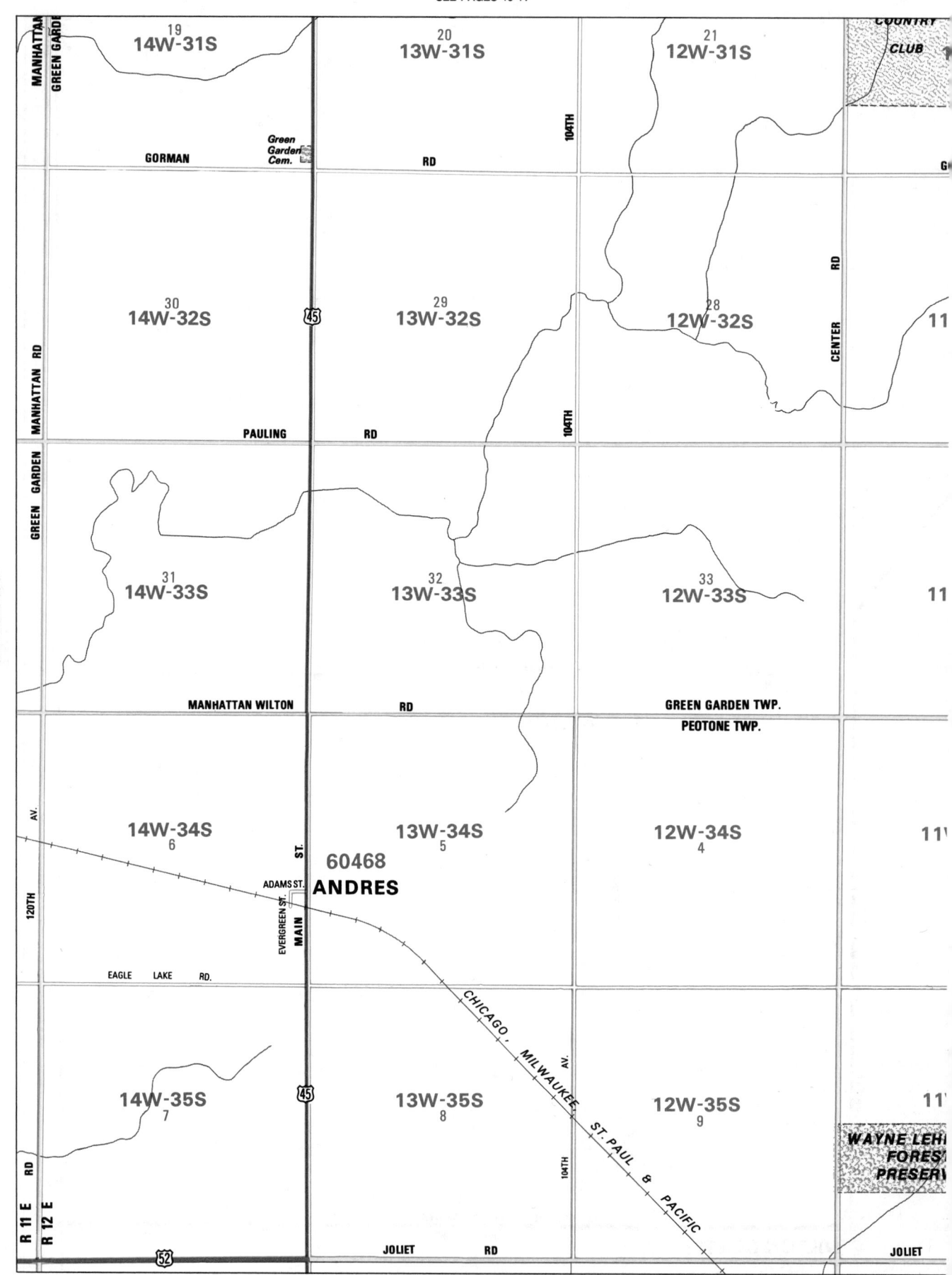

SEE PAGES 50-51

SEE PAGES 64-65

SEE PAGES 40-41

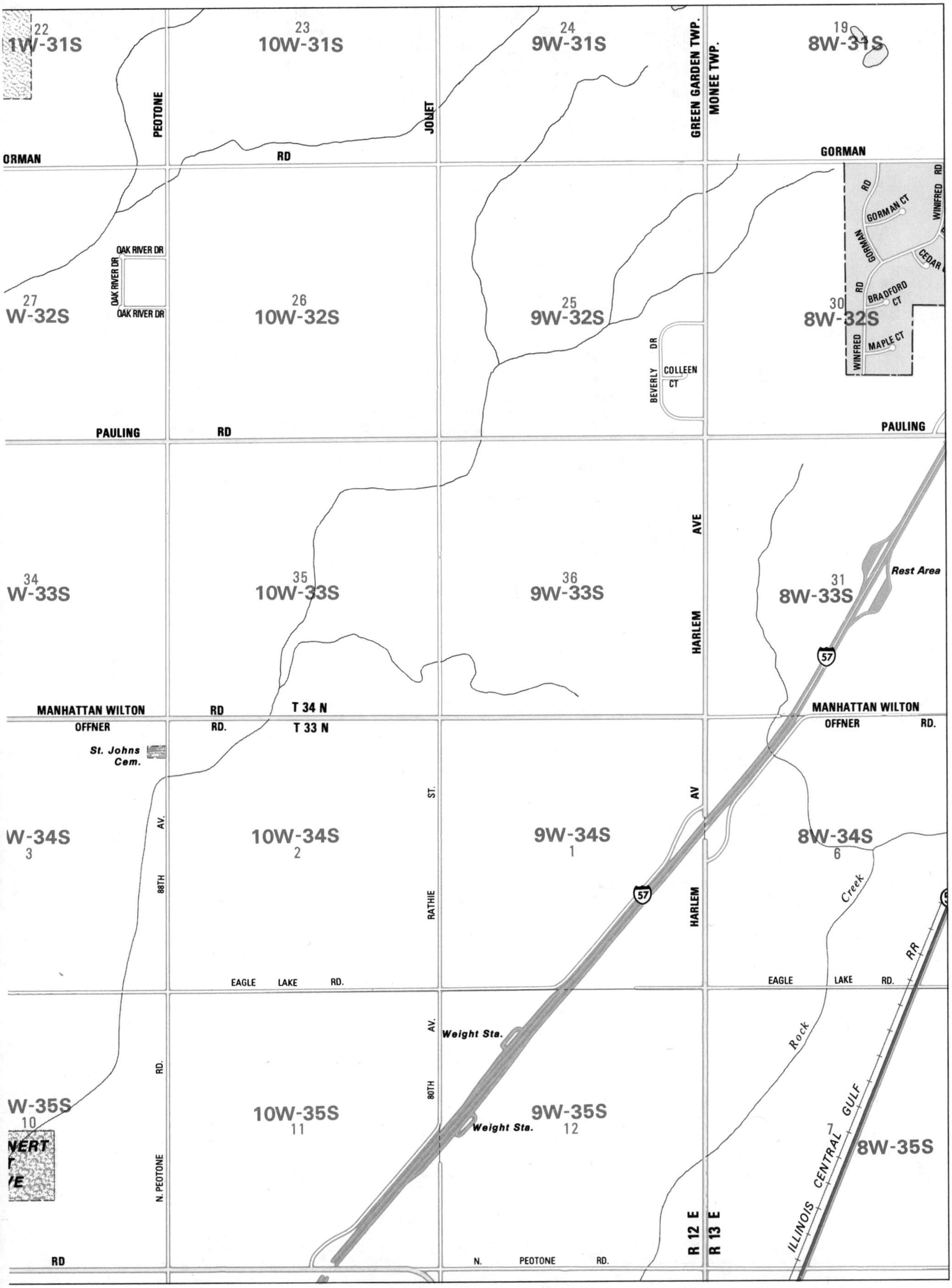

22
1W-31S

23
10W-31S

24
9W-31S

19
8W-31S

PEOTONE

JOLIET

GREEN GARDEN TWP.

MONEE TWP.

ORMAN

GORMAN

RD

GORMAN

RD

WINFRED RD

GORMAN CT

CEDAR

27
W-32S

OAK RIVER DR

OAK RIVER DR

26
10W-32S

25
9W-32S

30
8W-32S

BRADFORD CT

WINFRED

MAPLE CT

BEVERLY DR

COLLEEN CT

PAULING

RD

PAULING

34
W-33S

35
10W-33S

36
9W-33S

AVE

HARLEM

Rest Area

31
8W-33S

57

SEE PAGES 54-55

MANHATTAN WILTON

OFFNER

RD.

T 34 N

T 33 N

RD.

St. Johns Cem.

MANHATTAN WILTON

OFFNER

RD.

AV.

88TH

ST.

RATHIE

W-34S
3

10W-34S
2

9W-34S
1

AV

HARLEM

57

8W-34S
6

Creek

EAGLE LAKE RD.

EAGLE LAKE RD.

AV.

80TH

Weight Sta.

Rock

ILLINOIS CENTRAL GULF

R.R.

RD.

W-35S
10

N. PEOTONE

10W-35S
11

Weight Sta.

9W-35S
12

R 12 E

R 13 E

7

8W-35S

NERT

E

RD

N. PEOTONE RD.

SEE PAGES 64-65

SEE PAGES 42-43

SEE PAGES 52-53

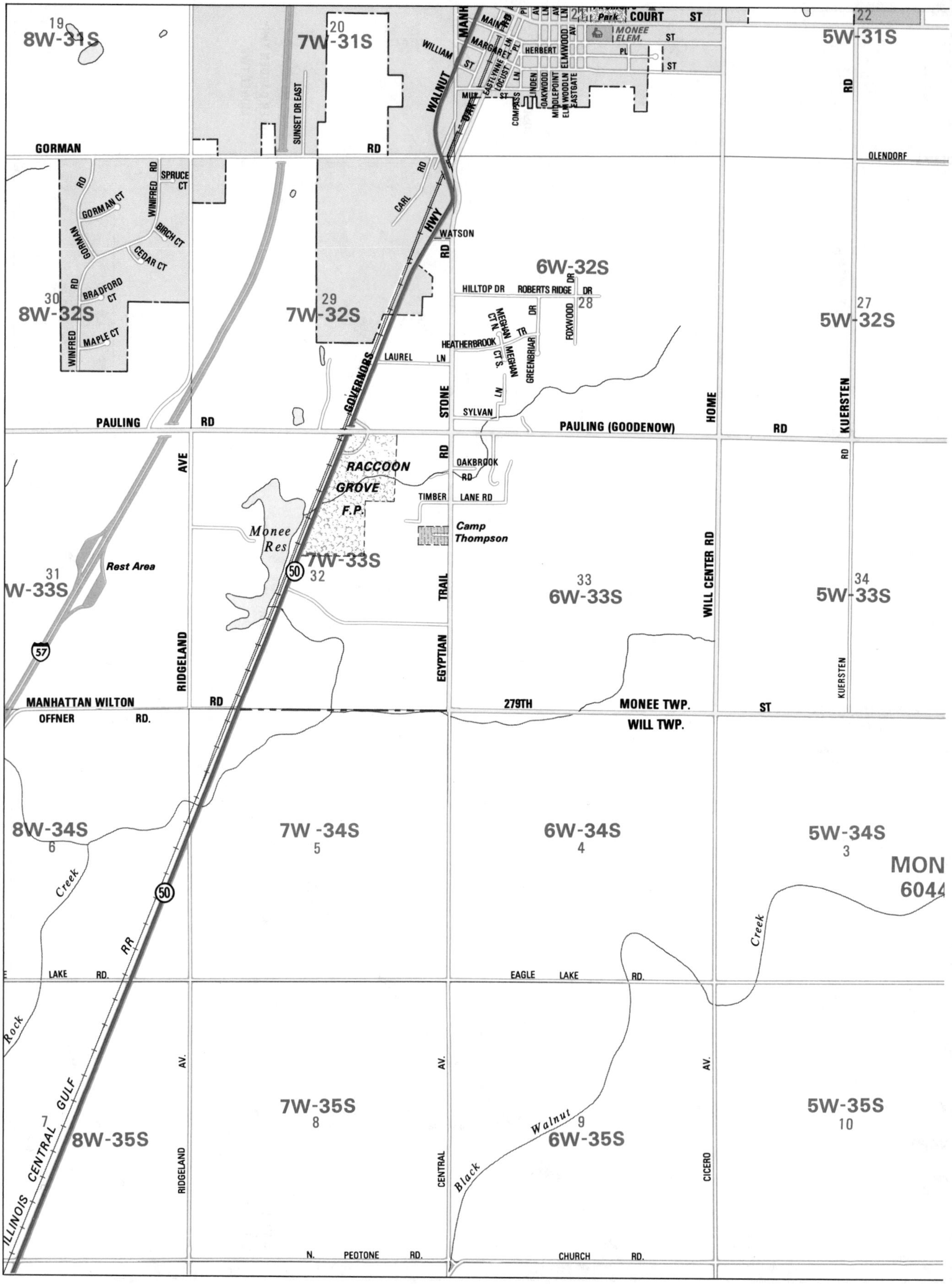

8W-31S

7W-31S

5W-31S

19

20

22

WILLIAM

WALNUT

MANH

MAIN

PB

MARGARET

ST

EASTLYNNE

LOCUST

LN

PL

HERBERT

LINDEN

OAKWOOD

MIDDLEPOINT

ELM WOODLN

ELMWOOD

EASTGATE

COMPASS

MILK

ST

Park

COURT ST

MONEE ELEM.

ST

PL

ST

ST

AV

GORMAN

OLENDORF

WINFRED RD

SPRUCE CT

GORMAN RD

GORMAN CT

BIRCH CT

CEDAR CT

GORMANN

BRADFORD CT

RD

MAPLE CT

WINFRED

CARL RD

HWY

WATSON

6W-32S

HILLTOP DR

ROBERTS RIDGE DR

FOXWOOD DR

MEGHAN CT N.

MEGHAN TR

GREENBRIAR DR

HEATHERBROOK

MEGHAN CT S.

8W-32S

30

7W-32S

29

5W-32S

27

STONE RD

LAUREL LN

SYLVAN

PAULING RD

PAULING (GOODENOW)

HOME RD

KUERSTEN RD

GOVERNORS

AVE

RACCOON GROVE F.P.

OAKBROOK RD

TIMBER

LANE RD

Monee Res

Camp Thompson

Rest Area

7W-33S

WILL CENTER RD

W-33S

31

32

EGYPTIAN

TRAIL

33

6W-33S

34

5W-33S

57

RIDGELAND

KUERSTEN RD

MANHATTAN WILTON RD.

OFFNER RD.

279TH

MONEE TWP.

WILL TWP.

ST

50

8W-34S

6

7W-34S

5

6W-34S

4

5W-34S

3

MON 6044

Creek

50

Creek

LAKE RD.

EAGLE LAKE RD.

Rock

AV.

AV.

Walnut

AV.

7W-35S

8

Black

6W-35S

9

5W-35S

10

ILLINOIS CENTRAL GULF

RR.

8W-35S

7

RIDGELAND

CENTRAL

CICERO

N. PEOTONE RD.

CHURCH RD.

SUNSET DR EAST

RD

SEE PAGES 66-67

SEE PAGES 56-57

23   CRETE   MONEE   RD   24

4W-31S   **3W-31S**   19   **2W-31S**   MONEE

RD   OLENDORF   RD

26   25   30
4W-32S   **3W-32S**   **2W-32S**

HARVEST   LN

OHLENDORF

PAULING (GOODENOW)   RD

NACKE

GOODENOW   RD

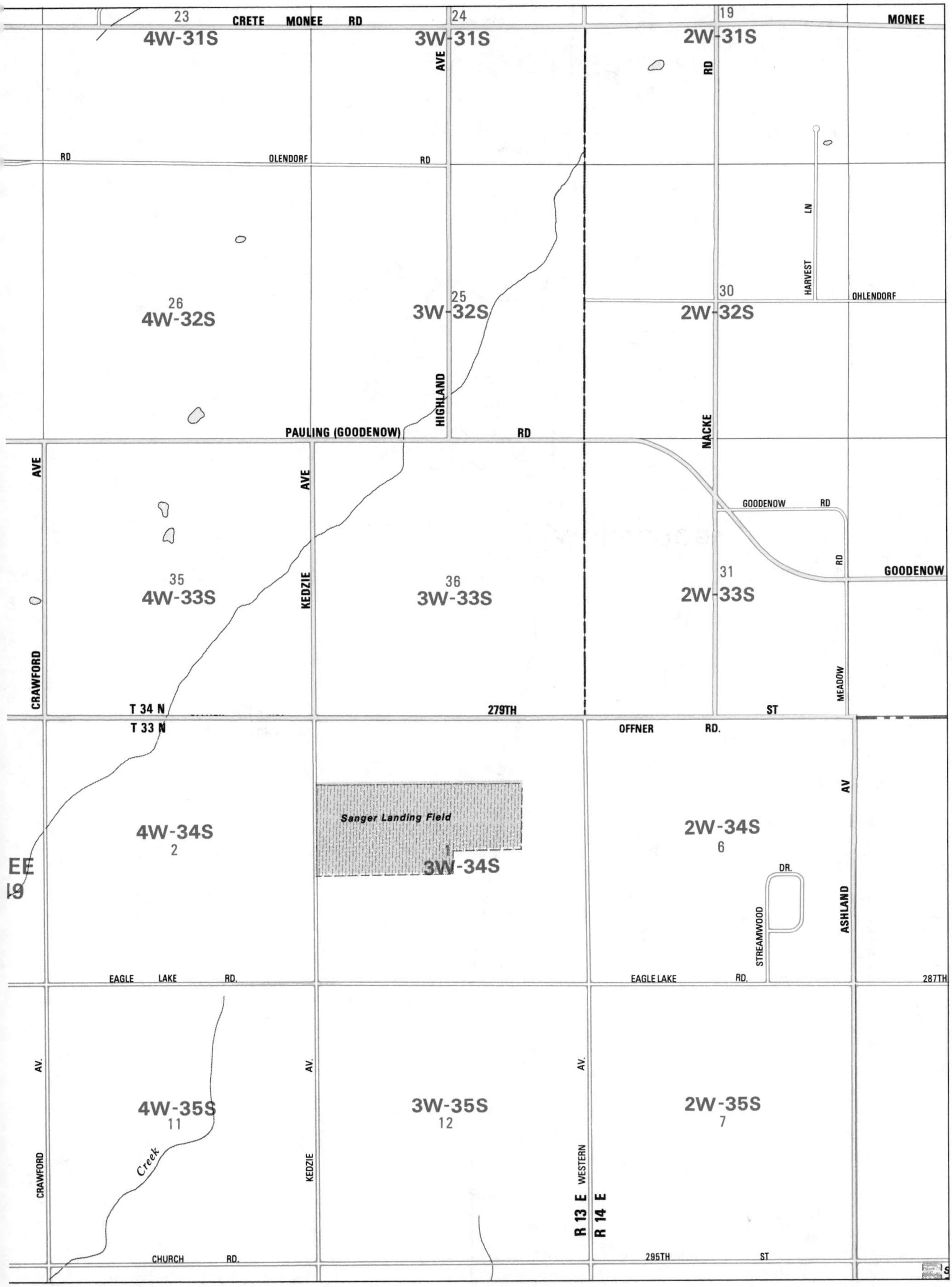

35   36   31
4W-33S   **3W-33S**   **2W-33S**   GOODENOW

RD

MEADOW

CRAWFORD AVE

KEDZIE AVE

HIGHLAND

T 34 N   279TH   ST
T 33 N

OFFNER   RD.

4W-34S   Sanger Landing Field   **2W-34S**
2   1   6

EE
9   **3W-34S**

AV

STREAMWOOD   DR.

ASHLAND

EAGLE   LAKE   RD.   EAGLE LAKE   RD.   287TH

AV.   AV.   AV.

4W-35S   **3W-35S**   **2W-35S**
11   12   7

CRAWFORD   KEDZIE   Creek

WESTERN

R 13 E   R 14 E

CHURCH   RD.   295TH   ST

SEE PAGES 44-45

MONEE RD 20

NEW MONEE RD

MONEE RD

1W-31S

0W-31S

0E-31S

PRIN
GREG

WINDSOR AV
COUNTESS LN
DUCHESS LN
AV

WINSTON AV
KENT AV

Balmoral

Racing

①

Club

COTTAGE

BEMES RD

ROOT DR

UNION PACIFIC

394

HARVEST LN

OHLENDORF 29 RD

1W-32S

DUNLEITH CT
ELMSCOURT 28

0W-32S

0E-32S

27

ARLINGTON LN

AUBURN LN
STANTON LN

TAM-O-SHANTER DR
TAM-O-SHANTER DR
TAM-O-SHANTER CT
BALMORAL DR
BALDWIN
BALMORAL DR
BUTLER CT
WOODS CT
PEBBLEBEACH

CALUMET EXPWY

RIDGELAND AV

BALMORAL WOODS

AV
COLUMBIA AV
AV
AV
AV

GOLF CLUB

SEE PAGES 54-55

BROADWAY
GOODENOW
GLENDALE
ROSEWOOD
HALSTED
GREENVIEW
JANICE
AV
AV

GOODENOW

FRANK ST
MAPLE ST

OW RD

RD

GOODENOW 32 RD

33

0E-33S

DUTTON 34

GOO

1W-33S

0W-33S

FOREST VIEW LN

CHESTNUT ST
HICKORY ST

MEADOW

①

CRETE TWP.
WASHINGTON TWP.

ASHLAND AV

1W-34S
5

HALSTEAD ST

0W-34S
4

RR

OE-34S
3

DR.

CHICAGO & EASTERN ILLINOIS
LOUISVILLE & NASHVILLE

287TH ST

EAGLE LAKE RD.

RR

SOUTHPARK AV.

CHICAGO, MILWAUKEE, ST. PAUL & PACIFIC RR

1W-35S
8

0W-35S
9

RD

0E-35S
10

HAHN'S

Creek

CHURCH RD

295TH ST

St. Pauls

COTTAGE GROVE AVE

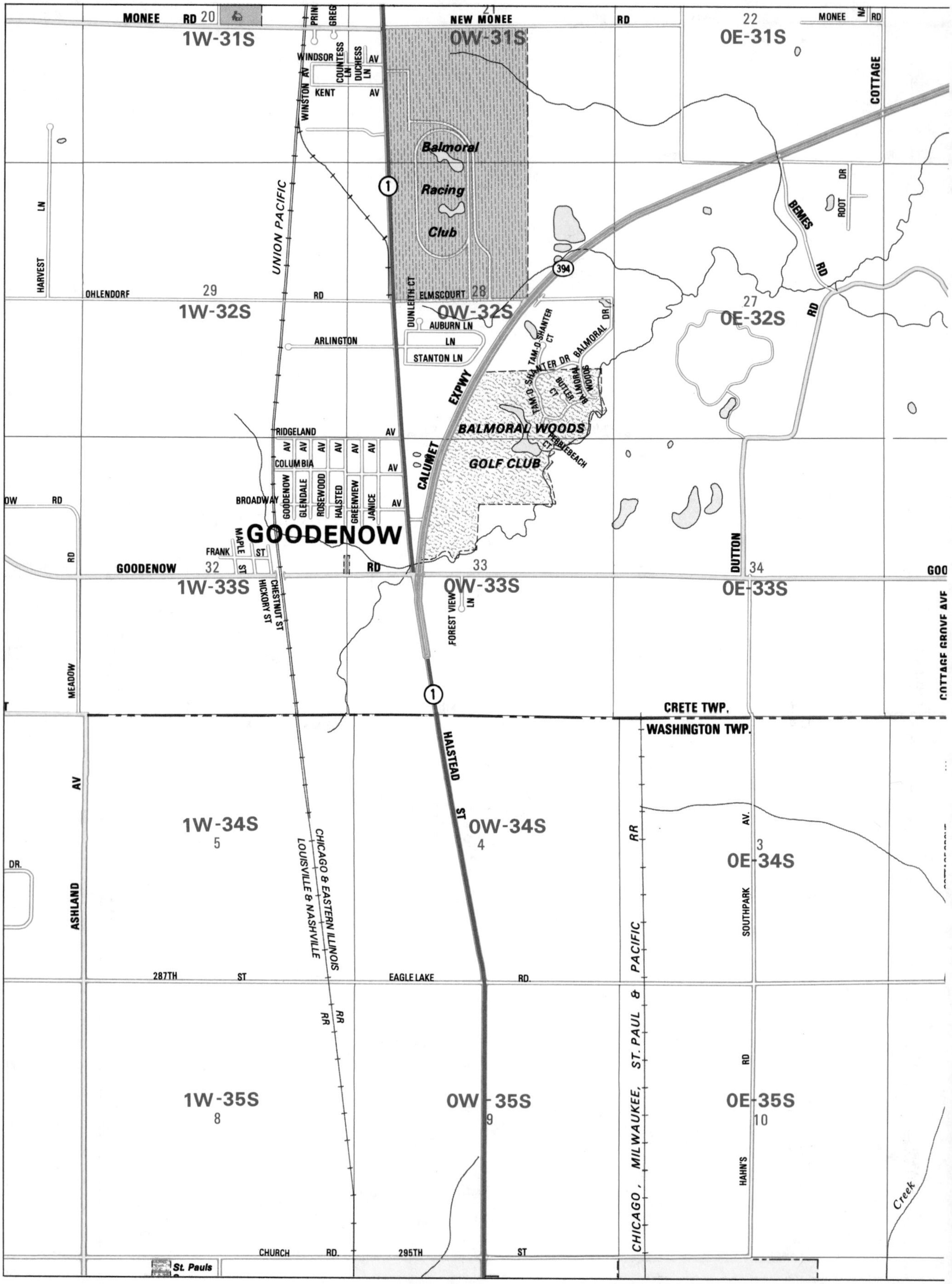

SEE PAGES 68-69

SEE PAGES 44-45

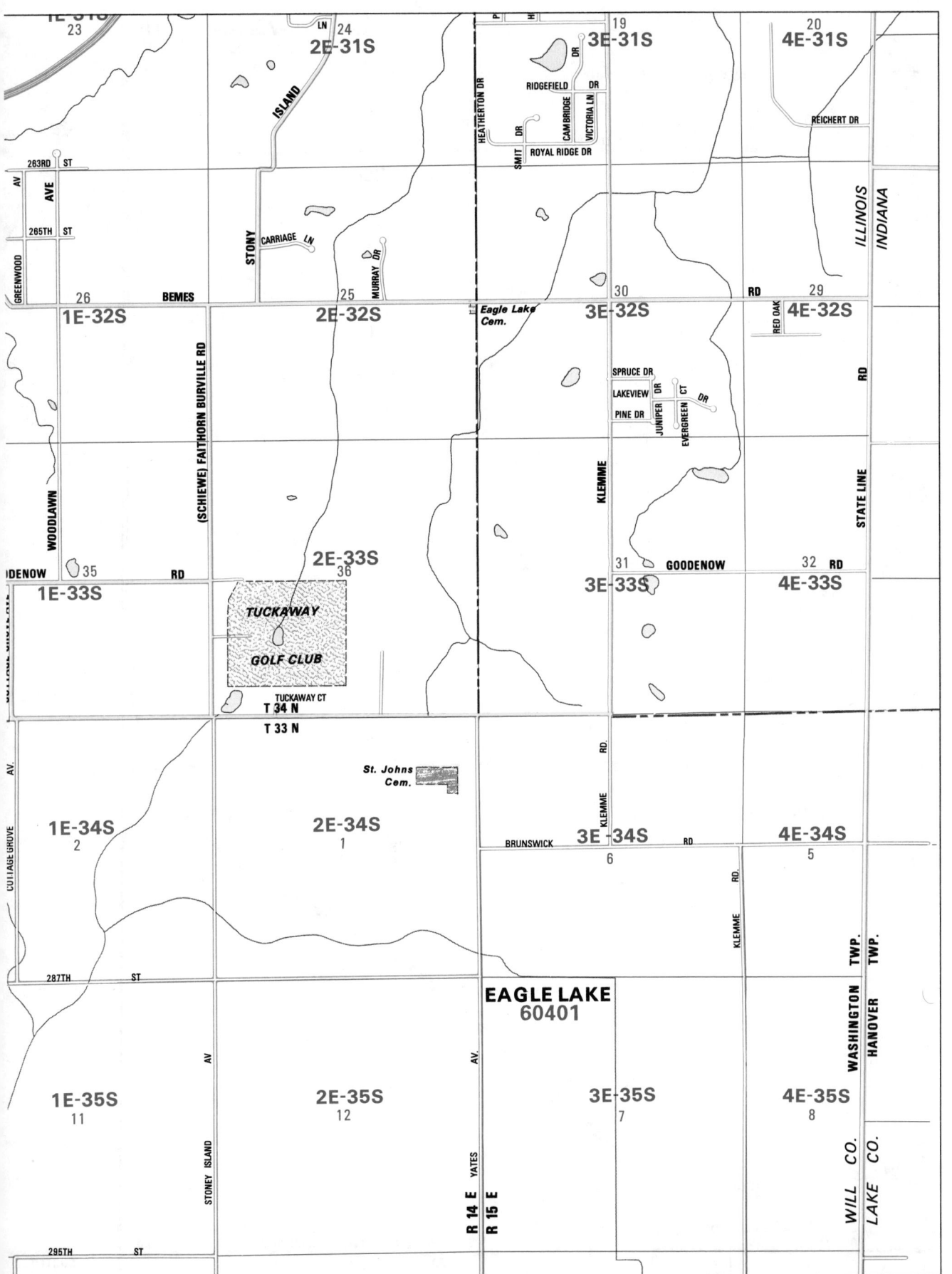

1E-31S
23

LN 24
2E-31S

19
3E-31S

20
4E-31S

HEATHERTON DR

RIDGEFIELD DR
CAM BRIDGE DR
VICTORIA LN

SMIT DR
ROYAL RIDGE DR

REICHERT DR

263RD ST

AVE

265TH ST

GREENWOOD

STONY ISLAND

CARRIAGE LN

MURRAY DR

BEMES

26
1E-32S

25
2E-32S

Eagle Lake Cem.

30
3E-32S

RD

29
4E-32S

RED OAK

ILLINOIS
INDIANA

(SCHIEWE) FAITHORN BURVILLE RD

WOODLAWN

SPRUCE DR
LAKEVIEW DR
PINE DR
JUNIPER DR
EVERGREEN CT DR

KLEMME

STATE LINE RD

ODENOW 35 RD

1E-33S

2E-33S
36

TUCKAWAY GOLF CLUB

TUCKAWAY CT
T 34 N
T 33 N

31
3E-33S

GOODENOW

32 RD
4E-33S

St. Johns Cem.

CUTTAGE GROVE AV

1E-34S
2

2E-34S
1

BRUNSWICK

KLEMME RD

3E-34S
6

RD

4E-34S
5

KLEMME RD.

287TH ST

AV

EAGLE LAKE
60401

AV

WASHINGTON TWP.
HANOVER TWP.

1E-35S
11

STONEY ISLAND

2E-35S
12

YATES
R 14 E
R 15 E

3E-35S
7

4E-35S
8

WILL CO.
LAKE CO.

295TH ST

SEE PAGES 68-69

SEE PAGES 46-47

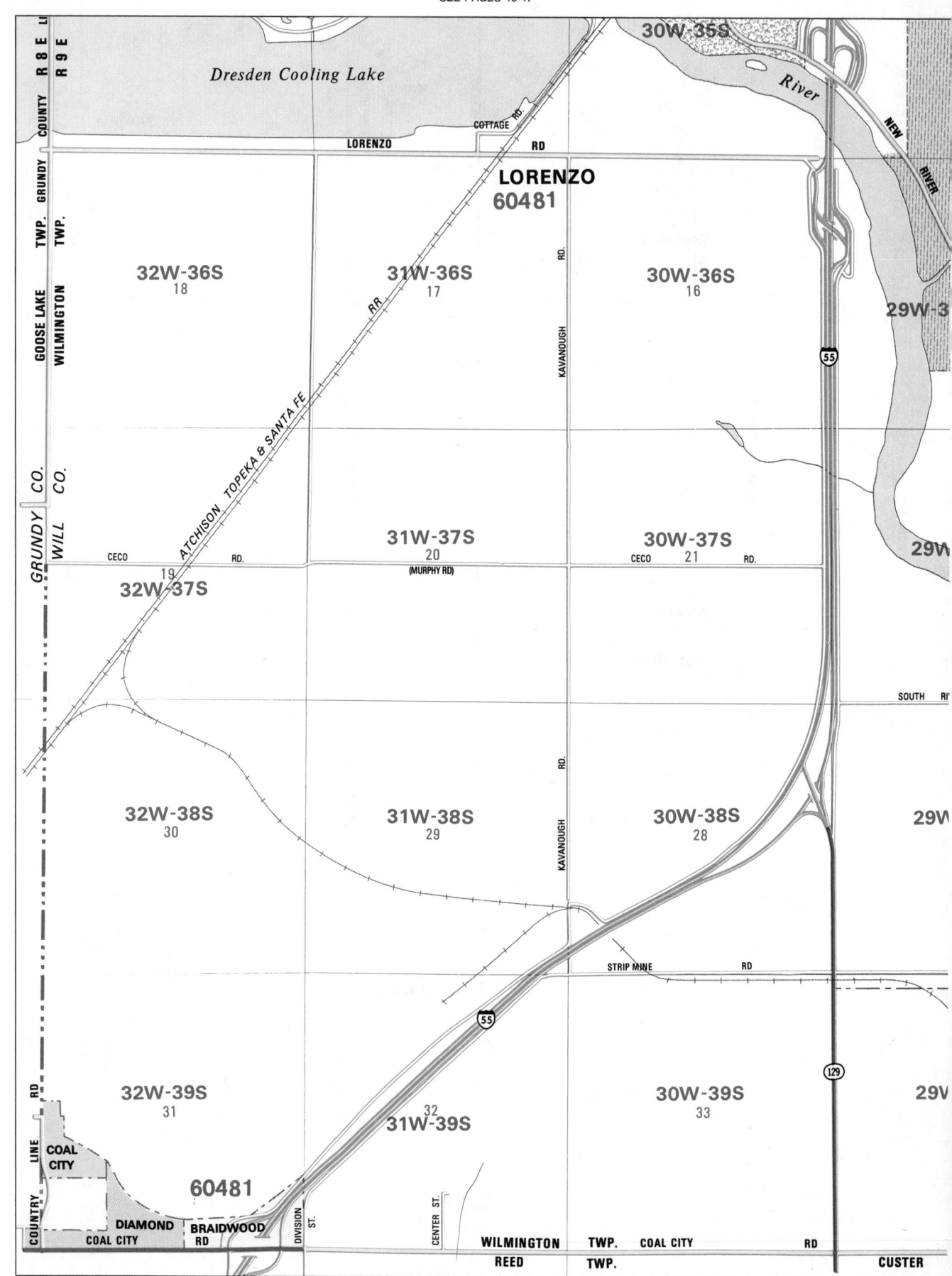

Dresden Cooling Lake

30W-35S

COUNTY LI

R 8 E

R 9 E

GRUNDY TWP.

WILMINGTON TWP.

GOOSE LAKE

COTTAGE RD.

LORENZO RD

LORENZO
60481

River

NEW RIVER

32W-36S
18

31W-36S
17

30W-36S
16

29W-3

KAVANOUGH RD.

55

GRUNDY CO.

WILL CO.

ATCHISON TOPEKA & SANTA FE

RR

CECO RD.

32W-37S
19

31W-37S
20
(MURPHY RD)

30W-37S
21
CECO RD.

29W

SOUTH RI

32W-38S
30

31W-38S
29

KAVANOUGH RD.

30W-38S
28

29W

STRIP MINE RD

55

129

32W-39S
31

31W-39S
32

30W-39S
33

29W

LINE RD

COAL CITY

60481

COUNTRY

DIAMOND

BRAIDWOOD RD

DIVISION ST.

CENTER ST.

COAL CITY

WILMINGTON TWP. **COAL CITY** RD

REED TWP.

CUSTER

SEE PAGES 70-71

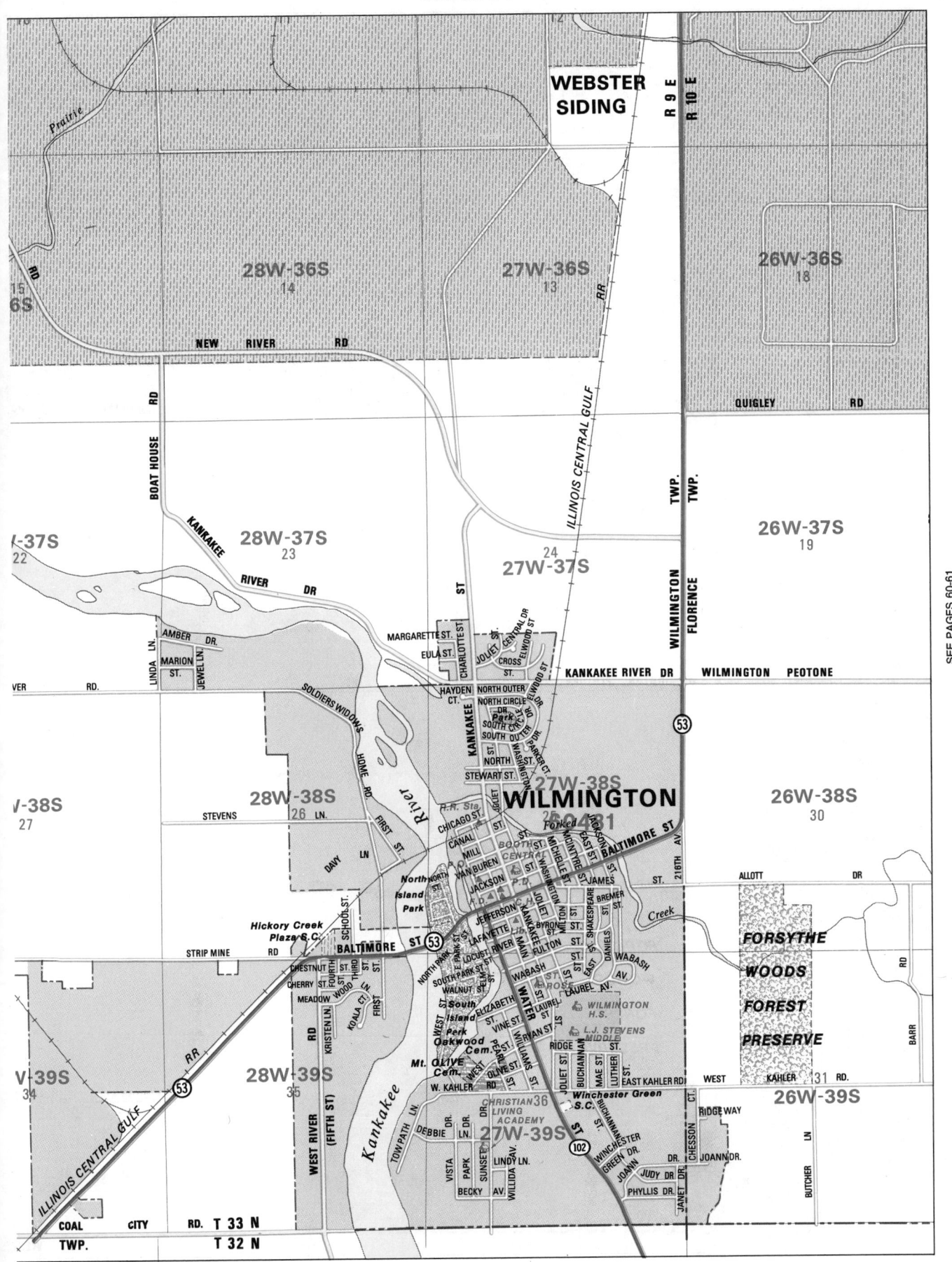

WEBSTER SIDING

R 9 E
R 10 E

Prairie

28W-36S
14

27W-36S
13

26W-36S
18

6S

RD

15

NEW RIVER RD

QUIGLEY RD

BOAT HOUSE RD

ILLINOIS CENTRAL GULF

TWP.
TWP.

KANKAKEE

37S
22

28W-37S
23

RIVER DR

27W-37S

26W-37S
19

WILMINGTON
FLORENCE

AMBER DR

LINDA LN
JEWEL LN

MARION ST

MARGARETTE ST.

EULA ST.

CHARLOTTE ST.

JOLIET ST
CROSS ST

CENTRAL DR
ELWOOD ST

RIVER RD.

HAYDEN CT.

NORTH OUTER DR

KANKAKEE RIVER DR

WILMINGTON PEOTONE

NORTH CIRCLE DR
Park
SOUTH CIRCLE DR
SOUTH OUTER DR

ELWOOD DR
PARKER CT

NORTH WASHINGTON

53

KANKAKEE ST

NORTH

SOLDIERS WIDOWS

HOME RD

38S
27

28W-38S
26

STEVENS LN.

River

STEWART ST.

R.R. Sta.

CHICAGO ST.

27W-38S

WILMINGTON
30 31

26W-38S
30

FIRST ST

DAVY LN

CANAL

MILL

VAN BUREN

JACKSON

Forked

JOLIET

BOOTH
CENTRAL
P.D.

MICHELLE ST
MONTANE ST

BALTIMORE ST

JAMES ST

216TH AV

ALLOTT DR

North
North
Island
Park

WASHINGTON

JOLIET

KANKAKEE

RIVER

ELSTER ST
HUSSON

Creek

RD

Hickory Creek
Plaza S.C.

SCHOOL ST.

BALTIMORE ST

53

JEFFERSON ST
LAFAYETTE ST
LOCUST ST

BYRON ST
MITTON ST

BREMER ST
SHAKESPEARE ST

FORSYTHE

STRIP MINE

CHESTNUT ST

NORTH PARK ST
SOUTH E PARK ST
SOUTH PARK ST

FULTON ST

DANIELS ST

WABASH

RD

CHERRY ST

KOHLA CT

WALNUT ST

ELM ST

WABASH

ST

LAUREL AV.

WOODS

MEADOW LN

FOURTH ST
THIRD ST

WEST
South
Island
Park
Oakwood
Cem.

ELIZABETH ST

ROSE

WATER

LAUREL

FOREST

WOOD

FIRST ST

VINE ST.

RYAN ST.

Wilmington
H.S.

BARR RD

KRISTEN LN

Mt. OLIVE
Cem.

WEST
JUDY

OLIVE ST

WILLIAMS ST

RIDGE

BUCHANNAN

L.J. STEVENS
MIDDLE
ST

PRESERVE

53

39S
34

ILLINOIS CENTRAL GULF

RR

28W-39S
35

WEST RIVER (FIFTH ST)

W. KAHLER RD

TOWPATH LN

DEBBIE

VISTA

PARK

Kankakee

JOLIET ST

MAE ST

Winchester Green
S.C.

EAST KAHLER RD

LUTHER

WEST

KAHLER

131 RD.

26W-39S

CHESSON CT

Christian
LIVING
ACADEMY

27W-39S

36

102

WINCHESTER
GREEN DR

RIDGE WAY

JANET DR

JOANN DR

BUTCHER LN

SUNSET DR

LINDY LN

BECKY AV.

WILLIDA

JOANN

JUDY DR

PHYLLIS DR

COAL CITY RD. T 33 N
TWP. T 32 N

SEE PAGES 60-61

SEE PAGES 48-49

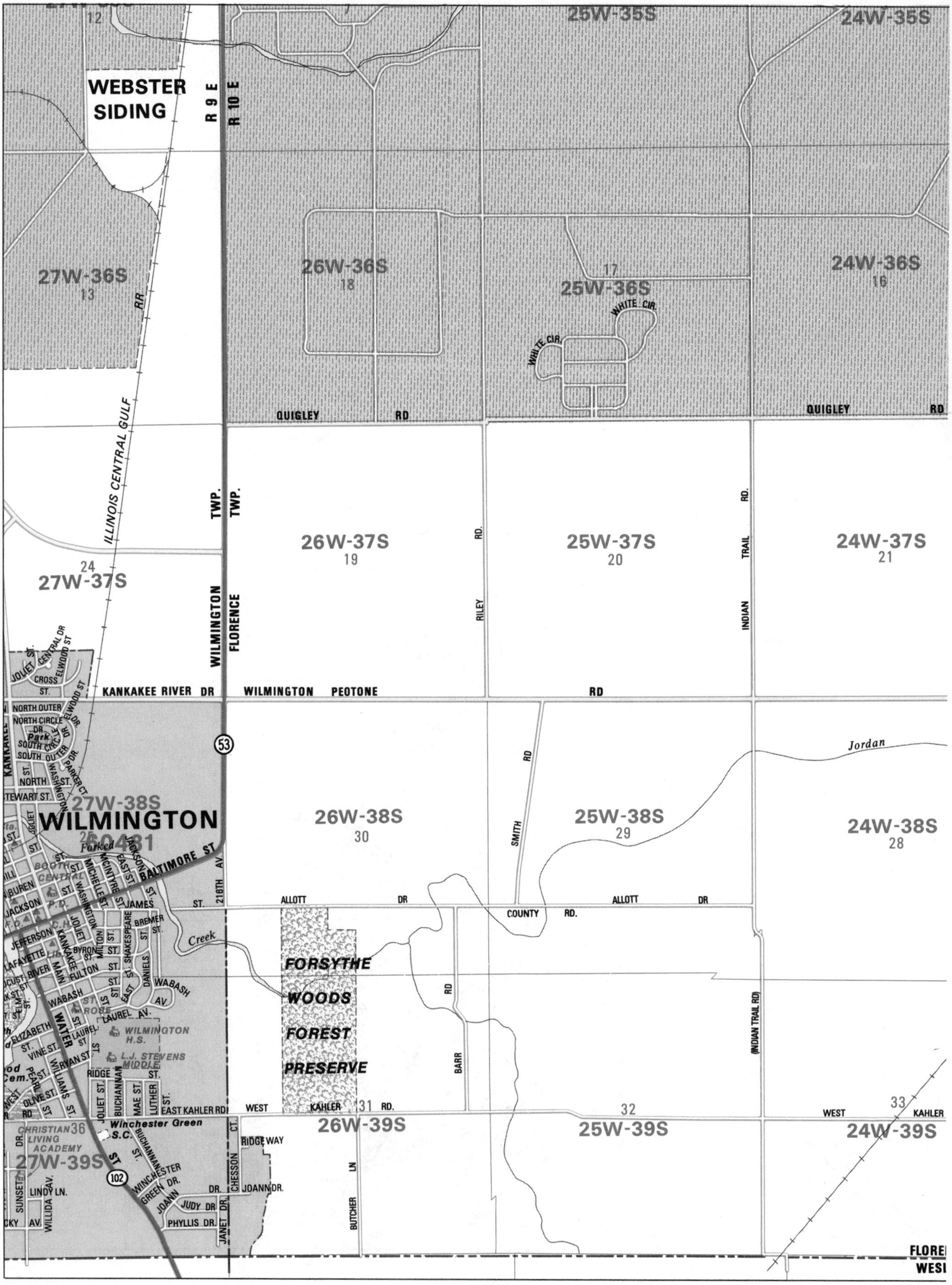

12

WEBSTER SIDING

R 9 E

R 10 E

25W-35S

24W-35S

27W-36S
13

RR

ILLINOIS CENTRAL GULF

26W-36S
18

17

25W-36S

WHITE CIR.

WHITE CIR.

24W-36S
16

QUIGLEY   RD

QUIGLEY   RD.

SEE PAGES 58-59

TWP.

TWP.

WILMINGTON

FLORENCE

26W-37S
19

RILEY   RD.

25W-37S
20

INDIAN   TRAIL   RD.

24W-37S
21

24

27W-37S

KANKAKEE RIVER DR

WILMINGTON   PEOTONE

RD

NORTH OUTER

NORTH CIRCLE DR

Park

SOUTH CIRCLE

SOUTH OUTER

NORTH

STEWART ST.

JOLIET CENTRAL DR

ELWOOD ST

CROSS ST.

WOOD ST

WASHINGTON

PARKER CT

53

SMITH   RD

Jordan

27W-38S
WILMINGTON
60481

Forked

BOOTH

CENTRAL

MCINTYRE

MICHELS

EAST ST.

JACKSON

BUREN

Creek

216TH   AV.

BALTIMORE ST.

26W-38S
30

25W-38S
29

INDIAN TRAIL RD.

24W-38S
28

JAMES   ST.

BREMER ST.

SHAKESPEARE

JEFFERSON

LAFAYETTE

RIVER   ST.

MAIN

KANKAKEE

JOLIET

MILTON

BYRON ST.

DANIELS   ST.

ALLOTT   DR

COUNTY   RD.

ALLOTT   DR

WABASH

AV.

AUGUST   ST.

ELM   ST.

WABASH ST.

FULTON

ROBE

ST.

FORSYTHE

RD

ELIZABETH ST.

VINE ST.

PEARL ST.

WATER   ST.

LAUREL

WILLIAMS ST.

RYAN ST.

LAUREL   AV.

WILMINGTON H.S.

RIDGE   ST.

L.J. STEVENS MIDDLE ST.

WOODS

FOREST

BARR   RD

PRESERVE

Cem.

OLIVE ST.

PEARL   RD.

JOLIET ST.

BUCHANNAN ST.

MAE ST.

LUTHER   ST.

RIDGE   ST.

EAST KAHLER RD

WEST

KAHLER

31

RD.

WEST   KAHLER

32

WEST   KAHLER

33

CHRISTIAN LIVING ACADEMY

36

Winchester Green S.C.

BUCHANNAN

RIDGE WAY

26W-39S

25W-39S

24W-39S

27W-39S

102

SUNSET   AV.

LINDY LN.

WILLIDA

CKY   AV.

WINCHESTER GREEN DR.

JOANN

JUDY DR.

PHYLLIS DR.

DR.

JOANN DR.

CHESSON   CT.

JANET   ST.

BUTCHER   LN

FLORE

WES

SEE PAGES 72-73

SEE PAGES 48-49

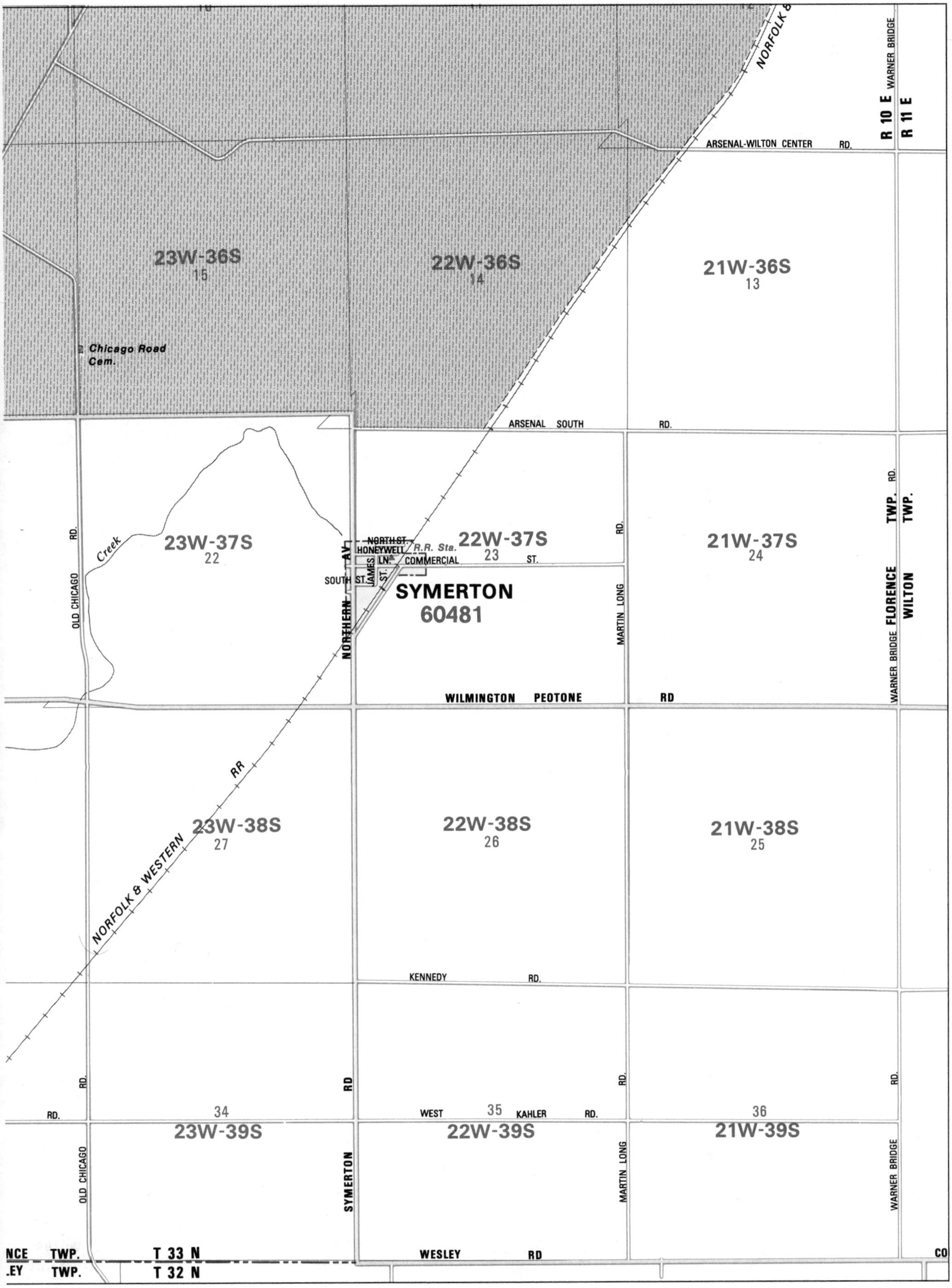

R 10 E

R 11 E

WARNER BRIDGE

ARSENAL-WILTON CENTER   RD.

NORFOLK &

23W-36S
15

22W-36S
14

21W-36S
13

Chicago Road
Cem.

ARSENAL   SOUTH   RD.

Creek

OLD CHICAGO   RD.

23W-37S
22

NORTH ST.
HONEYWELL
LN.   R.R. Sta.
SOUTH ST.
JAMES ST.

NORTHERN   AV.

COMMERCIAL

SYMERTON
60481

22W-37S
23

ST.

RD.

MARTIN LONG

21W-37S
24

WARNER BRIDGE

FLORENCE TWP.

WILTON TWP.

SEE PAGES 62-63

WILMINGTON   PEOTONE   RD

NORFOLK & WESTERN   RR

23W-38S
27

22W-38S
26

21W-38S
25

KENNEDY   RD.

RD.

SYMERTON   RD

RD.

RD.

34

WEST   35   KAHLER   RD.

36

MARTIN LONG   RD.

WARNER BRIDGE   RD.

RD.

23W-39S

OLD CHICAGO   RD.

SYMERTON   RD

22W-39S

MARTIN LONG   RD.

21W-39S

NCE   TWP.   **T 33 N**

.EY   TWP.   **T 32 N**

WESLEY   RD

CO

SEE PAGES 72-73

SEE PAGES 50-51

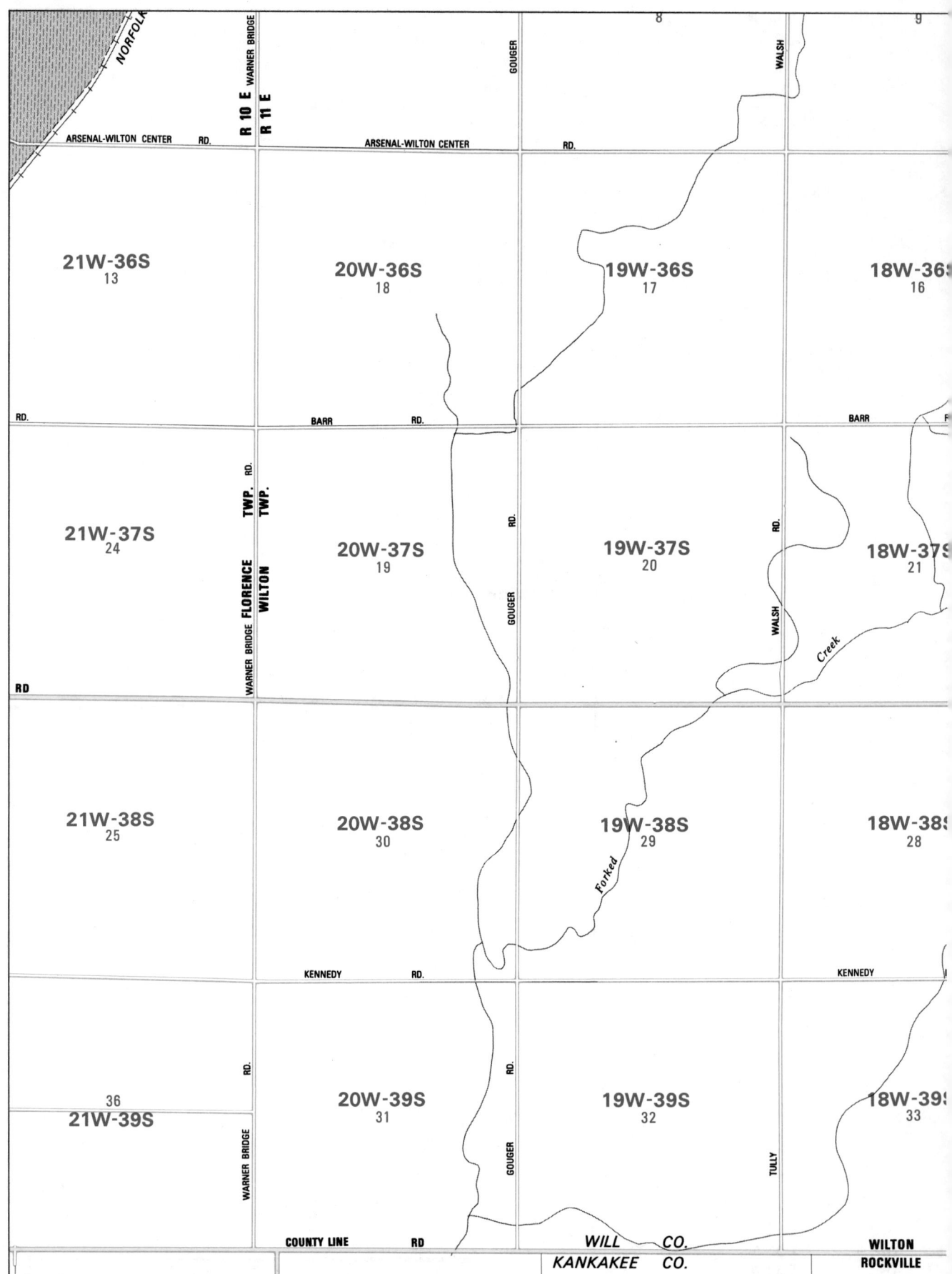

SEE PAGES 60-61

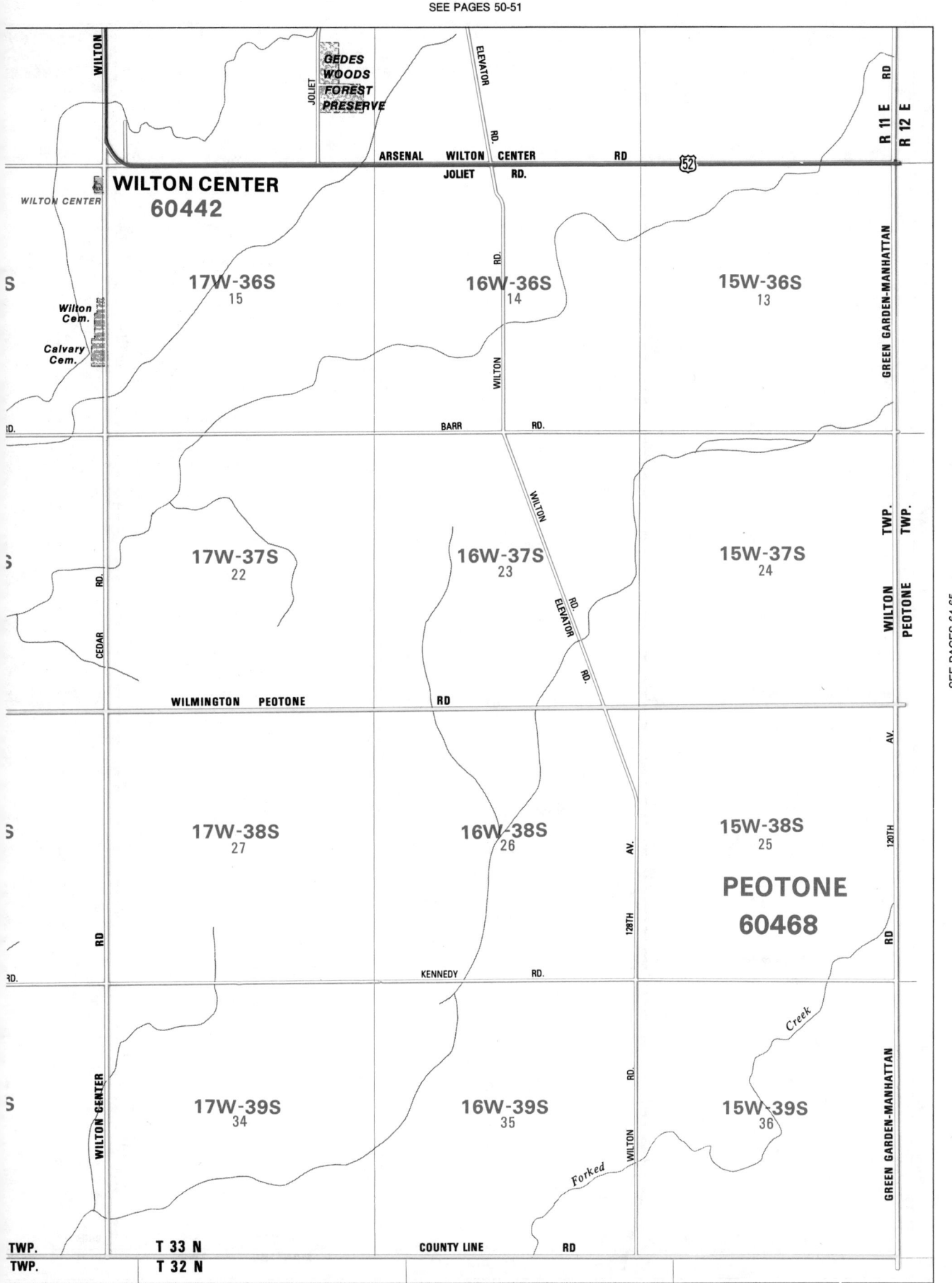

WILTON

JOLIET

GEDES
WOODS
FOREST
PRESERVE

ELEVATOR

RD.

R 11 E

R 12 E

RD

ARSENAL   WILTON   CENTER   RD   52

JOLIET   RD.

WILTON CENTER
60442

WILTON CENTER

Wilton
Cem.

Calvary
Cem.

RD.

17W-36S
15

RD.

WILTON

16W-36S
14

15W-36S
13

GREEN GARDEN-MANHATTAN

BARR   RD.

WILTON

RD.

CEDAR   RD.

17W-37S
22

16W-37S
23

ELEVATOR

RD.

15W-37S
24

TWP.

TWP.

WILTON

PEOTONE

SEE PAGES 64-65

WILMINGTON   PEOTONE   RD

AV.

RD.

17W-38S
27

16W-38S
26

AV.

120TH

15W-38S
25

128TH

PEOTONE
60468

KENNEDY   RD.

Creek

RD.

WILTON CENTER

17W-39S
34

RD.

16W-39S
35

WILTON

15W-39S
36

Forked

GREEN GARDEN-MANHATTAN

TWP.
TWP.

T 33 N

T 32 N

COUNTY LINE   RD

SEE PAGES 52-53

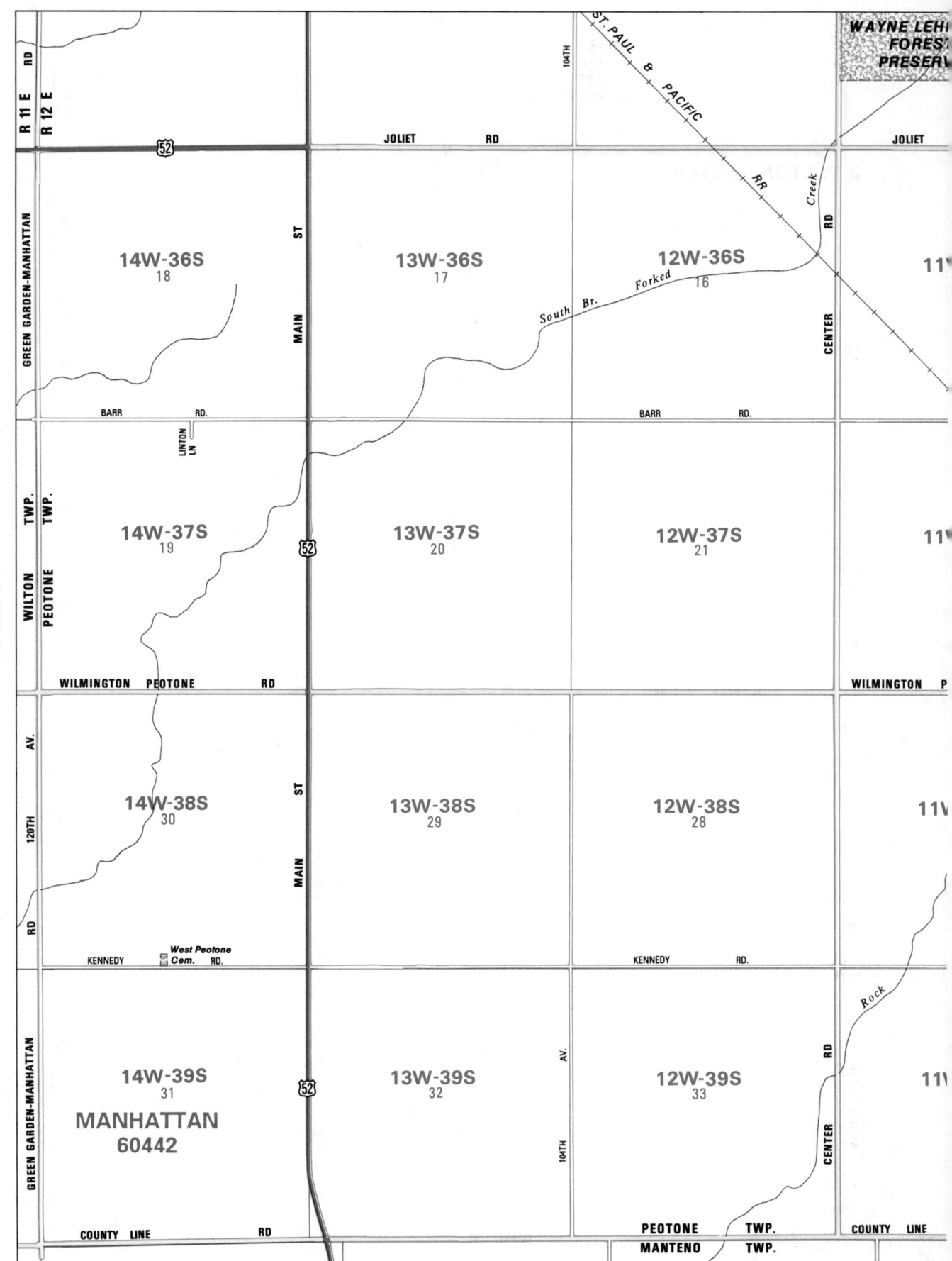

SEE PAGES 62-63

8W-35S

10W-36S
14

9W-36S
13

8W-36S
18

W-36S
15

R 12 E

R 13 E

ILLINOIS CENTRAL

Creek

N. PEOTONE RD.

N. PEOTONE

RD

Rock

JOLIET

RD.

HARLEM

AV.

57

JOLIET RD.

50

60468

PEOTONE

9W-37S

H.A. Rathje Mill Hist. Site

10W-37S

CORNING 23 ST.

OAK ST.
ST. ST.

DIVISION ST.

GLENVIEW

MILL ST.

PEOTONE

AV.

WEST ST.

PENNY LN.

SUMMER AV.

LINCOLN ST.

CRAWFORD ST.

Lib

WASHINGTON ST.

2ND ST.

Pk.

NORTH ST.

WOOD AV.

MAIN

ST.

P.O.

WALNUT ST.

HAVERT ST.

SIXTH

HAWTHORNE

BARTON LN.

HICKORY LN.

Cem.

ST.

CORNING ST.

CORNING ST.

8W-37S

19

W-37S
22

N. PEOTONE RD.

GARFIELD AV.

MAPLE ST.

LOCUST LN.

LOUISE LN.

LARK ST.

BONNIE

MEADOW LN.

JEAN LN.

RATHJE

ETHEL ST.

DEER CT.

HANS BRINKER DR.

CROWN LN.

MANOR DR.

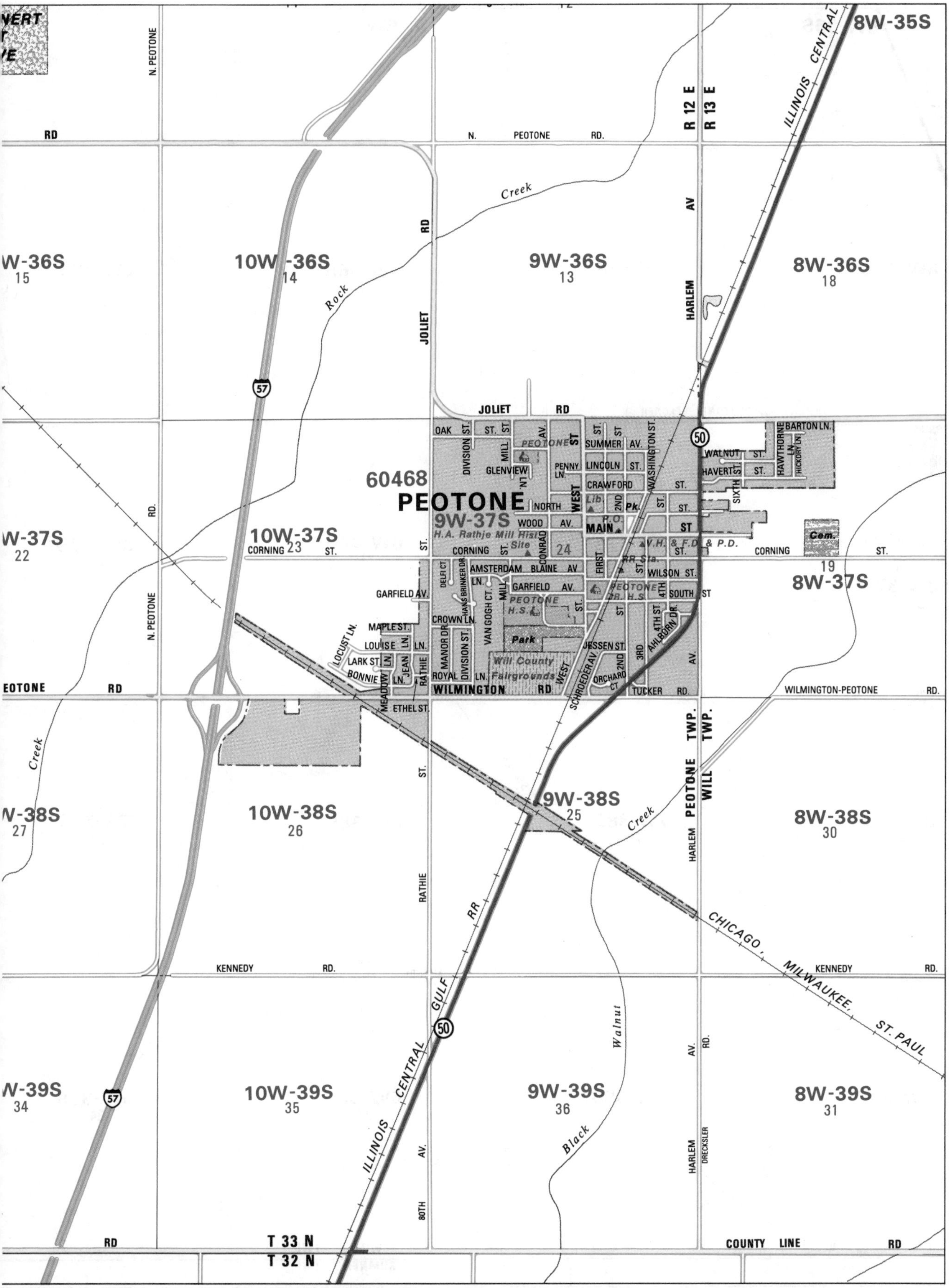

ROYAL LN.

DIVISION ST.

AMSTERDAM BLAINE AV.

VAN GOGH CT.

MILL ST.

GARFIELD AV.

PEOTONE H.S.

Park

Will County Fairgrounds

PEOTONE DR.

PEOTONE H.S.

CONRAD ST.

FIRST ST.

WILSON ST.

RR Sta.

V.H. & F.D. & P.D.

3RD

4TH ST.

SOUTH ST.

2ND

ORCHARD CT.

JESSEN ST.

AHLBORN DR.

TUCKER RD.

SCHROEDER AV.

WEST

RD.

WILMINGTON RD.

WILMINGTON-PEOTONE RD.

PEOTONE TWP.

WILL TWP.

HARLEM

10W-38S
26

9W-38S
25

Creek

8W-38S
30

ILLINOIS CENTRAL GULF RR

RATHJE ST.

50

KENNEDY RD.

Walnut

CHICAGO, MILWAUKEE, ST. PAUL

KENNEDY RD.

N-38S
27

Creek

57

10W-39S
35

9W-39S
36

Black

8W-39S
31

ILLINOIS CENTRAL

AV.

80TH

HARLEM

DRECKSLER

AV.

RD.

N-39S
34

RD

T 33 N

T 32 N

COUNTY LINE RD

SEE PAGES 54-55

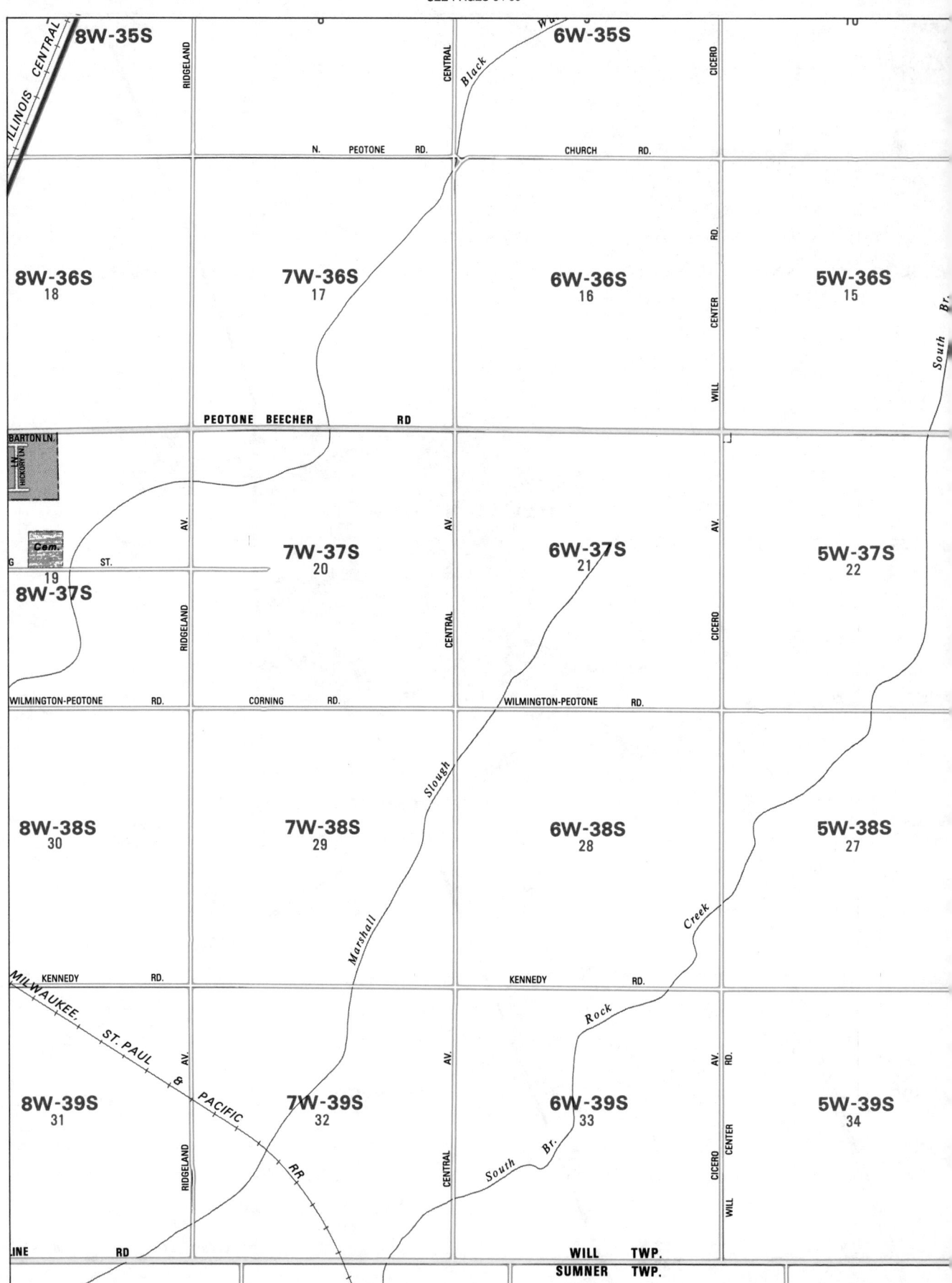

SEE PAGES 64-65

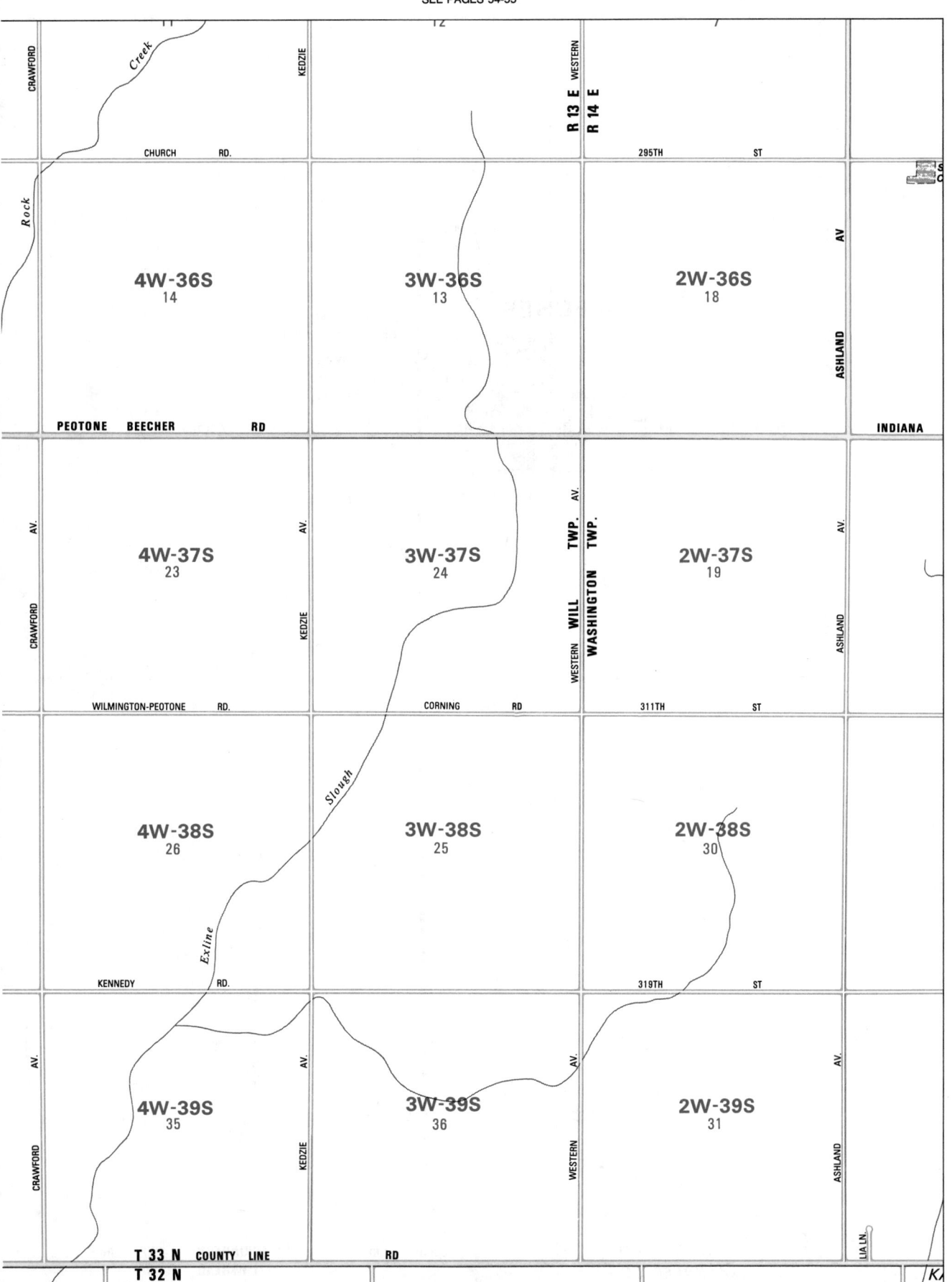

SEE PAGES 54-55

SEE PAGES 68-69

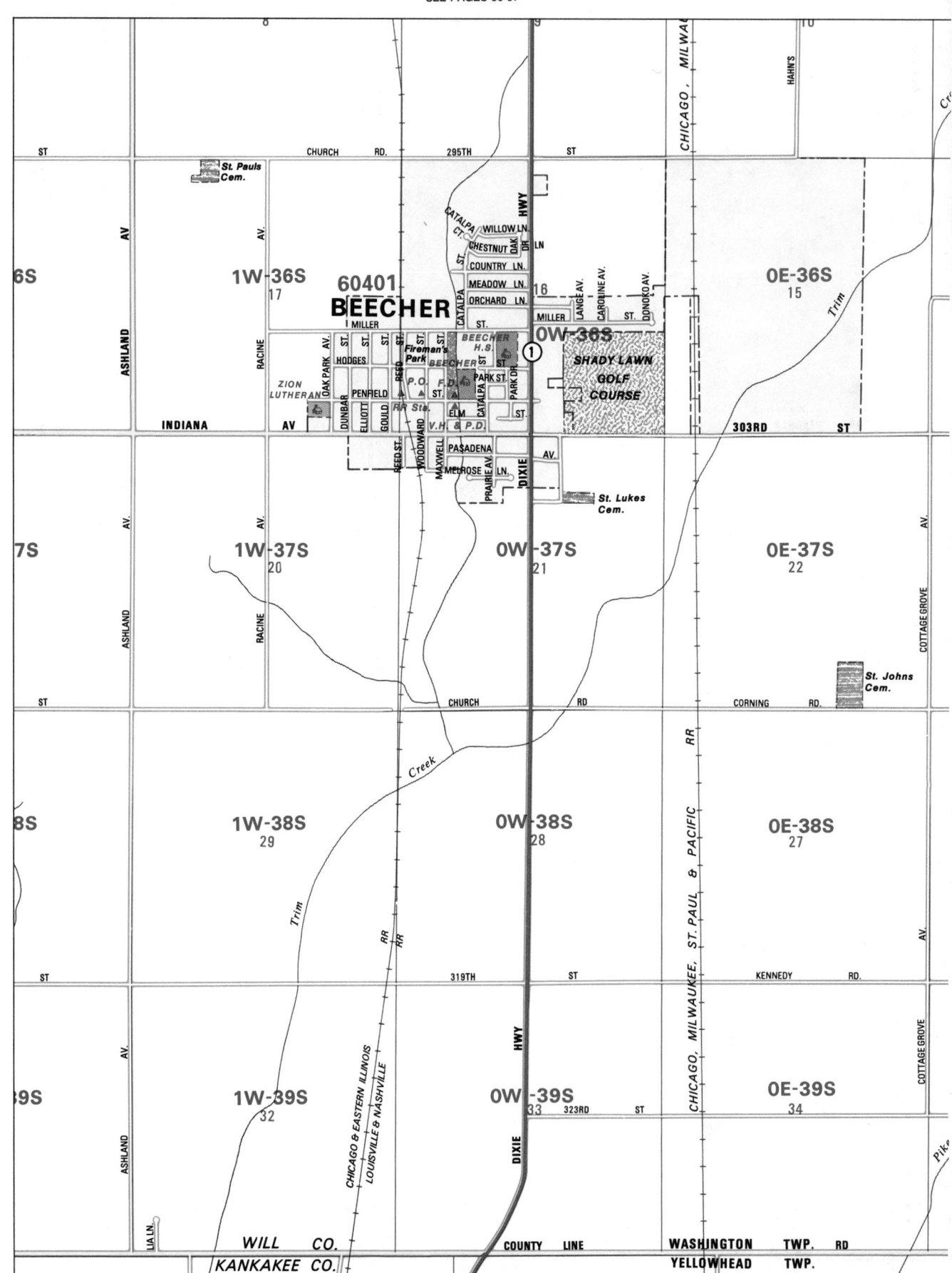

SEE PAGES 66-67

WILL CO.

LAKE CO.

STONEY ISLAND

YATES

R 14 E

R 15 E

295TH ST

1E-36S
14

2E-36S
13

3E-36S
18

4E-36S
17

STATE LINE

RD

303RD ST

WASHINGTON TWP.

WEST CREEK TWP.

AV

AV

1E-37S
23

2E-37S
24

3E-37S
19

4E-37S
20

KLEMME
RD.

STONEY ISLAND

YATES

311TH ST

311TH ST

CORNING RD.

Creek

ILLINOIS

INDIANA

1E-38S
26

(Dixie)

2E-38S
25

3E-38S
30

4E-38S
29

319TH ST

319TH ST

KENNEDY RD.

Creek

AV

AV

RD.

RD

1E-39S
35

2E-39S
36

3E-39S
31

4E-39S
32

323RD ST

323RD ST

STONEY ISLAND

YATES

KLEMME

STATE LINE

T 33 N

T 32 N

COUNTY LINE RD

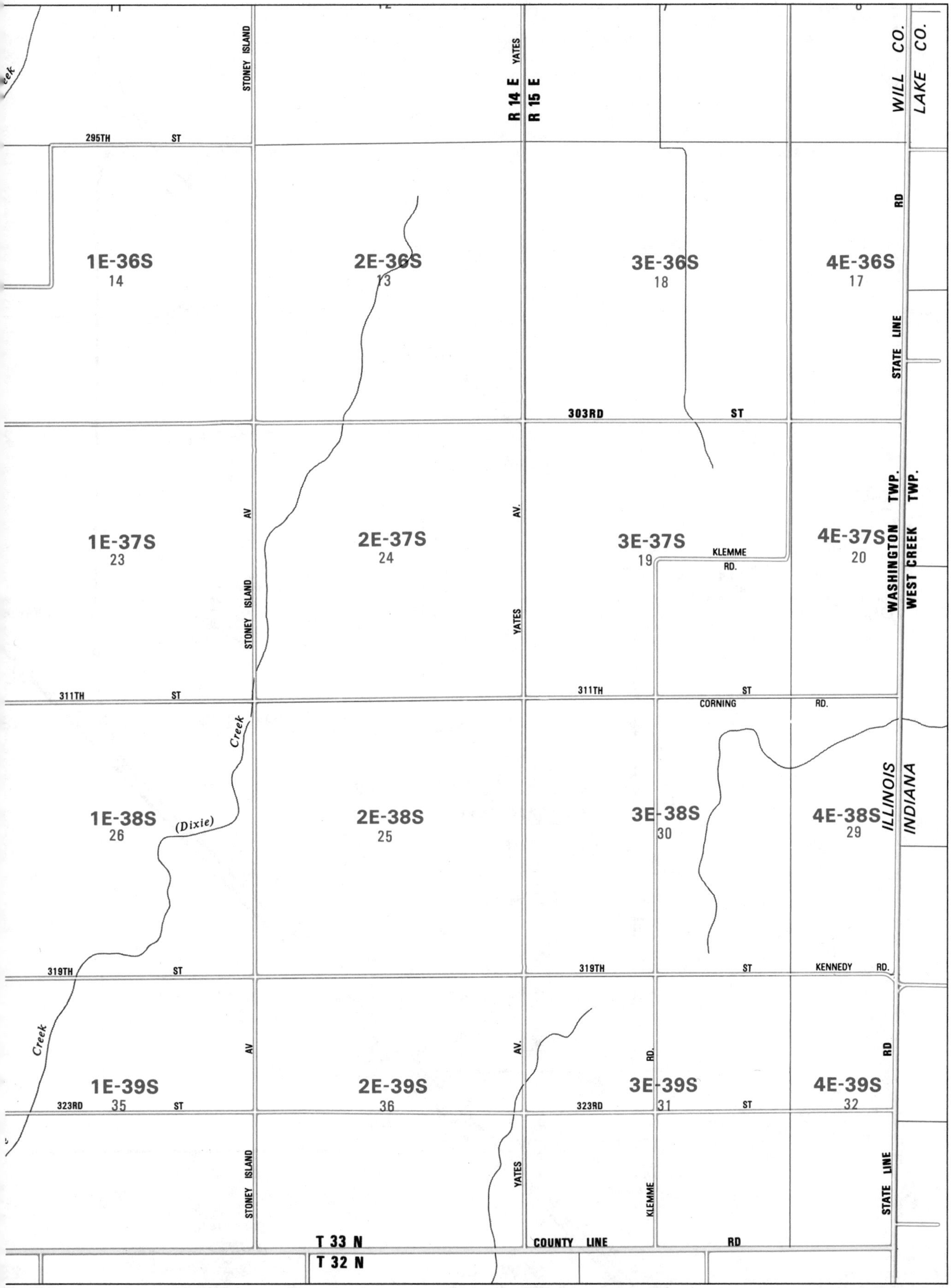

SEE PAGES 58-59

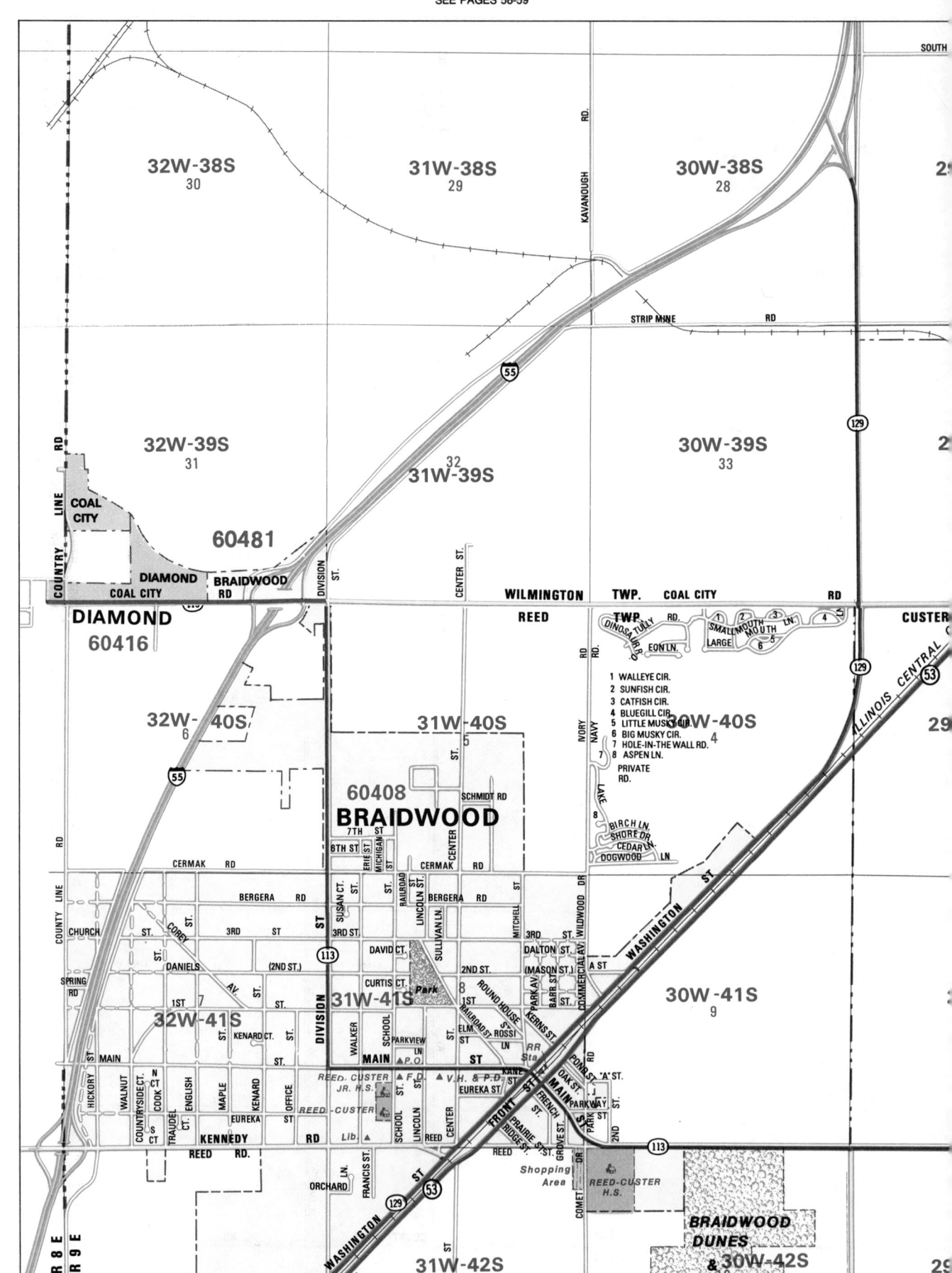

SEE PAGES 74-75

RIVER RD.
LINDA
MARION ST.
JEWEL

KANKAKEE RIVER DR
WILMINGTON PEOTONE

CHAR
CROSS ST
WOOD ST

HAYDEN CT.
NORTH OUTER DR.
NORTH CIRCLE DR.
Park
SOUTH CIRCLE DR
SOUTH OUTER ST
PARKER CT DR

53

9W-38S
27

28W-38S
26 LN.
STEVENS

SOLDIERS WIDOWS
HOME RD.
FIRST ST
DAVY LN

KANKAKEE
Island
NORTH ST
STEWART ST.

27W-38S
WILMINGTON
60481

26W-38S
30

River

H.R. Sta.
CHICAGO ST
P.O.
CANAL ST
MILL ST
VAN BUREN ST
BOOTH ST
CENTRAL ST
JACKSON ST
P.D.
NORTH ISLAND ST

MCKINLEY ST
MICHELLE
FORREST ST
JAMES ST
BREMER ST
BALTIMORE ST.
216TH AV.
ALLOTT DR

STEVENS
North Island Park
JEFFERSON ST
LAFAYETTE ST
KANKAKEE ST
BYRON
MAIN ST
FULTON ST
SHAKESPEARE ST
DANIELS ST
WABASH
AV.

Creek
RD

Hickory Creek Plaza S.C.
SCHOOL ST.
Baltimore
53
CHESTNUT ST
CHERRY ST
FOURTH ST
THIRD ST
WOOD CT
KOALA CT
FIRST ST
MEADOW

FORSYTHE WOODS FOREST PRESERVE

WEST RIVER RD (FIFTH ST)
KRISTEN LN
South Park St
WABASH ST
ROSE ST
South Island Park
Oakwood Cem.
ELIZABETH ST
VINE ST
WILLIAMS ST
LAUREL ST
BRYAN ST
VIVIAN ST
LAUREL AV.
WILMINGTON H.S.
L.J. STEVENS MIDDLE
RIDGE
BUCHANNAN
MAE ST
LUTHER ST
EAST KAHLER RD
WEST KAHLER RD.
JOLIET ST

28W-39S
35
Mt. OLIVE Cem.
OLIVE ST
W. KAHLER RD
36
CHRISTIAN LIVING ACADEMY

Winchester Green S.C.
BUCHANNAN ST
RIDGE WAY
WEST KAHLER RD.

26W-39S

9W-39S
34
53
RR
ILLINOIS CENTRAL GULF

COAL CITY RD. T 33 N
TWP. T 32 N

27W-39S
TOW PATH LN
DEBBIE LN
VISTA PARK DR
SUNSET DR
LINDY LN.
BECKY AV.
WILLIDA AV.
102
WINCHESTER GREEN DR.
JOANN DR.
JUDY DR
PHYLLIS DR.
JANET DR
CHESSON CT.
JOANN DR.
BUTCHER LN

Kankakee

GULF

KOERNER CT.
DORSET DR
VISTA DR
HAMILTON ST
ALMA DR
MAPLE ST
JOHN ST
MEADOWVIEW LN

9W-40S
3

28W-40S
2

WEST RIVER RD

Kankakee

CHURCH ST
ALBERT LN
BASS ST
ROBERT DR
TROUT ST
BRUNING DR
SUMAC ST
27W-40S
LAKEWOOD SHORES
60481
102

26W-40S
6

BALLOU RD

WOOD ST
WOODVIEW DR.
GROVE ST
CHARLES ST
HINTZE ST
LAKEWOOD DR
HINTZE ST

Forked

29W-41S
10
SHENK RD (PVT)

28W-41S
11

27W-41S
12

RIVERSIDE DR.
SHORT ST
NORTH
GREENWOOD DR
ROSEWOOD DR.
WILDWOOD ST.
WILDWOOD DR.
MAPLE ST
LAKESIDE DR
WILLOW DR.
EVERGREEN ST.
ELMWOOD AV.
R 9 E
R 10 E
REST HAVEN RD

26W-41S
7

River
ERICKSON LN (PVT)
POPLAR ST
ORCHARD AV.
WALNUT AV.
GRAND AV.
PINE GROVE AV.
JUNE WAY AV.
STEWART AV.
HICKORY AV.
OAK ST
DAVY LN (PVT)

W-42S

28W-42S

27W-42S

26W-42S

SEE PAGES 60-61

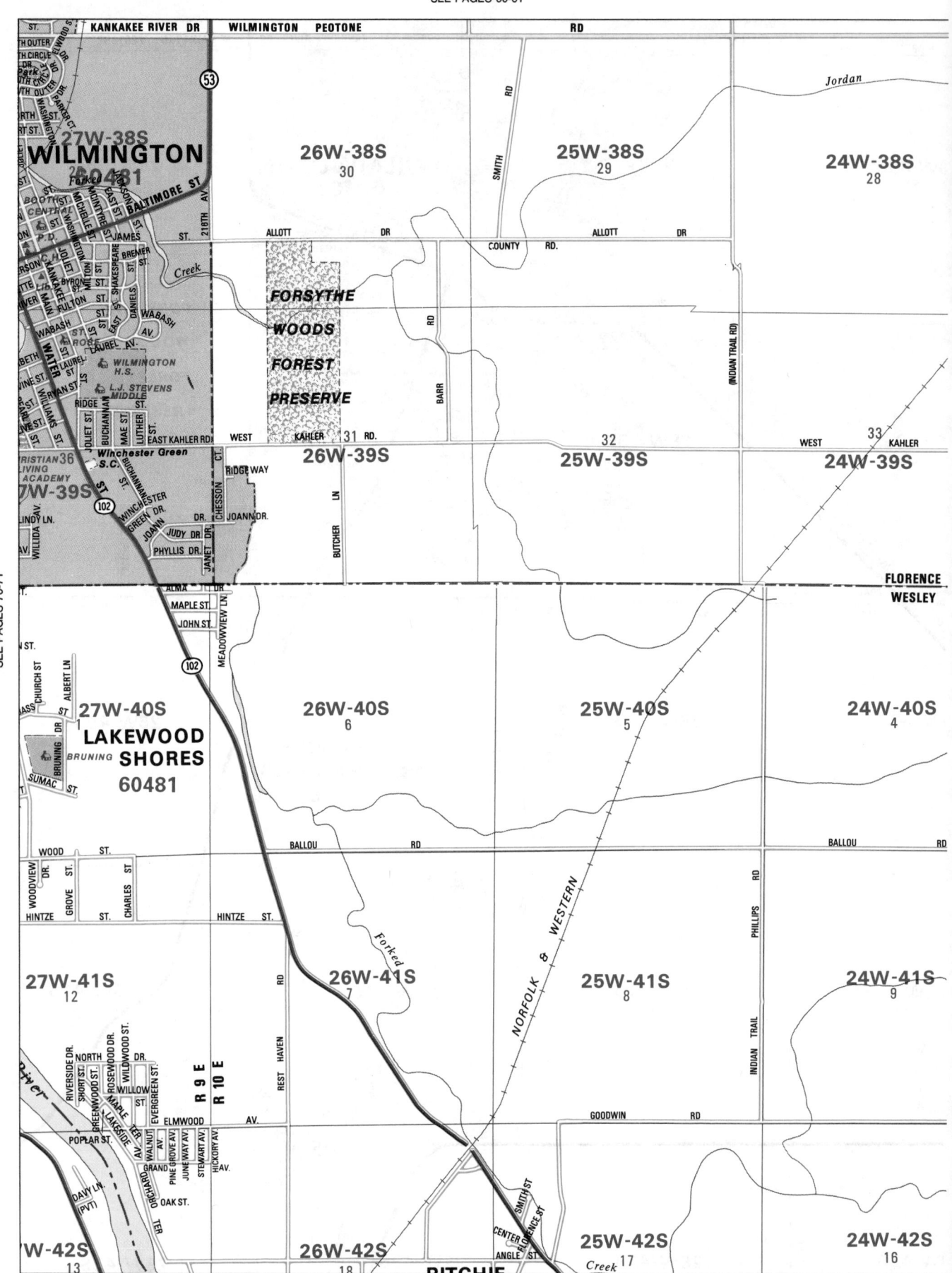

SEE PAGES 70-71

SEE PAGES 76-77

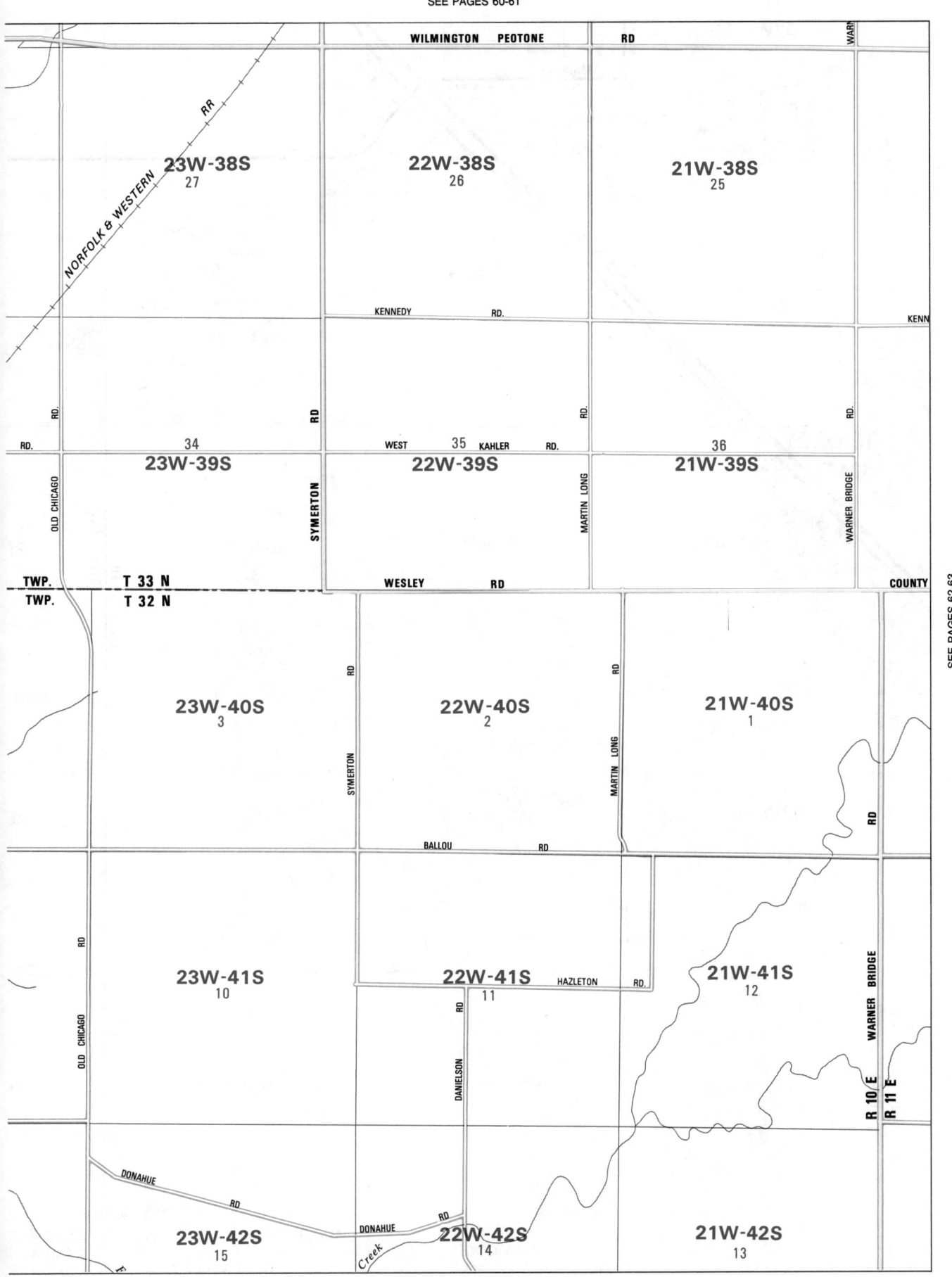

WILMINGTON   PEOTONE   RD

23W-38S
27

22W-38S
26

21W-38S
25

NORFOLK & WESTERN   RR

KENNEDY   RD.

KENN

OLD CHICAGO   RD.

RD.

34

23W-39S

SYMERTON   RD

WEST   35   KAHLER   RD.

22W-39S

RD.

MARTIN LONG

36

21W-39S

WARNER BRIDGE   RD.

TWP.   T 33 N
TWP.   T 32 N

WESLEY   RD

COUNTY

23W-40S
3

SYMERTON   RD

22W-40S
2

MARTIN LONG   RD

21W-40S
1

RD

BALLOU   RD

OLD CHICAGO   RD

23W-41S
10

22W-41S
11

DANIELSON   RD

HAZLETON   RD.

21W-41S
12

WARNER BRIDGE   RD

R 10 E   R 11 E

DONAHUE   RD

23W-42S
15

DONAHUE   RD

Creek

22W-42S
14

21W-42S
13

SEE PAGES 70-71

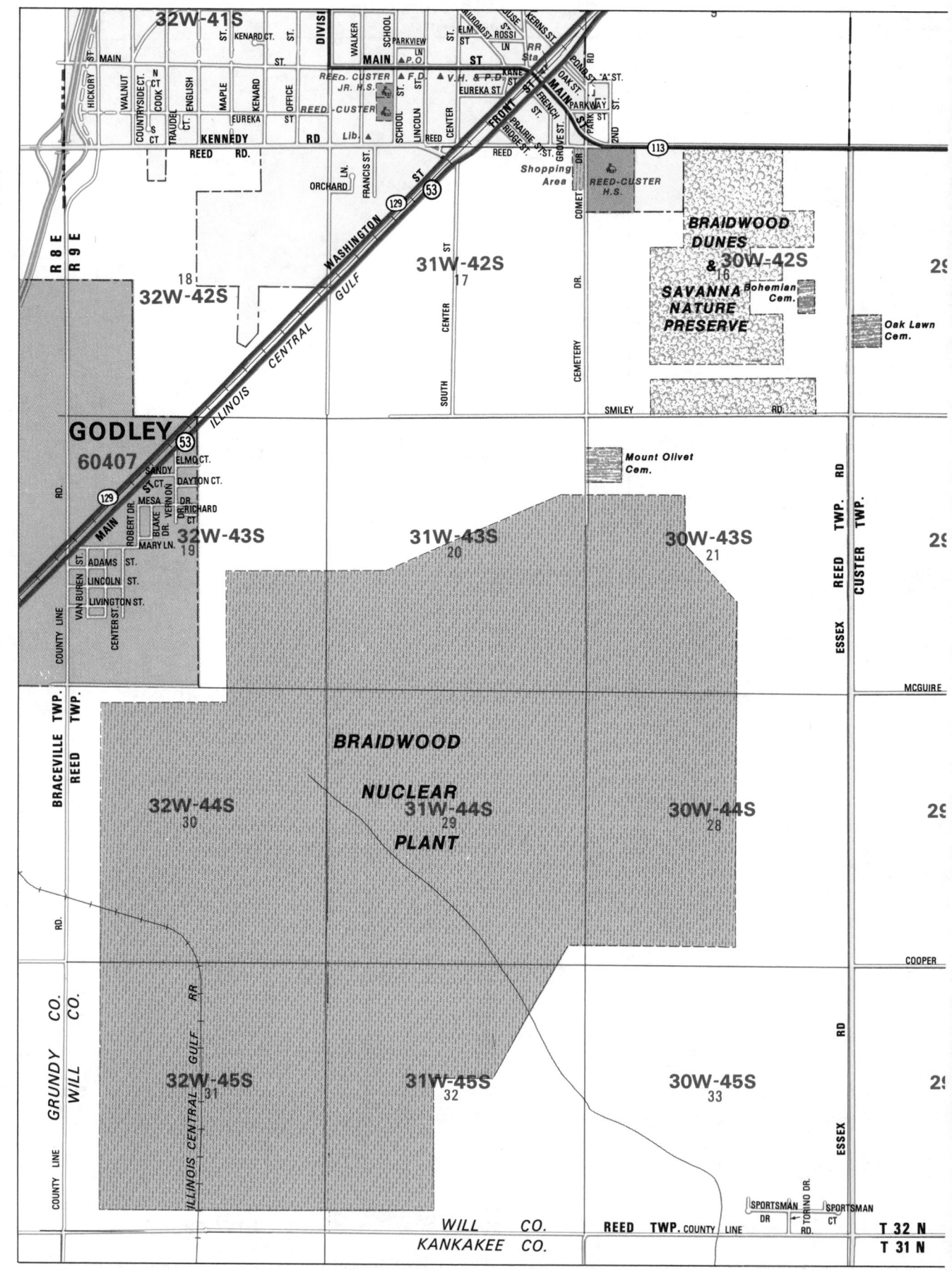

32W-41S

KENARD CT. ST.

MAIN ST.

HICKORY ST. MAIN

WALNUT

COUNTRYSIDE CT.

COOK CT. N

TRAUDEL CT. S CT.

ENGLISH

MAPLE

KENARD

OFFICE

WALKER

SCHOOL

PARKVIEW LN.

MAIN ST.

ELM ST.

LOUISE ST.

RAILROAD ST.

ROSSI

KERNS ST.

POND ST.

OAK ST.

"A" ST.

RR Sta.

KENNEDY RD.

REED RD.

ORCHARD LN.

FRANCIS ST.

REED-CUSTER JR. H.S.

SCHOOL ST.

LINCOLN ST.

REED

CENTER ST.

F.D.

P.O.

V.H. & P.D.

EUREKA ST.

KANE ST.

MAIN ST.

FRENCH ST.

PRAIRIE ST.

RIDGE ST.

GROVE ST.

2ND ST.

PARK ST.

PARKWAY

REED

129   **53**

WASHINGTON ST.

ILLINOIS CENTRAL GULF

Shopping Area

Comet Dr.

REED-CUSTER H.S.

113

31W-42S
17

CENTER ST.

SOUTH

CEMETERY DR.

SMILEY

BRAIDWOOD DUNES & 30W-42S
16

SAVANNA NATURE PRESERVE

Bohemian Cem.

Oak Lawn Cem.

RD.

29

GODLEY

60407

129   **53**

ELMO CT.

SANDY CT.

DAYTON CT.

MESA DR.

VERNON DR.

BLAKE DR.

ROBERT DR.

MARY LN.

RICHARD CT.

MAIN

Mount Olivet Cem.

ESSEX

REED TWP. RD.

CUSTER TWP.

32W-43S
19

ADAMS ST.

LINCOLN ST.

LIVINGTON ST.

VAN BUREN ST.

CENTER ST.

COUNTY LINE RD.

31W-43S
20

30W-43S
21

29

MCGUIRE

BRACEVILLE TWP.

REED TWP.

GRUNDY CO.

WILL CO.

ILLINOIS CENTRAL GULF RR

BRAIDWOOD

NUCLEAR

PLANT

32W-44S
30

31W-44S
29

30W-44S
28

29

COOPER

ESSEX RD.

32W-45S
31

31W-45S
32

30W-45S
33

29

COUNTY LINE

SPORTSMAN DR

TORINO DR.

SPORTSMAN CT

RD.

WILL   CO.

KANKAKEE   CO.

REED   TWP. COUNTY LINE

T 32 N

T 31 N

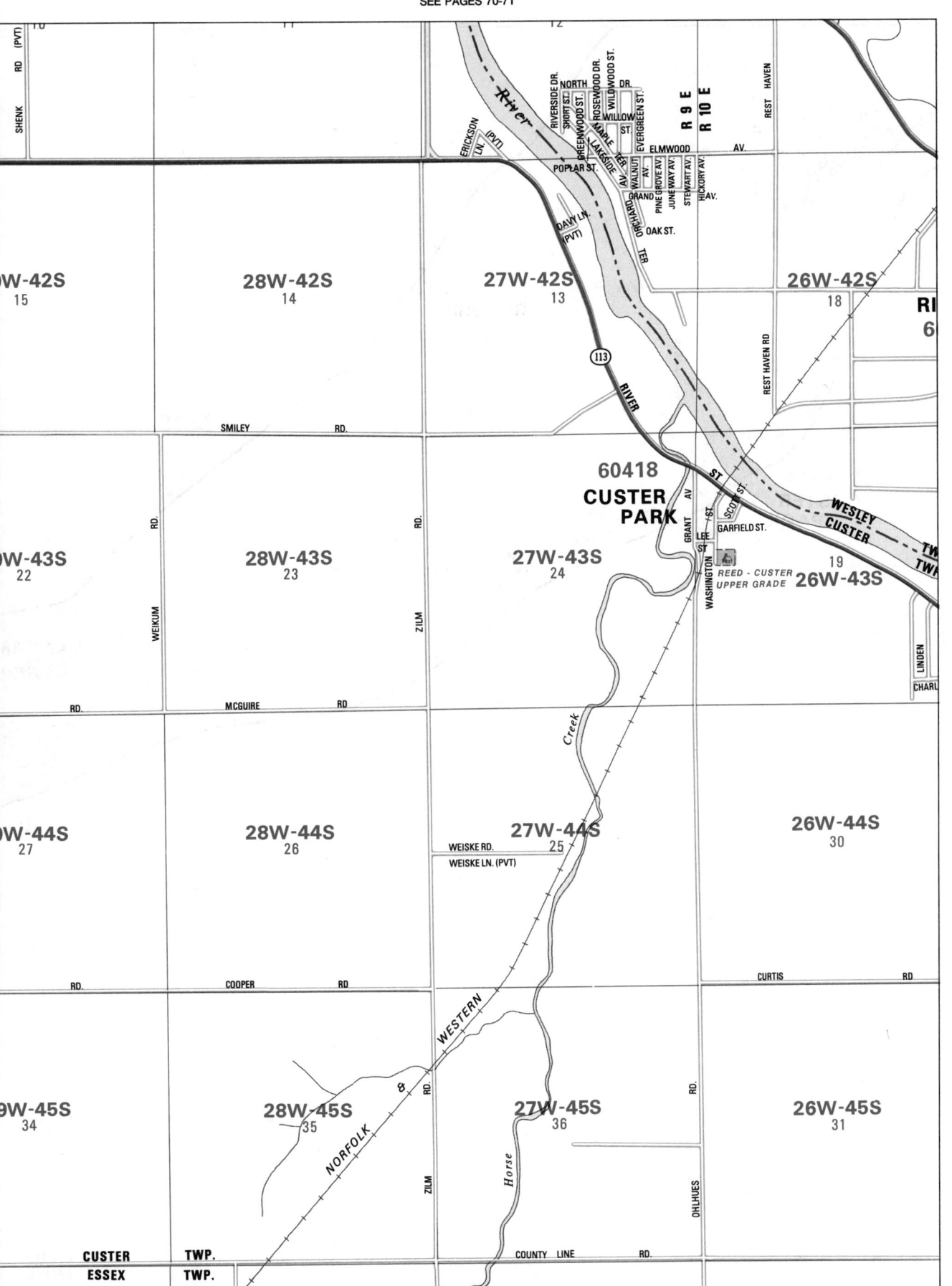

SHENK   RD. (PVT)

River

ERICKSON LN. (PVT)

RIVERSIDE DR.
SHORT ST.
NORTH
GREENWOOD ST.
LAKESIDE
MAPLE TER.
ROSEWOOD DR.
WILDWOOD ST.
WILDWOOD DR.
WILLOW ST.
EVERGREEN ST.
ELMWOOD   AV.
WALNUT AV.
PINE AV.
GROVE AV.
JUNE WAY AV.
STEWART AV.
HICKORY AV.
GRAND TER.
ORCHARD TER.
POPLAR ST.
OAK ST.
DAVY LN. (PVT)

R 9 E
R 10 E

REST HAVEN

9W-42S
15

28W-42S
14

27W-42S
13

26W-42S
18

RI
6

113

RIVER

SMILEY   RD.

REST HAVEN RD

60418
CUSTER PARK

WESLEY
CUSTER
TW
TW

RD.

WEIKUM

ZILM   RD.

GRANT AV
SCOTT ST.
LEE ST.
GARFIELD ST.
WASHINGTON

REED - CUSTER
UPPER GRADE

9W-43S
22

28W-43S
23

27W-43S
24

26W-43S
19

LINDEN
CHARL

RD.   MCGUIRE   RD

Creek

9W-44S
27

28W-44S
26

27W-44S
25

WEISKE RD.
WEISKE LN. (PVT)

26W-44S
30

RD.   COOPER   RD

CURTIS   RD.

WESTERN

Horse

NORFOLK
&
ZILM RD.

RD.

OHLHUES

9W-45S
34

28W-45S
35

27W-45S
36

26W-45S
31

CUSTER   TWP.
ESSEX   TWP.

COUNTY   LINE   RD.

SEE PAGES 72-73

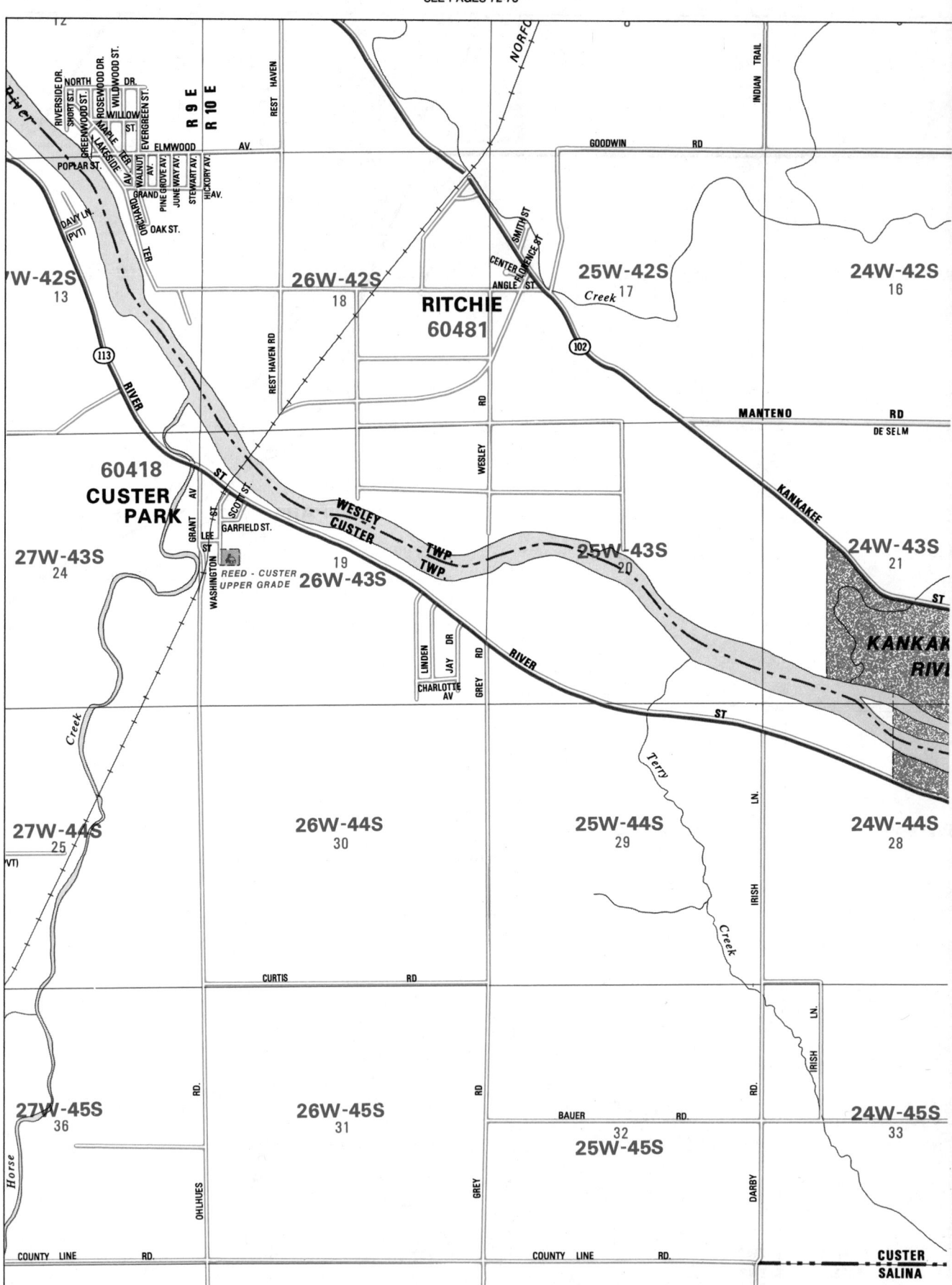

SEE PAGES 74-75

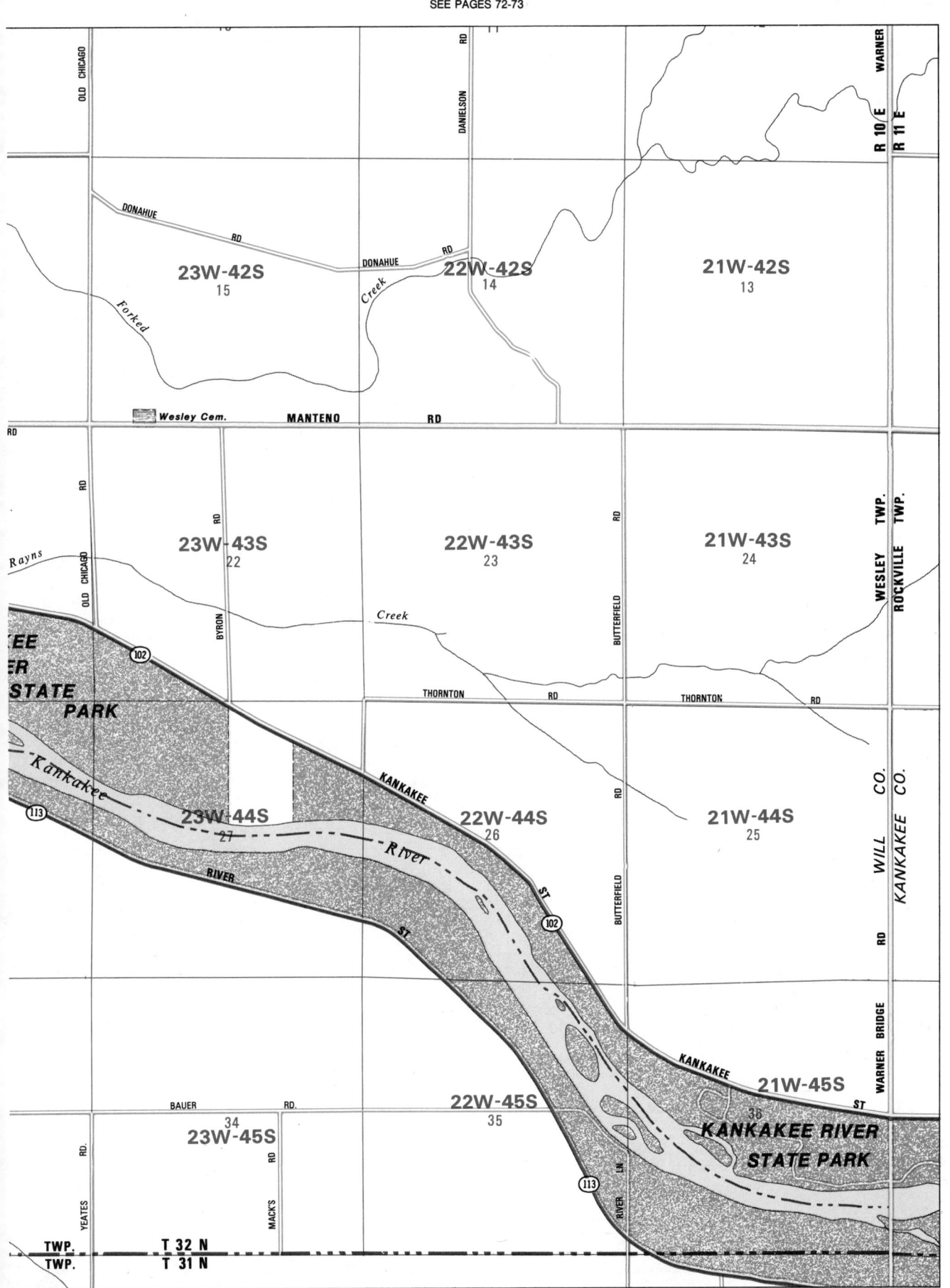

OLD CHICAGO

WARNER

DANIELSON RD

R 10 E
R 11 E

DONAHUE RD

23W-42S
15

DONAHUE RD

Creek

22W-42S
14

21W-42S
13

Forked

RD

RD

Rayns

OLD CHICAGO RD

BYRON RD

Wesley Cem.

MANTENO RD

23W-43S
22

Creek

22W-43S
23

BUTTERFIELD RD

21W-43S
24

WESLEY TWP.
ROCKVILLE TWP.

KEE
ER
STATE
PARK

102

Kankakee

113

THORNTON RD

THORNTON RD

KANKAKEE

23W-44S
27

River

RIVER

ST

ST

22W-44S
26

102

ST

RD

BUTTERFIELD RD

21W-44S
25

WILL CO.
KANKAKEE CO.

WARNER BRIDGE RD

KANKAKEE

BAUER RD.

34
23W-45S

RD.

MACK'S RD

22W-45S
35

113

RIVER LN

KANKAKEE

ST

21W-45S
36

**KANKAKEE RIVER
STATE PARK**

YEATES RD.

TWP.
TWP.

T 32 N
T 31 N

Street Index

Beecher

Street	Grid	Page
Caroline Av. N&S 0W-36S		68
Catalpa Ct. NW&SE . 0W-36S		68
Catalpa St. N&S 0W-36S		68
Chestnut Ln. E&W .. 0W-36S		68
Country Ln. E&W 0W-36S		68
Dixie Hwy. N&S		
.... 0W-37S,0W-36S		68
Donoko Av. N&S 0W-36S		68
Dunbar St. N&S 1W-36S		68
Elliott St. N&S 1W-36S		68
Elm St. E&W 1W-36S		68
Gould St. N&S 1W-36S		68
Hodges St. E&W		
.... 1W-36S,0W-36S		68
Indiana Av. E&W		
.... 1W-36S,0W-36S		68
Lange Av. N&S 1W-36S		68
Maxwell St. N&S		
.... 0W-37S,0W-36S		68
Meadow Ln. E&W .. 0W-36S		68
Melrose Ln. E&W 0W-37S		68
Miller St. E&W		
.... 1W-36S,0W-36S		68
Oak Dr. N&S 0W-36S		68
Oak Park Av. N&S .. 1W-36S		68
Orchard Ln. N&S 0W-36S		68
Park Dr. N&S 0W-36S		68
Park St. E&W 0W-36S		68
Pasadena Av. E&W .. 0W-37S		68
Penfield St. E&W		
.... 1W-36S,0W-36S		68
Prairie Av. N&S 0W-37S		68
Reed St. N&S		
.... 0W-37S,0W-36S		68
Rt. 1 N&S .. 0W-37S,0W-36S		68
Willow Ln. E&W 0W-36S		68
Woodward St. N&S		
.... 0W-37S,0W-36S		68
295th St. E&W		
.... 0W-36S,0E-36S		68
303rd St. E&W		
.... 0W-36S,0E-36S		68

Bolingbrook*
(Also see Du Page County)

Street	Grid	Page
Aberdeen Ct. N&S .. 22W-10S		7
Adams St. NE&SW		
.... 24W-12S		6
Alcester Cir. E&W .. 24W-10S		6
Alex Dr. E&W 26W-11S		6
Algonquin Ct. N&S.. 24W-11S		6
Allegheny Cir. 23W-11S		7
Allison Ct. NE&SW.. 22W-11S		7
Andover Ct. N&S .. 22W-12S		7
Anne Ct. NE&SW .. 21W-10S		7
Anne Dr. NE&SW .. 21W-10S		7
Anne Ln. E&W 21W-10S		7
Annerino Dr. N&S .. 22W-12S		7
Antelope Ct. NW&SE		
.... 26W-12S		6
Apache Ct. N&S .. 24W-11S		6
Applewood Dr. NE&SW		
.... 25W-11S		6
Appling Ln. E&W .. 23W-11S		7
Arcadia Ct. N&S.... 23W-11S		7
Ardglass Ct. NE&SW		
.... 22W-11S		7
Aristocrat Dr. N&S.. 25W-11S		6
Arrowhead Ln. E&W		
.... 23W-11S		7
Ashbury Av. N&S		
.... 22W-12S,22W-9S		7
Ashbury Ct. N&S		
.... 22W-10S		7
Ashbury Ln. N&S		
.........22W-11S,22W-10S		7
Ashley Rd. NE&SW		
.... 21W-10S		7
Aspen Dr. NE&SW .. 23W-11S		7
Assembly Ct. NW&SE		
.... 22W-10S		7
Assembly Dr. E&W 22W-10S		7
Bakersmill Dr. N&S 24W-12S		6
Balewood Dr. N&S.. 24W-12S		6
Balmoral Dr. N&S .. 22W-11S		7
Barbers Corner Rd. NS&EW		
.... 22W-11S		7
Barnwood Dr. NS&EW		
.... 25W-11S		6
Beaconridge Dr. N&S		
.... 22W-11S		7
Bear Creek Dr. NS&EW		
.... 26W-11S		6
Beaver Creek Dr. E&W		
.... 26W-11S		6
Bedford Dr. N&S .. 22W-11S		7
Beech Dr. E&W 24W-12S		6
Blackberry Ct. E&W		
.... 24W-12S		6
Blackberry Dr. N&S		
.... 24W-12S		6
Blackfoot Dr. N&S .. 26W-12S		6

Street	Grid	Page
Blackhawk Ln. E&W		
.... 24W-11S		6
Blair Ln. E&W 23W-11S		7
Bolingbrook Dr. N&S		
.... 23W-12S,23W-10S		7
Bonnie Brae Ct. NW&SE		
.... 22W-10S		7
Bonnie Brae Dr. NE&SW		
.... 22W-10S		7
Bothwell Ln. E&W .. 25W-11S		6
Boughton Rd. NE&SW		
.... 23W-11S,21W-10S		7
Boulder Ridge Ct. E&W		
.... 26W-11S		6
Boulder Ridge Dr. E&W		
.... 26W-11S		6
Bowie Ct.* N&S .. 24W-11S		6
Bowman Ct. N&S .. 22W-11S		7
Bradford Pl. N&S .. 25W-12S		6
Braemar Glen NS&EW		
.... 22W-12S		7
Bramble Av. E&W .. 25W-12S		6
Bramble Ct. NW&SE		
.... 25W-12S		6
Brandon Ct. N&S .. 23W-11S		7
Briarcliff Rd. E&W		
.... 24W-11S,26W-12S		6
Brighton Ln. N&S .. 25W-11S		6
Bristol Way NW&SE		
.... 24W-12S		6
Brittany Ln. E&W .. 25W-11S		6
Brompton Cir. NS&EW		
.... 22W-10S		7
Brompton Ct. NW&SE		
.... 25W-11S		6
Brook Ct. E&W 23W-11S		7
Brookwood Ln. N&S		
.... 25W-11S		6
Brookwood Ln. E. NS&EW		
.... 25W-11S		6
Brookwood Ln. W. NS&EW		
.... 25W-11S		6
Bryant Way E&W .. 24W-11S		6
Buck Eye E&W 24W-11S		6
Buckboard Ct. NE&SW		
.... 26W-13S		10
Buckboard Rd. NW&SE		
.... 26W-13S		10
Bunker Hill Dr. N&S		
.... 24W-12S		6
Bunting Cir. NW&SE		
.... 21W-10S		7
Butte Ln. NE&SW .. 26W-12S		6
Butte View Dr. NE&SW		
.... 26W-12S		6
Butternut Ct. E&W .. 24W-12S		6
Butternut Dr. N&S .. 24W-12S		6
Cades St. NE&SW.. 22W-10S		7
Calico Ct. NE&SW .. 26W-13S		10
Calisades Rd. N&S 21W-13S		11
Callery Ct. N&S 25W-12S		6
Callery Dr. N&S 25W-12S		6
Calumet Ct. N&S .. 24W-11S		6
Camelot Ct. NE&SW		
.... 22W-11S		7
Camelot Way E&W 22W-11S		7
Campbell Dr. N&S .. 25W-11S		6
Canterbury NE&SW		
.... 23W-12S		7
Canyon Dr. N&S		
.... 26W-12S,26W-11S		6
Capitol Dr. NW&SE		
.... 21W-11S		7
Carlyle Ln. E&W .. 24W-12S		6
Carmel Ln. E&W .. 23W-12S		7
Carol Ct. NE&SW .. 25W-11S		6
Carol Ln. NE&SW .. 25W-11S		6
Carriage Ln. E&W 26W-11S		6
Cascadia Ct. NW&SE		
.... 21W-13S		11
Castle Cir. E&W .. 21W-11S		7
Cedarwood Ct. N&S		
.... 22W-12S		7
Chantileed Av. E&W		
.... 25W-12S		6
Charleston Ct. N&S		
.... 22W-10S		7
Charleston Dr. E&W		
.... 22W-10S		7
Charlotte St. N&S .. 22W-10S		7
Charlotte Ln. E&W .. 22W-10S		7
Chase Ct. E&W 22W-10S		7
Cherokee Dr. N&S .. 24W-11S		6
Cherry Ct. E&W 23W-11S		7
Chesapeak Dr. NS&EW		
.... 21W-11S		7
Cheshire Ct. NE&SW		
.... 22W-11S		7
Cheyenne Ct. NW&SE		
.... 24W-11S		6
Chippewa Ct. N&S .. 24W-11S		6
Churchill Ct. E&W .. 25W-11S		6
Cinema Dr. E&W 22W-13S		11
Clear Dr. E&W 26W-12S		6
Cleveland Dr. E&W 24W-10S		6

Street	Grid	Page
Clifton Ln. NE&SW 25W-11S		6
Clover Ln. NS&EW 23W-10S		7
Cobb Cir. E&W 23W-11S		7
Cochise Cir. E&W .. 24W-11S		6
Collingwood Ln. NW&SE		
.... 25W-11S		6
Colony Ct. N&S 22W-12S		7
Columbus Ct. E&W		
.... 26W-11S		6
Comanche Dr. E&W		
.... 26W-12S		6
Commonwealth Ct. NE&SW		
.... 25W-11S		6
Commonwealth Dr. N&S		
.... 25W-11S		6
Concord Ln. NE&SW		
.... 23W-10S		7
Copper Ct. NE&SW		
.... 25W-11S		6
Coral Ct. E&W 23W-11S		7
Cornwall Ct. NW&SE		
.... 23W-10S		7
Cortez Dr. N&S 23W-11S		7
Cottonwood Cir. E&W		
.... 24W-11S		6
Cougar Dr. NW&SE		
.... 26W-12S		6
Country Ridge Ln. NS&EW		
.... 25W-11S		6
Creekside Dr. N&S.. 24W-12S		6
Crestwood Ln. NW&SE		
.... 25W-11S		6
Crossing Rd. NE&SW		
.... 22W-12S		7
Cumberland Ln. E&W		
.........24W-12S,23W-12S		6
Cypress Dr. E&W .. 23W-12S		7
Dakota Ct. N&S 24W-11S		6
Davinci Dr. E&W .. 22W-11S		7
Davis Ct. N&S...... 23W-11S		7
Davis Ln. N&S...... 23W-11S		7
Dean Cir. E&W 23W-11S		7
Deer Crossing Ln. NW&SE		
.... 22W-10S		7
Deerfield Ct. E&W .. 22W-12S		7
Degas Cir. E&W .. 22W-11S		7
Delaware Cir. NS&EW		
.... 23W-11S		7
Delaware Ln. E&W.. 23W-11S		7
Delmar Ct. E&W 23W-12S		7
Denver Yellow Pine Dr. N&S		
.... 24W-11S		6
Derbyshire Ct. NE&SW		
.... 23W-10S		7
Derbyshire Rd. NE&SW		
.... 23W-10S		7
Devon Ct. N&S 22W-10S		7
Devon Ln. NE&SW 22W-10S		7
Devonshire Ct. E&W		
.... 22W-12S		7
Diane Ln. NW&SE .. 22W-10S		7
Donovan Rd. E&W .. 22W-13S		11
Dorchester Ct. N&S		
.... 22W-10S		7
Dorchester Dr. N&S		
.... 22W-10S		7
Douglass Way NW&SE		
.... 22W-10S		7
Drake Av. E&W .. 25W-12S		6
Drake Ct. N&S 25W-12S		6
DuPage Dr. N&S .. 24W-11S		6
Dubois Cir.* N&S .. 23W-10S		7
Dunham Ct. NE&SW		
.... 21W-10S		7
Dunham Ln. NS&EW		
.... 21W-10S		7
Dunmore Ct. NW&SE		
.... 23W-10S		7
Eagle Ln. E&W 24W-11S		6
Edgehill Dr. E&W .. 23W-11S		7
Edinburgh Ct. E&W		
.... 22W-10S		7
Elkhorn Ct. N&S.... 24W-11S		6
Elm Ct. N&S 22W-11S		7
Emerson Cir. E&W.. 24W-11S		6
Ennis Ln. E&W 22W-10S		7
Eric Way NS&EW .. 22W-10S		7
Essex Ct. E&W 24W-12S		6
Essington Rd. N&S		
.... 28W-13S		9
Evergreen Ct. E&W		
.... 22W-12S		7
Fairwood Ct. N&S		
.... 23W-12S		7
Fairwood Dr. E&W .. 23W-12S		7
Falcon Ridge Way E&W		
.........22W-11S,21W-11S		7
Falmore Ln. N&S .. 23W-11S		7
Far Hills Dr. N&S .. 24W-11S		6
Farm Gate Ln. N&S		
.... 25W-12S		6
Faversham Ct. E&W		
.... 22W-11S		7
Fawn Ct. E&W 26W-12S		6

Street	Grid	Page
Feather Sound Dr. NS&EW		
.... 21W-10S		7
Feathersound Dr. NS&EW		
.... 21W-10S		7
Fernwood Dr. NS&EW		
.... 23W-11S		7
Flagstaff Dr. N&S .. 24W-11S		6
Fleetwood Ct. E&W		
.... 22W-12S		7
Forest Ct. N&S 21W-11S		7
Forest Way E&W .. 21W-11S		7
Foxhead Ct. E&W .. 24W-11S		6
Foxwood Ln. N&S .. 26W-12S		6
Friars Ct. NE&SW .. 22W-11S		7
Fuller Ln.* NS&EW .23W-10S		7
Galahad Rd. E&W .. 22W-11S		7
Gannet Ln. NE&SW		
.... 21W-10S		7
Garden Dr. N&S .. 24W-12S		6
Gary E&W 21W-11S		7
Gehrig Cir. E&W .. 23W-11S		7
Genesee Ct. E&W .. 23W-11S		7
Gibraltar NW&SE .. 22W-12S		7
Glen Lake Dr. NS&EW		
.... 25W-11S		6
Giengary Dr. E&W .. 23W-12S		7
Grace Ct. N&S 23W-11S		7
Grady Dr. NE&SW .. 25W-12S		6
Granada Ct. E&W .. 23W-11S		7
Grand Canyon E&W		
.... 23W-11S		7
Grant Dr. E&W 23W-11S		7
Green Mountain Ct. E&W		
.... 21W-11S		7
Green Mountain Dr. NS&EW		
.... 21W-11S		7
Greencrest Dr. E&W		
.... 22W-11S		7
Greentree Ct. N&S .. 22W-12S		7
Greentree Ln. N&S 22W-12S		7
Greenwood Ct. E&W		
.... 22W-12S		7
Grove Ct. NE&SW .. 24W-12S		6
Grove Ln. NW&SE .. 24W-12S		6
Gull Point Ct. N&S .. 21W-10S		7
Hadleigh Rd. E&W .. 22W-11S		7
Hallmark Ln. N&S .. 22W-12S		7
Hampoon Ct. N&S .. 24W-12S		6
Hampshire Ct. E&W		
.... 22W-10S		7
Hampshire Ln. NE&SW		
.... 22W-10S		7
Hanover Ct. N&S .. 22W-12S		7
Harris Dr. E&W 24W-11S		6
Hartford Ln. N&S .. 25W-11S		6
Harvest Ct. NE&SW		
.... 26W-13S		10
Harvest Dr. N&S.... 26W-13S		10
Harwood Dr. N&S .. 23W-12S		7
Hawthorne Ct. E&W		
.... 24W-12S		6
Heritage Ln. E&W .. 26W-11S		6
Heron Cir. E&W .. 21W-10S		7
Hickory Oaks Ct. NW&SE		
.... 26W-11S		6
Hickory Oaks Dr. N&S		
.... 26W-11S		6
Hidden Valley Ct. NE&SW		
.... 26W-11S		6
Hidden Valley Dr. E&W		
.... 26W-11S		6
Home Pl. E&W .. 25W-12S		6
Homer E&W 22W-11S		7
Homestead Dr. NS&EW		
.... 21W-11S		7
Homewood Dr. N&S		
.... 21W-11S		7
Hudson Ct. N&S.... 24W-11S		6
Hughes Pl. N&S 22W-10S		7
Huntington Way N&S		
.........22W-11S,22W-10S		7
Hywood Ct. N&S .. 23W-11S		7
Idlewood Ct. E&W .. 22W-12S		7
Independence Ct. N&S		
.... 22W-11S		7
Independence Dr. N&S		
.... 22W-11S		7
Ingleside Dr. N&S .. 25W-12S		6
International E&W		
.... 22W-12S,21W-12S		7
Ironbark Way NS&EW		
.... 22W-12S		7
Jameson Ct. NE&SW		
.... 22W-11S		7
Jamestown Ln. E&W		
.... 21W-12S		7
Janes Av. N&S		
.........21W-11S,21W-10S		7
Jeffrey Ln.* E&W .. 23W-10S		7
Jennifer Ln. NS&EW		
.... 25W-11S		6
Jerome Ln.* N&S .. 23W-10S		7
Jill Ln.* E&W 23W-10S		7
Joliet Rd. NE&SW .. 21W-11S		7
Jordan Way N&S .. 22W-10S		7

Street	Grid	Page
Joy Dr. N&S		
.........22W-11S,22W-10S		7
Julie Rd. E&W 22W-11S		7
Juniper Ln. E&W .. 23W-11S		7
Justine Av. NE&SW		
.... 22W-11S		7
Karen Cir. N&S 21W-11S		7
Kedvale Ct. N&S .. 24W-11S		6
Kenilworth Dr. NS&EW		
.... 22W-11S		7
Kenmore Dr. N&S .. 22W-12S		7
Keystone Ct. NE&SW		
.... 22W-10S		7
Keystone Dr. N&S .. 22W-10S		7
Kildeer Dr. N&S 21W-10S		7
King Arthur Ct. N&S		
.... 21W-11S		7
King Arthur Way E&W		
.... 21W-11S		7
Kings Rd. N&S		
.........27W-14S,27W-13S		9
Kingston Rd. N&S .. 22W-12S		7
Kirkwood Cir. E&W		
.... 23W-11S		7
Knight Ct. N&S 22W-11S		7
Lacross N&S 23W-11S		7
Lafayette E&W .. 23W-11S		7
Lakeside Dr. N&S .. 23W-11S		7
Lakewood Farms Rd. NS&EW		
.... 26W-13S		10
Lancaster Dr. N&S		
.........23W-11S,23W-12S		7
Lancelot Ct. N&S .. 22W-11S		7
Lancelot Way N&S ..22W-11S		7
Langford Ct. N&S .. 22W-10S		7
Langford Dr. E&W .. 22W-10S		7
Larchmont Ln. N&S .. 24W-12S		6
Laurie Cir. NE&SW 21W-11S		7
Lavina Ct. NE&SW .. 24W-10S		6
Lawton Ln. N&S .. 22W-11S		7
Lee Ln. E&W 23W-11S		7
Leicester Ct. NW&SE		
.... 22W-11S		7
Leslie Ln. E&W 22W-10S		7
Lexington Dr. NS&EW		
.... 23W-11S		7
Liberty Ct. N&S 22W-10S		7
Liberty Dr. E&W .. 22W-10S		7
Lily Cache Ln. E&W		
.... 25W-12S		6
Lincolnshire Ct. E&W		
.... 23W-11S		7
Linconshire Ln. N&S		
.... 23W-11S		7
Linden Ct. E&W .. 24W-12S		6
Lindsey Ct. E. E&W		
.... 25W-11S		6
Lindsey Ct. W. E&W		
.... 25W-11S		6
Lindsey Ln. NE&SW		
.... 25W-11S		6
Lyons Rd. N&S 23W-11S		7
Malibu Ct. N&S 22W-11S		7
Malibu Dr. N&S 22W-11S		7
Mallview Ln. N&S .. 22W-11S		7
Manor Ct. NW&SE .. 22W-12S		7
Maplewood Dr. N&S		
.... 23W-12S		7
Marian Ct. N&S 22W-11S		7
Marquette Ct. E&W		
.... 22W-10S		7
Mayfield Dr. N&S .. 23W-11S		7
Mayflower Ln. N&S		
.... 23W-11S		6
Meadowbrook Dr. NS&EW		
.... 22W-11S		7
Medical Center Dr. E&W		
.... 23W-13S		11
Melissa Dr. NS&EW		
.... 21W-11S		7
Mellbrook Ln. N&S.. 22W-12S		7
Merlin Ct. N&S 22W-11S		7
Merrill Ct. N&S 25W-11S		6
Mesa Dr. E&W 26W-12S		6
Michael Dr. NE&SW		
.... 26W-11S		6
Michelangelo NE&SW		
.... 22W-11S		7
Millsteam Dr. N&S .. 24W-12S		6
Miner Dr. E&W 22W-10S		7
Monet E&W 22W-11S		7
Monroe Rd. NW&SE		
.... 22W-10S		7
Monterey Dr. N&S .. 22W-11S		7
Monticello Cir. NS&EW		
.... 23W-11S		7
Moore NW&SE 22W-11S		7
Musial Ct. E&W .. 23W-11S		7
Mutwood Ct. NE&SW		
.... 22W-11S		7
Naperville Rd. NW&SE		
.... 25W-12S		6
Nausau Av. N&S.... 22W-11S		7
Navajo Ct. E&W .. 23W-11S		7

	Grid	Page
New Castle Ln. NE&SW	23W-12S	7
Newberry Ln. NS&EW	23W-12S	7
Newport Dr. N&S	23W-12S	7
Newton Ct. NE&SW	22W-11S	7
Niagara Dr. E&W	21W-11S	7
Nicholas Dr. NE&SW	26W-11S	6
Norman Way NW&SE	22W-10S	7
Northridge Dr. E&W	22W-11S	7
Northridge Ct. N&S	22W-11S	7
Nottingham Dr. N&S	23W-11S	7
Oakwood Dr. N&S	23W-12S	7
Old Chicago Dr. E&W	23W-13S	11
Old Elm Dr. N&S	23W-11S	7
Old Stone Rd. E&W	24W-12S	6
Olive Pl. NE&SW	22W-10S	7
Olympic Dr. E&W	22W-11S	7
Orchard Dr. NW&SE	24W-12S	6
Ottawa Dr. N&S	22W-11S	7
Overland Dr. N&S	24W-13S	10
Oxford Rd. NE&SW	22W-12S	7
Ozark Dr. N&S	23W-11S	7
Pacific Dr. N&S	22W-11S	7
Pamela Dr. E&W	22W-10S	7
Park Ct. E&W	21W-11S	7
Park Lawn Ct. NE&SW	22W-11S	7
Park Meadow Ln. NE&SW	22W-11S	6
Parkside Ct. NE&SW	26W-13S	10
Parkside Dr. N&S	26W-13S	10
Parliament N&S	22W-10S	7
Partridge Av. E&W	25W-12S	6
Pebble Ct. NW&SE	25W-11S	6
Pembrook Ct. E&W	25W-11S	6
Pennsbury Ct. E&W	25W-11S	6
Penny Ct. NW&SE	25W-11S	6
Penny Ln. NW&SE	25W-11S	6
Penwood Ln. NW&SE	25W-11S	6
Pepperwood Dr. N&S	24W-12S	6
Pierce Dr. N&S	23W-11S	7
Pilgrim Ct. N&S	21W-12S	7
Pinecrest Ct. E&W	22W-10S	7
Pinecrest Rd. N&S	22W-12S,22W-10S	7
Pinto Dr. E&W	24W-11S	6
Planeview Ln. NS&EW	24W-12S	6
Plymouth Ln. N&S	23W-11S	7
Ponderosa Dr. E&W	24W-12S	6
Pontiac Ln. E&W	24W-11S	6
Poplar Ln. E&W	24W-12S	6
Porter Ln. E&W	22W-10S	7
Portsmouth Ln. E&W	25W-12S	6
Potomac Dr. E&W	23W-11S	7
Prairie Cir. NW&SE	24W-12S	6
Preston Dr. N&S	21W-10S	7
Princeton Dr. E&W	24W-12S	6
Providence Ct. N&S	22W-12S	7
Prudentail Corp. Cr. NW&SE	22W-11S	7
Pueblo Dr. N&S	24W-11S	6
Puffin Ct. NW&SE	21W-10S	7
Purchase Ct. E&W	22W-10S	7
Quadangle Dr. N&S	23W-12S	7
Quail Run E&W	25W-12S	6
Quarry N&S	25W-12S	6
Queenswood Av. NS&EW	21W-11S	7
Racine Ln.* E&W	23W-10S	7
Radcliff Dr. N&S	24W-12S	6
Raphael E&W	22W-11S	7
Red Tr. NS&EW	26W-13S	10
Redspier Ct. NW&SE	25W-12S	6
Redwing Ct. N&S	24W-11S	6
Redwood Rd. E&W	21W-11S	7
Regent Ct. E&W	24W-12S	6
Remington Blvd. E&W	25W-13S,24W-13S	10
Revere Ct. NW&SE	22W-11S	7
Richmond Dr. N&S	24W-12S	6
Ridgewood Ct. E&W	25W-12S	6
Ridgewood Dr. NS&EW	25W-12S	6
Roanoke Ct. N&S	23W-11S	7
Roberts Rd.* E&W	23W-10S	7
Robinhood Ct. E&W	22W-11S	7
Robinhood Way NS&EW	23W-11S,22W-11S	7

	Grid	Page
Rockhurst Rd. E&W	22W-11S,21W-11S	7
Rocklyn Ct. E&W	23W-12S	7
Roman Cir. NS&EW	21W-11S	7
Rose Hill Dr. E&W	24W-11S	6
Rothbury Dr. E&W	22W-10S	7
Round Hill Ct. NE&SW	21W-11S	7
Royal Crest Ct. NW&SE	22W-10S	7
Ruth Cir. E&W	23W-11S	7
Rye Ct. E&W	22W-10S	7
Salem Sq. NS&EW	22W-12S	7
Sandalwood Ct. NE&SW	25W-11S	6
Sandalwood Ln. NW&SE	25W-11S	6
Scmidt Rd N&S	24W-13S,24W-11S	10
Scott Dr. NE&SW	22W-10S	7
Seabury Rd. NS&EW	22W-11S,21W-11S	7
Seminole Ln. E&W	23W-11S	7
Seneca Ln. N&S	24W-11S	6
Seneca Ln. E&W	23W-11S	7
Seneca Way N&S	24W-11S	6
Shady Ln. N&S	24W-12S	6
Sharon Dr. NW&SE	26W-12S	6
Shawnee N&S	24W-11S	6
Shenandoah Ct. N&S	23W-11S	7
	22W-12S	7
Sherington Ct. N&S	22W-11S	7
Sherwood Ct. N&S	22W-11S	7
Shield Ct. N&S	22W-11S	7
Shilling Ct. N&S	22W-10S	7
Sierra Ln. E&W	23W-11S	7
Sioux E&W	23W-11S	7
Skokie Ln. E&W	23W-11S	7
Somerset Ln. E&W	22W-12S	7
Spangler Dr. N&S	23W-15S	10
Springhill Dr. N&S	24W-12S	6
Springleaf Dr. N&S	24W-12S	6
Springwood Way NS&EW	25W-11S	6
Spruce Rd. E&W	21W-11S	7
St. Andrews Dr. E&W	22W-10S	7
Stafford Ct. N&S	22W-11S	7
Stafford Way N&S	22W-11S	7
Standish Ct. N&S	22W-11S	7
Staunton Ct. N&S	22W-11S	7
Steck Ct. NW&SE	24W-10S	6
Stonegate Ct. N&S	25W-11S	6
Stoneham Ct. NE&SW	22W-11S	7
Sullivan E&W	22W-11S	7
Summit Ln. NW&SE	24W-12S	6
Sundance Dr. E&W	24W-12S	6
Sunset Ln. N&S	22W-11S	7
Sunshine Ct. E&W	26W-11S	6
Sunshine Dr. N&S	26W-11S	6
Sunshine E. Dr. E&W	26W-11S	6
Swan Cir. N&S	21W-10S	7
Swinford Ct. NW&SE	23W-10S	7
Sword Way N&S	21W-11S	7
Sylvan Rd. N&S	22W-11S	7
Tahoe Ct. E&W	25W-12S	6
Talcot Ln. N&S	22W-11S	7
Tam O'Shanter Ct. N&S	22W-10S	7
Tam O'Shanter Dr. N&S	22W-10S	7
Tamarack Dr. N&S	24W-12S	6
Tamms Ln. E&W	24W-12S	6
Tarrington Way NW&SE	22W-11S	7
Taylor Ct. N&S	25W-12S	6
Tecumseh Dr. N&S	26W-12S	6
Tellemesh Dr. N&S	26W-12S	6
Territorial Dr. E&W	25W-13S	10
Thackery Dr. E&W	25W-12S	6
Thistle Av. NS&EW	25W-12S	6
Thomas Rd. N&S	22W-10S	7
Thornhurst Dr. NS&EW	22W-11S,21W-11S	7
Tilden Ln. NS&EW	25W-12S	6
Timber Ct. E&W	24W-12S	7
Tomahawk Ct. N&S	24W-11S	6
Towner Ct. N&S	25W-12S	6
Towner Ln. N&S	25W-12S	6
Tracy Way NS&EW	22W-10S	7
Tudor Ct. N&S	22W-11S	7
Tuttle Dr. E&W	27W-11S	5
Upton Rd. N&S	21W-11S	7
Valley Dr. N&S	24W-12S	6
Van Gogh Cir. N&S	22W-12S	7
Venetian Dr. NE&SW	22W-12S	7
Vermont Dr. E&W	22W-10S	7
Vernon Rd. E&W	22W-11S,21W-11S	7

	Grid	Page
Victoria Ct. N&S	22W-10S	7
Vigilance Ct. E&W	22W-10S	7
Vincent Dr. N&S	26W-11S	6
Vista Dr. N&S	26W-11S	6
Wagon Wheel Ct. NW&SE	26W-13S	10
Walden Ct. E&W	25W-11S	6
Walker Dr. E&W	22W-10S	7
Walnut Cir. NS&EW	23W-11S	7
Warwick Ct. E&W	21W-11S	7
Waterman Dr. N&S	24W-12S	6
Weber Rd. NE&SW	25W-12S	6
Wedgewood Way E&W	22W-11S	7
Wellington Ln. N&S	22W-12S	7
West Wind Dr. E&W	24W-12S	6
Westchester Ct. NE&SW	22W-10S	7
Westchester Ln. N&S	22W-10S	7
Wethersfield Ln. N&S	22W-12S	7
Whitbury Ct. N&S	22W-10S	7
White Oak Rd. E&W	21W-11S	7
Whitwater Dr. N&S	24W-11S	6
Wildwood Ln. E&W	23W-11S	7
Williamsburg N&S	22W-12S	7
Willow Ct. E&W	23W-11S	7
Willoway Dr. N&S	21W-11S	7
Wimbledon Ct. NW&SE	22W-10S	7
Windsor Dr. N&S	24W-11S	6
Winston Dr. NE&SW	22W-11S	7
Wobbler Dr. N&S	21W-10S	7
Wood Ct. NE&SW	25W-11S	6
Woodcreek Rd. N&S	23W-13S,23W-12S	11
Wright Ct. NE&SW	22W-11S	7
Wyeth N&S	22W-11S	7
Yellow Pine Dr. N&S	24W-11S	6
York St. N&S	23W-11S	7
Yorkshire Sq. NS&EW	22W-12S	7
107th St. N&S	25W-12S,21W-12S	6
111th St. E&W	23W-13S,23W-12S	11
119th St. E&W	27W-13S	9

Braidwood

	Grid	Page
'A' St. E&W	30W-41S	70
'B' St. N&S	31W-41S	70
Barr St. N&S	31W-41S	70
Bergera Rd. E&W	32W-41S,31W-41S	70
Center St. N&S	31W-41S,31W-40S	70
Church St. E&W	32W-41S,31W-41S	70
Coal City Rd. E&W	32W-39S	70
Commercial Av. N&S	31W-41S	70
Cook St. N&S	32W-41S	70
Corey Av. NW&SE	32W-41S	70
Countryside Ct. N&S	32W-41S	70
County Line Rd. N&S	32W-41S	70
Curtis Ct. E&W	31W-41S	70
Dalton St. E&W	31W-41S	70
Daniels (2nd St.) E&W	32W-41S,31W-41S	70
David St. E&W	31W-41S	70
Division St. (Hwy. 113) N&S	32W-41S,32W-39S	70
Elm St. E&W	31W-42S	74
English St. N&S	32W-41S	70
Erie St. N&S	31W-40S	70
Eureka St. E&W	32W-41S,31W-41S	70
Francis St.	31W-42S	74
French St. N&S	31W-41S	70
Front St. (Il. 53) NE&SW	31W-41S	70
Grove St. N&S	31W-41S	70
Hall St. E&W	32W-41S,31W-41S	70
Hickory St. N&S	31W-41S	70
Highway 113 EW&NS	32W-39S,30W-41S	70
Interstate 55 NE&SW	32W-41S,32W-39S	70
Kane St. N&S	31W-41S	70
Kenard Ct. E&W	31W-41S	70
Kennedy Rd. (Reed Rd.) E&W	32W-41S,31W-41S	70
Kerns St. NW&SE	31W-41S	70
Lincoln St. N&S	31W-41S	70
Main St. E&W	32W-41S,31W-41S	70
Maple St. N&S	31W-41S	70
Mason (2nd) St. E&W	31W-41S	70
Michigan St. N&S	31W-40S	70
Mitchell St. N&S	31W-41S	70

	Grid	Page
Oak St. NW&SE	31W-41S	70
Office St. N&S	32W-41S	70
Orchard Ln. NS&EW	31W-42S	74
Park Av. N&S	31W-41S	70
Park Rd. N&S	31W-41S	70
Parkview Ln. E&W	31W-41S	70
Parkway St. E&W	31W-41S,30W-41S	70
Pond St. NW&SE	31W-41S	70
Prairie St. NW&SE	31W-41S	70
Railroad St. NW&SE	31W-40S	70
Reed Rd. E&W	32W-41S	70
Reed St. N&S	31W-41S	70
Ridge St. NW&SE	31W-41S	70
Rossi Ln. E&W	31W-41S	70
Round House St. NW&SE	31W-41S	70
Route 53 NE&SW	31W-42S,30W-41S	74
Route 129 NE&SW	31W-42S,30W-41S	74
Schmidt Rd. E&W	31W-40S	70
School St. N&S	31W-41S,31W-40S	70
Small St. E&W	31W-40S	70
Spring Rd. E&W	32W-41S	70
Sullivan St. N&S	31W-41S	70
Susan Ct. N&S	31W-41S	70
Traudel St. N&S	32W-41S	70
Walker St. N&S	31W-41S	70
Walnut St. N&S	32W-41S	70
Washington St. NE&SW	31W-42S,30W-41S	74
2nd St. (Mason St.) E&W	32W-41S,31W-41S	70
3rd St. E&W	31W-41S	70
4th St. E&W	31W-41S	70
6th St. E&W	31W-40S	70
7th St. E&W	31W-40S	70

Channahon

	Grid	Page
Andrew Av. E&W	28W-27S	27
Belles Ln. N&S	29W-27S	26
Birch St. E&W	31W-30S	34
Blackberry St. N&S	31W-31S,31W-30S	34
Bluebird Ct. N&S	30W-29S	34
Bluff Rd. E&W	31W-29S,30W-29S	34
Bluff Ridge Ct. NW&SE	30W-29S	34
Bluff St. N&S	31W-30S	34
Bridge St. E&W	30W-30S,31W-30S	34
Bud St. N&S	29W-27S	26
Buell St. NE&SW	31W-30S	34
Burr Ct. N. NW&SE	30W-29S	34
Burr Ct. S. NE&SW	30W-29S	34
Burr Rd. N&S	30W-29S	34
Canal St. NW&SE	31W-30S	34
Cardinal Dr. N&S	30W-29S	34
Carrie Av. N&S	28W-27S	27
Center St. N&S	31W-30S	34
Channahon Dr. E&W	31W-30S	34
Cherokee Tr. E&W	31W-30S	34
Cherry St. N&S	31W-30S	34
Chipwood Dr. NS&EW	31W-29S	34
Deal Av. N&S	28W-27S	27
Devon Dr. NE&SW	29W-27S	26
Donna Av. N&S	28W-27S	27
Eames St. N&S	32W-30S,29W-28S	34
Edwin C. Dr. N&S	30W-29S	34
Edwin Dr. N&S	30W-29S	34
Egret Dr. NE&SW	30W-29S	34
Elsie Av. N&S	31W-30S	34
Essex Dr. NE&SW	29W-27S	26
Ford Rd. N&S	32W-29S	34
Frances Ct. NE&SW	29W-27S	26
Frances Way NW&SE	29W-27S	26
Freyer St. NW&SE	31W-30S	34
Goldfinch Ct. N&S	30W-29S	34
Green Heron Dr. N&S	30W-29S	34
Harriet St. N&S	29W-27S	26
Hawthorne Av. N&S	29W-27S	34
Heritage Valley Dr. NE&SW	31W-29S	34
Hummingbird Dr. E&W	30W-29S	34
Interstate 55 N&S	30W-29S,29W-27S	34
Janet Ln. E&W	30W-29S	34
Jessup St. E&W	31W-30S	34
Joliet St. N&S	31W-30S	34
Joliet-Channahon Rd. NE&SW	29W-28S,30W-28S	34
Joseph Av. N&S	28W-27S	27
Junior Av. N&S	30W-29S	34
Karl Av. E&W	28W-27S	27
Kathey Dr. E&W	29W-27S	26

	Grid	Page
Kent Rd. NW&SE	29W-27S	26
Kevins Dr. NW&SE	32W-31S	34
Knapps St. E&W	31W-30S	34
Kurt Rd. E&W	29W-27S	26
Lois Av. E&W	31W-30S	34
Majestic Dr. N&S	31W-29S	34
Manor Ln. N&S	29W-28S	34
Maple St. E&W	31W-30S	34
Mary Lou Av. E&W	28W-27S	27
Matthew Ct. E&W	30W-29S	34
May Ct. N&S	31W-31S	34
May St. E&W	32W-31S,31W-31S	34
McClintock Rd. NE&SW	29W-27S,28W-27S	26
McCowan St. E&W	31W-30S	34
McKinley Rd. N&S	32W-32S	46
Meadowlark Dr. E&W	30W-29S	34
Michael Dr. NE&SW	29W-27S	26
Moorman Av. NE&SW	30W-29S	34
North St. NW&SE	30W-29S	34
North St. E&W	31W-30S	34
Northern Illinois Dr NW&SE	30W-29S	34
Oakview Ct. E&W	32W-31S	34
Parkside Dr. N&S	32W-31S	34
Patricia Ln. NW&SE	29W-27S	26
Peak Dr. E&W	30W-29S	34
Pine St. E&W	31W-30S	34
Plum St. N&S	31W-30S	34
Pottawatomie Tr. NE&SW	32W-30S	34
Quail Dr. E&W	30W-29S	34
River Tr. N&S	32W-30S	34
Roberts Av. E&W	30W-29S	34
Robin Ct. N&S	30W-29S	34
Rock Ridge Ct. N&S	30W-29S	34
Sage St. N&S	31W-30S	34
Sandpiper Dr. N&S	30W-29S	34
South St. E&W	31W-30S	34
Spruce St. E&W	31W-30S	34
Steve Ritoff Dr. NW&SE	30W-29S	34
Story St. NE&SW	31W-30S	34
Summit Rd. E&W	30W-29S	34
Sunset Dr. N&S	30W-29S	34
Sussex Dr. E&W	29W-27S	26
Teal Dr. NS&EW	30W-29S	34
Tonka Dr. NE&SW	30W-29S	34
Tree Ridge Ct. NW&SE	30W-29S	34
Tryon St. N&S	31W-30S,31W-29S	34
Turnstone Blvd E&W	30W-29S	34
U.S. Route 6 N&SE	32W-30S,29W-27S	34
Valley Dr. NS&EW	30W-29S	34
Willard St. N&S	31W-30S,31W-29S	34

Channahon Township

	Grid	Page
Arsenal Rd. E&W	30W-31S,27W-30S	34
Bell Rd. N&S	32W-30S,32W-29S	34
Blodget Rd. E&W	32W-33S	46
Bluff Rd. E&W	31W-30S,29W-30S	34
Bonnell Ct. NW&SE	32W-33S	46
Canal Rd. NE&SW	30W-28S	34
Center St. N&S	30W-29S	34
Channahon Rd. NE&SW	30W-28S,29W-28S	34
Cottonwood Ct. NE&SW	32W-28S	34
Cottonwood Ln. N&S	32W-28S	34
Dirst Rd. N&S	32W-28S,32W-29S	34
Durkee Rd. E&W	30W-31S	34
Eames St. E&W	32W-30S,29W-28S	34
Ford Rd. N&S	32W-29S	34
Front St. E&W	31W-30S,30W-30S	34
Grande Dr. N&S	32W-28S	34
Graystone Ct. NW&SE	32W-33S	46
Grundy County Line Rd. N&S	32W-33S	46
Hickory Ct. E&W	32W-28S	34
Hickory Ln. N&S	32W-28S	34
Ingolsby Rd. N&S	32W-28S	34
Interstate 55 N&S	30W-30S	34
Joliet Rd. E&W	31W-30S,30W-30S	34
Joliet-Channahon Rd. NE&SW	30W-28S,30W-28S	35
Kankakee St. N&S	27W-33S,27W-31S	47
Lakeview Dr. NS&EW	32W-28S	34
Marble Ct. NW&SE	31W-29S	34

Street	Grid	Page
Marble Rd. NE&SW	31W-29S	34
McEvilly Rd. E&W	32W-28S	34
McKinley Rd. N&S	32W-32S,32W-31S	46
Millsdale Ext. Rd. E&W	28W-29S,27W-29S	35
Millsdale Rd. E&W	27W-29S	35
North St. NW&SE	30W-29S	34
Olin Rd. NS&EW	28W-29S	35
Paradise Dr. NE&SW	32W-31S	34
Simotes Ln. NS&EW	32W-33S	46
Sunset Ct. E&W	32W-29S	34
Timberwood Ct. E&W	32W-29S	34
Timberwood Dr. N&S	32W-29S	34
Tow Path Ln. E&W	31W-29S	34
U.S. Route 6 NE&SW	32W-30S,29W-28S	34
Willow Ln. N&S	32W-28S	34

Crest Hill

Street	Grid	Page
Alma Dr. N&S	25W-21S	20
Arbor Ln. N&S	25W-21S	20
Ardaugh Av. E&W	26W-21S	20
Ardaugh Dr. N&S	26W-21S	20
Aster Dr. N&S	27W-21S	19
August St. E&W	25W-21S	20
Barthelone Av. N&S	25W-21S	20
Bicentennial Av. NW&SE	26W-22S	20
Birkey Av. E&W	24W-21S	20
Brian Dr. E&W	25W-21S	20
Broadway St. N&S	24W-21S	20
Burry Cir Dr. NS&EW	25W-21S	20
Burry Ct. E&W	25W-21S	20
Buttercup Ln.* N&S	27W-21S	19
Carmel Ct. E&W	25W-21S	20
Carnation Dr. NW&SE	27W-21S	19
Caton Farm Rd. E&W	25W-20S	20
Cedarwood Dr. NE&SW	27W-22S	19
Center St. N&S	24W-21S	20
Chaney Av. E&W	24W-21S	20
Circle St. NW&SE	26W-21S	20
Clement St. N&S	24W-21S	20
Connie Dr. N&S	26W-21S	20
Cora St. N&S	24W-21S	20
Cowing Ln. NS&EW	25W-21S	20
Crestwood Dr. N&S	26W-21S	20
Crystal Ct. NW&SE	25W-21S	20
Daffodil Dr. NW&SE	27W-21S	19
Daisy Ln.* E&W	27W-21S	19
Dearborn St. N&S	24W-21S	20
Donmaur St. E&W	26W-21S	20
Dorothy Ct. E&W	24W-21S	20
Durness Ct. NW&SE	26W-20S	20
Eldorado Dr. N&S	26W-21S	20
Elrose Ct. E&W	25W-21S	20
Elsie Av. E&W	24W-21S	20
Fern St. NE&SW	26W-21S	20
Frederick St. NE&SW	25W-21S	20
Garden Dr. NW&SE	27W-21S	19
Gaylord N&S	27W-21S	19
Gladiola Ln.* N&S	27W-21S	19
Grandview Av. N&S	26W-21S	20
Green St. NW&SE	26W-21S	20
Greengold St. NW&SE	26W-21S	20
Hawthorn Av. N&S	24W-21S	20
Heiden Av. N&S	25W-21S	20
Hess Dr. E&W	26W-21S	20
Hickory St. N&S	24W-21S	20
Highland St. N&S	24W-21S	20
Hoffman St. N&S	24W-21S	20
Hosmer Ln. NS&EW	25W-21S	20
Imperial Dr. N&S	26W-21S	20
Inner Ct. Dr. NS&EW	25W-21S	20
Iris Ln.* E&W	27W-21S	19
Jasmine Dr. N&S	27W-21S	19
Jonquil Ln.* NE&SW	27W-21S	19
Kelly Av. N&S	24W-21S	20
Knapp Dr. NW&SE	26W-21S,25W-21S	20
Leness Ln. N&S	26W-21S	20
Lincoln Av. E&W	26W-21S	20
Live Oak Dr. NW&SE	27W-22S	19
Lotus Ln.* E&W	27W-21S	19
Ludwig Av. E&W	25W-21S,24W-21S	20
Lynwood St. NW&SE	26W-21S	20
Macher St. N&S	26W-21S	20
Manico Dr. NE&SW	26W-21S	20
Marietta Av. N&S	25W-21S	20
Marlboro Dr. N&S	25W-21S	20
Marlboro Ln. E&W	25W-21S	20
Meadows Ct. E&W	25W-21S	20
Morning Glory Ln.* E&W	27W-21S	19
Nicholson St. N&S	24W-21S	20
Noonan St. N&S	26W-21S	20
Oakland Av. N&S	24W-21S	20
Oakland St. N&S	24W-21S	20
Orchid Ln.* N&S	27W-21S	19
Parkrose St. NW&SE	26W-21S	20
Pasadena Av. E&W	24W-21S	20
Patio Dr. NW&SE	27W-21S	19
Pioneer Rd. N&S	26W-22S	20
Plainfield Rd. NW&SE	26W-21S	20
Pleasant Dr. E&W	25W-21S	20
Poppy Ln.* N&S	27W-21S	19
Post Oak Ct. NE&SW	27W-22S	19
Prairie Av. N&S	25W-21S	20
Rahill Ct. E&W	25W-21S	20
Raynor Av. N&S	25W-21S	20
Red Oak Dr. NS&EW	27W-22S	19
Rock Run Dr. N&S	26W-22S	20
Root St. NW&SE	26W-21S	20
Rose Av. E&W	24W-21S	20
Rosel Ct. E&W	25W-21S	20
Route 7 E&W	25W-21S	20
Route 53 N&S	24W-21S	20
Sak Dr. E&W	25W-21S	20
Stern Av. E&W	24W-21S	20
Sullivan Ct. E&W	24W-21S	20
Surry Ct. E&W	25W-21S	20
Susanna Way E&W	26W-21S	20
Sweetbriar St. E&W	26W-21S	20
Sybil Rd. N&S	26W-21S	20
Sycamore St. NE&SW	26W-21S	20
Theodore St. E&W	25W-21S	20
Tulip Ln.* E&W	27W-21S	19
U.S. Route 30 NW&SE	26W-21S	20
University St. NW&SE	26W-21S	20
Violet Ln.* NE&SW	27W-21S	19
Waverly Ct. NE&SW	25W-21S	20
Webb St. NE&SW	26W-21S	20
Weber Rd. N&S	26W-21S	20
Wilcox St. N&S	24W-21S	20
William Dr. N&S	26W-21S	20
Willowbridge Rd. N&S	25W-21S	20

Crete

Street	Grid	Page
Aberdeen Dr. E&W	0E-28S	44
Alpine Ln. E&W	0E-28S	44
Amanda Ct. N&S	1W-30S	44
Amber Ln. N&S	1W-29S	44
Arbon Ct. E&W	0E-28S	44
Arthur Ct. NE&SW	1E-28S	45
Arthur Rd. N&S	1E-28S	45
Ashley Ct. N&S	1W-29S	44
Austin Av. N&S	0W-30S	44
Ayr Ln. N&S	0E-28S	44
Barbara Ct. N&S	0W-29S	44
Barry Ln. E&W	0W-30S	44
Beacon Pl. NE&SW	0E-29S	44
Beckwith Ln. N&S	0E-28S	44
Belfast Ter. NE&SW	0E-28S	44
Benton St. N&S	0W-30S	44
Berk Ln. E&W	0W-28S	44
Brackman Av. E&W	0W-30S	44
Bradford Pl. E&W	0W-28S	44
Brenner Ct. E&W	0E-28S	44
Briconock Pl. N&S	0E-29S	44
Bristol Dr. NE&SW	0E-28S	44
Carol Ct. N&S	1E-28S	45
Carol Ln. E&W	1E-28S	45
Cass St. E&W	0W-29S	44
Chalet Ct. E&W	0E-28S	44
Chalet Dr. E&W	0E-28S	44
Chalet Ln. N&S	0E-28S	44
Charles Ct. N&S	1E-28S	45
Charles St. E&W	1E-28S	45
Chester Ter. E&W	0W-29S	44
Cheviot Pl. E&W	0W-28S	44
Columbia Av. N&S	1W-28S	44
Columbia St. N&S	1W-29S	44
Cornwall Dr. E&W	0E-28S	44
Cottage Grove Av. N&S	0E-28S	44
Craig Av. E&W	1W-30S	44
Crete Rd. NE&SW	0E-29S	44
Dairy Ln. N&S	0W-29S	44
Dale Dr. N&S	1E-28S	45
Denell Ct. NE&SW	0W-28S	44
Denell Dr. E&W	0W-28S	44
Derby Pl. NE&SW	0E-29S	44
Dewar Ter. E&W	0E-29S	44
Division St. E&W	1W-29S,0W-29S	44
Donavan Ct. E. E&W	1E-28S	45
Donavan Dr. N&S	1E-28S	45
Doncaster Dr. E&W	0E-28S	44
Donovan Ct., West NW&SE	1E-28S	45
Douglas Ln. N&S	0W-30S,0W-29S	44
Dover Ter. E&W	0E-28S	44
Dunbar Ter. NE&SW	0E-28S	44
Durham Dr. N&S	0W-28S	44
East St. N&S	0W-30S,0W-29S	44
East Thames Dr. N&S	0W-29S	44
Edgewood Av. NE&SW	0W-29S	44
Edward Dr. N&S	1E-28S	45
Elizabeth St. N&S	0W-29S	44
Enterprize Av. N&S	0W-30S	44
Essex Ter. E&W	0W-28S,0E-28S	44
Exchange St. E&W	1W-29S,0E-29S	44
Fairway Ct. E&W	0W-28S	44
Fairway Dr. N&S	0W-28S	44
Fairway Ln. N&S	0E-29S	44
Faithorn Av. N&S	0W-30S	44
Faithorn Rd. E&W	0W-30S	44
Falmouth Ter. N&S	0W-28S	44
Farm St. E&W	1W-30S	44
Geneva Ct. N&S	0E-28S	44
Gilbert Ct. N&S	1E-28S	45
Gloucester Dr. N&S	0E-29S	44
Gordon Dr. E&W	1W-30S	44
Grant Blvd. E&W	0E-29S	44
Greenbriar Dr. N&S	0E-28S	44
Greenwood Ter. E&W	0E-29S	44
Griffith Ln. N&S	0E-29S	44
Halsted St. N&S	1W-28S	44
Haweswood Ct. E&W	0E-28S	44
Haweswood Dr. N&S	0E-28S	44
Hereford Dr. E&W	0W-28S	44
Herman St. E&W	1W-30S	44
Hewes St. N&S	0W-30S	44
Honey Ln. E&W	1E-28S	45
Hubbard St. E&W	0W-30S	44
Huntington Ter. N&S	0W-28S	44
Huntley Ct. NW&SE	1E-28S	45
Huntley Ter. NE&SW	0E-28S,1E-28S	44
Innsbruck Ln. N&S	0E-28S	44
Inverness Ln. N&S	0E-29S,0E-28S	44
Islay Ln. N&S	0E-29S,0E-28S	44
Jaqueline Dr. E&W	1E-28S	45
Jefferson St. E&W	1W-29S	44
Jennifer Ln. N&S	1E-28S	45
Joe Orr Ct. N&S	0W-29S	44
Joe Orr Rd. E&W	0W-29S	44
Juliet Ln. E&W	1E-28S	45
Kent Pl. N&S	0E-28S	44
Lakeshore Dr. N&S	0E-29S	44
Lancaster Ter. NS&EW	0W-29S	44
Laurel Ct. N&S	0W-30S	44
Lincoln Av. N&S	1W-29S	44
Lincoln Dr. E&W	0E-29S	44
Lincolnshire Ter. E&W	0E-28S	44
Linden St. E&W	1W-29S	44
Ludbury Ln. NE&SW	0E-29S	44
Lyle Ln. N&S	1E-28S	45
Maeburn Ter. E&W	0E-29S	44
Main St. N&S	0W-30S,0W-29S	44
Melanie Ct., North E&W	1E-28S	45
Melanie Ct., South E&W	1E-28S	45
Melanie Ln. N&S	1E-28S	45
Merioneth Dr. N&S	0E-29S	44
Michael Av. E&W	1W-30S	44
Milburn Av. E&W	0W-29S	44
Mill St. N&S	1W-29S	44
Monmouth Pl. NE&SW	0E-29S	44
Montgomery Dr. N&S	0E-29S	44
Moray Ln. E&W	0E-28S	44
Moray Ter. E&W	0E-28S	44
Naoma Dr. N&S	1W-30S	44
Nicolette Dr. N&S	1E-28S	45
Nilts Pl. N&S	0W-28S	44
Norfolk Dr. E&W	0W-28S	44
North St. E&W	0W-29S	44
Oak Leaf Ct. N&S	0W-29S	44
Oak Ln. N&S	0E-29S	44
Oakwood Dr. E&W	0W-29S	44
Old Monee Rd. E&W	1W-30S	44
Orchard Dr. E&W	1W-30W	38
Oriole St. E&W	0W-30S	44
Park St. N&S	0W-30S	44
Patricia Ln. E&W	1E-28S	45
Pembrook Ter. E&W	0W-29S	44
Peoria Av. N&S	1W-29S	44
Perry Av. N&S	0W-30S,0W-29S	44
Radnor Ln. E&W	0W-28S	44
Richton Rd. E&W	0E-28S	44
Robert Ln. E&W	1E-28S	45
Rohe Ln. N&S	0W-29S	44
Ronald Rd. N&S	1E-28S	45
Rosedale Ter. NE&SW	0E-28S	44
Rt. 1 N&S	0W-30S,0W-29S	44
Salop Pl. NW&SE	0E-28S	44
Sangamon St. N&S	1W-30S	44
Selleck St. N&S	0W-30S	44
Sheridan Av. E&W	0W-29S	44
Somerset Ln. N&S	0W-28S	44
South Wold Dr. E&W	0W-28S	44
St. Andrews Dr. NS&EW	0E-28S	44
Stafford Pl. E&W	0W-28S	44
State St. N&S	0W-28S	44
Stewart Ter. E&W	0E-29S	44
Surry Ln. E&W	0E-28S	44
Sussex Ter. E&W	0E-28S	44
Tee Ct., N. E&W	0W-28S	44
Tee Ct., S. E&W	0W-28S	44
Tiverton Ln. N&S	0W-28S	44
Troon Ln. N&S	0E-29S	44
Union St. N&S	1W-30S,1W-29S	44
Vardon Ln. N&S	0E-29S	44
Vincennes St. N&S	1W-30S	44
Wales Pl. E&W	0W-28S	44
Warwick Dr. N&S	0W-28S	44
Washington Av. N&S	1W-29S	44
Wayne Dr. N&S	1W-30S	44
Williams Ln. N&S	0E-29S	44
Willow Ln. N&S	0E-28S	44
Willow Ln. N&S	1W-30S	44
Wilton Ln. N&S	0W-28S	44
Wood St. N&S	0W-29S	44
Worchester Dr. E&W	0W-28S	44
York Pl. NE&SW	0W-29S	44
Zurich Ct. N&S	0E-28S	44
1st St. E&W	1W-29S	44
2nd St. E&W	1W-29S	44
3rd St. E&W	1W-29S	44
4th St. E&W	1W-29S	44
5th St. E&W	1W-29S	44

Crete Township

Street	Grid	Page
Accorn Ct. NE&SW	2W-28S	43
Anne Ln. E&W	2E-30S	45
Arlington Ln. E&W	1W-32S,0W-32S	56
Ashland Av. N&S	2W-29S,1W-28S	43
Auburn Ln. E&W	0W-32S	56
Balmoral Dr. N&S	0W-32S	56
Balmoral Woods N&S	0W-32S	56
Bemes Rd. E&W	0E-32S,4E-32S	56
Bending Creek Tr. E&W	3E-30S	45
Bremer Ct. NE&SW	4E-29S	45
Bremer Tr. NW&SE	4E-29S	45
Broadview Av. E&W	1W-29S	44
Broadway Av. E&W	1W-33S	56
Brookside Ct. N&S	3E-29S	45
Brookview Dr. NS&EW	2W-28S	43
Buried Oak Dr. E&W	3E-29S,4E-29S	45
Burville Rd. E&W	1E-30S,3E-30S	45
Butler Ct. N&S	0W-32S	56
Calumet Expwy. NE&SW	0W-33S,1E-29S	56
Cambridge Dr. N&S	3E-31S	45
Carriage Ln. E&W	2E-32S	57
Cedar Ln. NW&SE	3E-30S	45
Challas Entrance Rd. E&W	4E-29S	45
Cherry Ln. E&W	3E-30S	45
Chestnut Ln. NW&SE	3E-30S	45
Chestnut St. N&S	1W-33S	56
Columbia Av. N&S	1W-28S	44
Columbia Av. E&W	1W-33S	56
Cottage Grove Av. N&S	1E-33S,0E-31S	57
Countess Ln. N&S	1W-31S	44
Creek Wood E&W	3E-29S	45
Creekside Ct. NW&SE	3E-30S	45
Cumberland Ct. N&S	4E-29S	45
Cumberland Ln. E&W	4E-29S	45
Danne Rd. E&W	3E-30S	45
Deer Run Ct. E&W	3E-29S	45
Deer Run Tr. N&S	3E-29S	45
Dixie Hwy. N&S	0W-31S	44
Dorothy Dr. N&S	2E-29S	45
Duchess Ln. N&S	0W-31S	44
Dunleith Ct. N&S	0W-32S	56
Dutton Rd. N&S	0E-33S	56
Elmscourt E&W	0W-32S	56
Equestrian Tr. NE&SW	3E-28S	45
Evergreen Ct. N&S	3E-32S	57
Exchange St. E&W	2W-29S,4E-29S	43
Faithorn Burville Rd. N&S	1E-33S,1E-32S	57
Faithorn Rd. E&W	0W-30S	44
Forest Ln. E&W	2W-29S	43
Forest View Ct. N&S	3E-29S	45
Forest View Dr. E&W	3E-29S,4E-29S	45
Forest View Ln. N&S	0W-33S	56
Frank St. E&W	1W-33S	56
Gaisor Ct. NW&SE	2E-29S	45
Gaisor Dr. E&W	2E-29S	45
Glendale Av. N&S	1W-33S	56
Goodenow Av. N&S	1W-33S	56
Goodenow Rd. E&W	2W-33S,4E-33S	55
Greenview Av. N&S	0W-33S	56
Greenwood Av. N&S	1E-32S	57
Gregory Ln. N&S	1W-31S	44
Halsted Av. N&S	1W-33S	56
Harvest Ln. N&S	2W-32S	55
Heatherton Dr. N&S	3E-31S	45
Hickory Ct. N&S	3E-30S	45
Hickory Ln. E&W	3E-30S	45
Hickory St. N&S	1W-33S	56
Hickory St. E&W	2E-30S	45
Hilltop Rd. N&S	3E-31S	45
Janice Av. N&S	0W-33S	56
Jonathan Ln. N&S	3E-28S	45
Juniper Dr. N&S	3E-32S	57
Kent Av. E&W	1W-31S	44
King Rd. NE&SW	2W-29S,2W-28S	43
Klemme Rd. N&S	3E-33S,3E-30S	57
Kurt Ln. N&S	3E-28S	45
Lakeside Ct. NW&SE	4E-29S	45
Lakeside Ter. N&S	4E-29S	45
Lakeview Dr. E&W	3E-32S	57
Laura Ln. N&S	2E-29S	45
Longwood Dr. E&W	3E-29S	45
Loomis St. N&S	1W-29S,1W-28S	44
Louella St. E&W	0E-31S	44
Maple Ln. NW&SE	2W-28S	43
Maple St. N&S	1W-33S	56
Maple St. E&W	2E-30S	45
Marmion Dr. E&W	2E-29S	45
Mary Ln. N&S	3E-31S	45
Meadow Rd. N&S	2W-33S	55
Meadowood Dr. N&S	2W-29S	43
Meadowood Rd. N&S	2W-29S	43
Michaels St. N&S	2E-30S	45
Monee Rd. E&W	2W-31S,0E-31S	43
Mulberry Ct. E&W	3E-29S	45
Mulberry Ln. N&S	3E-29S	45
Munz Rd. E&W	0E-30S	44
Murray Dr. N&S	2E-32S	57
Nacke Rd. N&S	2W-33S,2W-31S	55
Nancy St. N&S	0E-31S	44
New Monee Rd. E&W	0W-31S	44
Norfolk Av. E&W	2W-28S	43
Norway Trail NW&SE	4E-29S	45
Notre Dame Av. E&W	2E-29S	45
Oak Ct. NE&SW	2W-28S	43
Oak Ln. N&S	4E-30S	45
Oak St. E&W	2E-29S	45
Ohlendorf Rd. E&W	1W-32S	56
Old Monee Rd. E&W	2W-30S	43
Old Post Rd. NW&SE	2E-29S,3E-29S	45
Palamino Dr. NS&EW	3E-28S	45
Park Av. N&S	1W-28S	44
Pebblebeach Ct. N&S	0W-32S	56
Pheasant Ct. NW&SE	4E-29S	45
Pheasant Tr. NE&SW	4E-29S	45
Pine Ct. E&W	3E-29S	45
Pine Dr. E&W	3E-32S	57
Plum Valley Dr. N&S	3E-29S	45
Poplar Ct. E&W	3W-29S	43
Poplar Ln. E&W	3E-29S	45
Prairieview Av. N&S	3W-31S	43
Pricilla Ln. E&W	3E-28S	45
Princess Ln. N&S	1W-31S	44
Rainbow Ln. E&W	2E-31S	45
Rana Dr. E&W	3E-31S	45
Raymond Dr. NE&SW	2E-28S,3E-28S	45
Red Oak N&S	4E-32S	57
Reichert Dr. E&W	4E-31S	45
Richton Rd. E&W	2W-28S,2E-28S	43
Ridgefield Dr. E&W	3E-31S	45
Ridgeland Av. E&W	1W-32S	56
Rietveld Dr. E&W	2E-29S	45
Root Dr. N&S	0E-32S	56
Rosewood Av. N&S	1W-33S	56
Round House Rd. N&S	0E-31S,0E-30S	44
Royal Ridge Dr. E&W	3E-31S	45

Grid   Page

Rt. 1 N&S .. 0W-33S,0W-30S 56
Sangamon St. N&S .. 1W-31S 44
Schiewe Rd. N&S
...........1E-33S,1E-32S 57
Schultze Rd. N&S....2W-28S 43
Shady Grove Ct. E&W
.... 3E-30S 45
Shagbark Tr. E&W ... 3E-30S 45
Sharon Ln. N&S...... 3E-31S 45
Smit Dr. N&S 3E-31S 45
Southfield Av. E&W ..1W-29S 44
Spruce Ct. NW&SE .. 3E-30S 45
Spruce Dr. E&W 3E-32S 57
Stanton Ln. E&W ... 0W-32S 56
State Line Rd. N&S
...........4E-33S,4E-29S 57
Stoneridge Ln. N&S .. 3E-29S 45
Stoney Island Av. N&S
...........2E-32S,2E-29S 57
Susan Ln. N&S 3E-31S 45
Tam-O-Shanter Ct. N&S
.... 0W-32S 56
Tam-O-Shanter Dr. NS&EW
.... 0W-32S 56
Terry Av. N&S 2E-29S 45
Thorn Creek Ln. E&W
.... 2W-29S 43
Timberidge Ct. NE&SW
.... 3E-29S 45
Timberline Ct. E&W .. 3E-29S 45
Timberline Tr. N&S .. 3E-29S 45
Torrence Av. N&S ... 2E-28S 45
Tuckaway Ct. N&S .. 2E-33S 57
Valleyview Ln. E&W ..2W-29S 43
Van Drunen Dr. E&W
.... 3E-31S 45
Vera E&W 3E-29S 45
Victoria N&S....... 3E-31S 45
Volbrecht Rd. N&S .. 1E-29S 45
Walden Ln. N&S..... 3E-29S 45
Walnut St. N&S
...........2E-30S,2E-29S 45
West Ravine Rd. NS&EW
.... 3E-29S 45
Whispering Oak Ln. N&S
.... 3E-30S 45
White Oaks Ln. N&S..2E-30S 45
Wildwood Ct. N&S...3E-29S 45
Wildwood Tr. N&S
...........3E-30S,3E-29S 45
Willow Brook Ct. E&W
.... 3E-30S 45
Willow Brook Tr. NS&EW
.... 3E-29S 45
Windsor Av. E&W....1W-31S 44
Winston Av. N&S 1W-31S 44
Wolpers Rd. E&W .. 2W-29S 43
Woodland Dr. N&S .. 1E-30S 45
Woodland Tr. N&S... 3E-29S 45.
Woodlawn Av. N&S .. 1E-33S 57
235th St. E&W 2W-28S 43
263rd St. E&W 1E-29S 45
265th St. E&W 1E-32S 57

Custer Township
Bauer Rd. E&W
.... 25W-45S,22W-45S 76
Charlotte Av. E&W ..26W-45S 76
Cooper Rd. E&W
.... 29W-44S,28W-44S 74
County Line Rd. E&W
.... 29W-45S,25W-45S 74
Curtis Rd. E&W 26W-44S 76
Darby Rd. N&S 25W-45S 76
Davy Ln. (Pvt.) NE&SW
.... 27W-42S 75
Erickson Ln. (Pvt.) N&S
.... 27W-41S 71
Essex Rd. N&S
.... 29W-45S,29W-42S 74
Garfield St. E&W .. 26W-43S 76
Grant Av. N&S .. 27W-43S 75
Grey Rd. N&S
.... 26W-45S,26W-43S 76
Irish Ln NS&EW
.... 25W-44S,24W-45S 76
Jay Dr. N&S........26W-43S 76
Johnson Rd. E&W
.... 29W-40S,28W-40S 70
Lee St. E&W 26W-43S 76
Linden N&S .. 26W-43S 76
Mack's Rd. E&W...23W-45S 77
McGuire Rd. N&S
.... 29W-43S,28W-43S 74
Ohlhues Rd. N&S
.... 27W-45S,26W-43S 75
River Ln. N&S......22W-45S 77
River St. NW&SE
.... 27W-45S,22W-45S 75
Route 53 NE&SW .. 29W-45S 70
Route 113 NW&SE
.... 29W-42S,22W-45S 74
Route 129 N&S 29W-40S 70
Scott St. N&S 26W-43S 76
Shenk Rd. (Pvt.) N&S
.... 29W-41S 70
Smiley Rd. E&W
.... 28W-42S,27W-42S 75
Washington St. N&S
.... 26W-43S 76
Weikum Rd. N&S .. 29W-43S 74
Weiske Ln. (Pvt.) E&W
.... 27W-44S 75

Grid   Page

Weiske Rd. E&W .. 27W-44S 75
West River Rd. NW&SE
.... 28W-40S,27W-41S 71
Yeates Rd. N&S ... 24W-45S 76
Zilm Rd. N&S
.... 28W-45S,28W-43S 75

Diamond
Coal City Rd. E&W 32W-39S 58
County Line Rd. N&S
.... 32W-39S 58

Du Page Township
Adlai E. Stevenson Expy.
E&W . 26W-15S,21W-11S 10
Arbor Dr. E&W 25W-14S 10
Balsam Ct. NW&SE
.... 26W-13S 10
Big Run Ln. N&S .. 21W-15S 11
Carlisle Dr. NE&SW
.... 23W-10S 7
Catalina Dr. NW&SE
.... 23W-10S 7
Crimson Ct. N&S .. 26W-15S 10
Donovan Dr. E&W
.... 22W-13S,21W-13S 11
Ferguson Rd. E&W
.... 26W-13S 10
Forest Ln. N&S .. 21W-15S 11
Fox Valley Expy. NW&SE
.... 26W-15S 10
Kings Rd. (proposed) E&W
.... 25W-13S,26W-12S 10
Magnolia Dr. N&S .. 26W-15S 10
Mulberry St. N&S .. 26W-15S 10
Naperville Rd. NW&SE
.... 24W-14S,26W-11S 10
Normantown Rd. E&W
.... 26W-14S,24W-15S 10
North Carrilon Dr. NS&EW
.... 26W-15S 10
Old Orchard Ln., E. N&S
.... 21W-15S 11
Old Orchard Ln., N. N&S
.... 21W-15S 11
Old Orchard Ln., W. N&S
.... 21W-15S 11
Orchid St. N&S .. 26W-15S 10
Recreation Dr. NS&EW
.... 25W-15S 10
Red Cedar Ln. NS&EW
.... 26W-15S 10
Redberry Cir. NE&SW
.... 26W-15S 10
Redberry Ct. NE&SW
.... 26W-15S 10
Remington Blvd. NE&SW
.... 23W-12S 7
Royce Rd. E&W
.... 25W-10S,23W-10S 6
Royce Rd. (proposed) E&W
.... 23W-10S,22W-10S 7
Schmidt Rd. N&S
.... 24W-13S,24W-12S 10
Silk Tree Cir. NE&SW
.... 26W-15S 10
Silverleaf Ct. N&S ..26W-15S 10
Silverleaf Dr. N&S .. 26W-15S 10
Smith Rd. N&S
.... 21W-16S,21W-15S 15
Sumac Ct. N&S .. 26W-13S 10
Tall Pines Ln. . 26W-15S 10
Weber Rd. N&S
.... 26W-15S,26W-12S 10
Williams Rd. N&S . 26W-14S 10
Windham Pkwy. E&W
.... 26W-14S 10
107th St. E&W .. 21W-12S 7
111th St. E&W 26W-12S 6
115th St. E&W .. 25W-13S 10
135th St. E&W
.... 25W-15S,21W-15S 10

Elwood
Beattie N&S .. 25W-31S 36
Bush Dr. NS&EW .. 25W-31S 36
Chicago Av. N&S
.... 25W-32S,25W-31S 48
Douglas St. NE&SW
.... 25W-32S 48
Gardner St. E&W .. 25W-32S 48
Jackson St. N&S .. 25W-32N 48
Lincoln St. N&S .. 25W-32S 48
Linebarger Ct. N&S
.... 25W-31S 36
Manhattan Rd. E&W
.... 25W-30S 36
Matteson St. N&S .. 25W-32S 48
Mississippi Av. E&W
.... 25W-32S 48
Morris St. E&W 25W-32S 48
North St. E&W 25W-32S 48
Parks St. E&W 25W-32S 48
Route 53 NE&SW .. 25W-32N 48
South St. E&W 25W-32S 48
Spencer St. N&S .. 25W-32S 48
St. Louis St. N&S .. 25W-32S 48
Tehle Rd. E&W 25W-31S 36
Wood Rd. E&W .. 25W-32S 48
Wood St. NE&SW .. 25W-32S 48

Florence Township
Grid   Page

Allott Dr. E&W
.... 26W-38S,25W-38S 60
Arsenal South Rd. E&W
.... 22W-37S,21W-37S 61
Arsenal-Wilton Center Rd.
E&W 21W-35S 49
Barr Rd. N&S
.... 26W-39S,26W-38S 60
Butcher Ln. N&S .. 26W-39S 60
Commercial St. E&W
.... 22W-37S 61
County Rd. . 26W-38S 60
Doyle Rd. E&W
.... 26W-34S,21W-34S 48
Indian Trail Rd. N&S
.... 25W-39S,25W-37S 60
Kennedy Rd. E&W
.... 22W-38S,21W-38S 61
Manhattan-Wilton Rd. E&W
.... 25W-33S,22W-33S 48
Martin Long Rd. E&W
.... 22W-39S,22W-37S 61
Northern Av. N&S
.... 23W-38S,23W-37S 61
Old Chicago Rd.
.... 24W-39S,24W-36S 60
Quigley Rd. E&W
.... 26W-36S,23W-36S 60
Riley Rd. NS&EW
.... 26W-37S,25W-34S 60
Route 53 N&S
.... 26W-38S,26W-34S 60
Smith Rd. NE&SW .. 25W-38S 60
South Townline Rd. E&W
.... 21W-33S 49
Symerton Rd. N&S 23W-39S 61
Warner Bridge Rd. N&S
.... 21W-39S,21W-35S 61
Wesley Rd. E&W
.... 22W-39S,21W-39S 61
West Kahler Rd. E&W
.... 21W-39S 61
White Cir. NS&EW .. 25W-36S 60
Wilmington-Peotone Rd. E&W
.... 26W-37S,21W-37S 60

Frankfort
Abbotsford Ln. NS&EW
.... 9W-26S 33
Aberdeen Rd. E&W
.... 10W-26S,9W-26S 33
Ash St. N&S 12W-26S 32
Ashely Ct. NW&SE 11W-26S 33
Ashford Av. E&W .. 13W-26S 32
Aspen St. N&S .. 12W-25S 32
Ayeshire Ct. NE&SW
.... 9W-26S 33
Azalea Rd. N&S 11W-26S 33
Bankview Dr. NS&EW
.... 12W-25S 32
Birchwood Dr. E&W
.... 12W-25S,12W-24S 32
Boulder Ct. E&W .. 12W-25S 32
Bowen St. E&W .. 12W-25S 32
Box Elder Rd. N&S
.... 12W-25S 32
Brookridge Creek Dr. E&W
.... 13W-25S 32
Brookridge Ct. N&S
.... 13W-25S 32
Brookridge Dr. NE&SW
.... 13W-25S 32
Brookside Ln. E&W
.... 12W-25S 32
Butternut Cir., N. N&S
.... 13W-25S 32
Butternut Cir., S. NS&EW
.... 13W-25S 32
Butternut Cir., W. N&S
.... 13W-25S 32
Butternut Ct. NE&SW
.... 12W-25S 32
Butternut Tr. N&S .. 13W-25S 32
California Av. E&W .. 12W-26S 32
Carpenter St. N&S .. 12W-25S 32
Cedar Rd. N&S 12W-26S 32
Center Rd. N&S .. 12W-25S 32
Chestnut Ct. N&S .. 11W-26S 33
Chestnut Dr. N&S .. 11W-26S 33
Church Ct. E&W....11W-26S 33
Church St. E&W .. 11W-26S 33
Colony Ln. NW&SE .. 9W-26S 33
Colorado Av. E&W ..12W-25S 32
Concord Dr. N&S .. 13W-25S 32
Corsair Rd. E&W .. 11W-27S 33
Cottonwood Rd. NW&SE
.... 12W-25S 32
Coventry Ln. E&W .. 9W-25S 33
Dover Cir. NW&SE ..10W-26S 33
Drummore Ln. E&W ..9W-26S 33
Durham Dr. E&W .. 9W-26S 33
East Brook Rd. N&S
.... 13W-25S 32
Eastern Av. N&S .. 12W-26S 32
Eisner Rd. N&S .. 13W-25S 32
Elm St. N&S
.... 12W-26S,12W-25S 32
Elmwood St. E&W .. 12W-25S 32
Evergreen St. N&S 11W-26S 33
Fairway Ln. NW&SE
.... 10W-26S 33

Grid   Page

Firth Ct. NE&SW .. 10W-26S 33
Fox Run Cir. NS&EW
.... 11W-25S 33
Fox Run Ct. N&S .. 11W-25S 33
Franklin Av. E&W
.... 12W-25S,11W-25S 32
Georgetown Commons E&W
.... 9W-25S 33
Georgetown Rd. N&S
.... 9W-25S 33
Glen Bard Ln. NE&SW
.... 9W-26S 33
Glen Eagles Ct. NW&SE
.... 9W-26S 33
Golf Club Ln. NW&SE
.... 9W-26S 33
Gulfstream Rd. N&S
.... 11W-27S 33
Hackberry Rd. NW&SE
.... 12W-25S 32
Harlem Av. N&S
.... 9W-27S,9W-26S 33
Hawthorne Dr. N&S
.... 10W-26S 33
Hawthorne Rd. NW&SE
.... 12W-25S 32
Heather Ln. N&S .. 10W-26S 33
Heritage Ct. NE&SW
.... 9W-26S 33
Heritage Dr. NE&SW
.... 13W-26S 32
Hickory Ct. N&S .. 13W-25S 32
Hickory St. N&S ... 12W-25S 32
Highland Rd. E&W .. 9W-26S 33
Hornbeam Ct. NW&SE
.... 13W-25S 32
Hunt Club Dr. N&S 11W-25S 33
Hunterwoods Dr. N&S
.... 9W-25S 33
Illinois Rd. E&W .. 12W-25S 32
Indiana Ct. E&W .. 12W-25S 32
Industry Av. E&W .. 11W-27S 33
Iowa Ct. E&W 12W-25S 32
Johnson Av. NS&EW
.... 12W-25S 32
Kansas St. E&W 12W-25S 32
Kean Av. N&S 12W-25S 32
Lambrecht Rd. N&S
.... 11W-27S 33
Laraway Rd. E&W
.... 12W-26S,9W-26S 32
Larch Ct. 'A' E&W ..11W-26S 33
Larch Ct. 'B' E&W .. 11W-26S 33
Larch Ln. N&S .. 11W-26S 33
Leslie Ln. N&S 12W-24S 32
Lexington Ln. E&W
.... 13W-25S 32
Lincoln Hwy. E&W
.... 11W-25S,9W-25S 32
Lincoln Ln. N&S
.... 13W-25S,12W-25S 32
Linden Ct. E&W .. 11W-26S 33
Linden Dr. N&S .. 11W-26S 33
Linden Dr. N&S .. 12W-25S 32
Locust St. N&S
.... 12W-26S,12W-25S 32
Luther Ln. N&S .. 11W-26S 33
Maple Ct. E&W 10W-26S 33
Maple Ridge Ct. NE&SW
.... 13W-25S 32
Maple St. N&S
.... 12W-26S,12W-25S 32
McDonald Ln. E&W
.... 12W-25S 32
Meadow Av. E&W .. 12W-25S 32
Michigan Rd. E&W .. 12W-25S 32
Michigan St. N&S .. 12W-25S 32
Milton Bridge Tr. NE&SW
.... 9W-26S 33
Mulberry Rd. N&S .. 12W-25S 32
Nebraska Av. E&W
.... 12W-26S,11W-26S 32
Nebraska Ct. 'A' N&S
.... 11W-26S 33
Nebraska Ct. 'B' N&S
.... 11W-26S 33
Nevada Ct. 'A' N&S
.... 11W-26S 33
Nevada Ct. 'B' N&S
.... 11W-26S 33
Nevada St. E&W .. 11W-26S 33
North St. E&W 12W-25S 32
Oak Ridge Ct. N&S
.... 13W-26S 32
Oak St. N&S 12W-26S 32
Oakton Av. NW&SE
.... 13W-26S 32
Oakton Ct. E&W....13W-26S 32
Oakwood Dr. N&S
.... 12W-25S,12W-24S 32
Ohio Rd. E&W 12W-25S 32
Old N. Church Rd. NS&EW
.... 9W-25S 33
Ontario St. E&W .. 11W-26S 33
Oregon St. N&S .. 12W-25S 32
Overlook Ct., E. E&W
.... 12W-24S 32
Overlook Ct., W. NW&SE
.... 12W-24S 32
Overlook Dr. N&S .. 12W-24S 32
Pacific St. E&W .. 12W-25S 32
Parkway Ln. N&S .. 11W-27S 33
Pembrook Ct. NW&SE
.... 13W-26S 32

Grid   Page

Pfaff Dr. E&W 11W-26S 33
Pfeiffer Rd. N&S
.... 11W-27S,11W-26S 33
Pheasant Trail NW&SE
.... 10W-26S 33
Plymouth Ct. N&S .. 10W-26S 33
Prestwick Dr. E&W
.... 10W-25S,9W-26S 33
Princeton Cir. NS&EW
.... 13W-26S 32
Princeton Ct. NE&SW
.... 13W-26S 32
Pueblo Ct. E&W ... 12W-25S 32
Red Bud Ct. E&W .. 13W-25S 32
Regency Dr. N&S .. 13W-26S 32
Rt. 30 E&W
.... 13W-25S,9W-25S 32
Rt. 45 N&S 12W-25S 32
Sager Rd. E&W 12W-25S 32
Sauk Trail Dr. NW&SE
.... 11W-26S,9W-26S 33
Smith St. N&S......12W-25S 32
Somerset Ct. N&S .. 13W-26S 32
Southwick Ct. E&W ..9W-27S 33
Southwick Dr. NS&EW
.... 9W-27S 33
Spruce Dr. N&S
.... 12W-25S,12W-24S 32
St. Andrew's Way E&W
.... 9W-26S 33
Steeplechase Dr. NS&EW
.... 9W-26S 33
Steger Rd. E&W .. 9W-27S 33
Stonebridge Rd. N&S
.... 13W-25S,13W-24S 32
Stuart Cir. N&S .. 13W-26S 32
Sycamore Ln. N&S 11W-26S 33
Tanglewood Ln. N&S
.... 12W-25S,12W-24S 32
Thorn Trail N&S 9W-26S 33
Timber Ct. NE&SW
.... 10W-26S 33
Troon Cir. N&S 9W-26S 33
Utah St. E&W 12W-25S 32
Vail Dr. N&S 12W-25S 32
Vermont Rd. N&S .. 12W-25S 32
Virginia Ct. E&W 9W-26S 33
Walnut Ridge Ct. NE&SW
.... 13W-26S 32
Walnut Ridge Ct., N. NE&SW
.... 13W-26S 32
Walnut Ridge Ct., S. NE&SW
.... 13W-26S 32
Walnut St. N&S .. 12W-26S 32
Warmke Ln. NE&SW
.... 12W-25S 32
Westbrook Rd. E&W
.... 13W-26S 32
Williamsburg Ct. N&S
.... 13W-26S 32
Williamsburg Tr. E&W
.... 13W-26S 32
Willow St. N&S .. 12W-25S 32
Wisconsin Rd. E&W
.... 12W-25S 32
Woods Dr. E&W .. 12W-24S 32
Yankee Ridge Ct. NW&SE
.... 12W-25S 32
Yankee Ridge Dr. NE&SW
.... 12W-25S 32
Yorktown Rd. E&W .. 9W-25S 33
84th Av. N&S 10W-25S 33
95th Av. N&S 11W-25S 33
96th Av. NE&SW
.... 13W-26S,12W-24S 32

Frankfort Township
Abbotsford Ln. E&W
.... 9W-26S 33
Acorn Ridge Dr. N&S
.... 9W-25S,9W-24S 33
Apple Ct. N&S......11W-23S 25
Arden Ln. N&S .. 14W-23S 25
Arran Dr. N&S .. 9W-23S 25
Ash Av. E&W .. 11W-23S 25
Aviemore Ln. N&S .. 9W-25S 33
Ayshire Ct. SW&NE .. 9W-26S 33
Barbara Ln. E&W .. 9W-24S 33
Baron Rd. N&S .. 14W-23S 24
Beechnut Dr. N&S .. 11W-23S 25
Benton Dr. NE&SW .. 9W-23S 25
Birch Av. E&W .. 11W-23S 25
Birchwood Ln. N&S .. 9W-25S 33
Braemar Ln. E&W .. 9W-23S 25
Briarwood Ln. N&S
.... 13W-25S 32
Brickstone Dr. E&W
.... 10W-24S 33
Camelot Av. E&W .. 14W-23S 24
Carrie Ct. NS&EW .. 9W-23S 25
Catulpa St. N&S...11W-23S 25
Cedar Rd. N&S .. 13W-26S 32
Center Rd. N&S .. 11W-27S 33
Cherry St. N&S .. 11W-23S 25
Cleveland Rd. E&W
.... 13W-25S,13W-24S 32
Cobblestone Ct. N&S
.... 9W-24S 33
Countryside Ct. E&W
.... 10W-24S 33
Countryside Dr. E&W
.... 10W-24S 33

WILL COUNTY 83

	Grid	Page
Countryside Ln. N&S	10W-24S	33
Dove Ln. E&W	9W-24S	33
Driftwood N&S	9W-24S	33
Drummore Ln. E&W	9W-26S	33
Duchess Av. E&W	14W-23S	24
Eastern Av. E&W	14W-26S	32
Edinburgh Ln. N&S	9W-23S	25
Elizabeth Ln. N&S	9W-25S	33
Ellen Ln. NS&EW	9W-25S	33
Elm Av. E&W	11W-23S	25
Elmwood Ct. E&W	13W-24S	32
Ethyl Ln. E&W	11W-22S	25
Evan Ter. E&W	9W-23S	25
Evergreen Dr. E&W	10W-24S	33
Fallbrook Ct. E&W	9W-24S	33
Farmington Dr. N&S	9W-23S	25
Francis Rd. NW&SE	14W-23S	24
Frankfort Square Rd. NS&EW	9W-25S,9W-23S	33
Franklin Av. E&W	11W-25S	33
Front St. NE&SW	14W-23S	24
Garden Ln. E&W	9W-23S	25
Garrett Ct. NE&SW	9W-23S	25
Glasgow Dr. N&S	9W-23S	25
Glen Eagles Ct. NW&SE	9W-26S	33
Glenbark Ln. NE&SW	9W-26S	33
Glenshire Dr. E&W	9W-25S	33
Graceland Ln. N&S	9W-24S	33
Grand Prairie Ln. N&S	9W-24S	33
Green Meadow Ln. N&S	9W-24S	33
Greenfield Ln. N&S	9W-24S	33
Hamilton Rd. E&W	14W-24S	32
Harlem Av. N&S	9W-27S,9W-26S	33
Hawthorn St. E&W	11W-23S	25
Hawthorne Dr. E&W	9W-25S	33
Hickory Creek Ct. N&S	9W-25S	33
Hickory Creek Dr. E&W	9W-25S	33
Hickory St. N&S	11W-23S	25
Highview Ln. NW&SE	9W-23S	25
Hill Av. E&W	10W-25S	33
Hillgate Rd. N&S	14W-24S	32
Hillside Rd. N&S	10W-25S	33
Hilltop Rd. NW&SE	13W-24S	32
Holly Ln. NW&SE	9W-24S	33
Humbolt Dr. NE&SW	9W-25S	33
Hunter Dr. N&S	9W-24S	33
Interstate 80 E&W	13W-22S,10W-22S	24
Inverary Dr. NE&SW	9W-25S	33
Inverness Ln. NS&EW	9W-23S	25
Ironwood Ct. E&W	13W-24S	32
Ironwood Dr. N&S	13W-25S,13W-24S	32
Ivy Path N&S	9W-24S	33
Jamison Dr. E&W	9W-24S	33
Jean Rd. N&S	14W-22S	24
Jessica Ln. NS&EW	9W-24S	33
Jonquil Ln. N&S	9W-24S	33
Josephine Dr. E&W	14W-22S	24
Juniper Ct. NW&SE	9W-24S	33
Kingston Ct. NW&SE	9W-24S	33
Kingston Dr. NE&SW	9W-24S	33
Kohlwood Dr. N&S	13W-24S	32
Laraway Rd. E&W	12W-26S,9W-26S	32
Laurel Dr. E&W	9W-24S	33
Leila Dr. N&S	14W-22S	24
Lenore Ln. N&S	11W-22S	25
Locust St. N&S	12W-25S,12W-24S	32
Magnolia Av. E&W	11W-23S	25
Magnolia Dr. E&W	10W-24S	33
Magnolia Ln. E&W	9W-24S	33
Mallory Dr. N&S	9W-24S	33
Maple (Marley) Rd. E&W	14W-22S	24
Marjorie Pkwy. N&S	14W-22S	24
Marley (Maple) Rd. E&W	14W-22S	24
Mary Therese Ln. N&S	10W-24S	33
Mathews Dr. NS&EW	9W-25S	33
Milton Bridge Tr. E&W	9W-26S	33
Moorefield Dr. E&W	9W-24S	33
New Castle Dr. E&W	9W-24S	33
Normandy Ln. N&S	14W-24S	32
North Av. E&W	10W-24S,9W-24S	33
North Ct. N&S	9W-24S	33
Norwood Dr. E&W	10W-24S	33
Nutwood Ct. E&W	9W-24S	33
Oak Ln. N&S	9W-25S,9W-24S	33
Old Coach Tr. N&S	9W-23S	25
Old Mill Rd. NS&EW	10W-24S	33
Orchard Dr. E&W	10W-24S	33
Osage Rd. N&S	14W-27S,14W-25S	32
Patricia Ln. N&S	9W-23S	25
Pebble Ln. E&W	9W-24S	33
Pfeiffer Rd. N&S	11W-27S,11W-25S	33
Pheasant Trail E&W	10W-26S	33
Pine Hill Rd. NW&SE	10W-24S,9W-24S	33
Pine St. E&W	11W-23S	25
Pleasant Av. E&W	11W-22S	25
Poplar St. N&S	11W-23S	25
Prestwick St. E&W	9W-26S	33
Queens Ct. NW&SE	14W-23S	24
Redwood Ln. N&S	11W-23S	25
River Rd. N&S	10W-25S	33
Riverton Ct. NE&SW	9W-23S	25
Roeland Ct. NE&SW	9W-23S	25
Rose Ln. N&S	14W-22S	24
Rosewood Dr. N&S	10W-24S	33
Rosewood Ln. E&W	12W-23S,11W-23S	24
Royce Ct. E&W	9W-24S	33
Rt. 30 E&W	14W-25S,9W-25S	32
Rt. 45 N&S	11W-24S,11W-23S	33
Ryeland Dr. E&W	10W-24S	33
Saint Andrews Way E&W	9W-26S	33
Sara Rd. N&S	14W-22S	24
School House Rd. N&S	13W-24S	32
Shetland Dr. N&S	9W-26S	33
Silo Dr. E&W	9W-24S	33
Skidmore Ct. N&S	9W-23S	25
Skye Dr. N&S	9W-23S	25
Spruce Ln. N&S	10W-24S	33
St. Francis Rd. E&W	11W-24S,9W-24S	33
Steger Rd. E&W	14W-27S,9W-27S	32
Stephaine Ln. E&W	14W-22S	24
Sterling Dr. NW&SE	9W-24S	33
Stuart Cir. N&S	9W-26S	33
Sycamore Dr. N&S	10W-24S	33
Sycamore St. N&S	11W-23S	25
Tack Ln. E&W	9W-24S	33
Tartan Rd. E&W	9W-23S	25
Thistle Ln. N&S	9W-23S	25
Thorndale Dr. N&S	10W-24S	33
Timberview Dr. N&S	13W-24S	32
Town Line Rd. E&W	14W-27S,14W-23S	32
Troon Cir. N&S	9W-26S	33
Troon Trail N&S	9W-26S	33
Victorian Dr. N&S	14W-23S	24
Virginia Pkwy. E&W	14W-22S	24
Walnut St. N&S	11W-23S	25
Westport Dr. N&S	10W-24S	33
White Fence Ct. N&S	9W-24S	33
Willow Ln. E&W	11W-23S	25
Windmill Dr. E&W	9W-24S	33
Wintergreen Dr. E&W	10W-24S	33
Wishing Well Dr. E&W	9W-24S	33
Wolf Rd. N&S	14W-27S,14W-24S	32
Woodland Ct. NE&SW	8W-25S	33
Woodlawn Dr. E&W	9W-25S	33
Woodruff Ct. N&S	9W-23S	25
Woodvale Av. E&W	10W-25S	33
42nd St. N&S	11W-22S	25
68th Av. N&S	9W-25S	33
76th Av. N&S	9W-22S	25
78th Av. N&S	9W-25S	33
79th Av. N&S	9W-25S	33
80th Av. N&S	10W-27S,10W-22S	33
82nd Av. N&S	10W-25S	33
84th Av. N&S	10W-25S,10W-24S	33
88th Av. N&S	11W-24S,11W-22S	33
92nd Av. N&S	11W-25S	33
93rd Av. N&S	11W-25S	33
94th Av. N&S	11W-25S	33
104th Av. N&S	13W-27S,13W-22S	32
113th Av. N&S	14W-22S	24
114th Av. N&S	14W-23S,14W-22S	24
115th Av. N&S	14W-23S	24
118th Av. N&S	14W-23S	24
187th St. E&W	14W-22S,10W-22S	24
189th St. E&W	14W-22S	24
191st St. E&W	14W-22S	24
192nd St. E&W	14W-23S	24
193rd St. E&W	14W-23S	24
194th St. E&W	14W-23S,9W-23S	24
195th St. E&W	14W-23S	24
203rd St. E&W	13W-24S	32
204th St. E&W	13W-24S	32

Godley

	Grid	Page
Adams St. E&W	32W-43S	74
Blake Dr. N&S	32W-43S	74
Center St. N&S	32W-43S	74
County Line Rd. N&S	32W-43S,32W-42S	74
Dayton Ct. E&W	32W-43S	74
Elmo Ct. E&W	32W-43S	74
Lincoln St. E&W	32W-43S	74
Livington St. E&W	32W-43S	74
Main St. NE&SW	32W-43S	74
Mary Ln. E&W	32W-43S	74
Mesa Dr. E&W	32W-43S	74
Richard Ct. E&W	32W-43S	74
Robert Dr. N&S	32W-43S	74
Route 53 NE&SW	32W-43S	74
Route 129 NE&SW	32W-43S	74
Sandy Ct. E&W	32W-43S	74
Van Buren St. N&S	32W-43S	74
Vernon Dr. N&S	32W-43S	74

Green Garden Township

	Grid	Page
Beverly Dr. NS&EW	9W-32S	53
Center Rd. N&S	12W-33S,12W-28S	52
Colleen Ct. E&W	9W-32S	53
Dralle Rd. E&W	14W-29S,9W-29S	40
Gorman Rd. E&W	14W-31S,9W-31S	40
Green Garden-Manhattan Rd N&S	14W-33S,14W-29S	52
Harlem Av. N&S	9W-33S,9W-28S	53
Harvest Hills Dr. NS&EW	12W-29S	40
Joliet Rd. N&S	10W-33S,10W-28S	53
Manhattan Monee Rd. E&W	14W-30S,9W-30S	40
Manhattan Wilton Rd. E&W	14W-33S,9W-33S	52
Pauling Rd. E&W	14W-32S,9W-32S	52
Peotone Rd. N&S	11W-33S,11W-28S	52
Prairie Creek Dr. E&W	12W-30S	40
Steger Rd. E&W	14W-28S,9W-28S	40
Stuenkel Rd. E&W	14W-28S,9W-28S	40
Town Line Rd. N&S	14W-28S	40
U.S. Route 45 N&S	14W-33S,13W-28S	52
104th Av. N&S	13W-33S,13W-28S	52

Homer Township

	Grid	Page
Acorn Ct. N&S	19W-17S	16
Acorn Dr. N&S	19W-17S	16
Acorn Ln. E&W	18W-16S	16
Acorn Ridge Dr. E&W	17W-19S	17
Alice Dr. NE&SW	17W-17S	17
Allen Ct. N&S	15W-16S	17
Allison Ln. N&S	15W-17S	17
Alpine Dr. E&W	15W-18S	17
Anand Brook Dr. E&W	16W-16S,15W-16S	17
Angelica Dr. N&S	17W-17S	17
Annico Dr. N&S	16W-18S	17
Appaloosa Ln. N&S	15W-17S	17
Apple Ct. NW&SE	17W-17S	17
Arabian Dr. E&W	16W-18S	17
Arapano Tr. N&S	16W-16S	17
Arboretum Dr. NS&EW	17W-17S	17
Archer Av. NE&SW	20W-16S,19W-16S	16
Arlene Dr. E&W	15W-18S	17
Arrowhead Dr. E&W	19W-21S	22
Ashley Ct. N&S	15W-20S	23
Aspen Ln. E&W	15W-20S	23
Atlantic Dr. E&W	16W-17S	17
Badger Ln. N&S	16W-18S	17
Basham Av. N&S	20W-16S	16
Baywood Ln. NE&SW	16W-17S	17
Beaver Ct. N&S	16W-16S	17
Beaver Den Tr. E&W	15W-19S	17
Beaver Lake Dr. NS&EW	16W-16S	17
Bedford Ln. E&W	20W-16S	16
Bell Rd. N&S	16W-18S,16W-16S	17
Beths Dr. NE&SW	19W-17S	16
Birchdale Dr. N&S	17W-17S	17
Birchwood Ct. NW&SE	17W-17S	17
Birchwood Dr. E&W	16W-17S	17
Black Forest Ct. E&W	15W-18S	17
Black Forest Tr. E&W	15W-18S	17
Black Pine Tr. NW&SE	15W-18S	17
Bosi Dr. E&W	19W-19S	16
Boula Av. N&S	20W-17S	16
Boulder Dr. N&S	17W-17S	17
Brads Way N&S	19W-17S	16
Bramble Ct. E&W	16W-18S	17
Brentwood Dr. E&W	15W-20S	23
Briarwood Dr. E&W	16W-17S	17
Bridle Ct. N&S	15W-17S	17
Bruce Cir. NS&EW	18W-21S,18W-20S	22
Bruce Rd. E&W	20W-20S,15W-20S	22
Burning Bush Ct. NE&SW	16W-16S	17
Burr Oak Dr. E&W	15W-20S	23
Buttercup Ct. E&W	16W-18S	17
Cannonade Ct. NW&SE	15W-16S	17
Canvasback Ct. N&S	15W-17S	17
Carefree Dr. E&W	17W-17S	17
Carlton Ln. NW&SE	15W-17S	17
Casey Ln. NS&EW	17W-17S	17
Catawba Ct. E&W	15W-19S	17
Catawba Ln. E&W	16W-16S	17
Catawba Rd. NW&SE	15W-19S	17
Catawba Tr. E&W	16W-16S	17
Cave Creek Ct. E&W	17W-17S	17
Cedar Creek Ct. N&S	16W-17S	17
Cedar Ct. N&S	18W-20S	22
Cedar Glen Dr. NW&SE	18W-18S	16
Cedar Rd. N&S	18W-16S	17
Cedar Rd. N&S	18W-21S,18W-18S	22
Cedar Ridge Dr. E&W	15W-17S	17
Cedarbend Dr. N&S	16W-17S	17
Cedarwood Dr. NS&EW	18W-19S	16
Chancellor Dr. E&W	19W-19S	16
Chelsea Ct. NE&SW	17W-19S	17
Cherokee Tr. NW&SE	16W-16S	17
Chestnut Ct. N&S	15W-16S	17
Chestnut Ln. N&S	15W-16S	17
Chicago-Bloomington Rd. NE&SW	17W-21S	23
Chicasaw Tr. N&S	16W-16S	17
Chicory Tr. N&S	18W-16S	16
Chippewa Tr. N&S	16W-16S	17
Choctaw Tr. E&W	16W-16S	17
Church St. N&S	18W-17S	16
Cinnamon Creek Ln. E&W	18W-18S	16
Citation Ct. NE&SW	15W-16S	17
Citation Dr. NS&EW	15W-16S	17
Citation Dr. Cir.	15W-16S	17
Clifton Ln. E&W	17W-17S	17
Clydesdale Ln. N&S	15W-17S	17
Coachman's Rd. N&S	17W-17S,17W-16S	17
Codo Dr. N&S	18W-19S	16
Coke Rd. (Pvt.) N&S	19W-16S	16
Colt Dr. E&W	16W-18S	17
Comandora Cir. NS&EW	15W-20S	23
Country Woods Dr. NW&SE	15W-18S	17
Court Connection NW&SE	17W-21S	23
Cranberry Ct. N&S	16W-18S	17
Creek Ct. N. NE&SW	18W-17S	16
Creek Ct. S. NW&SE	18W-17S	16
Creek View Dr. NW&SE	18W-16S	16
Creekside Ct. NW&SE	16W-16S	17
Creekside Dr. NS&EW	16W-16S	17
Creekview Dr. NW&SE	18W-16S	16
Creekwood Ct. N&S	16W-19S	17
Creekwood Dr. N&S	16W-19S	17
Creekwood Ln. E&W	16W-19S	17
Creme Rd. N&S	18W-17S	16
Crestwood Dr. E&W	18W-16S	16
Cricketwood Ct. NE&SW	17W-17S	17
Cricketwood Dr. NS&EW	17W-17S	17
Crystal Lake Ct. E&W	16W-21S	23
Crystal Lake Dr. N&S	16W-21S,16S-20S	23
Dalphon Ct. NW&SE	15W-16S	17
Deerpath Dr. N&S	15W-20S	23
Deervalley Dr. E&W	17W-17S	17
Derby Dr. NE&SW	15W-16S	17
Derby Ln. NS&EW	15W-16S	17
Derby Rd. N&S	17W-16S	17
Dixon Ln. N&S	17W-17S	17
Doede Ln. E&W	16W-20S	23
Dogwood Dr. N&S	15W-18S	17
Dokter Pl. E&W	16W-19S	17
Doonaree Ct. NW&SE	17W-16S	17
Driftwood Ln. NE&SW	17W-21S	23
Dublin Ct. E&W	17W-16S	17
Dublin Dr. E&W	17W-16S	17
E. Glen Dr. N&S	15W-17S	17
Eagle Bend NS&EW	19W-21S	22
East Abbott Rd. N&S	15W-17S	17
East St. N&S	19W-16S	16
Edgewood Dr. E&W	15W-18S	17
Edinburgh Ct. E&W	18W-17S	16
Edinburgh Ct. E. N&S	18W-17S	16
Edinburgh Ct. W. N&S	18W-17S	16
Edmonds (Meador) St. N&S	17W-21S	23
Elizabeth Ln. N&S	17W-17S	17
Elm St. N&S	19W-16S	16
Erin Ct. N&S	16W-17S	17
Erin Dr. N&S	17W-16S	17
Evergreen Ter. N&S	15W-20S	23
Fairway View Av. E&W	17W-16S	17
Farmview St. E&W	16W-18S	17
Fieldview Ln. N&S	20W-18S	16
Filly Dr. E&W	16W-18S	17
Finch Ct. E&W	16W-18S	17
Flint Ct. E&W	15W-17S	17
Flint Dr. E&W	15W-17S	17
Forest Ct. N&S	19W-18S	16
Forestview Dr. NS&EW	15W-16S	17
Foxboro Ln. N&S	17W-21S	23
Gadwall Ct. NW&SE	15W-17S	17
Garavogue Av. N&S	15W-16S	17
Gatses Estates E&W	18W-17S,17W-17S	16
Gatses Ln. E&W	18W-17S,17W-17S	16
Glen Crest Ln. E&W	18W-18S	16
Glen Dale Ln. N&S	18W-18S	16
Glen Entrance E&W	16W-21S	23
Glenview Cir. E&W	18W-18S	16
Glenwood Ct. N&S	18W-18S	16
Glenwood Ln. E&W	18W-18S	16
Golden Oak Dr. N&S	16W-17S,16W-16S	17
Gombis Dr. E&W	18W-21S	22
Gougar Rd. N&S	20W-21S,20W-18S	22
Graceland Dr. N&S	15W-18S	17
Gunner Ct. E&W	15W-19S	17
Hadley Rd. NE&SW	16W-20S	23
Haggard St. E&W	16W-18S	17
Hampton Ct. N&S	19W-16S	16
Hank Ct., E. E&W	15W-17S	17
Hank Ct., W. E&W	15W-17S	17
Hass Rd. N&S	16W-21S	23
Hawk Haven E&W	19W-16S	16
Heather Ct. N&S	15W-17S	17
Heather Ln. N&S	15W-17S	17
Heatherwood Ct. N&S	17W-17S	17
Heatherwood Dr. NS&EW	17W-17S	17
Hedge Apple Dr. E&W	16W-16S	17
Heritage Dr. N&S	18W-20S	22
Hiawatha Ct. N&S	16W-16S	17
Hiawatha Dr. E&W	16W-16S	17
Hickory Av. NS&EW	18W-16S,17W-6S	16
Hickory Ct. E&W	17W-16S	17
Hidden Valley Dr. NS&EW	16W-19S	17

84 WILL COUNTY

	Grid	Page
Hidden Valley Tr. N&S	16W-19S	17
Hillary Ln. N&S	19W-17S	16
Hillcrest Rd. N&S	18W-17S	16
Hillside Ct. N&S	16W-18S	17
Hilltop Dr. E&W	17W-21S	23
Holm Ct. N&S	15W-17S	17
Holm Dr. E&W	15W-17S	17
Hunter Ct. N&S	20W-16S	16
Ignace Ct. NW&SE	15W-19S	17
Ironwood Cir. Cir.	17W-17S	17
Iroquis Tr. E&W	16W-16S	17
Ivy Ct. N&S	15W-17S	17
James Ct. E&W	19W-21S	22
Janas Dr. E&W	19W-16S	16
Joan Marie Dr. E&W	15W-18S	17
Juniper Ln. E&W	15W-20S	23
Kelly Ct. N&S	20W-18S	16
Kensington Dr. N&S	17W-19S	17
Kerry Dr. NE&SW	17W-16S	17
Kerry Ln. NE&SW	17W-16S	17
Kickapoo Tr. N&S	16W-16S	17
King Rd. N&S	18W-17S,18W-16S	16
Kiowa Ct. N&S	16W-16S	17
Kundrat Ln. E&W	19W-16S	16
Lacrosse Dr. N&S	19W-21S	22
Lady Bar Av. N&S	15W-16S	17
Lady Bar Dr. N&S	15W-16S	17
Lady Bar Ln. E&W	15W-16S	17
Lakeview Ct. N&S	15W-16S	17
Lakeview Dr. E&W	15W-16S	17
Lakeview Pl. N&S	15W-16S	17
Lakeview Tr. E&W	15W-18S	17
Lancaster Ct. E&W	15W-16S	17
Larkspur Ct. N&S	16W-21S	23
Larkspur Dr. NE&SW	17W-21S	23
Lauffer Rd. NW&SE	16W-20S	23
Laura Ln. N&S	17W-17S	17
Leach Dr. N&S	19W-19S	16
Lemont Rd. N&S	19W-16S	16
Linsey Ln. E&W	20W-21S	22
Lisa Ln. E&W	17W-17S	17
Little Creek Dr. NS&EW	16W-16S	17
Little Creek Dr. Cir.	16W-16S	17
Long Run Ct. E&W	16W-16S	17
Long Run Dr. N&S	18W-16S	16
Longmeadow Ln. E&W	15W-20S	23
Longrun Dr. N&S	18W-16S	16
Longview Dr. N&S	19W-16S	16
Loretta Ln. E&W	17W-17S	17
Lost Boy Ln. N&S	16W-16S	17
Mackinac Ct. NW&SE	15W-19S	17
Mackinac Rd. NE&SW	15W-19S	17
Magnolia Dr. E&W	15W-18S	17
Mallard Dr. N&S	15W-17S,15W-16S	17
Mallard Lake Dr. (Pvt.) N&S	19W-18S	16
Mallard Ln. N&S	15W-17S	17
Mallard Ln. (Pvt.) NS&EW	19W-18S	16
Manitou Ct. NE&SW	15W-19S	17
Manitou Rd. NS&EW	15W-19S	17
Manor Ln. N&S	18W-18S	16
Maple Av. N&S	18W-16S	16
Maple Ct. N&S	16W-18S	17
Maramel Dr. N&S	16W-19S	17
Marilyn Ln. N&S	17W-17S	17
Marti Rd. N&S	15W-21S	23
Martingale Ln. E&W	15W-17S	17
Maverick Tr. E&W	16W-18S	17
McCarron Rd. N&S	18W-21S,18W-20S	22
McDonald Dr. E&W	20W-18S	16
McKenzie Av. E&W	20W-18S	16
Meador Rd. (Edmonds St.) N&S	17W-21S	23
Meadow View Ln. E&W	16W-18S	17
Meadowcrest Dr. N&S	15W-20S	23
Meadowland Dr. NE&SW	15W-17S	17
Meadowlark Ct. E&W	16W-18S	17
Meath Cir. N&S	17W-16S	17
Meath Ct. N&S	17W-16S	17
Meath Dr. E&W	17W-16S	17
Monaghan Dr. N&S	17W-16S	17
Monaghan Rd. N&S	17W-16S	17
Morgan Ct. NE&SW	15W-17S	17
Mormann Ln. N&S	18W-16S	16
Morning Glory Ct. E&W	16W-18S	17

	Grid	Page
Muir Dr. N&S	20W-18S	16
Mulberry Ct. N&S	16W-18S	17
Mulraney Dr. NS&EW	16W-17S	17
Murvey Ct. NW&SE	15W-16S	17
Murvey Dr. E&W	15W-16S	17
Mustang Dr. N&S	15W-17S	17
N. Hawthorne Ct. N&S	19W-17S	16
Navajo Tr. E&W	16W-16S	17
Navajo Tr. E&W	17W-16S	17
Needles Ct. E&W	15W-16S	17
Nolan Ct. E&W	19W-16S	16
O'Connell Dr. E&W	18W-21S	22
Oak Av. E&W	17W-16S	17
Oak Ct. E&W	16W-17S	17
Oak Ct. NW&SE	16W-19S	17
Oak Meadow Ct. NW&SE	16W-20S	23
Oak Ridge Ct. NE&SW	16W-16S	17
Oak Ridge Dr. N&S	16W-16S	17
Oak Ridge Ln. NW&SE	16W-16S	17
Oak St. N&S	19W-16S	16
Oak St. E&W	21W-21S	21
Oak Tr. Cir.	16W-17S	17
Oak Valley Ct. E&W	16W-19S	17
Oak Valley Tr. NS&EW	16W-19S	17
Oakridge Ct. N&S	16W-16S	17
Oakridge Dr. N&S	16W-16S	17
Oakridge Ln. NW&SE	16W-16S	17
Oakview Ct. E&W	16W-19S	17
Oakwood Dr. E&W	16W-17S	17
Onondaga NW&SE	16W-16S	17
Orchard Ln. E&W	18W-16S	16
Paddock Ln. NE&SW	16W-18S	17
Palomino Ct. N&S	15W-17S	17
Pantigo Ln. E&W	19W-19S	16
Parker Rd. N&S	16W-21S,17W-17S	23
Parkside Dr. N&S	16W-18S	17
Parnell Cir. NE&SW	18W-21S	22
Partridge Ln. E&W	15W-16S	17
Pawnee Ln. NE&SW	16W-16S	17
Pebble Creek Ct. N&S	17W-17S	17
Pebble Creek Dr. N&S	17W-17S	17
Peppermill Tr. N&S	19W-19S	16
Pheasant Cir. NS&EW	18W-16S	16
Pheasant Ct. N&S	16W-19S	17
Pheasant Ln. N&S	15W-17S,15W-16S	17
Pin Oak Ct. E&W	16W-16S	17
Pin Oak Ct. N&S	16W-19S	17
Pin Oak Dr. E&W	16W-16S	17
Pine Grove Dr. NS&EW	16W-17S	17
Pine Hill Dr. N&S	17W-19S	17
Pinegrove Ct. E&W	16W-17S	17
Pineview Dr. N&S	15W-20S	23
Pinto St. NE&SW	16W-18S	17
Pleasant Cir. Cir.	18W-16S	16
Pleasant Ct. E&W	19W-16S	16
Potawatamie Ct. N&S	16W-16S	17
Potawatomi Tr. N&S	16W-16S	17
Prairie Ct. E&W	15W-17S	17
Prairie Dr. NW&SE	15W-17S	17
Prairie Hill Dr. N&S	19W-16S	16
Purley Ct. NE&SW	19W-16S	16
Quail Ln. E&W	15W-16S	17
Quail Run Ct. N&S	19W-16S	16
Rado Dr. NS&EW	16W-16S	17
Rambling Rd. (Pvt.) E&W	15W-17S	17
Red Cedar Tr. N&S	15W-18S	17
Red Oak Ct. E&W	16W-19S	17
Regents Rd. N&S	19W-19S	16
Reiter Dr. E&W	18W-20S	22
Renee Ln. N&S	17W-17S	17
Rickerman Rd. N&S	20W-17S	16
Ridge Ct. N&S	20W-16S	16
Ridgewood Dr. NS&EW	16W-19S	17
Robert Emmett Dr. N&S	18W-21S	22
Rockbluff Wy. NS&EW	17W-17S	17
Roger Rd. N&S	20W-16S	16
Ron Ct. E&W	18W-20S	22
Rosarie Dr. N&S	16W-18S	17
Rosewood Dr. E&W	15W-17S	17
Rt. 7 E&W	19W-19S,15W-19S	16
Rt. 171 NE&SW	20W-16S	16
Rt. 355 (proposed) N&S	18W-21S,20W-16S	22
Russ Ln. N&S	16W-16S	17

	Grid	Page
Rycon Dr. E&W	16W-21S	23
Rynberk Ct. N&S	20W-16S	16
S. Bell Rd. N&S	15W-20S	23
S. Hawthorne Ct. N&S	19W-17S	16
S. Wedgewood Dr. N&S	15W-20S	23
Saddle Brooke Ln. NS&EW	17W-17S	17
Sage Ct. E&W	16W-18S	17
Samuel Dr. NS&EW	16W-18S	17
Sandi Ln. E&W	20W-18S	16
Sandstone Dr. E&W	17W-17S	17
Schroeder Dr. E&W	19W-19S	16
Scott Ct. NW&SE	20W-18S	16
Scott Dr. N&S	20W-18S	16
Sea Biscut Ct. N&S	15W-16S	17
Secretariat Ln. NS&EW	15W-16S	17
Seminole Tr. E&W	17W-16S,16W-16S	17
Shady Ln. NS&EW	17W-17S	17
Shady Ln. NW&SE	19W-18S	16
Shaeffer Ct. N&S	20W-16S	16
Shannon Dr. NS&EW	17W-16S	17
Sharon Dr. N&S	15W-18S	17
Shetland Ct. NE&SW	20W-18S	16
Shetland Dr. N&S	15W-17S	17
Shoshoni Dr. E&W	16W-16S	17
Shoshoni Tr. E&W	17W-16S	17
Silo Ct. NE&SW	16W-18S	17
Silo Dr. N&S	16W-18S	17
Somerset Ct. N&S	15W-16S	17
Somerset Rd. NE&SW	15W-16S	17
Sonya Ln. E&W	17W-17S	17
South Janas Pkwy. NS&EW	19W-16S	16
Spaniel Dr. Cir.	15W-19S	17
Sparrow Ct. E&W	16W-18S	17
Split Rail Ct. NE&SW	16W-16S	17
Split Rail Dr. N&S	16W-16S	17
Spring Creek Rd. E&W	18W-16S,17W-16S	16
Spring Ct. E&W	16W-18S	17
Spring Lake Dr. NE&SW	15W-17S	17
Starr Ln. (Pvt.) N&S	20W-16S	16
State St. N&S	19W-17S	16
Stately Oaks Cir. NS&EW	19W-17S	16
Stately Oaks Dr. E&W	19W-17S	16
Sterling Dr. NE&SW	20W-16S	16
Stirrup Ct. N&S	15W-17S	17
Stone Creek Ct. NE&SW	16W-17S	17
Stone Oak Wy. E&W	17W-17S	17
Stonegate Dr. NS&EW	16W-17S	17
Stonehaven Ln. E&W	18W-17S	16
Stonewood Dr. E&W	16W-17S	17
Sulky Ct. NW&SE	16W-18S	17
Sulky Dr. N&S	16W-18S	17
Sunset Ct. N&S	17W-17S	17
Surrey Ln. E&W	15W-17S	17
Swaps Ct. E&W	15W-16S	17
Sweetwater Dr. E&W	16W-17S	17
Tamarack Ln. E&W	15W-20S	23
Tameling Dr. E&W	20W-16S	16
Tara Dr. NW&SE	17W-16S	17
Teakwood Dr. N&S	16W-16S	17
Teal Dr. N&S	15W-17S	17
Teal Ln. (Pvt.) N&S	19W-18S	16
Terrier Ct. E&W	15W-19S	17
Thicket Ct. N&S	16W-16S	17
Thistle Ct. E&W	16W-18S	17
Thorn Apple Dr. E&W	15W-20S	23
Thornbush Dr. NS&EW	16W-16S	17
Thornwood Ln. E&W	19W-17S	16
Tim Tam Ct. NE&SW	15W-16S	17
Timber Lane Ct. NW&SE	17W-17S	17
Trails End Ct. NE&SW	16W-16S	17
Trails End Dr. N&S	16W-16S	17
Trailside Dr. N&S	16W-18S	17
Turf Ln. N&S	15W-17S	17
Turnberry Dr. N&S	15W-16S	17
Twin Creek Ln. N&S	18W-17S	16
Twin Lakes Dr. N&S	15W-18S	17

	Grid	Page
Twin Oaks Ct. N&S	16W-19S	17
Twin Oaks Ct. N&S	18W-16S	16
Valley View Dr. E&W	16W-19S	17
Vaysee Ct. E&W	19W-16S	16
Venetion Ct. NE&SW	15W-16S	17
Venetion Way NS&EW	15W-16S	17
Victoria Ct. E&W	15W-17S	17
Walnut Av. NW&SE	18W-16S	16
Walnut Ct. N&S	16W-18S	17
Weather Vane Ln. N&S	16W-18S	17
Wedgewood Dr. NE&SW	15W-20S	23
Welter Ct. N&S	16W-17S	17
West Abbott Rd. N&S	15W-17S	17
West St. N&S	19W-16S	16
Wheatstone Dr. E&W	19W-19S	16
Whirlaway Ct. N&S	15W-16S	17
White Oak St. NW&SE	16W-19S	17
White Pines Ct. NE&SW	15W-18S	17
White Pines Tr. NW&SE	15W-18S	17
Wilco Dr. NS&EW	15W-17S	17
Will Cook Rd. N&S	15W-18S,15W-16S	17
Willow Oak Ct. N&S	16W-18S	17
Winchester Ct. N&S	15W-16S	17
Winchester Dr. NS&EW	15W-16S	17
Windmill Dr. N&S	19W-19S	16
Wingate Dr. N&S	15W-17S	17
Wirlway Ct. N&S	15W-16S	17
Wood Duck Ct. N&S	15W-17S	17
Woodcrest Av. N&S	15W-17S	17
Woodduck Ln. E&W	18W-16S	16
Woodland Dr. E&W	16W-18S	17
Woodmill Ct. E&W	17W-17S	17
Wooly Hill Ct. E&W	15W-16S	17
Wooly Hill Dr. N&S	15W-16S	17
Yorkshire Dr. E&W	15W-17S	17
Zuck Ct. E&W	18W-19S	16
7th St. E&W	20W-19S	16
135th Av. N&S	20W-17S	16
135th St. E&W	18W-16S	16
136th Av. N&S	20W-17S	16
136th St. E&W	18W-16S	16
138th St. E&W	20W-16S,19W-16S	16
139th St E&W	19W-16S,18W-16S	16
140th Pl. E&W	20W-16S	16
140th St. E&W	20W-16S	16
141th St. E&W	20W-16S,19W-16S	16
143rd St. E&W	18W-16S,16W-16S	16
144th Pl. E&W	20W-17S	16
145th Pl. E&W	20W-17S	16
146th St. E&W	20W-17S	16
147th St. E&W	19W-17S,18W-17S	16
151st St. E&W	18W-17S,16W-17S	16
159th St. E&W	20W-18S,16W-18S	16
163rd St. E&W	19W-19S,18W-19S	16
167th St. E&W	20W-19S,15W-19S	16
175th St. E&W	18W-20S	22
179th St. E&W	17W-21S,15W-21S	23

Jackson Township

	Grid	Page
Alice Ct. NE&SW	22W-30S	37
Arsenal Rd. E&W	26W-30S,25W-30S	36
Bernhard Rd. E&W	22W-28S	37
Bicentennial Dr. NS&EW	22W-30S	37
Brandon Rd. NS&EW	25W-29S,25W-28S	36
Breen Rd. E&W	23W-29S,22W-29S	37
Bridge Rd. N&S	24W-29S	36
Brown Rd. (Sweedler) E&W	23W-31S,21W-31S	37
Bush Rd. N&S	25W-30S	36
Cherry Hill Rd. N&S	21W-32S,21W-28S	49
Chicago Rd. N&S	25W-31S,25W-30S	36
Clair Ct. NW&SE	22W-30S	37
Coldwater Rd. N&S	24W-32S,24W-30S	48
Craig Rd. E&W	26W-29S,24W-28S	36
Creek Dr. N&S	24W-30S	36
Diagonal Rd. NE&SW	26W-31S,25W-31S	36
Eaton Av. E&W	24W-29S	36
Gary Ray Dr. N&S	24W-29S	36
Gas Plant Rd. E&W	26W-29S,25W-29S	36
Gladys Av. E&W	24W-29S	36
Gurney Rd. (Joliet Arsenal) E&W	23W-32S,21W-32S	49
Hemphill Dr. NS&EW	22W-30S	37
Jeffrey Ct. N&S	24W-30S	36
Keith Allen Dr. NE&SW	24W-29S	36
Kinder Rd. E&W	22W-30S	37
Lois Ln. N&S	24W-30S	36
Manhattan-Wilton Rd. E&W	25W-33S,22W-33S	48
Marian Av. N&S	21W-30S	37
Millsdale Ext. Rd. E&W	26W-29S,25W-29S	36
Millsdale Rd. NE&SW	26W-29S	36
Mississippi Av. E&W	26W-32S	48
Noel Rd. E&W	24W-29S	36
North Town Line Rd. E&W	23W-28S,21W-28S	37
Old Wilmington Rd. NE&SW	26W-32S	48
Patterson Rd. N&S	26W-29S	36
Pheasant Dr. N&S	24W-30S	36
Quail Ct. E&W	24W-30S	36
Raymond Dr. NS&EW	24W-29S	36
Rich Ct. E&W	21W-30S	37
Ridge Rd. N&S	22W-32S,22W-28S	49
Rob Av. E&W	21W-30S	37
Ron Ct. E&W	21W-30S	37
Rondorey Rd. N&S	24W-30S	36
Rowell Av. N&S	23W-29S,23W-28S	37
Rowell Rd. N&S	23W-32S,23W-30S	49
Rt. 53 NE&SW	24W-29S,24W-28S	36
Sharp Rd. E&W	24W-28S	36
South Townline Rd. E&W	21W-33S	49
Spangler Rd. E&W	21W-29S	37
Stone Rd. E&W	21W-28S	37
Sweitzer Rd. E&W	24W-28S	36
Sycamore St. N&S	24W-30S	36
Tanglewood Dr. E. E&W	24W-30S	36
Tanglewood Dr. W. E&W	24W-30S	36
Tehle Rd. NE&SW	24W-31S,24W-30S	36
Timber Ct. NW&SE	24W-30S	36
Timber Dr. N&S	24W-30S	36
Timber Dr. N. E&W	24W-30S	36
Timber Dr. S. E&W	24W-30S	36
Walnut St. N&S	24W-30S	36
Wilmington Rd. N&S	26W-29S,26W-28S	36
Wood Rd. E&W	24W-32S,24W-31S	48

Joliet

	Grid	Page
A Wilhelmi Dr. E&W	28W-21S	19
Abe St. N&S	23W-23S	21
Ada St. E&W	22W-22S	21
Addleman St. N&S	28W-22S	19
Adella Av. N&S	22W-24S	29
Agnes Av. E&W	25W-22S	20
Akin Av. N&S	23W-24S	29
Alamosa Ct. E&W	19W-22S	22
Alamosa St. N&S	19W-22S	22
Alann Dr. N&S	26W-22S	20
Albert Av. E&W	23W-24S	29
Albert D'Ottavio Dr. NE&SW	28W-22S	19
Aldrich Dr. N&S	29w-21S	18
Alessio Dr. E&W	23W-25S	29
Alexander Dr. E&W	27W-23S	19
Allen St. E&W	24W-24S	28
Amber Ct. NW&SE	28W-20S	19
Ambrose Ln. N&S	29W-21S	18
Ann Ct. N&S	26W-23S	20
Ann St. N&S	26W-23S	20
Antram Av. N&S	22W-22S	21
Apollo Dr. N&S	27W-23S	19
Arch Ct. N&S	23W-23S	21
Arden Ln. E&W	26W-23S	20
Arden Pl. E&W	27W-23S,26W-23S	19
Arizona Av. N&S	22W-24S	29
Arthur Av. N&S	22W-22S	21
Ashbury Cir. Dr. E&W	26W-22S	20
Aspen Ct. N&S	28W-23S	19
Audrey Av. E&W	25W-24S	28
Avalon Av. E&W	26W-23S	20

	Grid	Page
Baker Av. N&S 23W-24S		29
Bankview Ln. NE&SW 28W-23S		19
Barber Ct. E&W ... 27W-23S		19
Barber Ln. NW&SE 27W-23S		19
Barber Ln. NE&SW 27W-22S		19
Barney Dr. N&S 27W-23S		19
Barthelme Av. N&S 25W-22S		20
Bartleson St. E&W ..23W-24S		29
Basswood Rd. E&W 21W-22S		21
Bataan St. NW&SE 27W-21S		19
Beach St. N&S 24W-24S		28
Bedford Dr. E&W .. 30W-21S		18
Beechwood Rd. E&W 21W-22S		21
Bellarmine Dr. E. N&S 26W-24S		28
Bellarmine Dr. W. N&S 26W-24S		28
Belle Av. NW&SE .. 22W-22S		21
Belmont Av. . 21W-23S		21
Benedict Av. E&W ..26W-24S		28
Bennette Av. N&S .. 23W-23S		29
Bentley Dr. N&S .. 29W-21S		18
Benton St. E&W 24W-23S,22W-23S		20
Bergstrom St. E&W 28W-22S		19
Berlin Av. E&W .. 24W-22S		20
Bethal Dr. N&S 27W-23S		19
Betula St. N&S 25W-24S		28
Big Timber Dr. E&W 28W-23S		19
Birchwood Ln. E&W 26W-22S		20
Bissel St. N&S 22W-23S		21
Black Rd. E&W 29W-22S,25W-22S		18
Blackman Av. N&S 24W-24S		28
Blandford Av. E&W 19W-22S		22
Bluff St. N&S 24W-24S,24W-22S		28
Bogden Ln. Cir. .. 21W-22S		21
Border Dr. N&S .. 26W-23S		20
Boston Av. E&W .. 25W-22S		20
Boulder Av. N&S .. 22W-24S		29
Bowen Pl. E&W 23W-24S		29
Brentwood Dr. N&S 26W-22S		20
Brentwood Pl. NE&SW 26W-22S		20
Briargate Av. E&W..25W-22S		20
Bridge St. N&S ..24W-23S		20
Brighton Ln. NE&SW 30W-21S		18
Broadway Av. N&S 24W-23S,24W-22S		20
Bronk Rd. N&S 29W-22S,29W-21S		18
Brooks Av. NE&SW 24W-23S		20
Brookwood Ct. NW&SE 26W-22S		20
Bruce St. E&W .. 23W-22S		20
Bryan Av. E&W .. 25W-22S		20
Buck Av. E&W .. 28W-22S		19
Buell Av. E&W .. 25W-22S		20
Buell Ct. NW&SE .. 25W-22S		20
Burr Oaks Rd. E&W 28W-23S		19
Burry St. N&S 25W-22S		20
Bush Rd. NW&SE .. 28W-26S		27
Bush St. N&S 23W-24S		29
Caddy Ln. E&W ... 27W-23S		19
Cagwin Av. N&S....24W-24S		28
California Av. N&S..22W-22S		21
Campbell St. E&W 27W-23S,24W-23S		19
Canal Rd. NE&SW .. 28W-27S		27
Canal St. N&S 24W-23S		20
Canterbury Ct. N&S 30W-21S		18
Canterbury St. E&W 30W-21S		18
Capri Av. E&W .. 26W-24S		28
Cardinal Ln. E&W .. 22W-22S		21
Carson Av. N&S....24W-23S		20
Cass St. E&W 24W-23S,21W-23S		20
Casseday Av. N&S 23W-23S		21
Cassie Dr. NE&SW 27W-23S		19
Cassie Dr. NW&SE 27W-22S		19
Catalpa St. E&W .. 23W-23S		21
Caterpillar Dr. N&S 27W-24S		27
Catherine St. N&S 25W-23S,25W-22S		20
Cathy Dr. NS&EW .. 28W-25S		27
Caton Av. E&W 26W-23S,25W-23S		20
Caton Farm Rd. E&W 32W-20S,27W-20S		18
Cecily Dr. N&S 27W-21S		19
Center St. N&S 24W-24S,24W-22S		28
Central Av. N&S .. 24W-24S		28
Channahon Rd. NE&SW 28W-27W,27W-26S		27
Charlesworth Av. N&S 23W-22S		21
Charlotte Av. NE&SW 24W-23S		20
Chase Av. E&W 23W-22S,22W-22S		21
Cherry St. N&S 24W-24S		28
Chestnut Hill Rd. N&S 30W-21S		18
Chicago St. N&S 24W-24S,24W-23S		28
Chippewa St. N&S .. 24W-25S		29
Christina Ln. NE&SW 25W-22S		20
Christine Ct. N&S .. 28W-25S		27
Clara Av. E&W 26W-22S,25W-22S		20
Clark St. N&S 22W-22S		21
Claudia Av. N&S .. 22W-24S		29
Clay St. N&S 23W-23S		21
Cleary Av. E&W .. 25W-22S		20
Cleary St. E&W .. 25W-22S		20
Clement St. N&S 24W-23S,24W-22S		20
Cleveland Av. E&W 23W-22S		21
Clinton St. E&W 24W-23S,22W-23S		20
Cochrane St. E&W ..25W-24S		28
Colburn Av. E&W .. 23W-24S		29
Collins St. N&S 23W-23S,23W-22S		21
Colorado Av. NW&SE 25W-22S		20
Columbia St. N&S 24W-23S,23W-23S		20
Comstock St. N&S..24W-24S		28
Connecticut Av. NE&SW 25W-22S		20
Copley St. N&S .. 28W-25S		27
Copperfield Av. E&W 22W-23S		21
Cora St. N&S 24W-22S		20
Cornelia St. N&S .. 25W-23S		20
Corregidor St. NW&SE 27W-21S		19
Cottage Pl. E&W .. 25W-24S		28
Cottonwood Rd. E&W 21W-22S		21
Country Glen Dr. E&W 29W-22S		18
County Farm Rd. E&W 28W-24S		27
Court St. N&S.....22W-24S		29
Cove Cir. E&W .. 29W-21S		18
Coventry Ct. NE&SW 28W-23S		19
Cowles Av. N&S .. 25W-23S		20
Crabapple Ct. N&S 27W-23S		19
Crabapple Dr. N&S 27W-23S		19
Craig Dr. E&W .. 28W-25S		27
Cranberry Ct. N&S..27W-23S		19
Creed St. N&S .. 23W-23S		21
Croghan St. E&W 25W-24S,24W-24S		28
Crowley Av. E&W 24W-23S,23W-23S		20
Curtis Av. E&W 25W-22S,24W-22S		20
Cutter Av. N&S .. 22W-22S		21
Cypress Ct. NE&SW 26W-22S		20
Cypress Ln. N&S .. 26W-22S		20
Cypress St. E&W .. 22W-22S		21
D. Hutchinson Av. E&W 28W-21S		19
Darcy Av. E&W ... 24W-24S		28
Dartmoor Dr. .26W-22S		20
David Av. E&W .. 23W-25S		29
Davis Av. N&S .. 25W-24S		28
Davis St. E&W .. 24W-24S		28
Dawes Av. N&S...25W-22S		20
DeKalb St. E&W....24W-24S		28
Dearborn St. N&S .. 24W-24S		28
Deerpath Dr. E&W ..27W-23S		19
Delmar Av. N&S 25W-22S,24W-22S		20
Delrose St. N&S 27W-21S		19
Demmond (Desmond) Ct. E&W 22W-22S		21
Dennis Ct. E&W .. 23W-22S		21
Des Plaines St. N&S 24W-24S,24W-23S		28
Desmond (Demmond) Ct. E&W 22W-22S		21
Devonshire Dr. NE&SW 26W-22S		20
Dewey Av. N&S .. 24W-24S		28
Dillman St. N&S .. 23W-22S		21
Division St. N&S .. 24W-23S		20
Dixon Av. N&S .. 24W-23S		20
Dogwood St. E&W 28W-23S		19
Donna St. N&S .. 25W-22S		20
Donnie St. N&S .. 25W-22S		20
Dora Av. N&S.... 22W-22S		21
Doris St. E&W .. 23W-25S		29
Douglas St. N&S 27W-23S,25W-23S		19
Dover St. N&S .. 23W-22S		21
Draper Av. N&S .. 22W-22S		21
Duncan St. E&W .. 24W-24S		28
Dwight Av. N&S .. 25W-24S		28
E. Bevan Dr. N&S .. 26W-23S		20
E. Palladium Dr. N&S 26W-23S		20
Eagle St. N&S 22W-23S		21
Earl Av. N&S 25W-24S		28
Eastern Av. N&S 23W-24S,23W-23S		29
Edgecreek Dr. E&W 19W-23S		22
Edgehill St. N&S ..22W-22S		21
Edgerton Dr. N&S .. 26W-22S		20
Edie Ln. N&S 27W-23S		19
Edward St. N&S .. 24W-25S		28
Elgin Av. E&W ... 22W-22S		21
Elizabeth Ct. N&S .. 26W-22S		20
Elizabeth St. N&S .. 26W-22S		20
Ella Av. E&W 23W-25S		29
Elm St. N&S 23W-23S		29
Elmwood Av. E&W .23W-24S		29
Elwood St. N&S .. 23W-24S		29
Emery St. N&S .. 26W-23S		28
Emmett St. NW&SE 24W-24S		28
Englewood Av. N&S 22W-22S		21
Erins Glen Dr. NW&SE 29W-22S		18
Essington Rd. N&S 28W-23S,28W-19S		19
Ewen Av. E&W .. 23W-23S		21
Ewing St. N&S .. 22W-22S		21
Executive Dr. E&W 28W-22S		19
Fairbanks N&S .. 22W-22S		21
Fairchild Av. E&W .. 22W-22S		21
Fairlane Dr. N&S...27W-23S		19
Fairmont Av. N&S .. 22W-22S		21
Fairview Av. N&S .. 22W-22S		21
Fairway Dr. NW&SE 27W-23S		19
Farragut Pl. E&W .. 25W-23S		20
Ferris Pl. E&W 26W-24S		28
Ferris St. NW&SE .. 25W-24S		28
Fetz Av. NE&SW .. 23W-24S		29
Fiday Rd. E&W .. 28W-21S		19
Fiesta Dr. N&S .. 21W-23S		21
Fitzpatrick Ct. N&S 28W-21S		19
Florence Av. E&W .. 23W-24S		29
Fouser Ct. E&W .. 23W-24S		29
Francine Av. E&W .. 26W-24S		28
Francis St. E&W 23W-22S,22W-22S		21
Frank Av. N&S 24W-22S		20
Franklin St. N&S...23W-23S		21
Fredrick St. N&S .. 25W-22S		20
Frontage Rd. N&S 28W-22S,28W-21S		19
Fuller Av. N&S....22W-22S		21
Gael Dr. N&S 27W-22S		19
Gage Ct. NW&SE .. 22W-22S		21
Gage St. N&S 22W-22S		21
Galway Rd. N&S .. 29W-22S		18
Gardner St. N&S 23W-25S,23W-24S		29
Garfield Av. N&S .. 24W-23S		20
Garland Ct. N&S .. 22W-22S		21
Garnsey Av. N&S .. 23W-23S		21
Garvin St. N&S .. 22W-22S		21
George Av. N&S 27W-22S,26W-22S		19
Georgia Av. NE&SW 25W-22S		20
Glen St. N&S 23W-22S		21
Glenwood Av. E&W 27W-23S,24W-23S		19
Glenwood Green Dr. E&W 26W-23S		20
Godfrey Pl. E&W .. 23W-23S		21
Golfview Dr. NE&SW 27W-21S		19
Gordon Pl. N&S .. 25W-23S		20
Gould St. N&S .. 22W-22S		21
Grand Blvd. NE&SW 25W-24S		28
Granite St. N&S .. 24W-22S		20
Grant Av. E&W 23W-24S,22W-24S		29
Gross St. E&W .. 23W-22S		21
Grover St. N&S .. 23W-24S		29
Hacker Av. N&S .. 23W-22S		21
Hague Ct. N&S .. 22W-22S		21
Hague St. N&S .. 22W-22S		21
Haldeman Av. NE&SW 25W-24S		28
Hammes Av. N&S 26W-24S,26W-23S		28
Hampton Av. .. 30W-21S		18
Harrison Av. .. 23W-22S		21
Harrogate Ln. E&W 26W-22S		20
Harwood St. N&S .. 22W-23S		21
Hebbard St. N&S .. 23W-24S		29
Hebbard St. N&S...25W-22S		20
Heiden Av. N&S .. 25W-24S		28
Helles Av. N&S .. 24W-24S		28
Henderson Av. N&S 23W-23S,23W-22S		21
Hennepin Av. 28W-20S,27W-20S		19
Hennessy Rd. .. 27W-23S		19
Henry St. NE&SW .. 23W-24S		29
Heritage Dr. .. 27W-23S		19
Herkimer St. N&S 23W-23S,23W-22S		21
Hickory Creek Dr. NW&SE 19W-23S		22
Hickory St. N&S 24W-24S,24W-22S		28
High St. N&S 22W-23S		21
Highland Av. N&S .. 24W-23S		20
Higley Av. N&S .. 22W-24S		29
Hill St. E&W 23W-22S		21
Hillside Rd. NW&SE 22W-24S		29
Hintze Ct. E&W .. 27W-22S		19
Hobbs Av. N&S .. 23W-24S		29
Hobbs Dr. N&S .. 25W-22S		20
Holly Berry Ct. N&S 26W-22S		20
Holly Berry Dr. N&S 26W-22S		20
Holly Ct. NE&SW .. 28W-20S		19
Holmes Av. N&S....22W-22S		21
Home St. E&W 23W-25S,22W-25S		29
Homestead Pl. N&S 27W-23S		19
Honeywood Ct. NE&SW 29W-21S		18
Horseshoe Dr. NS&EW 25W-22S		20
Hosmer St. N&S .. 25W-22S		20
Houbolt Av. N&S 28W-26S,28W-24S		27
Hugh St. N&S 22W-22S		21
Hunter Av. N&S 24W-24S		28
Hutchins St. E&W 25W-22S,24W-22S		20
Hyde Park Av. N&S 24W-24S		28
Ida St. N&S 25W-24S		28
Illinois St. N&S 24W-24S		28
Indian Wood Ln. E&W 28W-23S		19
Industry Av. E&W .. 24W-22S		20
Ingalls Av. N&S 28W-22S,24W-22S		19
Interstate 55 N&S 29W-24S,28W-19S		26
Interstate 80 N&S 29W-26S,23W-24S		26
Inwood Dr. NS&EW 27W-23S		19
Iowa Av. N&S......23W-24S		29
Irene St. E&W 24W-24S		28
Iroquois St. E&W .. 22W-25S		29
Irving St. E&W 23W-23S		21
Ivy Pl. E&W 26W-24S		28
Ivy St. N&S 26W-24S		28
Jackson St. E&W 23W-23S,22W-23S		21
Jaguar St. N&S .. 28W-23S		19
Jane Ct. NW&SE .. 28W-22S		19
Jasper St. E&W 25W-24S,24W-24S		28
Jefferson St. E&W 29W-24S,23W-23S		26
Jeffery Dr. E&W 28W-23S,27W-23S		19
Jerald Dr. NE&SW .. 28W-25S		27
Jerome Av. E&W .. 23W-23S		21
Jersey Av. N&S .. 24W-23S		20
John St. E&W 26W-23S,24W-23S		20
Joliet St. N&S 24W-25S,24W-23S		28
Jonathon Simpson Dr. E&W 28W-23S		19
Jones St. E&W 26W-23S,25W-23S		20
Joyce Rd. N&S 26W-24S		28
Judy Dr. N&S .. 28W-22S		19
Julia St. NW&SE .. 22W-22S		21
June St. N&S 25W-22S		20
Junie Ct. N&S 25W-22S		20
Junior College Dr. N. E&W 28W-22S		27
Junior College Dr. S. E&W 28W-22S		27
Juniper Av. E&W .. 28W-22S		19
Juniper St. E&W....22W-22S		21
Karen Ct. E&W .. 28W-25S		27
Karen Dr. N&S .. 28W-25S		27
Kathlyn Pl. N&S .. 26W-24S		28
Katie Ct. N&S 27W-21S		19
Kelly Av. N&S 24W-22S		20
Kenilworth Av. N&S 25W-22S		20
Kenmore Dr. E&W .. 25W-22S		20
Kerry Ln. NW&SE .. 29W-22S		18
Key St. N&S 27W-23S		19
Kiep St. N&S 28W-25S		27
Kingswood Ct. NE&SW 28W-23S		19
Kinsey Av. N&S .. 25W-24S		28
Knowlton Av. E&W 23W-23S		21
Knox Pl. N&S 25W-23S		20
Krakar Av. N&S .. 22W-22S		21
Krings Ln. N&S .. 25W-22S		20
Kronmeyer Av. E&W 23W-22S		21
Kungs Way N&S ..27W-23S		19
Lafayette St. E&W ..24W-24S		28
Lakelawn Ct. N&S .. 25W-22S		20
Lakeview Av. E&W 25W-24S		28
Lakewook Dr. N&S 28W-22S		19
Lambert St. E&W .. 22W-22S		21
Lambeth Ln. N&S .. 19W-22S		22
Landau Av. N&S ..23W-23S		21
Larkin Av. N&S 26W-24S,26W-22S		28
Laurel Av. N&S ... 24W-24S		28
Lavinia Ln. N&S .. 26W-23S		20
Lawrence Av. N&S..24W-22S		20
Lawton Av. NE&SW 24W-25S		28
Leach N&S 22W-23S		21
Leawood Ct. NE&SW 28W-25S		27
Leawood Dr. N&S .. 28W-25S		27
Lehman St. N&S...23W-24S		29
Lemore Av. E&W .. 25W-22S		20
Lemorr Av. N&S....25W-22S		20
Liberty St. E&W 23W-23S		21
Lilac Ln. N&S 26W-22S		20
Lime St. NW&SE .. 24W-23S		20
Lincoln Av. E&W ..23W-24S		29
Linden Av. E&W....23W-24S		29
Linne Av. N&S....23W-25S		29
Little Big Horn Dr. E&W 27W-21S		19
Lloyds Av. N&S .. 23W-22S		21
Logan Av. N&S .. 22W-24S		29
Lois Pl. N&S 26W-23S,26W-22S		20
Longford Dr. NW&SE 28W-25S		27
Longwood Dr. N&S 21W-23S,21W-22S		21
Loral Av. E&W 25W-22S		20
Loren Dr. N&S 28W-25S		27
Loretta Av. NE&SW 25W-22S		20
Louis St. E&W 23W-23S		21
Louise Ln. N&S .. 28W-25S		27
Lucas St. E&W .. 24W-24S		28
Lucie Dr. N&S....29W-24S		26
Luther Av. N&S .. 22W-22S		21
M. Hershman Dr. N&S 28W-21S		19
Mack St. N&S .. 25W-23S		20
Macomber St. N&S 23W-24S		29
Madeline St. E&W .. 24W-24S		28
Madison St. N&S .. 26W-23S		20
Madonna Av. NE&SW 26W-24S		28
Magnolia Av. N&S .. 22W-22S		21
Magnolia St. N&S .. 27W-22S		19
Mall Loop Dr. E&W 28W-20S		19
Manor Av. N&S .. 25W-24S		28
Manor Ct. NE&SW .. 25W-24S		28
Maple Rd. NE&SW 22W-23S		21
Maple St. N&S .. 23W-23S		21
Marble St. NW&SE 24W-23S		20
Margaret St. N&S .. 26W-24S		28
Marion St. E&W 25W-24S,23W-24S		28
Market St. N&S .. 24W-24S		28
Marmion Av. NE&SW 26W-24S		28
Marquette Ct. NW&SE 25W-23S		20
Marquette Rd. NS&EW 26W-23S,25W-23S		20
Maserati Dr. NW&SE 27W-21S		19
Mason Av. E&W 28W-23S,24W-23S		19
Massachusetts Av. NE&SW 25W-22S		20
Maude Av. E&W....23W-25S		29
May St. N&S 25W-24S,25W-22S		28
Mayfield Av. E&W 26W-23S,25W-23S		20
McDonald Av. N&S 29W-24S		26
McDonough St. E&W 28W-24S,24W-24S		27
McKay St. N&S 22W-22S		21
McKinley Av. N&S .. 24W-25S		28
Mcfarlane Dr. E&W 26W-22S		20
Mcroberts St. N&S..23W-23S		21
Meadowwood Dr. N&S 28W-23S		19
Meda Av. N&S 23W-24S		29
Meeker Av. 23W-22S,22W-22S		21
Melchoir Ct. N&S .. 23W-23S		29
Memenas Pl. E&W..24W-24S		28
Menlo Ct. N&S 23W-23S		27
Michigan St. N&S .. 23W-23S		21
Midhurst Ct. E&W .. 26W-22S		20
Midhurst Ln. NE&SW 26W-22S		20
Midland Ct. E&W .. 25W-24S		28
Midland St. N&S 25W-24S,23W-24S		28
Miles Av. N&S .. 23W-24S		29
Millboro St. N&S .. 25W-24S		28
Miller Av. N&S....23W-24S		29
Mills Rd. E&W 23W-25S,22W-25S		29
Mississippi Av. N&S 23W-24S		29
Moran St. E&W 25W-22S,24W-22S		20

	Grid	Page
Morgan St. E&W		
..........25W-24S,24W-24S	28
Morris St. N&S 26W-24S	28
Mosby Ct. N&S 28W-25S	27
Mound Rd. N&S		
..........29W-26S,27W-25S	26
Mound St. N&S 24W-24S	29
Mulberry Rd. E&W ..21W-22S	21
Mulford Ln. N&S....28W-25S	27
Munroe St. E&W .. 24W-24S	28
Natoma Ct. N&S....28W-25S	27
Neal Av. N&S 23W-25S	29
Newton Pl. E&W ...23W-24S	29
Nicholson St. N&S		
..........24W-23S,24W-22S	20
Norley Av. N&S 25W-22S	20
Norley Dr. E&W .. 26W-22S	20
Normandy St. E&W		
..........28W-20S	19
North Av. N&S 25W-22S	20
Norton Av. E&W ...23W-23S	21
Notre Dame Av. N&S		
..........26W-24S	28
O'Niel St. N&S 25W-24S	28
Oak Leaf St. E&W ..26W-24S	28
Oakland Av. N&S		
..........24W-23S,24W-22S	20
Oakley Av. N&S 26W-24S	28
Oakview Av. N&S .. 22W-24S	29
Oakwood Dr. E&W 28W-23S	19
Odell Av. NS&EW .. 23W-24S	29
Ohio St. N&S 23W-23S	21
Old Meadow Ct. E&W		
..........29W-21S	18
Olde Mill Crossing N&S		
..........30W-21S	18
Ole Farm Rd. NE&SW		
..........29W-21S	18
Olive Pl. E&W 23W-23S	21
Oneida St. E&W		
..........26W-23S,24W-23S	20
Orchard Ln. E&W .. 26W-22S	20
Oregon Av. NE&SW		
..........25W-22S	20
Osgood St. E&W .. 23W-24S	29
Ottawa St. N&S		
..........24W-24S,24W-23S	28
Page Av. N&S 22W-23S	21
Pamela Ct. NE&SW		
..........28W-23S	19
Pandola Av. E&W .. 28W-22S	19
Park Ct. E&W 25W-24S	28
Park Dr. N&S 25W-24S	28
Parks Av. N&S 23W-23S	21
Parkview Av. E&W ..25W-24S	28
Parkwood Ct. E&W		
..........21W-22S	21
Parkwood Dr. E&W		
..........21W-23S,21W-22S	21
Pearl St. N&S 24W-23S	20
Pearson Dr. N&S .. 26W-22S	20
Pembrook Ct. N&S 30W-21S	18
Pennsylvania Av. NE&SW		
..........25W-22S	20
Pequot St. N&S .. 22W-25S	29
Pine St. NE&SW....24W-23S	20
Pioneer Ct. N&S .. 28W-25S	27
Plainfield Rd. NW&SE		
..........28W-19S,24W-19S	13
Plaza Dr. E&W....25W-22S	20
Pleasant Knoll Dr. N&S		
..........27W-22S	19
Pleasant St. N&S .. 24W-24S	28
Ponderosa Ct. N&S		
..........28W-23S	19
Poplar Av. N&S .. 23W-24S	29
Prairie Av. N&S		
..........25W-24S,24W-22S	28
Prestige St. N&S .. 27W-23S	19
Preston Pl. N&S ...22W-25S	29
Price Ct. E&W....28W-25S	27
Prieboy Av. E&W .. 28W-20S	19
Prior St. E&W 24W-24S	28
Prospect Pl. NE&SW		
..........25W-24S	28
Purdue Ct. N&S .. 26W-24S	28
Putter Ct. N&S....27W-23S	19
Quincy Ct. E&W .. 28W-25S	27
Railroad St. NE&SW		
..........24W-24S	28
Randy Rd. E&W .. 28W-23S	19
Raub St. N&S 24W-23S	20
Raymond St. N&S ..28W-22S	19
Raynor Av. N&S		
..........25W-24S,25W-22S	28
Rebbeca Rd. N&S ..27W-23S	19
Redwood Dr. N&S ..25W-24S	28
Reed St. N&S		
..........25W-24S,25W-22S	28
Reeves Rd. NE&SW		
..........27W-26S	27
Regency Ridge Dr. E&W		
..........25W-24S	28
Reichman St. E&W ..25W-24S	28
Reid St. N&S 23W-22S	21
Republic Av. N&S .. 26W-23S	20
Richards St. N&S		
..........23W-24S,24W-24S	29
Richmond Cir. E&W		
..........26W-23S	20
Richmond St. E&W		
..........27W-23S,24W-23S	19
Rick Ct. N&S 28W-22S	19

	Grid	Page
Ridgemont Av. NW&SE		
..........22W-22S	21
Ridgewood Av. E&W		
..........22W-23S	21
River Rd. N&S....31W-22S	18
River St. N&S		
..........24W-25S,24W-24S	28
Riverside Dr. NW&SE		
..........29W-31S	34
Robert Ct. N&S .. 28W-23S	19
Robin Ln. N&S .. 22W-22S	21
Rock Run Dr. E&W		
..........29W-24S,28W-24S	26
Rogina Ct. E&W .. 28W-22S	19
Romayne Av. N&S ..26W-24S	28
Ronnie Ct. N&S .. 25W-22S	20
Rooney Av. N&S . 27W-22S	19
Roosevelt Av. N&S		
..........25W-22S	20
Rosemont Dr. E&W		
..........25W-23S	19
Rosemont Rd. E&W		
..........22W-23S	19
Ross St. E&W		
..........25W-22S,24W-22S	20
Rossiter Ct. E&W .. 29W-21S	18
Rossiter Pkwy. N&S		
..........29W-21S	18
Royce Av. N&S .. 23W-22S	21
Rt. 7 N&S		
..........26W-24S,26W-22S	28
Rt. 53 N&S		
..........24W-25S,24W-22S	28
Rt. 59 N&S		
..........30W-22S,30W-21S	18
Ruby St. E&W		
..........24W-22S,23W-22S	20
Russell St. E&W .. 24W-22S	20
Ryehill Dr. NS&EW..29W-22S	18
Saint Joseph Av. N&S		
..........26W-24S	28
Saint Louis St. N&S		
..........23W-24S	29
Salem Dr. N&S .. 25W-24S	28
Schriber St. N&S .. 25W-22S	20
Scott St. N&S 23W-23S	21
Scribner St. N&S .. 22W-23S	21
Sesser St. N&S .. 24W-24S	28
Sharen Ct. NW&SE		
..........28W-22S	19
Sheila Dr. N&S .. 27W-22S	19
Shelby St. E&W .. 24W-24S	28
Sheridan Av. E&W ..24W-24S	28
Sherman St. E&W .. 23W-24S	29
Sherman St. N&S .. 23W-24S	29
Sherwood Pl. E&W 25W-23S	20
Shiloh St. N&S .. 28W-25S	27
Siegmund St. N&S..22W-24S	29
Smith Av. E&W .. 24W-22S	20
Somerset St, N&S .. 19W-22S	22
Somme Ct. NW&SE		
..........27W-21S	19
Somme St. NW&SE		
..........27W-21S	19
Sonny Ln. N&S .. 26W-22S	20
South St. E&W 24W-24S	28
Spring St. E&W .. 24W-23S	20
Springfield Av. N&S		
..........26W-23S	20
Squirrel St. N&S....22W-23S	21
St. Francis Av. E&W		
..........26W-24S	28
St. Jude Av. NE&SW		
..........26W-24S	28
Stadium St. N&S .. 27W-23S	19
Star Ln. N&S 27W-23S	19
State St. N&S 23W-23S	21
Stateville Rd. E&W..27W-20S	19
Stefani Ct. NE&SW		
..........19W-22S	22
Steven Smith St. N&S		
..........28W-22S	19
Stevens Av. NE&SW		
..........22W-23S	21
Stirling Av. E&W ..22W-23S	21
Stone St. NW&SE . 24W-23S	20
Stonehurst Ct. E&W		
..........28W-25S	27
Stratford Ct. N&S		
..........26W-23S,25W-23S	20
Strong Av. E&W .. 23W-24S	29
Stryker Av. N&S		
..........25W-24S,25W-23S	28
Summit St. N&S		
..........24W-23S,24W-22S	20
Sunset Strip N&S . 27W-23S	19
Surrey St. NE&SW 28W-25S	27
Susan St. N&S 23W-23S	19
Swing Ct. NW&SE . 27W-23S	19
T. Hickey Av. E&W		
..........28W-21S,27W-21S	19
Tamarack Dr. E&W		
..........21W-23S	21
Tana Ln. N&S......27W-23S	19
Taney Butler Av. .. 24W-24S	28
Tara St. NW&SE .. 29W-22S	18
Taylor St. E&W		
..........26W-23S,24W-23S	20
Terrace Av. NE&SW		
..........26W-23S	20
Terrence Ferry Dr. N&S		
..........28W-22S	19
Terry Dr. N&S		
..........27W-26S,27W-23S	27

	Grid	Page
Texas Av. NE&SW ..25W-22S	20
Theodore St. E&W		
..........32W-21S,27W-21S	18
Timber Oaks Dr. E&W		
..........28W-23S	19
Timbercrest Dr. NW&SE		
..........28W-23S	19
Tonti NE&SW		
..........28W-20S,27W-20S	19
Traymore Dr. E&W ..28W-22S	19
Twin Oaks Dr. NS&EW		
..........28W-23S,27W-23S	19
U.S. Route 6 NE&SW		
..........28W-27S,22W-23S	27
U.S. Route 30 NW&SE		
..........27W-20S,22W-23S	19
U.S. Route 52 E&W		
..........29W-24S,24W-24S	26
Union St. N&S....23W-24S	29
Vail Ct. E&W 19W-22S	22
Valley Av. N&S .. 23W-22S	21
Valley Forge Dr. E&W		
..........27W-21S	19
Van Buren St. E&W		
..........24W-23S,23W-24S	20
Verdun Dr. E&W .. 27W-21S	19
Vernon Av. E&W		
..........25W-22S,24W-22S	20
Vimy Ridge Dr. E&W		
..........27W-21S	19
Vine St. N&S 24W-23S	20
Vintage Dr. N&S .. 28W-22S	19
Virginia St. E&W		
..........23W-22S,22W-23S	21
Vonesch Rd. NE&SW		
..........28W-20S,28W-19S	19
Voyager Dr. NE&SW		
..........27W-20S	19
W. Allen St. E&W .. 26W-24S	28
W. Bevan Av. N&S..26W-23S	20
W. Palladium Dr. N&S		
..........26W-23S	20
W. Park Av. NE&SW		
..........25W-24S	28
W. Parkfront E&W . 25W-24S	28
Wabash Av. N&S .. 23W-22S	21
Wagner Av. N&S .. 25W-22S	20
Wake Island Dr. NE&SW		
..........27W-21S	19
Wallace St. E&W .. 24W-24S	28
Walnut St. N&S		
..........22W-23S	21
Walsh Av. N&S .. 24W-23S	20
Walsh Way E&W .. 27W-23S	19
Ward Av. E&W .. 23W-23S	21
Washington St. E&W		
..........23W-24S,22W-24S	29
Water St. N&S		
..........24W-25S,24W-24S	28
Waverly St. N&S....25W-22S	20
Webster Av. NW&SE		
..........26W-24S	28
Webster St. E&W		
..........24W-23S,23W-23S	20
Wee Ct. E&W 22W-23S	21
Wenberg St. E&W . 22W-23S	21
West Acres Rd. E&W		
..........26W-23S,25W-23S	20
West Ridge Rd. NE&SW		
..........28W-23S	19
Western Av. E&W		
..........25W-23S,24W-23S	20
Westfield Rd. E&W 26W-22S	20
Westminister Rd. N&S		
..........23W-26S,26W-22S	20
Westport Dr. N&S . 28W-23S	19
Westshire Dr. E&W		
..........25W-23S,25W-22S	20
Westwood Av. E&W		
..........25W-24S	28
Wheeler Av. N&S .. 25W-24S	28
White Av. N&S .. 22W-24S	29
Whitley Av. E&W .. 23W-24S	29
Whitney Av. NW&SE		
..........24W-23S	20
Whittier Av. N&S .. 24W-23S	20
Wilcox St. N&S		
..........24W-24S,24W-22S	28
Wild Oaks Ct. E&W		
..........28W-23S	19
Wildwood Dr. E&W		
..........28W-23S	19
Willard St. N&S .. 23W-23S	21
William St. N&S		
..........25W-24S,25W-23S	28
Williamson Av. N&S		
..........23W-22S,22W-22S	21
Willow Av. N&S .. 24W-24S	28
Wilson Av. N&S .. 23W-24S	29
Windsor Ln. N&S .. 28W-25S	27
Winston Av. NE&SW		
..........23W-24S	29
Winthrop Av. N&S ..26W-22S	20
Wisconsin St. N&S		
..........22W-22S	21
Woodland Ct. NE&SW		
..........25W-23S	28
Woodland Dr. NE&SW		
..........25W-24S	28
Woodlawn Av. N&S		
..........25W-23S	20
Woodruff Rd. E&W		
..........23W-22S,22W-22S	21
Woods Av. E&W....25W-24S	28

	Grid	Page
Woodward Av. NS&EW		
..........22W-22S	21
Woodworth Av. N&S		
..........24W-23S	20
Workman Av. E&W		
..........24W-24S	28
Wyoming Av. N&S ..25W-22S	20
York Av. N&S 23W-24S	29
Youngs Av. N&S .. 23W-23S	21
1st Av. E&W		
..........23W-24S,22W-24S	29
2nd Av. E&W		
..........23W-24S,22W-24S	29
3rd Av. E&W		
..........23W-24S,22W-24S	29
4th Av. E&W 23W-24S	29
5th Av. E&W 23W-24S	29

Joliet Township

	Grid	Page
Adelia Av. N&S .. 22W-24S	29
Alessio Dr. E&W .. 23W-25S	29
Algonquin St. N&S..21W-23S	21
Anderson Av. N&S..21W-23S	29
Argyle Av. N&S		
..........21W-24S,21W-23S	29
Ashley Ct. N&S .. 24W-26S	28
Aspen Dr. NW&SE..21W-26S	29
Barr Elms Av. N&S 21W-24S	29
Bartel Rd. N&S 21W-26S	29
Bates Rd. NW&SE..23W-25S	29
Belmont Av. E&W .. 21W-23S	21
Berkley Av. N&S....21W-24S	29
Beverly St. E&W .. 24W-26S	28
Blackhawk St. N&S		
..........21W-23S	21
Bond Rd. N&S 21W-23S	21
Bradford Rd. E&W..23W-26S	29
Brandon Rd. N&S		
..........25W-27S,25W-25S	28
Breen St. N&S 24W-25S	28
Briggs St. N&S		
..........22W-26S,22W-23S	29
Brookside Dr. E&W		
..........23W-25S	29
Brown Av. N&S 22W-23S	21
Buffalo St. N&S .. 22W-25S	29
Burger Av. N&S .. 22W-26S	29
Burke Dr. N&S .. 21W-24S	29
Caledonian St. N&S		
..........21W-24S,21W-23S	29
Camilla Av. E&W .. 24W-26S	28
Carey St. E&W .. 21W-25S	29
Caroline Av. E&W .. 22W-26S	29
Carriage Ln. N&S .. 21W-24S	29
Cayuga St. N&S 21W-23S	21
Cecelia Av. N&S		
..........24W-26S,24W-25S	28
Champlain Av. E&W		
..........24W-25S	28
Champlain St. E&W		
..........24W-25S	28
Channahon Rd. NE&SW		
..........26W-25S,25W-25S	28
Charity Av. NE&SW		
..........23W-25S	29
Charles St. N&S....22W-26S	29
Cherry Hill Rd. N&S		
..........21W-27S,21W-23S	29
Cherry Ln. NE&SW		
..........23W-25S	29
Chicago St. N&S		
..........23W-26S,23W-23S	29
Clairmont Av. E&W		
..........21W-24S,21W-23S	29
Cleveland Av. N&S		
..........21W-22S	21
Commonwealth Av. E&W		
..........22W-26S	29
Copperfield Av. E&W		
..........21W-23S	21
Cornelius Av. N&S..23W-25S	29
Crestwood Dr. E&W		
..........21W-23S	21
Dallas Pl. NE&SW .. 23W-26S	29
David Av. E&W 21W-25S	29
Davison St. N&S....21W-25S	29
Deep Haven Dr. N&S		
..........21W-22S	21
Des Plaines St. N&S		
..........24W-26S,24W-25S	28
Diamond 'K' Ln. N&S		
..........22W-25S	29
Dickens St. E&W .. 22W-25S	29
Doris Av. E&W 23W-25S	29
E. Circle Dr. N&S .. 21W-24S	29
East Holm Dr. N&S		
..........21W-22S	21
East Port Av. N&S ..21W-22S	21
Edison Rd. E&W....23W-26S	29
Elgin Av. E&W....21W-23S	21
Ellen Av. E&W 22W-24S	29
Ellsworth Av. NE&SW		
..........24W-25S	28
Elm St. N&S 23W-25S	29
Elmer Av. N&S .. 23W-24S	29
Elmwood Ln. N&S..21W-26S	29
Emerson Pl. E&W .. 23W-26S	29
Erie St. N&S 22W-25S	29
Eunice Av. E&W .. 22W-26S	29
Farrel Rd. N&S .. 21W-26S	29
Faust Av. N&S .. 21W-26S	29
Fernwood Av. N&S		
..........22W-22S	21

	Grid	Page
Fox St. N&S 21W-23S	21
Fuller St. E&W .. 22W-24S	29
Gage St. N&S 22W-22S	21
Geneva St. E&W .. 24W-26S	28
Girard Blvd. NE&SW		
..........23W-26S	29
Greenfield Av. N&S		
..........22W-25S	29
Greenlaw St. N&S .. 23W-24S	29
Gregory Av. N&S .. 22W-22S	21
Grinton St. N&S 23W-23S	21
Hadsell Av. N&S....24W-25S	28
Hamilton St. E&W .. 24W-26S	28
Hanover Av. E&W .. 21W-24S	29
Harlow St. N&S .. 22W-25S	29
Hartford Pl. N&S .. 21W-24S	29
Haviland St. N&S .. 24W-26S	28
Hawthorne Av. NW&SE		
..........23W-25S	29
Hebbard Av. N&S .. 24W-25S	29
Helen Av. N&S .. 23W-25S	29
Herman Ln. N&S .. 22W-25S	29
Highland Pkwy. N&S		
..........21W-23S	21
Hillcrest Rd. NE&SW		
..........21W-23S	21
Hills St. N&S .. 22W-25S	29
Hillside Rd. E&W .. 23W-26S	29
Home St. E&W .. 21W-23S	29
Houston Av. N&S .. 23W-26S	29
Howliston Ct. E&W		
..........22W-26S	29
Howliston St. N&S .. 22W-26S	29
Ignatius St. N&S		
..........24W-26S,24W-25S	28
Independence Av. N&S		
..........21W-24S	29
Industry Av. E&W .. 24W-22S	20
Interstate 80 E&W		
..........23W-24S,21W-24S	29
Jackson St. NW&SE		
..........22W-23S	21
Jacob Av. E&W .. 24W-25S	28
Jeffrey St. N&S .. 22W-22S	21
Jessie St. N&S		
..........21W-25S,21W-24S	29
Joliet Rd. E&W		
..........26W-27S,25W-27S	28
Joliet St. N&S		
..........24W-26S,24W-25S	28
Judge Ct. E&W .. 23W-25S	29
Julia Av. E&W .. 24W-26S	28
Karner Dr. N&S .. 21W-25S	29
Kenmore Av. N&S .. 24W-26S	29
Keuka St. N&S .. 24W-26S	29
Kildare Av. N&S .. 24W-26S	29
Kinmonth Dr. N&S..21W-25S	29
Laraway Rd. E&W		
..........26W-26S,21W-26S	28
Leana Av. N&S .. 24W-25S	28
Lincoln Hwy. E&W ..21W-23S	21
Lind St. N&S 22W-25S	21
Little Dorritt St. N&S		
..........21W-25S	29
Logan Av. NS&EW 22W-24S	29
Logan Berry Ln. N&S		
..........21W-26S	29
Longwood Ct. N&S		
..........21W-25S	29
Lorette Av. N&S .. 24W-25S	28
Lorraine Av. E&W .. 21W-24S	29
Louis Rd. NW&SE . 23W-26S	29
Luana Rd. NW&SE 23W-26S	29
Mable Rd. E&W .. 21W-26S	29
Manhattan Rd. NW&SE		
..........23W-25S,22W-25S	29
Maple Rd. NE&SW		
..........22W-22S,21W-22S	21
Maplecrest Av. N&S		
..........22W-22S	21
Maplewood Ln. N&S		
..........21W-25S,21W-24S	29
Marcella Av. N&S .. 24W-26S	28
Marie Av. E&W .. 22W-22S	21
Marigold Dr. N&S		
..........21W-25S,21W-24S	29
Markgrafs St. E&W		
..........24W-25S	28
Maude Av. N&S .. 21W-25S	29
McKinley Av. N&S		
..........24W-26S,24W-25S	28
Melicent Av. E&W .. 24W-26S	28
Melrose Av. N&S .. 22W-25S	29
Melrose St. E&W .. 22W-25S	29
Miami St. N&S....21W-23S	21
Middletree Rd. NS&EW		
..........23W-26S,23W-25S	29
Milan Av. N&S....22W-25S	29
Mills Rd. E&W		
..........22W-25S,21W-25S	29
Minton Rd. N&S .. 24W-26S	28
Moen Av. E&W .. 26W-25S	28
Mohawk St. N&S .. 21W-23S	21
Moore Av. N&S .. 23W-26S	29
Morningside Ct. N&S		
..........21W-23S	21
New Lenox Rd. E&W		
..........22W-24S,21W-24S	29
Newton Av. E&W .. 22W-25S	29
North Town Line Rd. E&W		
..........22W-27S,21W-27S	29
Nowell Av. N&S .. 22W-25S	29
Oak Av. N&S 22W-23S	21
Oakview Av. N&S . 22W-24S	29

	Grid	Page
Oakwood Ct. NW&SE	21W-26S	29
Oakwood Ln. E&W	21W-26S	29
Old Elm Rd. E&W	23W-25S	29
Old Manhattan Rd. NW&SE	21W-27S,21W-26S	29
Oliver Pl. N&S	21W-24S,21W-23S	29
Ontario Av. E&W	24W-25S	28
Ontario St. E&W	24W-25S	28
Osage St. N&S	21W-23S	21
Oscar Av. E&W	23W-24S	29
Ottawa St. N&S	24W-26S,24W-25S	28
Park Rd. N&S	21W-24S,21W-23S	29
Parker Av. E&W	22W-23S	21
Patterson Rd. NE&SW	25W-26S,24W-25S	28
Pawnee St. N&S	22W-25S	29
Peale St. N&S	21W-24S	29
Pearl St. N&S	21W-25S	29
Penfield Av. N&S	22W-25S	29
Phesant Run Rd. E&W	23W-26S	29
Pickwick St. N&S	22W-23S	21
Pico Ct. N&S	24W-25S	28
Pico St. E&W	24W-25S	28
Pineridge Dr. N&S	23W-25S	29
Pinewood Ln. E&W	21W-26S	29
Ponderosa Ct. E&W	21W-26S	29
Pontiac St. N&S	21W-23S	21
Porter St. N&S	22W-22S	21
Quarry St. NE&SW	23W-25S	29
Quarry St. E&W	24W-25S	28
Rachel Av. E&W	24W-26S	28
Randall Ct. N&S	23W-24S	29
Raymay Dr. N&S	21W-24S	29
Red Bud Dr. E&W	23W-25S	29
Retta Ct. NW&SE	23W-25S	29
Richard St. N&S	23W-26S	29
Rickey Dr. N&S	21W-25S	29
River St. N&S	24W-25S	28
Rose Av. E&W	24W-25S	28
Rowell Av. N&S	23W-24S,23W-27S	29
Rt. 53 N&S	24W-27S,24W-25S	28
Sandall Ct. N&S	23W-24S	29
Schorie Av. N&S	21W-24S	29
Schuster St. E&W	21W-25S	29
Sehring St. NS&EW	24W-25S	28
Seneca St. E&W	24W-25S	28
Shawnita St. E&W	21W-23S	21
Sherman St. NW&SE	23W-25S	29
Spencer Rd. E&W	22W-25S,20W-25S	29
Spencer St. NS&EW	22W-24S	29
Springgreen Dr. N&S	21W-25S	29
Starr Dr. N&S	23W-25S	29
Steinley Av. E&W	22W-26S	29
Stirling Av. E&W	22W-23S,21W-23S	21
Suffolk Rd. NS&EW	21W-23S	21
Sugar Creek Dr. NS&EW	23W-26S,23W-25S	29
Sugar Creek Rd. N&S	23W-25S	29
Sugar Maple Ln. N&S	23W-27S	29
Sugar Valley Rd. N&S	23W-25S	29
Sugarcreek Ct. N&S	23W-26S	29
Sugarford Way NW&SE	23W-25S	29
Sunrise Ln. N&S	23W-26S	29
Superior Av. N&S	24W-26S,24W-25S	28
Superior St. E&W	24W-25S	28
Sweitzer Rd. E&W	24W-27S,23W-27S	28
Thayer Av. N&S	22W-23S	21
Tulare St. E&W	24W-26S	28
U.S. Route 6 NE&SW	26W-25S,21W-22S	28
U.S. Route 30 E&W	21W-23S	21
U.S. Route 52 NW&SE	23W-24S,21W-27S	29
Ullian Av. NW&SE	24W-26S	28
Valley Plwy. N&S	23W-26S	29
Valleyview Pl. E&W	22W-22S,21W-22S	21
Violetta Av. N&S	21W-22S	21
Voyager Dr. NE&SW	27W-20S	19
W. Circle Dr. N&S	21W-24S	29
Walden Rd. E&W	23W-26S	29
Walnut St. N&S	22W-23S	21
Washington St. E&W	21W-24S	29
Wedgewood Dr. N&S	24W-26S	28
White Av. N&S	22W-26S	29
Whitehall Ct. N&S	23W-26S	29
Wildwood Ln. E&W	23W-25S	29
Wilhelm Ct. E&W	23W-24S	29
Wilmot Av. E&W	24W-26S,23W-26S	28
Winifred St. N&S	24W-25S	28
Wisconsin Av. NE&SW	25W-25S,24W-25S	28
Woodlawn Av. E&W	23W-25S	29
Woodruff Av. NE&SW	25W-25S,24W-25S	28
Yew Ct. E&W	21W-26S	29
Zarley Av. E&W	24W-25S	28
Zarley Blvd. E&W	24W-25S	28
Zarley Rd. E&W	23W-25S	29
Zipp Rd. E&W	23W-27S	29
Zurich Rd. E&W	24W-26S	28
2nd Av. E&W	21W-24S	29
3rd Av. E&W	21W-24S	29
4th Av. E&W	22W-24S,21W-24S	29
5th Av. E&W	22W-24S,21W-24S	29

Lockport

	Grid	Page
Adams St. NE&SW	22W-19S,21W-19S	15
Ames St. NE&SW	22W-19S	15
Amy Ct. NW&SE	21W-19S	15
Anand Pkwy. NS&EW	21W-18S	15
Arbor Ter. NW&SE	20W-18S	15
Archer Rd. NE&SW	22W-18S,21W-18S	15
Ashley Ct. E&W	21W-19S	15
Banberry Cir. N&S	20W-18S	15
Banyon Ct. N&S	20W-18S	15
Basin Dr. E&W	21W-18S	15
Bianco Villa Dr. N&S	21W-17S	15
Boehme St. NE&SW	22W-20S	21
Broadway St. NE&SW	22W-18S,22W-17S	15
Broadway St. N&S	23W-18S	15
Bruce Ct. N&S	22W-20S	21
Buck Dr. E&W	20W-17S	15
Buckeye Ct. N&S	20W-18S	15
Canal St. NE&SW	22W-19S	15
Chandra Dr. NW&SE	21W-18S	15
Charles St. E&W	22W-20S	21
Chestnut Ct. NE&SW	20W-18S	15
Cleveland St. N&S	21W-19S	15
Clinton St. NE&SW	22W-19S	15
Connor Av. N&S	22W-20S	21
Coral Dr. E&W	21W-18S	15
Country Ln. N&S	22W-18S	15
Cove Av. N&S	21W-18S	15
Daggett St. NW&SE	22W-20S	21
Darin Dr. E&W	21W-19S	15
Daviess Dr. NW&SE	22W-19S	15
Daviess St. NE&SW	22W-19S	15
Delpark Av. E&W	23W-20S,22W-20S	21
Denver Hill Dr. NE&SW	22W-20S	21
Dharam Dr. NE&SW	21W-18S	15
Diane Ct. E&W	22W-20S	21
Division St. E&W	22W-19S,21W-19S	15
Doe Tr. E&W	20W-17S	15
Douglas Pkwy. NS&EW	20W-18S	15
East St. N&S	22W-20S	21
Edgewood Dr. N&S	20W-18S	15
Farrel Rd. N&S	21W-18S,21W-29S	15
Fawn Path E&W	20W-17S	15
Frances St. E&W	23W-20S	21
Franklin St. NW&SE	22W-18S	15
Garfield St. N&S	22W-19S	15
Georges Dr. E&W	22W-20S	21
Gloria St. E&W	21W-19S	15
Grandview St. NS&EW	22W-19S,21W-19S	15
Hamilton St. NE&SW	22W-20S,22W-18S	21
Harbor Ct. E&W	21W-18S	15
Heritage Ln. NW&SE	22W-19S	15
Holly Ct. E&W	21W-19S	15
Illini Dr. E&W	21W-19S	15
Industry Dr. NW&SE	22W-19S	15
Isabella Av. N&S	21W-19S	15
Jacquays St. N&S	21W-19S	15
Janice Av. E&W	22W-20S	21
Jefferson St. NE&SW	22W-20S,22W-18S	21
Jennifer Ct. NE&SW	21W-19S	15
Joann Ct. E&W	22W-20S	21
Johnson St. N&S	22W-20S	21
Joseph Dr. N&S	22W-20S	21
Kathrine St. E&W	21W-19S	15
Key West Dr. E&W	21W-18S	15
Lawrence Av. N&S	22W-20S	21
Lejeune Av. N&S	21W-19S	15
Lincoln St. N&S	22W-19S	15
Lock Ln. NW&SE	21W-19S	15
Lone Star Dr. N&S	22W-20S	21
Lorraine Ct. E&W	22W-20S	21
Madison St. NE&SW	22W-20S,22W-19S	21
Maitland Dr. E&W	21W-19S	15
Maple Ct. E&W	21W-19S	15
Market St. NW&SE	22W-18S	15
Mary Ann Ln. N&S	22W-20S	21
Maryknoll Dr. NW&SE	22W-18S,21W-18S	15
May St. E&W	23W-20S	21
Mayo St. NW&SE	22W-20S	21
McGregor St. N&S	21W-18S	15
McKinley Ct. NS&EW	22W-19S	15
McKinley St. NS&EW	22W-19S	15
Michael St. E&W	21W-19S	15
Mikan Ct. E&W	22W-20S	21
Mikan Dr. E&W	22W-20S	21
Milne Dr. NS&EW	21W-19S	15
Milne St. NW&SE	22W-19S	15
Morgan St. NW&SE	22W-18S	15
Muelh St. N&S	21W-19S	15
New Av. NE&SW	22W-18S	15
North St. E&W	21W-18S	15
Northern Dr. NW&SE	21W-18S	15
Parkview Ln. E&W	23W-20S,22W-20S	21
Peachtree Ln. NW&SE	21W-20S	21
Porter Pl. N&S	21W-19S	15
Prairie St. NE&SW	23W-20S	21
Primrose Ln. NW&SE	20W-18S	15
Princess Ct. N&S	22W-20S	21
Prodehl Dr. E&W	21W-19S	15
Putnam Dr. E&W	21W-19S	15
Read St. N&S	21W-19S	15
Redwood Ct. NE&SW	20W-18S	16
Reef Rd. N&S	21W-18S	15
Rhonda Dr. N&S	21W-19S	15
Rio Grande Dr. NE&SW	22W-20S	21
Rodney St. N&S	21W-19S	15
Rosanne Ln. E&W	22W-20S	21
Rt. 7 E&W	23W-19S,21W-19S	15
Rt. 53 N&S	23W-19S,23W-18S	15
Runyan Dr. E&W	21W-19S	15
Shasi St. NW&SE	21W-18S	15
Shoals Dr. NE&SW	21W-18S	15
Silver Spur Ln. NS&EW	22W-20S	21
Sisson St. N&S	22W-20S	21
South St. E&W	22W-20S	21
State St. NE&SW	22W-20S,22W-18S	21
Stephanie Dr. N&S	21W-19S	15
Strawberry Hill Dr. NS&EW	21W-20S	21
Sujada Ct. N&S	21W-18S	15
Summit Dr. N&S	21W-20S,21W-19S	21
Sunshine Ct. NE&SW	21W-20S	21
Sycamore Ct. NW&SE	20W-18S	15
Table St. NE&SW	22W-18S	15
Theodore Ct. N&S	21W-18S	15
Thornton Av. E&W	22W-18S,21W-19S	15
Tonelli Tr. NW&SE	21W-19S	15
Valley Dr. NW&SE	21W-19S	15
Vine St. NE&SW	22W-19S	15
Warren St. NW&SE	22W-18S	15
Washburn Way NS&EW	21W-19S	15
Washington St. NE&SW	22W-20S,22W-18S	21
Water St. NE&SW	22W-19S	15
West St. N&S	22W-20S	21
Whelan St. E&W	22W-20S	21
White Tail Way E&W	20W-17S	15
Will Dr. E&W	21W-19S	15
2nd St. NW&SE	22W-19S,21W-19S	15
3rd St. NW&SE	22W-19S,21W-19S	15
4th St. NW&SE	22W-19S	15
5th St. NW&SE	22W-19S	15
6th St. NW&SE	22W-19S	15
7th St. E&W	22W-19S,21W-19S	15
8th St. E&W	22W-19S,21W-19S	15
9th St. NW&SE	23W-19S,21W-19S	15
10th St. NW&SE	22W-19S	15
11th St. NW&SE	22W-19S	15
12th St. NW&SE	22W-19S	15
13th St. NW&SE	22W-19S	15
14th St. NW&SE	22W-19S	15
15th St. NW&SE	22W-20S,22W-19S	21
16th St. NW&SE	22W-20S	21
17th St. NW&SE	22W-20S	21
18th St. NW&SE	22W-20S	21
19th St. NW&SE	22W-20S	21
20th Ct. E&W	22W-20S	21
151st St. E&W	21W-18S	15
167th St. E&W	20W-20S	22

Lockport Township

	Grid	Page
Aberdeen Dr. E&W	26W-19S	14
Airport Rd. E&W	26W-17S,23W-17S	14
Albert Dr. N&S	25W-19S	14
Alicia Ct. NW&SE	26W-19S	14
Alpine Ct. NE&SW	26W-16S	14
Alpine Rd. NW&SE	26W-16S	14
Amherst Av. N&S	23W-21S	21
Archer Rd. NE&SW	22W-18S,21W-17S	15
Argo Ln. E&W	21W-18S	15
Barret St. N&S	23W-21S	21
Barry Av. E&W	23W-21S,22W-21S	21
Bayberry Rd. NE&SW	26W-16S	14
Bellwood Av. E&W	23W-21S,22W-21S	21
Birch St. E&W	26W-20S	20
Blossom St. E&W	26W-20S	20
Blue Blossom Ct. NE&SW	26W-16S	14
Blue Blossom Ln. NW&SE	26W-16S	14
Bluff St. E&W	22W-18S	15
Bonnie Dr. N&S	21W-18S	15
Brassel St. N&S	22W-21S	21
Briggs St. N&S	22W-21S,22W-18S	21
Bristolcone Dr. N&S	26W-16S	14
Broadway St. NE&SW	24W-21S,23W-19S	20
Brown St. N&S	22W-20S	21
Bruce Rd. E&W	23W-20S,21W-20S	21
California Av. N&S	22W-21S	21
Cameron Av. E&W	23W-20S,22W-20S	21
Carillon Dr., South NW&SE	26W-16S	14
Central Park N&S	23W-21S	21
Cherry Ln. N&S	26W-16S	14
Circle E. N&S	21W-18S	15
Circle W. N&S	21W-18S	15
Cliff Av. N&S	23W-21S	21
Connor Av. E&W	23W-20S,22W-20S	21
Cornell Av. E&W	23W-21S	21
Cottonwood Ln. NE&SW	26W-16S	14
Country Ln. N&S	22W-18S,22W-17S	15
Dartmouth Av. NW&SE	23W-21S	21
Dawson Av. E&W	24W-17S	14
Disposal Rd. E&W	23W-16S	15
Division St. E&W	25W-19S,21W-19S	14
Dundee Dr. N&S	21W-18S	15
Earl St. N&S	21W-18S	15
East Rd. N&S	22W-20S	21
East St. N&S	22W-21S,21W-18S	21
Edinburg Dr. N&S	21W-18S	15
Elm St. E&W	21W-18S	15
Englewood Av. N&S	22W-21S	21
Exhibition Pl. N&S	23W-21S	21
Exter Ct. N&S	23W-21S	21
Fairmont Av. E&W	23W-21S,22W-21S	21
Fairmount Av. N&S	22W-21S	21
Farrel Rd. N&S	21W-21S,21W-18S	21
Foley Dr. E&W	26W-19S	14
Forest Ln. N&S	21W-17S	15
Gaylord Rd. N&S	26W-20S,26W-19S	20
Giesler Dr. N&S	21W-18S	15
Godfrey Av. E&W	22W-21S	21
Green Garden Pl. N&S	23W-21S,23W-20S	21
Hacker Dr. N&S	26W-20S	20
Hartman Dr. N&S	23W-17S	15
Harvard St. E&W	23W-21S	21
Hawthorn Pl. N&S	23W-21S	21
Hazelnut Ln. NS&EW	26W-16S	14
Hemlock St. E&W	26W-20S	20
Highland Av. E&W	21W-18S	15
Hopkins Rd. N&S	24W-18S	14
Hughes Av. E&W	23W-20S,22W-20S	21
Jacquie Av. E&W	24W-17S,23W-17S	14
Janet Av. E&W	24W-17S,23W-17S	14
John Ct. N&S	25W-19S	14
John Kirkham Dr. E&W	24W-17S,23W-17S	14
Joliet Rd. N&S	23W-17S,23W-16S	15
LaSalle Ct. NE&SW	23W-21S	21
Laurie Dr. N&S	21W-18S	15
Lawrence Av. N&S	22W-20S	21
Lejeune Av. N&S	21W-19S	15
Lewis Dr. N&S	24W-17S	14
Lockport Rd. NE&SW	23W-21S,22W-16S	21
Lois Av. N&S	22W-20S	21
Luther Av. N&S	22W-21S	21
Magnolia Dr. N&S	26W-16S	14
Mahogany Ln. NW&SE	26W-16S	14
Mary Dr. E&W	25W-19S	14
Maryknoll Dr. NS&EW	21W-18S	15
Maryville Dr. NE&SW	21W-19S	15
May St. N&S	22W-21S	21
McCameron Av. NS&EW	21W-18S	15
McGilvray Dr. E&W	26W-19S	14
McGregor Rd. N&S	21W-18S	15
McGregor St. N&S	21W-18S	15
McIntosh Rd. N&S	22W-20S	21
McIntyre Dr. N&S	21W-18S	15
Miramor St. N&S	21W-19S	15
Neil Ct. N&S	23W-21S	21
Nobes Av. E&W	23W-21S,22W-21S	21
North Av. E&W	23W-21S	21
North St. E&W	22W-21S	21
North St. E&W	22W-18S,21W-18S	15
Northern Dr. NW&SE	21W-18S	15
Oak Av. E&W	23W-21S,21W-21S	21
Oak St. N&S	21W-18S	15
Orange Blossom Ln. NW&SE	26W-16S	14
Oxford Pl. E&W	23W-21S	21
Pendleton St. N&S	21W-19S	15
Pilcher Rd. E&W	26W-16S,25W-16S	14
Pine Av. N&S	23W-20S	21
Prairie Av. E&W	22W-21S	21
Princeton St. E&W	23W-21S,22W-21S	21
Rankin Dr. N&S	26W-19S	14
Redbud Ln. N&S	26W-16S	14
Renwick Rd. E&W	26W-18S,23W-19S	14
Riley Av. E&W	23W-21S,22W-21S	21
Riverview Av. E&W	23W-21S,22W-21S	21
Robson Dr. N&S	21W-18S	15
Rosalind St. E&W	22W-21S,21W-21S	21
Rosemarie Ln. N&S	26W-19S	14
Rt. 53 N&S	23W-19S,23W-16S	15
Rt. 171 NE&SW	21W-17S	15
S. Carillon Dr. NW&SE	26W-16S	14
School House Ln. N&S	21W-17S	15
Sheffield Av. NW&SE	23W-21S	21
Smith Rd. N&S	21W-16S	15
South St. E&W	23W-21S,22W-21S	21
State St. NE&SW	23W-20S,22W-18S	21
Stateville Rd. E&W	26W-19S,23W-19S	14
Stuart Ct. NW&SE	21W-18S	15
Stuart Rd. N&S	21W-18S	15
Taylor Rd. E&W	26W-16S,24W-16S	14
Terminal Ct. N&S	23W-17S	15
Thomas Dr. E&W	25W-19S	14
Thorndale Ct. E&W	21W-18S	15
Thornton Av. E&W	22W-18S,21W-18S	15
Volz Av. N&S	22W-21S	21
W.143rd St. E&W	21W-16S	15
Weber Rd. N&S	26W-20S,26W-16S	20
Weslyan St. E&W	23W-21S	21
West Bluff St. N&S	21W-18S	15
Woodbrook Ln. E&W	21W-17S	15
Yale Av. E&W	23W-21S,22W-21S	21
1st St. E&W	21W-19S	15
135th St. E&W	21W-16S	15
144th St. E&W	21W-17S	15
145th St. E&W	21W-17S	15
147th St. E&W	21W-17S	15

Manhattan

	Grid	Page
Brett Ct. NE&SW	18W-31S	38
Brett Dr. NS&EW	18W-31S	38
Cochran St. E&W	19W-31S	38
Eastern Av. N&S	19W-31S,19W-30S	38
Eberhart St. NE&SW	19W-30S	38
Fairview Dr. NS&EW	19W-30S	38
Front St. NE&SW	19W-31S	38
Henry St. E&W	19W-31S	38
Jan Av. N&S	18W-31S	38
Jessie St. E&W	19W-31S,18W-31S	38
Kay Av. N&S	18W-31S	38
Lee St. E&W	18W-31S	38
Lincoln St. N&S	19W-31S	38
Madison St. N&S	19W-31S	38
Manhattan-Monee Rd. E&W	18W-30S	38
Marion St. NW&SE	19W-30S	38
May St. NS&EW	19W-31S	38
McClure Av. NE&SW	19W-31S	38
Morgan Ct. NE&SW	19W-30S	38
North St. E&W	19W-30S,18W-31S	38
Park Av. N&S	19W-31S,19W-30S	38
Prairie Av. E&W	19W-30S	38
Sharp Dr. N&S	19W-31S,19W-30S	38
Shelia Rd. N&S	19W-31S	38
Smith Rd. E&W	19W-29S	38
State St. NW&SE	18W-31S,19W-30S	38
Terry Ct. E&W	18W-31S	38
Thelma Av. N&S	18W-31S	38
Trask St. E&W	19W-31S	38
U.S. Rt. 52 NW&SE	18W-31S,19W-30S	38
Western Av. E&W	19W-30S	38
Whitson St. E&W	19W-31S,18W-31S	38
Woodrow Av. NE&SW	19W-30S	38
1st Ct. N&S	18W-31S	38
1st St. E&W	19W-31S,18W-31S	38
2nd St. E&W	19W-31S,18W-31S	38
3rd St. E&W	19W-31S,18W-31S	38

Manhattan Township

	Grid	Page
Baker Rd. E&W	20W-28S,15W-28S	38
Barton Rd. N&S	15W-33S	51
Brett Dr. E&W	18W-31S	38
Brian Ct. NE&SW	18W-31S	38
Brown St. E&W	20W-31S,19W-31S	38
Brynn Dr. N&S	18W-31S	38
Carriage Ln. E&W	17W-29S	39
Cedar Rd. N&S	18W-33S,18W-28S	50
Creek Dr. NE&SW	20W-29S	38
Delaney Rd. E&W	20W-28S,15W-28S	38
Diane Way NS&EW	20W-29S	38
Eastern Av. N&S	19W-33S,19W-28S	50
Elwood Manhattan Rd. E&W	20W-30S	38
Gallager (Haley) Rd. E&W	20W-31S,15W-31S	38
Gougar Rd. N&S	20W-33S,20W-31S	50
Green Garden-Manhattan Rd N&S	15W-33S,15W-28S	51
Haley (Gallager) Rd. EW&NS	17W-31S,16W-31S	39
Highland Dr. N&S	15W-28S	39
Hoff Rd. E&W	20W-32S,16W-32S	50
Julianne Ct. NW&SE	18W-31S	38
Julianne Dr. NS&EW	18W-31S	38
Kankakee St. N&S	17W-33S,17W-28S	51
Kathryn Av. E&W	20W-29S	38
Koehler Rd. N&S	16W-32S,16W-28S	51
Malibu Dr. N&S	19W-28S	38
Manhattan-Monee Rd. E&W	18W-30S,15W-30S	38
Old Manhattan Rd. NW&SE	20W-29S,20W-28S	38
Ottawa Dr. E&W	16W-28S	39
Prairie Estates Dr. E&W	16W-28S	39
Sauk Dr. N&S	16W-28S	39
Scheer Rd. N&S	15W-28S,15W-27S	39
Schoolhouse Rd. N&S	15W-32S,15W-29S	51
Seltzer Rd. N&S	20W-30S	38
Sioux Ct. E&W	16W-28S	39
Smith Rd. E&W	19W-29S,15W-29S	38
State St. NW&SE	18W-31S	38
Stuenkel Rd. E&W	15W-29S	39
Susan Ln. NE&SW	20W-29S	38
U.S. Rt. 52 NW&SE	18W-33S,20W-28S	50
Watkins Rd. E&W	18W-32S,15W-32S	50
1st St. E&W	18W-31S	38

Minooka

	Grid	Page
Bell Rd. N&S	32W-29S	34
Colina Calle Ct. E&W	32W-29S	34
Dirst Rd. NS&EW	32W-28S,32W-29S	34
El Dorado Ct. N&S	32W-29S	34
Grande Dr. E&W	32W-28S	34
Jardin St. NW&SE	32W-29S	34
Manuel Ct. E&W	32W-29S	34
Maria Ct. E&W	32W-29S	34
McEvilly Rd. E&W	32W-29S	34
Miguel Ct. E&W	32W-29S	34
Rio Roco Av. E&W	32W-29S	34
San Carlos NS&EW	32W-29S	34
San Carlos Rd. NS&EW	32W-29S	34
Santos Av. N&S	32W-29S	34
Valle Vista Ct. N&S	32W-29S	34
Vista Ct. N&S	32W-28S	34
Zapata Rd. NW&SE	32W-29S	34

Mokena

	Grid	Page
Aileen Av. NE&SW	13W-23S	24
Alta Vista Way E&W	14W-26S	32
Anna St. N&S	13W-24S	32
Anne Ln. N&S	13W-24S,13W-23S	32
Ashbury Ct. E&W	14W-26S	32
Ashford Cir. N&S	14W-24S	32
Bormet Dr. E&W	11W-23S	25
Boyer Ct. E&W	13W-24S	32
Brett Dr. N&S	13W-23S	24
Bryant Ct. N&S	13W-23S	24
Bryant Rd. E&W	13W-23S	32
Bryn Mawr Way E&W	14W-26S	32
Burr Oak Ln. N&S	14W-24S	32
Cambridge Dr. NS&EW	12W-23S	24
Camden Bay NW&SE	13W-22S	24
Camden Ct. NS&EW	13W-22S	24
Catherine Ln. N&S	13W-24S,13W-23S	32
Center St. N&S	13W-24S,13W-23S	32
Cleveland St. E&W	13W-22S	24
Creekview Cir. NW&SE	14W-22S	24
Creekview Ct. NW&SE	14W-22S	24
Creekview Ln. N&S	14W-22S	24
Cross St. NE&SW	13W-23S	24
Daniel Tr. E&W	13W-23S	24
Darla Trail N&S	13W-23S	24
Denny Av. NE&SW	13W-23S	24
Derby Ct. NE&SW	12W-23S	24
Derby Ln. NS&EW	12W-23S	24
Derek Way NW&SE	13W-23S	24
Diana Ct. NE&SW	14W-26S	32
Dickens Bay NE&SW	13W-22S	24
Dilber Bay N&S	13W-22S	24
Division St. E&W	13W-23S	24
East Brightway Dr. N&S	14W-24S	32
Edgewood Dr. NE&SW	14W-24S	32
English Bay Ln. N&S	13W-22S	24
Enterprise Dr. E&W	11W-23S	25
Fiona Av. NW&SE	13W-23S	24
Forest Ct. N&S	14W-26S	32
Front St. NE&SW	14W-23S,13W-23S	24
Frontage Rd. NW&SE	11W-22S	25
Giles Dr. E&W	12W-23S	24
Glennell Av. N&S	14W-23S	24
Glennell Ln. N&S	14W-23S	24
Greenview Av. N&S	14W-23S	24
Halligan Way E&W	13W-23S	24
Hamilton Rd. E&W	14W-24S	32
Hastings Ln. E&W	12W-23S	24
Henry Dr. N&S	11W-23S	25
Hiawatha Blvd. NE&SW	13W-23S	24
Hickory Cr. Dr. NS&EW	11W-22S	25
Hillside Ln. E&W	14W-22S	24
Hummingbird Ln. E&W	14W-22S	24
Irwin Dr. E&W	11W-22S	25
Jacob Dr. E&W	13W-22S	24
Jomar Ct. E&W	13W-23S	24
Kerry Ct. E&W	13W-23S	24
Kevin Ct. NE&SW	12W-23S	24
Kimberly Tr. NS&EW	13W-23S	24
Kingston Way N&S	14W-26S	32
Kirkstone Way NW&SE	12W-23S	24
Kluth Dr. E&W	14W-24S	32
LaPorte Rd. E&W	12W-23S	24
Leetrim Ct. NW&SE	13W-23S	24
Lincoln Hwy. E&W	14W-25S	32
Lindsay Ln. NE&SW	12W-23S	24
London Bay NE&SW	13W-22S	24
London Ln. N&S	13W-22S	24
Mannheim Rd. N&S	12W-23S,12W-22S	24
Marilyn Way E&W	14W-26S	32
Maurita Ct. N&S	13W-23S	24
McGovney St. NE&SW	13W-23S	24
Meadow Creek Dr. N&S	14W-22S	24
Meadowview Dr. N&S	14W-22S	24
Michael Ct. N&S	13W-23S	24
Midland Av. N&S	13W-23S	24
Mokena St. N&S	13W-24S,13W-23S	32
N. Woodland Cir. E&W	14W-24S	24
North Brightway Dr. NE&SW	14W-24S	32
O'Connell Av. NE&SW	13W-23S	24
Old Castle Dr. NE&SW	14W-24S	32
Osage Rd. N&S	14W-25S	32
Oxford Ct. E&W	12W-23S	24
Park Dr. N&S	11W-23S	25
Park Way Ln. N&S	14W-22S	24
Parker St. E&W	13W-23S	24
Patricia Ln. N&S	13W-24S,13W-23S	32
Pheasant Ct. N&S	14W-23S	24
Pheasant Ln. N&S	14W-23S	24
Pickran Rd. NW&SE	15W-23S	23
Plattner Ct. N&S	14W-24S	32
Plattner Dr. E&W	14W-24S	32
Revere Rd. E&W	13W-23S	24
Ridge Rd. N&S	11W-23S	25
Robert St. E&W	14W-23S	24
Rose N&S	14W-24S	32
Royal Ct. NE&SW	12W-23S	24
Rt. 45 N&S	12W-23S,12W-22S	24
S. Woodland Cir. E&W	14W-24S	32
Sandra Ct. NW&SE	14W-26S	32
Sarris Rd. NW&SE	15W-23S	23
Scarth Ln. N&S	13W-23S	24
Scott St. N&S	13W-24S	32
Sevan Rd. NW&SE	15W-23S	23
South Brightway Dr. E&W	14W-24S	32
St. Albert Ct. N&S	14W-26S	32
St. John Dr. E&W	14W-26S	32
St. Joseph Ct. E&W	13W-23S	24
Stafford Ct. NE&SW	12W-23S	24
Stephan Ct. E&W	13W-23S	24
Stephanie Ln. E&W	14W-22S	24
Stonehenge Dr. N&S	12W-23S	24
Sussex Ct. E&W	12W-23S	24
Swan Dr. N&S	14W-22S	24
Swanberg Ln. N&S	13W-23S	24
Swinford Ct. NW&SE	14W-24S	32
Swinford Ln. NE&SW	14W-24S	32
Therese St. N&S	13W-24S,13W-23S	32
Thomas Dr. NW&SE	13W-23S	24
Union Av. N&S	13W-23S	24
Union St. NW&SE	13W-23S	24
Vincent Ct. E&W	14W-26S	32
W. Woodland Cir. N&S	14W-24S	32
Weber Ct. N&S	13W-23S	24
Weber Rd. N&S	13W-23S	24
Westport Way E&W	13W-23S	24
Wild Field Ct. NW&SE	13W-23S	24
Willow Av. NE&SW	13W-23S	24
Willow Av. E&W	13W-23S	24
Willow Ln. E&W	14W-23S	24
Wolf Rd. N&S	14W-24S,14W-22S	32
1st St. NE&SW	13W-23S	24
2nd St. NE&SW	13W-23S	24
3rd St. NE&SW	13W-23S	24
82nd St. N&S	10W-22S	25
97th St. N&S	12W-23S	24
104th Av. N&S	13W-22S	24
114th Av. N&S	14W-23S	24
114th St. N&S	14W-24S	32
116th Av. N&S	14W-23S	24
188th St. E&W	10W-22S	25
189th St. E&W	10W-22S	25
191st St. E&W	14W-22S,10W-22S	24
193rd St. E&W	14W-23S	24
194th St. E&W	14W-23S,12W-23S	24
195th St. E&W	14W-23S,12W-23S	24
196th St. E&W	12W-23S	24
197th St. E&W	14W-23S,12W-23S	24

Monee

	Grid	Page
Astor Pl. E&W	6W-31S	42
Birch Ct. NW&SE	8W-32S	53
Bradford Ct. E&W	8W-32S	53
Carl Rd. NE&SW	7W-32S	54
Cedar Ct. NW&SE	8W-32S	53
Compass Ln. N&S	6W-31S	42
Court St. E&W	7W-31S,6WS-31S	42
Eastgate Av. N&S	6W-31S	42
Eastlynne Ln. NE&SW	6W-31S	42
Elmwood Ln. N&S	6W-31S	42
Gorman Ct. E&W	8W-32S	53
Gorman Rd. E&W	8W-31S,7W-31S	41
Gorman Rd. N&S	8W-32S	53
Goverors Hwy. NE&SW	7W-32S	54
Herbert Pl. E&W	6W-31S	42
Home Av. N&S	6W-31S	42
Interstate 57 N&S	7W-31S	42
Linden Av. N&S	6W-31S	42
Locust Pl. NE&SW	6W-31S	42
Main St. E&W	6W-31S	42
Manhattan Rd. N&S	6W-31S	42
Manhattan-Monee Rd. E&W	8W-30S	41
Maple Ct. E&W	8W-32S	53
Margaret St. E&W	6W-31S	42
Middlepoint Av. N&S	6W-31S	42
Mill St. E&W	6W-31S	42
Oak Rd. NE&SW	6W-31S	42
Oakwood Ln. N&S	6W-31S	42
Pinewood Ln. N&S	6W-31S	42
Ridgeland Av. N&S	8W-31S	41
Roosevelt St. E&W	7W-31S	42
Rt. 50 N&S	7W-32S,6W-31S	54
Spruce Ct. E&W	8W-32S	53
Sunset Dr. East N&S	7W-31S	42
Walnut St. NE&SW	7W-31S,6W-31S	42
Watson E&W	7W-32S	54
William St. NW&SE	7W-31S,6W-31S	42
Wilson St. E&W	7W-31S,6W-31S	42
Winfred Rd. N&S	8W-32S	53

Monee Township

	Grid	Page
Bong Rd. E&W	8W-30S	41
Boyington Ln. NE&SW	8W-30S	41
Central Av. N&S	7W-28S	42
Central Park Av. N&S	4W-28S	43
Chennault Av. N&S	8W-30S	41
Cicero Av. N&S	6W-29S	42
Crawford Av. N&S	5W-33S	54
Crete Monee Rd. E&W	4W-31S,3W-31S	43
Doolittle Rd. NE&SW	8W-30S	41
Dralle Rd. E&W	7W-29S,5W-30S	42
Egypitian Trail Rd. N&S	7W-33S	54
Foss Rd. E&W	8W-30S	41
Foxwood Dr. N&S	6W-32S	54
Gabreski Ln. E&W	8W-30S	41
Goodenow (Pauling) Rd. E&W	6W-32S,3W-32S	54
Gorman Rd. E&W	8W-31S	41
Governors Hwy. NE&SW	7W-32S	54
Greenbrier Dr. N&S	6W-32S	54
Hamilton Av. N&S	7W-30S	42
Hamilton Rd. E&W	4W-30S	43
Hansan Ct. NW&SE	8W-30S	41
Hawthorne Av. E&W	4W-28S	43
Heatherbrook Tr. E&W	6W-32S	54
Highland Av. N&S	3W-32S,3W-31S	55
Hilltop Dr. E&W	6W-32S	54
Home Av. N&S	6W-32S,6W-28S	54
Interstate 57 NE&SW	8W-33S,7W-28S	53
J.L. Smith Ln. E&W	8W-30S	41
Johnson Ct. E&W	8W-30S	41
Kedzie Av. N&S	4W-33S	55
Kuersten Rd. N&S	5W-33S,5W-31S	54
Laurel Ln. E&W	7W-32S	54
Manhattan Wilton Rd. E&W	8W-33S	53
Manhatten Monee Rd. E&W	8W-30S	41
McCampbell Rd. E&W	8W-30S	41
McNamara Rd. E&W	6W-29S	42
Meghan Ct. N. N&S	6W-32S	54
Meghan Ct. S. N&S	6W-32S	54
Murphy Ln. N&S	8W-30S	41
O'Hare Ct. N&S	8W-30S	41
Oak Hill Dr. E&W	4W-29S	43
Oakbrook Rd. E&W	6W-33S	54
Olendorf Rd. E&W	5W-31S,3W-31S	42
Pauling (Goodenow) Rd. E&W	8W-32S,3W-32S	53
Pinewood Dr. NE&SW	4W-31S	43
Pinewood Ln. N&S	4W-31S	43
Ridgeland Av. N&S	8W-33S,8W-28S	53
Roberts Ridge Dr. E&W	6W-32S	54
Rt. 50 NE&SW	7W-33S,6W-28S	54
Sandy Ct. E&W	5W-31S	42
Smith Rd. E&W	6W-29S	42
Spruce Av. E&W	4W-28S	43
Steger Monee Rd. N&S	5W-30S,4W-30S	42
Stone Rd. N&S	7W-32S	54
Stuenkel Rd. E&W	8W-28S,4W-28S	41
Sylvan Ln. E&W	6W-32S	54
Timber Lane Rd. N&S	7W-33S,6W-33S	54
Wagner Dr. E&W	8W-30S	41
Welch Ct. NW&SE	8W-30S	41
Western Av. N&S	3W-29S,3W-28S	43
Whispering Hill Ln. E&W	8W-31S	41
Will Center Rd. N&S	6W-33S	54
Willow Creek Ln. N&S	8W-31S	41
279th St. E&W	6W-33S,3W-33S	54

Naperville*
(Also see Du Page County)

	Grid	Page
Alamosa Dr. E&W	27W-10S	5
Albert Hall Ct. E&W	28W-11S	5
Alexandria Dr. NE&SW	25W-10S	6
Allegany Dr. NE&SW	25W-10S	6
Alyssa Ct.* NW&SE	26W-10S	6
Alyssa Dr.* NW&SE	26W-10S	6
Amherst Ct. NE&SW	25W-10S	6
Amy Ct. E&W	26W-10S	6
Apple River Dr. NS&EW	25W-10S	6
Ardley Ct. NW&SE	26W-10S	6
Ariel Ct. N&S	28W-12S	5
Arlington Av. E&W	26W-10S,25W-10S	6
Ashbury Dr. N&S	28W-11S	5
Ashtonlee Ct. N&S	26W-10S	6
Aster Ct. N&S	24W-10S	6
Augustana Ct. N&S	24W-10S	6
Augustana Dr. E&W	25W-10S,24W-10S	6
Austin Ct. N&S	28W-11S	5
Austin St. N&S	28W-11S	5
Aviara Ct. E&W	30W-10S	4
Avon Ct. E&W	28W-11S	5
Bards Av. E&W	28W-11S	5
Barkdoll Ct. N&S	25W-10S	6
Barkdoll Rd. N&S	25W-10S	6
Barnes Ln. N&S	28W-11S	5
Barth Dr. N&S	25W-10S	6
Beaconsfield Av. E&W	27W-10S	5
Becket Rd. E&W	28W-11S	5
Bennett Dr. N&S	28W-11S	5
Bennington Ct. N&S	27W-10S	5
Black Walnut Ct. E&W	24W-10S	6
Blodgett Ct. NW&SE	27W-10S	5
Bluebell Ct. E&W	24W-10S	6
Bluemont Ct. N&S	25W-10S	6
Book Rd. NE&SW	29W-13S	9
Boswell Ln. NE&SW	28W-11S	5
Boulder Ct. N&S	27W-10S	5
Braddock Dr. N&S	25W-10S	6

	Grid	Page
Branford Ln. NS&EW	28W-11S	5
Breckenridge Ln. NS&EW	27W-10S	5
Briarhill Ct. N&S	27W-10S	5
Briarhill Dr. EW&NS	27W-10S	5
Brockton Cir. NS&EW	27W-10S	5
Bronte Av. E&W	28W-11S	5
Bronte Ct. N&S	28W-11S	5
Brook Crossing Ct. E&W	28W-10S	5
Brossman N&S	28W-11S	5
Brown Ct. N&S	24W-10S	6
Bundle Flower Ct. N&S	29W-13S	9
Burbany Ct. N&S	28W-11S	5
Caine Dr. N&S	28W-11S	5
Calico Av. E&W	28W-10S	5
Calvert Ct. NW&SE	25W-10S	6
Candeur Dr. N&S	28W-11S	5
Carlyle Rd. E&W	28W-11S	5
Carrboro Ct. NW&SE	25W-10S	6
Cassin Rd. E&W	25W-10S	6
Catclaw Ct. N&S	29W-13S	9
Celeste Ln. NS&EW	28W-11S	5
Centenary Ct. N&S	25W-10S	6
Chaucer Ct. E&W	24W-10S	6
Cherokee Ct. NE&SW	28W-12S	5
Cheyenne Dr. N&S	27W-10S	5
Chilvers Ct. NE&SW	25W-10S	6
Choctaw Cir. Cir.	28W-12S	5
Christie Ct. E&W	28W-11S	5
Cimarron Ct. N&S	25W-10S	6
Clearwater Dr. NS&EW	28W-12S	5
Cobblestone Ct. E&W	28W-10S	5
Colgate Ct. E&W	25W-10S	6
Colorado Ct. E&W	27W-10S	5
Conan Doyle Rd. E&W	28W-11S	5
Coneflower Dr. NE&SW	29W-13S	9
Connecticut Dr. E&W	27W-10S	5
Copperfield Ct.* N&S	26W-10S	6
Copperfield Dr. N&S	27W-10S	5
Coralberry Cir. E&W	29W-13S	9
Cordula Ln. NS&EW	28W-11S	5
Covington Ct. E&W	25W-10S	6
Crestfield Ct.* NW&SE	26W-10S	6
Crimson Ct. E&W	28W-10S	5
Cromwell Ln. NS&EW	28W-11S	5
Crooked Tree Ct. NE&SW	26W-10S	6
Cumberland Ct. E&W	25W-10S	6
Daisy Ln. E&W	28W-10S	5
Danbury Dr. E&W	26W-10S	6
Dandelion Ct. NE&SW	29W-13S	8
Darwin St. N&S	28W-11S	5
DeFoe Ct. N&S	28W-11S,28W-12S	5
DeLaSalle Av. E&W	27W-10S	5
Delaware Ct. N&S	28W-12S	5
Dewhurst St. N&S	28W-10S	5
Diamond Ct. N&S	28W-12S	5
Dilorenzo Dr. E&W	25W-10S	6
Dorchester Ct. E&W	27W-10S	5
Dryden Ct. E&W	28W-11S	5
Du Pahze St. NS&EW	27W-10S,26W-10S	5
Dublin Dr. NE&SW	28W-12S	5
Eliot Ln. NS&EW	28W-11S	5
Emory Av. E&W	25W-10S	6
Emporia Ct. N&S	28W-12S	5
English Ct. E&W	28W-11S	5
Esquire Cir. NS&EW	28W-12S	6
Falkner Dr. N&S	28W-11S	5
Fieldstone Ct. E&W	28W-10S	5
Fireside Ct. E&W	28W-10S	5
Flambeau Dr. N&S	28W-10S	5
Fleetwood Ct. NS&EW	25W-10S	6
Fox River Ct. N&S	26W-10S	6
Fox River Dr. NS&EW	26W-10S,25W-10S	6
Frost St. NS&EW	28W-11S	5
Garnette Ct. N&S	28W-12S,28W-11S	5
Gateshead Dr. NS&EW	28W-11S,26W-10S	5
Ginger Ln. E&W	24W-10S	6
Glen Echo Rd. E&W	25W-10S	6

	Grid	Page
Gleneagles Dr. N&S	27W-10S	5
Goldtown Ct. E&W	28W-12S	5
Greenfield Ct. E&W	28W-10S	5
Haddassah Ct. E&W	27W-10S	5
Haddassah Dr. NW&SE	26W-10S	6
Halifax Ct. NE&SW	28W-12S	5
Hamlet Rd. E&W	28W-11S	5
Hampshire Ct. E&W	26W-10S	6
Hanover Ct. N&S	26W-10S	6
Hartford Ct. N&S	27W-10S	5
Harvest Ct. E&W	28W-10S	5
Hawley Ct. NW&SE	27W-10S	5
Hearthside Ct. E&W	28W-10S	5
Heirloom Ct. N&S	28W-12S	5
Hempstead Av. E&W	27W-10S	5
Heritage Ct. NE&SW	25W-10S	6
High Meadow Dr. E&W	29W-13S	9
Hobbes Dr. N&S	28W-11S	5
Hollenback Ct. N&S	27W-10S	5
Hollingswood Av. E&W	28W-11S	5
Homestead Dr. N&S	28W-10S	5
Hoyer Ct. NW&SE	25W-10S	6
Hudson Ct. NE&SW	26W-10S	6
Jubilant Ct. N&S	28W-11S	5
Junegrass Dr. N&S	29W-13S	8
Kalamazo River Dr. N&S	26W-10S	6
Kaskaskia Ct. N&S	26W-10S	6
Kayak Dr. E&W	28W-12S	5
Keats Dr. E&W	28W-11S	5
Keim Rd. N&S	25W-10S	6
Keystone Ct. N&S	27W-10S	5
Kingsley Dr. N&S	26W-10S	6
Knock Knoll Rd. NE&SW	26W-10S	6
Kubek Rd. E&W	28W-10S	5
Lagoon Ct. E&W	28W-12S	5
Lagoon Ln. NE&SW	28W-12S	5
Lakeside Ct. NW&SE	28W-10S	5
Landore Dr. N&S	28W-11S	5
Lawrence Dr. N&S	28W-12S,28W-11S	5
Leach Dr. NS&EW	28W-10S	5
Leawood Ct. E&W	28W-10S	5
Lehigh Cir. NS&EW	25W-10S	6
Lemmon Dr. NE&SW	26W-10S	6
Leverenz Dr. NS&EW	27W-10S	5
Lindenwood Ln. N&S	24W-10S	6
Lindholm Ct. NW&SE	27W-10S	5
Litchfield Ct. N&S	27W-10S	5
Longacre Dr. N&S	28W-10S	5
Lotus Ct. N&S	24W-10S	6
Mackinaw Ct. E&W	26W-10S	6
Madera Ln. N&S	24W-10S	6
Magenta Ct. E&W	28W-11S	5
Malvina Ct. N&S	28W-11S	5
Mandeville Ln. NS&EW	28W-12S,28W-11S	5
Marengo Ct. E&W	28W-10S	5
Marshall Ct. N&S	25W-10S	6
Massachusetts Av. EW&NS	27W-10S	5
Mayapple Ct. E&W	24W-10S	6
McCartny Ln. NE&SW	26W-10S	6
Meadow Green Dr. NE&SW	26W-10S	6
Mecan Dr. N&S	28W-10S	5
Mercer Ct. N&S	24W-10S	6
Mesa Dr. NE&SW	27W-10S	5
Meyer Ct. NW&SE	28W-10S	5
Midhurst Ct. E&W	26W-9S	6
Midland Dr. N&S	28W-10S	5
Milkweed Dr. N&S	29W-13S	9
Modaff Rd. N&S	27W-10S,27W-8S	5
Montrose Ct. NE&SW	25W-10S	6
Morningstar Ct. E&W	28W-10S	5
Nantucket Dr. NE&SW	27W-10S	5
Naper Blvd. NE&SW	25W-10S	6
Neskola Ct. NE&SW	28W-10S	5
Newgate Av. NW&SE	26W-10S	6
Newman Ct. E&W	28W-11S	5
Newport Dr. NS&EW	27W-10S	5

	Grid	Page
Norfolk Ct. N&S	25W-10S	6
Northhampton Ct. E&W	27W-10S	5
Norwich Ct. N&S	27W-10S	5
Nottingham Dr. NS&EW	27W-10S	5
Oak Bluff Ct. NE&SW	26W-10S	6
Oberlin Ct. N&S	25W-10S	6
Oconto Ct. E&W	28W-10S	5
Old Mill Ct. E&W	28W-10S	5
Old Wehrli Rd. N&S	24W-10S	6
Orwell Rd. E&W	28W-11S	5
Othello Dr. N&S	28W-11S	5
Papaw Dr. N&S	29W-13S	9
Parador Dr. N&S	28W-11S	5
Pebblestone Rd. NW&SE	26W-10S	6
Penny Royal Dr. N&S	29W-13S	8
Peshtigo Av. E&W	28W-10S	5
Petra Ct. N&S	28W-11S	5
Petworth Ct. N&S	26W-10S	6
Piedmont Cir. NS&EW	25W-10S	6
Piper Av. E&W	26W-10S	6
Poet Ct. N&S	28W-11S	5
Pontiac Cir. NE&SW	26W-10S	6
Potomac Av. E&W	25W-10S	6
Pradel Dr. N&S	29W-13S	9
Prairie Sage Dr. E&W	29W-13S	9
Prarie Dock Rd. E&W	29W-13S	9
Prelude Ct. E&W	28W-11S	5
Prestige Ct. N&S	28W-11S	5
Princess Rd. E&W	28W-12S	5
Providence Ct. NW&SE	27W-10S	5
Pueblo Ct. E&W	27W-10S	5
Putnam Av. NE&SW	26W-10S	6
Red River Ct. E&W	26W-10S	6
Relstar Ct. N&S	28W-11S	5
Remington Dr. N&S	25W-10S	6
Richwood Ct. N&S	28W-12S	5
Rio Grande Ct. NE&SW	25W-10S	6
Rivanna Ct. E&W	25W-10S	6
River Woods Ct. NE&SW	26W-10S	6
Rivermist Ct. NS&EW	26W-10S	6
Roanoke Ct. E&W	25W-10S	6
Robert Dr. E&W	28W-12S	5
Rock Ct. N&S	27W-10S	5
Rock River Cts. E&W	26W-10S	6
Rock Spring Ct. E&W	27W-10S	5
Rock Spring Rd. E&W	27W-10S	5
Rode St. N&S	31W-11S	4
Rosada Dr. N&S	28W-12S,28W-11S	5
Rosehill Ct. N&S	25W-10S	6
Roxbury Dr. E&W	27W-10S	5
Royce Rd. E&W	25W-10S,24W-10S	6
Royce Woods Ct. NS&EW	25W-10S	6
Ryan Ct. N&S	28W-10S	5
Saddlebrook Cir. Cir.	30W-10S	4
Salisbury Dr. NS&EW	27W-10S	5
Salt River Ct. NE&SW	26W-10S	6
Saltmeadow Rd. E&W	29W-13S	9
San Luis Ct. N&S	27W-10S	5
Sandlilly Dr. N&S	29W-13S	9
Saratoga Rd. E&W	28W-11S	5
Saw Mill Rd. NE&SW	26W-10S	6
Saxon Ln. NS&EW	28W-11S	5
Schultz St. E&W	31W-11S	4
Scottsbridge Rd. NW&SE	26W-10S	6
Sebastian Ct. E&W	28W-12S	5
Sebastian Dr. N&S	28W-12S	5
Seiler Dr. N&S	27W-10S	5
Settlers Ct. NE&SW	26W-10S	6
Settlers Dr. NE&SW	26W-10S	6
Seville Av. E&W	27W-10S	5
Shakespeare Ln. E&W	28W-11S	5
Shaw Av. E&W	28W-11S	5
Sherwood Ct. NW&SE	27W-10S	5
Shimer Ct. E&W	25W-10S	6
Shire Ct. N&S	28W-11S	5
Sierra Av. NE&SW	26W-10S	6
Sierra Ct.* NE&SW	26W-10S	6
Silver Spur Ct. NE&SW	27W-10S	5

	Grid	Page
Snapdragon Rd. E&W	29W-13S	9
Spencer Ct. N&S	28W-11S	5
Spinner Ct. N&S	27W-10S	5
St. Croix Av. E&W	28W-10S	5
Starlite Ct. NS&EW	28W-10S	5
Steamboat Rd. E&W	27W-10S	5
Stillwater Ct. NE&SW	25W-10S	6
Sweetbroom Ct. E&W	29W-13S	9
Sweetbroom Rd. E&W	29W-13S	8
Switchgrass Dr. N&S	29W-13S	9
Thackery Ct. E&W	28W-11S	5
Thackery Ln. NW&SE	28W-11S	5
Thames Ct. NE&SW	28W-12S	5
Thatcher Dr. N&S	29W-13S	9
Thaxton Ct. N&S	25W-10S	6
Thurman Av. E&W	25W-10S	6
Tiara Ct. E&W	28W-12S	5
Tiffany Ct.* NW&SE	26W-11S	6
Townsend Dr. NE&SW	26W-10S	6
Traee Cir. Cir.	28W-12S	5
Trappers Ct. NE&SW	26W-10S	6
Treesdale Ct. NE&SW	30W-10S	4
Trillium Ln. N&S	24W-10S	6
Tuscany Dr. NS&EW	28W-11S	5
Tussell St. N&S	28W-11S	5
University Ct. N&S	25W-10S	6
Velvet Bent Ct. N&S	29W-13S	9
Vermillion Ct. NW&SE	25W-10S	6
Vesper Ct. N&S	28W-12S	5
Wabash Ct. NW&SE	25W-10S	6
Waterbury Ct. E&W	27W-10S	5
Watford Dr. NW&SE	28W-10S	5
Watson Av. E&W	28W-12S,28W-11S	5
Waupaca Ct. N&S	28W-10S	5
Wayland Ln. N&S	26W-10S	6
Waynesburg St. N&S	26W-10S	6
Weatherford Ln. NS&EW	26W-10S	6
Wedgewood Dr. N&S	27W-10S	5
Wendy Rd. N&S	27W-10S	5
Wentworth Ct. E&W	26W-10S	6
Wesley Av. NW&SE	26W-10S	6
West Branch Ct. NE&SW	26W-10S	6
Westbrook Cir. NS&EW	26W-10S	6
Wicklow Way Rd. E&W	28W-12S	5
Wild Timothy Dr. E&W	29W-13S	9
Wildrose Ct. E&W	24W-10S	6
Windemere Av. E&W	28W-10S	5
Winter Park Dr. N&S	27W-10S	5
Wirestem Ct. N&S	29W-13S	9
Wishing Well Dr. NW&SE	28W-10S	5
Wolfe Ct. N&S	28W-11S	5
Woodland Cir. NS&EW	26W-10S,26W-9S	6
Woodview Cir. NE&SW	26W-10S	6
Woodview Ct. NE&SW	26W-10S	6
Wooster Ct. N&S	24W-10S	6
Worthing Dr. NS&EW	26W-10S	6
Xavier Ct. N&S	24W-10S	6
Yellowstar Ct. E&W	29W-13S	9
Yellowstar Ln. E&W	29W-13S	9
Yorktown Cir. Cir.	28W-12S	5
Yorktown Ct. NE&SW	28W-12S	5
91st St. E&W	24W-10S	6

New Lenox

	Grid	Page
Abraham Dr. E&W	18W-23S	22
Aeronca Ct. E&W	18W-25S	30
Andrea Dr. NS&EW	19W-26S	30
Argyle Dr. E&W	17W-25S	31
Argyle Ln. E&W	19W-27S	30
Armstrong Ln. E&W	18W-25S	30
Ash St. E&W	17W-24S	31
Aspen Dr. NS&EW	17W-24S	31
Auburn Ct. NE&SW	20W-24S	30
Barbara Ln. E&W	18W-23S	22
Barnside Rd. N&S	17W-25S	31
Beech Ln. E&W	18W-25S	30
Beechcraft Dr. N&S	18W-25S	30
Bentley Rd. N&S	17W-25S	31
Blandford Av. NE&SW	19W-22S	22
Boeing Dr. N&S	18W-25S	30
Bon Terre Rd. NW&SE	17W-25S	31
Briarcrest Ln. NW&SE	19W-26S	30
Carlton Ct. E&W	19W-27S	30
Cedar Rd. N&S	18W-25S,18W-22S	30
Central Rd. NW&SE	18W-25S	30
Cessna Ct. E&W	18W-25S	30
Cherry Hill Rd. N&S	21W-24S	29
Cherrywood Ln. NE&SW	19W-26S	30
Church St. NS&EW	17W-24S	31
Cimarron Dr. NW&SE	19W-26S	30
Circlegate Rd. NE&SW	17W-26S	31
Clinton St. N&S	18W-23S,18W-22S	22
Cooper St. N&S	17W-26S,17W-25S	31
Copper St. N&S	17W-24S	31
Corsair Ct. E&W	18W-25S	30
Country Creek Dr. E&W	18W-26S	30
Coventry Rd. E&W	19W-23S	22
Cricket Av. NE&SW	19W-22S	22
Daniel Lewis Dr. NE&SW	19W-27S	30
Dawn Wy. N&S	17W-25S	31
Delmar Dr. NW&SE	19W-26S	30
Digby Dr. NE&SW	19W-27S	30
Dorset Ln. E&W	20W-24S	30
Doxbury Ln. N&S	20W-24S	30
Eagle Vista Dr. NW&SE	19W-26S	30
Earl Ct. N&S	17W-25S	31
Eastgate Ct. E&W	17W-25S	31
Elm St. E&W	17W-24S	31
Elm St. N&S	18W-24S,18W-23S	30
Essex Ln. N&S	20W-24S	30
Fernwood Ter. N&S	19W-26S	30
Ferro Dr. E&W	19W-24S	30
Fir St. E&W	17W-24S	31
First Av. E&W	18W-25S	30
Forest St. E&W	18W-24S	30
Fourth Av. E&W	18W-25S,17W-25S	30
Foxwood Dr. N&S	18W-26S	30
Francis Rd. E&W	18W-23S,17W-23S	22
Gall Ln. E&W	17W-24S	31
Garnet Dr. NS&EW	19W-24S	30
Gear Dr. N&S	18W-25S	30
Gifford Pl. E&W	20W-24S	30
Gina Dr. E&W	18W-25S	30
Ginger Ln. E&W	19W-27S	30
Glenn Dr. N&S	18W-25S	30
Golf Av. NE&SW	19W-22S	22
Gougar Rd. N&S	19W-26S,19W-22S	30
Grandview Dr. NE&SW	19W-26S	30
Greenbriar Dr. N&S	18W-24S	30
Grissom Ln. E&W	18W-25S	30
Grumman Ct. E&W	18W-25S	30
Gum St. E&W	17W-24S	31
Haines St. E&W	18W-24S	30
Hauser Ct. E&W	17W-24S	31
Hawthorn Ln. E&W	17W-23S	23
Heatherway Ln. N&S	19W-26S	30
Hickory St. E&W	18W-24S	30
Honey Ln. NS&EW	19W-24S	30
Inner Ct. E&W	18W-23S	22
Interstate 80 NE&SW	19W-24S,18W-23S	30
Ione Dr. E&W	19W-23S	22
Jackson Branch Dr. N&S	19W-27S	30
John St. E&W	18W-24S	30
Joliet Hwy. E&W	18W-25S,17W-25S	30
Keithland Ct. E&W	17W-24S	31
Kerry Winde Dr. N&S	19W-27S	30
Kimber Dr. E&W	18W-24S	30
Kingston Dr. N&S	20W-24S	30
Knollside Rd. N&S	17W-25S	31
Kris Dr. NS&EW	19W-24S,18W-24S	30
Lake Rd. N&S	17W-25S	31
Lambeth Ln. N&S	19W-23S	22

Street	Grid	Page
Laraway Rd. E&W	19W-26S,18W-26S	30
Lear Ln. E&W	18W-25S	30
Lewis Ln. NE&SW	19W-23S	22
Lincoln Hwy. E&W	17W-24S	31
Livingston Dr. E&W	18W-24S	30
Manor Ct. NE&SW	18W-24S	30
Manor Dr. NW&SE	18W-24S	30
Maple Ln. E&W	18W-23S	22
Maple St. NW&SE	18W-24S,17W-24S	30
Markev Ln. E&W	18W-24S	30
Marla Ln. N&S	18W-23S	22
McDivitt Dr. N&S	18W-25S	30
McDonough St. E&W	20W-24S	30
Meadow Ridge Ln. E&W	18W-26S	30
Misty Creek Dr. NW&SE	18W-25S	30
Moss Ln. E&W	20W-24S	30
Mustang Ln. E&W	18W-25S	30
Nelson Rd. N&S	19W-26S,19W-24S	30
New Lenox Rd. E&W	19W-24S,18W-24S	30
Northgate Rd. N&S	17W-25S	31
Oak Dr. N&S	18W-24S	30
Oak St. E&W	17W-24S	31
Oakview Dr. E&W	17W-24S	31
Ogden Rd. N&S	18W-25S	30
Old New Lenox Rd. NW&SE	18W-24S	30
Oxford Ct. N&S	18W-25S	30
Park Ln. NW&SE	18W-24S	30
Pine Grove Ln. N&S	19W-23S	22
Pine Pl. N&S	18W-24S	30
Pine St. N&S	18W-25S,18W-23S	30
Piper Dr. N&S	18W-25S	30
Poplar Ln. E&W	17W-24S	31
Prairie Av. N&S	17W-25S	31
Root St. N&S	17W-24S	31
Rossford Ln. N&S	20W-24S	30
Roy St. NS&EW	17W-25S	31
Rt. 355 (proposed) NW&SE	18W-22S	22
Sanford Av. NS&EW	19W-27S	30
Scenic Ct. E&W	18W-24S	30
Schoolgate Ct. N&S	17W-25S	31
Schoolgate Rd. N&S	17W-26S	31
Second Av. E&W	18W-25S,17W-25S	30
Shagbark Ct. N&S	19W-26S	30
Shagbark Rd. E&W	19W-26S,18W-26S	30
Sheffield Dr. NW&SE	20W-24S	30
Southgate Rd. N&S	17W-26S,17W-25S	31
Southwest Hwy. E&W	19W-22S,18W-22S	22
Spencer Rd. E&W	18W-25S,17W-25S	30
Springcreek St. N&S	19W-22S	22
Stafford Ct. E&W	18W-25S	30
Stafford Dr. N&S	18W-25S	30
Stone Ct. NW&SE	19W-24S	30
Stonebridge Ct. E&W	19W-26S	30
Stonebridge Dr. NW&SE	19W-26S	30
Stonegate Rd. NS&EW	17W-26S,17W-25S	31
Sunset Tr. E&W	17W-25S	31
Surf Dr. E&W	18W-24S	30
Third Av. E&W	18W-25S	30
Timber Ct. N&S	19W-26S	30
Timber Pl. E&W	19W-26S,18W-26S	30
Twilight Ln. NE&SW	17W-25S	31
U.S. Route 30 E&W	19W-24S,17W-24S	30
Vine St. N&S	18W-24S	30
Wallace St. E&W	18W-24S	30
Warren Av. N&S	18W-25S	30
Washington St. E&W	20W-24S	30
Western Av. N&S	18W-25S	30
White Ln. E&W	18W-25S	30
Wildwood Dr. NE&SW	18W-25S	30
Willow Ln. E&W	18W-23S	22
Wimborne Av. NW&SE	19W-22S	22
Winsor Ln. N&S	19W-23S	22
Winter Park Dr. E&W	18W-26S	30
Wintree Ln. N&S	20W-24S	30
Wood St. E&W	18W-24S,17W-24S	30
Woodlawn Rd. E&W	18W-25S,17W-25S	30
York Pl. NE&SW	20W-24S	30
Young Dr. N&S	18W-25S	30

New Lenox Township

Street	Grid	Page
Abbington Ln. N&S	16W-27S	31
Airway Ct. E&W	16W-26S	31
Alan Dr. E&W	17W-25S	31
Alison Tr. E&W	15W-24S	31
Amherst Ct. N&S	20W-24S	30
Anderson Dr. N&S	16W-26S	31
Anderson Rd. N&S	16W-25S	31
Arrowhead Rd. NE&SW	15W-25S	31
Ashington Ct. E&W	16W-24S	31
Banbury Ln. N&S	19W-24S	30
Barbara Ln. E&W	18W-23S	22
Bellchase Rd. N&S	17W-26S	31
Bernard Ct. N&S	15W-22S	23
Betty Ln. E&W	17W-22S	23
Beverly Blvd. E&W	18W-23S,17W-23S	22
Bittersweet Dr. E&W	16W-24S	31
Bjork Dr. N&S	19W-23S	22
Blodgett Rd. N&S	17W-22S	23
Branchaw Blvd. E&W	18W-23S	22
Briarcliff Dr. NS&EW	16W-26S	31
Brisbane Rd. N&S	16W-24S	31
Bruce Ct. NE&SW	15W-22S	23
Bryan Ter. NS&EW	15W-24S	31
Buckboard Dr. E&W	17W-23S	23
Burl Ct. NS&EW	20W-24S	30
Calmer Rd. E&W	20W-24S	30
Carol Rd. E&W	18W-23S	22
Carrie Ct. NW&SE	16W-22S	23
Carrington Ct. NW&SE	16W-24S	31
Cedar Rd. N&S	18W-27S,18W-22S	30
Charlotte Ct. E&W	17W-25S	31
Chatfield Rd. NS&EW	16W-24S	31
Chelsae Ct. E&W	16W-25S	31
Chelsae St. E&W	16W-25S	31
Chessington Ln. N&S	16W-27S	31
Clearing Ct. E&W	16W-26S	31
Clinton St. N&S	19W-23S	22
Conlee Dr. N&S	16W-22S	23
Constance Dr. N&S	16W-24S	31
Constitution Rd. E&W	16W-25S	31
Cooper St. N&S	17W-25S	31
Corrie Ln. E&W	16W-26S	31
Covey Ct. N&S	15W-25S	31
Covey Dr. E&W	15W-25S	31
Craig Ct NW&SE	15W-22S	23
Cresent Pl. N&S	20W-24S	30
Cynthia Ln. NS&EW	17W-22S	23
Debbie Ln. E&W	17W-22S	23
Delaney Rd. E&W	20W-27S,15W-27S	30
Donald Ct. N&S	15W-23S	23
Dougall Rd. NW&SE	20W-24S,20W-23S	30
E. Circle Dr. NW&SE	15W-25S,15W-24S	31
Edmonds St. N&S	17W-24S,17W-23S	31
Edward Pkwy. NE&SW	16W-22S	23
Elizabeth Ln. E&W	17W-22S	23
Elm St. N&S	18W-23S	22
Emiley Ln. E&W	15W-24S	31
Erskine Rd. NW&SE	20W-24S,20W-23S	30
Farm View Rd. N&S	16W-27S	31
Florence Rd. E&W	16W-22S	23
Ford Dr. E&W	18W-26S	30
Fourth Av. E&W	17W-25S	31
Francis Rd. E&W	19W-23S,15W-23S	22
Fredonia St. E&W	17W-26S	31
Garfield Av. N&S	15W-25S,15W-24S	31
Giddington Ct. NW&SE	16W-24S	31
Golf Rd. E&W	20W-22S	22
Gordon St. N&S	18W-23S	22
Gougar Rd. N&S	20W-26S,20W-22S	30
Green St. N&S	18W-24S,18W-23S	30
Haley Rd. N&S	15W-22S	23
Harmoni Ln. E&W	16W-24S	31
Hass Rd. N&S	16W-22S	23
Hawkshead Dr. N&S	15W-27S	31
Hazelwood Dr. E&W	17W-26S	31
Hempstead Pl. N&S	20W-24S	30
Herr Dr. E&W	16W-24S	31
Hickory St. N&S	15W-22S	23
Hillside Dr. NW&SE	16W-23S	23
Hillside Rd. E&W	18W-24S	30
Hoberg Dr. NE&SW	20W-23S	22
Holly Hill Cir. N&S	19W-24S	30
Howell Dr. NS&EW	16W-26S	31
Hunter Tr. NS&EW	16W-23S	23
Illini Dr. E&W	16W-27S	31
Indian Mound Dr. NE&SW	15W-25S,15W-24S	31
Interstate 80 NE&SW	20W-24S,15W-22S	30
Jennifer Ct. NE&SW	15W-24S	31
Joey Ct. N&S	15W-23S	23
Joliet Hwy. E&W	17W-25S,16W-25S	31
Josephine Dr. E&W	15W-22S	23
Kalarama Dr. E&W	17W-23S	23
Kankakee St. N&S	16W-25S	31
Knollwood Pl. N&S	20W-24S	30
Lakeside Dr. NE&SW	16W-23S	23
Lancaster Dr. NW&SE	20W-24S	30
Laraway Rd. E&W	20W-26S,15W-26S	30
Laura Ln. E&W	18W-23S	22
Leanne Ct. E&W	17W-22S	23
Leanne Ln. N&S	17W-22S	23
Lenox St. E&W	17W-23S	23
Leven Av. N&S	15W-27S	31
Lila's Ct. E&W	17W-23S	23
Lincoln Hwy. NW&SE	16W-24S,15W-25S	31
Lincolnway Dr. N&S	16W-24S	31
Lincolnway Rd. N&S	16W-27S	31
Locust Ln. E&W	17W-24S	31
London Rd. N&S	16W-23S	23
Lynn Pkwy. NW&SE	16W-22S	23
Main St. NE&SW	15W-22S	23
Maple St. E&W	15W-22S	23
Maray Av. N&S	17W-25S	31
Marley Rd. NE&SW	16W-24S,16W-23S	31
Marshall Ct. NW&SE	15W-22S	23
Mary Ct. N&S	16W-23S	23
Maureen Ct. E&W	15W-24S	31
Maureen Dr. N&S	15W-24S	31
Mayfair Ct. N&S	20W-23S	22
Mayfair Dr. NW&SE	20W-23S	22
Melrose St. E&W	16W-25S	31
Menno Ct. NW&SE	18W-23S	22
Menno Dr. NE&SW	18W-23S	22
Michael Ln. E&W	18W-23S	22
Michigan Rd. E&W	18W-25S,17W-25S	30
Montieth Av. N&S	20W-24S	30
Morrcambe Bay Dr. E&W	15W-27S	31
Nelson Rd. N&S	19W-27S,19W-25S	30
Oak Rail Dr. N&S	16W-27S	31
Oakdale Av. NE&SW	20W-23S	22
Oakview Ct. NW&SE	15W-24S	31
Ogden Av. N&S	18W-25S	30
Old New Lenox Rd. NW&SE	19W-24S,18W-24S	30
Parker Rd. N&S	16W-23S,16W-22S	23
Pembroke Av. N&S	20W-24S,20W-23S	30
Pennington Ct. E&W	16W-24S	31
Pheasant Ln. E&W	16W-27S	31
Pine St. N&S	18W-25S	30
Pioneer Pl. NW&SE	15W-25S	31
Plaza Dr. E&W	16W-24S	31
Pleasant St. N&S	16W-24S	31
Pottawatamie Ln. NW&SE	15W-25S	31
Prairie Rd. N&S	17W-25S	31
Quails Roost Dr. E&W	15W-25S	31
Rachel Dr. N&S	17W-22S	23
Raymond Dr. E&W	15W-22S	23
Redwood Av. E&W	17W-23S	23
Regan Rd. NW&SE	16W-23S	23
Regan Rd. N&S	17W-23S	23
Regent St. E&W	16W-25S	31
Richard Av. N&S	16W-23S,16W-22S	23
Riivendell Dr. E&W	15W-24S	31
Robert St. N&S	17W-25S	31
Rodger Ct. NE&SW	15W-22S	23
Ross Dr. N&S	17W-22S	23
Rt. 355 (proposed) NW&SE	17W-23S,18W-22S	23
Ruth Rd. NS&EW	16W-22S	23
Rydal St. N&S	15W-27S	31
Scheer Rd. N&S	15W-27S	31
Schmule Rd. N&S	16W-26S,16W-23S	31
Scott Ct. NW&SE	15W-24S	31
Shelly Ln. N&S	15W-24S	31
Shirley Pkwy. E&W	16W-24S	31
Somerset Ct. N&S	16W-25S	31
Somerset Pl. NE&SW	16W-25S	31
Somerset St. NS&EW	16W-25S	31
Southwest Hwy. E&W	20W-22S,15W-22S	22
Spector Rd. N&S	18W-23S	22
Spencer Rd. NS&EW	19W-25S,16W-27S	30
Star-Lite Dr. NW&SE	20W-24S	30
Suffolk Rd. E&W	20W-23S	22
Summerfield Dr, E&W	17W-22S	23
Sunset Ln. N&S	20W-24S	30
Sycamore Ln. E&W	18W-24S	30
Tammy Dr. NS&EW	16W-22S	23
Tara St. E&W	17W-26S	31
Terry Ln. N&S	17W-25S	31
Terryellen Ln. E&W	18W-23S	22
Tessington Ct. NW&SE	16W-24S	31
Thomas Ct. E&W	15W-22S	23
Thomas Ln. N&S	17W-23S	23
Thornhouse Crescent SW&NE	19W-24S	30
Timothy Ln. NS&EW	16W-22S	23
Tomahawk Ridge E&W	16W-27S	31
Ton-Ell Av. N&S	17W-25S	31
Tracy St. N&S	17W-22S	23
Tudor Ln. N&S	16W-26S	31
U.S. Route 6 E&W	20W-22S,15W-22S	22
U.S. Route 30 E&W	20W-23S,15W-25S	22
U.S. Route 52 NW&SE	20W-27S	30
Valley St. E&W	15W-22S	23
Vine St. N&S	18W-24S,18W-23S	30
Virginia Ln. N&S	16W-22S	23
W. Circle Dr. N&S	15W-25S,15W-24S	31
Wagon Dr. E&W	17W-23S	23
Walona Av. N&S	17W-25S	31
Walter Ct. NW&SE	15W-22S	23
Walter Dr. E&W	15W-24S	31
Warren Dr. NS&EW	15W-22S	23
Washington St. E&W	20W-24S	30
Westmorland Av. N&S	15W-27S	31
Whitehall Rd. N&S	17W-26S	31
William St. N&S	17W-25S	31
Willow Rd. West E&W	19W-24S,18W-24S	30
Willow St. E&W	18W-24S	30
Wisconsin Av. N&S	18W-25S	30
Woodedge Ln. N&S	15W-23S	23
Woodlawn Rd. E&W	18W-25S	30
Woodside Dr. N&S	16W-23S	23
2nd. Av. E&W	17W-25S	31
184th Pl. E&W	16W-22S	23
185th St. E&W	16W-22S	23

Park Forest*

(Also see Cook County)

Street	Grid	Page
Braeburn Rd. SW&NE	3W-28S	43
Brookside Dr. SW&NE	3W-28S	43
E. Sycamore Dr. E&W	2W-28S	43
Glen St. N&S	3W-28S	43
Monee Ct. NW&SE	3W-28S	43
Monee Rd. NE&SW	3W-28S,3W-27S	43
Nanti St. NE&SW	4W-28S,3W-28S	43
Natoma St. N&S	4W-28S	43
Nauvoo St. E&W	4W-28S	43
Navajo St. N&S	4W-28S	43
Neosho N&S	4W-28S	43
Niagara St. N&S	3W-28S,3W-27S	43
North Arbor Trail N&S	2W-28S	43
South Arbor Trail N&S	2W-28S	43
Tahoe St. N&S	4W-28S	43
Talala St. NS&EW	4W-28S	43
Tamarack St. E&W	4W-28S	43
Tampa St. E&W	4W-28S	43
Thorn Creek Dr. N&S	3W-28S	43
Tioga St. E&W	4W-28S	43
Titonka St. N&S	4W-28S	43
Tomahawk St. N&S	4W-28S	43
Topeka St. E&W	4W-28S	43
Towanda Ct. NW&SE	4W-28S	43
Wold Ct. N&S	3W-28S	43
Woodland E&W	3W-28S	43

Peotone

Street	Grid	Page
Ahlborn Dr. NE&SW	9W-37S	65
Amsterdam Ln. E&W	9W-37S	65
Barton Ln. E&W	8W-37S	65
Blaine Av. E&W	9W-37S	65
Bonnie Ln. E&W	10W-37S	65
Conrad Av. N&S	9W-37S	65
Corning St. E&W	9W-37S	65
Crawford St. E&W	9W-37S	65
Crown Ln. E&W	9W-37S	65
Delfi Ct. N&S	9W-37S	65
Division St. N&S	9W-37S	65
Ethel St. E&W	10W-37S	65
First St. N&S	9W-37S	65
Garfield Av. E&W	10W-37S,9W-37S	65
Glenview Ln. NS&EW	9W-37S	65
Hans Brinker Ct. E&W	9W-37S	65
Hans Brinker Dr. N&S	9W-37S	65
Harlem Av. N&S	9W-37S	65
Havert St. E&W	8W-37S	65
Hawthorne Ln. N&S	8W-37S	65
Hickory Ln. N&S	8W-37S	65
Jean Ln. N&S	10W-37S	65
Jessen St. E&W	9W-37S	65
Joliet Rd. E&W	9W-37S	65
Lark St. E&W	10W-37S	65
Lincoln St. E&W	9W-37S	65
Louise Ln. E&W	10W-37S	65
Main St. E&W	9W-37S	65
Manor Dr. N&S	9W-37S	65
Maple St. E&W	10W-37S	65
Meadow Ln. N&S	10W-37S	65
Mill St. N&S	9W-37S	65
North St. E&W	9W-37S	65
Oak St. E&W	9W-37S	65
Orchard Ct. E&W	9W-37S	65
Penny Ln. E&W	9W-37S	65
Peotone-Beecher Rd. E&W	8W-37S	65
Rathie St. N&S	9W-37S	65
Royal Ln. E&W	9W-37S	65
Rt. 50 N&S	9W-37S	65
Schroeder Av. NE&SW	9W-37S	65
Sixth St. N&S	8W-37S	65
South St. E&W	9W-37S	65
Summer Av. E&W	9W-37S	65
Tucker Rd. E&W	9W-37S	65
Van Gough Ct. N&S	9W-37S	65
Walnut St. E&W	8W-37S	65
West St. N&S	9W-37S	65
Wilmington Rd. E&W	10W-37S,9W-37S	65
Wilson St. E&W	9W-37S	65
Wood Av. E&W	9W-37S	65
2nd St. N&S	9W-37S	65
3rd St. N&S	9W-37S	65

Peotone Township

Street	Grid	Page
Adams St. E&W	14W-34S	52
Barr Rd. E&W	14W-36S,11W-36S	64
Center Rd. N&S	12W-39S,12W-34S	64
Corning St. E&W	10W-37S	65
County Line Rd. E&W	14W-39S,11W-39S	64
Eagle Lake Rd. E&W	14W-34S,9W-34S	52
Evergreen St. N&S	14W-34S	52
Green Garden-Manhattan Rd N&S	14W-39S,14W-34S	64
Harlem Av. N&S	9W-39S,9W-34S	65
Interstate 57 NE&SW	11W-39S,9W-34S	64
Joliet Rd. NS&EW	13W-35S,9W-36S	52
Kennedy Rd. E&W	14W-38S,9W-38S	64
Linton Av. N&S	14W-37S	64
Locust Ln. NE&SW	10W-37S	65
Main St. N&S	14W-38S,14W-34S	64
Manhattan-Wilton Rd. E&W	14W-34S,9W-34S	52
N. Peotone Rd. N&S	11W-37S,11W-34S	64
N. Peotone Rd. E&W	9W-35S	53
Offner Rd. E&W	11W-34S,9W-34S	52
Rathie St. N&S	10W-38S,10W-34S	65
Rt. 50 NE&SW	10W-39S,9W-38S	65
U.S. Rt. 45 N&S	14W-35S,14W-34S	52

	Grid	Page
U.S. Rt. 52 NS&EW		
	14W-39S,14W-35S	64
Wilmington-Peotone Rd. E&W		
	14W-37S,10W-37S	64
80th Av. N&S		
	10W-39S,10W-35S	65
88th Av. N&S	11W-34S	52
104th Av. N&S		
	13W-39S,13W-34S	64
120th Av. N&S		
	14W-38S,14W-34S	64

Plainfield

	Grid	Page
Amboy St. E&W	29W-17S	12
Amherst Ct. N. N&S		
	30W-16S	12
Amherst Ct. S. N&S		
	30W-16S	12
Arnold St. NW&SE	30W-17S	12
Arrowhead Ct. N&S		
	31W-18S	12
Ash St. NW&SE	29W-17S	12
Asther Ct. E&W	30W-15S	8
Bartlett Av. N&S	29W-17S	12
Buttercup Ct. N&S	31W-18S	12
Center St. N&S	29W-17S	12
Chicago St. E&W		
	30W-18S,29W-18S	12
Clover Ct. N&S	31W-18S	12
Coilplus Dr. N&S	30W-16S	12
College St. NE&SW		
	29W-17S	12
Columbia Cir. E&W		
	30W-15S	8
Commercial St. E&W		
	30W-18S	12
Corbin St. NW&SE	29W-18S	12
Creekside Dr. NW&SE		
	31W-18S	12
Daisy Ct. N&S	31W-18S	12
DesPlaines St. N&S		
	30W-18S,30W-17S	12
Dillman St. E&W	29W-18S	12
Division St. N&S		
	30W-18S,30W-16S	12
DuPage Dr. NW&SE		
	29W-16S	12
Eastern Av. N&S	29W-17S	12
Evans St. E&W	29W-17S	12
Fort Beggs Dr. E&W		
	30W-18S	12
Foxriver St. N&S	30W-18W	12
Golden Meadow Ct. NW&SE		
	30W-15S	8
Golden Meadow Dr. N&S		
	30W-15S	8
Hartong St. N&S	29W-18S	12
Hawthorne Cir. NS&EW		
	29W-17S	12
Hazelcrest Dr. E&W		
	30W-16S	12
Heather Ct. E&W	30W-15S	8
Highview Ct. E&W	30W-18S	12
Hillcrest Dr. E&W	30W-16S	12
Homestead Ct. E&W		
	31W-18S	12
Huron St. NW&SE	29W-17S	12
Illini Dr. E&W	31W-18S	12
Illinois St. N&S		
	30W-18S,30W-17S	12
James St. NE&SW	30W-18S	12
Joliet Rd. NW&SE	29W-18S	12
Joseph Av. E&W	30W-16S	12
Kelly Av. N&S	30W-16S	12
Lakeside Dr. N&S	29W-18S	12
Lincoln Hwy. N&S	30W-16S	12
Lockport St. E&W		
	30W-17S,29W-17S	12
Lynn St. E&W	29W-19S	12
Main St. NE&SW		
	30W-17S,29W-18S	12
Mallard Dr. N. N&S	30W-16S	12
Mallard Dr. S. N&S	30W-16S	12
Maple Ct. N&S	29W-17S	12
Marybrook Dr. N&S		
	30W-16S	12
Meadow Ln. N&S	31W-18S	12
Meadowbrook N&S		
	30W-16S	12
Michigan St. NW&SE		
	29W-17S	12
Milk Farm Rd. E&W		
	29W-20S	18
Mill St. NW&SE	29W-17S	12
New Kirk Dr. N&S	30W-18S	12
Norman Av. E&W	29W-18S	12
Oak St. NE&SW	30W-17S	12
Old Indian Boundary Line		
NE&SW	31W-18S	12
Ottawa St. E&W		
	30W-18S,29W-18S	12
Parklane Rd. NE&SW		
	30W-16S	12
Partridge Dr. NW&SE		
	30W-16S	12
Pawnees St. N&S	31W-18S	12
Penn Rd. NW&SE	30W-17S	12
Pheasant Chase Dr. E&W		
	30W-16S	12
Plainfield-Naperville Rd.		
NE&SW	29W-17S,28W-15S	12
Plainsman Cir. NS&EW		
	29W-20S	18

	Grid	Page
Plainsman Ct. N&S	29W-20S	18
Plaza Dr. E&W	30W-18S	12
Poppy Ln. N&S	31W-18S	12
Prairie Dr. N&S	31W-18S	12
Pratt Ln. N&S	30W-18S	12
Quail Ct. E&W	30W-16S	12
Quail Ct., North E&W		
	30W-16S	12
Quail Run Dr. NS&EW		
	30W-16S	12
River Dr. NE&SW	30W-18S	12
River Rd. NE&SW	30W-18S	12
Robin Ct. N&S	31W-18S	12
Rt. 59 N&S		
	30W-19S,30W-16S	12
Rt. 126 E&W		
	31W-18S,30W-17S	12
Sandbank Dr. E&W		
	31W-18S	12
Spring Hill Ln. E&W		
	29W-17S	12
St. Joseph Av. E&W		
	30W-16S	12
Sundance Ct. E&W		
	31W-18S	12
Sunshine Cir. NS&EW		
	31W-18S	12
Sunshine Ct. E&W	31W-18S	12
Teal Ct. N&S	30W-16S	12
Timber Dr. NW&SE		
	31W-18S	12
U.S. Route 30 E&W		
	30W-17S,29W-18S	12
Union Ct. N&S	29W-18S	12
Union St. E&W		
	30W-18S,29W-18S	12
Vandyke Rd. N&S		
	30W-17S,30W-16S	12
Vista Dr. N&S	30W-18S	12
Whiskey Rd. E&W	30W-16S	12
Willow Dr. E&W	31W-18S	12
Willow Dr. E. N&S	31W-18S	12
Willow Rd. N&S	31W-18S	12

Plainfield Township

	Grid	Page
Addelman St. N&S	28W-21S	19
Anderson St. E&W	28W-19S	13
Ann Ct. E&W	31W-20S	18
Ann Dr. N&S	31W-20S	18
Ardith St. E&W	27W-20S	19
Ash St. NW&SE	29W-17S	12
Beckwith Ct. N&S	29W-21S	18
Benson Av. N&S	29W-18S	12
Bently Dr. E&W	29W-21S	18
Berry St. E&W	27W-20S	19
Biltmore St. E&W	27W-20S	19
Blossom St. NW&SE		
	27W-20S	19
Brittany St. E&W	30W-19S	12
Bronk Rd. N&S	29W-21S	18
Brown St. E&W		
	29W-18S,28W-18S	12
Budler Rd. N&S		
	27W-17S,27W-16S	13
Burbank St. E&W	27W-20S	19
Bussey Dr. E&W	28W-19S	13
Byrum Blvd. N&S	28W-20S	19
Caitlin Ct. N&S	30W-20S	18
Caroline Dr. NS&EW		
	27W-19S	13
Carrier Cir. NS&EW		
	29W-21S	18
Castle Ct. E&W	28W-21S	19
Catherine Ct. N&S	31W-20S	18
Catherine Dr. NS&EW		
	31W-20S	18
Caton Farm Rd. E&W		
	32W-20S,27W-20S	18
Cecily Dr. N&S	29W-19S	12
Central Dr. N&S	27W-20S	19
Chapman Ct. N&S	29W-21S	18
Chapman Dr. E&W	29W-21S	18
Charlotte Rd. NW&SE		
	28W-19S	13
Chestnut St. E&W	27W-20S	19
Chevy Chase Dr. NW&SE		
	27W-20S	19
Clark St. E&W	28W-19S	13
Coghill St. N&S	27W-20S	19
Colerain St. E&W	28W-19S	13
Colinia St. N&S	28W-19S	13
Collins Dr. NE&SW	29W-18S	12
Commonwealth Av. N&S		
	27W-20S	19
Conant Av. NW&SE		
	27W-20S	19
Copper Dr. N&S	29W-17S	12
Cosmic Dr. N&S	28W-20S	19
County Line Rd. N&S		
	32W-21S,32W-18S	18
Crystal Dr. NW&SE		
	27W-20S	19
Culver Ct. N&S	29W-21S	18
Culver St. E&W	29W-19S	12
Dan Ireland Dr. E&W		
	27W-19S	13
Dan O'Connell Dr. N&S		
	29W-19S	12
Davy Ct. E&W	29W-19S	12
Debbie Ct. NE&SW		
	27W-20S	19
Delia Av. N&S	32W-20S	18
Dina Ct. E&W	31W-20S	18

	Grid	Page
Division St. N&S		
	30W-18S,30W-16S	12
Dixmoor St. N&S	27W-20S	19
Drauden Rd. N&S		
	31W-20S,31W-19S	18
DuPage Dr. NW&SE		
	29W-16S	12
Dupage Country Club Dr.		
NW&SE	29W-20S	18
Earl Dr. E&W	28W-21S	19
Eastline Dr. N&S	28W-20S	19
Easy St. E&W	30W-19S	12
Edgewood Dr. NE&SW		
	28W-19S	13
Ellna Ct. N&S	28W-19S	13
Emlong St. NW&SE		
	27W-20S	19
Emyvale Dr. E&W	30W-20S	18
Enid St. N&S	27W-20S	19
Essington Rd. N&S		
	28W-20S,27W-20S	19
Evergreen Dr. E&W		
	27W-19S	13
Ewing Ct. N&S	28W-19S	13
Ewing St. E&W	28W-19S	13
Farmingdale Dr. NE&SW		
	30W-19S	12
Farmington Av. N&S		
	27W-20S	19
Feeney Dr. E&W	28W-19S	13
Fern St. E&W	28W-20S	19
Fiday Rd. E&W	28W-21S	19
Flagar Ct. E&W	29W-21S	18
Flanders Ct. N&S	29W-21S	18
Flossmoor St. N&S	27W-20S	19
Flower St. E&W	27W-20S	19
Foster Ct. E&W	29W-21S	18
Francis Dr. N&S	27W-18S	13
Fraser Rd. E&W	30W-19S	12
Frederick Av. N&S	29W-18S	12
Frontage Rd. N&S	28W-21S	19
Gael Dr. NE&SW	27W-19S	13
Gagne Ln. NW&SE	29W-20S	18
Galena St. N&S	27W-19S	13
Garden Ct. N&S	27W-20S	19
Garden Dr. NE&SW		
	29W-17S	12
Garden St. E&W	27W-20S	19
Gaylord Rd. N&S		
	27W-20S,27W-19S	19
George Ct. N&S	28W-19S	13
Getson Av. N&S	29W-18S	12
Gibson Ct. N&S	29W-21S	18
Glasglow St. E&W	27W-20S	19
Glenwoodie St. NE&SW		
	27W-20S	19
Golden Ln. N&S	29W-17S	12
Golf St. E&W	27W-20S	19
Golfmoor St. N&S	27W-20S	19
Golfview Dr. E&W	27W-20S	19
Goodhue Ln. E&W	29W-21S	18
Grape St. E&W	27W-20S	19
Greenview Cir. NW&SE		
	27W-20S	19
Greenview Dr. N&S		
	27W-21S	19
Greenway St. N&S	27W-20S	19
Greglawn Av. N&S	27W-20S	19
Grinton Dr. E&W	28W-19S	13
Harmony Dr. NE&SW		
	30W-19S	12
Harris Dr. E&W	28W-21S	19
Heather Ln. N&S	28W-21S	19
Heathercreek Dr. NE&SW		
	29W-20S	18
Hel-Mar St. N&S	28W-20S	19
Helen Dr. E&W	27W-18S	13
Hickory St. E&W	28W-19S	13
Hollylynn Ln. E&W	28W-20S	19
Honora Dr. NE&SW		
	30W-20S	18
Howard Av. N&S	28W-19S	13
Interstate 55 N&S		
	29W-21S,27W-16S	18
Ivy Ln. NE&SW	28W-19S	13
Jamestown Rd. E&W		
	32W-21S	18
Joann Dr. E&W	28W-21S	19
Joliet Rd. NW&SE	29W-18S	12
Judith Dr. N&S	28W-19S	13
Julie Av. N&S	29W-21S	18
June St. N&S	31W-20S	18
Kay Ct. N&S	31W-20S	18
Kay Dr. NE&SW	31W-20S	18
Kellog St. N&S	27W-20S	19
Kenwood Av. N&S	27W-20S	19
Kenyon E&W	29W-17S	12
Kerns Dr. N&S	29W-17S	12
Kieth St. E&W	27W-20S	19
Lake Dr. NS&EW	29W-16S	12
Lavida Blvd. NE&SW		
	27W-20S	19
Lawrence St. NW&SE		
	28W-19S	13
Leach Dr. E&W	28W-19S	13
Lee St. E&W		
	29W-18S,28W-19S	12
Legion St. N&S	30W-19S	12
Lenwood Dr. N&S	29W-16S	13
Lewood Dr. N&S	30W-19S	12
Lily Cache Rd. NE&SW		
	29W-18S,28W-19S	12
Lincoln Hwy. N&S		
	31W-17S,31W-16S	12

	Grid	Page
Link Ln. E&W	28W-19S	13
Lockner Blvd. NS&EW		
	28W-20S	19
Lockport Rd. E&W		
	28W-17S,27W-17S	13
Lorraine Av. NW&SE		
	28W-19S	13
Margaret Ct. E&W	28W-19S	13
Mary Ct. NE&SW	31W-20S	18
Mary Ln. N&S	31W-20S	18
Massey Av. N&S	29W-18S	12
Mathews St. E&W	29W-18S	12
McAllister Rd. N&S	28W-19S	13
McClellan Av. N&S	28W-19S	13
McGrath Dr. N&S	28W-19S	13
McKenna Dr. N&S	30W-20S	18
Meridan Dr. N&S	28W-20S	19
Metcalf Ct. E&W	29W-21S	18
Michael Dr. N&S	27W-18S	13
Mink Farm Rd. E&W		
	29W-20S	18
Nuclear Dr. N&S	28W-20S	19
Oak St. E&W	28W-19S	13
Old Indian Boundary Line		
NE&SW	32W-19S,31W-18S	12
Olympia St. NE&SW		
	27W-20S	19
Patterson St. E&W	28W-19S	13
Pauline Dr. E&W	31W-20S	18
Peacedale NW&SE	30W-19S	12
Peach St. NW&SE	27W-20S	19
Pecan St. E&W	27W-20S	19
Peerless Dr. N&S	28W-19S	13
Pennington Ln. E&W		
	29W-17S	12
Penny Ln. N&S	29W-17S	12
Peterson Dr. NW&SE		
	28W-19S	13
Pilcher Rd. E&W		
	32W-16S,27W-16S	12
Pincrest Dr. N&S	27W-19S	13
Pine Cone Ln. E&W		
	31W-19S	12
Plainfield Rd. E&W		
	28W-19S,27W-20S	13
Plainfield-Naperville Rd.		
NE&SW	29W-17S,29W-16S	12
Plantation Dr. E&W		
	32W-21S	18
Pleasant Ct. NW&SE		
	30W-19S	12
Plum St. E&W	27W-20S	19
Poplar St. E&W	27W-20S	19
Pratt Ct. E&W	29W-21S	18
Raymond St. N&S	28W-21S	19
Regan Rd. NS&EW		
	28W-20S	19
Renwick Park Dr. N&S		
	30W-19S	12
Renwick Rd. E&W		
	32W-18S,27W-18S	12
Richmond Rd. NS&EW		
	32W-21S	18
River Ct. NS&EW	30W-18S	12
River Rd. N&S	31W-19S	12
Robert Av. E&W	29W-18S	12
Rowley Rd. E&W	31W-18S	12
Rt. 59 N&S		
	29W-21S,30W-19S	18
Rt. 126 E&W		
	32W-17S,27W-16S	12
Ruben St. E&W	29W-19S	12
S. Mayleon NE&SW		
	29W-19S,28W-19S	12
Saddle Dr. N&S	27W-19S	13
Sanborn Cir. NW&SE		
	29W-21S	18
Satellite Dr. N&S	28W-20S	19
September Dr. E&W		
	28W-21S	19
Serenity St. N&S	30W-19S	12
Sergeant Ln. N&S	29W-21S	18
Silver Ln. N&S	29W-17S	12
South End Rd. NW&SE		
	27W-19S	13
Spangle Rd. NE&SW		
	28W-19S	13
State Rd. E&W	27W-19S	13
Stateville Rd. E&W	29W-18S	12
Steiner Rd. N&S	32W-17S	12
Surrey Ln. NW&SE	28W-19S	13
Surrey Ln. E&W	32W-21S	18
Sussex Rd. E&W	32W-21S	18
Teeway St. NE&SW		
	27W-20S	19
Thomas Ct. N&S	27W-18S	13
Timberview Dr. NE&SW		
	28W-19S	13
Tyler Dr. E&W	28W-21S	19
U.S. Route 30 NW&SE		
	31W-16S,27W-20S	12
Updyke Rd. NS&EW		
	28W-16S	12
Vandyke Rd. N&S		
	30W-17S,30W-16S	12
Vanhorn Ln. NW&SE		
	29W-20S	18
Vermette St. N&S	29W-19S	18
Vesta Dr. N&S	28W-20S	19
Virginia Ct. E&W	32W-21S	18
Vonesch Rd. NE&SW		
	30W-20S	18
Wattles Ct. E&W	29W-21S	18
Wayne Av. E&W	27W-20S	19

	Grid	Page
Wedgewood Ct. E&W		
	30W-19S	12
Weller Dr. E&W	29W-16S	12
Westgate Ln. E&W	27W-19S	13
Westline Dr. N&S	28W-20S	19
Whiskey Rd. E&W		
	32W-16S,30W-16S	12
Wilkins Ln. N&S	29W-17S	12
Wilkins Pl. NW&SE	29W-17S	12
Willardshire Rd. E&W		
	28W-20S	19
William Ct. NW&SE		
	29W-16S	12
Williamsburg Dr. N&S		
	32W-21S	18
Wilshire Blvd. E&W		
	27W-20S	19
Winding Creek Rd. NE&SW		
	28W-19S	13
Winston Av. E&W	29W-21S	18
143rd St. E&W		
	28W-16S,27W-16S	13

Reed Township

	Grid	Page
'A' St. E&W	30W-41S	70
Aspen Ln. E&W	30W-40S	70
Big Musky Cir. NE&SW		
	30W-40S	70
Birch Ln. E&W	30W-40S	70
Bluegill Cir. NS&EW		
	30W-40S	70
Catfish Cir. NS&EW		
	30W-40S	70
Cedar Ln. NS&EW	30W-40S	70
Cemetery Rd. N&S	31W-42S	74
Center St. N&S		
	31W-42S,31W-40S	74
Coal City Rd. E&W		
	32W-40S,30W-40S	70
Comet St. E&W	31W-42S	74
County Line Rd. NS&EW		
	32W-40S,30W-45S	70
Dogwood Ln. NS&EW		
	30W-40S	70
Eon Ln. E&W	30W-40S	70
Essex Rd. N&S		
	30W-45S,30W-42S	74
Fifth St. E&W	32W-40S	70
Hole-in-the-Wall Rd. NS&EW		
	30W-40S	70
Interstate 55 NE&SW		
	32W-40S	70
Ivory Rd. N&S	30W-40S	70
Kennedy Rd. E&W	32W-42S	74
Lake Shore Dr. NS&EW		
	30W-40S	70
Large Mouth Ln. E&W		
	30W-40S	70
Little Musky Cir. E&W		
	30W-40S	70
Navy Rd. N&S	30W-40S	70
Park St. E&W	30W-41S	70
Rt. 53 NE&SW		
	32W-42S,30W-40S	74
Rt. 113 NS&EW		
	32W-40S,30W-41S	70
Rt. 129 NE&SW		
	32W-42S,30W-40S	74
Small Mouth Ln. E&W		
	30W-40S	70
Smiley Rd. E&W		
	31W-42S,30W-42S	74
Sportsman Ct. E&W		
	30W-45S	74
Sportsman Dr. E&W		
	30W-45S	74
Sunfish Cir. NS&EW		
	30W-40S	70
Torino Dr. N&S	30W-45S	74
Tully Rd. NE&SW	30W-40S	70
Walleye Cir. NS&EW		
	30W-40S	70
1st St. N&S	30W-41S	70
2nd St. N&S	30W-41S	70

Rockdale

	Grid	Page
Anita St. E&W	25W-25S	28
Bellevue Av. E&W		
	26W-25S,25W-25S	28
Brandon Rd. NW&SE		
	25W-25S	28
Central Av. N&S		
	25W-25S,25W-24S	28
Channahon Rd. E&W		
	25W-25S	28
Connell St. N&S	25W-25S	28
Davis Av. N&S		
	25W-25S,25W-24S	28
Emery St. N&S		
	26W-25S,26W-24S	28
Ferris St. N&S	25W-25S	28
Fisher Av. N&S	25W-25S	28
Hays St. N&S	25W-25S	28
Howard St. N&S		
	25W-25S,25W-24S	28
Jewett St. N&S	25W-25S	28
Kinsey Av. E&W		
	25W-25S,25W-24S	28
Lakeview Av. E&W		
	26W-24S,25W-24S	28
Larkin Rd. N&S		
	26W-25S,26W-24S	28

	Grid	Page
Margaret St. N&S	26W-25S,26W-24S	28
Meadow Av. E&W	26W-25S,25W-25S	28
Midland Av. N&S	25W-25S,25W-24S	28
Moen Av. E&W	26W-25S,25W-25S	28
Morris St. N&S	26W-25S,26W-24S	28
Mound Rd. NE&SW	26W-25S,25W-25S	28
Otis Av. E&W	26W-25S,25W-25S	28
Railroad St. NE&SW	25W-25S,24W-25S	28
Ravine Av. NW&SE	25W-25S,25W-24S	28
Raynor Av. NE&SW	25W-25S	28
Rock Island Av. NE&SW	25W-25S	28
Stillwell Av. N&S	25W-25S	28
Stryker Av. N&S	26W-25S,26W-24S	28
Thorne St. N&S	25W-25S	28
Warren St. N&S	25W-25S,25W-24S	28
Wheeler Av. N&S	25W-25S	28

Romeoville

	Grid	Page
Abbeywood Dr. E&W	23W-14S	11
Airport Rd. E&W	24W-17S,23W-18S	14
Alexander Cir. NS&EW	24W-15S	10
Allenhurst Ct.* E&W	23W-13S	11
Ambassador Dr. NS&EW	23W-14S	11
Amherst Av. N&S	23W-15S	11
Anderson Dr. NS&EW	23W-15S	11
Anton Dr. E&W	23W-15S	11
Arbor Dr. E&W	25W-14S	10
Arcadia Dr. E&W	24W-14S	10
Arlington Dr. E&W	24W-15S,23W-15S	10
Arnold Av. N&S	24W-15S	10
Arsenal Rd. E&W	23W-16S	15
Ashton Av. N&S	24W-16S	14
Avalon Av. E&W	24W-16S	14
Baker Cir. Cir.	23W-14S	11
Beacon Av. E&W	23W-14S	11
Beechwood Rd.* E&W	24W-14S	10
Belmont Ct. E&W	24W-15S	10
Belmont Dr. N&S	24W-16S,24W-15S	14
Berkshire Av. N&S	24W-15S	10
Beverly J. Griffin Dr. E&W	24W-15S	10
Birchwood Dr.* E&W	24W-14S	10
Bluff Dr. N&S	23W-14S	11
Bluff Rd. E&W	22W-13S	11
Briarwood Av. N&S	24W-16S	14
Bristol Av. N&S	24W-16S	14
Broadway St. N&S	23W-18S	15
Bull Run Dr. E&W	23W-15S	11
Camden Av. N&S	23W-15S	11
Cascade Dr. NW&SE	23W-14S	11
Caton Farm Rd. E&W	25W-20S,24W-20S	20
Cedar Ct.* N&S	24W-14S	10
Cedarbend Dr.* E&W	24W-14S	10
Civic Center Dr. E&W	23W-15S	11
Clifton Dr. NS&EW	24W-15S	10
Concord Av. N&S	23W-15S	11
Corona E&W	24W-14S	10
Cranbrook Ct.* E&W	24W-13S	10
Dalhart Av. N&S	23W-15S	11
DeLaSalle Dr. E&W	23W-18S	15
Delta Av. N&S	23W-14S	11
Devon Ln.* E&W	24W-13S	10
Devonwood Av. E&W	23W-15S	11
Dexter Av. N&S	23W-14S	11
Disposal Rd. E&W	23W-16S	15
Dover Av. N&S	23W-15S	11
Driftwood Av. N&S	24W-14S	11
Dunbridge Ln.* E&W	24W-13S	10
Eaton Av. N&S	24W-16S	14
Echo Av. N&S	23W-14S	11
Elgin Av. N&S	23W-14S	11
Elmwood Rd.* E&W	23W-13S	11
Emery Av. N&S	24W-16S	14
Enterprise Dr. N&S	23W-14S,23W-13S	11
Erie Dr. E&W	24W-15S	10
Essex Av. N&S	24W-14S	11
Everette Av. N&S	24W-16S	14
Evergreen Ct. N&S	24W-16S	14
Fairfax Av. N&S	24W-16S	14

	Grid	Page
Farmbrook Ct.* E&W	24W-14S	10
Farragut Av. NS&EW	24W-14S	10
Fenton Av. N&S	24W-15S	10
Fernwood Ct. N&S	23W-14S	11
Flambeau Ct. N&S	24W-15S	10
Forestwood Dr. NE&SW	23W-14S	11
Fremont Av. N&S	24W-16S	14
Front St. N&S	23W-16S,23W-15S	15
Front St. N&S	23W-16S	15
Frontage Rd. E&W	25W-14S	10
Gardner Ct.* E&W	24W-14S	10
Garland Ave. N&S	24W-15S	10
Gavin Av. N&S	24W-16S	14
Geneva Av. NS&EW	24W-14S	10
Glen Av. N&S	24W-15S	10
Gordon Av. N&S	23W-16S	15
Gordon Dr. E&W	23W-16S	15
Greenwood Av. E&W	23W-15S	11
Halde Av. N&S	24W-15S	14
Hale Av. N&S	24W-15S	14
Haller Av. E&W	24W-16S,23W-16S	14
Halstead Ct. E&W	24W-15S	10
Hamerick Av. E&W	24W-16S	14
Hamrick Av. E&W	24W-16S	14
Hanson Cir. Cir.	23W-16S	15
Harris Dr.* E&W	24W-13S	11
Hayes Av. E&W	24W-16S,23W-16S	14
Healy Av. E&W	24W-16S,23W-16S	14
Hemlock Av. N&S	24W-16S	14
Hickory Av. N&S	24W-16S	14
Hickory Dr. E&W	23W-14S	11
High Rd. N&S	22W-15S	11
Hillcrest Dr. NS&EW	24W-14S	10
Homer Av. N&S	24W-16S	14
Honeybear Ln.* E&W	23W-14S	11
Honeytree Dr. N&S	23W-14S	11
Hopkins Rd. N&S	24W-18S	14
Hudson Av. N&S	24W-15S,24W-14S	10
Huron Dr. N&S	24W-15S	10
Independence Blvd. N&S	23W-18S,23W-17S	15
Iola Av. N&S	24W-16S	15
Jackson Av. N&S	22W-16S	15
Jackson St. N&S	22W-16S	15
Joliet Rd. N&S	23W-17S,23W-14S	15
Jordon Av. N&S	24W-15S	10
Juliet Av. N&S	23W-15S	11
Karen Av. E&W	24W-16S,23W-16S	14
Kent Av. N&S	24W-16S	14
Kenyon Av. N&S	24W-15S	10
Kingston Dr. N&S	24W-15S	10
Kirman Av. N&S	23W-16S	15
Lakeside Ct. N&S	23W-14S	11
Lakeside Dr. N&S	23W-14S,23W-13S	11
Lakeview Ct. E&W	25W-14S	10
Lakeview Dr. E&W	25W-14S	10
Larson Cir. Cir.	23W-15S	11
Laurel Av. N&S	24W-15S	10
Lemont Rd. N&S	22W-15S	11
Linden Av. N&S	24W-16S	14
Lynn Dr. N&S	24W-15S	10
Macon Av. N&S	24W-16S	14
Marquette Dr. E&W	23W-13S	11
Mc Kool Av. E&W	24W-16S,23W-16S	14
Mendota Ln. N&S	24W-15S	10
Michigan Dr. N&S	23W-14S	11
Mikan Ln. E&W	23W-15S	11
Montrose Dr. NS&EW	24W-15S	10
Murphy Dr. E&W	24W-16S	14
Naperville Av. E&W	23W-14S,23W-13S	11
Naperville Rd. N&S	23W-14S	11
Nelson Av. N&S	24W-15S	10
Newland Av. E&W	24W-15S	10
Newman Ct. N&S	24W-15S	10
Nippert Av. N&S	24W-16S	14
Normantown Rd. E&W	25W-15S,23W-15S	10
Oakton Av. N&S	24W-15S	10
Ontario Dr. E&W	24W-15S	10
Palmer Av. N&S	24W-15S	10
Parkwood Av. N&S	23W-15S,23W-14S	11
Pell Av. N&S	24W-16S,23W-16S	14
Phelps Av. N&S	23W-15S	11
Pinetree Ct.* N&S	24W-14S	10
Poplar Av. N&S	24W-16S	14
Poplar Blvd. N&S	24W-16S	14
Ridgewood Av. E&W	23W-14S	11
Riveredge Dr. N&S	23W-15S	11
Robbin Dr.* N&S	23W-14S	11
Rocbaar Dr. N&S	23W-14S	11
Rogers Rd. N&S	24W-14S	10

	Grid	Page
Romeo Rd. E&W	23W-16S,22W-16S	15
Romeoville Rd. E&W	23W-16S,22W-16S	15
Roof Av. N&S	24W-16S	14
Rt. 7 NE&SW	24W-20S,23W-19S	20
Rt. 53 N&S	23W-18S,23W-16S	15
Saybrook Ct.* E&W	24W-14S	10
Scarbury Ct.* E&W	24W-14S	10
Sherman Rd. NS&EW	22W-13S	11
Sherwood* N&S	24W-14S	10
Sinde Cir. NS&EW	23W-16S	15
Six Pines Dr.* E&W	24W-14S	10
Spangler Av. N&S	23W-15S	11
St. Claire Dr. E&W	24W-15S	10
Stateville Rd. E&W	25W-19S,23W-19S	14
Superior Dr. N&S	24W-15S	10
Tallman Av. E&W	24W-16S,23W-16S	14
Techni Rd. N&S	24W-17S	14
Troxel Av. N&S	24W-16S	14
Union Av. E&W	24W-16S	14
Waldon Ct.* N&S	24W-14S	10
Washington Av. E&W	22W-16S	15
Washington St. E&W	22W-16S	15
Weber Rd. N&S	25W-20S	20
Wildwood Ct.* E&W	24W-14S	10
Willow Ct.* E&W	24W-14S	10
Windham Pkwy. E&W	25W-14S	10
Windsor Ln.* E&W	24W-13S	10
Winnebago Ct. E&W	24W-15S	10
Woodlawn Av. N&S	25W-15S	10
Yates Av. N&S	24W-15S	10
135th St. E&W	24W-16S	14
135th St. E&W	24W-15S	10

Shorewood

	Grid	Page
'A' Ct. E&W	30W-24S	26
Abbeywood Dr. N&S	31W-24S	26
Amendodge Dr. E&W	30W-24S	26
Arrowhead Dr. N&S	30W-23S	18
Bass St. E&W	30W-24S	26
Birch Dr. E&W	30W-23S	18
Black Rd. E&W	30W-22S	18
Brook Forest Av. N&S	30W-23S,30W-22S	18
Brookshore Ct. NE&SW	30W-24S	26
Brookshore Dr. N&S	30W-24S,30W-23S	26
Buckinghan Pl. N&S	31W-24S	26
Ca-Crest Dr. N&S	30W-23S	18
Cambridge Ln. NW&SE	30W-24S,31W-24S	26
Caprice Dr. NE&SW	30W-24S	26
Cardinal Pl. N&S	30W-23S	18
Carla Dr. N&S	30W-23S	18
Cedar Ln. N&S	30W-23S	18
Channahon St. NE&SW	29W-24S	26
Clark Dr. NE&SW	30W-24S	26
Cobblestone Ct. E&W	30W-23S	18
Collingwood Ct. E&W	30W-23S	18
Colonade Rd. N&S	30W-24S	26
Colony Ct. NE&SW	30W-23S	18
Conquest Ct. NE&SW	31W-24S	26
Cortina Ct. N&S	30W-24S	26
Cottage St. N&S	30W-24S	26
Country Rd. NW&SE	30W-23S	18
D Ct. NS&EW	30W-24S	26
Danny Dr. N&S	30W-23S	18
Dantes St. N&S	29W-24S	26
David Dr. N&S	30W-23S	18
Deerwood Dr. E&W	30W-23S	18
Diana Av. N&S	29W-24S	26
DuPage St. N&S	30W-24S	26
Earl Rd. N&S	30W-24S	26
Edgebrook Dr. N&S	30W-23S	18
Franklin Ln. N&S	30W-24S	26
Garden Ter. N&S	30W-24S	26
Geneva St. E&W	30W-24S	26
Greenfield Rd. N&S	30W-23S	18
Grove St. N&S	30W-24S	26
Harvard Ln. NW&SE	30W-24S	26
Harvest Dr. NW&SE	30W-23S	18

	Grid	Page
Helene St. N&S	29W-24S	26
Highland Dr. E&W	30W-23S	18
Hummingbird Dr. E&W	30W-23S	18
Hunter Dr. E&W	31W-24S	26
Huntington Ct. N&S	30W-24S	26
Huntington Dr. E&W	30W-24S	26
Interstate 55 NE&SW	29W-24S,29W-23S	26
Jefferson St. N&S	30W-23S,29W-24S	18
Kensington St. E&W	30W-24S	26
Lancer Ct. E&W	30W-23S	18
Lark Ln. E&W	30W-23S	18
Laura Av. NS&EW	29W-24S	26
Lynwood Ct. NE&SW	30W-24S	26
Madeira Dr. NE&SW	30W-24S	26
Maple Ln. N&S	30W-23S	18
Mary St. N&S	30W-24S	26
Marya Ln. NE&SW	30W-24S	26
Meadow Dr. E&W	30W-23S	18
Meyers St. E&W	30W-24S	26
Moorland Av. NW&SE	30W-23S	18
Oak Ct. NW&SE	29W-24S	26
Oak Dr. E&W	30W-23S	18
Oakwood Dr. E&W	30W-23S	18
Olde Ivey Rd. N&S	31W-24S	26
Oxford Ln. E&W	30W-24S	26
Parkshore Dr. N&S	30W-23S	18
Picnic St. N&S	30W-24S	26
Princeton Ct. E&W	30W-24S	26
Ranchwood Dr. E&W	30W-23S	18
Raven Rd. N&S	30W-24S	26
Ravinia Ct. N&S	30W-24S	26
Ravinia Dr. N&S	30W-24S	26
Regency Ct. NW&SE	30W-24S	26
Ridge Dr. E&W	30W-23S	18
Ridge Rd. N&S	30W-24S	26
Riffles E&W	30W-24S	26
River Rd. N&S	31W-23S	18
Riverfront Dr. N&S	30W-24S	26
Robin Hill Dr. NS&EW	30W-23S	18
Rock Rd. E&W	30W-24S	26
Rome Ct. E&W	30W-23S	18
Rosewood St. E&W	30W-24S	26
Rt. 59 N&S	30W-24S,30W-22S	26
Rushwood Dr. E&W	30W-23S	18
Sarah Av. E&W	29W-24S	26
School Rd. E&W	29W-24S	26
Sean Dr. N&S	30W-23S	18
Seil Rd. NE&SW	30W-24S	26
Shady Ln. N&S	30W-23S	18
Shore Dr. NE&SW	29W-24S	26
Shorewood Dr. N&S	30W-23S	18
Shorewood Ln. E&W	30W-24S	26
Stanford Dr. E&W	30W-24S	26
Summer St. N&S	30W-24S	26
Sunrise Dr. E&W	30W-23S	18
Sunset Dr. NW&SE	30W-23S	18
Sunset Ln. N&S	30W-24S	26
Trout St. N&S	30W-24S	26
Turtle St. N&S	30W-24S	26
U.S. Route 52 E&W	30W-24S,29W-24S	26
Valencia Dr. NE&SW	30W-24S	26
Wales Ct. E&W	30W-23S	18
West Shore Dr. NE&SW	30W-24S	26
Westwood Ct. N&S	30W-24S	26
Westwood Dr. N&S	30W-24S	26
Wheatland Dr.* NW&SE	30W-23S	18
Whipperwill Ct. NW&SE	30W-23S	18
William St. E&W	29W-24S	26
Windsor Dr. N&S	30W-24S	26
Woodbrook Ct. E&W	30W-23S	18
Yale Ct. NW&SE	30W-24S	26

Steger*
(Also see Cook County)

	Grid	Page
Ashland Av. N&S	1W-28S	44
Barbara Ln. E&W	1W-28S	44
Berk Ln. E&W	0W-28S	44
Butler Av. N&S	0W-28S	44
Camden Pl. NE&SW	0W-28S	44
Carol Dr. N&S	1W-28S	44
Carpenter St. N&S	1W-28S	44
Cedarwood Dr. E&W	1W-28S	44
Christine Ct. E&W	1W-28S	44
Cornwall Dr. N&S	0W-28S	44
Craig Ct. E&W	1W-28S	44

	Grid	Page
Crescent Dr. NE&SW	1W-28S	44
Dorsetshire Dr. N&S	0W-28S	44
Durham Dr. N&S	0W-28S	44
Emerald Av. N&S	1W-28S	44
Evergreen Av. E&W	1W-28S	44
Florence Av. N&S	0W-28S	44
Grace St. E&W	1W-28S	44
Green St. N&S	1W-28S	44
Halsted St. N&S	1W-28S	44
Hereford Dr. E&W	0W-28S	44
Holeman Av. N&S	0W-28S	44
Hopkins Av. N&S	0W-28S	44
Jennifer St. N&S	1W-28S	44
John St. N&S	1W-28S	44
Knollwood Dr. E&W	0W-28S	44
Lewis Av. N&S	0W-28S	44
Loverock Av. N&S	0W-28S	44
McKinley Av. E&W	1W-28S	44
Morgan Ct. N&S	1W-28S	44
Morgan St. N&S	1W-28S	44
Park Av. N&S	1W-28S	44
Peoria St. N&S	1W-28S	44
Phillips Av. N&S	0W-28S	44
Richton Rd. E&W	1W-28S,0W-28S	44
Route 1 N&S	1W-28S	44
Sally Dr. N&S	0W-28S	44
Sangamon St. N&S	1W-28S	44
Stella Bl. N&S	1W-28S	44
Susan Ln. NS&EW	1W-28S	44
Tinbury Pl. NW&SE	0W-28S	44
Tiverton Ct. E&W	0W-28S	44
Union Av. N&S	1W-28S	44
Valleywood Dr. N&S	1W-28S	44
Wallace Av. N&S	0W-28S	44
Williams St. N&S	1W-28S	44
30th St* E&W	0W-28S	44
34th Pl. E&W	1W-28S	44
34tn St. E&W	1W-28S	44
35th Pl. E&W	1W-28S	44
35th St. E&W	1W-28S,0W-28S	44
36th Pl. E&W	1W-28S	44
36th St. E&W	1W-28S,0W-28S	44
37th Pl. E&W	1W-28S	44
37th St. E&W	1W-28S,0W-28S	44

Symerton

	Grid	Page
Commercial St. E&W	23W-37S,22W-37S	61
Honeywell Ln. E&W	22W-37S	61
James St. N&S	22W-37S	61
North St. E&W	22W-37S	61
Northern Av. N&S	22W-37S	61
South St. E&W	22W-37S	61
Traverse St. E&W	22W-37S	61

Tinley Park*
(Also see Cook County)

	Grid	Page
East Hillside Dr. N&S	9W-23S	25
Interstate 80 E&W	9W-22S	25
Rt. 43 N&S	9W-22S	25
76th Av. N&S	9W-22S	25
187th St. E&W	10W-22S	25

Troy Township

	Grid	Page
Addelman Dr.	28W-22S	19
Aston Martin Dr. NE&SW	27W-23S	19
Bailey Dr. N&S	29W-22S	18
Bankview Ln. NE&SW	28W-23S	19
Barber Ln. N&S	27W-22S	19
Baskin Dr. N&S	29W-22S	18
Beckett Dr. N&S	27W-23S	19
Black Rd. E&W	32W-22S,27W-22S	18
Bluemont Dr. E&W	30W-25S	26
Bramble Hill Rd. N&S	30W-27S	26
Bridge St. N&S	30W-26S	26
Bronk Rd. N&S	29W-22S	18
Burr Ridge Dr. N&S	30W-26S	26
Bush Rd. NW&SE	28W-26S	27
Camelot Ct. NE&SW	30W-27S	26
Camelot Rd. E&W	30W-27S	26
Campbell Dr. E&W	27W-23S	19
Canal Rd. NE&SW	30W-27S	26
Cannock Chase E&W	30W-26S	26
Canterbury St. N&S	30W-25S	26
Cassie Dr. N&S	27W-22S	19
Chancer Dr. E&W	27W-23S	19
Clarage Ct. E&W	30W-25S	26
Cleary Av. E&W	28W-22S	19
Colleen Ct. NW&SE	30W-26S	26
County Farm Rd. E&W	32W-24S,29W-24S	26
County Line Rd. N&S	30W-27S,32W-22S	26
Dover Cloud Ln. N&S	27W-23S	19

	Grid	Page
East Cliff Dr. NE&SW		
	30W-27S	26
Fairway St. N&S	28W-23S	19
Francis Ct. E&W	30W-25S	26
Frank Dr. E&W	29W-22S	18
Frontage Rd. N&S		
	30W-27S,29W-22S	26
George Av. E&W	27W-22S	19
Glenwood Av. E&W		
	27W-23S	19
Greenmeadow Ln. E&W		
	28W-23S	19
Guinevere Ln. NW&SE		
	30W-27W	26
Hickory Walk Ct. E&W		
	30W-27S	26
Ingalls Av. E&W	27W-22S	19
Ingolsby Rd. N&S		
	32W-27S,32W-26S	26
Interstate 55 N&S		
	30W-27S,29W-22S	26
Interstate 80 E&W		
	32W-27S,27W-25S	26
Jensen St. E&W 30W-25S		26
John Lee Ct. NE&SW		
	30W-26S	26
Joliet Rd. E&W		
	28W-27S,27W-27S	27
Juniper St. E&W 28W-22S		19
Kelly Ann Ln. N&S . 30W-26S		26
Kings Ct. NW&SE . 30W-27S		26
Kipling Ct. E&W 30W-26S		26
Knightwood Dr. N&S		
	30W-27S	26
Lakewood Ln. NE&SW		
	30W-25S	26
Lancelot Ln. NW&SE		
	30W-27S	26
Leominster Av. E&W		
	28W-22S	19
Mantry Dr. N&S 27W-23S		19
Marlene Ct. E&W . 30W-25S		26
Mattox Ct. N&S ... 30W-26S		26
McDonald Ct. E&W		
	30W-25S	26
Merlin Ct. NW&SE . 30W-27S		26
Mound Ct. NW&SE		
	28W-26S	27
Mound Rd. N&S		
	32W-25S,30W-25S	26
Murphy Dr. N&S ... 27W-22S		19
Nadia Dr. N&S....... 30W-26S		26
Oak Dr. E&W 30W-26S		26
Park River Ln. NS&EW		
	30W-25S	26
Pellinore Dr. E&W . 30W-27S		26
Pine Dr. N&S 30W-25S		26
Queens Ct. NW&SE		
	30W-27S	26
Red Oak Ln. E&W . 30W-26S		26
River Rd. N&S		
	31W-27S,31W-22S	26
Rock Rd. E&W 30W-24S		26
Rock Run Dr. NE&SW		
	29W-25S,29W-24S	26
Rollingwood Ct. E&W		
	28W-23S	19
Rollingwood Ln. NE&SW		
	28W-23S	19
Rolls Royce Dr. E&W		
	27W-23S	19
Ron Lee Ct. N&S . 30W-25S		26
Ronnie Ct. E&W ... 30W-25S		26
Rosary Ln. N&S ... 27W-22S		19
Rt. 59 N&S 29W-22S		18
Sarver Dr. N&S ... 29W-24S		26
Schock Dr. N&S ... 29W-22S		18
Schubert Dr. E&W . 30W-26S		26
Shepley Rd. E&W		
	32W-27S,30W-27S	26
Sir Gallahad Dr. N&S		
	30W-27S,30W-26S	26
Sir Gawain Dr. NE&SW		
	30W-27S,30W-26S	26
Sterling Dr. E&W .. 27W-23S		19
Theodore St. E&W		
	32W-22S,30W-22S	18
Timberline Ct. N&S 28W-23S		19
Timberline Dr. N&S		
	28W-23S,28W-22S	19
Troy Rd. E&W		
	32W-23S,31W-23S	18
Twin Oaks Dr. E&W		
	27W-23S	19
U.S. Route 52 E&W		
	32W-24S,29W-23S	26
West Ridge NW&SE		
	28W-23S	19
Woodridge Way E&W		
	30W-26S	26

University Park

	Grid	Page
Abbott Ln. E&W 3W-30S		43
Amherst Ln. E&W...3W-30S		43
Ann Ln. E&W.......3W-30S		43
Anvil Ct. N&S......3W-30S		43
Ashridge Ln. NE&SW		
	3W-30S	43
Barbara Ct. N&S 3W-29S		43
Barrington Ct. E&W . 3W-30S		43
Barrow Ct. N&S 3W-30S		43
Blackhawk Dr. N&S . .3W-30S		43

	Grid	Page
Bond St. E&W		
	7W-30S,6W-30S	42
Bradfield Ct. N&S .. 3W-30S		43
Brampton Ct. NW&SE		
	3W-30S	43
Brockhurst Ct. N&S . 3W-30S		43
Buckley Ct. E&W ... 3W-30S		43
Burnham Dr. N&S 3W-30S		43
Carmel Ct.* NE&SW 4W-29S		43
Central Av. N&S 7W-30S		42
Circle Dr. E&W 3W-29S		43
Clover Ln. E&W 3W-30S		43
Cordoba St. N&S 3W-30S		43
Crawford Av. N&S .. 5W-29S		42
Cypress Ct.* E&W .. 4W-29S		43
Delmar Ct.* NW&SE 4W-29S		43
Doral Ter. NE&SW .. 3W-30S		43
Dralle Rd. E&W 7W-29S		42
Driftwood Ct.* E&W . .4W-29S		43
Exchange St. E&W		
	4W-29S,3W-29S	43
Farmview Ct. NE&SW		
	3W-30S	43
Farmview Rd. E&W .. 3W-30S		43
Fiesta Dr. E&W 3W-30S		43
Gail Av. E&W.......7W-30S		42
Gordon Ter. E&W ... 3W-29S		43
Governors Hwy. NE&SW		
	5W-28S	42
Hamilton Av. N&S...7W-30S		42
Hamilton Rd. E&W .. 3W-30S		43
Harvest Ln. NS&EW 3W-30S		43
Hickock Av. E&W ... 3W-30S		43
Hickock Dr. E&W ... 3W-30S		43
Home Av. N&S 6W-30S		42
Horner Av. E&W 6W-30S		42
Interstate 57 N&S...7W-30S		42
Irving Pl. E&W 3W-30S		43
Kedzie Av. N&S 4W-30S		43
King Rd. N&S 2W-30S		43
Landau Rd. E&W ... 3W-30S		43
Love Ln. E&W 3W-29S		43
Marina Ct.* N&S ... 4W-30S		43
Mendicino Ct.* NE&SW		
	4W-30S	43
Mesa Ct. E&W 3W-30S		43
Mildred Ln. N&S 3W-30S		43
Mission Dr. E&W ... 3W-30S		43
Monterey Ct. N&S .. 3W-30S		43
Monterey Ct.* E&W . .4W-30S		43
Morningside Dr. NW&SE		
	3W-30S	43
Nathan Rd. E&W ... 3W-30S		43
Oakside Ln. E&W ... 3W-29S		43
Old Forge Ct. N&S . 3W-30S		43
Old Monee Rd. E&W		
	2W-30S	43
Pacifica Ct.* NE&SW		
	4W-30S	43
Palmer Av. E&W ... 7W-30S		42
Pebble Beach Ct. N&S		
	2W-30S	43
Pebble Ct.* N&S ... 4W-29S		43
Pine Oak Ln. NW&SE		
	3W-30S	43
Pine Trace Ct. E&W . .4W-29S		43
Pinewood Ct. E&W .. 4W-29S		43
Redwood Ct.* NE&SW		
	4W-29S	43
Regent Rd. E&W ... 3W-30S		43
Reigel Farm Rd. N&S		
	3W-30S	43
Rt. 50 N&S..........6W-30S		42
Samson Ct. N&S ... 3W-30S		43
Samson Dr. N&S ... 3W-30S		43
Sandpiper Ct.* N&S . .4W-29S		43
Sandra Dr. E&W ... 3W-29S		43
Sierra Ct.* E&W 4W-30S		43
Sierra Dr. N&S 3W-30S		43
Spring Ct.* N&S 4W-29S		43
Stegar Monee Rd. NW&SE		
	4W-29S	43
Stuenkel Rd. E&W .. 5W-28S		42
Sullivan Ln. E&W ... 3W-30S		43
Sunset Ct.* NW&SE 4W-30S		43
Surrey Ln. E&W 3W-30S		43
Union Dr. N&S 3W-30S		43
Western Av. N&S ... 2W-30S		43
Wildwood Ct.* E&W . .4W-29S		43
Wright Rd. N&S ,... 3W-29S		43

Washington Township

	Grid	Page
Ashland Av. N&S		
	2W-39S,2W-34S	67
Brunswick Rd. E&W		
	3E-34S,4E-34S	57
Church Rd. E&W ... 0W-37S		68
Church Rd. E&W ... 1W-35S		56
Corning Rd. E&W		
	0E-37S,4E-37S	68
Cottage Grove Av. N&S		
	0E-39S,1E-36S	68
County Line Rd. E&W		
	2W-39S,4E-39S	67
Dixie Hwy. N&S		
	0W-39S,0W-36S	68
Eagle Lake Rd. E&W		
	2W-34S,0E-34S	55
Hahn's Rd. N&S......0E-35S		56
Halstead St. N&S ... 0W-34S		56
Indiana Av. N&S ... 1W-36S		68
Kennedy Rd. E&W		
	0E-38S,4E-38S	68

	Grid	Page
Klemme Rd. NS&EW		
	3E-39S,3E-34S	69
Lia Ln. N&S 1W-39S		68
Oak Park Av. N&S . . 1W-36S		68
Racine Av. N&S		
	1W-37S,1W-36S	68
Rt. 1 N&S .. 0W-39S,0W-34S		68
Southpark Av. N&S .. 0E-34S		56
State Line Rd. N&S		
	4E-39S,4E-34S	69
Stony Island Av. N&S		
	1E-39S,1E-34S	69
Streamwood Dr. NS&EW		
	2W-34S	55
Yates Av. N&S		
	2E-39S,2E-34S	69
287th St. E&W		
	1W-34S,2E-34S	56
295th St. E&W		
	2W-35S,1E-35S	55
303rd St. E&W		
	0E-37S,4E-37S	68
311th St. E&W		
	2W-37S,4E-37S	67
319th St. E&W		
	2W-38S,4E-38S	67
323rd St. E&W		
	0W-39S,4E-39S	68

Wesley Township

	Grid	Page
Albert Ln. N&S 27W-40S		71
Alma Dr. E&W		
	27W-40S,26W-40S	71
Angle St. E&W 25W-42S		76
Ballou Rd. E&W		
	26W-40S,21W-40S	72
Bass St. E&W 27W-40S		71
Bruning Dr. N&S.... 27W-40S		71
Butterfield Rd. N&S		
	25W-45S,22W-43S	76
Byron Rd. N&S 23W-43S		77
Center NW&SE .. 25W-42S		76
Charles St. N&S 27W-40S		71
Church St. N&S 27W-40S		71
Danielson Rd. N&S		
	22W-42S,22W-41S	77
De Selm Rd. E&W . 24W-43S		76
Donahue Rd. E&W		
	23W-42S,22W-42S	77
Dorset Dr. E&W 27W-40S		71
Elmwood Av. E&W		
	27W-41S,26W-41S	71
Evergreen St. N&S..27W-41S		71
Florence St. NE&SW		
	25W-42S	76
Goodwin Rd. E&W		
	25W-41S,24W-41S	72
Grand Av. E&W 27W-42S		75
Greenwood St. N&S		
	27W-41S	71
Grove St. N&S 27W-41S		71
Hamilton St. E&W .. 27W-40S		71
Hazelton Rd. E&W		
	22W-41S,21W-41S	73
Hickory Av. N&S...27W-42S		75
Hintze St. E&W		
	27W-41S,26W-41S	71
Indian Trail-Phillips Rd. N&S		
	25W-41S,25W-40S	72
John St. E&W		
	27W-40S,26W-40S	71
Juneway Av. N&S .. 27W-42S		75
Kankakee St. NW&SE		
	24W-43S,22W-44S	76
Koerner Ct. E&W .. 27W-40S		71
Lakeside Av. NW&SE		
	27W-41S	75
Lakewood Dr. N&S		
	27W-40S	71
Manteno Rd. E&W		
	25W-42S,21W-42S	76
Maple St. E&W		
	27W-40S,26W-40S	71
Maple Ter. E&W		
	27W-42S,27W-41S	75
Martin Long Rd. N&S		
	22S-40S	73
Meadowview Ln. N&S		
	26W-40S	72
North Dr. E&W 27W-41S		71
Oak St. E&W 27W-42S		75
Old Chicago Rd. N&S		
	24W-43S,24W-40S	76
Orchard Ter. NW&SE		
	27W-42S	75
Pine Grove Av. N&S		
	27W-42S	75
Poplar St. NE&SW .. 27W-41S		71
Rest Haven Rd. N&S		
	26W-42S,26W-41S	76
Riverside Dr. N&S .. 27W-41S		71
Robert St. N&S 27W-40S		71
Rosewood Dr. N&S		
	27W-41S	71
Rt. 102 NW&SE		
	27W-40S,22S-44S	71
Short St. N&S 27W-42S		75
Smith St. NE&SW .. 25W-42S		76
Stewart Av. N&S ... 27W-42S		75
Sumac St. NW&SE 27W-40S		71
Symerton Rd. N&S 23W-40S		73
Thornton Rd. E&W		
	22W-43S,21W-43S	77

	Grid	Page
Trout St. NE&SW .. 27W-40S		71
Vista Dr. N&S 27W-40S		71
Walnut Av. N&S.... 27W-42S		75
Warner Bridge Rd. N&S		
	21W-44S,21W-40S	77
Wesley Rd. N&S		
	26W-43S,26W-42S	76
Wildwood St. N&S .. 27W-41S		71
Willow St. E&W 27W-41S		71
Wood St. E&W 27W-40S		71
Woodview Dr. N&S		
	27W-41S	71

Wheatland Township

	Grid	Page
Aero Ct. E&W 29W-14S		9
Aero Dr. N&S 29W-14S		9
Ainsdale Ct. E&W .. 30W-12S		4
Alago Rd. N&S 28W-10S		5
Alamosa Ct. E&W .. 27W-10S		5
Alan Dr. N&S 28W-10S		5
Alice Ln. E&W...... 30W-12S		4
Alma Ln. N&S....... 28W-10S		5
Andermann Dr. E&W		
	29W-11S	5
Andrew Dr. N&S .. 29W-14S		9
Anvil Ct. N&S 30W-13S		8
Arrowhead Dr. N&S		
	30W-10S	4
Ascot Ct. E&W 30W-12S		4
Ashbury Rd. N&S .. 28W-11S		5
Asther Ct. E&W 30W-15S		8
Book Rd. N&S		
	29W-14S,29W-11S	9
Boughton Rd. NE&SW		
	27W-12S	5
Brancaster Dr. NE&SW		
	30W-12S	4
Brenner Pl. N&S... 30W-12S		4
Bridle Path Ct. N&S		
	29W-12S	5
Burnham Ct. N&S .. 30W-12S		4
Carl Rd. N&S		
	31W-11S,31W-10S	4
Castellan Ln. NS&EW		
	29W-10S	5
Castle Dr. N&S 29W-10S		5
Cathy Ln. NS&EW .. 29W-15S		9
Chapins Rd. N&S		
	32W-14S,29W-14S	8
Charles Ct. N&S 27W-10S		5
Christine Ct. NE&SW		
	29W-13S	9
Cindy Dr. NW&SE . 32W-13S		8
Clow Creek Rd. N&S		
	31W-11S	4
Clow Ct. E&W......29W-13S		9
Columbia Cir. E&W		
	30W-15S	8
Condor Dr. N&S		
	29W-12S,29W-11S	5
Cotswald Dr. E&W . 29W-14S		9
Country Club Dr. NW&SE		
	30W-10S	4
Countryview Dr. E&W		
	28W-10S	5
Courtney Rd. NE&SW		
	32W-13S	8
Curtis Ln. E&W 28W-10S		5
Darcum Cir. N&S .. 28W-11S		5
Davids Ct. E&W 30W-12S		4
Davidson Dr. E&W 28W-10S		5
Dawn Av. N&S 30W-12S		4
De John Blvd. E&W		
	30W-12S	4
Deer Chase Ln. NS&EW		
	29W-12S	5
Denise St. N&S 29W-15S		9
Dickens St. N&S...28W-11S		5
Douglas Dr. E&W .. 29W-15S		9
Douglas Rd. E&W .. 28W-10S		5
Douglas Rd. E&W .. 29W-15S		9
Drawbridge Rd. E&W		
	29W-10S	5
Elaine Dr. N&S 32W-15S		8
Elizabeth Dr. N&S . 32W-15S		8
Essington Rd. N&S		
	28W-13S	9
Fairview Ct. NW&SE		
	29W-13S	9
Falcon Rd. E&W ... 29W-12S		5
Ferguson Rd. E&W		
	29W-13S,28W-13S	9
Fox Valley Expy. E&W		
	28W-14S,27W-14S	9
Ganton Ct. E&W.... 30W-12S		4
Giddings Dr. EW&NS		
	29W-13S	9
Golden Meadow Ct. NW&SE		
	30W-15S	8
Golden Meadow Dr. N&S		
	30W-15S	8
Gordon Ct. E&W .. 28W-10S		5
Graver Ct. E&W 28W-13S		9
Graver Ln. E&W 30W-13S		8
Greenburg Ct. E&W		
	30W-13S	8
Grey Fox Tr. N&S .. 29W-12S		5
Gristmill Ct. E&W .. 29W-10S		5
Grommon Rd. E&W		
	28W-11S	5
Hadfield St. N&S .. 32W-15S		8
Hanslick Ct. N&S .. 31W-10S		4
Hartman Dr. E&W . 29W-11S		5

	Grid	Page
Heggs Rd. N&S		
	32W-14S,32W-10S	8
Helene Av. N&S 30W-12S		4
High Meadow Dr. N&S		
	29W-13S	9
Highland Dr. N&S . 30W-13S		8
Hillview Dr. E&W . 28W-10S		5
Honey Locust Dr. NS&EW		
	29W-13S	9
Hummingbird St. N&S		
	29W-12S	5
Hunt Club Ct. E&W		
	29W-12S	5
Industrial Dr. E. N&S		
	29W-14S	9
Industrial Dr. N. E&W		
	29W-14S	9
Industrial Dr. S. E&W		
	29W-14S	9
Jane Ct. E&W...... 29W-11S		5
Jester Ct. E&W 29W-11S		5
Joyce Ln. E&W 28W-10S		5
Juanita Dr. E&W .. 28W-10S		5
Karen Dr. NS&EW . 32W-13S		8
Kelly Ct. E&W 32W-13S		8
Kennemer Ct. E&W		
	30W-12S	4
Killarney Ct. E&W . 30W-12S		4
Kings Rd. N&S 27W-14S		9
Kings Rd. (proposed) NW&SE		
	27W-13S,27W-11S	9
Kipling Ct. E&W 27W-10S		5
Kristal Rd. NS&EW 29W-13S		9
Kristi Dr. NW&SE . 32W-13S		8
Lakeside Dr. N&S . 29W-10S		5
Lakeview Cir. E&W		
	28W-10S	5
Laurel Ln. E&W 28W-10S		5
Leona Dr. NW&SE . 29W-14S		9
Leverenz Rd. E&W 28W-10S		5
Lillian Ct. N&S...... 29W-11S		5
Lincoln Hwy. NW&SE		
	32W-13S	8
Linda Ln. E&W 31W-10S		4
Lois Ln. N&S 29W-13S		9
Mallard Ln. NS&EW		
	29W-11S	5
Marie Ct. E&W 29W-13S		9
Mark Dr. E&W...... 29W-11S		5
Marquette Av. NE&SW		
	27W-10S	5
Marte Ct. NW&SE . 29W-13S		9
Mayfair Meadows Dr. NS&EW		
	27W-10S	5
Mayre Av. N&S 30W-12S		4
McGrath Ln. N&S .. 29W-11S		5
Meadow Dr. E&W . 27W-10S		5
Meadow Lake Dr. E&W		
	29W-12S	5
Meadow Lily Ct. NW&SE		
	30W-15S	8
Meadow Ln. N&S . 29W-15S		9
Merlin Cir. NE&SW . 29W-10S		5
Muirfield St. N&S . 30W-12S		4
Nancy Ct. E&W 28W-10S		5
Naperville Rd. N&S		
	28W-14S,28W-13S	9
Newmarket Dr. N&S		
	30W-12S	4
Nighthawk Rd. N&S		
	29W-11S	5
Normantown Rd. N&S		
	31W-10S,31W-14S	4
Old Joliet Gravel Rd. E&W		
	32W-10S,31W-10S	4
Old Wheaton Dr. E&W		
	27W-10S	5
Oswego Rd. NE&SW		
	32W-11S,31W-11S	4
Overman Ct. NW&SE		
	29W-15S	9
Owl Rd. N&S 29W-12S		5
Oxford Ln. N&S 29W-14S		9
Palmer Cir. NS&EW		
	30W-10S	4
Park Ln. N&S 28W-10S		5
Partlow Dr. N&S .. 29W-11S		5
Paul Cir. N&S 29W-15S		9
Paula Ct. NE&SW . 29W-15S		9
Perkins Ct. E&W....28W-10S		5
Pheasant Run Ct. E&W		
	29W-12S	5
Pilchers Rd. E&W		
	32W-15S,28W-15S	8
Pine Ridge Ct. NW&SE		
	29W-12S	5
Plainfield-Naperville Rd. N&S		
	28W-15S,28W-13S	9
Plainview Dr. E&W . 28W-10S		5
Plantation Ct. E&W		
	30W-10S	4
Poet Cir. N&S 28W-11S		5
Quail Ridge Ct. N&S		
	29W-12S	5
Ramm Ct. N&S 31W-10S		4
Ramm Rd. N&S .. 31W-10S		4
Rebecca Ct. N&S .. 29W-11S		5
Red Oak Rd. NS&EW		
	29W-13S	9
Red Stable Dr. NW&SE		
	29W-12S	5
Rhea Dr. N&S 29W-13S		9
Richard St. E&W .. 28W-11S		5
Ridge Rd. N&S 27W-10S		5

	Grid	Page
River Dr. N&S	27W-11S	5
Rivercrest Dr. NE&SW	29W-15S	9
Riverside Ct. NW&SE	29W-13S	9
Robert Av. E&W	21W-11S	7
Robert Dr. N&S	28W-12S,28W-11S	5
Robin Ln. E&W	31W-11S	4
Royal Birkdale Dr. E&W	30W-12S	4
Royal County Down Dr. NW&SE	30W-12S	4
Royal Lytham Dr. N&S	30W-12S	4
Royal Porthcawl Dr. NS&EW	30W-12S	4
Royal Portrush Dr. E&W	30W-12S	4
Schillinger Dr. NS&EW	29W-11S	5
Sharon Ct. E&W	29W-13S	9
Sharon Rd. NS&EW	29W-13S	9
Shell Lake Dr. E&W	29W-10S	5
Sheri St. N&S	27W-11S	5
Shoger Dr. NS&EW	31W-10S	4
Snowbird St. N&S	29W-11S	5
South Point Rd. N&S	29W-13S	9
Southfield Dr. N&S	29W-13S	9
Southpoint Dr. N&S	29W-13S	9
Spalding School Dr. N&S	29W-14S	9
Sparrow Ln. E&W	29W-12S	5
Springbrook Dr. N&S	27W-10S	5
Springdale Cir. N&S	28W-10S	5
Springlake Dr. E&W	29W-10S	5
Staffeldt Av. E&W	29W-11S	5
Stevenson Dr. N&S	28W-11S	5
Stratford Dr. E&W	29W-14S	9
Sunningdale Ct. N&S	30W-12S	4
Sunridge Dr. N&S	31W-11S	4
Swandyke Ct. N&S	27W-10S	5
Tern Ln. E&W	29W-13S	9
Thatcher Ct. NW&SE	29W-13S	9
Thatcher Dr. NW&SE	29W-13S	9
Thelma Ct. NW&SE	29W-15S	9
Timothy Av. E&W	27W-11S	5
Treyburn Rd. NW&SE	30W-10S	4
Turnberry Ct. NE&SW	30W-12S	4
Vandyke Rd. N&S	30W-15S	8
Vicky St. N&S	29W-15S	9
Wagner Rd. E&W	29W-11S,28W-11S	5
Walter Ln. NW&SE	29W-11S	5
Walton Heath Dr. N&S	30W-12S	4
Webster Ln. NW&SE	29W-11S	5
Wentworth Dr. N&S	30W-12S	4
West St. NS&EW	27W-11S	5
White Eagle Dr. NS&EW	30W-10S	4
White Eagle Dr., S.E. NW&SE	30W-10S	4
Whittington Ln. N&S	29W-10S,29W-11S	5
Wild Flower Dr. N&S	29W-13S	9
Wilde Av. E&W	28W-11S	5
Wileden Ln. N&S	28W-10S	5
Windjammer Ln. N&S	29W-10S	5
Winter Park Ct. E&W	27W-10S	5
Wolf Rd. N&S	29W-14S	9
Wolfs Crossing Rd. NE&SW	32W-11S,31W-11S	4
Woodhall Ct. E&W	30W-12S	4
Worcester Dr. N&S	27W-10S	5
91st St. E&W	31W-10S,30W-10S	4
95th Av. E&W	30W-11S,27W-11S	5
95th St. E&W	31W-10S,28W-10S	4
96th Ct. E&W	27W-11S	5
103rd St. E&W	31W-11S,29W-11S	4
104th St. E&W	28W-12S	5
111th St. E&W	32W-12S,26W-12S	4
119th St. E&W	32W-13S,27W-13S	8
127th St. E&W	31W-14S,30W-14S	8
128th St. E&W	32W-15S	8
129th St. E&W	32W-15S	8
131st St. E&W	32W-15S	8
135th St. E&W	30W-15S,29W-15S	8
248th St. N&S	31W-14S,31W-11S	8

Will Township

	Grid	Page
Central Av. N&S	7W-39S,7W-34S	66
Church Rd. E&W	6W-35S,3W-35S	54
Cicero Av. N&S	6W-39S,6W-34S	66
Corning Rd. E&W	7W-37S,3W-37S	66
Corning St. E&W	8W-37S,7W-37S	65
County Line Rd. E&W	8W-39S,3W-39S	65
Crawford Av. N&S	5W-39S,5W-34S	66
Drecksler Rd. N&S	8W-39S	65
Eagle Rd. E&W	8W-34S,3W-34S	53
Harlem Av. N&S	8W-39S,8W-34S	65
Joliet Rd. E&W	8W-35S,3W-35S	53
Kedzie Av. N&S	4W-39S,4W-34S	67
Kennedy Rd. E&W	8W-38S,3W-38S	65
Manhattan-Wilton Rd. E&W	8W-34S,7W-34S	53
N. Peotone Rd. E&W	8W-35S,7W-35S	53
Offner Rd. E&W	8W-34S,3W-34S	53
Peotone-Beecher Rd. E&W	3W-36S,3W-36S	67
Ridgeland Av. N&S	8W-39S,8W-34S	65
Rt. 50 NE&SW	8W-37S,7W-34S	65
Western Av. N&S	3W-39S,3W-34S	67
Will Center Rd. N&S	5W-39S,5W-36S	66
Wilmington-Peotone Rd. E&W	8W-37S,3W-37S	65
219th St. E&W	6W-34S,3W-34S	54

Wilmington

	Grid	Page
Amber Dr. NW&SE	28W-37S	59
Baltimore St. (IL Hwy 53) NE&SW	28W-38S,27W-38S	59
Becky Av. E&W	27W-39S	59
Bremer St. E&W	27W-38S	59
Buchannan St. N&S	27W-39N	59
Byron St. E&W	27W-38S	59
Canal St. NE&SW	27W-38S	59
Central Dr. NE&SW	27W-37S	59
Charlotte St. N&S	27W-37S	59
Cherry St. E&W	28W-39S	59
Chesson Ct. N&S	26W-39S	60
Chestnut St. E&W	28W-39S	59
Chicago St. NE&SW	27W-38S	59
Cross St. E&W	27W-37S	59
Daniels St. N&S	27W-39S,27W-38S	59
Debbie Ln. E&W	28W-39S,27W-39S	59
E. Park St. N&S	27W-39S,27W-38S	59
East Kahler Rd. E&W	27W-39S	59
East St. E&W	27W-39S,27W-38S	59
Elizabeth St. NE&SW	27W-39S	59
Elm St. N&S	27W-39S	59
Elwood St. NE&SW	27W-38S,27W-37S	59
Eula St. E&W	27W-37S	59
Fifth St. (West River Rd.) N&S	28W-39S	59
First St. N&S	28W-39S,28W-38S	59
Fourth St. N&S	28W-39S	59
Fulton St. NE&SW	27W-39S,27W-38S	59
Hayden Ct. E&W	27W-38S	59
Hwy. 53 NS&EW	28W-39S,27W-38S	59
Jackson St. NS&EW	27W-38S	59
James St. E&W	27W-38S	59
Janet Dr. N&S	27W-39S	59
Jefferson St. N&S	27W-38S	59
Jewel Ln. N&S	28W-37S	59
JoAnn Dr. E&W	27W-39S,26W-39S	59
Joliet St. NE&SW	27W-37S	59
Joliet St. N&S	27W-39S,27W-38S	59
Judy Dr. E&W	27W-39S	59
Kankakee River Dr. E&W	27W-37S	59
Kankakee St. N&S	27W-39S,27W-37S	59
Kirsten Ln. N&S	28W-39S	59
Koala Ct. NE&SW	28W-39S	59
Lafayette St. NE&SW	27W-38S	59
Laurel Av. E&W	27W-39S	59
Laurel St. NE&SW	27W-39S	59
Linda Ln. N&S	28W-37S	59
Lindy Ln. E&W	27W-39S	59
Locust St. NE&SW	27W-39S	59
Luther St. N&S	27W-39S	59
Mae St. N&S	27W-39S	59
Main St. NW&SE	27W-39S,27W-38S	59
Margarette St. E&W	27W-37S	59
Marion St. E&W	28W-37S	59
McIntyre St. NW&SE	27W-38S	59
Meadow Wood Ln. E&W	28W-39S	59
Michelle St. NW&SE	27W-38S	59
Mill St. NE&SW	27W-38S	59
Milton St. N&S	27W-38S	59
North Circle Dr. E&W	27W-38S	59
North Outer Dr. NS&EW	27W-38S	59
North Park St. NE&SW	27W-39S,27W-38S	59
North St. E&W	27W-38S	59
North St. NS&EW	27W-39S	59
Olive St. NE&SW	27W-39S	59
Park Dr. N&S	27W-39S	59
Parker Ct. NW&SE	27W-39S	59
Pearl St. NW&SE	27W-39S	59
Phyllis Dr. E&W	27W-39S	59
Ridge St. E&W	27W-39S	59
Ridge Way E&W	26W-39S	60
River St. NE&SW	27W-38S	59
Rt. 102 NW&SE	8W-39S,27W-38S	59
Ryan St. NE&SW	27W-39S	59
School St. N&S	28W-38S	59
Shakespeare St. N&S	27W-38S	59
Soldiers Widows Home Rd. NW&SE	28W-38S	59
South Circle Dr. NS&EW	27W-38S	59
South Outer Dr. NS&EW	27W-38S	59
South Park St. NE&SW	27W-39S	59
Stevens Ln. E&W	28W-38S	59
Stewart St. E&W	27W-38S	59
Strip Mine Rd. E&W	29W-38S,28W-39S	58
Sunset Dr. N&S	27W-39S	59
Third St. N&S	27W-38S	59
Tow Path Ln. NE&SW	28W-39S	59
Van Buren St. NE&SW	27W-38S	59
Vine St. NE&SW	27W-39S	59
Vista Dr. N&S	27W-39S	59
W. Kahler Rd. E&W	27W-39S	59
Wabash Av. NS&EW	27W-39S	59
Wabash St. NE&SW	27W-39S	59
Walnut St. E&W	27W-39S	59
Washington St. NW&SE	27W-39S	59
Water St. NE&SW	27W-39S,27W-38S	59
West River Rd. (Fifth St.) N&S	28W-39S	59
West St. NS&EW	27W-39S	59
Williams St. NW&SE	27W-39S	59
Willida Av. N&S	27W-39S	59
Winchester Green Dr. NE&SW	27W-38S	59
216th Av. N&S	27W-38S	59

Wilmington Township

	Grid	Page
Ash Ln. N&S	32W-34S	46
Blodgett Rd. E&W	32W-34S,30W-34S	46
Boat House Rd. N&S	29W-37S,29W-36S	58
Carp Dr. NS&EW	32W-34S	46
Catfish Dr. E&W	32W-34S	46
Ceco Rd. (Murphy Rd.) E&W	32W-37S,30W-37S	58
Cedar Ln. N&S	32W-34S	46
Center St. N&S	31W-38S	58
Cherry Ln. NE&SW	32W-34S	46
Coal City Rd. (IL Hwy. 113) E&W	32W-39S,28W-39S	58
Cottage Rd. N&S	32W-35S,30W-35S	46
County Line Rd. N&S	32W-39S	58
Crappie Dr. E&W	32W-34S	46
Division St. N&S	32W-39S	58
Elm Dr. NE&SW	32W-34S	46
Fir Ln. N&S	32W-34S	46
Grundy County Line Rd. E&W	32W-37S,32W-34S	58
Holly Hock NW&SE	32W-34S	46
Hwy 53 NE&SW	29W-39S,27W-35S	58
Hwy 129 N&S	30W-39S,30W-38S	58
Interstate 55 N&S	32W-39S,30W-34S	58
Kankakee River Dr. NW&SE	28W-37S,27W-37S	59
Kankakee St. N&S	27W-37S,27W-33S	59
Kavanough Rd. N&S	31W-38S,31W-36S	58
Locust Ln. NS&EW	32W-34S	46
Lorenzo Rd. E&W	32W-35S,30W-35S	46
Murphy Rd. E&W	31W-37S	58
New River Rd. NW&SE	29W-35S,27W-37S	46
North River Rd. NW&SE	30W-35S,30W-34S	46
School St. N&S	28W-38S	59
South River Rd. E&W	29W-37S	58
Stevens Ln. E&W	28W-38S	59
Strip Mine Rd. E&W	30W-38S,28W-38S	58
Townline Rd. E&W	29W-34S,27W-34S	46
Walnut Dr. Ln. N&S	32W-34S	46
Willow Ct. E&W	32W-34S	46
Willow Dr. NW&SE	32W-34S	46
Willow Ln. NW&SE	32W-34S	46
Willow St. N&S	32W-34S	46

Wilton Township

	Grid	Page
Arsenal Rd. E&W	20W-36S,15W-36S	62
Arsenal-Wilton Center Rd. E&W	20W-35S,15W-35S	50
Cedar Rd. N&S	18W-37S	62
County Line Rd. E&W	20W-39S,15W-39S	62
Doyle Rd. E&W	20W-34S,18W-34S	50
Draves Rd. E&W	17W-34S	51
Eagle Lake Rd. E&W	16W-34S,15W-34S	51
Eastern Av. N&S	19W-37S,19W-34S	62
Elevator Rd. N&S	16W-37S,16W-34S	63
Gouger Rd. E&W	20W-39S,20W-34S	62
Green Garden-Manhatten Rd N&S	15W-39S,15W-34S	63
Grove St. E&W	17W-34S,16W-34S	51
Hwy 52 NS&EW	18W-34S,15W-35S	50
Joliet Rd. (St.) N&S	17W-35S,17W-34S	51
Kennedy Rd. E&W	20W-38S,15W-38S	62
Manhattan-Wilton Rd. E&W	20W-34S,15W-34S	50
Offner Rd. E&W	20W-34S,15W-34S	50
Olive St. E&W	17W-34S	51
Tully Rd. N&S	19W-39S,19W-38S	62
Washington St. N&S	17W-34S	51
Wilmington-Peotone Rd. E&W	20W-37S,15W-37S	62
Wilton Center Rd. N&S	18W-39S,18W-34S	63
Wilton Rd. N&S	16W-39S,16W-36S	63
120th Av. N&S	15W-38S,15W-34S	63
128th Av. N&S	16W-38S	63

Woodridge*
(Also see Du Page County)

	Grid	Page
Davey Rd. E&W	21W-12S,20W-12S	7
Marmon Dr. N&S	21W-12S	7